30 *Years*
Previous Solved Papers

GATE 2025

Electrical Engineering

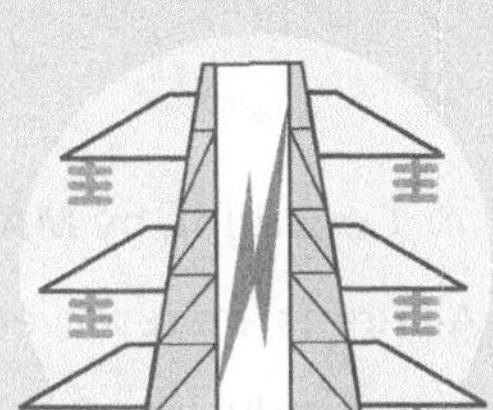

- Fully solved with explanations
- Analysis of previous papers
- Topicwise presentation
- Thoroughly revised & updated

MADE EASY PUBLICATIONS

www.madeeasy.in

MADE EASY Publications Pvt. Ltd.
Corporate Office: 44-A/4, Kalu Sarai (Near Hauz Khas Metro Station), New Delhi-110016 | **Ph. :** 9021300500
Email : infomep@madeeasy.in | **Web :** www.madeeasypublications.org

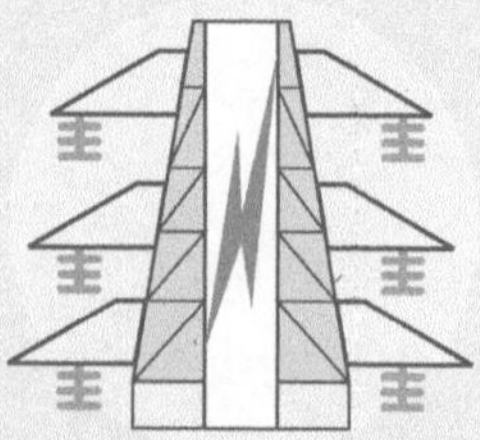

GATE - 2025

Electrical Engineering

Topicwise Previous GATE Solved Papers (1995-2024)

Editions

1st Edition	:	2008
2nd Edition	:	2009
3rd Edition	:	2010
4th Edition	:	2011
5th Edition	:	2012
6th Edition	:	2013
7th Edition	:	2014
8th Edition	:	2015
9th Edition	:	2016
10th Edition	:	2017
11th Edition	:	2018
12th Edition	:	2019
13th Edition	:	2020
14th Edition	:	2021
15th Edition	:	2022
16th Edition	:	2023
17th Edition	**:**	**2024**

Preface

B. Singh (Ex. IES)

Over the period of time the GATE examination has become more challenging due to increasing number of candidates. Though every candidate has ability to succeed but competitive environment, in-depth knowledge, quality guidance and good source of study is required to achieve high level goals.

The new edition of **GATE 2025 Solved Papers : Electrical Engineering** has been fully revised, updated and edited. The whole book has been divided into topicwise sections.

At the beginning of each subject, analysis of previous papers are given to improve the understanding of subject.

I have true desire to serve student community by way of providing good source of study and quality guidance. I hope this book will be proved an important tool to succeed in GATE examination. Any suggestions from the readers for the improvement of this book are most welcome.

B. Singh (Ex. IES)
Chairman and Managing Director
MADE EASY Group

GATE-2025

Electrical Engineering

CONTENTS

Electric Circuits

UNIT

I

CONTENTS

Electric Circuits

Syllabus

Network Elements: Ideal Voltage and Current Sources, Dependent Sources, R, L, C, M Elements;
Network Solution Methods: KCL, KVL, Node and Mesh analysis;
Network Theorems: Thevenin's, Norton's, Superposition and Maximum Power Transfer theorem; Transient response of DC and AC networks, sinusoidal steady-state analysis, resonance, two port networks, balanced three phase circuits, star-delta transformation, complex power and power factor in AC circuits.

Analysis of Previous GATE Papers

Exam Year	1 Mark Ques.	2 Marks Ques.	5 Marks Ques.	Total Marks
1995	2	–	–	**2**
1996	3	1	–	**5**
1997	4	6	2	**26**
1998	3	2	3	**28**
1999	4	5	2	**24**
2000	1	3	1	**12**
2001	5	1	1	**12**
2002	1	7	3	**30**
2003	3	6	–	**15**
2004	1	7	–	**15**
2005	4	7	–	**18**
2006	2	6	–	**14**
2007	–	7	–	**14**
2008	2	6	–	**14**
2009	2	6	–	**14**
2010	–	3	4	**11**
2011	–	3	5	**13**

Exam Year	1 Mark Ques.	2 Marks Ques.	Total Marks
2012	5	6	**17**
2013	2	3	**8**
2014 Set-1	2	2	**6**
2014 Set-2	3	2	**7**
2014 Set-3	3	3	**9**
2015 Set-1	4	3	**10**
2015 Set-2	3	3	**9**
2016 Set-1	4	5	**14**
2016 Set-2	5	4	**13**
2017 Set-1	2	3	**8**
2017 Set-2	2	2	**6**
2018	3	4	**11**
2019	1	3	**7**
2020	2	2	**6**
2021	4	5	**14**
2022	5	1	**7**
2023	2	4	**10**
2024	3	2	**7**

1 Basics

1.1 In the circuit shown in figure, X is an element which always absorbs power. During a particular operation, it sets up a current of 1 amp in the direction shown and absorbs a power P_x. It is possible that X can absorb the same power P_x for another current i. Then the value of this current is

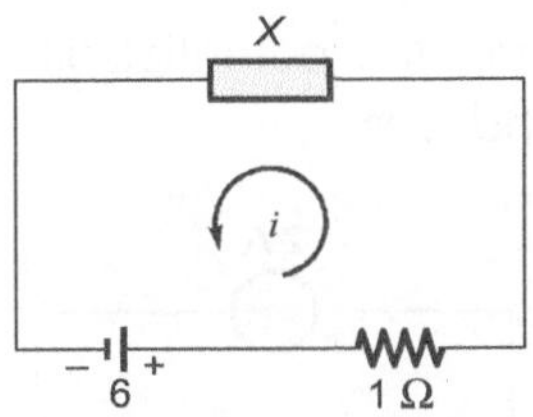

(a) $(3-\sqrt{14})$ Amps (b) $(3+\sqrt{14})$ Amps
(c) 5 Amps (d) None of these

[1996 : 1 M]

1.2 A practical current source is usually represented by
(a) a resistance in series with an ideal current source.
(b) a resistance in parallel with an ideal current source.
(c) a resistance in parallel with an ideal voltage source.
(d) None of these [1997 : 1 M]

1.3 A 10 V battery with an internal resistance of 1 Ω is connected across a non-linear load whose V-I characteristic is given by $7I = V^2 + 2V$. The current delivered by the battery is .... A.

[1997 : 2 M]

1.4 The value of E and I for the circuit shown in the figure, are .... V and .... A.

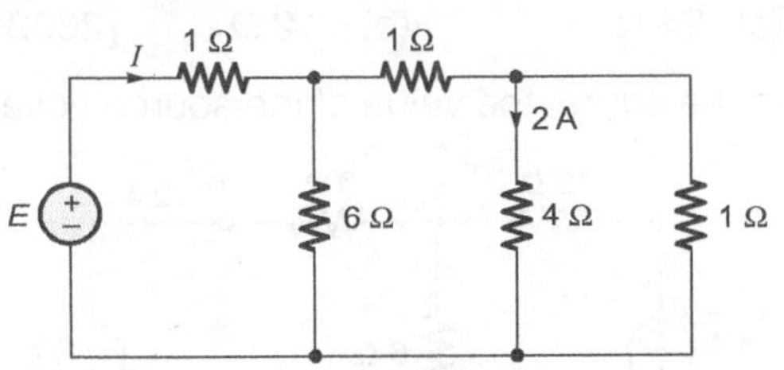

[1997 : 2 M]

1.5 The voltage and current waveforms for an element are shown in the figure. The circuit element is _____ and its value is _______.

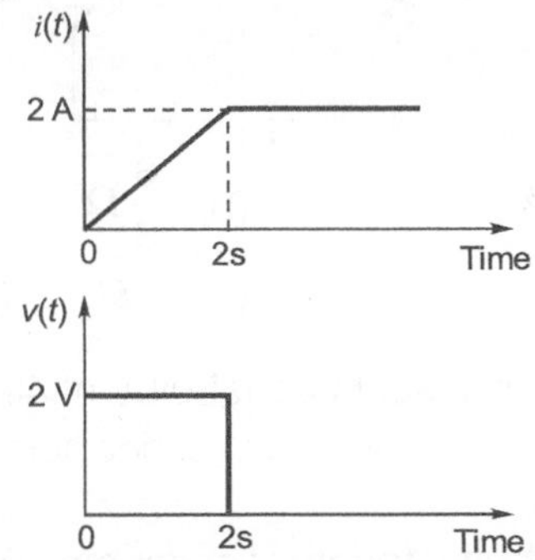

[1997 : 2 M]

1.6 For the circuit shown in the figure, the capacitance measured between terminals B and Y will be

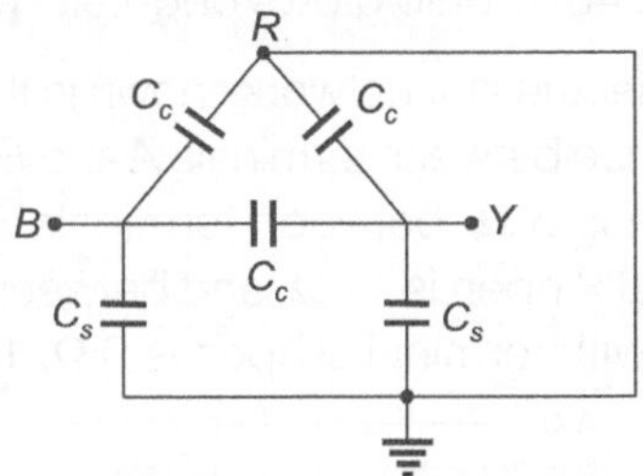

(a) $C_c + \left(\frac{C_s}{2}\right)$ (b) $C_s + \left(\frac{C_c}{2}\right)$
(c) $\frac{(C_s + 3C_c)}{2}$ (d) $3C_c + 2C_s$

[1999 : 1 M]

1.7 When a resistor R is connected to a current source, it consumes a power of 18 W. When the same R is connected to a voltage source having the same magnitude as the current source, the power absorbed by R is 4.5 W. The magnitude of the current source and the value of R are
(a) $\sqrt{18}$ A and 1 Ω (b) 3 A and 2 Ω
(c) 1 A and 18 Ω (d) 6 A and 0.5 Ω

[1999 : 2 M]

1.8 When a periodic triangular voltage of peak amplitude 1 V and frequency 0.5 Hz is applied to a parallel combination of 1 Ω resistance and 1 F capacitance, the current through the voltage source has waveform.

(a) 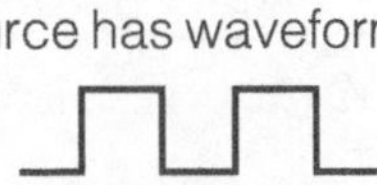 (b) 
(c)  (d) 

[1999 : 2 M]

1.9 The circuit shown in the figure is equivalent to a load of

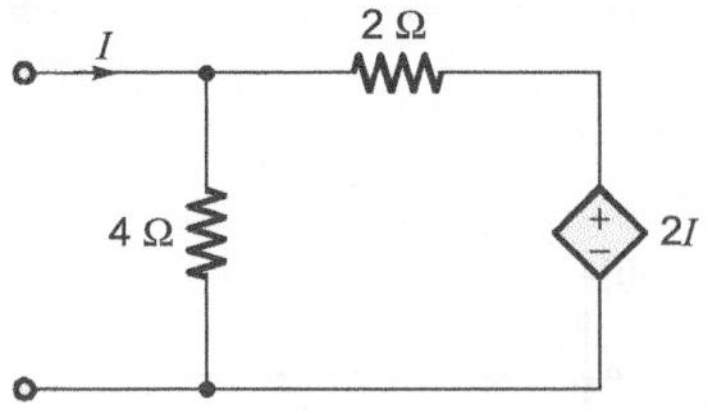

(a) $\frac{4}{3}\ \Omega$ (b) $\frac{8}{3}\ \Omega$

(c) $4\ \Omega$ (d) $2\ \Omega$ **[2000 : 2 M]**

1.10 Two incandescent light bulbs of 40 W and 60 W ratings are connected in series across the mains. Then

(a) the bulbs together consume 100 W.
(b) the bulbs together consume 50 W.
(c) the 60 W bulb glows brighter.
(d) the 40 W bulb glows brighter. **[2001 : 1 M]**

1.11 Consider the star network shown in the figure. The resistance between terminals *A* and *B* with terminal *C* open is 6 Ω, between terminals *B* and *C* with terminal *A* open is 11 Ω, and between terminals *C* and A with terminal *B* open is 9 Ω. Then

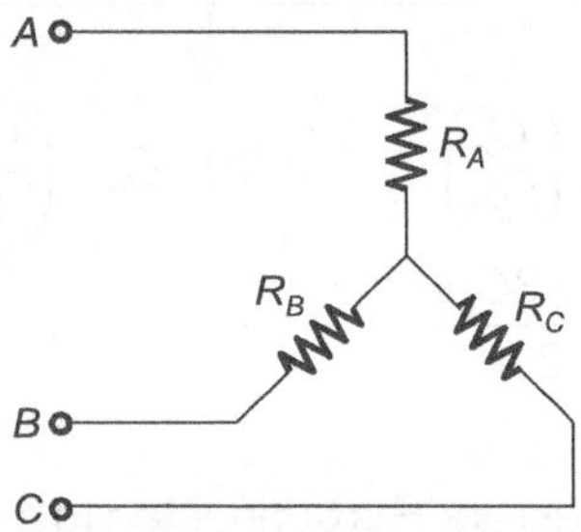

(a) $R_A = 4\ \Omega$, $R_B = 2\ \Omega$, $R_C = 5\ \Omega$
(b) $R_A = 2\ \Omega$, $R_B = 4\ \Omega$, $R_C = 7\ \Omega$
(c) $R_A = 3\ \Omega$, $R_B = 3\ \Omega$, $R_C = 4\ \Omega$
(d) $R_A = 5\ \Omega$, $R_B = 1\ \Omega$, $R_C = 10\ \Omega$ **[2001 : 2 M]**

1.12 A segment of a circuit is shown in the figure $V_R = 5$ V, $V_C = 4 \sin 2t$. The voltage V_L is given by

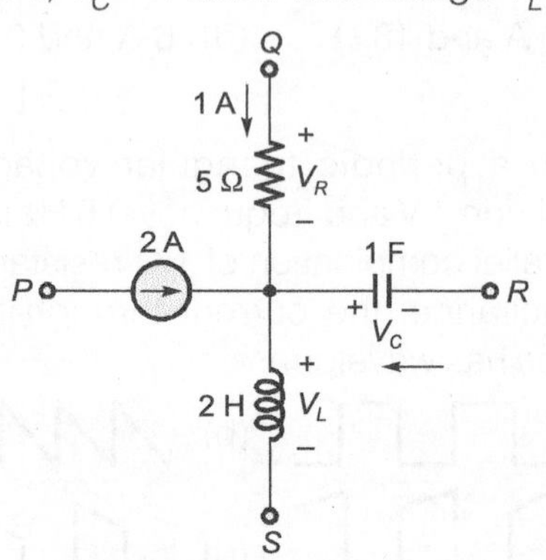

(a) $3 - 8 \cos 2t$ (b) $32 \sin 2t$

(c) $16 \sin 2t$ (d) $16 \cos 2t$ **[2003 : 1 M]**

1.13 The figure shows the waveform of the current passing through an inductor of resistance 1 Ω and inductance 2 H. The energy absorbed by the inductor in the first four seconds is

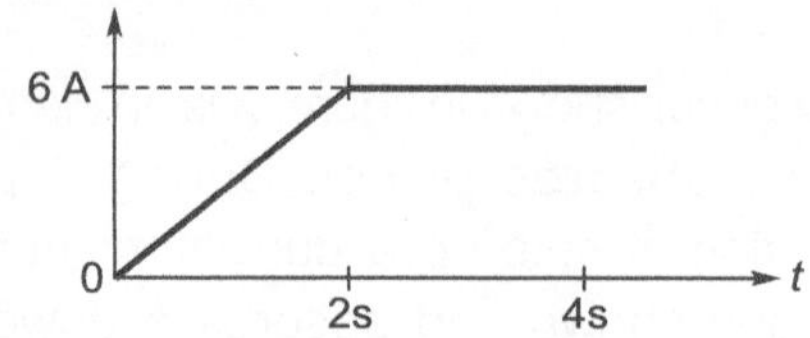

(a) 144 J (b) 98 J

(c) 132 J (d) 168 J **[2003 : 1 M]**

1.14 In the figure, the potential difference between points *P* and *Q* is

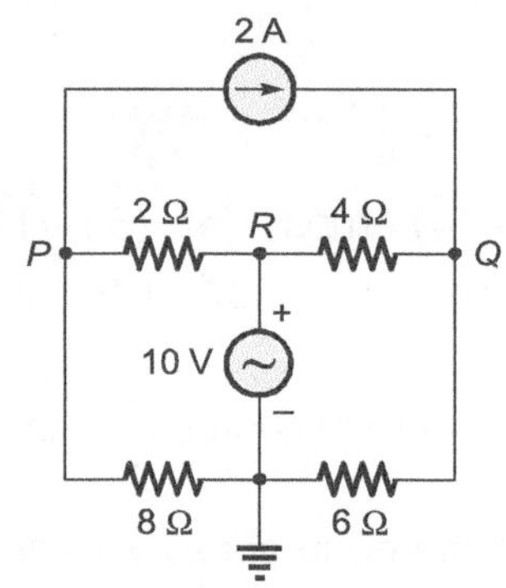

(a) 12 V (b) 10 V

(c) –6 V (d) 8 V **[2003 : 2 M]**

1.15 In the figure, the value of *R* is

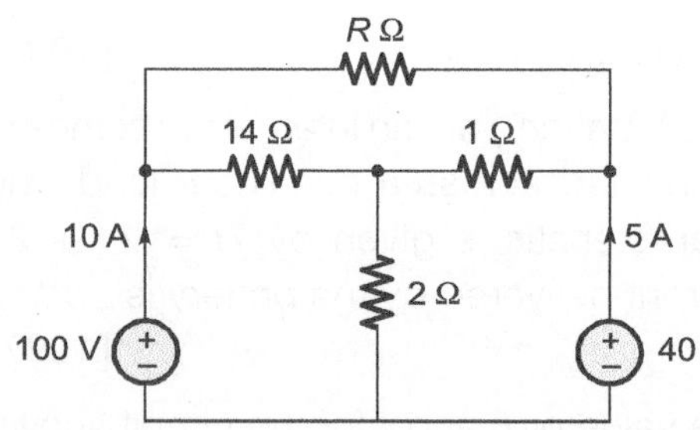

(a) 10 Ω (b) 18 Ω

(c) 24 Ω (d) 12 Ω **[2003 : 2 M]**

1.16 In the figure, the value of the source voltage is

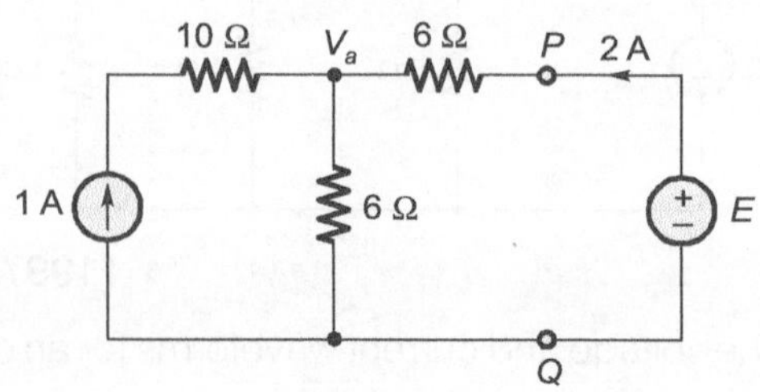

(a) 12 V (b) 24 V

(c) 30 V (d) 44 V **[2004 : 2 M]**

1.17 In the figure, the value of resistance R in Ω is

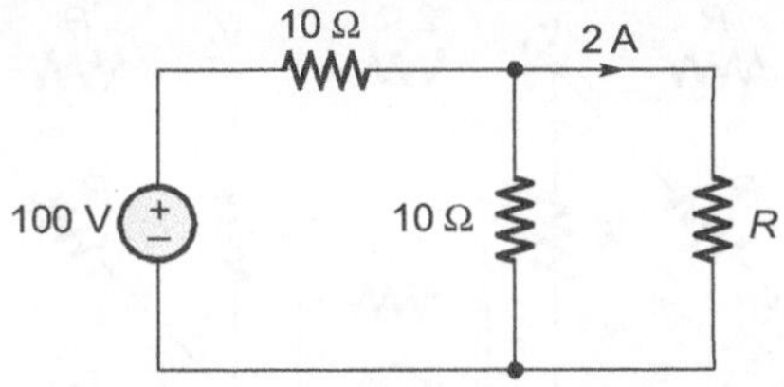

(a) 10 (b) 20
(c) 30 (d) 40 [2004 : 2 M]

1.18 In the figure, R_a, R_b and R_c are 20 Ω, 10 Ω and 10 Ω respectively. The resistances R_1, R_2 and R_3 in Ω of an equivalent star-connection are

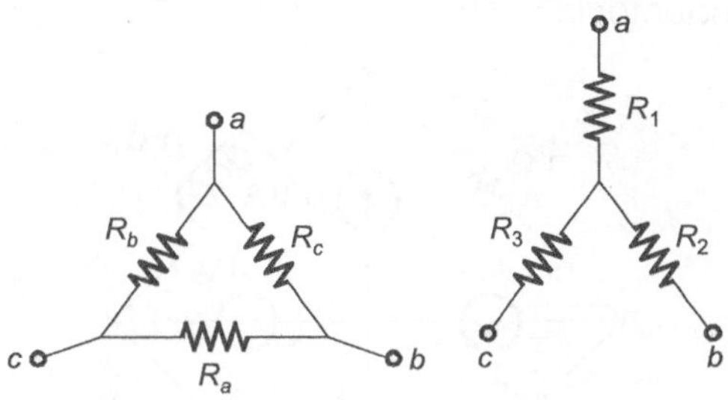

(a) 2.5, 5, 5 (b) 5, 2.5, 5
(c) 5, 5, 2.5 (d) 2.5, 5, 2.5 [2004 : 2 M]

1.19 The rms value of the current in a wire which carries a d.c. current of 10 A and a sinusoidal alternating current of peak value 20 A is

(a) 10 A (b) 14.14 A
(c) 15 A (d) 17.32 A [2004 : 2 M]

1.20 In the figure given below the value of R is

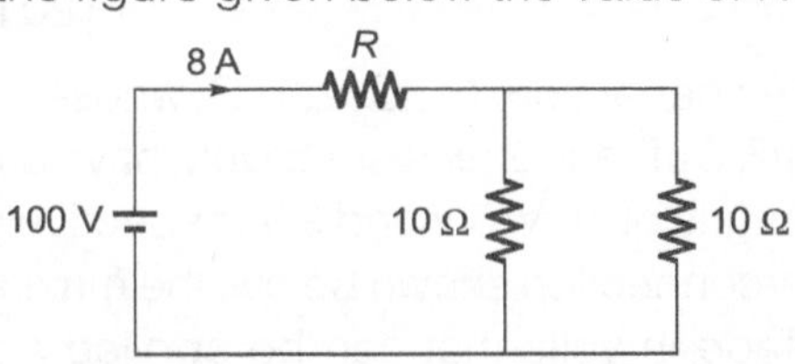

(a) 2.5 Ω (b) 5.0 Ω
(c) 7.5 Ω (d) 10.0 Ω [2005 : 1 M]

1.21 A 3 V dc supply with an internal resistance of 2 Ω supplies a passive non-linear resistance characterized by the relation $V_{NL} = I_{NL}^2$. The power dissipated in the non linear resistance is

(a) 1.0 W (b) 1.5 W
(c) 2.5 W (d) 3.0 W [2007 : 2 M]

1.22 Assuming ideal elements in the circuit shown below, the voltage V_{ab} will be

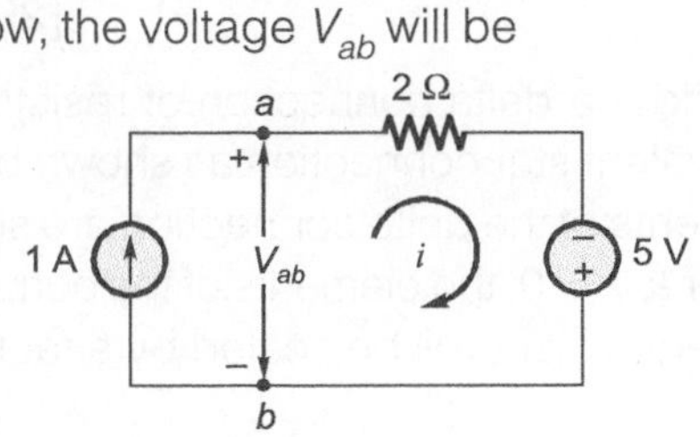

(a) –3 V (b) 0 V
(c) 3 V (d) 5 V [2008 : 2 M]

1.23 In the circuit shown in the figure, the value of the current i will be given by

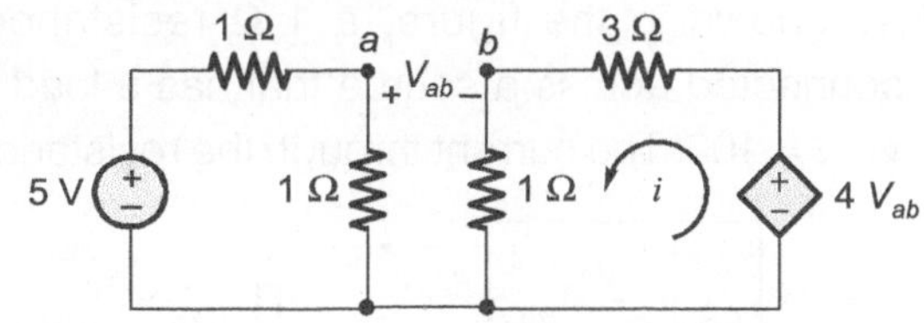

(a) 0.31 A (b) 1.25 A
(c) 1.75 A (d) 2.5 A [2008 : 2 M]

1.24 The current through the 2 kΩ resistance in the circuit shown is

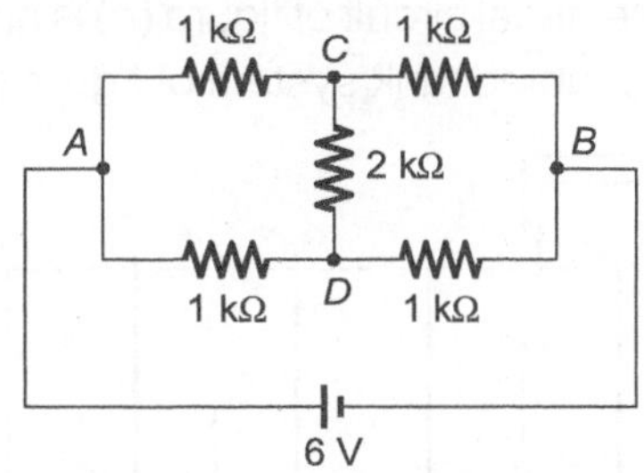

(a) 0 mA (b) 1 mA
(c) 2 mA (d) 6 mA [2009 : 1 M]

1.25 How many 200 W/220 V incandescent lamps connected in series would consume the same total power as a single 100 W/220 V incandescent lamp?

(a) Not possible (b) 4
(c) 3 (d) 2 [2009 : 1 M]

1.26 The equivalent capacitance of the input loop of the circuit shown is

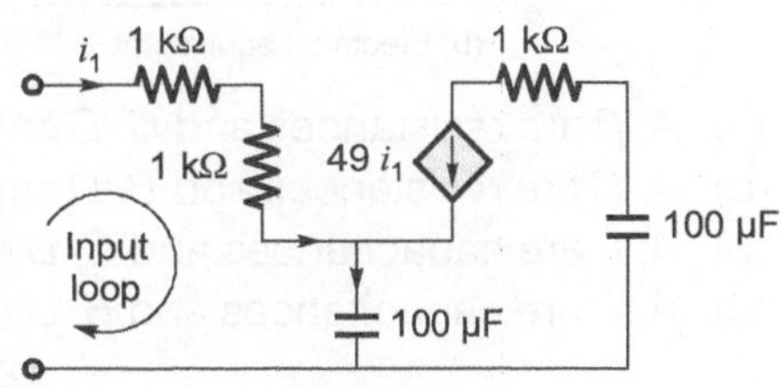

(a) 2 µF (b) 100 µF
(c) 200 µF (d) 4 µF [2009 : 2 M]

1.27 For the circuit shown, find out the current flowing through the 2 Ω resistance. Also identify the changes to be made to double the current through the 2 Ω resistance.

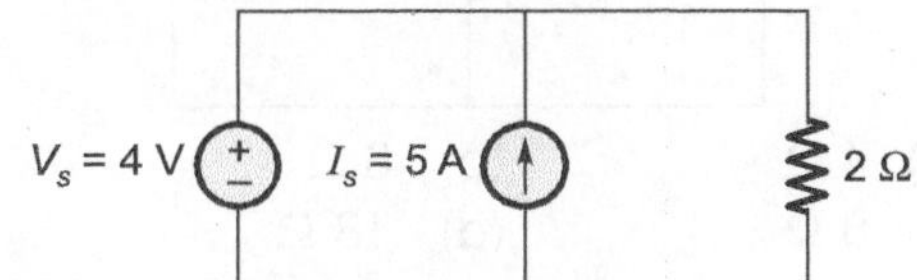

(a) (5 A; Put $V_s = 20$ V)
(b) (2 A; Put $V_s = 8$ V)
(c) (5 A; Put $I_s = 10$ A)
(d) (7 A; Put $I_s = 12$ A) [2009 : 2 M]

1.28 As shown in the figure, a 1 Ω resistance is connected across a source that has a load line $v + i = 100$. The current through the resistance is

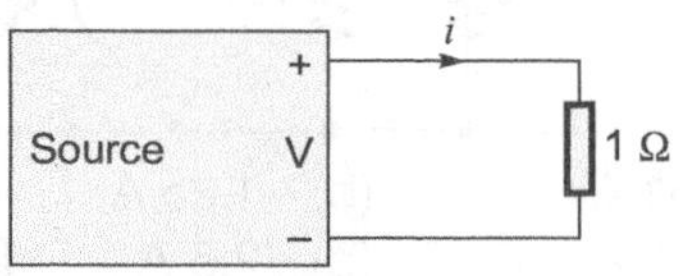

(a) 25 A (b) 50 A
(c) 100 A (d) 200 A [2010 : 1 M]

1.29 If the electrical circuit of figure (b) is an equivalent of the coupled tank system of figure (a), then

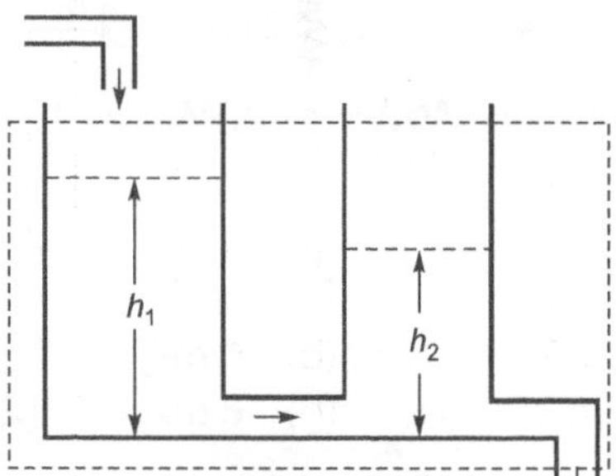

(a) Coupled tank

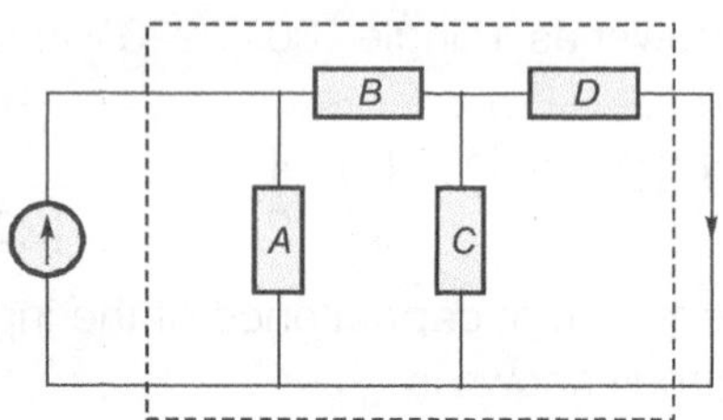

(b) Electrical equivalent

(a) *A*, *B* are resistances and *C*, *D* capacitances
(b) *A*, *C* are resistances and *B*, *D* capacitances
(c) *A*, *B* are capacitances and *C*, *D* resistances
(d) *A*, *C* are capacitances and *B*, *D* resistances [2010 : 1 M]

1.30 If the 12 Ω resistor draws a current of 1 A as shown in the figure, the value of resistance *R* is

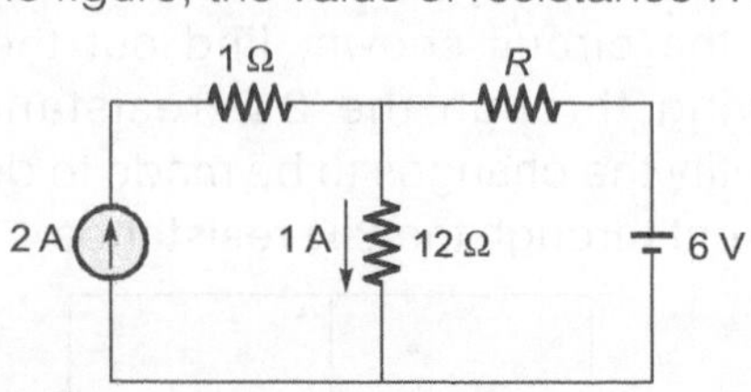

(a) 4 Ω (b) 6 Ω
(c) 8 Ω (d) 18 Ω [2010 : 2 M]

1.31 If $V_A - V_B = 6$ V, then $V_C - V_D$ is

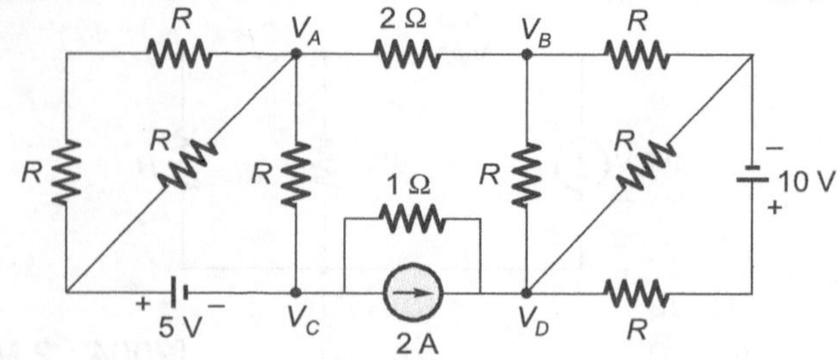

(a) –5 V (b) 2 V
(c) 3 V (d) 6 V [2012 : 2 M]

1.32 In the circuit shown below, the current through the inductor is

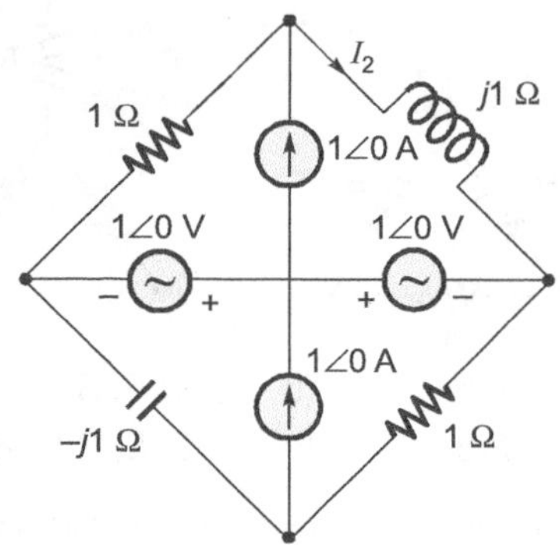

(a) $\dfrac{2}{1+j}$ A (b) $\dfrac{-1}{1+j}$ A

(c) $\dfrac{1}{1+j}$ A (d) 0 A

[2012 : 1 M]

1.33 Three capacitors C_1, C_2 and C_3 whose values are 10 μF, 5 μF, and 2 μF respectively, have breakdown voltages of 10 V, 5 V and 2 V respectively. For the interconnection shown below, the maximum safe voltage in volts that can be applied across the combination, and the corresponding total charge in μC stored in the effective capacitance across the terminals are, respectively

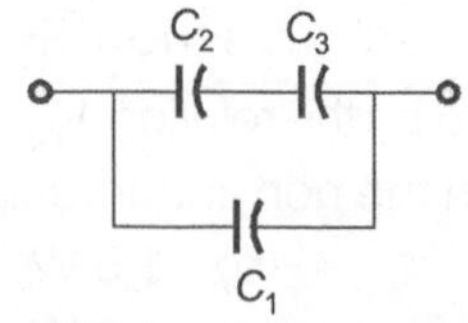

(a) 2.8 and 36 (b) 7 and 119
(c) 2.8 and 32 (d) 7 and 80 [2013 : 2 M]

1.34 Consider a delta connection of resistors and its equivalent star connection as shown below. If all elements of the delta connection are scaled by a factor *k*, $k > 0$, the elements of the corresponding star equivalent will be scaled by a factor of

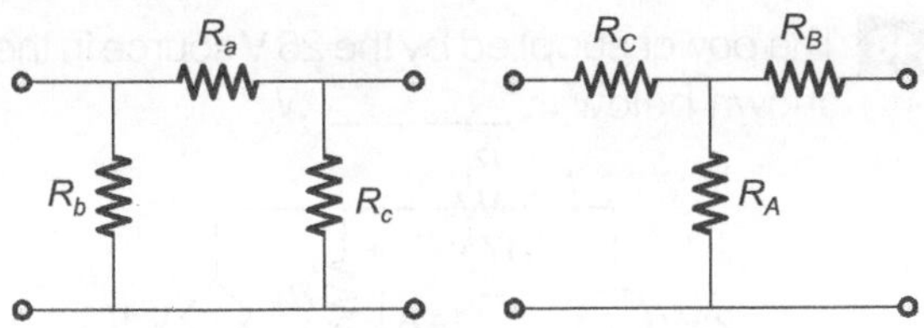

(a) k^2 (b) k

(c) $1/k$ (d) $\sqrt{k}$ [2013 : 1 M]

1.35 The three circuit elements shown in the figure are part of an electric circuit. The total power absorbed by the three circuit elements in watts is ______ .

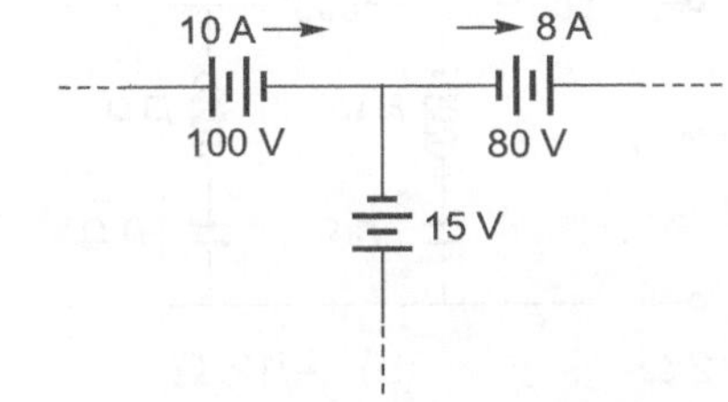

[2014 : 1 M, Set-1]

1.36 An incandescent lamp is marked 40 W, 240 V. If resistance at room temperature (26°C) is 120 Ω, and temperature coefficient of resistance is 4.5×10^{-3}/°C, then its 'ON' state filament temperature in °C is approximately ______. [2014 : 2 M, Set-1]

1.37 In the figure, the value of resistor R is $\left(25 + \frac{I}{2}\right)$ ohms, where I is the current in amperes. The current I is ______.

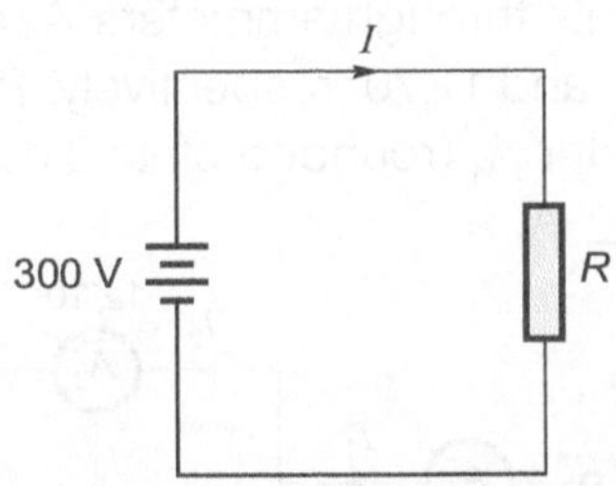

[2014 : 2 M, Set-1]

1.38 The power delivered by the current source, in the figure, is ______ .

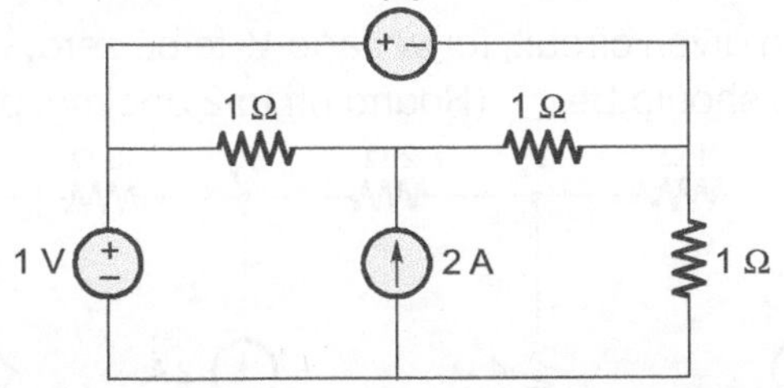

[2014 : 2 M, Set-3]

1.39 The voltages developed across the 3 Ω and 2 Ω resistors shown in the figure are 6 V and 2 V respectively, with the polarity as marked. What is the power (in Watt) delivered by the 5 V voltage source?

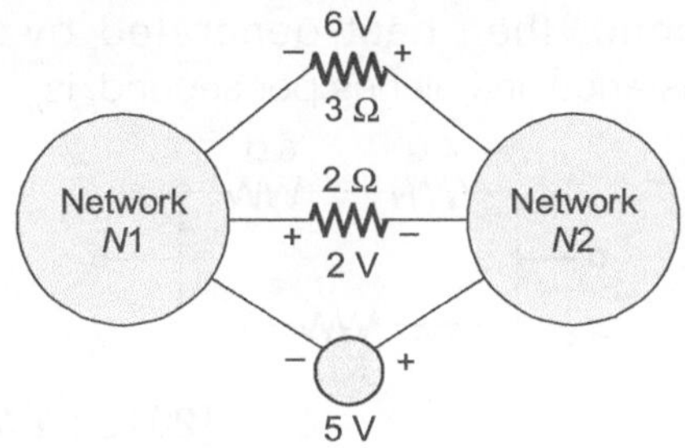

(a) 5 (b) 7

(c) 10 (d) 14 [2015 : 1 M, Set-1]

1.40 In the given circuit, the parameter k is positive, and the power dissipated in the 2 Ω resistor is 12.5 W. The value of k is ____.

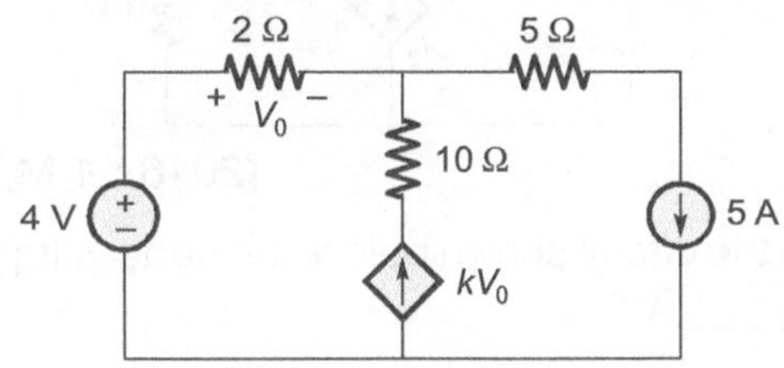

[2015 : 2 M, Set-1]

1.41 The current i(in Ampere) in the 2 Ω resistor of the given network is _____.

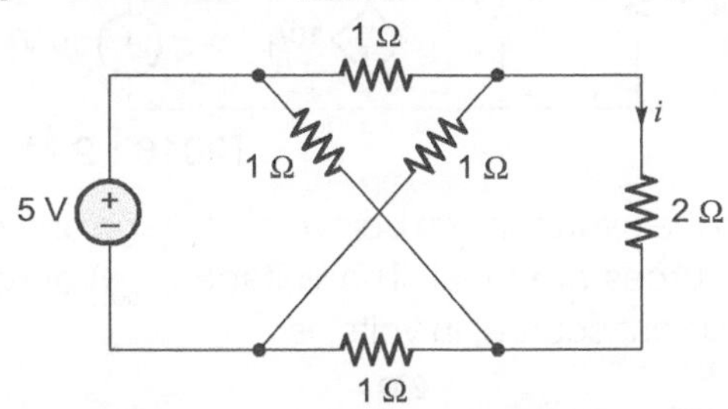

[2015 : 1 M, Set-2]

1.42 R_A and R_B are the input resistances of circuits as shown below. The circuits extend infinitely in the direction shown. Which one of the following statements is TRUE?

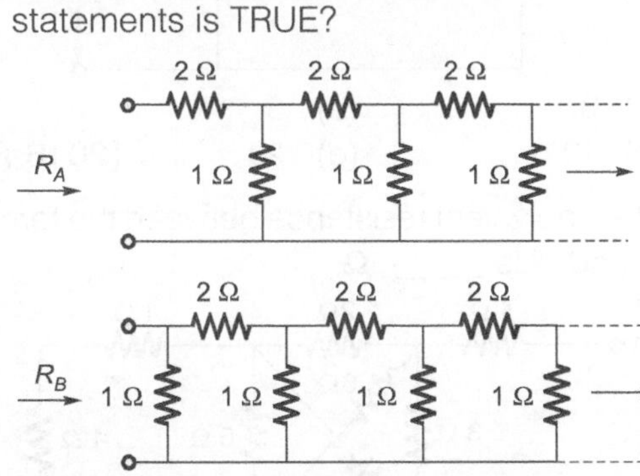

(a) $R_A = R_B$ (b) $R_A = R_B = 0$

(c) $R_A < R_B$ (d) $R_B = R_A/(1 + R_A)$

[2016 : 1 M, Set-1]

1.43 In the portion of a circuit shown, if the heat generated in 5 Ω resistance is 10 calories per second, then heat generated by the 4 Ω resistance, in calories per second, is _____.

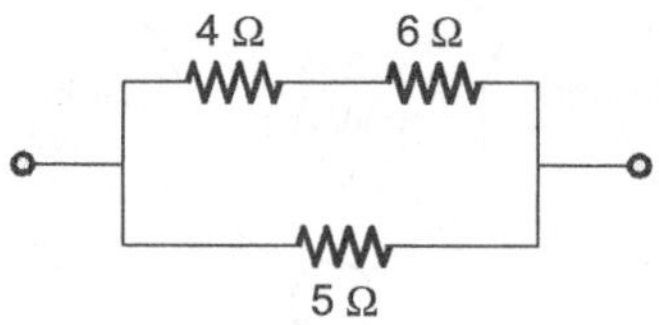

[2016 : 1 M, Set-1]

1.44 In the given circuit, the current supplied by the battery (in Ampere) is ________.

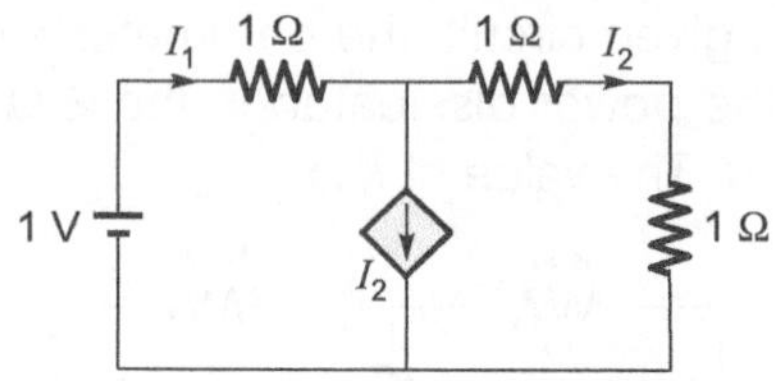

[2016 : 1 M, Set-1]

1.45 In the circuit shown below, the node voltage V_A is ______ V.

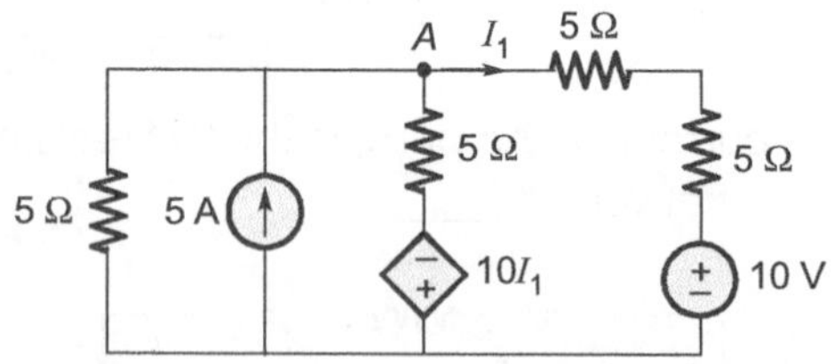

[2016 : 2 M, Set-1]

1.46 In the circuit shown below, the voltage and current sources are ideal. The voltage (V_{out}) across the current source, in volts, is

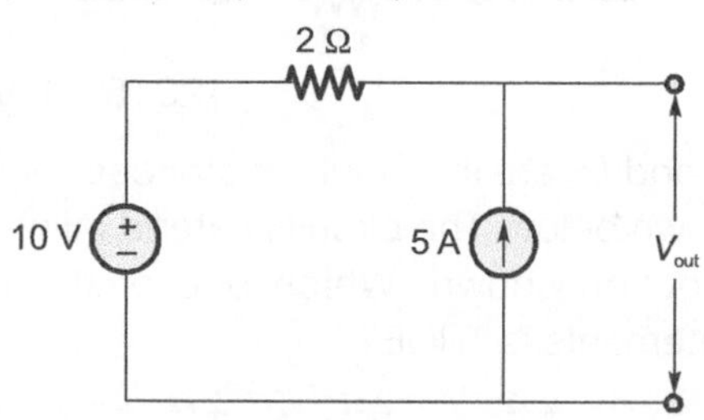

(a) 0 (b) 5
(c) 10 (d) 20 [2016 : 1 M]

1.47 The equivalent resistance between the terminals *A* and *B* is ______ Ω.

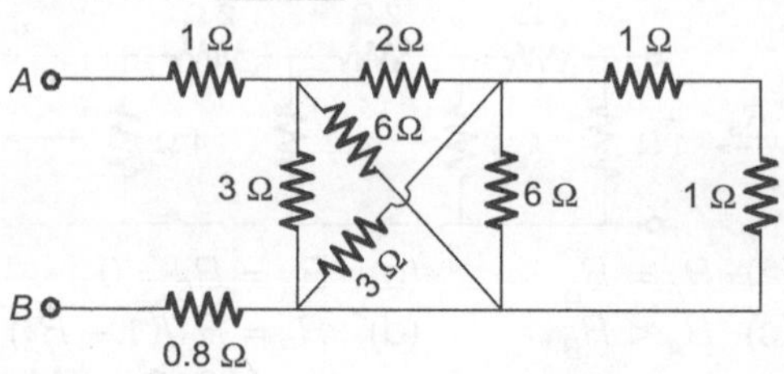

[2017 : 1 M, Set-1]

1.48 The power supplied by the 25 V source in the figure shown below is ________ W.

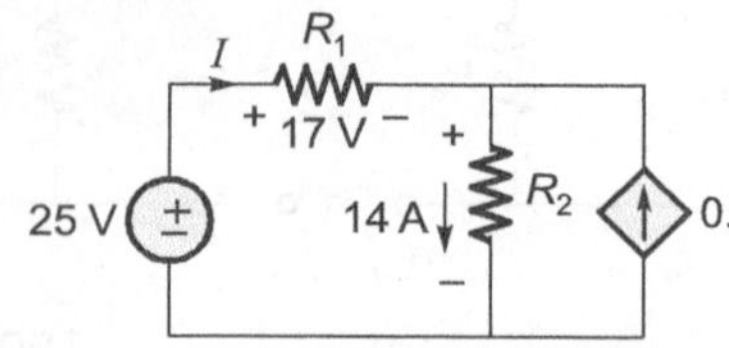

[2017 : 1 M, Set-1]

1.49 The equivalent impedance Z_{eq} for the infinite ladder circuit shown in the figure is

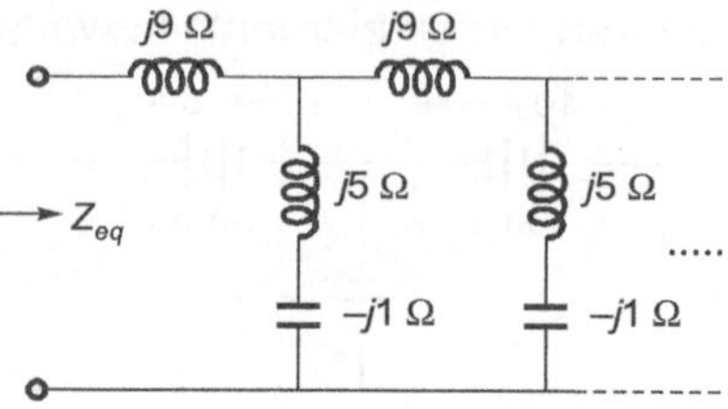

(a) $j12\ \Omega$ (b) $-j12\ \Omega$
(c) $j13\ \Omega$ (d) $13\ \Omega$ [2018 : 2 M]

1.50 The current *I* flowing in the circuit shown below in Amperes (round off to one decimal place) is ____.

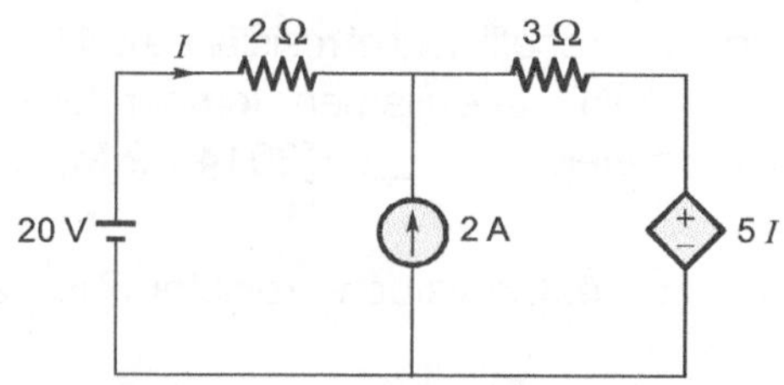

[2019 : 1 M]

1.51 Currents through ammeters A_2 and A_3 in fig.are $1\angle 10°$ and $1\angle 70°$ respectively. The reading of the ammeter A_1 (rounded off to 3 decimal places) is ______ A.

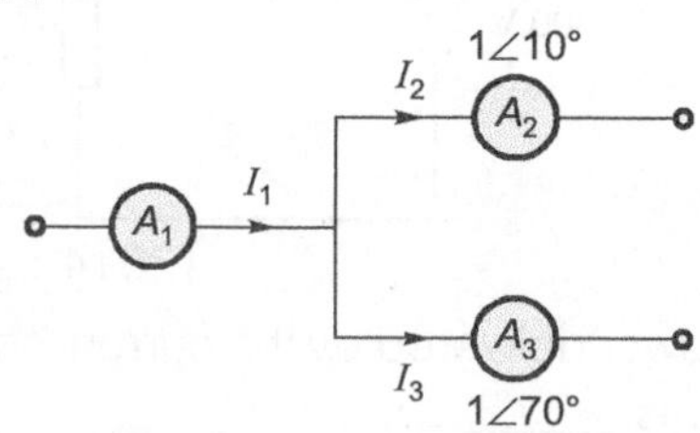

[2020 : 1 M]

1.52 In given circuit, for voltage V_y to be zero, value of β should be __. (Round off to 2 decimal places).

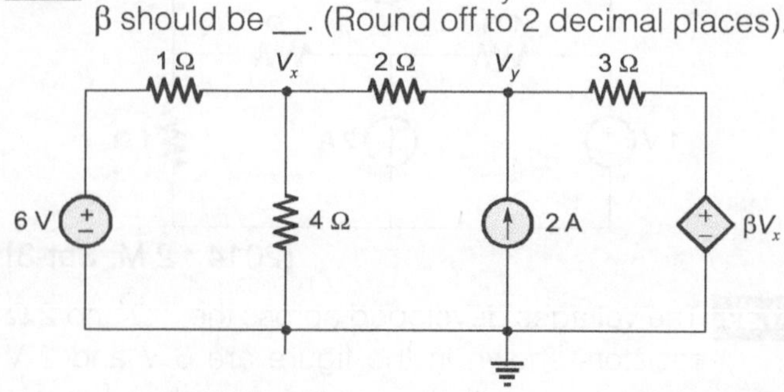

[2021 : 1 M]

1.53 In circuit shown below, the magnitude of voltage V_1 in volts, across the 8 kΩ resistor is _____.

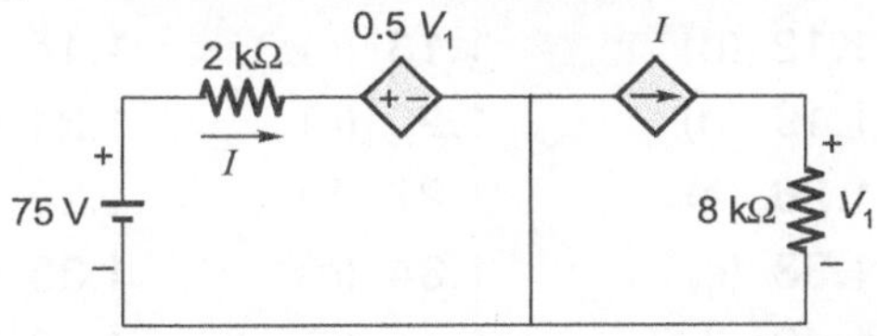

[2022 : 2 M]

1.54 The value of parameters of the circuit shown in the figure are :
R_1 = 2 Ω, R_2 = 2 Ω, R_3 = 3 Ω, L = 10 mH, C = 100 μF

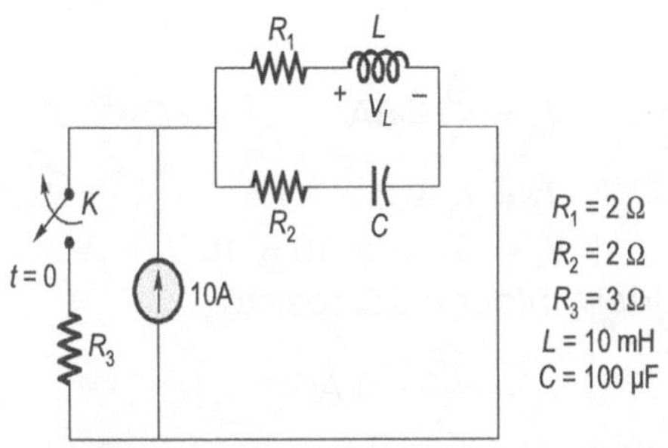

For time $t < 0$, the circuit is at steady state with the switch 'K' in closed condition. If the switch is opened at $t = 0$, the value of the voltage across the inductor (V_L) at $t = 0^+$ in Volts is _____ (Round off to 1 decimal place).

[2023 : 1 M]

1.55 For the circuit shown in the figure, V_1 = 8 V, DC and I_1 = 8 A, DC. The voltage V_{ab} in Volts is _________ (Round off to 1 decimal place).

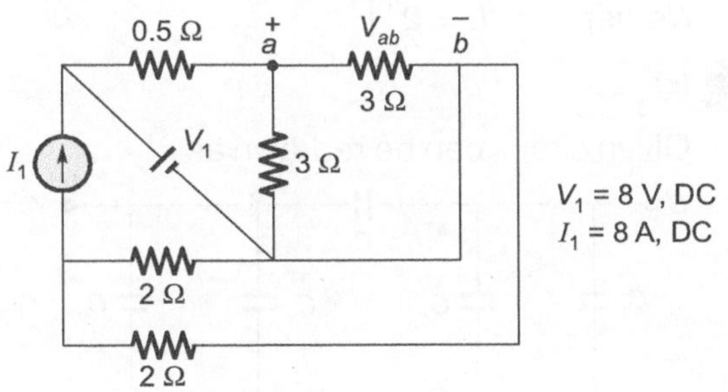

[2023 : 1 M]

1.56 The number of junctions in the circuit is

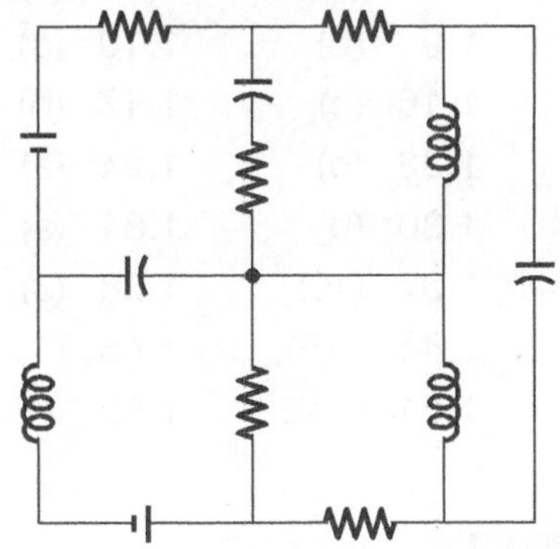

(a) 8 (b) 6
(c) 7 (d) 9

[2024 : 1 M]

1.57 All the elements in the circuit are ideal. The power delivered by the 10 V source in watts is

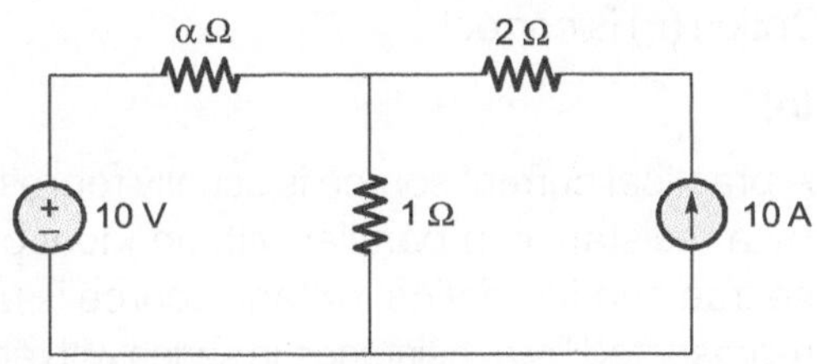

(a) 100
(b) 0
(c) dependent on the value of α
(d) 50

[2024 : 1 M]

■■■■

Answers Basics

1.1 (c)	1.2 (b)	1.3 (5)	1.4 (31)	1.5 (2)	1.6 (c)	1.7 (b)
1.8 (d)	1.9 (b)	1.10 (d)	1.11 (b)	1.12 (b)	1.13 (c)	1.14 (c)
1.15 (d)	1.16 (c)	1.17 (b)	1.18 (a)	1.19 (d)	1.20 (c)	1.21 (a)
1.22 (a)	1.23 (b)	1.24 (a)	1.25 (d)	1.26 (a)	1.27 (b)	1.28 (b)
1.29 (d)	1.30 (b)	1.31 (a)	1.32 (c)	1.33 (c)	1.34 (b)	1.35 (330)
1.36 (2470.44°)	1.37 (10)	1.38 (3)	1.39 (a)	1.40 (0.5)	1.41 (0)	1.42 (d)
1.43 (2)	1.44 (0.5)	1.45 (11.42)	1.46 (d)	1.47 (3)	1.48 (250)	1.49 (a)
1.50 (1.4)	1.51 (1.732)	1.52 (–3.25)	1.53 (100)	1.54 (8)	1.55 (6)	1.56 (b)
1.57 (b)						

Explanations Basics

1.1 (c)

$$P_x = P_{6\,V} - P_{1\,\Omega} = 6\times 1 - 1^2\times 1 = 5 \text{ W}$$

By putting the options, it can be concluded that for $i = 5$ A.

$$P_x = (6\times 5) - (5^2\times 1) = 5 \text{ W}$$

Option (c) is correct.

1.2 (b)

A practical current source is usually represented by a resistance in parallel with an ideal current source and a practical voltage source is usually represented by a resistance in series with an ideal voltage source.

1.3 Sol.

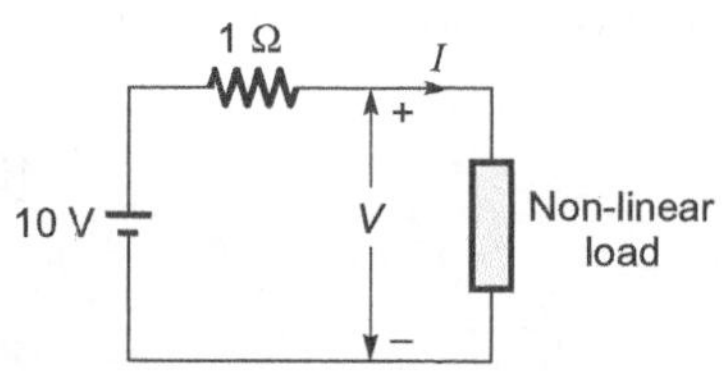

Using KVL, $V + I = 10$ ...(i)

Given, $7I = V^2 + 2V$ ...(ii)

On solving equation (i) and equation (ii) we get,

$$V = 5\text{ V}, \quad I = 5\text{ A}$$

1.4 Sol.

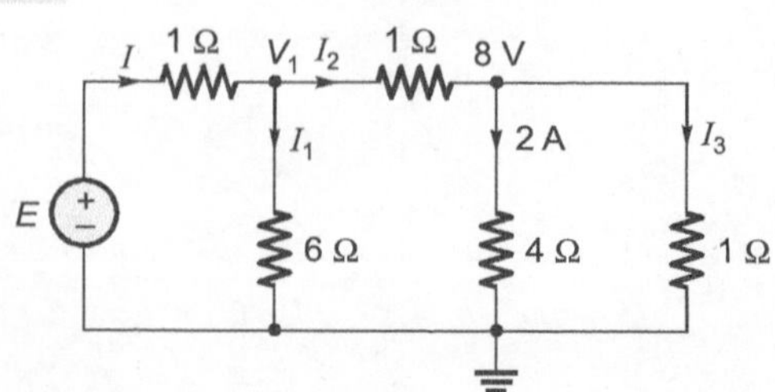

Voltage across 4 Ω resistor = 4 × 2 = 8 V

Current through 1 Ω resistor,

$$I_3 = \frac{8}{1} = 8\text{A}$$

$$I_2 = I_3 + 2 = 10 \text{ A}$$

$$V_1 = 8 + 1\times 10 = 18 \text{ V}$$

Current through 6 Ω resistor,

$$I_1 = \frac{18}{6} = 3 \text{ A}$$

Current through 1 Ω resistor,

$$I = I_1 + I_2 = 3 + 10 = 13 \text{ A}$$

$$E = V_1 + I\cdot 1 = 18 + 13\times 1 = 31 \text{ V}$$

1.5 Sol.

For the given waveforms,

$$v(t) = 2\frac{di(t)}{dt}$$

Comparing it with $v(t) = L\frac{di(t)}{dt}$

We get, $L = 2$ H

1.6 (c)

Given circuit can be redrawn as:

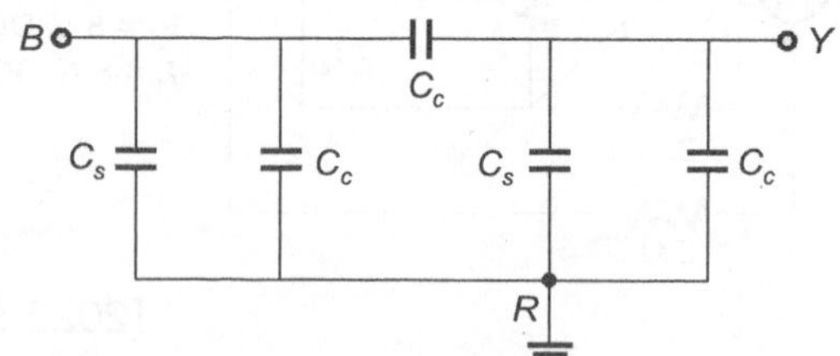

$$C_{BY} = \frac{C_s + C_c}{2} + C_c = \frac{C_s + 3C_c}{2}$$

1.7 (b)

When resistor R is connected to a current source,

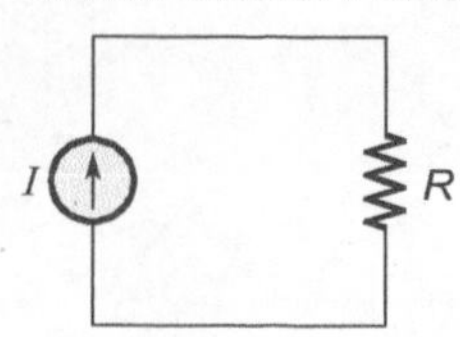

$$P = I^2 R = 18 \text{ W}$$

When resistor R is connected to a voltage source,

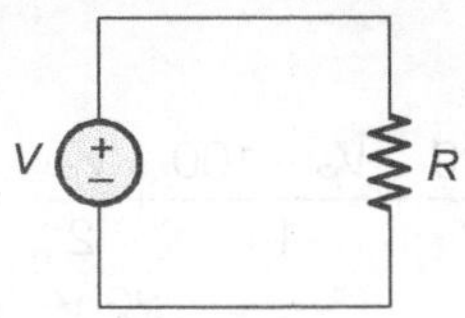

$$P = \frac{V^2}{R} = 4.5\text{ W}$$

Given, $V = I$ (in magnitude)

$$\Rightarrow \quad I^2R = 18 \quad \text{...(i)}$$

$$\frac{I^2}{R} = 4.5 \quad \text{...(ii)}$$

On solving these two equations, we get,

$$I = 3\text{ A}; \quad R = 2\,\Omega$$

1.8 (d)

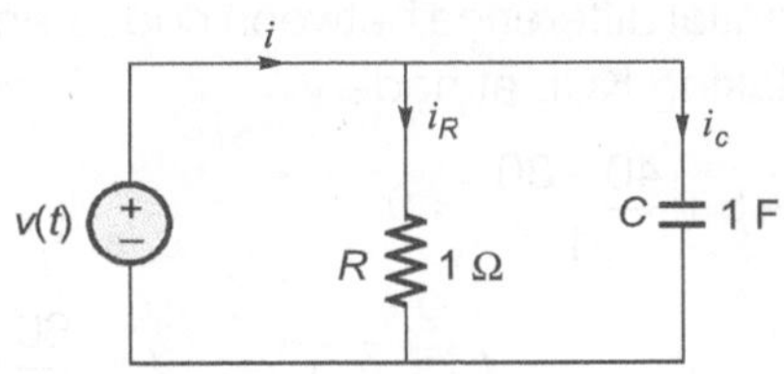

$$i = i_R + i_c = \frac{v(t)}{1} + 1\cdot\frac{dv(t)}{dt}$$

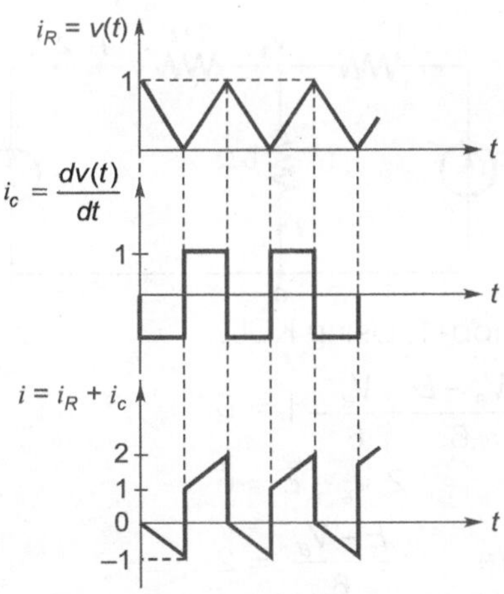

1.9 (b)

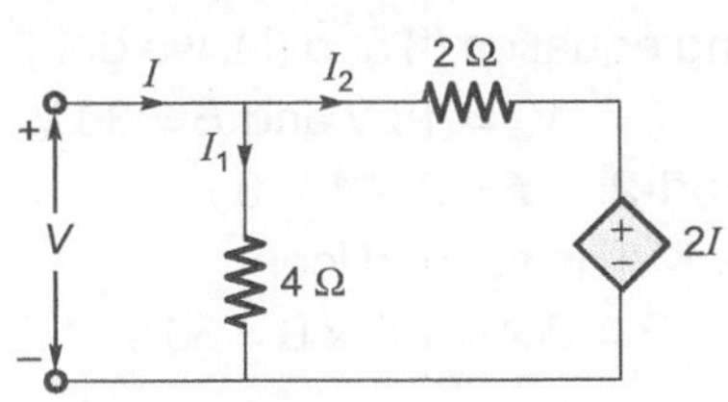

Current through 4 Ω resistor, $I_1 = \frac{V}{4}$

Current through 2 Ω resistor, $I_2 = \frac{V - 2I}{2}$

Total current, $I = I_1 + I_2 = \frac{V}{4} + \frac{V - 2I}{2}$

$$\Rightarrow \quad I = \frac{V}{4} + \frac{V}{2} - I \Rightarrow 2I = \frac{3}{4}V$$

$$\text{Load} = \frac{V}{I} = \frac{8}{3}\,\Omega$$

1.10 (d)

$$\because \quad P \propto \frac{1}{R}$$

Therefore resistance of 40 W bulb > resistance of 60 W bulb.

For series connection, current through both the bulbs will be same $P = I^2R$ (for series connection).

Power consumed by 40 W bulb > power consumed by 60 W bulb.

Hence, the 40 W bulb glows brighter.

1.11 (b)

When C is open, $R_{AB} = R_A + R_B = 6\,\Omega$

When B is open, $R_{AC} = R_A + R_C = 9\,\Omega$

When A is open, $R_{BC} = R_B + R_C = 11\,\Omega$

On solving above equations

$R_A = 2\,\Omega$, $R_B = 4\,\Omega$ and $R_C = 7\,\Omega$

1.12 (b)

By KCL,

$$I_P + I_Q + I_C + I_L = 0$$

$$2 + 1 + I_C + I_L = 0$$

But, $$I_C = C\times\frac{dv}{dt}$$

$$= 1\times\frac{d}{dt}(4\sin 2t) = (8\cos 2t)$$

$$\therefore \quad I_L = -(2 + 1 + 8\cos 2t)$$

$$= -3 - 8\cos 2t$$

$$\therefore \quad V_L = L\left(\frac{di}{dt}\right) = 2\times 2\times 8\sin 2t$$

$$= 32\sin 2t$$

Note: KCL is based on the law of conservation of charges.

1.13 (c)

For $0 < t < 2s$ current varies linearly with time and given as $i(t) = 3t$ and for $2s < t < 4s$ current is constant, $i(t) = 6$ A.

The energy absorbed by the inductor (Resistance neglected) in the first 2 sec,

$$E_L = \int_0^T Li\frac{di}{dt}dt = E_{L_1} + E_{L_2}$$

$$E_{L_1} = \int_0^2 Li\left(\frac{di}{dt}\right)dt = \int_0^2 2\times 3t\times 3\,dt$$

$$= 18\int_0^2 t\,dt = 18 \times \frac{t^2}{2}\Big|_0^2 = 18\times\left[\frac{4}{2} - 0\right] = 36 \text{ J}$$

The energy absorbed by the inductor in (2 s < t < 4 s) second,

$$E_{L_2} = \int_2^4 Li\left(\frac{di}{dt}\right) dt = \int_2^4 2\cdot 6 \cdot 0\, dt = 0 \text{ J}$$

A pure inductor does not dissipate energy but only stores it. Due to resistance, some energy is dissipated in the resistor. Therefore, total energy absorbed by the inductor is the sum of energy stored in the inductor and the energy dissipated in the resistor.

The energy dissipated by the resistance in 4 sec.

$$E_R = \int_0^T i^2 R\,dt = \int_0^2 (3t)^2 \times 1\,dt + \int_2^4 6^2 \times 1\,dt$$

$$= \int_0^2 (9t^2)\,dt + 36\int_2^4 1\,dt$$

$$= 9\times\frac{t^3}{3}\Big|_0^2 + 36t\Big|_2^4 = 9\times\left(\frac{8}{3}\right) + 36\times 2$$

$$= 24 + 72 = 96 \text{ J}$$

The total energy absorbed by the inductor in 4 sec = 96 J + 36 J = 132 J

1.14 (c)

Given: $V_R = 10$ V

By KCL,

$$\frac{V_P - 10}{2} + 2 + \frac{V_P}{8} = 0 \quad \text{...(i)}$$

$$\frac{V_Q - 10}{4} - 2 + \frac{V_Q}{6} = 0 \quad \text{...(ii)}$$

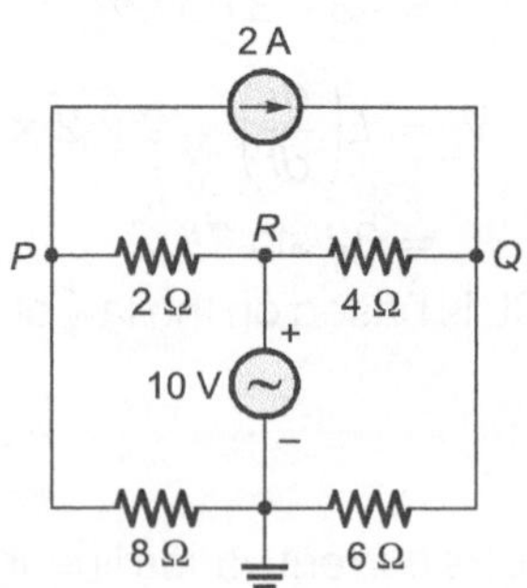

From equation (i),

$$4(V_P - 10) + 2\times 8 + V_P = 0$$

$$4V_P - 40 + 16 + V_P = 0$$

$$5V_P - 24 = 0 \Rightarrow V_P = 4.8$$

From equation (ii),

$$6(V_Q - 10) - 2\times 4\times 6 + 4V_Q = 0$$

$$10V_Q - 108 = 0$$

$$\therefore \quad V_Q = 10.8$$

$$\therefore \quad V_P - V_Q = -6 \text{ V}$$

1.15 (d)

By KCL,

$$\therefore \quad \frac{V_P - 40}{1} + \frac{V_P - 100}{14} + \frac{V_P}{2} = 0$$

$$22V_P = 660$$

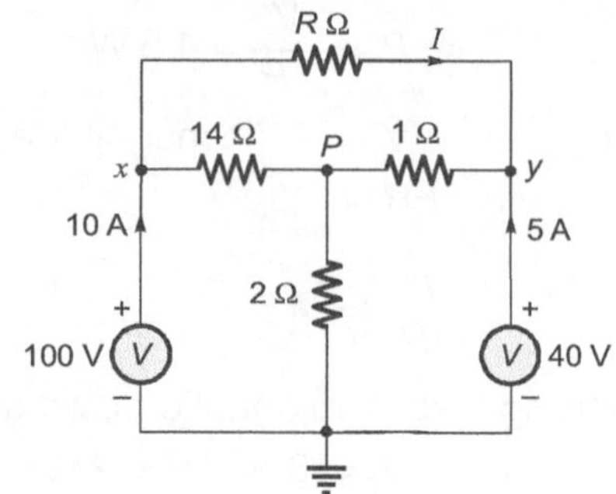

$$\therefore \quad V_P = 30 \text{ V}$$

Potential difference between node x and y = 60 V

By taking KCL at node y

$$-I - 5 + \frac{40 - 30}{1} = 0$$

$$\therefore \quad I = 5 \text{ A} \Rightarrow I = \frac{60}{5} = 12\ \Omega$$

1.16 (c)

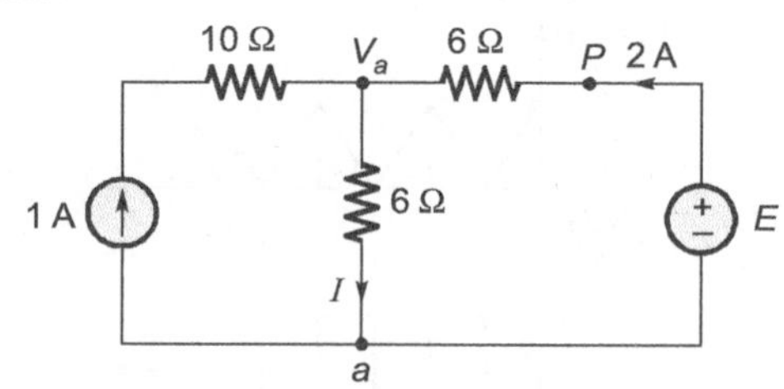

Method-1: Using KCL,

$$\frac{V_a - E}{6} + \frac{V_a}{6} - 1 = 0$$

$$\Rightarrow \quad 2V_a - E = 6 \quad \text{...(i)}$$

Where, $\frac{E - V_a}{6} = 2$

$$\Rightarrow \quad E - V_a = 12 \quad \text{...(ii)}$$

Solving equation (i) and (ii), we get

$$V_a = 18 \text{ V and } E = 30 \text{ V}$$

Method-2: $I = 2 + 1 = 3$ A

Apply KVL in second loop,

$$E = 2\times 6 + 3\times 6 = 30 \text{ V}$$

1.17 (b)

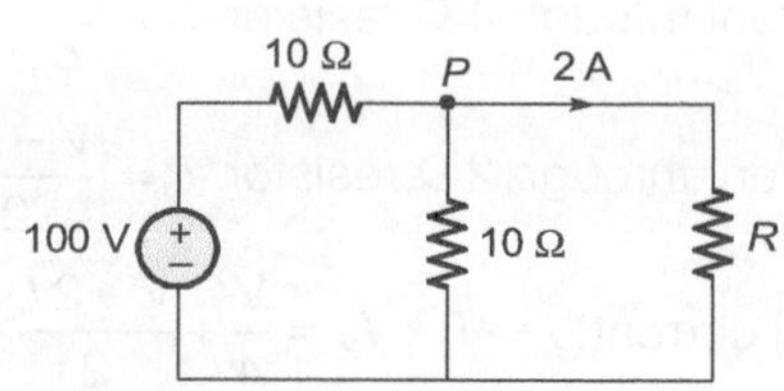

$$\frac{V_P - 100}{10} + \frac{V_P}{10} + 2 = 0$$

$$2V_P - 100 + 20 = 0$$

$$\therefore \quad V_P = \frac{80}{2} = 40 \text{ V}$$

$$\therefore \quad R = \frac{V_P}{2} = \frac{40}{2} = 20\,\Omega$$

1.18 (a)

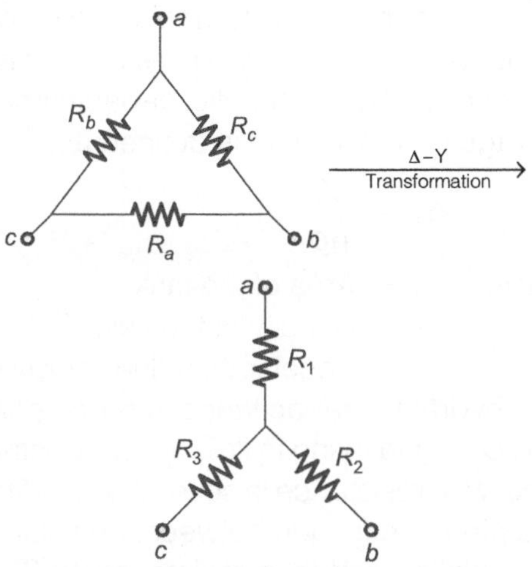

Given: $R_a = 20\,\Omega$

$R_b = 10\,\Omega$ and $R_c = 10\,\Omega$

$$R_1 = \frac{R_bR_c}{R_a + R_b + R_c} = \frac{10 \times 10}{20 + 10 + 10} = 2.5\,\Omega$$

$$R_2 = \frac{R_cR_a}{R_a + R_b + R_c} = \frac{10 \times 20}{20 + 10 + 10} = 5\,\Omega$$

$$R_3 = \frac{R_aR_b}{R_a + R_b + R_c} = \frac{20 \times 10}{20 + 10 + 10} = 5\,\Omega$$

Remember: If all the branches of Δ-connection has same impedance Z, then impedance of branch of Y-connection be $Z/3$.

1.19 (d)

R.M.S value of d.c current = 10 A = I_{dc}

R.M.S value of sinusoidal current = $\left(\frac{20}{\sqrt{2}}\right)$ A $= I_{ac}$

R.M.S value of resultant,

$$I_R = \sqrt{I_{dc}^2 + I_{ac}^2} = \sqrt{10^2 + \left(\frac{20}{\sqrt{2}}\right)^2} = 17.32 \text{ A}$$

1.20 (c)

The Resultant (R) when viewed from voltage source

$$= \frac{100}{8} = 12.5$$

$$R + 10\,||\,10 = 12.5\,\Omega$$

$$\therefore R = 12.5 - 10\,||\,10 = 12.5 - 5 = 7.5\,\Omega$$

1.21 (a)

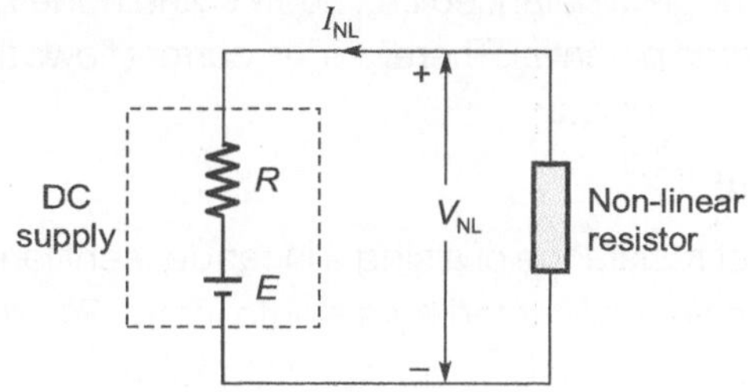

$$V_{NL} = I_{NL}^2 \quad \text{...(i)}$$

$$V_{NL} = E - I_{NL}R \text{ where, } E = 3 \text{ V}$$

and $\quad R = 2\,\Omega$

$$V_{NL} = 3 - 2I_{NL} = I_{NL}^2$$

$$I_{NL}^2 + 2I_{NL} - 3 = 0$$

$$I_{NL} = -3 \text{ A or } 1 \text{ A}$$

–3 A is rejected, because the non-linear resistor is passive and the only active element in the circuit is 3 V DC supply. Which is supplying the power to the resistor. So, $\quad I_{NL} = 1$ A

Power dissipated in the non-linear resistor

$$= V_{NL}I_{NL} = I_{NL}^2 I_{NL} = I_{NL}^3 = 1^3 = 1 \text{W}$$

1.22 (a)

$$i = 1 \text{ A}$$

Applying KVL, we get

$$V_{ab} - 2i + 5 = 0$$

$$V_{ab} = -5 + 2i = -5 + 2 \times 1 = -3 \text{ V}$$

Note: KVL is based on the conservation of energy.

1.23 (b)

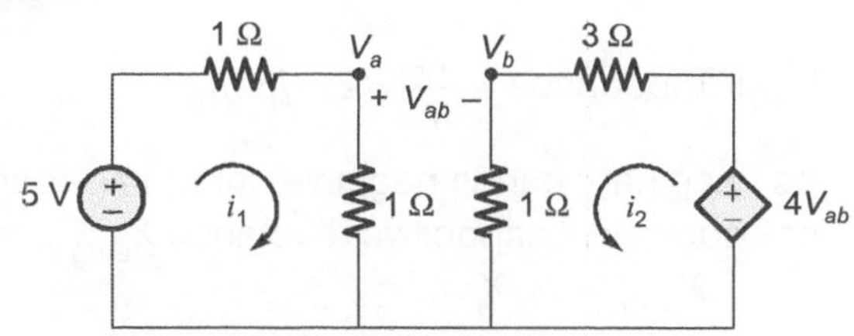

By KVL in Loop-1,

$$5 - i_1 - i_1 = 0$$

$$i_1 = \frac{5}{2} = 2.5 \text{ A}$$

$$\therefore \quad V_a = 2.5 \text{ V}$$

By KVL in Loop-2,

$$4V_{ab} = 3i_2 + i_2$$

$$i_2 = \frac{4V_{ab}}{4} = V_{ab}$$

$$\therefore \quad V_b = 1 \times i_2 = V_{ab}$$

$$V_b = V_a - V_b = \frac{V_a}{2} = \frac{2.5}{2} = 1.25 \text{ V}$$

$$i_2 = V_{ab} = V_b \Rightarrow i_2 = 1.25 \text{ A}$$

1.24 (a)

Bridge is balanced i.e., node *C* and node *D* are at same potential. Therefore, no current flows through 2 kΩ resistor.

1.25 (d)

Let resistance of a single incandescent lamp = *R*.
Power consumed by a single lamp, *P* = 200 W.
When connected across voltage, *V* = 220 V.

$$\text{So, } P = \frac{V^2}{R} \Rightarrow 200 = \frac{220^2}{R} \Rightarrow R = 242 \text{ W}$$

Let, *n* number of lamps are connected in series across voltage *V* = 200 V.
So total resistance of lamps,

$$R_{eq.} = nR = 242\,n$$

Total power consumed,

$$P = \frac{V^2}{R_{eq.}} \Rightarrow 100 = \frac{220^2}{242n} \Rightarrow n = 2$$

1.26 (a)

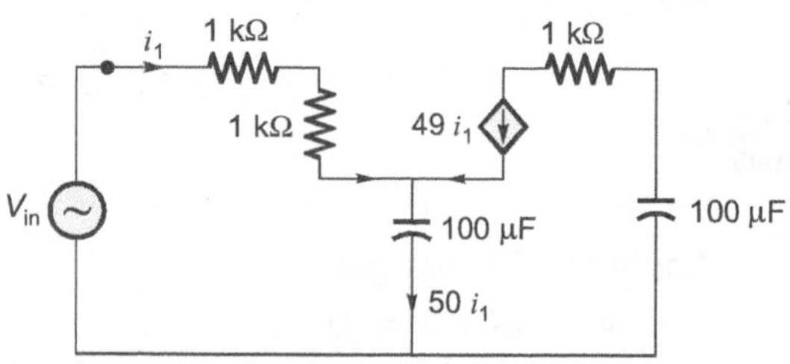

Applying KVL, we get

$$V_{in} - i_1(1 + 1) - 50\,i_1(-jX_C) = 0$$

$$\Rightarrow \quad V_{in} = i_1[2 - j50\,X_C]$$

$$\text{Input impedance} = \frac{V_{in}}{i_1} = 2 - j50X_C$$

As imaginary part is negative, input impedance has equivalent capacitive reactance $X_{C\,eq.}$

$$X_{C\,eq.} = 50\,X_C$$

$$\frac{1}{\omega C_{eq.}} = \frac{50}{\omega C} = \frac{50}{\omega \times 100} = \frac{1}{2\omega}$$

$$C_{eq.} = 2\,\mu\text{F}$$

1.27 (b)

Voltage across 2 Ω resistance = V_s = 4 V

$$\text{Current through 2 Ω resistance} = \frac{V_s}{R} = \frac{4}{2} = 2 \text{ A}$$

Current source has no effect, when connected across voltage source.
So, to double current though 2 Ω resistance, voltage source is doubled, i.e.,

$$V_s = 8 \text{ V}$$

1.28 (b)

A resistor has linear characteristics
i.e., $V = Ri \Rightarrow V = i$
Load line, $V + i = 100$
$i + i = 100$

$$\text{Current through resistance, } i = \frac{100}{2} = 50 \text{ A}$$

1.29 (d)

In such system, volumetric flow rate *C* is analogous to current and pressure is analogous to voltage. The hydraulic capacitance due to storage in gravity field is defined as,

$$C = \frac{A}{\rho g}$$

where, *A* = Area of the tank
ρ = Density of the fluid
g = Acceleration due to gravity

The hydraulic capacitance is represented by *A* and *C*. Liquid trying to flow out of a container, can meet with resistance in several ways. If the outlet is a pipe, the friction between the liquid and the pipe walls produces resistance to flow. Such resistance is represented by *B* and *D*.

1.30 (b)

Assuming voltage of the node V_a

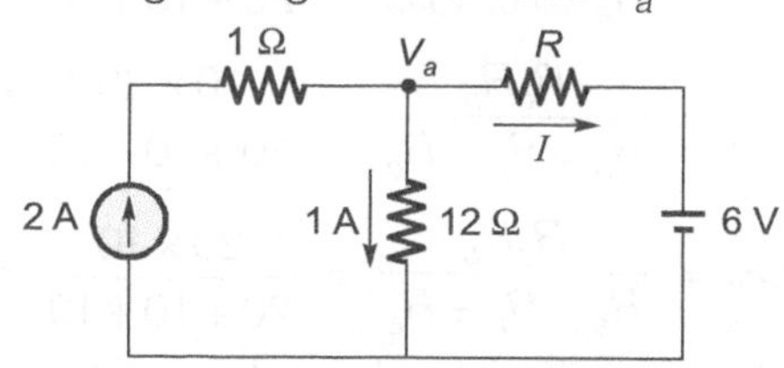

$$V_a = 1 \times 12 = 12 \text{ V}$$

Applying KCL, we get

$$-2 + 1 + I = 0 \Rightarrow I = 1 \text{ A}$$

$$I = \frac{V_a - 6}{R} = \frac{12 - 6}{R} = \frac{6}{R}$$

$$\Rightarrow \quad I = \frac{6}{R} \Rightarrow R = 6\,\Omega$$

1.31 (a)

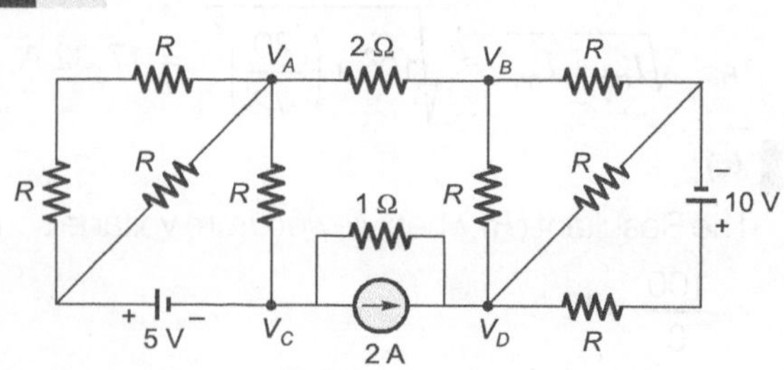

⇓

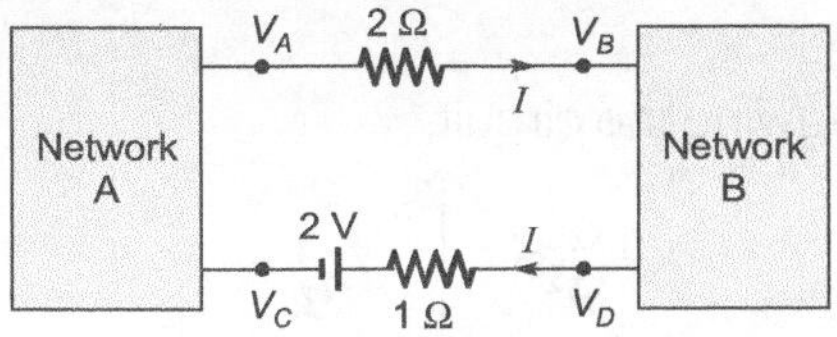

$$V_A - V_B = 2I \Rightarrow 2I = 6 = 3\text{ A}$$
$$V_C + 2 + 1 \times I = V_D$$
$$V_C - V_D = -2 - 3 = -5\text{ V}$$

1.32 (c)

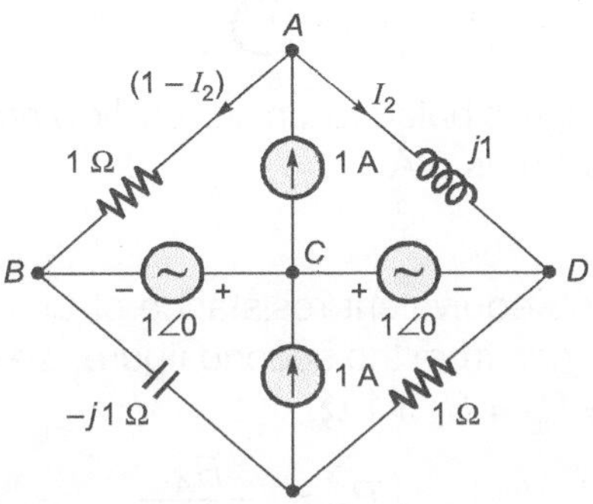

Apply KCL node at 'A',
so, current flowing through 1 Ω is $(1 - I_2)$
Applying KVL in $ABCD$ loop,

$$1\angle 0 - 1\angle 0 + 1(1 - I_2) - jI_2 = 0$$

$$I_2 = \frac{1}{1+j}$$

1.33 (c)

$$Q = CV$$
$$Q_1 = C_1V_1 = 10 \times 10^{-6} \times 10 = 100\ \mu\text{C}$$
$$Q_2 = C_2V_2 = 5 \times 10^{-6} \times 5 = 25\ \mu\text{C}$$
$$Q_3 = C_3V_3 = 2 \times 10^{-6} \times 2 = 4\ \mu\text{C}$$

Capacitors C_2 and C_3 are in series.
In series charge is same.
So, the maximum charge on C_2 and C_3 will be minimum of (Q_2, Q_3) = min(25 μC, 4 μC) = 4 μC = Q_{23}.
In series the equivalent capacitance of C_2 and C_3 is

$$C_{23} = \frac{C_2C_3}{C_2 + C_3} = \frac{5 \times 2}{5 + 2} = \frac{10}{7}\mu\text{F}$$

So, the equivalent voltage,

$$V_{23} = \frac{Q_{23}}{C_{23}} = \frac{4 \times 10^{-6}}{\frac{10}{7} \times 10^{-6}} = \frac{28}{10} = 2.8\text{ V}$$

In parallel, the voltage is same.

$$V_1 = V_{23} = 2.8\text{ V}$$

Change in capacitor C_1

$$Q_1 = C_1V_1 = 10 \times 10^{-6} \times 2.8 = 28\ \mu\text{C}$$

In parallel, the total charge

$$Q = Q_1 + Q_{23} = 4 + 28 = 32\ \mu\text{C}$$

1.34 (b)

$$R_A = \frac{R_bR_c}{R_a + R_b + R_c}$$
$$R'_a = kR_a;\ R'_b = kR_b;\ R'_c = kR_c$$
$$R'_A = \frac{kR_b \cdot kR_c}{kR_a + kR_b + kR_c} = \frac{k^2R_bR_c}{k(R_a + R_b + R_c)}$$
$$= k \times \frac{R_bR_c}{R_a + R_b + R_c} = kR_A$$

1.35 Sol.

Given electrical circuit is shown below:

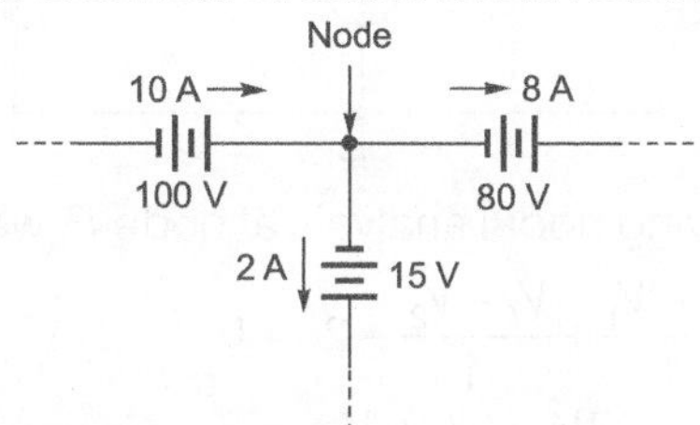

Applying KCL at node, current through 15 V voltage source = 2 A.
∴ Power absorbed by 100 V voltage source = 10 × 100 = 1000 Watt.
Power absorbed by 80 V voltage source = −(80 × 8) = −640 Watts and power absorbed by 15 V voltage source = −(15 × 2) = −30 Watt.
∴ Total power absorbed by the three circuit elements = (100 − 640 − 30) Watts = 330 Watts

1.36 Sol.

Let the resistance of incandescent lamp

$$= R_T = \frac{V^2}{P} = \frac{(240)^2}{40} = 1440\ \Omega$$

Given, $R_0 = 120\ \Omega$, $\alpha = 4.5 \times 10^{-3}/°\text{C}$
Let, R_T be the resistance of the filament in ON state at temperature T.

Then, $$R_T = R_0[1 + \alpha\Delta T]$$

or, $$[1 + \alpha\Delta T] = \frac{R_T}{R_0} = \frac{1440}{120} = 12$$

or, $$\alpha\Delta T = 11 \text{ or } \Delta T = 2444.44°\text{C}$$

or, $$T = 2444.44° + 26° = 2470.44°\text{C}$$

∴ ON state temperature of filament = 2470.44°C

1.37 Sol.

Given, $$R = \left(25 + \frac{I}{2}\right)\Omega \text{ or } I = (2R - 50)$$

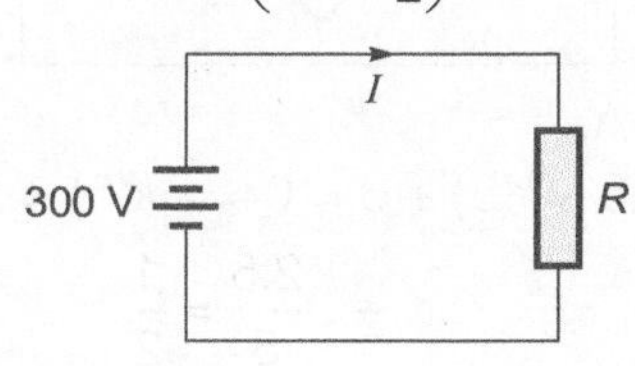

Applying KVL in given loop, we have:

$$300 = IR \text{ or } 300 = (2R - 50) \times R$$

we get, $R = 30\ \Omega$ or $-5\ \Omega$

Since resistance can't be negative. Therefore,

$$R = 30\ \Omega$$

Hence, $I = (2R - 50) = (2 \times 30 - 50)\text{A} = 10\text{ A}$

1.38 Sol.

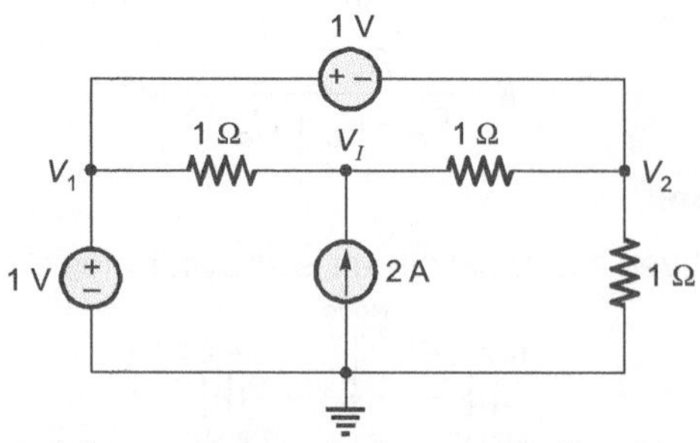

Applying nodal analysis at node P, we have:

$$\frac{V_I - V_1}{1} + \frac{V_I - V_2}{1} - 2 = 0$$

or, $$2V_I - (V_1 + V_2) = 2$$

or, $$V_I = \left[\frac{2 + (V_1 + V_2)}{2}\right] \quad ...(i)$$

Also, $V_1 - V_2 = 1$ volt and $V_1 = 1$ volt

$\therefore$ $V_2 = V_1 - 1 = 1 - 1 = 0$ volt

Putting values of V_1 and V_2 in equation (i), we get:

$$V_I = \left[\frac{2 + (1 + 0)}{2}\right] = \frac{3}{2}\text{ volt}$$

$\therefore$ Power delivered by the current source

$$= V_I \cdot I = \frac{3}{2} \times 2 = 3\text{ Watt} \qquad [\because I = 2\text{ A (given)}]$$

1.39 (a)

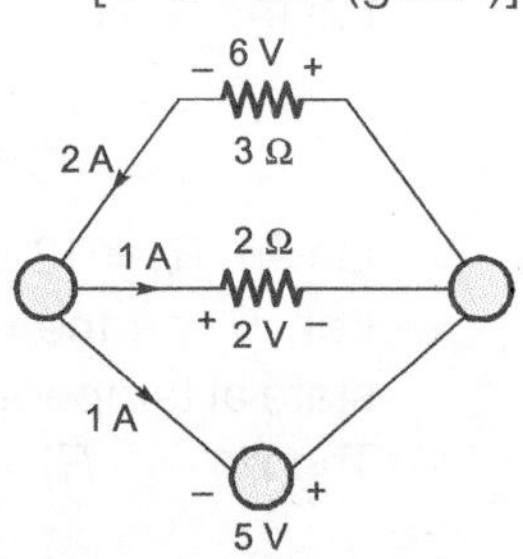

Power $= 5 \times 1$

$= 5$ Watt

1.40 Sol.

$$\Rightarrow \quad \frac{V_0^2}{2} = 12.5$$

$$\Rightarrow \quad V_0^2 = 12.5 \times 2 \Rightarrow V_0 = 5$$

$$I_0 = \frac{5}{2} = 2.5\text{ A}$$

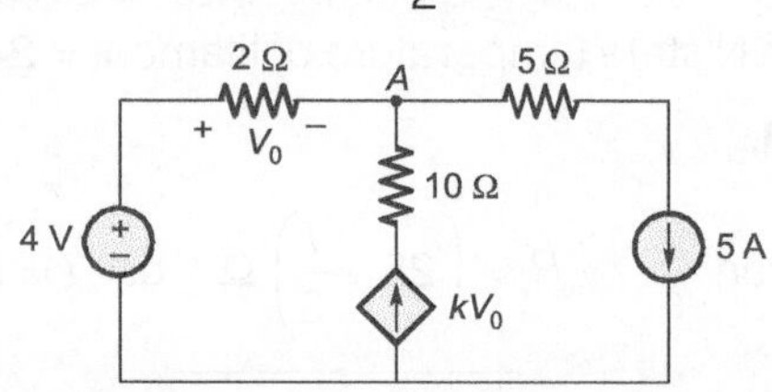

KCL at A,

$\Rightarrow$ $-2.5 - k(5) + 5 = 0 \Rightarrow k(5) = 2.5$

$$\Rightarrow \quad k = \frac{2.5}{5} = \frac{1}{2}$$

1.41 Sol.

Redrawing the circuit,

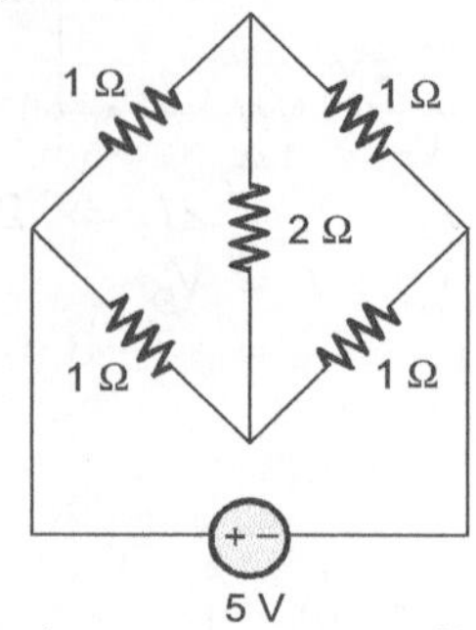

Bridge is balance, so current flowing through 2 Ω resistor is 0 A.

1.42 (d)

If the equivalent resistance of the first figure is R_A then from the second figure, we can see that the $R_B = R_A \parallel 1\ \Omega$.

$$R_B = \frac{R_A}{R_A + 1}$$

1.43 Sol.

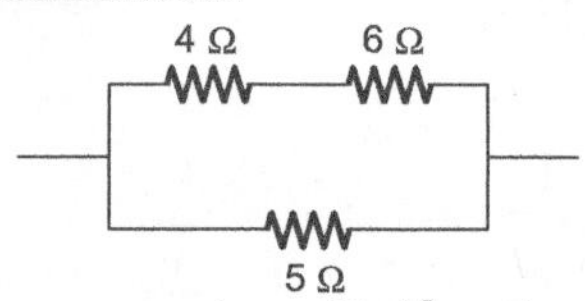

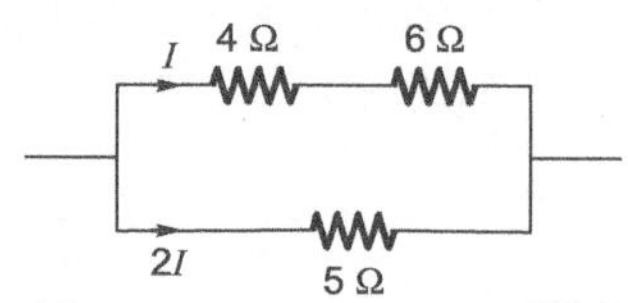

and $\quad (2I)^2 \times 5 = 10$

$$\Rightarrow \quad I^2 = \frac{10}{5 \times 4} = \frac{2.5}{5} = 0.5$$

So, $\quad I^2 \times 4 = 0.5 \times 4 = 2$ cal/sec.

1.44 Sol.

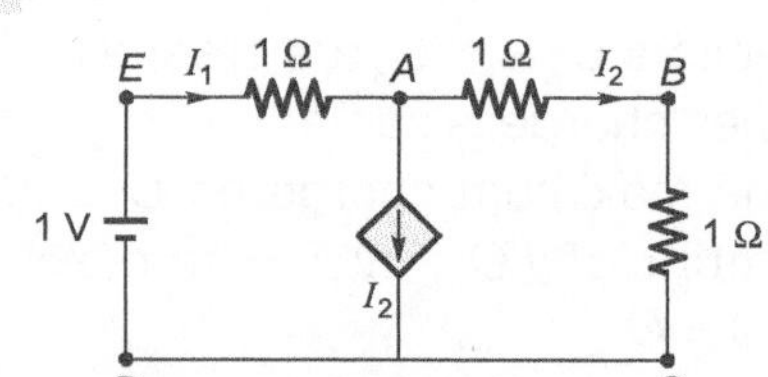

Applying KCL at node A,

$$-I_1 + I_2 + I_2 = 0 \Rightarrow 2I_2 = I_1 \quad ...(i)$$

and applying KVL in loop ABCD,

$$1 - I_1 - I_2 - I_2 = 0 \Rightarrow I_1 + 2I_2 = 1 \quad ...(ii)$$

From equation (i) and (ii),

$$\Rightarrow \quad 2I_2 + 2I_2 = 1 \Rightarrow 4I_2 = 1$$

$$\Rightarrow \quad I_2 = \frac{1}{4}\text{ A and } I_1 = 2 \times \frac{1}{4} = \frac{1}{2}\text{ A}$$

1.45 Sol.

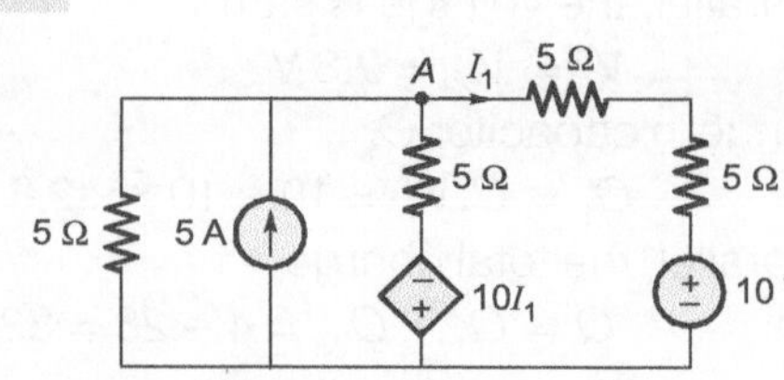

Applying KCL at node A, we get

$$\frac{V_A}{5}+\frac{V_A-10}{10}+\frac{V_A+10I_1}{5} = 5$$

So, $2V_A + V_A - 10 + 2V_A + 20I_1 = 5$

$$5V_A + 20I_1 = 60$$

Since, $$I_1 = \frac{V_A-10}{10}$$

So, $5V_A + 2V_A - 20 = 60 \Rightarrow 7V_A = 80$

$$V_A = \frac{80}{7} = 11.42 \text{ volt}$$

1.46 (d)

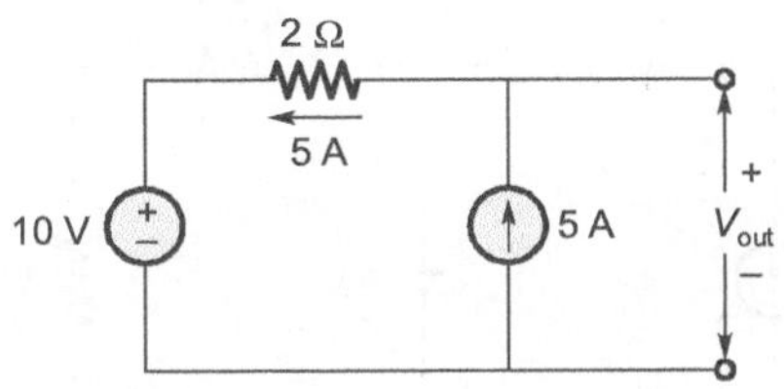

So, $$V_{out} = (5 \times 2) + 10 = 20 \text{ V}$$

1.47 Sol.

Consider the following circuit diagram,

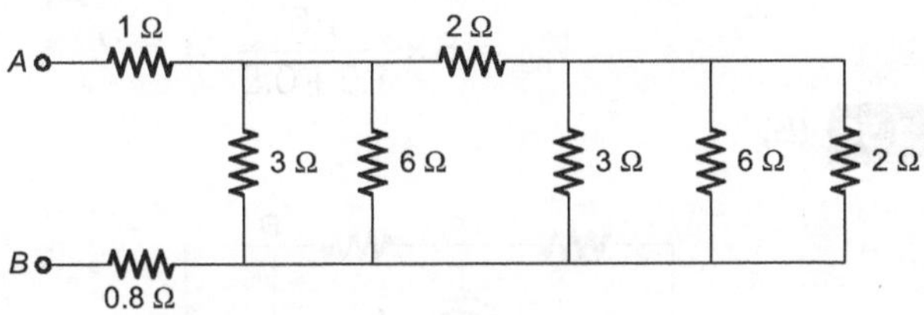

After rearrangement we get,

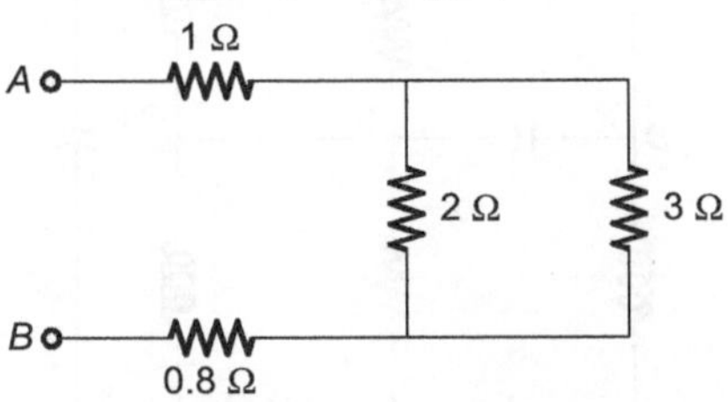

Now, $$R_{AB} = 1+\frac{6}{5}+0.8 = 3\ \Omega$$

1.48 Sol.

Using KCL at node, we get

$$I + 0.4I = 14 \text{ or } I = 10 \text{ A}$$

Now, power supplied, $P = 25 \times 10 = 250$ W

1.49 (a)

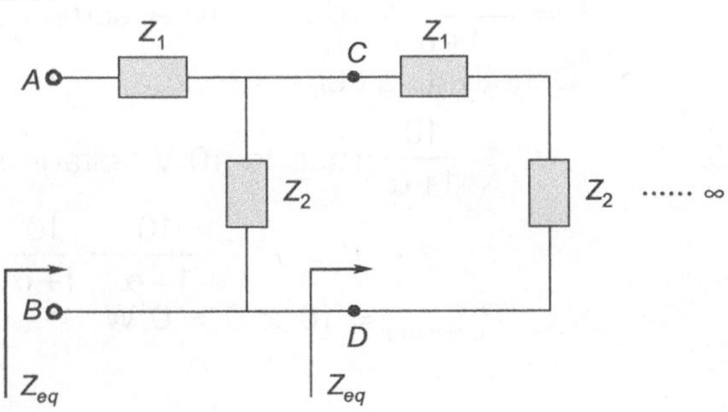

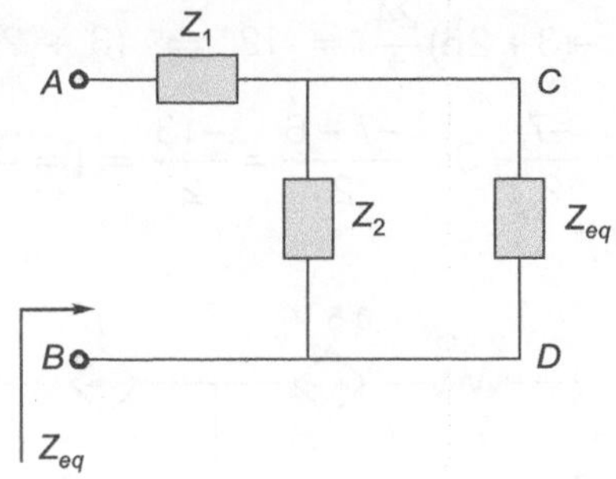

$$Z_1 = j9;\quad Z_2 = j5 - j1 = j4$$

$$Z_{eq} = Z_1+\frac{Z_2Z_{eq}}{Z_2+Z_{eq}}$$

By solving above equation, $Z_{eq} = j12$

1.50 Sol.

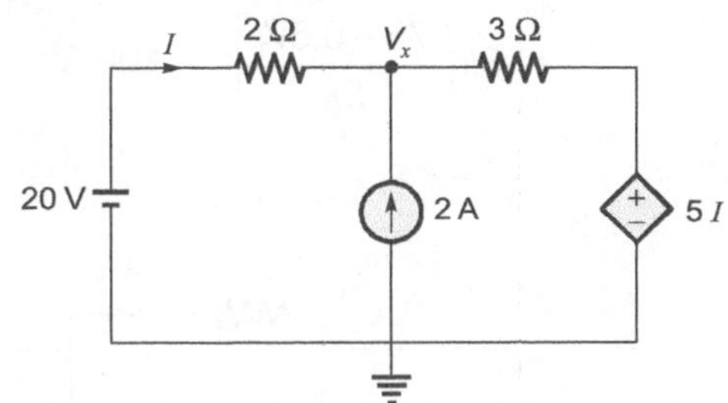

Applying nodal at node x,

$$-I-2+\frac{V_x-5I}{3} = 0$$

$$-3I - 6 + V_x - 5I = 0 \Rightarrow 8I = V_x - 6 \quad \text{...(i)}$$

As, $$I = \frac{20-V_x}{2} \Rightarrow V_x = 20 - 2I \quad \text{...(ii)}$$

Substituting (ii) by (i),

$8I = 20 - 2I - 6 \Rightarrow 10I = 14 \Rightarrow I = 1.4$ A

1.51 Sol.

$$I = 1\angle 10° + 1\angle 70° = 1.732\angle 40°$$

The reading of ammeter is 1.732 A.

1.52 (-3.25)

By KCL at node x,

$$\frac{V_x-6}{1}+\frac{V_x}{4}+\frac{V_x-V_y}{2} = 0$$

$$4V_x - 24 + V_x + 2V_x - 2V_y = 0$$

$$7V_x - 2V_y = 24 \quad \text{...(i)}$$

If $V_y = 0$

$$\Rightarrow \quad 7V_x = 24 \Rightarrow V_x = \frac{24}{7} \text{ V}$$

By KCL at node y,

$$\frac{V_y-V_x}{2}+\frac{V_y-\beta V_x}{3} = 2$$

$$3V_y - 3V_x + 2V_y - 2\beta V_x = 12$$

$$5V_y - (3+2\beta)V_x = 12 \quad \text{...(ii)}$$

$$V_y = 0$$

$$-(3+2\beta)\frac{24}{7} = 12 \Rightarrow (3+2\beta) = \frac{-7}{2}$$

$$2\beta = \frac{-7}{2} - 3 = \frac{-7-6}{2} = \frac{-13}{2} \Rightarrow \beta = \frac{-13}{4} = -3.25$$

1.53 (100)

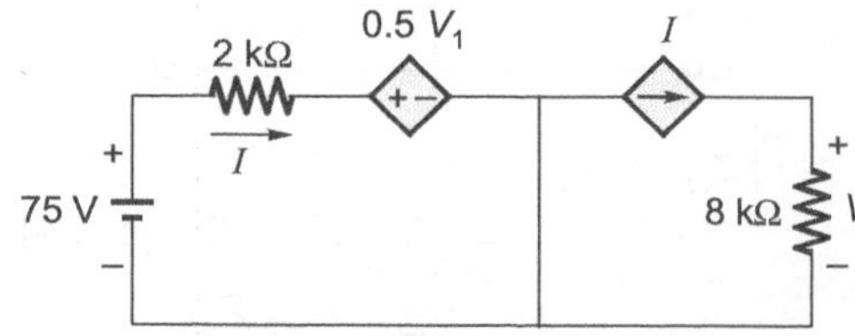

The voltage across 8 kΩ is : $V_1 = 8\ kI$

Write kVL equation in first loop.

$$75 = 2kI + 0.5V_1 \Rightarrow I = \frac{75 - 0.5V_1}{2k}$$

$$V_1 = 8k\left(\frac{75-0.5V_1}{2k}\right) = 100 \text{ V}$$

1.54 (8)

For $t = 0^-$

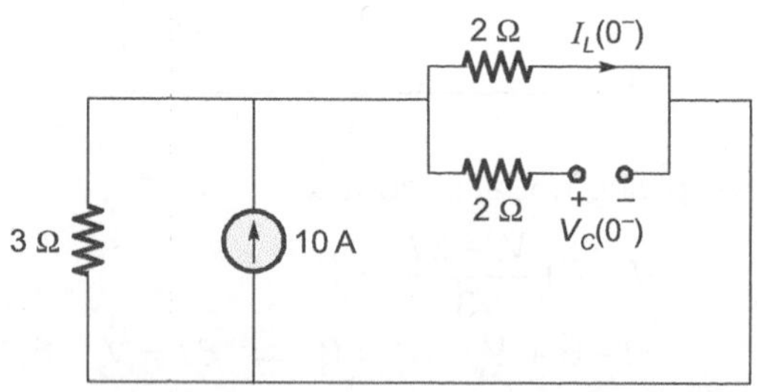

$$I_L(0^-) = 10 \times \frac{3}{3+2} = 6 \text{ A}$$

$$V_C(0^-) = 6 \times 2 = 12 \text{ V}$$

For $t = 0^+$

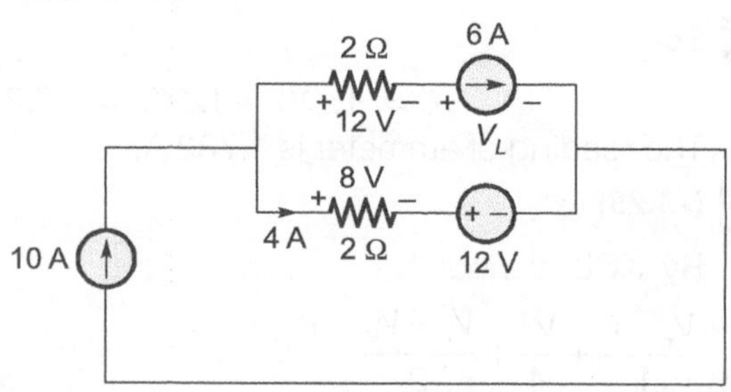

$$20 = 12 + V_L \Rightarrow V_L = 8 \text{ V}$$

1.55 (6)

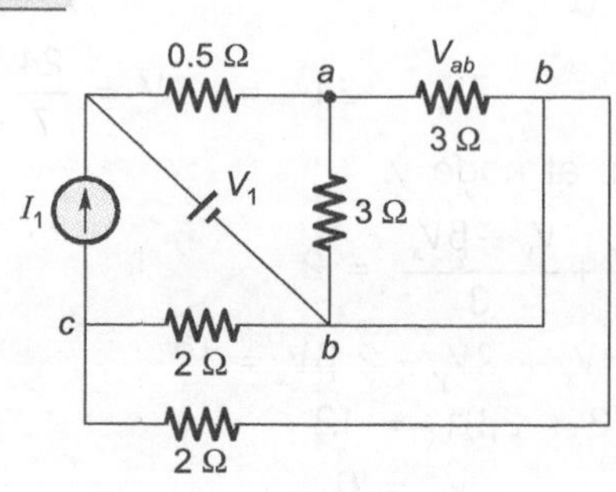

⇒

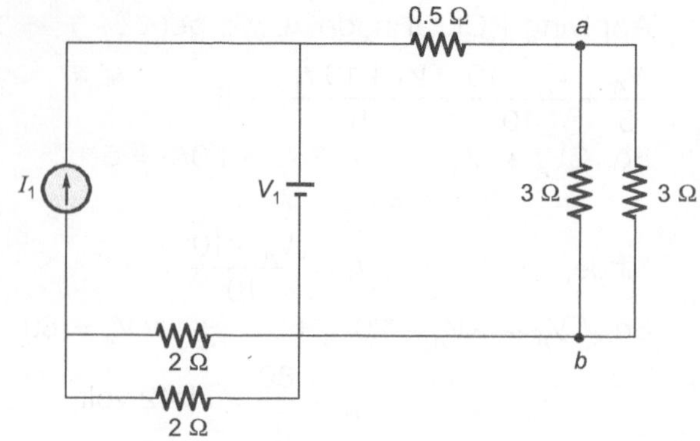

Circuit can be redrawn

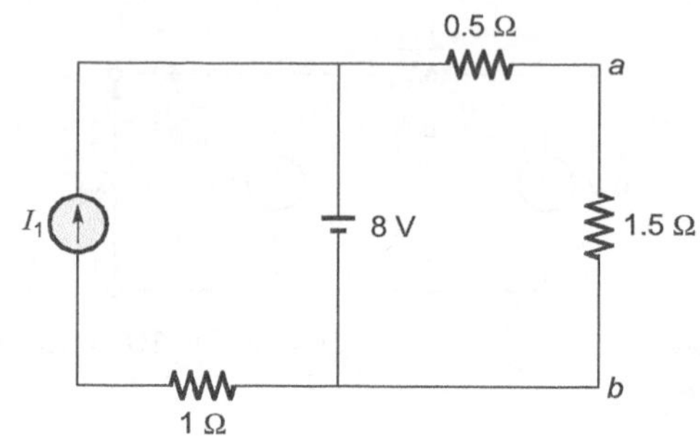

By applying voltage division rule, we can get

$$V_{ab} = 8 \times \frac{1.5}{1.5+0.5} = 6 \text{ V}$$

1.56 (b)

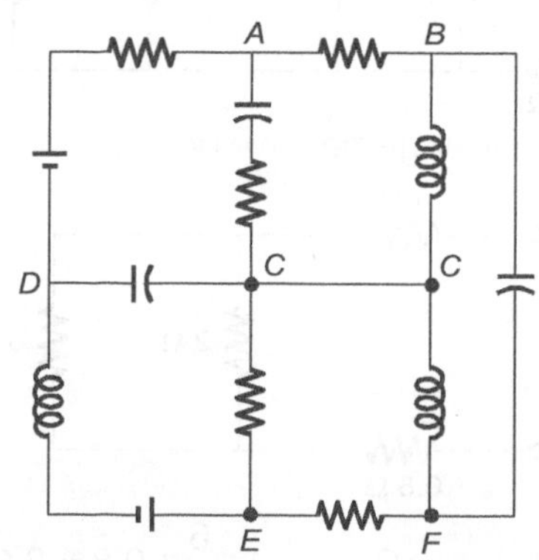

A point at which more than two elements are joints together is called Junction.

A, B, C, D, E and F are junction.

1.57 (b)

- Deactivating voltage source,

$$I' = \frac{-(10\times 1)}{1+\alpha} \text{ (due to 10 A current source)}$$

- Deactivating current source,

$$I'' = \frac{10}{1+\alpha} \text{ (due to 10 V voltage source)}$$

$$I = I' + I'' = \frac{10}{1+\alpha} - \frac{10}{1+\alpha} = 0 \text{ A}$$

$$P_{\text{delivered}} = 10 \times 0 = 0 \text{ W}$$

2 Steady State AC Analysis

2.1 In the circuit shown in figure, ammeter A_2 reads 12 A and A_3 reads 9 A. A_1 will read _____ A.

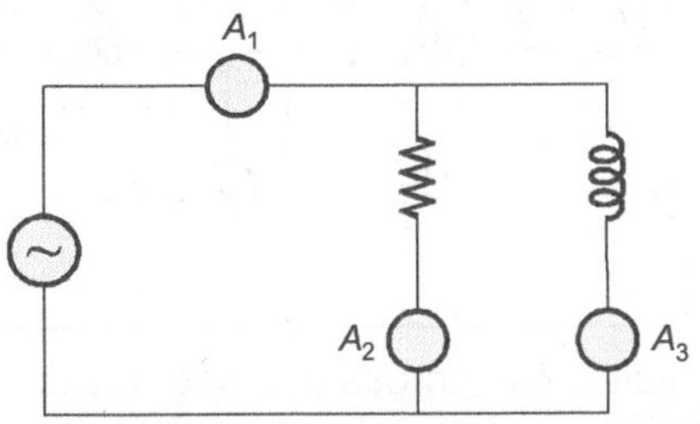

[1995 : 1 M]

2.2 Energy stored in the capacitor over a cycle, when excited by an AC source is
(a) the same as that due to a DC source of equivalent magnitude.
(b) half of that due to a DC source of equivalent magnitude.
(c) zero.
(d) None of the above [1997 : 1 M]

2.3 The rms value of a half wave rectified symmetrical square wave current of 2 A is
(a) $\sqrt{2}$ A (b) 1 A
(c) $\frac{1}{\sqrt{2}\text{ A}}$ (d) $\sqrt{3}$ A [1999 : 1 M]

2.4 Current I_1, I_2 and I_3 meet at a junction (node) in a circuit. All currents are marked as entering the node. If $I_1 = -6\sin(\omega t)$ mA and $I_2 = 8\cos(\omega t)$ mA, then I_3 will be
(a) $10\cos(\omega t + 36.87°)$ mA
(b) $14\cos(\omega t + 36.87°)$ mA
(c) $-14\sin(\omega t + 36.87°)$ mA
(d) $-10\cos(\omega t + 36.87°)$ mA [1999 : 2 M]

2.5 In the figure, the admittance values of the elements in Siemens are $Y_R = 0.5 + j0$, $Y_L = 0 - j1.5$, $Y_C = 0 + j0.3$ respectively. The value of I as a phasor when the voltage, E across the elements is $10\angle 0°$ V is

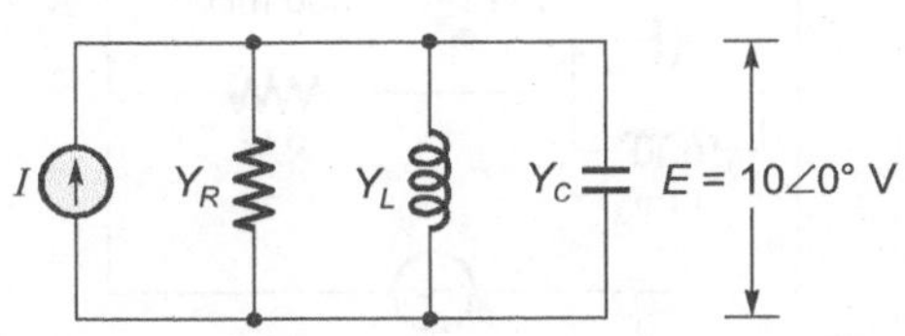

(a) $1.5 + j0.5$ (b) $5 - j18$
(c) $0.5 + j1.8$ (d) $5 - j12$ [2004 : 2 M]

2.6 The RMS value of the voltage $u(t) = 3 + 4\cos(3t)$ is
(a) $\sqrt{17}$ V (b) 5 V
(c) 7 V (d) $(3 + 2\sqrt{2})$ V
[2005 : 1 M]

2.7 The RL circuit of the figure is fed from a constant magnitude, variable frequency sinusoidal voltage source V_{IN}. At 100 Hz, the R and L elements each have a voltage drop u_{RMS}. If the frequency of the source is changed to 50 Hz, then new voltage drop across R is

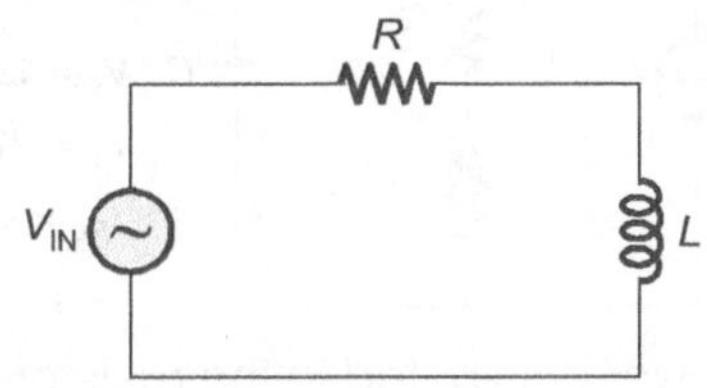

(a) $\sqrt{\frac{5}{8}}\,u_{rms}$ (b) $\sqrt{\frac{2}{3}}\,u_{rms}$
(c) $\sqrt{\frac{8}{5}}\,u_{rms}$ (d) $\sqrt{\frac{3}{2}}\,u_{rms}$ [2005 : 2 M]

2.8 An energy meter connected to an immersion heater (resistive) operating on an AC 230 V, 50 Hz, AC single phase source reads 2.3 units (kWh) in 1 hour. The heater is removed from the supply and now connected to a 400 V peak to peak square wave source of 150 Hz. The power in kW dissipated by the heater will be
(a) 3.478 (b) 1.739
(c) 1.540 (d) 0.870 [2006 : 2 M]

2.9 The rms value of the current $i(t)$ in the circuit shown below is

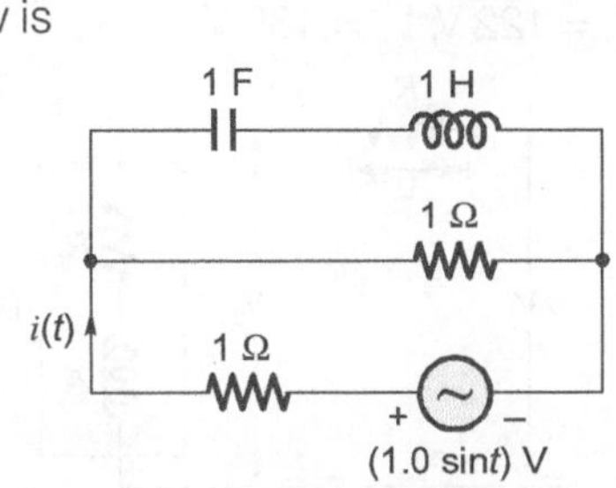

(a) $\frac{1}{2}$ A (b) $\frac{1}{\sqrt{2}}$ A
(c) 1 A (d) $\sqrt{2}$ A [2011 : 1 M]

2.10 The voltage applied to a circuit is $100\sqrt{2}\cos(100\pi t)$ volts and the circuit draws a current of $10\sqrt{2}\sin(100\pi t + \pi/4)$ Amperes. Taking the voltage as the reference phasor, the phasor representation of the current in Amperes is

(a) $10\sqrt{2}\angle -\frac{\pi}{4}$ (b) $10\angle -\frac{\pi}{4}$

(c) $10\angle +\frac{\pi}{4}$ (d) $10\sqrt{2}\angle +\frac{\pi}{4}$

[2011 : 1 M]

Common Data for Questions (2.11 and 2.12):

An RLC circuit with relevant data is given below.

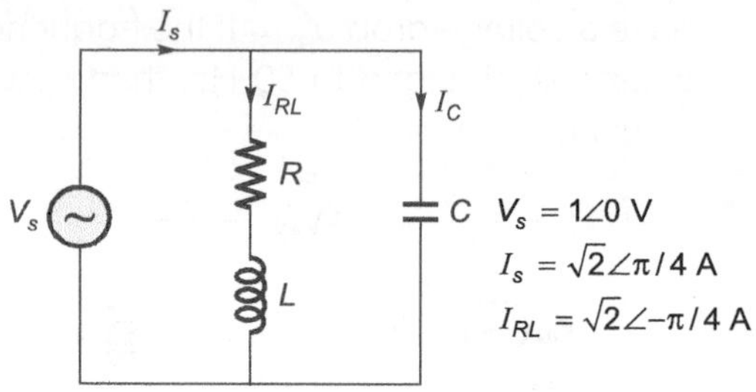

2.11 The power dissipated in the resistor R is

(a) 0.5 W (b) 1 W

(c) $\sqrt{2}$ W (d) 2 W [2011 : 2 M]

2.12 The current I_C in the figure above is

(a) $-j2$ A (b) $-j\frac{1}{\sqrt{2}}$ A

(c) $+j\frac{1}{\sqrt{2}}$ A (d) $+j2$ A [2011 : 2 M]

2.13 The average power delivered to an impedance $(4 - j3)\ \Omega$ by a current $5\cos(100\pi t + 100)$ A is

(a) 44.2 W (b) 50 W

(c) 62.5 W (d) 125 W [2012 : 1 M]

Statement for Linked Answer Questions (2.14 and 2.15):

In the circuit shown, the three voltmeter readings are $V_1 = 220$ V, $V_2 = 122$ V, $V_3 = 136$ V.

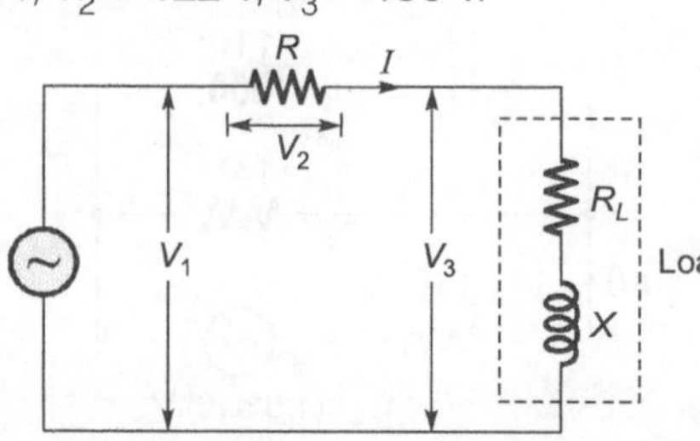

2.14 The power factor of the load is

(a) 0.45 (b) 0.50

(c) 0.55 (d) 0.60 [2012 : 2 M]

2.15 If $R_L = 5\ \Omega$, the approximate power consumption in the load is

(a) 700 W (b) 750 W

(c) 800 W (d) 850 W [2012 : 2 M]

2.16 The total power dissipated in the circuit, shown in the figure, is 1 kW.

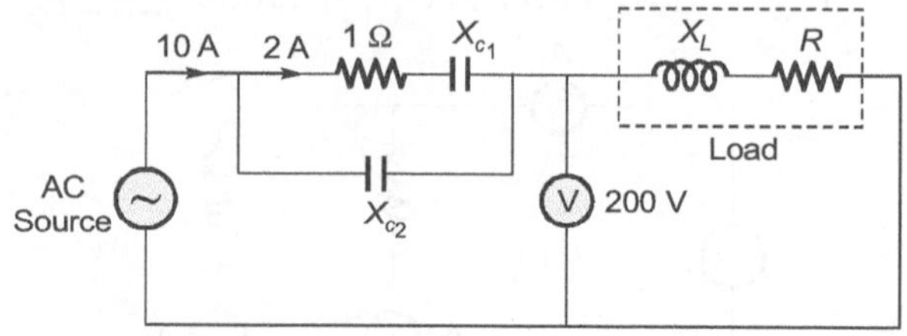

The voltmeter, across the load, reads 200 V. The value of X_L is ______ . [2014 : 2 M, Set-2]

2.17 The voltage (V) and current (A) across a load are as follows : $V(t) = 100\sin(\omega t)$

$i(t) = 10\sin(\omega t - 60°) + 2\sin(3\omega t) + 5\sin(5\omega t)$.

The average power consumed by the load, in W, is ______. [2016 : 1 M, Set-2]

2.18 A resistance and a coil are connected in series and supplied from a single phase, 100 V, 50 Hz AC source as shown in the figure below. The rms values of possible voltages across the resistance (V_R) and coil(V_C) respectively, in volts, are

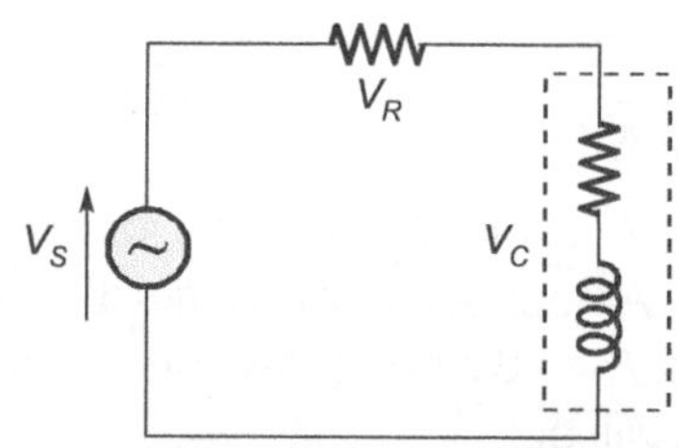

(a) 65, 35 (b) 50, 50

(c) 60, 90 (d) 60, 80 [2016 : 1 M]

2.19 In the circuit shown below, the supply voltage is $10\sin(1000t)$ volts. The peak value of steady state current through the 1 Ω resistor, in Amp, is ____.

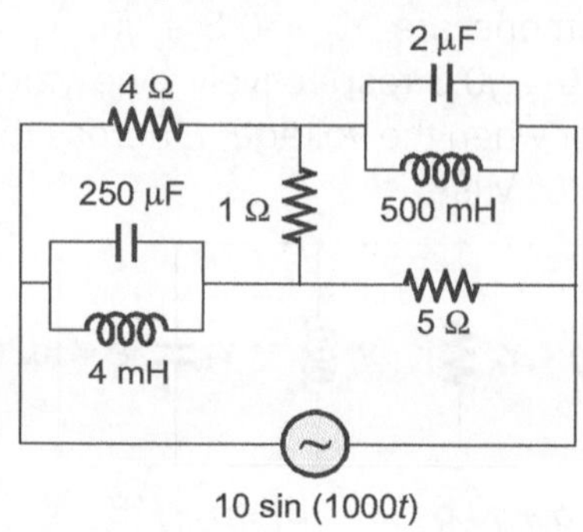

[2016 : 2 M, Set-1]

2.20 In the figure, the voltages are $v_1(t) = 100\cos(\omega t)$, $v_2(t) = 100\cos\left(\omega t + \frac{\pi}{18}\right)$ and $v_3(t) = 100\cos\left(\omega t + \frac{\pi}{36}\right)$. The circuit is in sinusoidal steady-state, and $R << \omega L$. P_1, P_2 and P_3 are the average power outputs. Which one of the following statements is true?

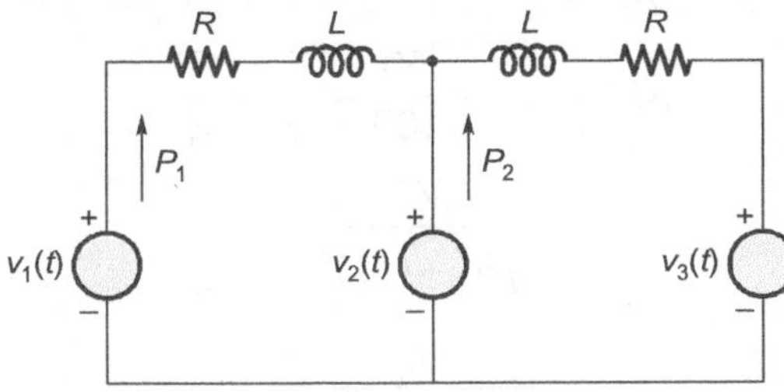

(a) $P_1 = P_2 = P_3 = 0$
(b) $P_1 < 0,\ P_2 > 0,\ P_3 > 0$
(c) $P_1 < 0,\ P_2 > 0,\ P_3 < 0$
(d) $P_1 > 0,\ P_2 < 0,\ P_3 > 0$ [2018 : 1 M]

2.21 The voltage across the circuit in the figure and the current through it, are given by the following expressions:

$$v(t) = 5 - 10\cos(\omega t + 60°)\text{ V}$$
$$i(t) = 5 + X\cos(\omega t)\text{ A}$$

where, $\omega = 100\pi$ radian/s. If the average power delivered to the circuit is zero, then the value of X(in Ampere) is _____ (upto 2 decimal places).

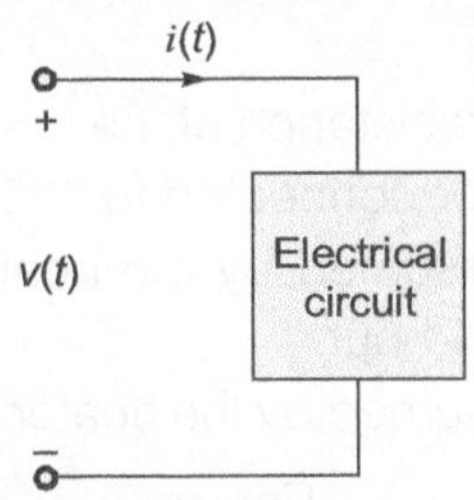

[2018 : 2 M]

2.22 A 0.1 μF capacitor charged to 100 V is discharged through a 1 kΩ resistor. The time in ms (round off to two decimal) required for the voltage across the capacitor to drop to 1 V is____. [2019 : 2 M]

2.23 A signal generator having a source resistance of 50 Ω is set to generate a 1 kHz sinewave. Open circuit terminal voltage is 10 V peak-to-peak. Connecting a capacitor across the terminals reduces the voltage to 8 V peak-to-peak. The value of this capacitor is ______ μF. (Round off to 2 decimal places). [2021 : 1 M]

2.24 In the given circuit, the value of capacitor C that makes current $I = 0$ is ______ μF.

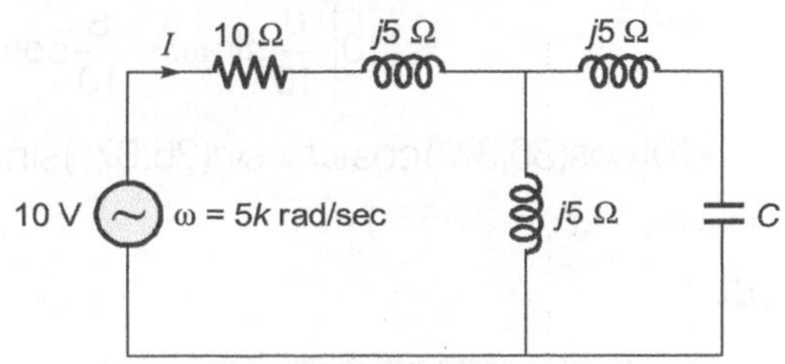

[2021 : 1 M]

Answers Steady State AC Analysis

2.1	(15)	2.2	(c)	2.3	(a)	2.4	(d)	2.5	(d)	2.6	(a)	2.7	(c)
2.8	(b)	2.9	(b)	2.10	(b)	2.11	(b)	2.12	(d)	2.13	(b)	2.14	(a)
2.15	(b)	2.16	(17.332)	2.17	(250)	2.18	(*)	2.19	(1)	2.20	(c)	2.21	(10)
2.22	(0.46)	2.23	(2.38)	2.24	(20)								

Explanations Steady State AC Analysis

2.1 Sol.

∵ Current in resistor and inductor will be in quadrature for same voltage across them.

$$\therefore \quad I_{A_1} = \sqrt{I_{A_2}^2 + I_{A_3}^2} = \sqrt{12^2 + 9^2}$$
$$= 15\text{ A}$$

2.2 (c)

When excited by an AC source, capacitor stores the energy in one half cycle and delivers that energy in another half cycle. Hence total energy stored in a capacitor over a complete cycle, when excited by an AC source is zero.

2.3 (a)

To find: RMS value of $i(t)$

we have,

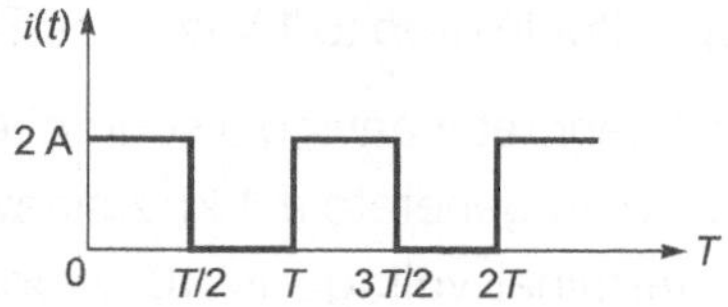

$$I_{rms} = \sqrt{\frac{1}{T}\int_0^T i^2(t)\,dt} = \sqrt{\frac{1}{T}\times 4\times\frac{T}{2}} = \sqrt{2}\text{ A}$$

2.4 (d)

$$I_1 + I_2 + I_3 = 0 \Rightarrow I_3 = -I_1 - I_2$$
$$= -[-6\sin(\omega t)] - 8\cos\omega t$$
$$= 10\left[\frac{6}{10}\sin\omega t - \frac{8}{10}\cos\omega t\right]$$
$$= -10[\cos(36.87°)\cos\omega t - \sin(36.87°)\sin\omega t]$$
$$= -10\cos(\omega t + 36.87°)\text{ mA}$$

2.5 (d)

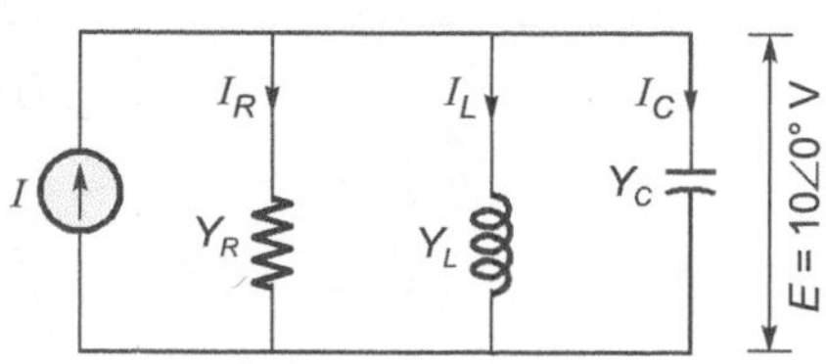

$$I_R = Y_R E = (0.5 + j0)\times 10\angle 0° = 5\text{ A}$$
$$I_Y = Y_L E = (0 - j1.5)\times 10\angle 0° = -j15\text{ A}$$
$$I_C = Y_C E = (0 + j0.3)\times 10\angle 0° = j3\text{ A}$$
$$I = I_R + I_Y + I_C = 5 + (-j15) + j3 = 5 - j12\text{ A}$$

2.6 (a)

R.M.S value of DC voltage = $V_{dc}^{(rms)} = 3$ V

R.M.S value of AC voltage = $V_{ac}^{(rms)} = \left(\frac{4}{\sqrt{2}}\right)$ V

∴ R.M.S value of the voltage

$$= \sqrt{3^2 + \left(\frac{4}{\sqrt{2}}\right)^2} = \sqrt{9+8} = \sqrt{17}\text{ V}$$

2.7 (c)

At, $f = 100$ Hz

$$|V_R| = |V_L|$$

as R and L are series connected, current through R and L is same, so

$$IR = IX_L = I\omega L \Rightarrow R = X_L = \omega L$$

$$I = \frac{V_{in}}{\sqrt{R^2 + X_L^2}} = \frac{V_{in}}{\sqrt{R^2 + R^2}} = \frac{V_{in}}{\sqrt{2}R}$$

$$V_R = u_{rms} = IR$$

$$V_R = \left(\frac{V_{in}}{\sqrt{2}R}\right)\times R = \frac{V_{in}}{\sqrt{2}}$$

$$\Rightarrow \quad V_{in} = \sqrt{2}\,u_{rms} \quad \text{...(i)}$$

At, $f = 50$ Hz

$$X_L \propto f$$

So, $$X_L' = X_L \times \frac{50}{100} = \frac{X_L}{2} = \frac{R}{2}$$

$$I' = \frac{V_{in}}{\sqrt{R^2 + (X_L')^2}} = \frac{V_{in}}{\sqrt{R^2 + \left(\frac{R}{2}\right)^2}} = \frac{2V_{in}}{\sqrt{5}R}$$

$$V_R' = I'R = \left(\frac{2V_{in}}{\sqrt{5}R}\right)R = \frac{2}{\sqrt{5}}V_{in}$$

From equation (i), we get

$$V_R' = \frac{2}{\sqrt{5}}\times(\sqrt{2}u_{rms}) = \frac{2\sqrt{2}}{\sqrt{5}}u_{rms} = \sqrt{\frac{8}{5}}u_{rms}$$

2.8 (b)

Assuming resistance of the heater = R

(i) When heater connected to 230 V, 50 Hz source, energy consumed by the heater = 2.3 units or 2.3 kWh in 1 hour.

Power consumed by the heater

$$= \frac{\text{Energy}}{\text{Time period}} = \frac{2.3\text{kWh}}{1\text{hour}}$$

$$P_1 = 2.3\text{ kW}$$

RMS value of the input voltage = V_{rms} = 230 V

$$P_1 = \frac{V_{rms}^2}{R} \Rightarrow 2.3\times 10^3 = \frac{230^2}{R} \Rightarrow R = 23\ \Omega$$

(ii) When heater connected to 400 V (peak to peak) square wave source of 150 Hz

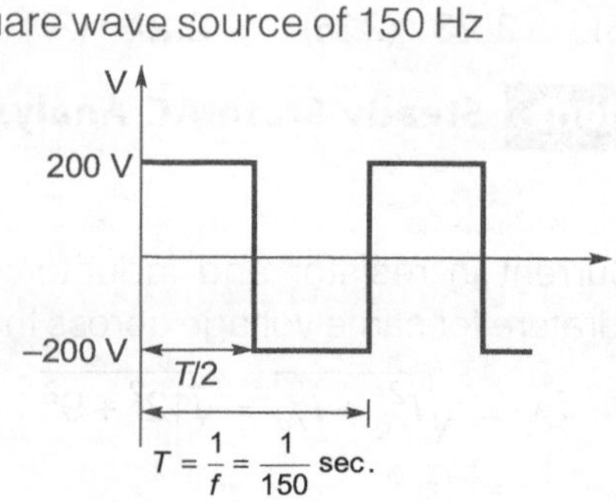

V_{rms} value of the input voltage

$$V_{rms} = \left[\frac{1}{T}\int_0^T V^2 dt\right]^{1/2}$$

$$= \left[\frac{1}{T}\left\{\int_0^{T/2} 200^2 dt + \int_{T/2}^{T} (-200)^2 dt\right\}\right]^{1/2}$$

$$V_{rms} = 200 \text{ V}$$

$$P_2 = \frac{V_{rms}^2}{R} = \frac{200^2}{23} \times 10^{-3} \text{ kW} = 1.739 \text{ kW}$$

2.9 (b)

$$V_s = 1 \sin t \equiv V_m \sin\omega t$$
$$V_m = 1 \text{ V and } \omega = 1 \text{ rad/sec}$$

Impedance of the branch containing inductor and capacitor,

$$Z = j(X_L - X_C) = j\left(\omega L - \frac{1}{\omega C}\right) = j\left(1 \times 1 - \frac{1}{1 \times 1}\right) = 0$$

So, this branch is short-circuit and the whole current flow through it

$$i(t) = \frac{1.0 \sin t}{1} = 1.0 \sin t$$

$$\text{RMS value of the current} = \frac{1}{\sqrt{2}} \text{ A}$$

2.10 (b)

$$V(t) = 100\sqrt{2}\cos(100\pi t)$$

Voltage represented in phasor form:

$$V_{ph} = V_{rms}\angle\phi = \frac{100\sqrt{2}}{\sqrt{2}}\angle 0°$$

$$i(t) = 10\sqrt{2}\sin\left(100\pi t + \frac{\pi}{4}\right)$$

$$i(t) = 10\sqrt{2}\cos\left(100\pi t + \frac{\pi}{4} - \frac{\pi}{2}\right)$$

$$I_{ph} = \frac{10\sqrt{2}}{\sqrt{2}}\angle\left(\frac{\pi}{4} - \frac{\pi}{2}\right) = 10\angle\left(-\frac{\pi}{4}\right) \text{ A}$$

2.11 (b)

Power supplied by the source = $V_s I_s \cos\phi$

Where, ϕ = Angle between V_s and $I_s = \pi/4$

Inductor and capacitor do not consume power.

Therefore, power dissipated in

R = Power supplied by the source

$$P_R = V_s I_s \cos\phi$$

$$= 1 \times \sqrt{2} \times \cos\frac{\pi}{4} = \sqrt{2} \times \frac{1}{\sqrt{2}} = 1 \text{ W}$$

2.12 (d)

Using KCL,

$$-I_s + I_{RL} + I_C = 0 \Rightarrow I_C = I_s - I_{RL}$$

$$= \sqrt{2}\angle\frac{\pi}{4} - \sqrt{2}\angle -\frac{\pi}{4} = 2\angle 90° = +j2 \text{ A}$$

2.13 (b)

$$i = 5\cos(100\pi t + 100°) \text{ A} = 5\angle 100° \text{ A}$$
$$Z = (4 - j3)\ \Omega = 5\angle -36.87°\ \Omega$$
$$v = iZ = 25\angle 63.13° \text{ V}$$

The average power is

$$P = \frac{1}{2}V_m I_m \cos\phi = \frac{1}{2} \times 25 \times 5 \times \cos(36.87°)$$

$$= \frac{1}{2} \times 25 \times 5 \times \frac{4}{5} = 50 \text{ W}$$

Alternate Method:

$$P = |I_{rms}|^2 R = \left(\frac{5}{\sqrt{2}}\right)^2 4 = 50 \text{ W}$$

2.14 (a)

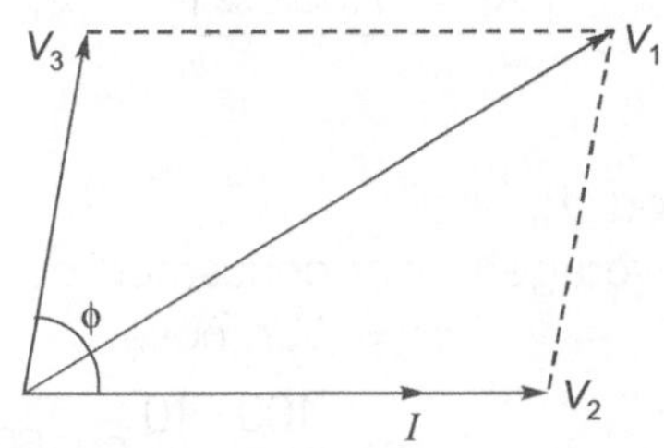

$$V_1^2 = V_2^2 + V_3^2 + 2V_2V_3\cos\phi$$

$$(220)^2 = (122)^2 + (136)^2 + 2 \times 122 \times 136 \times \cos\phi$$

$$\Rightarrow \cos\phi = 0.45$$

2.15 (b)

Given, $R_L = 5\ \Omega$

$$\therefore \quad \cos\phi = \frac{R_L}{|Z|}$$

$$\Rightarrow \quad |Z| = \frac{5}{0.45} = 11.11$$

Power consumed by load,

$$P_L = \left(\frac{V_3}{|Z|}\right)^2 R_L = \left(\frac{136}{11.11}\right)^2 \times 5 = 749.1$$

$$\simeq 750 \text{ W}$$

2.16 Sol.

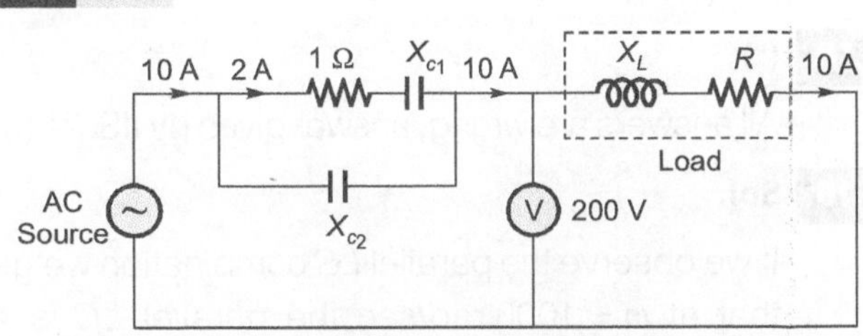

Given, total power dissipated in the circuit

= 1 kW = 1000 Watt

$\therefore \; 2^2 \times 1 + 10^2 \times R = 1000$

or, $\quad R = \dfrac{998}{100} = 9.98\ \Omega$

Also, voltage drop across R;

$$V_R = IR = 10 \times 9.98 = 99.8 \text{ volt}$$

Voltage drop across load,

$$V = 200 \text{ volt} = \sqrt{V_R^2 + V_{X_L}^2}$$

$\therefore$ Voltage drop across inductor,

$$V_{X_L} = \sqrt{V^2 - V_R^2} = \sqrt{(200)^2 - (99.8)^2} = 173.32 \text{ volt}$$

Now, $V_{X_L} = IX_L$ or $X_L = \dfrac{V_{X_L}}{I}$

$$= \frac{173.32}{10} = 17.332\ \Omega$$

$\therefore \quad \boxed{X_L = 17.332\ \Omega}$

2.17 Sol.

Method 1:

The average power conusmed by the load,

$$P = V_1 I_1 \cos\phi_1 = \frac{100}{\sqrt{2}} \cdot \frac{10}{\sqrt{2}} \cos 60^\circ = 250\text{ W}$$

Method 2:

$V(t) = 100 \sin(\omega t)$

$i(t) = 10 \sin(\omega t - 60^\circ) + 2 \sin(3\omega t) + 5 \sin(5\omega t)$

$P = V(t)\, i(t)$

$= 1000 \sin\omega t \cdot \sin(\omega t - 60^\circ) + 200 \sin\omega t \cdot \sin 3\omega t + 500 \sin\omega t \sin 5\omega t$

$= 500[\cos(\omega t - \omega t + 60^\circ) - \cos(\omega t + \omega t - 60^\circ)] + 100[\cos(\omega t - 3\omega t) - \cos(\omega t + 3\omega t)] + 250[\cos(\omega t - 5\omega t) - \cos(\omega t + 500\,\omega t]$

$= 500[\cos 60^\circ - \cos(2\omega t - 60^\circ)] + 100(\cos(-2\omega t) - \cos 4\omega t] + 250[\cos(-4\omega t) - \cos(6\omega t)]$

$= 500 \cos 60^\circ = 250$ W

Average value of $\cos(2\omega t - 60^\circ)$, $\cos(2\omega t)$, $\cos(4\omega t)$, $\cos(6\omega t)$ will be zero.

2.18 (*)

All answers are wrong, answer given by IISc is (c).

2.19 Sol.

If we observe the parallel LC combination we get that at $\omega = 1000$ rad/sec the parallel LC is at resonance thus it is open circuited.

The circuit given in question can be redrawn as

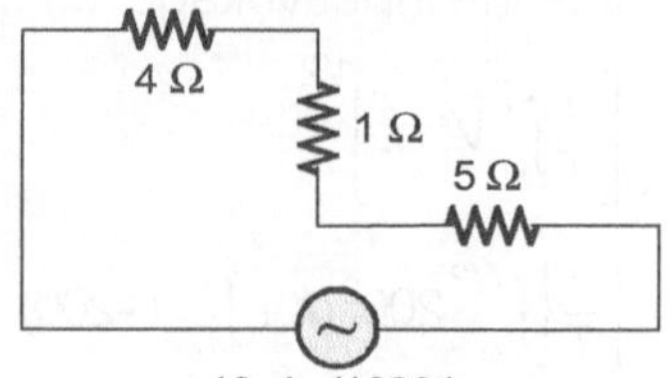

10 sin (1000t)

So, $\quad I = \dfrac{10 \sin 1000t}{10} = \sin 100t$

So, peak value is 1 Amp.

2.20 (c)

$V_2 : \dfrac{\pi}{18} = \dfrac{180^\circ}{18} = 10^\circ$

$V_3 : \dfrac{\pi}{36} = \dfrac{180^\circ}{36} = 5^\circ$

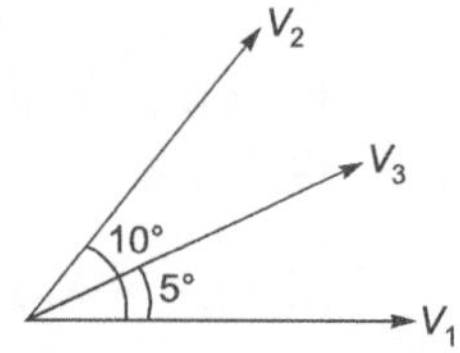

V_2 leads V_1 and V_3.

So, V_2 is a source, V_1 and V_3 are absorbing.

Hence, $\quad P_2 > 0, \quad P_1, P_3 < 0$

2.21 Sol.

Given that,

$v(t) = 5 - 10 \cos(\omega t + 60^\circ)$

$i(t) = 5 + X \cos(\omega t - 0^\circ)$

$P_{req} = 0$

$$0 = 5 \times 5 + \frac{1}{2}[(-10)(X)\cos(60^\circ)]$$

$$-25 = \frac{1}{2}[(-10)(X)\cos(60^\circ)]$$

$X = 10$

2.22 Sol.

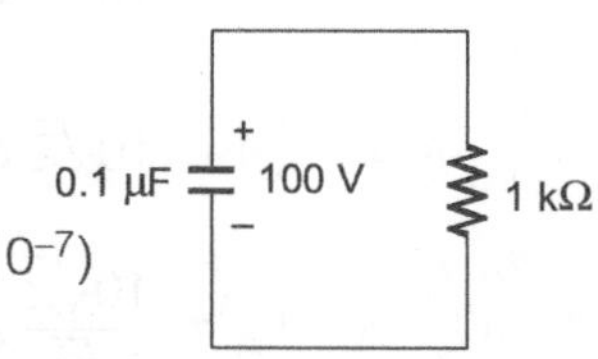

$v_c(t) = V_0\, e^{-t/\tau}$

$V_0 = 100$ V

$\tau = RC = (10^3)(10^{-7}) = 10^{-4}$ sec

$\therefore \; v_c(t) = 100 e^{-10^4 t}$ V

Let the time required by the voltage across the capacitor to drop to 1 V is t_1,

$\therefore \quad v_c(t_1) = 100 e^{-10^4 t_1} \qquad v_C(t_1) = 1\text{V}$

$\Rightarrow \quad 1 = 100 e^{-10^4 t_1}$

$e^{-10^4 t_1} = 0.01$

$t_1 = 0.46$ msec

Alternate solution:

$C = 0.1\ \mu$F, $\quad R = 1$ kΩ

$E = 100$ V, $\quad v = 1$ V

$$R = \frac{0.4343 \times t}{C \times \log(E/v)}$$

$$\Rightarrow \qquad 10^3 = \frac{0.4343 \times t}{0.1 \times 10^{-6} \times \log\left(\frac{100}{1}\right)}$$

$$t = \frac{10^{-4} \times \log(100)}{0.4343}$$

$$= 4.60511 \times 10^{-4} = 0.46 \text{ msec}$$

2.23 (2.38)

$$\frac{V_0(j\omega)}{V_i(j\omega)} = \frac{1}{1 + j\omega RC}$$

$$\frac{8}{10}\angle\theta = \frac{1}{1 + j\omega RC}$$

$$1 + j\omega RC = 1.25\angle\theta$$

$$\omega RC = \sqrt{1.25^2 - 1^2}$$

$$= 0.75$$

$$C = \frac{0.75}{2000\pi \times 50}$$

$$= 2.38\ \mu\text{F}$$

2.24 (20)

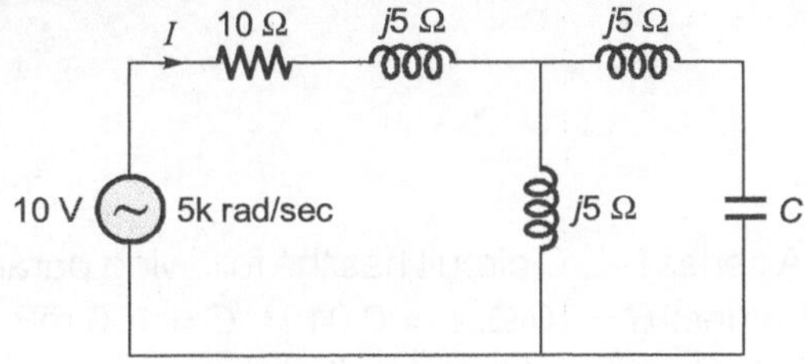

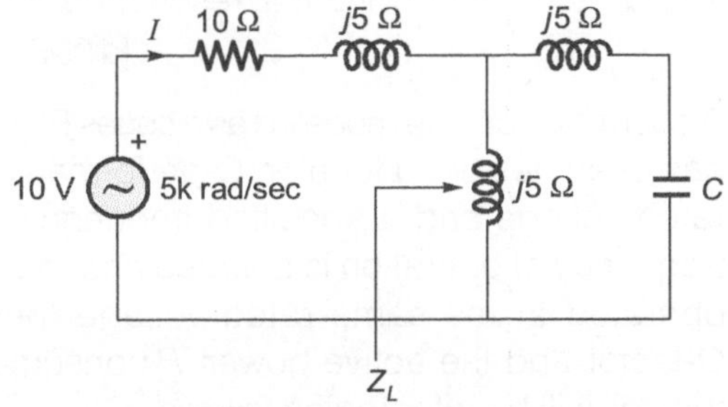

$$Z_L = (j5) \parallel (j5 - jX_c)$$

$$\frac{(j5)(j5 - jX_c)}{j5 + j5 - jX_c} = \infty$$

$$j5 + j5 - jX_c = 0 \Rightarrow jX_c = j10 \Rightarrow X_c = 10\ \Omega$$

$$X_c = \frac{1}{\omega C} = 10 \Rightarrow C = \frac{1}{10 \times 5 \times 10^3}$$

$$C = \frac{1}{5 \times 10^4} \times \frac{10^2}{10^2} = \frac{10^2}{5 \times 10^6} = 20\ \mu\text{F}$$

■■■■

3 Resonance and Locus Diagrams

3.1 A series R-L-C circuit has the following parameter values: $R = 10\ \Omega$, $L = 0.01$ H, $C = 100$ mF. The Q-factor of the circuit at resonance is _____.

[1995 : 1 M]

3.2 A coil (which can be modeled as a series RL circuit) has been designed for high Q-performance at a rated voltage and a specified frequency. If the frequency of operation is doubled and the coil is operated at the same rated voltage then the Q-factor and the active power, P consumed by the coil will be affected as follows.

(a) P is doubled, Q is halved
(b) P is halved, Q is doubled
(c) P remains constant, Q is doubled
(d) P decreased 4 times, Q is doubled

[1996 : 2 M]

3.3 A sinusoidal source of voltage V and frequency, f is connected to a series circuit of variable resistance R and a fixed reactance X. The locus of the tip of the current phasor I as R is varied from 0 to ∞ is

(a) a semicircle with a diameter of V/X.
(b) a straight line with a slope of R/X.
(c) an ellipse with V/R as major axis.
(d) a circle of radius R/X and origin at $(0, V/2)$.

[1998 : 1 M]

3.4 A circuit with a resistor, inductor and capacitor in series is resonant at f_0 Hz. If all the component values are now doubled, the new resonant frequency is

(a) $2 f_0$ (b) still f_0
(c) $\frac{f_0}{4}$ (d) $\frac{f_0}{2}$ [1998 : 1 M]

3.5 A fixed capacitor of reactance $-j0.02\ \Omega$ is connected in parallel across a series combination of a fixed inductor of reactance $j0.01\ \Omega$ and a variable resistance R. As R is varied from zero to infinity, the locus diagram of the admittance of this R–L–C circuit will be

(a) a semi-circle of diameter $j100$ and center at zero.
(b) a semi-circle of diameter $j50$ and center at zero.
(c) a straight line inclined at an angle.
(d) a straight line parallel to the x-axis.

[1999 : 2 M]

3.6 A series R-L-C circuit when excited by a 10 V sinusoidal voltage source of variable frequency, exhibits resonance at 100 Hz and has a 3 dB bandwidth of 5 Hz. The voltage across the inductor L at resonance is

(a) 10 V (b) $10\sqrt{2}$ V
(c) $\frac{10}{\sqrt{2}\text{ V}}$ (d) 200 V

[1999 : 1 M]

3.7 The current in the circuit shown in the figure is

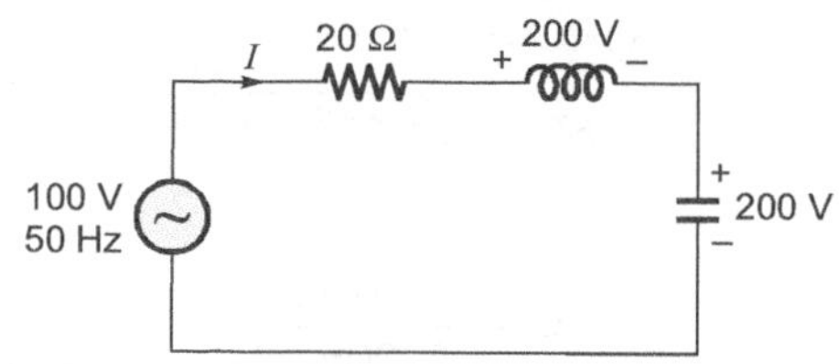

(a) 5 A (b) 10 A
(c) 15 A (d) 25 A

[1999 : 1 M]

3.8 In a series RLC circuit at resonance, the magnitude of the voltage developed across the capacitor

(a) is always zero.
(b) can never be greater than the input voltage.
(c) can be greater than the input voltage, however it is 90° out of phase with the input voltage.
(d) can be greater than the input voltage, and is in phase with the input voltage.

[2001 : 1 M]

3.9 In the circuit shown in the figure, what value of C will cause a unity power factor at the ac source?

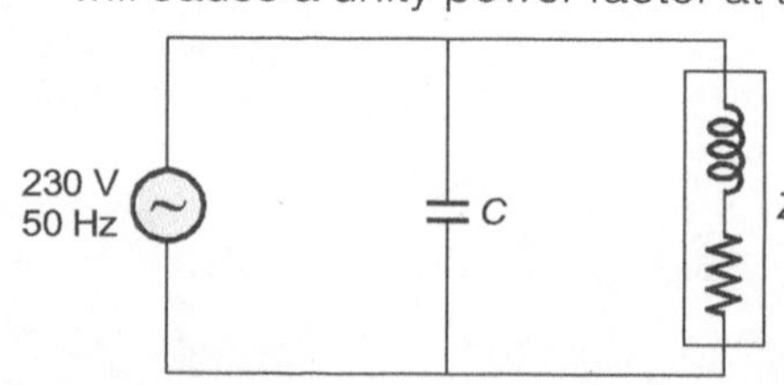

(a) 68.1 μF (b) 165 μF
(c) 0.681 μF (d) 6.81 μF [2002 : 2 M]

3.10 A series RLC circuit has $R = 50\ \Omega$; $L = 100\ \mu$H and $C = 1\ \mu$F. The lower half power frequency of the circuit is

(a) 30.55 kHz (b) 3.055 kHz
(c) 51.92 kHz (d) 1.92 kHz [2002 : 2 M]

3.11 In the circuit of figure, the magnitudes of V_L and V_C are twice that of V_R. Given that, $f = 50$ Hz, the inductance of the coil is

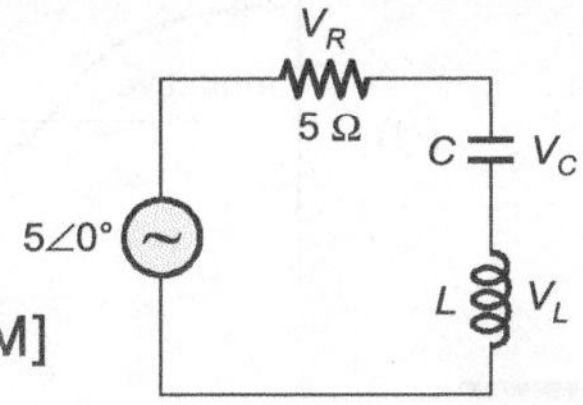

(a) 2.14 mH
(b) 5.30 H
(c) 31.8 mH
(d) 1.32 H [2003 : 2 M]

3.12 The value of Z in the figure which is most appropriate to cause parallel resonance at 500 Hz is

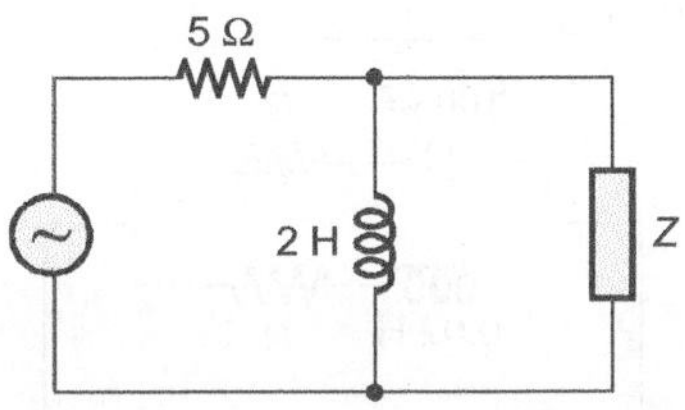

(a) 125.00 mH (b) 304.20 μF
(c) 2.0 μF (d) 0.05 μF [2004 : 1 M]

3.13 The circuit shown in the figure is energized by a sinusoidal voltage source V_1 at a frequency which causes resonance with a current of I.

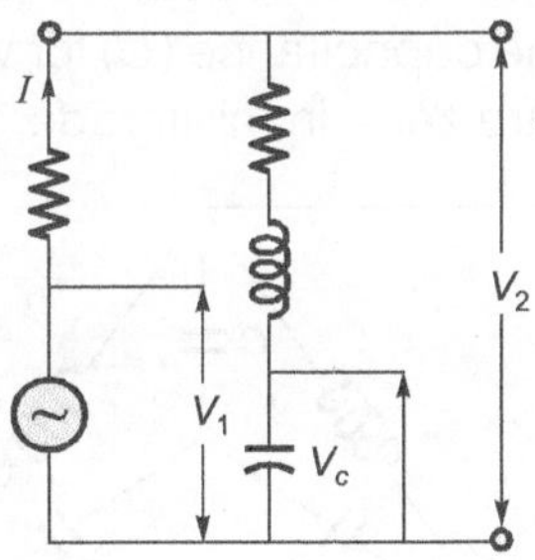

The phasor diagram which is applicable to this circuit is

(a)
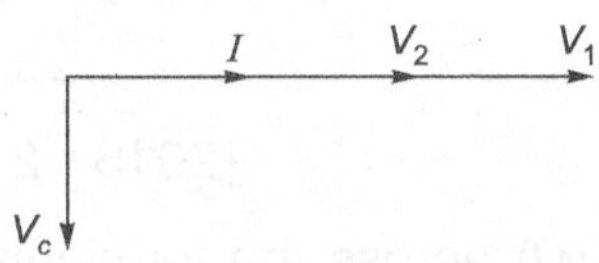

(b)
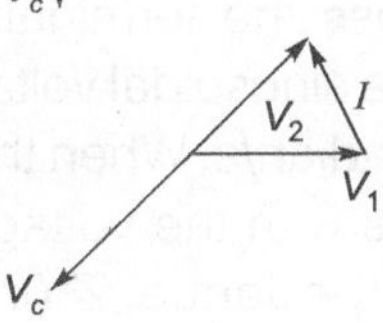

(c)
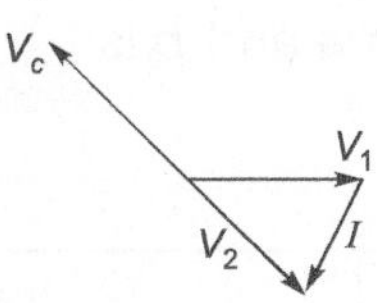

(d)
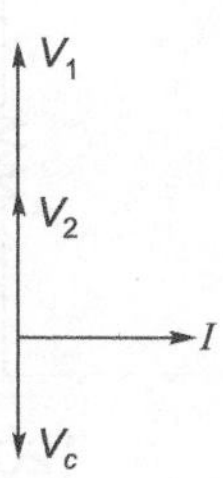

[2006 : 2 M]

3.14 The R-L-C series circuit shown is supplied from a variable frequency voltage source. The admittance-locus of the R-L-C network at terminals AB for increasing frequency ω is

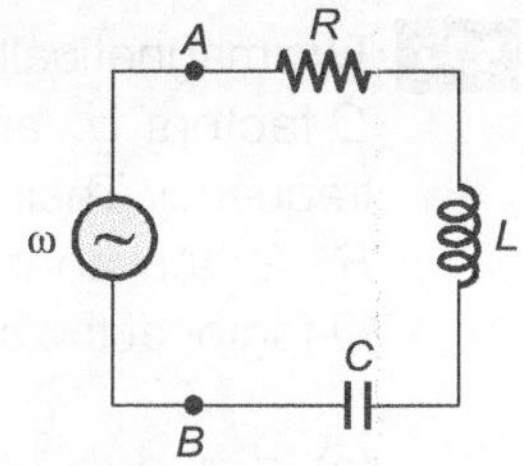

(a)
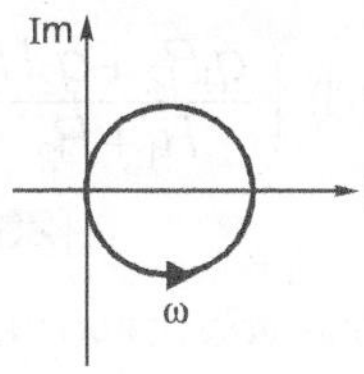

(b)
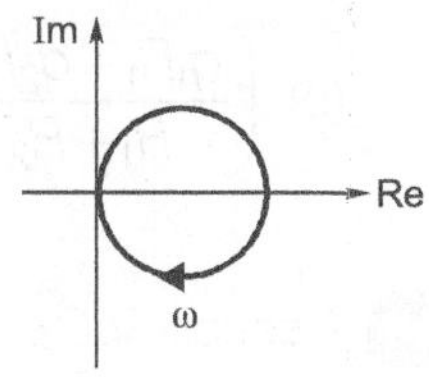

(c)
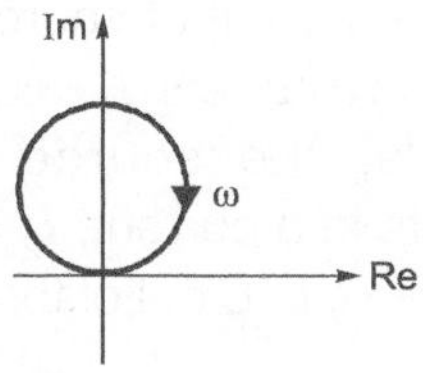

(d)
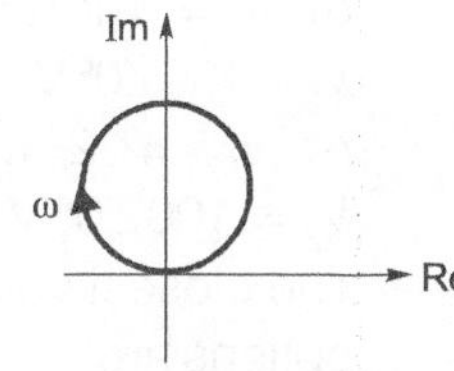

[2007 : 2 M]

3.15 In the figure given below all phasors are with reference to the potential at point "O". The locus of voltage phasor V_{YX} as R is varied from zero to infinity is shown by

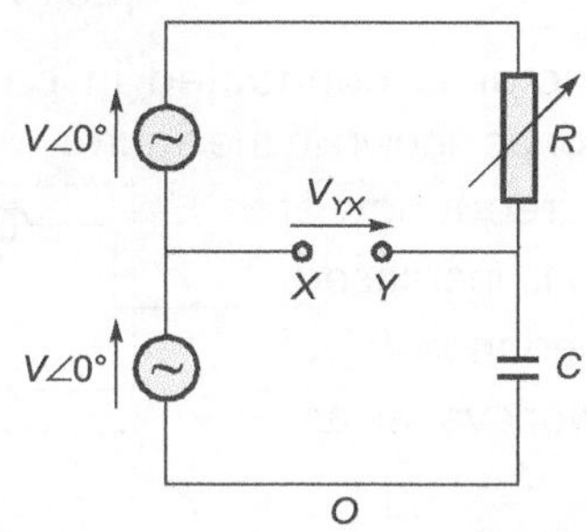

(a)
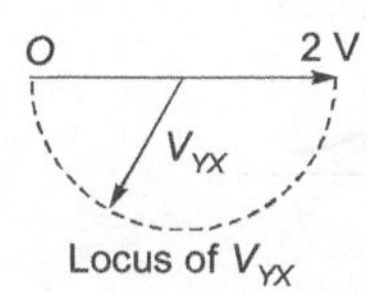

(b)
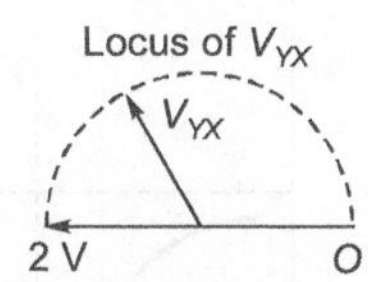

(c)
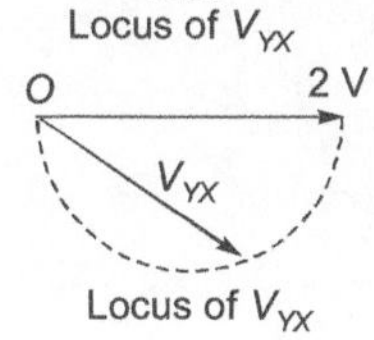

(d)
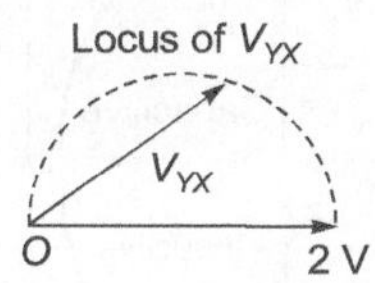

[2007 : 2 M]

3.16 The resonant frequency for the given circuit will be

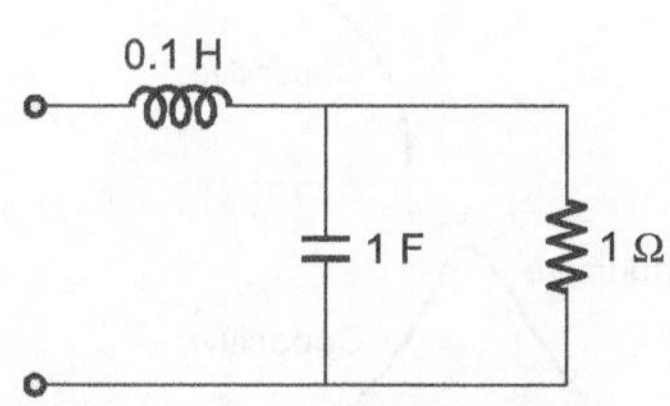

(a) 1 rad/s (b) 2 rad/s
(c) 3 rad/s (d) 4 rad/s [2008 : 2 M]

3.17 Two magnetically uncoupled inductive coils have Q-factors q_1 and q_2 at the chosen operating frequency. Their respective resistances are R_1 and R_2. When connected in series, their effective Q-factor at the same operating frequency is

(a) $q_1 + q_2$

(b) $\left(\frac{1}{q_1}\right)+\left(\frac{1}{q_2}\right)$

(c) $\left(\frac{q_1R_1+q_2R_2}{R_1+R_2}\right)$

(d) $\left(\frac{q_1R_2+q_2R_1}{R_1+R_2}\right)$

[2013 : 2 M]

3.18 A series RLC circuit is observed at two frequencies. At ω_1 = 1 k rad/s, we note that source voltage $V_1 = 100\angle 0°$ V results in a current, $I_1 = 0.03\angle 31°$ A. At ω_2 = 2 k rad/s, the source voltage, $V_2 = 100\angle 0°$ V results in a current, $I_2 = 2\angle 0°$ A. The closest values for R, L, C out of the following options are

(a) R = 50 Ω; L = 25 mH; C = 10 μF

(b) R = 50 Ω; L = 10 mH; C = 25 μF

(c) R = 50 Ω; L = 50 mH; C = 5 μF

(d) R = 50 Ω; L = 5 mH; C = 50 μF

[2014 : 2 M Set-3]

3.19 An inductor is connected in parallel with a capacitor as shown in the figure.

As the frequency of current i is increased, the impedance (z) of the network varies as

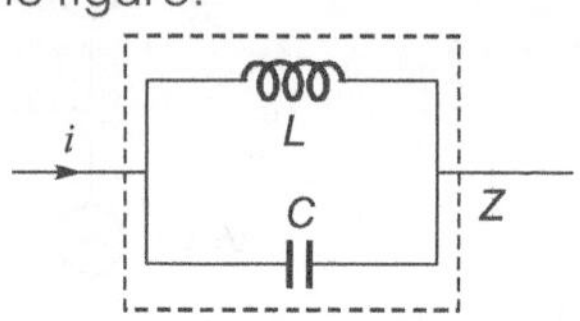

(a)

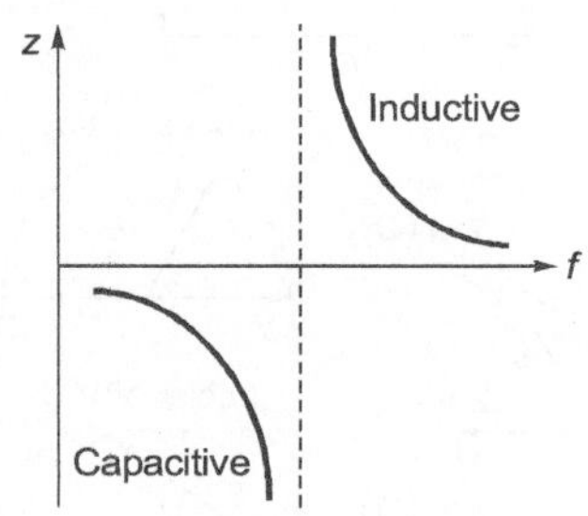

(b)

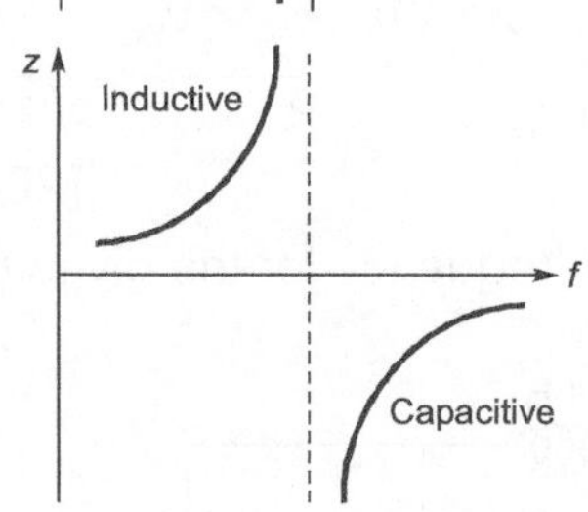

(c)

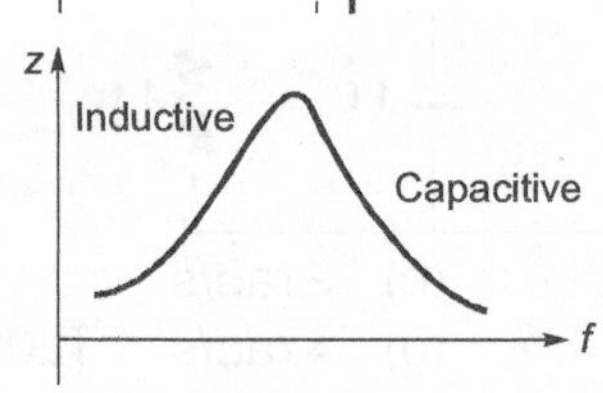

(d)

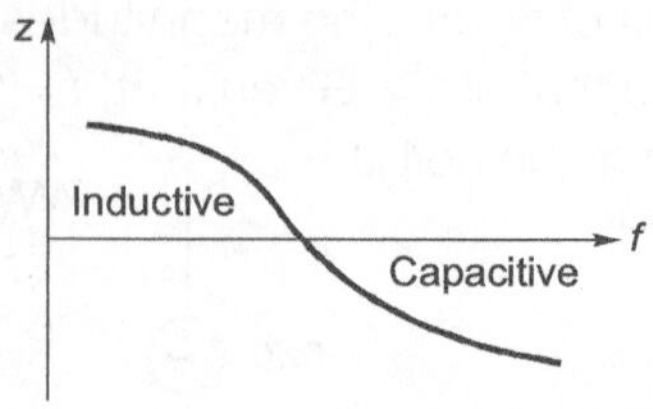

[2015 : 1 M, Set-1]

3.20 The circuit below is excited by a sinusoidal source. The value of R, in Ω, for which the admittance of the circuit becomes a pure conductance at all frequencies is ______.

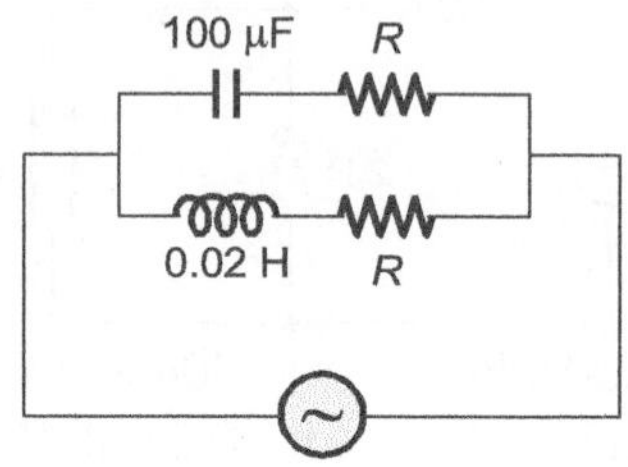

[2016 : 2 M, Set-1]

3.21 In the balanced 3-phase, 50 Hz, circuit shown below, the value of inductance (L) is 10 mH. The value of the capacitance (C) for which all the line currents are zero, in millifarads, is _____.

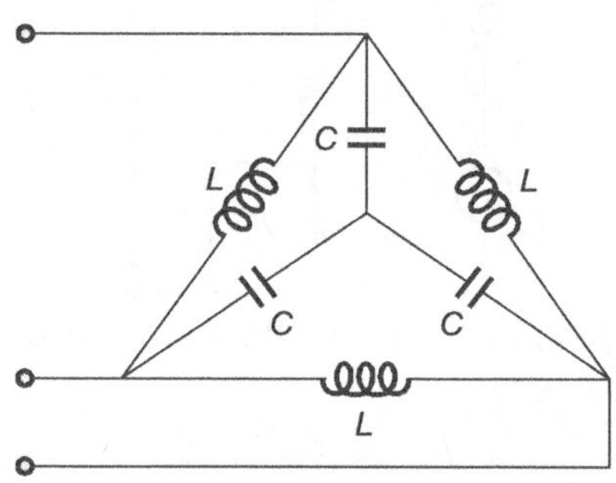

[2016 : 2 M, Set-2]

3.22 The voltage $v(t)$ across the terminals a and b shown in the figure, is a sinusoidal voltage having a frequency ω = 100 radian/s. When the inductor current $i(t)$ is in phase with the voltage $v(t)$, the magnitude of the impedance Z (in Ω) seen between the terminals a and b is _____ (upto 2 decimal places).

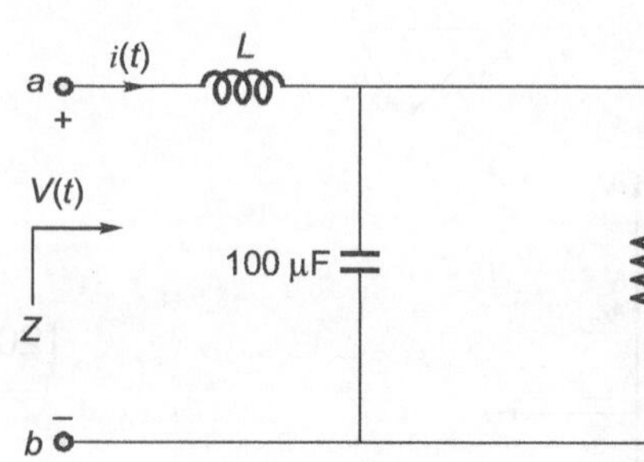

[2018 : 2 M]

3.23 A DC voltage source is connected to a series L-C circuit by turning on the switch S at time $t = 0$ as shown in the figure. Assume : $i(0) = 0$, $v(0) = 0$. Which one of the following circular loci represents the plot of $i(t)$ versus $v(t)$?

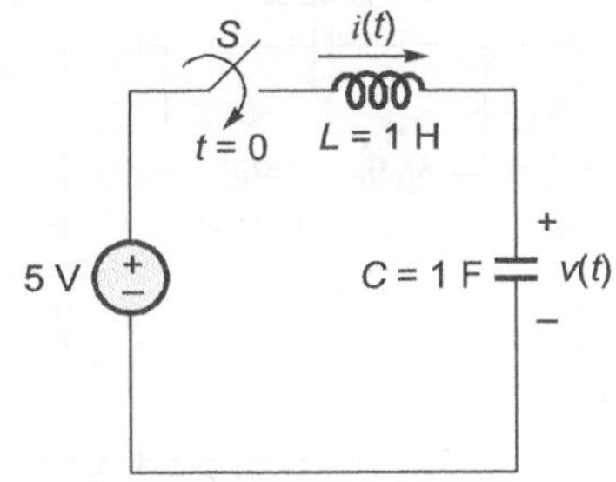

(a)
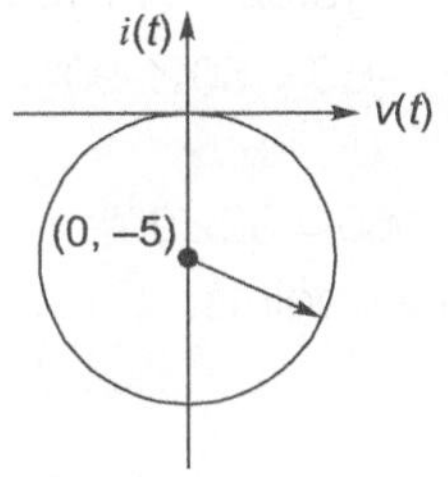

(b)
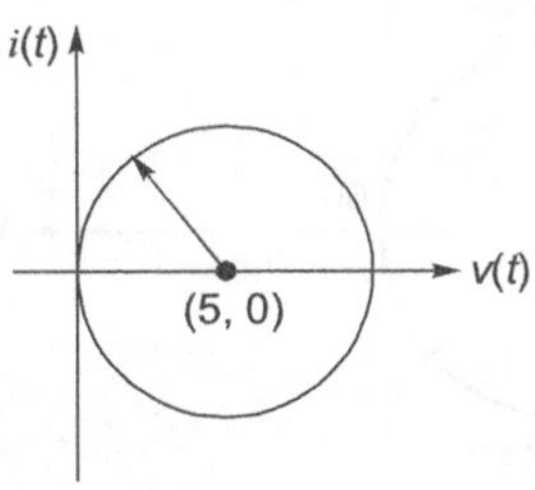

(c)
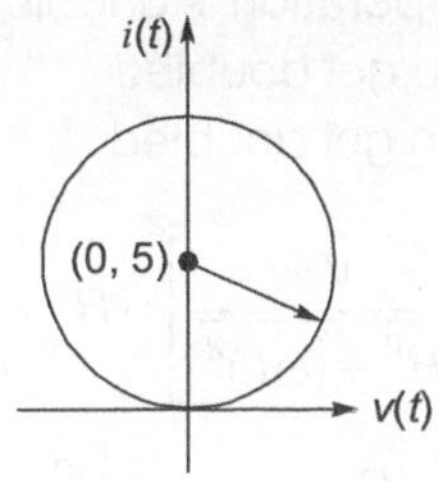

(d)
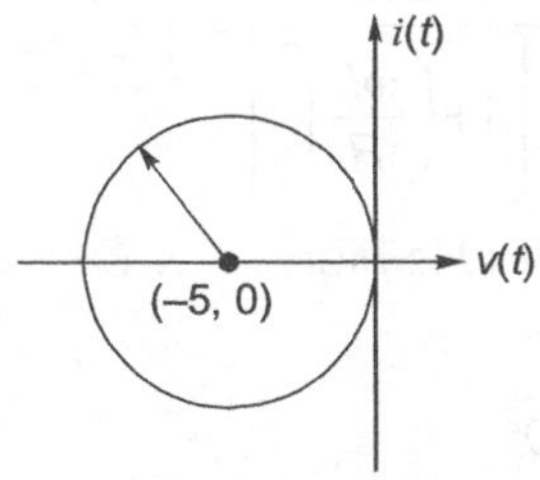

[2018 : 2 M]

3.24 An inductor having a Q-factor of 60 is connected in series with a capacitor having a Q-factor of 240. The overall Q-factor of the circuit is _____. (round off to nearest integer). [2022 : 1 M]

3.25 The network shown below has a resonant frequency of 150 kHz and a bandwidth of 600 Hz. The Q-factor of the network is ____. (round off to nearest integer).

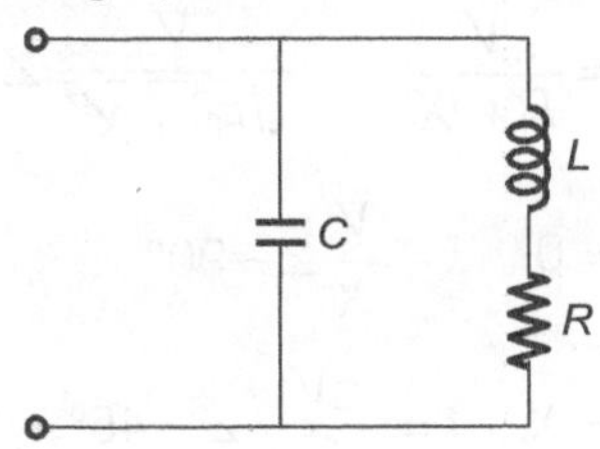

[2022 : 1 M]

■■■■

Answers Resonance and Locus Diagrams

3.1	(0.032)	3.2	(d)	3.3	(a)	3.4	(d)	3.5	(a)	3.6	(d)	3.7	(a)
3.8	(c)	3.9	(a)	3.10	(b)	3.11	(c)	3.12	(d)	3.13	(a)	3.14	(b)
3.15	(a)	3.16	(c)	3.17	(c)	3.18	(b)	3.19	(b)	3.20	(14.14)	3.21	(3.03)
3.22	(50)	3.23	(b)	3.24	(48)	3.25	(250)						

Explanations Resonance and Locus Diagrams

3.1 Sol.

$$Q_0 = 0.032$$

For series RLC circuit,

$$\text{Q-factor at resonance} = \frac{\omega_0 L}{R}.$$

$$\omega_0 = \frac{1}{\sqrt{LC}} = \frac{1}{\sqrt{(0.01)\times(100\times10^{-3})}}$$

$$= 10\sqrt{10} \text{ rad/sec.}$$

$$Q = \omega_0\frac{L}{R} = \frac{10\sqrt{10}\times0.01}{10} = 0.032$$

3.2 (d)

$$Q = \frac{\omega L}{R}$$

When frequency of operation is doubled,

$\omega = 2\pi f$, also get doubled

Consequently, Q also get doubled

$$P = I^2R\left[\frac{V}{\sqrt{R^2+(\omega L)^2}}\right]^2 \cdot R$$

$$= \frac{V^2 \cdot R}{R^2\left[1+\left(\frac{\omega L}{R}\right)^2\right]} = \frac{V^2}{R(1+Q^2)}$$

$\because$ It is given that Q is high. $\therefore Q^2 >> 1$

$$\Rightarrow \quad P \simeq \frac{V^2}{RQ^2}$$

$\because$ Q is doubled. $\therefore$ P decreased 4 times.

3.3 (a)

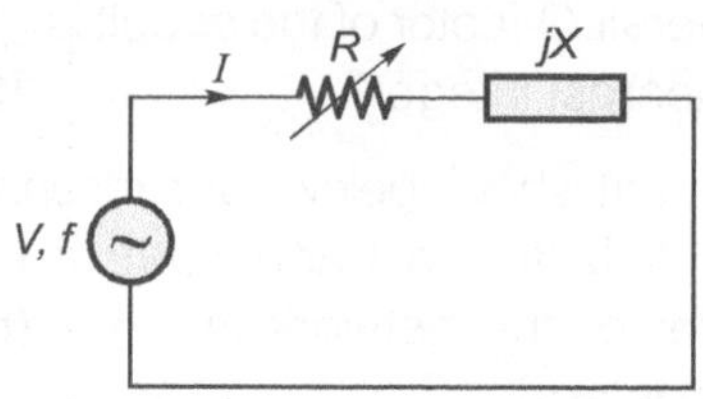

$$I = \frac{V}{R+jX} = \frac{V}{\sqrt{R^2+X^2}}\angle -\tan^{-1}\left(\frac{X}{R}\right)$$

For $R = 0$, $I = \frac{V}{X}\angle -90°$

For $R = X$, $I = \frac{V}{\sqrt{2}X}\angle -45°$

For $R = \infty$, $I = 0\angle 0°$

On plotting these three points we get,

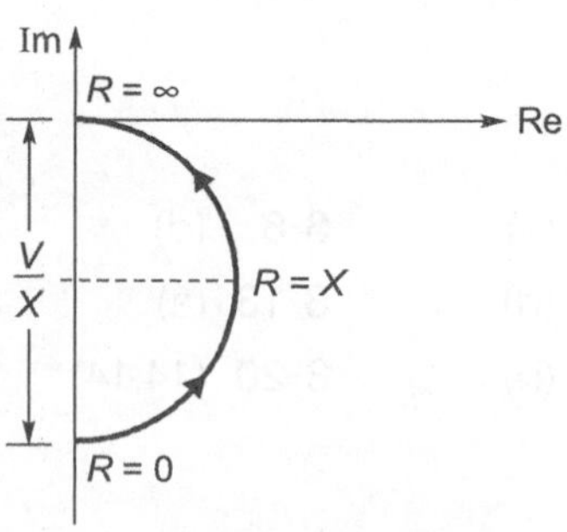

Hence locus of $\vec{I}$ is a semi-circle having diameter of V/X.

3.4 (d)

$$f_0 = \frac{1}{2\pi\sqrt{LC}} \quad \text{(for series RLC resonance)}$$

$$f_{new} = \frac{1}{2\pi\sqrt{2L \times 2C}}$$

(when all the components' values are doubled)

Hence, $f_{new} = \frac{f_0}{2}$

3.5 (a)

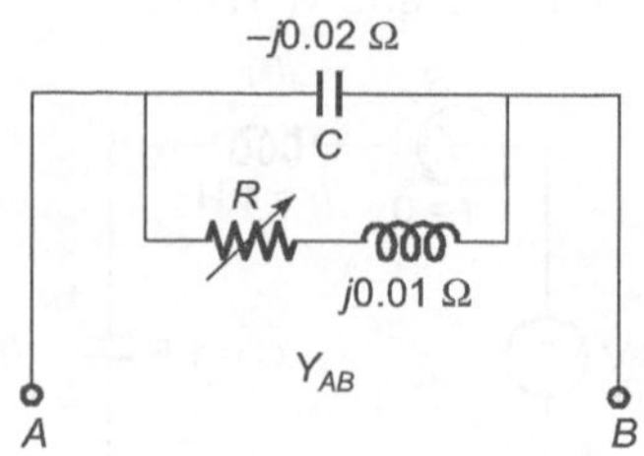

$$Y_{AB} = \frac{1}{-j0.02} + \frac{1}{R+j0.01}$$

For $R = 0$, $Y_{AB} = -j50 = 50\angle -90°$

For $R = 0.01$, $Y_{AB} = 50$

For $R = \infty$, $Y_{AB} = j50 = 50\angle 90°$

On plotting these three points,

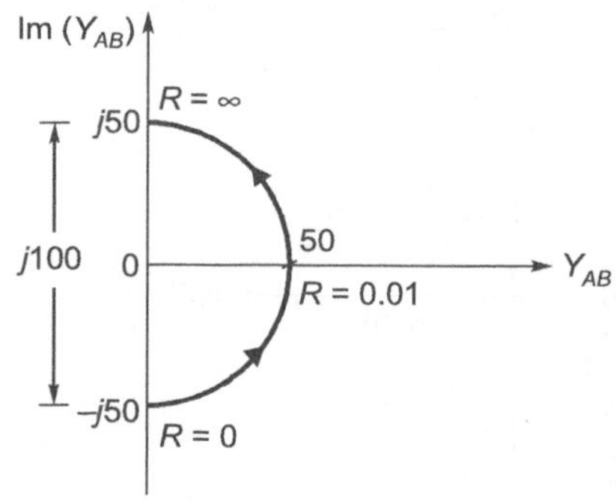

Hence, locus of $\vec{Y}_{AB}$ is a semicircle of diameter $j100$ and center at zero.

3.6 (d)

$$Q = \frac{\text{Resonance frequency}}{\text{Bandwidth}} = \frac{f_0}{\Delta f} = \frac{100}{5} = 20$$

$\because$ At resonance,

$$|V_L| = |V_C| = Q \cdot |V_{source}|$$

$$\therefore \quad |V_L| = 20 \times 10 = 200 \text{ V}$$

3.7 (a)

$\because$ $V_L = -V_c$ (given)

So, this is a case of RLC series resonance.

Hence, $I = \frac{V}{R}$ (at resonance) $= \frac{100}{20} = 5$ A

3.8 (c)

In a series RLC circuit, at resonance

$V_L = jQV_{source}$ and $V_c = -jQV_{source}$

Also for $Q > 1$, $|V_c| > |V_{source}|$.

Hence, option (c) is correct.

3.9 (a)

$$Y = j\omega C + \frac{1}{30\angle 40^\circ} = j\omega C + 0.0255 - j0.0214$$
$$= 0.0255 + j(\omega C - 0.0214)$$
$$= \text{Real}(Y) + j\,\text{Img}(Y)$$

To have a unity power factor at AC source i.e. resonance condition,

$$\text{Img}(Y) = 0$$
$$\Rightarrow \quad \omega C - 0.0214 = 0$$
$$\because \quad \omega = 2\pi \times 50$$
$$\therefore \quad C = \frac{0.0214}{100\pi} = 68.1\,\mu\text{F}$$

3.10 (b)

$$\omega_o = \frac{1}{\sqrt{LC}} = \frac{1}{\sqrt{100\times10^{-6}\times10^{-6}}} = 10^5 \text{ r/s}$$

$$\Delta\omega = \frac{R}{L} = \frac{50}{100\times10^{-6}} = 50\times10^4 \text{ r/s}$$

$$\omega_{lower} = \left[\sqrt{\omega_0^2 + \left(\frac{\Delta\omega}{2}\right)^2} - \frac{\Delta\omega}{2}\right]$$

$$= \left[\sqrt{(10^5)^2 + \left(\frac{5\times10^5}{2}\right)^2} - \frac{5\times10^5}{2}\right]$$

$$= 10^5\left[\sqrt{1+6.25} - 2.5\right] = 0.193 \times 10^5 \text{ r/s}$$

$$f_{lower} = \frac{\omega_{lower}}{2\pi} = \frac{0.193\times10^5}{2\pi}$$
$$= 3065 \text{ Hz} \simeq 3.055 \text{ kHz}$$

3.11 (c)

$$V = V_R + i(V_L - V_C)$$

Since, $|V_L| = |V_C|$ and $|V_L| = 2|V_R|$

Therefore, the circuit is at resonance and

$$V_R = V$$

$$\text{Quality factor} = \frac{V_L}{V} = \frac{V_L}{V_R} = \frac{2V_R}{V_R} = 2$$

As we know,

$$Q = \frac{\omega_0 L}{R} \Rightarrow 2 = \frac{2\pi f \times L}{5} \Rightarrow L = 31.8 \text{ mH}$$

3.12 (d)

At resonance, the circuit should be in unity power factor.

∴ Hence 'Z' should be capacitive.

Admittance of the parallel circuit,

$$Y = \frac{1}{jL\omega} + \frac{1}{1/jC\omega} = 0$$

$$\frac{-1}{L\omega} + C\omega = 0$$

$$\therefore C = \frac{1}{L\times\omega^2} = \frac{1}{2\times(2\pi\times500)^2} = 0.05\,\mu\text{F}$$

3.13 (a)

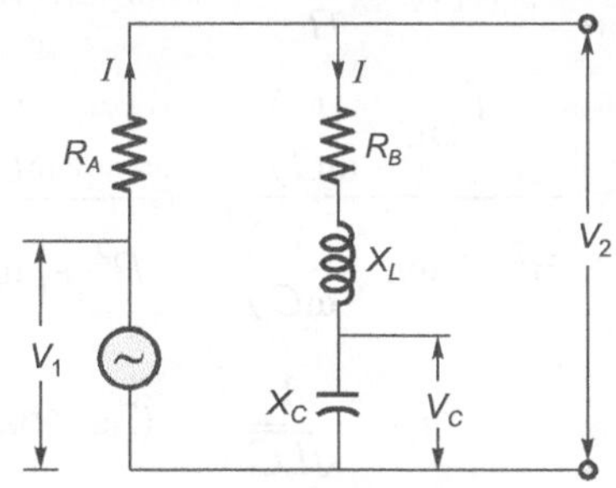

$$Z = R_A + R_B + j(X_L - X_C)$$

At resonance, $X_L = X_C$

So, $Z = R_A + R_B$

Therefore, input impedance is purely resistive, is minimum, and the input voltage and output current are in phase. So, V_1 and I are in phase

$$V_2 = \frac{V_1}{R_A + R_B + j(X_L - X_C)} \times [R_B + j(X_L - X_C)]$$

But, $X_L = X_C$

$$V_2 = \frac{V_1}{R_A + R_B} \times R_B$$

Therefore, V_2 is in phase with V_1 and $V_2 < V_1$

Voltage across the capacitor,

$$V_C = I \times X_C = I \times \frac{1}{j\omega C} = \frac{I}{\omega C}\angle -90^\circ$$

So, V_C lags the current by 90°.

The phasor diagram on the basis of above analysis.

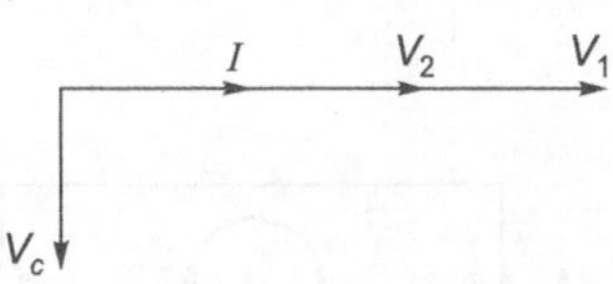

3.14 (b)

Admittance of the series connected RLC,

$$Y = \frac{1}{R + j\left(\omega L - \frac{1}{\omega C}\right)} = \frac{R - j\left(\omega L - \frac{1}{\omega C}\right)}{R^2 + \left(\omega L - \frac{1}{\omega C}\right)^2}$$

{By rationalization}

Separating, real and imaginary part of admittance.

$$\text{Re}[Y] = \frac{R}{R^2 + \left(\omega L - \frac{1}{\omega C}\right)^2}$$

For any value of ω, the real part of always positive.

When, $\omega L = \dfrac{1}{\omega C}$

At, $\omega_0 = \dfrac{1}{\sqrt{LC}}$ (Resonance)

$\text{Re}[Y] = \dfrac{1}{R}$ (Maximum value)

$$I_m(Y) = \frac{-\left(\omega L - \frac{1}{\omega C}\right)}{R^2 + \left(\omega L - \frac{1}{\omega C}\right)^2} = \frac{\left(\frac{1}{\omega C} - \omega L\right)}{R^2 + \left(\omega L - \frac{1}{\omega C}\right)^2}$$

At, $\omega_0 = \dfrac{1}{\sqrt{LC}}$ (Resonance)

Imaginary part of zero

$\Rightarrow$ $\text{Im}(Y) = 0$

For, $0 < \omega < \omega_0$

$\dfrac{1}{\omega C} > \omega L$

Therefore, $\text{Im}[Y] > 0$

For, $\omega_0 < \omega < \infty$

$\dfrac{1}{\omega C} < \omega L$

Therefore, $\text{Im}[Y] < 0$

On the basis of above analysis, the admittance locus is

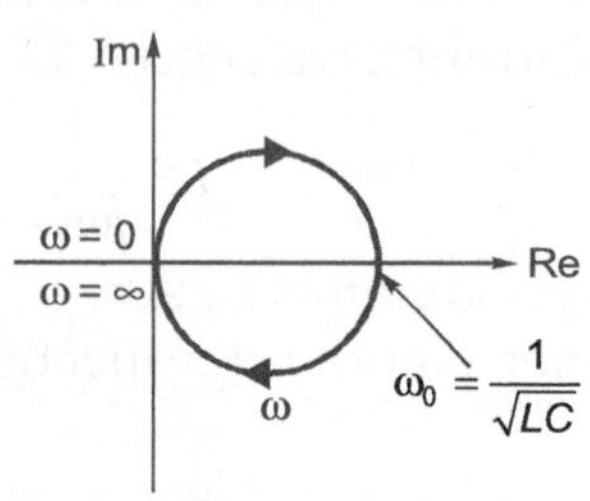

3.15 (a)

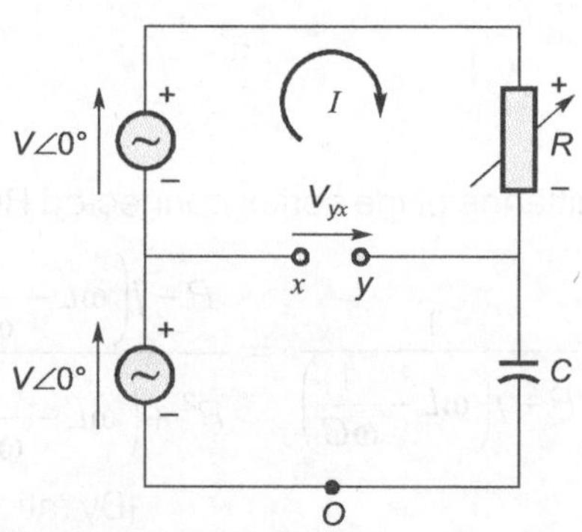

Let capacitive reactance = X_C

$$I = \frac{V\angle 0° + V\angle 0°}{R - jX_C} = \frac{2V}{R - jX_C}$$

Using KVL,

$$V_{YX} + IR - V = 0 \Rightarrow V_{YX} = V - IR$$

$$V_{YX} = V - \left(\frac{2V}{R - jX_C}\right)R = \frac{V(R - jX_C) - 2VR}{R - jX_C}$$

$$= -\frac{V(R + jX_C)}{(R - jX_C)}$$

Method-1:

$$V_{YX} = -V\left[\frac{R + jX_C}{R - jX_C}\right]$$

When, $R = 0$

$$V_{YX} = -V\left[\frac{0 + jX_C}{0 - jX_C}\right] = V$$

$$\Rightarrow \quad V_{YX} = -V \times \left[\frac{1 + j\frac{X_C}{R}}{1 - j\frac{X_C}{R}}\right]$$

When $R \to \infty$: $V_{YX} = -V$

Method-2:

$$V_{YX} = -V\left[\frac{R + jX_C}{R - jX_C}\right]$$

$$= V\angle 180° \times \left(\frac{\sqrt{R^2 + X_C^2}\angle \tan^{-1}\left(\frac{X_C}{R}\right)}{\sqrt{R^2 + X_C^2}\angle \tan^{-1}\left(\frac{-X_C}{R}\right)}\right)$$

$$= V\angle\left(180° + 2\tan^{-1}\left(\frac{X_C}{R}\right)\right)$$

Magnitude of $V_{YX} = V$

So, option (c) and (d) can not be correct, as magnitude is 2 V in these two options.

Angle of $V_{YX} = 180° + 2\tan^{-1}\left(\dfrac{X_C}{R}\right)$

When $R = 0$: $\angle V_{YX} = 180° + 2\tan^{-1}(\infty)$
$= 180° + 2 \times 90° = 360°$

When $R = \infty$: $\angle V_{YX} = 180° + 2\tan^{-1}(0) = 180°$

On the basis of above analysis, the locus of V_{YX} is drawn below:

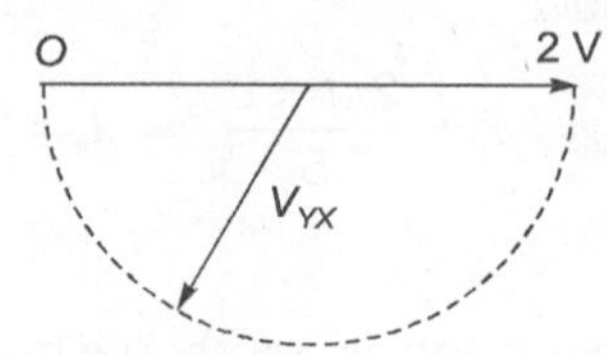

Locus of V_{YX}

3.16 (c)

Input impedance,

$$z = j\omega L + R \,\Big\|\, \frac{1}{j\omega C} = j\omega L + \frac{R}{1 + j\omega RC}$$

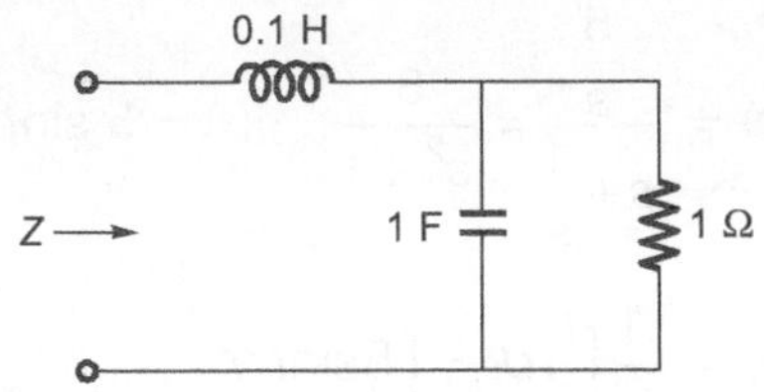

$$\therefore\ z = j0.1\omega + \frac{1}{1+j\omega}\times\frac{1-j\omega}{1-j\omega} = j0.1\omega + \frac{1-j\omega}{1+\omega^2}$$

$$= \frac{1}{1+\omega^2} + j\times\left(0.1\omega - \frac{\omega}{1+\omega^2}\right)$$

At resonance, imaginary part must be zero.

$$0.1\omega - \frac{\omega}{1+\omega^2} = 0 \Rightarrow 0.1 = \frac{1}{1+\omega^2}$$

$$\omega^2 + 1 = 10 \Rightarrow \omega^2 = 9 = 3 \text{ rad/sec}$$

3.17 (c)

$$q_1 = \frac{\omega L_1}{R_1}$$

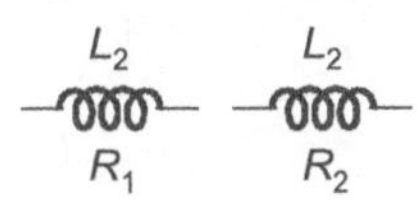

$$\Rightarrow \quad L_1 = \frac{q_1 R_1}{\omega}$$

Similarly, $L_2 = \dfrac{q_2 R_2}{\omega}$

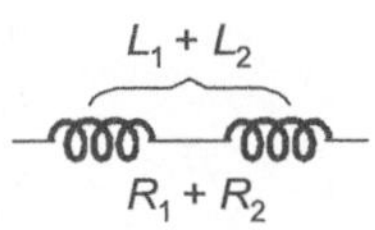

$$Q = \frac{\omega(L_1+L_2)}{R_1+R_2} = \frac{q_1 R_1 + q_2 R_2}{R_1+R_2}$$

3.18 (b)

Given,

At $\omega_1 = 1$ k rad/s,

$V_1 = 100\angle 0°$ V, $I_1 = 0.03\angle 31°$ A

At $\omega_2 = 2$ k rad/s,

$V_2 = 100\angle 0°$ V, $I_2 = 2\angle 0°$ A

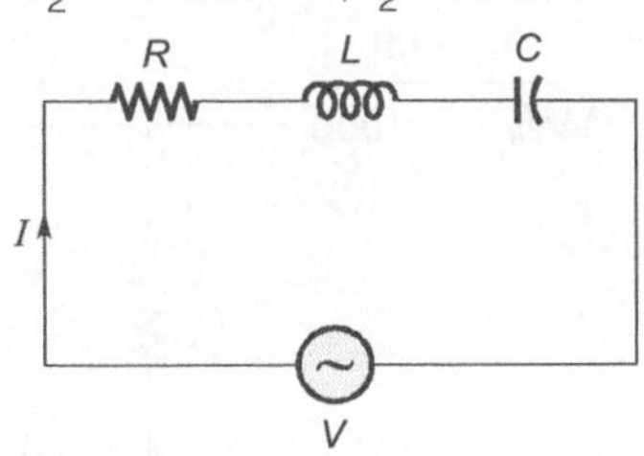

At $\omega_2 = 2$ k rad/s, voltage and current are in phase. Thus, it is a case of series resonance,

$$X_{L_{\omega_2}} = X_{C_{\omega_2}}$$

$$\therefore \quad Z = R = \frac{V_2}{I_2} = \frac{100\angle 0°}{2\angle 0°} = 50\ \Omega$$

$\therefore$ Resistance of circuit, $R = 50\ \Omega$

Now, at $\omega_1 = 1$ k rad/s,

$$Z = \frac{V_1}{I_1} = \frac{100\angle 0°}{0.03\angle 31°} = \frac{100}{0.03}\angle -31°\ \Omega \quad ...(i)$$

Also, $Z = |Z|\cdot\angle\tan^{-1}\left[\dfrac{X_L - X_C}{R}\right]$ ...(ii)

(at $\omega_1 = 1$ k rad/s)

Comparing equations (i) and (ii), we have:

$$-31° = \tan^{-1}\left[\frac{X_L - X_C}{R}\right]$$

or, $\tan(-31°) = \dfrac{X_L - X_C}{R}$

or, $X_L - X_C = R\tan(-31°) = 50 X - 0.6 = -30$

$\therefore X_{L_{\omega_1}} - X_{C_{\omega_1}} = -30$ ...(iii)

(At $\omega_1 = 1$ k rad/s)

Also, at $\omega_2 = 2$ k rad/s,

$$X_{L_{\omega_2}} = X_{C_{\omega_2}} \text{ or } \omega_2 L = \frac{1}{\omega_2 C}$$

or, $L = \dfrac{1}{\omega_2^2 C}$ ...(iv)

From equation (iii), $\omega_1 L - \dfrac{1}{\omega_1 C} = -30$

or, $\omega_1\left(\dfrac{1}{\omega_2^2 C}\right) - \dfrac{1}{\omega_1 C} = -30$ [Using (iv)]

or, $\dfrac{\omega_1}{\omega_2^2 C} - \dfrac{1}{\omega_1 C} = -30$

or, $\dfrac{1\times 10^3}{4\times 10^6 C} - \dfrac{1}{10^3 C} = -30$

or, $\left(\dfrac{10^{-3}}{4C} - \dfrac{10^{-3}}{C}\right) = -30$

or, $\dfrac{-3}{4C}\times 10^{-3} = -30$ or $C = \dfrac{3\times 10^{-3}}{4\times 30}$

or, $C = 25\times 10^{-6}$ F $= 25$ μF

Substituting the value of C in equation (iv), we get:

$$L = \frac{1}{\omega_2^2 C} = \frac{1}{(2\times 10^3)^2\times 25\times 10^{-6}} = \frac{1}{100} = 10 \text{ mH}$$

Therefore, values are:

$R = 50\ \Omega$, $L = 10$ mH, $C = 25$ μF

3.19 (b)

$$\Rightarrow \quad Z = j\omega L \,||\, \frac{1}{j\omega C} = \frac{L/C}{j\omega L + \dfrac{1}{j\omega C}}$$

$$= \frac{(L/C)\,j\omega C}{-\omega^2 LC + 1}$$

$$Z = \frac{j\omega L}{1-\omega^2 LC}$$

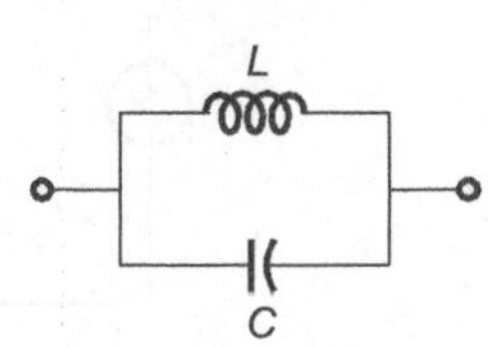

$$|Z| = \frac{\omega L}{1-\omega^2 LC}$$

For, $1 > \omega^2 LC$; $Z = +\text{ve}$

For, $1 < \omega^2 LC$; $Z = -\text{ve}$

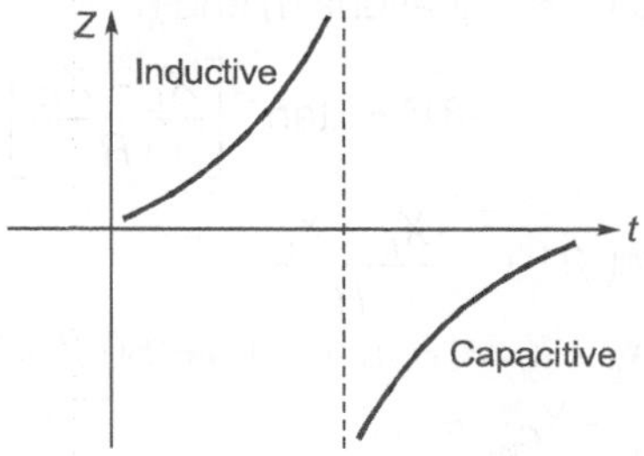

3.20 Sol.

The resonant frequency for the circuit is

$$\omega_0 = \frac{1}{\sqrt{LC}}\sqrt{\frac{R_L^2 - L/C}{R_C^2 - L/C}}$$

Since $(R_L = R_C = R)$

So, the circuit will have zero real part of admittance when,

$$R = \sqrt{\frac{L}{C}} = \sqrt{\frac{0.02}{100\,\mu\text{F}}} = 14.14\,\Omega$$

3.21 Sol.

Using star to delta conversion,

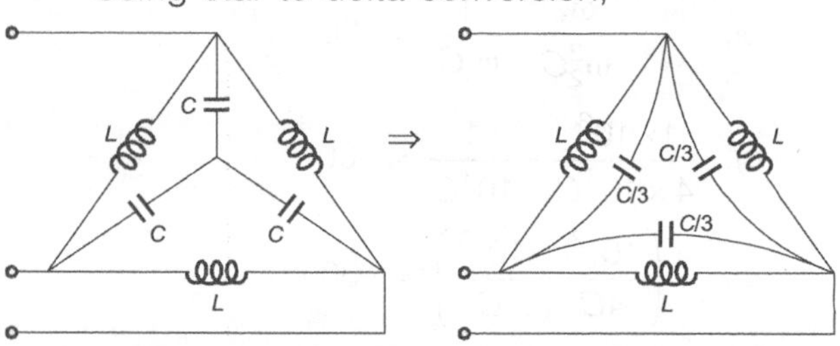

Line current will be zero when the parallel pair of induction-capacitor is resonant at $f = 50$ Hz.

So, $50 \times 2\pi = \frac{1}{\sqrt{LC/3}} \Rightarrow 100\pi = \frac{1}{\sqrt{LC/3}}$

Since $L = 10$ mH, C will be 3.03 mF.

3.22 Sol.

At resonance imaginary part of $Z_{eq} = 0$

$$\text{Real of } Z_{eq} = \frac{R_1 X_c^2}{R_1^2 + X_c^2} = \frac{100\times100\times100}{100^2+100^2} = 50\,\Omega$$

3.23 (b)

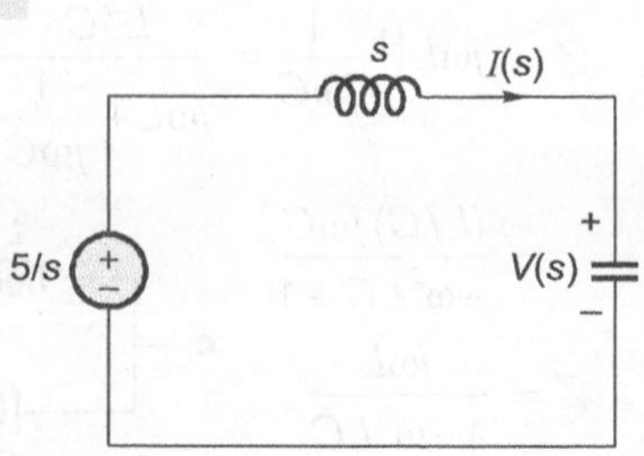

$$I(s) = \frac{\frac{5}{s}}{s+\frac{1}{s}} = \frac{5}{s^2+1};\quad i(t) = 5\sin t$$

$$v(t) = \frac{1}{C}\int_0^t i\,dt = \int_0^t 5\sin t\,dt$$

$$v(t) = 5[-\cos t]_0^t = 5[-\cos t + 1] = 5 - 5\cos t$$

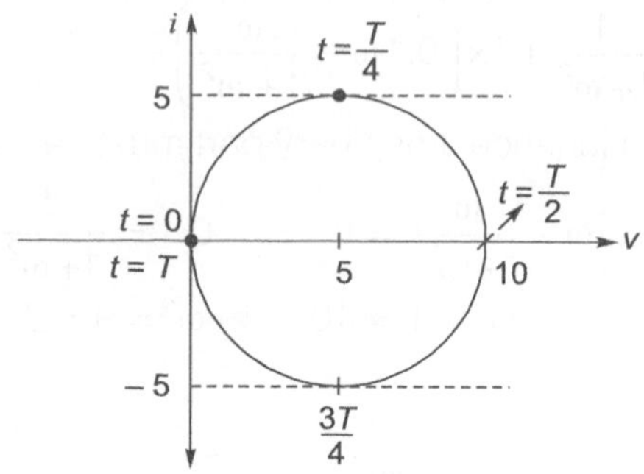

t	$i(t)$	$v(t)$
0	0	0
$\frac{T}{4}$	5	5
$\frac{T}{2}$	0	10
$\frac{3T}{4}$	–5	5
T	0	0

3.24 (48)

The overall Q-factor of the circuit is shown in below figure.

$$Q = \frac{Q_L \cdot Q_C}{Q_L + Q_C} = \frac{60\times240}{60+240} = 48$$

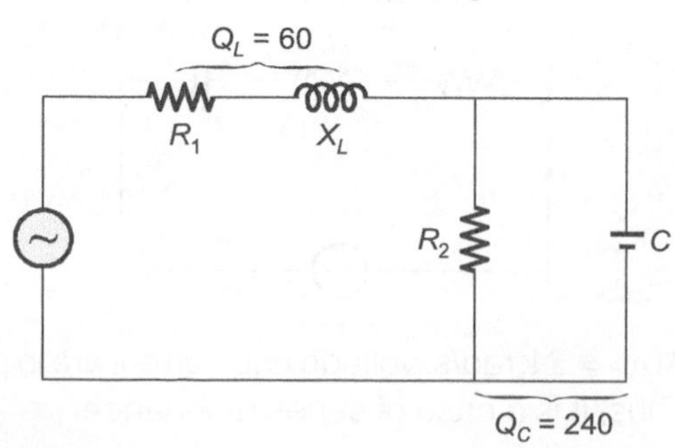

3.25 (250)

Resonance frequency, $f_o = 150$ kHz

Bandwidth width, B.W. = 600 Hz

The Q-factor of the network

$$Q = \frac{f_o}{\text{Bandwidth}} = \frac{150k}{600} = 250$$

■■■■

4 Network Theorems

4.1 For the circuit shown in the figure. The Norton equivalent source current value is....... A and its resistance is.......Ohm.

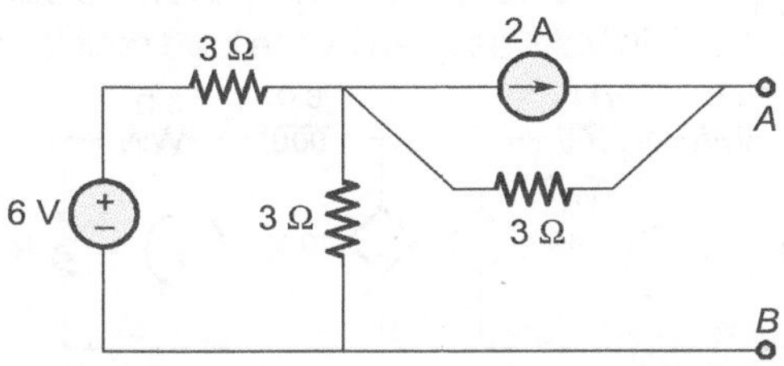

[1997 : 2 M]

4.2 Viewed from the terminals *A* and *B*, the following circuit shown in the figure can be reduced to an equivalent circuit of a single voltage source in series with a single resistor with the following parameters.

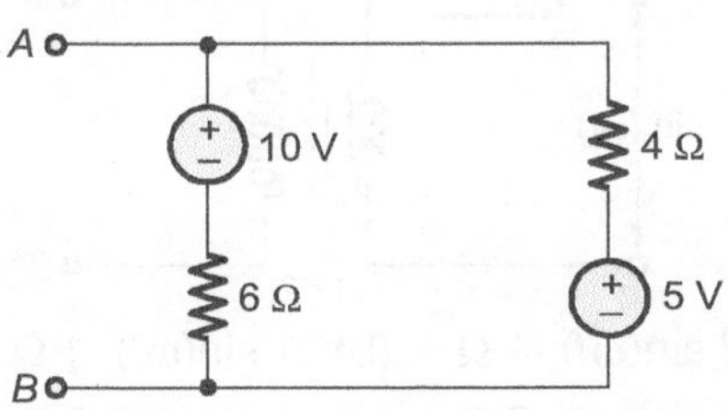

(a) 10 volt source in series with 10 Ω resistor.
(b) 7 volt source in series with 2.4 Ω resistor.
(c) 15 volt source in series with 2.4 Ω resistor.
(d) 1 volt source in series with 10 Ω resistor.

[1998 : 2 M]

4.3 In the figure, $Z_1 = 10\angle{-60°}$, $Z_2 = 10\angle 60°$, $Z_3 = 50\angle 53.13°$. Thevenin impedance seen from *X-Y* is

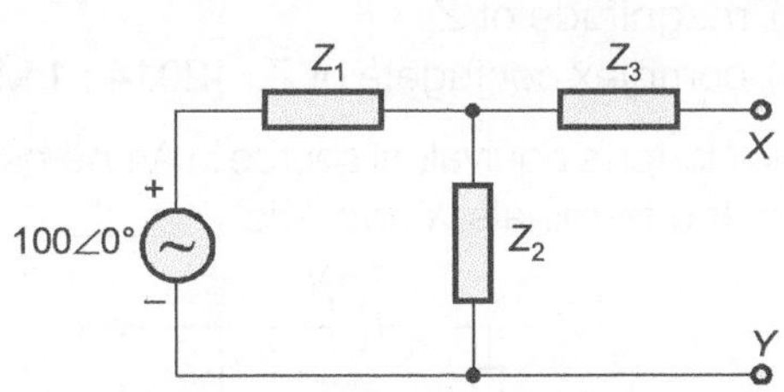

(a) $56.66\angle 45°$ (b) $60\angle 30°$
(c) $70\angle 30°$ (d) $34.4\angle 65°$ [2003 : 1 M]

4.4 Two AC sources feed a common variable resistive load as shown in the figure. Under the maximum power transfer condition, the power absorbed by the load resistance R_L is

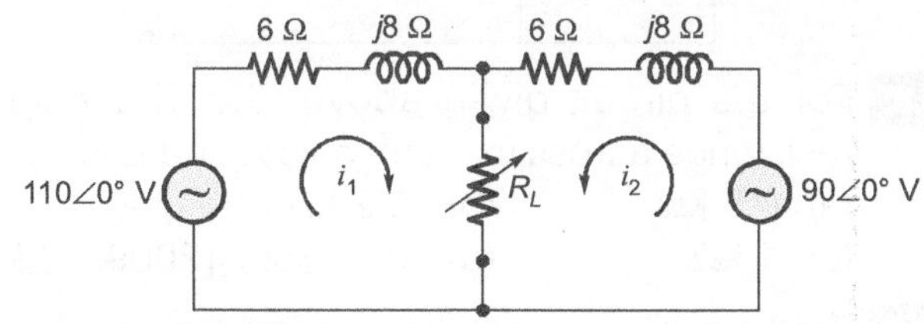

(a) 2200 W (b) 1250 W
(c) 1000 W (d) 625 W [2003 : 2 M]

4.5 In the given figure, the Thevenin's equivalent pair (voltage, impedance), as seen at the terminals *P-Q*, is given by

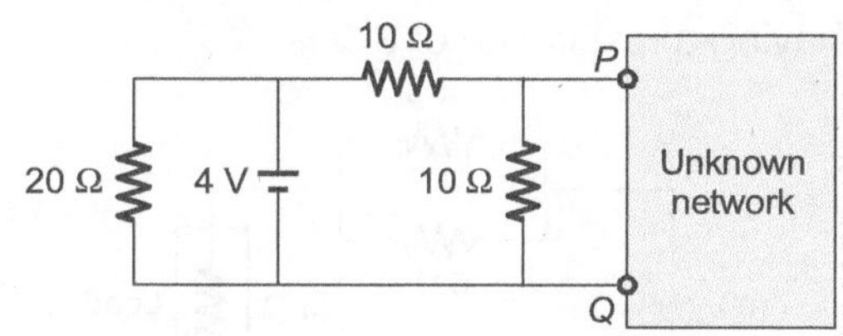

(a) (2 V, 5 Ω) (b) (2 V, 7.5 Ω)
(c) (4 V, 5 Ω) (d) (4 V, 7.5 Ω) [2005:2 M]

4.6 In the figure the current source is $1\angle 0$ A, $R = 1\ \Omega$, the impedances are $Z_C = -j\ \Omega$, and $Z_L = 2j\ \Omega$. The Thevenin equivalent looking into the circuit across *X-Y* is

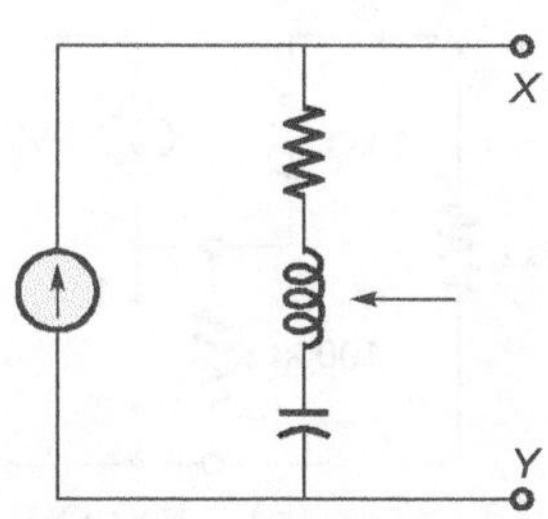

(a) $5\sqrt{2}\angle 0°$ V, $(1+2j)\ \Omega$
(b) $2\angle 45°$ V, $(1-2j)\ \Omega$
(c) $2\angle 45°$ V, $(1+j)\ \Omega$
(d) $\sqrt{2}\angle 45°$ V, $(1+j)\ \Omega$ [2006 : 1 M]

4.7 The Thevenin's equivalent of a circuit operating at ω = 5 rad/s, has $V_{oc} = 3.71\angle{-15.9°}$ V and $Z_0 = 2.38 - j0.667\ \Omega$. At this frequency, the minimal realization of Thevenin's impedance will have a

(a) resistor and a capacitor and an inductor
(b) resistor and a capacitor
(c) resistor and an inductor
(d) capacitor and an inductor [2008 : 1 M]

Statement for Linked Answer Questions (4.8 and 4.9):

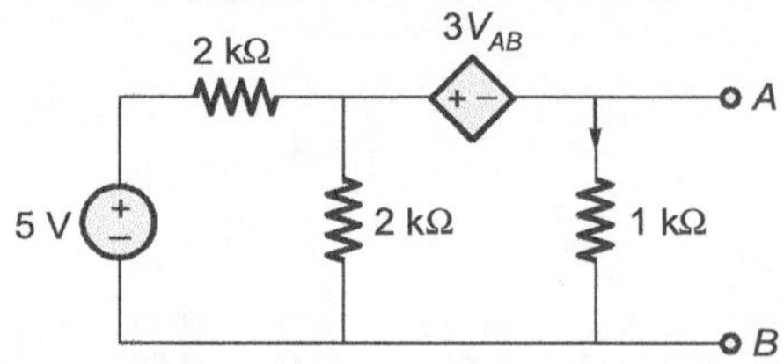

4.8 For the circuit given above, the Thevenin's resistance across the terminals *A* and *B* is
(a) 0.5 kΩ (b) 0.2 kΩ
(c) 1 kΩ (d) 0.11 kΩ [2009 : 2 M]

4.9 For the circuit given above, the Thevenin's voltage across the terminals *A* and *B* is
(a) 1.25 V (b) 0.25 V
(c) 1 V (d) 0.5 V [2009 : 2 M]

4.10 In the circuit given below, the value of *R* required for the transfer of maximum power to the load having a resistance of 3 Ω is

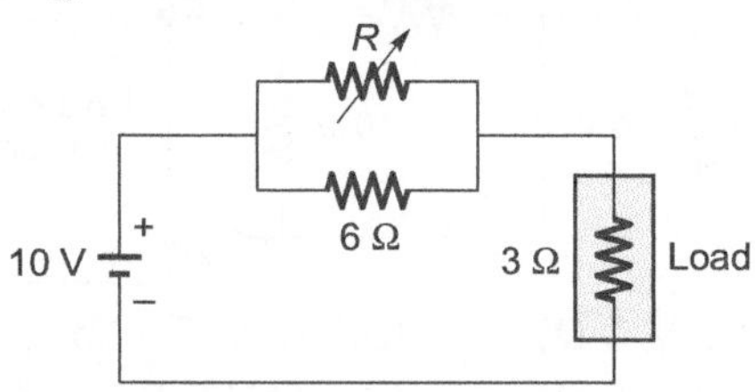

(a) zero (b) 3 Ω
(c) 6 Ω (d) infinity [2011 : 1 M]

4.11 The impedance looking into nodes 1 and 2 in the given circuit is

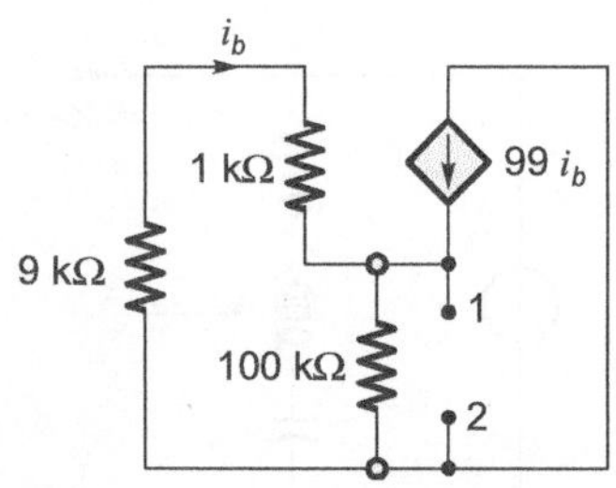

(a) 50 Ω (b) 100 Ω
(c) 5 kΩ (d) 10.1 kΩ [2012 : 1 M]

4.12 Assuming both the voltage sources are in phase, the value of *R* for which maximum power is transferred from circuit *A* to circuit *B* is

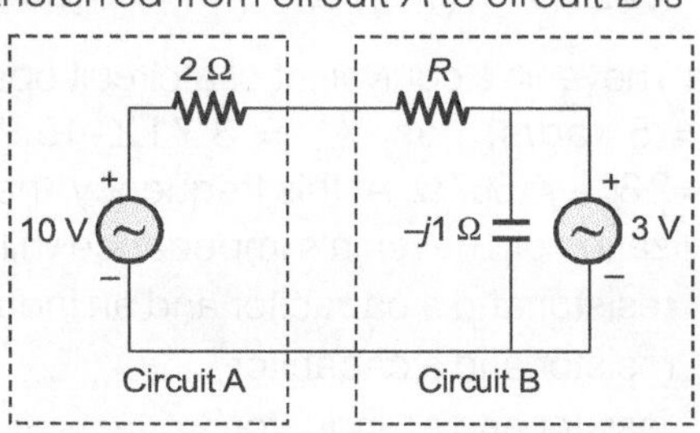

(a) 0.8 Ω (b) 1.4 Ω
(c) 2 Ω (d) 2.8 Ω [2012 : 2 M]

4.13 A source $v_s(t) = V\cos 100\pi t$ has an internal impedance of $(4 + j3)\ \Omega$. If a purely resistive load connected to this source has to extract the maximum power out of the source, its value in Ω should be
(a) 3 (b) 4
(c) 5 (d) 7 [2013 : 1 M]

4.14 In the circuit shown below, if the source voltage $V_s = 100\angle 53.13°$V then the Thevenin's equivalent voltage in Volts as seen by the load resistance R_L is

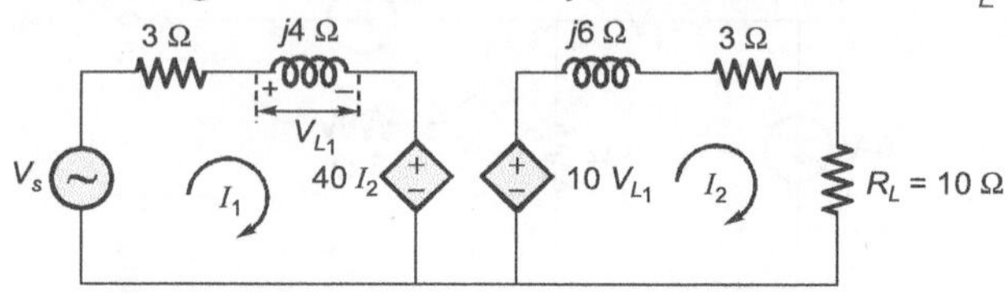

(a) $100\angle 90°$ V (b) $800\angle 0°$ V
(c) $800\angle 90°$ V (d) $100\angle 60°$ V [2013 : 2 M]

4.15 Assuming an ideal transformer, the Thevenin's equivalent voltage and impedance as seen from the terminals *x* and *y* for the circuit in figure are

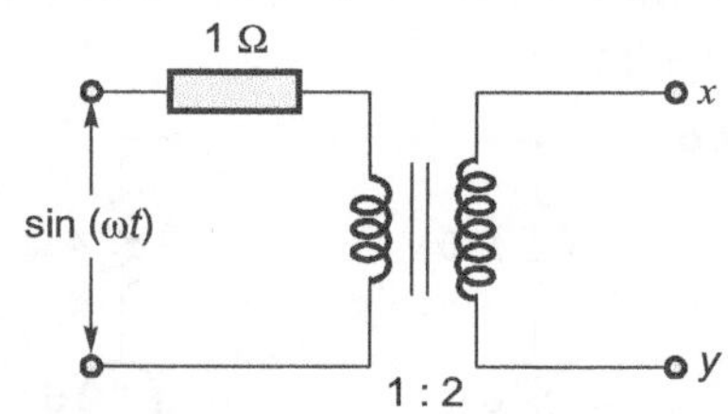

(a) 2 sin(ω*t*), 4 Ω (b) 1 sin(ω*t*), 1 Ω
(c) 1 sin(ω*t*), 2 Ω (d) 2 sin(ω*t*), 0.5 Ω
[2014 : 1 M, Set-2]

4.16 A non-ideal voltage source V_s has an internal impedance of Z_s. If a purely resistive load is to be chosen that maximizes the power transferred to the load, its values must be
(a) 0
(b) real part of Z_s
(c) magnitude of Z_s
(d) complex conjugate of Z_s [2014 : 1 M, Set-3]

4.17 The Norton's equivalent source in Amperes as seen into the terminals *X* and *Y* is ______.

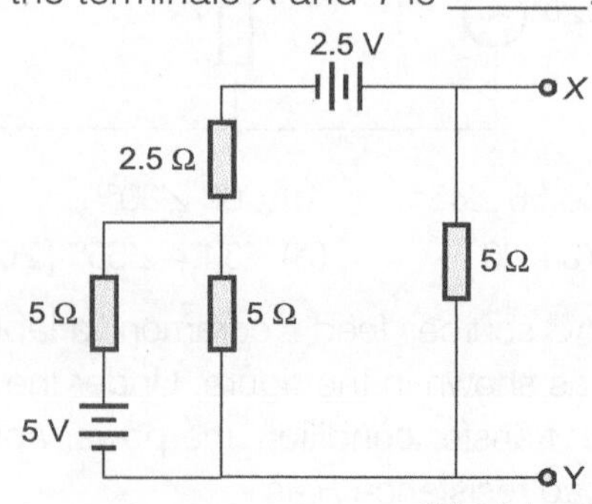

[2014 : 2 M, Set-3]

4.18 The voltage across the capacitor, as shown in the figure, is expressed as:

$$V_c(t) = A_1 \sin(\omega_1 t - \theta_1) + A_2 \sin(\omega_2 t - \theta_2)$$

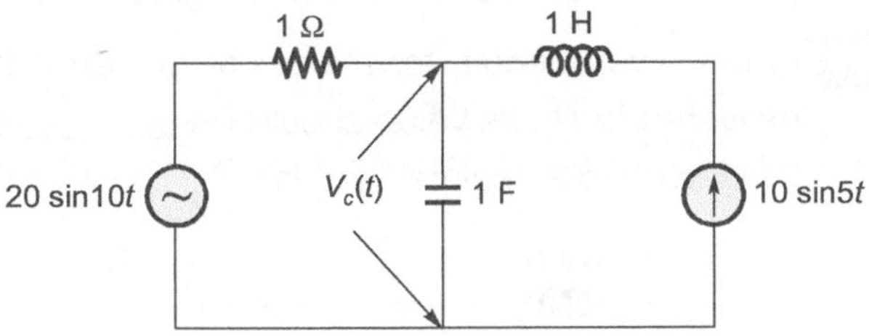

The values of A_1 and A_2 respectively, are

(a) 2.0 and 1.98 (b) 2.0 and 4.20
(c) 2.5 and 3.50 (d) 5.0 and 6.40

[2014 : 2 M, Set-2]

4.19 For the given circuit, the Thevenin equivalent is to be determined. The Thevenin voltage, V_{Th} (in Volt), seen from terminal AB is ____.

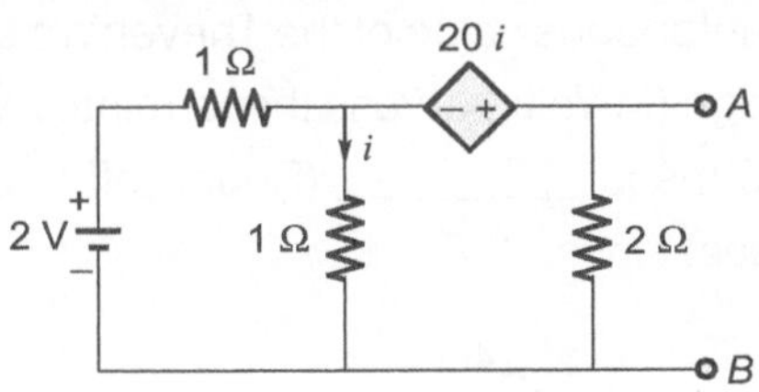

[2015 : 1 M, Set-1]

4.20 The circuit shown in the figure has two sources connected in series. The instantaneous voltage of the AC source (in Volt) is given by $V(t) = 12 \sin t$. If the circuit is in steady state. then the rms value of current (in Ampere) flowing in the circuit is ___.

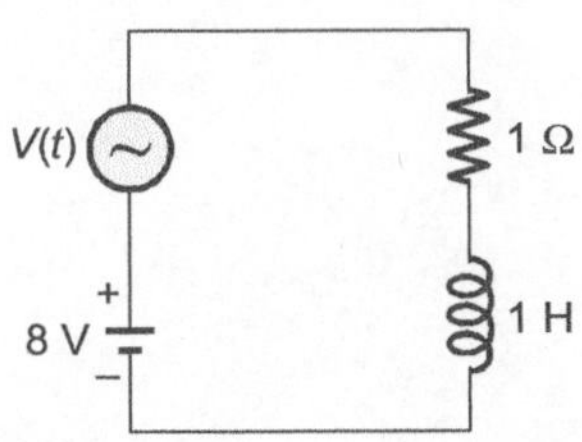

[2015 : 2 M, Set-1]

4.21 In a linear two-port network, when 10 V is applied to Port 1, a current of 4 A flows through Port 2 when it is short-circuited. When 5 V is applied to Port 1, a current of 1.25 A flows through a 1 Ω resistance connected across Port 2. When 3 V is applied to Port 1, the current (in Ampere) through a 2 Ω resistance connected across Port 2 is ____.

[2015 : 2 M, Set-1]

4.22 In the circuit shown below, the maximum power transferred to the resistor R is ______W.

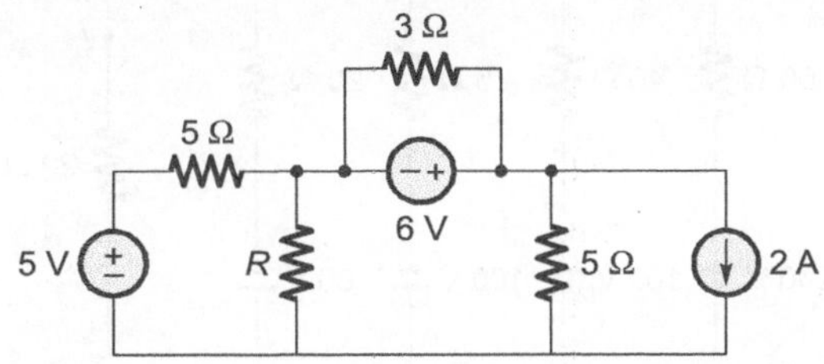

[2017 : 2 M, Set-1]

4.23 In the circuit shown below, the value of capacitor C required for maximum power to be transferred to the load is

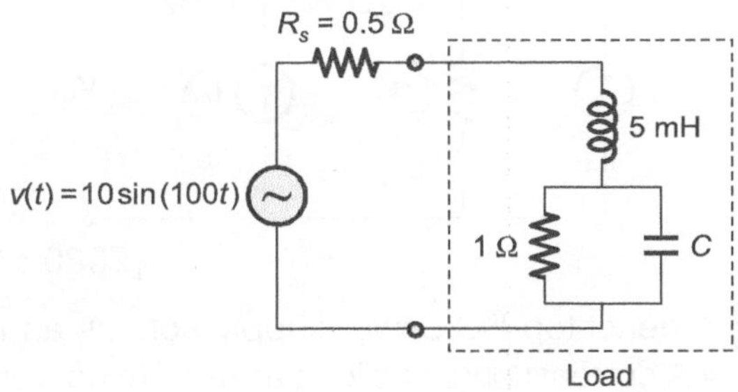

(a) 1 nF (b) 1 μF
(c) 1 mF (d) 10 mF

[2017 : 2 M, Set-2]

4.24 For the network given in the figure below, the Thevenin's voltage V_{ab} is

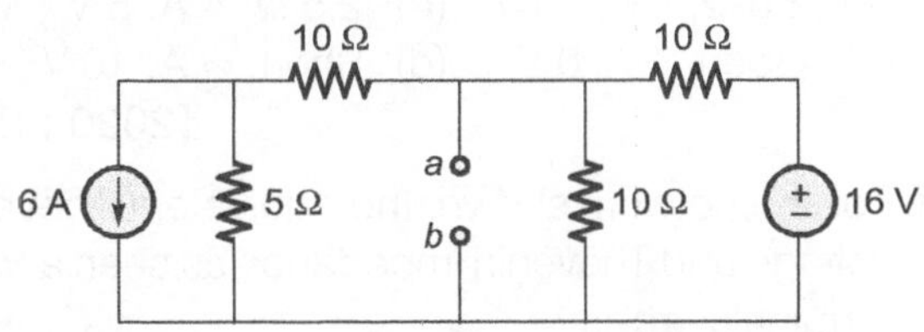

(a) −1.5 V (b) −0.5 V
(c) 0.5 V (d) 1.5 V

[2017 : 2 M, Set-2]

4.25 For the given 2-port network, the value of transfer impedance Z_{21} in ohms is ______.

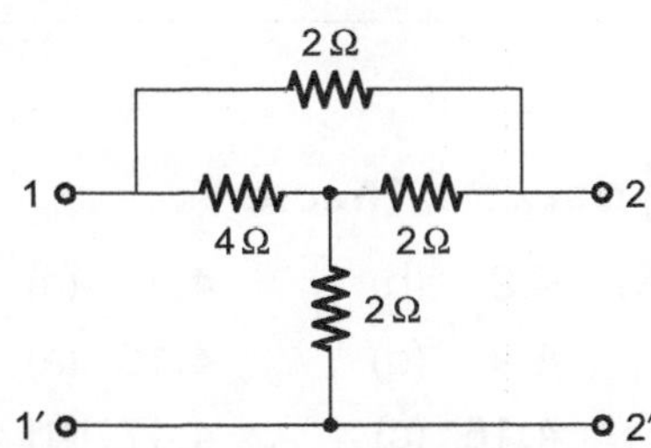

[2017 : 1 M, Set-2]

4.26 The current I flowing in the circuit shown below in Ampere is _____.

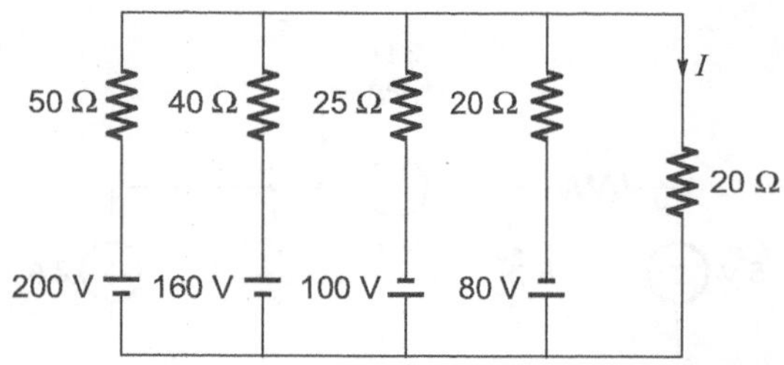

[2019 : 2 M]

4.27 The Thevenin equivalent voltage, V_{TH}, in V (rounded off to 2 decimal places) of the network shown below, is _______ .

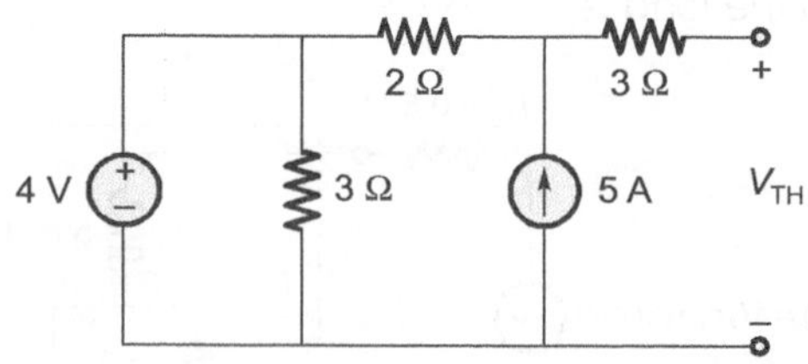

[2020 : 1 M]

4.28 A benchtop DC power supply acts as an ideal 4 A current source as long as its terminal voltage is below 10 V. Beyond this point, it begins to behave as an ideal 10 V voltage source for all load currents going down to 0 A. When connected to an ideal rheostat, find the load resistance value at which maximum power is transferred, and the corresponding load voltage and current.

(a) 2.5 Ω, 4 A, 10 V (b) 2.5 Ω, 4 A, 5 V
(c) Open, 4 A, 0 V (d) Short, ∞ A, 10 V

[2020 : 2 M]

4.29 For the network shown, the equivalent Thevenin voltage and Thevenin impedance as seen across terminals '*ab*' is

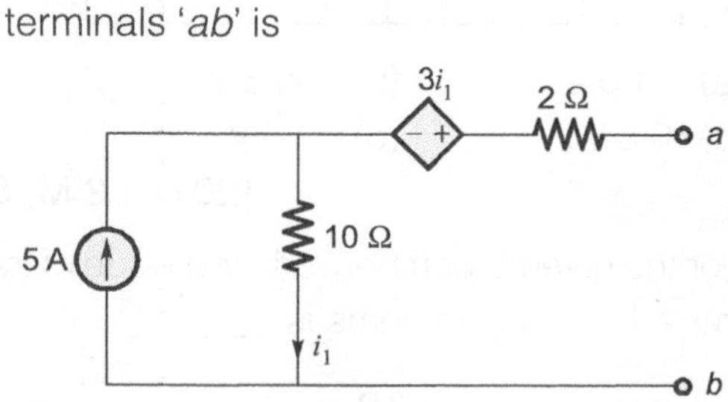

(a) 10 V in series with 12 Ω
(b) 35 V in series with 2 Ω
(c) 50 V in series with 2 Ω
(d) 65 V in series with 15 Ω [2021 : 1 M]

4.30 In the given circuit, for maximum power to be delivered to R_L, its value should be _______ Ω. (Round off to 2 decimal places)

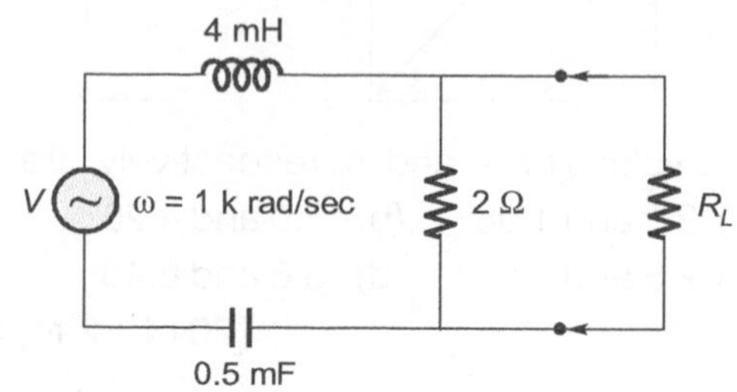

[2021 : 2 M]

4.31 For the circuit shown, if $i = \sin 1000t$, the instantaneous value of the Thevenin's equivalent voltage (in Volts) across the terminals *a*-*b* at time $t = 5$ ms is _______ (Round off to 2 decimal places).

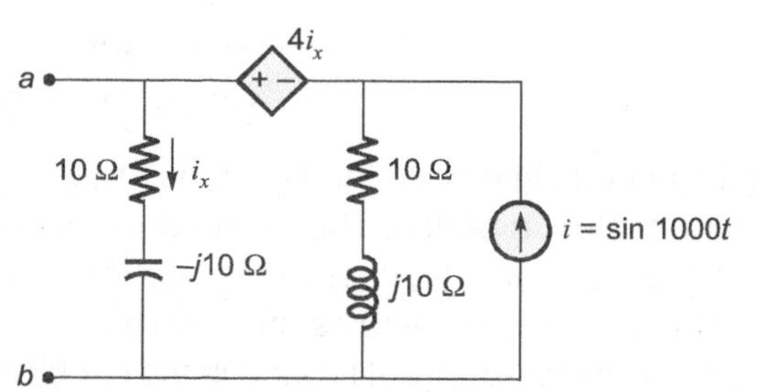

[2023 : 2 M]

Answers Network Theorems

4.1	(2, 4.5)	4.2	(b)	4.3	(a)	4.4	(d)	4.5	(a)	4.6	(d)	4.7	(b)
4.8	(b)	4.9	(d)	4.10	(a)	4.11	(a)	4.12	(a)	4.13	(c)	4.14	(c)
4.15	(a)	4.16	(c)	4.17	(1)	4.18	(a)	4.19	(3.36)	4.20	(10)	4.21	(0.545)
4.22	(3.025)	4.23	(d)	4.24	(a)	4.25	(3)	4.26	(0)	4.27	(14)	4.28	(a)
4.29	(d)	4.30	(1.41)	4.31	(–11.98)								

Explanations Network Theorems

4.1 Sol.

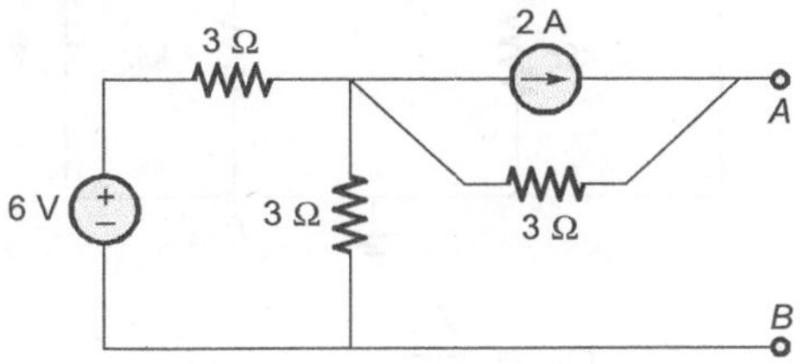

Using source transformation

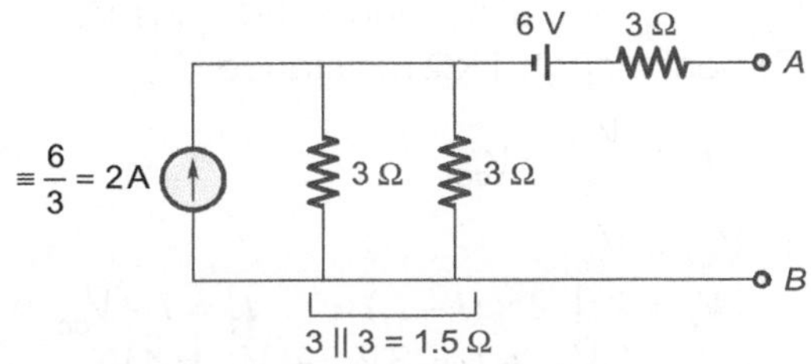

Using source transformation

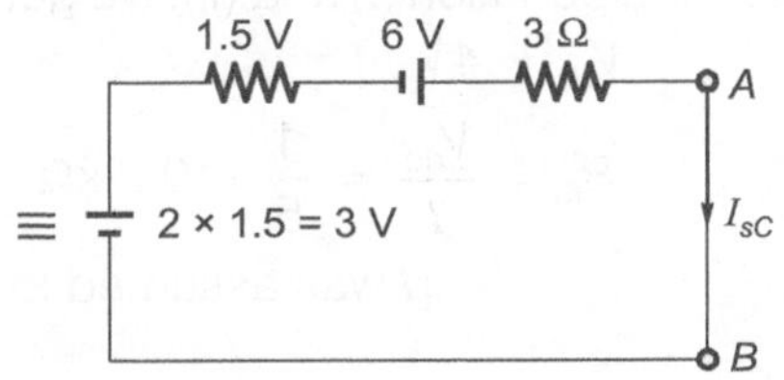

$$I_{sc} = \frac{3+6}{1.5+3} = 2\text{ A}; \quad R_N = 1.5 + 3 = 4.5\ \Omega$$

4.2 (b)

Using Millman's theorem

$$V = \frac{V_1Y_1 + V_2Y_2}{Y_1 + Y_2} = \frac{\frac{10}{6}+\frac{5}{4}}{\frac{1}{6}+\frac{1}{4}} = 7\text{V}$$

$$R = \frac{1}{Y_1 + Y_2} = \frac{1}{\frac{1}{6}+\frac{1}{4}} = 2.4\ \Omega$$

4.3 (a)

By Thevenin's theorem,

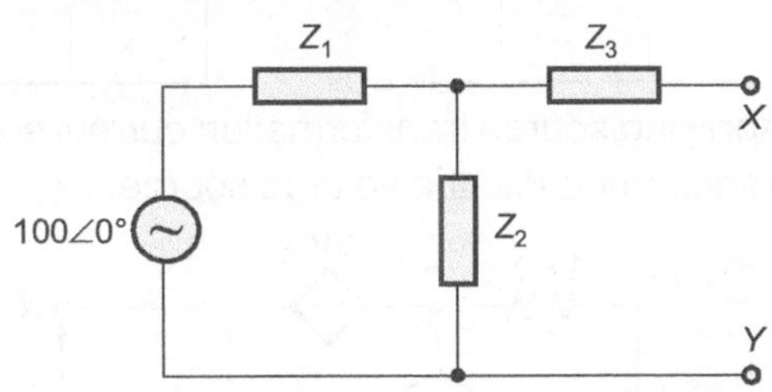

$$Z_{Th} = Z_{X-Y} = Z_1 \parallel Z_2 + Z_3 = \frac{Z_1 \times Z_2}{(Z_1 + Z_2)} + Z_3$$

$$= \frac{10\angle{-60} \times 10\angle 60}{(10\angle{-60} + 10\angle 60)} + (50\angle 53.13) = 56.66\angle 45°\ \Omega$$

4.4 (d)

For obtaining power absorbed by R_L under maximum power transfer condition. We find Thevenin's equivalent circuit across R_L.

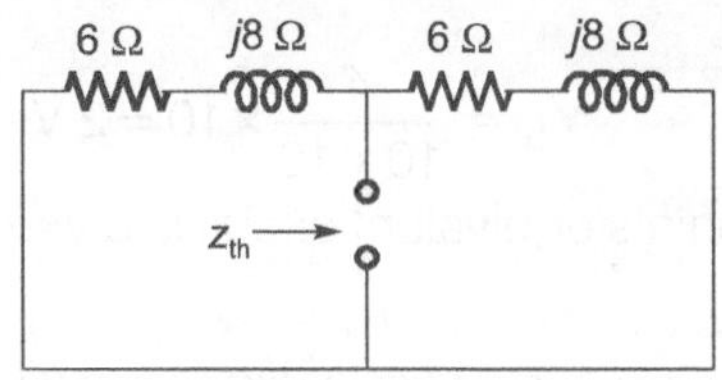

Z_{th} is calculated by short-circuting the voltage sources.

$$Z_{th} = (6 + j8) \,||\, (6 + j8) = 3 + j4\ \Omega$$

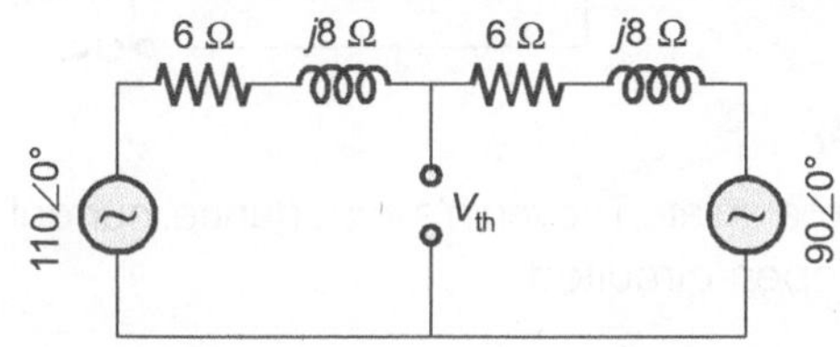

$$\frac{V_{th} - 110\angle 0°}{6 + j8} + \frac{V_{th} - 90\angle 0°}{6 + j8} = 0$$

$$V_{th} = 100\ \angle 0°\text{ V}$$

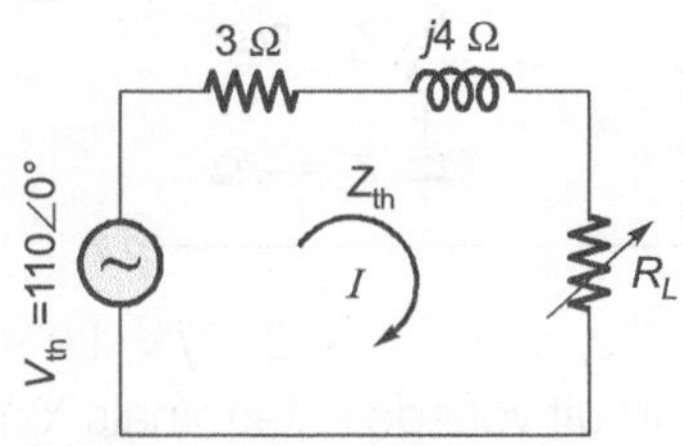

For maximum power transfer,

$$R_L = \sqrt{R_{th}^2 + X_{th}^2} = \sqrt{3^2 + 4^2} = 5\ \Omega$$

$$I = \frac{V_{th}}{(3 + j4) + R_L} = \frac{100}{8 + j4}$$

$$= 11.18\angle{-26.56°}\text{ A}$$

Power absorbed by R_L (max.)

$$= I^2R_L = 11.18^2 \times 5 = 625\text{ W}$$

4.5 (a)

To calculate R_{th}, (seen at terminals P-Q), voltage source is short-circuit

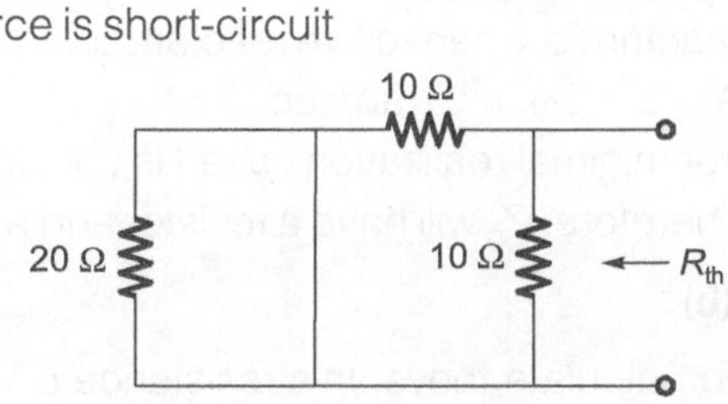

$$R_{th} = 10 \,||\, 10 = 5\ \Omega$$

V_{th} = Open-circuit voltage at the terminals, P-Q.

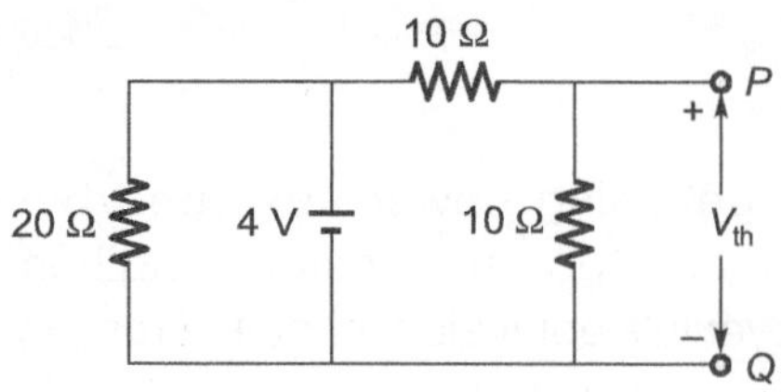

$$V_{th} = \frac{4}{10+10} \times 10 = 2\text{ V}$$

Thevenin's equivalent circuit is given below.

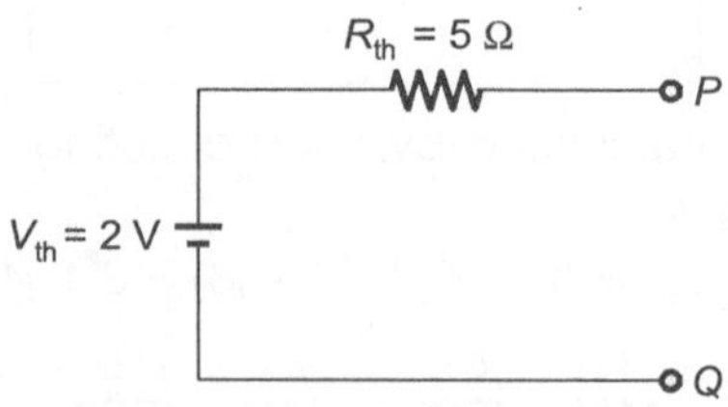

4.6 (d)

To calculate Thevenin's impedance, current-source is open-circuited

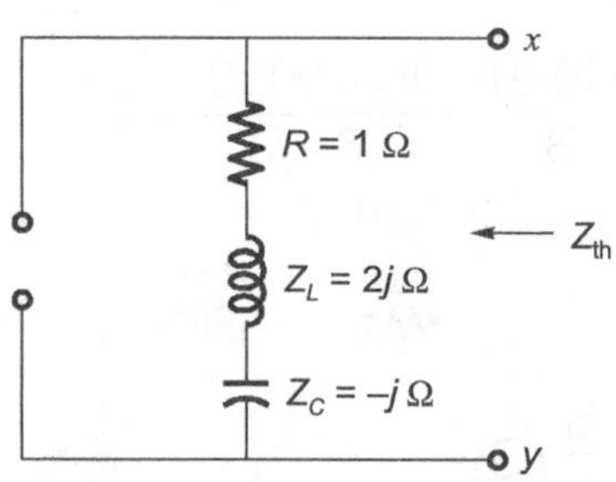

$Z_{th} = R + Z_L + Z_C = 1 + 2j - j = 1 + j\ \Omega$

Open-circuit voltage at terminals X-Y

$= I \times Z_{th} = 1\angle 0 \times (1 + j) = \sqrt{2}\angle 45°$ volts

4.7 (b)

Thevenin's Impedance:

$$Z_0 = 2.38 - j0.667\ \Omega$$

As real part is not zero, so Z_0 has resistor

$$\text{Img}[Z_0] = -j0.667$$

Case-I:

Z_0 has capacitor (as Img$[Z_0]$ is negative)

Case-II:

Z_0 has both capacitor and inductor, but inductive reactance < capacitive reactance.

At, $\omega = 5$ rad/sec

For minimal realization case-I is considered.

Therefore, Z_0 will have a resistor and a capacitor.

4.8 (b)

To calculate thevenin's resistance 5 V source is short-circuited and V_{dc} source is connected at terminals A and B

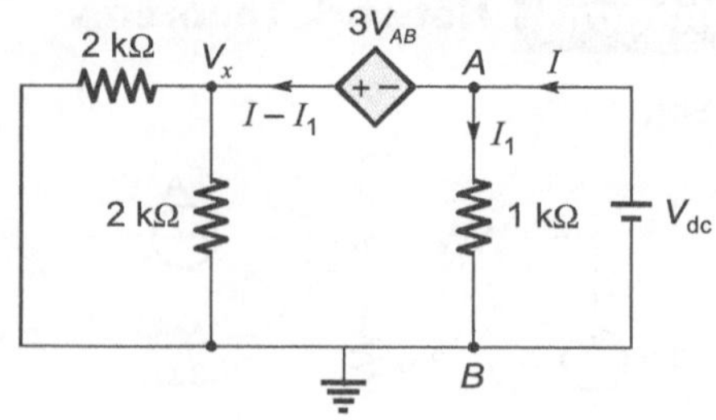

$$R_{th} = \frac{V_{dc}}{I}$$

$V_{AB} = V_{dc}$ and assuming I and I_1 in mA.

Current though 1 kΩ resistance

$$I_1 = \frac{V_{dc}}{1} = V_{dc}$$

$$V_{AB} = V_{dc}$$

$$V_x = 2 || 2 \times (I - I_1) = I - I_1 = I - V_{dc} \quad \text{...(i)}$$

$$V_x = 3V_{AB} + I_1 = 3V_{dc} + V_{dc} = 4V_{dc} \quad \text{...(ii)}$$

Comparing equation (i) and (ii), we get

$$I - V_{dc} = 4V_{dc}$$

$$R_{th} = \frac{V_{dc}}{I} = \frac{1}{5} = 0.2\text{ k}\Omega$$

[I was assumed to be in mA]

4.9 (d)

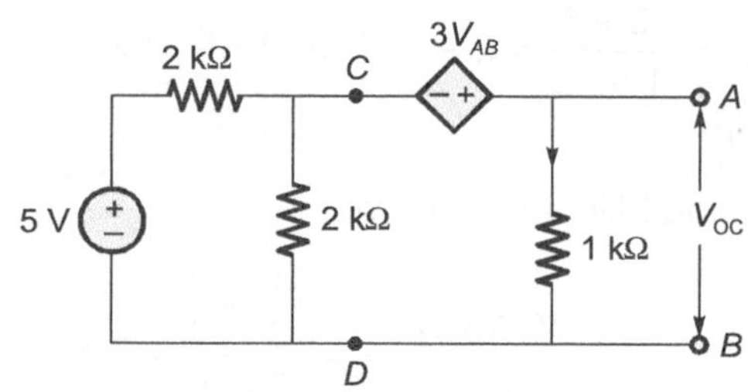

To calculate, Thevenin's voltage, terminals A-B are kept open.

Applying source transformation, voltage source is transformed into current source.

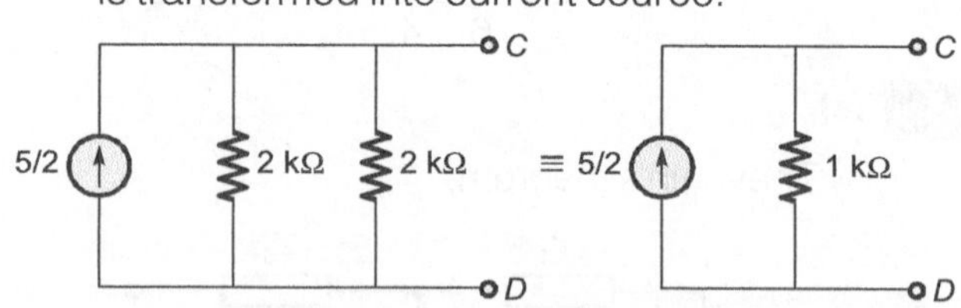

Applying source transformation current source is transformed into the voltage source.

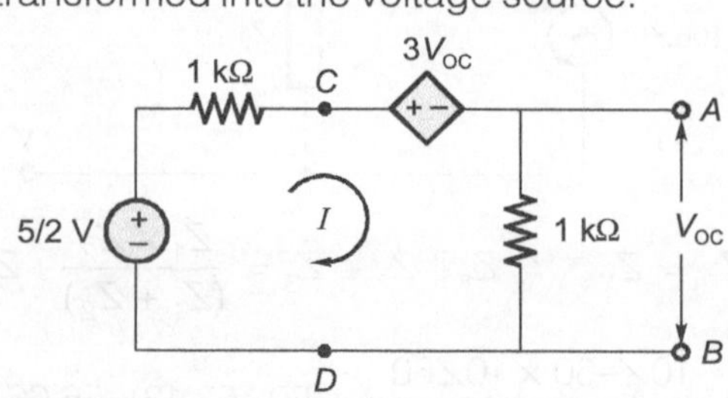

Applying KVL, {I is assumed to be in mA}

$$\frac{5}{2} - I - 3V_{OC} - I = 0 \qquad \text{...(i)}$$

$$I = \frac{V_{OC}}{I} = V_{OC} \qquad \text{...(ii)}$$

Put value of I in equation (i), we get

$$\frac{5}{2} - 5V_{OC} = 0 \Rightarrow V_{OC} = 0.5 \text{ V}$$

4.10 (a)

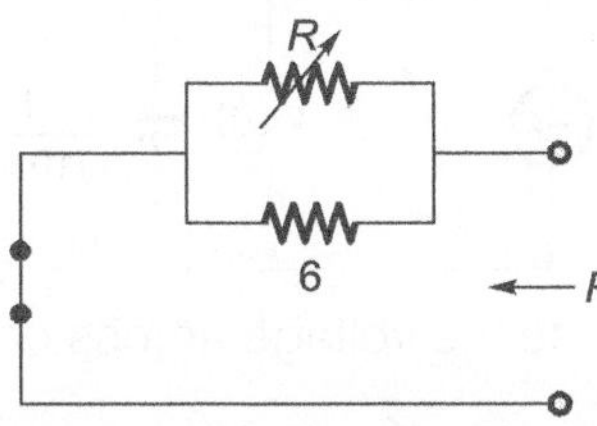

For maximum power transfer to the load, the source resistance must be minimum, i.e., zero.

So, $R = 0\ \Omega$

4.11 (a)

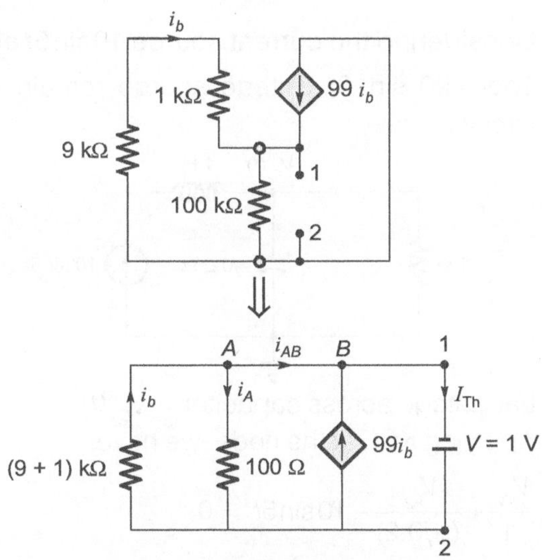

To find Thevenin impedance across node 1 and 2. Connect a 1 V source and find the current through voltage source.

Then, $Z_{Th} = \dfrac{1}{I_{Th}}$

By applying KCL at node B and A

$$i_{AB} + 99i_b = I_{Th}$$

$$i_b = i_A + i_{AB}$$

$$\Rightarrow \quad i_b - i_A + 99i_b = I_{Th}$$

$$\Rightarrow \quad 100i_b - i_A = I_{Th} \qquad \text{...(i)}$$

By applying KVL in outer loop

$$10 \times 10^3\, i_b = 1 \Rightarrow i_b = 10^{-4}\text{A}$$

and $10 \times 10^3\, i_b = -100\, i_A$

$$\Rightarrow \quad i_A = -100\, i_b$$

$\therefore$ From equation (i), we get

$$100i_b + 100i_b = I_{Th}$$

$$\Rightarrow \quad I_{Th} = 200i_b = 200 \times 10^{-4} = 0.02$$

$$\therefore \quad Z_{Th} = \frac{1}{I_{Th}} = \frac{1}{0.02} = 50\ \Omega$$

4.12 (a)

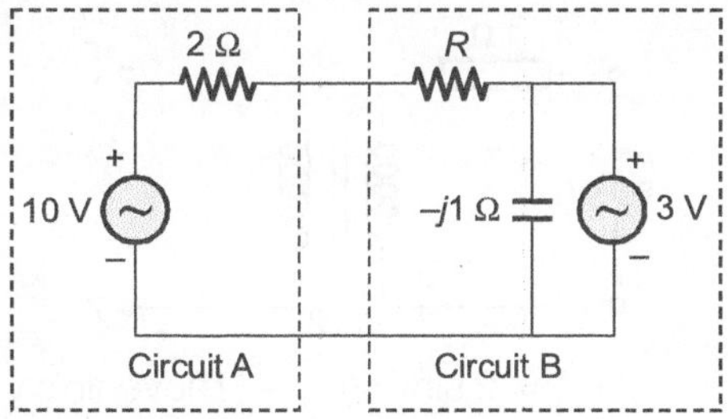

Thevenin equivalent circuit

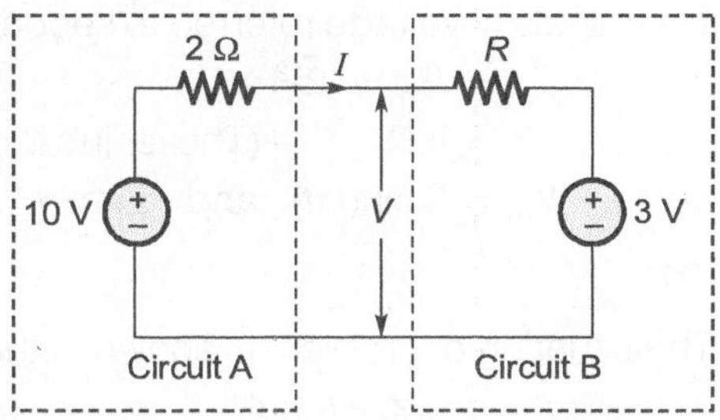

$$I = \frac{7}{R+2} \quad \text{and} \quad V = 10 - 2I$$

$$= 10 - \frac{14}{R+2} = \frac{10R+6}{R+2}$$

Power transferred from circuit A to circuit B

$$P = VI = \frac{10R+6}{R+2} \times \frac{7}{R+2}$$

For P to be maximum $\dfrac{dP}{dR} = 0$

$$(R + 2)^2 (10) - (10R + 6) \times 2(R + 2) = 0$$

$$5R^2 + 20R + 20 - 10R^2 + 26R + 12 = 0$$

$$5R^2 + 6R = 8 \Rightarrow R = 0.8\ \Omega$$

4.13 (c)

Using maximum power transfer theorem,

$$R_L = |Z| = |4 - j3| = \sqrt{4^2 + 3^2} = 5\ \Omega$$

4.14 (c)

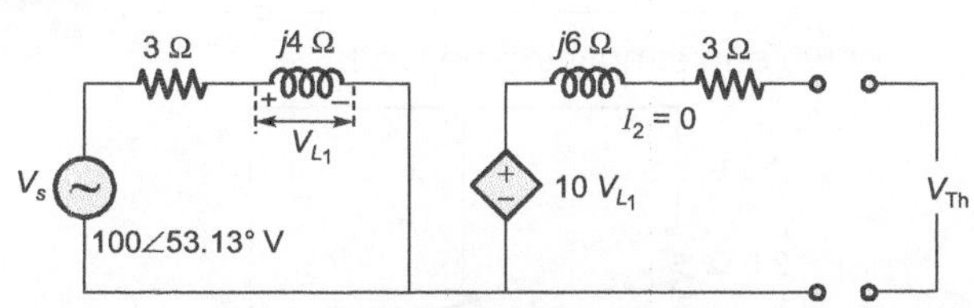

$$V_{L_1} = \left(\frac{j4}{3 + j4}\right) 100\angle 53.13° = 80\angle 90° \text{ V}$$

$$V_{Th} = 10\, V_{L_1} = 800\angle 90° \text{ V}$$

4.15 (a)

Thevenin's equivalent voltage = Voltage referred to secondary.

We have: $\dfrac{\sin \omega t}{V_{\text{Th}}} = \dfrac{1}{2}$

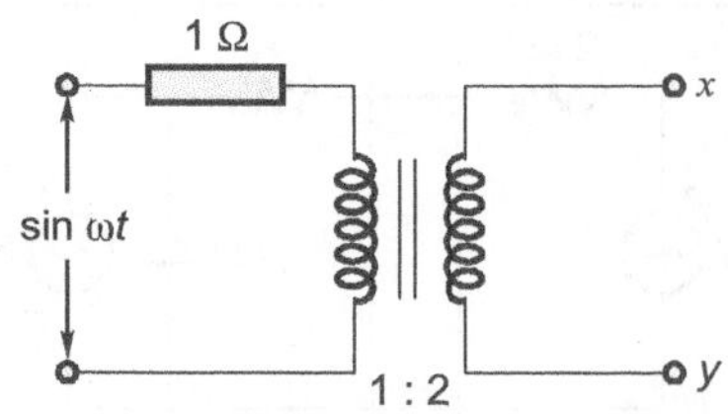

or, $V_{\text{Th}} = 2\sin(\omega t)$ ... (Thevenin's voltage)

Also, Thevenin's impedance seen from the x and y terminals = Voltage referred to secondary side

$\therefore$ $Z_{\text{Th}} = R_{\text{Th}} = (2)^2 \times 1$

$= 4\,\Omega$ ...(Thevenin's impedance)

So, $V_{\text{Th}} = 2\sin(\omega t)$ and $Z_{\text{Th}} = R_{\text{Th}} = 4\,\Omega$

4.16 (c)

The situation of problem is shown in the figure:

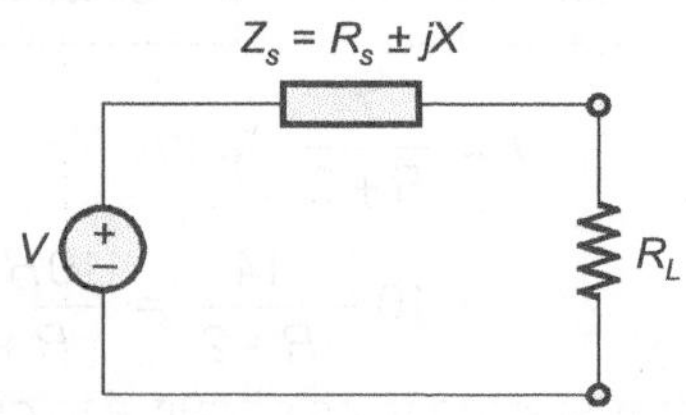

For the transfer of maximum power from source to load,

$$R_L = \sqrt{R_s^2 + X^2} = |Z|$$

Hence, option (c) is correct.

4.17 Sol.

Using source transformation theorem,

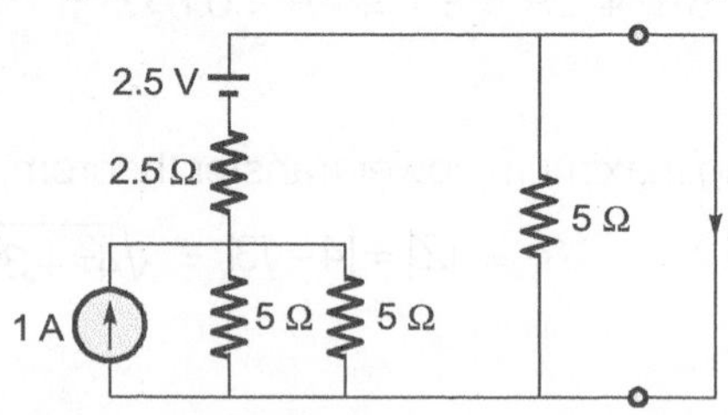

or we can simply the network,

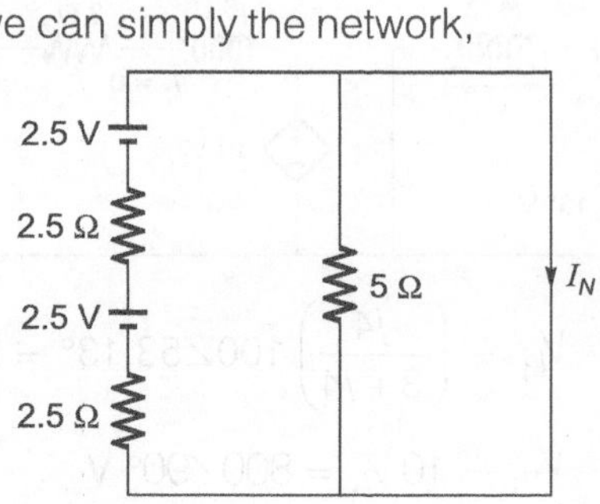

Now from the circuit, we get

$$I_N = \frac{5}{5} = 1\text{ A}$$

4.18 (a)

Let us apply superposition theorem,

Considering the voltage source 20 sin10*t* alone:

Then, 10 sin5*t* remain open circuited.

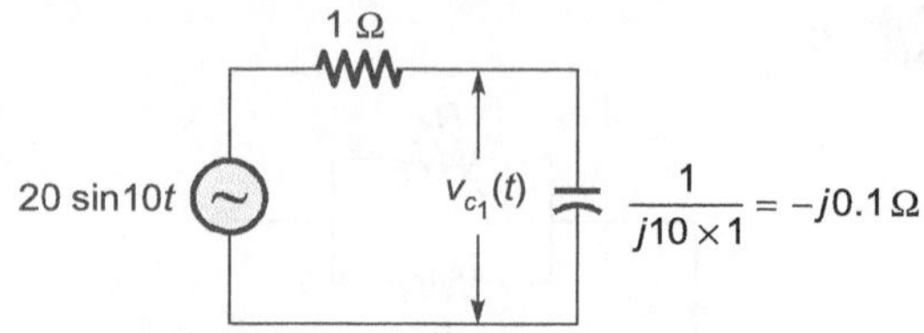

Let, $V_{C_1}(t)$ be the voltage across capacitor,

$$\therefore \quad V_{C_1}(t) = \left(\frac{-j0.1}{1-j0.1}\right) \times 20\sin 10t$$

$$= (1.99\angle{-84.28°})\sin 10t$$

$$\therefore \quad V_{C_1}(t) \approx 2\sin(10t - 84.28°) \quad \text{...(i)}$$

Considering the current source 10sin5*t* alone:

Then, 20 sin10*t* voltage source remain short circuited

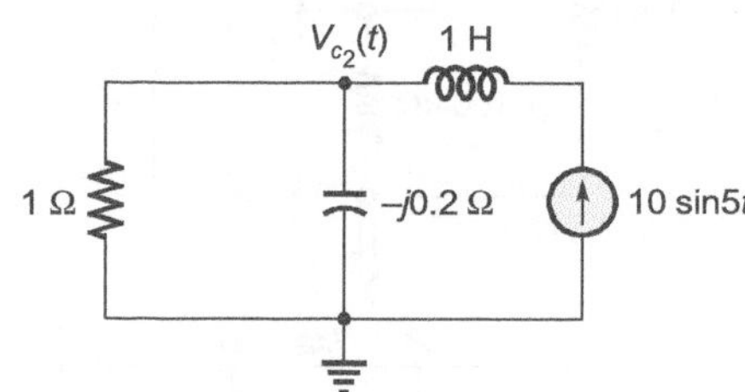

Let voltage across capacitor = $V_{C_2}(t)$

Applying KCL at the node, we have:

$$\frac{V_{C_2}}{1} + \frac{V_{C_2}}{(-j0.2)} - 10\sin 5t = 0$$

or, $$V_{C_2}(t) = \frac{10\sin 5t}{(1+j5)}$$

or, $$V_{C_2}(t) = 1.97\sin(5t - 78.69°) \quad \text{...(ii)}$$

Using superposition theorem, voltage across capacitor is

$$V_c(t) = V_{C_1}(t) + V_{C_2}(t) = 2\sin(10t - 84.28°) + 1.97\sin(5t - 78.69°)$$

$$\therefore V_c(t) = 2\sin(10t - 84.28°) + 1.97\sin(5t - 78.69°) \quad \text{...(iii)}$$

Given : $V_c(t) = A_1\sin(\omega_1 t - \theta_1) + A_2\sin(\omega_2 t - \theta_2)$ ...(iv)

Comparing equations (iii) and (iv), we have:

$\boxed{A = 2 \text{ and } B = 1.97 \approx 1.98}$ (closest answer).

4.19 Sol.

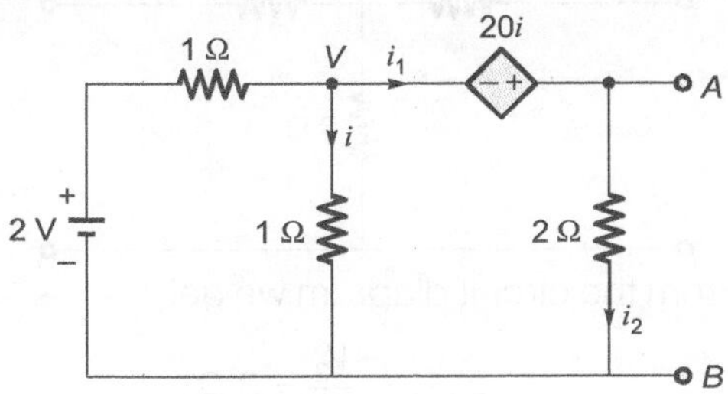

$$2-1(i_1+i)-i = 0$$
$$2-i_1-2i = 0$$
$$2i+i_1 = 2 \quad ...(i)$$
$$\frac{V-2}{1}+\frac{V}{1}+\frac{V-(-20i)}{2} = 0$$
$$2(V-2)+2V+V+20i = 0$$
$$4V-4+V+20i = 0 \Rightarrow 5V+20i = 4$$

where, $\frac{V}{1} = i$

$$25i = 2 \Rightarrow i = \frac{4}{25}\text{ A}$$
$$i_1 = 2-2\times\frac{4}{25} = 1.68\text{ A}$$
$$V_{AB} = 1.68\times 2 = 3.36\text{ V}$$

4.20 Sol.

$$i(t) = \frac{8}{1}+\frac{12\sin t}{\sqrt{1+1}} = 8+6\sqrt{2}\sin t$$
$$i_{\text{rms}}(t) = \sqrt{8^2+\frac{1}{2}(6\sqrt{2})^2} = 10\text{ A}$$

4.21 Sol.

$V_1 = 10$ V, $I_2 = 4$ A, $V_2 = 0$ Cond. ...(*i*)

$V_1 = 5$ V, $I_2 = 1.25$ A, $V_2 = 1.25\times 1 = 1.25$ V Cond. ...(*ii*)

$V_1 = 3$ V, $I_2 = ?$, $R = 2\ \Omega$ Cond. ...(*iii*)

As we know from ABCD parameter,

$$V_1 = AV_2 - BI_2; \quad I_1 = CV_2 - DI_2$$

From condition (*i*),

$$10 = A(0)-B(4); \quad B = -\frac{10}{4}$$

From condition (*ii*),

$$5 = A(1.25)-\left(-\frac{10}{4}\right)\times(1.25)$$
$$A = \frac{\left(5-\frac{12.5}{4}\right)}{1.25} = 1.5$$

From condition (*iii*), we get

$$3 = 1.5(2I)-\left(-\frac{10}{4}\right)\times I = 3I+2.5I = 5.5I$$
$$I = 0.545\text{ A}$$

4.22 Sol.

To get R_{th} and V_{th}, consider the following steps.

Case-1: For R_{th}

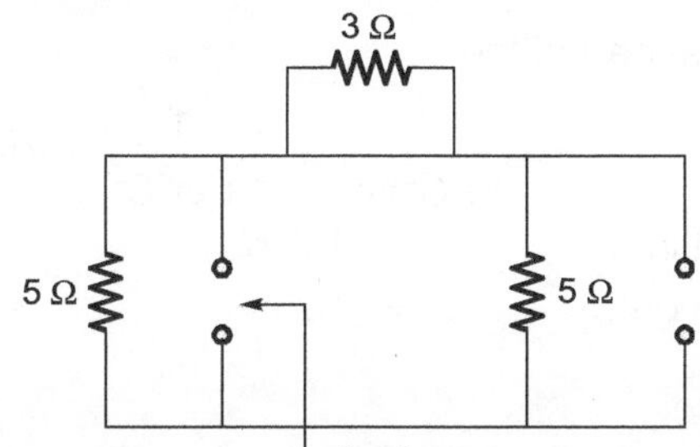

$$R_{Th} = \frac{5\times 5}{5+5} = 2.5\ \Omega$$
$$R_{th} = 2.5\ \Omega$$

Case-2: For V_{th}

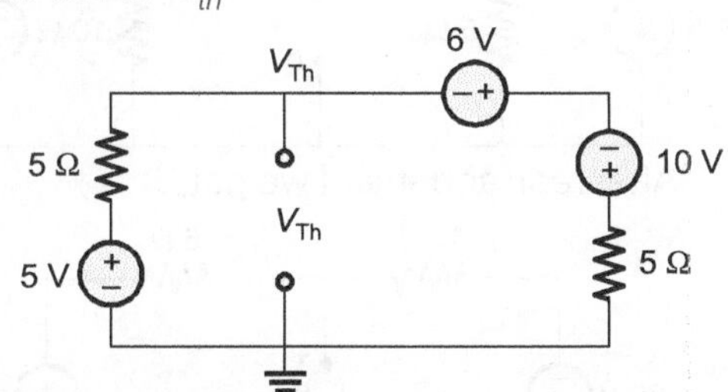

Applying KCL at node,

$$\frac{V_{th}-5}{5}+\frac{V_{th}+16}{5} = 0$$
$$2V_{th} = -11 = -5.5\text{ V}$$

Maximum power transferred,

$$P_{\max} = \frac{V_{th}^2}{4R_L} = 3.025\text{ W}$$

4.23 (d)

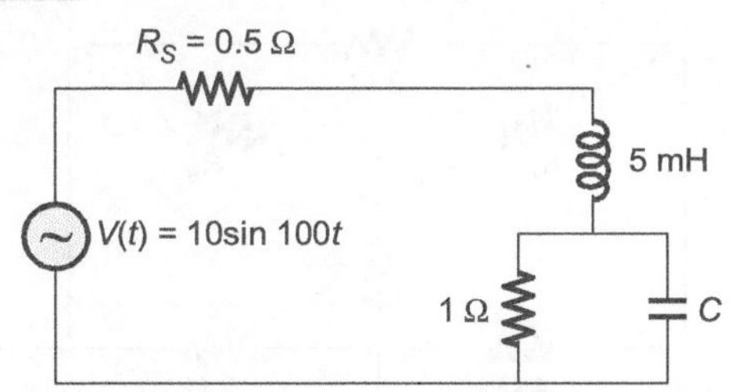

The frequency at which the load is resistive and it is equal to 0.5 Ω i.e. The load is resistive means, the imaginary part of the load is equal to zero and real part is equal to 0.5 Ω.

$$Z_{\text{load}} = \frac{1\times\frac{1}{Cs}}{1+\frac{1}{Cs}}+Ls = \frac{1}{1+Cs}+Ls$$

$$= \frac{(1-Cs)}{1-C^2s^2} + Ls$$

Put $s = j\omega$;

$$Z_{load} = \frac{1-j\omega C}{1+C^2\omega^2} + j\omega L$$

$$= \frac{1}{1+C^2\omega^2} + j\omega\left(L - \frac{C}{(1+C^2\omega^2)}\right)$$

Real part of the

$$Z_{load} = \frac{1}{1+C^2\omega^2} = \frac{1}{1+\omega^2C^2} = 0.5$$

Putting, $\omega = 100$ r/s

We get, $C = 10$ mF

4.24 (a)

Consider the following circuit,

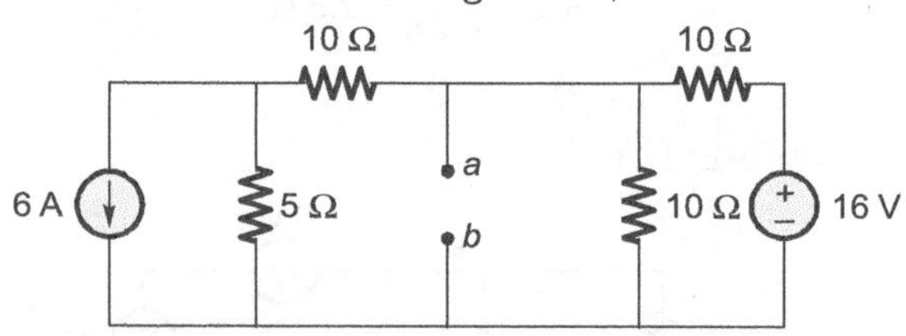

After rearrangement we get,

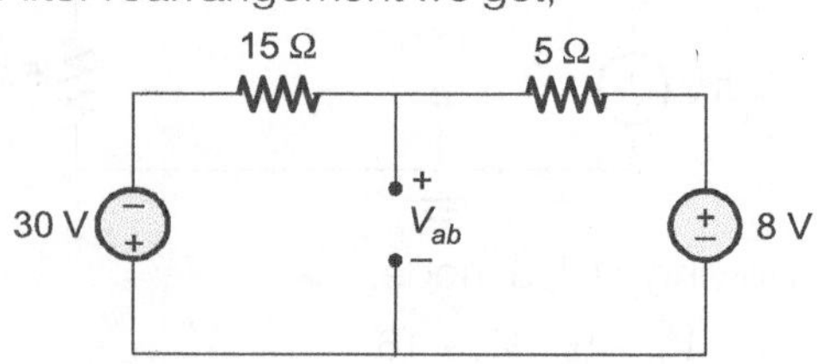

From circuit using KCL,
Voltage,

$$\frac{V_{ab}+30}{15} + \frac{V_{ab}-8}{5} = 0$$

$$V_{ab} + 30 + 3V_{ab} - 24 = 0 \Rightarrow V_{ab} = -1.5 \text{ V}$$

4.25 Sol.

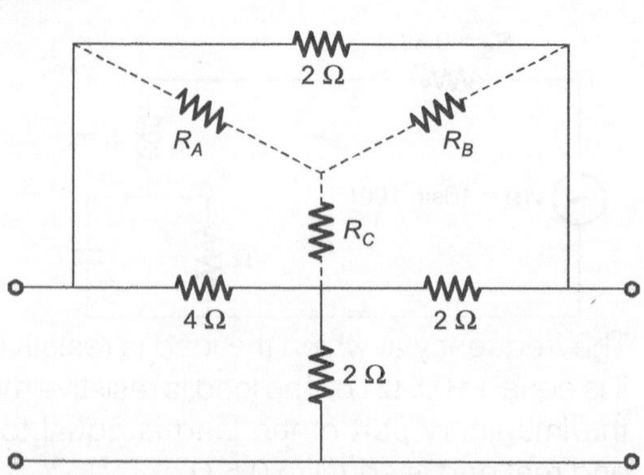

where, $R_A = 1\ \Omega$; $R_B = 1\ \Omega$

$$R_C = \frac{1}{2}\ \Omega$$

After rearrangement consider the following circuit,

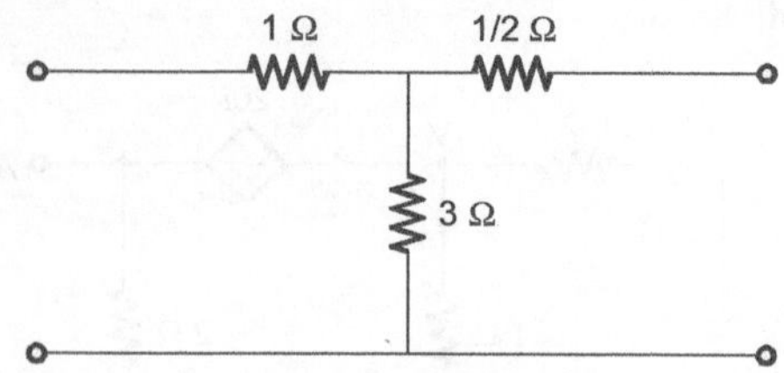

From the circuit diagram we get,

$$Z_{21} = \frac{V_2}{I_1} = 3\ \Omega$$

4.26 Sol.

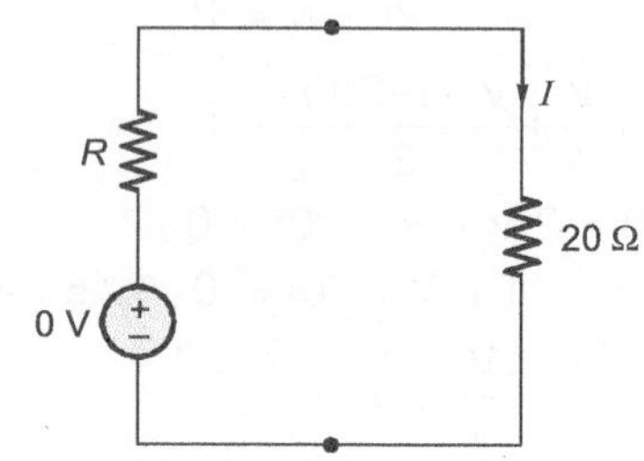

By Millman's theorem,

$$E = \frac{\frac{200}{50} + \frac{160}{40} - \frac{100}{25} - \frac{80}{20}}{\frac{1}{50} + \frac{1}{40} + \frac{1}{25} + \frac{1}{20}} = 0 \text{ V}$$

$$\frac{1}{R} = \frac{1}{50} + \frac{1}{40} + \frac{1}{25} + \frac{1}{20}$$

Simplified circuit,

$\therefore$ $I = 0$ A

4.27 Sol.

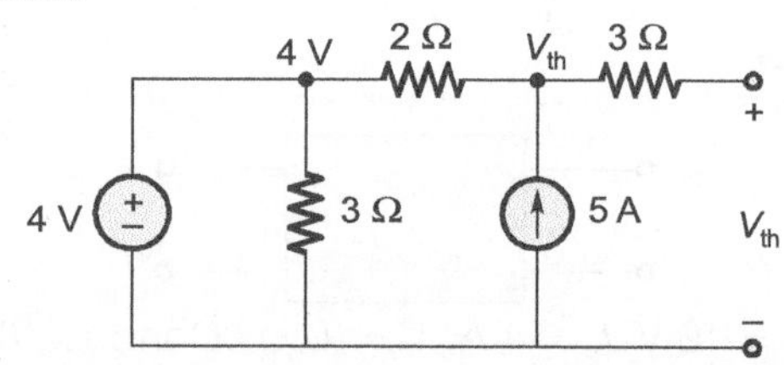

$$\frac{V_{th}-4}{2} = 5 \Rightarrow V_{th} = 14 \text{ V}$$

4.28 (a)

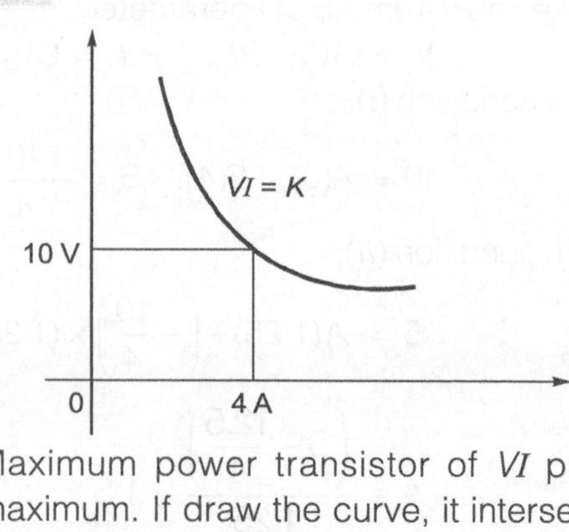

Maximum power transistor of *VI* product is maximum. If draw the curve, it intersect (10, 4) that will give maximum power.

The terminal voltage is 10 V (Load voltage) and current is 4 A (Load current).

Load resistance is $\frac{10}{4} = 2.5\ \Omega$.

4.29 (d)

Given circuit can be resolved as shown below,

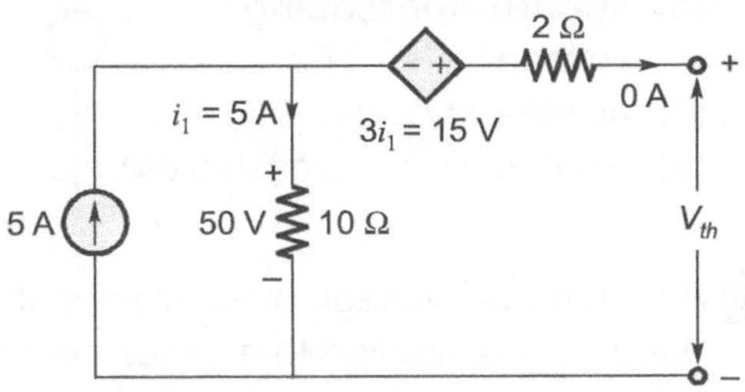

$$V_{TH} = 15 + 50 = 65\ \text{V}$$

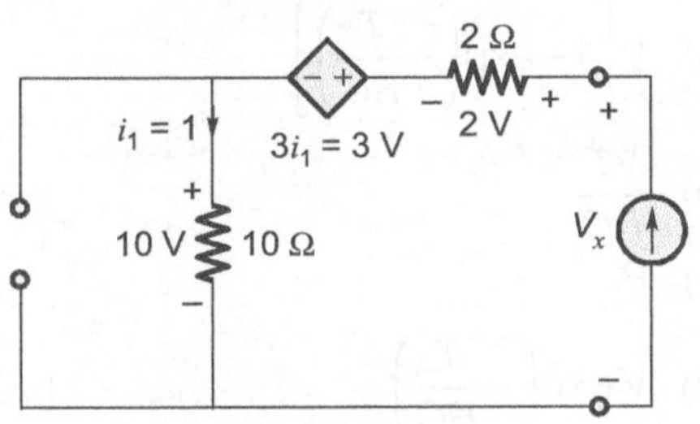

$$V_x = 2 + 3 + 10 = 15\ \text{V}$$

$$R_{TH} = \frac{V_x}{1} = 15\ \Omega$$

4.30 (1.41)

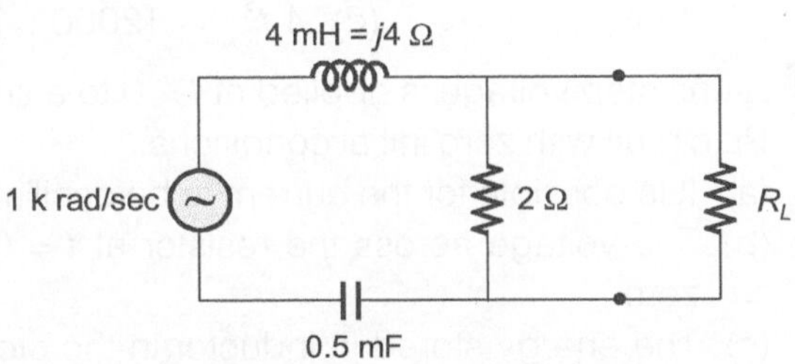

$$= \frac{-j}{\omega C} = \frac{-j}{1000 \times 0.5 \times 10^{-3}} = -j2\Omega$$

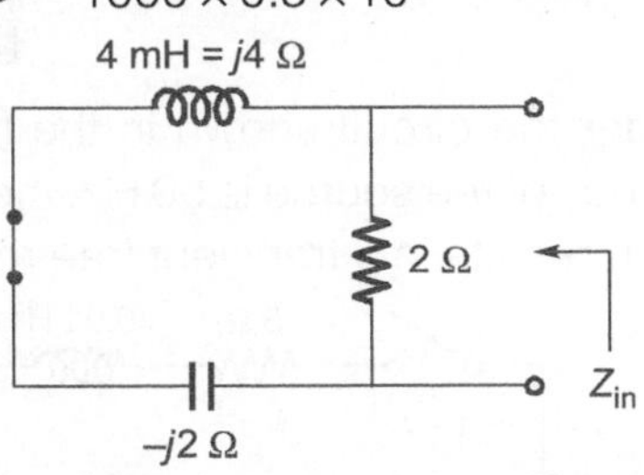

$$Z_{in} = 2 \parallel j2 = \frac{j4}{2 + j2} = \frac{j2}{1 + j1}$$

For maximum power transfer,

$$R_L = |Z_{TH}| = \frac{2}{\sqrt{2}} = \sqrt{2} = 1.414\ \Omega$$

4.31 (-11.98)

Let $\sin 1000t = 1\angle 0°$ A

Applying source transformation,

$$V_s = (10 + j10) \times i = (10 + j10) \times 1\angle 0° = (10 + j10)\ \text{Volts}$$

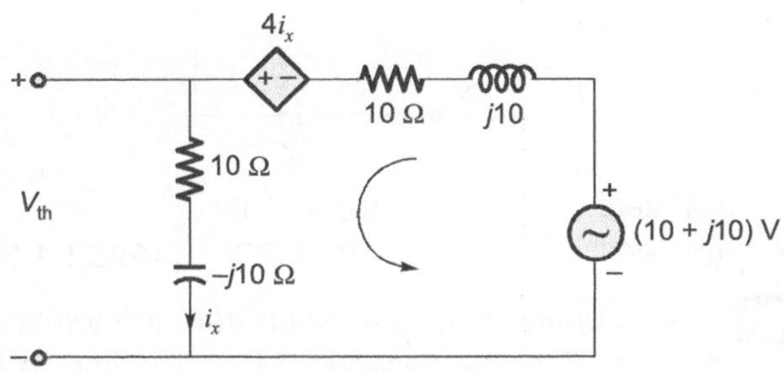

$$V_{th} = i_x(10 - j10)$$

Applying KVL

$$10 + j10 = (10 + j10)i_x - 4i_x + (10 - j10)i_x$$

$$i_x = \frac{10 + j10}{16}$$

$$V_{th} = i_x(10 - j10) = \frac{100 + 100}{16} = 12.5\angle 0°\ \text{Volts}$$

$$V_{th} = 12.5 \sin 1000t$$

At t = 5 ms, $V_{th} = 12.5 \sin 1000 \times 5 \times 10^{-3} = -11.98$ V

■■■■

5 Transients and Steady State Response

5.1 In the series RC circuit shown in the figure the voltage across C starts increasing when the DC source is switched ON. The rate of increase of voltage across C at the instant just after the switch is closed (i.e., at $t = 0^+$) will be

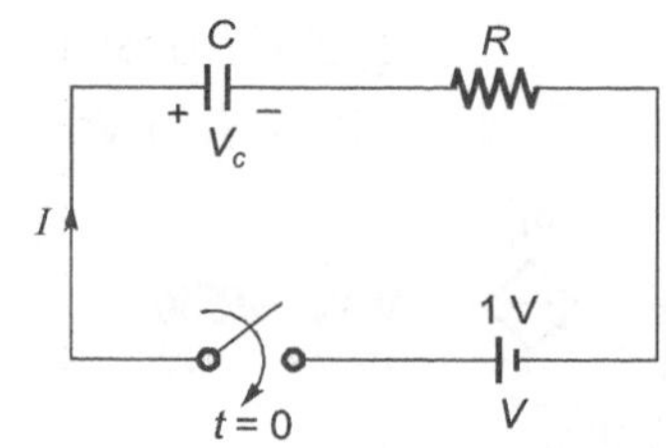

(a) zero (b) infinity
(c) RC (d) $1/RC$ [1996 : 1 M]

5.2 The v-i characteristic as seen from the terminal pair (A, B) of the network of Fig. 1 is shown in the Fig. (2). If an inductance of value 6 mH is connected across the terminal – pair (A, B), the time constant of the system will be

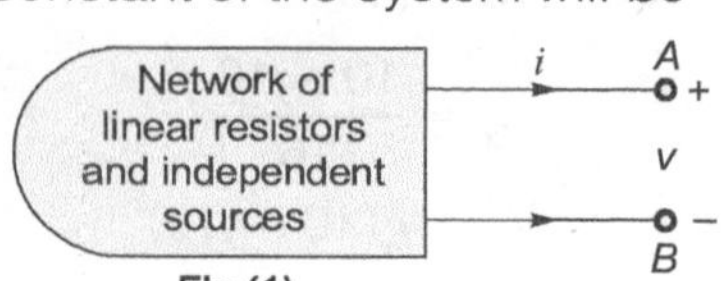

Fig.(1)

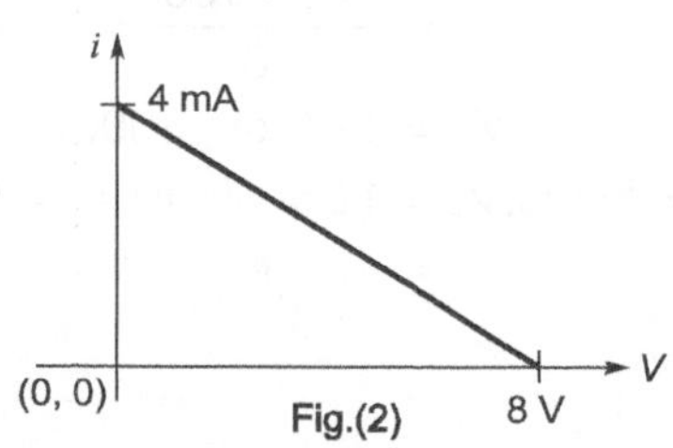

Fig.(2)

(a) 3 μsec
(b) 12 μsec
(c) 32 μsec
(d) unknown, unless the actual network is specified [1996 : 1 M]

5.3 An ideal voltage source will charge an ideal capacitor
(a) in infinite time (b) exponentially
(c) instantaneously (d) none of these
[1997 : 1 M]

5.4 In the circuit shown in figure, it is desired to have a constant direct current $i(t)$ through the ideal inductor L. The nature of the voltage source $v(t)$ must be

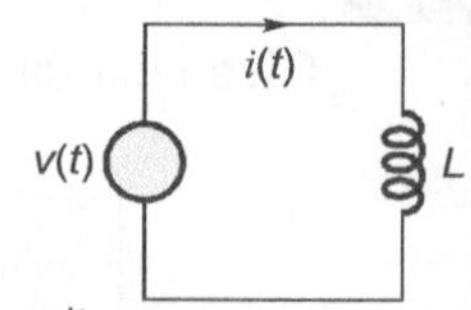

(a) constant voltage
(b) linearly increasing voltage
(c) an ideal impulse
(d) exponentially increasing voltage
[1998 : 1 M]

5.5 A rectangular voltage pulse of magnitude V and duration T is applied to a series combination of resistance R and capacitance C. The maximum voltage developed across the capacitor is

(a) $V\left[1-\exp\left(-\frac{T}{RC}\right)\right]$

(b) $\frac{VT}{RC}$

(c) V

(d) $V\exp\left(-\frac{T}{RC}\right)$ [1999 : 2 M]

5.6 A voltage waveform $v(t) = 12\,t^2$ is applied across a 1 H inductor for $t \geq 0$, with initial current through it being zero. The current through the inductor for $t \geq 0$ is given by
(a) $12\,t$ (b) $24\,t$
(c) $12\,t^3$ (d) $4\,t^3$ [2000 : 1 M]

5.7 A unit step voltage is applied at $t = 0$ to a series RL circuit with zero initial conditions.
(a) It is possible for the current to be oscillatory.
(b) The voltage across the resistor at $t = 0^+$ is zero.
(c) The energy stored in inductor in the steady state is zero.
(d) The resistor current eventually falls to zero.
[2001 : 1 M]

5.8 Consider the circuit shown in the figure. If the frequency of the source is 50 Hz, then a value of t_0 which results in a transient free response is

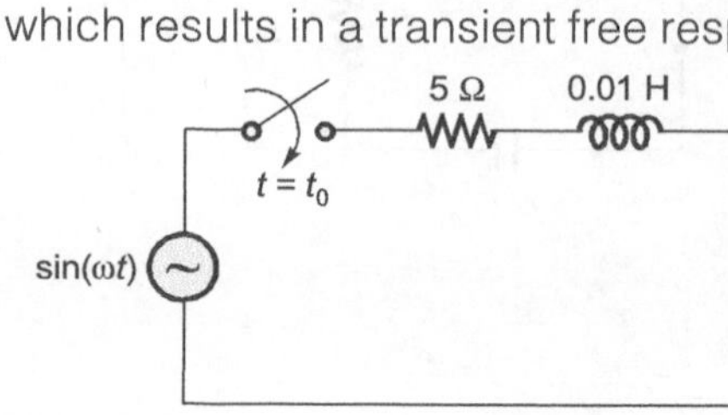

(a) 0 ms (b) 1.78 ms
(c) 2.71 ms (d) 2.91 ms [2002 : 2 M]

5.9 An 11 V pulse of 10 μs duration is applied to the circuit shown in the figure. Assuming that the capacitor is completely discharged prior to applying the pulse, the peak value of the capacitor voltage is

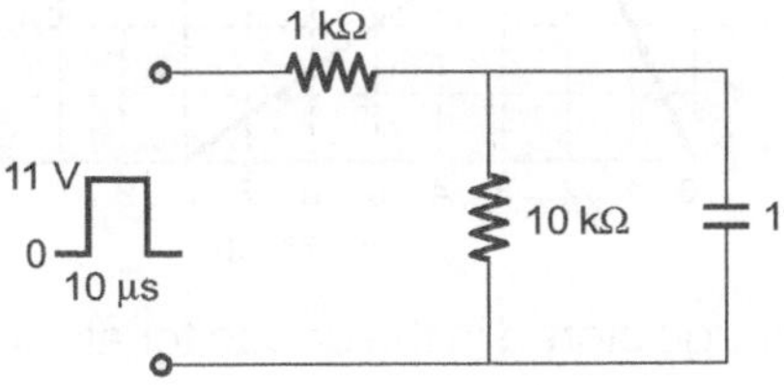

(a) 11 V (b) 5.5 V
(c) 6.32 V (d) 0.96 V **[2002 : 2 M]**

5.10 In the circuit shown in the figure, the switch S is closed at time (t = 0). The voltage across the inductor at $t = 0^+$, is

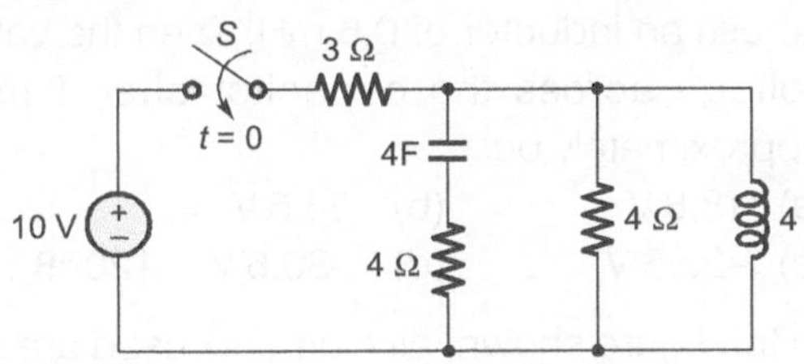

(a) 2 V (b) 4 V
(c) –6 V (d) 8 V **[2003 : 2 M]**

5.11 In the figure, the capacitor initially has a charge of 10 Coulomb. The current in the circuit one second after the switch S is closed will be

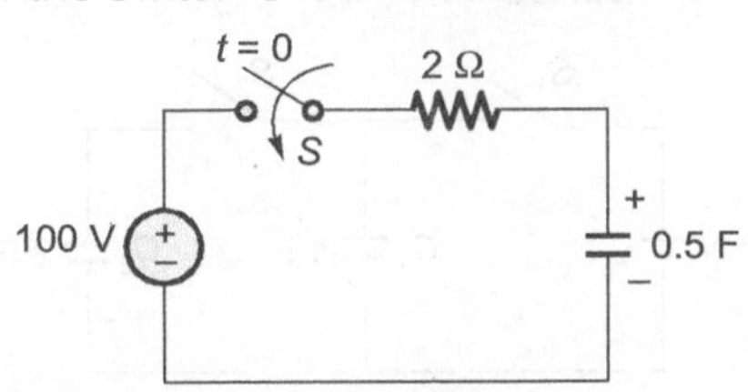

(a) 14.7 A (b) 18.5 A
(c) 40.0 A (d) 50.0 A **[2004 : 2 M]**

5.12 In the figure given, for the initial capacitor voltage is zero. The switch is closed at t = 0. The final steady-state voltage across the capacitor is

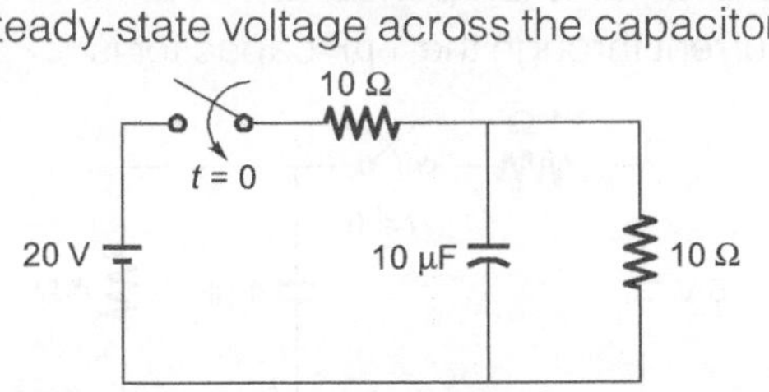

(a) 20 V (b) 10 V
(c) 5 V (d) 0 V **[2005 : 1 M]**

5.13 The circuit shown in the figure is in steady state, when the switch is closed at t = 0. Assuming that the inductance is ideal, the current through the inductor at $t = 0^+$ equals

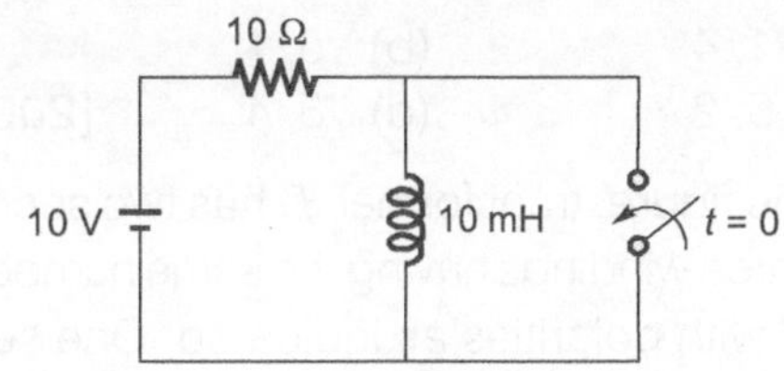

(a) 0 A (b) 0.5 A
(c) 1 A (d) 2 A **[2005 : 2 M]**

Statement for Linked Answer Questions (5.14 and 5.15):

A coil of inductance 10 H and resistance 40 Ω is connected as shown in the figure. After the switch S has been in contact with point 1 for a very long time, it is moved to point 2 at, t = 0.

5.14 If, at $t = 0^+$, the voltage across the coil is 120 V, the value of resistance R is

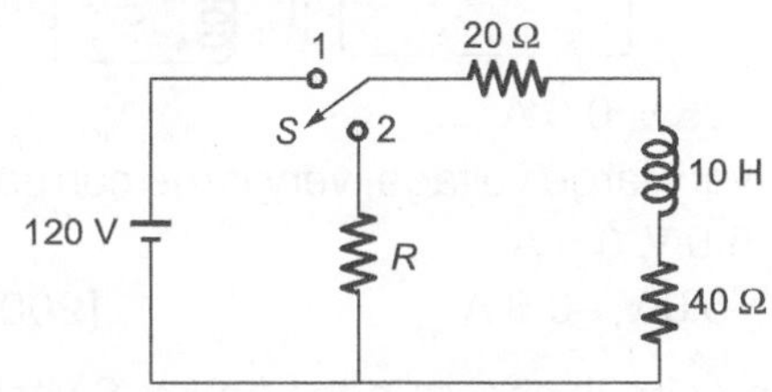

(a) 0 Ω (b) 20 Ω
(c) 40 Ω (d) 60 Ω **[2005 : 2 M]**

5.15 For the value of resistance obtained in (a), the time taken for 95% of the stored energy to be dissipated is close to

(a) 0.10 sec (b) 0.15 sec
(c) 0.50 sec (d) 1.0 sec **[2005 : 2 M]**

5.16 An ideal capacitor is charged to a voltage V_0 and connected at t = 0 across an ideal inductor L. (The circuit now consists of a capacitor and inductor alone). If we let $\omega_0 = \frac{1}{\sqrt{LC}}$, the voltage across the capacitor at time $t > 0$ is given by

(a) V_0 (b) $V_0 \cos(\omega_0 t)$
(c) $V_0 \sin(\omega_0 t)$ (d) $V_0 e^{-\omega_0 t} \cos(\omega_0 t)$
[2006 : 2 M]

5.17 In the circuit shown in the figure, the current source I = 1 A, the voltage source V = 5 V, $R_1 = R_2 = R_3$ = 1 Ω, $L_1 = L_2 = L_3$ = 1 H, $C_1 = C_2$ = 1 F. The currents (in A) through R_3 and the voltage source V respectively will be

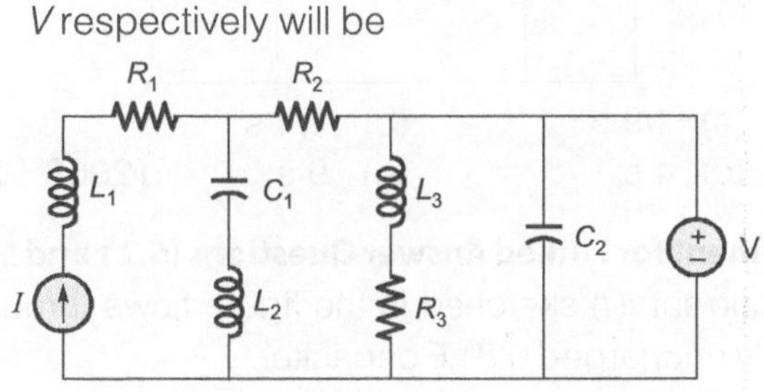

(a) 1, 4 (b) 5, 1
(c) 5, 2 (d) 5, 4 [2006 : 2 M]

5.18 In the figure, transformer T_1 has two secondaries, all three windings having the same number of turns and with polarities as indicated. One secondary is shorted by a 10 Ω resistor R, and the other by a 15 mF capacitor. The switch SW is opened ($t = 0$) when the capacitor is charged to 5 V with the left plate as positive. At ($t = 0^+$) the voltage V_p and current I_R are

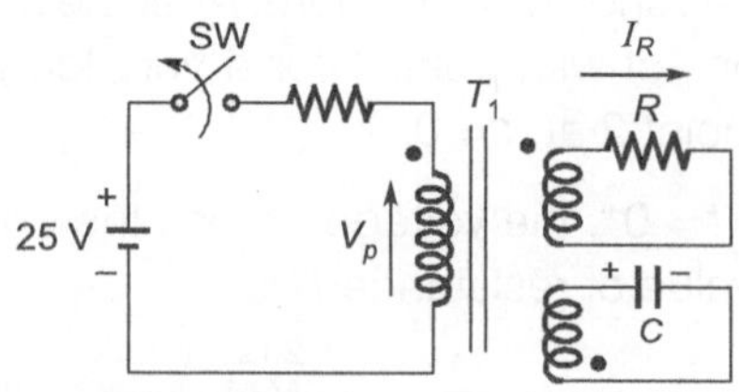

(a) –25 V, 0.0 A
(b) very large voltage, very large current
(c) 5.0 V, 0.5 A
(d) –5.0 V, –0.5 A [2007 : 2 M]

5.19 In the circuit shown in the figure. Switch SW_1 is initially closed and SW_2 is open. The inductor L carries a current of 10 A and the capacitor charged to 10 V with polarities as indicated. SW_2 is closed at $t = 0$ and SW_1 is opened at $t = 0$. The current through C and the voltage across L at ($t = 0^+$) is

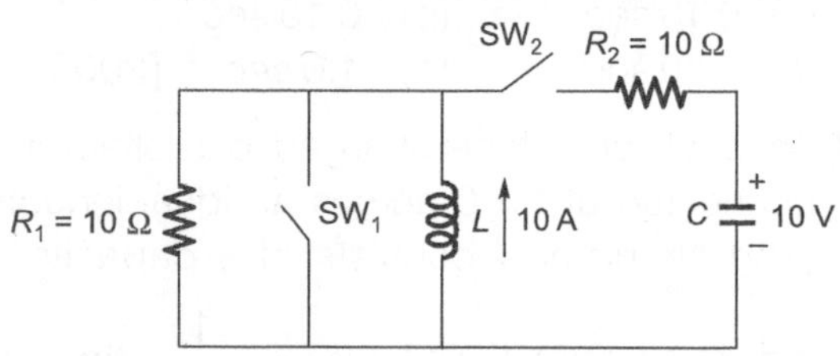

(a) 55 A, 4.5 V (b) 5.5 A, 45 V
(c) 45 A, 5.5 A (d) 4.5 A, 55 V
[2007 : 2 M]

5.20 The time constant for the given circuit will be

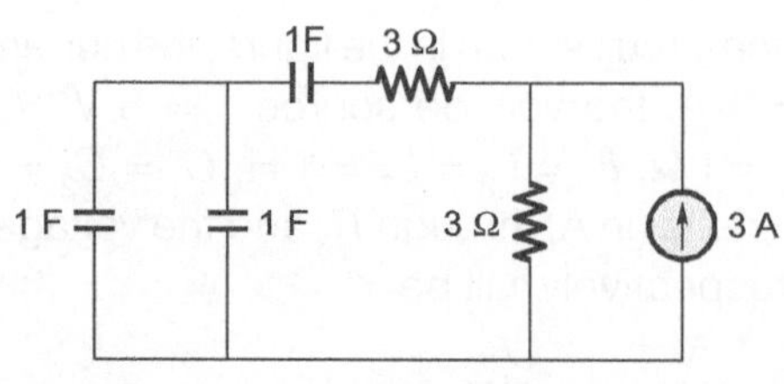

(a) 1/9 s (b) 1/4 s
(c) 4 s (d) 9 s [2008 : 2 M]

Statement for Linked Answer Questions (5.21 and 5.22):
The current $i(t)$ sketched in the figure flows through a initially uncharged 0.3 nF capacitor.

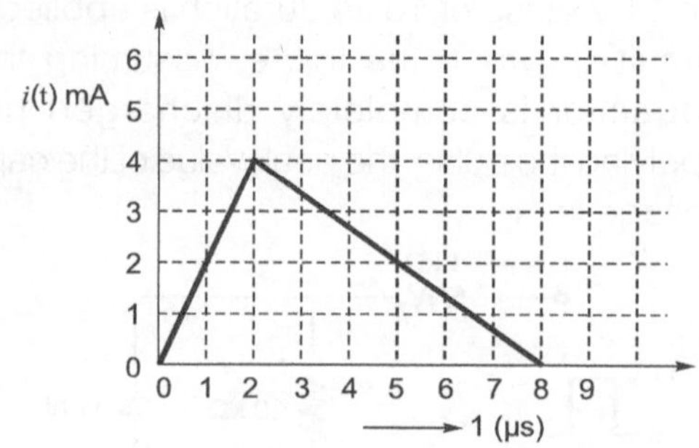

5.21 The charge stored in the capacitor at $t = 5$ μs, will be
(a) 8 nC (b) 10 nC
(c) 13 nC (d) 16 nC [2008 : 2 M]

5.22 The capacitor charged upto 5 μs, as per the current profile given in the figure, is connected across an inductor of 0.6 mH. Then the value of voltage across the capacitor after 1 μs will approximately be
(a) 18.8 V (b) 23.5 V
(c) –23.5 V (d) –30.6 V [2008 : 2 M]

5.23 In the figure shown, all elements used are ideal. For time $t < 0$, S_1 remained closed and S_2 open. At $t = 0$, S_1 is opened and S_2 is closed. If the voltage V_{C_2} across the capacitor C_2 at $t = 0^-$ is zero, the voltage across the capacitor combination at $t = 0^+$ will be

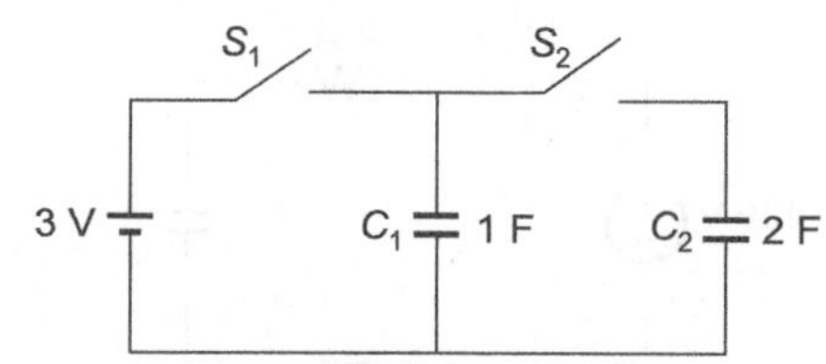

(a) 1 V (b) 2 V
(c) 1.5 V (d) 3 V [2009 : 2 M]

5.24 The switch in the circuit has been closed for a long time. It is opened at $t = 0$. At $t = 0^+$, the current through the 1 μF capacitor is

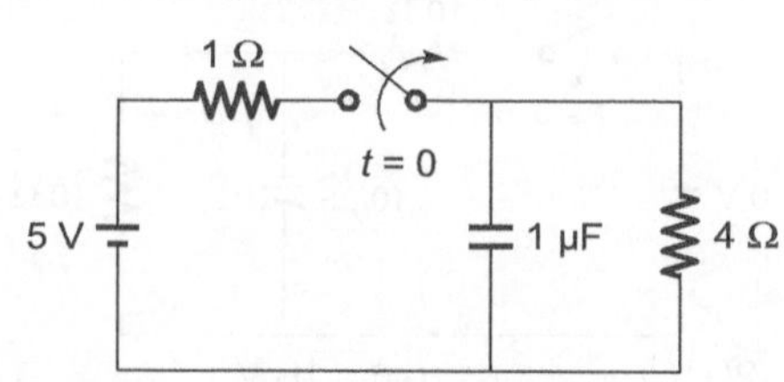

(a) 0 A (b) 1 A
(c) 1.25 A (d) 5 A [2010 : 1 M]

5.25 The L-C circuit shown in the figure has an inductance $L = 1$ mH and a capacitance $C = 10$ μF.

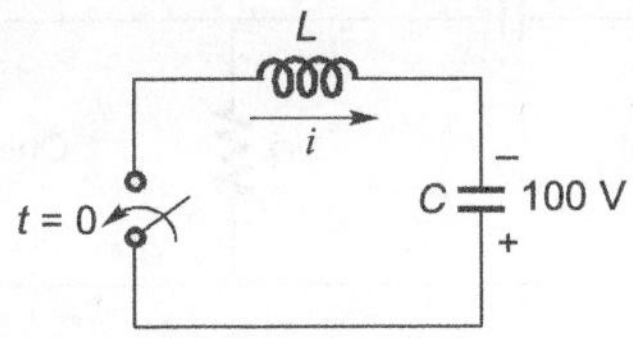

The initial current through the inductor is zero, while the initial capacitor voltage is 100 V. The switch is closed at $t = 0$. The current 'i' through the circuit is

(a) $5 \cos(5 \times 10^3 t)$ A
(b) $5 \sin(10^4 t)$ A
(c) $10 \cos(5 \times 10^3 t)$ A
(d) $10 \sin(10^4 t)$ A [2010 : 2 M]

5.26 In the following figure, C_1 and C_2 are ideal capacitors. C_1 has been charged to 12 V before the ideal switch S is closed at $t = 0$. The current $i(t)$ for all t is

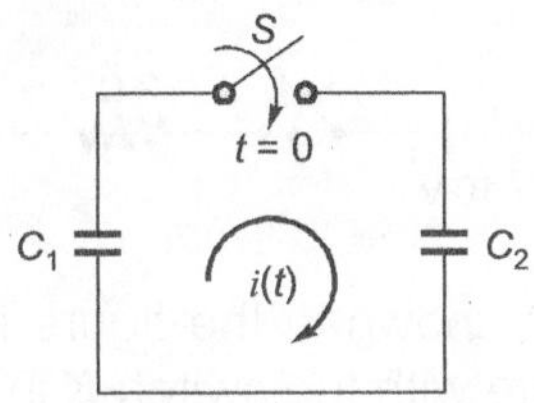

(a) zero
(b) a step function
(c) an exponentially decaying function
(d) an impulse function [2012 : 1 M]

5.27 A combination of 1 μF capacitor with an initial voltage $v_c(0) = -2$ V in series with a 100 Ω resistor is connected to a 20 mA ideal DC current source by operating both switches at $t = 0$ is as shown in the figure. Which of the following graphs shown in the options approximates the voltage v_s across the current source over the next few seconds?

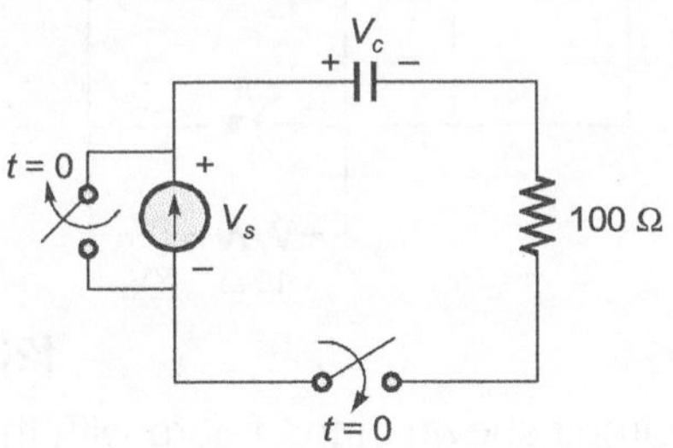

(a)
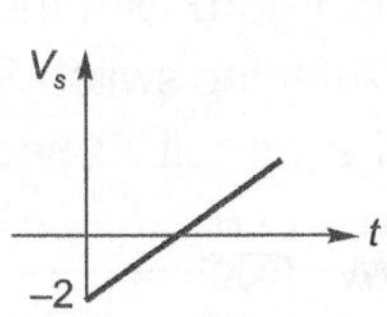

(b)
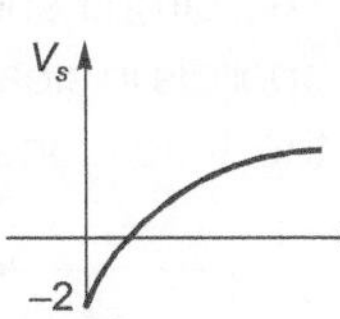

(c)
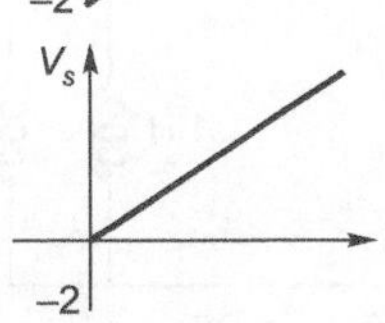

(d)
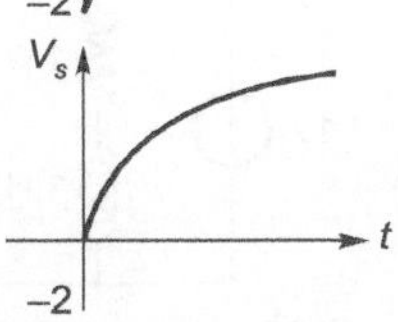

[2014 : 1 M, Set-1]

5.28 The switch SW shown in the circuit is kept at position '1' for a long duration. At $t = 0^+$, the switch is moved to position '2'. Assuming $|V_{02}| > |V_{01}|$, the voltage $v_c(t)$ across the capacitor is

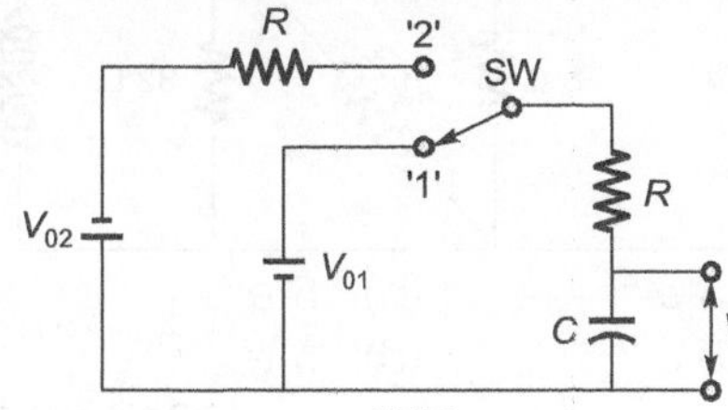

(a) $v_c(t) = V_{02}(1 - e^{-t/2RC}) - V_{01}$
(b) $v_c(t) = V_{02}(1 - e^{-t/2RC}) + V_{01}$
(c) $v_c(t) = -(V_{02} + V_{01})(1 - e^{-t/2RC}) - V_{01}$
(d) $v_c(t) = (V_{02} - V_{01})(1 - e^{-t/2RC}) + V_{01}$

[2014 : 1 M, Set-2]

5.29 A series RL circuit is excited at $t = 0$ by closing a switch as shown in the figure. Assuming zero initial conditions, the value of $\frac{d^2I}{dt^2}$ at $t = 0^+$ is

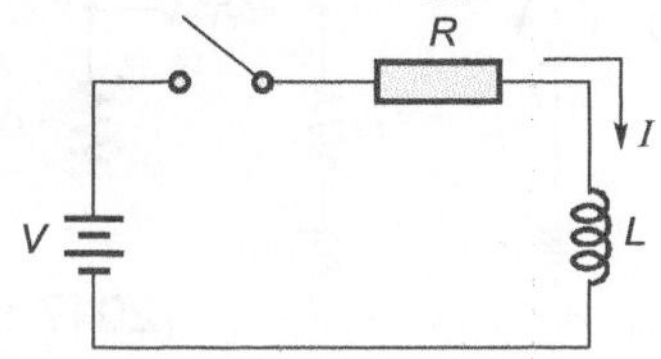

(a) $\frac{V}{L}$ (b) $\frac{-V}{R}$

(c) 0 (d) $\frac{-RV}{L^2}$ [2015 : 1 M, Set-2]

5.30 In the circuit shown, switch S_2 has been closed for a long time. At time $t = 0$ switch S_1 is closed. At $t = 0^+$, the rate of change of current through the inductor, in Amperes per second, is _______.

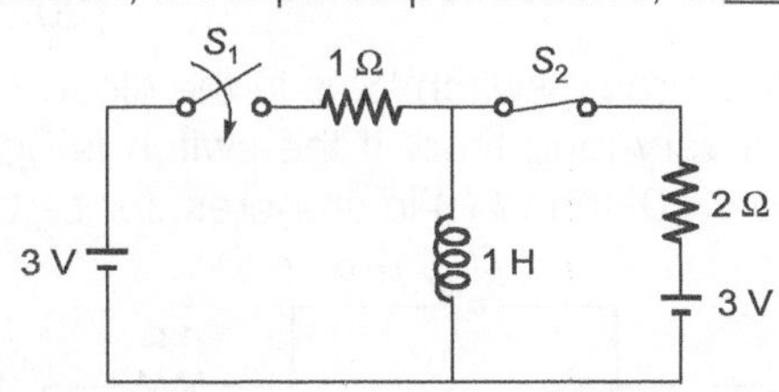

[2016 : 2 M, Set-1]

5.31 In the circuit shown below, the initial capacitor voltage is 4 V. Switch S_1 is closed at $t = 0$. The charge (in μC) lost by the capacitor from $t = 25$ μs to $t = 100$ μs is_______.

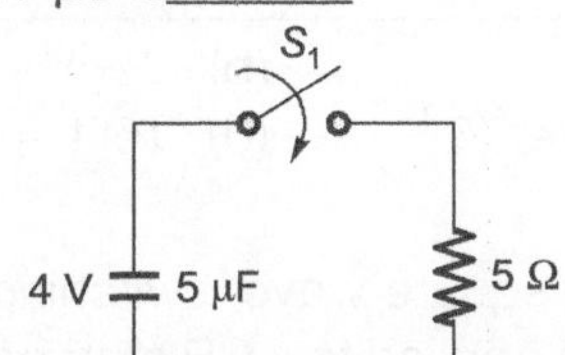

[2016 : 2 M, Set-2]

5.32 The switch in the figure below was closed for a long time. It is opened at $t = 0$. The current in the inductor of 2 H for $t \geq 0$, is

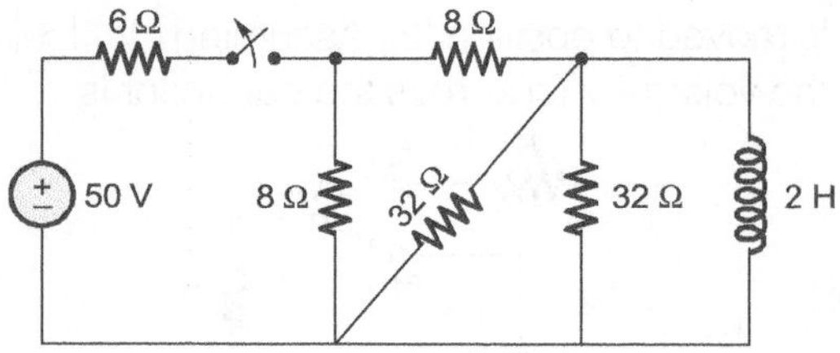

(a) $2.5\,e^{-4t}$ (b) $5\,e^{-4t}$
(c) $2.5\,e^{-0.25t}$ (d) $5\,e^{-0.25t}$

[2017 : 2 M, Set-1]

5.33 The initial charge in the 1 F capacitor present in the circuit shown is zero. The energy in joules transferred from the DC source until steady state condition is reached equals _____. (Give the answer upto one decimal place.)

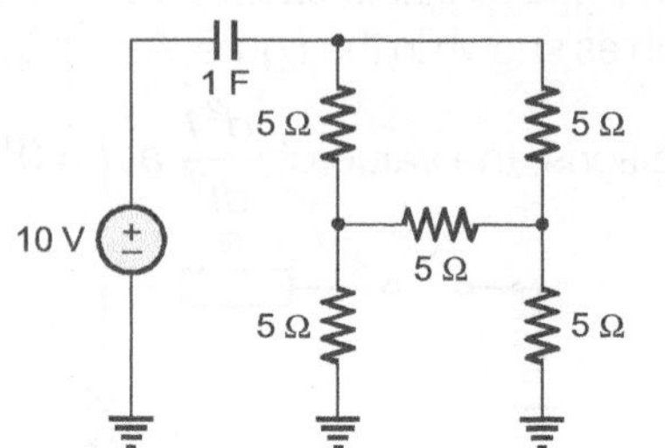

[2017 : 1 M, Set-2]

5.34 A resistor and a capacitor are connected in series to a 10 V DC supply through a switch. The switch is closed at $t = 0$, and the capacitor voltage is found to cross 0 V at $t = 0.4\tau$, where τ is the circuit time constant. The absolute value of percentage change required in the initial capacitor voltage if the zero crossing has to happen at $t = 0.2\tau$ is _______ (rounded off to 2 decimal places).

[2020 : 2 M]

5.35 In the circuit, switch '*S*' is in the closed position for a very long time. If the switch is opened at time $t = 0$, then $i_L(t)$ in amperes, for $t \geq 0$ is

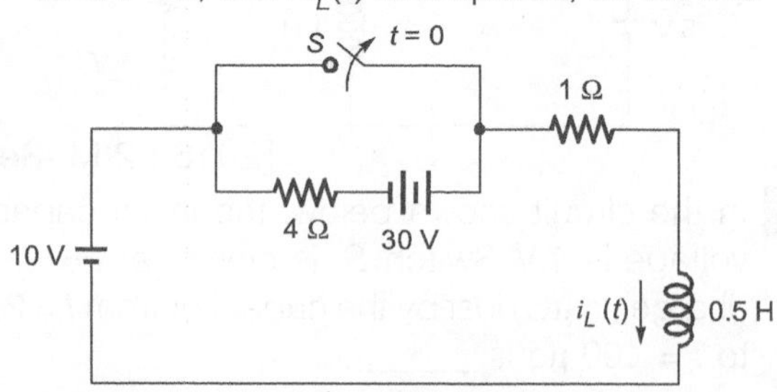

(a) 10 (b) $8e^{-10t}$
(c) $8 + 2e^{-10t}$ (d) $10(1 - e^{-2t})$

[2021 : 2 M]

5.36 A 100 Hz square wave, switching between 0 V and 5 V, is applied to a CR high-pass filter circuit as shown, The output voltage waveform across the resistor is 6.2 V peak-to-peak, If the resistance *R* is 820 Ω. then the value *C* is _____ uF. (Round off to 2 decimal places.)

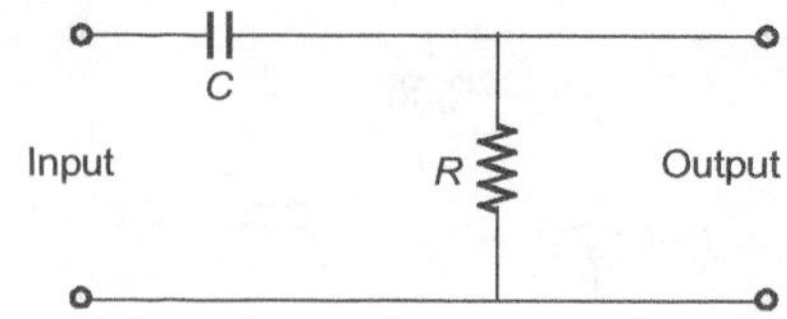

[2021 : 2 M]

5.37 In the circuit shown below, the switch *S* is closed at $t = 0$. The magnitude of the steady state voltage, in volts, across the 6 Ω resistor is ____. (round off to two decimal places).

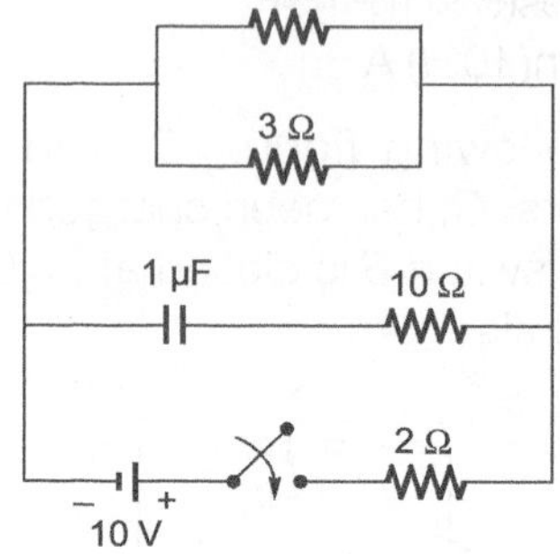

[2022 : 1 M]

5.38 The circuit shown in the figure is initially in the steady state with the switch *K* in open condition and $\bar{K}$ in closed condition. The switch *K* is closed and $\bar{K}$ is opened simultaneously at the instant $t = t_1$, where $t_1 > 0$. The minimum value of t_1 in milliseconds, such that there is no transient in the voltage across the 100 µF capacitor, is __________ (Round off to 2 decimal places).

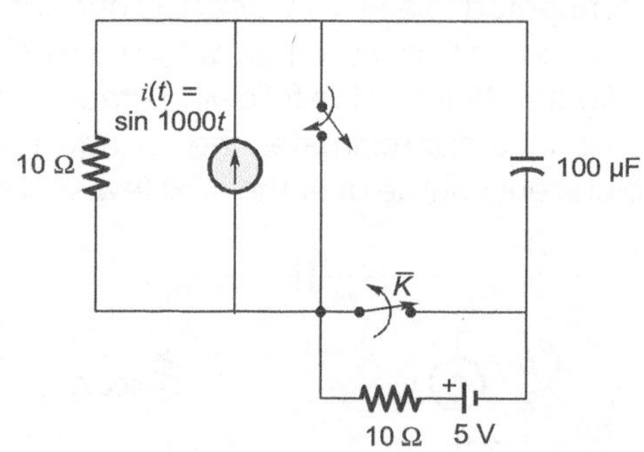

[2023 : 2 M]

5.39 The circuit shown in the figure with the switch *S* open, is in steady state. After the switch *S* is closed, the time constant of the circuit in seconds is

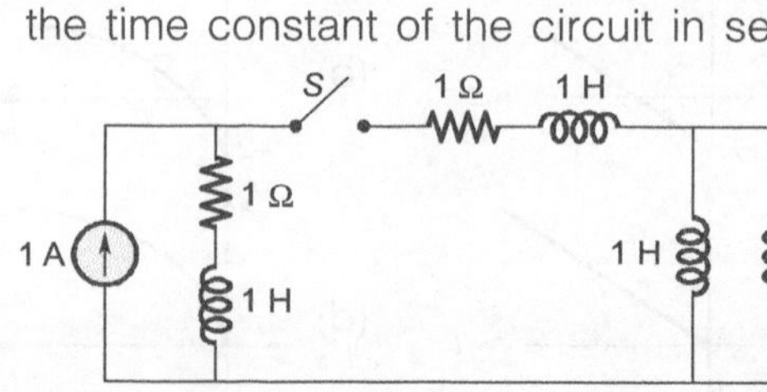

(a) 1.5 (b) 1.25
(c) 0 (d) 1

[2024 : 1 M]

■■■■

Answers Transients and Steady State Response

5.1 (d)	5.2 (a)	5.3 (c)	5.4 (c)	5.5 (a)	5.6 (d)	5.7 (b)
5.8 (b)	5.9 (c)	5.10 (b)	5.11 (a)	5.12 (b)	5.13 (c)	5.14 (c)
5.15 (b)	5.16 (b)	5.17 (d)	5.18 (d)	5.19 (d)	5.20 (c)	5.21 (c)
5.22 (d)	5.23 (a)	5.24 (b)	5.25 (d)	5.26 (d)	5.27 (c)	5.28 (*)
5.29 (d)	5.30 (2)	5.31 (6.99)	5.32 (a)	5.33 (100)	5.34 (54.99)	5.35 (c)
5.36 (12.46)	5.37 (5)	5.38 (1.57)	5.39 (b)			

Explanations Transients and Steady State Response

5.1 (d)

When switch is closed, current through capacitor,

$$I = C\frac{dV_c(t)}{dt}$$

$$\therefore \quad V = RI + V_c(t)$$

$$1 = RC\frac{dV_c(t)}{dt} + V_c(t)$$

$\because$ At $t = 0^+$, $V_c(0^+) = 0$

$$\therefore \quad 1 = RC\frac{dV_c(0^+)}{dt} + 0$$

Hence, $\dfrac{dV_c(0^+)}{dt} = \dfrac{1}{RC}$

5.2 (a)

$$R = \frac{V_{oc}}{I_{sc}} = \frac{8}{4\times 10^{-3}} = 2\text{ k}\Omega$$

$$L = 6\text{ mH}$$

Time constant,

$$\tau = \frac{L}{R} = \frac{6\times 10^{-3}}{2\times 10^{3}} = 3\ \mu\text{sec.}$$

5.3 (c)

$\because$ Ideal voltage has zero internal resistance,

$\therefore$ Time constant $\tau = RC = 0$

Hence, capacitor will charge instantaneously.

5.4 (c)

$$V = L\frac{di(t)}{dt}$$

For $\quad i(t) = I\,u(t)$

$$V = L\frac{d}{dt}I\,u(t)$$

$$= LI\,\delta(t) \text{ viz an ideal impulse function}$$

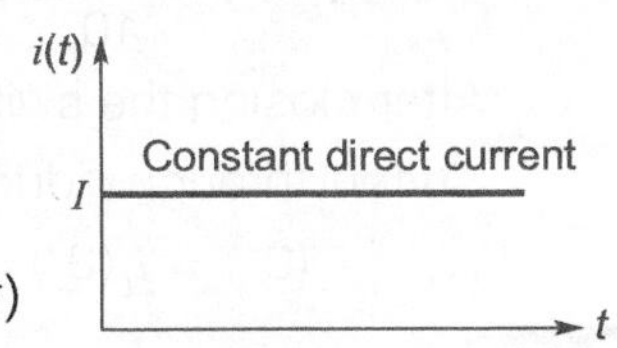

5.5 (a)

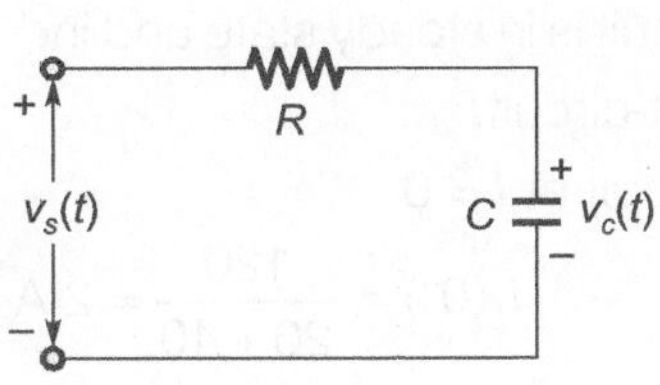

Given,

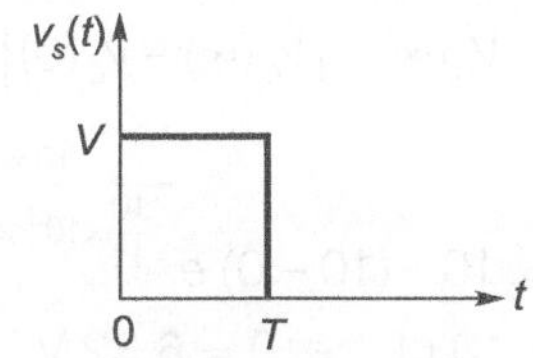

i.e., $\quad v_s = V[u(t) - u(t-T)]$

$\because \quad v_c(t) = v_s(t)\cdot(1 - e^{-t/RC})$

$\quad = V(1 - e^{-t/RC})\cdot[u(t) - u(t-T)]$

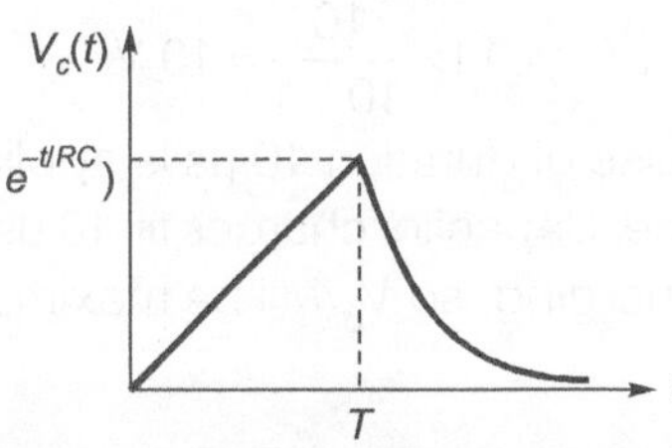

i.e., capacitor charges till $t = T$ and then discharges.

Hence, $\boxed{v_c(\max) = V(1 - e^{-t/RC})}$

5.6 (d)

Current through inductor,

$$i(t) = \frac{1}{L}\int_{-\infty}^{t} v(t)\,dt = \frac{1}{1}\int_{0}^{t} 12t^2\,dt \text{ for } t \ge 0 = 4t^3\text{ A}$$

5.7 (b)

At $t = 0^+$ inductor works as open circuit, hence complete source voltage drops across it and consequently, current through the resistor R is zero. Hence, voltage across the resistor at $t = 0^+$ is zero, and further with time it rises according to $V_R(t) = (1 - e^{-Rt/L})\,u(t)$.

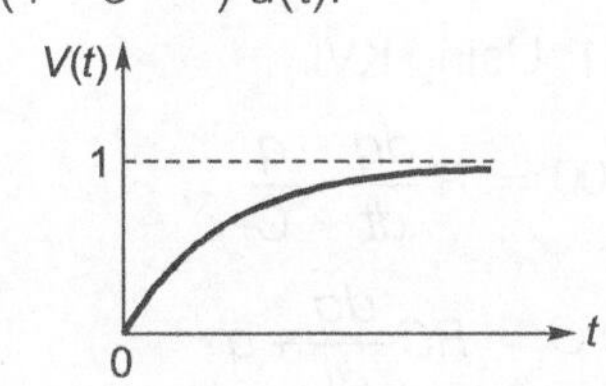

5.8 (b)

For transient free response,

$$\tan(\omega t_0) = \frac{\omega L}{R}$$

$$\tan(2\pi\times 50\times t_0) = \frac{2\pi\times 50\times 0.01}{5}$$

$$2\pi\times 50\times t_0 = \tan^{-1}\left(\frac{\pi}{5}\right)$$

$$= 32.14^\circ = 0.561 \text{ rad}$$

$$t_0 = \frac{0.561}{100\pi} = 1.786 \text{ ms}$$

5.9 (c)

$$V_c(t) = V_c(\infty) - [V_c(\infty) - V_c(0)]e^{-t/RC}$$

$$V_c(\text{peak}) = 10 - (10-0)\, e^{-\frac{10\times 10^{-6}}{\frac{10}{11}\times 10^3\times 11\times 10^{-9}}}$$

$$= 10\,(1 - e^{-1}) = 6.32 \text{ V}$$

$$\left[\text{where } R_{net} = 10\,||\,1 = \frac{10}{11}\text{ k}\Omega\right]$$

$$V_c(\infty) = 11\times\frac{10}{10+1} = 10 \text{ V}$$

∵ Pulse of duration 10 μs is applied.
Hence, Capacitor charges till 10 μs and then starts discharging, so V_c will be maximum at $t = 10$ μs.

5.10 (b)

Before closing the switch, the circuit was not energized, therefore, current through inductor and voltage across capacitor is zero.
After closing the switch, at $t = 0^+$ inductor acts as open-circuit and capacitor acts as short-circuit.
Equivalent circuit at $t = 0^+$

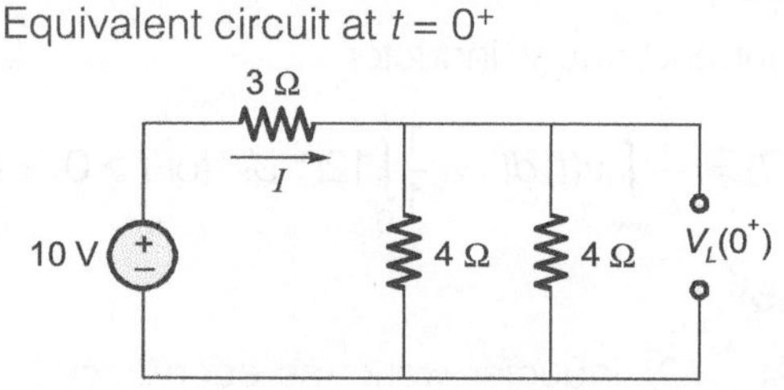

$$I = \frac{10}{3 + 4\,||\,4} = 2 \text{ A}$$

$$V_L(0^+) = I\times(4\,||\,4) = 2\times 2 = 4 \text{ V}$$

5.11 (a)

Method-1: Using KVL,

$$100 = R\frac{dq}{dt} + \frac{q}{C}$$

$$100\,C = RC\frac{dq}{dt} + q$$

$$\int_{q_0}^{q}\frac{dq}{100C - q} = \frac{1}{RC}\int_0^t dt$$

$$100\,C - q = (100\,C - q_0)\, e^{-t/RC}$$

$$i = \frac{dq}{dt} = \frac{(100C - q_0)}{RC}\, e^{-1/RC}$$

$$e^{-t/RC} = 40\, e^{-1} = 14.7 \text{ A}$$

Method-2: At ($t = 0$) switch is closed

$$V_c(0^-) = V_c(0^+) = \frac{Q}{C} = \frac{10}{0.5} = 20 \text{ V}$$

$$V_c(\infty) = 100 \text{ V}$$

$$V_c(t) = V_c(\infty) + \{V_c(0^+) - V_c(\infty)\}e^{-t/RC}$$

$$V_c(t) = 100 + (20 - 100)\, e^{-t/1}$$

$$\therefore \quad i_c(t) = C\frac{dV_c(t)}{dt}$$

$$= 0.5\times -80\times(-e^{-t}) = 40\, e^{-t}$$

$$i_c(t)|_{t=1} = 40\, e^{-1} = 14.71 \text{ A}$$

5.12 (b)

At ($t \to 0^+$), the capacitor act as short-circuit. At ($t \to \infty$), the capacitor will become open circuit.

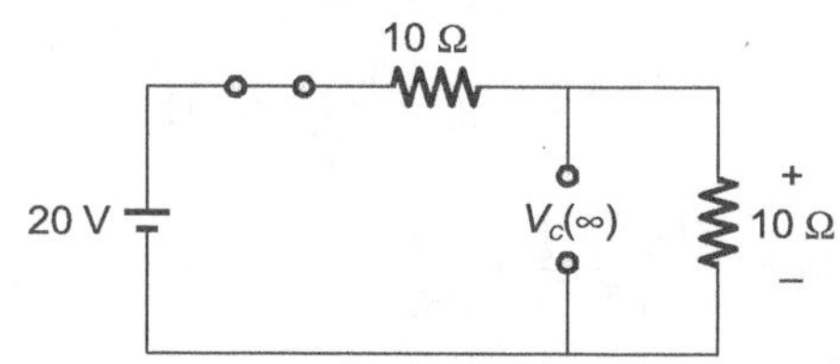

∴ Voltage across capacitor

$$= \frac{20}{10+10}\times 10 = 10 \text{ V}$$

5.13 (c)

Before closing the switch, at $t = 0^-$, the circuit is in steady state. So, inductor behaves as short-circuit

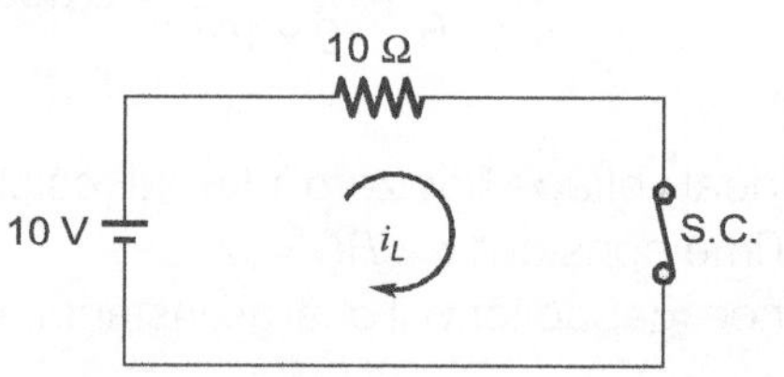

$$i_L(0^-) = \frac{10}{10} = 1 \text{ A}$$

After closing the switch, at $t = 0^+$
Current through inductor can not change abruptly.

$$\therefore \quad i_L(0^+) = i_L(0^-) = 1 \text{ A}$$

5.14 (c)

Before moving the switch, at $t = 0^-$
The circuit is in steady state and inductor behaves as short-circuit.
The circuit at $t = 0^-$

$$i_L(0^-) = \frac{120}{20+40} = 2 \text{ A}$$

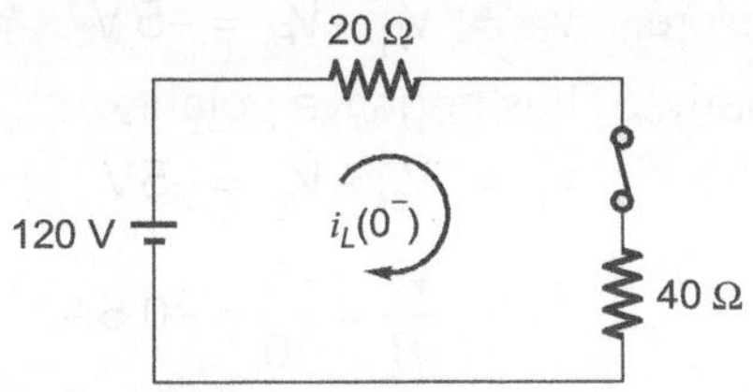

After moving the switch,
At $t = 0^+$
Current through inductor can not change abruptly.
So, $i_L(0^+) = i_L(0^-) = 2$ A

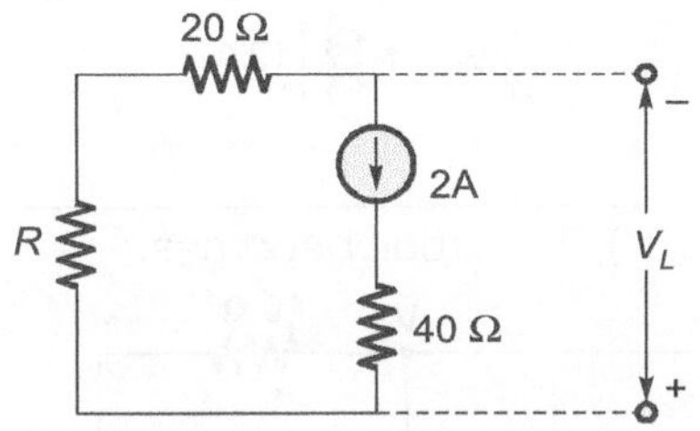

$$V_L = i_L(0^+) \times \{20 + R\}$$
$$120 = 2 \times (20 + R)$$
$$R = 40\ \Omega$$

5.15 (b)

The circuit (in s-domain)

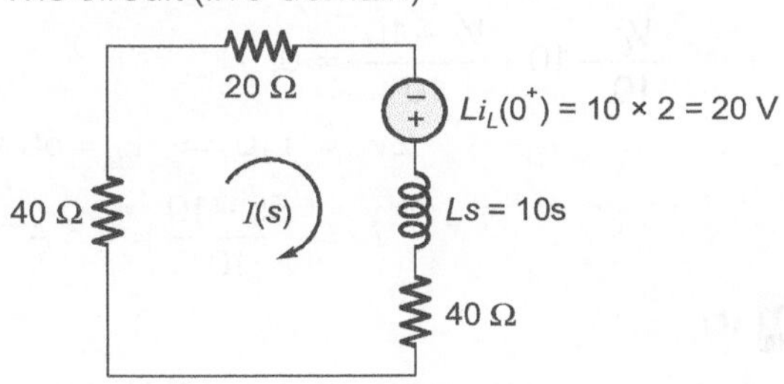

$$I(s) = \frac{20}{(20+40+40)+10s}$$
$$= \frac{20}{10s+100} = \frac{2}{s+10}$$
$$i(t) = \mathcal{L}^{-1}[I(s)] = \mathcal{L}^{-1}\left[\frac{2}{s+10}\right] = 2e^{-10t}$$

or,
$$i(t) = i_L(0^+)e^{-\frac{R_{eff}}{L}t}$$
$$= 2e^{-\frac{(20+40+40)}{10}t} = 2e^{-10t}$$

Initial stored energy in inductor
$$W_0 = \frac{1}{2}Li_L^2(0^+) = \frac{1}{2}\times 10 \times 2^2 = 20 \text{ Joules}$$

Remaining energy in inductor
$$W_1 = 0.05\ W_0 = 0.05 \times 20 = 1 \text{ Joule}$$
$$\frac{1}{2}Li_1^2 = 1$$
$$\frac{1}{2}\times 10 \times i_1^2 = 1 \Rightarrow i_1 = \frac{1}{\sqrt{5}} = 0.4472 \text{ A}$$

$$i = 2e^{-10T}$$

Let at $t = T$, current decrease to i_1
$$0.4472 = 2e^{-10T}$$
$$T \approx 0.15 \text{ sec.}$$

5.16 (b)

Voltage across capacitor will discharge through inductor upto voltage across the capacitor becomes zero. During this period, electrostatic energy stored in capacitor is transferred into magnetic energy which is stored in inductor.

Now inductor will start charging capacitor, magnetic energy in inductor is converted into electrostatic energy in capacitor.

Expression for $V_c(t)$ can be obtained in s-domain.

As capacitor is charged initially to voltage V_0, then representation of capacitor in s-domain

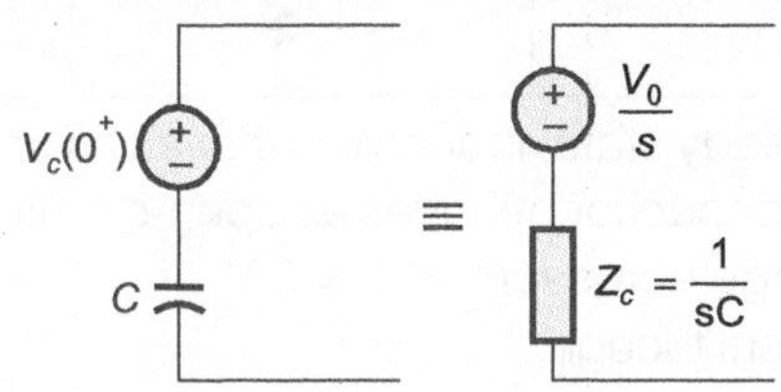

As current though the inductor is zero at $t = 0$, then

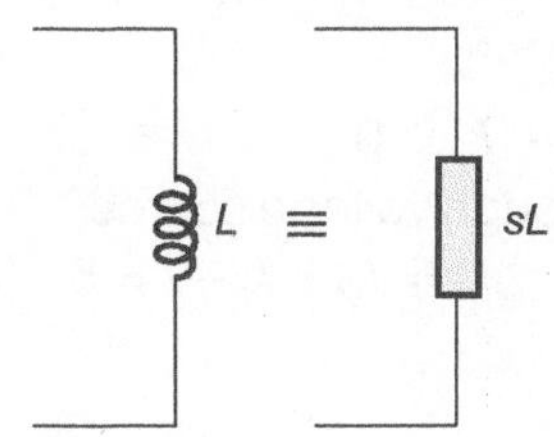

The circuit at $t > 0$

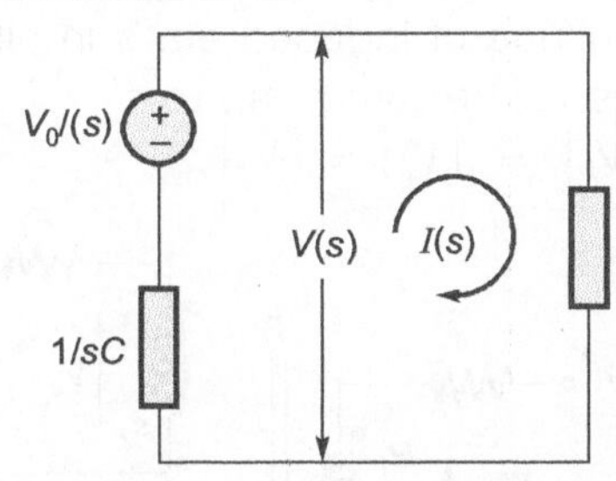

$$I(s) = \frac{V_0/s}{\frac{1}{sC}+sL} = \frac{V_0}{s^2L+\frac{1}{C}} = \frac{V_0}{L}\left(\frac{1}{s^2+\frac{1}{LC}}\right)$$

Voltage across capacitor = Voltage across inductor = $V(s)$,
$$V(s) = I(s) \times (sL)$$
$$= \frac{V_0}{L}\left[\frac{1}{s^2+\frac{1}{LC}}\right]\times(sL) = V_0\left[\frac{s}{s^2+\frac{1}{LC}}\right]$$

As, $\omega_0 = \frac{1}{\sqrt{LC}}$

$$V(s) = V_0\left[\frac{s}{s^2+\omega_0^2}\right]$$

Voltage across the capacitor

$$= V(t) = \mathcal{L}^{-1}[V(s)] = \mathcal{L}^{-1}\left[\frac{V_0 s}{s^2+\omega_0^2}\right]$$

$$V(t) = V_0\cos\omega_0 t \quad \text{where,} \quad \omega_0 = \frac{1}{\sqrt{LC}}$$

5.17 (d)

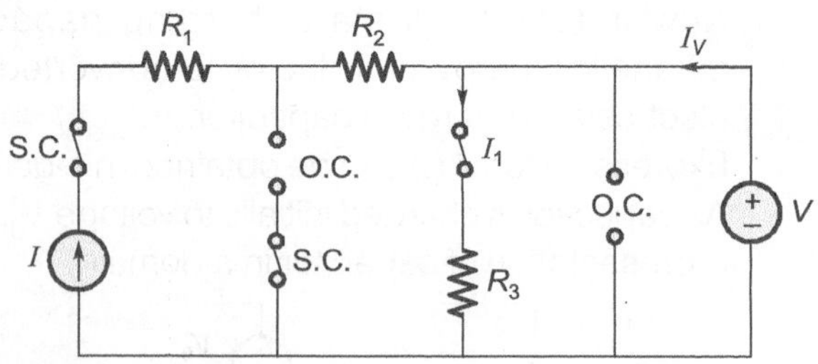

In steady state, inductor behaves as short-circuit and capacitor behaves as open-circuit

Voltage across, $R_3 = V = 5$ V

Current though,

$$R_3 = I_1 = \frac{V}{R_3} = \frac{5}{1} = 5\text{ A}$$

Apply KCL,

$$-I + I_1 - I_V = 0$$

Current though voltage source

$$= I_V = I_1 - I = 5 - 1 = 4\text{ A}$$

5.18 (d)

All the three windings has same number of turns, so magnitude of induced emf's in all the three windings will be same, i.e.,

$$|V_P| = |V_s| = |V_T|$$

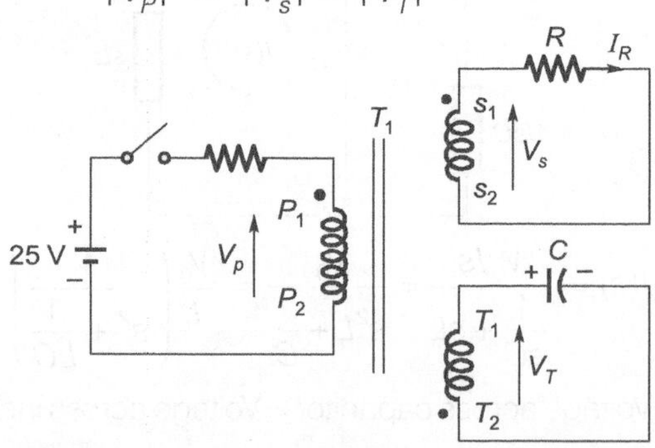

Polarity of the windings is decided on the basis of dot-convention. As capacitor is charged to 5 V with left plate as positive.

So, T_1 is positive w.r.t. T_2

$$V_T = V_{T_1} - V_{T_2} = 5\text{ V}$$

As T_2 has negative polarity. So, P_1 has negative polarity.

Therefore, $V_P = V_{P_1} - V_{P_2} = -5$ V

Similarly, S_1 has negative polarity

So, $V_s = V_{S_1} - V_{S_2} = -5$V

$$I_R = \frac{V_s}{R} = \frac{-5}{10} = -0.5\text{ A}$$

5.19 (d)

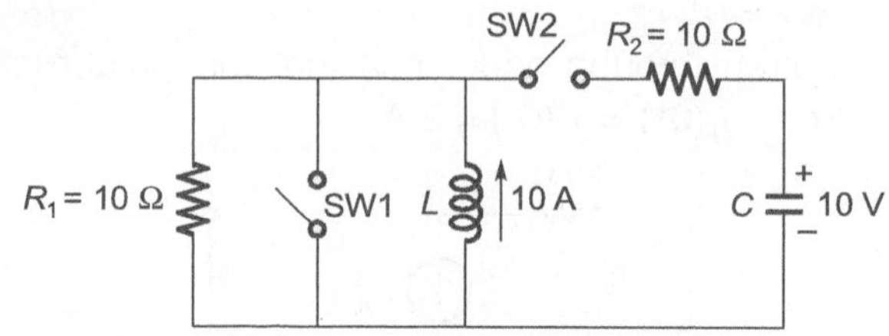

at ($t = 0^+$), the circuit becomes,

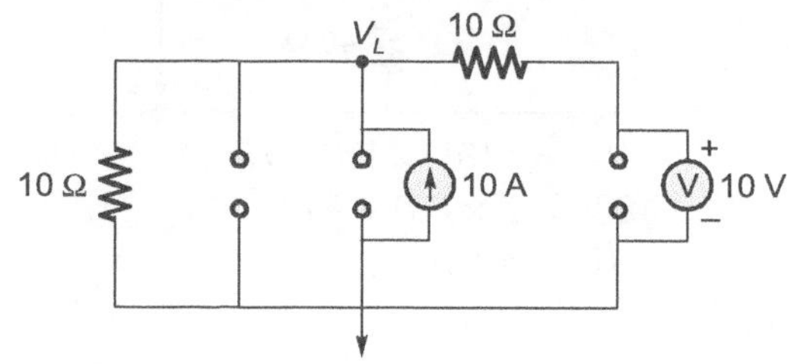

By KCL,

$$\frac{V_L}{10} - 10 + \frac{V_L - 10}{10} = 0$$

$$\therefore \quad 2V_L = 110 \Rightarrow V_L = 55\text{ V}$$

$$I_C = \frac{55-10}{10} = 4.5\text{ A}$$

5.20 (c)

For finding time const, we neglect current source as a open circuit. ∴ circuit becomes

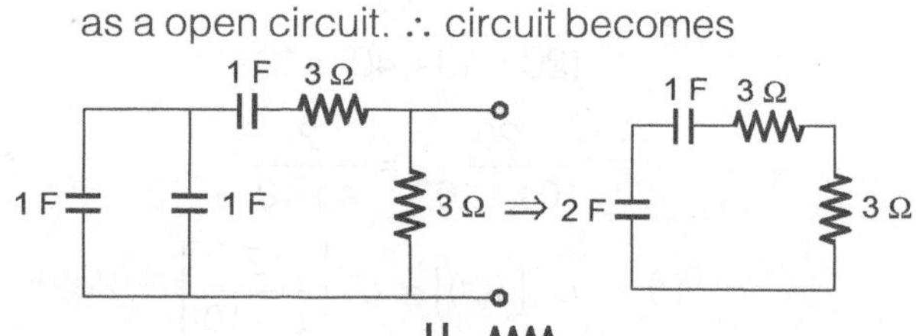

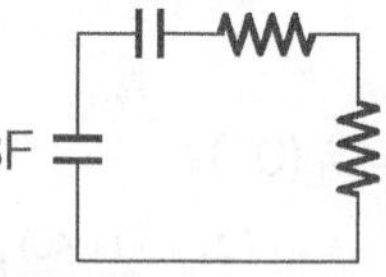

∴ $C_{eq} = 2/3$F $6\ \Omega = R_{eq}$

∴ Time constant $= R_{eq}C_{eq} = 6 \times \frac{2}{3} = 4$ sec

5.21 (c)

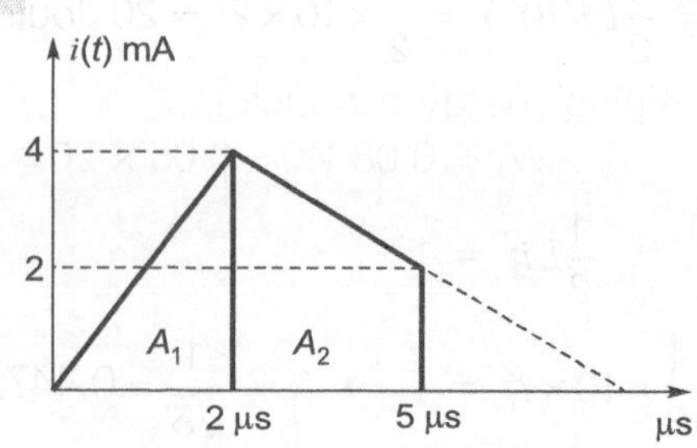

Charged stored in the capacitor = Area under $i-t$ curve

$$Q = A_1 + A_2$$

$$= \frac{1}{2}(2\times10^{-6})\times(4\times10^{-3})$$

$$+\frac{1}{2}(4+2)\times10^{-3}\times(5-2)\times10^{-6}$$

$$= \left[4+\frac{6\times3}{2}\right]\times10^{-9} = 13 \text{ nC}$$

5.22 (d)

Capacitor charged upto 5 μs, so total charge stored in capacitor = Q = 13 nC.

Voltage across the capacitor before connecting to inductor

$$V_0 = \frac{Q}{C} = \frac{13\times10^{-9}}{0.3\times10^{-9}} = 43.33 \text{ V}$$

Voltage across the capacitor at time t

$V_c(t)$ at t = 1 μs

$$V_c(t)\Big|_{t=1\mu s} = [V_0 \cos\omega_0 t]_{t=1\,\mu s}$$

$$\omega_0 t = \frac{1}{\sqrt{0.6\times10^{-3}\times0.3\times10^{-9}}}\times1\times10^{-6}$$

$$= 2.357 \text{ rad} = 135°$$

$$V_c(t)\Big|_{t=1\mu s} = 43.33\times\cos 135° \approx -30.6 \text{ V}$$

5.23 (a)

At $t = 0^-$, S_1 is closed, S_2 is open

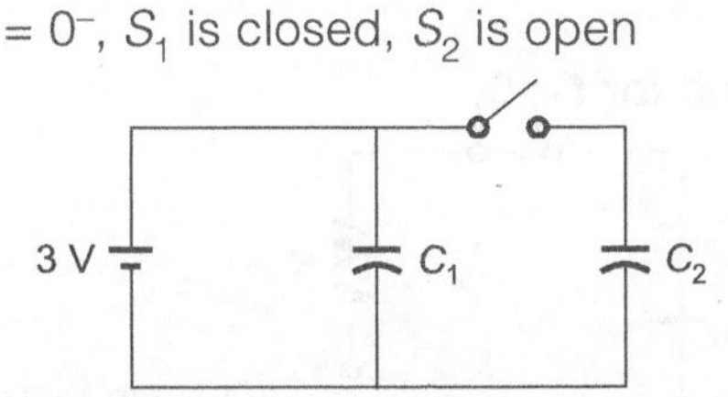

C_1 gets charged upto 3 V. Charge stored in C_1

$$Q_0 = C_1 V = 1\times3 = 3\ C$$

Voltage across C_2 is zero at $t = 0^-$, so no charge is stored in C_2. At $t > 0$, S_1 is open and S_2 is closed.

Charge stored (Q_0) initially in C_2 gets redistributed between C_1 and C_2

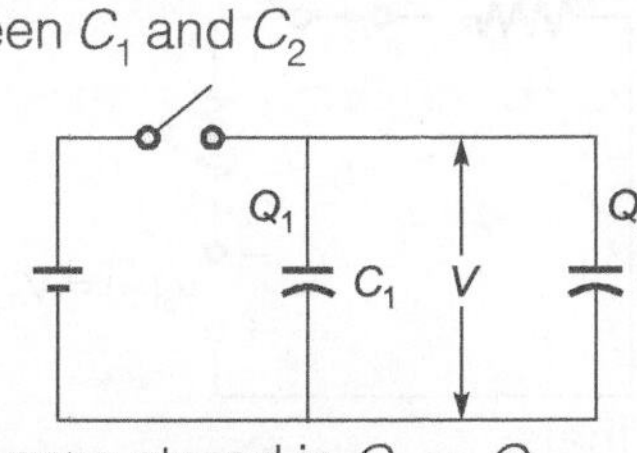

Let charge stored in $C_1 = Q_1$

Charge stored in $C_2 = Q_1$

According to conservation of charge

$$Q_1 + Q_2 = Q_0 = 3 \qquad \text{...(i)}$$

Voltage across C_1 = Voltage across C_2

$$\frac{Q_1}{C_1} = \frac{Q_2}{C_2} \Rightarrow \frac{Q_1}{1} = \frac{Q_2}{2}$$

$$Q_2 = 2\,Q_1 \qquad \text{...(ii)}$$

Solving equation (i) and (ii) we get

$$Q_1 = 1\ C \text{ and } Q_2 = 2C$$

$$\text{Voltage across combination} = \frac{Q_1}{C_1} = \frac{1}{1} = 1\text{ V}$$

Alternate method:

$$V_{c1} = V_{c2} = \frac{V_{c1}C_1 + V_{c2}C_2}{C_1 + C_2}$$

5.24 (b)

As the switch has been closed for a long time, the circuit is in steady state. At steady state, capacitor is open circuit,

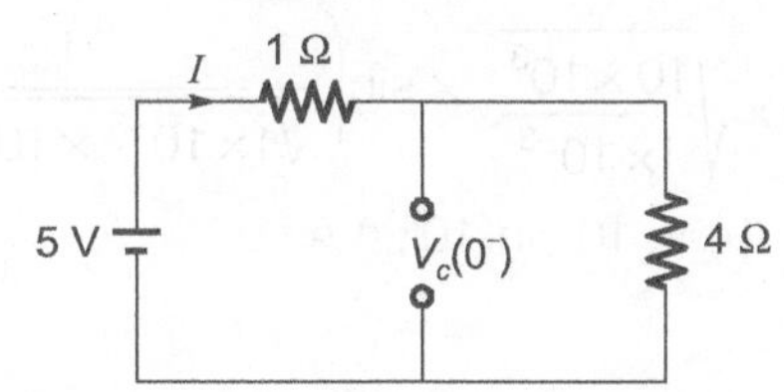

Circuit at $t = 0^-$

Using KVL,

$$5 - I - 4I = 0 \Rightarrow I = 1 \text{ A}$$

$$V_C(0^-) = 4\times1 = 4 \text{ V}$$

As the voltage across capacitor can not change abruptly,

So, $\quad V_C(0^+) = V_C(0^-) = 4$ V

Circuit at $t = 0^+$

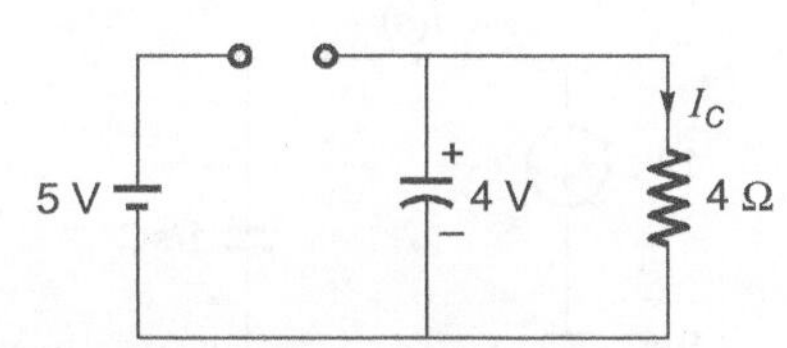

Current through capacitor at $t = 0^+$

$$I_C(0^+) = \frac{4}{4} = 1 \text{ A}$$

5.25 (d)

Initial current through the inductor is zero and capacitor voltage is charged upto voltage

$$V_c(0^-) = 100 \text{ V}$$

As current through inductor and voltage across capacitor can not change abruptly.

So, after closing the switch,

$$i_L(0^+) = i_L(0^-) = 0$$

and $\quad V_c(0^+) = V_c(0^-) = 100$ V

The circuit is s-domain

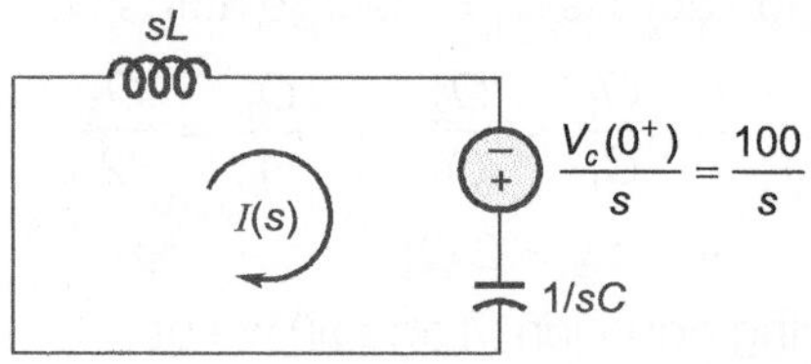

$$I(s) = \frac{100/s}{\left(sL + \frac{1}{sC}\right)} = \frac{100}{L}\left(\frac{1}{s^2 + \frac{1}{LC}}\right)$$

$$= 100\sqrt{\frac{C}{L}}\left(\frac{1/\sqrt{LC}}{s^2 + (1/\sqrt{LC})^2}\right)$$

Taking inverse Laplace transform

$$i(t) = \mathcal{L}^{-1}[I(s)] = 100\sqrt{\frac{C}{L}}\ \sin\frac{1}{\sqrt{LC}}t$$

$$= 100 \times \sqrt{\frac{10\times10^3}{1\times10^{-3}}} \times \sin\left(\frac{1}{\sqrt{1\times10^{-3}\times10\times10^{-6}}}t\right)$$

$$i(t) = 10\sin(10^4 t)\text{ A}$$

5.26 (d)

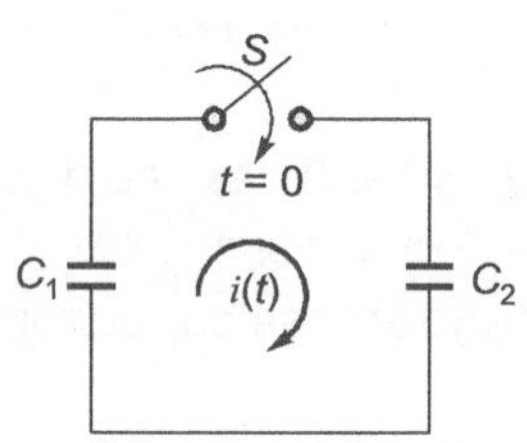

Circuit is s-domain,

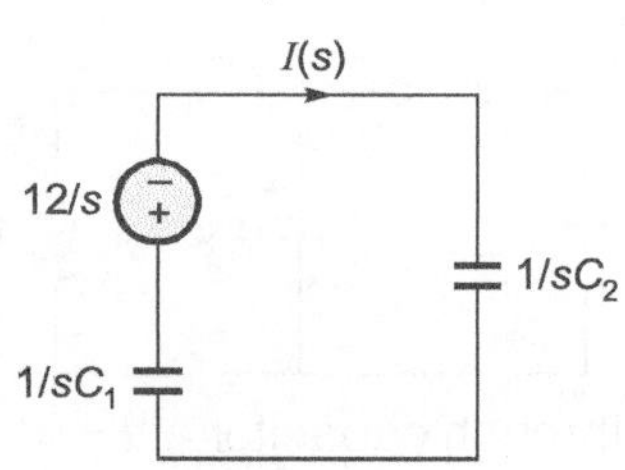

By applying kVL,

$$\frac{12}{s} + \frac{I(s)}{s}\left(\frac{1}{C_1} + \frac{1}{C_2}\right) = 0$$

$$I(s) = -\frac{12C_1C_2}{C_1 + C_2} = k\text{ (constant)}$$

$$\Rightarrow \quad i(t) = k\delta(t)$$

$\therefore$ Current $i(t)$ is an impulse function.

5.27 (c)

Given, $C = 1\ \mu\text{F}$, $v_c(0) = -2$ V, $R = 100\ \Omega$, $I = 20$ mA.

Circuit for the given condition at time $t > 0$ is shown below:

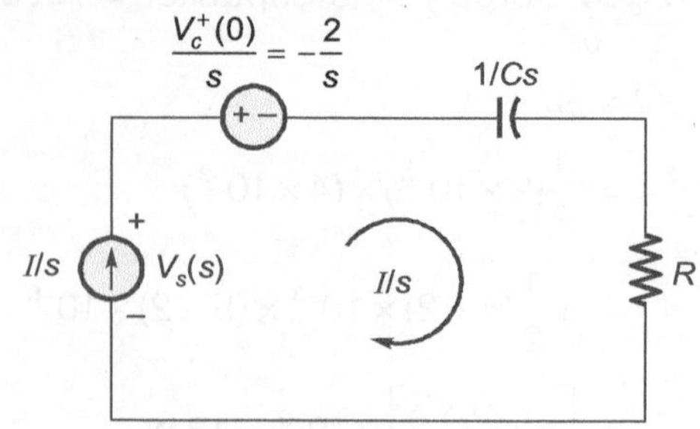

Applying KVL, we have,

$$V_s(s) = \left(\frac{-2}{s}\right) + \frac{I}{s}\left(R + \frac{I}{Cs}\right)$$

$$= \frac{1}{s}\left[-2 + IR + \frac{I}{Cs}\right] = \frac{1}{s}\left[(IR - 2) + \frac{I}{Cs}\right]$$

Putting values of R, C and I, we get,

$$V_s(s) = \frac{1}{s}\left[(20\times10^{-3}\times200 - 2) + \left(\frac{20\times10^{-3}}{10^{-6}}\right)\times\frac{1}{s}\right]$$

$$= \frac{1}{s}\left[(2-2) + 20\times10^3\times\frac{1}{s}\right] = \frac{20\times10^3}{s^2}$$

$$\therefore \quad V_s(s) = \frac{20\times10^3}{s^2}$$

or, $\quad V_s(t) = 20000t\,u(t)$

$\therefore \quad V_s(s) = (20000)t\,u(t)...$

which is the equation of a straight line passing through origin. Hence, option (c) is correct.

5.28 (*)

Circuit for $t < 0$,

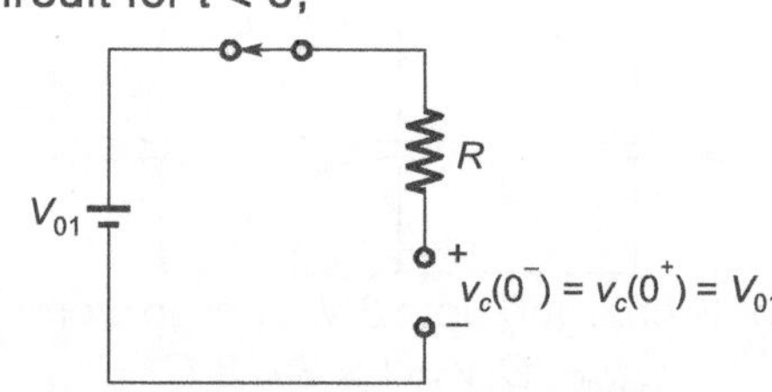

Circuit for $t = \infty$

In steady state capacitor becomes open circuit

$$\therefore \quad v_c(\infty) = -V_{02}$$

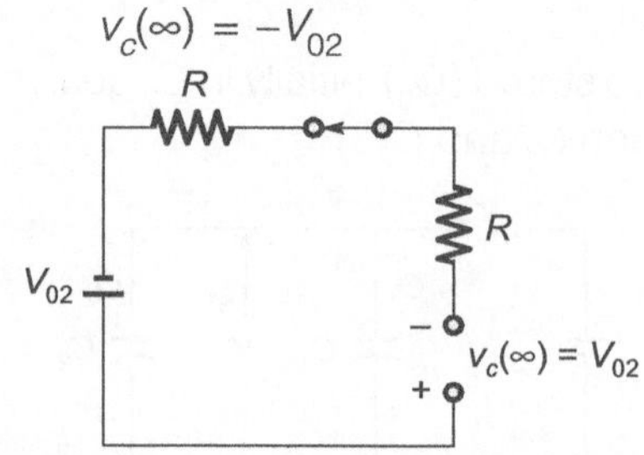

We know that:

$$v_c(t) = v_c(\infty) - \left[v_c(\infty) - v_c(0^+)\right]e^{-t/\tau}$$

$$\tau = \text{Time constant of given circuit} = 2RC$$

$$\therefore \quad v_c(t) = -V_{02} - (-V_{02} - V_{01})\,e^{-t/2RC}$$

$= (V_{02} + V_{01}) - (V_{02} - V_{01})\, e^{-t/2RC} - V_{01}$

or, $\boxed{V_c(t) = (V_{02} + V_{01})\,(e^{-t/2RC} - 1) + V_{01}}$

5.29 (d)

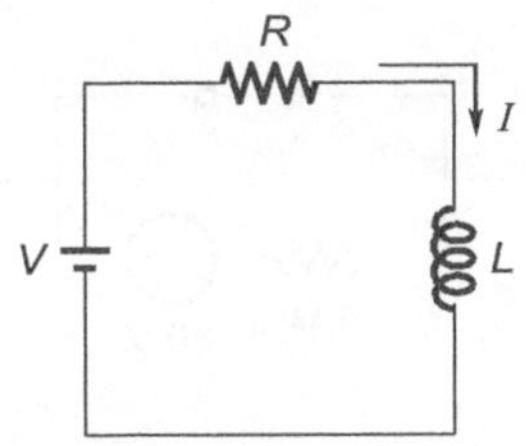

Initially ($t = 0^-$) the inductor would be uncharged.

So, $I(0^+) = 0$

The KVL in the loop will be

$$V = RI + L\frac{dI}{dt}$$

At $t = 0^+$, $V = RI(0^+) + L\frac{dI}{dt}(0^+)$

Since, $I(0^+) = 0$

So, $\frac{dI}{dt}(0^+) = \frac{V}{L}$

Now, lets differentiate the above equation

So, $$\frac{dV}{dt} = R\frac{dI}{dt} + L\frac{d^2I}{dt^2}$$

$$0 = R\frac{dI}{dt} + L\frac{d^2I}{dt^2}$$

At $t = 0^+$

$$0 = R\frac{dI}{dt}(0^+) + L\frac{dI^2}{dt^2}(0^+)$$

So, $\frac{d^2I}{dt^2}(0^+) = \left\{-\frac{R}{L^2}\cdot V\right\}$

5.30 Sol.

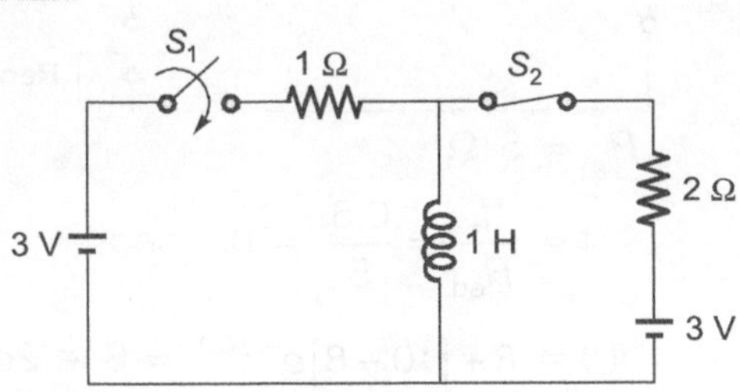

At $t = 0^-$,

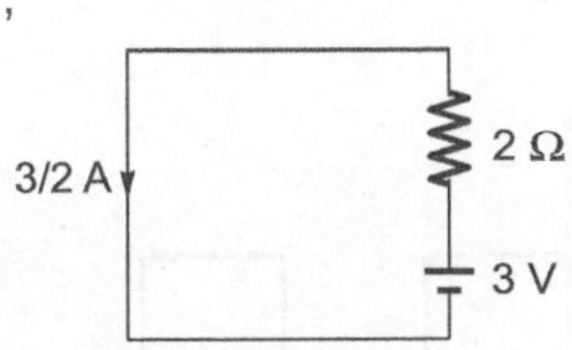

$$i_L(0^+) = i_L(0^-) = 1.5\text{ A}$$

At $t = 0^+$,

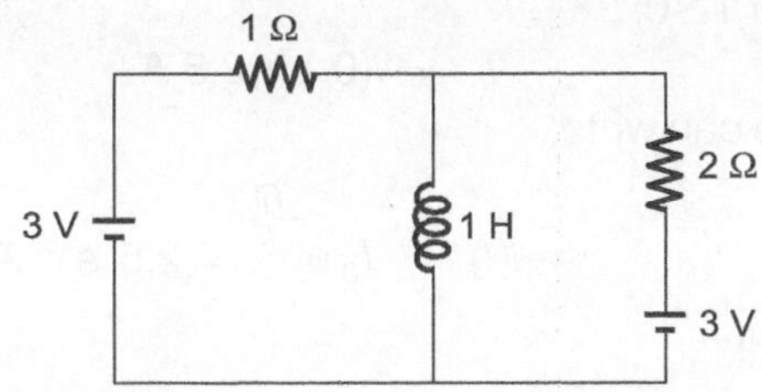

KCL at node A,

$$\frac{V_A - 3}{1} + \frac{3}{2} + \frac{V_A - 3}{2} = 0$$

$$2(V_A - 3) + 3 + (V_A - 3) = 0$$

$$3\,V_A = 6, \quad V_A = 2$$

$$V_A = L\frac{di(0^+)}{dt} = 2$$

$$\frac{di(0^+)}{dt} = \frac{2}{L} = \frac{2}{1} = 2\text{ A/ sec}$$

5.31 Sol.

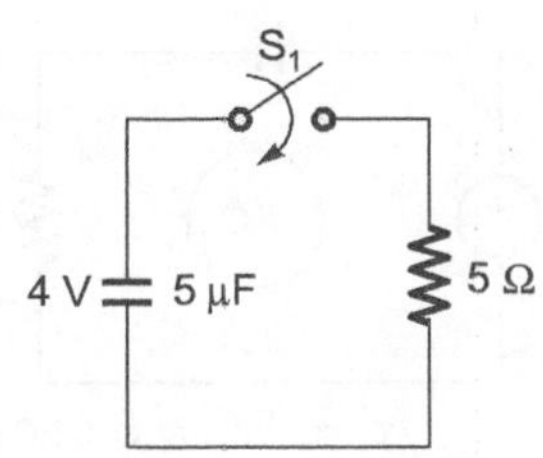

$$i(t) = \left(\frac{4}{5}e^{-t/\tau}\right)$$

$$\tau = RC = 25 \times 10^{-6}\text{ sec}$$

Change lost by capacitor from t = 25 μs to 100 μs is

$$\int_{25\,\mu\text{sec}}^{100\,\mu\text{sec}} i(t)dt = 6.99 \times 10^{-6}\text{ C}$$

5.32 (a)

From the given circuit, consider the following circuit diagram,

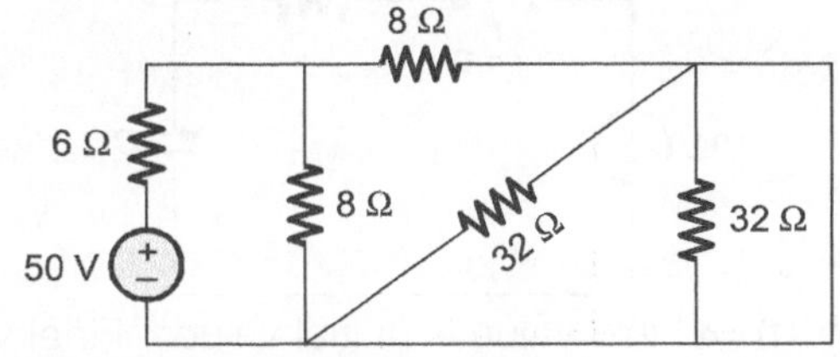

After rearrangement,

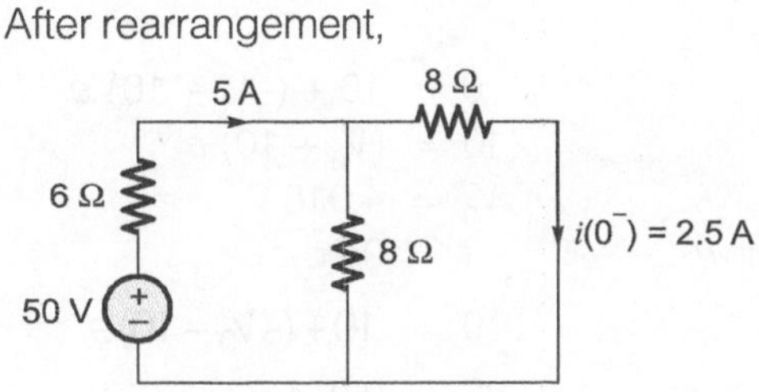

For $t \geq 0$;

$$I_0 = i(0^-) = 2.5 \text{ A}$$

we can write,

$$i(t) = I_0 e^{-\frac{Rt}{L}} = 2.5\, e^{-4t} A$$

5.33 Sol.

Consider the following circuit diagram,

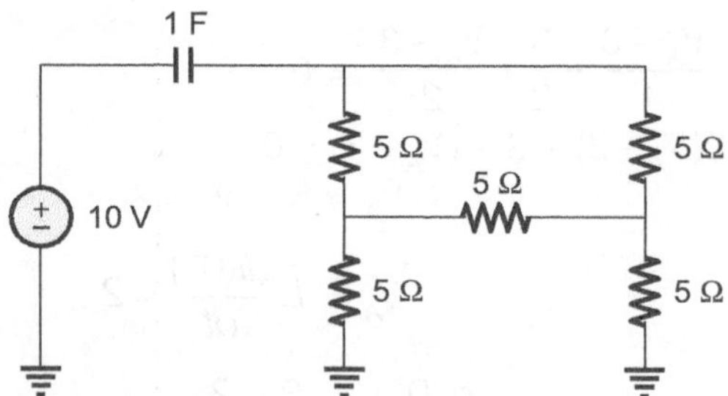

After minimizing circuit elements we can have the following circuit,

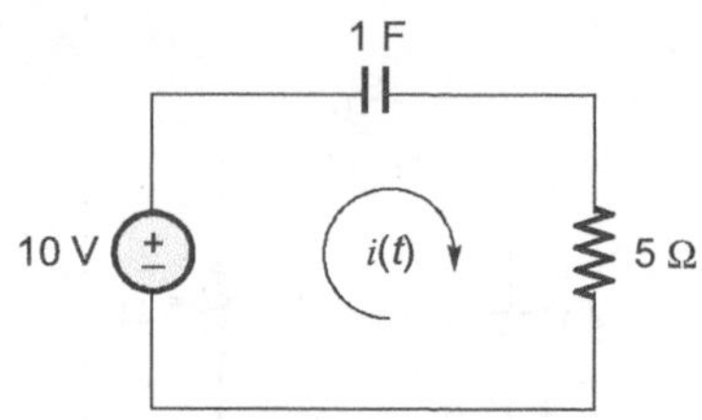

Here, $\tau = RC = 5$ sec.

Now current,

$$i(t) = \frac{V}{R}e^{-t/\tau} = \frac{10}{5}e^{-t/5} = 2e^{-0.2t}$$

Energy supplied by the source,

$$E = \int_0^{\infty} 10 \times 2e^{-0.2t}\, dt = 100 \text{ J}$$

5.34 Sol.

If initial charge polarities on the capacitor is opposite to the supply voltage then only the capacitor voltage crosses the zero line.

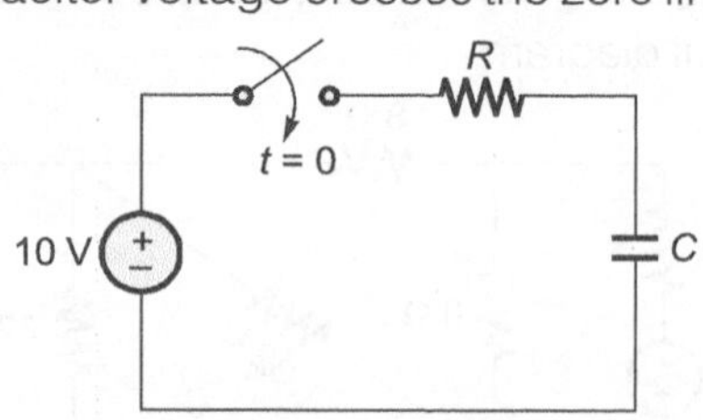

$V_c(t) \Rightarrow$ Final value + (Initial value – Final value) $e^{-t/\tau}$

$$0 = 10 + (-V_0 - 10)\, e^{-0.4}$$
$$10 = (V_0 + 10)\, e^{-0.4}$$
$$V_0 = 4.918 \text{ V}$$

Now, $t = 0.2\tau$

$$0 = 10 + (-V_0' - 10)e^{-0.2}$$
$$V_0' = 2.214$$

%change in voltage

$$= \frac{4.918 - 2.214}{4.918} \times 100\% = 54.99\%$$

5.35 (c)

At $t = 0^-$

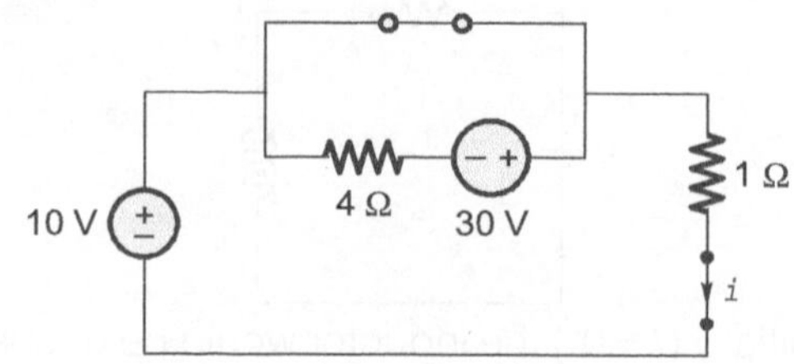

$$i_L(0^-) = \frac{10}{1} = 10 \text{ A}$$

For $t > 0$

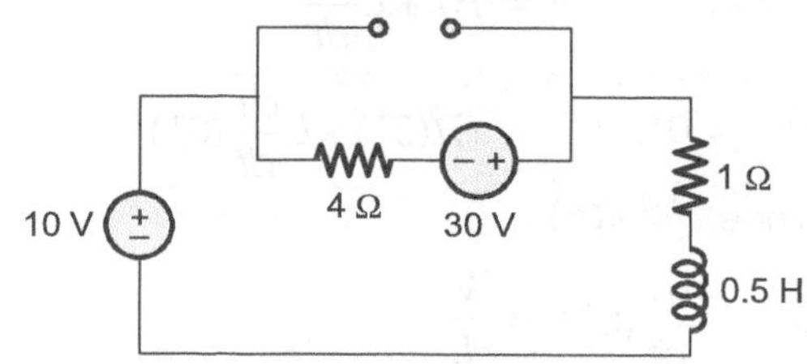

At $t = \infty$

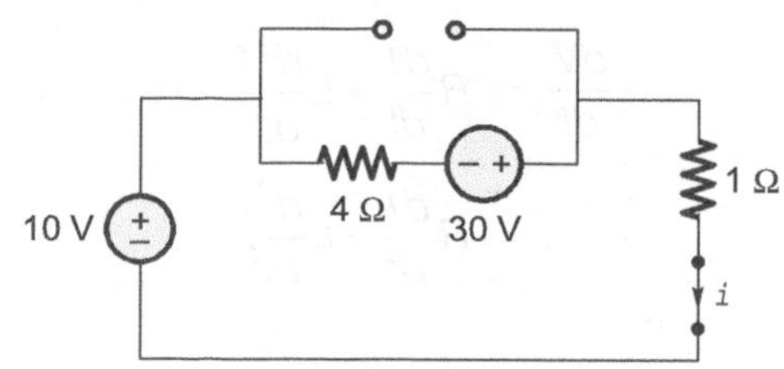

$$i(\infty) = \frac{40}{5} = 8 \text{ A}$$

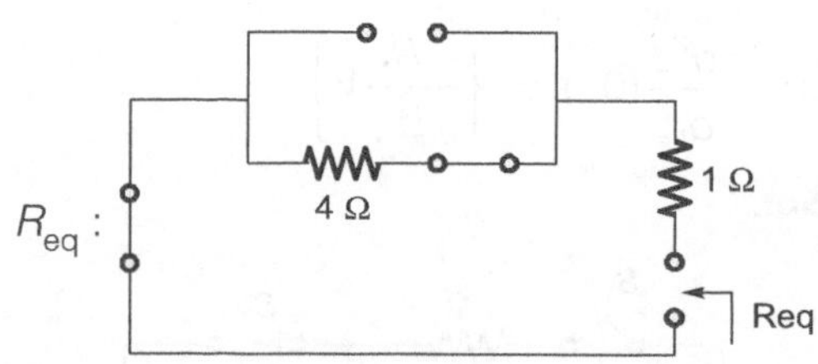

$$R_{eq} = 5\ \Omega$$

$$\tau = \frac{L}{R_{eq}} = \frac{0.5}{5} = 0.1 \text{ sec}$$

$$i(t) = 8 + [10 - 8]e^{-t/0.1} = 8 + 2e^{-10t} \text{ A}$$

5.36 (12.46)

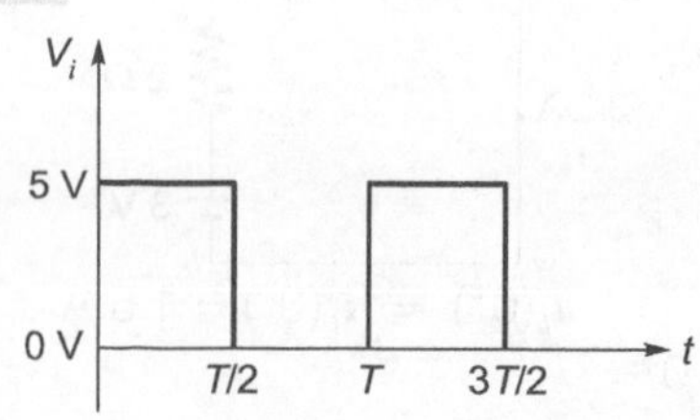

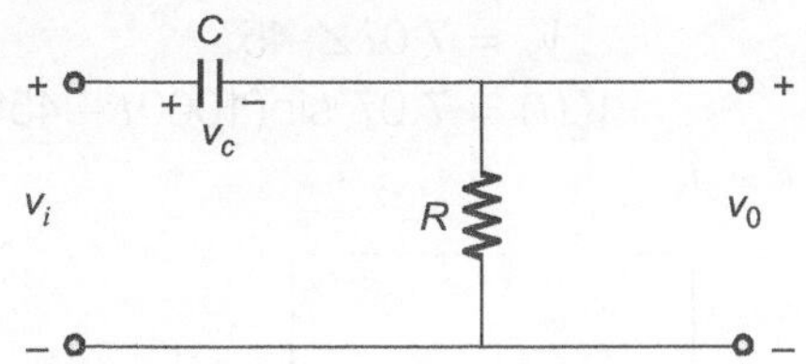

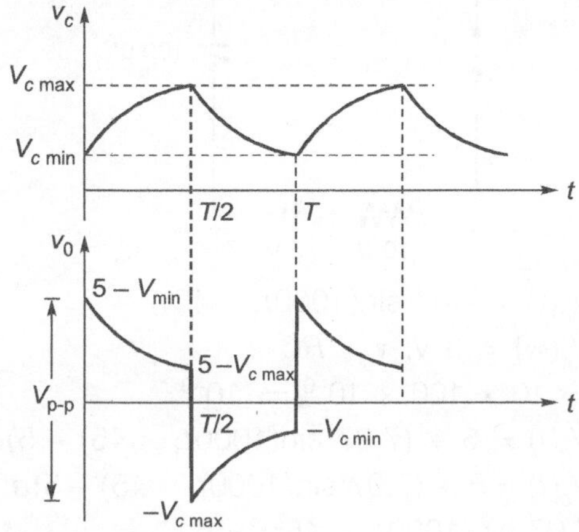

$$V_0 = V_i - V_c$$

For 1st half cycle, $V_0 = 5 - V_c$

For 2nd half cycle, $V_0 = -V_c$

$$V_{p\text{-}p} = (5 - V_{c\,min}) - (-V_{c\,max})$$

$$6.2 = 5 + V_{c\,max} - V_{c\,min}$$

$$\Rightarrow V_{c\,max} - V_{c\,min} = 1.2 \qquad ...(\alpha)$$

For first half cycle, i.e., $0 < t < \frac{T}{2}$

$$v_c(0^+) = v_c(0) = v_c(0^-) = v_{c\,min}$$

$$v_c(\infty) = 5\text{ V}$$

$$\therefore\ v_c(t) = v_c(\infty) + \left[v_c(0^+) - v_c(\infty)\right]e^{-t/\tau}$$

$$v_c(t) = 5 + [V_{cmin} - 5]e^{-t/2\tau} = V_{cmax}$$

$$\Rightarrow V_{c\,max} = 5\left[1 - e^{-T/2\tau}\right] + V_{cmin}e^{-T/2\tau} \qquad ...(i)$$

For $\frac{T}{2} < t < T$

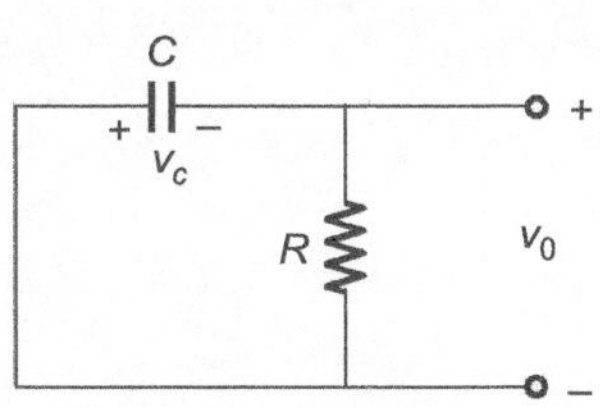

$$v_c(t) = v_c\left(\frac{T}{2}\right)e^{-t(t-T/2)\tau}$$

$$\therefore\quad v_c(t) = V_{cmax}e^{-(t-T/2)\tau}$$

at $t = T$, $\quad v_c = V_{c\,min}$

$$\Rightarrow\quad V_{c\,min} = V_{cmax}e^{-T/2\tau} \qquad ...(ii)$$

As $V_{c\,max} - V_{c\,min} = 1.2$ [From (α)]

$$\therefore V_{cmax} - V_{c\,max}\ e^{-T/2t} = 1.2$$

$$V_{c\,max} = \frac{1.2}{1 - e^{-T/2\tau}} \qquad ...(iii)$$

$$\Rightarrow\quad V_{c\,max} = \frac{1.2}{1 - e^{-T/2\tau}}$$

$$= 5\left[1 - e^{-T/2\tau}\right] + V_{cmin}e^{-2\tau}$$

From (ii),

$$V_{c\,max} = 5\left[1 - e^{-T/2\tau}\right] + \left(V_{cmax}e^{-T/2\tau}\right)e^{-T/2\tau}$$

$$V_{c\,max} = 5\left[1 - e^{-T/2\tau}\right] + V_{cmax}e^{-T/\tau}$$

$$\Rightarrow V_{cmax}\left[1 - e^{-T/\tau}\right] = 5\left[1 - e^{-T/2\tau}\right]$$

$$V_{c\,max} = \frac{5\left[1 - e^{-T/2\tau}\right]}{\left[1 + e^{-T/2\tau}\right]\left[1 - e^{-T/2\tau}\right]}$$

Using equation (iii),

$$\frac{1.2}{1 - e^{-T/2\tau}} = \frac{5}{1 + e^{-T/2\tau}}$$

$$\Rightarrow 1.2 + 1.2\ e^{-T/2\tau} = 5 - 5\ e^{-T/2t}$$

$$\Rightarrow\quad 6.2\ e^{-T/2\tau} = 3.8$$

$$e^{-T/2\tau} = \frac{3.8}{6.2} = 0.6129$$

$$\frac{T}{2\tau} = 0.4895$$

as $\quad T = \frac{1}{f} = \frac{1}{100}\text{ sec}$

and $\quad \tau = RC = 820\text{ C}$

$$\Rightarrow \frac{1}{(100)(2)(820)C} = 0.4895$$

$$\therefore\quad C = 12.46\ \mu\text{F}$$

5.37 (5)

In steady state capactiro acts as an open circuit for DC supply.

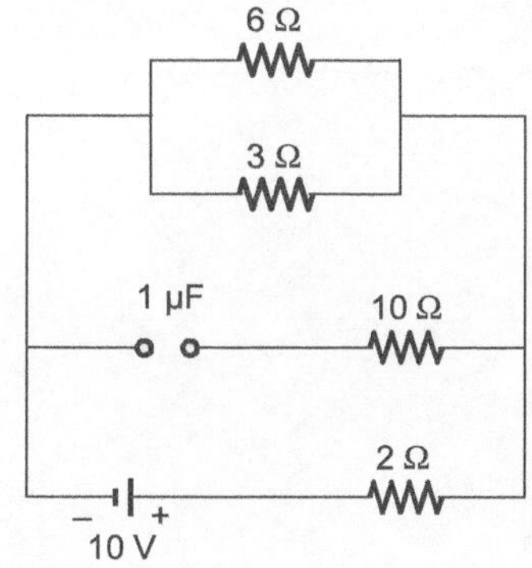

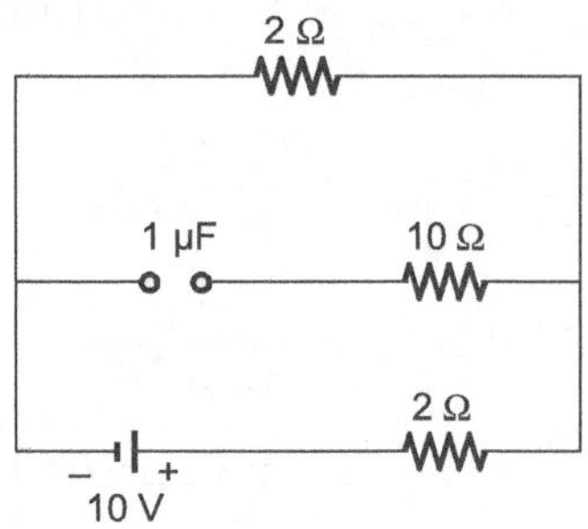

The voltage across 6 Ω is

$$V_o = 10 \times \frac{2}{2+2} = 5 \text{ V}$$

5.38 (1.57)

$t = 0^-$

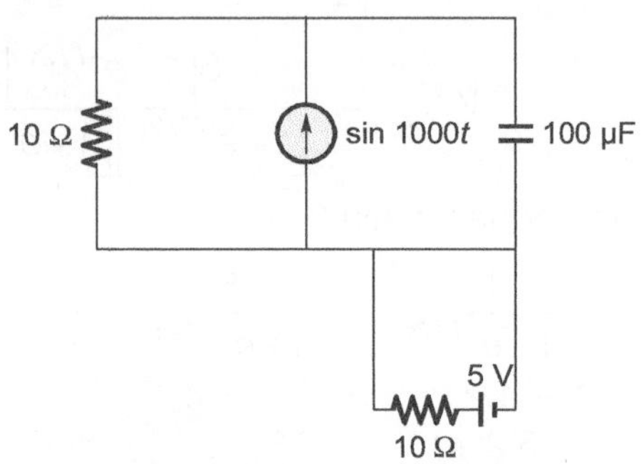

$$X_C = \frac{1}{\omega C} = \frac{1}{1000 \times 100 \times 10^{-6}} \Rightarrow 10\ \Omega$$

$$V_C = 1\angle 0° \times \frac{10}{10 - j10} * -j10$$

$$= \frac{(10)(-j10)}{10 - j10} \times \frac{10 + j10}{10 + j10} \Rightarrow 5 - j5$$

$$V_c = 7.07\angle -45°$$

$$V_c(t) = 7.07 \sin(1000t - 45°)$$

At $t = t_1$

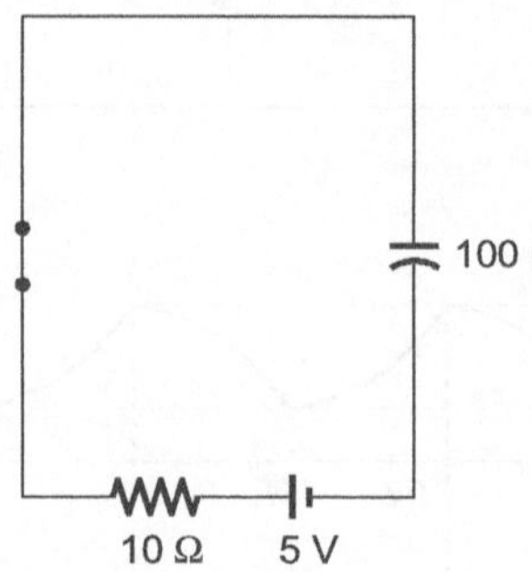

$V_c(t) = 7.07 \sin(1000t_1 - 45°)$

$V_c(\infty) = 5$ V; $\tau = RC$

$\Rightarrow 10 \times 100 \times 10^{-6} \Rightarrow 10^{-3}$

$V_c(t) = 5 + (7.07 \sin(1000t_1 - 45) - 5)e^{-(t/10^{-3})}$

$V_c(t) = 5 + (7.07 \sin(1000t_1 - 45) - 5)e^{-1000(t-t_1)}$

$7.07 \sin(1000t_1 - 45°) = 5$

$$\sin(1000t_1 - 45°) = \frac{5}{7.07}$$

$$1000t_1 - 45° = 45°$$

$$t_1 = 1.57 \text{ msec}$$

5.39 (b)

$$L_{eq} = 1 + 1 + (1 \| 1) = 2.5 \text{ H}$$

$$R_{eq} = 2\ \Omega$$

Time constant, $\tau = \frac{L_{eq}}{R_{eq}} = \frac{2.5}{2} = 1.25$

■■■■

6 Two Port Network and Network Functions

6.1 For the two-port network shown in the figure, the admittance matrix is

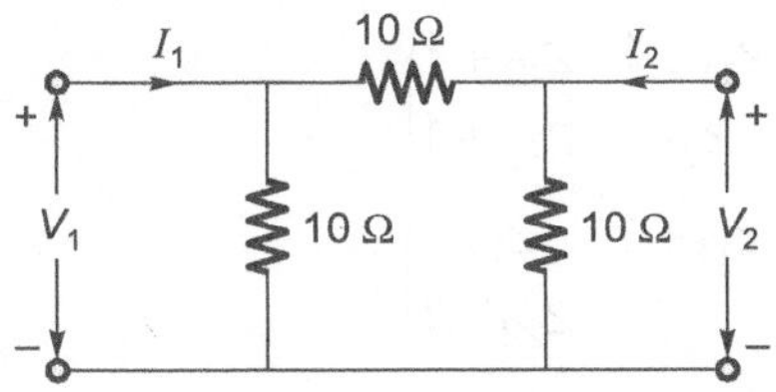

[1997 : 2 M]

6.2 A two-port device is defined by the following pair of equations:

$i_1 = 2v_1 + v_2$ and $i_2 = v_1 + v_2$

Its impedance parameters $\begin{bmatrix} Z_{11} & Z_{12} \\ Z_{21} & Z_{22} \end{bmatrix}$ are given by

(a) $\begin{bmatrix} 2 & 1 \\ 1 & 1 \end{bmatrix}$ (b) $\begin{bmatrix} 1 & -1 \\ -1 & 2 \end{bmatrix}$

(c) $\begin{bmatrix} 1 & 1 \\ 1 & 2 \end{bmatrix}$ (d) $\begin{bmatrix} 2 & -1 \\ -1 & 1 \end{bmatrix}$

[2000 : 2 M]

6.3 A passive two-port network is in a steady state. Compared to its input, the steady state output can never offer

(a) higher voltage (b) lower impedance

(c) greater power (d) better regulation

[2001 : 1 M]

6.4 A two-port network, shown in figure, is described by the following equations:

$I_1 = Y_{11} E_1 + Y_{12} E_2$

$I_2 = Y_{21} E_1 + Y_{22} E_2$

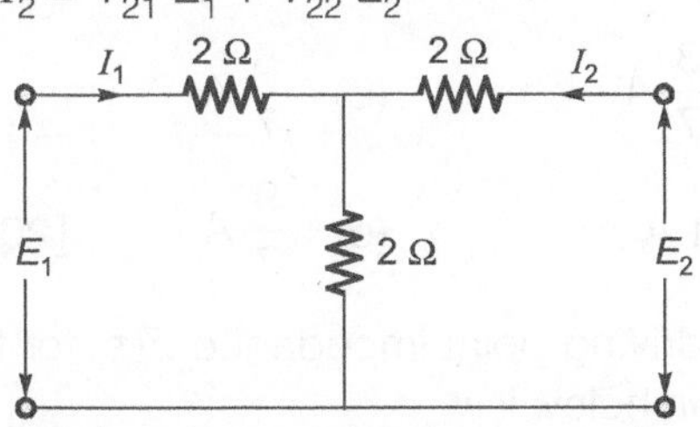

The admittance parameters, Y_{11}, Y_{12}, Y_{21} and Y_{22} for the network shown are

(a) 0.5 mho, 1 mho, 2 mho and 1 mho respectively.

(b) $\frac{1}{3}$ mho, $-\frac{1}{6}$ mho, $-\frac{1}{6}$ mho and $\frac{1}{3}$ mho resp.

(c) 0.5 mho, 0.5 mho, 1.5 mho and 2 mho resp.

(d) $-\frac{2}{5}$ mho, $-\frac{3}{7}$ mho, $\frac{3}{7}$ mho and $\frac{2}{5}$ mho resp.

[2002 : 2 M]

6.5 The h-parameters for a two-port network are defined by $\begin{bmatrix} E_1 \\ I_2 \end{bmatrix} = \begin{bmatrix} h_{11} & h_{12} \\ h_{21} & h_{22} \end{bmatrix} \begin{bmatrix} I_1 \\ E_2 \end{bmatrix}$.

For the two-port network shown in the figure, the value of h_{12} is given by

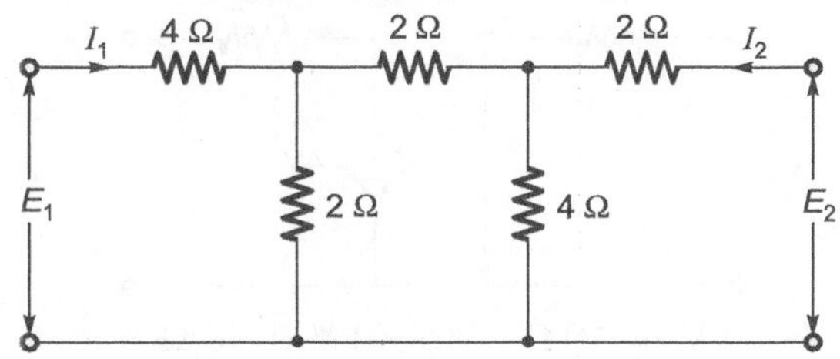

(a) 0.125 (b) 0.167

(c) 0.625 (d) 0.25

[2003 : 2 M]

6.6 The Z-matrix of a 2-port network is given by $Z = \begin{bmatrix} 0.9 & 0.2 \\ 0.2 & 0.6 \end{bmatrix}$.

The element Y_{22} of the corresponding Y matrix of the same network is given by

(a) 1.2 (b) 0.4

(c) –0.4 (d) 1.8

[2004 : 2 M]

6.7 For the two port network shown in the figure the z-matrix is given by

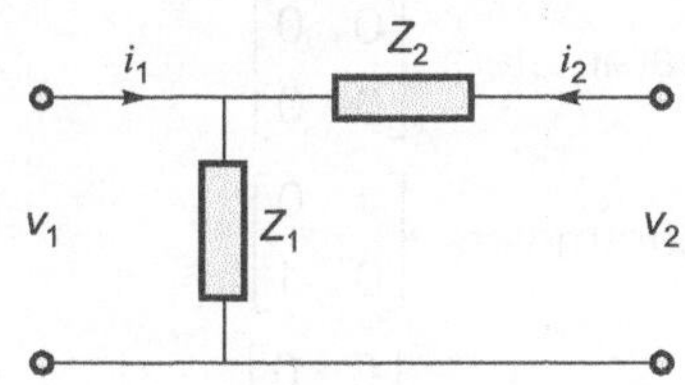

(a) $\begin{bmatrix} Z_1 & Z_1 + Z_2 \\ Z_1 + Z_2 & Z_2 \end{bmatrix}$

(b) $\begin{bmatrix} Z_1 & Z_1 \\ Z_1 + Z_2 & Z_2 \end{bmatrix}$

(c) $\begin{bmatrix} Z_1 & Z_2 \\ Z_2 & Z_1 + Z_2 \end{bmatrix}$

(d) $\begin{bmatrix} Z_1 & Z_1 \\ Z_1 & Z_1 + Z_2 \end{bmatrix}$

[2005 : 1 M]

6.8 Two networks are connected in cascade as shown in the figure. With the usual notations the equivalent *A*, *B*, *C* and *D* constants are obtained. Given that, $C = 0.025\angle45°$, the value of Z_2 is

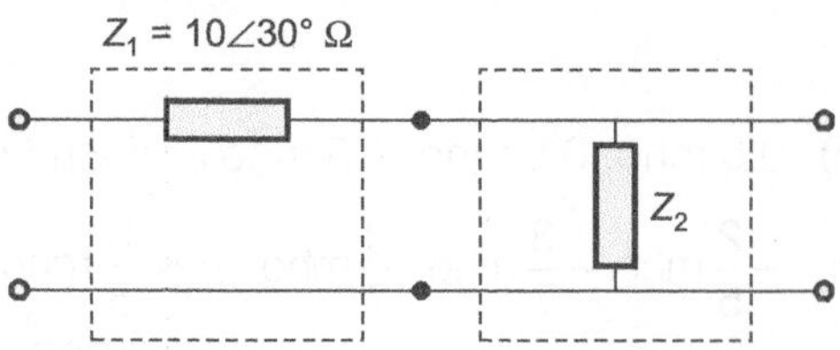

(a) $10\angle30°\ \Omega$ (b) $40\angle{-45°}\ \Omega$
(c) $1\ \Omega$ (d) $0\ \Omega$ [2005 : 2 M]

6.9 The parameters of the circuit shown in the figure are $R_i = 1\ \text{M}\Omega$, $R_0 = 10\ \Omega$, $A = 10^6$ V/V. If $V_i = 1\ \mu$V, the output voltage, input impedance and output impedance respectively are

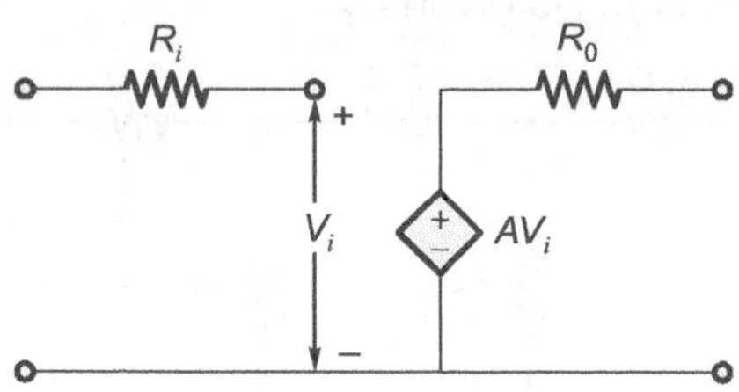

(a) 1 V, ∞, 10 Ω (b) 1 V, 0, 10 Ω
(c) 1 V, 0, ∞ (d) 10 V, ∞, 10 Ω
[2006 : 2 M]

6.10 The parameter type and the matrix representation of the relevant two port parameters that describe the circuit shown are

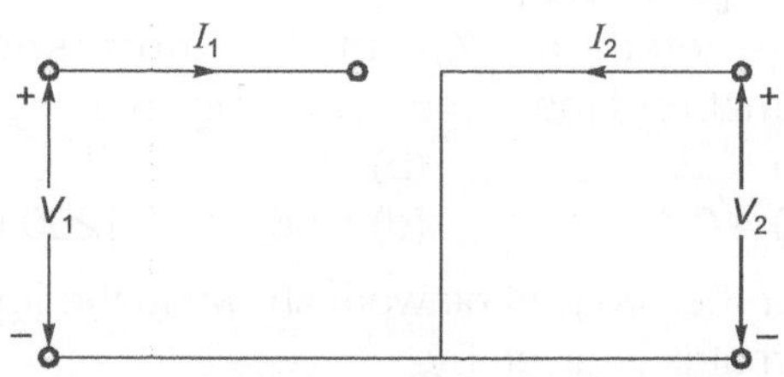

(a) z-parameters, $\begin{bmatrix} 0 & 0 \\ 0 & 0 \end{bmatrix}$

(b) h-parameters, $\begin{bmatrix} 1 & 0 \\ 0 & 1 \end{bmatrix}$

(c) g-parameters, $\begin{bmatrix} 0 & 0 \\ 0 & 0 \end{bmatrix}$

(d) z-parameters, $\begin{bmatrix} 1 & 0 \\ 0 & 1 \end{bmatrix}$ [2006 : 2 M]

6.11 The two-port network *P* shown in the figure has port 1 and 2, denoted by terminals (*a*, *b*) and (*c*, *d*), respectively. It has an impedance matrix *Z* with parameters denoted by Z_{ij}. A 1 Ω resistor is connected in series with the network at port 1 as shown in the figure. The impedance matrix of the modified two-port network (shown as a dashed box) is

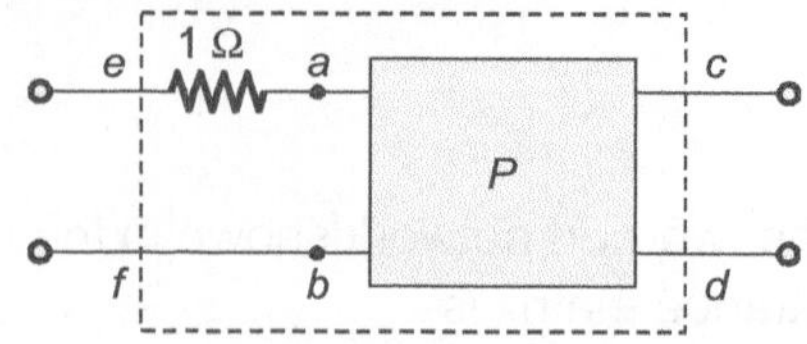

(a) $\begin{pmatrix} Z_{11}+1 & Z_{12}+1 \\ Z_{21} & Z_{22}+1 \end{pmatrix}$

(b) $\begin{pmatrix} Z_{11}+1 & Z_{12} \\ Z_{21} & Z_{22}+1 \end{pmatrix}$

(c) $\begin{pmatrix} Z_{11}+1 & Z_{12} \\ Z_{21} & Z_{22} \end{pmatrix}$

(d) $\begin{pmatrix} Z_{11}+1 & Z_{12} \\ Z_{21}+1 & Z_{22} \end{pmatrix}$ [2010 : 2 M]

Common Data for Questions (6.12 and 6.13):

With 10 V DC connected at port *A* in the linear non-reciprocal two-part network shown below, the following were observed:

(i) 1 Ω connected at port *B* draws a current of 3 A

(ii) 2.5 Ω connected at port *B* draws a current of 2 A

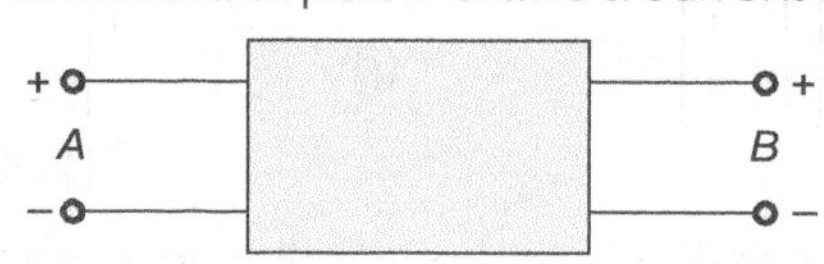

6.12 For the same network with 6 V DC connected at port A, 1 Ω connected at port B draws 7/3 A. If 8 V DC is connected to port A, the open-circuit voltage at port *B* is

(a) 6 V (b) 7 V
(c) 8 V (d) 9 V
[2012 : 2 M]

6.13 With 10 V DC connected at port *A*, the current drawn by 7 Ω connected at port *B* is

(a) $\frac{3}{7} A$ (b) $\frac{5}{7} A$

(c) 1 A (d) $\frac{9}{7} A$ [2012 : 2 M]

6.14 The driving point impedance *Z*(*s*) for the circuit shown below is

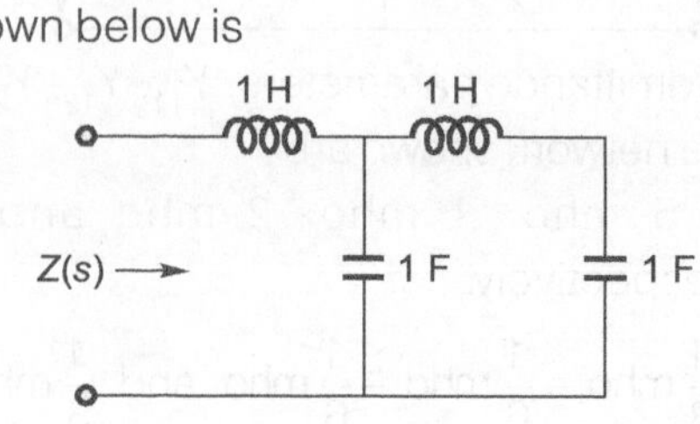

(a) $\dfrac{s^4+3s^2+1}{s^3+2s}$ (b) $\dfrac{s^4+2s^2+4}{s^2+2}$

(c) $\dfrac{s^2+1}{s^4+s^2+1}$ (d) $\dfrac{s^3+1}{s^4+s^2+1}$

[2014 : 1 M, Set-3]

6.15 The z-parameters of the two port network shown in the figure are $z_{11} = 40\ \Omega$, $z_{12} = 60\ \Omega$, $z_{21} = 80\ \Omega$ and $z_{22} = 100\ \Omega$. The average power delivered to $R_L = 20\ \Omega$, in watts, is______.

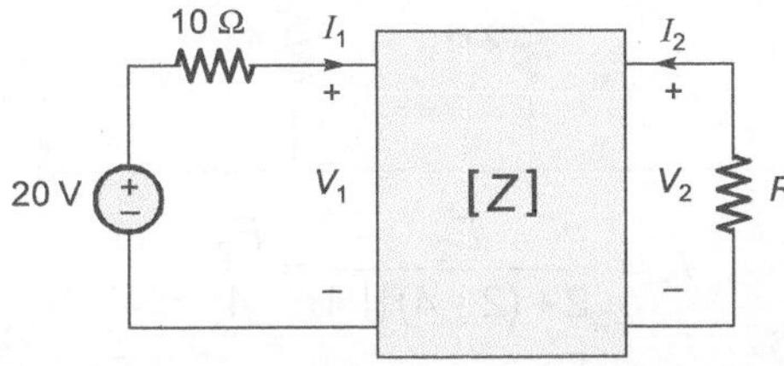

[2016 : 2 M, Set-2]

6.16 The driving point input impedance seen from the source V_S of the circuit shown below, in Ω, is ___.

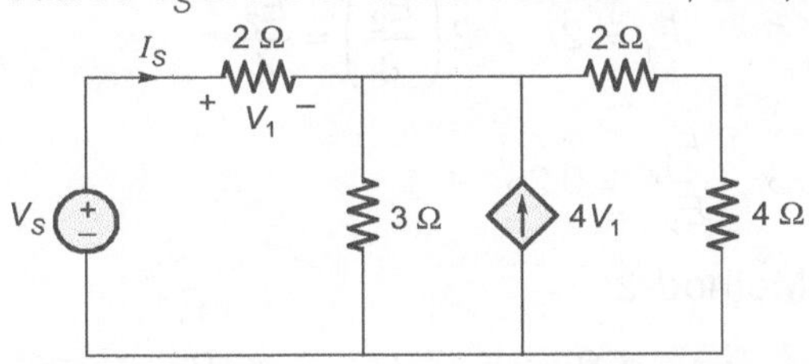

[2016 : 2 M, Set-2]

6.17 Two passive two-port networks are connected in cascade as shown in the figure. A voltage source is connected at port 1.

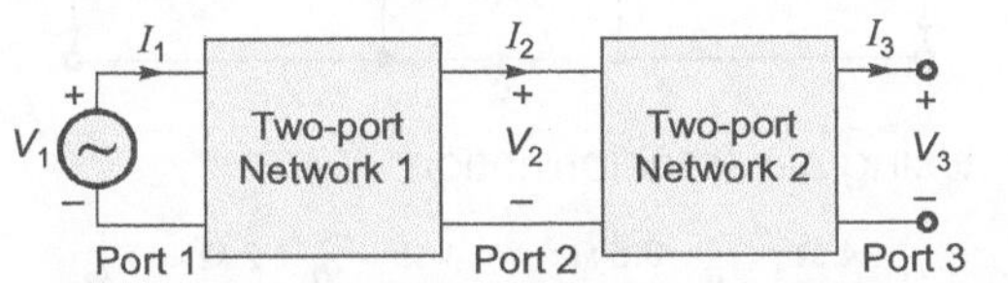

Given : $V_1 = A_1V_2 + B_1I_2$; $I_1 = C_1V_2 + D_1I_2$

$V_2 = A_2V_3 + B_2I_3$; $I_2 = C_2V_3 + D_2I_3$

A_1, B_1, C_1, D_1, A_2, B_2, C_2 and D_2 are the generalized circuit constants. If the Thevenin equivalent circuit at port 3 consists of a voltage source V_T and an impedance Z_T, connected in series, then

(a) $V_T = \dfrac{V_1}{A_1A_2}$, $Z_T = \dfrac{A_1B_2 + B_1D_2}{A_1A_2 + B_1C_2}$

(b) $V_T = \dfrac{V_1}{A_1A_2 + B_1C_2}$, $Z_T = \dfrac{A_1B_2 + B_1D_2}{A_1A_2}$

(c) $V_T = \dfrac{V_1}{A_1 + A_2}$, $Z_T = \dfrac{A_1B_2 + B_1D_2}{A_1 + A_2}$

(d) $V_T = \dfrac{V_1}{A_1A_2 + B_1C_2}$, $Z_T = \dfrac{A_1B_2 + B_1D_2}{A_1A_2 + B_1C_2}$

[2017 : 2 M, Set-1]

6.18 In the two-port network shown, the h_{11} parameter $\left(\text{where, } h_{11} = \dfrac{V_1}{I_i}, \text{ when } V_2 = 0\right)$ in ohms is ______ (up to 2 decimal places).

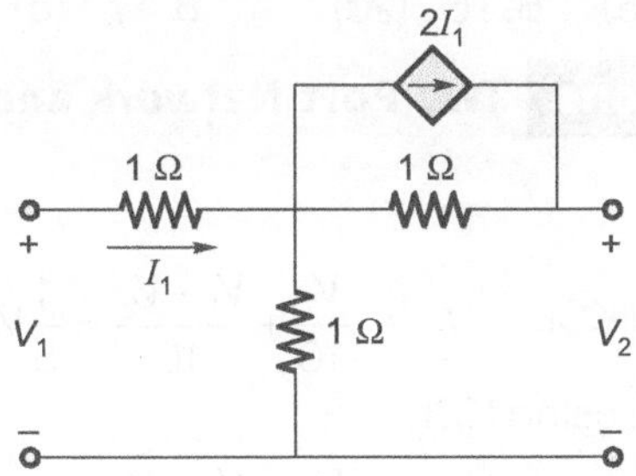

[2018 : 1 M]

6.19 The admittance parameters of the passive resistive two-port network shown in the figure are :

$y_{11} = 5\text{S}$, $y_{22} = 1\text{S}$, $y_{12} = y_{21} = -2.5\text{S}$

The power delivered to the load resistor R_L in Watt is _______ (Round off to 2 decimal places).

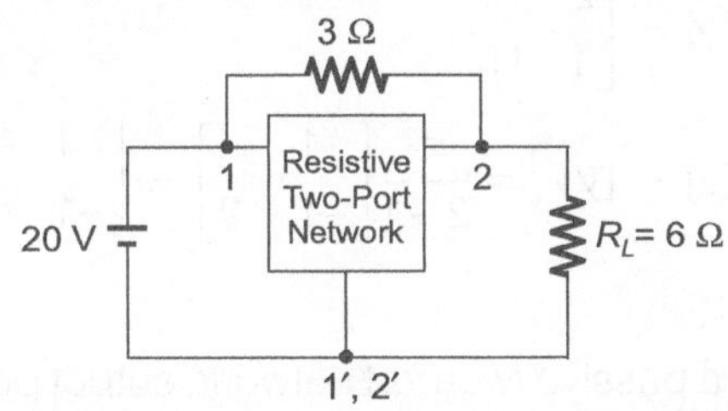

[2023 : 2 M]

6.20 Two passive two-port network P and Q are connected as shown in the figure. The impedance matrix of network P is $Z_P = \begin{bmatrix} 40\Omega & 60\Omega \\ 80\Omega & 100\Omega \end{bmatrix}$. The admittance matrix of network Q is $Y_Q = \begin{bmatrix} 5S & -2.5S \\ -2.5S & 1S \end{bmatrix}$. Let the ABCD matrix of the two-port network R in the figure be $\begin{bmatrix} \alpha & \beta \\ \gamma & \delta \end{bmatrix}$. The value of β in Ω is _______ . (rounded off to 2 decimal places)

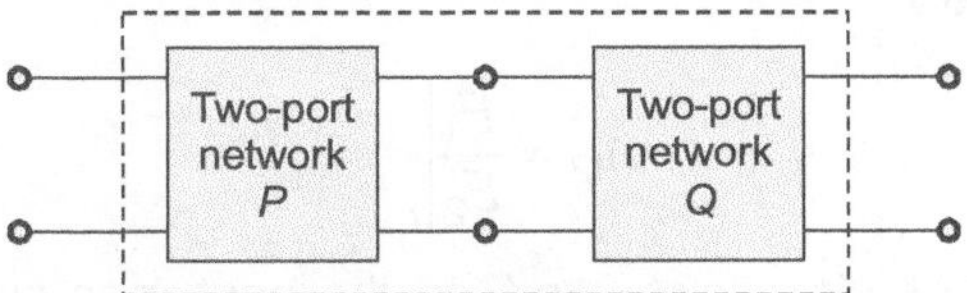

Two-port network R

[2024 : 2 M]

■■■■

Answers Two Port Network and Network

6.1 (Sol.)	6.2 (b)	6.3 (c)	6.4 (b)	6.5 (d)	6.6 (d)	6.7 (d)
6.8 (b)	6.9 (a)	6.10 (c)	6.11 (c)	6.12 (b)	6.13 (c)	6.14 (a)
6.15 (35.55)	6.16 (20)	6.17 (d)	6.18 (0.5)	6.19 (238)	6.20 (–19.80)	

Explanations Two Port Network and Network

6.1 Sol.

Using KCL, $I_1 = \frac{V_1}{10} + \frac{V_1 - V_2}{10} = \frac{1}{5}V_1 - \frac{1}{10}V_2$

Again using KCL,

$$I_2 = \frac{V_2}{10} + \frac{V_2 - V_1}{10} = -\frac{1}{10}V_1 + \frac{1}{5}V_2$$

Hence, $[y] = \begin{bmatrix} 0.2 & -0.1 \\ -0.1 & 0.2 \end{bmatrix}$

6.2 (b)

$$[y] = \begin{bmatrix} 2 & 1 \\ 1 & 1 \end{bmatrix}$$

$$[z] = [y]^{-1} = \frac{1}{2-1}\begin{bmatrix} 1 & -1 \\ -1 & 2 \end{bmatrix} = \begin{bmatrix} 1 & -1 \\ -1 & 2 \end{bmatrix}$$

6.3 (c)

For a passive two-port network, output power can never be greater than input power.

6.4 (b)

Using KVL, $E_1 = 2I_1 + 2(I_1 + I_2)$

Again using KVL, $E_2 = 2I_2 + 2(I_1 + I_2)$

$$\Rightarrow \quad [z] = \begin{bmatrix} 4 & 2 \\ 2 & 4 \end{bmatrix}$$

$$[y] = [z]^{-1}$$

$$= \frac{1}{(4 \times 4) - (2 \times 2)} \cdot \begin{bmatrix} 4 & -2 \\ -2 & 4 \end{bmatrix}$$

$$\begin{bmatrix} Y_{11} & Y_{12} \\ Y_{21} & Y_{22} \end{bmatrix} = \begin{bmatrix} 1/3 & -1/6 \\ -1/6 & 1/3 \end{bmatrix}$$

6.5 (d)

$$h_{12} = \left.\frac{E_1}{E_2}\right|_{I_1=0}$$

or h_{12} is ratio of E_1 to E_2 for the input open-circuited condition.

Two methods are provided to solve the problem.

Method-1:

Assuming,

$$I_1 = 0$$

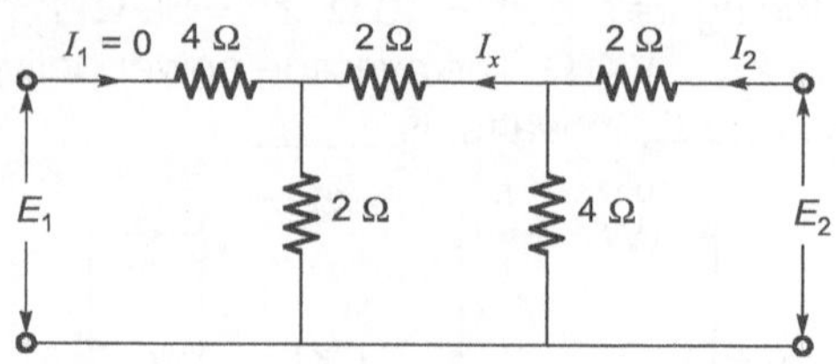

$$I_2 = \frac{E_2}{2 + (2+4)\,||\,4} = \frac{E_2}{4}$$

$$I_x = \frac{I_2}{(2+2)+4} \times 4 = \frac{I_2}{2} = \frac{1}{2}\left(\frac{E_2}{4}\right) = \frac{E_2}{8}$$

$$E_1 = 2I_x = 2\left(\frac{E_2}{8}\right) = \frac{E_2}{4}$$

$$\Rightarrow \quad \frac{E_1}{E_2} = 0.25$$

Method-2:

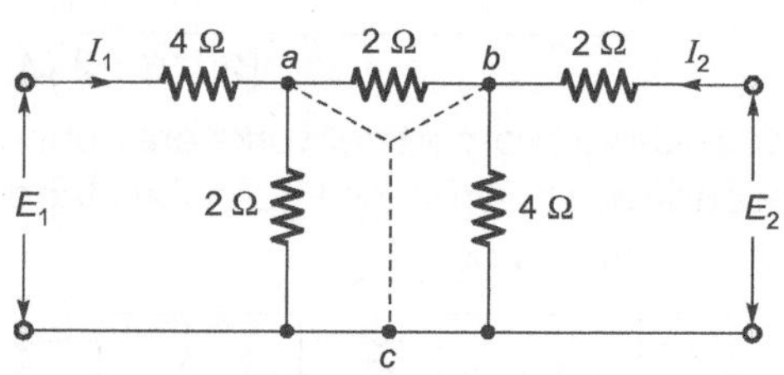

Using Δ-Y transformation:

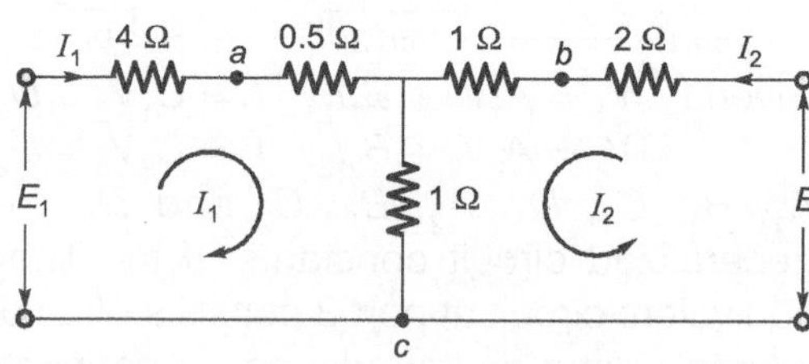

$$E_1 = (4 + 0.5)\,I_1 + 1(I_1 + I_2) = 5.5\,I_1 + I_2 \quad \text{...(i)}$$

$$E_2 = (2 + 1)\,I_2 + 1(I_1 + I_2) = I_1 + 4I_2 \quad \text{...(ii)}$$

Put, $I_1 = 0$ in equation (i) and (ii),

$$E_1 = I_2 \quad \text{and} \quad E_2 = 4I_2$$

$$h_{12} = \left.\frac{E_1}{E_2}\right|_{I_1=0} = 0.25$$

6.6 (d)

$$[z] = \begin{bmatrix} 0.9 & 0.2 \\ 0.2 & 0.6 \end{bmatrix}$$

$$[y] = [z]^{-1} = \frac{\begin{bmatrix} 0.6 & -0.2 \\ -0.2 & 0.9 \end{bmatrix}}{[0.9 \times 0.6 - 0.04]} = \begin{bmatrix} 1.2 & -0.4 \\ -0.4 & 1.8 \end{bmatrix}$$

$y_{22} = 1.8$

6.7 (d)

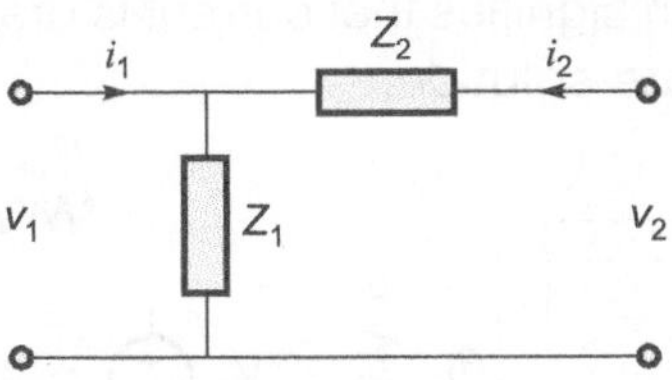

Two methods are provided to solve the problem.

Method-1:

When $i_2 = 0$: $V_1 = V_2 = i_1 Z_1$

so, $$Z_{11} = \left.\frac{V_1}{i_1}\right|_{i_2=0} = Z_1 \quad ...(i)$$

$$Z_{21} = \left.\frac{V_2}{i_1}\right|_{i_2=0} = Z_1 \quad ...(ii)$$

when, $i_1 = 0$; $V_1 = i_2 Z_1$

$$\Rightarrow \quad Z_{12} = \left.\frac{V_1}{i_2}\right|_{i_1=0} = Z_1 \quad ...(iii)$$

$$V_2 = i_2(Z_1 + Z_2)$$

$$\Rightarrow \quad Z_{22} = \left.\frac{V_2}{i_2}\right|_{i_1=0} = Z_1 + Z_2 \quad ...(iv)$$

Method-2:

$$V_1 = (i_1 + i_2) Z_1 = Z_1 i_1 + Z_1 i_2 \quad ...(v)$$

$$V_2 = Z_2 i_1 + Z_1(i_1 + i_2) = Z_1 i_1 + (Z_1 + Z_2) i_2 \quad ...(vi)$$

From equation (i) to (iv)

$$\text{z-matrix} = \begin{bmatrix} Z_1 & Z_1 \\ Z_1 & Z_1 + Z_2 \end{bmatrix}$$

6.8 (b)

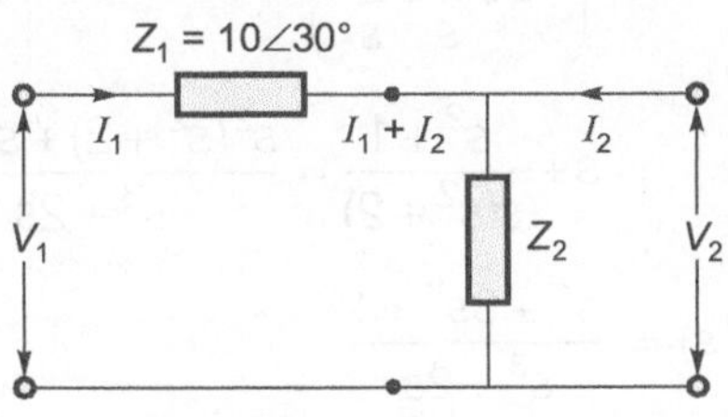

$$V_1 = AV_2 + BI_2$$

$$I_1 = CV_2 + DI_2$$

$$V_2 = Z_2(I_1 + I_2) \quad ...(i)$$

$$C = \left.\frac{I_1}{V_2}\right|_{I_2=0}$$

Putting, $I_2 = 0$ in equation (i)

$$V_2 = Z_2 I_1$$

$$\Rightarrow \quad Z_2 = \left.\frac{V_2}{I_1}\right|_{I_2=0} = \frac{1}{C} = \frac{1}{0.025\angle 45°}$$

$$Z_2 = 40\angle -45°\ \Omega$$

6.9 (a)

Output voltage $= V_0 = AV_i$

$$V_0 = 10^6 \times 1 \times 10^{-6} = 1\text{ V}$$

[Given: $A = 10^6$ V/V, $V_i = 1\ \mu$V]

To calculate input impedance, V_{DC} source is connected at input port

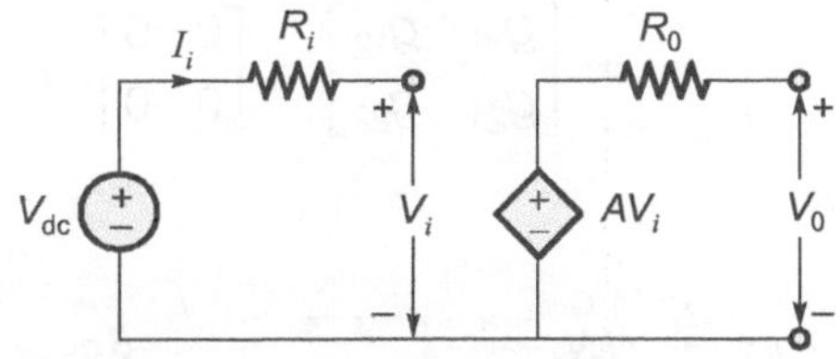

Input impedance, $Z_i = \dfrac{V_{DC}}{I_i}$

As loop is not closed, $I_i = 0$

So, $$Z_i = \frac{V_{DC}}{0} \to \infty$$

To calculate output impedance, V_{DC} source is connected at output port.

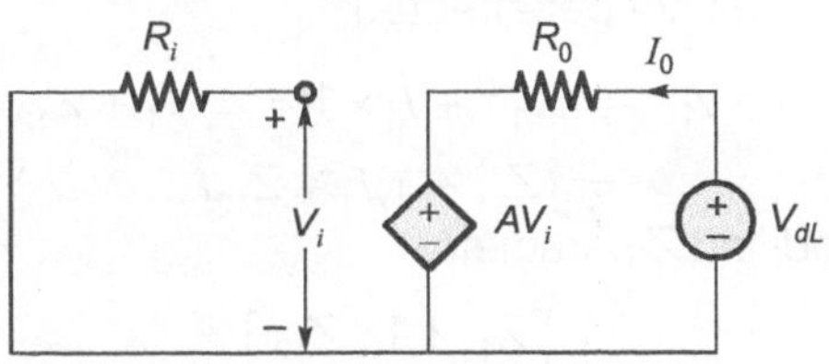

Output impedance,

$$Z_0 = \frac{V_{DC}}{I_0} = \frac{I_0 R_0 + AV_i}{I_0}$$

As $V_i = 0$, $$Z_0 = \frac{I_0 R_0 + A \times 0}{I_0} = R_0 = 10\ \Omega$$

6.10 (c)

$$I_1 = g_{11} V_1 + g_{12} I_2$$

$$V_2 = g_{21} V_1 + g_{22} I_2$$

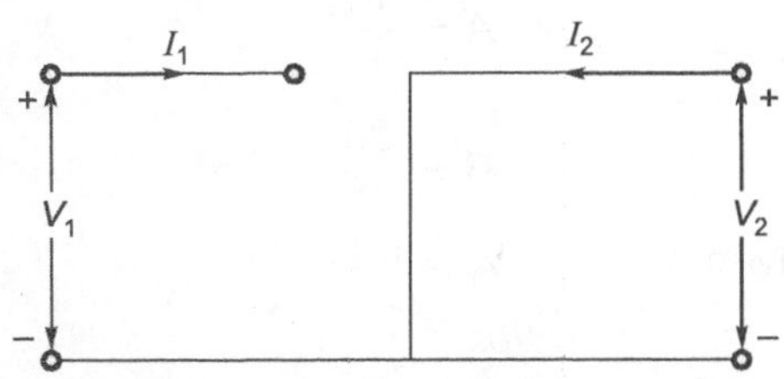

Since, port-1 is open-circuit, $I_1 = 0$

Port-2 is short-circuit, $V_2 = 0$

$$g_{11} = \left.\frac{I_1}{V_1}\right|_{I_2=0} = \frac{0}{V_1} = 0$$

$$g_{12} = \left.\frac{I_1}{I_2}\right|_{V_1=0} = \frac{0}{I_2} = 0$$

$$g_{21} = \left.\frac{V_2}{V_1}\right|_{I_2=0} = \frac{0}{V_1} = 0$$

$$g_{22} = \left.\frac{V_2}{I_1}\right|_{V_1=0} = \frac{0}{I_2} = 0$$

So, g-parameters,

$$= \begin{bmatrix} g_{11} & g_{12} \\ g_{21} & g_{22} \end{bmatrix} = \begin{bmatrix} 0 & 0 \\ 0 & 0 \end{bmatrix}$$

6.11 (c)

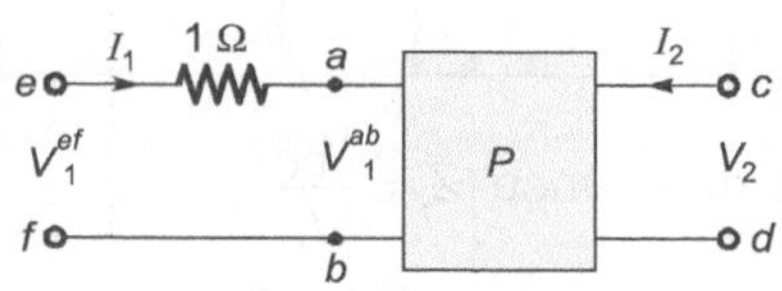

$$V_1^{ab} = Z_{11}I_1 + Z_{12}I_2 \quad \text{...(i)}$$
$$V_2 = Z_{21}I_1 + Z_{22}I_2 \quad \text{...(ii)}$$

As 1 Ω resistor is connected in series with the network at port-1.

V_2 does not get affected,

$$V_1^{ef} = V_1^{ab} + I_1 \times 1 = Z_{11}I_1 + Z_{12}I_2 + I_1$$
$$= (Z_{11}+1)I_1 + Z_{12}I_2$$

Modified Z-parameter

$$= \begin{bmatrix} Z_{11}+1 & Z_{12} \\ Z_{21} & Z_{22} \end{bmatrix}$$

6.12 (b)

(i) $V_1 = 10$ V; $V_2 = 3$ V

$I_2 = -3$ A $\quad V_1 = AV_2 - BI_2$

$10 = 3A + 3B$ ...(i)

(ii) $V_2 = 5$ V

$I_2 = -2$ A

$10 = 5A + 2B$ ...(ii)

$$A = \frac{10}{9} \quad \text{...(iii)}$$

$$B = \frac{20}{9} \quad \text{...(iv)}$$

Given, $V_1 = 8$ V

$(V_2)_{OC} = ?$

$I_2 = 0$

$V_1 = AV_2 - BI_2$

$8 = A(V_2)_{OC} - 0$

$$(V_2)_{OC} = \frac{8}{A} = \frac{8}{10/9} = 7.2 \text{ V}$$

6.13 (c)

Given, $V_1 = 10$ V; $\quad V_2 = (-7I_2)$

$V_1 = AV_2 - BI_2$

$$10 = -7I_2A - BI_2 = -\frac{70}{9}I_2 - \frac{20}{9}I_2$$

$I_2 = -1$ A

–ve sign signifies that current is drawn.

Alternate solution:

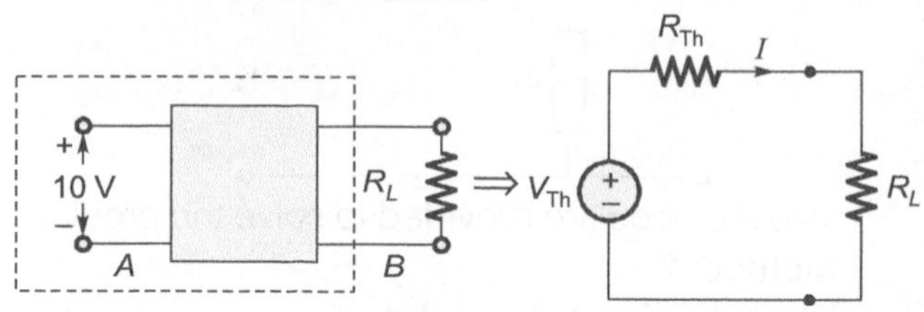

(i) $$3 = \frac{V_{Th}}{1+R_{Th}}$$

(ii) $$2 = \frac{V_{Th}}{2.5+R_{Th}}$$

$$\Rightarrow \quad \frac{3}{2} = \frac{R_{Th}+2.5}{R_{Th}+1}$$

$$\Rightarrow \quad 2R_{Th} + 5 = 3R_{Th} + 3$$

$$\Rightarrow \quad R_{Th} = 2\ \Omega \quad \text{and} \quad V_{Th} = 9 \text{ V}$$

$$I = \frac{9}{7+2} = 1 \text{ Amp.}$$

6.14 (a)

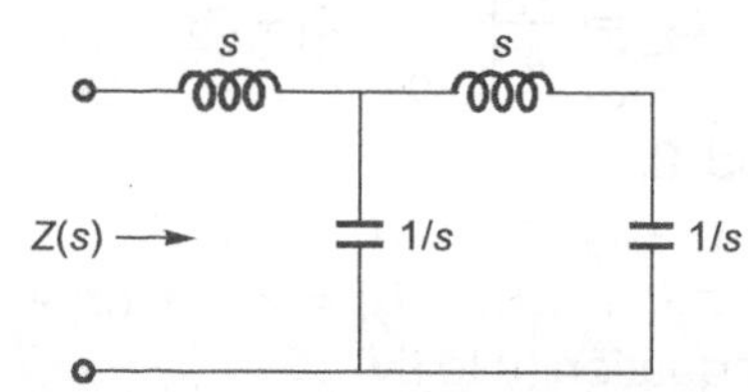

Driving point impedance, $Z(s)$ is

$$Z(s) = s + \left(\frac{\left(s+\frac{1}{s}\right)\times\frac{1}{s}}{s+\frac{1}{s}+\frac{1}{s}}\right) = s + \left(\frac{s^2+1}{s^2}\right)\times\frac{s}{(s^2+2)}$$

$$= s + \frac{s^2+1}{s(s^2+2)} = \frac{s^2(s^2+2)+s^2+1}{s^3+2s}$$

$$\text{or, } Z(s) = \frac{s^4+3s^2+1}{s^3+2s}$$

6.15 Sol.

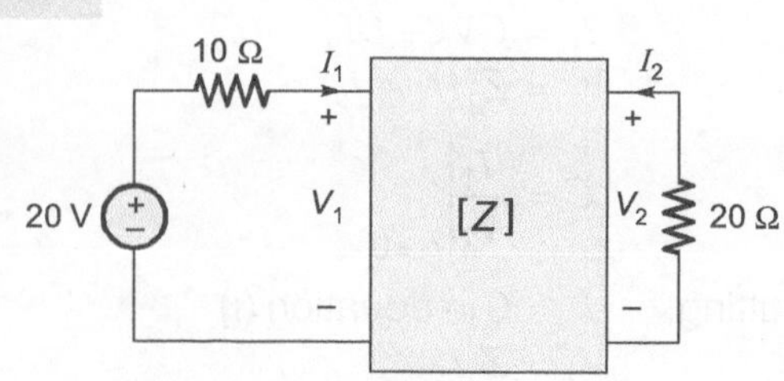

Given, $Z_{11} = 40\ \Omega,\ Z_{12} = 60\ \Omega$

$Z_{21} = 80\ \Omega,\ Z_{22} = 100\ \Omega$

From the figure,

$$V_2 = -20\, I_2 \qquad \text{...(i)}$$

and
$$V_1 = 40I_1 + 60I_2 \qquad \text{...(ii)}$$

$$V_2 = 80I_1 + 100I_2 \qquad \text{...(iii)}$$

From equation (i) and (iii), we get

So,
$$-20I_2 = 80I_1 + 100I_2$$

$$\Rightarrow \quad I_2 = -\frac{2}{3}I_1 \qquad \text{...(iv)}$$

Using equation (ii) and (iv), we get

So, $V_1 = 40I_1 + 60I_2 = 40I_1 + 60\left(\frac{-2}{3}I_1\right)$

$$V_1 = 0$$

From the figure, we have

$$20 = 10I_1 + V_1$$

Since $V_1 = 0$, $I_1 = 2$ A

So, $I_2 = -\frac{4}{3}$ A

Power dissipated in

$$P_{R_L} = I_2^2 R_L = \left(\frac{4}{3}\right)^2 \times 20$$

$$= \frac{16}{9} \times 20 = 35.55\text{ W}$$

6.16 Sol.

To find impedance seen by V_s

$$Z_s = \frac{V_s}{I_s};\quad V_1 = 2I_s$$

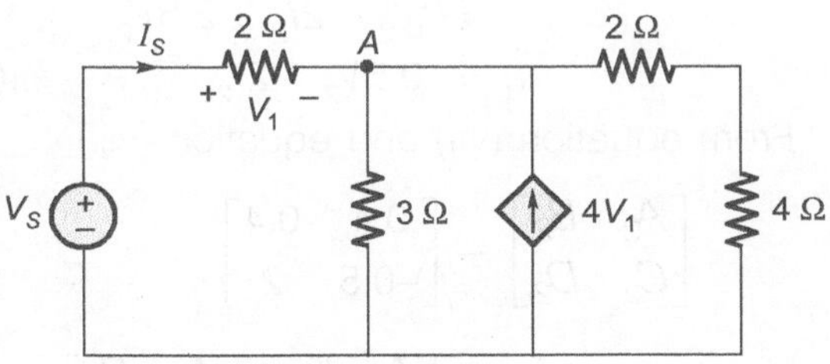

Applying KCL at node A,

$$I_s + 4\,V_1 = \frac{V_A}{3} + \frac{V_A}{6}$$

$$V_A = V_s - V_1 \text{ and } V_1 = 2I_s$$

So,
$$I_s + 8I_s = \frac{V_s - 2I_s}{3} + \frac{V_s - 2I_s}{6}$$

$$\Rightarrow \quad 54\,I_s = 2V_s - 4I_s + V_s - 2I_s$$

$$\Rightarrow \quad 3V_s = 60I_s$$

$$\frac{V_s}{I_s} = 20\ \Omega$$

6.17 (d)

For two port networks we can write,

$$\begin{bmatrix} A & B \\ C & D \end{bmatrix} = \begin{bmatrix} A_1 & B_1 \\ C_1 & D_1 \end{bmatrix}\begin{bmatrix} A_2 & B_2 \\ C_2 & D_2 \end{bmatrix}$$

or
$$A = A_1A_2 + B_1C_2 \qquad \text{...(i)}$$

$$B = A_1B_2 + B_1D_2 \qquad \text{...(ii)}$$

or
$$V_1 = AV_2 - BI_2 \qquad \text{...(iii)}$$

To get, V_T ($I_2 = 0$); from equation (iii),

$$V_2 = V_T = \frac{V_1}{A} = \frac{V_1}{A_1A_2 + B_1C_2}$$

To get Z_T ($V_1 = 0$); from equation (*iii*),

$$V_1 = AV_2 - BI_2$$

$$0 = AV_2 - BI_2$$

$$Z_T = \frac{V_2}{I_2} = \frac{B}{A} = \frac{A_1B_2 + B_1D_2}{A_1A_2 + B_1C_2}$$

6.18 Sol.

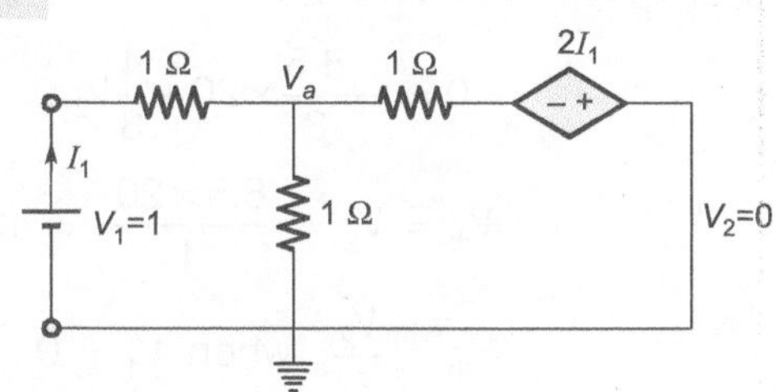

By KCL,

$$\frac{V_a - 1}{1} + \frac{V_a}{1} + \frac{V_a + 2I_1}{1} = 0$$

$$3V_a + 2I_1 = 1 \qquad \text{...(i)}$$

$$I_1 = \frac{1 - V_a}{1} \qquad \text{...(ii)}$$

Substitute equation (ii) in equation (i), we get

$$V_a = -1$$

$$I_1 = \frac{1 - V_a}{1} = \frac{1-(-1)}{1} = 2$$

$$h_{11} = \frac{V_1}{I_1} = \frac{1}{2} = 0.5\Omega$$

6.19 (238)

$$Y = [Y]_A + [Y]_B$$

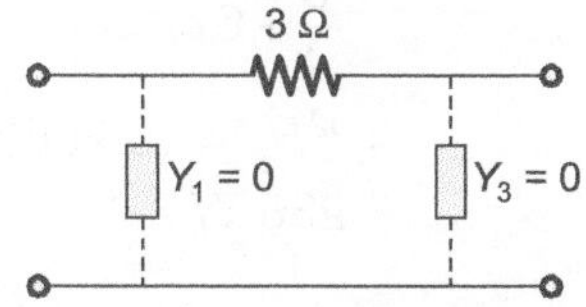

$$Y_A = \begin{bmatrix} Y_1 + Y_2 & -Y_2 \\ -Y_2 & Y_2 + Y_3 \end{bmatrix}$$

$$Y_A = \begin{bmatrix} \frac{1}{3} & -\frac{1}{3} \\ -\frac{1}{3} & \frac{1}{3} \end{bmatrix}$$

$$Y_B = \begin{bmatrix} 5 & -2.5 \\ -2.5 & 1 \end{bmatrix} s$$

$$Y = \begin{bmatrix} \frac{1}{3} & -\frac{1}{3} \\ -\frac{1}{3} & \frac{1}{3} \end{bmatrix} + \begin{bmatrix} 5 & -2.5 \\ -2.5 & 1 \end{bmatrix}$$

$$= \begin{bmatrix} \frac{16}{3} & -\frac{8.5}{3} \\ -\frac{8.5}{3} & \frac{4}{3} \end{bmatrix}$$

$$I_1 = \frac{16}{3}V_1 - \frac{8.5}{3}V_2 \quad ...(1)$$

$$I_2 = -\frac{8.5}{3}V_1 + \frac{4}{3}V_2 \quad ...(2)$$

Put $I_2 = 0$ and $V_1 = 20$ V

$$0 = -\frac{8.5}{3} \times 20 + \frac{4}{3}V_2$$

$$V_{th} = V_2 = \frac{8.5 \times 20}{4} = 42.5 \text{ V}$$

$$R_{th} = \frac{V_2}{I_2}, \text{ when } V_1 = 0$$

$$I_2 = \frac{4}{3}V_2$$

$$\frac{V_2}{I_2} = \frac{3}{4}\Omega$$

Equivalent circuit

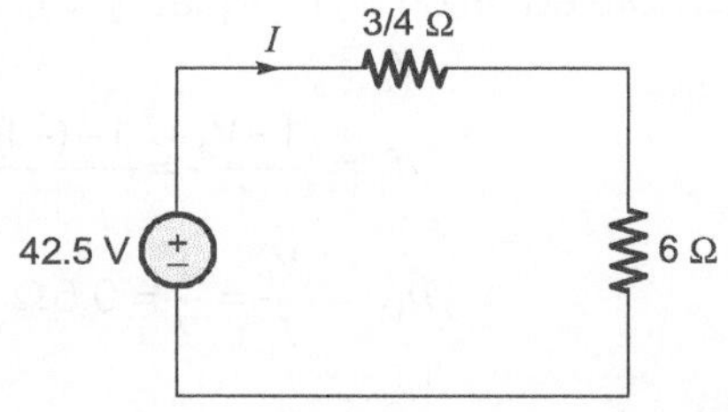

$$I = \frac{42.5}{\frac{3}{4}+6} = 6.296 \text{ A}$$

$$P = I^2R_L = (6.296)^2 \times 6$$
$$= 238 \text{ W}$$

6.20 (-19.80)

$$Z_P = \begin{bmatrix} 40 & 60 \\ 80 & 100 \end{bmatrix}; \quad Y_Q = \begin{bmatrix} 5S & -2.5S \\ -2.5S & 1S \end{bmatrix}$$

From Z_P matrix

$$V_1 = 40I_1 + 60I_2 \quad ...(i)$$
$$V_2 = 80I_1 + 100I_2 \quad ...(ii)$$

Convert Z-parameters in to ABCD parameters

$$V_1 = AV_2 - BI_2$$
$$I_1 = CV_2 - DI_2$$

From equation (ii),

$$I_1 = \frac{V_2}{80} - \frac{100}{80}I_2$$

$$I_1 = \frac{V_2}{80} - \frac{5}{4}I_2 \quad ...(iii)$$

Sub equation (iii) in equation (i),

$$V_1 = 40\left(\frac{V_2}{80} - \frac{5}{4}I_2\right) + 60I_2$$

$$V_1 = \frac{V_2}{2} + 10I_2 \quad ...(iv)$$

From equation (iii) and (iv),

$$\begin{bmatrix} A_1 & B_1 \\ C_1 & D_1 \end{bmatrix} = \begin{bmatrix} 0.5 & -10 \\ \frac{1}{80} & \frac{5}{4} \end{bmatrix}$$

$$[Y]_Q = \begin{bmatrix} 5 & -2.5 \\ -2.5 & 1 \end{bmatrix}$$

$$I_1 = 5V_1 - 2.5V_2 \quad ...(v)$$
$$I_2 = -2.5V_1 + V_2 \quad ...(vi)$$

Convert Y-parameters to ABCD parameters

From equation (vi),

$$V_1 = \frac{V_2}{2.5} - \frac{I_2}{2.5}$$

$$V_1 = 0.4V_2 - 0.4I_2 \quad ...(vii)$$

Sub equation (vii) in equation (v),

$$I_1 = 5(0.4V_2 - 0.4I_2) - 2.5V_2$$
$$= 2V_2 - 2I_2 - 2.5V_2$$
$$I_1 = -0.5V_2 - 2I_2 \quad ...(viii)$$

From equation (vii) and equation (viii),

$$\begin{bmatrix} A_2 & B_2 \\ C_2 & D_2 \end{bmatrix} = \begin{bmatrix} 0.4 & 0.4 \\ -0.5 & 2 \end{bmatrix}$$

$$T = \begin{bmatrix} A_1 & B_1 \\ C_1 & D_1 \end{bmatrix} * \begin{bmatrix} A_2 & B_2 \\ C_2 & D_2 \end{bmatrix}$$

$$= \begin{bmatrix} 0.5 & -10 \\ \frac{1}{80} & \frac{5}{4} \end{bmatrix} * \begin{bmatrix} 0.4 & 0.4 \\ -0.5 & 2 \end{bmatrix}$$

$$\beta = 0.5 \times 0.4 + (-10) \times 2$$
$$= -19.8 \ \Omega$$

■■■■

7 Magnetically Coupled Circuits, Network Topology/ Graph Theory & Filters

7.1 Two identical coils of negligible resistance when connected in series across a 200 V, 50 Hz source draws a current of 10 A. When the terminals of one of the coils are reversed, then current drawn is 8 A. The coefficient of coupling between the two coils is ______. [1997 : 2 M]

7.2 A major advantage of active filters is that they can be realized without using

(a) op-amps (b) inductors
(c) resistors (d) capacitors

[1997 : 1 M]

7.3 The effective inductance of the circuit across the terminals A, B in the figure shown below is

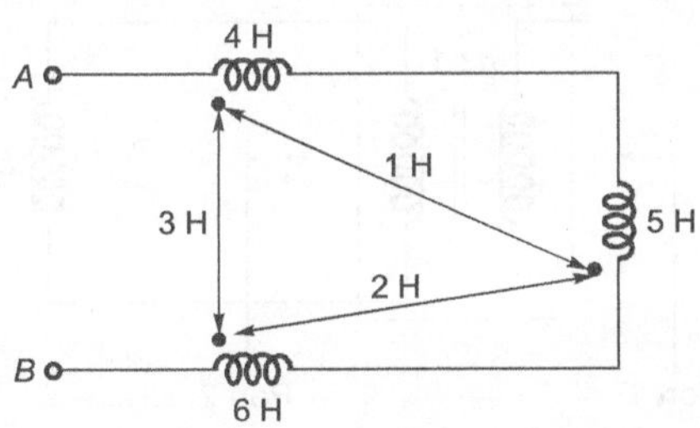

(a) 9 H (b) 21 H
(c) 11 H (d) 6 H [1998 : 2 M]

7.4 The impedance seen by the source in the circuit in the figure is given by

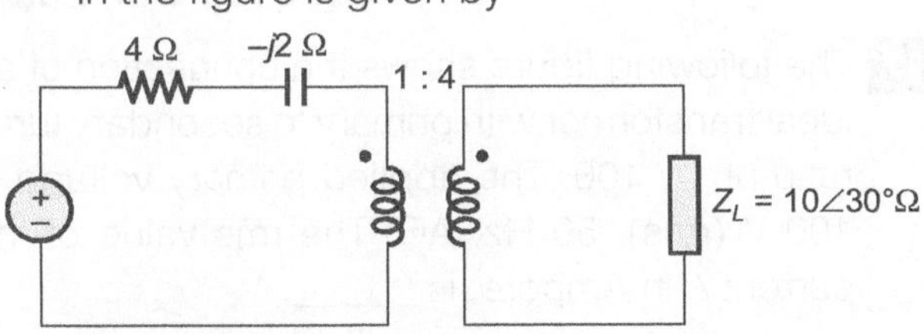

(a) $(0.54 + j0.313)$ ohms
(b) $(4 - j2)$ ohms
(c) $(4.54 - j1.69)$ ohms
(d) $(4 + j2)$ ohms [2000 : 2 M]

7.5 Given, two coupled inductors L_1 and L_2, their mutual inductance M satisfies

(a) $M = \sqrt{L_1^2 + L_2^2}$ (b) $M > \frac{(L_1 + L_2)}{2}$

(c) $M > \sqrt{L_1 L_2}$ (d) $M \le \sqrt{L_1 L_2}$

[2001 : 1 M]

7.6 In the circuit shown in the figure it is found that the input AC voltage (v_i) and current i are in phase. The coupling coefficient is $K = \frac{M}{\sqrt{L_1 L_2}}$, where M is the mutual inductance between the two coils. The value of K and the dot polarity of the coil P-Q are

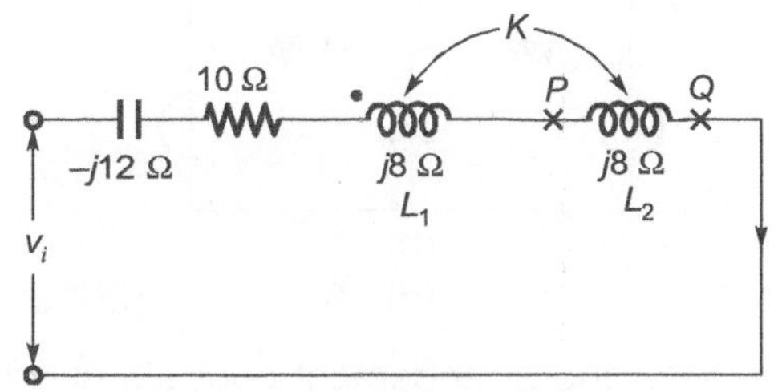

(a) $K = 0.25$ and dot at P
(b) $K = 0.5$ and dot at P
(c) $K = 0.25$ and dot at Q
(d) $K = 0.5$ and dot at Q [2002 : 2 M]

7.7 The graph of an electrical network has N nodes and B branches. The number of links L, with respect to the choice of a tree, is given by

(a) $B - N + 1$ (b) $B + N$
(c) $N - B + 1$ (d) $N - 2B - 1$

[2002 : 1 M]

7.8 A first order, low pass filter is given with $R = 50\ \Omega$ and $C = 5\ \mu F$. What is the frequency at which the gain of the voltage transfer function of the filter is 0.25?

(a) 4.92 kHz (b) 0.49 kHz
(c) 2.46 kHz (d) 24.6 kHz

[2002 : 2 M]

7.9 The matrix A given below is the node incidence matrix of a network. The columns correspond to branches of the network while the rows correspond to nodes. Let $V[v_1 v_2 \ldots v_6]^T$ denote the vector of branch voltages while $I = [i_1\, i_2 \ldots i_6]^T$ that of branch currents. The vector $E = [e_1\ e_2\ e_3\ e_4]^T$ denotes the vector of node voltages relative to a common ground.

$$A = \begin{bmatrix} 1 & 1 & 1 & 0 & 0 & 0 \\ 0 & -1 & 0 & -1 & 1 & 0 \\ -1 & 0 & 0 & 0 & -1 & -1 \\ 0 & 0 & -1 & 1 & 0 & 1 \end{bmatrix}$$

Which of the following statements is true?

(a) The equations: $V_1 - V_2 + V_3 = 0$, $V_3 + V_4 - V_5 = 0$ are KVL equations for the network for some loops.
(b) The equations: $V_1 - V_3 - V_6 = 0$, $V_4 + V_5 - V_6 = 0$ are KVL equations for the network for some loops.
(c) $E = AV$
(d) $AV = 0$ are KVL equations for the network.

[2007 : 2 M]

7.10 The number of chords in the graph of the given circuit will be

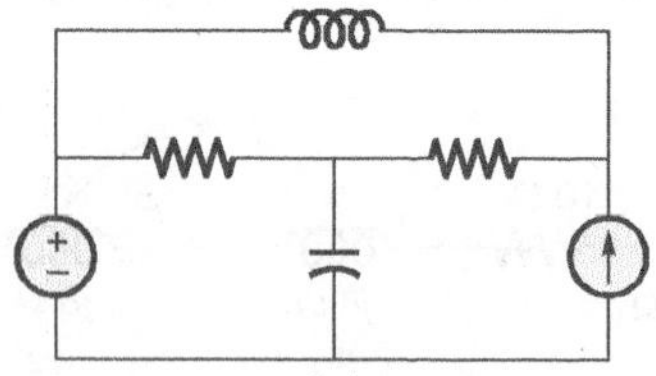

(a) 3 (b) 4
(c) 5 (d) 6

[2008 : 1 M]

7.11 The following arrangement consists of an ideal transformer and an attenuator which attenuates by a factor of 0.8. An AC voltage $V_{WX1} = 100$ V is applied across WX to get an open circuit voltage V_{YZ1} across YZ. Next, an AC voltage $V_{YZ2} = 100$ V is applied across YZ to get an open circuit voltage V_{WX2} across WX. Then, V_{YZ1}/V_{WX1}, V_{WX2}/V_{YZ2} are respectively,

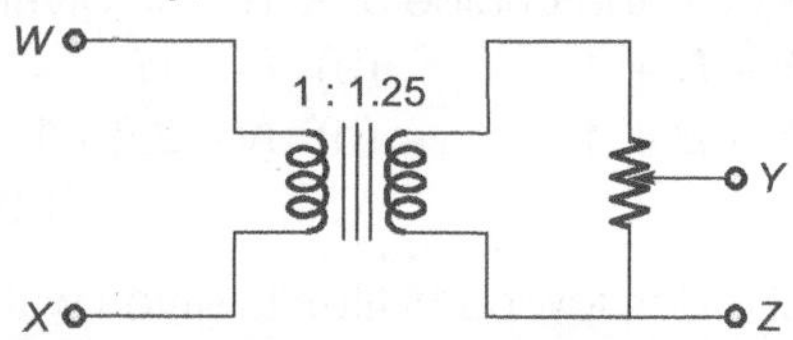

(a) $\frac{125}{100}$ and $\frac{80}{100}$ (b) $\frac{100}{100}$ and $\frac{80}{100}$
(c) $\frac{100}{100}$ and $\frac{100}{100}$ (d) $\frac{80}{100}$ and $\frac{80}{100}$

[2013 : 2 M]

7.12 Two identical coupled inductors are connected in series. The measured inductances for the two possible series connections are 380 μH and 240 μH. Their mutual inductance in μH is ______.

[2014 : 1 M, Set-2]

7.13 Find the transformer ratios a and b such that the impedance (Z_{in}) is resistive and equals 2.5 Ω when the network is excited with a sine wave voltage of angular frequency of 5000 rad/s.

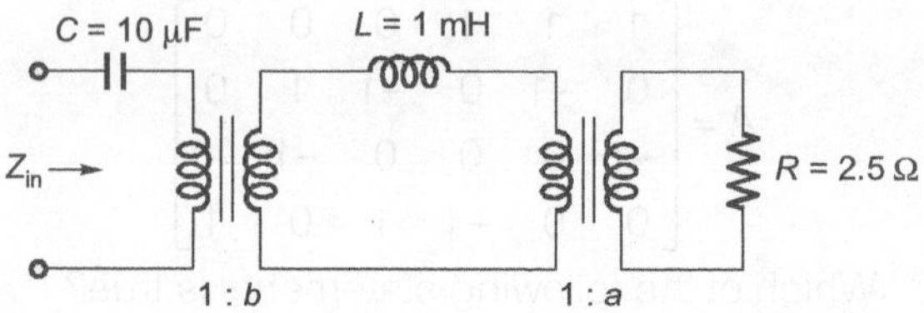

(a) $a = 0.5$, $b = 2.0$ (b) $a = 2.0$, $b = 0.5$
(c) $a = 1.0$, $b = 1.0$ (d) $a = 4.0$, $b = 0.5$

[2015 : 1 M, Set-2]

7.14 Two identical coils each having inductance L are placed together on the same core. If an overall inductance of aL is obtained by interconnecting these two coils, the minimum value of a is ______.

[2015 : 2 M, Set-2]

7.15 The graph associated with an electrical network has 7 branches and 5 nodes. The number of independent KCL equations and the number of independent KVL equations, respectively, are
(a) 2 and 5 (b) 5 and 2
(c) 3 and 4 (d) 4 and 3

[2016 : 1 M, Set-2]

7.16 If an ideal transformer has an inductive load element at port 2 as shown in the figure below, the equivalent inductance at port 1 is

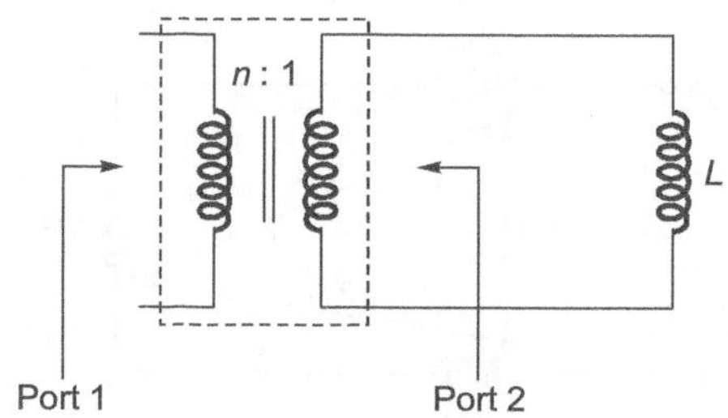

(a) nL (b) n^2L
(c) $\frac{n}{L}$ (d) $\frac{n^2}{L}$

[2016 : 1 M, Set-1]

7.17 The following figure shows the connection of an ideal transformer with primary to secondary turns ratio of 1 : 100. The applied primary voltage is 100 V (rms), 50 Hz, AC. The rms value of the current I, in Ampere, is ______.

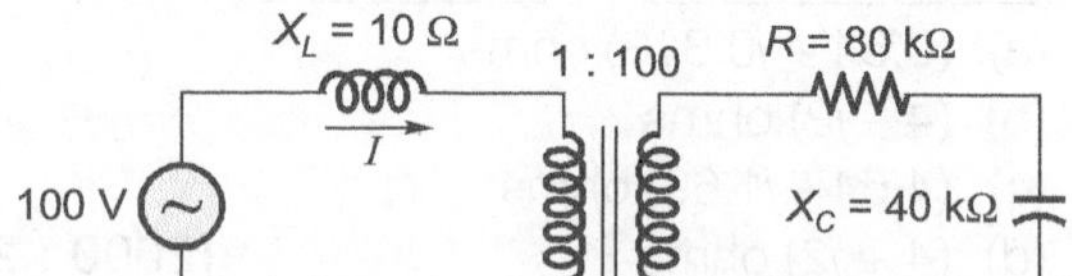

[2016 : 1 M, Set-2]

7.18 The graph of a network has 8 nodes and 5 independent loops. The number of branches of the graph is
(a) 11 (b) 12
(c) 13 (d) 14

[2018 : 1 M]

7.19 The input impedance $Z_{in}(s)$, for the network shown in

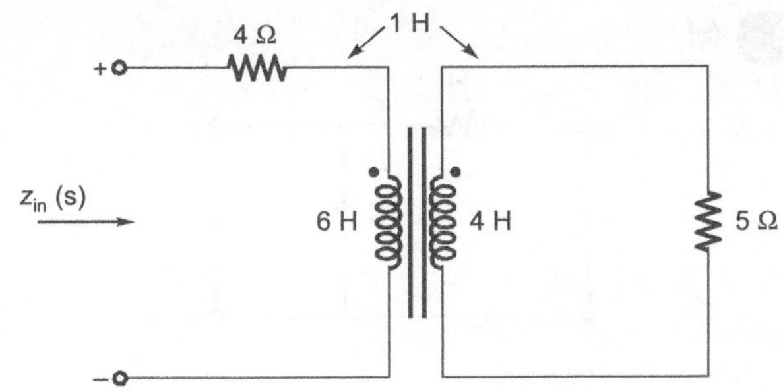

(a) $7s + 4$ (b) $\frac{23s^2 + 46s + 20}{4s + 5}$

(c) $\frac{25s^2 + 46s + 20}{4s + 5}$ (d) $6s + 4$ [2021 : 2 M]

7.20 An air-core radio-frequency transformer as shown has a primary winding and a secondary winding. The mutual inductance M between the windings of the transformer is ______ μH, (Round off to 2 decimal places).

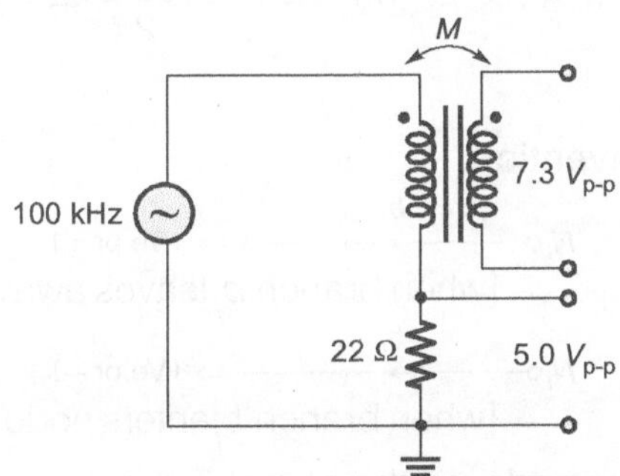

[2021 : 2 M]

7.21 The transfer function of a real system, $H(s)$, is given as :

$$H(s) = \frac{As + B}{s^2 + Cs + D},$$

where A, B, C and D are positive constants. This system cannot operate as

(a) low pass filter
(b) high pass filter
(c) band pass filter
(d) an integrator [2022 : 1 M]

7.22 For the circuit shown in the figure, the source frequency is 5000 rad/sec. The mutual inductance between the magnetically coupled inductors is 5 mH with their self inductances being 125 mH and 1 mH. The Thevenin's impedance. Z_{th}, between the terminals P and Q in Ω is ____ (rounded off to 2 decimal places).

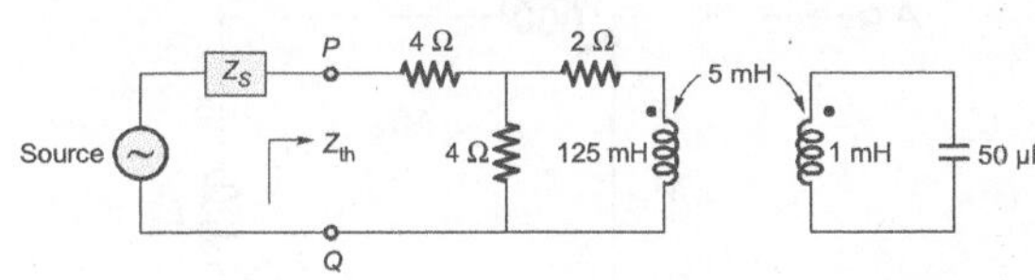

[2024 : 2 M]

Answers Magnetically Coupled Circuits, Network Topology/Graph Theory & Filters

7.1 (1/9)	7.2 (b)	7.3 (c)	7.4 (c)	7.5 (d)	7.6 (c)	7.7 (a)
7.8 (c)	7.9 (b)	7.10 (a)	7.11 (b)	7.12 (35)	7.13 (b)	7.14 (0)
7.15 (d)	7.16 (b)	7.17 (10)	7.18 (b)	7.19 (b)	7.20 (51.10)	7.21 (b)
7.22 (5.33)						

Explanations Magnetically Coupled Circuits, Network Topology/Graph Theory & Filters

7.1 Sol.

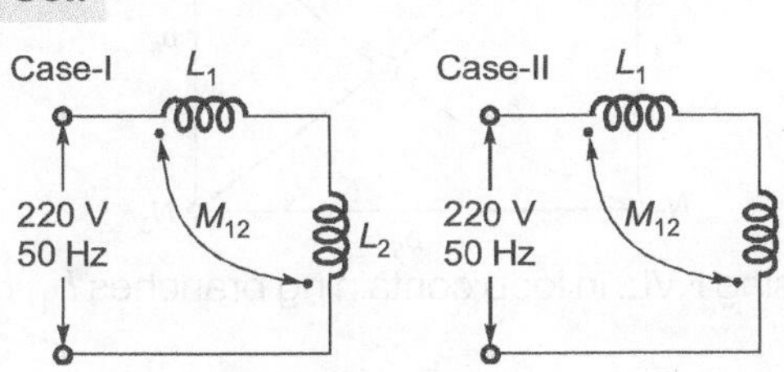

$$|I_1| = \frac{V}{\omega(L_1 + L_2 - 2M_{12})}$$

$$|I_2| = \frac{V}{\omega(L_1 + L_2 + 2M_{12})}$$

From above expressions, it is clear that

$$|I_1| > |I_2|$$

∴ Taking, $|I_1| = 10$ A and $|I_2| = 8$ A

and putting, $V = 220$ V

and, $L_1 = L_2 = L$

We have,

$$10 = \frac{200}{\omega(2L-2M_{12})} \quad ...(i)$$

$$8 = \frac{200}{\omega(2L+2M_{12})} \quad ...(ii)$$

On solving equation (i) and equation (ii),

We get, $M_{12} = \frac{1}{9}L$

Which can be written as:

$$M_{12} = \frac{1}{9}\sqrt{L \cdot L} = \frac{1}{9}\sqrt{L_1 \cdot L_2}$$

Hence, coefficient of coupling $= \frac{1}{9}$.

7.2 (b)

Inductive coils are bulky in nature.

7.3 (c)

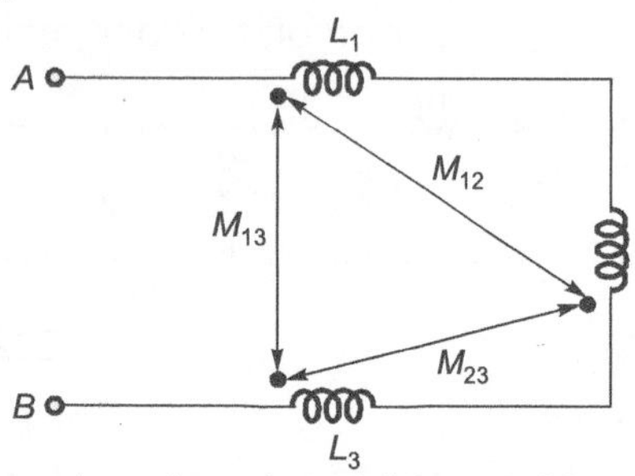

$$L = L_1 + L_2 + L_3 - 2M_{12} + 2M_{23} - 2M_{13}$$
$$= 4+5+6-(2\times1)+(2\times2)-(2\times3) = 11 \text{ H}$$

7.4 (c)

$$Z = (4-j2)+\left(\frac{1}{4}\right)^2 \times 10\angle30^\circ = (4.54 - j1.69)\ \Omega$$

7.5 (d)

$$M = K\sqrt{L_1L_2}$$

Where, K = Coefficient of coupling

$\because \ 0 < K < 1$

$$\therefore \quad M \le \sqrt{L_1L_2}$$

7.6 (c)

Input AC voltage and current will be in phase only at resonance condition, i.e., $X_C = X_L$

$$|-j12| = \left|j8+j8+2k\sqrt{(j8)\times(j8)}\right|$$

$$12 = 8+8+16k$$

$$\Rightarrow \quad k = -\frac{4}{16} = -\frac{1}{4} = -0.25$$

Hence coupling will be opposite.

$\therefore$ Dot will be at Q.

7.7 (a)

Number of links $= B-(N-1) = B-N+1$

7.8 (c)

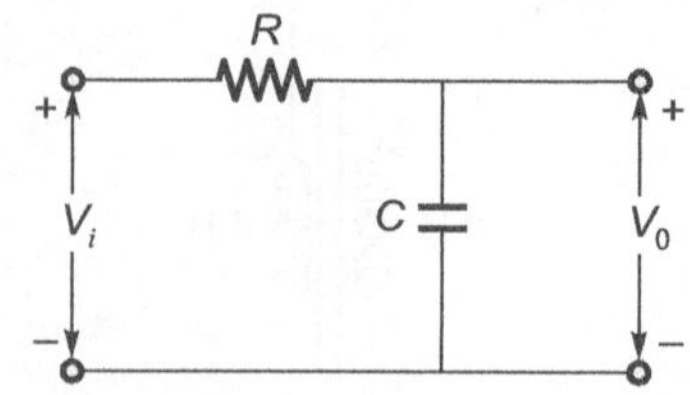

$$\text{T.F.} = \frac{V_0}{V_i} = \frac{\frac{1}{j\omega C}}{R+\frac{1}{j\omega C}} = \frac{1}{1+j\omega CR}$$

$$\therefore \quad |\text{Gain}| = \frac{1}{\sqrt{1+(\omega CR)^2}}$$

$$0.25 = \frac{1}{\sqrt{1+(\omega\times5\times10^{-6}\times50)^2}}$$

On solving, we get

$$\omega = 15.49 \times 10^3 \text{ r/s} \Rightarrow f = \frac{15.49}{2\pi} = 2.46 \text{ kHz}$$

7.9 (b)

Convention:

N_i o——b_j——→ $\Rightarrow$ +Ve or +1

[when branch b_j leaves away node N_i]

N_i o——b_j——→ $\Rightarrow$ +Ve or −1

[when branch b_j enters node N_i]

Given node incidence matrix

$A =$

Nodes ↓ \ Branches →	b_1	b_2	b_3	b_4	b_5	b_6
N_1	1	1	1	0	0	0
N_2	0	−1	0	−1	1	0
N_3	−1	0	0	0	−1	−1
N_4	0	0	−1	1	0	1

eg.: Branch b_1 enters N_3 and leaves N_1

The oriented network graph

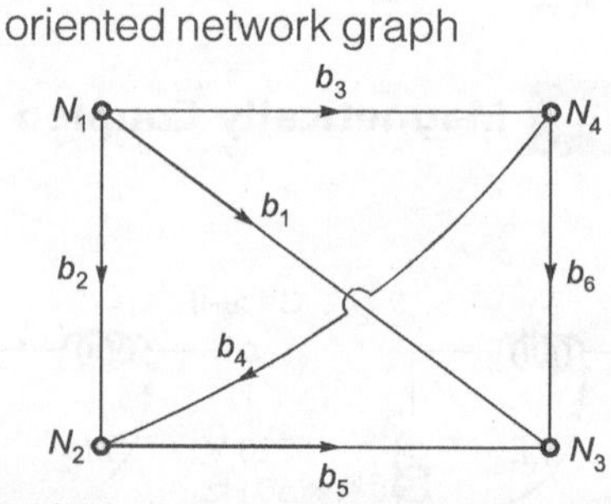

Using KVL, in loop containing branches b_1, b_3 and b_6

$$V_1 - V_6 - V_3 = 0$$

In loop containing branches, b_4, b_5 and b_6

$$V_4 + V_5 - V_6 = 0$$

7.10 (a)

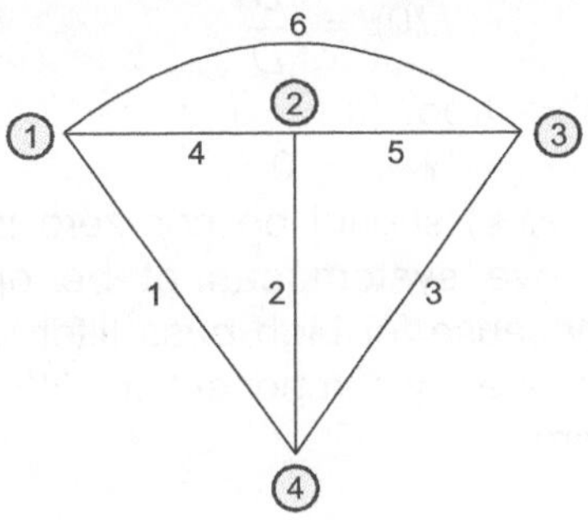

No. of branches = $b = 6$

No. of nodes = $n = 4$

No. of chords = $b-(n-1) = 6-(4-1) = 3$

7.11 (b)

$$V_{YZ_1} = 100 \times 1.25 \times 0.8 = 100 \text{ V}$$

In second case, when 100 V is applied at YZ terminals, this whole 100 V will appear across the secondary winding.

Hence, $V_{WX_2} = \dfrac{100}{1.25} = 80 \text{ V} \Rightarrow \dfrac{V_{YZ_1}}{V_{WX_1}} = \dfrac{100}{100}$

$$\frac{V_{YZ_2}}{V_{WX_2}} = \frac{80}{100}$$

7.12 Sol.

The two possible series connection are shown below:

Let the mutual inductance be M

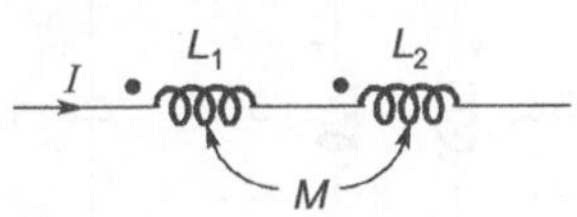

(i) Additive connection,

$$L_{eqv} = L_1 + L_2 + 2M = 380 \ \mu\text{H}$$

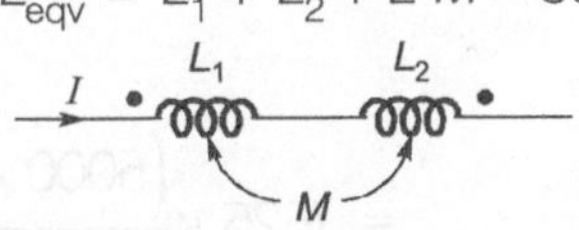

(ii) Subtractive connection,

$$L_{eqv} = L_1 + L_2 - 2M = 240 \ \mu\text{H}$$

Thus, $L_1 + L_2 + 2M = 380 \ \mu\text{H}$ ...(i)

and $L_1 + L_2 - 2M = 240 \ \mu\text{H}$ ...(ii)

Solving equations (i) and (ii), 23 get:

$4M = 140 \ \mu\text{H}$ or $M = 35 \ \mu\text{H}$

$\therefore$ Mutual inductance, $M = 35 \ \mu\text{H}$

7.13 (b)

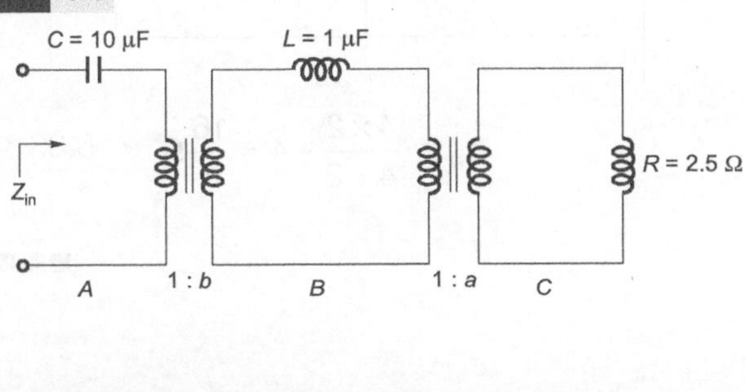

$$(R_{eq})_B = \frac{2.5}{a^2}$$

$$(Z_{eq})_B = \left[\frac{2.5}{a^2} + j\omega(1 \times 10^{-3})\right]$$

$$(Z_{eq})_A = \left[\frac{2.5}{a^2} + j\omega(1 \times 10^{-3})\right] \Big/ b^2$$

$$Z_{in} = \frac{2.5}{a^2b^2} + \frac{j\omega(10^{-3})}{b^2} + \frac{1}{j\omega(10 \times 10^{-5})}$$

$$Z_{in} = \frac{2.5}{a^2b^2} + j\left[\frac{5 \times 10^3 \times 10^{-3}}{b^2} - \frac{1}{5 \times 10^3 \times 10 \times 10^{-6}}\right] \quad \text{...(i)}$$

From problem, $Z_{in} = 2.5$ ...(ii)

From equation (i) and (ii), $a^2 b^2 = 1$ ...(iii)

$$\frac{5}{b^2} - \frac{1}{5 \times 10^{-2}} = 0 \Rightarrow 5 \times 5 \times 10^{-2} = \beta^2$$

$\Rightarrow$ $b = 0.5$ and $a = 2$

7.14 Sol.

Case-I : $L_{eff.} = L_1 + L_2 = 2L, \ a = 2$

Case-II : $L_{eff.} = \dfrac{L_1L_2}{L_1+L_2} = \dfrac{L^2}{2L} = \dfrac{L}{2}, \quad a = 0.5$

Case-III : If both are differentially coupled then

$$L_{eff.} = 0$$

Minimum value = 0

7.15 (d)

No. of KCL equations = $n - 1 = 5 - 1 = 4$

Number of KVL equations

$= b - (n-1) = 7 - (5-1) = 3$

7.16 (b)

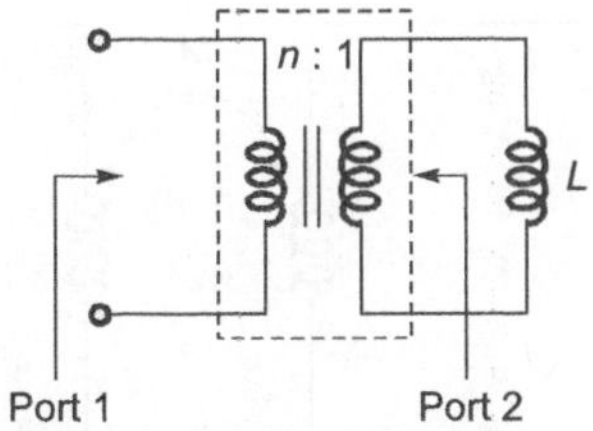

At port 1, i.e., high voltage side impedance will be high and current will be low, So n^2L.

7.17 Sol.

The below circuit can be drawn by transferring secondary circuit to primary side.

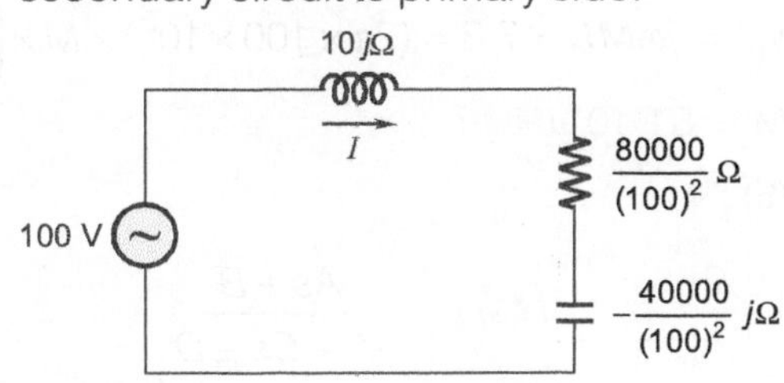

$$I = \frac{100\text{ V}}{(8+10j-4j)\Omega} = \frac{100V}{(8+6j)\Omega}$$

So, the RMS value of I will be 10 A.

7.18 (b)

$$\text{Loops} = b-(N-1)$$
$$5 = b-(8-1) = b-7 = 12$$

7.19 (b)

Circuit in s-domain,

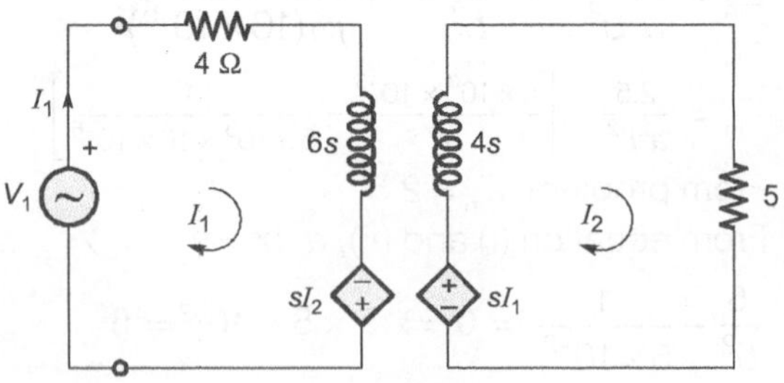

By KVL,

$$-sI_1 + (4s+5)I_2 = 0$$

$$\Rightarrow \quad I_2 = \frac{s}{4s+5}I_1$$

$$V_1 = (4+6s)I_1 - \frac{s^2}{4s+5}I_1$$

$$\frac{V_1}{I_1} = \frac{(4+6s)(4s+5)-s^2}{4s+5}$$

$$= \frac{24s^2+30s+16s+20-s^2}{4s+5}$$

$$Z_{in} = \frac{23s^2+46s+20}{4s+5}$$

7.20 (51.10)

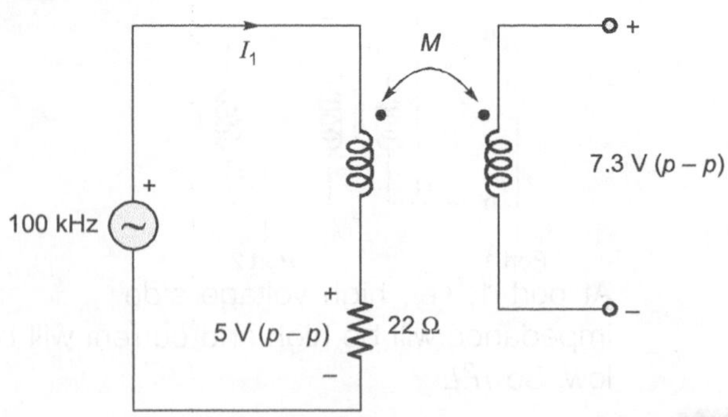

$$I_1 = \frac{5}{22}(p-p)$$

$$V_0 = j\omega MI_1 = 7.3 = (2\pi\times 100\times 10^3)\times M\times\left(\frac{5}{22}\right)$$

$$M = 51.10\ \mu\text{H}$$

7.21 (b)

$$H(s) = \frac{As+B}{s^2+Cs+D}$$

At low = freq. ($s = 0$)

$$H(0) = \frac{B}{D}$$

At high – freq. ($s = \infty$)

$$H(\infty) = 0$$

Since, $H(\infty)$ should be non-zero for HPF. So, the above system cannot be operated as HPF because for high pass filter should pass high frequency component i.e., $H(\infty)$ should be non-zero.

7.22 (5.33)

Given: ω = 5000 rad/sec

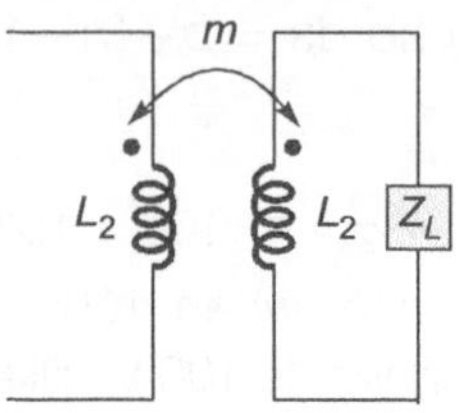

$$Z_{in} = jX_{L1} + \frac{(\omega m)^2}{jX_{L2}+Z_L}$$

$$jX_{L1} = 125\text{ mH}\times 5000 = j625\ \Omega$$

$$\frac{-j}{\omega C} = \frac{-j}{5000\times 50\times 10^{-6}} = -j4\Omega$$

$$jX_{L2} = 1\text{ mH}\times 5000 = +j5\ \Omega$$

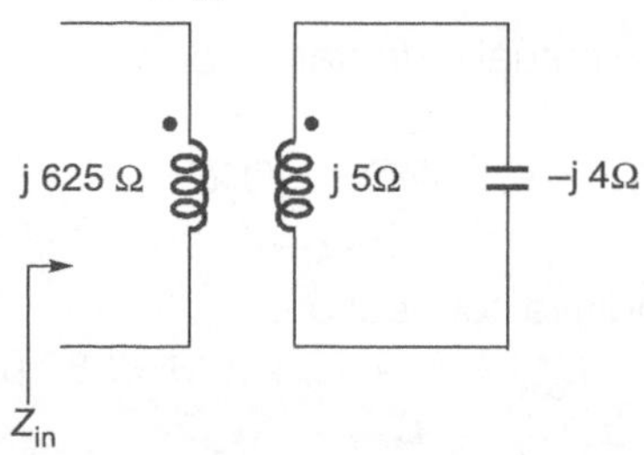

$$Z_{in} = j625 + \frac{\left(5000\times 5\times 10^{-3}\right)^2}{j5-j4}$$
$$= j625 - j625 = 0\ \Omega$$

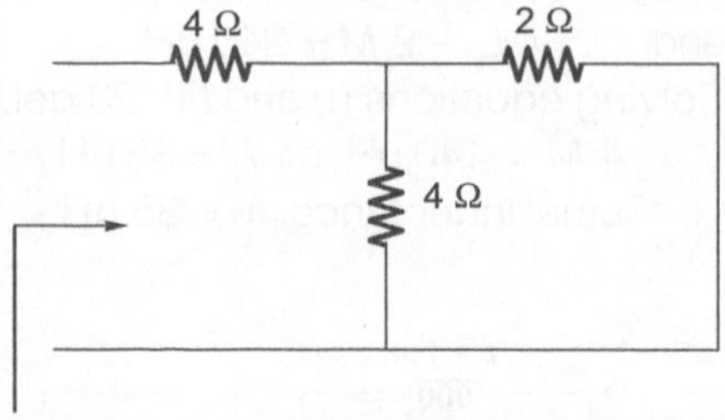

$$Z_m = \frac{4\times 2}{4+2}+4 = \frac{16}{3}\Omega = 5.33\ \Omega$$

■■■■

8 Three-Phase Circuits

8.1 For the three-phase circuit shown in the figure the ratio of the currents $I_R : I_Y : I_B$ is given by

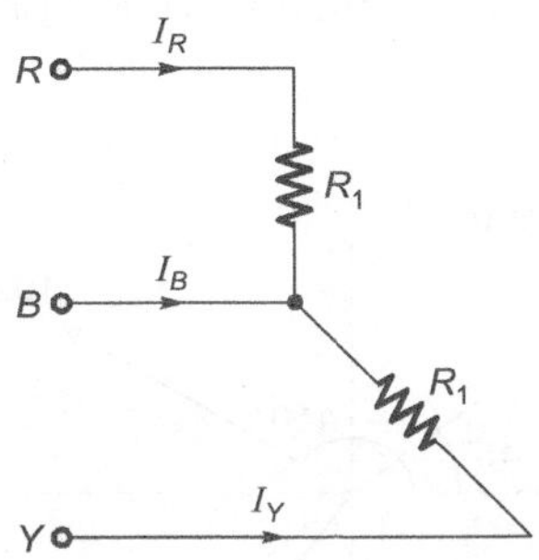

(a) $1 : 1 : \sqrt{3}$ (b) $1 : 1 : 2$
(c) $1 : 1 : 0$ (d) $1 : 1 : \sqrt{3/2}$

[2005 : 2 M]

8.2 A two-phase load draws the following phase current:

$$i_1(t) = I_m \sin(\omega t - \phi_1);\ i_2(t) = I_m \cos(\omega t - \phi_2)$$

These currents are balanced if ϕ_1 is equal to

(a) $-\phi_2$ (b) ϕ_2
(c) $\left(\frac{\pi}{2} - \phi_2\right)$ (d) $\left(\frac{\pi}{2} + \phi_2\right)$ [2012 : 1 M]

8.3 The line *A* to neutral voltage is $10\angle 15°$ V for a balanced three-phase star connected load with phase sequence *ABC*. The voltage of line *B* with respect to line *A* is given by

(a) $10\sqrt{3}\angle 105°$ V (b) $10\angle 105°$ V
(c) $10\sqrt{3}\angle -75°$ V (d) $-10\sqrt{3}\angle 90°$ V

[2014 : 1 M, Set-3]

8.4 In the given network $V_1 = 100\angle 0°$ V, $V_2 = 100\angle -120°$ V, $V_3 = 100\angle +120°$ V. The phasor current *i* (in Ampere) is

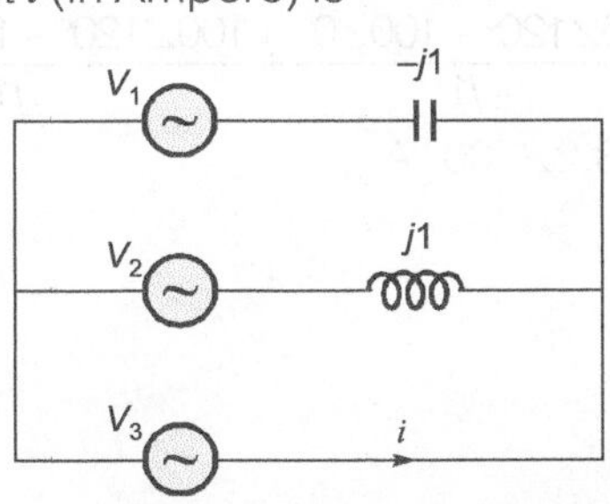

(a) $173.2\angle -60°$ (b) $173.2\angle 120°$
(c) $100.0\angle -60°$ (d) $100.0\angle 120°$

[2015 : 2 M, Set-2]

8.5 The line currents of a three-phase four wire system are square waves with amplitude of 100 A. The three currents are phase shifted by 120° wish respect to each other. The rms value of neutral current is

(a) 300 A (b) 0 A
(c) 100 A (d) $\frac{100}{\sqrt{3}}$ A [2019 : 2 M]

8.6 In the circuit shown below, a three-phase star-connected unbalanced load is connected to a balanced three-phase supply of $100\sqrt{3}$ V with phase sequence *ABC*. The star connected load has $Z_A = 10\ \Omega$ and $Z_B = 20\angle 60°\ \Omega$. The value of Z_C in Ω, for which the voltage difference across the nodes *n* and *n′* is zero, is

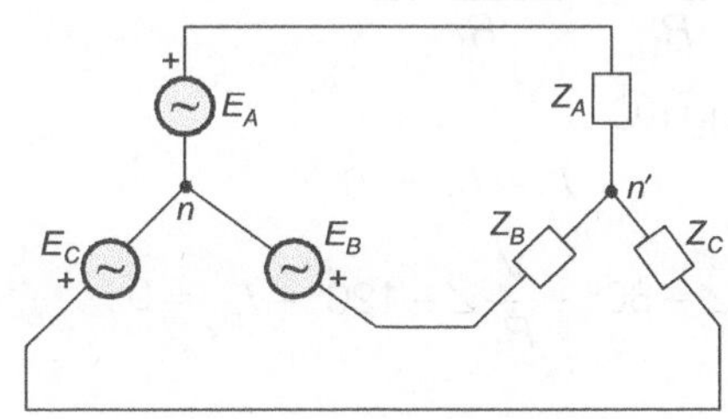

(a) $20\angle -30°$ (b) $20\angle 30°$
(c) $20\angle -60°$ (d) $20\angle 60°$ [2022 : 1 M]

8.7 A balanced delta connected load consisting of the series connection of one resistor (R = 15 Ω) and a capacitor (C = 212.21 μF) in each phase is connected to three-phase, 50 Hz, 415 V supply terminals through a line having an inductance of L = 31.83 mH per phase, as shown in the figure. Considering the change in the supply terminal voltage with loading to be negligible, the magnitude of the voltage across the terminals V_{AB} in Volts is ______ (Round off to the nearest integer).

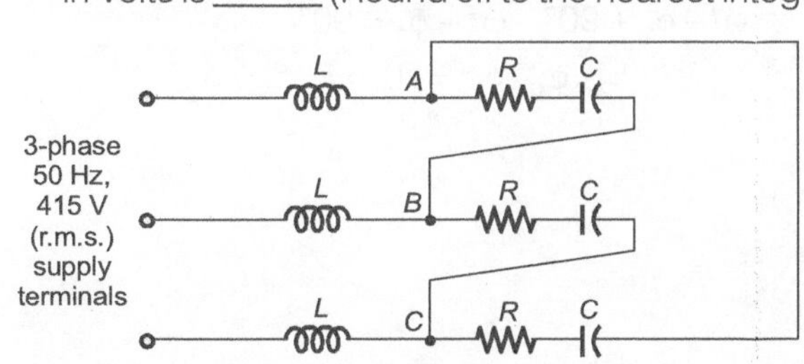

R = 15 Ω, L = 31.783 mH, C = 212.21 μF

[2023 : 2 M]

Answers Three-Phase Circuits

8.1 (a) 8.2 (b) 8.3 (c) 8.4 (b) 8.5 (c) 8.6 (c) 8.7 (415)

Explanations Three-Phase Circuits

8.1 (a)

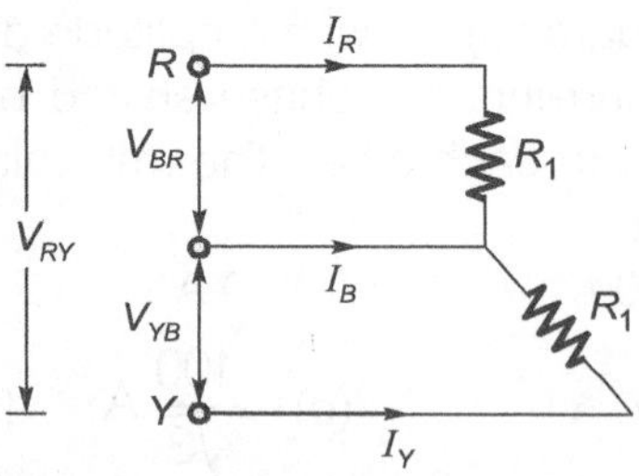

Assuming phase-sequence to be *RYB*

Taking V_{RY} as the reference,

$V_{RY} = V\angle 0°$; $V_{YB} = V\angle -120°$; $V_{BR} = V\angle -240°$

$$I_R = \frac{V_{RB}}{R_1} = -\frac{V_{BR}}{R_1} = -\frac{V\angle -240°}{R_1} = \frac{V}{R_1}\angle -60°$$

$$I_Y = \frac{V_{YB}}{R_1} = \frac{V\angle -120°}{R_1}$$

Using KCL,

$$I_R + I_Y + I_B = 0$$

$$\frac{V}{R_1}\angle -60° + \frac{V}{R_1}\angle -120° + I_B = 0$$

$$\Rightarrow \quad I_B = \sqrt{3}\frac{V}{R_1}\angle 90°$$

So, $I_R : I_Y : I_B = \frac{V}{R_1} : \frac{V}{R_1} : \sqrt{3}\frac{V}{R_1} = 1 : 1 : \sqrt{3}$

8.2 (b)

In two phase, currents are balanced if phase difference is 90°.

$i_1(t) = I_m \sin(\omega t - \phi_1)$; $i_2(t) = I_m \sin(\omega t - \phi_2 + 90°)$

Phase difference is 90°,

$$\omega t - \phi_2 + 90° - (\omega t - \phi_1) = 90°$$

$$\omega t - \phi_2 + 90° - \omega t + \phi_1 = 90°$$

$$\phi_1 - \phi_2 = 0 \Rightarrow \phi_1 = \phi_2$$

8.3 (c)

Given, $V_{AN} = 10\angle 15°$ volt

As the system is balanced and phase sequence is *ABC*,

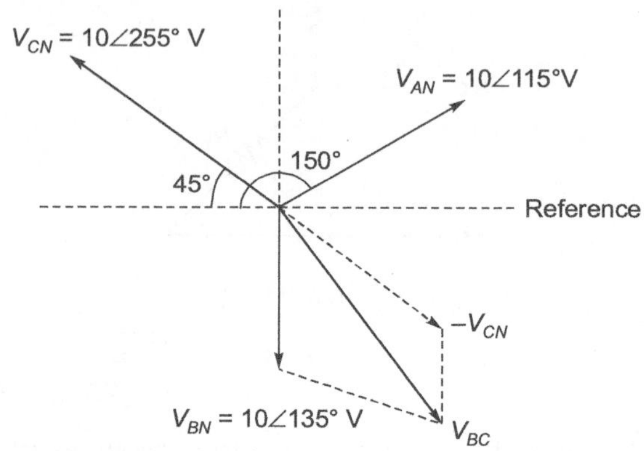

Therefore, $V_{AN} = 10\angle 15°$

$V_{BN} = 10\angle 135°$ V and $V_{CN} = 10\angle 255°$ V

$\therefore$ Voltage of line w.r.t. line *C* is

$$V_{BC} = V_{BN} - V_{CN} = 10\angle 255° - 10\angle 135°$$

or, $$V_{BC} = 10\sqrt{3}\angle -75° \text{ Volt}$$

8.4 (b)

'i' is taken as output,

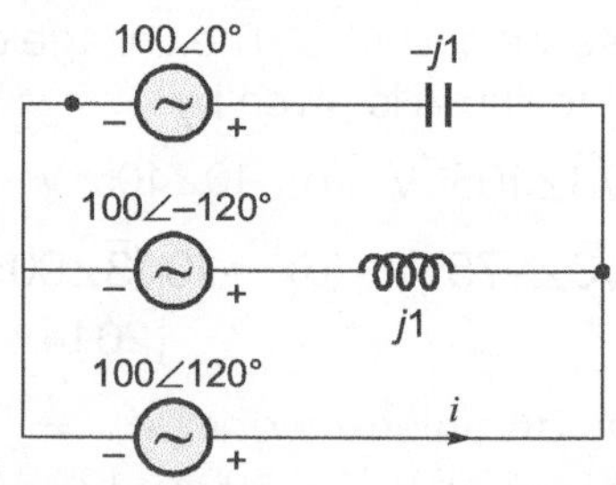

Apply KCL at node. Node voltage is $100\angle 120°$

$$i = \frac{100\angle 120° - 100\angle 0°}{-j1} + \frac{100\angle 120° - 100\angle -120°}{j1}$$

$i = 173.2\angle -60°$ A

8.5 (c)

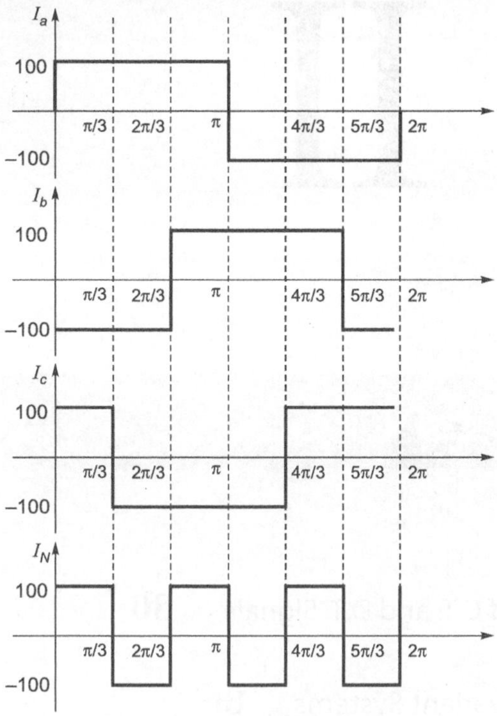

$$I_N = I_a + I_b + I_c$$
$$(I_N)_{rms} = 100\text{ A}$$

8.6 (c)

n and n' are same potential,

So, $I_A + I_B + I_C = 0$

$$\frac{E_A}{Z_A} + \frac{E_B}{Z_B} + \frac{E_C}{Z_C} = 0$$

$$\frac{100\angle 0°}{10} + \frac{100\angle -120°}{20\angle 60°} + \frac{100\angle 120°}{Z_C} = 0$$

$$10 + 5\angle -180° + \frac{100\angle 120°}{Z_C} = 0$$

$$Z_C = 20\angle -60°\ \Omega$$

8.7 (415)

$$X_L = 2\pi fL$$
$$= 2\pi \times 50 \times 31.83 \times 10^{-3}$$
$$= 10\ \Omega$$

$$X_C = \frac{1}{2\pi fC}$$
$$= \frac{1}{2\pi \times 50 \times 212.21 \times 10^{-6}}$$
$$= 15\ \Omega$$

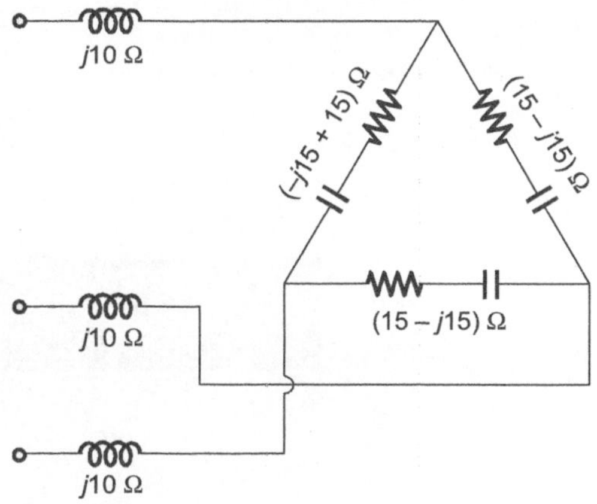

Delta to Star $\Rightarrow Z_Y = \frac{Z_\Delta}{3} = \frac{15 - j15}{3}$

$$= (5 - j5)\ \Omega$$

The per phase equivalent circuit

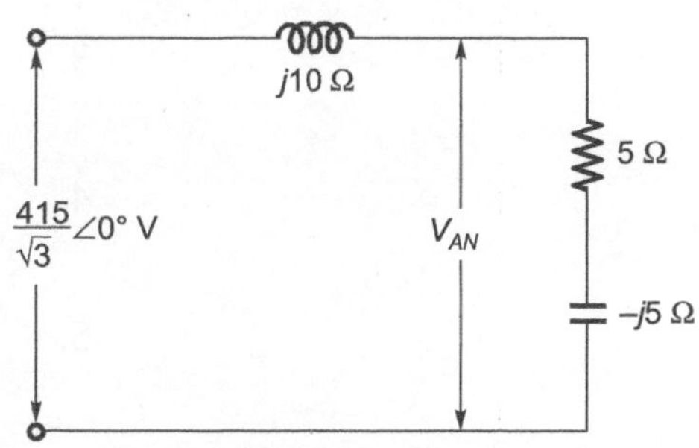

Per Phase Equivalent Circuit

$$V_{AN} = \frac{(5 - j5)}{(5 - j5 + j10)} \times \frac{415}{\sqrt{3}} = \frac{(5 - j5)}{(5 + j5)} \times \frac{415}{\sqrt{3}}$$

$$|V_{AN}| = \frac{415}{\sqrt{3}}\text{ V}$$

So, $V_{AB} = 415$ V

Signals and Systems

UNIT

II

CONTENTS

Signals and Systems

Syllabus

Representation of Continuous and Discrete Time Signals, Shifting and Scaling Properties, Linear Time Invariant and Causal Systems, Fourier Series Representation of Continuous and Discrete Time Periodic Signals, Sampling Theorem, Applications of Fourier Transform for Continuous and Discrete Time Signals, Laplace Transform and Z-transform. R.M.S. Value, Average Value Calculation for any general Periodic Waveform

Analysis of Previous GATE Papers

Exam Year	1 Mark Ques.	2 Marks Ques.	3 Marks Ques.	5 Marks Ques.	Total Marks
1995	4	–	–	–	**4**
1996	1	1	–	–	**3**
1997	–	–	–	2	**10**
1998	1	1	–	–	**3**
1999	–	1	–	–	**2**
2000	1	–	–	–	**1**
2001	–	1	–	1	**7**
2002	5	–	–	1	**10**
2003	–	–	–	–	**–**
2004	1	2	–	–	**5**
2005	–	4	–	–	**8**
2006	3	4	–	–	**11**
2007	2	5	–	–	**12**
2008	2	7	–	–	**16**
2009	1	3	–	–	**7**
2010	2	4	–	–	**10**
2011	2	2			**6**

Exam Year	1 Mark Ques.	2 Mark Ques.	Total Marks
2012	2	4	**10**
2013	6	1	**8**
2014 Set-1	2	3	**8**
2014 Set-2	2	2	**6**
2014 Set-3	3	3	**9**
2015 Set-1	2	2	**6**
2015 Set-2	1	4	**9**
2016 Set-1	4	3	**10**
2016 Set-2	5	2	**9**
2017 Set-1	4	2	**8**
2017 Set-2	3	2	**7**
2018	1	3	**7**
2019	4	1	**6**
2020	4	3	**10**
2021	2	4	**10**
2022	–	4	**8**
2023	5	3	**11**
2024	3	4	**11**

1 Introduction of C.T. and D.T. Signals

1.1 The rms value of the periodic waveform $e(t)$, shown in the figure is

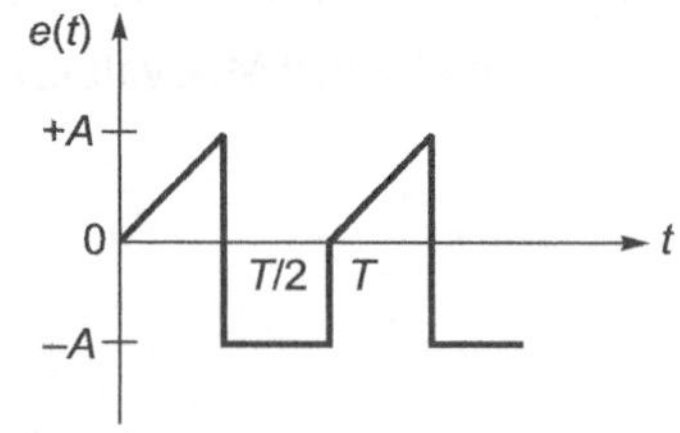

(a) $\sqrt{\frac{3}{2}}A$ (b) $\sqrt{\frac{2}{3}}A$

(c) $\sqrt{\frac{1}{3}}A$ (d) $\sqrt{2}A$ **[1995 : 1 M]**

1.2 A current impulse of $5\delta(t)$, is forced through a capacitor C. The voltage $v_c(t)$, across the capacitor is given by

(a) $5t$ (b) $5\,u(t) - C$

(c) $\frac{5}{C}t$ (d) $\frac{5u(t)}{C}$ **[2002 : 1 M]**

1.3 What is the rms value of the voltage waveform shown in the figure?

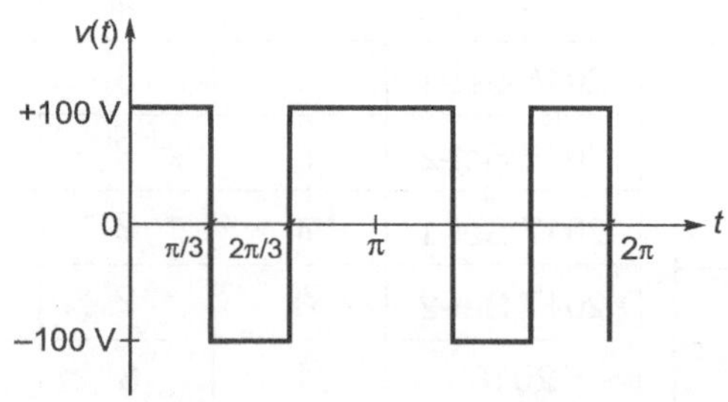

(a) $\frac{200}{\pi}$ V (b) $\frac{100}{\pi}$ V

(c) 200 V (d) 100 V **[2002 : 1 M]**

1.4 Which of the following is true?

(a) A finite amplitude signal is always time bounded.

(b) A time bounded signal always possesses finite energy.

(c) A time bounded signal is always zero outside the interval $[-t_0, t_0]$ for some t_0.

(d) A time bounded signal is always finite in amplitude. **[2006 : 1 M]**

1.5 A continuous-time system is described by $y(t) = e^{-|x(t)|}$, where $y(t)$ is the output and $x(t)$ is the input. $y(t)$ is bounded

(a) only when $x(t)$ is bounded

(b) only when $x(t)$ is non-negative

(c) only for $t \leq 0$ if $x(t)$ is bounded for $t \geq 0$

(d) even when $x(t)$ is not bounded **[2006 : 2 M]**

1.6 Given a sequence $x[n]$, to generate the sequence $y[n] = x[3 - 4n]$, which one of the following procedures would be correct?

(a) First delay $x[n]$ by 3 samples to generate $z_1[n]$, then pick every 4th sample of $z_1[n]$ to generate $z_2[n]$, and then finally time reverse $z_2[n]$ to obtain $y[n]$.

(b) First advance $x[n]$ by 3 samples to generate $z_1[n]$, then pick every 4th sample of $z_1[n]$ to generate $z_2[n]$, and then finally time reverse $z_2[n]$ to obtain $y[n]$.

(c) First pick every fourth sample of $x[n]$ to generate $v_1[n]$, time-reverse $v_1[n]$ to obtain $v_2[n]$, and finally advance $v_2[n]$ by 3 samples to obtain $y[n]$.

(d) First pick every fourth sample of $x[n]$ to generate $v_1[n]$, time-reverse $v_1[n]$ to obtain $v_2[n]$, and finally delay $v_2[n]$ by 3 samples to obtain $y[n]$. **[2008 : 2 M]**

1.7 The function shown in the figure can be represented as

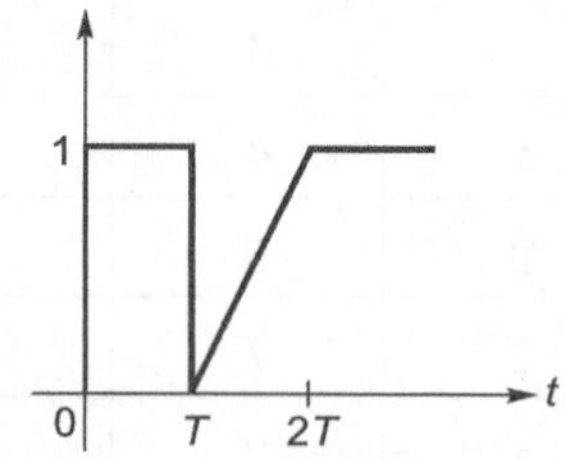

(a) $u(t) - u(t-T) + \frac{(t-T)}{T}u(t-T) - \frac{(t-2T)}{T}u(t-2T)$

(b) $u(t) + \frac{t}{T}u(t-T) - \frac{t}{u}u(t-2T)$

(c) $u(t) - u(t-T) + \frac{(t-T)}{T}u(t) - \frac{(t-2T)}{T}u(t)$

(d) $u(t) + \frac{(t-T)}{T}u(t-T) - 2\frac{(t-2T)}{T}u(t-2T)$

[2014 : 2 M, Set-1]

1.8 The value of $\int_{-\infty}^{+\infty} e^{-t}\delta(2t-2)dt$, where $\delta(t)$ is the Dirac delta function, is

(a) $\frac{1}{2e}$ (b) $\frac{2}{e}$

(c) $\frac{1}{e^2}$ (d) $\frac{1}{2e^2}$

[2016 : 1 M, Set-1]

1.9 Consider the system with following input-output relation $y[n] = (1 + (-1)^n)\, x[n]$, where $x[n]$ is the input and $y[n]$ is the output. The system is

(a) invertible and time invariant
(b) invertible and time varying
(c) non-invertible and time invariant
(d) non-invertible and time varying

[2017 : 1 M, Set-1]

1.10 The mean square value of the given periodic waveform $f(t)$ is _____.

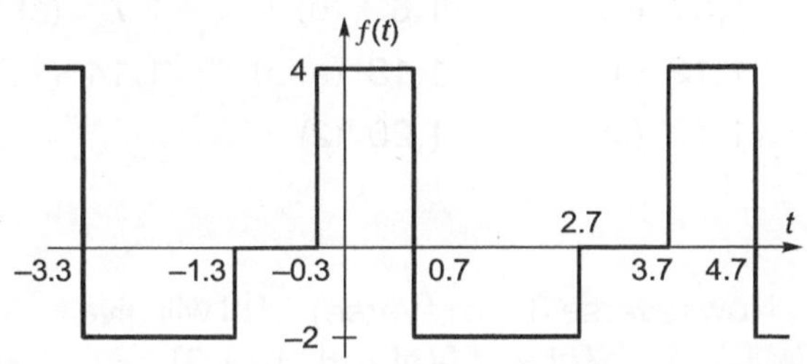

[2017 : 1 M, Set-2]

1.11 The signal energy of the continuous-time signal, $x(t) = [(t-1)\,u(t-1)] - [(t-2)\,u(t-2)] - [(t-3)\,u(t-3)] + (t-4)\,u(t-4)]$ is

(a) $\frac{11}{3}$ (b) $\frac{7}{3}$

(c) $\frac{1}{3}$ (d) $\frac{5}{3}$

[2018 : 2 M]

1.12 The symbol, a and T, represent positive quantities, and $u(t)$ is the unit step function. Which one of the following impulse responses is NOT the output of a causal linear time-invariant system?

(a) $1 + e^{-at}\,u(t)$ (b) $e^{+at}\,u(t)$
(c) $e^{-a(t-T)}\,u(t)$ (d) $e^{-a(t+T)}\,u(t)$

[2019 : 1 M]

1.13 x_R and x_A are, respectively, the rms and average values of $x(t) = x(t-T)$, and similarly, y_R and y_A are, respectively, the rms and average values of $y(t) = kx(t) \cdot k$, T are independent of t. Which of the following is true?

(a) $y_A = kx_A$; $y_R = kx_R$
(b) $y_A = kx_A$; $y_R \neq kx_R$
(c) $y_A \neq kx_A$; $y_R \neq kx_R$
(d) $y_A \neq kx_A$; $y_R = kx_R$

[2020 : 1 M]

1.14 The waveform shown in solid line is obtained by clipping a full-wave rectified sinusoid (shown dashed). The ratio of the RMS value of the full-wave rectified waveform to the RMS value of the clipped waveform is ______. (Round off to 2 decimal places.)

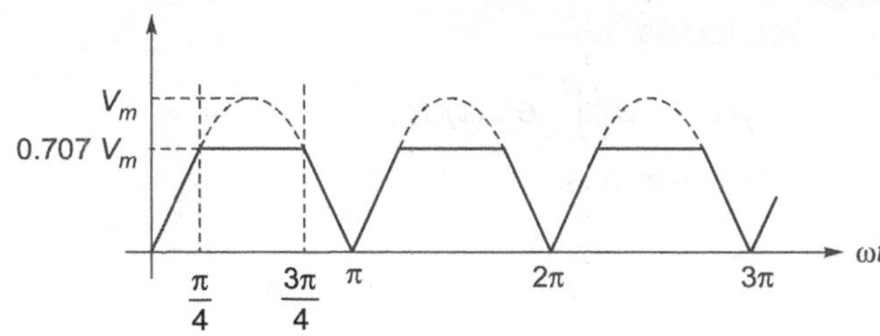

[2021 : 2 M]

1.15 Which of the following statement(s) is/are true?

(a) If an LTI system is causal, it is stable.
(b) A discrete time LTI system is causal if and only if its response to a step input $u[n]$ is 0 for $n < 0$.
(c) If a discrete time LTI system has an impulse response $h[n]$ of finite duration the system is stable.
(d) If the impulse response $0 < |h[n]| < 1$ for all n, then the LTI system is stable.

[2023 : 1 M]

1.16 The period of the discrete-time signal $x[n]$ described by the equation below is N = _____ (Round off to the nearest integer).

$$x[n] = 1 + 3\sin\left(\frac{15\pi}{8}n + \frac{3\pi}{4}\right) - 5\sin\left(\frac{\pi}{3}n - \frac{\pi}{4}\right)$$

[2023 : 2 M]

1.17 Suppose signal $y(t)$ is obtained by the time-reversal of signal $x(t)$, i.e., $y(t) = x(-t)$, $-\infty < t < \infty$. Which one of the following options is always true for the convolution of $x(t)$ and $y(t)$?

(a) It is an odd signal
(b) It is an even signal
(c) It is an anti-causal signal
(d) It is a causal signal

[2024 : 1 M]

1.18 Consider the discrete-time system T_1 and T_2 defined as follows :

$$\{T_1x\}[n] = x[0] + x[1] + \ldots + x[n]$$

$$\{T_2x\}[n] = x[0] + \frac{1}{2}x[1] + \cdots + \frac{1}{2^n}x[n]$$

Which one of the following statement is true?

(a) T_1 and T_2 are not BIBO stable.
(b) T_1 is BIBO stable but T_2 is not BIBO stable.
(c) T_1 is not BIBO stable but T_2 is BIBO stable.
(d) T_1 and T_2 are BIBO stable.

[2024 : 2 M]

1.19 The input $x(t)$ and the output $y(t)$ of a system are related as

$$y(t) = e^{-t}\int_{-\infty}^{t} e^{\tau}x(\tau)d\tau, \quad -\infty < t < \infty$$

The system is

(a) nonlinear
(b) linear but not time-invariant
(c) linear and time-invariant
(d) noncasual

[2024 : 2 M]

1.20 If the energy of a continuous-time signal $x(t)$ is E and the energy of the signal $2x(2t-1)$ is cE, then c is _____ (rounded off to 1 decimal place).

[2024 : 2 M]

Answers Introduction of C.T. and D.T. Signals

1.1 (b)	1.2 (d)	1.3 (d)	1.4 (c)	1.5 (d)	1.6 (b)	1.7 (a)
1.8 (a)	1.9 (d)	1.10 (6)	1.11 (d)	1.12 (a)	1.13 (a, b)	1.14 (1.21)
1.15 (b)	1.16 (48)	1.17 (b)	1.18 (c)	1.19 (c)	1.20 (2)	

Explanations Introduction of C.T. and D.T. Signals

1.1 (b)

$$\text{RMS value} = \left[\frac{1}{T}\left\{\int_0^{T/2}\left(\frac{2A}{T}t\right)^2 dt + \int_{T/2}^{T} A^2 dt\right\}\right]^{1/2}$$

$$= \sqrt{\frac{2}{3}}A$$

1.2 (d)

$$V_c(t) = \frac{1}{C}\int_{-\infty}^{t} i(t)dt = \frac{1}{C}\int_{-\infty}^{t} 5\delta(t)dt = \frac{5}{C}u(t)$$

1.3 (d)

RMS value

$$= \left[\frac{1}{\pi}\left\{\int_0^{\pi/3}(100)^2 dt + \int_{\pi/3}^{2\pi/3}(-100)^2 dt + \int_{2\pi/3}^{\pi}(100)^2 dt\right\}\right]^{12}$$

$$= \left[\frac{1}{\pi}\cdot 100^2\pi\right]^{1/2} = 100\text{ V}$$

1.6 (b)

$$y[n] = x[3-4n] = x[-4n+3]$$

So to obtain $y[n]$ we first advance $x[n]$ by 3 unit.

i.e., $z_1[n] = x[n+3]$

Now we will take every fourth sample of $z_1[n]$

i.e., $z_2[n] = z_1[4n] = x[4n+3]$

Now reverse (time reverse) $z_2[n]$ will give

$$y[n] = z_2[-n] = x[-4n+3]$$

1.7 (a)

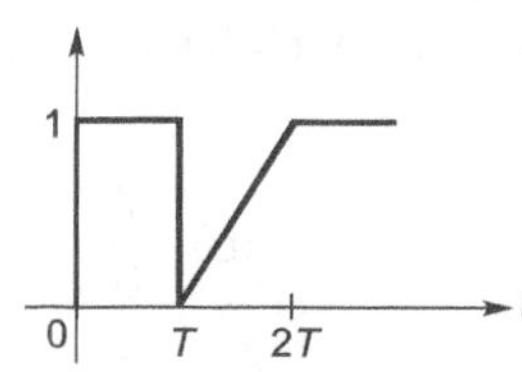

The given function can be realized as follows:

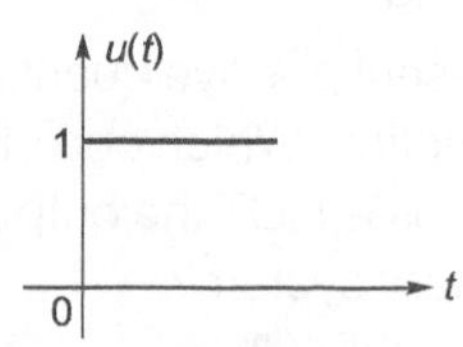

$u(t)$
(Unit step function)
(i)

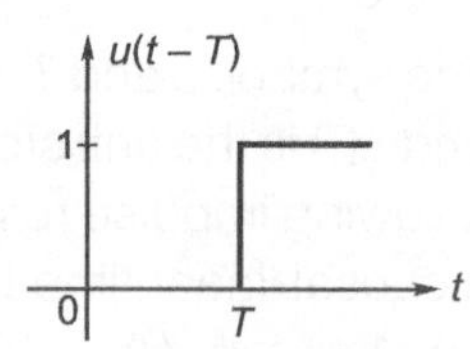

$u(t-T)$
(Unit step function right shifted)
(ii)

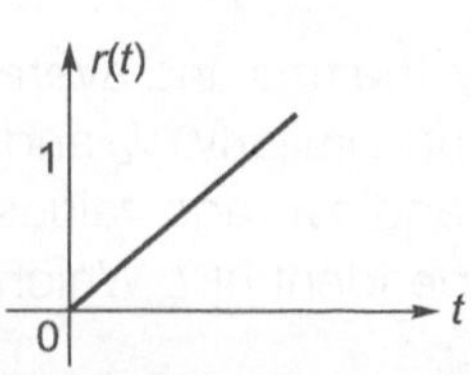

$r(t) = tu(t)$
(Ramp function)

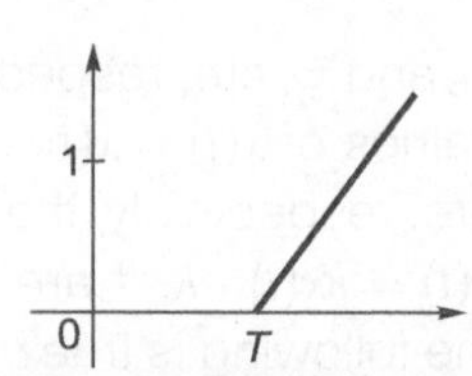

$\frac{(t-T)}{T}u(t-T)$
(Right shifted unit ramp function)

(iii)

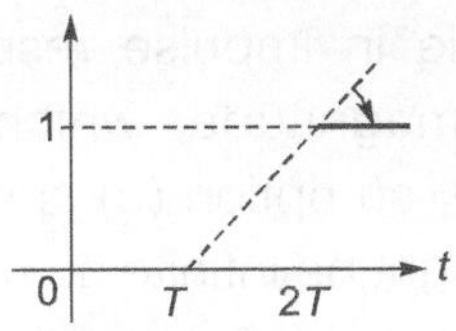

$\frac{(t-2T)}{T}u(t-2T)$

(iv)

Combining the functions (i), (ii), (iii) and (iv), we get the given function

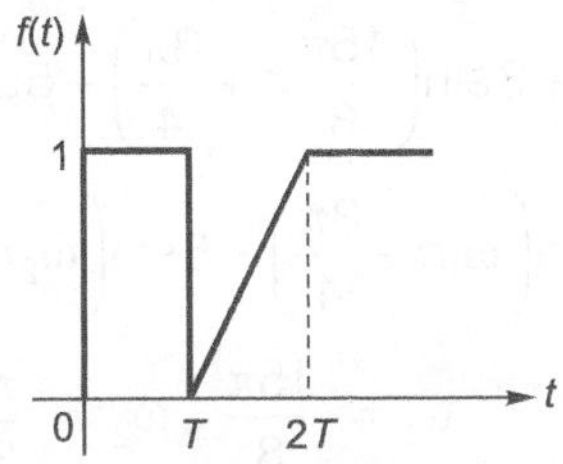

Therefore, $f(t) = u(t) - u(t-T) + \frac{(t-T)}{T}u(t-T)$

$$-\frac{(t-2t)}{T}u(t-2T)$$

$$f(t) = u(t) - u(t-T) + \frac{1}{T}r(t-T) - \frac{1}{T}r(t-2T)$$

or

$$f(t) = u(t) - u(t-T) + \frac{1}{T}(t-T)\,u(t-T) - \frac{1}{T}(t-2T)u(t-2T)$$

1.8 (a)

To find value of $\int_{-\infty}^{\infty} e^{-t}\delta(2t-2)dt$

Since, $\delta(2t-2) = \frac{1}{2}\delta(t-1)$

Above integral can be written as

$$\int_{-\infty}^{\infty} e^{-t}\frac{1}{2}\delta(t-1)dt = \frac{1}{2}e^{-1} = \frac{1}{2e}$$

1.9 (d)

Given relationship,

$$y(n) = [1 + (-1)^n]\,x(n)$$

Time invariance test:

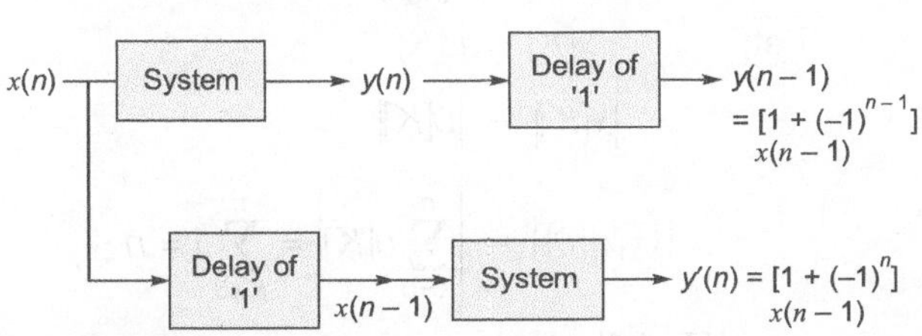

Since, $y(n-1) \neq y'(n)$

So, the system is time variant.

Invertibility test:

$x(n)$	$y(n)$
$\delta(n-1)$	$[1+(-1)^n]\,\delta(n-1) = [1+(-1)^1]\,\delta(n-1) = 0\cdot\delta(n-1) = 0$
$2\delta(n-1)$	$[1+(-1)^n]\,2\delta(n-1) = [1+(-1)^1]\,2\delta(n-1) = 0.2\delta(n-1) = 0$

Thus, we are getting many to one mapping between input and output. So, the system is non-invertible.

1.10 Sol.

Mean square value = Power of $f(t)$

$$\text{Mean square value} = \frac{1}{T_0}\int_{T_0}|f(t)|^2\,dt$$

$$= \frac{1}{4}[4^2 \times 1 + 2^2 \times 2] = \frac{16+8}{4} = \frac{24}{4} = 6$$

1.11 (d)

$$x(t) = r(t-1) - r(t-2) - r(t-3) + r(t-4)$$

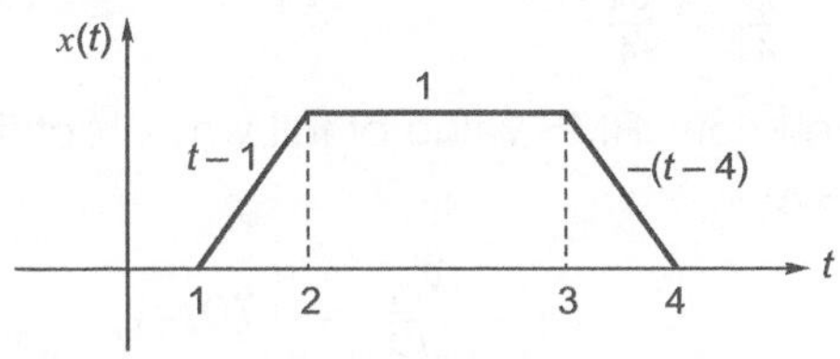

$$Ex(t) = \int_{-\infty}^{\infty}|x(t)|^2\,dt$$

$$= \int_1^2\left|(t-1)^2\right|dt + \int_2^3 1^2\,dt + \int_3^4\{-(t-4)\}^2\,dt$$

$$= \frac{1}{3} + 1 + \frac{1}{3} = \frac{2}{3} + 1 = \frac{5}{3}$$

1.12 (a)

a and T represents positive quantities.

$u(t)$ is unit step function.

$h(t) = 1 + e^{-at}\,u(t)$, is non-causal

Here '1' is a constant and two sided so the system will be non-causal, because for causal system.

$$h(t) = 0, \quad t < 0$$
$$\neq 0, \quad t > 0$$

1.13 (a, b)

Given, $y(t) = kx(t)$ ...(1)

then, Average of $y(t) = k \times$ Average of $x(t)$

$\Rightarrow \quad y_A = kx_A$

From equation (1),

Power of $y(t) = |k|^2 \cdot$ power of $x(t)$

$\Rightarrow \quad y_R^2 = |k|^2 \cdot x_R^2 \quad$ [$\because$ Power = Rms^2]

$\Rightarrow \quad y_R = |k| \cdot x_R$

Case (i): When k is real and positive then,

$$|k| = k \quad \text{and} \quad y_R = kx_R$$

Thus option (a) is satisfied.

Case (ii): When k is imaginary or complex or real and negative then,

$$|k| \neq k \quad \text{and} \quad y_R \neq kx_R$$

Thus option (b) is satisfied, option (a) and (b) both satisfies the given condition.

1.14 (1.21)

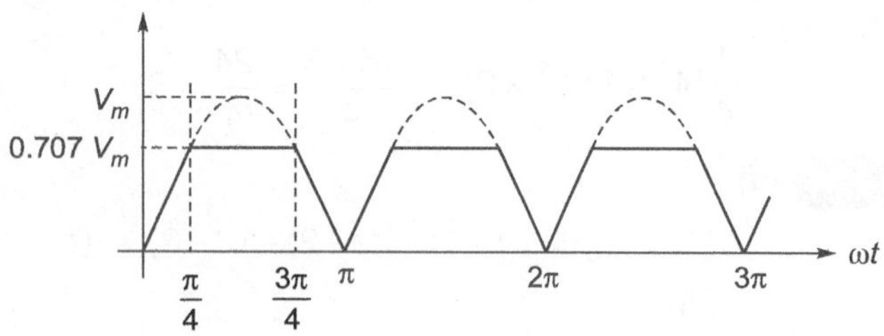

We know, RMS value of full wave rectified sine wave

$$= \frac{V_m}{\sqrt{2}} = 0.707\ V_m$$

For clipped wave form

As $0 \rightarrow \frac{\pi}{4}$ and $\frac{3\pi}{4} \rightarrow \pi$ wave form are identical

rms value of clipped wave

$$= \sqrt{\frac{1}{\pi}\left[2\int_0^{\pi/4} V_m^2 \sin^2 \omega t + \int_{\pi/4}^{3\pi/4} (0.707V_m)^2 d\omega t\right]}$$

$$= \sqrt{\frac{1}{\pi}\left(\int_0^{\pi/4} \frac{2V_m^2}{2}[1-\cos 2\omega t]d\omega t + \frac{1}{2}V_m^2 \times \frac{\pi}{2}\right)^{1/2}}$$

$$= \left[\frac{1}{\pi}\left[V_m^2\left(\frac{\pi}{4} - \frac{1}{2}\sin\frac{\pi}{2}\right) + \frac{V_m^2}{4}\pi\right]\right]^{1/2}$$

$$= \left[\frac{V_m^2}{\pi}\left[\frac{\pi}{4} - \frac{1}{2}\right] + \frac{V_m^2}{4}\right]^{1/2} = 0.5838\ V_m$$

$$\text{Ratio} = \frac{0.707}{0.5838} = 1.21$$

1.15 (b)

As we know for causal LTI-system :

$$h(n) = 0,\ n < 0$$

If input is $u(n)$, then step-response $S(n)$ will start either from $n = 0$ or right-side of $n = 0$.

i.e., $\quad S(n) = 0,\ n < 0$

- A sample in impulse response may have infinite magnitude, which makes system unstable, so option (c) is incorrect.
- There might be infinite number of samples in impulse response, so option (d) is incorrect.
- Some causal LTI system may become unstable, so option (a) is not correct.

1.16 (48)

$$x[n] = 1 + 3\sin\left(\frac{15\pi}{8}n + \frac{3\pi}{4}\right) - 5\sin\left(\frac{\pi}{3}n - \frac{\pi}{4}\right)$$

$$= 1 + 3\sin\left(\omega_1 n + \frac{3\pi}{4}\right) - 5\sin\left(\omega_2 n - \frac{\pi}{4}\right)$$

where, $\quad \omega_1 = \frac{15\pi}{8},\ \omega_2 = \frac{\pi}{3}$

Now, $\quad N_1 = \frac{2\pi}{\omega_1}K_1 = \frac{2\pi}{\frac{15\pi}{8}}K_1 = \frac{16}{15}K_1$

$$= 16 \text{ (for } K_1 = 15)$$

$$N_2 = \frac{2\pi}{\omega_2}K_2 = \frac{2\pi}{\frac{\pi}{3}}K_2 = 6K_2 = 6 \text{ (for } K_2 = 1)$$

Time-period of $x(n)$ is :

$N = \text{LCM}[N_1, N_2] = \text{LCM}[16, 6] = 48$

1.17 (b)

$$z(t) = x[t] * y[t]$$

$$z(t) = x[t] * x[-t]$$

Replace t with $-t$

$$z[-t] = x[-t] * x[t]$$

$$z[-t] = x[t] * x[-t]$$

$$z[-t] = z[t]$$

even function.

1.18 (c)

$$\{T_1 x\}[n] = x[0] + x[1] + \dots + x[n]$$

$$\{T_1 x\}[n] = \sum_{K=0}^{n} x[K]$$

Let, $\quad x[K] = u[K]$

$$|x[K]| = |u[K]|$$

$$|\{T_1 x\}[n]| = \left|\sum_{K=0}^{n} u[K]\right| = \sum_{K=0}^{n} 1 = n$$

$\because \quad |[T_1 x(n)]|_{n=\infty} = \infty$; $\quad \therefore$ system is not BIBO stable

$$\{T_2 x\}[n] = x[0] + \frac{1}{2}x[1] + \dots + \frac{1}{2^n}x[n]$$

$$= \sum_{K=0}^{n} \left(\frac{1}{2}\right)^K x(K)$$

Let, $x(K) = u[K]$

$$|x[K]| = |u[K]|$$

$$\{T_2x\}[n] = \sum_{K=0}^{n} \left|\left(\frac{1}{2}\right)^K u(K)\right|$$

$$\{T_2x\}[n] = \sum_{K=0}^{n} \left(\frac{1}{2}\right)^K$$

$$|[T_2x(n)|_{n=\infty} = \sum_{K=0}^{\infty} \left(\frac{1}{2}\right)^K = \frac{1}{1-\frac{1}{2}} = 2$$

(stable)

T_1 is not BIBO stable but T_2 is BIBO stable.

1.19 (c)

- System will follow law of superposition. Therefore, system will be linear.
- Time-invariance check:

$$y(t-t_o) = e^{-(t-t_o)} \int_{-\infty}^{t-t_o} e^{\tau} \cdot x(\tau)d\tau \quad \text{... (i)}$$

For input $x(t - t_o)$, output will be

$$y'(t) = e^{-t} \int_{-\infty}^{t} e^{\tau} \cdot x(\tau - t_o)d\tau$$

Put $\tau - t_o = k$, then $d\tau = dk$ and for $\tau = t$, $k = t - t_o$

$$= e^{-t} \int_{-\infty}^{t-t_o} e^{k+t_o} \cdot x(k)dk$$

$$= e^{-t} \times e^{t_o} \times \int_{-\infty}^{t-t_o} e^{k} \cdot x(k)dk$$

$$= e^{-(t-t_o)} \times \int_{-\infty}^{t-t_o} e^{k} \cdot x(k)dk$$

$$= e^{-(t-t_o)} \times \int_{-\infty}^{t-t_o} e^{\tau} \cdot x(\tau)d\tau \quad \text{... (ii)}$$

Since, $y'(t) = y(t - t_o)$

So, the system is time-invariant.

1.20 (2)

$$x(t) \rightarrow E \text{ (energy)}$$

$$x(t-1) \rightarrow E$$

$$x(2t-1) \rightarrow \frac{E}{2}$$

$$2x(2t-1) \rightarrow (2)^2(E/2) = 2E = CE$$

$$C = 2$$

■■■■

2 Linear Time Invariant Systems

2.1 The convolution of the functions $f_1(t) = e^{-2t}\, u(t)$ and $f_2(t) = e^{t}\, u(t)$ is equal to ____. [1995 : 1 M]

2.2 The impulse response of an initially relaxed linear system is $e^{-2t}\, u(t)$. To produce a response of $t\, e^{-2t}\, u(t)$, the input must be equal to

(a) $2\, e^{-t}\, u(t)$ (b) $\frac{1}{2} e^{-2t}\, u(t)$

(c) $e^{-2t}\, u(t)$ (d) $e^{-t}\, u(t)$ [1995 : 1 M]

2.3 The unit impulse response of a system is given as:

$$c(t) = -4e^{-t} + 6e^{-2t}$$

The step response of the same system for $t \geq 0$ is equal to

(a) $-3e^{-2t} - 4e^{-t} + 1$ (b) $-3e^{-2t} + 4e^{-t} - 1$

(c) $-3e^{-2t} - 4e^{-t} - 1$ (d) $3e^{-2t} + 4e^{-t} - 1$

[1996 : 2 M]

2.4 Let $s(t)$ be the step response of a linear system with zero initial conditions. Then the response of this system to an input $u(t)$ is

(a) $\int_0^t s(t-\tau)u(\tau)d\tau$

(b) $\frac{d}{dt}\left[\int_0^t s(t-\tau)u(\tau)d\tau\right]$

(c) $\int_0^t s(t-\tau)\left[\int_0^t u(\tau_1)d\tau_1\right]d\tau$

(d) $\int_0^t s(t-\tau)^2 u(\tau)d\tau$ [2002 : 1 M]

2.5 $x[n] = 0;\ n < -1,\ n > 0,\ x[-1] = -1,\ x[0] = 2$ is the input and $y[n] = 0;\ n < -1,\ n > 2,\ y[-1] = -1 = y[1],\ y[0] = 3,\ y[2] = -2$ is the output of a discrete-time LTI system.

The system impulse response $h[n]$ will be

(a) $h[n] = 0;\ n < 0,\ n > 2,\ h[0] = 1,\ h[1] = h[2] = -1$

(b) $h[n] = 0;\ n < -1,\ n > 1, h[-1] = 1,\ h[0] = h[1] = 2$

(c) $h[n] = 0;\ n < 0,\ n > 3,\ h[0] = -1,\ h[1] = 2,\ h[2] = 1$

(d) $h[n] = 0;\ n < -2,\ n > 1,\ h[-2] = h[1] = h[-1] = -h[0] = 3$ [2006 : 2 M]

2.6 If $u(t)$, $r(t)$ denote the unit step and unit ramp functions respectively and $u(t) * r(t)$ their convolution, then the function $u(t+1) * r(t-2)$ is given by

(a) $(1/2)\,(t-1)\,(t-2)$

(b) $(1/2)\,(t-1)\,(t+2)$

(c) $(1/2)\,(t-1)^2\, u(t-1)$

(d) None of the above

2.7 Let a signal $a_1 \sin(\omega_1 t + \phi_1)$ be applied to a stable linear time invariant system. Let the corresponding steady state output be represented as $a_2\, F(\omega_2 t + \phi_2)$. Then which of the following statements is true?

(a) F is not necessarily a "sine" ,or "cosine" function but must be periodic with $\omega_1 = \omega_2$.

(b) F must be a "sine" or "cosine" function with $a_1 = a_2$.

(c) F must be a "sine" function with $\omega_1 = \omega_2$ and $\phi_1 = \phi_2$.

(d) F must be a "sine" or "cosine" function with $\omega_1 = \omega_2$. [2007 : 1 M]

2.8 The impulse response of a causal linear time-invariant system is given as $h(t)$. Now consider the following two statements:

Statement (I): priniciple of superposition holds

Statement (II): $h(t) = 0$ for $t < 0$.

Which one of the following statements is correct?

(a) Statement (I) is correct and statement (II) is wrong.

(b) Statement (II) is correct and statement (I) is wrong.

(c) Both Statement (I) and Statement (II) are wrong.

(d) Both Statement (I) and Statement (II) are correct. [2008 : 1 M]

2.9 A signal $e^{-\alpha t} \sin(\omega t)$ is the input to a real Linear Time Invariant system. Given K and ϕ are constants, the output of the system will be of the from $Ke^{-\beta t} \sin(\nu t + \phi)$ where

(a) β need not be equal to α but v equal to ω.

(b) V need not be equal to ω but β equal to α.

(c) β equal to α and ν equal to ω.

(d) β need not be equal to α and ν need not be equal to ω. [2008 : 1 M]

2.10 A system with $x(t)$ and output $y(t)$ is defined by the input-output relation:

$$Y(t) = \int_{-\infty}^{-2t} x(t)\, d\tau$$

The system will be
(a) casual, time-invariant and unstable
(b) causal, time-invariant and stable
(c) non-causal, time-invariant and unstable
(d) non-causal, time-variant and unstable

[2008 : 2 M]

2.11 A Linear Time Invariant system with an impulse response $h(t)$ produces output $y(t)$ when input $x(t)$ is applied. When the input $x(t-\tau)$ is applied to a system with impulse response $h(t-\tau)$, the output will be
(a) $y(\tau)$
(b) $y(2(t-\tau))$
(c) $y(t-\tau)$
(d) $y(t-2\tau)$ [2009 : 1 M]

2.12 A cascade of 3 Linear Time Invariant systems is causal and unstable. From this, we conclude that
(a) each system in the cascade is individually causal and unstable.
(b) at least one system is unstable and at least one system is causal.
(c) at least one system is causal and all systems are unstable.
(d) the majority are unstable and the majority are causal. [2009 : 2 M]

2.13 The system represented by the input-output relationship $y(t) = \int_{-\infty}^{5t} x(\tau)\, d\tau, t > 0$ is
(a) Linear and causal
(b) Linear but not causal
(c) Causal but not linear
(d) Neither linear nor causal [2010 : 1 M]

2.14 Given the finite length input $x[n]$ and the corresponding finite length output $y[n]$ of an LTI system as shown below, the impulse response $h[n]$ of the system is

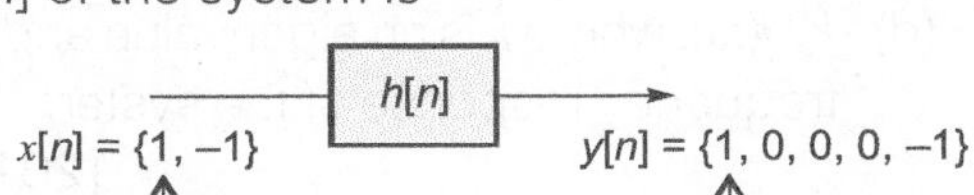

(a) $h[n] = \{\underset{\uparrow}{1}, 0, 0, 1\}$ (b) $h[n] = \{\underset{\uparrow}{1}, 0, 1\}$

(c) $h[n] = \{\underset{\uparrow}{1}, 1, 1, 1\}$ (d) $h[n] = \{\underset{\uparrow}{1}, 1, 1\}$

[2010 : 2 M]

2.15 Given two continuous time signals $x(t) = e^{-t}$ and $y(t) = e^{-2t}$ which exist for $t > 0$, the convolution $z(t) = x(t) * y(t)$ is
(a) $e^{-t} - e^{-2t}$
(b) e^{-3t}
(c) e^{+t}
(d) $e^{-t} + e^{-2t}$

[2011 : 1 M]

2.16 Let $y[n]$ denote the convolution of $h[n]$ and $g[n]$, where $h[n] = (1/2)^n\, u[n]$ and $g[n]$ is a causal sequence. If $y[0] = 1$ and $y[1] = 1/2$, then $g[1]$ equals
(a) 0
(b) 1/2
(c) 1
(d) 3/2 [2012 : 2 M]

2.17 The input $x(t)$ and output $y(t)$ of a system are related as $y(t) = \int_{-\infty}^{t} x(\tau)\cos(3\tau)\, d\tau$. The system is
(a) time-invariant and stable
(b) stable and not time-invariant
(c) time-invariant and not stable
(d) not time-invariant and not stable

[2012 : 2 M]

2.18 The impulse response of a system is $h(t) = t\,u(t)$. For an input $u(t-1)$, the output is
(a) $\frac{t^2}{2}u(t)$
(b) $\frac{t(t-1)}{2}u(t-1)$
(c) $\frac{(t-1)^2}{2}u(t-1)$
(d) $\frac{t^2-1}{2}u(t-1)$

[2013 : 1 M]

2.19 Two systems with impulse responses $h_1(t)$ and $h_2(t)$ are connected in cascade. Then the overall impulse response of the cascaded system is given by
(a) product of $h_1(t)$ and $h_2(t)$
(b) sum of $h_1(t)$ and $h_2(t)$
(c) convolution of $h_1(t)$ and $h_2(t)$
(d) subtraction of $h_2(t)$ from $h_1(t)$ [2013 : 1 M]

2.20 The impulse response of a continuous time system is given by $h(t) = \delta(t-1) + \delta(t-3)$. The value of the step response at $t = 2$ is
(a) 0
(b) 1
(c) 2
(d) 3 [2013 : 2 M]

2.21 $x(t)$ is non-zero only for $T_x < t < T'_x$, and similarly, $y(t)$ is non-zero only for $T_y < t < T'_y$. Let $z(t)$ be convolution of $x(t)$ and $y(t)$. Which one of the following statements is TRUE?
(a) $z(t)$ can be non-zero over an unbounded interval.

(b) $z(t)$ is non-zero for $t < T_x + T_y$.

(c) $z(t)$ is zero outside of $T_x + T_y < t < T_x' + T_y'$.

(d) $z(t)$ is non-zero for $t > T_x' + T_y'$.

[2014 : 1 M, Set-1]

2.22 Consider an LTI system with impulse response $h(t) = e^{-5t}\,u(t)$. If the output of the system is $y(t) = e^{-3t}\,u(t) - e^{-5t}\,u(t)$ then the input, $x(t)$ is given by

(a) $e^{-3t}\,u(t)$ (b) $2\,e^{-3t}\,u(t)$

(c) $e^{-5t}\,u(t)$ (d) $2\,e^{-5t}\,u(t)$

[2014 : 1 M, Set-2]

2.23 The impulse response $g(t)$ of a system, G, is as shown in the figure (a). What is the maximum value attained by the impulse response of two cascaded blocks of G as shown in the figure (b)?

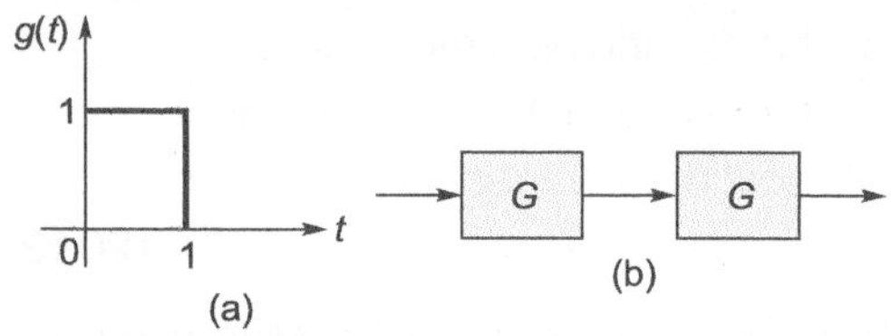

(a) $\frac{2}{3}$ (b) $\frac{3}{4}$

(c) $\frac{4}{5}$ (d) 1 [2015 : 1 M, Set-1]

2.24 A moving average function is given by $y(t) = \frac{1}{T}\int_{t-T}^{t} u(\tau)d\tau$. If the input u is a sinusoidal signal of frequency $\frac{1}{2T}$ Hz, then in steady state, the output y will lag u (in degree) by ______.

[2015 : 1 M, Set-1]

2.25 For linear time invariant systems, that are Bounded Input Bounded Output stable, which one of the following statements is TRUE?

(a) The impulse response will be integrable, but may not be absolutely integrable.

(b) The unit impulse response will have finite support.

(c) The unit step response will be absolutely integrable.

(d) The unit step response will be bounded.

[2015 : 2 M, Set-2]

2.26 Consider a continuous-time system with input $x(t)$ and output $y(t)$ given by : $y(t) = x(t)\cos(t)$
This system is

(a) linear and time-invariant

(b) non-linear and time-invariant

(c) linear and time-varying

(d) non-linear & time-varying [2016 : 1 M, Set-1]

2.27 The output of a continuous-time, linear time-invariant system is denoted by $T\{x(t)\}$ where $x(t)$ is the input signal. A signal $z(t)$ is called eigen-signal of the system T, when $T\{z(t)\} = y\,z(t)$, where y is a complex number, in general, and is called an eigen value of T. Suppose the impulse response of the system T is real and even. Which of the following statements is TRUE?

(a) $\cos(t)$ is an eigen-signal but $\sin(t)$ is not.

(b) $\cos(t)$ and $\sin(t)$ are both eigen-signals but with different eigen values.

(c) $\sin(t)$ is an eigen-signal but $\cos(t)$ is not.

(d) $\cos(t)$ and $\sin(t)$ are both eigen-signals with identical eigen values.

[2016 : 2 M, Set-1]

2.28 Consider a causal LTI system characterized by differential equation $\frac{dy(t)}{dt} + \frac{1}{6}y(t) = 3x(t)$. The response of the system to the input $x(t) = 3e^{-t/3}\,u(t)$, where $u(t)$ denotes the unit step function, is

(a) $9\,e^{-t/3}\,u(t)$

(b) $9\,e^{-t/6}\,u(t)$

(c) $9\,e^{-t/3}\,u(t) - 6e^{-t/6}\,u(t)$

(d) $54\,e^{-t/6}\,u(t) - 54\,e^{-t/3}\,u(t)$ [2016 : 1 M]

2.29 Let $z(t) = x(t) * y(t)$, where "$*$" denotes convolution. Let c be a positive real-valued constant. Choose the correct expression for $z(ct)$.

(a) $c \cdot x(ct) * y(ct)$ (b) $x(ct) * y(ct)$

(c) $c \cdot x(t) * y(ct)$ (d) $c \cdot x(ct) * y(t)$

[2017 : 1 M, Set-1]

2.30 A continuous-time input signal $x(t)$ is an eigen function of an LTI system, if the output is

(a) $k\,x(t)$, where k is an eigenvalue.

(b) $ke^{j\omega t}x(t)$, where k is an eigenvalue and $e^{j\omega t}$ is a complex exponential signal.

(c) $x(t)\,e^{j\omega t}$, where $e^{j\omega t}$ is a complex exponential signal.

(d) $k\,H(\omega)$, where k is an eigenvalue and $H(\omega)$ is a frequency response of the system.

[2018 : 1 M]

2.31 Suppose for input $x(t)$ a linear time-invariant system with impulse response $h(t)$ produces output $y(t)$, so that $x(t) * h(t) = y(t)$. Further, if $|x(t)| * |h(t)| = z(t)$, which of the following statements is true?

(a) For some but not all $t \in (-\infty, \infty)$, $z(t) \le y(t)$

(b) For all $t \in (-\infty, \infty)$, $z(t) \ge y(t)$

(c) For all $t \in (-\infty, \infty)$, $z(t) \le y(t)$

(d) For some but not all $t \in (-\infty, \infty)$, $z(t) \ge y(t)$

[2020 : 2 M]

2.32 If the input $x(t)$ and output $y(t)$ of a system are related as $y(t) = \max(0, x(t))$, then the system is

(a) non-linear and time variant

(b) linear and time-variant

(c) linear and time-invariant

(d) non-linear and time-invariant

[2021 : 1 M]

2.33 Two discrete-time linear time-invariant systems with impulse response $h_1[n] = \delta[n-1] + \delta[n+1]$ and $h_2[n] = \delta[n] + \delta[n-1]$ are connected in cascade, where $\delta[n]$ is the Kronecker delta. The impulse response of the cascaded system is

(a) $\delta[n-2] + \delta[n+1]$

(b) $\delta[n-2] + \delta[n-1] + \delta[n] + \delta[n+1]$

(c) $\delta[n-1]\,\delta[n] + \delta[n+1]\,\delta[n-1]$

(d) $\delta[n]\,\delta[n-1] + \delta[n-2]\,\delta[n+1]$

[2021 : 1 M]

2.34 Consider the system as shown below :

where $y(t) = x(e^t)$. The system is

(a) linear and causal.

(b) linear and non-causal.

(c) non-linear and causal.

(d) non-linear and non-causal. [2022 : 2 M]

2.35 Let a causal LTI system be governed by the following differential equation $y(t) + \frac{1}{4}\frac{dy}{dt} = 2x(t)$, where $x(t)$ and $y(t)$ are the input and output respectively. Its impulse response is

(a) $2e^{-\frac{1}{4}t}u(t)$ (b) $2e^{-4t}u(t)$

(c) $8e^{-\frac{1}{4}t}u(t)$ (d) $8e^{-4t}u(t)$

[2022 : 2 M]

■■■■

Answers Linear Time Invariant Systems

2.1 (Sol.)	2.2 (c)	2.3 (b)	2.4 (b)	2.5 (a)	2.6 (c)	2.7 (d)
2.8 (d)	2.9 (c)	2.10 (d)	2.11 (d)	2.12 (b)	2.13 (b)	2.14 (c)
2.15 (a)	2.16 (a)	2.17 (d)	2.18 (c)	2.19 (c)	2.20 (b)	2.21 (c)
2.22 (b)	2.23 (d)	2.24 (90)	2.25 (d)	2.26 (c)	2.27 (d)	2.28 (d)
2.29 (a)	2.30 (a)	2.31 (b)	2.32 (d)	2.33 (b)	2.34 (b)	2.35 (d)

Explanations Linear Time Invariant Systems

2.1 Sol.

$$\because \quad f_1(t) = e^{-2t}u(t) \Rightarrow F_1(s) = \frac{1}{s+2}$$

and $$f_2(t) = e^{t}u(t) \Rightarrow F_2(s) = \frac{1}{s-1}$$

$$F_1(s)\,F_2(s) = \frac{1}{(s+2)(s-1)} = \frac{-1/3}{s+2} + \frac{1/3}{s-1}$$

$\because F_1(s)\cdot F_2(s) = f_1(t) * f_2(t)$

Multiplication in s-domain results in convolution in time domain.

$$\therefore \quad f_1(t) * f_2(t) = \mathcal{L}^{-1}\left[\frac{-1/3}{s+2} + \frac{1/3}{s-1}\right]$$

$$= \left[-\frac{1}{3}e^{-2t} + \frac{1}{3}e^{t}\right]u(t)$$

2.2 (c)

$$H(s) = \mathcal{L}[e^{-2t}u(t)] = \frac{1}{s+2}$$

$$Y(s) = \mathcal{L}[t\,e^{-2t}u(t)] = \frac{1}{(s+2)^2}$$

$$\Rightarrow \quad X(s) = \frac{Y(s)}{X(s)} = \frac{1}{s+2}$$

Hence, input $x(t) = \mathcal{L}^{-1}[X(s) = e^{-2t}u(t)$

2.3 (b)

Impulse response $= -4e^{-t} + 6e^{-2t}$

$$\text{Step response} = \int_{-\infty}^{t} c(t)dt = \int_0^t -4e^{-t}dt + \int_0^t 6e^{-2t}dt$$

$$= 4e^{-t} - 3e^{-2t} - 1$$

Hence, option (b) is correct.

2.5 (a)

For finite duration convolution,
$x[n]$ have M terms
$h[n]$ have N terms
then, $y[n]$ should have terms $(M + N - 1)$

Here, $x[n] = \{-1, \underset{\uparrow}{2}\}$

$$y[n] = \{-1, \underset{\uparrow}{3}, -1, -2\}$$

So, $h[n]$ should have only 3 terms and $h[n]$ have value starting from origin or $[n = 0]$ only because $y[n]$ have value start from $[n = -1]$

$\therefore$ Consider option (I)

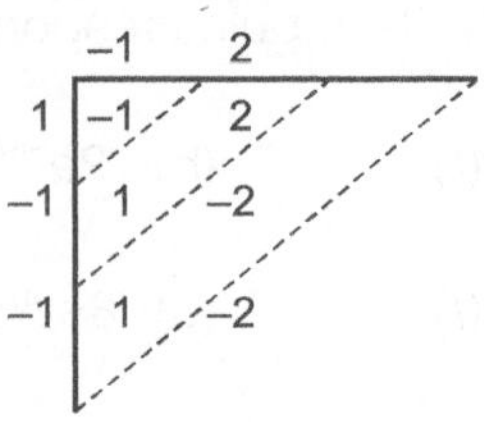

$$y[n] = \{-1, \underset{\uparrow}{3}, -1, -2\}$$

So option (a) is correct.

2.6 (c)

$$u(t+1) * r(t-2) = \int_{-\infty}^{\infty} r(\tau - 2)\, u(t+1-\tau)\, d\tau$$

$$z(t) = \int_{-\infty}^{(t+1)} r(\tau - 2)\, d\tau$$

$$z(t) = \int_{2}^{(t+1)} (\tau - 2)\, d\tau$$

For, $(t + 1) \geq 2$

$$z(t) = \left(\frac{\tau^2}{2} - 2\tau\right)_2^{(t+1)} = \frac{1}{2}[\tau^2 - 4\tau]_2^{(t+1)}$$

For, $t \geq 1 = \frac{1}{2}[(t+1)^2 - 4(t+1) - 4 + 8]$

$$z(t) = \frac{1}{2}[t^2 + 2t + 1 - 4t - 4 + 4]$$

For, $t \geq 1$

$$z(t) = \frac{1}{2}(t-1)^2$$

For $t \geq 1$, $z(t) = \frac{1}{2}(t-1)^2 u(t-1)$

2.7 (d)

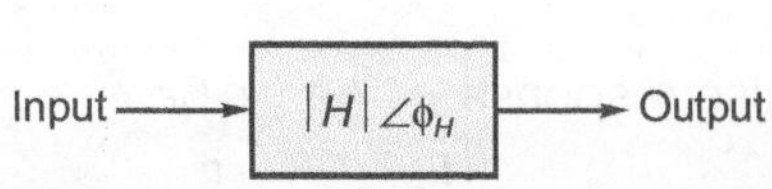

The effect of linear time invariant system is best understood by considering that, at any given input frequency ω_1, the output is sinusoidal at same frequency, as the input applied, with amplitude given by output = $|H|$ × input and phase equal to $\phi_2 = \phi_H + \phi_1$, where $|H|$ is the magnitude of the frequency response and ϕ_H is its phase angle. Both $|H|$ and ϕ_H are functions of frequency.

2.8 (d)

Since system is causal therefore its impulse response will be zero for $t < 0$.
So statement-II is correct.
$\Rightarrow$ since system is linear it also satisfies the principle of superposition.
So both the statement are correct.

2.9 (c)

$$e^{-\alpha t}\sin(\omega_0 t) \xrightarrow[h(t)\,=\,\text{real signal}]{\text{LTI system}} Ke^{-\beta t}\sin(vt + \phi)$$

For real $h(t)$: $h(t) = h^*(t)$

Applying Laplace transform, $H(s) = H^*(s^*)$

$$\Rightarrow \quad H(s + j\omega) = H^*(\sigma - j\omega) \quad [\text{Put } s = \sigma + j\omega]$$

$$\omega = \omega_0, \quad \sigma = -a$$

$$\Rightarrow \quad H(-\alpha + j\omega_0) = H^*(-\alpha - j\omega_0)$$

$$\Rightarrow \quad |H(-\alpha + j\omega_0)|\, e^{j\angle H(-\alpha + j\omega_0)}$$

$$= |H(-\alpha - j\omega_0)|\, e^{-j\angle H(-\alpha - j\omega_0)}$$

By comparison,

$$|H(-\alpha + j\omega_0)| = |H(-\alpha - j\omega_0)| \quad \text{...(i)}$$

and $\angle H(-\alpha + j\omega_0) = -\angle H(-\alpha - j\omega_0)$ ...(ii)

Now system input is,

$$x(t) = e^{-\alpha t}\sin\omega_0 t = \frac{1}{2j}e^{-\alpha t}(e^{j\omega_0 t} - e^{-j\omega_0 t})$$

$$= \frac{1}{2j}\left[e^{t(-\alpha + j\omega_0)} - e^{t(-j\omega_0 - \alpha)}\right] = \frac{1}{2j}[x_1(t) - x_2(t)]$$

where, $x_1(t) = e^{t(-\alpha + j\omega_0)}$ and $x_2(t) = e^{t(-\alpha - j\omega_0)}$

For $x_1(t)$ system output is,

$$y_1(t) = H(s)\big|_{s = -\alpha + j\omega_0} \cdot e^{(-\alpha + j\omega_0)t}$$

$$= H(-\alpha + j\omega_0)\, e^{(-\alpha + j\omega_0)t}$$

Similarly, for $x_2(t)$ system output is,

$$y_2(t) = H(s)\big|_{s = -\alpha - j\omega_0} \cdot e^{(-\alpha - j\omega_0)t}$$

$$= H(-\alpha - j\omega_0)\, e^{(-\alpha - j\omega_0)t}$$

Thus, the overall output $y(t)$ is,

$$y(t) = \frac{1}{2j}[y_1(t) - y_2(t)]$$

[$\because$ By using property of linear system]

$$= \frac{1}{2j}\left[H(-\alpha + j\omega_0)e^{(-\alpha + j\omega_0)t} - H(-\alpha - j\omega_0)e^{(-\alpha - j\omega_0)t}\right]$$

$$= \frac{e^{-at}}{2j}\left[|H(-\alpha + j\omega_0)|\, e^{j\angle H(-\alpha + j\omega_0)} \cdot e^{j\omega_0 t}\right.$$

$$-|H(-\alpha - j\omega_0)|e^{j\angle H(-\alpha-j\omega_0)} \cdot e^{-j\omega_0 t}\Big]$$

$$= \frac{e^{-at}}{2j}|H(-\alpha + j\omega_0)|\Big[e^{j\angle H(-\alpha+j\omega_0)}\, e^{j\omega_0 t}$$

$$-e^{-j\angle H(-\alpha+j\omega_0)} \cdot e^{-j\omega_0 t}\Big]$$

[By using equation (i) and (ii)]

$$= e^{-at}|H(-\alpha + j\omega_0)|$$

$$\frac{1}{2j}\Big[e^{j}\{\omega_0 t + \angle H(-\alpha + j\omega_0)\}$$

$$- e^{-j}\{\omega_0 t + \angle H(-\alpha + j\omega_0)\}\Big]$$

$$= e^{-at}|H(-\alpha + j\omega_0)|\sin[\omega_0 t + \angle H(-\alpha + j\omega_0)]$$

$$= Ke^{-\beta t}\sin(vt + \alpha)$$

2.10 (d)

First of all we will check for causality,

$$y(t) = \int_{-\infty}^{-2t} x(\tau)d\tau$$

So for $t = -2$

$$y(-2) = \int_{-\infty}^{4} x(\tau)d\tau$$

So output depends on future values of input along with past and present values of input so system is non-causal.

Let us find output for shifted input $x(t - t_0)$

$$y'(t) = \int_{-\infty}^{2t} x(\tau - t_0)d\tau$$

$$y'(t) = \int_{-\infty}^{(2t-t_0)} x(\tau)d\tau \quad \text{...(i)}$$

Now shift the output by t_0 then,

$$y(t - t_0) = \int_{-\infty}^{2(t-t_0)} x(\tau)d\tau \quad \text{...(ii)}$$

So from eq. (i) and (ii),

$$y'(t) \neq y(t - t_0)$$

Therefore system is time variant.

Non-causal and time variant is present only in option (d).

2.11 (d)

Case 1: $Y(s) = H(s) \cdot X(s)$

Case 2: input $= x(t - \tau) \Rightarrow X(s) \cdot e^{-s\tau}$

impulse response $= h(t - \tau) \Rightarrow H(s) \cdot e^{-s\tau}$

Output: $Y'(s) = X(s) \cdot e^{-s\tau} \cdot H(s)\, e^{-s\tau}$

$$Y'(s) = X(s) \cdot H(s) \cdot e^{-s2t} = Y(s)\, e^{-2sT}$$

$$\Rightarrow \quad y'(t) = y(t - 2\tau)$$

2.12 (b)

Since in cascade overall impulse response

$$h(t) = h_1(t) * h_2(t) * h_3(t)$$

$h_1(t)$, $h_2(t)$, $h_3(t)$ are impulse responses of individual systems.

Since initial point where $h(t)$ is nonzero is $t \geq 0$ and since in convolution initial point

$$= t_1 + t_2 + t_3$$

where t_1, t_2, t_3 are initial points of $h_1(t)$, $h_2(t)$, $h_3(t)$ respectively.

So for it to be greater than or equal to zero at least one of them t_1, t_2 or t_3 must be +ve i.e. greater than zero so at least one of them must be causal. Similarly if one (at least) of the systems becomes unstable then overall system will become unstable.

2.13 (b)

Integrator is always a linear system.

Since, $\quad y(t) = \int_{-\infty}^{5t} x(\tau)d\tau \qquad t > 0$

For $t = 1$, $\quad y(t) = \int_{-\infty}^{5} x(\tau)d\tau$

here value at $t = 1$ depends on future values like at $t = 2, 3, \ldots$ of input $x(t)$.

So, it is a noncausal system.

2.14 (c)

$$x[n] = \{\underset{\uparrow}{1}, -1\}, M = 2$$

$$y[n] = \{\underset{\uparrow}{1}, 0, 0, 0, -1\}, N_1 = 5$$

Since output has → number of elements

$$N_1 = M + N - 1$$

where, N is number of elements in impulse response $h[n]$

$$\therefore \quad 5 = 2 + N - 1 \;\Rightarrow\; N = 4$$

Let it be a_1, a_2, b_1, b_2

$$y[n] = [\underset{\uparrow}{a_1}, (a_2 - a_1), (b_1 - a_2), (b_2 - b_1), -b_2]$$

Comparing with

$$y[n] = [\underset{\uparrow}{1}, 0, 0, 0, -1]$$

$$a_1 = 1; \qquad a_2 = a_1 = 1$$

$$b_1 = a_1 = 1; \; b_2 = b_1 = 1$$

$$\therefore \quad h[n] = \{\underset{\uparrow}{1}, 1, 1, 1\}$$

2.15 (a)

$$x(s) = \frac{1}{s+1}, \qquad y(s) = \frac{1}{s+2}$$

$$z(s) = x(s) \cdot y(s)$$

$$= \frac{1}{s+1} \times \frac{1}{s+2} = \frac{1}{s+1} - \frac{1}{s+2}$$

$$z(t) = e^{-t} - e^{-2t}$$

2.16 (a)

Given, $g[n]$ is a causal sequence

$\therefore$ $g[n]$ will be zero for $n < 0$

$$y[n] = \sum_{k=-\infty}^{\infty} g[k]\, h[n-k]$$

and $h[n] = \left(\frac{1}{2}\right)^n u[n]$

$$h[n] = \begin{cases} \left(\frac{1}{2}\right)^n & ; \ n \geq 0 \\ 0 & ; \ n < 0 \end{cases}$$

$$h[0] = 1, \quad h[1] = \frac{1}{2}, \quad h[2] = \frac{1}{4}$$

$$y[0] = g[0]\, h[0] + g[1]\, h[-1] + \cdots$$

$$1 = g[0] \times 1 + 0 + 0 + \cdots$$

$$g[0] = 1$$

$$y[1] = g[0]\, h[1] + g[1]\, h[0] + g[2]\, h[-1] + \cdots$$

$$\frac{1}{2} = 1 \cdot \frac{1}{2} + g[1] \cdot 1 + 0 \Rightarrow g[1] = 0$$

2.17 (d)

$$y = \int_{-\infty}^{t} x(\tau) \cos(3\tau)\, d\tau$$

$$y(t - t_0) = \int_{-\infty}^{t-t_0} x(\tau) \cos(3\tau)\, d\tau$$

$y'(t)$ for input $x(t - t_0)$ is

$$y'(t) = \int_{-\infty}^{t} x(\tau - t_0) \cos 3\tau\, d\tau$$

$$y'(t) = \int_{-\infty}^{(t-t_0)} x(\tau) \cos 3(\tau + t_0)\, d\tau$$

$y'(t) \neq y(t - t_0)$ so system is not time invariant for input $x(\tau) = \cos(3\tau)$ bounded input

$$y(t) = \int_{-\infty}^{t} \cos^2(3\tau)\, d\tau \to \infty \text{ as } t \to \infty$$

So, for bounded input, output is not bounded therefore system is not stable.

2.18 (c)

$$h(t) = t\, u(t)$$

$$H(s) = \frac{1}{s^2}$$

$$\Rightarrow \quad \frac{Y(s)}{U(s)} = \frac{1}{s^2}$$

$$Y(s) = \frac{1}{s^2} \frac{e^{-s}}{s}$$

$$\Rightarrow \quad y(t) = \frac{(t-1)^2}{2} u(t-1)$$

2.19 (c)

In cascade connection,

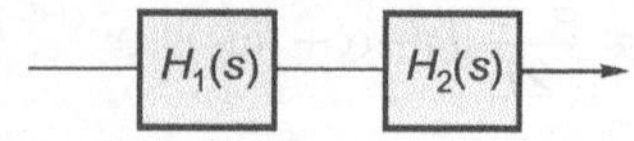

$$H(s) = H_1(s) \cdot H_2(s) \Rightarrow h(t) = h_1(t) * h_2(t)$$

2.20 (b)

For step response, impulse response can be integrated.

$$\Rightarrow \quad y(t) = u(t-1) + u(t-3)$$

$$y(2) = u(1) + u(-1) = 1 + 0 = 1$$

2.21 (c)

$x(t)$ is non-zero for $T_x < t < T_x'$

and $y(t)$ is non-zero for $T_y < t < T_y'$

$$\because \quad z(t) = x(t) \otimes y(t)$$

then the limits of the resultant signal is

$$T_x + T_y < t < T_x' + T_y'$$

i.e. $z(t)$ is non-zero for $T_x + T_y < t < T_x' + T_y'$

2.22 (b)

Impulse response of an LTI system

$$= \text{transfer function} = \frac{Y(s)}{X(s)} = H(s)$$

where, $y(t) = e^{-3t} u(t) - e^{-5t} u(t)$

$$\therefore \quad Y(s) = \frac{1}{(s+3)} - \frac{1}{(s+5)} = \frac{2}{(s+3)(s+5)}$$

Also, $h(t) = e^{-5t} u(t)$

$$\therefore H(s) = \frac{1}{(s+5)}$$

$$\therefore X(s) = \frac{Y(s)}{H(s)} = \frac{2}{(s+3)(s+5)} \times (s+5) = \frac{2}{(s+3)}$$

$\therefore$ Input $= x(t) = 2e^{-3t} u(t)$

2.23 (d)

$$g(t) = u(t) - u(t-1)$$

$$G(s) = \frac{1}{s} - \frac{e^{-s}}{s}$$

or, $\quad G(s) \times G(s) = g(t) * g(t)$

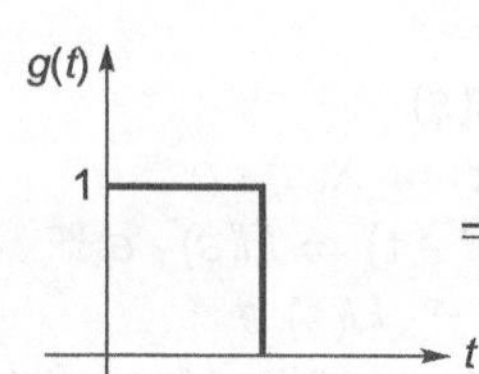

$\Rightarrow$

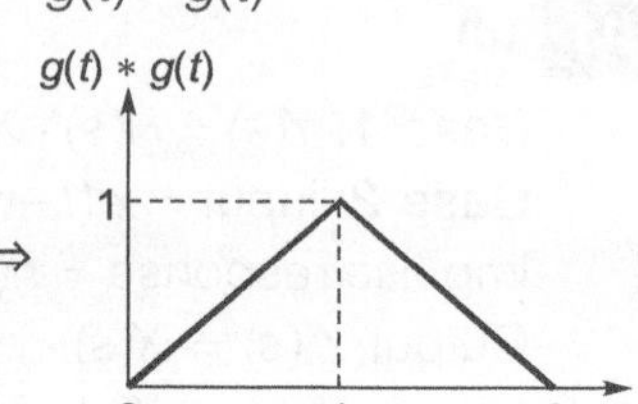

Maximum value = 1

2.24 (90)

System input: $\sin\omega_0 t$

$$f_0 = \frac{1}{2T}\text{ Hz}$$

$$\therefore\ \omega_0 = 2\pi f_0 = \frac{2\pi}{2T} = \frac{\pi}{T}\text{ rad/sec.}$$

$$y(t) = \frac{1}{T}\int_{t-T}^{t}\sin\omega_0\tau\, d\tau = \frac{1}{T}[-\cos\omega_0 t]_{t-T}^{t}$$

$$= -\frac{1}{T}[\cos\omega_0 t - \cos[\omega_0(t-T)]]\ \left[\because \omega_0 = \frac{t}{T}\right]$$

$$= -\frac{1}{T}\left[\cos\omega a t - \cos\left(\frac{\pi t}{T} - \pi\right)\right]$$

$$= -\frac{1}{T}[\cos\omega_0 t + \cos\omega_0 t] = -\frac{2}{T}\cos\omega_0 t$$

$$= \frac{2}{T}\sin(\omega_0 t - 90°)$$

= Output is lagging by 90° w.r.t. input.

2.26 (c)

$$y(t) = x(t)\cos(t)$$

To check linearity,

$y_1(t) = x_1(t)\cos(t)$ [$y_1(t)$ is output for $x_1(t)$]

$y_2(t) = x_2(t)\cos(t)$ [$y_2(t)$ is output for $x_2(t)$]

So the output for $(x_1(t) + x_2(t))$ will be

$y(t) = [x_1(t) + x_2(t)]\cos(t) = y_1(t) + y_2(t)$

So the system is linear to check time invariance.

The delayed output,

$$y(t - t_0) = x(t - t_0)\cos(t - t_0)$$

The output for delayed input,

$$y(t, t_0) = x(t - t_0)\cos(t)$$

Since $y(t - t_0) \neq y(t, t_0)$. System is time varying.

2.27 (d)

Given that impulse response is real and even, Thus $H(j\omega)$ will also be real and even.

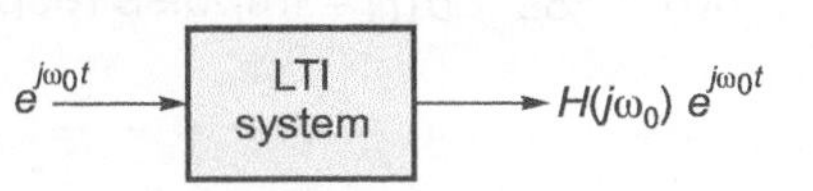

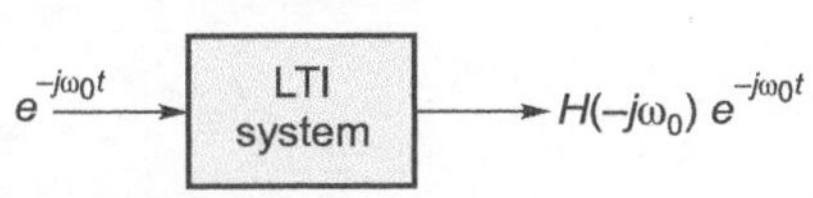

Since, $H(j\omega)$ is real and even thus

$$H(j\omega_0) = H(-j\omega_0)$$

Now, $\cos(t)$ is input, i.e., $\dfrac{e^{jt} + e^{-jt}}{2}$ is input

Output will be

$$\frac{H(j1)e^{jt} + H(-j1)e^{-jt}}{2} = H(j1)\left[\frac{e^{jt} + e^{-jt}}{2}\right]$$

$$= H(j1)\cos(t)$$

If $\sin(t)$ is input, i.e., $\dfrac{e^{jt} - e^{-jt}}{2}$ is input/

Output will be

$$\frac{H(j1)e^{jt} - H(-j1)e^{-jt}}{2j} = H(j1)\left[\frac{e^{jt} - e^{-jt}}{2j}\right]$$

$$= H(j1)\sin t$$

So, $\sin(t)$ and $\cos(t)$ are eigen signals with same eigen values.

2.28 (d)

The differential equation, $\dfrac{dy(t)}{dt} + \dfrac{1}{6}y(t) = 3\,x(t)$

So, $sY(s) + \dfrac{1}{6}Y(s) = 3\,X(s)$

$$Y(s) = \frac{3\,X(s)}{\left(s+\frac{1}{6}\right)};\quad X(s) = \frac{9}{\left(s+\frac{1}{3}\right)}$$

So, $Y(s) = \dfrac{9}{\left(s+\frac{1}{3}\right)\left(s+\frac{1}{6}\right)} = \dfrac{54}{\left(s+\frac{1}{6}\right)} - \dfrac{54}{\left(s+\frac{1}{3}\right)}$

So, $y(t) = (54e^{-1/6t} - 54e^{-1/3t})\,u(t)$

2.29 (a)

Time scaling property of convolution.

If, $x(t) * y(t) = z(t)$

Then, $x(ct) * y(ct) = \dfrac{1}{c}z(ct)$

$\Rightarrow\quad z(ct) = c \times x(ct) * y(ct)$

2.30 (a)

Eigen-function is a type of input for which output is constant times of input.

i.e.,

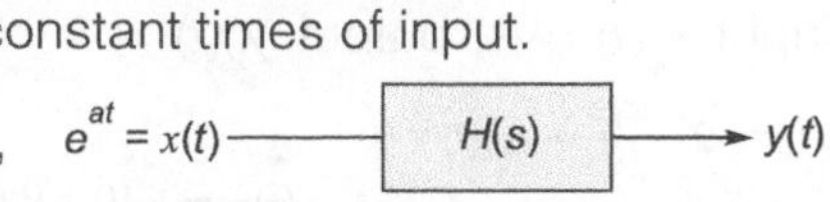

where, $x(t)$ = system input = eigen-function

$H(s)$ = transfer function of system

$y(t)$ = system output

Here, $y(t) = H(s)|_{s=a}\cdot e^{at} = K\cdot x(t)$

where, K = eigen-value = $H(s)|_{s=a}$

$x(t)$ = eigen-function input = $H(s)|_{s=a}$

2.31 (b)

Since, $y(t) = x(t) + h(t)$ and $z(t) = |x(t)| \times |h(t)|$

Case-1:

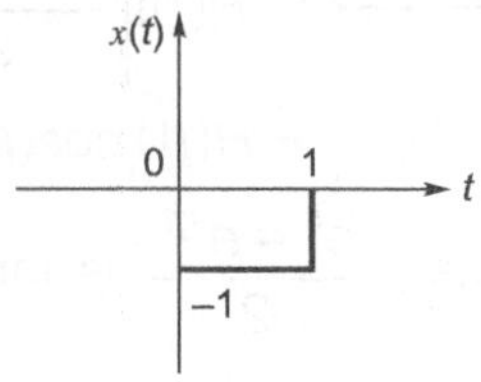

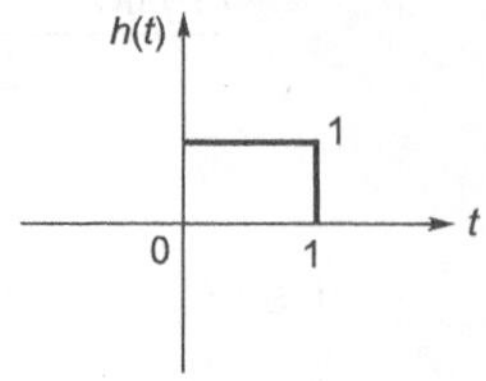

then, $y(t)$ and $z(t)$

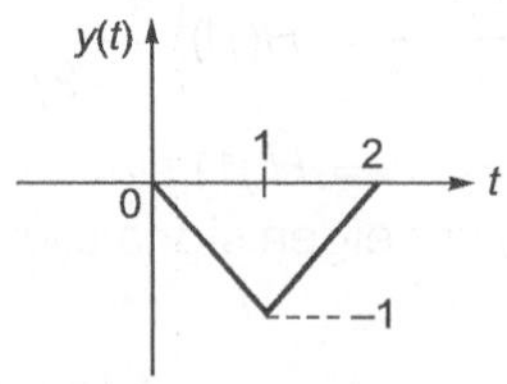

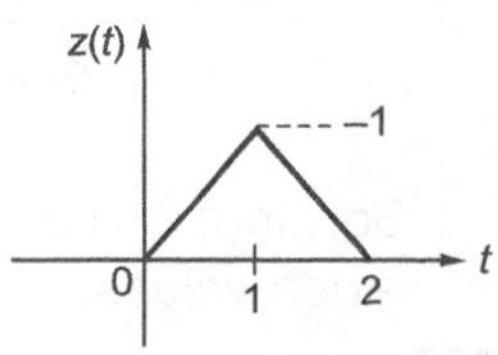

Case-2:

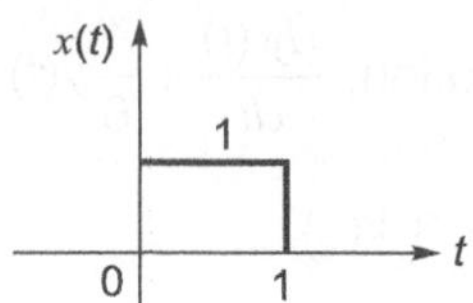

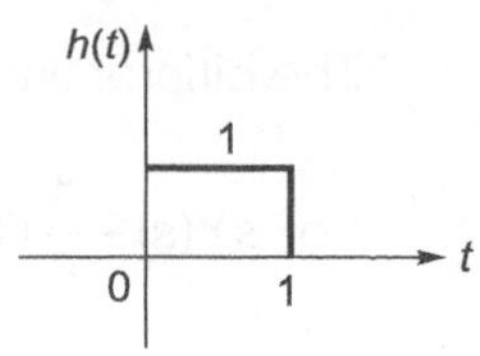

then, $y(t) = z(t)$

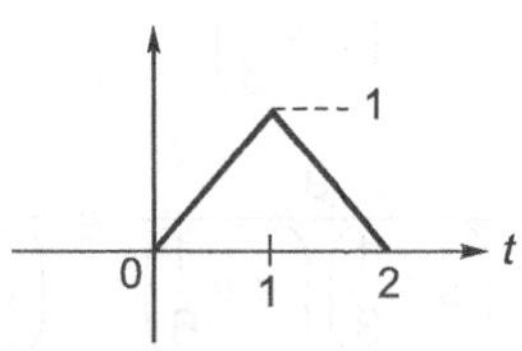

Thus, $z(t) \geq y(t)$, for all 't'

2.32 (d)

$$y(t) = \max(0, x(t)) = \begin{cases} 0, & x(t) < 0 \\ x(t), & x(t) > 0 \end{cases}$$

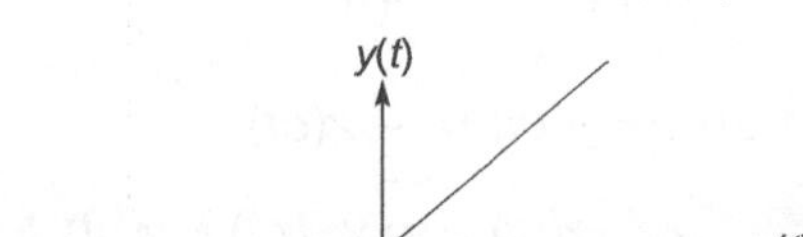

Linearity check :

at input $x_1(t) = -2$, output $y_1(t) = 0$

at input $x_2(t) = 1$, output $y_2(t) = 1$

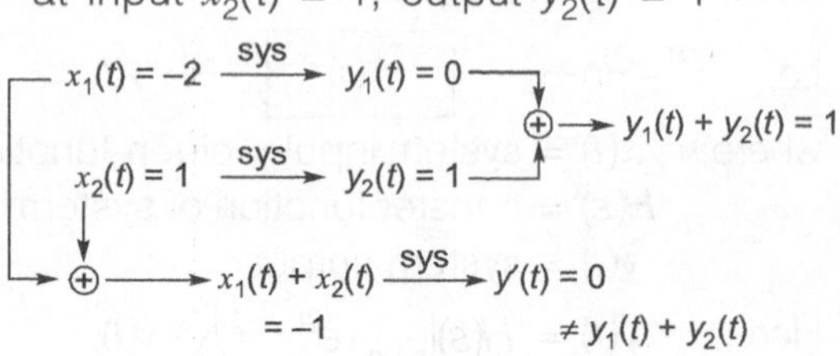

$\therefore$ system is non-linear because it violates law of additivity.

Check for time-invariance :

Delayed O/P :

$$y(t - t_o) = \begin{cases} x(t-t_o), & x(t-t_o) > 0 \\ 0 & x(t-t_o) < 0 \end{cases}$$

O/P of system when input is $x(t - t_o) = f(t)$

$$y_1(t) = \begin{cases} f(t), & f(t) > 0 \\ 0, & f(t) < 0 \end{cases} = \begin{cases} x(t-t_o), & x(t-t_o) > 0 \\ 0, & x(t-t_o) < 0 \end{cases}$$

Therefore, system is time-invariant.

2.33 (b)

$h(n)$ = Resultant impulse response = $h_1(n) * h_2(n)$

By applying z-transform

$$H(z) = H_1(z).H_2(z) = (z + z^{-1})(1 + z^{-1})$$
$$= z + z^{-1} + 1 + z^{-2}$$

By applying inverse ZT,

$$h(n) = \delta(n + 1) + \delta(n - 1) + \delta(n) + \delta(n - 2)$$

2.34 (b)

Part-1 : The given system will satisfy law of superposition. So the system will be linear.

Part-2 : $y(t) = x(e^t)$

Put, $t = 0$, $y(0) = x(1)$

Since output is depending on future value of input. So the system is non-causal.

2.35 (d)

By applying Laplace-transform on given differential equation,

$$Y(s) + \frac{1}{4}s.Y(s) = 2X(s)$$

$$\Rightarrow \frac{Y(s)}{X(s)} = \frac{2}{1+\frac{s}{4}} = \frac{8}{s+4} \Rightarrow H(s) = \frac{8}{s+4}$$

By taking inverse LT,

$$h(t) = 8e^{-4t}u(t) = \text{Impulse response of system.}$$

■■■■

3 Fourier Series

3.1 A periodic rectangular signal, $x(t)$ has the waveform as shown in the figure. Frequency of the fifth harmonic of its spectrum is

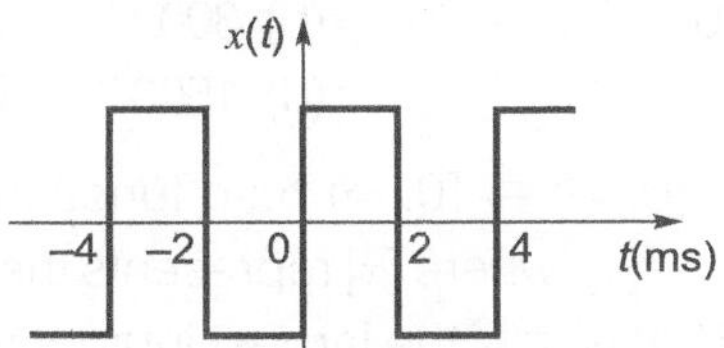

(a) 40 Hz (b) 200 Hz
(c) 250 Hz (d) 1250 Hz **[1996 : 1 M]**

3.2 If an AC voltage wave is corrupted with an arbitrary number of harmonics, then the overall voltage waveform differs from its fundamental frequency component in terms of

(a) only the peak values
(b) only the rms values
(c) peak and rms values
(d) all the three measures (peak, rms and average values) **[2000 : 1 M]**

3.3 Fourier Series for the waveform, $f(t)$ shown in figure is

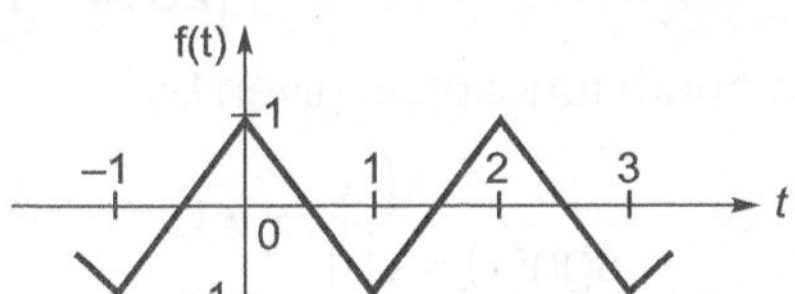

(a) $\frac{8}{\pi^2}\left[\sin(\pi t)+\frac{1}{9}\sin(3\pi t)+\frac{1}{25}\sin(5\pi t)+.....\right]$

(b) $\frac{8}{\pi^2}\left[\sin(\pi t)-\frac{1}{9}\cos(3\pi t)+\frac{1}{25}\sin(5\pi t)+.....\right]$

(c) $\frac{8}{\pi^2}\left[\cos(\pi t)+\frac{1}{9}\cos(3\pi t)+\frac{1}{25}\cos(5\pi t)+.....\right]$

(d) $\frac{8}{\pi^2}\left[\cos(\pi t)-\frac{1}{9}\sin(3\pi t)+\frac{1}{25}\sin(5\pi t)+.....\right]$

[2002 : 1]

3.4 The rms value of the periodic waveform given in the figure is

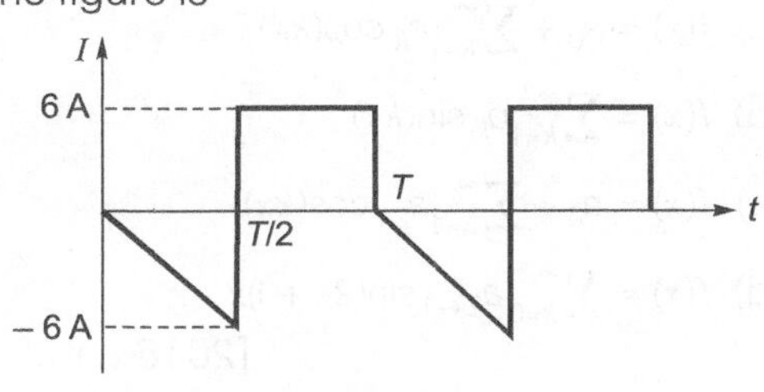

(a) $2\sqrt{6}$ A (b) $6\sqrt{2}$ A
(c) $\sqrt{4/3}$ A (d) 1.5 A **[2004 : 2 M]**

3.5 The RMS value of the resultant current in a wire which carries a DC current of 10 A and a sinusoidal alternating current of peak value 20 A is

(a) 14.1 A (b) 17.3 A
(c) 22.4 A (d) 30.0 A **[2004 : 2 M]**

3.6 For the triangular wave form shown in the figure, the RMS value of the voltage is equal to

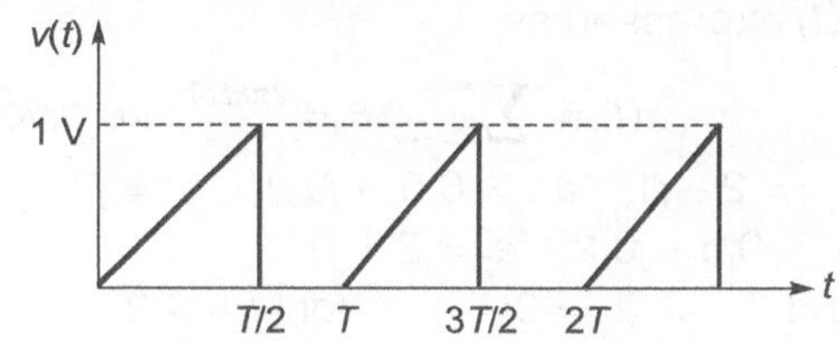

(a) $\sqrt{\frac{1}{6}}$ V (b) $\sqrt{\frac{1}{3}}$ V

(c) $\frac{1}{3}$ V (d) $\sqrt{\frac{2}{3}}$ V **[2005 : 2 M]**

3.7 The Fourier series for the function $f(x) = \sin^2 x$ is

(a) $\sin x + \sin 2x$ (b) $1 - \cos 2x$
(c) $\sin 2x + \cos 2x$ (d) $0.5 - 0.5 \cos 2x$

[2005 : 2 M]

3.8 $x(t)$ is a real valued function of a real variable with period T. Its trigonometric Fourier Series expansion contains no terms of frequency $\omega = 2\pi\,(2k)/T$; $k = 0, 1, 2, ...$ Also, no sine terms are present. Then $x(t)$ satisfies the equation

(a) $x(t) = -x(t - T)$
(b) $x(t) = x(T - t) = -x(-t)$
(c) $x(t) = x(T - t) = -x\left(t-\frac{T}{2}\right)$
(d) $x(t) = x(t - T) = x\left(t-\frac{T}{2}\right)$ **[2006 : 1 M]**

3.9 A signal $x(t)$ is given by

$$x(t) = \begin{cases} 1, -T/4 < t \le 3T/4 \\ -1, 3T/4 < t \le 7T/4 \\ -x(t+T) \end{cases}$$

Which among the following gives the fundamental Fourier term of $x(t)$?

(a) $\frac{4}{\pi}\cos\left(\frac{\pi t}{T}-\frac{\pi}{4}\right)$ (b) $\frac{\pi}{4}\cos\left(\frac{\pi t}{2T}+\frac{\pi}{4}\right)$

(c) $\frac{4}{\pi}\sin\left(\frac{\pi t}{T}-\frac{\pi}{4}\right)$ (d) $\frac{\pi}{4}\sin\left(\frac{\pi t}{2T}+\frac{\pi}{4}\right)$

[2007 : 2 M]

3.10 Let $x(t)$ be a periodic signal with time period T, Let $y(t) = x(t-t_0) + x(t+t_0)$ for some t_0. The Fourier Series coefficients of $y(t)$ are denoted by b_k. If $b_k = 0$ for all odd K. Then t_0 can be equal to

(a) $\frac{T}{8}$ (b) $\frac{T}{4}$

(c) $\frac{T}{2}$ (d) $2T$ [2008 : 2 M]

3.11 The Fourier Series coefficients, of a periodic signal $x(t)$ expressed as,

$$x(t) = \sum_{k=-\infty}^{\infty} a_k e^{j2\pi kt/T} \text{ are given by}$$

$a_{-2} = 2 - j1$; $a_{-1} = 0.5 + j0.2$; $a_0 = j2$;
$a_1 = 0.5 - j0.2$; $a_2 = 2 + j1$;
and $a_k = 0$; for $|k| > 2$

Which of the following is true?

(a) $x(t)$ has finite energy because only finitely many coefficients are non-zero
(b) $x(t)$ has zero average value because it is periodic.
(c) The imaginary part of $x(t)$ is constant.
(d) The real part of $x(t)$ is even. [2009 : 2 M]

3.12 The second harmonic component of the periodic waveform given in the figure has an amplitude of

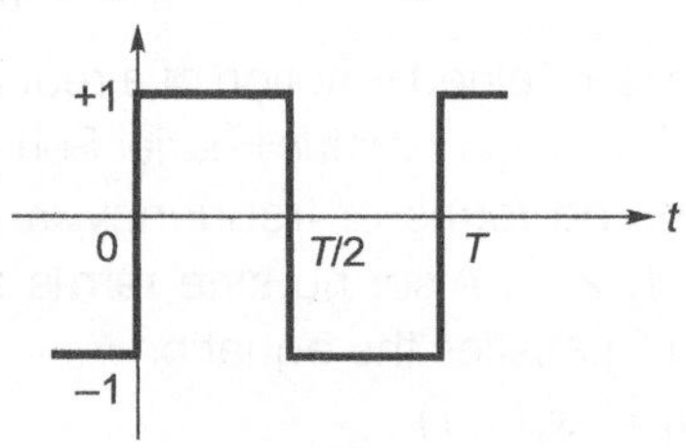

(a) 0 (b) 1

(c) $\frac{2}{\pi}$ (d) $\sqrt{5}$ [2010 : 1 M]

3.13 The Fourier expansion:

$$f(t) = a_0 + \sum_{n=1}^{\infty} a_n \cos n\omega t + b_n \sin n\omega t \quad \text{of the}$$

periodic signal shown below will contain the following non zero terms:

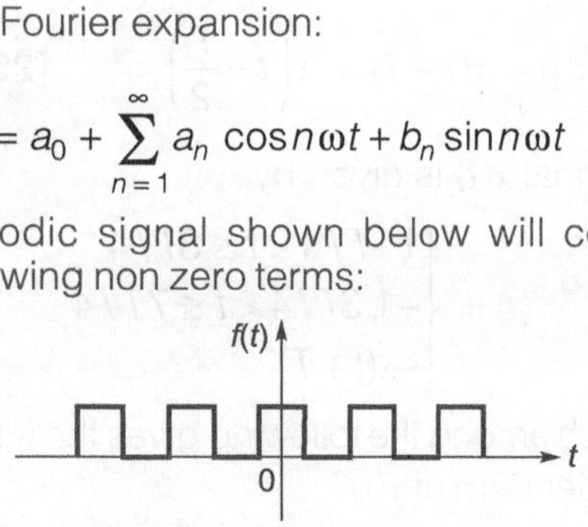

(a) a_0 and b_n, $n = 1, 3, 5, ... \infty$
(b) a_0 and a_n, $n = 1, 2, 3, ... \infty$
(c) a_0, a_n and b_n, n = 1, 2, 3, ... ∞
(d) a_0 and a_n, $n = 1, 3, 5, ... \infty$ [2011 : 1 M]

3.14 For a periodic signal

$$v(t) = 30\sin 100t + 10\cos 300t + 6\sin\left(500t + \frac{\pi}{4}\right),$$

the fundamental frequency in rad/s is

(a) 100 (b) 300
(c) 500 (d) 1500 [2013 : 1 M]

3.15 Let $g : [0, \infty) \rightarrow [0, \infty)$ be a function defined by $g(x) = x - [x]$, where $[x]$ represents the integer part of x. (That is, it is the largest integer which is less than or equal to x). The value of the constant term in the Fourier series expansion of $g(x)$ is ______ .

[2014 : 2 M, Set-1]

3.16 For a periodic square wave, which one of the following statements is TRUE?

(a) The Fourier series coefficients do not exist.
(b) The Fourier series coefficient exist but the reconstruction converges at most point.
(c) The Fourier series coefficient exist and the reconstruction converges at no points.
(d) The Fourier series coefficient exist and the reconstruction converges at every point.

[2014 : 1 M, Set-1]

3.17 The signum function is given by

$$\text{sgn}(x) = \begin{cases} \frac{x}{|x|}; & x \neq 0 \\ 0; & x = 0 \end{cases}$$

The Fourier series expansion of sgn(cos(t)) has

(a) only sine terms with all harmonics.
(b) only cosine terms with all harmonics.
(c) only sine terms with even numbered harmonics.
(d) only cosine terms with odd numbered harmonics. [2015 : 2 M, Set-1]

3.18 Let $f(x)$ be a real, periodic function satisfying $f(-x) = -f(x)$. The general form of its Fourier series representation would be

(a) $f(x) = a_0 + \sum_{k=1}^{\infty} a_k \cos(kx)$

(b) $f(x) = \sum_{k=1}^{\infty} b_k \sin(kx)$

(c) $f(x) = a_0 + \sum_{k=1}^{\infty} a_{2k} \cos(kx)$

(d) $f(x) = \sum_{k=0}^{\infty} a_{2k+1} \sin(2k+1)x$

[2016 : 1 M, Set-2]

3.19 Let the signal $x(t) = \sum_{k=-\infty}^{+\infty} (-1)^k \delta\left(t - \frac{k}{2000}\right)$ be passed through an LTI system with frequency response $H(\omega)$, as given in the figure.

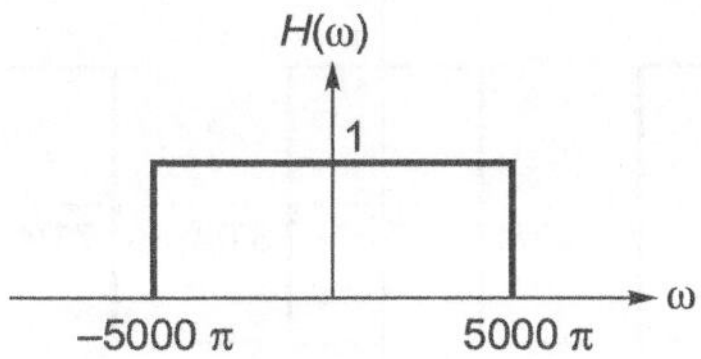

The Fourier series representation of the output is given as

(a) $4000 + 4000\cos(2000\pi t) + 4000\cos(4000\pi t)$
(b) $2000 + 2000\cos(2000\pi t) + 2000\cos(4000\pi t)$
(c) $4000\cos(2000\pi t)$
(d) $2000\cos(2000\pi t)$ **[2017 : 2 M, Set-1]**

3.20 Consider $g(t) = \begin{cases} t - \lfloor t \rfloor, & t \geq 0 \\ t - \lceil t \rceil, & \text{otherwise} \end{cases}$, where $t \in R$.
Here $\lfloor t \rfloor$ represents the largest integer less than or equal to t and $\lceil t \rceil$ denotes the smallest integer greater than or equal to t. The coefficient of the second harmonic component of the Fourier series representing $g(t)$ is ____ . **[2017 : 1 M, Set-1]**

3.21 A periodic function $f(t)$, with a period of 2π, is represented as its Fourier series,

If, $f(t) = a_0 + \sum_{n=1}^{\infty} a_n \cos nt + \sum_{n=1}^{\infty} b_n \sin nt$

$$f(t) = \begin{cases} A\sin t, & 0 \leq t \leq \pi \\ 0, & \pi < t < 2\pi \end{cases}$$

the Fourier series coefficients a_1 and b_1 of $f(t)$ are:

(a) $a_1 = 0;\ b_1 = \frac{A}{\pi}$ (b) $a_1 = \frac{A}{\pi};\ b_1 = 0$
(c) $a_1 = 0;\ b_1 = \frac{A}{2}$ (d) $a_1 = \frac{A}{2};\ b_1 = 0$

[2019 : 2 M]

3.22 The discrete time Fourier series representation of a signal $x[n]$ with period N is written as $x[n] = \sum_{k=0}^{N-1} a_k e^{j(2kn\pi/N)}$. A discrete time periodic signal with period $N = 3$, has the non-zero Fourier series coefficients: $a_{-3} = 2$ and $a_4 = 1$. The signal is

(a) $2 + 2e^{-\left(j\frac{2\pi}{6}n\right)} \cos\left(\frac{2\pi}{6}n\right)$
(b) $1 + 2e^{\left(j\frac{2\pi}{6}n\right)} \cos\left(\frac{2\pi}{6}n\right)$
(c) $1 + 2e^{\left(j\frac{2\pi}{3}n\right)} \cos\left(\frac{2\pi}{6}n\right)$
(d) $2 + 2e^{\left(j\frac{2\pi}{6}n\right)} \cos\left(\frac{2\pi}{6}n\right)$ **[2022 : 2 M]**

Answers Fourier Series

3.1 (d)	3.2 (c)	3.3 (c)	3.4 (a)	3.5 (b)	3.6 (a)	3.7 (d)
3.8 (c)	3.9 (a)	3.10 (b)	3.11 (c)	3.12 (a)	3.13 (d)	3.14 (a)
3.15 (0.5)	3.16 (b)	3.17 (d)	3.18 (b)	3.19 (c)	3.20 (Sol.)	3.21 (c)
3.22 (b)						

Explanations Fourier Series

3.1 (d)

$\because$ Fundamental frequency $= \frac{1}{4 \times 10^{-3}}$ Hz

$\therefore$ Freq. of fifth harmonic $= \frac{5}{4 \times 10^{-3}} = 1250$ Hz

3.3 (c)

$\because f(t)$ is an even function with half wave symmetry,
$\therefore$ DC term as well as sine terms = 0
Only the cosine terms with odd harmonics will be present. $\therefore$ Option (c) is correct.

3.4 (a)

$$I_{rms} = \left[\frac{1}{T}\int_0^T i^2(t)\right]^{1/2}$$

$$I_{rms}^2 = \left[\frac{1}{T}\int_0^{T/2} \left(-\frac{12t}{T}\right)^2 dt + \frac{1}{T}\int_{T/2}^{T} 6^2 dt\right]$$

$$= \left[\frac{1}{T}\int_0^{T/2} \frac{144(t)^2}{T^2} dt + \frac{36}{T}\int_{T/2}^{T} dt\right]$$

$$= \frac{144}{T^3}\left[\frac{t^3}{3}\right]_0^{T/2} + \frac{36}{T}\left[\frac{T}{2}\right]$$

$$= \frac{144}{8\times 3}+\frac{36}{2} = 24$$

$$\therefore\ I_{rms} = 2\sqrt{6}\text{ A}$$

3.5 (b)

Rms value of D.C. current = 10 A
Rms value of alternating current

$$= 20/\sqrt{2} = 10\sqrt{2}\text{ A}$$

∴ Rms value of resultant current

$$= \sqrt{10^2+(10\sqrt{2})^2}\text{ A} = 17.32\text{ A}$$

3.6 (a)

From the wave symmetry,

$$V_{rms} = \left[\frac{1}{T}\left[\int_0^{T/2}\left(\frac{2t}{T}\right)^2 dt + \int_{T/2}^{T} 0\, dt\right]\right]^{1/2}$$

$$V_{rms}^2 = \frac{1}{T}\int_0^{T/2}\frac{4t^2}{T^2} = \frac{4}{3T^3}\left[\frac{T^3}{8}\right] = \frac{1}{6}$$

$$\therefore\quad V_{rms} = \sqrt{\frac{1}{6}}\text{ V}$$

3.7 (d)

$$f(n) = \sin^2 x$$

for finding the fourier series expansion

$$f(x) = \sin^2 x = \frac{1-\cos 2x}{2} = 0.5 - 0.5\cos 2x$$

3.8 (c)

Since trigonometric fourier series has no sine terms and has only cosine terms therefore this will be an even signal, i.e., it will satisfy

$$x(t) = x(-t)$$

or we can write, $x(t-T) = x(-t+T)$
but signal is periodic with period T.
Therefore, $x(t-T) = x(t)$
therefore, $x(t) = x(T-t)$ ...(i)
Now since signal contains only odd harmonics i.e. no terms of frequency

$\omega = \frac{2\pi\times 2k}{T}$, k = 0, 1, 2, 3, 4, i.e., no even harmonics.

This means signal contains half wave symmetry this implies that,

$$x(t) = -x\left(t-\frac{T}{2}\right) \quad ...(ii)$$

From (i) and (ii), we get

$$x(t) = x(T-t) = -x\left(t-\frac{T}{2}\right)$$

3.9 (a)

According to definition of signal given in question the $x(t)$ will be as

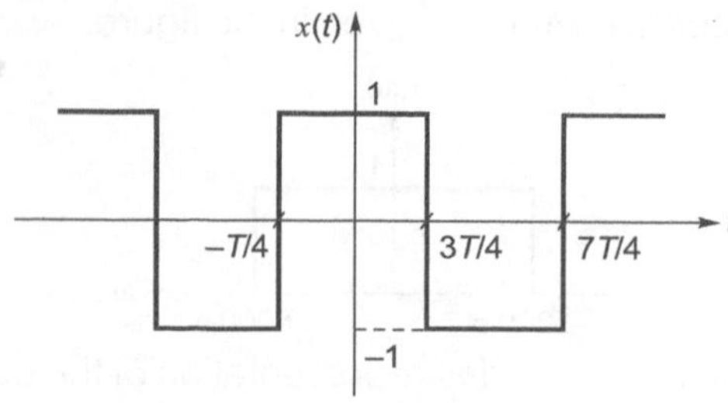

So it is periodic with period,

$$T_0 = \frac{7T}{4}-\left(-\frac{T}{4}\right) = 2T$$

∴ Fundamental angular frequency

$$\omega_0 = \frac{2\pi}{T_0} = \frac{2\pi}{2T} = \frac{\pi}{T}$$

Now, C_1 = (exponential series coeff. for k = 1)

$$C_1 = \frac{1}{T_0}\int_{T_0} x(t)e^{-j\omega_0 t}\, dt$$

$$C_1 = \frac{1}{T_0}\left[\int_{-T/4}^{3T/4} e^{-j\omega_0 t}dt + \int_{3T/4}^{7T/4} -1e^{-j\omega_0 t}dt\right]$$

$$C_1 = \frac{1}{j2\pi}[2e^{j\pi/4}+2e^{j\pi/4}]$$

$$C_1 = \frac{2}{j\pi}e^{j\pi/4} = \frac{2}{\pi}e^{-j\pi/4} = \frac{2}{\pi}\left[\frac{1}{\sqrt{2}}-\frac{j}{\sqrt{2}}\right]$$

Comparing it with $C_1 = \frac{a_1}{2}-\frac{jb_1}{2}$

$$a_1 = \frac{4}{\sqrt{2}\pi},\ b_1 = \frac{4}{\sqrt{2}\pi}$$

∴ Fundamental Fourier term will be

$$= a_1\cos\omega_0 t + b_1\sin\omega_0 t$$

$$= \frac{4}{\sqrt{2}\pi}\cos\left(\frac{\pi}{T}t\right)+\frac{4}{\sqrt{2}\pi}\sin\left(\frac{\pi}{T}t\right)$$

$$= \frac{4}{\pi}\cos\left[\frac{\pi}{T}t-\frac{\pi}{4}\right]$$

3.10 (b)

$$y(t) = x(t-t_0) + x(t+x_0)$$

Since, $x(t)$ is periodic with period T.
Therefore, $x(t-t_0)$ and $x(t+t_0)$ will also be periodic with period T.

$$\therefore\quad b_k = a_k e^{-jk\omega_0 t_0} + a_k e^{jk\omega_0 t_0} \text{ (By property)}$$

a_k is Fourier series coefficient of signal $x(t)$ therefore,

$$b_k = a_k\left[e^{-jk\omega_0 t_0} + e^{jk\omega_0 t_0}\right] = 2a_k \cos k\omega_0 t_0$$

Since, $b_k = 0$ for odd k

i.e. $k\omega_0 t_0 = k\dfrac{\pi}{2}$ where $k = \pm1, \pm3, \pm5, ...$

$$\Rightarrow \quad \omega_0 t_0 = \frac{\pi}{2} \Rightarrow t_0 = \frac{\pi}{2}\times\frac{T}{2\pi} = \frac{T}{4}$$

3.11 (c)

$$\begin{aligned}x(t) &= a_{-2}e^{-j2\omega_0 t} + a_{-1}e^{-j\omega_0 t} + a_0 \\ &\quad + a_1 e^{j\omega_0 t} + a_2 e^{j2\omega_0 t} \\ &= (2-j)e^{-j2\omega_0 t} + (2+j)e^{j2\omega_0 t} \\ &\quad + (0.5+j0.2)e^{-j\omega_0 t} + (0.5-j0.2)e^{j\omega_0 t} + 2j \\ &= 2(e^{-j2\omega_0 t} + e^{j2\omega_0 t}) + j[e^{j2\omega_0 t} - e^{-j2\omega_0 t}] \\ &\quad + 0.5(e^{j\omega_0 t} + e^{-j\omega_0 t}) - j0.2(e^{j\omega_0 t} - e^{-j\omega_0 t}) + 2j \\ &= 4\cos 2\omega_0 t - 2\sin 2\omega_0 t + \cos\omega_0 t \\ &\quad + 0.4\sin\omega_0 t + 2j \\ &= \text{Real}[x(t)] + j\,\text{Img}[x(t)]\end{aligned}$$

where,

$$\begin{aligned}\text{Real}[x(t)] &= 4\cos 2\omega_0 t - 2\sin 2\omega_0 t + \cos\omega_0 t \\ &\quad + 0.4\sin\omega_0 t \\ &= \text{Neither even nor odd signal} \\ \text{Img}[x(t)] &= 2 = \text{constant}\end{aligned}$$

3.12 (a)

The given signal is odd as well as having half wave symmetry.

So it has only sine terms with odd harmonics. So second harmonic component = 0.

3.13 (d)

Let, $x(t)$ = Even and HWS

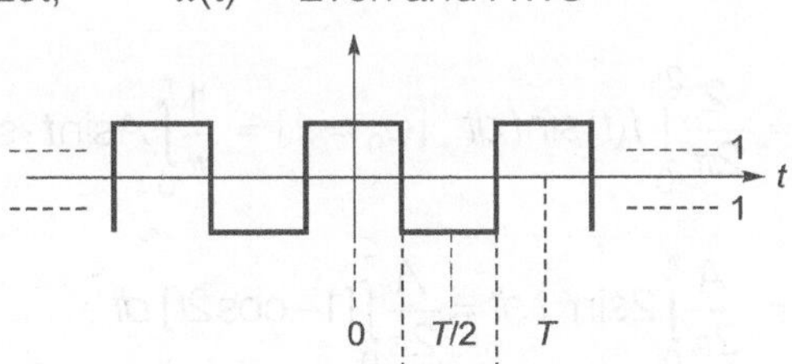

Fourier Series expansion of $x(t)$ contains cos terms with odd harmonics.

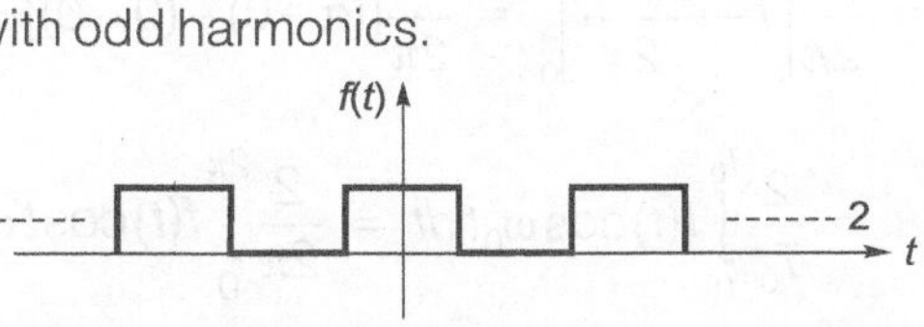

Now, $f(t) = 1 + x(t)$

Fourier Series of $f(t)$ contains dc and cos terms with odd harmonics.

3.14 Sol.

ω_0 = GCD (100, 300, 500) = 100 rad/s

3.15 (0.5)

Given, function $g(x) = x - [x]$ where $[x]$ is a integer part of x. hen function $g(x)$ will be

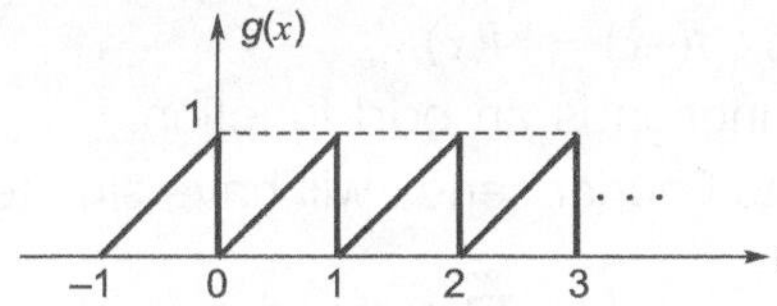

The value of the constant term (or) DC term in the Fourier series expansion of $g(x)$ is

$$a_0 = \frac{1}{T}\int_0^T f(x)\,dx = \frac{\text{Area in one period}}{\text{one period}} = \frac{1}{2}\times1\times1 = 0.5$$

3.16 (b)

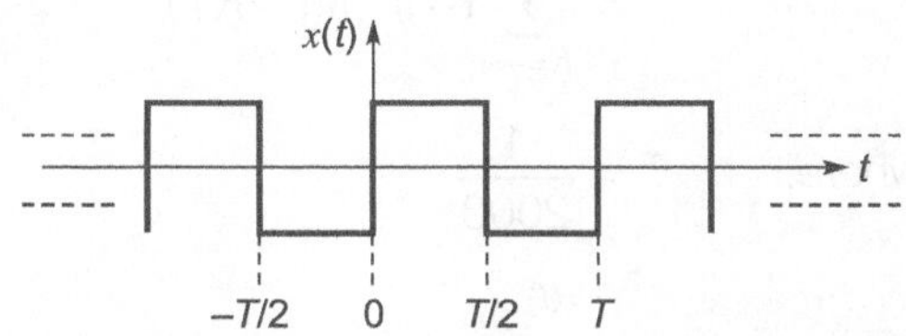

Reconstruction of signal by its Fourier series coefficient is not possible at those points where signal is discontinuous.

In the above figure, at integer multiples of '$T/2$', signal recovery is not possible by using its coefficient.

Therefore, reconstruction of $x(t)$ by using its coefficient is possible at most of the points except those instants where $x(t)$ is discontinuous.

3.17 (d)

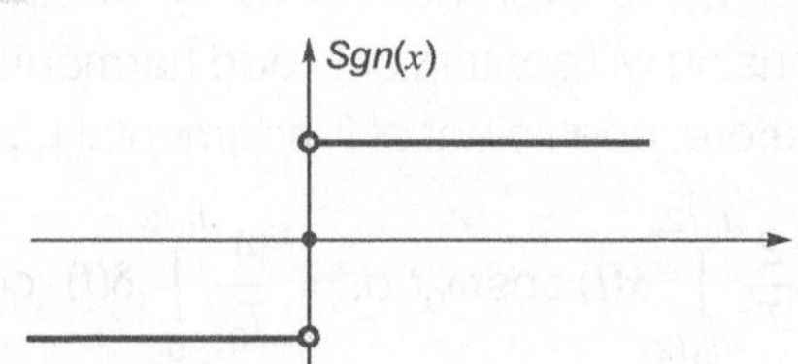

So, cos(t) is

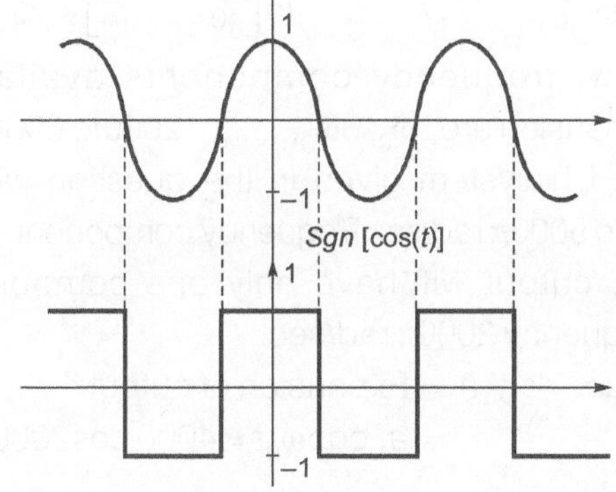

So, *sgn* (cos*t*) is a rectangular signal which is even and has half wave symmetry. So, Fourier series will have only cosine terms with odd harmonics only.

3.18 (b)

Given : $f(-x) = -f(x)$
So, function is an odd function.
So, the Fourier series will have sine terms only.

So, $$f(x) = \sum_{k=1}^{\infty} b_x \sin(kx)$$

3.19 (c)

$$x(t) = \sum_{K=-\infty}^{\infty} (-1)^K \delta\left(t - \frac{K}{2000}\right)$$

$$= \sum_{K=-\infty}^{\infty} (-1)^K \delta(t - KT)$$

where, $T = \dfrac{1}{2000}$

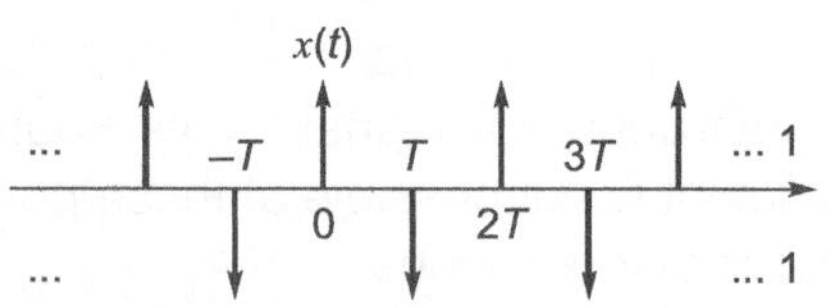

Time period:

$$T_0 = 2T = \frac{1}{1000}$$

$$\omega_0 = \frac{2\pi}{T_0} = 2000\pi \text{ rad}$$

Since, $x(t)$ is even-half wave symmetric. So, its expansion will contain only odd harmonics of cos. Therefore, coefficient of fundamental harmonic is

$$a_1 = \frac{2}{T}\int_{-T_0/2}^{T_0/2} x(t)\cos\omega_0 t\, dt = \frac{4}{T_0}\int_0^{T_0/2} \delta(t)\cdot\cos\omega_0 t\, dt$$

$$= \frac{4}{T_0}\int_0^{T_0/2} \delta(t)\cdot\cos 0\, dt = \frac{4}{T_0}\left[\int_0^{T_0/2}\delta(t)\, dt\right] = \frac{4}{T_0} = 4000$$

Now, frequency components available in expansion are : $\omega_0, 3\omega_0, \ldots\ldots$; $2000\pi, 6000\pi, \ldots\ldots$
As, LTI system given in the question will pass upto 5000π rad/sec frequency component of input. So, output will have only one component of frequency 2000π rad/sec.

Thus, $y(t)$ = Expansion of output
$= a_1 \cos\omega_0 t = 4000 \cos 2000\pi t$

3.20 Sol.

Given that, $$g(t) = \begin{cases} t - \lfloor t \rfloor, & t \geq 0 \\ t - \lceil t \rceil, & \text{otherwise} \end{cases}$$

where,

$\lfloor t \rfloor$ = Greatest integer less than or equal to '*t*'.

$\lceil t \rceil$ = Smallest integer greater than or equal to '*t*'.

Now,

$t \geq 0$			$t < 0$		
$g(t)$	t	$\lfloor t \rfloor$	$g(t)$	t	$\lceil t \rceil$
$t - 0$	0, 1	0	$t + 0$	–1, 0	0
$t - 1$	1, 2	1	$t + 1$	–2, –1	–1
$t - 2$	2, 3	2	$t + 2$	–3, –2	–2
$t - 3$	3, 4	3	$t + 3$	–4, –3	–3
⋮			⋮		

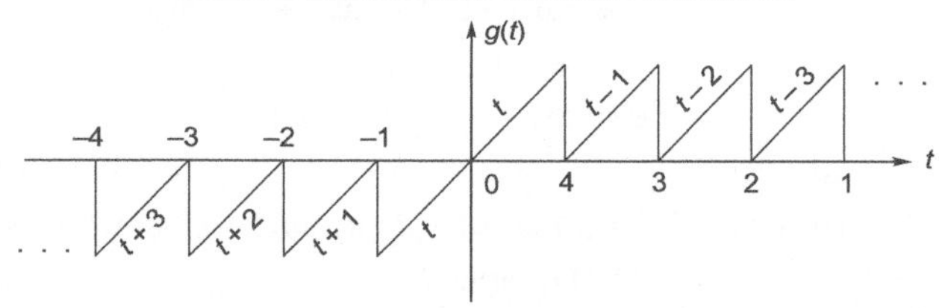

Since, $g(t)$ is non-periodic. So, there is no Fourier series expansion of this signal and hence no need to calculate harmonic here.

3.21 (c)

$$T_0 = 2\pi \Rightarrow \omega_0 = \frac{2\pi}{T_0} = 1$$

Now, $$b_1 = \frac{2}{T_0}\int_0^{T_0} f(t)\sin\omega_0 t\, dt$$

$$= \frac{2}{2\pi}\int_0^{2\pi} f(t)\sin t\, dt \quad [\omega_0 = 1] = \frac{1}{\pi}\int_0^{\pi} A\sin t\cdot\sin t\, dt$$

$$= \frac{A}{2\pi}\int_0^{\pi} 2\sin^2 t\, dt = \frac{A}{2\pi}\int_0^{\pi}[1 - \cos 2t]\, dt$$

$$= \frac{A}{2\pi}\left[t - \frac{\sin 2t}{2}\right]_0^{\pi} = \frac{A}{2\pi}[(\pi - 0) - (0 - 0)] = \frac{A}{2}$$

$$a_1 = \frac{2}{T_0}\int_0^{T_0} f(t)\cos\omega_0 t\, dt = \frac{2}{2\pi}\int_0^{2\pi} f(t)\cos t\, dt$$

$$= \frac{1}{\pi}\int_0^{\pi} A\sin t\cdot\cos t\, dt = \frac{A}{2\pi}\int_0^{\pi}\sin 2t\, dt = 0$$

3.22 (b)

Given : $x[n] = \sum_{K=0}^{N-1} a_K e^{j\left(\frac{2\pi}{N}Kn\right)}$

$$N = 3; \quad a_{-3} = 2; \quad a_K = a_{K+N}$$
$$K = -3; \; a_{-3} = a_{-3+N} = a_{-3+3} = a_0$$
$$a_{-3} = a_0; \; a_4 = 1; \; a_4 = a^*_{-4}$$
$$a_{-4} = 1; \quad K = 4; \quad a_4 = a_{4+3} = a_7$$
$$a_{-4} = a_{-4+3} = a_{-1}; \; a_{-1} = a_1 = 1$$
$$a_1 = a_4 = 1$$

$$x[n] = \sum_{K=0}^{3-1} a_k e^{j\left(\frac{2\pi}{3}kn\right)} = \sum_{K=0}^{2} a_k e^{j\left(\frac{2\pi}{3}kn\right)}$$

$$= a_0 + a_1 e^{j\left(\frac{2\pi}{3}n\right)} + a_2 e^{j\left(\frac{2\pi}{3}2n\right)} = 2 + 1\cdot e^{j\left(\frac{2\pi}{3}n\right)} + 0$$

$$x[n] = 2 + e^{j\left(\frac{2\pi}{3}n\right)} = 1 + e^{j\left(\frac{2\pi}{3}n\right)} + 1$$

$$= 1 + e^{\frac{j2\pi}{3}n} + e^{-\frac{j2\pi}{6}n} \cdot e^{\frac{j2\pi}{6}n}$$

$$= 1 + e^{\frac{j2\pi}{6}n}\left[e^{\frac{j2\pi}{6}n} + e^{-\frac{j2\pi}{6}n}\right]$$

$$= 1 + 2e^{\frac{j2\pi}{6}n}\left[\frac{e^{\frac{j2\pi}{6}n} + e^{\frac{-j2\pi}{6}n}}{2}\right]$$

$$= 1 + 2e^{\frac{j2\pi}{6}n}\cos\left(\frac{2\pi}{6}n\right)$$

■■■■

4 Fourier Transform

4.1 Let, $x(t) = \text{rect}\left(t - \frac{1}{2}\right)$ [where, rect(t) =1 for $-\frac{1}{2} \le x \le \frac{1}{2}$ and zero otherwise]. Then $\text{sinc}(x) = \frac{\sin(\pi x)}{\pi x}$, the Fourier Transform of $x(t) + x(-t)$ will be given by

(a) $\text{sinc}\left(\frac{\omega}{2\pi}\right)$

(b) $2\,\text{sinc}\left(\frac{\omega}{2\pi}\right)$

(c) $2\,\text{sinc}\left(\frac{\omega}{2\pi}\right)\cos\left(\frac{\omega}{2}\right)$

(d) $\text{sinc}\left(\frac{\omega}{2\pi}\right)\sin\left(\frac{\omega}{2}\right)$ **[2008 : 2 M]**

4.2 A signal $x(t) = \text{sinc}(\alpha t)$ where α is a real constant $\left(\text{sinc}\,(x) = \frac{\sin(\pi x)}{\pi x}\right)$ is the input to a Linear time invariant system whose impulse response $h(t) = \text{sinc}(\beta t)$, where β is a real constant. If min (α, β) denotes the minimum of α and β and similarly max (α, β) denotes the maximum of α and β, and K is a constant, which one of the following statements is true about the output of the system?

(a) It will be of the form $K\,\text{sinc}(\gamma t)$ where $\gamma = \min(\alpha, \beta)$.

(b) It will be of the $K\,\text{sinc}(\gamma t)$ where $\gamma = \max(\alpha, \beta)$.

(c) It will be of the form $K\,\text{sinc}(\alpha t)$.

(d) It can't be a sinc type of signal. **[2008 : 2 M]**

4.3 A continuous time LTI system with system function $H(\omega)$ has the following pole-zero plot. For this system, which of the alternatives is TRUE?

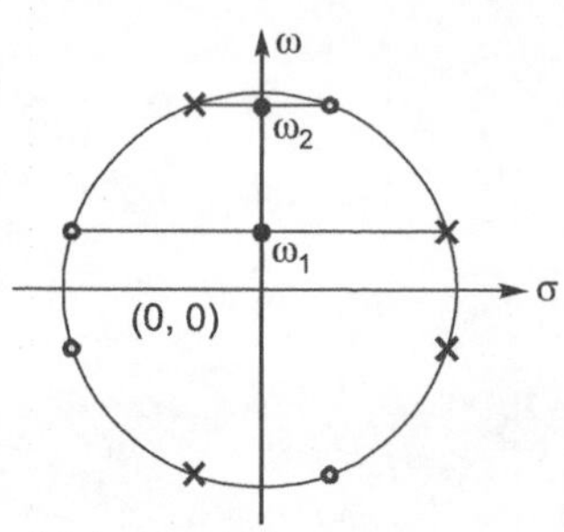

(a) $|H(0)| > |H(\omega)|; |\omega| > 0$

(a) $|H(\omega)|$ has multiple maxima, at ω_1 and ω_2

(c) $|H(0)| < |H(\omega)|; |\omega| > 0$

(d) $|H(\omega)|$ = constant; $-\infty < \omega < \infty$

[2014 : 2 M, Set-3]

4.4 $x(t)$ is a positive rectangular pulse from $t = -1$ to $t = +1$ with unit height as shown in the figure. The value of $\int_{-\infty}^{\infty} |X(\omega)|^2 d\omega$ {where $X(\omega)$ is the Fourier transform of $x(t)$} is

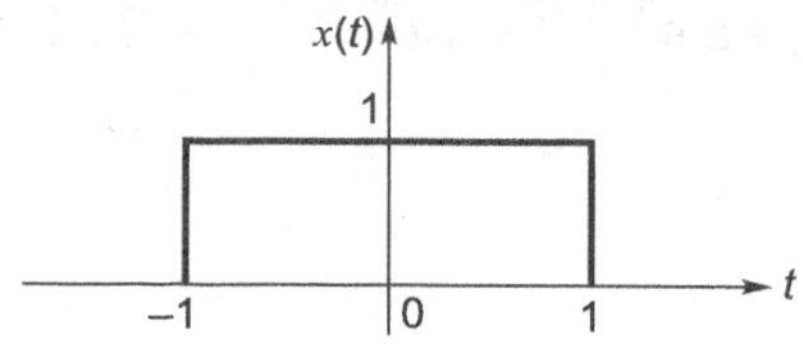

(a) 2 (b) 2π

(c) 4 (d) 4π **[2010 : 2 M]**

4.5 The Fourier transform of a signal $h(t)$ is $H(j\omega) = (2\cos\omega)(\sin 2\omega)/\omega$. The value of $h(0)$ is

(a) $\frac{1}{4}$ (b) $\frac{1}{2}$

(c) 1 (d) 2 **[2012 : 2 M]**

4.6 Let $f(t)$ be a continuous time signal and let $F(\omega)$ be its Fourier Transform defined by

$$F(\omega) = \int_{-\infty}^{\infty} f(t)e^{-j\omega t}\,dt$$

and $g(t)$ is defined by

$$g(t) = \int_{-\infty}^{\infty} F(u)e^{-jut}\,du$$

What is the relationship between $f(t)$ and $g(t)$?

(a) $g(t)$ would always be proportional to $f(t)$.

(b) $g(t)$ would be proportional to $f(t)$ if $f(t)$ is an even function.

(c) $g(t)$ would be proportional to $f(t)$ only if $f(t)$ is a sinusoidal function.

(d) $g(t)$ would never be proportional to $f(t)$.

[2014 : 2 M, Set-1]

4.7 A function $f(t)$ is shown in the figure.

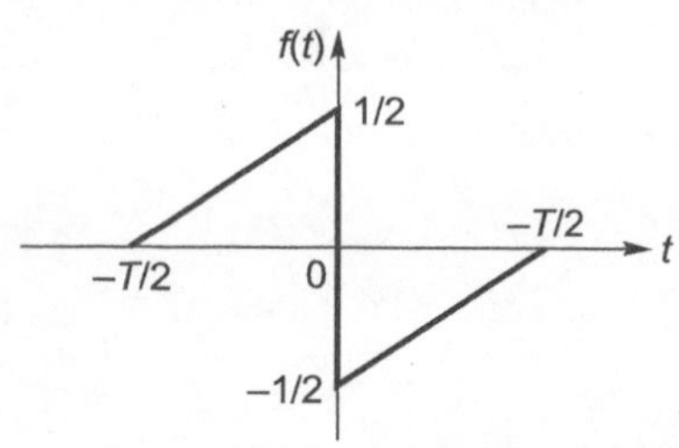

The Fourier transform $F(\omega)$ of $f(t)$ is
(a) real and even function of ω
(b) real and odd function of ω
(c) imaginary and odd function of ω
(d) imaginary and even function of ω
[2014 : 1 M, Set-3]

4.8 A signal is represented by

$$x(t) = \begin{cases} 1 & |t| < 1 \\ 0 & |t| > 1 \end{cases}$$

The Fourier transform of the convolved signal $y(t) = x(2t) * x\left(\frac{t}{2}\right)$ is

(a) $\frac{4}{\omega^2}\sin\left(\frac{\omega}{2}\right)\sin(2\omega)$ (b) $\frac{4}{\omega^2}\sin\left(\frac{\omega}{2}\right)$

(c) $\frac{4}{\omega^2}\sin(2\omega)$ (d) $\frac{4}{\omega^2}\sin^2\omega$

[2014 : 1 M, Set-3]

4.9 A differentiable non constant real even function $x(t)$ has a derivative $y(t)$, and their respective Fourier Transforms are $X(\omega)$ and $Y(\omega)$. Which of the following statements is TRUE?
(a) $X(\omega)$ and $Y(\omega)$ are both real
(b) $X(\omega)$ is real and $Y(\omega)$ is imaginary
(c) $X(\omega)$ and $Y(\omega)$ are both imaginary
(d) $X(\omega)$ is imaginary and $Y(\omega)$ is real
[2014 : 2 M, Set-3]

4.10 Consider a signal defined by

$$x(t) = \begin{cases} e^{j10t} & \text{for } |t| \le 1 \\ 0 & \text{for } |t| > 1 \end{cases}$$

Its Fourier Transform is

(a) $\frac{2\sin(\omega-10)}{\omega-10}$ (b) $2e^{j10}\frac{\sin(\omega-10)}{\omega-10}$

(c) $\frac{2\sin\omega}{\omega-10}$ (d) $e^{j10\omega}\frac{2\sin\omega}{\omega}$

[2015 : 2 M, Set-2]

4.11 Suppose $x_1(t)$ and $x_2(t)$ have the Fourier transforms as shown below.

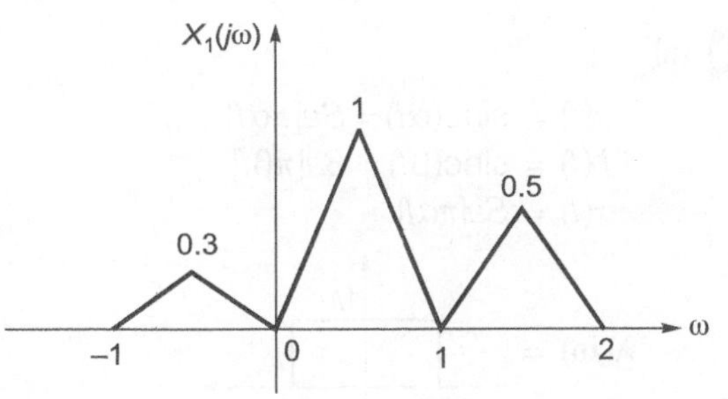

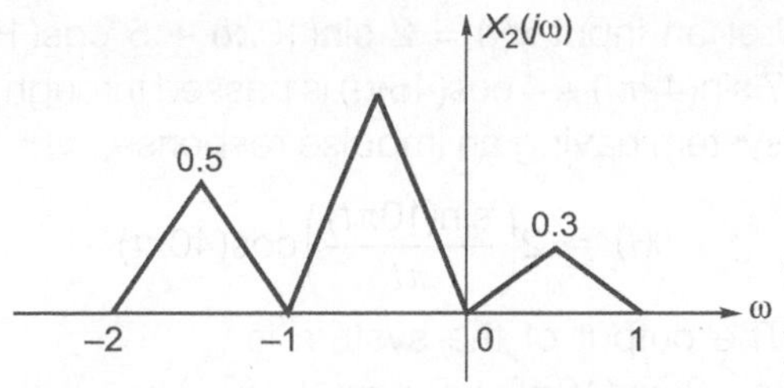

Which one of the following statements is TRUE?
(a) $x_1(t)$ and $x_2(t)$ are complex and $x_1(t)\,x_2(t)$ is also complex with non zero imaginary part.
(b) $x_1(t)$ and $x_2(t)$ are real and $x_1(t)\,x_2(t)$ is also real.
(c) $x_1(t)$ and $x_2(t)$ are complex but $x_1(t)\,x_2(t)$ is real.
(d) $x_1(t)$ and $x_2(t)$ are imaginary but $x_1(t)\,x_2(t)$ is real.
[2016 : 2 M, Set-1]

4.12 Suppose the maximum frequency in a band-limited signal $x(t)$ is 5 kHz. Then, the maximum frequency in $x(t)\cos(2000\pi t)$, in kHz, is ____.
[2016 : 1 M, Set-2]

4.13 The value of the integral $2\int_{-\infty}^{\infty}\left(\frac{\sin 2\pi t}{\pi t}\right)dt$ is equal to
(a) 0 (b) 0.5
(c) 1 (d) 2 [2016 : 2 M, Set-2]

4.14 The Fourier transform of a continuous time signal $x(t)$ is given by

$$X(\omega) = \frac{1}{(10+j\omega)^2}\cdot -\infty < \omega < \infty,$$

where $j = \sqrt{-1}$ and ω denotes frequency. Then the value of $|\ln x(t)|$ at $t = 1$ is ___ (upto 1 decimal place). (ln denotes the logarithm to base e).
[2018 : 2 M]

4.15 Lef $f(t)$ be an even function, i.e. $f(-t) = f(t)$ for all t. Let the Fourier transform of $f(t)$ be defined as $F(\omega) = \int_{-\infty}^{\infty} f(t)e^{-j\omega t}dt$. Suppose $\frac{dF(\omega)}{d\omega} = -\omega F(\omega)$ for all ω, and $F(0) = 1$. Then
(a) $f(0) = 0$ (b) $f(0) = 1$
(c) $f(0) > 1$ (d) $f(0) < 1$ [2021 : 2 M]

4.16 Consider a continuous-time signal $x(t)$ defined by $x(t) = 0$ for $|t| > 1$, and $x(t) = 1-|t|$ for $|t| \le 1$. Let the Fourier transform of $x(t)$ be defined as $X(\omega) = \int_{-\infty}^{\infty} x(t)e^{-j\omega t}dt$. The maximum magnitude of $X(\omega)$ is ______.
[2021 : 2 M]

4.17 Let an input $x(t) = 2\sin(10\pi t) + 5\cos(15\pi t) + 7\sin(42\pi t) + 4\cos(45\pi t)$ is passed through an LTI system having an impulse response,

$$h(t) = 2\left(\frac{\sin(10\pi t)}{\pi t}\right)\cos(40\pi t)$$

The output of the system is

(a) $2\sin(10\pi t) + 5\cos(15\pi t)$
(b) $5\cos(15\pi t) + 7\sin(42\pi t)$
(c) $7\sin(42\pi t) + 4\cos(45\pi t)$
(d) $2\sin(10\pi t) + 4\cos(45\pi t)$ [2022 : 2 M]

4.18 The Fourier transform $X(\omega)$ of the signal $x(t)$ is given by

$$X(\omega) = 1, \quad \text{for } |\omega| < W_o$$
$$= 0, \quad \text{for } |\omega| > W_o$$

Which one of the following statements is true?

(a) $x(t)$ tends to be an impulse as $W_0 \to \infty$.
(b) $x(0)$ decreases as W_0 increases.
(c) At $t = \frac{\pi}{2W_0}, x(t) = -\frac{1}{\pi}$
(d) At $t = \frac{\pi}{2W_0}, x(t) = \frac{1}{\pi}$ [2023 : 1 M]

4.19 The discrete-time Fourier transform of a signal $x[n]$ is $X(\Omega) = (1 + \cos\Omega)e^{-j\Omega}$. Consider that $x_p[n]$ is a periodic signal of period $N = 5$ such that

$$x_p[n] = x[n], \text{ for } n = 0, 1, 2$$
$$= 0, \text{ for } n = 3, 4$$

Note that $x_p[n] = \sum_{k=0}^{N-1} a_k e^{j\frac{2\pi}{N}kn}$. The magnitude of the Fourier series coefficient a_3 is _______ (Round off to 3 decimal places).

[2023 : 2 M]

4.20 Let $X(\omega)$ be the Fourier transform of the signal, $x(t) = e^{-t^4}\cos t, -\infty < t < \infty$.
The value of the derivative of $X(\omega)$ at $\omega = 0$ at is ______ (rounded off to 1 decimal place)

[2024 : 1 M]

■■■■

Answers Fourier Transform

4.1 (c)	4.2 (a)	4.3 (d)	4.4 (d)	4.5 (c)	4.6 (b)	4.7 (c)
4.8 (a)	4.9 (b)	4.10 (a)	4.11 (c)	4.12 (6)	4.13 (d)	4.14 (10)
4.15 (d)	4.16 (1)	4.17 (c)	4.18 (a)	4.19 (0.038)	4.20 (0)	

Explanations Fourier Transform

4.1 (c)

$$x(t) = \text{rect}\,[t - 1/2]$$
$$x(t) = 1 \qquad 0 \le t \le 1$$
$$= 0 \qquad \text{otherwise}$$

$$F[x(t)] = \int_{-\infty}^{\infty} x(t)\, e^{-j\omega t} dt$$

$$= \int_0^1 1 e^{-j\omega t}\, dt = \frac{1}{j\omega}(1 - e^{-j\omega})$$

$$= \frac{2}{\omega}\left[\frac{e^{j\omega/2} - e^{-j\omega/2}}{2j}\right] x\, e^{-j\omega/2} = \frac{\sin\omega/2}{\omega/2} \cdot e^{-j\omega/2}$$

$$F[x(t) + x(-t)] \xrightarrow{\text{F.T.}} X(\omega) + X(-\omega)$$

$$= \frac{\sin\omega/2}{\omega/2}\,[e^{-j\omega/2} + e^{j\omega/2}] = \frac{\sin\omega/2}{\omega/2}\, 2\cos\frac{\omega}{2}$$

$$= 2\,\text{sinc}\left(\frac{\omega}{2\pi}\right)\cos\left(\frac{\omega}{2}\right)$$

4.2 (a)

$$x(t) = \text{sinc}(\alpha t) = Sa[\pi\alpha t]$$
$$h(t) = \text{sinc}(\beta t) = Sa[\pi\beta t]$$
$$x(t) = Sa[\pi\alpha t]$$

$\therefore \; X(j\omega) =$

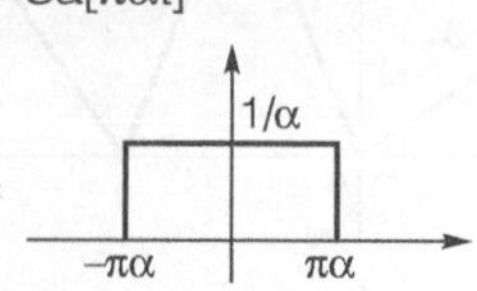

$$h(t) = Sa[\pi\beta t]$$

$\Rightarrow H(j\omega) =$

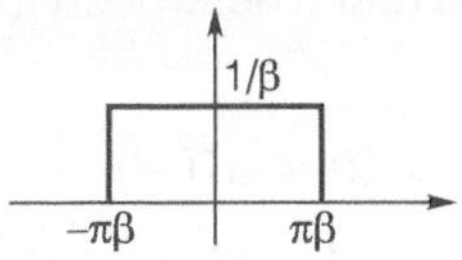

$$y(t) = h(t) * x(t)$$

$$\Rightarrow Y(j\omega) = X(j\omega)\,H(j\omega)$$

$Y(j\omega) =$ if $\alpha < \beta$

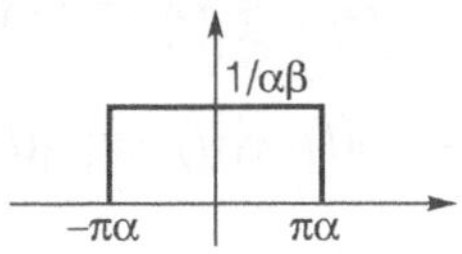

or, $Y(j\omega) =$ if $\beta < \alpha$

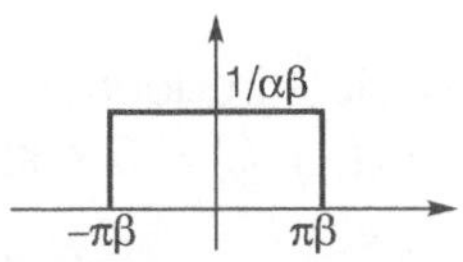

so, $y(t) = \dfrac{1}{\beta}Sa(\pi\alpha t)$ if $\alpha < \beta$

$y(t) = \dfrac{1}{\beta}\text{sinc}(\alpha t)$ if $\alpha < \beta$

or, $y(t) = \dfrac{1}{\alpha}\text{sinc}(\beta t)$ if $\beta < \alpha$

So output is of the form $k\,\text{sinc}(\gamma t)$

where, $\gamma = \min(\alpha, \beta)$

4.3 (d)

$\Rightarrow$ Symmetrically located pole and zero.

$\Rightarrow$ All pass filter.

$\Rightarrow$ (constant magnitude) $(-\infty \le \omega \le \infty)$

4.4 (d)

By using Parseval's theorem,

$$\int_{-\infty}^{\infty}|X(\omega)|^2\,d\omega = 2\pi\int_{-1}^{1}1\,dt = 2\pi \times 2$$

$$\int_{-\infty}^{\infty}|X(\omega)|^2\,d\omega = 4\pi$$

4.5 (c)

$$H(j\omega) = \cos\omega \cdot \frac{2\sin 2\omega}{\omega} = \frac{e^{j\omega}+e^{-j\omega}}{2}\cdot\frac{2\sin 2\omega}{\omega}$$

$$= \frac{1}{2}\left[e^{j\omega}\cdot\left(\frac{2\sin 2\omega}{\omega}\right) + e^{-j\omega}\left(\frac{2\sin 2\omega}{\omega}\right)\right]$$

Let, $X(\omega) = \dfrac{2\sin 2\omega}{\omega}$

Then, $x(t) = \begin{cases} 1 \;;\; -2 < t < 2 \\ 0 \;;\; \text{otherwise} \end{cases}$

By time shifting property

$$h(t) = \frac{1}{2}(x(t+1) + x(t-1))$$

$$\therefore\; h(0) = \frac{1}{2}[x(1) + x(-1)] = \frac{1}{2}[1+1] = 1$$

4.6 (b)

Given that : $f(t) \rightleftharpoons F(\omega)$

and $g(t) = \displaystyle\int_{-\infty}^{\infty} F(u)e^{-jut}\,du$

Now, $f(t) = \dfrac{1}{2\pi}\displaystyle\int_{-\infty}^{\infty} F(\omega)e^{j\omega t}\,d\omega$

$\Rightarrow\; 2\pi f(t) = \displaystyle\int_{-\infty}^{\infty} F(\omega)e^{j\omega t}\,d\omega$

$\omega \to u,\; 2\pi f(t) = \displaystyle\int_{-\infty}^{\infty} F(u)e^{jut}\,du$

$t \to -t,\; 2\pi f(-t) = \displaystyle\int_{-\infty}^{\infty} F(u)e^{-jut}\,du = g(t)$

$\Rightarrow\; g(t) = 2\pi f(-t) = 2\pi f(t)$

[If $f(t)$ is even then, $f(t) = f(-t)$]

$\Rightarrow\; g(t)\;\alpha\; f(t)$ if $f(t)$ is even.

4.7 (c)

Given signal $f(t)$ is an odd signal. Hence, $F(\omega)$ is imaginary and odd function of ω.

4.8 (a)

Given signal can be drawn as

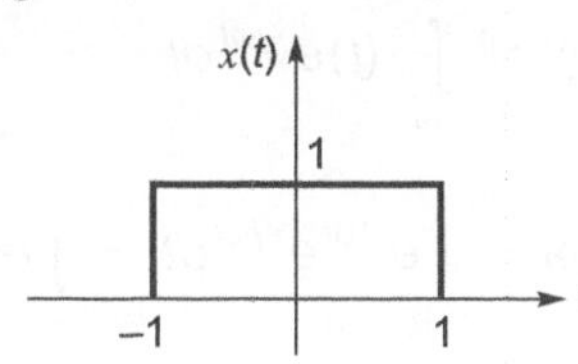

For signal,

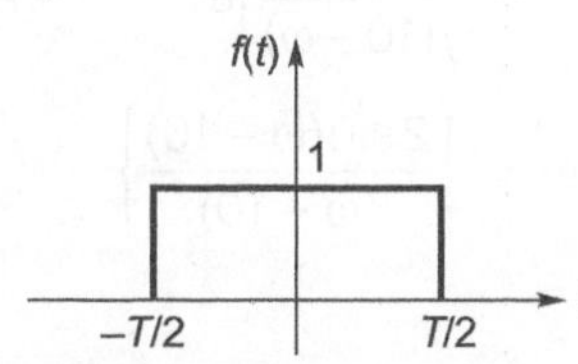

$F(\omega)$ is given by

$$F(\omega) = T\;Sa\left(\omega\left(\frac{T}{2}\right)\right)$$

(T: Area of gate pulse; $\frac{T}{2}$: $\frac{\text{Gate width}}{2}$)

$\therefore\; x(t) \leftrightarrow X(\omega) = 2\,Sa(\omega)$

Now, $x(t) \leftrightarrow X(\omega)$

Then by time scaling,

$$x(at) \longleftrightarrow \frac{1}{|a|}X(\omega/a)$$

$\therefore \quad x(2t) \longleftrightarrow Sa\left(\frac{\omega}{2}\right)$ ...(i)

and $x\left(\frac{t}{2}\right) \longleftrightarrow 4\,Sa(2\omega)$ ...(ii)

Now, $\quad y(t) = x(2t) * x(t/2)$

Convolution in time domain multiplication in frequency domain,

$$Y(\omega) = 4\,Sa\left(\frac{\omega}{2}\right) Sa(2\omega) = \frac{4\sin\left(\frac{\omega}{2}\right)}{\frac{\omega}{2}} \frac{\sin(2\omega)}{2\omega}$$

$$= \frac{4}{\omega^2}\sin\left(\frac{\omega}{2}\right)\sin(2\omega)$$

4.9 (b)

For real even function $x(t)$ the Fourier transform $X(\omega)$ is always real even. $y(t)$ is a derivative of $x(t)$ which is a real odd function because derivative of even function is an odd function and hence, Fourier transform $Y(\omega)$ is imaginary odd.

4.10 (a)

Since, $x(t) = \begin{cases} e^{j10t} & \text{for } |t| \le 1 \\ 0 & \text{for } |t| > 1 \end{cases}$

as $\quad X(\omega) = \int_{-\infty}^{\infty} x(t)e^{-j\omega t}dt$

so, $\quad X(\omega) = \int_{-1}^{1} e^{j10t}e^{-j\omega t}dt = \int_{-1}^{1} e^{jt(10-\omega)}dt$

$$= \frac{1}{j(10-\omega)}[e^{j(10-\omega)} - e^{-j(10-\omega)}]$$

$$X(\omega) = \left\{\frac{2\sin(\omega-10)}{(\omega-10)}\right\}$$

4.11 (c)

By observing $X_1(j\omega)$ and $X_2(j\omega)$ we can say that they are not conjugate symmetric. Since the fourier transform is not conjugate symmetric the signal will not be real. So $x_1(t), x_2(t)$ are not real. Now the fourier transform of $x_1(t) \cdot x_2(t)$ will be $\frac{1}{2\pi}X_1(j\omega) * X_2(j\omega)$ and by looking at $X_1(j\omega)$ and $X_2(j\omega)$ we can say that $X_1(j\omega) * X_2(j\omega)$ will be conjugate symmetric and thus $x_1(t) \times x_2(t)$ will be real.

By observing $X_1(\omega)$ and $X_2(\omega)$, we can say

$$X_2(\omega) = X_1(-\omega)$$
$$\downarrow \text{IFT}$$
$$x_2(t) = x_1(-t) \qquad \text{...(i)}$$

Now, $X_1(\omega)$ is real. Therefore, $x_1(t)$ will be conjugate symmetric.

i.e., $\quad x_1(t) = x_1^*(-t)$

$$\downarrow t = -t$$

$$x_1(-t) = x_1^*(t)$$

Now, $\quad x_1(t) \cdot x_2(t) = x_1(t) \cdot x_1(-t)$ [From (i)]

$$= x_1(t) \cdot x_1^*(t) = |x_1(t)|^2 = \text{real function}$$

4.12 (6)

Maximum possible frequency of $x(t)$ $(2000\pi t)$

$$= (f_1 + f_2) = 5 \text{ kHz} + 1 \text{ kHz} = 6 \text{ kHz}$$

4.13 (d)

The Fourier transform of

$$\frac{2\sin(t\tau/2)}{t} \longrightarrow 2\pi\,\text{rect}\left(\frac{\omega}{\tau}\right)$$

$$\frac{\sin(2\pi t)}{\pi t} \longrightarrow \text{rect}\left(\frac{\omega}{4\pi}\right)$$

So, $\int_{-\infty}^{\infty} \frac{\sin(2\pi t)}{\pi t} e^{-j\omega t}dt = \text{rect}\left(\frac{\omega}{4\pi}\right)$

Putting $\omega = 0$ in above equation

$$\int_{-\infty}^{\infty} \frac{\sin(2\pi t)}{\pi t}dt = 1; \qquad 2\int_{-\infty}^{\infty} \frac{\sin(2\pi)}{\pi t}dt = 2$$

4.14 (10)

$$x(t) \rightleftharpoons X(j\omega) = \frac{1}{(10+j\omega)^2}$$

By taking inverse Fourier transform,

$$x(t) = te^{-10t}u(t)$$

Now, $\quad x(t)_{t=1} = 1 \times e^{-10} \times 1 = e^{-10}$

Thus, $\quad |\ln\{x(t)\}| = |\ln(e^{-10})| = |-10| = 10$

4.15 (d)

$$f(t) \rightleftharpoons F(\omega) = \int_{-\infty}^{\infty} f(t)e^{-j\omega t}dt$$

The following informations are given about $f(t) \rightleftharpoons F(\omega)$.

(i) $f(t) = f(-t)$

(ii) $F(\omega)|_{\omega=0} = 1$

(iii) $\frac{dF(\omega)}{d\omega} = -\omega F(\omega)$

From (iii), $\frac{dF(\omega)}{d\omega} + \omega F(\omega) = 0$

By solving the above linear differential equations, (by mathematics)

$$\text{In } F(\omega) = -\frac{\omega^2}{2} \Rightarrow F(\omega) = K \cdot e^{-\omega^2/2} \quad ...(iv)$$

Put $\omega = 0$, $F(0) = K \Rightarrow 1 = K$ (from info. (ii)

From (iv), $F(\omega) = e^{-\omega^2/2}$

As we know,

$$e^{-at^2}, \quad a > 0 \rightleftharpoons \sqrt{\frac{\pi}{a}} e^{-\omega^2/4a}$$

$$\text{At } a = \frac{1}{2}, e^{-t^2/2} \rightleftharpoons \sqrt{\frac{\pi}{1/2}} \cdot e^{-\omega^2/2}$$

$$f(t) = \frac{1}{\sqrt{2\pi}} e^{-t^2/2} \rightleftharpoons e^{-\omega^2/2} = F(\omega)$$

$$\text{Thus,} \quad f(t) = \frac{1}{\sqrt{2\pi}} e^{-t^2/2}$$

$$\text{At } t = 0, \quad f(0) = \frac{1}{\sqrt{2\pi}} < 1$$

4.16 (1)

Fourier transform,

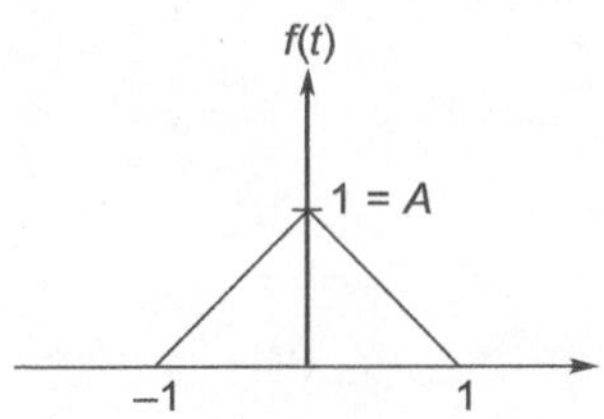

$$F(\omega) = A\tau Sa^2\left(\frac{\omega\tau}{2}\right)$$

As $A = 1$, $\tau = 1$,

$$F(\omega) = Sa^2\left(\frac{\omega}{2}\right)$$

$$F(\omega)\big|_{peak} = F(0) = Sa^2(0) = 1$$

∵ Peak value of sampling function occurs at $\omega = 0$, Peak value = 1

4.17 (c)

Given that :

$$x(t) = 2\sin(10\pi t) + 5\cos(15\pi t) + 7\sin(42\pi t) + 4\cos(45\pi t)$$

$$h(t) = 2\left[\frac{\sin(10\pi t)}{\pi t}\right]\cos(40\pi t)$$

$$\frac{\sin 10\pi t}{\pi t} \rightarrow$$

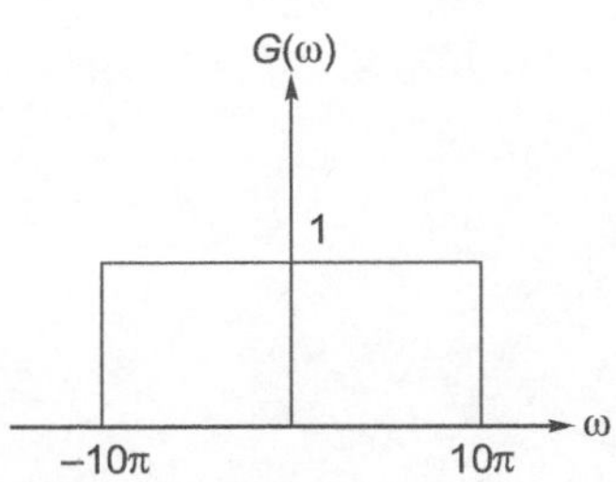

$$h(t) = 2g(t)\cos(40\pi t)$$

Apply Fourier transform

$$H(\omega) = 2\left[\frac{G(\omega - 40\pi) + G(\omega + 40\pi)}{2}\right]$$

$$= G(\omega - 40\omega) + G(\omega + 40\omega)$$

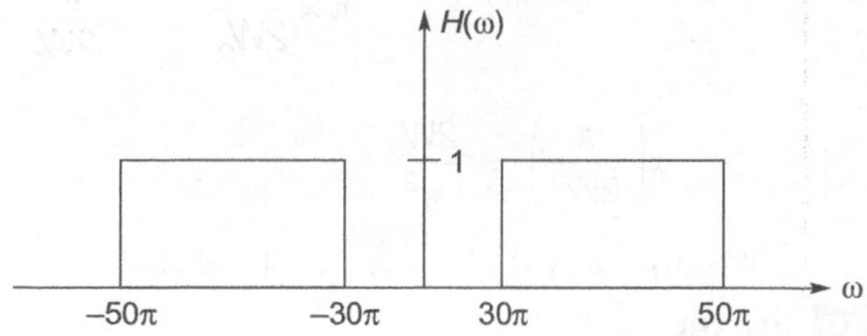

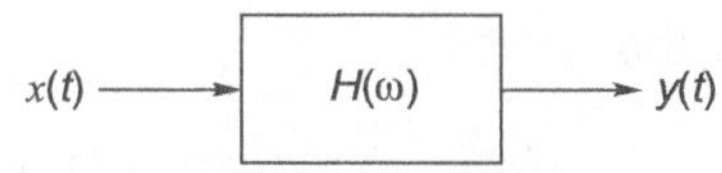

$$2\sin(10\pi t) + 5\cos(15\pi t) + 7\sin(42\pi t) + 4\cos(45\pi t)$$

$\omega_1 = 10\pi \quad \omega_2 = 15\pi \quad \omega_3 = 42\pi \quad \omega_4 = 45\pi$

The output of LTI system is

$$y(t) = 7\sin(42\pi t) + 4\cos(45\pi t)$$

4.18 (a)

We know that :

$A_oS_a(kt) \rightleftharpoons$

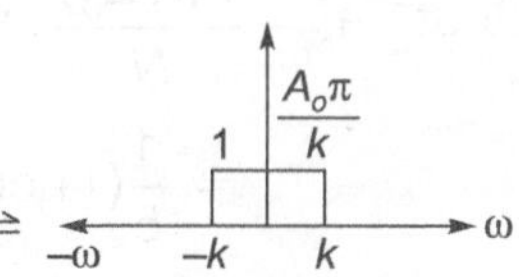

Comparing with :

$x(t) \rightleftharpoons$

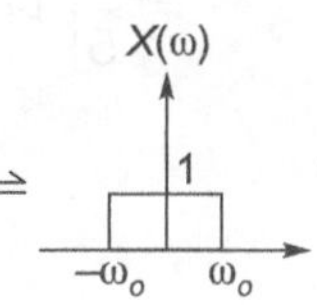

$$k = \omega_o$$

$$\frac{A_o\pi}{k} = 1$$

$$A_o = \frac{k}{\pi} = \frac{W_o}{\pi}$$

$$\therefore x(t) = \frac{W_o}{\pi} Sa(W_o t) = \frac{W_o}{\pi} \cdot \frac{\sin(W_o t)}{W_o t}$$

$$x(t) = \frac{\sin(W_o t)}{\pi t}$$

Case 1 : if $W_o = \infty$

then $X(\omega) = 1$

then $x(t) = \delta(t)$

Thus, option (a) is correct.

Case 2 : $x(0) =$

$$\lim_{t\to 0} \frac{\sin(W_o t)}{\pi t} = \lim_{t\to 0} \frac{W_o\cos(W_o t)}{\pi} = \frac{W_o}{\pi}$$

$$x(0) \propto W_0$$

Thus, option (b) is incorrect.

Case 3 : $x\left(\frac{\pi}{2W_o}\right) = \dfrac{\sin\left(W_o \times \frac{\pi}{2W_o}\right)}{\pi \times \frac{\pi}{2W_o}} = \dfrac{1}{\frac{\pi^2}{2W_o}}$

$$x\left(\frac{\pi}{2W_o}\right) = \frac{2W_o}{\pi^2}$$

Thus, option (c) and (d) are incorrect.

4.19 (0.038)

Given that :

$$x(n) \Longleftrightarrow X(e^{j\Omega}) = (1 + \cos\Omega)e^{-j\Omega}$$

and $x_p(n)$ = Periodic signal $\Longleftrightarrow a_K$ = DFS-coefficient with $N = 5$

where $x_p(n) = \begin{cases} x(n), & \text{for } n = 0,1,2 \\ 0, & \text{for } n = 3, 4 \end{cases}$

Shortcut : $a_K = \dfrac{X(K\Omega_o)}{N}$, where $\Omega_o = \dfrac{2\pi}{N} = \dfrac{2\pi}{5}$

$$= \frac{1}{5}(1+\cos K\Omega_o)\cdot e^{-jK\Omega_o}$$

$$= \frac{1}{5}\left[1+\cos\frac{2\pi}{5}K\right]\cdot e^{-j\frac{2\pi}{5}K}$$

Now, $|a_K| = \dfrac{1}{5}\left[1+\cos\dfrac{2\pi}{5}K\right]$

Put $K = 3$; $|a_3| = \dfrac{1}{5}\left[1+\cos\dfrac{6\pi}{5}\right]$

$$= \frac{1}{5}(1-0.809) = 0.038$$

4.20 (0)

$$tx(t) \rightleftarrows j\frac{dX(\omega)}{d\omega}$$

$$f(t) = \frac{tx(t)}{j} \rightleftarrows \frac{dX(\omega)}{d\omega} = F(\omega)$$

$$\left.\frac{dX(\omega)}{d\omega}\right|_{\omega=0} = F(\omega)\Big|_{\omega=0} = \text{area of } f(t)$$

$$= \int_{-\infty}^{\infty} \frac{tx(t)}{j}dt$$

= area of odd-function = 0

$$\begin{bmatrix} \because x(t) = \text{even-function} \\ \therefore \frac{t}{j}x(t) = \text{odd-function} \end{bmatrix}$$

■■■■

5 Laplace Transform

5.1 The Laplace transformation of $f(t)$ is $F(s)$.

Given: $F(s) = \dfrac{\omega}{s^2 + \omega^2}$, the final value of $f(t)$ is

(a) infinity (b) zero
(c) one (d) None of these

[1995 : 1 M]

5.2 The Laplace transform of $(t^2 - 2t)\,u(t-1)$ is

(a) $\dfrac{2}{s^3}e^{-s} - \dfrac{2}{s^2}e^{-s}$ (b) $\dfrac{2}{s^3}e^{-2s} - \dfrac{2}{s^2}e^{-s}$

(c) $\dfrac{2}{s^3}e^{-s} - \dfrac{1}{s}e^{-s}$ (d) none of the above

[1998 : 2 M]

5.3 The output of a linear time invariant control system is $c(t)$ for a certain input $r(t)$. If $r(t)$ is modified by passing it through a block whose transfer function is e^{-s} and then applied to the system, the modified output of the system would be

(a) $\dfrac{r(t)}{1+e^{t}}$ (b) $\dfrac{r(t)}{1+e^{-t}}$

(c) $c(t-1)$ (d) $r(t)\,u(t-1)$

[1998 : 1 M]

5.4 A rectangular current pulse of duration T and magnitude 1 has the Laplace transform

(a) $\dfrac{1}{s}$ (b) $\left(\dfrac{1}{s}\right)\exp(-Ts)$

(c) $\left(\dfrac{1}{s}\right)\exp(Ts)$ (d) $\left(\dfrac{1}{s}\right)[1-\exp(-Ts)]$

[1999 : 2 M]

5.5 Given the relationship between the input $u(t)$ and the output $y(t)$ to be

$$y(t) = \int_0^t (2+t-\tau)\,e^{-3(t-\tau)}\,u(\tau)\,d\tau$$

the transfer function $Y(s)/U(s)$ is

(a) $\dfrac{2e^{-2s}}{s+3}$ (b) $\dfrac{s+2}{(s+3)^2}$

(c) $\dfrac{2s+5}{s+3}$ (d) $\dfrac{2s+7}{(s+3)^2}$ [2001 : 2 M]

5.6 Let $Y(s)$ be the Laplace transformation of the function $y(t)$, then the final value of the function is

(a) $\lim_{s\to 0} Y(s)$ (b) $\lim_{s\to \infty} Y(s)$

(c) $\lim_{s\to 0} sY(s)$ (d) $\lim_{s\to \infty} sY(s)$

[2002 : 1 M]

5.7 Consider the function, $F(s) = \dfrac{5}{s(s^2+3s+2)}$ where $F(s)$ is the Laplace transform of the function $f(t)$. The initial value of $f(t)$ is equal to

(a) 5 (b) $\dfrac{5}{2}$

(c) $\dfrac{5}{3}$ (d) 0 [2004 : 1 M]

5.8 The Laplace transform of a function $f(t)$ is

$$F(s) = \frac{5s^2 + 23s + 6}{s(s^2 + 2s + 2)}$$

As $t \to \infty$, $f(t)$ approaches

(a) 3 (b) 5

(c) $\dfrac{17}{2}$ (d) ∞ [2005 : 2 M]

5.9 The integrator, given by $y(t) = \int_{-\infty}^{t} x(\alpha)\,d\alpha$

(a) has no finite singularities in its double sided Laplace Transform $H(s)$.
(b) produces a bounded output for every causal bounded input.
(c) produces a bounded output for every anticausal bounded input.
(d) has no finite zeroes in its double sided Laplace Transform $H(s)$. [2006 : 2 M]

Common Data Questions (5.10 and 5.11):

Given: $f(t)$ and $g(t)$ as shown below:

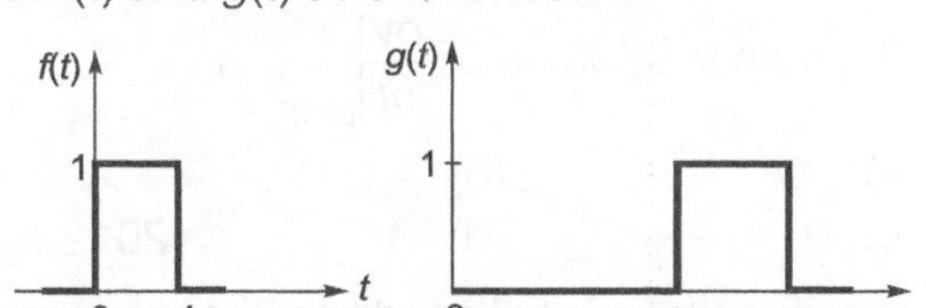

5.10 $g(t)$ can be expressed as:

(a) $g(t) = f(2t-3)$ (b) $g(t) = f\left(\dfrac{t}{2} - 3\right)$

(c) $g(t) = f\left(2t - \frac{3}{2}\right)$ (d) $g(t) = f\left(\frac{t}{2} - \frac{3}{2}\right)$

[2010 : 2 M]

5.11 The Laplace transform of $g(t)$ is

(a) $\frac{1}{s}(e^{3s} - e^{5s})$ (b) $\frac{1}{s}(e^{-5s} - e^{-3s})$

(c) $\frac{e^{-3s}}{s}(1 - e^{-2s})$ (d) $\frac{1}{s}(e^{5s} - e^{3s})$

[2010 : 2 M]

5.12 Let the Laplace transform of a function $f(t)$ which exists for $t > 0$ be $F_1(s)$ and the Laplace transform of its delayed version $f(t - \tau)$ be $F_2(s)$. $F_1^*(s)$ be the complex conjugate of $F_1(s)$ with the Laplace variable set as $s = \sigma + j\omega$. If $G(s) = \frac{F_2(s)\,F_1^*(s)}{|F_1(s)|^2}$, then the inverse Laplace transform of $G(s)$ is

(a) an ideal impulse $\delta(t)$
(b) an ideal delayed impulse $\delta(t - \tau)$
(c) an ideal step function $u(t)$
(d) an ideal delayed step function $u(t - \tau)$

[2011 : 2 M]

5.13 The unilateral Laplace transform of $f(t)$ is $\frac{1}{s^2 + s + 1}$. The unilateral Laplace transform of $t\, f(t)$ is

(a) $-\frac{s}{(s^2 + s + 1)^2}$ (b) $-\frac{2s + 1}{(s^2 + s + 1)^2}$

(c) $\frac{s}{(s^2 + s + 1)^2}$ (d) $\frac{2s + 1}{(s^2 + s + 1)^2}$

[2012 : 1 M]

5.14 Consider the differential equation:

$$\frac{d^2 y(t)}{dt^2} + 2\frac{dy(t)}{dt} + y(t) = \delta(t) \text{ with}$$

$$y(t)\Big|_{t = 0^-} = -2 \text{ and } \frac{dy}{dt}\Big|_{t = 0^-} = 0$$

The numerical value of $\frac{dy}{dt}\Big|_{t = 0^+}$ is

(a) –2 (b) –1
(c) 0 (d) 1 [2012 : 2 M]

5.15 Assuming zero initial condition, the response $y(t)$ of the system given below to a unit step input $u(t)$ is

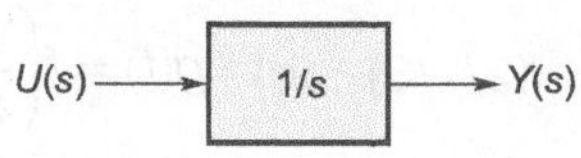

(a) $u(t)$ (b) $t\, u(t)$

(c) $\frac{t^2}{2}u(t)$ (d) $e^{-t}u(t)$ [2013 : 1 M]

5.16 Which one of the following statements is NOT TRUE for a continuous time causal and stable LTI system?

(a) All the poles of the system must lie on the left side of the $j\omega$-axis.
(b) Zeros of the system can lie anywhere in the s-plane.
(c) All the poles must lie within $|s| = 1$.
(d) All the roots of the characteristic equation must be located on the left side of the $j\omega$ -axis.

[2013 : 1 M]

5.17 Consider an LTI system with transfer function

$$H(s) = \frac{1}{s(s + 4)}$$

If the input to the system is $\cos(3t)$ and the steady state output is $A \sin(3t + \alpha)$, then the value of A is

(a) $\frac{1}{30}$ (b) $\frac{1}{15}$

(c) $\frac{3}{4}$ (d) $\frac{4}{3}$ [2014 : 1 M, Set-2]

5.18 The Laplace transform of $f(t) = 2\sqrt{t/\pi}$ is $s^{-3/2}$. The Laplace transform of $g(t) = \sqrt{1/\pi t}$ is

(a) $\frac{3s^{-5/2}}{2}$ (b) $s^{-1/2}$

(c) $s^{1/2}$ (d) $s^{3/2}$

[2015 : 1 M, Set-2]

5.19 The Laplace Transform of $f(t) = e^{2t}\sin(5t)\,u(t)$ is

(a) $\frac{5}{s^2 - 4s + 29}$ (b) $\frac{5}{s^2 + 5}$

(c) $\frac{s - 2}{s^2 - 4s + 29}$ (d) $\frac{5}{s + 5}$

[2016 : 1 M, Set-1]

5.20 The transfer function of a system is $\frac{Y(s)}{R(s)} = \frac{s}{s + 2}$. The steady state output $y(t)$ is $A\cos(2t + \phi)$ for the input $\cos(2t)$. The values of A and ϕ, respectively are?

(a) $\frac{1}{\sqrt{2}}, -45°$ (b) $\frac{1}{\sqrt{2}}, +45°$

(c) $\sqrt{2}, -45°$ (d) $\sqrt{2}, +45°$

[2016 : 1 M, Set-1]

5.21 Consider a causal LTI system characterized by differential equation $\frac{dy(t)}{dt}+\frac{1}{6}y(t)=3x(t)$. The response of the system to the input $x(t) = 3e^{-\frac{t}{3}}u(t)$, where $u(t)$ denotes the unit step function, is

(a) $9e^{-\frac{t}{3}}u(t)$

(b) $9e^{-\frac{t}{6}}u(t)$

(c) $9e^{-\frac{t}{3}}u(t)-6e^{-\frac{t}{6}}u(t)$

(d) $54e^{-\frac{t}{6}}u(t)-54e^{-\frac{t}{3}}u(t)$ **[2016 : 1 M, Set-2]**

5.22 The solution of the differential equation, for $t > 0$, $y''(t) + 2y'(t) + y(t) = 0$ with initial conditions $y(0) = 0$ and $y'(0) = 1$, is [$u(t)$ denotes the unit step function],

(a) $te^{-t}u(t)$ (b) $(e^{-t} - te^{-t})u(t)$
(c) $(-e^{-t} + te^{-t})u(t)$ (d) $e^{-t}u(t)$

[2016 : 1 M, Set-2]

5.23 Consider a linear time-invariant system with transfer function $H(s)=\frac{1}{(s+1)}$. If the input is $\cos(t)$ and the steady state output is $A\cos(t+\alpha)$, then the value of A is_______. **[2016 : 1 M, Set-2]**

5.24 The output response of a system is denoted as $y(t)$, and its Laplace transform is given by

$$Y(s)=\frac{10}{s(s^2+s+100\sqrt{2})}$$

The steady state value of $y(t)$ as

(a) $100\sqrt{2}$ (b) $\frac{1}{10\sqrt{2}}$

(c) $10\sqrt{2}$ (d) $\frac{1}{100\sqrt{2}}$ **[2019 : 1 M]**

5.25 A system transfer function is

$$H(s)=\frac{a_1s^2+b_1s+c_1}{a_2s^2+b_2s+c_2}$$

If $a_1 = b_1 = 0$, and all other coefficients are positive, the transfer function represents a

(a) high pass filter (b) low pass filter
(c) notch filter (d) band pass filter

[2019 : 1 M]

5.26 The inverse Laplace transform of $H(s)=\frac{s+3}{s^2+2s+1}$ for $t \geq 0$ is

(a) $2t\,e^{-t} + e^{-t}$ (b) $3e^{-t}$
(c) $3t\,e^{-t} + e^{-t}$ (d) $4t\,e^{-t} + e^{-t}$

[2019 : 1 M]

5.27 Consider a linear time-invariant system whose input $r(t)$ and output $y(t)$ are related by the following differential equation,

$$\frac{d^2y(t)}{dt^2}+4y(t)=6\,r(t)$$

The poles of this system are at

(a) $+4j, -4j$ (b) $+4, -4$
(c) $+2, -2$ (d) $+2j, -2j$ **[2020 : 1 M]**

5.28 Which of the following statements is true about the two sided Laplace transform?

(a) It has no poles for any bounded signal that is non-zero only inside a finite time interval.
(b) If a signal can be expressed as a weighted sum of shifted one sided exponentials, then its Laplace Transform will have no poles.
(c) It exists for every signal that may or may not have a Fourier transform.
(d) The number of finite poles and finite zeroes must be equal. **[2020 : 1 M]**

5.29 The causal realization of a system transfer function $H(s)$ having poles at (2, –1), (–2, 1) and zeroes at (2, 1), (–2, –1) will be

(a) unstable, complex, all pass
(b) unstable, real, high pass
(c) stable, complex, low pass
(d) stable, real, all pass **[2020 : 2 M]**

5.30 A continuous-time system that is initially at rest is described by

$$\frac{dy(t)}{dt}+3y(t)=2x(t)$$

where $x(t)$ is the input voltage and $y(t)$ is the output voltage. The impulse response of the system is

(a) $3e^{-2t}$ (b) $\frac{1}{3}e^{-2t}u(t)$

(c) $2e^{-3t}u(t)$ (d) $2e^{-3t}$ **[2023 : 1 M]**

5.31 If $u(t)$ is the unit step function, then the region of convergence (ROC) of the Laplace transform of the signal

$$x(t) = e^{t^2}[u(t-1) - u(t-10)]$$

is

(a) Re $(s) \geq 10$ (b) $-\infty <$ Re $(s) < \infty$
(c) Re $(s) \leq 1$ (b) $1 \leq$ Re $(s) \leq 10$

[2024 : 1 M]

Answers Laplace Transform

5.1 (d)	5.2 (c)	5.3 (c)	5.4 (d)	5.5 (d)	5.6 (c)	5.7 (d)
5.8 (a)	5.9 (d)	5.10 (d)	5.11 (c)	5.12 (b)	5.13 (d)	5.14 (d)
5.15 (b)	5.16 (c)	5.17 (b)	5.18 (b)	5.19 (a)	5.20 (b)	5.21 (d)
5.22 (a)	5.23 (0.707)	5.24 (b)	5.25 (b)	5.26 (a)	5.27 (d)	5.28 (a)
5.29 (a)	5.30 (c)	5.31 (b)				

Explanations Laplace Transform

5.1 (d)

$$\because \quad F(s) = \frac{\omega}{s^2+\omega^2}$$

$\Rightarrow$ $f(t) = \sin(\omega t)\, u(t)$ which is a periodic signal so final value theorem is not applicable on it.

Final value of $f(t)$ may be any value between –1 and +1.

5.2 (c)

$$\mathcal{L}(t^2 - 2t)\, u(t-1)$$

$$= \mathcal{L}[(t-1)^2 u(t-1) - u(t-1)] = \frac{2e^{-s}}{s^3} - \frac{e^{-s}}{s}$$

5.3 (c)

$$r(t) \xrightarrow{\text{LTI system}} c(t)$$

$$r(t) \xrightarrow[e^{-s}]{\text{Unit delay block}} r(t-1) \xrightarrow{\text{LTI system}} c(t-1)$$

5.4 (d)

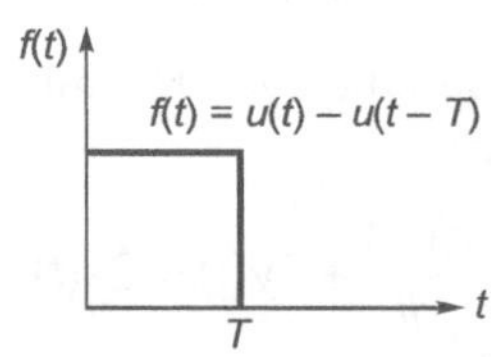

$$\mathcal{L}[u(t) - u(t-T)] = \frac{1}{s} - \frac{e^{-Ts}}{s} = \frac{1-e^{-Ts}}{s}$$

5.5 (d)

$$y(t) = \int_0^t (2+t-\tau)e^{-3(t-\tau)}\, u(\tau)\, d\tau$$

$$y(t) = [(2+t)\, e^{-3t}] * u(t)$$

$$\therefore \quad H(s) = \frac{2}{s+3} + \frac{1}{(s+3)^2}$$

$$\therefore \quad \text{T.F.} = H(s) = \frac{2s+7}{(s+3)^2}$$

5.6 (c)

Final value $= \lim\limits_{s\to 0} sY(s)$

5.7 (d)

$$F(s) = \frac{5}{s(s^2+3s+2)}$$

By initial value theorem,

$$\lim_{t\to 0} f(t) = \lim_{s\to\infty} sF(t) = \lim_{s\to\infty} \cdot \frac{5\times s}{s(s^2+3s+2)} = 0$$

5.8 (a)

$$F(s) = \frac{5s^2+23s+6}{s(s^2+2s+2)}$$

As $F(s)$ has only left hand poles.

By final value theorem,

$$\lim_{t\to\infty} F(t) = \lim_{s\to 0} sF(s) = \frac{6}{2} = 3$$

5.10 (d)

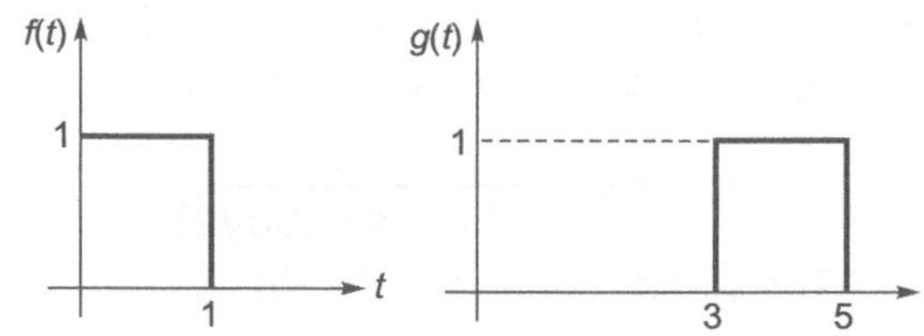

Since, $g(t)$ has width of 2-unit and $f(t)$ has 1 unit therefore we have to first expand $f(t)$ by 2 unit and for this we have to a scale $f(t)$ by 1/2.

i.e., $\quad f_1(t) = f\left(\frac{1}{2}t\right)$

Now shift it by three unit to get $g(t)$,

$$g(t) = f_1(t-3)$$

$$g(t) = f\left(\frac{t-3}{2}\right) = f\left(\frac{t}{2} - \frac{3}{2}\right)$$

5.11 (c)

$$g(t) = \begin{cases} 1 & 3 < t < 5 \\ 0 & t < 3, t > 5 \end{cases}$$

$$\Rightarrow \quad g(t) = u(t-3) - u(t-5)$$

$$L\{g(t)\} = L\{u(t-3)\} - L\{u(t-5)\}$$

$$= \frac{1}{s}e^{-3s} - \frac{1}{s}e^{-5s} = \frac{1}{s}(e^{-3s} - e^{-5s})$$

$$= \frac{e^{-3s}}{s}(1-e^{-2s})$$

5.12 (b)

$$F_2(t) = L\{f(t-\tau)\} = e^{-s\tau} F_1(s)$$

$$G(s) = \frac{e^{-s\tau} F_1(s) F_1^*(s)}{|F_1(s)|^2} = e^{-s\tau}$$

$$G(t) = \delta(t-\tau)$$

5.13 (d)

If $\quad L[f(t)] = \bar{f}(s)$

then, $\quad L[t^n f(t)] = (-1)^n \frac{d^n}{ds^n}\bar{f}(s)$

In this problem,

Given, $\quad L[f(t)] = \frac{1}{s^2+s+1} = \bar{f}(s)$

We need,

$$L[t\,f(t)] = (-1)^1 \frac{d^1}{ds^1}\bar{f}(s) = -\frac{d}{ds}\bar{f}(s)$$

$$= -\frac{d}{ds}\left[\frac{1}{s^2+s+1}\right] = -\left[\frac{-1}{(s^2+s+1)^2}\right] \times (2s+1)$$

$$= \left[\frac{2s+1}{(s^2+s+1)^2}\right]$$

5.14 (d)

$$\frac{d^2y(t)}{dt^2} + 2\frac{dy(t)}{dt} + y(t) = \delta(t)$$

$$\Rightarrow s^2Y(s) - sy(0) - y'(0) + 2[sY(s) - y(0)] + Y(s) = 1$$

$$\Rightarrow Y(s)\,[s^2 + 2s + 1] - (s \times -2) - y'(0) - (2 \times -2) = 1$$

$$\Rightarrow Y(s)\,[s^2 + 2s + 1] = -2s - 3$$

$$\Rightarrow \quad Y(s) = \frac{-2s-3}{s^2+2s+1} = \frac{-2}{s+1} - \frac{1}{(s+1)^2}$$

$$\Rightarrow \quad y(t) = -2e^{-t} - te^{-t}$$

$$\frac{dy(t)}{dt} = 2e^{-t} - (-t\,e^{-t} + e^{-t}) = e^{-t} + te^{-t}$$

$$\left.\frac{dy(t)}{dt}\right|_{t=0^+} = 1 + 0 = 1$$

5.15 (b)

$$Y(s) = \frac{1}{s}U(s) = \frac{1}{s^2}$$

$$y(t) = t\,u(t)$$

5.16 (c)

All poles must lie within $|Z| = 1$

5.17 (b)

Given, $\quad H(s) = \frac{1}{s(s+4)}$

$$r(t) = \text{input} = \cos(3t) = \cos\omega t$$

$\therefore \quad \omega = 3$ rad/s

$$H(j\omega) = \frac{1}{(j\omega)(j\omega+4)}$$

$$\text{Now, } |H(j\omega)| = \frac{1}{\omega\sqrt{\omega^2+4^2}} = \frac{1}{3\sqrt{25}} = \frac{1}{15} \quad (\text{at } \omega = 3)$$

$$\angle H(j\omega) = -90° - \tan^{-1}\frac{\omega}{4}$$

$$= -90 - \tan^{-1}\left(\frac{3}{4}\right) = -126.86° \quad (\text{at } \omega = 3)$$

$$\therefore \quad c(t) = \frac{1}{15}\cos(3t - 126.86°)$$

$$= \frac{1}{15}\sin(3t - 36.86°) \quad \text{...(i)}$$

$$\text{or,} \quad c(t) = A\sin(3t + \alpha) \quad \text{...(ii)}$$

Comparing eqn. (i) and (ii), we have $A = \frac{1}{15}$

5.18 (b)

Given : $\quad f(t) = 2\sqrt{\frac{t}{\pi}} \Longleftrightarrow F(s) = s^{-3/2}$

By using property of differentiation in time,

$$\frac{df(t)}{dt} \Longleftrightarrow sF(s)$$

$$\Rightarrow \quad \frac{2}{\sqrt{\pi}} \cdot \frac{1}{2} t^{-1/2} \Longleftrightarrow sF(s)$$

$$\Rightarrow \quad \frac{1}{\sqrt{\pi t}} \Longleftrightarrow s \cdot s^{-3/2} \quad \Rightarrow \quad \frac{1}{\sqrt{\pi t}} \Longleftrightarrow s^{-1/2}$$

5.19 (a)

Laplace transform of $\sin 5t\,u(t) \rightarrow \frac{5}{s^2+25}$

$$e^{2t}\sin 5t\,u(t) \rightarrow \frac{5}{(s-2)^2+25} = \frac{5}{s^2-4s+29}$$

5.20 (b)

$$\frac{Y(s)}{R(s)} = \frac{s}{s+2}$$

$$y(t) = A\cos(2t + \phi),\; r(t) = \cos 2t$$

$$\because \quad H(s) = \frac{s}{(s+2)}$$

$$H(j\omega) = \frac{j\omega}{j\omega+2}; |H(j\omega)| = \frac{\omega}{\sqrt{\omega^2+4}};$$

$$\angle H(j\omega) = 90° - \tan^{-1}\left(\frac{\omega}{2}\right)$$

$\because$ $\omega = 2$ (given)

$$|H(j\omega)| = \frac{2}{\sqrt{4+4}} = \frac{2}{2\sqrt{2}} = \frac{1}{\sqrt{2}}$$

$$|H(j\omega)| = 90° - \tan^{-1}(1) = 45°$$

$\because$ Hence, $A = 1 \times |H(j\omega)|_{\omega=2} = 1 \times \frac{1}{\sqrt{2}} = 0.707$

and $\phi = +45°$

5.21 (d)

The differential equation,

$$\frac{dy(t)}{dt} + \frac{1}{6}y(t) = 3x(t)$$

So, $sY(s) + \frac{1}{6}Y(s) = 3X(s)$

$$Y(s) = \frac{3X(s)}{\left(s+\frac{1}{6}\right)}; \quad X(s) = \frac{3}{\left(s+\frac{1}{3}\right)}$$

So, $Y(s) = \frac{9}{\left(s+\frac{1}{3}\right)\left(s+\frac{1}{6}\right)} = \frac{54}{\left(s+\frac{1}{6}\right)} - \frac{54}{\left(s+\frac{1}{3}\right)}$

So, $y(t) = (54e^{-t/6} - 54e^{-t/3})\,u(t)$

5.22 (a)

The differential equation is

$y''(t) + 2y'(t) + y(t) = 0$

So, $(s^2Y(s) - sy(0) - y'(0)) + 2[sY(s) - y(0)] + Y(s) = 0$

So, $$Y(s) = \frac{sy(0) + y'(0) + 2y(0)}{(s^2+2s+1)}$$

Given that, $y'(0) = 1$, $y(0) = 0$

So, $$Y(s) = \frac{1}{(s+1)^2}$$

So, $y(t) = te^{-t}\,u(t)$

5.23 Sol.

$$H(s) = \frac{1}{(s+1)}$$

Put $s = j\omega$, $H(j\omega) = \frac{1}{j\omega+1}$

$$|H(j\omega)| = \frac{1}{\sqrt{\omega^2+1}}$$

$\because$ Input $x(t) = \cos(t)$

Here $\omega = 1$ rad/sec and $|x(t)| = 1$

Hence, steady state output

$$y(t) = |H(j\omega)|_{\omega=1}\cos(t + \angle H(j\omega)|_{\omega=1})$$

$$A = |H(j\omega)|_{\omega=1} = \frac{1}{\sqrt{2}} = 0.707$$

5.24 (b)

Steady state value of $y(t) = \underset{s\to 0}{Lt}\, sY(s)$

$$= \underset{s\to 0}{Lt}\,\frac{10s}{s(s^2+s+100\sqrt{2})} = \frac{10}{100\sqrt{2}} = \frac{1}{10\sqrt{2}}$$

5.25 (b)

$$H(s) = \frac{c_1}{a_2s^2+b_2s+c_2} \quad \text{as } a_1 = b_1 = 0$$

$$= \frac{c_1}{(1+s\tau_1)(1+s\tau_2)}$$

Put $s = 0$, $H(0) = \frac{c_1}{c_2}$

Put $s = \infty$, $H(\infty) = 0$

At s = 0	At s = ∞	Type of filter
Constant	0	LPF
0	Constant	HPF
0	0	BPF
1	1	BSF

which represents 2nd order low pass filter.

5.26 (a)

$$\mathcal{L}^{-1}\left(\frac{s+3}{s^2+2s+1}\right) = \mathcal{L}^{-1}\left(\frac{s+1+2}{(s+1)^2}\right)$$

$$= \mathcal{L}^{-1}\left(\frac{1}{s+1} + \frac{2}{(s+1)^2}\right) = e^{-t} + 2t\,e^{-t}$$

5.27 (d)

$$\frac{dy(t)}{dt^2} + 4y(t) = 6\,r(t)$$

$$[s^2+4]\,Y(s) = 6\,R(s)$$

$$\frac{Y(s)}{R(s)} = \frac{6}{s^2+4}$$

Poles: $s^2 + 4 = 0 \Rightarrow s = \pm j2$

5.29 (a)

Pole locations: (2, –1) and (–2, 1)
Zero locations: (2, 1) and (–2, 1)

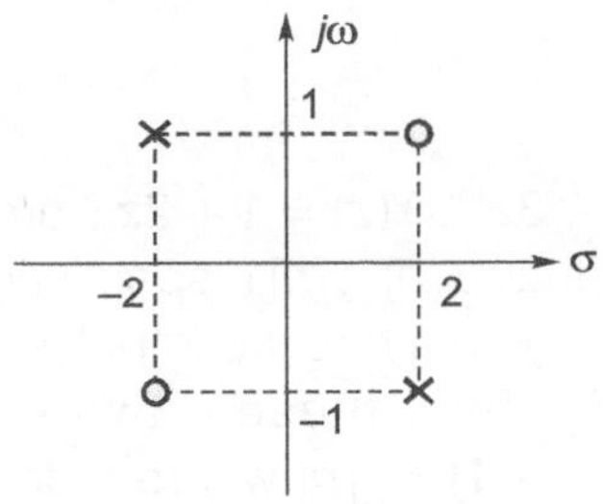

Filter is all pass since for each pole, there is a mirror image zero.

System is unstable because for stability of causal system all poles should lie in the L.H.S. of s-plane and here one pole is lying in the R.H.S.

$$H(s) = \frac{[s-(2+j)][s-(-2-j)]}{[s-(2+j)][s-(-2+j)]}$$

$$= \frac{[s-(2+j)][s+(2+j)]}{[s-(2-j)][s+(-2-j)]}$$

$$= \frac{s^2-(2+j)^2}{s^2-(2-j)^2} = \frac{s^2-(4-1+4j)}{s^2-(4-1-4j)}$$

$$= \frac{s^2-3-4j}{s^2-3+4j}$$

i.e., transfer function is complex.

5.30 (c)

$$\frac{dy(t)}{dt}+3y(t) = 2x(t) \quad ...(1)$$

For impulse response

$$x(t) = \delta(t)$$

$$\therefore \quad y(t) = h(t)$$

So, taking laplace transform of equation (1)

$$sH(s) + 3H(s) = 2(1)$$

$$H(s) = \frac{2}{s+3}$$

$$\Updownarrow$$

$$h(t) = 2e^{-3t}u(t)$$

5.31 (b)

Since $x(t)$ is finite duration signal. So ROC will be $-\infty < \sigma < \infty$.

6 Z-Transform

6.1 If $u(t)$ is the unit step and $\delta(t)$ is the unit impulse function, the inverse z-transform of $F(z) = \frac{1}{z+1}$ for $k > 0$ is

(a) $(-1)^k \delta(k)$ (b) $\delta(k) - (-1)^k u(k)$
(c) $(-1)^k u(k)$ (d) $u(k) - (-1)^k \delta(k)$

[2005 : 2 M]

6.2 A discrete real all pass system has a pole at $z = 2\angle 30°$; it, therefore

(a) also has a pole at $0.5\angle 30°$
(b) has a constant phase response over the z-plane: $\arg|H(z)|$ = constant
(c) is stable only if it anticausal
(d) has a constant phase response over the unit circle: $\arg|H(e^{j\Omega})|$ = constant [2006 : 1 M]

6.3 The discrete-time signal

$$x[n] \leftrightarrow X(z) = \sum_{n=0}^{\infty} \frac{3^n}{2+n} z^{2n}$$

where $\leftrightarrow$ denotes a transform-pair relationship, is orthogonal to the signal

(a) $y_1[n] \leftrightarrow Y_1(z) = \sum_{n=0}^{\infty} \left(\frac{2}{3}\right)^n z^{-n}$

(b) $y_2[n] \leftrightarrow Y_2(z) = \sum_{n=0}^{\infty} (5^n - n) z^{-(2n+1)}$

(c) $y_3[n] \leftrightarrow Y_3(z) = \sum_{n=-\infty}^{\infty} 2^{-|n|} z^{-n}$

(d) $y_4[n] \leftrightarrow Y_4(z) = 2z^{-4} + 3z^{-2} + 1$

[2006 : 2 M]

6.4 Consider the discrete-time system shown in the figure where the impulse response of $G(z)$ is $g(0) = 0$, $g(1) = g(2) = 1$, $g(3) = g(4) = \ldots = 0$

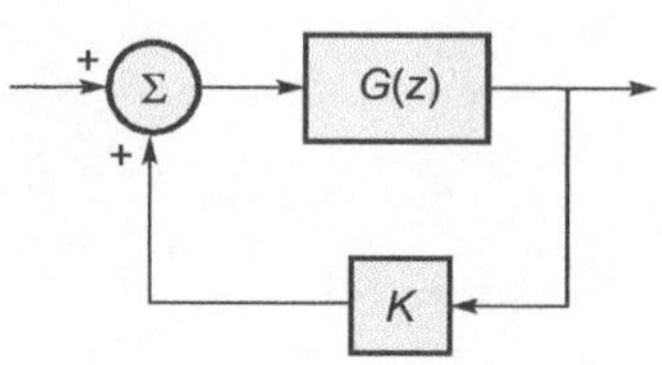

This system is stable for range of values of K

(a) $\left(-1, \frac{1}{2}\right)$ (b) $(-1, 1)$

(c) $\left(-\frac{1}{2}, 1\right)$ (d) $\left(-\frac{1}{2}, 2\right)$ [2007 : 2 M]

6.5 $X(z) = 1 - 3z^{-1}$, $Y(z) = 1 + 2z^{-2}$ are Z-transforms of two signals $x[n]$, $y[n]$ respectively. A linear time invariant system has the impulse response $h[n]$ defined by these two signals as $h[n] = x[n-1] * y[n]$ where $*$ denotes discrete time convolution. Then the output of the system for the input $\delta[n-1]$

(a) has Z-transform $z^{-1} X(z) Y(z)$
(b) equals $\delta[n-2] - 3\delta[n-3] + 2\delta[n-4] - 6\delta[n-5]$
(c) has Z-transform $1 - 3z^{-1} + 2z^{-2} - 6z^{-3}$
(d) does not satisfy any of the above three

[2007 : 2 M]

Linked Answer Questions (6.6 and 6.7):

6.6 A signal is processed by a causal filter with transfer function $G(s)$. For a distortion free output signal waveform, $G(s)$ must

(a) provide zero phase shift for all frequency.
(b) provide constant phase shift for all frequency.
(c) provide linear phase shift that is proportional to frequency.
(d) provide a phase shift that is inversely proportional to frequency. [2007 : 2 M]

6.7 $G(z) = \alpha z^{-1} + \beta z^{-3}$ is a digital band stop filter with a phase characteristics same as that of the above question if

(a) $\alpha = \beta$ (b) $\alpha = -\beta$
(c) $\alpha = \beta^{(1/3)}$ (d) $\alpha = \beta^{(-1/3)}$ [2007 : 2 M]

6.8 $H(z)$ is a transfer function of a real system when a signal $x[n] = (1 + j)^n$ is the input to such a system, the output is zero.

Further, the Region of convergence (ROC) of $\left(1 - \frac{1}{2}z^{-1}\right) H(z)$ is the entire Z-plane (except $z = 0$). It can then be inferred that $H(z)$ can have a minimum of

(a) one pole and one zero
(b) one pole and two zeros
(c) two poles and one zero
(d) two poles and two zeros [2008 : 2 M]

6.9 Given $X(z) = \frac{z}{(z-a)^2}$ with $|z| > a$, the residue of $X(z)z^{n-1}$ at $z = a$ for $n \geq 0$ will be

(a) a^{n-1} (b) a^n
(c) na^n (d) na^{n-1} [2008 : 2 M]

6.10 The z-transform of a signal $x[n]$ is given by $4z^{-3} + 3z^{-1} + 2 - 6z^2 + 2z^3$. It is applied to a system, with a transfer function $H(z) = 3z^{-1} - 2$. Let the output be $y(n)$. Which of the following is true?
(a) $y(n)$ is non causal with finite support
(b) $y(n)$ is causal with infinite support
(c) $y(n) = 0; |n| > 3$
(d) $\text{Re}[Y(z)]_{z=e^{j\theta}} = -\text{Re}[Y(z)]_{z=e^{-j\theta}}$;
$\text{Im}[Y(z)]_{z=e^{j\theta}} = \text{Im}[Y(z)]_{z=e^{-j\theta}}$; $-\pi \le \theta < \pi$
[2009 : 2 M]

6.11 If $x[n] = \left(\frac{1}{3}\right)^{|n|} - \left(\frac{1}{2}\right)^n u[n]$, then the region of convergence (ROC) of its z-transform in the Z-plane will be
(a) $\frac{1}{3} < |z| < 3$ (b) $\frac{1}{3} < |z| < \frac{1}{2}$
(c) $\frac{1}{2} < |z| < 3$ (d) $\frac{1}{3} < |z| < 2$ [2012 : 1 M]

6.12 Let $X(z) = \frac{1}{1 - z^{-3}}$ be the z-transform of a causal signal $x[n]$. Then, the values of $x[2]$ and $x[3]$ are
(a) 0 and 0 (b) 0 and 1
(c) 1 and 0 (d) 1 and 1
[2014 : 2 Marks, Set-1]

6.13 A discrete system is represented by the difference equation

$$\begin{bmatrix} X_1(k+1) \\ X_2(k+1) \end{bmatrix} = \begin{bmatrix} a & a-1 \\ a+1 & 1 \end{bmatrix} \begin{bmatrix} X_1(k) \\ X_2(k) \end{bmatrix}$$

It has initial conditions $X_1(0) = 1$; $X_2(0) = 0$. The pole locations of the system for $a = 1$, are
(a) $1 + j0$ (b) $-1 - j0$
(c) $-1 + j0$ (d) $0 + j1$
[2014 : 2 M, Set-2]

6.14 An input signal $x(t) = 2 + 5 \sin(100\pi t)$ is sampled with a sampling frequency of 400 Hz and applied to the system whose transfer function is represented by

$$\frac{Y(z)}{X(z)} = \frac{1}{N}\left(\frac{1 - z^{-N}}{1 - z^{-1}}\right)$$

where, N represents the number of samples per cycle. The output $y(n)$ of the system under steady state is
(a) 0 (b) 1
(c) 2 (d) 5
[2014 : 2 M, Set-2]

6.15 Consider a discrete time signal given by

$$x[n] = (-0.25)^n u[n] + (0.5)^n u[-n-1]$$

The region of convergence of its Z-transform would be
(a) the region inside the circle of radius 0.5 and centered at origin.
(b) the region outside the circle of radius 0.25 and centered at origin.
(c) the annular region between the two circles, both centered at origin and having radii 0.25 and 0.5.
(d) the entire Z-plane.
[2015 : 2 M, Set-1]

6.16 The z-Transform of a sequence $x[n]$ is given as $X(z) = 2z + 4 - 4/z + 3/z^2$. If $y[n]$ is the first difference of $x[n]$, then $Y[z]$ is given by
(a) $2z + 2 - 8/z + 7/z^2 - 3/z^3$
(b) $-2z + 2 - 6/z + 1/z^2 - 3/z^3$
(c) $-2z - 2 + 8/z - 7/z^2 + 3/z^3$
(d) $4z - 2 - 8/z - 1/z^2 + 3/z^3$
[2015 : 2 M, Set-2]

6.17 Let $S = \sum_{n=0}^{\infty} n\alpha^n$ where $|\alpha| < 1$. The value of α in the range $0 < \alpha < 1$, such that $S = 2a$ is _____.
[2016 : 2 M, Set-1]

6.18 Consider a causal and stable LTI system with rational transfer function $H(z)$, whose corresponding impulse response begins at $n = 0$. Furthermore, $H(1) = \frac{5}{4}$. The poles of $H(z)$ are

$$p_k = \frac{1}{\sqrt{2}} \exp\left(j\frac{(2k-1)\pi}{4}\right) \text{ for } k = 1, 2, 3, 4$$

The zeros of $H(z)$ are all at $z = 0$. Let $g[n] = j^n h[n]$. The value of $g[8]$ equals _______ .
(Give the answer up to three decimal places.)
[2017 : 2 M, Set-1]

6.19 The pole-zero plots of three discrete-time systems P, Q and R on the z-plane are shown below.

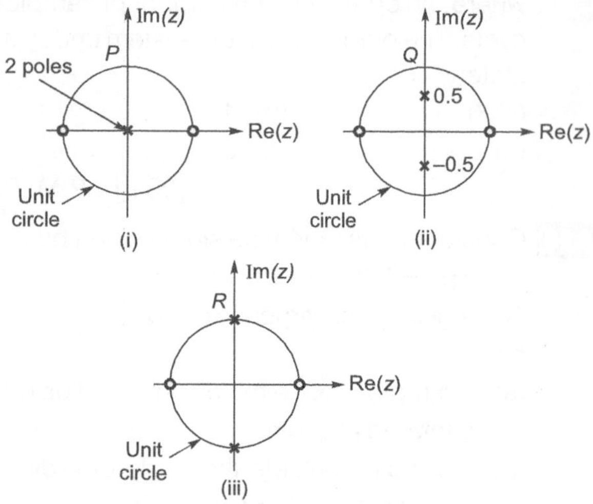

Which one of the following is TRUE about the frequency selectivity of these systems?

(a) All three are high-pass filters.
(b) All three are band-pass filters.
(c) All three are low-pass filters.
(d) P is a low-pass filter, Q is a band-pass filter and R is a high-pass filter.

[2017 : 1 M, Set-2]

6.20 A cascade system having the impulse responses $h_1(n) = \{\underset{\uparrow}{1}, -1\}$ and $h_2(n) = \{\underset{\uparrow}{1}, 1\}$ is shown in the figure below, where symbol - denotes the time origin.

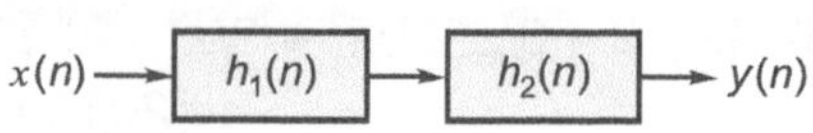

The input sequence $x(n)$ for which the cascade system produces an output sequence $y(n) = \{\underset{\uparrow}{1}, 2, 1, -1, -2, -1\}$ is

(a) $x(n) = \{\underset{\uparrow}{1}, 2, 1, 1\}$
(b) $x(n) = \{\underset{\uparrow}{1}, 1, 2, 2\}$
(c) $x(n) = \{\underset{\uparrow}{1}, 1, 1, 1\}$
(d) $x(n) = \{\underset{\uparrow}{1}, 2, 2, 1\}$

[2017 : 2 M, Set-2]

6.21 Consider a signal $x[n] = \left(\frac{1}{2}\right)^n 1[n]$, where $1[n] = 0$ if $n < 0$, and $1[n] = 1$ if $n \geq 0$. The z-transform of $x[n-k]$, $k > 0$ is $\dfrac{z^{-k}}{1-\frac{1}{2}z^{-1}}$ with region of convergence being

(a) $|z| < \frac{1}{2}$
(b) $|z| > \frac{1}{2}$
(c) $|z| < 2$
(d) $|z| > 2$

[2020 : 1 M]

6.22 Which of the following options is true for a linear time-invariant discrete time system that obeys the difference equation:

$$y[n] - ay[n-1] = b_0x[n] - b_1x[n-1]$$

(a) When $x[n] = 0$, $n < 0$, the function $y[n]$; $n > 0$ is solely determined by the function $x[n]$.
(b) The system is necessarily causal.
(c) $y[n]$ is unaffected by the values of $x[n-k]$; $k > 2$.
(d) The system impulse response is non-zero at infinitely many instants.

[2020 : 2 M]

6.23 The causal signal with z-transform $z^2(z-a)$ is ($u[n]$ is the unit step signal)

(a) $n^{-1} a^n u[n]$
(b) $(n+1)a^n u[n]$
(c) $n^2 a^n u[n]$
(d) $a^{2n} u[n]$

[2021 : 2 M]

6.24 The Z-transform of a discrete signal $x[n]$ is

$$X(z) = \frac{4z}{\left(z-\frac{1}{5}\right)\left(z-\frac{2}{3}\right)(z-3)} \text{ with ROC } = R$$

Which one of the following statements is true?

(a) Discrete-time Fourier transform of $x[n]$ converges if R is $|z| > 3$
(b) Discrete-time Fourier transform of $x[n]$ converges if R is $\frac{2}{3} < |Z| < 3$
(c) Discrete-time Fourier transform of $x[n]$ converges if R is such that $x[n]$ is a left-sided sequence
(d) Discrete-time Fourier transform of $x[n]$ converges if R is such that $x[n]$ is a right-sided sequence

[2023 : 1 M]

6.25 If the Z-transform of a finite-duration discrete-time signal $x[n]$ is $X(z)$, then the Z-transform of the signal $y[n] = x[2n]$ is

(a) $Y(z) = X(z^2)$
(b) $Y(z) = \frac{1}{2}\left[X(z^2) + X(-z^2)\right]$
(c) $Y(z) = \frac{1}{2}\left[X(z^{-1/2}) + X(-z^{-1/2})\right]$
(d) $Y(z) = \frac{1}{2}\left[X(z^{1/2}) + X(-z^{1/2})\right]$

[2024 : 2 M]

Answers Z-Transform

6.1 (b)	6.2 (c)	6.3 (b)	6.4 (a)	6.5 (b)	6.6 (c)	6.7 (a)
6.8 (d)	6.9 (d)	6.10 (a)	6.11 (c)	6.12 (b)	6.13 (a)	6.14 (c)
6.15 (c)	6.16 (a)	6.17 (0.29)	6.18 (0.097)	6.19 (b)	6.20 (d)	6.21 (b)
6.22 (d)	6.23 (b)	6.24 (b)	6.25 (d)			

Explanations Z-Transform

6.1 (b)

$$F(z) = 1-\frac{z}{z+1} = 1-\frac{1}{1+z^{-1}} = \delta(k)-(-1)^k\,u(k)$$

6.2 (c)

For causal system, If all the poles are inside the Unit Circle then system is stable, and converse in true for anti-causal system.

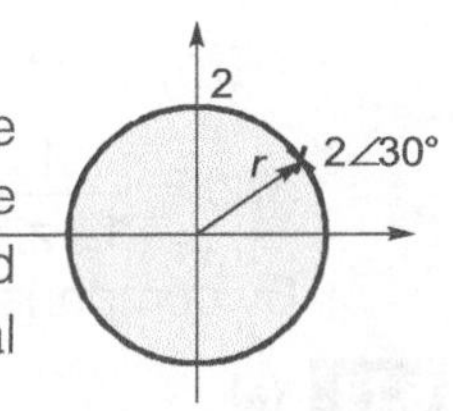

6.3 (b)

$$x(n) \rightleftharpoons X(z) = \sum_{n=0}^{\infty} \frac{3^n}{2+n} z^{2n}$$

In $X(z)$, power of z are even. Therefore, samples in $x(n)$ are available at even instants of time.

By observing all the options.

Option (b):

$$y_2(n) \rightleftharpoons Y_2(z) = \sum_{n=0}^{\infty} (5^n - n)\, z^{-(2n+1)}$$

In $Y_2(z)$, powers of z are odd. Therefore, samples in $y_2(n)$ are available only at odd instants of time.

Hence, $$\sum_{n=-\infty}^{\infty} x(n)\cdot y_2(n) = 0$$

Thus, $x(n)$ is orthogonal to $y_2(n)$.

6.4 (a)

Given: $g(1) = g(2) = 1$, otherwise 0

i.e., $g[n] = \delta[n-1] + \delta[n-2]$

therefore, $G(z) = z^{-1} + z^{-2}$

therefore overall transfer function of closed loop system,

$$T(z) = \frac{G(z)}{1-kG(z)} = \frac{z^{-1}+z^{-2}}{1-k(z^{-1}+z^{-2})}$$

or, $$T(z) = \frac{z+1}{z^2-k(z+1)}$$

So the system will be stable if it's outer most pole will lie inside the unit circle.

Location of poles,

$$z = \frac{+k \pm \sqrt{k^2+4k}}{2}$$

$$\frac{k+\sqrt{k^2+4k}}{2} < 1$$

$$k+\sqrt{k^2+4k} < 2$$

$$\sqrt{k^2+4k} < (2-k)$$

$$\Rightarrow \quad k^2 + 4\,k < (2-k)^2$$

$$\Rightarrow \quad k^2 + 4\,k < 4 + k^2 - 4\,k$$

$$\Rightarrow \quad 8\,k < 4 \;\Rightarrow\; k < \frac{1}{2}$$

This condition is satisfied in only option (a).

6.5 (b)

$$X(z) = (1-3z^{-1})$$
$$Y(z) = (1+2z^{-2})$$
$$h[n] = x(n-1) * y[n]$$
$$\Rightarrow H[z] = z^{-1}\,X(z)\cdot Y(z)$$
$$\Rightarrow H[z] = z^{-1}\,(1-3z^{-1})\,(1+2z^{-2})$$
$$\Rightarrow H[z] = z^{-1}(1+2z^{-2}-3z^{-1}-6z^{-3})$$
$$\Rightarrow H[z] = (z^{-1}-3z^{-2}+2z^{-3}-6z^{-4})$$

when input,

$I(n) = \delta[n-1]$ then, $I[z] = z^{-1}$

$\therefore$ Output, $P(n) = h[n] * I[n]$

$$\Rightarrow P(z) = H(z)\,I(z)$$
$$P(z) = (z^{-1}-3z^{-2}+2z^{-3}-6z^{-4})\times z^{-1}$$
$$\Rightarrow P(z) = z^{-2}-3z^{-3}+2z^{-4}-6z^{-5}$$
$$\therefore P(n) = \delta[n-2]-3\delta[n-3]+2\delta[n-4]-6\delta[n-5]$$

6.7 (a)

$$G(z) = \alpha z^{-1} + \beta z^{-3}$$

For, low frequency ($z = 1$) $\rightarrow$ $G(1) = \alpha + \beta$ ...(i)

For, high freq. ($z = 1$) $\rightarrow$ $G(-1) = -(\alpha + \beta)$ ...(ii)

For now, $g(n) = \{\underset{\uparrow}{0}, \alpha, 0, \beta\}$

For linear phase FIR filter, g(n) should be either even symmetric or odd symmetric about virtual y-axis.

To, satisfy the above condition.

Case (i): $\alpha = -\beta$

(odd symmetric about virtual y-axis)

But in this case,

$$G(1) = G(-1) = 0$$

i.e., at low frequency $\Rightarrow G(z) = 0$
and at high frequency $\Rightarrow G(z) = 0$
Thus in this case filter will be band-pass.

Case (ii): $\alpha = \beta$
(even symmetric about virtual y-axis)

In this case,
At low frequency $(z = 1) \Rightarrow G(1) = 2\alpha$
At low frequency $(z = -1) \Rightarrow G(-1) = -2\alpha$
Filter will be either all pass or band stop.

Now, $G(e^{j\omega}) = \alpha(e^{-j\omega} + e^{-j3\omega})$

$$= \alpha e^{-j2\omega}(e^{j\omega} + e^{-j\omega}) = 2ae^{-j2\omega}\cos\omega$$

6.9 (d)

$$x(z) = \frac{z}{(z-a)^2}$$

$$\therefore \quad z^{n-1}X(z) = \frac{z^n}{(z-a)^2}$$

Since, $z = a$ is a pole of second order therefore residue at $z = a$

$$= \frac{1}{\lfloor 1}\left[\frac{d}{dz}\left\{(z-a)^2 \cdot \frac{z^n}{(z-a)^2}\right\}\right]_{at\ z=a}$$

$$= [nz^{n}-1]_{z=a} = na^{n-1}$$

6.10 (a)

$$y[n] = x[n] * H[n]$$
$$y[z] = x[z]\,H[z]$$
$$y[z] = (4z^{-3} + 3z^{-1} + 2 - 6z^2 + 2z^3)(3z^{-1} - 2)$$
$$= 12z^{-4} + 9z^{-2} + 6z^{-1} - 18z + 6z^2$$
$$-8z^{-3} - 6z^{-1} - 4 + 12z^2 - 4z^3$$
$$= -4z^3 + 18z^2 - 18z + 12z^{-4} + 9z^{-2} - 4 - 8z^{-3}$$

$y[n] \neq 0$ for $n < 0$
Therefore, it is non-causal with finite support.

6.11 (c)

Let, $x_1[n] = \left(\frac{1}{3}\right)^{|n|}$ and $x_2[n] = \left(\frac{1}{2}\right)^n u[n]$

$$\Rightarrow \quad x_1[n] = \left(\frac{1}{3}\right)^n u[n] + \left(\frac{1}{3}\right)^{-n} u[-n-1]$$

$$\left(\frac{1}{3}\right)^n u[n] \xleftrightarrow{z} \frac{1}{1-\frac{1}{3}z^{-1}};\ \text{ROC}: |z| > \frac{1}{3}$$

$$\left(\frac{1}{3}\right)^{-n} u[-n-1] \xleftrightarrow{z} \frac{-1}{1-\left(\frac{1}{3}\right)^{-1} z^{-1}};\ \text{ROC}: |z| < 3$$

and $x_2[n] = \left(\frac{1}{2}\right)^n u[n] \xleftrightarrow{z} \frac{1}{1-\frac{1}{2}z^{-1}};\ \text{ROC}: |z| > \frac{1}{2}$

$\therefore$ ROC is $\frac{1}{2} < |z| < 3$.

6.12 (b)

$$X(z) = \frac{1}{1-z^{-3}}$$

From z-transform definition,

$$X(z) = \sum_{n=-\infty}^{\infty} x[n]\, z^{-n}$$

$$\begin{array}{r|l} & 1 + z^{-3} + \ldots \\ 1 - z^{-3} & 1 \\ & 1 - z^{-3} \\ \hline & z^{-3} \\ & z^{-3} \end{array}$$

$$= x(0) + x(1)\,z^{-1} + x(2)\,z^{-2} + x(3)\,x^{-3} \ldots$$

$$X(z) = \frac{1}{1-z^{-3}} = 1 + 0\cdot z^{-1} + 0\cdot z^{-2} + 1\cdot z^{-3} \ldots$$

By comparison $x(2) = 0$ and $x(3) = 1$

6.13 (a)

Given that,

$$\begin{bmatrix} X_1(K+1) \\ X_2(K+1) \end{bmatrix} = \begin{bmatrix} a & a-1 \\ a+1 & 1 \end{bmatrix}\begin{bmatrix} X_1(K) \\ X_2(K) \end{bmatrix}$$

With initial conditions,

$$X_1(0) = 1;\ X_2(0) = 0$$

For $a = 1$ we can write

$$\begin{bmatrix} X_1(K+1) \\ X_2(K+1) \end{bmatrix} = \begin{bmatrix} 1 & 0 \\ 2 & 1 \end{bmatrix}\begin{bmatrix} X_1(K) \\ X_2(K) \end{bmatrix}$$

$$\Rightarrow \quad X_1(K+1) = X_1(K) \quad \ldots(i)$$
$$X_2(K+1) = 2X_1(K) + X_2(K) \quad \ldots(ii)$$

Applying z-transform on equation (i),

$$zX_1(z) - zX_1(0) = X_1(z)$$
$$\Rightarrow \quad X_1(z)[z-1] = zX_1(0)$$
$$\Rightarrow X_1(z) = \frac{z}{z-1} \quad [\text{Given, } X_1(0) = 1] \quad \ldots(iii)$$

From equation (ii),

$$zX_2(z) - zX_2(0) = 2X_1(z) + X_2(z)$$
$$\Rightarrow \quad X_2(z)[z-1] = 2X_1(z) + zX_2(0)$$

[Given, $X_2(0) = 0$]

$$\Rightarrow \quad \frac{X_2(z)}{X_1(z)} = \frac{2}{z-1}$$

Thus, transfer function, $H(z) = \frac{2}{z-1}$

Therefore, pole location is $z = 1$.

6.14 (c)

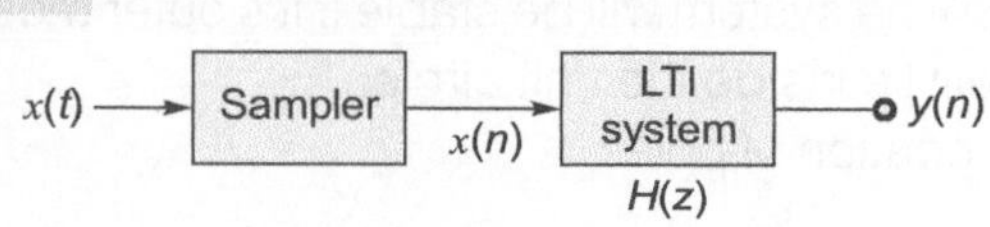

$$f_s = 400 \text{ Hz}$$
$$x(t) = 2 + 5\sin(100\pi t)$$
$$t \to xT_s = \frac{n}{f_s} = \frac{n}{400}$$
$$x(n) = 2 + 5\sin\left(100\pi \cdot \frac{n}{400}\right) = 2 + 5\sin\frac{\pi}{4}n$$

Here, $\omega_0 = \frac{\pi}{4}$

Therefore, N = time-period of $x(n) = \frac{2\pi}{\omega_0} = \frac{2\pi}{\pi/4} = 8$

Now, $H(z) = \frac{1}{N}\left[\frac{1-z^{-N}}{1-z^{-1}}\right]$

Applying, $z = e^{j\omega}$

$$H(e^{j\omega}) = \frac{1}{N}\left[\frac{1-e^{-j\omega N}}{1-e^{-j\omega}}\right]$$

For sinusoidal part of $x(n)$:

$$H(e^{j\omega_0}) = \frac{1}{N}\left[\frac{1-e^{-j\omega_0 N}}{1-e^{-j\omega_0}}\right]$$

$$\left[\text{Note} : \omega_0 N = \frac{\pi}{4} \times 8 = 2\pi\right]$$

$$= \frac{1}{N}\left[\frac{1-e^{-j2\pi N}}{1-e^{-j\pi/4}}\right] = 0$$

Therefore for sinusoidal part of input, system output is zero. For DC part of input,

$$y(n) = \text{output} = H(e^{j0}) \times \text{input DC value}$$

$$\left[\because H(e^{j\omega})\Big|_{\omega=0} = 1\right]$$

$$= 1 \times 2 = 2$$

Thus, steady state output = 2.

6.15 (c)

$$x[n] = (-0.25)^n u(n) + (0.5)^n u(-n-1)$$

Signal $x[n]$ is sum of two signals, one is right sided $[(-0.25)^n u(n)]$ and other is left sided $[(0.5)^n u(-n-1)]$. The right sided signal will have pole at location with magnitude 0.25. So, ROC is $|z| > 0.25$. The left sided signal will have pole at location with magnitude 0.5. So, ROC is $|z| < 0.5$. So, ROC of $X(z)$ (Z transform of $x(n)$ will be) $0.25 < |z| < 0.5$.

6.16 (a)

$y(n)$ is first difference of $x(n)$

so, $y(n) = x(n) - x(n-1)$

so, $Y(z) = X(z) - Z^{-1} X(z)$

$$Y(z) = (2z + 4 - 4z^{-1} + 3z^{-2})$$
$$-(2 + 4z^{-1} - 4z^{-2} + 3z^{-3})$$
$$Y(z) = 2z + 2 - 8z^{-1} + 7z^{-2} - 3z^{-3}$$

6.17 Sol.

The Z-transform of

$$\alpha^n u(n) \longrightarrow \frac{1}{(1-\alpha Z^{-1})}$$

and

$$n\alpha^n u(n) \longrightarrow \frac{\alpha Z^{-1}}{(1-\alpha Z^{-1})^2}$$

so,

$$\frac{\alpha Z^{-1}}{(1-\alpha Z^{-1})^2} = \sum_{n=0}^{\infty} n\alpha^n Z^{-n}$$

If we put $Z = 1$ in above equation we get

$$\frac{\alpha}{(1-\alpha)^2} = \sum_{n=0}^{\infty} n\alpha^n$$

Since, $\sum_{n=0}^{\infty} n\alpha^n = 2\alpha = \frac{\alpha}{(1-\alpha)^2}$

So, $2 = \frac{1}{(1-\alpha)^2} \Rightarrow \alpha = 0.29$

6.18 Sol.

Pole location of $H(z)$ are given as,

$$z = \frac{1}{\sqrt{2}} e^{j(2K-1)\frac{\pi}{4}}; \quad K = 1, 2, 3, 4$$

$$z_1 = \frac{1}{\sqrt{2}} e^{j\pi/4} = \frac{1}{2}(1+j)$$

$$z_2 = \frac{1}{\sqrt{2}} e^{j3\pi/4} = \frac{1}{2}[-1+j]$$

$$z_3 = \frac{1}{\sqrt{2}} e^{j5\pi/4} = \frac{1}{2}[-1-j]$$

$$z_4 = \frac{1}{\sqrt{2}} e^{j7\pi/4} = \frac{1}{2}[1-j]$$

Now,

$$H(z) = \frac{K \cdot z^4}{(z-z_1)(z-z_2)(z-z_3)(z-z_4)} \quad \text{...(i)}$$

[As $h(n)$ is causal & it starts from $n = 0$, so numerator will have same order as denominator is having]. By solving equation (i),

$$H(z) = \frac{Kz^4}{z^4 + \frac{1}{4}} \quad \text{...(ii)}$$

Given : $H(1) = \frac{5}{4}$

From equation (ii),

$$H(1) = \frac{K}{1+\frac{1}{4}} \Rightarrow \frac{5}{4} = \frac{K}{5/4}$$

$$\Rightarrow \quad K = \frac{25}{16}$$

$$H(z) = \frac{\frac{25}{16}z^4}{z^4 + \frac{1}{4}}$$

By using division rule,

$$H(z) = \frac{25}{16}\left[1 - \frac{1}{4}z^{-4} + \frac{1}{16}z^{-8} + ...\right]$$

Thus, $h(8) = \frac{1}{16} \times \frac{25}{16} = 0.097$

It is given that,

$$g(n) = j^n h(n)$$
$$g(8) = j^8 h(8) = h(8) = 0.097$$

6.19 (b)

For the figure (i), we have

$$H(z) = \frac{K(z^2 - 1)}{z^2}$$

At low frequency, i.e., $z = 1$

$$|H(1)| = 0$$

At high frequency, i.e., $z = -1$

$$|H(-1)| = 0$$

Hence filter type is BPF.

For the figure (ii), we have

$$H(z) = \frac{K(z^2 - 1)}{z^2 + 0.25}$$

At low frequency, i.e., $z = 1$

$$|H(1)| = 0$$

At high frequency, i.e., $z = -1$

$$|H(-1)| = 0$$

Hence filter type is BPF.

For the figure (iii), we have

$$H(z) = \frac{z^2 - 1}{z^2 + 1}$$

At low frequency, i.e., $z = 1$

$$|H(1)| = 0$$

At high frequency, i.e., $z = -1$

$$|H(-1)| = 0$$

Hence filter type is BPF.

Alternate solution:

Since all 3 pole-zero plots have zeros at $z = 1$ and $z = -1$.

So filter will not pass low frequency and high frequency components.

Therefore all are bandpass filter.

Note: In digital filter

For low frequency $\Rightarrow z = 1$

For high frequency $\Rightarrow z = -1$

6.20 (d)

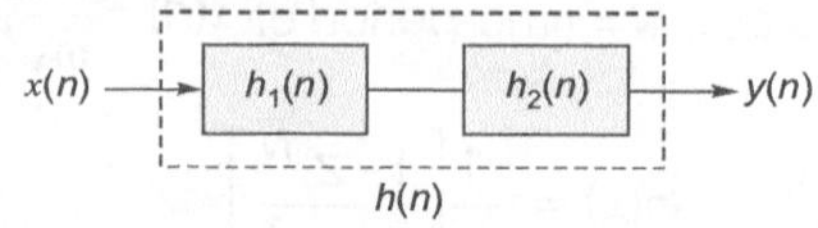

Now, $h(n)$ = overall system impulse response

$$= h_1(n) * h_2(n) = \{\underset{\uparrow}{1}, -1\} * \{\underset{\uparrow}{1}, 1\}$$

$$= \{\underset{\uparrow}{1}, 0, -1\} \rightleftharpoons H(z) = 1 - z^{-2}$$

As we know,

$$H(z) = \frac{Y(z)}{X(z)} \Rightarrow X(z) = \frac{Y(z)}{H(z)}$$

$$= \frac{1 + 2z^{-1} + z^{-2} - z^{-3} - 2z^{-4} - z^{-5}}{1 - z^{-2}}$$

$$= 1 + 2z^{-1} + 2z^{-2} + z^{-3}$$

$$\Rightarrow \quad x(n) = \{1, 2, 2, 1\}$$

6.21 (b)

$$x(n) = \left(\frac{1}{2}\right)^n u(n),$$

ROC of $x(n)$: $|z| > \frac{1}{2}$

$$x(n-k) \rightleftharpoons X(z) = \frac{z^{-k}}{1 - \frac{1}{2}z^{-1}},$$

ROC of $x(n-k)$: $|z| > \frac{1}{2}$

For $x(n-k)$ ROC will be $|z| > \frac{1}{2}$.

6.22 (d)

$$y(n) - ay(n-1) = b_0 x(n) - b_1 x(n-2)$$

By applying ZT,

$$Y(z) - az^{-1} Y(z) = b_0 X(z) - b_1 z^{-1} X(z)$$

$$\Rightarrow \quad H(z) = \frac{Y(z)}{X(z)} = \frac{b_0 - b_1 z^{-1}}{1 - az^{-1}}$$

By taking right-sided inverse ZT,

$$h(n) = b_0 a^n u(n) - b_1 a^{n-1} u(n-1)$$

By taking left-sided inverse ZT,

$$h(n) = -b_0 a^n u(-n-1) + b_1 a^{n-1} u(-n)$$

Thus system is not necessarily causal.

The impulse response is non-zero at infinitely many instants.

6.23 (b)

As we know,

$$n.a^n u(n) \rightleftharpoons \frac{az}{(z-a)^2}$$

$$\frac{1}{a}.n.a^n u(n) \rightleftharpoons \frac{z}{(z-a)^2}$$

$$f(n) = n.a^{n-1} u(n) \rightleftharpoons \frac{z}{(z-a)^2} = F(z)$$

Time-shifting property,

$$f(n) \rightleftharpoons F(z)$$

$$f(n+1) \rightleftharpoons z.F(z)$$

$$x(n) = (n+1)a^n u(n+1) \rightleftharpoons \frac{z^2}{(z-a)^2} = X(z)$$

Thus,
$$x(n) = (n+1)a^n u(n+1)$$
$$= (n+1)a^n.u(n)$$
$$[\because (n+1)u(n+1) = (n+1)u(n)]$$

6.24 (b)

If ROC is $\frac{2}{3} < |Z| < 3$ then it is including $Z = 1$ circle or unit-circle. So, DTFT will converge.

6.25 (d)

$$Y(z) = \sum_{n=-\infty}^{\infty} y(n) z^{-n} = \sum_{n=-\infty}^{\infty} x(2n) z^{-n}$$

Let $\quad 2n = m$

then
$$n = \frac{m}{2} = \sum_{m=\text{even-int.}} x(m) z^{-m/2}$$

$$= \sum_{m=-\infty}^{\infty} \left[\frac{x(m) + (-1)^m x(m)}{2}\right] z^{-m/2}$$

$$= \frac{1}{2}\left[\sum_{m=-\infty}^{\infty} x(m) z^{-m/2} + \sum_{m=-\infty}^{\infty} (-1)^m x(m) z^{-m/2}\right]$$

$$= \frac{1}{2}\left[\sum_{m=-\infty}^{\infty} x(m)(z^{1/2})^{-m} + \sum_{m=-\infty}^{\infty} x(m)(-z^{1/2})^{m}\right]$$

$$= \frac{1}{2}\left[x(z^{1/2}) + x(-z^{1/2})\right]$$

■■■■

7 Sampling

7.1 The frequency spectrum of a signal is shown in the figure. If this is ideally sampled at intervals of 1 ms, then the frequency spectrum of the sampled signal will be

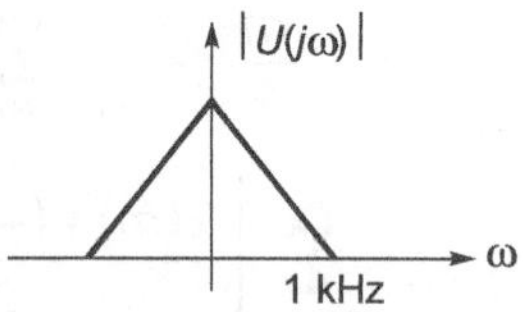

(a) 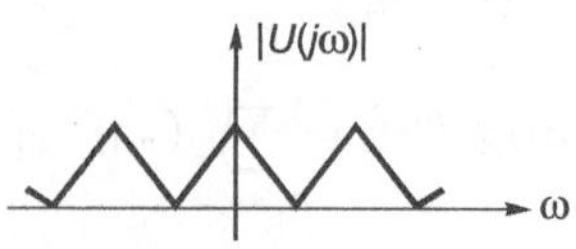

(b) 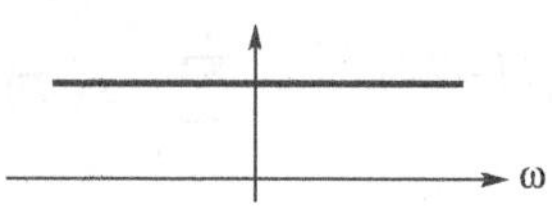

(c) 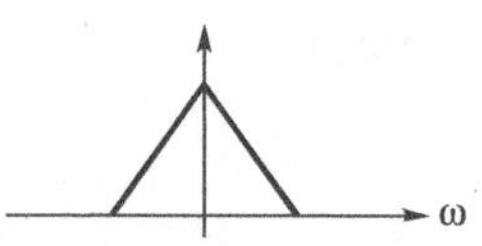

(d) 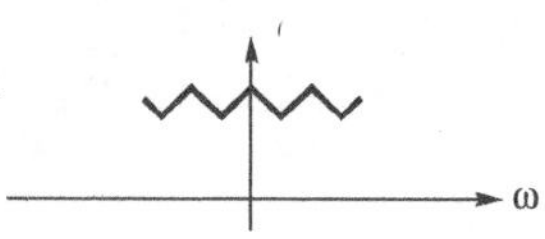

[2007 : 1 M]

7.2 A band-limited signal with a maximum frequency of 5 kHz is to be sampled. According to the sampling theorem, the sampling frequency in kHz which is not valid is

(a) 5 (b) 12
(c) 15 (d) 20 [2013 : 1 M]

7.3 A sinusoid $x(t)$ of unknown frequency is sampled by an impulse train of period 20 ms. The resulting sample train is next applied to an ideal low-pass filter with a cut-off at 25 Hz. The filter output is seen to be a sinusoid of frequency 20 Hz. This means that $x(t)$ has a frequency of

(a) 10 Hz (b) 60 Hz
(c) 30 Hz (d) 90 Hz

[2014 : 2 M, Set-3]

7.4 For the signal $f(t) = 3\sin 8\pi t + 6\sin 12\pi t + \sin 14\pi t$, the minimum sampling frequency (in Hz) satisfying the Nyquist criterion is ____.

[2014 : 1 M, Set-3]

7.5 Let $x_1(t) \leftrightarrow X_1(\omega)$ and $x_2(t) \leftrightarrow X_2(\omega)$ be two signals whose Fourier Transforms are as shown in the figure below. In the figure, $h(t) = e^{-2|t|}$ denotes the impulse response.

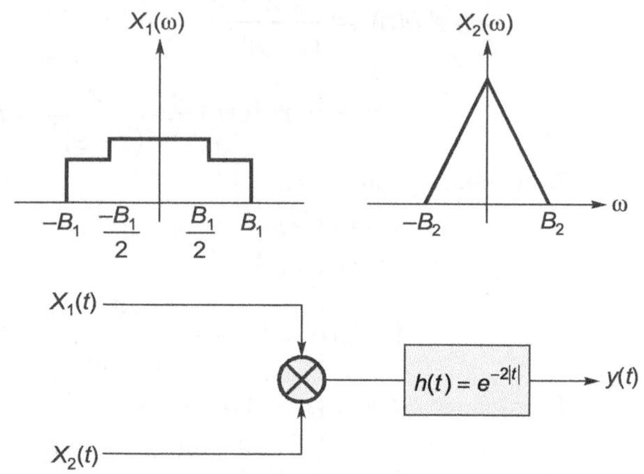

For the system shown above, the minimum sampling rate required to sample $y(t)$, so that $y(t)$ can be uniquely reconstructed from its samples, is

(a) $2B_1$ (b) $2(B_1 + B_2)$
(c) $4(B_1 + B_2)$ (d) ∞

[2016 : 2 M, Set-2]

7.6 The output $y(t)$ of the following system is to be sampled, so as to reconstruct it from its samples uniquely. The required minimum sampling rate is

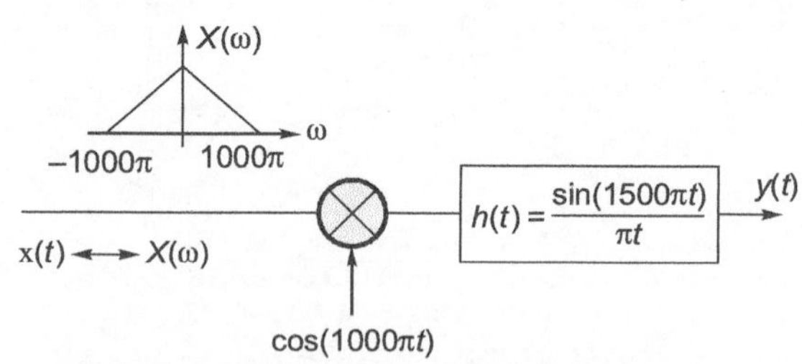

(a) 1000 samples/s (b) 1500 samples/s
(c) 2000 samples/s (d) 3000 samples/s

[2017 : 2 M]

7.7 Consider the two continuous-time signals defined below:

These signals are sampled with a sampling period of $T = 0.25$ seconds to obtain discrete-time signals $x_1[n]$ and $x_2[n]$, respectively. Which one of the following statements is true?

(a) The energy of $x_1[n]$ is greater than the energy of $x_2[n]$.
(b) The energy of $x_2[n]$ is greater than the energy of $x_1[n]$.
(c) $x_1[n]$ and $x_2[n]$ have equal energies.
(d) Neither $x_1[n]$ nor $x_2[n]$ is a finite energy signal.

[2018 : 2 M]

7.8 For the signals $x(t)$ shown in the figure, $z(t) = x(t) * y(t)$ is maximum at $t = T_1$. Then T_1 in seconds is __________ (Round off to the nearest integer).

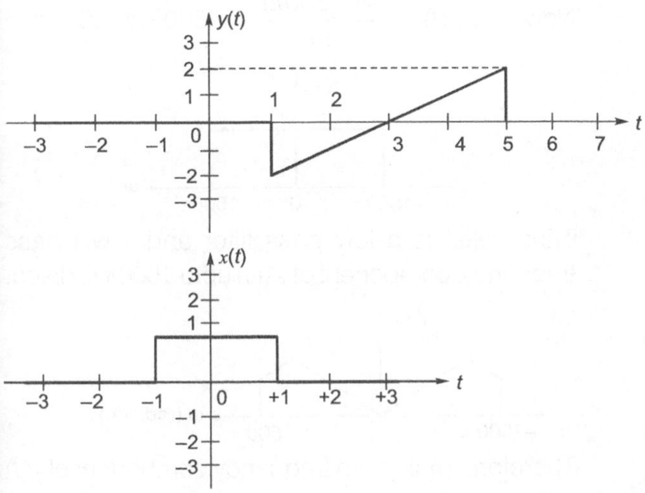

[2023 : 1 M]

7.9 A signal $x(t) = 2\cos(180\pi t)\cos(60\pi t)$ is sampled at 200 Hz and then passed through an ideal low pass filter having cut-off frequency of 100 Hz. The maximum frequency present in the filtered signal in Hz is __________ (Round off to the nearest integer).

[2023 : 2 M]

■■■■

Answers Sampling

7.1 (b) 7.2 (a) 7.3 (c) 7.4 (14) 7.5 (b) 7.6 (b) 7.7 (a)
7.8 (4) 7.9 (80)

Explanations Sampling

7.1 (b)

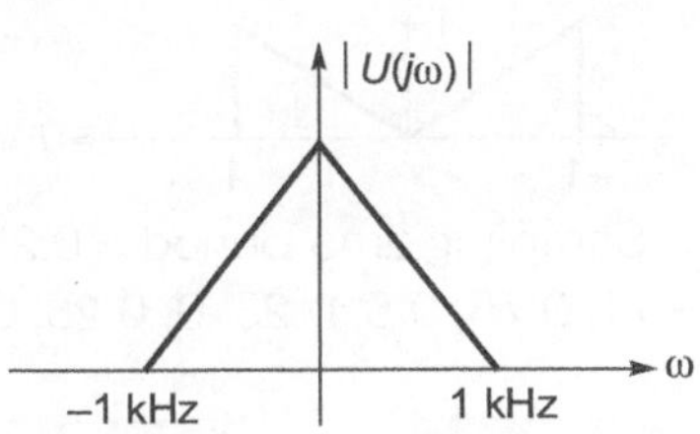

Given that, sampling interval = 1 msec

i.e., $T_s = 1 \text{ msec} = 10^{-3}$ sec

Therefore sampling frequency

$$f_s = \frac{1}{T_s} = \frac{1}{10^{-3}} = 1 \text{ kHz}$$

after sampling new signal in frequency domain

$$U_T(f) = \frac{1}{T_s}\sum_{n=-\infty}^{\infty} U(f - nf_s)$$

∴ Spectrum of sampled signal will be

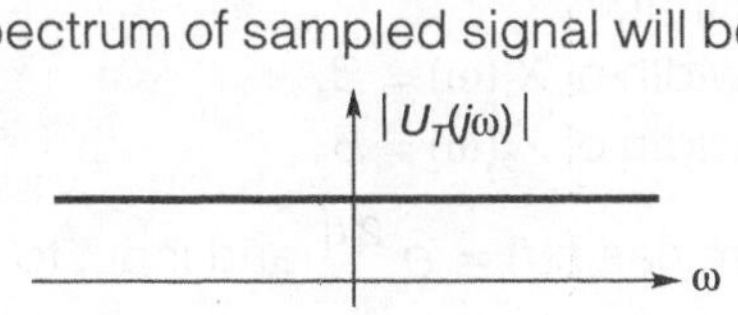

7.2 (a)

$$(f_s)_{\min} = 2f_m = 2 \times 5 = 10 \text{ kHz}$$

7.3 (c)

Given, impulse train of period 20 ms.

Then, sampling frequency $= \dfrac{1}{20 \times 10^{-3}} = 50$ Hz

If the input signal $x(t) = \cos\omega_m(t)$ having spectrum

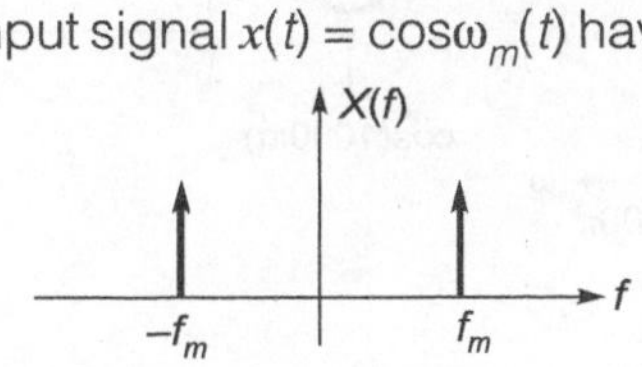

The filtered out sinusoidal signal has 20 Hz frequency the sampling must be under sampling. The output signal which is an under sampled signal with sampling frequency 50 Hz is

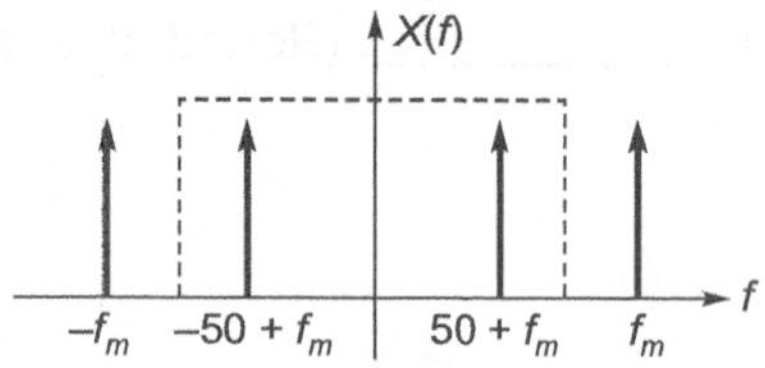

and $\quad 50 - f_m = 20 \text{ Hz} \Rightarrow f_m = 30 \text{ Hz}$

7.4 (14)

Method-I:

$$f_{m_1} = 4 \text{ Hz}; \quad f_{m_2} = 6 \text{ Hz}; \quad f_{m_3} = 7 \text{ Hz}$$

Then the minimum sampling frequency satisfying the Nyquist criterion is = 14 Hz.

Method-II:

Sampling frequency = 2 * [Highest frequency component of $f(t)$].

Given, $f(t) = 3 \sin 8\pi t + 6 \sin 12\pi t + \sin 14\pi t$

Different frequency components are

$$\omega_1 = 8\pi \qquad \therefore f_1 = \frac{8\pi}{2\pi} = 4 \text{ Hz}$$

$$\omega_2 = 12\pi \qquad \therefore f_2 = \frac{12\pi}{2\pi} = 6 \text{ Hz}$$

$$\text{and } \omega_2 = 14\pi \quad \therefore f_3 = \frac{14\pi}{2\pi} = 7 \text{ Hz}$$

Clearly, highest frequency component of the signal,

$$f(t) = f_3 = 7 \text{ Hz}$$

$\therefore$ Minimum sampling frequency,

$$f_s = 2 \times f_3 = 2 \times 7 = 14 \text{ Hz}$$

7.5 (b)

Given that,

Bandwidth of $X_1(\omega) = B_1$

Bandwidth of $X_2(\omega) = B_2$

System has $h(t) = e^{-2|t|}$ and input to the system is $x_1(t) \cdot x_2(t)$

The bandwidth of $x_1(t) \cdot x_2(t)$ is $B_1 + B_2$.

The bandwidth of output will be $B_1 + B_2$.

So, sampling rate will be $2(B_1 + B_2)$.

7.6 (b)

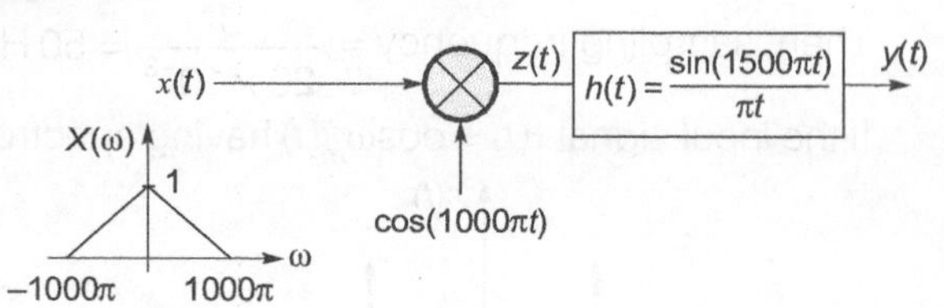

From the above block diagram,

$$Z(t) = x(t) \cos 1000\pi t$$

By using modulation property of fourier transform,

$$Z(\omega) = \frac{1}{2}[X(\omega + 1000\pi) + X(\omega - 1000\pi)]$$

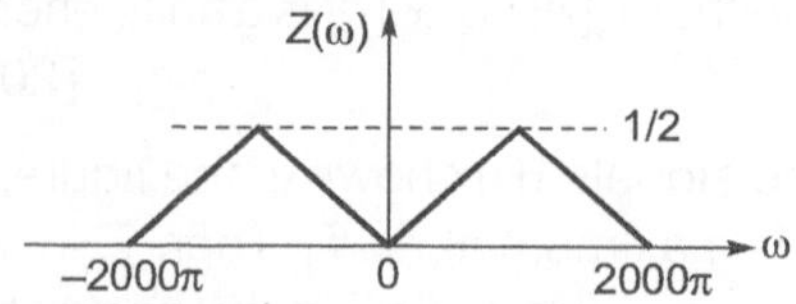

Now, $\quad h(t) = \frac{\sin 1500\pi t}{\pi t} = 1500\, sa(1500\pi t)$

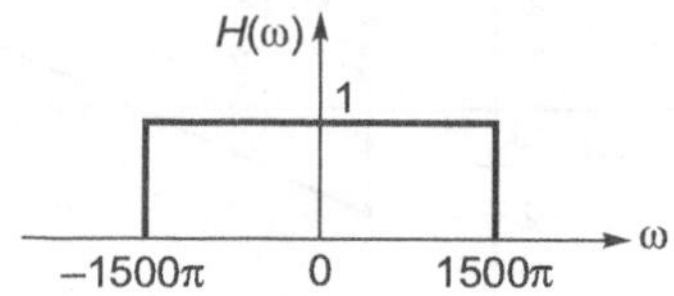

Thus, $H(\omega)$ is a low pass filter and it will pass frequency, component of $Z(\omega)$ upto 1500π rad/sec.

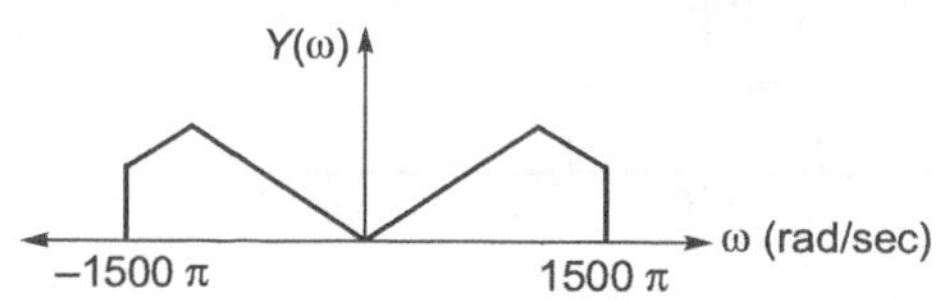

Therefore, maximum frequency component of $y(t)$ is

$$\omega_m = 1500\pi \text{ rad/sec} \text{ or } f_m = 750 \text{ Hz}$$

So, the minimum sampling rate for $y(t)$ is

$$f_{s\,min} = 2 f_m = 2 \times 750 = 1500 \text{ Hz}$$
$$= 1500 \text{ samples/sec}$$

7.7 (a)

$$x_1(t) = \begin{cases} |t|, -1 \le t \le 1 \\ 0, \text{ otherwise} \end{cases}$$

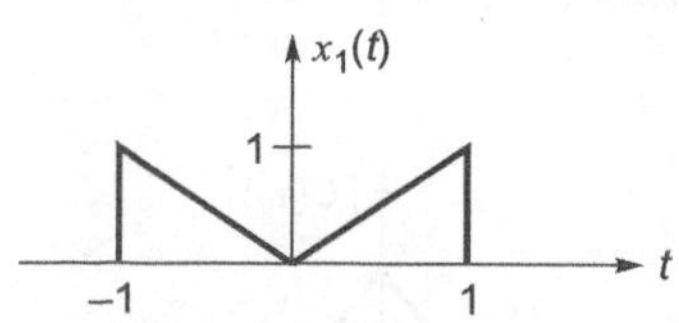

T_s = Sampling time-period = 0.25 sec

$x_1(n) = \{1, 0.75, 0.5, 0.25, 0, 0.25, 0.5, 0.75, 1\}$

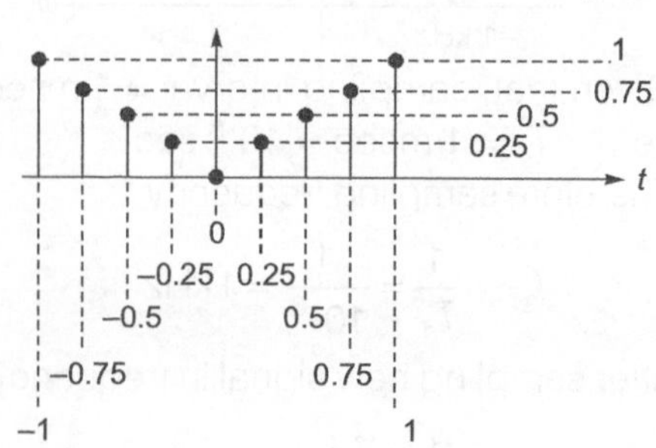

Now, $x_2(t) = \begin{cases} 1-|t|, -1 \le t \le 1 \\ 0, \text{ otherwise} \end{cases}$

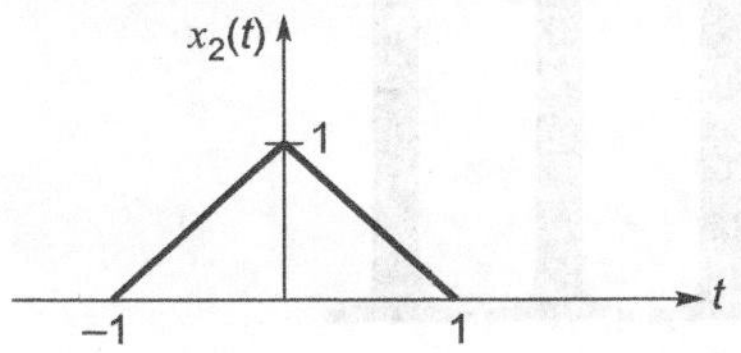

$x_2(n) = \{0, 0.25, 0.5, 0.75, 1, 0.75, 0.5, 0.25, 0\}$

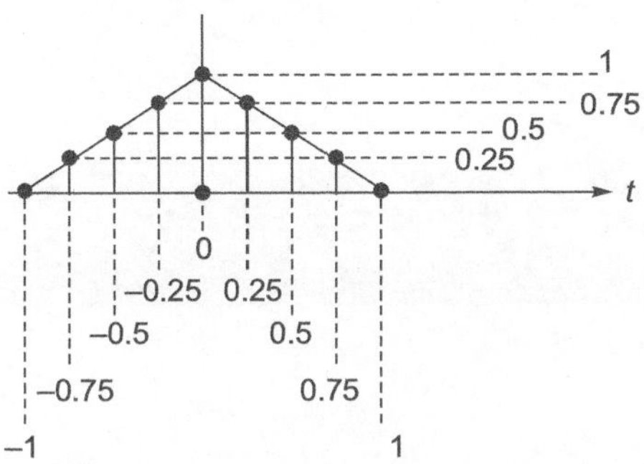

Since $x_1(n)$ is having one more non-zero sample of amplitude '1' as compared to $x_2(n)$. Therefore, energy of $x_1(n)$ is greater than energy of $x_2(n)$.

7.8 (4)

$z(t) = y(t) * x(t) = y(t) * [u(t + 1) - u(t - 1)]$

$= y(t) * u(t) * [\delta(t + 1) - \delta(t - 1)]$

[Property : $f(t) * \delta(t - t_o) = f(t - t_o)$]

$= u(t) * [y(t) * \delta(t + 1) - y(t) * \delta(t - 1)]$

$= u(t) * [y(t + 1) - y(t - 1)]$

$= u(t) * f(t)$ where $f(t) = y(t + 1) - y(t - 1)$

$$\Rightarrow \quad z(t) = \int_{-\infty}^{t} f(\tau)d\tau$$

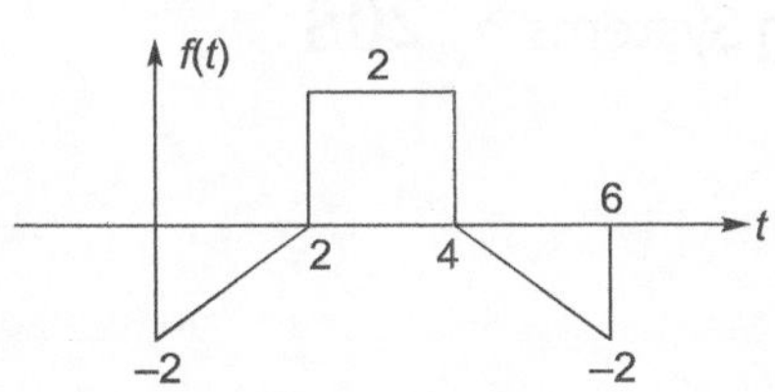

Now, $z(t)|_{t=0} = 0$

$z(t)|_{t=2}$ = Area of $f(t)$ upto "$t = 2$"

$= -\left(\frac{1}{2}\times 2\times 2\right) = -2$

$z(t)|_{t=4}$ = Area of $f(t)$ upto "$t = 4$" $= -2 + 4 = 2$

$z(t)|_{t=6}$ = Area of $f(t)$ upto "$t = 6$" $= 0$

$\therefore$ $z(t)$ is maximum at $t = 4$.

7.9 (80)

Given that :

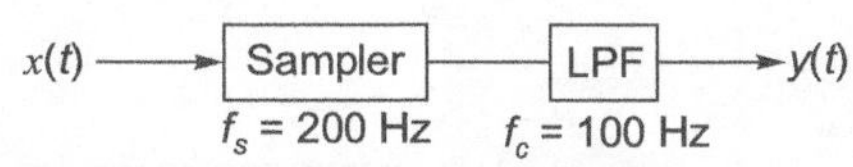

$x(t) = 2\cos(180\pi t)\cos(60\pi t)$

$= \cos(240\pi t) + \cos(120\pi t)$

where $f_1 = 120$ Hz, $f_2 = 60$ Hz

Frequency components present at sampler output

: $f_1, f_s \pm f_1, 2f_s \pm f_1, ...$

$f_2, f_s \pm f_2, 2f_s \pm f_2, ...$

: $120, 200 \pm 120, ...$

$60, 200 \pm 60, ...$

: $120, 80, 320, ...$

$60, 140, 260, ...$

Now, LPF will pass only 60 Hz and 80 Hz because these frequencies are less than cut-off frequency 100 Hz, i.e., LPF-output : 60 Hz, 80 Hz.

So, maximum frequency available at LPF o/p is 80 Hz.

Electrical Machines

UNIT III

CONTENTS

Electrical Machines

Syllabus

Single Phase Transformer: Equivalent Circuit, Phasor Diagram, Open Circuit and Short Circuit Tests, Regulation and Efficiency;

Three Phase Transformers: Connections, Vector Groups, Parallel Operation; Auto-Transformer, Electromechanical Energy Conversion Principles;

DC Machines: Separately Excited, Series and Shunt, Motoring and Generating Mode of Operation and their Characteristics, Speed Control of DC Motors;

Three Phase Induction Machines: Principle of Operation, Types, Performance, Torque-Speed Characteristics, No-Load and Blocked-Rotor Tests, Equivalent Circuit, Starting and Speed Control; Operating Principle of Single-Phase Induction Motors;

Synchronous Machines: Cylindrical and Salient Pole Machines, Performance and Characteristics, Regulation and Parallel Operation of Generators, Starting of Synchronous Motors; Types of Losses and Efficiency Calculations of Electric Machines

Analysis of Previous GATE Papers

Exam Year	1 Mark Ques.	2 Marks Ques.	3 Marks Ques.	5 Marks Ques.	Total Marks
1995	5	1	–	1	**14**
1996	11	1	–	–	**13**
1997	–	3	–	7	**38**
1998	5	3	–	2	**21**
1999	4	8	–	2	**30**
2000	4	8	–	2	**30**
2001	7	4	–	4	**35**
2002	2	3	–	4	**28**
2003	7	11	–	–	**29**
2004	6	13	–	–	**32**
2005	4	10	–	–	**24**
2006	4	12	–	–	**28**
2007	7	7	–	–	**21**
2008	4	10	–	–	**24**
2009	3	8	–	–	**19**
2010	2	4	–	–	**10**
2011	3	3	–	–	**9**

Exam Year	1 Mark Ques.	2 Mark Ques.	Total Marks
2012	1	3	**7**
2013	3	2	**7**
2014 Set-1	3	4	**11**
2014 Set-2	4	4	**12**
2014 Set-3	4	5	**14**
2015 Set-1	2	5	**12**
2015 Set-2	2	0	**2**
2016 Set-1	4	4	**12**
2016 Set-2	1	4	**9**
2017 Set-1	2	5	**12**
2017 Set-2	2	6	**14**
2018	3	4	**11**
2019	3	5	**13**
2020	3	4	**11**
2021	2	3	**8**
2022	2	4	**10**
2023	4	3	**10**
2024	2	2	**6**

1 D.C. Machines

1.1 A differentially compounded DC motor with interpoles and with brushes on the neutral axis is to be driven as a generator in the direction with the same polarity of the terminal voltage. It will then

(a) be a cumulatively compounded generator but the interpole coil connections are to be reversed.
(b) be a cumulatively compounded generator without reversing the interpole coil connections.
(c) be a differentially compounded generator without reversing the interpole coil connections.
(d) be a differentially compounded generator but the interpole coil connections are to be reversed. **[1995 : 1 M]**

1.2 A 4-pole dynamo with wave wound armature has 51 slots containing 20 conductors in each slot. The induced emf is 357 volts and the speed is 8500 rpm. The flux per pole will be

(a) 3.5 mWb (b) 1.2 mWb
(c) 14 mWb (d) 21 mWb **[1995 : 2 M]**

1.3 The torque speed characteristic of a Repulsion motor resembles with which of the following DC motor characteristic?

(a) separately excited (b) shunt
(c) series (d) compound
[1996 : 1 M]

1.4 A 200 V, 10 kW lap-wound DC generator has 10 poles and 500 conductors on its armature. If the pole face covers 80% of the pole pitch, the pole face conductors required to fully compensate for armature reaction will be ____ conductors/pole.
[1997 : 1 M]

1.5 A 240 V DC shunt motor with an armature resistance of 0.5 Ω has a full load current of 40 A. Find the ratio of the stalling torque to the full load torque when a resistance of 1 Ω is connected in series with the armature?

(a) 4 (b) 12
(c) 6 (d) None of the above
[1998 : 2 M]

1.6 A 4-pole lap-wound DC generator has a developed power of P watts and voltage of E volts. Two adjacent brushes of the machine are removed as they are worn out. If the machine operates with the remaining brushes, the developed voltage and power that can be obtained from the machine are

(a) E, P (b) $E/2$, $P/2$
(c) E, $P/4$ (d) E, $P/2$ **[1999 : 1 M]**

1.7 Match the following:

	List-I		List-II
A.	No load and Blocked rotor	1.	Transformer
B.	Sumpner's test	2.	Induction machine
C.	Swinburn's test	3.	Synchronous machine
		4.	D.C. machine

Codes:

	A	B	C
(a)	2	1	4
(b)	1	2	3
(c)	3	2	1
(d)	4	1	2

[1999 : 2 M]

1.8 A DC shunt motor is running at 1200 rpm, when excited with 220 V DC. Neglecting losses and saturation, the speed of the motor when connected to a 175 V d.c. supply is

(a) 750 rpm (b) 900 rpm
(c) 1050 rpm (d) 1200 rpm **[1999 : 2 M]**

1.9 The compensating winding in a DC machine

(a) is located in armature slots for compensation of the armature reaction.
(b) is located on commutating poles for improving the commutation.
(c) is located on pole shoes for avoiding the flashover at the commutator surface.
(d) is located on pole shoes to avoid the sparking at the brushes. **[2000 : 1 M]**

1.10 A 240 V DC series motor takes 40 A when giving its rated output at 1500 rpm. Its resistance is 0.3 Ω. The value of resistance which must be added to obtain rated torque at 1000 rpm is

(a) 6 ohms (b) 5.7 ohms
(c) 2.2 ohms (d) 1.9 ohms **[2000 : 2 M]**

1.11 A permanent magnet DC commutator motor has a no load speed of 6000 rpm when connected to a 120 V DC supply. The armature resistance is 2.5 ohms and other losses may be neglected. The

speed of the motor with supply voltage of 60 V developing a torque of 0.5 Nm, is ___ (in rpm)
(a) 3000 (b) 2673
(c) 2836 (d) 5346 [2000 : 2 M]

1.12 An electric motor with "constant output power" will have a torque speed characteristic in the form of a
(a) straight line through the origin
(b) straight line parallel to the speed axis
(c) circle about the origin
(d) rectangular hyperbola [2001 : 2 M]

1.13 A DC series motor fed from rated supply voltage is overloaded and its magnetic circuit is saturated. The torque-speed characteristic of this motor will be approximately represented by which curve of figure.

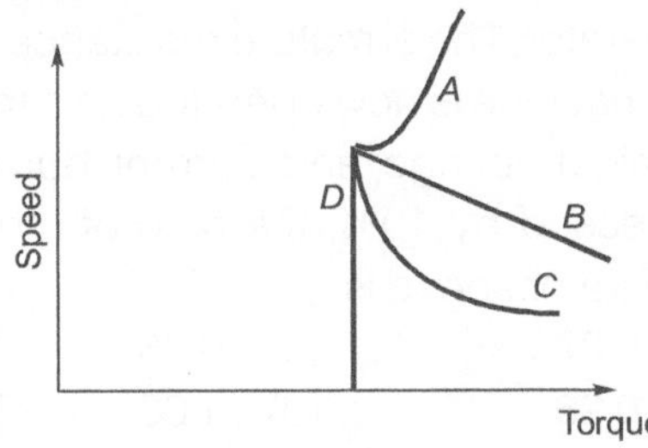

(a) Curve A (b) Curve B
(c) Curve C (d) Curve D [2002 : 1 M]

1.14 A 200 V, 2000 rpm, 10 A, separately excited DC motor has an armature resistance of 2 Ω. Rated DC voltage is applied to both the armature and field winding of the motor. If the armature draws 5 A from the source, the torque developed by the motor is ______. [2002 : 2 M]

1.15 The speed/torque regimes in a DC motor and the control methods suitable for the same are given respectively in **List-I** and **List-II**:

List-I	List-II
A. Field control	1. Below base speed
B. Armature Control	2. Above base speed
	3. Above base torque
	4. Below base torque

Codes:

	A	B
(a)	1	3
(b)	2	4
(c)	2	3
(d)	1	4

[2003 : 1 M]

1.16 To conduct load test on a dc shunt motor, it is coupled to a generator which is identical to the motor. The field of the generator is also connected to the same supply source as the motor. The armature of generator is connected to a load resistance. The armature resistance is 0.02 p.u. Armature reaction and mechanical losses can be neglected. With rated voltage across the motor, the load resistance across the generator is adjusted to obtain rated armature current in both motor and generator. The p.u. value of this load resistance is
(a) 1.0 (b) 0.98
(c) 0.96 (d) 0.94 [2003 : 2 M]

1.17 Following are some of the properties of rotating electrical machines
P : Stator winding current is DC, rotor winding current is AC.
Q : Stator winding current is AC, rotor winding current is DC.
R : Stator winding current is AC, rotor winding current is AC.
S : Stator has salient poles and rotor has commutator.
T : Rotor has salient poles and sliprings and stator is cylindrical
U : Both stator and rotor have poly-phase windings

DC machines, Synchronous machines and Induction machines exhibit some of the above properties as given in the following table. Indicate the correct combination from this table

	DC Machines	Synchronous Machines	Induction Machines
(a)	P.S.	Q.T.	R.U.
(b)	Q.U.	P.T.	R.S.
(c)	P.S.	R.U.	Q.T.
(d)	R.S.	Q.U.	P.T.

[2003 : 2 M]

1.18 A DC series motor driving an electric train faces a constant power load. It is running at rated speed and rated voltage. If the speed has to be brought down to 0.25 p.u. the supply voltage has to be approximately brought down to
(a) 0.75 p.u. (b) 0.5 p.u.
(c) 0.25 p.u. (d) 0.125 p.u. [2003 : 2 M]

1.19 The armature resistance of a permanent magnet dc motor is 0.8 Ω. At no load, the motor draws 1.5 A from a supply voltage of 25 V and runs at 1500 rpm. The efficiency of the motor while it is operating on load at 1500 rpm drawing a current of 3.5 A from the same source will be
(a) 48.0% (b) 57.1%
(c) 59.2% (d) 88.8% [2004 : 2 M]

1.20 A 8 pole, DC generator has a simplex wave-wound armature containing 32 coils of 6 turns each. Its flux per pole is 0.06 Wb. The machine is running at 250 rpm. The induced armature voltage is
(a) 96 V (b) 192 V
(c) 384 V (d) 768 V [2004 : 2 M]

1.21 Two magnetic poles revolve around a stationary armature carrying two coil $(c_1 - c_1', c_2 - c_2')$ as shown in the figure. Consider the instant when the poles are in a position as shown. Identify the correct statement regarding the polarity of the induced emf at this instant in coilsides c_1 and c_2.

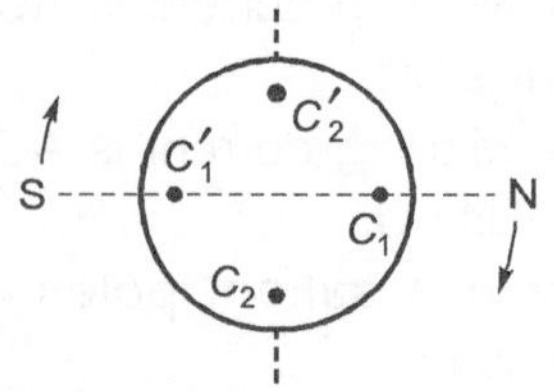

(a) ⊙ in c_1, no emf in c_2
(b) ⊗ in c_1, no emf in c_2
(c) ⊙ in c_2, no emf in c_1
(d) ⊗ in c_2, no emf in c_1 [2005 : 2 M]

1.22 A 50 kW DC shunt motor is loaded to draw rated armature current at any given speed. When driven
(i) At half the rated speed by armature voltage control and
(ii) At 1.5 times the rated speed by field control, the respective output powers delivered by the motor are approximately.
(a) 25 kW in (i) and 75 kW in (ii)
(b) 25 kW in (i) and 50 kW in (ii)
(c) 50 kW in (i) and 75 kW in (ii)
(d) 50 kW in (i) and 50 kW in (ii) [2005 : 2 M]

1.23 In relation to DC machines, match the following and choose the correct combination

List-I (Performance Variables)
A. Armature emf (E)
B. Developed torque (T)
C. Developed power (P)

List-II (Proportional to)
1. Flux (ϕ), speed (ω) and armature current (I_a)
2. ϕ and ω only
3. ϕ and I_a only
4. I_a and ω only
5. I_a only

Codes:

	A	B	C
(a)	3	3	1
(b)	2	5	4
(c)	3	5	4
(d)	2	3	1

[2005 : 2 M]

1.24 In a DC machine, which of the following statement is true?
(a) Compensating winding is used for neutralizing armature reaction while interpole winding is used for producing residual flux.
(b) Compensating winding is used for neutralizing armature reaction while interpole winding is used for improving commutation.
(c) Compensating winding is used for improving commutation while interpole winding is used for neutralizing armature reaction.
(d) Compensating winding is used for improving commutation while interpole winding is used for producing residual flux. [2006 : 1 M]

1.25 A 220 V DC machine supplies 20 A at 200 V as a generator. The armature resistance is 0.2 ohm. If the machine is now operated as a motor at same terminal voltage and current but with the flux increased by 10%, the ratio of motor speed to generator speed is
(a) 0.87 (b) 0.95
(c) 0.96 (d) 1.06 [2006 : 2 M]

1.26 The DC motor, which can provide zero speed regulation at full load without any controller is
(a) series
(b) shunt
(c) cumulative compound
(d) differential compound [2007 : 1 M]

Statement for Linked Answer Questions (1.27 and 1.28):

A 240 V, DC shunt motor draws 15 A while supplying the rated load at a speed of 80 rad/s. The armature resistance is 0.5 Ω and the field winding resistance is 80 Ω.

1.27 The net voltage across the armature resistance at the time of plugging will be
(a) 6 V (b) 234 V
(c) 240 V (d) 474 V [2008 : 2 M]

1.28 The external resistance to be added in the armature circuit to limit the armature current to 125% of its rated value is
(a) 31.1 Ω (b) 31.9 Ω
(c) 15.1 Ω (d) 15.9 Ω [2008 : 2 M]

1.29 Figure shows the extended view of a 2 pole DC machine with 10 armature conductors. Normal brush positions are shown by A and B, placed at the interpolar axis. If the brushes are now shifted, in the direction of rotation, to A' and B' as shown, the voltage waveform $V_{A'B'}$ will resemble

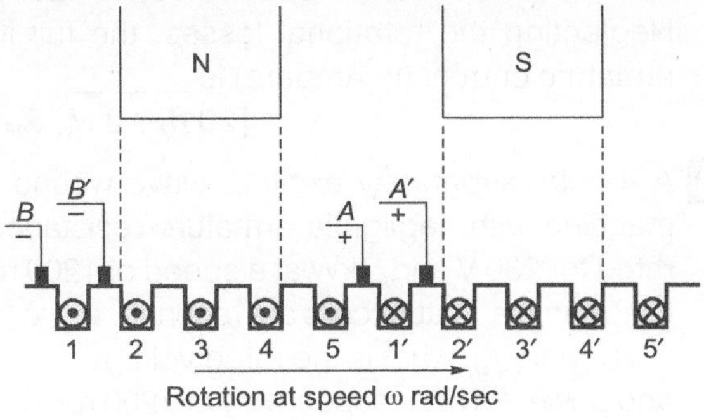

(a)

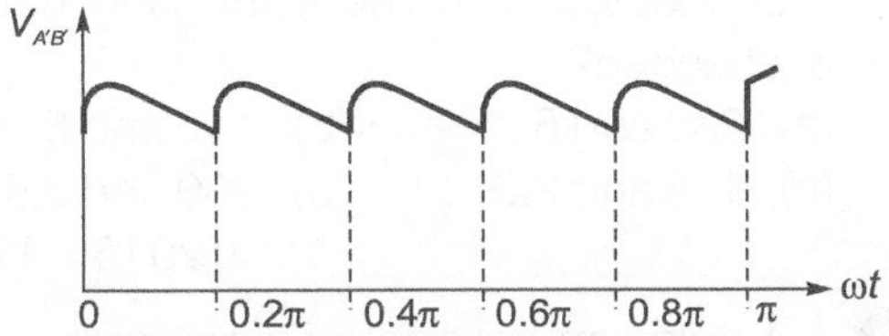

(b)

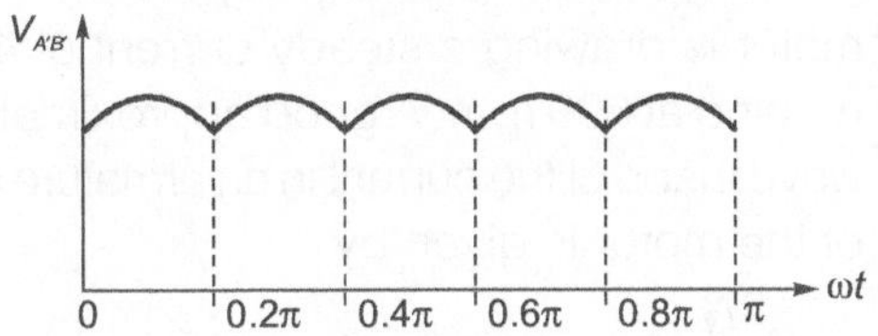

(c)

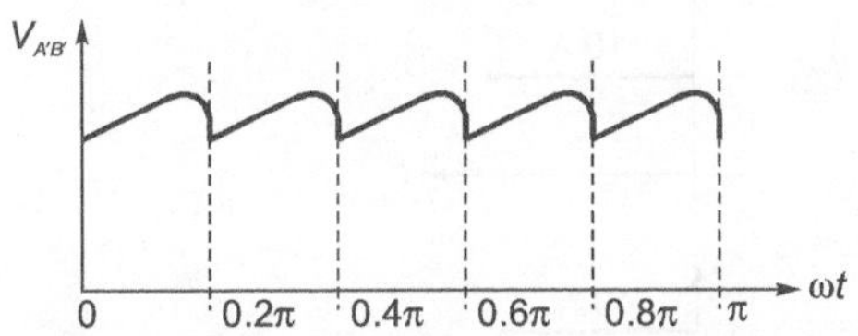

(d)

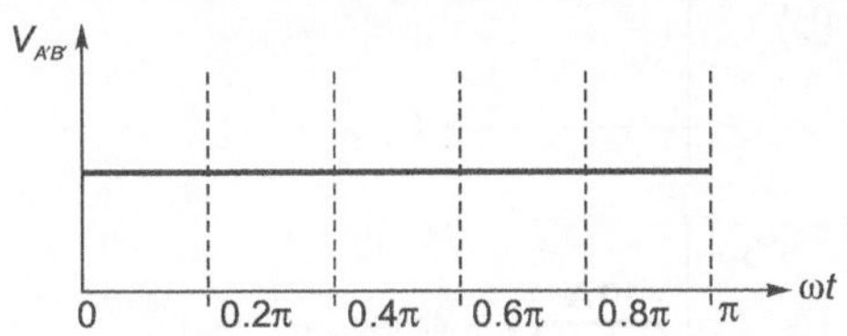

[2009 : 2 M]

Common Data for Questions (1.30 and 1.31):

A separately excited DC motor runs at 1500 rpm under no-load with 200 V applied to the armature. The field voltage is maintained at its rated value. The speed of the motor, when it delivers a torque of 5 Nm, is 1400 rpm as shown in the figure. The rotational losses and armature reaction are neglected.

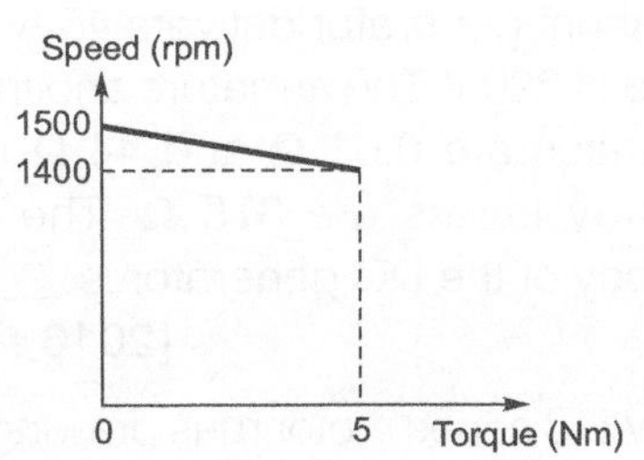

1.30 The armature resistance of the motor is,

(a) 2 Ω (b) 3.4 Ω
(c) 4.4 Ω (d) 7.7 Ω [2010 : 2 M]

1.31 For the motor to deliver a torque of 2.5 Nm at 1400 rpm, the armature voltage to be applied is

(a) 125.5 V (b) 193.3 V
(c) 200 V (d) 241.7 V [2010 : 2 M]

1.32 A 4-point starter is used to start and control the speed of a

(a) DC shunt motor with armature resistance control
(b) DC shunt motor with field weakening control
(c) DC series motor
(d) DC compound motor [2011 : 1 M]

1.33 A 220 V, DC shunt motor is operating at a speed of 1440 rpm. The armature resistance is 1.0 Ω and armature current is 10 A. If the excitation of the machine is reduced by 10%, the extra resistance to be put in the armature circuit to maintain the same speed and torque will be

(a) 1.79 Ω (b) 2.1 Ω
(c) 3.1 Ω (d) 18.9 Ω [2011 : 2 M]

1.34 A 220 V, 15 kW, 1000 rpm shunt motor with armature resistance of 0.25 Ω, has a rated line current of 68 A and a rated field current of 2.2 A. The change in field flux required to obtain a speed of 1600 rpm while drawing a line current of 52.8 A and a field current of 1.8 A is

(a) 18.18% increase (b) 18.18% decrease
(c) 36.36% increase (d) 36.36% decrease

[2012 : 2 M]

1.35 A 15 kW, 230 V dc shunt motor has armature circuit resistance of 0.4 Ω and field circuit resistance of 230 Ω. At no load and rated voltage, the motor runs at 1400 rpm and the line current drawn by the motor is 5 A. At full load, the motor draws a line current of 70 A. Neglect armature reaction. The full load speed of the motor in rpm is ______.

[2014 : 2 M, Set-1]

1.36 A 250 V DC shunt machine has armature circuit resistance of 0.6 Ω and field circuit resistance of 125 Ω. The machine is connected to 250 V supply mains. The motor is operated as a generator and then as a motor separately. The line current of the machine in both the cases is 50 A. The ratio of the speed as a generator to the speed as a motor is ______.

[2014 : 2 M, Set-2]

1.37 For a single-phase, two winding transformer, the supply frequency and voltage are both increased by 10%. The percentage changes in the hysteresis loss and eddy current loss, respectively, are

(a) 10 and 21 (b) –10 and 21
(c) 21 and 10 (d) –21 and 10

[2014 : 2 M, Set-2]

1.38 The no load speed of a 230 V separately excited dc motor is 1400 rpm. The armature resistance drop and the brush drop are neglected. The field current is kept constant at rated value. The torque of the motor in Nm for an armature current of 8 A is ______ . [2014 : 1 M, Set-3]

1.39 A separately excited 300 V DC shunt motor under no load runs at 900 rpm drawing an armature current of 2 A. The armature resistance is 0.5 Ω and leakage inductance is 0.01 H. When loaded, the armature current is 15 A. Then the speed in rpm is ______ . [2014 : 2 M, Set-3]

1.40 A separately excited DC generator has an armature resistance of 0.1 Ω and negligible armature inductance. At rated field current and rated rotor speed, its open-circuit voltage is 200 V. When this generator is operated at half the rated speed, with half the rated field current, an un-charged 1000 μF capacitor is suddenly connected across the armature terminals. Assume that the speed remains unchanged during the transient. At what time (in microsecond) after the capacitor is connected with the voltage across it reach 25 V?

(a) 62.25 (b) 69.3
(c) 73.25 (d) 77.3

[2015 : 1 M, Set-1]

1.41 A DC motor has the following specifications: 10 hp, 37.5 A, 230 V; flux/pole = 0.01 Wb, number of poles = 4, number of conductors = 666, number of parallel paths = 2. Armature resistance = 0.267 Ω. The armature reaction is negligible and rotational losses are 600 Ω. The motor operates from a 230 V DC supply. If the motor runs at 1000 rpm, the output torque produced (in Nm) is ______. [2015 : 2 M, Set-1]

1.42 A separately excited DC motor runs at 1000 rpm on no load when its armature terminals are connected at a 200 V DC source and the rated voltage is applied to the field winding. The armature resistance of this motor is 1 Ω. The no-load armature current is negligible. With the motor developing its full load torque, the armature voltage is set so that the rotor speed is 500 rpm. When the load torque is reduced to 50% of the full load value under the same armature voltage conditions, the speed rises to 520 rpm. Neglecting the rotational losses, the full load armature current (in Ampere) is ______.

[2015 : 2 M, Set-1]

1.43 A 4-pole, separately excited, wave wound DC machine with negligible armature resistance is rated for 230 V and 5 kW at a speed of 1200 rpm. If the same armature coils are reconnected to form a lap winding, what is the rated voltage (in volts) and power (in kW) respectively at 1200 rpm of the reconnected machine if the field circuit is left unchanged?

(a) 230 and 5 (b) 115 and 5
(c) 115 and 2.5 (d) 230 and 2.5

[2015 : 1 M, Set-2]

1.44 A 4-pole, lap-connected, separately excited DC motor is drawing a steady current of 40 A while running at 600 rpm. A good approximation for the waveshape of the current in an armature conductor of the motor is given by

(a)
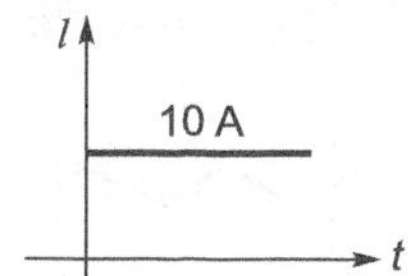

(b)
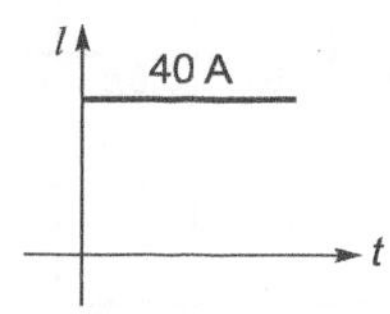

(c)
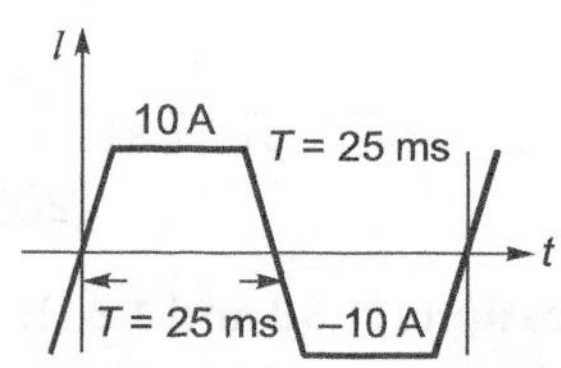

(d)
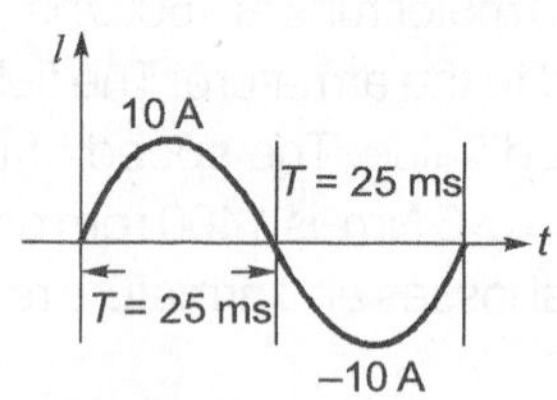

[2016 : 1 M, Set-1]

1.45 A DC shunt generator delivers 45 A at a terminal voltage of 220 V. The armature and the shunt field resistances are 0.01 Ω and 44 Ω respectively. The stray losses are 375 Ω. The percentage efficiency of the DC generator is ______.

[2016 : 2 M, Set-1]

1.46 A 220 V DC series motor runs drawing a current of 30 A from the supply. Armature and field circuit resistance are 0.4 Ω and 0.1 Ω respectively. The load torque varies as the square of the speed. The flux in the motor may be taken as being proportional to the armature current. To reduce the speed of the motor by 50%, the resistance in ohms that should be added in series with the armature is _____. (Give answer upto two decimal places) [2017 : 2 M, Set-1]

1.47 A separately excited DC generator supplies 150 A to a 145 V DC grid. The generator is running at 800 rpm. The armature resistance is 0.1 Ω. If the speed of the generator is increased to 1000 rpm, the current in amperes supplied by the generator to the DC grid is ________ . (Give the answer up to one decimal place.) [2017 : 2 M, Set-1]

1.48 A 120 V DC shunt motor takes 2 A at no load. It takes 7 A on full load while running at 1200 rpm. The armature resistance is 0.8 Ω, and the shunt field resistance is 240 Ω. The no load speed, in rpm, is ______. [2017 : 2 M, Set-2]

1.49 A 220 V, 10 kW, 900 rpm separately excited DC motor has an armature resistance R_a = 0.02 Ω. When the motor operates at rated speed and with rated terminal voltage, the electromagnetic torque developed by the motor is 70 Nm. Neglecting the rotational losses of the machine, the current drawn by the motor from the 220 V supply is
(a) 34.2 A (b) 30 A
(c) 22 A (d) 4.84 A
[2017 : 2 M, Set-2]

1.50 A separately excited DC motor has an armature resistance R_a = 0.05 Ω. The field excitation is kept constant. At an armature voltage of 100 V, the motor produces a torque of 500 Nm at zero speed. Neglecting all mechanical losses, the no-load speed of the motor (in radian/s) for an armature voltage of 150 V is _____ (upto 2 decimal places). [2018 : 1 M]

1.51 A 200 V DC series motor, when operating from rated voltage while driving a certain load, draws 10 A current and runs at 1000 rpm. The total series resistance is 1 Ω. The magnetic circuit is assumed to be linear. At the same supply voltage, the load torque is increased by 44%. The speed of the motor in rpm (rounded to the nearest integer) is _______. [2018 : 2 M]

1.52 A 220 V DC shunt motor takes 3 A at no-load. It draws 25 A when running at full-load at 1500 rpm. The armature and shunt resistance are 0.5 Ω and 220 Ω, respectively. The no-load speed in rpm (round off to two decimal places) is ______. [2019 : 2 M]

1.53 A 250 V DC shunt motor has an armature resistance of 0.2 Ω and a field resistance of 100 Ω. When the motor is operated on no-load at rated voltage. It draws an armature current of 5 A and runs at 1200 rpm. When a load is coupled to the motor, it draws total line current of 50 A at rated voltage, with a 5 % reduction in the air-gap flux due to armature reaction. Voltage drop across the brushes can be taken as 1 V per brush under all operating conditions. The speed of the motor, in rpm, under this loaded condition, is closest to:
(a) 900 (b) 1200
(c) 1000 (d) 1220 [2020 : 2 M]

1.54 A belt-driven DC shunt generator running at 300 RPM delivers 100 kW to a 200 V DC grid. It continues to run as a motor when the belt breaks, taking 10 kW from the DC grid. The armature resistance is 0.025 Ω, field resistance is 50 Ω, and brush drop is 2 V. Ignoring armature reaction, the speed of the motor is _______ RPM. (Round off 2 decimal places) [2021 : 2 M]

1.55 A 280 V, separately excited DC motor with armature resistance of 1 Ω and constant field excitation drives a load. The load torque is proportional to the speed. The motor draws a current of 30 A when running at a speed of 1000 rpm. Neglect frictional losses in the motor. The speed, in rpm, at which the motor will run, if an additional resistance of value 10 Ω is connected in series with the armature, is ____. (round off to nearest integer). [2022 : 2 M]

1.56 A separately excited DC motor rated 400 V, 15 A, 1500 RPM drives a constant torque load at rated speed operating from 400 V DC supply drawing rated current. The armature resistance is 1.2 Ω. If the supply voltage drops by 10% with field current unaltered then the resultant speed of the motor in RPM is ___________ (Round off to the nearest integer). [2023 : 1 M]

1.57 A 5 kW, 220 V DC shunt motor has 0.5 Ω armature resistance including brushes. The motor draws a no-load current of 3 A. The field current is constant at 1 A. Assuming that the core and rotational losses are constant and independent of the load, the current (in amperes) drawn by the motor while delivering the rated load, for the best possible efficiency, is ________ (rounded off to 2 decimal places). [2024 : 2 Marks]

■■■■

Answers D.C. Machines

1.1 (b)	1.2 (b)	1.3 (c)	1.4 (4)	1.5 (a)	1.6 (c)	1.7 (a)
1.8 (d)	1.9 (c)	1.10 (d)	1.11 (b)	1.12 (d)	1.13 (b)	1.14 (4.299)
1.15 (b)	1.16 (c)	1.17 (a)	1.18 (b)	1.19 (a)	1.20 (c)	1.21 (a)
1.22 (b)	1.23 (d)	1.24 (b)	1.25 (a)	1.26 (d)	1.27 (d)	1.28 (a)
1.29 (a)	1.30 (b)	1.31 (b)	1.32 (b)	1.33 (a)	1.34 (d)	
1.35 (1240.63)	1.36 (1.271)	1.37 (a)	1.38 (12.55)	1.39 (880.43)	1.40 (b)	1.41 (57.8)
1.42 (8)	1.43 (b)	1.44 (c)	1.45 (86.84%)	1.46 (10.75)	1.47 (550)	
1.48 (1241.81)	1.49 (b)	1.50 (600)	1.51 (825)	1.52 (1579.33)	1.53 (d)	
1.54 (275.18)	1.55 (483)	1.56 (1343)	1.57 (27.28)			

Explanations D.C. Machines

1.1 (b)

Since, the direction of armature current will be reversed, the nature will be cumulative. This reverse current will also change the nature of interpole flux. Hence no need to reverse the connection.

1.2 (b)

$$E_b = \frac{\phi PZN}{60A} \Rightarrow 357 = \frac{\phi \times 4 \times 51 \times 20 \times 8500}{60 \times 2}$$

$$\phi = 1.235 \text{ mWb} \approx 1.2 \text{ mWb}$$

1.4 Sol.

Compensating conductor/pole

$$= \frac{Z}{AP} \times 0.8 = \frac{500}{10 \times 10} \times 0.8 = 4$$

1.5 (a)

At stall condition,

$$E_b = 0 \qquad \text{(As } N = 0\text{)}$$

$$V = I_{stall} \cdot R_a$$

$$240 = I_{stall} \cdot (0.5\ \Omega + 1\ \Omega)$$

$$I_{stall} = 160 \text{ A}$$

$$\frac{T_{stall}}{T_{start}} = \frac{I_{stall}}{I_{start}} = \frac{160}{40} = 4$$

1.6 (c)

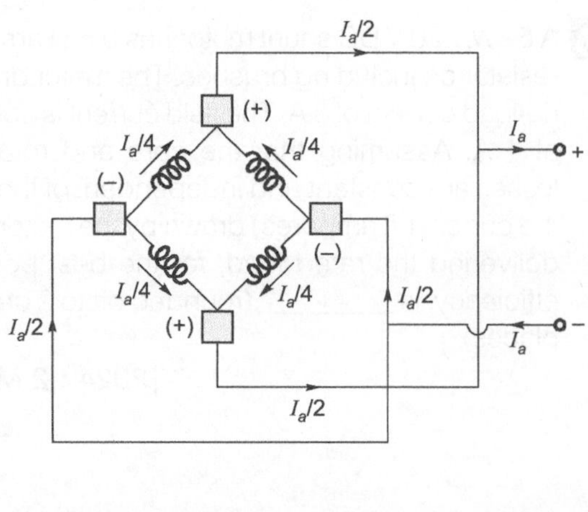

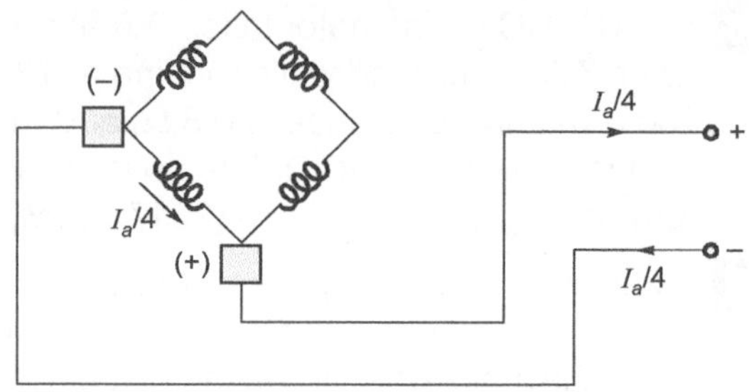

$$E' = E; \qquad I'_a = \frac{I_a}{4},$$

$$\therefore \qquad P' = E\frac{I_a}{4} = \frac{P}{4}$$

1.8 (d)

For DC shunt motor,

$$I_f = \frac{V}{R_f}$$

$$\Rightarrow \qquad \phi = K'V$$

Also, $V = E_b$ (As lossless motor)

$$V = K\phi N$$

$$N = \frac{V}{K\phi} = \frac{V}{KK'V} = \frac{1}{KK'}$$

Hence, N = Constant, irrespective of V.

1.10 (d)

As torque is constant, both armature current and flux remains constant,

$$V = E_b + I_a R_a \Rightarrow 240 = E_b + 40 \times 0.3$$

$$E_b = 228 \text{ V}$$

$$E_{b_2} = \frac{N_2}{N_1} E_{b_1} \qquad \text{(As } \phi = \text{constant)}$$

$$= \frac{1000}{1500} \times 228 = 152 \text{ V}$$

Now, $240 = 152 + 40(0.3 + R_{ext})$

$$R_{ext} = 1.9\ \Omega$$

1.11 (b)

$$V = E_b + I_a R_a \quad \text{(At no load } I_a = 0\text{)}$$

$$\Rightarrow \quad E_{b_1} = 120\text{ V}$$

$$E_{b_2} = \frac{N_2 \times 120}{6000} = 0.02\, N_2$$

$$\text{or,} \quad N_2 = 50\, E_{b_2} \quad \text{...(i)}$$

$$\text{Similarly,} \quad 60 = E_{b_2} + I_a \times 2.5$$

$$\Rightarrow \quad E_{b_2} = (60 - I_a \times 2.5) \quad \text{...(ii)}$$

$$\text{Also,} \quad E_{b_2} I_a = T \times \omega$$

$$\text{or,} \quad \frac{E_{b_2}}{2.5}(60 - E_{b_2}) = \frac{0.5 \times 2\pi \times N_2}{60}$$

$$\text{or,} \quad E_{b_2}(60 - E_{b_2}) = \frac{2.5\pi}{60} \cdot E_{b_2} \times 50$$

$$\Rightarrow \quad E_{b_2} = 53.455\text{ V}$$

$$\text{and} \quad N_2 = 2672.75\text{ rpm}$$

1.12 (d)

$$P = \frac{2\pi N}{60} \cdot T = \text{Constant}$$

$$xy = C \quad \text{(Rectangular Hyperbola)}$$

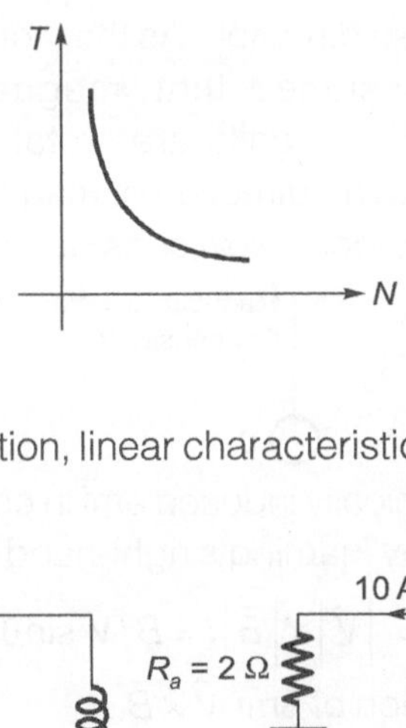

1.13 (b)

At saturation, linear characteristics are observed.

1.14 Sol.

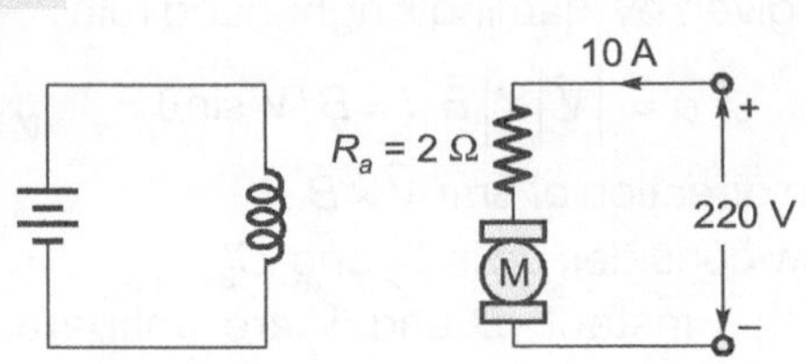

$$(E_b)_{\text{rated}} = 200 - 2 \times 10 = 180\text{ V}$$

$$\text{Rated power} = 180 \times 10 = 1800\text{ W}$$

$$\text{Rated torque} = \frac{1800 \times 60}{2\pi \times 2000} = 8.598\text{ Nm}$$

Torque at 5 A,

$$\frac{T_2}{T_1} = \frac{I_{a_2}}{I_{a_2}}$$

$$\Rightarrow \quad T_2 = \frac{5}{10} \times 8.598 = 4.299\text{ Nm}$$

1.15 (b)

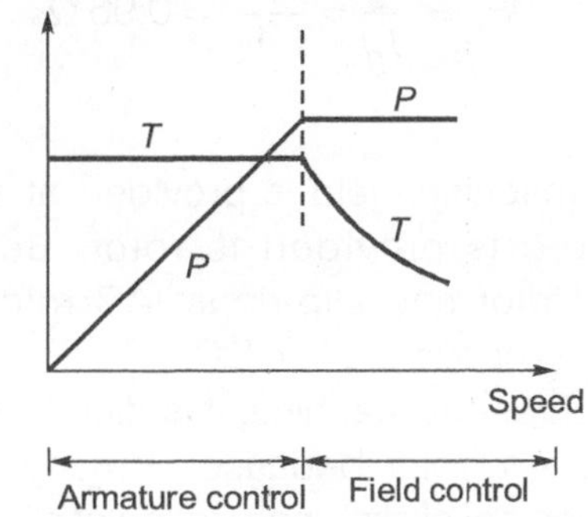

Field Control: Field control provides constant power drive. Since the speed is inversely proportional to flux/pole $\left(\omega \propto \frac{1}{\phi}\right)$ field can not be increased; it can only be weakened. So this control is suitable for speed control above base speed as torque is directly proportional to flux/pole ($T \propto \alpha$) for a given armature current. So this control provides torque below base torque.

Armature control: It provides constant-torque drive. By keeping the field current at maximum value full load torque can be obtained at full load armature current at all speeds. This control provides speeds below base speed.

1.16 (c)

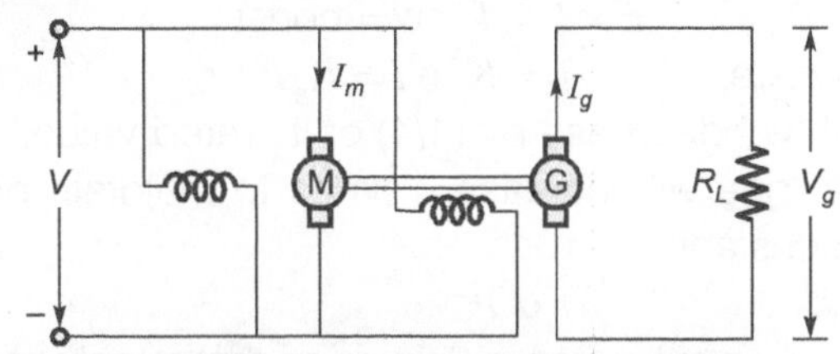

All calculations are done in p.u. system.

Motor is connected across rated voltage $V = 1$ pu.

Rated armature current flows in both motor and generator

$$I_g = I_m = 1\text{ pu}$$

Armature resistance of both motor and generator

$$R_m = R_g = 0.02\text{ pu}$$

Back emf in motor

$$= E_b = V - I_m R_m = 1 - 1 \times 0.02 = 0.98\text{ pu}$$

Mechanical power output of the motor

$$= E_b I_m = 0.98 \times 1 = 0.98\text{ pu}$$

This power is given to the generator and Armation reaction and mechanical losses are neglected.

Input power of generator = Output power of motor

$$E_g I_g = E_b I_m$$

$$E_g = E_b = 0.98\text{ pu}$$

Terminal voltage of the generator

$$V_g = E_g - I_g R_g$$

$$= 0.98 - 1 \times 0.02 = 0.96 \text{ pu}$$

$$R_L = \frac{V_g}{I_g} = \frac{0.96}{1} = 0.96\ \Omega$$

1.17 (a)

In DC machine, field is provided at stator and armature is provided is rotor. Because of commutator and slip-rings, AC rotor winding current is converted into DC.

In synchronous machine, it is difficult to provide armature in rotor because of many factor like insulation no. of slip rings, rating of slip ring, high rating of armature. Therefore, armature is provided in stator and field in rotor.

Induction machines are singly fed machine i.e. they draw excitation from the same supply. So no external field supply is required.

1.18 (b)

In series motor,

Back emf, $E = V - I(R_a + R_{se})$

If, $(R_a + R_{se}) = R$

then, $E = V - IR$ where, $E = K_n \phi N$

$$\therefore \quad N = \frac{V - IR}{K_n \phi}$$

In series motor, $\phi \propto I$

$\therefore$ At constant power load

$$E \times I = T \times w = \text{const}$$

where, $T = K_a \phi I = K_a I^2$

If 'w_n' decrease by (1/4) of its rated value, then torque will increase 4 times to maintain power constant.

$$\therefore \quad T \propto I^2$$

$\therefore$ Current increase 2 times so that supply voltage approximately brought down to 0.5 pu.

1.19 (a)

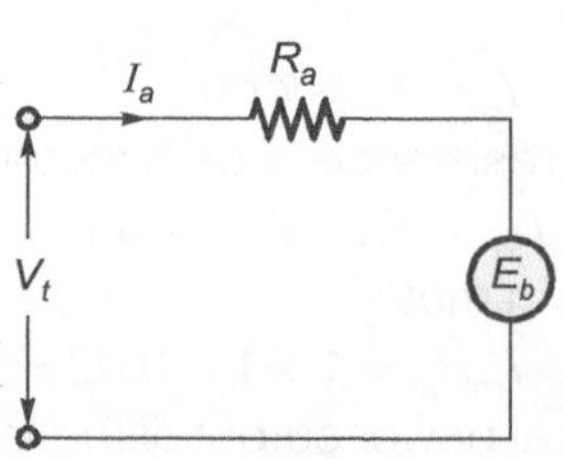

At no load, $I_{a0} = 1.5$ A

Supply voltage $= V_t = 250$ V

Back emf at no load,

$E_{bo} = V_t - I_{a0} R_a = 25 - 1.5 \times 0.8 = 23.8$ V

Power developed by the motor

$P_0 = E_{b0} I_{a0} = 23.8 \times 1.5 = 35.7$ W

As the motor is working at no load, power P_0 supplies no load losses.

When the motor is on load,

$$I_{a1} = 3.5 \text{ A}$$

$$E_{a1} = V_t - I_{a1} R_a = 25 - 3.5 \times 0.8 = 22.2 \text{ V}$$

Power developed by the motor

$$P_1 = E_{a1} I_{a1} = 22.2 \times 3.5 = 77.7 \text{ W}$$

Power available at shaft

$P_s = P_1 -$ no load losses $= 77.7 - 35.7 = 42$ W

$$\text{Efficiency} = \frac{\text{Output power}}{\text{Input Power}} \times 100$$

$$= \frac{P_0}{V_t I_{a_1}} \times 100 = \frac{42}{25 \times 3.5} \times 100 = 48\%$$

1.20 (c)

No. of coils = 32. Each coil has 6 turns

Total no. of turns = 32 × 6 = 192

Each turn has 2 conductors

So total no. of conductors = Z = 192 × 2 = 384

For wave would armature no. of parallel paths = A = 2.

Expression of induced emf,

$$E = \frac{PZ}{60A} \phi N = \frac{8 \times 384}{60 \times 2} \times 0.06 \times 250 = 384 \text{ V}$$

1.21 (a)

Magnetic poles revolves around stationary coils in clockwise direction. As the motion is relative, it can be assumed that, magnetic poles are stationary and coils are rotating in counter-clockwise. The direction of magnetic flux is from N-pole to S-pole. Now consider coils C_1 and C_i

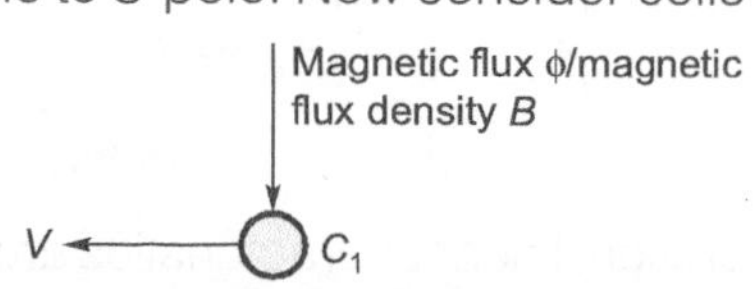

The dynamically induced emf in conductor C_1 can be given by Flaming's right-hand rule

$$e = |\vec{V}| \times |\vec{B}| l = BlV \sin\theta$$

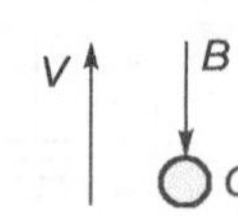

The direction of emf $\vec{V} \times \vec{B}$

Now consider, coils C_2 and C_2'

At this instant, B and V are antiparallel, i.e., $\theta = 180°$. So, $\sin 180° = 0$. Hence, no emf is induced in C_2.

1.22 (b)

Armature control provides constant-torque drive. In the shunt motor case by keeping the field current at maximum value full load torque can be obtained at full load armature current at all speeds.

$$T = \text{Constant}$$

$$T\omega = \text{Output Power } (P_0)$$

$$\Rightarrow \quad \frac{P_0}{\omega} = \text{Constant}$$

$$(P_0)_1 = 50 \text{ kW}$$

$$\omega_2 = \frac{\omega_1}{2}$$

$$\frac{(P_0)_1}{\omega_1} = \frac{(P_0)_2}{\omega_2}$$

$$\Rightarrow \quad \frac{50}{\omega_1} = \frac{(P_0)_2}{\omega_1/2}$$

$$(P_0)_2 = 25 \text{ kW}$$

(Output power delivered at half the rated speed by armature control).

Field control provides constant-power drive. Since the speed is inversely proportional to the flux/pole $\left(\omega \propto \frac{1}{\phi}\right)$ while the torque is directly proportional to flux/pole ($T \propto \phi$) for a given armature current.

$$\text{Output power} = T\omega = \left(\frac{k_1}{\phi}\right)(k_2\phi)$$

$$= k_1 k_2 = \text{Constant}$$

Output power delivered by the motor remains constant, i.e., 50 kW

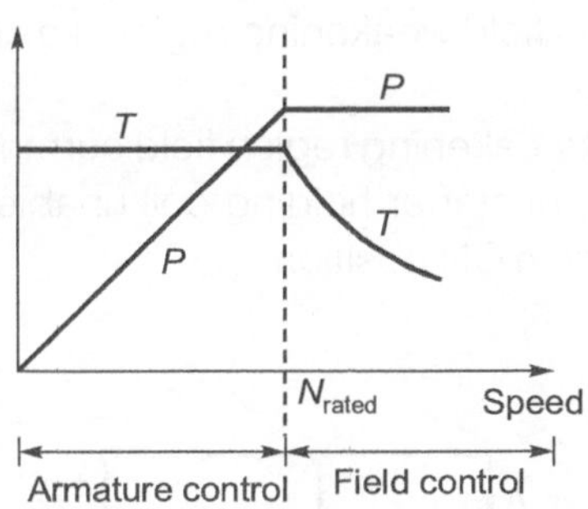

1.23 (d)

As we know,

(i) Armature emf, $E = k_a \phi \omega$
Where k_a is a constant
So, E depends on ϕ and ω only.

(ii) Developed torque, $T = k_a \phi I_a$
T depends on ϕ and I_a only.

(iii) Developed power, $P = EI_a = (k_a \phi \omega) I_a$
or, $P = T\omega = (k_a \phi I_a)\omega$
So, P depends on ϕ, ω and I_a.

1.24 (b)

Compensating winding is placed in slots cut out in pole faces such that the axis of this winding coincides with the brush axis. The compensating winding neutralizes the armature mmf directly under the pole while in the interpolar region, there is incomplete neutralization.

To speed up the commutation process, the reactance voltage must be neutralized by injecting a suitable polarity dynamical (speed) voltage into the commutating coil. In order that this injection is restricted to commutating coils, narrow interpolar are provided in the interpolar region.

1.25 (a)

For generator,
$E = V + I_a R_a = 200 + 20 \times 0.2 = 204$
For motor, $V = E + I_a R_a$
$E = 200 - 20 \times 0.2 = 196$

$$\frac{E_1}{E_2} = \frac{N_1}{N_2} \times \frac{\phi_1}{\phi_2} \Rightarrow \frac{204}{196} = \frac{N_1}{N_2} \times \frac{1}{1.1}$$

$$\therefore \quad \frac{N_2}{N_1} = \frac{196}{204 \times 1.1} = 0.87$$

1.26 (d)

Speed-current characteristics of DC motors
It can seen from the characteristics, speed regulation can be zero at full load in the case of differential compound DC motor.

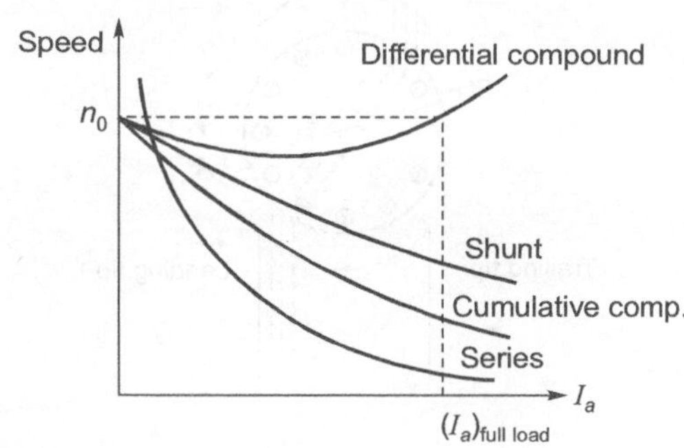

1.27 (d)

Before plugging

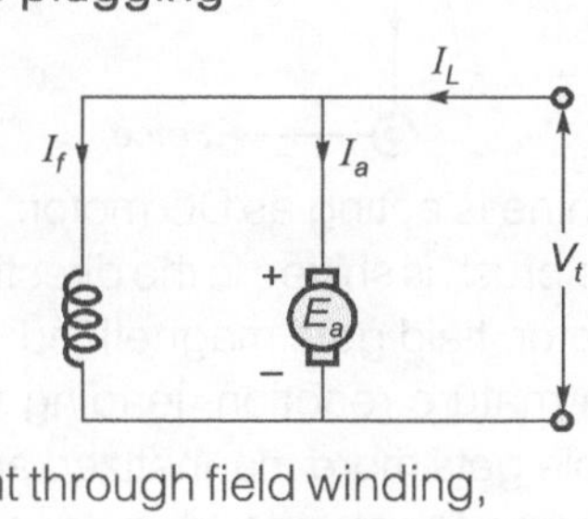

Current through field winding,

$$I_f = \frac{V_t}{R_f} = \frac{240}{80} = 3 \text{ A}$$

$$I_L = I_f + I_a$$

Current through armature

$$I_a = I_L - I_f = 15 - 3 = 12 \text{ A}$$

$$\text{Back emf} = E_a = V_t - I_a R_a$$
$$= 240 - 12 \times 0.5 = 234 \text{ V}$$

In plugging, the connection of armature winding is reversed. A strong braking torque is achieved by maintaining the supply voltage to the armature with connections reversed.

So, the net voltage across the armature resistance at the time of plugging

$$E_a + V = 240 + 234 = 474 \text{ V}$$

1.28 (a)

Rated armature current = 12 A

Armature current during braking

$$I_b = 1.25 \times 12 = 15 \text{ A}$$

Total armature resistance including external resistance during braking

$$R_b = R_a + R_{ext}$$

$$R_b = \frac{E+V}{I_b} = \frac{474}{15} = 31.6\ \Omega$$

$$= R_a + R_{ext} = 31.6\ \Omega$$

$$0.5 + R_{ext} = 31.6\ \Omega \Rightarrow R_{ext} = 31.1\ \Omega$$

1.29 (a)

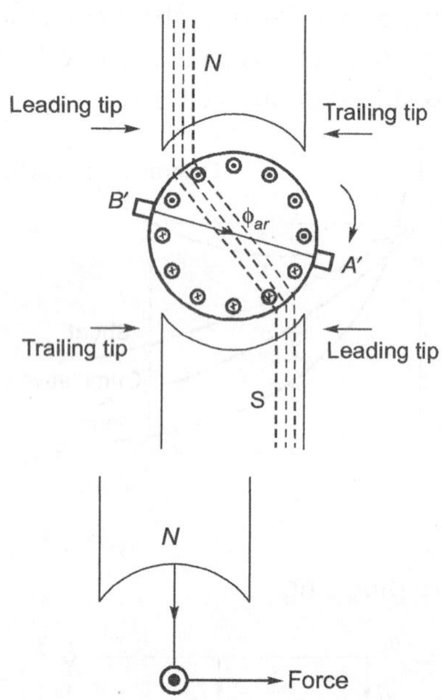

DC machine is acting as DC motor.

When the brush is shifted in the direction of rotation in DC motor, field gets magnetized.

Due to armature reaction, leading tip of N-pole and S-pole gets more magnetized and trailing tip of N-pole and S-pole gets demagnetized.

But increase in flux density is more than the decrease in flux density.

Due to magnetizing effect of armature reaction waveform of $V_{A'B'}$ will be as given option (a).

1.30 (b)

At no load $I_0 = 0$, $N_0 = 1500$ rpm,

$$E_a = V = 200 \text{ V}$$

Assuming $N \propto E_a$,

at $N = 1400$ rpm

$$\Rightarrow E_a = 200 \times \frac{1400}{1500} = 186.67 \text{ V}$$

$$I_a = \frac{T\omega}{E_a} = \frac{5 \times 1400 \times \frac{2\pi}{60}}{186.67} = 3.925 \text{ A}$$

$$I_a R_a = V - E = 200 - 186.67 = 13.33 \text{ V}$$

$$R_a = \frac{13.33}{3.925} = 3.39\ \Omega$$

1.31 (b)

For $N = 1400$ rpm, $E_a = 186.67$ V

Required 'V' $= E_a + I_a R_a = 186.67 + I_a(3.39)$

where I_a is function of torque,

To develop 5 N-m it requires 3.925 A

$$I_a = 3.925\left(\frac{2.5}{5}\right)$$

$$I_a = \frac{3.925}{2} = 1.9625 \text{ A}$$

$$V_{applied} = 186.67 + (1.9625)(3.39) = 193.3 \text{ V}$$

1.32 (b)

4-point starter is used to control the speed of shunt motor in field weakening region. i.e.; above rated speeds.

In field weakening region field current will reduce in 3-point starter holding coil unable to hold the plunger in ON position.

1.33 (a)

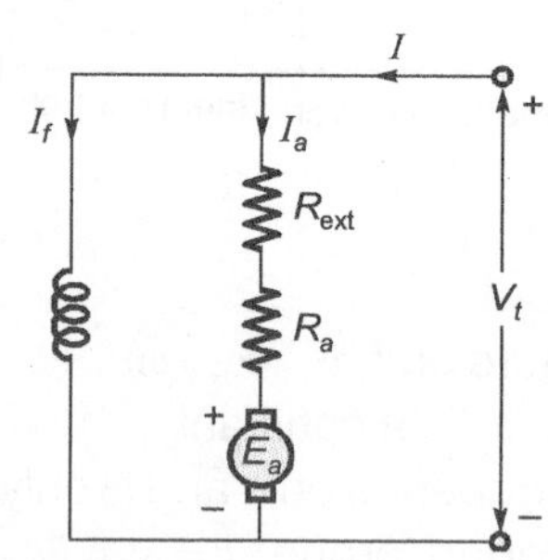

Case (i):

$$R_a = 1\ \Omega;\quad I_{a1} = 10 \text{ A}$$

$$N_1 = 1440 \text{ rpm};\quad V_t = 220 \text{ V}$$

$$E_{a1} = V_t - I_{a1}R_a = 210 \text{ V} \qquad \{R_{ext} = 0\}$$

Case (ii):

Back emf,

$$E_a \propto \phi N;\quad \phi_2 = 0.9\,\phi_1 \text{ and } N_2 = N_1$$

As speed is constant

$$E_a \propto \phi$$

$$E_{a2} = \frac{\phi_2}{\phi_1} E_{a_1} = \frac{0.9\phi_1}{\phi_1} \times 210 \Rightarrow E_{a2} = 189 \text{ V}$$

Torque $\propto \phi I_a$

As torque is constant

$$T_1 = T_2$$
$$\phi I_{a1} = \phi_2 I_{a2}$$
$$\phi_1 \times 10 = 0.9\phi_1 I_{a2}$$
$$I_{a2} = 11.11 \text{ A}$$
$$E_{a2} = V_t - I_{a2}(R_a + R_{ext})$$
$$189 = 220 - 11.11(1 + R_{ext})$$

External resistance = R_{ext} = 1.79 Ω

1.34 (d)

Given: $r_a = 0.25\ \Omega$

$$I_{f1} = 2.2 \text{ A}; \quad I_{f2} = 1.8 \text{ A}$$
$$I_{a1} = 68 - 2.2 = 65.8 \text{ A}$$
$$I_{a2} = 52.8 - 1.8 = 51 \text{ A}$$
$$E_a = V_t - I_a r_a$$
$$E_{a1} = 220 - 65.8 \times 0.25 = 203.55 \text{ V}$$
$$E_{a2} = 220 - 51 \times 0.25 = 207.25 \text{ V}$$
$$E_a = K\phi\omega_m$$
$$\frac{E_{a1}}{E_{a2}} = \frac{\phi_1}{\phi_2} \cdot \frac{\omega_{m1}}{\omega_{m2}}$$
$$\frac{203.55}{207.25} = \frac{\phi_1}{\phi_2} \times \frac{1000}{1600}$$
$$\frac{\phi_1}{\phi_2} = 1.571$$

i.e., field flux reduces

∴ % change in flux

$$= \frac{\phi_2 - \phi_1}{\phi_1} = -36.36\%$$

⇒ Decreases in flux.

1.35 Sol.

Given, $R_a = 0.4\ \Omega$, $R_f = 230\ \Omega$, $V = 230$ volt

At no-load, speed, $N_0 = 1400$ rpm, $I_{L0} = 5$ A

Field current, $I_f = \frac{V}{R_f} = \frac{230}{230} = 1 \text{ A}$

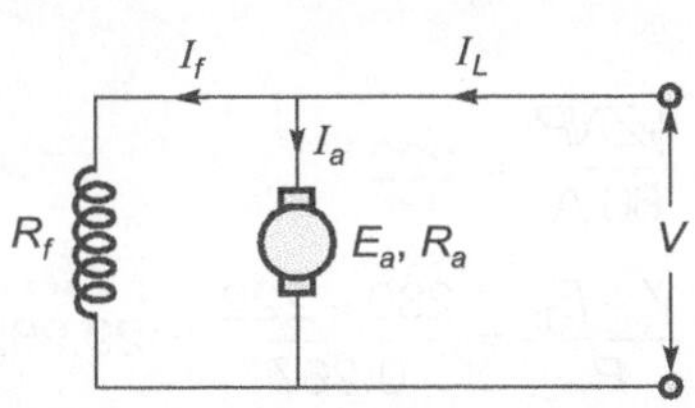

∴ No-load armature current,

$$I_{a0} = 5 - 1 = 4 \text{ A}$$

∴ Induced emf at no-load,

$$E_{a0} = V - I_{a0}R_a = 230 - 4 \times 0.4 = 228.4 \text{ volt}$$

Also, at full load,

$$I_{Lfl} = 70 \text{ A}$$

∴ $I_{afl} = I_{Lfl} - I_f = 70 - 1 = 69$ A

∴ Induced emf at full-load,

$$E_{afl} = V - I_{afl} R_a$$

or, $E_{afl} = 230 - 69 \times 0.4 = 202.4$ volt

We know that,

$$E_a \propto \phi\omega_m$$
$$\alpha\ \phi N \quad \text{(where, } N = \text{Speed in rpm)}$$

or, $E_a \propto N$

(As flux, ϕ = Constant in shunt motor)

$$\therefore \quad \frac{E_{afl}}{E_{a0}} = \frac{N_{fl}}{N_0} \text{ or } N_{fl} = \left(\frac{E_{afl}}{E_0}\right) \times N_0$$
$$= \left(\frac{202.4}{228.4}\right) \times 1400$$

Therefore, full-load speed of motor = 1240.63 rpm

1.36 Sol.

Given, $V = 250$ volt, $R_a = 0.6\ \Omega$, $R_f = 125\ \Omega$

$I_L = 50$ A both for motor as well as generator action.

Field current, $I_f = \frac{V}{R_f} = \frac{250}{125} = 2 \text{ A}$

For motor action:

$$E_{am} = V - I_a R_a$$
$$I_a = I_L - I_f = 50 - 2 = 48 \text{ A}$$
$$\therefore \quad E_{am} = 250 - 48 \times 0.6 = 221.2 \text{ volt}$$

For generator action:

$$E_{ag} = V + I_a R_a$$
$$I_a = I_L + I_f = 50 + 2 = 52 \text{ A}$$
$$\therefore \quad E_{ag} = 250 + 52 \times 0.6 = 281.2 \text{ volt}$$

For a DC shunt machine, $E_a \propto \phi\omega_m$

or, $E_a \propto N$(speed) $(\because \phi = \text{constant})$

$$\therefore \quad \frac{E_{ag}}{E_{am}} = \frac{N_g}{N_m} \text{ or } \frac{N_g}{N_m} = \frac{281.2}{221.2} = 1.271$$

∴ Ratio of the speed as a generator to the speed of motor = 1.271.

1.37 (a)

Eddy current loss, $P_e \propto B_m^2 f^2$

and Hysteresis loss,

$$P_h \propto B_m^x f \propto B_m^{1.6} f \quad (\text{for } x = 1.6)$$

When, $\frac{V}{f}$ = constant, $P_e \propto V^2$

and $P_h \propto \frac{V^{1.6}}{f^{0.6}}$

Given, $V_2 = 1.1\,V_1$ and $f_2 = 1.1\,f_1$

$$\therefore \quad \frac{P_{e_2}}{P_{e_1}} = \left(\frac{V_2}{V_1}\right)^2 = (1.1)^2 = 1.21$$

$$\text{or,} \quad \frac{P_{e_2} - P_{e_1}}{P_{e_1}} = 0.21$$

$\therefore$ % increase in eddy current loss

$$= \left[\frac{P_{e_2} - P_{e_1}}{P_{e_1}}\right] \times 100 = 21\%$$

$$\text{Also,} \quad \frac{P_{h_2}}{P_{h_1}} = \left(\frac{V_2}{V_1}\right)^{1.6} \times \left(\frac{f_1}{f_2}\right)^{0.6}$$

$$= (1.1)^{1.6} \times \left(\frac{1}{1.1}\right)^{0.6} = 1.1$$

$$\text{or,} \quad \left(\frac{P_{h_2} - P_{h_1}}{P_{h_1}}\right) = 0.1$$

$\therefore$ % increase in hysteresis loss

$$= \left(\frac{P_{h_2} - P_{h_1}}{P_{h_1}}\right) \times 100 = 10\%$$

Thus, hysteresis loss increases by 10% and eddy current loss by 21%.

1.38 Sol.

Given, $I_a = 8$ A

No-load speed,

$$N_0 = 1400 \text{ rpm}, V = 230 \text{ volt}$$

$$\therefore \quad \omega_0 = \frac{2\pi}{60} \times N_0 = \frac{2\pi}{60} \times 1400 = 146.6 \text{ rad/s}$$

Power developed in the motor,

$$P = E_a \cdot I_a$$

Also, Power, P = Torque $\times\, \omega_0$

$$\text{or, torque,} \quad T = \frac{P}{\omega_0} = \frac{E_a I_a}{\omega_0}$$

Since, armature resistance drop and brush drop are neglected therefore,

$$E_a \simeq V \approx 230 \text{ volt}$$

$$\therefore \text{ Torque,} \quad T = \frac{230 \times 8}{146.6} = 12.55 \text{ N-m}$$

1.39 Sol.

Given, $V = 300$ volt, No-load speed $= 900$ rpm, No-load current,

$$I_{a0} = 2 \text{ A}$$

$\therefore$ No-load induced emf,

$$E_{a0} = V - I_{a0}R$$

$$= 300 - 2 \times 0.5 = 299 \text{ volt}$$

Also, full-load armature current,

$$I_{af} = 15 \text{ A}$$

$\therefore$ Induced emf in armature at load,

$$E_{af} = V - I_{af}R_a = 300 - 15 \times 0.5 = 292.5 \text{ volt}$$

Let the speed of motor during loaded condition be N.

We know that:

$$E_a \propto \phi N \text{ or } E_a \propto N \quad (\phi = \text{constant})$$

$$\therefore \quad \frac{N}{N_0} = \frac{E_{af}}{E_{a0}} \text{ or } N = \frac{E_{af}}{E_{a0}} \cdot N_0 = \left(\frac{292.5}{299}\right) \times 900$$

$$N = 880.43 \text{ rpm}$$

1.40 (b)

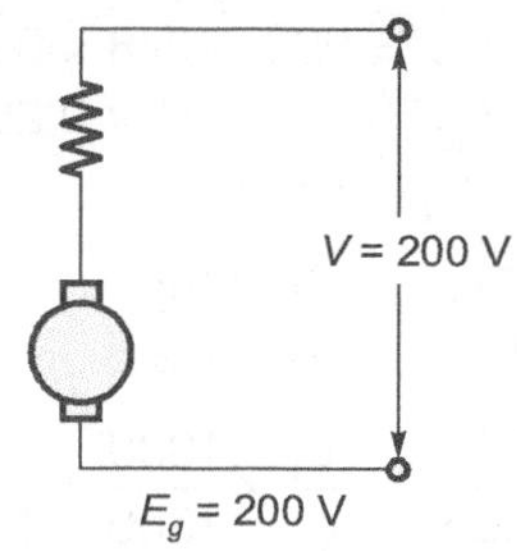

$R_a = 0.1\ \Omega$, $I_{f\,\text{rated}}$, ω_{rated}

$$V_{OC} = 200 \text{ V}$$

$$I_f' = \frac{I_f}{2};\ \omega' = \frac{\omega_{\text{rated}}}{2}$$

$$E_g = K\phi\omega$$

$$E_g' = K\frac{\phi}{2} \times \frac{\omega}{2} = \frac{200}{4} = 50 \text{ V}$$

$$V = V(\infty) - [V(\infty) - V(0)]\, e^{-t/T} = 50 - [50 - 0]\, e^{-t/T}$$

$$25 = 50\,[1 - e^{-t/RC}]$$

$$\frac{1}{2} = 1 - e^{-t/100 \times 10^{-6}}$$

$$e^{-t/10^{-4}} = \frac{1}{2} \Rightarrow t = 0.693 \times 10^{-4} = 69.3\ \mu\text{s}$$

1.41 Sol.

$$E_b = \frac{\phi ZNP}{60\,A} = 222 \text{ V}$$

$$I_a = \frac{V - E_b}{P_a} = \frac{230 - 222}{0.267} = 29.9625 \text{ A}$$

$$T = \frac{60}{2\pi(1000)}\,[222 \times 29.96 - 600] = 57.8 \text{ N-m}$$

1.42 Sol.

$V_t = 200$ V, $N = 1000$ rpm, $I_{a0} = 0$ A ...(1)

$V_t = V$ (say), $N = 500$ rpm, $T_d = T_{fL}$ ...(2)

$V_t = V$ (said), $N = 520$ rpm,

$$T_L' = \frac{T_L}{2} = \frac{T_{FL}}{2} \quad ...(3)$$

$$P_{losses} = 0 \quad \text{(Neglected)}$$

$$R_a = 1\ \Omega$$

From condition (1),

$$E_b = V_t - I_a R_a \quad \text{(Neglected at no load)}$$

$$E_{b_1} = 200\text{ V},\ N = 1000\text{ rpm}$$

Since from (2),

$$N_2 = 500\text{ rpm}$$

$$E_{b_2} = \frac{500 \times 200}{1000} = 100\text{ V}$$

$$\because \quad E_{b_2} = V - I_{fL}\cdot R_a;$$

$$100 = V - I_{fL}\cdot R_a;$$

$$100 = V - I_{fL} \quad ...(4)$$

Again (3), $N_3 = 520$ rpm,

$$E_{b_3} = \frac{520 \times 200}{1000} = 104\text{ V}$$

As we know, $104 = V - \frac{I_{fL}}{2} \times R = V - \frac{I_{fL}}{2}$ ...(5)

From eqn. (4) and (5), $100 + I_{fL} = 104 + \frac{I_{fL}}{2}$

On solving, we get $I_{fL} = 8$ A

1.43 (b)

$A = 2$, separately excited DC machine

$R_a \approx 0$, 230 V, 5 kW, $N = 1200$ rpm

For wave winding,

$$230 = \frac{\phi \times 4 \times 1200 \times Z_1}{60 \times 2} \quad ...(i)$$

For lap winding,

$$V = \frac{\phi \times 4 \times 1200 \times Z_1}{60 \times 4} \quad ...(ii)$$

Dividing equation (*i*) by equation (*ii*),

$$\frac{230}{V} = \frac{4}{2} \Rightarrow V = 115\text{ V}$$

For power:

Wave:

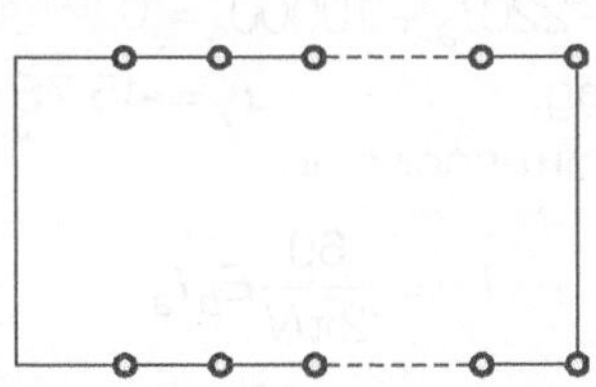

Total conductor in single path $= \frac{Z}{2}$

Let, resistance of $\left(\frac{Z}{2}\right)$ conductor $= R$

Total resistance, $P_1 = \frac{V_1^2}{R_1} = \frac{V_1^2}{(R/2)}$ ...(*i*)

Lap:

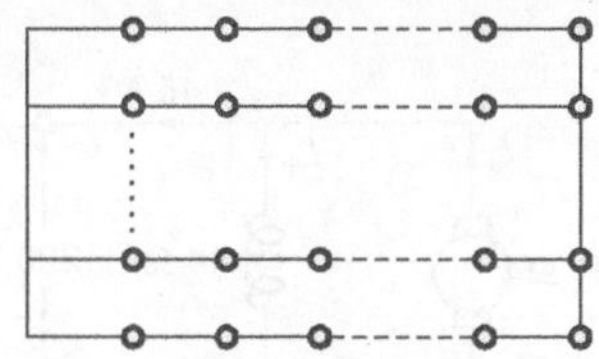

Total conductor in single path $= \frac{Z}{4}$

Let resistance of single path $= \frac{R}{2}$

Total resistance $= \frac{R/2}{4} = \frac{R}{8}$

$$P_2 = \frac{V_2^2}{R_2} = \frac{V_2^2}{(R/8)} \quad ...(ii)$$

By equation (*i*) and (*ii*), we get

$$\frac{P_1}{P_2} = \frac{V_1^2/(R/2)}{V_2^2/(R/8)} = \frac{(230)^2}{R/2} \times \frac{R/8}{(715)^2} = 1$$

1.44 (c)

4 pole, lap - connected, separately excited dc motor, $N = 600$ rpm

$$I_{total} = 40\text{ A}$$

Parallel path = 4 = no. of poles

Current in each parallel path $= \frac{40}{4}$ 10 A

Current in armature conductor = 10 A

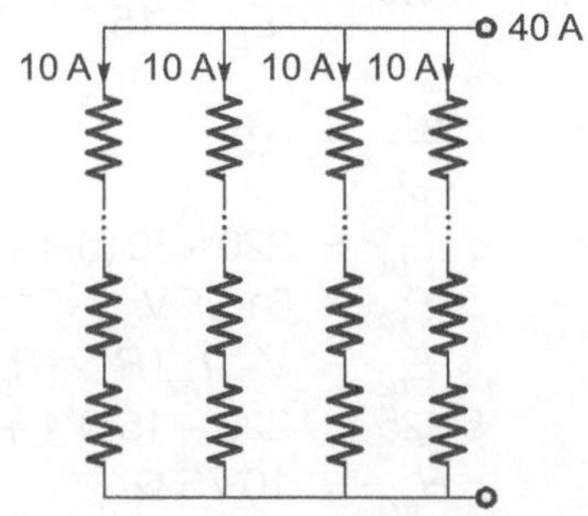

$$\text{Speed} = 600\text{ rpm}$$

$$N = \frac{120 \times f}{4} = 600$$

$$f = \frac{600 \times 4}{120} = 20\text{ Hz}$$

Time period, $T = \frac{1}{f} = \frac{1}{20} = 50$ msec

So,

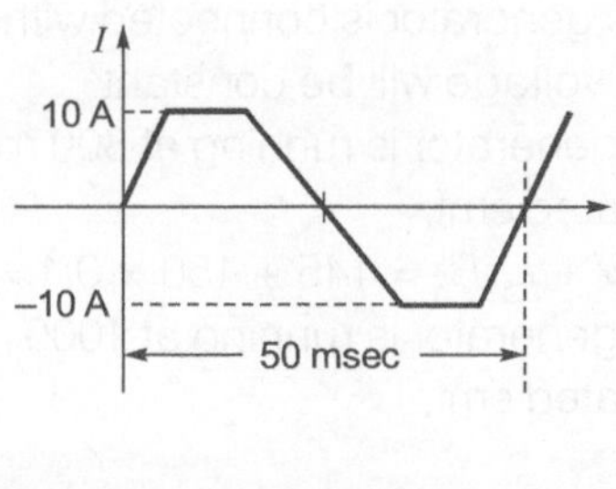

1.45 Sol.

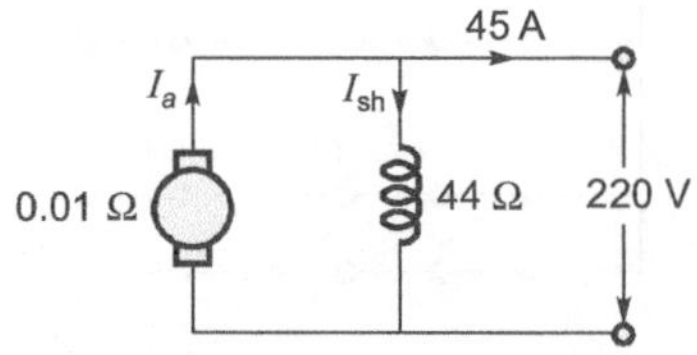

Stay losses = 375 Watt

Total copper losses = $I_a^2R_a + I_{sh}^2R_{sh}$

$= 50^2 \times (0.01) + (5)^2 \times 44 = 1125$ Watt

$$h = \frac{\text{Output}}{\text{Output + losses}}$$

$$= \frac{220 \times 45}{220 \times 45 + (1125) + 375} = 0.86842 \text{ or } 86.84\%$$

1.46 Sol.

$$T \propto N^2 \quad \phi \propto I_a \quad T \propto I_a^2$$

$$\therefore \quad N^2 \propto I_a^2$$

$$\therefore \quad \frac{N_2}{N_1} = \frac{I_{a2}}{I_{a1}} \Rightarrow 0.5 = \frac{I_{a2}}{30}$$

$$\Rightarrow \quad I_{a2} = 15 \text{ A}$$

For series motor,

$$\frac{N_2}{N_1} = \frac{E_{b2}}{E_{b1}} \times \frac{I_{a1}}{I_{a2}}$$

$$\therefore \quad 0.5 = \frac{E_{b2}}{E_{b1}} \times \frac{30}{15}$$

$$\therefore \quad \frac{E_{b2}}{E_{b1}} = 0.25$$

$$E_{b1} = 220 - 30\,(0.4 + 0.1) = 205 \text{ V}$$

$$\therefore \quad E_{b2} = 51.25 \text{ V}$$

$$\therefore \quad E_{b2} = V - I_{a2}(R_a + R_{se} + R_{ext})$$

$$51.25 = 220 - 15(0.4 + 0.1 + R_{ext})$$

$$\therefore \quad R_{ext} = 10.75\ \Omega$$

1.47 Sol.

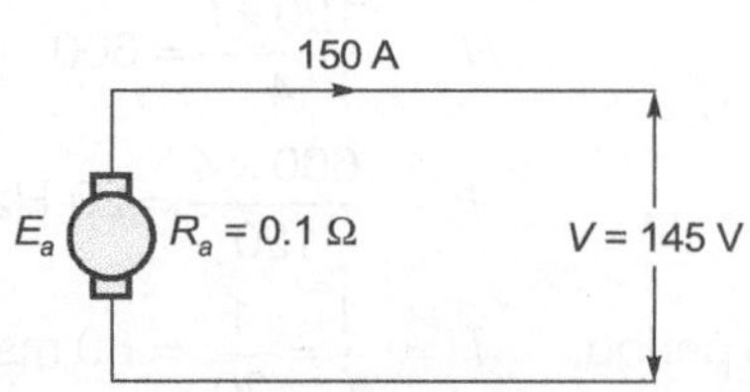

Given, $\quad R_a = 0.1\ \Omega$

As, DC generator is connected with DC grid, the output voltage will be constant.

When generator is running at 800 rpm,

Generated emf,

$E_{a1} = V + I_{a1}R_a = 145 + 150 \times 0.1 = 160$ V

When generator is running at 1000 rpm,

Generated emf,

$$E_{a2} = E_{a1} \times \frac{N_2}{N_1} = 160 \times \frac{1000}{800} = 200 \text{ V}$$

So, new armature current will be,

$$I_{a2} = \frac{E_{a2} - V}{R_a} = \frac{200 - 145}{0.1} = \frac{55}{0.1} = 550 \text{ A}$$

1.48 Sol.

Shunt motor take 7 A on full load and runs @ 1200 rpm.

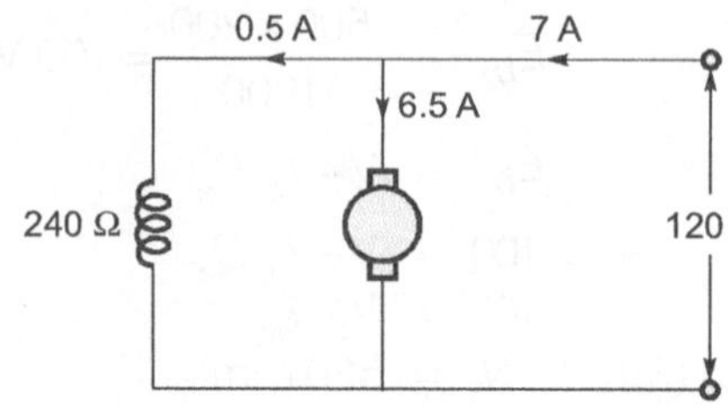

$$\therefore \quad E_{b1} = 120 - 65(0.8) = 114.8 \text{ V}$$

$$N_1 = 1200 \text{ rpm}$$

Shunt motor takes 2 A at no load

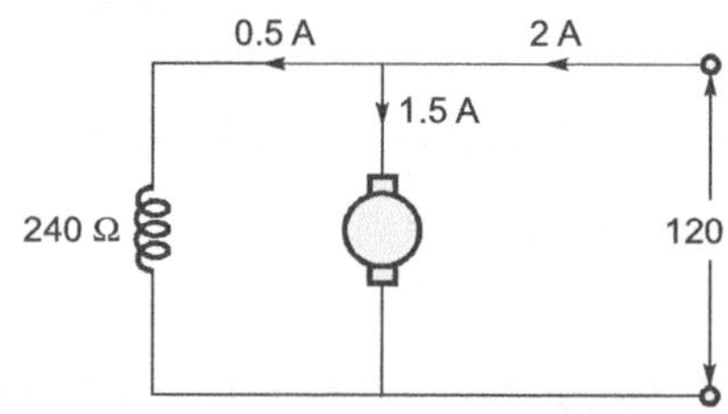

$$E_{b2} = 120 - 1.5(0.8) = 118.8$$

For shunt motor,

$$\frac{N_2}{N_1} = \frac{E_{b2}}{E_{b1}} \Rightarrow N_2 = \frac{118.8}{114.8} \times 1200 = 1241.81 \text{ rpm}$$

1.49 (b)

Neglecting rotor loss,

$$E_bI_a = 10 \text{ kW}$$

$$(V - I_aR_a)\,I_a = 10 \text{ kW}$$

$$[220 - I_a(0.02)]\,I_a = 10000$$

$$220I_a - 0.02I_a^2 - 10000 = 0$$

$$0.02I_a^2 - 220I_a + 10000 = 0$$

By solving, $\quad I_a = 45.75$ A

Torque corresponding,

$$T = \frac{60}{2\pi N}E_bI_a$$

$$\therefore \quad T = \frac{60}{2\pi(900)}10000 = 106.15 \text{ N-m}$$

Separately excited motor, $T \propto I_a$

For a torque 106.15 N-m, $I_a = 45.75$ A

For a torque 70 N-m, $I_a = ?$

New, $\quad I_a = \dfrac{70 \times 45.75}{106.15} = 30.16$ A

1.50 Sol.

Given, separately initiated DC motor.
Field excitation is constant,

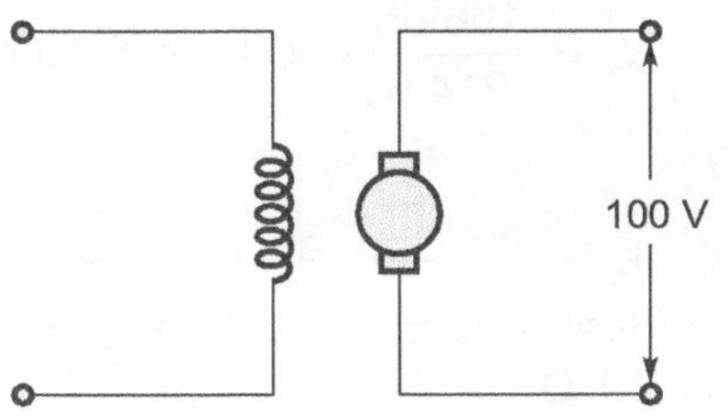

Producing a torque of 500 N-m
At zero speed,

$$\because \quad N = 0, \quad E_b = 0$$

$$E_b = V - I_a R_a$$

$$\therefore \quad 0 = 100 - I_a(0.05)$$

$$\therefore \quad I_{a1} = \frac{100}{0.05} = 2000 \text{ A}$$

$$T = \frac{60}{2\pi N} E_b I_a = \frac{60}{2\pi N} \frac{\phi ZNP}{60\,A} \cdot I_a$$

Let, $T_1 = \frac{1}{2\pi} \frac{ZP}{A} \phi I_{a1}$ N-m

Let, $T_1 = k I_{a1}$

where, $k = \frac{1}{2\pi} \frac{ZP\phi}{A}$

$$T_1 = 500 \text{ N-m}, I_{a1} = 2000 \text{ A}$$

$$k = \frac{500}{2000} = 0.25$$

When motor runs on no-load given all mechanical losses neglected. No-load current is negligible and the voltage drop at no-load can be neglegible.

$$\therefore \quad E_b \simeq V = 150 \text{ V}$$

$$E_b = \frac{\phi ZNP}{60A}$$

Already solved,

$$k = \frac{1}{2\pi}\left(\frac{ZP\phi}{A}\right)$$

$$\therefore \quad \frac{ZP\phi}{A} = 2\pi(k)$$

$$\therefore \quad E_b = 2\pi(k) \cdot \frac{N}{60}$$

$$\therefore \quad E_b = k \times \omega$$

(ω : No-load speed in rad/sec.)

$$\therefore \quad \omega = \frac{E_b}{k}$$

$$\therefore \quad \omega = \frac{150}{0.25} = 600 \text{ rad/sec.}$$

1.51 Sol.

From circuit diagram,

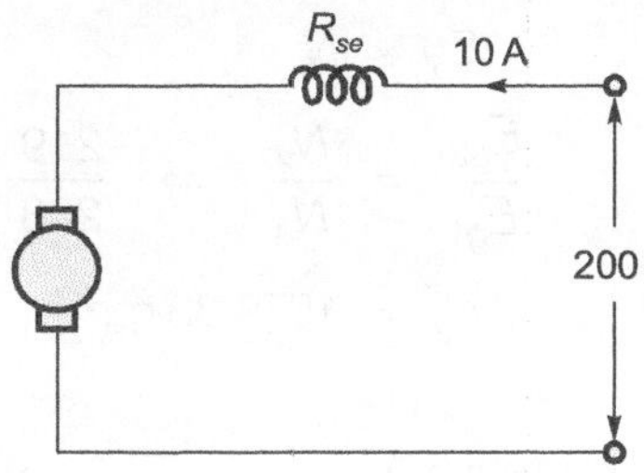

$$E_{b1} = V - I_a(R_a + R_{se}) = 200 - 10(1) = 190 \text{ V}$$

$$N_1 = 1000 \text{ rpm}$$

Load torque increased by 44% ($T \propto I_a^2$)

$$\therefore \quad \frac{T_2}{T_1} = \left(\frac{I_{a2}}{I_{a1}}\right)^2 \Rightarrow \frac{1.44 T_1}{T_1} = \left(\frac{I_{a2}}{10}\right)^2$$

$$\therefore \quad I_{a2}^2 = 144 \Rightarrow I_{a2} = 12 \text{ A}$$

At same voltage,

$$E_{b2} = V - I_{a2}(R_a + R_{se})$$

$$\therefore \quad E_{b2} = 200 - 12(1) = 188 \text{ V}$$

For series motor,

$$N \propto \frac{E_b}{\phi} (\phi \propto I_a)$$

$$\frac{N_2}{N_1} = \frac{E_{b2}}{E_{b1}} \times \frac{I_{a1}}{I_{a2}}$$

$$\therefore \quad \frac{N_2}{1000} = \frac{188}{190} \times \frac{10}{12} = 824.56 \text{ rpm} \simeq 825 \text{ rpm}$$

1.52 Sol.

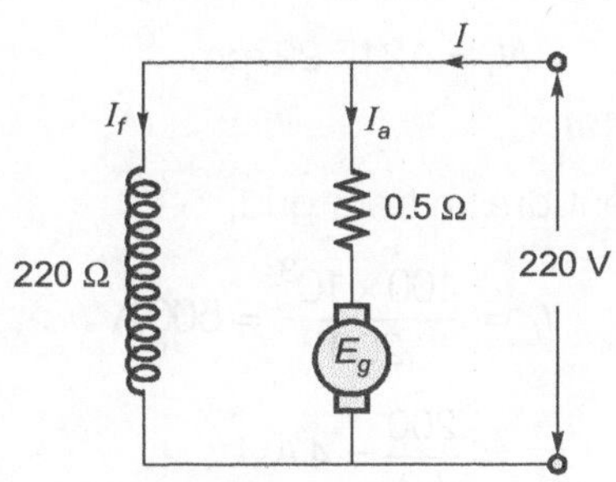

At full load condition,

$$N_1 = 1500 \text{ rpm}; \quad I_1 = 25 \text{ A}$$

$$I_f = \frac{220}{220} = 1 \text{ A}$$

$$I_{a1} = I_1 - I_f = 24 \text{ A}$$

$$E_{g1} = V - I_{a1} R_a = 220 - 24 \times 0.5 = 208 \text{ V}$$

At no-load condition,

$$I_2 = 3 \text{ A}; \quad I_f = 1 \text{ A}$$

$$I_{a2} = I_2 - I_f = 2 \text{ A}$$

$$E_{g2} = 220 - 2 \times 0.5 = 219 \text{ V}$$

$$E_g \propto \phi N$$

Flux is constant,

$$E_g \propto N$$

$$\frac{E_{g2}}{E_{g1}} = \frac{N_2}{N_1} \Rightarrow \frac{219}{208} = \frac{N_2}{1500}$$

$$N_2 = 1579.33 \text{ rpm}$$

1.53 (d)

No load current 5 A

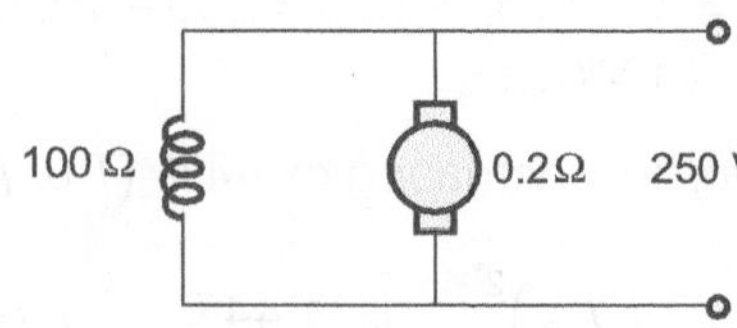

$$\text{B.R.D} = 1 \text{ V per brush}$$

$$\text{Loaded, } I_L = 50 \text{ A}$$

$$R_{sh} = 100\ \Omega$$

$$I_{sh} = \frac{250}{100} = 2.5 \text{ A}$$

$$I_{a0} = 2.5 \text{ A}$$

$$I_{aL} = 47.5 \text{ A}$$

$$V = E_b + I_a R_a + \text{B.R.D}$$

$$E_{b\,\text{no load}} = V - I_{a0} R_a - \text{B.R.D}$$

$$= 250 - 2.5(0.2) - 1 \times 2 = 247.5 \text{ V}$$

$$E_{b\,\text{load}} = 250 - 47.5(0.2) - 1 \times 2 = 238.5 \text{ V}$$

$$\frac{N_2}{N_1} = \frac{E_{b2}}{E_{b1}} \times \frac{\phi_1}{\phi_2}$$

$$\frac{N_2}{1200} = \frac{238.5}{247.5} \times \frac{\phi_1}{0.95\phi_1}$$

$$N_2 = 1217.22 \text{ rpm}$$

1.54 (275.18)

Current drawn from grid,

$$I_L = \frac{100 \times 10^3}{200} = 500 \text{A}$$

$$I_{sh} = \frac{200}{50} = 4 \text{A}$$

$$I_a = 504 \text{ A}$$

$$E_q = V + I_a R_a + \text{Brush drop}$$

$$= 200 + 504\ (0.025) + 2\text{V} = 214.6 \text{ V}$$

In motoring case: $VI = 10$ kW, $V = 200$ V

$$\therefore \quad I = \frac{10000}{200} = 50A$$

Field current, $I_f = 4$A, $I_a = I_L - I_f = 46$ A

$$E_b = V - I_a R_a - \text{Brush Drop}$$

$$= 200 - 46\ (0.025) - 2 = 196.85 \text{ V}$$

$$\frac{N_m}{N_g} = \frac{E_b}{E_g} \Rightarrow N_m = \frac{E_b}{E_g} \times N_g$$

$$\therefore \quad N_m = \frac{196.85}{214.6} \times 300 = 275.18 \text{ rpm}$$

1.55 (483)

Given: $V = 280$, Sep. excited motor, $R_a = 1\ \Omega$, ϕ : Constant, $T_a \propto I_a = 30$ A, $N = 1000$ rpm, $R_{ext} = 10\ \Omega$

$$T \propto \phi I_a$$

$\therefore$ Flux constant, $T \propto I_a$

$$\frac{T_2}{T_1} = \frac{I_{a2}}{I_{a1}}$$

Given: $T \propto N$, $\therefore \frac{T_2}{T_1} = \frac{N_2}{N_1}$

$$\therefore \quad \frac{I_{a2}}{I_{a1}} = \frac{N_2}{N_1}$$

Let, $I_{a1} = 30$ A, $N_1 = 1000$ rpm

$$\therefore \quad I_{a2} = \frac{30}{1000}(N_2) = 0.03\ N_2$$

$$N \propto \frac{E_f}{\phi}$$

$$\frac{N_2}{N_1} = \frac{E_{b2}}{E_{b1}} \Rightarrow \frac{N_2}{N_1} = \frac{V - I_{a2}(R_a + R_{ext})}{V - I_{a1} R_a}$$

$$\frac{N_2}{1000} = \frac{280 - 0.03 N_2 (1 + 10)}{280 - 30(1)}$$

$$= \frac{280 - 0.33 N_2}{250}$$

$$250 N_2 = 280 \times 1000 - 330 N_2 = 482.76 \text{ rpm}$$

1.56 (1343)

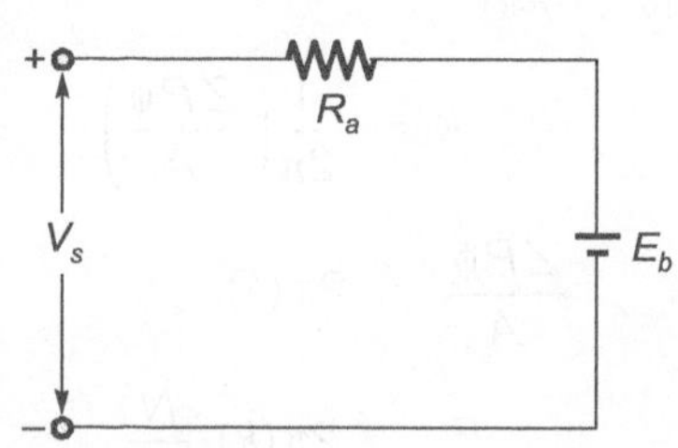

Given torque is constant, $T = K\phi I_a$ = Constant

Field is also unchanged, ϕ = Constant

So, I_a = Constant

The back emf with rated voltage

$$E_{b1} = V_s - I_a R_a = 400 - 1.2 \times 15 = 382 \text{ V}$$

and back emf with 10% drop in supply voltage

$$E_{b2} = V_{s2} - I_aR_a = 400 \times 0.90 - 1.2 \times 15$$

$$E_{b2} = 342 \text{ V}$$

We know $E_b \propto N$ (for separately excited motor)

So, $$\frac{E_{b2}}{E_{b1}} = \frac{N_2}{N_1}$$

$$N_2 = \frac{342}{382} \times 1500 = 1342.93 \approx 1343 \text{ rpm}$$

1.57 (27.28)

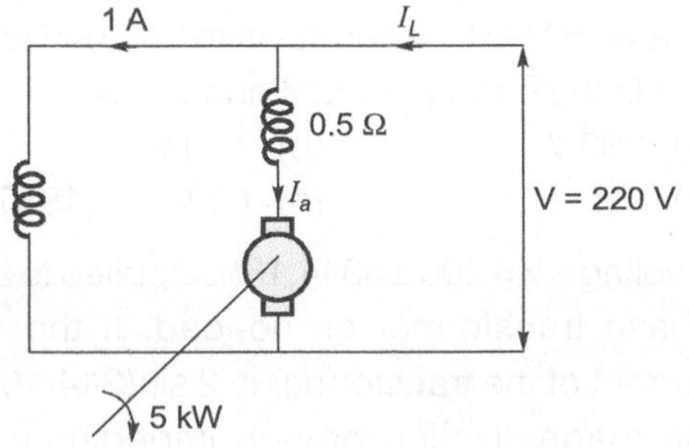

At no load, $E_b = V - I_aR_a$

$$= 220 - (3 - 1)0.5$$

$$E_b = 219 \text{ V}$$

No load power developed

$$= \text{Mechanical loss} = E_b \cdot I_a$$

$$= 219 \times 2 = 438 \text{ W}$$

At rated load :

Given : $P_{out} = 500$ W

$$E_b' I_a' = 5000 + \text{Mechanical Loss}$$

$$(V - I_a'R_a)(I_a') = 5438 \text{ W}$$

$$-I_a'^2 + 0.5 + V.I_a' = 5438$$

$$0.5I_a'^2 - 220I_a' + 5438 = 0 \Rightarrow I_a' = 26.28 \text{ A}$$

Current drawn by motor

$$I_L = I_a' + 1 = 27.28 \text{ A}$$

■■■■

2 Transformers

2.1 The percentage impedance of a 100 kVA, 11 kV/400 V, delta/wye, 50 Hz transformer is 4.5%. For the circulation of half the full load current during short circuit test, with low voltage terminals shorted, the applied voltage on the high voltage side will be
(a) 200 V (b) 247.5 V
(c) 250 V (d) 230 V **[1995 : 1 M]**

2.2 An induction motor is fed from a balanced three-phase supply at rated voltage and frequency through a bank of three single phase transformers connected in delta-delta. One unit of the bank develops fault and is removed. Then
(a) single phasing will occur and the machine fails so start.
(b) single phasing will not occur but the motor terminal voltages will become unbalanced and the machine can be loaded to the extent of 57.7% of its rating.
(c) the machine can be loaded to the extent of 57.7% of its rating with balanced supply at its terminals.
(d) the machine can be loaded to the extent of $66\frac{2}{3}$% with balanced supply at its terminals.
[1995 : 1 M]

2.3 The function of oil in a transformer is
(a) to provide insulation and cooling
(b) to provide protection against lightning
(c) to provide protection against short-circuit
(d) to provide lubrication **[1996 : 1 M]**

2.4 Auto transformer is used in transmission and distribution
(a) when operator is not available
(b) when iron losses are to be reduced
(c) when efficiency considerations can be ignored
(d) when the transformation ratio is small
[1996 : 1 M]

2.5 Keeping in view the requirement of parallel operation, which of the 3-phase connections given below are possible?
(a) delta-delta to delta-star
(b) delta-delta to star-delta
(c) delta-star to delta-star
(d) delta-star to star-delta **[1996 : 1 M]**

2.6 The low voltage winding of a 400/230 V, single-phase, 50 Hz transformer is to be connected to a 25 Hz the supply voltage should be
(a) 230 V (b) 460 V
(c) 115 (d) 65 V **[1997 : 2 M]**

2.7 A voltage $V = 400 \sin 314.16t$ is applied to a single-phase transformer on no-load. If the no-load current of the transformer is $2 \sin(314.16t - 85°)$, the magnetization branch impedance will be approximately
(a) $141\angle 90°$ (b) $200\angle -85°$
(c) $200\angle 85°$ (d) $282\angle -80°$
[1997 : 2 M]

2.8 3-phase transformer bank consists of three identical 2300/230 V, 15 kVA single-phase transformers connect in delta/delta. The bank supplies a 20 kVA, unity p.f. 3-phase load. If one of the single-phase transformer develops a fault, and is removed, the load carried by each of the two transformers now operating in open delta will be ______ kVA.
(a) 10 kVA (b) 20 kVA
(c) 15 kVA (d) None
[1997 : 2 M]

2.9 The laws of electromagnetic induction (Faraday's and Lenz's law) are summarized in the following equation
(a) $e = iR$ (b) $e = L\frac{di}{dt}$
(c) $e = -\frac{d\Psi}{dt}$ (d) None of these
[1998 : 1 M]

2.10 The efficiency of a 100 kVA transformer is 0.98 at full as well as at half load. For this transformer at full load the copper loss
(a) is less than core loss
(b) is equal to core loss
(c) is more than core loss
(d) None of the above **[1998 : 1 M]**

2.11 The magnetizing current in a transformer is rich in
(a) 3rd harmonic (b) 5th harmonic
(c) 7th harmonic (d) 13th harmonic
[1998 : 1 M]

2.12 A 50 Hz transformer having equal hysteresis and eddy current losses at rated excitation is operated at 45 Hz at 90% of its rated voltage. Compared to rated operating point, the core losses under this condition.
(a) reduce by 10% (b) reduce by 19%
(c) reduce by 14.5% (d) remain unchanged
[1998 : 2 M]

2.13 The windings of a Q kVA, V_1/V_2 volt, three-phase delta connected core type transformer are reconnected to work as a single-phase transformer. The maximum voltage and the power ratings of the new configuration are
(a) $\frac{V_1}{V_2}, Q$ (b) $\frac{V_1}{V_2}, Q/3$
(c) $\frac{\sqrt{3}\,V_1}{V_2}, 2Q$ (d) $\frac{2V_1}{V_2}, \frac{2Q}{3}$
[1999 : 2 M]

2.14 A 10 kVA, 400 V/200 V single-phase transformer with 10% impedance draws a steady short-circuit line current of
(a) 50 A (b) 150 A
(c) 250 A (d) 350 A [1999 : 2 M]

2.15 The percentage resistance and percentage reactance of a 10 kVA, 400 V/200 V, 3-phase transformer are 2% and 10% respectively. If the constant losses in the machine are 1%, the maximum possible percentage efficiency of the transformer is
(a) 98.32 (b) 97.25
(c) 96.85 (d) 96.12 [1999 : 2 M]

2.16 A 400 V/100 V, 10 kVA two winding transformer is reconnected as an auto transformer across a suitable voltage source. The maximum rating of such an arrangement could be
(a) 50 kVA (b) 15 kVA
(c) 12.5 kVA (d) 8.75 kVA [1999 : 2 M]

2.17 A 10 kVA, 400 V/200 V, single-phase transformer with a percentage resistance of 3% and percentage reactance of 6% is supplying a current of 50 A to a resistive load. The value of the load voltage is
(a) 194 V (b) 390 V
(c) 192 V (d) 196 V [1999 : 2 M]

2.18 Two single-phase transformers *A* and *B* have the following parameters.
Transformer A: 400 V/200 V, 10 kVA, percentage resistance and percentage reactance are 3% and 4% respectively.
Transformer B: 5 kVA, 400 V/200 V, percentage resistance and percentage reactance are 4% and 3% respectively. These two transformers are connected in parallel and they share a common load of 12 kW at a power factor of 0.8 lag. Determine the active and reactive power delivered by transformer A. [1999 : 2 M]

2.19 A 3-phase delta/star transformer is supplied 6000 V on delta connected side. The terminal voltage on the secondary side when supplying full load at 0.8 lagging power factor is 415 V. The equivalent resistance and reactance drops for the transformer are 1% and 5% respectively. The turns ratio of the transformer is
(a) 14 (b) 24
(c) 42 (d) 20 [2000 : 2 M]

2.20 A single-phase transformer is to be switched to the supply to have minimum inrush current. The switch should be closed at
(a) maximum supply voltage
(b) zero supply voltage
(c) $\frac{1}{\sqrt{2}}$ maximum supply voltage
(d) $\frac{1}{2}$ maximum supply voltage [2001 : 1 M]

2.21 The core flux of a practical transformer with a resistive load
(a) is strictly constant with load changes
(b) increases linearly with load
(c) increases as the square root of the load
(d) decreases with increased load [2001 : 1 M]

2.22 In the protection of transformers, harmonic restraint is used to guard against
(a) magnetizing inrush current
(b) unbalanced operation
(c) lightning
(d) switching over-voltages [2001 : 1 M]

2.23 A 3-phase transformer has rating of 20 MVA, 220 kV (Star)/33 kV (Delta) with leakage reactance 12%. The transformer reactance (in ohms) referred to each phase of the L.V. delta connected side is
(a) 23.5 (b) 19.6
(c) 18.5 (d) 8.7 [2001 : 2 M]

2.24 In a single-phase 3 winding transformer. The turns ratio for primary : secondary : tertiary windings is 20 : 4 : 1 with the lagging currents of 50 A at a power factor of 0.6 in the tertiary winding find the primary current and power factor. [2002 : 1 M]

2.25 A 1 kVA, 230 V/100 V single-phase 50 Hz transformer having negligible winding resistance and leakage inductance is operating under saturation, while 250 V, 50 Hz, sinusoidal supply is connected to the high voltage winding. A resistive load is connected to the low voltage winding which draws rated current. Which one of the following quantities will not be sinusoidal?

(a) voltage induced across the low voltage winding
(b) core flux
(c) load current
(d) current drawn from source [2002 : 2 M]

2.26 A 400 V/200 V/200 V, 50 Hz three winding transformer is connected is shown in figure. The reading of the voltmeter 'V' will be

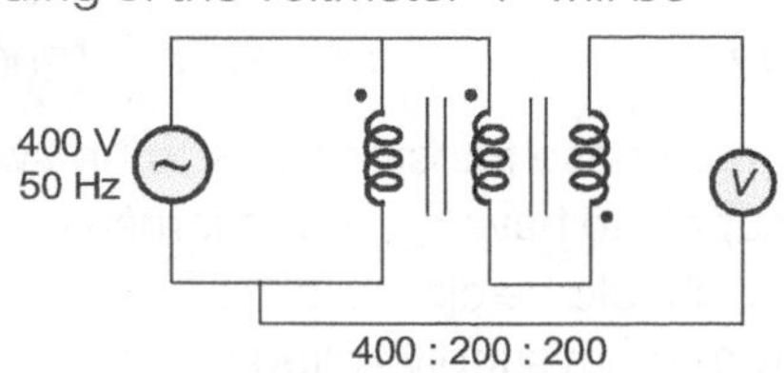

(a) 0 V (b) 400 V
(c) 600 V (d) 800 V [2002 : 2 M]

2.27 A single phase transformer has a maximum efficiency of 90% at full load and unity power factor. Efficiency at half load, at the same power factor is

(a) 86.7% (b) 88.26%
(c) 88.9% (d) 87.8% [2003 : 1 M]

2.28 Figure shows an ideal single-phase transformer. The primary and secondary coils are wound on the core as shown. Turns ratio $(N_1/N_2) = 2$. The correct phasors of voltages E_1, E_2, currents I_1, I_2 and core flux Φ are as shown in

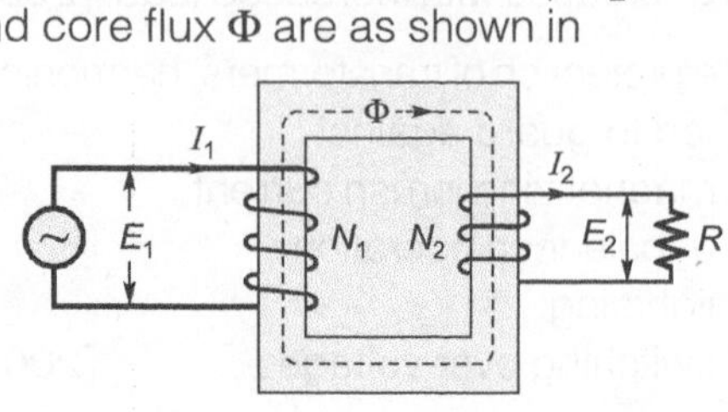

(a)
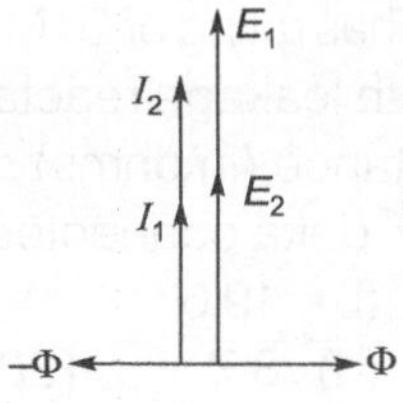

(b)
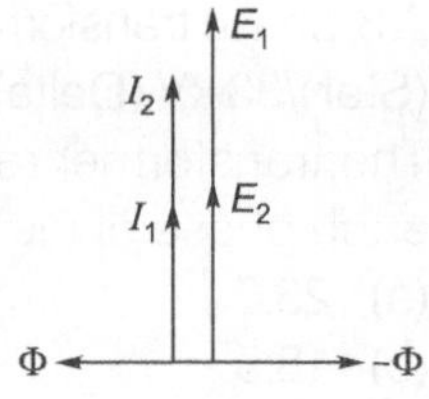

(c)
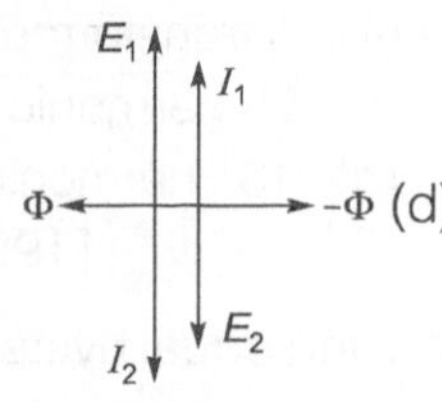

(d)
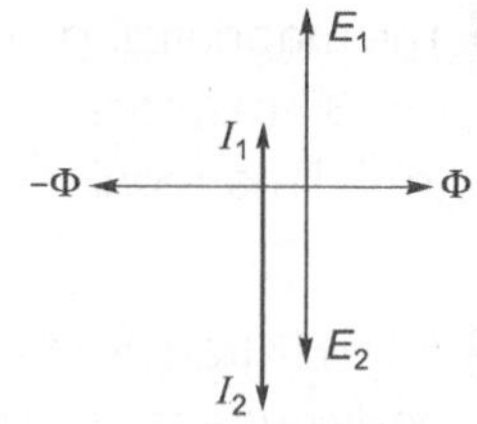

[2003 : 2 M]

2.29 The figure shows a Δ-Y connected 3-phase distribution transformer used to step down the voltage from 11000 V to 415 V line-to-line. It has two switches S_1 and S_2. Under normal conditions S_1 is closed and S_2 is open. Under certain superior conditions S_1 is open and S_2 is closed. In such a case the magnitude of the voltage across the *LV* terminals *a* and *c* is

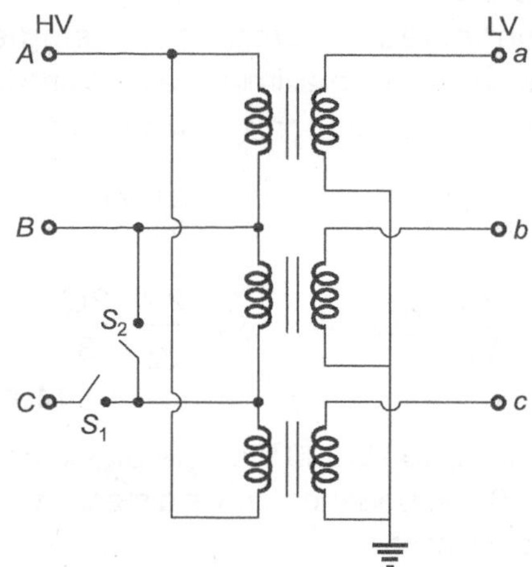

(a) 240 V (b) 480 V
(c) 415 V (d) 0 V [2003 : 2 M]

2.30 The figure shows an ideal three winding transformer windings are wound on the same core as shown. The turns ratio $N_1 : N_2 : N_3$ is 4 : 2 : 1. A resistor of 10 Ω is connected across winding-2. A capacitor of reactance 2.5 Ω is connected across winding-3. Winding-1 is connected across a 400 V, as supply. If the supply voltage phasor $V_1 = 400\angle 0°$ V, the supply current phasor I_1 is given by

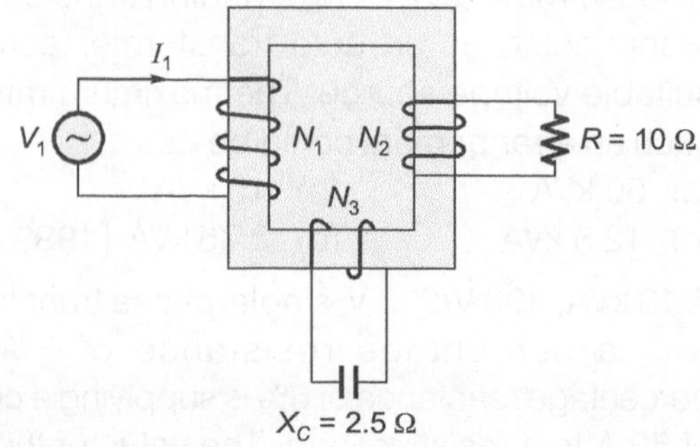

(a) $(-10 + j10)$ A (b) $(-10 - j10)$ A
(c) $(10 + j10)$ A (d) $(10 - j10)$ A

[2003 : 2 M]

2.31 A 500 kVA, 3-phase transformer has iron losses of 300 W and full load copper losses of 600 W. The percentage load at which the transformer is expected to have maximum efficiency is
(a) 50.0% (b) 70.7%
(c) 141.4% (d) 200.0%
[2004 : 1 M]

2.32 A 50 kVA, 3300/230 V single-phase transformer is connected as an autotransformer shown in figure. The nominal rating of the autotransformer will be

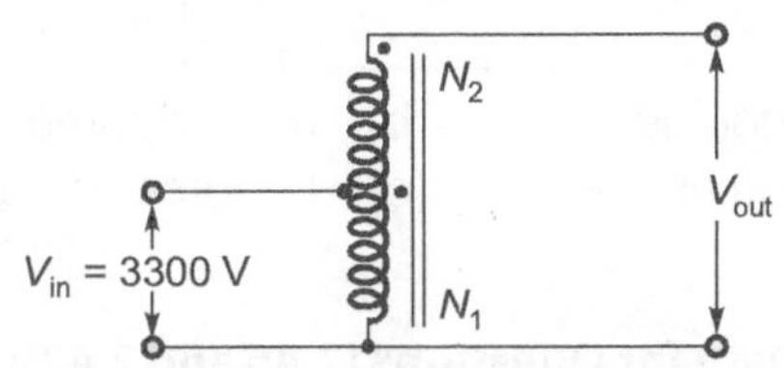

(a) 50.0 kVA (b) 53.5 kVA
(c) 717.4 kVA (d) 767.4 kVA
[2004 : 2 M]

2.33 The resistance and reactance of a 100 kVA, 11000/400 V, Δ-Y distribution transformer are 0.02 and 0.07 pu respectively. The phase impedance of the transformer referred to the primary is
(a) $(0.02 + j0.07)\,\Omega$
(b) $(0.55 + j1.925)\,\Omega$
(c) $(15.125 + j52.94)\,\Omega$
(d) $(72.6 + j254.1)\,\Omega$ [2004 : 2 M]

2.34 The equivalent circuit of a transformer has leakage reactances X_1, X_2', and magnetizing reactance X_M. Their magnitudes satisfy
(a) $X_1 >> X_2' >> X_M$ (b) $X_1 << X_2' << X_M$
(c) $X_1 \approx X_2' >> X_M$ (d) $X_1 \approx X_2' << X_M$
[2005 : 1 M]

2.35 Which three-phase connection can be used in a transformer to introduce a phase difference of 30° between its output and corresponding input line voltages?
(a) Star-Star (b) Star-Delta
(c) Delta-Delta (d) Delta-Zigzag
[2005 : 1 M]

2.36 In transformers, which of the following statements is valid?
(a) In an open circuit test, copper losses are obtained while in short circuit test, core losses are obtained.
(b) In an open circuit test, current is drawn at high power factor.
(c) In a short circuit test, current is drawn at zero power factor.
(d) In an open circuit test, current is drawn at low power factor. [2006 : 1 M]

2.37 Two transformers are to be operated in parallel such that they share load in proportion to their kVA ratings. The rating of the first transformer is 500 kVA and its pu leakage impedance is 0.05 pu. If the rating of second transformer is 250 kVA, its pu leakage impedance is
(a) 0.20 (b) 0.10
(c) 0.05 (d) 0.025 [2006 : 2 M]

Data for Questions (2.38 and 2.39):
A 300 kVA transformer has 95% efficiency at full load 0.8 pf lagging and 96% efficiency at half load, unity pf.

2.38 The iron loss (P_i) and copper loss (P_c) in kW, under full load operation are
(a) $P_c = 4.12$, $P_i = 8.51$
(b) $P_c = 6.59$, $P_i = 9.21$
(c) $P_c = 8.51$, $P_i = 4.12$
(d) $P_c = 12.72$, $P_i = 3.07$ [2006 : 2 M]

2.39 What is the maximum efficiency (in%) at unity pf load?
(a) 95.1 (b) 96.2
(c) 96.4 (d) 98.1 [2006 : 2 M]

2.40 In a transformer, zero voltage regulation at full load is
(a) not possible
(b) possible at unity power factor load
(c) possible at leading power factor load
(d) possible at lagging power factor load
[2007 : 1 M]

2.41 A single-phase 50 kVA, 250 V/500 V two winding transformer has an efficiency of 95% at full load, unity power factor. If it is reconfigured as a 500 V/750 V autotransformer, its efficiency at its new rated load at unity power factor will be
(a) 95.752% (b) 97.851%
(c) 98.276% (d) 99.241% [2007 : 2 M]

2.42 Three single-phase transformers are connected to form a 3-phase transformer bank. The transformers are connected in the following manner:

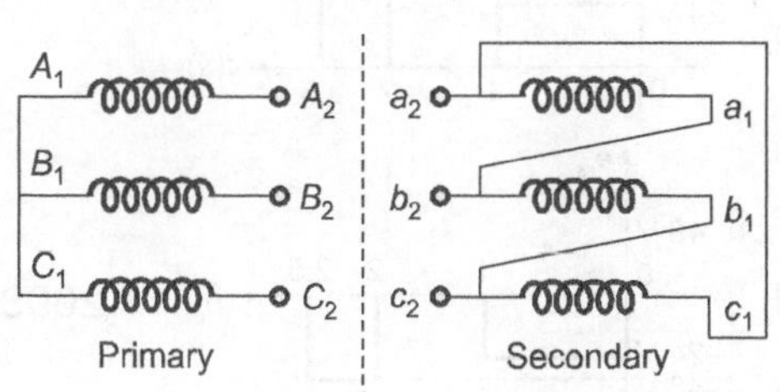

The transformer connection will be represented by

(a) Yd0 (b) Yd1
(c) Yd6 (d) Yd11 [2008 : 1 M]

2.43 It is desired to measure parameters of 230 V/115 V, 2 kVA, single-phase transformer. The following wattmeters are available in a laboratory:

W_1 : 250 V, 10 A, Low Power Factor
W_2 : 250 V, 5 A, Low Power Factor
W_3 : 150 V, 10 A, High Power Factor
W_4 : 150 V, 5 A, High Power Factor

The wattmeters used in open circuit test and short circuit test of the transformer will respectively be

(a) W_1 & W_2 (b) W_2 & W_4
(c) W_1 & W_4 (d) W_2 & W_3 [2008 : 1 M]

2.44 The core of a two-winding transformer is subjected to a magnetic flux variation as indicated in the figure.

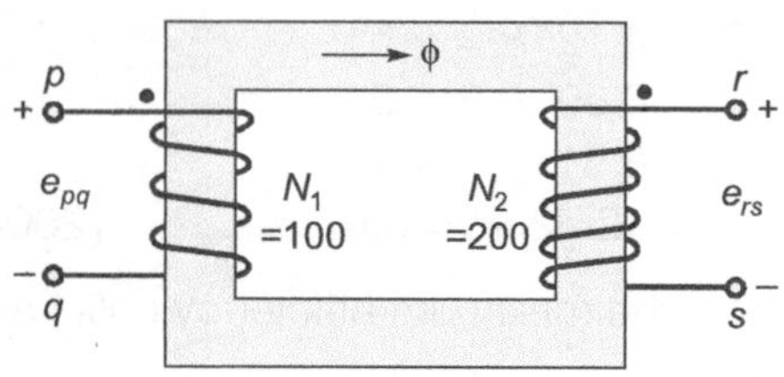

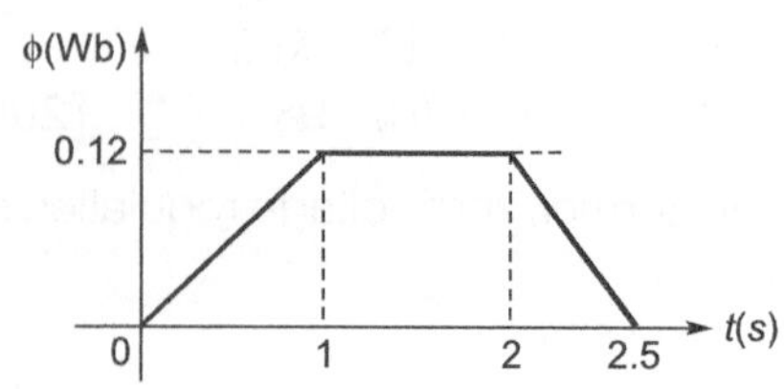

The induced emf (e_{rs}) in the secondary winding as a function of time will be of the form

(a)

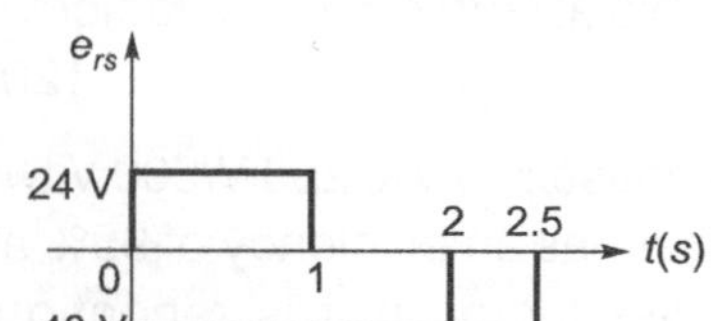

(b)

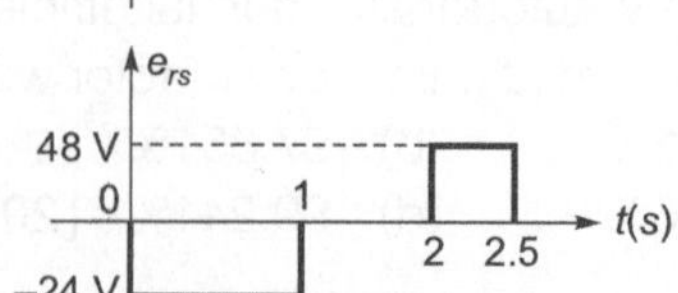

(c)

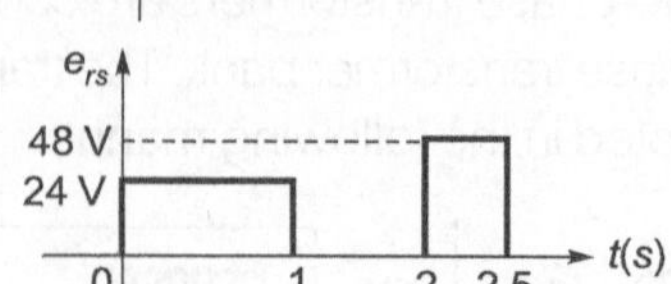

(d)

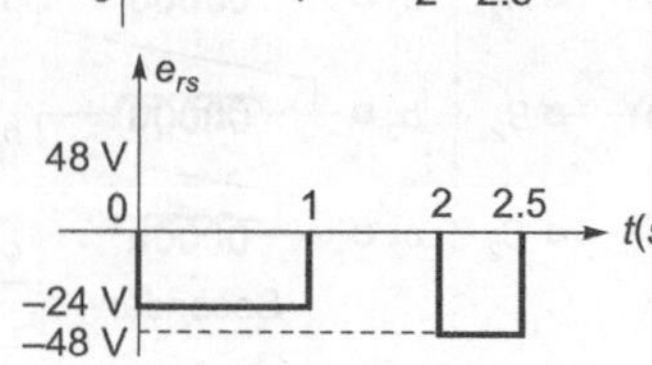

[2008 : 2 M]

2.45 The single-phase, 50 Hz iron core transformer in the circuit has both the vertical arms of cross sectional area 20 cm^2 and both the horizontal arms of cross sectional area 10 cm^2. If the two windings shown were wound instead on opposite horizontal arms, the mutual inductance will

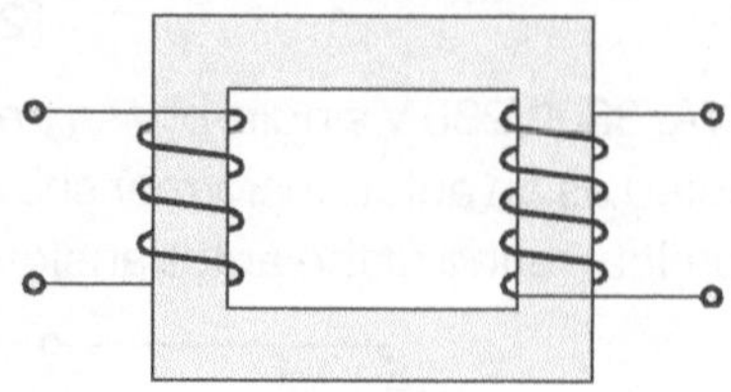

(a) double (b) remain same
(c) be halved (d) become one quarter
[2009 : 1 M]

Common Data for Questions (2.46 and 2.47):

The star-delta transformer shown below is excited on the star side with balanced, 4-wire, 3-phase, sinusoidal voltage supply of rated magnitude. The transformer is under no load condition.

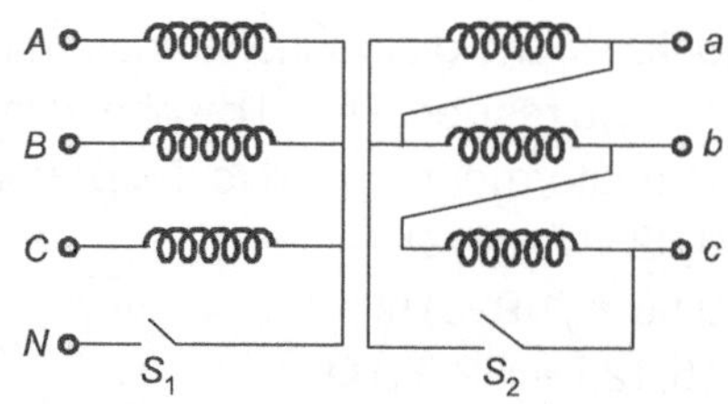

2.46 With both *S*1 and *S*2 open, the core flux waveform will be

(a) a sinusoid at fundamental frequency
(b) flat-topped with third harmonic
(c) peaky with third-harmonic
(d) none of these [2009 : 2 M]

2.47 With *S*2 closed and *S*1 open, the current waveform in the delta winding will be

(a) a sinusoid at fundamental frequency
(b) flat-topped with third harmonic
(c) only third-harmonic
(d) none of these [2009 : 2 M]

Common Data for Questions (2.48 and 2.49):

The circuit diagram shows a two-winding, lossless transformer with no leakage flux, excited from a current source, $i(t)$, whose waveform is also shown. The transformer has a magnetizing inductance of $400/\pi$ mH.

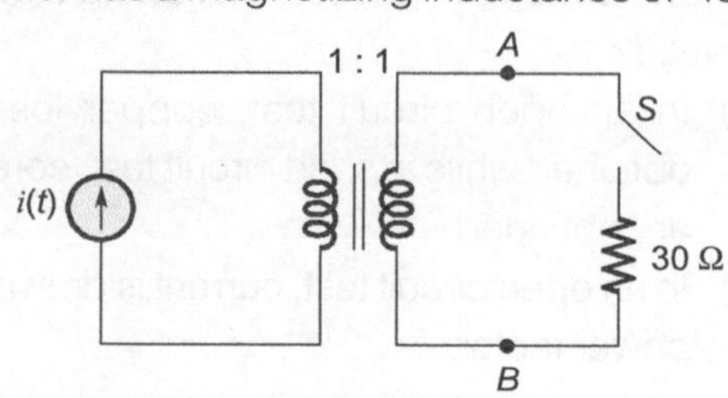

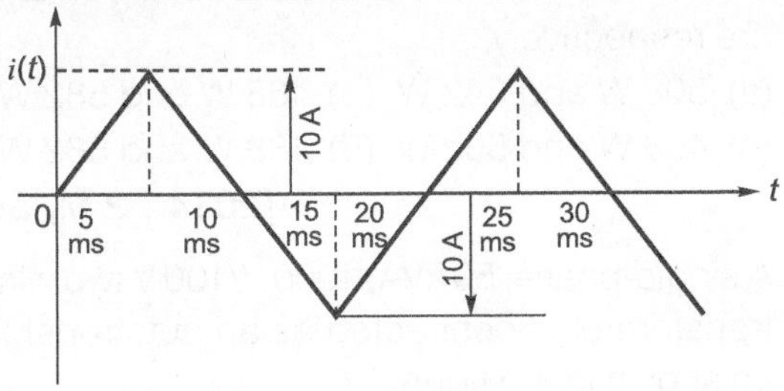

2.48 The peak voltage across *A* and *B*, with *S* open is

(a) $\frac{400}{\pi}$ V (b) 800 V

(c) $\frac{4000}{\pi}$ V (d) $\frac{800}{\pi}$ V [2009 : 2 M]

2.49 If the waveform of $i(t)$ is changed to $i(t) = 10\sin(100\pi t)$ A, the peak voltage across *A* and *B* with *S* closed is

(a) 400 V (b) 240 V

(c) 320 V (d) 160 V [2009 : 2 M]

Statement for Linked Answer Questions (2.50 and 2.51):

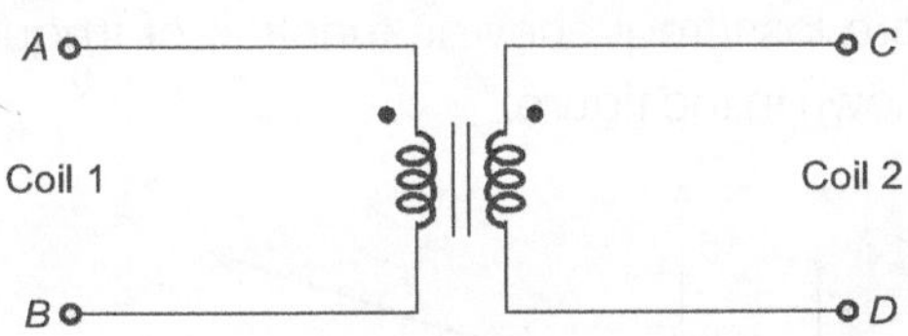

The figure above shows coils 1 and 2, with dot markings as shown, having 4000 and 6000 turns respectively. Both the coils have a rated current of 25 A. Coil 1 is excited with single phase, 400 V, 50 Hz supply.

2.50 The coils are to be connected to obtain a single-phase, $\frac{400}{1000}$ V, auto-transformer to drive a load of 10 kVA. Which of the options given should be exercised to realize the required auto-transformer?

(a) Connect A and D; Common B
(b) Connect B and D; Common C
(c) Connect A and C; Common B
(d) Connect A and C; Common D [2009 : 2 M]

2.51 In the autotransformer obtained in **Q. 2.56**, the current in each coil is

(a) Coil-1 is 25 A and Coil-2 is 10 A
(b) Coil-1 is 10 A and Coil-2 is 25 A
(c) Coil-1 is 10 A and Coil-2 is 15 A
(d) Coil-1 is 15 A and Coil-2 is 10 A

[2009 : 2 M]

2.52 A single-phase transformer has a turns ratio of 1 : 2, and is connected to a purely resistive load as shown in the figure. The magnetizing current drawn is 1 A, and the secondary current is 1 A. If core losses and leakage reactances are neglected, the primary current is

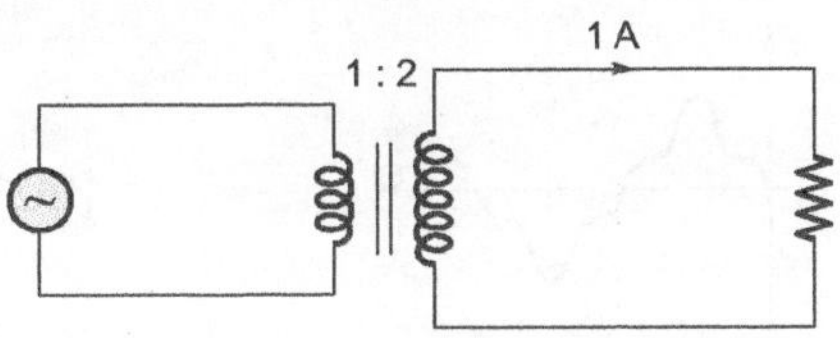

(a) 1.41 A (b) 2 A
(c) 2.24 A (d) 3 A [2010 : 1 M]

2.53 A balanced star-connected and purely resistive load is connected at the secondary of a star-delta transformer as shown in the figure. The line-to-line voltage rating of the transformer is 110 V/200 V. Neglecting the non-idealities of the transformer, the impedance '*Z*' of the equivalent star-connected load, referred to the primary side of the transformer, is

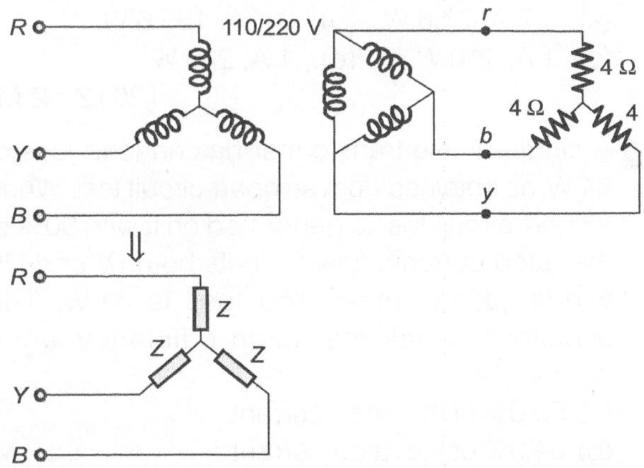

(a) $(3 + j0)\ \Omega$ (b) $(0.866 - j0.5)\ \Omega$
(c) $(0.866 + j0.5)\ \Omega$ (d) $(1 + j0)\ \Omega$ [2010 : 2 M]

2.54 A single-phase air core transformer, fed from a rated sinusoidal supply, is operating at no load. The steady state magnetizing current drawn by the transformer from the supply will have the waveform

(a)

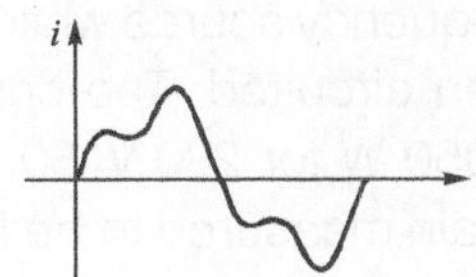

(b)

(c) 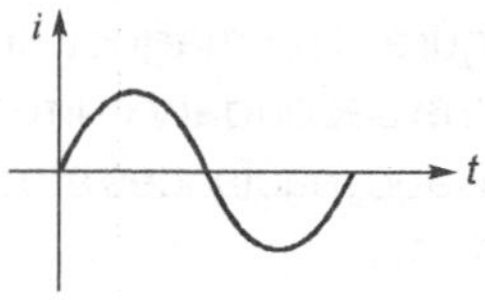

(d) 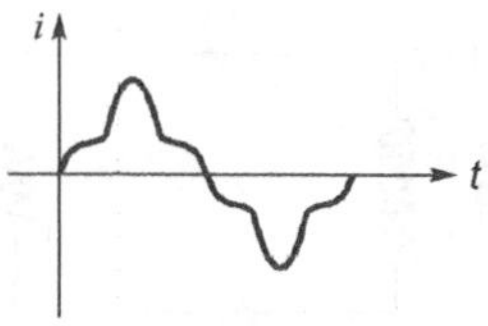

[2011 : 1 M]

2.55 A single phase 10 kVA, 50 Hz transformer with 1 kV primary winding draws 0.5 A and 55 W, at rated voltage and frequency, on no load. A second transformer has a core with all its linear dimensions $\sqrt{2}$ times the corresponding dimensions of the first transformer. The core material and lamination thickness are the same in both transformers. The primary windings of both the transformers have the same number of turns. If a rated voltage of 2 kV at 50 Hz is applied to the primary of the second transformer, then the no load current and power, respectively, are

(a) 0.7 A, 77.8 W (b) 0.7 A, 155.6 W
(c) 1 A, 110 W (d) 1 A, 220 W

[2012 : 2 M]

2.56 A single-phase transformer has no-load loss of 64 W, as obtained from an open-circuit test. When a short-circuit test is performed on it with 90% of the rated currents flowing in its both LV and HV windings, the measured loss is 81 W. The transformer has maximum efficiency when operated at

(a) 50.0% of the rated current
(b) 64.0% of the rated current
(c) 80.0% of the rated current
(d) 88.8% of the rated current [2013 : 1 M]

2.57 For a specified input voltage and frequency, if the equivalent radius of the core of a transformer is reduced by half, the factor by which the number of turns in the primary should change to maintain the same no load current is

(a) 1/4 (b) 1/2
(c) 2 (d) 4 [2014 : 1 M, Set-1]

2.58 The core loss of a single-phase, 230/115 V, 50 Hz power transformer is measured from 230 V side by feeding the primary (230 V side) from a variable voltage variable frequency source while keeping the secondary open circuited. The core loss is measured to be 1050 W for 230 V, 50 Hz input. The core loss is again measured to be 500 W for 138 V, 30 Hz input. The hysteresis and eddy current losses of the transformer for 230 V, 50 Hz input are respectively,

(a) 508 W and 542 W (b) 468 W and 582 W
(c) 498 W and 552 W (d) 488 W and 562 W

[2014 : 2 M, Set-1]

2.59 A single-phase, 50 kVA, 1000 V/100 V two winding transformer is connected as an autotransformer as shown in the figure.

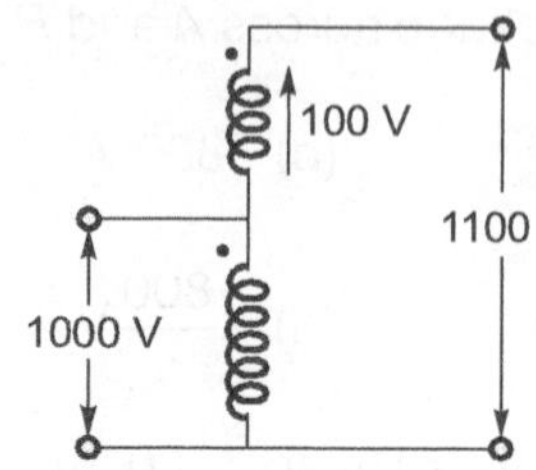

The kVA rating of the autotransformer is _____ .

[2014 : 1 M, Set-2]

2.60 An open circuit test is performed on 50 Hz transformer, using variable frequency source and keeping V/f ratio constant, to separate its eddy current and hysteresis losses. The variation of core loss/frequency as function of frequency is shown in the figure.

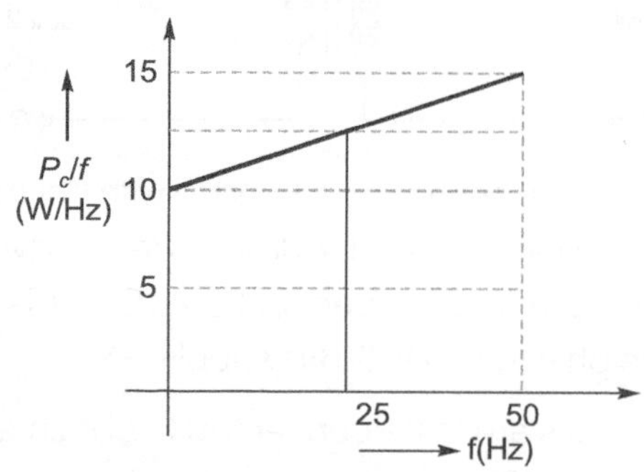

The hysteresis and eddy current losses of the transformer at 25 Hz respectively are

(a) 250 W and 2.5 W
(b) 250 W and 62.5 W
(c) 312.5 W and 62.5 W
(d) 312.5 W and 250 W [2014 : 2 M, Set-3]

2.61 The load shown in the figure absorbs 4 kW at a power factor of 0.89 lagging.

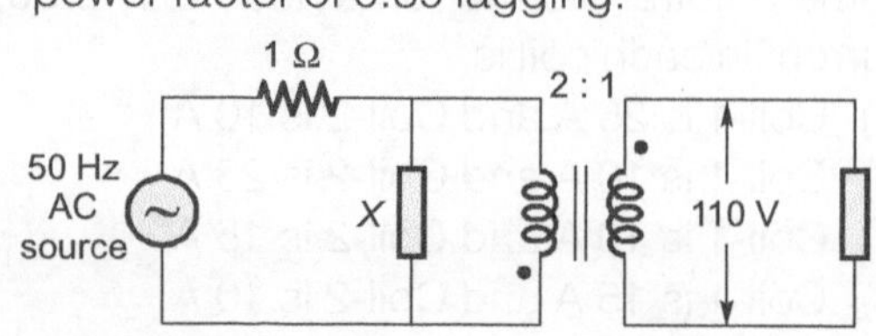

Assuming the transformer to be ideal, the value of the reactance X to improve the input power factor to unity is _____. [2014 : 2 M, Set-3]

2.62 The parameters measured for a 220 V/110 V, 50 Hz, single-phase transformer are:
Self inductance of primary winding = 45 mH.
Self inductance of secondary winding = 30 mH.
Mutual inductance between primary and secondary windings = 20 mH.
Using the above parameters, the leakage (L_{l1}, L_{l2}) and magnetizing (L_m) inductances as referred to primary side in the equivalent circuit respectively, are
(a) 5 mH, 20 mH and 40 mH
(b) 5 mH, 80 mH and 40 mH
(c) 25 mH, 10 mH and 20 mH
(d) 45 mH, 30 mH and 20 mH
[2014 : 2 M, Set-3]

2.63 The primary mmf is least affected by the secondary terminal conditions in a
(a) power transformer
(b) potential transformer
(c) current transformer
(d) distribution transformer [2015 : 1 M, Set-1]

2.64 Two signal-phase transformer T_1 and T_2 each rated at 500 kVA are operated in parallel. Percentage impedances of T_1 and T_2 are $(1 + j6)$ and $(0.8 + j4.8)$, respectively. To share a load of 1000 kVA at 0.8 lagging power factor, the contribution of T_2 (in kVA) is _______.
[2015 : 2 M, Set-1]

2.65 A 200/400 V, 50 Hz, two-winding transformer is rated at 20 kVA. Its windings are connected as an auto-transformer of rating 200/600 V. A resistive load of 12 Ω is connected to the high voltage (600 V) side of the auto-transformer. The value of equivalent load resistance (in Ohm) as seen from low voltage side is _____. [2015 : 2 M, Set-1]

2.66 The self inductance of the primary winding of a single phase, 50 Hz, transformer is 800 mH, and that of the secondary winding is 600 mH. The mutual inductance between these two windings is 480 mH. The secondary winding of this transformer is short circuited and the primary winding is connected to a 50 Hz, single phase, sinusoidal voltage source. The current flowing in both the windings is less than their respective rated currents. The resistance of both windings can be neglected. In this condition, what is the effective inductance (in mH) seen by the source?
(a) 416 (b) 440
(c) 200 (d) 920
[2015 : 1 M, Set-1]

2.67 A single-phase 400 V, 50 Hz transformer has an iron loss of 5000 W at the rated condition. When operated at 200 V, 25 Hz, the iron loss is 2000 W. When operated at 416 V, 52 Hz, the value of the hysteresis loss divided by the eddy current loss is _____.
[2016 : 2 M, Set-1]

2.68 A single-phase, 22 kVA, 2200 V/ 220 V, 50 Hz, distribution transformer is to be connected as an auto-transformer to get an output voltage of 2420 V. Its maximum kVA rating as an autotransformer is
(a) 22 (b) 24.2
(c) 242 (d) 2420
[2016 : 2 M, Set-1]

2.69 A single-phase, 2 kVA, 100/200 V transformer is reconnected as an auto-transformer such that its kVA rating is maximum. The new rating, in kVA, is _______.
[2016 : 2 M, Set-2]

2.70 Three single-phase transformers are connected to form a delta-star three-phase transformer of 110 kV/11 kV. The transformer supplies at 11 kV a load of 8 MW at 0.8 p.f. lagging to a nearby plant. Neglect the transformer losses. The ratio of phase currents in delta side to star side is
(a) $1:10\sqrt{3}$ (b) $10\sqrt{3}:1$
(c) 1 : 10 (d) $\sqrt{3}:10$
[2016 : 2 M, Set-2]

2.71 If an ideal transformer has an inductive load element at port 2 as shown in the figure below, the equivalent inductance at port 1 is

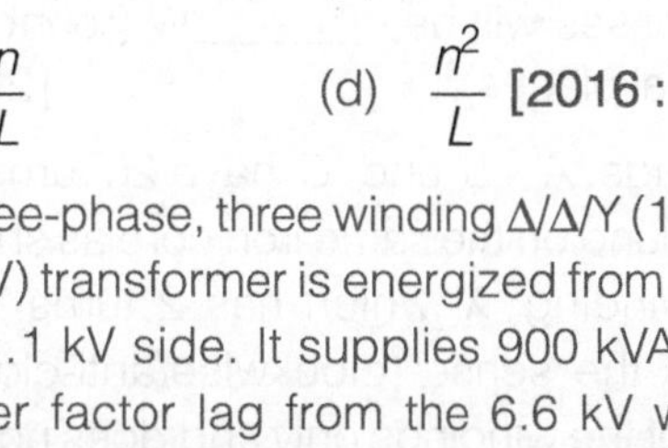

(a) nL (b) n^2L
(c) $\frac{n}{L}$ (d) $\frac{n^2}{L}$ [2016 : 1 M, Set-1]

2.72 A three-phase, three winding Δ/Δ/Y (1.1 kV/6.6 kV/ 400 V) transformer is energized from AC mains at the 1.1 kV side. It supplies 900 kVA load at 0.8 power factor lag from the 6.6 kV winding and 300 kVA load at 0.6 power factor lag from the 400 V winding. The RMS line current in ampere drawn by the 1.1 kV winding from the mains is _____. (Give the answer up to one decimal place.)
[2017 : 2 M, Set-1]

2.73 If the primary line voltage rating is 3.3 kV(Y-side) of a 25 kVA, Y-Δ transformer (the per phase turns ratio is 5 : 1), then the line current rating of the secondary side (in Ampere) is______.
[2017 : 2 M, Set-2]

2.74 A single-phase 100 kVA, 1000 V/100 V, 50 Hz transformer has a voltage drop of 5% across its series impedance at full load. Of this, 3% is due to resistance. The percentage regulation of the transformer at full load with 0.8 lagging power factor is
(a) 4.8 (b) 6.8
(c) 8.8 (d) 10.8 [2018 : 1 M]

2.75 A three-phase, 900 kVA, $3\,kV/\sqrt{3}\,kV$ (Δ/Y), 50 Hz transformer has primary (high voltage side) resistance per phase of 0.3 Ω and secondary (low voltage side) resistance per phase of 0.02 Ω. Iron loss of the transformer is 10 kW. The full load % efficiency of the transformer operated at unity power factor is ______ (upto 2 decimal places).
[2018 : 2 M]

2.76 A 5 kVA, 50 V/100 V, single-phase transformer has a secondary terminal voltage of 95 V when loaded. The regulation of the transformer is
(a) 5% (b) 1%
(c) 4.5% (d) 9% [2019 : 1 M]

2.77 A single-phase transformer of rating 25 kVA, supplied a 12 kW load at power factor of 0.6 lagging. The additional load at unity power factor in kW (round off to two decimal places) that may be added before this transformer exceeds its rated kVA is _____. [2019 : 2 M]

2.78 A single-phase, 4 kVA, 200 V/100 V, 50 Hz transformer with laminated CRGO steel core has rated no-load loss of 450 W. When high-voltage winding is excited with 160 V, 40 Hz sinusoidal AC supply, no-load losses are found to be 320 W. When high-voltage winding of same transformer is supplied from a 100 V, 25 Hz sinusoidal AC source, no-load losses will be _______W (rounded off to 2 decimal places). [2020 : 1 M]

2.79 Windings '*A*', '*B*' and '*C*' have 20 turns each and are wound on the same iron core as shown, along with winding '*X*' which has 2 turns. The figure shows the sense (clockwise/anti-clockwise) of each of the windings only and does not reflect the exact number of turns, If windings '*A*', '*B*' and '*C*' are supplied with balanced 3-phase voltages at 50 Hz and there is no core saturation, the no-load RMS voltage (in V, rounded off to 2 decimal places) across winding '*X*' is ________ .

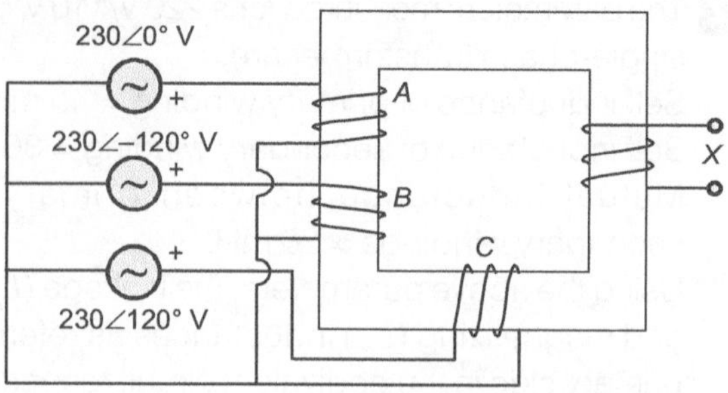

[2020 : 2 M]

2.80 In a single-phase transformer, the total iron loss is 2500 W at nominal voltage of 440 V and frequency 50 Hz. The total iron loss is 850 W at 220 V and 25 Hz. Then, at nominal voltage and frequency, the hysteresis loss and eddy current loss respectively are
(a) 250 W & 600 W (b) 900 W & 1600 W
(c) 1600 W & 900 W (d) 600 W & 250 W
[2021 : 2 M]

2.81 When the winding *c-d* of the single-phase, 50 Hz, two winding transformer is supplied from an AC current source of frequency 50 Hz, the rated voltage of 200 V (rms), 50 Hz is obtained at the open-circuited terminals *a-b*. The cross sectional area of the core is 5000 mm^2 and average core length traversed by the mutual flux is 500 mm. The maximum allowable flux density in the core is B_{max} = 1 Wb/m^2 and the relative permeability of the core material is 5000. The leakage impedance of winding *a-b* and winding *c-d* at 50 Hz are $(5 + j100\pi \times 0.16)$ Ω and $(11.25 + j100\pi \times 0.36)$ Ω, respectively. Considering the magnetizing characteristics to be linear and neglecting core loss, the self-inductance of the winding *a-b* in millihenry is ______ (Round off to 1 decimal place).

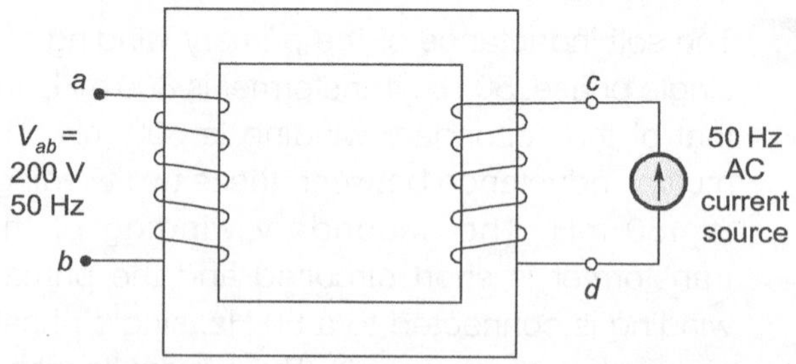

[2023 : 2 M]

2.82 Which one of the following options represents possible voltage polarities in a single phase two winding transformer? Here, V_p is the applied primary voltage, E_p is the induced primary voltage, V_s is the open circuit secondary voltage, and E_s is the induced secondary voltage.

(a)
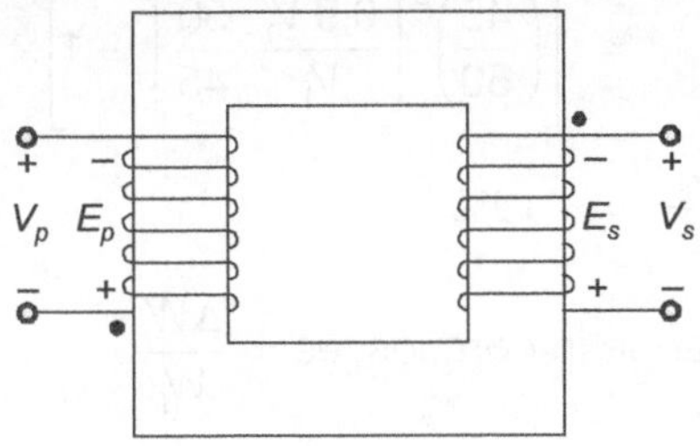

(b)
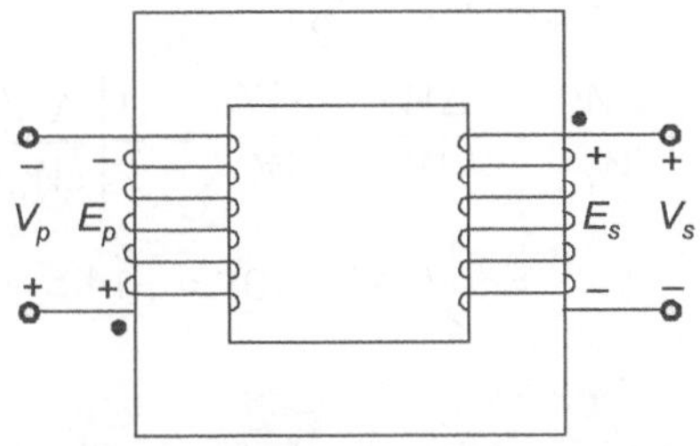

(c)
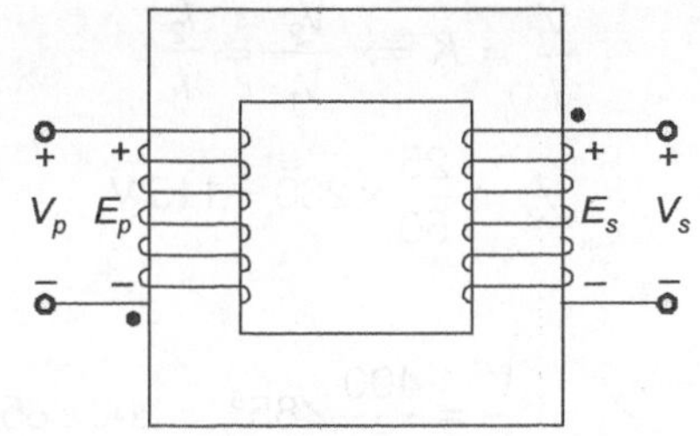

(d)
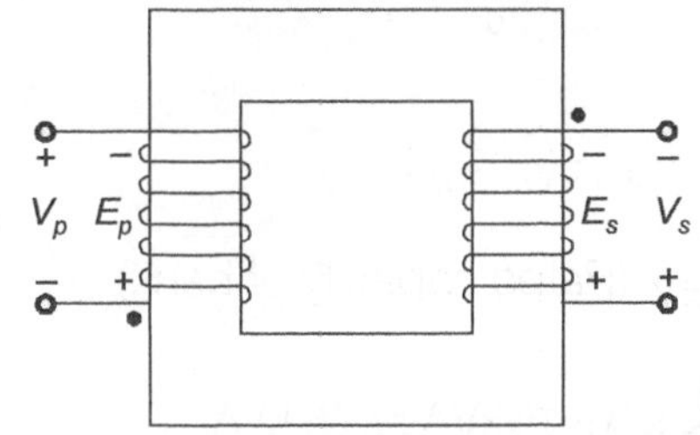

[2024 : 1 M]

Answers Transformers

2.1 (b)	2.2 (c)	2.3 (a)	2.4 (d)	2.5 (c, d)	2.6 (c)	2.7 (c)
2.8 (a)	2.9 (c)	2.10 (c)	2.11 (a)	2.12 (c)	2.13 (a)	2.14 (c)
2.15 (b)	2.16 (a)	2.17 (a)	2.18 (Sol.)	2.19 (b)	2.20 (a)	2.21 (d)
2.22 (a)	2.23 (b)	2.24 (Sol.)	2.25 (d)	2.26 (a)	2.27 (d)	2.28 (a)
2.29 (b)	2.30 (c)	2.31 (b)	2.32 (d)	2.33 (d)	2.34 (d)	2.35 (b)
2.36 (d)	2.37 (c)	2.38 (c)	2.39 (b)	2.40 (c)	2.41 (c)	2.42 (b)
2.43 (d)	2.44 (a)	2.45 (b)	2.46 (b)	2.47 (c)	2.48 (d)	2.49 (b)
2.50 (a)	2.51 (d)	2.52 (c)	2.53 (d)	2.54 (c)	2.55 (b)	2.56 (c)
2.57 (b)	2.58 (a)	2.59 (550)	2.60 (b)	2.61 (23.61)	2.62 (b)	2.63 (c)
2.64 (Sol.)	2.65 (1.33)	2.66 (a)	2.67 (1.4423)	2.68 (c)	2.69 (6)	2.70 (a)
2.71 (b)	2.72 (Sol.)	2.73 (37.88)	2.74 (a)	2.75 (97.36%)	2.76 (a)	2.77 (7.21)
2.78 (162.50)	2.79 (*)	2.80 (b)	2.81 (2218.43)	2.82 (b)		

Explanations Transformers

2.1 (b)

For half of the rated current, the voltage should be half of that voltage which will give full load current.

$$\text{i.e.,} \quad \frac{4.5\times 11\times 10^3}{100} = 495 \text{ V}$$

For half of the rated current,

$$V = \frac{495}{2} = 247.5 \text{ V}$$

2.2 (c)

This is a case of open delta connection which gives balanced supply voltage with reduced rating.

2.5 (c, d)

Transformers *i* the same phase group can be operated in parallel. The transformer with +30° and –30° angle can be operated in parallel by reversing the primary and secondary phase sequence of one transformer.

2.6 (c)

In order to have same value of core flux (V/f) ratio is to be kept constant. Also in a transformer, core flux is constant.

$$V \propto B_m f \propto \frac{\phi f}{A}$$

$$\Rightarrow \quad \frac{V}{f} = K \Rightarrow \frac{V_2}{V_1} = \frac{f_2}{f_1}$$

$$\Rightarrow \quad V_2 = \frac{25}{50} \times 230 = 115 \text{ V}$$

2.7 (c)

$$Z_0 = \frac{V}{I_0} = \frac{400}{2}\angle 85^\circ = 200\angle 85^\circ\ \Omega$$

2.8 (a)

Maximum kVA supplied

$$= \frac{1}{\sqrt{3}} \times \text{(Rated capacity of bank)}$$

$$= \frac{1}{\sqrt{3}} \times 3 \times 15 \text{ kVA} \simeq 26 \text{ kVA}$$

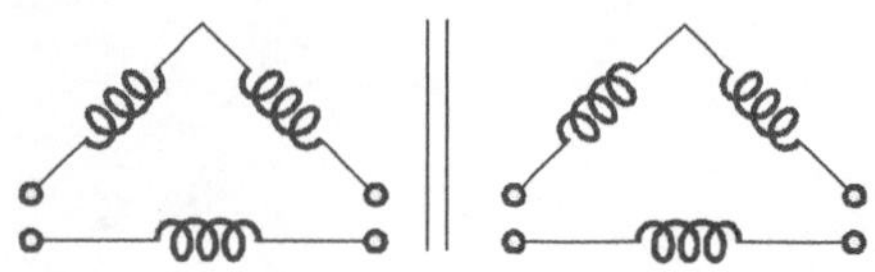

In case of bank of similar or identical transformers, each transformer generally share the load in equal proportion.

Hence load shared by one transformer

$$= \frac{20}{2} \text{ kVA} = 10 \text{ kVA}$$

2.10 (c)

$$0.98 = \frac{\text{kVA}}{\text{kVA} + P_i + P_c} \quad \text{...(i)}$$

$$\text{and} \quad 0.98 = \frac{0.5 \text{ kVA}}{0.5 \text{ kVA} + P_i + 0.25 P_c}$$

$$= \frac{\text{kVA}}{\text{kVA} + 2P_i + 0.5 P_c} \quad \text{...(ii)}$$

From equation (i) and (ii),

$P_i + P_c = 2\,P_i + 0.5\,P_c \Rightarrow 0.5\,P_c = P_i \Rightarrow P_c = 2\,P_i$

2.12 (c)

Percentage change in hysteresis loss

$$= \left[\frac{f_2}{f_1}\left(\frac{V_2}{V_1}\cdot\frac{f_1}{f_2}\right)^x - 1\right] \times 100$$

$$= \left[\frac{45}{50}\left(\frac{0.9\,V_1}{V_1}\cdot\frac{50}{45}\right)^x - 1\right] \times 100$$

$$\frac{\Delta W_h}{W_h} = -10\%$$

Percentage change in eddy current loss

$$= \left[\left(\frac{f_2}{f_1}\right)^2 \cdot \left(\frac{V_2}{V_1}\cdot\frac{f_1}{f_2}\right)^2 - 1\right] \times 100$$

$$= \left[\left(\frac{45}{50}\right)^2 \left[\frac{0.9\,V_1}{V_1}\cdot\frac{50}{45}\right]^2 - 1\right] \times 100$$

$$\frac{\Delta W_e}{W_e} = -19\%$$

$$\% \text{ change in core losses} = \frac{\Delta W_i}{W_i}$$

$$\text{Initially,} \quad W_h = W_e = \frac{W_i}{2}$$

$$\Rightarrow \quad \frac{\Delta W_i}{W_i} = \frac{\Delta W_h}{W_i} + \frac{\Delta W_c}{W_i} = \frac{1}{2}\left[\frac{\Delta W_h}{W_h} + \frac{\Delta W_e}{W_e}\right]$$

$$= \frac{1}{2}(-10 - 19) = -14.5\%$$

2.13 (a)

- If we neglect the effect of saturation and failure of insulation (as maximum voltage is being asked) the voltage ratings will be three times of their previous values and same will be the kVA ratings.
- The practical problems like failure of insulation can not be incorporated as nothing is mentioned in the question.

2.14 (c)

$$\text{Primary current} = \frac{10 \times 1000}{400} = 25 \text{ A}$$

$$\text{Short circuit current} = \frac{25 \text{ A}}{10/100} = 250 \text{ A}$$

2.15 (b)

$$x = \sqrt{\frac{\text{Iron loss}}{\text{F.L. copper loss}}} = \sqrt{\frac{0.01}{0.02}} = 0.707$$

$$\eta_{max} = \frac{0.707 \times 1}{0.707 \times 1 + 0.01 + 0.01} \times 100$$

$$= 97.248\% \approx 97.25\%$$

2.16 (a)

The auto transformer connections can be made as below:

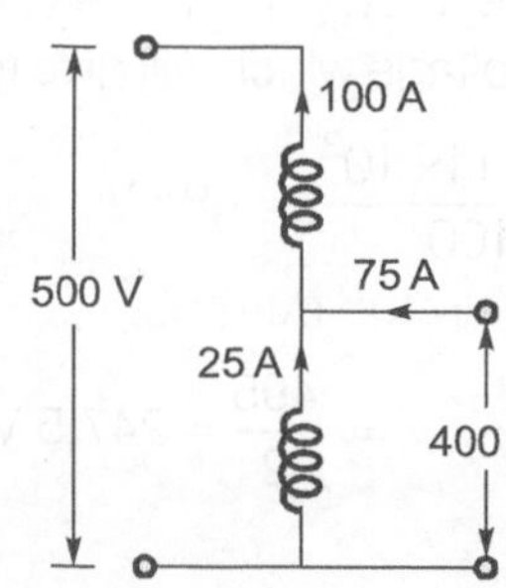

$$I_p = \frac{10 \times 1000}{400} = 25 \text{ A}$$

$$I_s = \frac{400}{100}\times 25 = 100 \text{ A}$$

$$\text{Rating} = 500 \text{ V} \times 100 \text{ A} = 50 \text{ kVA}$$

2.17 (a)

$$\%\text{ regulation} = \frac{(IR\cos\phi + IX\sin\phi)}{I}\times 100$$

$$= (R\cos\phi + X\sin\phi)\times 100$$

Being a resistive load, $\phi = 0$

$$\%\text{ regulation} = (3\times\cos 0° + 6\times\sin 0°) = 3\%$$

$= \frac{3\times 200}{100} = 6$ V. ∴ terminal voltage = 200 – 6 = 194 V

2.18 Sol.

$$S_A^* = S_L^*\left(\frac{Z_b}{Z_a + Z_b}\right)$$

$$= \left(\frac{12}{0.8}\angle{-36.87}\right)\left(\frac{8+j6}{3+j4+8+j6}\right)$$

$$= (7.466 - j6.787) \text{ kVA}$$

2.19 (b)

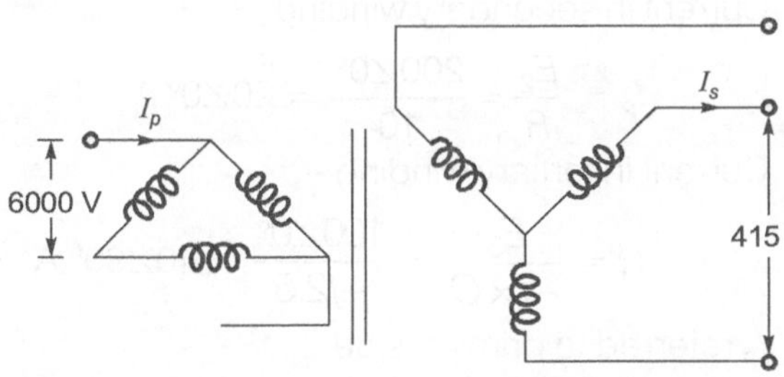

$$\text{Percentage voltage regulation} = (R\cos\phi + X\sin\phi) = (1\times 0.8 + 5\times 0.6) = 3.8\%$$

Also voltage regulation

$$= \frac{\text{Induced voltage} - \text{Terminal voltage}}{\text{Terminal voltage}}$$

$$\Rightarrow \frac{3.8}{100} = \frac{\text{I.V.} - 415}{415}$$

$$\text{I.V.} = 415\times\left(1 + \frac{3.8}{100}\right) = 430.77 \text{ V}$$

$$(\text{I.V.})_{\text{per phase}} = \frac{430.77}{\sqrt{3}} = 248.71 \text{V}$$

$$\text{Turns ratio} = \frac{6000}{248.71} = 24.12 \simeq 24$$

2.20 (a)

When the value of input voltage is maximum, rate of change of core flux is minimum; as both are 90° out of phase in case of sinusoidal input.

2.23 (b)

$$[(Z)_{\text{base}}]_{\text{per phase}} = \frac{(V_{\text{phase}})^2}{[S_{\text{base}}]_{\text{phase}}} = \frac{(33 \text{ kV})^2}{\frac{20}{3}\text{ MVA}}$$

$$Z_{\text{actual}} = Z_{\text{base}}\times Z_{\text{pu}} = \frac{3\times(33)^2}{20}\times\frac{12}{100} = 19.6\ \Omega$$

2.24 Sol.

$$\frac{I_1}{I_3} = \left(\frac{N_3}{N_1}\right)$$

$$I = \left(\frac{1}{20}\right)\cdot 50\angle -\cos^{-1}(0.6)$$

$$= 2.5\angle{-\cos^{-1}(0.6)}$$

2.26 (a)

The two 200 turns windings are connected in additive polarity. Hence the output voltage will be 400 V. The difference of this 400 V and the voltage induced in first windings (i.e. 400 turns) is same.

2.27 (d)

$$\text{Efficiency} = \frac{\text{Output}}{\text{Input}} = \frac{\text{Output}}{\text{Output} + \text{Losses}}$$

$$\text{Output} = xS\cos\phi$$

where, S = Rating of the machine

x = % of the full load

$\cos\phi$ = Power factor

P_i = Iron or core loss

P_{cu} = Full load copper loss

For maximum efficiency,

$$P_i = x^2 P_{Cu}$$

$$\text{Efficiency} = \eta = \frac{xS\cos\phi}{xS\cos\phi + P_i + x^2 P_{Cu}}$$

Max efficiency,

$$\eta_{max} = \frac{xS\cos\phi}{xS\cos\phi + 2P_i} \Rightarrow 0.9 = \frac{1\times S\times 1}{1\times S\times 1 + 2P_i}$$

$$P_i = 0.055 \text{ S}$$

$$\text{Full load copper loss} = 1^2 P_{Cu} = P_i$$

$$P_{cu} = 0.055 \text{ S}$$

Efficiency of half load, upf

$$x = \frac{1}{2}$$

$$\eta = \frac{\frac{1}{2}\times S\times 1}{\frac{1}{2}\times S\times 1 + 0.055S + \left(\frac{1}{2}\right)^2\times 0.055S}\times 100 \approx 87.8\%$$

2.28 (a)

As per Lenz's law, $E_1 = N_1 \frac{d\phi}{dt}$

The positive direction of this emf opposes the positive current direction.

If, $\phi = \phi_m \sin \omega t$

$$E_1 = N_1 \frac{d}{dt}(\phi_m \sin\omega t) = N_1 \phi_m \omega \cos \omega t$$

E_1 leads the flux by 90°.

Similarly, $E_2 = N_2 \frac{d\phi}{dt}(\phi_m \sin\omega t)$

and E_2 leads the flux by 90°.

So E_1 and E_2 are in phase.

But, $\frac{E_1}{E_2} = \frac{N_1}{N_2} = 2$

$\Rightarrow$ $E_1 = 2E_2$ or $E_1 > E_2$

$$I_2 = \frac{E_2}{R}$$

Because of resistive load, I_2 is in phase with E_2.

$$\frac{I_1}{I_2} = \frac{N_2}{N_1} = \frac{1}{2} \Rightarrow I_1 = \frac{I_2}{2} \Rightarrow I_1 < I_2$$

I_1 and I_2 are in phase, because

$$I_1 = I_2 + \text{Exciting current } (I_m)$$

But I_m is very small. On the basis of above analysis, phasor diagram is

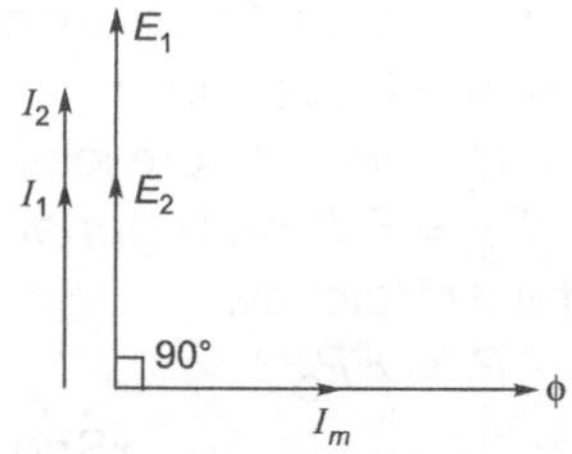

2.29 (b)

As the transformer is Δ-Y connected, it means line-to-line voltage on primary side induces line to neutral for phase voltage on secondary side due to transformer.

$$\text{Turn ratio} = \frac{N_1}{N_2} = a = \frac{11\text{kV}}{415/\sqrt{3}} = \frac{11\sqrt{3}}{415} \times 10^3$$

$$V_{AB} = aV_a,\ V_{BC} = aV_b \text{ and } V_{AC} = aV_c$$

When S_2 is closed, B and C are at same potential.

So, $V_{BC} = V_B - V_C = 0$

Applying KVL in, Δ connection

$V_{AB} + V_{BC} + V_{AC} = 0$

$\Rightarrow$ $V_{AC} = -V_{AB}$

Let, $V_{AB} = 11 \times 10^3 \angle 0°$

$V_{AC} = -11 \times 10^3 \angle 0° = 11 \times 10^3 \angle 180°$

Voltage across terminals a and c,

$$V_{ac} = V_a - V_c$$

$$= \frac{V_{AB}}{a} - \frac{V_{AC}}{a} = \frac{1}{a}[V_{AB} - V_{AC}]$$

$$= \frac{415}{11\sqrt{3} \times 10^3}[11 \times 10^3 \angle 0° - 11 \times 10^3 \angle 180°]$$

$$= \frac{415}{11\sqrt{3} \times 10^3}[11 \times 10^3 \times 2] \approx 480 \text{ V}$$

2.30 (c)

Turns ratio, $N_1 : N_2 : N_3$ is 4 : 2 : 1

Induced emf in winding 1 = $E_1 \approx V_1 = 400\angle 0°$V

As we know,

$$\frac{E_1}{E_2} = \frac{N_1}{N_2} \Rightarrow E_2 = \frac{N_2}{N_1}E_1 = \left(\frac{2}{4}\right)400\angle 0°$$

$$= 200\angle 0° \text{ V}$$

and $\frac{E_1}{E_3} = \frac{N_1}{N_3}$

$$\Rightarrow E_3 = \left(\frac{N_3}{N_1}\right)E_1 = \left(\frac{1}{4}\right)400\angle 0° = 100\angle 0° \text{ V}$$

Current in secondary winding

$$I_2 = \frac{E_2}{R} = \frac{200\angle 0°}{10} = 20\angle 0° \text{ A}$$

Current in tertiary winding

$$I_3 = \frac{E_3}{-j \times C} = \frac{100\angle 0°}{-j2.5} = 40\angle 90° \text{ A}$$

I_2 referred to primary side

$$I_2' = \left(\frac{N_2}{N_1}\right)I_2 = \left(\frac{2}{4}\right) \times (20\angle 0°) = 0\angle 0° \text{A}$$

I_3 reffered to primary side

$$I_3' = \left(\frac{N_3}{N_1}\right)I_3 = \left(\frac{1}{4}\right)(40\angle 90°) = 10\angle 90°$$

The supply current

$$I_1 = I_2' + I_3' = 10\angle 0° + 10\angle 90° = (10 + j10) \text{ A}$$

2.31 (b)

Copper loss at any load

$$P_{cu} = x^2 \times [\text{Full load copper loss}]$$
$$= x^2 \times P_{fl,\,cu}$$

where, x = fraction of load

Maximum occurs when

Copper loss = Iron loss

$$x^2 \times P_{fl,\,Cu} = P_i \Rightarrow x^2 \times 600 = 300$$
$$x = 0.707 = 70.7\%$$

2.32 (d)

Consider two-winding transformer

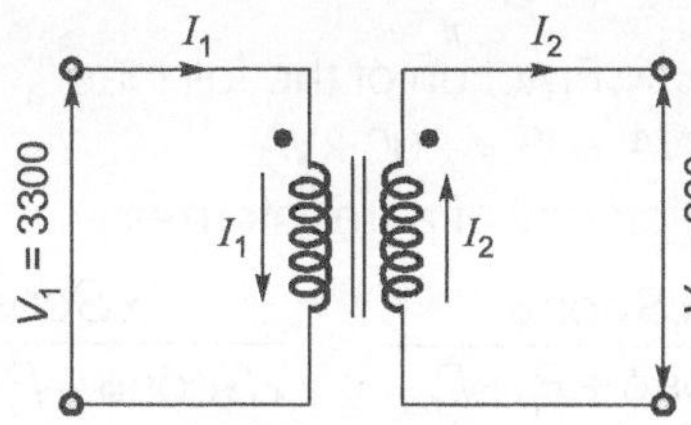

$$I_1 = \frac{50 \times 10^3}{3300} = 15.15 \text{ A}$$

$$I_2 = \frac{50 \times 10^3}{230} = 217.34 \text{ A}$$

When this transformer is connected as step-up auto transformer

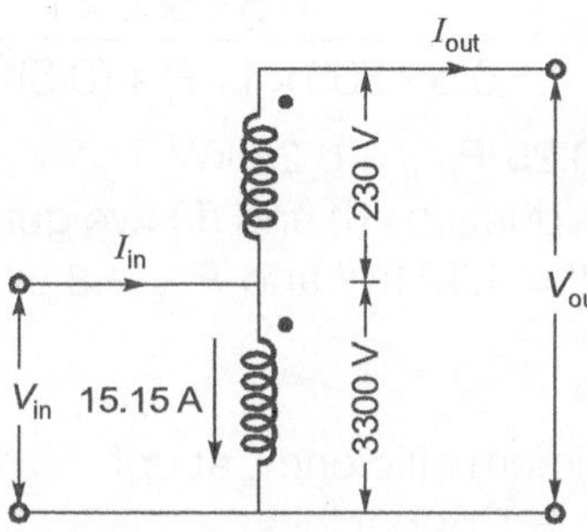

$V_{in} = 3300; \quad V_{out} = 3300 + 230 = 3530$ V

$I_{out} = I_2 = 217.34$ A

$I_{in} = I_{out} + 15.15 = 232.49$ A

kVA rating of the auto transformer

$S_{auto} = V_{in} I_{in} = 3300 \times 232.49 \times 10^{-3} \approx 767.4$ kVA

2.33 (d)

Rated kVA = 100 kVA

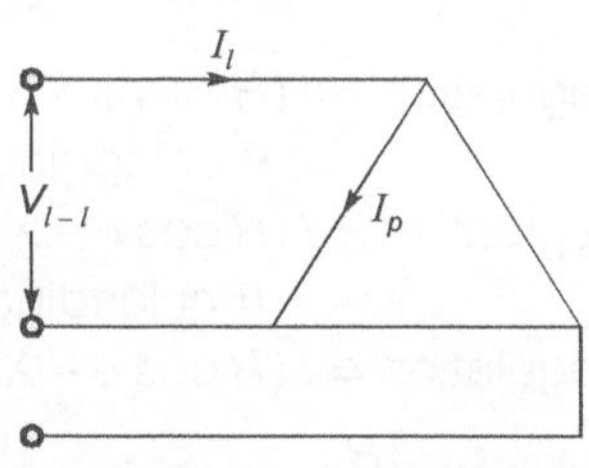

Rated primary line to line voltage

$V_{l-l} = 11000$ V

Rated primary phase voltage

$V_P = V_{l-l} = 11000$ V

Rated primary side line current

$$I_l = \frac{\text{Rated kVA} \times 10^3}{\sqrt{3} V_{l-l}} = \frac{100 \times 10^3}{\sqrt{3} \times 11000} \text{ A}$$

Rated primary phase current,

$$I_P = \frac{I_1}{\sqrt{3}} = \frac{100 \times 10^3}{3 \times 11000}$$

Base phase impedance,

$$Z_B = \frac{V_P}{I_P} = \frac{11000}{\left(\frac{100 \times 10^3}{3 \times 11000}\right)} = 3630\ \Omega$$

p.u. phase impedance = $0.02 + j0.07$ pu

Phase impedance = Base phase impedance × $(0.02 + j0.07)$

$= 3630 \times (0.02 + j0.07)\ \Omega = (72.6 + j254.1)\ \Omega$

2.34 (d)

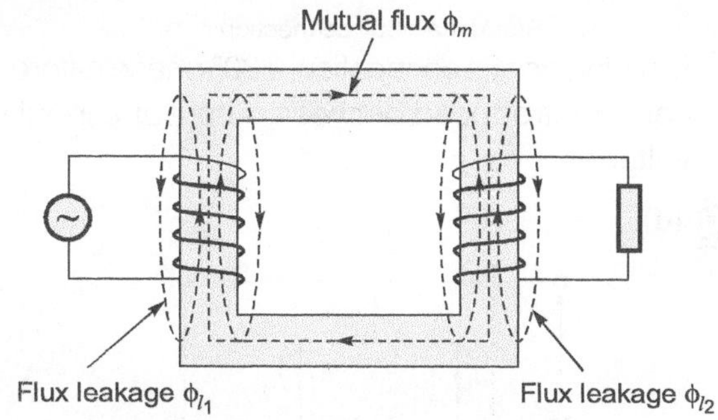

A major part of the total flux is confined to the core as mutual flux ϕ_m linking both the primary and secondary, a small amount of flux does leak through paths which lie mostly in air and link separating the individual windings.

Since, reactance ∝ amount of the flux linkage

$\therefore \quad X_m >> X_1 \approx X_2'$

2.35 (b)

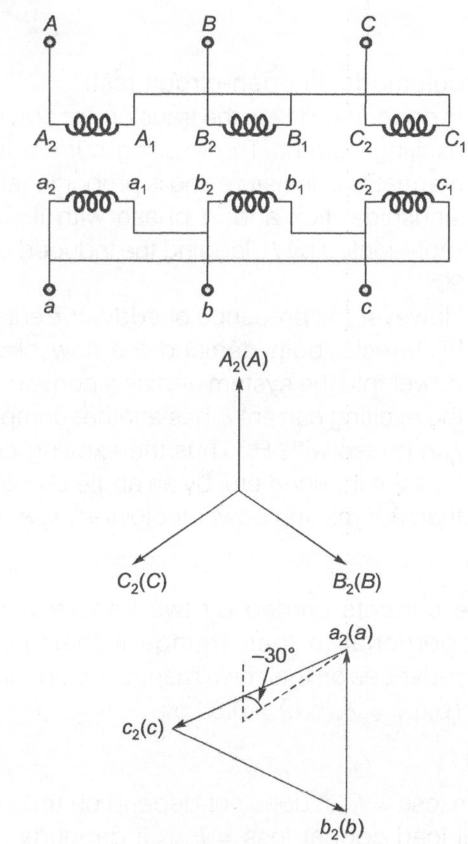

Star/Delta –30° connection

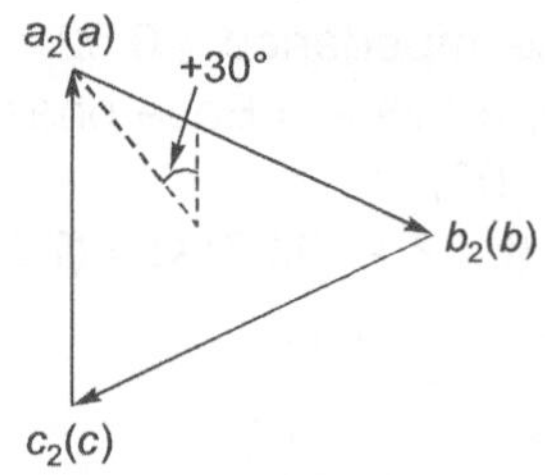

Star/delta +30° connection

By using above connection, ±30° phase difference can be introduced between output and input line voltages.

2.36 (d)

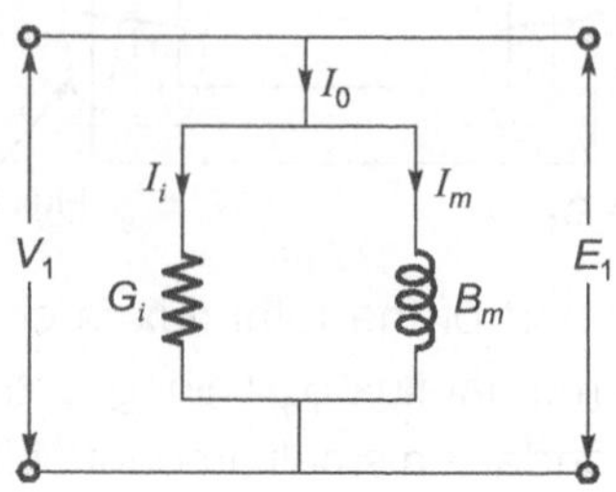

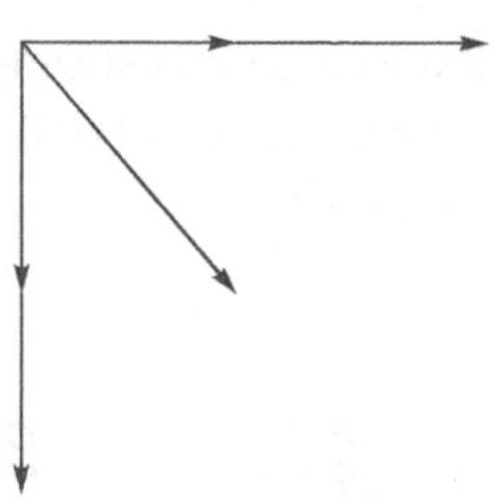

Circuit model in open-circuit test:

- In open-circuit test, the transformer draws only exciting current. The exciting current is only magnetizing in nature and is proportional to the sinusoidal flux and in phase with it. This is represented by I_m lagging the induced emf by 90°.
- However the presence of eddy-currents, and hysteresis, both demand the flow of active power into the system and as a consequence the exciting current I_0 has another component I_i in phase with E_1. Thus the exciting current lags the induced emf by an angle slightly less than 90° making power factor very low.

2.37 (c)

The currents carried by two transformers are proportional to their ratings if their per-unit impedances on their own ratings are equal.

Z_2(p.u.) = Z_1(p.u) = 0.05 p.u.

2.38 (c)

Iron loss = P_i, it does not depend on the load.
Full load copper loss = $P_{c,\,fl}$ it depends on the load

$$P_{cu} = x^2 P_{c,\,fl}$$

where, x = Fraction of the full load
Rated KVA = S = 300 kVA
η =% efficiency of a transformer

$$= \frac{xS\cos\phi}{xS\cos\phi + P_i + P_{cu}} = \frac{xS\cos\phi}{xS\cos\phi + P_i + x^2 P_{c,fl}}$$

(i) $\eta = 95\%,\ x = 1,\ \cos\phi = 0.8$

$$0.95 = \frac{1\times 300\times 0.8}{1\times 300\times 0.8 + P_i + 1^2 \times P_{cfl}}$$

$$\Rightarrow P_i + P_{c,\,fl} = 12.63 \text{ kW} \quad \text{...(i)}$$

(ii) $\eta = 96\%,\ x = 0.5\ ,\ \cos\phi = 1$

$$0.96 = \frac{0.5\times 300\times 1}{0.5\times 300\times 1 + P_i + (0.5)^2 \times P_{c,fl}}$$

$$\Rightarrow P_i + 0.25\ P_{c,\,fl} = 6.25 \text{ kW} \quad \text{...(ii)}$$

Solving equations (i) and (ii), we get

$$P_i = 4.12 \text{ kW and } P_{c,fl} = 8.51 \text{ kW}$$

2.39 (b)

For maximum efficiency, at upf
Iron loss = Copper loss

$$P_i = x^2 P_{c,\,fl}$$

$$4.12 = x^2\, 8.51 \ \Rightarrow\ x = 0.696$$

$$\eta_{max} = \frac{xS\cos\phi}{xS\cos\phi + 2P_i} \times 100$$

$$= \frac{0.696\times 300\times 1}{0.696\times 300\times 1 + 2\times 4.12}\times 100 = 96.2\%$$

2.40 (c)

Voltage regulation = $I(R\cos\phi + X\sin\phi)$
(For lagging power factor)
Voltage regulation = $I(R\cos\phi - X\sin\phi)$
(For leading power factor)
Voltage regulation = $I(R\cos\phi - X\sin\phi) = 0$

$$\tan\phi = \frac{R}{X}$$

So zero voltage regulation is possible for leading power factor.

When, $\tan\phi = \dfrac{R}{X}$

2.41 (c)

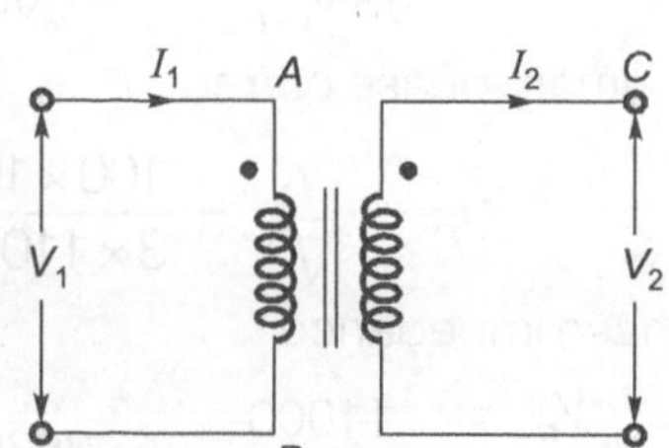

In two winding transformer

$V_1 = 500$ V; $V_2 = 250$ V

$$I_1 = \frac{50\times 10^3}{500} = 100\text{ A}$$

and $$I_2 = \frac{50\times 10^3}{250} = 200\text{ A}$$

$$\eta = \frac{xS\cos\phi}{xS\cos\phi + losses}$$

Full load,

$$x = 1,\ \text{upf},\ \cos\phi = 1,\ S = 50$$

$$0.95 = \frac{1\times 50\times 1}{1\times 50\times 1 + \text{losses}}$$

Losses = 2.63

Losses = Iron loss + Copper loss

When two winding transformer is connected as step up autotransformer

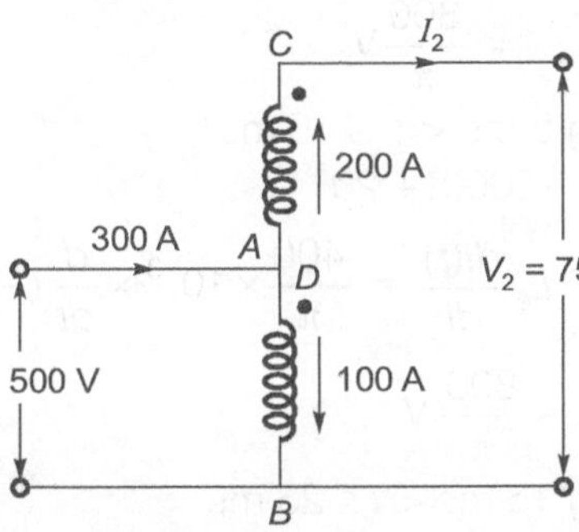

$I_2 = 200$; $V_2 = 750$

kVa rating of auto transformer

$$S_{auto} = V_2 I_2 = 750 \times 200 = 150\text{ kVA}$$

As current through windings and voltage across windings are equal in two-winding transformer and autotransformer. Losses remain same at full load efficiency at upf and full load

$$\%\eta = \frac{xS_{auto}\cos\phi}{xS_{auto}\cos\phi + \text{losses}} \times 100$$

$$= \frac{1\times 150\times 1}{1\times 150\times 1 + 2.63}\times 100 = 98.276\%$$

2.42 (b)

Phasor diagram of primary and secondary voltage.

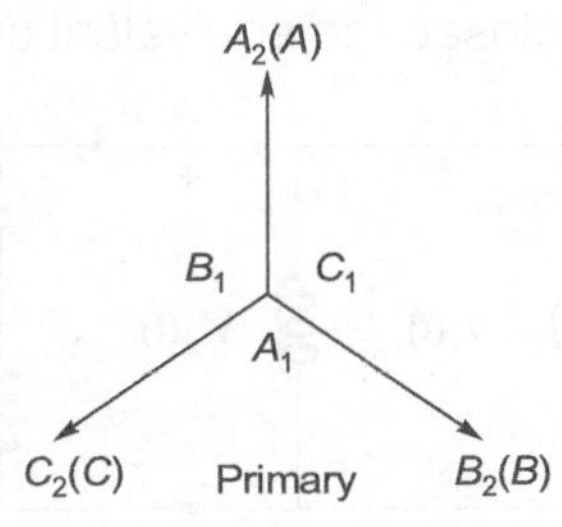

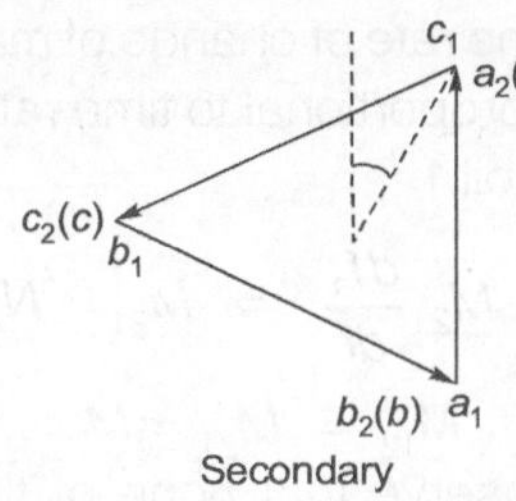

It is observed from the phasor diagram that phase a to neutral voltage (equivalent star basis) on the delta side lags by –30° to the phase to neutral voltage on the star side. This connection is known as –30° connection or Yd1 connection.

2.43 (d)

Open-circuit Test: In open-circuit test, no-load current I_0 is very small (it is usually 2-6% of the rated current).

No-load current I_0 lags E by slightly less than 90°, so power factor is very low.

Therefore, wattmeter W_2 is suitable for the open-circuit test.

Short-circuit Test: In short-circuit test, V_{sc} needed to circulate the full-load current is very low.

Under these conditions, I_0 is only about 0.1 to 0.5% of full load current. As I_0 is highly lagging but it is very small as compared to full load current, therefore the power factor is high.

So, wattmeter W_3 is suitable for the short-circuit test.

2.45 (b)

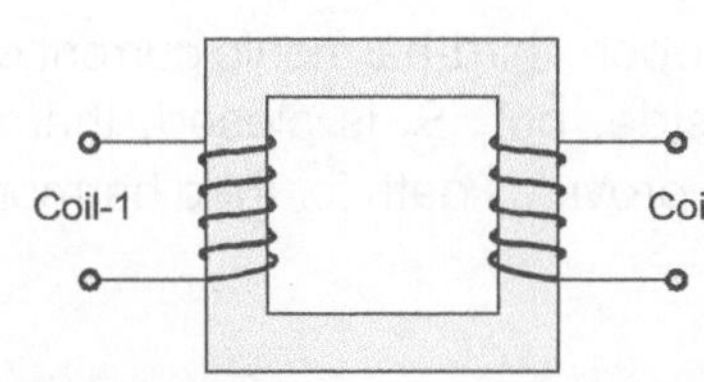

Mutual inductance is where the flux of two or more coils are linked so that voltage is induced in one coil proportional to the rate of change of current in another.

The mutual inductance that exists between the two coils can be greatly increased by positioning them on a common soft iron core or by increasing the number of turns of either coil.

Consider:

The changing flux in coil 1 is proportional to changing current in coil 2.

$$N_1\frac{d\phi_{12}}{dt} = M_{12}\frac{dI_2}{dt} \Rightarrow M_{12} = N_1\frac{\phi_{12}}{I_2}$$

Similarly time rate of change of magnetic flux ϕ_{21} in coil 2 is proportional to time rate of change of current in coil 1

$$N_2 \frac{d\phi_{21}}{dt} = M_{21}\frac{dI_1}{dt} \Rightarrow M_{21} = N_2 \frac{\phi_{21}}{I_1}$$

Also, $\quad M_{12} = M_{21} = M$

We can observe that none of the parameters changes due to placing of coils on horizontal arms hence mutual inductance will remain unchanged.

2.46 (b)

As S_2 is open, third harmonic current cannot flow in delta winding. S_1 is open, neutral gets isolated, therefore no third harmonic current in star side. Since third harmonic current cannot flow in either side, the magnetizing current I_m is almost sinusoidal.

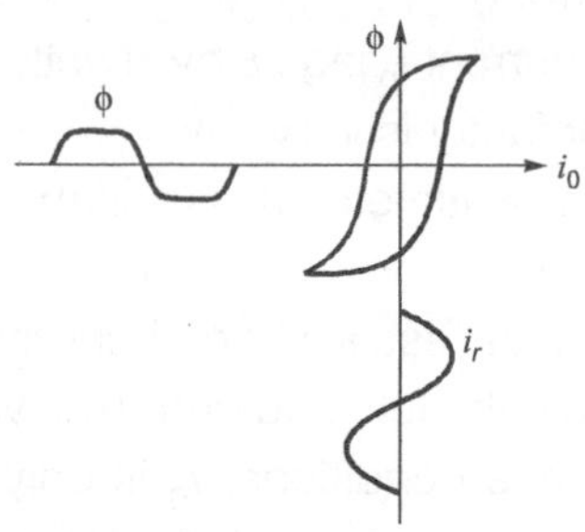

T_0 satisfies the B-H curve, the core, the core flux must then be non-sinusoidal it is a flat-topped wave. This can be verified by assuming a sinusoidal i_m and then finding out the ϕ wave shape from the $\phi - i_m$ relationship.

2.47 (c)

As S_1 is open, third-harmonic current cannot flow in star side, but, S_2 is closed, therefore delta winding provides path for third harmonic current to flow.

2.48 (d)

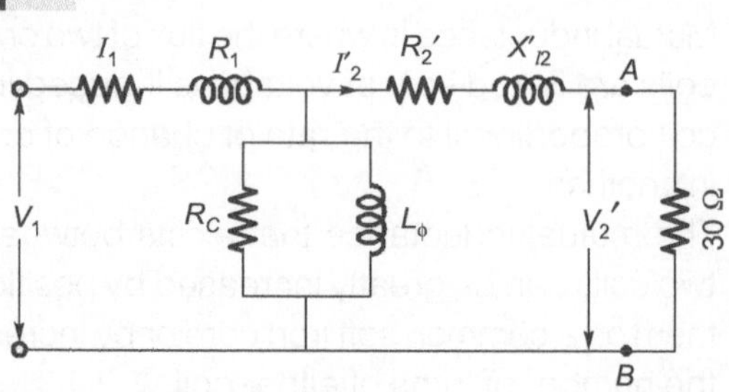

No leakage flux $\Rightarrow X_{l_1}$ and X'_{l_2} are neglected.

Transformer is lossless $\Rightarrow R_1$, R_2' and R_c are neglected. L_ϕ = magnetizing inductances = $\frac{400}{\pi}$ mH.

Therefore, the above equivalent circuit can be redrawn with S open.

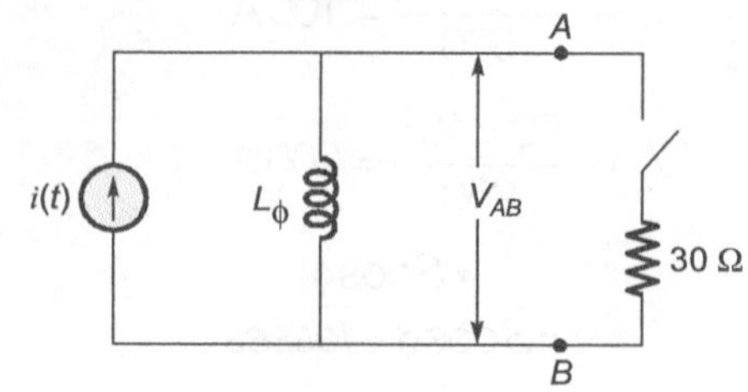

$$V_{AB} = L_\phi \frac{di(t)}{dt}$$

(i) During $0 < t \le 5$ ms, $i(t)$ increases linearly with time

$$i(t) = \frac{10}{5\times10^{-3}}t = 2000t$$

$$V_{AB} = L_\phi \frac{di(t)}{dt} = \frac{400}{\pi}\times10^{-3}\times\frac{d(2000t)}{dt}$$

$$= \frac{800}{\pi}\text{V}$$

(ii) During 5 ms $< t \le 15$ ms

$$i(t) = -2000t + 20$$

$$V_{AB} = L_\phi \frac{di(t)}{dt} = \frac{400}{\pi}\times10^{-3}\times\frac{d}{dt}(-2000t+20)$$

$$= -\frac{800}{\pi}\text{V}$$

(iii) During 15 ms $< t \le 25$ ms

$$i(t) = 2000\,t - 40$$

$$V_{AB} = L_\phi \frac{di(t)}{dt} = \frac{400}{\pi}\times10^{-3}\times\frac{d}{dt}(2000t-40)$$

$$= \frac{800}{\pi}\text{V}$$

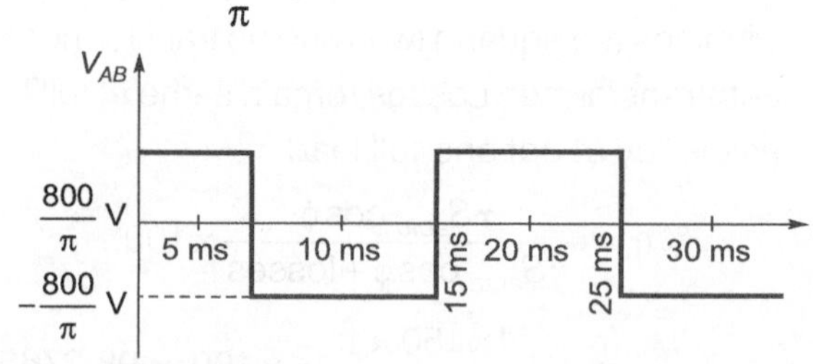

Therefore, peak voltage across A and $B = \frac{800}{\pi}$V

2.49 (b)

When s is closed the equivalent circuit,

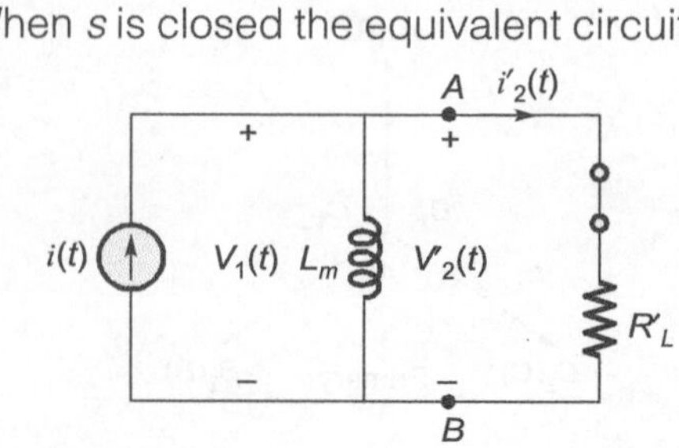

Here, $R_L' = \left(\frac{N_1}{N_2}\right)^2 R_L$

$$= \left(\frac{1}{1}\right)^2 \times 30 = 30\ \Omega$$

Load current referred to primary side,

$$i_2'(t) = \frac{j\omega L_m}{j\omega L_m + R_L'} \times i(t)$$

Load voltage referred to primary side,

$$v_2' = i_2'(t) R_L'$$

$$\Rightarrow \quad V_2'(t) = \frac{j\omega L_m R_L'}{R_L' + j\omega L_m} \times i(t)$$

So, secondary voltage across A and B

$$V_2(t) = \frac{N_2}{N_1} V_2'(t) = \frac{1}{1} \times V_2'(t)$$

$$\Rightarrow \quad V_2(t) = \frac{j\omega L_m R_L'}{R_L' + j\omega L_m} \times i'(t)$$

Given, $i(t) = 10\cos 100\pi t$

$\therefore$ Frequency,

$$\omega = 100\pi$$

$$V_2(t) = \frac{j100\pi \times \frac{400}{\pi} \times 10^{-3} \times 30}{30 + j100\pi \times \frac{400}{\pi} \times 10^{-3}} \times 10\sin 100\pi t$$

$$= V_m \cos(100\pi t + \phi)$$

Peak value of secondary voltage,

$$V_m = \left|\frac{j12000}{30 + j40}\right| V = 240\text{ V}$$

2.50 (a)

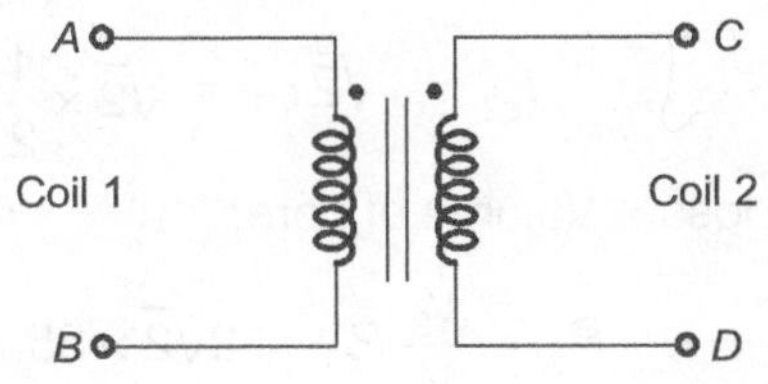

No. of turns $N_1 = 4000$ and $N_2 = 6000$

$$\frac{E_1}{E_2} = \frac{N_1}{N_2} \Rightarrow E_2 = \frac{N_2}{N_1} E_1$$

$$= \frac{6000}{4000} \times 400 = 600\text{ V}$$

The two coils are connected to obtain step up autotransformer $\frac{400\text{ V}}{1000\text{ V}}$.

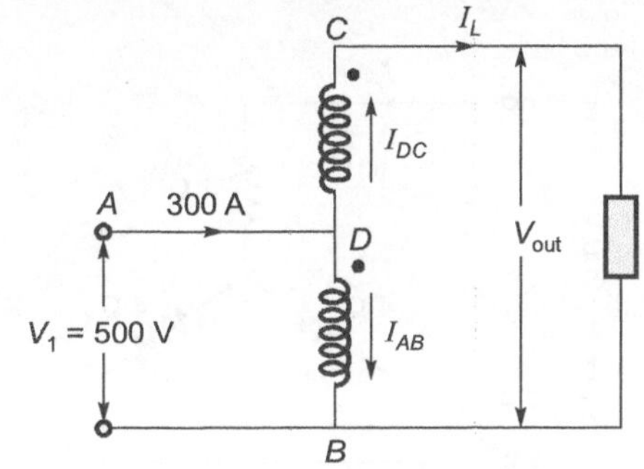

$$V_{in} = 400\text{ V}$$

$$V_{out} = V_{AB} + V_{CD}$$

$$= 400 + 600 = 1000\text{ V}$$

2.51 (d)

Refer to circuit-diagram of autotransformer drawn in previous solution,

load connected,

$$S = 10\text{ kVA}$$

$$I_L = \text{Load current}$$

$$= \frac{S}{V_{out}} = \frac{10 \times 10^3}{1000} = 10\text{ A}$$

Current through coil 2

$$I_{DC} = I_L = 10\text{ A}$$

Input current of autotransformer

$$I_{in} = \frac{S}{V_{in}} = \frac{10 \times 10^3}{400}$$

$$\Rightarrow \quad I_{in} = 25\text{ A}$$

Using KCL, $I_{in} = I_{DC} + I_{AB}$

$$\Rightarrow \quad 25 = 10 + I_{AB}$$

Current through coil 1 $= I_{AB} = 15$ A

2.52 (c)

Secondary current,

$$I_2 = 1\angle 0°\text{ A}$$

Secondary current referred to primary side,

$$I_2' = \left(\frac{N_2}{N_1}\right) I_2 = \frac{2}{1} \times 1\angle 0° = 2\angle 0°\text{ A}$$

As the core losses are neglected, magnetizing current (I_m) will be in phase with flux (ϕ). Therefore I_m lags the induced emf by 90°

$$I_m = 1\angle -90°\text{ A}$$

Primary current,

$$I_1 = I_m + I_2'$$

$$= 1\angle -90° + 2\angle 0°$$

$$\approx 2.24\angle -26.56°\text{ A}$$

2.53 (d)

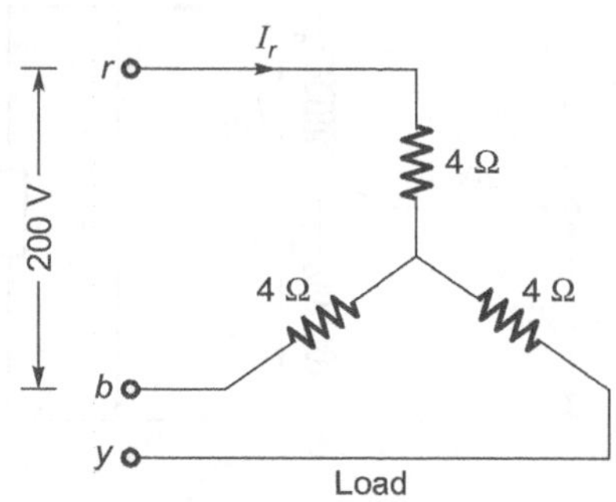

Let VA rating of the transformer S.

Rated voltage (line to line) on secondary side = 200 V

$$\text{rated current} = \frac{S}{\sqrt{3}\,200}$$

Base impedance

$$= \frac{\text{Rated phase to neutral voltage}}{\text{Rated current}}$$

$$Z_B = \frac{200/\sqrt{3}}{S/200\sqrt{3}} = \frac{200^2}{S}$$

pu impedance on secondary side

$$= \frac{4}{Z_B} = \frac{4}{200^2/S} = \frac{4S}{200^2}$$

pu impedance in primary side and secondary side are equal.

Therefore *pu* impedance on primary side

$$Z_{1\,pu} = \frac{4S}{200^2}$$

Rated phase to neutral voltage on primary side,

$$V_{p1} = \frac{100}{\sqrt{3}}$$

Rated primary current (phase)

$$= \frac{S}{\sqrt{3}V_{l\text{-}l}} = \frac{S}{\sqrt{3}100}$$

Base impedance

$$= \frac{\text{rated phase to neutral voltatge}}{\text{rated current}}$$

$$= \frac{100/\sqrt{3}}{S/100\sqrt{3}} = \frac{100^2}{S}$$

Load impedance reffered to primary side

Z_1 = base impedance (primary side) × *pu* impedance

$$= \frac{100^2}{S} \times \frac{4S}{200^2} = (1 + j0)\ \Omega$$

2.54 (c)

The transformer is fed from sinusoidal supply, so flux (ϕ) set up in the transformer is also sinusoidal. As the transformer is air-cored, there will be no saturation and hysteresis effect in transformer. So B-H curve is linear, due to which magnetizing current is also sinusoidal.

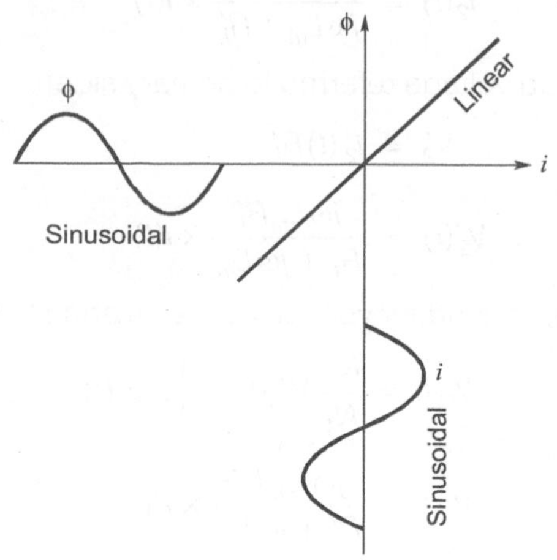

2.55 (b)

We know, $E = \sqrt{2}\pi N f \phi_m$

as N and f are constant,

$$\frac{E_1}{E_2} = \frac{\phi_{m1}}{\phi_{m2}} = \frac{1}{2}$$

$$\phi_{m2} = 2\phi_{m1}$$

and also, $\phi_m = \frac{\mu N i A}{d}$

$$\Rightarrow \quad \frac{\phi_{m1}}{\phi_{m2}} = \frac{i_{e1}A_1 d_2}{i_{e2}A_2 d_1} = \frac{\sqrt{2}I_{e1}}{2i_{e2}} = \frac{1}{2}$$

$$\Rightarrow \quad i_{e2} = \sqrt{2}i_{e1} = \sqrt{2}\times\frac{1}{2} = 0.7\text{ A}$$

Core loss ∝ Volume of core,

$$\therefore \quad P_{c2} = \frac{V_2}{V_1}P_{c1} = 2\sqrt{2}\times 55 = 155.6\text{ W}$$

2.56 (c)

For 90% current,

$$P_{cu} = (0.9)^2 P_{fl\,cu}$$

$$P_{fl\,cu} = \frac{81}{0.81} = 100\text{ W}$$

$$x = \sqrt{\frac{P_{core}}{P_{fl\,cu}}} = \sqrt{\frac{64}{100}} = 0.8 = 80\%$$

2.57 (b)

Reluctance will be increased by two times. By considering flux density as constant flux should be reduced to 25% as area is reduced to 25%.

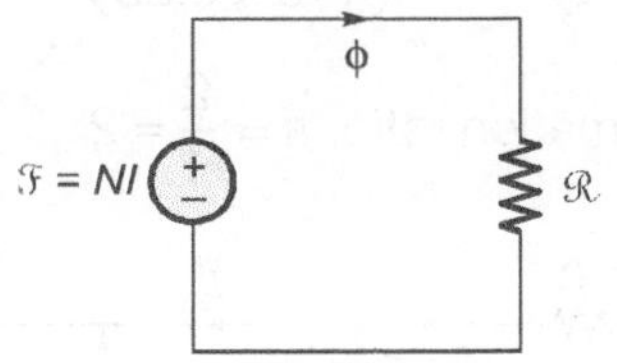

$$\mathcal{F} = \phi\mathcal{R}$$

$$\mathcal{F}' = \frac{\phi}{4}\cdot 2\mathcal{R} = \frac{\mathcal{R}\phi}{2}$$

For given current, number of turns should be reduced by half.

2.58 (a)

Given, at 230 V, 50 Hz, core loss = 1050 W and at 138 V, 30 Hz, core loss = 500 W

Let, $V_1 = 230$ volt, $f_1 = 50$ Hz,

Eddy current loss $= P_{e_1}$,

Hysteresis loss $= P_{h_1}$

$V_2 = 138$ volt, $f_2 = 30$ Hz,

Eddy current loss $= P_{e_2}$,

Hysteresis loss $= P_{h_2}$

We know that,

eddy current loss, $P_e \propto B_m^2 f^2$

and hysteresis loss, $P_h \propto B_m^x \cdot f \propto B_m^{1.6} f$

(Typical value of $x = 1.6$)

The transformer induced emf equation is:

$$V_1 = \sqrt{2}\pi f \phi_m \cdot N$$

or, $$V_1 = \sqrt{2}\pi f \cdot B_m \cdot A \cdot N$$

when, $\frac{V}{t}$ = constant

i.e. flux = constant (to avoid over fluxing)

Eddy current loss, $P_e \propto V^2$

and hysteresis loss, $P_h \propto \frac{V^{1.6}}{f^{0.6}}$

$$\therefore \quad \frac{P_{e_2}}{P_{e_1}} = \left(\frac{V_2}{V_1}\right)^2 = \left(\frac{138}{230}\right)^2 = 0.36$$

or, $$P_{e_2} = 0.36\,P_{e_1} \quad \text{...(i)}$$

Also, $$\frac{P_{h_2}}{P_{e_1}} = \left(\frac{V_2}{V_1}\right)^{1.6} \times \left(\frac{f_1}{f_2}\right)^{0.6}$$

$$= \left(\frac{138}{230}\right)^{1.6} \times \left(\frac{50}{30}\right)^{0.6}$$

$$= (0.6)^{1.6} \times \frac{1}{(0.6)^{0.6}} = 0.6$$

$$\therefore \quad P_{h_2} = 0.6\,P_{h_1} \quad \text{...(ii)}$$

Given, at 230 V, 50 Hz,

Core loss $= P_{e_1} + P_{h_1} = 1050$ W ...(iii)

at 138 V, 30 Hz,

Core loss $= P_{e_2} + P_{h_2} = 500$ W

or, $0.36\,P_{e_1} + 0.6 P_{h_1} = 500$ W ...(iv)

Solving equations (iii) and (iv), we get

$$P_{h_1} = 508 \text{ W and } P_{e_1} = 542 \text{ W}$$

Therefore, hysteresis and eddy current losses at 230 V, 50 Hz are 508 W and 542 W respectively.

2.59 Sol.

Method-I:

Given, 50 kVA, 1000 V/100 V, 2-winding transformer

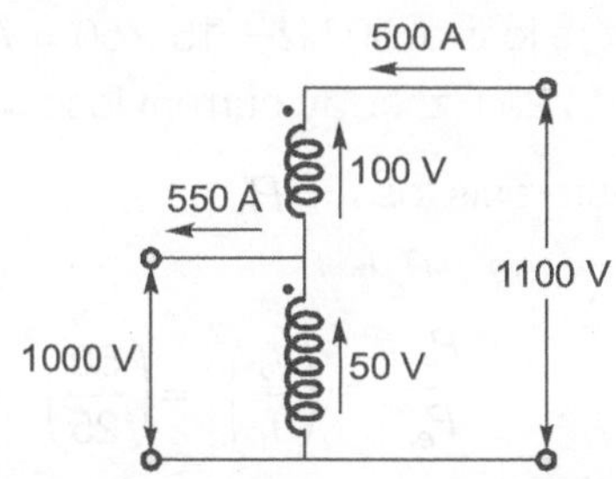

Current in HV side (winding) $= \frac{50 \times 10^3}{1000} = 50$ A

Current in LV winding $= \frac{50 \times 10^3}{100} = 500$ A

The direction and magnitude of the currents in HV and LV windings according to the dot polarity is shown in figure.

Therefore, kVA rating of the autotransformer

$1000 \times 550 = 550$ kVA

or $500 \times 1100 = 550$ kVA

Method-II:

From the figure, autotransformation ratio,

$$a_{\text{auto}} = \frac{1100}{1000} = \frac{11}{10}$$

We know that:

$$\frac{S_{\text{auto}}}{S_{2\text{ winding}}} = \left(\frac{a_{\text{auto}}}{a_{\text{auto}-1}}\right) = \left(\frac{\frac{11}{10}}{\frac{11}{10} - 1}\right)$$

$$= \frac{11}{10} \times \frac{10}{1} = 11$$

$\therefore S_{\text{auto}} = 11 \times S_2$ winding $= 11 \times 50$ kVA $= 550$ kVA

2.60 (b)

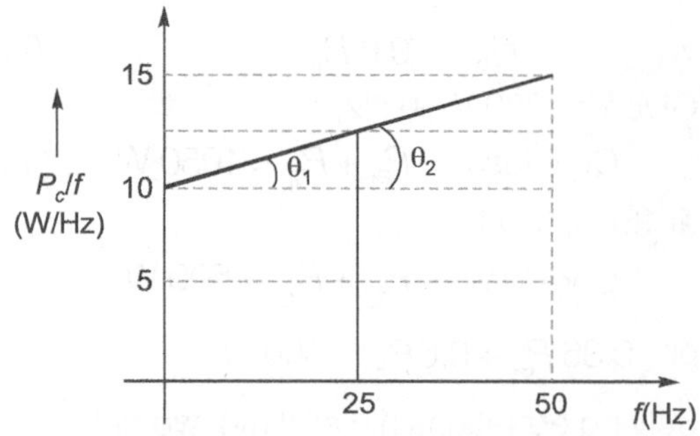

Let core loss at 25 Hz be P_0 per Hz.

Then, equating the two slopes of the figure, we have :

$$\left(\frac{P_0 - 10}{25}\right) = \frac{5}{50}$$

or, $P_0 = 2.5 + 10 = 12.5$ W/Hz

∴ Core loss at 25 Hz = 12.5 × 25 = 312.5 Watt

Also, core loss at 50 Hz = 15 × 50 = 750 Watt

Let at (f_1) 25 Hz, eddy current loss = P_{e_1}

and hysteresis loss = P_{h_1}

We know that, $P_e \propto f^2$

$$\therefore \quad \frac{P_{e_2}}{P_{e_1}} = \left(\frac{f_2}{f_1}\right)^2 = \left(\frac{50}{25}\right)^2 = 4$$

$$\therefore \quad P_{e_2} = 4P_{e_1} \quad \text{...(i)}$$

Also, $P_h \propto f$

$$\therefore \quad \frac{P_{h_2}}{P_{h_1}} = \frac{f_2}{f_1} = \frac{50}{25} = 2$$

$$\therefore \quad P_{h_2} = 2P_{h_1} \quad \text{...(ii)}$$

Now, at $f_1 = 25$ Hz,

$$P_{e_1} + P_{h_1} = 312.5 \quad \text{...(iii)}$$

and at $f_2 = 50$ Hz

$$P_{e_2} + P_{h_2} = 750 \text{ W}$$

or, $4P_{e_1} + 2P_{h_1} = 750$ W

or, $2P_{e_1} + P_{h_1} = 375$ W ...(iv)

(Using equation (i) and (ii))

Solving equations (iii) and (iv), we get

$P_{e_1} = 62.5$ Watt and $P_{h_1} = 250$ Watt

Thus, at 25 Hz, hysteresis loss = 250 W and eddy current loss = 62.5 W.

2.61 Sol.

Method-I:

As transformed is ideal, induced emf in secondary winding = load voltage

i.e., $E_2 = V_2 = 110$ Volt

Load current/secondary winding current,

$$I_2 = \left(\frac{4000}{110 \times 0.89}\right) = 40.86 \text{ A}$$

Transformation ratio, $a = \frac{2}{1} = 2$

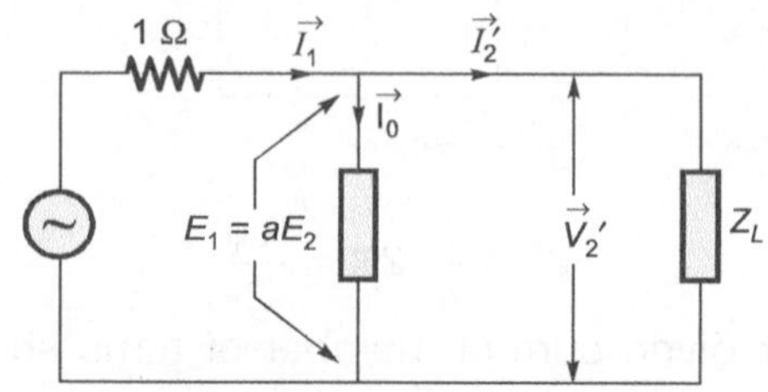

(Equivalent circuit offered to primary)

Given, cosϕ = 0.89 or, ϕ = 27.12°

Secondary current referred to primary,

$$I_2' = \frac{I_a}{a} = 20.43 \text{ A}$$

$$\therefore \quad \vec{I_2} = 20.43\angle{-27.12°} \text{ A},$$

$$E_1 = aE_2 = 2 \times 110 = 220 \text{ volt}$$

Now, $\vec{I_1} = \vec{I_0} + \vec{I_2'}$

Since, input pf required is unity and the load is inductive (lagging), therefore, x must be capacitive reactance.

$$\therefore \quad \vec{I_1}\angle 0° = \frac{E_1}{-jX} + \vec{I_2'} = \frac{jE_1}{X} \times \vec{I_2'}$$

$$= \left(\frac{j220}{X}\right) + (20.43\angle - 27.12°)$$

$$= \frac{j220}{X} + (18.18 - j9.31)$$

or, $$\vec{I_1}\angle 0° = \sqrt{(18.18)^2 + \left(\frac{220}{X} - 9.31\right)^2}\angle\theta° \quad \text{...(i)}$$

where, $$\theta = \tan^{-1}\left[\frac{\frac{220}{X} - 9.31}{18.18}\right]$$

Equating angle of equation (i) on both sides, we get

$$\tan^{-1}\left(\frac{\frac{220}{X} - 9.31}{18.18}\right) = 0°$$

or, $\frac{220}{X} - 9.31 = 0$ or $X = 23.63\ \Omega$

Method-II:

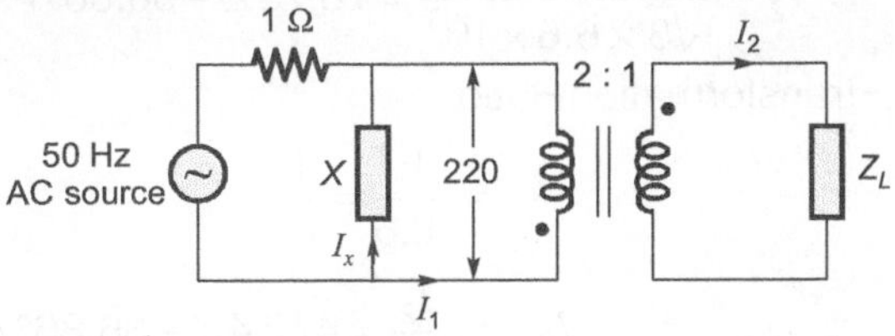

$$I_2 = \frac{4\times10^3}{110\times0.89} = 40.86\angle-27.13^\circ$$

$$I_1 = \frac{40.86}{2}\angle-27.13^\circ = 20.43\angle-27.73^\circ$$

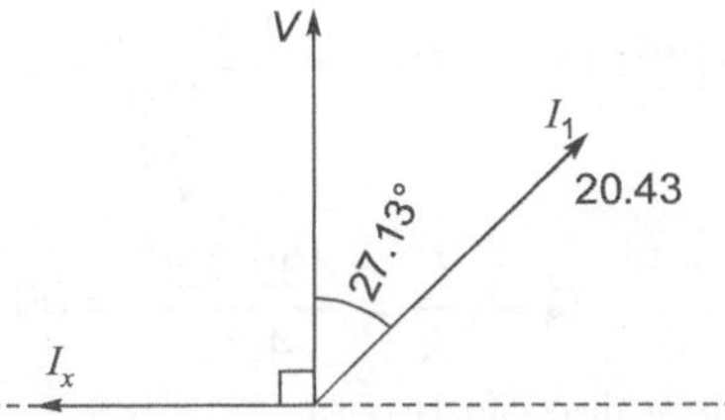

For unity pf horizontal component of (I_1) should be cancelled with (I_x).

So, $20.43\sin(27.13^\circ) = \dfrac{220}{X} \Rightarrow X = 23.61\ \Omega$

2.62 (b)

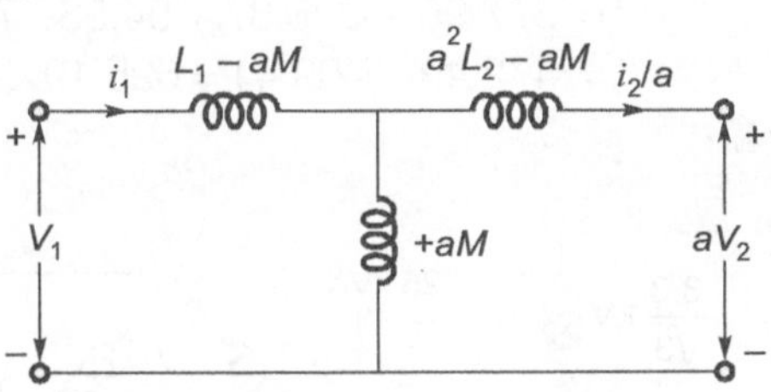

$L_1 = 45$ mH, $L_2 = 30$ mH, $M = 20$ mH,
$a = 2$ (220/110 V)

$$L_1 - aM = 45 - 2\times20 = 5 \text{ mH}$$
$$a^2L_2 - aM = 2^2\times30 - 2\times20 = 80 \text{ mH}$$
$$aM = 2\times20 = 40 \text{ mH}$$

2.64 Sol.

$$S_{T_2} = \frac{1+j6}{(1+j6)+(0.8+j4.8)}\times1000\angle\cos^{-1}(0.8)$$

$$= \frac{(1+j6)}{(1.8+j10.8)}1000\angle\cos^{-1}(0.8)$$

$$= 0.55\angle0\times100\angle\cos^{-1}(0.8) = 555.55\angle36.89^\circ$$

2.65 Sol.

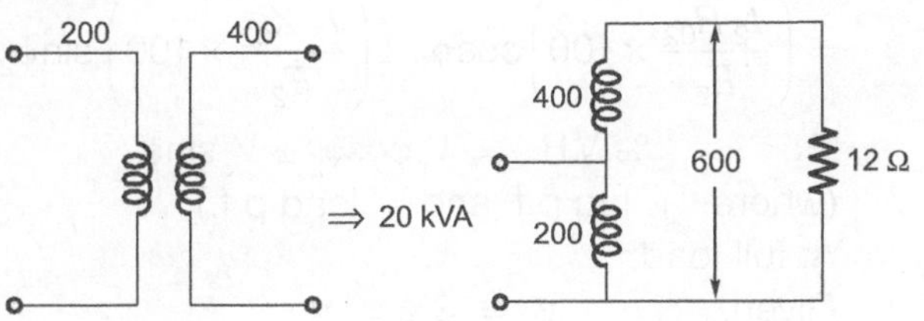

$$K = \frac{200}{600} = \frac{1}{3}$$

$$R_{eq} = 12\times\frac{1}{9} = 1.33\ \Omega$$

2.66 (a)

Self inductance of primary coil is 500 mH.
Self inductance of secondary coil is 600 mH.
Mutual inductance is 480 mH.
So, $L_1 = 800$ mH; $L_2 = 600$ mH; $M = 480$ mH

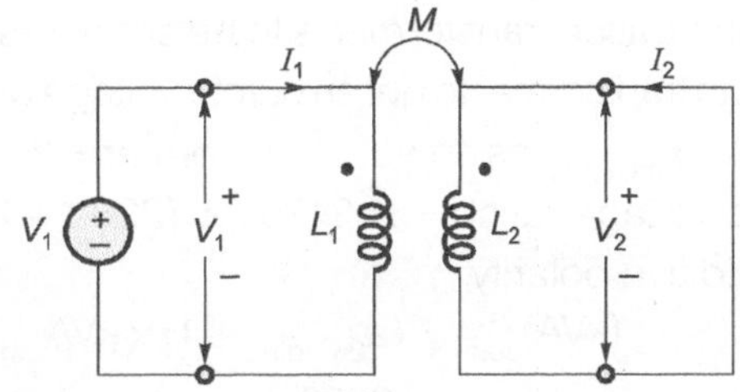

The voltage source V_1 is sinusoidal with frequency 50 Hz.

So, $V_2 = L_2\dfrac{dI_2}{dt} + \dfrac{MdI_1}{dt} \Rightarrow V_2 = 0$

so, $j\omega L_2 I_2 = -j\omega M I_1$

so, $I_2 = -I_1\cdot\dfrac{M}{L_2} = -\dfrac{4}{5}I_1$

Now, $V_1 = L_1\dfrac{dI_1}{dt} + \dfrac{MdI_2}{dt}$

so,
$$V_1 = j\omega L_1 I_1 + j\omega M I_2$$
$$V_1 = j\omega[L_1I_1 + MI_2]$$
$$V_1 = j\omega[800 \text{ mH} - 384 \text{ mH}]\,I_1$$
$$\frac{V_1}{I_1} = j\omega\,[416 \text{ mH}]$$

So, equivalent inductance is 416 mH.

2.67 Sol.

400 V, 50 Hz transformer,

$$Pi = 5000 \text{ Watt}$$

When, 200 V, 25 Hz ⇒ $Pi = 2000$ Watt

416 V, 52 Hz ⇒ $\dfrac{P_h}{P_e} = ?$

$$Pi = P_h + P_e$$
$$P_h \propto f B_m^{\ x}$$
$$P_e \propto f^2 B_m^{\ 2}$$

As in the problem

$$\frac{V}{f} = \frac{400}{50} = \frac{200}{25} = \frac{416}{52} = 8 = \text{constant}$$

$$P_h = Af \text{ and } P_e = Bf^2$$

From given data,

$$2000 = (P_{25Hz})_i = A(25) + B(25)^2 \quad ...(i)$$

$$5000 = (P_{50\text{ Hz}})_i = A(50) + B(50)^2 \quad \text{...(ii)}$$

Solving equation (i) and (ii),

$A = 60$, $B = 0.8$,

$$(P_h)_{52\text{ Hz}} = Af = 60 \times 52 = 3120 \text{ Watt}$$

$$(P_e)_{52\text{ Hz}} = Bf^2 = 0.8 \times (52)^2 = 2163.2 \text{ Watt}$$

$$\frac{(P_h)_{52\text{ Hz}}}{(P_e)_{52\text{ Hz}}} = \frac{3120}{2163.2} = 1.4423$$

2.68 (c)

22 KVA, 2200 V/220 V, 50 Hz

Distribution transformer is to be connected as on auto transformer to get an output voltage of 2420 V.

$(\text{kVA})_{\text{maximum}}$ as an auto transformer = ?

As voltage rating = is 2420 i.e. (2200 + 220) V additive polarity,

$$(\text{kVA})_{\text{auto}} = (a_{2\text{winding}} + 1) \times \text{kVA}_{2\text{winding}}$$

where, $$a_{2\text{winding}} = \frac{2200}{220} = 10$$

$$(\text{kVA})_{\text{auto}} = (10 + 1) \times 22 = 242$$

2.69 Sol.

2 kVA, 100/200 V transformer,

$$a_{2\text{ winding}} = \frac{200}{100} = 2$$

$$[(\text{kVA})_{\text{auto}}]_{\max} = (a_{2\text{winding}} + 1)(\text{kVA})_{2\text{winding}} = (2 + 1) \times 2 = 6$$

2.70 (a)

At 11 kV, load is 8 MW, 0.8 p.f. lagging

$$\Rightarrow \quad \frac{(V_{ph})_\Delta}{(V_{ph})_Y} = \frac{(I_{ph})_Y}{(I_{ph})_\Delta}$$

$$\Rightarrow \quad (I_{ph})_\Delta = (I_{ph})_Y \times \frac{(V_{ph})_Y}{(V_{ph})_\Delta}$$

$$\frac{(I_{ph})_\Delta}{(V_{ph})_Y} = \frac{11/\sqrt{3}}{110} = 1 : 10\sqrt{3}$$

2.71 (b)

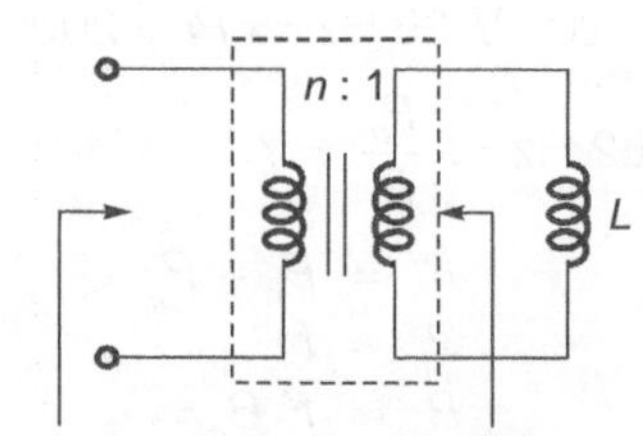

At port 1 which is high voltage side, so impedance will be high and current will be low.

So, the correct answer is n^2L.

2.72 Sol.

Load on 6.6 kV winding = 900 kVA @ 0.8 pf lag

$$\therefore \; I_2 = \frac{900 \times 10^3}{\sqrt{3} \times 6.6 \times 10^3} = 78.73\angle -36.86° \text{ A}$$

Transformation Ratio,

$$k = \frac{1.1}{6.6} = \frac{1}{6}$$

$$\therefore \quad I_2' = \frac{I_2}{k} = 472.4\angle -36.86° \text{ A}$$

Load on 400 V winding = 300 kVA at 0.6 p.f.

$$\therefore \; I_3 = \frac{300 \times 10^3}{\sqrt{3} \times 400} = 433.025\angle -53.13° \text{ A}$$

$$\text{Transformation ratio} = \frac{1.1 \times 1000}{\frac{400}{\sqrt{3}}} = 4.76$$

$$I_3' = \frac{I_3}{k} = \frac{433.025}{4.76} = 90.97\angle -53.13°$$

So line value of current

$$= \sqrt{3} \times 90.97\angle -53.13° = 157.56\angle -53.13° \text{ A}$$

Current drawn by 1.1 kV winding will be

$$I_1 = I_2' + I_3'$$
$$= 472.4\angle -36.86 + 157.56\angle -53.13°$$
$$= 377.96 - j283.37 + 94.53 - j126.04$$
$$= 472.49 - j409.41 = 625.19\angle -40.9° \text{ A}$$

2.73 Sol.

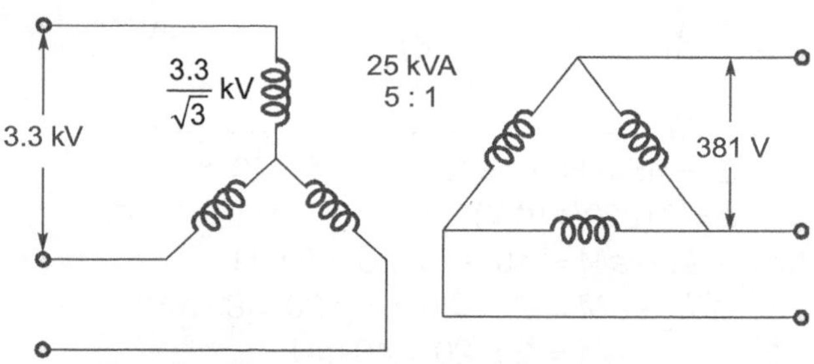

From turn ratio,

$$V_2 = 381 \text{ V/phase}$$

$$V_{2\text{ line}} = 381 \text{ V}$$

$$\sqrt{3} V_2 I_2 = 25000$$

$$\therefore \quad I_2 = \frac{25000}{\sqrt{3} \times 381} = 37.88 \text{ A}$$

2.74 (a)

Percent voltage regulation

$$= \left(\frac{I_2 R_{02}}{E_2} \times 100\right) \cos\phi_2 \pm \left(\frac{I_2 X_{02}}{E_2} \times 100\right) \sin\phi_2$$

$$\% \text{ V.R.} = V_r \cos\phi_2 \pm V_x \sin\phi_2$$

(where, '+' lag p.f. and '–' lead p.f.)

At full load:

Given, $V_r = 3\%$

Impedance drop, $V_z = 5\%$

$\therefore$ Reluctance drop,

$$V_x = \sqrt{5^2 - 3^2} = 4\%$$

Voltage regulation at full load at 0.8 p.f. lagging

V.R. = 3(0.8) + 4(0.6) = 2.4 + 2.4 = 4.8%

2.75 Sol.

900 kVA, Δ/Y, 3-phase transformer,

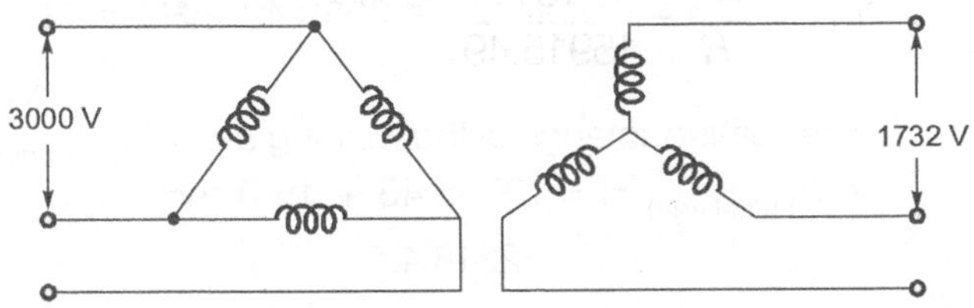

Given : $R_1 = 0.3\ \Omega/\text{ph};\ R_2 = 0.02\ \Omega/\text{ph}$

Iron loss = 10 kW

Primary line current,

$$I_{l_1} = \frac{900 \times 1000}{\sqrt{3} \times 3000} = 173.2 \text{ A}$$

Primary phase current,

$$I_{ph_1} = 100 \text{ A}$$

Secondary line current,

$$I_{l_2} = \frac{900 \times 1000}{\sqrt{3} \times 1732} = 300 \text{ A}$$

Secondary phase current,

$$I_{ph_2} = 300 \text{ A}$$

Cu loss of H.V. side

$$= 3 \times I_1^2 R_1 = 3(100)^2 (0.3) = 9000 \text{ Watts}$$

Cu loss of L.V. side

$$= 3 \times I_2^2 R_2 = 3(300)^2 (0.02) = 5400 \text{ Watts}$$

Full load efficiency @ u.p.f.

$$\eta = \frac{1 \times 900 \times 1000 \times 1}{1 \times 900 \times 1000 + 9000 + 5400 + 10000}$$

$$= \frac{900000}{924400} = 0.9736 \text{ (or) } 97.36\%$$

2.76 (a)

$$\text{Voltage regulation} = \frac{V_{NL} - V_{FL}}{V_{NL}} \times 100$$

$$= \frac{100 - 95}{100} \times 100 = 5\%$$

2.77 Sol.

For a 12 kW, 0.6 pf lagging load,

$P_L = 12$ kW

$$Q_L = \frac{12}{0.6} \times \sin(\cos^{-1} 0.6) = 16 \text{ kVAR}$$

Transformer rating,

$S = 25$ kVA

Let us assume load that can be added is P kW then,

$$S^2 = (P + P_L)^2 + Q_L^2$$

$$25^2 = (P + 12)^2 + 16^2$$

$$P = 7.21 \text{ kW}$$

2.78 Sol.

200 V, 50 Hz, P_c = 450 Watt

160 V, 40 Hz, P_c = 320 Watt

100 V, 25 Hz, P_c = ? Watt

$$\frac{V}{f} = \text{constant} = \frac{200}{50} = \frac{160}{40} = \frac{100}{25}$$

So, $P_c = Af + Bf^2$

$$450 = A \times (50) + B \times (50)^2 \quad \text{...(i)}$$

$$320 = A \times (40) + B \times (40)^2 \quad \text{...(ii)}$$

From equations (i) and (ii),

$$\frac{450}{50} = A + B(50) \quad \text{...(iii)}$$

$$\frac{320}{40} = A + B(40) \quad \text{...(iv)}$$

Equation (iii) – (iv),

$$(9 - 8) = B(10)$$

$$B = \frac{1}{10}$$

and $A = 9 - \frac{1}{10} \times 50 = 4$

Now at 100 V, 25 Hz,

$$P_c = 4 \times 25 + \frac{1}{10} \times (25)^2$$

$$= 100 + 62.5 = 162.50 \text{ Watt}$$

2.79 (*)

[Data given in question is ambiguous leading to dead short-circuit of supply, marks to all was awarded for this question].

2.80 (b)

W_{i1} = 2500 W at 440 V, 50 Hz

W_{i2} = 850 W at 220 V, 25 Hz

$W_{i3} = R_{e_3} + P_{h_3}$ at 440 V, 50 Hz

$R_{e_3} = ?,\ P_{h_3} = ?,\ \frac{V}{f} = \text{constant}$

$\Rightarrow$ $2500 = Af + Bf^2$

$$\left\{\frac{400}{50} = \frac{220}{25} = \text{Constant}\right\}$$

or, $\frac{2500}{f} = A + Bf$

or, $\dfrac{2500}{50} = A + B(50)$ ...(i)

and $\dfrac{850}{25} = A + B\ (25)$ ...(ii)

Solving (i) and (ii), we get

$$25B = \frac{2500}{50} - \frac{850}{25} = \frac{2500-1700}{50} = \frac{800}{50} = 16$$

$$B = \frac{16}{25}$$

and from (i), $A = 50 - \dfrac{16}{25} \times 50 = 50 - 32 = 18$

So, at 50 Hz

$$P_h = Af = 18 \times 50 = 900 \text{ W}$$

$$P_e = Bf^2 = \left(\frac{16}{25}\right) \times (50)^2 = 1600 \text{ W}$$

2.81 (2218.43)

Given : $l = 500$ mm $= 0.5$ m

$A = 5000$ mm^2 $= 5 \times 10^{-3}$ m^2

$\mu_r = 5000$; $E = 200$ V

So, $R = \dfrac{l}{\mu_o \mu_x A}$

$$= \frac{0.5}{4\pi \times 10^{-7} \times 5000 \times 5 \times 10^{-3}} = 15915.49$$

and $E = 4.44 f N B_m A$

$$N = \frac{E}{4.44 f B_m A} = \frac{200}{4.44 \times 50 \times 1 \times 5 \times 10^{-3}}$$

$$N \cong 181$$

$$L_{12} = \frac{N^2}{R} = \frac{181}{15915.49} = 2058.43 \text{ mH}$$

The self-inductance of the winding a-b = $L_{12(\text{linkage})}$ + $L_{11(\text{leakage})}$ H = (2058.43 + 160) mH

= 2218.43

2.82 (b)

- E_p is induced so as to oppose V_p.
- Polarity of voltage at "Dots" is equal; this decides polarity of E_s.
- E_s and V_s oppose each other.

■■■■

3 Three Phase Induction Machines

3.1 An induction motor runs under stable condition with constant torque load at 1250 rpm off a 50 Hz supply. Its number of poles is ___.

[1995 : 1 M]

3.2 When started by means of an auto transformer with 50% tapping, supply current at start of an induction motor is reduced to _____ of that when started by means of a star-delta starter.

[1995 : 2 M]

3.3 Unbalanced supply voltage given to a 3-phase, delta-connected induction motor will cause
(a) zero sequence currents
(b) less heating of the rotor
(c) negative sequence component current
(d) all of these

[1996 : 1 M]

3.4 When the supply voltage to an induction motor is reduced by 10%, the maximum torque will decrease by approximately
(a) 5% (b) 10%
(c) 20% (d) 40% [1996 : 1 M]

3.5 In an induction motor, if the air gap is increased
(a) speed will reduce
(b) efficiency will improve
(c) power factor will be lower
(d) breakdown torque will reduce [1996 : 1 M]

3.6 An induction motor having full load torque of 60 Nm when delta connected, develops a starting torque of 120 N-m. For the same supply voltage, if the motor is changed to star connection the starting torque developed will be
(a) 40 Nm (b) 60 Nm
(c) 90 Nm (d) 120 Nm [1996 : 2 M]

3.7 Out of the several characteristics shown in figure. Identify the appropriate ones to match the following for a 3-phase induction motor.

	x-axis	y-axis	
(a)	Load	Efficiency	(P) Curve I
(b)	Speed	Current	(Q) Curve II
(c)	Speed	Power factor	(R) Curve III
			(S) Curve IV
			(T) Curve V

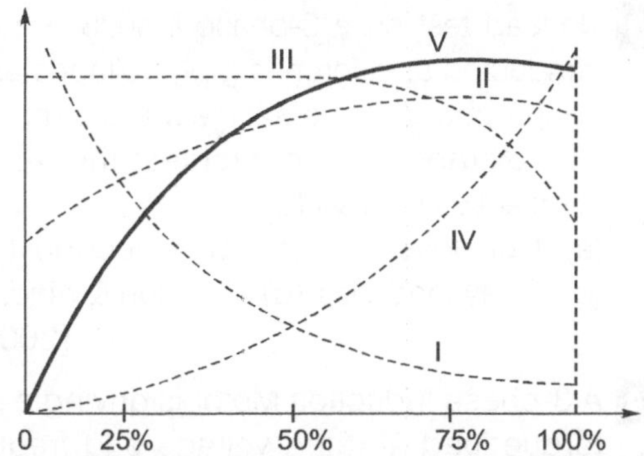

[1996 : 2 M]

3.8 A 3-phase squirrel cage induction motor has a full load efficiency of 0.8 and a maximum efficiency of 0.9 it is operated at a slip of 0.6 by applying a reduce voltage. The efficiency of the rotor at this operating points is
(a) less than 4
(b) greater than 0.6
(c) in the range of 0.8 ± 0.1
(d) none of the above [1998 : 1 M]

3.9 The following starting method for an induction motor is inferior in view of the poor starting torque per ampere of the line current drawn:
(a) Direct on line starting
(b) Auto transformer method of starting
(c) Series inductor method of starting
(d) Star-Delta method of starting [1999 : 1 M]

3.10 The power input to a 415 V, 50 Hz, 6-pole, 3-phase induction motor running at 975 rpm is 40 kW. The stator losses are 1 kW and friction and windage losses total 2 kW. The efficiency of the motor is
(a) 92.5% (b) 90%
(c) 91% (d) 88% [2000 : 2 M]

3.11 If a 400 V, 50 Hz, star connected, 3-phase squirrel cage induction motor is operated from a 400 V, 75 Hz supply, the torque that the motor can now provide while drawing rated current from the supply?
(a) decreases
(b) increases
(c) remains same
(d) increase or decrease depending on rotor resistance

[2002 : 1 M]

3.12 The rotor of a three phase, 5 kW, 400 V, 50 Hz, slip ring induction motor is wound for 6 poles while its stator is wound for 4 poles. The approximate average no load steady state speed when this motor is connected to 400 V, 50 Hz supply is

[2002 : 1 M]

3.13 No-load test on a 3-phase induction motor was conducted at different supply voltages and a plot of input power versus voltage was drawn. This curve was extrapolated to intersect the *y*-axis. This intersection point yields

(a) Core loss (b) Stator copper loss
(c) Stray load loss (d) Friction & windage loss

[2003 : 1 M]

3.14 A 3-phase Induction Motor is driving a constant torque load at rated voltage and frequency. If both voltage and frequency are halved, following statements relate to the new condition if stator resistance, leakage reactance and core loss are ignored

1. The difference between synchronous speed and actual speed remains same
2. The airgap flux remains same
3. The stator current remains same
4. The p.u. slip remains same

Among the above, correct statements are

(a) All (b) 1, 2 and 3
(c) 2, 3 and 4 (d) 1 and 4 [2003 : 2 M]

3.15 An AC induction motor is used for a speed control application. It is driven from an inverter with a constant V/f control. The motor name-plate details are as follows (number of poles = 2)

V : 415 V, Ph : 3 *f* : 50 Hz N : 2850 rpm

The motor is run with the inverter output frequency set at 40 Hz, and with half the rated slip. The running speed of the motor is

(a) 2400 rpm (b) 2280 rpm
(c) 2340 rpm (d) 2790 rpm [2003 : 2 M]

3.16 The direction of rotation of a 3-phase induction motor is clockwise when it is supplied with 3-phase sinusoidal voltage having phase sequence A-B-C. For counterclockwise rotation of the motor, the phase sequence of the power supply should be

(a) B-C-A (b) C-A-B
(c) A-C-B (d) B-C-A or C-A-B

[2004 : 1 M]

3.17 The synchronous speed for the seventh space harmonic mmf wave of a 3-phase, 8 pole, 50 Hz induction machine is

(a) 107.14 rpm in forward direction
(b) 107.14 rpm in reverse direction
(c) 5250 rpm in forward direction
(d) 5250 rpm in reverse direction [2004 : 2 M]

3.18 A 400 V, 15 kW, 4 pole, 50 Hz, Y-connected induction motor has full load slip of 4%. The output torque of the machine at full load is

(a) 1.66 Nm (b) 95.50 Nm
(c) 99.47 Nm (d) 624.73 Nm

[2004 : 2 M]

3.19 On the torque/speed curve of induction motor shown in the figure four points of operation are marked as *W*, *X*, *Y* and *Z*. Which one of them represents the operation at a slip greater than 1?

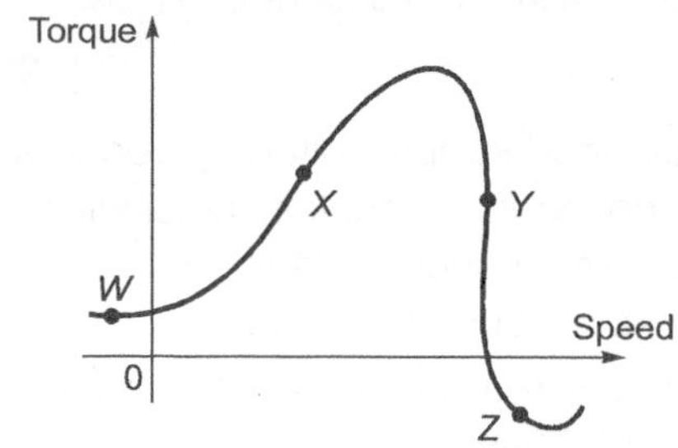

(a) *W* (b) *X*
(c) *Y* (d) *Z* [2005 : 1 M]

3.20 For an induction motor, operating at a slip *s*, the ratio of gross power output to air gap power is equal to

(a) $(1-s)^2$ (b) $(1-s)$
(c) $\sqrt{(1-s)}$ (d) $(1-\sqrt{s})$ [2005 : 1 M]

3.21 Under no load condition, if the applied voltage to an induction motor is reduced from the rated voltage to half the rated value,

(a) the speed decreases and the stator current increases.
(b) both the speed and the stator current decrease.
(c) the speed and the stator current remain practically constant.
(d) there is negligible change in the speed but the stator current decreases. [2005 : 2 M]

3.22 A three-phase cage induction motor is started by direct-on-line (DOL) switching at the rated voltage. If the starting current drawn is 6 times the full load current, and the full load slip is 4%, then ratio of the starting developed torque to the full load torque is approximately equal to

(a) 0.24 (b) 1.44
(c) 2.40 (d) 6.00

[2005 : 2 M]

3.23 **Assertion (A):** Under *V/f* control of induction motor, the maximum value of the developed torque remains constant over a wide range of speed in the sub-synchronous region.

Reason (R): The magnetic flux is maintained almost constant at the rated value by keeping the ratio *V/f* constant over the considered speed range.

(a) Both [A] and [R] are true and [R] is the correct reason for [A].
(b) Both [A] and [R] are true but [R] is not the correct reason for [A].
(c) Both [A] and [R] are false.
(d) [A] is true but [R] is false.

[2005 : 2 M]

3.24 The speed of a 4-pole induction motor is controlled by varying the supply frequency while maintaining the ratio of supply voltage to supply frequency (*V/f*) constant. At rated frequency of 50 Hz and rated voltage of 400 V its speed is 1440 rpm. Find the speed at 30 Hz, if the load torque is constant

(a) 882 rpm (b) 864 rpm
(c) 840 rpm (d) 828 rpm

[2006 : 2 M]

3.25 A 3 phase, 4 pole, 400 V, 50 Hz, star connected induction motor has following circuit parameters

$r_1 = 1.0\ \Omega,\ r_2' = 0.5\ \Omega,\ x_1 = x_2' = 1.2\ \Omega,\ x_m = 35\ \Omega$

The starting torque when the motor is started direct-on-line is (use approximate equivalent circuit model)

(a) 63.6 Nm (b) 74.3 Nm
(c) 190.8 Nm (d) 222.9 Nm

[2006 : 2 M]

3.26 A 3 phase, 10 kW, 400 V, 4 pole, 50 Hz, star connected induction motor draws 20 A on full load. Its no load and blocked rotor test data are given below.

No Load Test: 400 V 6 A 1002 W
Blocked Rotor Test: 90 V 15 A 762 W

Neglecting copper loss in No Load Test and core loss in Blocked Rotor Test, estimate motor's full load efficiency.

(a) 76% (b) 81%
(c) 82.4% (d) 85%

[2006 : 2 M]

3.27 The electromagnetic torque T_e of a drive and its connected load torque T_L are as shown below. Out of the operating points *A*, *B*, *C* and *D*, the stable ones are

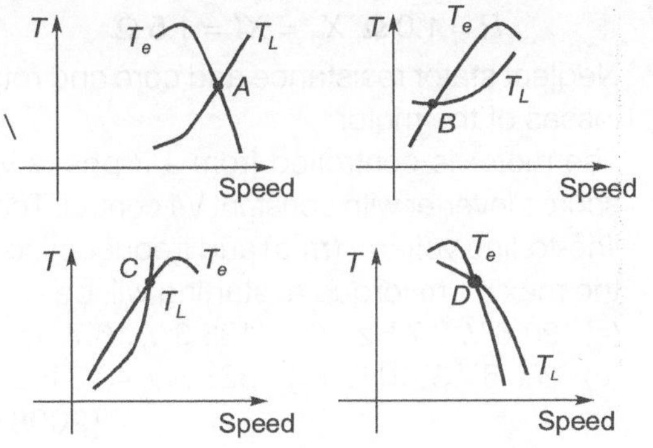

(a) *A*, *C*, *D* (b) *B*, *C*
(c) *A*, *D* (d) *B*, *C*, *D* [2007 : 1 M]

3.28 A three-phase squirrel cage induction motor has a starting torque of 150% and a maximum torque of 300% with respect to rated torque at rated voltage and rated frequency. Neglect the stator resistance and rotational losses. The Value of slip for maximum torque is

(a) 13.48% (b) 16.42%
(c) 18.92% (d) 26.79% [2007 : 2 M]

Common Data for Questions (3.29 to 3.30):

A three phase squirrel cage induction motor has a starting current of seven times the full load current and full load slip of 5%.

3.29 If an autotransformer is used for reduced voltage starting to provide 1.5 per unit starting torque, the autotransformer ratio(%) should be

(a) 57.77% (b) 72.56%
(c) 78.25% (d) 81.33% [2007 : 2 M]

3.30 If a star-delta starter is used to start this induction motor, the per unit starting torque will be

(a) 0.607 (b) 0.816
(c) 1.225 (d) 1.616 [2007 : 2 M]

3.31 If a starting torque of 0.5 per unit is required then the per unit starting current should be

(a) 4.65 (b) 3.75
(c) 3.16 (d) 2.13 [2007 : 2 M]

3.32 A 400 V, 50 Hz, 30 hp, three-phase induction motor is drawing 50 A current at 0.8 power factor lagging. The stator and rotor copper losses are 1.5 kW and 900 W respectively. The friction and windage losses are 1050 W and the core losses are 1200 W. The air-gap power of the motor will be

(a) 23.06 kW (b) 24.11 kW
(c) 25.01 kW (d) 26.21 kW [2008 : 2 M]

3.33 A 400 V, 50 Hz, 4 pole, 1400 rpm, star connected squirrel cage induction motor has the following parameters referred to the stator:

$R = 1.0\ \Omega,\ X_s = X_r' = 1.5\ \Omega$

Neglect stator resistance and core and rotational losses of the motor.

The motor is controlled from a 3-phase voltage source inverter with constant V/f control. The stator line-to-line voltage (rms) and frequency to obtain the maximum torque at starting will be:

(a) 20.6 V, 2.7 Hz (b) 133.3 V, 16.7 Hz
(c) 266.6 V, 33.3 Hz (d) 323.3 V, 40.3 Hz

[2008 : 2 M]

Statement for Linked Answer Questions (3.34 and 3.35):

A 3-phase, 440 V, 50 Hz, 4 pole, slip ring induction motor is feed from the rotor side through an auto transformer and the stator is connected to a variable resistance as shown in the figure.

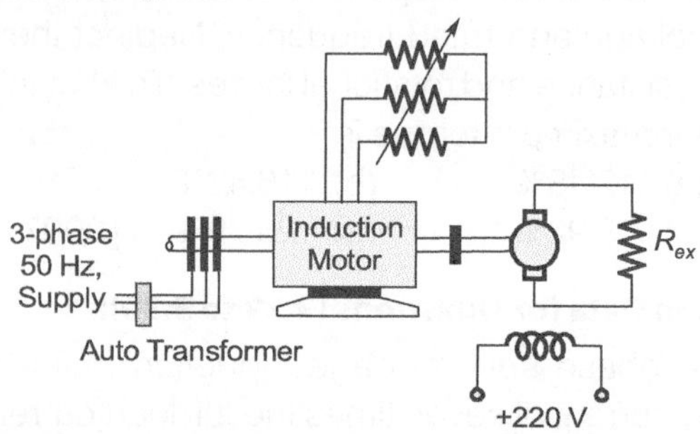

The motor is coupled to a 220 V, separately excited DC generator feeding power to fixed. Resistance of 10 Ω. Two wattmeter method in used to measure the input power to induction motor. The variable resistance is adjusted such that motor runs at 1410 rpm and the following reading were recorded, $W_1 = 1800$ W, $W_2 = -200$ W.

3.34 The speed of rotation of stator magnetic field with respect to rotor structure will be

(a) 90 rpm in the direction of rotation
(b) 90 rpm in the opposite direction of rotation
(c) 1500 rpm in the direction of rotation
(d) 1500 rpm in the opposite direction of rotation

[2008 : 2 M]

3.35 Neglecting all losses of both the machines, the DC generator power output and the current through resistance (R_{ex}) will respectively be

(a) 96 W, 3.10 A (b) 120 W, 3.46 A
(c) 1504 W, 12.26 A (d) 1880 W, 13.71 A

[2008 : 2 M]

3.36 A balanced three-phase voltage is applied to a star-connected induction motor, the phase to neutral voltage being V. The stator resistance, rotor resistance referred to the stator, stator leakage reactance, rotor leakage reactance referred to the stator, and the magnetizing reactance are denoted by r_s, r_r, x_s, x_r and X_m, respectively. The magnitude of the starting current of the motor is given by:

(a) $\dfrac{V}{\sqrt{(r_s + r_r)^2 + (x_s + x_r)^2}}$

(b) $\dfrac{V}{\sqrt{r_s^2 + (x_s + X_m)^2}}$

(c) $\dfrac{V}{\sqrt{(r_s + r_r)^2 + (X_m + x_r)^2}}$

(d) $\dfrac{V}{\sqrt{r_s^2 + (X_m + x_r)^2}}$

[2010 : 1 M]

3.37 A separately excited dc machine is coupled to a 50 Hz, three-phase, 4-pole induction machine as shown in the figure. The DC machine is energized first and the machines rotate at 1600 rpm. Subsequently the induction machine is also connected to a 50 Hz, three-phase source, the phase sequence being consistent with the direction of rotation. In steady state,

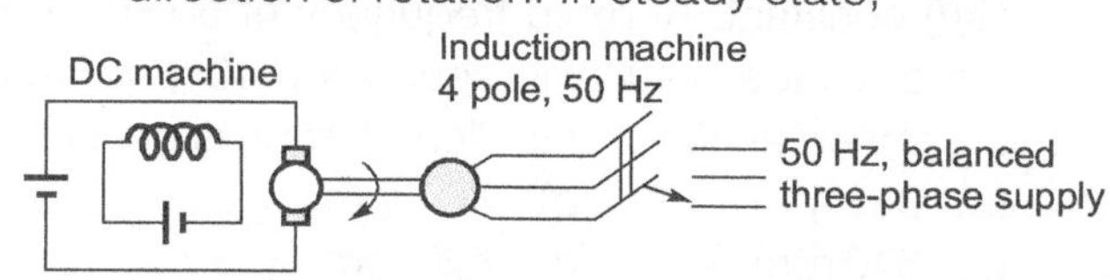

(a) both machines act as generators
(b) the dc machine acts as a generator, and the induction machine acts as a motor
(c) the dc machine acts as a motor, and the induction machine acts as a generator
(d) both machines act as motors

[2010 : 2 M]

3.38 A three-phase 400 V, 6 pole, 50 Hz, squirrel cage induction motor is running at a slip of 5%. The speed of stator magnetic field with respect to rotor magnetic field and speed of rotor with respect to stator magnetic field are

(a) zero, –50 rpm
(b) zero, 950 rpm
(c) 1000 rpm, –50 rpm
(d) 1000 rpm, 950 rpm

[2011 : 2 M]

3.39 The slip of an induction motor normally does not depend on

(a) rotor speed (b) synchronous speed
(c) shaft torque (d) core-loss component

[2012 : 1 M]

3.40 The locked rotor current in a 3-phase, star connected 15 kW, 4-pole, 230 V, 50 Hz induction motor at rated conditions is 50 A. Neglecting losses and magnetizing current, the approximate locked rotor line current drawn when the motor is connected to a 236 V, 57 Hz supply is

(a) 58.5 A (b) 45.0 A
(c) 42.7 A (d) 55.6 A

[2012 : 2 M]

3.41 Leakage flux in an induction motor is
(a) flux that leaks through the machine
(b) flux that links both stator and rotor windings
(c) flux that links none of the windings
(d) flux that links the stator winding or the rotor winding but not both [2013 : 1 M]

3.42 A 4-pole induction motor, supplied by a slightly unbalanced three-phase 50 Hz source, is rotating at 1440 rpm. The electrical frequency in Hz of the induced negative sequence current in the rotor is
(a) 100 (b) 98
(c) 52 (d) 48 [2013 : 2 M]

3.43 An 8-pole, 3-phase, 50 Hz induction motor is operating at speed of 700 rpm. The frequency of the rotor current of the motor in Hz is ______.
[2014 : 1 M, Set-1]

3.44 A 3-phase, 50 Hz, six pole induction motor has a rotor resistance of 0.1 Ω and reactance of 0.92 Ω. Neglect the voltage drop in stator and assume that the rotor resistance is constant. Given that the full load slip is 3%, the ratio of maximum torque to full load torque is
(a) 1.567 (b) 1.712
(c) 1.948 (d) 2.134 [2014 : 2 M]

3.45 A three-phase, 4-pole, self excited induction generator is feeding power to a load at a frequency f_1. If the load is partially removed, the frequency becomes f_2. If the speed of the generator is maintained at 1500 rpm in both the cases, then
(a) f_1, f_2 > 50 Hz and $f_1 > f_2$
(b) f_1 < 50 Hz and f_2 > 50 Hz
(c) f_1, f_2 < 50 Hz and $f_2 > f_1$
(d) f_1 > 50 Hz and f_2 < 50 Hz [2014 : 1 M, Set-2]

3.46 A single-phase induction motor draws 12 MW power at 0.6 lagging power. A capacitor is connected in parallel to the motor to improve the power factor of the combination of motor and capacitor to 0.8 lagging. Assuming that the real and reactive power drawn by the motor remains same as before, the reactive power delivered by the capacitor in MVAR is ______. [2014 : 1 M, Set-2]

3.47 In a constant *V*/*f* control of induction motor, the ratio *V*/*f* is maintained constant from 0 to base frequency, where V is the voltage applied to the motor at fundamental frequency *f*. Which of the following statements relating to low frequency operation of the motor is TRUE?
(a) At low frequency, the stator flux increases from its rated value.
(b) At low frequency, the stator flux decreases from its rated value.
(c) At low frequency, the motor saturates.
(d) At low frequency, the stator flux remains unchanged at its rated value.
[2014 : 1 M, Set-2]

3.48 A three-phase slip-ring induction motor, provided with a commutator winding, is shown in the figure. The motor rotates in clockwise direction when the rotor windings are closed.

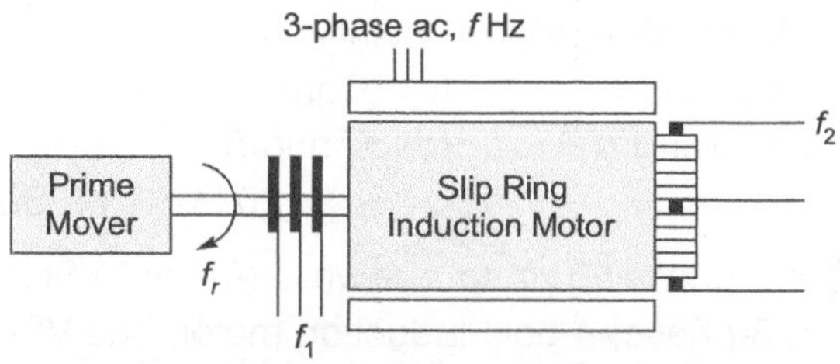

If the rotor winding is open-circuited and the system is made to run at rotational speed f_r with the help of prime-mover in anti-clockwise direction, then the frequency of voltage across slip rings is f_1 and frequency of voltage across commutator brushes is f_2. The values of f_1 and f_2 respectively are
(a) $f + f_r$ and f (b) $f - f_r$ and f
(c) $f - f_r$ and $f + f_r$ (d) $f + f_r$ and $f - f_r$
[2014 : 2 M, Set-2]

3.49 A single-phase induction motor is provided with capacitor and centrifugal switch in series with auxiliary winding. The switch is expected to operate at a speed of 0.7 N_s, but due to malfunctioning the switch fails to operate. The torque-speed characteristic of the motor is represented by

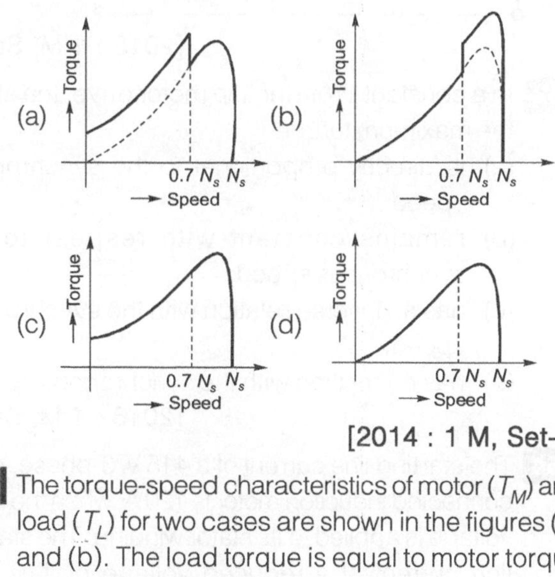

[2014 : 1 M, Set-3]

3.50 The torque-speed characteristics of motor (T_M) and load (T_L) for two cases are shown in the figures (a) and (b). The load torque is equal to motor torque at points *P*, *Q*, *R* and *S*.

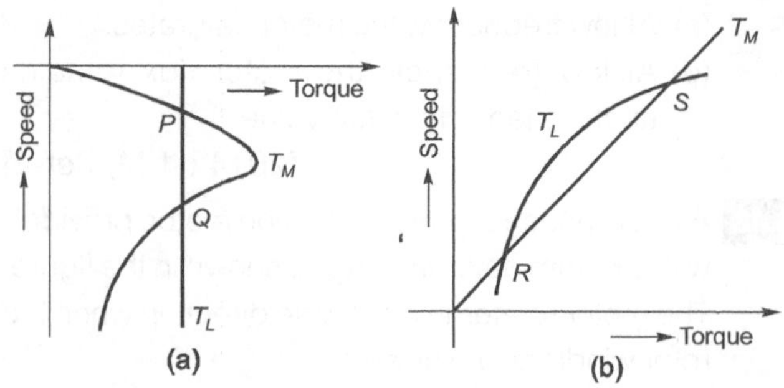

The stable operating points are

(a) P and R (b) P and S
(c) Q and R (d) Q and S

[2014 : 1 M, Set-3]

3.51 A 3-phase 50 Hz square wave (6-step) VSI feeds a 3-phase, 4 pole induction motor. The VSI line voltage has a dominant 5th harmonic component. If the operating slip of the motor with respect to fundamental component voltage is 0.04, the slip of the motor with respect to 5th harmonic component of voltage is ______.

[2015 : 2 M, Set-1]

3.52 The figure shows the per-phase equivalent circuit of a two-pole three-phase induction motor operating at 50 Hz. The "air-gap" voltage, V_g across the magnetizing inductance, is 210 V rms, and the slip, s, is 0.05. The torque (in Nm) produced by the motor is ________.

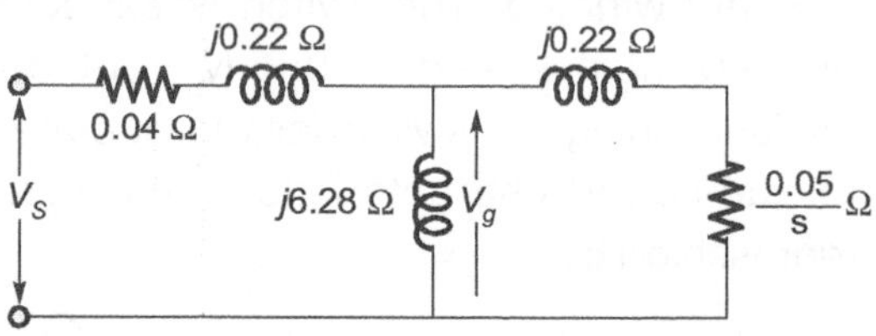

[2015 : 1 M, Set-2]

3.53 In a constant V/f induction motor drive, the slip at the maximum torque

(a) is directly proportional to the synchronous speed.
(b) remains constant with respect to the synchronous speed.
(c) has an inverse relation with the synchronous speed.
(d) has no relation with the synchronous speed.

[2016 : 1 M, Set-1]

3.54 The starting line current of a 415 V, 3-phase, delta connected induction motor is 120 A, when the rated voltage is applied to its stator winding. The starting line current at a reduced voltage of 110 V, in Ampere, is ______. [2016 : 2 M, Set-2]

3.55 A 4-pole induction machine is working as an induction generator. The generator supply frequency is 60 Hz. The rotor current frequency is 5 Hz. The mechanical speed of the rotor in RPM is

(a) 1350 (b) 1650
(c) 1950 (d) 2250

[2017 : 1 M, Set-1]

3.56 A 3-phase, 4-pole, 400 V, 50 Hz squirrel-cage induction motor is operating at a slip of 0.02. The speed of the rotor flux in mechanical rad/sec, sensed by a stationary observer, is closest to

(a) 1500 (b) 1470
(c) 157 (d) 154

[2017 : 1 M, Set-2]

3.57 A star-connected, 12.5 kW, 208 V (line), 3-phase, 60 Hz squirrel cage induction motor has following equivalent circuit parameters per phase referred to the stator:

$R_1 = 0.3\ \Omega$, $R_2 = 0.3\ \Omega$, $X_1 = 0.41\ \Omega$, $X_2 = 0.41\ \Omega$.

Neglect shunt branch in the equivalent circuit. The starting current (in Ampere) for this motor when connected to an 80 V (line), 20 Hz, 3-phase AC source is ______. [2017 : 2 M, Set-2]

3.58 The parameter of an equivalent circuit of a three-phase induction motor affected by reducing the rms value of the supply voltage at the rated frequency is

(a) stator resistance
(b) rotor leakage reactance
(c) rotor resistance
(d) magnetizing reactance [2019 : 1 M]

3.59 A delta-connected, 3.7 kW, 400 V(line), 3-phase, 4-pole, 50 Hz squirrel-cage induction motor has the following equivalent circuit parameters per phase referred to stator. $R_1 = 5.39\ \Omega$, $R_2 = 5.72\ \Omega$, $X_1 = X_2 = 8.22\ \Omega$. Neglect shunt branch in the equivalent circuit. The starting line current in amperes (round off to two decimal places) when it is connected to a 100 V(line), 10 Hz, three-phase AC source is ______. [2019 : 2 M]

3.60 A three-phase, 50 Hz, 4-pole induction motor runs at no-load with a slip of 1%. With full load, the slip increases to 5 %. The % speed regulation of the motor (rounded off to 2 decimal places) is ______.

[2020 : 1 M]

3.61 The power input to a 500 V, 50 Hz. 6-pole. 3-phase induction motor running at 975 RPM is 40 kW. The total stator losses are 1 kW. If the total friction and windage losses are 2.025 kW. then the efficiency is _____ %. [2021 : 1 M]

3.62 An 8-pole, 50 Hz. three-phase, slip-ring induction motor has an effective rotor resistance of 0.08 Ω per phase. Its speed at maximum torque is 650 RPM. The additional resistance per phase that must be inserted in the rotor to achieve maximum torque at start is _____Ω. (Round off to 2 decimal places). Neglect magnetizing current and stator leakage impedance. Consider equivalent circuit parameters referred to stator. [2021 : 2 M]

3.63 The frequencies of the stator and rotor currents flowing in a three-phase 8-pole induction motor are 40 Hz and 1 Hz, respectively. The motor speed, in rpm, is _____. (round off to nearest integer). [2022 : 1 M]

3.64 A 4-pole induction motor with inertia of 0.1 kg-m^2 drives a constant load torque of 2 Nm. The speed of the motor is increased linearly from 1000 rpm to 1500 rpm in 4 seconds as shown in the figure below. Neglect losses in the motor. The energy, in joules, consumed by the motor during the speed change is _____. (round off to nearest integer).

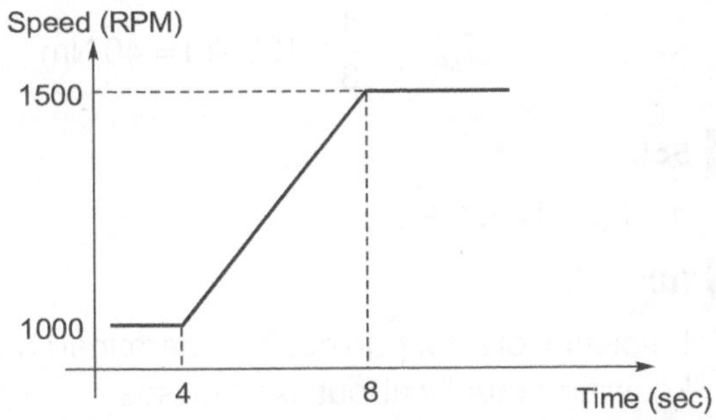

[2022 : 2 M]

3.65 A 3-phase, 415 V, 50 Hz induction motor draws 5 times the rated current at rated voltage at starting. It is required to bring down the starting current from the supply to 2 times of the rated current using a 3-phase autotransformer. If the magnetizing impedance of the induction motor and no load current of the autotransformer is neglected, then the transformation ratio of the autotransformer is given by ____. (round off to two decimal places). [2022 : 2 M]

3.66 The following columns present various modes of induction machine operation and the ranges of slip

(A) : Mode of Operation
(a) Running in generator mode
(b) Running in motor mode
(c) Plugging in motor mode

(B) : Range of Slip
(p) From 0.0 to 1.0
(q) From 1.0 to 2.0
(r) From −1.0 to 0.0

The correct matching between the elements in column **A** with those of column **B** is
(a) a-r, b-p, and c-q (b) a-r, b-q, and c-p
(c) a-p, b-r, and c-q (d) a-q, b-p, and c-r
[2023 : 1 M]

3.67 A 10-pole, 50 Hz, 240 V, single phase induction motor runs at 540 RPM while driving rated load. The frequency of induced rotor currents due to backward field is
(a) 100 Hz (b) 95 Hz
(c) 10 Hz (d) 5 Hz
[2023 : 1 M]

3.68 A three phase 415 V, 50 Hz, 6-pole, 960 RPM, 4 HP squirrel cage induction motor drives a constant torque load at rated speed operating from rated supply and delivering rated output. If the supply voltage and frequency are reduced by 20%, the resultant speed of the motor in RPM (neglecting the stator leakage impedance and rotational losses) is _________ (Round off to the nearest integer).
[2023 : 2 M]

3.69 A three phase, 50 Hz, 6 pole induction motor runs at 960 rpm. The stator copper loss, core loss, and the rotational loss of the motor can be neglected. the percentage efficiency of the motor is
(a) 96 (b) 98
(c) 92 (d) 94
[2024 : 1 M]

3.70 A 3-phase star connected slip ring induction motor has the following parameters referred to the stator:

$$R_s = 3\ \Omega,\quad X_s = 2\ \Omega,$$
$$X_r' = 2\ \Omega,\quad R_r' = 2.5\ \Omega,$$

The per phase stator to rotor effective turns ratio is 3 :1. The rotor winding is also star connected. The magnetizing reactance and core loss of the motor can be neglected. To have maximum torque at starting, the value of the extra resistance in ohms (referred to the rotor side) to be connected in series with each phase of the rotor winding is ______ (rounded off to 2 decimal places).
[2024 : 2 M]

■■■■

Answers Three Phase Induction Machines

3.1 (4)	3.2 (0.433)	3.3 (c)	3.4 (c)	3.5 (c, d)	3.6 (a)	3.7 (Sol.)
3.8 (d)	3.9 (c)	3.10 (b)	3.11 (a)	3.12 (Sol.)	3.13 (d)	3.14 (b)
3.15 (c)	3.16 (c)	3.17 (a)	3.18 (c)	3.19 (a)	3.20 (b)	3.21 (d)
3.22 (b)	3.23 (a)	3.24 (c)	3.25 (a)	3.26 (b)	3.27 (a)	3.28 (d)
3.29 (c)	3.30 (b)	3.31 (c)	3.32 (c)	3.33 (b)	3.34 (d)	3.35 (c)
3.36 (a)	3.37 (c)	3.38 (a)	3.39 (d)	3.40 (b)	3.41 (d)	3.42 (b)
3.43 (3.33)	3.44 (c)	3.45 (c)	3.46 (7.01)	3.47 (b)	3.48 (a)	3.49 (c)
3.50 (b)	3.51 (1.2)	3.52 (401.64)	3.53 (c)	3.54 (31.807)	3.55 (c)	3.56 (c)
3.57 (70.05)	3.58 (d)	3.59 (14.95)	3.60 (4.21)	3.61 (90)	3.62 (0.52)	3.63 (585)
3.64 (1733)	3.65 (0.63)	3.66 (a)	3.67 (b)	3.68 (760)	3.69 (a)	3.70 (0.28)

Explanations Three Phase Induction Machines

3.1 Sol.

$$N_s > N \quad \text{(For stable θ running)}$$

$$\frac{120 \times 50}{P} > 1250 \Rightarrow P < \frac{120 \times 50}{1250}$$

$$P < 4.8$$

The value of P will always be an integer nearest to this range the value is '4'.

3.2 Sol.

$$I_{auto} = (0.5)^2 \times I_{sc}$$

$$I_{Y/\Delta} = \frac{1}{\sqrt{3}} I_{sc}$$

$$\frac{I_{auto}}{I_{Y/\Delta}} = (0.5)^2 \times \sqrt{3} = 0.433$$

3.3 (c)

Any unbalance in supply system, will cause negative sequence component.

3.4 (c)

$$(T_e)_{max} \propto V^2$$

$$\Rightarrow \quad T_2 = \left(\frac{V_2}{V_1}\right)^2 \cdot T_1 = \left(\frac{0.9V_1}{V_1}\right)^2 T_1 = 0.81\ T_1$$

$$\%\ \text{change} = \frac{T_2 - T_1}{T_1} \simeq -19\%$$

$\Rightarrow$ Decreases by 19%.

3.5 (c, d)

If air gap increases, leakage flux increases. This increased leakage flux will cause the motor current more lagging in nature and hence p.f. gets deteriorated. Also as breakdown torque is inversely proportional to leakage reactance, it will also reduce.

3.6 (a)

When the connection is delta $(I_L)_\Delta = \sqrt{3}(I_P)_\Delta$

When the connection is star $(I_L)_Y = (I_P)_Y$

Also, $T_{st} \propto (I_{st})^2$

$$\Rightarrow \quad \frac{(T_{st})_Y}{(T_{st})_\Delta} = \left[\frac{(I_P)}{\sqrt{3} I_P}\right]^2$$

$$\Rightarrow \quad (T_{st})_Y = \frac{1}{3} \times 120\,\text{Nm} = 40\ \text{Nm}$$

3.7 Sol.

a - T, b - R, c – P

3.8 (d)

Efficiency of rotor can not be determined without knowing actually about fixed losses.

3.9 (c)

Except this method, all other methods do not alter the value of current. In this method both starting voltage as well as starting current gets reduced.

3.10 (b)

Air gap power = 40 kW – 1 kW = 39 kW

Synchronous speed

$$= \frac{120 \times 50}{6} = 1000\ \text{rpm}$$

$$\text{Slip} = \frac{1000 - 975}{1000} = 0.025$$

Gross power = $(1 - s) \times 39$ kW = 38.025 kW

Net power = 38.025 kW – 2 kW = 36.025 kW

$$\text{Efficiently} = \frac{36.025}{(40)} \times 100 \simeq 90\%$$

3.11 (a)

$$T = 3I^2\frac{R_2}{s}\cdot\frac{60}{2\pi N_s}$$

As the motor is drawing rated current from the supply,

$$T \propto \frac{1}{N_s}$$

or, $$T \propto \frac{1}{\text{frequency}} \qquad \left[\because N_s = \frac{120f}{P}\right]$$

Hence as frequency increases, torque decreases.

3.12 Sol.

Induction motor will not work in this condition.

3.13 (d)

Power input at no-load (P_0) provided losses only as the shaft output is zero. These losses comprise. P_i (Iron/core loss) and P_{wf} (windage and friction loss).

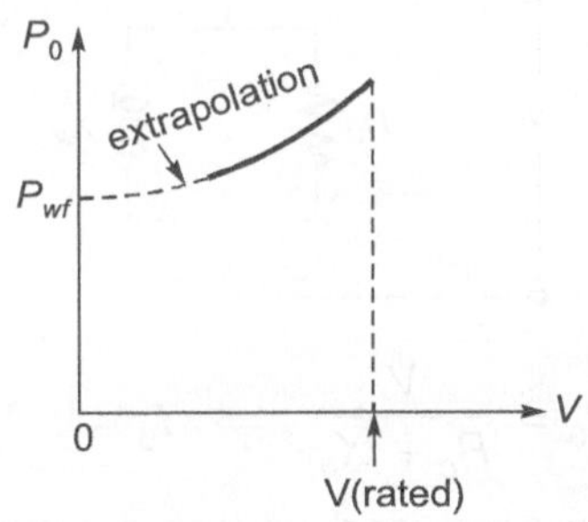

As the voltage is reduced below the rated value, the core-loss decreases as the square of voltage. Since the slip does not increase significantly, the windage and friction loss remains almost constant. The plot of P_0 versus V is extrapolated to $V = 0$ which gives, P_{wf} as $P_i = 0$ at zero voltage.

3.14 (b)

$$V \propto f\phi$$

In induction machine

$$V_2 = \frac{V_1}{2} \text{ and } f_2 = \frac{f_1}{2}$$

$$\frac{V_1}{V_2} = \frac{f_1\phi_1}{f_2\phi_2}$$

$$\phi_2 = \left(\frac{V_2}{V_1}\right)\left(\frac{f_1}{f_2}\right)\phi_1 = \left(\frac{V_1/2}{V_1}\right)\left(\frac{f_1}{f_1/2}\right)\phi_1 = \phi_1$$

So, air-gap flux remains same.

$(X_2')_1$ = nominal rotor stanstill reactance referred to stator (at frequency f_1)

$$(X_2')_2 = \left(\frac{f_2}{f_1}\right)(X_2')_1 = \left(\frac{f_1/2}{f_1}\right)(X_2')_1 = \frac{(X_2')_1}{2}$$

$$(\omega_s)_2 = \left(\frac{f_2}{f_1}\right)(\omega_s)_1 = \left(\frac{f_1/2}{f_1}\right)(\omega_s)_1 = \frac{(\omega_s)_1}{2}$$

Torque developed in induction motor

$$T = \frac{3}{\omega_s}\frac{V^2R_2'/s}{(R_2'/s)^2 + X_2'^2}$$

For normal steady state operating region, slip is very low

$$\frac{R_2'}{s} >> X_2'$$

$$T = \frac{3}{\omega_s}\frac{V^2R_2'/s}{(R_2'/s)^2} = \frac{3}{\omega_s}\frac{sV^2}{R_2'}$$

As the load torque is constant

$$T_1 = T_2$$

$$\frac{s_1V_1^2}{(\omega_s)_1R_2'} = \frac{s_2V_2^2}{(\omega_s)_2R_2'}$$

$$\Rightarrow s_2 = \frac{(\omega_s)_2}{(\omega_s)_1}\times\left(\frac{V_1}{V_2}\right)^2 s_1 = \frac{(\omega_s)_1/2}{(\omega_s)_1}\times\left(\frac{V_1}{V_1/2}\right)^2 s_1$$

$$s_2 = \frac{1}{2}\times 4\times s_1 = 2s_1$$

Difference between synchronous speed and actual speed

$$\omega_s - \omega_r = s\omega_s$$

$$(\omega_s)_1 - (\omega_r)_2 = s_2(\omega_s)_2$$

$$= (2s_1)\times\frac{(\omega_s)_1}{2} = s_1(\omega_s)_1$$

Therefore, the difference between synchronous speed and actual speed remains same.

Stator current

$$|I_s| = \frac{V}{\sqrt{(R_2'/s)^2 + X_2'^2}}$$

as $R_2'/s >> X_2'$

$$|I_s| = \frac{sV}{R_2'};\quad |I_{s1}| = s_1\frac{V_1}{R_2'};\quad |I_{s2}| = s_2\frac{V_2}{R_2'}$$

$$|I_{s2}| = (2s_1)\times\frac{(V_1/2)}{R_2'} = \frac{s_1V_1}{R_2'}$$

$$|I_{s1}| = |I_{s2}|$$

Therefore, stator current remains same.

3.15 (c)

N_{S1} = 3000 rpm (synchronous speed at 50 Hz)

N_1 = 2850 rpm (motor speed at 50 Hz)

$$S_{FL} = \frac{3000 - 2850}{3000} = 0.05$$

(rated slip at 50 Hz)

Where, by (V/f) control,

$$N_{s2} = 3000\left(\frac{40}{50}\right) = 2400 \text{ rpm}$$

(synchronous speed at 40 Hz)

$\therefore$ N_2 = New running speed of motor

$$N_{s2}\left(1-\frac{S_{FL}}{2}\right) = 2400\left(1-\frac{0.05}{2}\right) = 2340 \text{ rpm}$$

3.16 (c)

To reverse the direction of rotation, phase sequence of the supply has to be reversed. For clockwise direction, the phase sequence was A-B-C for counter-clockwise direction, the phase sequence has to be A-C-B.

3.17 (a)

Synchronous speed

$$N_s = \frac{120f}{P} = \frac{120\times 50}{8} = 750 \text{ rpm}$$

mmf wave contain space harmonics e.g. fifth and seventh harmonics which correspond to poles five and seven times that of the fundamental.

Since the space-phase different between fundamental poles of the winding phase is (0°, 120°, 240°), this difference is (0°, 240°, 120°) for fifth harmonic poles and (0°, 120°, 240°) for the seventh. Hence the fifth harmonic poles rotate backwards with synchronous speed $N_s/5$ and the seventh harmonic poles rotate forward at

$$\frac{N_s}{7} = \frac{750}{7} = 107.14 \text{ rpm}$$

3.18 (c)

$$N_s \text{ (syn. speed)} = \frac{120\times f}{P}$$

$$= \frac{120\times 50}{4} = 1500 \text{ rpm}$$

$$\text{Torque developed} = T = \frac{P}{w_s(1-s)}$$

$$= \frac{15\times 10^3\times 60}{2\pi\times 1440} = 99.47 \text{ Nm}$$

3.19 (a)

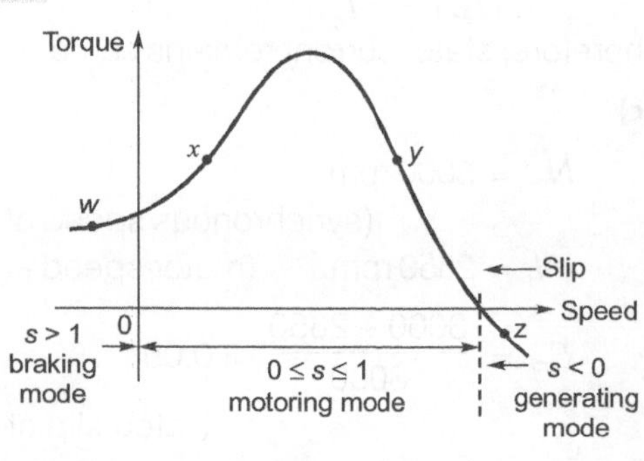

In braking mode, the motor runs in opposite direction to rotating field i.e. N_r (rotor speed) is negative,

$$s = \frac{N_s-(N_r)}{N_s} = \frac{N_s+N_r}{N_s} > 1$$

3.20 (b)

Let, P_g = Air gap power

P_{cu} = Rotor copper loss = sP_g

P_m = Gross power output = Air gap power – Rotor copper loss

$$P_m = P_g - sP_g = P_g(1-s)$$

$$\Rightarrow \frac{P_m}{P_g} = (1-s)$$

3.21 (d)

Let no-load current = I_0

Equivalent-circuit model of induction motor at no-load is drawn.

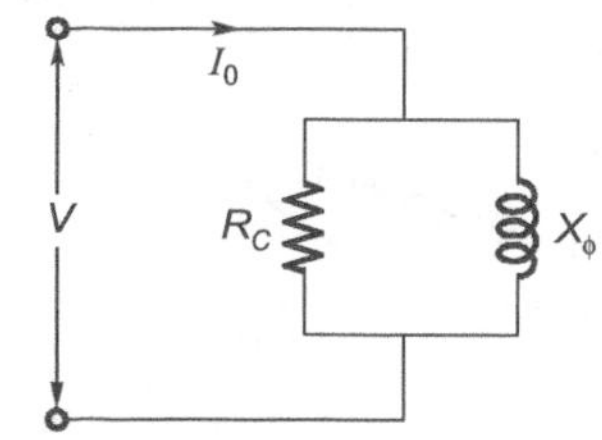

$$I_0 = \frac{V}{R_c + jX_\phi} \Rightarrow I_0 \propto V$$

If voltage is reduced, no-load current (I_0) reduces. Slip is extremely law (of the order 0.001) at no-load, even if voltage is reduced, there is negligible change in the speed.

3.22 (b)

Torque in induction motor is given by,

$$T = \frac{3}{\omega_s}(I_2')\frac{R_2'}{s}$$

Full load torque

$$T_{fl} = \frac{2}{\omega_s}(I_{2fl}')^2\frac{R_2'}{s_{fl}}$$

At starting, $s = 1$

Starting torque

$$T_{st} = \frac{3}{\omega_s}(I_{2st}')^2 R_2'$$

$$\frac{T_{st}}{T_{fl}} = \left(\frac{I_{2,st}'}{I_{2,fl}'}\right)^2 s_{fl} = 6^2\times 0.04 = 1.44$$

3.23 (a)

The resultant air-gap flux per pole is given by

$$\phi_r = \frac{1}{4.44k_{\omega_1}N_{pn_1}}\left(\frac{V}{f}\right) \Rightarrow \phi_r \propto \left(\frac{V}{f}\right)$$

Therefore, in order to avoid saturation in stator and rotor cores which would cause sharp increase in magnetization current, the flux f_p must be kept constant. To achieve this, $\frac{V}{f}$ ratio must be kept constant.

Let, V_0 = Nominal voltage

f_0 = Nominal frequency

X_{20}' = nominal rotor standstill reactance reffered to stator then at any frequency f,

Then at any frequency f,

$$V = \left(\frac{f}{f_0}\right)V_0$$

$$X_2' = \left(\frac{f}{f_0}\right)X_{20}'$$

$$\omega_s = \left(\frac{f}{f_0}\right)\omega_{20}$$

$$T_{max} = \frac{3}{\left(\frac{f}{f_0}\right)\omega_{s0}}\left[\frac{0.5\left(\frac{f}{f_0}\right)^2 V_0^2}{\left(\frac{f}{f_0}\right)X'_{20}}\right] = \frac{3}{\omega_{s0}}\frac{0.5V_0^2}{X'_{20}}$$

= constant, independent of f

3.24 (c)

$$N_{s_1} = \frac{120f_1}{P} = \frac{120\times 50}{4} = 1500 \text{ rpm}$$

$$s_1 = \frac{N_{s_1} - N_{R_1}}{N_{s_1}} = \frac{1500-1440}{1500} = 0.04$$

As $\frac{V}{f}$ = Constant

$$\frac{V_1}{f_1} = \frac{V_2}{f_2}$$

$$\Rightarrow \quad V_2 = V_1\left(\frac{f_2}{f_1}\right) = 400\times\frac{30}{50} = 240 \text{ V}$$

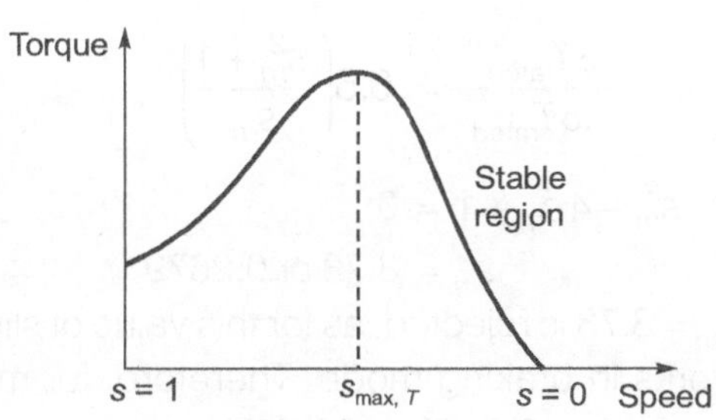

Torque developed in induction motor

$$T = \frac{3}{\omega_s}\frac{V^2R'_2/s}{(R_2'/s)^2 + X_2'^2}$$

in stable region, slip is very low, so

$$\frac{R_2'}{s} >> X_2'$$

So, T can be approximated as

$$T = \frac{3}{\omega_s}\frac{V^2R_2'/s}{(R_2'/s)^2} = \frac{3}{\omega_s}\frac{8V^2}{R_2'}$$

$$T \propto \frac{sV^2}{f}$$

$$\frac{T_1}{T_2} = \frac{s_1}{s_2}\left(\frac{V_1}{V_2}\right)^2 \times\left(\frac{f_2}{f_1}\right)$$

as load torque is constant, $T_1 = T_2$

$$s_2 = s_1\left(\frac{V_1}{V_2}\right)^2 \times\left(\frac{f_2}{f_1}\right)$$

$$= 0.04\times\left(\frac{400}{240}\right)^2\times\left(\frac{30}{50}\right) = 0.067$$

$$N_{s2} = \frac{120f_2}{P} = \frac{120\times 30}{4} = 900\text{rpm}$$

Rotor speed,

$N_{r2} = N_{s2}(1 - s_2) = 900\,(1 - 0.067) \approx 840$ rpm

3.25 (a)

Rated line-to-line voltage, $V_{l\text{-}l} = 400$ V

Phase to neutral voltage, $V_p = \frac{V_{l\text{-}l}}{\sqrt{3}} = \frac{400}{\sqrt{3}}$ V

Synchronous speed at 50 Hz

$$N_s = \frac{120f}{P} = \frac{120\times 50}{4} = 1500 \text{ rpm}$$

$$\omega_s = \frac{2\pi N_s}{60} = \frac{2\pi\times 1500}{60} = 157.08 \text{ rad/sec}$$

s = 1, at starting

Starting torque,

$$T_s = \frac{3}{\omega_s}\frac{V_P^2 r_2'}{(r_1 + r_2')^2 + (x_1 + x_2')^2}$$

$$\Rightarrow \quad T_s = \frac{3}{157.08}\times\frac{(400/\sqrt{3})^2\times 0.5}{(1+0.5)^2 + (1.2+1.2)^2}$$

$\approx$ 63.6 N-m

3.26 (b)

No-load losses, P_1 = 1002 W = 1.002 kW (does not depend on load)

Copper loss at 15 A, P_{cu} = 762 W

Full load current = 20 A

So, copper loss on full load

$$P_{cu,\,fl} = (P_{cu})\left(\frac{20}{15}\right)^2 = 762\times\left(\frac{20}{15}\right)^2 = 1.354 \text{ kW}$$

Total losses,

$$P_1 + P_{cu,fl} = 1.002 + 1.354 = 2.356 \text{ kW}$$

Power available at shaft or output power on full load

$$P_0 = 10 \text{ kW}$$

Input power = Output power + losses

$$= 10 + 2.356 = 12.356 \text{ kW}$$

$$\% \text{ efficiency} = \frac{\text{Output power}}{\text{Input power}}\times 100$$

$$= \frac{10}{12.356}\times 100 \approx 81\%$$

3.27 (a)

Accelerating torque = $T_a = T_e - T_L$

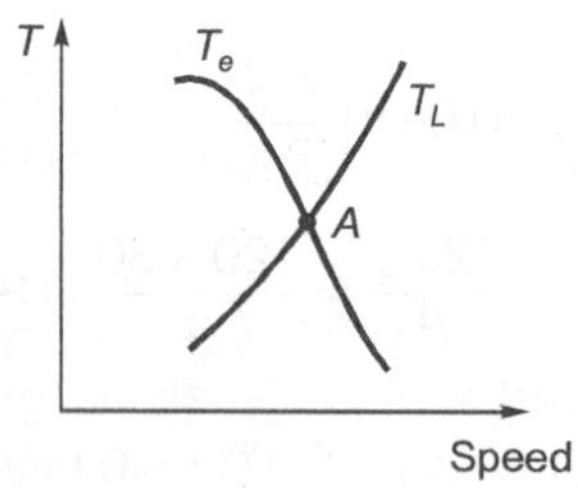

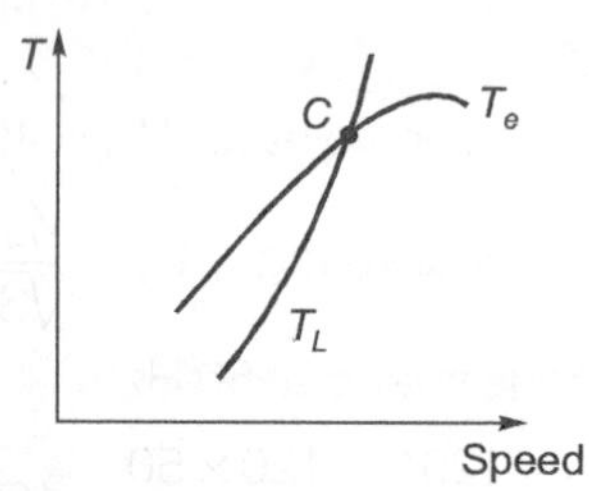

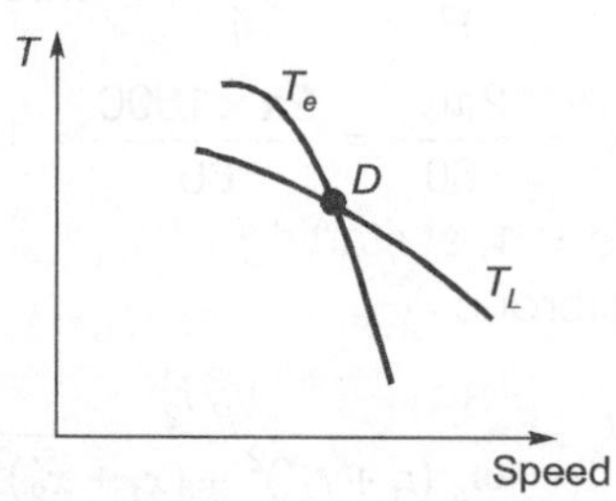

At points A, C, D

$$T_L = T_e \Rightarrow T_a = T_e - T_L = 0$$

it due to some disturbance speed decreases

$$T_e > T_L \Rightarrow T_a = T_e - T_L > 0$$

so, rotor accelerates and speed increases.

It speed increases due to some disturbances

$$T_e < T_L \Rightarrow T_a = T_e - T_L < 0$$

So, rotor decelerates and speed decreases.

So, point A, C, D are stable.

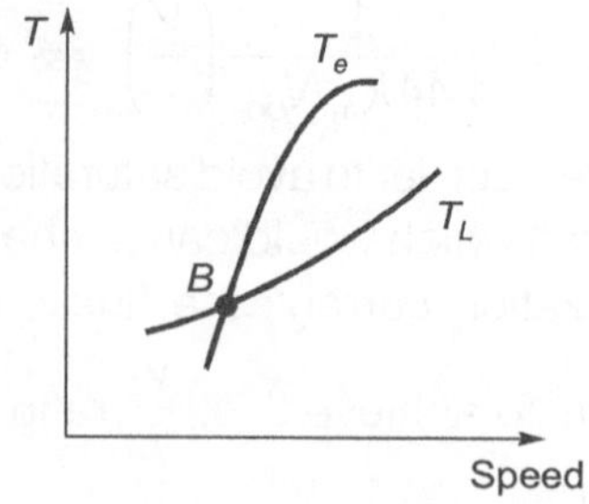

At point B,

$$T_e = T_L$$
$$T_a = T_e - T_L = 0$$

If speed decreases, due to some disturbance

$$T_e < T_L$$
$$T_a = T_e - T_L < 0$$

So rotor decelerates, and rotor speed keeps on decreasing.

If speed increases,

$$T_e > T_L$$
$$T_a = T_e - T_L > 0$$

Rotor accelerates, and rotor speed keeps on increasing. So, point B is unstable.

3.28 (d)

Assuming rated torque = T_{rated}

Starting torque

$$T_s = 150\% \text{ of } T_{rated} = 1.5\, T_{rated}$$

Maximum torque,

$$T_m = 300\% \text{ of } T_{rated} = 3.0\, T_{rated}$$

As we know,

$$T_s = \frac{3}{\omega_s}\frac{V^2 R_2'}{R_2'^2 + X_2'^2}$$

and
$$T_m = \frac{3}{\omega_s}\frac{0.5V^2}{X_2'}$$

$$\frac{T_m}{T_s} = \frac{0.5}{X_2'}\frac{R_2'^2 \times X_2'^2}{R_2'} = 0.5\left[\frac{(R_2/X_2')^2 + 1}{(R_2'/X_2')}\right]$$

As we know, maximum torque occurs at

$$s_m = \frac{R_2'}{X_2'}$$

$$\frac{T_m}{T_s} = 0.5\left(\frac{s_m^2 + 1}{s_m}\right)$$

$$\frac{3T_{rated}}{1.5T_{rated}} = 0.5\left(\frac{s_m^2+1}{s_m}\right)$$

$$s_m^2 - 4s_m + 1 = 0$$

$$s_m = 3.73 \text{ or } 0.2679$$

$s_m = 3.73$ is rejected, as for this value of slip motor works in braking mode. Therefore, for motoring mode $s_m = 0.2679$ or 26.79%

3.29 (c)

In case of auto transformer, the starting torque can be expressed as ratio of the full load

$$\frac{T_s}{T_{fl}} = x^2\left(\frac{I_{sc}}{I_{fl}}\right)^2 s_{fl} \qquad ...(i)$$

where, x = Autotransformer ratio
$T_s = 1.5\ T_{fl}$ (given)
$I_{sc} = 7\ I_{fl}$ (given)
$s_{fl} = 0.05$

Substituting the above values in equation (i),

$$\frac{1.5T_{fl}}{T_{fl}} = x^2\left(\frac{7I_{fl}}{I_{fl}}\right)^2 \times 0.05 \Rightarrow x^2 = 0.6122$$

Transformer ratio, $x = 0.7825$ or 78.25%

3.30 (b)

In case of star-delta starter, the ratio of starting torque to full load torque,

$$\frac{T_s}{T_{fl}} = \frac{1}{3}\left(\frac{I_{sc}}{I_{fl}}\right)^2 \times s_{fl} = \frac{1}{3}\left(\frac{7I_{fl}}{I_{fl}}\right)^2 \times 0.05$$

$$T_s = 0.816\ T_{fl}$$

Per unit starting torque, $T_s = 0.816$ pu

3.31 (c)

$$T_s = 0.5\ T_{fl}$$

and $s_{fl} = 0.05$ (given)

In the case of direction line starting,

$$\frac{T_s}{T_{fl}} = \left(\frac{I_{sc}}{I_{fl}}\right)^2 \times s_{fl}$$

$$\frac{0.5T_{fl}}{T_{fl}} = \left(\frac{I_{sc}}{I_{fl}}\right)^2 \times 0.05$$

$$I_{sc} = 3.16\ I_{fl}$$

Per unit starting current, $I_{sc} = 3.16$ pu

3.32 (c)

Line to line supply voltage = V_{l-l} = 400 V

Current drawn by the motor,

$$I_L = 50 \text{ A at } 0.8 \text{ pf}$$

Input power,

$$P = \sqrt{3}V_{l-l}I_L\cos\phi$$

$$= \sqrt{3}\times 400\times 50\times 0.8 = 27.71 \text{ kW}$$

Air-gap power (P_g) in the power that is transferred from the stator to the rotor via the air-gap magnetic field.

P_g = Input power – Core loss – Copper loss in stator = 27.71 – 1.5 – 1.2 = 25.01 kW

3.33 (b)

For max. torque slip = $S_m = \dfrac{R'_r}{x_{sm} + x'_{rm}}$

For starting torque, $S_m = 1$

$$x_{sm} + x'_{rm} = R'_r$$

$$2\pi f_m L_s + 0.2\pi f_m L'_r = 1$$

(frequency at maximum torque) f_m

$$= \frac{1}{2\pi(L_s + L_r)}$$

$$L_s = \frac{x_s}{2\pi\times 50} = \frac{1.5}{2\pi\times 50}$$

$$L'_r = \frac{1.5}{2\pi\times 50}$$

$$\therefore \quad f_m = \frac{1}{1.5/50 + 1.5/50} = \frac{50}{3} = 16.7 \text{ Hz}$$

In const V/f control method, $\dfrac{V_1}{f_1} = \dfrac{400}{50} = 8$

$$\therefore \quad V_2/f_2 = 8 \Rightarrow V_2 = f_2 \times 8$$

$$= 16.7 \times 8 = 133.3 \text{ volts}$$

3.34 (d)

If the three-phase supply is given to the rotor windings through an autotransformer, the three-phase rotor current will generate a rotating field in the air-gap, rotating at the synchronous speed with respect to rotor. If the rotor is kept stationary, this rotating field will also rotate in the gap at the synchronous speed voltage and current will be induced in the stator winding and a torque will be developed. If the rotor is allowed to move, it will rotate as per the Lenz's law, opposing the rotation of the rotating field decreasing the induced voltage in the stator winding. Thus at a particular speed, the frequency of the stator circuit will correspond to the slip speed.

Synchronous speed,

$$N_s = \frac{120f}{P} = \frac{120\times 50}{4} = 1500 \text{rpm}$$

Rotor speed,

$$N_r = 1410 \text{ rpm}$$

$$\text{Slip} = \frac{N_s - N_r}{N_s} = \frac{1500 - 1410}{1500} = 0.06$$

Slip frequency induced in stator

$$sf = 0.06 \times 50 = 3 \text{ Hz}$$

$$\text{Slip speed} = \frac{120\times 3}{4} = 90 \text{ rpm}$$

As stator magnetic field rotates 90 rpm in the opposite direction of the rotation of rotor, therefore, speed of the stator field with respect to rotor

$$= 1410 + 90 = 1500 \text{ rpm}$$

3.35 (c)

$$\text{Slip} = \frac{90}{1500} = 0.06$$

Total power input to induction motor within

$$= 1800 - 200 = 1600 \text{ W}$$

Power output of induction motor

$$P_0 = (1 - s)\,P_{in} = 1600(1 - 0.06) = 1504 \text{ W}$$

Since all the losses are neglected.

Induction motor power output = d.c. generator input = d.c. generator output = 1504 W

$$\therefore \quad I^2R_{ex} = 1504 \Rightarrow I = \sqrt{\frac{1504}{10}} = 12.26 \text{ A}$$

3.36 (a)

Starting current in induction motor can be as large as 5 to 6 times the full load current. As compared to starting current (I_{st}), exciting current (I_m) is very small. $\{I_m << I_{st}\}$

Therefore, shunt branch can be neglected in the circuit model of the motor.

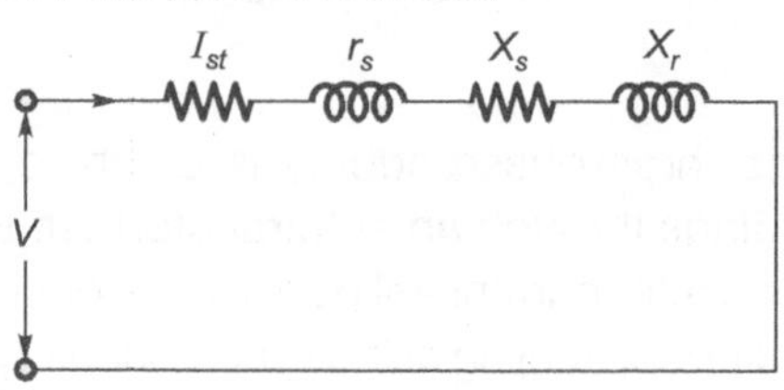

$$I_{st} = \frac{V}{(r_s + r_r) + j(X_s + X_r)}$$

$$|I_{st}| = \frac{V}{\sqrt{(r_s + r_r)^2 + (X_s + X_r)^2}} \text{ A}$$

3.37 (c)

As both the machines are coupled, rotor of both the machines rotate at same speed.

Rotor speed of DC machine = 1600 rpm

Rotor speed of induction machine =

N_r = rotor speed of DC machine

N_r = 1600 rpm

Speed of stator field in induction machine = N_s

$$N_s = \frac{120f}{P} = \frac{120 \times 50}{4} = 1500 \text{ rpm}$$

Slip of rotor w.r.t. stator field in induction machine,

$$s = \frac{N_s - N_r}{N_s} = \frac{1500 - 1600}{1500} = -0.067$$

As slip is negative, induction machine acts as a generator and the dc machine acts as a motor.

3.38 (a)

Stator and rotor magnetic fields rotates at same speed. So difference in speed is zero. Speed of stator magnetic field is N_s (synchronous speed)

$$N_s = \frac{120f}{p} = \frac{120 \times 50}{6} = 1000 \text{ rpm}$$

Rotor speed = $N_s(1 - s)$ = 950 rpm

So, speed of rotor with respect to stator magnetic field

$$= N_r - N_s = 950 - 1000 = -50 \text{ rpm}$$

3.39 (d)

$$\text{Slip} = \frac{N_s - N_r}{N_s}$$

⇒ From above formula slip depends upon:

(1) Synchronous speed (N_s)

(2) Rotor speed (N_r)

As the shaft torque depends upon rotor speed therefore the slip also depends on shaft torque. And core-losses are independent of slip.

3.40 (b)

At standstill, the rotor current is,

$$I_2 = \frac{E_2}{\sqrt{R^2 + X^2}}$$

As losses are zero,

$$\Rightarrow \quad I_2 = \frac{E_2}{X} = \frac{E_2}{j\omega L}$$

$$\therefore \quad \frac{I_2}{I_2'} = \frac{E_2}{E_2'} \times \frac{\omega'}{\omega}$$

$$\Rightarrow \quad I_2' = \frac{236 \times 50}{230 \times 57} \times 50 = 45 \text{ A}$$

3.42 (b)

$$N_s = \frac{120 \times 50}{14} = 1500 \text{ rpm}$$

$$s = \frac{1500 - 1440}{1500} = 0.04$$

For negative sequence component,

$$s' = 2 - s = 2 - 0.04 = 1.96$$

$$f' = s'f = 1.96 \times 50 = 98.00 \text{ Hz}$$

3.43 Sol.

Given, $P = 8$, $f = 50$ Hz, $N_r = 700$ rpm

Synchronous speed,

$$N_s = \frac{120f}{P} = \frac{120 \times 50}{8} = 750 \text{ rpm}$$

$$\therefore \quad \text{Slip, } S = \left(\frac{N_s - N_r}{N_s}\right) = \left(\frac{750 - 700}{750}\right)$$

$$= \frac{50}{750} = \frac{1}{15}$$

∴ Frequency of rotor current,

$$f_2 = sf_1 = \frac{1}{15} \times 50 = 3.33 \text{ Hz}$$

Hence, answer is 3.33 Hz.

3.44 (c)

Given, $R_2' = 0.1\,\Omega$, $X_2' = 0.92\,\Omega$, $S_{fl} = 3\% = 0.03$

Slip at maximum torque,

$$S_m = \frac{R_2'}{X_2'} = \frac{0.1}{0.92} = 0.1087$$

We know that, $$\frac{T_{fl}}{T_{em}} = \frac{2}{\left(\frac{S_m}{S_{fl}} + \frac{S_{fl}}{S_m}\right)}$$

$$\therefore \quad \frac{T_{fl}}{T_{em}} = \frac{2}{\left(\frac{0.1087}{0.03} + \frac{0.03}{0.1087}\right)}$$

$$= \frac{2}{(3.623 + 0.275)} = \frac{2}{3.898}$$

or, $$\frac{T_{em}}{T_{fl}} = \frac{3.898}{2} = 1.949$$

Therefore, ratio of maximum torque to full-load

torque $= \dfrac{T_{em}}{T_{fl}} = 1.949$

Option (c) is close to the answer.

3.45 (c)

An induction generator always operates at negative slip

i.e., $$\frac{N_s - N_r}{N_s} = S < 0 \text{ or } N_s < N_r$$

As the generator speed is maintained at a constant speed of 1500 rpm therefore, the synchronous speed will be less than 1500 rpm in both the cases (i.e. when at full load as well as reduced load).

$N_s = 1500$ rpm for $P = 4$ and $f = 50$ Hz.

Thus, for $N_s < 1500$ rpm, $f_1, f_2 < 50$ Hz.

$$\left(\because f = \frac{N_s P}{120} \text{ or } N_s = \frac{120 f}{P}\right)$$

Also, when load decreases then, frequency increases.

Hence, $f_2 > f_1$. Therefore, option (c) is correct.

3.46 Sol.

Given, motor draws 12 MW at 0.6 pf lag i.e. $\cos\phi = 0.6$ ∴ $\sin\phi = 0.8$.

∴ $P = 12$ MW

and $$Q = \frac{P}{\cos\phi} \times \sin\phi = \frac{12}{0.6} \times 0.8 = 16 \text{ MVAR}$$

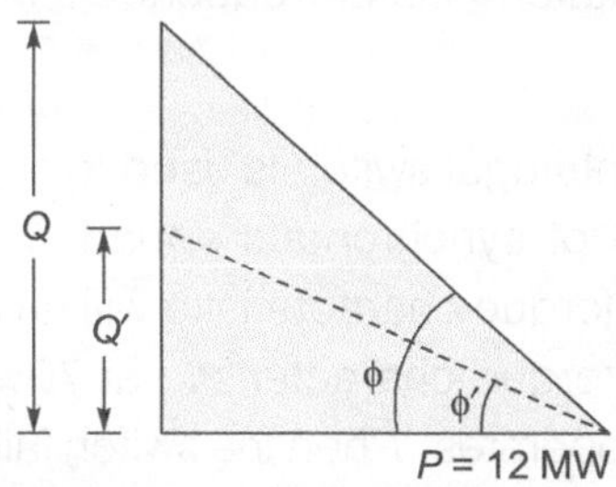

For improving the power factor to 0.8 lag, Q should be reduced, i.e., ϕ reduces,

$$\phi' = (16 - Q) \text{ MVAR}$$

Where, Q is the reactive power delivered by the capacitor.

Now power factor, $\cos\phi' = 0.8$

or $\phi' = \cos^{-1}(0.8) = 36.86°$

from power triangle,

$$\tan\phi' = \frac{Q'}{P} \text{ or, } Q' = P\tan\phi'$$

$$= 12 \tan 36.86°$$

$$Q' = 8.99 \text{ MVAR}$$

or, $16 - Q = 8.99$

or $Q = 16 - 8.99 = 7.01$ MVAR

∴ Reactive power delivered by the capacitor;

$Q = 7.01$ MVAR.

3.47 (b)

In a 3-ϕ induction motor, induced e.m.f. equation per phase is given by

$$E_{ph_1} = K_w \sqrt{2}\pi f \phi N_{ph} \approx V_1$$

where, ϕ = flux per pole = $1.5\,\phi_m$

or, $$\phi \propto \frac{V}{f}$$

Since V/f ratio is constant (to avoid over fluxing), therefore,

$$\left(\frac{V}{f}\right)_{\text{rated frequency}} > \left(\frac{V}{f}\right)_{\text{reduced frequency}}$$

$$\therefore \quad \boxed{\phi_{\text{rated}} > \phi_{\text{reduced}}}$$

3.48 (a)

Frequency of voltage across slip ring = Frequency of voltage induced in the rotor winding.

Let speed corresponding to stator frequency,

$f = N_s$.
and speed corresponding to $f_r = N_r$
Then, frequency of rotor induced emf

$$= f + f_r = f_1\left(\text{corresponding to } f_1 = \frac{(N_s + N_r)P}{120}\right)$$

Also, frequency of induced voltage across commutator = stator frequency = f.

3.49 (c)

The centrifugal switch is used to snaps the load at 70% of synchronous speed. Therefore, the speed-torque characteristic will have kink in the speed-torque characteristic at 70% of N_s if the switch operates. When the switch fails to operate, 1-ϕ induction motor will have torque-speed characteristic similar to a three-phase induction motor.

3.50 (b)

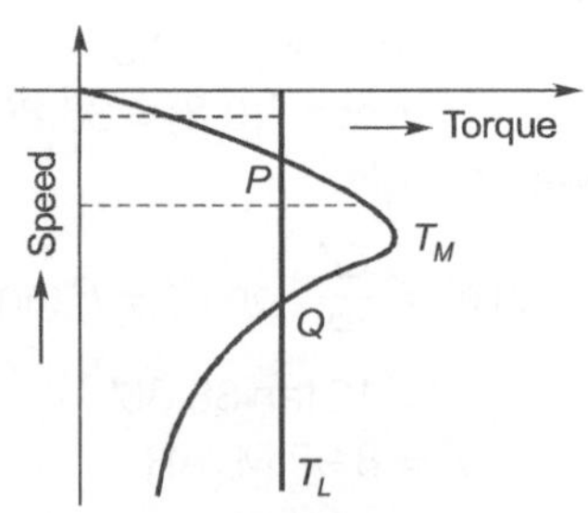

(a)

$$S = \left(\frac{N_s - N_r}{N_s}\right)$$

For given system,
Slip = (2 – s) (speed = negative)
Below point P, $T_M - T_L < 0$
⇒ Speed decreases
∴ S increases ∴ slip = (2 – s) decreases
Above point P, $T_M - T_L > 0$
⇒ Speed increases
∴ S decreases and (2 – s) increases.
Thus, system is stable near point P.

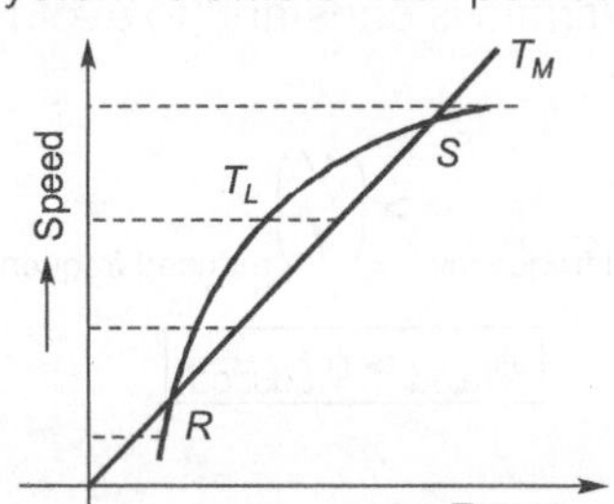

(b)

POINT R:
Below point R: $T_M - T_L < 0 \Rightarrow$ speed decreases
Above point R: $T_M - T_L > 0 \Rightarrow$ Speed increases
Thus, system can't regain stability at point R.
∴ R is unstable point.

POINT S:
Below point S: $T_M - T_L > 0 \Rightarrow$ speed increases
Above point S: $T_M - T_L < 0 \Rightarrow$ speed decreases
Thus, system can regain stability near point S.

3.51 Sol.

Given, $\quad \text{Slip} = \dfrac{N_s - N_R}{N_s} = 0.04$

$$N_R = 0.96\, N_s$$

Speed of 5th time harmonic = 5 N_s

$$\text{Slip} = \frac{5N_s + N_R}{5N_s} = 1.192$$

3.52 Sol.

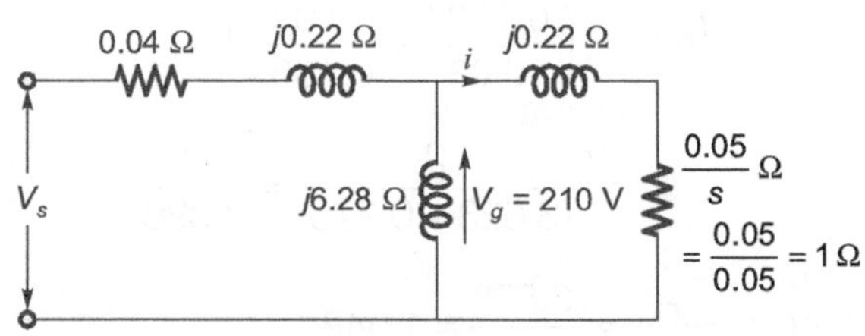

$$i = \frac{210}{1 + j0.22} = 205.09\angle{-12.40°}$$

Active power = $(205.09)^2 \times 1 = P_g$

$$P_m = (1 - 0.05) \times P_g = 39958.81 \text{ Watt}$$

$$\text{Torque} = \frac{39958.81 \times 60}{2\pi \times 2850} \times 3$$

$$= 133.88 \times 3 = 401.64 \text{ N-m}$$

3.53 (c)

$$f_0 = \text{Nominal frequency}$$

$$\omega_s = \left(\frac{f}{f_0}\right)\omega_{s_0} \quad \text{...(i)}$$

$$S_{max,T} = \left(\frac{f_0}{f}\right)\cdot\left(\frac{R_2}{X_{20}^1}\right) \quad \text{...(ii)}$$

From (i) and (ii),

$$S_{max,T} = \left(\frac{\omega_{s_0}}{\omega_s}\right)\cdot\left(\frac{R_2}{X_{20}^1}\right)$$

$$S_{max,T} = \left(\frac{\omega_{s_0}}{\omega_s}\right)\cdot\left(\frac{R_2}{X_{20}^1}\right)$$

$$S_{max,T} \propto \frac{1}{\omega_s}$$

3.54 Sol.

415 V, 3-phase, Δ connected induction motor $(I_{st})_{line}$ = 120 A at rated voltage.
at, V = 110 V, i.e. reduced voltage

$$I_{st} = x(I_{st})_{rated}$$

where, $$x = \frac{V_{reduced}}{V_{rated}} = \frac{110}{415}$$

$$(I_{st})_{at\ 110\ V} = \left(\frac{110}{415}\right) \times 120 = 31.807\ A$$

3.55 (c)

Induction Generator Mode :

$$N > N_s \quad \text{i.e. } (s \text{ is negative})$$

$$\because \quad P = 4,\ f = 60\ Hz,$$

$$N_s = 1800\ rpm$$

Rotor frequency given = 5 Hz ⇒ $(s \times f)$

$$\because \quad f = 60\ Hz$$

$$\therefore \quad s = \frac{5}{60} = 0.0833$$

∵ It is induction generator slip will be – 0.0833.

$$s = \frac{1800 - N}{1800} = -0.0833$$

$$\therefore \quad N = 1950\ rpm$$

3.56 (c)

$$N_s = \frac{120 \times 50}{4} = 1500\ rpm$$

$$s = 0.02$$

As it was specified rotor rmf with respect to stationary part/stator.
Rotor rmf w.r.t. stator is always at N_s.
∴ Speed of rotor flux in mechanical radian/sec.

$$= \frac{2 \times 3.14 \times 1500}{60} = 157\ rad/sec.$$

3.57 Sol.

Equivalent circuit representation during starting – according to question.

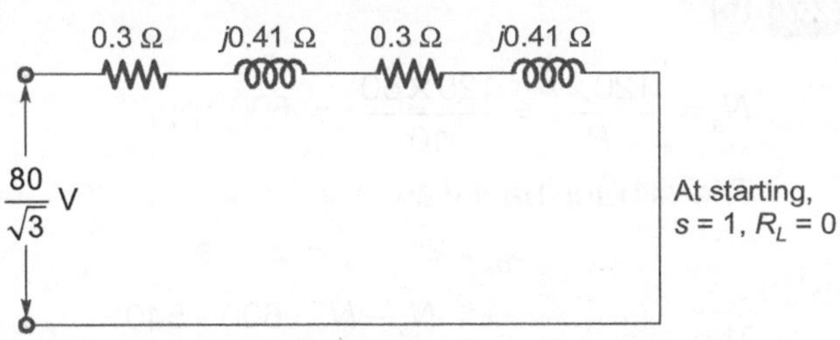

Total resistance reference to stator = 0.6 Ω
Total reactance reference to stator = 0.82 Ω
Reactance for 60 Hz → 0.82 Ω
Reactance for 20 Hz → ?

$\because X_e \propto F$ Reactance for 20 Hz ⇒ $\frac{20}{60} \times 0.82$

∴ X for 20 Hz frequency = 0.273 Ω
∴ Starting current

$$\Rightarrow \quad I_{st} = \frac{80\sqrt{3}}{\sqrt{(0.6)^2 + (0.273)^2}} = 70.05\ A$$

3.59 (14.95)

$$I_L = \sqrt{3}\, I_{ph}$$

At f = 10 Hz,

$$X_1 = X_2 = 8.22 \times \frac{10}{50} = 1.644\ \Omega$$

$$I_{ph} = \frac{V_{ph}}{\sqrt{(R_1 + R_2)^2 + (X_1 + X_2)^2}}$$

$$= \frac{100}{\sqrt{(5.39 + 5.72)^2 + (1.644 + 1.644)^2}}$$

$$= \frac{100}{11.586} = 8.63\ A$$

$$I_L = 14.95\ A$$

3.60 Sol.

4 pole, 50 Hz I.M has no load slip 1%
4 pole, 50 Hz I.M has full load slip 5%

$$N_S = 1500\ rpm$$

$$N_0 = N_s(1 - s) = 1500(1 - 0.01) = 1485$$

$$N = N_s(1 - s) = 1500(1 - 0.01) = 1425$$

Speed regulation is

$$\%S.R. = \frac{N_0 - N}{N} \times 100$$

$$= \frac{1485 - 1425}{1425} \times 100 = 4.21\%$$

3.61 (90)

$P_{i/p}$ = 40 kW, stator loss = 1 kW, F and W = 2.025 kW

$$\text{Stator o/p} = 40 - 1 = 39\ kW$$

$$\text{Slip} = \frac{1000 - 975}{1000} = 0.025$$

$$\text{Rotor o/p} = \text{Rotor i/p} \times (1 - s)$$

$$= 39\ (1 - 0.025) = 38.025\ kW$$

$$\text{Motor o/p} = 38.025 - 2.025 = 36\ kW$$

$$\eta = \frac{\text{Motor output}}{\text{Motor input}} = \frac{36}{40} \times 100 = 90\%$$

3.62 (0.52)

$$N_{max} = 650\ rpm,\ P = 8,\ 50\ Hz$$

$$s_m = \frac{750-650}{750} = 0.1333$$

$$s_m = \frac{R_2}{X_2}$$

$$\therefore \quad X_2 = \frac{R_2}{s_m} = \frac{0.08}{0.133} = 0.601\,\Omega$$

$$R_2 = 0.08\,\Omega,\ X_2 = 0.601\,\Omega$$

Condition for maximum T_{st}

$$\Rightarrow \quad R_2 = X_2$$

$$\therefore \quad R_2 + R_{ext} = X_2$$

$$\therefore \quad R_{ext} = 0.601 - 0.08 = 0.521\,\Omega$$

3.63 (585)

Stator frequency= 40 Hz;

Rotor frequency = 1 Hz

Poles = 8, N = ?

$$sf = 1$$

$$\Rightarrow \quad s = \frac{1}{40} = 0.025$$

$$N_s = \frac{120\times 40}{8} = 600$$

$$N = N_s(1-s)$$

$$\therefore \quad N = 600(1-0.025) = 585 \text{ rpm}$$

3.64 (1733)

Equation of speed vs. time from 4 sec to 8 sec.

$$N = 125t + 500$$

$$\Rightarrow \quad 4 \text{ sec} \le t \le 8 \text{ sec}$$

Now, $J\dfrac{d\omega}{dt} = T_e - T_L$

or $$T_e = J\frac{d\omega}{dt} + T_L$$

or $$\omega T_e = J\omega\frac{d\omega}{dt} + \omega T_L$$

or $$P_e = J\omega\frac{d\omega}{dt} + \omega T_L$$

or $$\frac{dE}{dt} = J\omega\frac{d\omega}{dt} + \omega T_L$$

or $$dE = J\omega d\omega + T_L\omega dt$$

or $$E = J\int \omega d\omega + T_L\int \omega dt$$

or $$E = J\left(\frac{2\pi}{60}\right)^2 \int_{1000}^{1500} N dN + T_L \frac{2\pi}{60}\int_4^8 N dt$$

$$= J\left(\frac{2\pi}{60}\right)^2 \left(\frac{N^2}{2}\right)_{1000}^{1500} + 2\cdot\frac{2\pi}{60}\int_4^8 (125t+500)dt$$

$$= 0.1\times\left(\frac{2\pi}{60}\right)^2\left(\frac{1500^2-1000^2}{2}\right) + 2\cdot\frac{2\pi}{60}\left[\frac{125t^2}{2}+500t\right]_4^8$$

$$= 685.389 + 1047.197 = 1732.5865 \text{ J}$$

3.65 (0.63)

Given: $I_{sc} = 5\,I_{rated} \Rightarrow I_{sc} = 5I_f$

It is required to make starting current to "$2I_f$" using auto transformer.

Line current with auto transformer,

$$I_L = x^2 I_{sc}$$

Given : $I_L = 2I_f \Rightarrow 2I_f = x^2 I_{sc}$

$$\frac{2I_f}{I_{sc}} = x^2$$

$$\Rightarrow \quad x^2 = 2\times\left(\frac{1}{5}\right) \Rightarrow x = \sqrt{\frac{2}{5}} = 0.63$$

3.66 (a)

Mode	Slip Range
Generator Mode	–1.0 to 0.0
Motor Mode	0.0 to 1.0
Plugging Mode	1.0 to 2.0

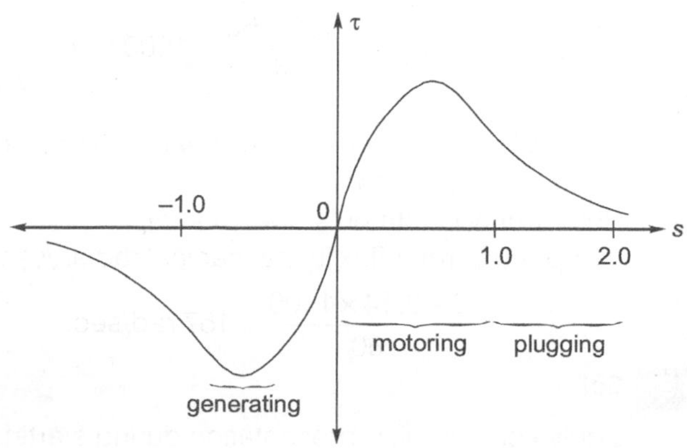

3.67 (b)

$$N_s = \frac{120\times f}{P} = \frac{120\times 50}{10} = 600 \text{ rpm}$$

The slip for backward slip,

$$s_b = 2 - s_f = 2 - s$$

The slip, $$s = \frac{N_s - N}{N_s} = \frac{600-540}{600} = 0.1$$

So, $$s_b = 2 - 0.1 = 1.9$$

So, frequency due to backward field,

$$s_b f = 1.9 \times 50 = 95 \text{ Hz}$$

3.68 (760)

Method 1 :

$$N_s = 120\times\frac{50}{6} = 1000 \text{ rpm}$$

$\Rightarrow$ Slip speed = 40 rpm

Here, V/f = Constant. So, slip speed remains constant.

For $f = 50 \times 0.80 = 40$ Hz, the synchronous speed

$$N_{s2} = 120\times\frac{40}{6} = 800 \text{ rpm}$$

So, motor speed $N_2 = N_{s2}$ – slip speed
$= 800 - 40 = 760$ rpm

Method 2 :

Torque, $T \propto \dfrac{sV^2}{f}$

$$\frac{T_2}{T_1} = \frac{s_2}{s_1}\left(\frac{V_2}{V_1}\right)^2\left(\frac{f_2}{f_1}\right)$$

$$1 = \frac{s_2}{0.04}(0.8)^2\times\frac{1}{0.8}$$

$$s_2 = 0.05$$

So, $N_2 = N_{s2}(1-s_2) = \dfrac{120\times f_2}{P}(1-s_2)$

$$N_2 = \frac{120\times 40}{6}(1-0.05) = 760 \text{ rpm}$$

3.69 (a)

$$N_s = \frac{120\times 50}{6} = 1000 \text{ rpm}$$

$$s = \frac{1000-960}{1000} = 0.04$$

$$\frac{\text{Rotor output}}{\text{Rotor input}} = 1 - s$$

$$\eta = 1 - 0.04 = 0.96 \text{ or } 96\%$$

3.70 (0.28)

Condition for the maximum torque

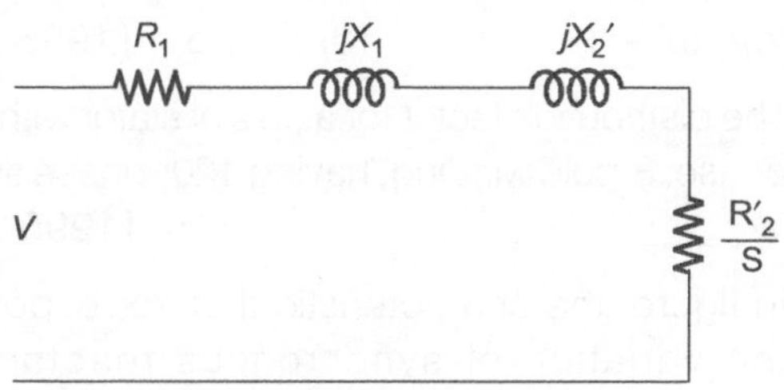

$$\frac{R_2'}{s} = \sqrt{3^2+4^2} = 5$$

Substituting, s = 1

As maximum torque at starting

$$R_2' + R_{ext}' = 5$$

$$R_{ext}' = 5 - 2.5 = 2.5\ \Omega \text{ (referred to stator)}$$

Referring to rotor side, $\dfrac{2.5}{3^2} = 0.277\ \Omega$

■■■■

4 Synchronous Machines

4.1 A synchronous motor on load draw a current at a leading power factor angle ϕ. If the internal power factor angle, which is the phase angle between the excitation emf and the current in time phasor diagram is Ψ. Then the air gap excitation mmf lags the armature mmf by

(a) Ψ (b) $\pi/2 + \Psi$
(c) $\pi/2 - \Psi$ (d) $\Psi + \phi$ [1995 : 1 M]

4.2 The distribution factor for a 36 slot stator with three-phase, 8-pole winding, having 120° phase spread, is _____ [1995 : 2 M]

4.3 In figure, the characteristic that corresponds to the variation of synchronous reactance of synchronous motor with field current is

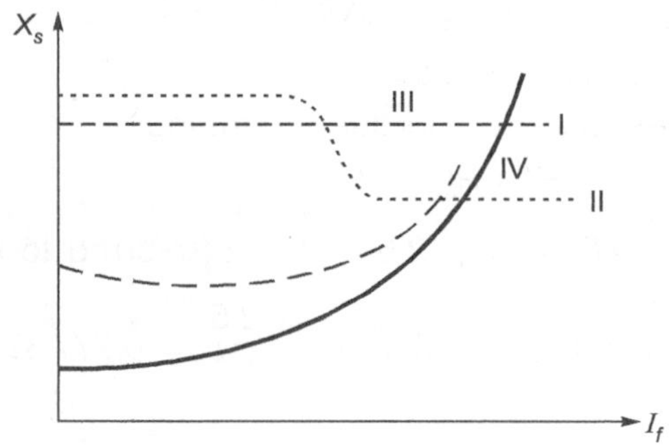

(a) Curve I (b) Curve II
(c) Curve III (d) Curve IV [1996 : 1 M]

4.4 During hunting of synchronous motor

(a) negative phase sequence currents are generated
(b) harmonics are developed in the armature circuit
(c) damper bar develops torque
(d) field excitation increases [1996 : 1 M]

4.5 The per unit voltage of two synchronous machines connected through a lossless line are $0.95\angle 10°$ and $1.0\angle 0°$. Match the two sides in the following.

List-I
(A) Real power of machine 1
(B) Reactive power of machine 1
(C) Power factor of machine 1

List-II
(P) Positive real power
(Q) Positive reactive power
(R) Negative real power
(S) Negative reactive power
(T) Leading power factor
(U) Lagging power factor [1996 : 2 M]

4.6 A single-phase alternator has a synchronous reactance of 2 ohms and negligible resistance. If it supplies 10 A to a purely capacitive load at 200 V, then generated emf will be ____ V and the regulation will be _____% [1997 : 2 M]

4.7 A synchronous generator connected to an infinite bus is overexcited.
Considering only the reactive power, from the point of the system the machine acts as

(a) a capacitor (b) an inductor
(c) a resistor (d) none of these [1998: 1 M]

4.8 The armature of a single phase alternator is completely wound with T single turn coils distributed uniformly. The induced voltage in each turn is 2 Volts(rms). The emf of the whole winding is

(a) $2T$ Volt (b) $1.11T$ Volt
(c) $1.414T$ Volt (d) $1.273T$ Volt [1998 : 2 M]

4.9 Higher synchronous reactance is preferred in the present day alternators, because one can have

(a) reduced sub-transient currents.
(b) reduced harmonic currents.
(c) reduced transient currents.
(d) higher voltage regulation with load. [1999 : 1 M]

4.10 A three-phase alternator is wound with a 60 degree phase-spread armature windings and develops 300 kVA. If the armature is reconnected utilizing all the coils for single-phase operation with a phase spread of 180 degrees, the new rating of the machine is

(a) 100 kVA (b) 200 kVA
(c) 250 kVA (d) 300 kVA [1999 : 2 M]

4.11 The phase sequence of a three-phase alternator will reverse if

(a) the field current is reversed keeping the direction of rotation same
(b) the field current remains the same but the direction of rotation is reversed

(c) the field current is reversed and the number of poles is doubled

(d) the number of poles is doubled without reversing the field current [2000 : 1 M]

4.12 A single-phase, 2000 V alternator has armature resistance and reactance of 0.8 ohms and 4.94 ohms respectively. The voltage regulation of the alternator at 100 A load at 0.8 leading power factor is

(a) 7% (b) –8.9%

(c) 14% (d) 0% [2000 : 2 M]

4.13 It is desirable to eliminate 5th harmonic voltage from the phase voltage of an alternator. The coils should be short-pitched by an electrical angle of

(a) 30° (b) 36°

(c) 72° (d) 18° [2001 : 1 M]

4.14 The figure shows the magnetization curves of an alternator at rated armature current, unity power factor and also at no-load. The magnetization curve for rated armature current, 0.8 power factor leading is given by

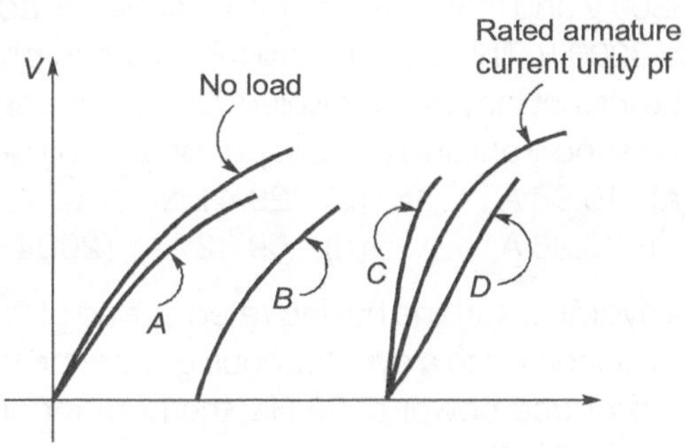

(a) Curve A (b) Curve B

(c) Curve C (d) Curve D [2001 : 1 M]

4.15 X_d, X_d' and X_d'' are steady state d-axis synchronous reactance, transient d-axis reactances and sub-transient d-axis reactance of a synchronous machine respectively. Which of the following statement is true?

(a) $X_d > X_d' > X_d''$ (b) $X_d'' > X_d' > X_d$

(c) $X_d' > X_d'' > X_d$ (d) $X_d > X_d'' > X_d'$

[2001 : 1 M]

4.16 A 50 kW synchronous motor is tested by driving it by another motor. When the excitation is not switched on, the driving motor takes 800 W. When the armature is short-circuited and the rated armature current of 10 A is passed through it, the driving motor requires 2500 W. On open-circuiting the armature with rated excitation, the driving motor takes 1800 W. Calculate the % efficiency of the synchronous motor at 50% load. Neglect the losses in the driving motor. [2001 : 2 M]

4.17 A star connected 440 V, 50 Hz alternator has per phase synchronous reactance of 10 Ω. It supplies a balanced capacitive load current of 20 A. As shown in the per phase equivalent circuit of figure. It is desirable to have zero voltage regulation. The load power factor should be

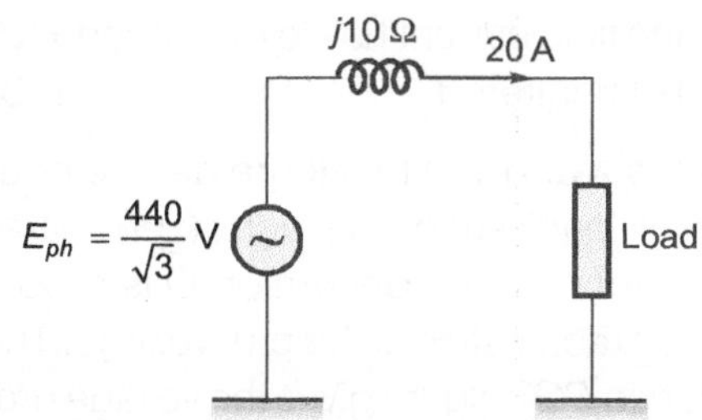

(a) 0.82 (b) 0.47

(c) 0.39 (d) 0.92 [2001 : 2 M]

4.18 The synchronous speed for the seventh space harmonic mmf wave of a 3-phase, 8 pole, 50 Hz induction machine is

(a) 107.14 rpm in forward direction

(b) 107.14 rpm in reverse direction

(c) 5250 rpm in forward direction

(d) 5250 rpm in reverse direction [2001 : 2 M]

4.19 The flux per pole in a synchronous motor with the field circuit ON and the stator disconnected from the supply is found to be 25 mWb. When the stator is connected to the rated supply with the field excitation unchanged, the flux per pole in the machine is found to be 20 mWb while the motor is running on no-load. Assuming no-load losses to be zero, the no-load current drawn by the motor from the supply

(a) lags the supply voltage

(b) leads the supply voltage

(c) is in phase with the supply voltage

(d) is zero [2002 : 2 M]

4.20 A round rotor generator with internal voltage E_1 = 2.0 p.u. and X = 1.1 p.u. is connected to a round rotor synchronous motor with internal voltage E_2 = 1.3 p.u. and X = 1.2 p.u. The reactance of the line connecting the generator to the motor is 0.5 p.u. When the generator supplies 0.5 p.u. power, the rotor angle difference between the machines will be

(a) 57.42° (b) 1°

(c) 32.58° (d) 122.58° [2003 : 1 M]

4.21 A stand alone engine driven Synchronous generator is feeding a partly inductive load. A capacitor is now connected across the load to completely nullify the inductive current. For this operating condition

(a) the field current and fuel input have to be reduced
(b) the field current & fuel input have to be increased
(c) the field current has to be increased & fuel input left unaltered
(d) the field current has to be reduced & fuel input left unaltered [2003 : 1 M]

4.22 Curves X and Y in the figure denote open circuit and full-load zero power factor (zpf) characteristics of a synchronous generator. Q is a point on the zpf characteristics at 1.0 p.u. voltage. The vertical distance PQ in figure gives the voltage drop across

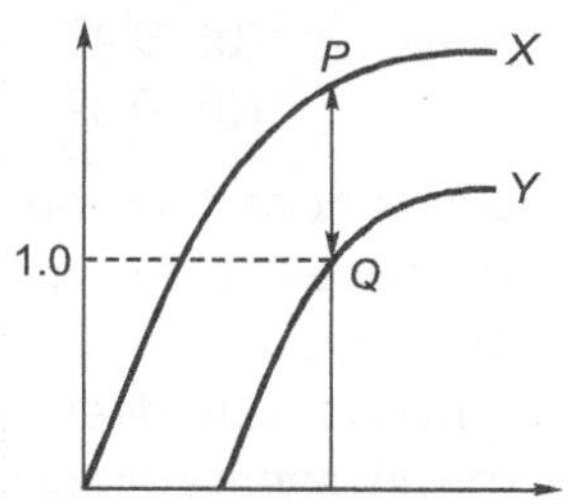

(a) Synchronous reactance
(b) Magnetizing reactance
(c) Potier reactance
(d) Leakage reactance [2003 : 1 M]

4.23 A 4-pole, 3-phase, double-layer winding is housed in a 36-slot stator for an ac machine with 60° phase spread. Coil span is 7 short pitched. Number of slots in which top and bottom layers belong to different phases is

(a) 24 (b) 18
(c) 12 (d) 0 [2003 : 2 M]

4.24 When Stator and Rotor windings of a 2-pole rotating electrical machine are excited, each would produce a sinusoidal mmf distribution in the airgap with peak values F_s and F_r respectively. The rotor mmf lags stator mmf by a space angle δ at any instant as shown in figure.

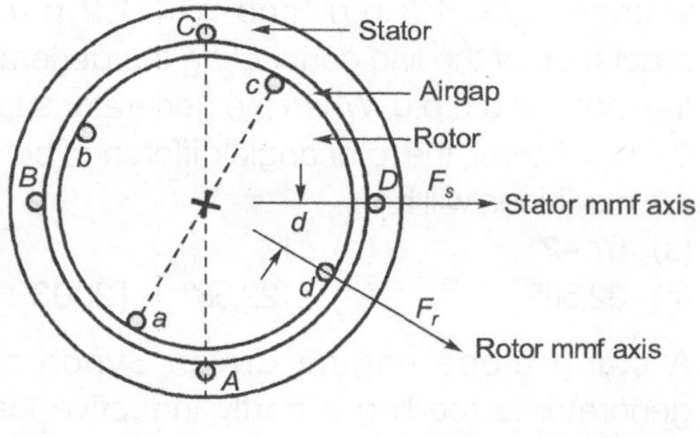

Thus, half of stator and rotor surfaces will form one pole with the other half forming the second pole. Further, the direction of torque acting on the rotor can be clockwise or counter-clockwise.

The following Table gives four sets of statements as regards poles and torque. Select the correct set corresponding to the mmf axis as shown in figure.

	Stator Surface ABC forms	Stator Surface CDA forms	Rotor Surface abc forms	Rotor Surface cda forms	Torque Surface
(a)	North Pole	South Pole	North Pole	South Pole	Clockwise
(b)	South Pole	North Pole	North Pole	South Pole	Counter Clockwise
(c)	North Pole	South Pole	South Pole	North Pole	Counter Clockwise
(d)	South Pole	North Pole	South Pole	North Pole	Clockwise

[2003 : 2 M]

4.25 Two 3-phase, Y-connected alternators are to be paralleled to a set of common busbars. The armature has a per phase synchronous reactance of 1.7 Ω and negligible armature resistance. The line voltage of the first machine is adjusted to 3300 V and that of the second machine is adjusted to 3200 V. The machine voltages are in phase at the instant they are paralleled. Under this condition, the synchronizing current per phase will be

(a) 16.98 A (b) 29.41 A
(c) 33.96 A (d) 58.82 A [2004 : 2 M]

4.26 A hydraulic turbine having rated speed of 250 rpm is connected to a synchronous generator. In order to produce power at 50 Hz, the number of poles required in the generator are

(a) 6 (b) 12
(c) 16 (d) 24 [2004 : 2 M]

4.27 A 400 V, 50 kVA, 0.8 pf leading Δ-connected, 50 Hz synchronous machine has a synchronous reactance of 2 Ω and negligible armature resistance. The friction and windage losses are 2 kW and the core loss is 0.8 kW. The shaft is supplying 9 kW load at a power factor of 0.8 leading. The line current drawn is

(a) 12.29 A (b) 16.24 A
(c) 21.29 A (d) 36.88 A [2004 : 2 M]

4.28 A 500 MW 3-phase Y-connected synchronous generator has a rated voltage of 21.5 kV at 0.85 pf. The line current when operating at full load rated conditions will be

(a) 13.43 kA (b) 15.79 kA
(c) 23.25 kA (d) 27.36 kA

[2004 : 2 M]

4.29 In relation to the synchronous machines, which one of the following statements is false?
(a) In salient pole machines, the direct-axis synchronous reactance is greater than the quadrature-axis synchronous reactance.
(b) The damper bars help the synchronous motor self start.
(c) Short circuit ratio is the ratio of the field current required to produce the rated voltage on open circuit to the rated armature current.
(d) The V-curve of a synchronous motor represents the variation in the armature current with field excitation, at a given output power.
[2005 : 2 M]

Data for Questions (4.30 and 4.31):

A 1000 kVA, 6.6 kV, 3-phase star connected cylindrical pole synchronous generator has a synchronous reactance of 20 Ω. Neglect the armature resistance and consider operation at full load and unity power factor.

4.30 The induced emf (line-to-line) is close to
(a) 5.5 kV (b) 7.2 kV
(c) 9.6 kV (d) 12.5 kV [2005 : 2 M]

4.31 The power (or torque) angle is close to
(a) 13.9° (b) 18.3°
(c) 24.6° (d) 33.0° [2005 : 2 M]

4.32 A synchronous generator is feeding a zero power factor (lagging) load at rated current. The armature reaction is
(a) magnetizing
(b) demagnetizing
(c) cross-magnetizing
(d) ineffective [2006 : 2 M]

4.33 A 3-phase, 400 V, 5 kW, star connected synchronous motor having an internal reactance of 10 Ω is operating at 50% load, unity pf. Now, the excitation is increased by 1%. What will be the new load in percent, if the power factor is to be kept same? Neglect all losses and consider linear magnetic circuit.
(a) 67.9% (b) 56.9%
(c) 51% (d) 50% [2006 : 2 M]

Data for Questions (4.34 to 4.36):

A 4 pole, 50 Hz, synchronous generator has 48 slots in which a double layer winding is used. Each coil has 10 turns and is short pitched by an angle to 36° electrical. The fundamental flux per pole is 0.025 Wb.

4.34 The line-to-line induced emf (in volts), for a three phase star connection is approximately
(a) 808 (b) 888
(c) 1400 (d) 1538 [2006 : 2 M]

4.35 The line-to-line induced emf (in volts), for a three phase connection is approximately
(a) 1143 (b) 1332
(c) 1617 (d) 1791 [2006 : 2 M]

4.36 The fifth harmonic component of phase emf (in volts), for a three phase star connection is
(a) 0 (b) 269
(c) 281 (d) 808 [2006 : 2 M]

4.37 A three-phase synchronous motor connected to ac mains is running at full load and unity power factor. If its shaft load is reduced by half, with field current held constant, its new power factor will be
(a) unity
(b) leading
(c) lagging
(d) dependent on machine parameters
[2007 : 2 M]

4.38 A 100 kVA, 415 V (line-line), star-connected synchronous machine generates rated open circuit voltage of 415 V at a field current of 15 A. The short circuit armature current at a field current of 10 A is equal to the rated armature current. The per unit saturated synchronous reactance is
(a) 1.731 (b) 1.5
(c) 0.666 (d) 0.577 [2007 : 2 M]

4.39 Distributed winding and short chording employed in AC machines will result in
(a) increase in emf and reduction in harmonics
(b) reduction in emf and increase in harmonics
(c) increase in both emf and harmonics
(d) reduction in both emf and harmonics
[2008 : 1 M]

Statement for Linked Answer (4.40 and 4.41):

A synchronous motor is connected to an infinite bus at 1.0 pu voltage and draws 0.6 pu current at unity power factor. Its synchronous reactance is 1.0 pu and resistance is negligible.

4.40 The excitation voltage (E) and load angle (δ) will respectively be
(a) 0.8 pu and 36.86° lag
(b) 08 pu and 36.86° lead
(c) 1.17 pu and 30.96° lead
(d) 1.17 pu and 30.96° lag [2008 : 2 M]

4.41 Keeping the excitation voltage same, the load on the motor is increased such that the motor current increases by 20%. The operating power factor will become

(a) 0.995 lagging (b) 0.995 leading
(c) 0.791 lagging (d) 0.848 leading

[2008 : 2 M]

4.42 A field excitation of 20 A in a certain alternator results in an armature current of 400 A in short circuit and a terminal voltage of 2000 V on open circuit. The magnitude of the internal voltage drop within the machine at a load current of 200 A is

(a) 1 V (b) 10 V
(c) 100 V (d) 1000 V [2009 : 1 M]

4.43 A three-phase, salient pole synchronous motor is connected to an infinite bus. It is operated at no load at normal excitation. The field excitation of the motor is first reduced to zero and then increased in the reverse direction gradually. Then the armature current

(a) increases continuously
(b) first increases and then decreases steeply
(c) first decreases and then increases steeply
(d) remains constant [2011 : 1 M]

4.44 The direct axis and quadrature axis reactances of a salient pole alternator are 1.2 p.u. and 1.0 p.u. respectively. The armature resistance is negligible. If this alternator is delivering rated kVA at upf and at rated voltage then its power angle is

(a) 30° (b) 45°
(c) 60° (d) 90° [2011 : 2 M]

4.45 The angle δ in the swing equation of a synchronous generator is the

(a) angle between stator voltage and current.
(b) angular displacement of the rotor with respect to the stator.
(c) angular displacement of the stator mmf with respect to a synchronously rotating axis.
(d) angular displacement of an axis fixed to the rotor with respect to a synchronously rotating axis. [2013 : 1 M]

4.46 A star connected 400 V, 50 Hz, 4 pole synchronous machine gave the following open circuit and short circuit test result:

Open circuit test: V_{oc} = 400 V (rms, line-to-line) at field current, I_f = 2.3 A

Short circuit test: I_{sc} = 10 A (rms, phase) at field current, I_f = 1.5 A

The value of per phase synchronous impedance in Ω at rated voltage is ______.

[2014 : 1 M, Set-1]

4.47 A three-phase synchronous generator is to be connected to the infinite bus. The lamps are connected as shown in the figure for the synchronization. The phase sequence of bus voltage is R-Y-B and that of incoming generator voltage is R′-Y′-B′.

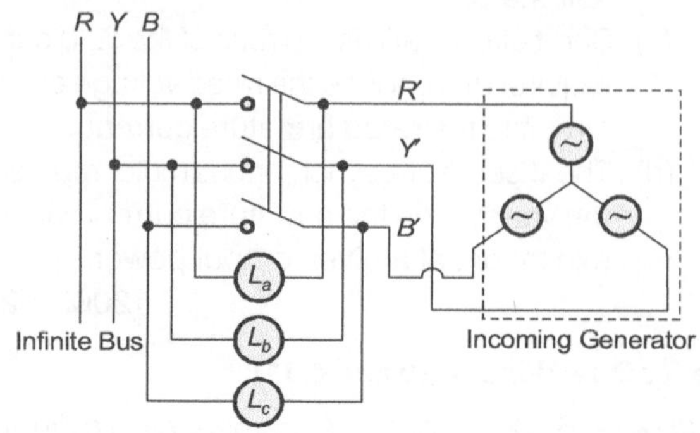

If was found that the lamps are becoming dark in the sequence L_a-L_b-L_c. It means that the phase sequence of incoming generator is

(a) opposite to infinite bus and its frequency is more than infinite bus.
(b) opposite to infinite bus but its frequency is less than infinite bus.
(c) same as infinite bus and its frequency is more than infinite bus.
(d) same as infinite bus and its frequency is less than infinite bus.

[2014 : 2 M, Set-1]

4.48 A 20 pole alternator is having 180 identical stator slots with 6 conductors in each slot. All the coils of a phase are in series. If the coils are connected to realize single-phase winding, the generated voltage is V_1. If the coils are reconnected to realize three-phase star connected winding, the generated phase voltage is V_2. Assuming full pitch, single-layer winding, the ratio V_1/V_2 is

(a) $\frac{1}{\sqrt{3}}$ (b) $\frac{1}{2}$
(c) $\sqrt{3}$ (d) 2 [2014 : 2 M, Set-2]

4.49 In a synchronous machine, hunting is predominantly damped by

(a) mechanical losses in the rotor
(b) iron losses in the rotor
(c) copper losses in the stator
(d) copper losses in the rotor [2014 : 1 M, Set-3]

4.50 A non-salient pole synchronous generator having synchronous reactance of 0.8 pu is supplying 1 pu power to a unity power factor load at a terminal voltage of 1.1 pu. Neglecting the armature resistance, the angle of the voltage behind the synchronous reactance with respect to the angle of the terminal voltage in degrees is ____.

[2014 : 2 M, Set-3]

4.51 Consider a system consisting of a synchronous generator working at a lagging power factor, a synchronous motor working at an overexcited condition and a directly grid-connected induction generator. Consider capacitive VAr to be a source and inductive VAr to be a sink of reactive power. Which one of the following statements is TRUE?

(a) Synchronous motor and synchronous generator are sources and induction generator is a sink of reactive power.
(b) Synchronous motor and induction generator are sources and synchronous generator is a sink of reactive power.
(c) Synchronous motor is a source and induction generator and synchronous generator are sinks of reactive power.
(d) All are sources of reactive power.

[2016 : 1 M, Set-1]

4.52 A three-phase, 50 Hz salient-pole synchronous motor has a per-phase direct-axis reactance (X_d) of 0.8 pu and a per-phase quadrature-axis reactance (X_q) of 0.6 pu. Resistance of the machine is negligible. It is drawing full-load current at 0.8 pf (leading). When the terminal voltage is 1 pu, per-phase induced voltage, in pu, is ______.

[2016 : 2 M, Set-1]

4.53 The power consumption of an industry is 500 kVA, at 0.8 p.f. lagging. A synchronous motor is added to raise the power factor of the industry to unity. If the power intake of the motor is 100 kW, the p.f. of the motor is _______.

[2016 : 2 M, Set-2]

4.54 A three-phase, 50 Hz, star-connected cylindrical-rotor synchronous machine is running as a motor. The machine is operated from a 6.6 kV grid and draws current at unity power factor (UPF). The synchronous reactance of the motor is 30 Ω per phase. The load angle is 30°. The power delivered to the motor in kW is ______ . (Give the answer up to one decimal place.) [2017 : 1 M, Set-1]

4.55 Two parallel connected, three-phase, 50 Hz, 11 kV, star-connected synchronous machines A and B, are operating as synchronous condensers. They together supply 50 MVAR to a 11 kV grid. Current supplied by both the machines are equal. Synchronous reactances of machine *A* and machine *B* are 1 Ω and 3 Ω, respectively. Assuming the magnetic circuit to be linear, the ratio of excitation current of machine A to that of machine *B* is _______ . (Give the answer up to two decimal places.) [2017 : 2 M, Set-1]

4.56 If a synchronous motor is running at a leading power factor, its excitation induced voltage (E_f) is

(a) equal to terminal voltage V_t
(b) higher than the terminal voltage V_t
(c) less than terminal voltage V_t
(d) dependent upon supply voltage V_t

[2017 : 1 M, Set-2]

4.57 A 3-phase, 50 Hz generator supplies power of 3 MW at 17.32 kV to a balanced 3-phase inductive load through an overhead line. The per phase line resistance and reactance are 0.25 Ω and 3.925 Ω respectively. If the voltage at the generator terminal is 17.87 kV, the power factor of the load is ____.

[2017 : 2 M, Set-2]

4.58 A 25 kVA, 400 V, Δ-connected, 3-phase, cylindrical rotor synchronous generator requires a field current of 5 A to maintain the rated armature current under short-circuit condition. For the same field current, the open-circuit voltage is 360 V. Neglecting the armature resistance and magnetic saturation, its voltage regulation (in % with respect to terminal voltage), when the generator delivers the rated load at 0.8 pf leading, at rated terminal voltage is ______. [2017 : 2 M, Set-2]

4.59 In a salient pole synchronous motor, the developed reluctance torque attains the maximum value when the load angle in electrical degree is

(a) 0 (b) 45°
(c) 60° (d) 90° [2018 : 1 M]

4.60 A three-phase synchronous motor draws 200 A from the line at unity power factor at rated load. Considering the same line voltage and load, the line current at a power factor of 0.5 leading is

(a) 200 A (b) 100 A
(c) 400 A (d) 300 A [2019 : 1 M]

4.61 A 220 V (line), three-phase, Y-connected, synchronous motor has a synchronous impedance of (0.25 + *j*2.5) Ω/phase. The motor draws the rated current of 10 A at 0.8 pf leading. The rms value of line-to-line internal voltage in volts (round off to two decimal places) is ____. [2019 : 2 M]

4.62 A three-phase cylindrical rotor synchronous generator has a synchronous reactance X_s and a negligible armature resistance. The magnitude of per phase terminal voltage is V_A and the magnitude of per phase induced emf is E_A. Considering the following two statements, *P* and *Q*.

P : For any three-phase balanced leading load connected across the terminals of this synchronous generator, V_A is always more than E_A.

Q : For any three-phase balanced lagging load connected across the terminals of this synchronous generator, V_A is always less than E_A.

Which of the following options is correct?

(a) P is true and Q is true.
(b) P is true and Q is false.
(c) P is false and Q is false.
(d) P is false and Q is true. **[2020 : 1 M]**

4.63 The figure below shows the per-phase Open Circuit Characteristics (measured in V) and Short Circuit Characteristics (measured in A) of a 14 kVA, 400 V, 50 Hz, 4-pole, 3-phase, delta connected alternator, driven at 1500 rpm. The field current, If is measured in A. Readings taken are marked as respective (x, y) coordinates in the figure. Ratio of the unsaturated and saturated synchronous impedances $(Z_{s(unsat)}/Z_{s(sat)})$ of the alternator is closest to

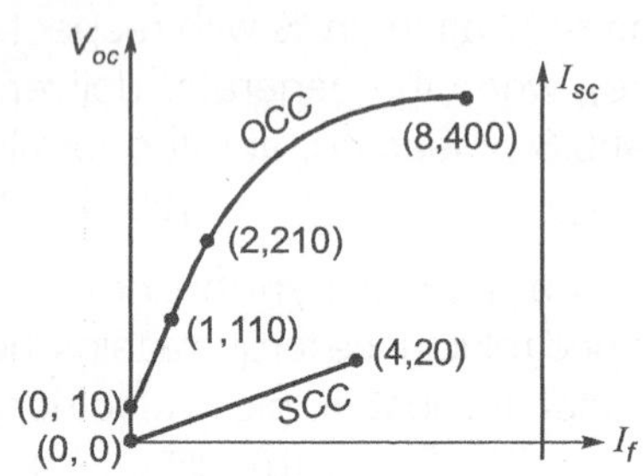

(a) 1.000 (b) 2.100
(c) 2.000 (d) 2.025 **[2020 : 2 M]**

4.64 A cylindrical rotor synchronous generator with constant real power output and constant terminal voltage is supplying 100 A current to a 0.9 lagging power factor load. An ideal reactor is now connected in parallel with the load, as a result of which the total lagging reactive power requirement of the load is twice the previous value while the real power remains unchanged. The armature current is now ______ A (rounded off to 2 decimal places). **[2020 : 2 M]**

4.65 Consider the table given:

Constructional feature	Machine type	Mitigation
(P) Damper bars	(S) Induction motor	(X) Hunting
(Q) Skewed rotor slots	(T) Transformer	(Y) Magnetic locking
(R) Compensating winding	(U) Synchronous machine	(Z) Armature reaction
	(V) DC machine	

The correct combination that relates the constructional feature, machine type and mitigation is

(a) P-U-X, Q-V-Y, R-T-Z
(b) P-V-X, Q-U-Z, R-T-Y
(c) P-U-X, Q-S-Y, R-V-Z
(d) P-T-Y, Q-V-Z, R-S-X **[2021 : 1 M]**

4.66 A star-connected 3-phase, 400 V, 50 kVA, 50 Hz synchronous motor has a synchronous reactance of 1 ohm per phase with negligible armature resistance. The shaft load on the motor is 10 kW while the power factor is 0.8 leading. The loss in the motor is 2 kW. The magnitude of the per phase excitation emf of the motor, in volts, is _____. (round off to nearest integer). **[2022 : 2 M]**

4.67 A three-phase synchronous motor with synchronous impedance of $0.1+j0.3$ per unit per phase has a static stability limit of 2.5 per unit. The corresponding excitation voltage in per unit is _______ (Round off to 2 decimal places). **[2023 : 2 M]**

4.68 A 3-phase. 11 kV. 10 MVA synchronous generator is connected to an inductive load of power factor $(\sqrt{3}/2)$ via a lossless line with a per-phase inductive reactance of 5 Ω. The per-phase synchronous reactance of the generator is 30 Ω with negligible armature resistance. If the generator is producing the rated current at the rated voltage, then the power factor at the terminal of the generator is

(a) 0.63 lagging (b) 0.87 lagging
(c) 0.63 leading (d) 0.87 leading

[2024 : 2 M]

■■■■

Answers Synchronous Machines

4.1 (c)	4.2 (0.844)	4.3 (b)	4.4 (c)	4.5 (Sol.)	4.6 (–10%)	4.7 (a)
4.8 (d)	4.9 (c)	4.10 (b)	4.11 (b)	4.12 (b)	4.13 (b)	4.14 (c)
4.15 (a)	4.16 (87.72%)	4.17 (d)	4.18 (a)	4.19 (b)	4.20 (c)	4.21 (d)
4.22 (a)	4.23 (a)	4.24 (c)	4.25 (a)	4.26 (d)	4.27 (c)	4.28 (b)
4.29 (c)	4.30 (b)	4.31 (c)	4.32 (b)	4.33 (a)	4.34 (c)	4.35 (c)
4.36 (a)	4.37 (b)	4.38 (c)	4.39 (d)	4.40 (d)	4.41 (a)	4.42 (d)
4.43 (b)	4.44 (b)	4.45 (d)	4.46 (15.06)	4.47 (a)	4.48 (d)	4.49 (d)
4.50 (33.49)	4.51 (a)	4.52 (1.606)	4.53 (0.316)	4.54 (838.3)	4.55 (0.745)	4.56 (b)
4.57 (0.8)	4.58 (–14.56%)	4.59 (b)	4.60 (c)	4.61 (245.36)	4.62 (d)	4.63 (*)
4.64 (125.29)	4.65 (c)	4.66 (245)	4.67 (1.58)	4.68 (a)		

Explanations Synchronous Machines

4.2 Sol.

$$m = \frac{36}{3\times 8} = 1.5$$

$$\frac{(\text{slots/ph})}{\text{poles}} = \frac{(36/3)}{8} = \frac{12}{8} = \frac{4\times 3}{4\times 2}$$

Which gives $S_k = 3$

$$K_d = \frac{\sin\left(\frac{120}{2}\right)}{3\sin\left(\frac{120}{2\times 3}\right)} = 0.844$$

4.5 Sol.

A - P, B - S, C - T.

4.6 Sol.

$$E_f\angle\delta = V\angle 0 + I_a\angle\phi\cdot Z_s\angle\phi$$
$$= 200 + 10\angle 90°\cdot 2\angle 90°$$
$$= 200 - 20 = 180\text{ V}$$

$$V.R. = \frac{|E_f| - |V|}{|V|} = \frac{180-200}{200}\times 100 = -10\%$$

4.7 (a)

It will absorb leading reactive power in this case or it will supply lagging reactive power.

4.8 (d)

$$\text{Distribution factor, } K_d = \frac{\sin\left(\frac{m\gamma}{2}\right)}{\left(\frac{m\gamma}{2}\right)}$$

$$= \frac{\sin(\pi/2)}{\pi/2} = \frac{2}{\pi}$$

$$E' = K_d\cdot N\cdot E$$

$$= \frac{2}{\pi}\cdot T\cdot 2 = 1.273T$$

4.9 (c)

They are used to limit fault and transient currents.

4.10 (b)

$$E' \propto K_d K_p Nf$$

$$K_{d_1} = \frac{\sin\left(\frac{\pi/3}{2}\right)}{\left(\frac{\pi/3}{2}\right)} = \frac{3}{\pi}$$

$$K_{d_2} = \frac{\sin(\pi/2)}{\pi/2} = \frac{2}{\pi}$$

Since, the armature conductors are same, the current carrying capability will be same.

kVA rating $\propto E' \propto K_d$

$$\Rightarrow \quad \frac{300}{(\text{kVA})_\pi} = \frac{3/\pi}{2/\pi} = \frac{3}{2}$$

$$\Rightarrow \quad (\text{kVA})_\pi = 200\text{ kVA}$$

4.12 (b)

$$E_f\angle\delta = V\angle 0° + (I_a\angle\phi\cdot Z_s\angle\theta)$$
$$= 2000\angle 0° + 100\angle\cos^{-1}(0.8)\cdot(0.8 + j4.94)$$
$$= 1822.316\angle 14.07°\text{ V}$$

$$V.R. = \frac{|E| - |V|}{|V|}\times 100$$

$$= \frac{1822.316 - 2000}{2000}\times 100 = -8.884\%$$

4.13 (b)

For fifth harmonics, $5\beta = 180° \Rightarrow \beta = 36°$

4.16 Sol.

The losses corresponding to 'without excitation' condition are frictional and windage losses,

$$W_f = 800\text{ W}$$

When armature is short circuited, losses will be corresponding to copper losses,

$$W_{cu} = 2500 - 800 = 1700 \text{ W}$$

When armature is open circuited with excitation, losses are corresponding to iron losses,

$$W_i = 1800 - 800 = 1000 \text{ W}$$

$$\eta = \frac{25,000}{25,000 + 800 + 1700 + 1000} = 87.72\%$$

4.17 (d)

$$E_f\angle\delta = V\angle 0 + I_a\angle\phi \cdot Z_s\angle\theta$$
$$= V\angle 0 + I_a Z_s\angle\theta + \phi \quad \text{...(i)}$$

Squaring both side, we get

$$|E_f|^2 = |V|^2 + |I_a|^2|Z_s|^2 + 2|V||I_a||Z_s|\cos(\theta + \phi)$$

For zero regulation, $|E_f| = |V|$

$$\Rightarrow \cos(\theta + \phi) = -\frac{|I_a||Z_s|}{2|V|} = -\frac{20 \times 10}{2 \times \frac{440}{\sqrt{3}}} = -0.3936$$

$$\theta + \phi = 113.178^\circ$$
$$\phi = 113.178^\circ - 90^\circ = 23.178^\circ$$
$$\cos\phi = 0.9192$$

4.18 (a)

$$N_s = \frac{120f}{P} = \frac{120 \times 50}{8} = 750 \text{ rpm}$$

Speed of 7th harmonic $= \frac{750}{7} = 107.14$ rpm

For 7th harmonic:

$7 \times 120^\circ = 840^\circ - 360^\circ - 360^\circ = 120^\circ$

$\Rightarrow$ Forward direction

4.19 (b)

Armature reaction is trying to deteriorate the main flux. So motor must be operating with leading power factor.

4.20 (c)

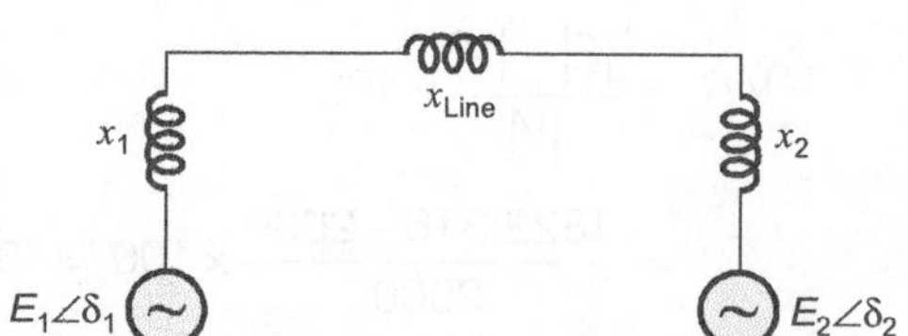

$$x_{eq} = x_1 + x_{line} + x_2 = 1.1 + 0.5 + 1.2 = 2.8 \text{ p.u.}$$
$$E_1 = 2.0 \text{ p.u. and } E_2 = 1.3 \text{ p.u.}$$

Rotor angle difference $= \delta_1 - \delta_2$

Power transferred from generator to motor

$$P = \frac{E_1 E_2}{x_{eq}}\sin(\delta_1 - \delta_2)$$

$$0.5 = \frac{2 \times 1.3}{2.8}\sin(\delta_1 - \delta_2)$$
$$\delta_1 - \delta_2 = 32.58^\circ$$

4.21 (d)

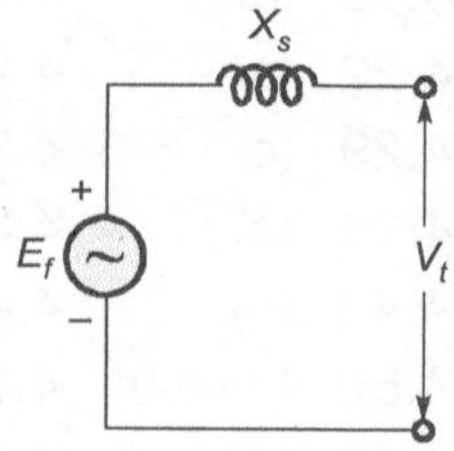

Assuming resistance of the armature to be zero.

In first case, the generator is feeding a partly inductive load. It means that generator is supplying lagging power. The generator supplies a lagging power factor current when it is overexcited which is represented by E_{f_1}.

In second case, a capacitor is connected across the load to completely nullify the inductive current. It means the generator supplies no reactive power and unity power factor current is drawn from the generator. The excitation (E_{f_2}) corresponding to unity power factor is known as normal excitation.

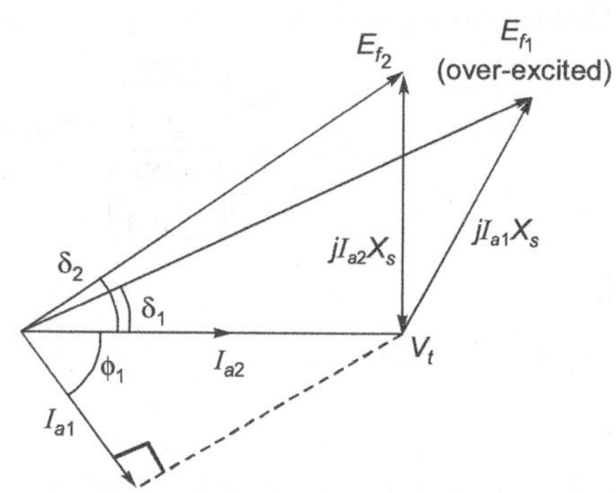

From the phasor diagram $E_{f_1} > E_{f_2}$ and as the field current approximately directly proportional to excitation, the field current has to be reduced. From phasor diagram $E_{f_1}\sin\delta_1 = E_{f_2}\sin\delta_2$ = constant.

$$P_e \text{(power delivered)} = \frac{E_f V_t}{X_s}\sin\delta$$

Fuel input remains unaltered.

4.22 (a)

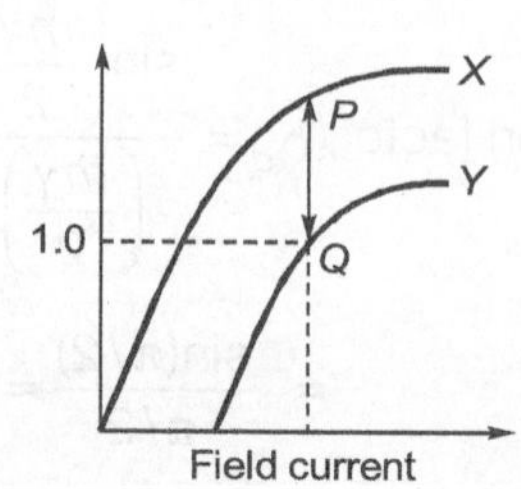

The vertical distance PQ between O.C. (curve X) and ZPF characteristic (curve Y) in above characteristic is synchronous reactance.

4.23 (a)

$$\text{Pole pitch} = \frac{36}{4} = 9 \text{ slots}$$

Coil span = 7 slots, Slot/pole/ph = 3

So that 3 slots in one phase, if it is chorded by 2 slot, then

Out of 3 → 2 have different phase

∴ Out of 36 → 24 have different phase

4.24 (c)

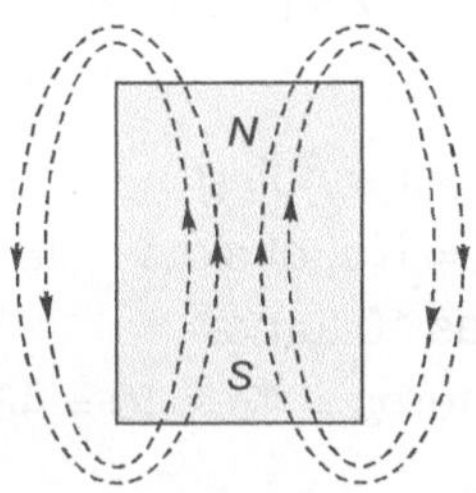

Inside a magnetic material, flux flows from S-pole to N-pole.

Outside a magnetic material flux flow from N-pole to S-pole.

When stator winding is excited it produces flux. If ABC is assumed to N-pole flux will link rotor. As we know, flux flows from N-pole to S-pole externally, abc forms S-pole.

Inside rotor, flux flows from surface *abc* to *cda* so, surface cda forms N-poles.

To identify direction of torque

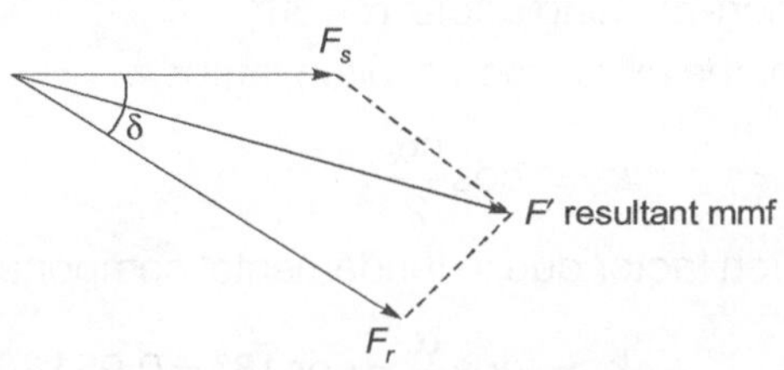

The torque acts in a direction to reduce δ i.e. to align the two fields. So, torque on the rotor acts in counter-clockwise direction to reduce δ.

4.25 (a)

Per phase synchronizing current

$$= \frac{|E_1 - E_2|}{Z_{s_1} + Z_{s_2}}$$

We have, $Z_{s_1} = Z_{s_2}$

E_1 and E_2 must be of phase quantities

$$\therefore \quad I_{sy} = \frac{\left|\frac{3300}{\sqrt{3}} - \frac{3200}{\sqrt{3}}\right|}{2 \times 1.7} = 16.98 \text{ A}$$

4.26 (d)

$$P = \frac{120 f}{N}$$

where, P = Number of poles

N = Rated speed = 250 rpm

f = 50 Hz

$$P = \frac{120 \times 50}{250} = 24$$

4.27 (c)

Power available at shaft = P_{shaft} = 9 kW

Losses = Core loss + Friction and windage loss + Copper loss

Core loss= 0.8 kW

Friction and windage loss = 2 kW

Armature resistance is negligible, so copper loss is neglected.

∴ Losses = 0.8 + 2 = 2.8 kW

Input power = $P = P_{shaft}$ + Losses

= 9 + 2.8 = 11.8 kW

Line current drawn

$$= \frac{P}{\sqrt{3} V_L \cos\phi} = \frac{11.8 \times 10^3}{\sqrt{3} \times 400 \times 0.8} = 21.29 \text{ A}$$

4.28 (b)

3-ϕ power = $\sqrt{3} V_L I_L \cos\phi$

Rated 3-ϕ power, P = 500 MW

Power factor, cos ϕ = 0.85

Rated line voltage, V_L = 21.5 kV

$$\text{Rated line current, } I_L = \frac{P}{\sqrt{3} V_L \cos\phi}$$

$$\Rightarrow \quad I_L = \frac{500 \times 10^6}{\sqrt{3} \times 21.5 \times 10^3 \times 0.85} = 15.79 \text{ kA}$$

4.29 (c)

(i) P_d, P_q = permeance of pole-arc oriented along the d-axis/q-axis.

As, $P_d > P_q$

∴ $X_d > X_q$

(ii) Damper bars not only minimize hunting to a tolerable level but also help the synchronous motor self-start.

(iii) SCR is defined as the ratio of the field current required to produce rated voltage on open-circuit to the field current required to produce rated armature current with the armature terminals shorted while the machine is mechanically run at synchronous speed.

4.30 (b)

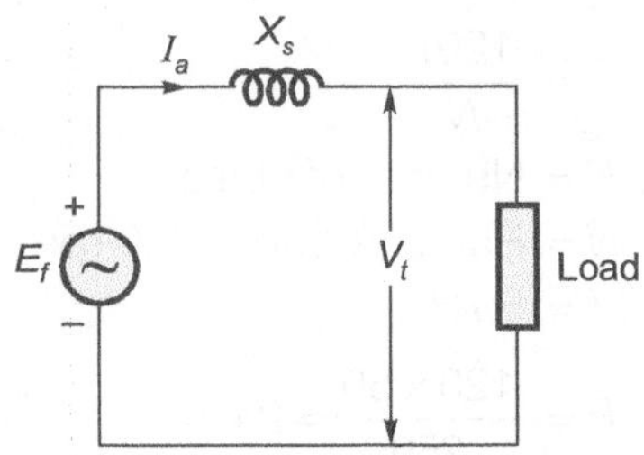

Rated output power, $S = 1000$ kVA

Rated line to line voltage, $V_{l-l} = 6.6$ kV

Rated current,

$$I_a = \frac{S}{\sqrt{3}\,V_{l-l}} = \frac{1000}{\sqrt{3}\times 6.6} = 87.477\,\text{A}$$

As generator is star connected.

Phase to neutral voltage,

$$V_t = \frac{V_{l-l}}{\sqrt{3}} = \frac{6.6}{\sqrt{3}} = 3.81\text{ kV}$$

Taking V_t as the reference,

$$V_t = 3.81\angle 0^\circ\text{ kV}$$

$$I_a = 87.477\angle 0^\circ\text{ A} \quad \text{[unity power factor]}$$

$$E_f = V_t + I_a X_s$$

$$= 3.81\angle 0^\circ + 87.477\angle 0^\circ \times (j20)\times 10^{-3}$$

$$= 4.19\angle 24.6^\circ\text{ kV}$$

Induced emf (line to line)

$$= \sqrt{3}E_f = \sqrt{3}\times 4.19$$

$$= 7.2\text{ kV}$$

4.31 (c)

From previous solution, $\delta = 24.6^\circ$

4.32 (b)

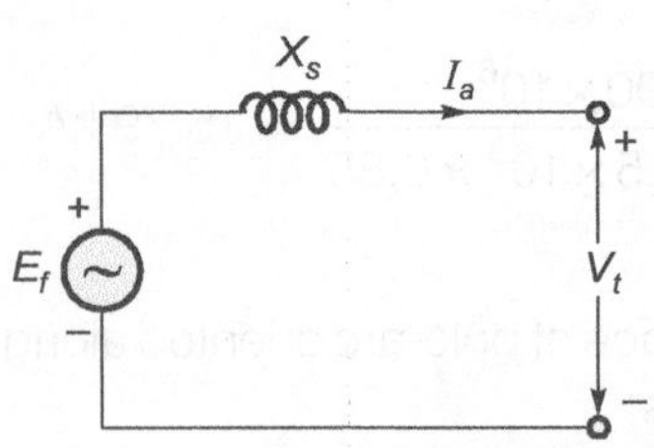

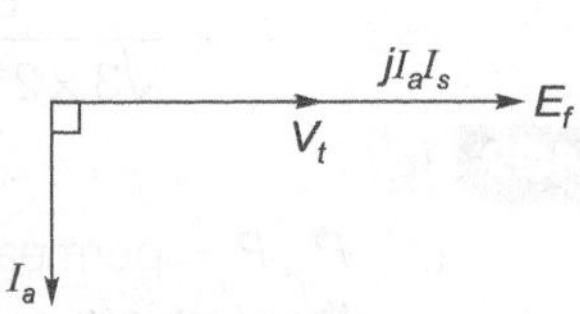

$E_f > V_t$, the generator is overexcited. Therefore armature-reaction is demagnetizing in nature, i.e., ϕ_{ar} opposes ϕ_f.

4.33 (a)

$$I_{fl} = \frac{5\times 10^3}{\sqrt{3}\times 400\times 1} = 7.21\text{ A}$$

For synchronous motor,

$$E^2 = (V\cos\theta - I_aR_a)^2 + (V\sin\theta - I_aX_s)^2$$

$$\therefore \quad E = \sqrt{(400/\sqrt{3})^2 + (10\times 3.6)^2}$$

$$= 233.729\text{ V}$$

If there is 1% increase in E_1, then E_1 becomes,

$$E_2 = 233.729\times 1.01 = 236\text{ V}$$

$$\therefore \quad I_aX = \sqrt{E_2^2 - V^2}$$

$$= \sqrt{236^2 - (400/\sqrt{3})^2} = 48.93$$

$$\therefore \quad I_a = 4.893\text{ A}$$

$$\therefore \quad \%\text{ load} = \frac{4.893}{7.21} = 67.86\%$$

4.34 (c)

In double layer winding

No. of coils = No. of slots = 48

Each coil has 10 turns

Total no. of turns = 48 × 10 = 480

$$T_{Ph} = \text{No. of turns/phase} = \frac{480}{3} = 160$$

$$\text{Slots/pole/phase, } m = \frac{48}{4\times 3} = 4$$

$$\gamma = \text{Slot angle} = \frac{180\times\text{poles}}{\text{slots}} = \frac{180\times 4}{48} = 15^\circ$$

Distribution factor,

$$k_d = \frac{\sin\frac{m\gamma}{2}}{m\sin\frac{\gamma}{2}} = \frac{\sin\left(\frac{4\times 15}{2}\right)}{4\sin\left(\frac{15}{2}\right)} = 0.957$$

Short-pitching angle, $\alpha = 36^\circ$

For the nth harmonic, pitch factor,

$$k_p = \cos\frac{n\alpha}{2}$$

pitch factor due to fundamental component,

$$k_p = \cos\frac{\alpha}{2} = \cos 18^\circ = 0.951$$

Winding factor,

$$k_w = k_d k_p = 0.957\times 0.951 = 0.910$$

Induced emf per phase,

$$E_{Ph} = 4.44\,k_w f\,T_{ph}\,\phi$$

$$= 4.44\times 0.910\times 50\times 160\times 0.025 \approx 808\text{ V}$$

$$\Rightarrow E_{line} = \sqrt{3}E_{Ph} = 1400\text{ V}$$

4.36 (a)

Pitch factor due to 5th harmonic component,

$$(k_p)_5 = \cos\left(\frac{5\times 36}{2}\right) = \cos 90 = 0$$

[as pitch factor is zero, induced emf due to 5th harmonic component is zero]

4.37 (b)

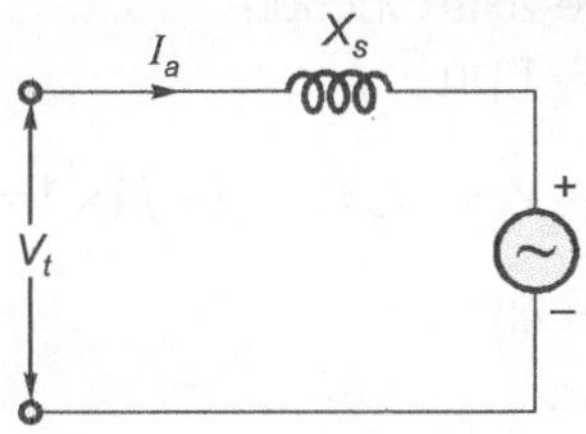

Neglecting armature resistance R_a,

$$V_t = E_f + jI_a X_s$$

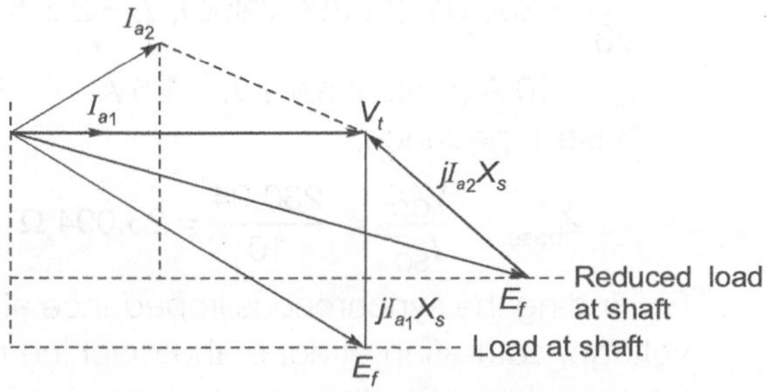

It can be seen from the phasor diagram that, when load is reduced at shaft keeping field current constant the new power factor becomes leading.

4.38 (c)

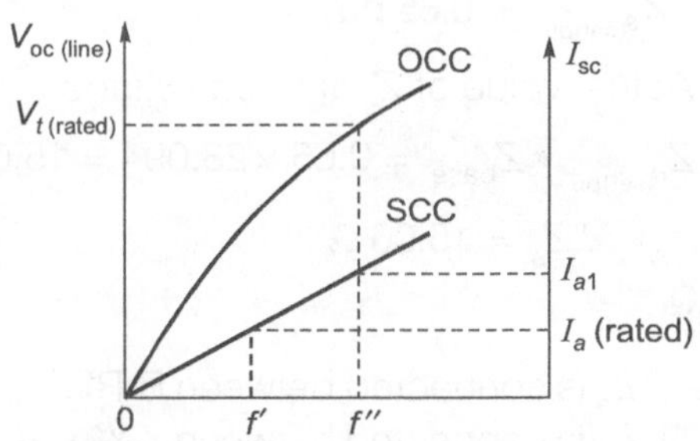

Rated open circuit voltage is obtained at

$$of' = 15 \text{ A}$$

$$\text{Rated armature current} = \frac{100 \times 1000}{\sqrt{3} \times 415} = 139.12 \text{ A}$$

Short-circuit characteristics is linear.

Short-circuit armature current at field current 15 A

$$I_{a_1} = I_a \times \frac{of''}{of'} = 139.12 \times \frac{15}{10} = 208.68 \text{ A}$$

Phase voltage,

$$V_p = \frac{V_{oc\,(line)}}{\sqrt{3}} = \frac{415}{\sqrt{3}}$$

Saturated synchronous reactance

$$= X_s = \frac{V_p}{I_{a_1}} = \frac{415/\sqrt{3}}{208.68} = 1.148\ \Omega$$

Base impedance,

$$Z_B = \frac{415^2}{100 \times 10^3} = 1.722\ \Omega$$

Per unit saturated synchronous reactance

$$= \frac{X_s}{Z_B} = \frac{1.148}{1.722} = 0.666 \text{ pu}$$

4.39 (d)

Distribution factor in distributed winding

$$k_d\,(n^{th}\text{ harmonic}) = \frac{\sin\left(\frac{mn\gamma}{2}\right)}{m\sin\left(\frac{n\gamma}{2}\right)}$$

k_d is always less than 1, so emf reduces.

If $mn\gamma/2$ is chosen in such a way that $\sin mn\gamma/2 = 0$, n^{th} harmonic gets eliminated.

Similarly,

$$\text{Pitch factor} = k_p(n^{th}\text{ harmonic}) = \cos\frac{n\alpha}{2}$$

k_p is always less than 1, so emf reduces.

If $\frac{n\alpha}{2}$ is chosen in such a way that

$\cos\frac{n\alpha}{2} = 0$, n^{th} harmonic gets eliminated.

4.40 (d)

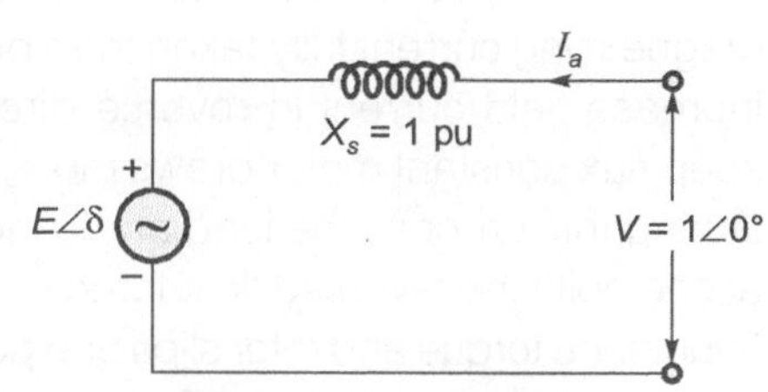

Motor draws 0.6 pu current at upf,

$$I_a = 0.6\angle 0°$$

$$E\angle\delta = V_t - jI_aX_s = 1\angle 0° - j\,0.6 \times 1$$

$$\approx 1.17\angle{-30.96°}$$

So, $|E| = 1.17$ pu and $\delta = 30.96°$ lag

4.41 (a)

Let power factor = $\cos\phi$ lagging

$$I_a = 0.6 \times 1.2 = 0.72 \text{ pu}$$

$$E\angle\delta = V_t - jI_a\angle{-\phi} \times X_s$$

$$E(\cos\delta + j\sin\delta)$$

$$= 1\angle 0° - jI_aX_s(\cos\phi - j\sin\phi)$$

$$= 1 - j\,0.72 \times 1(\cos\phi - j\sin\phi)$$

$$= (1 - 0.72\sin\phi) - j\,0.72\cos\phi$$

$$E^2 = (1 - 0.72\sin\phi)^2 + (0.72\cos\phi)^2$$

$$= 1 + 0.72^2\sin^2\phi - 1.44\sin\phi + 0.72^2\cos^2\phi$$

$$1.17^2 = 1 + 0.72^2 - 1.44\sin\phi$$

$$\sin\phi = 0.1038 \Rightarrow \phi = 5.96°$$

$$\cos\phi \approx 0.995 \text{ lagging}$$

4.42 (d)

If R_a is neglected,

$$\vec{E}_f = \vec{V}_t + j\vec{I}_a X_s$$

For a given field current under short-circuit condition,

$$I_a = I_{sc};\ V_t = 0$$

$$X_s = \frac{E_f}{I_{sc}}$$

But, $E_f = V_{oc}$

(open-circuit voltage i.e. $I_a = 0$ with the same field current).

Then with the linearity assumption

$$X_s = \left.\frac{V_{oc}}{I_{sc}}\right|_{I_f = \text{constant}} = \frac{2000}{400} = 5\ \Omega$$

Internal voltage drop

$$I_a X_s = 200 \times 5 = 1000\text{ V}$$

4.43 (b)

- As field current reduces the flux will start reducing, to keep this flux constant the armature draws high current from bus.
- At zero field current the motor acts as synchronous reluctance motor in this case the magnetizing current fully taken from bus. If we increase field current in reverse direction to keep flux constant motor draws more current. During this process the load angle increases at one point the reversed field force dominates reluctance torque and rotor slips one pole pitch and align to opposite pole. The instant aligns to opposite pole the flux will be very high to reduce this flux current drops steeply to synchronous motor value. Here torque is reluctance + synchronous motor torque.

4.44 (b)

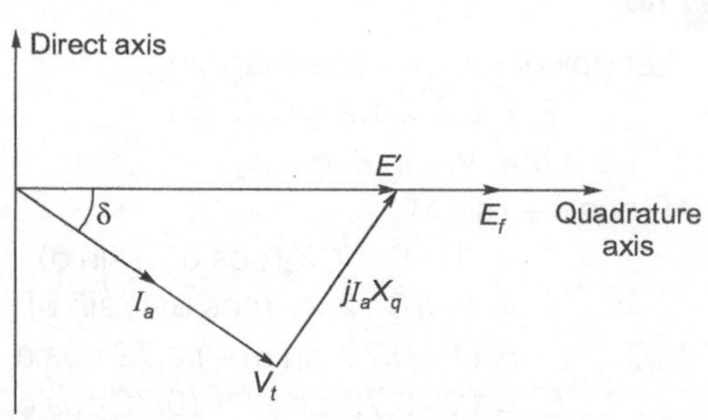

Power angle is the angle between E_f and V_t but as the E' and E_f are in phase, angle between E' and V_t = power-angle = δ

Rated power is delivered at upf, so I_a and V_t are in phase,

$$I_a = 1\angle 0^\circ \text{ pu}$$

Quadrature-axis function,

$$X_q = 1 \text{ pu}$$

$$E' = V_t + jI_aX_q = 1 + j1\times 1 = \frac{1}{\sqrt{2}}\angle 45^\circ$$

$$\delta = 45^\circ$$

4.46 Sol.

Given, for $V_{OC} = 400$ V(rms)

$$= \frac{400}{\sqrt{3}} = 230.94 \text{ (Phase value)},\ I_f = 2.3\text{ A}$$

for $I_{sc} = 10$ A (rms, phase), $I_f = 1.5$ A

∴ Base impedance,

$$Z_{\text{base}} = \frac{V_{OC}}{I_{SC}} = \frac{230.94}{10} = 23.094\ \Omega$$

For finding the synchronous impedance at rated voltage, saturation effect of the machine has to be considered (i.e. the adjusted value of Z_s) short circuit ratio,

$$\text{SCR} = \frac{I_f \text{ for } V_{OC}}{I_f \text{ for } I_{SC}} = \frac{1}{Z_{S(\text{sat})\text{pu}}} = \frac{2.3}{1.5}$$

or, $Z_{S(\text{sat})\text{pu}} = 0.65$ pu

∴ Actual value of Z_s at rated voltage

$$= Z_{S(\text{sat})\text{pu}} * Z_{\text{base}} = 0.65 \times 23.094 = 15.06\ \Omega$$

$$\therefore \quad Z_s = 15.06\ \Omega$$

4.47 (a)

As L_a is connected between R-R′

L_b is connected between Y-Y′

L_c is connected between B-B′

Therefore, the given method of synchronization is called "synchronization by dark lamp method."

In this method of synchronisation, correct phase sequence of the incoming generator is obtained when all the three lamps become dark together and bright together. Since lamps are becoming dark in the sequence L_a-L_b-L_c therefore, the phase sequence of incoming generator is opposite to the infinite bus.

4.49 (d)

Hunting is a phenomenon occurring in case of synchronous generator and motor where the machine oscillates about the equilibrium point. To reduce this phenomenon, damper bars are used which are embedded in the rotor pole faces of the synchronous machines. Thus, these damper bars causes additional copper loss in the rotor for damping the undesirable phenomenon of hunting.

4.50 Sol.

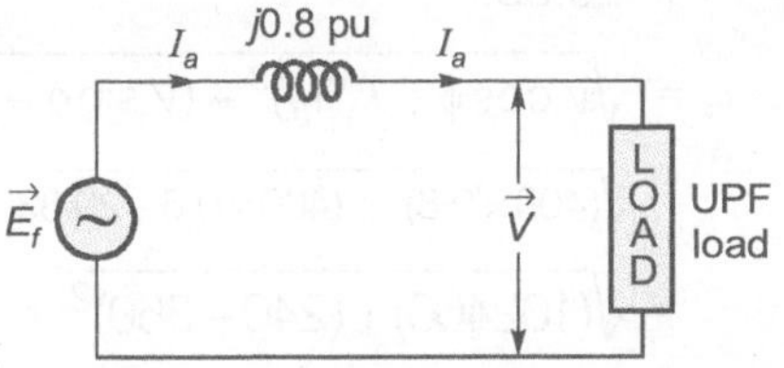

Given, $X_s = 0.8$ pu, $\cos\phi = 1$,

$V = 1.1\angle 0°$ pu (reference voltage)

$P = 1$ pu

Let, $E_f\angle\delta$ be the excitation voltage

Now, power supplied = 1 pu

i.e., $VI_a\cos\phi = 1$ or $I_a = \dfrac{1}{V\cos\phi} = \dfrac{1}{1.1\times 1}$

$= 0.91$ pu

Now, $\vec{E}_f\angle\delta = \vec{V} + j\vec{I}_a X_s$

$= 1.1\angle 0° + j(0.91)\times(0.8)$

$= 1.1 + j0.728 = 1.319\angle 33.49°$ pu

$\therefore$ $\boxed{E_f\angle\delta = 1.319\angle 33.49° \text{ pu}}$

Hence, the angle of the voltage behind the synchronous reactance = Power angle (δ)

or, $\delta = 33.49$

4.51 (a)

Synchronous generator working at a lagging power factor, will supply active power and lagging reactive power, so this is source.

Synchronous motor working at an overexcited condition i.e. leading p.f. operation so it will take active power and leading reactive power in other words taking active power and supplying lagging reactive power.

Induction generator generates active power but as there is no dc excitation in rotor so it takes lagging reactive power.

4.52 Sol.

Synchronous motor at leading p.f.

$X_d = 0.8, \quad \phi = 36.86$

$X_q = 0.6, \quad R_q = 0$

$$\tan\psi = \frac{V\sin\phi + I_a\cdot X_a}{V\cos\phi} = 56.30$$

For synchronous motor at leading p.f.

$\psi = \phi + \delta$

$\Rightarrow$ $\delta = 19.70°$

Now, $E = V\cos\delta + I_d X_d$

$I_d = I_a \sin f = 0.831$

$E = (1)\cos(19.7) + (0.831)(0.8) = 1.606$

4.53 Sol.

$P_1 = 500\times 0.8 = 400$ kW

$Q_1 = 500\times 0.6 = 300$ kVAR

The power factor is to be raised to unity

The motor has to supply 300 kVAR

The motor rating is 100 kW, 300 kVAR

$$\phi_m = \tan^{-1}\left(\frac{Q}{P}\right)$$

$$\phi_m = \tan^{-1}\left(\frac{300}{100}\right) = 71.56$$

Power factor of motor

$= \cos\phi_m = \cos 71.56 = 0.316$

4.54 Sol.

For synchronous motor;

$$\bar{V}_t = \bar{E}_f + j\bar{I}_a X_s$$

As current is drawn at unity power factors.

Therefore,

$$E_f\cos\delta = V_t$$

or $$E_f = \frac{V_t}{\cos\delta} = \frac{6.6\text{ kV}}{\sqrt{3}}\times\frac{1}{\cos 30°}$$

$$E_f = \frac{6.6\times 10^3}{\sqrt{3}\times\sqrt{3}}\times 2\text{ V}$$

$$V_t = \frac{6.6\times 10^3}{\sqrt{3}}\text{ V}$$

$\therefore$ $$P_e = \frac{3V_{ph}.E_{ph}}{X_s}\sin\delta$$

$$= \frac{3\times\dfrac{6.6\times 10^3}{\sqrt{3}}\times\dfrac{6.6\times 10^3}{\sqrt{3}\times\sqrt{3}}\times 2}{30}\times\sin 30°$$

$= 838.3$ kW

4.55 Sol.

Synchronous machine A:

11 kV, 50 Hz, Y connected, $X_s = 1\ \Omega$

Synchronous machine B:

11 kV, 50 Hz, Y connected, $X_s = 3\ \Omega$

Given both supplying only reactive power 50 MVAR to 11 kV grid equally

$\therefore$ $Q_A = \sqrt{3}\,VI_A\sin\phi = 25$ MVAR

$Q_B = \sqrt{3}\,VI_B\sin\phi = 25$ MVAR

$\therefore$ $$I_A = \frac{25\times 10^6}{\sqrt{3}\times 11\times 10^3} = 1312.2\text{ A}$$

$\therefore$ $I_B = 1312.2$ A

$E_A = V - jI_A X_s$

$$= \frac{11\times10^3}{\sqrt{3}} - j1312.2\angle 90\times 1$$

$$E_B = V - jI_B X_s = \frac{11\times10^3}{\sqrt{3}} - j1312.2\angle 90\times 3$$

Given magnetic circuit to be linear,

$$\therefore \quad E \propto I_f$$

$$\therefore \quad \frac{I_{f_A}}{I_{f_B}} = \frac{E_A}{E_B}$$

$$E_A = 7663.05 \text{ V}$$

$$E_B = 10287.45 \text{ V}$$

$\therefore$ Ratio of excitation current of machine A to machine B

$$\frac{E_A}{E_B} = 0.745$$

4.56 (b)

Synchronous motor-leading p.f. over excited.

$$\therefore \quad E_f > V$$

4.57 Sol.

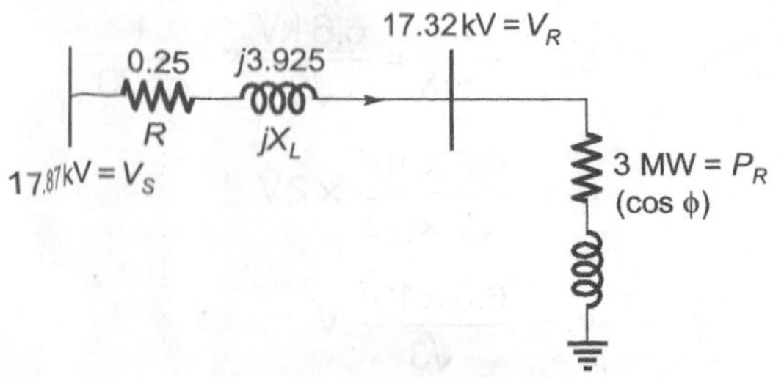

$$V_S - V_R = I(R\cos\phi + X_L\sin\phi)$$

$$I = \frac{P_R}{V_R\cos\phi}$$

$$V_S - V_R = \frac{P_R}{V_R\cos\phi}(R\cos\phi + X_L\sin\phi)$$

$$\tan\phi = \left(\frac{V_S V_R - V_R^2 - RP_R}{X_L P_R}\right)$$

$$\phi = \tan^{-1}\left[\frac{V_S V_R - V_R^2 - RP_R}{X_L P_R}\right] \Rightarrow \text{Substitute all values}$$

$$\text{Power factor} = \cos\phi = 0.8 \text{ lag}$$

4.58 Sol.

25 kVA, 400 V, Δ-synchronous generator

$$\therefore I_{\text{rated}} = \frac{25000}{\sqrt{3}\times 400} = 36.085 \text{ A}$$

$$I_{\text{phase}} = \frac{36.085}{\sqrt{3}} = 20.83 \text{ A}$$

$$X_s = \left.\frac{V_{OC}}{I_{a_{\text{rated}}}}\right|_{I_f \text{ same}}$$

$$\therefore \quad X_s = \frac{360}{20.83} = 17.28\ \Omega/\text{phase}$$

$$E = \sqrt{(V\cos\phi + I_a R_a)^2 + (V\sin\phi - I_a X_s)^2}$$

$$= \sqrt{(400\times0.8)^2 + (400\times0.6 - 20.83(17.28))^2}$$

$$= \sqrt{(102400) + (240-360)^2} = 341.76 \text{ V}$$

$$\therefore \%\text{V.R.} = \frac{E-V}{V}\times 100$$

$$= \frac{341.76-400}{400}\times 100 = -14.56\%$$

4.59 (b)

Salient pole synchronous motor power and torque relations per phase:

$$P = \frac{E_b V}{X_t}\cdot\sin\delta + \frac{V^2}{2}\left(\frac{X_d - X_q}{X_d X_q}\right)\sin(2\delta) \text{ Watts}$$

$$T = \frac{60}{2\pi N_S}\left[\frac{E_b V}{X_d}\sin\delta + \frac{V^2}{2}\left(\frac{X_d - X_q}{X_d X_q}\right)\sin 2\delta\right] \text{N-m}$$

The second term is reluctance power or reluctance torque, which is directly proportional to sin2δ.

$\therefore$ Reluctance torque will be maximum.

When, $\sin 2\delta = 90° \Rightarrow \delta = 45°$

4.60 (c)

$$P = VI\cos\phi$$

For same voltage and load,

$$I\cos\phi = \text{constant}$$

$$I_1\cos\phi_1 = I_2\cos\phi_2$$

at unity power factor,

$$I_1 = 200 \text{ A}$$

$$200\times 1 = I_2\times 0.5 \Rightarrow I_2 = 400 \text{ A}$$

4.61 Sol.

For synchronous motor,

$$E_g = V_t - IZ$$

$$V_t = \frac{220}{\sqrt{3}} \text{ V (phase)}$$

$$Z = (0.25 + j2.5)\ \Omega$$

$$I = 10\angle\cos^{-1}0.8 \text{ A}$$

$$E_g = \frac{220}{\sqrt{3}} - (0.25 + j2.5)\times 10\angle\cos^{-1}0.8$$

$$E_g = 141.658\angle{-8.728°} \text{ V (phase)} = 245.36 \text{ V (line)}$$

4.63 (*)

At 400 V, $\quad I_f = 8$ A

So, air gap line equation will be like,

$$y = mx + C = 100x + 10 \quad \text{...(i)}$$

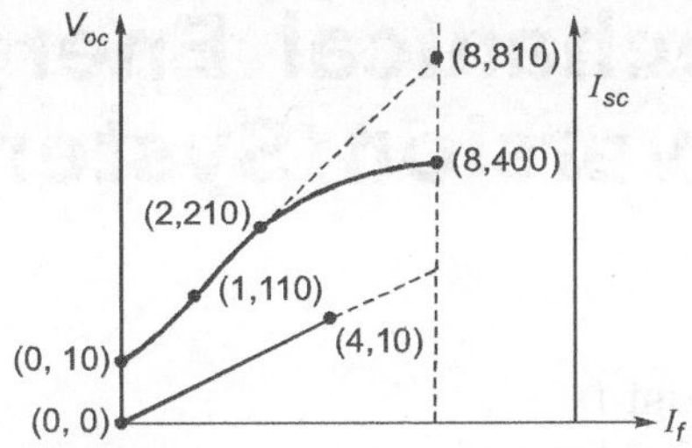

At $I_f = 8$ A,

Unsaturated voltage = 100 × 8 + 10 = 810

$$Z_{sunsat} = \frac{810}{20} = \frac{81}{2} \quad \text{...(ii)}$$

and $$Z_{saturated} = \frac{400}{20} = \frac{40}{2} \quad \text{...(iii)}$$

From equation (ii) and (iii),

$$\frac{Z_{unsaturated}}{Z_{saturated}} = \frac{81/2}{40/2} = 2.025$$

Note: As, (0,10), (1, 110) and (2, 210) are colinear so, air gap line is an shown in the figure.

∗ Data given is not realistic because at $V_{(residual)}$, $I_{(SC)}$ should have some non-zero value. Gate has given answer as (b) option i.e. (2.1).

4.64 Sol.

At $P_{constant}$,

$$I_{a1}\cos\phi_1 = I_{a2}\cos\phi_2$$

$$\cos\phi_1 = 0.9$$

$$\tan\phi_1 = 0.484 = \frac{Q}{P}$$

$$\Rightarrow \quad \frac{2Q}{P} = 0.9686 = \tan\phi_2$$

$$\cos\phi_2 = 0.7182$$

$$\therefore \quad 100 \times 0.9 = I_{a2} \times 0.7182 \Rightarrow I_{a2} = 125.29 \text{ A}$$

4.66 (245)

Y-connected, 3 phase, 400V, 50 kVA, Synch motor, X_s = 1 Ω/ph, Shaft load P_{sh} = 10 kW, p.f. = 0.8 loading losses = 2 kW, R_a: Negligible

Find excitation EMF (E_b):

$$E_b = \sqrt{(V\cos\phi - I_aR_a)^2 + (V\sin\phi \pm I_aX_s)^2}$$

(+ for lead and – for lagg)

To find I_a: Current drawn, we need to find Input.

Input = Output + Losses + P_{sh} + Losses

= 10 + 2 = 12 kW

$$\sqrt{3}V_LI_L\cos\phi = 12 \text{ kW}$$

$$I_L = \frac{12000}{\sqrt{3}\times 400\times 0.8} = 21.65A = I_a$$

$$E_b = \sqrt{\left(\frac{400}{\sqrt{3}}\times 0.8\right)^2 + \left[\frac{400}{\sqrt{3}}\times 0.6 + 21.65(1)\right]^2}$$

$$E_b = \sqrt{(V\cos\phi - I_aR_a)^2 + (V\sin\phi + I_aX_s)^2}$$

$$= \sqrt{34135.336 + (138.568 + 21.65)^2}$$

$$E_b = 244.55 \text{ V per phase}$$

4.67 (1.58)

Given : $Z_s = (0.1 + j0.3)$

$$= \sqrt{0.1^2 + 0.3^2}\angle\tan^{-1}\left(\frac{0.3}{0.1}\right)$$

$$Z = 0.3162\angle 71.56° \text{ pu}$$

The developed power will be maximum, if δ = θ

$$P_m = \frac{E_fV}{Z_s} - \frac{E_f^2}{Z_s}\cos\theta$$

$$2.5 = \frac{E_f(1)}{\frac{1}{\sqrt{10}}} - \frac{E_f^2}{\frac{1}{\sqrt{10}}}\cos\left(\tan^{-1}\frac{0.3}{0.1}\right)$$

$$E_f^2 - \sqrt{10}E_f + 2.5 = 0 \Rightarrow E_f = 1.5811 \text{ pu}$$

4.68 (a)

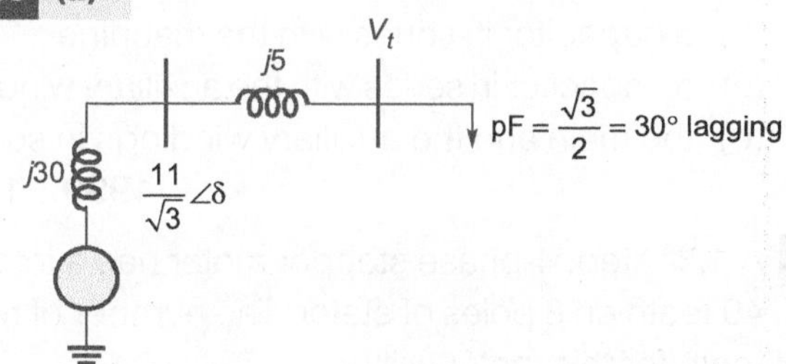

$$\text{pF} = \frac{\sqrt{3}}{2} = 30° \text{ lagging}, \; I_{rated} = \frac{10\times 10^6}{\sqrt{3}\times 11\times 10^3}$$

= 524.86 A

$$E = \sqrt{(V\cos 30 + IR)^2 + (V\sin 30° + IX)^2}$$

$$\frac{11K}{\sqrt{3}} = \sqrt{\left(V\cdot\frac{\sqrt{3}}{2}\right)^2 + \left(\frac{V}{2} + 524.86(5)\right)^2}$$

$$\left(\frac{11K}{\sqrt{3}}\right)^2 = V^2\cdot\frac{3}{4} + \left(\frac{V}{2} + 2624.319\right)^2$$

$$\left(\frac{11K}{\sqrt{3}}\right)^2 = \frac{3}{4}V^2 + \frac{V^2}{4} + 2624.319V + 2624.319^2$$

$$V_t = 4618.105 \text{ V}$$

$$E\angle\delta = V\angle 0° + I_{rated} \times Z$$

$$= 4618.105 + 524.86\angle{-30°} \times (j5)$$

$$E\angle\delta = 6350.84\angle 20.97$$

$$\phi_{net} = 20.97 - (-30)$$

$$\cos\phi_{net} = 0.629 \text{ lagging}$$

■■■■

5 Single Phase Induction Motors, Special Purpose Machines and Electromechanical Energy Conversion Systems

5.1 Which type of motor is most suitable for computer printer drive?
(a) Reluctance motor
(b) Hysteresis motor
(c) Shaded pole motor
(d) Stepper motor [1996 : 1 M]

5.2 In case of a split phase motor, the phase shift between currents in the two windings is around
(a) 30° (b) 70°
(c) 90° (d) 120° [1996 : 1 M]

5.3 Starting torque can be obtained in the case of a single-phase induction motor with identical main and auxiliary windings by connecting
(a) a capacitor across the mains
(b) a capacitor in series with the machine
(c) a capacitor in series with the auxiliary winding
(d) the main and the auxiliary windings in series [1999 : 1 M]

5.4 A 1.8° step, 4-phase stepper motor has a total of 40 teeth on 8 poles of stator. The number of rotor teeth for this motor will be
(a) 40 (b) 50
(c) 100 (d) 80 [2000 : 1 M]

5.5 Ratio of the rotor reactance X to the rotor resistance R for a two-phase servomotor
(a) is equal to that of a normal induction motor.
(b) is less than that of a normal induction motor.
(c) is greater than that of a normal induction motor.
(d) may be less or greater than that of a normal induction motor. [2000 : 1 M]

5.6 The hysteresis loop of a magnetic material has an area of 5 cm with the scales given as 1 cm = 2 AT and 1 cm = 50 mWb. At 50 Hz the total hysteresis loss is
(a) 15 W (b) 20 W
(c) 25 W (d) 50 W [2001 : 2 M]

5.7 **List-I** lists different applications and **List-II** lists the motors for these applications. Match the application with the most suitable motor and choose the right combination among the choices given these after

List-I
A. Food mixer
B. Cassette tape recorder
C. Domestic water pump
D. Escalator

List-II
1. Permanent magnet dc motor
2. Single-phase induction motor
3. Universal motor
4. Three-phase induction motor
5. DC series motor
6. Stepper motor

Codes:

	A	B	C	D
(a)	3	6	4	5
(b)	1	3	2	4
(c)	3	1	2	4
(d)	3	2	1	4

[2003 : 1 M]

5.8 A single phase induction motor with only the main winding excited would exhibit the following response at synchronous speed
(a) Rotor current is zero.
(b) Rotor current is non-zero and is at slip frequency.
(c) Forward and backward rotating fields are equal.
(d) Forward rotating field is more than the backward rotating field. [2003 : 2 M]

5.9 For a given stepper motor, the following torque has the highest numerical value.
(a) Detent torque (b) Pull-in torque
(c) Pull-out torque (d) Holding torque [2004 : 1 M]

5.10 The following motor definitely has a permanent magnet rotor
(a) DC commutator motor
(b) Brushless dc motor
(c) Stepper motor
(d) Reluctance motor [2004 : 1 M]

5.11 The type of single-phase induction motor having the highest power factor at full load is
(a) shaded pole type
(b) split-phase type
(c) capacitor-start type
(d) capacitor-run type [2004 : 1 M]

5.12 For a linear electromagnetic circuit, the following statement is true.
(a) Field energy is equal to the co-energy
(b) Field energy is greater than the co-energy
(c) Field energy is lesser than the co-energy
(d) Co-energy is zero [2004 : 1 M]

5.13 A rotating electrical machine having its self-inductances of both the stator and the rotor windings, independent of the rotor position will be definitely not develop ____ torque.
(a) starting (b) synchronizing
(c) hysteresis (d) reluctance [2004 : 2 M]

5.14 A single-phase, 230 V, 50 Hz, 4 pole, capacitor-start induction motor has the following stand-still impedances:
Main winding, $Z_m = 6.0 + j4.0\ \Omega$
Auxiliary winding, $Z_a = 8.0 + j6.0\ \Omega$
The value of the starting capacitor required to produce 90° phase difference between the currents in the main and auxiliary windings will be
(a) 176.84 μF (b) 187.24 μF
(c) 265.26 μF (d) 280.86 μF [2004 : 2 M]

5.15 For a 1.8°, 2-phase bipolar stepper motor, the stepping rate is 100 steps/second. The rotational speed of the motor in rpm is
(a) 15 (b) 30
(c) 60 (d) 90 [2004 : 2 M]

5.16 In a single phase induction motor driving a fan load, the reason for having a high resistance rotor is to achieve
(a) low staring torque
(b) quick acceleration
(c) high efficiency
(d) reduced size [2005 : 2 M]

5.17 For a single phase capacitor start induction motor which of the following statements is valid?
(a) The capacitor is used for power factor improvement.
(b) The direction of rotation can be changed by reversing the main winding terminals.
(c) The direction of rotation cannot be changed.
(d) The direction of rotation can be changed by interchanging the supply terminals.
[2006 : 1 M]

5.18 A 3-ϕ, 3-stack, variable reluctance step motor has 20 poles on each rotor and stator stack. The step angle of this step motor is
(a) 3° (b) 6°
(c) 9° (d) 18° [2007 : 2 M]

5.19 In a stepper motor, the detent torque means
(a) minimum of the static torque with the phase winding excited.
(b) maximum of the static torque with the phase winding excited.
(c) minimum of the static torque with the phase winding unexcited.
(d) maximum of the static torque with the phase winding unexcited. [2008 : 1 M]

5.20 A 230 V, 50 Hz, 4 pole, single -phase induction motor is rotating in the clockwise (forward) direction at a speed of 1425 rpm. If the rotor resistance at standstill is 7.8 Ω, then the effective rotor resistance in the backward branch of the equivalent circuit will be
(a) 2 Ω (b) 4 Ω
(c) 78 Ω (d) 156 Ω [2008 : 2 M]

5.21 A 200 V, 50 Hz, single-phase induction motor has the following connection diagram and winding orientations shown. MM′ is the axis of the main stator winding (M_1M_2) and AA' is that of the auxiliary winding (A_1A_2). Directions of the winding axis indicate direction of flux when currents in the windings are in the directions shown. Parameters of each winding are indicated. When switch S is closed, the motor

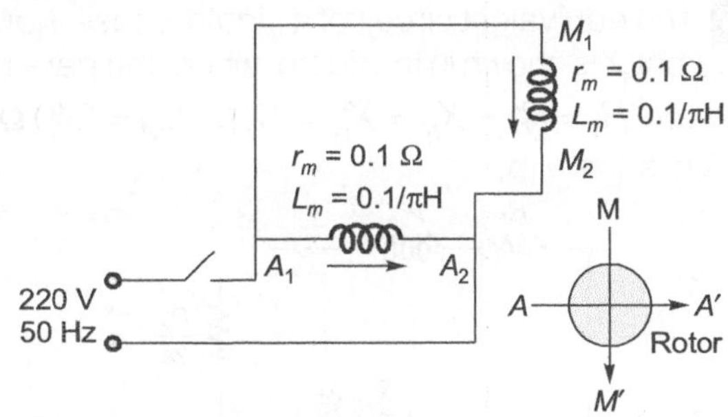

(a) rotates clockwise
(b) rotates anticlockwise
(c) does not rotate
(d) rotates momentarily and comes to a halt
[2009 : 2 M]

5.22 The direction of rotation of a single-phase capacitor run induction motor is reversed by
(a) interchanging the terminals of the AC supply.
(b) interchanging the terminals of the capacitor.
(c) interchanging the terminals of the auxiliary winding.
(d) interchanging the terminals of both the windings. [2016 : 1 M, Set-2]

5.23 The flux linkage (λ) and current (i) relation for an electromagnetic system is $\lambda = (\sqrt{i})/g$. When i = 2 A and g(air-gap length) = 10 cm, the magnitude of mechanical force on the moving part, in N, is ____. [2016 : 1 M, Set-2]

5.24 A 375 W, 230 V, 50 Hz capacitor start single-phase induction motor has the following constants for the main the auxiliary windings (at starting):

Z_m = (12.50 + *j*15.75) Ω (main winding). Z_a = (24.50 + *j*12.75) Ω (auxiliary winding). Neglecting the magnetizing branch, the value of the capacitance (in μF) to be added in series with the auxiliary winding to obtain maximum torque at starting is _____ . [2017 : 2 M, Set-1]

5.25 A transformer with toroidal core of permeability μ is shown in the figure. Assuming uniform flux density across the circular core cross-section of radius *r* << *R*, and neglecting any leakage flux, the best estimate for the mean radius *R* is

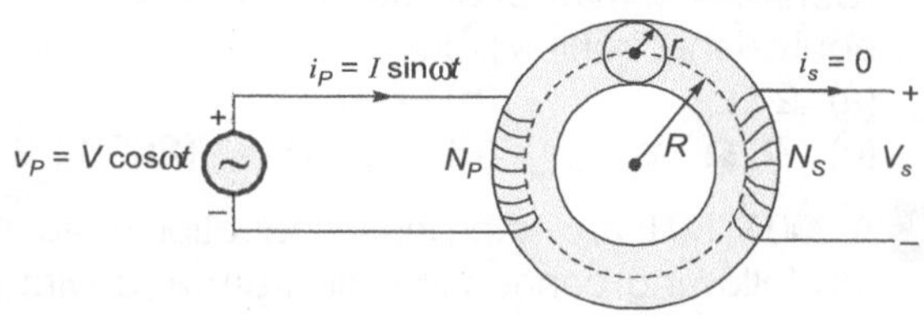

(a) $\dfrac{\mu V r^2 N_P^2 \omega}{I}$ (b) $\dfrac{\mu I r^2 N_P N_S \omega}{V}$

(c) $\dfrac{\mu V r^2 N_P^2 \omega}{2I}$ (d) $\dfrac{\mu I r^2 N_P^2 \omega}{2V}$

[2018 : 2 M]

5.26 The equivalent circuit of a single-phase induction motor is shown in the figure, where the parameters are $R_1 = R_2' = X_{l_1} = X_{l_2}' = 12\ \Omega$, $X_M = 240\ \Omega$ and *s* is the slip.

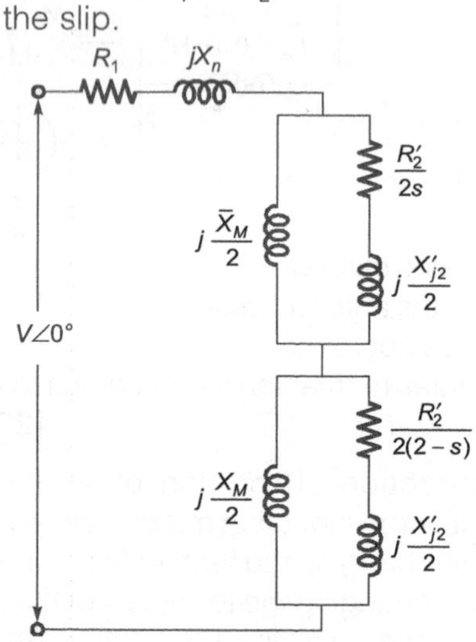

At no-load the motor speed can be approximated to be the synchronous speed. The no-load lagging power factor of the motor is _____ (upto 3 decimal places). [2018 : 2 M]

5.27 The magnetic circuit shown below has uniform cross-sectional area and air gap of 0.2 cm. The mean path length of the core is 40 cm. Assume that leakage and fringing fluxes are negligible. When the core relative permeability is assumed to be infinite, the magnetic flux density computed in the air gap is 1 tesla. With same Ampere-turns, if core relative permeability is assumed to be 1000 (linear), flux density in tesla (round off to 3 decimal places) calculated in the air gap is _______.

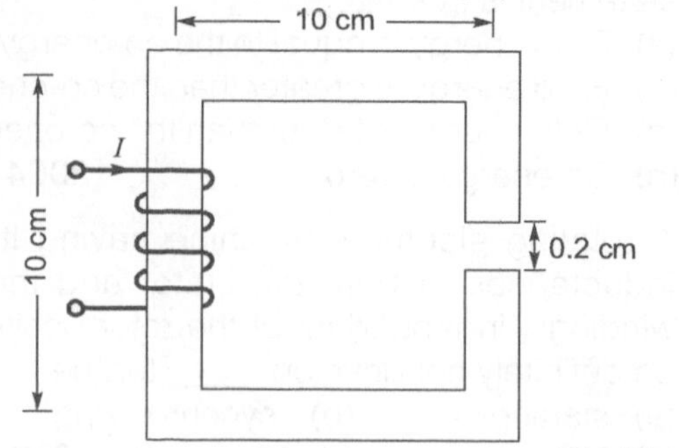

[2019 : 2 M]

5.28 The type of single-phase induction motor, expected to have the maximum power factor during steady state running condition, is

(a) split phase (resistance start).
(b) shaded pole.
(c) capactior start.
(d) capacitor start, capacitor run. [2022 : 1 M]

5.29 The four stator conductors (*A*, *A*′, *B* and *B*′) of a rotating machine are carrying DC currents of the same value, the directions of which are shown in the figure (i). The rotor coils *a*-*a*′ and *b*-*b*′ are formed by connecting the back ends of conductors '*a*' and '*a*′' and '*b*' and '*b*′', respectively, as shown in figure (ii). The e.m.f. induced in coil *a*-*a*′ and coil *b*-*b*′ are denoted by $E_{a\text{-}a'}$ and $E_{b\text{-}b'}$, respectively. If the rotor is rotated at uniform angular speed ω rad/s in the clockwise direction then which of the following correctly describes the $E_{a\text{-}a'}$ and $E_{b\text{-}b'}$?

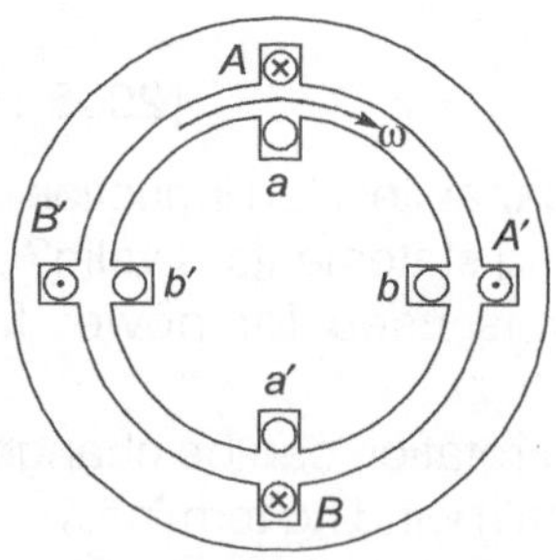

fig. (i) cross-section view

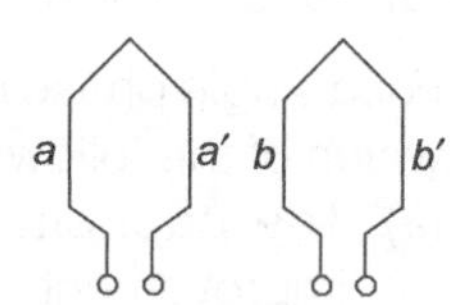

fig. (ii) rotor winding connection diagram

(a) $E_{a\text{-}a'}$ and $E_{b\text{-}b'}$ have finite magnitudes and are in the same phase.
(b) $E_{a\text{-}a'}$ and $E_{b\text{-}b'}$ have finite magnitudes with $E_{b\text{-}b'}$ leading $E_{a\text{-}a'}$
(c) $E_{a\text{-}a'}$ and $E_{b\text{-}b'}$ have finite magnitudes with $E_{a\text{-}a'}$ leading $E_{b\text{-}b'}$
(d) $E_{a\text{-}a'} = E_{b\text{-}b'} = 0$

[2023 : 1 M]

Answers — Single Phase Induction Motors, Special Purpose Machines and Electromechanical Energy Conversion Systems

5.1 (d)	5.2 (a)	5.3 (c)	5.4 (b)	5.5 (b)	5.6 (c)	5.7 (c)
5.8 (d)	5.9 (c)	5.10 (b)	5.11 (d)	5.12 (a)	5.13 (d)	5.14 (a)
5.15 (b)	5.16 (b)	5.17 (b)	5.18 (b)	5.19 (d)	5.20 (a)	5.21 (b)
5.22 (c)	5.23 (188.56)	5.24 (149.51)	5.25 (d)	5.26 (0.106)	5.27 (0.833)	5.28 (d)
5.29 (d)						

Explanations — Single Phase Induction Motors, Special Purpose Machines and Electromechanical Energy Conversion Systems

5.3 (c)

As the speed gets build up, auxiliary winding is disconnected along with starting capacitor.

5.4 (b)

$$\text{Step angle} = \frac{360°}{\text{No. of phases} \times \text{No. of rotor teeth}}$$

$$\Rightarrow \text{No. of rotor teeths} = \frac{360°}{4 \times 1.8°} = 50$$

5.6 (c)

The area under the hysteresis wave is a means to measure the hysteresis loss per cycle.

Area under the curve = 5 cm^2

$$= 5 \times (2\text{AT} \times 50 \text{ mWb}) = 500 \text{ mW}$$

For a cycle of 50 Hz,

$$\text{Loss} = 500 \text{ mW} \times 50 \text{ Hz} = 25 \text{ W}$$

5.7 (c)

Universal motors are used where light weight is important, as in vacuum cleaners and portable tools e.g. food mixer which usually operate at high speeds (1500 – 15000 rpm). Domestic water pumps are usually of low rating. So single phase induction motor can be used for such application. Three-phase induction motor is suitable for escalator as it has high starting torque.

5.8 (d)

A 1-ϕ induction motor with only the main winding excited would have forward rotating field is more than the backward rotating field at synchronous speed.

5.9 (c)

Detent Torque: It is amount of torque that the motor produces when it is not energized. No current is flowing through the windings.

Holding Torque: It is amount of torque that the motor produces when it has rated current flowing through the windings but motor is at rest.

Pull-in-Torque: Shows the maximum value of torque at given speeds that the motor can start, stop or reverse in synchronism with the input pulses.

Pull-out Torque: Shows the maximum value of torque at given speeds that the motor can generate while running in synchronism. If the motor is run outside of this curve, it will stall.

5.10 (b)

Brushless dc motor definitely has a permanent magnet.

5.11 (d)

Capacitor start type motor has high starting torque and therefore is used for hard starting loads such as compressors, conveyors, pumps etc.

Capacitor-run type motor has better running power factor and efficiency and a quiter and smoother operation. It is used for both easy and hard to start loads. Moder applications of such motors are ceiling fans, blowers etc.

5.12 (a)

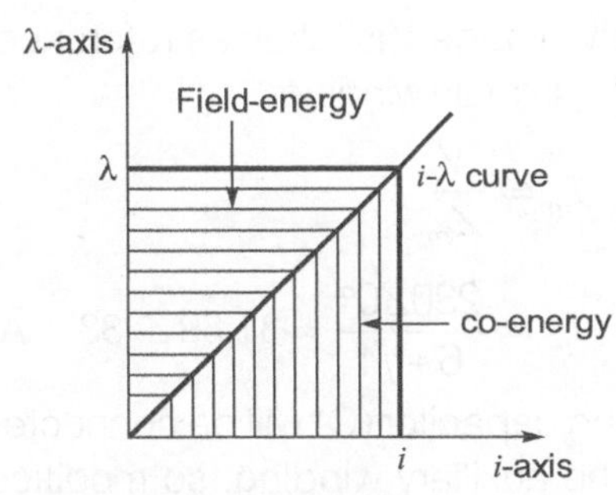

Where, $\lambda = N\phi$ = Flux linkage

Field energy is the energy absorbed by the magnetic system to establish flux ϕ.

For a linear electromagnetic circuit

$$\text{Field energy} = \text{Co-energy} = \frac{1}{2}\lambda i$$

5.13 (d)

$$L_s = \frac{N_s^2}{Rl_s} \text{ and } L_r = \frac{N_r^2}{Rl_r}$$

where,

L_s & L_r are self-inductances of stator and rotor. N_s and N_r are no. of turns energized in stator and rotor.

Rl_s = Reluctance seen by the stator flux.

Rl_r = Reluctance seen by the rotor flux.

As L_s and L_r are independent of the rotor position, the reluctances Rl_s and Rl_r does not vary with the rotor movement. Therefore, reluctance torque terms.

$\frac{1}{2}I_s^2\frac{dls}{d\theta}$ and $\frac{1}{2}I_r^2\frac{dL_r}{d\theta}$ become zero.

5.14 (a)

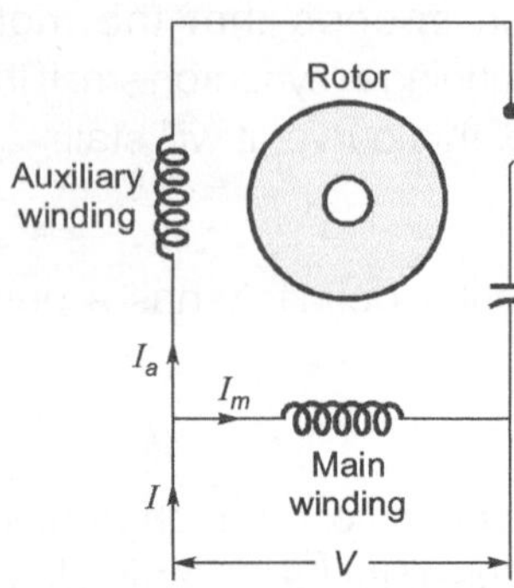

$Z_m = 6 + j4.0\ \Omega$

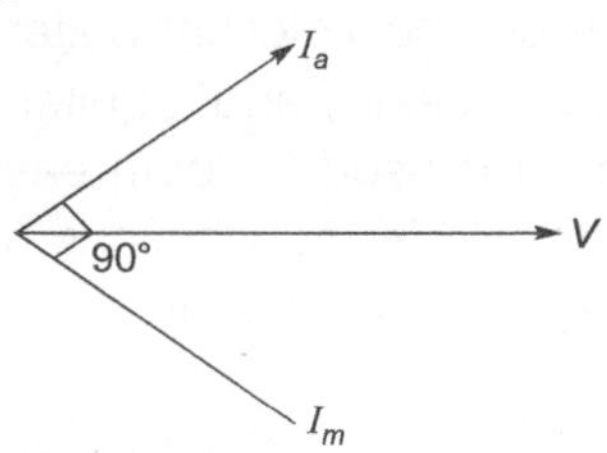

Supply voltage V is taken as reference.

Current in main winding

$$I_m = \frac{V}{Z_m}$$

$$\Rightarrow \quad I_m = \frac{230\angle 0^\circ}{6+j4} = 31.89\angle{-33.7^\circ}\text{A}$$

Starting capacitor (C) will be connected in series with the auxiliary winding, so modified auxiliary winding impedance.

$$Z_m = 8 + j6 - \frac{j}{\omega C}$$

$$= 8 + j\left(6 - \frac{1}{\omega C}\right) = 8 - j\left(\frac{1}{\omega C} - 6\right)$$

Current in auxiliary winding,

$$I_a = \frac{V}{Z_a} = \frac{230\angle 0^\circ}{8 - j\left(\frac{1}{\omega C} - 6\right)}$$

$$= |I_a|\angle \tan^{-1}\left(\frac{\frac{1}{\omega C} - 6}{8}\right)$$

Capacitor is provided to make angle between I_a and I_m 90°.

So, $\angle I_m + \angle I_a = 90^\circ \Rightarrow \angle I_a = 90^\circ - 33.7^\circ = 56.3^\circ$

$$\tan^{-1}\left(\frac{\frac{1}{\omega C} - 6}{8}\right) = 56.3^\circ$$

$$\frac{1}{\omega C} = 18$$

$$\omega = 2\pi f = 2\pi \times 50 \text{ rad/sec}$$

$$C \approx 176.84\ \mu\text{F}$$

5.15 (b)

Stepping rate is 100 steps/sec.

Step required for one revolution $= \frac{360^\circ}{1.8}$

= 200 steps.

∴ Time required for one revolution = 2 seconds

rev/sec = 0.5 rps ⇒ rev/minute = 30 rpm

5.16 (b)

In 1-ϕ induction motor, total torque developed is given by

$$T_{total} = T_f - T_b$$

where,

T_f = Torque developed due to forward field

T_b = Torque developed due to backward field

$$T_{total} = T_f - T_b = \frac{I_m^2 R_2}{2\omega_s}\left(\frac{1}{s} - \frac{1}{2-s}\right)$$

$$T_{total} \propto R_2$$

Higher is the rotor resistance (R_2), more is the total torque developed (T_{total}) which results in quick acceleration.

5.17 (b)

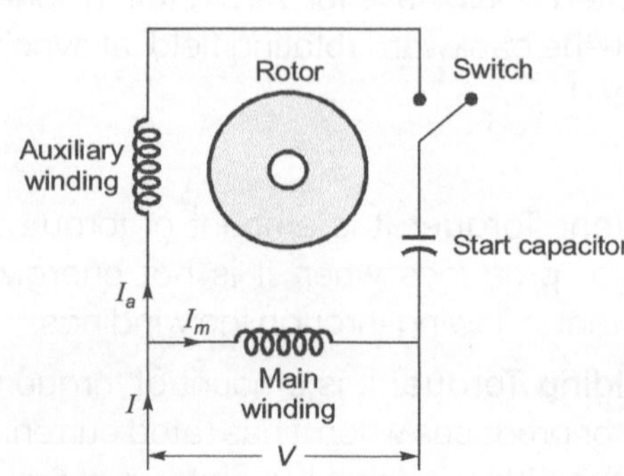

Start capacitor is used to provide 90° phase difference between I_m and I_a.

If supply terminals are interchanged, I_a and I_m will flow in the opposite direction, so torque will act in the same direction. Therefore, the direction of rotation will remain same.

5.18 (b)

No. of teeth = No. of poles = 20

Step angle,

$$\alpha = \frac{360°}{(\text{No. of stacks})\times(\text{No. of teeth})} = \frac{360°}{3\times 20} = 6°$$

5.19 (d)

Detent Torque: It is amount of torque that the motor produces when it is not energized. No current is flowing through the windings.

5.20 (a)

Rotor resistance at stand still, $R = 7.8\ \Omega$

Synchronous speed,

$$N_s = \frac{120f}{P} = \frac{120\times 50}{4} = 1500 \text{ rpm}$$

The slip (s) of rotor with respect to forward field

$$s = \frac{N_s - N_r}{N_s} = \frac{1500-1425}{1500} = 0.05$$

The slip of rotor with respect to backward field

$$= 2 - s = 2 - 0.05 = 1.95$$

Effective rotor resistance

$$= \frac{R_2}{2(2-s)} = 7.8 \Rightarrow \frac{7.8}{2\times 1.95} = 2\ \Omega$$

5.21 (b)

$f = 50$ Hz

Impedance of main winding

$$Z_m = r_m + j2\pi f L_m = 0.1 + j2\pi\times 50\times\frac{0.1}{\pi}$$

$$\Rightarrow \quad Z_m = 0.1 + j\,10\ \Omega$$

Impedance of auxiliary winding,

$$Z_a = r_a + j2\pi f L_a = 1 + j2\pi\times 50\times\frac{10}{\pi}$$

$$\Rightarrow \quad Z_a = 1 + j1000\ \Omega$$

Current through main winding

$$I_m = \frac{V_s}{Z_m} = \frac{220}{0.1+j10} \Rightarrow I_m = 22\angle{-89.427°}\text{ A}$$

Current through auxiliary winding

$$I_a = \frac{V_s}{Z_a} = \frac{220}{1+j1000} = 0.22\angle{-89.942}$$

Taking V_s as the reference

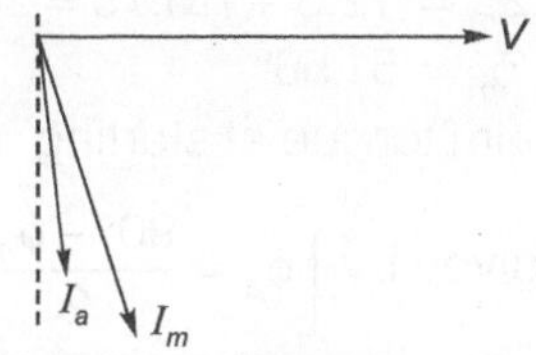

I_m leads I_a the fields created by the two current also have same difference thereby constituting an unbalanced field system. The result is the production of the starting torque.

Space-orientation of the field

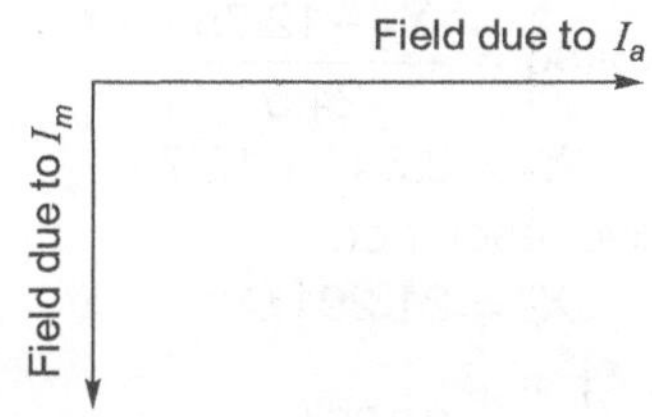

The motor rotates in the direction of leading phase to lagging phase. In this case, the motor rotates anti-clockwise.

5.22 (c)

Inter changing the terminals of the auxiliary winding.

5.23 Sol.

Energy in magnetic system,

$$W_f = \int_0^{\lambda} i(\lambda)\, d\lambda$$

$$\because \quad i = \lambda^2 \cdot g^2 \quad \text{(given)}$$

$$W_f = \int_0^{\lambda} \lambda^2 \cdot g^2 \cdot d\lambda = g^2 \cdot \frac{\lambda^3}{2}$$

Now, mechanical force,

$$F_F = -\frac{\partial W_f(\lambda, g)}{\partial g}$$

$$= -\frac{\partial}{\partial g}\left(\frac{g^2\cdot\lambda^3}{3}\right) = -\frac{2}{3}\lambda^3\cdot g$$

$$\because \quad i = 2\text{ A},\ g = 10\text{ cm},\ \lambda = 10\sqrt{2}$$

$$\text{Hence, } |F_F| = \frac{2}{3}\times(10\sqrt{2})^3\times 0.1 = 188.56\text{ N}$$

5.24 Sol.

Auxiliary winding impedance = $(24.5 + j12.75)\ \Omega$

Let X_c be the reactance of the capacitor connected in auxiliary winding

$$Z_a = (24.5 + j12.75 - jX_c)\ \Omega$$

$$X_a = (X_c - 12.75),\ R_a = 24.5\ \Omega$$

$$\therefore \quad \phi_a = \tan^{-1}\frac{X_c - 12.75}{24.5} \qquad ...(i)$$

$$Z_m = 12.5 + j15.75 = 20.10\angle 51.56°\ \Omega$$
$$\phi_m = 51.56°$$

For maximum torque at starting, condition on ϕ_a and ϕ_m is given by $\left[\phi_a = \dfrac{90° - \phi_m}{2}\right]$

$$\phi_a = \frac{90 - 51.56}{2} = 19.22° \quad ...(ii)$$

Comparing equation (*i*) and (*ii*),

$$\tan(19.22°) = \frac{X_c - 12.75}{24.5}$$
$$0.3486 = \frac{X_c - 12.75}{24.5}$$
$$X_c = 8.541 + 12.75$$

Capacitive reactance,

$$X_c = 21.291\ \Omega$$
$$\frac{1}{2\pi fc} = 21.291$$
$$\Rightarrow \quad C = \frac{1}{2\pi \times 50 \times 21.291} = 149.51\ \mu F$$

5.25 (d)

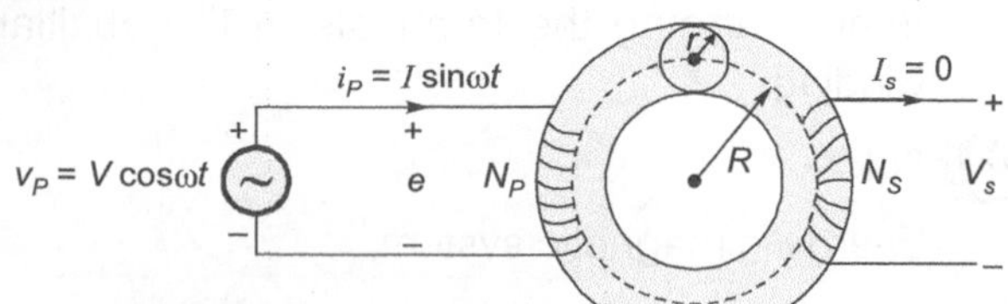

Since we know reluctance of core:

$$\Re = \frac{l_c}{\mu a}$$

Here, l_c = mean core length: a = area of core

$$\text{Reluctance,} \quad \Re = \frac{2\pi R}{\mu \times \pi r^2} = \frac{2R}{\mu r^2}$$

[where, R = mean radius of core]

$$\text{or} \quad R = \frac{\Re \cdot \mu r^2}{2}$$

$$\text{and} \quad \Re = \frac{\text{mmf}}{\text{flux}} = \frac{N_p I_p}{\phi}$$

$$\text{Here,} \quad \phi = -\frac{1}{N_p}\int_0^t e\,dt$$

[where e is primary generated voltage]

$$\text{or,} \quad \phi = -\frac{1}{N_p}\int_0^t -(V_p)\,dt \quad [\because \text{since, } V_p = -e]$$

$$\phi = \frac{1}{N_p}\int_0^t V\cos\omega t\,dt = \frac{V}{\omega N_p}\sin\omega t$$

Now reluctance,

$$\Re = \frac{N_p . I\sin\omega t}{\dfrac{V}{\omega N_p}\sin\omega t} = \frac{N_p^2 . \omega I}{V}$$

So mean radius,

$$R = \frac{N_p^2 \times \omega I \times \mu r^2}{2V}$$

5.26 Sol.

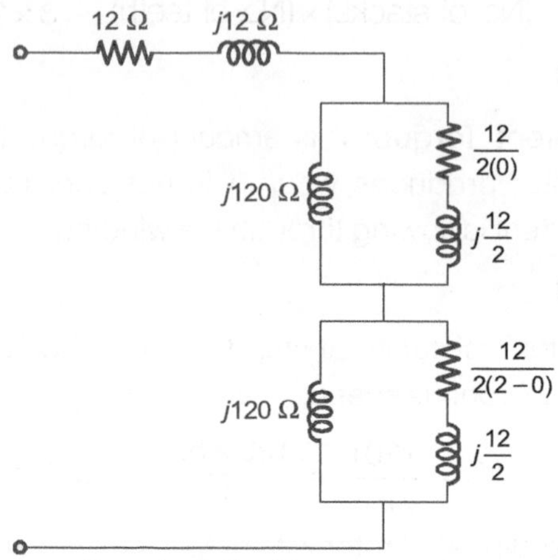

On no-load, N ; N_s
slip, $s \cong 0$

⇓

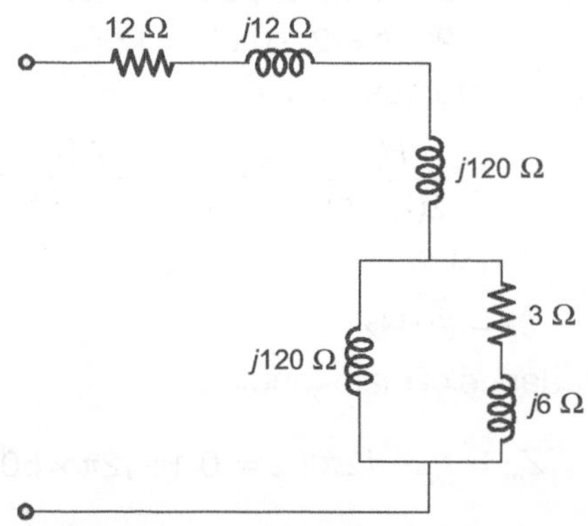

Simplifying the above circuit into a simple R-L circuit,

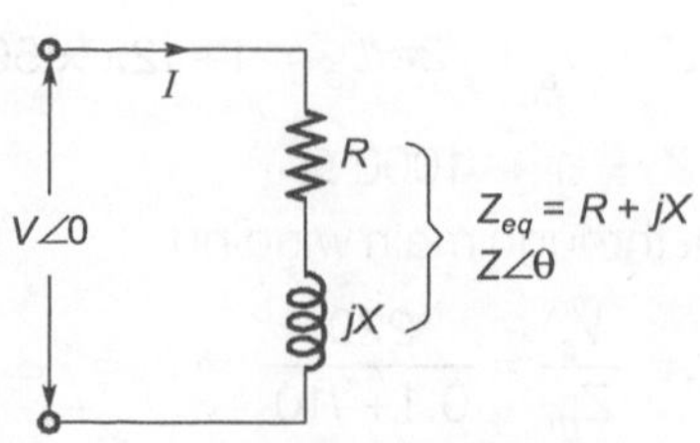

$$Z_{eq} = \frac{(3 + j6)(j120)}{(3 + j126)} + [12 + j132]$$
$$= 138.563\angle 83.9\ \Omega$$

Current drawn by motor is

$$I = \frac{V\angle 0}{Z\angle\theta}$$

{∴ θ : impedance angle will be p.f. angle}

∴ No-load lagging p.f. of motor is (cosθ),

cos(83.9) = 0.106 lagging power factor

5.27 Sol.

Given: $l_{ag} = 0.2$ cm, $l_m = 40$ cm

$B_0 = 1$ Tesla, $\phi = \frac{\text{mmf}}{\Re}$

For same mmf, $\phi \propto \frac{1}{\Re}$

In case-1: $\Re_1 = \frac{l_m}{\mu_0 A} = \frac{l_{ag}}{\mu_0 A} + \frac{l_m}{\mu_0\mu_r A}$

As $\mu_r = 0$

$$\therefore \quad \Re_1 = \frac{0.2\times 10^{-2}}{\mu_0 A}$$

In case-2:

$$\Re_2 = \frac{l_{ag}}{\mu_0 A} + \frac{l_m}{\mu_0\mu_r A}$$

$$= \frac{0.2\times 10^{-2}}{\mu_0 A} + \frac{40\times 10^{-2}}{1000\mu_0 A} = \frac{0.24\times 10^{-2}}{\mu_0 A}$$

As, flux, $\phi \;\alpha\; B$ for uniform cross section area.

$$\therefore B_2 = \frac{B_1\times\Re_1}{\Re_2} = \frac{1\times 0.2\times 10^{-2}/\mu_0 A}{0.24\times 10^{-2}/\mu_0 A} = 0.833 \text{ T}$$

5.29 (d)

Since the coil span of both coils *a–a'* and *b–b'* is $Y_s = 2Y_p$, i.e., twice the pole pitch, the resultant magnitudes of $E_{a\text{-}a'} = E_{b\text{-}b'} = E + (-E) = 0$.

■■■■

Power Systems

UNIT

IV

CONTENTS

Power Systems

Syllabus

Basic Concepts of Electrical Power Generation, AC and DC Transmission Concepts, Models and Performance of Transmission Lines and Cables, Economic Load Dispatch (with and without considering transmission losses), Series and Shunt Compensation, Electric Field Distribution and Insulators, Distribution Systems, Per Unit Quantities, Bus Admittance Matrix, Gauss- Seidel and Newton-Raphson Load Flow Methods, Voltage and Frequency Control, Power Factor Correction, Symmetrical Components, Symmetrical and Unsymmetrical Fault Analysis, Principles of Over Current, Differential, Directional and Distance Protection; Circuit Breakers, System Stability Concepts, Equal Area Criterion.

Analysis of Previous GATE Papers

Exam Year	1 Mark Ques.	2 Marks Ques.	3 Marks Ques.	5 Marks Ques.	Total Marks
1995	7	1	–	–	**9**
1996	6	3	–	–	**12**
1997	4	1	–	1	**11**
1998	3	6	–	3	**30**
1999	5	6	–	2	**27**
2000	4	7	–	2	**28**
2001	1	6	–	3	**28**
2002	4	4	–	1	**17**
2003	4	11	–	–	**26**
2004	6	9	–	–	**24**
2005	4	7	–	–	**18**
2006	4	10	–	–	**24**
2007	4	9	–	–	**22**
2008	3	8	–	–	**19**
2009	2	5	–	–	**12**
2010	3	5	–	–	**13**
2011	5	4	–	–	**13**

Exam Year	1 Mark Ques.	2 Mark Ques.	Total Marks
2012	3	2	**7**
2013	1	3	**7**
2014 Set-1	3	5	**13**
2014 Set-2	2	4	**10**
2014 Set-3	3	3	**9**
2015 Set-1	1	4	**9**
2015 Set-2	2	3	**8**
2016 Set-1	2	4	**10**
2016 Set-2	4	3	**10**
2017 Set-1	3	4	**11**
2017 Set-2	3	3	**9**
2018	4	4	**12**
2019	4	4	**12**
2020	3	3	**9**
2021	4	4	**12**
2022	4	3	**10**
2023	2	2	**6**
2024	2	4	**10**

1 Performance of Transmission Lines, Line Parameters and Corona

1.1 The surge impedance of a 400 km long overhead transmission line is 400 Ω. For a 200 km length of the same line, the surge impedance will be
(a) 200 Ω (b) 800 Ω
(c) 400 Ω (d) 100 Ω [1995 : 1 M]

1.2 For a 500 Hz frequency excitation, a 50 km short power line will be modeled as
(a) Short line (b) Medium line
(c) Long line (d) Data insufficient
[1996 : 1 M]

1.3 For equilateral spacing of conductors of an untransposed 3-phase line, we have
(a) balanced receiving end voltage and no communication interference.
(b) unbalanced receiving end voltage and no communication interference.
(c) balance receiving end voltage and communication interference.
(d) unbalanced receiving end voltage and communication interference. [1996 : 2 M]

1.4 Corona losses are minimized when
(a) Conductor size is reduced
(b) Smoothness of conductor is reduced
(c) Sharp points are provided in the line hardware
(d) Current density in conductors is reduced
[1999 : 1 M]

1.5 A 220 kV, 20 km long, 3-phase transmission line has the following: *A*, *B*, *C*, *D* constants. $A = D = 0.96\angle 3°$, $B = 55\angle 65°$ Ω/phase, $C = 0.5 \times 10^{-4}\angle 90°$ S/phase. Its correct charging current/phase is
(a) $\frac{11}{\sqrt{3}}$ A (b) $11\sqrt{3}$ A
(c) 11 A (d) $\frac{220}{\sqrt{3}}$ A [1999 : 2 M]

1.6 For a single-phase overhead line having solid copper conductors of diameter 1 cm, spaced 60 cm between centres, the inductance in mH/km is
(a) 0.05 + 0.2 ln 60
(b) 0.2 ln 60
(c) 0.05 + 0.2 ln (60/0.5)
(d) 0.2 ln (60/0.5) [1999 : 2 M]

1.7 The corona loss on a particular system at 50 Hz is 1 kW/km per phase. The corona loss at 60 Hz would be
(a) 1 kW/km per phase
(b) 0.83 kW/km per phase
(c) 1.2 kW/km per phase
(d) 1.13 kW/km per phase [2000 : 2 M]

1.8 The conductors of a 10 km long, single-phase, two wire line are separated by a distance of 1.5 m. The diameter of each conductor is 1 cm. If the conductors are of copper, the inductance of the circuit is
(a) 50.0 mH (b) 45.3 mH
(c) 23.8 mH (d) 19.6 mH [2001: 2 M]

1.9 Consider a long, two-wire line composed of solid round conductors. The radius of both conductors is 0.25 cm and the distance between their centres is *l* meters. If this distance is doubled, then the inductance per unit length.
(a) Doubles
(b) Halves
(c) Increase but does not double
(d) Decrease but does not halve [2002 : 1 M]

1.10 A long wire composed of a smooth round conductor runs above parallel to the ground (Assumed to be a large conductive plane). A high voltage exists between the conductor and the ground. The maximum electric stress occurs at
(a) The upper surface of the conductor
(b) The lower surface of the conductor
(c) The ground surface
(d) Midway between the conductor and the ground
[2002 : 1 M]

1.11 Bundled conductors are mainly used in high voltage overhead transmission lines to
(a) reduce transmission line losses
(b) increase mechanical strength of the line
(c) reduce corona
(d) reduce sag [2003 : 1 M]

1.12 The ABCD parameters of a 3-phase overhead transmission line are: $A = D = 0.9\angle 0°$; $B = 200\angle 90°$ Ω and $C = 0.95 \times 10^{-3}\angle 90°$ S. At no-load condition a shunt inductive reactor is connected at the receiving end of the line to limit the receiving-end voltage to be equal to the sending-end voltage. The ohmic value of the reactor is

(a) $\infty\ \Omega$ (b) $2000\ \Omega$
(c) $105.26\ \Omega$ (d) $1052.6\ \Omega$

[2003 : 2 M]

1.13 The generalized circuit constants of a 3-phase, 220 kV rated voltage, medium length transmission line are

$$A = D = 0.936 + j0.016 = 0.936\angle 0.98^\circ$$
$$B = 33.5 + j138 = 142.0\angle 76.4^\circ\ \Omega$$
$$C = (-5.18 + j914)\times 10^{-6}\ \mho$$

If the load at the receiving end is 50 MW at 220 kV with a power factor of 0.9 lagging, then magnitude of line to line sending end voltage should be

(a) 133.23 kV (b) 220.00 kV
(c) 230.78 kV (d) 246.30 kV [2004 : 2 M]

1.14 The concept of an electrically short, medium and long line is primarily based on the

(a) nominal voltage of the line
(b) physical length of the line
(c) wavelength of the line
(d) power transmitted over the line [2006 : 1 M]

1.15 The *A*, *B*, *C*, *D* constants of a 220 kV line are: $A = D = 0.94\angle 10$, $B = 130\angle 730$, $C = 0.001\angle 900$. If the sending end voltage of the line for a given load delivered at nominal voltage is 240 kV, the % voltage regulation of the line is

(a) 5 (b) 9
(c) 16 (d) 21 [2006 : 2 M]

1.16 A single phase transmission line and a telephone line are both symmetrically strung one below the other, in horizontal configurations, on a common tower. The shortest and longest distances between the phase and telephone conductors are 2.5 m and 3 m respectively. The voltage (volt/km) induced in the telephone circuit, due to 50 Hz current of 100 amps in the power circuit is

(a) 4.81 (b) 3.56
(c) 2.29 (d) 1.27 [2006 : 2 M]

1.17 Consider a bundled conductor of an overhead line consisting of three identical sub-conductors placed at the corners of an equilateral triangle as shown in the figure. If we neglect the charges on the other phase conductors and ground, and assume that spacing between sub-conductors is much larger than their radius, the maximum electric field intensity is experienced at

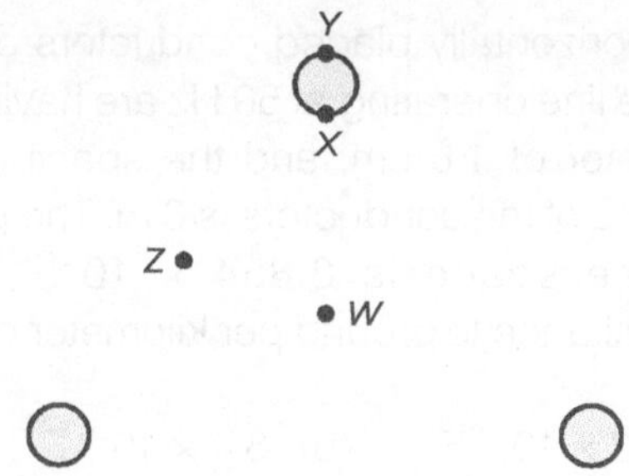

(a) Point X (b) Point Y
(c) Point Z (d) Point W [2007 : 1 M]

1.18 The total reactance and total susceptance of a lossless overhead EHV line, operating at 50 Hz, are given by 0.045 pu and 1.2 pu respectively. If the velocity of wave propagation is 3×10^5 km/s, then the approximate length of the line is

(a) 122 km (b) 172 km
(c) 222 km (d) 272 km [2007 : 2 M]

1.19 An extra high voltage transmission line of length 300 km can be approximate by a lossless line having propagation constant $\beta = 0.00127$ radians per km. Then the percentage ratio of line length to wavelength will be given by

(a) 24.24% (b) 12.12%
(c) 19.05% (d) 6.06% [2008 : 1 M]

1.20 For a fixed value of complex power flow in a transmission line having a sending end voltage *V*, the real power loss will be proportional to

(a) V (b) V^2
(c) $\frac{1}{V^2}$ (d) $\frac{1}{V}$ [2009 : 1 M]

1.21 A 50 Hz synchronous generator is initially connected to a long lossless transmission line which is open circuited at the receiving end. With the field voltage held constant, the generator is disconnected from the transmission line. Which of the following may be said about the steady state terminal voltage and field current of the generator?

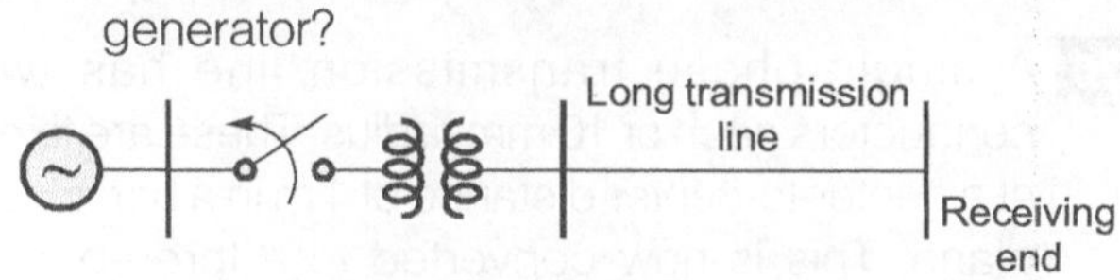

(a) The magnitude of terminal voltage decreases, and the field current does not change.
(b) The magnitude of terminal voltage increases, and the field current does not change.
(c) The magnitude of terminal voltage increases, and the field current increases.
(d) The magnitude of terminal voltage does not change and the field current decreases.

[2010 : 2 M]

1.22 The horizontally placed conductors of a single-phase line operating at 50 Hz are having outside diameter of 1.6 cm, and the spacing between centers of the conductors is 6 m. The permittivity of free space is 8.854×10^{-12} F/m. The capacitance to ground per kilometer of each line is

(a) 4.2×10^{-9} F (b) 8.4×10^{-9} F
(c) 4.2×10^{-12} F (d) 8.4×10^{-12} F

[2014 : 2 M, Set-2]

1.23 In a long transmission line with r, l, g and c are the resistance, inductance, shunt conductance and capacitance per unit length, respectively, the condition for distortionless transmission is

(a) $rc = lg$ (b) $r = \sqrt{l/c}$
(c) $rg = lc$ (d) $g = \sqrt{c/l}$

[2014 : 1 M, Set-3]

1.24 For a 400 km long transmission line, the series impedance is $(0.0 + j0.5)$ Ω/km and the shunt admittance is $(0.0 + j5.0)$ μmho/km. The magnitude of the series impedance (in Ω) of the equivalent π circuit of the transmission line is ______.

[2014 : 2 M, Set-3]

1.25 A composite conductor consists of three conductors of radius R each. The conductors are arranged as shown below. The geometric mean radius (GMR) (in cm) of the composite conductor is kR. The value of k is ______.

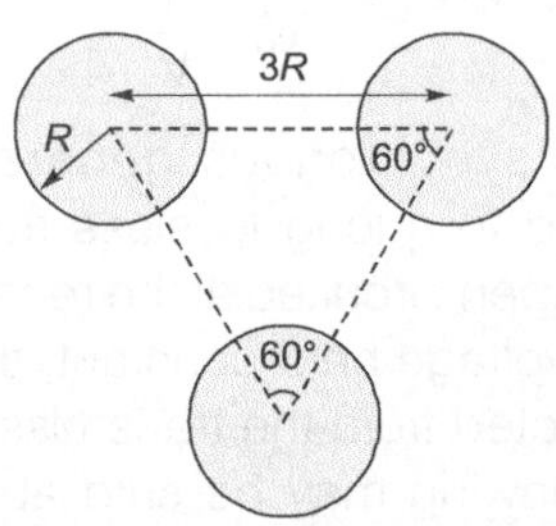

[2015 : 2 M, Set-2]

1.26 A single-phase transmission line has two conductors each of 10 mm radius. These are fixed at a center-to-center distance of 1 m in a horizontal plane. This is now converted to a three-phase transmission line by introducing a third conductor of the same radius. This conductor is fixed at an equal distance D from the two single-phase conductors. The three-phase line is fully transposed. The positive sequence inductance per phase of the three-phase system is to be 5% more than that of the inductance per conductor of the single-phase system. The distance D, in meters, is _____.

[2016 : 2 M, Set-2]

1.27 At no load condition, a 3-phase, 50 Hz, lossless power transmission line has sending-end and receiving-end voltages of 400 kV and 420 kV respectively. Assuming the velocity of traveling wave to be the velocity of light, the length of the line, in km, is ______.

[2016 : 2 M, Set-2]

1.28 A source is supplying a load through a 2-phase, 3-wire transmission system as shown in the figure below. The instantaneous voltage and current in phase-a are $v_{an} = 220 \sin(100\pi t)$ V and $i_a = 10 \sin(100\pi t)$ A, respectively. Similarly for phase-b, the instantaneous voltage and current are $v_{bn} = 220 \cos(100\pi t)$ V and $i_b = 10 \cos(100\pi t)$ A, respectively.

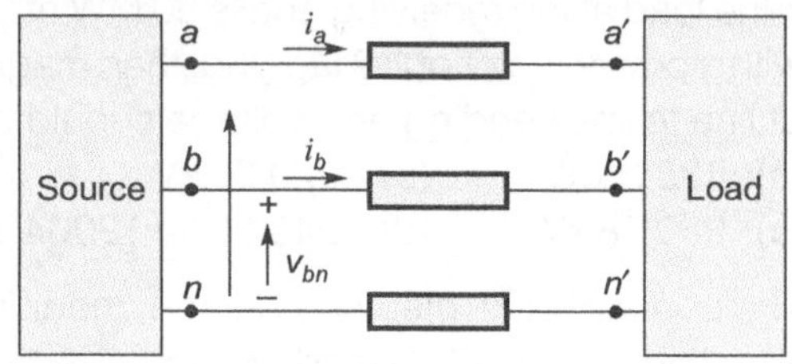

The total instantaneous power flowing from the source to the load is

(a) 2200 W
(b) $2200 \sin^2(100\pi t)$ W
(c) 4400 W
(d) $2200 \sin(100\pi t) \cos(100\pi t)$ W

[2017 : 1 M, Set-1]

1.29 The nominal-π circuit of a transmission line is shown in the figure.

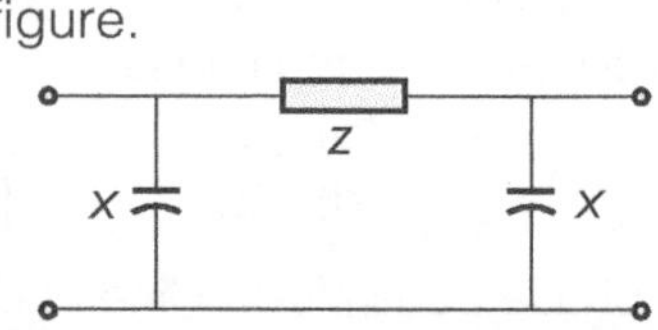

Impedance, $Z = 100\angle 80°$ Ω and reactance $X = 3300$ Ω. The magnitude of the characteristic impedance of the transmission line, in Ω, is ___. (Give the answer up to one decimal place.)

[2017 : 1 M, Set-2]

1.30 Consider an overhead transmission line with 3-phase. 50 Hz balanced system with conductors located at the vertices of an equilateral triangle of length $D_{ab} = D_{bc} = D_{ca} = 1$ m as shown in fig.

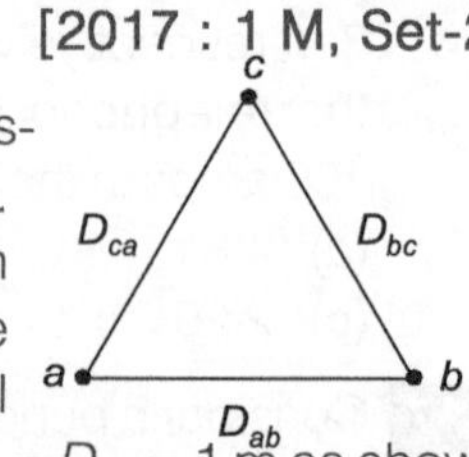

The resistances of the conductors are neglected. The geometric mean radius (GMR) of each conductor is 0.01 m. Neglecting the effect of ground, the magnitude of positive sequence reactance in Ω/km (rounded off to three decimal places) is ______ .

[2017 : 2 M, Set-2]

1.31 For the balanced Y-Y connected 3-phase circuit shown in the figure below, the line-line voltage is 208 V rms and the total power absorbed by the load is 432 Ω at a power factor of 0.6 leading.

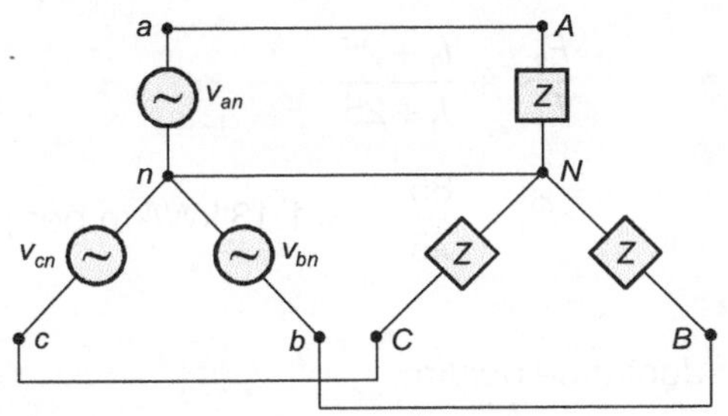

The approximate value of the impedance Z is

(a) $33\angle-53.1°\,\Omega$ (b) $60\angle 53.1°\,\Omega$

(c) $60\angle-53.1°\,\Omega$ (d) $180\angle-53.1°\,\Omega$

[2017 : 2 M, Set-2]

1.32 A three-phase load is connected to a three-phase balanced supply as shown in the figure. If $V_{an} = 100\angle 0°$ V. $V_{bn} = 100\angle-120°$ V and $V_{cn} = 100\angle-240°$ V (angles are considered positive in the anti-clockwise direction), the value of R for zero current in the neutral wire is _____ Ω (upto 2 decimal places).

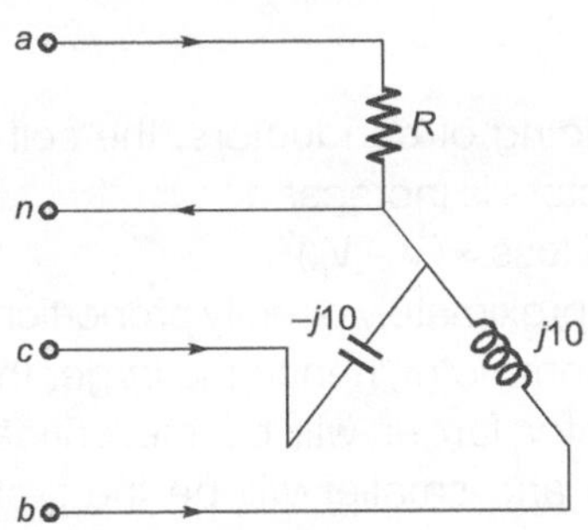

[2018 : 2 M]

1.33 A three-phase 50 Hz, 400 kV transmission line is 300 km long. The line inductance is 1 mH/km per phase, and the capacitance is 0.01 μF/km per phase. The line is under open circuit condition at the receiving end and energized with 400 kV at the sending end, the receiving end line voltage in kV (round off to two decimal places) will be _____.

[2019 : 2 M]

1.34 The geometric mean radius of a conductor, having four equal strands with each strand of radius r, as shown in the figure below, is

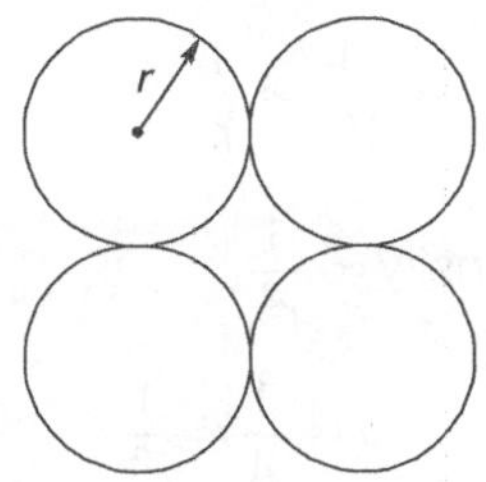

(a) $4r$ (b) $1.414r$

(c) $2r$ (d) $1.723r$

[2022 : 1 M]

Answers Performance of Transmission Lines, Line Parameters and Corona

1.1 (c)	1.2 (c)	1.3 (d)	1.4 (d)	1.5 (a)	1.6 (c)	1.7 (d)
1.8 (c)	1.9 (c)	1.10 (b)	1.11 (c)	1.12 (b)	1.13 (c)	1.14 (b)
1.15 (c)	1.16 (c)	1.17 (b)	1.18 (c)	1.19 (d)	1.20 (c)	1.21 (a)
1.22 (b)	1.23 (a)	1.24 (187)	1.25 (1.913)	1.26 (1.427)	1.27 (294.59)	1.28 (a)
1.29 (406.2)	1.30 (0.289)	1.31 (c)	1.32 (5.77)	1.33 (418.85)	1.34 (d)	

Explanations Performance of Transmission Lines, Line Parameters and Corona

1.1 (c)

Surge impedance is not a function of length of the line.

1.2 (c)

$\Rightarrow$ Length of line × frequency

$= 50 \times 500 = 25000 > 10000$

then the line is modelled as long line.

1.4 (d)

Corona loss $\propto (V - V_C)^2$

Also, $V_C \propto r\,ln\left(\frac{d}{r}\right)$

$\Rightarrow \quad V_C \propto r \quad \text{and} \quad V_C \propto -lnr$

$\because \quad e^x > x$ (this relation always holds for $x > 0$)

As r being radius of conductor, $r > 0$

$$\Rightarrow \quad e^r > r$$

$$r > ln\,r$$

with this we can conclude that

$$V_C \propto r$$

$$\text{Corona loss} \propto \frac{1}{V_C} \propto \frac{1}{r} \quad \text{...(i)}$$

$$\text{Current density} \propto \frac{1}{r^2} \quad \text{...(ii)}$$

$$J \propto \frac{1}{A} \propto \frac{1}{r^2}$$

Corona loss $\propto$ current density.

Methods of option (a), (b) and (c) will increase corona losses, if applied.

V_C = critical disruptive voltage

1.5 (a)

$$\text{Charging current/phase, } I_C = \frac{V_{phase}}{X_c}$$

$$\left(X_C = \frac{1}{Y}\right) = \frac{220 \times 10^3}{\sqrt{3} \times \left[\frac{1}{0.5 \times 10^{-4}} \angle -90°\right]}$$

$$= \frac{220 \times 10^3 \times 0.5 \times 10^{-4}}{\sqrt{3}} \angle 90° = \frac{11}{\sqrt{3}} \angle 90° \text{A}$$

1.6 (c)

Inductance,

$$L = 0.2\,ln\left(\frac{d}{r'}\right) \text{mH/km} = 0.2\,ln\left(\frac{d}{e^{1/4}r}\right) \text{mH/km}$$

$$= 0.2ln\left(\frac{d}{r}\right) \text{mH/km} - 0.2ln(e^{-1/4}) \text{ mH/km}$$

$$= 0.2ln\left(\frac{d}{r}\right) \text{mH/km} + \frac{0.2}{4} \text{mH/km}$$

$$= \left[0.05 + 0.2ln\left(\frac{d}{r}\right)\right] \text{mH/km}$$

$$L = \left[0.05 + 0.2\,ln\left(\frac{60}{0.5}\right)\right] \text{mH/km}$$

1.7 (d)

Corona loss,

$$P_{corona} = 2.41 \times \sqrt{\frac{V}{d}}\left(\frac{f+25}{8}\right)(V - V_0)^2 \text{ W/km/ph}$$

$$\Rightarrow \quad \frac{P_2}{P_1} = \frac{f_2 + 25}{f_1 + 25}$$

$$P_2 = \frac{85}{75} \times 1 = 1.13 \text{ kW/km per phase}$$

1.8 (c)

Inductance per km,

$$L_{ab} = 0.4\,ln\left(\frac{d}{r'}\right) \text{mH/km}$$

$$= 0.4ln\left(\frac{150}{0.7788 \times \frac{1}{2}}\right) \text{mH/km} = 2.3815 \text{ mH/km}$$

For 10 km long line, L_{ab} = 23.815 mH

1.9 (c)

$$L \propto ln\left(\frac{d}{r'}\right) \quad \Rightarrow ln\,(2d) > ln\,d$$

1.11 (c)

- By bonding of conductors, the self GMD of the conductors is increased.
- Corona loss $\propto (V - V_0)^2$
  V_0 is approximately directly proportional to the size of the conductor, hence the larger the size of the conductor larger will be the critical disruptive voltage and smaller will be the factor $(V - V_0)^2$ and hence, smaller will be the corona loss.

1.12 (b)

Reactance,

$$X_{L,\,SH} = \frac{B}{1-A} = \frac{200\angle 90°}{1-0.9} = 2000\angle 90°\ \Omega$$

1.13 (c)

$$A = 0.936\,\angle 0.98° \text{ and } B = 142\angle 76.4°\ \Omega$$

$$= \frac{220}{\sqrt{3}} \text{kV} = 127\angle 0° \text{ kV}$$

Receiving end current,

$$I_R = \frac{50 \times 10^6}{\sqrt{3} \times 220 \times 10^3 \times 0.9} = 145.8 \text{ A}$$

$$I_R = 0.1458\,\angle{-25.84°} \text{ kA}$$

(Power factor = 0.9 lagging)

V_s(sending end phase voltage) = $AV_R + BI_R$

$$V_s = (0.936\,\angle 0.98°) \times (127\angle 0°) + (142\angle 76.4°)$$

$$\times (0.1458\angle{-25.84°})$$
$$= 133.23\angle 7.77° \text{ kV}$$

Line to line sending end voltage

$$= |V_s|_{l-l} = \sqrt{3} \times 133.23 \text{ kV} = 230.78 \text{ kV}$$

1.14 (b)

L = length of line

If $L \le 80$ km

The line is short line.

If $80 \text{ km} < L \le 250$ km

The line is medium line.

If $L > 250$ km

The line is long line.

1.15 (c)

Sending end voltage = V_s = 240 kV

Full-load receiving end voltage

$$= (V_R)_{FL} = 220 \text{ kV}$$

$$V_{s(\text{per phase})} = \frac{240}{\sqrt{3}} \text{ kV}$$

As we know,

$$V_{s(\text{per phase})} = AV_{R(\text{per phase})} + BI_R$$

At no-load,

$$I_R = 0$$

No-load receiving end voltage (per phase)

$$= (V_R)_{NL\,(\text{per phase})} = \frac{V_{s(\text{per phase})}}{A}$$

$$= \frac{240}{\sqrt{3} \times 0.94} = \frac{255.32}{\sqrt{3}} \text{ kV}$$

No-load receiving end voltage (line to line)

$$= (V_R)_{NL} = 255.32 \text{ kV}$$

$$\% \text{ voltage regulation} = \frac{(V_R)_{NL} - (V_R)_{FL}}{(V_R)_{FL}} \times 100$$

$$= \frac{255.32 - 220}{220} \times 100 \approx 16\%$$

1.16 (c)

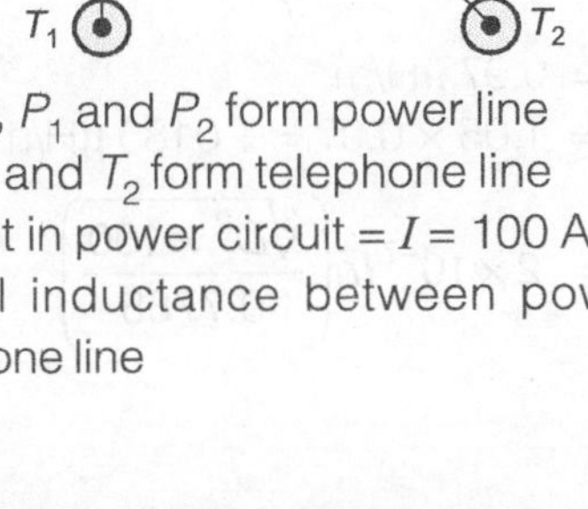

Where, P_1 and P_2 form power line and T_1 and T_2 form telephone line

Current in power circuit = I = 100 A

Mutual inductance between power line and telephone line

$$= M_{PT} = 4 \times 10^{-7} \ln\left(\frac{D_2}{D_1}\right) \text{ H/m}$$

$$\Rightarrow \quad M_{PT} = 4 \times 10^{-7} \times \ln\left(\frac{3}{2.5}\right) = 0.73 \times 10^{-7} \text{ H/m}$$

Voltage induced in the telephone line

$$= |V_T| = \omega M_{PT} I$$

$$= 2\pi \times 50 \times 0.73 \times 10^{-7} \times 100$$

$$\Rightarrow \quad |V_T| = 2.29 \times 10^{-3} \text{ V/m} = 2.29 \text{ V/km}$$

1.17 (b)

Electric field intensity at various points are shown as follows.

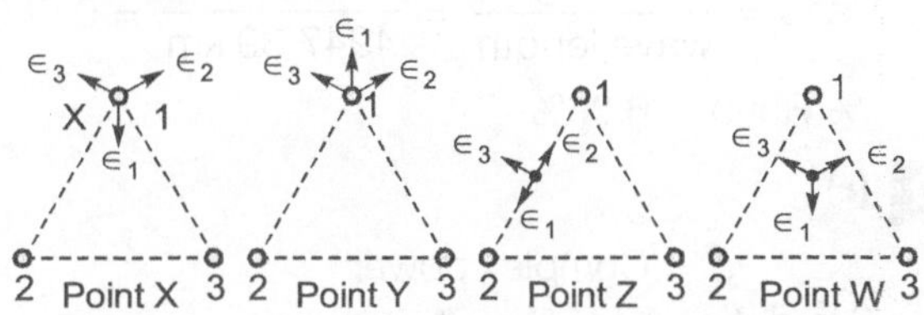

It is clear from the above diagrams that minimum cancellation of vector occurs at the point Y. Hence maximum electric field intensity.

1.18 (c)

For loss less line,

$$\beta = \sqrt{\omega L \cdot \omega C}$$

$$\beta = \sqrt{0.045 \times 1.2} = 0.2324$$

$$\beta = \frac{2\pi f l}{v}$$

$$l = \frac{0.2324 \times 3 \times 10^5}{2\pi \times 50} = 222 \text{ km}$$

Alternative solution:

Reactance = X (in p.u.) = 0.045 p.u.

Suspectance = Y (in p.u.) = 1.2 p.u.

Let base impedance = Z_B

$$X \text{ (in } \Omega) = 0.045\, Z_B$$

$$Y \text{ (in } ℧) = \frac{1.2}{Z_B} ℧$$

Assuming impedance of line = L H/km

$$X = \omega l L$$

Capacitance of line = C F/km

$$Y = \omega l C$$

Where, l = Length of the line

$$L = \frac{X}{\omega l} = \frac{0.045\, Z_B}{\omega l}$$

and $$C = \frac{Y}{\omega l} = \frac{1.2}{\omega l\, Z_B}$$

Velocity of propagation is given by

$$v_c = \frac{1}{\sqrt{LC}}$$

$$V_c = \frac{1}{\sqrt{\left(\frac{0.045 Z_B}{\omega l}\right)\left(\frac{1.2}{\omega l\, Z_B}\right)}} = \frac{\omega l}{\sqrt{0.45 \times 1.2}}$$

$$3 \times 10^5 = \frac{2 \times \pi \times 50 l}{\sqrt{0.045 \times 1.2}} \quad \Rightarrow \quad l \approx 222 \text{ km}$$

1.19 (d)

Wave length = λ, propagation constant= β

$$\lambda = \frac{2\pi}{\beta} = \frac{2\pi}{0.00127} = 4947.39 \text{ km}$$

$$\text{Ratio} = \frac{\text{line length}}{\text{wave length}} = \frac{300 \text{ km}}{4947.39 \text{ km}} = 0.0606$$

% Ratio = 6.06%

1.20 (c)

S = Complex power

S is defined as, $S = VI$

$$I = \frac{S}{V}$$

If R is the resistance of transmission line

$$\text{Real power loss} = I^2 R = \left(\frac{S}{V}\right)^2 R = \frac{S^2 R}{V^2}$$

Since, S and R are fixed

$$\text{Real power loss} \propto \frac{1}{V^2}$$

1.21 (a)

- As field voltage is held constant, so field current does not change.
- When the generator is connected with open-circuit transmission line, line draws charging current.

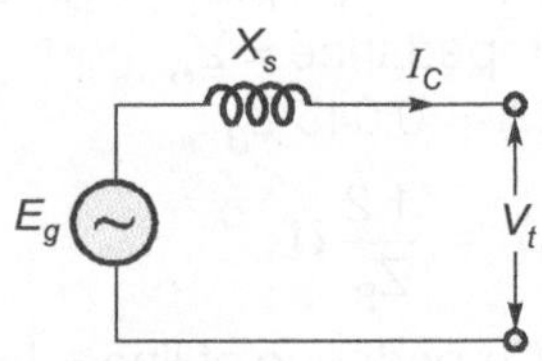

Therefore V_t is higher than E_g, i.e., $V_t > E_g$

But when the generator is disconnected from the line, no charging current is delivered by the generator i.e. $I_C = 0$. In this $V_t = E_g$

So, terminal voltage decreases.

1.22 (b)

Capacitance w.r.t. ground per kilometre

$$C = \frac{2\pi\epsilon_0}{ln\left(\frac{d}{r}\right)} = \frac{2\pi \times \epsilon_0}{ln\left(\frac{2 \times 6}{1.6 \times 10^{-2}}\right)}$$

$$= 8.4 \times 10^{-12} \text{ F/m} = 8.4 \times 10^{-9} \text{ F/km}$$

1.23 (a)

The condition for a transmission line to be distortionless is

$$\boxed{\frac{r}{g} = \frac{l}{c}}$$

1.24 Sol.

$$z = (0 + j0.5)\ \Omega/\text{km},$$
$$Z = zl = (0 + j200)\ \Omega$$
$$y = (0 + j5)\ \mu\mho/\text{km},$$
$$Y = yl = (0 + j5) \times 10^{-6} \times 400$$
$$= (0 + j2 \times 10^{-3})\ \mho$$

$$\gamma l = \sqrt{ZY} = \sqrt{(200 \times 2 \times 10^{-3})}$$

$$\gamma l = j0.6324 = \alpha l + j\beta l = 0 + j\beta l$$

$$Z_c = \sqrt{\frac{Z}{Y}} = \sqrt{\frac{200}{2 \times 10^{-3}}} = 316.22\ \Omega$$

Equivalent π-network:

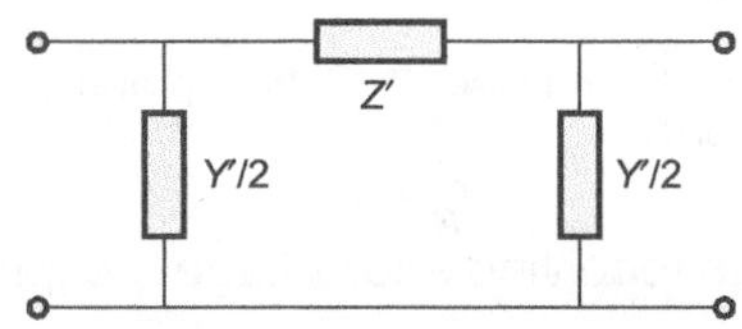

$$Z = Z_C \sinh \gamma l = Z_C \sinh(\alpha l + j\beta l)$$
$$= Z_C(\sinh \alpha l \cdot \cos\beta l + j\cosh \alpha l \ \ \sin\beta l)$$
$$= 316.22\,[\sinh(0) \times \cos\beta l + j\cosh(0)$$

$$\sin\left(0.6324 \times \frac{180}{\pi}\right) = 187\ \Omega$$

1.25 Sol.

Geometric mean radius,

$$= \left[(0.7788R \times 3R \times 3R)^3\right]^{1/3^2} = 1.913\,R = KR$$

$K = 1.913$

1.26 Sol.

In first case,

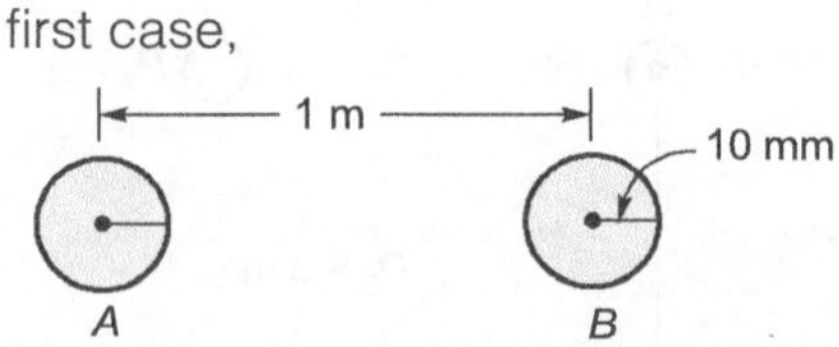

$$L_1 = 2 \times 10^{-7}\, ln\,\frac{D}{r} = 2 \times 10^{-7} ln\left(\frac{100}{0.7788}\right)$$

$$L_1 = 0.97\ \mu\text{H/m}$$
$$L_2 = 1.05 \times 0.97 = 1.0185\ \mu\text{H/m}$$

$$L_2 = 2 \times 10^{-7} ln\left(\frac{\sqrt[3]{D^2 \times 100}}{0.7788}\right)$$

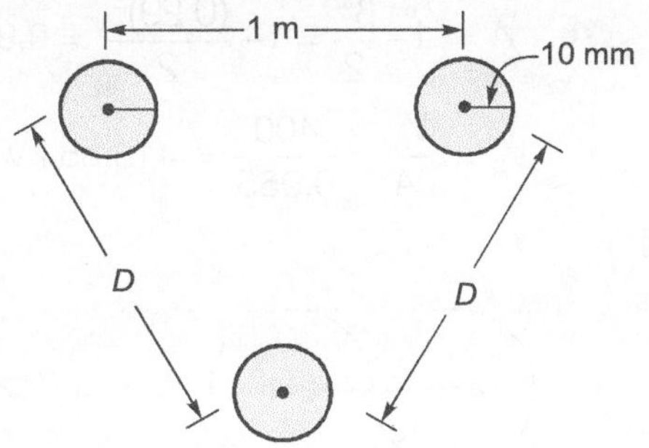

$$ln\left(\frac{\sqrt[3]{D^2\times 100}}{0.7788}\right) = \frac{1.0185\times 10^{-6}}{2\times 10^{-7}} = 5.0925$$

$$e^{5.0925} = \frac{\sqrt[3]{100D^2}}{0.7788} \Rightarrow D = 142.7 \text{ cm} = 1.427 \text{ m}$$

1.27 Sol.

At no load,

$$V_s = AV_R$$
$$400 = A420$$
$$A = \frac{400}{420} = 0.9524$$
$$A = 1+\frac{YZ}{2} = 1+\frac{(r+j\omega L)(g+j\omega C)}{2}$$

For lossless line $r = 0$, $g = 0$

Then, $$A = 1-\frac{(\omega C)(\omega L)}{2}$$

$$\beta l = \sqrt{\omega L\,\omega C}$$
$$A = 0.9524 = 1-\frac{\beta^2 l^2}{2}$$
$$\beta l = 0.3085$$
$$\beta = \frac{0.3085}{l}$$
$$\frac{V}{f} = \frac{2\pi}{\beta}$$
$$\frac{3\times 10^5}{50} = \frac{2\pi}{\left(\frac{0.3085}{l}\right)}$$
$$l = 294.59 \text{ km}$$

1.28 (a)

$$V_{an} = 220\sin(100\pi t)\text{ V}$$
$$i_a = 10\sin(100\pi t)\text{ A}$$
$$V_{bn} = 220\cos(100\pi t)\text{ V}$$
$$i_b = 10\cos(100\pi t)\text{ A}$$
$$p = V_{an}\cdot i_a + V_{bn}\cdot i_b$$
$$\Rightarrow \quad = 2200 \text{ W}$$

1.29 Sol.

$$Z = 100\angle 80^\circ\ \Omega$$
$$X = 3300\ \Omega$$
$$\frac{y}{2} = \frac{1}{X} \Rightarrow \frac{1}{3300}$$

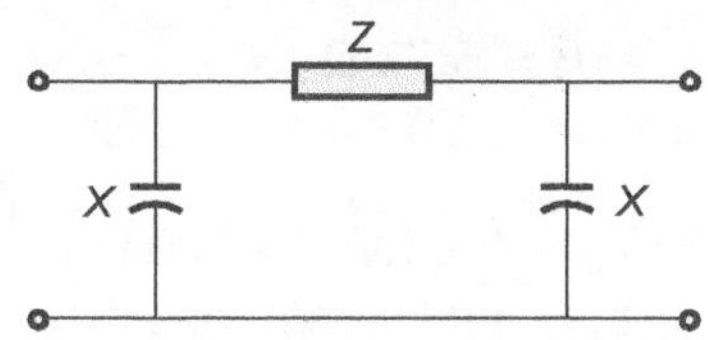

$$y = \frac{2}{3300} = 6.06\times 10^{-4}$$
$$Z_C = \sqrt{\frac{Z}{y}} = \sqrt{\frac{100}{6.06\times 10^{-4}}} = 406.2\ \Omega$$

1.30 Sol.

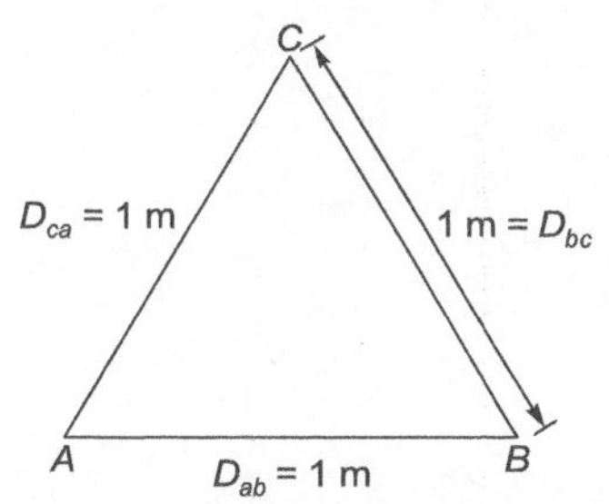

$$X_L = 2\pi fL = 2\pi f\times 2\times 10^{-7} ln\left(\frac{D_m}{D_s}\right)$$
$$= 2\pi\times 50\times 2\times 10^{-7} ln\left(\frac{1}{0.01}\right)$$
$$= 2.89\times 10^{-4}\ \Omega/\text{m} \quad \Rightarrow \quad X_L = 0.289\ \Omega/\text{km}$$

1.31 (c)

For star connection,

$$P_{3\text{-}\phi} = \left|\frac{V_L^2}{Z}\right|.\cos\phi$$
$$Z = \frac{V_L^2}{P_{3\text{-}\phi}}\cdot\cos\phi = 60\ \Omega = \frac{208^2}{430}\times 0.6$$
$$\cos\phi = 0.6 \text{ lead} \Rightarrow \phi = 53.13^\circ$$
$$\therefore \quad Z = 60\angle{-53.13^\circ}\ \Omega$$

1.32 Sol.

From the given voltages,

$$I_R = \frac{V_{RN}}{R} = \frac{100\angle 0^\circ}{R}$$
$$I_Y = \frac{V_{YN}}{jX_L} = \frac{100\angle -120^\circ}{j10} = 10\angle{-210^\circ}$$

$$I_B = \frac{V_{BN}}{-jX_C} = \frac{100\angle-240°}{-j10} = 10\angle-150°$$

For $I_N = 0$

$$I_R + I_Y + I_B = 0$$

$$\frac{100}{R} + 10\angle-210° + 10\angle-150° = 0$$

$$R = 5.77\ \Omega$$

1.33 Sol.

Given : $V_s = 400$ kV; $l = 300$ km

$L_1 = 1$ mH/km/phase

$C_1 = 0.01$ μF/km/phase

$$v = \frac{1}{\sqrt{L_1C_1}} = \frac{1}{\sqrt{1\times10^{-3}\times0.01\times10^{-6}}}$$

$$= 3.16 \times 10^5 \text{ km/sec}$$

$$\beta' = \frac{2\pi fl}{v} = \frac{2\pi\times50\times300}{3.16\times10^5} = 0.29$$

$$A = 1-\frac{\beta^2}{2} = 1-\frac{(0.29)^2}{2} = 0.955$$

$$V_R = \frac{V_S}{A} = \frac{400}{0.955} = 418.85 \text{ kV}$$

1.34 (d)

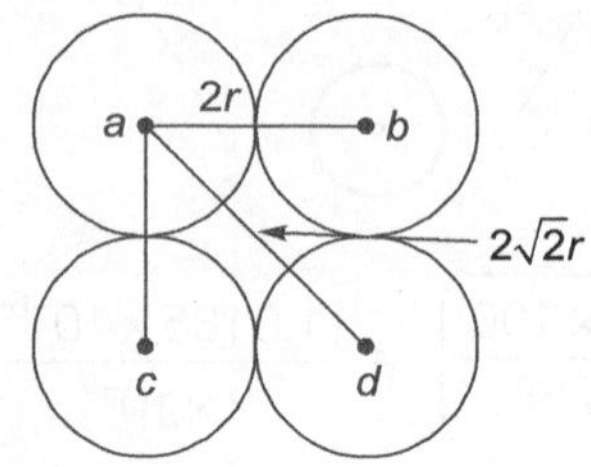

$$\text{GMR} = [D_{aa}\ D_{ab}\ D_{ac}\ D_{ad}]^{1/4}$$

$$= \left[e^{-1/4}r\cdot2r\cdot2r\cdot2\sqrt{2}r\right]^{1/4} = 1.723r$$

■■■■

2 Compensation Techniques and Voltage Profile Control

2.1 A load draws 100 kW at 0.7 p.f. lagging from a 3-phase, 11 kV supply. It is desired to raise the p.f. to 0.95 lagging using shunt capacitors. Calculate the rating of the capacitor required. [1997 : 2 M]

2.2 Series capacitive compensation in EHV transmission lines is used to
(a) Reduce the line loading
(b) Improve the stability of the system
(c) Reduce the voltage profile
(d) Improve the protection of line [1998 : 1 M]

2.3 A shunt reactor of 100 MVAR is operated at 98% of its rated voltage and at 96% of its rated frequency. The reactive power absorbed by the reactor is _____ (in MVAR)
(a) 98 (b) 104.02
(c) 96.04 (d) 100.04 [1998 : 2 M]

2.4 The voltage phasor of a circuit is $10\angle 15°$ V and the current phasor is $2\angle -45°$ A. The active and the reactive powers in the circuit are
(a) 10 W and 17.32 VAR
(b) 5 W and 8.66 VAR
(c) 20 W and 60 VAR
(d) $20\sqrt{2}$ W and $10\sqrt{2}$ VAR [1999 : 1 M]

2.5 A 3-phase, 11 kV, 50 Hz, 200 kW load has a power factor of 0.8 lag. A delta connected 3-phase capacitor is used to improve the power factor to unity. The capacitance per-phase of the capacitor in micro-farad is
(a) 3.948 (b) 1.316
(c) 0.439 (d) 11.844 [1999 : 2 M]

2.6 A transmission line has equal voltages at the two ends, maintained constant by two sources. A third source is to be provided to maintain constant voltage (equal to end voltages) at either the midpoint of the line or at 75% of the distances from the sending end. Then the maximum power transfer capabilities of the line in the original case and the other two cases respectively will be in the following ratios.
(a) 1 : 1 : 1 (b) 1 : 2 : 1/0.75
(c) 1 : 2 : 4 (d) 1 : 4 : 6 [2000 : 2 M]

2.7 A lossless radial transmission line with surge impedance loading
(a) Taking negative VAR at sending end and zero VAR at receiving end.
(b) Takes positive VAR at sending end and zero VAR at receiving end.
(c) Has flat voltage profile and unity power factor at all points along it.
(d) Has sending end voltage higher than receiving end voltage and unity power factor at sending end. [2001 : 1 M]

2.8 Consider the model shown in the figure of a transmission line with a series capacitor at its mid-point. The maximum voltage on the line is at the location

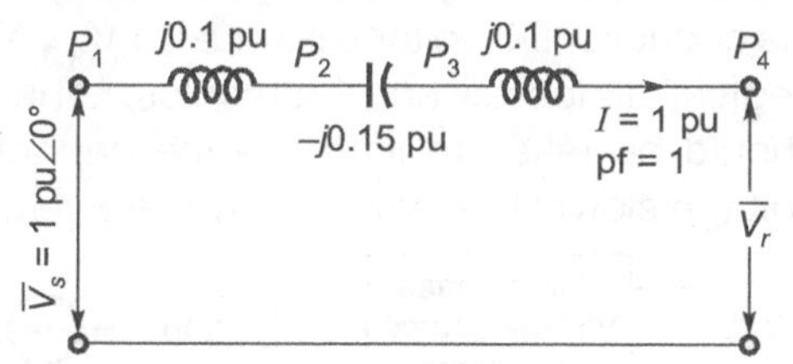

(a) P_1 (b) P_2
(c) P_3 (d) P_4
[2001 : 2 M]

2.9 A 3-phase 11-kV generator feeds power to a constant power unity power factor load of 100 MW through a 3-phase transmission line. The line-to-line voltage at the terminals of the machine is maintained constant at 11 kV. The per unit positive sequence impedance of the line based on 100 MVA and 11 kV is $j0.2$. The line to line voltage at the load terminals is measured to be less than 11 kV. The total reactive power to be injected at the terminals of the load to increase the line-to-line voltage at the load terminals to 11 kV is _____ (in MVAR)
(a) 100 (b) 10.1
(c) –100 (d) –10.1 [2003 : 2 M]

2.10 A balanced delta connected load of $(8 + j6)\ \Omega$ per phase is connected to a 400 V, 50 Hz, 3-phase supply lines. If the input power factor is to be improved to 0.9 by connecting a bank of star connected capacitors the required kVAR of the bank is
(a) 42.7 (b) 10.2
(c) 28.8 (d) 38.4 [2003 : 2 M]

2.11 At an industrial sub-station with a 4 MW load, a capacitor of 2 MVAR is installed to maintain the load power factor at 0.97 lagging. If the capacitor goes out of service, the load power factor becomes
(a) 0.85
(b) 1.00
(c) 0.80 lag
(d) 0.90 lag [2005 : 2 M]

2.12 A 400 V, 50 Hz, three phase balanced source supplies power to a star connected load whose rating is $12\sqrt{3}$ KVA, 0.8 pf (lag). The rating (in KVAR) of the delta connected (capacitive) reactive power bank necessary to bring the pf to unity is
(a) 28.78
(b) 21.60
(c) 16.60
(d) 12.47 [2006 : 2 M]

2.13 The figure below shows a three phase self-commutated voltage source converter connected to a power system. The converter's dc bus capacitor is marked as C in the figure. The circuit in initially operating in steady state with $\delta = 0$ and the capacitor DC voltage is equal to $V_{dc0.}$ You may neglect all losses and harmonics. What action should be taken to increase the capacitor DC voltage slowly to a new steady state value?

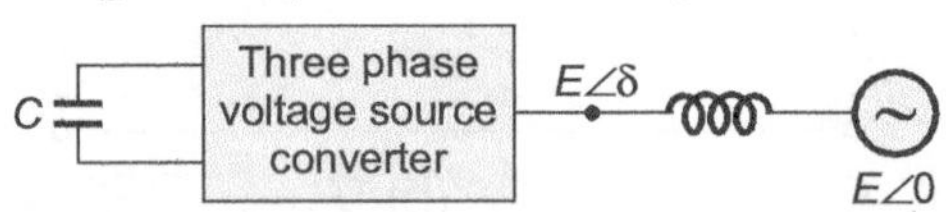

(a) Make δ positive and maintain it at a positive value.
(b) Make δ positive and return it to its original value.
(c) Make δ negative and maintain it at a negative value.
(d) Make δ negative and return it to its original value. [2007 : 2 M]

2.14 A 230 V (Phase), 50 Hz, three-phase,4-wire, system has a sequence ABC. A unity power-factor load of 4 kW is connected between phase A and neutral N. It is desired to achieve zero neutral current through the use of a pure inductor and pure capacitor in the other two phases. The value of inductor and capacitor is
(a) 72.95 mH in phase C & 139.02 μF in phase B.
(b) 72.95 mH in Phase B & 139.02 μF in phase C.
(c) 42.12 mH in Phase C & 240.79 μF in phase B.
(d) 42.12 mH in Phase b & 240.79 μF in phase C. [2007 : 2 M]

2.15 A lossless transmission line having Surge Impedance Loading (SIL) of 2280 MW is provided with a uniformly distributed series capacitive compensation of 30%. Then, SIL of the compensated transmission line will be
(a) 1835 MW
(b) 2280 MW
(c) 2725 MW
(d) 3257 MW [2008 : 2 M]

2.16 Match the items in **List-I (To)** with the items in **List-II (Use)** and select the correct answer using the codes given below the lists.

List-I
A. Improve power factor
B. Reduce the current ripples
C. Increase the power flow in line
D. Reduce the Ferranti effect

List-II
1. Shunt reactor
2. Shunt capacitor
3. Series capacitor
4. Series reactor

Codes:

	A	B	C	D
(a)	2	3	4	1
(b)	2	4	3	1
(c)	4	3	1	2
(d)	4	1	3	2

[2009 : 2 M]

2.17 For enhancing the power transmission in a long EHV transmission line, the most preferred method is to connect a
(a) series inductive compensator in the line.
(b) shunt inductive compensator at the receiving end.
(c) series capacitive compensator in the line.
(d) shunt capacitive compensator at the sending end. [2011 : 1 M]

2.18 For the system shown below, S_{D1} and S_{D2} are complex power demands at bus 1 and bus 2 respectively. If $|V_2| = 1$ pu, the VAR rating of the capacitor (Q_{G2}) connected at bus 2 is

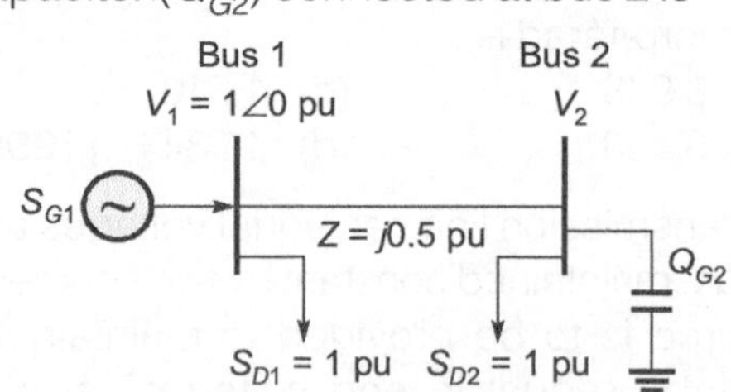

(a) 0.2 pu
(b) 0.268 pu
(c) 0.312 pu
(d) 0.4 pu [2012 : 2 M]

2.19 Shunt reactors are sometimes used in high voltage transmission systems to
(a) limit the short circuit current through the line.
(b) compensate for the series reactance of the line under heavily loaded condition.
(c) limit over voltages at the load side under lightly loaded condition.

(d) compensate for the voltage drop in the line under heavily loaded condition.

[2014 : 1 M, Set-2]

2.20 The complex power consumed by a constant voltage load is given by $(P_1 + jQ_1)$, where $1 \text{ kW} \le P_1 \le 1.5 \text{ kW}$ and $0.5 \text{ kVAR} \le Q_1 \le 1 \text{ kVAR}$. A compensating shunt capacitor is chosen such that $|Q| \le 0.25$ kVAR, where Q is the net reactive power consumed by the capacitor load combination. The reactive power (in kVAR) supplied by the capacitor is _______ .

[2014 : 2 M, Set-3]

2.21 The inductance and capacitance of a 400 kV, 3-ϕ, 50 Hz lossless transmission line are 1.6 mH/km/phase and 10 nF/km/phase respectively. The sending end voltage is maintained at 400 kV. To maintain a voltage of 400 kV at the receiving end, when the line is delivering 300 MW load, the shunt compensation required is

(a) capacitive (b) inductive
(c) resistive (d) zero

[2016 : 1 M, Set-2]

2.22 A load is supplied by a 230 V, 50 Hz source. The active power P and the reactive power Q consumed by the load are such that $1 \text{ kW} \le P \le 2 \text{ kW}$ and $1 \text{ kVAR} \le Q \le 2 \text{ kVAR}$. A capacitor connected across the load for power factor correction generates 1 kVAR reactive power. The worst case power factor after power factor correction is

(a) 0.447 lag (b) 0.707 lag
(c) 0.894 lag (d) 1

[2017 : 2 M, Set-1]

2.23 Consider the two bus power system network with given loads as shown in the figure. All the values shown in the figure are in per unit. The reactive power supplied by generator G_1 and G_2 are Q_{G1} and Q_{G2} respectively. The per unit values of Q_{G1}, Q_{G2}, and line reactive power loss (Q_{loss}) respectively are

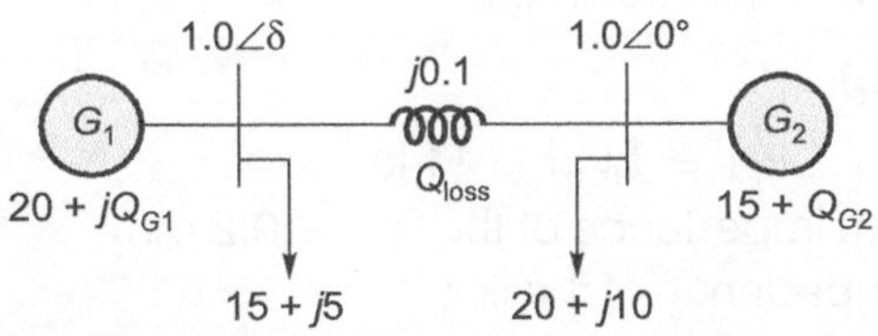

(a) 5.00, 12.68, 2.68 (b) 6.34, 10.00, 1.34
(c) 6.34, 11.34, 2.68 (d) 5.00, 11.34, 1.34

[2018 : 2 M]

2.24 Bus 1 with voltage magnitude $V_1 = 1.1$ p.u. is sending reactive power Q_{12} towards bus 2 with voltage magnitude $V_2 = 1$ p.u. through a lossless transmission line of reactance X. Keeping the voltage at bus 2 fixed at 1 p.u., magnitude of voltage at bus 1 is changed, so that the reactive power Q_{12} sent from bus 1 is increased by 20%. Real power flow through the line under both the conditions is zero. The new value of the voltage magnitude, V_1, in p.u. (rounded off to 2 decimal places) at bus 1 is _______.

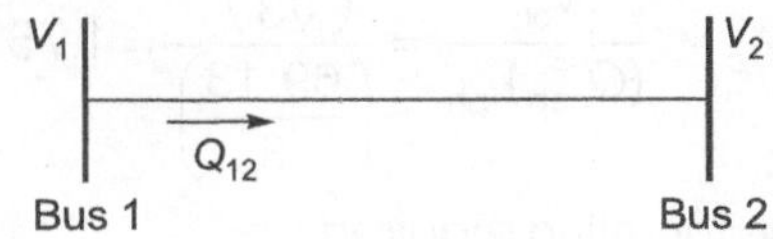

[2020 : 2 M]

2.25 For the three-bus lossless power network shown in the figure, the voltage magnitudes at all the buses are equal to 1 per unit (pu), and the differences of the voltage phase angles are very small. The line reactances are marked in the figure, where α, β, γ and x are strictly positive. The bus injections P_1 and P_2 are in pu. If $P_1 = mP_2$, where $m > 0$, and the real power flow from bus 1 to bus 2 is 0 pu. then which one of the following options is correct?

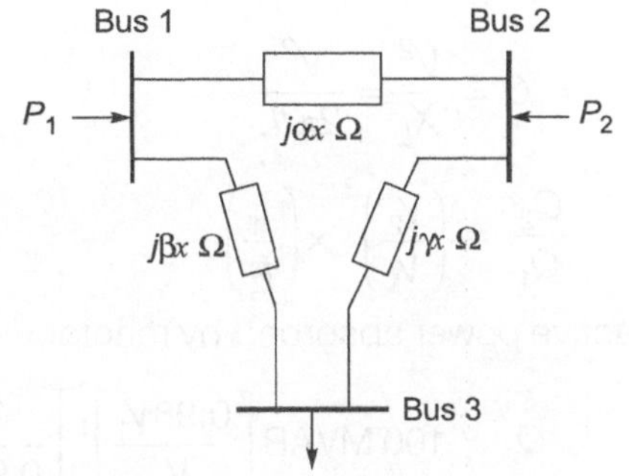

(a) $\alpha = m\beta$ (b) $\beta = m\gamma$
(c) $\gamma = m\beta$ (d) $\alpha = m\gamma$

[2024 : 2 M]

■■■■

Answers Compensation Techniques and Voltage Profile Control

2.1 (1.82 μF)	2.2 (b)	2.3 (d)	2.4 (a)	2.5 (b)	2.6 (b)	2.7 (c)
2.8 (c)	2.9 (b)	2.10 (b)	2.11 (c)	2.12 (d)	2.13 (d)	2.14 (b)
2.15 (c)	2.16 (b)	2.17 (c)	2.18 (b)	2.19 (c)	2.20 (0.75)	2.21 (b)
2.22 (b)	2.23 (c)	2.24 (1.12)	2.25 (c)			

Explanations Compensation Techniques and Voltage Profile Control

2.1 Sol.

$$Q_{old} = 100\tan(\cos^{-1}0.7) = 102 \text{ kVAR}$$
$$Q_{new} = 100\tan(\cos^{-1}0.95) = 32.87 \text{ kVAR}$$
$$\therefore Q_{cap} = Q_{old} - Q_{new} = 69.13 \text{ kVAR}$$

$$X_c = \frac{V_{ph}^2}{(Q_{cap})_{1ph}} = \frac{\left(\frac{11}{\sqrt{3}}\right)^2}{\left(\frac{69.13}{3}\right)} = 1.75 \text{ kW}$$

Capacitor rating required

$$\therefore C = \frac{1}{2\pi f X_C} = \frac{1}{2\pi\times1.75\times10^3\times50} = 1.82\ \mu\text{F}$$

2.2 (b)

Synchronizing power,

$$P_{syn} = \frac{3EV}{X}\cos\delta$$

As total reactance decreases the synchronizing power increases as $P_{syn} \propto \frac{1}{X}$.

2.3 (d)

$$Q = \frac{V^2}{X_L} = \frac{V^2}{2\pi fL}$$

$$\frac{Q_2}{Q_1} = \left(\frac{V_2}{V_1}\right)^2\times\left(\frac{f_1}{f_2}\right)$$

Reactive power absorbed by reactor,

$$Q_2 = 100\text{ MVAR}\left[\frac{0.98V_1}{V_1}\right]^2\left[\frac{f_1}{0.96f_1}\right]$$
$$= 100.042 \text{ MVAR}$$

2.4 (a)

Complex power = VI^*

$$P + jQ = 10\angle15^\circ\times2\angle45^\circ = 20\angle60^\circ$$
$$= (10 + j10\sqrt{3}) = (10 + j17.32)$$

$\Rightarrow$ $P = 10$ W and $Q = 17.32$ VAR

2.5 (b)

Line current,

$$I_L = \frac{200\times10^3}{\sqrt{3}\times11\times10^3\times0.8}\angle-\cos^{-1}(0.8)$$
$$= 10.4973 - j7.8729$$

The reactive part of the line current is to be supplied by capacitor bank in order to have unity pf.

$$I_{C(\text{phase})} = 7.8729/\sqrt{3} \qquad (\Delta \text{ connection})$$

$$\frac{V}{X_c} = \frac{7.8729}{\sqrt{3}}$$

$$X_c = \frac{11000\times\sqrt{3}}{7.8729} = 2420.0178\ \Omega$$

Capacitance per phase,

$$C = \frac{1}{2\pi\times50\times2420.02} = 1.31532\ \mu\text{F}$$

2.6 (b)

$$P_{max} = \frac{V^2}{x} \qquad \text{...(i)}$$

With compensator at mid point,

$$P' = \frac{V^2}{x/2}\sin\frac{\delta}{2}$$

$$P'_{max} = 2P_{max} \qquad \text{...(ii)}$$

With compensator at 75% distance $x' = 0.75x$

$$P''_{max} = \frac{V^2}{0.75x} = \frac{1}{0.75}P_{max} \qquad \text{...(iii)}$$

From (i), (ii) and (iii), the ratio is $1:2:\frac{1}{0.75}$.

2.8 (c)

$$V_{P_2} = 1\text{ pu} - 1\text{ pu}\times j0.1\text{ pu}$$
$$V_{P_3} = 1\text{ pu} + 1\text{ pu}\times j0.05\text{ pu}$$
$$V_{P_4} = 1\text{ pu} - 1\text{ pu}\times j0.05\text{ pu}$$

$\Rightarrow$ P_3 is maximum.

2.9 (b)

$$|V_s| = |V_R| = 11\text{ kV}$$

p.u impedance of the line = j0.2 p.u.

Impedance of the line

= p.u. impedance × Base impedance

$$x = j0.2\times\frac{11^2}{100} = j0.242\Omega$$

Active power at receiving end = $P_R = 100$ MW

$$P_R = \frac{|V_s||V_R|}{x}\sin\delta \Rightarrow 100 = \frac{11\times 11}{0.242}\sin\delta$$

$\delta = 11.54°$

Reactive power at receiving end

$$Q_R = \frac{|V_s||V_R|}{X}\cos\delta - \frac{|V_R|^2}{x}$$

$$= \frac{11\times 11}{0.242}\cos 11.54 - \frac{11^2}{0.242}$$

$= -10.1$ MVAR

Since, power factor = 1

Load reactive power should be zero.

Therefore, reactive power to be injected at terminals

$= -Q_R = -(-10.1) = 10.1$ MVAR

2.10 (b)

KVAR of capacitor bank,

$$Q_C = \frac{P_1(\tan\phi_1 - \tan\phi_2)}{3(2\pi f V_{ph}^2)} \quad ...(I)$$

$$P_1 = \frac{3\cdot V_L^2}{Z}\cos\phi_1 \text{ for } \Delta \text{ load}$$

Star capacitor bank,

$$V_{ph} = \frac{V_L}{\sqrt{3}} \quad \text{where, } V_L = 400 \text{ V}$$

$Z = 8 + j6, |Z| = 10$

$\phi_1 = \tan^{-1}(6/8)$

$\cos\phi_2 = 0.9$

Substitute in equation (i),

$Q_C = 10.2$ kVAR

2.11 (c)

$Q_C = P_1(\tan\phi_1 - \tan\phi_2)$

$2 = 4(\tan\phi_1 - \tan(\cos^{-1} 0.97))$

Load power factor,

$\cos\phi_1 = 0.8$ lag

2.12 (d)

Rating of the load $12\sqrt{3}$ kVA at 0.8 pf lag

Active power demand of the load = P_L

$= 12\sqrt{3}\times 0.8 = 16.63$ kW

$\cos\theta = 0.8$

$\Rightarrow \sin\theta = 0.6$

Reactive power demand of the load = Q_L

$= 12\sqrt{3}\sin\theta = 12\sqrt{3}\times 0.6$

$\Rightarrow Q_L = 12.47$ kVAR

For unity pf, the total reactive power is zero.

Assuming kVAR rating of capacitor bank = Q_C

$Q_L + Q_C = 0$

$Q_C = -Q_L = -12.47$ kVAR

So, capacitor banks supplies 12.47 kVAR reactive power to the load.

2.13 (d)

- To increase the capacitor DC voltage, energy has to be supplied by the generator to capacitor through converter.
- Power flow occurs from higher angle to lower angle. As angle of induced emf of generator is 0, so δ should be made negative to make the power flow in required direction.

2.14 (b)

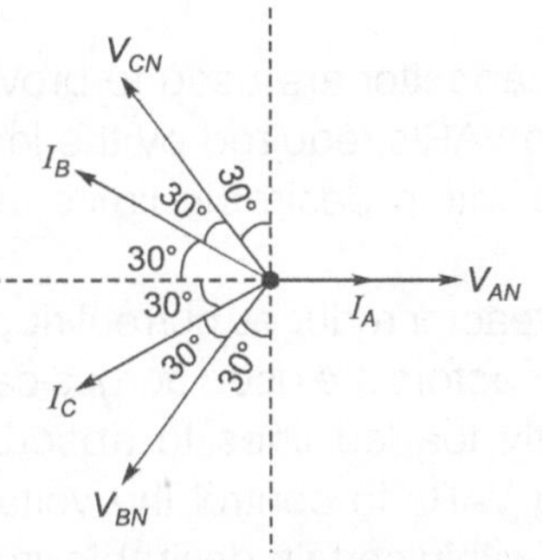

Taking V_{AN} as the reference, phase voltage

$V_{AN} = 230\angle 0°$V

$V_{BN} = 230\angle -120°$V

$V_{CN} = 230\angle 120°$V

$$I_A = \frac{P}{V_{AN}\cos\phi} = \frac{4\times 10^3}{230\times 1} = 17.4\angle 0°$$

- If a pure inductor is present in phase B, then I_B lags V_{BN} by 90°.
- If a pure capacitor is present in phase C, then I_C leads V_{CN} by 90°.

If current through neutral is to be zero.

$I_B \sin 30° = I_C \sin 30°$

$\Rightarrow I_B = I_C = I$

$I_A = I_B\cos 30° + I_C\cos 30°$

$= 2I\cos 30° = \sqrt{3}I$

$$|I_B| = |I_C| = |I| = \frac{|I_A|}{\sqrt{3}}$$

$$= \frac{17.04}{\sqrt{3}} = 10.05 \text{ A}$$

$$|I_B| = \frac{|V_{BN}|}{\omega L}$$

$$\Rightarrow 10.05 = \frac{230}{2\pi\times 50\times L}$$

$L = 72.847$ mH

$$|I_C| = |V_{CN}| \cdot \omega C$$

$$\Rightarrow 10.05 = 230 \times 2 \times \pi \times 50 \times C$$

$$C = 139.087\ \mu F$$

2.15 (c)

Surge impedance loading,

$$SIL = 2280\ MW$$

$$Z_c = \text{Surge impedance}$$

$$SIL_2 = \frac{SIL_1}{\sqrt{1-\gamma_{se}}}$$

$$SIL_2 = \frac{2280}{\sqrt{1-0.3}} = 2725.12\ MW$$

2.16 (b)

- Shunt capacitor are used to provide part of the reactive VAR's required by the load to keep the voltage within desirable limits and to improve factor.
- Series reactor reduces current ripple.
- Shunt reactors are used across capacitive loads or lightly loaded lines to absorb some of the leading VARs to control the voltage across the load to within certain desirable limit.
- Series capacitor compensation reduces the series impedance of the line. Power flow in line $\propto \frac{1}{X_L}$, power flow in line increases, as X_L decreases.

2.17 (c)

Series capacitance compensation reduces reactance of the line.

$$\text{Power flow in line} \propto \frac{1}{\text{reactance of the line}}$$

As reactance of line decreases, power flow in the line C_a increases.

2.18 (b)

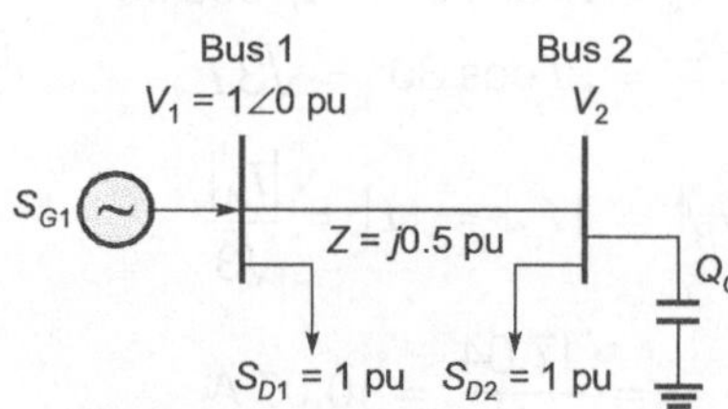

$$\text{Real power, } P_r = \frac{|V_S||V_R|}{|X|}\sin\delta$$

$$1 = \frac{1.0 \times 1.0}{0.5}\sin\delta$$

$$\Rightarrow \quad \delta = \sin^{-1}(0.5) = 30°$$

Reactive power,

$$Q_r = \frac{|V_S||V_r|}{|X|}\cos\delta - \frac{|V|^2}{|X|}$$

$$= \frac{1.0 \times 1.0}{0.5}\cos 30 - \frac{1^2}{0.5} = \frac{\left(\frac{\sqrt{3}}{2}\right)}{\left(\frac{1}{2}\right)} - 2$$

$$\Rightarrow \quad 1.732 - 2 = -0.268$$

But $Q_r + Q_C = 0$

$$\Rightarrow \quad Q_C = -Q_r \Rightarrow 0.268 \text{ p.u}$$

2.19 (c)

Shunt reactor is basically an inductor having no-core or core is replaced by cement slab to avoid saturation problem. These are widely used to reduce the fault level or the short circuit current through the line.

2.20 Sol.

Given, $S_{load} = P_1 + jQ_1$

$1\text{ kW} \le P_1 \le 1.5\text{ kW}$ and

$0.5\text{ kVAR} \le Q_1 \le 1\text{ kVAR}$

Shunt capacitor will only supply reactive power and the total reactive power consumed by the load and the capacitor together will be less than 0.25 kVAR i.e. $|Q| \le 0.25$ kVAR.

∴ Reactive power supplied by capacitor will be $Q_c = (1 - 0.25) = 0.75$ kVAR (lagging)

2.21 (b)

$$Z_n = \sqrt{\frac{L}{C}} = \sqrt{\frac{1.6 \times 10^{-3}}{10 \times 10^{-9}}} = 400\ \Omega$$

$$SIL = \frac{400 \times 400}{400} = 400\ MW$$

In the second case SIL decreases means Z_n increases.

Z_n increases with increase in inductance 'L'.

So, it is inductive.

Load < SIL means, line behaves capacitive to compensate it inductor to be placed.

2.22 (b)

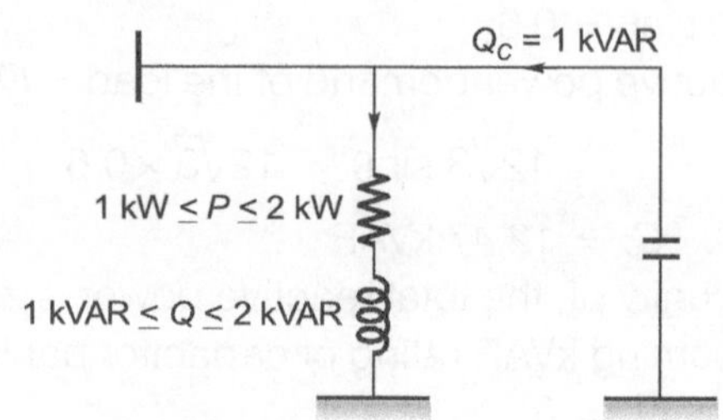

Load power factor,

$$\cos\phi_1 = \cos\left[\tan^{-1}\frac{Q}{P}\right] = \cos\left[\tan^{-1}\frac{1}{1}\right]$$

$$= 0.707 \text{ lag}$$

$$Q_c = P[\tan\phi_1 - \tan\phi_2]$$

$$1 = 1(\tan\cos^{-1}0.707 - \tan\phi_2)$$

$$\phi_2 = 45° \Rightarrow \cos\phi_2 = 0.707 \text{ lag}$$

2.23 (c)

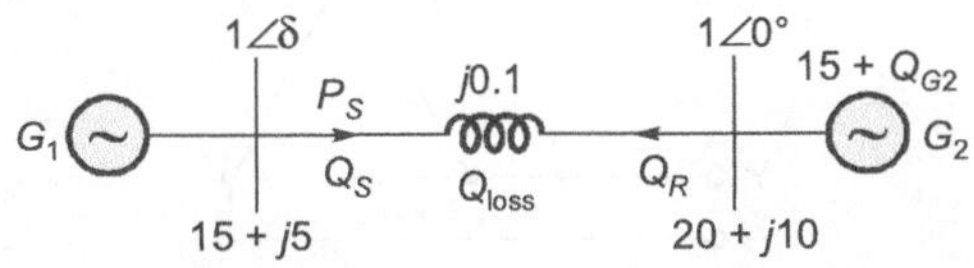

At G_2 load demand is 20 pu, G_2 supplies only 15 pu. Remaining supplied by G_1 through transmission line.

Real power,

$$P_s = \left|\frac{V_S V_R}{X_L}\right|\sin\delta$$

$$5 = \frac{1\times1}{0.1}\sin\delta$$

$$\delta = 30°$$

$$Q_S = \frac{V_S^2}{X_L} - \frac{V_S V_R}{X_L}\cos\delta$$

$$= \frac{1^2}{0.1} - \frac{1\times1}{0.1}\cos30° = 1.34 \text{ p.u.}$$

$$Q_R = \left|\frac{V_S V_R}{X_L}\right|\cos\delta - \left|\frac{V_R^2}{X_L}\right|$$

$$= \left|\frac{1\times1}{0.1}\right|\cos30° - \frac{1^2}{0.1} = -1.34 \text{ p.u.}$$

$$Q_{loss} = Q_S - Q_R$$

$$= 1.34 - (-1.34) = 2.68 \text{ p.u.}$$

At G_1: $Q_{G1} = Q_{load} + Q_S$

$$= 5 + 1.34 = 6.34 \text{ p.u.}$$

At G_2: $Q_{G2} = Q_{load} + (-Q_R)$

$$= 10 - (-1.34) = 11.34 \text{ p.u.}$$

2.24 Sol.

With real power zero, load angle $\delta = 0$

With initial values, $V_1 = 1.1$, $V_2 = 1$

$$Q_{12} = \frac{V_1^2}{X} - \frac{V_1V_2}{X}\sin\delta$$

$$= \frac{(1.1)^2}{X} - \frac{1.1\times1}{X}\sin0 = \frac{0.11}{X}$$

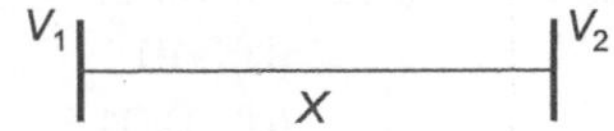

With increased value of voltage,

new value of $Q_{12} = 1.2Q_{12}$, $V_2 = 1$

$$1.2\,Q_{12} = \frac{V_1^2}{X} - \frac{V_1\times1}{X} = 1.2\times\frac{0.11}{X}$$

$$V_1^2 - V_1 - 0.132 = 0$$

$$V_1 = 1.12, \quad -0.118$$

Hence the practical value in per unit, $V_1 = 1.12$ p.u.

2.25 (c)

$$P_1 = mP_2$$

$$P_1 = \frac{1\times1}{\beta x}\sin\theta$$

$$P_2 = \frac{1\times1}{\gamma x}\sin\theta$$

$$\frac{1\times1}{\beta x}\sin\theta = m\frac{1\times1}{\gamma x}\sin\theta$$

$$\frac{1}{\beta} = \frac{m}{\gamma} \Rightarrow \gamma = m\beta$$

■■■■

3 Distribution Systems, Cables and Insulators

3.1 The insulation resistance of a cable of length 10 km is 1 MΩ. For a length of 100 km of the same cable, the insulation resistance will be ___ (in MΩ)

(a) 1 (b) 10
(c) 0.1 (d) 0.01

[1995 : 1 M]

3.2 A cable has the following characteristics. $L = 0.201$ μH/m and $C = 196.2$ pF/m. The velocity of wave propagation through the cable is

(a) 32 m/s (b) 159.24 m/ms
(c) 0.0312 m/s (d) 159.24 m/(μ-s)

[1998 : 2 M]

3.3 A single phase AC distributor supplies two single phase loads as shown in figure. The voltage drop from A to C is

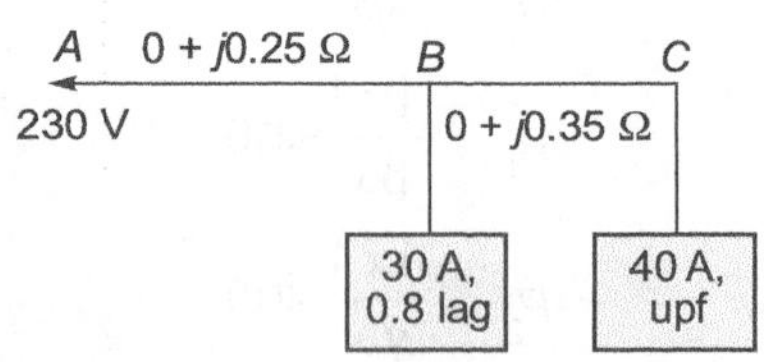

(a) 4.5 V (b) 31.5 V
(c) 30.34 V (d) 20 V

[1999 : 2 M]

3.4 A dc distribution system is shown in figure with load currents as marked. The two ends of the feeder are fed by voltage sources such that $V_P - V_Q = 3$ V. The value of the voltage V_P for a minimum voltage of 220 V at any point along the feeder is

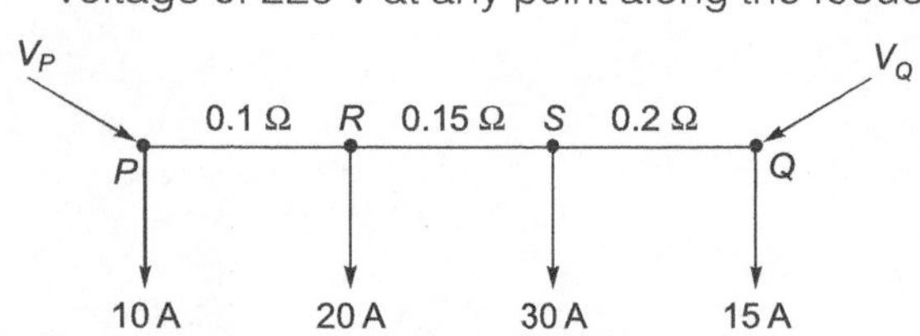

(a) 225.89 V (b) 222.89 V
(c) 220.0 V (d) 228.58 V

[2003 : 2 M]

3.5 The rated voltage of a 3-phase power system is given as

(a) rms phase voltage
(b) peak phase voltage
(c) rms line to line voltage
(d) peak line to line voltage

[2004 : 1 M]

3.6 The phase sequence of the 3-phase system shown in the figure is

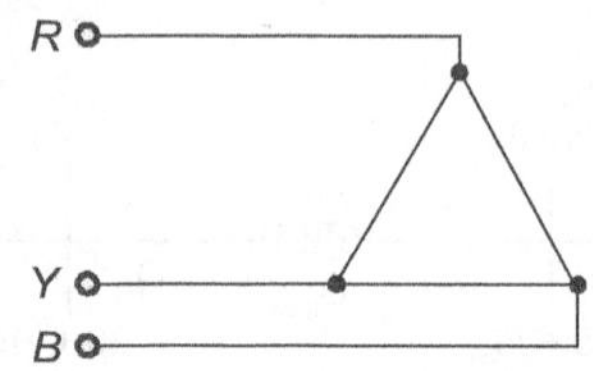

(a) RYB (b) RBY
(c) BRY (d) YBR

[2004 : 1 M]

3.7 A 110 kV, three core coaxial, XLPE insulated power cable delivering power at 50 Hz, has a capacitance of 125 nF/km. If the dielectric loss tangent of XLPE is 2×10^{-4}, then dielectric power loss in this cable in W/km is

(a) 5.0 (b) 31.7
(c) 37.8 (d) 189.0

[2004 : 2 M]

3.8 Single line diagram of a 4-bus single source distribution system is shown below. Branches e_1, e_2, e_3 and e_4 have equal impedances. The load current values indicated in the figure are in per unit.

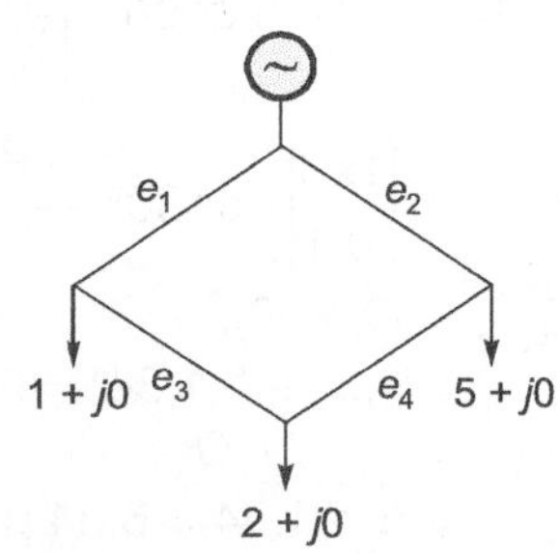

Distribution company's policy requires radial system operation with minimum loss. This can be achieved by opening of the branch

(a) e_1 (b) e_2
(c) e_3 (d) e_4

[2008 : 2 M]

3.9 Consider a three-phase, 50 Hz, 11 kV distribution system. Each of the conductors is suspended by an insulator string having two identical porcelain insulators. The self capacitance of the insulator is 5 times the shunt capacitance between the link and the ground, as shown in the figures. The voltage across the two insulators are

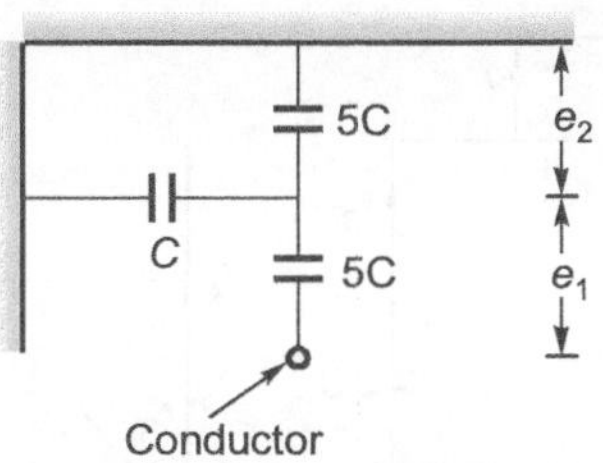

(a) e_1 = 3.74 kV, e_2 = 2.61 kV
(b) e_1 = 3.46 kV, e_2 = 2.89 kV
(c) e_1 = 6.0 kV, e_2 = 4.23 kV
(d) e_1 = 5.5 kV, e_2 = 5.5 kV [2010 : 2 M]

3.10 Consider a three-core, three-phase, 50 Hz, 11 kV cable whose conductors are denoted as *R*, *Y* and *B* in the figure. The inter-phase capacitance (C_1) between each pair of conductors is 0.2 μF and the capacitance between each line conductor and the sheath is 0.4 μF. The per-phase charging current is

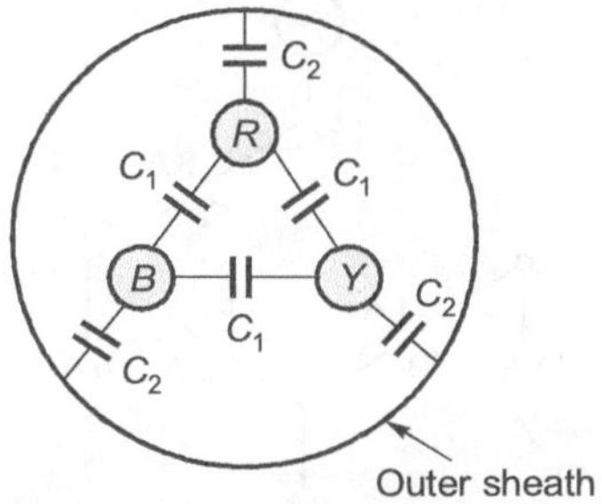

(a) 2.0 A (b) 2.4 A
(c) 2.7 A (d) 3.5 A [2010 : 2 M]

3.11 The undesirable property of an electrical insulating material is
(a) high dielectric strength
(b) high relative permittivity
(c) high thermal conductivity
(d) high insulation resistivity [2014 : 1 M, Set-1]

3.12 A distribution feeder of 1 km length having resistance, but negligible reactance, is fed from both the ends by 400 V, 50 Hz balanced sources. Both voltage sources S_1 and S_2 are in phase. The feeder supplies concentrated loads of unity power factor as shown in the figure.

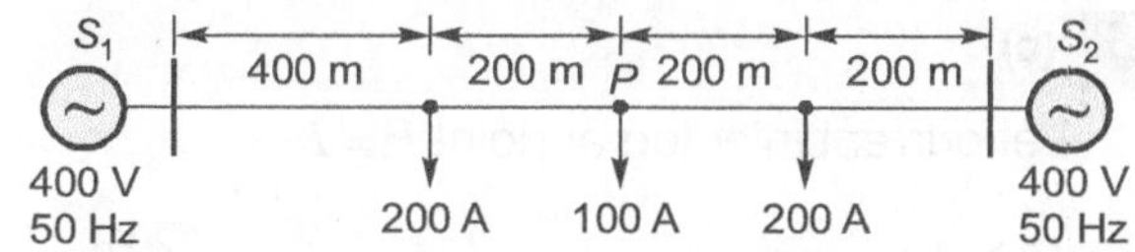

The contribution of S_1 and S_2 in 100 A current supplied at location P respectively, are
(a) 75 A and 25 A (b) 50 A and 50 A
(c) 25 A and 75 A (d) 0 A and 100 A
[2014 : 2 M, Set-1]

3.13 A three-phase cable is supplying 800 kW and 600 kVAr to an inductive load. It is intended to supply an additional resistive load of 100 kW through the same cable without increasing the heat dissipation in the cable, by providing a three-phase bank of capacitors connected in star across the load. Given the line voltage is 3.3 kV, 50 Hz, the capacitance per phase of the bank, expressed in microfarads, is _______. [2016 : 2 M, Set-1]

3.14 Two single-core power cables have total conductor resistances of 0.7 Ω and 0.5 Ω, respectively, and their insulation resistances (between core and sheath) are 600 MΩ and 900 MΩ. respectively, When the two cables are joined in series, the ratio of insulation resistance to conductor resistance is _______ × 10^6. [2021 : 1 M]

3.15 If only 5% of the supplied power to a cable reaches the output terminal, the power loss in the cable, in decibels, is _____. (round off to nearest integer). [2022 : 1 M]

Answers Distribution Systems, Cables and Insulators

3.1 (c)	3.2 (d)	3.3 (c)	3.4 (a)	3.5 (c)	3.6 (b)	3.7 (b)
3.8 (d)	3.9 (b)	3.10 (a)	3.11 (c)	3.12 (d)	3.13 (47.96)	3.14 (300)
3.15 (13)						

Explanations Distribution Systems, Cables and Insulators

3.1 (c)

Insulation resistance,

$$R_{ins} = \frac{\rho}{2\pi l} ln\left(\frac{R}{r}\right) \Omega \Rightarrow \frac{R_2}{R_1} = \left(\frac{l_1}{l_2}\right)$$

$$R_2 = 1\text{M}\Omega\left(\frac{10}{100}\right) = 0.1\text{M}\Omega$$

3.2 (d)

Velocity of wave propagation,

$$v = \frac{1}{\sqrt{LC}} = \frac{1}{\sqrt{0.201\times10^{-6}\times196.2\times10^{-12}}}$$

$$= \frac{1000}{\sqrt{0.201\times196.2}\,\mu\text{s/m}} = 159.24 \text{ m/μs}$$

3.3 (c)

$V_C - (40 \times j0.35) - (40 + 30\angle -\cos^{-1}(0.8)) \times j0.25 = V_A$

Voltage drop,

$$V_{CA} = 16.6207\angle 74.29° + 14\angle 90°$$
$$= 30.34\ \angle 81.468°\ \text{V}$$

3.4 (a)

Let current injected at point $P = I_P$

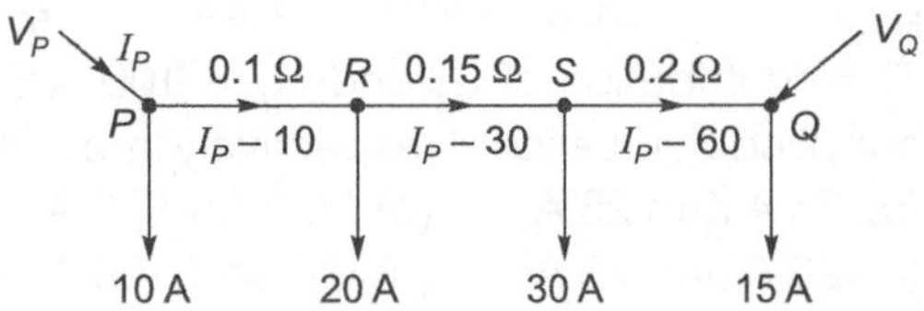

$$V_P - V_Q = 0.1(I_P - 10) + 0.15(I_P - 30) + 0.2(I_P - 60)$$
$$\Rightarrow \quad 3 = 0.45\,I_P - 17.5$$
$$\Rightarrow \quad I_P = 45.55\ \text{A}$$

Direction of current will be in opposite in section *SQ*. So, points will have minimum voltage,

$$V_s = 220\ \text{V}$$
$$V_p = V_s + 0.1(I_P - 10) + 0.15(I_P - 30)$$
$$= 220 + 0.1 \times (45.55 - 10) + 0.15(45.55 - 30)$$
$$\approx 225.89\ \text{V}$$

3.5 (c)

- Generators are specified in 3-ϕ MVA, line to line voltage and per-phase reactance (equivalent star).
- Transformers are specified in 3-ϕ MVA, line to line transformation ratio and per phase (equivalent star) impedance on one side.
- Load are specified in 3-ϕ MW, line to line voltage and power factor.

3.6 (b)

- The phase sequence of the given figure in question is RBY.
- RYB, BRY and YBR represent the same phase sequence of the figure (given below).

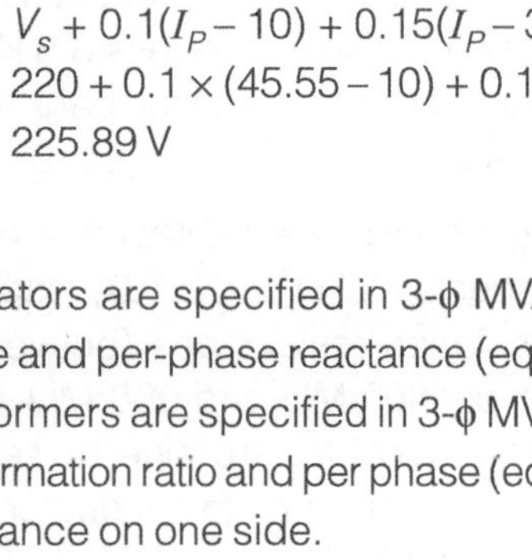

3.7 (b)

$$\text{V(Phase voltage)} = \frac{110}{\sqrt{3}}\ \text{kV}$$
$$C = 125\ \text{nF/km}$$
$$\tan\delta = 2 \times 10^{-4}$$

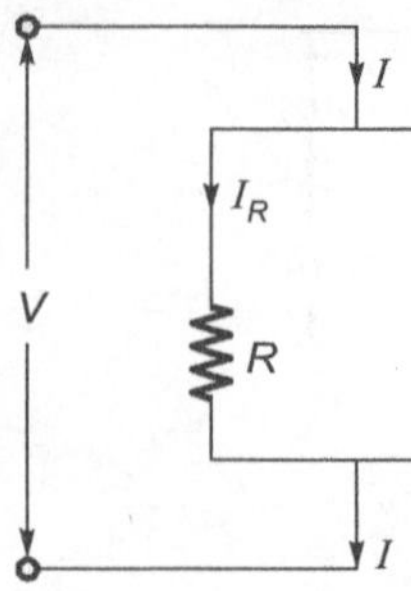

Equivalent of a cable

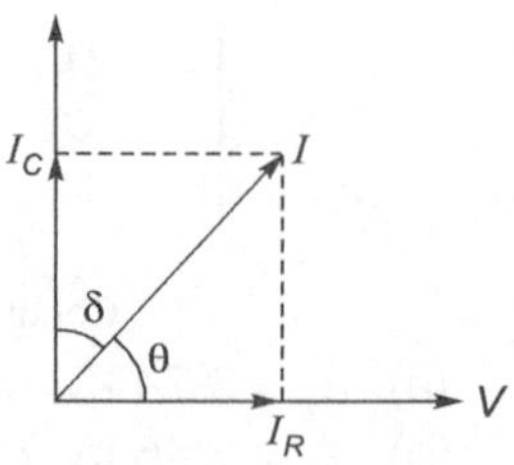

Phasor diagram

Dielectric power loss in cable = $P = V^2\omega C\tan\delta$

$$P = \left(\frac{110 \times 10^3}{\sqrt{3}}\right)^2 \times 2 \times \pi \times 50 \times 125 \times 10^{-9} \times 2 \times 10^{-4}$$
$$\approx 31.7\ \text{W/km}$$

3.8 (d)

Assuming impedance of each branch *R*

(i) if e_1 is opened

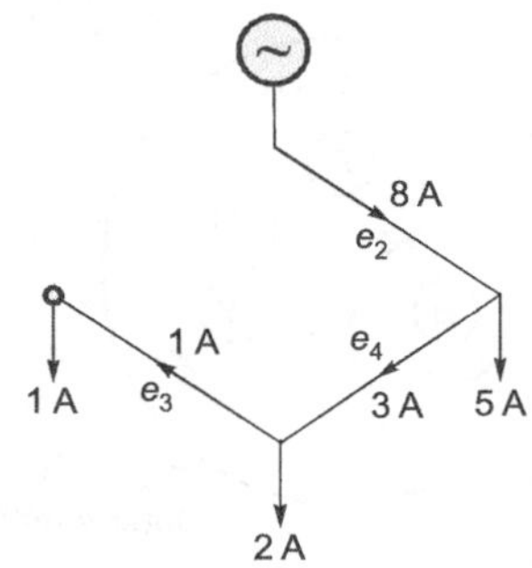

Total losses = $8^2R + 3^2R + 1^2R = 74R$

(ii) If e_2 is opened

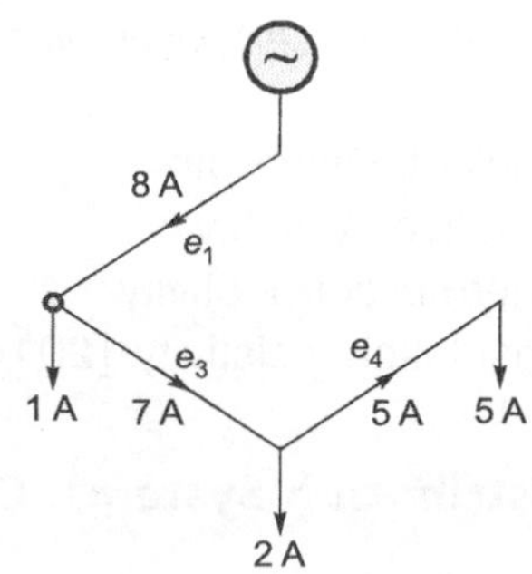

Total losses = $8^2R + 7^2R + 5^2R = 138R$

(iii) If e_3 is opened

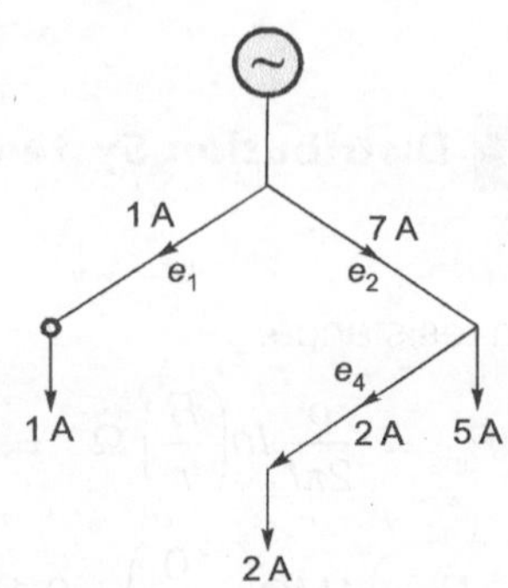

Total losses = $1^2R + 7^2R + 2^2R = 54R$

(iv) If e_4 is opened

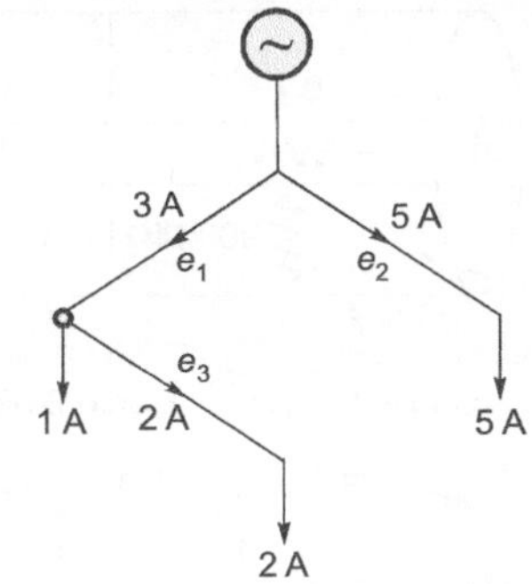

Total losses = $5^2R + 3^2R + 2^2R = 38R$

Operation with minimum loss can be achieved by opening line e_4.

3.9 (b)

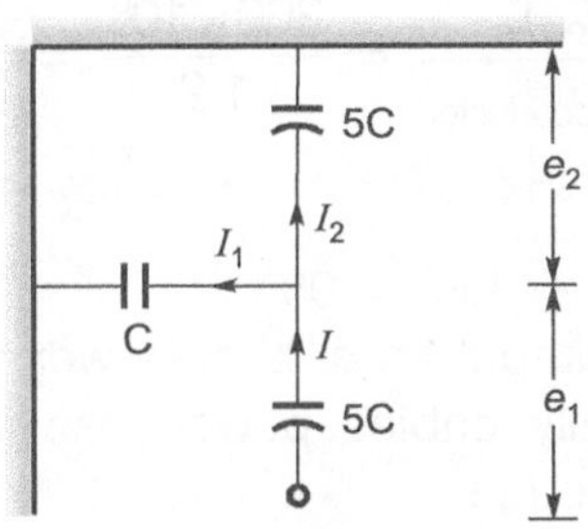

Line to line voltage = $V_{l-l} = 11$ kV

$$V_P = \text{Phase to ground voltage}$$

$$= \frac{V_{l-l}}{\sqrt{3}} \Rightarrow V_P = \frac{11}{\sqrt{3}}\text{ kV}$$

$$e_1 + e_2 = \frac{11}{\sqrt{3}}\text{ kV} = 6.35\text{ kV} \quad \ldots\text{(i)}$$

$$I = I_1 + I_2$$

$$(j\omega 5C)\times e_1 = (j\omega C)\times e_2 + (j\omega 5C)\times C_2$$

$$5e_1 = 6e_2 \quad \ldots\text{(ii)}$$

Solving equation (i) and (ii), we get

$$e_1 = 3.46\text{ kV and } e_2 = 2.89\text{ kV}$$

3.10 (a)

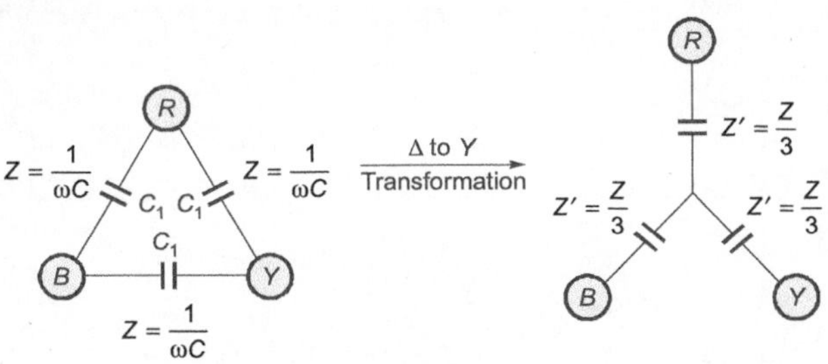

$$Z' = Z/3$$

$$\Rightarrow \quad \frac{1}{\omega C'} = \frac{1}{3\omega C}$$

$$C' = 3C$$

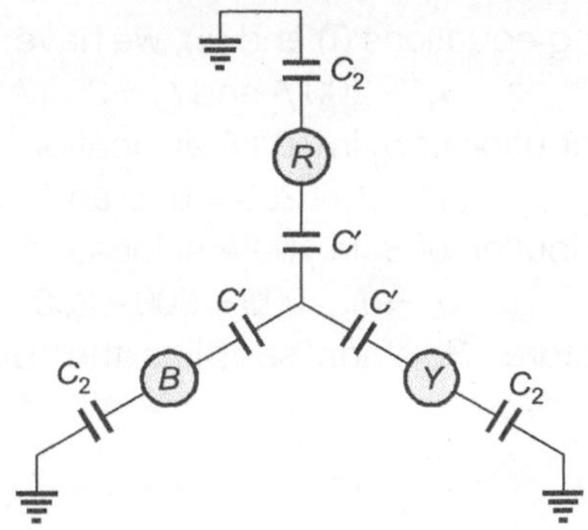

Equivalent capacitance (C_{eq}) between a phase and ground

$$C_{eq} = C' + C_2 = 3C_1 + C_2$$

$$C_{eq} = 3 \times 0.2 + 0.4 = 1\ \mu\text{F}$$

Per phase voltage = $V_P = \frac{11}{\sqrt{3}}$ kV

Per phase charging current

$$I_C = j\omega C_{eq} V_P$$

$$|I_C| = \omega C_{eq} V_P$$

$$= 2\pi \times 50 \times 1 \times 10^{-6} \times \frac{11}{\sqrt{3}} \times 10^3 \approx 2\text{ A}$$

3.12 (d)

Let the resistance of whole length of feeder be $R\,\Omega$, length of feeder = 1000 m.

$\therefore$ Resistance per unit length = $\frac{R}{1000}\ \Omega$ / meter

$\therefore$ Resistance of 400 m length

$$= \frac{R}{1000} \times 400 = \frac{2R}{5}\ \Omega$$

$\therefore$ Resistance of 200 m length = $\frac{R}{5}\ \Omega$

Let the current supplied by the sources be I_1 and I_2.

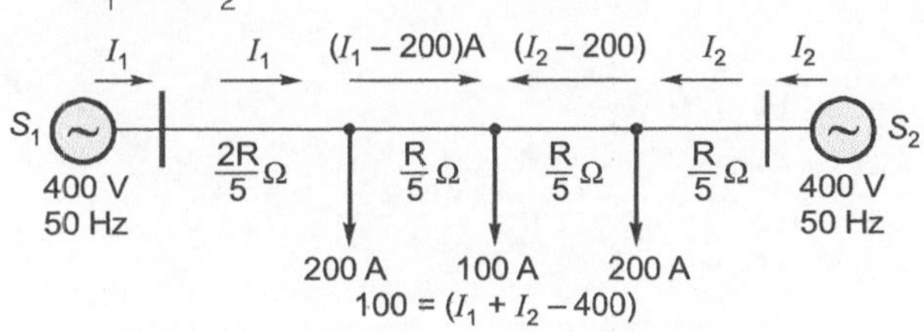

Applying KVL from source S_1 to S_2, we have:

$$400 = \frac{2R}{5}I_1 + \frac{R}{5}(I_1 - 200) - \frac{R}{5}(I_2 - 200) - \frac{R}{5}I_2 + 400$$

$$\text{or, } \frac{2}{5}I_1 + \frac{I_1 - 200}{5} + \frac{200 - I_2}{5} - \frac{I_2}{5} = 0$$

$$\text{or, } 2I_1 + I_1 - 200 + 200 - I_2 - I_2 = 0$$

$$\text{or, } 3I_1 - 2I_1 = 0 \text{ or } I_1 = \frac{2}{3}I_2 \quad \ldots\text{(i)}$$

Also, $I_1 + I_2 - 400 = 100$

or, $I_1 + I_2 = 500$ ...(ii)

Solving equations (i) and (ii), we have:

$$I_1 = 200 \text{ A and } I_2 = 300 \text{ A}$$

Contribution of S_1 in 100 A at location

$$P = I_1 - 200 = 0 \text{ A} \text{ and}$$

Contribution of S_2 in 100 A at location

$$P = I_2 - 200 = 300 - 200 = 100 \text{ A}$$

Therefore, S_2 alone supplies the total load at location P.

3.13 Sol.

$$\text{KVA}_1 = \sqrt{800^2 + 600^2} = 1000 \text{ KVA}$$

Without excessive heat dissipation means current should be constant (i.e.) KVA erating must be constant.

In second case active power,

$$P = 800 + 100 = 900 \text{ KW}$$

Reactive power in second case,

$$Q_2 = \sqrt{1000^2 - 900^2} = 435.889 \text{ KVAR}$$

Reactive power supplied by the three phase bank

$$= 600 - 435.889 = 164.11 \text{ KVAR}$$

$$Q_{bank}/\text{ph} = \frac{164.11}{3} = 54.7 \text{ KVAR}$$

$$V/\text{ph} = \frac{3.3}{\sqrt{3}} = 1.9052 \text{ KV}$$

$$Q_c/\text{ph} = \frac{(V/ph)^2}{X_C}$$

$$X_C = \frac{(1.9052 \times 10^3)^2}{54.7 \times 10^3} = 66.36 \text{ W}$$

$$C = \frac{1}{2\pi f X_C} = \frac{1}{2\pi \times 50 \times 66.36}$$

$$= 47.96 \,\mu\text{F}$$

3.14 (300)

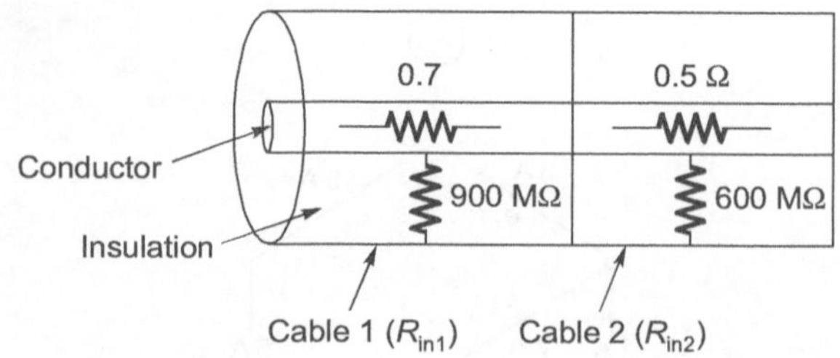

$$(R_{eq})_{conductor} = 0.7 + 0.5 = 1.2 \ \Omega$$

$$(R_{eq})_{in} = \frac{1}{(R_{eq})_{in}} = \frac{1}{(R_{in})_1} + \frac{1}{(R_{in})_2}$$

$$= \frac{1}{900 \times 10^6} + \frac{1}{600 \times 10^6}$$

$$(R_{eq})_{in} = 360 \times 10^6 \ \Omega$$

$$\frac{(R_{eq})_{in}}{(R_{eq})_{conductor}} = \frac{360 \times 10^6}{1.2} = 300 \times 10^6.$$

3.15 (13)

% power loss = 95%

Power output as a % of power is 5%

Practically cables power loss in decibels is calculated as

Power loss in decibels

$$= 10\log\left(\frac{95}{5}\right) \approx 12.78 \approx 13 \text{ db}$$

■■■■

4 Economic Power Generation and Load Dispatch

4.1 The incremental cost characteristic of two generators delivering 200 MW are as follows:

$$\frac{dF_1}{dP_1} = 20 + 0.1P_1; \quad \frac{dF_2}{dP_2} = 16 + 0.2\,P_2$$

For economic operation, the generation P_1 and P_2 should be

(a) $P_1 = P_2 = 100$ MW
(b) $P_1 = 80$ MW, $P_2 = 120$ MW
(c) $P_1 = 200$ MW, $P_2 = 0$ MW
(d) $P_1 = 120$ MW, $P_2 = 80$ MW **[2000 : 2 M]**

4.2 In a power system, the fuel inputs per hour of plants 1 and 2 are given as

$$F_1 = 0.20\,P_1^2 + 30\,P_1 + 100 \text{ Rs. per hour.}$$

$$F_2 = 0.25\,P_2^2 + 40\,P_2 + 80 \text{ Rs. per hour.}$$

The limits of generators are

$20 \le P_1 \le 80$ MW; $40 \le P_2 \le 200$ MW

Find the economic operating schedule of generation, if the load demand is 130 MW, neglecting transmission losses. **[2000 : 2 M]**

4.3 A power system has two generators with the following cost curves.

Generator 1: $C_1(P_{G1}) = 0.006P_{G1}^2 + 8P_{G1} + 350$
(Thousand Rupees/Hour)

Generator 2: $C_2(P_{G2}) = 0.006P_{G2}^2 + 7P_{G2} + 400$
(Thousand Rupees/Hour)

The generator limits are :

$100 \text{ MW} \le P_{G1} \le 650 \text{ MW}$
$50 \text{ MW} \le P_{G2} \le 500 \text{ MW}$

A load demand of 600 MW is supplied by the generators in an optimal manner. Neglecting losses in the transmission network, determine the optimal generation of each generator. **[2001 : 2 M]**

4.4 Consider a power system with three identical generators. The transmission losses are negligible. One generator ($G1$) has a speed governor which maintains its speed constant at the rated value, while the other generators ($G2$ and $G3$) have governors with a droop of 5%. If the load of the system is increased, then in steady state.

(a) Generation of $G2$ and $G3$ is increased equally while generation of $G1$ is unchanged.
(b) Generation of $G1$ alone is increased while generation of $G2$ and $G3$ is unchanged.
(c) Generation of $G1$, $G2$ and $G3$ is increased equally.
(d) Generation of $G1$, $G2$ and $G3$ is increased in the ratio 0.5 : 25 : 0.25. **[2002 : 1 M]**

4.5 Incremental fuel costs (in some appropriate unit) for a power plant consisting of three generating units are :

$IC_1 = 20 + 0.3\,P_1$, $IC_2 = 30 + 0.4\,P_2$, $IC_3 = 30$

where P_1 is the power in MW generated by unit i, for $i = 1$, 2 and 3.

Assume that all the three units are operating all the time. Minimum and maximum loads on each unit are 50 MW and 300 MW respectively. If the plant is operating on economic load dispatch to supply the total power demand of 700 MW, the power generated by each unit is

(a) $P_1 = 242.86$ MW; $P_2 = 157.14$ MW; and $P_3 = 300$ MW
(b) $P_1 = 157.14$ MW; $P_2 = 242.86$ MW; and $P_3 = 300$ MW
(c) $P_1 = 300.0$ MW; $P_2 = 300.0$ MW; and $P_3 = 100$ MW
(d) $P_1 = 233.3$ MW; $P_2 = 233.3$ MW; and $P_3 = 233.4$ MW **[2003 : 2 M]**

4.6 A load centre is at an equidistant from the two thermal generating stations G_1 and G_2 as shown in the figure. The fuel cost characteristics of the generating stations are given by

$F_1 = a + bP_1 + cP_1^2$ Rs/hour
$F_2 = a + bP_2 + 2cP_2^2$ Rs/hour

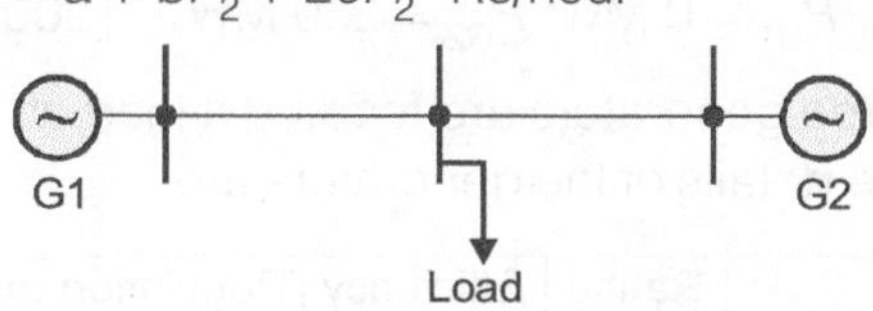

where P_1 and P_2 are the generation in MW of G_1 and G_2, respectively. For most economic generation to meet 300 MW of load, P_1 and P_2, respectively, are

(a) 150, 150 (b) 100, 200
(c) 200, 100 (d) 175, 125 **[2005 : 2 M]**

4.7 The incremental cost curves in Rs/ MWhr for two generators supplying a common load of 700 MW are shown in the figures. The maximum and minimum generation limits are also indicated. The optimum generation schedule is:

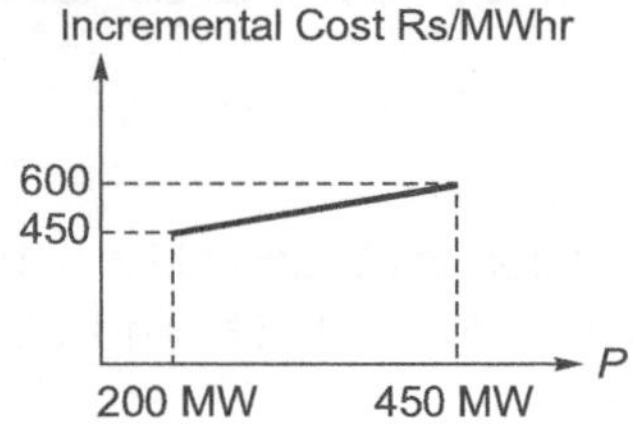

GENERATOR A

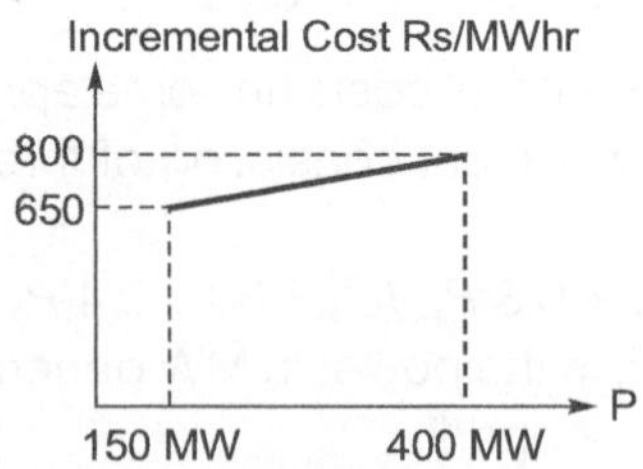

GENERATOR B

(a) Generator A : 400 MW, Generator B : 300 MW
(b) Generator A : 350 MW, Generator B : 350 MW
(c) Generator A : 450 MW, Generator B : 250 MW
(d) Generator A : 425 MW, Generator B: 275 MW

[2007 : 1 M]

4.8 A lossless power system has to serve a load of 250 MW. There are two generators ($G1$ and $G2$) in the system with cost curves C_1 and C_2 respectively defined as follows:

$C_1(P_{G1}) = P_{G1} + 0.055 \times P_{G1}^2$

$C_2(P_{G2}) = 3\ P_{G2} + 0.03 \times P_{G2}^2$

Where P_{G1} and P_{G2} are the MW injections from generator $G1$ and $G2$ respectively. Thus, the minimum cost dispatch will be

(a) P_{G1} = 250 MW; P_{G2} = 0 MW
(b) P_{G1} = 150 MW; P_{G2} = 100 MW
(c) P_{G1} = 100 MW; P_{G2} = 150 MW
(d) P_{G1} = 0 MW; P_{G2} = 250 MW [2008 : 2 M]

4.9 Three generators are feeding a load of 100 MW. The details of the generators are

	Rating (MW)	Efficiency (%)	Regulation (pu) on 100 MVA base
Generator-1	100	20	0.02
Generator-2	100	30	0.04
Generator-3	100	40	0.03

In the event of increased load power demand, which of the following will happen?

(a) All the generators will share equal power
(b) Generator-3 will share more power compared to Generator-1
(c) Generator-1 will share more power compared to Generator-2
(d) Generator-2 will share more power compared to Generator-3

[2009 : 2 M]

4.10 A load center of 120 MW derives power from two power stations connected by 220 kV transmission lines of 25 km and 75 km as shown in the figure below. The three generators $G1$, $G2$ and $G3$ are of 100 MW capacity each and have identical fuel cost characteristics. The minimum loss generation schedule for supplying the 120 MW load is

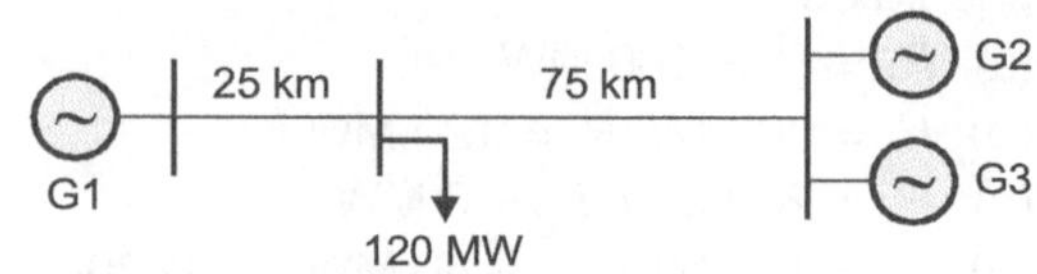

(a) $P1$ = 80 MW + losses
$P2$ = 20 MW
$P3$ = 20 MW

(b) $P1$ = 60 MW
$P2$ = 30 MW + losses
$P3$ = 30 MW

(c) $P1$ = 40 MW
$P2$ = 40 MW
$P3$ = 40 MW + losses

(d) $P1$ = 30 MW + losses
$P2$ = 45 MW
$P3$ = 45 MW

[2011 : 2 M]

4.11 The figure shows a two-generator system supplying a load of P_D = 40 MW, connected at bus 2.

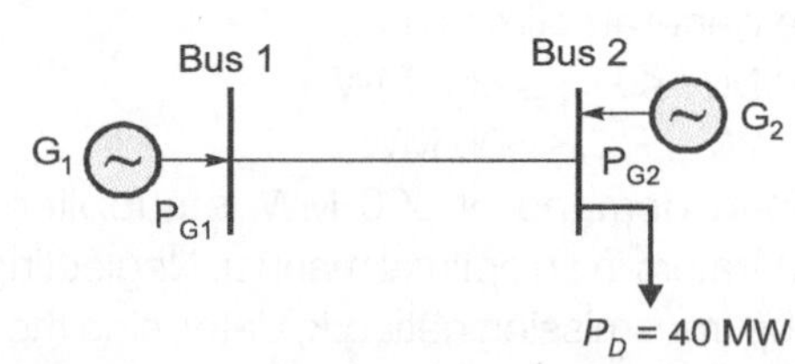

The fuel cost of generators G_1 and G_2 are:

$IC_1(P_{G1})$ = 10,000 Rs/MWh and

$IC_2(P_{G2})$ = 12,500 Rs/MWh

and the loss in the line is $P_{loss(pu)} = 0.5P_{G1(pu)}^2$, where the loss coefficient is specified in pu on a 100 MVA base. The most economic power generation schedule in MW is

(a) P_{G1} = 20, P_{G2} = 22
(b) P_{G1} = 22, P_{G2} = 20
(c) P_{G1} = 20, P_{G2} = 20
(d) P_{G1} = 0, P_{G2} = 40 [2012 : 1 M]

4.12 The fuel cost functions of two power plants are:

Plant P_1 : $C_1 = 0.05Pg_1^2 + A\,Pg_1 + B$

Plant P_2 : $C_2 = 0.10Pg_2^2 + 3A\,Pg_2 + 2B$

where, Pg_1 and Pg_2 are the generated powers of two plants, and A and B are the constants. If the two plants optimally share 1000 MW load at incremental fuel cost of 100 Rs/MWh, the ratio of load shared by plants P_1 and P_2 is

(a) 1 : 4 (b) 2 : 3
(c) 3 : 2 (d) 4 : 1

[2014 : 2 M, Set-1]

4.13 Consider the economic dispatch problem for a power plant having two generating units. The fuel costs in Rs/MWh along with the generation limits for the two units are given below:

$C_1(P_1) = 0.01\,P_1^{\,2} + 30P_1 + 10$;
$100 \text{ MW} \le P_1 \le 150 \text{ MW}$

$C_2(P_2) = 0.05\,P_2^{\,2} + 10P_2 + 10$;
$100 \text{ MW} \le P_2 \le 180 \text{ MW}$

The incremental cost (in Rs/MWh) of the power plant when it supplies 200 MW is ______.

[2015 : 2 M, Set-1]

4.14 The incremental costs (in Rs/MWh) of operating two generating units are function of their respective powers P_1 and P_2 in MW, and are given by

$$\frac{dC_1}{dP_1} = 0.2P_1 + 50; \quad \frac{dC_2}{dP_2} = 0.24P_2 + 40$$

where, $20 \text{ MW} \le P_1 \le 150 \text{ MW}$
$20 \text{ MW} \le P_2 \le 150 \text{ MW}$

For a certain load demand, P_1 and P_2 have been chosen such that $\frac{dC_1}{dP_1}$ = 76 Rs/MWh and $\frac{dC_2}{dP_2}$ = 68.8 Rs/MWh. If the generations are rescheduled to minimize the total cost, then P_2 is ______.

[2015 : 2 M, Set-2]

4.15 Two generating units rated 300 MW and 400 MW have governor speed regulation of 6% and 4% respectively from no load to full load. Both the generating units are operating in parallel to share a load of 600 MW. Assuming free governor action, the load shared by the larger units is ______ MW.

[2017 : 2 M, Set-2]

4.16 Two generators have cost functions F_1 and F_2. Their incremental-cost characteristics are

$$\frac{dF_1}{dP_1} = 40 + 0.2P_1; \quad \frac{dF_2}{dP_2} = 32 + 0.4P_2$$

They need to deliver a combined load of 260 MW. Ignoring the network losses, for economic operation, the generation P_1 and P_2 (in MW) are

(a) $P_1 = 160, P_2 = 100$ (b) $P_1 = P_2 = 130$
(c) $P_1 = 120, P_2 = 140$ (d) $P_1 = 140, P_2 = 120$

[2021 : 1 M]

4.17 The fuel cost functions in rupees/hour for two 600 MW thermal power plants are given by

Plant 1 : $C_1 = 350 + 6P_1 + 0.004P_1^2$

Plant 2 : $C_2 = 450 + aP_2 + 0.003P_2^2$

where P_1 and P_2 are power generated by plant 1 and plant 2, respectively, in MW and a is constant. The incremental cost of power (λ) is 8 rupees per MWh. The two thermal power plants together meet a total power demand of 550 MW. The optimal generation of plant 1 and plant 2 in MW, respectively, are

(a) 200, 350 (b) 250, 300
(c) 325, 225 (d) 350, 200 [2022 : 2 M]

4.18 The expressions of fuel cost of two thermal generating units as a function of the respective power generation P_{G1} and P_{G2} are given as :

$F_1(P_{G1}) = 0.1aP_{G1}^2 + 40P_{G1} + 120$ Rs/hour
$\rightarrow 0 \text{ MW} \le P_{G1} \le 350 \text{ MW}$

$F_2(P_{G2}) = 0.2P_{G2}^2 + 30P_{G2} + 100$ Rs/hour
$\rightarrow 0 \text{ MW} \le P_{G2} \le 300 \text{ MW}$

where a is a constant. For a given value of a, optimal dispatch requires the total load of 290 MW to be shared as P_{G1} = 175 MW and P_{G2} = 115 MW. With the load remaining unchanged, the value of a is increased by 10% and optimal dispatch is carried out. The changes in P_{G1} and the total cost of generation, F (= F_1 + F_2) in Rs/hour will be as follows :

(a) P_{G1} will decrease and F will increase.
(b) Both P_{G1} and F will increase.
(c) P_{G1} will increase and F will decrease.
(d) Both P_{G1} and F will decrease.

[2023 : 1 M]

4.19 The incremental cost curves of two generators (Gen A and Gen B) in a plant supplying a common load are shown in the figure. If the incremental cost of supplying the common load is Rs. 7400 per MWh, then the common load in MW is ______ (rounded off to the nearest ingeger).

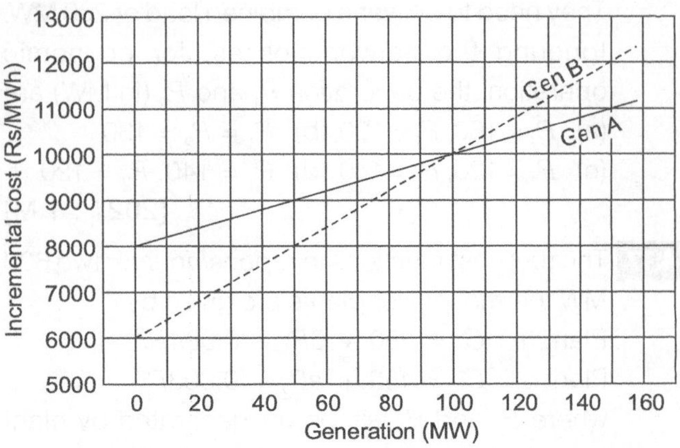

[2024 : 1 M]

4.20 Which of the following option is/are correct for the Automatic Generation Control (AGC) and Automatic Voltage Regulator (AVR) installed with synchronous generators?

(a) AGC regulates the generator's mechanical power input while AVR regulates the field current of the synchronous generator.

(b) AGC response has a local effect on frequency while AVR response lias a global effect on voltage.

(c) AGC response has a global effect on frequency while AVR response has a local effect on voltage.

(d) AGC regulates the field current of the synchronous generator while AVR regulates the generator's mechanical power input.

[2024 : 2 M]

■■■■

Answers Economic Power Generation and Load Dispatch

4.1	(d)	4.2	(Sol.)	4.3	(Sol.)	4.4	(b)	4.5	(a)	4.6	(c)	4.7	(c)
4.8	(c)	4.9	(c)	4.10	(a)	4.11	(a)	4.12	(d)	4.13	(52)		
4.14	(136.36)	4.15	(400)	4.16	(a)	4.17	(b)	4.18	(a)	4.19	(35)	4.20	(a, c)

Explanations Economic Power Generation and Load Dispatch

4.1 (d)

$$P_1 + P_2 = 200 \text{ MW} \quad \text{...(i)}$$

For economic operations

$$20 + 0.1P_1 = 16 + 0.2P_2$$

$$4 = 0.2P_2 - 0.1P_1$$

$$2P_2 - P_1 = 40 \text{ MW} \quad \text{...(ii)}$$

From (i) and (ii), we get

$$P_1 = 120 \text{ MW}; P_2 = 80 \text{ MW}$$

4.2 Sol.

Incremental fuel costs by differentiating fuel input equation,

$$\frac{dF_1}{dP_1} = 0.4P_1 + 30 \quad \text{...(i)}$$

$$\frac{dF_2}{dP_2} = 0.5P_2 + 40 \quad \text{...(ii)}$$

For economic generation,

$$0.4P_1 + 30 = 0.5P_2 + 40$$

$$\text{or } 4P_1 - 5P_2 = 100 \quad \text{...(A)}$$

$$\text{and } P_1 + P_2 = 130 \quad \text{...(B)}$$

Solving we get,

$$P_1 = 83.33 \text{ MW}; \quad P_2 = 46.67 \text{ MW}$$

Which is exceeding the upper limit of first generator

$$\therefore \quad P_1 = 80 \text{ MW and } P_2 = 50 \text{ MW}$$

4.3 Sol.

$$\frac{dC_1}{dP_{G_1}} = 0.012P_{G_1} + 8$$

$$\frac{dC_2}{dP_{G_2}} = 0.012P_{G_2} + 7$$

For economic generation,

$$\frac{dC_1}{dP_{G_1}} = \frac{dC_2}{dP_{G_2}}$$

$$0.012P_{G_1} + 8 = 0.012P_{G_2} + 7$$

$$P_{G_2} - P_{G_1} = 83.33 \quad \text{...(i)}$$

$$P_{G_1} + P_{G_2} = 600 \quad \text{...(ii)}$$

Solving (i) and (ii), we get

$$P_{G_1} = 258.335 \text{ MW}$$

$$P_{G_2} = 341.665 \text{ MW}$$

Both the limits are in the respective specified ranges.

4.5 (a)

When P_1 is minimum i.e. $P_1 = 50$ MW

$IC_1 = 20 + 0.3 \times 50 = 35$

When P_2 is minimum i.e. $P_2 = 50$ MW

$IC_2 = 30 + 0.4 \times 50 = 50$

For minimum value of P_1 and P_2

IC_1 and $IC_2 > IC_3$

Therefore, $P_3 = 300$ MW

[maximum load is assigned to unit 3]

Remaining power is to shared by unit 1 and 2.

So, $P_1 + P_2 = 700 - 300 = 400$ MW ...(i)

For optimal operation,

$$IC_1 = IC_2$$

$$20 + 0.3\,P_1 = 30 + 0.4\,P_2$$

$$0.3P_1 - 0.4\,P_2 = 10 \quad ...(ii)$$

Solving equation (i) and (ii), we get

$$P_1 = 242.86 \text{ MW and } P_2 = 157.14 \text{ MW}$$

4.6 (c)

Fuel-cost curve of generating station 1

$$F_1 = a + bP_1 + cP_1^2 \text{ Rs/hour}$$

Fuel-cost curve of generating station 2

$$F_2 = a + bP_2 + 2cP_2^2$$

The slope of the fuel-cost curve, i.e., $\dfrac{dF_i}{dP_i}$ is called the incremental fuel-cost (IC) and is expressed in Rs/MWh.

$$IC_1 = \frac{dF_1}{dP_1} = b + 2cP_1$$

$$IC_2 = \frac{dF_2}{dP_2} = b + 4cP_2$$

For economic generation,

$$IC_1 = IC_2$$

$$b + 2cP_1 = b + 4cP_2$$

$$P_1 = 2P_2 \quad ...(i)$$

To meet 300 MW of load

$$P_1 + P_2 = 300 \quad ...(ii)$$

Solving equation (i) and (ii)

$P_1 = 200$ MW and $P_2 = 100$ MW

4.7 (c)

Maximum incremental cost in Rs/Mwhr for generator $A = 600$ (at 450 MW)

Minimum incremental cost in Rs/Mwhr for generator $B = 650$ (at 150 MW)

As maximum value of incremental cost of A is less than minimum value of B,

∴ Generator 'A' will operate at its maximum (o/p) 450 MW and B at $(700 - 450) = 250$ MW

4.8 (c)

Given cost curves,

$$C_1(P_{G1}) = P_{G1} + 0.055P_{G1}^2$$

Differentiating, $$\frac{dC_1}{P_{G1}} = 1 + 0.11\,P_{G1}$$

Similarly, $$C_2(P_{G2}) = P_{G2} + 0.03P_{G2}^2$$

$$\frac{dC_2}{P_{G2}} = 3 + 0.06\,P_{G2}$$

For minimum cost analysis,

$$\frac{dC_1}{P_{G1}} = \frac{dC_2}{P_{G2}}$$

$$1 + 0.11\,P_{G1} = 3 + 0.06\,P_{G2} \quad ...(i)$$

$$P_{G1} + P_{G2} = 250 \text{ MW} \quad ...(ii)$$

Solving equation (i) and (ii),

we get, $P_{G1} = 100$ MW

and $P_{G2} = 150$ MW

4.9 (c)

Let x_1, x_2 and x_3 are reactances of generator-1 generator-2 and generator-3 respectively.

Neglecting armature resistance of all the three generators,

$VR_1 = 0.02$ p.u.

$VR_2 = 0.04$ p.u and

$VR_3 = 0.03$ p.u.

$VR_1 < VR_3 < VR_2$

Since, voltage regulation (VR) ∝ reactance of generator,

$$X_1 < X_3 < X_2$$

Power shared by a generator

$$\propto \frac{1}{\text{reactance of generator}}$$

Power shared by

Generator 1 > Generator 2 > Generator 3

4.10 (a)

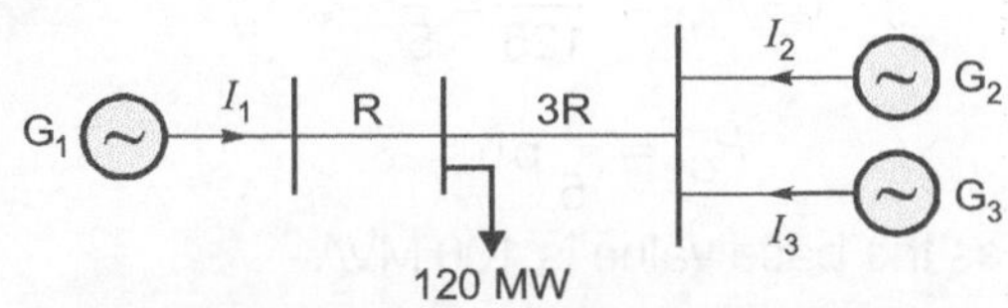

Let load center is connected at point x

If r is resistance/km of line

Resistance of section between 1 and x

$$R_1 = r \times 25 = R$$

Resistance of section between x and 2,

$$R_2 = r \times 75 = 3R$$

Current (I) fed by generator ∝ power produced by the generator.

$I_1 \propto P_1, I_2 \propto P_2$ and $I_3 \propto P_3$

Transmission losses = P_L

$$= I_1^2 R + I_2^2 3R + I_3^2 3R$$

$$P_L \propto (P_1^2 + 3P_2^2 + 3P_3^2)$$

$$P_L = K[P_1^2 + 3P_2^2 + 3P_3^2]$$

Option (a):

$$P_1 = 80\text{ MW};\ P_2 = 20\text{ MW};\ P_3 = 20\text{ MW}$$
$$P_L = K[80^2 + 3 \times 20^2 + 3 \times 20^2]$$
$$= 8800\text{ K MW}$$

Option (b):

$$P_1 = 60\text{ MW};\ P_2 = 30\text{ MW};\ P_3 = 30\text{ MW}$$
$$P_L = K[60^2 + 3 \times 30^2 + 3 \times 30^2]$$
$$= 9000\text{ K MW}$$

Option (c):

$$P_1 = 40\text{ MW};\ P_2 = 40\text{ MW};\ P_3 = 40\text{ MW}$$
$$P_L = K[40^2 + 3 \times 40^2 + 3 \times 40^2]$$
$$= 11200\text{ K MW}$$

Option (d):

$$P_1 = 30\text{ MW};\ P_2 = 45\text{ MW};\ P_3 = 45\text{ MW}$$
$$P_L = K[30^2 + 3 \times 45^2 + 3 \times 45^2]$$
$$= 13050\text{K MW}$$

So, option (a) gives minimum losses.

4.11 (a)

From exact cordination equation,

$$\frac{dF_n}{dP_{Gn}} \cdot \frac{1}{1 - \dfrac{\partial P_L}{\partial P_{Gn}}} = \lambda$$

Given: $P_L = 0.5P_{G1}^2$

$$\frac{\partial P_L}{\partial P_{G1}} = P_{G1} \text{ and } \frac{\partial P_L}{\partial P_{G2}} = 0$$

$$\Rightarrow \frac{dF_1}{dP_{G1}} \cdot \frac{1}{1 - P_{G1}} = \frac{dF_2}{dP_{G2}} \cdot \frac{1}{1 - 0}$$

$$\Rightarrow 10000 \times \frac{1}{1 - P_{G1}} = 12500$$

$$1 - P_{G1} = \frac{100}{125} = \frac{4}{5}$$

$$P_{G1} = \frac{1}{5}\text{ p.u.}$$

as the base value is 100 MVA

$$P_{G1} = \frac{1}{5} \times 100 = 20\text{ MW}$$

$$\Rightarrow \quad P_L = 0.5P_{G1}^2 = 0.5\left(\frac{1}{5}\right)^2 = 0.02\text{ p.u.}$$

$$\Rightarrow \quad P_L = 0.02 \times 100 = 2\text{ MW}$$
$$P_D = P_{G1} + P_{G2} - P_L$$
$$40 = 20 + P_{G2} - 2$$
$$\Rightarrow \quad P_{G2} = 22\text{ MW}$$

4.12 (d)

Given, $P_{g_1} + P_{g_2} = 1000$ ...(i)

Also, incremental cost of production of

$$g_1 = \frac{dC_1}{dP_{g_1}} = 0.1P_{g_1} + A \quad \text{...(ii)}$$

and incremental cost of production of

$$g_2 = \frac{dC_2}{dP_{g_2}} = 0.2P_{g_2} + 3A \quad \text{...(iii)}$$

For optimum load sharing,

$$\frac{dC_1}{dP_{g_1}} = \frac{dC_2}{dP_{g_2}}$$

or, $0.1P_{g_1} + A = 0.2P_{g_2} + 3A$

or, $0.1P_{g_1} - 0.2P_{g_2} = 2A$

or, $P_{g_1} - 2P_{g_2} = 20A$ ..(iv)

Given, $\dfrac{dC_1}{dP_{g_1}} = \dfrac{dC_2}{dP_{g_2}} = 100\text{ Rs/MWh}$

Putting this value in equation (ii), we have

$$100 = 0.1P_{g_1} + A$$

or, $A = (100 - 0.1P_{g_1})$ ...(v)

Using equations (iv) and (v), we have:

$$P_{g_1} - 2P_{g_2} = 20(100 - 0.1P_{g_1})$$

or, $3P_{g_1} - 2P_{g_2} = 20004$ ...(vi)

Solving equations (i) and (vi), we get:

$$P_{g_1} = 800\text{ MW and } P_{g_2} = 200\text{ MW}$$

$$\therefore \quad P_{g_1} : P_{g_2} = 800 : 200 = 4 : 1$$

4.13 Sol.

Fuel cost curve 1 :

$$C_1 = 0.01P_1^2 + 30P_1 + 10$$

$$\frac{dC_1}{dP_1} = 2 \times 0.01P_1 + 30 = 0.02\,P_1 + 30$$

Fuel cost curve 2 :

$$C_2 = 0.05P_2^2 + 10P_2 + 10$$

$$\frac{dC_2}{dP_2} = 0.1\,P_2 + 10$$

For optimum incremental cost,

$$\frac{dC_1}{dP_1} = \frac{dC_2}{dP_2}$$

$$0.02P_1 + 30 = 0.1P_2 + 10$$

$$0.02P_1 - 0.1P_2 = -20 \quad ...(i)$$

Given, $P_1 + P_2 = 200$ ...(ii)

Solving equation (*i*) and (*ii*), we get

$$P_2 = 200,\ P_1 = 0$$

But the range of P_1 and P_2 are specified as

$$100 \text{ MW} \le P_1 \le 150 \text{ MW}$$

$$100 \text{ MW} \le P_2 \le 180 \text{ MW}$$

Hence, $P_2 = 200$ MW is not valid

So, both P_1 and P_2 are operating at

$P_1 = 100,\ P_2 = 100 \Rightarrow P_1 + P_2 = 200$

$$IC_1 = \frac{dC_1}{dP_1} = 0.02 \times P_1 + 30 = 32 \text{ Rs/MW}$$

$$IC_2 = \frac{dC_2}{dP_2} = 0.1 \times P_2 + 10 = 20 \text{ Rs/MW}$$

4.14 Sol.

Given incremental costs respectively

$$0.2P_1 + 50 = 76$$

$$P_1 = 130 \text{ mW}$$

$$0.24\,P_2 + 40 = 68.8$$

$$P_2 = 120 \text{ mW}$$

$$P_1 + P_2 = 250 \quad ...(i)$$

For minimized schedule,

$$0.2P_1 + 50 = 0.24P_2 + 40$$

$$\Rightarrow 0.2P_1 - 0.24P_2 = -10 \quad ...(ii)$$

From equation (*i*) and (*ii*),

$$P_1 = 113.63 \text{ MW};\ P_2 = 136.36 \text{ MW}$$

4.15 Sol.

Given : Speed regulation of generating unit is from no load to full load.

Hence the rated frequency (50 Hz) will be common for both units at no load and will droop to lower value as loaded to full load.

Let, P_1 = Load shared by 400 MW machine

P_2 = Load shared by 300 MW machine

For 400 MW machine drop in frequency from no load to full load $= \dfrac{50 \times 4}{100} = 2$ Hz.

New frequency of operation of 400 MW machine,

$$f_1 = 50 - \frac{2}{400}P_1.$$

For 300 MW machine drop in frequency from no load to full load $= \dfrac{50 \times 6}{100} = 3$ Hz.

New frequency of operation of 300 MW machine,

$$f_2 = 50 - \frac{3}{300}P_2.$$

We also know,

$$P_1 + P_2 = 600$$

...(Load shared between two units)

$$P_2 = 600 - P_1$$

$$f_1 = f_2 = f$$

...(Units operation is in parallel)

$$50 - \frac{2}{400}P_1 = 50 - \frac{3}{300}(600 - P_1)$$

$$\frac{P_1}{200} = \frac{600 - P_1}{100}$$

$$100P_1 = 200(600 - P_1)$$

$$P_1 = 1200 - 2P_1$$

$$3P_1 = 1200$$

$$P_1 = 400 \text{ MW}$$

4.16 (a)

$$IC_1 = IC_2$$

$$40 + 0.2P_1 = 32 + 0.4P_2$$

$$0.4P_2 - 0.2P_1 = 8 \quad ...(i)$$

$$P_2 + P_1 = 260 \quad ...(ii)$$

Solving equation (i) and (ii),

$$P_1 = 160 \text{ MW};\quad P_2 = 100 \text{ MW}$$

4.17 (b)

$$IC_1(P_1) = 0.008P_1 + 6$$

$$IC_2(P_2) = 0.006P_2 + a$$

$$I_{C1} = I_{C2} = \lambda = 8$$

$$\Rightarrow I_{C1} = 8$$

$$0.008P_1 + 6 = 8$$

$$P_1 = 250 \text{ MW},$$

$$P_2 = 300 \text{ MW}$$

4.18 (a)

Cost function : $C_1(P_{G1}) = 0.1aP^2_{G1} + 40P_{G1} + 120$ Rs/hr0 MW $\le P_{G1} \le 350$ MW

$C_2(P_{G2}) = 0.2P^2_{G2} + 30P_{G2} + 100$ Rs/hr0 MW $\le P_{G2} \le 300$ MW

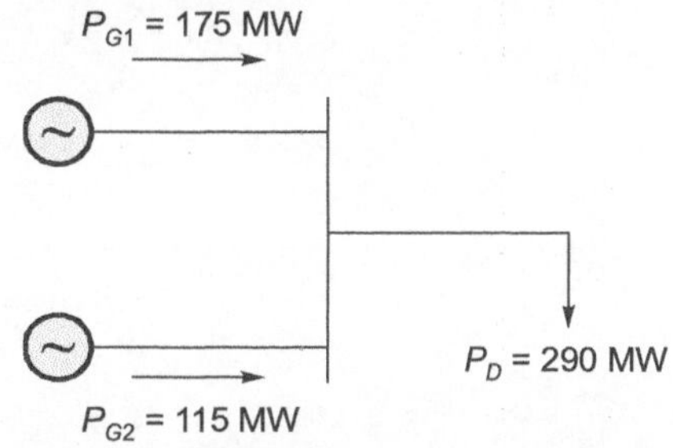

Incremental costs :

$$IC_1(P_{G1}) = (0.2aP_{G1} + 40) \text{ Rs/Mwhr}$$

$$IC_2(P_{G2}) = (0.4P_{G2} + 30) \text{ Rs/Mwhr}$$

For economic scheduling of P_o = 290 MW, it is given that

$P_{G1} = 175$ MW; $P_{G2} = 115$ MW

Hence, $I_{C1}(P_{G1} = 175) = I_{C2}(P_{G2} = 115)$

$\Rightarrow 0.2a \times 175 + 40 = 0.4 \times 115 + 30$

$a = 1.02857$

$\Rightarrow F = C_1(P_{G1} = 175) + C_2(P_{G2} = 115)$

$= 10270 + 6195 = 16465$ Rs/hr.

Now if a is increased by 10%

$a' = 1.1a = 1.1314$

For economic scheduling of same demand

$P_D = 290$ MW

$P_{G1} + P_{G2} = 290$ ...(1)

$I_{C1} = I_{C2}$

$0.2 \times 1.1314 \times P_{G1} + 40 = 0.4P_{G2} + 30$

$0.2262P_{G1} - 0.4P_{G2} = -10$ ...(2)

Solving eqn. (1) and (2)

$P_{G1} = 169.25$ MW; $P_{G2} = 120.75$ MW

We can see P_{G1} has decreased.

$F = C_1(P_{G1} = 169.27) + C_2(P_{G2} = 120.70)$

$= 10170.22 + 6638.61 = 16808.83$ Rs/hr.

We can see total cost has increased.

So, correct answer is option (a).

4.19 (35)

Given, $\lambda = 7400$ Rs/MWhr

$$I_{CA}(P_{GA}) = \frac{2000}{100}P_{GA} + 8000$$

$$I_{CB}(P_{GB}) = 40P_{GB} + 6000$$

$$\lambda = I_{CA} = I_{CB} = 7400$$

$$200P_{GA} + 8000 = 7400$$

$$P_{GA} = -600$$

$$P_{GA} < 0$$

$$P_{GB} = \frac{7400 - 6000}{40} = 35$$

$$P_{GA} + P_{GB} = 35 \text{ MW}$$

■■■■

5 Fault Analysis

5.1 For an unbalanced fault, with paths for zero sequence currents, at the point of fault
(a) the negative and zero sequence voltages are minimum.
(b) the negative and zero sequence voltages are maximum.
(c) the negative sequence voltage is minimum and zero sequence voltage is maximum.
(d) the negative sequence voltage is maximum and zero sequence voltage is minimum.
[1996 : 1 M]

5.2 For a fault at the terminals of a synchronous generator, the fault current is maximum for a
(a) 3-phase fault
(b) 3-phase to ground fault
(c) line-to ground fault
(d) line-to-line fault [1997 : 1 M]

5.3 For the network shown in the figure, zero sequence reactances in p.u. are indicated. The zero sequence driving point reactance of the node 3 is

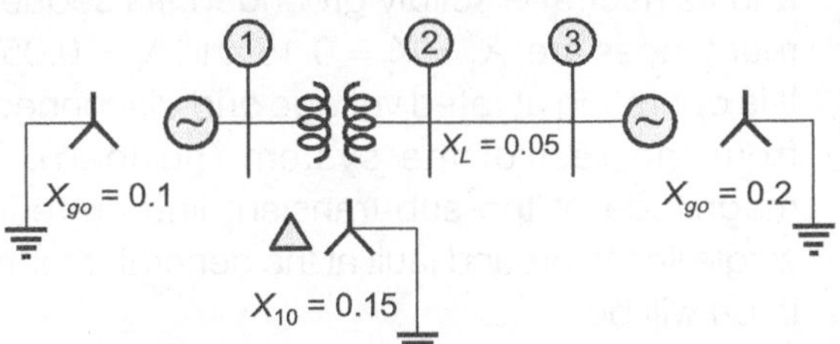

(a) 0.12 p.u. (b) 0.30 p.u.
(c) 0.10 p.u. (d) 0.20 p.u. [1998 : 2 M]

5.4 Determine the magnitudes of the symmetrical components (I_{R1}, I_{R2} and I_{R0}) of the currents in a 3-phase (RYB) three wire system, when a short circuit occurs between R and Y phase wires, the fault current being 100 A. [1999 : 2 M]

5.5 For the configuration shown in figure the breaker connecting a large system to bus 2 is initially open. The system 3-phase fault level at bus 3 under this condition is not known. After closing the system breaker, the 3-phase fault level at bus 1 was found to be 5.0 p.u. What will be the new 3-phase fault level at system bus 3 after the interconnection? All per unit values are on common base. Prefault load currents are neglected and prefault voltages are assumed to be 1.0 p.u. at all buses.

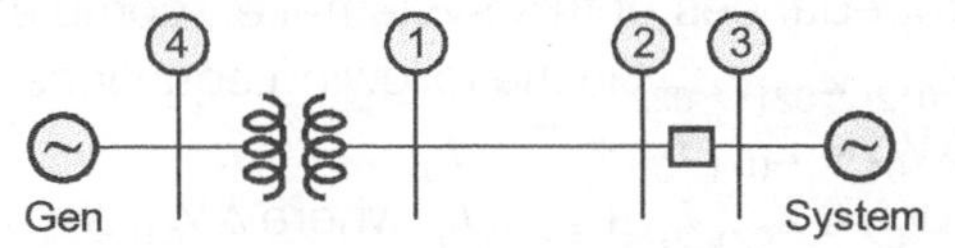

E_g = 1.0 p.u., X_T = 0.2 p.u., X_{Line} = 0.3 p.u., X_d = 0.2 p.u.

[2000 : 2 M]

5.6 The severity of line-to-ground and 3-phase faults at the terminals of an unloaded synchronous generator is to be same. If the terminal voltage is 1.0 p.u. and $X_1 = X_2 = j0.1$ p.u, $X_0 = j0.05$ p.u., for the alternator, then the required inductive reactance for neutral grounding is
(a) 0.0166 p.u. (b) 0.05 p.u.
(c) 0.1 p.u. (d) 0.15 p.u. [2000 : 2 M]

5.7 A 50 Hz alternator is rated 500 MVA, 20 kV, with X_d = 1.0 per unit and X_d'' = 0.2 per unit. It supplies a purely resistive load of 400 MW at 20 kV. The load is connected directly across the generator terminals when a symmetrical fault occurs at the load terminals. The initial rms current in the generator in per unit is
(a) 7.22 (b) 5.05
(c) 3.22 (d) 2.25 [2001 : 2 M]

5.8 A generator is connected to a transformer which feeds another transformer through a short feeder. The zero sequence impedance values are expressed in p.u. on a common base and are indicated in figure. The Thevenin equivalent zero sequence impedance at point B (in p.u.) is

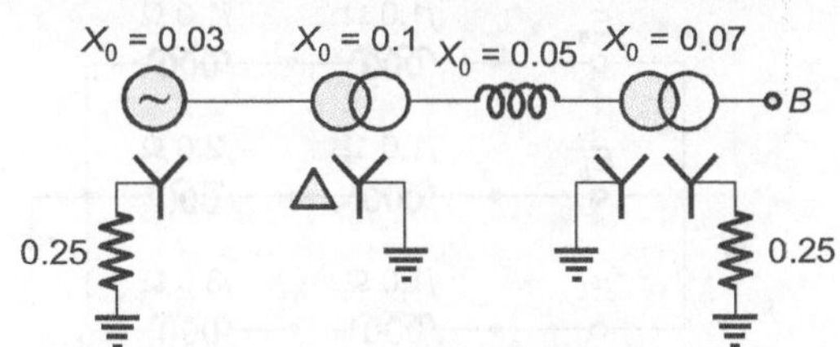

(a) 0.8 + j0.6 (b) 0.75 + j0.22
(c) 0.75 + j0.25 (d) 1.5 + j0.25
[2002 : 2 M]

5.9 Two transposed 3-phase lines run parallel to each other. The equation describing the voltage drop in both lines given below.

$$\begin{bmatrix} \Delta V_{a1} \\ \Delta V_{b1} \\ \Delta V_{c1} \\ \Delta V_{a2} \\ \Delta V_{b2} \\ \Delta V_{c2} \end{bmatrix} = \begin{bmatrix} 0.15 & 0.05 & 0.05 & 0.04 & 0.04 & 0.04 \\ 0.05 & 0.15 & 0.05 & 0.04 & 0.04 & 0.04 \\ 0.05 & 0.05 & 0.15 & 0.04 & 0.04 & 0.04 \\ 0.04 & 0.04 & 0.04 & 0.15 & 0.05 & 0.05 \\ 0.04 & 0.04 & 0.04 & 0.05 & 0.015 & 0.05 \\ 0.04 & 0.04 & 0.04 & 0.05 & 0.05 & 0.15 \end{bmatrix} \begin{bmatrix} I_{a1} \\ I_{b1} \\ I_{c1} \\ I_{a2} \\ I_{b2} \\ I_{c2} \end{bmatrix}$$

Compute the self and mutual zero sequence impedances of this system, i.e., complete Z_{011}, Z_{012}, Z_{021}, Z_{022} in the following equations.

$\Delta V_{01} = Z_{011} I_{01} + Z_{012} I_{02}$

$\Delta V_{02} = Z_{021} I_{01} + Z_{022} I_{02}$ where $\Delta V_{01}, \Delta V_{02}, I_{01}, I_{02}$ are the zero sequence voltage drops and currents for the two lines respectively. **[2002 : 2 M]**

5.10 A 20-MVA, 6.6-kV, 3-phase alternator is connected to a 3-phase transmission line. The per unit positive-sequence, negative-sequence and zero-sequence impedances of the alternator are $j0.1$, $j0.1$ and $j0.04$ respectively. The neutral of the alternator is connected to ground through an inductive reactor of $j0.05$ p.u. The per unit +ve, –ve and zero-sequence impedances of the transmission line are $j0.1$, $j0.1$ and $j0.3$, respectively. All per unit values are based on the machine ratings. A solid ground fault occurs at one phase of the far end of the transmission line. The voltage of the alternator neutral with respect to ground during the fault is

(a) 513.8 V (b) 889.9 V
(c) 1112.0 V (d) 642.2 V **[2003 : 2 M]**

5.11 A three-phase alternator generating unbalanced voltages is connected to an unbalanced load through a 3-phase transmission line as shown in figure. The neutral of the alternator and the star point of the load are solidly grounded. The phase voltages of the alternator are $E_a = 10\angle 0°$V, $E_b = 10\angle -90°$ V, $E_c = 10\angle 120°$ V. The +ve sequence component of the load current is

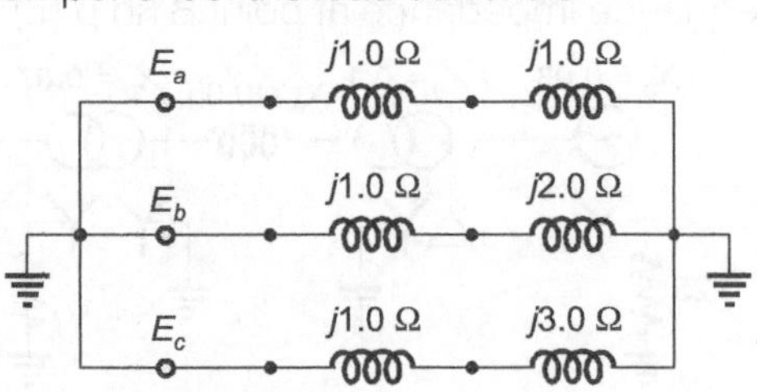

(a) $1.310\angle -107°$ A (b) $0.332\angle -120°$ A
(c) $0.996\angle -120°$ A (d) $3.510\angle -81°$ A

[2003 : 2 M]

5.12 A 3-phase generator rated at 110 MVA, 11 kV is connected through circuit breakers to a transformer. The generator is having direct axis sub-transient reactance $X''_d = 19\%$, transient reactance $X'_d = 26\%$ and synchronous reactance = 130%. The generator is operating at no load and rated voltage when a three phase short circuit fault occurs between the breakers and the transformer. The magnitude of initial symmetrical rms current in the breakers will be

(a) 4.44 kA (b) 22.20 kA
(c) 30.39 kA (d) 38.45 kA **[2004 : 2 M]**

5.13 A 3-phase transmission line supplies Δ-connected load Z. The conductor 'c' of the line develops an open circuit fault as shown in figure. The currents in the lines are as shown on the diagram. The +ve sequence current component in line 'a' will be

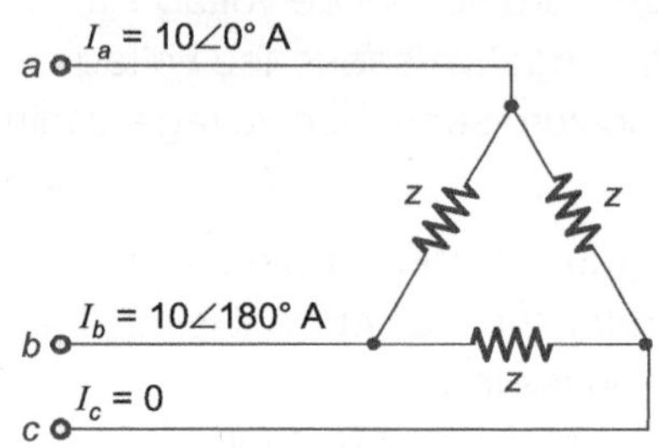

(a) $5.78\angle -30°$ A (b) $5.78\angle 90°$ A
(c) $6.33\angle 90°$ A (d) $10.00\angle -30°$ A

[2004 : 2 M]

5.14 A500 MVA, 50 Hz, 3-phase turbo-generator produces power at 22 kV. Generator is Y-connected and its neutral is solidly grounded. Its sequence reactances are $X_1 = X_2 = 0.15$ and $X_0 = 0.05$ pu. It is operating at rated voltage and disconnected from the rest of the system (no load). The magnitude of the sub-transient line current for single line to ground fault at the generator terminal in pu will be

(a) 2.851 (b) 3.333
(c) 6.667 (d) 8.553 **[2004 : 2 M]**

Data for Questions (5.15 and 5.16):

At a 220 kV substation of a power system, it is given that the three-phase fault level is 4000 MVA and single-line to ground fault level is 5000 MVA. Neglecting the resistance and the shunt susceptances of the system.

5.15 The positive sequence driving point reactance at the bus is

(a) 2.5 Ω (b) 4.033 Ω
(c) 5.5 Ω (d) 12.1 Ω **[2005 : 2 M]**

5.16 The zero sequence driving point reactance at the bus is

(a) 2.2 Ω (b) 4.84 Ω
(c) 18.18 Ω (d) 22.72 Ω

[2005 : 2 M]

5.17 The parameters of transposed overhead transmission line are given as:
Self reactance X_s = 0.4 Ω/km and Mutual reactance X_m = 0.1 Ω/km
The positive sequence reactance x_1 and zero sequence reactance x_0, respectively, in Ω/km are
(a) 0.3, 0.2 (b) 0.5, 0.2
(c) 0.5, 0.6 (d) 0.3, 0.6 [2005 : 2 M]

5.18 Three identical star connected resistors of 1.0 p.u. are connected to an unbalanced 3 phase supply. The load neutral is isolated. The symmetrical components of the line voltages in p.u. are:

$$V_{ab1} = X\angle\theta_1,\ V_{ab2} = Y\angle\theta_2$$

If all the p.u. calculations are with the respective base values, the phase to neutral sequence voltages are
(a) $V_{an1} = X\angle(\theta_1 + 30°)$,
$V_{an2} = Y\angle(\theta_2 - 30°)$
(b) $V_{an1} = X\angle(\theta_1 - 30°)$,
$V_{an2} = Y\angle(\theta_2 + 30°)$
(c) $V_{an1} = \frac{1}{\sqrt{3}} X\angle(\theta_1 - 30°)$,
$V_{an2} = \frac{1}{\sqrt{3}} Y\angle(\theta_2 + 30°)$
(d) $V_{an1} = \frac{1}{\sqrt{3}} X\angle(\theta_1 - 60°)$,
$V_{an2} = \frac{1}{\sqrt{3}} Y\angle(\theta_2 - 60°)$ [2006 : 2 M]

5.19 A three phase balanced star connected voltage source with frequency ω rad/s is connected to a star connected balanced load which is purely inductive. The instantaneous line currents and phase to neutral voltages are denoted by (i_a, i_b, i_c) and (V_{an}, V_{bn}, V_{cn}) respectively, and their rms values are denoted by V and I.

If $R = [V_{an}\ V_{bn}\ V_{cn}]\begin{bmatrix} 0 & \frac{1}{\sqrt{3}} & -\frac{1}{\sqrt{3}} \\ -\frac{1}{\sqrt{3}} & 0 & \frac{1}{\sqrt{3}} \\ \frac{1}{\sqrt{3}} & -\frac{1}{\sqrt{3}} & 0 \end{bmatrix}\begin{bmatrix} i_a \\ i_b \\ i_c \end{bmatrix}$,

then the magnitude of R is
(a) 3 VI (b) VI
(c) 0.7 VI (d) 0 [2007 : 2 M]

5.20 Suppose we define a sequence transformation between "a-b-c" and "p-n-o" variables as follows:

$$\begin{bmatrix} f_a \\ f_b \\ f_c \end{bmatrix} = K\begin{bmatrix} 1 & 1 & 1 \\ \alpha^2 & \alpha & 1 \\ \alpha & \alpha^2 & 1 \end{bmatrix}\begin{bmatrix} f_p \\ f_n \\ f_0 \end{bmatrix}$$

where $\alpha = e^{j\frac{2\pi}{3}}$ and k is a constant now, if it is given that:

$$\begin{bmatrix} V_p \\ V_n \\ V_0 \end{bmatrix} = \begin{bmatrix} 0.5 & 0 & 0 \\ 0 & 0.5 & 0 \\ 0 & 0 & 2.0 \end{bmatrix}\begin{bmatrix} i_p \\ i_n \\ i_0 \end{bmatrix}$$

and $\begin{bmatrix} V_a \\ V_b \\ V_c \end{bmatrix} = z\begin{bmatrix} i_a \\ i_b \\ i_c \end{bmatrix}$ then,

(a) $z = \begin{bmatrix} 1.0 & 0.5 & 0.75 \\ 0.75 & 1.0 & 0.5 \\ 0.5 & 0.75 & 1.0 \end{bmatrix}$

(b) $z = \begin{bmatrix} 1.0 & 0.5 & 0.5 \\ 0.5 & 1.0 & 0.5 \\ 0.5 & 0.5 & 1.0 \end{bmatrix}$

(c) $z = 3k^2\begin{bmatrix} 1.0 & 0.75 & 0.5 \\ 0.5 & 1.0 & 0.75 \\ 0.75 & 0.5 & 1.0 \end{bmatrix}$

(d) $z = \frac{K^2}{3}\begin{bmatrix} 1.0 & -0.5 & -0.5 \\ -0.5 & 1.0 & -0.5 \\ -0.5 & -0.5 & 1.0 \end{bmatrix}$ [2007 : 2 M]

5.21 A 3-phase transmission line is shown in the figure:

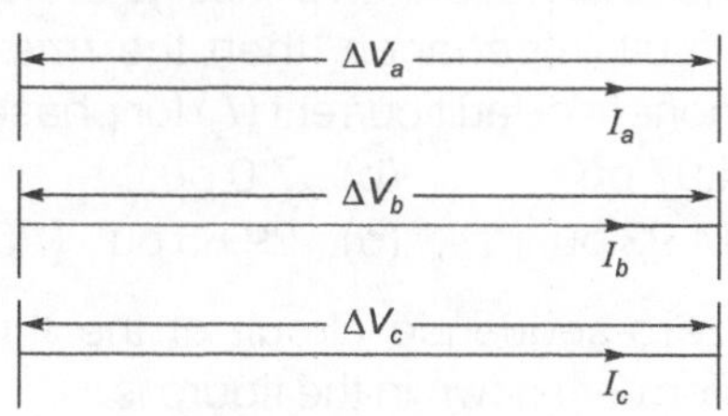

Voltage drop across the transmission line is given by the following equation:

$$\begin{bmatrix} \Delta V_a \\ \Delta V_b \\ \Delta V_c \end{bmatrix} = \begin{bmatrix} Z_s & Z_m & Z_m \\ Z_m & Z_s & Z_m \\ Z_m & Z_m & Z_s \end{bmatrix}\begin{bmatrix} I_a \\ I_b \\ I_c \end{bmatrix}$$

Shunt capacitance of the line can be neglect. If the line has positive sequence impedance of 15 Ω and zero sequence impedance of 48 Ω, then the values of Z_s and Z_m will be

(a) $Z_s = 31.5\ \Omega$; $Z_m = 16.5\ \Omega$
(b) $Z_s = 26\ \Omega$; $Z_m = 11\ \Omega$
(c) $Z_s = 16.5\ \Omega$; $Z_m = 31.5\ \Omega$
(d) $Z_s = 11\ \Omega$; $Z_m = 26\ \Omega$ [2008 : 1 M]

Data for Questions (5.22 to 5.24):

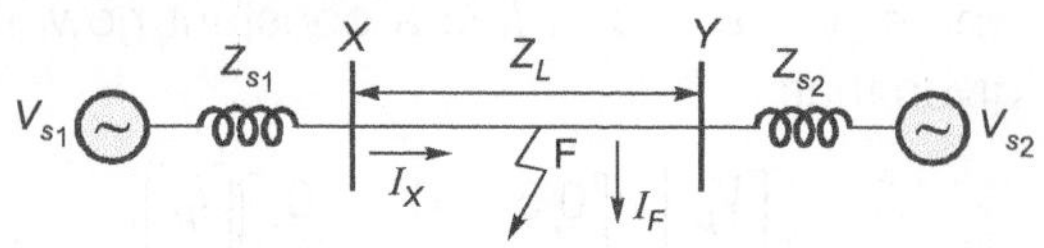

Given that: $V_{s1} = V_{s2} = 1 + j0$ p.u, +ve sequence impedance are $Z_{s1} = Z_{s2} = 0.001 + j0.01$ p.u and $Z_L = 0.006 + j0.06$ p.u, 3-ϕ. Base MVA = 100, voltage base = 400 kV($L-L$).

Nominal system frequency = 50 Hz. The reference voltage for phase 'a' is defined as $V(t) = V_m \cos(\omega t)$. A symmetrical 3-$\phi$ fault occurs at centre of the line, i.e. at point 'F' at time 'to' the +ve sequence impedance from source S_1 to point 'F' equals $(0.004 + j0.04)$ p.u. The wave form corresponding to phase 'a' fault current from bus X reveals that decaying d.c. offset current is –ve and in magnitude at its maximum initial value. Assume that the negative sequence are equal to +ve sequence impedances and the zero sequence (Z) are 3 times +ve sequence (Z).

5.22 The instant (t_0) of the fault will be
(a) 4.682 ms (b) 9.667 ms
(c) 14.667 ms (d) 19.667 ms [2008 : 2 M]

5.23 The rms value of the ac component of fault current (I_x) will be
(a) 3.59 kA (b) 5.07 kA
(c) 7.18 kA (d) 10.15 kA [2008 : 2 M]

5.24 Instead of the three phase fault, if a single line to ground fault occurs on phase 'a' at point 'F' with zero fault impedance, then the rms of the ac component of fault current (I_x) for phase 'a' will be
(a) 4.97 pu (b) 7.0 pu
(c) 14.93 pu (d) 29.85 pu [2008 : 2 M]

5.25 The zero-sequence circuit of the three phase transformer shown in the figure is

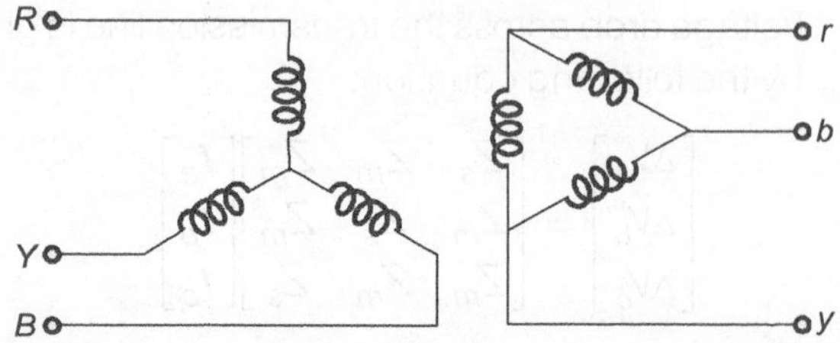

(a)
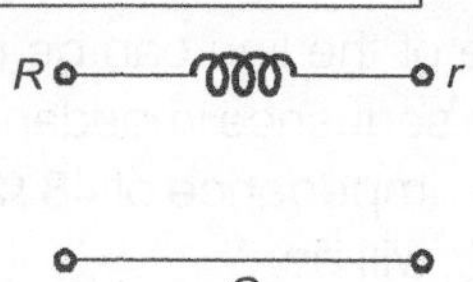

(b)
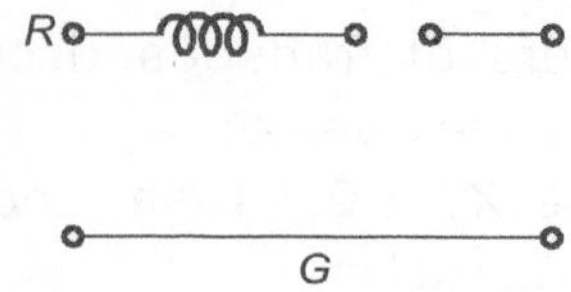

(c)
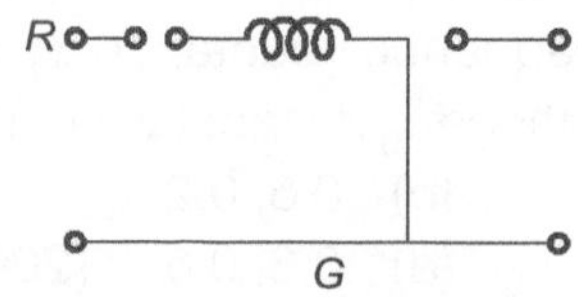

(d)
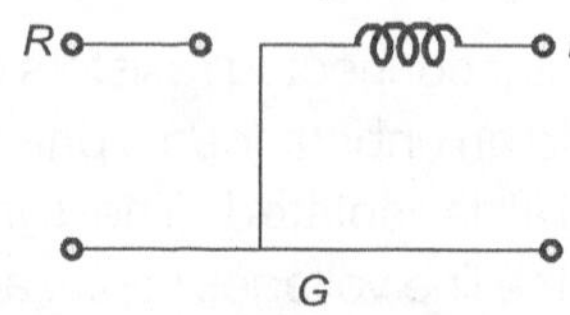

[2010 : 2 M]

Statement for Linked Answer Questions (5.26 and 5.27):
Two generator units $G1$ and $G2$ are connected by 15 kV line with a bus at the mid-point as shown below,

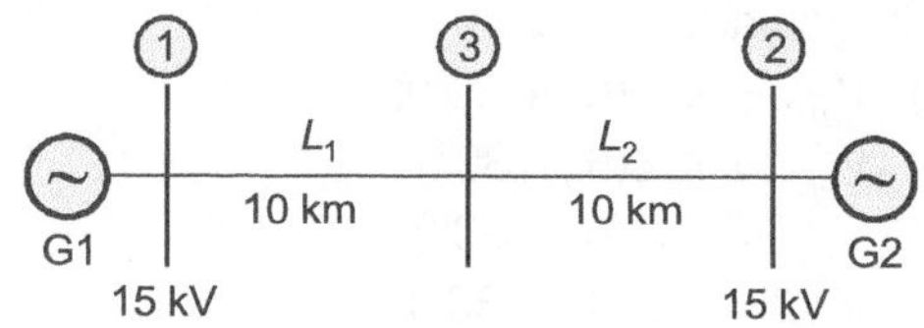

$G1$ = 250 MVA, 15 kV, positive sequence reactance X = 25% on its own base.
$G2$ = 100 MVA, 15 kV, positive sequence reactance X = 10% on its own base.
L_1 and L_2 = 10 km, positive sequence reactance $X = 0.225\ \Omega$/km.

5.26 For the above system, the positive sequence diagram with the p.u. values on the 100 MVA common base is

(a)
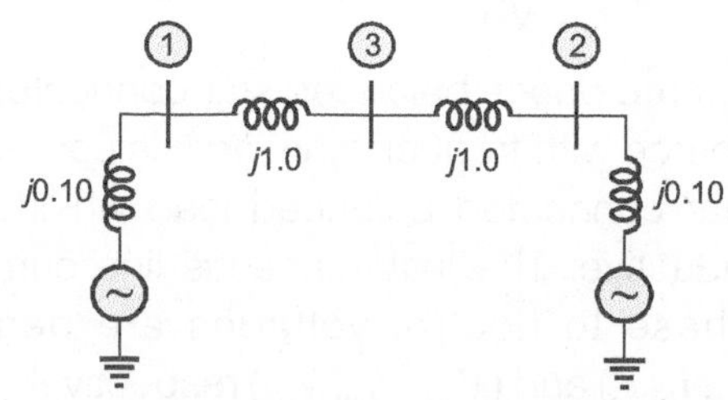

(b)
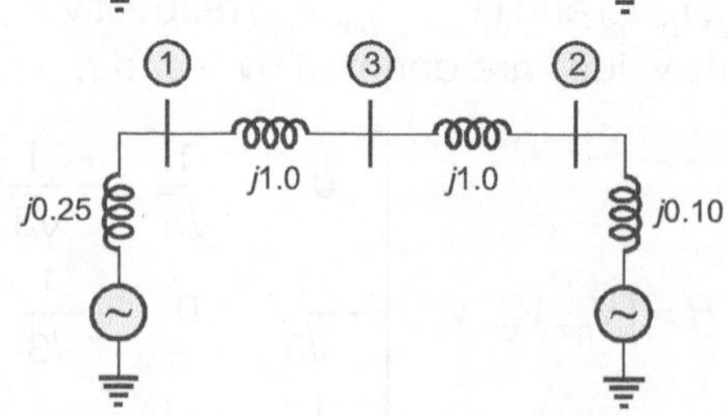

(c)
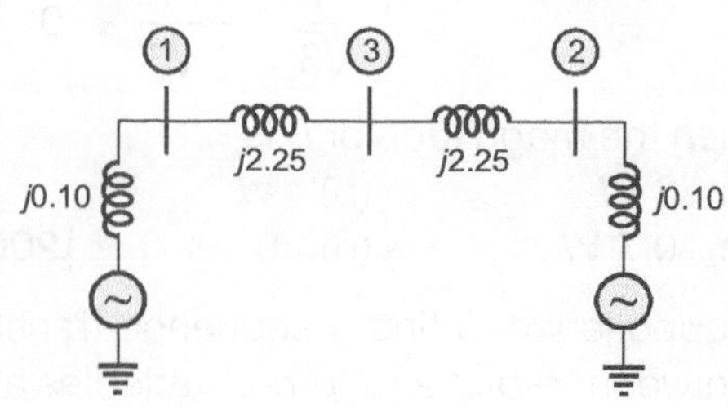

(d) 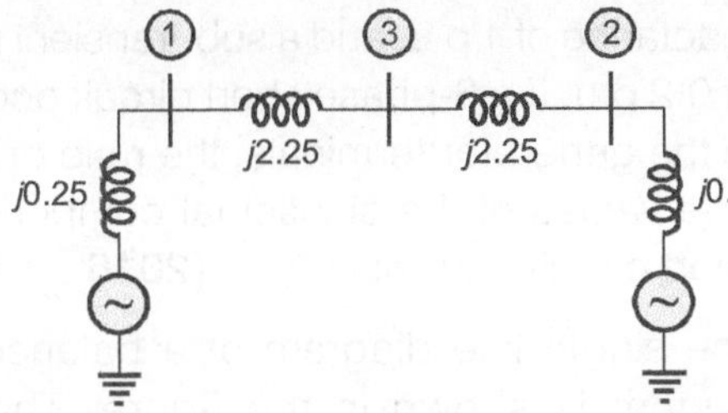

[2011 : 2 M]

5.27 In the above system, the three-phase fault MVA at the bus 3 is
(a) 82.55 MVA (b) 85.11 MVA
(c) 170.91 MVA (d) 181.82 MVA

[2011 : 2 M]

5.28 The sequence components of the fault current are as follows: $I_{\text{positive}} = j1.5$ pu, $I_{\text{negative}} = -j0.5$ pu, $I_{\text{zero}} = -j1$ pu. The type of fault in the system is
(a) LG (b) LL
(c) LLG (d) LLLG [2012 : 1 M]

5.29 Three-phase to ground fault takes place at locations F_1 and F_2 in the system shown in the figure.

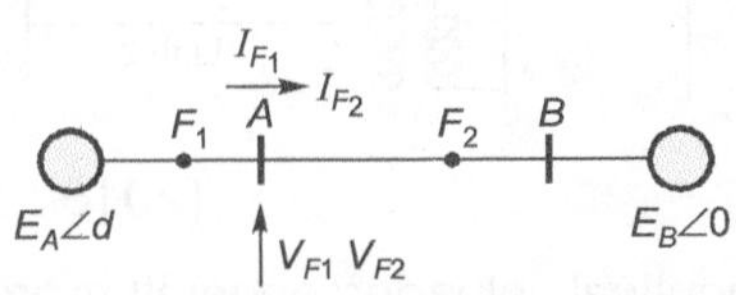

If the fault takes place at location F_1, then the voltage and the current at bus A are V_{F1} and I_{F1} respectively. If the fault takes place at location F_2, then the voltage and the current at bus A are V_{F2} and I_{F2} respectively. The correct statement about voltages and currents during faults at F_1 and F_2 is
(a) V_{F1} leads I_{F1} and V_{F2} leads I_{F2}
(b) V_{F1} leads I_{F1} and V_{F2} lags I_{F2}
(c) V_{F1} lags I_{F1} and V_{F2} leads I_{F2}
(d) V_{F1} lags I_{F1} and V_{F2} lags I_{F2}

[2014 : 1 M, Set-1]

5.30 A 2-bus system and corresponding zero sequence network are shown in the figure.

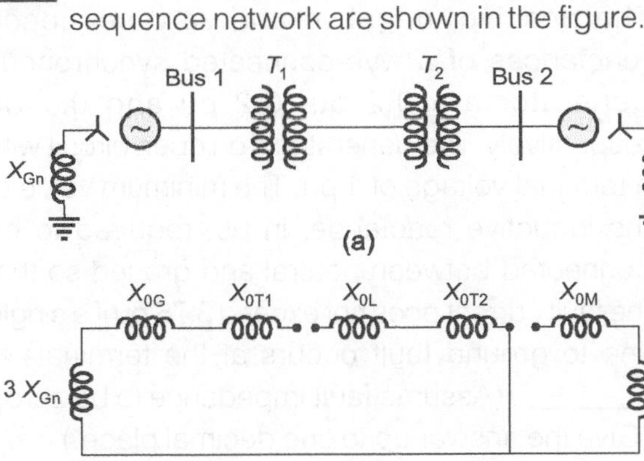

(a)

(b)

The transformer T_1 and T_2 are connected as
(a) ⅄ ⅄ and ⅄ △
(b) ⏚⅄ ⅄ and ⏚⅄ △
(c) ⏚⅄ △ and △ ⏚⅄
(d) △ ⏚⅄ and ⅄ ⏚⅄ [2014 : 1 M, Set-1]

5.31 In an unbalanced three-phase system, phase current $I_a = 1\angle(-90°)$ pu, negative sequence current $I_{b2} = 4\angle(-150°)$ pu, zero sequence current $I_{c0} = 3\angle 90°$ pu. The magnitude of phase current I_b in pu is
(a) 1.00 (b) 7.81
(c) 11.53 (d) 13.00

[2014 : 2 M, Set-1]

5.32 A three-phase, 100 MVA, 25 kV generator has solidly grounded neutral. The positive, negative and the zero sequence reactances of the generator are 0.2 pu, 0.2 pu and 0.05 pu, respectively, at the machine base quantities. If a bolted single phase to ground fault occurs at the terminal of the unloaded, generator, the fault current in amperes immediately after the fault is ________.

[2014 : 2 M, Set-2]

5.33 A sustained three-phase fault occurs in the power system shown in the figure. The current and voltage phasors during the fault (on a common reference), after the natural transients have died down, are also shown. Where is the fault located?

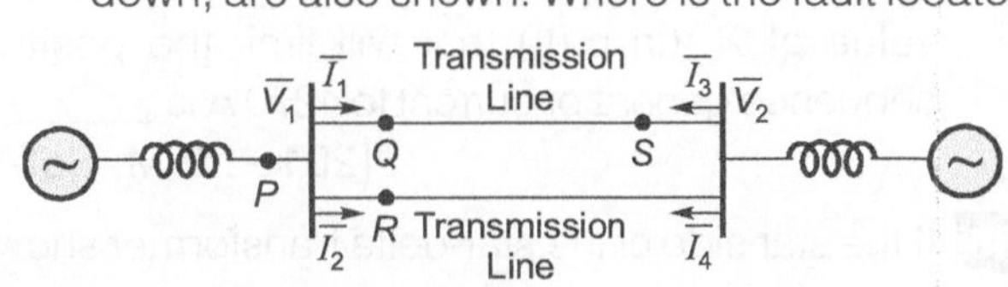

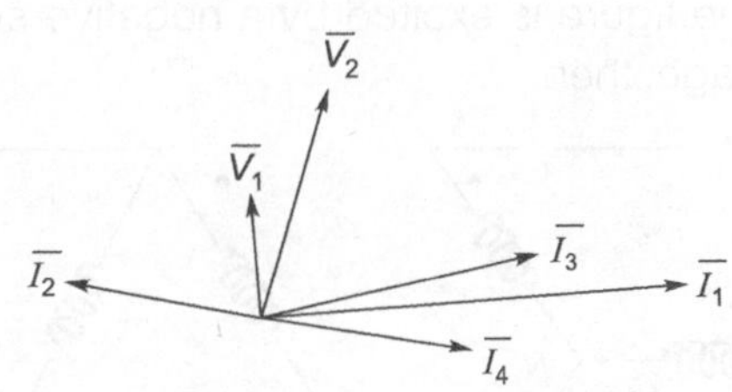

(a) Location P (b) Location Q
(c) Location R (d) Location S

[2015 : 2 M, Set-1]

5.34 For a power system network with n nodes, Z_{33} of its bus impedance matrix is $j0.5$ per unit. The voltage at node 3 is $1.3\angle -10°$ per unit. If a capacitor having reactance of $-j3.5$ per unit is now added to the network between node 3 and the reference node, the current drawn by the capacitor per unit is

(a) $0.325\angle-100°$ (b) $0.325\angle80°$
(c) $0.371\angle-100°$ (d) $0.433\angle80°$

[2013 : 2 M]

5.35 For a fully transposed transmission line
(a) positive, negative and zero sequence impedances are equal.
(b) positive and negative sequence impedances are equal.
(c) zero and positive sequence impedances are equal.
(d) negative and zero sequence impedances are equal. [2014 : 1 M, Set-3]

5.36 The magnitude of three-phase fault currents at buses *A* and *B* of a power system are 10 pu and 8 pu, respectively. Neglect all resistances in the system and consider the pre-fault system to be unloaded. The pre-fault voltage at all buses in the system is 1.0 pu. The voltage magnitude at bus *B* during a three-phase fault at bus *A* is 0.8 pu. The voltage magnitude at bus *A* during a three-phase fault at bus *B*, in pu, is _________?

[2016 : 1 M, Set-1]

5.37 A 30 MVA, 3-phase, 50 Hz, 13.8 kV, star-connected synchronous generator has positive, negative and zero sequence reactances, 15%, 15% and 5% respectively. A reactance (X_n) is connected between the neutral of the generator and ground. A double line to ground fault takes place involving phases '*b*' and '*c*', with a fault impedance of j0.1 p.u. The value of X_n (in p.u.) that will limit the positive sequence generator current to 4270 A is _______.

[2016 : 2 M, Set-1]

5.38 If the star side of the star-delta transformer shown in the figure is excited by a negative sequence voltage, then

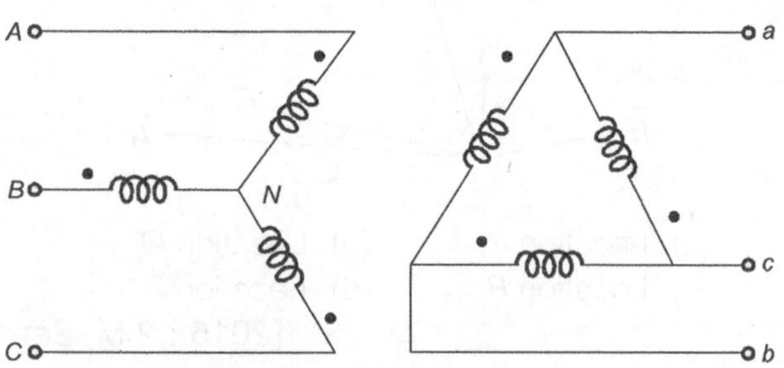

(a) V_{AB} leads V_{ab} by 60°
(b) V_{AB} lags V_{ab} by 60°
(c) V_{AB} leads V_{ab} by 30°
(d) V_{AB} lags V_{ab} by 30° [2016 : 2 M, Set-1]

5.39 A 50 MVA, 10 kV, 50 Hz, star-connected, unloaded three-phase alternator has a synchronous reactance of 1 p.u. and a sub-transient reactance of 0.2 p.u. If a 3-phase short circuit occurs close to the generator terminals, the ratio of initial and final values of the sinusoidal component of the short circuit current is ____. [2016 : 1 M, Set-2]

5.40 The single line diagram of a balanced power system is shown in the figure. The voltage magnitude at the generator internal bus is constant and 1.0 p.u. The p.u. reactances of different components in the system are also shown in the figure. The infinite bus voltage magnitude is 1.0 p.u. A three phase fault occurs at the middle of line 2. The ratio of the maximum real power that can be transferred during the pre-fault condition to the maximum real power that can be transferred under the faulted condition is ______.

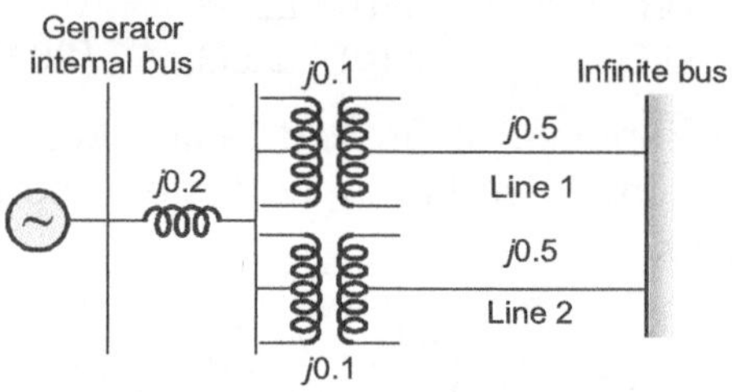

[2016 : 2 M, Set-2]

5.41 Two identical unloaded generators are connected in parallel as shown in the figure.

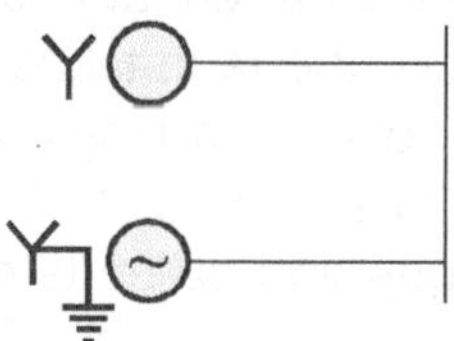

Both the generators are having positive, negative and zero sequence impedance of *j*0.4 p.u, *j*0.3 p.u. and j0.15 p.u., respectively. If the pre-fault voltage is 1 p.u., for a line-to-ground (L-G) fault at the terminals of the generators, the fault current, in p.u., is ______. [2016 : 2 M, Set-2]

5.42 The positive, negative and zero sequence reactances of a wye-connected synchronous generator are 0.2 pu, 0.2 pu and 0.1 pu respectively. The generator is on open circuit with a terminal voltage of 1 pu. The minimum value of the inductive reactance, in pu, required to be connected between neutral and ground so that the fault current does not exceed 3.75 pu if a single line to ground fault occurs at the terminals is _______. (Assume fault impedance to be zero). (Give the answer up to one decimal place.)

[2017 : 2 M, Set-1]

5.43 The series impedance matrix of a short three-phase transmission line in phase coordinates is

$\begin{bmatrix} Z_s & Z_m & Z_m \\ Z_m & Z_s & Z_m \\ Z_m & Z_m & Z_s \end{bmatrix}$. If the positive sequence impedance is $(1 + j10)\ \Omega$, and the zero sequence is $(4 + j31)\ \Omega$, then the imaginary part of Z_m (in Ω) is _____(up to 2 decimal places). [2018 : 1 M]

5.44 The positive, negative and zero sequence impedances of a three-phase generator are Z_1, Z_2 and Z_0 respectively. For a line-to-line fault with fault impedance Z_f, the fault current is $I_{f1} = kI_f$, where I_f is the fault current with zero fault impedance. The relation between Z_f and k is

(a) $Z_f = \dfrac{(Z_1 + Z_2)(1-k)}{k}$

(b) $Z_f = \dfrac{(Z_1 + Z_2)(1+k)}{k}$

(c) $Z_f = \dfrac{(Z_1 + Z_2)\,k}{1-k}$

(d) $Z_f = \dfrac{(Z_1 + Z_2)\,k}{1+k}$ [2018 : 2 M]

5.45 The positive, negative and zero sequence impedances of a 125 MVA, three-phase, 15.5 kV, star-grounded, 50 Hz generator are $j0.1$ pu, $j0.05$ pu and $j0.01$ pu respectively on the machine rating base.
The machine is unloaded and working at the rated terminal voltage. If the grounding impedance of the generator is $j0.01$ pu, then the magnitude of fault current for a b-phase to ground fault (in kA) is _____ (upto 2 decimal places). [2018 : 1 M]

5.46 In the circuit shown below, the switch is closed at $t = 0$. The value of θ in degrees which will give the maximum value of DC offset of the current at the time of switching is

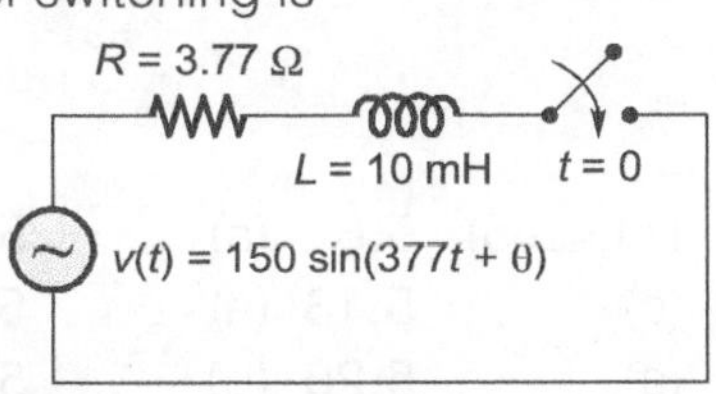

(a) –30 (b) 60
(c) –45 (d) 90 [2019 : 1 M]

5.47 Five alternators each rated 5 MVA, 13.2 kV with 25% of reactance on its own base are connected in parallel to a busbar. The short-circuit level in MVA at the busbar is _____. [2019 : 1 M]

5.48 A 30 kV, 50 Hz, 50 MVA generator has the positive, negative and zero sequence reactances of 0.25 pu, 0.15 pu, 0.05 pu, respectively. The neutral of the generator is grounded with a reactance so that the fault current for a bolted LG fault and that of a bolted three-phase fault at the generator terminal are equal. The value of grounding reactance in ohms (round off to one decimal place) is ______. [2019 : 2 M]

5.49 A cylindrical rotor synchronous generator has steady state synchronous reactance of 0.7 pu and subtransient reactance of 0.2 pu. It is operating at $(1+ j0)$ pu terminal voltage with an internal emf of $(1 + j0.7)$ pu. Following a three-phase solid short circuit fault at the terminal of the generator, the magnitude of the subtransient internal emf (rounded off to 2 decimal places) is _______ pu. [2020 : 2 M]

5.50 A three-phase balanced voltage is applied to the load shown. The phase sequence is RYB. The ratio $\dfrac{|I_B|}{|I_R|}$ is ______ .

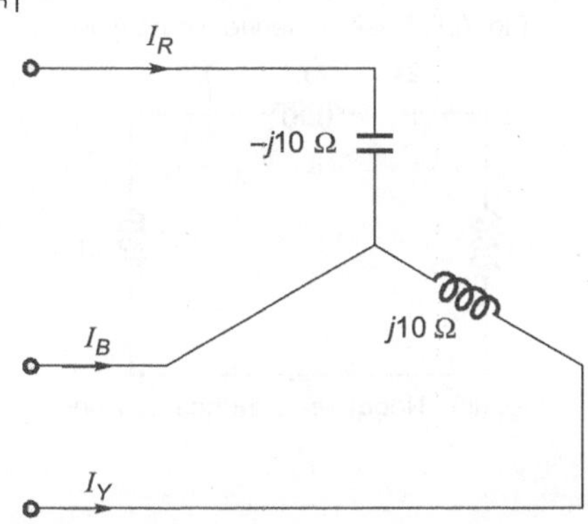

[2021 : 2 M]

5.51 Suppose I_A, I_B and I_C are a set of unbalanced current phasors in a three-phase system. The phase-B zero sequence current $I_{B0} = 0.1\angle 0°$ p.u. If phase-A current $I_A = 1.1\angle 0°$ p.u. and phase-C current $I_c = (1\angle 120° + 0.1)$ p.u., then I_B in p.u. is

(a) $1.1\angle 240° - 0.1\angle 0°$ (b) $1\angle 240° - 0.1\angle 0°$
(c) $1\angle -120° + 0.1\angle 0°$ (d) $1.1\angle -120° + 0.1\angle 0°$ [2021 : 2 M]

5.52 The valid positive, negative and zero sequence impedances (in p.u.), respectively, for a 220 kV, fully transposed three-phase transmission line, from the given choices are

(a) 1.1, 0.15 and 0.08
(b) 0.15, 0.15 and 0.35
(c) 0.2, 0.2 and 0.2
(d) 0.1, 0.3 and 0.1 [2022 : 1 M]

5.53 The two-bus power system shown in figure (i) has one alternator supplying a synchronous motor load through a Y-Δ transformer. The positive, negative and zero-sequence diagrams of the system are shown in figures (ii), (iii) and (iv), respectively. All reactances in the sequence diagrams are in p.u. For a bolted line-to-line fault (fault impedance = zero) between phases '*b*' and '*c*' at bus 1, neglecting all pre-fault currents, the magnitude of the fault current (from phase '*b*' to '*c*') in p.u. is ________ (Round off to 2 decimal places).

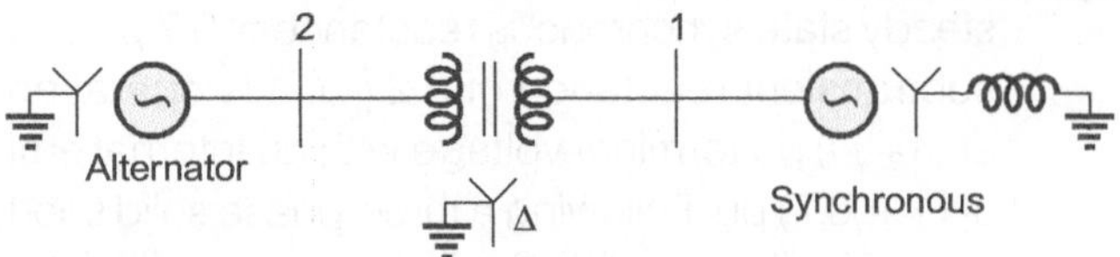

Fig. (i) : Single-line diagram of the power system

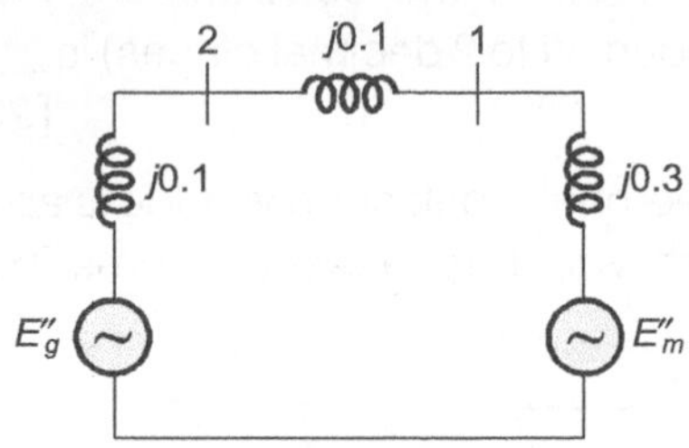

Fig. (ii) : Positive-sequence network

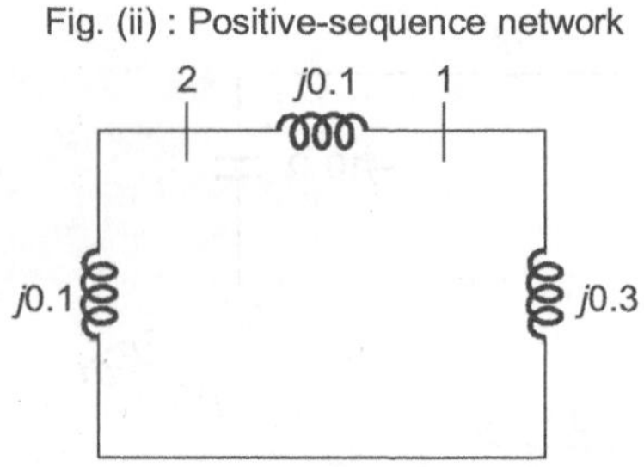

Fig. (iii) : Negative-sequence network

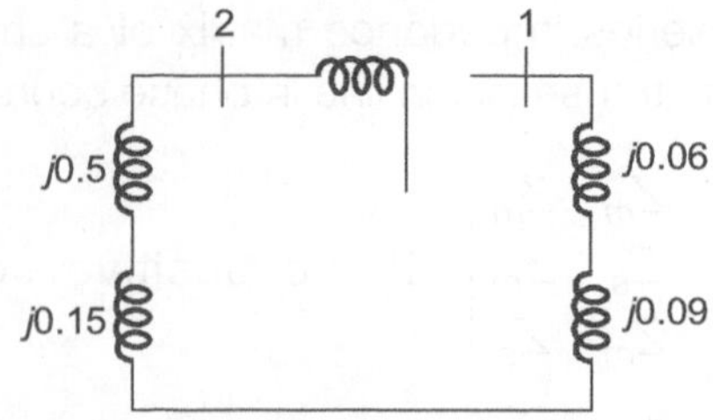

Fig. (iv) : Zero-sequence network

[2023 : 2 M]

5.54 For a two-phase network, the phase voltages V_p and V_q are to be expressed in terms of sequence voltages V_α and V_β as $\begin{bmatrix} V_p \\ V_q \end{bmatrix} = S \begin{bmatrix} V_\alpha \\ V_\beta \end{bmatrix}$. The possible option(s) for matrix S is/are

(a) $\begin{bmatrix} 1 & 1 \\ 1 & 1 \end{bmatrix}$ (b) $\begin{bmatrix} 1 & 1 \\ 1 & -1 \end{bmatrix}$

(c) $\begin{bmatrix} 1 & 1 \\ 1 & 0 \end{bmatrix}$ (d) $\begin{bmatrix} -1 & 1 \\ 1 & 1 \end{bmatrix}$

[2024 : 2 M]

Answers Fault Analysis

5.1	(b)	5.2	(c)	5.3	(c)	5.4	(57.74)	5.5	(11.4285)	5.6	(a)	5.7	(b)
5.8	(b)	5.9	(Sol.)	5.10	(d)	5.11	(d)	5.12	(c)	5.13	(a)	5.14	(d)
5.15	(d)	5.16	(b)	5.17	(d)	5.18	(b)	5.19	(a)	5.20	(b)	5.21	(b)
5.22	(a)	5.23	(a)	5.24	(c)	5.25	(c)	5.26	(a)	5.27	(d)	5.28	(c)
5.29	(c)	5.30	(b)	5.31	(c)	5.32	(15393)	5.33	(b)	5.34	(d)	5.35	(b)
5.36	(0.84)	5.37	(1.07)	5.38	(d)	5.39	(5.0)	5.40	(2.286)	5.41	(6)	5.42	(0.1)
5.43	(7)	5.44	(a)	5.45	(73.52)	5.46	(c)	5.47	(100)	5.48	(1.8)	5.49	(1.02)
5.50	(1)	5.51	(c)	5.52	(b)	5.53	(7.21)	5.54	(*)				

Explanations Fault Analysis

5.1 (b)

Unbalanced fault → Negative sequence dominate.
Ground fault → Zero sequence dominates.

5.3 (c)

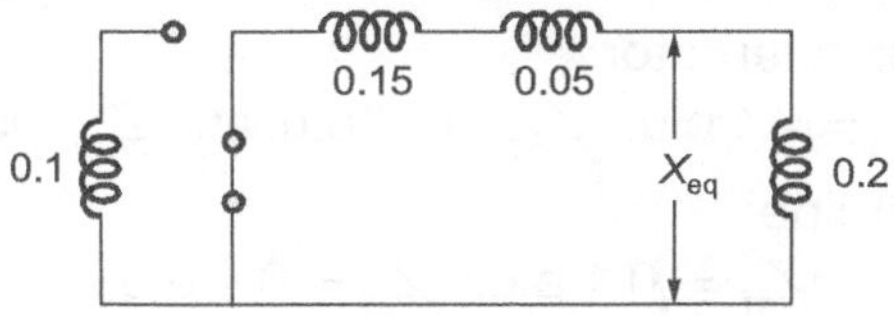

$$X_{eq} = 0.2 \parallel 0.2 = 0.1 \text{ p.u.}$$

5.4 Sol.

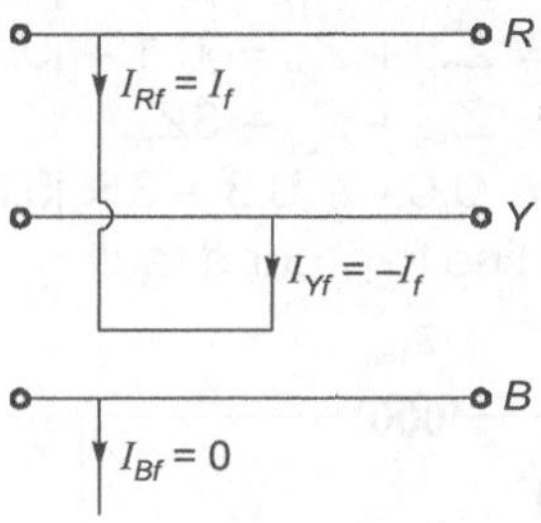

$$I_{B_f}^{(0)} = \frac{1}{3}(I_{R_f} + I_{Y_f} + I_{B_f}) = \frac{1}{3}(I_f - I_f + 0) = 0$$

$$I_{B_f}^{(0)} = I_{R_f}^{(0)} = I_{Y_f}^{(0)} = 0; \quad I_{R_f}^{(0)} = 0$$

$$I_{B_f} = I_{B_f}^{(0)} + I_{B_f}^{(1)} + I_{B_f}^{(2)} = 0 \Rightarrow I_{B_f}^{(2)} = I_{B_f}^{(1)} \quad \text{...(1)}$$

Fault current,

$$I_f = I_{R_f} = I_{R_f}^{(0)} + I_{R_f}^{(1)} + I_{R_f}^{(2)} = 0 + \alpha^2 I_{B_f}^{(1)} + \alpha I_{B_f}^{(2)}$$

$$I_f = (\alpha^2 - \alpha) I_{B_f}^{(1)} = \sqrt{3}\angle -90\ I_{B_f}^{(1)}$$

$$\left|I_{B_f}^{(1)}\right| = \frac{|I_f|}{\sqrt{3}} = \frac{100}{\sqrt{3}} \text{ A}; \quad \left|I_{R_f}^{(1)}\right| = \left|I_{B_f}^{(1)}\right| = \frac{100}{\sqrt{3}}$$

From equation (1),

$$\left|I_{R_f}^{(2)}\right| = \left|I_{B_f}^{(2)}\right| = \left|I_{B_f}^{(1)}\right| = \frac{100}{\sqrt{3}} \text{ A}$$

$$\left|I_{R_f}^{(0)}\right| = 0$$

$$\left|I_{R_f}^{(1)}\right| = \left|I_{R_f}^{(2)}\right| = \frac{100}{\sqrt{3}} \text{ A}$$

5.5 Sol.

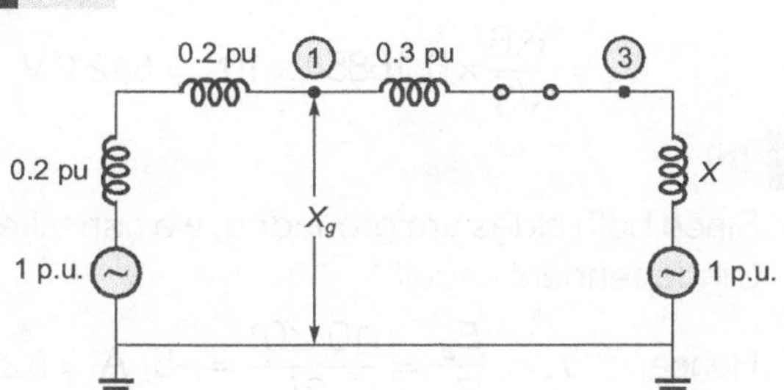

$$I_{3\text{-}\phi} = \frac{E_f}{X_1}$$

At bus (1),

$$X_1 = \frac{(0.4)(0.3 + X)}{(0.4) + (0.3 + X)}$$

$$\frac{E_f}{X_1} = 5 \text{ pu} \quad \text{(given)}$$

or $X + 0.7 = 2(0.3 + X)$

⇒ $X = 0.1$ pu

Now at bus (3)

$$X_{1\,eq} = \frac{(0.2+0.2+0.3)\times 0.1}{(0.2+0.2+0.3+0.1)} = \frac{0.7\times 0.1}{0.8}$$

∴ Fault level input,

$$I_f = \frac{1 \text{ pu}}{X_{eq}} = \frac{0.8}{0.1\times 0.7} = 11.4285 \text{ pu}$$

5.6 (a)

The fault current in case of line to ground fault is given by

$$I_{L\text{-}G} = \frac{3E_f}{X_1 + X_2 + X_0 + 3X_n}$$

Also, $I_{3\text{-}\phi} = \dfrac{E_f}{X_1}$

According to question,

$$I_{L\text{-}G} = I_{3\text{-}\phi}$$

$$\frac{3E_f}{X_1 + X_2 + X_0 + 3X_n} = \frac{E_f}{X_1}$$

$$2X_1 = X_2 + X_0 + 3X_n$$

$$2(0.1) = 0.1 + 0.05 + 3X_n$$

$$X_n = 0.0167 \text{ pu}$$

5.7 (b)

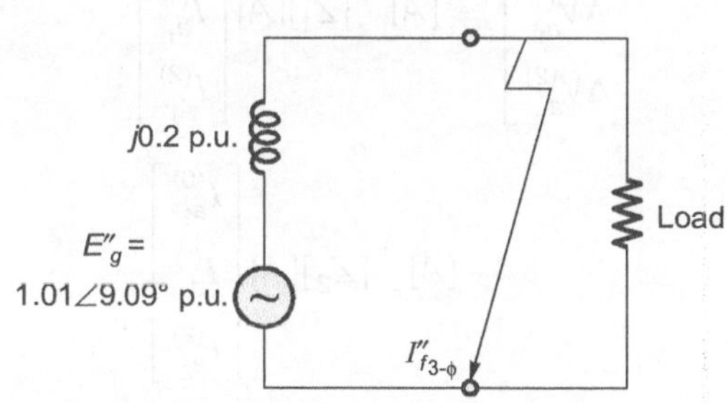

$$P_{3\text{-}\phi\,(\text{p.u.})} = V_{\text{p.u.}}\, I_{\text{p.u.}} \cos\theta$$

$$P_{\text{p.u.}} = \frac{400}{500} = 0.8 \text{ p.u.}$$

(∵ Base MVA = 500 MVA)
(P_{actual} = 400 MW)

Since, $V_{\text{p.u.}} = 1\angle 0°$ p.u.

$\therefore \quad I_{p.u.} = 0.8$

$$E_g'' = 1\angle 0° + j0.2\,(0.8\angle 0°)$$
$$= 1.01\angle 9.09° \text{ p.u.}$$

Initial rms current,

$$I_f'' = \frac{1.01}{0.2} = 5.05 \text{ p.u.}$$

5.8 (b)

Zero sequence impedance diagram can be drawn as :

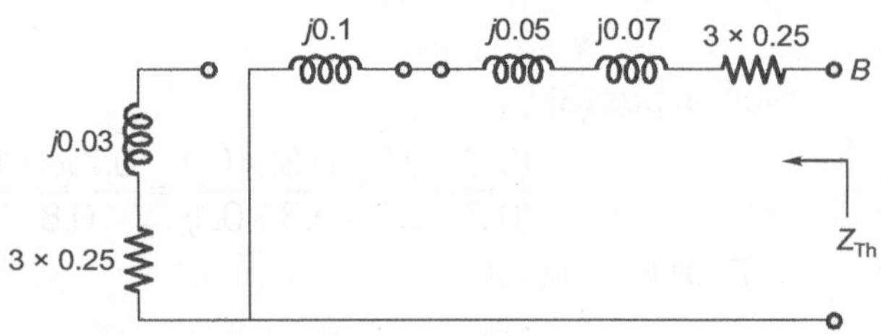

$$Z_{Th} = 0.75 + j0.22$$

5.9 Sol.

$$\begin{bmatrix}\Delta V_{a_1}\\ \Delta V_{b_1}\\ \Delta V_{c_1}\end{bmatrix} = \begin{bmatrix}0.15 & 0.05 & 0.05\\ 0.05 & 0.15 & 0.05\\ 0.05 & 0.05 & 0.15\end{bmatrix}\begin{bmatrix}I_{a_1}\\ I_{b_1}\\ I_{c_1}\end{bmatrix} + \begin{bmatrix}0.04 & 0.04 & 0.04\\ 0.04 & 0.04 & 0.04\\ 0.04 & 0.04 & 0.04\end{bmatrix}\begin{bmatrix}I_{a_2}\\ I_{b_2}\\ I_{c_2}\end{bmatrix}$$

$$[A]\begin{bmatrix}\Delta V_{a_1}^{(0)}\\ \Delta V_{a_1}^{(1)}\\ \Delta V_{a_1}^{(2)}\end{bmatrix} = [Z_1][A]\begin{bmatrix}I_{a_1}^{(0)}\\ I_{a_1}^{(1)}\\ I_{a_1}^{(2)}\end{bmatrix} + [Z_2][A]\begin{bmatrix}I_{a_2}^{(0)}\\ I_{a_2}^{(1)}\\ I_{a_2}^{(2)}\end{bmatrix}$$

$$\begin{bmatrix}\Delta V_{a_1}^{(0)}\\ \Delta V_{a_1}^{(1)}\\ \Delta V_{a_1}^{(2)}\end{bmatrix} = [A]^{-1}[Z_1][A]\begin{bmatrix}I_{a_1}^{(0)}\\ I_{a_1}^{(1)}\\ I_{a_1}^{(2)}\end{bmatrix} + [A]^{-1}[Z_2][A]\begin{bmatrix}I_{a_2}^{(0)}\\ I_{a_2}^{(1)}\\ I_{a_2}^{(2)}\end{bmatrix}$$

$$= \begin{bmatrix}0.25 & 0 & 0\\ 0 & 0.1 & 0\\ 0 & 0 & 0.1\end{bmatrix}\begin{bmatrix}I_{a_1}^{(0)}\\ I_{a_1}^{(1)}\\ I_{a_1}^{(2)}\end{bmatrix} + \begin{bmatrix}0.12 & 0 & 0\\ 0 & 0 & 0\\ 0 & 0 & 0.1\end{bmatrix}\begin{bmatrix}I_{a_2}^{(0)}\\ I_{a_2}^{(1)}\\ I_{a_2}^{(2)}\end{bmatrix}$$

$$\Delta V_{a_1}^{(0)} = 0.25 I_{a_1}^{(0)} + 0.12 I_{a_2}^{(0)}$$

$\Rightarrow \quad Z_{011} = 0.25, Z_{012} = 0.12$

Similarly,

$$Z_{021} = 0.12, Z_{022} = 0.25$$

$$\begin{bmatrix}Z_{011} & Z_{012}\\ Z_{021} & Z_{022}\end{bmatrix} = \begin{bmatrix}0.25 & 0.12\\ 0.12 & 0.25\end{bmatrix}$$

5.10 (d)

For alternator:

$Z_{1g} = j0.1$ p.u.; $Z_{2g} = j0.1$ p.u. and $Z_{0g} = j0.04$ p.u.

For line:

$$Z_{1l} = j0.1 \text{ p.u.};\ Z_{2l} = j0.1 \text{ p.u.}$$
$$Z_{0l} = j0.3 \text{ p.u.};\ Z_n = j0.05 \text{ p.u.}$$

Equivalent Sequence Impedances:

$$Z_{1eq} = Z_{1g} + Z_{1l} = j0.1 + j0.1 = j0.2 \text{ p.u.}$$
$$Z_{2eq} = Z_{2g} + Z_{2l} = j0.1 + j0.1 = j0.2 \text{ p.u.}$$
$$Z_{0eq} = Z_{0g} + Z_{0l} + 3Z_n$$
$$= j0.04 + j0.3 + 3 \times j0.05 = j0.49 \text{ p.u.}$$

For single line to ground fault

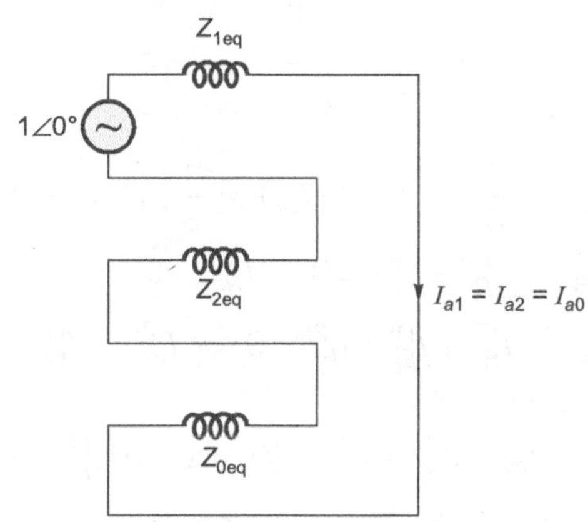

$$I_{a1} = I_{a2} = I_{a0} = \frac{1\angle 0}{Z_{1eq} + Z_{2eq} + Z_{0eq}}$$

$$= \frac{1\angle 0}{j(0.2 + 0.2 + 0.49)} = -j1.1236 \text{ p.u.}$$

Fault current $= I_f = 3I_{a1} = -j3.3708$ p.u.

Voltage of neutral w.r.t. ground

$$V_n = I_f \times Z_n = -j3.3708 \times j0.05$$
$$= 0.16854 \text{ p.u.}$$

Per phase base voltage $= V_p$

$$= \frac{6.6}{\sqrt{3}}\text{kV} = \frac{6.6}{\sqrt{3}} \times 10^3 \text{ V}$$

$$V_n = \frac{6.6}{\sqrt{3}} \times 0.16854 \times 10^3 = 642.2 \text{ V}$$

5.11 (d)

Since both sides are grounded, we can take it as 3 independent circuit.

Hence, $\quad I_a = \dfrac{E_a}{Z_a} = \dfrac{10\angle 0°}{2j} = -5j \text{ A} = 5\angle -90°$

$$I_b = \frac{E_b}{Z_b} = \frac{10\angle -90}{3j}$$
$$= 3.33\angle{-180} \text{ A}$$
$$I_c = \frac{E_c}{Z_c} = \frac{10\angle 120}{4j} = 2.5\angle 30° \text{A}$$
$$I_{a1} = 1/3(I_a + \alpha I_b + \alpha^2 I_c)$$
$$= 1/3[5\angle{-90} + 3.33\angle(-180+120) + 2.5\angle(240+30)]$$
$$= 1/3(5\angle 90 + 3.33\angle{-60} + 2.5\angle 270)$$
$$= 3.510\angle{-81°} \text{ A}$$

5.12 (c)

The generator is operating at no-load and rated voltage

So, $E_g = 1.0\angle 0°$ p.u.

Initial symmetrical current

$$= I = \frac{E_g}{X_d''} = \frac{1\angle 0°}{j0.19}$$
$$I = 5.26\angle{-90°} \text{ p.u.}$$
$$\text{Base current} = \frac{(\text{MVA})_{\text{Base}}}{\sqrt{3}\,(\text{kV})_{\text{Base}}} = \frac{100}{\sqrt{3}\times 11} \text{kA}$$
$$|I| = 5.26 \times \text{Base current}$$
$$= 5.26 \times \frac{100}{\sqrt{3}\times 11} \text{kA} = 30.37 \text{ kA}$$

5.13 (a)

$I_a = 10\angle 0°$ A; $I_b = 10\angle 180°$

and $I_c = 0$ A

$\alpha = 1\angle 120°$

Positive sequence current component in line '*a*'

$$I_{a1} = \frac{1}{3}[I_a + \alpha I_b + \alpha^2 I_C]$$
$$= \frac{1}{3}[10\angle 0° + (1\angle 120°)\times(10\angle 180°) + (1\angle 240°)\times 0]$$
$$I_{a1} = 5.77\angle{-30°}\text{A}$$

5.14 (d)

For single-line to ground fault,

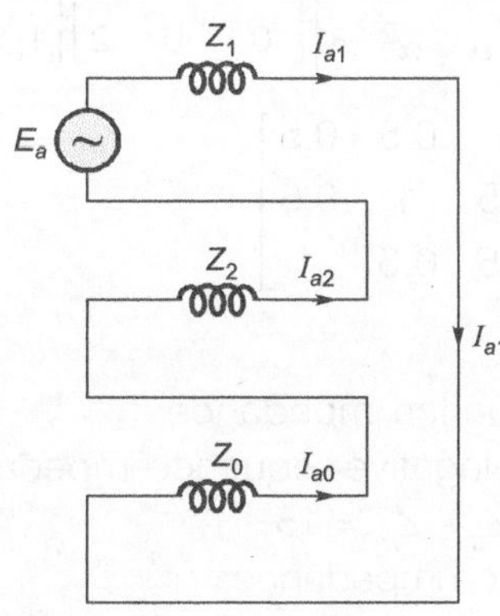

Since, the generator is operating at no-load and rated voltage

$E_a = 1\angle 0$ p.u.

$Z_1 = jX_1 = j0.15$ p.u.

$Z_2 = jX_2 = j0.15$ p.u.

and $Z_0 = jX_0 = j0.05$ p.u.

Sequence currents,

$$I_{a1} = I_{a2} = I_{a0} = \frac{E_a}{Z_1 + Z_2 + Z_0}$$
$$= \frac{1\angle 0°}{j0.15 + j0.15 + j0.05}$$
$$= 2.857\angle 90° \text{ pu}$$

Fault current $= 3I_{a1} = 3 \times 2.857 = 8.57$ pu

5.15 (d)

Fault level (in MVA) = $\sqrt{3}$ x prefault line voltage (kV) × fault current (in kA)

Pre-fault line voltage = V_{prefault} = 220 kV

3-ϕ fault

3-ϕ fault level = 4000 MVA

$\sqrt{3} V_{\text{prefault}} I_f$ = 4000 MVA

$$I_f = \frac{4000\times 10^6}{\sqrt{3}\times 220\times 10^3} = 10.5 \text{ kA}$$

Let positive sequence driving point reactance $= X_1$

$$X_1 = \frac{V_{\text{prefault}}/\sqrt{3}}{I_f} = \frac{220/\sqrt{3}}{10.5} = 12.1\Omega$$

5.16 (b)

Let negative sequence reactance = X_2

Assuming, $X_2 = X_1$

and zero sequence reactance = X_0

In single-line to ground fault,

Fault level (MVA) = 5000 MVA

$$\text{Fault current} = I_f = \frac{\text{Fault-level}}{\sqrt{3}\times V_{\text{prefault}}}$$
$$= \frac{5000\times 10^6}{\sqrt{3}\times 220\times 10^3} = 13.12 \text{ kA}$$

In SLG fault

Positive sequence current

$$= I_{a1} = \frac{I_f}{3} = \frac{13.12}{3} = 4.37 \text{ kA}$$
$$I_{a1} = \frac{V_{\text{prefault}}(\text{Per phase})}{X_1 + X_2 + X_0}$$
$$4.37 = \frac{220/\sqrt{3}}{2X_1 + X_0} = \frac{127}{2\times 12.1 + X_0}$$

Zero sequence reactance, $X_0 \approx 4.84\ \Omega$

5.17 (d)

For transmission line

Positive sequence impedance,

$$X_1 = X_s - X_m = 0.4 - 0.1 = 0.3\ \Omega/\text{km}$$

Zero sequence impedance,

$$X_0 = X_s + 2X_m = 0.4 + 0.2 = 0.6\ \Omega/\text{km}$$

5.18 (b)

Positive (negative) sequence component of line voltage is $\sqrt{3}$ times of positive (negative) sequence component of phase voltage and leads (lags) by 30° in actual values but in p.u. on respective base values they are equal,

$$V_{an}^{(1)} = X\angle\theta_1 - 30°; \quad V_{an}^{(2)} = Y\angle\theta_2 + 30°$$

5.19 (a)

$$R = [V_{an}\ V_{bn}\ V_{cn}]\begin{bmatrix} 0 & \frac{1}{\sqrt{3}} & -\frac{1}{\sqrt{3}} \\ -\frac{1}{\sqrt{3}} & 0 & \frac{1}{\sqrt{3}} \\ \frac{1}{\sqrt{3}} & -\frac{1}{\sqrt{3}} & 0 \end{bmatrix}\begin{bmatrix} i_a \\ i_b \\ i_c \end{bmatrix} \quad \text{...(A)}$$

Assuming phase sequence *abc*, taking V_{an} as the reference,

$$V_{an} = V\angle 0°;$$
$$V_{bn} = V\angle -120°;$$
$$V_{cn} = V\angle 120° \quad \text{...(B)}$$

V is the rms value of V_{an}, V_{bn} and V_{cn}

I is the rms value of i_a, i_b and i_c

From equation (B) we can say

$$V_{an} = \sqrt{2}V\sin\omega t;$$

$$\left.\begin{aligned} V_{bn} &= \sqrt{2}V\sin(\omega t - 120°) \\ V_{cn} &= \sqrt{2}V\sin(\omega t + 120°) \end{aligned}\right\} \quad \text{...(C)}$$

and similarly due to lagging load,

$$i_a = \sqrt{2}I\sin(\omega t - 90°)$$
$$i_b = \sqrt{2}I\sin(\omega t - 90 - 120°)$$
$$i_c = \sqrt{2}I\sin(\omega t - 90 + 120°)$$

From equation (C) by considering line values of voltage,

$$V_{ab} = \sqrt{6}V\sin(\omega t + 30°)$$
$$V_{bc} = \sqrt{6}V\sin(\omega t - 90°)$$
$$V_{ca} = \sqrt{6}V\sin(\omega t + 150°)$$

From equation (A) we have

$$= -\left[\frac{V_{bn} - V_{cn}}{\sqrt{3}}\ \frac{V_{cn} - V_{an}}{\sqrt{3}}\ \frac{V_{an} - V_{bn}}{\sqrt{3}}\right]\begin{bmatrix} i_a \\ i_b \\ i_c \end{bmatrix}$$

$$= -\left[(\sqrt{2}V\sin(\omega t - 90°))(\sqrt{2}I\sin(\omega t - 90°) + \sqrt{2}V\sin(\omega t + 150°)\cdot \sqrt{2}I\sin(\omega t + 150°) + \sqrt{2}V\sin(\omega t + 30°)\cdot\sqrt{2}I\sin(\omega t + 30°)\right]$$

$$\because \quad \left(\sin A\sin B = \frac{\cos(A-B) - \cos(A+B)}{2}\right)$$

$$= -[VI - VI\cos(2\omega t + 180°) + VI - VI\cos(2\omega t + 300°) + VI - VI\cos(2\omega t + 60)]$$

$$= -[3\,VI - VI(\cos(2\omega t - 180°) + \cos(2\omega t + 300°) + \cos(2\omega t + 60))]$$

$$= R = -3\,VI = |R| = 3VI$$

5.20 (b)

$$\text{Here,} \quad \begin{bmatrix} V_p \\ V_n \\ V_o \end{bmatrix} = \begin{bmatrix} 0.5 & 0 & 0 \\ 0 & 0.5 & 0 \\ 0 & 0 & 2 \end{bmatrix}\begin{bmatrix} I_p \\ I_n \\ I_0 \end{bmatrix}$$

$$\text{where,} \quad \begin{bmatrix} V_a \\ V_b \\ V_c \end{bmatrix} = [Z]\begin{bmatrix} I_a \\ I_b \\ I_c \end{bmatrix} \quad \text{...(i)}$$

$$\begin{bmatrix} I_a \\ I_b \\ I_c \end{bmatrix} = K\begin{bmatrix} 1 & 1 & 1 \\ \alpha^2 & \alpha & 1 \\ \alpha & \alpha^2 & 1 \end{bmatrix}\begin{bmatrix} I_p \\ I_n \\ I_0 \end{bmatrix} = KA\begin{bmatrix} I_p \\ I_n \\ I_0 \end{bmatrix}$$

Similarly, $V_{ph} = K[A]\,V_{sequence}$

$$= K[A]\begin{bmatrix} 0.5 & 0 & 0 \\ 0 & 0.5 & 0 \\ 0 & 0 & 2 \end{bmatrix}[I_{sequence}]$$

$$= K[A]\begin{bmatrix} 0.5 & 0 & 0 \\ 0 & 0.5 & 0 \\ 0 & 0 & 2 \end{bmatrix}\frac{[A^{-1}]}{K}I_{ph} \quad \text{...(ii)}$$

Comparing equations (i) with (ii),

$$\therefore \quad Z = A\begin{bmatrix} 0.5 & 0 & 0 \\ 0 & 0.5 & 0 \\ 0 & 0 & 2 \end{bmatrix}[A^{-1}]$$

$$= \frac{1}{3}\begin{bmatrix} 1 & 1 & 1 \\ \alpha^2 & \alpha & 1 \\ \alpha & \alpha^2 & 1 \end{bmatrix}\begin{bmatrix} 0.5 & 0 & 0 \\ 0 & 0.5 & 0 \\ 0 & 0 & 2 \end{bmatrix}\begin{bmatrix} 1 & \alpha & \alpha^2 \\ 1 & \alpha^2 & \alpha \\ 1 & 1 & 1 \end{bmatrix}$$

$$= \begin{bmatrix} 1 & 0.5 & 0.5 \\ 0.5 & 1 & 0.5 \\ 0.5 & 0.5 & 1 \end{bmatrix}$$

5.21 (b)

Positive sequence impedance

= Negative sequence impedance

$= Z_s - Z_m = 15$ ...(i)

Zero-sequence impedance

$$= Z_s + 2Z_m = 48 \quad ...(ii)$$

Solving equation (i) and (ii), we get

$$Z_s = 26\ \Omega \text{ and } Z_m = 11\ \Omega$$

5.22 (a)

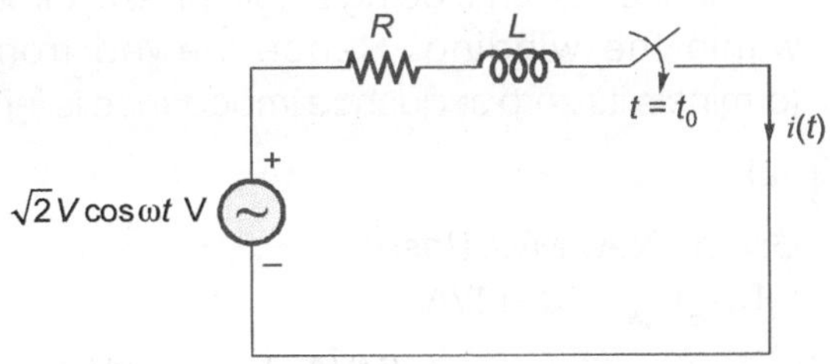

To find $i(t)$ for $t(0^+)$ we can convert the above circuit in following way

Let, $t' = t - t_0 \Rightarrow t = t' + t_0$

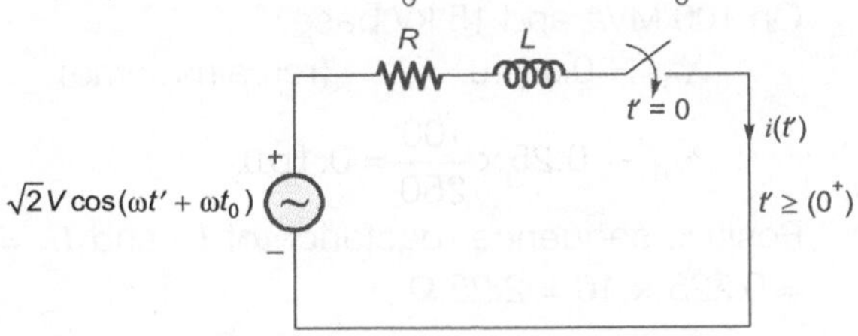

$$i(t') = \frac{\sqrt{2}V}{|z|}\cos(\omega t' + \omega t_0 - \alpha) + A\,e^{-\frac{t'}{\tau}} \quad (t' \geq t_0^+)$$

Symmetrical short circuit current,

$$|Z| = \sqrt{R^2 + (\omega L)^2}$$

$$\alpha = \tan^{-1}\frac{\omega L}{R}$$

$$\tau = \frac{L}{R}$$

Now, $i(t' = 0^-) = i(t' = 0^+) = 0$

$$\Rightarrow \frac{\sqrt{2}V}{|Z|}\cos(\omega t_0 - \alpha) + A = 0$$

$$A = -\frac{\sqrt{2}V}{|Z|}\cos(\omega t_0 - \alpha)$$

Now change,

$$t' = t - t_0$$

$$i(t) = \frac{\sqrt{2}V}{|Z|}\cos(\omega t - \alpha)$$

$$-\frac{\sqrt{2}V}{|Z|}\cos(\omega t_0 - \alpha)\,e^{-\frac{(t-t_0)R}{L}} \quad t \geq t_0^+$$

Now, if the initial value of dc off set (A) is $-\frac{\sqrt{2}V}{|Z|}$

it requires that $\cos(\omega t_0 - \alpha) = 1$

$$\Rightarrow \omega t_0 - \alpha = 0 \Rightarrow t_0 = \frac{\alpha}{\omega}$$

$$t_0 = \frac{\tan^{-1}\frac{\omega L}{R}}{\omega}$$

In problem given the circuit for 3-ϕ fault can be drawn as:

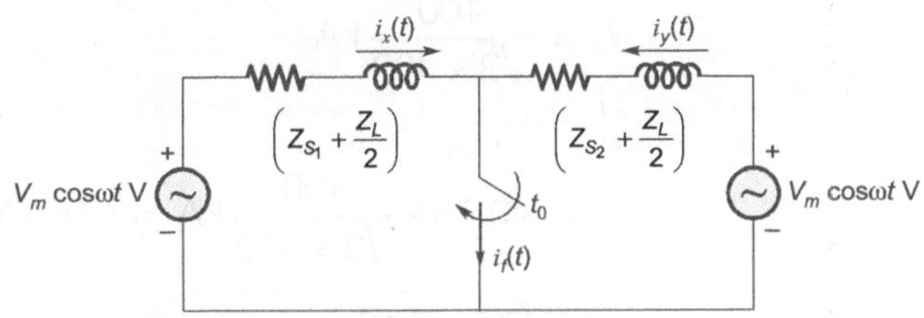

According to the derivation:

For initial value of dc off set to be $-\frac{\sqrt{2}V}{|Z|}$ we required

$$t_0 = \frac{\tan^{-1}\frac{\omega L}{R}}{\omega} = \frac{\tan^{-1}\frac{0.04}{0.004}}{2\times 3.14\times 50} = 4.68 \text{ msec.}$$

5.23 (a)

Fault occurs at mid point 'F'

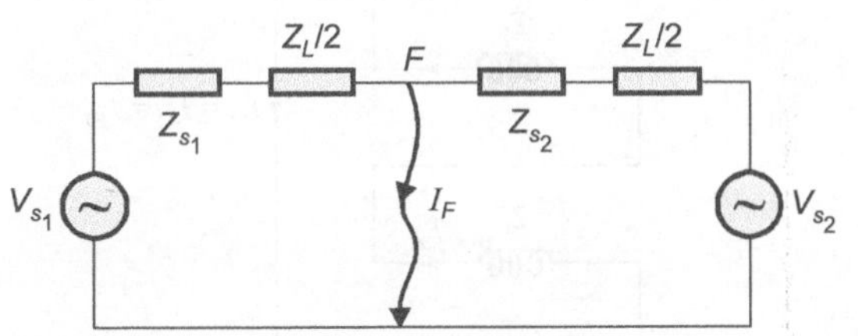

Thevenins equivalent,

$$V_s = V_{s_1} = V_{s_2} = 1.0\angle 0 \text{ pu}$$

$$Z_1 = \left(Z_{s_1} + \frac{Z_L}{2}\right) \Big|\Big| \left(Z_{s_2} + \frac{Z_L}{2}\right)$$

as $Z_{s_1} = Z_{s_2} = 0.001 + j0.01$ pu

and $Z_L = 0.006 + j0.06$

$$Z_1 = \frac{1}{2}\left(Z_{s_1} + \frac{Z_L}{2}\right)$$

$$= \frac{1}{2}\left(0.001 + j0.01 + \frac{0.006 + j0.06}{2}\right)$$

$$Z_1 = 0.002 + j0.02$$

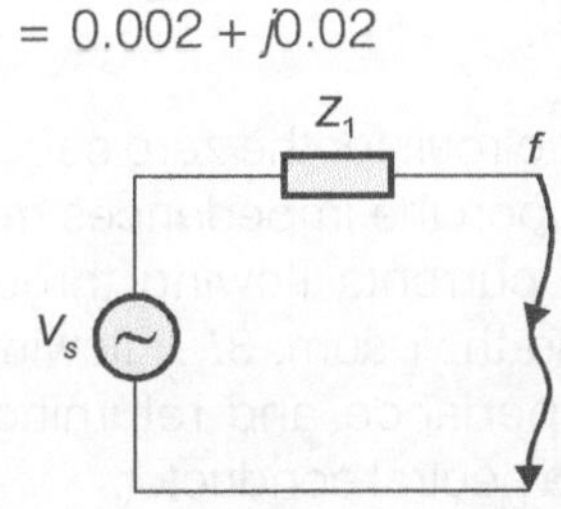

Fault current, $I_f = \frac{V_s}{Z_1} = \frac{1\angle 0°}{0.002 + j0.02}$

$$= 49.75\angle{-84.29} \text{ p.u.}$$

$$|I_f| \text{ p.u.} = 49.75$$

$$\text{Base current} = \frac{\text{Base MVA}}{\sqrt{3} \times \text{Voltage Base}} \text{ kA}$$

$$I_B = \frac{100}{\sqrt{3} \times 400} \text{ kA}$$

$$|I_f| = |I_f|_{\text{p.u.}} \times I_B$$

$$= 49.75 \times \frac{100}{\sqrt{3} \times 400} \text{ kA} = 7.18 \text{ kA}$$

$$I_x^f = \frac{|I_f|}{2} = \frac{7.18}{2} \text{ kA} = 3.59 \text{ kA}$$

5.24 (c)

$$Z_1 = 0.002 + j0.02 \text{ pu}$$
$$Z_2 = Z_1 = 0.002 + j0.02 \text{ pu}$$
$$Z_0 = 3Z_1 = 0.006 + j0.006 \text{ pu}$$

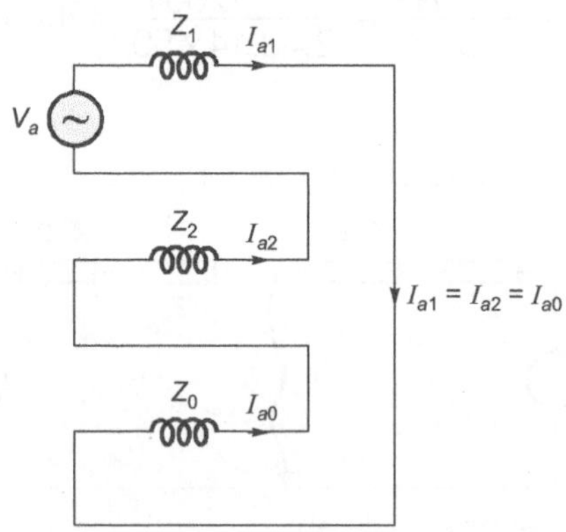

$$I_{a0} = I_{a1} = I_{a2} = \frac{V_s}{Z_1 + Z_2 + Z_0}$$

$$= \frac{V_s}{Z_1 + Z_2 + 3Z_1}$$

$$= \frac{V_s}{5Z_1} = \frac{1\angle 0^\circ}{5 \times (0.002 + j0.02)}$$

$$= 9.95\angle{-84.29} \text{ p.u.}$$

$$\text{Fault current} = |I_f|_{\text{p.u.}} = 3\,|I_{a0}|$$

$$= 3 \times 9.95 = 29.85 \text{ p.u.}$$

$$I_x^f = \frac{|I_f|_{\text{p.u.}}}{2} = \frac{29.85}{2} = 14.93 \text{ p.u.}$$

5.25 (c)

- Equivalent circuit for the zero sequence network depends upon the impedances met by the zero sequence currents flowing through the three phases and their sum, $3I_{a0}$, flowing through the neutral impedance and returning through the ground or a neutral conductor.
- If these is no complete path for zero sequence currents in a circuit, the zero sequence impedance is infinite.

Thus a Y-connected circuit with ungrounded neutral has infinite impedance to zero sequence currents.

- A delta-connected circuit provides no path for zero sequence currents flowing in the line. The zero-sequence currents being single phase, circulate within the winding. Hence viewed from its terminals its zero sequence impedance is infinite.

5.26 (a)

Given : New MVA Base,
$(MVA_B)_{new} = 100$ MVA

$$Z_{\text{p.u.(new)}} = Z_{\text{p.u.(old)}} \times \frac{(MVA_B)_{new}}{(MVA_B)_{old}} \times \frac{(kV_B)^2_{old}}{(kV_B)^2_{new}}$$

$$(kV_B)_{old} = (kV_B)_{new} = 15 \text{ kV}$$

On 100 MVA and 15 kV base,

$$X_{G2} = 0.1 \text{ p.u.} \qquad \text{(remains same)}$$

$$X_{G1} = 0.25 \times \frac{100}{250} = 0.1 \text{ p.u.}$$

Positive sequence reactance of L_1 and $L_2 = X_L$ $= 0.225 \times 10 = 2.25\ \Omega$

$$\text{Base impedance} = Z_B = \frac{(kV_B)^2_{new}}{(MVA_B)_{new}}$$

$$= \frac{15^2}{100} = 2.25\ \Omega$$

$$X_L(\text{in pu}) = \frac{2.25}{2.25} = 1 \text{ p.u.}$$

{p.u. reactance of section L_1 and L_2}

Hence, option (a) is correct.

5.27 (d)

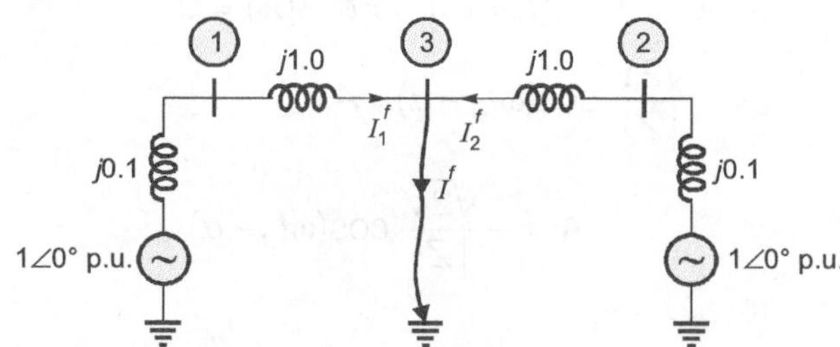

$$\text{Fault current} = I^f = I_1^f + I_2^f$$

$$I_1^f = I_2^f = \frac{1\angle 0^\circ}{j1.0 + j0.1} = -j0.9091 \text{ p.u.}$$

$$I^f = -j1.8182 \text{ p.u.}$$

Fault level = 1 × 1.8182 = 1.8182 p.u.

Fault level (in MVA) = 1.8182 × 100 = 181.82 MVA

5.28 (c)

Only 'LLG' fault has a unique property that the sum of all sequence components of the fault current is zero.

5.29 (c)

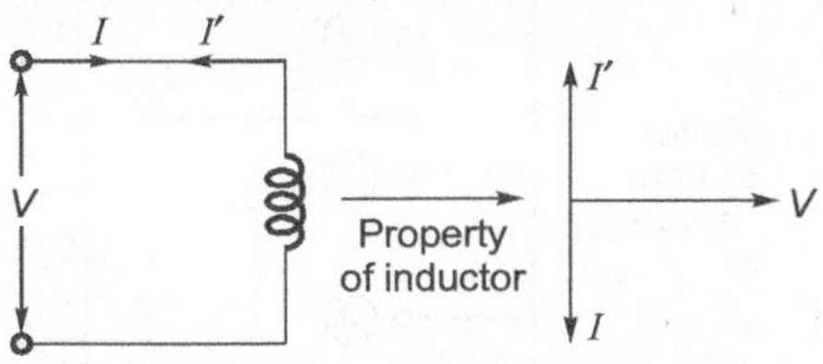

For the fault at location F_1 the equivalent circuit is shown below.

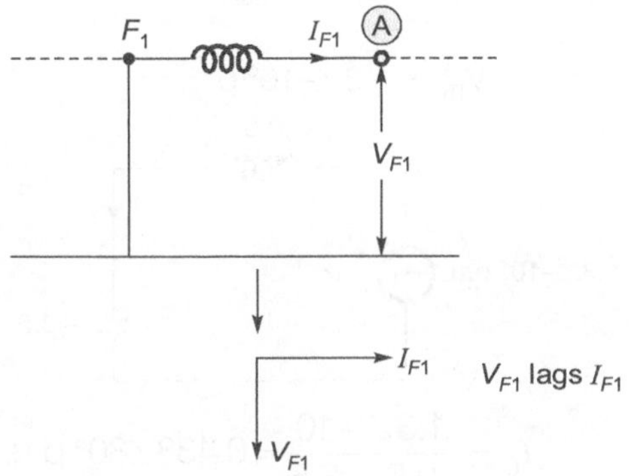

For the fault at location F_2 the equivalent circuit is shown below.

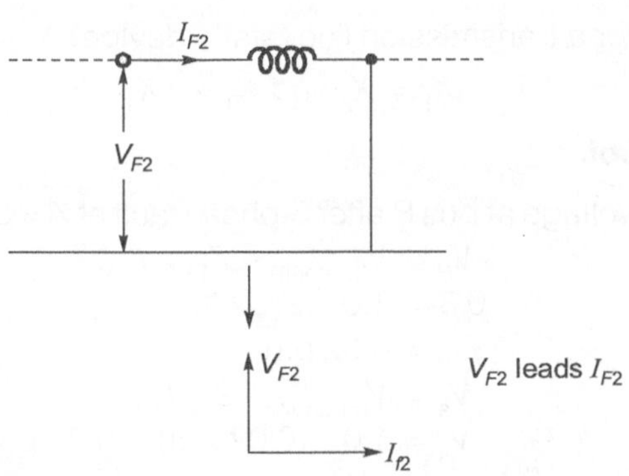

5.30 (b)

We have to use switch diagram for the zero-sequence network of transformer.

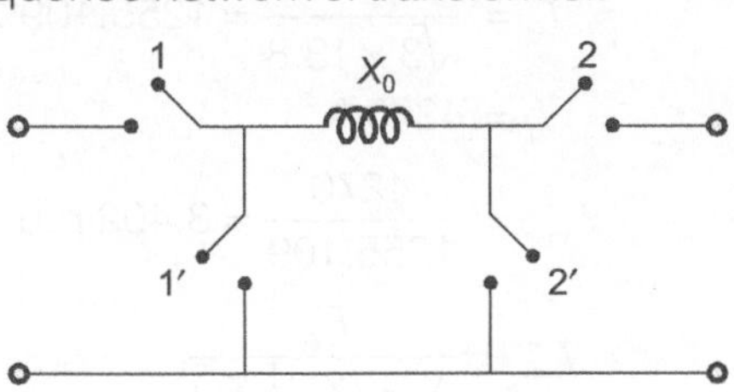

- Switch 1 or 2 is closed when winding in Y and neutral grounded.
- Switch 1′ or 2′ is closed when winding is in Δ.

Given, for transformer T_1, switch 1 is closed and 2 is open.

Therefore, transformer will be connected in primary with grounded neutral and secondary in star with isolated neutral (i.e. grounded Y – Y).

For transformer T_2, switch 1 is closed and 2′ is closed, therefore, primary winding is star connected with neutral grounded whereas secondary will be delta connected (i.e. grounded Y – Δ).

5.31 (c)

Given : $I_a = 1\angle(-90°)$ pu

$I_{b_2} = 4\angle(-150°)$ pu

$I_{c_0} = 3\angle(90°)$ pu

As zero sequence current in all the three phases of unbalanced 3-ϕ system are equal, therefore,

$$I_{a_0} = I_{b_0} = I_{c_0} = 3\angle 90° \text{ pu}$$

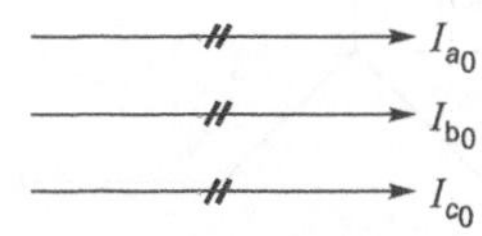

Zero sequence currents

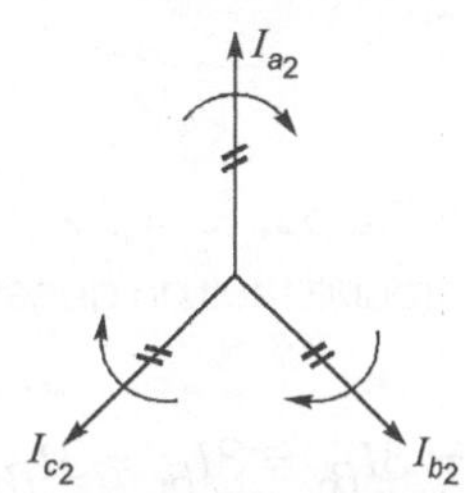

Negative sequence currents

Also, $I_{b_2} = 4\angle -150°$

= Negative phase sequence current of phase *b*.

Therefore, referring to the negative phase sequence phasor,

$$I_{a_2} = 4\angle(-150° - 120°) = 4\angle -270° \text{ pu}$$

We know that, phase current vectors in matrix form are given by

$$\begin{bmatrix} I_a \\ I_b \\ I_c \end{bmatrix} = \begin{bmatrix} 1 & 1 & 1 \\ 1 & a^2 & a \\ 1 & a & a^2 \end{bmatrix} \begin{bmatrix} I_{a_0} \\ I_{a_1} \\ I_{a_2} \end{bmatrix}$$

(where, operator $a = 1\angle 120°$)

Here, $I_a = I_{a_0} + I_{a_1} + I_{a_2}$

Given : $I_a = 1\angle -90°$ and

$I_{a_0} = 3\angle 90°$ pu and

$I_{a_2} = 4\angle -270°$ pu (as obtained above)

Therefore,

$$I_{a_1} = I_a - (I_{a_0} + I_{a_2})$$

$$= (1\angle -90°) - (3\angle 90° + 4\angle 270°)$$

or, $I_{a_1} = 8\angle -90°$ pu

Therefore,

$$I_b = I_{a_0} + a^2 I_{a_1} + a I_{a_2}$$

$$= (3\angle 90°) + (1\angle 120°)^2 (8 - \angle 90°) + (1\angle 120°)(4\angle -270°)$$

$$= 3\angle 90° + 8\angle 150° + 4\angle -150°$$

$$I_b = 11.53\angle 154.3° \text{ pu}$$

∴ Magnitude of phase current,

$$I_b = 11.53 \text{ pu}$$

5.32 Sol.

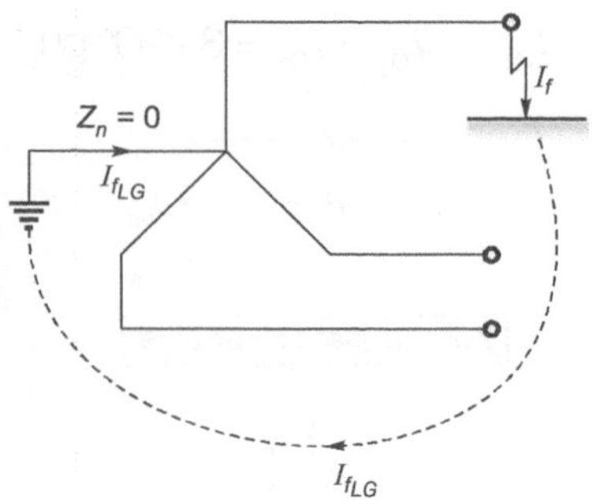

Given, $X_1 = 0.2$ pu, $X_2 = 0.2$ pu, $X_3 = 0.05$ pu

For a line-to-ground fault on generator, fault current is given by

$$I_{f_{LG}} = 3I_{R_0} = 3I_{R_1} = 3I_{R_2}$$

$$= \frac{3E}{Z_1 + Z_2 + Z_0}$$

$$I_{f_{LG}} = \frac{3\times 1}{0.2+0.2+0.05} = \frac{3}{0.45} = \frac{20}{3} \text{ pu}$$

Also, base MVA = 100

Base KVA = 25

∴ Base current,

$$I_b = \frac{100\times 10^6}{\sqrt{3}\times 25\times 10^3} = 2.309 \text{ kA}$$

∴ Actual fault current in ampere

$$= (I_{f_{LG}})_{pu} \times I_b = \frac{20}{3}\times 2.309\times 10^3$$

∴ $I_f = 15393$ Ampere

5.33 (b)

From the phasor I_2 and I_4 are having 180° phase shift. $|I_1|$ is higher than all other. Hence fault occurred at point Q.

5.34 (d)

Thevenin's equivalent circuit between bus (3) and reference bus = $Z_{33} = j0.5$

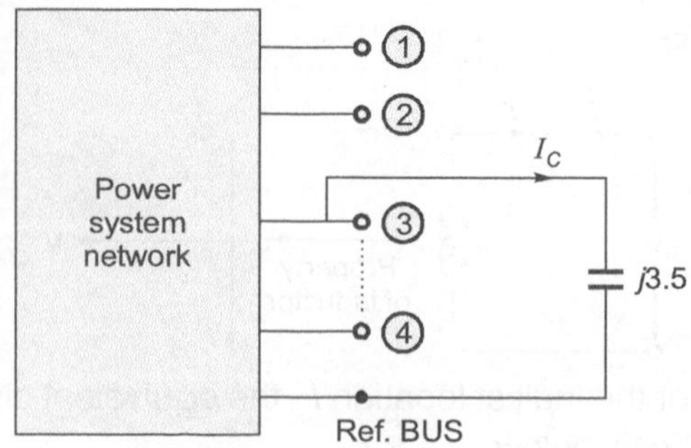

Thevenin equivalent voltage between 3rd bus and ref

$$V_{Th} = 1.3\angle -10° \text{ p.u.}$$

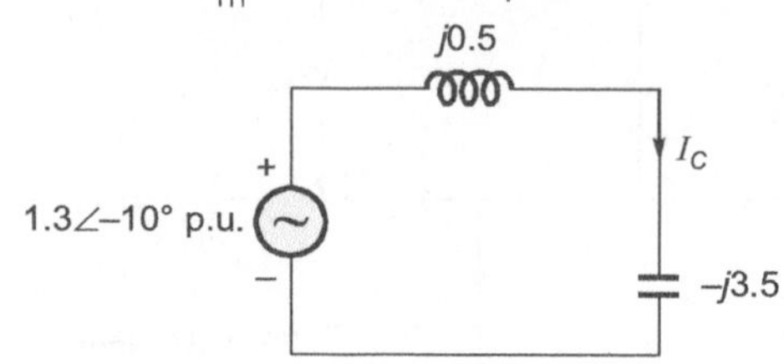

$$I_C = \frac{1.3\angle -10}{j0.5 - j3.5} = 0.433\angle 80° \text{ p.u.}$$

5.35 (b)

For a transmission line (static device),

$$X_1 = X_2 \text{ and } X_0 >> X_1$$

5.36 Sol.

Voltage at bus B after 3-phase fault at $A = 0.8$ p.u.

$$V_B = V_{(prefault)} - Z_{12}\times I_{f(A)}$$

$$0.8 = 1.0 - Z_{12}\times 10$$

$$Z_{12} = 0.02 \text{ p.u.}$$

$$V_A = V_{(prefault)} - Z_{12} I_{f(B)}$$

$$V_A = 1.0 - (0.02\times 8) = 0.84 \text{ p.u.}$$

5.37 Sol.

Base current,

$$I_B = \frac{30\times 10^3}{\sqrt{3}\times 13.8} = 1255.109 \text{ A}$$

$$I_f = 4270 \text{ A}$$

$$I_{p.u.} = \frac{4270}{1255.109} = 3.402 \text{ p.u.}$$

$$I_{g1} = \frac{E_a}{X_1 + (X_2 || X_0)}$$

where, $X_0 = X_0 + 3(Z_n + Z_f)$

$$X_0 = 3Z_n + 0.35$$

$$3.402 = \frac{1.0}{0.15 + \dfrac{0.15\times(3Z_n + 0.35)}{0.15 + 3Z_n + 0.35}}$$

By solving the equation,

$$Z_n = 1.07 \text{ p.u.}$$

5.38 (d)

According to negative sequence phasors.

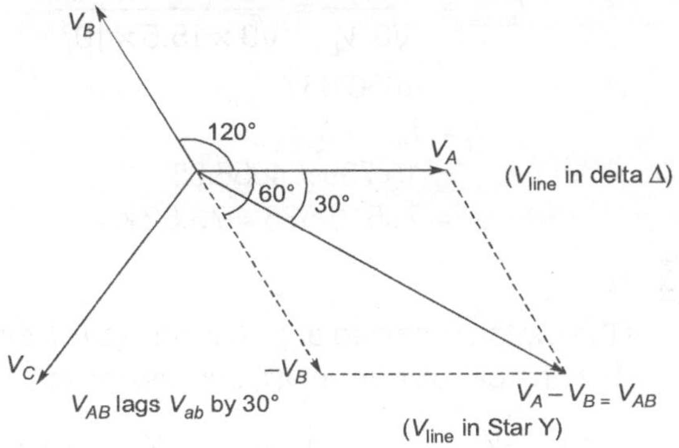

5.39 Sol.

$$I'' = \frac{E_g}{X''_d} \quad ; \quad I = \frac{E_g}{X}$$

$$\frac{I''}{I} = \frac{X}{X''_d} = \frac{1.0}{0.2} = 5.0 \text{ p.u.}$$

5.40 Sol.

$$\text{The ratio} = \frac{P_1}{P_2} = \frac{X_2}{X_1}$$

$X_1 =$

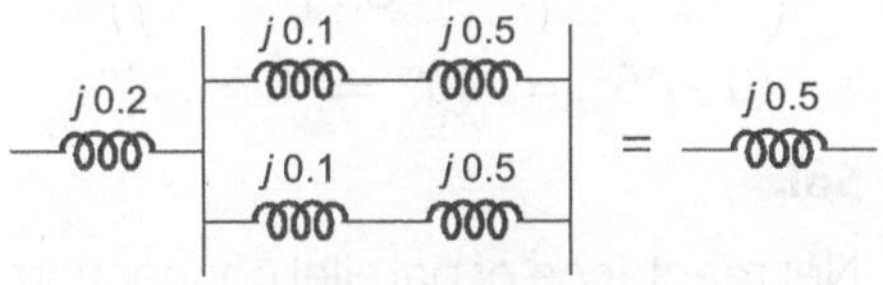

$X_2 =$

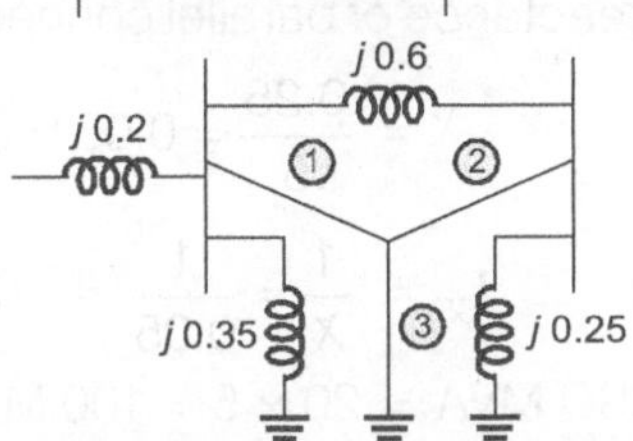

Equivalent star values will be

1. $\Rightarrow \quad \dfrac{j0.35 \times j0.6}{j1.2} = j0.175$

2. $\Rightarrow \quad \dfrac{j0.6 \times j0.25}{j1.2} = j0.125$

3. $\Rightarrow \quad \dfrac{j0.35 \times j0.25}{j1.2} = j0.0729$

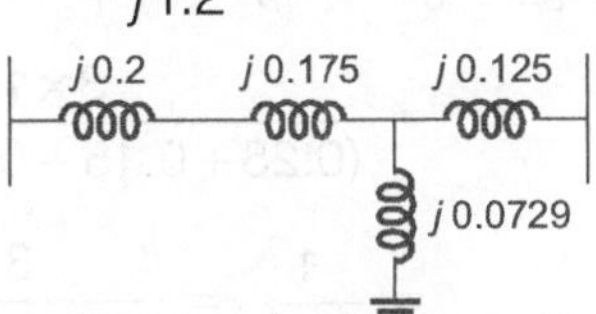

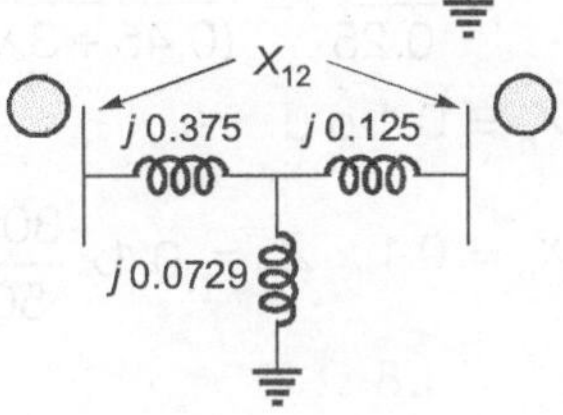

$$X_{12} = X_2$$

$$= j0.375 + j0.125 + \frac{j0.375 \times j0.125}{j0.0729}$$

$$= j1.1430 \text{ p.u.}$$

$$\frac{P_1}{P_2} = \frac{X_2}{X_1} = \frac{j1.1430}{j0.5} = 2.286$$

5.41 Sol.

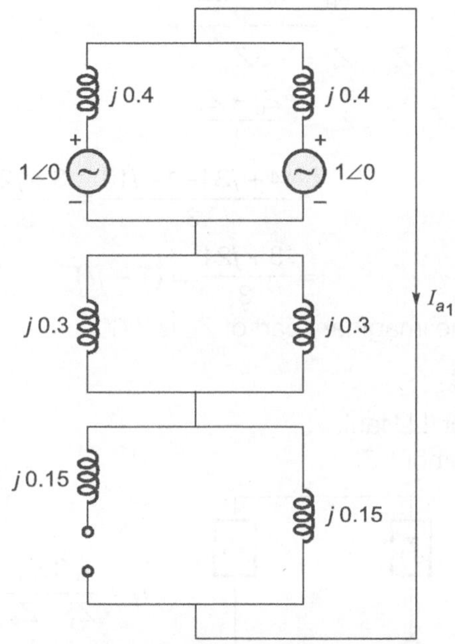

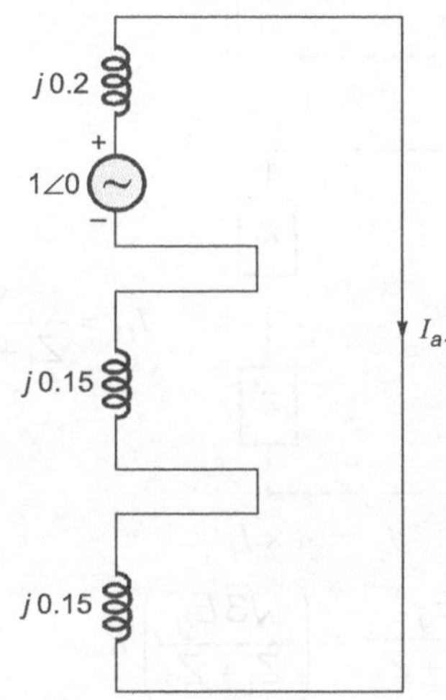

$$I_f = 3I_{a_1} = \frac{3}{0.2 + 0.15 + 0.15} = 6 \text{ p.u.}$$

5.42 Sol.

$$Z_1 = 0.2; \; Z_2 = 0.2; \; Z_0 = 0.1$$

$$V_{Th} = 1; \; I_f = 3.75$$

For LG fault;

$$I_f = \frac{3V_{Th}}{Z_1 + Z_2 + Z_0 + 3Z_n}$$

$$3.75 = \frac{3 \times 1}{0.2 + 0.2 + 0.1 + 3Z_n}$$

$$Z_n = 0.1 \text{ pu.}$$

5.43 Sol.

$$Z_1 = (1 + j10)\ \Omega$$
$$Z_0 = (4 + j31)\ \Omega$$

We know, $Z_1 = Z_s - Z_m$

$$Z_0 = Z_s + 2Z_m$$

$$Z_1 - Z_0 = -3Z_m$$

$$Z_m = \frac{Z_0 - Z_1}{3}$$

$$= \frac{4 + j31 - 1 - j10}{3} = \frac{3 + j21}{3}$$

$$= \frac{3 + j21}{3} = (1 + j7)$$

The imaginary part of Z_m is 7.00.

5.44 (a)

For 'LL' fault:

Without Z_f,

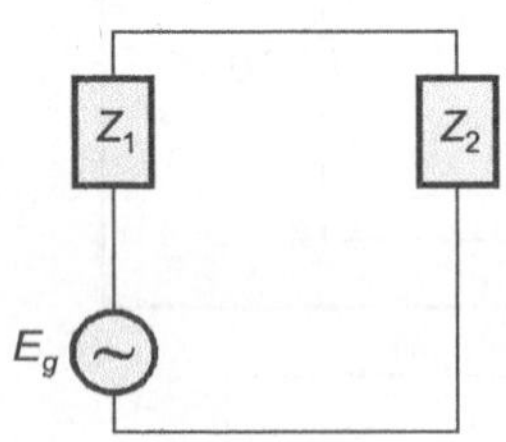

$$I_f = \frac{\sqrt{3} E_g}{Z_1 + Z_2}$$

With Z_f,

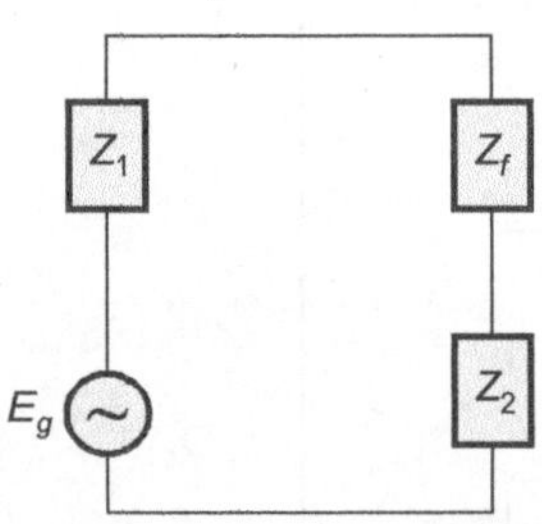

$$I_{f1} = \frac{\sqrt{3} E_g}{Z_1 + Z_2 + Z_f}$$

Given, $I_{f1} = k \times I_f$

$$\frac{\sqrt{3} \cdot E_g}{Z_1 + Z_2 + Z_f} = \left(\frac{\sqrt{3} E_g}{Z_1 + Z_2}\right) k$$

$$Z_1 + Z_2 = k(Z_1 + Z_2 + Z_f)$$

$$Z_f = \frac{(Z_1 + Z_2)(1 - k)}{k}$$

5.45 Sol.

For LG fault,

$$I_{f\text{p.u.}} = \frac{3.V_{Th}}{z_1 + z_2 + z_0 + 3z_n}$$

$$= \frac{3 \times 1}{0.1 + 0.05 + 0.01 + 3(0.01)}$$

$$= 15.789 \text{ p.u.}$$

$$I_{base} = \frac{P}{\sqrt{3}.V_L} = \frac{125 \times 10^6}{\sqrt{3} \times 15.5 \times 10^3}$$

$$= 4656.05 \text{ A}$$

$$I_f = I_{f\text{p.u.}} \cdot I_{base}$$
$$= 15.789 \times 4656.05$$
$$= 73516.538 = 73.52 \text{ kA}$$

5.46 (c)

If the switch is closed at $t = 0$ in series R-L circuit. Then the circuit current $i(t)$ expression is

$$i(t) = \left\{\frac{-V_m}{\sqrt{R^2 + X_L^2}} \sin(\theta - \phi)\right\} e^{-t/\tau} + \frac{V_m}{\sqrt{R^2 + X_L^2}} \sin(\omega t - \phi)$$

The first term of expression indicates DC offset current.

For maximum value of DC offset current, the angle should be 90°

$$-\left(\theta - \tan^{-1}\left(\frac{\omega L}{R}\right)\right) = 90°$$

$$-\left(\theta - \tan^{-1}\left(\frac{377 \times 10 \times 10^{-3}}{3.77}\right)\right) = 90°$$

$$(\theta - 45°) = -90° \Rightarrow \theta = -45°$$

5.47 Sol.

Net reactance of parallel connection,

$$X = \frac{0.25}{5} = 0.05 \text{ p.u.}$$

$$I_{SC} = \frac{1}{X} = \frac{1}{0.05} = 20 \text{ p.u.}$$

$$\text{SC MVA} = 20 \times 5 = 100 \text{ MVA}$$

5.48 Sol.

Given sequence reactances,

$X_1 = 0.25$ p.u.; $X_2 = 0.15$ p.u.; $X_0 = 0.05$ p.u.

Fault current,

$$I_{f(LG)} = I_{f(3\text{-}\phi)}$$

$$\frac{3V_{pu}}{(X_1 + X_2 + X_0 + 3X_n)} = \frac{V_{pu}}{X_1}$$

$$= \frac{3 \times 1}{(0.25 + 0.15 + 0.05 + 3X_n)}$$

$$= \frac{1}{0.25} = \frac{3}{(0.45 + 3X_n)} = \frac{1}{0.25}$$

$$\Rightarrow \quad X_n = 0.1 \text{ p.u.}$$

$$X_n = 0.1 \times Z_B = 0.1 \times \frac{30^2}{50}$$

$$= 1.8\ \Omega$$

5.49 Sol.

Prefault current,

$$I_0 = \frac{E_f - V_t}{jX_d} = \frac{1 + j0.7 - 1}{j0.7} = 1$$

Subtransient induced emf,

$$E_f'' = V_0 + jX_d'' I_0$$
$$= 1 + j0.2 \times 1 = 1 + j0.2$$
$$|E_f''| = \sqrt{1^2 + 0.2^2} = 1.02$$

5.50 (1)

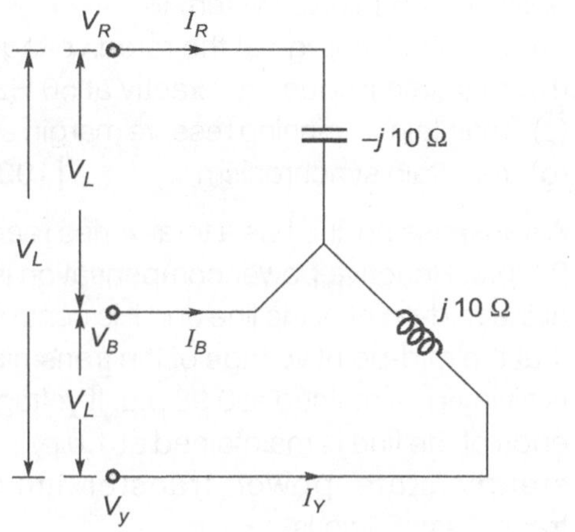

$$I_R = \frac{V_{RB}}{-j10} = \frac{V_L\angle -60°}{-j10} = \frac{V_L}{10}\angle 30°$$

$$I_Y = \frac{V_{YB}}{j10} = \frac{V_L\angle -120°}{-j10} = \frac{V_L}{10}\angle 150°$$

and $\quad I_B = -(I_R + I_Y) = \dfrac{V_L}{10}\angle -90°$

$$\therefore \quad \left|\frac{I_B}{I_R}\right| = 1$$

5.51 (c)

$$I_{B0} = \frac{1}{3}(I_A + I_B + I_C)$$

$$0.1 = \frac{1}{3}(1.1\angle 0 + I_B + 1\angle 120° + 0.1)$$

$$I_B = 0.3 - 1.1\angle 0 - 0.1 - 1\angle 120°$$
$$I_B = -0.9 - 1\angle 120°$$
$$I_B = 0.1 + 1\angle 240°$$
$$I_B = 1\angle -120° + 0.1\angle 0°$$

5.52 (b)

For 3-ϕ full transferred transmission line,

$$X_1 \cong X_1 < X_0$$

So suitable answer from given options is option (b).

$$X_1 = X_2 = 0.15$$
$$X_0 = 0.35$$

5.53 (7.21)

$$(Z_{1eq})_{bus\ 1} = j0.2 \parallel j0.3 = \frac{j0.2 \times 0.3}{0.5} = j0.12 \text{ pu}$$

$$(Z_{2eq})_{bus\ 1} = j0.2 \parallel j0.3 = j0.12 \text{ pu}$$

The per phase equivalent circuit for LL fault.

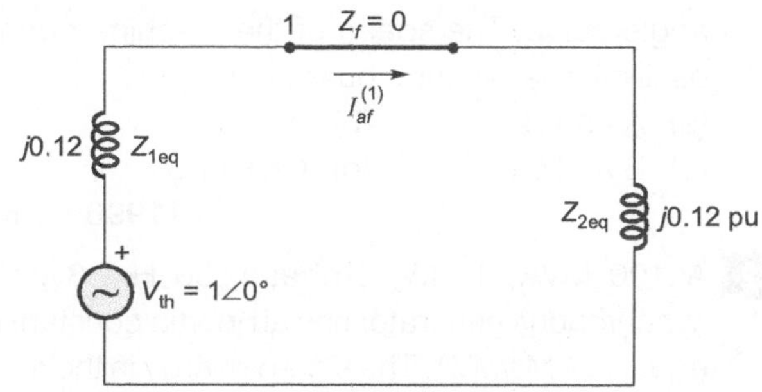

$$I_{af}^{(1)} = \frac{1\angle 0°}{j0.12 + j0.12} = 4.16 \text{ pu}$$

So, fault current, $I_F = \sqrt{3} \times I_{af}^{(1)} = 7.21$ pu

5.54 (*)

For a 2-phase system :

In a two phase balanced system, phase difference between two phase voltages is 90°.

Assume : $V_p = V\angle 0°$

$V_q = V\angle -90°$

By using Fortescue's Theorem :

Sequence 1 : **Sequence 2 :**

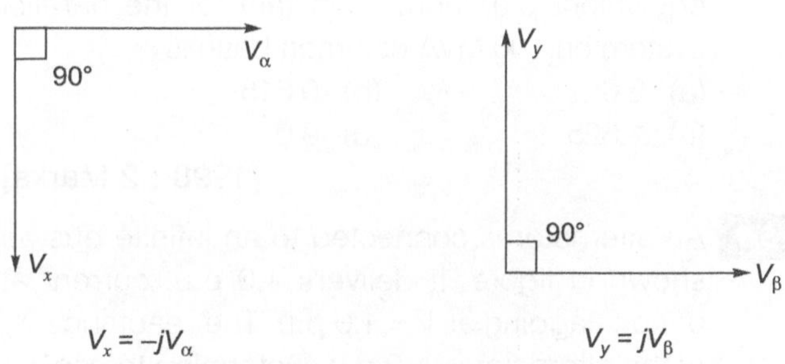

V_p and V_q can be restored in terms of sequence components as

$$V_p = V_\alpha + V_\beta$$
$$V_q = V_x + V_y = -jV_\alpha + jV_\beta$$

$$\begin{bmatrix} V_p \\ V_q \end{bmatrix} = \begin{bmatrix} 1 & 1 \\ -j & j \end{bmatrix}\begin{bmatrix} V_\alpha \\ V_\beta \end{bmatrix} = [S]\begin{bmatrix} V_\alpha \\ V_\beta \end{bmatrix}$$

where, $\quad S = \begin{bmatrix} 1 & 1 \\ -j & j \end{bmatrix}$

■■■■

6 Power System Stability

6.1 During a disturbance on a synchronous machine, the rotor swings from A to B, before finally settling down to a steady state at point C on the power angle curve. The speed of the machine during oscillation is synchronous at point(s).

(a) A and B (b) A and C
(c) B and C (d) Only at C

[1996 : 1 M]

6.2 A 100 MVA, 11 kV, 3-phase, 50 Hz, 8 pole synchronous generator has an inertia constant H equal to 4 MJ/MVA. The stored energy in the rotor of the generator at synchronous speed will be

(a) 100 MJ (b) 400 MJ
(c) 800 MJ (d) 12.5 MJ [1997 : 1 M]

6.3 The use high speed circuit breakers

(a) reduces the short circuit current
(b) improves system stability
(c) decreases system stability
(d) increase the short-circuit current

[1997 : 1 M]

6.4 A power station consists of two synchronous generators A and B of ratings 250 MVA and 500 MVA with inertia 1.6 p.u. and 1.0 p.u., respectively on their own base MVA ratings. The equivalent p.u. inertia constant for the parallel system on 100 MVA common base is

(a) 2.6 (b) 0.615
(c) 1.625 (d) 9.0

[1998 : 2 Marks]

6.5 An alternator is connected to an infinite bus as shown in figure. It delivers 1.0 p.u. current at 0.8 p.f. lagging at $V = 1.0$ p.u. The reactance X_d of the alternator is 1.2 p.u. Determine the active power output and the steady state power limit. Keeping the active power fixed, if the excitation is reduced, find the critical excitation corresponding to operation at stability limit.

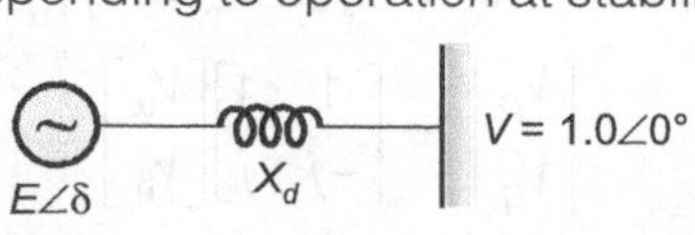

[1998 : 2 M]

6.6 The load carrying capability of a long AC transmission line is

(a) always limited by the conductor size.
(b) limited by stability considerations.
(c) reduced at low ambient temperature.
(d) decreased by the use of bundled conductors of single conductors. [1999 : 1 M]

6.7 Steady state stability of a power system is the ability of the power system to

(a) maintain voltage at the rated voltage level
(b) maintain frequency exactly at 50 Hz
(c) maintain a spinning reserve margin at all times
(d) maintain synchronism [1999 : 1 M]

6.8 A transmission line has a total series reactance of 0.2 p.u. Reactive power compensation is applied at the mid-point of the line and it is controlled such that the mid-point voltage of the transmission line is always maintained at 0.98 p.u. If voltage at both ends of the line is maintained at 1.0 p.u., then the steady state power transfer limit of the transmission line is

(a) 9.8 p.u. (b) 4.9 p.u.
(c) 19.6 p.u. (d) 5 p.u. [2002 : 2 M]

6.9 A generator delivers power of 1.0 p.u. to an infinite bus through a purely reactive network. The maximum power that could be delivered by the generator is 2.0 p.u. A three-phase fault occurs at the terminals of the generator which reduces the generator output to zero. The fault is cleared after t_c second. The original network is then restored. The maximum swing of the rotor angle is found to be δ_{max} = 110 electrical degree. Then the rotor angle in electrical degrees at $t = t_c$ is

(a) 55 (b) 70
(c) 69.14 (d) 72.4 [2003 : 2 M]

6.10 A 800 kV transmission line is having per phase line inductance of 1.1 mH/km and per phase line capacitance of 11.68 nF/km. Ignoring the length of the line, its ideal power transfer capability in MW is

(a) 1204 MW (b) 1504 MW
(c) 2085 MW (d) 2606 MW

[2004 : 2 M]

6.11 A new generator having $E_g = 1.4\angle 30^\circ$ pu [equivalent to (1.212 + j 0.70) pu] and synchronous reactance 'X_s' of 1.0 pu on the system base, is to be connected to a bus having voltage V_t in the existing power system. This existing power system can be represented by Thevenin's voltage $E_{th} = 0.9\angle 0°$ pu in series with Thevenin's impedance $Z_{th} = 0.25\angle 90°$ pu. The magnitude of the bus voltage V_t of the system in pu will be

(a) 0.990 (b) 0.973
(c) 0.963 (d) 0.900 [2004 : 2 M]

6.12 A 50 Hz, 4-pole, 500 MVA, 22 kV turbo-generator is delivering rated MVA at 0.8 power factor. Suddenly a fault occurs reducing is electric power output by 40%. Neglect losses and assume constant power input to the shaft. The accelerating torque in the generator in MNm at the time of the fault will be

(a) 1.528 (b) 1.018
(c) 0.848 (d) 0.509 [2004 : 2 M]

6.13 An 800 kV transmission line has a maximum power transfer capacity of *P*. If it is operated at 400 kV with the series reactance unchanged, then new maximum power transfer capacity is approximately

(a) *P* (b) 2*P*
(c) *P*/2 (d) *P*/4 [2005 : 1 M]

6.14 A generator with constant 1.0 p.u. terminal voltage supplies power through a step-up transformer of 0.12 p.u. reactance and a double-circuit line to an infinite bus bar as shown in the figure. The infinite bus voltage is maintained at 1.0 p.u. Neglecting the resistances and susceptances of the system, the steady state stability power limit of the system is 6.25 p.u. If one of the double-circuit is tripped, then resulting steady state stability power limit in p.u. will be

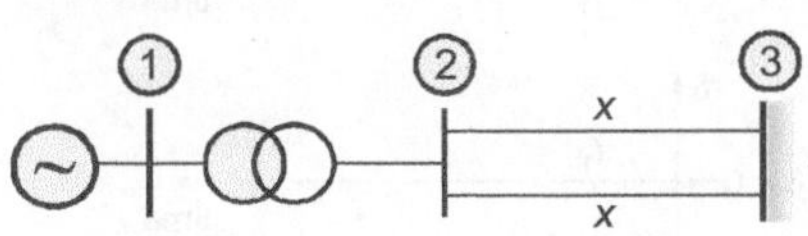

(a) 12.5 p.u. (b) 3.125 p.u.
(c) 10.0 p.u. (d) 5.0 p.u.

[2005 : 2 M]

Statement for Common Data Questions (6.15 and 6.16):

A generator feeds power to an infinite bus through a double circuit transmission line. A 3-phase fault occurs at the middle point of one of the lines. The infinite bus voltage is 1 pu, the transient internal voltage of the generator is 1.1 pu and the equivalent transfer admittance during fault is 0.8 pu. The 100 MVA generator has an inertia constant of 5 MJ/MVA and it was delivering 1.0 pu power prior of the fault with rotor power angle of 30°. The system frequency is 50 Hz.

6.15 The initial accelerating power (in pu) will be

(a) 1.0 (b) 0.6
(c) 0.56 (d) 0.4 [2006 : 2 M]

6.16 If the initial accelerating power is X pu, the initial acceleration in elect deg/sec, and the inertia constant in MJ-sec/elect deg respectively will be

(a) 31.4 X, 18 (b) 1800 X, 0.056
(c) X/1800, 0.056 (d) X/31.4, 18

[2006 : 2 M]

6.17 Consider a synchronous generator connected to an infinite bus by two identical parallel transmission lines. The transient reactance X' of the generator is 0.1 pu and the mechanical power input to it is constant at 1.0 pu. Due to some previous disturbance, the rotor angle (δ) is undergoing an undamped oscillation, with the maximum value of δ(t) equal to 130°. One of the parallel lines trip due to relay maloperation at an instant when δ(*t*) = 130° as shown in the figure. The maximum value of the per unit line reactance, *X* such that the system does not lose synchronism subsequent to this tripping is

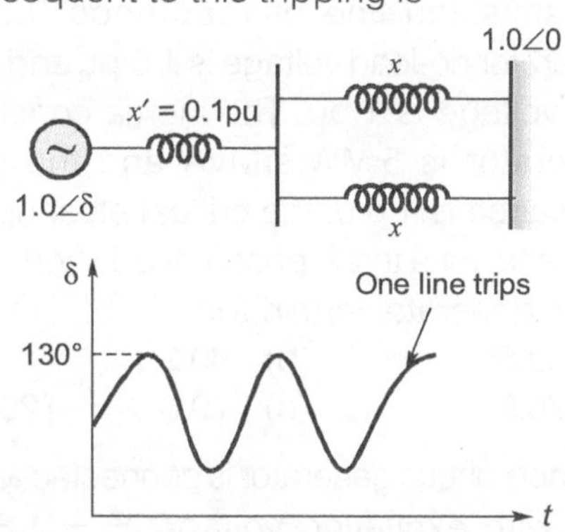

(a) 0.87 (b) 0.74
(c) 0.67 (d) 0.54 [2007 : 2 M]

6.18 An isolated 50 Hz synchronous generator is rated at 15 MW which is also the maximum continuous power limit of its prime mover. It is equipped with a speed governor with 5% droop. Initially, the generator is feeding three loads of 4 MW each at 50 Hz. One of these loads is programmed to trip permanently if the frequency falls below 48 Hz. If an additional load of 3.5 MW is connected then the frequency will settle down to

(a) 49.417 Hz (b) 49.917 Hz
(c) 50.083 Hz (d) 50.583 Hz [2007 : 2 M]

6.19 A lossless single machine infinite bus power system is shown below:

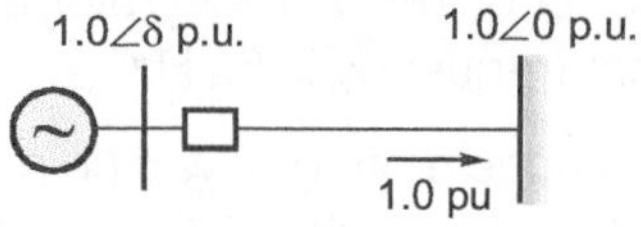

The synchronous generator transfers 1.0 per unit of power to the infinite bus. The critical clearing time of circuit breaker is 0.28 s. If another identical synchronous generator is connected in parallel to the existing generator and each generator is scheduled to supply 0.5 per unit of power, then the critical clearing time of the circuit breaker will

(a) reduce to 0.14 s
(b) reduce but will be more than 0.14 s
(c) remain constant at 0.28 s
(d) increase beyond 0.28 s **[2008 : 2 M]**

6.20 A 500 MW, 21 kV, 50 Hz, 3-phase, 2-pole synchronous generator having a rated p.f. = 0.9, has a moment of inertia of 27.5×10^3 kg-m^2. The inertia constant (H) will be

(a) 2.44 MJ/MVA (b) 2.71 MJ/MVA
(c) 4.88 MJ/MVA (d) 5.42 MJ/MVA

[2009 : 2 M]

6.21 A cylindrical rotor generator delivers 0.5 pu power in the steady-state to an infinite bus through a transmission line of reactance 0.5 pu. The generator no-load voltage is 1.5 pu and the infinite bus voltage is 1 pu. The inertia constant of the generator is 5 MW-s/MVA and the generator reactance is 1 p.u. the critical clearing angle, in degrees, for a three-phase dead short circuit fault at the generator terminal is

(a) 53.5 (b) 60.2
(c) 70.8 (d) 79.6 **[2012 : 2 M]**

6.22 A synchronous generator is connected to an infinite bus with excitation voltage E_f = 1.3 pu. The generator has a synchronous reactance of 1.1 pu and is delivering real power (P) of 0.6 pu to the bus. Assume the infinite bus voltage to be 1.0 pu. Neglect stator resistance. The reactive power (Q) in pu supplied by the generator to the bus under this condition is _____. **[2014 : 2 M, Set-2]**

6.23 There are two generators in a power system. No load frequencies of the generators are 51.5 Hz and 51 Hz, respectively, and both are having droop constant of 1 Hz/MW. Total load in the system is 2.5 MW. Assuming that the generators are operating under their respective droop characteristics, the frequency of the power system in Hz in the steady state is _____.

[2014 : 2 M, Set-2]

6.24 A 50 Hz generating unit has *H*-constant of 2 MJ/MVA. The machine is initially operating in steady state at synchronous speed, and producing 1 pu of real power. The initial value of the rotor angle δ is 5°, when a bolted three phase to ground short circuit fault occurs at the terminal of the generator. Assuming the input mechanical power to remain at 1 pu, the value of δ in degrees, 0.02 second after the fault is _____.

[2015 : 2 M, Set-1]

6.25 The synchronous generator shown in the figure is supplying active power to an infinite bus via two short, lossless transmission lines, and is initially in steady state. The mechanical power input to the generator and the voltage magnitude *E* are constant. If one line is tripped at time t_1 by opening the circuit breakers at the two ends (although there is no fault), then it is seen that the generator undergoes a stable transient. Which one of the following waveforms of the rotor angle d shows the transient correctly?

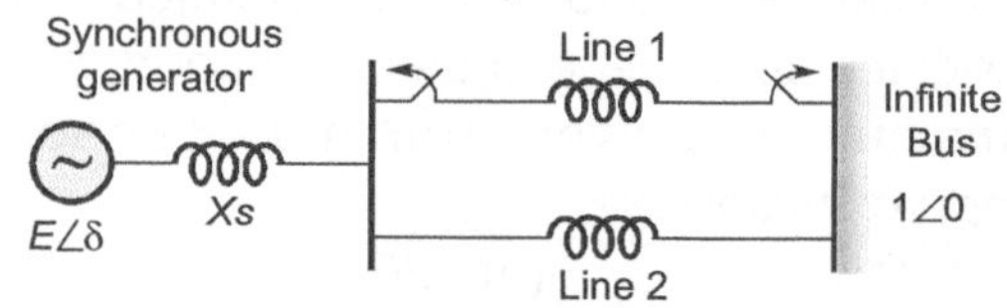

(a)
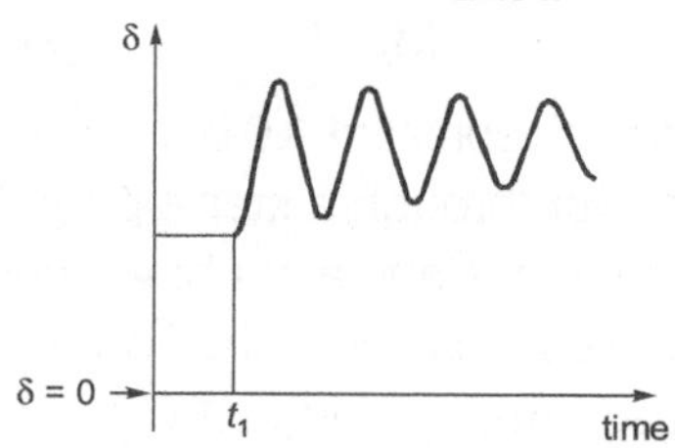

(b)
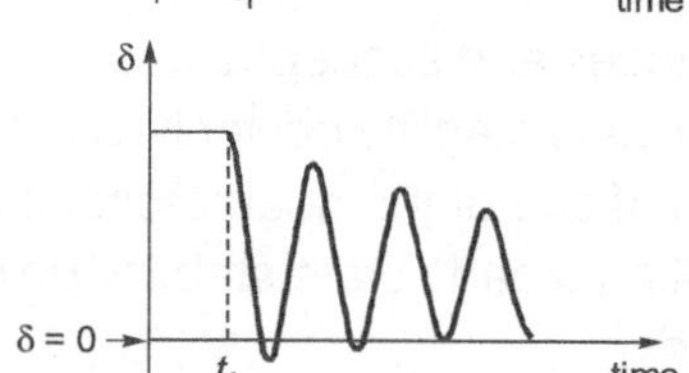

(c)
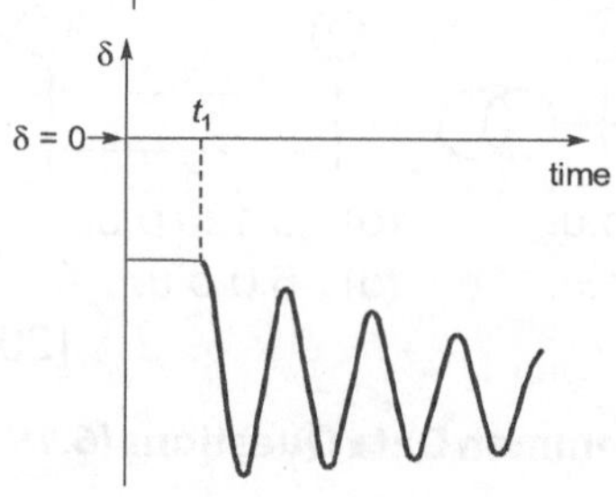

(d)
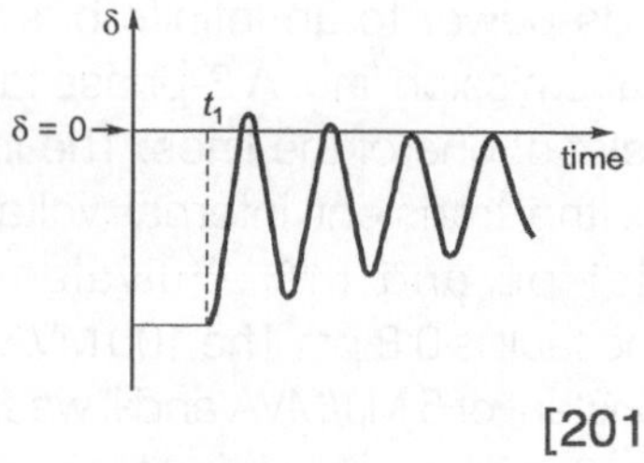

[2015 : 1 M, Set-2]

6.26 The figure shows the single line diagram of a single machine infinite bus system.

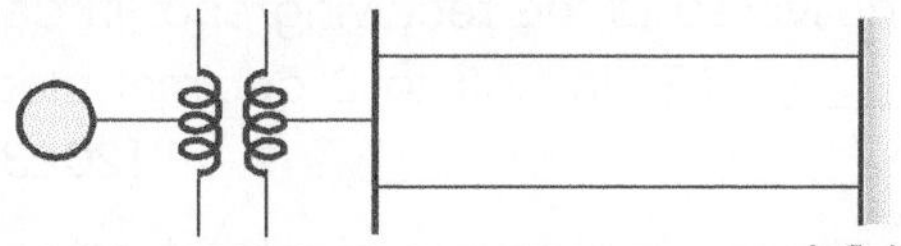

The inertia constant of the synchronous generator $H = 5$ MW-s/MVA. Frequency is 50 Hz. Mechanical power is 1 pu. The system is operating at the stable equilibrium point with rotor angle δ equal to 30°. A three-phase short-circuit fault occurs at a certain location on one of the circuits of the double circuit transmission line. During fault, electrical power in pu is $P_{max} \sin \delta$. If the values of δ and $d\delta/dt$ at the instant of fault clearing are 45° and 3.762 radian/s respectively, then P_{max} (in pu) is ____.

[2014 : 2 M, Set-3]

6.27 The figure shows the single line diagram of a power system with a double circuit transmission line. The expression for electrical power is 1.5 sinδ, where δ is the rotor angle. The system is operating at the stable equilibrium point with mechanical power equal to 1 pu. If one of the transmission line circuits is removed the maximum value of δ, as the rotor swings is 1.221 radian. If the expression for electrical power with one transmission line circuit removed is $P_{max} \sin \delta$, the value of P_{max}, in pu is____. (Give answer upto three decimal places.)

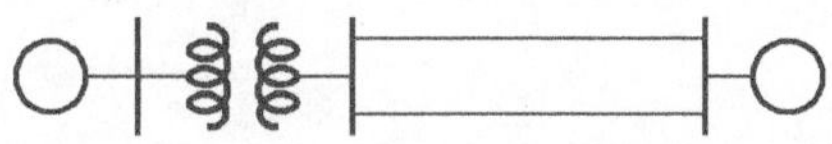

[2017 : 2 M, Set-1]

6.28 A 3-phase, 2-pole, 50 Hz, synchronous generator has a rating of 250 MVA, 0.8 pf lagging. The kinetic energy of the machine at synchronous speed is 1000 MJ.
The machine is running steadily at synchronous speed and delivering 60 MW power at a power angle of 10 electrical degrees. If the load is suddenly removed, assuming the acceleration is constant for 10 cycles, the value of the power angle after 5 cycles is_____electrical degrees.

[2017 : 2 M, Set-1]

6.29 Consider a lossy transmission line with V_1 and V_2 as the sending and receiving end voltages, respectively. Z and X are the series impedance and reactance of the line, respectively. The steady-state stability limit for the transmission line will be

(a) greater than $\left|\frac{V_1 V_2}{X}\right|$

(b) less than $\left|\frac{V_1 V_2}{X}\right|$

(c) equal to $\left|\frac{V_1 V_2}{X}\right|$

(d) equal to $\left|\frac{V_1 V_2}{Z}\right|$

[2018 : 1 M]

6.30 In the single machine infinite bus system shown below, the generator is delivering the real power of 0.8 pu at 0.8 power factor lagging to the infinite bus. The power angle of the generator in degrees (round off to one decimal place) is _____.

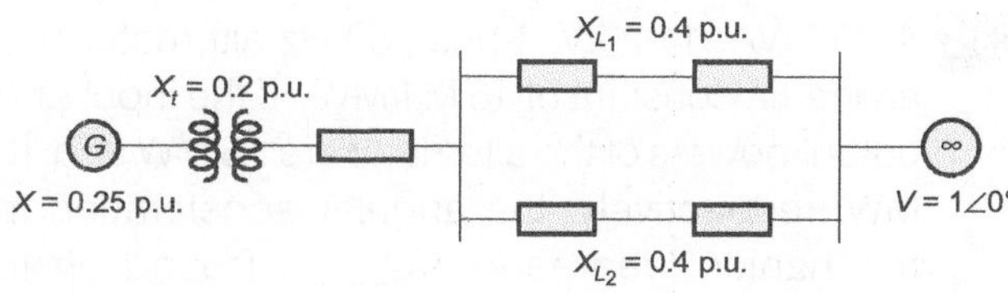

[2019 : 2 M]

6.31 A single 50 Hz synchronous generator on droop control was delivering 100 MW power to a system. Due to increase in load, generator power had to be increased by 10 MW. as a result of which, system frequency dropped to 49.75 Hz. Further increase in load in the system resulted in a frequency of 49.25 Hz. At this condition, the power in MW supplied by the generator is _______ (rounded off to 2 decimal places). [2020 : 1 M]

6.32 An alternator with internal voltage of $1\angle\delta_1$ p.u. and synchronous reactance of 0.4 p.u. is connected by a transmission line of reactance 0.1 p.u. to a synchronous motor having synchronous reactance 0.35 p.u. and internal voltage of $0.85\angle\delta_2$ p.u. If the real power supplied by the alternator is 0.866 p.u, then $(\delta_1 - \delta_2)$ is _____ degrees. (Round off to 2 decimal places)
(Machines are of non-salient type. Neglect resistances)

[2021 : 1 M]

6.33 In the figure shown, self-impedances of the two transmission lines are 1.5j p.u each, and Z_m = 0.5j p.u is the mutual impedance. Bus voltages shown in the figure are in p.u. Given that $\delta > 0$, the maximum steady-state real power that can be transferred in p.u from Bus-1 to Bus-2 is

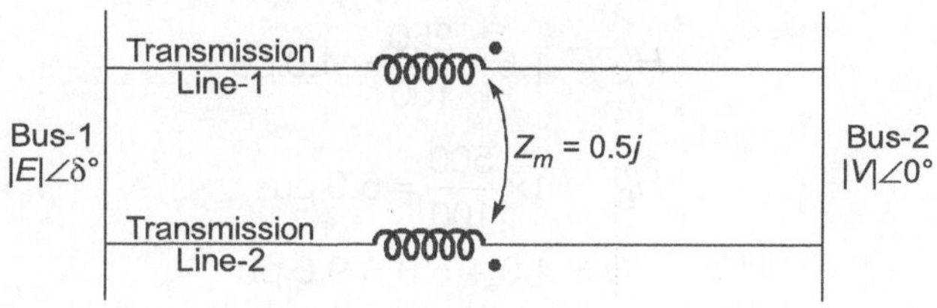

(a) $\frac{|E||V|}{2}$ (b) $\frac{3|E||V|}{2}$

(c) $|E||V|$ (d) $2|E||V|$

[2021 : 2 M]

6.34 Two generating units rated for 250 MW and 400 MW have governor speed regulations of 6% and 6.4%, respectively, from no load to full load. Both the generating units are operating in parallel to share a load of 500 MW. Assuming free governor action, the load shared in MW, by the 250 MW generating unit is __. (round off to nearest integer).

[2022 : 2 M]

6.35 A 20 MVA, 11.2 kV, 4-pole, 50 Hz alternator has an inertia constant of 15 MJ/MVA. If the input and output powers of the alternator are 15 MW and 10 MW, respectively, the angular acceleration in mechanical degree/s^2 is _____. (round off to nearest integer).

[2022 : 2 M]

6.36 A 50 Hz, 275 kV line of length 400 km has the following parameters:

Resistance : R = 0.035 Ω/km;

Inductance : L = 1 mH/km;

Capacitance : C = 0.01 μF/km;

The line is represented by the nominal-π model. With the magnitudes of the sending end and the receiving end voltages of the line (denoted by V_S and V_R, respectively) maintained at 275 kV, the phase angle difference (θ) between V_S and V_R required for maximum possible active power to be delivered to the receiving end, in degree is ________ (Round off to 2 decimal places).

[2023 : 1 M]

6.37 The single line diagram of a lossless system is shown in the figure. The system is operating in steady-state at a stable equilibrium point with the power output of the generator being P_{max} sin δ. where δ is the load angle and the mechanical power input is 0.5 P_{max}. A fault occurs on line 2 such that the power output of the generator is less than 0.5.P_{max} during the fault. After the fault is cleared by opening line 2. The power output of the generator is $\{P_{max}/\sqrt{2}\}\sin\delta$. If the critical fault clearing angle is π/2 radians, the accelerating area on the power angle curve is _______ times P_{max} (rounded off to 2 decimal places).

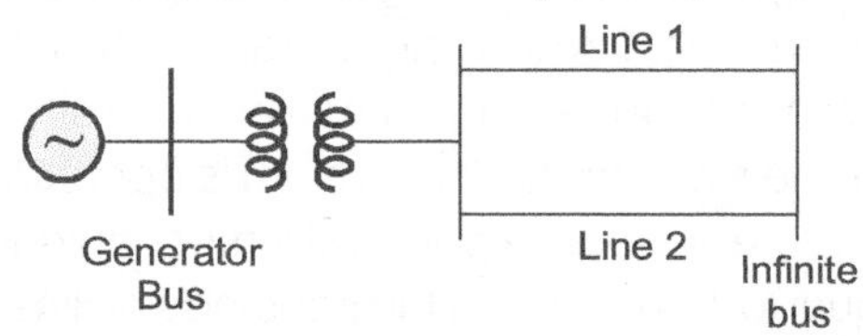

[2024 : 2 M]

Answers Power System Stability

6.1 (a)	6.2 (b)	6.3 (b)	6.4 (d)	6.5 (0.96)	6.6 (b)	6.7 (d)
6.8 (a)	6.9 (c)	6.10 (c)	6.11 (b)	6.12 (b)	6.13 (d)	6.14 (d)
6.15 (c)	6.16 (b)	6.17 (c)	6.18 (c)	6.19 (d)	6.20 (a)	6.21 (d)
6.22 (0.11)	6.23 (50)	6.24 (5.9)	6.25 (a)	6.26 (0.23)	6.27 (1.220)	6.28 (12.7)
6.29 (b)	6.30 (20.5)	6.31 (130)	6.32 (60)	6.33 (c)	6.34 (200)	6.35 (75)
6.36 (83.64)	6.37 (0.1)					

Explanations Power System Stability

6.2 (b)

Energy stored = 4 × 100 = 400 MJ

6.4 (d)

$$H'_A = 1.6\times\frac{250}{100} = 4.0 \text{ pu}$$

$$H'_B = 1\times\frac{500}{100} = 5.0 \text{ pu}$$

$$H = 4.0 + 5.0 = 9.0 \text{ pu}$$

6.5 Sol.

$$E_f\angle\delta = V\angle 0 + I\angle\phi\, Z_s\angle\theta$$
$$= 1\angle 0 + 1\angle -\cos^{-1}(0.8)\times 1.2\angle 90^\circ$$
$$= 1.969777\angle 29.1676^\circ \text{ pu}$$

Active power output = $VI\cos\phi$

= 1 pu × 1 pu × 0.8 = 0.8 pu

Steady state power limit

$$= \frac{|E||V|}{X} = \frac{1.96977\times 1}{1.2} = 1.64$$

$$\frac{|E||V|}{X} = 0.8 \text{ pu}$$

or $$\frac{|E| \times 1 \text{ pu}}{1.2 \text{ pu}} = 0.8 \text{ pu}$$

$$|E| = 0.96 \text{ pu}$$

6.8 (a)

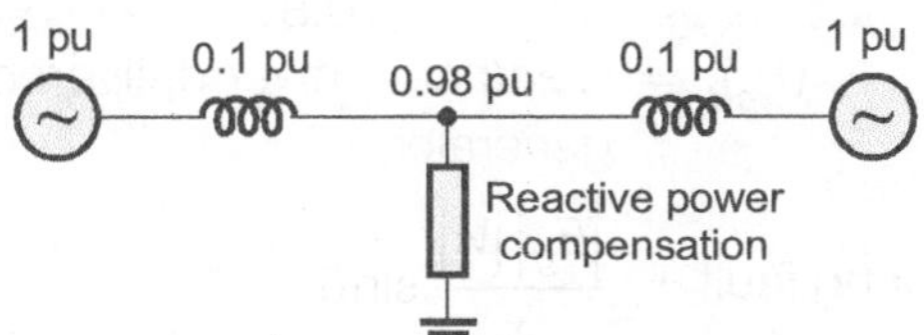

$$P_{ss} = \frac{1 \times 0.98}{0.1} = 9.8 \text{ pu}$$

6.9 (c)

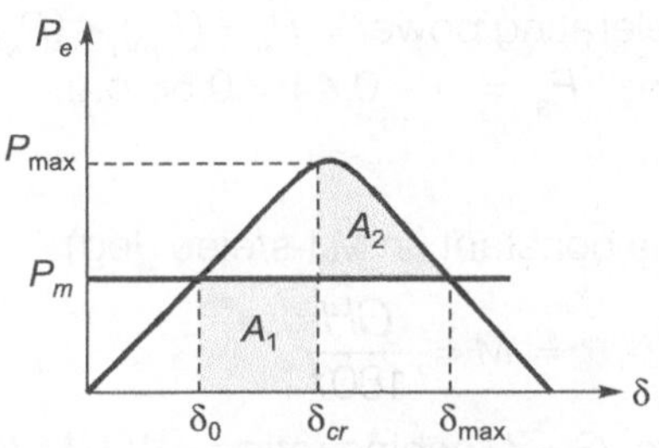

Power generated by the generator,

$$P_e = P_{max} \sin\delta$$

At δ_0, $P_e = 1$ p.u.

$$P_{max} = 2 \text{ p.u.}$$

$$\delta_{cr} = \text{rotor angle at } t = t_c$$

$$P_e = P_{max} \sin\delta$$

$$1 = 2 \sin\delta_0$$

$$\delta_0 = 30^\circ$$

- During fault, P_e become zero and the fault is cleared at δ_{cr}. Mechanical input to the generator remains constant, $P_m = 1$ p.u.
- Applying equal areal criterion,

$$A_1 = A_2$$

$$A_1 = \int_{\delta_0}^{\delta_{cr}} (P_m - 0)\, d\delta = P_m(\delta_{cr} - \delta_0) \quad \text{...(i)}$$

$$A_2 = \int_{\delta_{cr}}^{\delta_{max}} (P_{max} \sin\delta - P_m)\, d\delta$$

$$= P_{max}(\cos\delta_{cr} - \cos\delta_{max}) - P_m(\delta_{max} - \delta_{cr}) \quad \text{...(ii)}$$

Equating equation (i) and (ii), we get

$$P_m(\delta_{cr} - \delta_0) = P_{max}(\cos\delta_{cr} - \cos\delta_{max}) - P_m(\delta_{max} - \delta_{cr})$$

$$\cos\delta_{cr} = \frac{P_m}{P_{max}}(\delta_{max} - \delta_0) + \cos\delta_{max}$$

$$\cos\delta_{cr} = \frac{1}{2} \times (110 - 30) \times \frac{\pi}{180} + \cos 110$$

$$\delta_{cr} = 69.14^\circ$$

6.10 (c)

Parameter of the line

Inductance = $L = 1.1$ mH/km

Capacitance = $C = 11.68$ nF/km

Surge impedance of the line

$$= Z_s = \sqrt{\frac{L}{C}} = \sqrt{\frac{1.1 \times 10^{-3}}{11.68 \times 10^{-9}}}$$

$$\Rightarrow \quad Z_s = 306.88\ \Omega$$

Ideal power transfer capability

$$= \frac{V_{l-l}^2}{Z_s} = \frac{(800 \times 10^3)^2}{306.88}$$

$$\approx 2085 \times 10^6 \approx 2085 \text{ MW}$$

6.11 (b)

Thevenin's equivalent of existing system,

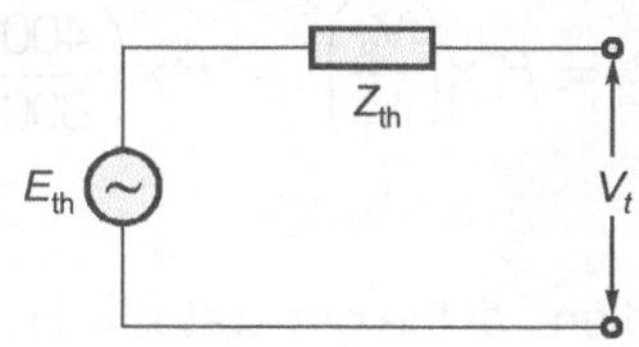

where,

$$Z_{th} = 0.25\angle 90^\circ \text{ p.u.}; \quad E_{th} = 0.9\angle 0^\circ \text{ p.u.}$$

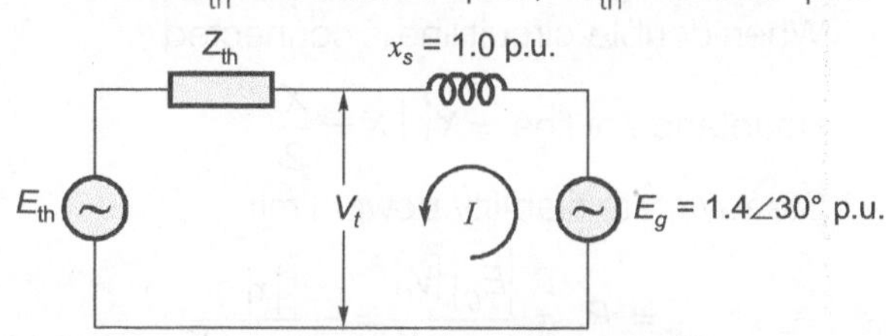

$$I = \frac{E_g - E_{th}}{jX_s + Z_{th}} = \frac{1.4\angle 30^\circ - 0.9\angle 0^\circ}{j1 + 0.25\angle 90^\circ}$$

$$= 0.56 - j0.25 \text{ p.u.}$$

$$V_t = E_g - jIx_s$$

$$= 1.4\angle 30^\circ\ (0.56 - j0.25) \times (j1)$$

$$= 0.973\angle 8.27^\circ \text{ p.u.}$$

$$|V_t| = 0.973 \text{ p.u.}$$

6.12 (b)

Before fault, mechanical input (P_m) = Electrical output (P_e)

$$= 500 \times 0.8 = 400 \text{ MW}$$

After fault, mechanical input remains same

$$= P_m = 400 \text{ MW}$$

While electrical output reduces by 40%.

So, electrical output at the time of fault

$$= (P_e)_f = 0.6 \times 400 = 240 \text{ MW}$$

Accelerating Power = $P_a = P_m - (P_e)_f$

$= 400 - 240 = 160$ MW

Synchronous speed (in rpm) = N_s

$$= \frac{120 \times f}{P} = \frac{120 \times 50}{4} = 1500 \text{ rpm}$$

Synchronous speed (in rad/sec) = ω_s

$$= \frac{2\pi N_s}{60} = \frac{2\pi \times 1500}{60} = 157 \text{ rad/sec}$$

$$P_a = T_a \omega_s$$

$$\Rightarrow T_a(\text{accelerating Torque}) = \frac{P_a}{\omega_s}$$

$$= \frac{160}{157} \text{ MN-m} = 1.018 \text{ MN-m}$$

6.13 (d)

Power transfer capacity $\propto V^2$

$$\frac{P_1}{P_2} = \frac{V_1^2}{V_2^2}$$

$$\Rightarrow \quad P_2 = P_1 \times \left(\frac{V_2}{V_1}\right)^2 = P \times \left(\frac{400}{800}\right)^2 = \frac{P}{4}$$

6.14 (d)

Induced emf of the generator = $|t_g| = 1$ p.u.

Terminal voltage = $|V_t| = 1$ p.u.

Reactance of transformer = $X_t = 0.12$ p.u.

When double-circuit line is connected

Reactance of line = $X || X = \frac{X}{2}$

Steady state stability power limit

$$= P_1 = \frac{|E_g||V_t|}{X_t + \frac{X}{2}} = \frac{|x|}{0.12 + \frac{X}{2}}$$

$$6.25 = \frac{1}{0.12 + \frac{x}{2}}$$

$X = 0.08$ p.u.

When one of the double circuit is tripled, steady state stability power limit

$$= P_2 = \frac{|E_g||V_t|}{X_t + X} = \frac{|x|}{0.12 + 0.08} = 5 \text{ p.u.}$$

6.15 (c)

Before fault

Electrical power delivered by the generator

$= P_e = 1.0$ p.u.

Mechanical input to the generator = P_m

At steady state,

$$P_m = P_e = 1.0 \text{ p.u.}$$

During fault:

Initial rotor angle = $\delta = 30°$

Internal voltage of the generator = $|E_g|$

$= 1.1$ p.u.

Voltage of the bus = $|V| = 1$ p.u.

Equivalent transfer admittance = 0.8

$$\text{Transfer reactance} = x = \frac{1}{0.8} = 1.25$$

$(P_e)_f$ = Electrical power delivered by the generator

$$\text{during fault} = \frac{|E_g||V_t|}{x} \sin\delta$$

$$(P_e)_f = \frac{1.1 \times 1}{1.25} \sin 30° = 0.44 \text{ p.u.}$$

Mechanical input do not change during fault

So, $(P_m)_f = 1$ p.u.

Accelerating power = $P_a = (P_m)_f - (P_e)_f$

$$\Rightarrow \quad P_a = 1 - 0.44 = 0.56 \text{ p.u.}$$

6.16 (b)

Inertia constant (in MJ-s/elec deg)

$$= M = \frac{GH}{180f}$$

where, G = Machine rating = 100 MVA

H = Inertia constant in MJ/MVA

= 5 MJ/MVA

$$M = \frac{GH}{180f} = \frac{5 \times 100}{180 \times 50}$$

$= 0.056$ MJ-s/elec deg

$$M_{(\text{in p.u.})} = \frac{H}{180f} = \frac{5}{180 \times 50}$$

α = Acceleration in elect deg/sec

$M\alpha = P_a$

P_a = Accelerating power = X

$$\alpha = \frac{P_a}{M} = \frac{X}{(5/180 \times 50)}$$

$= 1800\, X$ elec deg/sec

6.17 (c)

The curve P_{e1} and P_{e2} represents power before the tripping of one lines and after the tripping respectively

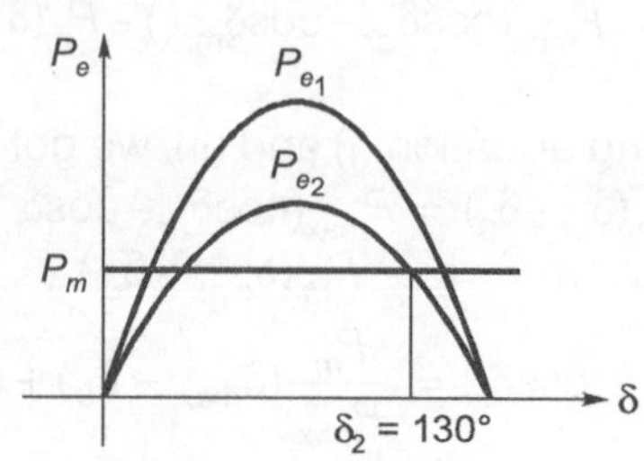

$$P_{e\max 2} = \frac{EV}{X_2}$$

$$X_2 = 0.1 + X$$

For max value of X system will not loose synchronization, if at δ_2

$$P_{e2} = P_m$$

$$\Rightarrow \frac{EV}{X_2} \sin \delta_2 = P_m$$

$$\frac{1 \times 1}{X_2} \sin 130° = 1$$

$$X_2 = 0.77 \Rightarrow X = 0.67$$

6.18 (a)

Here, change in (frequency) w.r.t. to power

$$\left(\frac{5}{15}\right) \times 3.5 = 1.16\%$$

$$\therefore \quad \Delta f = 0.583$$

$$\therefore \quad f = 50 - 0.583 = 49.417 \text{ Hz}$$

6.19 (d)

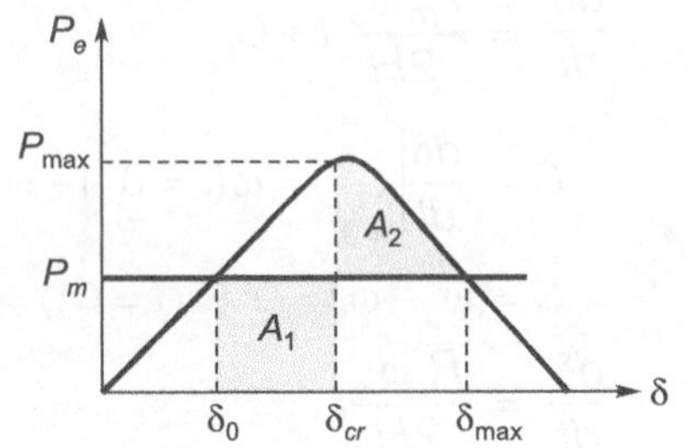

For such faults

δ_{cr} is given by

$$\delta_{cr} = \cos^{-1}[(\pi - 2\delta_0) \sin \delta_0 - \cos \delta_0] \quad \text{...(i)}$$

and t_{cr} = critical cleaning time

$$= \left(\frac{2H(\delta_{cr} - \delta_0)}{\pi f P_m}\right)^{1/2} \quad \text{...(ii)}$$

Case 1:

When only one generator is connected.

Mechanical input to the generator (P_{m_1}) = Electrical power delivered by the generator (P_{e_1})

$$\Rightarrow \quad P_{m_1} = P_{e_1} = 1 \text{ p.u.}$$

Case 2:

When two generator connected in parallel.

Electrical power delivered by each generator

$$(P_{e_2}) = 0.5 \text{ p.u.}$$

Mechanical input to each generator

$$(P_{m_2}) = P_{e_2}$$

$$P_{m_2} = P_{e_2} = 0.5 \text{ p.u.} = \frac{P_{m_1}}{2}$$

Assuming δ_0 in each case same,

$$\delta_{0_1} = \delta_{0_2} = \delta_0$$

So, δ_{cr} will also be same in both cases, (from equation (i)),

$$t_{cr} \propto \frac{1}{\sqrt{P_m}} \quad (\delta_{cr}, H \text{ and } \delta_0 \text{ are same})$$

$$\therefore \quad t_{cr_2} = \sqrt{\frac{P_{m_1}}{P_{m_2}}} \times t_{cr_1}$$

$$t_{cr_2} = \sqrt{\frac{P_{m_1}}{P_{m_1}/2}} \times 0.28$$

$$= \sqrt{2} \times 0.28 = 0.396 \text{ sec.}$$

6.20 (a)

J = moment of inertia = 27.5×10^3 kg-m^2

Synchronous speed = N_s

$$= \frac{120f}{P} = \frac{120 \times 50}{2} = 300 \text{ rpm}$$

Kinetic energy of the rotor

$$= \frac{1}{2} J\omega_s^2 = \frac{1}{2} \times 27.5 \times 10^3 \times (314.16)^2$$

$$\text{K.E.} = 1357 \text{ MJ}$$

$$G = \text{Machine rating} = \frac{500}{\text{pf}} = \frac{500}{0.9}$$

$$= 555.55 \text{ MVA}$$

Inertial constant

$$= H = \frac{\text{K.E.}}{G} = \frac{1357}{555.55} = 2.44 \text{ MJ/MVA}$$

6.21 (d)

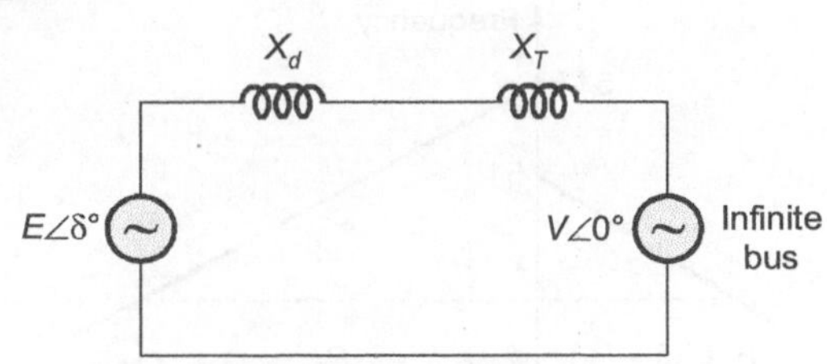

$$P_e = \frac{|E||V|}{(X_d + X_T)} \sin\delta_0$$

$$0.5 = \frac{1.5 \times 1}{(1 + 0.5)} \sin\delta_0$$

$$\delta_0 = \sin^{-1}(0.5) = 30° = \frac{\pi}{6}$$

Critical clearing angle

$$\delta_{cr} = \cos^{-1}[(\pi - 2\delta_0) \sin\delta_0 - \cos\delta_0]$$

$$= \cos^{-1}\left[\left(\pi - \frac{\pi}{3}\right)\times\frac{1}{2} - \frac{\sqrt{3}}{2}\right]$$

$$= \cos^{-1}(0.18) = 79.56°$$

6.22 Sol.

Given, $E_f = 1.3$ pu, $X_s = 1.1$ pu, $V = 1$ pu

$P = 0.6$ pu

Neglecting stator resistance i.e. $\theta_s = 90°$.

Output power delivered,

$$P = \frac{VE_f}{X_s}\sin\delta$$

or, $$\sin\delta = \frac{PX_s}{VE_f} = \frac{0.6\times 1.1}{1\times 1.3} = \frac{6.6}{13} \approx 0.5$$

∴ $$\delta = 30° = \text{Power angle}$$

∴ $$\cos\delta = \cos 30° = \frac{\sqrt{3}}{2}$$

Reactive power is given by,

$$Q = \left[\frac{VE_f}{X_s}\cos\delta - \frac{V^2}{X_s}\right]$$

$$= \left[\frac{1\times 1.3}{1.1}\times\frac{\sqrt{3}}{2} - \frac{1^2}{1.1}\right]$$

or, $$Q = (1.023 - 0.909) = 0.114 \text{ pu} \approx 0.11 \text{ pu}$$

∴ Reactive power (Q) supplied by the generator to bus = 0.11 pu.

6.23 Sol.

Let no-load frequency of generator-1 be 51.5 Hz and no-load frequency of generator-2 be 51 Hz.

Given, total load = 2.5 MW

and droop of both machines = 1 Hz/MW

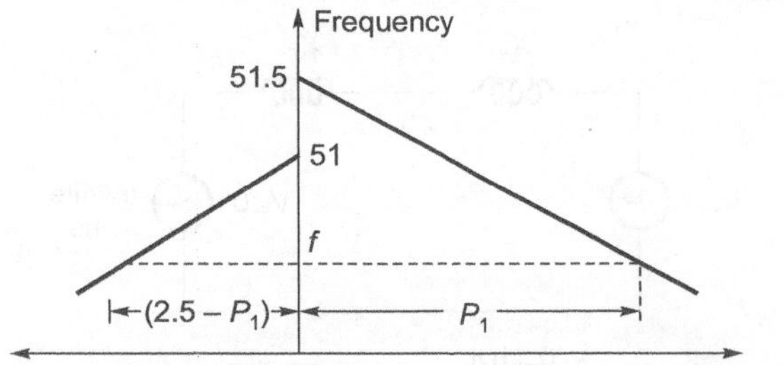

(Load frequency curve)

Let machine-1 shares a load of P_1 MW then, machine-2 will share a load of $(2.5 - P_1)$ MW

Let the steady state frequency of the system be f.

Now, $$\frac{51.5 - f}{P_1} = 1 = \text{droop}$$

or, $$P_1 = (51.5 - f) \quad \text{...(i)}$$

Also, $$\left(\frac{51 - f}{2.5 - P_1}\right) = 1 = \text{droop}$$

or, $$51 - f = 2.5 - P_1$$

or, $$P_1 = f - 48.5 \quad \text{...(ii)}$$

Equating equations (i) and (ii), we get:

$$51.5 - f = f - 48.5$$

or $$2f = 51.5 + 48.5 = 100$$

or $$f = \frac{100}{2} = 50 \text{ Hz}$$

Hence, steady state frequency,

$$f = 50 \text{ Hz}$$

6.24 Sol.

Let 3-ϕ fault occurs at $t = 0$

⇒ $P_e = 0$ for $t \geq 0^+$

Now, swing equation for $t \geq 0^+$

$$\frac{2H}{\omega_s}\frac{d^2\delta}{dt^2} = P_m$$

$$\frac{d^2\delta}{dt^2} = \frac{P_m\omega_s}{2H}$$

$$\frac{d\delta}{dt} = \frac{P_m\omega_s}{2H}t + C$$

$$C = \left.\frac{d\delta}{dt}\right|_{t=0^+} = \omega(t = 0^+) - \omega_s$$

$$C = \omega_s - \omega_s = 0 \quad (\omega(t = 0^+) = \omega_s)$$

$$\frac{d\delta}{dt} = \frac{P_m\omega_s}{2H}t$$

⇒ $$\delta = \frac{P_m\omega_s}{4H}t^2 + C'$$

$$C' = \delta|_{t=0^+} = \delta_0$$

$$\delta = \frac{P_m\omega_s}{4H}t^2 + \delta_0$$

Now in problem,

$\delta(t = 0.02)$

$$= \frac{1\times 2\times 180\times 50\times 0.02^2 + 5°}{4\times 2} = 5.9°$$

6.25 (a)

Initial value of $\delta = \delta_0$ at $t = t_1$ with both lines are in service of one of the line circuit breakes opened then δ is increased for same time and stabilizes to new value of δ.

6.26 Sol.

$$\frac{2H}{\omega_s}\frac{d^2\delta}{dt^2} = (P_m - P_e)$$

Let, $\omega_r = \dfrac{d\delta}{dt}$

$$\frac{2H}{\omega_s}\frac{d\omega_r}{dt} = (P_m - P_e)$$

Multiplying with ω_r in both side,

$$\frac{H}{\omega_s}\left(2\omega_r\frac{d\omega_r}{dt}\right) = (P_m - P_e)\frac{d\delta}{dt}$$

$$\Rightarrow \quad \frac{H}{\omega_s}2d(\omega_r^2) = (P_m - P_e)\,d\delta$$

$$\Rightarrow \quad \frac{H}{\omega_s}\int_{\omega_{r_1}}^{\omega_{r_2}} d(\omega_r^2) = \int_{\delta_1}^{\delta_2}(P_m - P_e)\,d\delta$$

where,

$$\omega_{r_1} = \left.\frac{d\delta}{dt}\right|_{\delta=\delta_1} = \omega(\delta=\delta_1) - \omega_s$$

$$\omega_{r_2} = \left.\frac{d\delta}{dt}\right|_{\delta=\delta_2} = \omega(\delta=\delta_2) - \omega_s$$

$$\frac{H}{\omega_s}(\omega_{r_2}^2 - \omega_{r_1}^2) = \int_{\delta_1}^{\delta_2}(P_m - P_e)\,d\delta$$

In problem,

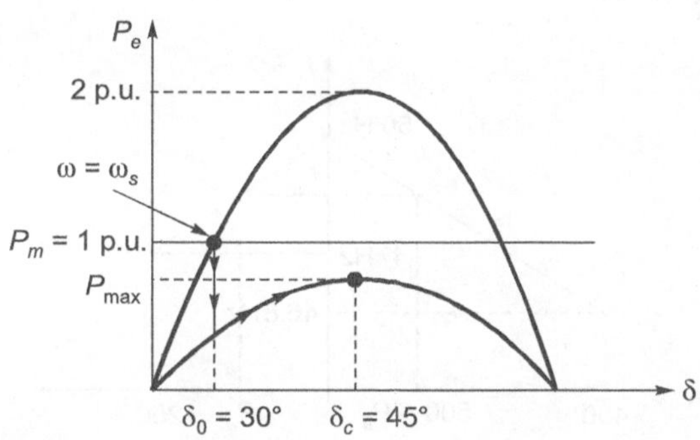

$$\frac{d\delta}{dt} = 0; \quad \frac{d\delta}{dt} = 3.762$$

$$\frac{5}{314}(3.762^2 - 0^2) = \int_{30^\circ}^{45^\circ}(1 - P_{max}\sin\delta)\,d\delta$$

$$\Rightarrow \quad P_{max} = 0.23 \text{ p.u.}$$

6.27 Sol.

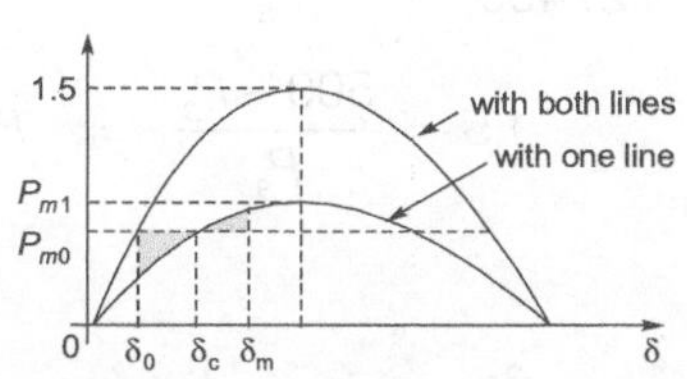

Using equal area criteria: $A_1 = A_2$

$$\int_{\delta_0}^{\delta_c}(P_{mo} - P_{max1}\sin\delta)\,d\delta = \int_{\delta_c}^{\delta_2}(P_{max1}\sin\delta - P_{mo})\,d\delta$$

By solving above integration,

$$P_{max1} = \frac{P_{mo}(\delta_2 - \delta_0)}{\cos\delta_0 - \cos\delta_2}$$

Given data:

$$P_{m0} = 1 \text{ pu}$$
$$P_e = 1.5\sin(\delta_0) = P_{m0} = 1 \text{ pu}$$
$$\delta_0 = 41.8^\circ = 0.7295 \text{ rad}$$
$$\delta_2 = 1.221 \text{ rad} = 69.95^\circ$$

Substitute above values in above equation,

$$P_{max1} = \frac{1(1.221 - 0.7295)}{\cos 41.8 - \cos 69.95} = 1.220 \text{ p.u.}$$

6.28 Sol.

$$S = 250 \text{ MVA},$$
$$\cos\phi = 0.8$$
$$\text{K.E.} = 1000 \text{ MJ}$$
$$P_e = 60 \text{ MW}$$
$$\delta_0 = 10^\circ$$
$$t = 10 \text{ cycles} = \frac{10}{50} = 0.2 \text{ sec}$$
$$t = 5 \text{ cycles} \Rightarrow 0.1 \text{ sec}$$

Load is removed,

$$P_e = 0$$
$$P_a = P_m - P_e$$
$$\Rightarrow \quad P_m = 60 \text{ MW},$$

$$M = \frac{HS}{180f} = \frac{K.E.}{180f} = \frac{1000}{180\times 50} = 0.111$$

$$\frac{d^2\delta}{dt^2} = \frac{P_a}{M} = 545.45 \text{ ele.degree/s}^2$$

Integrating the above equation, we get

$$\int\left(\frac{d^2\delta}{dt^2}\right)dt = \int\frac{P_a}{M}dt = 545.45\times t$$

Integrating it once again with dt

$$= \int 545.45\times t(dt)$$

$$\delta = \frac{545.45\times t^2}{2} = 2.7^\circ$$

So, new value of power angle

$$= 10^\circ + 2.7^\circ = 12.7^\circ$$

6.29 (b)

With only x:

$$P_{max} = \left|\frac{V_1 V_2}{x}\right|$$

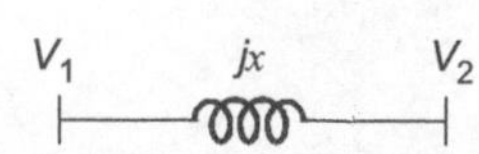

With Lossy Tr. Line

$$P = \left|\frac{V_1 V_2}{z}\right|\cos(\beta - \delta) - \left|\frac{AV_2^2}{z}\right|\cos(\beta - \alpha)$$

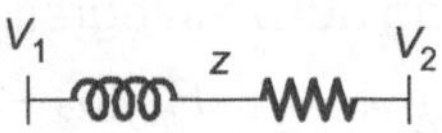

$\therefore$ With Lossy Line $P_{max} < \left|\frac{V_1 V_2}{x}\right|$

6.30 Sol.

$$X = 0.25 + 0.2 + 0.4 || 0.4$$
$$= 0.45 + 0.2 = 0.65 \text{ pu}$$
$$P = V_{pu} \times I_{pu} \cos\phi$$
$$0.8 = 1 \times I_{pu} \times 0.8$$
$$I_{pu} = 1 \text{ pu}$$
$$E = V + jI_a X_s = 1 + 1\angle{-36.86°} \times j0.65$$
$$= 1.484\angle 20.51° \text{ pu}$$
$$\delta = 20.51°$$

6.31 Sol.

Assumed full load frequency is 50 Hz

$$\tan\theta = \frac{50 - 49.75}{110 - 100} = \frac{49.75 - 49.25}{(x - (110))}$$

$$\frac{0.25}{10} = \frac{0.5}{(x - 110)}$$

$$x - 110 = \frac{0.5 \times 10}{0.25}$$

$$x = 110 + 20 = 130 \text{ MW}$$

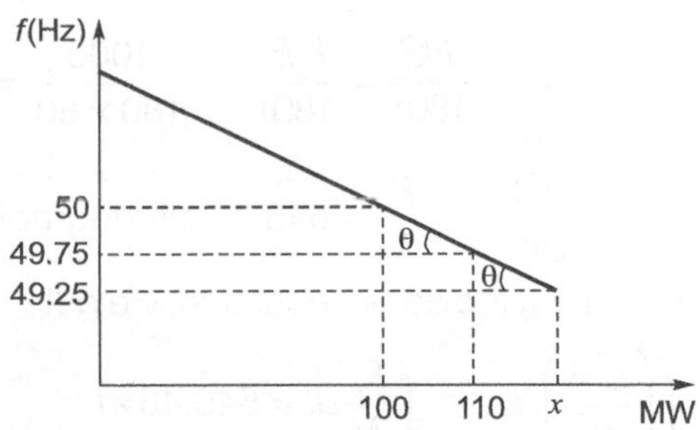

6.32 (60)

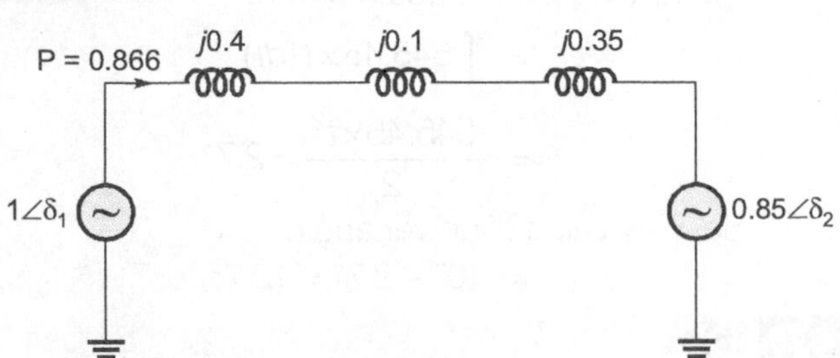

Given real power in (p.u.),

$$P = 0.866$$

$$P = \frac{EV\sin(\delta_1 - \delta_2)}{X_{eq}}$$

$$= \left|\frac{1 \times 0.85}{0.4 + 0.1 + 0.35}\right| \sin(\delta_1 - \delta_2)$$

$$(\delta_1 - \delta_2) = 60°$$

6.33 (c)

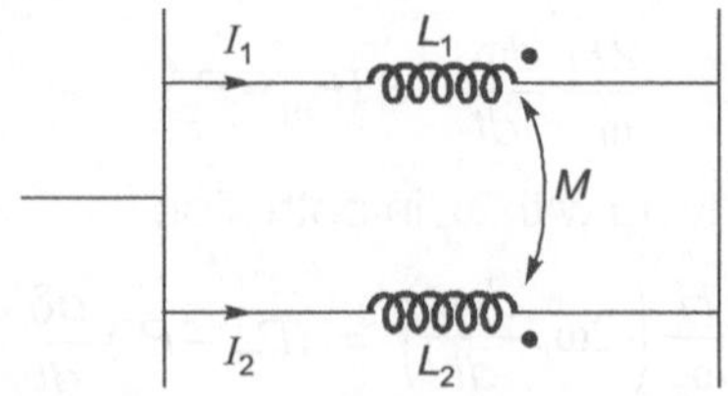

$\Rightarrow$

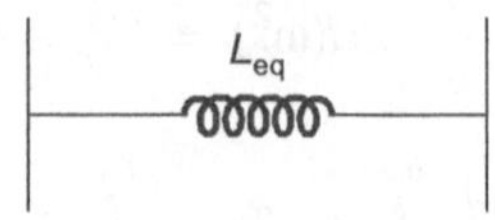

$$L_{eq} = \frac{L_1 L_2 - M^2}{(L_1 + L_2 - 2M)}$$

$$X_{eq} = \frac{1.5 \times 1.5 - 0.5^2}{1.5 + 1.5 - 2 \times 0.5} = 1 \text{ p.u.}$$

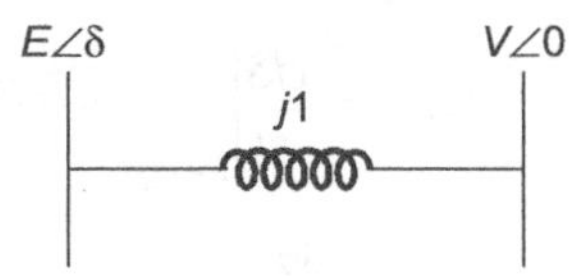

$$P_{max} = \frac{|E||V|}{1} = |E||V|$$

6.34 (200)

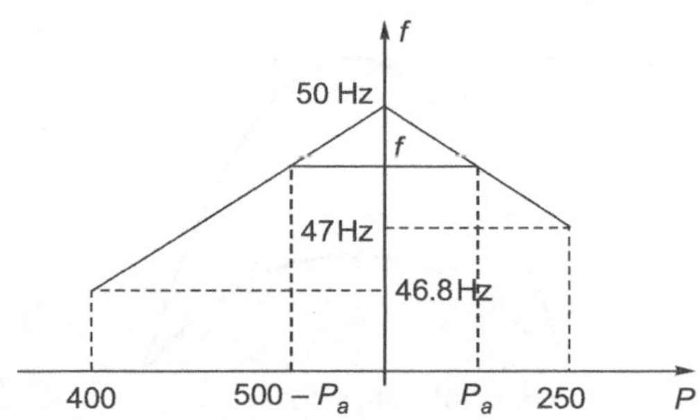

$$\frac{3}{250} = \frac{50 - f}{P_a} \quad \text{...(i)}$$

$$\frac{3.2}{400} = \frac{50 - f}{500 - P_a} \quad \text{...(ii)}$$

Equation (i) divide by equation (ii),

$$\frac{3/250}{3.2/400} = \frac{500 - P_a}{P_a}$$

$$1.5 = \frac{500 - P_a}{P_a} \Rightarrow P_a = 200 \text{ MW}$$

6.35 (75)

$$M\frac{d^2\delta}{dt^2} = P_a$$

$$P_a = 15 - 10 = 5$$

So, $m\alpha = P_a \Rightarrow \alpha = \dfrac{P_a}{m}$

$$m = \frac{Hs}{180f} = \frac{15 \times 20}{180 \times 50} = \frac{1}{30}$$

So, $$\alpha = \frac{5}{1/30}$$

$$= 150 \text{ electrical degree/sec}^2$$

$$\alpha = 150 \times \frac{2}{P} \text{ mech. degree/sec}^2$$

$$\alpha = 75 \text{ mechanical degree}$$

6.36 (83.64)

For nominal π model :

$B = Z = [R + j(2\pi fL)]l$

$= [0.035 + j(2\pi \times 50 \times 1 \times 10^{-3})] \times 400$

$= 126.44\angle 83.642° = |B|\angle\beta$

$V_S\angle\theta = V_S\angle\delta = 275\angle\delta$ $\qquad V_R\angle 0° = 275\angle 0°$

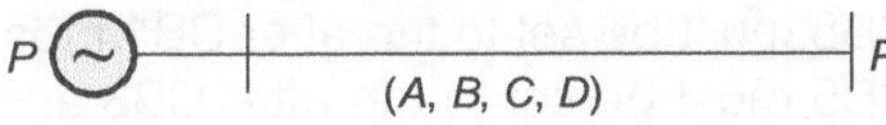

$$P_R = \left|\frac{V_S V_R}{B}\right|\cos(\beta - \delta) - \left|\frac{AV_R^2}{B}\right|\cos(\beta - \alpha)$$

$$\frac{\partial P_R}{\partial \delta} = 0 \Rightarrow \left|\frac{V_S V_R}{B}\right| \times (-\sin(\beta - \delta) \times (-1) = 0;\ \delta = \beta$$

$$\text{At } \delta = \beta \Rightarrow (P_R)_{max} = \left|\frac{V_S V_R}{B}\right| - \left|\frac{AV_R^2}{B}\right|\cos(\beta - \alpha)$$

Maximum P_R occurs at

$$\delta = \beta = 83.64° = \theta$$

6.37 (0.1)

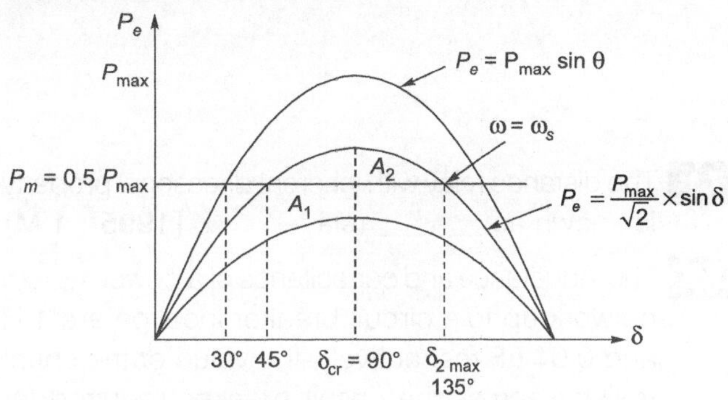

At critical clearing :

Accelerating area = Decelerating area

$$A_1 = A_2$$

$$\int_{90°}^{135°}\left(\frac{P_{max}}{\sqrt{2}}\sin\delta - 0.5P_{max}\right)d\delta = 0.107P_{max}$$

■■■■

7 Switch Gear and Protection

7.1 The distance relay with inherent directional property is known as _______ relay. [1995 : 1 M]

7.2 The inductance and capacitance of a power system network up to a circuit breaker location are 1 H and 0.01 μF respectively, the value of the shunt resistor across the circuit breaker, required for critical damping of the restriking voltage is ____. [1995 : 1 M]

7.3 The insulation level of a 400 kV, EHV over head transmission line is decided on the basis of
(a) Lightning over voltage
(b) Switching over voltage
(c) Corona inception voltage
(d) Radio and TV interference [1995 : 1 M]

7.4 Type of Relay most suited for

(a) Buchholz relay	(P) Feeder
(b) Translay relay	(Q) Transformer
(c) Carrier current, Phase comparison relay	(R) Radial distributed
(d) Directional over current relay	(S) Generator
(e) Negative sequence relay	(T) Ring main distributor
	(U) Long overload transmission line application

[1995 : 2 M]

7.5 If the fault current is 2000 A, the relay setting is 50% and CT ratio is 400 : 5, the plug setting multiplier will be
(a) 25 A (b) 15 A
(c) 50 A (d) 10 A [1996 : 1 M]

7.6 Resistance switching is normally employed in
(a) All breakers
(b) Bulk oil breakers
(c) Minimum oil breakers
(d) Air blast circuit breakers [1996 : 1 M]

7.7 Reactance relay is normally preferred for protection against
(a) Earth faults (b) Phase faults
(c) Open-circuit faults (d) None of these [1997 : 1 M]

7.8 The neutral of 10 MVA, 11 kV alternator is earthed through a resistance of 5 ohms. The earth fault relay is set to operate at 0.75 A. The CT's have a ratio of 1000/5. What percentage of the alternator winding is protected?
(a) 85% (b) 88.2%
(c) 15% (d) 11.8% [1998 : 2 M]

7.9 Three sections of a feeder are provided with circuit breakers CB1, CB2, CB3, CB4, CB5 and CB6. For a fault F as indicated in figure.

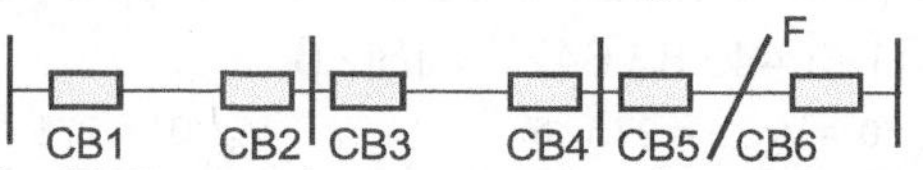

(a) CB5 must be set to trip after CB1 trips
(b) CB5 must be set to trip after CB3 and CB4 trip
(c) CB5 must be set to trip after CB2 trips
(d) CB5 must be set to trip before CB1, CB2, CB3 and CB4 trip [1999 : 1 M]

7.10 In an inverse definite minimum time, electromagnetic type over-current relay, the minimum time feature is achieved because of
(a) Saturation of the magnetic circuit
(b) Proper mechanical design
(c) Appropriate time delay element
(d) Electromagnetic damping [2000 : 1 M]

7.11 In a 3 step distance protection, the reach of the 3 zones of the relay at the beginning of the first line typically extends upto
(a) 100% of the first line, 50% of the second line and 20% of the third line.
(b) 80% of the first line, 50% of the second line and 20% of the third line.
(c) 80% of the first line, 20% of second line and 10% the third line.
(d) 50% of first line, 50% of second line and 20% of the third line. [2000 : 1 M]

7.12 The plug setting of a negative sequence relay is 0.2 A. The current transformer ratio is 5 : 1. The minimum value of line to line fault current for the operation of the relay is
(a) 1 A (b) $\frac{1}{1.732}$ A
(c) 1.732 A (d) $\frac{0.2}{1.732}$ A
[2000 : 2 M]

7.13 Consider the problem of relay coordination for the distance relays R_1 and R_2 on adjacent lines of a transmission system figure. The zone 1

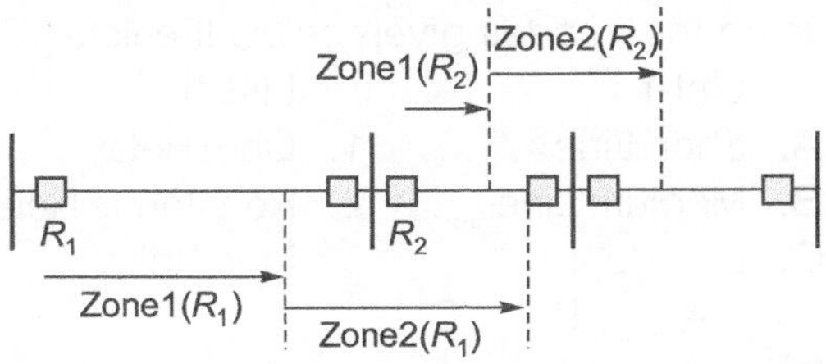

and zone 2 settings for both the relays are indicated on the diagram. Which of the following indicates the correct time setting for the zone 2 of relays R_1 and R_2.

(a) $TZ2_{R1} = 0.6s$, $TZ2_{R2} = 0.3s$
(b) $TZ2_{R1} = 0.3s$, $TZ2_{R2} = 0.6s$
(c) $TZ2_{R1} = 0.3s$, $TZ2_{R2} = 0.3s$
(d) $TZ2_{R1} = 0.1s$, $TZ2_{R2} = 0.3s$ [2002 : 1 M]

7.14 The interrupting time of a circuit breaker is the period between the instant of

(a) initiation of short circuit and the arc extinction on an opening operation.
(b) energizing of the trip circuit and the arc extinction on an opening operation.
(c) initiation of short circuit and the parting of primary arc contacts.
(d) energizing of the trip circuit and the parting of primary arc contacts. [2003 : 1 M]

7.15 A list of relays and the power system components protected by the relays are given in **List-I** and **List-II** respectively. Choose the correct match from the four choices given below:

List-I	List-II
A. Distance relay	1. Transformers
B. Under frequency relay	2. Turbines
C. Differential relay	3. Busbars
D. Buchholz relay	4. Shunt capacitors
	5. Alternators
	6. Transmission lines

Codes:

	A	B	C	D
(a)	6	5	3	1
(b)	4	3	2	1
(c)	5	2	1	6
(d)	6	4	5	3

[2003 : 2 M]

7.16 The transmission line distance protection relay having the property of being inherently directional is

(a) impedance relay
(b) MHO relay
(c) OHM relay
(d) reactance relay [2004 : 1 M]

7.17 A lightning stroke discharges impulse current of 10 kA (peak) on a 400 kV transmission line having surge impedance of 250 Ω. The magnitude of transient over-voltage travelling waves in either direction assuming equal distribution from the point of lightning strike will be

(a) 1250 kV (b) 1650 kV
(c) 2500 kV (d) 2900 kV [2004 : 2 M]

7.18 Keeping in view the cost and overall effectiveness, the following circuit breaker is best suited for capacitor bank switching

(a) vacuum (b) air blast
(c) SF_6 (d) oil [2006 : 1 M]

7.19 In a biased differential relay the bias is defined as a ratio of

(a) number of turns of restraining and operating coil
(b) operating coil current and restraining coil current
(c) fault current and operating coil current
(d) fault current and restraining coil current

[2006 : 1 M]

7.20 Consider the transformer connections in a part of a power system shown in the figure. The nature of transformer connections and phase shifts are indicated for all but one transformer.

Which of the following connections, and the corresponding phase shift θ, should be used for the transformer between A and B?

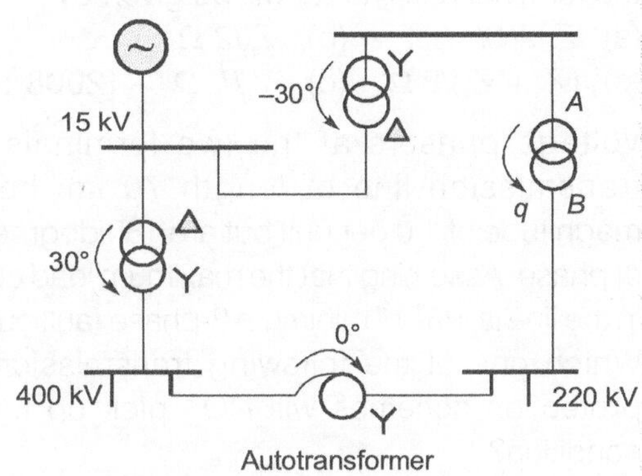

(a) Star - Star (θ = 0°)
(b) Star - Delta (θ = –30°)
(c) Delta - Star (θ = 30°)
(d) Star - Zigzag (θ = 30°) [2007 : 1 M]

7.21 Consider the protection system shown in the figure below. The circuit breakers numbered from 1 to 7 are of identical type. A single line to ground fault with zero fault impedance occurs at the midpoint of the line (at point F), but circuit breaker 4 fails to operate ("Stuck breaker"). If the relays are coordinated correctly, a valid sequence of circuit breaker operation is

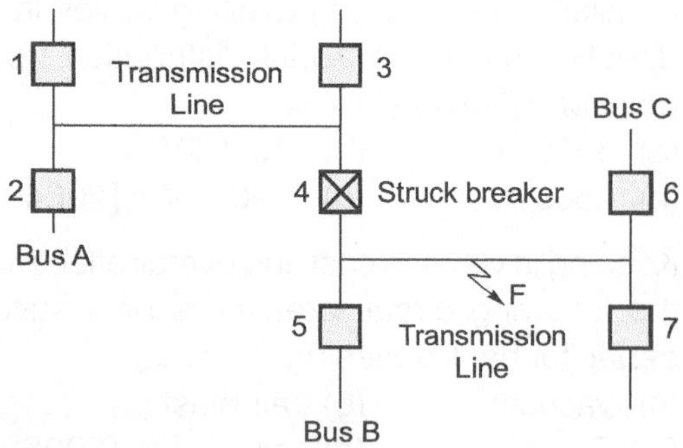

(a) 1, 2, 6, 7, 3, 5 (b) 1, 2, 5, 5, 7, 3
(c) 5, 6, 7, 3, 1, 2 (d) 5, 1, 2, 3, 6, 7

[2007 : 2 M]

7.22 A two machine power system in shown below. Transmission line *XY* has positive sequence impedance of $Z_1\,\Omega$ and zero sequence impedance of $Z_0\,\Omega$.

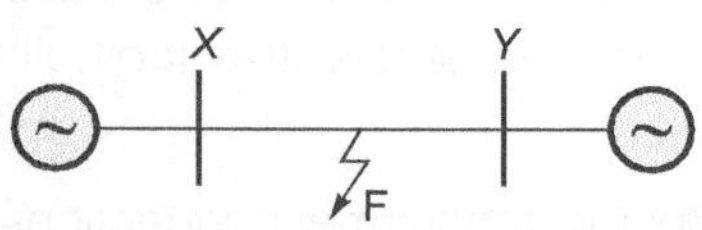

An '*a*' phase to ground fault with zero fault impedance occurs at the centre of he transmission line. Bus voltage at *X* and line current from *X* to *F* for the phase '*a*', are given by V_a Volts and I_a Amperes, respectively. Then, the impedance measured by the ground distance relay located at the terminal X of line XY will be given by

(a) $Z_1/2\,\Omega$ (b) $Z_0/2\,\Omega$
(c) $(Z_0 + Z_1)/2\,\Omega$ (d) $V_a/I_a\,\Omega$ [2008 : 1 M]

7.23 Voltage phasors at the two terminals of a transmission line of length 70 km have a magnitude of 1.0 per unit but are 180 degrees out of phase. Assuming that the maximum load current in the line is 1/5th of minimum 3-phase fault current. Which one of the following transmission line protection schemes will NOT pick up for this condition?

(a) Distance protection using mho relays with zone-1 set to 80% of the line impedance.
(b) Directional overcurrent protection set to pick up at 1.25 times the maximum load current
(c) Pilot relaying system with directional comparison scheme
(d) Pilot relaying system with segregated phase comparison scheme. [2008 : 2 M]

7.24 Match the items in **List-I** (Type of transmission line) with the items in **List-II** (Type of distance relay preferred) and select the correct answer using the codes given below the lists.

List-I	List-II
A. Short Line	1. Ohm Relay
B. Medium Line	2. Reactance Relay
C. Long Line	3. Mho Relay

	A	B	C
(a)	2	1	3
(b)	3	2	1
(c)	1	2	3
(d)	1	3	2

[2009 : 2 M]

7.25 A three-phase, 33 kV oil circuit breaker is rated 1200 A, 2000 MVA, 3s. The symmetrical breaking current is

(a) 1200 A (b) 3600 A
(c) 35 kA (d) 104.8 kA [2010 : 2 M]

7.26 Consider a stator winding of an alternator with an internal high-resistance ground fault. The currents under the fault condition are as shown in the figure. The winding is protected using a differential current scheme with current transformers of ratio 400/5 A as shown. The current through the operating coil is

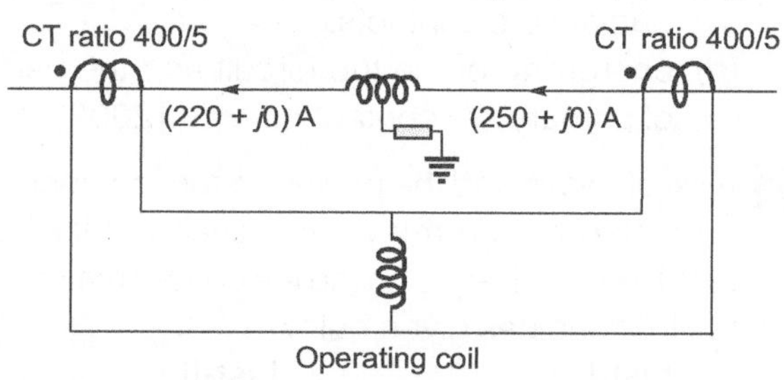

(a) 0.1875 A (b) 0.2 A
(c) 0.375 A (d) 60 kA [2010 : 2 M]

7.27 A nuclear power station of 500 MW capacity is located at 300 km away from a load center. Select the most suitable power evacuation transmission configuration among the following options

(a)
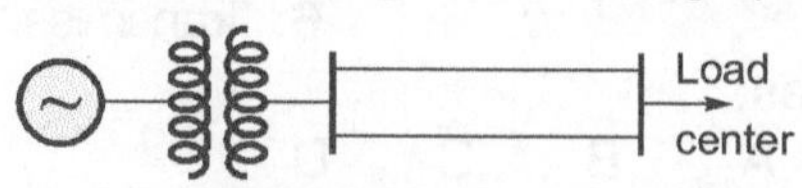

132 kV, 300 km double circuit

(b)
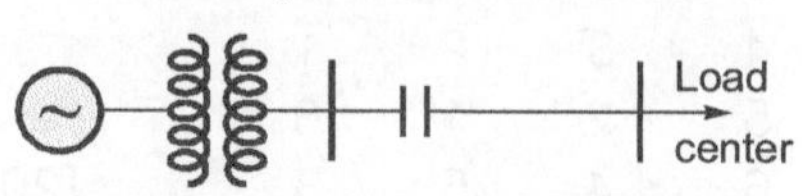

132 kV, 300 km single circuit with 40% series capacitor compensation

(c)
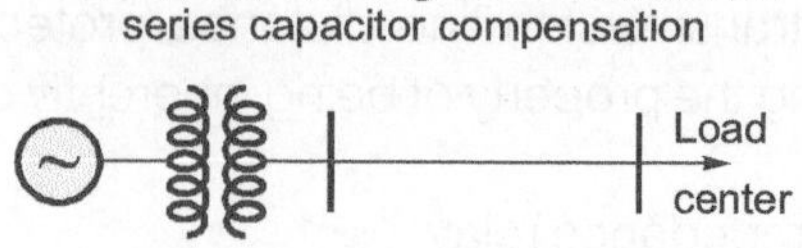

400 kV, 300 km single circuit

(d) 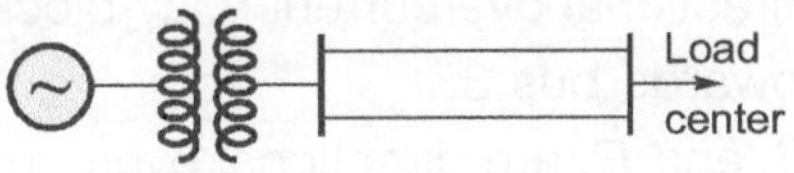

400 kV, 300 km double circuit

[2011 : 1 M]

7.28 A negative sequence relay is commonly used to protect
(a) an alternator
(b) a transformer
(c) a transmission line
(d) a bus bar [2011 : 1 M]

7.29 The over current relays for the line protection and loads connected at the buses are shown in the figure.

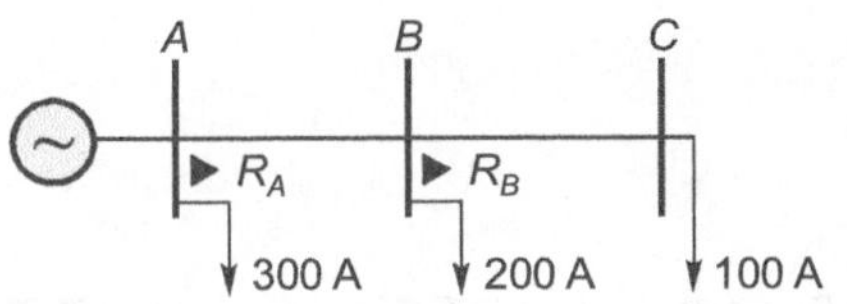

The relays are IDMT in nature having the characteristic

$$t_{op} = \frac{0.14 \times \text{Time Multiplier Setting}}{(\text{Plug Setting Multiplier})^{0.02} - 1}$$

The maximum and minimum fault currents at bus B are 2000 A and 500 A respectively. Assuming the time multiplier setting and plug setting for relay R_B to be 0.1 and 5 A respectively, the operating time of R_B (in seconds) is ______.

[2014 : 2 M, Set-1]

7.30 A 3-phase transformer rated for 33 kV/11 kV is connected in delta/star as shown in figure. The current transformers (CTs) on low and high voltage sides have a ratio of 500/5. Find the currents i_1 and i_2, if the fault current is 300 A as shown in the figure.

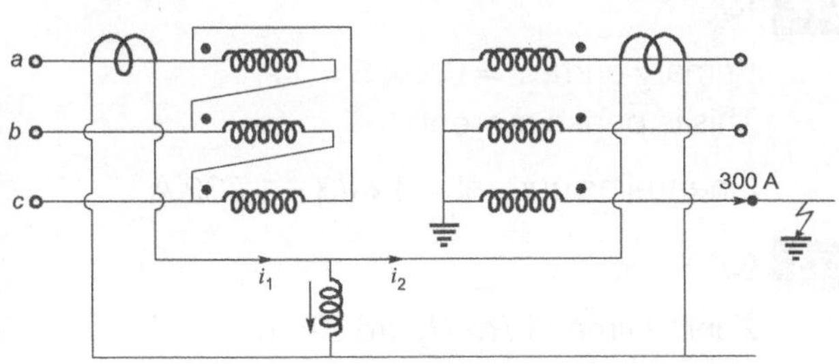

(a) $i_1 = \frac{1}{\sqrt{3}}$ A, $i_2 = 0$ A
(b) $i_1 = 0$ A, $i_2 = 0$ A
(c) $i_1 = 0$A, $i_2 = \frac{1}{\sqrt{3}}$ A
(d) $i_1 = \frac{1}{\sqrt{3}}$ A, $i_2 = \frac{1}{\sqrt{3}}$ A [2015 : 2 M, Set-2]

7.31 A power system with two generators is shown in the figure below. The system (generators, buses and transmission lines) is protected by six overcurrent relays R_1 to R_6. Assuming a mix of directional and nondirectional relays at appropriate locations, the remote backup relays for R_4 are

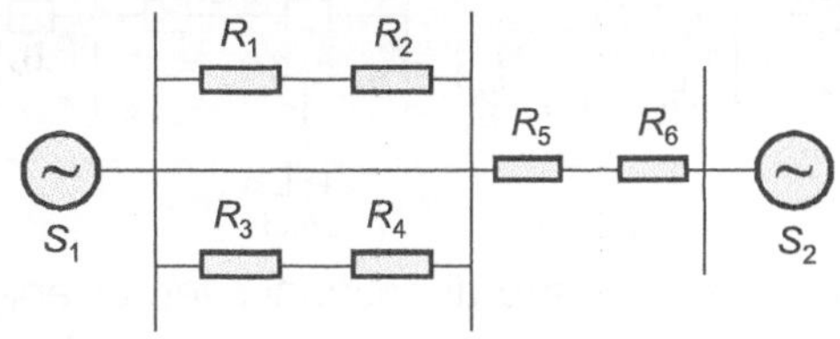

(a) R_1, R_2 (b) R_2, R_6
(c) R_2, R_5 (d) R_1, R_6

[2016 : 1 M, Set-2]

7.32 The total impedance of the secondary winding, leads and burden of a 5 A CT is 0.01 Ω. If the fault current is 20 times the rated primary current of the CT. The VA output of the CT is ______.

[2019 : 1 M]

7.33 In a 132 kV system, the series inductance upto the point of circuit breaker location is 50 mH. The shunt capacitance at the circuit breaker terminal is 0.05 μF. The critical value of resistance in ohms required to be connected across the circuit breaker contacts which will give no transient oscillation is ______.

[2019 : 2 M]

7.34 A lossless transmission line with 0.2 pu reactance per phase uniformly distributed along the length of the line, connecting a generator bus to a load bus, is protected up to 80% of its length by a distance relay placed at the generator bus. The generator terminal voltage is 1 pu. There is no generation at the load bus. The threshold pu current for operation of the distance relay for a solid three phase-to-ground fault on the transmission line is closest to:
(a) 1.00 (b) 5.00
(c) 3.61 (d) 6.25 [2020 : 1 M]

7.35 The most commonly used relay, for the protection of an alternator against loss of excitation, is
(a) offset Mho relay. (b) over current relay.
(c) differential relay. (d) Buchholz relay.

[2022 : 1 M]

7.36 For the three-bus power system shown in the figure, the trip signals to the circuit breakers B_1 to B_9 are provided by overcurrent relays R_1 to R_9, respectively, some of which have directional properties also. The necessary condition for the system to be protected for short circuit fault at any part of the system between bus 1 and the R-L loads with isolation of minimum portion of the network using minimum number of directional relays is

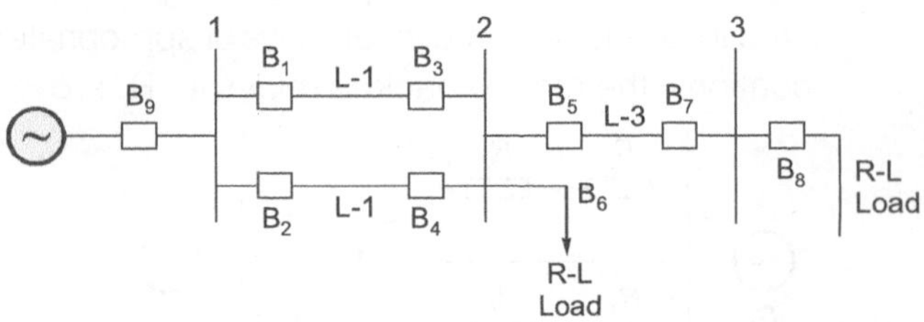

(a) R_3 and R_4 are directional overcurrent relays blocking faults towards bus 2.
(b) R_3 and R_4 are directional overcurrent relays blocking faults towards bus 2 and R_7 is directional overcurrent relay blocking faults towards bus 3.
(c) R_3 and R_4 are directional overcurrent relays blocking faults towards Line 1 and Line 2, respectively, R_7 is directional overcurrent relay blocking faults towards Line 3 and R_5 is directional overcurrent relay blocking faults towards bus 2.
(d) R_3 and R_4 are directional overcurrent relays blocking faults towards Line 1 and Line 2, respectively.

[2023 : 1 M]

Answers Switch Gear and Protection

7.1	(Sol.)	**7.2**	(5)	**7.3**	(b)	**7.4**	(Sol.)	**7.5**	(d)	**7.6**	(d)	**7.7**	(a)
7.8	(b)	**7.9**	(d)	**7.10**	(a)	**7.11**	(b)	**7.12**	(c)	**7.13**	(c)	**7.14**	(b)
7.15	(a)	**7.16**	(b)	**7.17**	(c)	**7.18**	(a)	**7.19**	(a)	**7.20**	(a)	**7.21**	(c)
7.22	(d)	**7.23**	(a)	**7.24**	(a)	**7.25**	(c)	**7.26**	(c)	**7.27**	(d)	**7.28**	(a)
7.29	(0.226)	**7.30**	(a)	**7.31**	(d)	**7.32**	(100)	**7.33**	(500)	**7.34**	(d)	**7.35**	(a)
7.36	(a)												

Explanations Switch Gear and Protection

7.1 Sol.

The distance relay with inherent directional property is known as **mho** relay.

7.2 Sol.

$$R_p = \frac{1}{2}\sqrt{\frac{L}{C}} = \frac{1}{2}\sqrt{\frac{1}{0.01\times10^{-6}}} = 5\text{ k}\Omega$$

7.3 (b)

Upto 345 kV, insulation level is decided based on lightning over voltages and above 345 kV, insulation level is decided based on switching over voltage.

7.4 Sol.

a — Q
b — P
c — U
d — T
e — S

7.5 (d)

Plug setting multiplier,

$$= \frac{\text{Fault current}}{\text{Relay current settings}} = \frac{2000/80}{0.5\times5} = 10$$

7.8 (b)

Percentage protected winding

$$= \left(1-\frac{IR}{V}\right)\times100\% = \left(1-\frac{5\times0.75\times1000/5}{11/\sqrt{3}\times1000}\right)\times100\%$$

$$= \left(1-\frac{0.75\times\sqrt{3}}{11}\right)\times100\%$$

$= 88.19\%$

7.12 (c)

Primary current = $0.2\times5 = 1$ A

This is phase current.

Line-to-line current = $1\times\sqrt{3} = 1.732$ A

7.13 (c)

Zone 2 time of R_1, R_2 are same.

7.14 (b)

Note: When a fault occurs in the protected circuit, the relay connected to the 'CT' actuates and closes its contact. Current flows from the battery in the trip circuit. As the trip coil of the breaker is energized the circuit breaker operating mechanism is actuated.

7.15 (a)

- Buchholz Relay is used to protect transformers which operates when the oil level in the conservator pipe of a transformer is lowered by the accumulation of gas caused by a poor connection or by an incipient breakdown of insulation (slowly developing fault).
- A differential relay is a suitably connected over current relay operating on the difference of currents at the two ends of a protected element.

7.16 (b)

MHO relay has the property of being inherently directional.

Note: $Q = K_3|V||I|\cos(\theta - \tau) - (-K_1)|V|^2$

For relay operation

$Q > 0$

$K_3|V||I|\cos(\theta-\tau) - (-K_1)|V|^2 > 0$

$$\frac{|V|}{|I|} = |z| < \frac{K_3}{-K_1}\cos(\theta-\tau) \quad ...(i)$$

The right had side of equation (i) is a circle with center located on the line determined by the parameter τ and passing through, the origin.

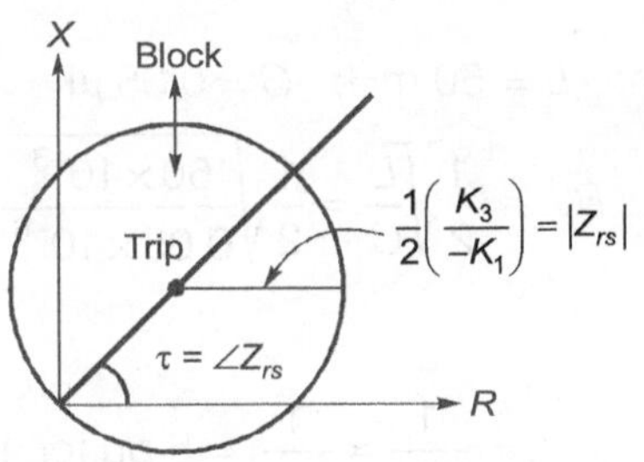

7.17 (c)

The impulse current of 10 kA (peak) will be equally divided in both directions since there is equal distribution on both sides. Then, magnitude of transient over voltage = 10000 A × 250

= 2500 kV (Assumed ideal impulse)

Impulse available for zero time.

7.18 (a)

The arc time constant is the least (a few μs) in vacuum circuit breakers as compared to other breaker types. The rapid building up of dielectric strength after final arc extinction (20 kV/μs) is unique feature of VCBs. These arc, therefore, ideally suited for capacitor switching.

7.19 (a)

$$N_0 k|I_1 - I_2| > \frac{N_r}{2}k|I_1 + I_2|$$

$$|I_1 - I_2| > \frac{N_r}{N_0}\left|\frac{I_1 + I_2}{2}\right|$$

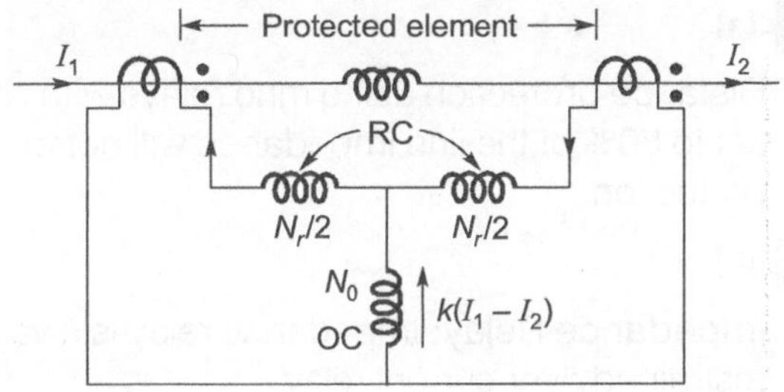

where,

N_0 = number of turns of the operating coil

N_r = number of turns of the restraining coil

The relay operates when the magnitude of the difference in the currents at the two ends of a protected element exceeds a certain percentage $((N_r / N_0)/100)$ of half the magnitude of their sum. Hence the name percentage differential relay.

7.20 (a)

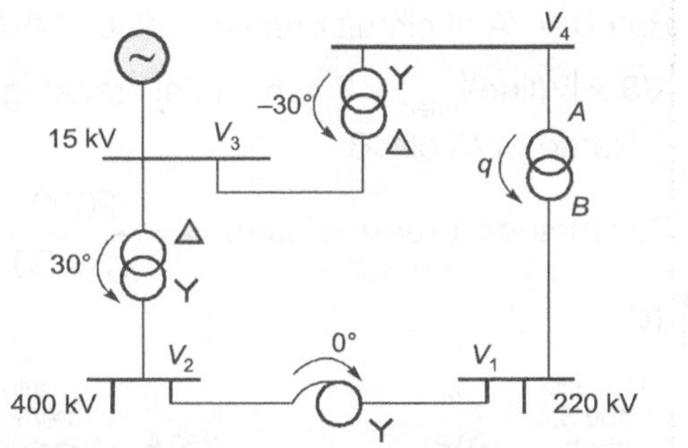

Taking V_1 as the reference

$V_1 = 220\angle 0°$ kV; $\angle\theta_1 = 0°$

Phase difference V_2 and V_1 is 0°.

So, $V_2 = 400\angle 0°$ kV

V_2 leads V_3 by 30°

So, $V_3 = 15\angle{-30°}$ kV; $\angle\theta_3 = -30°$

V_3 lags V_4 by 30°

$\angle\theta_3 = \angle\theta_4 - 30°$

$\angle\theta_4 = \angle\theta_3 + 30° = -30° + 30° = 0°$

Phase difference between V_1 and V_4 is θ

$\theta = \angle\theta_1 - \angle\theta_4 = 0 - 0 = 0°$

So, option (a) is correct.

7.21 (c)

If the relays are co-ordinated correctly, due to fault in a particular section, relay in that section must operate first, then relays in near by section if first fails, it is back up protection. Thus, the sequence is [5, 6, 7, 3, 1, 2].

7.22 (d)

As the ground distance relay is located at terminal impedance measured by the relay

$$= \frac{\text{Bus voltage}}{\text{current}} = \frac{V_a}{I_a}$$

7.23 (a)

Distance protection using mho relays with zone-I set to 80% of the line impedance will not provide protection.

7.24 (a)

Impedance Relay: Impedance relay is a voltage restrained over-current relay.

$$T = k_1 I^2 - k_2 V^2$$

Reactance Relay: Reactance relays is an over-current relay with directional restraint.

$$T = k_1 I^2 - k_3 VI \sin\theta$$

Mho Relay: mho relay is a voltage restrained direction relay.

$$T = k_3 VI \cos(\theta - t) - k_2 V^2$$

mho relay is inherently directional.

7.25 (c)

Rated MVA of circuit breaker = 2000 MVA

$\sqrt{3} \times |V(\text{line})|_{\text{rated}} \times$ Symmetrical breaking current = Rated MVA of CB

$$\text{Symmetrical breaking current} = \frac{2000}{\sqrt{3} \times 33} = 35 \text{ kA}$$

7.26 (c)

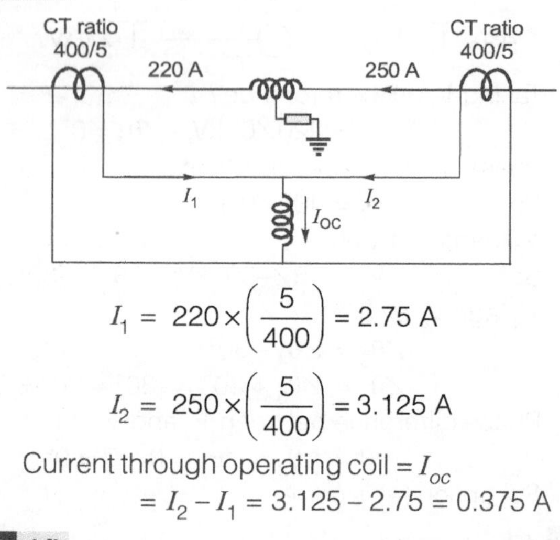

$$I_1 = 220 \times \left(\frac{5}{400}\right) = 2.75 \text{ A}$$

$$I_2 = 250 \times \left(\frac{5}{400}\right) = 3.125 \text{ A}$$

Current through operating coil = I_{oc}

$$= I_2 - I_1 = 3.125 - 2.75 = 0.375 \text{ A}$$

7.27 (d)

- For transmission of bulk power of very long distance high voltage (400 kV) is used.
- To increase reliability, double circuit is used.

7.29 Sol.

Minimum and maximum fault currents are

$$I_{f\min} = 500 \text{ A}, \quad I_{f\max} = 2000 \text{ A}$$

For relay R_B: TMS = 0.1, plug setting = 5 A

$$\text{PSM} = \frac{I_f}{I_{pk} \times \text{C.T. ratio}}$$

$$\text{CT ratio} = \frac{\text{Full load current}}{I_{pk}} = \frac{100}{5}$$

At relay R_B:

$$\text{With } I_{f\min} \Rightarrow \text{PSM}_1 = \frac{500}{\frac{100}{5} \times 5} = 5$$

$$\text{With } I_{f\max} \Rightarrow \text{PSM}_2 = \frac{2000}{\frac{100}{5} \times 5} = 20$$

Operating time with TMS = 0.1 and from IDMT characteristics

$$t_{op_2} = \frac{0.14 \times 0.1}{(20)^{0.02} - 1} = 0.226 \text{ sec}$$

7.30 (a)

$i_2 = 0$ since entire current flows through fault

$$i_1 = \frac{1}{\sqrt{3}} \text{ A}$$

7.32 Sol.

$$I_{\text{sec}} = 5 \times 20 = 100 \text{ A}$$

$$V = I_{\text{sec}} R = 100 \times 0.01 = 1 \text{ V}$$

VA output of the CT = VI_{sec} = 100 × 1 = 100 VA

7.33 Sol.

Given: $L = 50$ mH; $C = 0.05\ \mu$F

$$R_{cr} = \frac{1}{2}\sqrt{\frac{L}{C}} = \frac{1}{2}\sqrt{\frac{50 \times 10^{-3}}{0.05 \times 10^{-6}}} = 500\ \Omega$$

7.34 (d)

$$I_f = \frac{1}{Z_{Th}} = \frac{1}{0.2} = 5 \text{ pu for 100\% of line}$$

Relay is operated for 80%

$$Z_f = 0.8\, Z_l \Rightarrow 0.8 \times 0.2 = 0.16 \text{ p.u.}$$

For 80% of line,

$$I_f = \frac{1}{0.16} = 6.25 \text{ p.u.}$$

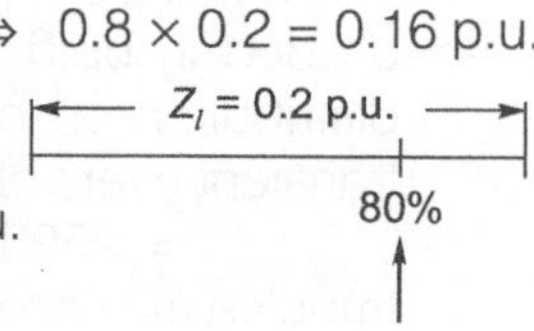

7.36 (a)

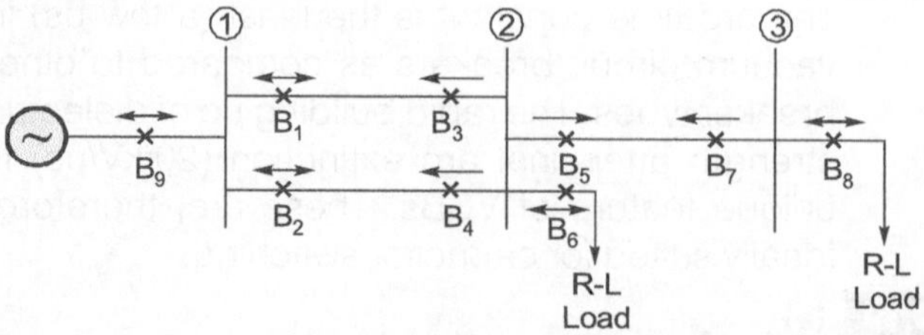

B_3, B_4 and B_5, B_7 are directional over current relays. All remaining are (B_1, B_2, B_6, B_8, B_9) non-directional over current relays.

■■■■

8 Generating Power Stations

8.1 Which material is used in controlling chain reaction in a nuclear reactor?
(a) Thorium (b) Heavy water
(c) Boron (d) Beryllium
[1996 : 2 M]

8.2 In a thermal power plant, the feed water coming to the economizer is heated using
(a) H.P. steam
(b) L.P. steam
(c) Direct heat in the furnace
(d) Flue gases [2000 : 1 M]

8.3 In thermal power plants, the pressure in the working fluid cycle is developed by
(a) condenser
(b) super heater
(c) feed water pump
(d) turbine [2004 : 1 M]

8.4 For harnessing low variable waterheads, the suitable hydraulic turbine with high percentage of reaction and runner adjustable vanes is
(a) Kaplan (b) Francis
(c) Pelton (d) Impeller [2004 : 1 M]

8.5 Out of the following plant categories
(i) Nuclear
(ii) Run-of-river
(iii) Pump Storage
(iv) Diesel
the base load power plants are
(a) (i) and (ii) (b) (ii) and (iii)
(c) (i), (ii) and (iv) (d) (i), (iii) and (iv)
[2009 : 1 M]

8.6 Base load power plants are
P. wind farms
Q. run-of-river plants
R. nuclear power plants
S. diesel power plants
(a) P, Q and S only (b) P, R and S only
(c) P, Q and R only (d) Q and R only
[2015 : 1 M, Set-1]

■■■■

Answers Generating Power Stations

8.1 (c) 8.2 (d) 8.3 (c) 8.4 (a) 8.5 (a) 8.6 (d)

Explanations Generating Power Stations

8.1 (c)

Boron and cadmium rods are used as control rods.

8.3 (c)

Note:
Condenser: The functions of condenser are
(i) To provide vacuum at outlet of steam turbine.
(ii) To condense the steam and pass on the condensate to boiler feed.

Super-heater is a device used to convert saturated steam or wet steam into dry steam used for power generation process.

8.4 (a)

Kaplan is used for run-of-river and poundage stations with heads of upto 70 m (low head). This type has an axial-flow rotor with variable-pitch blades.

8.5 (a)

Pumped storage plants and diesel stations supply power-during peak loads.
Note: Base load-plant
(i) Low operating cost
(ii) Capability of working continuously long period.

■■■■

9 Load Flow Studies

9.1 In load-flow analysis, a voltage controlled bus is treated as a load bus in subsequent iteration for a reactive power limit is violated (True/False).
[1995 : 1 M]

9.2 If the reference bus is changed in two load flow runs with same system data and power obtained for reference bus taken as specified P and Q in the later run

(a) the system losses will be unchanged but complex bus voltages will change.
(b) the system losses will change but complex bus voltages remain unchanged.
(c) the system losses as well as complex bus voltage will change.
(d) the system losses as well as complex bus voltage will be unchanged. [1996 : 1 M]

9.3 The per unit voltages of two synchronous machines connected through a lossless line are $0.95\angle 10°$ and $1.0\angle 0°$. Match the two sides in the following:

(a) Real power of of machine 1	(P) Positive real power
(b) Reactive power of machine 1	(Q) Positive reactive power
(c) Power factor of machine 1	(R) Negative real power
	(S) Negative reactive power
	(T) Leading power factor
	(U) Lagging power factor

[1996 : 2 M]

9.4 A power system consists of 2 areas (Area 1 and Area 2) connected by a single tie-line. It is required to carry out a load flow study on this system. While entering the network data, the tie-line data (connectivity and parameters) is inadvertently left out. If the load flow program is run with this incomplete data.

(a) The load flow will converge only if the slack bus is specified in area 1.
(b) The load flow will converge only if the slack bus is specified in area 2.
(c) The load flow will converge if the slack bus is specified in either area 1 or area 2.
(d) The load flow will not converge if only one slack bus is specified. [2002 : 2 M]

9.5 A power system consist of 300 buses out of which 20 buses are generator bus, 25 buses are ones with reactive power support and 15 buses are the ones with fixed shunt capacitors. All the other buses are load buses. It is proposed to perform a load flow analysis in the system using Newton-Raphson method. The size of the Newton Raphson Jacobian matrix is

(a) 553×553 (b) 540×540
(c) 555×555 (d) 554×554 [2003 : 1 M]

9.6 The bus impedance matrix of a 4-bus power system is given by

$$Z_{bus} = \begin{bmatrix} j0.3435 & j0.2860 & j0.2723 & j0.2277 \\ j0.2860 & j0.3408 & j0.2586 & j0.2414 \\ j0.2723 & j0.2586 & j0.2791 & j0.2209 \\ j0.2277 & j0.2414 & j0.2209 & j0.2791 \end{bmatrix}$$

A branch having an impedance of $j0.2\ \Omega$ is connected between bus 2 and the reference. Then the values of $Z_{22,\ new}$ and $Z_{23,\ new}$ of the bus impedance matrix of the modified network are respectively

(a) $j0.5408\ \Omega$ and $j0.4586\ \Omega$
(b) $j0.1260\ \Omega$ and $j0.0956\ \Omega$
(c) $j0.5408\ \Omega$ and $j0.0956\ \Omega$
(d) $j0.1260\ \Omega$ and $j0.1630\ \Omega$ [2003 : 2 M]

9.7 The network shown in the given figure has impedances in p.u. as indicated. The diagonal element Y_{22} of the bus admittance matrix Y_{BUS} of the network is

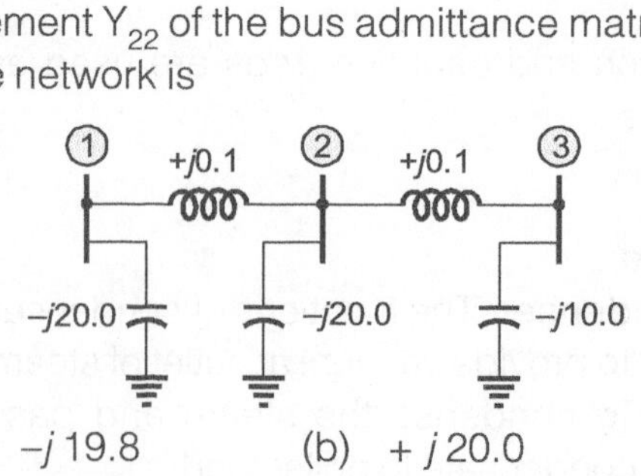

(a) $-j19.8$ (b) $+j20.0$
(c) $+j0.2$ (d) $-j19.95$ [2005 : 2 M]

9.8 The Gauss Seidel load flow method has following disadvantages. Tick the incorrect statement.

(a) Unreliable convergence
(b) Slow convergence

(c) Choice of slack bus affects convergence
(d) A good initial guess for voltages is essential for convergence [2006 : 2 M]

Statement for Common Data Questions (9.9 and 9.10)

For a power system the admittance and impedance matrices for the fault studies are as follows.

$$Y_{bus} = \begin{bmatrix} -j8.75 & j1.25 & j2.50 \\ j1.25 & -j6.25 & j2.50 \\ j2.50 & j2.50 & -j5.00 \end{bmatrix}$$

$$Z_{bus} = \begin{bmatrix} j0.16 & j0.08 & j0.12 \\ j0.08 & j0.24 & j0.16 \\ j0.12 & j0.16 & j0.34 \end{bmatrix}$$

The pre-fault voltages are 1.0 p.u. at all the buses. The system was unloaded prior to the fault. A solid 3 phase fault takes place at bus 2.

9.9 The post fault voltages at buses 1 and 3 in per unit respectively are
(a) 0.24, 0.63 (b) 0.31, 0.76
(c) 0.33, 0.67 (d) 0.67, 0.33 [2006 : 2 M]

9.10 The per unit fault feeds from generators connected to buses 1 and 2 respectively are
(a) 1.20, 2.51 (b) 1.55, 2.61
(c) 1.66, 2.50 (d) 5.00, 2.50 [2006 : 2 M]

9.11 Consider the two power systems shown in figure (a) below, which are initially not interconnected, and are operating in steady state at the same frequency. Separate loadflow solutions are computed individually for the two systems, corresponding to this scenario. The bus voltage phasors so obtained are indicated on figure (a). These two isolated systems are now interconnected by a short transmission line as shown in figure (b), and it is found that $P_1 = P_2 = Q_1 = Q_2 = 0$.

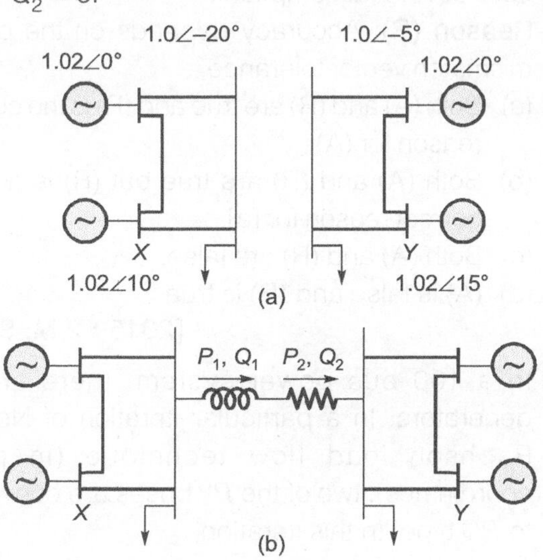

The bus voltage phase angular difference between generator bus X and generator bus Y after the interconnection is
(a) 10° (b) 25°
(c) –30° (d) 30° [2007 : 2 M]

9.12 For the Y-bus matrix of a 4-bus system given in per unit, the buses having shunt elements are

$$Y_{BUS} = j\begin{bmatrix} -5 & 2 & 2.5 & 0 \\ 2 & -10 & 2.5 & 4 \\ 2.5 & 2.5 & -9 & 4 \\ 0 & 4 & 4 & -8 \end{bmatrix}$$

(a) 3 and 4 (b) 2 and 3
(c) 1 and 2 (d) 1, 2 and 4 [2009 : 2 M]

9.13 Consider two buses connected by an impedance of $(0 + j5)\ \Omega$. The bus 1 voltage is $100\angle 30°$ V, and bus 2 voltage is $100\angle 0°$ V. The real and reactive power supplied by bus 1, respectively are
(a) 1000 W, 268 VAR
(b) –1000 W, –134 VAR
(c) 276.9 W, –56.7 VAR
(d) –276.9 W, 56.7 VAR [2010 : 1 M]

9.14 A three-bus network is shown in the figure below indicating the p.u. impedance of each element.

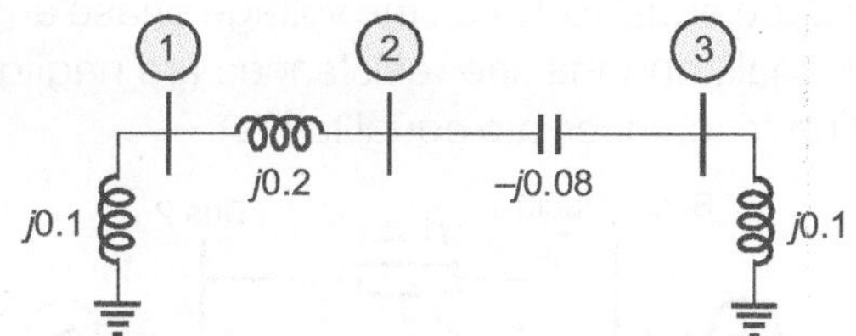

The bus admittance matrix, Y-bus, of the network is

(a) $j\begin{bmatrix} 0.3 & -0.2 & 0 \\ -0.2 & 0.12 & 0.08 \\ 0 & 0.08 & 0.02 \end{bmatrix}$

(b) $j\begin{bmatrix} -15 & 5 & 0 \\ 5 & 7.5 & -12.5 \\ 0 & -12.5 & 2.5 \end{bmatrix}$

(c) $j\begin{bmatrix} 0.1 & 0.2 & 0 \\ 0.2 & 0.12 & -0.08 \\ 0 & -0.08 & 0.10 \end{bmatrix}$

(d) $j\begin{bmatrix} -10 & 5 & 0 \\ 5 & 7.5 & 12.5 \\ 0 & 12.5 & -10 \end{bmatrix}$ [2011 : 2 M]

9.15 The bus admittance matrix of a three-bus three-line system is

$$Y = j\begin{bmatrix} -13 & 10 & 5 \\ 10 & -18 & 10 \\ 5 & 10 & -13 \end{bmatrix}$$

If each transmission line between the two buses is represented by an equivalent π-network, the magnitude of the shunt susceptance of the line connecting bus 1 and 2 is

(a) 4 (b) 2
(c) 1 (d) 0 [2012 : 1 M]

9.16 A single-phase load is supplied by a single-phase voltage source. If the current flowing from load to the source is $10\angle-150°$ A and if the voltage at the load terminals is $100\angle 60°$ V, then the

(a) load absorbs real power and delivers reactive power.
(b) load absorbs real power and absorbs reactive power.
(c) load delivers real power and delivers reactive power.
(d) load delivers real power and absorbs reactive power. [2013 : 1 M]

Statement for Linked Answer Questions (9.17 and 9.18):
In the following network, the voltage magnitudes at all buses are equal to 1 p.u., the voltage phase angles are very small, and the line resistances are negligible. All the line reactances are equal to $j1\ \Omega$.

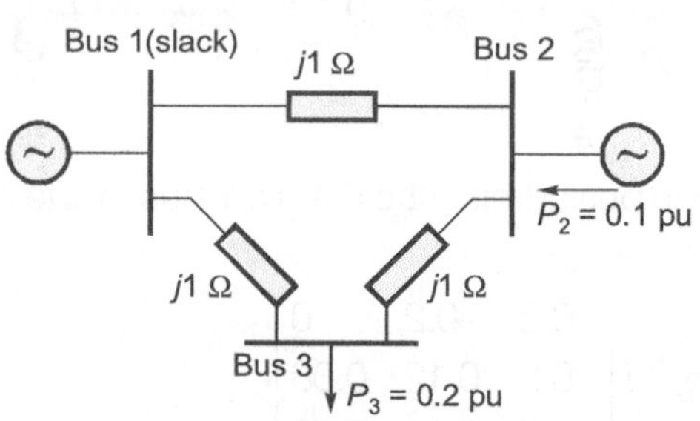

9.17 The voltage phase angles in rad at buses 2 and 3 are

(a) $\theta_2 = -0.1,\ \theta_3 = -0.2$
(b) $\theta_2 = 0,\ \theta_3 = -0.1$
(c) $\theta_2 = 0.1,\ \theta_3 = 0.1$
(d) $\theta_2 = 0.1,\ \theta_3 = 0.2$ [2013 : 2 M]

9.18 If the base impedance and the line-to-line base voltage are 100 Ω and 100 kV, respectively, then the real power in MW delivered by the generator connected at the slack bus is

(a) –10 (b) 0
(c) 10 (d) 20 [2013 : 2 M]

9.19 A two bus power system shown in the figure supplies load of $1.0 + j0.5$ pu.

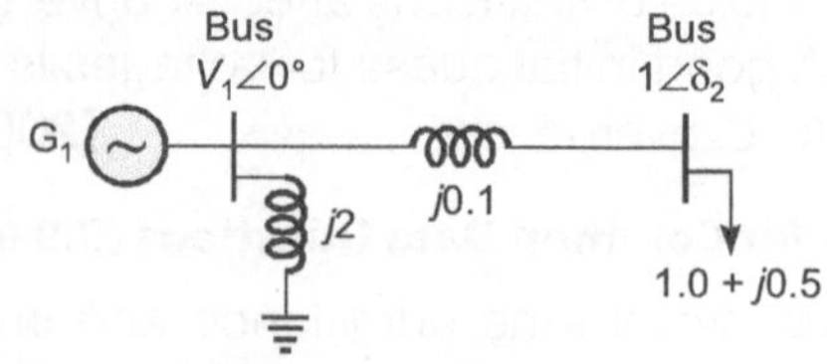

The values of V_1 in pu and δ_2 respectively are

(a) 0.95 and 6.00° (b) 1.05 and –5.44°
(c) 1.1 and –6.00° (d) 1.1 and –27.12°
[2014 : 2 M, Set-1]

9.20 A 183-bus power system has 150 PQ buses and 32 PV buses. In the general case, to obtain the load flow solution using Newton-Raphson method in polar co-ordinates, the minimum number of simultaneous equations to be solved is ______ .
[2014 : 1 M, Set-3]

9.21 A 3-bus power system network consists of 3 transmission lines. The bus admittance matrix of the uncompensated system is

$$\begin{bmatrix} -j6 & j3 & j4 \\ j3 & -j7 & j5 \\ j4 & j5 & -j8 \end{bmatrix} \text{pu}$$

If the shunt capacitance of all transmission lines is 50% compensated, the imaginary part of the 3rd row 3rd column element (in pu) of the bus admittance matrix after compensation is

(a) $-j7.0$ (b) $-j8.5$
(c) $-j7.5$ (d) $-j9.0$
[2015 : 1 M, Set-2]

9.22 Determine the correctness or otherwise of the following Assertion [A] and the Reason [R].
Assertion (A): Fast decoupled load flow method gives approximate load flow solution because it uses several assumptions.
Reason (R): Accuracy depends on the power mismatch vector tolerance.

(a) Both (A) and (R) are true and (R) is the correct reason for (A).
(b) Both (A) and (R) are true but (R) is not the correct reason for [a].
(c) Both (A) and (R) are false.
(d) (A) is false and (R) is true.
[2015 : 2 M, Set-1]

9.23 In a 100 bus power system, there are 10 generators. In a particular iteration of Newton Raphson load flow technique (in polar coordinates), two of the *PV* buses are converted to *PQ* type. In this iteration,

(a) the number of unknown voltage angles increases by two and the number of unknown voltage magnitudes increases by two.
(b) the number of unknown voltage angles remains unchanged and the number of unknown voltage magnitudes increases by two.
(c) the number of unknown voltage angles increases by two and the number of unknown voltage magnitudes decreases by two.
(d) the number of unknown voltage angles remains unchanged and the number of unknown voltage magnitudes decreases by two.

[2016 : 1 M, Set-1]

9.24 A power system has 100 buses including 10 generator buses. For the load flow analysis using Newton-Raphson method in polar coordinates, the size of the Jacobian is

(a) 189×189
(b) 100×100
(c) 90×90
(d) 180×180

[2016 : 1 M, Set-2]

9.25 A 10-bus power system consists of four generator buses indexed at $G1$, $G2$, $G3$, $G4$ and six load buses indexed as $L1$, $L2$, $L3$, $L4$, $L5$, $L6$. The generator-bus $G1$ is considered as slack bus, and the load buses $L3$ and $L4$ are voltage controlled buses. The generator at bus $G2$ cannot supply the required reactive power demand, and hence it is operating at its maximum reactive power limit. The number of non-linear equations required for solving the load flow problem using Newton-Raphson method in polar form is ______.

[2017 : 1 M, Set-1]

9.26 A 3-bus power system is shown in the figure below, where the diagonal elements of Y-bus matrix are: $Y_{11} = -j12$ pu, $Y_{22} = -j15$ pu and $Y_{33} = -j7$ pu.

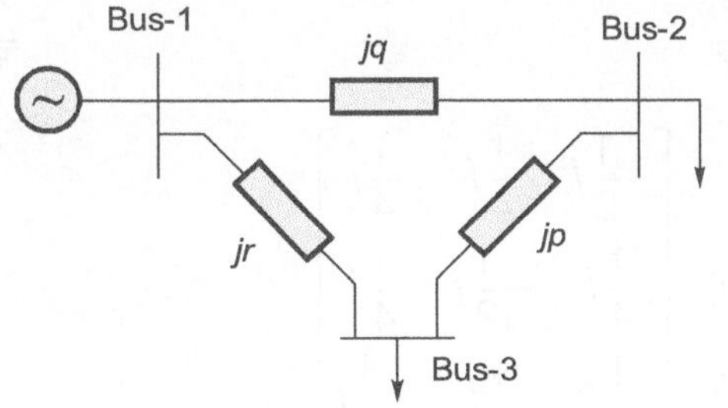

The per unit values of the line reactances p, q and r shown in the figure are

(a) $p = -0.2$, $q = -0.1$, $r = -0.5$
(b) $p = 0.2$, $q = 0.1$, $r = 0.5$
(c) $p = -5$, $q = -10$, $r = -2$
(d) $p = 5$, $q = 10$, $r = 2$

[2017 : 1 M, Set-1]

9.27 The bus admittance matrix for a power system network is

$$\begin{bmatrix} -j39.9 & j20 & j20 \\ j20 & -j39.9 & j20 \\ j20 & j20 & -j39.9 \end{bmatrix} \text{p.u.}$$

There is a transmission line, connected between buses 1 and 3, which is represented by the circuit shown in figure.

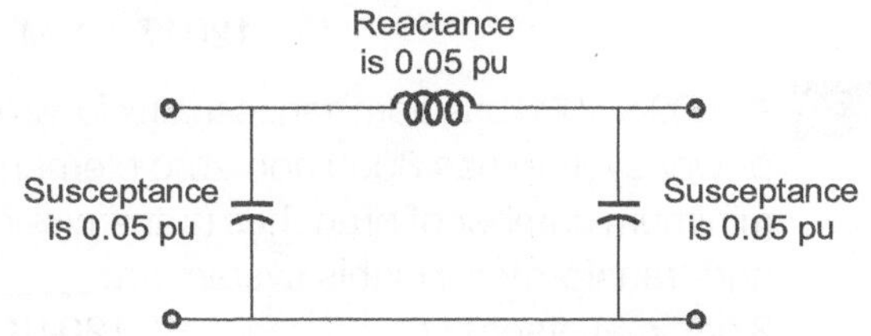

If this transmission line is removed from service, what is the modified bus admittance matrix?

(a) $\begin{bmatrix} -j19.9 & j20 & 0 \\ j20 & -j39.9 & j20 \\ 0 & j20 & -j19.9 \end{bmatrix}$ p.u.

(b) $\begin{bmatrix} -j39.95 & j20 & 0 \\ j20 & -j39.9 & j20 \\ 0 & j20 & -j39.95 \end{bmatrix}$ p.u.

(c) $\begin{bmatrix} -j19.95 & j20 & 0 \\ j20 & -j39.9 & j20 \\ 0 & j20 & -j19.95 \end{bmatrix}$ p.u.

(d) $\begin{bmatrix} -j19.95 & j20 & j20 \\ j20 & -j39.9 & j20 \\ j20 & j20 & -j19.95 \end{bmatrix}$ p.u.

[2017 : 2 M, Set-1]

9.28 In a load flow problem solved by Newton-Raphson method with polar coordinates, the size of the Jacobian is 100×100. If there are 20 PV buses in addition to PQ buses and a slack bus, the total number of buses in the system is______.

[2017 : 1 M, Set-2]

9.29 The figure shows the per-phase representation of a phase-shifting transformer connected between buses 1 and 2, where a is a complex number with non-zero real and imaginary parts.

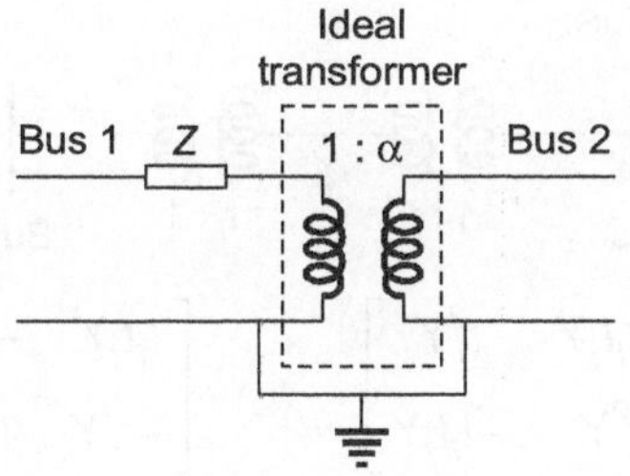

For the given circuit Y_{BUS} and Z_{BUS} are bus admittance matrix and bus impedance matrix, respectively, each of size 2 × 2. Which one of the following statements is true?

(a) Both Y_{BUS} and Z_{BUS} are symmetric.
(b) Y_{BUS} is symmetric and Z_{BUS} is unsymmetric.
(c) Y_{BUS} is unsymmetric and Z_{BUS} is symmetric.
(d) Both Y_{BUS} and Z_{BUS} are unsymmetric.

[2017 : 1 M, Set-2]

9.30 A 1000 × 1000 bus admittance matrix for an electric power system has 8000 non-zero elements. The minimum number of branches (transmission lines and transformers) in this system are _____ (upto 2 decimal places). [2018 : 1 M]

9.31 The per unit power output of a salient pole generator which is connected to an infinite bus, is given by the expression, $P = 1.4 \sin\delta + 0.15 \sin 2\delta$, where δ is the load angle. Newton-Raphson method is used to calculate the value of δ for $P = 0.8$ pu. If the initial guess is 30°, then its value (in degree) at the end of the first iteration is

(a) 15° (b) 28.48°
(c) 28.74° (d) 31.20° [2018 : 2 M]

9.32 The Y_{bus} matrix of a two-bus power system having two identical parallel lines connected between them in pu is given as,

$$Y_{bus} = \begin{bmatrix} -j8 & j20 \\ j20 & -j8 \end{bmatrix}$$

The magnitude of the series reactance of each line in pu (round off up to one decimal place) is _______ . [2019 : 1 M]

9.33 Out of the following options, the most relevant information needed to specify the real power (P) at the PV buses in a load flow analysis is

(a) solution of economic load dispatch
(b) base power of the generator
(c) rated power output of the generator
(d) rated voltage of the generator [2020 : 1 M]

9.34 Two buses, i and j, are connected with a transmission line of admittance Y, at the two ends of which there are ideal transformers with turns ratios as shown. Bus admittance matrix for the system is:

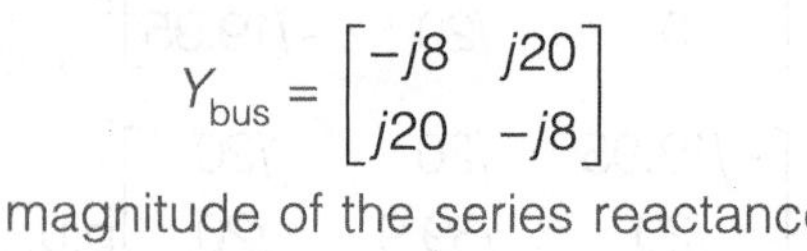

(a) $\begin{bmatrix} -t_it_jY & t_j^2Y \\ t_i^2Y & -t_it_jY \end{bmatrix}$ (b) $\begin{bmatrix} t_it_jY & -t_j^2Y \\ -t_i^2Y & t_it_jY \end{bmatrix}$

(c) $\begin{bmatrix} t_it_jY & -(t_i - t_j)^2Y \\ -(t_i - t_j)^2Y & t_it_jY \end{bmatrix}$

(d) $\begin{bmatrix} t_i^2Y & -t_it_jY \\ -t_it_jY & t_j^2Y \end{bmatrix}$ [2020 : 2 M]

9.35 Consider a power system consisting of N number of buses. Buses in this power system are categorized into slack bus, PV buses and PQ buses for load flow study. The number of PQ buses is N_L. The balanced Newton-Raphson method is used to carry out load flow study in polar form. H, S, M, and R are sub-matrices of the Jacobian matrix J as shown below:

$$\begin{bmatrix} \Delta P \\ \Delta Q \end{bmatrix} = J \begin{bmatrix} \Delta\delta \\ \Delta V \end{bmatrix}, \text{ where } J = \begin{bmatrix} H & S \\ M & R \end{bmatrix}$$

The dimensions of the sub-matrix is

(a) $N_L \times (N - 1 + N_L)$
(b) $N_L \times (N - 1)$
(c) $(N - 1) \times (N - 1 + N_L)$
(d) $(N - 1) \times (N - 1 - N_L)$ [2021 : 1 M]

9.36 A 3 Bus network is shown. Consider generators as ideal voltage sources. If rows 1, 2 and 3 of the Y_{Bus} matrix correspond to Bus 1, 2 and 3 respectively, then Y_{Bus} of the network is

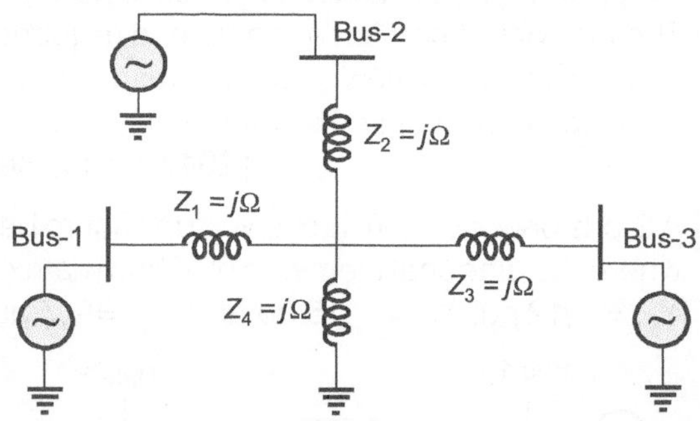

(a) $\begin{bmatrix} -\frac{1}{2}j & \frac{1}{4}j & \frac{1}{4}j \\ \frac{1}{4}j & -\frac{1}{2}j & \frac{1}{4}j \\ \frac{1}{4}j & \frac{1}{4}j & -\frac{1}{2}j \end{bmatrix}$

(b) $\begin{bmatrix} -\frac{3}{4}j & \frac{1}{4}j & \frac{1}{4}j \\ \frac{1}{4}j & -\frac{3}{4}j & \frac{1}{4}j \\ \frac{1}{4}j & \frac{1}{4}j & -\frac{3}{4}j \end{bmatrix}$

(c) $\begin{bmatrix} -4j & j & j \\ j & -4j & j \\ j & j & -4j \end{bmatrix}$

(d) $\begin{bmatrix} -4j & 2j & 2j \\ 2j & -4j & 2j \\ 2j & 2j & -4j \end{bmatrix}$ [2021 : 2 M]

9.37 The bus admittance (Y_{bus}) matrix of a 3-bus power system is given below.

$$\begin{array}{c|ccc} & 1 & 2 & 3 \\ \hline 1 & -j15 & j10 & j5 \\ 2 & j10 & -j13.5 & j4 \\ 3 & j5 & j4 & -j8 \end{array}$$

Considering that there is no shunt inductor connected to any of the buses, which of the following can NOT be true?

(a) Line charging capacitor of finite value is present in all three lines.
(b) Line charging capacitor of finite value is present in line 2-3 only.
(c) Line charging capacitor of finite value is present in line 2-3 only and shunt capacitor of finite value is present in bus 1 only.
(d) Line charging capacitor of finite value is present in line 2-3 only and shunt capacitor of finite value is present in bus 3 only.

[2023 : 1 M]

9.38 The three-bus power system shown in the figure has one alternator connected to bus 2 which supplies 200 MW and 40 MVAr power. Bus 3 is infinite bus having a voltage of magnitude $|V_3| = 1.0$ p.u. and angle of $-15°$. A variable current source, $|I|\angle\phi$ is connected at bus 1 and controlled such that the magnitude of the bus 1 voltage is maintained at 1.05 p.u. and the phase angle of the source current, $\phi = \theta_1 \pm \frac{\pi}{2}$, where θ_1 is the phase angle of the bus 1 voltage. The three buses can be categorized for load flow analysis as

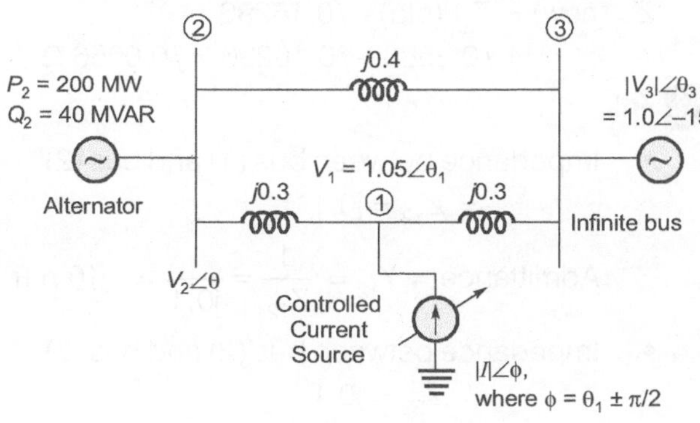

(a) Bus 1 Slack Bus
Bus 2 $P-|V|$ Bus
Bus 3 $P-Q$ Bus

(b) Bus 1 $P-|V|$ Bus
Bus 2 $P-|V|$ Bus
Bus 3 Slack Bus

(c) Bus 1 $P-Q$ Bus
Bus 2 $P-Q$ Bus
Bus 3 Slack Bus

(d) Bus 1 $P-|V|$ Bus
Bus 2 $P-Q$ Bus
Bus 3 Slack Bus

[2023 : 2 M]

9.39 The figure shows the single line diagram of a 4-bus power network, Branches b_1, b_2, b_3 and b_4 have impedances $4z$, z, $2z$ and $4z$ per-unit (pu), respectively, where $z = r + jx$, with $r > 0$ and $x > 0$. The current drawn from each load bus (marked as arrows) is equal to I pu, where $I \neq 0$. If the network is to operate with minimum loss, the branch that should be opened is

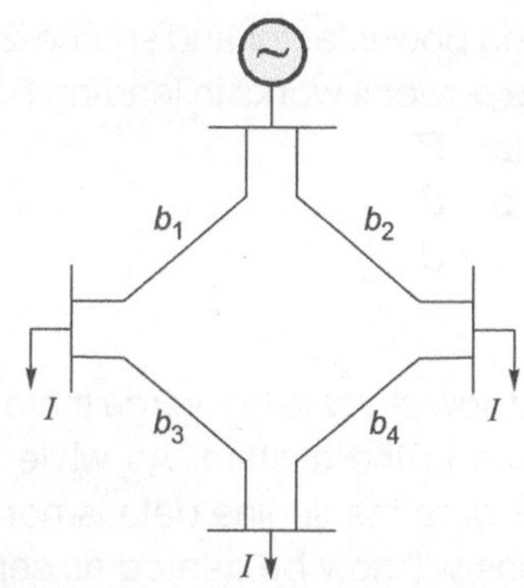

(a) b_3
(b) b_1
(c) b_4
(d) b_2

[2024 : 1 M]

Answers Load Flow Studies

9.1 (T)	9.2 (a)	9.3 (Sol.)	9.4 (d)	9.5 (d)	9.6 (b)	9.7 (d)
9.8 (d)	9.9 (d)	9.10 (c)	9.11 (a)	9.12 (c)	9.13 (a)	9.14 (b)
9.15 (b)	9.16 (b)	9.17 (b)	9.18 (c)	9.19 (b)	9.20 (332)	9.21 (b)
9.22 (d)	9.23 (b)	9.24 (a)	9.25 (14)	9.26 (b)	9.27 (c)	9.28 (61)
9.29 (d)	9.30 (3500)	9.31 (c)	9.32 (0.1)	9.33 (a)	9.34 (d)	9.35 (b)
9.36 (b)	9.37 (a, c)	9.38 (d)	9.39 (a)			

Explanations Load Flow Studies

9.2 (a)

The system losses are a function of system properties which are independent of choosing a type of bus. Where as the complex bus voltage is a function of parameters which are being calculated.

9.3 Sol.

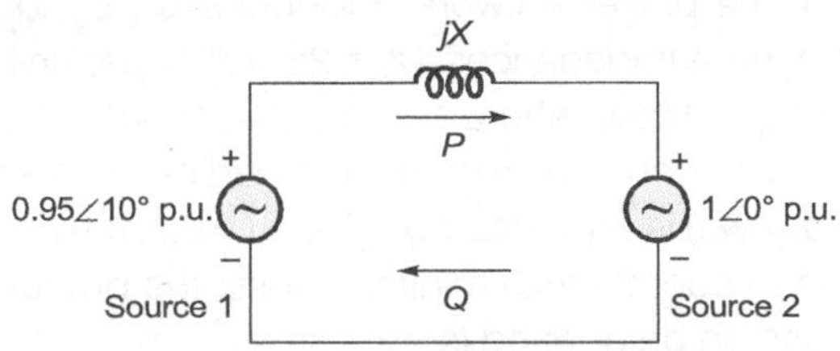

Since source-1 is drawing reactive power it works in lagging power factor and source-2 is supplying reactive power it works in leading power factor.

a – P
b – S
c – U

9.4 (d)

For load flow study to converge there must be one slack bus in the system. As while entering the network data the tie-line data is not considered, two areas will now be treated as separate areas hence two slack bus (one in each area) is required.

9.5 (d)

Total buses = 300

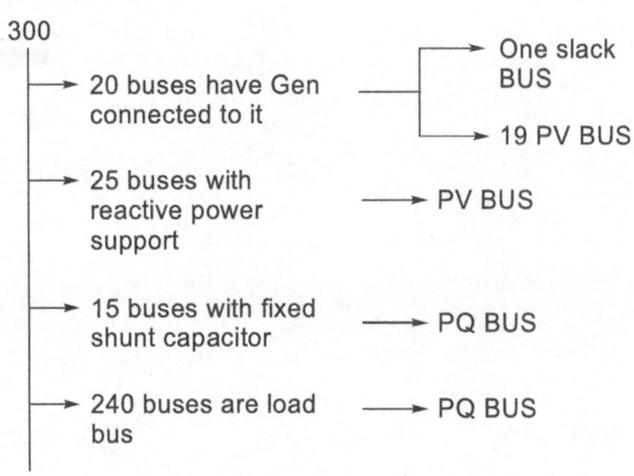

Size of jacobian matrix

$$= \text{No. of unknown bus voltage}$$
$$= 2 \times 300 - 2 - 19 - 25 = 554$$

9.6 (b)

$$Z_B(\text{new}) = Z_B(\text{old}) - \frac{1}{Z_{jj} + Z_b}\begin{bmatrix} Z_{1j} \\ \vdots \\ Z_{nj} \end{bmatrix}[Z_{j1} \,..\, Z_{jn}]$$

New elements (Z_b) is connected between j^{th} and reference bus.

Here, $j = 2$, $b = 4$

$$\therefore \frac{1}{Z_{ij} + Z_b}\begin{bmatrix} Z_{12} \\ Z_{22} \\ Z_{32} \\ Z_{42} \end{bmatrix}[Z_{21} \quad Z_{22} \quad Z_{23} \quad Z_{24}]$$

$$= \frac{1}{(j\,0.3408 + j\,0.2)}\begin{bmatrix} j\,0.2860 \\ j\,0.3408 \\ j\,0.2586 \\ j\,0.2414 \end{bmatrix} \times$$

$$[j\,0.2860 \quad j\,0.3408 \quad j\,0.2586 \quad j\,0.2414]$$

We are required only changes in Z_{22} and Z_{23}

$$\begin{bmatrix} - & - & - & - \\ - & j\,0.2147 & j\,0.16296 & - \\ - & - & - & - \\ - & - & - & - \end{bmatrix}$$

$$Z_{22}\,(\text{new}) = Z_{22}\,(\text{old}) - j\,0.2147 = j\,0.126$$
$$Z_{23}\,(\text{new}) = Z_{23}\,(\text{old}) - j\,0.16296$$
$$= j\,0.2586 - j\,0.16296 = j\,0.0956\ \Omega$$

9.7 (d)

- Impedance between bus (1) and bus (2)

$$= Z_{12} = j0.1$$

$$\text{Admittance} = Y_{12} = \frac{1}{Z_{12}} = \frac{1}{j0.1} = -j10 \text{ p.u.}$$

- Impedance between bus (2) and bus (3)

$$= Z_{23} = j0.1$$

$$\text{Admittance} = Y_{23} = \frac{1}{Z_{23}} = \frac{1}{j0.1} = -j10 \text{ p.u.}$$

- Impedance between bus (2) and ground

$$= Z_{20} = -j20 \text{ p.u.}$$

$$\text{Admittance} = Y_{20} = \frac{1}{Z_{20}} = \frac{1}{-j20} = j0.05 \text{ p.u.}$$

$$Y_{22} = Y_{20} + Y_{12} + Y_{23}$$
$$= j0.05 - j10 - j10 = -j19.95 \text{ p.u.}$$

9.8 (d)

The G.S. method has following disadvantage:
- unreliable convergence
- slow convergence
- sensitivity to the choice of slack bus

So the correct solution is (d).

9.9 (d)

$$\text{Pre fault } V_{BUS} = \begin{bmatrix} V_1^0 \\ V_2^0 \\ V_3^0 \end{bmatrix} = \begin{bmatrix} 1\angle 0° \\ 1\angle 0° \\ 1\angle 0° \end{bmatrix}$$

$$\text{Fault current, } I^f = \frac{V_r^0}{Z_{rr} + Z_f}$$

Solid 3-ϕ fault occurs at bus 2, $r = 2$ and

$$Z_F = 0$$
$$Z_{rr} = Z_{22} = j0.24,$$
$$V_r^0 = V_1^0 = 1\angle 0°$$

$$I^f = \frac{V_2^0}{Z_{22} + Z_f} = \frac{1\angle 0°}{j0.24 + 0} = -j4.167 \text{ p.u.}$$

Post fault voltage at bus i is given by

$$V_i^f = V_i^0 - Z_{ir} I^f$$

Post fault voltage at bus, $i = 1$

$$V_1^f = V_1^0 - Z_{12} I^f$$
$$= 1\angle 0° - j0.08 \times (-j4.167) = 0.67 \text{ p.u.}$$

Post fault voltage at bus, $i = 3$

$$V_3^f = V_3^0 - Z_{32} I^f$$
$$= 1\angle 0° - j0.16 \times (-j4.167) \approx 0.33 \text{ p.u.}$$

9.10 (c)

Y_{ii} = Sum of the admittance directly connected to i^{th} bus

$$Y_{iL} = y_{i0} + y_{i1} \dots y_{in} \quad (y_{ii} \text{ excluded})$$
$$Y_{ik} = -y_{ik}$$

So,
$$Y_{11} = y_{10} + y_{12} + y_{13}$$
$$Y_{12} = -y_{12}$$
$$Y_{13} = -y_{13}$$
$$y_{10} = Y_{11} + Y_{12} + y_{13}$$
$$= -j8.75 + j1.25 + j2.5$$
$$\Rightarrow \quad y_{10} = -j5 \text{ p.u.}$$

Similarly,

$$y_{20} = Y_{22} + Y_{21} + Y_{23}$$
$$y_{20} = -j6.25 + j1.25 + j2.5$$
$$\Rightarrow \quad y_{20} = -j2.5 \text{ p.u.}$$

$$Z_{10} = \frac{1}{y_{10}} = \frac{1}{-j5} = j0.2$$

$$Z_{20} = \frac{1}{y_{20}} = \frac{1}{-j2.5} = j0.4$$

$$I_{Gi}^f = \frac{E_{Gi} - V_i^f}{Z_{i0}}$$

[Current fed by generator i during fault]

$$E_{G1} = E_{G2} = 1\angle 0°$$

[The system was unloaded]

Current fed by generator 1

$$I_{G1}^f = \frac{E_{G1} - V_1^f}{Z_{10}} = \frac{1\angle 0° - 0.67\angle 0°}{j0.2}$$
$$= 1.65\angle -90° \text{ p.u.}$$

Current fed by generator 2,

$$I_{G2}^f = \frac{E_{G2} - V_2^f}{Z_{20}} = \frac{1\angle 0° - 0}{j0.4}$$
$$= 2.5\angle -90° \text{ p.u.}$$

9.11 (a)

Flow of power in transmission line depends on bus voltage magnitude and phase difference between bus voltages.

After interconnecting the two system by a line there is no power flow in the line hence bus voltage magnitude and phase difference between buses will remain same as before interconnecting the two systems.

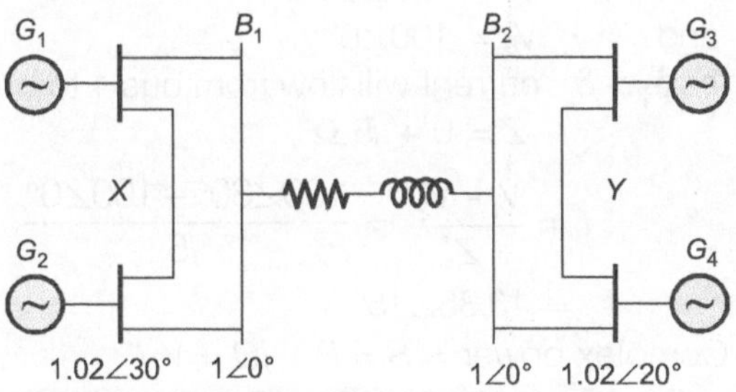

Phase difference between bus X and bus Y

$$= 30 - 20 = 10°$$

9.12 (c)

$$Y_{BUS} = \begin{bmatrix} Y_{11} & Y_{12} & Y_{13} & Y_{14} \\ Y_{21} & Y_{22} & Y_{23} & Y_{24} \\ Y_{31} & Y_{32} & Y_{33} & Y_{34} \\ Y_{41} & Y_{42} & Y_{43} & Y_{44} \end{bmatrix}$$

Here,

$$Y_{11} = y_{10} + y_{12} + y_{13} + y_{14} = -5j \quad \text{... (i)}$$
$$Y_{22} = y_{20} + y_{22} + y_{23} + y_{24} = -10j \quad \text{... (ii)}$$
$$Y_{33} = y_{30} + y_{32} + y_{33} + y_{34} = -9j \quad \text{... (iii)}$$
$$Y_{44} = y_{40} + y_{42} + y_{43} + y_{44} = -8j \quad \text{... (iv)}$$
$$Y_{12} = Y_{21} = -y_{12} = 2j$$
$$Y_{13} = Y_{31} = -y_{13} = 2.5j$$
$$Y_{14} = Y_{41} = -y_{14} = 0j$$
$$Y_{23} = Y_{32} = -y_{23} = 2.5j$$
$$Y_{24} = Y_{42} = -y_{24} = 4j$$
$$Y_{34} = Y_{43} = -y_{34} = 4j$$

From equation (i), we get

$$y_{10} = Y_{11} - Y_{12} - y_{13} - y_{14}$$
$$= -5j + 2j + 2.5j + 0j = -0.5j$$

From equation (ii), we get

$$y_{20} = Y_{22} - y_{12} - y_{23} - y_{24}$$
$$= -10j + 2j + 2.5j + 4j = -1.5j$$

From equation (iii), we get

$$y_{30} = Y_{33} - y_{31} - y_{32} - y_{34}$$
$$= -9j + 2.5j + 2.5j + 4j = 0$$

From equation (iv), we get

$$y_{40} = Y_{44} - y_{41} - y_{42} - y_{43}$$
$$= -8j - 0 + 4j + 4j = 0$$

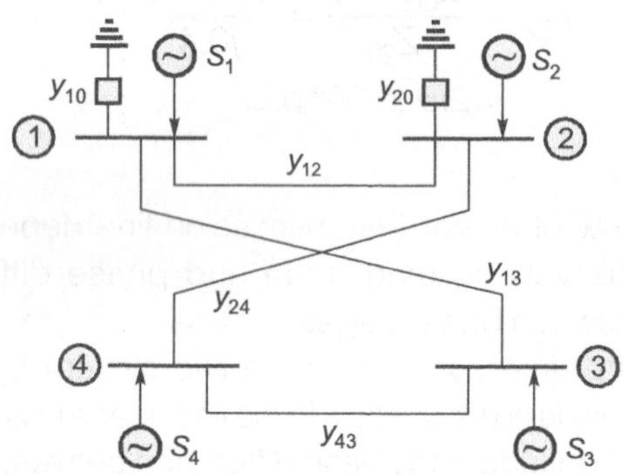

9.13 (a)

$$V_1 = 100\angle 30° \text{ V}$$
and $$V_2 = 100\angle 0° \text{ V}$$
as $\delta_1 > \delta_2$, current will flow from bus 1 to bus 2

$$Z = 0 + j5\ \Omega$$
$$I = \frac{V_1 - V_2}{Z} = \frac{100\angle 30° - 100\angle 0°}{j5}$$
$$= 10.35\angle 15° \text{ A}$$

Complex power $= S = P + jQ = V_1 I^*$

$$P + jQ = 100\angle 30° \times (10.35\angle 15°)^* \approx 1000 + j268$$
$$P = 1000 \text{ W and } Q = 268 \text{ VAR}$$

9.14 (b)

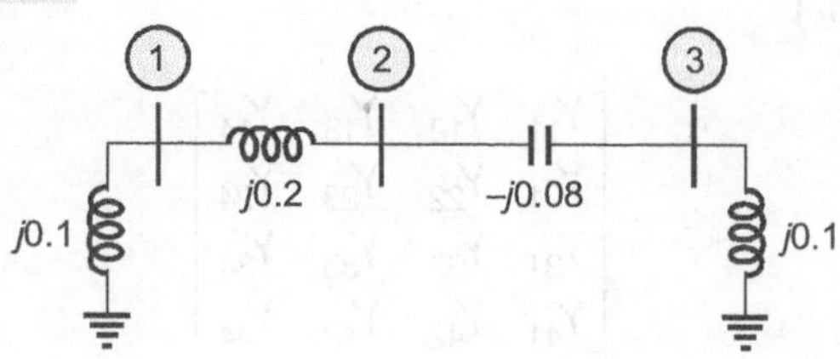

$$Z_{10} = j0.1$$
$$\Rightarrow y_{10} = \frac{1}{Z_{10}} = \frac{1}{j0.1} = -j10 \text{ p.u.}$$
$$Z_{12} = j0.2$$
$$\Rightarrow y_{12} = \frac{1}{Z_{12}} = \frac{1}{j0.2} = -j5 \text{ p.u.}$$
$$Z_{23} = -j0.8$$
$$\Rightarrow y_{23} = \frac{1}{Z_{23}} = \frac{1}{-j0.08} = j12.5 \text{ p.u.}$$
$$Z_{30} = j0.1$$
$$\Rightarrow y_{30} = \frac{1}{Z_{30}} = \frac{1}{j0.1} = -j10 \text{ p.u.}$$
$$Y_{11} = y10 + y_{12} = -j10 - j5 = -j15 \text{ p.u.}$$
$$Y_{12} = Y_{21} = -y_{12} = -(-j5) = j5 \text{ p.u.}$$
$$Y_{31} = Y_{13} = -y_{13} = 0$$
$$Y_{22} = y_{12} + y_{23} = -j5 + j12.5 = j7.5 \text{ p.u.}$$
$$Y_{33} = y_{30} + y_{23} = -j10 + j12.5 = j2.5 \text{ p.u.}$$
$$Y_{BUS} = j\begin{bmatrix} -15 & 5 & 0 \\ 5 & 7.5 & -12.5 \\ 0 & -12.5 & 2.5 \end{bmatrix}$$

9.15 (b)

y_{ik} = Series admittance of the line connecting buses i and k

$\frac{y'_{ik}}{2}$ = Half line charging admittance in bus admittance matrix

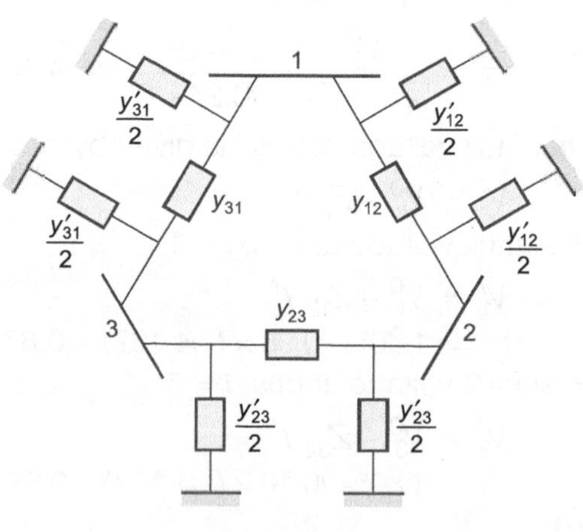

$$Y_{11} = \frac{y'_{12}}{2} + \frac{y'_{31}}{2} + y_{12} + y_{31} = -j13$$
$$y_{12} = -y_{12} = \text{j}10$$
$$y_{23} = -y_{23} = \text{j}10; \quad y_{31} = -y_{31} = \text{j}5$$
$$\therefore y'_{12} + y'_{31} = 2[-j13 + j10 + j5] = j4$$

Similarly,

$$y'_{12} + y'_{23} = 2[-j18 + j10 + j10] = j4$$
$$y'_{23} + y'_{31} = 2[-j13 + j10 + j5] = j4$$
$$\Rightarrow \quad y'_{12} = j6 - j4 = j2$$

9.16 (b)

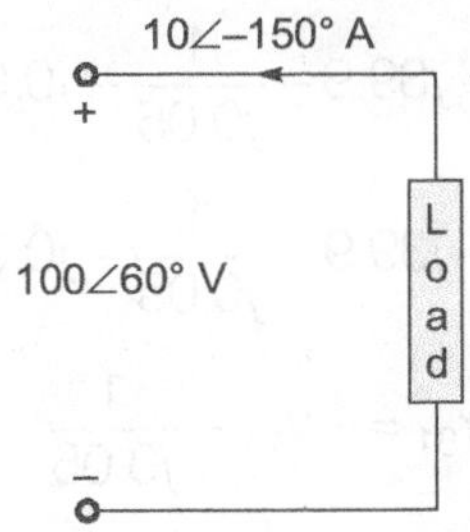

Complex power supplied by load

$= (100\angle 60°)(10\angle -150°)^* = -866.025 - j500$ VA

As supplied active and reactive power are negative. Load absorbs active and reactive power both.

9.17 (b)

$$P_1 + P_2 + P_3 = 0$$
$$P_1 + j0.1 - j0.2 = 0$$
$$P_1 = j0.1 \text{ p.u}$$

No power flow in between bus 1 and 2.

$$P_{12} \propto \sin(\theta_1 - \theta_2) = 0$$
$$\theta_1 - \theta_2 = 0$$

As bus 1 is a slack bus

$$\theta_1 = 0 \Rightarrow \theta_2 = 0$$

Now, $$P_{23} = \frac{V_2 V_3}{X}\sin(\theta_2 - \theta_3)$$

or $$0.1 = \frac{1\times 1}{1}\sin(\theta_2 - \theta_3)$$

$$\Rightarrow \quad \theta_2 - \theta_3 = \sin^{-1}(0.1) = 0.10016 \simeq 0.1$$
$$\theta_3 = -0.1$$

9.18 (c)

$$P_1 = 0.1 \text{ pu} = 0.1\times\frac{(100\text{ kV})^2}{100} = 10\text{ MW}$$

Note: IIT Bombay has given marks to all students in questions **9.19** and **9.20**.

9.19 (b)

$$P_2 + jQ_2 = (1.0 + j0.5) \text{ p.u.}$$
$$P_2 = 1 = \frac{|V_1||V_2|}{|X|}\sin(\delta_1 - \delta_2)$$
$$1 = \frac{V_1}{0.1}\sin(-\delta_2)$$
$$-\frac{0.1}{\sin\delta_2} = V_1 \quad \text{...(i)}$$
$$Q_2 = 0.5 = \frac{|V_2|}{|X|}(|V_1|\cos\delta_2 - |V_2|)$$
$$0.5 = \frac{1}{0.1}(V_1\cos\delta_2 - 1)$$
$$\frac{1.05}{\cos\delta_2} = V_1 \quad \text{...(ii)}$$

From equation (i) and (ii), we get

$$\tan\delta_2 = \frac{-0.1}{1.05} \Rightarrow \delta_2 = -5.44°$$

Substituting δ_2 value in equation (ii), we get

$$V_1 = 1.05 \text{ p.u.}$$

9.20 Sol.

$$n = 183$$
$$\text{No. of equations} = 2n - m - 2$$
$$= 2\times 183 - 32 - 2 = 332$$

where, n = Total number of buses

m = PV bus or reactive power support bus excluding slack bus

One bus already has taken as stack bus.

9.21 (b)

$$Y_{BUS} = \begin{bmatrix} y_{11}+y_{12}+y_{13} & -y_{12} & -y_{13} \\ -y_{21} & y_{21}+y_{22}+y_{23} & -y_{23} \\ -y_{31} & -y_{32} & y_{31}+y_{32}+y_{33} \end{bmatrix}$$

$$= \begin{bmatrix} -j6 & j3 & j4 \\ j3 & -j7 & j5 \\ j4 & j5 & -j8 \end{bmatrix}$$

By comparing for Y_{33} element

$$y_{31} + y_{32} + y_{30} = -j8$$
$$(-j4) + (-j5) + y_{30} = -j8$$
$$y_{30} = j1$$

After compensating with 50% of shunt capacitance $= j1 \times 0.5 = j0.5$

Hence, new value $= -j8 - j0.5 = -j8.5$

9.22 (d)

FDLF method gives approximate solution not because it uses several assumptions but due to the fixed number of iterations.

9.23 (b)

The voltage angle is the common variable for both *PV* and *PQ* buses so it remains unchanged. Where as voltage magnitude is specified in *PV* bus and not specified in *PQ* bus, so it increases by two.

9.24 (a)

Size of the Jacobian matrix is,

$$(2n - m - 1)\times(2n - m - 1)$$

Given that 10 generator buses, we need to assume with in the 10 buses one bus as slack bus then,

$$(2\times 100 - 10 - 1)\times(2\times 100 - 10 - 1) = 189\times 189$$

9.25 Sol.

Total number of buses ≡ 10

G_1 ≡ slack bus

G_2 ≡ PQ bus (reactive power limit is reached)

$\left.\begin{matrix} G_3 \\ G_4 \end{matrix}\right\}$ PV bus

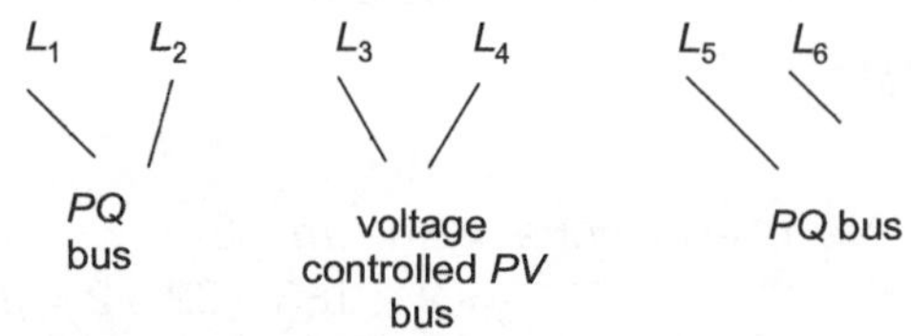

Minimum number of non linear equations to be solved

= Number of unknown bus voltage variables

= (2 × 10 – 4 – 2) = 14

9.26 (b)

Given : $Y_{11} = -j12$ p.u.; $Y_{22} = -j15$ p.u.

$Y_{33} = -j7$ p.u.

We know that,

$$Y_{11} = y_{12} + y_{13} = -j12 \text{ p.u} \quad ...(i)$$

$$Y_{22} = y_{12} + y_{23} = -j15 \text{ p.u} \quad ...(ii)$$

$$Y_{33} = y_{13} + y_{23} = -j7 \text{ p.u} \quad ...(iii)$$

From equation (i) and (ii),

$$y_{13} - y_{23} = j3 \text{ p.u}$$

$$y_{13} + y_{23} = -j7 \text{ p.u.} \quad ...(iv)$$

$$y_{13} = -j2 \text{ p.u}$$

$$y_{23} = -j5 \text{ p.u}$$

$$y_{12} = -j10 \text{ p.u}$$

The p.u. values of line reactances p, q and r are

$$jr = \frac{1}{-j2} = \text{j}0.5 \text{ p.u.}$$

$$jp = \frac{1}{-j5} = \text{j } 0.2 \text{ p.u.}$$

$$jq = \frac{1}{-j10} = \text{j } 0.1 \text{ p.u.}$$

∴ $p = 0.2$, $q = 0.1$, $r = 0.5$

9.27 (c)

$$Y_{bus} = \begin{bmatrix} -j39.9 & j20 & j20 \\ j20 & -j39.9 & j20 \\ j20 & j20 & -j39.9 \end{bmatrix}$$

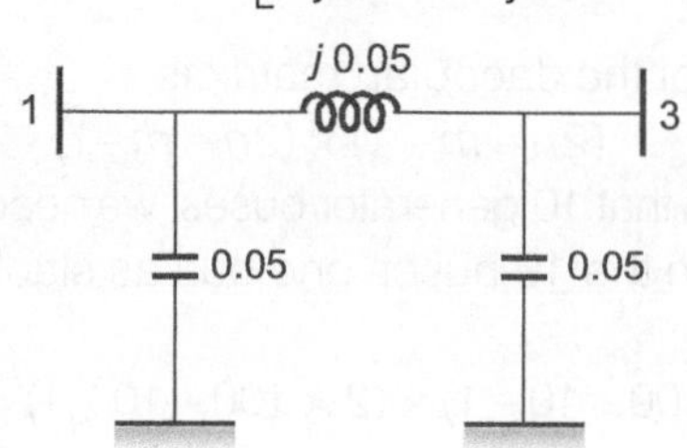

Reactance is eliminated in between Bus 1 and 3.

$$Y_{11\,(new)} = -j39.9 - \frac{1}{j0.05} - j0.05 = -j19.95$$

$$Y_{33\,(new)} = -j39.9 - \frac{1}{j0.05} - j0.05 = -j19.95$$

$$Y_{13\,(new)} = Y_{31} = -y_{13} - \frac{1}{j0.05}$$

$$Y_{31\,(new)} = -j20 - \frac{1}{j0.05} = 0$$

So, modified admittance matrix is,

$$= \begin{bmatrix} -j19.95 & j20 & 0 \\ j20 & -j39.9 & j20 \\ 0 & j20 & -j19.95 \end{bmatrix} \text{pu}$$

9.28 Sol.

$$[J] = [2n - m - 2]$$

$$100 = [2n - 20 - 2]$$

$$n = \frac{100 + 22}{2} = 61$$

Total number of buses = n = 61

9.29 (d)

Both Y_{BUS} and Z_{BUS} are unsymmetrical with transformer.

9.30 Sol.

Number of buses = 1000

No. of non-zero element = 8000

% sparsity = %x

$$= \frac{\text{No. of zero element}}{\text{Total no. of elements}} = \frac{10^6 - 8000}{10^6} = 0.992$$

$$\text{No. of Tr-lines} = \frac{N^2(1-x) - N}{2}$$

$$= \frac{1000^2(1-0.992) - 1000}{2} = 3500$$

9.31 (c)

$$P(\delta) = 1.4 \sin\delta + 0.15 \sin 2\delta = 0.8$$

$$= 1.4 \sin\delta + 0.15 \sin 2\delta - 0.8 = 0$$

$$P'(\delta) = \frac{d}{d\delta}(P(\delta)) = 1.4 \cos\delta + 0.30 \cos 2\delta$$

Given, $\delta_0 = 30°$

By using Newton Raphson method for single variable,

$$\Delta\delta = \delta_1 - \delta_0 = -\frac{f(\delta_0)}{f'(\delta_0)} = -\frac{P(\delta_0)}{P'(\delta_0)}$$

$$\delta_1 - 30° = -\frac{1.4 \sin 30° + 0.15 \sin 60° - 0.8}{1.4 \cos 30° + 0.3 \cos 60°}$$

$$\delta_1 - 30° = -0.0219 \text{ rad} = -1.26°$$
$$\delta_1 = 28.74°$$

9.32 (0.1)

$$Y_{12} = -(y_{12}) = -j20$$

Series admittance of each line

$$= \frac{Y_{12}}{2} = \frac{-j20}{2} = -j10$$

$$\text{Series reactance of each line} = \frac{1}{-j10} = j0.1 \text{p.u.}$$

9.34 (d)

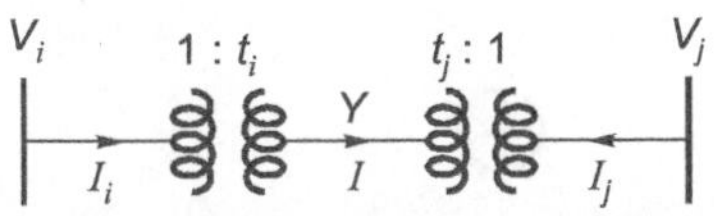

$$I = Y(t_iV_i - V_jt_j)$$
$$I_i = t_iI = t_i^2YV_i - t_it_jYV_j$$
$$I_j = -t_jI = -I_it_jYV_i + t_j^2YV_j$$

$$\begin{bmatrix} I_i \\ I_j \end{bmatrix} = \begin{bmatrix} t_i^2Y & -t_it_jY \\ -t_it_jY & t_j^2Y \end{bmatrix}\begin{bmatrix} V_i \\ V_j \end{bmatrix}$$

9.35 (b)

$$\begin{bmatrix} \Delta P \\ \Delta Q \end{bmatrix} = J\begin{bmatrix} \Delta\delta \\ \Delta V \end{bmatrix} \text{ where } J = \begin{bmatrix} H & S \\ M & R \end{bmatrix}$$

For size of M

Row = No. of unknown variables of $Q = N_L$

Column = No. of variable which has $\delta = N_L + (N - 1 - N_L) = N - 1$

So, size of $M = N_L \times (N - 1)$

9.36 (b)

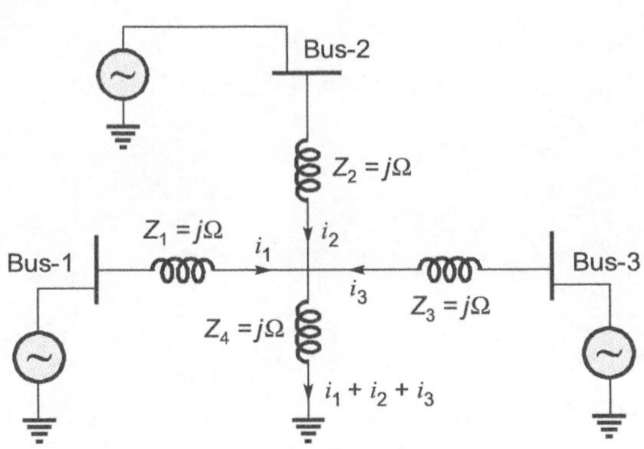

$$V_1 = j1i_1 + j1(i_1 + i_2 + i_3)$$
$$V_1 = 2ji_1 + ji_2 + ji_3$$

and
$$V_2 = ji_1 + 2ji_2 + ji_3$$
$$V_3 = ji_1 + ji_2 + 2ji_3$$

$$\therefore \quad \begin{bmatrix} V_1 \\ V_2 \\ V_3 \end{bmatrix} = \begin{bmatrix} 2j & j & j \\ j & 2j & j \\ j & j & 2j \end{bmatrix}\begin{bmatrix} i_1 \\ i_2 \\ i_3 \end{bmatrix}$$

$$Z_{bus} = \begin{bmatrix} 2j & j & j \\ j & 2j & j \\ j & j & 2j \end{bmatrix}$$

$$Y_{bus} = \begin{bmatrix} \frac{-3j}{4} & \frac{j}{4} & \frac{j}{4} \\ \frac{j}{4} & \frac{-3j}{4} & \frac{j}{4} \\ \frac{j}{4} & \frac{j}{4} & \frac{-3j}{4} \end{bmatrix}$$

9.37 (a, c)

$$[Y_{Bus}] = j\begin{bmatrix} -15 & 10 & 5 \\ 10 & -13.5 & 4 \\ 5 & 4 & -8 \end{bmatrix} \begin{matrix} \text{Sum} \\ = 0 \\ = j0.5 \\ = j1 \end{matrix}$$

No shunt branch is present at 1st bus.

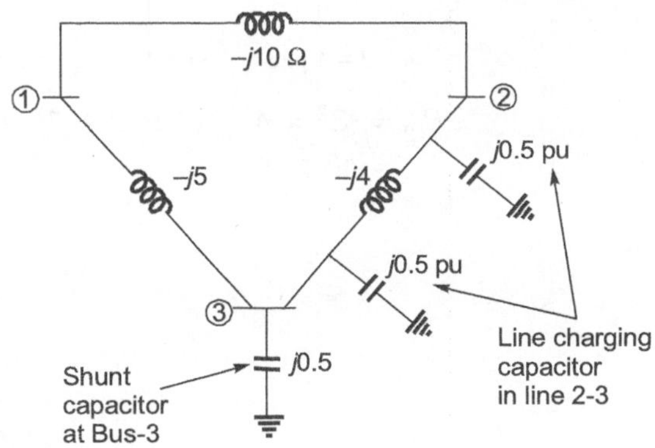

9.38 (d)

At Bus (1), voltage magnitude is maintained also specified active power is zero. So, it is a PV Bus.

At Bus (2), P_G and Q_G are specified and $|V|$, δ are unknown. Hence, it is PQ Bus.

At Bus (3), $|V|$ and δ are specified. So it is slack bus.

9.39 (a)

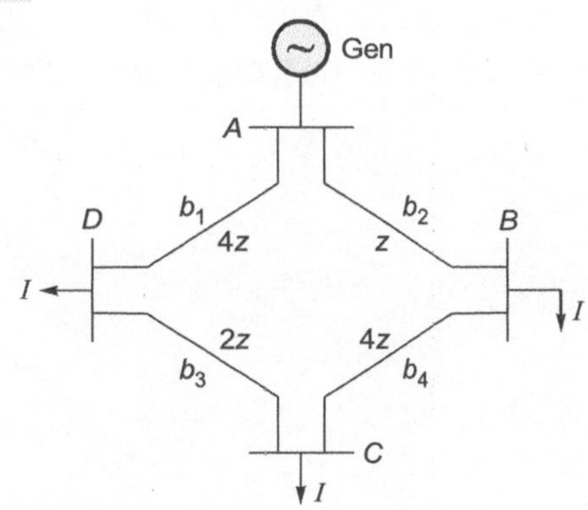

Assume, $|z| = |r + jX| = 1$ p.u.,

$I = 1$ p.u.

By opening branches ring becomes radial

open b_1 :

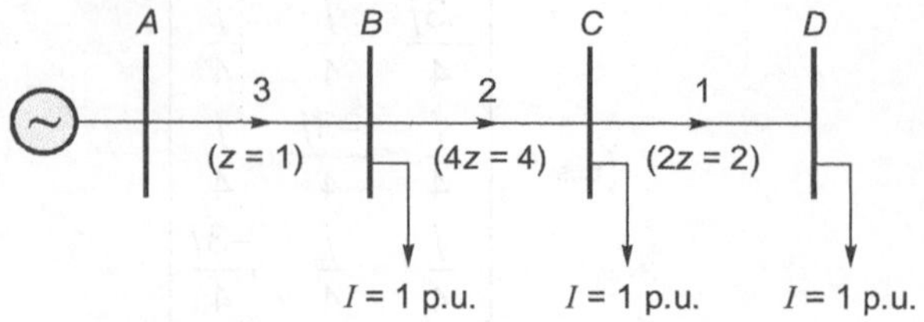

Total power loss,

$$P_{L1} = 1^2 \times 2 + 2^2 \times 4 + 3^2 \times 1$$
$$= 27 \text{ p.u.}$$

Open b_2 :

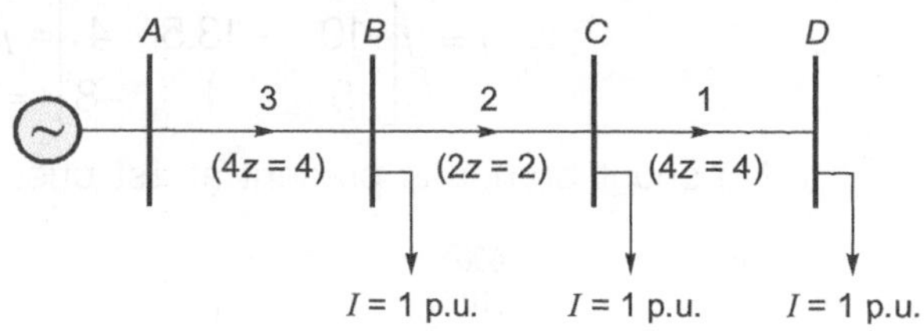

$$P_{L2} = (3^2 \times 4 + 2^2 \times 2 + 1^2 \times 4)$$
$$= 48 \text{ p.u.}$$

Open b_3 :

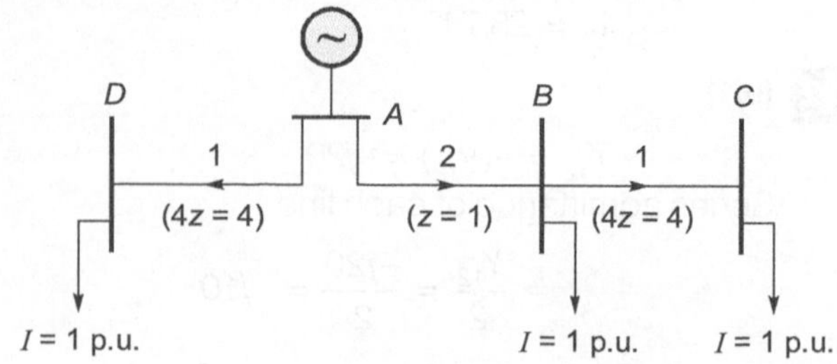

$$P_{L3} = (2^2 \times 1 + 1^2 \times 4 + 1^2 \times 4)$$
$$= 12 \text{ p.u.} \leftarrow \text{mininum power loss}$$

Open b_4 :

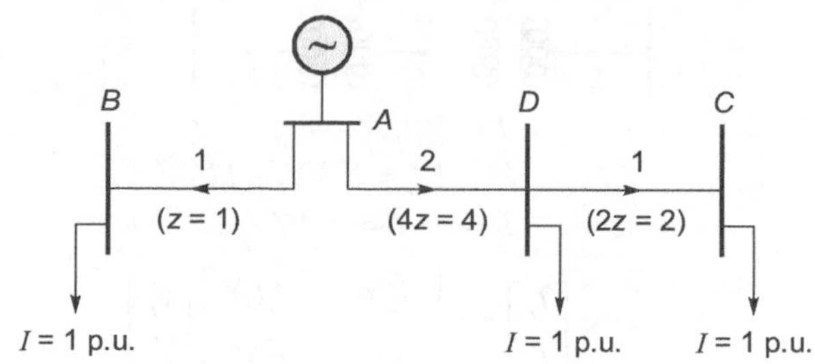

$$P_{L4} = (1^2 \times 1 + 2^2 \times 4 + 1^2 \times 2)$$
$$= 19 \text{ p.u.}$$

By opening branch b_3 Power loss is minimum.

■■■■

10 High Voltage DC Transmission

10.1 Bulk power transmission over long HVDC lines are preferred, on account of
(a) low cost of HVDC terminals
(b) no harmonics problems
(c) minimum line power losses
(d) simple protection **[1998 : 1 M]**

10.2 Out of the considerations (i) to (iv) listed below:
(i) No distance limitation related to steady state stability.
(ii) No reactive power requirement from the system at the two terminals.
(iii) No substantial effect on fault level of the two systems at the terminals inspite of the inter connection.
(iv) No corona problems.
The considerations which constitute advantages of HVDC transmission are
(a) All of the above (b) (i) and (iii)
(c) (iii) and (iv) (d) (ii) and (iv)
[2000 : 1 M]

10.3 Choose the appropriate auxiliary components of HVDC transmission system from the following
1. D.C. line inductor
2. A.C. line inductor
3. Reactive power sources
4. Distance relays on D.C. line
5. Series capacitance on A.C. line
(a) 1 and 2 (b) 1 and 3
(c) 2 and 4 (d) 4 and 5 **[2003 : 1 M]**

10.4 High Voltage DC (HVDC) transmission is mainly used for
(a) bulk power transmission over very long distances
(b) inter-connecting two systems with the same nominal frequency
(c) eliminating reactive power requirement in the operation
(d) minimizing harmonics at the converter stations
[2005 : 1 M]

10.5 An HVDC link consists of rectifier, inverter transmission line and other equipments. Which one of the following is true for this link?
(a) The transmission line produces/supplies reactive power.
(b) The rectifier consumes reactive power and the inverter supplies reactive power from/to the respective connected AC systems.
(c) Rectifier supplies reactive power and the inverter consumes reactive power to/from the respective connected AC systems.
(d) Both the converters (rectifier and inverter) consume reactive power from the respective connected AC systems. **[2006 : 1 M]**

10.6 Two regional systems, each having several synchronous generators and loads are inter-connected by an ac line and a HVDC link as shown in the figure. Which of the following statements is true in the steady state?

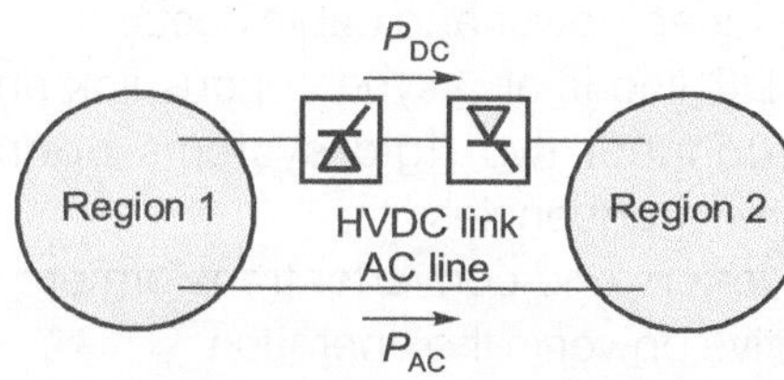

(a) Both regions need not have the same frequency.
(b) The total power flow between the regions (P_{AC} + P_{DC}) can be changed by controlling the HVDC converters alone.
(c) The power sharing between the ac line and the HVDC link can be changed by controlling the HVDC converters alone.
(d) The directions of power flow in the HVDC link(Pdc) cannot be reversed. **[2007 : 1 M]**

10.7 Power is transferred from system A to system B by an HVDC link as shown in the figure. If the voltages V_{AB} and V_{CD} are as indicated in the figure, and $I > 0$, then

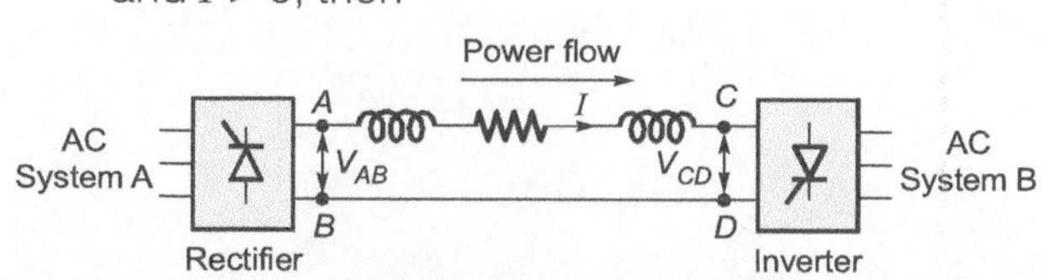

(a) $V_{AB} < 0$, $V_{CD} < 0$, $V_{AB} > V_{CD}$
(b) $V_{AB} > 0$, $V_{CD} > 0$, $V_{AB} < V_{CD}$
(c) $V_{AB} > 0$, $V_{CD} > 0$, $V_{AB} > V_{CD}$
(d) $V_{AB} > 0$, $V_{CD} < 0$ **[2010 : 1 M]**

Answers High Voltage DC Transmission

10.1 (c)	10.2 (b)	10.3 (b)	10.4 (a)	10.5 (b)	10.6 (c)	10.7 (c)

Explanations High Voltage DC Transmission

10.3 (b)

- D.C. line inductor, reduces the harmonic current in the DC line and possible transient over currents.
- Converters and converter transformers consume reactive power.

The reactive power may be supplied from
(i) AC filter
(ii) Shunt Capacitors
(iii) AC network
(iv) Static compensators
(v) Synchronous condensers.

10.4 (a)

- HVDC system are economical for long distance bulk power transmission by overhead lines, due to reduced power and cable costs.
- The DC line is an asynchronous link and it can interconnect two rigid systems operating at different frequencies.
- Converters and converter transformers requires reactive power in the operation.
- The AC/DC converters are sources of harmonics on AC as well as DC side.

10.5 (b)

The rectifier consumers reactive power and inverter supplies reactive power from/to the respective connected AC systems because no reactive power can be transmitted over a DC link.

10.6 (c)

Both regions are connected by HVDC link as well as AC line. So AC link is possible when both regions have same frequency.
By changing firing angle (α) of converter, we can change the power sharing, between the AC line and HVDC link.

10.7 (c)

Power flows in the indicated direction when $V_{AB} > 0$, $V_{CD} > 0$ and $V_{AB} > V_{CD}$.
Since current flows from high voltage to low voltage.

■■■■

11 Per Unit System

11.1 A 75 MVA, 10 kV synchronous generator has $X_d = 0.4$ p.u. The X_d value (in p.u.) to a base of 100 MVA 11 kV is

(a) 0.578 (b) 0.279
(c) 0.412 (d) 0.44 **[2001 : 2 M]**

11.2 The p.u. parameters for a 500 MVA machine on its own base are:

Inertia, $M = 20$ p.u.; Reactance, $X = 2$ p.u.

The p.u. values of inertia and reactance on 100 MVA common base, respectively, are

(a) 4, 0.4 (b) 100, 10
(c) 4, 10 (d) 100, 0.4 **[2005 : 1 M]**

11.3 A generator is connected through a 20 MVA, 13.8/138 kV step down transformer, to a transmission line. At the receiving end of the line a load is supplied through a step down transformer of 10 MVA, 138/69 kV rating. A 0.72 p.u. load, evaluated on load side transformer ratings as base values, is supplied from the above system. For system base values of 10 MVA and 69 kV in load circuit, the value of the load (in per unit) in generator circuit will be

(a) 36 (b) 1.44
(c) 0.72 (d) 0.18 **[2006 : 2 M]**

11.4 For the power system shown in the figure below, the specifications of the components are the following:

G1 : 25 kV, 100 MVA, X = 9%
G2 : 25 kV, 100 MVA, X = 9%
T1 : 25 kV/220 kV, 90 MVA, X = 12%
T2 : 220 kV/25 kV, 90 MVA, X = 12%
Line 1 : 200 kV, X = 150 ohms

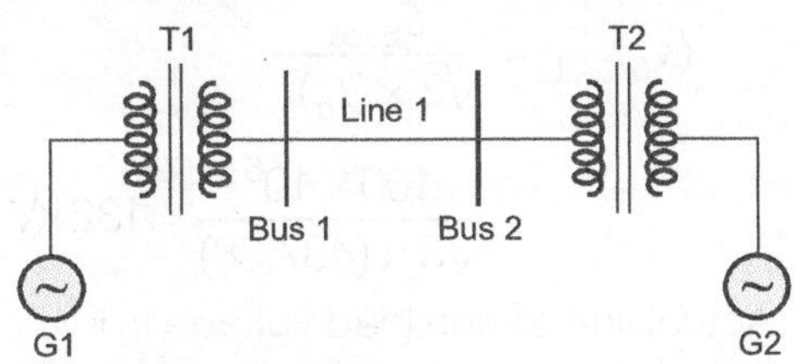

Choose 25 kV as the base voltage at the generator G1, and 200 MVA as the MVA base. The impedance diagram is

(a)
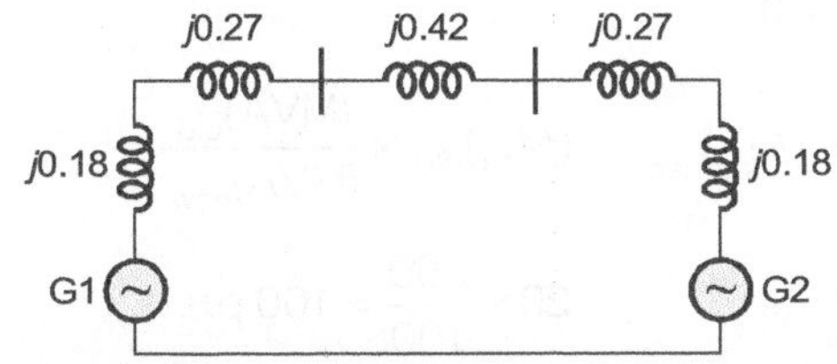

(b)
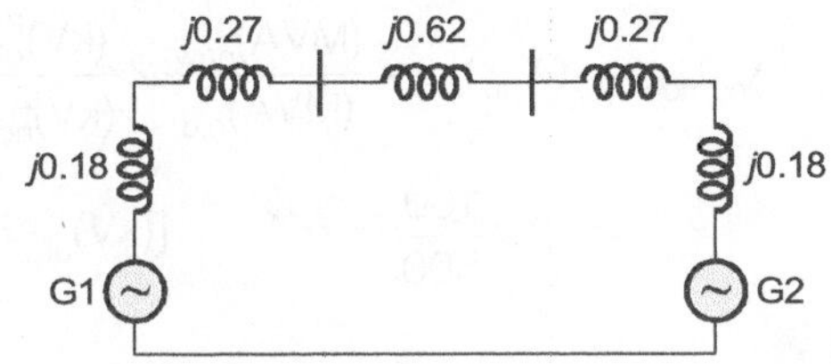

(c)
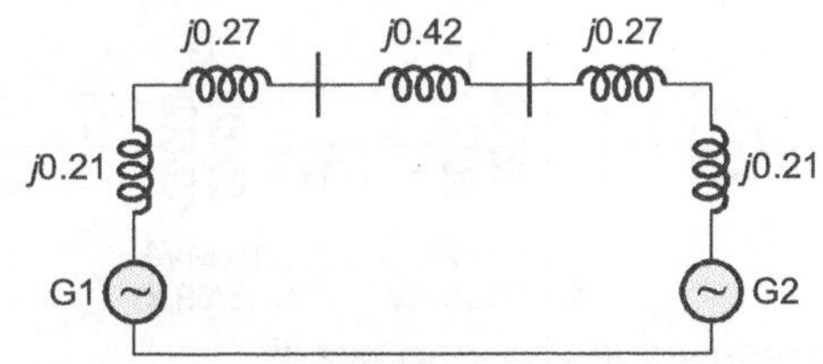

(d)
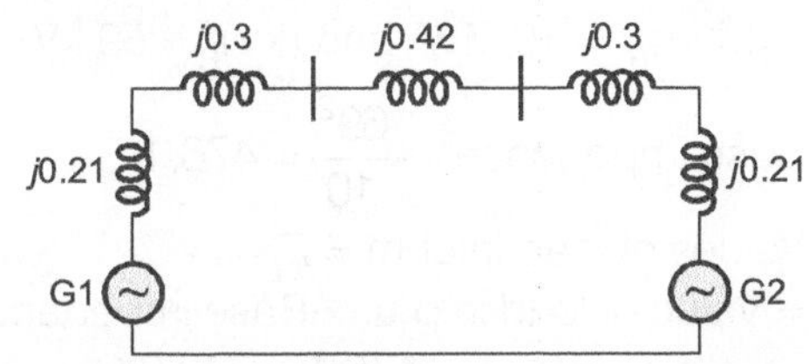

[2010 : 2 M]

11.5 A 3-phase star connected load is drawing power at a voltage of 0.9 pu and 0.8 power factor lagging. The three-phase base power and base current are 100 MVA and 437.38 A respectively. The line-to-line load voltage in kV is _____.

[2014 : 1 M, Set-2]

■■■■

Answers Per Unit System

11.1 (d) 11.2 (d) 11.3 (a) 11.4 (b) 11.5 (118.8)

Explanations Per Unit System

11.1 (d)

$$(X_d)_{new} = 0.4 \times \left(\frac{100}{75}\right)\left(\frac{10}{11}\right)^2 = 0.44077$$

11.2 (d)

$$(M_{pu})_{new} = (M_{pu})_{old} \times \frac{(MVA)_{old}}{(MVA)_{new}}$$

$$= 20 \times \frac{500}{100} = 100 \text{ p.u.}$$

$$(X_{pu})_{new} = (X_{pu})_{old} \times \frac{(MVA)_{new}}{(MVA)_{old}} \times \frac{(kV)^2_{old}}{(kV)^2_{new}}$$

$$= 2 \times \frac{100}{500} = 0.4 \qquad [(kV)_{old} = (kV)_{new}]$$

11.3 (a)

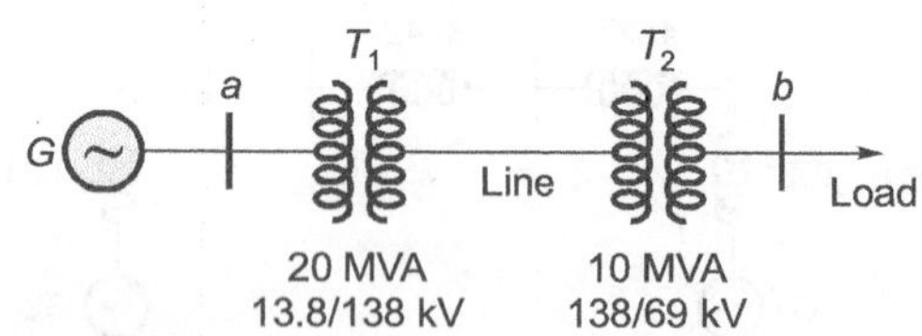

Base values on load circuit,

$(MVA)_B = 10$ MVA and $(kV)_B = 69$ kV

$$\text{Base impedance} = \frac{69^2}{10} = 476.1$$

Values of load in ohm = Z_L

= Value of load in p.u. × Base impedance

$\Rightarrow \quad Z_L = 0.72 \times 476.1 = 342.8\ \Omega$

Base values in generator circuit,

$(MVA)_B = 20$ MVA

$(kV)_B = 13.8$ kV

$$\text{Base impedance} = Z_B = \frac{(kV)_B^2}{(MVA)_B} = \frac{13.8^2}{20}$$

$$\Rightarrow \quad Z_B = 9.522\ \Omega = \frac{Z_L}{Z_B} = \frac{342.8}{9.522} = 36 \text{ p.u.}$$

11.4 (b)

$$Z_{pu(new)} = Z_{pu(old)} \times \frac{(MVA_B)_{new}}{(MVA_B)_{old}} \times \frac{(kV_B)^2_{old}}{(kV_B)^2_{new}}$$

$(MVA_B)_{new}$ = 200 MVA and $(kV_B)_{new} = 25$ kV

$$\text{Base impedance} = Z_B = \frac{(kV_B)^2}{(MVA_B)}$$

(kV_B) in line = 220 kV

$$Z_B = \frac{220^2}{200} = 242\ \Omega$$

$$X_{l\text{-pu}} = \frac{X_l}{Z_B} = \frac{150}{242} = 0.62 \text{ p.u.}$$

G_1: 25 kV, 100 MVA,

$X_{G1} = 0.09$ pu

$$X_{G1}(\text{On base 200 MVA}) = 0.09 \times \frac{200}{100} = 0.18 \text{ p.u.}$$

$\{(kV_B)_{old} = (kV_B)_{new}\}$

G_2: 2 kV, 100 MVA,

$X_{G2} = 0.09$ p.u.

$$X_{G2}(\text{on base 200 MVA}) = 0.09 \times \frac{200}{100} = 0.18 \text{ p.u.}$$

T_1: 25 kV/220 kV, 90 MVA,

$X_{T1} = 0.12$ p.u.

$$X_{T1}(\text{on base 200 MVA}) = 0.12 \times \frac{200}{90} = 0.27 \text{ p.u.}$$

T_2: 220 kV/25 kV, 90 MVA,

$X_{T2} = 0.12$ p.u.

$$X_{T2}(\text{on base 200 MVA}) = 0.12 \times \frac{200}{90} = 0.27 \text{ p.u.}$$

11.5 Sol.

Given, $V = 0.9$ pu, $\cos\phi = 0.8$

$S_{base} = 100$ MVA

$I_{base} = 437.38$ A

For a 3-ϕ star connected system,

$$S_{base} = \sqrt{3}(V_b)_{L-L}(I_b)_L$$

$$\therefore \quad (V_b)_{L-L} = \frac{S_{base}}{\sqrt{3} \times (I_b)_L}$$

$$= \frac{100 \times 10^6}{\sqrt{3} \times (437.38)} = 132 \text{ kV}$$

∴ Actual line-to-line load voltage in kV

= 0.9 × 132 kV = 118.8 kV

■■■■

12 Power System Transients

12.1 The reflection coefficient for the transmission line shown in the figure at ρ is

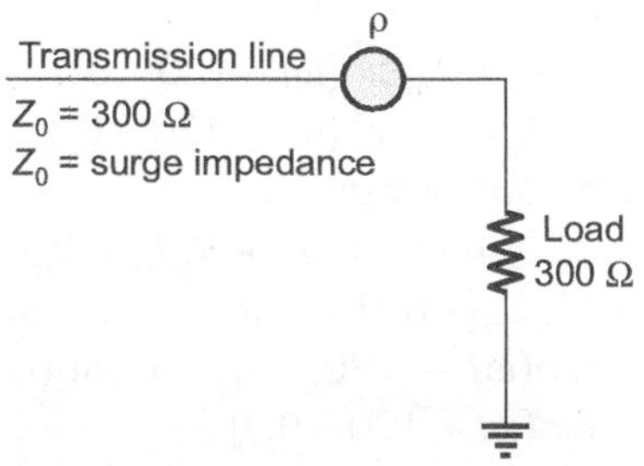

(a) +1 (b) –1
(c) 0 (d) 0.5 **[1998 : 1 M]**

12.2 An overhead line having a surge impedance of 400 Ω is connected in series with an underground cable having a surge impedance of 100 Ω. If a surge of 50 kV travels from the line towards the cable junctions, the value of the transmitted voltage wave at the junction is

(a) 30 kV (b) 20 kV
(c) 80 kV (d) –30 kV **[1999 : 1 M]**

12.3 A surge of 20 kV magnitude travels along a lossless cable towards its junction with two identical lossless overhead transmission lines. The inductance and the capacitance of the cable are 0.4 mH and 0.5 μF per km. The inductance and capacitance of the overhead transmission lines are 1.5 mH and 0.015 μF per km. The magnitude of the voltage at the junction due to surge is

(a) 36.72 kV (b) 18.36 kV
(c) 6.07 kV (d) 33.93 kV **[2003 : 2 M]**

12.4 Total instantaneous power supplied by a 3-phase AC supply to a balanced R-L load is

(a) zero
(b) constant
(c) pulsating with zero average
(d) pulsating with non-zero average

[2004 : 1 M]

12.5 The insulation strength of an EHV transmission line is mainly governed by

(a) load power factor
(b) switching over-voltages
(c) harmonics
(d) corona **[2005 : 1 M]**

12.6 Consider a step voltage wave of magnitude 1 pu travelling along a lossless transmission line that terminates in a reactor. The voltage magnitude across the reactor at the instant the travelling wave reaches the reactor is

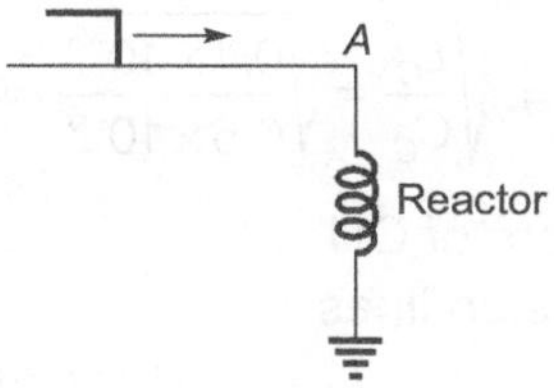

(a) –1 pu (b) 1 pu
(c) 2 pu (d) 3 pu

[2010 : 1 M]

Answers Power System Transients

12.1 (c) 12.2 (b) 12.3 (d) 12.4 (b) 12.5 (b) 12.6 (c)

Explanations Power System Transients

12.1 (c)

$$\rho_R = \frac{V_R}{V} = \left(\frac{Z_L - Z_0}{Z_L + Z_0}\right)$$

$\because \quad Z_0 = Z_L; \quad \therefore \rho_R = 0$

12.2 (b)

$$V_T = \left[\frac{2Z_C}{Z_L + Z_C}\right] V = \left[\frac{2 \times 100}{400 + 100}\right] \times 50 \text{ kV} = 20 \text{ kV}$$

12.3 (d)

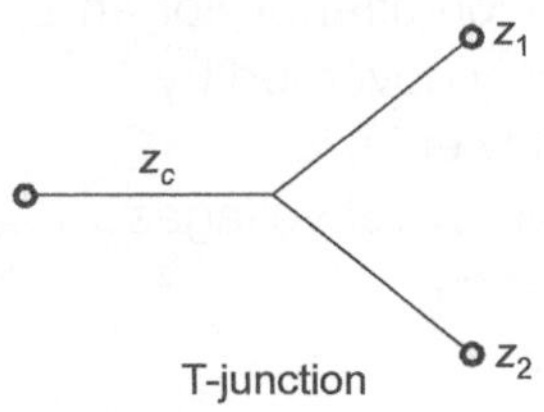

Where, $\quad Z_1 = Z_2$

Parameter of cable:

Inductance, $\quad L_C = 0.4$ mH/km and

Capacitance $\quad C_C = 0.5$ µF/km

Surge impedance of the cable = Z_C

$$= \sqrt{\frac{L_C}{C_C}} = \sqrt{\frac{0.4 \times 10^{-3}}{0.5 \times 10^{-6}}} = 28.28\ \Omega$$

Parameters of OH Transmission lines

Inductance, $\quad L_1 = 1.5$ mH/km and

Capacitance, $\quad C_1 = 0.015$ µF/km

Surge impedance of the lines = $Z_1 = Z_2$

$$= \sqrt{\frac{L_1}{C_1}} = \sqrt{\frac{1.5 \times 10^{-3}}{0.015 \times 10^{-6}}} = 316.22\ \Omega$$

$E = 20$ kV

Junction voltage = E_j

$$= \frac{2E \times \frac{1}{Z_C}}{\frac{1}{Z_C} + \frac{1}{Z_1} + \frac{1}{Z_2}} = \frac{2 \times 20 \times \frac{1}{28.28}}{\frac{1}{28.28} + \frac{2}{316.22}} = 33.93 \text{ kV}$$

12.4 (b)

- Impedance of the load

$$= Z_L = R + j\omega L = |Z| \angle\theta_L$$

where, $\quad \theta_L = \tan^{-1}\left(\frac{\omega L}{R}\right)$

- Voltages of 3-ϕ supply

$$V_a = V_m \sin \omega t$$
$$V_b = V_m \sin(\omega t - 120°)$$
$$V_c = V_m \sin(\omega t + 120°)$$
$$I_a = \frac{V_a}{Z_L} = \frac{V_m \sin\omega t}{|Z|\angle\theta_L} = I_m \sin(\omega t - \theta_L)$$

where, $\quad I_m = \frac{V_m}{|Z|}$

Similarly,

$$I_b = I_m \sin(\omega t - 120 - \theta_L)$$

and $\quad I_c = I_m \sin(\omega t + 120 - \theta_L)$

Instantaneous power

$$= P = V_a I_a + V_b I_b + V_c I_c$$

$$P = V_m I_m [\sin\omega t \cdot \sin(\omega t - \theta_L) + \sin(\omega t - 120°) \sin(\omega t - 120 - \theta_L) + \sin(\omega t + 120°) \cdot \sin(\omega t + 120 - \theta_L)]$$

$$= \frac{V_m I_m}{2} [\{\cos\theta_L - \cos(2\omega t - \theta_L)\}]$$
$$+ \{\cos\theta_L - \cos(2\omega t - 240 - \theta_L)\}$$
$$+ \{\cos\theta_L - \cos(2\omega t + 240 - \theta_L)\}]$$

$$\Rightarrow P = \frac{3V_m I_m}{2} \cos\phi = \text{constant}$$

12.5 (b)

Note: At transmission line voltages upto around 230 kV, the insulation level is dictated by the requirement of protection against lightning. For voltages from 230 kV to 700 kV, both switching transients and lightning overvoltages must be accounted for in deciding the insulation levels. In EHV (> 700 kV) switching surges cause overvoltages than lightning and are therefore mainly responsible for insulation level decision.

12.6 (c)

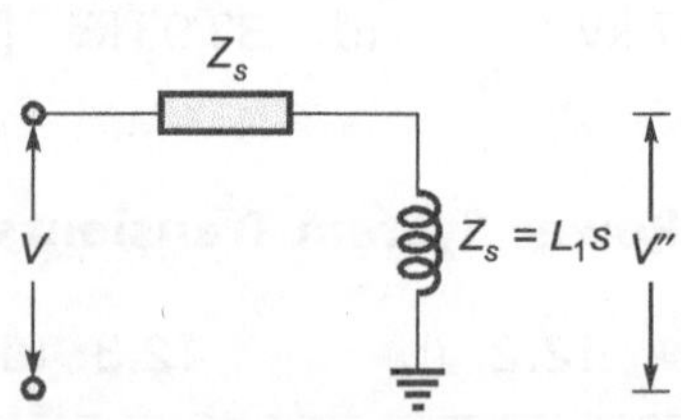

If a step voltage of magnitude 'V' travels through transmission line which is termianted with a inductive load $Z_L = L_1 s$ then voltage transmitted (Induced) across the $Z_L = L_1 s$

$$V''(t) = 2Ve^{-Z_s t / L_1}$$

At time $t = 0$, $\quad V''(t) = 2$ V

if $V = 1$, $\quad V'' = 2$ p.u.

Control Systems

UNIT

V

CONTENTS

Control Systems

Syllabus

Mathematical Modeling and Representation of Systems, Feedback Principle, Transfer Function, Block Diagrams and Signal Flow Graphs, Transient and Steady State Analysis of Linear Time Invariant Systems, Stability Analysis using Routh-Hurwitz and Nyquist Criteria, Bode Plots, Root Loci, Lag, Lead and Lead Lag Compensators; P, PI and PID Controllers; State Space Model, Solution of State Equations of LTI Systems.

Analysis of Previous GATE Papers

Exam Year	1 Mark Ques.	2 Marks Ques.	4 Marks Ques.	5 Marks Ques.	Total Marks
1995	5	–	–	1	**10**
1996	5	1	–	–	**7**
1997	2	2	–	1	**11**
1998	2	2	–	1	**11**
1999	–	–	–	–	**–**
2000	3	2	–	2	**17**
2001	3	–	–	2	**13**
2002	1	4	–	1	**14**
2003	3	7	–	–	**17**
2004	2	9	–	–	**20**
2005	3	6	–	–	**15**
2006	1	3	–	–	**7**
2007	1	7	–	–	**15**
2008	1	8	–	–	**17**
2009	4	5	–	–	**14**
2010	2	3	–	–	**8**
2011	3	3	–	–	**9**

Exam Year	1 Mark Ques.	2 Marks Ques.	Total Marks
2012	1	4	**9**
2013	2	4	**10**
2014 Set-1	2	2	**6**
2014 Set-2	2	3	**8**
2014 Set-3	2	3	**8**
2015 Set-1	2	4	**10**
2015 Set-2	2	3	**8**
2016 Set-1	1	4	**9**
2016 Set-2	1	4	**9**
2017 Set-1	3	5	**13**
2017 Set-2	2	3	**8**
2018	2	2	**6**
2019	2	3	**8**
2020	2	4	**10**
2021	2	3	**8**
2022	2	3	**8**
2023	3	3	**9**
2024	3	3	**9**

1 Mathematical Models of Physical Systems

1.1 The closed loop transfer function of a control system is given by

$$\frac{C(s)}{R(s)} = \frac{2(s-1)}{(s+2)(s+1)}$$

for a unit step input the output is

(a) $-3e^{-2t} + 4e^{-t} - 1$
(b) $-3e^{-2t} - 4e^{-t} + 1$
(c) zero
(d) infinity **[1995 : 1 M]**

1.2 For block diagram shown in figure $C(s)/R(s)$ is given by

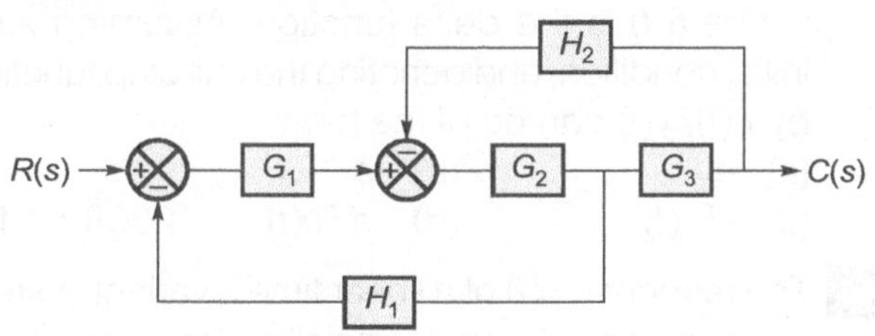

(a) $\dfrac{G_1G_2G_3}{1+H_2G_2G_3+H_1G_1G_2}$

(b) $\dfrac{G_1G_2G_3}{1+G_1G_2G_3H_1H_2}$

(c) $\dfrac{G_1G_2G_3}{1+G_1G_2G_3H_1+G_1G_2G_3H_2}$

(d) $\dfrac{G_1G_2G_3}{1+G_1G_2G_3H_1}$ **[1998 : 2 M]**

1.3 A linear time invariant system initially at rest, when subjected to a unit step input, gives a response $y(t) = te^{-t}, t > 0$. The transfer function of the system is

(a) $\dfrac{1}{(s+1)^2}$ (b) $\dfrac{1}{s(s+1)^2}$

(c) $\dfrac{s}{(s+1)^2}$ (d) $\dfrac{1}{s(s+1)}$ **[2000 : 1 M]**

1.4 The transfer function of the system described by $\dfrac{d^2y}{dt^2}+\dfrac{dy}{dt}=\dfrac{du}{dt}+2u$ with u as input and y as output is

(a) $\dfrac{(s+2)}{(s^2+s)}$ (b) $\dfrac{(s+1)}{(s^2+s)}$

(c) $\dfrac{2}{(s^2+s)}$ (d) $\dfrac{2s}{(s^2+s)}$ **[2002 : 2 M]**

1.5 A control system is defined by the following mathematical relationship

$$\frac{d^2x}{dt^2} + 6\frac{dx}{dt} + 5x = 12(1-e^{-2t})$$

The response of the system as $t \to \infty$ is

(a) $x = 6$ (b) $x = 2$
(c) $x = 2.4$ (d) $x = -2$ **[2003 : 1 M]**

1.6 A control system with certain excitation is governed by the following mathematical equation

$$\frac{d^2x}{dt^2} + \frac{1}{2}\frac{dx}{dt} + \frac{1}{18}x = 10 + 5e^{-4t} + 2e^{-5t}$$

The natural time constants of the response of the system are

(a) 2s and 5s (b) 3s and 6s
(c) 4s and 5s (d) 1/3s and 1/6s
[2003 : 2 M]

1.7 The block diagram of a control system is shown in figure. The transfer function $G(s) = Y(s)/U(s)$ of the system is

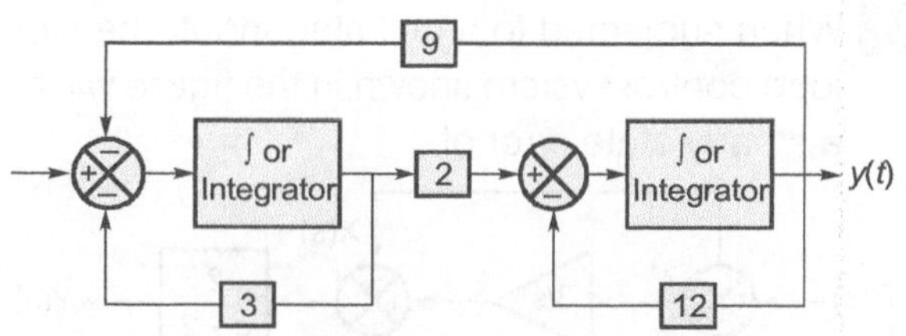

(a) $\dfrac{1}{18\left(1+\dfrac{s}{12}\right)\left(1+\dfrac{s}{3}\right)}$

(b) $\dfrac{1}{27\left(1+\dfrac{s}{6}\right)\left(1+\dfrac{s}{9}\right)}$

(c) $\dfrac{1}{27\left(1+\dfrac{s}{12}\right)\left(1+\dfrac{s}{9}\right)}$

(d) $\dfrac{1}{27\left(1+\dfrac{s}{9}\right)\left(1+\dfrac{s}{3}\right)}$ **[2003 : 2 M]**

1.8 For a tachometer, if $\theta(t)$ is the rotor displacement in radians, $e(t)$ is the output voltage and K_t is the tachometer constant in V/rad/sec, then the transfer function, $\frac{E(s)}{Q(s)}$ will be

(a) $K_t s^2$ (b) $\frac{K_t}{s}$

(c) $K_t s$ (d) K_t [2004 : 1 M]

1.9 For the block diagram shown in figure, the transfer function $\frac{C(s)}{R(s)}$ is equal to

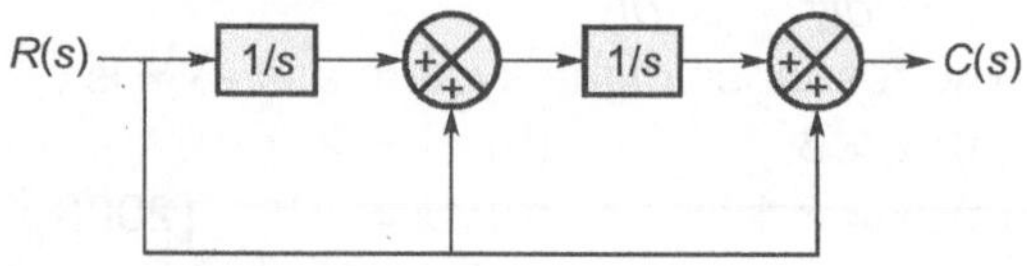

(a) $\frac{s^2+1}{s^2}$ (b) $\frac{s^2+s+1}{s^2}$

(c) $\frac{s^2+s+1}{s}$ (d) $\frac{1}{s^2+s+1}$ [2004 : 2 M]

1.10 The unit impulse response of a second order-under damped system starting from rest is given by $c(t) = 12.5\, e^{-6t} \sin 8t,\ t \geq 0$. The steady-state value of the unit step response of the system is equal to

(a) 0 (b) 0.25

(c) 0.5 (d) 1.0 [2004 : 2 M]

1.11 When subjected to a unit step input, the closed loop control system shown in the figure will have a steady state error of

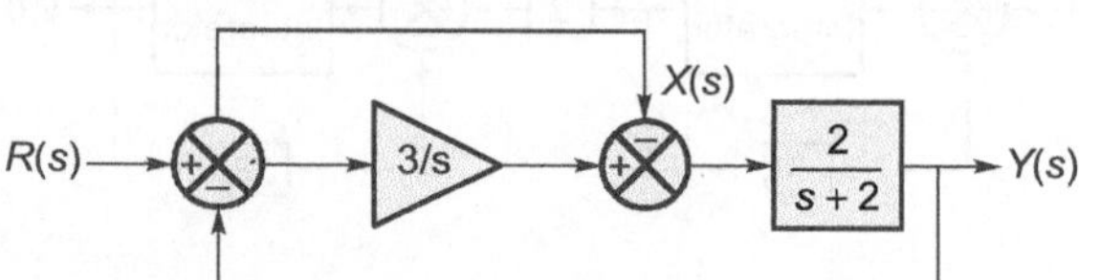

(a) −1.0 (b) −0.5

(c) 0 (d) 0.5 [2005 : 2 M]

1.12 The system shown in figure below

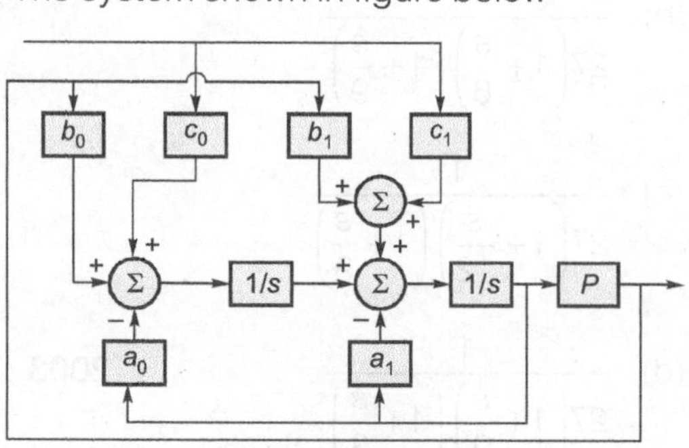

can be reduced to the form

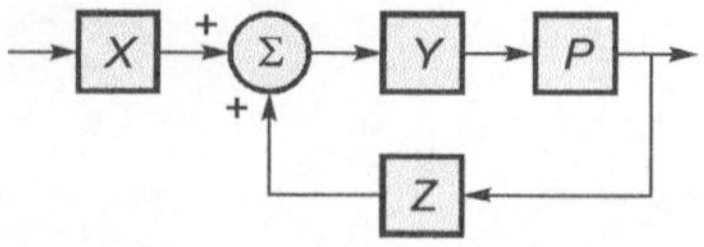

with

(a) $X = c_0 s + c_1,\ Y = 1/(s^2 + a_0 s + a_1),$
$Z = b_0 s + b_1$

(b) $X = 1,\ Y = (c_0 s + c_1)/(s^2 + a_0 s + a_1),$
$Z = b_0 s + b_1$

(c) $X = c_1 s + c_0,\ Y = (b_1 s + b_0)/(s^2 + a_1 s + a_0),$
$Z = 1$

(d) $X = c_1 s + c_0,\ Y = 1/(s^2 + a_1 s + a_0),$
$Z = b_1 s + b_0$ [2007 : 2 M]

1.13 A function $y(t)$ satisfies the following differential equation:

$$\frac{dy(t)}{dt} + y(t) = \delta(t)$$

where $\delta(t)$ is the delta function. Assuming zero initial condition, and denoting the unit step function by $u(t)$, $y(t)$ can be of the form

(a) e^t (b) e^{-t}

(c) $e^t u(t)$ (d) $e^{-t} u(t)$ [2008 : 1 M]

1.14 The response $h(t)$ of a linear time invariant system to an impulse $\delta(t)$, under initially relaxed condition is $h(t) = e^{-t} + e^{-2t}$. The response of this system for a unit step input $u(t)$ is

(a) $u(t) + e^{-t} + e^{-2t}$

(b) $(e^{-t} + e^{-2t})\, u(t)$

(c) $(1.5 - e^{-t} - 0.5\, e^{-2t})\, u(t)$

(d) $e^{-t}\delta(t) + e^{-2t} u(t)$ [2011 : 2 M]

1.15 The transfer function $\frac{V_2(s)}{V_1(s)}$ of the circuit shown below is

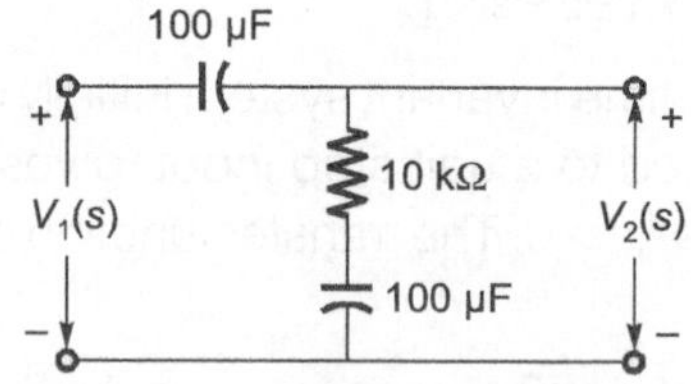

(a) $\frac{0.5s+1}{s+1}$ (b) $\frac{3s+6}{s+2}$

(c) $\frac{s+2}{s+1}$ (d) $\frac{s+1}{s+2}$ [2013 : 1 M]

1.16 The signal flow graph for a system is given below. The transfer function $Y(s)/U(s)$ for this system is

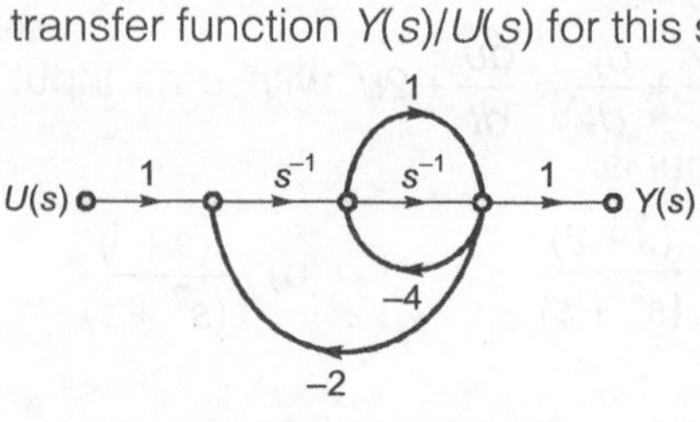

(a) $\dfrac{s+1}{5s^2+6s+2}$ (b) $\dfrac{s+1}{s^2+6s+2}$

(c) $\dfrac{s+1}{s^2+4s+2}$ (d) $\dfrac{1}{5s^2+6s+2}$

[2013 : 2 M]

1.17 The signal flow graph of a system is shown below. $U(s)$ is the input and $C(s)$ is the output.

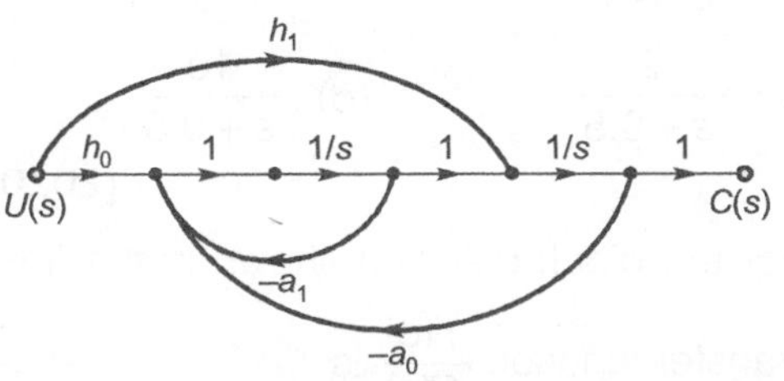

Assuming, $h_1 = b_1$ and $h_0 = b_0 - b_1a_1$, the input-output transfer function, $G(s) = \dfrac{C(s)}{U(s)}$ of the system is given by

(a) $G(s) = \dfrac{b_0s+b_1}{s^2+a_0s+a_1}$

(b) $G(s) = \dfrac{a_1s+a_0}{s^2+b_1s+b_0}$

(c) $G(s) = \dfrac{b_1s+b_0}{s^2+a_1s+a_0}$

(d) $G(s) = \dfrac{a_0s+a_1}{s^2+b_0s+b_1}$ [2014 : 1 M, Set-3]

1.18 For the signal-flow graph shown in the figure, which one of the following expressions is equal to the transfer function $\left.\dfrac{Y(s)}{X_2(s)}\right|_{X_1(s)=0}$?

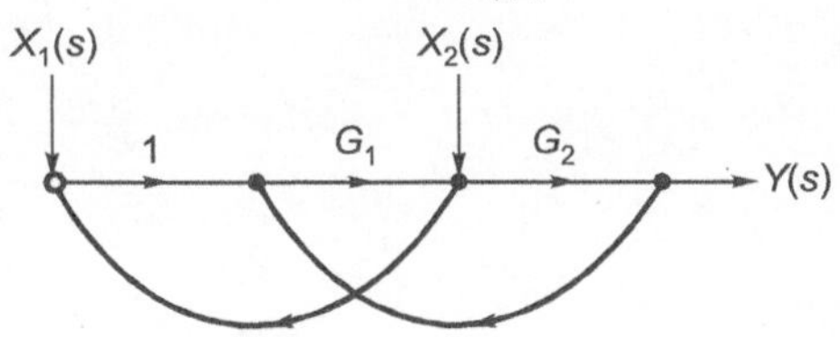

(a) $\dfrac{G_1}{1+G_2(1+G_1)}$ (b) $\dfrac{G_2}{1+G_1(1+G_2)}$

(c) $\dfrac{G_1}{1+G_1G_2}$ (d) $\dfrac{G_2}{1+G_1G_2}$

[2015 : 1 M, Set-1]

1.19 Find the transfer function $Y(s)/X(s)$ of the system given below:

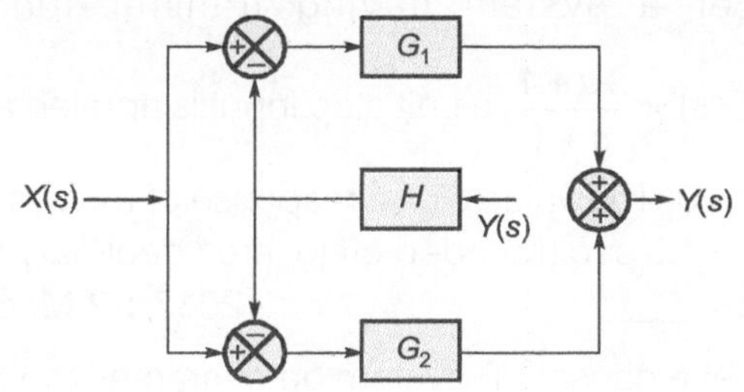

(a) $\dfrac{G_1}{1-HG_1}+\dfrac{G_2}{1-HG_2}$ (b) $\dfrac{G_1}{1+HG_1}+\dfrac{G_2}{1+HG_2}$

(c) $\dfrac{G_1+G_2}{1+H(G_1+G_2)}$ (d) $\dfrac{G_1+G_2}{1-H(G_1+G_2)}$

[2015 : 2 M, Set-1]

1.20 For the system governed by the set of equations:

$$\frac{dx_1}{dt} = 2x_1 + x_2 + u$$

$$\frac{dx_2}{dt} = -2x_1 + u$$

$$y = 3x_1$$

the transfer function $\dfrac{Y(s)}{U(s)}$ is given by

(a) $\dfrac{3(s+1)}{(s^2-2s+2)}$ (b) $\dfrac{3(2s+1)}{(s^2-2s+1)}$

(c) $\dfrac{(s+1)}{(s^2-2s+1)}$ (d) $\dfrac{3(2s+1)}{(s^2-2s+2)}$

[2015 : 2 M, Set-2]

1.21 In the system whose signal flow graph is shown in the figure, $U_1(s)$ and $U_2(s)$ are inputs. The transfer function $Y(s)/U_1(s)$ is

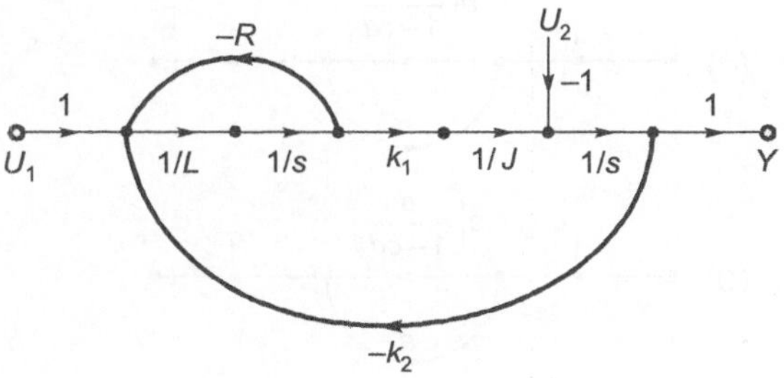

(a) $\dfrac{k_1}{JLs^2+JRs+k_1k_2}$

(b) $\dfrac{k_1}{JLs^2-JRs-k_1k_2}$

(c) $\dfrac{k_1-U_2(R+sL)}{JLs^2+(JR-U_2L)s+k_1k_2-U_2R}$

(d) $\dfrac{k_1-U_2(sL-R)}{JLs^2-(JR+U_2L)s-k_1k_2+U_2R}$

[2017 : 2 M, Set-1]

1.22 For a system having transfer function, $G(s) = \dfrac{-s+1}{s+1}$, a unit step input is applied at time $t = 0$. The value of the response of the system at $t = 1.5$ sec (rounded off to three decimal places) is ______. **[2017 : 2 M, Set-1]**

1.23 Let a causal LTI system be characterized by the following differential equation, with initial rest condition,

$$\frac{d^2y}{dt^2} + 7\frac{dy}{dt} + 10y(t) = 4x(t) + 5\frac{dx(t)}{dt}$$

Where, $x(t)$ and $y(t)$ are the input and output respectively. The impulse response of the system is ($u(t)$ is the unit step function)

(a) $2e^{-2t}u(t) - 7e^{-5t}u(t)$
(b) $-2e^{-2t}u(t) + 7e^{-5t}u(t)$
(c) $7e^{-2t}u(t) - 2e^{-5t}u(t)$
(d) $-7e^{-2t}u(t) + 2e^{-5t}u(t)$ **[2017 : 2 M, Set-1]**

1.24 Which of the options is an equivalent representation of the signal flow graph shown here?

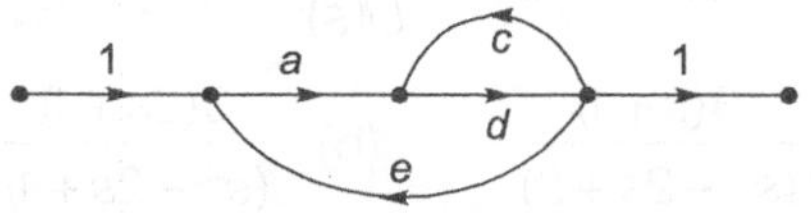

(a) 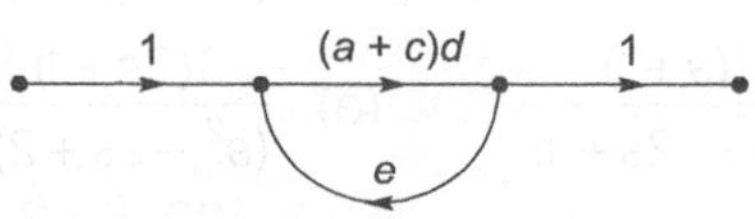

(b) 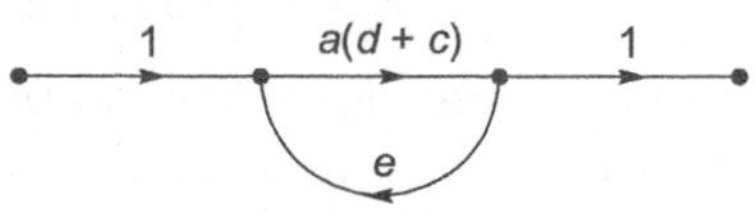

(c) 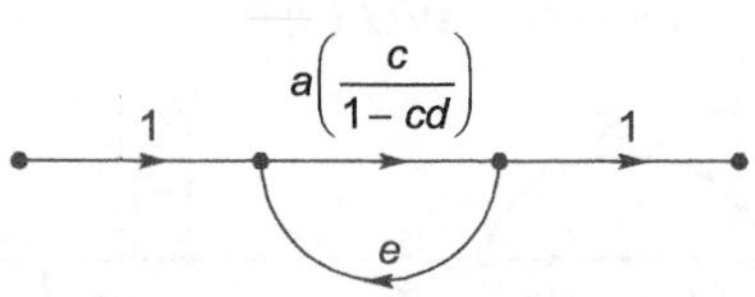

(d) 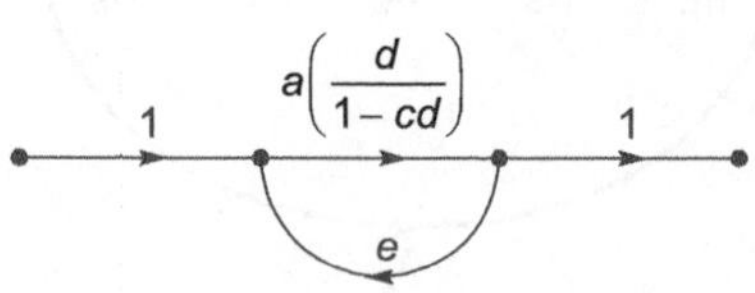

[2020 : 2 M]

1.25 Consider a permanent magnet dc (PMDC) motor which is initially at rest. At $t = 0$. a DC voltage of 5 V is applied to the motor. Its speed monotonically increases from 0 rad/s to 6.32 rad/s in 0.5 s and finally settles to 10 rad/s. Assuming that the armature inductance of the motor is negligible, the transfer function for the motor is

(a) $\dfrac{2}{0.5s+1}$ (b) $\dfrac{10}{0.5s+1}$

(c) $\dfrac{2}{s+0.5}$ (d) $\dfrac{10}{s+0.5}$

[2020 : 2 M]

1.26 For the block diagram shown in the figure, the transfer function $\dfrac{Y(s)}{R(s)}$ is

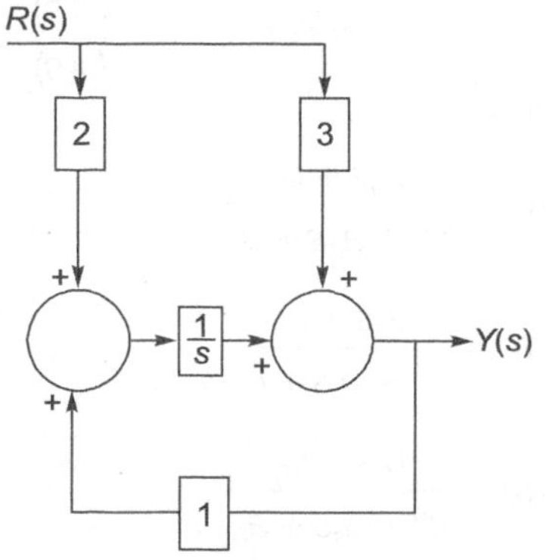

(a) $\dfrac{2s+3}{s+1}$ (b) $\dfrac{3s+2}{s-1}$

(c) $\dfrac{s+1}{3s+2}$ (d) $\dfrac{3s+2}{s+1}$

[2023 : 1 M]

'*' Not in the GATE-2020 syllabus.

Answers Mathematical Models of Physical Systems

1.1	(a)	1.2	(a)	1.3	(c)	1.4	(a)	1.5	(c)	1.6	(b)	1.7	(b)
1.8	(c)	1.9	(b)	1.10	(d)	1.11	(c)	1.12	(d)	1.13	(d)	1.14	(c)
1.15	(d)	1.16	(a)	1.17	(c)	1.18	(b)	1.19	(c)	1.20	(a)	1.21	(a)
1.22	(0.554)	1.23	(b)	1.24	(d)	1.25	(a)	1.26	(b)				

Explanations Mathematical Models of Physical Systems

1.1 (a)

$$\frac{C(s)}{1/s} = \frac{2(s-1)}{(s+2)(s+1)}$$

$$C(s) = \frac{-1}{s} + \frac{-3}{s+2} + \frac{4}{s+1}$$

$$\therefore \quad C(t) = -1 - 3e^{-2t} + 4e^{-t}$$

1.2 (a)

Number of forward paths, P_K = one $(G_1G_2G_3)$

Number of loops = two $(-G_1G_2H_1, -G_2G_3H_2)$

Number of non-touching loops = 0, $\Delta_K = 1$

$$\Delta = 1 + G_1G_2H_1 + G_2G_3H_2$$

$$\therefore \text{ T.F.} = \frac{C(s)}{R(s)} = \frac{P_K\Delta_K}{\Delta} = \frac{G_1G_2G_3}{1+G_1G_2H_1+G_2G_3H_2}$$

1.3 (c)

Corresponding Laplace transforms are :

$$X(s) = \mathcal{L}u(t) = \frac{1}{s}$$

$$Y(s) = \mathcal{L}te^{-t} = \frac{1}{(s+1)^2}$$

$$\therefore \quad \text{T.F.} = \frac{Y(s)}{X(s)} = \frac{s}{(s+1)^2}$$

1.4 (a)

$$\frac{d^2y}{dt^2} + \frac{dy}{dt} = \frac{du}{dt} + 2u$$

$$\Rightarrow \quad s^2Y(s) + sY(s) = sU(s) + 2U(s)$$

$$\therefore \quad \frac{Y(s)}{U(s)} = \frac{s+2}{(s^2+s)}$$

1.5 (c)

Taking Laplace transform on both sides

$$(s^2+6s+5)\,X(s) = 12\left(\frac{1}{s} - \frac{1}{s+2}\right)$$

$$= \frac{24}{s(s+2)}$$

$$X(s) = \frac{24}{s(s+2)(s+1)(s+5)}$$

Response at $t \to \infty$

Using final value theorem

$$\underset{t\to\infty}{\text{Lt}}\ x(t) = \underset{s\to 0}{\text{Lt}}\ sX(s)$$

$$= \underset{s\to 0}{\text{Lt}}\ \frac{s\times 24}{s(s+1)(s+2)(s+5)} = 2.4$$

1.6 (b)

Natural time constant of the response depends only on poles of the system.

$$T(s) = \frac{C(s)}{R(s)} = \frac{1}{s^2 + s/2 + 1/18}$$

$$= \frac{18}{18s^2+9s+1} = \frac{1}{(6s+1)(3s+1)}$$

This is in the form $\dfrac{1}{(1+sT_1)(1+sT_2)}$

$\therefore\ T_1, T_2$ = 6 sec, 3 sec.

1.7 (b)

Integrator are represented as 1/s in s-domain

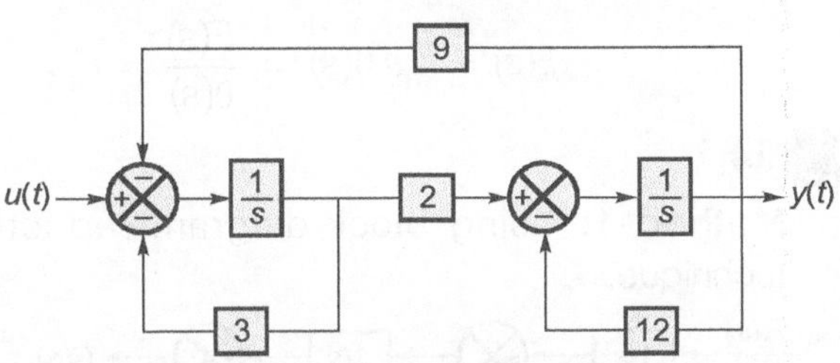

As per the block diagram, the corresponding signal flow graph is drawn

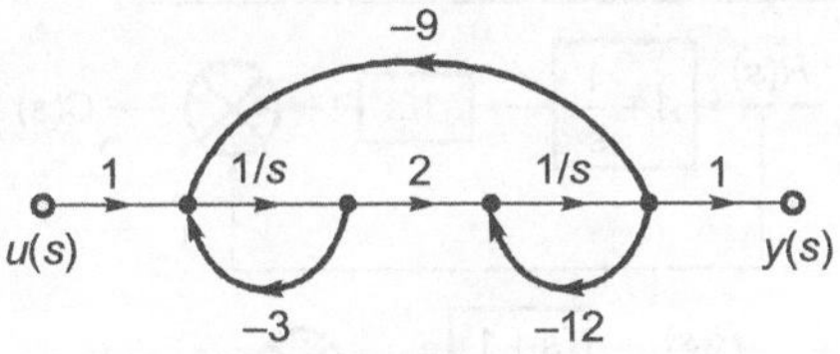

One forward path $P_1 = 2/s^2$

The individual loops are,

$$L_1 = -\frac{3}{s},\ L_2 = -\frac{12}{s} \text{ and } L_3 = -\frac{18}{s^2}$$

L_1 and L_2 are non-touching loops

$$L_1L_2 = \frac{36}{s^2}$$

The loops touches the forward path $\Delta_1 = 1$

The graph determinant is

$$\Delta = 1 - (L_1 + L_2 + L_3) + L_1 L_2$$

$$= 1 + \frac{3}{s} + \frac{12}{s} + \frac{18}{s^2} + \frac{36}{s^2}$$

Applying mason's gain formula,

$$G(s) = \frac{Y(s)}{U(s)} = \frac{P_1\Delta_1}{\Delta}$$

$$= \frac{2/s^2}{1 + \frac{3}{s} + \frac{12}{s} + \frac{18}{s^2} + \frac{36}{s^2}}$$

$$= \frac{2}{s^2 + 15s + 54} = \frac{2}{(s+9)(s+6)}$$

$$= \frac{1}{27\left(1 + \frac{s}{9}\right)\left(1 + \frac{s}{6}\right)}$$

1.8 (c)

$\theta(t)$ = rotor displacement in radians

$$\omega(t) = \frac{d\theta}{dt} = \text{angular speed in rad/sec}$$

Output voltage; $e(t) = K_t\omega(t) = K_t \frac{d\theta}{dt}$.

Taking Laplace transform on both sides

$$E(s) = K_t s\,\theta(s) \Rightarrow \frac{E(s)}{\theta(s)} = K_t s$$

1.9 (b)

Method-1: Using block-diagram reduction technique.

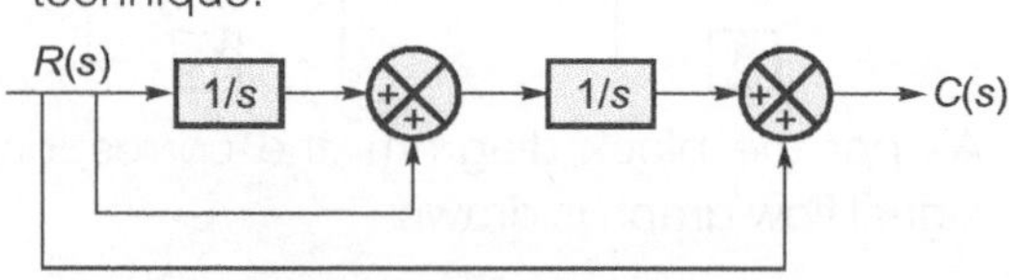

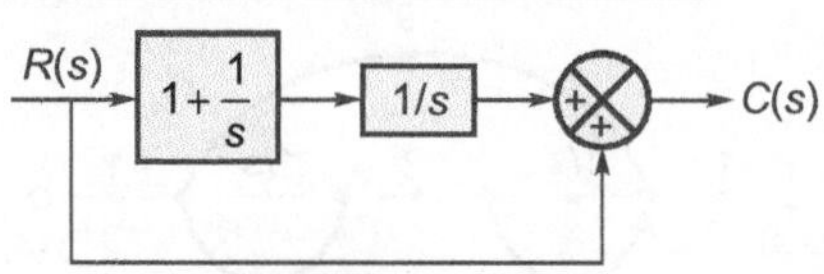

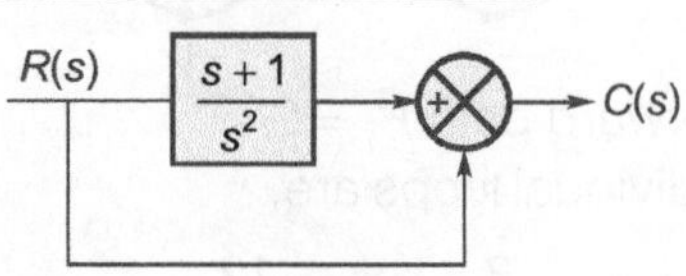

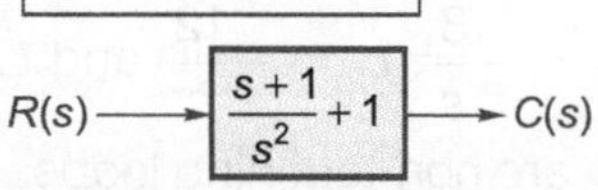

So, transfer function

$$= \frac{C(s)}{R(s)} = \frac{s^2 + s + 1}{s^2}$$

Method-2: Using signal flow graph

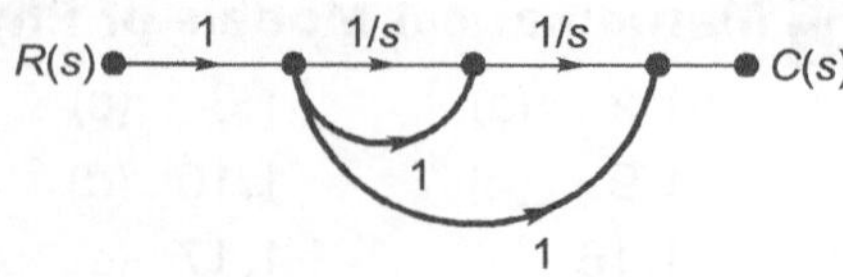

Three forward paths,

$$P_1 = \frac{1}{s}\frac{1}{s} = \frac{1}{s^2},\; P_2 = 1 \cdot \frac{1}{s} = \frac{1}{s} \text{ and } P_3 = 1$$

The number of individual loap = 0

So graph determinant = $\Delta = 1$

and $\Delta_1 = \Delta_2 = \Delta_3 = 1$

Applying Mason's gain formula

$$G(s) = \frac{C(s)}{R(s)} = \frac{P_1\Delta_1 + P_2\Delta_2 + P_3\Delta_3}{\Delta}$$

$$= \frac{\frac{1}{s^2} \cdot 1 + \frac{1}{s} \cdot 1 + 1 \cdot 1}{1} = \frac{s^2 + s + 1}{s^2}$$

1.10 (d)

Transfer function of a system is the unit impulse response of the system.

$$\text{Transfer function} = \frac{C(s)}{R(s)} = \mathcal{L}[12.5e^{-6t}\sin 8t]$$

$$= 12.5 \times \frac{8}{(s+6)^2 + 8^2} = \frac{100}{(s+6)^2 + 8^2}$$

when input is unit step, $R(s) = \frac{1}{s}$

$$C(s) = \frac{1}{s} \cdot \frac{100}{(s+6)^2 + 8^2}$$

Steady-state value of response, using final value theorem

$$C_{\text{steady-state}} = \underset{t\to\infty}{\text{Lt}}\; c(t) = \underset{s\to 0}{\text{Lt}}\; sC(s)$$

$$= \underset{s\to 0}{\text{Lt}}\; s\left[\frac{1}{s} \cdot \frac{100}{(s+6)^2 + 8^2}\right] = 1$$

1.11 (c)

Using signal flow graph,

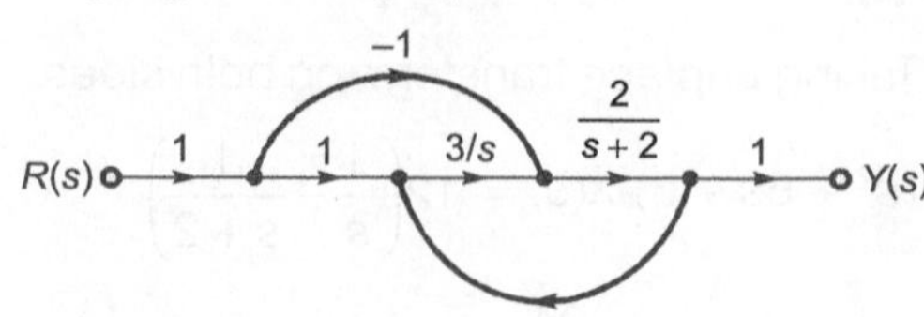

Forward path gains

$$P_1 = (-1) \times \frac{2}{s+2} = \frac{-2}{s+2}$$

$$\text{and} \quad P_2 = \frac{3}{s} \cdot \frac{2}{s+2} = \frac{6}{s(s+2)}$$

Individual loop,

$$L_1 = -1\times\frac{3}{s}\times\frac{2}{s+2} = \frac{-6}{s(s+2)}$$

Loop touches forward paths, therefore,

$$\Delta_1 = 1 \text{ and } \Delta_2 = 1$$

$$D = 1 - L_1$$

$$= 1+\frac{6}{s(s+2)} = \frac{s(s+2)+6}{s(s+2)}$$

Using Mason's gain formula,

$$\frac{Y(s)}{R(s)} = \frac{P_1\Delta_1 + P_2\Delta_2}{\Delta}$$

$$= \frac{-\frac{2}{s+2}\times 1+\frac{6}{s(s+2)}\times 1}{\frac{s(s+2)+6}{s(s+2)}}$$

$$\frac{Y(s)}{R(s)} = \frac{6-2s}{s^2+2s+6}$$

For unit step input, $R(s) = \frac{1}{s}$

$$Y(s) = R(s)\cdot\left(\frac{6-2s}{s^2+2s+6}\right)$$

$$\text{Error} = E(s) = R(s) - Y(s)$$

$$= R(s) - R(s)\left(\frac{6-2s}{s^2+2s+6}\right)$$

$$= R(s)\left[1-\frac{6-2s}{s^2+2s+6}\right]$$

$$E(s) = \frac{1}{s}\left[\frac{s^2+4s}{s^2+2s+6}\right]$$

Steady state value of error, using final value theorem,

$$e_{ss} = \underset{s\to 0}{\text{Lt}}\ sE(s) = \underset{s\to 0}{\text{Lt}}\ s\left[\frac{1}{s}\cdot\frac{s^2+4s}{s^2+2s+6}\right]$$

$$= \underset{s\to 0}{\text{Lt}}\left[\frac{s^2+4s}{s^2+2s+6}\right] = 0$$

1.12 (d)

The block-diagram can be redrawn as

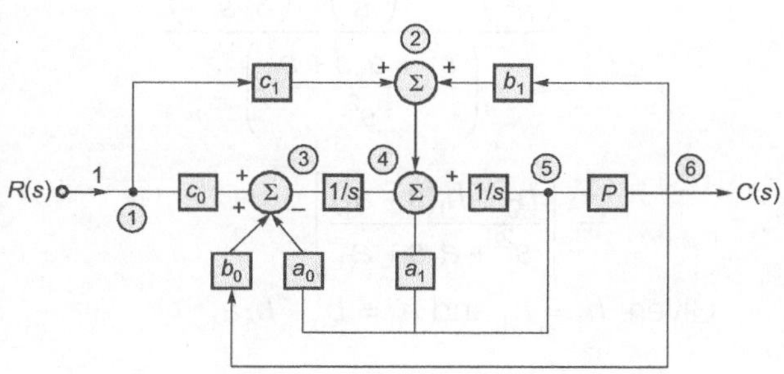

Signal flow graph of the block-diagram,

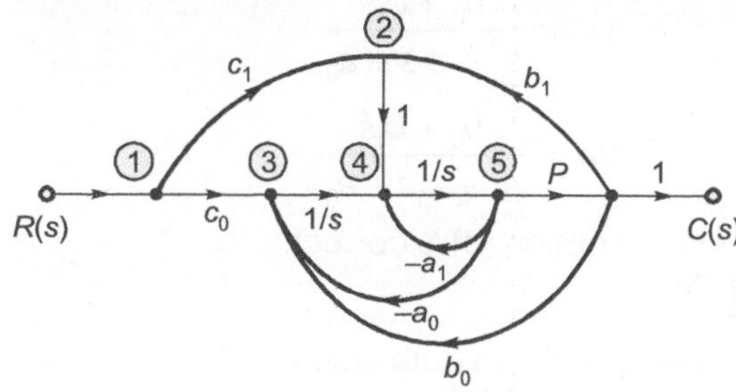

There are two forward paths:

$$P_1 = 1\times c_0\times\frac{1}{s}\times\frac{1}{s}\times P = \frac{c_0P}{s^2}$$

$$P_2 = 1\times c_1\times 1\times\frac{1}{s}\times P = \frac{c_1P}{s}$$

These are four individual loops

$$L_1 = -a_1\times\frac{1}{s} = -\frac{a_1}{s}$$

$$L_2 = \frac{1}{s}\times\frac{1}{s}\times -a_0 = -\frac{a_0}{s^2}$$

$$L_3 = \frac{1}{s}\times\frac{1}{s}\times P\times b_0 = \frac{b_0P}{s^2}$$

$$L_4 = b_1\times P\times\frac{1}{s} = \frac{b_1P}{s}$$

All the loops touch forward paths

$$\Delta_1 = \Delta_2 = 1$$

$$\Delta = 1-(L_1+L_2+L_3+L_4)$$

$$= 1+\frac{a_0}{s^2}+\frac{a_1}{s}-\frac{b_0P}{s^2}-\frac{b_1P}{s}$$

Using Mason's gain formula :

$$\frac{C(s)}{R(s)} = \frac{P_1\Delta_1+P_2\Delta_2}{s}$$

$$= \frac{\frac{c_0P}{s^2}+\frac{c_1P}{s}}{1+\frac{a_1}{s}+\frac{a_0}{s^2}-\frac{b_0P}{s^2}-\frac{b_1P}{s}}$$

$$\frac{C(s)}{R(s)} = \frac{c_0P+c_1P_s}{s^2+(a_0-b_1)s+(a_0-b_0P)}$$

$$\frac{C(s)}{R(s)} = \frac{P(c_0+c_1s)/(s^2+a_1s+a_0)}{1-(b_0+b_1s)P/(s^2+a_1s+a_0)} \quad \text{...(i)}$$

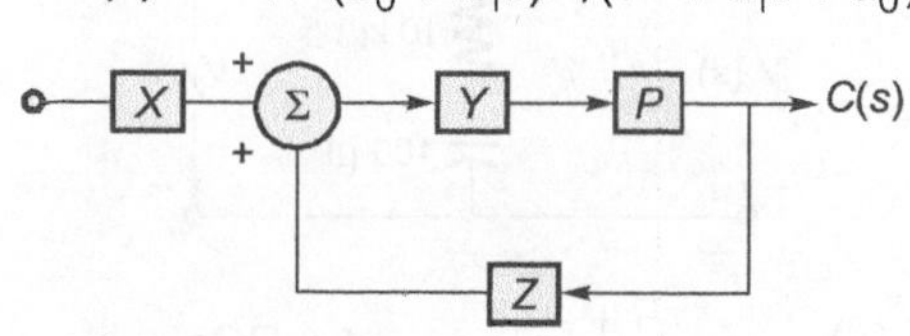

$$\frac{C(s)}{R(s)} = \frac{xyP}{1-yzP} \quad \text{...(ii)}$$

Comparing equation (i) and (ii), we get

$$xy = \frac{c_0 + c_1 s}{s^2 + a_1 s + a_0}$$

$$yz = \frac{b_0 + b_1 s}{s^2 + a_1 s + a_0}$$

Hence option (d) is correct.

1.13 (d)

Taking (L.T.) on both sides

$$Y(s)\,(s+1) = 1$$

$$\therefore \quad Y(s) = \frac{1}{s+1}$$

Taking inverse laplace transform

$$y(t) = e^{-t}\,u(t)$$

1.14 (c)

Transfer function of system is impulse response of the system with zero initial conditions.

Transfer function,

$$H(s) = \mathcal{L}(e^{-t} + e^{-2t}) = \frac{1}{s+1} + \frac{1}{s+2}$$

$$H(s) = \frac{C(s)}{R(s)} = \left(\frac{1}{s+1} + \frac{1}{s+2}\right)$$

$$R(s) = \frac{1}{s} \text{ (step input)}$$

$$C(s) = R(s)\cdot H(s) = \frac{1}{s}\left(\frac{1}{s+1} + \frac{1}{s+2}\right)$$

$$= \frac{1}{s(s+1)} + \frac{1}{s(s+2)}$$

$$C(s) = \left(\frac{1}{s} - \frac{1}{s+1}\right) + \frac{1}{2}\left(\frac{1}{s} - \frac{1}{s+2}\right)$$

$$= \frac{1.5}{s} - \frac{1}{s+1} - \frac{0.5}{s+2}$$

$$\text{Response} = c(t) = \mathcal{L}^{-1}[C(s)]$$

$$= \mathcal{L}^{-1}\left[\frac{1.5}{s} - \frac{1}{s+1} - \frac{0.5}{s+2}\right]$$

$$c(t) = (1.5 - e^{-t} - 0.5e^{-2t})\,u(t)$$

1.15 (d)

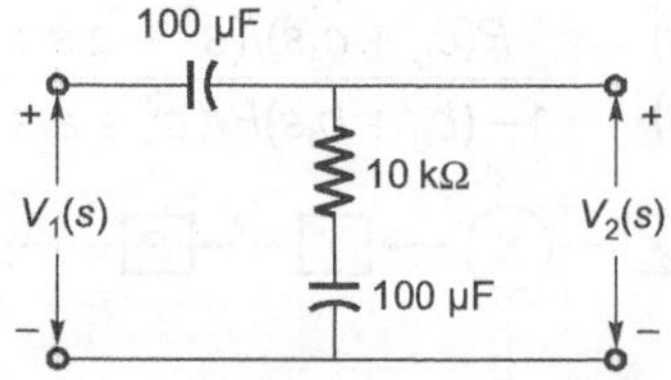

$$\frac{V_2(s)}{V_1(s)} = \frac{R + \frac{1}{Cs}}{\frac{1}{Cs} + R + \frac{1}{Cs}} = \frac{1 + RCs}{2 + RCs}$$

$$= \frac{1 + 10\times10^3 \times 100\times10^{-6} s}{2 + 10\times10^3 \times 100\times10^{-6} s} = \frac{1+s}{2+s}$$

1.16 (a)

Using Mason's gain formulae

$$\Delta = 1 - [-2s^{-1} - 2s^{-2} - 4 - 4s^{-1}]$$

$$\Delta = 1 + \frac{2}{s} + \frac{2}{s^2} + 4 + \frac{4}{s} = \frac{5s^2 + 6s + 2}{s^2}$$

$$P_1 = s^{-2} = \frac{1}{s^2}; \quad P_2 = s^{-1} = \frac{1}{s}$$

$$\Delta_1 = 1, \quad \Delta_2 = 2$$

$$\frac{Y(s)}{u(s)} = \frac{\sum P_k \Delta_k}{\Delta} = \frac{\frac{1}{s^2}\times 1 + \frac{1}{s}\times 1}{\frac{5s^2 + 6s + 2}{s^2}}$$

$$\frac{Y(s)}{u(s)} = \frac{s+1}{5s^2 + 6s + 2}$$

1.17 (c)

Using Mason gain formula,

Transfer function,

$$G(s) = \frac{C(s)}{U(s)} = \frac{\Sigma P_K \Delta_K}{\Delta}$$

Forward paths are:

$$P_1 = \frac{h_0}{s^2}, P_2 = \frac{h_1}{s}$$

Individual loops are:

$$L_1 = \frac{-a_1}{s}, L_2 = \frac{-a_0}{s^2}$$

Non-touching loops = zero

$$\therefore \quad \Delta_1 = 1,$$

$$\Delta_2 = 1 - \left(\frac{-a_1}{s}\right) = \left(1 + \frac{a_1}{s}\right) = \left(\frac{s + a_1}{s}\right)$$

$$\Delta = 1 - \left(\frac{-a_1}{s} - \frac{a_0}{s^2}\right) = \frac{s^2 + a_1 s + a_0}{s^2}$$

$$\therefore \; G(s) = \frac{C(s)}{U(s)} = \frac{P_1\Delta_1 + P_2\Delta_2}{\Delta}$$

$$= \frac{\left(\frac{h_0}{s^2}\right)\times 1 + \left(\frac{h_1}{s}\right)\times\left(\frac{s + a_1}{s}\right)}{\left(\frac{s^2 + a_1 s + a_0}{s^2}\right)}$$

$$= \left[\frac{h_0 + h_1(s + a_1)}{s^2 + a_1 s + a_0}\right]$$

Given: $h_1 = b_1$ and $h_0 = b_0 - b_1 a_1$

Thus, $G(s) = \dfrac{(b_0 - b_1 a_1) + b_1(s + a_1)}{(s^2 + a_1 s + a_0)}$

$$= \left(\frac{b_1 s + b_0}{s^2 + a_1 s + a_0}\right)$$

$$G(s) = \left(\frac{b_1 s + b_0}{s^2 + a_1 s + a_0}\right)$$

1.18 (b)

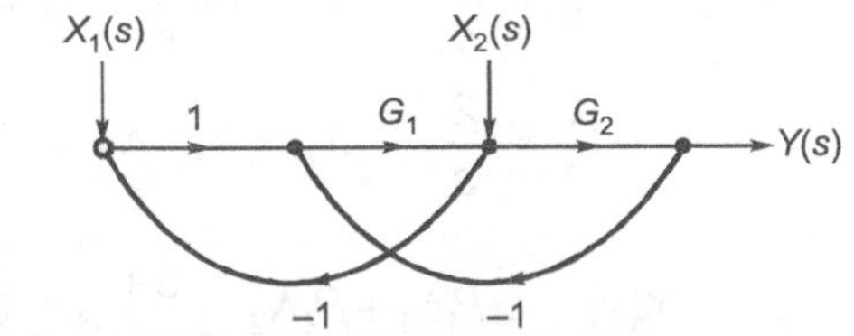

$$\left.\frac{Y(s)}{X_2(s)}\right|_{x_1(s)=0} = \frac{G_2}{1-(-G_1 G_2 - G_1)} = \frac{G_2}{1+G_1(G_2+1)}$$

1.19 (c)

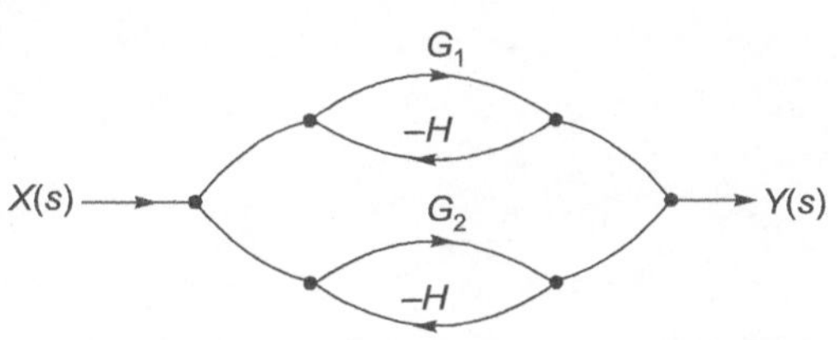

$$\frac{Y(s)}{X(s)} = \frac{G_1 + G_2}{1-(-G_1 H - G_2 H)} = \frac{G_1 + G_2}{1+G_1 H + G_2 H}$$

1.20 (a)

$$\frac{dx_1}{dt} = 2x_1 + x_2 + u \quad \text{...(i)}$$

$$\frac{dx_2}{dt} = -2x_1 + 4 \quad \text{...(ii)}$$

From equation (i),

$$sX_1 = 2X_1 + X_2 + u$$

$$sX_2 = -2X_1 + u$$

$$X_2 = \left(-\frac{2X_1 + u}{s}\right)$$

$$sX_1 = 2X_1 + \left(-\frac{2X_1 + u}{s}\right) + u$$

$$sX_1 - 2X_1 + \frac{2X_1}{s} = \frac{u}{s} + u$$

$$X_1\left(s - 2 + \frac{2}{s}\right) = u\left(1 + \frac{1}{s}\right)$$

$$\frac{X_1}{u} = \frac{\left(1+\frac{1}{s}\right)}{\left(s-2+\frac{2}{s}\right)}$$

$$\Rightarrow \quad \frac{y}{u} = \frac{3X_1}{u} = \frac{3\left(1+\frac{1}{s}\right)}{\left(s-2+\frac{2}{s}\right)}$$

$$\Rightarrow \quad \frac{y}{u} = \frac{3(s+1)}{(s^2 - 2s + 2)}$$

1.21 (a)

Transfer function,

$$\frac{Y(s)}{U_1(s)} = \frac{\frac{K_1}{LJs^2}[1]}{1-\left[-\frac{R}{Ls} - \frac{K_1 K_2}{LJs^2}\right]}$$

$$= \frac{K_1}{LJs^2 + RJs + K_1 K_2}$$

1.22 (0.554)

$$G(s) = \frac{-s+1}{s+1}$$

System output,

$$Y(s) = G(s)\cdot\frac{1}{s} = \frac{-s+1}{s+1}\cdot\frac{1}{s} = \frac{1}{s} - \frac{2}{s+1}$$

$$y(t) = u(t) - 2e^{-t}u(t)$$

$$y(1.5) = 1 - 2e^{-1.5} = 1 - 0.446 = 0.554$$

1.23 (b)

$$\frac{d^2y}{dt^2} + 7\frac{dy}{dt} + 10y = 4x + 5\frac{dx}{dt}$$

$$(s^2 + 7s + 10)\,Y(s) = (4 + 5s)\,X(s)$$

$$\frac{Y(s)}{X(s)} = \frac{5s+4}{s^2 + 7s + 10}$$

Impulse response $= \mathcal{L}^{-1}$ (Transfer function)

$$= \mathcal{L}^{-1}\frac{5s+4}{(s+2)(s+5)}$$

By partial fraction,

$$= \mathcal{L}^{-1}\left[-\frac{2}{s+2} + \frac{7}{s+5}\right] = -2e^{-2t}u(t) + 7e^{-5t}u(t)$$

1.24 (d)

Simplifying given signal flow graph:

Step-1:

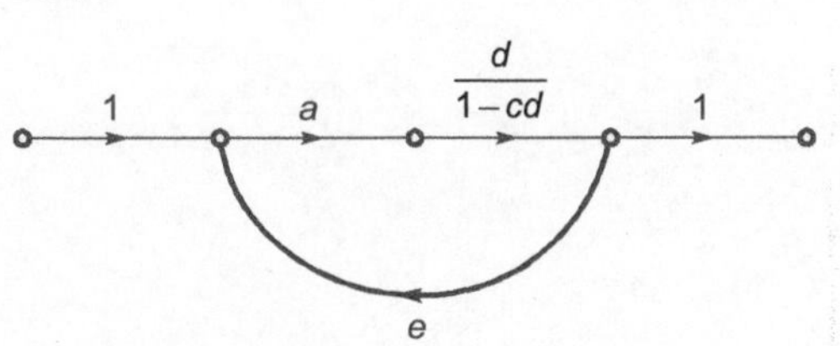

Step-2:

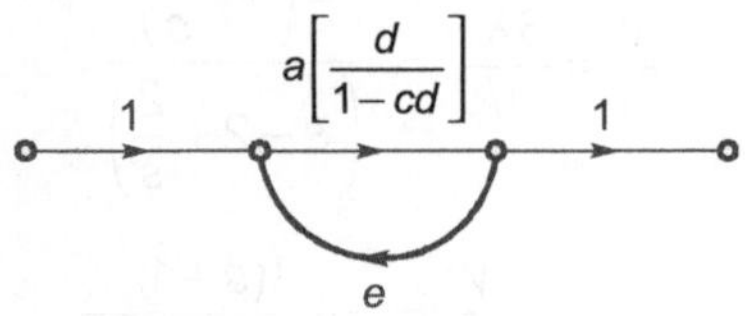

1.25 (a)

Input = 5 V

$$\Rightarrow \quad R(s) = \frac{5}{s}$$

$$\frac{C(s)}{R(s)} = \text{T.F.} = \frac{K}{1+Ts}$$

$$C(s) = \frac{5K}{s(1+Ts)}$$

$$\underset{s\to 0}{Lt}\, sC(s) = 5K = 10 \Rightarrow K = 2$$

Steady state speed = 10 rad/sec. (Given)

$$C(s) = \frac{2}{1+0.5s} \times \frac{5}{s}$$

$$\Rightarrow \quad \underset{s\to 0}{Lt}\, sC(s) = 10$$

1.26 (b)

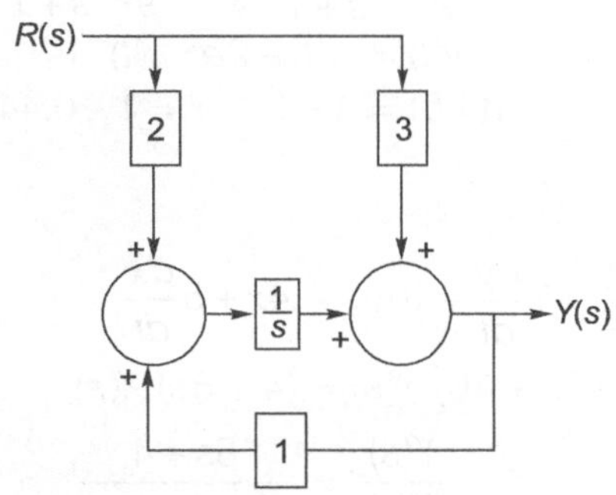

Constructing the signal flow graph :

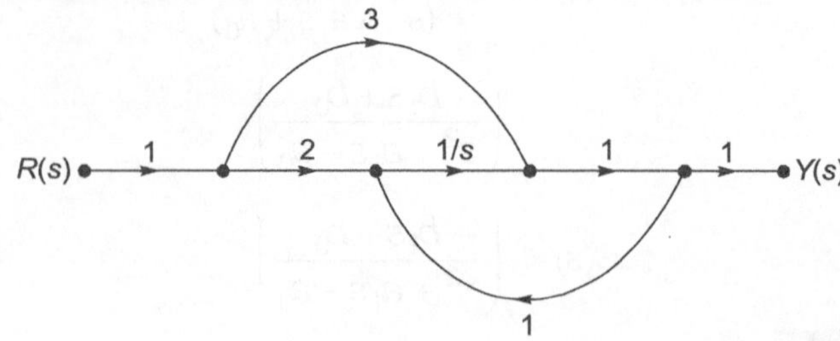

$$L_1 = \frac{1}{s}; \quad \Delta = 1-\frac{1}{s}; \quad P_1 = 3$$

$$P_2 = \frac{2}{s}; \quad \Delta_1 = 1; \quad \Delta_2 = 1$$

$$\frac{Y(s)}{R(s)} = \frac{P_1\Delta_1 + P_2\Delta_2}{\Delta} = \frac{3+\frac{2}{s}}{1-\frac{1}{s}} = \frac{3s+2}{s-1}$$

■■■■

2 Feedback Characteristics of Control Systems

2.1 The unit impulse response of a unit feedback control system is given by: $c(t) = -te^{-t} + 2e^{-t}$, $(t \geq 0)$ the open loop transfer function is equal to

(a) $\dfrac{s+1}{(s+2)^2}$ (b) $\dfrac{2s+1}{s^2}$

(c) $\dfrac{s+1}{(s+1)^2}$ (d) $\dfrac{s+1}{s^2}$ [1996 : 1 M]

2.2 Feedback control systems are

(a) insensitive to both forward and feedback path parameter changes.

(b) less sensitive to feedback path parameter changes than to forward path parameter changes.

(c) less sensitive to forward path parameter changes than to feedback path parameter changes.

(d) equally sensitive to forward and feedback path parameter changes. [2000 : 1 M]

2.3 As shown in the figure, a negative feedback system has an amplifier of gain 100 with ±10% tolerance in the forward path, and an attenuator of value 9/100 in the feedback path. The overall system gain in approximately:

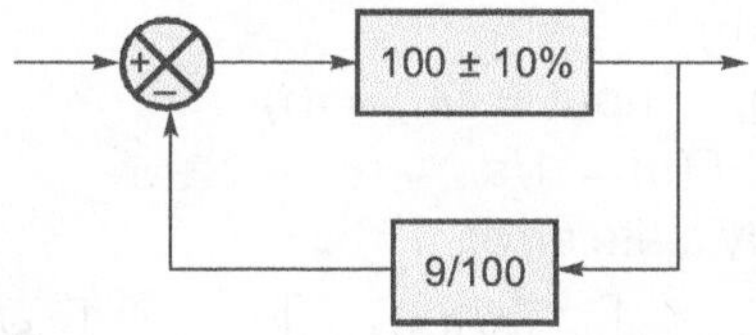

(a) 10 ± 1% (b) 10 ± 2%

(c) 10 ± 5% (d) 10 ± 10% [2010 : 1 M]

2.4 The open-loop transfer function of a DC motor is given as $\dfrac{\omega(s)}{V_a(s)} = \dfrac{10}{1+10s}$. When connected in feedback as shown below, the approximate value of K_a that will reduce the time constant of the closed loop system by one hundred times as compared to that of the open-loop system is

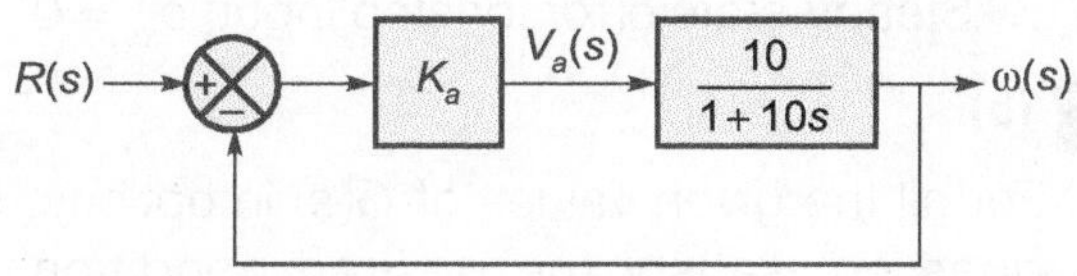

(a) 1 (b) 5

(c) 10 (d) 100 [2013 : 2 M]

2.5 The closed-loop transfer function of a system is $T(s) = \dfrac{4}{(s^2 + 0.4s + 4)}$. The steady state error due to unit step input is ______. [2014 : 1 M, Set-2]

2.6 The block diagram of a system is shown in the figure.

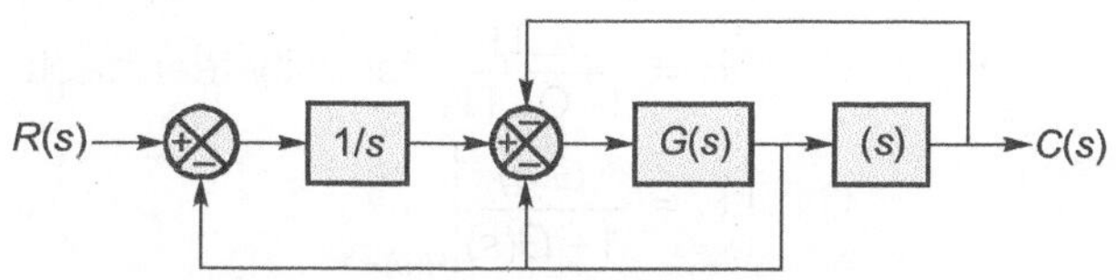

If the desired transfer function of the system is $\dfrac{C(s)}{R(s)} = \dfrac{s}{s^2+s+1}$ then $G(s)$ is

(a) 1 (b) s

(c) $\dfrac{1}{s}$ (d) $\dfrac{-s}{s^3+s^2-s-2}$

[2014 : 2 M, Set-3]

2.7 For the closed-loop system shown, the transfer function $\dfrac{E(s)}{R(s)}$ is

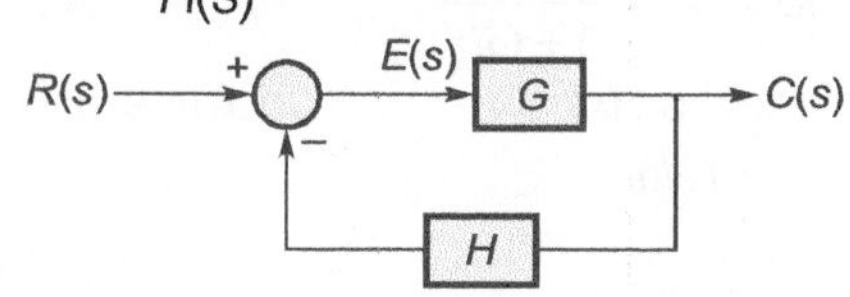

(a) $\dfrac{1}{1+G}$ (b) $\dfrac{1}{1+GH}$

(c) $\dfrac{GH}{1+GH}$ (d) $\dfrac{G}{1+GH}$ [2021 : 1 M]

2.8 For the block-diagram shown in the figure, the transfer function $\dfrac{C(s)}{R(s)}$ is

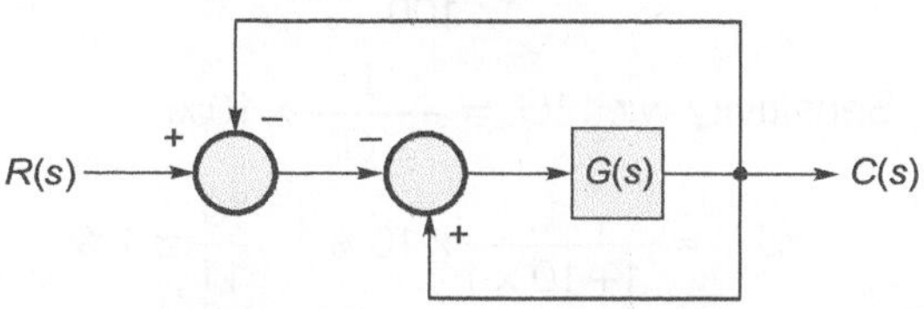

(a) $\dfrac{G(s)}{1-2G(s)}$ (b) $-\dfrac{G(s)}{1+2G(s)}$

(c) $-\dfrac{G(s)}{1-2G(s)}$ (d) $\dfrac{G(s)}{1+2G(s)}$

[2024 : 1 M]

Answers Feedback Characteristics of Control Systems

2.1 (b) 2.2 (c) 2.3 (a) 2.4 (c) 2.5 (0) 2.6 (b) 2.7 (b)
2.8 (c)

Explanations Feedback Characteristics of Control Systems

2.1 (b)

Close loop T.F. = $\mathcal{L}$(impulse response)

$$= \mathcal{L}(-te^{-t} + 2e^{-t}) = \frac{-1}{(s+1)^2} + \frac{2}{(s+1)}$$

$$\text{C.L.T.F.} = \frac{2s+1}{(s+1)^2}$$

$$\because \quad \text{C.L.T.F.} = \frac{\text{O.L.T.F.}}{1+\text{O.L.T.F.}} \text{ for unity feedback}$$

$$\text{i.e.} \quad \text{C.L.T.F.} = \left.\frac{G(s)}{1+G(s)}\right|_{H(s)=1}$$

$$\therefore \quad G(s) = \left.\frac{\text{C.L.T.F.}}{1-\text{C.L.T.F.}}\right|_{H(s)=1}$$

On putting the value of C.L.T.F.

We get, $G(s) = \dfrac{2s+1}{s^2}$

2.2 (c)

$$\because \quad S_G^{T.F.} = \frac{1}{1+GH}$$

$$|S_H^{T.F.}| = \frac{GH}{1+GH}$$

$\Rightarrow$ Less sensitive to 'change in G' than 'change in H'.

2.3 (a)

$$G = 100 \pm 10\%$$

$$\frac{\Delta G}{G} = 10\% \text{ or } 0.1; \quad H = \frac{9}{100}$$

$$\text{Overall gain,} \quad T = \frac{G}{1+GH} \quad \text{...(i)}$$

$$T = \frac{100}{1+100\times\frac{9}{100}} = 10$$

$$\text{Sensitivity w.r.t. 'G'} = \frac{1}{1+GH}\times 10\%$$

$$= \frac{1}{1+10\times 1}\times 10\% = \frac{10}{11} \cong 1\%$$

2.4 (c)

$$\frac{\omega(s)}{V_a(s)} = \frac{10}{1+10s} = \frac{10}{1+\tau s}$$

$$\tau = 10$$

Closed loop system,

$$\frac{\omega(s)}{R(s)} = \frac{\frac{10K_a}{1+10s}}{1+\frac{10K_a}{1+10s}} = \frac{10K_a}{1+10K_a+10s}$$

$$= \frac{\frac{10K_a}{1+10K_a}}{1+\frac{10}{1+10K_a}s}$$

According to question,

$$\frac{10}{1+10K_a} = \frac{1}{100}\times 10$$

$$1 + 10\,K_a = 100 \Rightarrow K_a = 9.9$$

2.5 (0)

Closed-loop transfer function,

$$T(s) = \frac{4}{(s^2+0.4s+4)}$$

$$\text{or,} \quad \frac{G(s)}{1+G(s)} = \left(\frac{4}{s^2+0.4s+4}\right) \quad [H(s)=1]$$

or, $(s^2 + 0.4s + 4) = 4 + 4\,G(s)$

$$\text{or,} \quad G(s) = \frac{4}{s(s+0.4)}$$

= Open loop transfer function (for unity feedback system)

Given, Input = $u(t) = r(t)$

$\therefore$ $R(s) = 1/s$

Steady state error,

$$e_{ss} = \lim_{s\to 0}\left[\frac{sR(s)}{1+G(s)H(s)}\right] = \lim_{s\to 0}\left[\frac{sR(s)}{1+G(s)}\right]$$

(For $H(s) = 1$)

$$= \lim_{s\to 0}\left[\frac{s\times\frac{1}{s}}{1+\frac{4}{s(s+0.4)}}\right] = \lim_{s\to 0}\left[\frac{s(s+0.4)}{s^2+0.4s+4}\right]$$

$$= \lim_{s\to 0}\left[\frac{s(s+0.4)}{s^2+0.4s+4}\right] = 0$$

$\therefore$ Steady-state error for step input, $e_{ss} = 0$.

2.6 (b)

For all the given values of $G(s)$ in options, $G(s)$ does not satisfy the desired condition. But option (b) is official answer given by IIT.

2.7 (b)

$$\frac{E(s)}{R(s)} = \frac{R(s) - H \times C(s)}{R(s)} = 1 - H \times \frac{C(s)}{R(s)}$$

$$= 1 - \frac{HG}{1+GH} = \frac{1+GH-GH}{1+GH} = \frac{1}{1+GH}$$

2.8 (c)

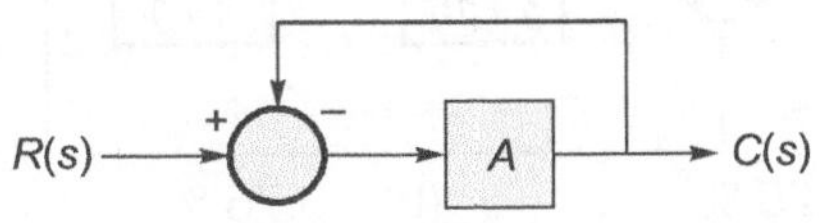

$$A = \frac{-G(s)}{1-G(s)}$$

Transfer fucntion :

$$\frac{C(s)}{R(s)} = \frac{A}{1+A}$$

$$= \frac{-G(s)}{1-G(s)-G(s)}$$

$$= \frac{-G(s)}{1-2G(s)}$$

3 Time Response Analysis

3.1 The steady state error due to a step input for type 1 system is ______. [1995 : 1 M]

3.2 Consider the unit step response of a unity feedback control system whose open loop transfer function is $G(s)=\dfrac{1}{s(s+1)}$. The maximum overshoot is equal to

(a) 0.143 (b) 0.153
(c) 0.163 (d) 0.173 [1996 : 1 M]

3.3 For a feedback control system of type 2, the steady state error for a ramp input is

(a) infinite (b) constant
(c) zero (d) indeterminate

[1996 : 1 M]

3.4 For the system shown in figure, with a damping ratio ξ of 0.7 and an undamped natural frequency ω_n of 4 rad/sec, the values of 'K' and 'a' are

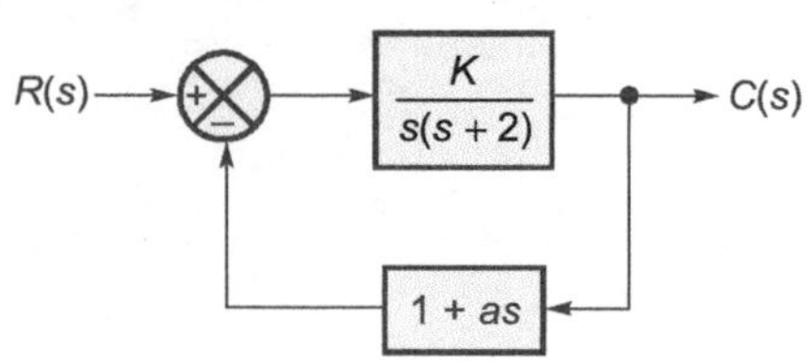

(a) $K = 4$, $a = 0.35$ (b) $K = 8$, $a = 0.455$
(c) $K = 16$, $a = 0.225$ (d) $K = 64$, $a = 0.9$

[1996 : 2 M]

3.5 A unity feedback system has open loop transfer function $G(s)$. The steady state error is zero for

(a) step input and type 1 $G(s)$
(b) ramp input and type 1 $G(s)$
(c) step input and type 0 $G(s)$
(d) ramp input and type 0 $G(s)$ [2000 : 1 M]

3.6 A unity feedback system has open loop transfer function $G(s)=\dfrac{25}{s(s+6)}$. The peak overshoot in the step-input response of the system is approximately equal to

(a) 5% (b) 10%
(c) 15% (d) 20% [2000 : 2 M]

3.7 The block diagram shown in figure gives a unity feedback closed loop control system. The steady state error in the response of the above system to unit step input is

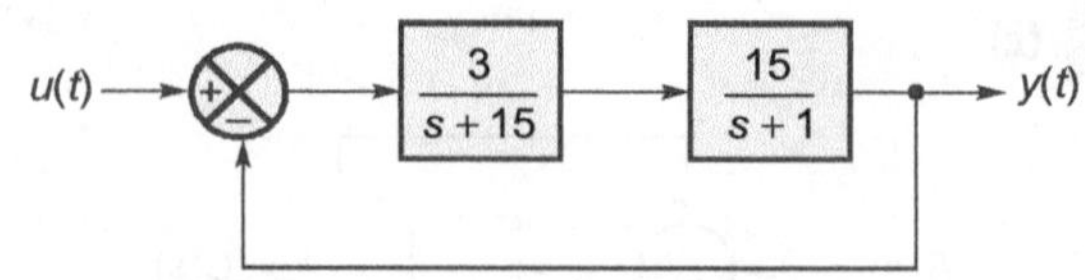

(a) 25 % (b) 0.75 %
(c) 6 % (d) 33 % [2003 : 2 M]

3.8 The block diagram of a closed loop control system is given by figure. The values of K and P such that the system has a damping ratio of 0.7 and an undamped natural frequency ω_n of 5 rad/sec, are respectively equal to

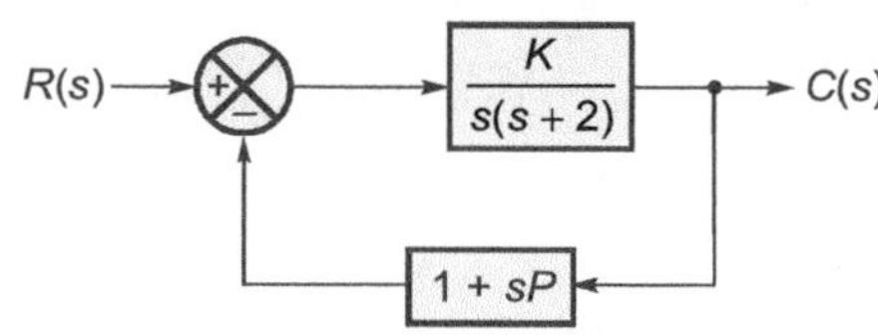

(a) 20 and 0.3 (b) 20 and 0.2
(c) 25 and 0.3 (d) 25 and 0.2 [2004 : 2 M]

3.9 Consider the feedback system shown below which is subjected to a unit step input. The system is stable and has following parameters $k_p = 4$, $k_i = 10$, $\omega = 500$ and $\xi = 0.7$. The steady state value of Z is

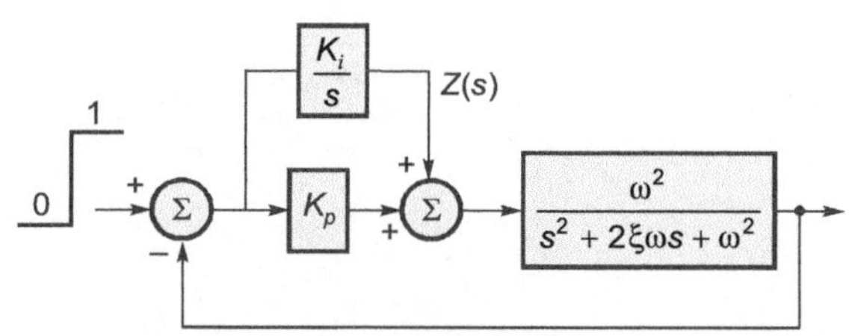

(a) 1 (b) 0.25
(c) 0.1 (d) 0 [2007 : 2 M]

Statement for Common Data Questions (3.10 and 3.11):

R-L-C circuit shown in figure,

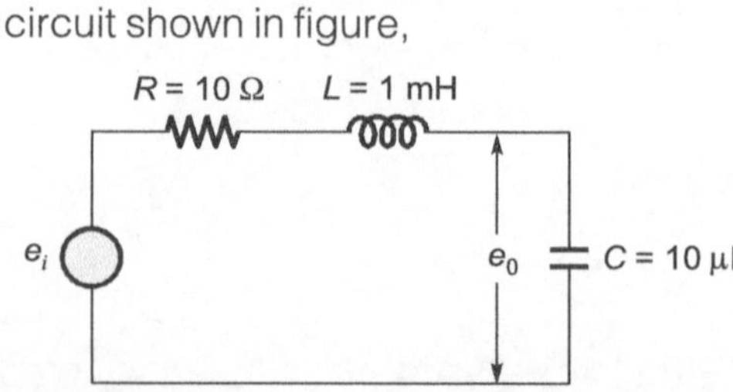

3.10 For a step-input e_i, the overshoot in the output e_0 will be

(a) 0, since the system is not under damped

(b) 5%
(c) 16%
(d) 48% [2007 : 2 M]

3.11 If the above step response is to be observed on a non-storage CRO, then it would be best have the e_i as a
(a) step function
(b) square wave of 50 Hz
(c) square wave of 300 Hz
(d) square wave of 2.0 KHz [2007 : 2 M]

3.12 The transfer function of a linear time invariant system is given as $G(s) = \frac{1}{s^2 + 3s + 2}$. The steady state value of the output of the system for a unit impulse input applied at time instant $t = 1$ will be
(a) 0 (b) 0.5
(c) 1 (d) 2 [2008 : 2 M]

3.13 The transfer function of a system is given as
$$\frac{100}{s^2 + 20s + 100}$$
The system is
(a) an overdamped system
(b) an underdamped system
(c) a critically damped system
(d) an unstable system [2008 : 2 M]

3.14 The unit-step response of a unity feedback system with open loop transfer function $G(s) = K/((s + 1)(s + 2))$ is shown in the figure. The value of K is

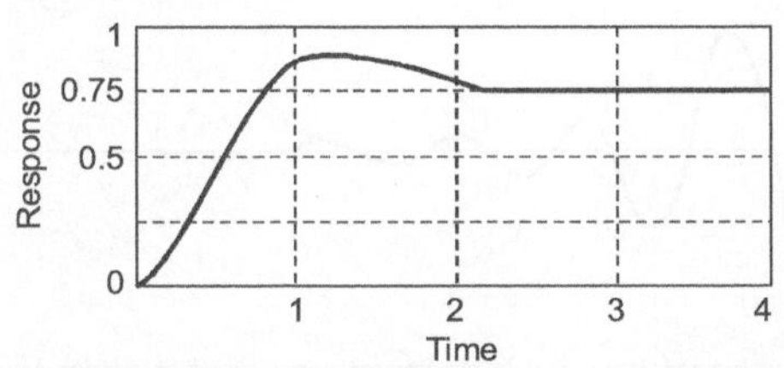

(a) 0.5 (b) 2
(c) 4 (d) 6 [2009 : 2 M]

3.15 For the system $\frac{2}{(s+1)}$, the approximate time taken for a step response to reach 98% of its final value is
(a) 1 s (b) 2 s
(c) 4 s (d) 8 s [2010 : 1 M]

3.16 The steady state error of a unity feedback linear system for a unit step input is 0.1. The steady state error of the same system, for a pulse input

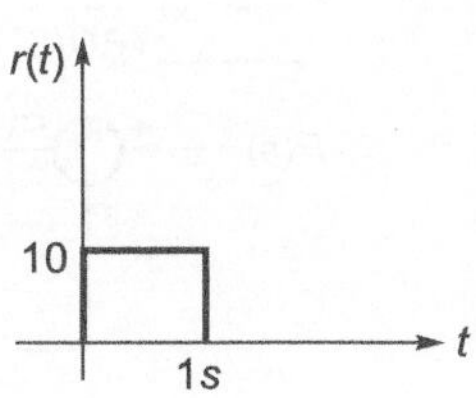

$r(t)$ having a magnitude of 10 and a duration of one second, as shown in the figure is

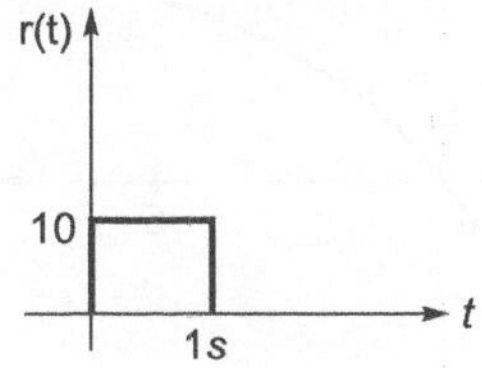

(a) 0 (b) 0.1
(c) 1 (d) 10 [2011 : 1 M]

3.17 A two-loop position control system is shown below.

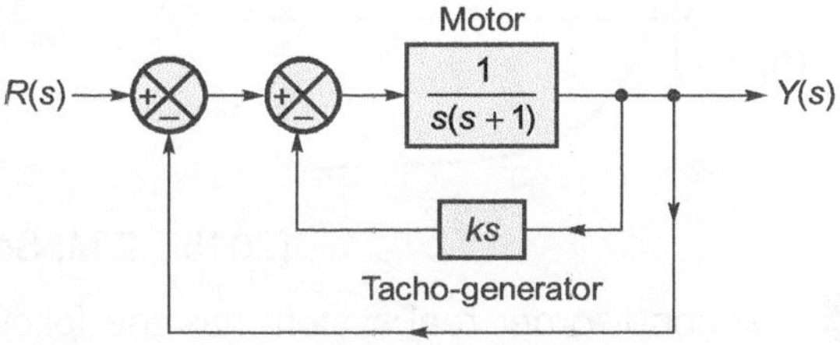

The gain k of the Tacho-generator influences mainly the
(a) peak overshoot
(b) natural frequency of oscillation
(c) phase shift of the closed loop transfer function at very low frequencies ($\omega \to 0$)
(d) phase shift of the closed loop transfer function at very high frequencies ($\omega \to \infty$) [2011 : 2 M]

3.18 An open loop control system results in a response of $e^{-2t}(\sin 5t + \cos 5t)$ for a unit impulse input. The DC gain of the control system is _______.
[2015 : 1 M, Set-2]

3.19 The unit step response of a system with the transfer function $G(s) = \frac{1-2s}{1+s}$ is given by which one of the following waveforms?

(a)
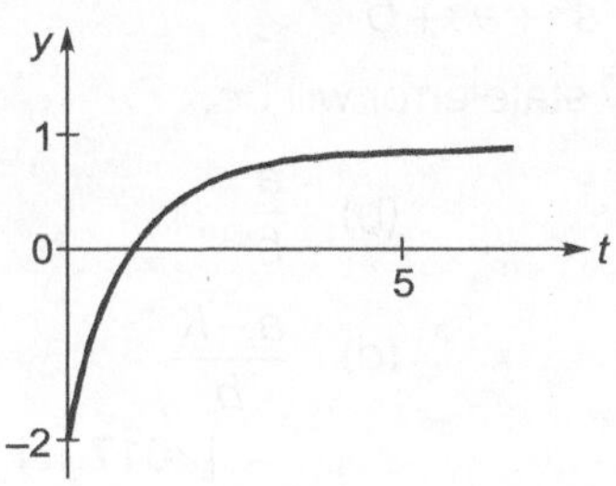

(b)
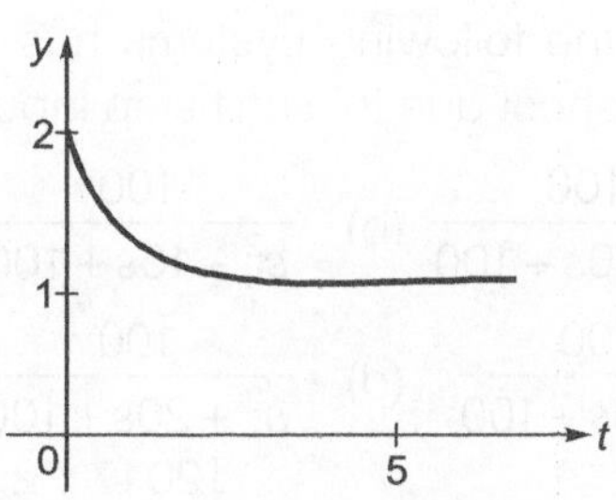

(c) 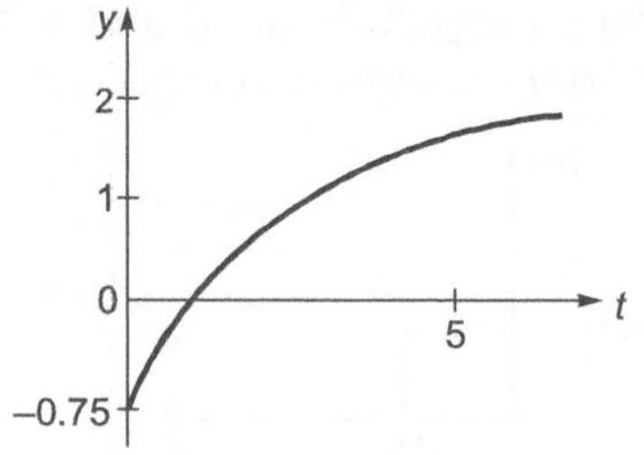

(d) 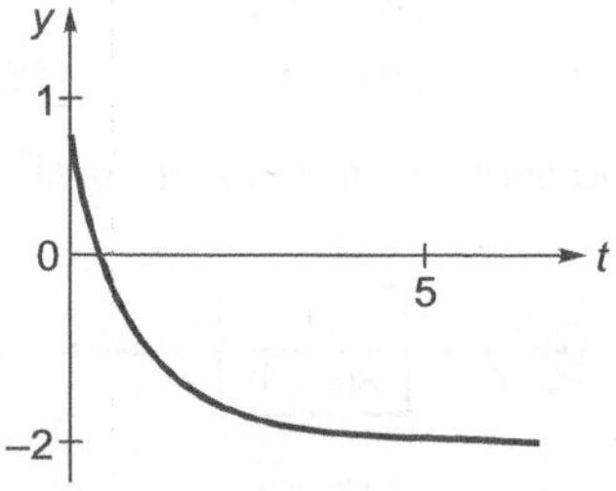

[2015 : 2 M, Set-2]

3.20 A second -order real system has the following properties :

(a) the damping ratio ζ = 0.5 and undamped natural frequency ω_n = 10 rad/s,

(b) the steady state value of the output, to a unit step input, is 1.02.

The transfer function of the system is

(a) $\dfrac{102}{s^2+5s+100}$ (b) $\dfrac{102}{s^2+10s+100}$

(c) $\dfrac{100}{s^2+10s+100}$ (d) $\dfrac{102}{s^2+5s+100}$

[2016 : 2 M, Set-2]

3.21 When a unit ramp input is applied to the unity feedback system having closed loop transfer function,

$$\frac{C(s)}{R(s)}=\frac{Ks+b}{s^2+as+b},(a>0,\ b>0,\ K>0),$$

the steady state error will be,

(a) 0 (b) $\dfrac{a}{b}$

(c) $\dfrac{a+K}{b}$ (d) $\dfrac{a-K}{b}$

[2017 : 1 M, Set-2]

3.22 Which of the following systems has maximum peak overshoot due to a unit step input?

(a) $\dfrac{100}{s^2+10s+100}$ (b) $\dfrac{100}{s^2+15s+100}$

(c) $\dfrac{100}{s^2+5s+100}$ (d) $\dfrac{100}{s^2+20s+100}$

[2017 : 2 M, Set-2]

3.23 Match the transfer functions of the second-order systems with the nature of the system given below.

Transfer functions	Nature of system
P. $\dfrac{15}{s^2+5s+15}$	I. Overdamped
Q. $\dfrac{25}{s^2+10s+25}$	II. Critically damped
R. $\dfrac{35}{s^2+18s+35}$	III. Underdamped

Codes:

(a) P-I, Q-II, R-III

(b) P-II, Q-I, R-III

(c) P-III, Q-II, R-I

(d) P-III, Q-I, R-II

[2018 : 1 M]

3.24 Consider a unity feedback system with forward transfer function given by

$$G(s)=\frac{1}{(s+1)(s+2)}$$

The steady-state error in the output of the system for a unit-step input is ____ (upto 2 decimal places.

[2018 : 1 M]

3.25 The unit step response $y(t)$ of a unity feedback system with open-loop transfer function

$G(s)H(s)=\dfrac{K}{(s+1)^2\,(s+2)}$ is shown in the figure.

The value of K is _____ (upto 2 decimal places).

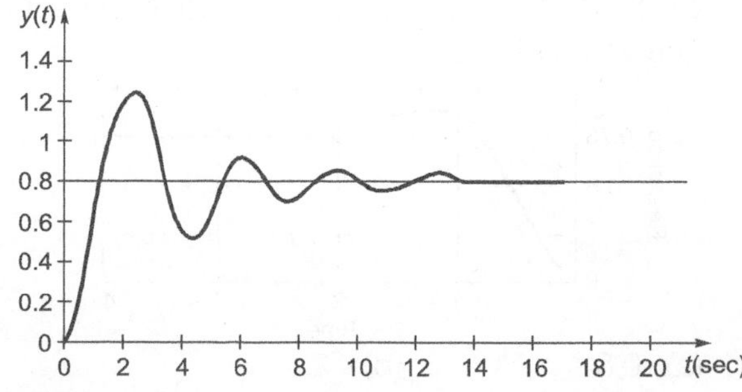

[2018 : 2 M]

3.26 Consider a closed-loop system as shown. $G_p(s)=\dfrac{14.4}{s(1+0.1s)}$ is the plant transfer function and $G_c(s)$ = 1 is the compensator. For a unit-step input, the output response has damped oscillations. The damped natural frequency is _______ rad/s. (Round off to 2 decimal places.)

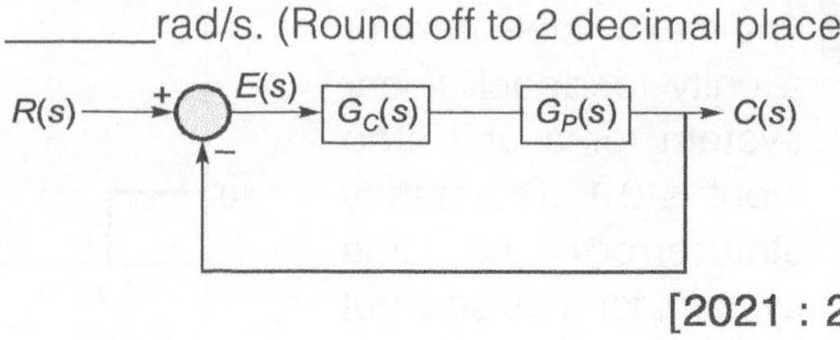

[2021 : 2 M]

3.27 The damping ratio and undamped natural frequency of a closed loop system as shown in the figure, are denoted as ξ and ω_n, respectively. The values of ξ and ω_n are

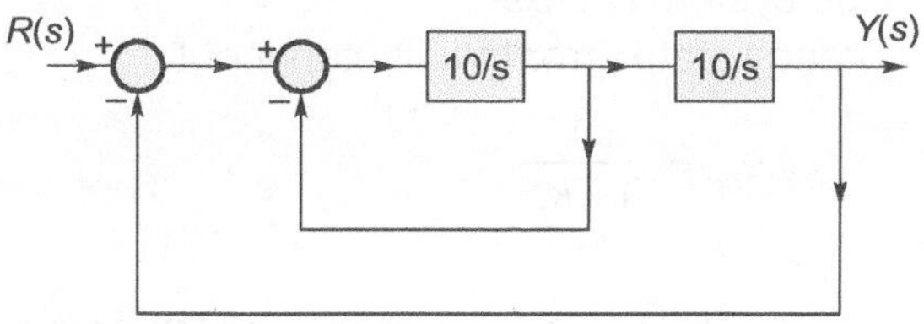

(a) $\xi = 0.5$ and $\omega_n = 10$ rad/s
(b) $\xi = 0.1$ and $\omega_n = 10$ rad/s
(c) $\xi = 0.707$ and $\omega_n = 10$ rad/s
(d) $\xi = 0.707$ and $\omega_n = 100$ rad/s

[2022 : 2 M]

3.28 Consider the cascaded system as shown in the figure. Neglecting the faster component of the trasient response, which one of the following options is a first-order pole-only approximation such that the steady-state values of the unit step responses of the original and the approximated system are same?

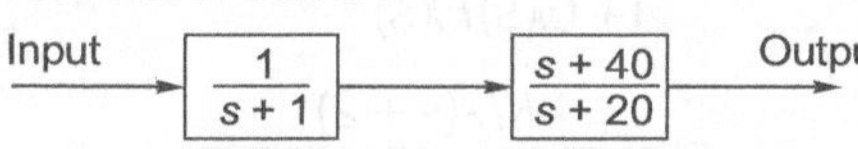

(a) $\dfrac{2}{s+20}$ (b) $\dfrac{2}{s+1}$

(c) $\dfrac{1}{s+20}$ (d) $\dfrac{1}{s+1}$

[2024 : 1 M]

3.29 Consider the standard second-order system of the form $\dfrac{\omega_n^2}{s^2+2\xi\omega_n s+\omega_n^2}$ with the poles p and p* having negative real parts. The pole locations are also shown in the figure. Now consider two such second-order systems as defined below :

System 1 : ω_n = 3 rad/sec and θ = 60°

System 2 : ω_n = 1 rad/sec and θ = 70°

Which one of the following statements is correct?

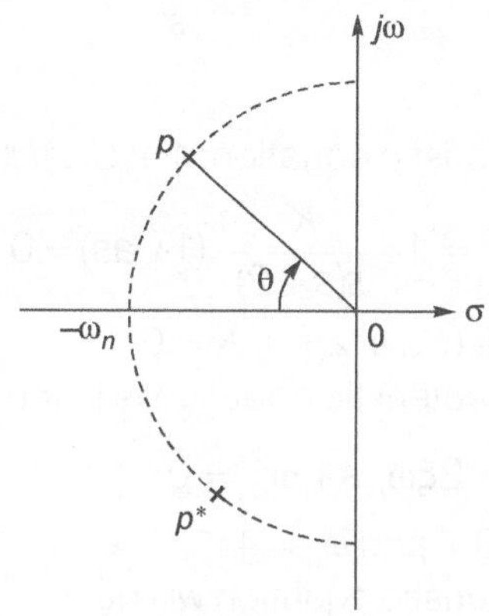

(a) Settling time cannot be computed from the given information.
(b) Settling time of System 1 is more than that of System 2.
(c) Settling times of both the systems are the same.
(d) Settling time of System 2 is more than that of System 1.

[2024 : 1 M]

Answers Time Response Analysis

3.1 (0)	3.2 (c)	3.3 (c)	3.4 (c)	3.5 (a)	3.6 (b)	3.7 (a)
3.8 (d)	3.9 (a)	3.10 (c)	3.11 (c)	3.12 (a)	3.13 (c)	3.14 (d)
3.15 (c)	3.16 (a)	3.17 (a)	3.18 (0.241)	3.19 (a)	3.20 (b)	3.21 (d)
3.22 (c)	3.23 (c)	3.24 (0.66)	3.25 (8)	3.26 (10.90)	3.27 (a)	3.28 (b)
3.29 (d)						

Explanations Time Response Analysis

3.1 (0)

$$\lim_{s\to 0} sE(s) = \lim_{s\to 0}\frac{s\cdot\frac{1}{s}}{1+\frac{G'(s)H(s)}{s}} = 0$$

3.2 (c)

$$G(s) = \frac{1}{s(s+1)}$$

∴ Characteristic equation = 1 + $G(s)$
$= s^2 + s + 1$

On comparing with $s^2 + 2\xi\omega_n s + \omega_n^2$

We have, $\xi = 0.5$

$$\therefore \quad M_p = e^{\left(\frac{-\xi\pi}{\sqrt{1-\xi^2}}\right)} = 0.1630$$

3.3 (c)

$$\lim_{s\to 0} sE(s) = \lim_{s\to 0} \frac{s\frac{1}{s^2}}{1+\frac{G'(s)H(s)}{s^2}} = 0$$

3.4 (c)

Characteristic equation: $1 + G(s)H(s) = 0$

$$= 1+\frac{K}{s(s+2)}\cdot(1+as) = 0$$

$\Rightarrow \quad s^2 + (Ka + 2)s + K = 0 \qquad ...(i)$

$\because$ Characteristic equation is in the form of

$$s^2 + 2\xi\omega_n s + \omega_n^2 = 0$$

For $\xi = 0.7$ and $\omega_n = 4$

Characteristic equation will be

$s^2 + 5.6s + 16 = 0 \qquad ...(ii)$

On comparing equation (i) and (ii),

we get, $K = 16$ and $a = 0.225$

3.5 (a)

Steady state error is zero for step input with type 1 $G(s)$.

3.6 (b)

$$G(s) = \frac{25}{s(s+6)}$$

Characteristic equation:

$$1 + G(s) = 0$$

$$1+\frac{25}{s(s+6)} = 0 \Rightarrow s^2 + 6s + 25 = 0$$

On comparing with

$$s^2 + 2\xi\omega_n s + \omega_n^2 = 0$$

we get, $\omega_n = 5$ and $\xi = 0.6$

$$\therefore \quad M_p = e^{\left(\frac{-\xi\pi}{\sqrt{1-\xi^2}}\right)} = 0.0948 \cong 10\%$$

3.7 (a)

$$G(s) = \left(\frac{3}{s+15}\right)\times\left(\frac{15}{s+1}\right)$$

$$= \frac{45}{(s+1)(s+15)} \text{ and } H(s) = 1$$

Open loop transfer function

$$= G(s)\cdot H(s) = \frac{45}{(s+1)(s+15)}$$

The system is type-0.

Steady state error to unit-step input

$$e_{ss} = \frac{1}{1+k_p}$$

where,

k_p = Position error constant = $\underset{s\to 0}{\text{Lt}}\, G(s)H(s)$

$$\Rightarrow \quad k_p = \underset{s\to 0}{\text{Lt}} \frac{45}{(s+1)(s+15)} = 3$$

$$e_{ss} = \frac{1}{1+k_p} = \frac{1}{1+3} = 0.25 \text{ or } 25\%$$

3.8 (d)

$$G(s) = \frac{k}{s(s+2)} \text{ and } H(s) = 1+sP$$

Closed-loop transfer function,

$$T(s) = \frac{G(s)}{1+G(s)H(s)}$$

$$= \frac{k/s(s+2)}{1+\frac{k}{s(s+2)}\cdot(1+sP)}$$

$$= \frac{k}{s(s+2)+k(1+sP)}$$

$$\Rightarrow \quad T(s) = \frac{k}{s^2+(2+kP)s+k}$$

So, characteristic equation

$$= s^2 + (2 + kP)\, s + P$$

Comparing with standard equation

$$= s^2 + 2\xi\omega_n s + \omega_n^2$$

$$k = \omega_n^2 = 5^2 = 25$$

(where ω_n = undamped natural frequency)

$$2\xi\omega_n = 2 + kP$$

$$\Rightarrow \quad 2\times 0.7\times 5 = 2 + 25P$$

(where ξ = Damping ratio)

$$P = 0.2$$

3.9 (a)

Step input $\Rightarrow R(s) = 1/s$

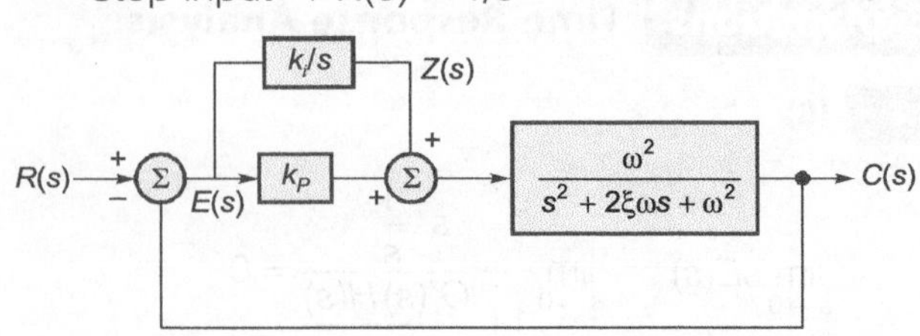

$$G(s) = \left(k_p + \frac{k_i}{s}\right)\left(\frac{\omega^2}{s^2 + 2\xi\omega s + \omega^2}\right)$$

and $H(s) = 1$

$$\frac{C(s)}{R(s)} = \frac{G(s)}{1+G(s)H(s)} = \frac{G(s)}{1+G(s)}$$

$$E(s) = R(s) - C(s)$$

$$= R(s)\left[1 - \frac{C(s)}{R(s)}\right]$$

$$= R(s)\left[1 - \frac{G(s)}{1+G(s)}\right] = \frac{R(s)}{1+G(s)}$$

$$Z(s) = \frac{k_i}{s}\cdot E(s) = \frac{\frac{k_i}{s}\cdot R(s)}{1+G(s)}$$

Steady state value of Z,

$$Z_{ss} = \underset{s\to 0}{\text{Lt}}\ sZ(s)$$

$$= \underset{s\to 0}{\text{Lt}} \frac{s\cdot\frac{k_i}{s}\cdot\frac{1}{s}}{1+\left(k_p + \frac{k_i}{s}\right)\left(\frac{\omega^2}{s^2 + 2\xi\omega s + \omega^2}\right)}$$

$$= \frac{k_i}{k_i\cdot\frac{\omega^2}{\omega^2}} = 1$$

3.10 (c)

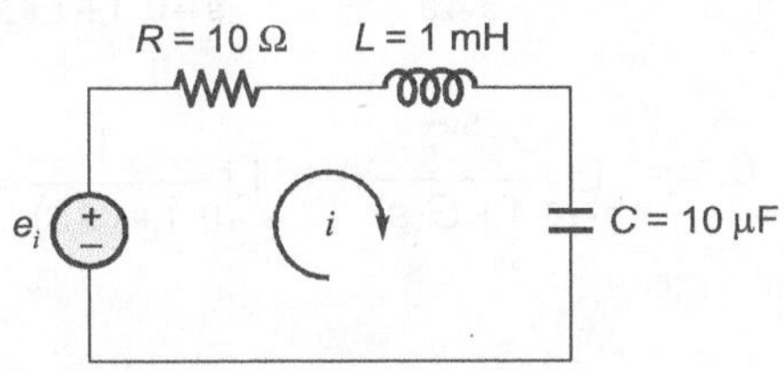

$$e_i = Ri + L\frac{di}{dt} + \frac{1}{C}\int i\,dt$$

Taking Laplace transform,

$$E_i(s) = \left(R + Ls + \frac{1}{Cs}\right)I(s)$$

$$I(s) = \frac{E_i(s)}{R + Ls + \frac{1}{Cs}}$$

$$e_0 = \frac{1}{C}\int i\,dt$$

$$\Rightarrow\quad E_0(s) = \frac{1}{Cs}I(s) = \frac{1}{Cs}\left[\frac{E_i(s)}{R + Ls + \frac{1}{Cs}}\right]$$

$$E_0(s) = \frac{E_i(s)}{RCs + LCs^2 + 1}$$

$$\frac{E_0(s)}{E_i(s)} = \frac{1}{LC\left(s^2 + \frac{R}{L}s + \frac{1}{LC}\right)}$$

Characteristic equation,

$$\Rightarrow\quad s^2 + \frac{R}{L}s + \frac{1}{LC} = 0$$

Comparing with,

$s^2 + 2\xi\omega_n s + \omega_n^2$

$$\omega_n = \frac{1}{\sqrt{LC}}$$

$$2\xi\omega_n = \frac{R}{L}$$

$$\xi = \frac{R}{L}\times\frac{1}{2\omega_n}$$

$$= \frac{R}{L}\times\frac{\sqrt{LC}}{2} = \frac{R}{2}\sqrt{\frac{C}{L}}$$

$$\xi = \frac{10}{2}\sqrt{\frac{10\times 10^{-6}}{1\times 10^{-3}}} = 0.5$$

$$\text{Overshoot} = P_P = e^{-\pi\xi/\sqrt{1-\xi^2}}$$

$$= e^{-\pi\times 0.5/\sqrt{1-0.5^2}} = 0.163 \text{ or } 16.3\%$$

3.11 (c)

$$\omega_n = \frac{1}{\sqrt{LC}}$$

$$= \frac{1}{\sqrt{1\times 10^{-3}\times 10\times 10^{-6}}} = 10^4 \text{ rad/sec}$$

$$\text{Settling time } (t_s) = \frac{4}{\xi\omega_n}$$

$$= \frac{4}{10^4\times 0.5} = 0.8 \text{ msec.}$$

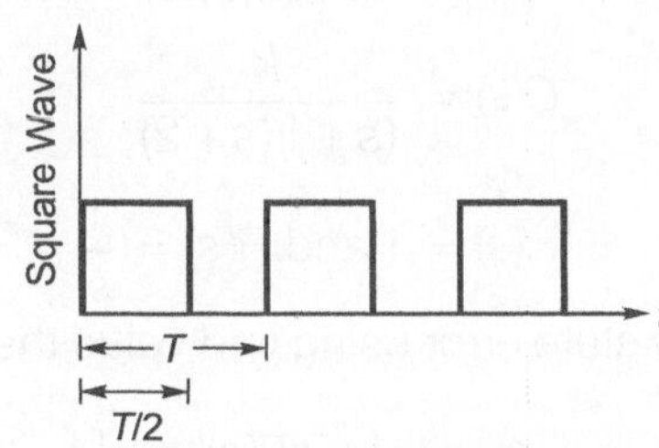

For a square wave $T/2$ should be greater than t_s

For $f_1 = 50$ Hz

$$\Rightarrow\quad \frac{T_1}{2} = \frac{1}{2\times 50} = 10 \text{ ms} >> t_s$$

For $f_2 = 300$ Hz

$$\Rightarrow\quad \frac{T_2}{2} = \frac{1}{2\times 300} = 1.67 \text{ ms} > t_s$$

For $f_3 = 2$ kHz

$$\Rightarrow \quad \frac{T_3}{2} = \frac{1}{2\times 2\times 10^3} = 0.25 \text{ ms} < t_s$$

Therefore, it would be best to have e_i as a square wave of 300 Hz.

3.12 (a)

$$r(t) = \text{unit impulse applied at } t = 1 = \delta(t-1)$$

$$R(s) = 1[r(t)] = e^{-s}$$

$$G(s) = \frac{C(s)}{R(s)} = \frac{1}{s^2+3s+2}$$

$$C(s) = R(s)G(s) = \frac{e^{-s}}{s^2+3s+2}$$

Steady state value of output, using final value theorem,

$$C_{ss} = \underset{s\to 0}{\text{Lt}}\, sC(s) = \underset{s\to 0}{\text{Lt}} \frac{se^{-s}}{s^2+3s+2} = 0$$

3.13 (c)

$$M(s) = \frac{100}{s^2+20s+100}$$

Comparing with standard form,

$$M(s) = \frac{\omega_n^2}{s^2+2\xi\omega_n s+\omega_n^2}$$

$\therefore \quad 2\xi\omega_n = 20; \quad \omega_n = 100; \quad \xi = 1$

$\therefore \quad \omega_n = 10$

$\therefore$ The system is critically damped.

3.14 (d)

Steady state value of response = 0.75

Input is unit-step. So steady state error

$$e_{ss} = 1 - 0.75 = 0.25$$

$$\text{Error} = E(s) = \frac{R(s)}{1+G(s)H(s)}$$

where, $\quad G(s) = \dfrac{k}{(s+1)(s+2)}$

$$H(s) = 1 \text{ and } R(s) = \frac{1}{s}$$

Steady state error using find value theorem,

$$e_{ss} = \underset{s\to 0}{\text{Lt}}\, sE(s) = \underset{s\to 0}{\text{Lt}} \frac{sR(s)}{1+G(s)H(s)}$$

$$= \underset{s\to 0}{\text{Lt}} \frac{s\cdot\frac{1}{s}}{1+\frac{k}{(s+1)(s+2)}} = \frac{1}{1+k/2}$$

$$\Rightarrow \quad 0.25 = \frac{1}{1+k/2} \Rightarrow 1+\frac{k}{2} = 4 \Rightarrow k = 6$$

3.15 (c)

$$\frac{C(s)}{R(s)} = \frac{2}{s+1}; \quad R(s) = \frac{1}{s} \text{ (step-input)}$$

$$C(s) = R(s)\left(\frac{2}{s+1}\right) = \frac{2}{s(s+1)} = 2\left[\frac{1}{s} - \frac{1}{s+1}\right]$$

$$c(t) = \mathcal{L}^{-1}[C(s)] = 2[-1 - e^{-t}]$$

Final value of $c(t) = C_{ss} = 2$

$$98\% \text{ of } C_{ss} = 0.98 \times 2 = 1.96$$

Let $t = T$, the response reaches 98% of its final values.

$$1.96 = 2[1 - e^{-T}]$$

$$T \approx 4 \text{ sec.}$$

3.16 (a)

Let the system is represented as

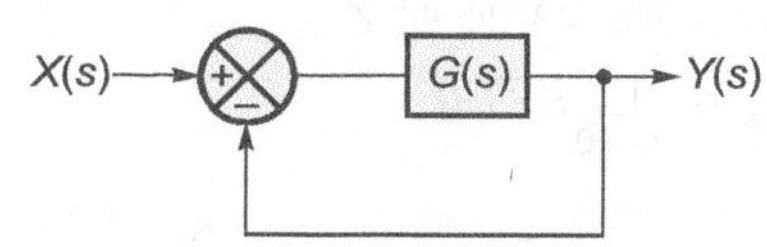

$$\frac{y(s)}{x(s)} = \frac{G(s)}{1+G(s)H(s)}$$

$$H(s) = 1 \text{ (unity feedback)}$$

$$\text{Error} = E(s) = \frac{x(s)}{1+G(s)}$$

Steady-state error for $x(s) = 1/s$, $e_{ss} = 0.1$

$$e_{ss} = \underset{s\to 0}{\text{Lt}}\, sE(s) = \underset{s\to 0}{\text{Lt}} \frac{sx(s)}{1+G(s)}$$

$$\Rightarrow \quad 0.1 = \underset{s\to 0}{\text{Lt}} \frac{s.\frac{1}{s}}{1+G(s)} \Rightarrow \underset{s\to 0}{\text{Lt}} \frac{1}{1+G(s)} = 0.1$$

When input,

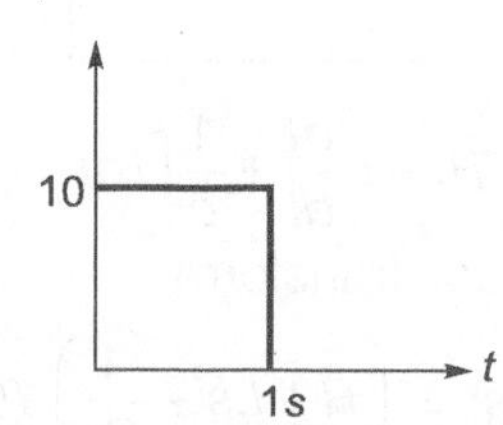

$$x(t) = 10[u(t) - u(t-1)]$$

$$x(s) = 10\left[\frac{1}{s} - \frac{e^{-s}}{s}\right] = \frac{10}{s}[1 - e^{-s}]$$

$$e_{ss} = \underset{s\to 0}{\text{Lt}}\, sE(s) = \underset{s\to 0}{\text{Lt}} \frac{sx(s)}{1+G(s)}$$

$$= \underset{s\to 0}{\text{Lt}} \frac{s\times\frac{10}{s}[1-e^{-s}]}{1+G(s)}$$

$$\Rightarrow \quad e_{ss} = \underset{s\to 0}{\text{Lt}} \frac{10(1-e^{-s})}{1+G(s)} = 0$$

3.17 (a)

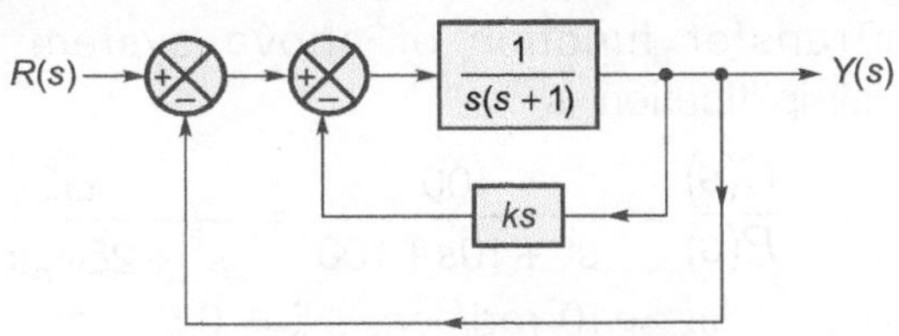

≡

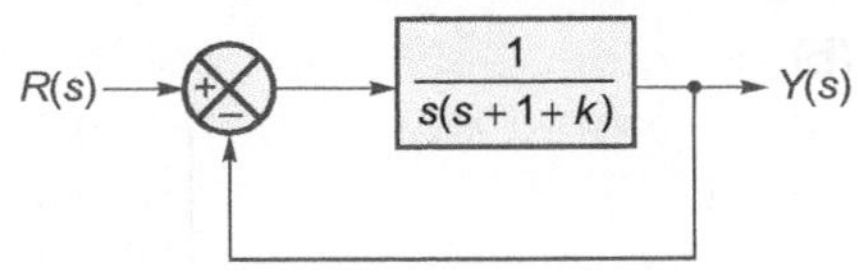

$$G(s) = \frac{1}{s(s+1+k)} \text{ and } H(s) = 1$$

Characteristic equation, $1 + G(s)H(s) = 0$

$$\Rightarrow \quad 1 + \frac{1}{s(s+1+k)} = 0$$

$\Rightarrow s(s + 1 + k) + 1 = 0 \Rightarrow s^2 + (k + 1)s + 1 = 0$

Comparing with, $s^2 + 2\xi\omega_n s + \omega_n^2 = 0$

Natural frequency $\omega_n = 1$

Remains constant and does not depend on k

$$2\xi\omega_n = k + 1 \quad \Rightarrow \quad \xi = \frac{k+1}{2}$$

Damping ratio depends on k

Peak overshoot = $m_p = e^{-\xi\pi/\sqrt{1-\xi^2}}$

Since m_p depends on ξ which depends on k.

Hence peak overshoot is influenced by gain (k) of the tacho-generator.

3.18 (0.241)

$$\frac{5}{(s+2)^2 + 5^2} + \frac{(s+2)}{(s+2)^2 + 5^2}$$

Putting $s = 0$,

$$= \frac{5}{2^2 + 5^2} + \frac{2}{2^2 + 5^2} = \frac{5}{29} + \frac{2}{29} = \frac{7}{29} = 0.241$$

3.19 (a)

$$\text{T.F.} = \frac{1-2s}{1+s}$$

$$C(s) = \frac{1-2s}{1+s} \cdot \frac{1}{s} = \frac{A}{1+s} + \frac{B}{s}$$

For A,

$$A = \lim_{s \to -1} (s+1) \cdot \frac{1-2s}{s(1+s)} = \frac{(1-2(-1))}{(-1)} = \frac{1+2}{-1} = -3$$

For B, $\quad B = \lim_{s \to 0} s \cdot \frac{1-2s}{(1+s)s} = 1$

$$C(s) = \frac{1}{s} - \frac{3}{1+s}; \quad C(t) = (1 - 3e^{-t})\, u(t)$$

3.20 (b)

Damping ratio, $\xi = 0.5$

Undamped natural frequency, $\omega_n = 10$ rad/sec

Steady state output to a unit step input,

$$C_{ss} = 1.02$$

Hence steady state error = 1.02 – 1.00

$$e_{ss} = 0.02$$

$\because$ Characteristics equation is,

$s^2 + 2\xi\omega_n s + \omega_n^2 = 0$

$s^2 + 2 \times 0.5 \times 10\, s + 100 = 0$

$s^2 + 10s + 100 = 0$

From options, if we take option (b)

then $\quad C_{ss} = \lim_{s \to 0} s \cdot C(s)$

$$= \lim_{s \to 0} s \times \frac{1}{s} \times \frac{102}{s^2 + 10s + 100}$$

$$C_{ss} = 1.02$$

Hence option (b) is correct answer.

3.21 (d)

Close loop transfer function $= \dfrac{Ks + b}{s^2 + as + b}$

Open loop transfer function

$$= G(s) = \frac{Ks + b}{s^2 + as + b - Ks - b}$$

$$G(s) = \frac{Ks + b}{s^2 + as - Ks} = \frac{Ks + b}{s(s + a - K)}$$

Steady state error for ramp input given to type-1 system = $1/K_V$ where, velocity error coefficient,

$$K_V = \underset{s \to 0}{Lt}\ s \cdot \frac{Ks + b}{s(s + a - K)} = \frac{b}{a - K}$$

Steady state error, $e_{ss} = \dfrac{a - K}{b}$

3.22 (c)

For maximum peak over shoot $M_P \propto \dfrac{1}{\xi}$

$\xi = 0.25$ for option (c) which is least among all options. Therefore correct option is (c).

3.23 (c)

$$P = \frac{15}{s^2 + 5s + 15}$$

$$\omega_n = \sqrt{15} = 3.872 \text{ rad/sec.}$$

$2\xi \times 3.872 = 5$

$$\xi = \frac{5}{2 \times 3.872} = 0.64 \quad \text{(Underdamped)}$$

$$Q = \frac{25}{s^2 + 10s + 25}$$

$$\omega_n = \sqrt{25} = 5 \text{ rad/sec.}$$

$$2\xi \times 5 = 10 \Rightarrow \xi = 1 \quad \text{(Critically damped)}$$

Observing all the options, option (c) is correct.

3.24 (0.66)

Steady-state error for type-0 and step input,

$$e_{ss} = \frac{1}{1+k_p}$$

$$k_p = \lim_{s \to 0} \frac{1}{(s+1)(s+2)} = \frac{1}{2}$$

$$e_{ss} = \frac{1}{1+\frac{1}{2}} = \frac{2}{3} = 0.66 \text{ units}$$

3.25 (8)

Closed loop transfer function,

$$\frac{C(s)}{R(s)} = \frac{\frac{K}{(s+1)^2(s+2)}}{1+\frac{K}{(s+1)^2(s+2)}}$$

$$\frac{C(s)}{R(s)} = \frac{K}{(s+1)^2(s+2)+K}$$

Given, $R(s) = \frac{1}{s}$

$$C(s) = \frac{K}{s\left[(s+1)^2(s+2)+K\right]}$$

$$\lim_{s \to 0} sC(s) = 0.8 \quad \text{(given)}$$

$$\frac{K}{2+K} = 0.8 \Rightarrow K = 8$$

3.26 (10.90)

$$q(s) = s^2 + 10s + 144 = 0$$

Natural frequency,

$$\omega_n = 12; \ \xi = \frac{5}{12}$$

Damped frequency,

$$\omega_d = \omega_n\sqrt{1-\xi^2} = 12\sqrt{\frac{119}{144}} = 10.90$$

3.27 (a)

Transfer function of above system after simplification is

$$\frac{C(s)}{R(s)} = \frac{100}{s^2+10s+100} = \frac{\omega_n^2}{s^2+2\xi\omega_n s+\omega_n^2}$$

$$\omega_n = 10 \text{ rad/sec}; \quad \xi = 0.5$$

3.28 (b)

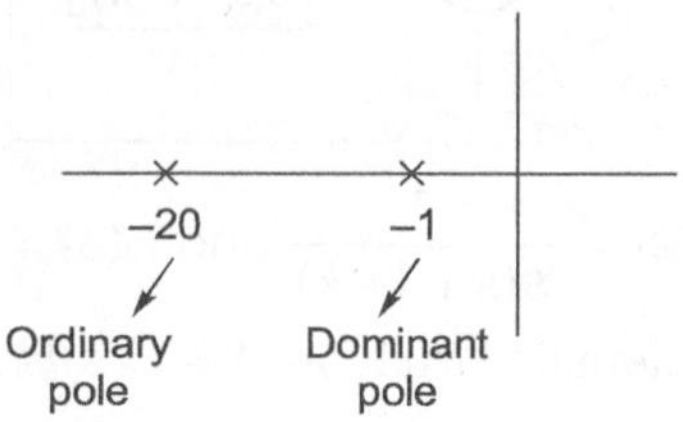

$$\text{Transfer function} \rightarrow \frac{40\left(1+\frac{s}{40}\right)}{20\left(1+\frac{s}{20}\right)(1+s)}$$

$$\text{Answer} = \frac{2}{s+1}$$

3.29 (d)

In system -1

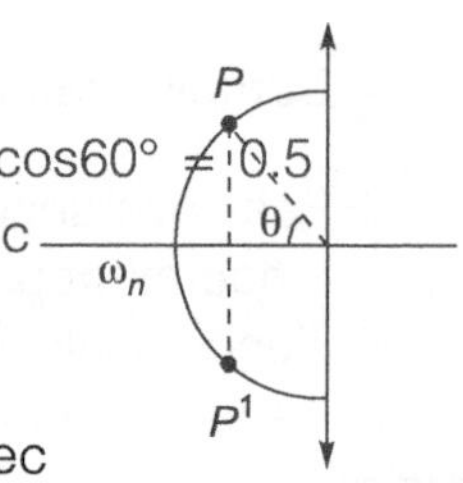

$$\xi_1 = \cos\theta = \cos 60^\circ = 0.5$$

$$\omega_{n1} = 3 \text{ rad/sec}$$

Settling time – t_{s1} =

$$\frac{4}{\xi\omega_n} = \frac{4}{0.5 \times 3} = \frac{8}{3} = 2.67 \text{ sec}$$

In system -2

$$\xi_2 = \cos\theta_2 = \cos 70^\circ$$

$$\omega_{n2} = 1 \text{ rad/sec}$$

Settling time – t_{s2} =

$$\frac{4}{\xi\omega_n} = \frac{4}{\cos 70^\circ \times 1} = 11.69 \text{ sec}$$

So, $t_{s2} > t_{s1}$

4 Concepts of Stability

4.1 Closed loop stability implies that $[1 + G(s) H(s)]$ has all the _____ in the left half of the s-plane. **[1995 : 1 M]**

4.2 Determine whether the system given by the block diagram of figure is stable

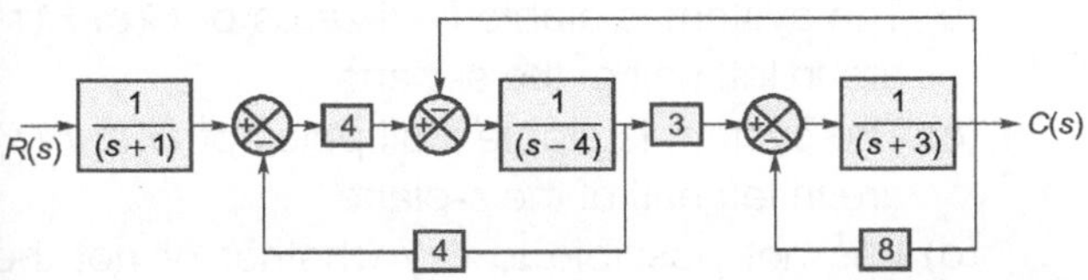

[1997 : 2 M]

4.3 The system represented by the transfer function

$$G(s) = \frac{s^2 + 10s + 24}{s^4 + 6s^3 - 39s^2 + 19s + 84}$$

has ____ pole(s) in the right-half of s-plane. **[1997 : 2 M]**

4.4 None of the poles of a linear control system lie in the right half of s-plane. For a bounded input, the output of this system

(a) is always bounded
(b) could be unbounded
(c) always tends to zero
(d) none of the above **[1998 : 1 M]**

4.5 The number of roots on the equation $2s^4 + s^3 + 3s^2 + 5s + 7 = 0$ that in lie in the right half of s-plane is

(a) zero (b) one
(c) two (d) three **[1998 : 2 M]**

4.6 The roots of the closed loop characteristic equation of the system shown above **(Q. 4.7)**.

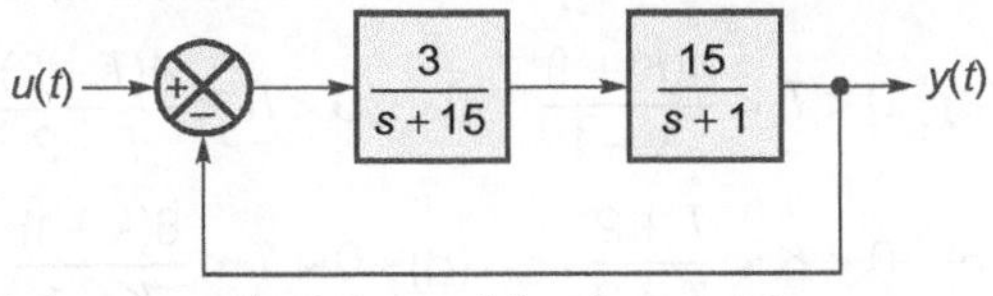

(a) –1 and –15 (b) 6 and 10
(c) –4 and –15 (d) –6 and –10 **[2003 : 2 M]**

4.7 The loop gain GH of a closed loop system is given by the following expression $\frac{K}{s(s+2)(s+4)}$. The value of K for which the system just becomes unstable is

(a) $K = 6$ (b) $K = 8$
(c) $K = 48$ (d) $K = 96$ **[2003 : 2 M]**

4.8 For the equation, $s^3 - 4s^2 + s + 6 = 0$ the number of roots in the left half of s-plane will be

(a) zero (b) one
(c) two (d) three **[2004 : 2 M]**

4.9 A unity feedback system, having an open loop gain $G(s)H(s) = \frac{k(1-s)}{(1+s)}$, becomes stable when

(a) $|k| > 1$ (b) $k > 1$
(c) $|k| < 1$ (d) $k < -1$ **[2005 : 2 M]**

4.10 The system shown in the figure is

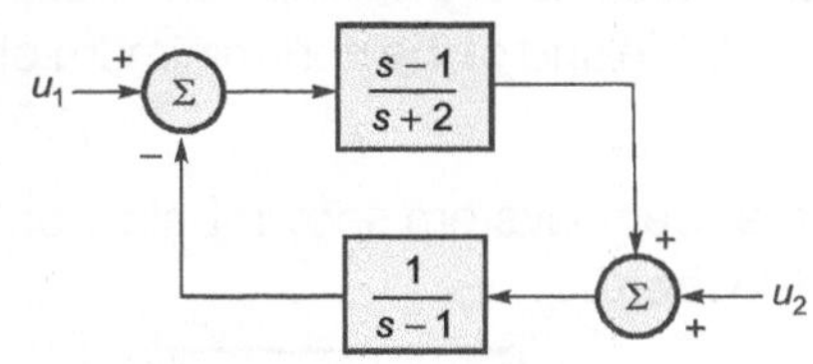

(a) Stable
(b) Unstable
(c) conditionally stable
(d) stable for input u_1, but unstable for input u_2 **[2007 : 1 M]**

4.11 If the loop gain k of a negative feedback system having a loop transfer function $k(s + 3)/(s + 8)^2$ is to be adjusted to induce a sustained oscillation then

(a) the frequency of this oscillation must be $4/\sqrt{3}$ rad/s.
(b) the frequency of this oscillation must be 4 rad/s.
(c) the frequency of this oscillation must be 4 or $4/\sqrt{3}$ rad/s.
(d) such a k does not exist. **[2007 : 2 M]**

4.12 The figures shows a feedback system where $k > 0$

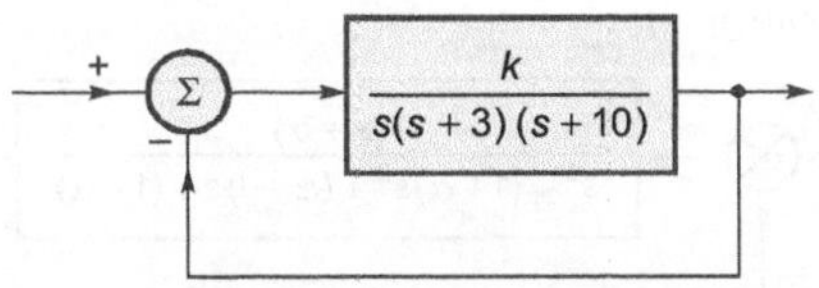

The range of k for which is stable will be given by

(a) $0 < k < 30$ (b) $0 < k < 39$
(c) $0 < k < 390$ (d) $k > 390$ **[2008 : 2 M]**

4.13 The first two rows of Routh's tabulation of a third order equation are as follows:

s^3 2 2

s^2 4 4

This means there are:

(a) two roots at $s = \pm j$ and one root in right half s-plane

(b) two roots at $s = \pm j2$ and one root in left half s-plane

(c) two roots at $s = \pm j2$ and one root in right half s-plane

(d) two roots at $s = \pm j$ and one root in left half s-plane [2009 : 1 M]

4.14 An open loop system represented by the transfer function is

$$G(s) = \frac{(s-1)}{(s+2)(s+3)}$$

(a) stable and of the minimum phase type

(b) stable and of the non-minimum phase type

(c) unstable and of the minimum phase type

(d) unstable and of the non-minimum phase type [2011 : 1 M]

4.15 The feedback system shown below oscillates at 2 rad/s when

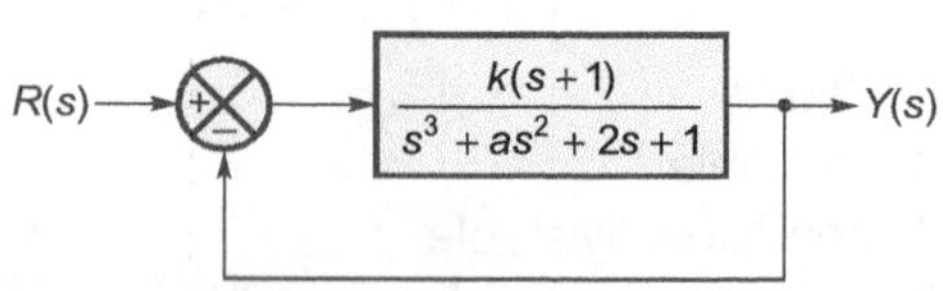

(a) $k = 2$ and $a = 0.75$

(b) $k = 3$ and $a = 0.75$

(c) $k = 4$ and $a = 0.5$

(d) $k = 2$ and $a = 0.5$ [2012 : 2 M]

4.16 In the formation of Routh-Hurwitz array for a polynomial, all the elements of a row have zero values. This premature termination of the array indicates the presence of

(a) only one root at the origin

(b) imaginary roots

(c) only positive real roots

(d) only negative real roots [2014 : 1 M, Set-1]

4.17 For the given system, it is desired that the system be stable. The minimum value of α for this condition is ______.

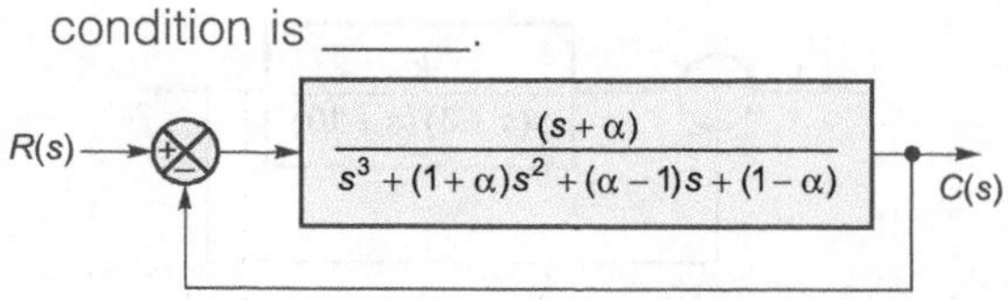

[2014 : 2 M, Set-1]

4.18 A system with the open loop transfer function:

$$G(s) = \frac{K}{s(s+2)(s^2+2s+2)}$$

is connected in a negative feedback configuration with a feedback gain of unity. For the closed-loop system to be marginally stable, the value of K is ______. [2014 : 2 M, Set-2]

4.19 A single-input single-output feedback system has forward transfer function $G(s)$ and feedback transfer function $H(s)$. It is given that $|G(s)H(s)| < 1$.

Which of the following is true about the stability of the system?

(a) The system is always stable.

(b) The system is stable if all zeros of $G(s)\,H(s)$ are in left half of the s-plane.

(c) The system is stable if all poles of $G(s)\,H(s)$ are in left half of the s-plane.

(d) It is not possible to say whether or not the system is stable from the information given. [2014 : 1 M, Set-3]

4.20 The transfer function of a second order real system with a perfectly flat magnitude response of unity has a pole at $(2 - j3)$. List all the poles and zeros.

(a) Poles at $(2 \pm j3)$, no zeros.

(b) Poles at $(\pm 2 - j3)$, one zero at origin.

(c) Poles at $(2 - j3)$, $(-2 + j3)$, zeroes at $(-2 - j3)$, $(2 + j3)$.

(d) Poles at $(2 \pm j3)$, zeros at $(-2 \pm j3)$. [2015 : 2 M, Set-1]

4.21 Given the following polynomial equation,

$$s^3 + 5.5s^2 + 8.5s + 3 = 0$$

The number of roots of the polynomial, which have real parts strictly less than −1, is______. [2016 : 2 M, Set-1]

4.22 The open loop transfer function of a unity feedback control system is given by

$$G(s) = \frac{K(s+1)}{s(1+Ts)(1+2s)}, K > 0, T > 0.$$

The closed loop system will be stable if,

(a) $0 < T < \frac{4(K+1)}{K-1}$ (b) $0 < K < \frac{4(T+2)}{T-2}$

(c) $0 < K < \frac{T+2}{T-2}$ (d) $0 < T < \frac{8(k+1)}{K-1}$

[2016 : 2 M, Set-2]

4.23 A closed loop system has the characteristic equation given by $s^3 + Ks^2 + (K+2)s + 3 = 0$. For this system to be stable, which one of the following conditions should be satisfied?

(a) $0 < K < 0.5$ (b) $0.5 < K < 1$

(c) $0 < K < 1$ (d) $K > 1$

[2017 : 1 M, Set-1]

4.24 The range of K for which all the roots of the equation $s^3 + 3s^2 + 2s + K = 0$ are in the left half of the complex s-plane is

(a) $0 < K < 6$ (b) $0 < K < 16$
(c) $6 < K < 36$ (d) $6 < K < 16$

[2017 : 2 M, Set-2]

4.25 The number of roots of polynomial, $s^7 + s^6 + 7s^5 + 14s^4 + 31s^3 + 73s^2 + 25s + 200$, in the left half of the complex plane is

(a) 3 (b) 4
(c) 5 (d) 6 [2018 : 2 M]

4.26 The characteristic equation of a linear time-invariant (LTI) system is given

$$\Delta(s) = s^4 + 3s^3 + 3s^2 + s + k = 0$$

The system is BIBO stable if

(a) $0 < k < \frac{12}{9}$ (b) $k > 3$
(c) $0 < k < \frac{8}{9}$ (d) $k > 6$ [2019 : 1 M]

4.27 Which of the following option is correct for the system shown below?

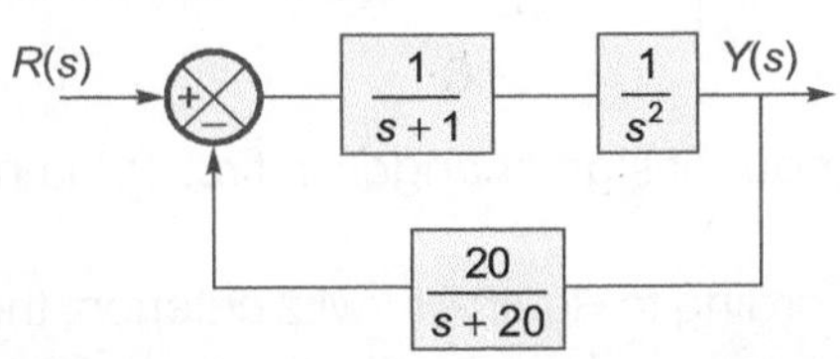

(a) 3rd order and stable
(b) 4th order and unstable
(c) 4th order and stable
(d) 3rd order and unstable [2020 : 2 M]

4.28 Consider a negative unity feedback system with the forward path transfer function $\frac{s^2+s+1}{s^3+2s^2+2s+K}$, where K is a positive real number. The value of K for which the system will have some of its poles on the imaginary axis is ______.

(a) 8 (b) 9
(c) 6 (d) 7 [2020 : 2 M]

4.29 The open loop transfer function of a unity gain negative feedback system is given by $G(s) = \frac{k}{s^2+4s-5}$. The range of k for which the system is stable, is

(a) $k > 3$ (b) $k < 3$
(c) $k > 5$ (d) $k < 5$ [2022 : 1 M]

■■■■

Answers Concepts of Stability

4.1 (Sol.)	4.2 (Sol.)	4.3 (Sol.)	4.4 (d)	4.5 (c)	4.6 (d)	4.7 (c)
4.8 (b)	4.9 (c)	4.10 (d)	4.11 (b)	4.12 (c)	4.13 (d)	4.14 (b)
4.15 (a)	4.16 (b)	4.17 (0.618)	4.18 (5)	4.19 (d)	4.20 (d)	4.21 (2)
4.22 (c)	4.23 (d)	4.24 (a)	4.25 (a)	4.26 (c)	4.27 (b)	4.28 (a)
4.29 (c)						

Explanations Concepts of Stability

4.2 Sol.

Because zeros of the $1 + G(s)H(s)$ represents closed loop transfer function poles which must lie in the left half of the s-plane.

4.3 Sol.

SFG:

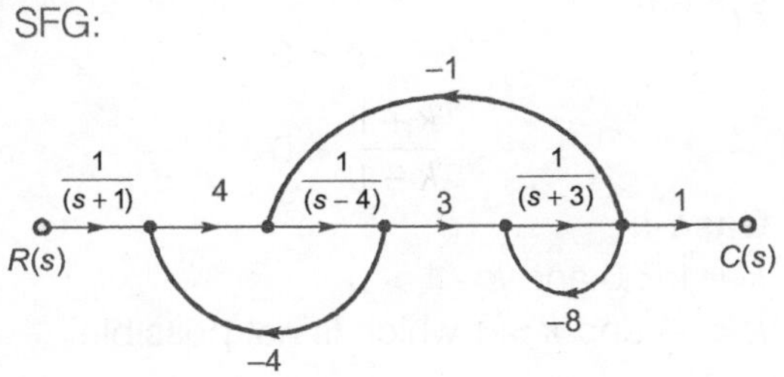

$$\therefore \text{ T.F.} = \frac{\frac{1}{s+1}\times 4\times\frac{1}{s-4}\times 3\times\frac{1}{s+3}\times[1]}{1-\left[\frac{-16}{s-4}-\frac{8}{s+3}-\frac{3}{(s+3)(s-4)}\right]+\left[\frac{16}{s-4}\times\frac{8}{s+3}\right]}$$

On solving,

$$\text{T.F.} = \frac{12}{s^3+24s^2+158s+135}$$

s^3	1	158
s^2	24	135
s^1	+ve	
s^0	+ve	

∵ No sign change in first column.

⇒ System is stable.

4.4 Sol.

Characteristic equation:

$s^4 + 6s^3 - 39s^2 + 19s + 84 = 0$

Routh Hurwitz Criteria:

s^4	1	−39	84
s^3	6	19	
s^2	−42	84	
s^1	31		
s^0	84		

∵ Two sign changes in first column.

∴ Two poles in the right-half of s-plane.

4.5 (d)

If the poles lie on imaginary axis and at the origin then output may be bounded or unbounded for a bounded input.

4.6 (c)

$2s^4 + s^3 + 3s^2 + 5s + 7 = 0$

s^4	2	3	7
s^3	1	5	0
s^2	−7	7	
s^1	6		
s^0	7		

∵ Two sign changes in first column.

∴ Two poles in the right half of s-plane.

4.7 (d)

$1 + GH = 0$ (characteristic equation),

$$1 + \frac{45}{(s+15)(s+1)} = 0$$

$$s^2 + 16s + 60 = 0$$

$$\therefore \quad (s+10)(s+6) = 0$$

∴ Roots are (−6, −10).

4.8 (c)

Characteristic equation = $1 + GH = 0$

$$\Rightarrow \quad 1 + \frac{K}{s(s+2)(s+4)} = 0$$

Characteristic eqn. ⇒ $s^3 + 6s^2 + 8s + K = 0$

The Routh-array is formed as follows:

s^3	1	8
s^2	6	K
s^1	$\frac{48-K}{6}$	
s^0	K	

According to Routh-Hurwitz criterion, for a stable system there should be no change of sign in the first column of routh array.

So, $\frac{48-K}{6} > 0$ and $K > 0 \Rightarrow 0 < K < 48$

Hence, for $K = 48$, the system just becomes unstable.

4.9 (b)

$s^3 - 4s^2 + s + 6 = 0$

The Routh-array is formed as follows:

s^3	1	1
s^2	−4	6
s^1	$\frac{-4 \times 1 - 6 \times 1}{-4} = 2.5$	
s^0	6	

Number of sign changes in first column of Routh array = 2.

According to Routh-Hurwitz criterion, the number of changes of sign in the first column gives the number of positive real part roots of the polynomial.

So, number of roots in RHS of s-plane = 2.

Total number of roots = 3

Hence, no. of roots in LHS of s-plane = 3 − 2 = 1

4.10 (c)

Characteristic equation

$$1 + G(s)H(s) = 0$$

$$\Rightarrow \quad 1 + \frac{k(1-s)}{1+s} = 0$$

$$\Rightarrow \quad s + 1 + k(1-s) = 0$$

$$\Rightarrow \quad s(1-k) + 1 + k = 0$$

$$s = \frac{k+1}{k-1}$$

For a stable system pole lies in left hand side of s-plane, it means and must be negative for stable system.

$$s < 0$$

$$\frac{k+1}{k-1} < 0$$

Case-I:

$k + 1 < 0$ and $k - 1 > 0$

$k < -1$ and $k > 1$ which is not possible.

Case-II:

$k + 1 > 0$ and $k - 1 < 0$

$k > -1$ and $k < 1$

$-1 < k < 1 \Rightarrow |k| < 1$

4.11 (d)

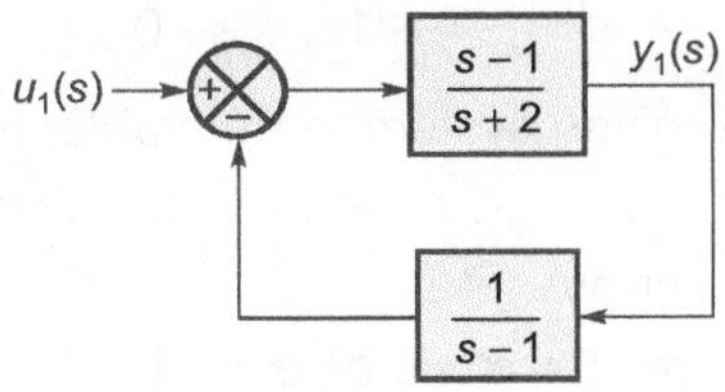

$$(T/F)_1 = \frac{\frac{(s-1)}{s+2}}{1+\frac{(s-1)}{(s+2)}\times\frac{1}{(s-1)}} = \frac{(s-1)}{(s+3)}$$

Pole is in LHS of s-plane, hence stable.

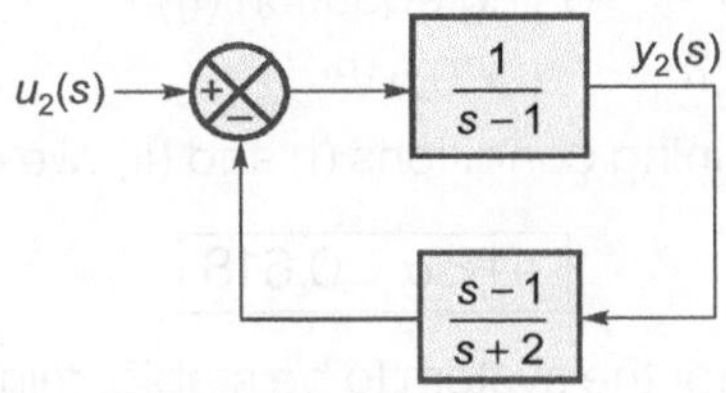

$$(T/F)_2 = \frac{\frac{1}{(s-1)}}{1+\frac{1}{(s-1)}\times\frac{(s-1)}{(s+2)}} = \frac{s+2}{(s-1)(s+3)}$$

Hence unstable as it has pole at right hand side of s-plane.

4.12 (b)

Open loop transfer function,

$$G(s)\,H(s) = k\frac{(s+3)}{(s+8)^2}$$

$\therefore$ Characteristic equation,

$$1 + G(s)\,H(s) = 0 \Rightarrow 1+k\frac{(s+3)}{(s+8)^2} = 0$$

$\Rightarrow s^2 + 16s + 64 + (s + 3) = 0$

$\Rightarrow s^2 + (16 + k)\,s + (64 + 3k) = 0$

Routh-Array,

s^2	1	$64+3k$
s^1	$16+k$	
s^0	$64+3k$	

For sustained oscillation,

$16 + k = 0 \Rightarrow k = -16$

Auxiliary equation,

$s^2 + 64 + 3k = 0$

$\Rightarrow s^2 + 64 + 3x(-16) = 0$

$s^2 + 16 = 0 \Rightarrow s = \pm j4$

So, frequency of oscillation is 4 rad/sec.

4.13 (c)

$$G(s) = \frac{k}{s(s+3)(s+10)} \text{ and } H(s) = 1$$

Characteristic equation,

$\Rightarrow \quad 1 + G(s)\,H(s) = 0$

$$1+\frac{k}{s(s+3)(s+10)} = 0$$

$\Rightarrow s(s + 3)(s + 10) + k = 0$

$\Rightarrow s^3 + 13s^2 + 30s + k = 0$

Routh-Array

s^3	1	30
s^2	13	k
s^1	$\frac{13\times30-k}{13}$	
s^0	k	

According to Routh-Hurwitz criterion.

For a stable system, signs of first column do not change

$$k > 0 \text{ and } \frac{13\times30-k}{13} > 0$$

Therefore system to be stable $0 < k < 390$.

4.14 (d)

Routh-Array :

s^3	2	2
s^2	4	4
s^1	$\frac{2\times4-4\times2}{4} = 0$	

The third row vanishes.

An auxiliary equation is formed using elements of 2nd row.

Auxiliary equation $A(s) = 4s^2 + 4 = 0 \Rightarrow s = \pm j$.

The derivative of this auxiliary equation is taken w.r.t. s and the coefficients of the differentiated equation are taken as the elements of 3rd row.

$$\frac{dA(s)}{ds} = 8s$$

Routh-Array

s^3	2	2
s^2	4	4
s^1	8	
s^0	4	

There is no root in RHS of s-plane.

Two roots of $s = \pm j$, so one root is in LHS of s-plane.

4.15 (b)

$$1+\frac{s-1}{(s+2)(s+3)}=0$$

$$s^2+6s+5=0$$

$$\begin{array}{c|cc} s^2 & 1 & 5 \\ s^1 & 6 & 0 \\ s^0 & 5 & 0 \end{array}$$

There is no root in RHS of s-plane.

4.16 (a)

The characteristic equation

$$1+G(s)\,H(s) = 0$$

$$\Rightarrow s^3+as^2+(k+2)s+(k+1)=0$$

Routh array for above equation,

$$\begin{array}{c|cc} s^3 & 1 & (k+2) \\ s^2 & a & (k+1) \\ s^1 & \dfrac{a(k+2)-(k+1)}{a} & 0 \\ s^0 & (k+1) & 0 \end{array}$$

$$as^2+(k+1) = 0$$

For $s = j\omega$ and $\omega = 2$

$$-a\omega^2+(k+1) = 0$$

$$k = 4a-1$$

and $a=\dfrac{k+1}{k+2}$ $\Rightarrow a = 0.75$ and $k = 2$

4.17 (b)

When all the elements of a row in Routh-Hurwitz array ends abruptly i.e. all elements of that row have a zero values then, the system will be either unstable or marginally stable. Marginally stable means it will have imaginary roots (two equal and opposite complex roots on imaginary axis of s-plane).

4.18 (0.618)

Given, $G(s)=\dfrac{(s+\alpha)}{s^3+(1+\alpha)s^2+(\alpha-1)s+(1-\alpha)}$

and $H(s) = 1$

Characteristic equation of given control system is given by

$$1+G(s)\,H(s)=0$$

or, $1+\left[\dfrac{s+\alpha}{s^3+(1+\alpha)s^2+(\alpha-1)s+(1-\alpha)}\right]=0$

or, $s^3+(1+\alpha)s^2+(\alpha-1)s+(1-\alpha)+s+\alpha=0$

or, $s^3+(1+\alpha)s^2+\alpha s+1=0$

Routh's array is

$$\begin{array}{c|cc} s^3 & 1 & \alpha \\ s^2 & (1+\alpha) & 1 \\ s^1 & \left(\dfrac{\alpha^2+\alpha-1}{1+\alpha}\right) & 0 \\ s^0 & 1 & 0 \end{array}$$

For the given system to be stable, there should not be any sign change in the first column of Routh's array.

Therefore, $1+\alpha>0$ or, $\alpha>-1$ ...(i)

Also, $\dfrac{\alpha^2+\alpha-1}{1+\alpha}>0$ or, $\alpha^2+\alpha-1>0$

or, $(\alpha+1.618)(\alpha-0.618)>0$

or, $\alpha<-1.618$ or, $\alpha>0.618$

As $\alpha>-1$ (Using equation (i))

therefore, $\alpha>0.618$ ...(ii)

Combining conditions (i) and (ii), we get

$$\boxed{-1<\alpha<0.618}$$

Thus, for the system to be stable, minimum value of $\alpha = 0.618$.

4.19 (5)

Given, $G(s)=\dfrac{K}{s(s+2)(s^2+2s+2)}$

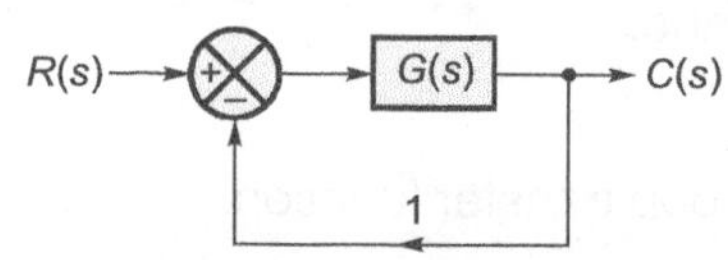

The characteristic equation of the given unity-negative feedback control system is given by

$$1+G(s)\,H(s)=0$$

or, $1+\dfrac{K}{s(s+2)(s^2+2s+2)}=0$

or, $s(s+2)(s^2+2s+2)+K=0$

or, $s^4+4s^3+6s^2+4s+K=0$

Forming Routh's array as shown below:

$$\begin{array}{c|ccc} s^4 & 1 & 6 & K \\ s^3 & 4 & 4 & 0 \\ s^2 & 5 & K & 0 \\ s^1 & \left(\dfrac{20-4K}{5}\right) & 0 & 0 \\ s^0 & K & 0 & 0 \end{array}$$

For stability of the system, $K>0$ and

$$\frac{20-4K}{5}>0 \text{ or } K<5$$

$\therefore$ For stability, $0 < K < 5$

For the given system to be marginally stable,

$$K = 5$$

4.21 (d)

Response of transfer function is unit for all ω.

$M = 1$; $P_1 = 2 - j3$

Second order system, hence number of poles = 2

Therefore, second pole $P_2 = 2 + j3$

Now for $M = 1$, and due to x-axis symmetry of root locus of transfer function, position of zeroes must be

$$Z_1 = -2 - j3 \quad \text{and} \quad Z_2 = -2 + j3$$

4.22 (2)

$s^3 + 5.5s^2 + 8.5\ s + 3 = 0$

Putting, $s = s_1 - 1$

$(s_1 - 1)^3 + 5.5(s_1 - 1)^2 + (8.5)(s_1 - 1) + 3 = 0$

$s_1^3 + 2.5s_1^2 + 0.5s_1 - 1 = 1$

s_1^3	1	0.5
s_1^2	2.5	−1
s_1	0.9	0
s_1^0	−1	

As there is one sign change hence, two roots of given polynomial will lie to the left of $s = -1$.

4.23 (c)

Open loop transfer function:

$$G(s) = \frac{K(s+1)}{s(1+Ts)(1+2s)};\ K > 0 \text{ and } T > 0$$

For closed loop system stability, characteristic equation is, $1 + G(s)H(s) = 0$

$$1+\frac{K(s+1)}{s(1+Ts)(1+2s)}\cdot 1 = 0$$

$s(1 + Ts)(1 + 2s) + K(s + 1) = 0$

$2Ts^3 + (2 + T)\ s^2 + (1 + K)s + K = 0$

Using Routh's criteria,

s^3	$2T$	$(1+K)$
s^2	$(2+T)$	K
s^1	$\frac{(2+T)(1+K)-2TK}{(2+T)}$	0
s^0	K	

For stability, $K > 0$

and $(2 + T)(1 + K) - 2TK > 0$

$K(2 + T - 2T) + (2 + T) > 0$

or

$-(T - 2)K + 2(2 + T) > 0$

$$-K > -\frac{(2+T)}{(T-2)} \text{ or } K < \frac{T+2}{(T-2)}$$

Hence for stability,

$$0 < K < \frac{T+2}{T-2}$$

4.24 (d)

Characteristic equation is,

$s^3 + Ks^2 + (K + 2)s + 3 = 0$

For this system to be stable, using Routh's criterion, we can write,

$$3 < K(K + 2)$$

or $K^2 + 2K - 3 > 0$

or $(K + 3)(K - 1) > 0$

Here, the valid answer will be out of all the options given, i.e., $K > 1$.

4.25 (a)

From the given equation,

$s^3 + 3s^2 + 2s + K = 0$

Using Routh's criterion, we get

$K < 6$ and $K > 0$ or $0 < K < 6$

4.26 (a)

Characteristic equation,

$s^7 + s^6 + 7s^5 + 14s^4 + 31s^3 + 73s^2 + 25s + 200 = 0$

s^7	1	7	31	25
s^6	1	14	73	200
s^5	−7	−42	−175	0
s^4	8	48	200	0
s^3	32	96	0	0
s^2	24	200	0	0
s^1	−170	0	0	0
s^0	200	0	0	0

Auxiliary equation,

$$A(s) = 8s^4 + 48s^2 + 200$$

$$\frac{d}{ds}A(s) = 32s^3 + 96s$$

Total number of poles = 7

Two sign change above auxiliary equation = 2 poles in RHS.

Two sign changes below auxiliary equation implies out of 4 symmetric roots about origin two poles are in LHS and two poles are in RHS.

$\therefore$ 3 poles in LHS and 4 poles in RHS.

4.27 (c)

Routh array is

$$\begin{array}{c|cc} s^4 & 1 & 3 \quad k \\ s^3 & 3 & 1 \\ s^2 & \frac{8}{3} & k \\ s^1 & \frac{\left(\frac{8}{3}-3k\right)}{8/3} & \\ s^0 & k & \end{array}$$

For BIBO stability,

$$\frac{\left(\frac{8}{3}-3k\right)}{\left(\frac{8}{3}\right)} > 0$$

$$\Rightarrow \qquad k < \frac{8}{9} \text{ and } k > 0$$

$$\therefore \qquad 0 < k < \frac{8}{9}$$

4.28 (b)

$$1+\frac{20}{s^2(s+1)(s+20)} = 0$$

$$(s^3 + s^2)(s + 20) + 20 = 0$$

$$s^4 + 20s^3 + s^3 + 20s^2 + 20 = 0$$

$$s^4 + 21s^3 + 20s^2 + 20 = 0$$

$$s^1 \text{ coefficient} = 0$$

Given system is fourth order system and unstable.

4.29 (a)

CE is $\qquad 1 + G(s)H(s) = 0$

$$\Rightarrow \qquad 1+\frac{s^2+s+1}{s^3+2s^2+2s+K} = 0$$

$$\Rightarrow \qquad s^3 + 3s^2 + 3s + (1 + K) = 0$$

R.H. criteria:

$$\begin{array}{c|cc} s^3 & 1 & 3 \\ s^2 & 3 & (1+K) \\ s^1 & 9-(1+K) & 0 \end{array}$$

For marginal stability,

$$9 - (1 + K) = 0 \Rightarrow K = 8$$

4.30 (c)

Characteristics equation,

$$1 + G(s) = 0$$

$$s^2 + 2s + (K - 5) = 0$$

By Routh table analysis, $k > 5$ and $k > 0$

For stable system $k > 5$

■■■■

5 Root Locus Techniques

5.1 In case of an armature controlled separately excited d.c. motor drive with closed loop speed control, an inner current loop is useful because it

(a) limits the speed of the motor to a safe value

(b) helps in improving the drive energy efficiency

(c) limits the peak current of the motor to the permissible value

(d) reduces the steady state speed error

[2001 : 1 M]

5.2 A unity feedback system has an open loop transfer function, $G(s) = \dfrac{K}{s^2}$. Its root locus plot will be

(a) 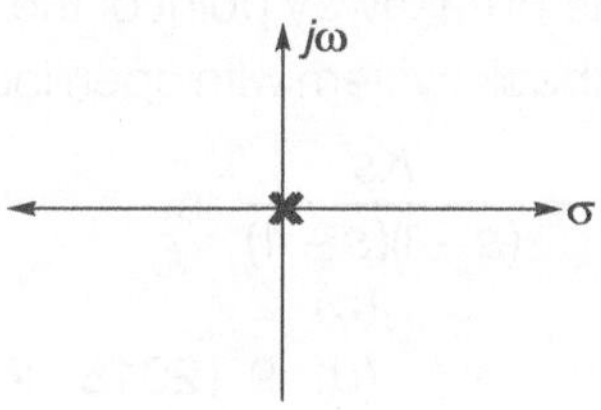

(b) 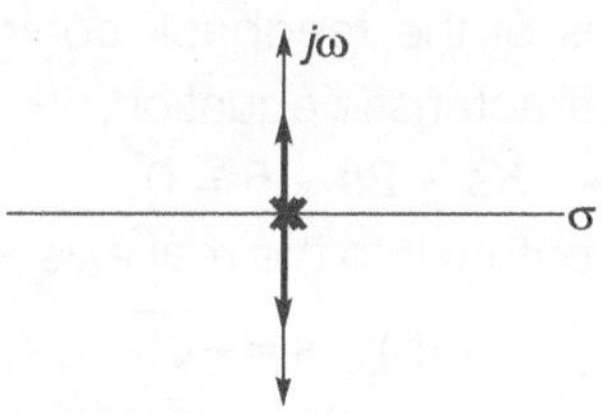

(c) 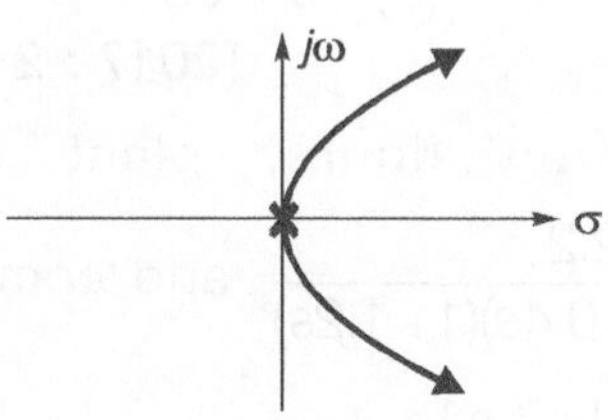

(d) 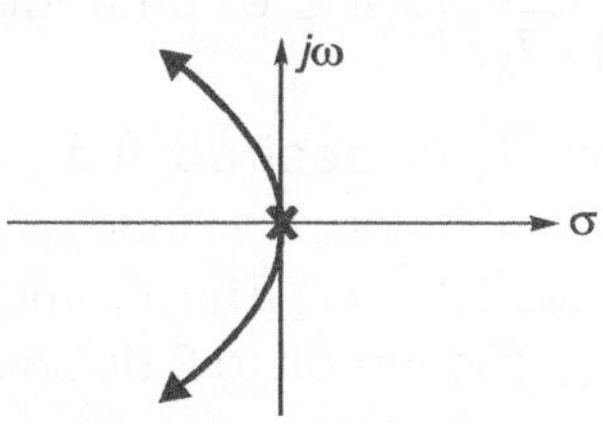

[2002 : 2 M]

5.3 The figure shows the root locus plot (location of poles not given) of a third order system whose open loop transfer function is

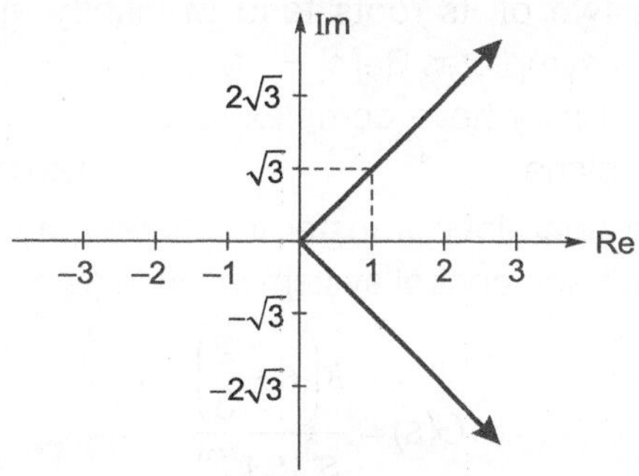

(a) $\dfrac{K}{s^3}$

(b) $\dfrac{K}{s^2(s+1)}$

(c) $\dfrac{K}{s(s^2+1)}$

(d) $\dfrac{K}{s(s^2-1)}$

[2005 : 1 M]

5.4 A closed-loop system has the characteristic function $(s^2 - 4)(s + 1) + K(s - 1) = 0$. Its root locus plot against K is

(a) 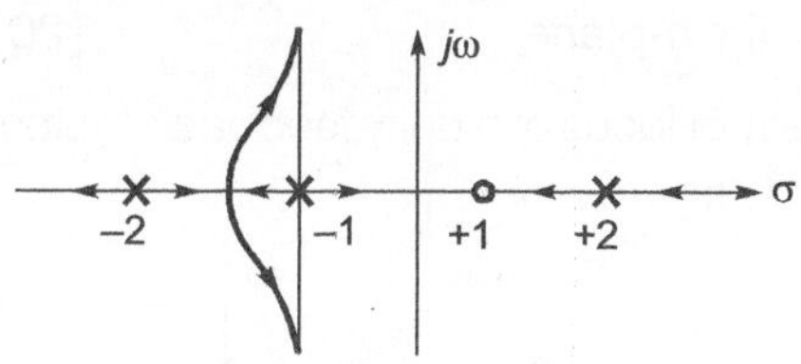

(b) 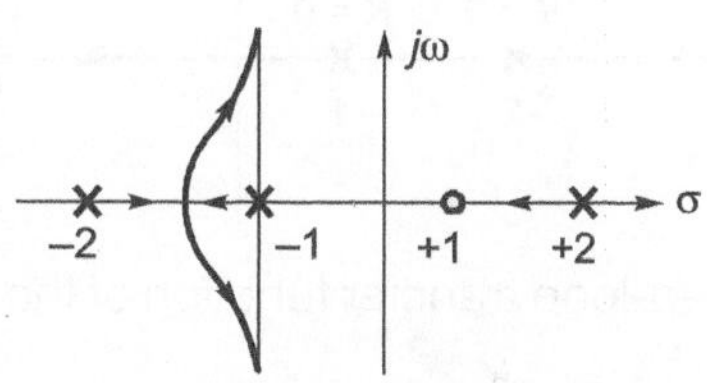

(c) 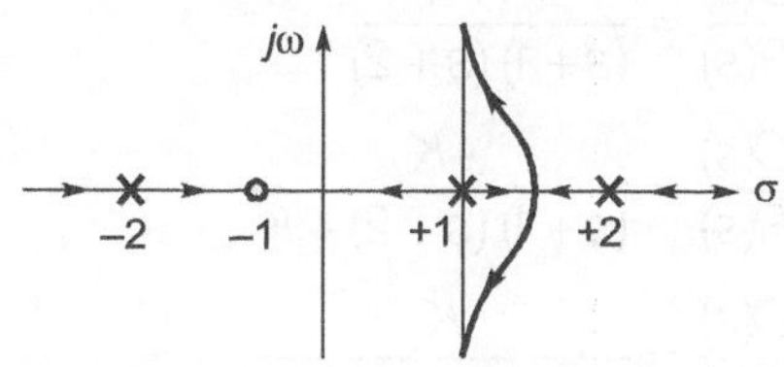

(d) 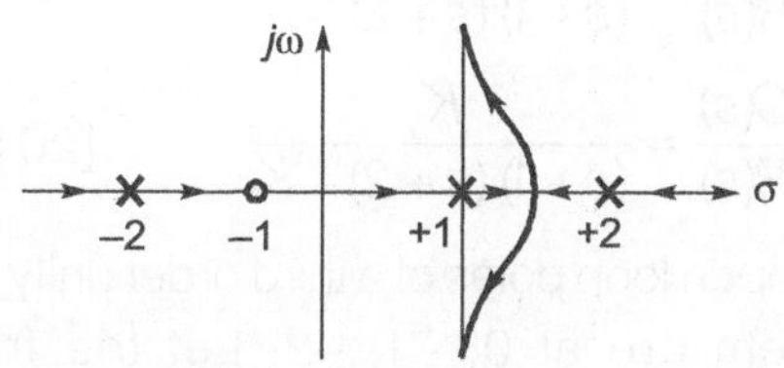

[2006 : 2 M]

5.5 The characteristic equation of a closed-loop system is $s(s+1)(s+3)+k(s+2)=0$, $k>0$. Which of the following statements is true?
(a) Its roots are always real.
(b) It cannot have a breakaway point in the range $-1 < \text{Re}[s] < 0$.
(c) Two of its roots tend to infinity along the asymptotes $\text{Re}[s] = -1$.
(d) It may have complex roots in the right half plane. [2010 : 2 M]

5.6 The open loop transfer function $G(s)$ of a unity feedback control system is given as,

$$G(s) = \frac{k\left(s+\frac{2}{3}\right)}{s^2(s+2)}$$

From the root locus, it can be inferred that when k tends to positive infinity
(a) three roots with nearly equal real parts exist on the left half of the s-plane.
(b) one real root is found on the right half of the s-plane.
(c) the root loci cross the $j\omega$ axis for a finite value of $k : k \neq 0$.
(d) three real roots are found on the right half of the s-plane. [2011 : 2 M]

5.7 The root locus of a unity feedback system is shown in the figure.

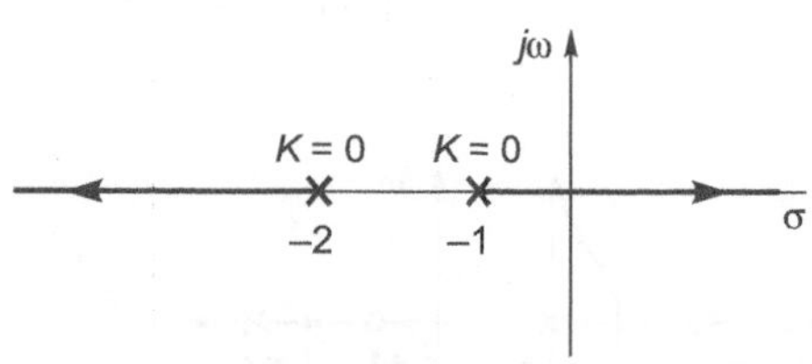

The closed-loop transfer function of the system is

(a) $\dfrac{C(s)}{R(s)} = \dfrac{K}{(s+1)(s+2)}$

(b) $\dfrac{C(s)}{R(s)} = \dfrac{-K}{(s+1)(s+2)+K}$

(c) $\dfrac{C(s)}{R(s)} = \dfrac{K}{(s+1)(s+2)-K}$

(d) $\dfrac{C(s)}{R(s)} = \dfrac{K}{(s+1)(s+2)+K}$ [2014 : 1 M]

5.8 The open loop poles of a third order unity feedback system are at 0, –1, –2. Let the frequency corresponding to the point where the root locus of the system transits to unstable region be K. Now suppose we introduce a zero in the open loop transfer function at –3, while keeping all the earlier open loop poles intact. Which one of the following is TRUE about the point where the root locus of the modified system transits to unstable region?
(a) It corresponds to a frequency greater than K
(b) It corresponds to a frequency less that K
(c) It corresponds to a frequency K
(d) Root locus of modified system never transits to unstable region [2015 : 2 M, Set-1]

5.9 An open loop transfer function $G(s)$ of a system is

$$G(s) = \frac{K}{s(s+1)(s+2)}$$

For a unity feedback system, the breakaway point of the root loci on the real axis occurs at,
(a) –0.42 (b) –1.58
(c) –0.42 and –1.58 (d) None of the above [2015 : 2 M, Set-2]

5.10 The gain at the breakaway point of the root locus of a unity feedback system with open loop transfer function $G(s) = \dfrac{Ks}{(s-1)(s-4)}$ is
(a) 1 (b) 2
(c) 5 (d) 9 [2016 : 2 M, Set-2]

5.11 The root locus of the feedback control system having the characteristic equation,

$$s^2 + 6Ks + 2s + 5 = 0$$

where $K > 0$, enters into the real axis at
(a) $s = -1$ (b) $s = -\sqrt{5}$
(c) $s = -5$ (d) $s = \sqrt{5}$ [2017 : 2 M, Set-2]

5.12 In the given figure, plant $G_p(s) = \dfrac{2.2}{(1+0.1s)(1+0.4s)(1+1.2s)}$ and compensator $G_c(s) = K\left(\dfrac{1+T_1 s}{1+T_2 s}\right)$. The external disturbance input is $D(s)$. It is desired that when the disturbance is a unit step, the steady-steady error should not exceed 0.1 unit. The minimum value of K is ________. (Round off to 2 decimal places)

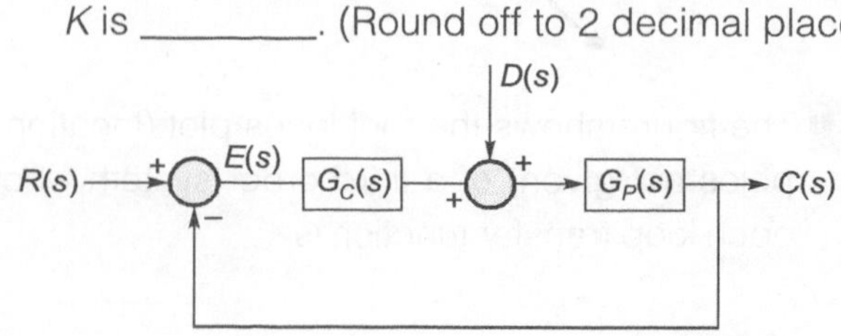

[2021 : 2 M]

5.13 Consider the closed-loop system shown in the figure with $G(s) = \dfrac{K(s^2 - 2s + 2)}{(s^2 + 2s + 5)}$. The root locus for the closed-loop system is to be drawn for $0 \le K < \infty$. The angle of departure (between 0° and 360°) of the root locus branch drawn from the pole $(-1 + j2)$, in degree, is ______ (rounded off to the nearest integer).

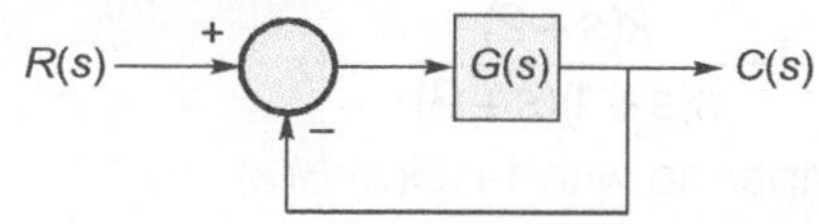

[2024 : 2 M]

Answers Root Locus Techniques

5.1	(c)	5.2	(b)	5.3	(a)	5.4	(b)	5.5	(c)	5.6	(a)	5.7	(c)
5.8	(d)	5.9	(a)	5.10	(a)	5.11	(b)	5.12	(9.54)	5.13	(7)		

Explanations Root Locus Techniques

5.1 (c)

Closed loop system limits the peak value.

5.2 (b)

$$G(s)H(s) = \frac{K}{s^2}$$

$$\therefore \quad P - Z = 2$$

$$\text{Centroid} = \frac{0}{2} = 0$$

Angle of asymptotes = 90°, 270°

∴ Option (b) is correct.

5.3 (a)

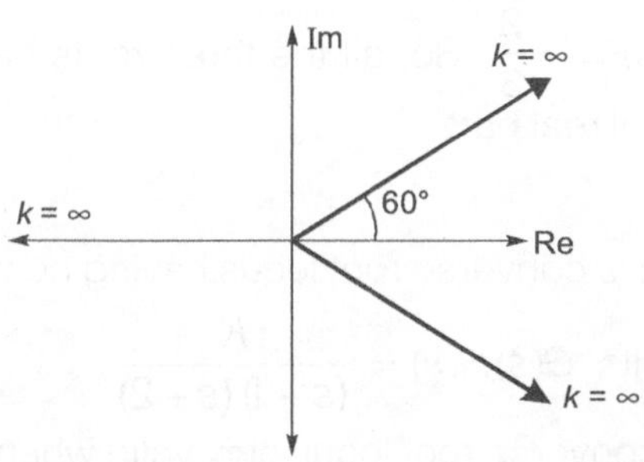

These are three asymptotes with angle 60°, 180° and 300°.

$$\text{Angle of asymptotes} = \frac{(2k+1)\times 180°}{P - Z}$$

where, $k = 0, 1, 2$ upto $(P - Z) - 1$

as angles are 60°, 180° and 300°

it means $P - Z = 3$

Intersection of asymptotes on real axis

$$x = \frac{\Sigma \text{poles} - \Sigma \text{ zeros}}{P - Z}$$

Since, system does not have zeros

$$x = \frac{\Sigma \text{poles}}{P}$$

As asymptotes intersect at origin, it means all the three poles are at origin. So, option (a) is correct.

5.4 (b)

Characteristic function

$$\Rightarrow \quad (s^2 - 4)(s + 1) + k(s - 1) = 0$$

$$\Rightarrow \quad 1 + \frac{k(s-1)}{(s^2 - 4)(s + 1)} \equiv 1 + G(s)H(s)$$

Open loop transfer function

$$= G(s)H(s) = \frac{k(s-1)}{(s^2 - 4)(s+1)}$$

Zero of OLTF $s = 1$; $z = 1$

Poles of OLTF $s = -1, -2, +2$, $P = 3$

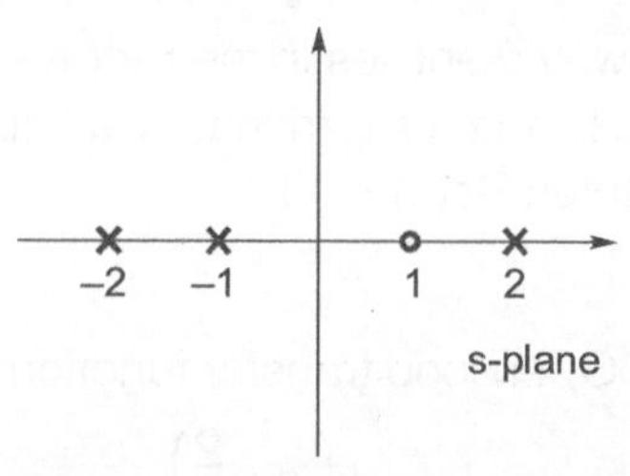

The root locus starts from open-loop poles and terminates either on open-loop zero or infinity. Root locus exist on a section of real axis it the sum of the open-loop poles and zeros to the right of the section is odd.

Number of branches terminating on infinity.

$$= P - Z = 3 - 1 = 2$$

Angle of asymptotes

$$= \frac{(2k+1)\times 180°}{P - Z} = \frac{(2k+1)\times 180°}{2}$$

$$= 90° \text{ and } 270°$$

Intersection of asymptotes on real axis (centroid)

$$= \frac{\Sigma \text{poles} - \Sigma \text{zeros}}{P - Z} = \frac{(-1 - 2 + 2) - (1)}{2} = -1$$

Option (b) is correct on the basic of above analysis.

5.5 (c)

Characteristic eqn., $s(s+1)(s+3)+k(s+2)=0$

$$1+\frac{k(s+2)}{s(s+1)(s+3)}=0$$

Comparing with $1+G(s)H(s)=0$

$G(s)H(s)$ = open-loop transfer function (OLTF)

$$=\frac{k(s+2)}{s(s+1)(s+2)}$$

Number of zeros, $Z=1$ (at -2)

Number of poles, $P=3$ (at 0, -1 and -3)

Number of branches terminating at infinity

$$=P-Z=3-1=2$$

Angle of asymptotes

$$=\frac{(2k+1)\times 180^\circ}{P-Z}=\frac{(2k+1)\times 180^\circ}{2}$$

$$=(2k+1)\times 90^\circ=90^\circ \text{ and } 270^\circ$$

$$\text{Centroid}=\frac{\Sigma\text{ poles}-\Sigma\text{ zeros}}{P-Z}$$

$$=\frac{0-1-3-(-2)}{2}=-1$$

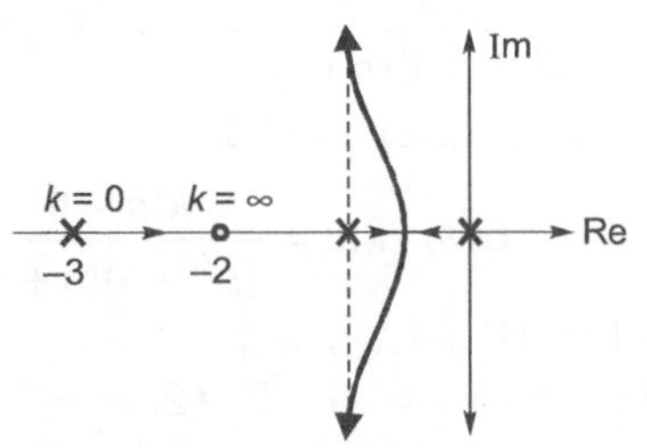

Breakaway point lies in the range $-1<\text{Re}[s]<0$ and two branches terminates at infinity along the asymptotes $\text{Re}(s)=-1$.

5.6 (a)

Given : Open loop transfer function

$$G(s)=\frac{k\left(s+\frac{2}{3}\right)}{s^2(s+2)} \quad \text{and} \quad H(s)=1$$

Characteristic equation, $1+G(s)H(s)=0$

$$\Rightarrow 1+\frac{k\left(s+\frac{2}{3}\right)}{s^2(s+2)}=0 \Rightarrow s^3+2s^2+k\left(s+\frac{2}{3}\right)=0$$

$$\Rightarrow s^3+2s^2+ks+\frac{2k}{3}=0$$

Routh Array:

s^3	1	k
s^2	2	$\frac{2k}{3}$
s^1	$\frac{2k-2k/3}{2}=\frac{2}{3}k$	
s^0	$\frac{2k}{3}$	

As $k>0$, there is no sign change in the 1st column of Routh array. So the system is stable and all the three roots lie on LHS of s-plane.

For $k>0$ $(k\neq 0)$, none of the row of Routh array becomes zero. So root loci does not cross the $j\omega$-axis.

Number of zero = $Z=1$

Number of poles = $P=3$

Number of branches terminating at infinity

$$=P-Z=3-1=2$$

$$\text{Angle of asymptotes}=\frac{(2k+1)\times 180^\circ}{P-Z}$$

$$=\frac{(2k+1)\times 180^\circ}{2}=(2k+1)\times 90^\circ=90^\circ \text{ and } 270^\circ$$

$$\text{Centroid}=\frac{\Sigma\text{poles}-\Sigma\text{zeros}}{P-Z}$$

$$=\frac{0+0-2-\left(-\frac{2}{3}\right)}{2}=-\frac{2}{3}$$

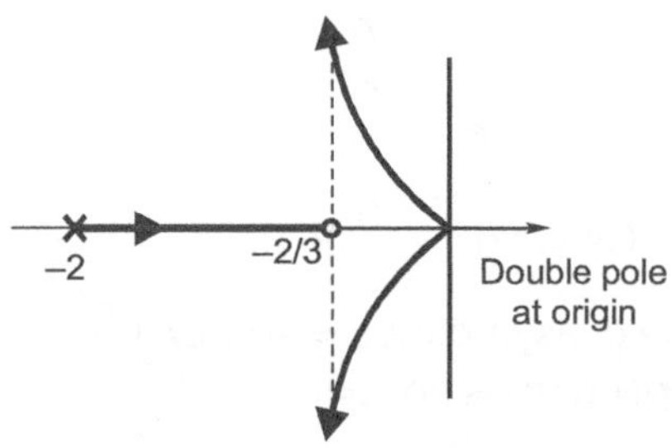

Since, all the three branches terminates at $R_e(s)=-\frac{2}{3}$. So, all the three roots have nearly equal real part.

5.7 (c)

This is converse root locus having no zero.

As, its $G(s)H(s)=\frac{K}{(s+1)(s+2)}$

Its converse root locus only valid when $K<0$

$$\text{So,} \quad \text{C.L.T.F.}=\frac{G(s)H(s)}{1-G(s)H(s)}$$

$$=\frac{\frac{K}{(s+1)(s+2)}}{1-\frac{K}{(s+1)(s+2)}}=\frac{K}{(s+1)(s+2)-K}$$

5.8 (d)

Open loop transfer function can be represented as $\frac{k(s+3)}{s(s+1)(s+2)}$

$P=3,\ Z=1 \Rightarrow P-Z=3-1=2$

Asymptotes angle = 90°, 180°

$$\text{Centroid} = \frac{(-3)-(-3)}{2} = 0$$

As root locus never cross asymptotes so, will remain in left side of x-axis.

5.9 (a)

$$G_1(s) = \frac{k}{s(s+1)(s+2)}$$

$$\Rightarrow 1+\frac{k}{s(s+1)(s+2)} = 0$$

$$\Rightarrow s(s^2+3s+2)+k = 0 \Rightarrow s^3+3s^2+2s+k = 0$$

$$\Rightarrow \qquad k = -s^3-3s^2-2s \qquad ...(i)$$

$$\Rightarrow \qquad \frac{dk}{ds} = 0$$

$$\frac{dk}{ds} = -3s^2-6s-2 = 0 \quad \Rightarrow \quad s = -0.422, -1.577$$

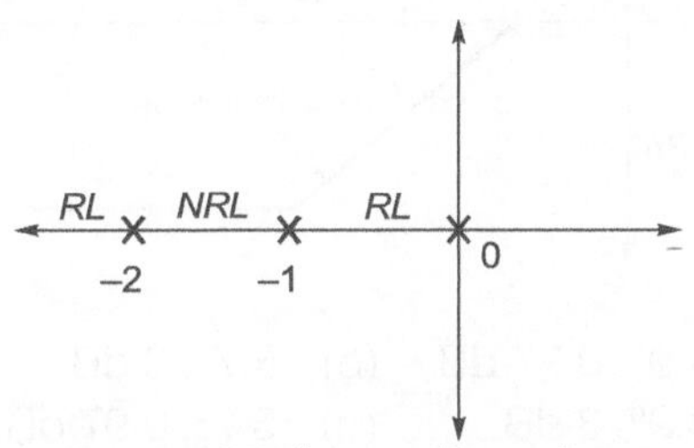

Only $s = -0.422$ lie on Root locus. Therefore breakaway point is $s = -0.42$.

5.10 (a)

$$\text{OLTF} \Rightarrow G(s) = \frac{Ks}{(s-1)(s-4)}$$

Now, characteristics equation,

$$1+G(s)H(s) = 0$$

$$\frac{Ks}{(s-1)(s-4)}+1 = 0$$

$$\Rightarrow \quad Ks+(s^2-5s+4) = 0$$

$$K = -\frac{(s^2-5s+4)}{s} = \left[s-5+\frac{4}{s}\right]$$

For break away point: $\dfrac{dK}{ds} = 0$

$$\frac{dK}{ds} = -\left[1-0-\frac{4}{s^2}\right] = 0$$

we get, $\quad s = \pm 2$

Therefore valid break away point is $s = 2$

Now gain at $s = 2$ is

$$\Rightarrow \quad K = \frac{\text{Product of distances from all the poles to break away point}}{\text{Product of distance from all the zeros to break away point}}$$

$$\text{Gain, } K = \frac{1\times 2}{2} = 1$$

5.11 (b)

Given, characteristic equation

$$1+G(s)H(s) = s^2+6Ks+2s+5 = 0$$

We can write,

$$GH(s) = \frac{6Ks}{s^2+2s+5}$$

The point at which root locus enters real axis (breakaway point) is given by

$$\frac{dK}{ds} = 0$$

$$K = -\frac{(s^2+2s+5)}{6s}$$

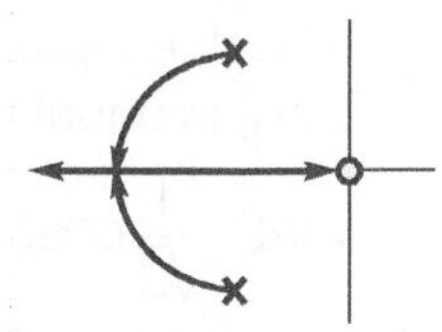

$$\frac{dK}{ds} = 0 \Rightarrow s = \pm\sqrt{5}$$

$$\therefore \qquad s = -\sqrt{5}$$

5.12 (9.54)

$$e_{ss} = \lim_{s\to 0}\left[\frac{sR}{1+G_CG_P} - \frac{sDG_P}{1+G_CG_P}\right]$$

$$R(s) = 0;\; D(s) = \frac{1}{s}$$

$$\therefore \qquad e_{ss} = \frac{2.2}{1+2.2K} = 0.1$$

If D is considered with +ve sign, then

$$K_{min} = -10.45$$

If D is considered with –ve sign, then

$$K_{min} = 9.54$$

(IITB has given answer as 9.54).

5.13 (7)

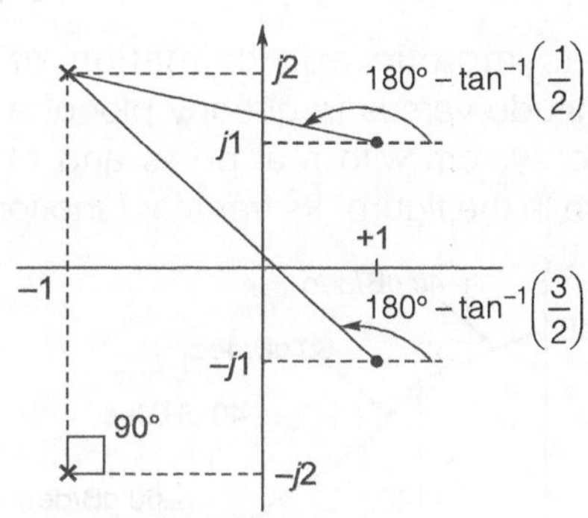

$$\phi_D = \pm(180+\phi)$$

$$\phi = \Sigma\phi_z - \Sigma\phi_p$$

$$\phi = \left\{\left[180-\tan^{-1}\left(\frac{1}{2}\right)\right]+\left[180-\tan^{-1}\left(\frac{3}{2}\right)\right]-90°\right\}$$

$$= 187°$$

$$\phi_D = \phi+180° = 367 = 7° \quad\Rightarrow\quad \phi_D = \pm 7°$$

6 Frequency Response Analysis

6.1 The closed loop transfer function of a control system is given by

$$\frac{C(s)}{R(s)} = \frac{1}{(1+s)}$$

For the input $r(t) = \sin t$, the steady state value of $c(t)$ is equal to

(a) $\frac{1}{\sqrt{2}}\cos t$ (b) 1

(c) $\frac{1}{\sqrt{2}}\sin t$ (d) $\frac{1}{\sqrt{2}}\sin\left(t - \frac{\pi}{4}\right)$

[1996 : 1 M]

6.2 The polar plot of a type-1, 3-pole, open-loop system is shown in the figure below. The closed loop system is

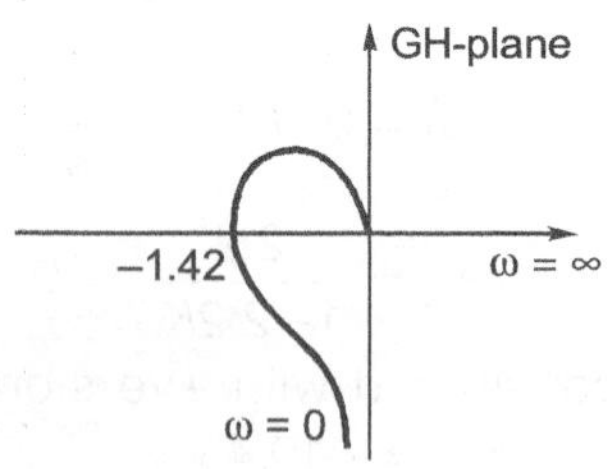

(a) always stable.
(b) marginally stable.
(c) unstable with one pole on the right half s-plane.
(d) unstable with two poles on the right half s-plane.

[2001 : 1 M]

6.3 The asymptotic approximation of the log-magnitude versus frequency plot of a minimum phase system with real poles and one zero is shown in the figure. Its transfer function is

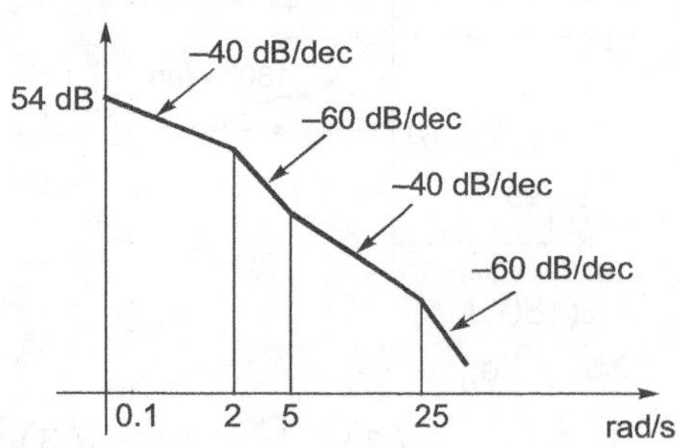

(a) $\frac{20(s+5)}{s(s+2)(s+25)}$ (b) $\frac{10(s+5)}{(s+2)^2(s+25)}$

(c) $\frac{20(s+5)}{s^2(s+2)(s+25)}$ (d) $\frac{50(s+5)}{s^2(s+2)(s+25)}$

[2001 : 1 M]

6.4 The asymptotic Bode plot of the transfer function $K/[1 + (s/a)]$ is given in figure. The error in phase angle and dB gain at a frequency of $\omega = 0.5$ a are respectively

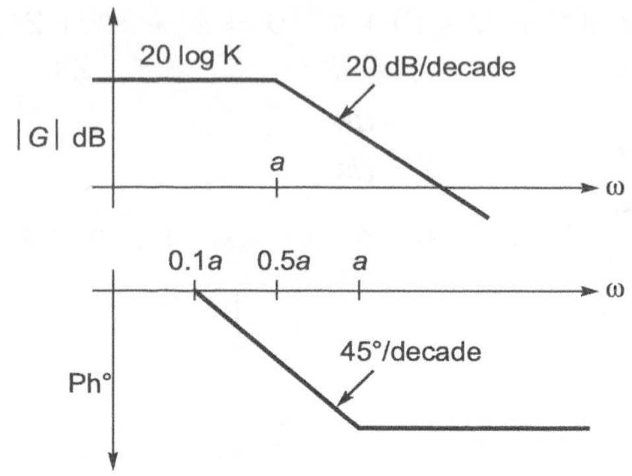

(a) 4.9°, 0.97 dB (b) 5.7°, 3 dB
(c) 4.9°, 3 dB (d) 5.7°, 0.97 dB

[2003 : 2 M]

6.5 The Nyquist plot of loop transfer function $G(s)\,H(s)$ of a closed loop control system passes through the point $(-1, j0)$ in the $G(s)\,H(s)$ plane. The phase margin of the system is

(a) 0° (b) 45°
(c) 90° (d) 180°

[2004 : 1 M]

6.6 In the system shown in figure, the input $x(t) = \sin t$. In the steady-state, the response $y(t)$ will be

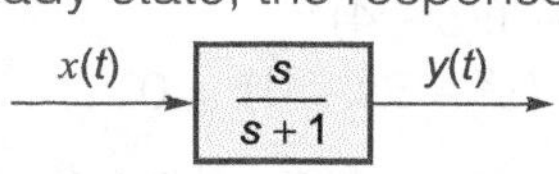

(a) $\frac{1}{\sqrt{2}}\sin(t - 45°)$

(b) $\frac{1}{\sqrt{2}}\sin(t + 45°)$

(c) $\sin(t - 45°)$
(d) $\sin(t + 45°)$

[2004 : 2 M]

6.7 The open loop transfer function of a unity feedback control system is given as $G(s) = \frac{as+1}{s^2}$. The value of 'a' to give a phase margin of 45° is equal to

(a) 0.141 (b) 0.441
(c) 0.841 (d) 1.141

[2004 : 2 M]

6.8 A system with zero initial conditions has the closed loop transfer function.

$$T(s) = \frac{s^2 + 4}{(s+1)(s+4)}$$

The system output is zero at the frequency

(a) 0.5 rad/sec (b) 1 rad/sec
(c) 2 rad/sec (d) 4 rad/sec [2005 : 1 M]

6.9 The gain margin of a unity feedback control system with the open loop transfer function is

$$G(s) = \frac{(s+1)}{s^2}$$

(a) 0 (b) $\frac{1}{\sqrt{2}}$
(c) $\sqrt{2}$ (d) ∞ [2005 : 1 M]

6.10 In the $GH(s)$ plane, the Nyquist plot of the loop transfer function $G(s)\,H(s) = \frac{\pi e^{-0.25s}}{s}$ passes through the negative real axis at the point.

(a) $(-0.25, j0)$ (b) $(-0.5, j0)$
(c) $(-1, j0)$ (d) $(-2, j0)$ [2005 : 2 M]

6.11 If the compensated system shown in the figure has a phase margin of 60° at the crossover frequency of 1 rad/sec, then value of the gain k is

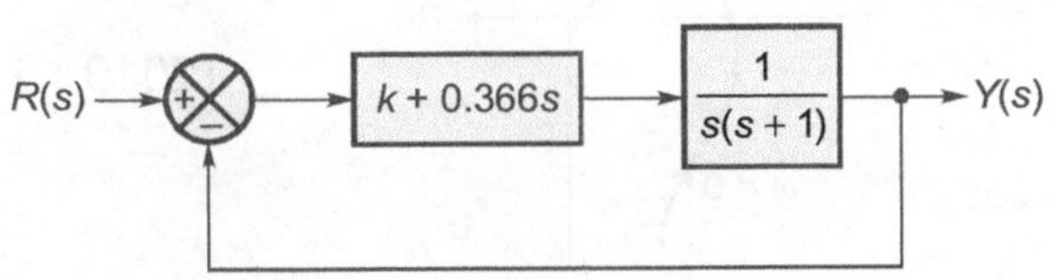

(a) 0.366 (b) 0.732
(c) 1.366 (d) 2.738 [2005 : 2 M]

6.12 The Bode magnitude plot of

$$H(j\omega) = \frac{10^4(1+j\omega)}{(10+j\omega)(100+j\omega)^2} \text{ is}$$

(a)

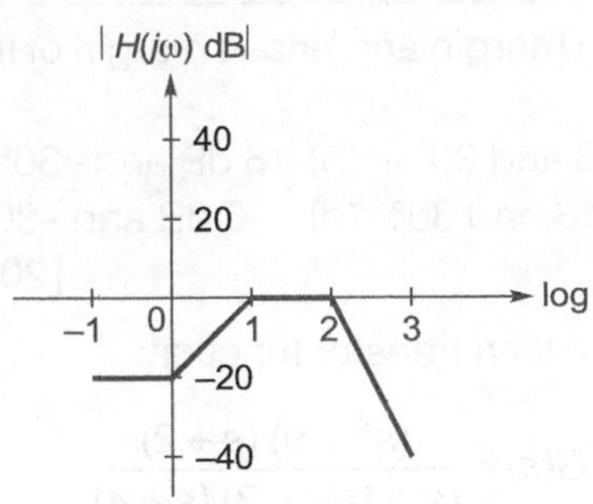

(b)

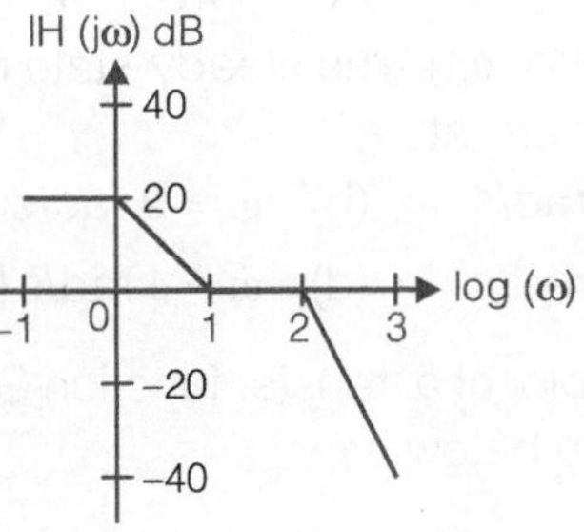

(c)

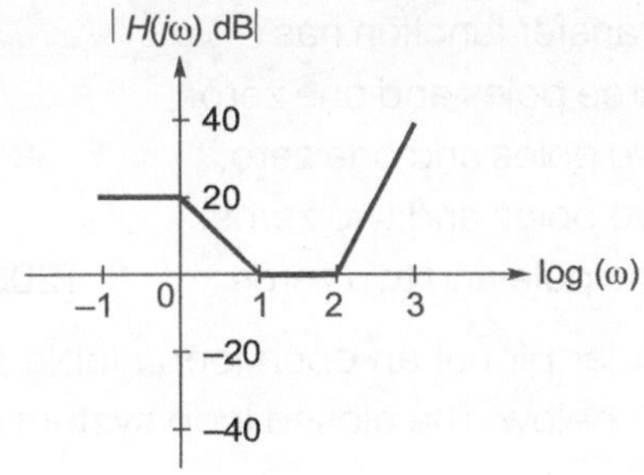

(d)

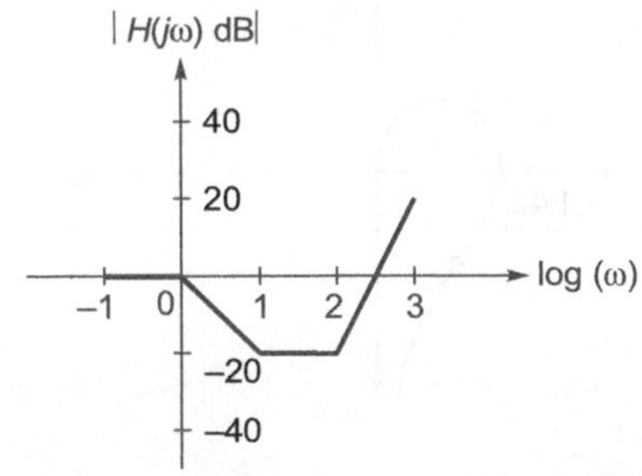

[2006 : 2 M]

6.13 Consider the following Nyquist plots of loop transfer functions over ω = 0 to ω = ∞. Which of these plots represents a stable closed loop system?

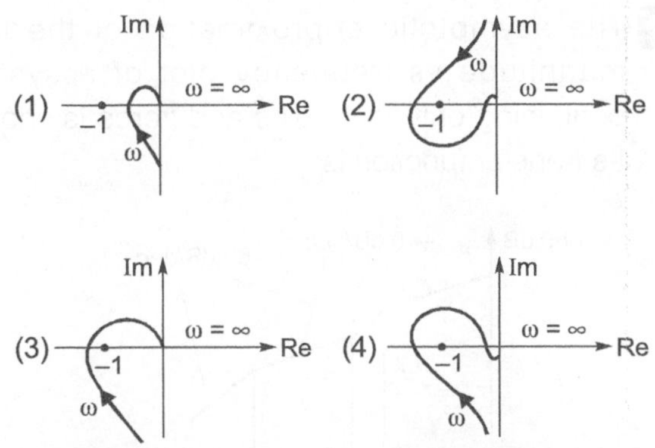

(a) (1) only (b) All, except (1)
(c) All, except (3) (d) (1) and (2) only [2006 : 2 M]

6.14 If x = Re $G(j\omega)$, and y = Im $G(j\omega)$ then for $\omega \to 0^+$, the Nyquist plot for $G(s) = 1/s(s+1)(s+2)$

(a) $x = 0$ (b) $x = -3/4$
(c) $x = y - 1/6$ (d) $x = y/\sqrt{3}$ [2007 : 2 M]

6.15 The asymptotic Bode magnitude plot of a min. phase transfer function is shown in the figure:

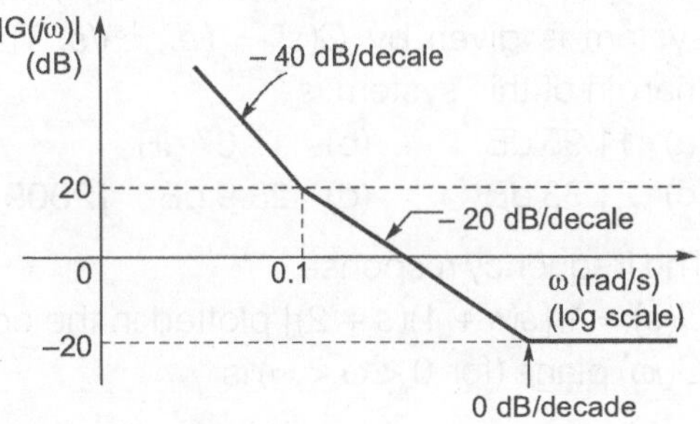

This transfer function has
(a) three poles and one zero.
(b) two poles and one zero.
(c) two poles and two zeros.
(d) one pole and two zeros. [2008 : 2 M]

6.16 The polar plot of an open loop stable system is shown below. The closed loop system is

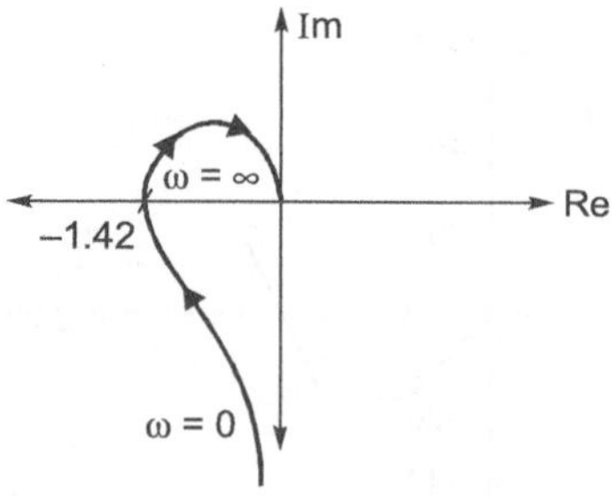

(a) always stable
(b) marginally stable
(c) unstable with one pole on the RHS s-plane
(d) unstable with two poles on the RHS s-plane
[2009 : 1 M]

6.17 The asymptotic approximation of the log-magnitude vs frequency plot of a system containing only real poles and zeros is shown. Its transfer function is

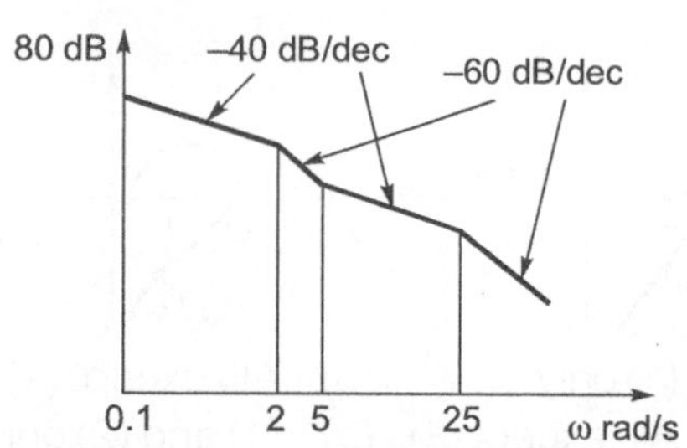

(a) $\dfrac{10(s+5)}{s(s+2)(s+25)}$ (b) $\dfrac{1000(s+5)}{s^2(s+2)(s+25)}$

(c) $\dfrac{100(s+5)}{s(s+2)(s+25)}$ (d) $\dfrac{80(s+5)}{s^2(s+2)(s+25)}$

[2009 : 1 M]

6.18 The open loop transfer function of a unity feedback system is given by $G(s) = (e^{-0.1s})/s$. The gain margin of this system is
(a) 11.95 dB (b) 17.67 dB
(c) 21.33 dB (d) 23.9 dB [2009 : 2 M]

6.19 The frequency response
$G(s) = 1/[s(s+1)(s+2)]$ plotted in the complex $G(j\omega)$ plane (for $0 < \omega < \infty$) is

(a)
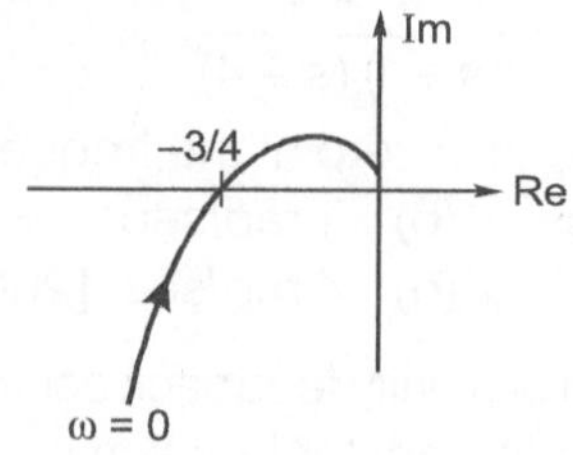

(b)
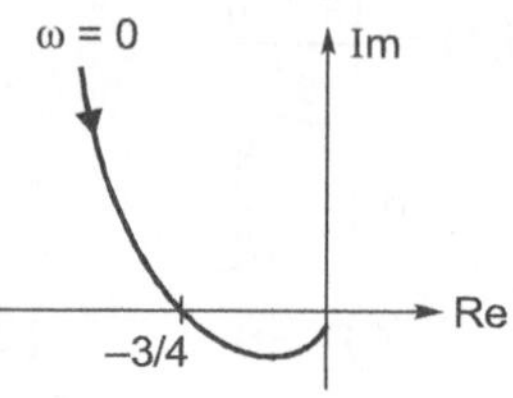

(c)
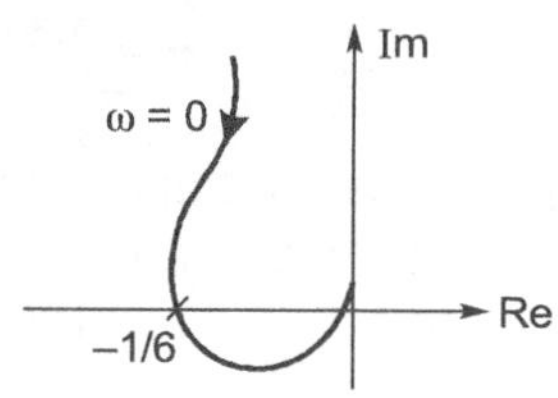

(d)
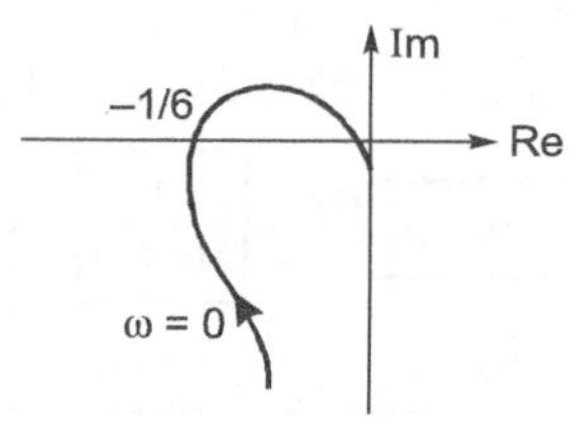

[2010 : 2 M]

6.20 The frequency response of a linear system $G(j\omega)$ is provided in the tabular form below.

$\lvert G(j\omega)\rvert$	1.3	1.2	1.0	0.8	0.5	0.3
$\angle G(j\omega)$	–130°	–140°	–150°	–160°	–180°	–200°

The gain margin and phase margin of the system are
(a) 6 dB and 30° (b) 6 dB and –30°
(c) –6 dB and 30° (d) –6 dB and –30°
[2011 : 1 M]

6.21 A system with transfer function:

$$G(s) = \frac{(s^2+9)(s+2)}{(s+1)(s+3)(s+4)}$$

is excited sin (ωt). The steady-state output of the system is zero at
(a) ω = 1 rad/s (b) ω = 2 rad/s
(c) ω = 3 rad/s (d) ω = 4 rad/s [2012 : 1 M]

6.22 The Bode plot of a transfer function G(s) is shown in the figure below:

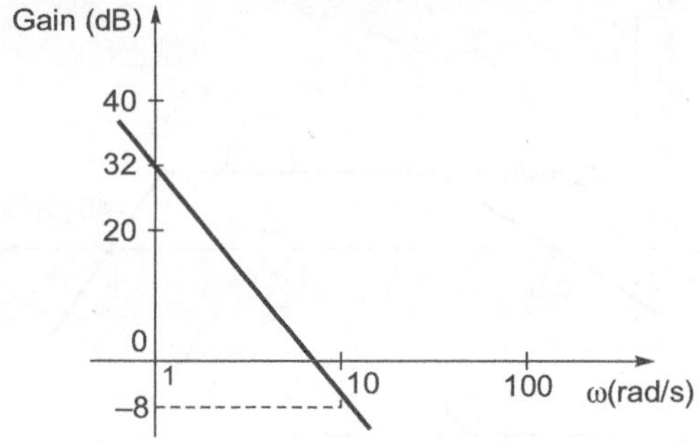

The gain ($20 \log |G(s)|$) is 32 dB and –8 dB at 1 rad/s and 10 rad/s respectively. The phase is negative for all ω. Then $G(s)$ is

(a) $\frac{39.8}{s}$ (b) $\frac{39.8}{s^2}$

(c) $\frac{32}{s}$ (d) $\frac{32}{s^2}$ [2013 : 1 M]

6.23 The Bode magnitude plot of the transfer function

$$G(s) = \frac{K(1+0.5s)(1+as)}{s\left(1+\frac{s}{8}\right)(1+bs)\left(1+\frac{s}{36}\right)}$$

is shown below.

Note that –6 dB/octave = –20 dB/decade. The value of a/bK is ______.

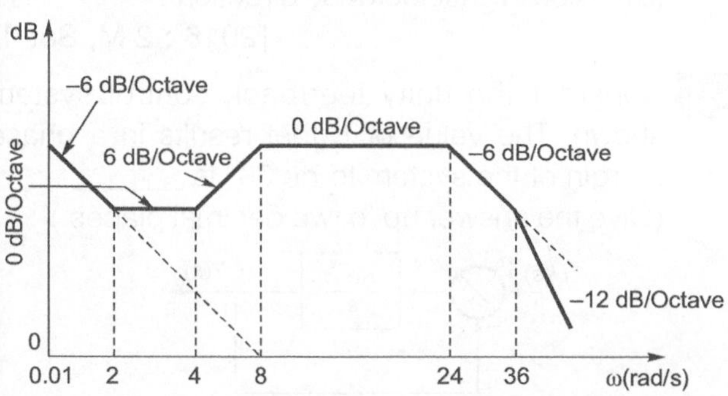

[2014 : 2 M, Set-1]

6.24 For the transfer function:

$$G(s) = \frac{5(s+4)}{s(s+0.25)(s^2+4s+25)}$$

The values of the constant gain term and the highest corner frequency of the Bode plot respectively are

(a) 3.2 and 5.0 (b) 16.0 and 4.0

(c) 3.2 and 4.0 (d) 16.0 and 5.0

[2014 : 2 M, Set-2]

6.25 The magnitude Bode plot of a network is shown in the figure

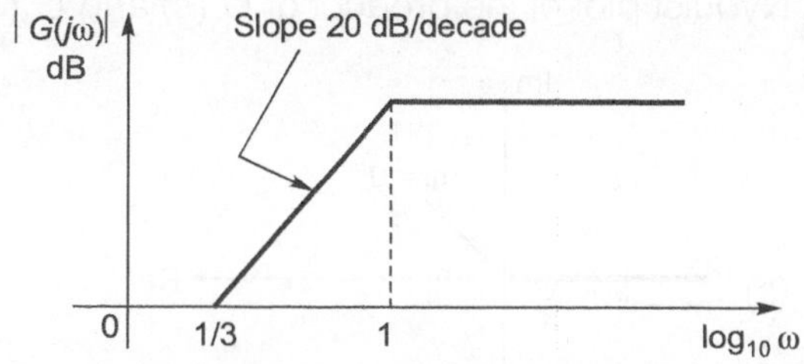

The maximum phase angle ϕ_m and the corresponding gain G_m respectively, are

(a) –30° and 1.73 dB (b) –30° and 4.77 dB

(c) +30° and 4.77 dB (d) +30° and 1.73 dB

[2014 : 2 M, Set-3]

6.26 A Bode magnitude plot for the transfer function $G(s)$ of a plant is shown in the figure. Which one of the following transfer functions best describes the plant?

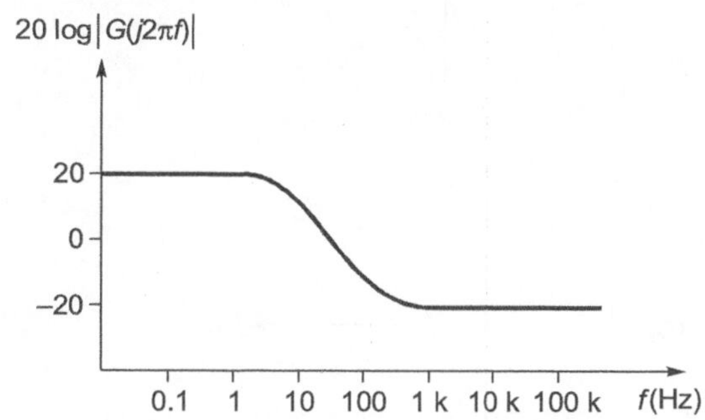

(a) $\frac{1000(s+10)}{s+1000}$ (b) $\frac{10(s+10)}{s(s+1000)}$

(c) $\frac{s+1000}{10s(s+10)}$ (d) $\frac{s+1000}{10(s+10)}$

[2015 : 1 M, Set-1]

6.27 Nyquist plots of two functions $G_1(s)$ and $G_2(s)$ are shown in figure.

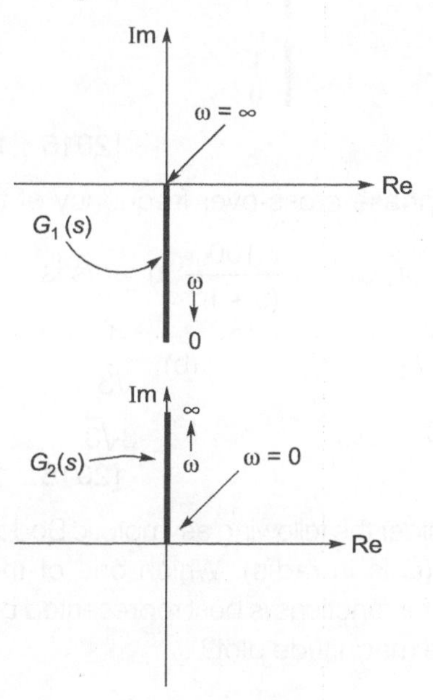

Nyquist plot of the product of $G_1(s)$ and $G_1(s)$ is

(a) 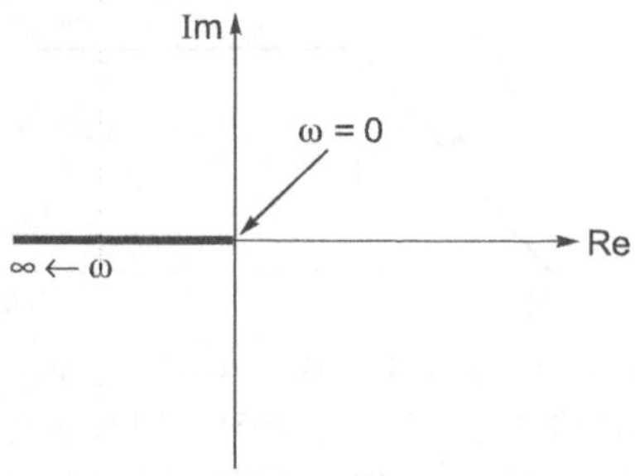

(b) 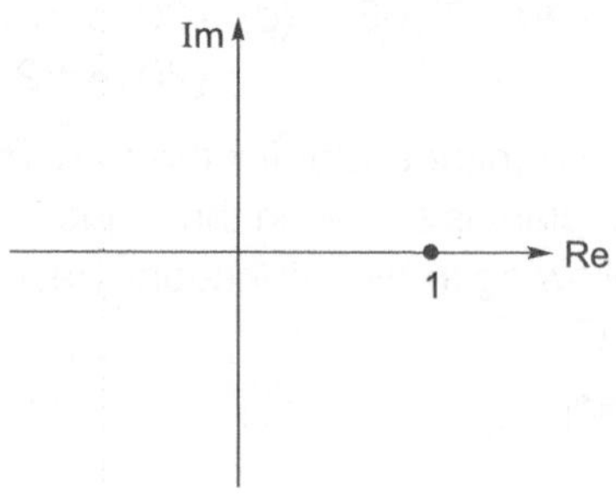

(c) 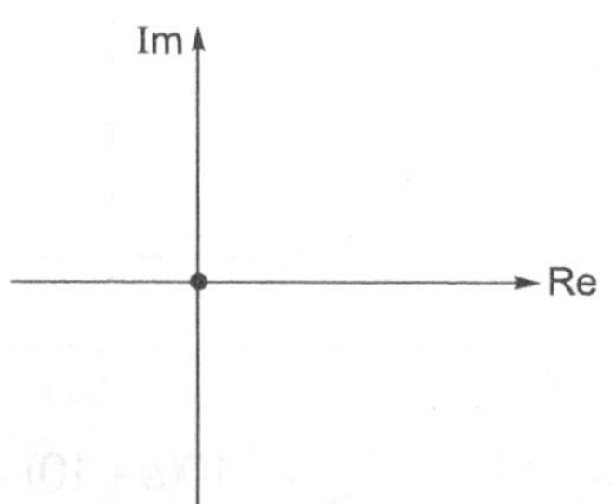

(d) 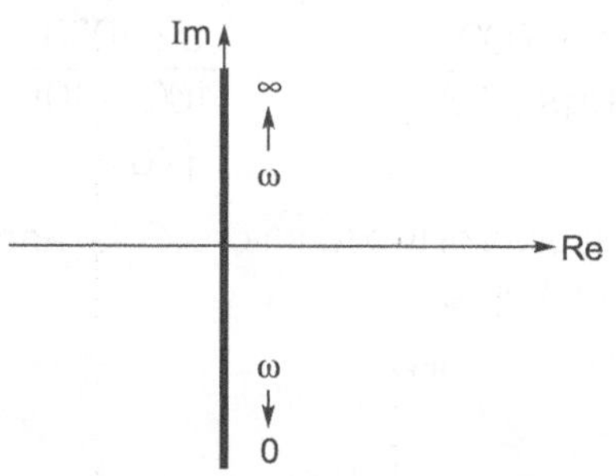

[2015 : 1 M, Set-2]

6.28 The phase cross-over frequency of the transfer function $G(s) = \dfrac{100}{(s+1)^3}$ in rad/s is

(a) $\sqrt{3}$ (b) $\dfrac{1}{\sqrt{3}}$

(c) 3 (d) $3\sqrt{3}$

[2016 : 1 M, Set-1]

6.29 Consider the following asymptotic Bode magnitude plot (ω is in rad/s). Which one of the following transfer functions is best represented by the above Bode magnitude plot?

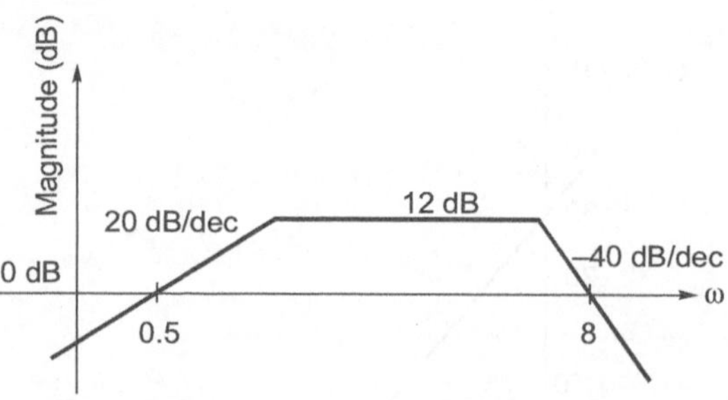

(a) $\dfrac{2s}{(1+0.5s)(1+0.25s)^2}$

(b) $\dfrac{4(1+0.25s)}{s(1+0.25s)}$

(c) $\dfrac{2s}{(1+2s)(1+4s)}$

(c) $\dfrac{4s}{(1+2s)(1+4s)^2}$

[2016 : 2 M, Set-1]

6.30 Loop transfer function of a feedback system is $G(s)H(s) = \dfrac{s+3}{s^2(s-3)}$. Take the Nyquist contour in the clockwise direction. Then, the Nyquist plot of $G(s)H(s)$ encircles $-1 + j0$

(a) once in clockwise direction
(b) twice in clockwise direction
(c) once in anticlockwise direction
(d) twice in anticlockwise direction

[2016 : 2 M, Set-1]

6.31 Consider the unity feedback control system shown. The value of K that results in a phase margin of the system to be 30° is ______ . (Give the answer up to two decimal places.)

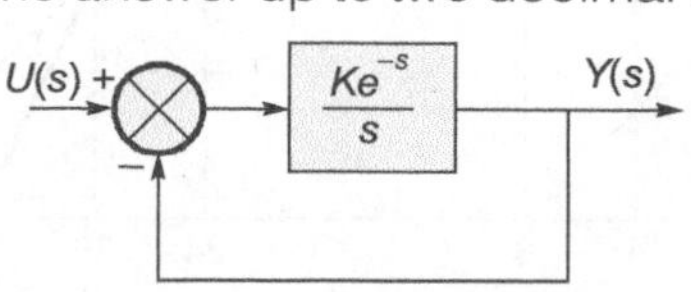

[2017 : 1 M, Set-1]

6.32 The transfer function of a system is given by, $\dfrac{V_0(s)}{V_i(s)} = \dfrac{1-s}{1+s}$. Let the output of the system be $v_0(t) = V_m \sin(\omega t + j)$ for the input, $v_i(t) = V_m \sin(\omega t)$. Then the minimum and maximum values of j (in radians) are respectively

(a) $\dfrac{-\pi}{2}$ and $\dfrac{\pi}{2}$ (b) $\dfrac{-\pi}{2}$ and 0

(c) 0 and $\dfrac{\pi}{2}$ (d) $-\pi$ and 0

[2017 : 1 M, Set-1]

6.33 The open loop transfer function of a unity feedback system is given by

$$G(s) = \frac{\pi e^{-0.25s}}{s}$$

In $G(s)$ plane, the Nyquist plot of $G(s)$ passes through the negative real axis at the point.

(a) $(-1.25, j0)$ (b) $(-0.5, j0)$
(c) $(-1.5, j0)$ (d) $(-0.75, j0)$ **[2019 : 1 M]**

6.34 The asymptotic Bode magnitude plot of minimum phase transfer function $G(s)$ is shown below.

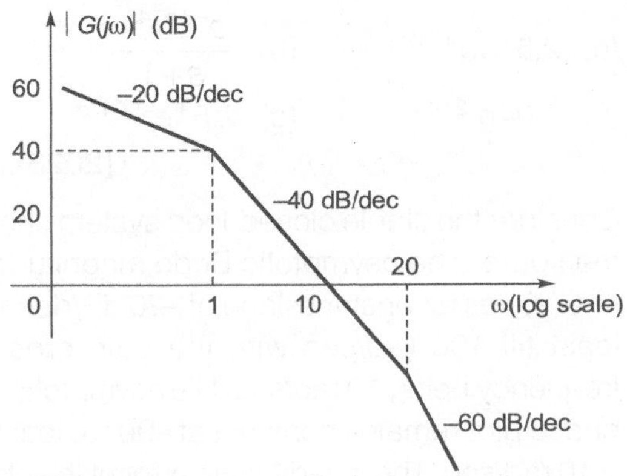

Consider the following statements:
Statement-1: Transfer function $G(s)$ has three poles and one zero.
Statement-2: At very high frequency ($\omega \to \infty$), the phase angle $\angle G(j\omega) = -\frac{3\pi}{2}$.
Which of the following options is correct?

(a) Statement 2 is true and statement 1 is false.
(b) Both the statements are false.
(c) Both the statements are true.
(d) Statement 1 is true and statement 2 is false.

[2019 : 2 M]

6.35 Consider a negative unity feedback system with forward path transfer function,

$$G(s) = \frac{K}{(s+a)(s-b)(s+c)}$$

where K, a, b, c are positive real numbers. For a Nyquist path enclosing the entire imaginary axis and right half of the s-plane in the clockwise direction, the Nyquist plot of $(1 + G(s))$, encircles the origin of $(1 + G(s))$ plane once in the clockwise direction and never passes through this origin for a certain value of K. Then, the number of poles of $G(s)/1 + G(s)$ lying in the open right half of the s-plane is ________. **[2020 : 1 M]**

6.36 A stable real linear time-invariant system with single pole at p, has a transfer function $H(s) = \frac{s^2+100}{s-p}$ with a dc gain of 5. The smallest positive frequency, in rad/s at unity gain is closed to:

(a) 11.08 (b) 78.13
(c) 8.84 (d) 122.87 **[2020 : 2 M]**

6.37 The Bode magnitude plot for the transfer function $\frac{V_0(s)}{V_i(s)}$ of the circuit is as shown. The value of R is ________ Ω. (Round off 2 decimal places)

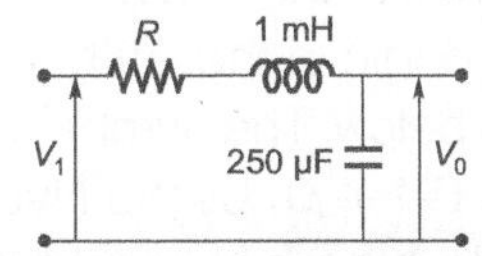

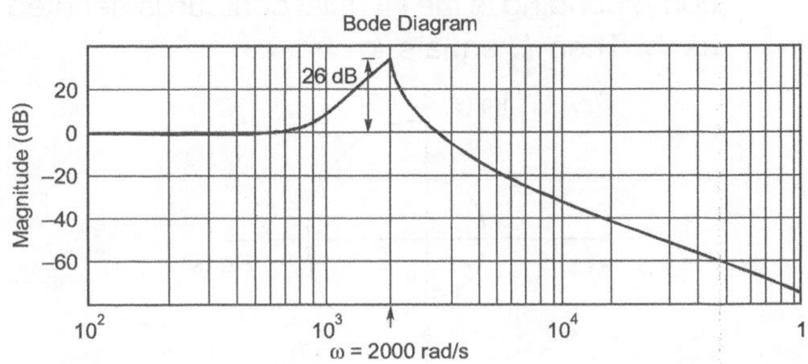

[2021 : 1 M]

6.38 The Bode magnitude plot of a first order stable system is constant with frequency. The asymptotic value of the high frequency phase, for the system, is –180°. This system has

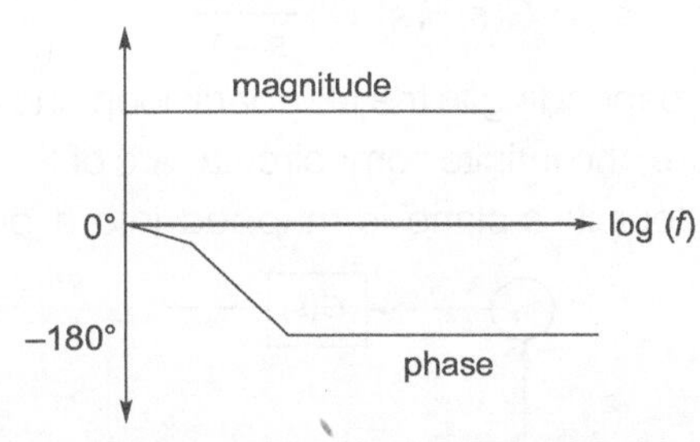

(a) one LHP pole and one RHP zero at the same frequency.
(b) one LHP pole and one LHP zero at the same frequency.
(c) two LHP poles and one RHP zero.
(d) two RHP poles and one LHP zero.

[2022 : 1 M]

6.39 An LTI system is shown in the figure where $G(s) = \frac{100}{s^2 + 0.1s + 100}$. The steady state output of the system, to the input $r(t)$, is given as $y(t) = a + b \sin(10t + \theta)$. The values of a and b will be

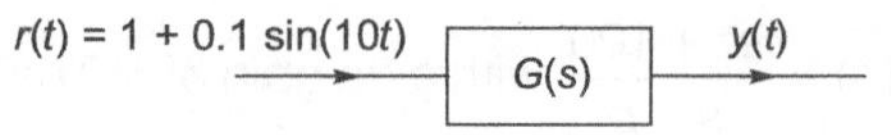

(a) $a = 1, b = 10$ (b) $a = 10, b = 1$
(c) $a = 1, b = 100$ (d) $a = 100, b = 1$

[2022 : 2 M]

6.40 The open loop transfer function of a unity gain negative feedback system is given as

$$G(s) = \frac{1}{s(s+1)}$$

The Nyquist contour in the s-plane encloses the entire right half plane and a small neighbourhood around the origin in the left half plane, as shown in the figure below. The number of encirclemenets of the point $(-1 + j0)$ by the Nyquist plot of $G(s)$, corresponding to the Nyquist contour, is denoted as N. Then N equals to

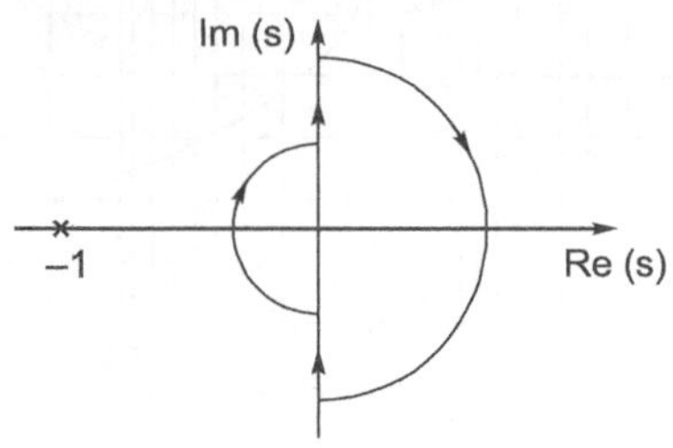

(a) 0 (b) 1
(c) 2 (d) 3 [2022 : 2 M]

6.41 In the Nyquist plot of the open-loop transfer function

$$G(s)H(s) = \frac{3s+5}{s-1}$$

corresponding to the feedback loop shown in the figure, the infinite semi-circular arc of the Nyquist contour in s-plane is mapped into a point at

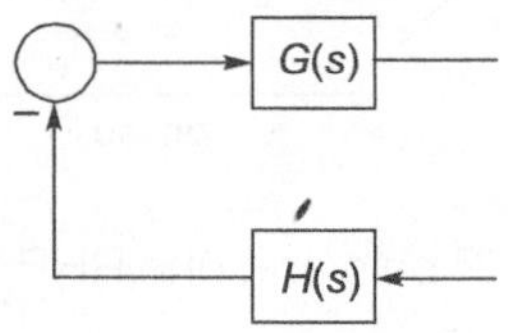

(a) $G(s)H(s) = \infty$ (b) $G(s)H(s) = 0$
(c) $G(s)H(s) = 3$ (d) $G(s)H(s) = -5$

[2023 : 1 M]

6.42 The magnitude and phase plots of an LTI system are shown in the figure. The transfer function of the system is

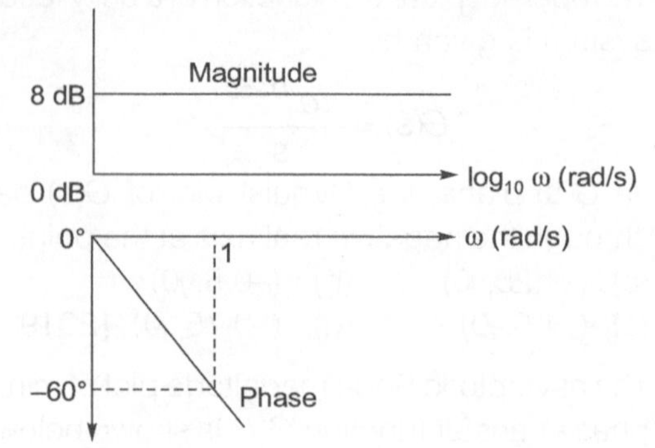

(a) $2.515e^{-0.032s}$ (b) $\frac{e^{-2.514s}}{s+1}$
(c) $1.04e^{-2.514s}$ (d) $2.51e^{-1.047s}$

[2023 : 2 M]

6.43 Consider the stable closed-loop system shown in the figure. The asymptotic Bode magnitude plot of $G(s)$ has a constant slope of –20 dB/decade at least till 100 rad/sec with the gain crossover frequency being 10 rad/sec. Hie asymptotic Bode phase plot remains constant at –90° at least till ω = 10 rad/sec. The steady-state error of the closed-loop system for a unit ramp input is ______ (rounded off to 2 decimal places).

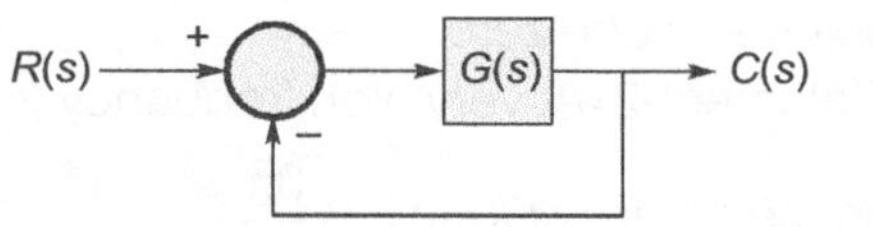

[2024 : 2 M]

6.44 Consider the stable closed-loop system shown in the figure. The magnitude and phase values of the frequency response of $G(s)$ are given in the table. The value of the gain $K_I (> 0)$ for a 50° phase margin is ______ (rounded off to 2 decimal places).

ω in rad/sec	Magnitude in dB	Phase in degrees
0.5	–7	–40
1.0	–10	–80
2.0	–18	–130
10.0	–40	–200

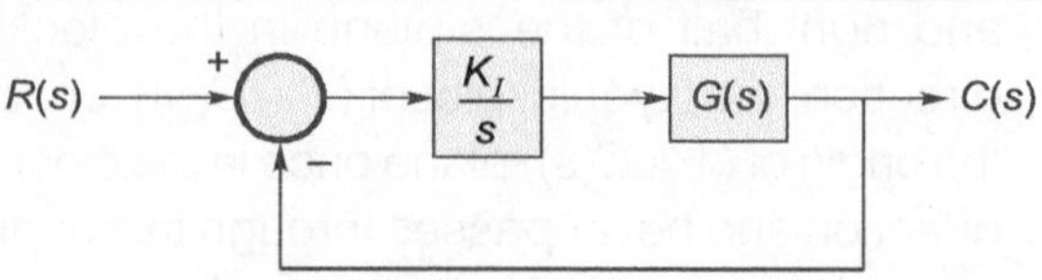

[2024 : 2 M]

Answers Frequency Response Analysis

6.1 (d)	6.2 (d)	6.3 (d)	6.4 (a)	6.5 (a)	6.6 (b)	6.7 (c)
6.8 (c)	6.9 (d)	6.10 (b)	6.11 (c)	6.12 (a)	6.13 (d)	6.14 (b)
6.15 (c)	6.16 (d)	6.17 (b)	6.18 (d)	6.19 (d)	6.20 (a)	6.21 (c)
6.22 (b)	6.23 (0.755)	6.24 (a)	6.25 (c)	6.26 (d)	6.27 (b)	6.28 (a)
6.29 (a)	6.30 (a)	6.31 (1.047)	6.32 (d)	6.33 (b)	6.34 (a)	6.35 (2)
6.36 (c)	6.37 (0.10)	6.38 (a)	6.39 (a)	6.40 (b)	6.41 (c)	6.42 (d)
6.43 (0.10)	6.44 (1.12)					

Explanations Frequency Response Analysis

6.1 (d)

$\because \quad r(t) = \sin t$

$\Rightarrow \quad \omega = 1$

$$\text{Gain} = \frac{1}{(1+s)} = \frac{1}{1+j\omega} = \frac{1}{1+j} = \frac{1}{\sqrt{2}}\angle -45^\circ$$

$$\therefore \text{Output} = \sin t \times \frac{1}{\sqrt{2}}\angle -45^\circ$$

$$= \frac{1}{\sqrt{2}}\sin(t-45^\circ)$$

6.2 (d)

Nyquist plot will be:

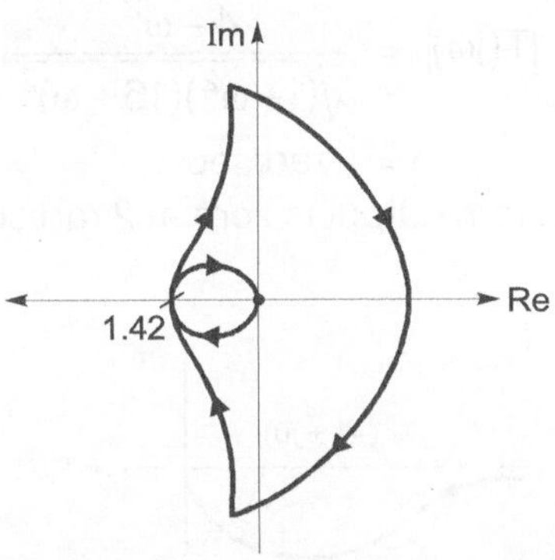

$\therefore$ Number of encirclement of $(-1, j0) = -2$ and number of right sided poles in open loop system = 0

$$N = P - Z$$

$$\therefore \quad -2 = 0 - 2 \Rightarrow Z = 2$$

$\therefore$ Closed loop system is unstable with two poles on the right half of s-plane.

6.3 (d)

Type: 2

Poles: 0, 0, 2, 25

Zeros: 5

$$\text{Gain} = m \log \omega + 20\log K$$

$$54 = -40 \log(0.1) + 20\log K \Rightarrow K = 5$$

$$\therefore \quad \text{T.F.} = \frac{5(1+0.2s)}{s^2(1+0.5s)(1+0.04s)}$$

$$= \frac{50(s+5)}{s^2(s+2)(s+25)}$$

6.4 (a)

$$\text{Transfer function} = G(s) = \frac{K}{1+\frac{s}{a}}$$

$$\text{Put } s = j\omega,\ G(j\omega) = \frac{K}{1+j\frac{\omega}{a}}$$

$$\text{Magnitude of } G(j\omega) = |G(j\omega)| = \frac{K}{\sqrt{1+\frac{\omega^2}{a^2}}}$$

$$\text{Phase of } G(j\omega)\ \angle G(j\omega) = -\tan^{-1}\left(\frac{\omega}{a}\right)$$

at $\omega = 0.5a$

$$|G(j\omega)|_{\omega=0.5a} = \frac{K}{\sqrt{1+\frac{(0.5a)^2}{a^2}}} = \frac{K}{\sqrt{1.25}}$$

in dB : $= 20 \log K - 20 \log (1.25)^{1/2}$

$= 20 \log K - 0.97$ dB

From the plot

Magnitude $= 20 \log K$

$\therefore$ error in dB gain = 0.97dB

$$\angle G(j\omega)|_{\omega=0.5a} = -\tan^{-1}\left(\frac{0.5a}{a}\right) = 26.56^\circ$$

Since it is -45° /dec (–ve slope) at $\omega = 0.5a$

phase $= -22.5^\circ$

Error in phase of $\angle G(j\omega)|_{\omega=0.5a}$

$= -22.5 - (-26.56)$

$= +4.06^\circ \approx +4.9^\circ$

6.5 (a)

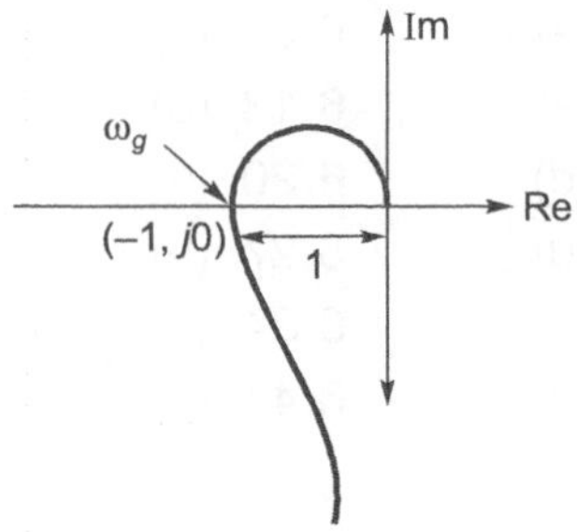

ω_g is gain cross over frequency at which the gain $|G(j\omega)H(j\omega)|$ becomes unity.

In this case, phase $\angle G(j\omega)\ H(j\omega)$ is $-180°$ at $\omega = \omega_g$.

$\phi = -180°$. So, phase margin $= 180° + \phi = 0°$.

6.6 (b)

$$x(t) = \sin t = 1 < 0; \quad \omega = 1 \text{ rad/sec}$$

Transfer function $= G(s) = \dfrac{s}{s+1}$

For sinusoidal input, $s = j\omega = j$

$$G(j\omega)\big|_{\omega=1} = \frac{j\omega}{j\omega+1} = \frac{j}{j+1} = \frac{1}{\sqrt{2}}\angle 45°$$

$$y(t) = \frac{1}{\sqrt{2}}\angle 45°.1\angle 0° = \frac{1}{\sqrt{2}}\angle 45°$$

$$y(t) = \frac{1}{\sqrt{2}}\sin(t+45°)$$

6.7 (c)

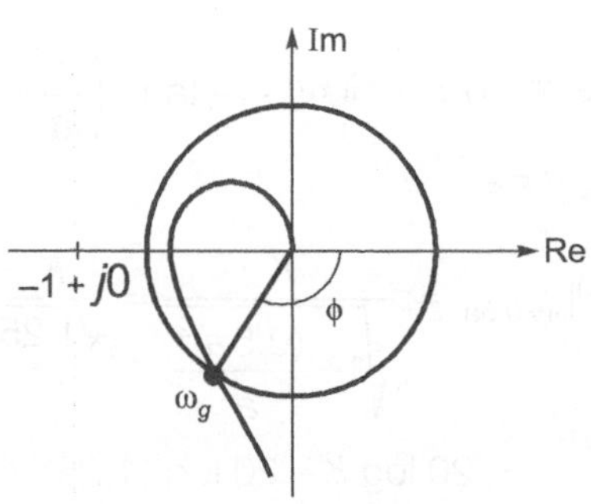

$$G(s) = \frac{as+1}{s^2}$$

ω_g = Gain-crossover frequency at which open gain is loop

$$\text{Phase margin} = 180 + \angle G(j\omega)$$

$$45° = 180° + \angle G(j\omega_g)$$

$$\angle G(j\omega_g) = -130°$$

$$\angle G(j\omega_g) = \frac{ja\omega_g + 1}{(j\omega_g)^2}$$

$$= -2 \times 90° + \tan^{-1} a\,\omega_g$$

$$\angle G(j\omega_g) = -180° + \tan^{-1} a\omega_g = -135°$$

$$\tan^{-1} a\,\omega_g = 45°$$

$$a\,\omega_g = 1$$

$$\omega_g = \frac{1}{a}$$

$$\text{Gain} = |G(j\omega_g)| = 1$$

$$\Rightarrow \quad \frac{\sqrt{1+a^2\omega_g^2}}{\omega_g^2} = 1 \Rightarrow \frac{\sqrt{1+a^2 \times \dfrac{1}{a^2}}}{1/a^2} = 1$$

$$\frac{1}{a^2} = \sqrt{2}$$

$$a = \frac{1}{2^{1/4}} = 0.841$$

6.8 (c)

$$T(s) = \frac{s^2+4}{(s+1)(s+4)}$$

For frequency response, put $s = j\omega$

$$T(j\omega) = \frac{(j\omega)^2+4}{(j\omega+1)(j\omega+4)} = \frac{4-\omega^2}{(1+j\omega)(4+j\omega)}$$

Magnitude of $T(j\omega)$,

$$|T(j\omega)| = \frac{4-\omega^2}{\sqrt{(1+\omega^2)(16+\omega)^2}}$$

$$|T(j\omega)| = \frac{4-\omega^2}{\sqrt{(1+\omega^2)(16+\omega)^2}} = 0$$

$$\omega = 2 \text{ rad/sec}$$

The system output is zero at 2 rad/sec.

6.9 (d)

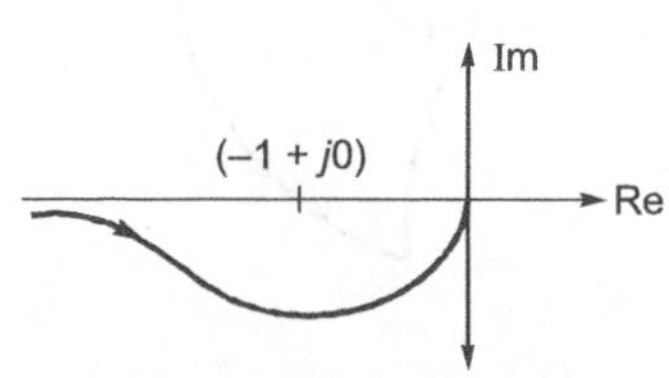

$-1 + j0$ is not enclosed, system is always stable.

$\therefore$ G.M. $= \infty$

6.10 (b)

$$G(s)\,H(s) = \frac{\pi e^{-0.25s}}{s}$$

Putting $s = j\omega$, for frequency response

$$G(j\omega)\,H(j\omega) = \frac{\pi e^{-j0.25\omega}}{j\omega}$$

Phase angle $= \angle G(j\omega)\,H(j\omega) = -90 - 0.25\omega$

at ω_{pc} Nyquist plot cuts negative real axis and this frequency $\angle G(j\omega_{pc})\,H(j\omega_{pc})$ is $-180°$.

$$\angle G(j\omega_{pc})\,H(j\omega_{pc}) = -90° - \frac{\omega}{4}$$

$$-180° = -90° - \frac{\omega}{4} = 360° \text{ or } 2\pi$$

$$|G(j\omega_{pc})H(j\omega_{pc})| = \frac{\pi e^{-j0.25\times2\pi}}{j2\pi}$$

$$= -\frac{j}{2}\left[\cos\frac{\pi}{2} - j\sin\frac{\pi}{2}\right] = -0.5 + j0$$

6.11 (c)

$$G(s) = \frac{k+0.366s}{s(s+1)} \text{ and } H(s) = 1$$

$$\text{OLTF} = G(s)H(s) = \frac{k+0.366s}{s(s+1)}$$

Phase margin = 60° at ω_{gc} (gain crossover frequency)

$$\Rightarrow \quad 180° + \angle G(j\omega_{pc})\,H(j\omega_{pc}) = 60°$$

$$\angle G(j\omega_{gc})\,H(j\omega_{gc}) = -120°$$

$$-90° - \tan^{-1}\omega_{gc} + \tan^{-1}\left(\frac{0.366\omega_{gc}}{k}\right) = -120°$$

$$\omega_{gc} = 1 \text{ rad/sec}$$

$$-90° - \tan^{-1}(1) + \tan^{-1}\left(\frac{0.366\times1}{k}\right) = -120°$$

$$= -90° - 45° + \tan^{-1}\left(\frac{0.366}{k}\right) = -120°$$

$$\Rightarrow \tan^{-1}\left(\frac{0.366}{k}\right) = 15° \Rightarrow k = \frac{0.366}{\tan 15°} = 1.366$$

6.12 (a)

$$H(j\omega) = \frac{10^4(1+j\omega)}{(10+j\omega)(100+j\omega)^2}$$

$$= \frac{(1+j\omega)}{10\times\left(1+\frac{j\omega}{10}\right)\times\left(1+\frac{j\omega}{100}\right)^2}$$

$$H(j\omega) = \frac{k(1+j\omega)}{\left(1+\frac{j\omega}{10}\right)\left(1+\frac{j\omega}{100}\right)^2}$$

where, $k = \frac{1}{10} = 0.1$

Corner frequencies

ω_1 = 1 rad/sc; ω_2 = 10 rad/sec

and ω_3 = 100 rad/sec

For frequency less than ω_1 i.e. $\omega < \omega_1$

Gain of the system is constant as there is no pole at origin.

Gain = 20 log k = 20 log 0.1 = –20 dB

At $\omega = \omega_1$ = 1 rad/sec or log ω_1 = log = 0

There is zero, so system gain increases with slope +20 dB/decade and system gain becomes 0 dB at ω = 10 rad/sec or

$$\log\omega|_{\omega=10} = \log_{10} = 1$$

At ω_2 = 10 or $\log\omega|_{\omega_2=10} = \log 10 = 1$

There is pole, so slope is –20 dB/decade.

Overall slope $\omega_2 < \omega < \omega_3$

= 20 dB/decade – 20 dB/decade

= 0 dB/decade

So, gain remains constant between

$\omega_2 < \omega < \omega_3$

or $1 < \log\omega < 2$

At $\omega = \omega_3$ = 100 rad/sec or

$$\log\omega|_{\omega_3=100} = \log 100 = 2$$

The double pole are present.

So, system gain decrease with –40 dB/decade.

6.14 (b)

$$G(s) = \frac{1}{s(s+1)(s+2)}$$

Put $s = j\omega$,

$$G(j\omega) = \frac{1}{j\omega(1+j\omega)(2+j\omega)}$$

$$= \frac{-j(1-j\omega)(2-j\omega)}{\omega(1+\omega^2)(4+\omega^2)}$$

$$= \frac{-j(2-3j\omega-\omega^2)}{\omega(1+\omega^2)(4+\omega^2)}$$

$$= \frac{-3\omega - j(2-\omega^2)}{\omega(1+\omega^2)(4+\omega^2)}$$

$$= \frac{-3}{(1+\omega^2)(4+\omega^2)} + \frac{j(\omega^2-2)}{\omega(1+\omega^2)(4+\omega^2)}$$

$$= x + iy$$

At $\omega \to 0$, $x \to -3/4$, $y \to -\infty$

6.15 (c)

Initial slope is –40 dB/decade, it means there are double pole at origin.

Slope changes from –40 dB/decade to –20 dB/decade. It means there is a zero.

Slope changes from –20 dB/decade to 0 dB/decade at some other frequency. It means there is one more zero.

Therefore transfer function has two poles and two zeros.

6.16 (d)

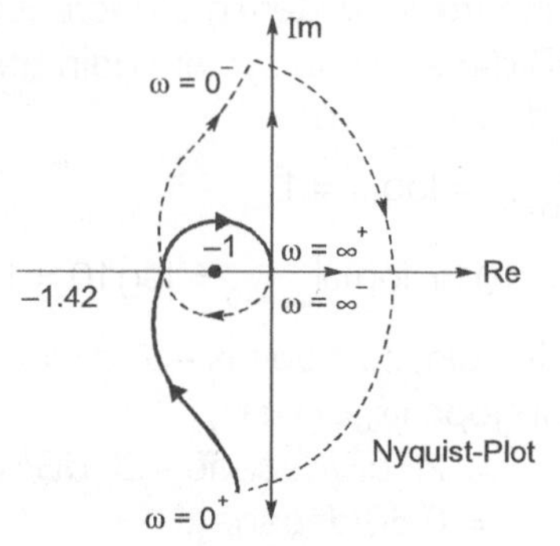

Two clockwise encirclement of 1 + *j*0.

$N = -2$

Open-loop system is stable $\Rightarrow P = 0$

$N = P - Z \Rightarrow -2 = 0 - Z$

Z = Number of closed loop poles in RHS of 8-planes. Hence the system is unstable.

6.17 (b)

Initial slope = −40 dB/dec.

Hence the system is type-2. So the corresponding term of the transfer function is $1/s^2$.

At ω = 2 rad/s, slope changes by −20 dB/dec. from −40 dB/dec to −60 dB/dec. Hence the corresponding term of the transfer function is

$$\frac{1}{\left(1+\frac{s}{2}\right)}.$$

At ω = 5 rad/s, slope changes by 20 dB/dec from −60 dB/dec to −40 dB/dec. Hence, the corresponding term of the transfer function is

$$\left(1+\frac{s}{5}\right).$$

At ω = 25 the slope changes by −20 dB/dec. from −40 dB/dec to −60 dB/dec. Hence, the corresponding term of the transfer function is

$$\frac{1}{1+\frac{s}{25}}.$$

$$G(s) = \frac{k\left(1+\frac{s}{5}\right)}{s^2\left(1+\frac{s}{2}\right)\left(1+\frac{s}{25}\right)}$$

At ω = 0.1 ($\omega \le 2$ rad/sec)

$$\text{Gain(dB)} = 20\log k - 40\log\omega$$

$$80 = 20\log k - 40\log 0.1$$

$$20\log k = 40 \Rightarrow k = 100$$

$$G(s) = \frac{100\left(1+\frac{s}{5}\right)}{s^2\left(1+\frac{s}{2}\right)\left(1+\frac{s}{25}\right)} = \frac{1000(s+5)}{s^2(s+2)(s+25)}$$

6.18 (d)

Open loop transfer function, $G(s) = \frac{e^{-0.1s}}{s}$

Put, $s = j\omega$, $G(j\omega) = \frac{e^{-j0.1\omega}}{j\omega}$

At phase crossover frequency (ω_{pc}), phase of OLTF is −180°

$$\angle G(j\omega)\big|_{\omega/\omega_{pc}} = -180° = -\pi$$

$$-\frac{\pi}{2} - 0.1\omega_{pc} = -\pi$$

$$-0.1\,\omega_{pc} = -\frac{\pi}{2} \Rightarrow \omega_{pc} = 5\pi$$

$\left|e^{-0.1\omega}\right|$ is always 1 for any value of ω

$$|G(j\omega)| = \frac{1}{\omega}$$

Gain at ω_{pc} (phase-cross frequency)

$$|G(j\omega)|_{\omega_{pc=5}} = \frac{1}{5\pi}$$

$$\text{Gain margin} = 20\log\frac{1}{|G(j\omega_{pc})|}$$

$$= 20\log 5\pi = 23.9 \text{ dB}$$

6.20 (a)

At gain across over frequency (ω_{gc}), magnitude of $G(j\omega)$ is 1.

$$|G(j\omega_{gc})| = 1$$

$$\text{Phase of } G(j\omega) = \angle G(j\omega_{gc}) = -150°$$

$$\text{Phase margin} = 180° + \angle G(j\omega_{gc})$$

$$= 180° - 150° = 30°$$

At phase cross frequency (ω_{pc}), phase of $G(j\omega)$ is −180°, $\angle G(j\omega_{pc}) = -180°$

M = magnitude of $G(j\omega)$ at

$$\omega_{pc} = |G(j\omega_{pc})| = 0.5$$

$$\text{Gain margin} = 20\log\frac{1}{M} = 20\log\frac{1}{0.5} = 6 \text{ dB}$$

6.21 (c)

For sinusoidal excitation,

$$s = j\omega$$

$$\therefore \quad G(j\omega) = \frac{(-\omega^2+9)(j\omega+2)}{(j\omega+1)(j\omega+3)(j\omega+4)}$$

For zero steady-state output $|G(j\omega)| = 0$

$$= \frac{(-\omega^2+9)\sqrt{\omega^2+4}}{(\sqrt{\omega^2+1})(\sqrt{\omega^2+9})(\sqrt{\omega^2+16})}$$

For zero steady-state output,

$\Rightarrow \quad \omega^2 = 9 \Rightarrow \omega = 3$ rad/sec

6.22 (b)

Let, $\quad G(s) = \dfrac{K}{s^n}$

$$G(j\omega) = \frac{K}{(j\omega)^n}$$

$\because$ Gain = 32 dB at ω = 1 rad/sec.

$$\therefore \quad 20\log\left|\frac{K}{1^n}\right| = 32$$

$\Rightarrow \quad 20 \log K = 32 \Rightarrow K = 39.8$

Now,

$\because$ Gain = –8 dB at ω = 10 rad/sec.

$$\therefore \quad 20\log\left|\frac{39.8}{10^n}\right| = -8 \Rightarrow 10^n = 100 \Rightarrow n = 2$$

6.23 Sol.

Given, $\quad G(s) = \dfrac{K(1+0.5s)(1+as)}{s\left(1+\dfrac{s}{8}\right)(1+bs)\left(1+\dfrac{s}{36}\right)}$

$(1 + as)$ is addition of zero to the transfer function whose contribution in slope = +20 dB/decade or –6 dB/octave.

$(1 + bs)$ is addition of pole to the transfer function whose contribution in slope = –20 dB/decade or –6 dB/octave.

Observing the change in the slope at different corner frequencies, we conclude that,

$$a = \frac{1}{4} \text{ rad/s and } b = \frac{1}{24} \text{ rad/s}$$

From ω = 0.01 rad/s to ω = 8 rad/s,

slope = –20 dB/decade

Let the vertical length in dB be y

$$\therefore \quad -20 = \left(\frac{0-y}{\log 8 - \log 0.01}\right)$$

$$\text{or,} \quad -20 = \frac{y}{\log 8 + 2} \quad \text{or } y = 58 \text{ dB}$$

Applying, $\quad y = mx + C$ at ω = 0.01 rad/s,

we have, $\quad 58 = -20 \log 0.01 + C$

or, $\quad C = 58 - 40 = 18$

Now, $\quad C = 20 \log K$

$$\text{or,} \quad \log K = \frac{18}{20} = 0.9$$

$$\therefore \quad K = \log^{-1}(0.9) = (10)^{0.9} = 7.94$$

$$\therefore \quad \frac{a}{bK} = \frac{\left(\frac{1}{4}\right)}{\left(\frac{1}{24}\right)\times 7.94} = \frac{24}{4\times 7.94} = 0.755$$

$$\therefore \quad \frac{a}{bK} = 0.755$$

6.24 (a)

Given, $\quad G(s) = \dfrac{5(s+4)}{s(s+0.25)(s^2+4s+25)}$

Reducing $G(s)$ in time constant form, we have:

$$G(s) = \frac{5\times 4\left(1+\frac{s}{4}\right)}{s\left[0.25\left(1+\frac{s}{0.25}\right)\right]\left[25\left(1+\frac{4}{25}s+\frac{1}{25}s^2\right)\right]}$$

$$= \frac{20(1+0.25s)}{s[0.25(1+4s)][25(1+0.16s+0.04s^2)]}$$

$$\text{or, } G(s) = \frac{3.2(1+0.25s)}{s(1+4s)(1+0.16s+0.04s^2)}$$

$\therefore$ The value of constant gain term = 3.2

$$\text{Also, } G(j\omega) = \frac{3.2\left(1+\frac{j\omega}{4}\right)}{(j\omega)\left(1+\frac{j\omega}{0.25}\right)\left(1+j0.16\omega-\frac{\omega^2}{25}\right)}$$

Corner frequencies are:

ω_1 = 4 rad/s; ω_2 = 0.25 rad/s

and $\quad \omega_3$ = 5 rad/s

$\therefore$ Highest corner frequency = ω_3 = 5 rad/s.

6.25 (c)

From the given Bode plot, it is clear that corner frequencies,

$$\omega_{c_1} = \frac{1}{3} \text{ and } \omega_{c_2} = 1$$

$\therefore$ Transfer function of given system is given by

$$T(s) = \frac{C(s)}{R(s)} = \left[\frac{1+3s}{1+s}\right] = \left[\frac{1+Ts}{1+\alpha Ts}\right]$$

Here, $\quad T = 3$ and $\alpha T = 1$ or, $\alpha = \dfrac{1}{3}$

As $\alpha < 1$, therefore the transfer function $T(s)$ represents a lead compensator having $\alpha = \dfrac{1}{3}$

$\therefore$ Maximum phase shift,

$$\phi_m = \tan^{-1}\left[\frac{1-\alpha}{2\sqrt{\alpha}}\right] = \tan^{-1}\left[\frac{1-\frac{1}{3}}{\frac{2}{\sqrt{3}}}\right]$$

$$= \tan^{-1}\left[\frac{2}{3}\times\frac{\sqrt{3}}{2}\right] = \tan^{-1}\left[\frac{1}{\sqrt{3}}\right] = 30°$$

$\therefore \quad \phi_m = 30°$

Also,

$$T(j\omega) = \frac{1+j3\omega}{1+j\omega}$$

$$\therefore \quad |T(j\omega)| = \frac{\sqrt{1+9\omega^2}}{\sqrt{1+\omega^2}}$$

$\therefore$ Gain G_m in dB $= 20\log|T(j\omega)|$

$$= 20\log_{10}\left[\frac{\sqrt{1+9\omega_m^2}}{1+\omega_m^2}\right]$$

Now, $\omega_m = \dfrac{1}{T\sqrt{\alpha}} = \dfrac{1}{\sqrt{3}}$

$$\therefore \quad G_m = 20\log_{10}\left[\sqrt{\frac{1+9\times\left(\frac{1}{\sqrt{3}}\right)^2}{1+\left(\frac{1}{\sqrt{3}}\right)^2}}\right]$$

$$= 20\log_{10}\left[\sqrt{\frac{1+3}{1+\frac{1}{3}}}\right] = 20\log_{10}\sqrt{3}$$

or, $GM = 4.77$ dB

6.26 (d)

From initial line equation,

$$\text{dB} = 20\log_{10}K \qquad ...(1)$$

We know, 20 dB = 20 $\log_{10} K$, i.e., $K = 10$

Now, from the given plot we know type of system is one.

Hence, $G(s) = \dfrac{K(1+sT_1)}{(1+sT_2)}$

$$T_1 = \frac{1}{10}; \; T_2 = \frac{1}{1000}$$

$$= \frac{10(1+0.001s)}{(1+0.1s)} = \frac{1}{10}\frac{(1000+s)}{(s+10)}$$

6.27 (b)

$$G_1(s) = \frac{1}{s}; \quad G_2(s) = s$$

$$G_1(s)\times G_2(s) = \frac{1}{s}\cdot s = 1$$

6.28 (a)

$$G(s) = \frac{100}{(s+1)^3}$$

$$G(j\omega) = \frac{100}{(1+j\omega)} = \frac{100}{1+(j\omega)^3+3.(j\omega)^2+3j\omega}$$

$$= \frac{100}{(1-3\omega^2)+j(3\omega-\omega^3)}$$

$$= \frac{100\left[(1-3\omega^2)-j\omega(3-\omega^2)^2\right]}{\left[(1-3\omega^2)+\omega^2(3-\omega^2)^2\right]}$$

For phase crossover frequency ω_{ph} Img $[G(j\omega)] = 0$;

Hence, $\omega(3-\omega^2) = 0$ or, $\omega = 0; \pm\sqrt{3}$,

therefore, $\omega_{ph} = \sqrt{3}$ rad/sec.

6.29 (a)

- From the given Bode plot, it is evident that there are 3 (three) poles in the transfer function, out of which there are double poles at corner frequency near but less than ω = 8 rad/sec and one pole is near but greater than ω = 0.5 rad/sec.
- The initial slope is +20 dB/dec. Therefore one zero exist at s = 0. So from all the given options, option (a) satisfies all the conditions. Therefore option (a) is correct.

6.30 (a)

Nyquist plot of $G(s)\,H(s) = \dfrac{s+3}{s^2(s-3)}$ is as shown below:

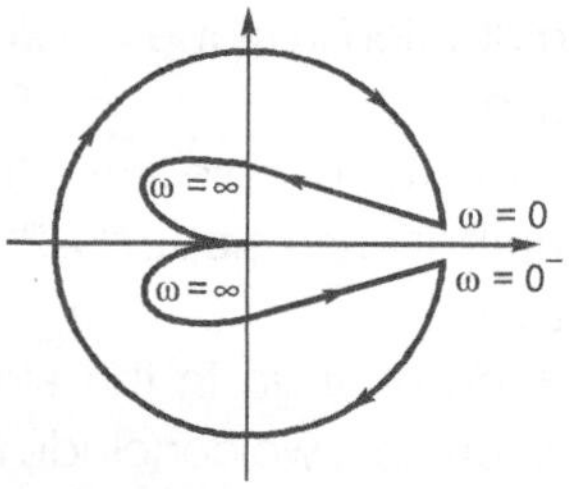

From the Nyquist plot $G(s)H(s)$ encircle $-1 + j0$ once in clockwise direction.

Alternate solution:

Characteristic equation,

$$1 + G(s)\,H(s) = 0$$

$$s^2(s-3) + (s+3) = 0$$

$$s^3 - 3s^2 + s + 3 = 0$$

Using Routh's array,

s^3	1	1
s^2	−3	3
s^1	2	0
s^0	3	

There are two sign changes, hence two poles in right half of s-plane exist.

$Z = 2, \; P = 1 \Rightarrow N = P - Z = -1$

One encirclement in clockwise direction.

6.31 (1.047)

Forward path transfer function,

$$G(s) = \frac{Ke^{-s}}{s}$$

Given : Phase margin = 30°

or Phase margin = 180° + ϕ

$$30° = 180° + \phi \Rightarrow \phi = -150°$$

$$\phi = \angle G(j\omega)_{\omega=\omega_{gc}}$$

[where, ω_{gc} is gain crossover frequency]

$$\angle G(j\omega) = -90° - 57.3\,\omega°$$

At $\omega = \omega_{gc}$ [gain crossover frequency]

$$|G(j\omega)| = 1$$

or $$\frac{K\times 1}{\omega} = 1 \Rightarrow \omega = \omega_{gc} = \text{K rad/sec.}$$

$$\therefore \angle G(j\omega)_{\omega=\omega_{gc}} = -90° - 57.3\text{ K}$$

$$-90° - 57.3\,K° = -150°$$

$$-57.3\,K° = -60° \Rightarrow K = \frac{60}{57.3} = 1.047$$

6.32 (d)

$$\frac{V_0(s)}{V_i(s)} = \frac{1-s}{1+s} = H(s)$$

For the minimum and maximum values of "ϕ".

$$H(j\omega) = \frac{1-j\omega}{1+j\omega} \Rightarrow \angle H(j\omega) = -2\tan^{-1}\omega$$

At $\omega = 0$; $\angle H(j\omega) = 0°$

At $\omega = \infty$; $\angle H(j\omega) = -\pi$

6.33 (b)

$$\angle G(j\omega) = -0.25\times\frac{180\omega_{pc}}{\pi} - 90° = -180°$$

$$-0.25\times\frac{180\omega_{pc}}{\pi} = -90°$$

$$\omega_{pc} = \frac{4\pi}{2} = 2\pi$$

$$|G(j\omega)|_{\omega=\omega_{pc}} = \left|\frac{\pi e^{-0.25s}}{s}\right|_{\omega=\omega_{pc}} = \frac{\pi}{2\pi} = 0.5$$

$\therefore$ Point is $(-0.5, j0)$.

6.34 (a)

$$G(s) = \frac{k}{s\left(1+\frac{s}{1}\right)\left(1+\frac{s}{20}\right)}$$

Transfer function shows 2 poles and no zeros.
So, statement 1 is wrong.

$$\angle G(j\omega) = -90 - \tan^{-1}\omega - \tan^{-1}\frac{\omega}{20}$$

$$\angle G(j\omega)|_{\omega\to\infty} = -270° = -\frac{3\pi}{2}\text{ rad}$$

So, statement 2 is correct.

6.35 Sol.

$$\text{O.L.T.F} = G(s) = \frac{K}{(s+a)(s-b)(s+c)}$$

$N = P - Z$; $P = 1$

$-1 = 1 - Z$; $N = -1$

$Z = 2$

6.36 (c)

$$H(s) = \text{T.F.} = \frac{s^2+100}{s-p}$$

$$\text{D.C. gain} = 5$$

$$\Rightarrow \frac{100}{-p} = 5 = p = -20$$

$$H(j\omega) = \frac{-\omega^2+100}{j\omega+20}; \quad |H(j\omega)| = \frac{-\omega^2+100}{\sqrt{\omega^2+400}}$$

$$\frac{-\omega^2+100}{\sqrt{\omega^2+400}} = 1 \Rightarrow \omega = 8.84\text{ rad/sec.}$$

6.37 (0.10)

From response plot, M_r = 26 dB = 20

$$\therefore \frac{1}{2\xi\sqrt{1-\xi^2}} = 20 \Rightarrow \xi = 0.025$$

From electrical network

$$\chi = \frac{R}{2}\sqrt{\frac{C}{L}} = 0.025 \Rightarrow R = 0.10\ \Omega$$

6.38 (a)

$$P = 1$$

Possibilities, $\frac{s-1}{s+1}$, $\frac{1-s}{1+s}$

Number of poles, $P = 1$

Number of zeros, $Z = 1$

6.39 (a)

$$B = A|G(j\omega)|$$

$$a = 1\times\left|\frac{100}{100-\omega^2+0.1j\omega}\right|_{\omega=0} = 1$$

$$b = 0.1\times\left|\frac{100}{100-\omega^2+0.1j\omega}\right|_{\omega=10} = 10$$

6.40 (b)

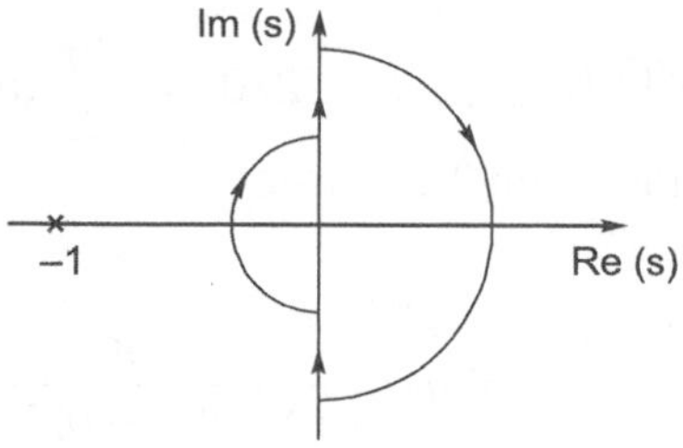

Number of Encirclement = 1

6.41 (c)

Given : OLTF, $GH = \dfrac{3s+5}{(s-1)}$

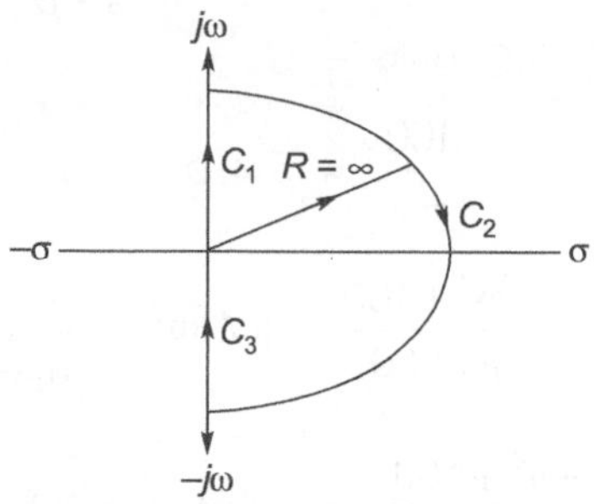

Here for mapping C_2,

$$s = \lim_{R\to\infty} Re^{j\theta};\ \theta : +\frac{\pi}{2} \text{ to } -\frac{\pi}{2}$$

$$GH = \lim_{R\to\infty} \frac{3Re^{j\theta}+5}{Re^{j\theta}-1}$$

$$= \lim_{R\to\infty} \frac{3e^{j\theta}+\dfrac{5}{R}}{e^{j\theta}-\dfrac{1}{R}} = \frac{3e^{j\theta}+0}{e^{j\theta}+0} = 3$$

$$GH = 3$$

6.42 (d)

The transfer function,

TF = Ke^{-sT_d} (transportation lag)

Given magnitude, M = 8 dB = 20 log [K]

K = 2.511

and angle at ω = 1 rad/sec = –60°

$$\text{Angle, } \phi = -\omega T_d \times \frac{180^\circ}{\pi}$$

$$60^\circ = -1\times T_d \times \frac{180^\circ}{\pi}$$

$$T_d = 1.047$$

So, required transfer function

$$TF = 2.511e^{-1.047s}$$

6.43 (0.1)

Initial slppe = –20 dB/sec

Open loop transfer function = $\dfrac{k}{s}$

At w = 1;20 $\log_{10}k$ – 20 log (1) = 20

k = 10

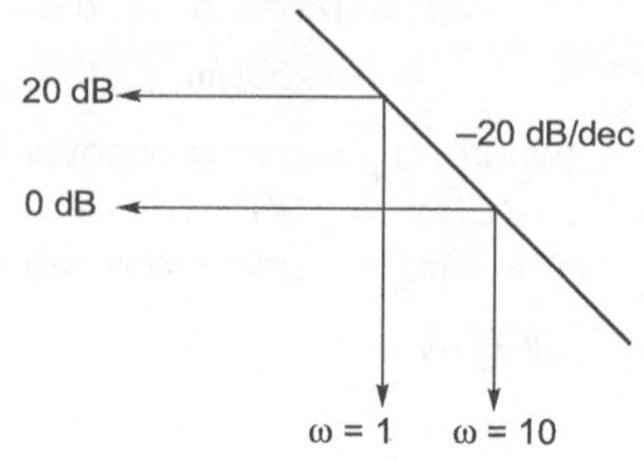

$$K_v = \lim_{s\to 0} s\times\frac{K}{s} = K = 10$$

e_{ss} for type -1 system for ramp input

$$e_{ss} = \frac{1}{k_v} = \frac{1}{10} = 0.1$$

6.44 (1.12)

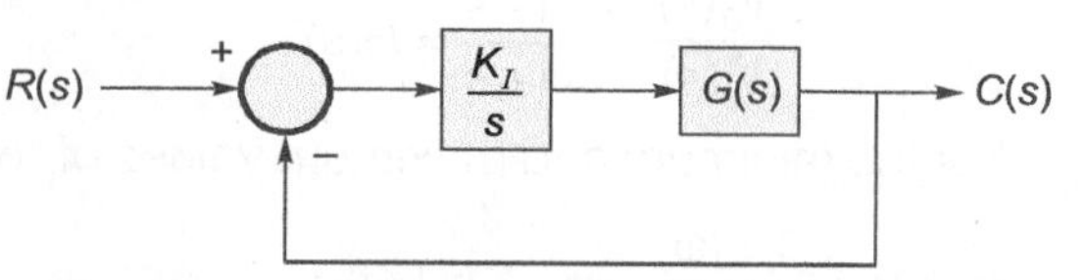

$$GH(s) = \frac{K_I}{s}G(s)$$

$$PM = 180^\circ + \angle GH(j\omega)_{\omega_{gc}} = 50^\circ$$

$$\Rightarrow 180^\circ + [-90^\circ + \angle G(j\omega)]_{\omega_{gc}} = 50^\circ$$

$$\therefore \quad \angle G(j\omega) = -40^\circ$$

at $\omega_{gc} \Rightarrow |GH(j\omega)| = 0$ dB

But at ω = 0.5;

$$|G(j\omega)| = -7 \text{ dB and } \angle G(j\omega) = -40^\circ$$

$$\therefore \quad \left|\frac{K_I}{s}\right|_{\omega=0.5} = +7 \text{ dB} = 10^{\left(\frac{7}{20}\right)} = 2.24$$

$$\therefore \quad \frac{K_I}{0.5} = 2.24$$

$$\therefore \quad K_I = 1.12$$

7 Design of Control Systems

7.1 Introduction of integral action in the forward path of a unity feedback system results in a
(a) marginally stability
(b) system with no steady state error
(c) system with increased stability margin
(d) system with better speed of response
[1997 : 1 M]

7.2 A differentiator has transfer function whose
(a) phase increases linearly with frequency
(b) amplitude remains constant
(c) amplitude increases linearly with frequency
(d) amplitude decreases linearly with frequency
[1997 : 1 M]

7.3 The phase lead compensations used to
(a) increase rise time and decrease overshoot.
(b) decrease both rise time and overshoot.
(c) increase both rise time and overshoot.
(d) decrease rise time and increase overshoot.
[1998 : 1 M]

7.4 $G(s) = \dfrac{(0.5s+1)}{(0.05s+1)}$ maximum phase lead of the compensator is
(a) 52 deg at 4 rad/sec
(b) 52 deg at 10 rad/sec
(c) 55 deg at 12 rad/sec
(d) none of these [2000 : 2 M]

7.5 A lead compensator used for a closed loop controller has the following transfer function

$$\frac{K\left(1+\dfrac{s}{a}\right)}{\left(1+\dfrac{s}{b}\right)}$$

For such a lead compensator
(a) $a < b$ (b) $b < a$
(c) $a > Kb$ (d) $a < Kb$ [2003 : 1 M]

7.6 The system $900/s(s+1)(s+9)$ is to be such that its gain-crossover frequency becomes same as its uncompensated phase crossover frequency and provides a 45° phase margin. To achieve this, one may use
(a) a lag compensator that provides an attenuation of 20 dB and a phase lag of 45° at the frequency of $3\sqrt{3}$ rad/s.
(b) a lead compensator that provides an amplification of 20 dB and a phase lead of 45° at the frequency of 3 rad/s.
(c) a lag-lead compensator that provides an amplification of 20 dB and a phase lag of 45° at the frequency of $\sqrt{3}$ rad/s.
(d) a lag-lead compensator that provides an attenuation of 20 dB and phase lead of 45° at the frequency of 3 rad/s. [2007 : 2 M]

7.7 The transfer functions of two compensators are given below:

$$C_1 = \frac{10(s+1)}{(s+10)},\ C_2 = \frac{s+10}{10(s+1)}$$

Which one of the following statements is correct?
(a) C_1 is lead compensator and C_2 is a lag compensator.
(b) C_1 is a lag compensator and C_2 is a lead compensator.
(c) Both C_1 and C_2 are lead compensator.
(d) Both C_1 and C_2 are lag compensator.
[2008 : 2 M]

Statement for Linked Answer Questions (7.8 and 7.9):
The transfer function of a compensator is given as

$$G_c(s) = \frac{s+a}{s+b}$$

7.8 $G_c(s)$ is a lead compensator if
(a) $a=1, b=2$ (b) $a=3, b=2$
(c) $a=-3, b=-1$ (d) $a=3, b=1$
[2012 : 2 M]

7.9 The phase of the above lead compensator is maximum at
(a) $\sqrt{2}$ rad/s (b) $\sqrt{3}$ rad/s
(c) $\sqrt{6}$ rad/s (d) $1/\sqrt{3}$ rad/s
[2012 : 2 M]

7.10 For the network shown in the figure below, the frequency (in rad/s) at which the maximum phase lag occurs is _______.

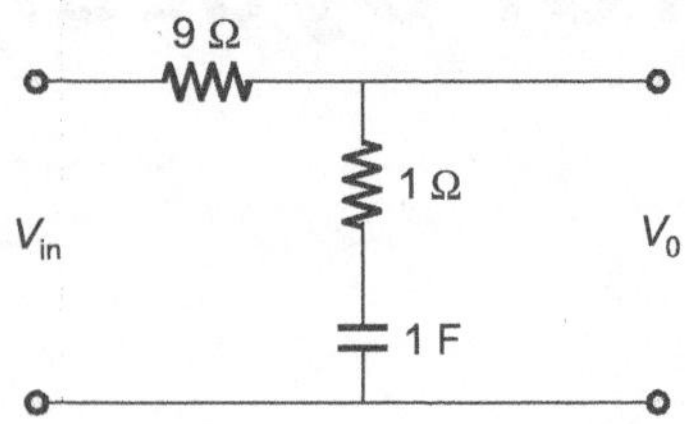

[2016 : 1 M, Set-2]

7.11 The transfer function $C(s)$ of a compensator is given below.

$$C(s) = \frac{\left(1+\frac{s}{0.1}\right)\left(1+\frac{s}{100}\right)}{(1+s)\left(1+\frac{s}{10}\right)}$$

The frequency range in which the phase (lead) introduced by the compensator reaches the maximum is

(a) $0.1 < \omega < 1$ (b) $1 < \omega < 10$
(c) $10 < \omega < 100$ (d) $\omega > 100$

[2017 : 1 M, Set-2]

7.12 The transfer function of a phase lead compensator is given by

$$D(s) = \frac{3\left(s+\frac{1}{3T}\right)}{\left(s+\frac{1}{T}\right)}$$

The frequency (in rad/sec), at which $\angle D(j\omega)$ is maximum is

(a) $\sqrt{\frac{3}{T^2}}$ (b) $\sqrt{3T}$
(c) $\sqrt{\frac{1}{3T^2}}$ (d) $\sqrt{3T^2}$ [2019 : 2 M]

7.13 Consider a unity-gain negative feedback system consisting of the plant $G(s)$ (given below) and a proportional-integral controller. Let the proportional gain and integral gain be 3 and 1, respectively. For a unit step reference input, the final values of the controller output and the plant output, respectively, are

$$G(s) = \frac{1}{s-1}$$

(a) ∞, ∞ (b) 1, 0
(c) 1, −1 (d) −1, 1

[2023 : 1 Mark]

7.14 Consider a lead compensator of the form

$$K(s) = \frac{1+\frac{s}{\alpha}}{1+\frac{s}{\beta\alpha}}, \beta > 1, \alpha > 0$$

The frequency at which this compensator produces maximum phase lead is 4 rad/s. At this frequency, the gain amplification provided by the controller, assuming asymptotic Bode-magnitude plot of $K(s)$, is 6 dB. The values of α, β, respectively, are

(a) 1, 16 (b) 2, 4
(c) 3, 5 (d) 2.66, 2.25

[2023 : 2 Marks]

■■■■

Answers Design of Control Systems

7.1 (b)	7.2 (c)	7.3 (b)	7.4 (d)	7.5 (a)	7.6 (d)	7.7 (a)
7.8 (a)	7.9 (a)	7.10 (0.316)	7.11 (a)	7.12 (c)	7.13 (d)	7.14 (b)

Explanations Design of Control Systems

7.1 (b)

Integral controller is widely used to eliminate steady state error.

7.2 (c)

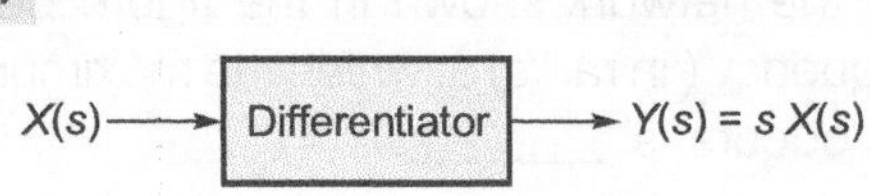

$$\therefore \quad H(s) = \frac{Y(s)}{X(s)} = s$$

Frequency response, $H(j\omega) = j\omega$

Hence, phase is constant at 90° for all ω but amplitude increases linearly with frequency.

7.3 (b)

Lead compensator increases bandwidth, hence decreases rise time. As it improves damping, so overshoot also decreases.

7.4 (d)

$$G(s) = \frac{0.5s+1}{0.05s+1}$$

On comparing with,

$$G(s) = \frac{1+aTs}{1+Ts}$$

we get, $T = 0.05$ and $a = 10$

$$\therefore \quad \phi_n = \sin^{-1}\left(\frac{a-1}{a+1}\right) = 55°$$

and $$\omega_m = \frac{1}{T\sqrt{a}} = 6.32 \text{ rad/sec.}$$

7.5 (a)

$$\text{Transfer function} = \frac{k\left(1+\frac{s}{a}\right)}{\left(1+\frac{s}{b}\right)}$$

Zero of TF = $-a$

Pole of TF = $-b$

For a lead-compensator, the zero is nearer to origin as compared to pole, hence the effect of zero is dominant, therefore, the lead-compensator when introduced in series with forward path of the transfer function the phase shift is increased.

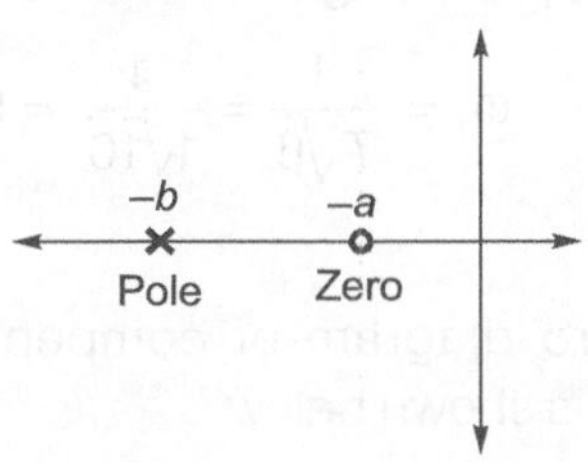

So, from pole-zero configuration of the compensator $a < b$.

7.6 (d)

Let uncompensated system,

$$T(s) = \frac{900}{s(s+1)(s+9)}$$

Phase crossover frequency of uncompensated system = (ω_{pc}), at this frequency phase of $T(j\omega)$ is $-180°$

Put $s = j\omega$ in $T(s)$

$$T(j\omega) = \frac{900}{j\omega(j\omega+1)(j\omega+9)}$$

$$\angle T(j\omega) = -90° - \tan^{-1}\omega - \tan^{-1}\left(\frac{\omega}{9}\right)$$

At, $(\omega_{pc})_1$, $\angle T(j\omega) = -180°$

$$-180° = -90° - \tan^{-1}(\omega_{pc})_1 - \tan^{-1}\left(\frac{(\omega_{pc})_1}{9}\right)$$

$$-90° = \tan^{-1}\frac{(\omega_{pc})_1 + \frac{(\omega_{pc})_1}{9}}{1-\frac{(\omega_{pc})_1^2}{9}}$$

$$\Rightarrow \quad 1-\frac{(\omega_{pc})_1^2}{9} = 0 \Rightarrow (\omega_{pc})_1 = 3 \text{ rad/ses}$$

Gain cross frequency of compensated system, $(\omega_{gc})_2$ = phase cross frequency of uncompensated system, $(\omega_{pc})_1$

$$\Rightarrow \quad (\omega_{gc})_2 = (\omega_{gc})_1 = 3 \text{ rad/sec}$$

$$\text{Phase-margin} = 180° + \angle T(j\omega)\big|_{\omega=(\omega_{gc})_2}$$

$$\Rightarrow \quad 45° = 180° + \angle T(j\omega)\big|_{\omega=(\omega_{pc})_2}$$

At $(\omega_{gc})_2 = 3$ rad/sec,

Phase angle of $\angle T(j\omega)$ is $-135°$.

and phase of uncompensated system is $-180°$ at 3 rad/sec.

Therefore, the compensator provides phase lead of 45° at the frequency of 3 rad/sec.

Let X dB is the gain provided by the compensator, so at gain cross frequency, $|T(j\omega)|_{com} = 1$ or 0 dB.

Gain of uncompensated system = $|T(j\omega)|_{un\text{-}com}$

$$= \frac{100}{\omega\sqrt{1+\omega^2}\sqrt{1+\left(\frac{\omega}{9}\right)^2}}$$

$|T(j\omega)|_{un\text{-}com}$ in dB

$$= 40 - 20\log\omega - 20\log\sqrt{1+\omega^2} - 20\log\sqrt{1+\left(\frac{\omega}{9}\right)^2}$$

Gain of compensated system,

$$|T(j\omega)|_{com} = X + |T(j\omega)_{un\text{-}com}|$$

$|T(j\omega)|_{com}$ must be zero at gain cross frequency $(\omega_{gc})_2$

$$|T(j\omega_{gc})_2|_{com} = X + 40 - 20\log(\omega_{gc})_2$$

$$-20\log\sqrt{1+(\omega_{gc})_2^2}$$

$$-20\log\sqrt{1+\frac{(\omega_{gc})^2}{9^2}} = 0$$

$$X + 40 - 20\log 3 - 20\log\sqrt{1+3^2}$$

$$-20\log\sqrt{1+\left(\frac{3}{9}\right)^2} = 0$$

$$X = -20 \text{ dB}$$

So, the compensator provides an attenuation of 20 dB. Hence option (d) is correct.

7.7 (a)

$$C_1 = \frac{10(s+1)}{(s+10)}$$

Zero at $s = -1$
Pole at $s = -10$

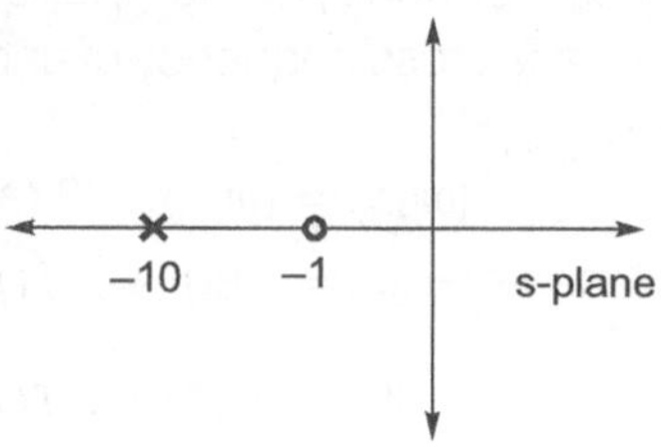

As zero is closer origin, zero dominates pole. Hence C_1 is lead compensator.

$$C_2 = \frac{s+10}{10(s+1)}$$

Zero at $s = -10$
Pole at $s = -1$

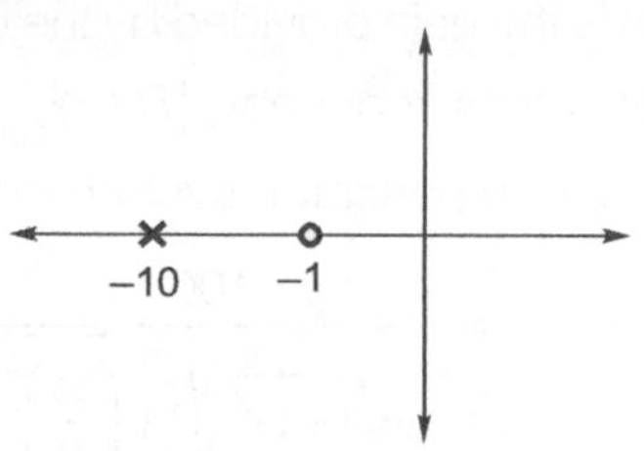

As pole is closer to origin, pole dominates zero. Hence C_2 is lag compensator.

7.8 (a)

$$G_c(s) = \frac{s+a}{s+b}$$

$$G_c(j\omega) = \frac{j\omega+a}{j\omega+b}$$

$$\angle G_c(j\omega) = \tan^{-1}\frac{\omega}{a} - \tan^{-1}\frac{\omega}{b}$$

$$= \tan^{-1}\left(\frac{\frac{\omega}{a}-\frac{\omega}{b}}{1+\frac{\omega^2}{ab}}\right)$$

For $G_c(s)$ to be a lead compensator

$$\angle G_c(j\omega) > 0$$

$$\frac{\omega}{a} > \frac{\omega}{b} \Rightarrow b > a$$

Option (a) satisfies the above equation.

7.9 (a)

$$\omega_m = \sqrt{\omega_{c_1} \times \omega_{c_2}} = \sqrt{1\times 2} = \sqrt{2} \text{ rad/ sec.}$$

7.10 (0.316)

Assuming,

$$R_1 = 9\ \Omega;\ R_2 = 1\ \Omega$$

We can write,

$$\frac{V_o(s)}{V_{in}(s)} = \frac{R_2 + \frac{1}{sC}}{R_1 + R_2 + \frac{1}{sC}} = \frac{1 + R_2 \cdot Cs}{1 + (R_1+R_2)Cs}$$

$$= \frac{1+R_2\cdot Cs}{1+\left(\frac{R_1+R_2}{R_2}\right)R_2 Cs}$$

Let, $R_2 C = T$

$$\frac{R_1+R_2}{R_2} = \beta$$

$$\therefore \quad \frac{V_o(s)}{V_{in}(s)} = \frac{1+Ts}{(1+\beta Ts)}$$

which represent a lag compensator

$\because$ Here, $T = R_2 C = 1.1 = 1$ sec

$$\beta = \frac{1+9}{1} = 10$$

Maximum phase lag occurs at frequency

$$\omega_n = \frac{1}{T\sqrt{\beta}} = \frac{1}{1\sqrt{10}} = 0.316 \text{ rad/sec.}$$

7.11 (a)

Pole zero diagram of compensator transfer function is shown below.

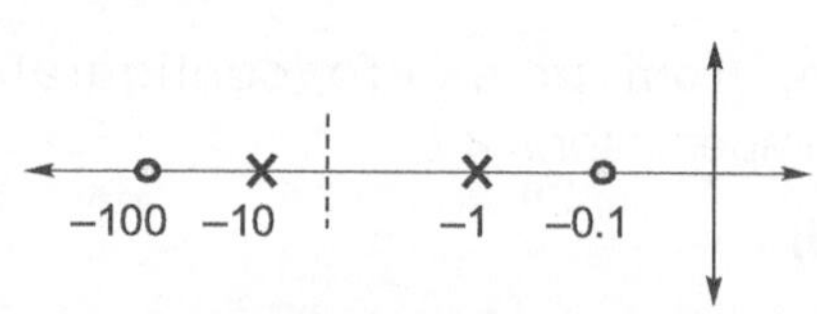

Maximum phase lead is between 0.1 and 1.

$$0.1 < \omega < 1$$

7.12 (c)

$$T(s) = \left(\frac{1+3Ts}{1+Ts}\right)$$

Frequency at which $\angle T(j\omega)$ is maximum,

$$(\omega_m) = \frac{1}{T\sqrt{\alpha}}$$

$$\alpha = \frac{1}{1/3} = 3$$

$$\omega_m = \frac{1}{T\sqrt{3}} = \sqrt{\frac{1}{3T^2}}$$

7.13 (d)

$$G(s) = \frac{1}{s-1}$$

$$H(s) = 1$$

Let the controller output be $C_1(s)$ and plant output be $C(s)$.

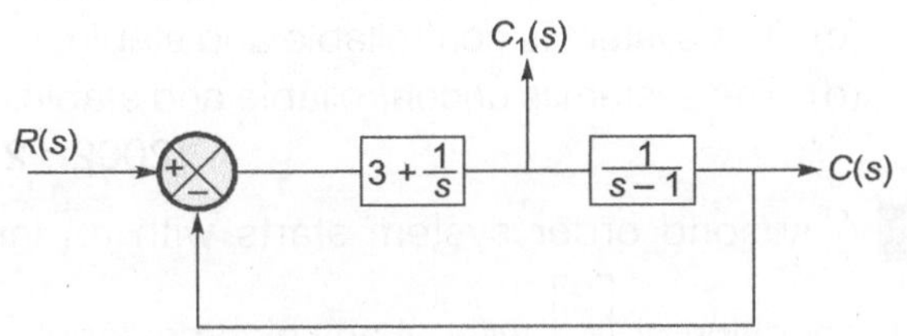

$$\frac{C(s)}{R(s)} = \frac{\frac{(3s+1)}{s}\times\frac{1}{(s+1)}}{1+\frac{(3s+1)\times 1}{s\times(s-1)}} = \frac{3s+1}{(s^2+3s-s+1)}$$

$$\frac{C(s)}{R(s)} = \frac{3s+1}{s^2+2s+1}$$

$$\because \quad R(s) = \frac{1}{s}$$

$$C(s) = \frac{1}{s}\times\frac{3s+1}{s^2+2s+1}$$

$$C_{ss} = \lim_{s\to 0} sC(s)$$

$$C_{ss} = \lim_{s\to 0} s\times\frac{3s+1}{s^2+2s+1}\times\frac{1}{s} = +1$$

As now, $$C(s) = C_1(s)\times\frac{1}{s-1}$$

$$C_1(s) = (s-1)C(s)$$

$$C_1(s) = (s-1)\times\left[\frac{3s+1}{s(s^2+2s+1)}\right]$$

$$(C_1)_{ss} = \lim_{s\to 0} sC_1(s) = -1$$

7.14 (b)

$$\text{TF} = \frac{1+\frac{s}{\alpha}}{1+\frac{s}{\alpha\beta}}, \alpha > 0, \beta > 1 = \frac{s+\alpha}{(s+\alpha\beta)}$$

Given : $\omega_m = \sqrt{\alpha(\alpha\beta)} = 4$

$$\alpha\sqrt{\beta} = 4 \qquad \text{...(1)}$$

Given : at $\omega = \omega_m$

$$M = \frac{1}{\sqrt{\beta}}$$

$$20 \log M = 6 \text{ dB}$$

$$20\log\frac{1}{\sqrt{\beta}} = 6$$

Amplification, $10 \log_{10} (\beta) = 6 \Rightarrow \beta = 4$

and $$\alpha = \frac{4}{\sqrt{4}} = 2$$

■■■■

8 State Variable Analysis

8.1 A system is described by the state equation $\dot{X} = AX + BU$. The output is given by $Y = CX$.

Where, $A = \begin{pmatrix} -4 & -1 \\ 3 & -1 \end{pmatrix}$ $B = \begin{pmatrix} 1 \\ 1 \end{pmatrix}$ $C = [1 \quad 0]$.

Transfer function $G(s)$ of the system is

(a) $\dfrac{s}{s^2+5s+7}$ (b) $\dfrac{1}{s^2+5s+7}$

(c) $\dfrac{s}{s^2+3s+2}$ (d) $\dfrac{1}{s^2+3s+2}$

[1995 : 1 M]

8.2 Given the matrix: $A = \begin{bmatrix} 0 & 1 & 0 \\ 0 & 0 & 1 \\ -6 & -11 & -6 \end{bmatrix}$

Its eigen values are ______ [1995 : 1 M]

8.3 Given the homogeneous state-space equation $\dot{X} = \begin{bmatrix} -3 & 1 \\ 0 & -2 \end{bmatrix} x$ the steady state value of $x_{ss} = \lim_{t\to\infty} x(t)$, given the initial state value of $x(t) = [10 - 10]^T$, is

(a) $x_{ss} = \begin{bmatrix} 0 \\ 0 \end{bmatrix}$ (b) $x_{ss} = \begin{bmatrix} -3 \\ -2 \end{bmatrix}$

(c) $x_{ss} = \begin{bmatrix} -10 \\ 10 \end{bmatrix}$ (d) $x_{ss} = \begin{bmatrix} \infty \\ \infty \end{bmatrix}$

[2001 : 1 M]

8.4 The state transition matrix for the system $\dot{X} = AX$ with initial state $X(0)$ is

(a) $(sI - A)^{-1}$

(b) $e^{At}X(0)$

(c) Laplace inverse of $[(sI - A)^{-1}]$

(d) Laplace inverse of $[(sI - A)^{-1} X(0)]$

[2002 : 1 M]

8.5 For the system

$$\dot{X} = \begin{bmatrix} 2 & 0 \\ 0 & 4 \end{bmatrix} X + \begin{bmatrix} 1 \\ 1 \end{bmatrix} u; y = [4 \quad 0]X,$$

with u as unit impulse and with zero initial state, the output y becomes

(a) $2e^{2t}$ (b) $4e^{2t}$

(c) $2e^{4t}$ (d) $4e^{4t}$ [2002 : 2 M]

8.6 For the system $X = \begin{bmatrix} 2 & 3 \\ 0 & 5 \end{bmatrix} X + \begin{bmatrix} 1 \\ 0 \end{bmatrix} u$, which of the following statement is true?

(a) The system is controllable but unstable.

(b) The system is uncontrollable and unstable.

(c) The system is controllable and stable.

(d) The system is uncontrollable and stable.

[2002 : 2 M]

8.7 A second order system starts with an initial condition of $\begin{bmatrix} 2 \\ 3 \end{bmatrix}$ without any external input. The state transition matrix for the system is given by $\begin{bmatrix} e^{-2t} & 0 \\ 0 & e^{-t} \end{bmatrix}$. The state of the system at the end of 1 second is given by

(a) $\begin{bmatrix} 0.271 \\ 1.100 \end{bmatrix}$ (b) $\begin{bmatrix} 0.135 \\ 0.368 \end{bmatrix}$

(c) $\begin{bmatrix} 0.271 \\ 0.736 \end{bmatrix}$ (d) $\begin{bmatrix} 0.135 \\ 1.100 \end{bmatrix}$ [2003 : 1 M]

8.8 The following equation defines a separately excited dc motor in the form of a differential equation

$$\frac{d^2\omega}{dt^2} + \frac{B}{J}\frac{d\omega}{dt} + \frac{K^2}{LJ}\omega = \frac{K}{LJ}V_a$$

The above equation may be organized in the state-space form as follows

$$\begin{bmatrix} \dfrac{d^2\omega}{dt^2} \\ \dfrac{d\omega}{dt} \end{bmatrix} = P \begin{bmatrix} \dfrac{d\omega}{dt} \\ \omega \end{bmatrix} + QV_a$$

where the P matrix is given by

(a) $\begin{bmatrix} -\dfrac{B}{J} & -\dfrac{K^2}{LJ} \\ 1 & 0 \end{bmatrix}$ (b) $\begin{bmatrix} -\dfrac{K^2}{LJ} & -\dfrac{B}{J} \\ 0 & 1 \end{bmatrix}$

(c) $\begin{bmatrix} 0 & 1 \\ -\dfrac{K^2}{LJ} & -\dfrac{B}{J} \end{bmatrix}$ (d) $\begin{bmatrix} 1 & 0 \\ -\dfrac{B}{J} & -\dfrac{K^2}{LJ} \end{bmatrix}$

[2003 : 2 M]

8.9 The state variable description of a linear autonomous system is, $\dot{X} = AX$, where X is the two dimensional state vector and A is the system matrix given by $A = \begin{bmatrix} 0 & 2 \\ 2 & 0 \end{bmatrix}$. The roots of the characteristic equation are

(a) –2 and +2 (b) $-j2$ and $+j2$

(c) –2 and –2 (d) +2 and +2

[2004 : 2 M]

Statement for common data Questions(8.10 and 8.11):

A state variable system $\dot{X}(t) = \begin{bmatrix} 0 & 1 \\ 0 & -3 \end{bmatrix} X(t) + \begin{bmatrix} 1 \\ 0 \end{bmatrix} U(t)$, with the initial condition $X(0) = [-1 \quad 3]^T$ and the unit step input $u(t)$ has

8.10 The state transition matrix

(a) $\begin{bmatrix} 1 & \frac{1}{3}(1-e^{-3t}) \\ 0 & e^{-3t} \end{bmatrix}$

(b) $\begin{bmatrix} 1 & \frac{1}{3}(e^{-t}-e^{-3t}) \\ 0 & e^{-t} \end{bmatrix}$

(c) $\begin{bmatrix} 1 & \frac{1}{3}(e^{-t}-e^{-3t}) \\ 0 & e^{-3t} \end{bmatrix}$

(d) $\begin{bmatrix} 1 & (1-e^{-t}) \\ 0 & e^{-t} \end{bmatrix}$

[2005 : 2 M]

8.11 The state transition equation:

(a) $X(t) = \begin{bmatrix} t-e^{-t} \\ e^{-t} \end{bmatrix}$

(b) $X(t) = \begin{bmatrix} t-e^{-t} \\ 3e^{-3t} \end{bmatrix}$

(c) $X(t) = \begin{bmatrix} t-e^{-3t} \\ 3e^{-3t} \end{bmatrix}$

(d) $X(t) = \begin{bmatrix} t-e^{-3t} \\ e^{-t} \end{bmatrix}$

[2005 : 2 M]

8.12 For a system with the transfer function:

$$H(s) = \frac{3(s-2)}{s^3+4s^2-2s+1}$$

the matrix A in the state space form $\dot{x} = Ax + Bu$ is equal to

(a) $\begin{bmatrix} 1 & 0 & 0 \\ 0 & 1 & 0 \\ -1 & 2 & -4 \end{bmatrix}$ (b) $\begin{bmatrix} 0 & 1 & 0 \\ 0 & 0 & 1 \\ -1 & 2 & -4 \end{bmatrix}$

(c) $\begin{bmatrix} 0 & 1 & 0 \\ 3 & -2 & 1 \\ 1 & -2 & 4 \end{bmatrix}$ (d) $\begin{bmatrix} 1 & 0 & 0 \\ 0 & 0 & 1 \\ -1 & 2 & -4 \end{bmatrix}$

[2006 : 1 M]

8.13 The state equation for the current I_1 shown in the network shown below in terms of the voltage V_x and the independent source V, is given by

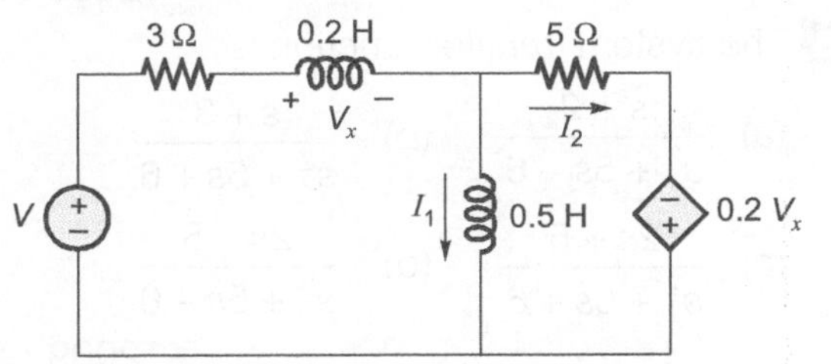

(a) $\frac{dI_1}{dt} = -1.4\,V_x - 3.75 I_1 + \frac{5}{4}V$

(b) $\frac{dI_1}{dt} = -1.4\,V_x - 3.75 I_1 - \frac{5}{4}V$

(c) $\frac{dI_1}{dt} = -1.4\,V_x + 3.75 I_1 + \frac{5}{4}V$

(d) $\frac{dI_1}{dt} = -1.4\,V_x + 3.75 I_1 - \frac{5}{4}V$ [2007 : 2 M]

Statement for Linked Answer Questions (8.14 and 8.15).

The sate space equation of a system is described by $\dot{x} = Ax + Bu$, $y = Cx$ where x is state vector, u is input, y is output and $A = \begin{bmatrix} 0 & 1 \\ 0 & -2 \end{bmatrix}$, $B = \begin{bmatrix} 0 \\ 1 \end{bmatrix}$, $C = [1 \quad 0]$

8.14 The transfer function $G(s)$ of this system will be

(a) $\frac{s}{(s+2)}$ (b) $\frac{s+1}{s(s-2)}$

(c) $\frac{s}{(s-2)}$ (d) $\frac{1}{s(s+2)}$ [2008 : 2 M]

8.15 A unity feedback is provided to the above system $G(s)$ to make it a closed loop system as shown in figure.

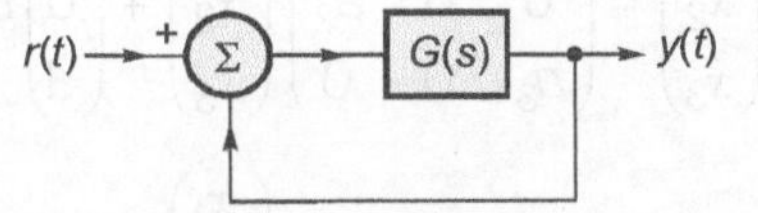

For a unit step input $r(t)$, the steady state error in the input will be

(a) 0 (b) 1
(c) 2 (d) ∞ [2008 : 2 M]

Statement for Common Data Questions (8.16 and 8.17):
A system is described by the following state and output equations:

$$\frac{dx_1(t)}{dt} = -3x_1(t) + x_2(t) + 2u(t)$$

$$\frac{dx_2(t)}{dt} = -2x_2(t) + u(t)$$

$$y(t) = x_1(t)$$

when $u(t)$ is the input and $y(t)$ is the output.

[2009 : 2 M]

8.16 The system transfer function is

(a) $\frac{s+2}{s^2+5s-6}$ (b) $\frac{s+3}{s^2+5s+6}$

(c) $\frac{2s+5}{s^2+5s+6}$ (d) $\frac{2s-5}{s^2+5s-6}$

[2009 : 2 M]

8.17 The state-transition matrix of the above system is

(a) $\begin{bmatrix} e^{-3t} & 0 \\ e^{-2t}+e^{-3t} & e^{-2t} \end{bmatrix}$

(b) $\begin{bmatrix} e^{-3t} & e^{-2t}-e^{-3t} \\ 0 & e^{-2t} \end{bmatrix}$

(c) $\begin{bmatrix} e^{-3t} & e^{-2t}+e^{-3t} \\ 0 & e^{-2t} \end{bmatrix}$

(d) $\begin{bmatrix} e^{3t} & e^{-2t}-e^{-3t} \\ 0 & e^{-2t} \end{bmatrix}$ [2009 : 2 M]

8.18 The system $\dot{X} = AX + BU$ and

$A = \begin{bmatrix} -1 & 2 \\ 0 & 2 \end{bmatrix}$, $B = \begin{bmatrix} 0 \\ 1 \end{bmatrix}$ is

(a) stable and controllable
(b) stable but uncontrollable
(c) unstable but controllable
(d) unstable and uncontrollable [2010 : 2 M]

8.19 The state variable description of an LTI system is given by

$$\begin{pmatrix} \dot{x}_1 \\ \dot{x}_2 \\ \dot{x}_3 \end{pmatrix} = \begin{pmatrix} 0 & a_1 & 0 \\ 0 & 0 & a_2 \\ a_3 & 0 & 0 \end{pmatrix} \begin{pmatrix} x_1 \\ x_2 \\ x_3 \end{pmatrix} + \begin{pmatrix} 0 \\ 0 \\ 1 \end{pmatrix} u$$

$$y = (1 \;\; 0 \;\; 0) \begin{pmatrix} x_1 \\ x_2 \\ x_3 \end{pmatrix}$$

where y is the output and u is the input. The system is controllable for

(a) $a_1 \neq 0, a_2 = 0, a_3 \neq 0$
(b) $a_1 = 0, a_2 \neq 0, a_3 \neq 0$
(c) $a_1 = 0, a_2 \neq 0, a_3 = 0$
(d) $a_1 \neq 0, a_2 \neq 0, a_3 = 0$ [2012 : 2 M]

Common Data for Questions (8.20 and 8.21):
The state variable formulation of a system is given as

$\begin{bmatrix} \dot{x}_1 \\ \dot{x}_2 \end{bmatrix} = \begin{bmatrix} -2 & 0 \\ 0 & -1 \end{bmatrix} \begin{bmatrix} x_1 \\ x_2 \end{bmatrix} + \begin{bmatrix} 1 \\ 1 \end{bmatrix} u$, $x_1(0) = 0$, $x_2(0) = 0$ and

$y = [1 \;\; 0] \begin{bmatrix} x_1 \\ x_2 \end{bmatrix}$.

8.20 The system is
(a) controllable but not observable
(b) not controllable but observable
(c) both controllable and observable
(d) both not controllable and not observable

[2013 : 2 M]

8.21 The response $y(t)$ to a unit step input is

(a) $\frac{1}{2} - \frac{1}{2}e^{-2t}$ (b) $1 - \frac{1}{2}e^{-2t} - \frac{1}{2}e^{-t}$

(c) $e^{-2t} - e^{-t}$ (d) $1 - e^{-t}$

[2013 : 2 M]

8.22 The state transition matrix for the system

$\begin{bmatrix} \dot{x}_1 \\ \dot{x}_2 \end{bmatrix} = \begin{bmatrix} 1 & 0 \\ 1 & 1 \end{bmatrix} \begin{bmatrix} x_1 \\ x_2 \end{bmatrix} + \begin{bmatrix} 1 \\ 1 \end{bmatrix} u$ is

(a) $\begin{bmatrix} e^t & 0 \\ e^t & e^t \end{bmatrix}$ (b) $\begin{bmatrix} e^t & 0 \\ t^2 e^t & e^t \end{bmatrix}$

(c) $\begin{bmatrix} e^t & 0 \\ te^t & e^t \end{bmatrix}$ (d) $\begin{bmatrix} e^t & te^t \\ 0 & e^t \end{bmatrix}$

[2014 : 1 M, Set-2]

8.23 Consider the system described by following state space equations:

$\begin{bmatrix} \dot{x}_1 \\ \dot{x}_2 \end{bmatrix} = \begin{bmatrix} 0 & 1 \\ -1 & -1 \end{bmatrix} \begin{bmatrix} x_1 \\ x_2 \end{bmatrix} + \begin{bmatrix} 0 \\ 1 \end{bmatrix} u$; $y = [1 \;\; 0] \begin{bmatrix} x_1 \\ x_2 \end{bmatrix}$

If u is unit step input, then the steady state error of the system is

(a) 0 (b) $\frac{1}{2}$

(c) $\frac{2}{3}$ (d) 1

[2014 : 2 M, Set-3]

8.24 In the signal flow diagram given in the figure, u_1 and u_2 are possible inputs whereas y_1 and y_2 are possible outputs. When would the SISO system derived from this diagram be controllable and observable?

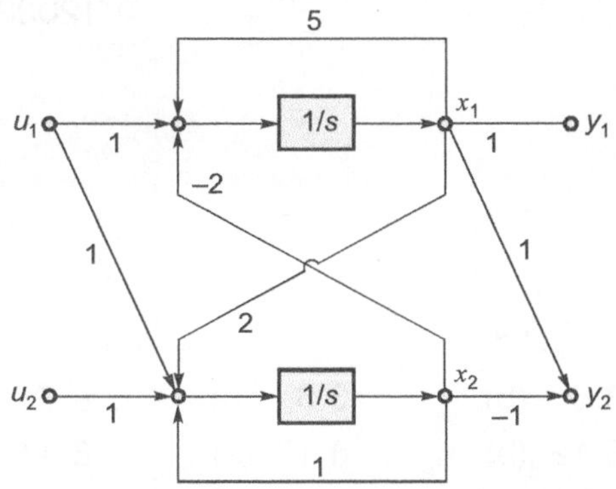

(a) When u_1 is the only input and y_1 is the only output.
(b) When u_2 is the only input and y_1 is the only output.
(c) When u_1 is the only input and y_2 is the only output.
(d) When u_2 is the only input and y_2 is the only output. [2015 : 2 M, Set-1]

8.25 Consider a linear time invariant system $\dot{x} = Ax$, with initial conditions $x(0)$ at $t = 0$. Suppose α and β are eigenvectors of (2×2) matrix A corresponding to distinct eigenvalues λ_1 and λ_2 respectively. Then the response $x(t)$ of the system due to initial condition $x(0) = \alpha$ is

(a) $\alpha e^{\lambda_1 t}$ (b) $\beta e^{\lambda_2 t}$
(c) $\alpha e^{\lambda_2 t}$ (d) $\alpha e^{\lambda_1 t} + \beta e^{\lambda_2 t}$

[2016 : 2 M, Set-2]

8.26 The second order dynamic system

$$\frac{dX}{dt} = PX + Qu$$
$$y = RX$$

has the matrices, P, Q and R as follows:

$$P = \begin{bmatrix} -1 & 1 \\ 0 & -3 \end{bmatrix} Q = \begin{bmatrix} 0 \\ 1 \end{bmatrix} R = [0 \quad 1]$$

The system has the following controllability and observability properties:
(a) Controllable and Observable
(b) Not controllable and Observable
(c) Controllable but not observable
(d) Not controllable and not observable

[2017 : 2 M, Set-2]

8.27 The transfer function of the system $Y(s)/U(s)$ whose state-space equations are given below is:

$$\begin{bmatrix} \dot{x}_1(t) \\ \dot{x}_2(t) \end{bmatrix} = \begin{bmatrix} 1 & 2 \\ 2 & 0 \end{bmatrix} \begin{bmatrix} x_1(t) \\ x_2(t) \end{bmatrix} + \begin{bmatrix} 1 \\ 2 \end{bmatrix} u(t)$$

$$y(t) = [1 \quad 0] \begin{bmatrix} x_1(t) \\ x_2(t) \end{bmatrix}$$

(a) $\dfrac{(s+2)}{(s^2 - 2s - 2)}$ (b) $\dfrac{(s-2)}{(s^2 + s - 4)}$

(c) $\dfrac{(s-4)}{(s^2 + s - 4)}$ (d) $\dfrac{(s+4)}{(s^2 - s - 4)}$

[2017 : 2 M, Set-1]

8.28 Consider the system described by the following state space representation:

$$\begin{bmatrix} \dot{x}_1(t) \\ \dot{x}_2(t) \end{bmatrix} = \begin{bmatrix} 0 & 1 \\ 0 & -2 \end{bmatrix} \begin{bmatrix} x_1(t) \\ x_2(t) \end{bmatrix} + \begin{bmatrix} 0 \\ 1 \end{bmatrix} u(t)$$

$$y(t) = [1 \quad 0] \begin{bmatrix} x_1(t) \\ x_2(t) \end{bmatrix}$$

If $u(t)$ is a unit step input and $\begin{bmatrix} x_1(0) \\ x_2(0) \end{bmatrix} = \begin{bmatrix} 1 \\ 0 \end{bmatrix}$, value of output $y(t)$ at $t = 1$ sec (rounded off to three decimal places) is_______.

[2017 : 2 M, Set-2]

8.29 Consider a state-variable model of a system:

$$\begin{bmatrix} \hat{x}_1 \\ \hat{x}_2 \end{bmatrix} = \begin{bmatrix} 0 & 1 \\ -\alpha & -2\beta \end{bmatrix} \begin{bmatrix} x_1 \\ x_2 \end{bmatrix} + \begin{bmatrix} 0 \\ \alpha \end{bmatrix} r$$

$$y = [1 \quad 0] \begin{bmatrix} x_1 \\ x_2 \end{bmatrix}$$

where y is the output, and r is the input. The damping ratio ξ and the undamped natural frequency ω_n (rad/sec) of the system are given by

(a) $\xi = \dfrac{\beta}{\sqrt{\alpha}}$; $\omega_n = \sqrt{\alpha}$

(b) $\xi = \sqrt{\alpha}$; $\omega_n = \dfrac{\beta}{\sqrt{\alpha}}$

(c) $\xi = \dfrac{\sqrt{\alpha}}{\beta}$; $\omega_n = \sqrt{\beta}$

(d) $\xi = \sqrt{\beta}$; $\omega_n = \sqrt{\alpha}$ [2019 : 2 M]

8.30 The state space representation of a first-order system is given as

$$\dot{x} = -x + u\,; \quad y = x$$

where, x is the state variable, u is the control input and y is the controlled output. Let

$u = -Kx$ be the control law, where K is the controller gain. To place a closed-loop pole at –2, the value of K is ______. [2021 : 2 M]

8.31 Consider the state-space description of an LTI system with matrices

$$A = \begin{bmatrix} 0 & 1 \\ -1 & -2 \end{bmatrix}, B = \begin{bmatrix} 0 \\ 1 \end{bmatrix}, C = [3 \quad -2], \quad D = 1$$

For the input, $\sin(\omega t)$, $\omega > 0$, the value of ω for which the steady-state output of the system will be zero, is _____ (Round off to the nearest integer). [2023 : 2 M]

■■■■

Answers State Variable Analysis

8.1 (a)	8.2 (–1,–2,–3)	8.3 (a)	8.4 (c)	8.5 (b)	8.6 (b)	8.7 (a)
8.8 (a)	8.9 (a)	8.10 (a)	8.11 (c)	8.12 (b)	8.13 (a)	8.14 (d)
8.15 (a)	8.16 (c)	8.17 (b)	8.18 (c)	8.19 (d)	8.20 (a)	8.21 (a)
8.22 (c)	8.23 (a)	8.24 (b)	8.25 (a)	8.26 (c)	8.27 (d)	8.28 (1.284)
8.29 (a)	8.30 (1)	8.31 (2)				

Explanations State Variable Analysis

8.1 (a)

$$\text{T.F.} = C[sI - A]^{-1} B + D$$

$$= [1 \quad 0] \cdot \begin{bmatrix} s+4 & 1 \\ -3 & s+1 \end{bmatrix}^{-1} \cdot \begin{bmatrix} 1 \\ 1 \end{bmatrix} + 0$$

$$= \frac{[1 \quad 0] \cdot \begin{bmatrix} s+1 & -1 \\ 3 & s+4 \end{bmatrix} \cdot \begin{bmatrix} 1 \\ 1 \end{bmatrix}}{(s+1)(s+4)+3}$$

$$= \frac{[s+1 \quad -1] \cdot \begin{bmatrix} 1 \\ 1 \end{bmatrix}}{s^2 + 5s + 4 + 3} = \frac{s}{s^2 + 5s + 7}$$

8.2 (–1, –2, –3)

$$|sI - A| = \begin{vmatrix} s & -1 & 0 \\ 0 & s & -1 \\ 6 & 11 & s+6 \end{vmatrix} = 0$$

$$s^3 + 6s^2 + 11s + 6 = 0$$

∴ Eigen values are $s = -1, -2, -3$.

8.3 (a)

$$[sI - A]^{-1} = \begin{bmatrix} s+3 & -1 \\ 0 & s+2 \end{bmatrix}^{-1}$$

$$= \begin{bmatrix} \frac{1}{s+3} & \frac{1}{(s+2)(s+3)} \\ 0 & \frac{1}{(s+2)} \end{bmatrix}$$

$$e^{At} = \mathcal{L}^{-1}[sI - A]^{-1}$$

$$= \begin{bmatrix} e^{-3t} & e^{-2t} - e^{-3t} \\ 0 & e^{-2t} \end{bmatrix}$$

$$x(t) = e^{At} X(0)$$

$$= \begin{bmatrix} e^{-3t} & e^{-2t} - e^{-3t} \\ 0 & e^{-2t} \end{bmatrix} \begin{bmatrix} 10 \\ -10 \end{bmatrix}$$

$$\therefore \quad x(t) = \begin{bmatrix} 20e^{-3t} - 10e^{-2t} \\ -10e^{-2t} \end{bmatrix}$$

$$\therefore \quad X_{ss} = \lim_{t \to \infty} x(t) = \begin{bmatrix} 0 \\ 0 \end{bmatrix}$$

8.4 (c)

$$e^{At} = \mathcal{L}^{-1}[sI - A]^{-1}$$

8.5 (b)

$$x(t) = e^{At} X(0) + \int_0^t e^{A(t-\tau)} BU(\tau)\, d\tau$$

$$\because \quad X(0) = 0 \quad \text{(given)}$$

$$\therefore \quad x(t) = 0 + \int_0^t \begin{bmatrix} e^{2(t-\tau)} & 0 \\ 0 & e^{4(t-\tau)} \end{bmatrix} \cdot \begin{bmatrix} 1 \\ 1 \end{bmatrix} \cdot \delta(t) dt$$

$$= \int_0^t \begin{bmatrix} e^{2(t-\tau)} \\ e^{4(t-\tau)} \end{bmatrix} \cdot \delta(t)\, dt = \begin{bmatrix} e^{2t} \\ e^{4t} \end{bmatrix}$$

$$\text{Output } y(t) = [4 \quad 0] \cdot x(t) = [4 \quad 0] \cdot \begin{bmatrix} e^{2t} \\ e^{4t} \end{bmatrix}$$

Hence, $y(t) = 4e^{2t}$

8.6 (b)

$$|Q_C| = \begin{vmatrix} 1 & 2 \\ 0 & 0 \end{vmatrix} = 0$$

⇒ Uncontrollable.

Characteristic equation:

$$|sI - A| = 0$$

$$\begin{vmatrix} s-2 & -3 \\ 0 & s-5 \end{vmatrix} = 0$$

$$s^2 - 7s + 13 = 0$$

⇒ Eigen values, $s = 3.5 \pm j0.866$, i.e., roots lies on right side of s-plane ⇒ unstable.

8.7 (a)

State transition matrix,

$$\phi(t) = \begin{bmatrix} e^{-2t} & 0 \\ 0 & e^{-t} \end{bmatrix}$$

Initial conditions,

$$x(0) = \begin{bmatrix} 2 \\ 3 \end{bmatrix}$$

Zero input response is given by

$$x(t) = \phi(t)\,x(0)$$

$$= \begin{bmatrix} e^{-2t} & 0 \\ 0 & e^{-t} \end{bmatrix}\begin{bmatrix} 2 \\ 3 \end{bmatrix} = \begin{bmatrix} 2e^{-2t} \\ 3e^{-t} \end{bmatrix}$$

State of the system at $t = 1$ s

$$x(t)\big|_{t=1} = \begin{bmatrix} 2e^{-2} \\ 3e^{-1} \end{bmatrix} = \begin{bmatrix} 0.271 \\ 1.100 \end{bmatrix}$$

8.8 (a)

Let, $x_1 = \dfrac{d\omega}{dt}$ and $x_2 = \omega$

$$\dot{x}_2 = x_1$$

$$\frac{d^2\omega}{dt^2} + \frac{B}{J}\frac{d\omega}{dt} + \frac{k^2}{LJ}\omega = \frac{k}{LJ}V_a$$

$$\frac{d^2\omega}{dt^2} = -\frac{B}{J}\frac{d\omega}{dt} - \frac{k^2}{LJ}\omega + \frac{k}{\omega}V_a$$

$$\Rightarrow \quad \dot{x}_1 = -\frac{B}{J}x_1 - \frac{k^2}{LJ}x_2 + \frac{k}{LJ}V_a$$

$$\dot{x}_2 = x_1$$

$$\begin{bmatrix} \dot{x}_1 \\ \dot{x}_2 \end{bmatrix} = P\begin{bmatrix} x_1 \\ x_2 \end{bmatrix} + QV_a$$

where,

$$P = \begin{bmatrix} -\dfrac{B}{J} & -\dfrac{k^2}{LJ} \\ 1 & 0 \end{bmatrix} \text{ and } Q = \begin{bmatrix} \dfrac{k}{LJ} \\ 0 \end{bmatrix}$$

8.9 (a)

$$\text{System a matrix} = A = \begin{bmatrix} 0 & 2 \\ 2 & 0 \end{bmatrix}$$

$$sI - A = \begin{bmatrix} s & 0 \\ 0 & s \end{bmatrix} - \begin{bmatrix} 0 & 2 \\ 2 & 0 \end{bmatrix} = \begin{bmatrix} s & -2 \\ -2 & s \end{bmatrix}$$

Characteristic equation ⇒ $|sI - A| = 0$

$$\begin{vmatrix} s & -2 \\ -2 & s \end{vmatrix} = 0 \Rightarrow s^2 - 4 = 0$$

Roots of the characteristic equation are –2 and + 2.

8.10 (a)

$$\dot{x}(t) = \begin{bmatrix} 0 & 1 \\ 0 & -3 \end{bmatrix}x(t) + \begin{bmatrix} 1 \\ 0 \end{bmatrix}V(t) \quad \text{...(i)}$$

$$\dot{x}(t) = Ax(t) + Bu(t) \quad \text{...(ii)}$$

Comparing equation (i) and (ii), we get

$$A = \text{System matrix} = \begin{bmatrix} 0 & 1 \\ 0 & -3 \end{bmatrix} \text{ and } B = \begin{bmatrix} 1 \\ 0 \end{bmatrix}$$

$$[sI - A] = \begin{bmatrix} s & 0 \\ 0 & s \end{bmatrix} - \begin{bmatrix} 0 & 1 \\ 0 & -3 \end{bmatrix} = \begin{bmatrix} s & -1 \\ s & s+3 \end{bmatrix}$$

$$[sI - A]^{-1} = \frac{\begin{bmatrix} s+3 & 1 \\ 0 & s \end{bmatrix}}{s(s+3) - 0x(-1)}$$

$$= \begin{bmatrix} \dfrac{1}{s} & \dfrac{1}{s(s+3)} \\ 0 & \dfrac{1}{s+3} \end{bmatrix}$$

$$[sI - A]^{-1} = \begin{bmatrix} \dfrac{1}{s} & \dfrac{1}{3}\left(\dfrac{1}{s} - \dfrac{1}{s+3}\right) \\ 0 & \dfrac{1}{s+3} \end{bmatrix}$$

State transition matrix = $\mathcal{L}^{-1}[(sI - A)^{-1}]$

$$\phi(t) = \begin{bmatrix} 1 & \dfrac{1}{3}(1 - e^{-3t}) \\ 0 & e^{-3t} \end{bmatrix}$$

8.11 (c)

ZIR (Zero Input Response) = $\phi(t) \times X(0)$

$$= \begin{bmatrix} 1 & \dfrac{1}{3}(1 - e^{-3t}) \\ 0 & e^{-3t} \end{bmatrix}\begin{bmatrix} -1 \\ 3 \end{bmatrix}$$

$$= \begin{bmatrix} -1 + 1 - e^{-3t} \\ 3\,e^{-3t} \end{bmatrix} = \begin{bmatrix} -e^{-3t} \\ 3e^{-3t} \end{bmatrix}$$

ZSR (Zero State Response)

$$= \mathcal{L}^{-1}[(sI - A)^{-1} BU(s)]$$

$$= \mathcal{L}^{-1}\left\{\begin{bmatrix} \frac{1}{s} & \frac{1}{s(s+3)} \\ 0 & \frac{1}{s+3} \end{bmatrix} \begin{bmatrix} 1 \\ 0 \end{bmatrix} \frac{1}{s}\right\}$$

$$= \mathcal{L}^{-1}\begin{bmatrix} 1/s^2 \\ 0 \end{bmatrix} = \begin{bmatrix} t \\ 0 \end{bmatrix}$$

∴ State transition equation

$$= ZIR + ZSR = \begin{bmatrix} -e^{-3t} \\ 3e^{-3t} \end{bmatrix} + \begin{bmatrix} t \\ 0 \end{bmatrix}$$

$$= \begin{bmatrix} t - e^{-3t} \\ 3e^{-3t} \end{bmatrix}$$

8.12 (b)

$$H(s) = \frac{Y(s)}{U(s)} = \frac{3(s-2)}{s^3 + 4s^2 - 2s + 1}$$

$$\frac{Y(s)}{X_1(s)} \cdot \frac{X_1(s)}{U(s)} = 3(s-2)\left(\frac{1}{s^3 + 4s^2 - 2s + 1}\right)$$

Let, $$\frac{X_1(s)}{U(s)} = \frac{1}{s^3 + 4s^2 - 2s + 1}$$

$$s^3x_1(s) + 4s^2x_1(s) - 2sx_1(s) + x_1(s) = u(s)$$

Replacing s by $\frac{d}{dt}$,

$$\frac{d^3x_1}{dt^2} + 4\frac{d^2x_1}{dt^2} - 2\frac{dx_1}{dt} + x_1 = u(t) \quad \text{...(i)}$$

Let, $$\frac{dx_1}{dt} = x_2 = \dot{x}_1$$

$$\frac{d^2x_1}{dt} = \dot{x}_2 = x_3$$

Replacing equation (i), we get

$$\dot{x}_3 + 4x_3 - 2x_2 + x_1 = u(t)$$

$$\dot{x}_3 = -x_1 + 2x_2 - 4x_3 + u(t)$$

$$\dot{x}_1 = x_2$$

$$\dot{x}_2 = x_3$$

$$\dot{x}_3 = -x_1 + 2x_2 - 4x_3 + u(t)$$

$$\begin{bmatrix} \dot{x}_1 \\ \dot{x}_2 \\ \dot{x}_3 \end{bmatrix} = \begin{bmatrix} 0 & 1 & 0 \\ 0 & 0 & 1 \\ -1 & 2 & -4 \end{bmatrix}\begin{bmatrix} x_1 \\ x_2 \\ x_3 \end{bmatrix} + \begin{bmatrix} 0 \\ 0 \\ 1 \end{bmatrix} u(t)$$

So, $$A = \begin{bmatrix} 0 & 1 & 0 \\ 0 & 0 & 1 \\ -1 & 2 & -4 \end{bmatrix}$$

8.13 (a)

Any state equation represents the dynamical behaviour of the given network. State equations usually follow a specific 'format' while being represented. On the left side of each state equation, the derivative of only one variable is used. On the right hand side a mathematical function is represented involving any or all the state variables and the sources.

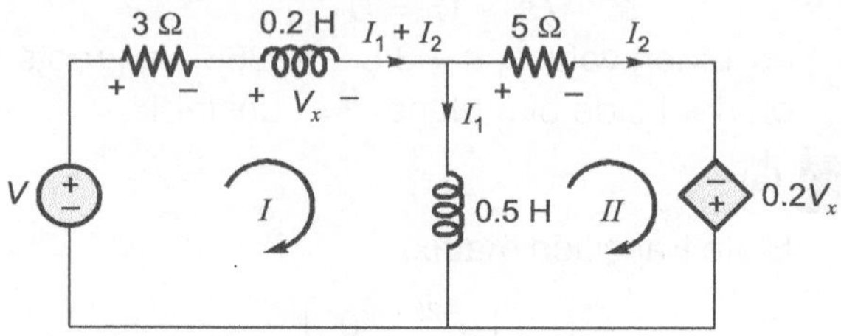

Using KVL in Loop-I,

$$V - 3(I_1 + I_2) - V_x - 0.5\frac{dI_1}{dt} = 0 \quad \text{...(i)}$$

Using KVL in Loop-II,

$$0.5\frac{dI_1}{dt} - 5I_2 + 0.2V_x = 0 \quad \text{...(ii)}$$

Eliminating I_2 from equation (i) and (ii), we get,

$$\frac{dI_1}{dt} = -1.4V_x - 3.75I_1 + \frac{5}{4}V$$

8.14 (d)

$$\dot{x} = Ax + Bu \quad \text{and} \quad y = Cx$$

where,

$$A = \begin{bmatrix} 0 & 1 \\ 0 & -2 \end{bmatrix}, B = \begin{bmatrix} 0 \\ 1 \end{bmatrix}; C = [1 \quad 0] \text{ and } D = 0$$

$$[sI - A] = \begin{bmatrix} s & 0 \\ 0 & s \end{bmatrix} - \begin{bmatrix} 0 & 1 \\ 0 & -2 \end{bmatrix} = \begin{bmatrix} s & -1 \\ 0 & s+2 \end{bmatrix}$$

$$[sI - A]^{-1} = \frac{1}{s(s+2)}\begin{bmatrix} s+2 & 1 \\ 0 & s \end{bmatrix}$$

Transfer function $= C[sI - A]^{-1} B + D$

$$= [1 \ 0]\frac{\begin{bmatrix} s+2 & 1 \\ 0 & s \end{bmatrix}\begin{bmatrix} 0 \\ 1 \end{bmatrix}}{s(s+2)}$$

$$= [1 \ 0]\frac{\begin{bmatrix} 1 \\ s \end{bmatrix}}{s(s+2)} = \frac{1}{s(s+2)}$$

8.15 (a)

$$G(s) = \frac{1}{s(s+2)}$$

and $H(s) = 1$

$r(t) = u(t)$

$$\Rightarrow \quad R(s) = \frac{1}{s}$$

$$\text{Error} = E(s) = \frac{R(s)}{1+G(s)H(s)}$$

$$= \frac{1/s}{1+\dfrac{1}{s(s+2)}}$$

$$\Rightarrow \quad E(s) = \frac{s+2}{s(s+2)+1}$$

Steady state error, using final value theorem

$$e_{ss} = \underset{s\to 0}{\text{Lt}}\; sE(s) = \underset{s\to 0}{\text{Lt}} \frac{s(s+2)}{s(s+2)+1} = 0$$

8.16 (c)

Selecting $x_1(t)$ and $x_2(t)$ as state variables.

$$\dot{x}_1(t) = \frac{dx_1(t)}{dt} = -3x_1(t) + x_2(t) + 2u(t)$$

$$\dot{x}_2(t) = \frac{dx_2(t)}{dt} = -2x_2(t) + u(t)$$

$$\begin{bmatrix} \dot{x}_1(t) \\ \dot{x}_2(t) \end{bmatrix} = \begin{bmatrix} -3 & 1 \\ 0 & -2 \end{bmatrix}\begin{bmatrix} x_1(t) \\ x_2(t) \end{bmatrix} + \begin{bmatrix} 2 \\ 1 \end{bmatrix} u(t)$$

$$\dot{x} = AX + BU$$

So, $A = \begin{bmatrix} -3 & 1 \\ 0 & -2 \end{bmatrix}$, $B = \begin{bmatrix} 2 \\ 1 \end{bmatrix}$

$$y(t) = x_1(t) = [1 \;\; 0]\begin{bmatrix} x_1(t) \\ x_2(t) \end{bmatrix}$$

$$y = CX + DU$$

So, $C = [1 \;\; 0]$ and $D = 0$

$$[sI - A] = \begin{bmatrix} s & 0 \\ 0 & s \end{bmatrix} - \begin{bmatrix} -3 & 1 \\ 0 & -2 \end{bmatrix} = \begin{bmatrix} s+3 & -1 \\ 0 & s+2 \end{bmatrix}$$

$$[sI - A]^{-1} = \frac{\begin{bmatrix} s+2 & 1 \\ 0 & s+3 \end{bmatrix}}{(s+2)(s+3)}$$

Transfer function $= C[sI - A]^{-1} B + D$

$$= [1 \;\; 0]\frac{\begin{bmatrix} s+2 & 1 \\ 0 & s+3 \end{bmatrix}}{(s+2)(s+3)}\begin{bmatrix} 2 \\ 1 \end{bmatrix} + 0$$

$$= \frac{[1 \;\; 0]\begin{bmatrix} 2(s+2)+1 \\ s+3 \end{bmatrix}}{(s+2)(s+3)}$$

$$= \frac{2s+5}{(s+2)(s+3)} = \frac{2s+5}{s^2+5s+6}$$

8.17 (b)

$$[sI - A]^{-1} = \frac{\begin{bmatrix} s+2 & 1 \\ 0 & s+3 \end{bmatrix}}{(s+2)(s+3)}$$

$$= \begin{bmatrix} \dfrac{1}{s+3} & \dfrac{1}{(s+2)(s+3)} \\ 0 & \dfrac{1}{s+2} \end{bmatrix}$$

$$= \begin{bmatrix} \dfrac{1}{s+3} & \dfrac{1}{s+2} - \dfrac{1}{s+3} \\ 0 & \dfrac{1}{s+2} \end{bmatrix}$$

State transition matrix $= \mathcal{L}^{-1}[(sI - A)^{-1}]$

$$= \mathcal{L}^{-1}\begin{bmatrix} \dfrac{1}{s+3} & \dfrac{1}{s+2} - \dfrac{1}{s+3} \\ 0 & \dfrac{1}{s+2} \end{bmatrix}$$

$$= \begin{bmatrix} e^{-3t} & e^{-2t} - e^{-3t} \\ 0 & e^{-2t} \end{bmatrix}$$

8.18 (c)

$$A = \begin{bmatrix} -1 & 2 \\ 0 & 2 \end{bmatrix} \text{ and } B = \begin{bmatrix} 0 \\ 1 \end{bmatrix}$$

$$[sI - A] = \begin{bmatrix} s & 0 \\ 0 & s \end{bmatrix} - \begin{bmatrix} -1 & 2 \\ 0 & 2 \end{bmatrix} = \begin{bmatrix} s+1 & -2 \\ 0 & s-2 \end{bmatrix}$$

$$[sI - A]^{-1} = \frac{\begin{bmatrix} s-2 & 2 \\ 0 & s+1 \end{bmatrix}}{(s+1)(s-2)} \quad \text{...(i)}$$

Transfer function $= C[sI - A]^{-1} B$

So denominator of equation (i) gives poles of the system.

$$(s+1)(s-2) = 0 \Rightarrow s = -1 \text{ and } 2$$

One pole lies in RHS of s-plane. Hence, the so, system is unstable.

For controllability, Q_c is defined as

$$Q_c = [B : AB]$$

$$AB = \begin{bmatrix} -1 & 2 \\ 0 & 2 \end{bmatrix}\begin{bmatrix} 0 \\ 1 \end{bmatrix} = \begin{bmatrix} 2 \\ 2 \end{bmatrix}$$

$$Q_c = [B : AB] = \begin{bmatrix} 0 & 2 \\ 1 & 2 \end{bmatrix}$$

$$|Q_c| \neq 0$$

Hence, the system is controllable.

8.19 (d)

$$A = \begin{bmatrix} 0 & a_1 & 0 \\ 0 & 0 & a_2 \\ a_3 & 0 & 0 \end{bmatrix}, \; B = \begin{bmatrix} 0 \\ 0 \\ 1 \end{bmatrix}, \; C = [1 \;\; 0 \;\; 0]^T$$

For system to be controllable, the metric Q_c must be non singular.

$$Q_c = \begin{bmatrix} B & AB & A^2B \end{bmatrix}$$

$$AB = [0 \;\; a_2 \;\; 0]^T$$

$$A^2 = \begin{bmatrix} 0 & 0 & a_1a_2 \\ a_2a_3 & 0 & 0 \\ 0 & a_1a_3 & 0 \end{bmatrix}$$

$$A^2B = [a_1a_2 \;\; 0 \;\; 0]^2$$

$$Q_c = \begin{bmatrix} 0 & 0 & a_1a_2 \\ 0 & a_2 & 0 \\ 1 & 0 & 0 \end{bmatrix}$$

$$|Q_c| = -a_1a_2^2$$

For Q_c to be nonsingular

$$|Q_c| \neq 0$$

$\therefore \quad a_1 \neq 0$ and $a_2 \neq 0$ and $a_3 \in R$

8.20 (a)

$$A = \begin{bmatrix} -2 & 0 \\ 0 & -1 \end{bmatrix}, \quad B = \begin{bmatrix} 1 \\ 1 \end{bmatrix}$$

$$AB = \begin{bmatrix} -2 & 0 \\ 0 & -1 \end{bmatrix}\begin{bmatrix} 1 \\ 1 \end{bmatrix} = \begin{bmatrix} -2 \\ -1 \end{bmatrix}$$

For controllability, $|B : AB| \neq 0$

or $\begin{vmatrix} 1 & -2 \\ 1 & -1 \end{vmatrix} = -1-(-2) = 1 \neq 0$

The system is controllable.

$$C^T = \begin{bmatrix} 1 \\ 0 \end{bmatrix}$$

$$A^TC^T = \begin{bmatrix} -2 & 0 \\ 0 & -1 \end{bmatrix}\begin{bmatrix} 1 \\ 0 \end{bmatrix} = \begin{bmatrix} -2 \\ 0 \end{bmatrix}$$

For observability, $|C^T : A^TC^T| \neq 0$

or $\begin{vmatrix} 1 & -2 \\ 0 & 0 \end{vmatrix} = 0$

The system is not observable.

8.21 (a)

$$[sI - A] = \begin{bmatrix} s & 0 \\ 0 & s \end{bmatrix} - \begin{bmatrix} -2 & 0 \\ 0 & -1 \end{bmatrix} = \begin{bmatrix} s+2 & 0 \\ 0 & s+1 \end{bmatrix}$$

$$[sI - A]^{-1} = \begin{bmatrix} \frac{1}{s+2} & 0 \\ 0 & \frac{1}{s+1} \end{bmatrix}$$

$$[sI - A]^{-1}[B] = \begin{bmatrix} \frac{1}{s+2} & 0 \\ 0 & \frac{1}{s+1} \end{bmatrix}\begin{bmatrix} 1 \\ 1 \end{bmatrix} = \begin{bmatrix} \frac{1}{s+2} \\ \frac{1}{s+1} \end{bmatrix}$$

$$C[sI - A]^{-1}[B] = [1 \;\; 0]\begin{bmatrix} \frac{1}{s+2} \\ \frac{1}{s+1} \end{bmatrix} = \frac{1}{s+2}$$

$$G(s) = \frac{1}{s+2}$$

$$\frac{Y(s)}{X(s)} = \frac{1}{s+2}$$

$$Y(s) = \frac{1}{s(s+2)} = \frac{1}{2}\left[\frac{1}{s} - \frac{1}{s+2}\right]$$

$$y(t) = \frac{1}{2}(1 - e^{-2t}) = \frac{1}{2} - \frac{1}{2}e^{-2t}$$

8.22 (c)

Given: $\begin{bmatrix} \dot{x}_1 \\ \dot{x}_2 \end{bmatrix} = \begin{bmatrix} 1 & 0 \\ 1 & 1 \end{bmatrix}\begin{bmatrix} x_1 \\ x_2 \end{bmatrix} + \begin{bmatrix} 1 \\ 1 \end{bmatrix}u$

or $\dot{x}(t) = Ax(t) + Bu$

Here, $[A] = \begin{bmatrix} 1 & 0 \\ 1 & 1 \end{bmatrix}, \; B = \begin{bmatrix} 1 \\ 1 \end{bmatrix}$

State transition matrix is given by,

$$e^{At} = \mathcal{L}^{-1}[sI - A]^{-1}$$

Now, $[sI - A] = \begin{bmatrix} s-1 & 0 \\ -1 & s-1 \end{bmatrix}$

$$\text{Adj}[sI - A] = \begin{bmatrix} s-1 & 0 \\ 1 & s-1 \end{bmatrix} \text{ and}$$

$$|sI - A| = (s-1)^2$$

$$\therefore \quad [sI - A]^{-1} = \frac{1}{|sI - A|}\text{Adj}[sI = A]$$

$$[sI - A]^{-1} = \frac{1}{(s-1)^2}\begin{bmatrix} s-1 & 0 \\ 1 & s-1 \end{bmatrix}$$

$$= \begin{bmatrix} \frac{1}{(s-1)} & 0 \\ \frac{1}{(s-1)^2} & \frac{1}{(s-1)} \end{bmatrix}$$

$$\therefore \mathcal{L}^{-1}[sI-A]^{-1} = \begin{bmatrix} e^t & 0 \\ te^t & e^t \end{bmatrix}$$

$$\therefore \text{ State transition matrix} = \begin{bmatrix} e^t & 0 \\ te^t & e^t \end{bmatrix}$$

8.23 (a)

Given: $A = \begin{bmatrix} 0 & 1 \\ -1 & -1 \end{bmatrix}$, $B = \begin{bmatrix} 0 \\ 1 \end{bmatrix}$, $C = [1 \;\; 0]$

Transfer function of the given system is given by

$$\frac{Y(s)}{U(s)} = C(sI-A)^{-1}B$$

$$\text{Now, } (sI-A) = \begin{bmatrix} s & 0 \\ 0 & s \end{bmatrix} - \begin{bmatrix} 0 & 1 \\ -1 & -1 \end{bmatrix} = \begin{bmatrix} s & -1 \\ 1 & s+1 \end{bmatrix}$$

$$(sI-A)^{-1} = \frac{1}{|sI-A|}[\text{Adj } A]$$

$$|sI-A| = s(s+1)+1 = (s^2+s+1)$$

$$\text{and } \text{Adj}[A] = \begin{bmatrix} s+1 & 1 \\ -1 & s \end{bmatrix}$$

$$\therefore (sI-A)^{-1} = \frac{1}{(s^2+s+1)}\begin{bmatrix} s+1 & 1 \\ -1 & s \end{bmatrix}$$

$\therefore$ Transfer function,

$$\frac{Y(s)}{U(s)} = \left(\frac{1}{s^2+s+1}\right)\left\{[1 \;\; 0]\begin{bmatrix} s+1 & 1 \\ -1 & s \end{bmatrix}\begin{bmatrix} 0 \\ 1 \end{bmatrix}\right\}$$

$$= \frac{1}{(s^2+s+1)}\left\{[(s+1) \;\; 1]\begin{bmatrix} 0 \\ 1 \end{bmatrix}\right\}$$

$$= \frac{1}{(s^2+s+1)} \times 1 = \frac{1}{(s^2+s+1)}$$

$$\therefore \frac{Y(s)}{U(s)} = \frac{1}{s^2+s+1}$$

Given, Input = unit step

$$\therefore U(s) = \frac{1}{s}$$

$$\therefore Y(s) = \frac{1}{s} \times \frac{1}{s^2+s+1} = \frac{1}{s(s^2+s+1)}$$

$$\therefore \text{ Final value} = \lim_{s\to 0}\{sY(s)\}$$

$$= \lim_{s\to 0}\left[\frac{1}{s^2+s+1}\right] = 1$$

$$\therefore \text{ Error} = \text{Final value} - \text{Initial value}$$

$$e_{ss} = 0$$

8.24 (b)

Equations from flow diagram,

$$\dot{x}_1 = 5x_1 - 2x_2 + u_1; \quad y_1 = x_1$$

$$\dot{x}_2 = 2x_1 + x_2 + u_1 + u_2; \quad y_2 = x_1 - x_2$$

$$\dot{x} = \underbrace{\begin{bmatrix} 5 & -2 \\ 2 & 1 \end{bmatrix}}_{A} x + \underbrace{\begin{bmatrix} 1 & 0 \\ 1 & 1 \end{bmatrix}}_{B} u; \; y = \underbrace{\begin{bmatrix} 1 & 0 \\ 1 & -1 \end{bmatrix}}_{C} x$$

Considering the SISO cases:

1. u_1 (input) and y_1 (output):

$$A \triangleq \begin{bmatrix} 5 & -2 \\ 2 & 1 \end{bmatrix}; B \triangleq \begin{bmatrix} 1 \\ 1 \end{bmatrix} \text{ and } C = [1 \;\; 0]$$

$$O \triangleq \begin{bmatrix} C \\ CA \end{bmatrix} = \begin{bmatrix} 1 & 0 \\ 5 & -2 \end{bmatrix}$$

$\Rightarrow$ Rank $(O) = 2 \Rightarrow$ observable

$$C \triangleq [B \;\; AB] = \begin{bmatrix} 1 & 3 \\ 1 & 3 \end{bmatrix}$$

$\Rightarrow$ Rank $(C) \neq 2 \Rightarrow$ non-controllable

2. u_2 (input) and y_1 (output):

$$A \triangleq \begin{bmatrix} 5 & -2 \\ 2 & 1 \end{bmatrix}; B \triangleq \begin{bmatrix} 0 \\ 1 \end{bmatrix} \text{ and } C \triangleq [1 \;\; -1]$$

$$O \triangleq \begin{bmatrix} 1 & 0 \\ 5 & -2 \end{bmatrix} \Rightarrow \text{Rank } (O) = 2 \Rightarrow \text{observable}$$

$$C \triangleq \begin{bmatrix} 0 & -2 \\ 1 & 1 \end{bmatrix} \Rightarrow \text{Rank } (C) = 2 \Rightarrow \text{controllable}$$

3. u_2 (input) and y_2 (output):
   Rank $(O) \neq 2 \Rightarrow$ non-observable
   Rank $(C) = 2 \Rightarrow$ controllable
4. u_1 (input) and y_2 (output):
   Neither observable and non-controllable

8.25 (a)

$$\dot{x} = Ax$$

Eigen values are λ_1 and λ_2. We can write,

$$\phi(t) = \begin{bmatrix} e^{\lambda_1 t} & 0 \\ 0 & e^{\lambda_2 t} \end{bmatrix}$$

Response due to initial conditions,

$$x(t) = \phi(t)\cdot x(0) \Rightarrow x(t) = \begin{bmatrix} e^{\lambda_1 t} & 0 \\ 0 & e^{\lambda_2 t} \end{bmatrix}\begin{bmatrix} \alpha \\ 0 \end{bmatrix} = \alpha\, e^{\lambda_1 t}$$

8.26 (c)

Given: $\dot{x}(t) = Px + Qu$, $y = Rx$

$$P = \begin{bmatrix} -1 & 1 \\ 0 & -3 \end{bmatrix}, \; Q = \begin{bmatrix} 0 \\ 1 \end{bmatrix}, \; R = [0 \; 1]$$

For controllability,

$$Q_c = [Q \;\; PQ]$$

$$[PQ] = \begin{bmatrix} -1 & 1 \\ 0 & -3 \end{bmatrix}\begin{bmatrix} 0 \\ 1 \end{bmatrix} = \begin{bmatrix} 1 \\ -3 \end{bmatrix}$$

$$\therefore \quad Q_c = \begin{bmatrix} 0 & 1 \\ 1 & -3 \end{bmatrix}$$

$\therefore \quad |Q_c| = -1 \neq 0 \cdots$ controllable

Also, for observability,

$$Q_0 = [R^T \quad P^T R^T]$$

$$[P^T R^T] = \begin{bmatrix} -1 & 0 \\ 1 & -3 \end{bmatrix}\begin{bmatrix} 0 \\ 1 \end{bmatrix} = \begin{bmatrix} 0 \\ -3 \end{bmatrix}$$

$$\therefore \quad Q_0 = \begin{bmatrix} 0 & 0 \\ 1 & -3 \end{bmatrix}$$

$\therefore \quad |Q_0| = (0 \times -3 - 0 \times 1) = 0 \cdots$ Non-observable

Since, $Q_c \neq 0$ and $Q_0 = 0$

$\therefore$ System is controllable but not observable.

8.27 (d)

Given: $\dot{X} = \begin{bmatrix} 1 & 2 \\ 2 & 0 \end{bmatrix} X + \begin{bmatrix} 1 \\ 2 \end{bmatrix} u$

and $Y = [1 \quad 0] X$

Transfer function $= C[sI - A]^{-1} B + D$

$$[sI - A]^{-1} = \begin{bmatrix} \frac{s}{s^2 - s - 4} & \frac{2}{s^2 - s - 4} \\ \frac{2}{s^2 - s - 4} & \frac{s-1}{s^2 - s - 4} \end{bmatrix}$$

$$(sI - A)^{-1} \times B = \begin{bmatrix} \frac{s+4}{s^2 - s - 4} \\ \frac{s-1}{s^2 - s - 4} \end{bmatrix}$$

$$\frac{Y(s)}{U(s)} = C[sI - A]^{-1} \times B = \frac{s+4}{s^2 - s - 4}$$

8.28 (1.284)

$$y(t) = [1 \quad 0]\, x(t)$$

$$x(t) = \mathcal{L}^{-1}[(sI - A)^{-1}]\, x(0) + \mathcal{L}^{-1}[(sI - A)^{-1}]\, BU(s)]$$

$$\mathcal{L}^{-1}[(sI - A)^{-1}]\, x(0) = \begin{bmatrix} 1 \\ 0 \end{bmatrix}$$

$$\mathcal{L}^{-1}[(sI - A)]^{-1}\, BU(s) = \begin{bmatrix} -0.25 + 0.5t + 0.25e^{-2t} \\ 0.5 - 0.5e^{-2t} \end{bmatrix}$$

$$\therefore \quad x(t) = \begin{bmatrix} 0.75 + 0.5t + 0.25e^{-2t} \\ 0.5 - 0.5e^{-2t} \end{bmatrix}$$

$$\therefore \quad y(t) = 0.75 + 0.5t + 0.25e^{-2t}$$
$$+ (0.5 \times 1) + 0.25(e^{-2})$$
$$y(t) = 0.75 \quad \therefore y(1) = 1.284$$

8.29 (a)

Characteristic equation is, $|sI - A| = 0$

$$|sI - A| = \begin{vmatrix} s & -1 \\ \alpha & s + 2\beta \end{vmatrix} = s^2 + 2s\beta + \alpha = 0$$

$$\therefore \quad \omega_n^2 = \alpha \Rightarrow \omega_n = \sqrt{\alpha}$$

$$2\xi\omega_n = 2\beta \Rightarrow \xi = \frac{\beta}{\sqrt{\alpha}}$$

8.30 (1)

$$\dot{x} = -x - Kx = x(-kI - I)$$

Characteristic equation,

$$|sI + KI + I| = 0 \Rightarrow |(s + 1 + K)I| = 0$$

$\therefore \quad s + 1 + K = 0$

$s = -1 - K \Rightarrow -2 = -1 - K \Rightarrow K = 1$

8.31 (2)

Given : $A = \begin{bmatrix} 0 & 1 \\ -1 & -2 \end{bmatrix}$; $B = \begin{bmatrix} 0 \\ 1 \end{bmatrix}$; $C = [3, -2]$ $D = 1$

The transfer function

$$TF = C[sI - A]^{-1} B + D$$

$$[sI - A]^{-1} = \begin{bmatrix} s & -1 \\ 1 & s+2 \end{bmatrix}^{-1} = \frac{1}{(s^2 + 2s + 1)}\begin{bmatrix} s+2 & 1 \\ -1 & s \end{bmatrix}$$

$$TF = \frac{1}{(s^2 + 2s + 1)}[3 \quad -2]\begin{bmatrix} s+2 & 1 \\ -1 & s \end{bmatrix}\begin{bmatrix} 0 \\ 1 \end{bmatrix} + 1$$

$$TF = \frac{1}{(s^2 + 2s + 1)}[3 \quad -2]\begin{bmatrix} 1 \\ s \end{bmatrix} + 1$$

$$TF = \frac{(3 - 2s)}{(s^2 + 2s + 1)} + 1 = \frac{s^2 + 4}{(s^2 + 2s + 1)}$$

$$H(j\omega) = \frac{4 - \omega^2}{(1 - \omega^2 + 2j\omega)}$$

The output will be zero, for $\omega^2 - 4 = 0$

$$\omega = 2 \text{ rad/sec}$$

■■■■

Electrical & Electronic Measurements

UNIT

VI

CONTENTS

Electrical & Electronic Measurements

Syllabus

Bridges and Potentiometers, Measurement of Voltage, Current, Power, Energy and Power Factor; Instrument Transformers, Digital Voltmeters and Multimeters, Phase, Time and Frequency Measurement; Oscilloscopes, Error Analysis.

Analysis of Previous GATE Papers

Exam Year	1 Mark Ques.	2 Marks Ques.	3 Marks Ques.	5 Marks Ques.	Total Marks
1995	5	–	–	–	**5**
1996	5	–	1	–	**8**
1997	3	–	–	–	**3**
1998	1	1	–	1	**8**
1999	1	2	–	1	**10**
2000	2	1	–	1	**9**
2001	2	1	–	–	**4**
2002	2	–	–	–	**2**
2003	3	8	–	–	**19**
2004	3	7	–	–	**17**
2005	3	5	–	–	**13**
2006	2	4	–	–	**10**
2007	1	1	–	–	**3**
2008	1	2	–	–	**5**
2009	2	2	–	–	**6**
2010	2	1	–	–	**4**
2011	3	1	–	–	**5**

Exam Year	1 Mark Ques.	2 Mark Ques.	Total Marks
2012	3	1	**5**
2013	2	1	**4**
2014 Set-1	2	2	**6**
2014 Set-2	2	2	**6**
2014 Set-3	2	2	**6**
2015 Set-1	2	1	**4**
2015 Set-2	3	7	**17**
2016 Set-1	0	0	**0**
2016 Set-2	1	2	**5**
2017 Set-1	2	0	**2**
2017 Set-2	2	2	**6**
2018	1	1	**3**
2019	–	2	**4**
2020	–	2	**4**
2021	1	–	**1**
2022	2	–	**2**
2023	–	1	**2**
2024	1	1	**3**

1 Characteristics of Instruments and Measurement Systems

1.1 Four ammeters M1, M2, M3 and M4 with the following specifications are available.

Instrument	Type	Full scale value (A)	Accuracy % of FS
M1	$3\frac{1}{2}$ digit dual slope	20	±0.10
M2	PMMC	10	±0.20
M3	Electro-dynamic	5	±0.50
M4	Moving-iron	1	±1.00

A current of 1 A is to be measured. To obtain minimum error in the reading, one should select meter

(a) M1 (b) M2
(c) M3 (d) M4 **[1995 : 1 M]**

1.2 A current of

$\lfloor 2+\sqrt{2}\sin(\sin 314t+30°)+2\sqrt{2}\cos(952t+45°)\rfloor$ A

is measured with a thermocouple type, 5 A full scale, class 1 meter. The meter reading would lie in the range,

(a) 5 A ± 1% (b) $(2+3\sqrt{2})$ A ± 1%
(c) 3 A ± 1.67% (d) 2 A ± 0.5%

[1999 : 2 M]

1.3 Resistance R_1 and R_2 have, respectively, nominal values of 10 Ω and 5 Ω, and tolerances of ±5% and ±10%. The range of values for the parallel combination of R_1 and R_2 is

(a) 3.077 Ω to 3.636 Ω
(b) 2.805 Ω to 3.371 Ω
(c) 3.237 Ω to 3.678 Ω
(d) 3.192 Ω to 3.435 Ω **[2001 : 2 M]**

1.4 A variable w is related to three other variables *x*, *y*, *z* as $w = xy/z$. The variables are measured with meters of accuracy ± 0.5% reading, ± 1% of full scale value and ± 1.5% reading. The actual readings of the three meters are 80, 20 and 50 with 100 being the full scale value for all three. The maximum limiting error in the measurement of *w* will be

(a) ±0.5% rdg (b) ±5.5% rdg
(c) ±6.7% rdg (d) ±7.0% rdg **[2006 : 2 M]**

1.5 The measurement system shown in the figure uses three sub-systems in cascade whose gains are specified as G_1, G_2 and $1/G_3$. The relative small errors associated with each respective subsystem G_1, G_2 and G_3 are ε_1, ε_2 and ε_3. The error associated with the output is:

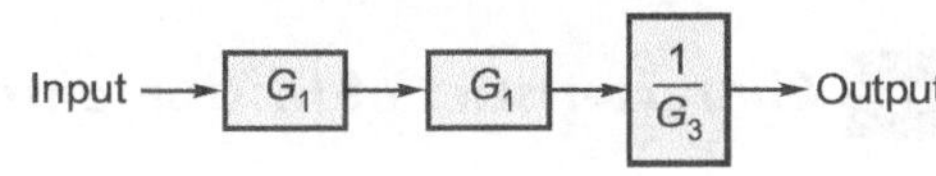

(a) $\varepsilon_1 - \varepsilon_2 + \varepsilon_3$ (b) $\frac{\varepsilon_1 . \varepsilon_2}{\varepsilon_3}$
(c) $\varepsilon_1 + \varepsilon_2 - \varepsilon_3$ (d) $\varepsilon_1 + \varepsilon_2 + \varepsilon_3$

[2009 : 1 M]

1.6 When the Wheatstone bridge shown in the figure is used to find the value of resistor R_X, the galvanometer *G* indicates zero current when R_1 = 50 Ω, R_2 = 65 Ω and R_3 = 100 Ω. If R_3 is known with ±5% tolerance on its nominal value of 100 Ω, what is the range of R_X in Ohms?

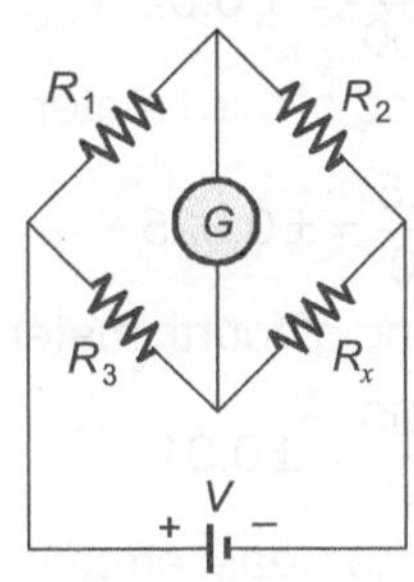

(a) [123.50, 136.50] (b) [125.89, 134.12]
(c) [117.00, 143.00] (d) [120.25, 139.75]

[2015 : 1 M, Set-1]

1.7 The following measurement are obtained on a single phase load: *V* = 220 V ±1%. *I* = 5.0 A ± 1% and *W* = 555 W ± 2%. If the power factor is calculated using these measurements, the worst case error in the calculated power factor in percent is _____. (Give answer up to one decimal place.)

[2017 : 1 M, Set-1]

1.8 Two resistors with nominal resistance values R_1 and R_2 have additive uncertainties ΔR_1 and ΔR_2 respectively. When these resistances are connected in parallel, the standard deviation of the error in the equivalent resistance *R* is

(a) $\pm\sqrt{\left\{\frac{\partial R}{\partial R_1}\Delta R_1\right\}^2 + \left\{\frac{\partial R}{\partial R_2}\Delta R_2\right\}^2}$

(b) $\pm\sqrt{\left\{\frac{\partial R}{\partial R_2}\Delta R_1\right\}^2+\left\{\frac{\partial R}{\partial R_1}\Delta R_2\right\}^2}$

(c) $\pm\sqrt{\left\{\frac{\partial R}{\partial R_1}\right\}^2\Delta R_2+\left\{\frac{\partial R}{\partial R_2}\right\}^2\Delta R_1}$

(d) $\pm\sqrt{\left\{\frac{\partial R}{\partial R_1}\right\}^2\Delta R_1+\left\{\frac{\partial R}{\partial R_2}\right\}^2\Delta R_2}$

[2017 : 1 M, Set-2]

Answers Characteristics of Instruments and Measurement Systems

1.1 (d) 1.2 (c) 1.3 (a) 1.4 (d) 1.5 (c) 1.6 (a) 1.7 (4)
1.8 (a)

Explanations Characteristics of Instruments and Measurement Systems

1.1 (d)

Error in reading of first meter

$$= \text{FSD} \times \text{accuracy} = 20\times\frac{\pm 0.1}{100} = \pm 0.02$$

Error in reading of second meter

$$= 10\times\frac{\pm 0.2}{100} = \pm 0.02$$

Error in reading of third meter

$$= 5\times\frac{\pm 0.5}{100} = \pm 0.025$$

Error in reading of fourth meter

$$= 1\times\frac{\pm 1.00}{100} = \pm 0.01$$

Fourth meter has least error.

1.2 (c)

Thermocouple type, instruments read rms value,

$$I_{\text{rms}} = \sqrt{2^2+\left(\frac{\sqrt{2}}{\sqrt{2}}\right)^2+\left(\frac{2\sqrt{2}}{\sqrt{2}}\right)^2}\ \text{A}$$

$$= \sqrt{4+1+4} = 3\ \text{A}$$

For class 1 meter, accuracy is 1% for 5 A range.

For 3 A, accuracy will be $\frac{5}{3}\% = 1.67\%$

1.3 (a)

$$\text{Range of } R_1 = 10 \pm 10\times\frac{5}{100} = 9.5\ \Omega \text{ to } 10.5\ \Omega$$

$$\text{Range of } R_2 = 5 \pm 5\times\frac{10}{100} = 4.5\ \Omega \text{ to } 5.5\ \Omega$$

$$\because \quad R_p = \frac{R_1 R_2}{R_1 + R_2}$$

$$\therefore \text{ Range of } R_p = \frac{9.5\times 4.5}{9.5+4.5} \text{ to } \frac{10.5\times 5.5}{10.5+5.5}$$

$$= 3.05\ \Omega \text{ to } 3.61\ \Omega$$

$\therefore$ Option (a) is closest.

1.4 (d)

Full scale reading of all three = 100

Readings of : $x = 80$; $y = 20$; $z = 50$

$$\delta x = \pm 0.5\% \text{ of reading} = \pm\frac{0.5\times 80}{100} = \pm 0.4$$

$$\delta y = \pm 1\% \text{ of full reading} = \pm\frac{1\times 100}{100} = \pm 1$$

$$\delta z = \pm 1.5\% \text{ of reading} = \pm\frac{1.5\times 50}{100} = \pm 0.75$$

Given, $\omega = \frac{xy}{z}$

Taking log, we get : $\log\omega = \log x + \log y - \log z$

Differentiating w.r.t. ω, we get

$$\frac{\delta\omega}{\omega} = \frac{\delta x}{x}+\frac{\delta y}{y}-\frac{\delta z}{z}$$

For maximum limiting error,

$$\frac{\delta\omega}{\omega} = \pm\left(\frac{0.4}{80}+\frac{1}{20}+\frac{0.75}{100}\right)\times 100 = \pm 7\%$$

1.5 (c)

$$\frac{dG_1}{G_1} = \epsilon_1,\ \frac{dG_2}{G_2} = \epsilon_2, \text{ and } \frac{dG_3}{G_3} = \epsilon_3$$

$$\text{Output } (y)_0 = \frac{G_1 G_3}{G_3}x$$

where, x = input

$$\ln y = \ln G_1 + \ln G_2 - \ln G_3 + \ln x$$

Differentiating both sides,

$$\frac{dy}{y} = \frac{dG_1}{G_1} + \frac{dG_2}{G_2} - \frac{dG_3}{G_3} + \frac{dx}{x}$$

No error is specified in input so, $\frac{dx}{x} = 0$

$$\frac{dy}{y} = \epsilon_1 + \epsilon_2 - \epsilon_3$$

1.6 (a)

$R_1 = 50\ \Omega$, $R_2 = 65\ \Omega$, $R_3 = 100 \pm 5\%\ \Omega$

The value of R_3 with ±5% tolerance,

$$R_3 = 100 \pm 5\%$$

$$= 100 + 100 \times \frac{5}{100} = 105\ \Omega$$

or,

$$= 100 - 100 \times \frac{5}{100} = 95\ \Omega$$

In both conditions, the bridge is balanced so under balance condition,

$$\frac{R_1}{R_3} = \frac{R_2}{R_x}$$

$$R_x = \frac{R_2 R_3}{R_1}$$

(i) When, $R_3 = 105\ \Omega$

$$\therefore \quad R_x = \frac{105 \times 65}{50} = 136.50\ \Omega$$

(ii) When, $R_3 = 95\ \Omega$

$$\therefore \quad R_x = \frac{95 \times 65}{50} = 123.50\ \Omega$$

1.7 Sol.

$$V = 220 \pm 1\%,$$

$$I = 5 \pm 1\%$$

$$W = 555 \pm 2\%$$

$$W = VI \cdot \cos(\phi)$$

$$\Rightarrow \quad \text{p.f} = \cos(\phi) = \frac{W}{V.I}$$

$$\text{p.f.} = \cos(\phi)$$

$$= \frac{555 \pm 2\%}{(220 \pm 1\%)(5 \pm 1\%)} = \frac{555}{220 \times 5} \pm 4\%$$

$$\text{p.f.} = \cos(\phi) = 0.5 \pm 4\%$$

1.8 (a)

$$\sigma_{res} = \sqrt{\left(\frac{\partial R}{\partial R_1}\right)^2 \sigma_1^2 + \left(\frac{\partial R}{\partial R_2}\right)^2 \sigma_2^2}$$

$$= \sqrt{\left(\frac{\partial R}{\partial R_1}\right)^2 \Delta R_1^2 + \left(\frac{\partial R}{\partial R_2}\right)^2 \Delta R_2^2}$$

■■■■

2 Galvanometers, Voltmeters and Ammeters

2.1 Two 100 V full scale PMMC type dc voltmeter having figure of merits (FOM) of 10 kΩ/V and 20 kΩ/V are connected in series. The series combination can be used to measure a maximum DC voltage of _____ V. [1995 : 1 M]

2.2 The scale of a voltmeter is uniform. Its type is
(a) moving iron
(b) induction
(c) moving coil permanent magnet
(d) moving coil dynamometer [1996 : 1 M]

2.3 An advantage of permanent magnet moving coil instrument is that is
(a) free from friction error.
(b) has high torque/weight ratio (of the moving parts).
(c) has low torque/weight ratio (of the moving parts).
(d) can be used on both AC and DC.
[1996 : 1 M]

2.4 For a given frequency, the deflecting torque of an induction ammeter is directly proportional to
(a) current^2 (b) current^3
(c) $\sqrt{\text{current}}$ (d) current [1996 : 1 M]

2.5 A DC voltmeter has a sensitivity of 1000 Ω/volt. When it measures half of full scale in 100 V range, the current through the voltmeter is
(a) 100 mA (b) 1 mA
(c) 0.5 mA (d) 50 mA [1998 : 1 M]

2.6 Two 100 μA full scale PMMC meters are employed to construct a 10 V and a 100 V full scale voltmeter. These meters will have figure of merits (sensitivities) as
(a) 10 kΩ/V and 10 kΩ/V
(b) 100 kΩ/V and 10 kΩ/V
(c) 10 kΩ/V and 100 kΩ/V
(d) 10 kΩ/V and 1 kΩ/V [1999 : 2 M]

2.7 A 100 μA ammeter has an internal resistance of 100 Ω. For extending its range to measure 500 μA, the shunt resistance required is of
(a) 20.0 Ω (b) 22.22 Ω
(c) 25.0 Ω (d) 50.0 Ω
[2001 : 1 M]

2.8 A Manganin swap resistance is connected in series with a moving coil ammeter consisting of a milli-ammeter and a suitable shunt in order to
(a) minimise the effect of temperature variation
(b) obtain large deflecting torque
(c) reduce the size of the meter
(d) minimise the effect of stray magnetic fields
[2003 : 1 M]

2.9 A rectifier type AC voltmeter consists of a series resistance R_s, an ideal full-wave rectifier bridge and a PMMC instrument as shown in the figure. The internal resistance of the instrument is 100 Ω and a full scale deflection is produced by a DC current of 1 mA. The value of R_s required to obtain full scale deflection with an AC voltage of 100 V (rms) applied to the input terminals is

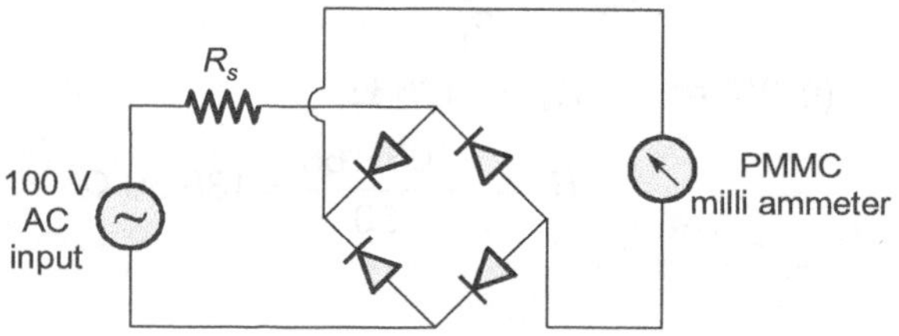

(a) 63.56 Ω (b) 69.93 Ω
(c) 89.93 kΩ (d) 141.3 kΩ [2003 : 2 M]

2.10 The inductance of a certain moving-iron ammeter is expressed as:
$L = 10 + 3\theta - (\theta^2/4)$ μH, where θ is the deflection in radians from the zero position. The control spring torque in 25×10^{-6} Nm/radian. The deflection of the pointer in radian when the meter carries a current of 5 A, is
(a) 2.4 (b) 2.0
(c) 1.2 (d) 1.0 [2003 : 2 M]

2.11 The effect of stray magnetic fields on the actuating torque of a portable instrument is maximum when the operating field of the instrument and the stray fields are
(a) perpendicular (b) parallel
(c) inclined at 60° (d) inclined at 30°
[2003 : 1 M]

2.12 A galvanometer with a full scale current of 10 mA has a resistance of 1000 Ω. The multiplying power

(the ratio of measured current to galvanometer current) of 100 Ω shunt with this galvanometer is
(a) 110 (b) 100
(c) 11 (d) 10 [2004 : 1 M]

2.13 A moving coil of a meter has 100 turns, and a length and depth of 10 mm and 20 mm respectively. It is positioned in a uniform radial flux density of 200 mT. The coil carries a current of 50 mA. The torque on the coil is
(a) 200 μNm (b) 100 μNm
(c) 2 μNm (d) 1 μNm [2004 : 2 M]

2.14 A moving iron ammeter produces a full scale torque of 240 μNm with a deflection of 120° at a current of 10 A. The rate of change of self inductance (μH/radian) of the instrument at full scale is
(a) 2.0 μH/radian (b) 4.8 μH/radian
(c) 12.0 μH/radian (d) 114.6 μH/radian
[2004 : 2 M]

2.15 A PMMC voltmeter is connected across a series combination of a DC voltage source V_1 = 2 V and and AC voltage source $V_2(t)$ = 3 sin(4t) V. The meter reads
(a) 2 V (b) 5 V
(c) $(2+\sqrt{3}/2)$ V (d) $(\sqrt{17}/2)$ V [2005 : 1 M]

2.16 A DC ammeter has a resistance of 0.1 Ω and its current range is 0-100 A. If the range is to be extended to 0-500 A, then meter requires the following shunt resistance
(a) 0.010 Ω (b) 0.011 Ω
(c) 0.025 Ω (d) 1.0 Ω [2005 : 2 M]

2.17 A 1000 V DC supply has two 1-core cables as its positive and negative leads: their insulation resistances to earth are 4 MΩ and 6 MΩ, respectively, as shown in the figure. A voltmeter with resistance 50 kΩ is used to measure the insulation of the cable. When connected between the positive core and earth, then voltmeter reads

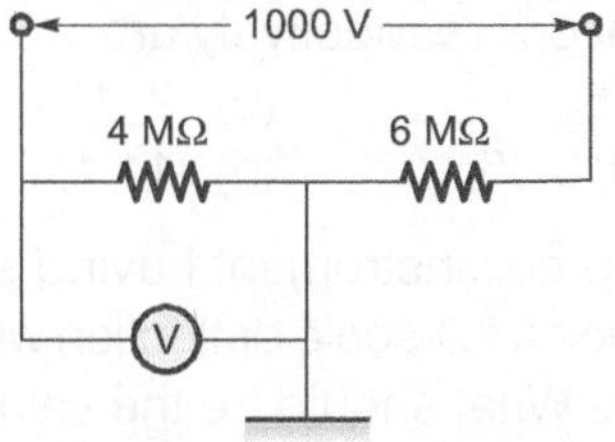

(a) 8 V (b) 16 V
(c) 24 V (d) 40 V [2005 : 2 M]

2.18 A current of $-8+6\sqrt{2}$ sin (ωt + 30°) A is passed through three meters. They are a centre zero PMMC meter, a true rms meter and a moving iron instrument. The respective readings (in A) will be
(a) 8, 6, 10 (b) 8, 6, 8
(c) –8, 10, 10 (d) –8, 2, 2 [2006 : 2 M]

2.19 An ammeter has a current range of 0 – 5 A, and its internal resistance is 0.2 Ω. In order to change the range to 0 – 25 A, we need to add a resistance of
(a) 0.8 Ω in series with meter
(b) 1.0 Ω in series with meter
(c) 0.04 Ω in parallel with meter
(d) 0.05 Ω in parallel with meter [2010 : 1 M]

2.20 A periodic voltage waveform observed on an oscilloscope across a load is shown. A permanent magnet moving coil (PMMC) meter connected across the same load reads

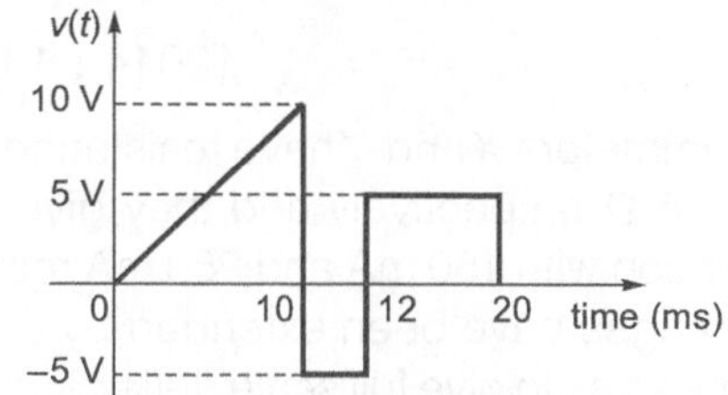

(a) 4 V (b) 5 V
(c) 8 V (d) 10 V [2012 : 1 M]

2.21 An analog voltmeter uses external multiplier settings. With a multiplier setting of 20 kΩ, it reads 440 V and with a multiplier setting of 80 kΩ, it reads 352 V. For a multiplier setting of 40 kΩ, the voltmeter reads
(a) 371 V (b) 383 A
(c) 394 V (d) 406 V [2012 : 2 M]

2.22 The input impedance of the permanent magnet moving coil (PMMC) voltmeter is infinite. Assuming that the diode shown in the figure below is ideal, the reading of the voltmeter in Volts is

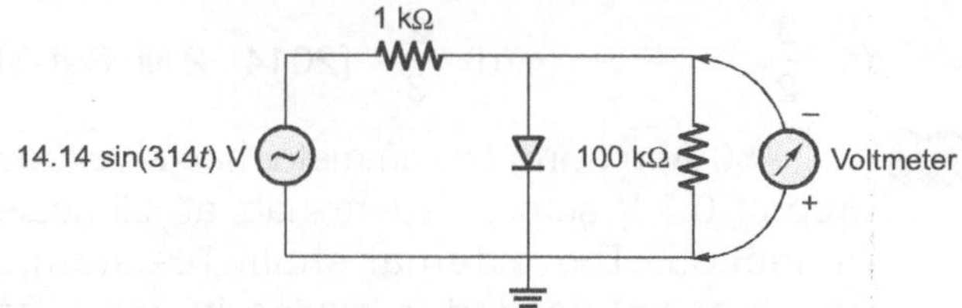

(a) 4.46 (b) 3.15
(c) 2.23 (d) 0 [2013 : 1 M]

2.23 The DC current flowing a circuit is measured by two ammeters, one PMMC and another electrodynamometer type, connected in series. The PMMC meter contains 100 turns in the coil, the flux density in the air gap is 0.2 Wb/m^2, and the area of this coil is 80 mm^2. The electrodynamometer ammeter has a change in mutual inductance with respect to deflection of 0.5 mH/rad. The spring constants of both the meters are equal. The value of current, at which the deflections of the two meters are same, is _____. [2014 : 2 M, Set-1]

2.24 The saw-tooth voltage waveform shown in the figure is fed to a moving iron voltmeter. Its reading would be close to _______.

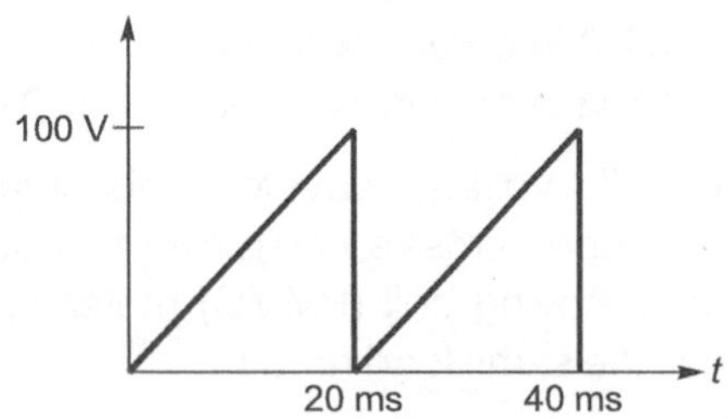

[2014 : 1 M, Set-2]

2.25 Two ammeters X and Y have resistances of 1.2 Ω and 1.5 Ω respectively and they give full scale deflection with 150 mA and 250 mA respectively. The ranges have been extended by connecting shunts so as to give full scale deflection with 15 A. The ammeters along with shunts are connected in parallel and then placed in a circuit in which the total current flowing is 15 A. The current in amperes indicated in ammeter X is _____. [2014 : 2 M, Set-2]

2.26 A periodic waveform observed across a load is represented by

$$V(t) = \begin{cases} 1+\sin\omega t & 0 \le \omega t < 6\pi \\ -1+\sin\omega t & 6\pi \le \omega t < 12\pi \end{cases}$$

The measured value, using moving iron voltmeter connected across the load, is

(a) $\sqrt{\dfrac{3}{2}}$ (b) $\sqrt{\dfrac{2}{3}}$

(c) $\dfrac{3}{2}$ (d) $\dfrac{2}{3}$ [2014 : 2 M, Set-3]

2.27 A (0 - 50 A) moving coil ammeter has a voltage drop of 0.1 V across its terminals at full scale deflection. The external shunt resistance (in milliohms) needed to extend its range to (0 - 500 A) is _____. [2015 : 1 M, Set-1]

2.28 Match **List-I (Instrument Type)** with **List-II (Used for)** the following:

List-I
A. Permanent magnet moving coil
B. Moving Iron connected through current transformer
C. Rectifier
D. Electrodynamometer

List-II
1. DC only
2. AC only
3. AC and DC

Codes:

	A	B	C	D
(a)	1	2	1	3
(b)	1	3	1	2
(c)	1	2	3	3
(d)	3	1	2	1

[2015 : 1 M, Set-2]

2.29 A capacitive voltage divider is used to measure the bus voltage V_{bus} in a high-voltage 50 Hz AC system as shown in the figure. The measurement capacitors C_1 and C_2 have tolerances of ±10% on their nominal capacitance values. If the bus voltage V_{bus} is 100 kV rms, the maximum rms output voltage V_{out} (in kV), considering the capacitor tolerance, is ________.

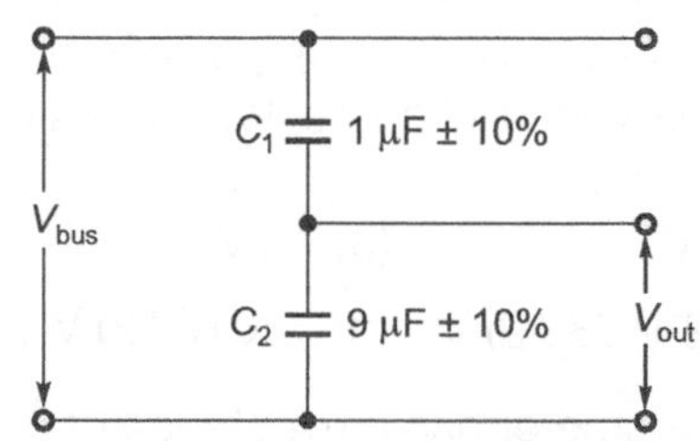

[2015 : 1 M, Set-2]

2.30 A DC voltage with ripple is given by $v(t) = [100 + 10 \sin(\omega t) - 5 \sin(3\omega)t)]$ volts. Measurements of this voltage $v(t)$, made by moving-coil and moving-iron voltmeters, show readings of V_1 and V_2 respectively. The value of $V_2 - V_1$, in volts, is _____. [2016 : 2 M, Set-1]

2.31 A 0 - 1 ampere moving iron ammeter has an internal resistance of 50 mΩ and inductance of 0.1 mH. A shunt coil is connected to extend its range to 0 - 10 A for all operating frequencies. The time constant in milliseconds and resistance in mΩ of the shunt coil respectively are

(a) 2, 5.55 (b) 2, 1
(c) 2.18, 0.55 (d) 11.1, 2 [2018 : 2 M]

2.32 A moving coil instrument having a resistance of 10 Ω, gives a full-scale deflection when the current is 10 mA. What should be the value of the series resistance, so that it can be used as a voltmeter for measuring potential difference up to 100 V?

(a) 9990 Ω (b) 990 Ω
(c) 9 Ω (d) 99 Ω

[2019 : 2 M]

2.33 Two balanced three-phase loads, as shown in the figure, are connected to a $100\sqrt{3}$ V, three-phase, 50 Hz main supply. Given $Z_1 = (18 + j24)\ \Omega$ and $Z_2 = (6 + j8)\ \Omega$. The ammeter reading, in amperes, is _____. (round off to nearest integer).

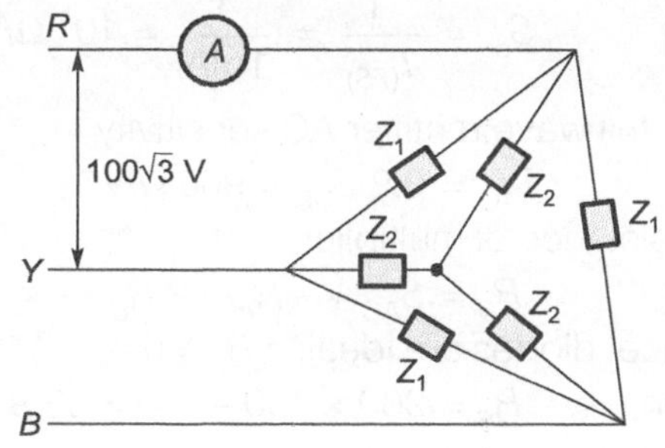

[2022 : 1 M]

Answers Galvanometers, Voltmeters and Ammeters

2.1 (150)	2.2 (c)	2.3 (b)	2.4 (a)	2.5 (c)	2.6 (a)	2.7 (c)
2.8 (a)	2.9 (c)	2.10 (c)	2.11 (b)	2.12 (c)	2.13 (a)	2.14 (b)
2.15 (a)	2.16 (c)	2.17 (a)	2.18 (c)	2.19 (d)	2.20 (a)	2.21 (d)
2.22 (a)	2.23 (3.2)	2.24 (57.73)	2.25 (10.28)	2.26 (a)	2.27 (0.22)	2.28 (c)
2.29 (11.95)	2.30 (0.312)	2.31 (a)	2.32 (a)	2.33 (20)		

Explanations Galvanometers, Voltmeters and Ammeters

2.1 Sol.

$$(I_{fsd})_1 = \frac{1}{10\,\text{k}\Omega/\text{V}} = 0.05\ \text{mA}$$

$$(I_{fsd})_2 = \frac{1}{20\,\text{k}\Omega/\text{V}} = 0.05\ \text{mA}$$

In series application, maximum current will be 0.05 mA

Resistance of M_1 = 10 kΩ/V × 100 V = 1 MΩ

Resistance of M_2 = 20 kΩ/V × 100 V = 2 MΩ

Hence, the maximum voltage rating

$$= I_{max} \cdot (R_{M_1} + R_{M_2})$$

$$= 0.05\ \text{mA} \times (1 + 2)\ \text{M}\Omega = 150\ \text{V}$$

2.2 (c)

In PMMC, $\theta \propto I_{av}$

2.4 (a)

In induction type meter, rms value is obtained.

2.5 (c)

$$s = \frac{1}{I_{fsd}} \Rightarrow I_{fsd} = \frac{1}{1\text{k}\Omega/\text{V}} = 1\text{mA}$$

For half scale, I_{hsd} = 0.5 mA

2.6 (a)

$$\text{Sensitivity} = \frac{1}{I_{fsd}} = \frac{1}{100 \times 10^{-6}} = 10\ \text{k}\Omega/\text{V for}$$

each

2.7 (c)

$$R_{sh} = \frac{R_a}{m-1} = \frac{100}{\frac{500}{100} - 1} = 25.0\ \Omega$$

2.8 (a)

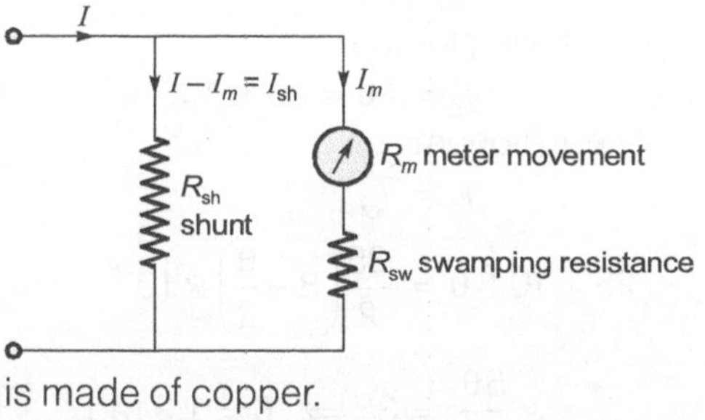

- Coil is made of copper.
  A swamping resistance (R_{sw}) of manganin (which has a negligible temperature coefficient) having a resistance 20 to 30 times the coil resistance is connected in series with the coil and a shunt of manganin is connected across this combination.
- Since copper forms a small fraction of the series combination, the proportion in which the currents would divide between the meter and the shunt would not change appreciably with the change in temperature.

2.9 (c)

$I_{(FS)}$ = Current required to produce full scale deflection

DC sensitivity is,

$$S_{dc} = \frac{1}{I_{(FS)}} = \frac{1}{1\,\text{mA}} = 10^3\ \Omega/\text{V}$$

For full wave rectifier AC sensitivity

$$S_{AC} = 0.9\ S_{DC} = 900\ \Omega/\text{V}$$

Resistance of multiplier,

$$R_S = S_{AC}\ V - R_m - 2\ R_d$$

Since, diodes are ideal, $R_d = 0$

then, $R_S = 900 \times 100 - 100 = 89.9\ \text{k}\Omega$

2.10 (c)

Deflecting torque in moving-iron ammeter,

$$T_d = \frac{1}{2} I^2 \frac{dL}{d\theta}$$

Inductance, $L = \left(10 + 3\theta - \frac{\theta^2}{4}\right)\ \mu\text{H}$

Rate of change of inductance with deflection,

$$\frac{dL}{d\theta} = \frac{d}{d\theta}\left(10 + 3\theta - \frac{\theta^2}{4}\right) = \left(3 - \frac{\theta}{2}\right)\ \mu\text{H/rad}$$

Current, $I = 5$ A

Deflecting torque,

$$T_d = \frac{1}{2} I^2 \frac{dL}{d\theta}$$

$$T_d = \frac{1}{2} \times 5^2 \times \left(3 - \frac{\theta}{2}\right) \times 10^{-6}$$

$$= \frac{25}{2}\left(3 - \frac{\theta}{2}\right) \times 10^{-6}\text{Nm}$$

Controlling torque,

$$T_c = k\theta = 25 \times 10^{-6}\ \theta$$

At equilibrium,

$$T_c = T_d$$

$$25 \times 10^{-6}\ \theta = \frac{25}{2}\left(3 - \frac{\theta}{2}\right) \times 10^{-6}$$

$$\frac{5\theta}{2} = 3 \Rightarrow \theta = 1.2\ \text{rad}$$

2.11 (b)

Due to stray magnetic field, torque is also produced which can affect the torque produced due to operating field. If both stray magnetic and operating field are parallel, torques due to both field become additive.

2.12 (c)

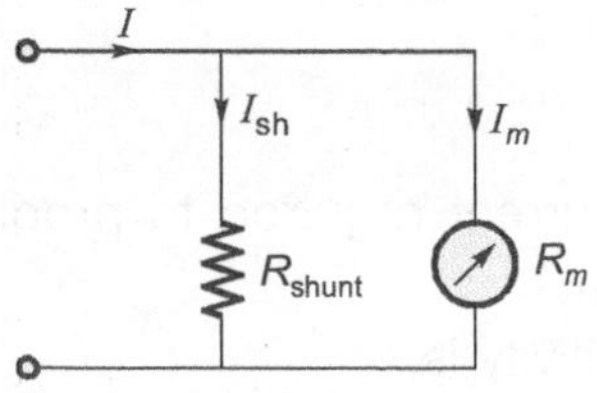

Full scale current of galvanometer

$$I_m = 10\ \text{mA}$$

Resistance of meter

$$R_m = 1000\ \Omega$$

Resistance of shunt

$$R_{sh} = 100\ \Omega$$

$$I_m = \frac{IR_{sh}}{R_{sh} + R_m}$$

Multiplying power

$$= \frac{I}{I_m} = \frac{R_{sh} + R_m}{R_m}$$

$$= \frac{100 + 1000}{100} = 11$$

2.13 (a)

T = Torque on the coil = NBAI

where,

N = Number of turns = 100

B = Flux density = 200 mT = 200×10^{-3} T

A = Area of the coil = length × depth

$= (10 \times 10^{-3}) \times (20 \times 10^{-3}) = 200 \times 10^{-6}\ \text{m}^2$

I = Current through coil = 50 mA = 50×10^{-3}

T = NBAI = $100 \times (200 \times 10^{-3}) \times (200 \times 10^{-6})$

$\times (50 \times 10^{-3}) = 2 \times 10^{-4}$ N-m = 200 μN-m

2.14 (b)

Torque produced,

$$T = \frac{1}{2} I^2 \frac{dL}{d\theta}$$

where, $I = 10$ A

and $T = 240\ \mu\text{N-m} = 240 \times 10^{-6}$ N-m

$$T = \frac{1}{2} I^2 \frac{dL}{d\theta}$$

$$240 \times 10^{-6} = \frac{1}{2} \times 10^2 \times \frac{dL}{d\theta}$$

Rate of change of self-inductance

$$= \frac{dL}{d\theta} = 4.8 \times 10^{-6}\ \text{H/radian}$$

$$\Rightarrow \frac{dL}{d\theta} = 4.8\ \mu\text{H/radian}$$

2.15 (a)

Total voltage across PMMC

$$= V_T = V_1 + V_2 = 2 + 3\sin(4t)\ \text{V}$$

PMMC reads average value

Average value of $V_1 = 2$ V

Average value of $V_2 = 0$

Average value of $V_T = 2$ V

So PMMC reads = 2 V

2.16 (c)

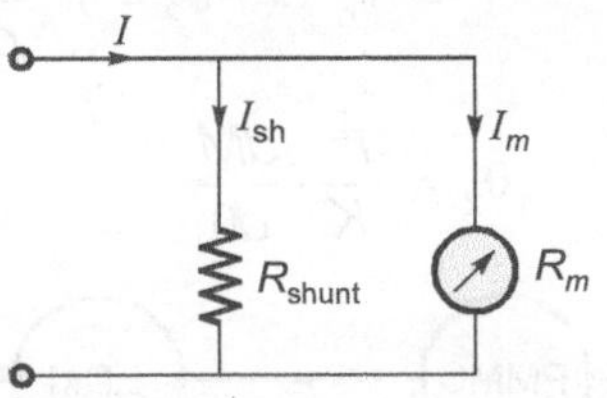

I_m = Full scale deflection current = 100 A

I = Current to be measured = 500 A

Multiplying power,

$$m = \frac{I}{I_m} = \frac{500}{100} = 5$$

$$R_m = 0.1\ \Omega$$

$$I_m = \frac{I R_{sh}}{R_{sh} + R_m} \Rightarrow m = 1 + \frac{R_m}{R_{sh}}$$

$$\Rightarrow \quad R_{sh} = \frac{R_m}{m-1} = \frac{0.1}{5-1} = 0.025\ \Omega$$

2.17 (a)

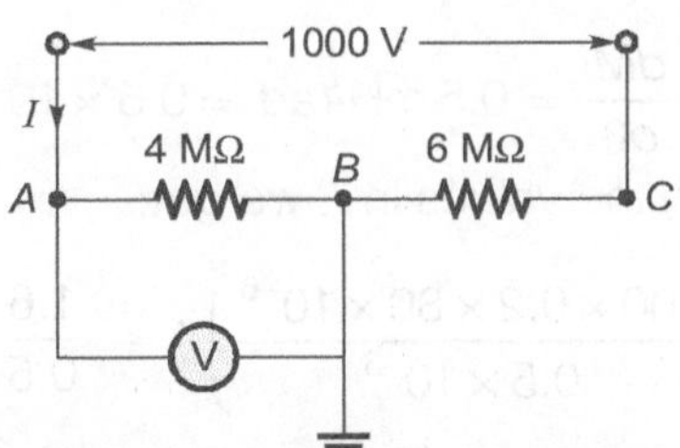

Resistance of voltmeter (R_v) appears parallel to 4 MΩ

Effective resistance between A and B,

$$R_{AB} = R_v \| 4\ \text{M}\Omega$$

$$R_v = 50\ \text{k}\Omega = 0.05\ \text{M}\Omega$$

$$R_{AB} = 0.05 \| 4\ \text{M}\Omega = 0.05\ \text{M}\Omega$$

$$R_{AC} = R_{AB} + R_{BC} = 0.05 + 6 = 6.05\ \text{M}\Omega$$

$$I = \frac{1000}{R_{AC}} = \frac{1000}{6.05}\ \mu\text{A}$$

Voltmeter reads,

$$IR_{AB} = \frac{1000}{6.05} \times 10^{-6} \times 0.05 \times 10^{6} \approx 8\ \text{V}$$

2.18 (c)

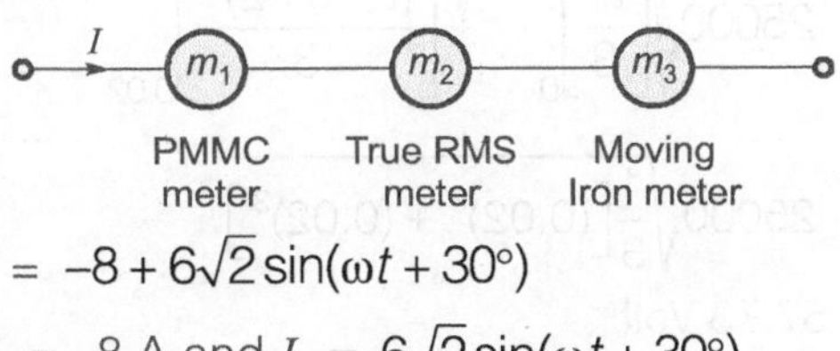

$$I = -8 + 6\sqrt{2}\sin(\omega t + 30°)$$

$$I_1 = -8\ \text{A and } I_2 = 6\sqrt{2}\sin(\omega t + 30°)$$

Average value of I_1 = –8 A

Average value of I_2 = 0

So, average value of I = –8 A

PMMC reads only average value of current

Therefore PMMC reads = –8 A

(Since it is centre zero type)

$$\text{RMS value of } I = \sqrt{(-8)^2 + \left(\frac{6\sqrt{2}}{\sqrt{2}}\right)^2} = 10\ \text{A}$$

RMS meter and moving iron meter both reads rms value of the current.

So, both m_2 and m_3 read 10 A.

2.19 (d)

To extend the range of the ammeter, a resistance R_{sh} is connected across the meter.

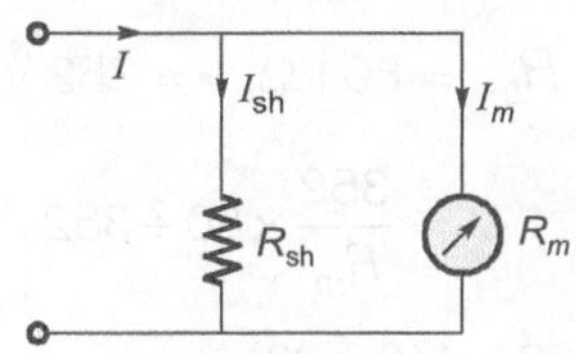

I_m = full scale deflection current = 5 A

I = 25 A

Multiplying power,

$$m = \frac{I}{I_m} = \frac{25}{5} = 5$$

$$I_m = \frac{IR_{sh}}{R_{sh} + R_m}$$

$$\Rightarrow \quad \frac{I}{I_m} = \frac{R_{sh} + R_m}{R_{sh}}$$

$$m = 1 + \frac{R_m}{R_{sh}}$$

$$R_{sh} = \frac{R_m}{m-1} = \frac{0.2}{5-1} = 0.05$$

2.20 (a)

As PMMC meter reads only DC value or average value and average value is equal to

$$V_{avg} = \frac{\text{Area under the curve}}{\text{Total time}}$$

$$V_{avg} = \frac{\left(\frac{1}{2} \times 10 \times 10\right) + (-5 \times 2) + (5 \times 8)}{20}$$

$$= \frac{80}{20} = 4\ \text{V}$$

2.21 (d)

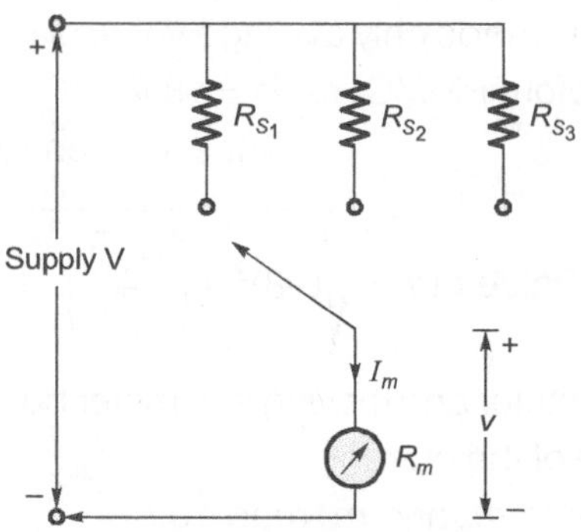

When, $R_{S_1} = 20\ k\Omega,\quad v = 440\ V$

$$V = \frac{440}{R_m} \times 20 + 440 \quad ...(i)$$

When, $R_{S_2} = 80\ k\Omega,\ v = 352\ V$

$$V = \frac{352}{R_m} \times 80 + 352 \quad ...(ii)$$

$$\Rightarrow \frac{352 \times 80 - 440 \times 20}{88} = R_m$$

$$R_m = 220\ k\Omega,\ V = 480\ V$$

When, $R_{S_3} = 40\ k\Omega$ and $V = 480\ V$

$$480 = \frac{v}{R_m} \times 40 + v$$

$$480 = \frac{13}{11} v \Rightarrow v = 406.15\ V$$

2.22 (a)

In the half cycle, D is ON

$\Rightarrow \quad V_0 = 0\ V$

In negative half cycle, D is OFF, PMMC voltmeter measures average value of V_0.

In case of half wave rectification,

$$V_{0\ avg.} = \frac{14.14}{\pi} \times \frac{100}{101} = 4.456\ V$$

2.23 Sol.

For a PMMC ammeter, at equilibrium, we have: controlling torque (T_c) = Deflecting torque (T_d)

or $\quad K\theta_1 = GI$ or $\theta_1 = \dfrac{GI}{K}$

where, $\quad G = nBA$

$$\therefore \quad \theta_1 = \frac{nBAI}{K} \quad ...(i)$$

For an electrodynamometer ammeter, at equilibrium,

Controlling torque = Deflecting torque

i.e., $\quad T_c = T_d$ or $K\theta_2 = I^2 \dfrac{dM}{d\theta}$

$$\text{or,} \quad \theta_2 = \frac{I^2}{K} \cdot \frac{dM}{d\theta} \quad ...(ii)$$

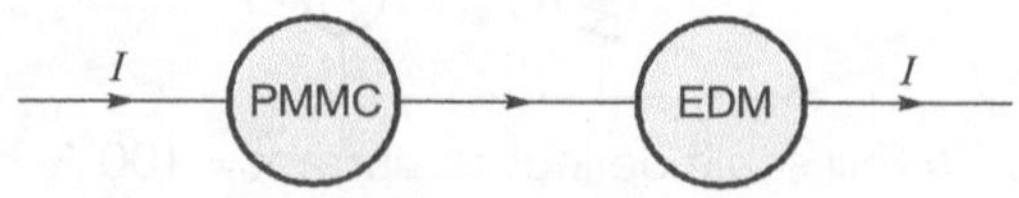

For the deflection in the two meters to be same,

$$\theta_1 = \theta_2$$

$$\text{or,} \quad \frac{nBAI}{K} = \frac{I^2}{K} \cdot \frac{dM}{d\theta}$$

$$\text{or,} \quad nBA = I\frac{dM}{d\theta} \text{ or } I = \frac{nBA}{\left(\frac{dM}{d\theta}\right)}$$

Given, $\quad n$ = Number of turns = 100,

$B = 0.2\ Wb/m^2$,

$A = 80\ mm^2 = 80 \times 10^{-6}\ m^2$

and $\quad \dfrac{dM}{d\theta} = 0.5\ mH/rad. = 0.5 \times 10^{-3}\ H/rad.$

Putting the values in I, we get:

$$I = \left(\frac{100 \times 0.2 \times 80 \times 10^{-6}}{0.5 \times 10^{-3}}\right) A = \frac{1.6}{0.5} A = 3.2\ A$$

$\therefore$ Required value of current,

$$I = 3.2\ \text{Amp.}$$

2.24 Sol.

Moving iron voltmeter reads rms value of voltage and current. From the given waveform of voltage,

$$V(t) = \begin{cases} 5000t, & 0 < t < 0.02s \\ 5000(t - 0.02), & 0.02 < t < 0.04s \end{cases}$$

$$\therefore V_{rms} = \sqrt{\frac{1}{T}\int_0^T [V(t)]^2 dt}$$

$$= \sqrt{\frac{1}{0.04}\int_0^{0.02} (5000t)^2 dt + \int_{0.02}^{0.04} \left[5000(t-0.02)^2\right] dt}$$

$$= 25000\sqrt{\left[\frac{t^3}{3}\right]_0^{0.02} + \left[\frac{(t-0.02)^3}{3}\right]_{0.02}^{0.04}}$$

$$= 25000\sqrt{\frac{1}{3}\left[(0.02)^3 + (0.02)^3\right]}$$

= 57.73 Volt

$\therefore$ Voltage reading by moving iron voltmeter = 57.73 Volts

2.25 Sol.

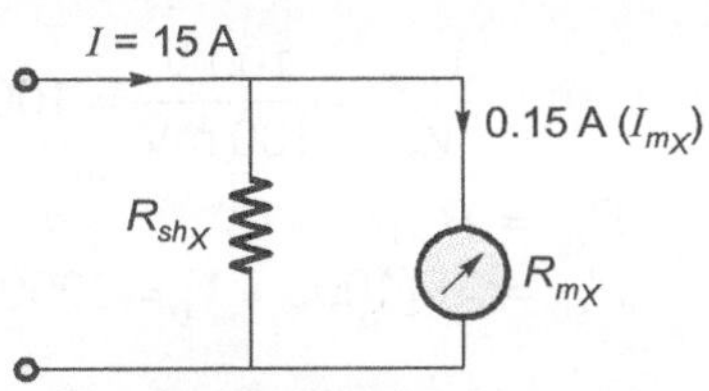

Ammeter-X

Given, $R_{mX} = 1.2\ \Omega,\ R_{mY} = 1.5\ \Omega$

$I_{mX} = 0.15$ A, $I_{mY} = 0.25$ A

and I = Full scale deflection current = 15 A

∴ Shunt multiplying factor for ammeter X is

$$m_X = \frac{I}{I_{mX}} = \frac{15}{0.15} = 100$$

and shunt multiplier resistance,

$$R_{shX} = \frac{R_{mX}}{(m_X - 1)} = \frac{1.2}{99} = 0.012\ \Omega$$

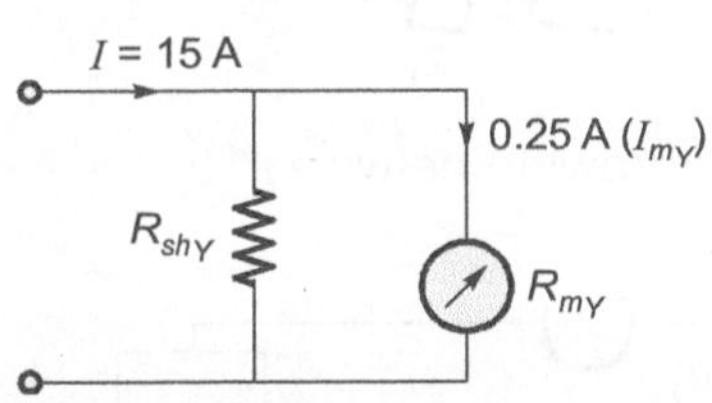

Also, shunt multiplying factor for ammeter Y is

$$m_Y = \frac{I}{I_{mY}} = \frac{15}{0.25} = 60$$

and shunt multiplier resistance,

$$R_{shY} = \frac{R_{mY}}{(m_Y - 1)} = \frac{1.5}{59} = 0.025\ \Omega$$

When these ammeters are connected in parallel as shown in the figure below:

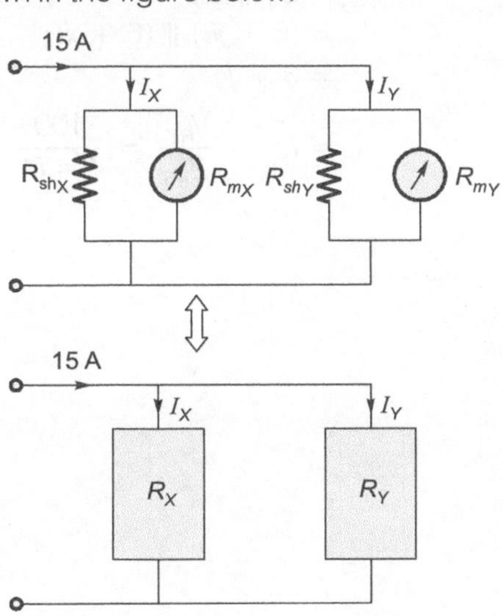

Let current in ammeter X be I_X then,

$$R_X = \left(\frac{R_{shX} \cdot R_{mX}}{R_{shX} + R_{mX}}\right) = \left(\frac{0.012 \times 1.2}{1.212}\right)$$

$= 0.011\ \Omega$ and

$$R_Y = \left(\frac{R_Y}{R_X + R_Y}\right) \times 15 = \frac{0.024}{0.035} \times 15 = 0.024\ \Omega$$

$$\therefore\ I_X = \left(\frac{0.024}{0.024 + 0.011}\right) \times 15$$

$$= \left(\frac{R_Y}{R_X + R_Y}\right) \times 15 = \frac{0.024}{0.035} \times 15 = 10.28 \text{ A}$$

∴ Current in ammeter X = 10.28 A

2.26 (a)

Since moving iron voltmeter reads rms value,

V_{rms}

$$= \sqrt{\frac{1}{2\pi}\left[\int_0^{6\pi}(1+\sin\omega t)^2 d\omega t + \int_{6\pi}^{12\pi}(-1+\sin\omega t)^2 d\omega t\right]}$$

$$= \sqrt{\frac{1}{2\pi}\left[\int_0^{6\pi}(1+\sin^2\omega t + 2\sin\omega t) d\omega t + \int_{6\pi}^{12\pi}(1+\sin^2\omega t - 2\sin\omega t) d\omega t\right]}$$

$$= \sqrt{\frac{1}{2\pi}\left[\int_0^{6\pi}\left\{1+\frac{(1-\cos 2\omega t)}{2} + 2\sin\omega t\right\} d\omega t + \int_{6\pi}^{12\pi}\left\{1+\frac{(1-\cos 2\omega t)}{2} - 2\sin\omega t\right\}\right]}$$

$$= \sqrt{\frac{1}{2\pi}\left[\int_0^{6\pi}\left(\frac{3}{2} - \frac{\cos 2\omega t}{2} + 2\sin\omega t\right) d\omega t + \int_{6\pi}^{12\pi}\left(\frac{3}{2} - \frac{\cos 2\omega t}{2} - 2\sin\omega t\right)\right]}$$

$$= \sqrt{\frac{1}{12\pi}\left[\frac{3}{2}(6\pi - 6) + \frac{3}{2}(12\pi - 6\pi)\right]} = \sqrt{\frac{3}{2}}$$

2.27 Sol.

$I_{fs} = 50$ A; $V_m = 0.1$ V

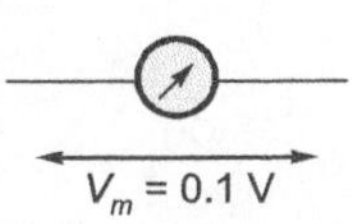

$$R_m = \frac{0.1}{50} = 2 \times 10^{-3}\ \Omega$$

$\because\ m = 10;$

$$R_{sh} = \frac{R_m}{(m-1)} = \frac{2 \times 10^{-3}}{9} = 0.22\ \text{m}\Omega$$

2.28 Sol.

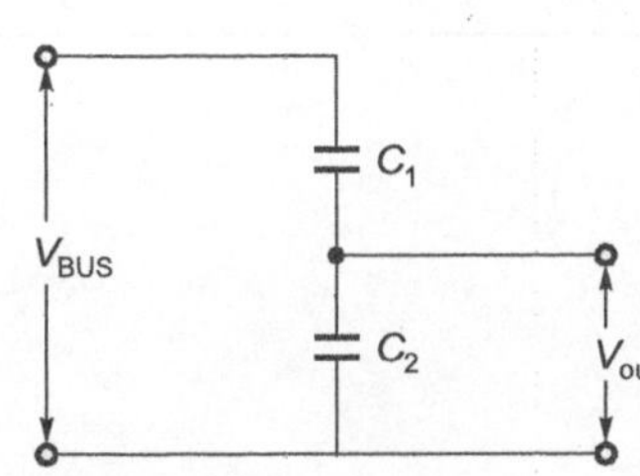

V_{BUS} is 100 kV_{rms}

$C_2 = 9\ \mu$F ± 10% and $C_1 = 1\ \mu$F ± 10%

To get maximum output voltage we need minimum C_2 and maximum C_1.

So, $C_2 = 8.1\ \mu F$ and $C_1 = 1.1\ \mu F$

so, $$V_{\text{out rms}} = \left(\frac{C_1}{C_1+C_2}\right) V_{\text{BUS}_{rms}} = 11.95 \text{ kV}$$

2.30 Sol.

$V(t) = 100 + 10 \sin(\omega t) - 5 \sin(3\omega t)$ volt

moving coil,

$$V_1 = V_{avg.} = 100 \text{ V}$$

Moving iron,

$$V_2 = V_{rms} = \sqrt{100^2 + \frac{1}{2}(10^2+5^2)} = 100.312$$

$$V_2 - V_1 = 0.312$$

2.31 (a)

Given, $I_m = 1$ A, $R_m = 50$ mΩ

$L_m = 0.1$ mH, $I = 10$ A

We know,

$$R_{sh} = \frac{R_m}{(m-1)};\quad m = \frac{I}{I_m}$$

Here, $$\frac{10}{1} = 10$$

$$\therefore R_{sh} = \frac{50\times10^{-3}}{10-1} = \frac{50\times10^{-3}}{9} = 5.55 \text{ m}\Omega$$

For all frequencies time constants of the shunt and meter arm should be equal.

i.e., $$\frac{\omega L_m}{R_m} = \frac{\omega L_{sh}}{R_{sh}}$$

or, $$\frac{L_m}{R_m} = \frac{L_{sh}}{R_{sh}}$$

or, $$\frac{L_m}{R_m} = \frac{0.1\times10^{-3}}{50\times10^{-3}} = 0.002 = 2 \text{ ms}$$

$\therefore$ Option (a) is correct.

2.32 (a)

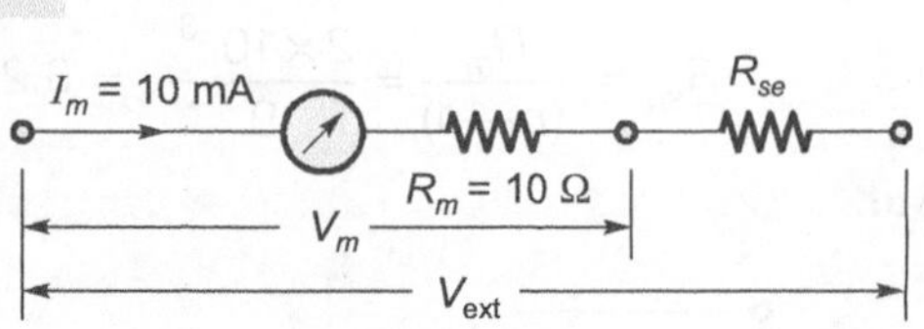

$V_m = I_m R_m = 10 \text{ mA} \times 10\ \Omega = 100$ mV

$(0 - 100 \text{ mV}) \Rightarrow (0 - 100 \text{ V})$

$$m = \frac{V_{ext}}{V_m} = \frac{100 \text{ V}}{100 \text{ mV}} = 1000$$

$$R_{se} = R_m[m-1]$$

$$R_{se} = 10[1000-1] = 9990\ \Omega$$

2.33 (20)

Convert Δ-Y_{Network}

$$Z'_1 = \frac{Z_1 \times Z_1}{Z_1+Z_1+Z_1} = \frac{Z_1}{3}$$

$$= \frac{18+j24}{3} = (6 + j8)\ \Omega$$

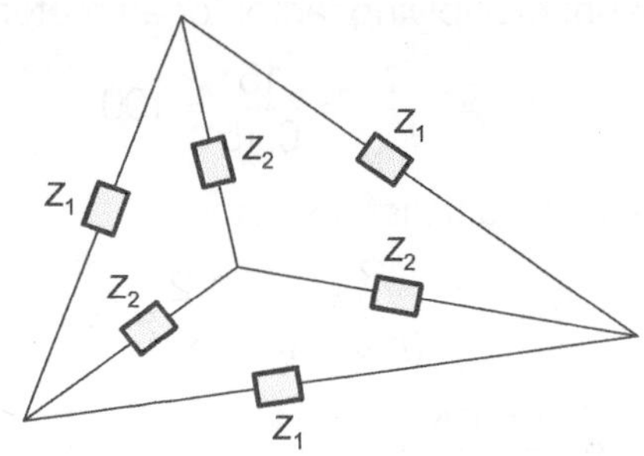

After redrawing network,

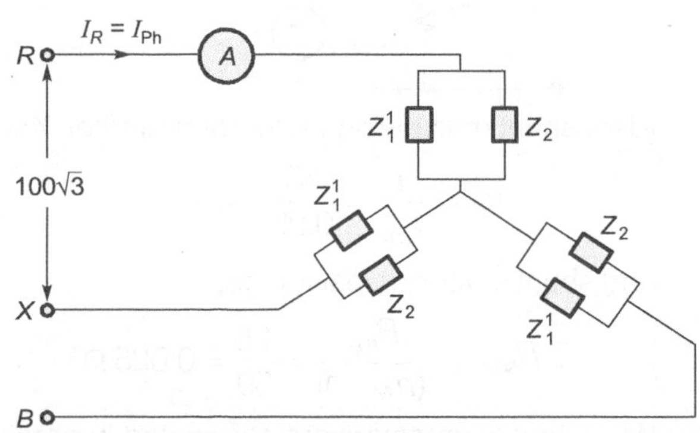

$$Z_{eq} = Z'_1 \parallel Z_2$$

$$= (6 + j8) \parallel (6 + j8)$$

$$= 3 + j4$$

$$I_R = I_{Ph} = \frac{V_{RY}}{Z_{eq}} = \frac{100}{3+j4}$$

$$= 20 \text{ A}$$

■■■■

3 Measurement of Resistance and Potentiometers

3.1 A Kelvin double bridge is best suited for the measurement of

(a) inductance (b) capacitance
(c) low resistance (d) high resistance

[1995 : 1 M]

3.2 A transfer instrument employed in the standardization of a polar type AC potentiometer is

(a) an electrostatic instrument
(b) a thermal instrument
(c) dynamometer instrument
(d) a moving coil instrument [1997 : 1 M]

3.3 A potentiometer is basically a

(a) deflection type instrument
(b) null type instrument
(c) deflection as well as null type instrument
(d) a digital instrument [1997 : 1 M]

3.4 The figure shows input attenuator of a multimeter. The meter reads full-scale with 12 V at M with the range switch at position B obtain full-scale deflection with the range switch position at D?

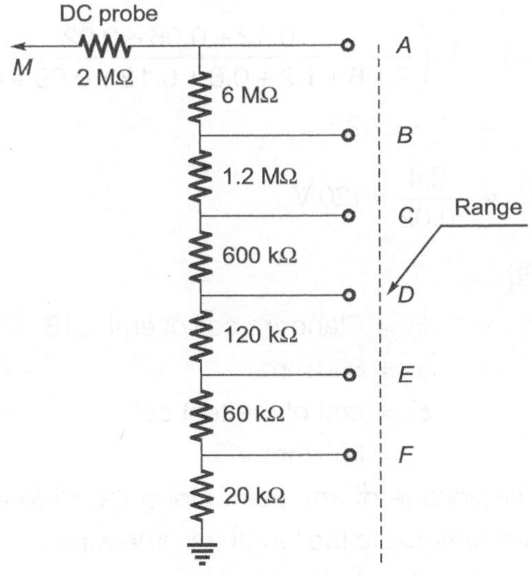

(a) 1 V (b) 150 V
(c) 120 V (d) 147 V [1998 : 2 M]

3.5 Kelvin double bridge is best suited for the measurement of

(a) resistance of very low value
(b) low value capacitance
(c) resistance of very high value
(d) high value capacitance [2002 : 1 M]

3.6 A DC potentiometer is designed to measure up to about 2 V with a slide wire of 800 mm. A standard cell of emf 1.18 V obtains balance at 600 mm. A test cell is seen to obtain balance at 680 mm. The emf of the test cell is

(a) 1.00 V (b) 1.34 V
(c) 1.50 V (d) 1.70 V [2004 : 1 M]

3.7 The set-up in the figure is used to measure resistance R. The ammeter and voltmeter resistances are 0.01 Ω and 2000 Ω, respectively. Their readings are 2 A and 180 V, respectively, giving a measured resistance of 90 Ω. The percentage error in the measurement is

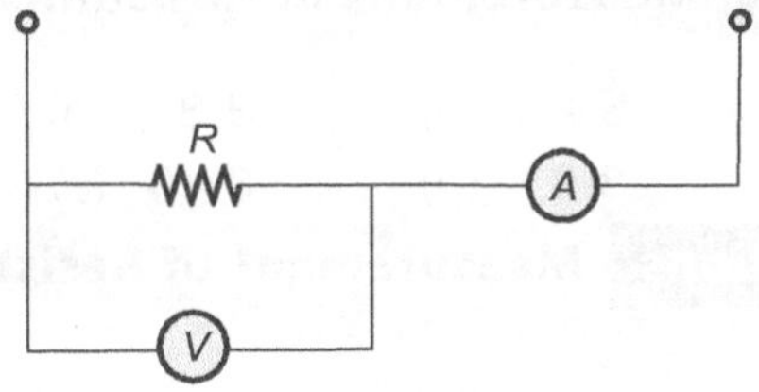

(a) 2.25 % (b) 2.35%
(c) 4.5% (d) 4.71% [2005 : 2 M]

3.8 R_1 and R_4 are the opposite arms of a Wheatstone bridge as are R_3 and R_2. The source voltage is applied across R_1 and R_3. Under balanced conditions which one of the following is true?

(a) $R_1 = R_3 R_4/R_2$ (b) $R_1 = R_2 R_3/R_4$
(c) $R_1 = R_2 R_4/R_3$ (d) $R_1 = R_2 + R_3 + R_4$

[2006 : 2 M]

3.9 Suppose that resistors R_1 and R_2 are connected in parallel to give an equivalent resistor R. If resistors R_1 and R_2 have tolerance of 1% each, the equivalent resistor R for resistors R_1 = 300 Ω and R_2 = 200 Ω will have tolerance of

(a) 0.5% (b) 1%
(c) 1.2% (d) 2%

[2014 : 2 M, Set-2]

3.10 An unbalanced DC Wheatstone bridge is shown in the figure. At what value of p will the magnitude of V_0 be maximum?

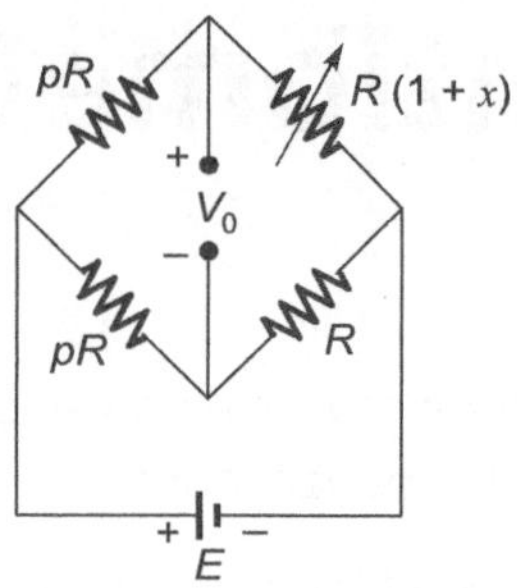

(a) $\sqrt{(1+x)}$ (b) $(1+x)$

(c) $\dfrac{1}{\sqrt{(1+x)}}$ (d) $\sqrt{(1-x)}$

[2015 : 2 M, Set-1]

3.11 A strain gauge forms one arm of the bridge shown in the figure below and has a nominal resistance without any load as $R_s = 300\ \Omega$. Other bridge resistances are $R_1 = R_2 = R_3 = 300\ \Omega$. The maximum permissible current through the strain gauge is 20 mA. During certain measurement when the bridge is excited by maximum permissible voltage and the strain gauge resistance is increased by 1% over the nominal value, the output voltage V_0 in mV is

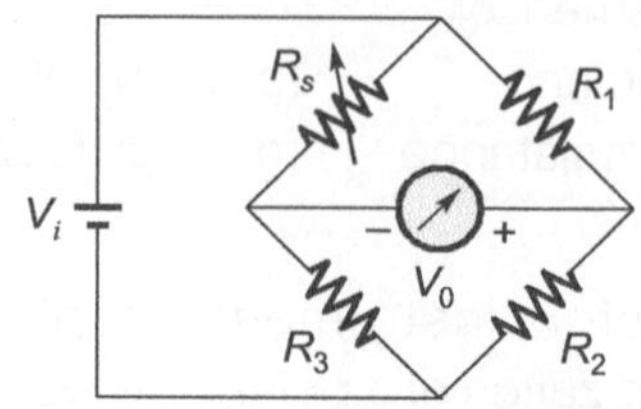

(a) 56.02 (b) 40.83
(c) 29.85 (d) 10.02 [2013 : 2 M]

Answers Measurement of Resistance and Potentiometers

3.1 (c)	3.2 (b)	3.3 (b)	3.4 (c)	3.5 (a)	3.6 (b)	3.7 (c)
3.8 (b)	3.9 (b)	3.10 (a)	3.11 (c)			

Explanations Measurement of Resistance and Potentiometers

3.1 (c)

It is used for the measurement of low resistance.

3.4 (c)

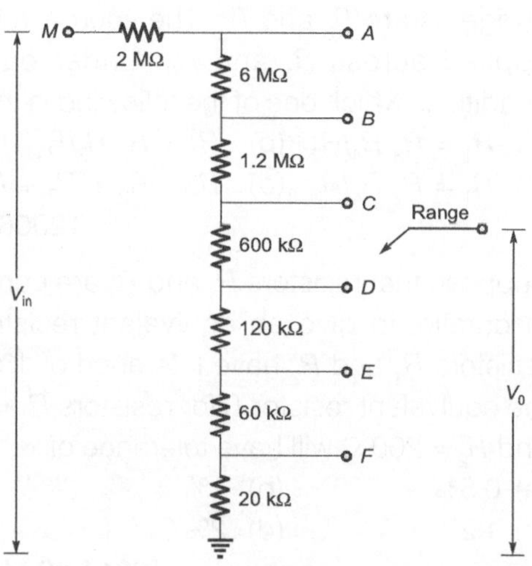

At B

$$V_0 = V_s\left(\frac{1.2+0.6+0.12+0.06+0.02}{2+6+1.2+0.6+0.12+0.06+0.02}\right)$$

$$= 12 \times 0.2 = 2.4\text{ V}$$

At D

$$2.4 = V_s\left(\frac{0.12+0.06+0.02}{2+6+1.2+0.6+0.12+0.06+0.02}\right)$$

$$= V_s \times 0.02$$

$$V_s = \frac{2.4}{0.02} = 120\text{ V}$$

3.6 (b)

E_1 = Standard call of emf 1.18
l_1 = 600 mm
E_2 = emf of the test cell
l_2 = 680 mm

The voltage of any point along the slide wire is proportional to the length of slide wire.

$$E \propto l$$

$$\frac{E_1}{E_2} = \frac{l_1}{l_2}$$

$$E_2 = \frac{l_2}{l_1}E_1 = \frac{680}{600}\times 1.18 \approx 1.34\text{ V}$$

3.7 (c)

Measured value of resistance = 90 Ω
Resistance of voltmeter, $R_V = 2000\ \Omega$
Voltage across voltmeter, $V = 180$

$$\text{Current through voltmeter} = \frac{V}{R_V}$$

$$= \frac{180}{2000} = 0.09\ \text{A}$$

Current through resistance,

$$(R) = I_R = 2 - I_v = 2 - 0.09$$

$$\Rightarrow \quad I_R = 1.91\ \text{A}$$

True value of resistance,

$$\frac{V}{I_R} = \frac{180}{1.91} = 94.24\,\Omega$$

$$\%\ \text{Error} = \frac{\text{Measured value} - \text{True value}}{\text{True value}} \times 100$$

$$= \frac{90 - 94.24}{94.24} \times 100 = -4.5\%$$

3.8 (b)

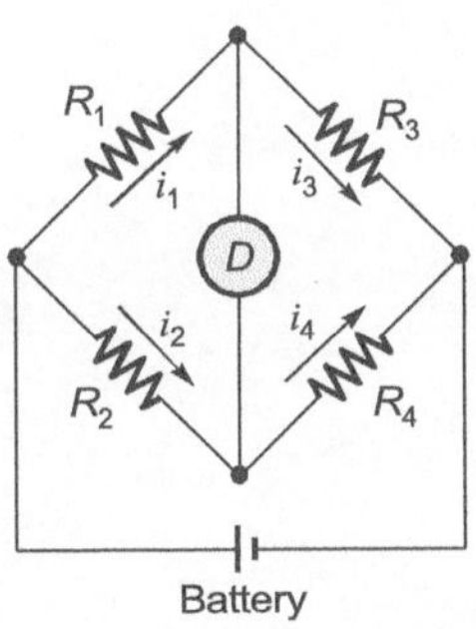

Adjustments are made in various arms of the bridge so that the voltage across the detector is zero and hence no current flows through it, when no current flows through detector the bridge is said to be balanced.

$$\text{Under conditions of balance} = R_1 = R_2 \frac{R_3}{R_4}$$

3.9 (b)

Given, $R_1 = 300\ \Omega$, $R_2 = 200\ \Omega$
Equivalent resistance of parallel combination of R_1 and R_2 is

$$R = \frac{R_1 R_2}{R_1 + R_2} = \frac{300 \times 200}{500} = 120\ \Omega = R_T$$

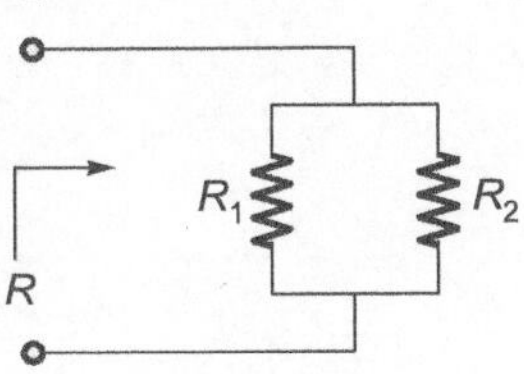

Given, tolerance of $R_1 = 1\%$

$$\therefore \quad R_1 = 300 \pm 3$$

$$R_1 = 303\ \Omega \text{ or } 297\ \Omega$$

Tolerance of $R_2 = 1\%$

$$\therefore \quad R_2 = 200 \pm 2 = 202\ \Omega \text{ or } 198\ \Omega$$

Equivalent resistance of the parallel combination of R_1 and R_2 in high range is:

$$R_H = \frac{303 \times 202}{303 + 202} = 121.2\ \Omega$$

∴ Tolerance for the equivalent resistance R in high range of R_1 and R_2 is

$$\varepsilon_R = \left(\frac{R_H - R}{R}\right) \times 100\%$$

$$= \left(\frac{121.2 - 120}{120}\right) \times 100 = \frac{1.2}{120} \times 100 = 1\%$$

Note: In low range of R_1 and R_2,

$$R_L = \left(\frac{297 \times 198}{495}\right) = 118.8\ \Omega$$

∴ Tolerance of equivalent resistance, R in low

$$\text{range} = \varepsilon_L = \left(\frac{R_L - R}{R}\right) \times 100\%$$

$$= \frac{118.8 - 120}{120} \times 100\% = -1\%$$

3.10 (a)

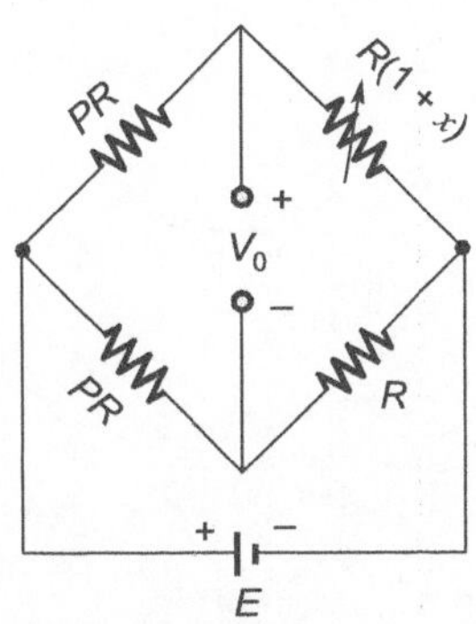

$$V_0 = \left[\frac{R(1+x)}{PR + R(1+x)} - \frac{R}{R + (PR)}\right] E$$

$$V_0 = \left[\frac{R(1+x)}{R + (1+x+P)} - \frac{R}{R + (1+P)}\right] E$$

$$V_0 = \left(\frac{1+x}{1+x+P} - \frac{1}{1+P}\right) E$$

$$V_0 = \left[\frac{(1+x)(1+P) - (1+x+P)}{(1+x+P)(1+P)}\right] E$$

$$V_0 = \left(\frac{1+x+Px+P-1-x-P}{(1+P+x+xP+P+P^2)}\right) E$$

$$= \left(\frac{Px}{(1+2P+P^2+x+xP)}\right)E$$

For maximum V_0, $\frac{dV_0}{dP} = 0$

$$\text{so, } \frac{(x)}{1+x+P(P+2+x)} - \frac{xP\{2P+x+2\}}{(1+x+P(2+P+x))^2} = 0$$

so, $[1 + x + P(2 + P + x)] = P(2P + x + 2)$

$\Rightarrow \quad 1 + x + 2P + P^2 + Px = 2P^2 + Px + 2P$

$\Rightarrow \quad 1 + x = P^2$

$\Rightarrow \quad P = \sqrt{1+x}$

3.11 (c)

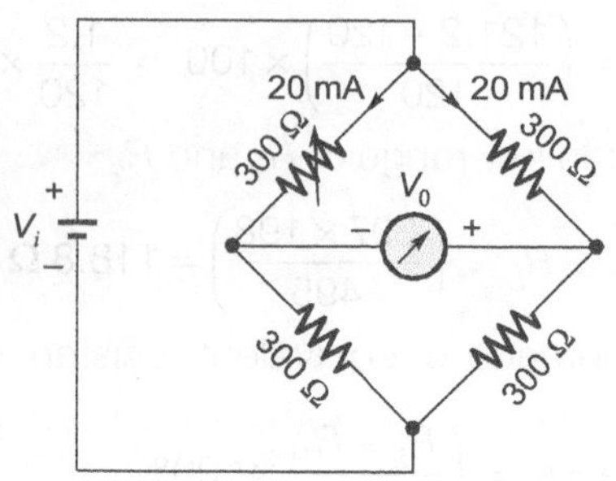

$$V_i = 20 \times 10^{-3} \times (300 + 300) = 12\text{ V}$$

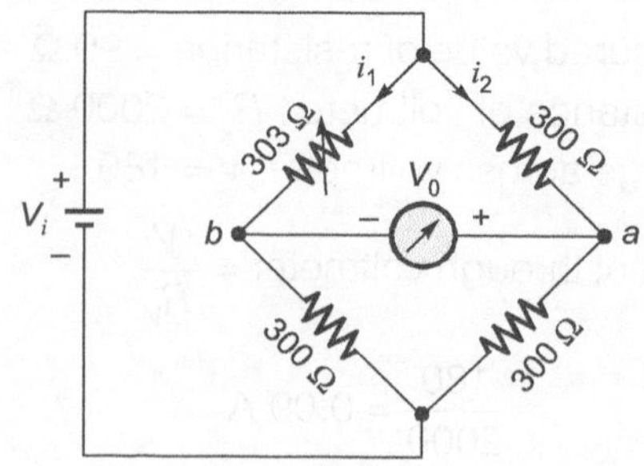

$$V_{ab} = V_a - V_b$$

Using voltage division rule,

$$V_{ab} = \left(\frac{300}{300+300} - \frac{300}{300+303}\right) \times 12$$

$$= 29.85\text{ mV}$$

4 A.C. Bridges

4.1 The items in **List-I** represent the various types of measurements to be made with a reasonable accuracy using a suitable bridge. The items in **List-II** represent the various bridges available for this purpose. Select the correct choice of the item in List-II for the corresponding item in List-I from the following:

List-I

A. Resistance in the milli-Ohm range
B. Low values of Capacitance
C. Comparison of resistances which are nearly equal
D. Inductance of a coil with a large time-constant

List-II

1. Wheatstone Bridge
2. Kelvin Double Bridge
3. Schering Bridge
4. Wien's Bridge
5. Hay's Bridge
6. Carey-Foster Bridge

Codes:

	A	B	C	D
(a)	2	3	6	5
(b)	2	6	4	5
(c)	2	3	5	4
(d)	1	3	2	6

[2003 : 2 M]

4.2 A bridge circuit is shown in the figure below. Which one of the sequence given below is most suitable for balancing the bridge?

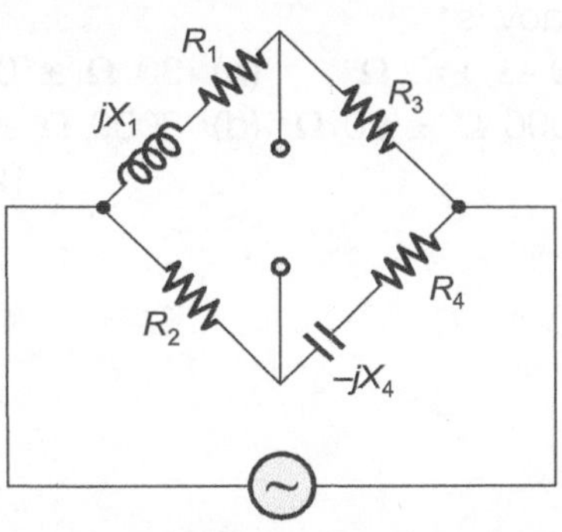

(a) First adjust R_4, and then adjust R_1
(b) First adjust R_2, and then adjust R_3
(c) First adjust R_2, and then adjust R_4
(d) First adjust R_4 and then adjust R_2

[2007 : 2 M]

4.3 The AC bridge shown in the figure is used to measure the impedance Z.

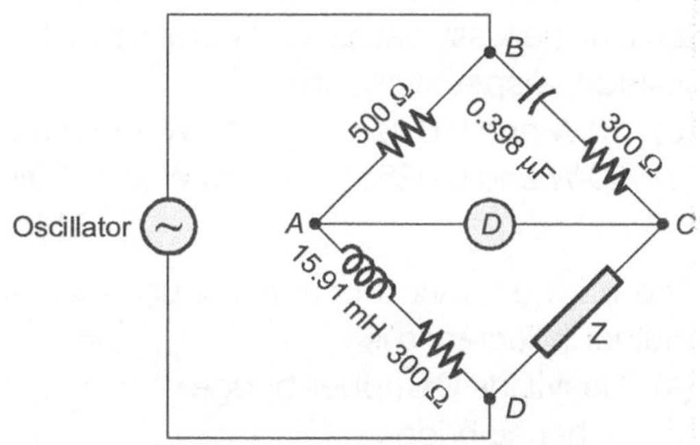

If the bridge is balanced for oscillator frequency $f = 2$ kHz, then the impedance Z will be

(a) $(260 + j0)\ \Omega$
(b) $(0 + j200)\ \Omega$
(c) $(260 - j200)\ \Omega$
(d) $(260 + j200)\ \Omega$

[2008 : 2 M]

4.4 The Maxwell's bridge shown in the figure is at balance. The parameters of the inductive coil are

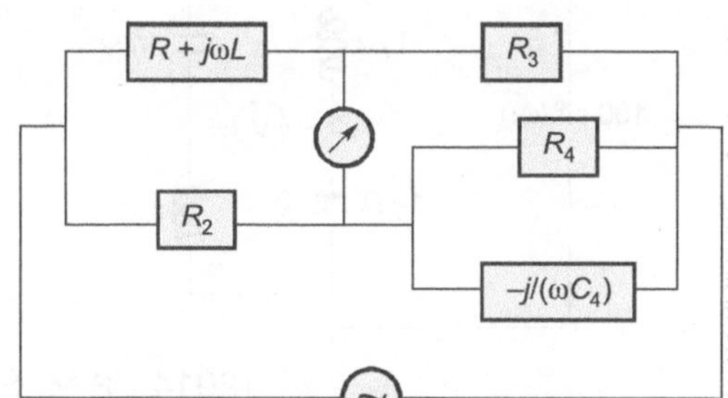

(a) $R = R_2R_3/R_4$, $L = C_4R_2R_3$
(b) $L = R_2R_3/R_4$, $R = C_4R_2R_3$
(c) $R = R_4/R_2R_3$, $L = 1/(C_4R_2R_3)$
(d) $L = R_4/R_2R_3$, $R = 1/(C_4R_2R_3)$

[2010 : 2 M]

4.5 The bridge circuit shown in the figure below is used for the measurement of an unknown element Z_x. The bridge circuit is best suited when Z_x is a

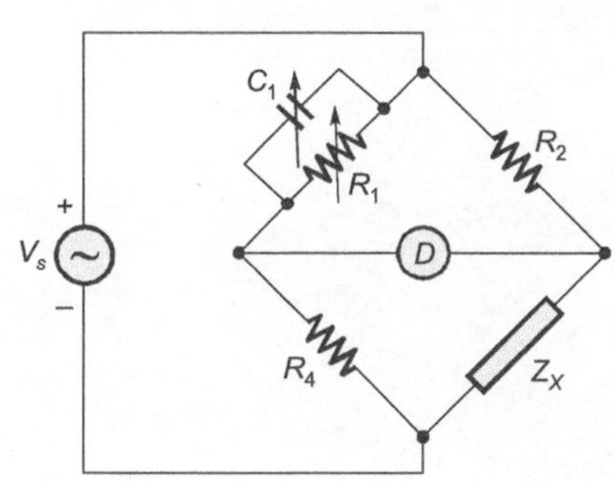

(a) low resistance (b) high resistance
(c) low Q inductor (d) lossy capacitor
[2011 : 1 M]

4.6 A lossy capacitor C_x, rated for operation at 5 kV, 50 Hz is represented by an equivalent circuit with an ideal capacitor C_P in parallel with a resistor R_P. The value of C_P is found to be 0.102 μF and the value of R_P = 1.25 MΩ. Then the power loss and tan δ of the lossy capacitor operating at the rated voltage, respectively, are
(a) 10 W and 0.0002 (b) 10 W and 0.0025
(c) 20 W and 0.025 (d) 20 W and 0.04
[2011 : 2 M]

4.7 The bridge method commonly used for finding mutual inductance is
(a) Heaviside Campbell bridge
(b) Schering bridge
(c) De Sauty bridge
(d) Wien bridge [2012 : 1 M]

4.8 The reading of the voltmeter (rms) in volts, for the circuit shown in the figure is _____.

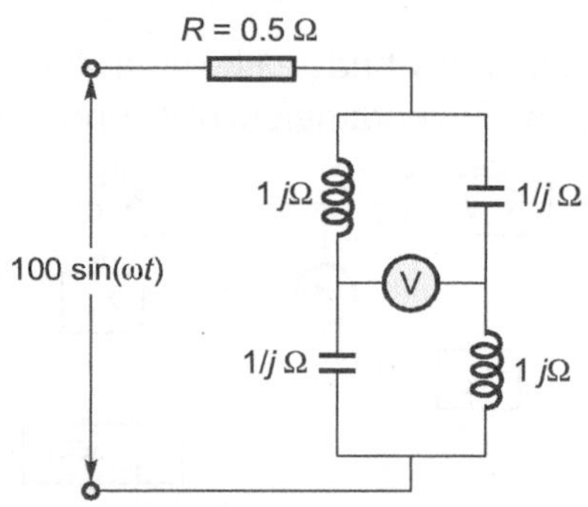

[2014 : 2 M, Set-1]

4.9 Three moving iron type voltmeters are connected as shown below. Voltmeter readings are V, V_1 and V_2, as indicated. The correct relation among the voltmeter readings is

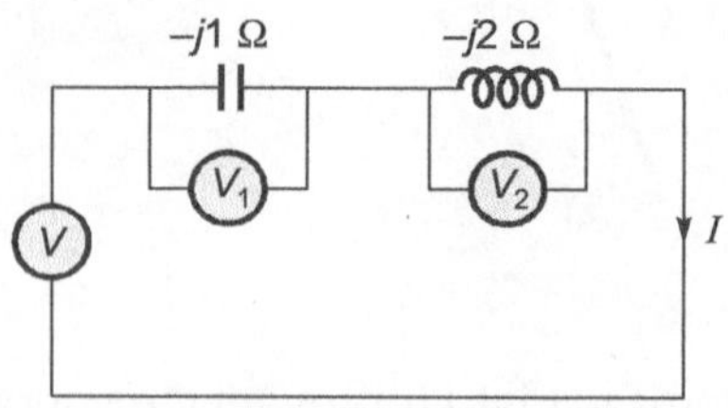

(a) $V = \frac{V_1}{\sqrt{2}} + \frac{V_2}{\sqrt{2}}$ (b) $V = V_1 + V_2$
(c) $V = V_1 V_2$ (d) $V = V_2 - V_1$
[2013 : 1 M]

4.10 In the bridge circuit shown, the capacitors are loss free. At balance, the value of capacitance C_1 in microfarad is _____.

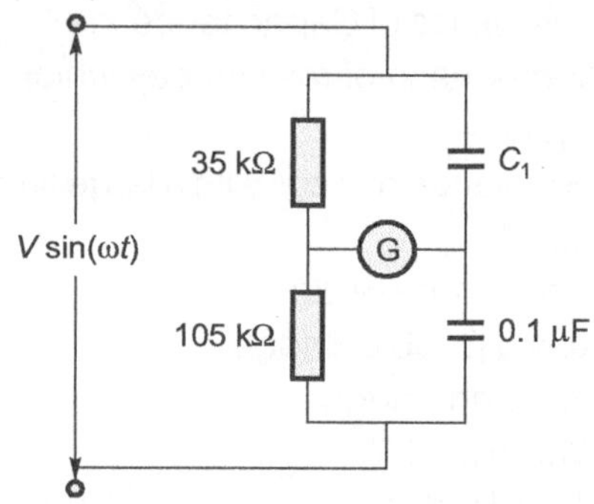

[2014 : 2 M, Set-3]

4.11 Inductance is measured by
(a) Maxwell bridge (b) Wien bridge
(c) Schering bridge (d) Kelvin bridge
[2021 : 1 M]

4.12 A balanced Wheatstone bridge $ABCD$ has the following arm resistances :
R_{AB} = 1 kΩ ± 2.1%; R_{BC} = 100 Ω ± 0.5%; R_{CD} is an unknown resistance;
R_{DA} = 300 Ω ± 0.4%. The value of R_{CD} and its accuracy is
(a) 30 Ω ± 3 Ω (b) 30 Ω ± 0.9 Ω
(c) 3000 Ω ± 90 Ω (d) 3000 Ω ± 3 Ω
[2022 : 1 M]

■■■■

Answers A.C. Bridges

4.1	(a)	4.2	(c)	4.3	(a)	4.4	(a)	4.5	(c)	4.6	(c)	4.7	(a)
4.8	(141.42)	4.9	(d)	4.10	(0.3)	4.11	(a)	4.12	(b)				

Explanations A.C. Bridges

4.1 (a)

- Wheat stone bridge is used for measurement of medium resistance.
- Kelvin double bridge is used for measurement of low resistance.
- Schering bridge is used for measurement of low value of capacitances.
- Wein's bridge is used for measurement of the frequency.
- Hay's bridge is used for measurement of inductance of a coil with a large time-constant.
- Carey-foster bridge is used for comparison of resistances which are nearly equal.

4.2 (c)

$$x_1 = \omega L_1 \text{ and } x_4 = \frac{1}{\omega C_4}$$

$$Z_1 = R_1 + jx_1 = R_1 + j\omega L_1$$

$$Z_2 = R_2$$

$$Z_3 = R_3$$

$$Z_4 = R_4 - jx_4 = R_4 - \frac{j}{\omega C_4}$$

Under balanced condition

$$Z_1 Z_4 = Z_2 Z_3$$

$$(R_1 + j\omega L_1)(R_4 - jx_4) = R_2 R_3$$

$$\left(R_1 R_4 + \frac{L_1}{C_4}\right) + j\left[\omega L_1 R_4 - \frac{R_1}{\omega C_4}\right] = R_2 R_3$$

Equating real and imaginary terms, we obtain

$$R_1 R_4 + \frac{L_1}{C_4} = R_2 R_3$$

$$\omega L_1 R_4 - \frac{R_1}{\omega C_4} = 0$$

Solving above equations, we get

$$L_1 = \frac{R_2 R_3 C_4}{1 + \omega^2 C_4^2 R_4^2}$$

and $$R_2 = \frac{\omega^2 R_2 R_3 R_4 C_4^2}{1 + \omega^2 C_4^2 R_4^2}$$

Q-factor of the coil $Q = \dfrac{\omega L_1}{R_1} = \dfrac{1}{\omega C_4 R_4}$

Therefore, $$L_1 = \frac{R_2 R_3 C_4}{1 + \left(\frac{1}{Q}\right)^2} \quad \text{...(i)}$$

and $$R_1 = \frac{\omega^2 R_2 R_3 R_4 C_4^2}{1 + \left(\frac{1}{Q}\right)^2} \quad \text{...(ii)}$$

For a value of a greater than 10, the term $(1/Q)^2$ will be smaller than 1/1000 and can be neglected Therefore equation (i) and (ii) reduces to

$$L_1 = R_2 R_3 C_4 \quad \text{...(iii)}$$

$$R_1 = \omega^2 R_2 R_3 C_4^2 \quad \text{...(iv)}$$

R_4 appears only in equation (iv) and R_2 appears in both equation (iii) and (iv).

So first R_2 is adjusted and then R_4 is adjusted.

4.3 (a)

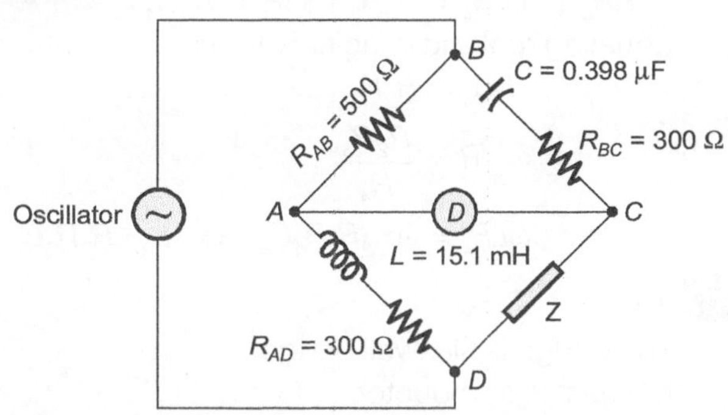

$$Z_{AB} = 500\ \Omega\ ; \quad Z_{CD} = Z$$

$$Z_{BC} = R_{BC} + \frac{1}{j\omega C}$$

$$= 300 - \frac{j}{2\pi \times 2 \times 10^3 \times 0.398 \times 10^{-6}}$$

$$Z_{BC} = 300 - j200\ \Omega$$

$$Z_{AD} = R_{AD} + j\omega L$$

$$Z_{AD} = 300 + j2\pi \times 2 \times 10^3 \times 15.91 \times 10^{-3}$$

$$= 300 + j\,200\ \Omega$$

At balance,

$$Z_{AB} \times Z_{CD} = Z_{BC} \times Z_{AD}$$

$$Z_{CD} = Z = \frac{Z_{BC} \times Z_{AD}}{Z_{AB}}$$

$$\Rightarrow \quad Z = \frac{(300 - j200) \times (300 + j200)}{500}$$

$$Z = (260 + j0)\ \Omega$$

4.4 (a)

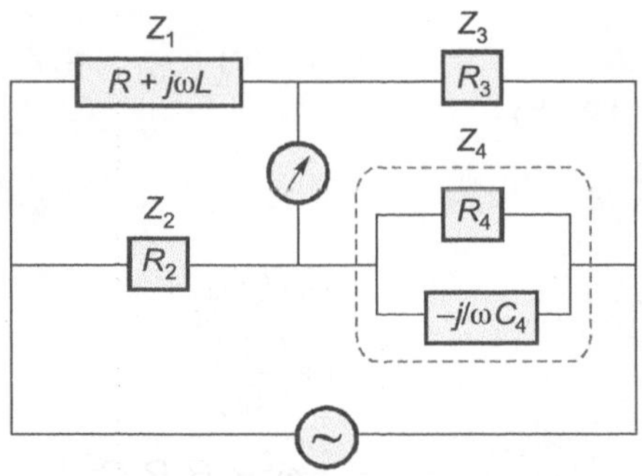

$$Z_1 = R + j\omega L;\ Z_2 = R_2;\ Z_3 = R_3$$

$$Z_4 = R_4 \left\| \left(\frac{-j}{\omega C_4}\right)\right.$$

$$= R_4 \left\| \left(\frac{1}{j\omega C_4}\right)\right. = \frac{R_4}{1 + j\omega C_4 R_4}$$

At balance,

$$Z_1 Z_4 = Z_2 Z_3$$

$$(R + j\omega L)\left(\frac{R_4}{1 + j\omega C_4 R_4}\right) = R_2 R_3$$

$$R_4(R + j\omega L) = R_2 R_3 (1 + j\omega C_4 R_4)$$

$$RR_4 + j\omega L R_4 = R_2 R_3 + j\omega R_2 R_3 R_4 C_4$$

Equating real and imaginary terms,

$$RR_4 = R_2 R_3$$

$$R = \frac{R_2 R_3}{R_4}$$

$$\omega L R_4 = \omega R_2 R_3 R_4 C_4 \Rightarrow L = R_2 R_3 C_4$$

4.5 (c)

The bridge is Maxwell bridge.

Element is an inductor.

Inductance = L_x effective resistance of the inductor = R_x

$$Q = \frac{\omega L_x}{R_x} = \omega C_1 R_1 \quad \text{...(i)}$$

The bridge is limited to measurement of low Q inductor ($1 < Q < 10$).

It is clear from equation (i) that the measurement of high Q coils demands a large value of resistance R_1, perhaps 10^5 or 10^6 Ω. The resistance boxes of such high values are very expensive. Thus for values of $Q > 10$, the bridge is unsuitable.

The bridge is also unsuited for coils with a very low values of Q (i.e. $Q < 1$).

4.6 (c)

$$C_P = 0.102\ \mu F$$

$$R_P = 1.25\ M\Omega$$

$$\text{Power loss} = \frac{V^2}{R_P} = \frac{(5\times10^3)^2}{1.25\times10^6} = \frac{25}{1.25} = 20\ W$$

$$\tan\delta = R/X = \frac{1}{\omega C_P R_P}$$

$$= \frac{1}{2\times\pi\times50\times0.102\times10^{-6}\times1.25\times10^6} = 0.025$$

4.7 (a)

Heaviside Campbell bridge method is commonly used for finding mutual inductance.

4.8 Sol.

Given bridge is shown in the figure:

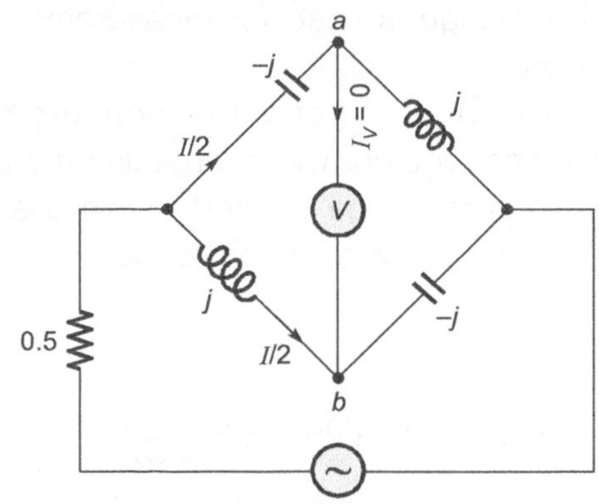

$v_i(t)$ = 100 sinωt

From the given bridge circuit, we see that product of opposite arm impedances are equal i.e., $(-j)\times(-j) = (j)(j)$ or $-1 = -1$

Hence, the bridge is balanced, i.e. no current flows through the voltage meter.

Now, $v_i(t) = 100 \sin \omega t = \frac{100}{\sqrt{2}}$ Volt (rms value)

∴ Net current supplied by the source is

$$I = \frac{100/\sqrt{2}}{(0.5 + Z_{eq})}\ A$$

$Z_{eq} = 0$ (i.e. net impedance of whole bridge = 0)

$$\therefore \quad I = \frac{100/\sqrt{2}}{0.5} = 141.42\ A$$

$$\therefore \quad \frac{I_2}{2} = 70.71\ A$$

Current flows through each parallel path of the bridge circuit therefore, $V = V_a - V_b$

$$= \left(\frac{I}{2}\times -j\right) - \left(\frac{I}{2}\times j\right)$$

$$= -jI = -j141.42 \text{ volt (rms value)}$$

∴ Reading of the voltmeter in volt (rms)

$$= V = V_{ab} = 141.42$$

4.9 (d)

$$V_1 = -|j1\Omega| \times I$$
$$V_2 = |j2\,\Omega| \times I$$
$$V = -j1\ \Omega \times I + j2\ \Omega \times I = V_2 - V_1$$

4.10 Sol.

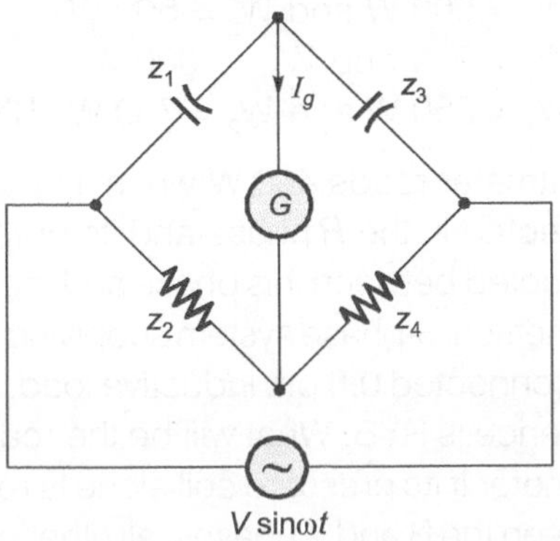

Redrawing the given bridge circuit, we have :

$$Z_1 = \frac{1}{j\omega C_1}\ \Omega,\ Z_2 = 35\ \text{k}\Omega,$$

$$Z_3 = \frac{10^6}{j0.1\,\omega}\ \Omega,\ Z_4 = 105\ \text{k}\Omega$$

At balance, current through galvanometer;

$$I_g = 0$$

and $|Z_1||Z_4| = |Z_2||Z_3|$

$$\therefore \left(\frac{1}{\omega C_1}\right) \times (105\ K) = (35\ K)\left(\frac{10^6}{0.1\omega}\right)$$

$$\text{or, } C_1 = \frac{105 \times 0.1}{35}\mu\text{F} = \frac{10.5}{35}\mu\text{F} = 0.3\ \mu\text{F}$$

$\therefore$ In μF, $C_1 = 0.3$

4.12 (b)

Under balanced condition,

$$R_{CD} = \frac{R_{BC} \times R_{DA}}{R_{AB}} = \frac{(100 \pm 0.5\%)(300 \pm 0.4\%)}{(1000 \pm 2.1\%)}$$

$$R_{CD} = \frac{100 \times 300}{1000} + (0.5\% + 0.4\% + 2.1\%)$$

$$R_{CD} = 30 \pm 3\%$$

% Error = ±3%

$$R_{CD} = 30 \pm 30 \times \frac{3}{100} = 30 \pm 0.9\ \Omega$$

■■■■

5 Measurement of Energy and Power

5.1 The moving coil in a dynamometer wattmeter is normally connected
(a) in series with the fixed coil
(b) across the supply
(c) in series with the load
(d) across the load [1996 : 1 M]

5.2 A water boiler at home is switched ON to the AC mains supplying power at 230 V/50 Hz. The frequency of instantaneous power consumed by the boiler is
(a) 0 Hz (b) 50 Hz
(c) 100 Hz (d) 150 Hz [1996 : 1 M]

5.3 A dynamometer type wattmeter responds to the
(a) average value of active power
(b) average value of reactive power
(c) peak value of active power
(d) peak value of reactive power [1997 : 1 M]

5.4 Electrodynamic type watt meters have large errors while measuring power in AC circuits at low power factor conditions, since the voltage across and the current through the
(a) current coil are not in phase
(b) current coil not in quadrature
(c) pressure coil are not in phase
(d) pressure coil are not in quadrature
[1999 : 1 M]

5.5 The minimum number of wattmeter(s) required to measure 3-phase, 3-wire balanced or unbalanced power is
(a) 1 (b) 2
(c) 3 (d) 4 [2001 : 1 M]

5.6 The line to line input voltage to the 3-phase. 50 Hz, AC circuit shown in figure is 100 V rms. Assuming that the phase sequence is RYB the wattmeter would read.

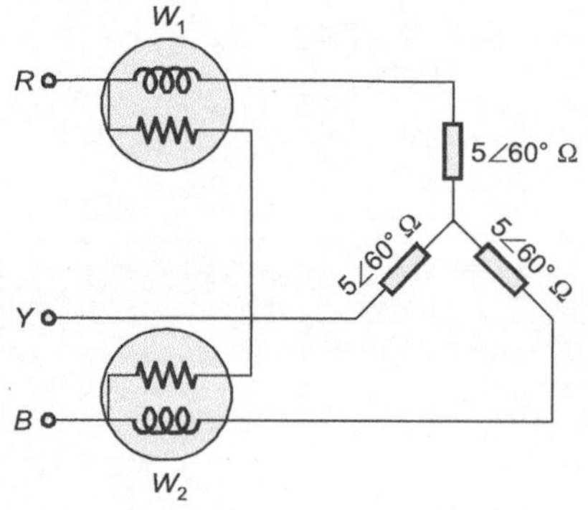

(a) W_1 = 886 W and W_2 = 896 W
(b) W_1 = 500 W and W_2 = 500 W
(c) W_1 = 0 W and W_2 = 1000 W
(d) W_1 = 250 W and W_2 = 750 W [2002 : 1 M]

5.7 A wattmeter reads 400 W when its current coil is connected in the *R* phase and its pressure coil is connected between this phase and the neutral of a symmetrical 3-phase system supplying a balanced star connected 0.8 p.f. inductive load. This phase sequence is RYB. What will be the reading of this wattmeter if its pressure coil alone is reconnected between the B and Y phases, all other connections remaining as before?
(a) 400.0 (b) 519.6
(c) 300.0 (d) 692.8 [2003 : 2 M]

5.8 The voltage-flux adjustment of a certain 1-phase 220 V induction watt-hour meter is altered so that the phase angle between the applied voltage and the flux due to it is 85° (instead of 90°). The errors introduced in the reading of this meter when the current is 5 A at power factors of unity and 0.5 lagging are respectively
(a) 3.8 mW, 77.4 mW
(b) –3.8 mW, –77.4 mW
(c) –4.2 W, –85.1 W
(d) 4.2 W, 85.1 W [2003 : 2 M]

5.9 The circuit in figure is used to measure the power consumed by the load. The current coil and the voltage coil of the wattmeter have 0.02 Ω and 1000 Ω resistances respectively. The measured power compared to the load power will be

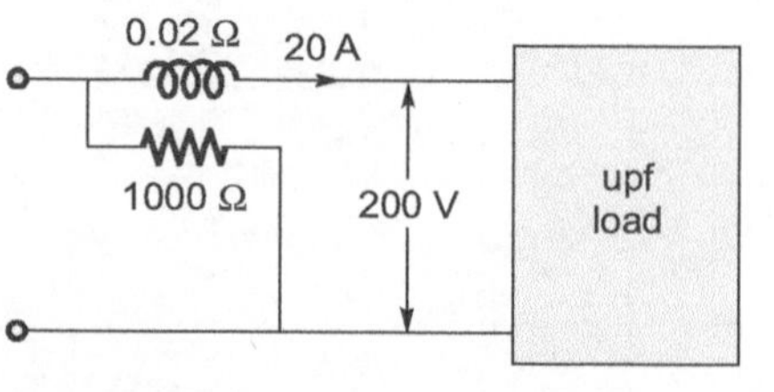

(a) 0.4% less (b) 0.2% less
(c) 0.2% more (d) 0.4% more
[2004 : 1 M]

5.10 A dc A-h meter is rated for 15 A, 250 V. The meter constant is 14.4 A-sec/rev. The meter constant at rated voltage may be expressed as

(a) 3750 rev/kWh (b) 3600 rev/kWh
(c) 1000 rev/kWh (d) 960 rev/kWh

[2004 : 2 M]

5.11 A single-phase load is connected between R and Y terminals of a 415 V, symmetrical, 3-phase, 4 wire system with phase sequence RYB. A wattmeter is connected in the system as shown in the figure. The power factor of the load is 0.8 lagging. The wattmeter will read

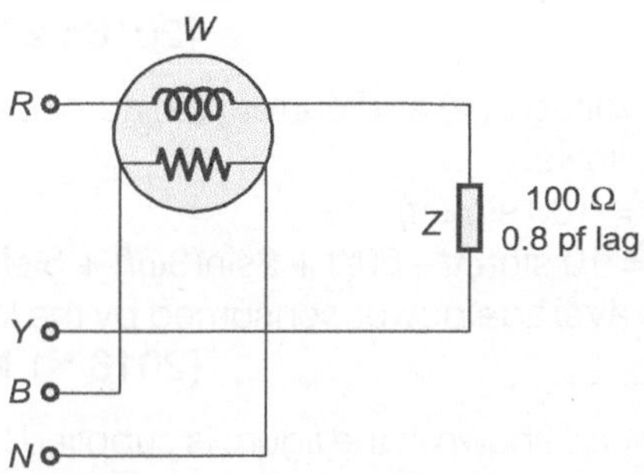

(a) –795 W (b) –597 W
(c) +597 W (d) +795 W [2004 : 2 M]

5.12 Two wattmeters, which are connected to measure the total power on a three-phase system supplying a balanced load, read 10.5 kW and –2.5 kW, respectively. The total power and the power factor respectively, are
(a) 13.0 kW, 0.334 (b) 13.0 kW, 0.684
(c) 8.0 kW, 0.52 (d) 8.0 kW, 0.334

[2005 : 2 M]

5.13 A sampling wattmeter (that computes power from simultaneously sampled values of voltage and current) is used to measure the average power of a load. The peak to peak voltage of the square wave is 10 V and the current is a triangular wave of 5 A p-p as shown in the figure. The period is 20 ms. The reading in W will be

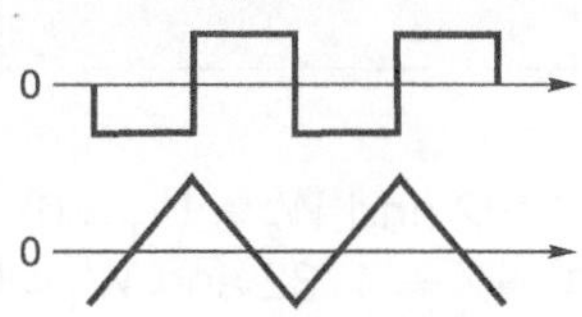

(a) 0 W (b) 25 W
(c) 50 W (d) 100 W [2006 : 1 M]

5.14 The pressure coil of a dynamometer type wattmeter is
(a) highly inductive
(b) highly resistive
(c) purely resistive
(d) purely inductive [2009 : 1 M]

5.15 The figure shows a three-phase delta connected load supplied from a 400 V, 50 Hz, 3-phase balanced source. The pressure coil (PC) and current coil (CC) of a wattmeter are connected to the load as shown, with the coil polarities suitably selected to ensure a positive deflection. The wattmeter reading will be

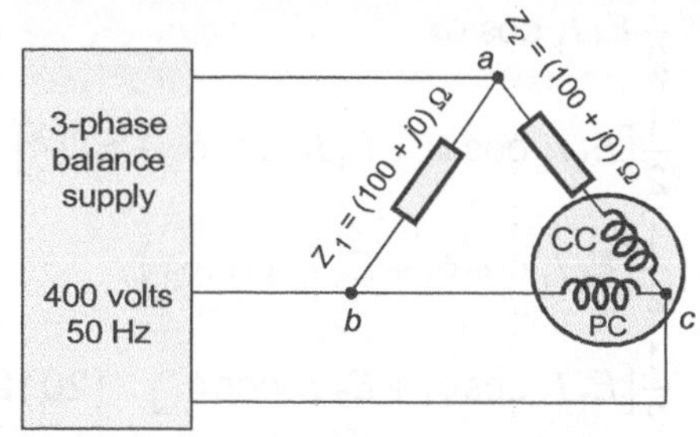

(a) 0 (b) 1600 Watt
(c) 800 Watt (d) 400 Watt

[2009 : 2 M]

5.16 A wattmeter is connected as shown in the figure. The wattmeter reads

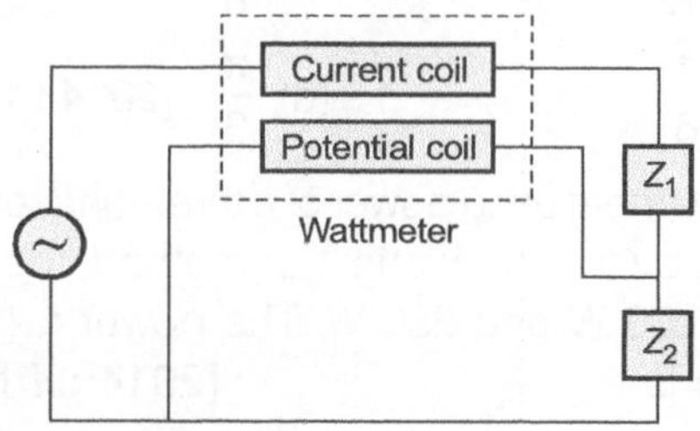

(a) Zero always
(b) Total power consumed by Z_1 and Z_2
(c) Power consumed by Z_1
(d) Power consumed by Z_2 [2010 : 1 M]

5.17 Consider the following statements:
(i) The compensating coil of a low power factor wattmeter compensates the effect of the impedance of the current coil.
(ii) The compensating coil of a low power factor wattmeter compensates the effect of the impedance of the voltage coil circuit.
(a) (i) is true but (ii) is false.
(b) (i) is false but (ii) is true.
(c) both (i) and (ii) are true.
(d) both (i) and (ii) are false. [2011 : 1 M]

5.18 For the circuit in the figure, the voltage and current expressions are:

$$v(t) = E_1 \sin(\omega t) + E_3 \sin(3\omega t)$$

and $$i(t) = I_1 \sin(\omega t - \phi_1) + I_3 \sin(3\omega t - \phi_3) + I_5 \sin(5\omega t)$$

The average power measured by the Wattmeter is

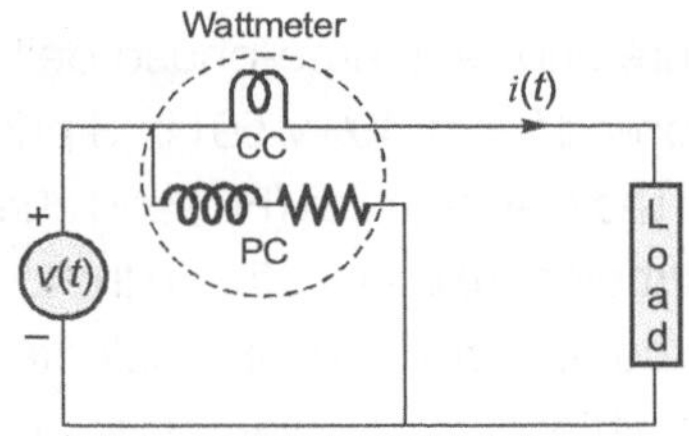

(a) $\frac{1}{2}E_1 I_1 \cos\phi_1$

(b) $\frac{1}{2}\left[E_1 I_1 \cos\phi_1 + E_1 I_3 \cos\phi_3 + E_1 I_5\right]$

(c) $\frac{1}{2}\left[E_1 I_1 \cos\phi_1 + E_3 I_3 \cos\phi_3\right]$

(d) $\frac{1}{2}\left[E_1 I_1 \cos\phi_1 + E_3 I_1 \cos\phi_1\right]$ [2012 : 1 M]

5.19 Power consumed by a balanced 3-phase, 3-wire load is measured by two wattmeter method. The first wattmeter reads twice that of the second. Then the load impedance angle in radians is

(a) $\frac{\pi}{12}$ (b) $\frac{\pi}{8}$

(c) $\frac{\pi}{6}$ (d) $\frac{\pi}{3}$ [2014 : 1 M, Set-1]

5.20 While measuring power of a three-phase balanced load by the two-wattmeter method, the readings are 100 W and 250 W. The power factor of the load is ______ . [2014 : 1 M, Set-2]

5.21 An LPF wattmeter of power factor 0.2 is having three voltage settings 300 V, 150 V and 75 V, and two current settings 5 A and 10 A. The full scale reading is 150. If the wattmeter is used with 150 V voltage setting and 10 A current setting, the multiplying factor of the wattmeter is ______ . [2014 : 1 M, Set-3]

5.22 A 3-phase balanced load which has a power factor of 0.707 is connected to a balanced supply. The power consumed by the load is 5 kW. The power is measured by the two-wattmeter method. The readings of the two wattmeters are

(a) 3.94 kW and 1.06 kW
(b) 2.50 kW and 2.50 kW
(c) 5.00 kW and 0.00 kW
(d) 2.96 kW and 2.04 kW [2015 : 1 M, Set-2]

5.23 The coils of a wattmeter have resistance 0.01 Ω and 1000 Ω; their inductances may be neglected. The wattmeter is connected as shown in the figure, to measure the power consumed by a load, which draws 25 A at power factor 0.8. The voltage across the load terminals is 30 V. The percentage error on the wattmeter reading is ______.

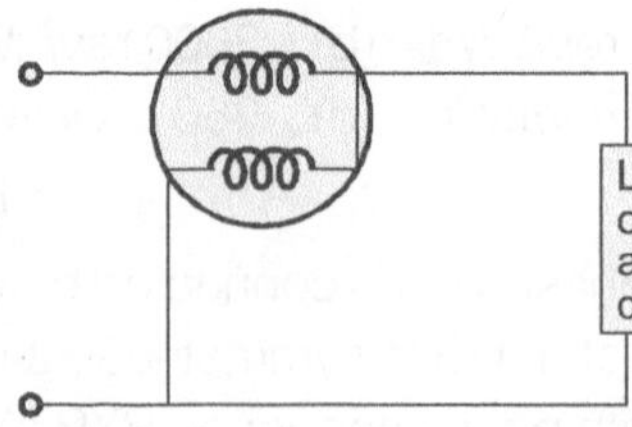

[2015 : 2 M, Set-2]

5.24 An energy meter, having meter constant of 1200 rev/kWh, makes 20 revolutions in 30 seconds for a constant load. The load, in kW, is ______. [2016 : 2 M, Set-2]

5.25 The voltage (V) and current (A) across a load are as follows:
$V(t) = 100 \sin(\omega t)$
$i(t) = 10 \sin(\omega t - 60°) + 2\sin(3\omega t) + 5 \sin(5\omega t)$.
The average power consumed by the load, in W, is______. [2016 : 1 M, Set-2]

5.26 The load shown in the figure is supplied by a 400 V (line-to-line), 3-phase source (RYB sequence). The load is balanced and inductive, drawing 3464 VA. When the switch S is in position N, the three wattmeters W_1, W_2 and W_3 read 577.35 W each. If the switch is moved to position Y, the readings of the wattmeters in watts will be:

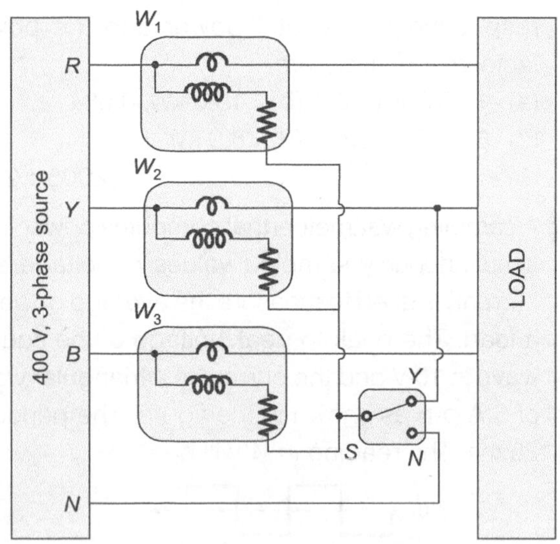

(a) $W_1 = 1732$ and $W_2 = W_3 = 0$
(b) $W_1 = 0$, $W_2 = 1732$ and $W_3 = 0$
(c) $W_1 = 866$ and $W_2 = 0$, $W_3 = 866$
(d) $W_1 = W_2 = 0$ and $W_3 = 1732$

[2017 : 2 M, Set-1]

5.27 Two wattmeter method is used for measurement of power in a balanced three-phase load supplied from a balanced three-phase system. If one of the wattmeters reads half of the other (both positive), then the power factor of the load is

(a) 0.532 (b) 0.632
(c) 0.707 (d) 0.866

[2018 : 1 M]

5.28 The voltage across and the current through a load are expressed as follows:

$$v(t) = -170\sin\left(377t - \frac{\pi}{6}\right) \text{ V}$$

$$i(t) = 8\cos\left(377t + \frac{\pi}{6}\right) \text{ A}$$

The average power in watts (round off to one decimal place) consumed by the load is _____.

[2019 : 2 M]

5.29 A 3-phase, star-connected, balanced load is supplied from a 3-phase, 400 V (rms), balanced voltage source with phase sequence R-Y-B, as shown in the figure. If the wattmeter reading is –400 W and the line current is I_R = 2 A (rms), then the power factor of the load per phase is

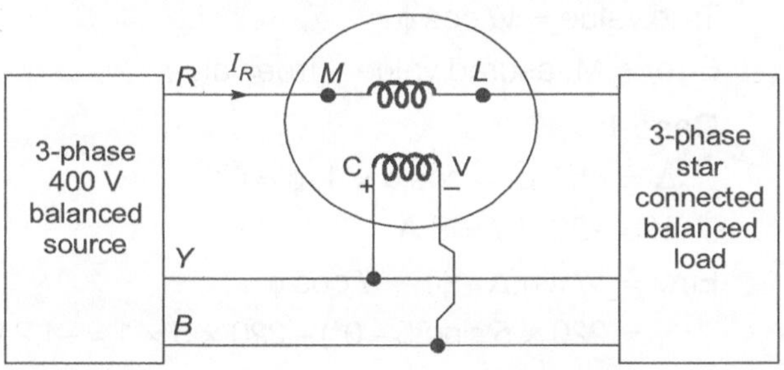

(a) Unity (b) 0.5 leading
(c) 0.866 leading (d) 0.707 lagging

[2023 : 2 Marks]

5.30 In the circuit shown, Z_1 = 50∠–90° and Z_2 = 200∠–30° Ω. It is supplied by a three phase 400 V source with the phase sequence being R-Y-B. Assume the watt meters W_1 and W_2 to be ideal. The magnitude of the difference between the readings of W_1 and W_2 in watts is _______ (rounded off to 2 decimal places).

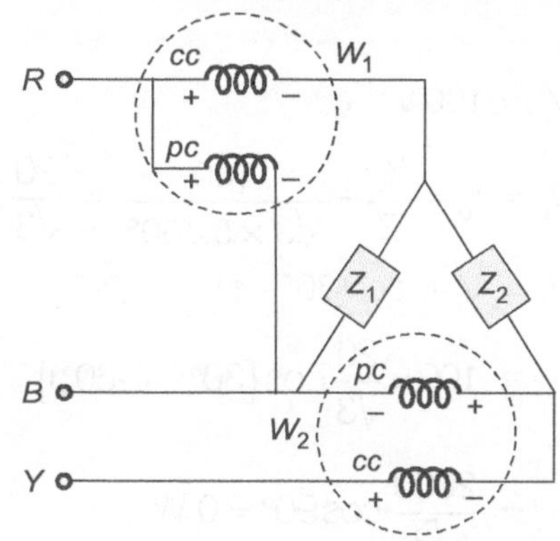

[2024 : 2 M]

Answers Measurement of Energy and Power

5.1 (b)	5.2 (c)	5.3 (a)	5.4 (c)	5.5 (b)	5.6 (c)	5.7 (b)
5.8 (c)	5.9 (c)	5.10 (c)	5.11 (b)	5.12 (d)	5.13 (a)	5.14 (b)
5.15 (c)	5.16 (d)	5.17 (b)	5.18 (c)	5.19 (c)	5.20 (0.8)	5.21 (2)
5.22 (a)	5.23 (0.15%)	5.24 (2)	5.25 (250)	5.26 (d)	5.27 (d)	5.28 (588.9)
5.29 (c)	5.30 (692.80)					

Explanations Measurement of Energy and Power

5.1 (b)

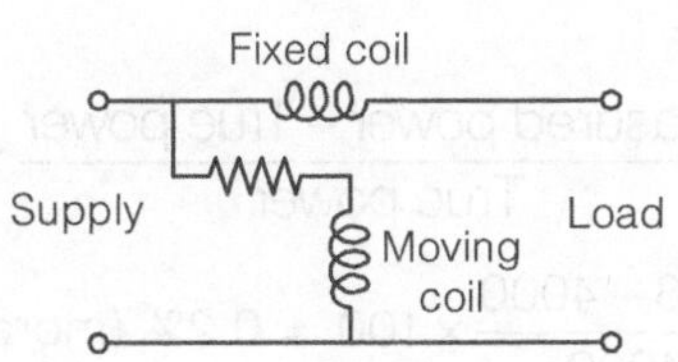

5.2 (c)

$$P_{inst} = V_m \sin\omega t \cdot \frac{V_m \sin\omega t}{R}$$

$$P_{inst} = \frac{V_m^2}{2R}(1 - \cos 2\omega t)$$

$$\therefore \quad f = 2 \times 50 = 100 \text{ Hz}$$

5.3 (a)

$T_D \propto VI\cos\phi$ (Active power)

Average is taken over a period.

5.5 (b)

Two wattmeter method can also take care of unbalance.

5.6 (c)

$$V_L = 100\text{ V}$$

$$I_L = I_p = \frac{V_p}{Z} = \frac{100}{\sqrt{3}\times 5\angle 60^\circ} = \frac{20}{\sqrt{3}}\angle -60^\circ\text{ A}$$

$$W_1 = V_L I_L \cos(30^\circ - \phi)$$

$$= 100\times\frac{20}{\sqrt{3}}\cos\{30^\circ - (-60^\circ)\}$$

$$= \frac{2000}{\sqrt{3}}\cos 90^\circ = 0\text{ W}$$

$$W_2 = V_L I_L \cos(30^\circ + \phi)$$

$$= 100\times\frac{20}{\sqrt{3}}\cos(30^\circ - 60^\circ) = 1000\text{ W}$$

5.7 (b)

Taking V_R as the reference and assuming phase to neutral voltage = V

$V_R = V\angle 0^\circ$; $V_Y = V\angle -120^\circ$ and $V_B = V\angle -240^\circ$

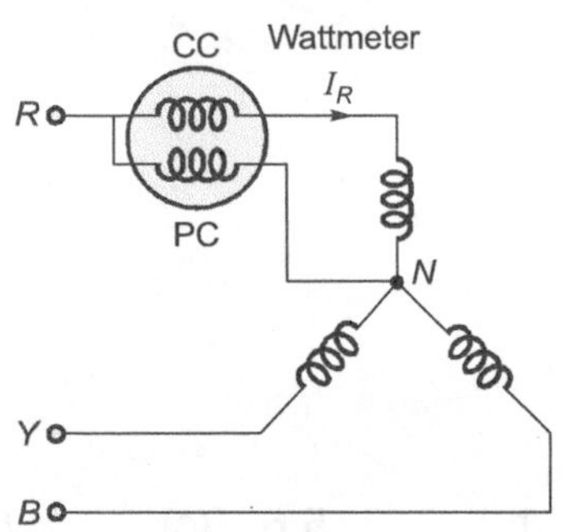

$$V_R I_R \cos\phi = 400\text{ W}$$

$$VI_R \times 0.8 = 400\text{ W}$$

$$VI_R = 400 \quad \text{...(i)}$$

$$I_R = I_R\angle -36.87^\circ$$

[p.f. = 0.8 lag, inductive load]

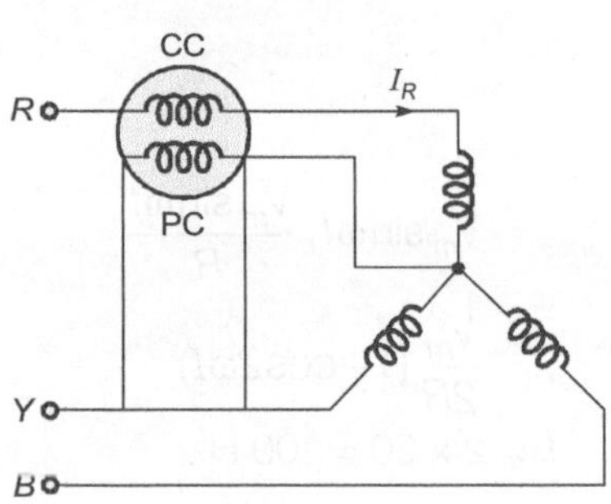

$$V_{YB} = V_{YB} - V_B = V\angle -120^\circ - V\angle -240^\circ$$

$$= \sqrt{3}\,V\angle -90^\circ\text{ V}$$

$$I_R = I_R\angle -36.87^\circ\text{ A}$$

Angle between V_{YB} and I_R

$$\theta = -90^\circ - (-36.87^\circ) = 53.13^\circ$$

As pressure coil connected between Y and B phases.

Reading of wattmeter = $V_{YB} I_R \cos\theta$

From equation (i),

$$= \sqrt{3}\,V \times I_R \times \cos(-53.13)$$

$$= \sqrt{3}\times 500\times 0.6 = 519.6\text{ W}$$

5.8 (c)

Measured value = $VI\sin(\Delta - \phi)$

Where,

Δ = Phase angle between voltage and flux

$\cos\phi$ = Power factor

True value = $VI\cos\phi$

Error = Measured value – True value

Case-I:

$\Delta = 85^\circ$, pf = $\cos\phi = 1$, $\phi = 0^\circ$

$V = 220$ V, $I = 5$ A

Error = $VI\sin(\Delta - \phi) - VI\cos\phi$

$= 220\times 5\sin(85 - 0^\circ) - 220\times 5\times 1 \approx -4.2$ W

Case-II:

$\Delta = 85^\circ$, pf = $\cos\phi = 0.5 \Rightarrow \phi = 60^\circ$

$V = 220$ V, $I = 5$ A

Error = $VI\sin(\Delta - \phi) - VI\cos\phi$

$= 220\times 5\times\sin(85 - 60^\circ) - 220\times 5\times 0.5 = -85.1$ W

5.9 (c)

Load power (true power) = $VI\cos\phi$

$= 200\times 20\times 1 = 4000$ W

Resistance of current coil = $R_{cc} = 0.02$

Current through $CC = I_c = 20$ A

Power consumed by current coil

$$= I_c^2 R_{cc} = 20^2\times 0.02 = 8\text{ W}$$

Measured power = Power consumed by load + Power consumed by current coil

Measured power = 4000 + 8 = 4008 W

% error

$$= \frac{\text{Measured power} - \text{True power}}{\text{True power}}\times 100$$

$$= \frac{4008 - 4000}{4000}\times 100 = 0.2\%\text{ (more)}$$

5.10 (c)

Meter constant, $m = 14.4$ A-sec/rev

To express meter constant in the unit rev/kWh

$$\text{Meter constant} = M' = \frac{1}{M} = \frac{1}{14.4}\frac{\text{rev}}{\text{A-sec}}$$

$$M' = \left(\frac{1}{14.4}\frac{\text{rev}}{\text{A-sec}}\right)\times\left(\frac{1}{250\,\text{V}}\right)\times 10^3 \times \frac{3600\,\text{sec}}{1\text{hr}}$$

$$= \frac{3600\times10^3}{14.4\times250} = 1000 \text{ rev/kWh}$$

5.11 (b)

Line to line voltage = 415 V

$$\text{Phase to neutral voltage} = \frac{415}{\sqrt{3}}\text{V}$$

Taking V_R as the reference,

$$V_{RN} = \frac{415}{\sqrt{3}}\angle 0^\circ\text{V};\quad V_{yN} = \frac{415}{\sqrt{3}}\angle{-120^\circ}\text{V}$$

$$\text{and}\quad V_{BN} = \frac{415}{\sqrt{3}}\angle{-240^\circ}\text{V}$$

$$V_{Ry} = V_{RN} - V_{yN}$$

$$= \frac{415}{\sqrt{3}}\angle 0^\circ\text{V} - \frac{415}{\sqrt{3}}\angle -120^\circ = 415\angle 30^\circ\text{ V}$$

Load current,

$$I_L = \frac{V_{Ry}}{z} = \frac{415\angle 30^\circ}{100\angle 36.87} = 4.15\angle{-6.87^\circ}\text{ A}$$

Current through current coil,

$$I_{cc} = I_L = 4.15\angle{-6.87^\circ}\text{ A}$$

Voltage across pressure coil,

$$V_{BN} = \frac{415}{\sqrt{3}}\angle{-240^\circ}\text{V}$$

Phase angle between V_{BN} and I_{cc},

$$\phi = -240^\circ - (-6.87^\circ) = 233.13^\circ$$

$$\text{Wattmeter reading} = |V_{BN}||I_L|\cos\phi$$

$$= \frac{415}{\sqrt{3}}\times 4.15\times\cos 233.13 \approx -597\text{ W}$$

5.12 (d)

$P_1 = 10.5$ kW and $P_2 = -2.5$ kW

Total power = $P_1 + P_2 = 10.5 - 2.5 = 8$ kW

$$\text{Power factor, } \cos\phi = \cos\left(\tan^{-1}\sqrt{3}\frac{P_1 - P_2}{P_1 + P_2}\right)$$

$$= \cos\left(\tan^{-1}\sqrt{3}\frac{10.5-(-2.5)}{10.5+(-2.5)}\right) = 0.334$$

5.13 (a)

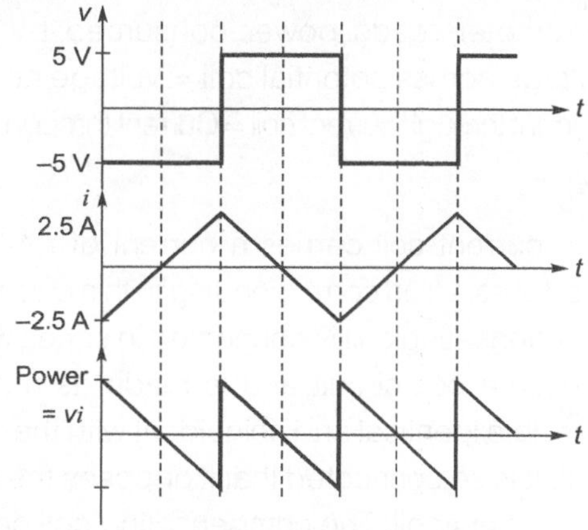

Positive power = Negative power

So average power = 0 W

5.14 (b)

It is difficult to have purely resistive pressure coil. The pressure coil has a small value of inductance, due to which error occurs in wattmeter readings.

5.15 (c)

Assuming phase sequence *abc*

Line to line voltage $V_{l-l} = 400$ V

Taking V_{ab} as the reference,

$$V_{ab} = V_{l-l}\angle 0^\circ = 400\angle 0^\circ\text{ V}$$

$$V_{bc} = 400\angle{-120^\circ}\text{ V}$$

$$V_{ca} = 400\angle{-240^\circ}\text{ V}$$

Current through current coil,

$$I_{cc} = \frac{V_{ca}}{Z_2} = \frac{400\angle -240^\circ}{100 + j0} = 4\angle{-240^\circ}\text{ A}$$

Voltage across pressure coil

$$V_{pc} = V_{bc} = 400\angle{-120^\circ}\text{ V}$$

$$\phi = \text{Angle between } I_{cc} \text{ and } V_{pc}$$

$$= -120 - (-240^\circ) = 120^\circ$$

Wattmeter reading,

$$|V_{pc}||I_{cc}|\cos\phi = 400\times4\times\cos(120^\circ) = -800\text{ W}$$

5.16 (d)

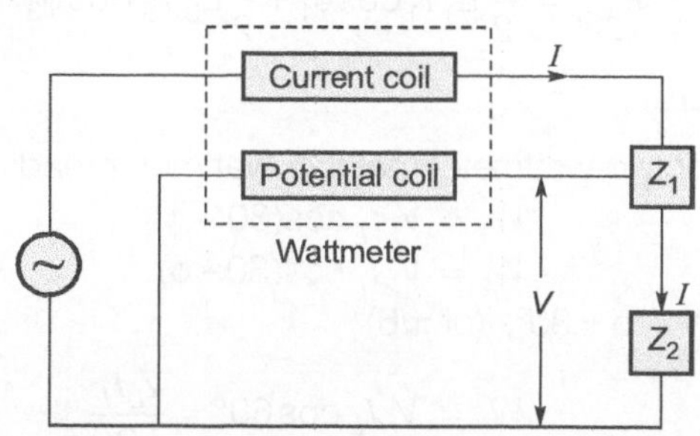

Potential coil draws negligible current, so current through z_1 and z_2 is same.

Current through current coil = $I_{cc} = I$

Voltage across potential coil = V_{pc}
Voltage across $z_2 = V_{pc} = V$
Wattmeter reads power consumed by z_2, as voltage across potential coil = Voltage across z_2
Current through current coil = Current through z_2

5.17 (b)

The current coil carries a current of $I + I_P$ and produces a filed corresponding to this current. The compensating coil is connected in series with the pressure coil circuit and is made as nearly as possible identical and coincident with the current coil. It is so connected that it opposes the field of the current coil. The compensating coil carries a current I_P and produces a field corresponding to this current. This field acts in opposition to the current coil field. Thus the resultant field is due to current I only. Hence the error caused by the pressure coil current flowing in the current coil is neutralized.

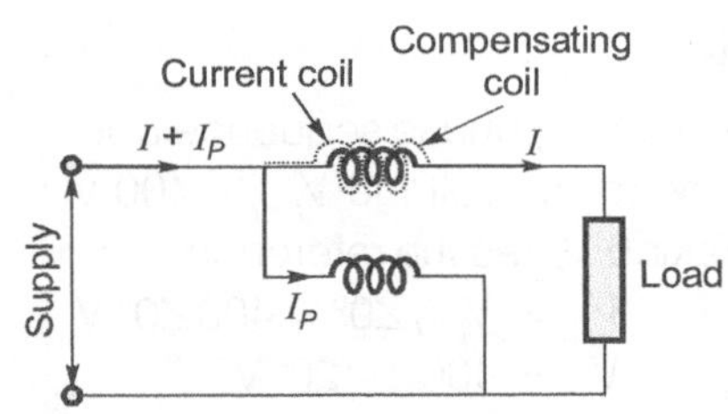

5.18 (c)

$$v(t) = E_1 \sin \omega t + E_3 \sin 3\omega t$$
$$i(t) = I_1 \sin(\omega t - \phi_1) + I_3 \sin(3\omega t - \phi_3) + I_5 \sin 5\omega t$$

Average power,

$$P_{avg} = \frac{1}{2\pi}\int_0^{2\pi} vi\, d(\omega t)$$

The product of different frequency terms have zero average value.

$$\therefore \quad P_{avg} = \frac{1}{2}E_1 I_1 \cos\phi_1 + \frac{1}{2}E_3 I_3 \cos\phi_3$$

5.19 (c)

For two-wattmeter method, wattmeter readings are

$$W_1 = V_L I_L \cos(30 + \phi)$$
$$W_2 = V_L I_L \cos(30 - \phi)$$

For $\phi = 30°$, (or $\pi/6$)

$$W_1 = V_L I_L \cos 60° = \frac{V_L I_L}{2}$$

and $\quad W_2 = V_L I_L \cos 0° = V_L I_L$

Therefore, $\quad W_2 = 2W_1$

5.20 Sol.

Given, $\quad W_1 = 250$ W, $\quad W_2 = 100$ W

Power factor angle is given by

$$\phi = \tan^{-1}\left[\frac{\sqrt{3}(W_1 - W_2)}{(W_1 + W_2)}\right]$$

$$= \tan^{-1}\left[\frac{\sqrt{3}(250 - 100)}{(250 + 100)}\right]$$

or, $$\phi = \tan^{-1}\left[\frac{\sqrt{3} \times 150}{350}\right]$$

$$= \tan^{-1}\left[\frac{3\sqrt{3}}{7}\right] = 36.58°$$

$\therefore$ Power factor of load,

$$\cos\phi = \cos 36.58° = 0.802 \approx 0.80$$

5.21 Sol.

Given, $\cos\phi = 0.2$ (LPF wattmeter)
Given, full scale reading of wattmeter,

$$W_m = 150 \text{ Watt}$$

Also, for $\quad V = 150$ volt, $I = 10$ A settings,
Wattmeter reading,

$$W = VI\cos\phi = 150 \times 10 \times 0.2 = 300$$

$\therefore$ Multiplying factor of wattmeter,

$$m = \frac{W}{W_m} = \frac{300}{150} = 2$$

5.22 (a)

p.f. = 0.707, $\phi = \cos^{-1}(0.707) = 45°$

$$P = 5 \text{ kW}$$
$$W_1 + W_2 = 5 \qquad \ldots(i)$$
$$\tan\phi = \frac{\sqrt{3}(W_1 - W_2)}{(W_1 + W_2)}$$
$$(W_1 + W_2) \times 1 = \sqrt{3}(W_1 - W_2)$$
$$W_1 - W_2 = \frac{5}{\sqrt{3}} \qquad \ldots(ii)$$

From equation (*i*) and (*ii*),

$$W_1 = 3.94 \text{ kW}; \quad W_2 = 1.06 \text{ kW}$$

5.23 Sol.

True value of power

$$= VI\cos\phi = 30 \times 25 \times 0.8 = 600 \text{ W}$$

In the given diagram,
Measured power = Power loss in pressure coil + Power in load

$$= \frac{V^2}{R_P} + VI\cos\phi = \frac{30^2}{1000} + 900 \text{ W} = 900.9 \text{ W}$$

∴ Percentage error in wattmeter reading

$$= \frac{\text{Measured value} - \text{True value}}{\text{True value}}$$

$$= \frac{600.9 - 600}{600} \times 100 = 0.15\%$$

5.24 Sol.

1200 rev/kWh, 20 rev, 30 sec

$$P_{\text{loss}} = \frac{20 \times 3600}{1200 \times 30} = 2 \text{ kW}$$

5.25 Sol.

Method 1:

The average power consumed by the load

$$P = V_1 I_1 \cos\phi_1 = \frac{100}{\sqrt{2}} \cdot \frac{10}{\sqrt{2}} \cos 60° = 250 \text{ W}$$

5.26 (d)

When switch is on Neutral side

$$W = W_1 + W_2 + W_3$$
$$= 577.35 + 577.35 + 577.35 = 1732 \text{ W}$$

VA load = 3464 = 3 $V_{ph} \cdot I_{ph}$

$\Rightarrow V_{ph} \cdot I_{ph} = 1154.66$

Each Ω meter reading

$$= V_{RN} \cdot I_R \cdot \cos(\angle V_{RN} \text{ and } I_R)$$
$$577.35 = V_{Ph} \cdot I_{ph} \cdot \cos(\phi)$$
$$\Rightarrow \quad \cos(\phi) = \frac{577.35}{1154.66} = 0.5$$
$$\phi = \cos^{-1}(0.5) \Rightarrow \phi = 60°$$

When switch is on Y-phase side

$$W_1 = V_{RY} \cdot I_R \cos(\angle V_{RY} \text{ and } I_R)$$
$$= \sqrt{3} V_{ph} \cdot I_{ph} \cdot \cos(30° + \phi)$$
$$W_1 = \sqrt{3} \times 1154.66 \times \cos(30° + 60°)$$
$$= 0 \text{ Watt}$$
$$W_3 = V_{BY} \cdot I_R \cdot \cos(\angle V_{BY} \text{ and } I_B)$$
$$= \sqrt{3} Y_{ph} \cdot I_{ph} \cdot \cos(30° - \phi)$$
$$W_3 = \sqrt{3} \times 1154.66 \times \cos(30° - 60°)$$
$$= 1732 \text{ Watts}$$
$$W_1 = 0, \quad W_2 = 0$$
$$W_3 = 1732 \text{ Watts}$$

5.27 (d)

In two Wattmeter method,

$$\tan\phi = \frac{\sqrt{3}(W_1 - W_2)}{(W_1 + W_2)}$$

Given, $W_2 = \dfrac{W_1}{2}$

$$\tan\phi = \frac{\sqrt{3}\left(W_1 - \frac{W_1}{2}\right)}{\left(W_1 + \frac{W_1}{2}\right)}$$

$$\phi = 30°$$
$$\cos\phi = \cos 30°$$
$$= 0.866$$

5.28 (588.9)

$$v(t) = -170 \sin\left(377t - \frac{\pi}{6}\right) \text{V} = v_{pc}$$

$$i(t) = 8\cos\left(377t + \frac{\pi}{6}\right) \text{A} = I_{cc}$$

$$i(t) = 8\sin\left(377t + \frac{2\pi}{3}\right) = I_{cc}$$

$$P_{\text{avg}} = \frac{1}{2} \times (-170)(8) \times \cos\left(-\frac{\pi}{6} - \frac{2\pi}{3}\right)$$

$$= \frac{1}{2} \times (-170)(8) \times \cos(150°)$$

$$= 588.9 \text{ W}$$

5.29 (c)

Wattmeter reading, $W = V_{YB} \times I_R \times \cos\phi'$

where ϕ' = phase angle between PC voltage and CC current, i.e., between V_{YB} and I_R.

$$-400 = 400 \times 2 \times \cos\phi'$$
$$\phi' = \cos^{-1}\left(-\frac{1}{2}\right)$$
$$= 120°$$

By above phasor diagram,

$$\phi = \phi' - 90° = 30°$$

and it is clearly seen that I_R leads V_R.

∴ Pf = cos ϕ = cos 30° = 0.866 leading

5.30 (692.8)

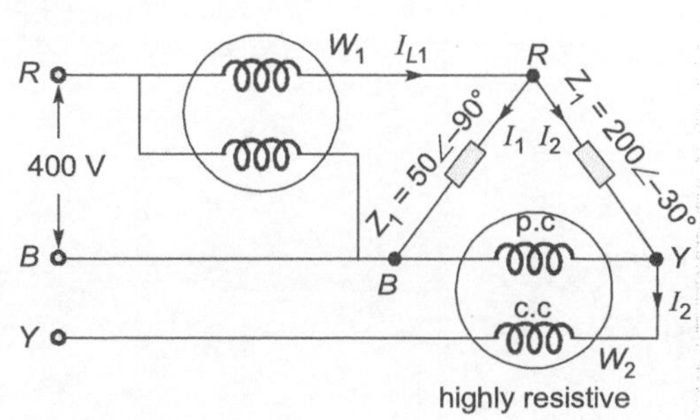

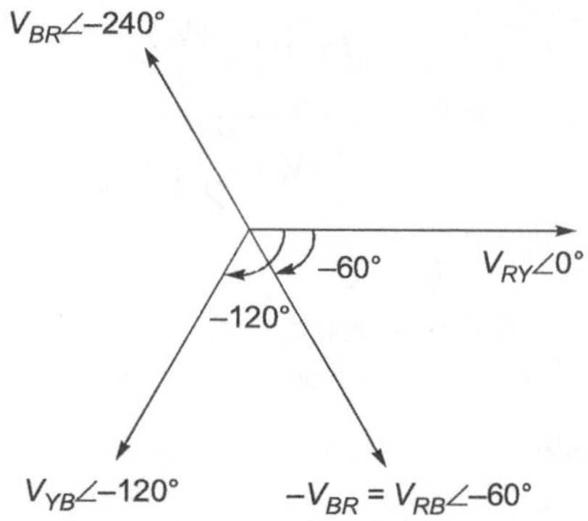

$$V_{RY} = 400\angle 0°$$
$$V_{YB} = 400\angle -120°$$
$$V_{BR} = 400\angle -240°$$
$$W(1) \Rightarrow \quad V_{PC} = V_{RB} = 400\angle -60°$$
$$I_1 = \frac{V_{RB}}{Z_1} = \frac{400\angle -60°}{50\angle -90°} = 8\angle 30°$$
$$I_2 = \frac{V_{RY}}{Z_2} = \frac{400\angle 0°}{200\angle -30°} = 2\angle 30° \Rightarrow I_{CC\,(2)}$$
$$I_{L1} = I_{CC\,(1)} = I_1 + I_2 = 8\angle 30° + 2\angle 30°$$
$$= 10\angle 30° \Rightarrow I_{CC\,(1)}$$
$$W_1 = V_{RB} \times I_{L1} \cos(\angle V_{RB} \text{ and } I_{L1})$$
$$W_1 = 400 \times 10 \times \cos(-60° - 30°) = 0 \text{ Watt}$$
$$W_2 = V_{YB} \times I_2 \cos(\angle V_{YB} \text{ and } I_2)$$
$$= 400 \times 2 \times \cos(-120° - 30°)$$
$$= -692.8 \text{ Watt}$$
$$W_1 - W_2 = 0 - (-692.8) = 692.8 \text{ Watt}$$

■■■■

6 CRO and Electronic Measurements

6.1 A certain oscilloscope with 4 cm screen has its own sweep output fed to its input. If the x and y sensitivities are same, the oscilloscope will display a

(a) triangular wave (b) diagonal line
(c) sine wave (d) circle **[1995 : 1 M]**

6.2 In an oscilloscope the input to the horizontal plates is a 100 Hz voltage signal. The lissajous pattern (A), (B) and (C) will be generated when different frequency voltage signals are applied to vertical plates. Match each lissajous pattern to the corresponding frequency by

A. 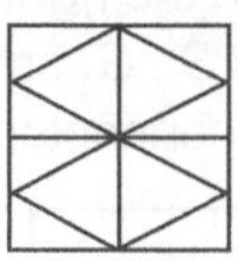 P. f_y = 50 Hz

Q. f_y = 66.66 Hz

B.  R. f_y = 150 Hz

S. f_y = 150 Hz

C. 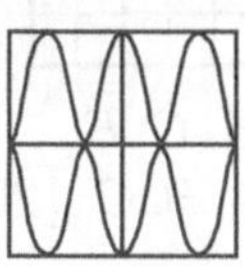 T. f_y = 200 Hz

U. f_y = 300 Hz

[1996 : 3 M]

6.3 Two in phase 50 Hz sinusoidal wave forms of unit amplitude are fed into channel-1 and channel-2 respectively of an oscilloscope. Assuming that the voltage scale, time scale and other settings are exactly the same for both the channels. What would be observed if the oscilloscope is operated in x-y mode?

(a) A circle of unit radius
(b) An ellipse
(c) A parabola
(d) A straight line inclined at 45° with respect to the x-axis **[2002 : 1 M]**

6.4 A reading of 120 is obtained when standard inductor was connected. The circuit of a Q-meter and the variable capacitor is adjusted to a value of 300 pF. A lossless capacitor of unknown value C_x is then connected in parallel with the variable capacitor and the same reading was obtained when the variable capacitor is readjusted to a value of 200 pF. The value of C_x in pF is

(a) 100 (b) 200
(c) 300 (d) 500 **[2003 : 1 M]**

6.5 The simplified block diagram of a 10-bit A/D converter of dual slope integrator type is shown in the figure. The 10-bit counter at the output is clocked by a 1 MHz clock. Assuming negligible timing overhead for the control logic, the maximum frequency of the analog signal that can be converted using this A/D converter is approximately

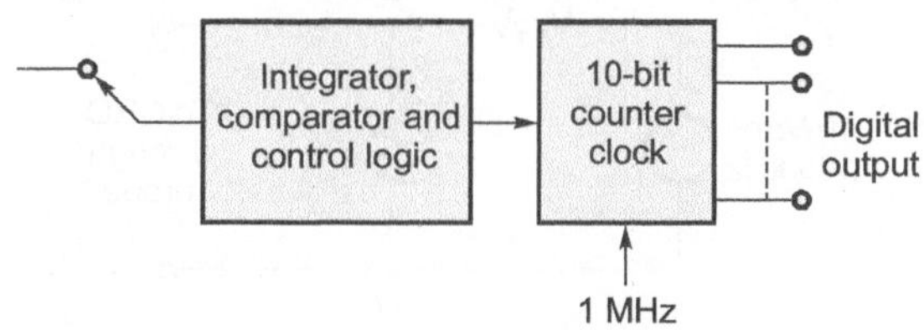

(a) 2 kHz (b) 1 kHz
(c) 500 Hz (d) 250 Hz **[2003 : 2 M]**

6.6 List-I represents the figures obtained on a CRO screen when the voltage signals $V_x = V_{xm} \sin\omega t$ and $V_y = V_{ym} \sin(\omega t + \Phi)$ are given to its X and Y plates respectively and Φ is changed. Choose the correct value of Φ from **List-I** to match with the corresponding figure of **List-II**.

List-I **List-II**

A. $\Phi = 0$ 1. 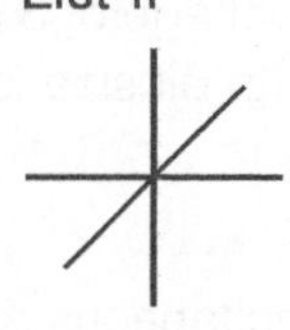

B. $\Phi = \pi/2$ 2. 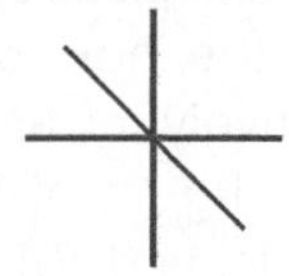

C. $\pi < \Phi < 3\pi/2$ 3. 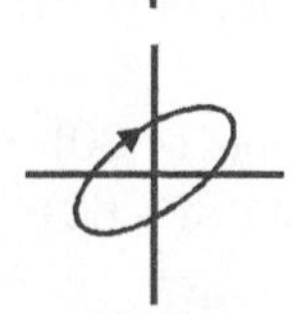

D. $\Phi = 3\pi/2$ 4. 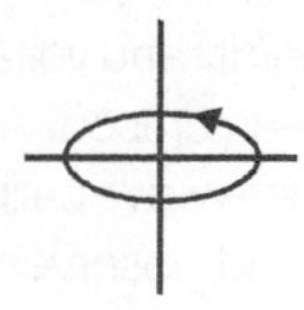

5. 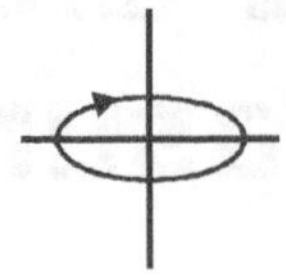

6. 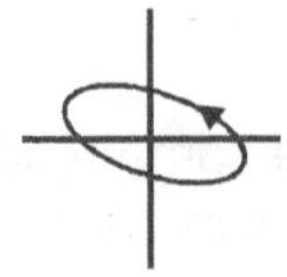

Codes:

	A	B	C	D
(a)	1	3	6	5
(b)	2	6	4	5
(c)	2	3	5	4
(d)	1	5	6	4

[2003 : 2 M]

6.7 A CRO probe has an impedance of 500 kΩ in parallel with a capacitance of 10 pF. The probe is used to measure the voltage between *P* and *Q* as shown in figure. The measured voltage will be

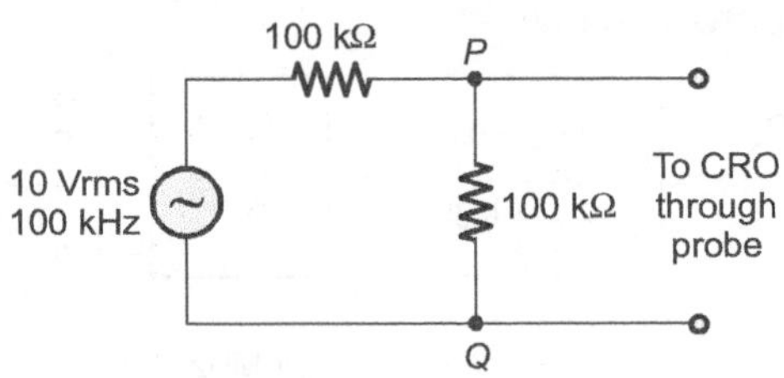

(a) 3.53 V (b) 4.37 V
(c) 4.54 V (d) 5.00 V [2004 : 2 M]

6.8 The Q-meter works on the principle of
(a) mutual inductance
(b) self inductance
(c) series resonance
(d) parallel resonance [2005 : 1 M]

6.9 A digital-to-analog converter with a full-scale output voltage of 3.5 V has a resolution close to 14 mV. Its bit size is
(a) 4 (b) 8
(c) 16 (d) 32 [2005 : 1 M]

6.10 The simultaneous application of signals $x(t)$ and $y(t)$ to the horizontal and vertical plates, respectively, of an oscilloscope, produces a vertical figure-of-8 display. If *P* and *Q* are constants, and $x(t) = P\sin(4t + 30)$, then $y(t)$ is equal to
(a) $Q\sin(4t - 30)$ (b) $Q\cos(2t + 30)$
(c) $Q\sin(8t + 60)$ (d) $Q\sin(4t + 30)$
[2005 : 2 M]

6.11 The time/div and voltage/div axes of an oscilloscope have been erased. A student connects a 1 kHz, 5 V p-p square wave calibration pulse to channel 1 of the scope and observes the screen to be as shown in the upper trace of the figure. An unknown signal is connected to channel 2 (lower trace) of the scope. It the time/div and V/div on both channels are the same, the amplitude (p-p) and period of the unknown signal are respectively

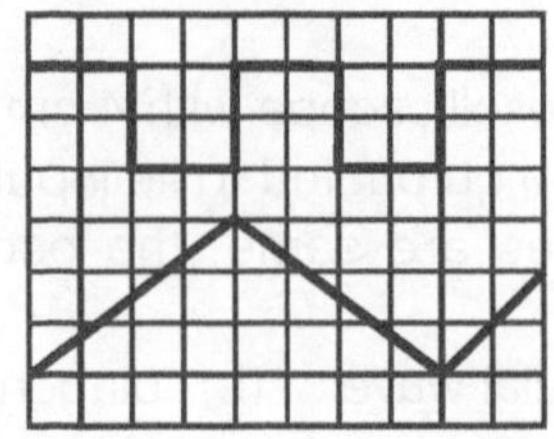

(a) 5 V, 1 ms (b) 5 V, 2 ms
(c) 7.5 V, 2 ms (d) 10 V, 1 ms [2006 : 1 M]

6.12 The probes of a non-isolated, two channel oscillocope are clipped to points *A*, *B* and *C* in the circuit of the adjacent figure. V_{in} is a square wave of a suitable low frequency. The display on Ch_1 and Ch_2 are as shown on the right. Then the "Signal" and "Ground" probes S_1, G_1 and S_2, G_2 of Ch_1 and Ch_2 respectively are connected to points:

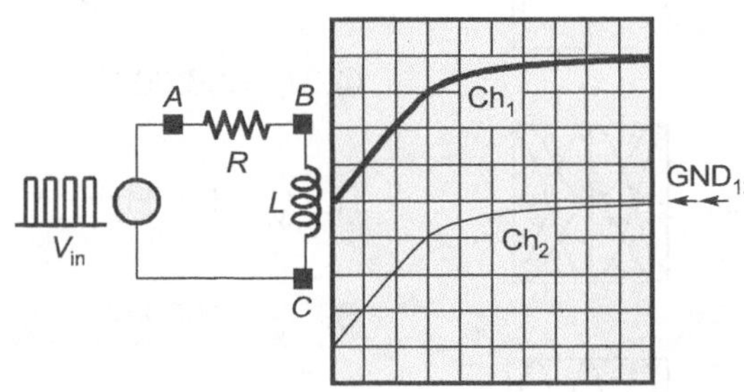

(a) *A, B, C, A* (b) *A, B, C, B*
(c) *C, B, A, B* (d) *B, A, B, C* [2007 : 1 M]

6.13 Two 8-bit ADCs, one of single slope integrating type and other of successive approximate type, take T_A and T_B times to convert 5 V analog input signal to equivalent digital output. If the input analog signal is reduced to 2.5 V, the approximate time taken two ADCs will respectively, be
(a) T_A, T_B (b) $T_A/2$, T_B
(c) T_A, $T_B/2$ (d) $T_A/2$, $T_B/2$ [2008 : 1 M]

6.14 Two sinusodial signals $\rho(\omega_1, t) = A\sin\omega_1 t$ and $q(\omega_2, t)$ are applied to *X* and *Y* inputs of a dual channel CRO. The Lissajous figure displayed on the screen is shown below:
The signal $q(\omega_2, t)$ will be represented as

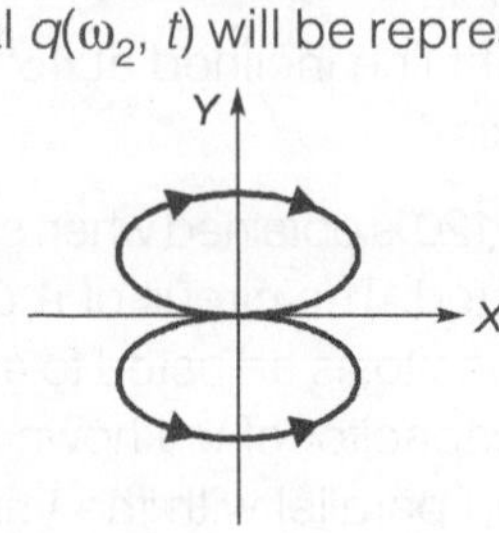

(a) $q(\omega_2 t) = A \sin \omega_2 t$, $\omega_2 = 2\omega_1$
(b) $q(\omega_2 t) = A \sin\omega_2 t$, $\omega_2 = \omega_1 /2$
(c) $q(\omega_2 t) = A \cos \omega_2 t$, $\omega_2 = 2\omega_1$
(d) $q(\omega_2 t) = A \cos \omega_2 t$, $\omega_2 = \omega_1 /2$ **[2008 : 2 M]**

6.15 The two inputs of a CRO are fed with two stationary periodic signals. In the *X-Y* mode, the screen shows a figure which changes from ellipse to circle and back to ellipse with its major axis changing orientation slowly and repeatedly. The following inference can be made from this.
(a) The signals are not sinusoidal.
(b) The amplitudes of the signals are very close but not equal.
(c) This signals are sinusoidal with their frequencies very close but not equal.
(d) There is a constant but small phase difference between the signals. **[2009 : 1 M]**

6.16 An average-reading digital multimeter reads 10 V when fed with a triangular wave, symmetric about the time-axis. For the same input an rms-reading meter will read

(a) $\frac{20}{\sqrt{3}}$ V (b) $\frac{10}{\sqrt{3}}$ V
(c) $20\sqrt{3}$ V (d) $10\sqrt{3}$ V **[2009 : 2 M]**

6.17 A dual trace oscilloscope is set to operate in the Alternate mode. The control input of the multiplexer use in the *y*-circuit is fed with a signal having a frequency equal to
(a) the highest frequency that the multiplexer can operate properly.
(b) twice the frequency of the time base (sweep) oscillator.
(c) the frequency of the time base (sweep) oscillator.
(d) half the frequency of the time base (sweep) oscillator. **[2011 : 1 M]**

6.18 A $4\frac{1}{2}$ digit DMM has the error specification as: 0.2% of reading + 10 counts. If a DC voltage of 100 V is read on its 200 V full scale, the maximum error that can be expected in the reading is
(a) ± 0.1% (b) ± 0.2%
(c) ± 0.3% (d) ± 0.4% **[2011 : 2 M]**

6.19 In an oscilloscope screen, linear sweep is applied at the
(a) vertical axis
(b) horizontal axis
(c) origin
(d) both horizontal and vertical axis
[2014 : 1 M, Set-1]

6.20 The two signals *S*1 and *S*2, shown in figure, are applied to *Y* and *X* deflection plates of an oscilloscope.

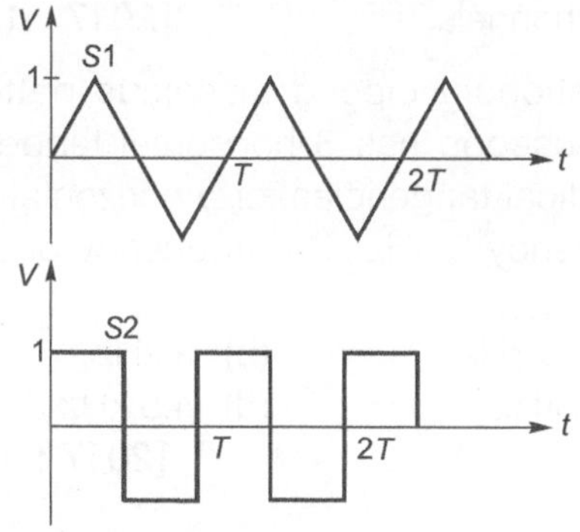

The waveform displayed on the screen is

(a)
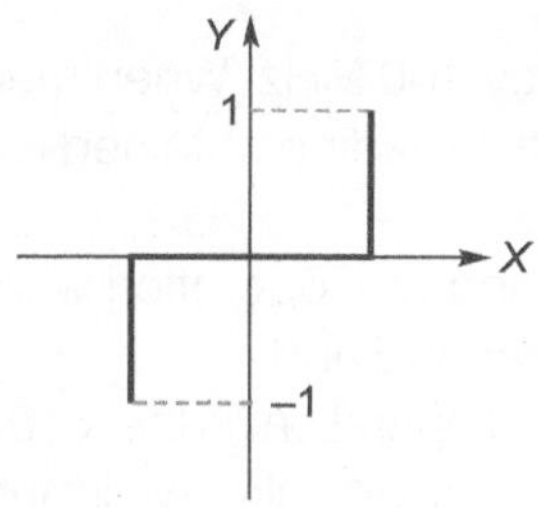

(b)
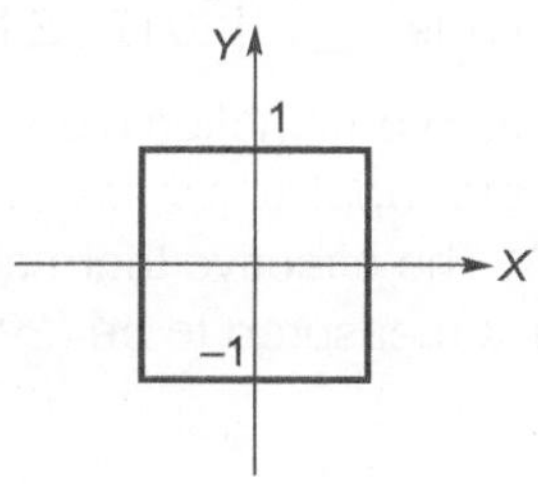

(c)
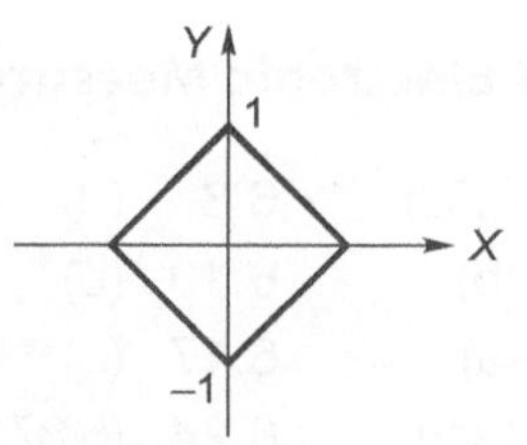

(d)
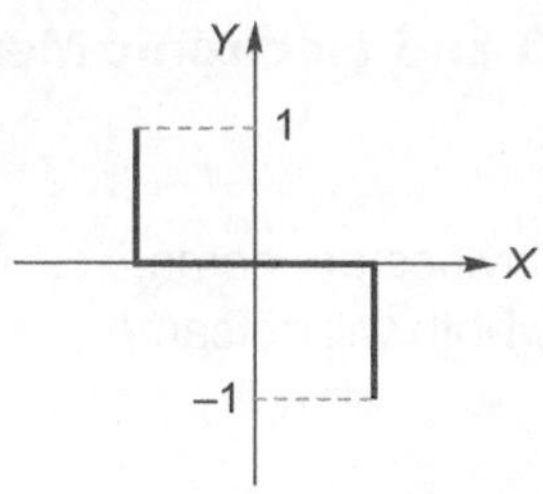

[2014 : 1 Mark, Set-3]

6.21 The slope and level detector circuit in a CRO has a delay of 100 ns. The start-stop sweep generator has a response time of 50 ns. In order to display correctly, a delay line of
(a) 150 ns has to be inserted into the *y*-channel.
(b) 150 ns has to be inserted into the *x*-channel.

(c) 150 ns has to be inserted into both x and y channels.
(d) 100 ns has to be inserted into both x and y channels. [2017 : 1 M, Set-1]

6.22 A stationary closed Lissajous pattern on an oscilloscope has 3 horizontal tangencies and 2 vertical tangencies for a horizontal input with, frequency 3 kHz. The frequency of the vertical input is,

(a) 1.5 kHz (b) 2 kHz
(c) 3 kHz (d) 4.5 kHz

[2017 : 1 M, Set-2]

6.23 A $10\frac{1}{2}$ digit timer counter possesses a base clock of frequency 100 MHz. When measuring a particular input, the reading obtained is the same in:

(i) Frequency mode of operation with a gating time of one second and
(ii) Period mode of operation (in the × 10 ns scale). The frequency of the unknown input (reading obtained) in Hz is____. [2017 : 2 M, Set-2]

6.24 A non-ideal Si-based pn-junction diode is tested by sweeping the bias applied across its terminals from –5 V to +5 V. The effective thermal voltage, V_T, for the diode is measured to be (29 ± 2) mV. The resolution of the voltage source in the measurement range is 1 mV. The percentage uncertainty (rounded off to 2 decimal plates) in the measured current at a bias voltage of 0.02 V is ______. [2020 : 2 M]

6.25 The temperature of the coolant oil bath for a transformer is monitored using the circuit shown. It contains a thermistor with a temperature-dependent resistance, $R_{thermistor} = 2(1 + \alpha T)$ kΩ. Where T is the temperature in °C. The temperature coefficient α, is –(4 ± 0.25) %/°C. Circuit parameters: R_1 = 1 kΩ, R_2 = 1.3 kΩ, R_3 = 2.6 kΩ. The error in the output signal (in V. rounded off to 2 decimal places) at 150° C is ______.

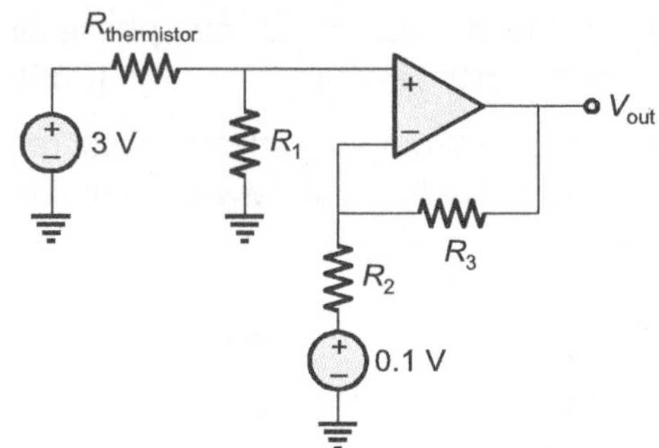

[2020 : 2 M]

Answers CRO and Electronic Measurements

6.1 (b)	6.2 (300)	6.3 (d)	6.4 (a)	6.5 (b)	6.6 (d)	6.7 (b)
6.8 (c)	6.9 (b)	6.10 (b)	6.11 (c)	6.12 (b)	6.13 (b)	6.14 (d)
6.15 (d)	6.16 (a)	6.17 (b)	6.18 (c)	6.19 (b)	6.20 (a)	6.21 (a)
6.22 (d)	6.23 (Sol)	6.24 (5.87)	6.25 (0.04)			

Explanations CRO and Electronic Measurements

6.1 (b)

∵ x and y sensitivities are same.
∴ Deflection by both will be same.

6.2 Sol.

(A) – (P):

$$\frac{f_x}{f_y} = \frac{2}{1} \Rightarrow f_y = \frac{100}{2} = 50 \text{ Hz}$$

(B) – (S):

$$\frac{f_x}{f_y} = \frac{2}{3} \Rightarrow f_y = 100 \times \frac{3}{2} = 150 \text{Hz}$$

(C) – (U):

$$\frac{f_x}{f_y} = \frac{1}{3} \Rightarrow f_y = 3 \times 100 = 300 \text{ Hz}$$

6.3 (d)

∵ Phase difference = 0° also $f_x = f_y$.

6.4 (a)

Q of the coil = 120

$$Q = \frac{\omega L}{R} = \frac{1}{\omega RC}$$

Case-I:
Q = 120 is obtained for C = 300 pF

Case-II:

For same reading i.e. $Q = 120$

Effective capacitance of the circuit should be 300 pF.

Effective capacitance of parallel capacitor,

$$C_{eff.} = C + C_x$$
$$C_x = C_{eff.} - C = 300 - 200 = 100 \text{ pF}$$

6.5 (b)

The wave form for dual slope integrater is shown. The maximum frequency can be attained when $T_2 - T_1 = 0$ and as

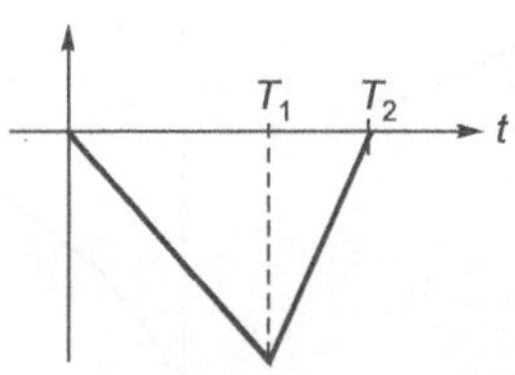

$$T_1 = 2^N T_C \qquad (T_C = \text{Clock period})$$

$$\therefore \quad f_{max} = \frac{1}{2^N \cdot T_C} = \frac{f_C}{2^N} = \frac{10^6}{1024} \approx 1 \text{ kHz}$$

6.6 (d)

When the two voltages are in phase i.e. [$\phi = 0°$ or $180°$], the trace is straight line. But for $\phi = 0°$, the trace has positive slope and for $\phi = 180°$, the trace has negative slope.

When the two voltages has phase displacement of 90°, the trace is a circle. If the direction of the trace is in the clockwise direction then the phase difference is $\phi = 90°$.

If the direction of the trace is anti-clockwise direction the phase difference is

$$(2\pi - \phi) = \left(2\pi - \frac{\pi}{2}\right) = 3\frac{\pi}{2}.$$

When the phase difference is not equal to 0° or 90°, the trace is an ellipse. For $\pi \le \phi \le \frac{3\pi}{2}$, the trace is an ellipse is anti-clockwise direction.

6.7 (b)

$$X_c = \frac{1}{2\pi fC}$$
$$= \frac{1}{2 \times \pi \times 100 \times 10^3 \times 10 \times 10^{-12}} = 159.15 \text{ k}\Omega$$

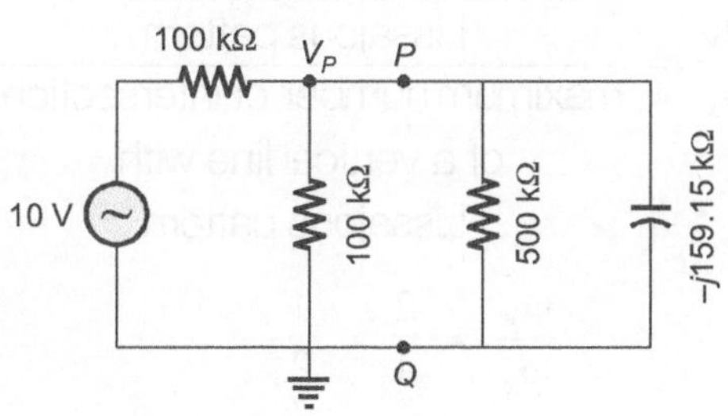

Taking supply voltage as the reference.

$$V_s = 10\angle 0 \text{ V}$$

Using KCL,

$$\frac{V_p - 10}{100 \times 10^3} + \frac{V_p}{100 \times 10^3} + \frac{V_p}{500 \times 10^3} + \frac{V_p}{-j159.15 \times 10^3} = 0$$
$$V_p = 4.37\angle -15.94 \text{ V}$$

6.8 (c)

$$Q = \frac{\omega_0 L}{R}$$

Where ω_0 is the resonant angular frequency, L is the inductance and R is the effective resistance of the coil.

The principle of working of Q-meter is based upon series resonance of R, L, C circuit.

6.9 (b)

Resolution of digital to analog converter

$$R = \frac{V_0}{2^N} = 14 \text{ mV}$$

where,

V_0 = Full scale output voltage = 3.5 V

N = Bit-size

$$\frac{V_0}{2^N} = 14 \text{ mV}$$

$$\Rightarrow \quad 2^N = \frac{3.5}{14 \times 10^{-3}} = 250 \Rightarrow N = 8$$

6.10 (b)

f_y = Frequency of signal applied to y plates

f_x = Frequency of signal applied to X plates

$$x(t) = P\sin(4t + 30)$$

So, $$f_x = \frac{4}{2\pi} \text{Hz}$$

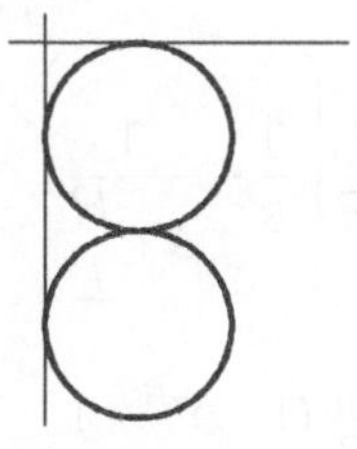

$$\frac{f_y}{f_x} = \frac{\text{No. of times tangent touches top or bottom}}{\text{No. of times tangent touches either side}} = \frac{1}{2}$$

$$f_y = \frac{f_x}{2} = \left(\frac{4}{2\pi}\right) \times \frac{1}{2} = \frac{2}{2\pi}$$

$$\omega_y = 2\pi f_y = 2\pi\left(\frac{2}{2\pi}\right) = 2 \text{ rad/sec.}$$

$Q\sin(2t + 15)$ has angular frequency

$\omega = 2$ rad/sec

6.11 (c)

Peak-peak (p-p) division of upper trace voltage = 2 and value of (p-p) voltage = 5 V

$$\therefore \quad \frac{\text{Voltage}}{\text{Divisions}} = 2.5 \text{ V}$$

Now it will be same for unknown voltage (p-p) division of unknown voltage = 3

$\therefore$ p-p voltage = 3 × 2.5 = 7.5 V

Frequency of upper trace = 1 kHz

$$\therefore \quad \text{Time period} = \frac{1}{10^3} = 1 \text{ ms}$$

Division of x-axis (upper) = 4

Division of x-axis (lower) = 8

$\therefore$ Period of unknown signal = 2 ms

6.12 (b)

Square wave is of low frequency. So, it can be assumed that time during which the wave forms are displayed on the screen, the voltage across R and L is V_{in}.

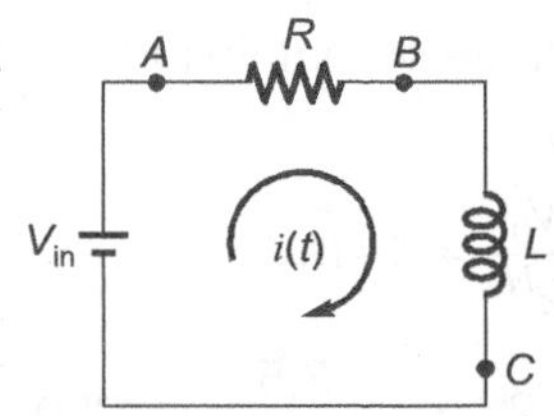

$$I(s) = \frac{V_{in}/s}{R+Ls} = \frac{V_{in}}{Ls\left(s+\frac{R}{L}\right)}$$

$$= \frac{V_{in}}{L}\left(\frac{1}{s} - \frac{1}{s+\frac{R}{L}}\right) \times \frac{L}{R}$$

$$I(s) = \frac{V_{in}}{R}\left(\frac{1}{s} - \frac{1}{s+\frac{R}{L}}\right)$$

$$i(t) = \frac{V_{in}}{R}(1-e^{-Rt/L})$$

V_{AB} = Voltage across resistance

$= R\,i(t) = V_{in}\,[1 - e^{-Rt/L}]$

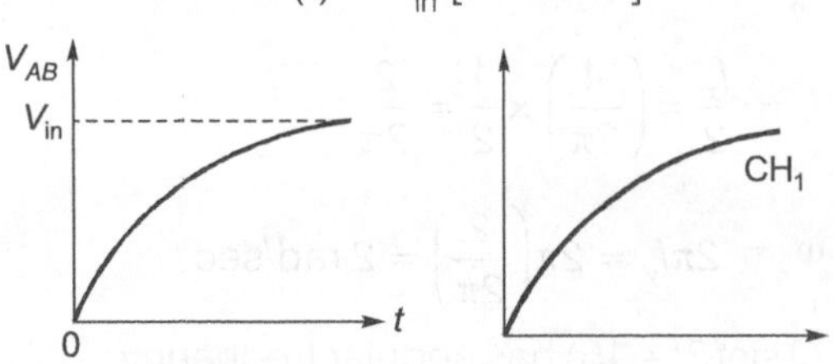

So, S_1 is connected to point A

and G_1 is connected to point B.

$$\text{Voltage across inductor} = V_{BC} = L\frac{di}{dt}$$

$$= L\frac{d}{dt}\left[\frac{V_{in}}{R}(1-e^{-Rt/L})\right]$$

$$V_{BC} = V_{in}e^{-Rt/L}$$

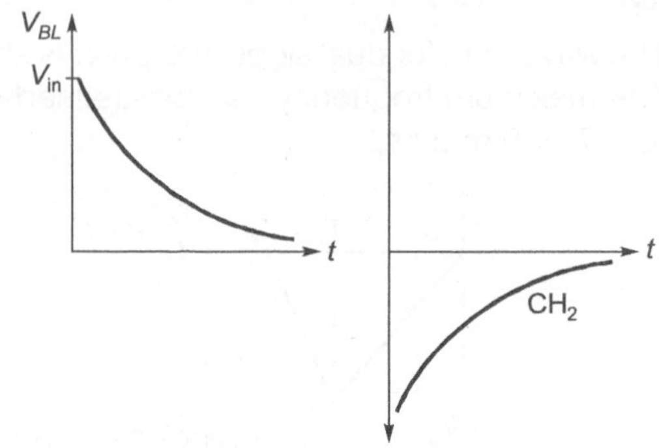

So, S_2 is connected to point C.

and G_2 is connected to point B.

6.13 (b)

- Single slope integrating type ADC utilizes digital counter techniques to measure time required for a voltage ramp to rise from zero to the input voltage.
- If conversion time for input voltage 5 V = T_A
  So, conversion time for input voltage 2.5 V = $T_A/2$
  Conversion time in successive type ADC does not depend on input voltage. So, conversion time for input voltage 2.5 V is also T_B.

6.14 (d)

Here, $\quad P(\omega_1 t) = A \sin \omega_1 t$

y_1 - line cut '4' times the Lissajous pattern

x_1 - line cut '2' times the Lissajous pattern

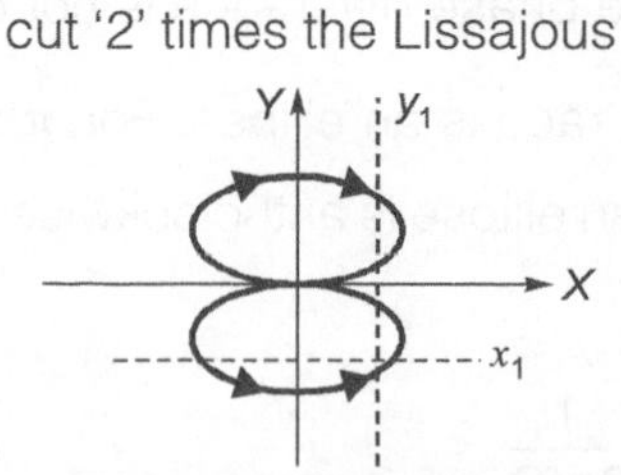

$$\frac{\omega_y}{\omega_x} = \frac{\text{maximum number of intersection of a horizontal line with Lissajous pattern}}{\text{maximum number of intersection of a vertical line with Lissajous pattern}}$$

$$\therefore \quad \frac{\omega_y}{\omega_x} = \frac{2}{4} = \frac{1}{2}$$

$$\therefore \qquad \omega_x = 2\omega_y$$
$$\omega_1 = 2\omega_2$$

and $q(\omega_2, t)$ will lead $p(\omega_1, t)$ by 90° as trace is a circle,

$$q(\omega_2, t) = A\sin(\omega_2 t + 90°)$$
$$q(\omega_2, t) = A\cos\omega_2 t$$

6.15 (d)

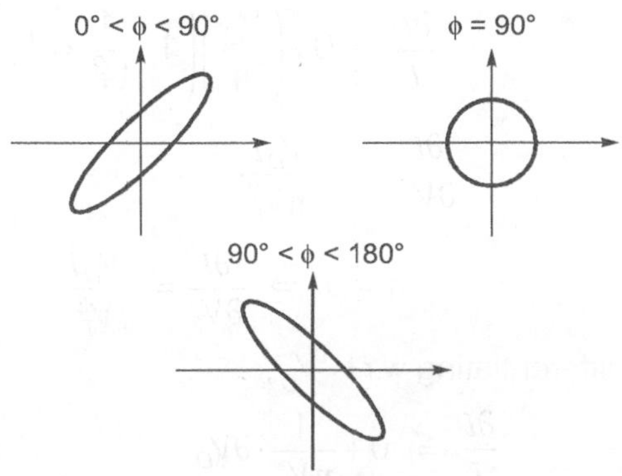

Because phase difference only patterns changes from ellipse to circle and back to ellipse.

6.16 (a)

For triangular wave,

$$\text{Average value} = \frac{V_m}{2}$$
$$\text{RMS value} = \frac{V_m}{\sqrt{3}}$$
$$\therefore \qquad \frac{V_m}{2} = 10\text{ V} \Rightarrow V_m = 20$$
$$\text{RMS value} = \frac{V_m}{\sqrt{3}} = \frac{20}{\sqrt{3}}\text{ V}$$

6.18 (c)

$4\frac{1}{2}$ digit

No. of full digits in case of $4\frac{1}{2}$ digit display = 4

So, maximum count with $4\frac{1}{2}$ digits display = 20,000

Full scale reading = 200 V

$$1\text{ count} = \frac{200\text{ V}}{20{,}000}$$

E_1 = Error corresponding to 10 counts

$$= 10 \times \frac{200}{20{,}000} = 0.1\text{V}$$

Reading = 100 V

Error corresponding to 0.2% of reading

$$E_2 = 0.2 \times \frac{100}{100} = 0.2\text{V}$$

$$\text{Total error} = E_1 + E_2 = 0.1 + 0.2$$
$$= 0.3\text{ V, i.e., } 0.3\%$$

6.19 (b)

A CRO uses a horizontal input voltage which is an internally generated ramp voltage (i.e. linear sweep voltage) called "Time Base of CRO". An oscilloscope is used to display a waveform which varies as a function of time. To reproduce the waveform accurately, the beam must have a constant horizontal velocity i.e. the deflecting voltage must increase linearly with time. This voltage is called a ramp voltage. Thus, a sawtooth waveform also called linear sweep voltage is applied to the horizontal deflecting system (or horizontal axis) which provides a time base to CRO.

6.20 (a)

The waveform pattern appearing on the screen of CRO is shown below. Here, S_1 is applied to vertical deflecting plate and S_2 to horizontal deflecting plate.

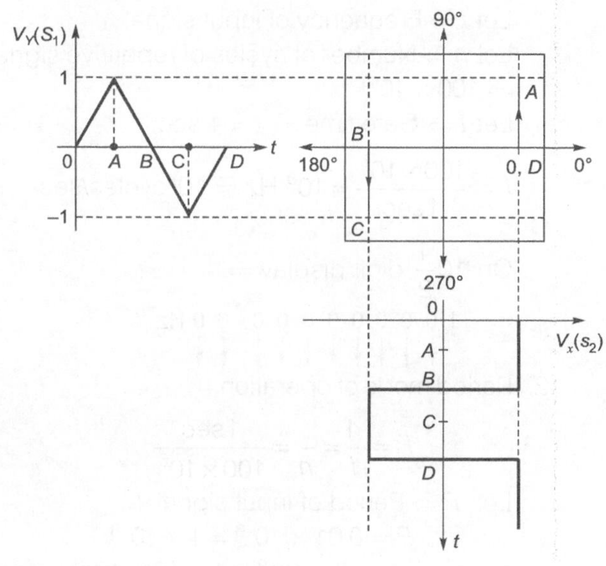

Points	$\lvert S_1\rvert$	$\lvert S_2\rvert$	$\theta = \tan^{-1}\left(\frac{S_1}{S_2}\right)$	$\sqrt{S_1^2 + S_2^2}$
0	$\lvert 0\rvert = 0$	$\lvert 1\rvert = 1$	0°	1
A	$\lvert 1\rvert = 1$	$\lvert 1\rvert = 1$	45°	$\sqrt{2}$
B	$\lvert 0\rvert = 0$	$\lvert -1\rvert = 1$	180°	1
C	$\lvert -1\rvert = 1$	$\lvert -1\rvert = 1$	225°	$\sqrt{2}$
D	$\lvert 0\rvert = 0$	$\lvert 1\rvert = 1$	0°	1

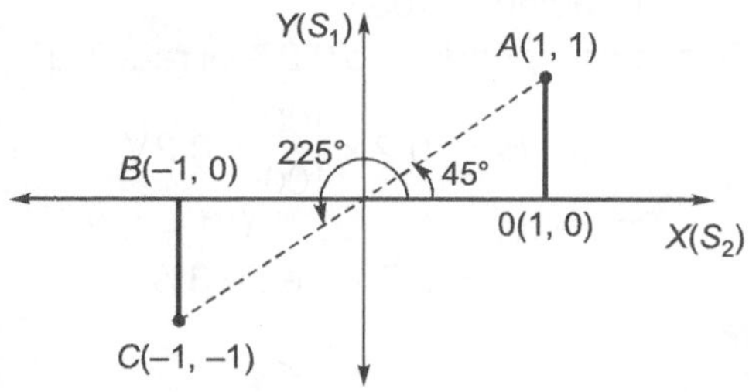

6.21 (a)

A delay line of 150 ns has to be inserted into the *y*-channel to get synchronous between sweep signal applied to *x*-plate and test signal applied to *y*-plate.

6.22 (d)

$$\frac{f_y}{f_x} = \frac{\text{Horizontal Tangencies}}{\text{Vertical Tangencies}}$$

$$\Rightarrow \quad \frac{f_y}{3} = \frac{3}{2} \quad \Rightarrow \quad f_y = 4.5 \text{ kHz}$$

6.23 Sol.

1. $10\frac{1}{2}$ digital timer counter:

Frequency mode of operation: $f = \dfrac{n}{t}$

Let $f \Rightarrow$ Frequency of input signal

Let $n \Rightarrow$ Number of cycles of repetitive signal $\Rightarrow 100 \times 10^6$

Let $t \Rightarrow$ Gate time $\Rightarrow t = 1$ sec.

$$f = \frac{100 \times 10^6}{1 \text{ sec}} = 10^8 \text{ Hz} = 10^8 \text{cycles/sec.}$$

On $10\frac{1}{2}$ digit display $\Rightarrow$

1 0 0 0 0 0 0 0 0 0 • 0 0 Hz
↓ ↓ ↓ ↓ ↓ ↓ ↓ ↓ ↓ ↓ ↓

2. Period mode of operation:

$$P = \frac{1}{f} = \frac{t}{n} = \frac{1 \text{ sec}}{100 \times 10^6}$$

Let, $P \Rightarrow$ Period of input signal

$$P = 0.01 \times 10^{-6} = 1 \times 10^{-8}$$
$$= 10 \times 10^{-9} = 1 \times 10 \times 10^{-9}$$
$$= 1 \times 10 \text{ n-sec.}$$

= 1 0 • 0 0 0 0 0 0 0 0 0 0 n-sec
↓ ↓ ↓ ↓ ↓ ↓ ↓ ↓ ↓ ↓ ↓

6.24 Sol.

$$V_T = (29 \pm 2) \text{ mV}$$

$$= \left(\underset{\downarrow \atop V_T}{0.029} \pm \underset{\downarrow \atop W_{V_T}}{(0.002)} \right) \text{V} \quad \Bigg| \quad \begin{matrix} V_D = 0.02 \text{ V} \\ W_{V_D} = 1 \text{mV} = 0.001 \text{ V} \end{matrix}$$

$$I_D = I = I_0 \cdot e^{\frac{V_D}{\eta V_T}}$$

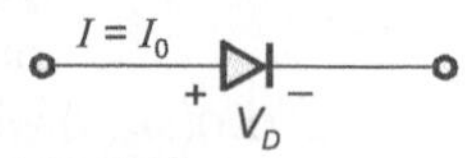

Applying log on both sides, we get

$$\ln(I) = \ln(I_0) + \frac{V_D}{\eta V_T}$$

Differentiating w.r.t. 'V_T',

$$\frac{\partial I}{I} = 0 + \left(\frac{V_D}{\eta}\right)\left(1 - \frac{1}{V_T^2} \times \partial V_T\right)$$

$$\Rightarrow \quad \frac{\partial I}{\partial V_T} = -\frac{V_D I}{\eta V_T^2}$$

For $\quad \eta = 1 \Rightarrow \dfrac{\partial I}{\partial V_T} = -\dfrac{V_D I}{V_T^2}$

Differentiating w.r.t. 'V_D',

$$\frac{\partial I}{I} = 0 + \frac{1}{\eta V_T} \cdot \partial V_D$$

$$\Rightarrow \quad \frac{\partial I}{\partial V_D} = -\frac{I}{\eta V_T}$$

$$\Rightarrow \quad \frac{\partial I}{\partial V_D} = \frac{I}{V_T}; \quad \text{for } \eta = 1$$

$$W_{\text{res}} = \sqrt{\left(\frac{\partial I}{\partial V_T}\right)^2 W_{V_T}^2 + \left(\frac{\partial I}{\partial V_D}\right)^2 W_{V_D}^2}$$

$$= \sqrt{\left(-\frac{V_D I}{V_T^2}\right)^2 W_{V_T}^2 + \left(\frac{I}{V_T}\right)^2 W_{V_D}^2}$$

$$= \sqrt{\frac{V_D^2 \cdot I^2}{V_T^4} \times W_{V_T}^2 + \left(\frac{I}{V_T}\right)^2 \times W_{V_D}^2}$$

$$W_{\text{res}} = \sqrt{\frac{I^2 \times (0.02)^2}{(0.029)^4} \times (0.002)^2 + \frac{I^2}{(0.029)^2} \times (0.001)^2}$$

$$= I \times \sqrt{2.262 \times 10^{-3} + 1.189 \times 10^{-3}}$$

$$= 0.0587 \times I$$

$$\% \frac{W_{\text{res}}}{I} = \pm 0.0587 \times 100$$

$$\Rightarrow \quad \% \frac{W_I}{I} = \pm 5.87\%$$

[Data given in question is incomplete, solution above has assumed $\eta = 1$]

6.25 Sol.

Given data,

$$R_{\text{thermistor}} = R_{\text{th}} = 2(1 + \alpha T) \text{ k}\Omega$$

$$\alpha = -(4 + 0.25)\% \text{ /°C}$$

$$= -(0.04 \pm 0.0025) \text{ °C}$$

$$\alpha_{\max} = -0.0425 \text{ /°C},$$

$$\alpha_{min} = -0.375/°C$$

Temperature, $T = 150°C$

$$R_1 = 1\ k\Omega,\ R_2 = 1.3\ k\Omega,\ R_3 = 2.6\ k\Omega$$

Considering, $\alpha = -0.04$

$$\Rightarrow \quad R_{th} = 2[1 - 0.04 \times 150] = -10\ k\Omega$$

$$V_0 = V_1 \times \frac{R_1}{R_1 + R_{th}}\left[1 + \frac{R_2}{R_3}\right]$$

$$= 3 \times \frac{1k}{1k + 10k}\left[1 + \frac{1.3k}{2.6k}\right] = 0.5\ V$$

Case-1:

Considering,

$$\alpha_{max} = -0.0425/°C$$

$$R_{thmax} = 2[1 + (-0.0425) \times 150]\ k\Omega$$

$$= -10.75\ k\Omega$$

$$V_0 = V_1 \times \frac{R_1}{R_1 + R_{th}}\left[1 + \frac{R_2}{R_3}\right]$$

$$= 3 \times \frac{1k}{1k - 10.75k}\left[1 + \frac{1.3k}{2.6k}\right]$$

$= -0.46$ volt $\Rightarrow$ For R_{thmax}

Case-2:

Considering,

$$\alpha_{min} = 0.0375/°C$$

$$R_{Th\,min} = 2[1 + (-0.0375) \times 150]\ k\Omega$$

$$= -9.25\ k\Omega$$

$$V_0 = 3 \times \frac{1k}{1k + (-9.25k)}\left[1 + \frac{1.3k}{2.6k}\right]$$

$= -0.54$ Volt = for R_{thmin}

Output voltage,

$$V_0 = 0.5 \pm 0.04 \Rightarrow \text{(Error)}$$

$$\text{Error} = 0.04$$

[Marks to all was awarded for this question]

■■■■

7 Instrument Transformers

7.1 Instrument transformers are known to introduce magnitude and phase errors in measurements. These are primarily due to
(a) improper connections on the primary side.
(b) measurement errors inherent in the meter connected to the transformer secondary.
(c) open and short-circuit parameters of the instrument transformers.
(d) None of these **[2000 : 1 M]**

7.2 In a constant voltage transformer (CVT), the output voltage remains constant due to
(a) capacitor (b) input inductor
(c) saturation (d) tapped windings
[2000 : 1 M]

7.3 A 500 A/5 A, 50 Hz current transformer has a bar primary. The secondary burden is a pure resistance of 1 Ω and it draws a current of 5 A. If the magnetic core requires 250 AT for magnetization, the percentage ratio error is
(a) 10.56 (b) –10.56
(c) 11.80 (d) –11.80 **[2003 : 2 M]**

7.4 A 50 Hz, bar primary CT has a secondary with 500 turns. The secondary supplies 5 A current into a purely resistive burden of 1 Ω. The magnetizing ampere-turns is 200. The phase angle between the primary and secondary current is
(a) 4.6° (b) 85.4°
(c) 94.6° (d) 175.4° **[2004 : 2 M]**

7.5 The core flux in the CT of Prob. **(Q.7.4)**, under the given operating conditions
(a) 0 (b) 45.0 μWb
(c) 22.5 mWb (d) 100.0 mWb
[2004 : 2 M]

7.6 A 200/1 Current transformer (CT) is wound with 200 turns on the secondary on a toroidal core. When it carries a current of 160 A on the primary, the ratio and phase errors of the CT are found to be –0.5% and 30 minutes respectively. If the number of secondary turns is reduced by 1 the new ratio error (%) and phase error (min) will be respectively
(a) 0.0, 30 (b) –0.5, 35
(c) –1.0, 30 (d) –1.0, 25 **[2006 : 2 M]**

7.7 The table lists two instrument transformers and their features :

Instrument Transformers	Features
	(P) Primary is connected in parallel to the grid
(X) Current Transformer (CT)	(Q) Open circuited secondary is not desirable
(Y) Potential Transformer (PT)	(R) Primary current is the line current
	(S) Secondary burden affects the primary current

The correct matching of the two column is
(a) X matches with P and Q; Y matches with R and S.
(b) X matches with Q and S; Y matches with P and R.
(c) X matches with Q and R; Y matches with P and S.
(d) X matches with P and R; Y matches with Q and S.

[2024 : 1 M]

■■■■

Answers Instrument Transformers

7.1 (d) 7.2 (d) 7.3 (b) 7.4 (d) 7.5 (b) 7.6 (a) 7.7 (c)

Explanations Instrument Transformers

7.3 (b)

Nominal ratio $= \frac{500}{5} = 100$

Number of turn in primary $N = 1$ (primary bar)

Magnetizing current,

$$I_m = \frac{\text{mmf}}{\text{No. of turns}} = \frac{250}{1} = 250\text{ A}$$

Secondary current,

$$I_s = 5\text{ A};\quad n = \frac{500}{5} = 100$$

Primary current,

$$I_P = \sqrt{(nI_s)^2 + I_m^2} = \sqrt{(100\times 5)^2 + (250)^2}$$

$\Rightarrow$ $I_P = 559.017$ A

$$\text{Actual ratio} = \frac{I_P}{5} = \frac{559.017}{5} = 111.803$$

Ratio error

$$= \frac{\text{Nominal Ratio} - \text{Actual ratio}}{\text{Actual ratio}}\times 100$$

$$= \frac{100 - 111.803}{111.803}\times 100 \approx -10.56\%$$

7.4 (d)

Phase angle between primary and secondary current

$$\theta = \left(\frac{180}{\pi}\right)\left(\frac{I_m}{nI_s}\right)\text{ degree}$$

No. of turns in primary, $N_p = 1$

No. of turns in secondary, $N_s = 500$

$$n = \text{Turn ratio} = \frac{N_s}{N_p} = 500$$

I_s = Secondary winding current = 5 A

Magnetizing ampere turn = 200

$$I_m = \frac{\text{Magnetizing ampere turn}}{N_p}$$

$$= \frac{200}{1} = 200\text{ A}$$

$$\theta = \left(\frac{180}{\pi}\right)\left(\frac{200}{500\times 5}\right) \approx 4.6°$$

7.5 (b)

Z_s = Burden = 1 Ω

Voltage induced in the secondary,

$$E_s = I_S \times Z_S = 5\times 1 = 5\text{ V}$$

$$E_s = 4.44\, f\phi_m N_s$$

$$\therefore\quad \phi_m = \frac{E_s}{4.44\, f N_s}$$

$$= \frac{5}{4.44\times 50\times 500} = 45\ \mu\text{Wb}$$

7.6 (a)

$$\text{Nominal ratio, } k_n = \frac{200}{1} = 200$$

Primary current, $I_P = 160$ A

$$\text{Actual ratio,}\quad R_1 = n_1\left(1 + \frac{I_e}{I_P}\right) \quad \ldots(i)$$

$$\%\text{ Ratio error} = \frac{k_n - R_1}{R_1}\times 100$$

$$-0.5 = \frac{200 - R_1}{R_1}\times 100$$

Actual ratio, $R_1 = 201$

n_1 = turn ratio = 200

$$R_1 = 200\left(1 + \frac{I_e}{I_P}\right) \quad \ldots(ii)$$

When number of secondary is reduced by 1,

$N_s = 199$, $n_2 = 199$

$$R_2 = n_2\left(1 + \frac{I_e}{I_P}\right) = 199\left(1 + \frac{I_e}{I_P}\right) \quad \ldots(iii)$$

Dividing equation (*ii*) and (*iii*),

$$\frac{R_1}{R_2} = \frac{200}{199} \Rightarrow R_2 = R_1\times\left(\frac{199}{200}\right)$$

$$= 201\times\frac{199}{200} = 199.995 \approx 200$$

$$\%\text{ Ratio error} = \frac{k_n - R_2}{R_2}\times 100$$

$$= \frac{200 - 200}{200}\times 100 = 0$$

$$\text{Phase angle error} = \theta = \frac{180}{\pi}\left(\frac{I_m}{I_P}\right)$$

Reduction of one or two turns of the secondary winding, no doubt, reduces the ratio error, but is has no effect on the phase angle error.

Analog Electronics

UNIT VII

CONTENTS

Analog Electronics

Syllabus

Simple Diode Circuits: Clipping, Clamping, Rectifiers

Amplifiers: Biasing, Equivalent Circuit and Frequency Response; Oscillators and Feedback Amplifiers

Operational Amplifiers: Characteristics and Applications; Single Stage Active Filters

Active Filters: Sallen Key, Butterwoth, VCOs and Timers

Analysis of Previous GATE Papers

Exam Year	1 Mark Ques.	2 Marks Ques.	3 Marks Ques.	5 Marks Ques.	Total Marks
1995	1	–	–	–	**1**
1996	4	–	–	–	**4**
1997	1	–	–	1	**6**
1998	1	2	–	2	**15**
1999	3	–	–	2	**13**
2000	2	2	–	1	**11**
2001	2	3	–	3	**23**
2002	2	2	–	2	**16**
2003	3	4	–	–	**11**
2004	3	6	–	–	**15**
2005	2	6	–	–	**14**
2006	2	3	–	–	**8**
2007	3	3	–	–	**9**
2008	1	6	–	–	**13**
2009	2	3	–	–	8
2010	3	1	–	–	5
2011	2	2	–	–	6
2012	2	2	–	–	6

Exam Year	1 Mark Ques.	2 Marks Ques.	Total Marks
2013	2	3	8
2014 Set-1	2	2	6
2014 Set-2	2	2	6
2014 Set-3	1	1	3
2015 Set-1	3	1	5
2015 Set-2	4	1	6
2016 Set-1	1	–	1
2016 Set-2	1	1	3
2017 Set-1	0	2	4
2017 Set-2	0	2	4
2018	1	1	3
2019	2	2	**6**
2020	2	2	**6**
2021	2	1	**4**
2022	3	2	**7**
2023	–	3	**6**
2024	–	3	**6**

1 Diodes and their Applications

1.1 The depletion region (or) space charge region (or) transition region in a semiconductor *p-n* junction diode has
(a) electrons and holes
(b) positive ions and electrons
(c) positive and negative ions
(d) negative ions and holes [1996 : 1 M]

1.2 As temperature is increased, the voltage across a diode carrying a constant current
(a) increases
(b) decreases
(c) remains constant
(d) may increase or decrease depending upon the doping levels in the junction
[1999 : 1 M]

1.3 The mobility of an electron in a conductor expressed in terms of
(a) cm^2/V-sec (b) cm/V-sec
(c) cm^2/V (d) cm^2/sec [1999 : 1 M]

1.4 A diode whose terminal characteristics are related as $i_D = I_s e^{V/V_T}$, where I_s is the reverse saturation current and V_T is thermal voltage (= 25 mV), is biased at i_D = 2 mA. Its dynamic resistance is
(a) 25 Ω (b) 12.5 Ω
(c) 50 Ω (d) 100 Ω [2000 : 2 M]

1.5 The forward resistance of the diode shown in the figure is 5 Ω and the remaining parameters are same as those of ideal diode. The DC components of the source current is

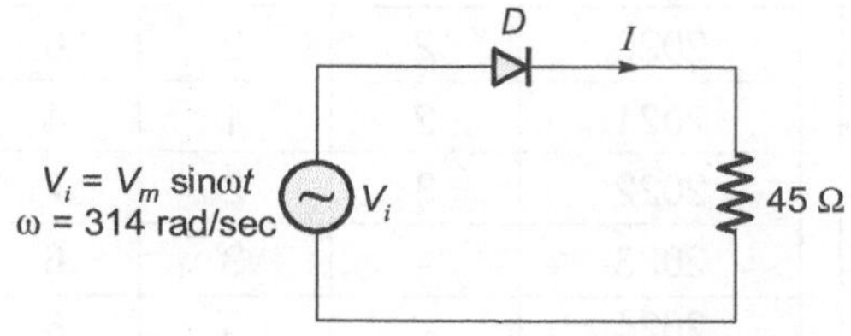

(a) $\frac{V_m}{50\pi}$ (b) $\frac{V_m}{50\pi\sqrt{2}}$
(c) $\frac{V_m}{100\pi\sqrt{2}}$ (d) $\frac{2V_m}{50\pi\sqrt{2}}$ [2002 : 1 M]

1.6 The cut-in voltage of both zener diode D_z and D shown in figure is 0.7 V, while breakdown voltage of the zener is 3.3 V and reverse break down of D is 5 V. The other parameters can be assumed to be the same as those of an ideal diode. The values of the peak output voltage (V_0) are
(a) 3.3 V in the positive half cycle and 1.4 V in the negative half cycle.
(b) 4 V in the positive half cycle and 5 V in the negative half cycle.
(c) 3.3 V in the both positive and negative half cycle.
(d) 4 V in the both positive and negative half cycle.

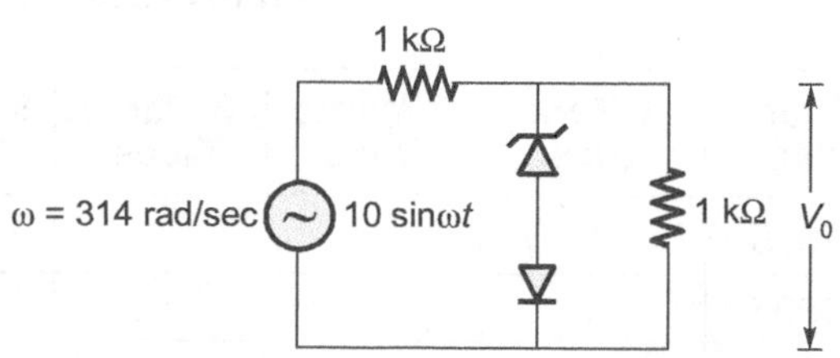

[2002 : 1 M]

1.7 In the single-phase diode bridge rectifier shown in the figure, the load resistor is R = 50 Ω. The source voltage is V = 200 sin ω*t*, where ω = 2π × 50 rad/s. The power dissipated in the load resistor R is

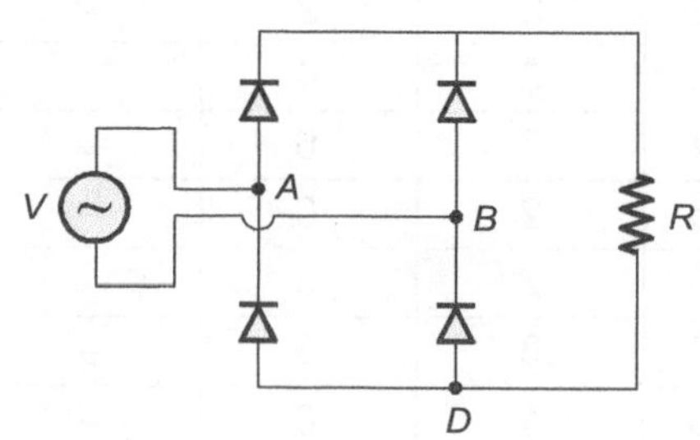

(a) $\frac{3200}{\pi}$ W (b) 400 W
(c) $\frac{400}{\pi}$ W (d) 800 W
[2002 : 2 M]

1.8 A voltage signal 10 sin ω*t* is applied to the circuit with ideal diodes, as shown in the figure. The maximum, and minimum values of the output waveform V_{out} of the circuit are respectively.

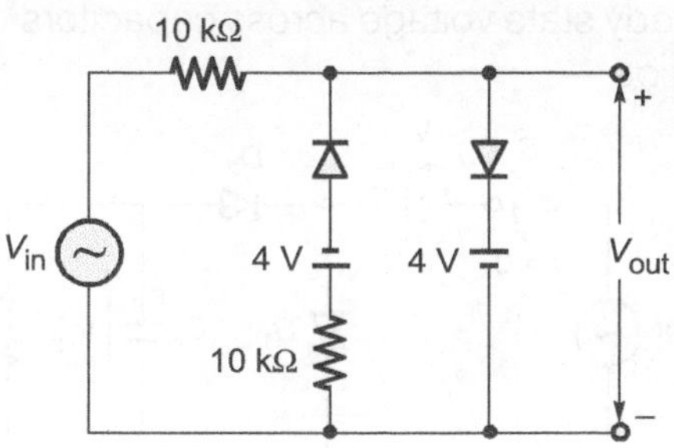

(a) + 10 V and – 10 V
(b) + 4 V and – 4 V
(c) + 7 V and – 4 V
(d) + 4 V and – 7 V [2003 : 2 M]

1.9 The current through the Zener diode in the figure is

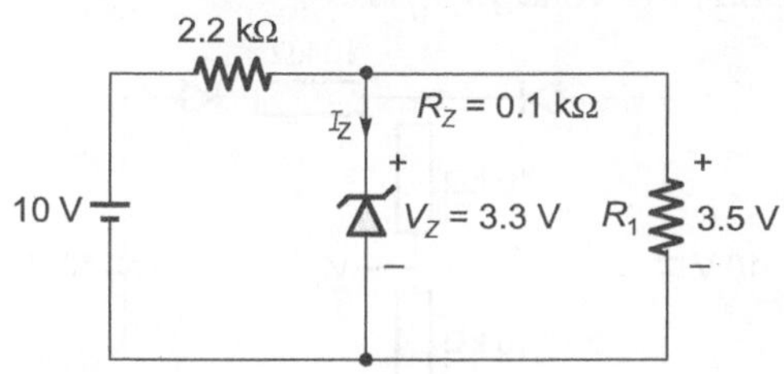

(a) 33 mA (b) 3.3 mA
(c) 2 mA (d) 0 mA [2004 : 1 M]

1.10 Assuming that the diodes are ideal in the figure, the current in diode D_1 is

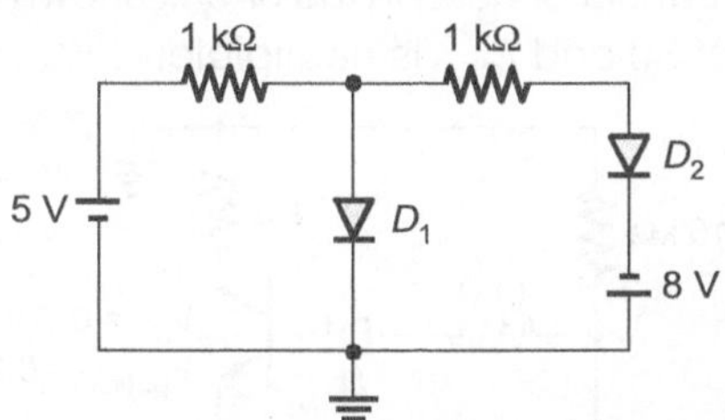

(a) 8 mA (b) 5 mA
(c) 0 mA (d) –3 mA [2004 : 2 M]

1.11 Assume that D_1 and D_2 in figure are ideal diodes. The value of current I is,

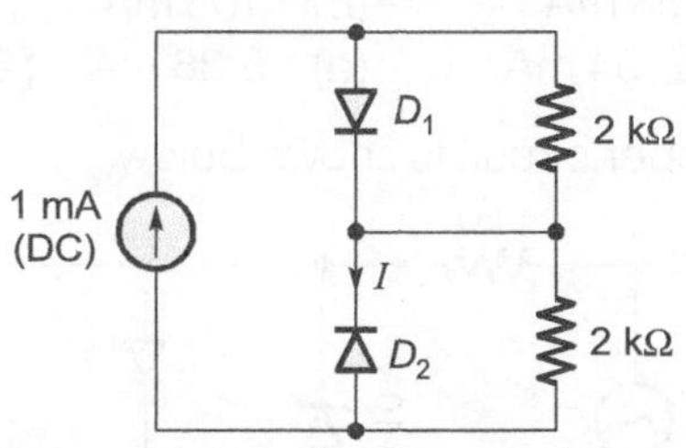

(a) 0 mA (b) 0.5 mA
(c) 1 mA (d) 2 mA [2005 : 1 M]

1.12 What are the states of the three ideal diodes of the circuit shown in the figure?

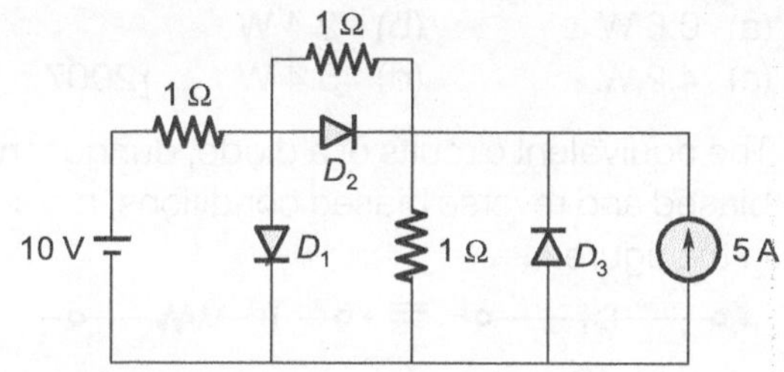

(a) D_1 ON, D_2 OFF, D_3 OFF
(b) D_1 OFF, D_2 ON, D_3 OFF
(c) D_1 ON, D_2 OFF, D_3 ON
(d) D_1 OFF, D_2 ON, D_3 ON [2006 : 1 M]

1.13 Assuming the diodes D_1 and D_2 of the circuit shown in the figure to be ideal ones, the transfer characteristics of the circuit will be

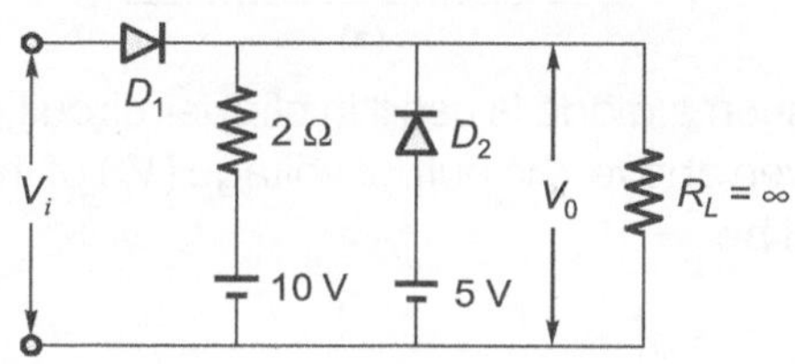

(a)
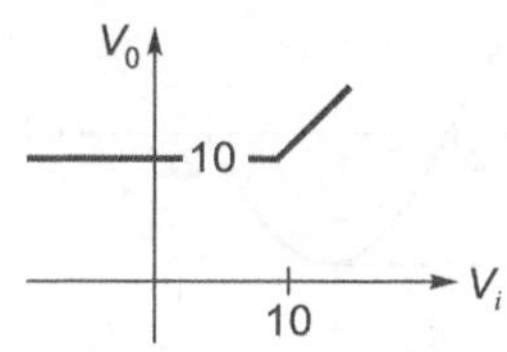

(b)
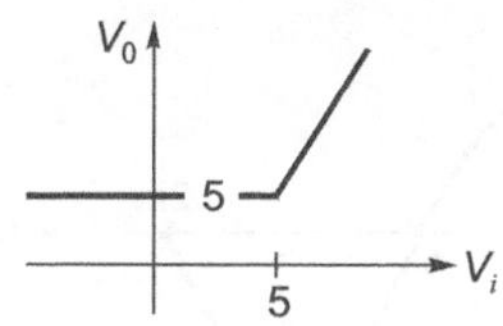

(c)
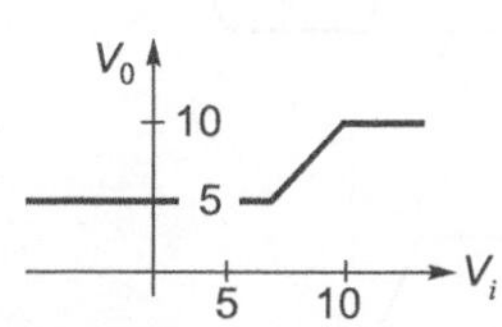

(d)
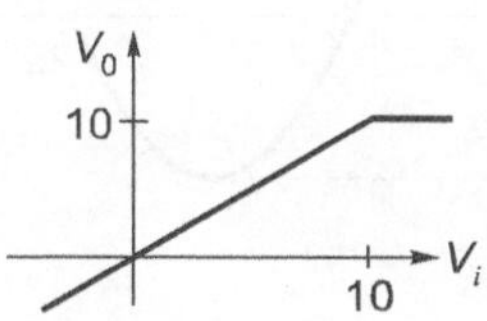

[2006 : 2 M]

1.14 The three-terminal linear voltage regulator is connected to a 10 Ω load resistor as shown in the figure. If V_{in} is 10 V, what is the power dissipated in the transistor?

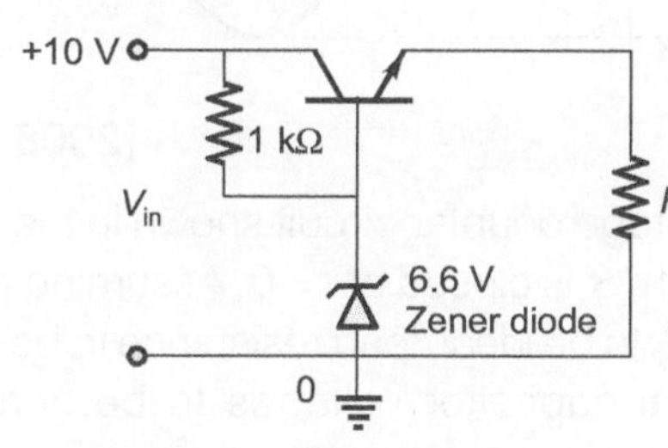

(a) 0.6 W (b) 2.4 W
(c) 4.2 W (d) 5.4 W [2007 : 1 M]

1.15 The equivalent circuits of a diode, during forward biased and reverse biased conditions, are shown in the figure.

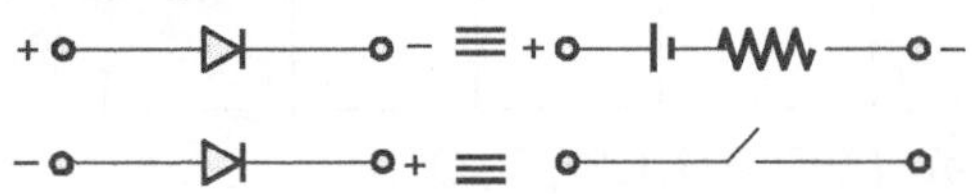

(a)

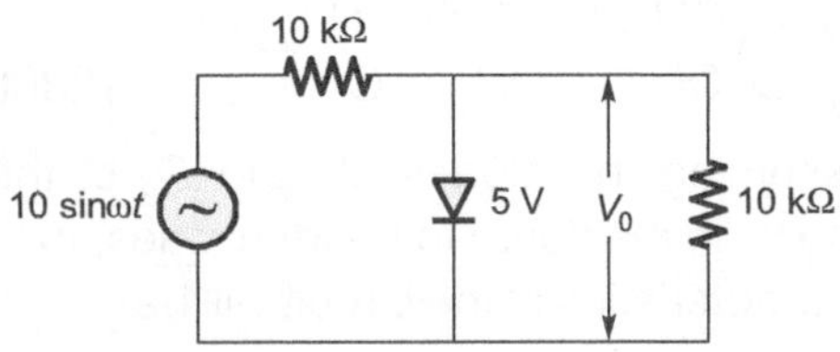

(b)

If such a diode is used in clipper circuit of figure given above, the output voltage (V_0) of the circuit will be

(a)

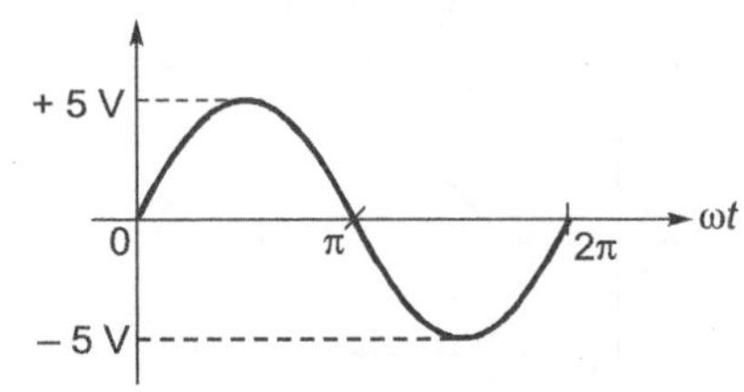

(b)

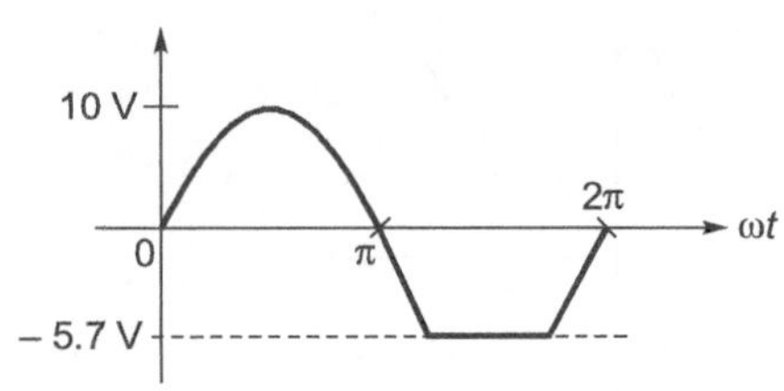

(c)

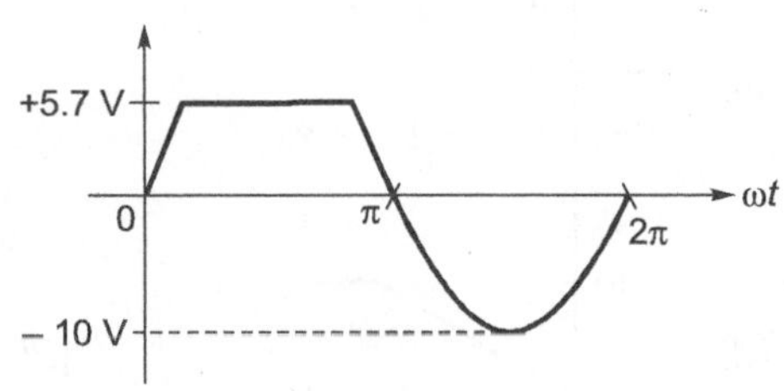

(d)

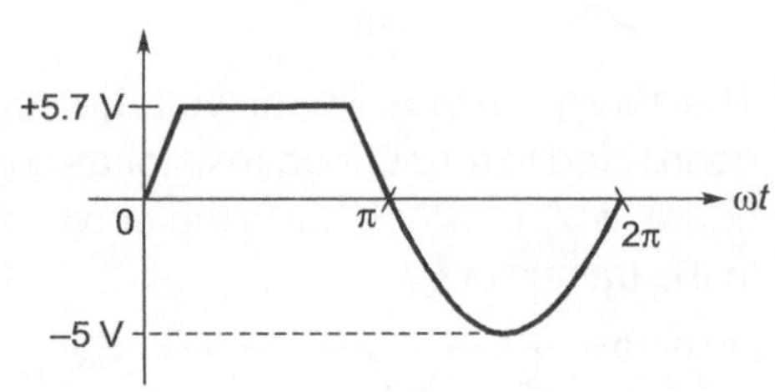

[2008 : 1 M]

1.16 In the voltage doubler circuit shown in the figure, the switch '*S*' is closed at $t = 0$. Assuming diodes D_1 and D_2 to be ideal, load resistance to be infinite and initial capacitor voltages to be zero. The steady state voltage across capacitors C_1 and C_2 will be

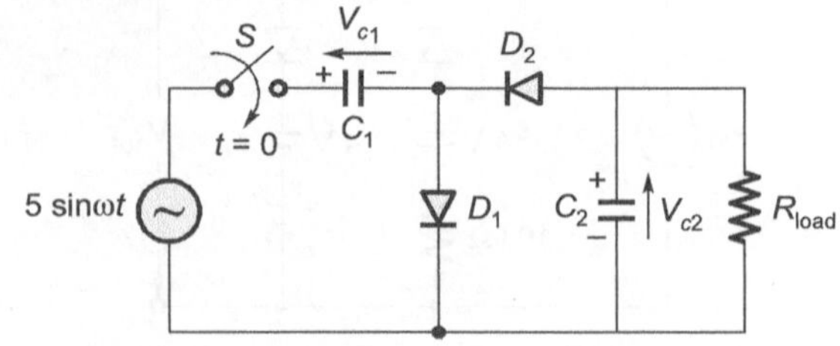

(a) $V_{c1} = 10$ V, $V_{c2} = 5$ V
(b) $V_{c1} = 10$ V, $V_{c2} = -5$ V
(c) $V_{c1} = 5$ V, $V_{c2} = 10$ V
(d) $V_{c1} = 5$ V, $V_{c2} = -10$ V [2008 : 2 M]

1.17 Assuming that the diodes in the given circuit are ideal, the voltage V_0 is

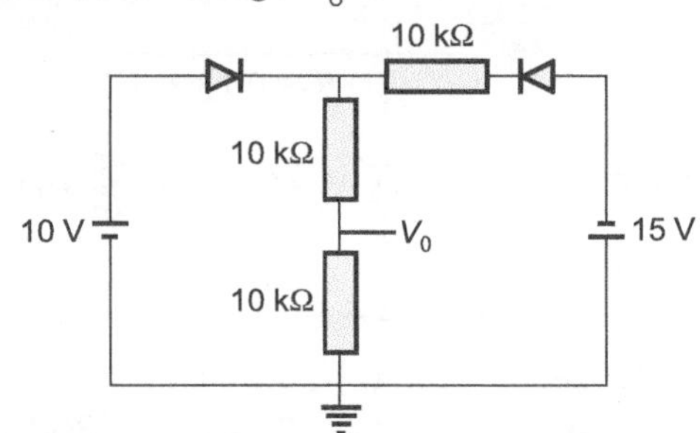

(a) 4 V (b) 5 V
(c) 7.5 V (d) 12.12 V [2010 : 1 M]

1.18 The transistor used in the circuit shown below has a β of 30 and I_{CBO} is negligible.

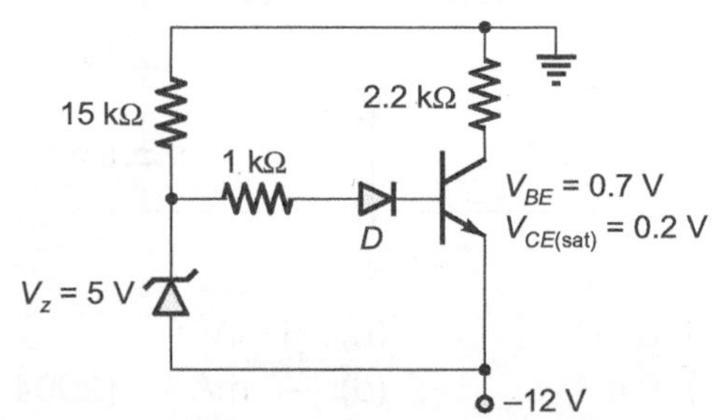

If the forward voltage drop of diode is 0.7 V, then the current through collector will be

(a) 168 mA (b) 108 mA
(c) 20.54 mA (d) 5.36 mA [2011 : 2 M]

1.19 A clipper circuit is shown below

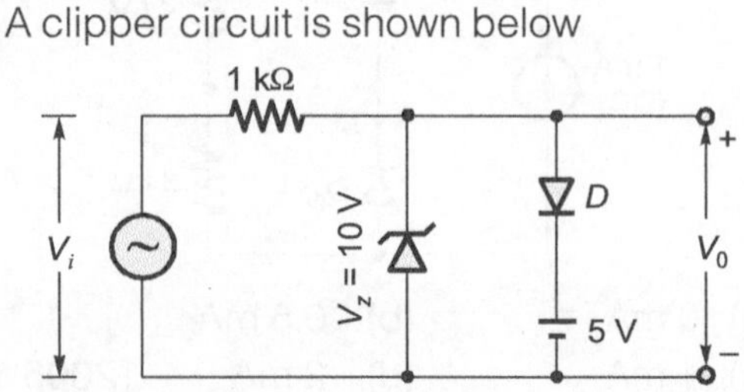

Assuming forward voltage drops of the diodes to be 0.7 V, the input-output transfer characteristics of the circuit is

(a) 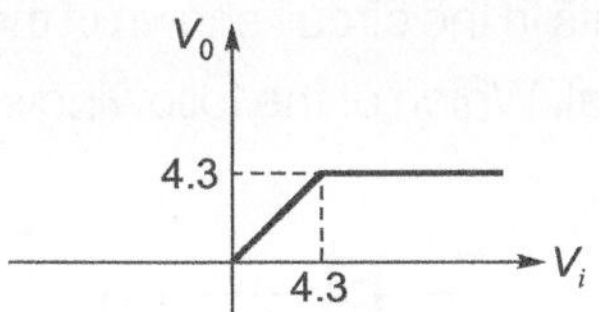

(b) 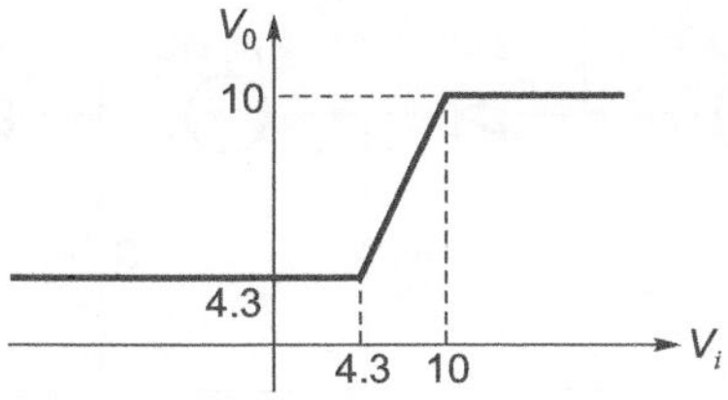

(c) 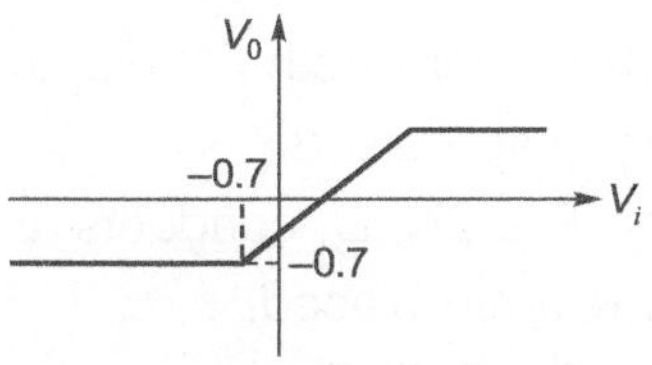

(d) 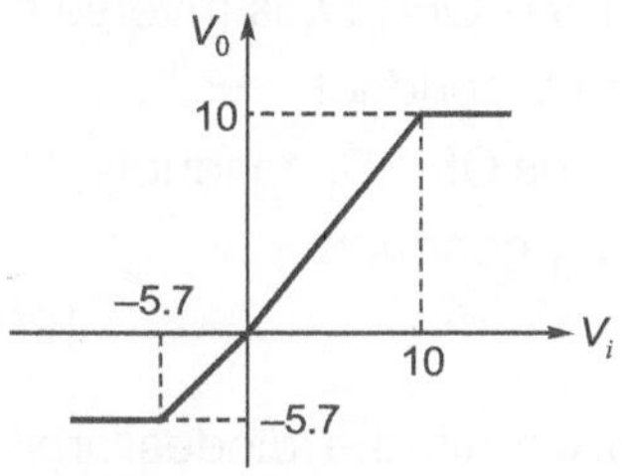

[2011 : 2 M]

1.20 The i-v characteristics of the diode in the circuit given below are:

$$i = \begin{cases} \dfrac{v-0.7}{500}\,\text{A}, & v \geq 0.7\text{ V} \\ 0\text{ A}, & v < 0.7\text{ V} \end{cases}$$

The current in the circuit is

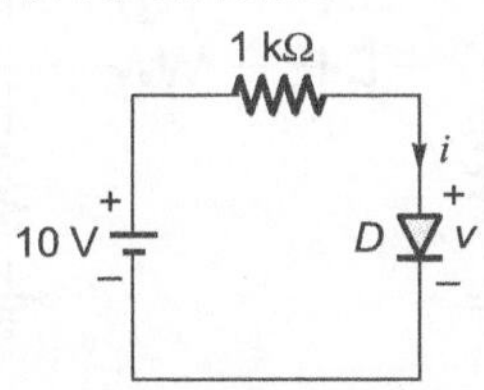

(a) 10 mA (b) 9.3 mA
(c) 6.67 mA (d) 6.2 mA [2012 : 1 M]

1.21 In the circuit shown below, the knee current of the ideal Zener diode is 10 mA. To maintain 5 V across R_L, the minimum value of R_L in Ω and the minimum power rating of the Zener diode in mW respectively are

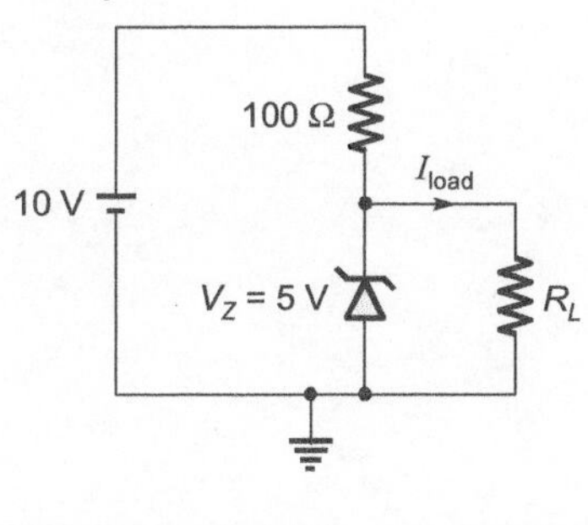

(a) 125 and 125 (b) 125 and 250
(c) 250 and 125 (d) 250 and 250

[2013 : 2 M]

1.22 A voltage 1000 sin ωt volts is applied across YZ. Assuming ideal diodes, the voltage measured across WX in Volts, is

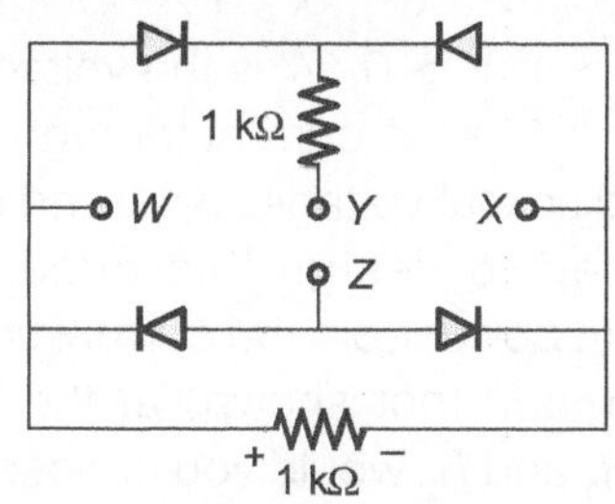

(a) sinωt
(b) $(\sin\omega t + |\sin\omega t|)/2$
(c) $(\sin\omega t - |\sin\omega t|)/2$
(d) 0 for all t [2013 : 2 M]

1.23 The sinusoidal ac source in the figure has an rms value of $\dfrac{20}{\sqrt{2}}$ V. Considering all possible values of R_L, the minimum value of R_s in Ω to avoid burnout of the Zener diode is ______.

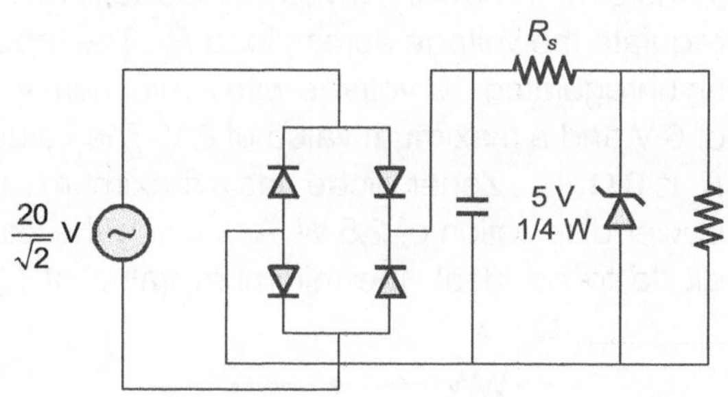

[2014 : 1 M, Set-2]

1.24 Assuming the diodes to be ideal in the figure, for the output to be clipped, the input voltage v_i must be outside the range

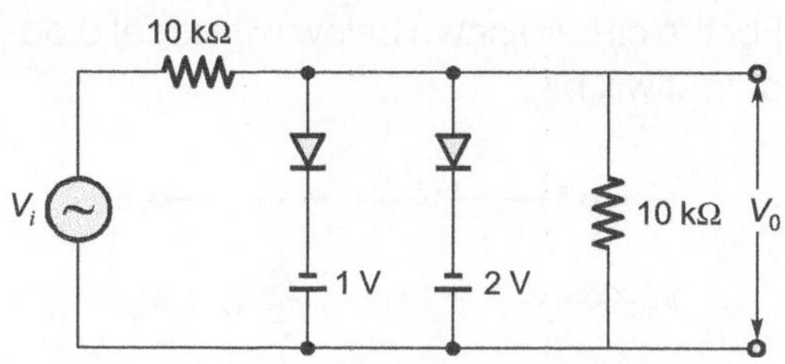

(a) –1 V to –2 V (b) –2 V to –4 V
(c) +1 V to –2 V (d) +2 V to –4 V

[2014 : 2 M, Set-2]

1.25 A non-ideal diode is biased with a voltage of –0.03 V, and a diode current of I_1 is measured. The thermal voltage is 26 mV and the ideality factor for the diode is 15/13. The voltage, in V, at which the measured current increases to $1.5I_1$ is closest to

(a) –4.50 (b) –0.09
(c) –0.02 (d) –1.50 [2020 : 2 M]

1.26 Consider the diode circuit shown below. The diode, D, obeys the current-voltage characteristic

$$I_D = I_S\left(\exp\left(\frac{V_D}{nV_T}\right) - 1\right),$$

where $n > 1$, $V_T > 0$, V_D is the voltage across the diode and I_D is the current through it. The circuit is biased so that voltage, $V > 0$ and current, $I < 0$. If you had to design this circuit to transfer maximum power from the current source (I_1) to a resistive load (not shown) at the output, what values R_1 and R_2 would you choose?

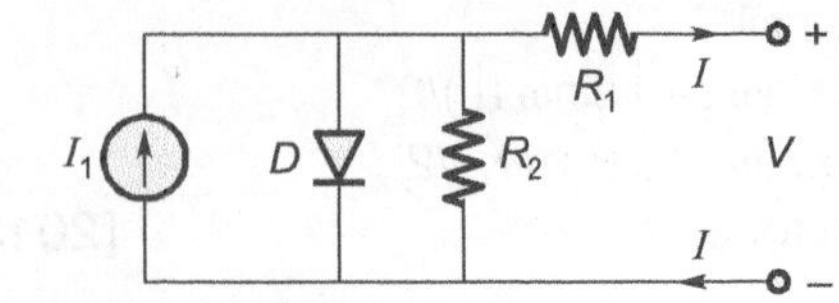

(a) Small R_1 and small R_2
(b) Large R_1 and large R_2
(c) Small R_1 and large R_2
(d) Large R_1 and small R_2 [2020 : 2 M]

1.27 In the circuit shown, a 5 V Zener diode is used to regulate the voltage across load R_0. The input is an unregulated DC voltage with a minimum value of 6 V and a maximum value of 8 V. The value of R_s is 6 Ω. The Zener diode has a maximum rated power dissipation of 2.5 W, Assuming the Zener diode to be ideal, the minimum value of R_0 is ______ Ω.

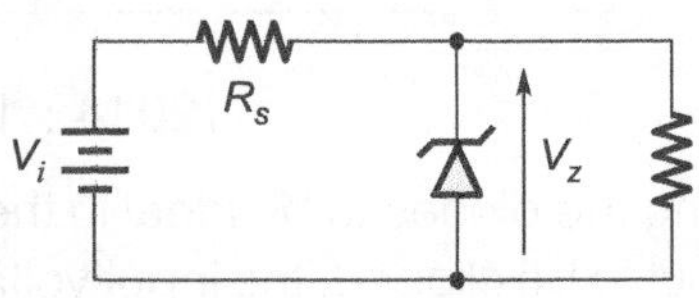

[2021 : 1 M]

1.28 For the circuit shown below with ideal diodes, the output will be

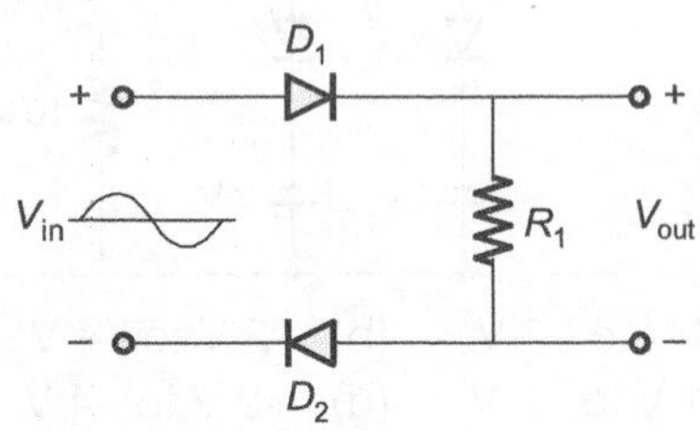

(a) $V_{out} = V_{in}$ for $V_{in} > 0$
(b) $V_{out} = V_{in}$ for $V_{in} < 0$
(c) $V_{out} = -V_{in}$ for $V_{in} > 0$
(d) $V_{out} = -V_{in}$ for $V_{in} < 0$ [2022 : 1 M]

1.29 All the elements in the circuit shown in the following figure are ideal. Which of the following statements is/are true?

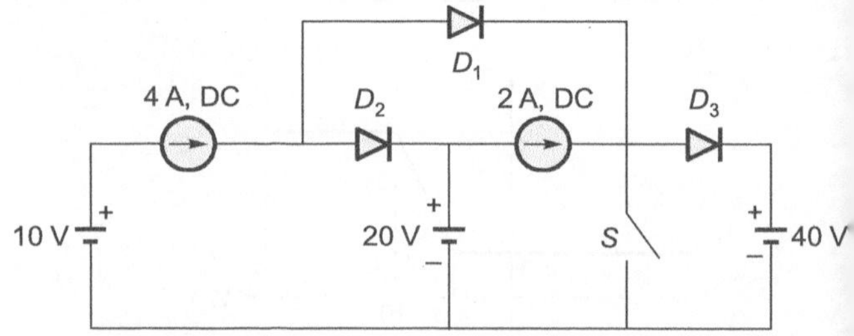

(a) When switch S is ON, both D_1 and D_2 conducts and D_3 is reverse biased.
(b) When switch S is ON, D_1 conducts and both D_2 and D_3 are reverse biased.
(c) When switch S is OFF, D_1 is reverse biased and both D_2 and D_3 conduct.
(d) When switch S is OFF, D_1 conducts, D_2 is reverse biased and D_3 conducts.

[2023 : 2 M]

1.30 In the given circuit, the diodes are ideal. The current I through the diode D_1 in milliamperes is ______ (rounded off to two decimal places).

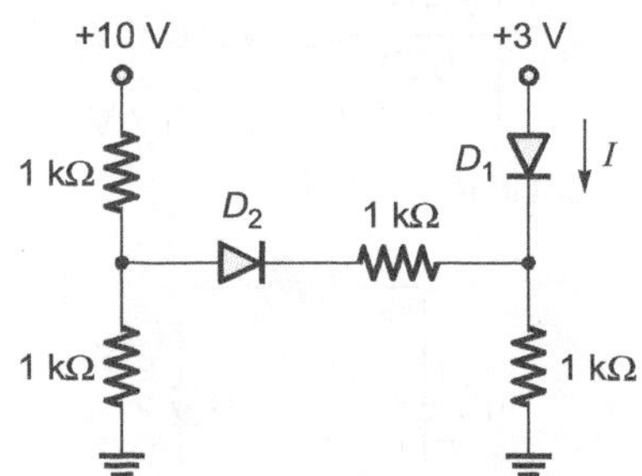

[2024 : 2 M]

Answers Diodes and their Applications

1.1 (c)	1.2 (b)	1.3 (a)	1.4 (b)	1.5 (a)	1.6 (b)	1.7 (b)
1.8 (d)	1.9 (c)	1.10 (c)	1.11 (a)	1.12 (a)	1.13 (a)	1.14 (b)
1.15 (a)	1.16 (d)	1.17 (b)	1.18 (d)	1.19 (c)	1.20 (d)	1.21 (b)
1.22 (d)	1.23 (300)	1.24 (b)	1.25 (b)	1.26 (c)	1.27 (30)	1.28 (a)
1.29 (b, c)	1.30 (1.67)					

Explanations Diodes and their Applications

1.1 (c)

Because of charge recombination occurs in depletion region, only positive and negative ions remains here.

1.2 (b)

$$\frac{dV_D}{dT} = -2.5 \text{ mV/°C}$$

1.3 (a)

$$\mu = \frac{\text{Drift velocity}}{\text{Electric field}} = \frac{\text{cm/s}}{\text{V/cm}} = \frac{\text{cm}^2}{V-s}$$

1.4 (b)

$$r_D = \frac{V_T}{I_D} = \frac{25 \text{ mV}}{2 \text{ mA}} = 12.5\ \Omega$$

1.5 (a)

$$V_{av} = \frac{V_m}{\pi}$$

$$I_{av} = \frac{V_{av}}{R} = \frac{V_{av}}{(5+45)} = \frac{V_{av}}{50} = \frac{V_m}{50\pi}$$

1.6 (b)

For positive half cycle, at peak input voltage

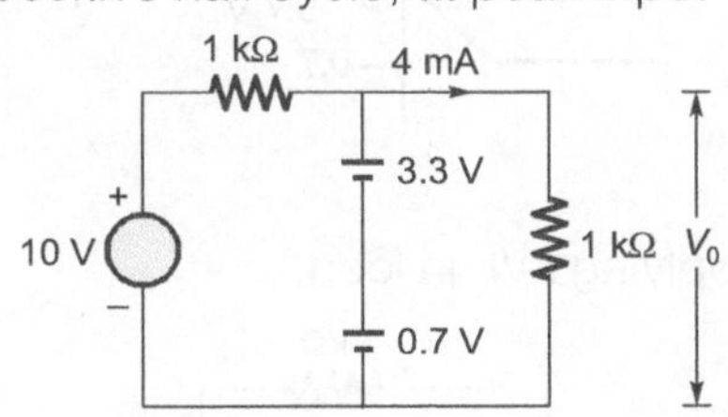

So, $V_o = 4$ V

For negative half cycle, at peak input voltage

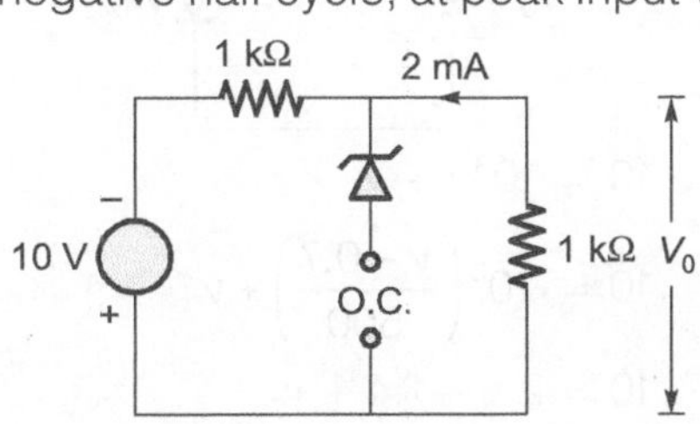

So, $V_o = 10 \times \left(\frac{1}{1+1}\right) = 5$ V

1.7 (b)

$$V_{rms} = \frac{V_m}{\sqrt{2}}$$

$$P = \frac{V_{rms}^2}{R} = \frac{V_m^2}{2R} = \frac{200^2}{2\times 50} = 400 \text{ W}$$

1.8 (d)

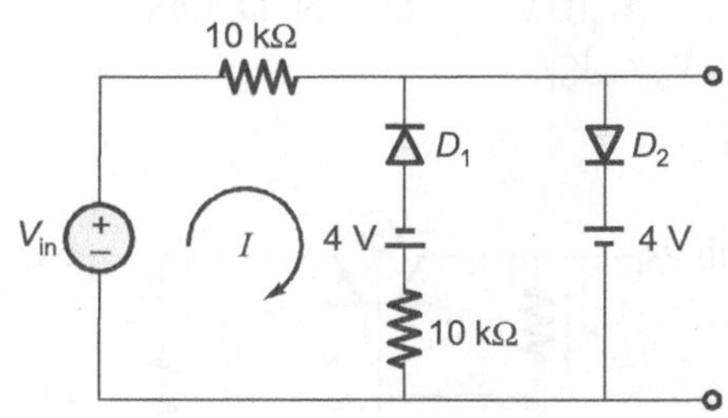

For +ve cycle, when $V_{in} < -4$ V

($V_{in} > 4$ V) D_2 ON, D_1 OFF, $V_{out} = 4$ V

where $V_{in} < 4$ V, then $V_{out} = V_{in}$ because D_2 also become (OFF).

For –ve cycle, when $V_{in} < -4$

D_1 ON, D_2 OFF

$$I = \frac{V_{in}+4}{20k}$$

But, V_{in} = –ve

$$\therefore \quad I = \frac{-V_{in}+4}{20} = \frac{-3}{10} \text{ mA}$$

($\because$ for minimum output $V_{in} = -10$ V)

$$V_{out} = \frac{-3}{10} \times 10k - 4 = -7 \text{ V}$$

1.9 (c)

$$R_Z = \frac{\Delta V_Z}{\Delta I_Z} = \frac{3.5-3.3}{\Delta I_z}$$

$$\therefore \quad \Delta I_z = \frac{0.2}{0.1k} = 2 \text{ mA}$$

1.10 (c)

From the circuit, if 'D_1' is 'ON' Kirchoff's law is not satisfied.

$\therefore$ D_1 is OFF and D_2 is ON

$\therefore$ $I = 0$ mA

1.11 (a)

Current will pass through its simplest path (or) low resistance path. D_1 become forward biased and D_2 is reverse biased

$I = 0$ mA

1.12 (a)

Analyzing the circuit, we can see that D_1 and D_2 are in forward bias and D_3 is in reverse bias. But no current flow through D_2, because current gets shortest path through D_1.

1.13 (a)

$V_i < 10$ V $\quad$ D_1 OFF, D_2 OFF

$V_o = 10$ V

$V_i > 10$ V $\quad$ D_1 ON, D_2 OFF

$V_o = V_{in}$

1.14 (b)

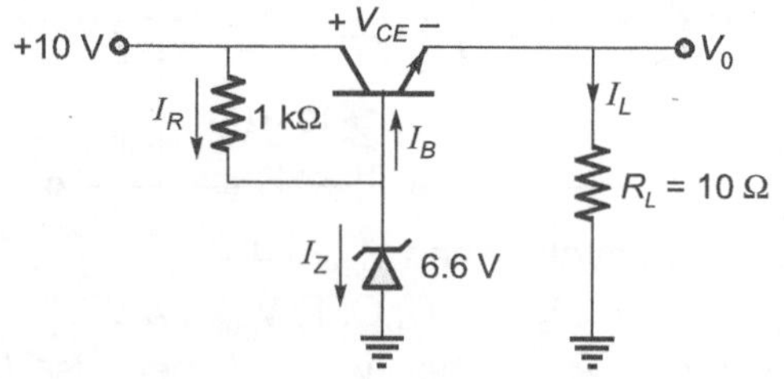

$$V_0 = V_z - V_{BE}$$
$$= 6.6\text{ V} - 0.7\text{ V} = 5.9\text{ V}$$
$$I_L = \frac{V_0}{R_L} = \frac{5.9}{10} = 0.59\text{ A} = I_C$$
$$V_{CE} = V_i - V_0 = 10 - 5.9 = 4.1\text{ V}$$
$$P_Q = V_{CE} \times I_C = 4.1 \times 0.59\text{ A} = 2.4\text{ W}$$

1.15 (a)

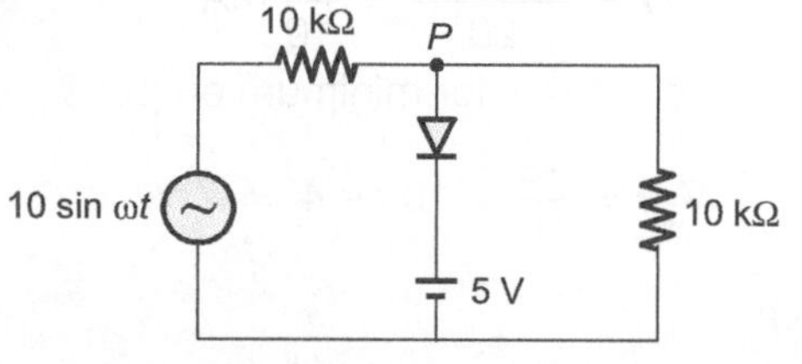

$$V_p = \frac{10}{10+10} \times 10\sin\omega t = 5\sin\omega t$$

Since, maximum voltage across at point P may be 5 V, hence voltage across the diode always will be less than or equal to zero. So it will be reversed always.

$$\therefore \quad V_0 = \frac{10}{10+10} \times 10\text{ Sin }\omega t = 5\text{ Sin }\omega t$$

1.16 (d)

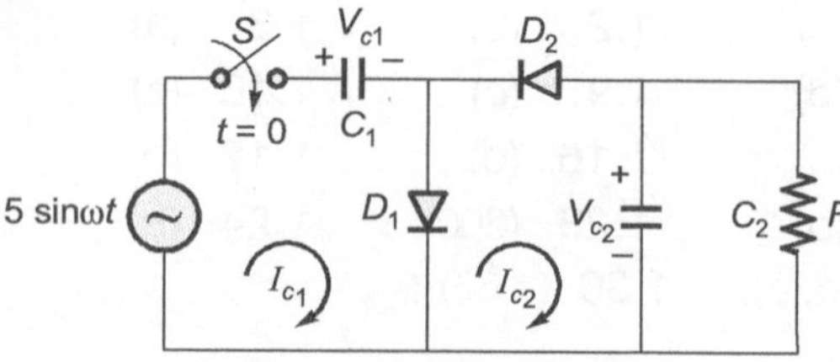

First capacitor charge through Diode (D_1) upto V_{max}(5 V) with shown polarities

$\therefore$ $V_{c1} = 5$ V

Now diode D_1 will be reversed biased and D_2 will be forward biased and capacitor C_2 will charge in reverse direction through diode D_2 upto 2 V_{max}

$\therefore$ $V_{c2} = -10$ V

1.17 (b)

Diode D_1 is on and diode D_2 is off.

So, $\quad V_0 = 10 \times \dfrac{10}{10+10} = 5$ V

1.18 (d)

$$I_{c(sat)} = \frac{0-(-12)-0.2}{R_c} = \frac{11.8}{2.2\text{ k}} = 5.3636\text{ mA}$$

This is the maximum current possible in any case.

1.19 (c)

For positive voltage the waveform clips at +5.7 V. For negative voltage at –0.7 V, the zener diode conducts and clips out.

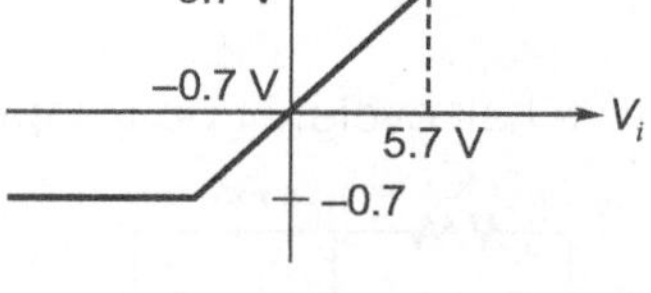

1.20 (d)

By applying kVL in loop,

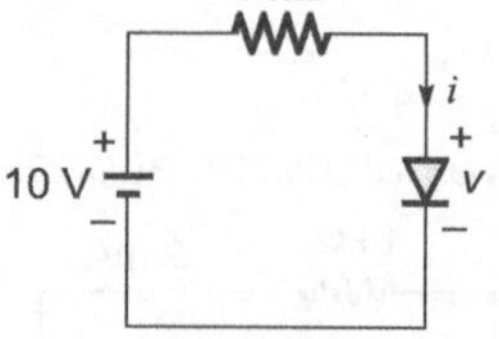

$$10 = 10^3\, i + v$$
$$10 = 10^3\left(\frac{v-0.7}{500}\right) + v$$
$$10 = 2v - 1.4 + v$$
$$3v = 11.4 \Rightarrow v = 3.8$$
$$\Rightarrow \quad i = (10 - 3.8) \times 10^{-3} = 6.2\text{ mA}$$

1.21 (b)

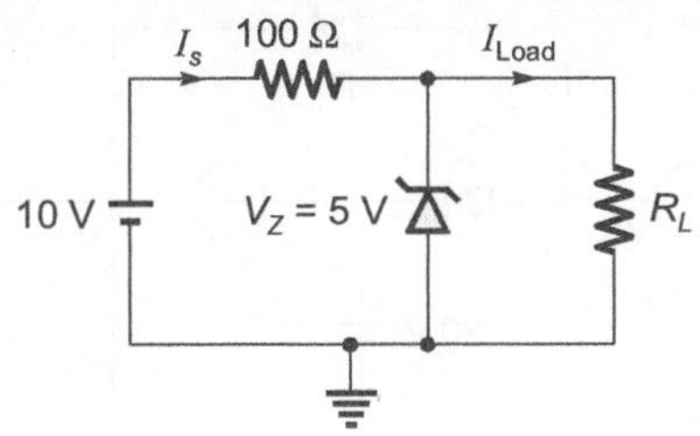

$$I_s = \frac{10-5}{100} = 50 \text{ mA}$$

$$I_{load} = 50 \text{ mA} - 10 \text{ mA} = 40 \text{ mA}$$

$$R_L = \frac{5\text{V}}{40 \text{ mA}} = 125\,\Omega$$

$$P_{min} = 50 \text{ mA} \times 5 \text{ V} = 250 \text{ mW}$$

1.22 (d)

For positive half cycle

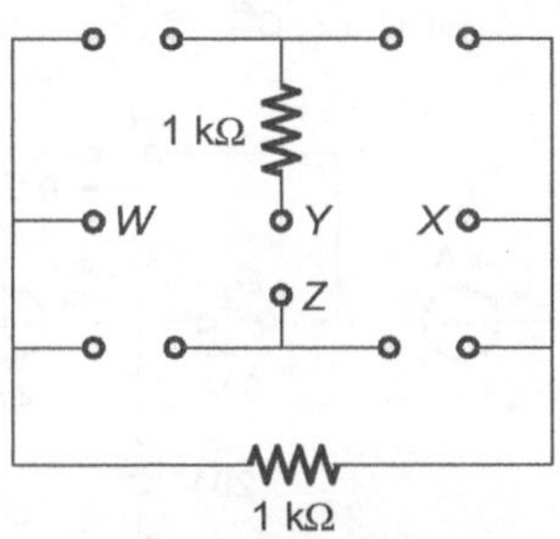

$$V_{WX} = 0$$

For negative half cycle,

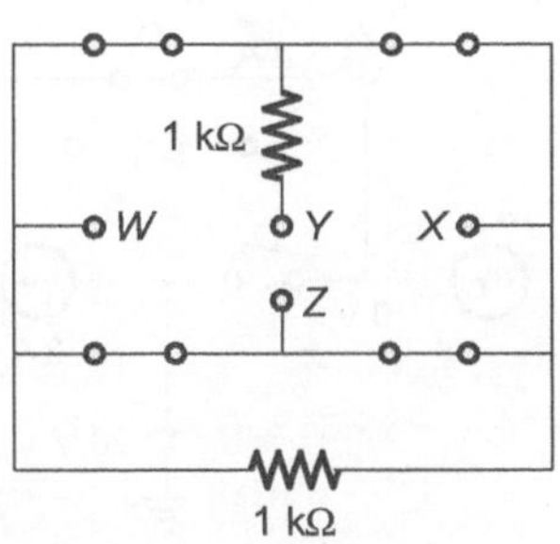

Short circuit condition, $V_{WX} = 0$

1.23 (300)

$$I_s \geq I_z + I_L$$

$$\frac{1}{20} \geq \frac{20-5}{R_s} \qquad \text{(Putting } I_L = 0\text{)}$$

$\Rightarrow$ $R_s \geq 300\,\Omega$. Hence, $R_{s_{min}} = 300\,\Omega$

1.24 (b)

Case-I: $V_i \leq -4$ V

$D_2 \to$ conducts; $D_1 \to$ OFF

So, $V_o = -2$ V

Case-II: $-4 \text{ V} \leq V_i \leq -2$ V

Both the diodes will be OFF.

$$V_o = V_1'$$

Case-III: $V_i \geq -2$ V

$D_1 \to$ conducts; $D_2 \to$ OFF

$V_o = -1$ V

1.25 (b)

$$I_1 = I_0\left[e^{\frac{0.03}{15/13 \times 26\text{ mV}}} - 1\right]$$

As, $V_D = -$ve '1' can not be neglected in diode current equation

$$I_1 = I_0[e^{-30 \text{ mV}/30 \text{ mV}} - 1]$$

$$= I_0[e^{-1} - 1] = -0.64\, I_0 \qquad \text{...(i)}$$

$$1.5 I_1 = I_0\left[e^{V_D/30 \text{ mV}} - 1\right]$$

$$-1.5 \times 0.64 I_0 = I_0\left[e^{V_D/30 \text{ mV}} - 1\right]$$

$$-0.96 = e^{V_{D2}/30 \text{ mV}} - 1$$

$$1 - 0.96 = e^{V_D/30 \text{ mV}} \Rightarrow 0.04 = e^{V_D/30 \text{ mV}}$$

$$30 \text{ mV} \ln(0.04) = V_D \Rightarrow V_D = -0.09 \text{ V}$$

1.26 (c)

R_1, low, R_2 high : $V_D = V \times \dfrac{R_2}{R_1 + R_2}$

If R_2 is large V_d (high). R_1 is less $V_D = V$

So for maximum power to deliver to load R_1 is small and R_2 is large.

1.27 (30)

To calculate $R_{0\,min}$, we must find $I_{L\,max}$

$$I_{s\,min} = I_z \text{ min} + I_{L\,max}$$

$$\frac{V_{i\,min} - V_z}{R_s} = I_z \text{ min} + I_{L\,max}$$

For ideal zener diode, $I_{z\,min} = 0$

$$\frac{V_{i\,min} - V_z}{R_s} = I_{L\,max} \Rightarrow \frac{6-5}{6} = I_{L\,max}$$

$$I_{L\,max} = \frac{1}{6}\text{A}$$

$$R_{0\,min} = \frac{V_z}{I_{L\,max}} = \frac{5}{1/6} = 30\,\Omega$$

1.28 (a)

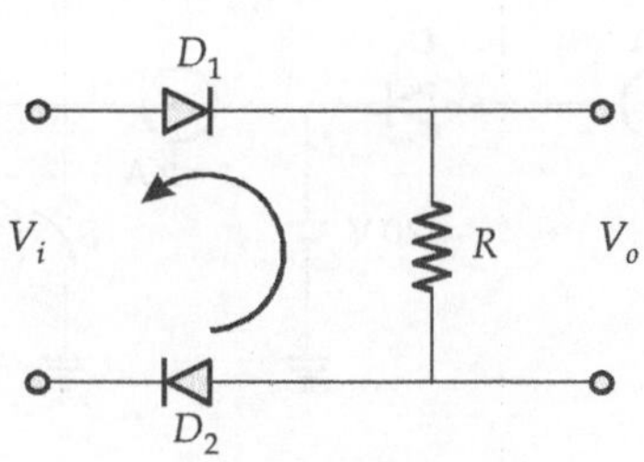

Positive half cycle

D_1 and D_2 will be ON.

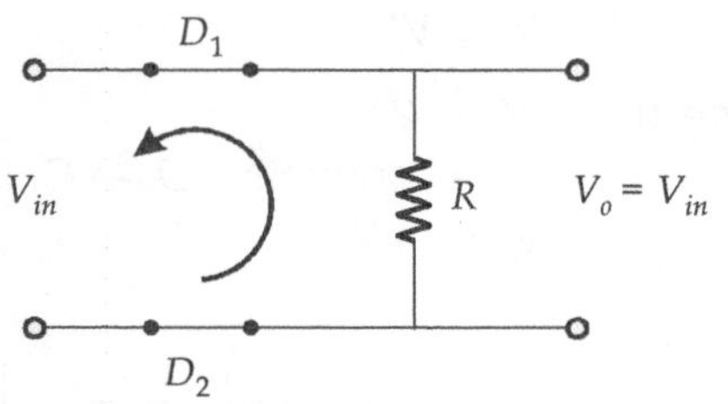

For Negative half cycle

D_1 and D_2 will be OFF.

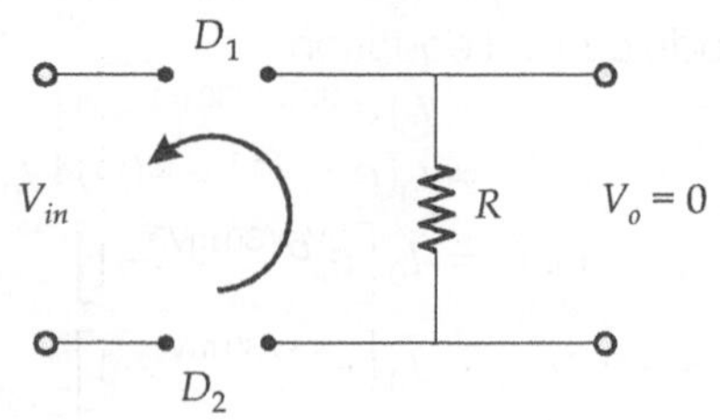

The output waveform

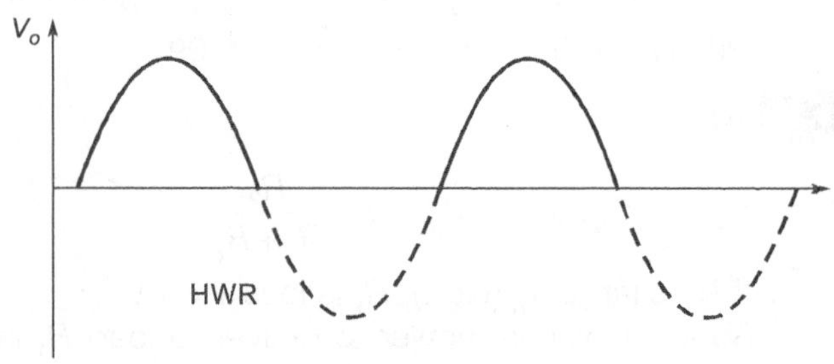

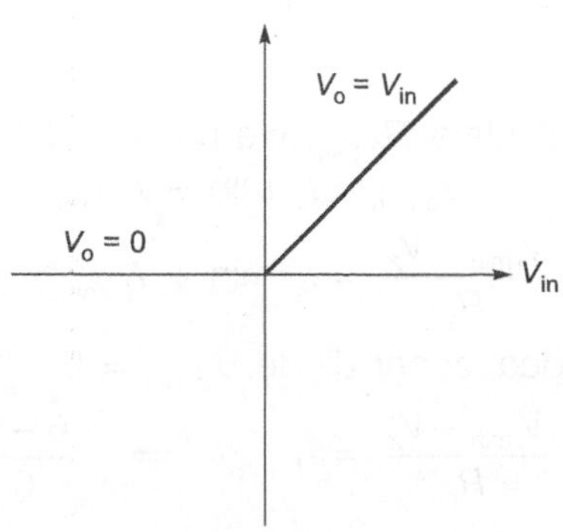

So, $\quad V_o = V_{in}$ for $V_{in} > 0$

1.29 (b, c)

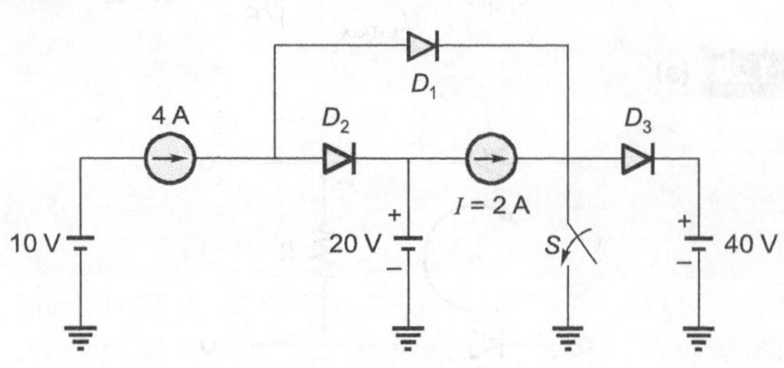

When switch on,

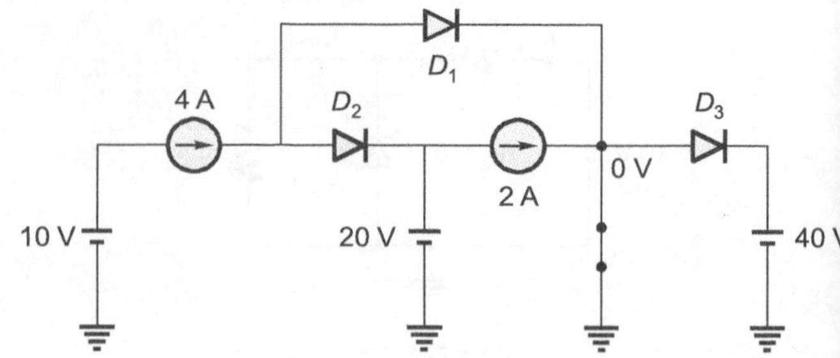

D_3 off by observation

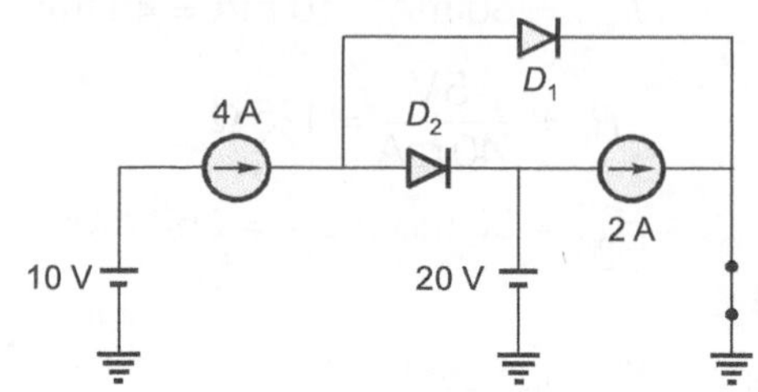

Assume D_1 and D_2 off

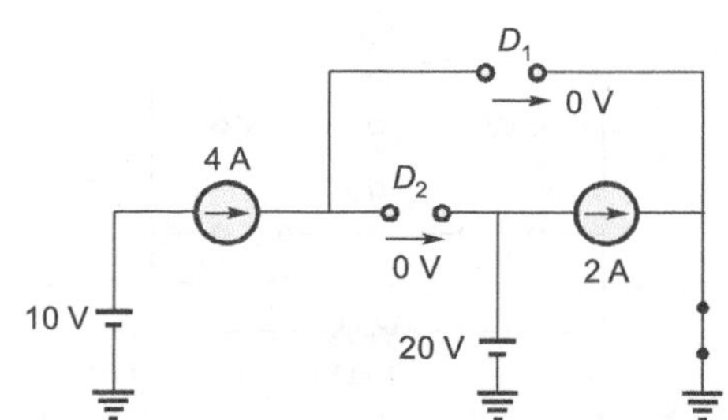

KCL is being violated. Assume D_1 on, D_2 off

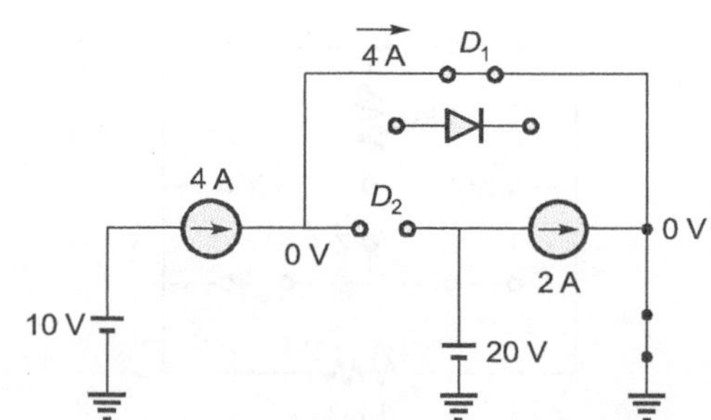

$\because$ D_2 is reverse biased, $\therefore$ our assumption is correct, i.e., D_1 = ON and D_2 = OFF and D_3 = OFF.

When switch is off,

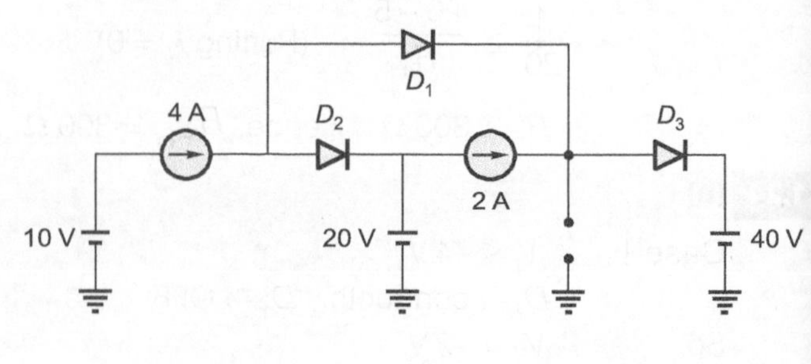

Assume D_1 off

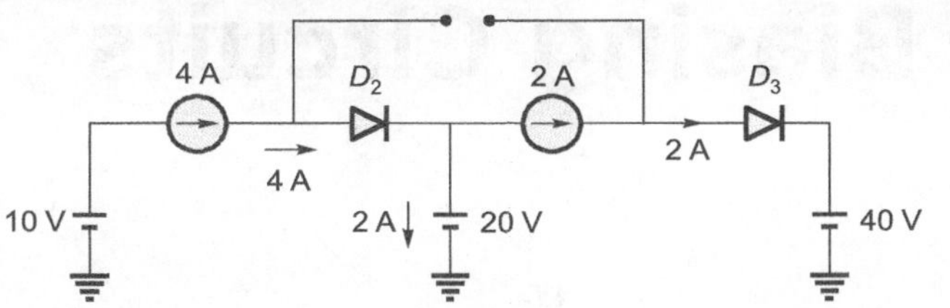

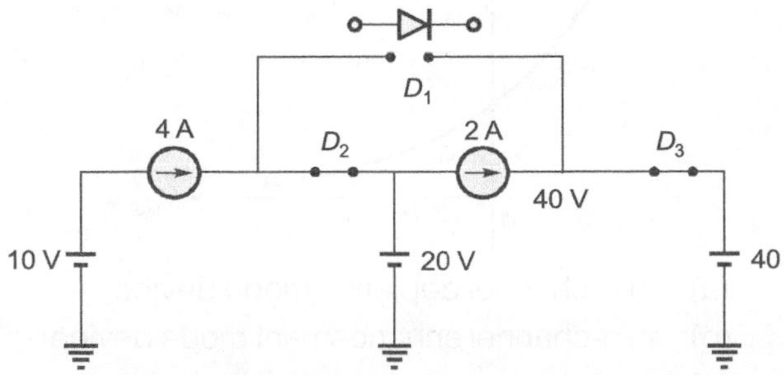

$\because$ D_1 is reverse biased, $\therefore$ D_1 is OFF
Switch off : D_1 off, D_2 and D_3 on.

1.30 (1.67)

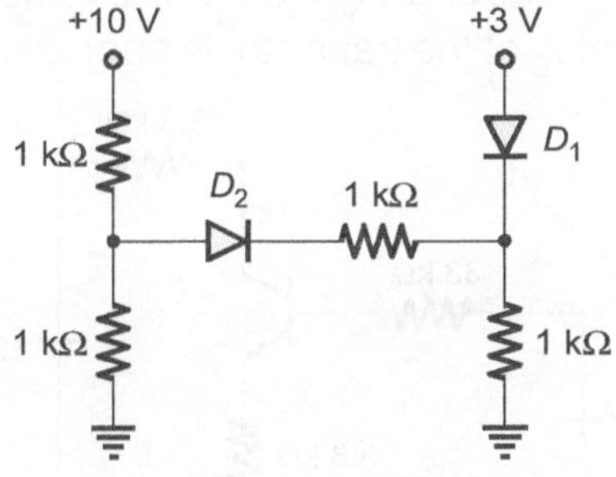

Assume D_2 OFF

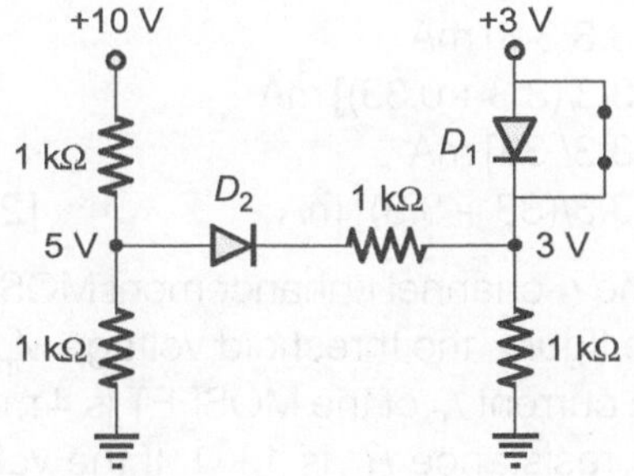

Tested : D_2 ON
Assume D_1 OFF

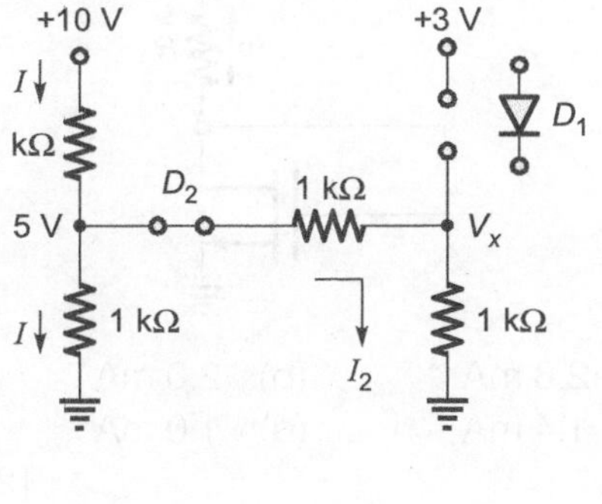

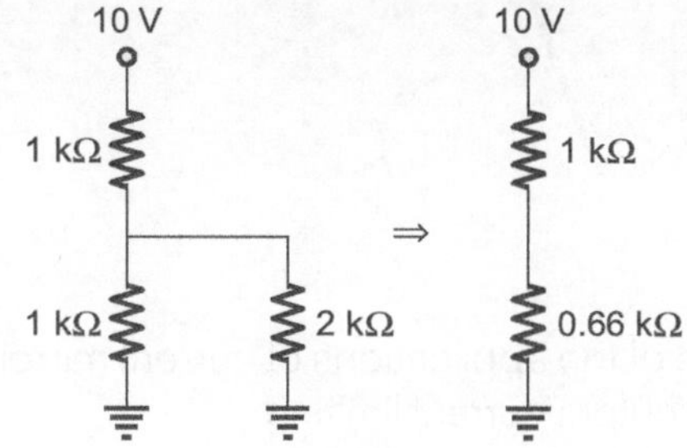

$$I = \frac{10}{1.66\,\text{K}} = 6.02\text{ mA}$$

$$I_2 = I \times \frac{1\text{k}\Omega}{3\text{k}\Omega} = 6.02\text{ mA} \times \frac{1}{3} = 2\text{ mA}$$

Tested : D_1 ON

$$V_X = 2\text{ mA} \times 1\text{ k}\Omega = 2\text{ V}$$

States: D_1 and D_2 ON

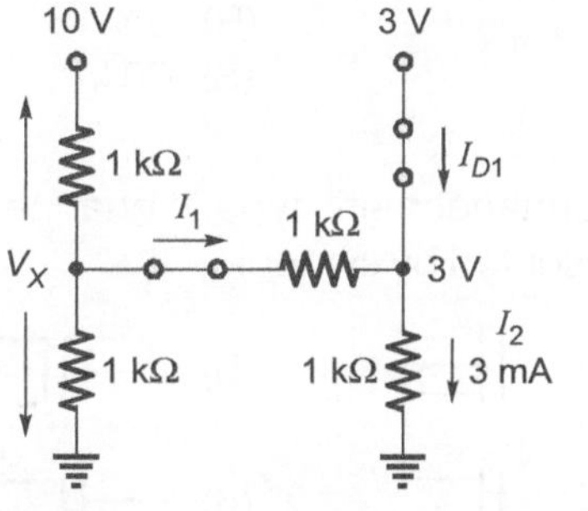

$$\frac{V_X - 10}{1\text{k}\Omega} + \frac{V_X - 3}{1\text{k}\Omega} + \frac{V_X}{1\text{k}\Omega} = 0$$

$$3V_X = 13$$

$$V_X = \frac{13}{3} = 4.33\text{ V}$$

$$I_1 + I_{D1} = I_2$$

$$\frac{4.33 - 3}{1\text{k}\Omega} + I_{D1} = 3\text{ mA}$$

$$I_{D1} = 3\text{ mA} - 1.33\text{ mA}$$
$$= 1.67\text{ mA}$$

■■■■

2 BJT, FET and their Biasing Circuits

2.1 One of the applications of current mirror is
(a) output current limiting
(b) obtaining a very high current gain
(c) current feedback
(d) temperature stabilized biasing

[1998 : 1 M]

2.2 A NPN Si transistor is meant for low-current audio amplification. Match its following characteristics against their values.

Characteristics	Values
(a) $V_{EB\max}$	(P) 0.7 V
(b) $V_{CB\max}$	(Q) 0.2 V
(c) $V_{CE\max}$	(R) 6 V
	(S) 50 V

[1998 : 2 M]

2.3 The enhancement type *n*-channel MOSFET is represented by symbol

(a) 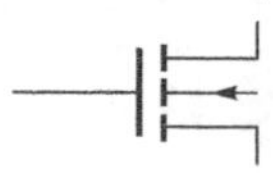
(b) 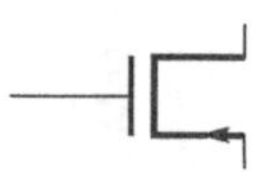
(c) 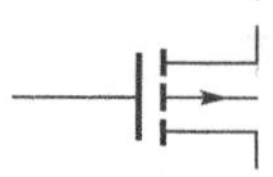
(d) 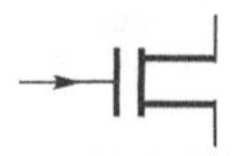

[1999 : 1 M]

2.4 In the circuit of figure, the value of the base current I_B will be

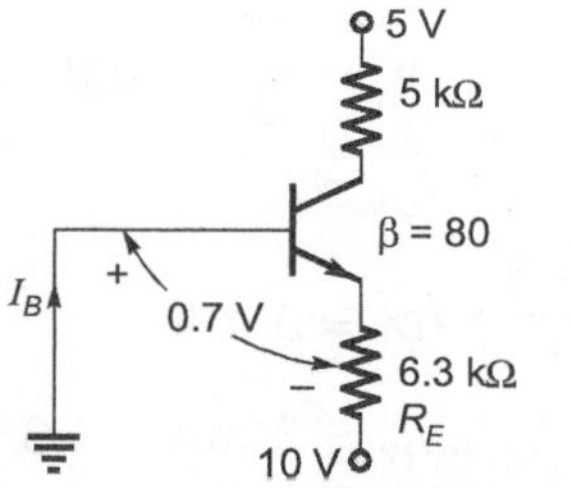

(a) 0 μA
(b) 18.2 μA
(c) 26.7 μA
(d) 40 μA

[2000 : 2 M]

2.5 An n-channel JFET having a pinch-off voltage (V_p) of –5 V shows a transconductance (g_m) of 1 mA/V when the applied Gate to source voltage (V_{GS}) is –3 V. Its maximum transconductance (in mA/V) is
(a) 1.5
(b) 2.0
(c) 2.5
(d) 3.0

[2001 : 2 M]

2.6 The variation of drain current with gate-to-source voltage ($I_D - V_{GS}$ characteristic) of a MOSFET is shown in the figure. The MOSFET is

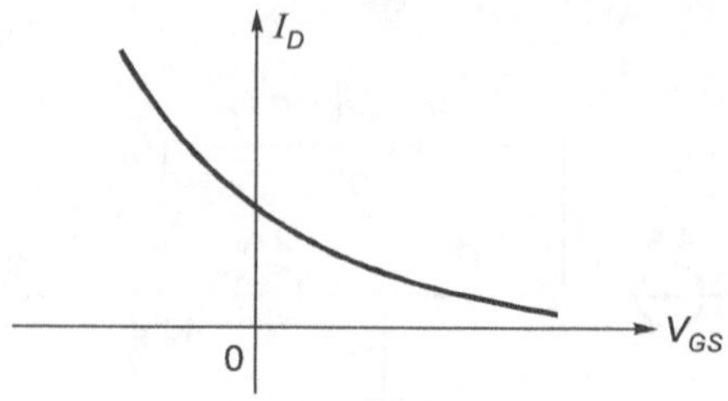

(a) an n-channel depletion mode device
(b) an n-channel enhancement mode device
(c) an p-channel depletion mode device
(d) an p-channel enhancement mode device

[2003 : 1 M]

2.7 In the circuit of figure, assume that the transistor has $h_{FE} = 99$ and $V_{BE} = 0.7$ V. The value of collector current I_C of the transistor is approximately

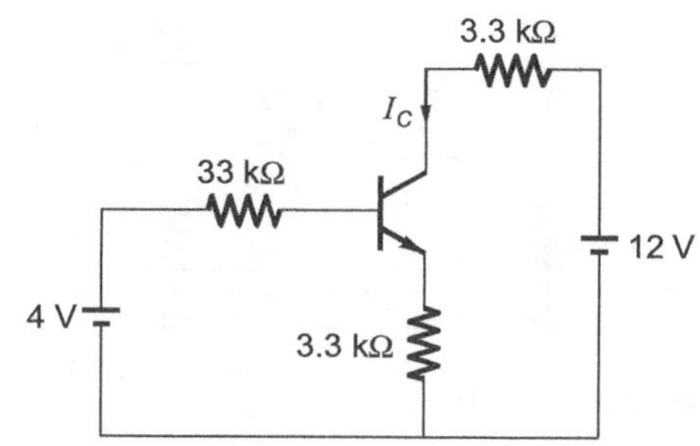

(a) [3.3/3.3] mA
(b) [3.3/(3.3 + 0.33)] mA
(c) [3.3/.33] mA
(d) [3.3/(33 + 3.3)] mA

[2003 : 1 M]

2.8 For the *n*-channel enhancement MOSFET shown in the figure, the threshold voltage $V_{th} = 2$ V. The drain current I_D of the MOSFET is 4 mA when the drain resistance R_D is 1 kΩ. If the value of R_D is increased to 4 kΩ, drain current I_D will become

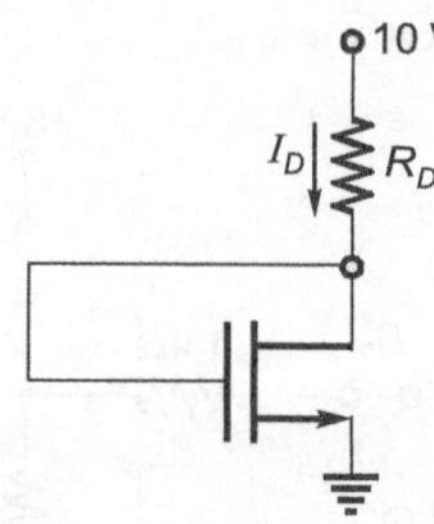

(a) 2.8 mA
(b) 2.0 mA
(c) 1.4 mA
(d) 1.0 mA

[2003 : 2 M]

2.9 Two perfectly matched silicon transistors are connected as shown in the figure. The value of the current I is

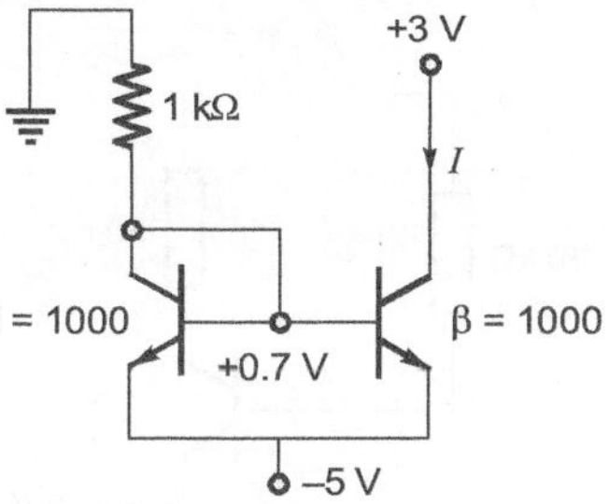

(a) 0 mA (b) 2.3 mA
(c) 4.3 mA (d) 7.3 mA [2004 : 1 M]

2.10 The transconductance g_m of the transistor shown in the figure is 10 mS. The value of the input resistance R_{IN} is

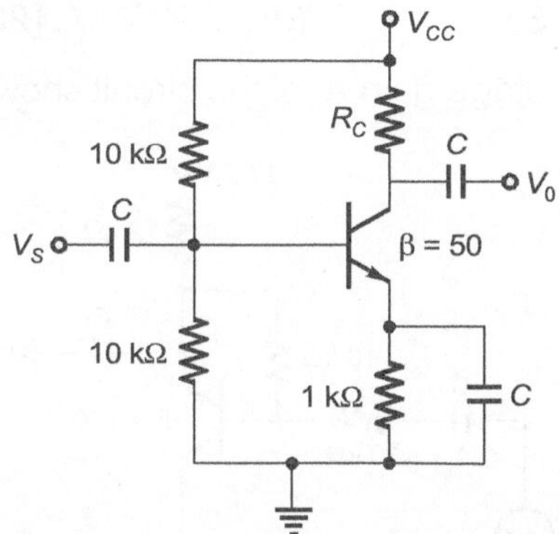

(a) 10.0 kΩ (b) 8.3 kΩ
(c) 5.0 kΩ (d) 2.5 kΩ [2004 : 2 M]

2.11 The value of R for which the PMOS transistor in figure will be biased in linear region is

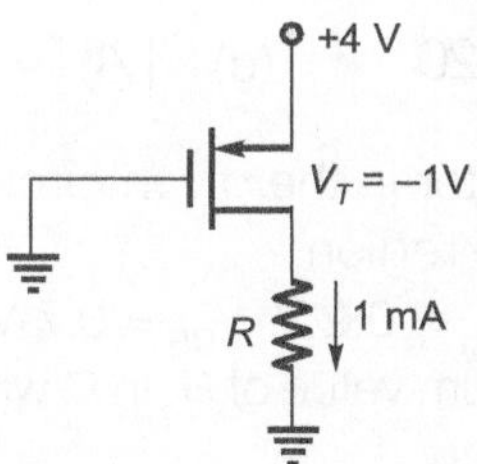

(a) 220 Ω (b) 470 Ω
(c) 680 Ω (d) 1200 Ω [2004 : 2 M]

2.12 Assume that the n-channel MOSFET shown in the figure is ideal, and that its threshold voltage is +1.0 V the voltage V_{ab} between nodes a and b is

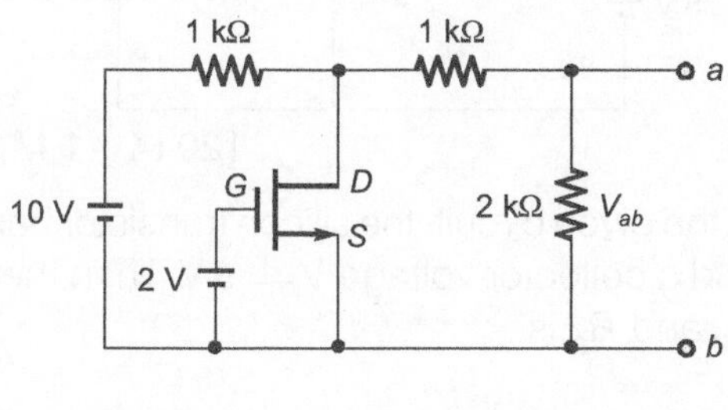

(a) 5 V (b) 2 V
(c) 1 V (d) 0 V [2005 : 1 M]

2.13 The common emitter amplifier shown in the figure is biased using a 1 mA ideal current source. The approximate base current value is

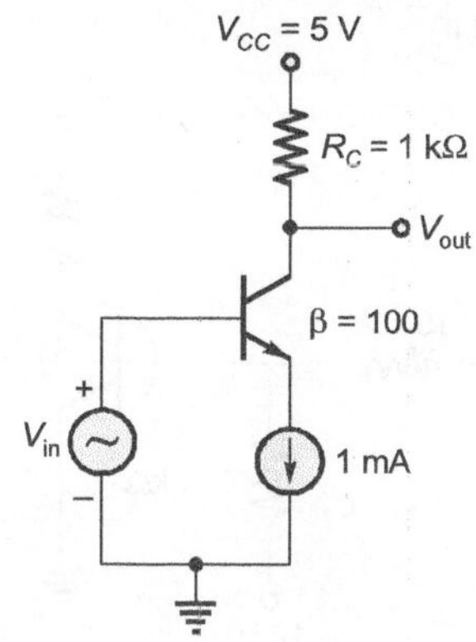

(a) 0 μA (b) 10 μA
(c) 100 μA (d) 1000 μA [2005 : 2 M]

Statement for common data Questions (2.14 and 2.15)

Assume that the threshold voltage of the n-channel MOSFET shown in figure is +0.75 V. The output characteristics of the MOSFET are also shown.

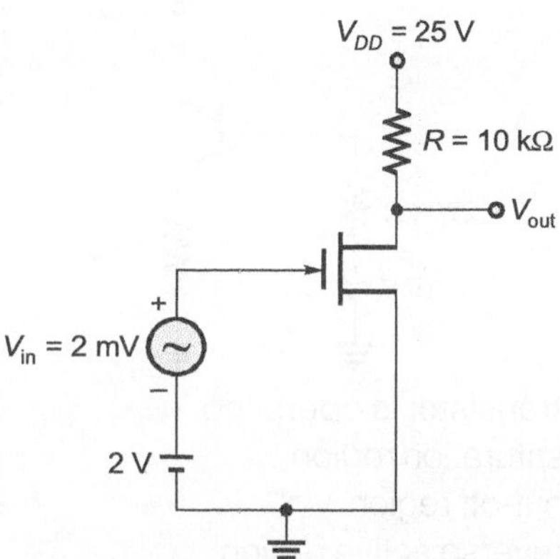

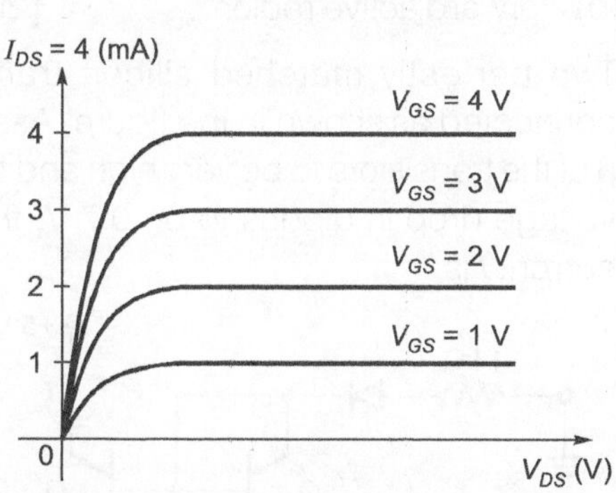

2.14 The transconductance of the MOSFET is
(a) 0.75 mS (b) 1 mS
(c) 2 mS (d) 10 mS
[2005 : 2 M]

2.15 The voltage gain of the amplifier is

(a) +5 (b) –7.5

(c) +10 (d) –10 [2005 : 2 M]

2.16 Consider the circuit shown in figure. If the β of the transistor is 30 and I_{CBO} is 20 nA and the input voltage is +5 V, the transistor would be operating in

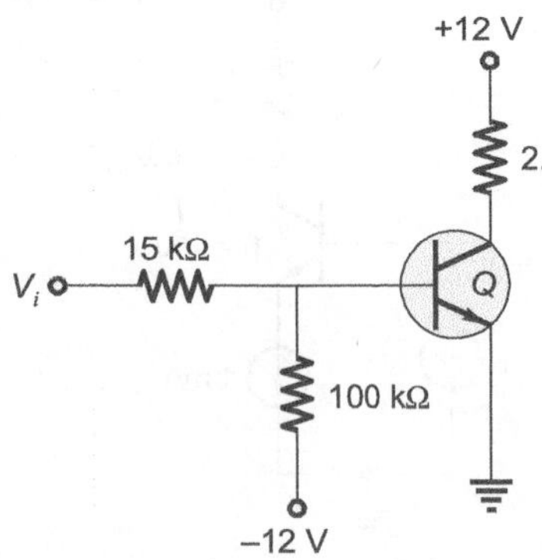

(a) saturation region (b) active region

(c) breakdown region (d) cut-off region

[2006 : 2 M]

2.17 The common emitter forward current gain of the transistor shown is $\beta_F = 100$.

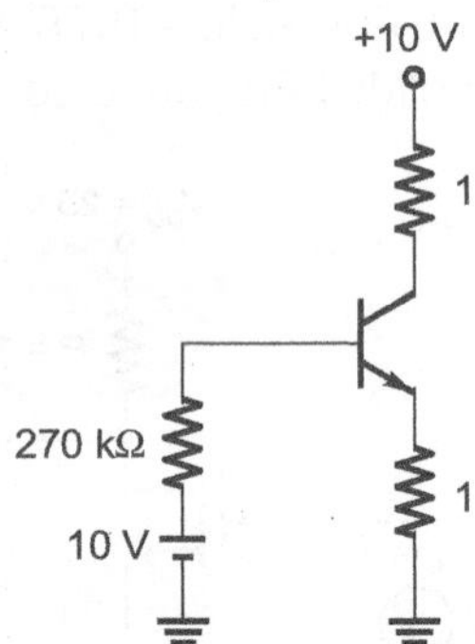

The transistor is operating in

(a) saturation region

(b) cut-off region

(c) reverse active region

(d) forward active region [2007 : 1 M]

2.18 Two perfectly matched silicon transistor are connected as shown in the figure. Assuming the β of the transistors to be very high and the forward voltage drop in diodes to be 0.7 V, the value of current I is

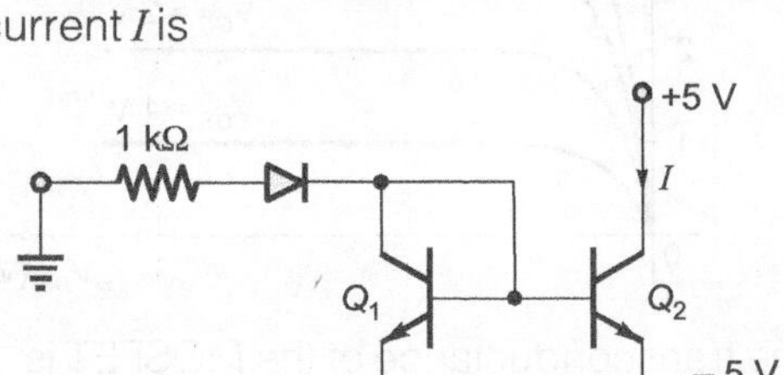

(a) 0 mA (b) 3.6 mA

(c) 4.3 mA (d) 5.7 mA [2008 : 2 M]

2.19 The transistor circuit shown uses a silicon transistor with $V_{BE} = 0.7$ V, $I_C \approx I_E$ and a DC current gain of 100. The value of V_0 is

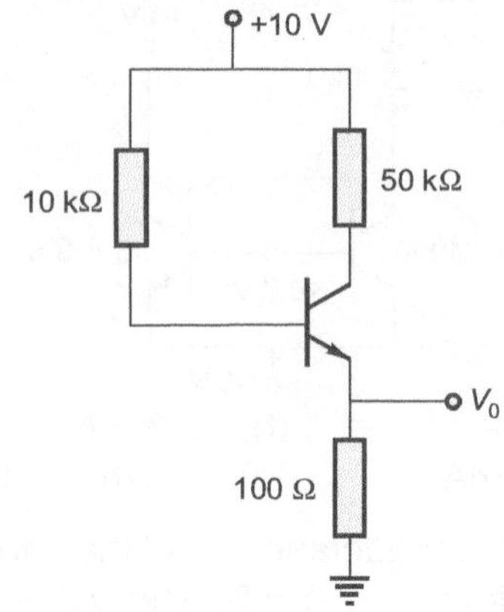

(a) 4.65 V (b) 5 V

(c) 6.3 V (d) 7.23 V [2010 : 2 M]

2.20 The voltage gain A_v of the circuit shown below is

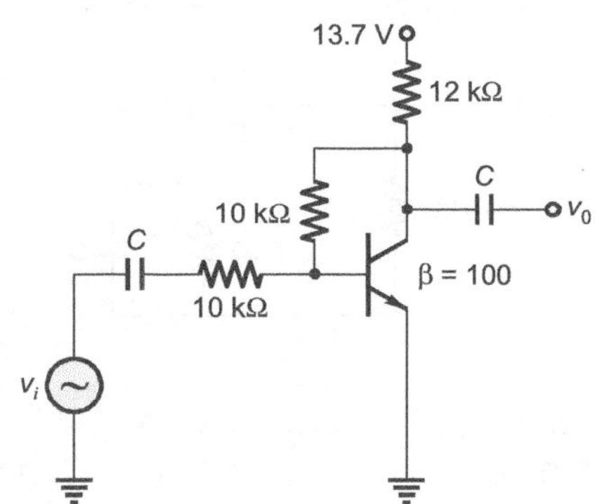

(a) $|A_v| \approx 200$ (b) $|A_v| \approx 100$

(c) $|A_v| \approx 20$ (d) $|A_v| \approx 10$ [2012 : 2 M]

2.21 The transistor in the given circuit should always be in active region.

Take $V_{CE(sat)} = 0.2$ V, $V_{BE} = 0.7$ V.

The maximum value of R_c in Ω which can be used, is ______.

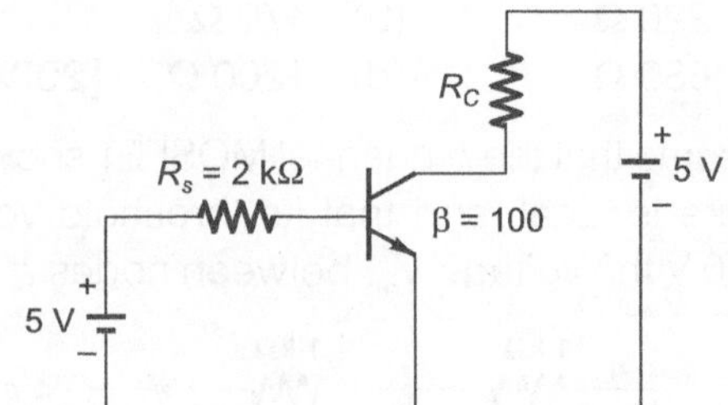

[2014 : 1 M, Set-2]

2.22 In the given circuit, the silicon transistor has β = 75 and a collector voltage $V_C = 9$ V. Then the ratio of R_B and R_C is

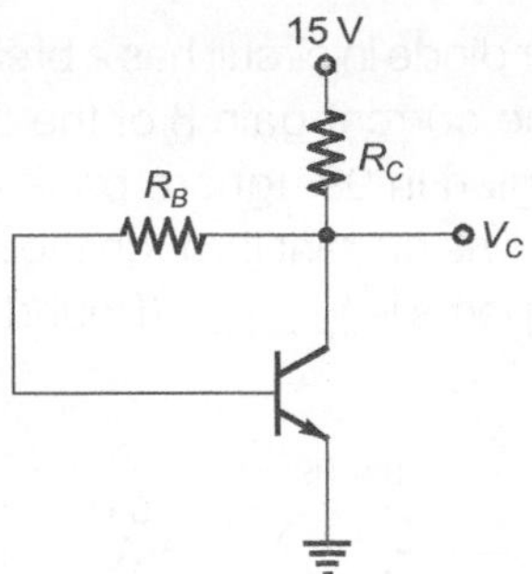

[2015 : 1 M, Set-1]

2.23 In the following circuit, the transistor is in active mode and $V_C = 2$ V. To get $V_C = 4$ V, we replace R_C with R'_C. Then the ratio R'_C / R_C is ______.

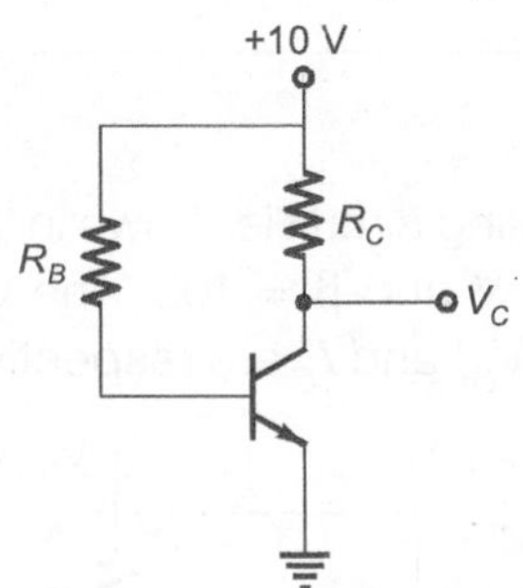

[2015 : 1 Mark, Set-2]

2.24 When a bipolar junction transistor is operating in the saturation mode, which one of the following statements is TRUE about the state of its collector-base (CB) and the base-emitter (BE) junctions?

(a) The CB junctions is forward biased and the BE junction is reverse biased.
(b) The CB junctions is reverse biased and the BE junction is forward biased.
(c) Both the CB and BE junctions are forward biased.
(d) Both the CB and BE junctions are reverse biased. [2015 : 1 M, Set-2]

2.25 A transistor circuit is given below. The Zener diode breakdown voltage is 5.3 V as shown below. Take base to emitter voltage drop to be 0.6 V. The value of the current gain β is ______.

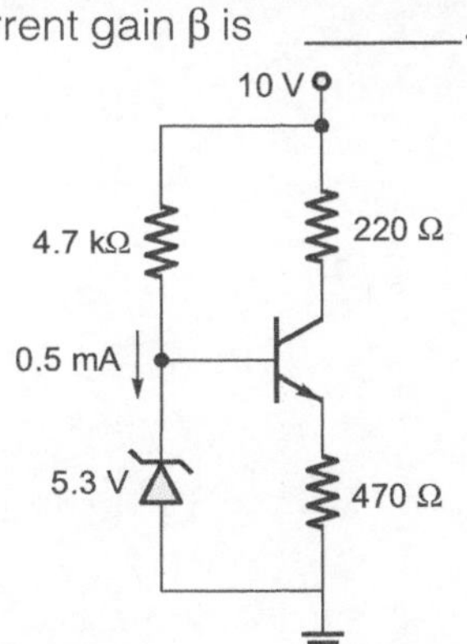

[2016 : 1 M, Set-1]

2.26 The circuit shown in the figure uses matched transistor with a thermal voltage $V_T = 25$ mV. The base currents of the transistors are negligible. The value of the resistance R in kΩ that is required to provide 1 mA bias current for the differential amplifier block shown is ______. (Give the answer up to one decimal place.)

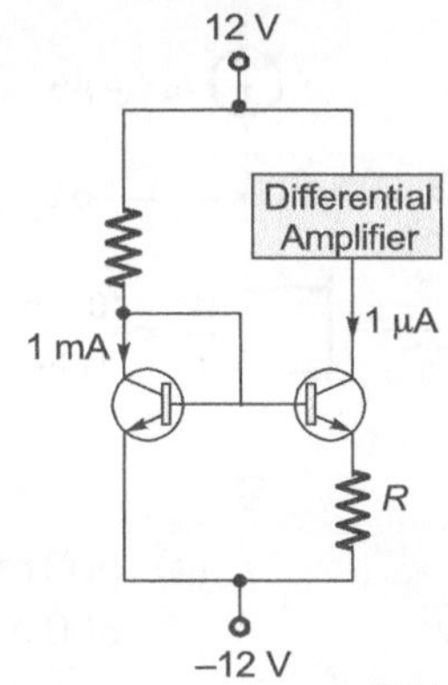

[2017 : 2 M, Set-1]

2.27 For the circuit shown in the figure below, it is given that $V_{CE} = \frac{V_{CC}}{2}$. The transistor has β = 29 and V_{BE} = 0.7 V when the B-E junction is forward biased.

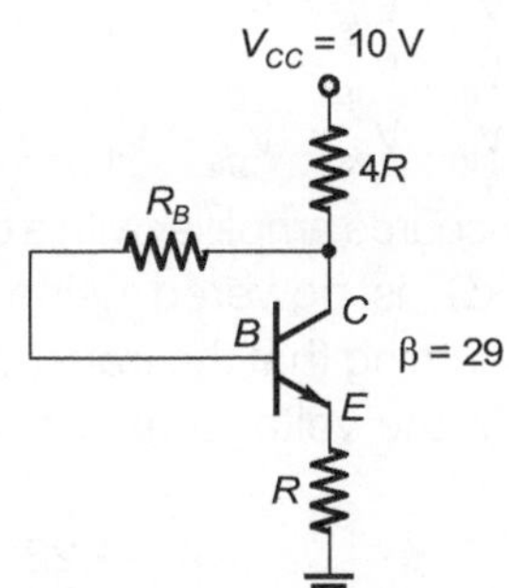

For this circuit, the value of R_B/R is

(a) 43 (b) 92
(c) 121 (d) 129

[2017 : 2 M, Set-2]

2.28 In the circuit shown in the figure, the bipolar junction transistor (BJT) has a current gain β = 100. The base-emitter voltage drop is a constant, V_{BE} = 0.7 V. The value of the Thevenin equivalent resistance R_{Th} (in Ω) as shown in the figure is _____ (upto 2 decimal places).

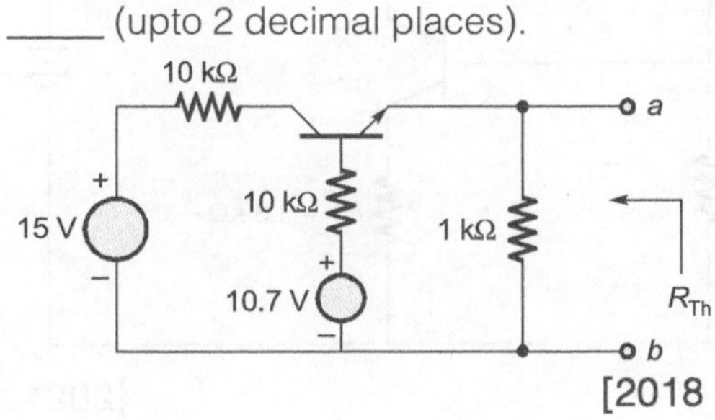

[2018 : 2 M]

2.29 The enhancement type MOSFET in the circuit below operates according to the square law. $\mu_n C_{ox}$ = 100 μA/V², the threshold voltage (V_T) is 500 mV. Ignore channel length modulation. The output voltage V_{out} is

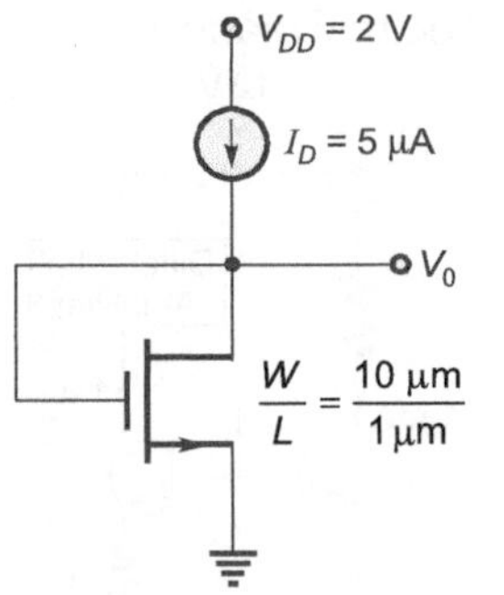

(a) 2 V (b) 500 mV
(c) 100 mV (d) 600 mV [2019 : 2 M]

2.30 Given, V_{gs} is the gate-source voltage, V_{ds} is the drain source voltage, and V_{th} is the threshold voltage of an enhancement type NMOS transistor, the conditions for transistor to be biased in saturation are

(a) $V_{gs} > V_{th}$; $V_{ds} \le V_{gs} - V_{th}$
(b) $V_{gs} < V_{th}$; $V_{ds} \ge V_{gs} - V_{th}$
(c) $V_{gs} > V_{th}$; $V_{ds} \ge V_{gs} - V_{th}$
(d) $V_{gs} < V_{th}$; $V_{ds} \le V_{gs} - V_{th}$ [2019 : 1 M]

2.31 A common-source amplifier with a drain resistance, R_D = 4.7 kΩ, is powered using a 10 V power supply. Assuming that the transconductance, g_m is 520 μA/V, the voltage gain of the amplifier is closest to:

(a) 1.22 (b) –1.22
(c) 2.44 (d) –2.44 [2020 : 1 M]

2.32 In the BJT circuit shown, beta of the PNP transistor is 100. Assume V_{BE} = –0.7 V.
The voltage across R_C will be 5 V when R_2 is ______ kΩ. (Round off 2 decimal places).

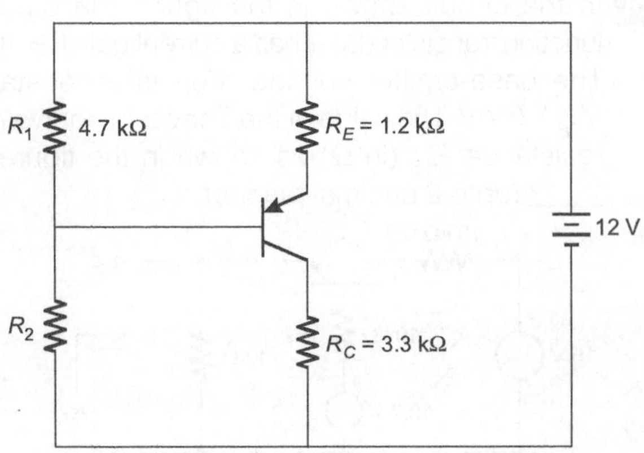

[2021 : 1 M]

2.33 The Zener diode in circuit has a breakdown voltage of 5 V. The current gain β of the transistor in the active region in 99. Ignore base-emitter voltage drop V_{BE}. The current through the 20 Ω resistance in milliamperes is ______(Round off to 2 decimal places).

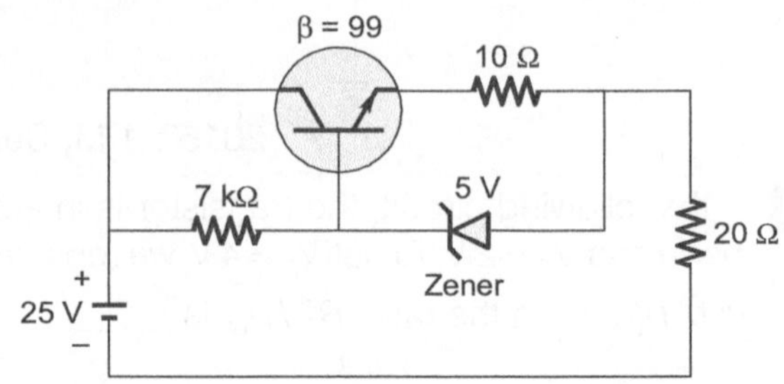

[2023 : 2 M]

2.34 A BJT biasing circuit is shown in the figure, where V_{BE} = 0.7 V and β = 100. The Quiescent Point values of V_{CE} and I_C are respectively

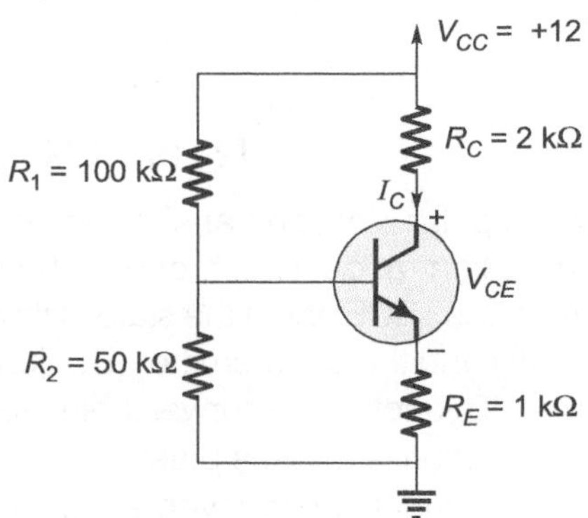

(a) 4.6 V and 2.46 mA
(b) 2.61 V and 3.13 mA
(c) 4.6 V and 3.13 mA
(d) 3.5 V and 2.46 mA

[2024 : 2 M]

Answers BJT, FET and their Biasing Circuits

2.1 (d)	2.2 (Sol.)	2.3 (a)	2.4 (b)	2.5 (c)	2.6 (c)	2.7 (b)
2.8 (c)	2.9 (c)	2.10 (d)	2.11 (d)	2.12 (d)	2.13 (b)	2.14 (b)
2.15 (d)	2.16 (b)	2.17 (d)	2.18 (b)	2.19 (a)	2.20 (d)	2.21 (22.32)
2.22 (105.13)	2.23 (0.75)	2.24 (c)	2.25 (19)	2.26 (172.7)	2.27 (d)	2.28 (90.09)
2.29 (d)	2.30 (c)	2.31 (d)	2.32 (17.06)	2.33 (250)	2.34 (a)	

Explanations BJT, FET and their Biasing Circuits

2.2 Sol.

(a) → (P)
(b) → (R)
(c) → (Q)

2.4 (b)

$$0 - 0.7 - I_b(\beta + 1) \times 6.3 = -10\text{ V}$$

$$I_b = \frac{10-0.7}{81 \times 6.3}\ \mu\text{A} = 18.452\ \mu\text{A}$$

2.5 (c)

$$g_m = -\frac{2I_{DSS}}{V_p}\left(1 - \frac{V_{GS}}{V_p}\right)$$

$$1\text{ mA/V} = \frac{-2I_{DSS}}{-5}\left(1 - \frac{-3}{-5}\right) = \frac{2}{5} \times \left(1 - \frac{3}{5}\right) I_{DSS}$$

$$\Rightarrow \quad I_{DSS} = \left(\frac{5}{2}\right)^2 \text{ mA}$$

$$(g_m)_{max} = -\frac{2I_{DSS}}{V_p} = -\frac{2 \times 25}{4 \times (-5)} = 2.5\text{ mA/V}$$

2.6 (c)

This is the characteristics of a *p*-channel depletion mode device.

2.7 (b)

Using KVL in Base-emitter Loop

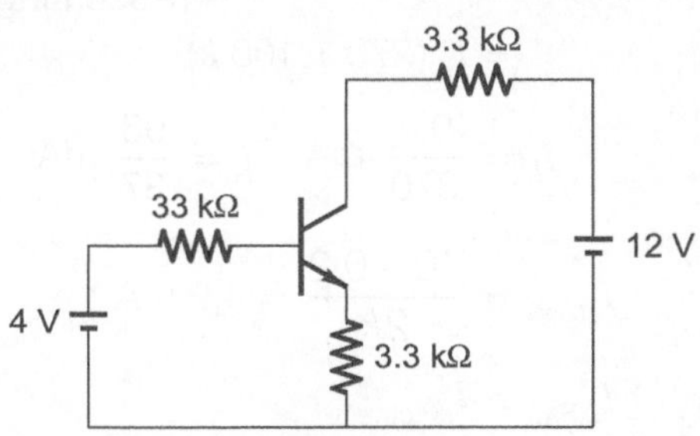

$$4 - 33 \times I_b - 0.7 - 3.3 \times 100\, I_b = 0 (I_c = \beta I_b)$$

$$\therefore \quad I_b = \frac{4 - 0.7}{(33 + 330)}$$

$$\therefore \quad I_c = \frac{3.3 \times 100}{33 + 330} = \frac{3.3}{(0.33 + 3.3)}\text{ mA}$$

2.8 (c)

$$I_D = K(V_{GS} - V_{th})^2$$
$$4 = K(6-2)^2$$
$$\therefore \quad K = \frac{1}{4}\text{ mA/V}^2$$
$$V_{GS} = 10 - 4 \times 1 = 6\text{ V}$$

when R_D is increased to 4 kΩ

$$V_{GS} = 10 - 4I_D$$
$$I_D = \frac{1}{4}(10 - 4I_D - 2)^2$$
$$4I_D = 16I_D^2 + 64 - 64I_D$$
$$16I_D^2 - 68I_D + 64 = 0$$
$$I_D = 2.84\text{ mA, } 1.4\text{ mA}$$

For MOSFET to be on, V_{GS} must be greater that V_{th} and this is possible only if

$$I_D = 1.4\text{ mA}$$

If $I_D = 2.84$ mA

then V_{GS} become –ve and less than V_{th} so transistor will be off for this value which is not possible since $V_{DG} = 0$; $V_{DG} > -V_{th}$.

2.9 (c)

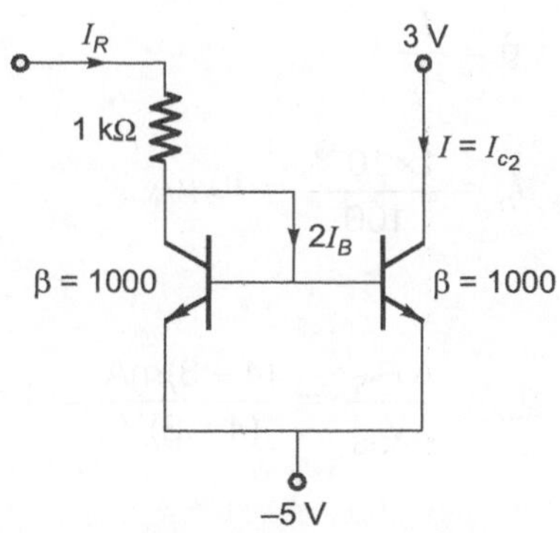

Both transistor are perfectly matched, hence

$$V_{BE2} = V_{BE1}$$

Therefore, $\dfrac{I_{c1}}{I_{c2}} = \exp\left[(V_{BE1} - V_{BE2})/V_T\right] = 1$

Also 'β' is same.

$$I_R = \frac{+5 - 0.7}{1\text{K}} = 4.3\text{ mA}$$

Writing KCL at node *B*,

$$I_c + 2I_b - I_R = 0$$

$$I_c = \beta I_b$$

$$\therefore \quad I_c = \left(\frac{\beta}{\beta+2}\right) I_R$$

$$\approx I_R \text{ (Because } \beta \text{ is very large)}$$

Hence, $I_{c1} = I_{c2} = I_R = 4.3$ mA

2.10 (d)

$$\text{Input Resistance} = \frac{h_{fe}}{g_m} = \frac{\beta}{g_m}$$

$$= \frac{50}{10 \times 10^{-3}} = 5 \text{ k}\Omega$$

But overall input resistance seen from source is

$$R_{IN} = 10 \parallel 10 \parallel 5 = 2.5 \text{ k}\Omega$$

2.11 (d)

Here, $V_S = +4$ V; $V_G = 0$ V; $V_T = -1$ V

Therefore, $V_{SG} = 4$ V

$$V_{SD} = V_S - V_D = 4 - I_a R$$

Now for linear region of operation

$$V_{SD} < (V_{SG} + V_T)$$

$$\Rightarrow \quad 4 - I_a R < (4-1) \Rightarrow I_a(R) > 1$$

$$\Rightarrow \quad 10^{-3} \times R > 1 \Rightarrow R > 1000 \ \Omega$$

2.12 (d)

MOSFET is N-channel. Gate through source is so connected that MOSFET will be in enhance mode and so conductivity of the channel will be increased very much and effectively 'β' terminal act as short circuited, So $V_{ab} = 0$ V.

2.13 (b)

$$\beta = \frac{I_c}{I_b}$$

$$I_b = \frac{1 \times 10^{-3}}{100} = 10 \ \mu\text{A}$$

2.14 (b)

$$g_m = \frac{\Delta I_{DS}}{\Delta V_{GS}} = \frac{(4-3)\text{mA}}{(4-3)\text{V}} = 1 \text{ mS}$$

2.15 (d)

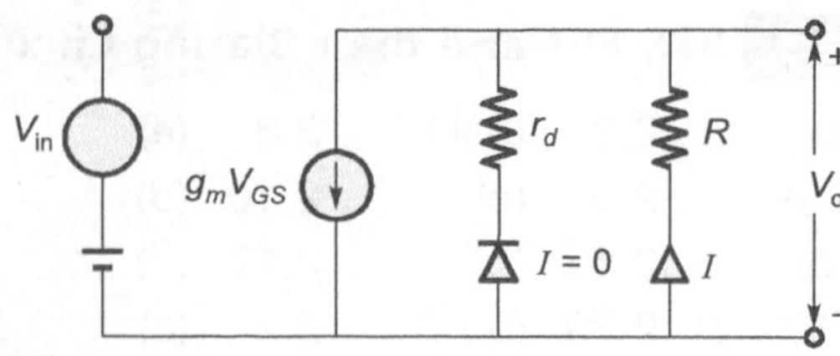

Since, $r_d >> R$

All current will pass through R

$$V_{out} = -g_m V_{GS} \cdot R$$

$$= -1 \times 10^{-3} \times 2 \times 10^{-3} \times 10 \times 10^{-3}$$

$$= -20 \text{ mV}$$

$$\text{Voltage gain} = \frac{V_{out}}{V_{in}} = \frac{20 \text{ mV}}{2 \text{ mV}} = -10$$

2.16 (b)

Assume BJT is in active Region and we neglect I_{CBO},

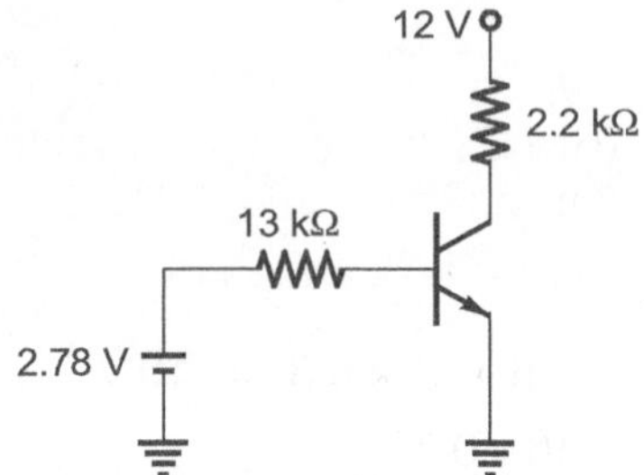

$$V_{th} = \frac{-12 \times 15}{115} + \frac{5 \times 100}{5} = 2.78 \text{ V}$$

$$R_{th} = (15 \,||\, 100) = \frac{15 \times 100}{115} = 13 \text{ k}\Omega$$

$$I_c = \beta I_b$$

$$\text{But,} \quad I_b = \frac{2.78 - 0.7}{13} = 0.16 \text{ mA}$$

$$I_c = 4.8 \text{ mA}$$

$$\text{But,} \quad I_{c(Sat)} = \frac{12 - 0.2}{2.2\text{k}} = 5.36 \text{ mA}$$

as $\quad I_{c(Sat)} > I_{c(active)}$

BJT is in active region.

2.17 (d)

We assume BJT is in active region, applying KVL in base emitter circuit

$$10 - 0.7 = 1 \text{ k}\Omega \times I_c + 270 \times I_b$$

(Assuming $I_C = I_E$)

$$= I_b(270 + 100\,k)$$

$$\therefore \quad I_b = \frac{9.3}{370} \text{ mA}; \quad I_c = \frac{93}{37} \text{ mA}$$

$$I_{c(Sat)} = \frac{10 - 0.2}{2k} = 4.9 \text{ mA}$$

$$I_{c(Sat)} > I_{c(active)}$$

∴ BJT is in active region.

2.18 (b)

Since both transistor are perfectly matched,

So, $\quad V_{BE1} = V_{BE2}$

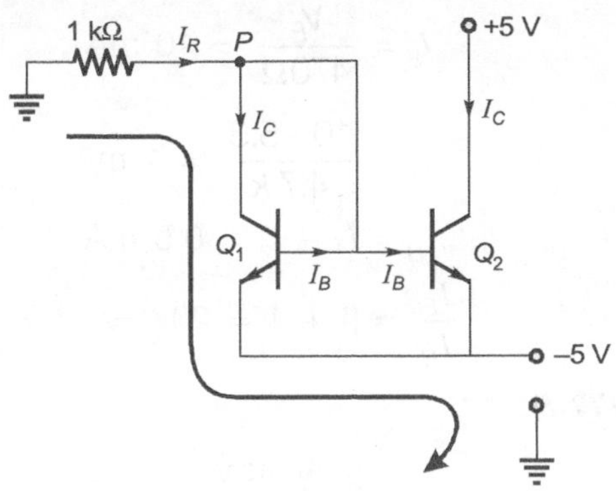

$$\frac{I_{c1}}{I_{c2}} = \exp\left[\frac{V_{BE1} - V_{BE2}}{V_T}\right] = e^0 = 1$$

Since β for both are same, therefore

$$I_{b1} = I_{b2} = I_b$$

Applying KVL to loop as shown,

$$I_R = \frac{0 - 0.7 - 0.7 - (-5)}{1\text{k}\Omega} = 3.6 \text{ mA}$$

By KCL,

$$I_R = I_c + 2\,I_b = I_c + 2\,\frac{I_c}{\beta}$$

$$\therefore \quad I_c = \frac{\beta}{\beta + 2} \times I_R$$

$$\approx I_R \quad \text{(Because β is very large)}$$

2.19 (a)

KVL ⇒

$$10 = 10\,kI_b + 0.7 + (100)\,(I_b)100$$

$$I_b = \frac{9.3}{20\text{k}} = 0.465 \text{ mA}$$

$$V_0 = 100 \times 0.465 \times 100 = 4.65 \text{ V}$$

2.20 (d)

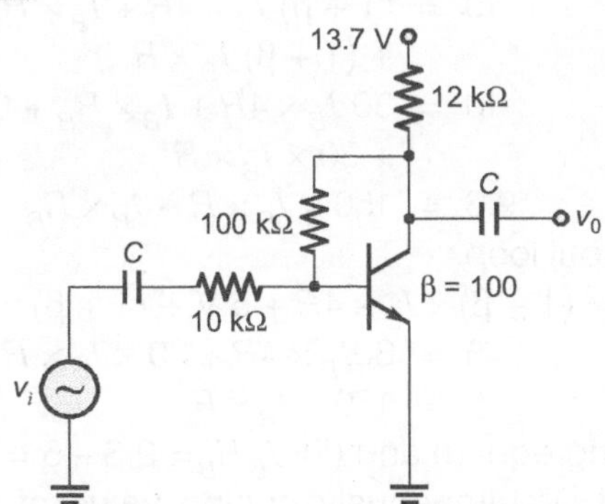

Equivalent A.C. model will be taking,

$$h_{ie} = 1\text{ k}\Omega\,;\ h_{fe} = \beta = 100$$

$$A_V = \frac{V_o}{V_{in}} = \frac{-h_{fe}i_b \times (100 \,||\, 12)\text{k}\Omega}{h_{ie} \times i_b}$$

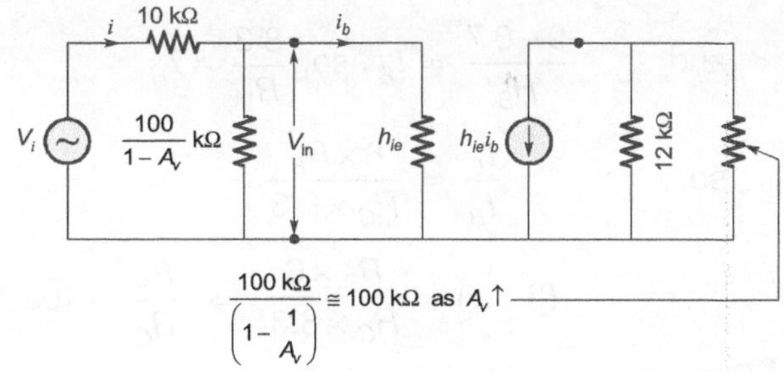

$$A_V = \frac{-100 \times 10.71\text{k}\Omega}{1\text{k}\Omega} = -1071.42$$

$$V_i = h_{ie}i_b + 10^4 i$$

$$= 10^3 i_b + 10^4\left[i_b + \frac{h_{ie} \times i_b}{\left(\frac{100\text{ k}\Omega}{1 - A_v}\right)}\right]$$

$$V_i = 10^3\,i_b + i_b\,[10^4 + (1072.42) \times 100]$$

$$A_{VS} = \frac{V_0}{V_i} = \frac{-h_{fe}\,i_b \times 10.7142 \times 10^3}{i_b[10^3 + 10^4 + 10^2 \times 1072.42]}$$

$$A_{VS} = \frac{-100 \times 10.7142 \times 10^3}{(10^3 + 10^4 + 10^2 \times 1072.42)}$$

$$A_{VS} = -9.06$$

$$|A_{VS}| = 9.06 \simeq 10$$

2.21 (22.32)

In input loop:

$$I_b = \frac{5 - 0.7}{2\text{ k}\Omega} = \frac{4.3}{2\text{ k}\Omega} = 2.15\text{ mA}$$

So, $\quad I_c = \beta \times I_b = 100 \times 2.15 \text{ mA} = 0.215 \text{ A}$

Now KVL in output loop:

$$V_{CE} = 5 - 0.215\,R_C$$

For active region $V_{CE} > 0.2$ V

$$\Rightarrow \quad 0.215\,R_C < 5 - 0.2$$

$$\Rightarrow \quad R_{C\max} = \frac{4.8}{0.215} \simeq 22.32\ \Omega$$

2.22 (105.13)

Consider the circuit shown in figure.

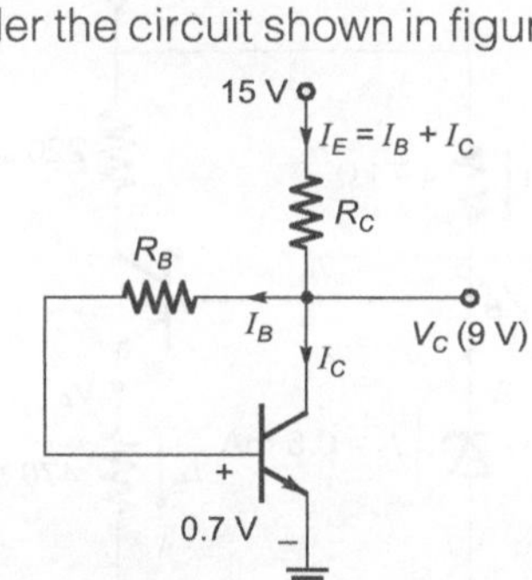

$$V_C = 9 \text{ V}$$

so, $\quad \dfrac{15 - 9}{R_C} = I_E \Rightarrow \dfrac{6}{R_C} = I_E$

and $\frac{9-0.7}{R_B} = I_B$ so, $\frac{8.3}{R_B} = I_B$

so, $\frac{I_E}{I_B} = \frac{6 \times R_B}{R_C \times 8.3}$

$$(\beta + 1) = \frac{R_B \times 6}{R_C \times 8.3} \Rightarrow \frac{R_B}{R_C} = 105.13$$

2.23 (0.75)

Case-I: $V_C = 2$ V

$$i_C = \frac{10-2}{R_C} \quad \text{...(i)}$$

$10 - i_B R_B - 0.7 = 0$

$$i_B = \frac{10-0.7}{R_B} \quad \text{...(ii)}$$

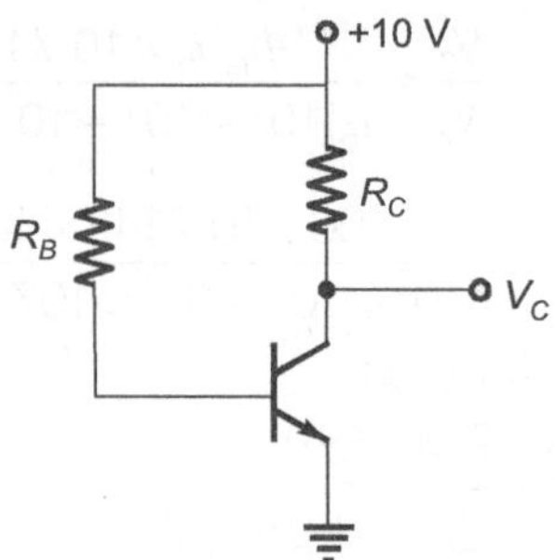

Case-II:

When, $V_C = 4$ V

$R_C \to R_C'$

$$i_C = \frac{10-4}{R_C'} \quad \text{...(iii)}$$

From equation (iii) and (i),

$$\frac{10-4}{R_C'} = \frac{10-2}{R_C} \Rightarrow \frac{R_C'}{R_C} = \frac{6}{8} = \frac{3}{4} = 0.75$$

2.25 (19)

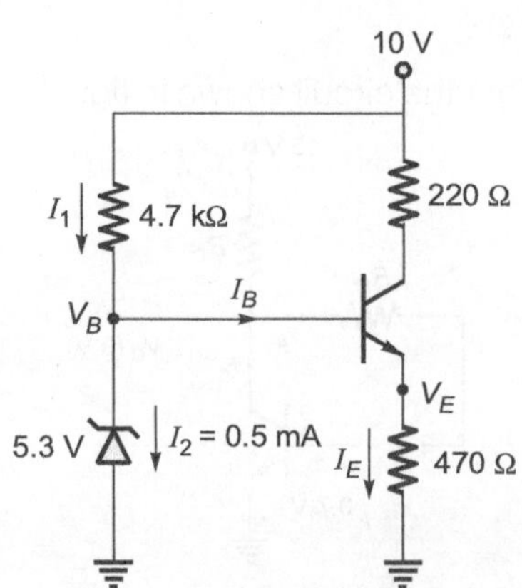

$V_B = 5.3$ V

$V_E = V_B - 0.6 = 4.7$ V

$$I_E = \frac{V_E}{470\Omega} = 10 \text{ mA}$$

$$I_1 = \frac{10-5.3}{4.7k} = 1 \text{ mA}$$

$$I_B = I_1 - I_2 = 0.5 \text{ mA}$$

$$\frac{I_E}{I_B} = \beta + 1 = 20 \Rightarrow \beta = 19$$

2.26 (172.7)

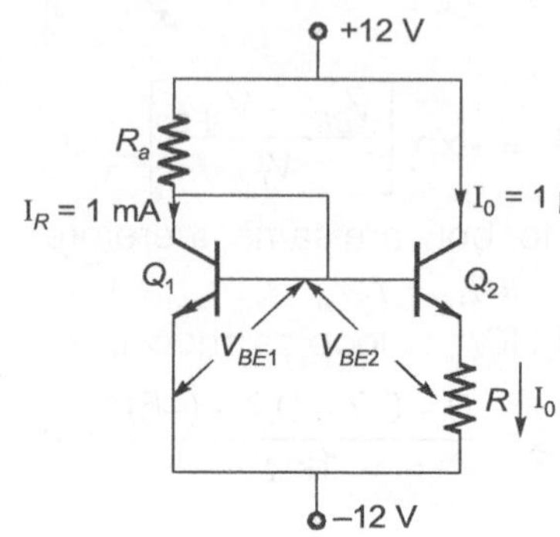

$$V_{BE1} = V_{BE2} + I_0 R$$

$$I_0 R = V_{BE1} - V_{BE2}$$

$$= V_I l n\left(\frac{I_R}{I_3}\right) - V_T l_n\left(\frac{I_0}{I_3}\right)$$

where, $I_s \to$ Reverse saturation current

$$R = \frac{V_T l_n\left(\frac{I_R}{I_0}\right)}{I_0} = \frac{0.025\, l n\left(\frac{1\text{mA}}{1\mu\text{A}}\right)}{1\mu\text{A}}$$

$$= \frac{0.025\, l n(10^3)}{1\mu\text{A}} = 172.7\text{ k}\Omega$$

2.27 (d)

In input loop,

$$10 = (1+\beta) I_B \times 4R + I_B \times R_B + 0.7 + (1+\beta) I_B \times R$$

$$10 = 30\, I_B \times 4R + I_B \times R_B + 0.7 + 30 \times I_B \times R$$

$$9.3 = 150 \times I_B \times R + I_B \times R_B \quad \text{...(i)}$$

Output loop,

$$10 = (1+\beta) \times I_B \times 4R + 5\text{ V} + (1+\beta) \times I_B \times R$$

$$5 = 30 I_B \times 4R + 30 \times I_B \times R$$

$$\Rightarrow \quad 5 = 150 \times I_B \times R \quad \text{...(ii)}$$

Using eqn. (i) and (ii), $I_B R_B = 9.3 - 5 = 4.3$ and simultaneously putting value of $I_B R$ from equation (ii) in equation (i),

$$9.3 = I_B R\left[150 + \frac{R_B}{R}\right] = \frac{5}{150}\left[150 + \frac{R_B}{R}\right]$$

$$279 = 150 + \frac{R_B}{R} \Rightarrow \frac{R_B}{R} = 129$$

2.28 (90.09)

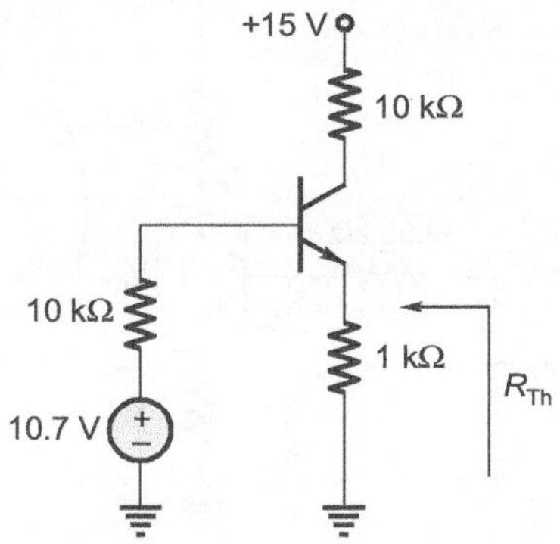

To calculate R_{Th} D.C. voltage should be short-circuited.

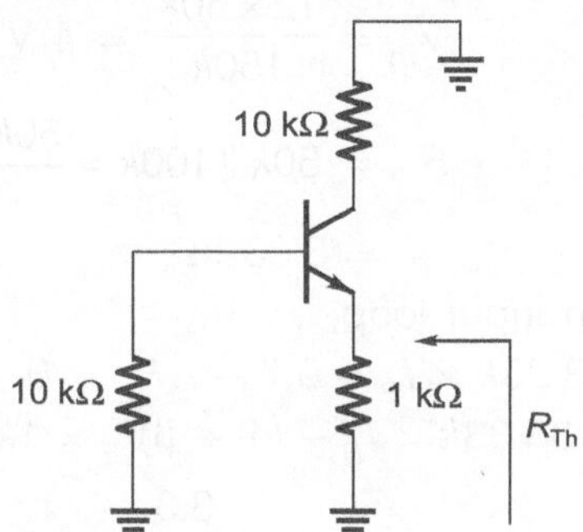

$$R_{Th} = 1\text{k}\Omega \,||\, \frac{10\text{ k}}{1+\beta}$$

$$= 1\text{k}\Omega \,||\, 99.0099 = 90.09\ \Omega$$

2.29 (d)

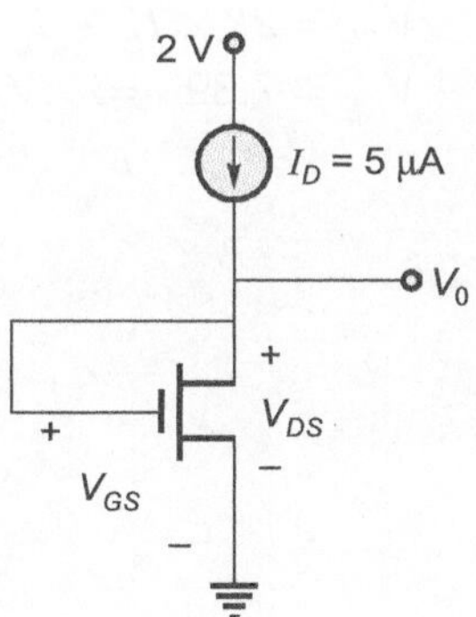

As, $\quad V_{DS} = V_{GS}$

MOSFET is in saturation,

$$I_D = \frac{1}{2}\mu_n C_{ox}\left(\frac{W}{L}\right)(V_{GS} - V_T)^2$$

$$5 \times 10^{-6} = \frac{1}{2} \times 100 \times 10^{-6} \times 10\ (V_{GS} - 0.5)^2$$

$\Rightarrow \quad V_{GS} = 0.6$ V

$\Rightarrow \quad V_0 = 600$ mV

2.30 (c)

For NMOS transistor to be in saturation the condition will be

$$V_{gs} > V_{th} \quad \text{and} \quad V_{ds} \geq V_{gs} - V_{th}$$

2.31 (d)

Given data: $R_D = 4.7$ kΩ, $\quad g_m = 520$ μA/V

Voltage gain of CS amplifier

$$= -g_m R_D$$

$$= -520\ \mu\text{AV} \times 4.7\ \text{k}\Omega = -2.44$$

2.32 (17.06)

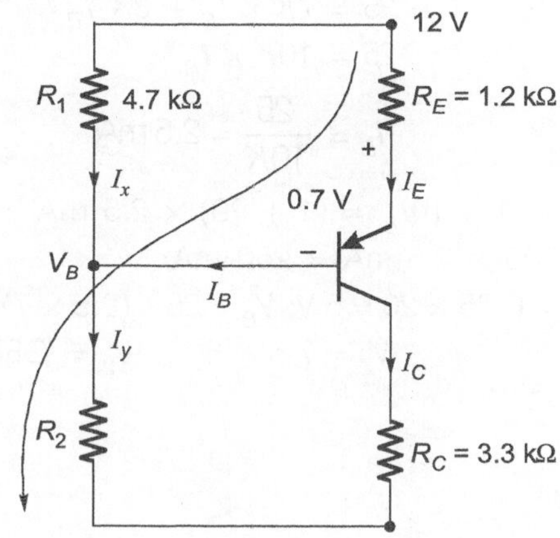

$$I_C = \frac{5\text{V}}{3.3\text{k}} = 1.515 \text{ mA}$$

$$I_E = 1.53 \text{ mA}$$

$$I_B = 0.0151 \text{ mA}$$

$$-12 + 1.2\text{k} \times 1.53\text{ m} + 0.7 + V_B = 0$$

$$V_B = 9.464 \text{ V}$$

$$I_x = \frac{12 - V_B}{4.7\text{k}} = \frac{12 - 9.464}{4.7\text{k}} = 0.539 \text{ mA}$$

$$I_x + I_B = I_y$$

$\Rightarrow \quad I_y = 0.5396 + 0.0151$

$$I_y = 0.5546 \text{ mA}$$

$$V_B = 0.5546\text{ m} \times R_2 = 9.464$$

$$R_2 = 17.06\ \text{k}\Omega$$

2.33 (250)

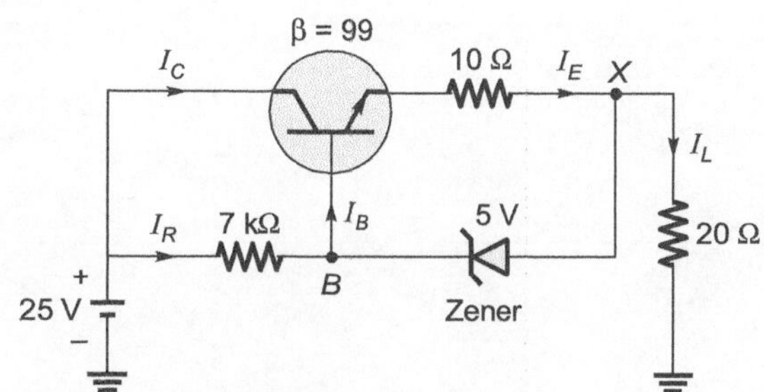

Assume zener off

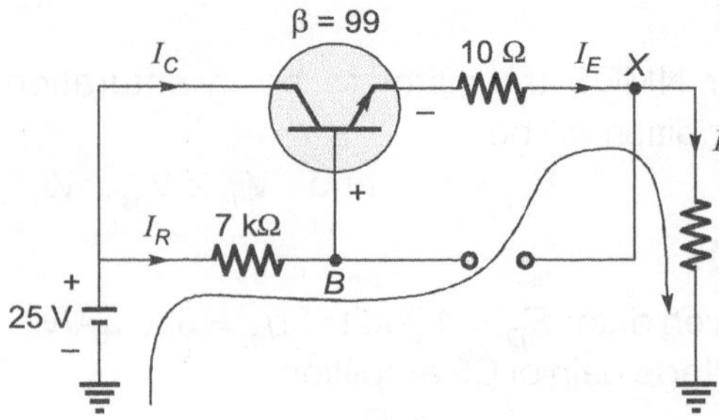

$$25 = 7K \times I_B + I_E\,(10\Omega + 20\Omega)$$
$$25 = 7K \times I_B + (1 + \beta)\, I_B\,(30\Omega)$$
$$25 = 7K \times I_B + 3000\Omega \times I_B$$
$$25 = 7K \times I_B + 3K\, I_B$$
$$25 = 10K \times I_B$$
$$I_B = \frac{25}{10K} = 2.5\,\text{mA}$$

$I_E = (1 + \beta)I_B = (1 + 99) \times 2.5$ mA

$= 100 \times 2.5$ mA $= 250$ mA

$V_X = 0.25 \times 20 = 5$ V; $V_B = 25 - (2.5 \times 7) = 7.5$ V

$\because \quad V_Z = 7.5 - 5, \ \therefore\ V_2 =$ OFF

2.34 (a)

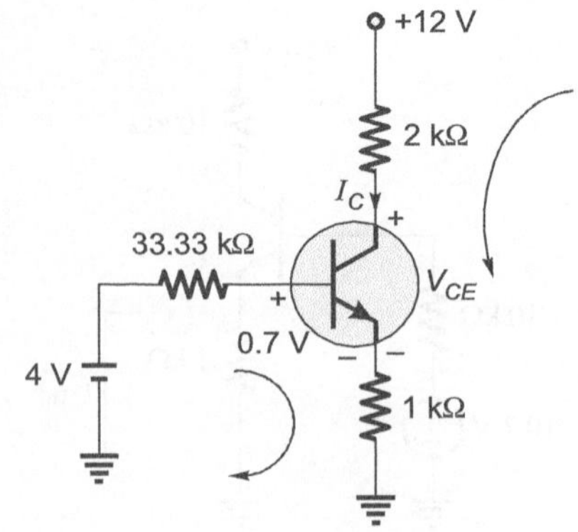

$\because \quad V_{BE} = 0.7$ V

$\therefore$ Transisotr is in ACTIVE region.

$$V_{Th} = \frac{12 \times 50k}{150k} = 4\text{ V}$$

$$R_{Th} = 50k \,\|\, 100k = \frac{50k \times 100k}{150k}$$
$$= 3.33\text{ k}\Omega$$

KVL in input loop,

$4 - 33.33k \times I_B - 0.7 - I_E R_E = 0$

$3.3 - 33.33k \times I_B - (1 + \beta)I_B \times 1k = 0$

$$\therefore \quad I_B = \frac{3.3}{33.33k + 101k} = 24.56\,\mu A$$

$$I_C = \beta I_B = 100 \times 24.56 \times 10^{-6}$$
$$= 2.46\text{ mA}$$
$$I_E = (1 + \beta)I_B = 2.47\text{ mA}$$

KVL in outer loop :

$12 - 2k \times I_C - V_{CE} - I_E R_E = 0$

$12 - V_{CE} = 2k \times I_C + I_E \times 1k$

$12 - V_{CE} = 7.39 \ \Rightarrow\ V_{CE} = 4.61$ V

3 Small Signal Analysis

3.1 In the transistor amplifier shown in figure, the ratio of small signal voltage gain, when the emitter resistor is bypassed by the capacitor 'C_e' to when it is not bypassed. (Assuming of simplified approximate h-parameter model for transistor) is

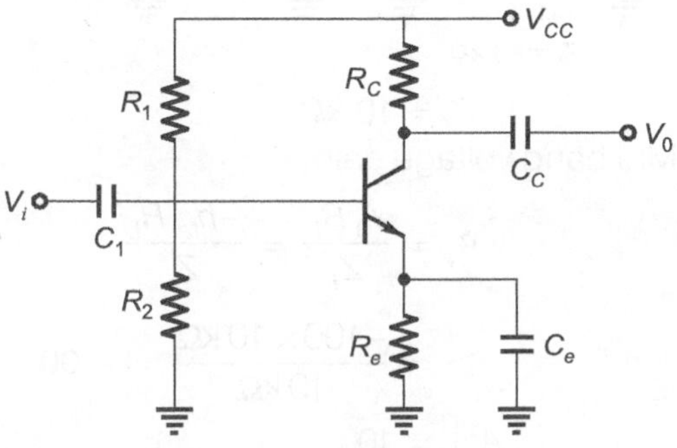

(a) 1
(b) h_{fe}
(c) $\dfrac{(1+h_{fe})R_e}{h_{ie}}$
(d) $1+\dfrac{(1+h_{fe})R_e}{h_{ie}}$

[1996 : 1 M]

3.2 In the single stage transistor amplifier circuit shown in figure, the capacitor C_E is removed then the AC small signal midband voltage gain of the amplifier

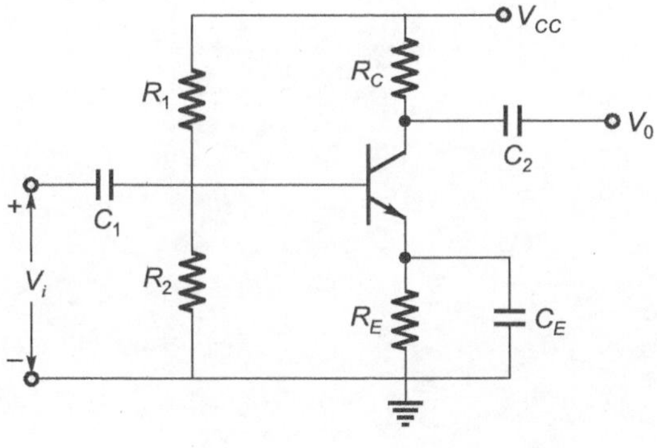

(a) increases
(b) decreases
(c) is unaffected
(d) drops to zero

[2001 : 1 M]

3.3 The magnitude of the mid-band voltage gain of the circuit shown in the figure is (Assuming h_{fe} of the transistor to be 100)

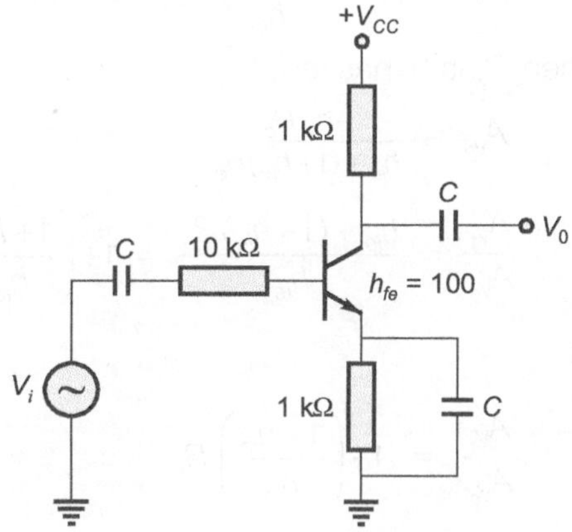

(a) 1
(b) 10
(c) 20
(d) 100

[2014 : 1 M, Set-1]

3.4 For an ideal MOSFET biased in saturation, the magnitude of the small signal current gain for a common drain amplifier is

(a) 0
(b) 1
(c) 100
(d) Infinite

[2022 : 1 M]

Answers Small Signal Analysis

3.1 (d) 3.2 (b) 3.3 (d) 3.4 (d)

Explanations Small Signal Analysis

3.1 (d)

When C_e is unbypassed,

$$A_{V_1} = -\frac{h_{fe} R_C}{h_{ie}}$$

When C_e is bypassed,

$$A_{V_2} = \frac{-h_{fe} R_c}{h_{ie} + (1+h_{fe})R_e}$$

$$\frac{A_{V_1}}{A_{V_2}} = \frac{h_{ie} + (1+h_{fe})R_e}{h_{ie}} = 1+\left(\frac{1+h_{fe}}{h_{ie}}\right)R_e$$

3.2 (b)

$$\frac{A_{V_1}}{A_{V_2}} = 1+\left(\frac{1+h_{fe}}{h_{ie}}\right)R_e$$

$$\frac{A_{V_1}}{A_{V_2}} > 1 \Rightarrow A_{V_2} < A_{V_1}$$

3.3 (d)

AC model,

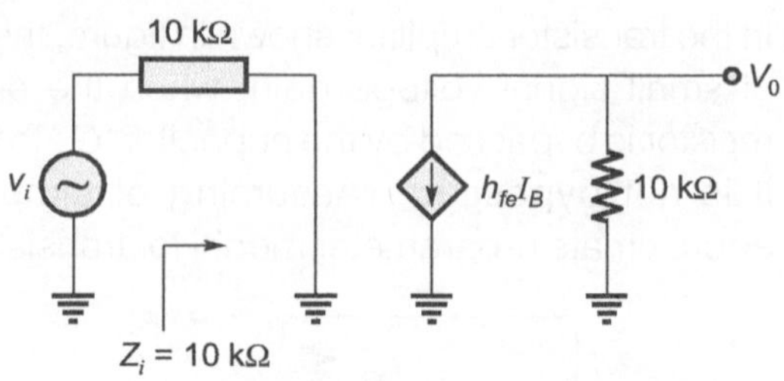

$$Z_i = 10\text{ k}\Omega$$

Mid band voltage gain

$$A_v = \frac{A_I R_L}{Z_i} = \frac{-h_{fe} R_L}{Z_i}$$

$$= \frac{-100 \times 10\text{ k}\Omega}{10\text{ k}\Omega} = -100$$

$$|A_v| = 100$$

3.4 (d)

Voltage gain $A_V = 1$

Current gain, $A_I = \frac{i_s}{i_G} = \frac{i_s}{0} = \infty$

■■■■

4 Operational Amplifiers

4.1 The common mode voltage of a unity gain (voltage follower) op-amp buffer in terms of its output voltage V is ______. [1995 : 1 M]

4.2 Let the magnitude of the gain in the inverting op-amp circuit shown be x with switch S_1 open. When the switch is closed, the magnitude of gain becomes.

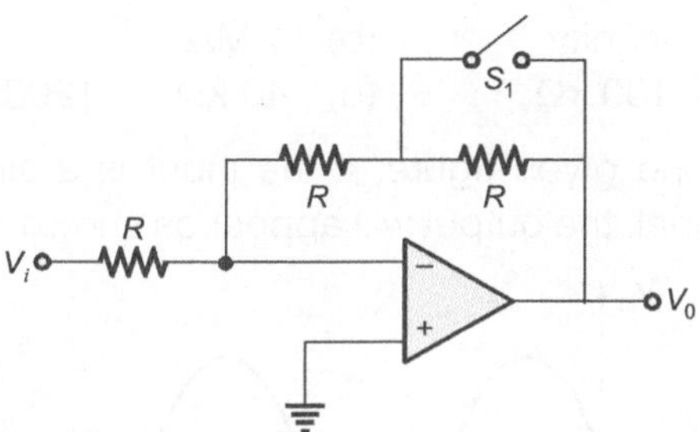

(a) $x/2$ (b) $-x$
(c) $2x$ (d) $-2x$ [1996 : 1 M]

4.3 A non-inverting op-amp amplifier is shown in the figure. The output voltage is

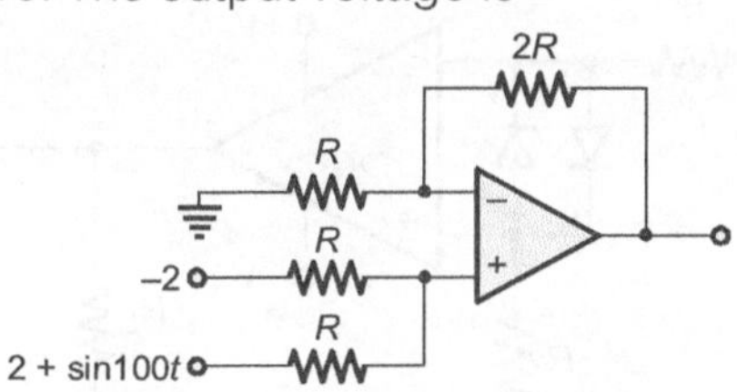

(a) $\frac{3}{2}$ sin100t (b) 3 sin100t
(c) 2 sin100t (d) None of these
[1996 : 1 M]

4.4 A major advantage of active filters is that they can be realized without using
(a) op-amps (b) inductors
(c) resistors (d) capacitors
[1997 : 1 M]

4.5 Match the following:
Circuits:

A.
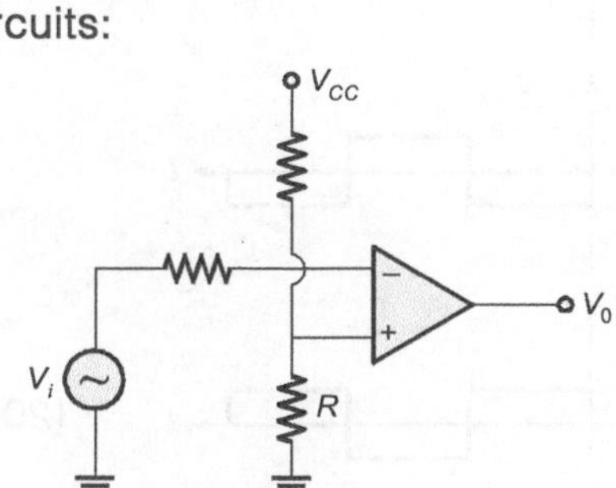

B.
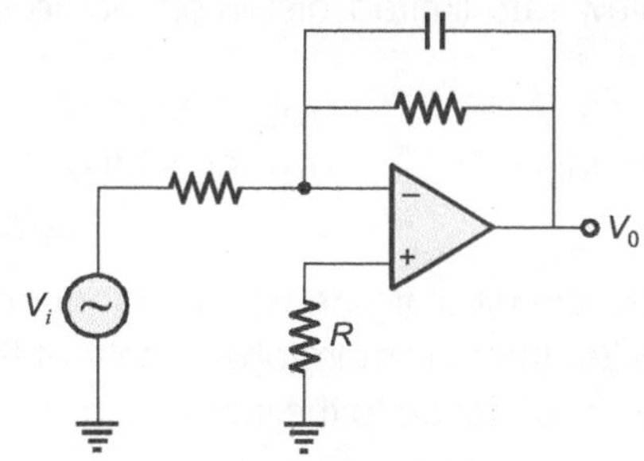

C.
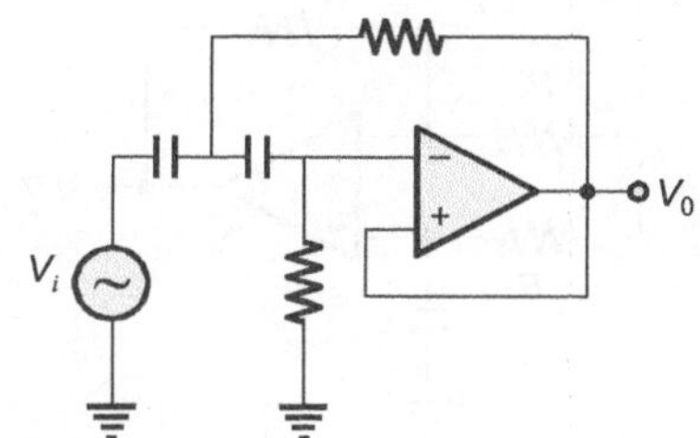

Functions:
P. High-pass filter Q. Amplifier
R. Comparator S. Low-pass filter
[1998 : 2 M]

4.6 The feedback factor for the circuit shown in the figure is

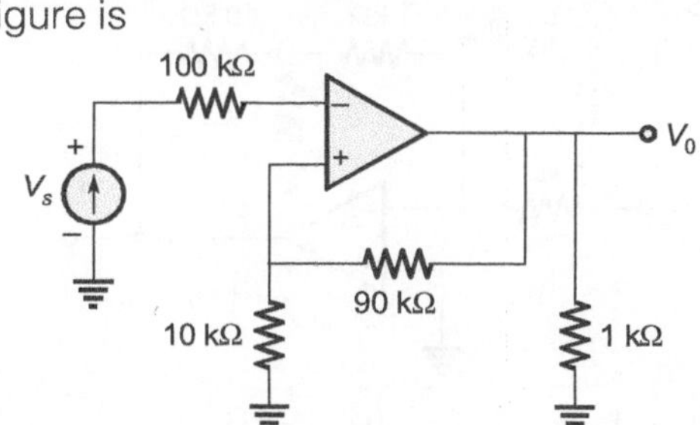

(a) $\frac{9}{100}$ (b) $\frac{9}{10}$
(c) $\frac{1}{9}$ (d) $\frac{1}{10}$ [2000 : 1 M]

4.7 The circuit shown in the figure uses an ideal op-amp working with +5 V and –5 V power supplies. The output voltage, V_0 is equal to

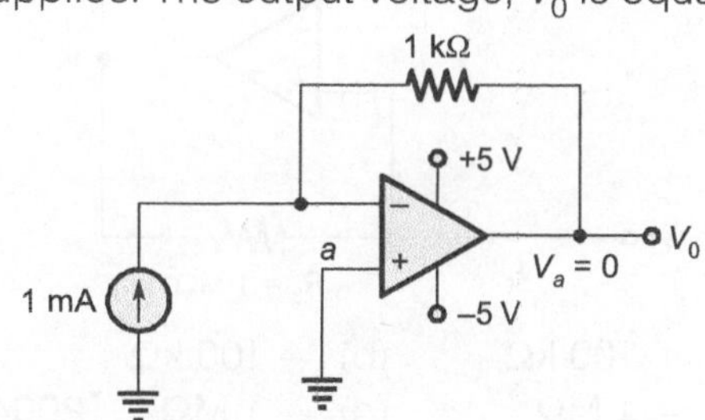

(a) +5 V (b) –5 V
(c) +1 V (d) –1 V [2000 : 1 M]

4.8 An op-amp having a slew rate of 62.8 V/μsec is connected in a voltage follower configuration. If the maximum amplitude of the input sinusoidal is 10 Volts, then the minimum frequency at which the slew rate limited distortion would set in the output is

(a) 1 MHz (b) 6.28 MHz
(c) 10 MHz (d) 62.8 MHz [2001 : 2 M]

4.9 For the circuit of figure with an ideal operational amplifier, the maximum phase shift of the output V_{out} with reference to the input V_{in} is

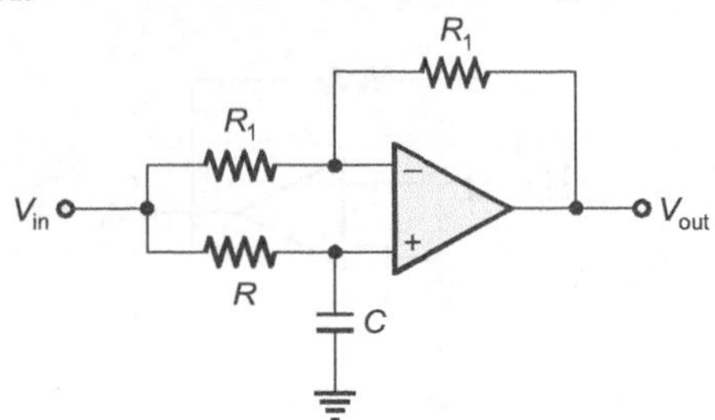

(a) 0° (b) –90°
(c) +90° (d) ±180° [2003 : 1 M]

4.10 Assuming the operational amplifier to be ideal, the gain V_{out}/V_{in} for the circuit shown in the figure is

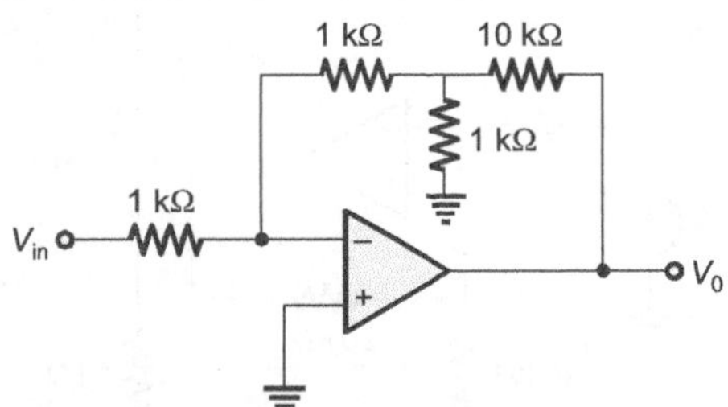

(a) – 1 (b) – 20
(c) – 100 (d) – 120 [2003 : 2 M]

4.11 The input resistance R_{IN} (= V_x/I_x) of the circuit in the figure is

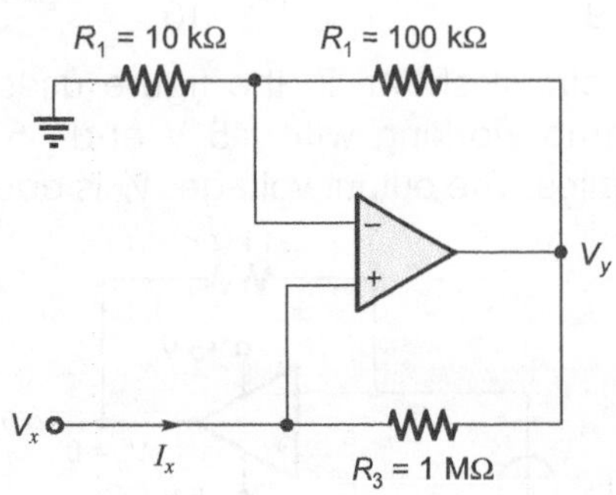

(a) + 100 kΩ (b) – 100 kΩ
(c) + 1 MΩ (d) – 1 MΩ [2004 : 2 M]

4.12 Consider the inverting amplifier, using an ideal operational amplifier shown in the figure. The designer wishes to realize the input resistance seen by the small-signal source to be as large as possible, while keeping the voltage gain between –10 and –25. The upper limit on R_F is 1 MΩ. The value of R_1 should be

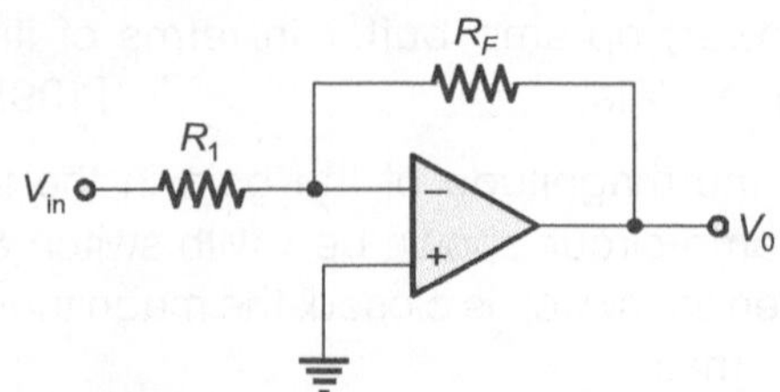

(a) Infinity (b) 1 MΩ
(c) 100 kΩ (d) 40 kΩ [2005 : 2 M]

4.13 In the given figure, if the input is a sinusoidal signal, the output will appear as shown in

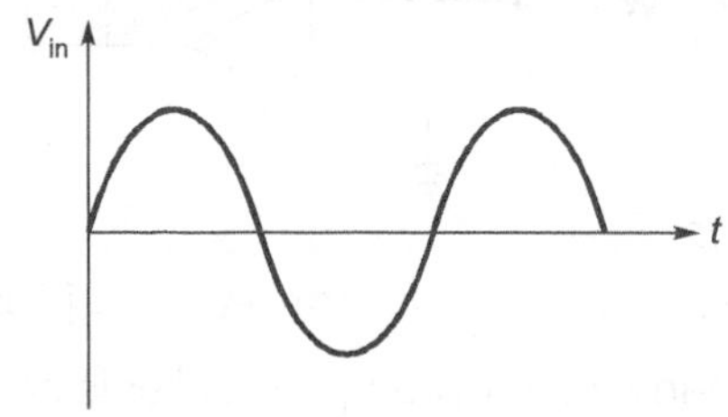

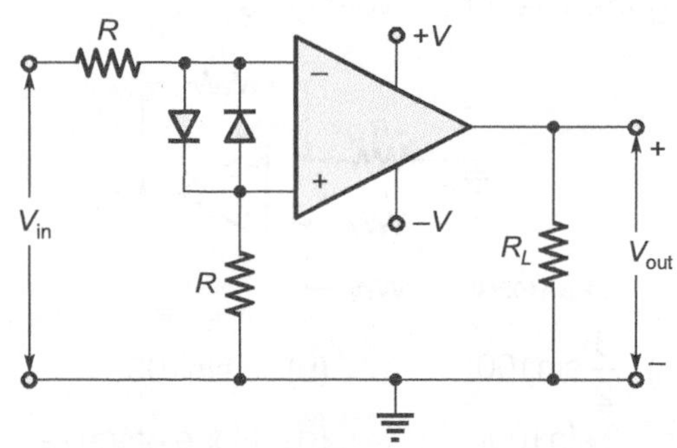

(a)
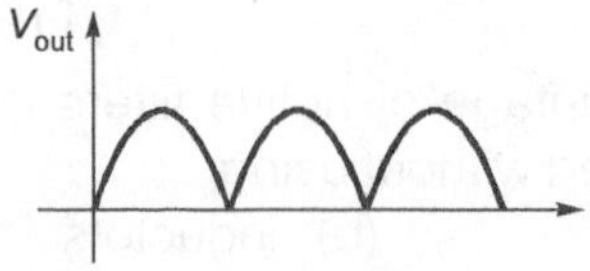

(b)
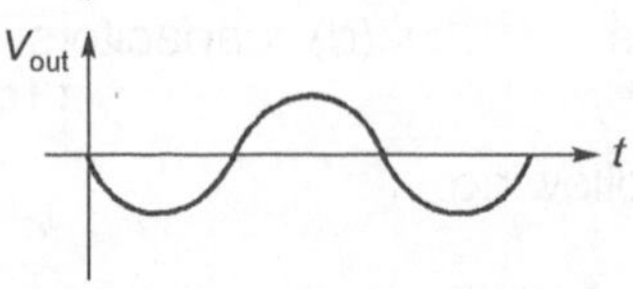

(c)
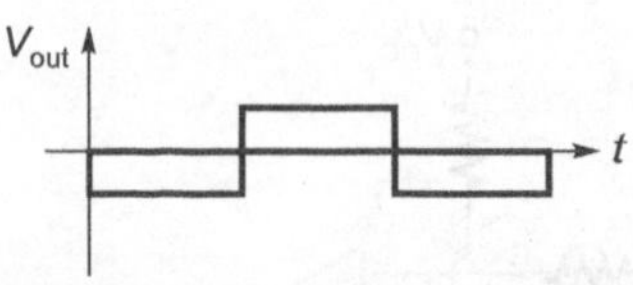

(d)
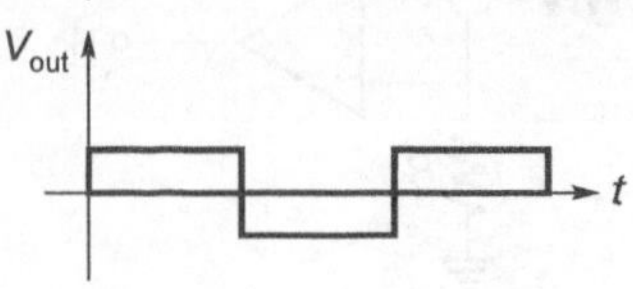

[2005 : 2 M

4.14 For a given sinusoidal input voltage, the voltage waveform at point P of the clamper circuit shown in the figure will be

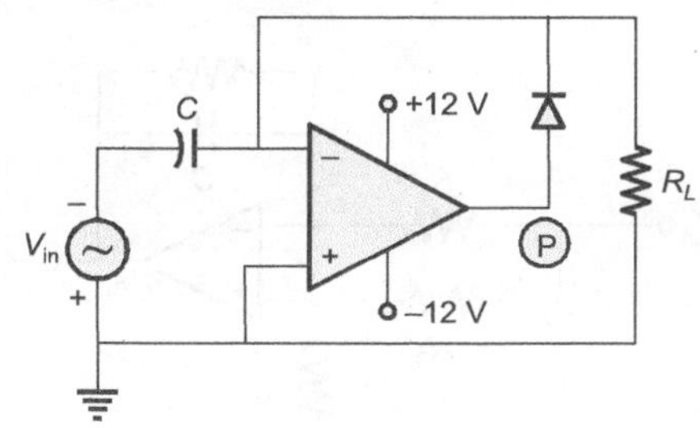

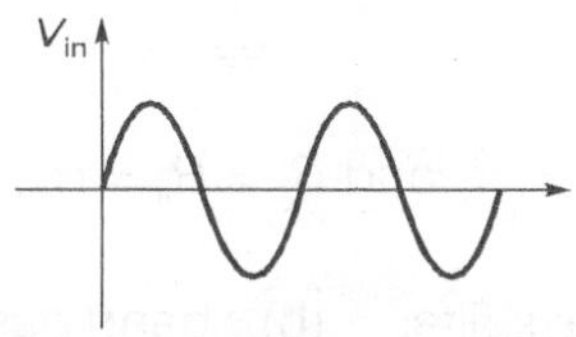

(a)

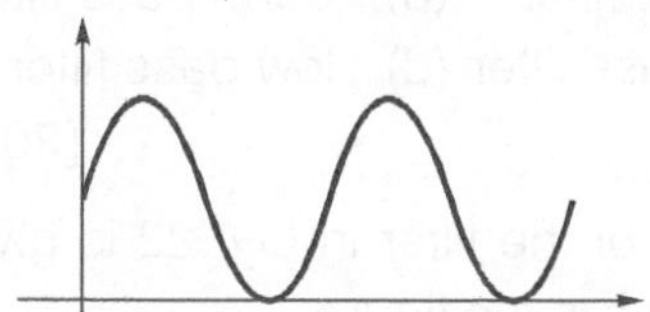

(b)

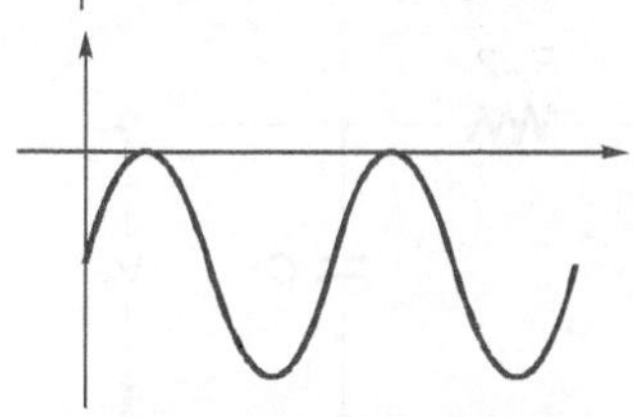

(c)

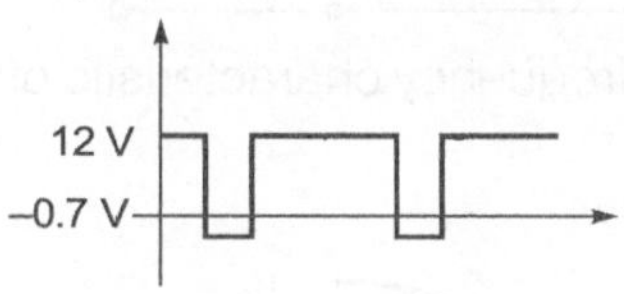

(d)

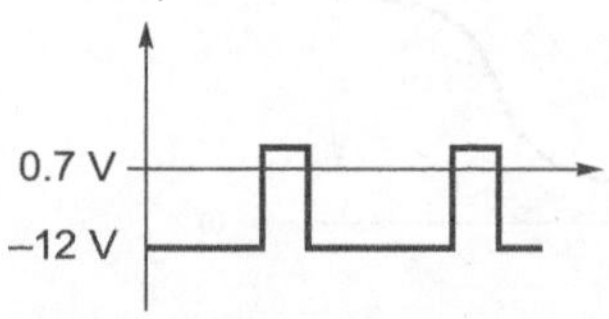

[2006 : 1 M]

4.15 A relaxation oscillator is made using OPAMP as shown in the figure. The supply voltages of the OPAMP are ± 12 V. The voltage waveform at point P will be

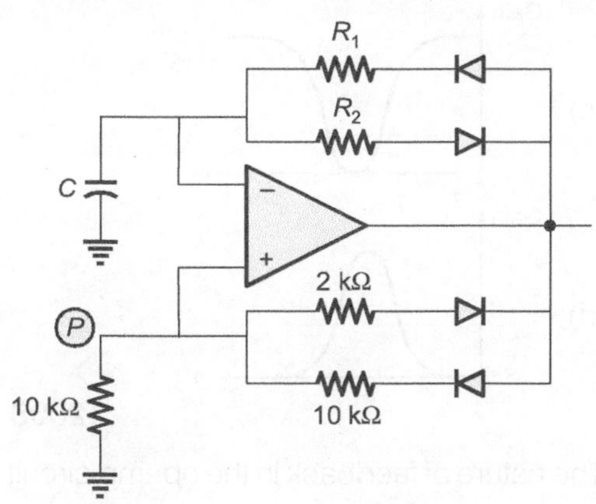

(a)

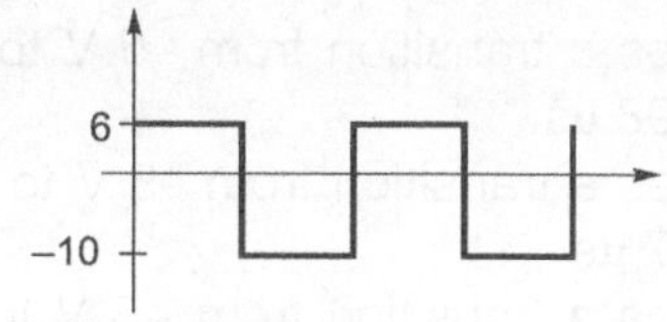

(b)

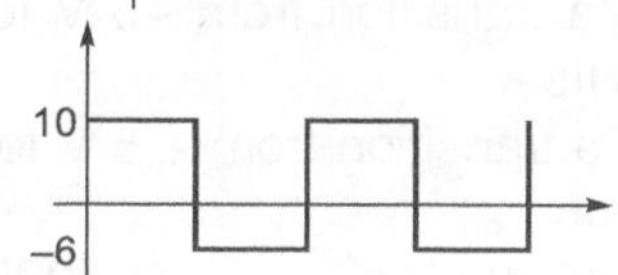

(c)

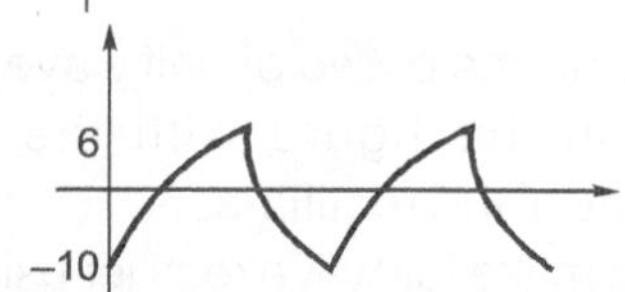

(d)

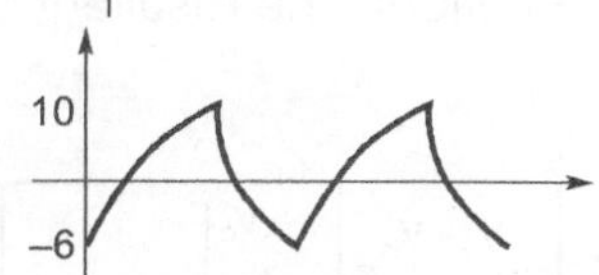

[2006 : 2 M]

4.16 The circuit shown in the figure is

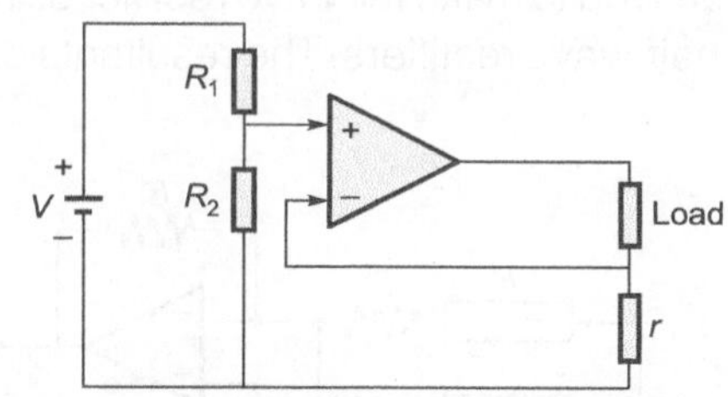

(a) a voltage source with voltage $\dfrac{rV}{R_1 \parallel R_2}$

(b) a voltage source with voltage $\dfrac{r \parallel R_2}{R_1} V$

(c) a current source with current $\dfrac{r \parallel R_2}{R_1 + R_2} \cdot \dfrac{V}{r}$

(d) a current source with current $\dfrac{R_2}{R_1 + R_2} \cdot \dfrac{V}{r}$

[2007 : 1 M]

4.17 The switch S in the circuit of the figure is initially closed, it is opened at time $t = 0$. You may neglect the Zener diode forward voltage drops. What is the behaviour of V_{OUT} for $t > 0$?

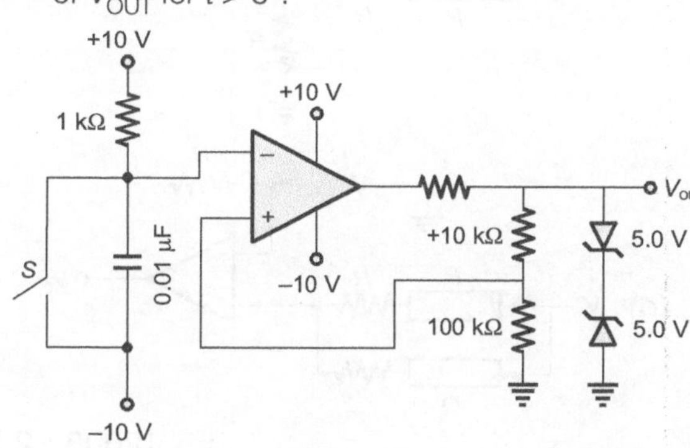

(a) It makes a transition from –5 V to +5 V at $t = 12.98\ \mu s$
(b) It makes a transition from –5 V to + 5 V at $t = 2.57\ \mu s$
(c) It makes a transition from +5 V to –5 V at $t = 12.98\ \mu s$
(d) It makes a transition from + 5 V to – 5 V at $t = 2.57\ \mu s$

[2007 : 2 M]

4.18 The block diagrams of two of half wave rectifiers are shown in the figure with the transfer characteristics of the rectifiers.
It is desired to make full wave rectifier using above two half-wave rectifiers. The resultant circuit will be

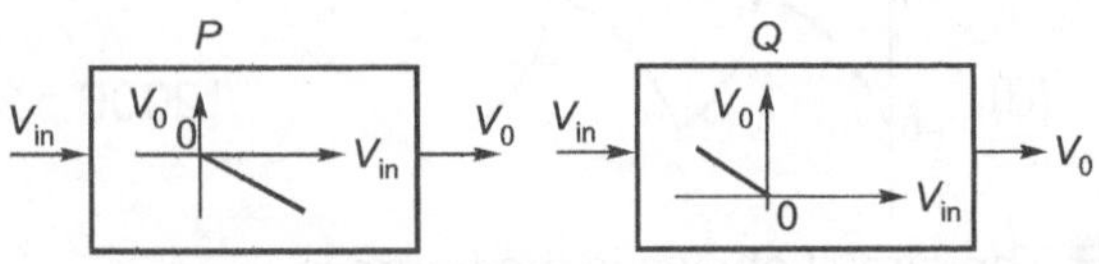

It is desired to make full wave rectifier using above two half-wave rectifiers. The resultants circuit will be

(a) 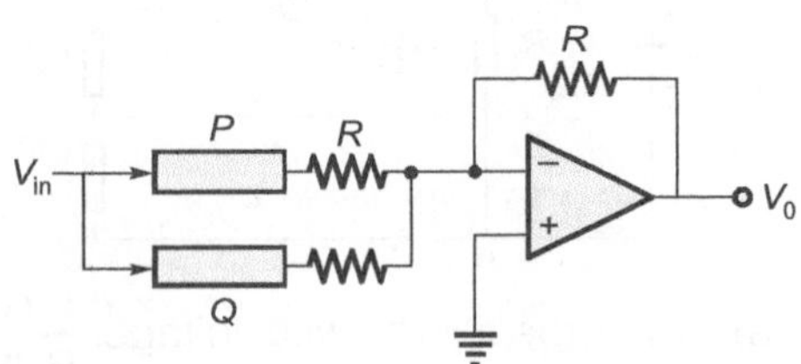

(b) 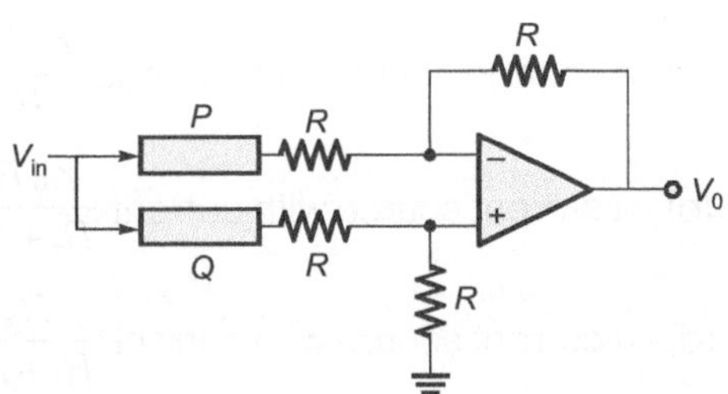

(c) 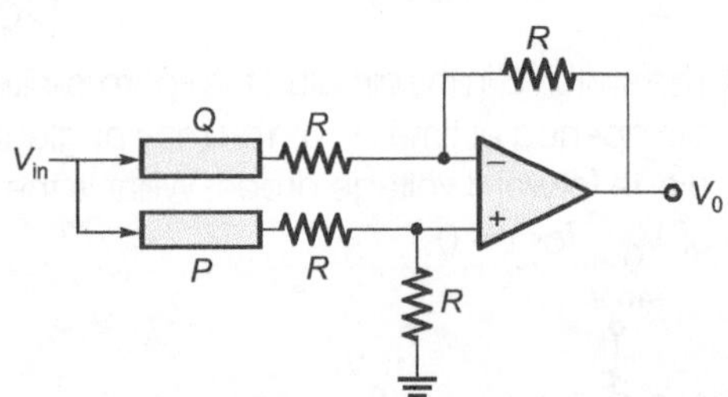

(d) 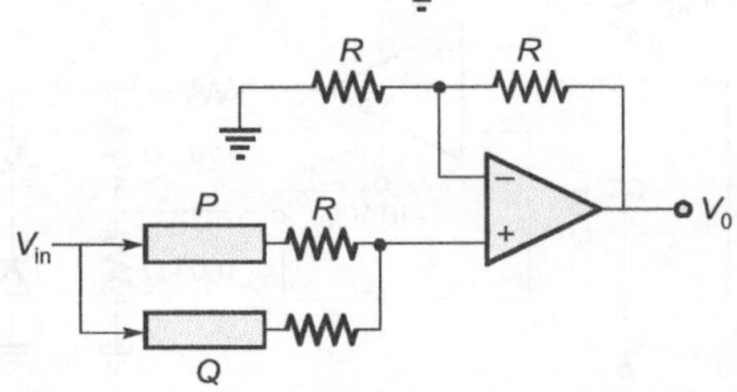

[2008 : 2 M]

Statement for Linked Answer Questions (4.19 and 4.20):

A general filter circuit is shown in the figure:

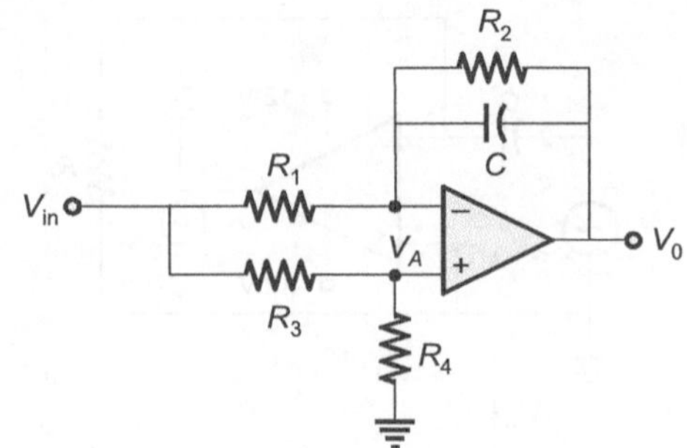

4.19 If $R_1 = R_2 = R_A$ and $R_3 = R_4 = R_B$, the circuit acts as a

(a) all pass filter (b) band pass filter
(c) high pass filter (d) low pass filter

[2008 : 2 M]

4.20 The output of the filter in Q.4.22 is given to the circuit shown in the figure.

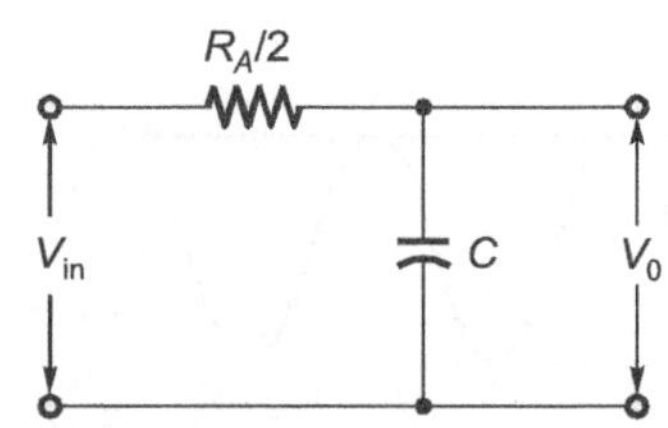

The gain vs frequency characteristic of the output (V_0) will be

(a) 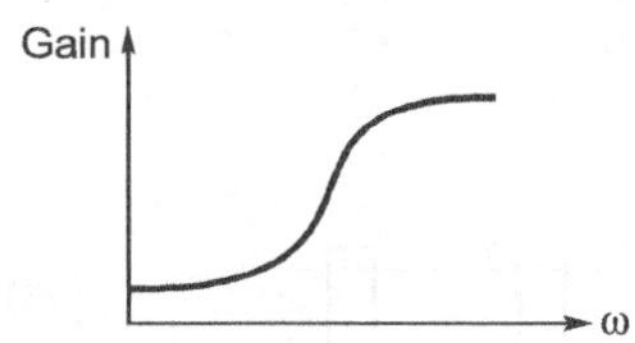

(b) 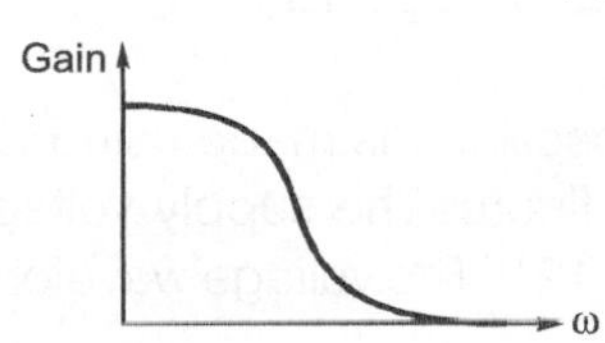

(c) 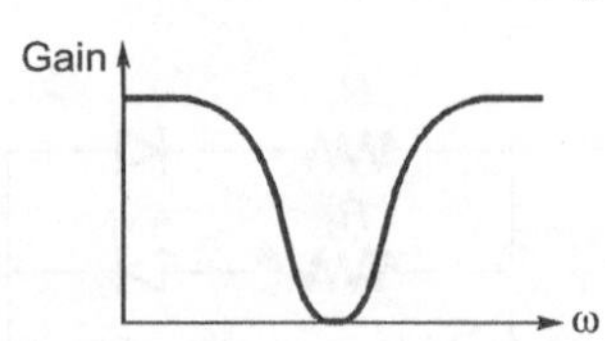

(d) 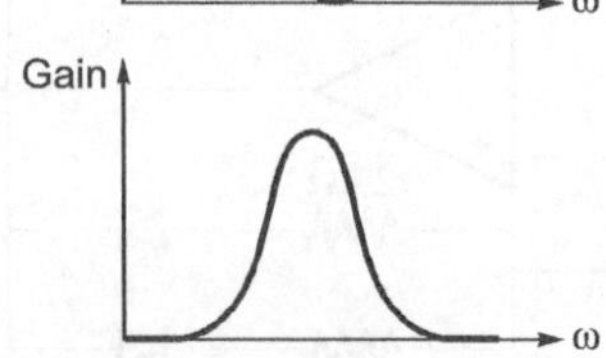

[2008 : 2 M]

4.21 The nature of feedback in the opamp circuit shown is

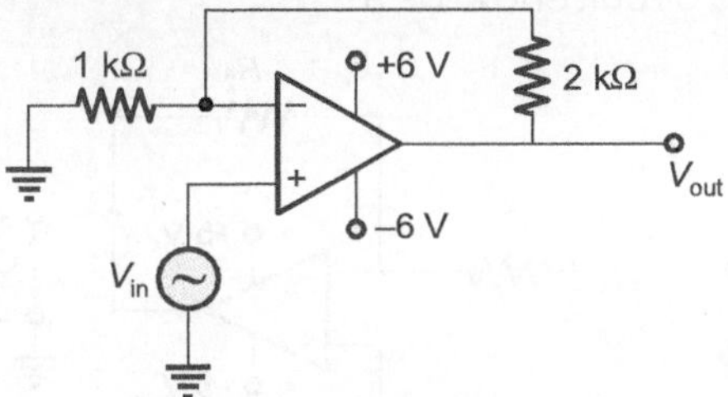

(a) Current – Current feedback
(b) Voltage – Voltage feedback
(c) Current – Voltage feedback
(d) Voltage – Current feedback [2009 : 1 M]

4.22 The following circuit has $R = 10\ k\Omega$, $C = 10\ \mu F$. The input voltage is a sinusoid at 50 Hz with an rms value of 10 V. Under ideal conditions, the current i_s from the source is

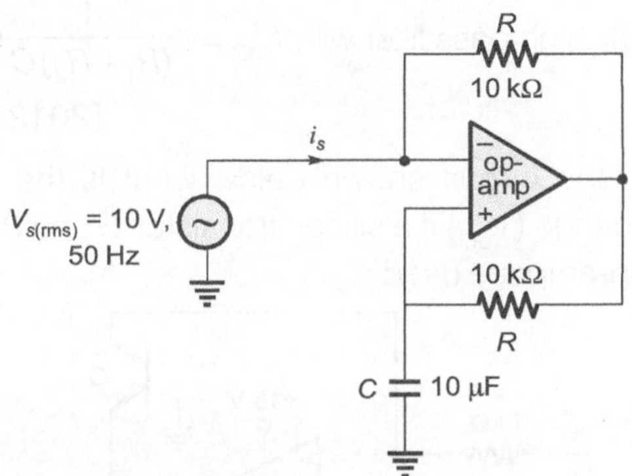

(a) 10π mA leading by 90°
(b) 20π mA leading by 90°
(c) 10 mA leading by 90°
(d) 10π mA lagging by 90° [2009 : 2 M]

4.23 Transformer and emitter follower can both be used for impedance matching at the output of an audio amplifier. The basic relationship between the input power P_{in} and output power P_{out} in both the cases is

(a) $P_{in} = P_{out}$ for both transformer and emitter follower
(b) $P_{in} > P_{out}$ for both transformer and emitter follower
(c) $P_{in} < P_{out}$ for transformer and $P_{in} = P_{out}$ for emitter follower
(d) $P_{in} = P_{out}$ for transformer and $P_{in} < P_{out}$ for emitter follower

[2009 : 2 M]

4.24 An ideal opamp circuit and its input waveform are shown in the figures. The output waveform of this circuit will be

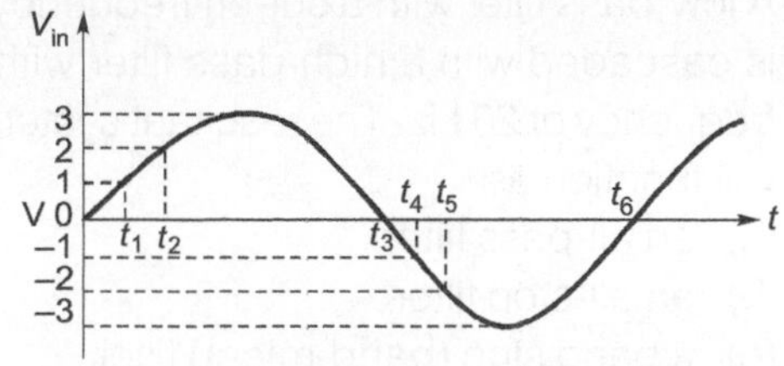

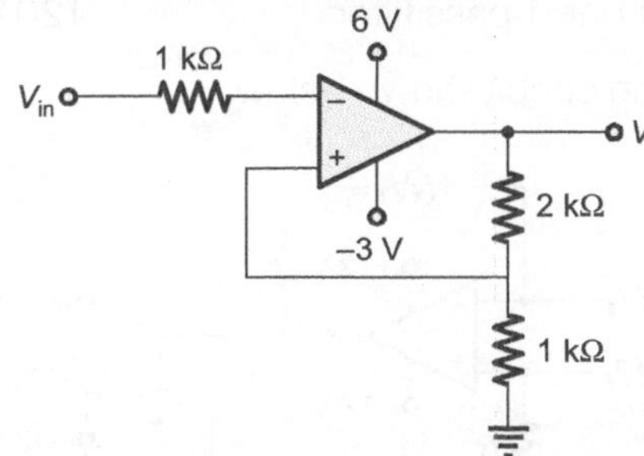

(a)
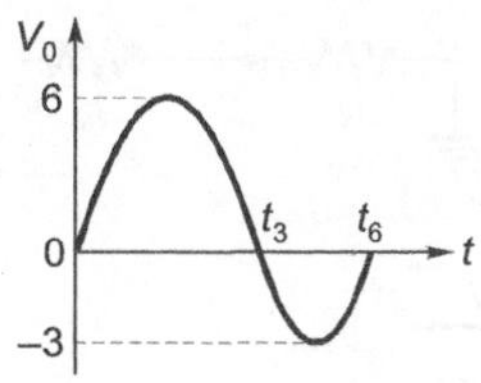

(b)
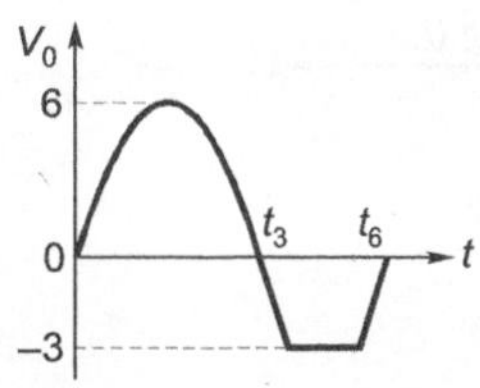

(c)
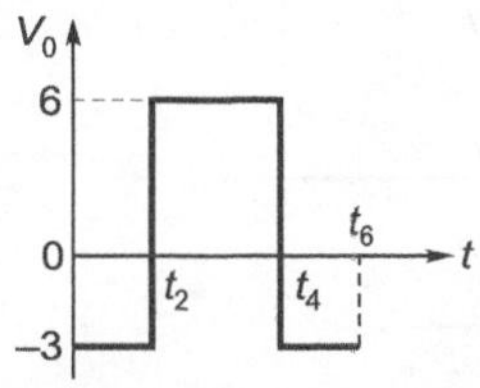

(d)
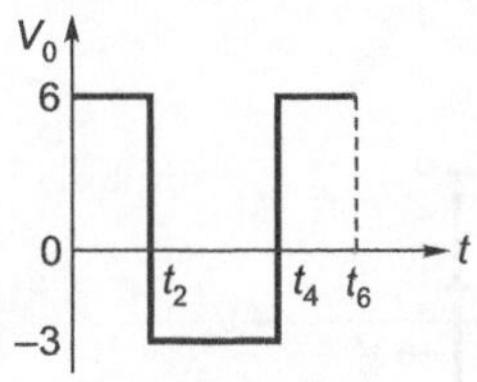

[2009 : 2 M]

4.25 Given that the op-amp is ideal, the output voltage V_0 is

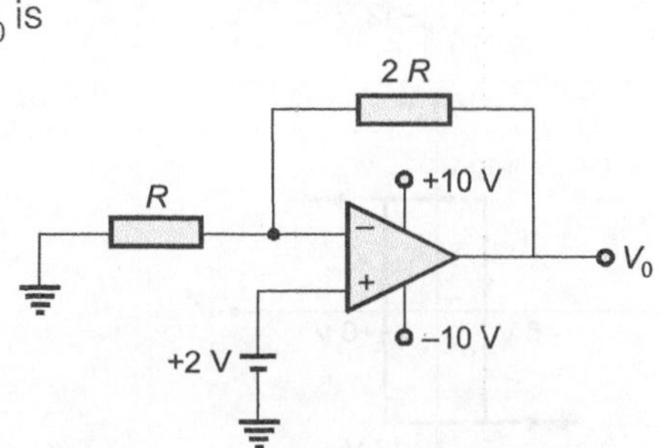

(a) 4 V (b) 6 V
(c) 7.5 V (d) 12.12 V [2010 : 1 M]

4.26 A low-pass filter with a cut-off frequency of 30 Hz is cascaded with a high-pass filter with a cut-off frequency of 20 Hz. The resultant system of filters will function as

(a) an all-pass filter
(b) an all-stop filter
(c) a band stop (band-reject) filter
(d) a band-pass filter [2011 : 1 M]

4.27 For the circuit shown below,

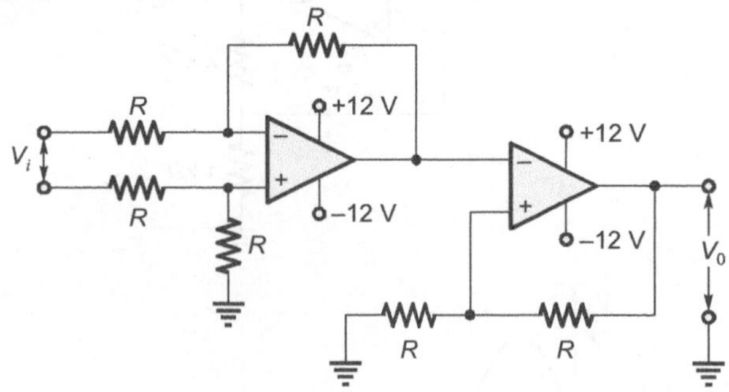

(a)

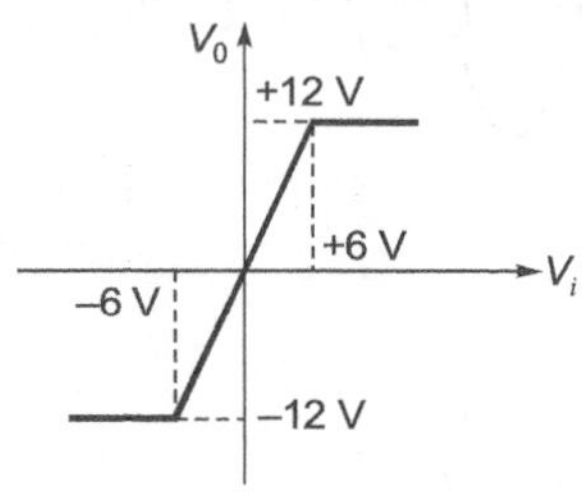

(b)

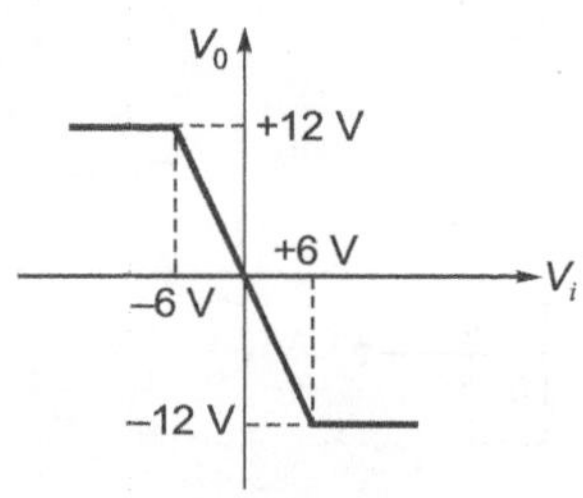

(c)

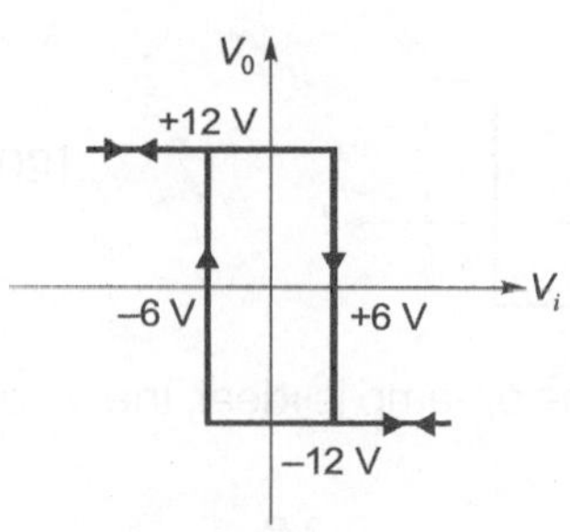

(d)

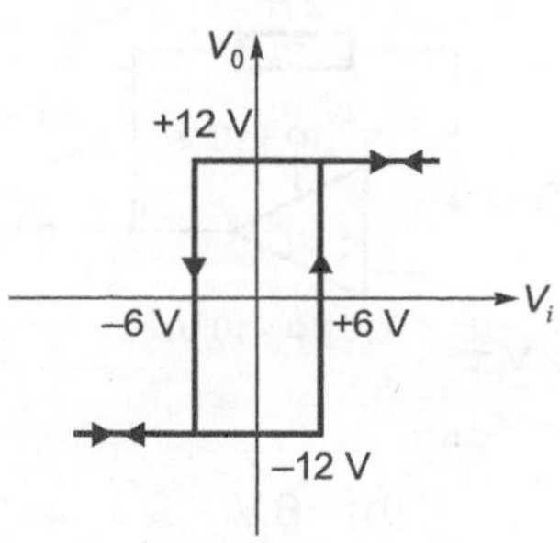

[2011 : 1 M]

4.28 The circuit shown is a

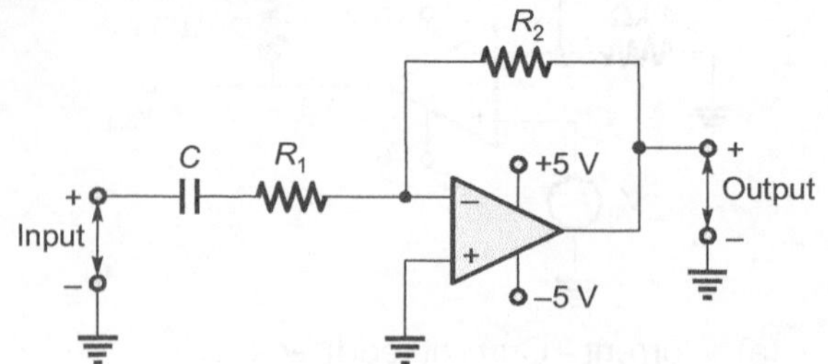

(a) low pass filter with $f_{3dB} = \dfrac{1}{(R_1+R_2)C}$ rad/s

(b) high pass filter with $f_{3dB} = \dfrac{1}{R_1 C}$ rad/s

(c) low pass filter with $f_{3dB} = \dfrac{1}{R_1 C}$ rad/s

(d) high pass filter with $f_{3dB} = \dfrac{1}{(R_1+R_2)C}$ rad/s

[2012 : 2 M]

4.29 In the circuit shown below what is the output voltage (V_{out}) if a silicon transistor Q and an ideal op-amp are used?

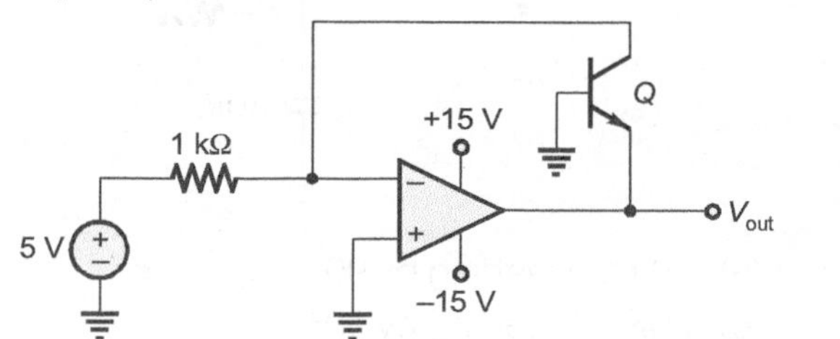

(a) –15 V (b) –0.7 V
(c) +0.7 V (d) +15 V [2013 : 1 M]

4.30 In the circuit shown below the op-amps are ideal. Then V_{out} in volts is

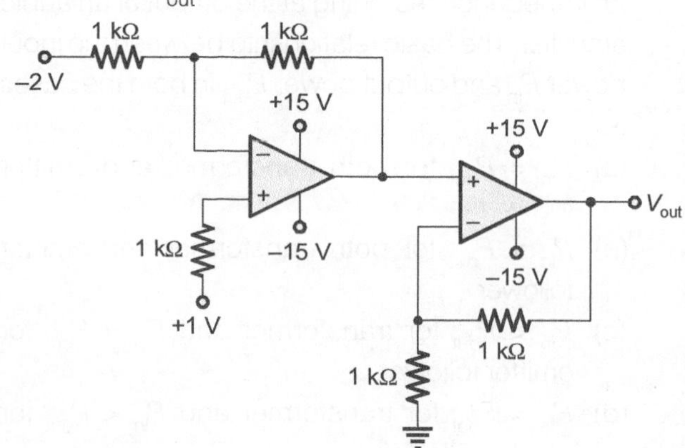

(a) 4 (b) 6
(c) 8 (d) 10 [2013 : 2 M]

4.31 Given that the op-amps in the figure are ideal, the output voltage V_0 is

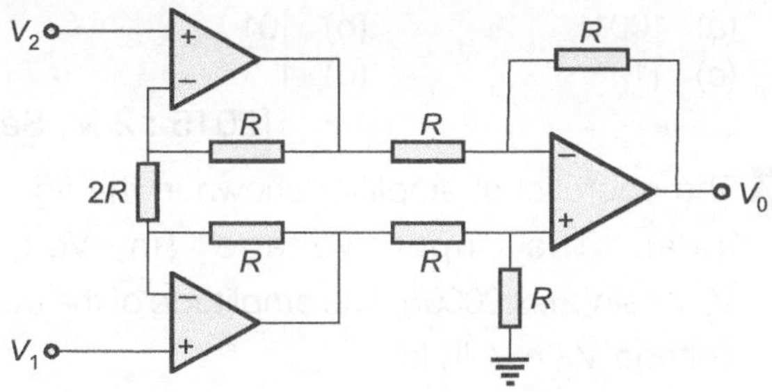

(a) $(V_1 - V_2)$ (b) $2(V_1 - V_2)$

(c) $\frac{(V_1 - V_2)}{2}$ (d) $(V_1 + V_2)$

[2014 : 2 M, Set-1]

4.32 In the figure shown, assume the op-amp to be ideal. Which of the alternatives gives the correct Bode plots for the transfer function $\frac{V_0(\omega)}{V_i(\omega)}$?

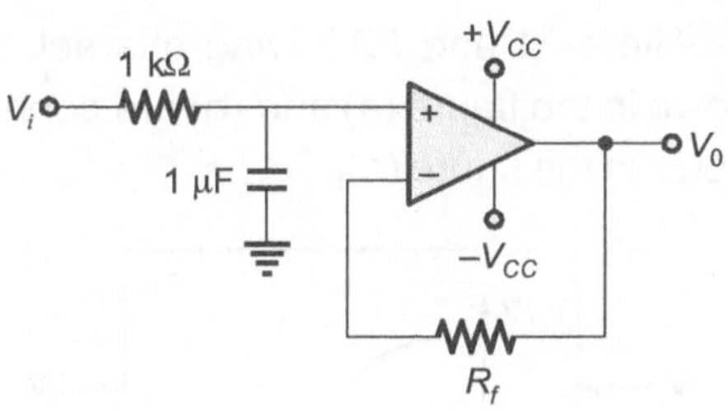

(a)

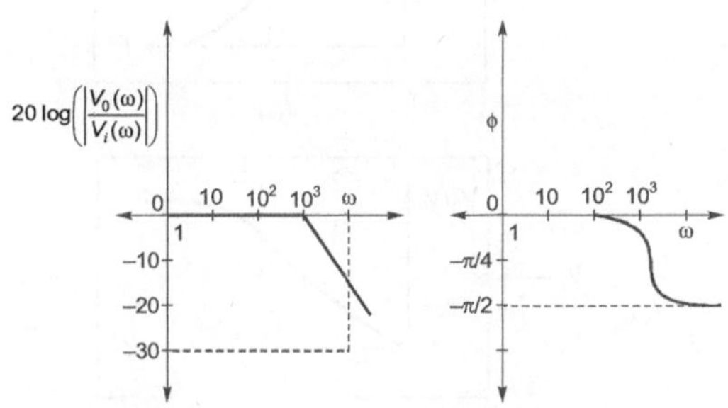

(b)

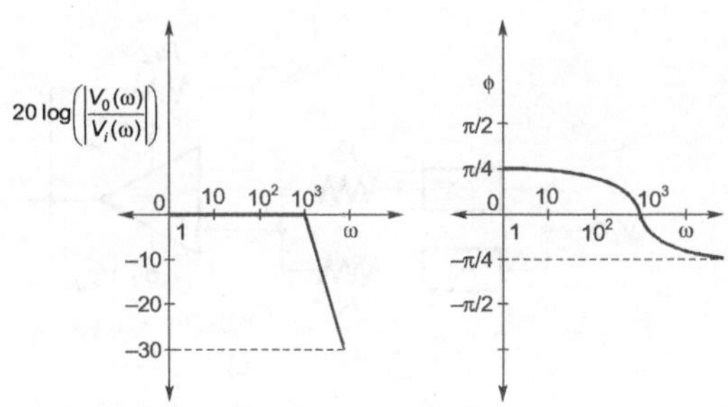

(c)

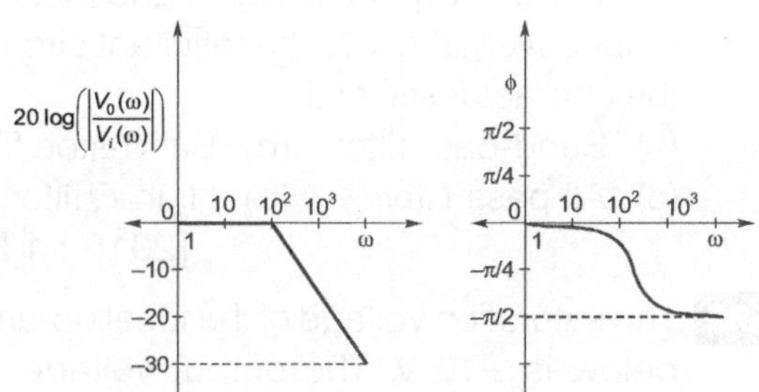

(d)

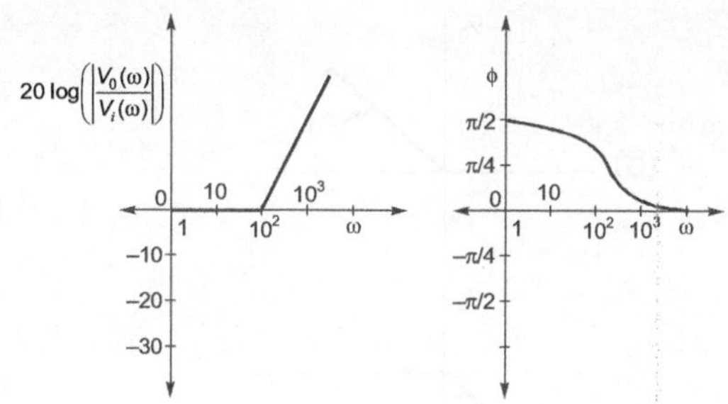

[2014 : 2 M, Set-1]

4.33 An operational amplifier circuits is shown in the figure.

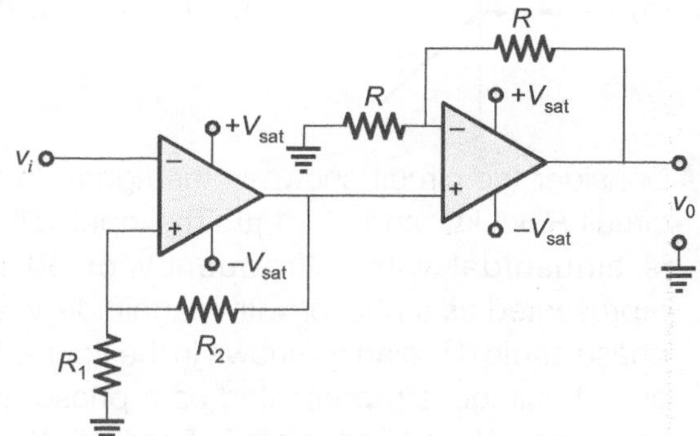

The output of the circuit for a given input v_i is

(a) $-\left(\frac{R_2}{R_1}\right)V_i$ (b) $-\left(1+\frac{R_2}{R_1}\right)V_i$

(c) $\left(1+\frac{R_2}{R_1}\right)V_i$ (d) $+V_{sat}$ or $-V_{sat}$

[2014 : 1 M, Set-3]

4.34 The transfer characteristic of the op-amp circuit shown in the figure is

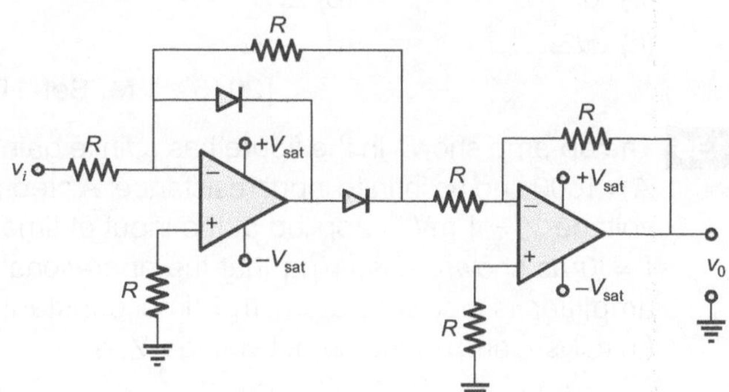

(a)

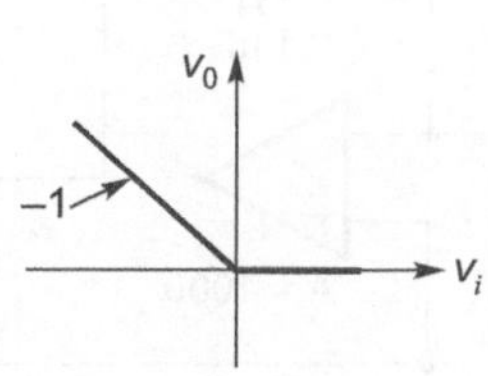

(b)
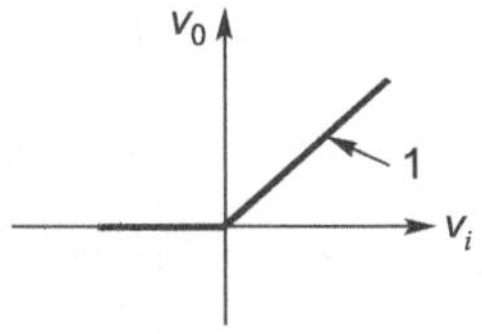

(c)
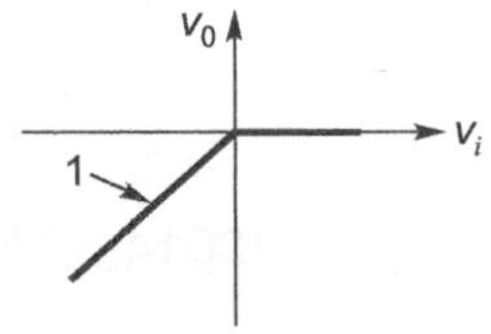

(d)
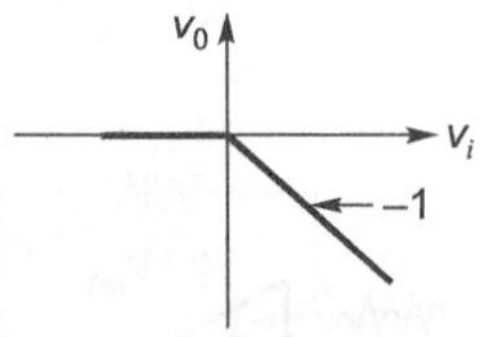

[2014 : 2 M, Set-3]

4.35 Consider the circuit shown in the figure. In this circuit $R = 1\ \text{k}\Omega$, and $C = 1\ \mu\text{F}$. The input voltage is sinusoidal with a frequency of 50 Hz, represented as a phasor with magnitude V_i and phase angle 0 radian as shown in the figure. The output voltage is represented as a phasor with magnitude V_0 and phase angle δ radian. What is the value of the output phase angle δ (in radian) relative to the phase angle of the input voltage?

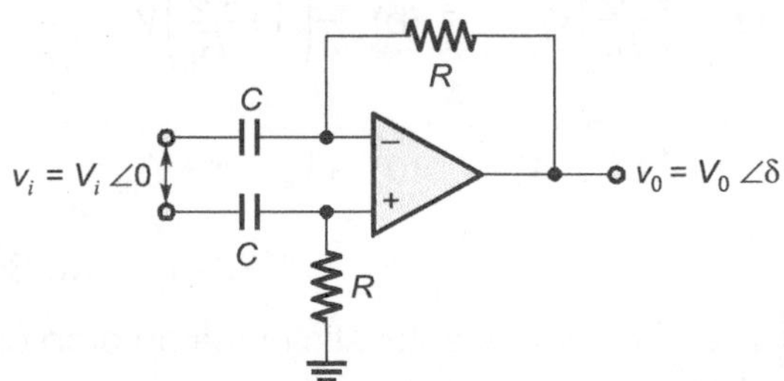

(a) 0 (b) π
(c) $\pi/2$ (d) $-\pi/2$

[2015 : 1 M, Set-1]

4.36 The op-amp shown in the figure has a finite gain $A = 1000$ and an infinite input resistance. A step-voltage $V_i = 1$ mV is applied at the input at time $t = 0$ as shown. Assuming that the operational amplifier is not saturated, the time constant (in millisecond) of the output voltage V_0 is

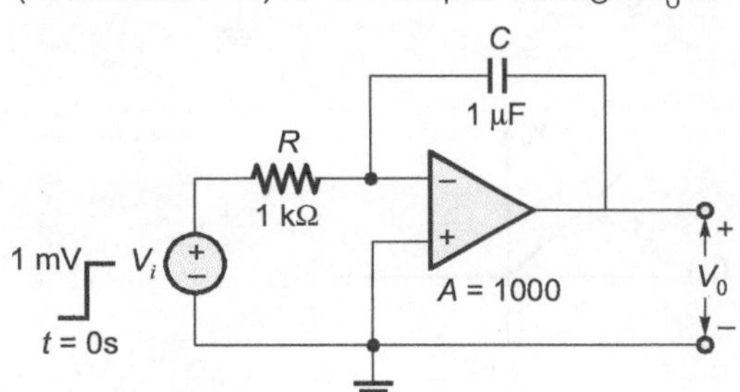

(a) 1001 (b) 101
(c) 11 (d) 1

[2015 : 2 M, Set-1]

4.37 The operational amplifier shown in the figure is ideal. The input voltage (in Volt) is $V_i = 2\sin(2\pi \times 2000t)$. The amplitude of the output voltage V_0(in Volt) is _____.

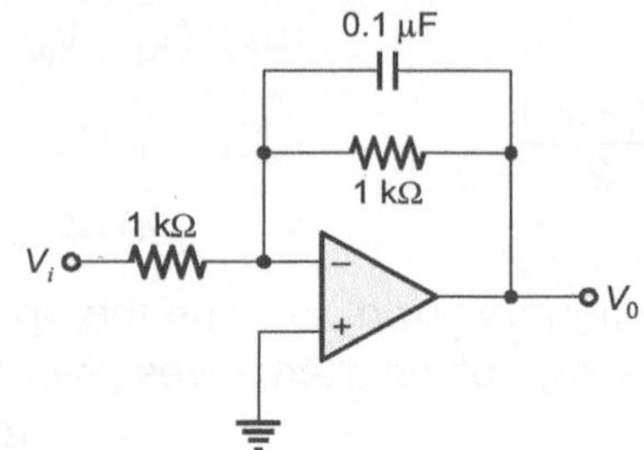

[2015 : 1 M, Set-2]

4.38 The filters $F1$ and $F2$ having characteristics as shown in the figure (a) and (b) are connected as shown in the figure (c).

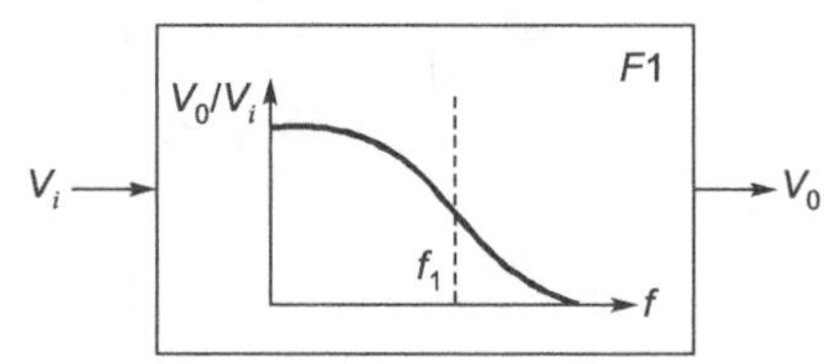

(a)

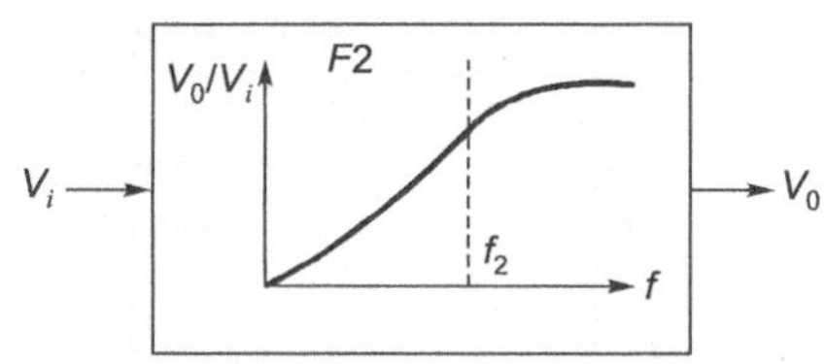

(b)

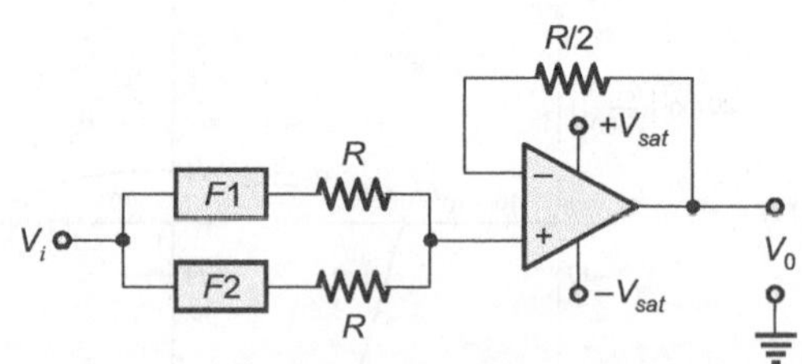

(c)

The cut-off frequencies of $F1$ and $F2$ are f_1 and f_2 respectively. If $f_1 < f_2$, the resultant circuit exhibits the characteristic of a

(a) Band-pass filter (b) Band-stop filter
(c) All pass filter (d) High-Q filter

[2015 : 1 M, Set-2]

4.39 The saturation voltage of the ideal op-amp shown below is ± 10 V. The output voltage V_0 of the following circuit in the steady-state is

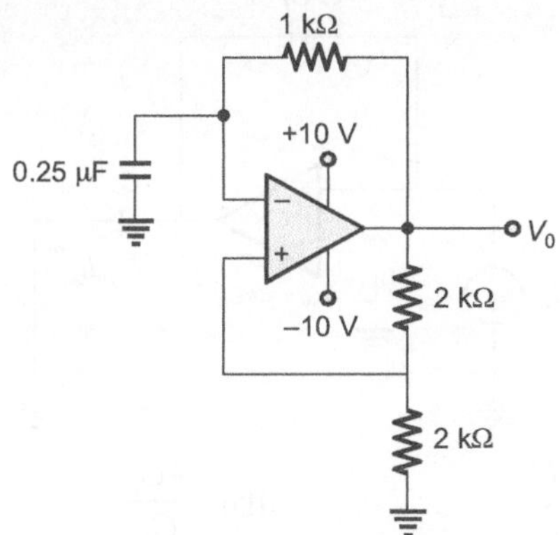

(a) square wave of period 0.55 ms
(b) triangular wave of period 0.55 ms
(c) square wave of period 0.25 ms
(d) triangular wave of period 0.25 ms

[2015 : 2 M, Set-2]

4.40 Of the four characteristics given below, which are the major requirements for an instrumentation amplifier?

P. High common mode rejection ratio
Q. High input impedance
R. High linearity
S. High output impedance

(a) P, Q and R only (b) P and R only
(c) P, Q and S only (d) Q, R and S only

[2015 : 1 M, Set-1]

4.41 The circuit shown below is an example of a

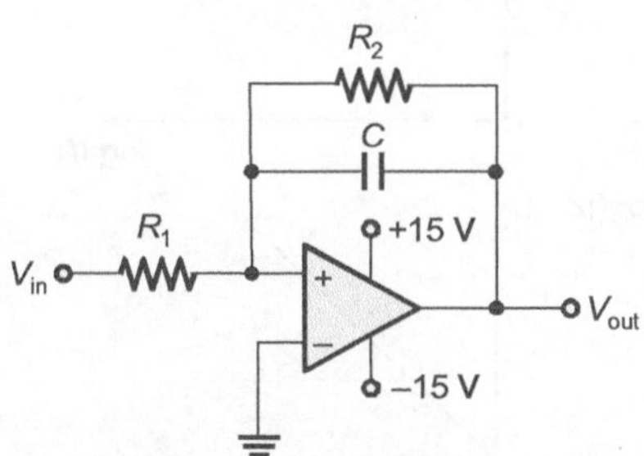

(a) low pass filter (b) band pass filter
(c) high pass filter (d) notch filter

[2016 : 1 M, Set-2]

4.42 For the circuit shown below, taking the opamp as ideal, the output voltage V_{out} in terms of the input voltages V_1, V_2 and V_3 is

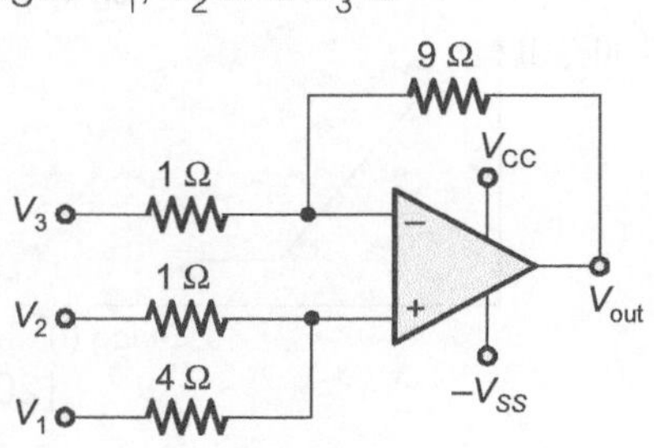

(a) $1.8\ V_1 + 7.2\ V_2 - V_3$
(b) $2\ V_1 + 8\ V_2 - 9\ V_3$
(c) $7.2\ V_1 + 1.8\ V_2 - V_3$
(d) $8\ V_1 + 2\ V_2 - 9\ V_3$

[2016 : 2 M, Set-2]

4.43 The approximate transfer characteristic for the circuit shown below with an ideal operational amplifier and diode will be

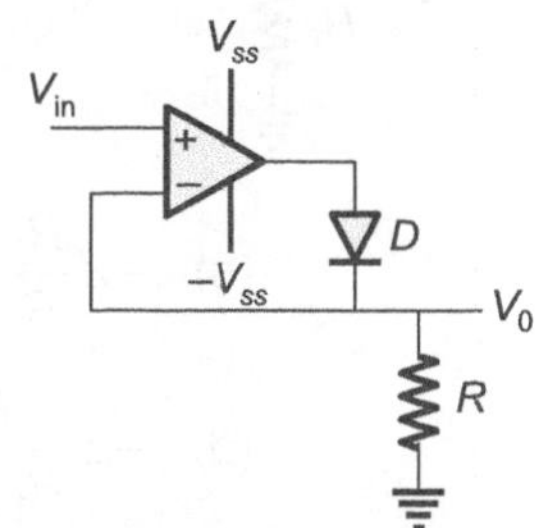

(a)

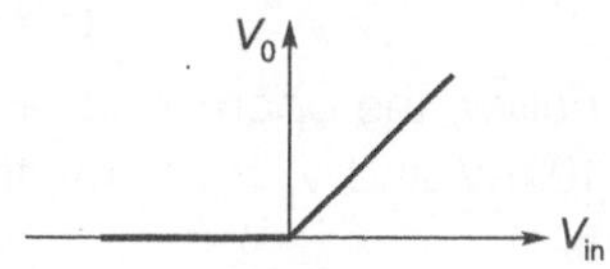

(b)

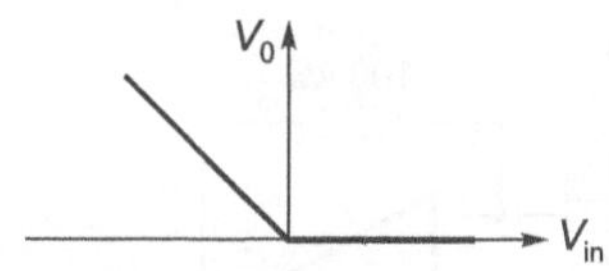

(c)

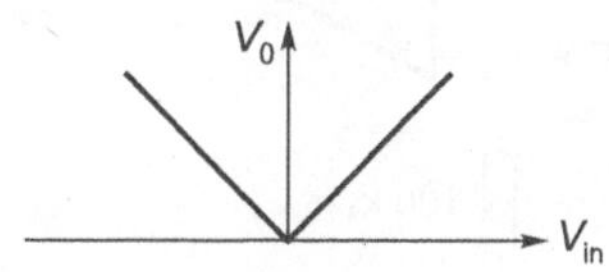

(d)

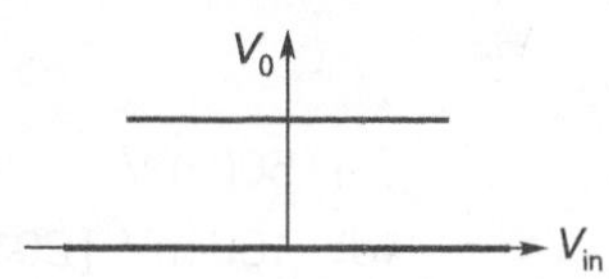

[2017 : 2 M, Set-1]

4.44 For the circuit shown below, assume that the OPAMP is ideal. Which one of the following is TRUE?

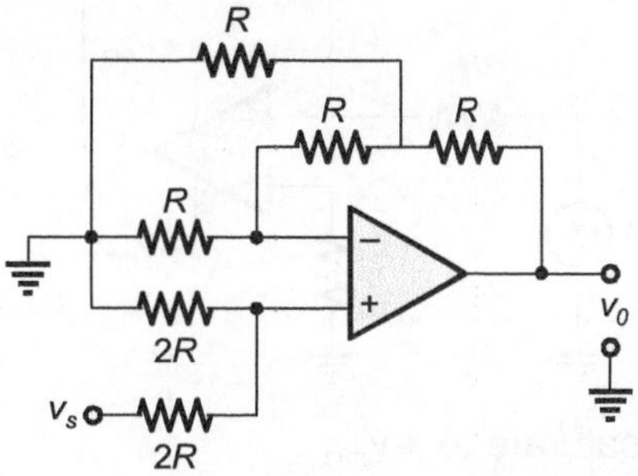

(a) $v_0 = v_s$ (b) $v_0 = 1.5\ v_s$
(c) $v_0 = 2.5\ v_s$ (d) $v_0 = 5\ v_s$

[2017 : 2 M, Set-2]

4.45 The op-amp shown in the figure is ideal. The input impedance V_{in}/i_{in} is given by

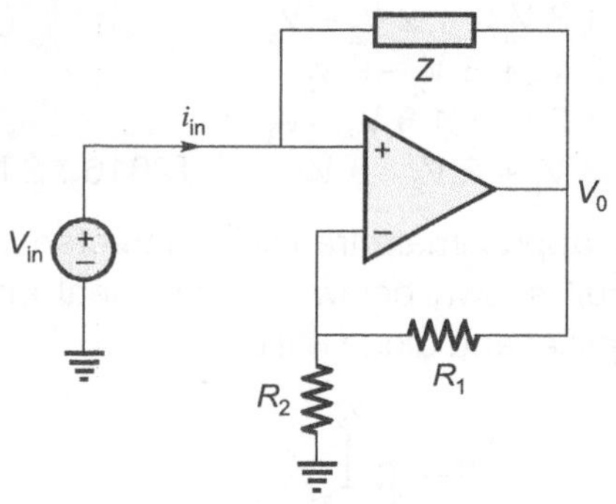

(a) $Z\dfrac{R_1}{R_2}$ (b) $-Z\dfrac{R_2}{R_1}$

(c) Z (d) $-Z\dfrac{R_2}{R_1+R_2}$

[2018 : 1 M]

4.46 In the circuit below, the operational amplifier is ideal. If V_1 = 10 mV and V_2 = 50 mV, the output voltage (V_{out}) is

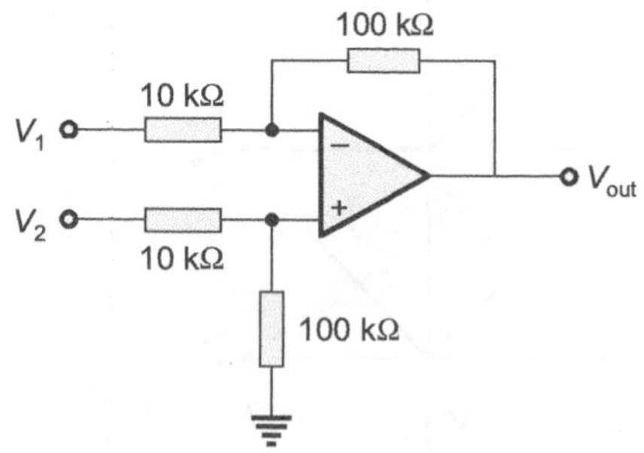

(a) 400 mV (b) 500 mV

(c) 600 mV (d) 100 mV [2019 : 2 M]

4.47 The steady state output (V_{out}), of the circuit shown below, will

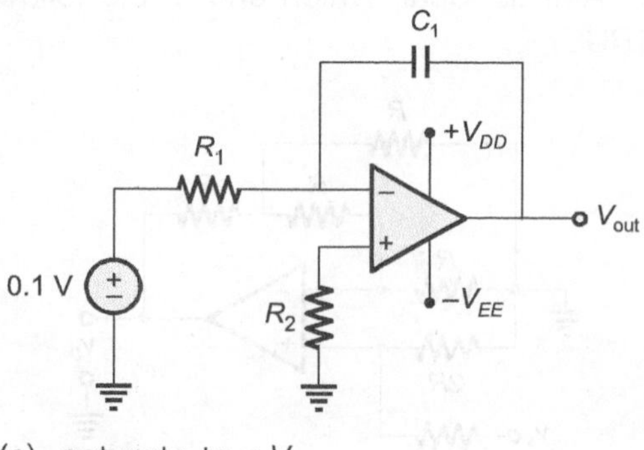

(a) saturate to $+V_{DD}$

(b) saturate to $-V_{EE}$

(c) become equal to 0.1 V

(d) become equal to –0.1 V [2022 : 1 M]

4.48 The current gain (I_{out}/I_{in}) in the circuit with an ideal current amplifier given below is

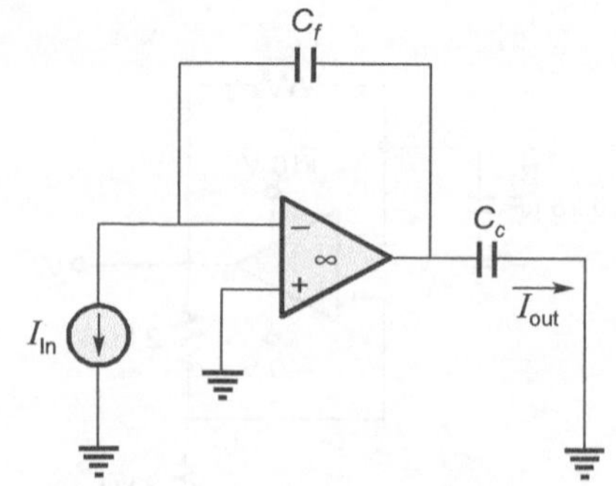

(a) $\dfrac{C_f}{C_c}$ (b) $\dfrac{-C_f}{C_c}$

(c) $\dfrac{C_c}{C_f}$ (d) $\dfrac{-C_c}{C_f}$ [2022 : 2 M]

4.49 The output impedance of a non-ideal operational amplifier is denoted by Z_{out}. The variation in the magnitude of Z_{out} with increasing frequency, f, in the circuit shown below, is best represented by

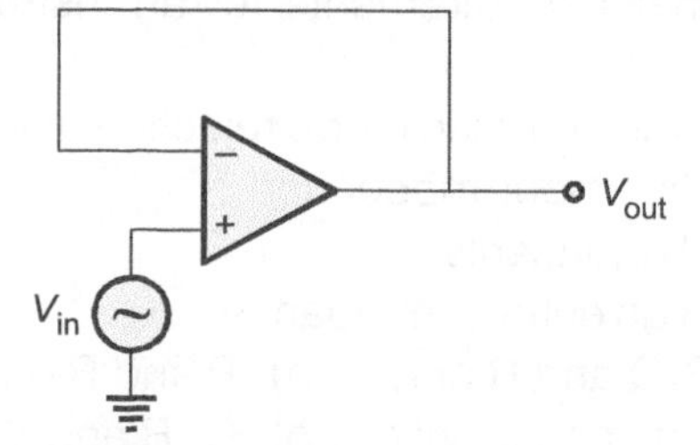

(a)

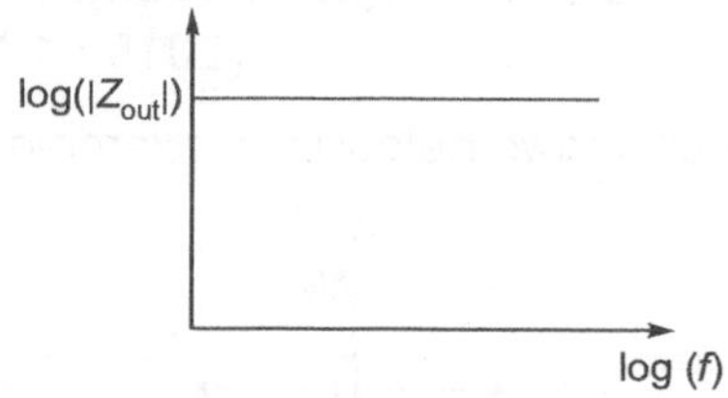

(b)

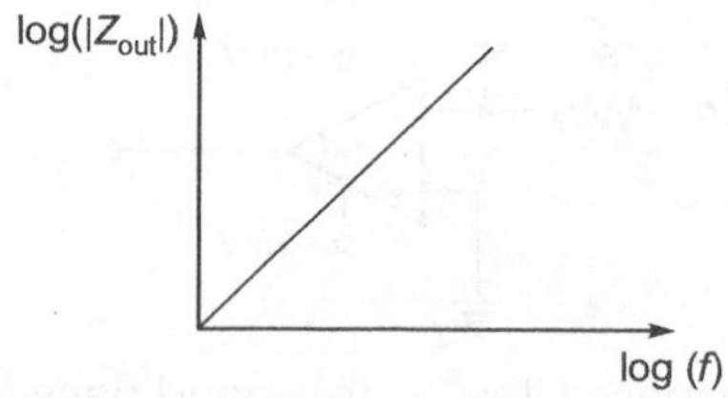

(c)

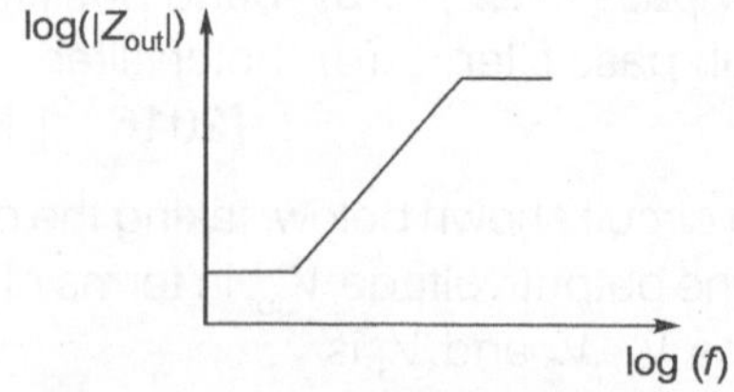

(d)

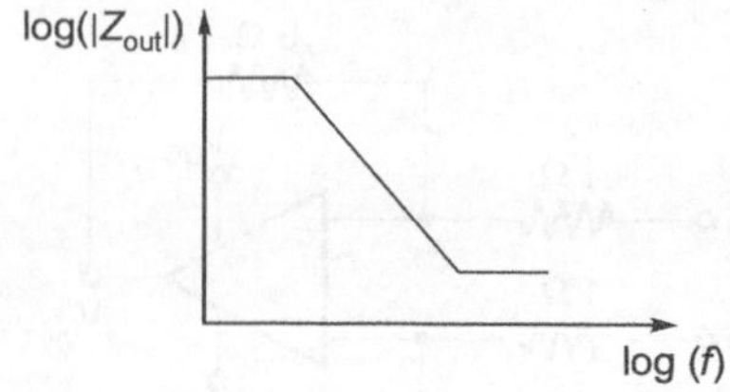

[2022 : 2 M]

4.50 Consider the OP AMP based circuit shown in the figure. Ignore the conduction drops of diodes D_1 and D_2. All the components are ideal and the breakdown voltage of the Zener is 5 V. Which of the following statements is true?

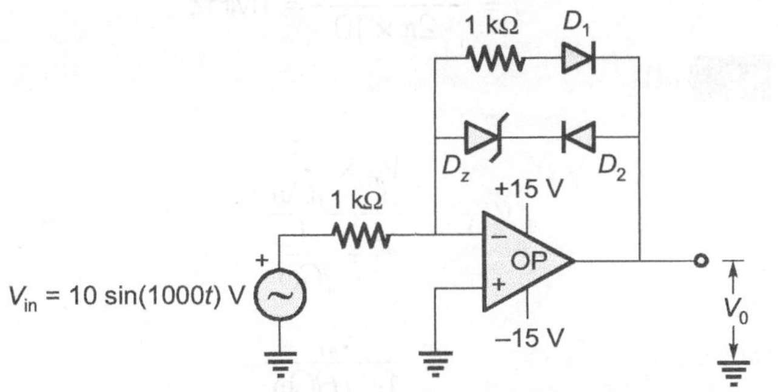

(a) The maximum and minimum values of the output voltage V_O are +15 V and –10 V, respectively.
(b) The maximum and minimum values of the output voltage V_O are +5 V and –15 V, resp..
(c) The maximum and minimum values of the output voltage V_O are +10 V and –5 V, resp..
(d) The maximum and minimum values of the output voltage V_O are +5 V and –10 V, resp.

[2023 : 2 M]

4.51 A difference amplifier is shown in the figure. Assume the op-amp to be ideal. The CMRR (in dB) of the difference amplifier is _______ (rounded off to 2 decimal places).

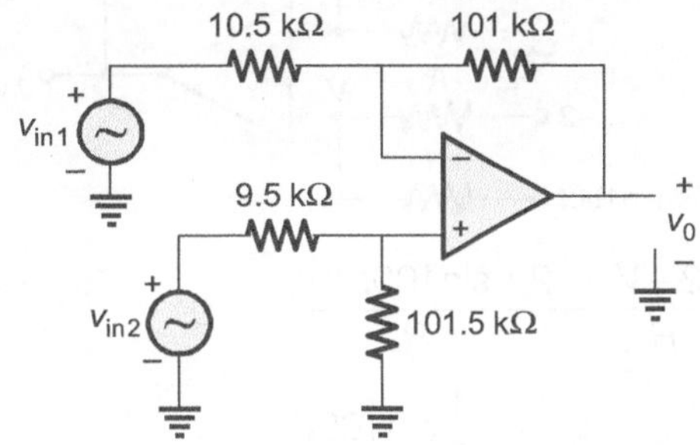

[2024 : 2 M]

Answers Operational Amplifiers

4.1 (Sol.)	4.2 (a)	4.3 (a)	4.4 (b)	4.5 (Sol.)	4.6 (d)	4.7 (d)
4.8 (a)	4.9 (d)	4.10 (d)	4.11 (b)	4.12 (d)	4.13 (c)	4.14 (d)
4.15 (a)	4.16 (d)	4.17 (c)	4.18 (b)	4.19 (c)	4.20 (d)	4.21 (b)
4.22 (d)	4.23 (d)	4.24 (d)	4.25 (b)	4.26 (d)	4.27 (d)	4.28 (b)
4.29 (b)	4.30 (c)	4.31 (b)	4.32 (a)	4.33 (d)	4.34 (c)	4.35 (d)
4.36 (d)	4.37 (1.245)	4.38 (b)	4.39 (a)	4.40 (a)	4.41 (a)	4.42 (d)
4.43 (a)	4.44 (c)	4.45 (b)	4.46 (a)	4.47 (b)	4.48 (c)	4.49 (c)
4.50 (d)	4.51 (40.52)					

Explanations Operational Amplifiers

4.1 Sol.

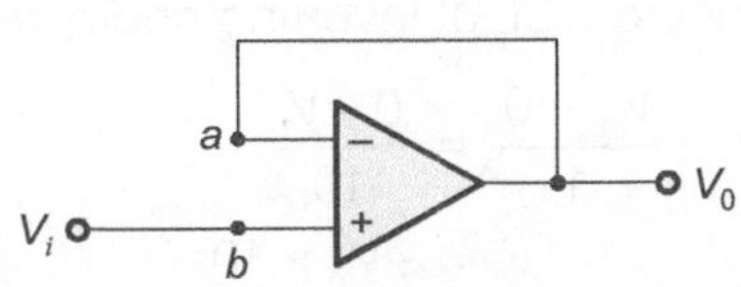

Here, $V_i = V_0$

$$V_{cm} = V_a + V_b = \frac{1}{2}(V_i + V_0) = V_i = V_0$$

4.2 (a)

$$x = -\frac{2R}{R} = -2$$

$$x' = -\frac{R}{R} = -1$$

$$\frac{x'}{x} = \frac{1}{2} \Rightarrow x' = \frac{1}{2}x$$

4.3 (a)

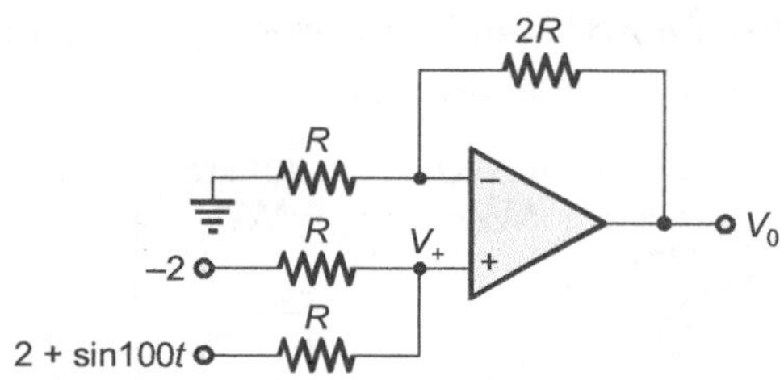

$$\frac{-2-V_+}{R}+\frac{2+\sin 100t-V_+}{R}=0$$

$$V_+ = \frac{1}{2}\sin 100t$$

$$\frac{0-V_+}{R} = \frac{V_+-V_o}{2R}$$

$$V_o = 3\,V_+ = \frac{3}{2}\sin 100t$$

4.5 Sol.

(A) → (R) : There is no feedback.

(B) → (S) : Input impedance is resistive and feedback is capacitive in nature.

(C) → (P) : Input impedance is capacitive and feedback is resistive in nature.

4.6 (d)

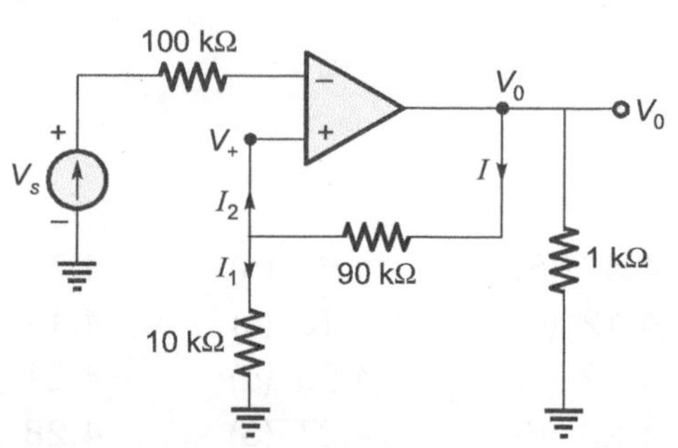

$I_2 = 0, \quad I_1 = I$

$V_+ = 10\ \text{k}\Omega \times I$

$V_o = (90 + 10)\ \text{k}\Omega \times I = 100\ \text{k}\Omega \times I$

$$\beta = \frac{V_+}{V_o} = \frac{10\ k\Omega \times I}{100\ k\Omega \times I} = \frac{1}{10}$$

4.7 (d)

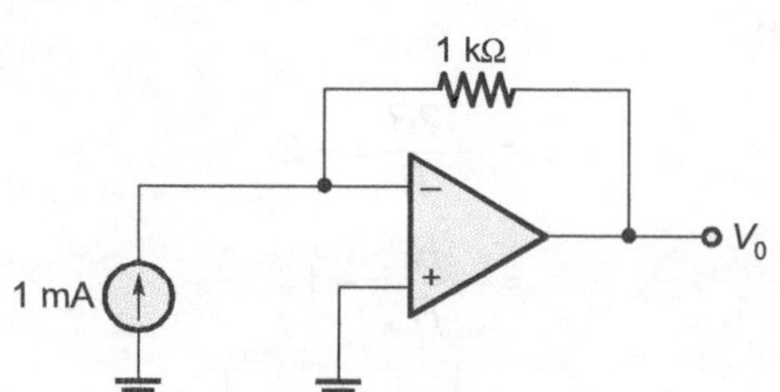

$V_0 = -1\ \text{mA} \times 1\ \text{k}\Omega = -1\ \text{V}$

4.8 (a)

$$\text{S.R.} = \omega V_m$$

$$62.8\ V/\mu s = 2\pi f \times 10$$

$$f = \frac{62.8\times 10^6}{2\pi\times 10} = 1\,\text{MHz}$$

4.9 (d)

$$V_{(+)} = \frac{V_{in}\times\frac{1}{jC\omega}}{R+\frac{1}{jC\omega}} = \frac{V_{in}}{1+jRC\omega}$$

where, $V_{(-)} = V_{(+)}$ [For ideal Op-amp]

$$\frac{V_{in}-V_{(-)}}{R_1} = \frac{V_{(-)}-V_0}{R_1}$$

$$V_o = 2\,V_{(-)} - V_{in} = 2V_{(+)} - V_{in} \qquad (V_{(-)} = V_{(+)})$$

$$= \left(\frac{2}{1+j\omega RC}-1\right)V_{in} = \left(\frac{1-j\omega RC}{1+j\omega RC}\right)V_{in}$$

$$\angle\left(\frac{V_0}{V_{in}}\right) = -2\tan^{-1}\omega RC$$

For $-90° \le \theta \le 90°$, maximum phase shift occurs (± 180°).

4.10 (d)

Using KCL at node 1 we have,

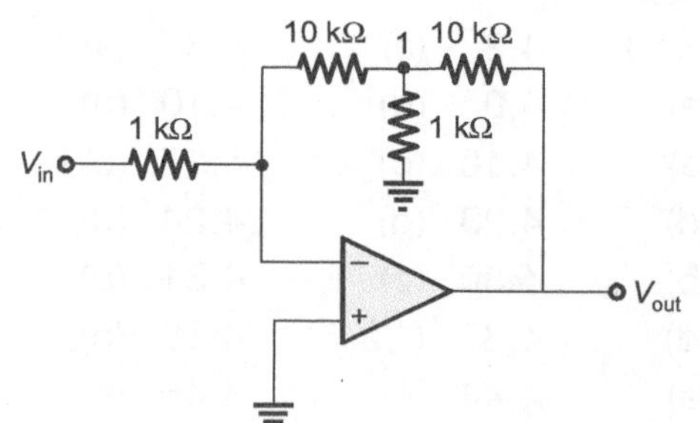

$$\frac{V_1-0}{10}+\frac{V_1}{1}+\frac{V_1-V_{out}}{10}=0$$

$$12\,V_1 = V_{out} \qquad \text{...(i)}$$

Also, using KCL at inverting node, we get

$$\frac{V_{in}-0}{1} = \frac{0-V_1}{10}$$

$$V_1 = -V_{in}\times 10 \qquad \text{...(ii)}$$

From equation (i) and (ii) we get,

$$\frac{V_{out}}{V_{in}} = -120$$

4.11 (b)

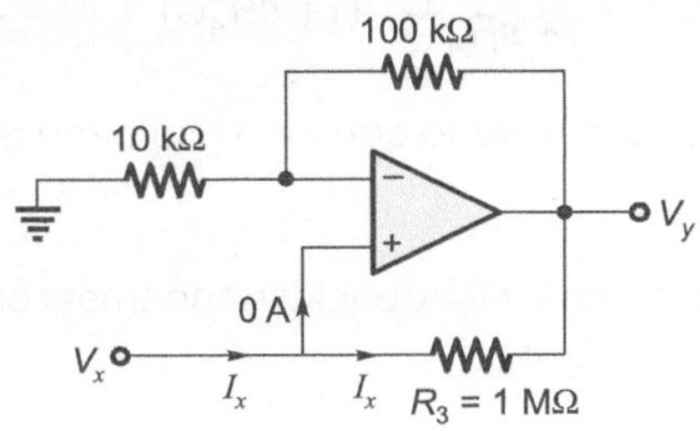

Here, $I_x = \dfrac{V_x - V_y}{R_3}$

But, $V_y = \left(1 + \dfrac{100K}{10K}\right) V_x = 11\ V_x$

$$I_x = \frac{-10V_x}{1M\Omega} \Rightarrow \frac{V_x}{I_x} = -100\ k\Omega$$

4.12 (d)

$$\frac{V_0}{V_{in}} = -\frac{R_f}{R_1}$$

If gain = –25 then, $R_1 = \dfrac{1000k}{25} = 40\ k\Omega$

If gain = –10 then,

$$R_1 = \frac{1000k}{10} = 100\ k\Omega$$

So, if we keep R_1 to be 100 kΩ then we never get the gain –25 for any R_F so we can keep R_1 to be 40 kΩ.

4.13 (c)

Output will be at its saturation values and it is having a phase difference of (–180°).
For +ve half of the input I[st] diode will be on making –ve terminal of op-amp to 0.7 V larger than the voltage at +ve input so output will be $-V_{sat.}$
For –ve half of input reverse of above will happen and output will go $+V_{sat}$.

4.14 (d)

For –ve half of the input +ve terminal of OPAMP is at higher potential than –ve terminal and output goes to $+V_{sat}$ but due to this high potential diode gets on and restricts the output to 0.7 V only.
And for +ve half of the input +ve terminal of OPAMP is at lower potential than –ve terminal's potential and output goes to $-V_{sat}$ and remains at $-V_{sat}$.

4.15 (a)

Output will be either at $+V_{sat}$ or $-V_{sat}$. When output will be at $+V_{sat}$ diode connected to 10 kΩ resistance will be on making voltage at point P equal to 6 V.
When output is at $-V_{sat}$ diode connected to 2 kΩ resistance will be on making voltage at point P equal to –10 V.

4.16 (d)

It behaves as current source because the output current (I_0) depends upon (V_{in}) and resistance only.

where, $$I_0 = \frac{\left(\dfrac{V \times R_2}{R_1 + R_2}\right)}{r} = \frac{R_2}{R_1 + R_2} \cdot \frac{V}{r}$$

4.17 (c)

Let the voltage at the non-inverting terminal of the opamp be V_a volts and the voltage at inverting terminal be V_b volts.
At $t = 0^+$ switch is open $V_0 = 5$ V and $V_b = -10$ V

$$\therefore \quad V_a = \frac{V_0 \times 10\,k}{100\,k + 10\,k} = \frac{50}{11}\ \text{volts}$$

As $t \to \infty$, $V_b \to 10$ V

$$V_b = V_f + (V_i - V_f)\, e^{-t/RC} \qquad \text{...(i)}$$

Putting, $V_b = \dfrac{50}{11}$ at $t = T_1$,

$V_i = -10$ V and $V_f = 10$ V in eq. (i)

We get, $T_1 = 12.98\ \mu$sec

$\therefore$ At $t = 12.98\ \mu$-sec V_0 changes from 5 V to –5 V

4.18 (b)

For ($V_{in} > 0$) : $P \to V_{01}$ = negative; $Q \to V_{02} = 0$

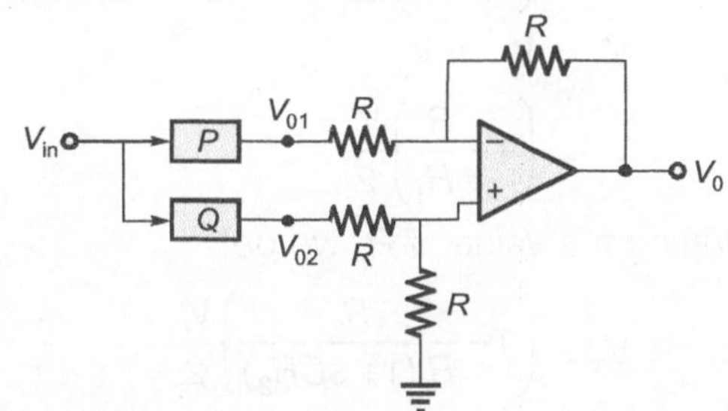

V_0 will be positive due to inverting action.
For ($V_{in} < 0$),

$P \to V'_{01} = 0$

$Q \to V'_{02}$ = positive

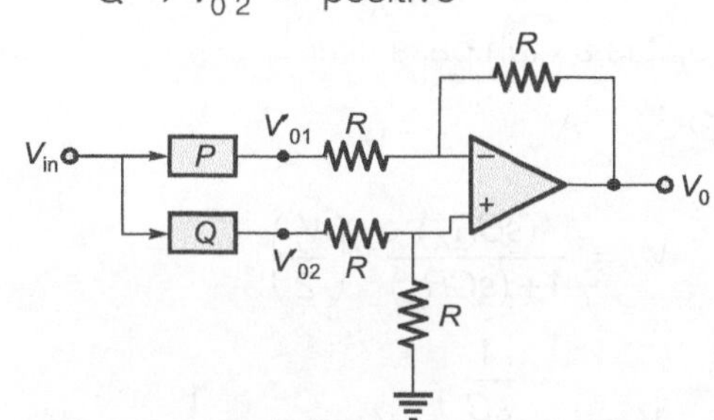

V_0 will be positive due to non-inverting action.
So, output is always rectified.

4.19 (c)

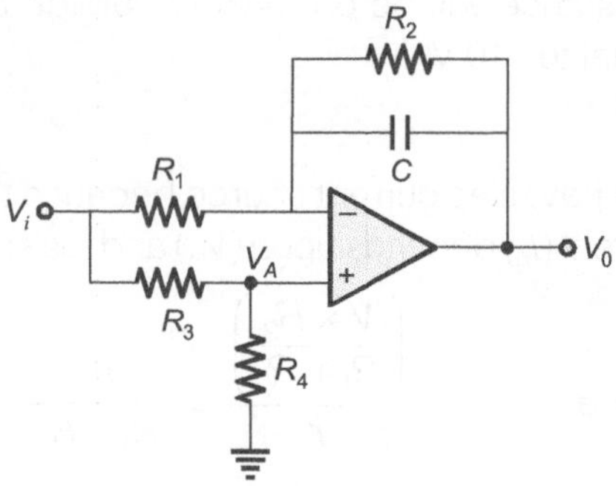

$$V_A = \frac{R_4}{R_4 + R_3} \times V_i,$$

$$V_A = \frac{R_B}{R_B + R_B} \times V_i$$

$$= \frac{V_i}{2}$$

Let, V_i only on inverting terminal

$$V_{01} = \frac{-R_f}{R_1} V_i$$

Here, $R_f = R_2 \,||\, \frac{1}{sC}$

$$= \frac{R_2}{1 + sCR_2}$$

Let, V_i only on non-inverting terminal

$$V_{02} = \left(1 + \frac{R_f}{R_1}\right)\frac{V_i}{2}$$

$$V_0 = V_{01} + V_{02}$$

$$= \left(1 - \frac{R_f}{R_1}\right)\frac{V_i}{2}$$

Putting the value of R_f, we get

$$V_0 = \left(1 - \frac{R_2}{R_1(1 + sCR_2)}\right)\frac{V_i}{2}$$

Here, $(R_1 = R_2 = R_A)$

$$\therefore \quad V_0 = \frac{1}{1 + \frac{1}{sCR_A}} \times \left(\frac{V_i}{2}\right)$$

So, it is a high pass filter.

4.20 (d)

$$V_{in} = \frac{(sCR_A)}{1 + (sCR_A)} \times \left(\frac{V_i}{2}\right)$$

$$\therefore \quad V_0 = \frac{\frac{1}{sC}}{\frac{1}{sC} + \frac{R_A}{2}} V_{in} = \frac{1}{\left(1 + sR_A \frac{C}{2}\right)} V_{in}$$

$$\therefore \quad V_0 = \frac{1}{1 + sR_A \frac{C}{2}} \cdot \frac{sR_AC}{(1 + sR_AC)} V_{in}$$

Which is similar to equation of a band pass filter.

4.21 (b)

Voltage-series feedback arrangement or voltage-voltage feedback.

4.22 (d)

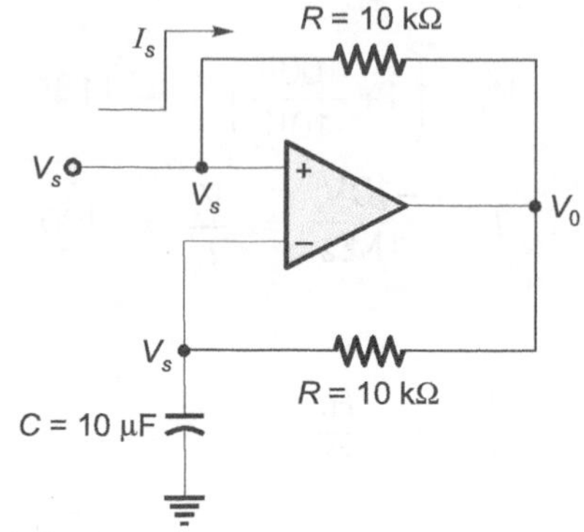

$$I_s = \frac{V_s - V_0}{R} \quad \text{...(1)}$$

$$V_s = \frac{\frac{1}{j\omega C}}{R + \frac{1}{j\omega C}} V_0 \quad \text{...(2)}$$

$$I_s = -V_s j\omega C$$

$$= -j(10 \times 2\pi \times 50 \times 10 \times 10^{-6})$$

$$= -j\,10\pi \text{ ... mA}$$

$$= 10\pi \text{ mA lagging by } 90°$$

4.23 (d)

For emitter follower,

$A_V \cong 1$; A_I is high

$\Rightarrow \quad A_p = A_V \cdot A_I$ is high $\Rightarrow P_{out} > P_{in}$

For transformer, $\quad P_{in} = P_{out}$

4.24 (d)

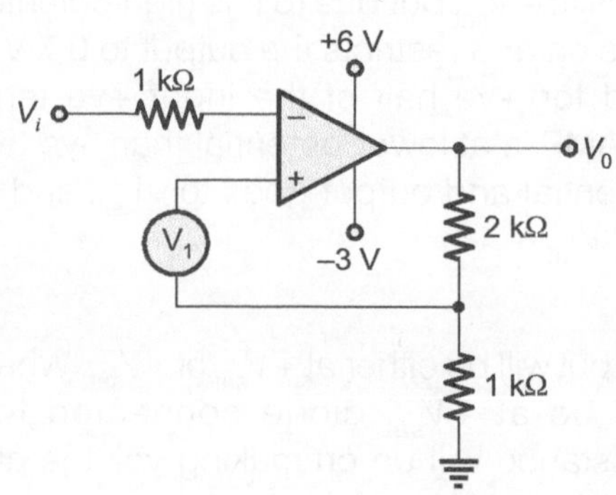

When $V_i < V_1$ (upto t_2) ; V_0 = +ve
When $V_i > V_1$ ($t_2 \le t \le t_4$) ; V_0 = –ve

$$V_1 = \frac{1}{1+2}V_0 = \frac{1}{3}V_0$$

4.25 (b)

$$V_0 = \left(1+\frac{2R}{R}\right)V = (1 + 2)\,(2) = 6\text{ V}$$

4.26 (d)

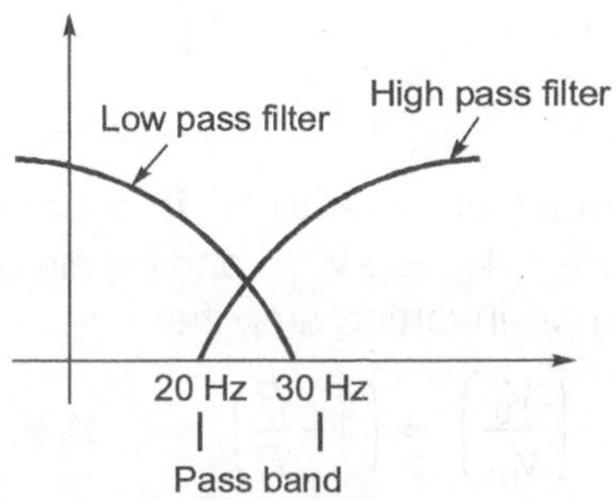

It is a band pass filter.

4.27 (d)

First section is differential amplifier having gain –1.

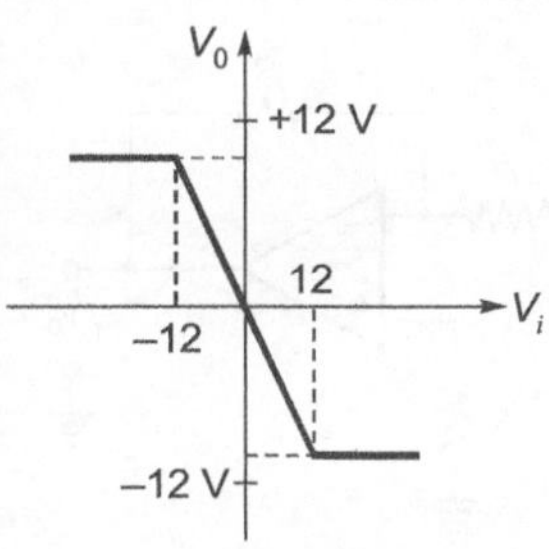

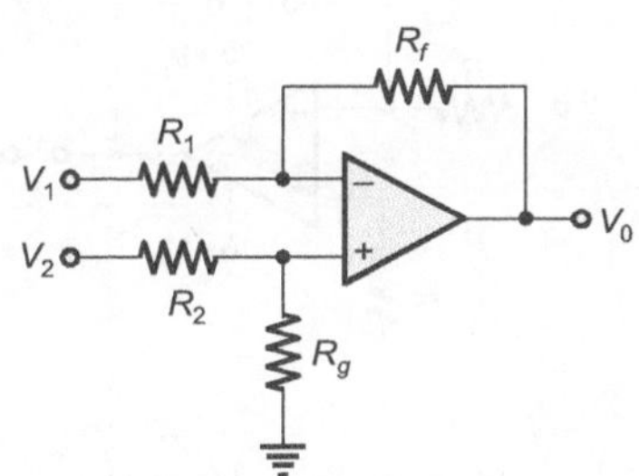

Output is

$$V_0 = \left(\frac{R_f + R_1}{R_g + R_2}\right)\frac{R_g}{R_1}V_2 - \frac{R_f}{R_1}V_1$$

$$V_0 = V_2 - V_1$$
$$= -V_i$$

Second stage-Schmitt trigger

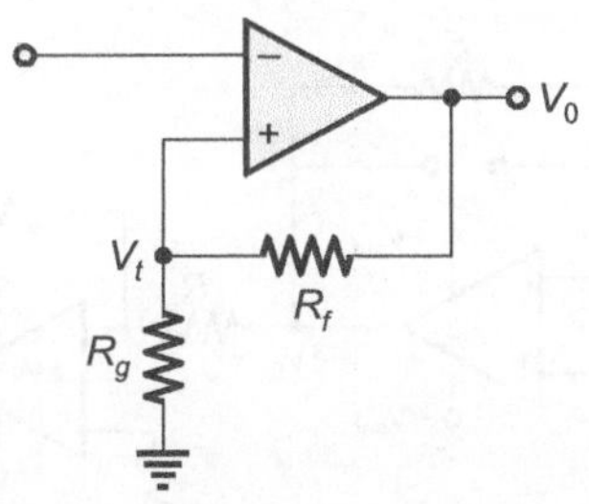

$$V_t = \pm\frac{R_g}{R_f + R_g}V_0 = \frac{V_0}{2} = \pm 6\,V$$

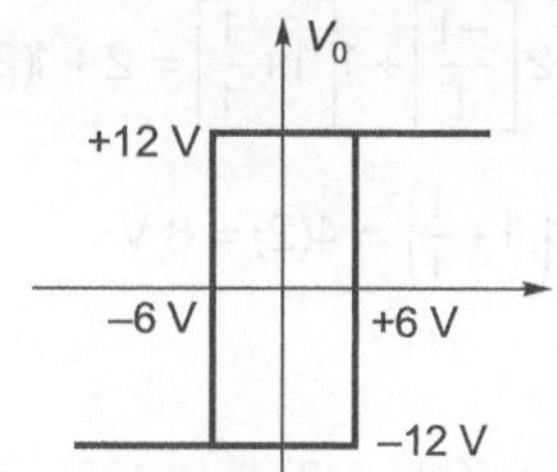

4.28 (b)

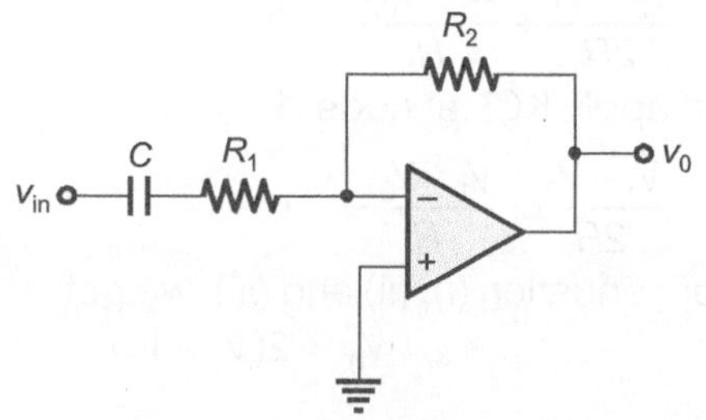

$$\frac{v_{in} - 0}{R_1 + \frac{1}{sC}} = \frac{0 - v_0}{R_2}$$

$$\frac{v_0}{v_i} = -\left(\frac{R_2}{R_1 + \frac{1}{sC}}\right)$$

It is a high pass filter with

$$f_{3\text{ dB}} = \frac{1}{R_1 C}\text{ rad/sec.}$$

4.29 (b)

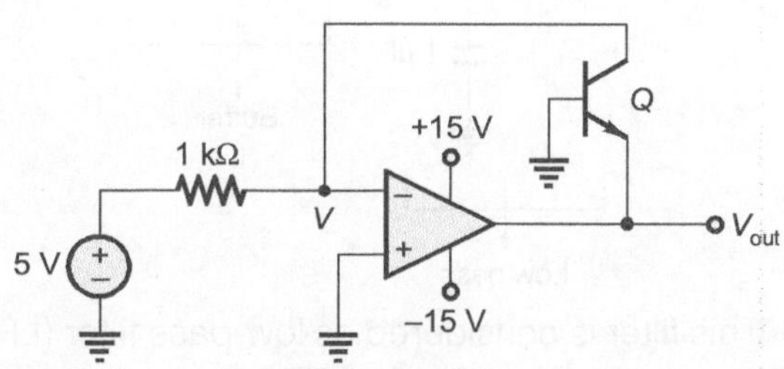

Using the concept of virtual ground, $V = 0$

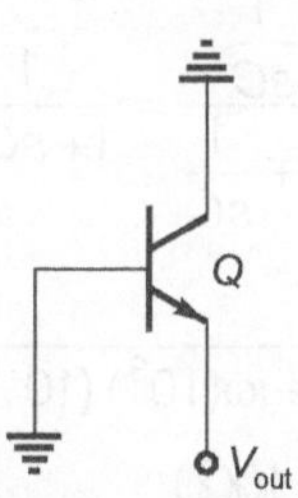

$$V_{out} = -0.7\text{ V}$$

4.30 (c)

Output of first op-amp

$$V_{out\,1} = -2\left[\frac{-1}{1}\right]+1\left[1+\frac{1}{1}\right] = 2+1(2) = 4\text{ V}$$

$$V_{out} = 4\left(1+\frac{1}{1}\right) = 4(2) = 8\text{ V}$$

4.31 (b)

Op-amp '3' circuit is a differential amplifier so,

$$V_o = V_{01} - V_{02} \quad \text{...(i)}$$

Now, apply KCL at node '2'

$$\frac{V_2 - V_1}{2R} + \frac{V_2 - V_{02}}{R} = 0 \quad \text{...(ii)}$$

and apply KCL at node '1'

$$\frac{V_1 - V_2}{2R} + \frac{V_1 - V_{01}}{R} = 0 \quad \text{...(iii)}$$

From equation (i), (ii) and (iii), we get

$$V_0 = 2(V_1 - V_2)$$

4.32 (a)

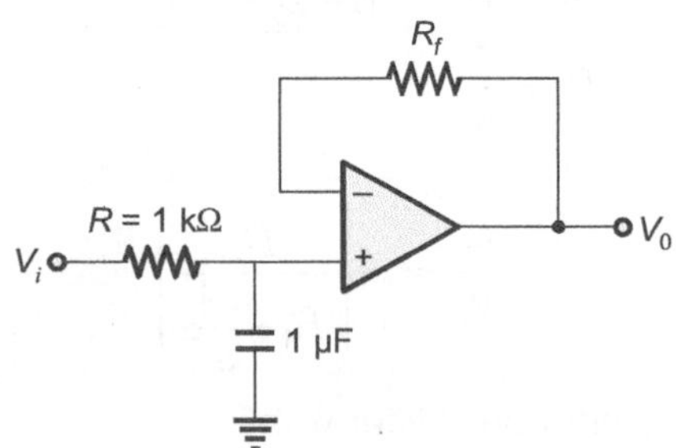

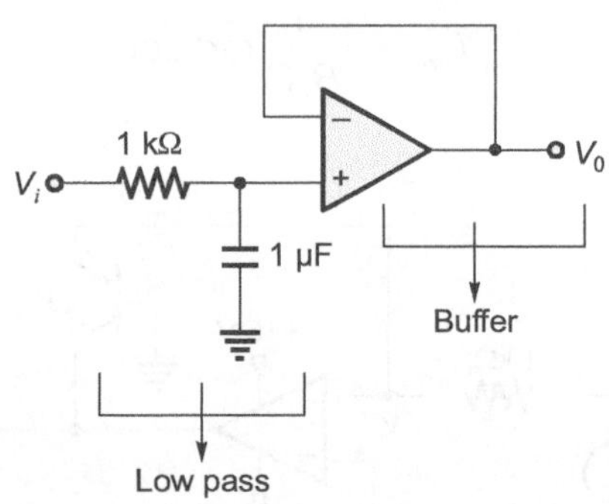

This filter is considered as low-pass filter (LPF)
The transfer function for LPF

$$\frac{V_o}{V_i} = \frac{\frac{1}{sC}}{R+\frac{1}{sC}} = \frac{1}{1+sCR}$$

$$= \frac{1}{1+j\omega(10^3)(10^{-6})} = \frac{1}{1+j\omega(10^{-3})}$$

$$= \frac{1000}{1000+j\omega} = \frac{1}{1+\frac{s}{1000}}$$

ω_c (corner frequency) = 10^3 rad/sec.
The gain at low frequencies,

$$A_v = \frac{1}{1+sCR} = \frac{1}{1} = 1$$

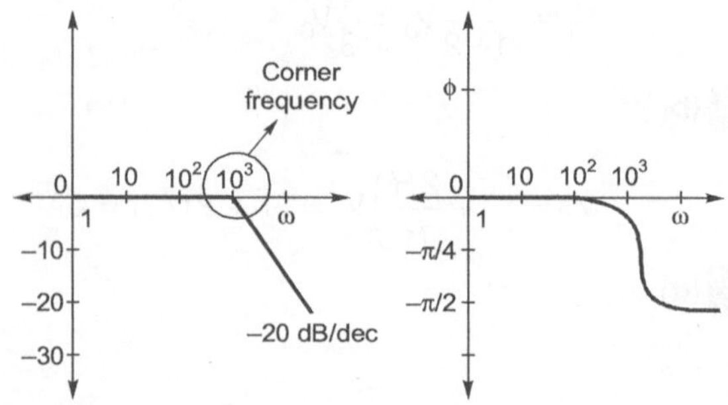

4.33 (d)

The circuit of op-amp '1' is a Schmitt trigger, therefore, $V_{01} = \pm V_{sat}$ and the circuit of op-amp '2' is a non-inverting amplifier

$$\therefore \quad \left(\frac{V_0}{V_{01}}\right) = \left(1+\frac{R}{R}\right) \Rightarrow V_0 = 2\,V_{01}$$

Where, $V_{01} = \pm V_{sat}$

Therefore, the answer is $V_0 = \pm V_{sat}$.

4.34 (c)

Case-I: $V_i > 0$, the circuit will look like

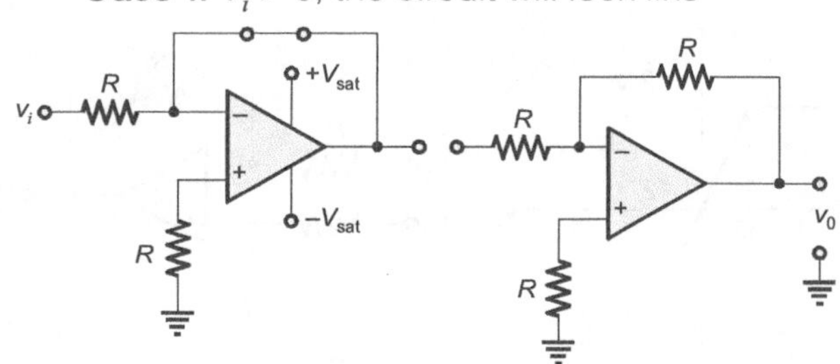

Hence $V_0 = 0$

Case-II: $V_i < 0$, the circuit will look like

$$\frac{V_{01}}{V_i} = -\frac{R}{R} \quad \text{...(i)}$$

$$V_{01} = -V_i \quad \text{...(ii)}$$

and

$$\frac{V_o}{V_{01}} = -\frac{R}{R}$$

$$V_0 = -V_{01} \quad \text{...(iii)}$$

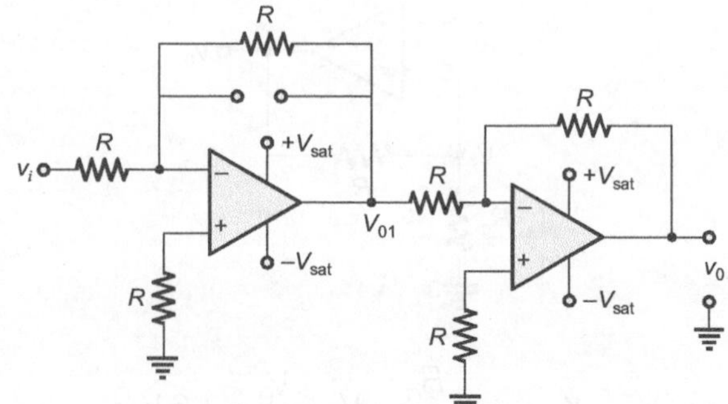

From equation (ii) and (iii), we get $V_0 = V_i$

4.35 (d)

The circuit is

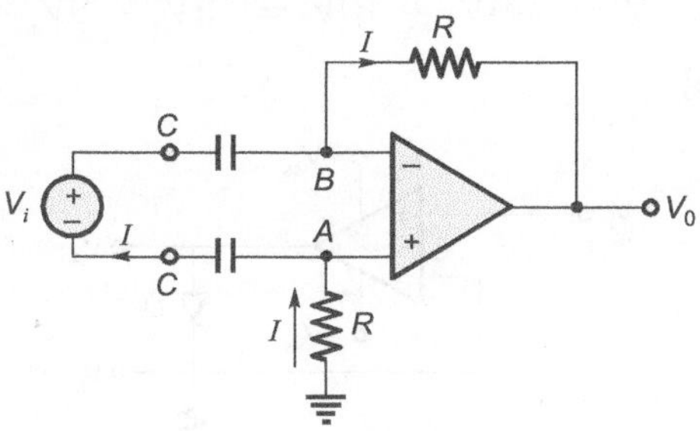

Since, circuit has negative feedback, so with help of virtual short $V_A = V_B$.

So, KVL in the loop from A to B given,

$$I\left(\frac{1}{sC}\right) - V_i + I\left(\frac{1}{sC}\right) = 0$$

So, $$I = \frac{V_i(sC)}{2}$$

so, $$V_A = -IR$$

$$V_A = -\frac{sCRV_i}{2} \quad \text{and} \quad V_A = V_B$$

so, $$V_B = -\frac{sCRV_i}{2}$$

Now, $$\frac{V_B - V_0}{R} = I$$

So, $$V_0 = V_B - IR$$

$$V_0 = -\frac{V_i(sCR)}{2} - \frac{V_i(sCR)}{2} = -V_i(sCR)$$

so, $$V_0 = -j\omega RCV_i$$

so, V_0 lag V_i by 90° or phase of V_0 w.r.t. V_i is –90°.

4.36 (d)

The circuit is

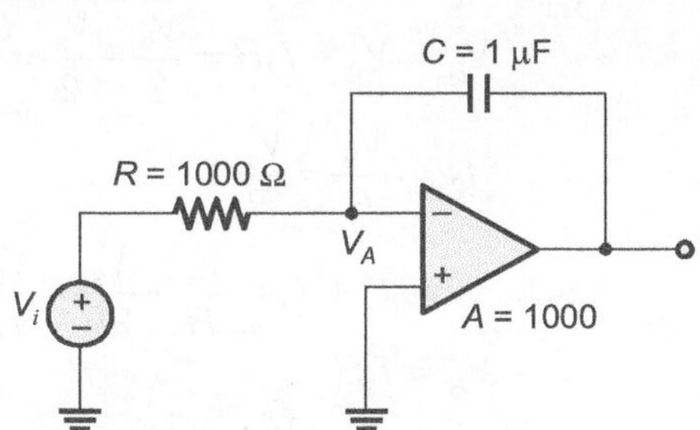

Now, the gain of op-amp is 1000

so, $$V_0 = 1000\,(V_+ - V_-)$$

since, $$V_+ = 0 \text{ (grounded)}$$

$$V^- = V_A$$

The above circuit can be redrawn as

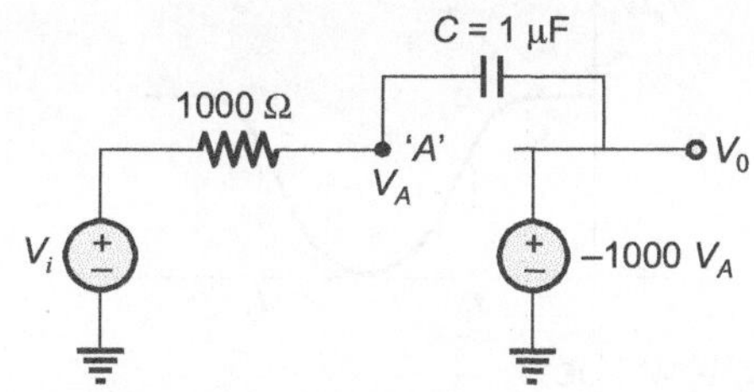

so, $$V_0 = -1000\text{ VA}$$

Now, KCL at node 'A'

$$\frac{V_A - V_i}{1000} = \left(\frac{V_A - V_0}{1/sC}\right)$$

$$\frac{V_A - V_i}{1000} = \frac{V_A - (-1000\,V_A)}{1/sC}$$

$$V_A - V_i = (1001\,V_A)\,sC$$

$$V_A - V_i = (1001) \times 1000\,V_A sC$$

$$V_A - V_i = s(1001000 \times 10^{-6})\,V_A$$

$$V_A - s(1.001)\,V_A = V_i$$

$$V_A = \left[\frac{V_i}{1 - s(1.001)}\right]$$

Since, $$V_0 = -1000\,V_A = -\frac{1000}{1 - s(1.001)}V_i$$

Since, pole is at (1/1.001), so time constant is approx '1'.

4.37 (1.245)

$$V_i = 2\sin(2\pi \times 2000t)$$

The transfer function of the system is

$$H(s) = \frac{1000}{(1000 \times 0.1 \times 10^{-6}s + 1)1000} = \frac{-1}{(10^{-4}s + 1)}$$

The input is $2\sin(2\pi \times 2000t)$

$$\omega = 4000\pi$$

$$V_0 = 2|H(j\omega)|_{\omega = 4000\pi} \times \sin(4000\pi t - \phi)$$

$$\text{Output} = 1.245\sin(4000\pi t - 51.46°)$$

So, amplitude of output is 1.245.

4.38 (b)

To check the type of system:

We apply a delta function at input $V_i = \delta(t)$

$$V(f) = 1 \forall f$$

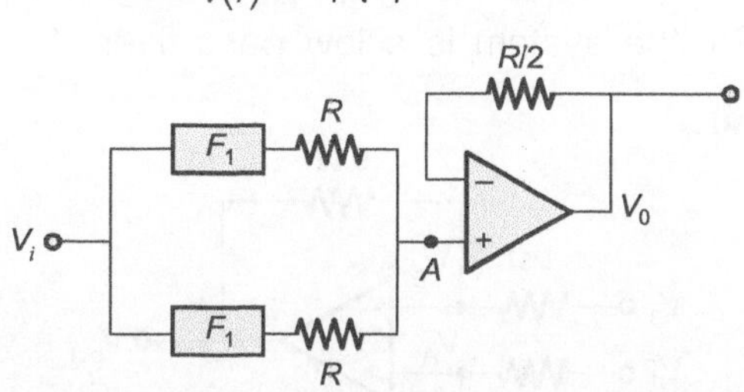

So, voltage at A will be same as voltage at output V_0. V_A will be equal to voltage due to F_1 + voltage due to F_2. Since, $f_1 < f_2$, so V_A/V_i will be

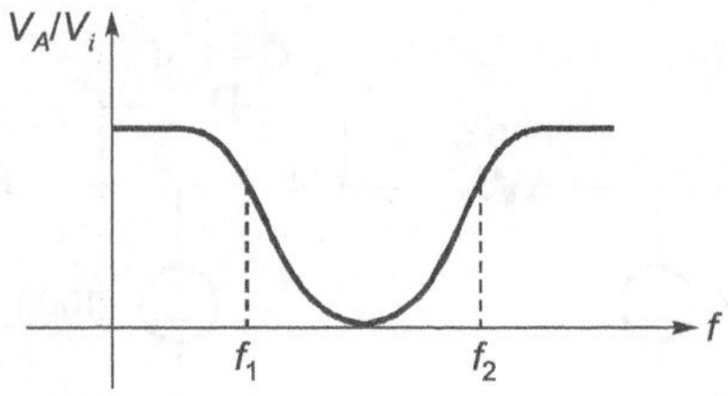

V_0/V_i will be

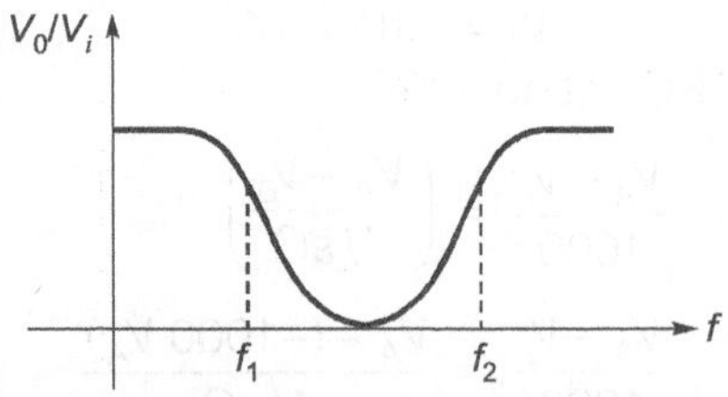

So, the system work as a band stop filter.

4.39 (a)

The given circuit in a astable multivibrators so output will be a periodic square wave and from the circuit $\beta = 0.5$.

So, time period will be $2\tau ln\left(\frac{1+\beta}{1-\beta}\right)$

$$T = 2 \times RC ln\left(\frac{1+0.5}{1-0.5}\right) = 0.55 \text{ ms}$$

4.41 (a)

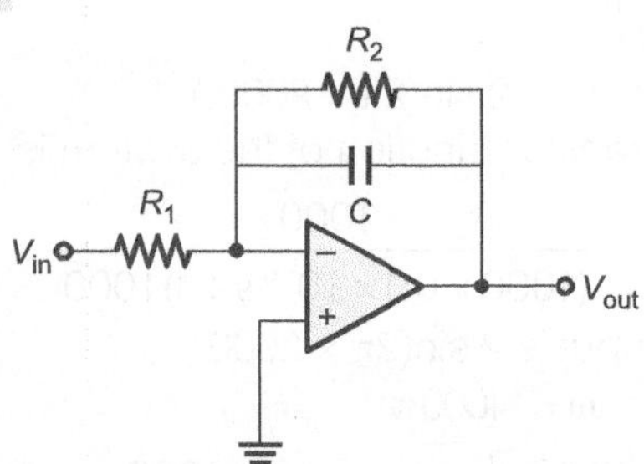

$$\frac{V_{out}}{V_{in}} = -\left[\frac{R_2 \cdot \frac{1}{j\omega C} \Big/ R_2 + \frac{1}{j\omega C}}{R_1}\right]$$

$$\frac{V_{out}}{V_{in}} = -\left[\frac{R_2}{R_1(R_2 j\omega C + 1)}\right]$$

So, the system is a low pass filter.

4.42 (d)

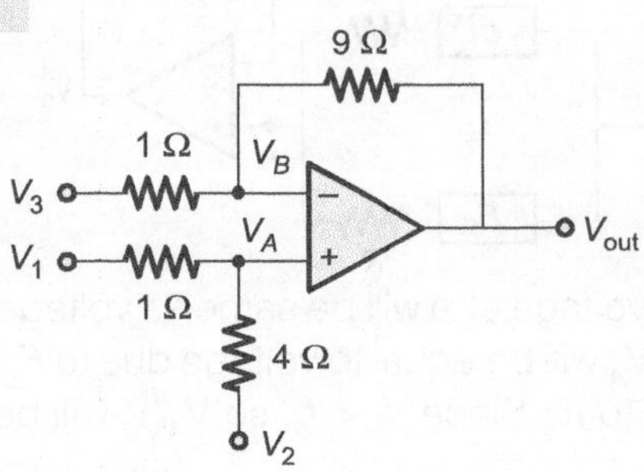

$$V_A = \left(\frac{4}{5}V_1 + \frac{1}{5}V_2\right)$$

$$V_{out} = -9V_3 + 10V_A = -9V_3 + 8V_1 + 2V_2$$

4.43 (a)

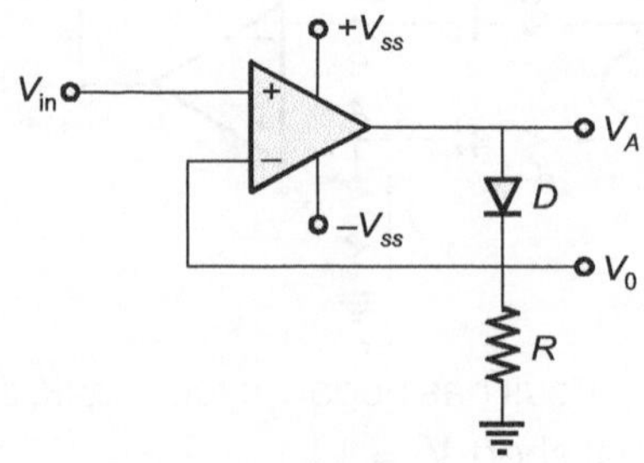

$V_{in} > 0$ $V_A = +V_{55}$, D on, $V_0 = V_{in}$
$V_{in} < 0$ $V_A = -V_{55}$, D off, $V_0 = 0$

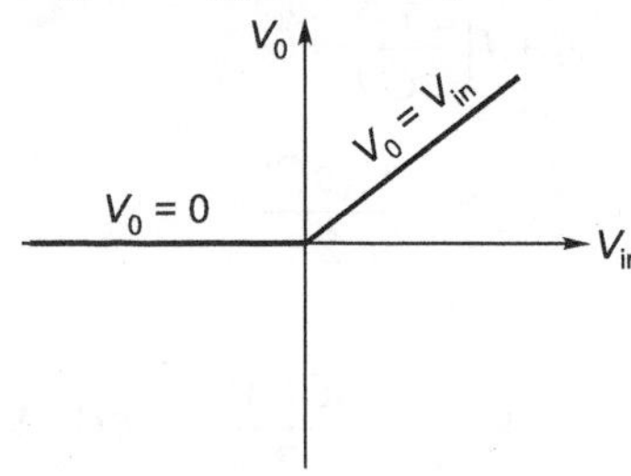

4.44 (c)

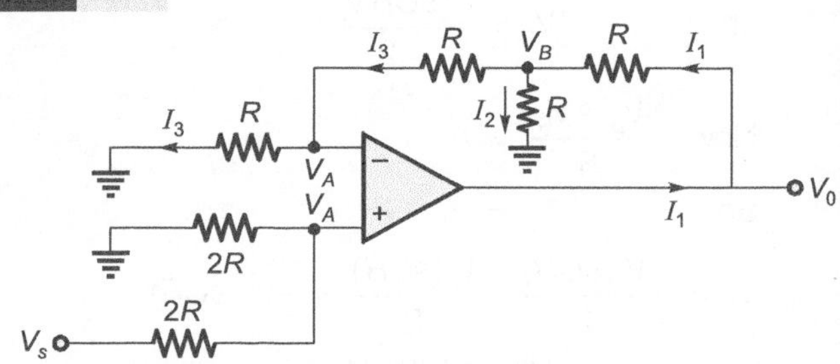

$$V_A = \frac{V_s}{2}$$

$$I_3 = \frac{V_A}{R} = \frac{V_s}{2R} \quad \text{...(i)}$$

$$V_B - V_A = I_3 R$$

$$V_B = V_A + I_3 R = \frac{V_s}{2} + \frac{V_s}{2} = V_s$$

$$I_2 = \frac{V_B}{R} = \frac{V_s}{R} \quad \text{...(ii)}$$

$$I_1 = I_2 + I_3 = \frac{V_s}{R} + \frac{V_s}{2R} = \frac{V_s}{R}[1.5]$$

$$V_0 - V_B = I_1 R$$

$$\Rightarrow \quad V_0 = V_B + \frac{V_s}{R}[1.5]R$$

$$= V_s + 1.5\,V_s = 2.5\,V_s$$

4.45 (b)

According to virtual ground,

$$V_A = V_B = V_{in}$$

At node A,

$$\frac{V_{in} - V_0}{Z} = i_{in} \quad \text{...(i)}$$

$$V_{in}\left[\frac{1}{R_2} + \frac{1}{R_1}\right] = \frac{V_0}{R_1}$$

$$V_{in}\left[\frac{R_1 + R_2}{R_1 R_2}\right] \times R_1 = V_0$$

$$V_0 = V_{in} \times \left[\frac{R_1 + R_2}{R_2}\right] \quad \text{...(ii)}$$

Equation (ii) in equation (i),

$$\frac{V_{in} - V_0}{Z} = i_{in}$$

$$\frac{V_{in} - V_{in}\left[\frac{R_1 + R_2}{R_2}\right]}{Z} = i_{in}$$

$$\frac{V_{in}}{i_{in}}\left[1 - \frac{R_1 + R_2}{R_2}\right] = Z$$

$$\frac{V_{in}}{i_{in}}\left[\frac{R_2 - R_1 + R_2}{R_2}\right] = Z$$

$$\frac{V_{in}}{i_{in}} = -Z \cdot \frac{R_2}{R_1}$$

4.46 (a)

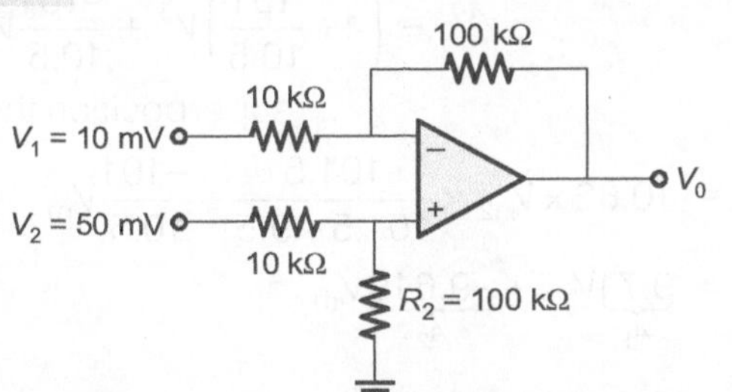

$$V_0 = \frac{R_2}{R_1}(V_2 - V_1) = \frac{100\text{k}}{10\text{k}}(50\text{ mV} - 10\text{ mV})$$

$$= 10\,(40\text{ mV}) = 400\text{ mV}$$

4.47 (b)

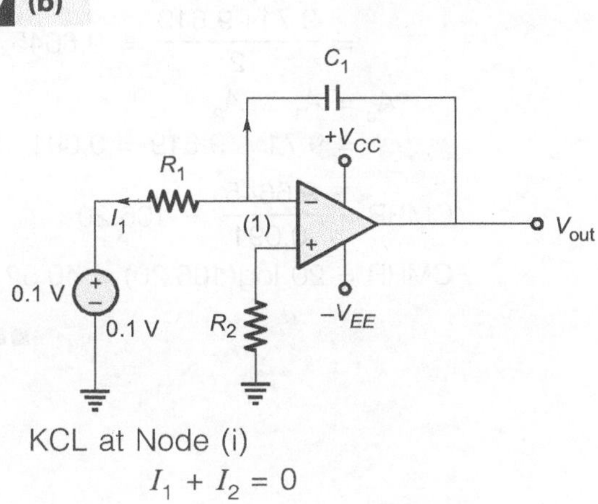

KCL at Node (i)

$$I_1 + I_2 = 0$$

$$\frac{V_1 - 0}{R_1} + C_1 \times \frac{d}{dt}(V_{out} - 0) = 0$$

$$\frac{dV_{out}}{dt} = -\frac{V_i}{R_1 C_1}$$

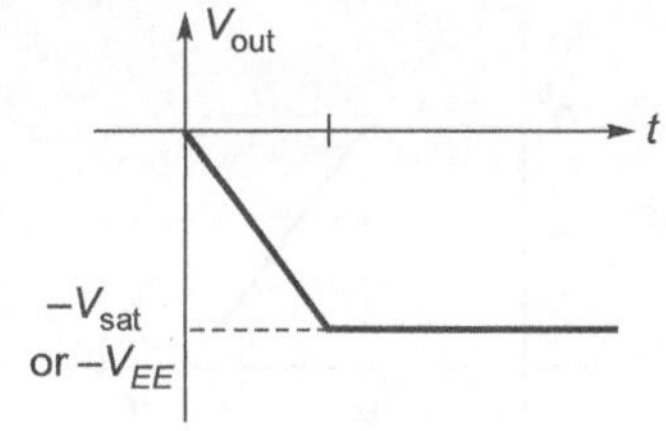

$$V_A = 0$$

$$V_{out} = -\frac{1}{R_1 C_1}\int V_i dt$$

$$V_i = 0.1\text{ V}$$

$$V_{out} = -\frac{1}{R_1 C_1}\int_t 0.1 \times dt = -\frac{0.1}{R_1 C_1} \times t$$

$$V_{out} = -k \times t$$

The output will be constant, i.e., $-V_{EE}$.

4.48 (c)

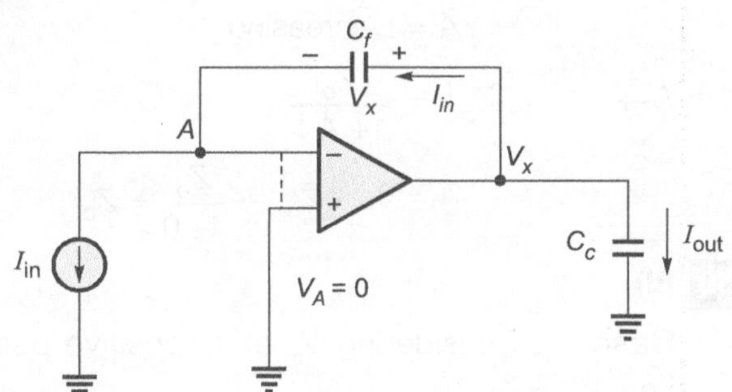

$$V_x = \frac{1}{C_f}\int I_{in} \times dt = \frac{I_{in}}{C_f} \times t$$

$$I_{out} = C_c \times \frac{dV_x}{dt} = C_c \times \frac{d}{dt}\left(\frac{I_{in}}{C_f} \times t\right) = C_c \times \frac{I_{in}}{C_f}$$

$$\frac{I_{out}}{I_{in}} = \frac{C_c}{C_f}$$

4.49 (c)

Analysis:

It is a voltage series feedback,

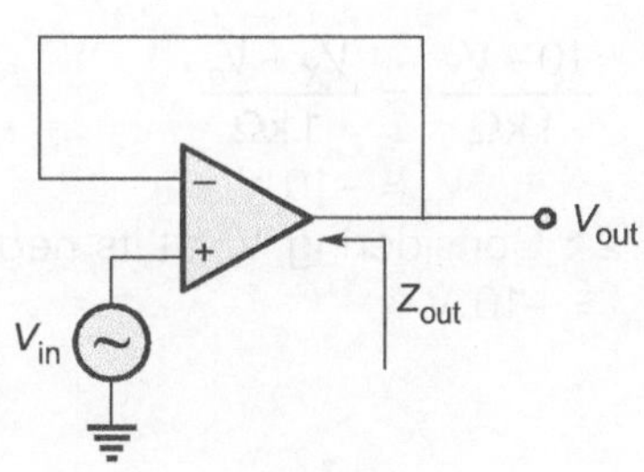

Voltage is a shunt connection,
Z_{out} feedback,

$$Z_{of} = \frac{Z_o}{1+A} = \frac{Z_o}{1+A} \text{ [Buffer, } \beta = 1]$$

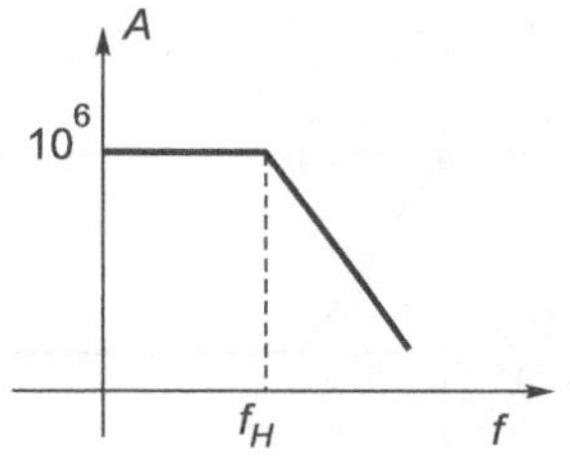

A = Open loop gain

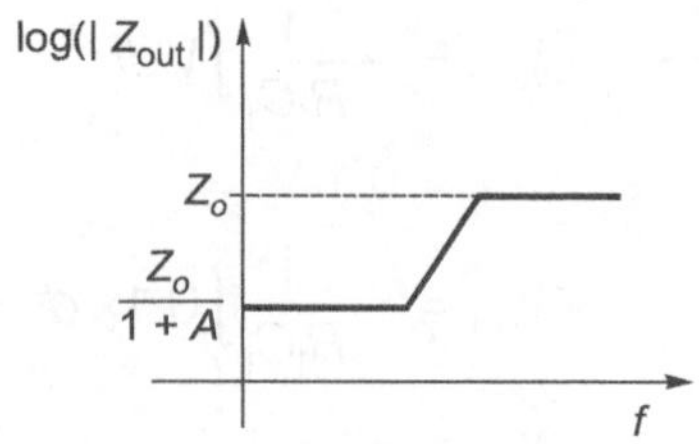

A = constant (low frequency)

$$Z_{of} = \frac{Z_o}{1+A} \text{[Constant]}$$

A = Decreasing

$$f \rightarrow \qquad Z_{of}\uparrow = \frac{Z_o}{1+A\downarrow}$$

$$A = 0; \quad Z_{of} = \frac{Z_o}{1+0} = Z_o$$

4.50 (d)

Case 1 : Considering V_{in} at its positive peak, i.e., V_{in} = 10 V

Diode D_1 = ON (F.B.); Diode D_2 = OFF (R.B.)

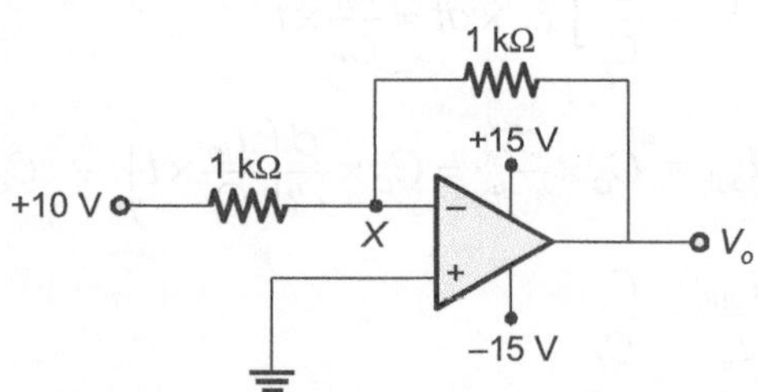

∵ Negative feedback is present. ∴ by virtual ground, $V_X = 0$.

By KCL at node X_i,

$$\frac{10 - V_X}{1\text{ k}\Omega} = \frac{V_X - V_o}{1\text{ k}\Omega}$$

$$V_o = -10 \text{ V}$$

Case 2 : Considering V_{in} at its negative peak, i.e., V_{in} = −10 V

Diode D_1 = OFF (R.B.); Diode D_2 = ON (F.B.)

Zener Diode, D_z= Breakdown

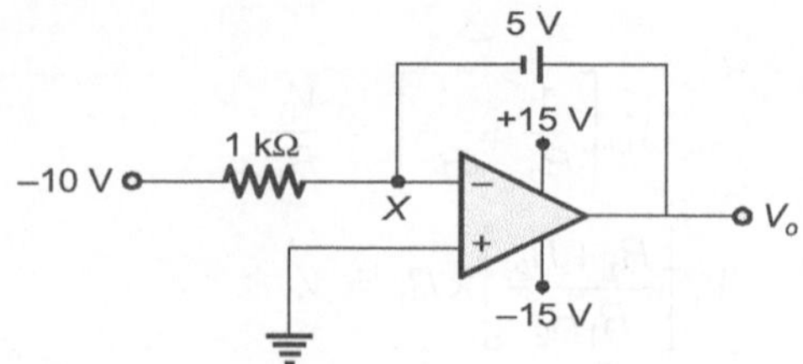

∵ Negative feedback is present. ∴ by virtual ground, $V_X = 0$.

By KVL, $-V_X - 5 + V_o = 0$

$$V_o = 5 \text{ V}$$

$$\therefore \qquad V_{o,max} = 5 \text{ V}; \quad V_{o,min} = -10 \text{ V}$$

4.51 (40.52)

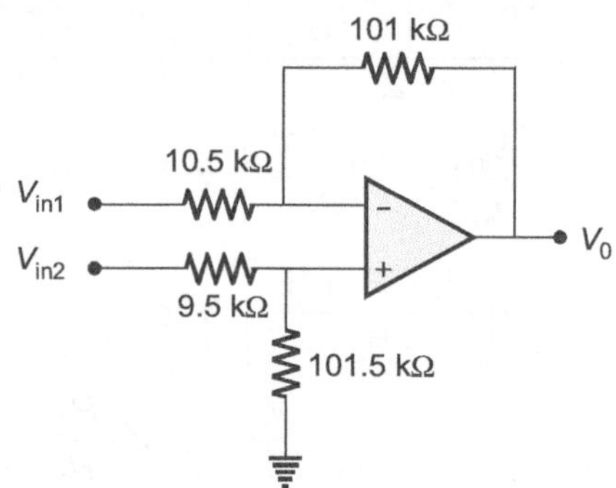

$$V_o = \left(1 + \frac{101}{10.5}\right)V^+ + \frac{-101}{10.5}V_{in1}$$

...(superposition theorem)

$$= 10.62 \times V_{in2} \times \frac{101.5}{101.5 + 9.5} + \frac{-101}{10.5}V_{in1}$$

$$= \underbrace{9.71}_{A_1}V_{in2} + \underbrace{-9.619}_{A_2}V_{in1}$$

$$\text{CMRR} = \frac{A_d}{A_c}$$

$$A_d = \frac{A_1 - A_2}{2}$$

$$= \frac{9.71 + 9.619}{2} = 9.6645$$

$$A_c = A_1 + A_2$$

$$= 9.71 - 9.619 = 0.091$$

$$\text{CMRR} = \frac{9.6645}{0.091} = 106.20$$

$$\text{CMRR} = 20 \log(106.20) = 40.52 \text{ dB}$$

■■■■

5 Oscillators and Feedback Amplifiers

5.1 An op-amp has an open-loop gain of 10^5 and an open loop upper cut-off frequency of 10 Hz. If this op-amp is connected as an amplifier with a closed loop gain of 100, then the new upper cut-off frequency is

(a) 10 Hz (b) 100 Hz
(c) 10 kHz (d) 100 kHz

[2001 : 1 M]

5.2 For the oscillator circuit shown in figure, the expression for the time period of oscillations can be given by (where $\tau = RC$)

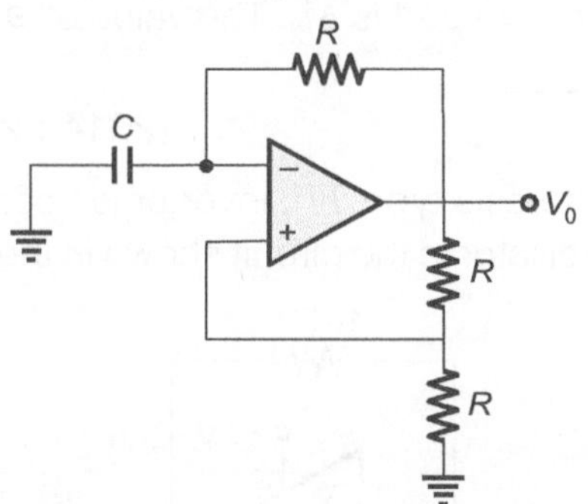

(a) $\tau \ln 3$ (b) $2\tau \ln 3$
(c) $\tau \ln 2$ (d) $2\tau \ln 2$ [2001 : 2 M]

5.3 The output voltage (V_0) of the schmitt trigger shown in figure swings between +15 V and –15 V. Assume that the operational amplifier is ideal. The output will change from +15 V to –15 V when the instantaneous value of the input sine wave is,

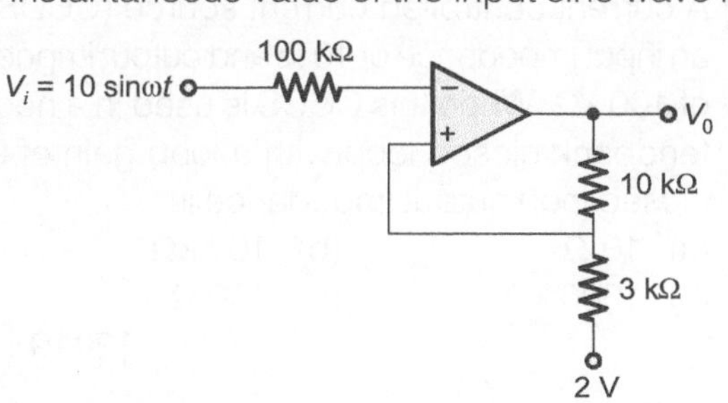

(a) 5 V in the positive slope only.
(b) 5 V in the negative slope only.
(c) 5 V in the positive and negative slopes.
(d) 3 V in the positive and negative slopes.

[2002 : 2 M]

5.4 The feedback used in the circuit shown in the figure can be classified as

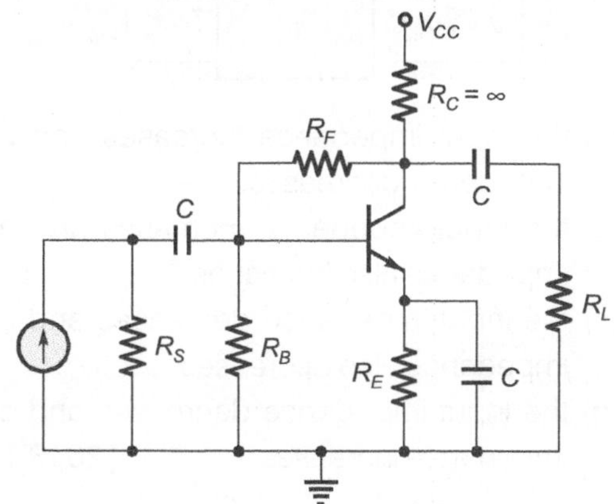

(a) shunt-series feedback
(b) shunt-shunt feedback
(c) series-shunt feedback
(d) series-series feedback

[2004 : 1 M]

5.5 The typical frequency response of a two-stage direct coupled voltage amplifier is as shown in

(a)
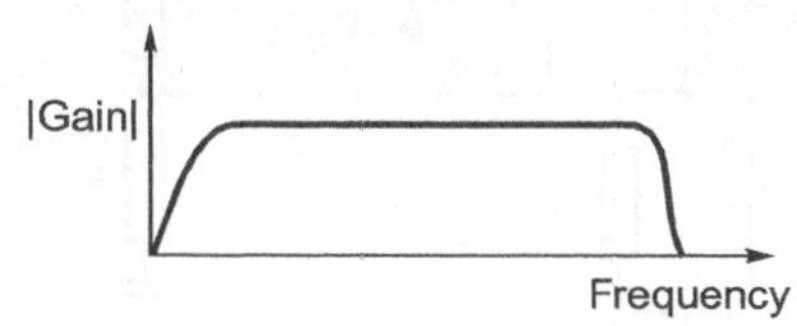

(b)
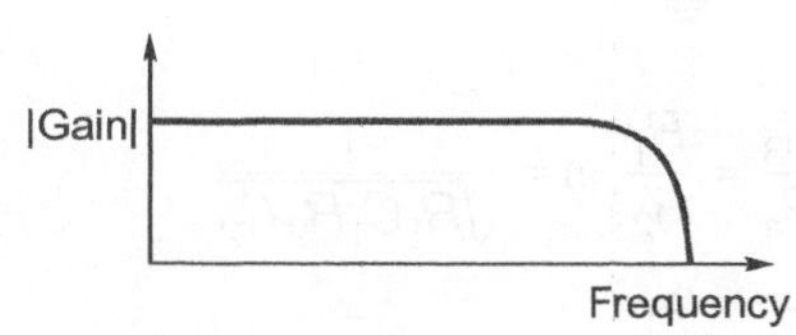

(c)
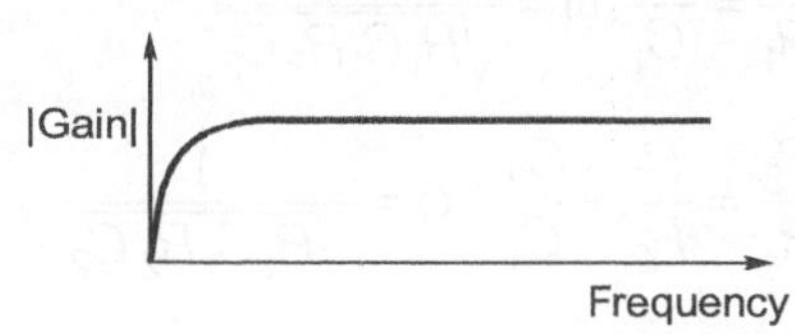

(d)
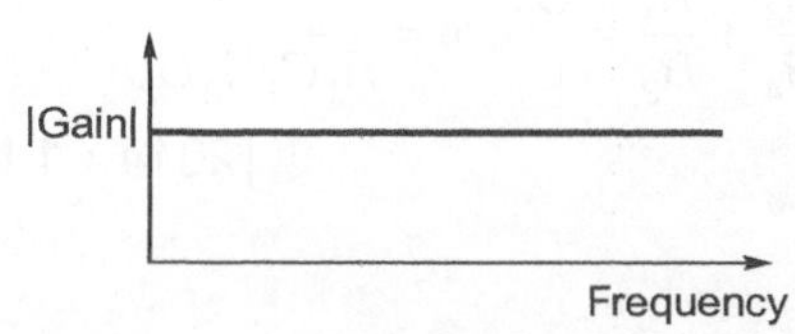

[2005 : 2 M]

5.6 In the feedback network shown below, if the feedback factor k is increased, then

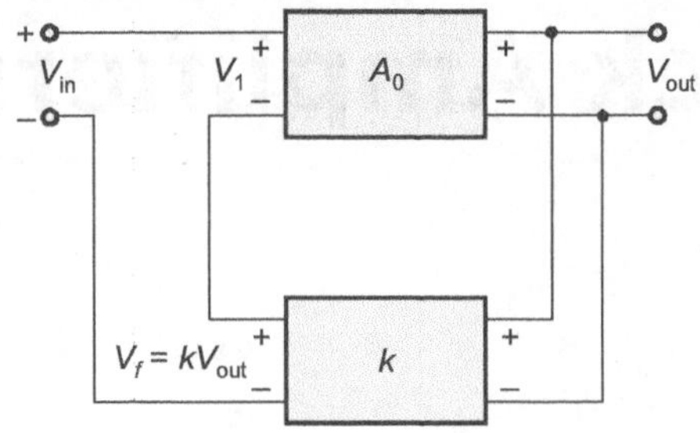

(a) the input impedance increases and output impedance decreases.
(b) the input impedance increases and output impedance also increases.
(c) the input impedance decreases and output impedance also decreases.
(d) the input impedance decreases and output impedance increases. [2013 : 1 M]

5.7 In the Wien Bridge oscillator circuit shown in the figure, the bridge is balanced when

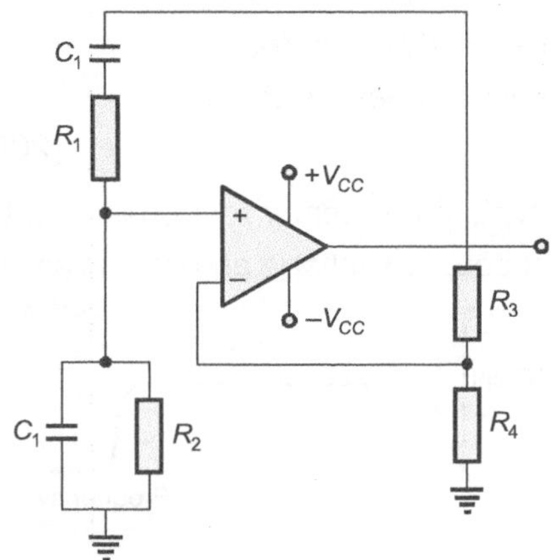

(a) $\frac{R_3}{R_4} = \frac{R_1}{R_2}, \omega = \frac{1}{\sqrt{R_1 C_1 R_2 C_2}}$

(b) $\frac{R_2}{R_1} = \frac{C_2}{C_1}, \omega = \frac{1}{\sqrt{R_1 C_1 R_2 C_2}}$

(c) $\frac{R_3}{R_4} = \frac{R_1}{R_2} + \frac{C_2}{C_1}, \ \omega = \frac{1}{\sqrt{R_1 C_1 R_2 C_2}}$

(d) $\frac{R_3}{R_4} + \frac{R_1}{R_2} = \frac{C_2}{C_1}, \omega = \frac{1}{R_1 C_1 R_2 C_2}$

[2014 : 1 M, Set-1]

5.8 An oscillator circuit using ideal op-amp and diodes is shown in the figure.

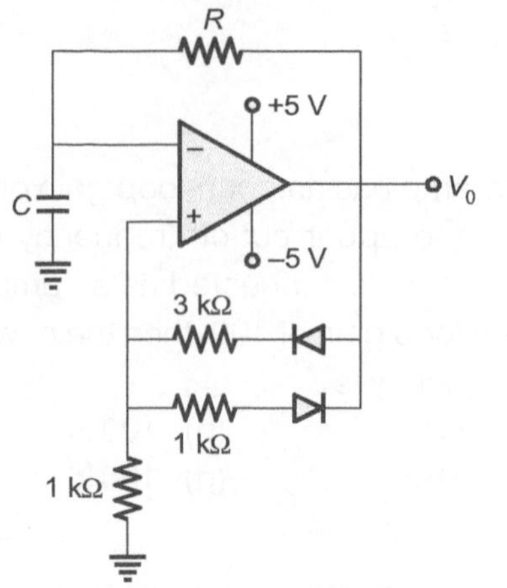

The time duration for +ve part of the cycle is Δt_1 and for −ve part is Δt_2. The value of $e^{\frac{\Delta t_1 - \Delta t_2}{RC}}$ will be ______ .

[2014 : 2 M, Set-2]

5.9 A hysteresis type *TTL* inverter is used to realize an oscillator in the circuit shown in the figure.

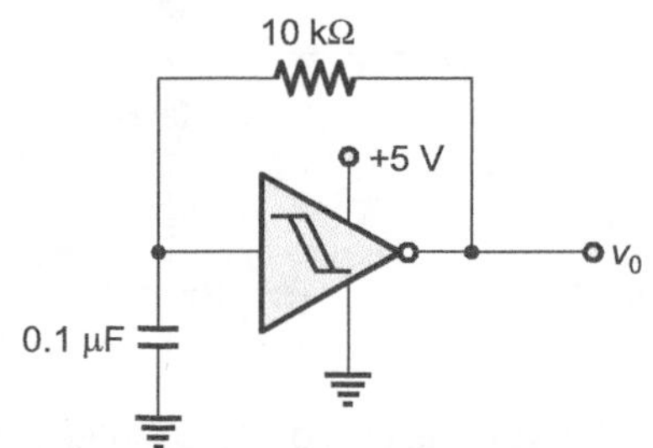

If the lower and upper trigger level voltages are 0.9 V and 1.7 V, the period (in ms), for which output is LOW, is ______. [2014 : 2 M, Set-3]

5.10 A current controlled current source (CCCS) has an input impedance of 10 Ω and output impedance of 100 kΩ. When this CCCS is used in a negative feedback closed loop with a loop gain of 9, the closed loop output impedance is
(a) 10 Ω
(b) 100 kΩ
(c) 1000 kΩ
(d) 100 Ω

[2019 : 1 M]

■■■■

Answers Oscillators and Feedback Amplifiers

5.1 (c)	5.2 (b)	5.3 (a)	5.4 (b)	5.5 (b)	5.6 (a)	5.7 (c)
5.8 (0.8)	5.9 (0.63)	5.10 (c)				

Explanations Oscillators and Feedback Amplifiers

5.1 (c)

$$A_{OL} = 10^5; \quad f = 10 \text{ Hz}$$

$$A_{CL} = \frac{A_{OL}}{1+\beta A_{OL}}$$

$$(1+\beta A_{OL}) = \frac{10^5}{10^2} = 10^3$$

$$f_2' = f_2(1+\beta A_{OL})$$

$$= 10 \times 10^3 \text{ Hz} = 10 \text{ kHz}$$

5.2 (b)

$$T = 2RC\, ln\, 3 = 2\tau\, ln 3$$

5.4 (b)

Equivalent circuit can be drawn with input voltage comparison and current feedback. It is shunt-shunt feedback.

5.5 (b)

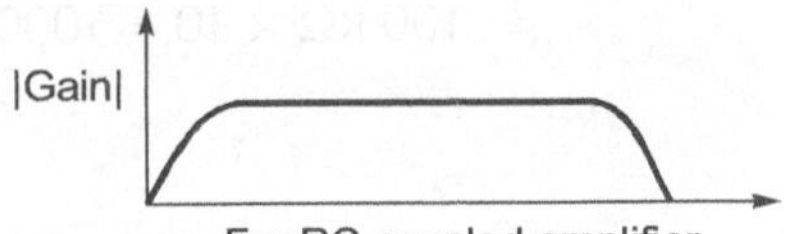

For RC coupled amplifier

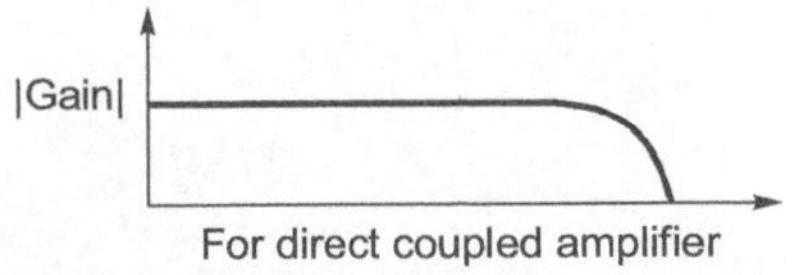

For direct coupled amplifier

5.6 (a)

The given configuration is a voltage-series feedback configuration.

So, the input impedance increases

$$R_{if} = R_i(1 + A_o k)$$

and, the output impedance decreases

$$R_{of} = \frac{R_o}{1+A_o k}$$

5.7 (c)

$$\frac{R_3}{R_4} = \frac{R_1}{R_2} + \frac{C_2}{C_1}$$

$$\omega = \frac{1}{\sqrt{R_1 R_2 C_1 C_2}}$$

5.8 (0.8)

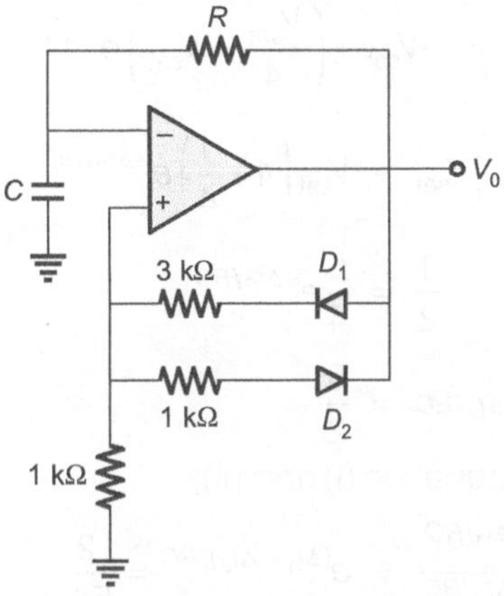

This circuit is astable multivibrator (or) free running oscillator.

When, $V_0 = +V_{sat}$

$$V_{UTP} = V_{sat} \times \frac{1\text{k}\Omega}{1\text{k}\Omega + 3\text{k}\Omega} = \frac{V_{sat}}{4}$$

When, $V_0 = -V_{sat}$

$$V_{LTP} = -V_{sat} \times \frac{1\text{k}\Omega}{1\text{k}\Omega + 1\text{k}\Omega} = \frac{-V_{sat}}{2}$$

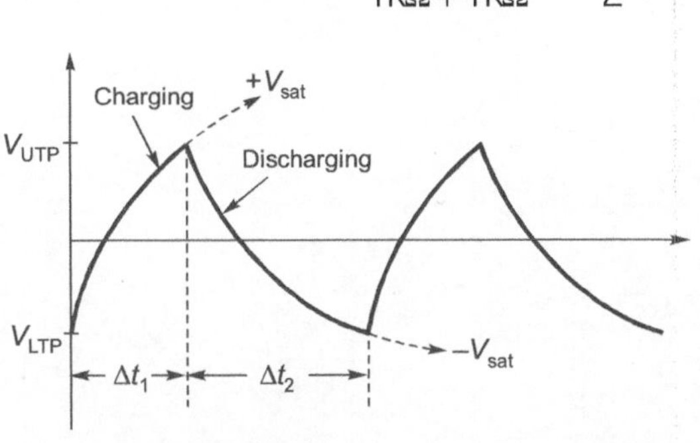

$$V_C = V_{final} + (V_{initial} - V_{final})e^{-t/RC}$$

In time $t = \Delta t_1$

$V_C = V_{UTP}$; $V_{initial} = V_{LTP}$; $V_{final} = +V_{sat}$

$$V_{UTP} = V_{sat} + (V_{LTP} - V_{sat})\, e^{-\Delta t_1/RC}$$

$$\frac{V_{sat}}{4} = V_{sat} + \left(-\frac{V_{sat}}{2} - V_{sat}\right) e^{-\Delta t_1/RC}$$

$$V_{sat}\left(-1+\frac{1}{4}\right) = -V_{sat}\left(\frac{1}{2}+1\right) e^{-\Delta t_1/RC}$$

$$\left(1-\frac{1}{4}\right) = \left(\frac{1}{2}+1\right) e^{-\Delta t_1/RC}$$

$$\frac{3}{4} = \frac{3}{2}e^{-\Delta t_1/RC}$$

$$e^{\Delta t_1/RC} = 2 \quad \text{...(i)}$$

In time $t = \Delta t_2$

$V_c = V_{LTP}$; $V_{initial} = V_{UTP}$; $V_{final} = -V_{sat}$

$$V_{LTP} = -V_{sat} + (V_{UTP} + V_{sat})e^{-\Delta t_2/RC}$$

$$-\frac{V_{sat}}{2} = -V_{sat} + \left(\frac{V_{sat}}{4} + V_{sat}\right)e^{-\Delta t_2/RC}$$

$$\left(1-\frac{1}{2}\right)V_{sat} = V_{sat}\left(1+\frac{1}{4}\right)e^{-\Delta t_2/RC}$$

$$\frac{1}{2} = \frac{5}{4}e^{-\Delta t_2/RC}$$

$$e^{\Delta t_2/RC} = \frac{5}{2} \quad \text{...(ii)}$$

From equation (i) and (ii),

$$\frac{e^{\Delta t_1/RC}}{e^{\Delta t_2/RC}} = e^{(\Delta t_1 - \Delta t_2)/RC} = \frac{2}{5/2} = \frac{4}{5} = 0.8$$

5.9 (0.63)

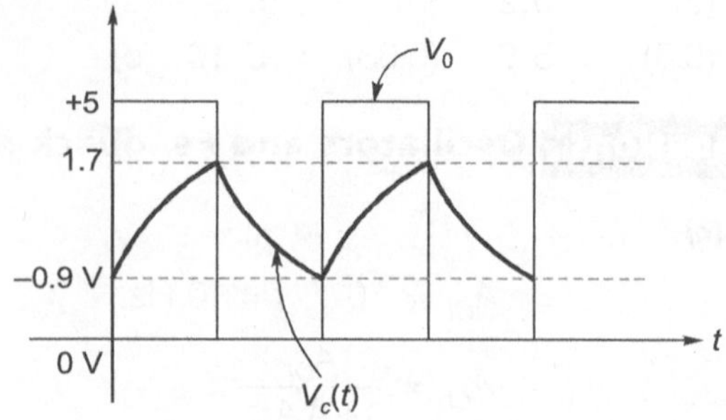

Discharging curve,

$$v_c(t) = 0 - (0 - 1.7)e^{-t/RC}$$

At, $t = T_2$

$$v_c(t) = 0.9 \text{ V}$$

$$0.9 = 1.7\, e^{-t/RC}$$

$$0.63 = T_2 \Rightarrow T_2 = 0.63 \text{ ms}$$

5.10 (c)

"CCCS" [Current Controlled Current Source Amplifier]

Given, $Z_0 = 100 \text{ k}\Omega$

Loop gain, $A\beta = 9$

$$Z_{0F} = Z_0[1 + A\beta]$$

[High impedance CS]

$$= 100 \text{ k}\Omega\,[1 + 9]$$

$$= 100 \text{ k}\Omega \times 10 = 1000 \text{ k}\Omega$$

6 Miscellaneous

6.1 The circuit of figure shows a 555 Timer IC connected as an astable multivibrator. The value of the capacitor C is 10 nF. The values of the resistors R_A and R_B for a frequency of 10 kHz and a duty cycle of 0.75 for the output voltage waveform are

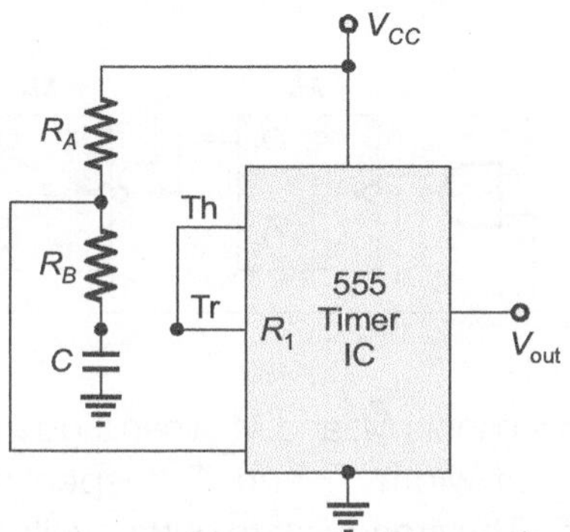

(a) R_A = 3.62 kΩ, R_B = 3.62 kΩ
(b) R_A = 3.62 kΩ, R_B = 7.25 kΩ
(c) R_A = 7.25 kΩ, R_B = 3.62 kΩ
(d) R_A = 7.25 kΩ, R_B = 7.25 kΩ [2003 : 2 M]

6.2* In the Schmitt trigger circuit shown in the figure, if $V_{CE(sat)}$ = 0.1 V, the output logic low level (V_{OL}) is

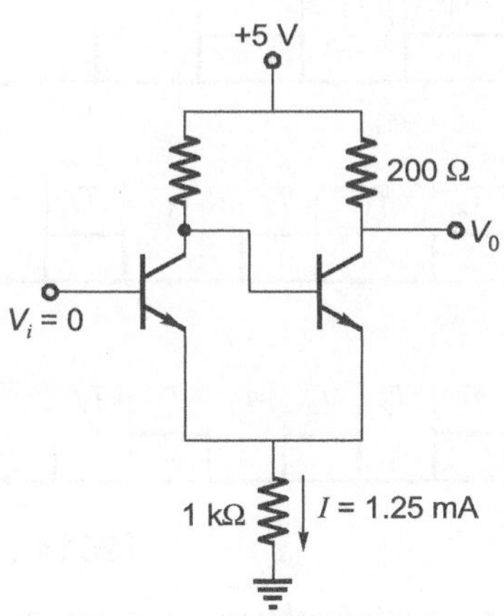

(a) 1.25 V (b) 1.35 V
(c) 2.50 V (d) 5.00 V [2004 : 2 M]

6.3* The input signal V_{in} shown in the figure is a 1 kHz square wave voltage that alternates between +7 V and –7 V with a 50% duty cycle. Both transistors have the same current gain, which is large. The circuit delivers power to the load resistor R_L. What is the efficiency of this circuit for the given input? Choose the closest answer.

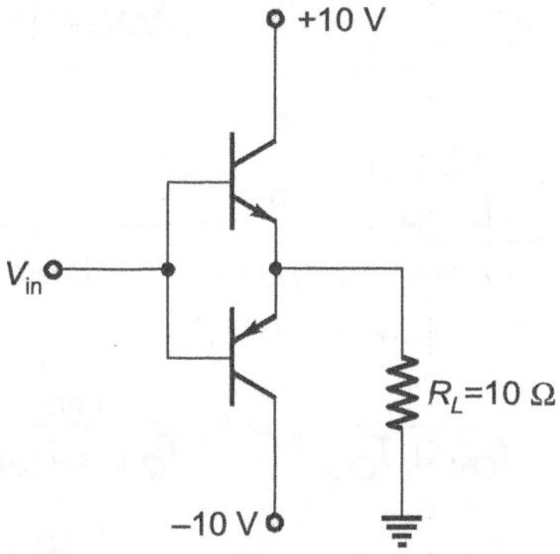

(a) 46% (b) 55%
(c) 63% (d) 92% [2007 : 2 M]

6.4 IC 555 in the figure is configured as an astable multivibrator. It is enabled to oscillate at t = 0 by applying a high input to pin 4. The pin description is: 1 and 8 – supply; 2-trigger; 4-reset; 6-threshold 7-discharge. The waveform appearing across the capacitor starting from t = 0, as observed on a storage CRO is

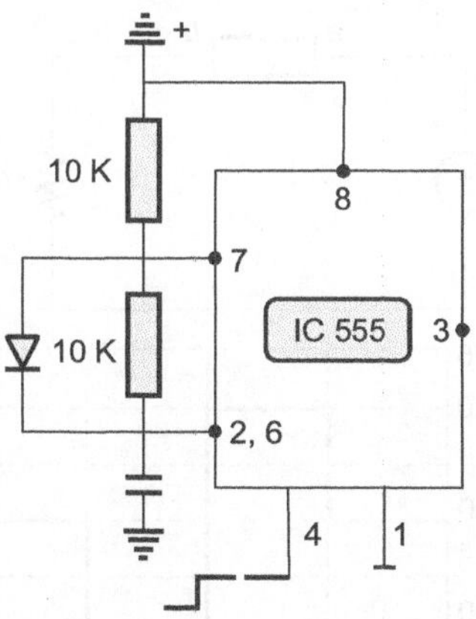

(a)
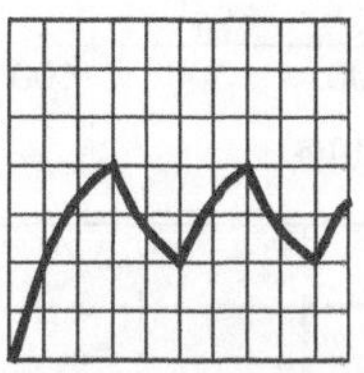
(b)
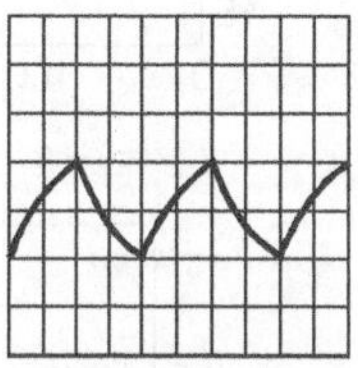
(c)
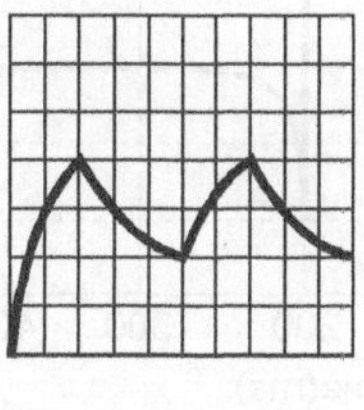
(d)
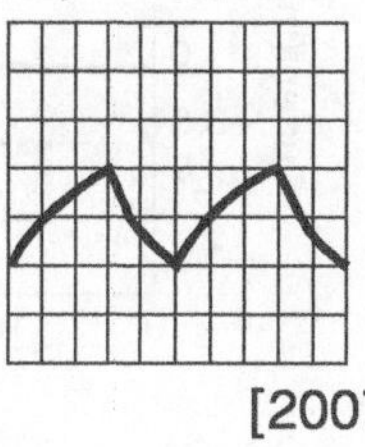

[2007 : 2 M]

6.5 The truth table of a monoshot shown in the figures is given in the table below.
Two monoshots, one positive edge triggered and other negative edge triggered, are connected shown in the figure. The pulse widths of the two monoshot outputs, Q_1 and Q_2 are and T_{ON1} and T_{ON2} respectively.

X	Y	Q	$\bar{Q}$
0	↑	⎍	⊔
↓	1	⎍	⊔

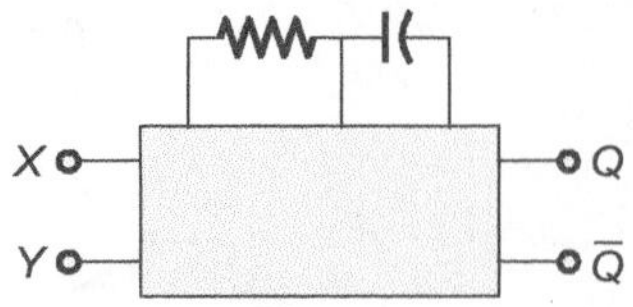

The frequency and the duty cycle of the signal at Q_1 will respectively be

(a) $f = \dfrac{1}{T_{ON_1} + T_{ON_2}}$, $D = \dfrac{T_{ON1}}{T_{ON_1} + T_{ON_2}}$

(b) $f = \dfrac{1}{T_{ON_1} + T_{ON_2}}$, $D = \dfrac{T_{ON2}}{T_{ON_1} + T_{ON_2}}$

(c) $f = \dfrac{1}{T_{ON_1}}$, $D = \dfrac{T_{ON_1}}{T_{ON_1} + T_{ON_2}}$

(d) $f = \dfrac{1}{T_{ON2}}$, $D = \dfrac{T_{ON1}}{T_{ON_1} + T_{ON_2}}$ [2008 : 2 M]

6.6 The following circuit has a source voltage V_s as shown in the graph. The current through the circuit is also shown.

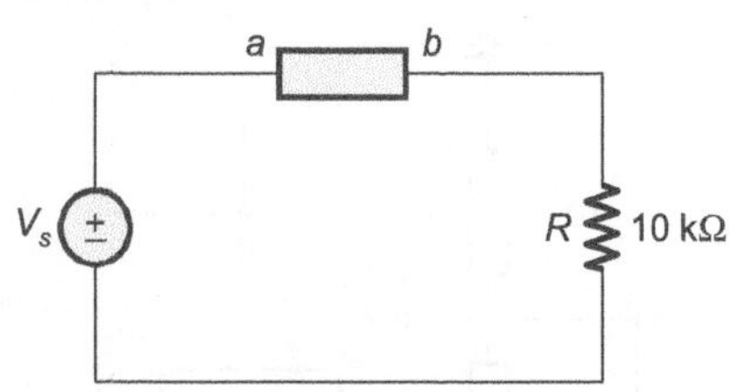

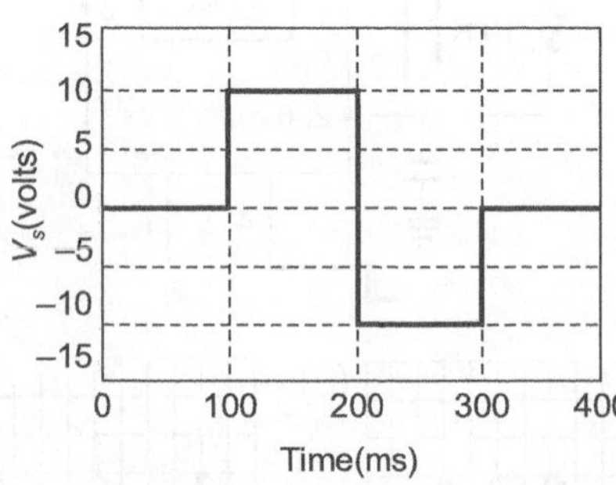

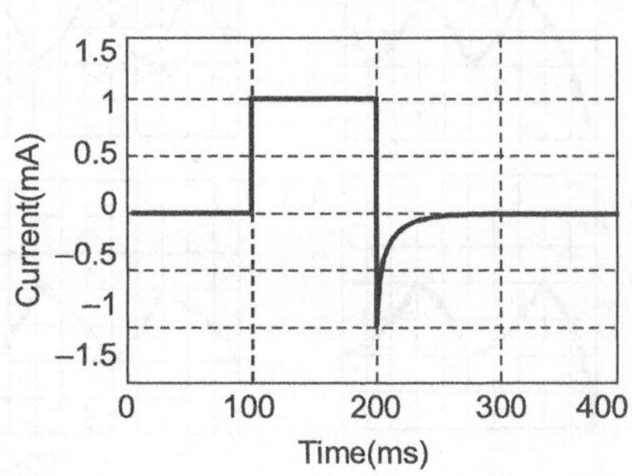

The element connected between a and b could be

(a) 

(b) 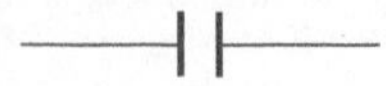

(c) 

(d) 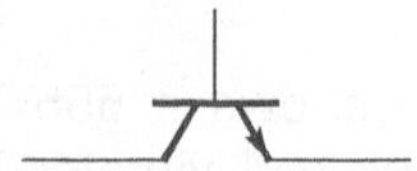

[2009 : 1 M]

6.7 Two monoshot multivibrators, one positive edge triggered (M_1) and another negative edge triggered (M_2), are connected as shown in figure.

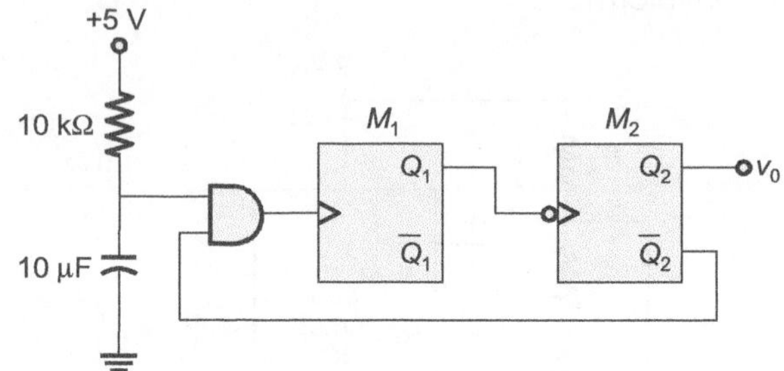

The monoshots M_1 and M_2 when triggered produce pulses of width T_1 and T_2 respectively, where $T_1 > T_2$. The steady state output voltage V_0 of the circuit is

(a) 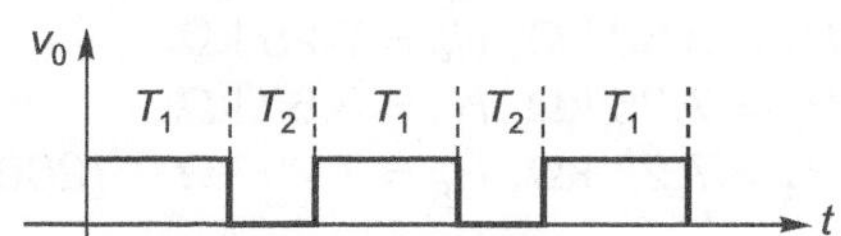

(b) 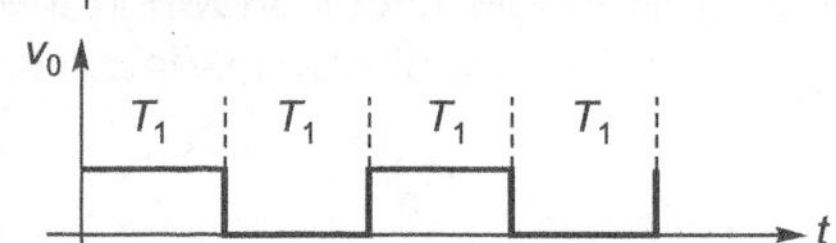

(c) 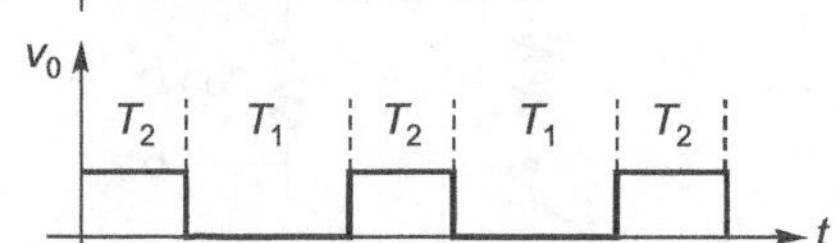

(d) 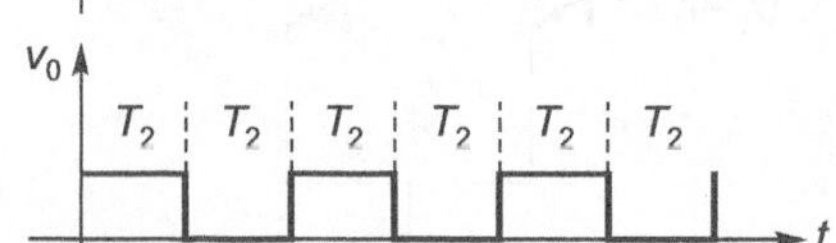

[2014 : 2 M, Set-3]

6.8 The cross-section of a metal-oxide-semiconductor structure is shown schematically. Starting from an uncharged condition, a bias of +3 V is applied to the gate contact with respect to the body contact. The charge inside the silicon dioxide layer is then measured to be $+Q$. The total charge contained within the dashed box shown, upon application of bias, expressed as a multiple of Q (absolute value in Coulombs, rounded off to the nearest integer) is ________.

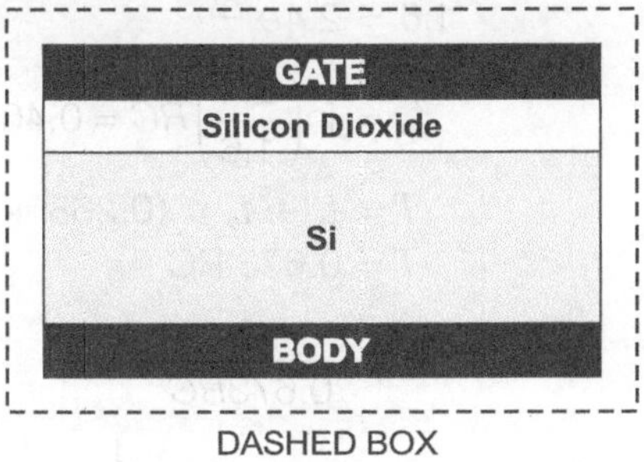

[2020 : 1 M]

6.9 A CMOS Schmitt-trigger inverter has a low output level of 0 V and a high output level of 5 V. It has input thresholds of 1.6 V and 2.4 V. The input capacitance and output resistance of the Schmitt-trigger are negligible. The frequency of the oscillator shown is _______ Hz. (Round off to 2 decimal places.)

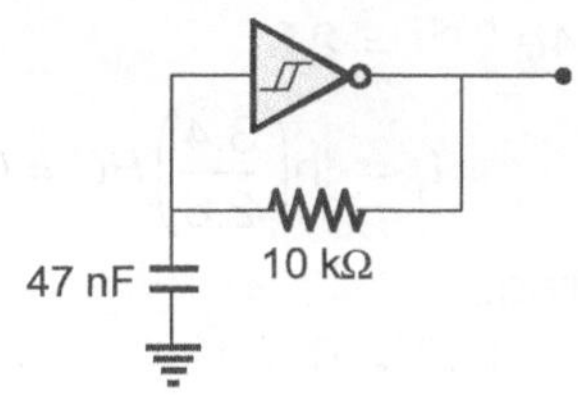

[2021 : 2 M]

Answers Miscellaneous

6.1 (c)	6.2 (b)	6.3 (d)	6.4 (a)	6.5 (a)	6.6 (a)	6.7 (c)
6.8 (0)	6.9 (3157.56)					

Explanations Miscellaneous

6.1 (c)

$$\text{Duty cycle} = \alpha = \frac{T_{ON}}{T}$$

$$\therefore \quad T_{ON} = \alpha T = .75 \times 10^{-4} = 75\ \mu\text{sec}$$

$$\frac{75 \times 10^{-6}}{0.7 \times 10^{-8}} = R_A + R_B$$

$$\therefore \quad R_A + R_B = 10.714\ \text{k}\Omega$$

$$\therefore \quad T_{OFF} = 0.7\ CR_B$$

$$\therefore \quad R_B = \frac{25 \times 10^{-6}}{0.7 \times 10^{-8}}$$

$$R_B = 3.57\ \text{k}\Omega,\ R_A = 7.14\ \text{k}\Omega$$

6.2 (b)

$V_i = 0$, then first transistor will be cut-off and current through left resistor will drive the second transistor into saturation.

Then, $V_0 = V_{CE,Sat} + 1.25 \times 10^{-3} \times 10^3 = 1.35$ V

6.3 (d)

It is a class-D amplifier, so η should be high η of class-D amplifier is 90% to 100%.

6.4 (a)

An astable multi-vibrator is providing pulses as given below:

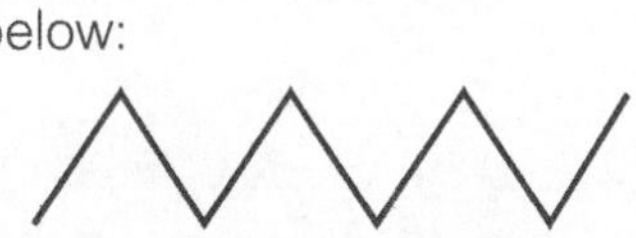

But in this case initial voltage at capacitor is zero so it starts from zero also charging time will be larger (normally) than discharging time but it is made equal by using a diode.

6.5 (a)

$$f = \frac{1}{T_{ON1} + T_{ON2}},\ D = \frac{T_{ON1}}{T_{ON1} + T_{ON2}}$$

6.6 (a)

Diode acts as a switch. When forward biased it is short circuited. But when suddenly reverse biased, current does not become zero instantly, initially the same current flow in opposite direction and after some time (turn off time) it will become zero.

6.8 Sol.

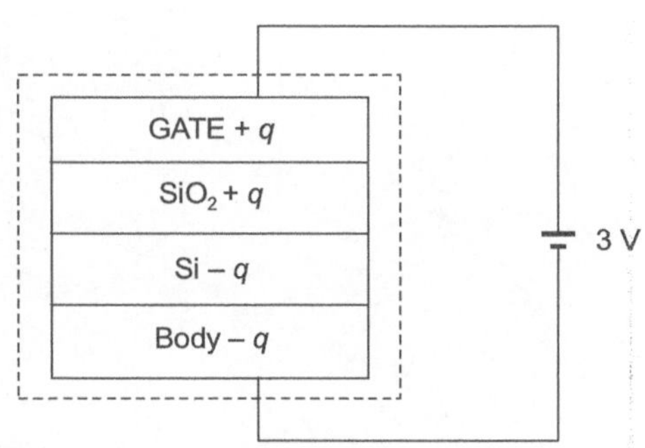

Overall charge in side the box $q + q - q - q = 0$ charge

6.9 (3157.56)

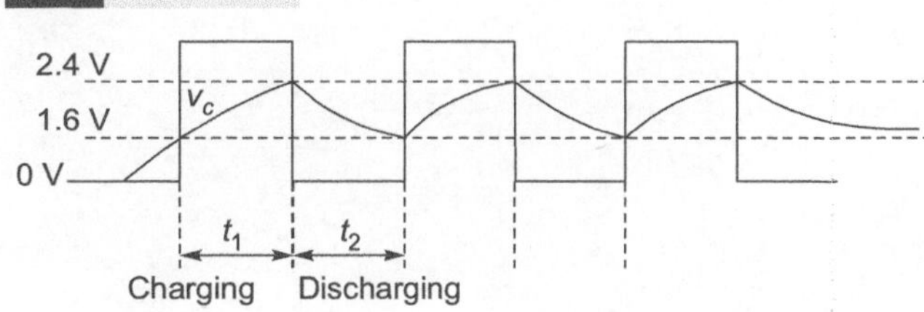

$$v_c(t) = v_{c\,\text{final}} + [v_{\text{initial}} - v_{c'\text{final}}]\, e^{-t/RC}$$

Charging, $v_c(t) = 5 + (1.6 - 5)e^{-t/RC}$

$$= 5 - 3.4\, e^{-t/RC}$$

$t = t_1$, $v_c(t) = 2.4$ V

$$2.4 = 5 - 3.4\, e^{-t/RC}$$

$$3.4e^{-t1/RC} = 2.6$$

$$t_1 = \ln\left(\frac{3.4}{2.6}\right) RC = 0.268 \times RC$$

Discharging,

$$v_c(t) = 0 + (2.4 - 0)\, e^{-t/RC}$$

$$= 2.4e^{-t/RC}$$

$t = t_2$, $v_c(t_a) = 1.6$ V

$$1.6 = 2.4e^{-t2/RC}$$

$$t_2 = \ln\left(\frac{2.4}{1.6}\right) RC = 0.405 \times RC$$

$$T = t_1 + t_2 = (0.268 + 0.405)\text{RC}$$

$$T = 0.673 \text{ RC}$$

$$f = \frac{1}{0.673RC}$$

$$= \frac{1}{0.673 \times 10^4 \times 47 \times 10^{-9}}$$

$$f = 3157.46 \text{ Hz}$$

■■■■

Digital Electronics

UNIT

VIII

CONTENTS

Digital Electronics

Syllabus

Combinatorial and Sequential Logic Circuits, Multiplexers, Demultiplexers, Schmitt Triggers, Sample and Hold Circuits, A/D and D/A Converters.

Analysis of Previous GATE Papers

Exam Year	1 Mark Ques.	2 Marks Ques.	3 Marks Ques.	5 Marks Ques.	Total Marks
1995	2	–	–	–	**2**
1996	1	–	1	–	**4**
1997	1	1	–	–	**8**
1998	2	1	–	–	**9**
1999	–	4	–	–	**8**
2000	1	2	–	–	**5**
2001	3	1	–	1	**10**
2002	3	1	–	2	**15**
2003	2	5	–	–	**12**
2004	2	4	–	–	**10**
2005	2	4	–	–	**10**
2006	–	5	–	–	**10**
2007	1	1	–	–	**3**
2008	–	3	–	–	**6**
2009	2	1	–	–	**4**
2010	–	4	–	–	**8**
2011	1	2	–	–	**5**

Exam Year	1 Mark Ques.	2 Mark Ques.	Total Marks
2012	2	1	**4**
2013	1	2	**5**
2014 Set-1	1	2	**5**
2014 Set-2	1	3	**7**
2014 Set-3	2	3	**8**
2015 Set-1	1	3	**7**
2015 Set-2	1	2	**5**
2016 Set-1	2	2	**6**
2016 Set-2	1	1	**3**
2017 Set-1	1	2	**5**
2017 Set-2	–	1	**2**
2018	1	2	**5**
2019	–	2	**4**
2020	1	1	**3**
2021	1	1	**3**
2022	2	–	**2**
2023	–	3	**6**
2024	3	–	**3**

1 Boolean Algebra & Minimization

1.1 The logic function $f=\overline{(x\cdot\bar{y})+(\bar{x}\cdot y)}$ is the same as

(a) $f=(x+y)(\bar{x}+\bar{y})$

(b) $f=\overline{(\bar{x}+\bar{y})(x+y)}$

(c) $f=\overline{(x\cdot y)}(\bar{x}\cdot\bar{y})$

(d) None of the above [1999 : 2 M]

1.2 The minimal product-of-sums function described by the K-map given in the figure, is

C \ AB	00	01	11	10
0	1	1	0	0
1	0	0	0	0

(a) $A'C'$ (b) $A'+C'$

(c) $A+C$ (d) AC [2000 : 2 M]

1.3 The boolean expression $\bar{X}Y\bar{Z}+\bar{X}\bar{Y}Z+XY\bar{Z}+X\bar{Y}Z+XYZ$ can be simplified to

(a) $X\bar{Z}+\bar{X}Z+YZ$ (b) $XY+\bar{Y}Z+Y\bar{Z}$

(c) $\bar{X}Y+YZ+XZ$ (d) $\overline{XY}+Y\bar{Z}+\bar{X}Z$

[2003 : 2 M]

1.4 The simplified form of the Boolean expression $Y=(\bar{A}BC+D)(\bar{A}D+\bar{B}\bar{C})$ can be written as

(a) $\bar{A}D+\bar{B}\bar{C}D$

(b) $AD+B\bar{C}D$

(c) $(\bar{A}+D)(\bar{B}C+\bar{D})$

(d) $A\bar{D}+BC\bar{D}$ [2004 : 2 M]

Statement for Linked Answer Questions (1.5 and 1.6):

The following Karnaugh map represents a function F.

X \ YZ (F)	00	01	11	10
0	1	1	1	0
1	0	0	1	0

1.5 A minimized form of the function F is

(a) $F=\bar{X}Y+YZ$ (b) $F=\bar{X}\bar{Y}+YZ$

(c) $F=\bar{X}\bar{Y}+Y\bar{Z}$ (d) $F=\bar{X}\bar{Y}+\bar{Y}Z$

[2010 : 2 M]

1.6 Which of the following circuits is a realization of the above function F?

(a)

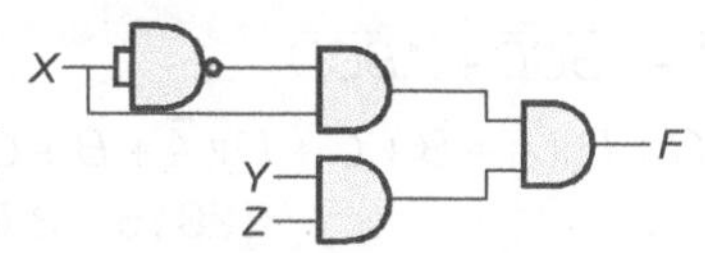

(b)

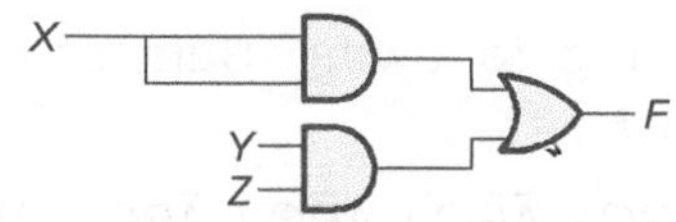

(c)

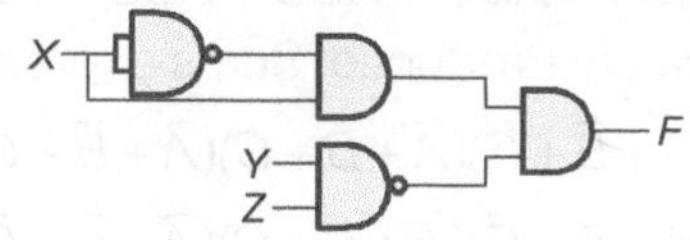

(d)

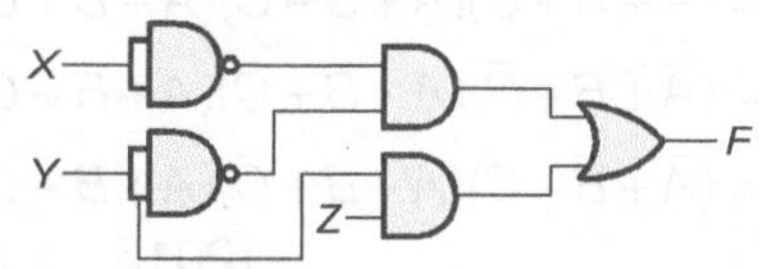

[2010 : 2 M]

1.7 In the sum of products function:

$$f(X, Y, Z)=\Sigma(2, 3, 4, 5)$$

the prime implicants are

(a) $\bar{X}Y+X\bar{Y}$

(b) $\bar{X}Y+X\bar{Y}\bar{Z}+X\bar{Y}Z$

(c) $\bar{X}Y\bar{Z}+\bar{X}YZ+X\bar{Y}$

(d) $\bar{X}Y\bar{Z}+\bar{X}YZ+X\bar{Y}\bar{Z}+X\bar{Y}Z$

[2012 : 1 M]

1.8 Which of the following is an invalid state in 8-4-2-1 Binary Coded Decimal counter

(a) 1 0 0 0 (b) 1 0 0 1

(c) 0 0 1 1 (d) 1 1 0 0

[2014 : 1 M, Set-2]

1.9 The SOP (sum of products) form of a Boolean function is $\Sigma(0, 1, 3, 7, 11)$, where inputs are A, B, C, D (A is MSB, and D is LSB). The equivalent minimized expression of the function is

(a) $(\bar{B}+C)(\bar{A}+C)(\bar{A}+\bar{B})(\bar{C}+D)$

(b) $(\bar{B}+C)(\bar{A}+C)(\bar{A}+\bar{C})(\bar{C}+D)$

(c) $(\bar{B}+C)(\bar{A}+C)(\bar{A}+\bar{C})(\bar{C}+\bar{D})$

(d) $(\bar{B}+C)(A+\bar{B})(\bar{A}+\bar{B})(\bar{C}+D)$

[2014 : 2 M, Set-2]

1.10 $f(A, B, C, D) = \Pi M(0, 1, 3, 4, 5, 7, 9, 11, 12, 13, 14, 15)$ is a maxterm representation of Boolean function $f(A, B, C, D)$ where A is the MSB and D is the LSB. The equivalent minimized representation of this function is

(a) $(A+\bar{C}+D)(\bar{A}+B+D)$

(b) $A\bar{C}D+\bar{A}BD$

(c) $\bar{A}C\bar{D}+A\bar{B}C\bar{D}+A\bar{B}\bar{C}\bar{D}$

(d) $(B+\bar{C}+D)(A+\bar{B}+\bar{C}+D)(\bar{A}+B+C+D)$

[2015 : 2 M, Set-1]

1.11 Consider the following Sum of Products expression, F.

$$F = ABC+\bar{A}\bar{B}C+A\bar{B}C+\bar{A}BC+\bar{A}\bar{B}\bar{C}$$

The equivalent Product of Sums expression is

(a) $F=(A+\bar{B}+C)(\bar{A}+B+C)(\bar{A}+\bar{B}+C)$

(b) $F=(A+\bar{B}+\bar{C})(A+B+C)(\bar{A}+\bar{B}+\bar{C})$

(c) $F=(\bar{A}+B+\bar{C})(A+\bar{B}+\bar{C})(A+B+C)$

(d) $F=(\bar{A}+\bar{B}+C)(A+B+\bar{C})(A+B+C)$

[2015 : 1 M, Set-2]

1.12 The output expression for the Karnaugh map shown below is

A \ BC	00	01	11	10
0	1	0	0	1
1	1	1	1	1

(a) $A+\bar{B}$ (b) $A+\bar{C}$

(c) $\bar{A}+\bar{C}$ (d) $\bar{A}+C$

[2016 : 1 M, Set-2]

1.13 The Boolean expression:

$$\overline{(a+\bar{b}+c+\bar{d})+(b+\bar{c})}$$

simplifies is

(a) 1   (b) $\overline{a\cdot b}$

(c) a, b (d) 0 [2016 : 2 M, Set-2]

1.14 The Boolean expression, $AB+A\bar{C}+BC$ simplifies to

(a) $BC+A\bar{C}$ (b) $AB+A\bar{C}+B$

(c) $AB+A\bar{C}$ (d) $AB+BC$

[2017 : 1 M, Set-1]

1.15 The output expression for the Karnaugh map shown below is

AB \ CD	00	01	11	10
00	0	0	0	0
01	1	0	0	1
11	1	0	1	1
10	0	0	0	0

(a) $B\bar{D}+BCD$ (b) $B\bar{D}+AB$

(c) $\bar{B}D+ABC$ (d) $B\bar{D}+ABC$

[2017 : 2 M, Set-1]

1.16 Digital input signals A, B, C with A as the MSB and C as the LSB are used to realize the Boolean function, $F = m_0 + m_2 + m_3 + m_5 + m_7$, where, m_i denotes the ith minterm. In addition, F has a don't care for m_1. The simplified expression for F is given by

(a) $\bar{A}\bar{C}+\bar{B}C+AC$ (b) $\bar{A}+C$

(c) $\bar{C}+A$ (d) $\bar{A}C+BC+A\bar{C}$

[2018 : 2 M]

1.17 The output expression for the Karnaugh map shown below is

RS \ PQ	00	01	11	10
00	0	1	1	0
01	1	1	1	1
11	1	1	1	1
10	0	0	0	0

(a) $QR+S$ (b) $Q\bar{R}+S$

(c) $Q\bar{R}+\bar{S}$ (d) $QR+\bar{S}$ [2019 : 2 M]

1.18 Simplified form of the Boolean function :

$F(P, Q, R, S) = \bar{P}\bar{Q}+\bar{P}QS+P\bar{Q}\bar{R}\bar{S}+P\bar{Q}R\bar{S}$

is

(a) $\bar{P}\bar{Q}+\bar{Q}\bar{S}$ (b) $P\bar{S}+Q\bar{R}$

(c) $\bar{P}Q+R\bar{S}$ (d) $\bar{P}S+\bar{Q}\bar{S}$

[2024 : 1 M]

■■■■

Answers Boolean Algebra & Minimization

1.1 (b)	1.2 (a)	1.3 (b)	1.4 (a)	1.5 (b)	1.6 (d)	1.7 (a)
1.8 (d)	1.9 (a)	1.10 (c)	1.11 (a)	1.12 (b)	1.13 (d)	1.14 (a)
1.15 (d)	1.16 (b)	1.17 (b)	1.18 (d)			

Explanations Boolean Algebra & Minimization

1.1 (b)

Given,

$$f = \overline{(x\cdot\overline{y})+(\overline{x}\cdot y)} = \overline{(x\cdot\overline{y})}\cdot\overline{(\overline{x}\cdot y)}$$

$$= (\overline{x}+y)\cdot(x+\overline{y}) = xy+\overline{x}\,\overline{y}$$

Option (b): $\overline{(\overline{x}+\overline{y})\,(x+y)} = \overline{(\overline{x}+\overline{y})}+\overline{(x+y)}$

$$= xy+\overline{x}\,\overline{y} = f$$

1.2 (a)

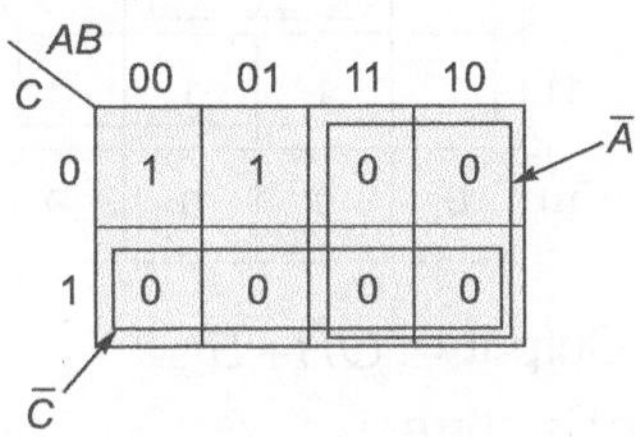

1.3 (b)

By K-map

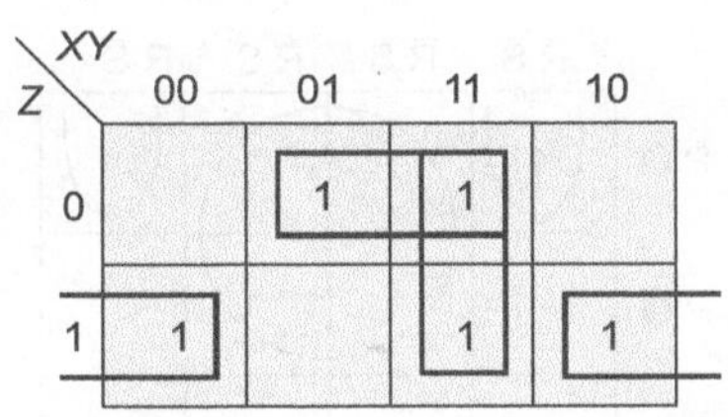

The simplified form is $XY + Y\overline{Z} + \overline{Y}Z$

1.4 (a)

$$Y = (\overline{A}BC + D)(\overline{A}D + \overline{B}\,\overline{C})$$

$$= \overline{A}BCD + \overline{A}D + \overline{B}\,\overline{C}D$$

$$= (\overline{A}D + \overline{B}\,\overline{C}D) \quad (A + 1 = 1)$$

1.5 (b)

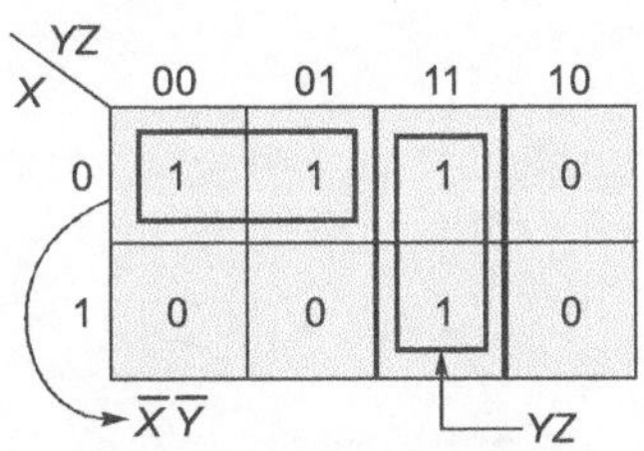

$$F = \overline{X}\,\overline{Y} + YZ$$

1.6 (d)

From the figure it is clear that, two NAND gates generate the $\overline{X}$ and $\overline{Y}$ and now two AND gates with i/ps $\overline{X}$ and $\overline{Y}$ and inputs Y and Z is used to generate two terms of SOP form and now OR gate is used to sum them and generate the F.

1.7 (a)

$$f(X, Y, Z) = \Sigma(2, 3, 4, 5)$$

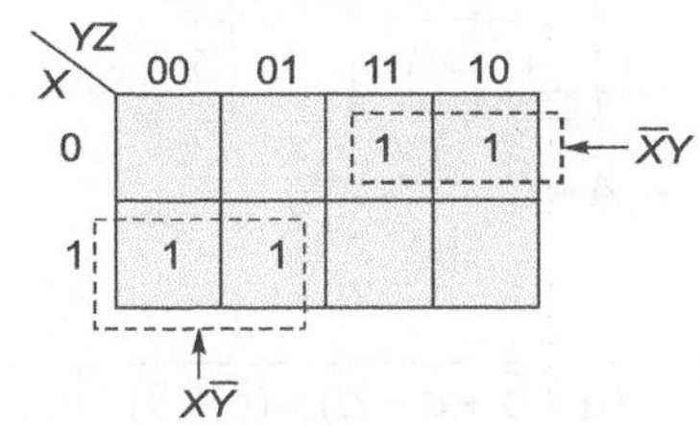

$$\therefore \quad f(X, Y, Z) = X\overline{Y} + \overline{X}Y$$

1.8 (d)

Binary coded decimal counter counts from 0 to 9. So, 1100 is an invalid state i.e. 12.

1.9 (a)

The 4 variable Boolean function is given in canonical sum of products form as,

$$f(A, B, C, D) = \Sigma(0, 1, 3, 7, 11)$$

As the options are given in the simplified product of sums form, we first convert the given function in canonical product of sums form, as under

$f(A, B, C, D) = \Pi(2, 4, 5, 6, 8, 9, 10, 12, 13, 14, 15)$

Now by plotting the above function on a 4 variable K-map (maxterms map), we obtain the simplified expression of the function

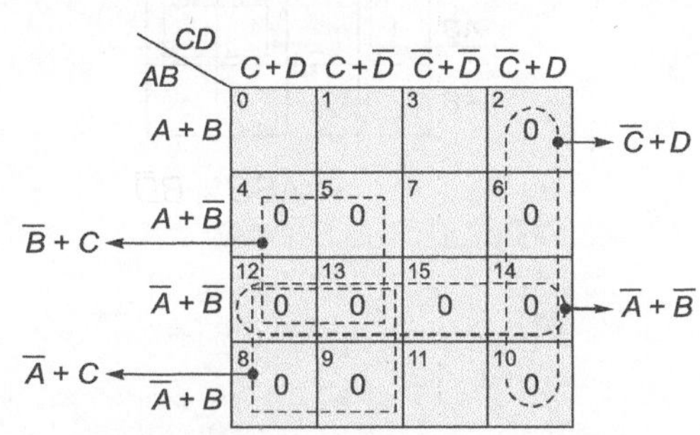

$$f = (\overline{B}+C)\cdot(\overline{A}+C)\cdot(\overline{A}+\overline{B})\cdot(\overline{C}+D)$$

1.11 (a)

The SOP form of F is (shown in K-map)

A \ BC	00	01	11	10
0	1	1	1	0
1	0	1	1	0

So, POS form can be formed using '0' from the K-map.

$$\text{POS} = (\bar{A} + B + C)(\bar{A} + \bar{B} + C)(A + \bar{B} + C)$$

1.12 (b)

A \ BC	00	01	11	10
0	1	0	0	1
1	1	1	1	1

$$F = A + \bar{C}$$

1.13 (d)

$$F = \overline{(a+\bar{b}+c+\bar{d})+(b+\bar{c})}$$

$$= \overline{(a+\bar{b}+c+\bar{d})}\cdot\overline{(b+\bar{c})}$$

$$= \bar{a}\cdot b\cdot \bar{c}\cdot d\cdot \bar{b}\cdot c$$

$$F = 0$$

1.14 (a)

	$\bar{B}\bar{C}$	$\bar{B}C$	BC	$B\bar{C}$
$\bar{A}$	(0)	(1)	1 (3)	(2)
A	1 (4)	(5)	1 (7)	1 (6)

$$A\bar{C} + BC$$

$$BC + A\bar{C}$$

1.15 (d)

	$\bar{C}\bar{D}$	$\bar{C}D$	CD	$C\bar{D}$
$\bar{A}\bar{B}$				
$\bar{A}B$	1			1
AB	1		1	1
$A\bar{B}$				

$$ABC + B\bar{D}$$

1.16 (b)

Given, $f = m_0 + m_2 + m_3 + m_5 + m_7$

and m_1 = don't care

	$\bar{B}\bar{C}$	$\bar{B}C$	BC	$B\bar{C}$	
$\bar{A}$	1 (0)	X (1)	1 (3)	1 (2)	$\bar{A} + C$
A	(4)	1 (5)	1 (7)	(6)	$f = \bar{A} + C$

1.17 (b)

RS \ PQ	00	01	11	10
00	0	1	1	0
01	1	1	1	1
11	1	1	1	1
10	0	0	0	0

$$\text{Output} = Q\bar{R} + S$$

Option (b) is correct.

1.18 (d)

	$\bar{R}\bar{S}$	$\bar{R}S$	RS	$R\bar{S}$
$\bar{P}\bar{Q}$	1	1	1	1
$\bar{P}Q$		1	1	
PQ				
$P\bar{Q}$	1			1

$$\text{Minimize function} = \bar{Q}\bar{S} + \bar{P}S$$

2 Logic Gates

2.1 The Boolean expression of the output of the logic circuit shown in figure is

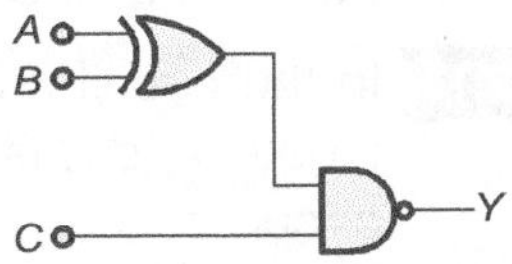

(a) $Y = \overline{A}\overline{B} + AB + \overline{C}$

(b) $Y = \overline{A}\overline{B} + AB + C$

(c) $Y = \overline{A}B + \overline{A}B + C$

(d) $Y = \overline{A}B + \overline{A}\overline{B} + \overline{C}$ [1996 : 1 M]

2.2 The output of a logic gate is "1" when all its inputs are at logic "0". The gate is either

(a) a NAND or an EX-OR gate

(b) a NOR or an EX-OR gate

(c) an AND or an EX-NOR gate

(d) a NOR or an EX-NOR gate [2001 : 1 M]

2.3 For the circuit shown in the figure the Boolean expression for the output *Y* in terms of inputs *P*, *Q*, *R* and *S* is

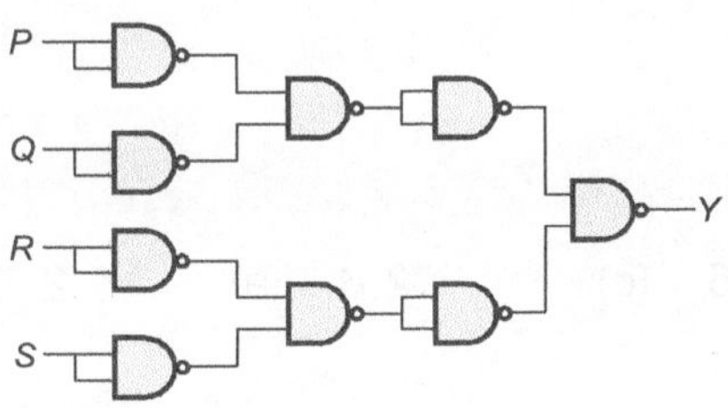

(a) $\overline{P} + \overline{Q} + \overline{R} + \overline{S}$ (b) $P + Q + R + S$

(c) $(\overline{P} + \overline{Q}) + (\overline{R} + \overline{S})$ (d) $(P + Q)(R + S)$

[2002 : 2 M]

2.4 *A*, *B*, *C*, and *D* are input, and *Y* is the output bit in the XOR gate circuit of the figure below. Which of the following statements about the sum *S* of *A*, *B*, *C*, *D* and *Y* is correct?

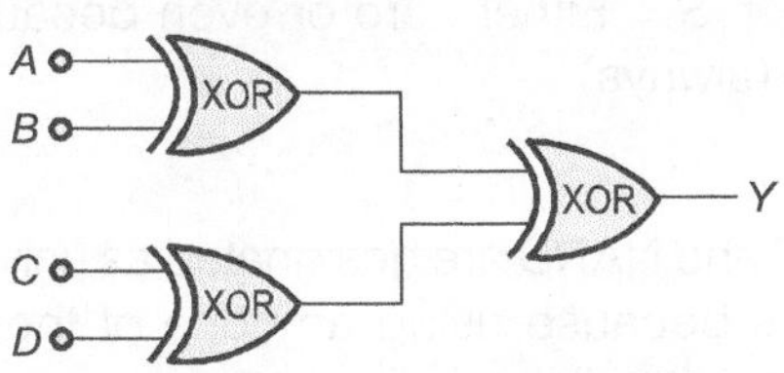

(a) *S* is always either zero or odd.

(b) *S* is always either zero or even.

(c) *S* = 1 only if the sum of *A*, *B*, *C* and *D* is even.

(d) *S* = 1 only if the sun of *A*, *B*, *C* and *D* is odd.

[2007 : 1 M]

2.5 The complete set of only those Logic Gates designated as Universal Gates is

(a) NOT, OR and AND Gates

(b) XNOR, NOR and NAND Gates

(c) NOR and NAND Gates

(d) XOR, NOR and NAND Gates [2009 : 1 M]

2.6 The output *Y* of the logic circuit given below is

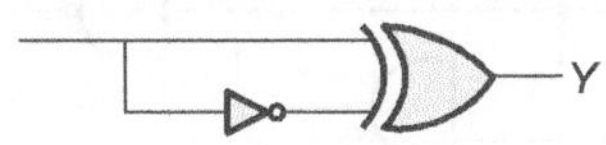

(a) 1 (b) 0

(c) x (d) $\overline{x}$ [2011 : 1 M]

2.7 A bulb in a staircase has two switches, one switch being at the ground floor and the other one at the first floor. The bulb can be turned ON and also can be turned OFF by any one of the switches irrespective of the state of the other switch. The logic of switching of the bulb resembles

(a) an AND gate (b) an OR gate

(c) an XOR gate (d) a NAND gate

[2013 : 1 M]

2.8 Which of the following logic circuits is a realization of the function *F* whose Karnaugh map is shown in the figure?

C \ AB	00	01	11	10
0	1	1		
1		1	1	

(a)

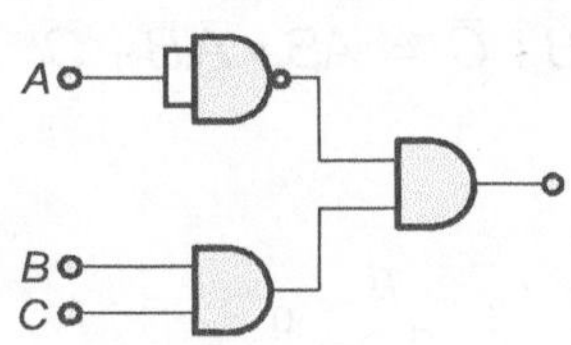

(b)

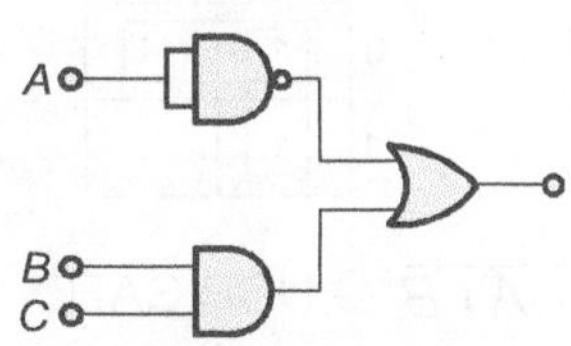

(c)

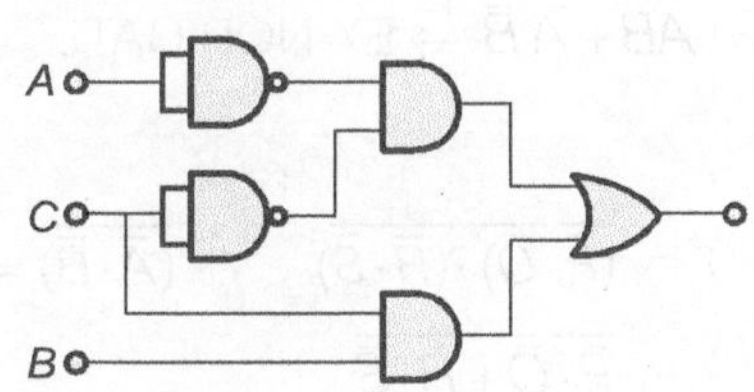

(d) 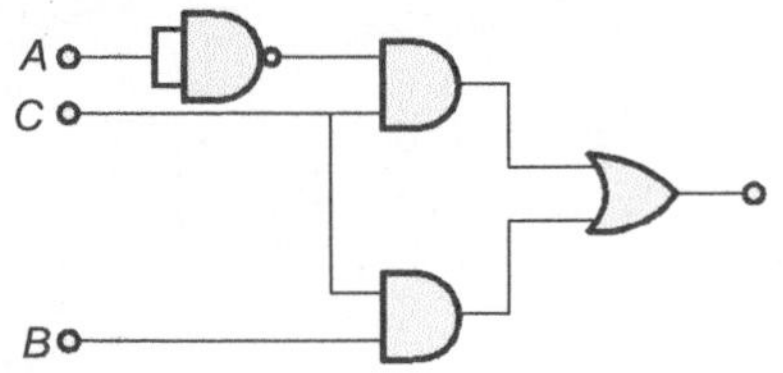

[2014 : 2 M, Set-1]

2.9 For a 3-input logic circuit shown below, the output Z can be expressed as

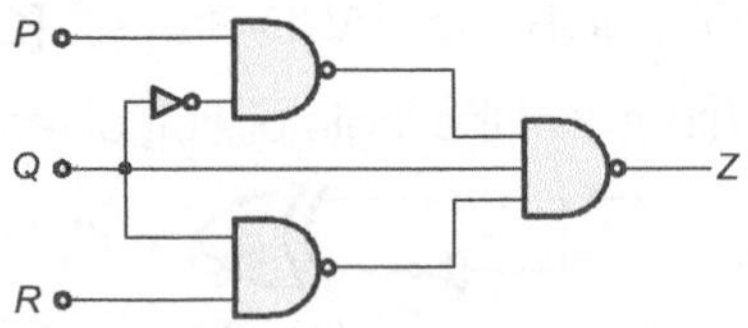

(a) $Q+\bar{R}$ (b) $P\bar{Q}+R$

(c) $\bar{Q}+R$ (d) $P+\bar{Q}+R$

[2017 : 1 M, Set-2]

2.10 In the logic circuit shown in the figure, Y is given by

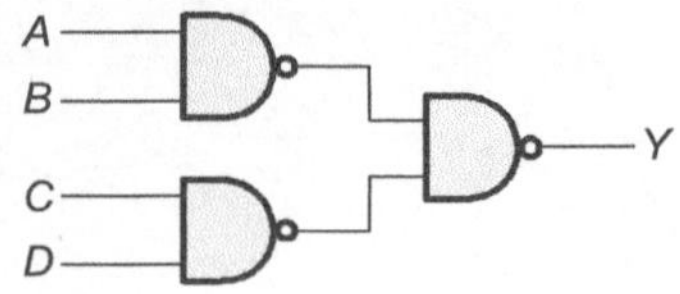

(a) $Y = ABCD$

(b) $Y = (A+B)(C+D)$

(c) $Y = A+B+C+D$

(d) $Y = AB+CD$ [2018 : 1 M, Set-2]

2.11 In the circuit shown below, X and Y are digital inputs, and Z is a digital output. The equivalent circuit is

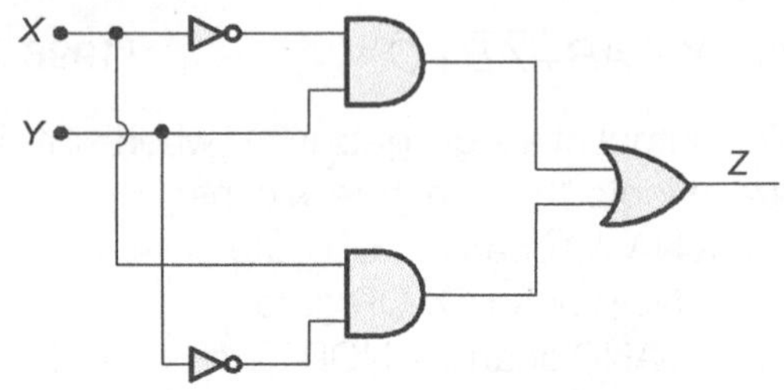

(a) XNOR gate (b) NOR gate

(c) NAND gate (d) XOR gate

[2019 : 2 M]

Answers Logic Gates

2.1 (a)	2.2 (d)	2.3 (b)	2.4 (b)	2.5 (c)	2.6 (a)	2.7 (c)
2.8 (c)	2.9 (c)	2.10 (d)	2.11 (d)			

Explanations Logic Gates

2.1 (a)

$$Y = \overline{(A\oplus B)\cdot C} = \overline{A\oplus B}+\bar{C}$$

$$= A\odot B+\bar{C} = AB+\bar{A}\bar{B}+\bar{C}$$

2.2 (d)

A	B	Y
0	0	1
0	1	X
1	0	X
1	1	X

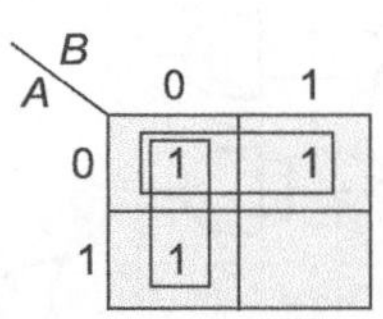

$Y = \bar{A}\bar{B} = \overline{A+B} \rightarrow$ NOR GATE

$Y = AB+\bar{A}\bar{B} \rightarrow$ EX-NOR GATE

2.3 (b)

$$Y = \overline{(\overline{P}\cdot\overline{Q})\cdot(\overline{R}\cdot\overline{S})} \quad \because \overline{(\bar{A}\cdot\bar{B})} = (A+B)$$

$$\therefore \quad Y = \overline{\overline{P}\cdot\overline{Q}}+\overline{\overline{R}\cdot\overline{S}}$$

$$\Rightarrow \quad Y = (P+Q+R+S)$$

2.4 (b)

$Y = A\oplus B\oplus C\oplus D$ from the given diagram.

We know that sum of any no. of bits is XOR of all bits.

So, $S = A\oplus B\oplus C\oplus D\oplus Y$

$S = Y\oplus Y$

$S =$ Either zero or even because LSB is zero (always)

2.5 (c)

NOR and NAND are designated as universal logic gates because using any one of them we can implement all the logic gates.

2.6 (a)

x	$\bar{x}$	y
1	0	1
0	1	1

$$Y = x\bar{\bar{x}}+\bar{x}\,\bar{x} = 1$$

2.7 (c)

Truth table of XOR gate is,

A	B	Y
0	0	0
0	1	1
1	0	1
1	1	0

So, from the XOR gate truth table it is clear that the bulb can be turned ON and also can be turned OFF by any one of the switches irrespective of the state of the other switch.

2.8 (c)

C \ AB	00	01	11	10
0	1	1		
1		1	1	

$$F = \bar{A}\bar{C} + BC$$

So, option (c) is correct.

2.9 (c)

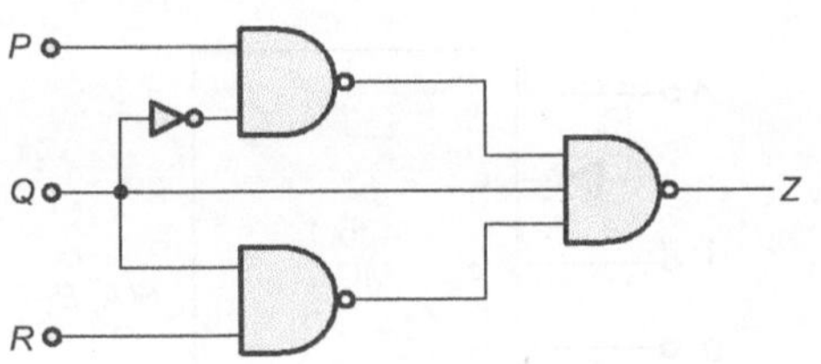

$$Z = \overline{\overline{P\bar{Q}}\cdot Q\cdot\overline{Q\cdot R}} = P\bar{Q}+\bar{Q}+\overline{\overline{Q\cdot R}}$$

$$= P\bar{Q}+\bar{Q}+QR = \bar{Q}(P+1)+QR$$

$$= \bar{Q}+QR$$

$$= (\bar{Q}+Q)\cdot(\bar{Q}+R) = \bar{Q}+R$$

2.10 (d)

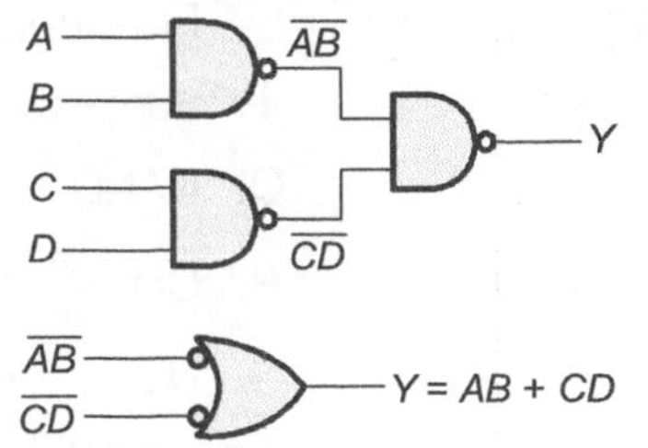

2.11 (d)

The Boolean expression for the output of the digital circuit is shown below.

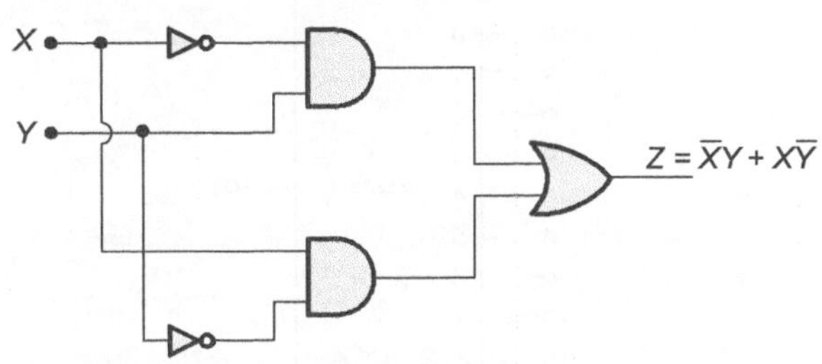

The above expression is of XOR gate.

3 Combinational Logic Circuits

3.1 Match the following:

	Logic		Function
A.	$\overline{X}+\overline{Y}$	P.	Sum
B.	XY	Q.	NAND
C.	$\overline{X}\,\overline{Y}$	R.	Carry
		S.	NOR

[1998 : 2 M]

3.2 The logic function $F = AC + ABD + ACD$ is to be realized using an 8 to 1 multiplexer shown in the figure, using A, C and D as control inputs.

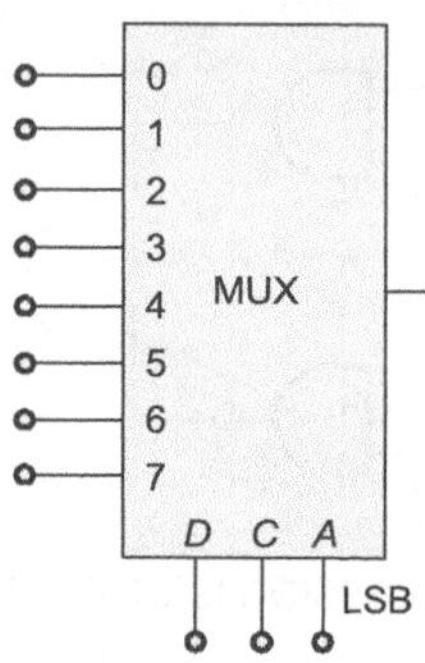

(a) Indicate the inputs to the applied at the terminals 0 to 7.
(b) Can the function be realize using a 4 to 1 multiplexer? [1999 : 2 M]

3.3 The output F of the 4-to-1 MUX shown in the figure is

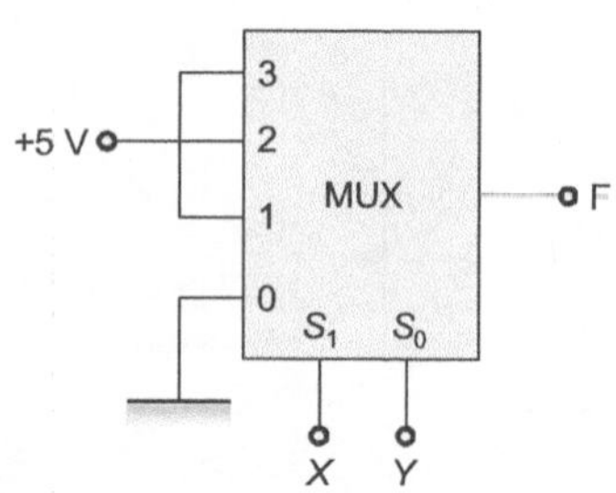

(a) $\overline{x\,y}+x$
(b) $x + y$
(c) $\overline{x}+\overline{y}$
(d) $xy+\overline{x}$ [2001 : 1 M]

3.4 The figure shows a 4 to 1 MUX to be used to implement the sum S of a 1-bit full adder with input bits P and Q and the carry input C_{in}. Which of the following combinations of inputs to I_0, I_1, I_2 and I_3 of the MUX will realize the sum S?

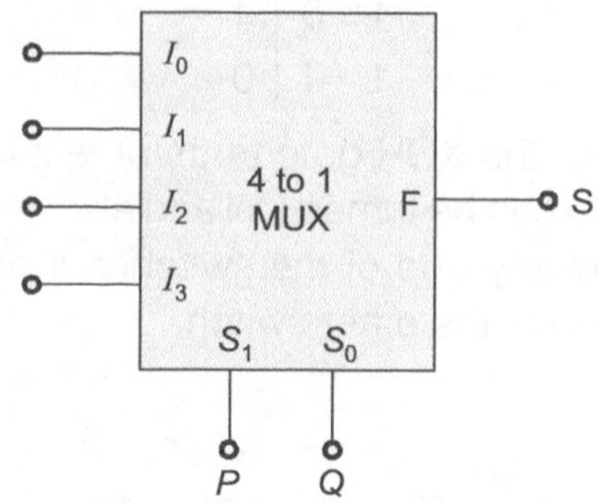

(a) $I_0 = I_1 = C_{in}$; $I_2 = I_3 = \overline{C}_{in}$
(b) $I_0 = I_1 = \overline{C}_{in}$; $I_2 = I_3 = C_{in}$
(c) $I_0 = I_3 = C_{in}$; $I_1 = I_2 = \overline{C}_{in}$
(d) $I_0 = I_3 = \overline{C}_{in}$; $I_1 = I_2 = C_{in}$ [2003 : 1 M]

3.5 A (4 × 1) MUX is used to implement a 3-input Boolean function as shown in the figure. The Boolean function $F(A, B, C)$ implemented is

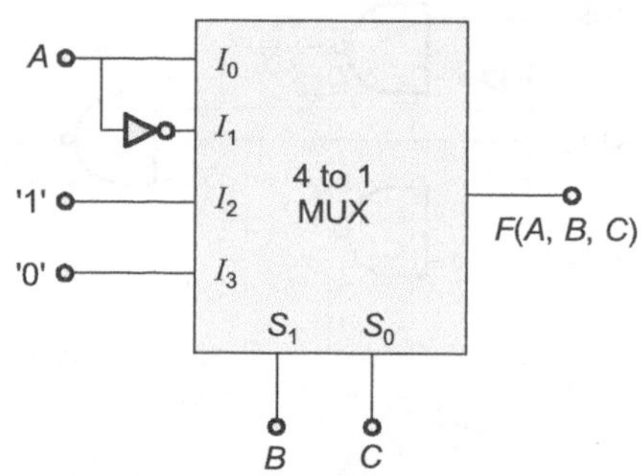

(a) $F(A, B, C) = \Sigma(1, 2, 4, 6)$
(b) $F(A, B, C) = \Sigma(1, 2, 6)$
(c) $F(A, B, C) = \Sigma(2, 4, 5, 6)$
(d) $F(A, B, C) = \Sigma(1, 5, 6)$ [2006 : 2 M]

3.6 A 3 line to 8 line decoder, with active low outputs, is used to implement a 3-variable Boolean function as shown in the figure.

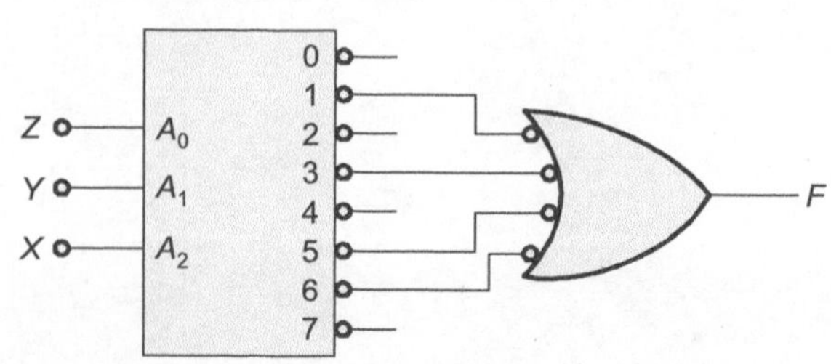

The simplified form of Boolean function $F(A, B, C)$ implemented in 'Product of Sum' form will be

(a) $(X+Z)\cdot(\bar{X}+\bar{Y}+\bar{Z})\cdot(Y+Z)$

(b) $(\bar{X}+\bar{Z})\cdot(X+Y+Z)\cdot(\bar{Y}+\bar{Z})$

(c) $(\bar{X}+\bar{Y}+Z)\cdot(\bar{X}+Y+Z)\cdot(X+\bar{Y}+Z)\cdot(X+Y+\bar{Z})$

(d) $(\bar{X}+\bar{Y}+Z)\cdot(\bar{X}+Y+\bar{Z})\cdot(X+\bar{Y}+Z)\cdot(X+\bar{Y}+\bar{Z})$

[2008 : 2 M]

3.7 The output Y of a 2-bit comparator is logic 1 whenever the 2-bit input A is greater than the 2-bit input B. The number of combinations for which the output is logic 1, is

(a) 4 (b) 6
(c) 8 (d) 10 [2012 : 1 M]

3.8 A 3-bit gray counter is used to control the output of the multiplexer as shown in the figure. The initial state of the counter is 000_2. The output is pulled high. The output of the circuit follows the sequence

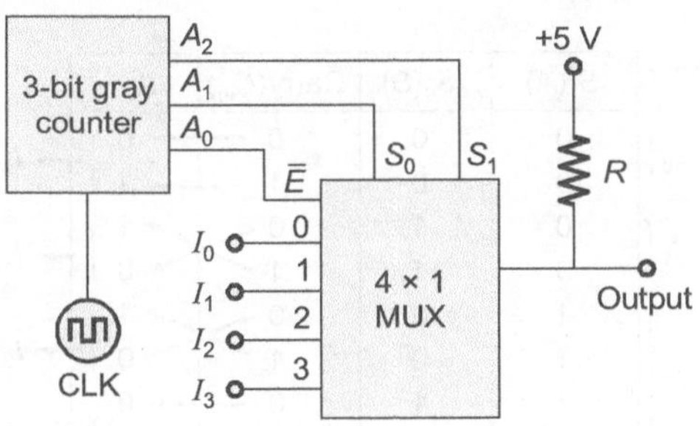

(a) I_0, 1, 1, I_1, I_3, 1, 1, I_2
(b) I_0, 1, I_1, 1, I_2, 1, I_2, 1
(c) 1, I_0, 1, I_1, I_2, 1, I_3, 1
(d) I_0, I_1, I_2, I_3, I_0, I_1, I_2, I_3 [2014 : 2 M, Set-3]

3.9 In the 4 × 1 multiplexer, the output F is given by $F = A \otimes B$. Find the required input '$I_3\, I_2\, I_1\, I_0$'

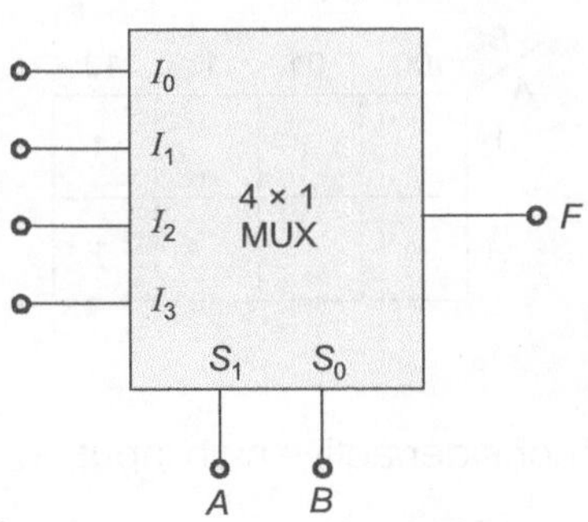

(a) 1010 (b) 0110
(c) 1000 (d) 1110

[2015 : 1 M, Set-1]

3.10 A Boolean function $f(A, B, C, D) = \Pi(1, 5, 12, 15)$ is to be implemented using an 8 × 1 multiplexer (A is MSB). The inputs ABC are connected to the select input $S_2\ S_1\ S_0$ of the multiplexer respectively.

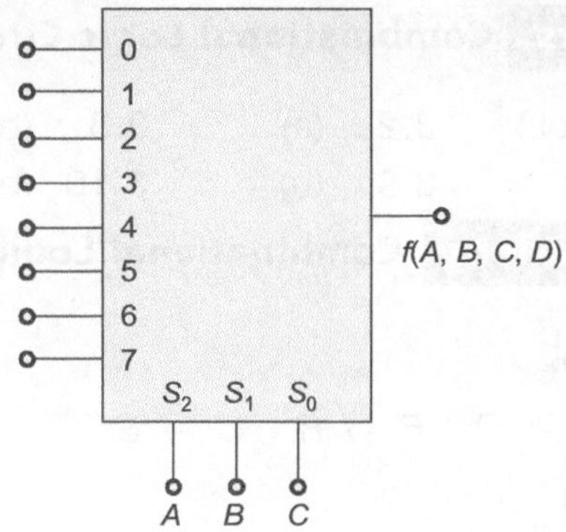

Which one of the following options gives the correct inputs to pins 0, 1, 2, 3, 4, 5, 6, 7 in order?

(a) $D, 0, D, 0, 0, 0, \bar{D}, D$

(b) $\bar{D}, 1, \bar{D}, 1, 1, 1, D, \bar{D}$

(c) $D, 1, D, 1, 1, 1, \bar{D}, D$

(d) $\bar{D}, 0, \bar{D}, 0, 0, 0, D, \bar{D}$ [2015 : 2 M, Set-2]

3.11 Consider the following circuit which uses a 2-to-1 multiplexer as shown in the figure below. The Boolean expression for output F in terms of A and B is?

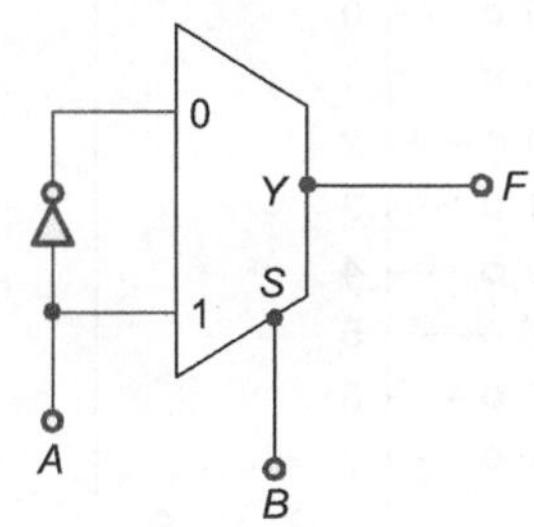

(a) $A \oplus B$ (b) $\overline{A+B}$
(c) $A + B$ (d) $\overline{A \oplus B}$

[2016 : 1 M, Set-1]

3.12 To obtain the Boolean function $F(X, Y) = X\bar{Y} + \bar{X}$, the inputs $PQRS$ in the figure should be

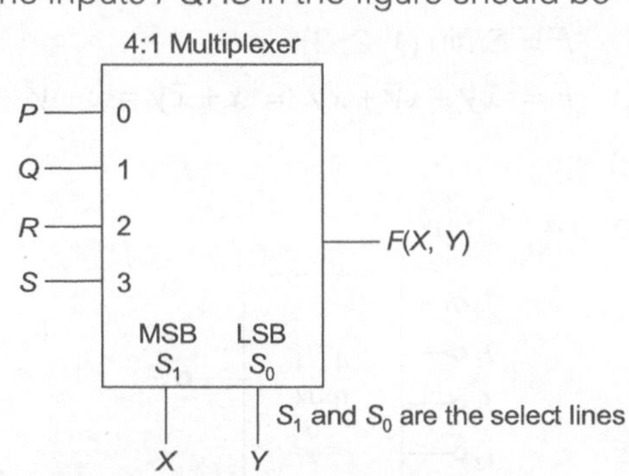

(a) 1110 (b) 1010
(c) 0110 (d) 0001

[2024 : 1 Mark]

Answers Combinational Logic Circuits

3.1	(Sol.)	3.2	(a)	3.3	(b)	3.4	(c)	3.5	(a)	3.6	(a)	3.7	(b)
3.8	(a)	3.9	(b)	3.10	(b)	3.11	(d)	3.12	(a)				

Explanations Combinational Logic Circuits

3.1 Sol.

$A \rightarrow Q, \quad B \rightarrow R, \quad C \rightarrow S.$

3.2 (a)

$$F = AC + ABD + ACD$$
$$= ACB + AC\bar{B} + ABDC + ABD\bar{C} + ACD\bar{B}$$
$$= ABC\bar{D} + A\bar{B}C\bar{D} + A\bar{B}CD + ABCD + AB\bar{C}D$$
$$= AC\bar{D}(\bar{B}+B) + ACD(B+\bar{B}) + A\bar{C}DB$$
$$= \underbrace{\bar{D}CA}_{3^{rd}\text{ line}}(B+\bar{B}) + \underbrace{DCA}_{7^{th}\text{ line}}(B+\bar{B}) + \underbrace{D\bar{C}A}_{5^{th}\text{ line}}B$$
$$= \bar{D}CA(1) + DCA(1) + D\bar{C}A\cdot(B)$$

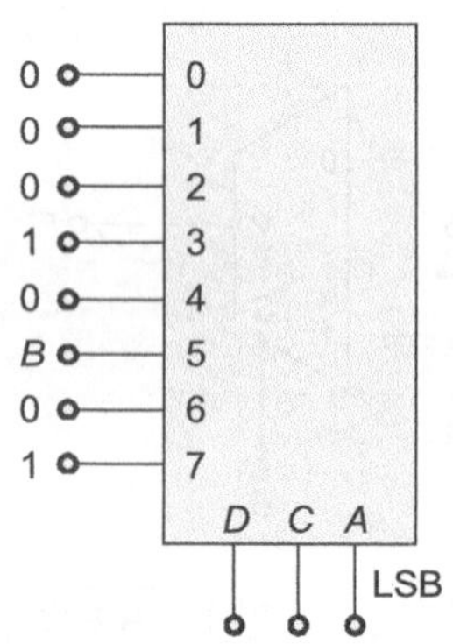

(b) It can not be realized with a 4 to 1 multiplexer as 5th and 7th lines are being used.

3.3 (b)

$\because \quad F = \Sigma\text{min}(1, 2, 3)$

$\therefore \quad F = \bar{x}y + x\bar{y} + xy = x + \bar{x}y = x + y$

3.4 (c)

For a 4 : 1 mux

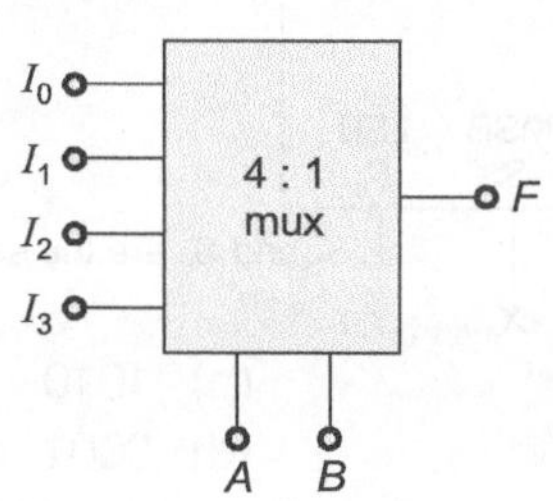

$$F = I_0\bar{A}\bar{B} + I_1\bar{A}B + I_2A\bar{B} + I_3AB$$

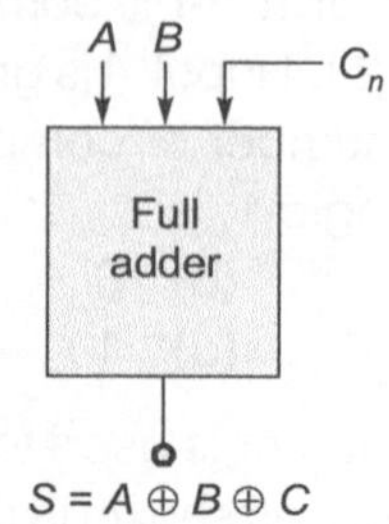

Where sum of full adder is $= A \oplus B \oplus C$

Truth table of Full adder

	$S_1(A)$	$S_0(B)$	Carry(C_{in})	Sum	
I_0	0	0	0	0	$I_0 = C_{in}$
	0	0	1	1	
I_1	0	1	0	1	$I_1 = \bar{C}_{in}$
	0	1	1	0	
I_2	1	0	0	1	$I_2 = \bar{C}_{in}$
	1	0	1	0	
I_3	1	1	0	0	$I_3 = C_{in}$
	1	1	1	1	

3.5 (a)

$$F(A, B, C) = A\bar{B}\bar{C} + \bar{A}\bar{B}C + B\bar{C}$$
$$= \Sigma(1, 2, 4, 6)$$

A \ BC	00	01	11	10
0		1		1
1	1			1

3.6 (a)

Let us consider active high input,

x \ yz	00	01	11	10
0	0	1	1	0
1	0	1	0	1

$$F = \Sigma(1, 3, 5, 6) = \Pi M(0, 2, 4, 7)$$
$$= (Y+Z)\cdot(X+Z)\cdot(\bar{X}+\bar{Y}+\bar{Z})$$

3.7 (b)

A B	No. of cases
01 00	1
10 00, 01	2
11 00, 01, 10	3
	Total = 6

3.8 (a)

For

A_2	A_1	A_0	S_0 (A_1)	S_1 (A_2)
0	0	0	0	0

MUX is enabled and output is I_0

For

A_2	A_1	A_0	S_0	S_1
0	0	1	0	0

MUX is disable and output is '1'
Similarly, for

A_2	A_1	A_0	S_0	S_1	$\bar{E}$ (A_0)	O/P
0	0	0	0	0	0	I_0
0	0	1	0	0	1	1
0	1	1	0	1	1	1
0	1	0	0	1	0	I_1
1	1	0	1	1	0	I_3
1	1	1	1	1	1	1
1	0	1	1	0	1	1
1	0	0	1	0	0	I_2

3.9 (b)

$$F = A \oplus B,\ I_3 I_2 I_1 I_0 = \bar{A}B + A\bar{B}$$

$$= \bar{A}\bar{B}I_0 + \bar{A}BI_1 + A\bar{B}I_2 + ABI_3$$

$$\Rightarrow \quad I_0 = 0, I_1 = 1, I_2 = 1, I_3 = 0$$

$$I_3 I_2 I_1 I_0 = 0110$$

3.10 (b)

Boolean function is $f(A, B, C, D) = \pi(1, 5, 12, 15)$ implemented with 8 × 1 MUX

	I_0	I_1	I_2	I_3	I_4	I_5	I_6	I_7
$\bar{D}$ –	(0)	(2)	(4)	(6)	(8)	(10)	12	(14)
D –	1	(3)	5	(7)	(9)	(11)	(13)	15
	$\bar{D}$	1	$\bar{D}$	1	1	1	D	$\bar{D}$

3.11 (d)

In the given multiplexer,

$$I_0 = \bar{A},\ I_1 = A$$

$$\text{Select} = B$$

$$F = \bar{B}I_0 + BI_1 = \bar{B}\bar{A} + AB = \overline{A \oplus B}$$

3.12 (a)

X	Y	F(X, Y)
0	0	1 (P)
0	1	1 (Q)
1	0	1 (R)
1	1	0 (S)

■■■■

4 Sequential Logic Circuits

4.1 For a J-K flip-flop, J input is tied to its own $\bar{Q}$ output and its K input is connected to its own Q output. If the flip-flop is fed with a clock of frequency 1 MHz, its Q output frequency will be ______. **[1995 : 1 M]**

4.2 For a flip-flop formed two NAND gates as shown in the figure. The unusable state corresponds to

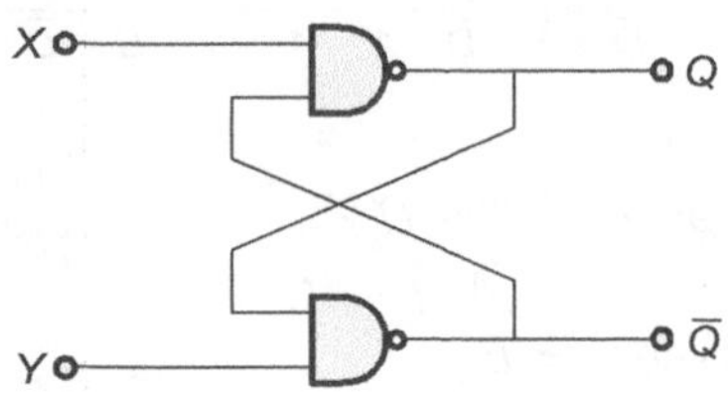

(a) $X = 0, Y = 0$ (b) $X = 0, Y = 1$
(c) $X = 1, Y = 0$ (d) $X = 1, Y = \infty$

[1999 : 2 M]

4.3 The frequency of the clock signal applied to the rising edge triggered D flip-flop shown in the figure is 10 kHz. The frequency of the signal available at Q is

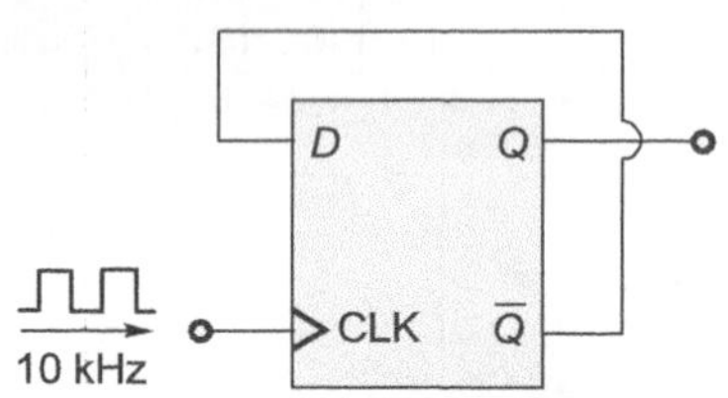

(a) 10 kHz (b) 2.5 kHz
(c) 20 kHz (d) 5 kHz **[2002 : 1 M]**

4.4 The shift register shown in figure is initially loaded with the bit pattern 1010. Subsequently the shift register is clocked, and with each clock pulse the pattern gets shifted by one bit position to the right. With each shift, the bit at the serial input is pushed to the left most position (msb). After how many clock pulses will the content of the shift register become 1010 again?

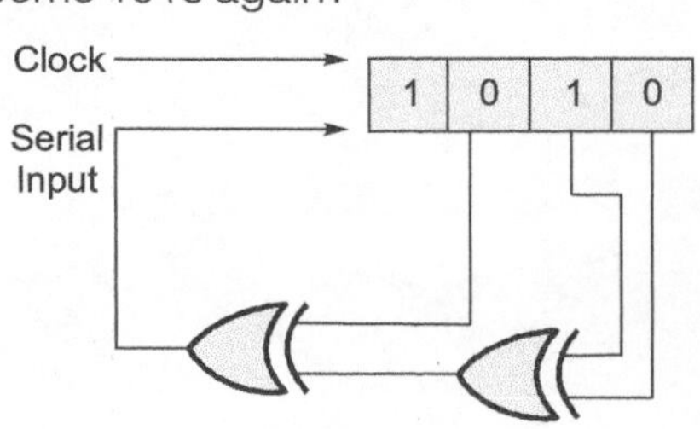

(a) 3 (b) 7
(c) 11 (d) 15 **[2003 : 2 M]**

4.5 An X-Y flip flop, whose Characteristic Table is given below is to be implemented using a J-K flip flop.

X	Y	Q_{n+1}
0	0	1
0	1	Q_n
1	0	$\bar{Q}_n$
1	1	0

This can be done by making

(a) $J = X, K = \bar{Y}$ (b) $J = \bar{X}, K = Y$
(c) $J = Y, K = \bar{X}$ (d) $J = \bar{Y}, K = X$

[2003 : 2 M]

4.6 The digital circuit using two inverters shown in the figure will act as

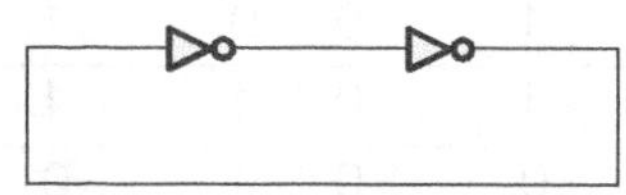

(a) a bistable multi-vibrator
(b) an astable multi-vibrator
(c) a monostable multi-vibrator
(d) an oscillator **[2004 : 1 M]**

4.7 A digital circuit which compares two numbers $A_3 A_2 A_1 A_0$, $B_3 B_2 B_1 B_0$ is shown in figure. To get output $Y = 0$, choose one pair of correct input numbers.

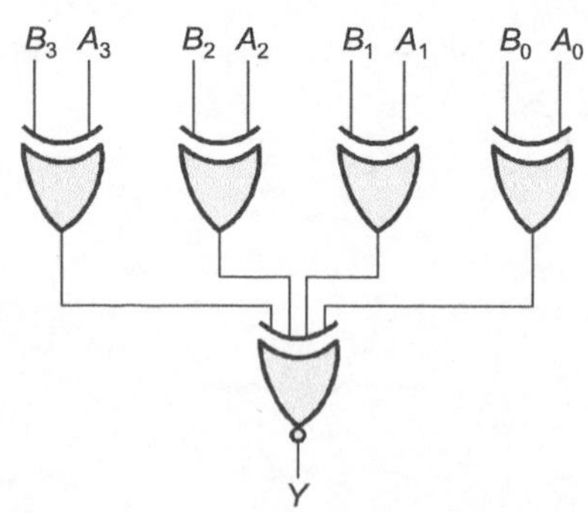

(a) 1010, 1010 (b) 0101, 0101
(c) 0010, 0010 (d) 1010, 1011 **[2004 : 2 M]**

4.8 The digital circuit shown in the figure generates a modified clock pulse at the output. Choose the correct output waveform from the options given below.

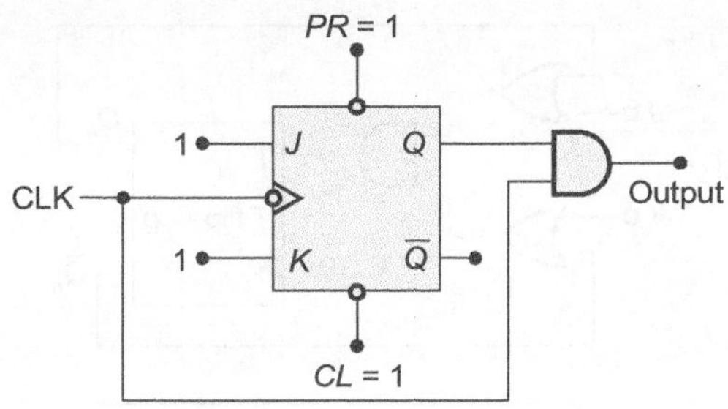

(a) 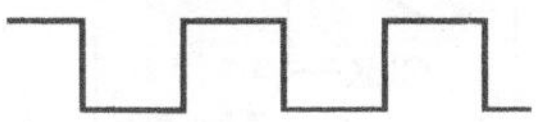

(b) 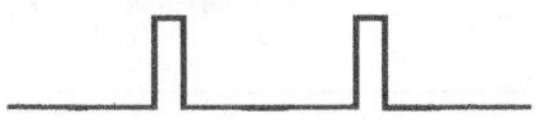

(c) 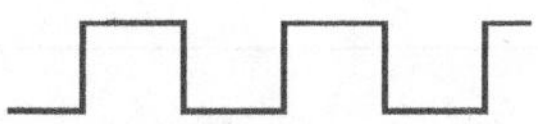

(d)  [2004 : 2 M]

4.9 The digital circuit shown in the figure works as

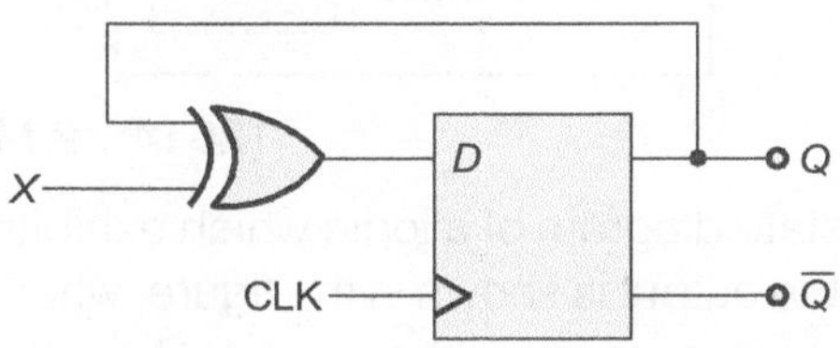

(a) JK flip-flop (b) Clocked RS flip-flop
(c) T flip-flop (d) Ring counter

[2005 : 1 M]

4.10 Select the circuit which will produce the given output Q for the input signals X_1 and X_2 given in the figure.

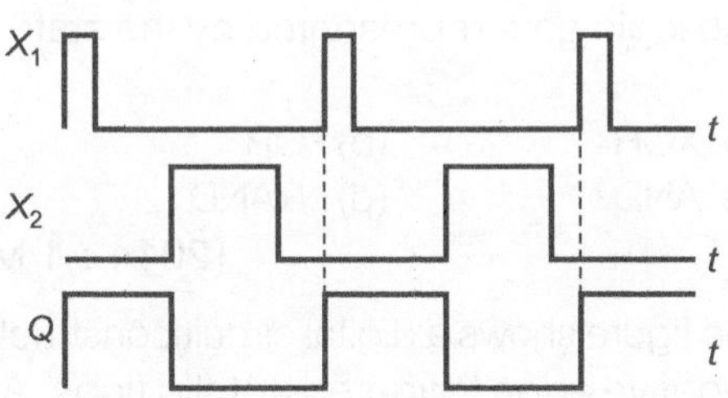

(a) 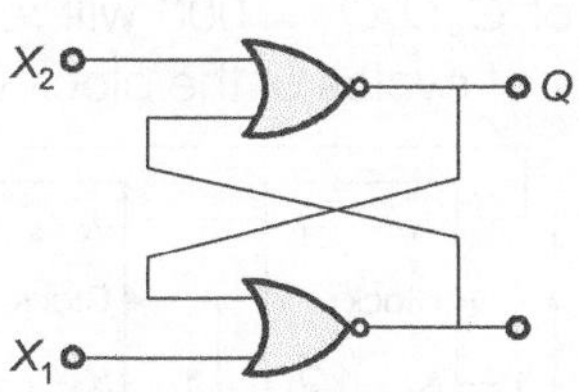

(b) 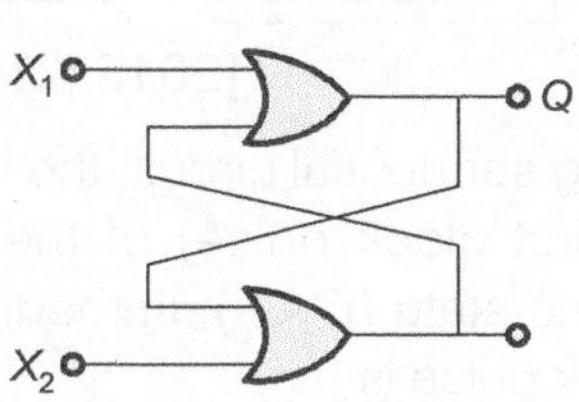

(c) 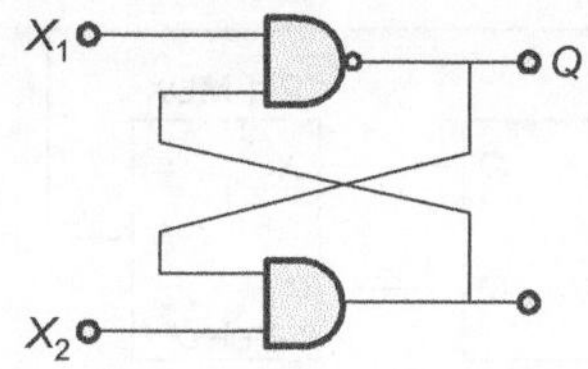

(d) 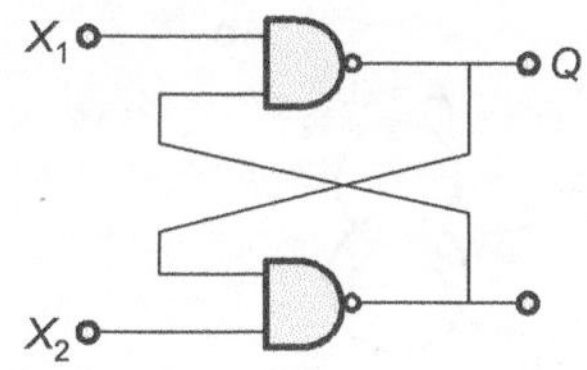

[2005 : 2 M]

4.11 In the figure, as long as $X_1 = 1$ and $X_2 = 1$, the output Q remains

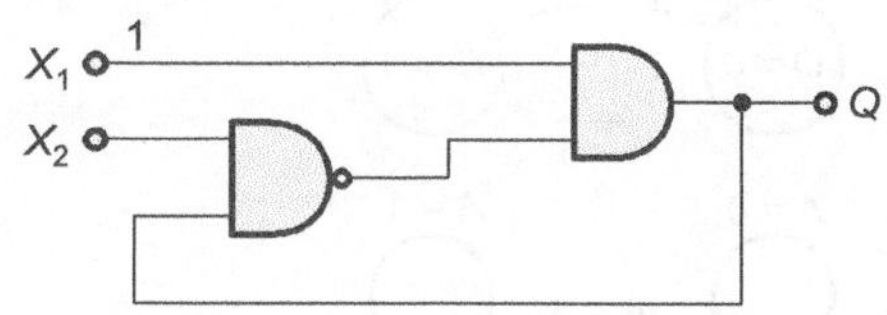

(a) at 1
(b) at 0
(c) at its initial value
(d) unstable

[2005 : 2 M]

4.12 A two-bit counter circuit is shown below.

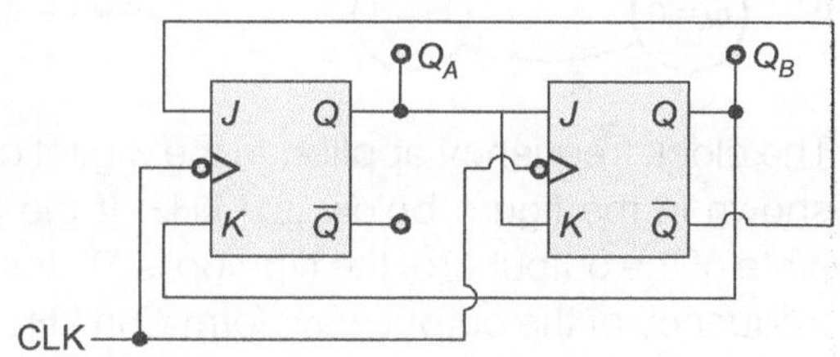

If the state Q_AQ_B of the counter at the clock time t_n is "10" then the state Q_AQ_B of the counter at $t_n + 3$ (after three cycles) will be

(a) 00 (b) 01
(c) 10 (d) 11 [2011 : 2 M]

4.13 Consider the given circuit:

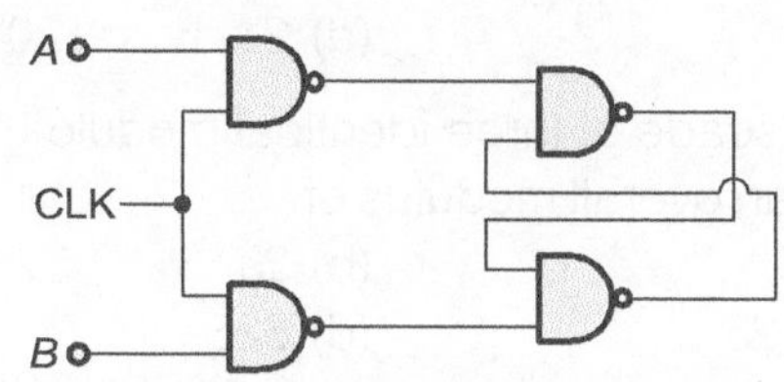

In this circuit, the race around
(a) does not occur
(b) occurs when CLK = 0
(c) occurs when CLK = 1 and $A = B = 1$
(d) occurs when CLK = 1 and $A = B = 0$

[2012 : 1 M]

4.14 The state transition diagram for the logic circuit shown is

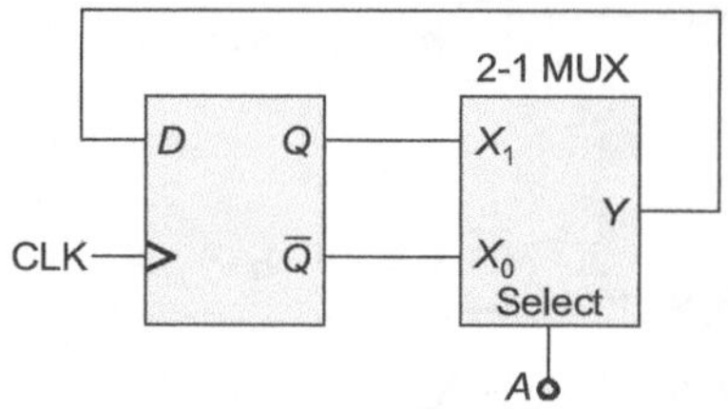

(a) 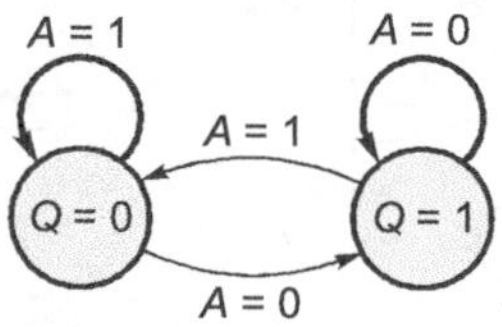

(b) 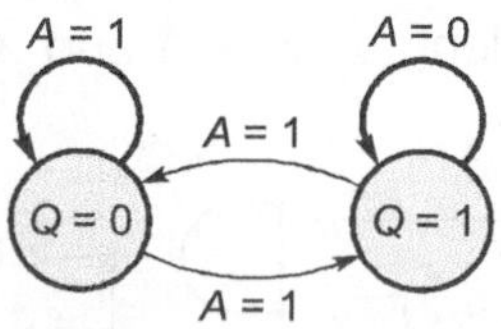

(c) 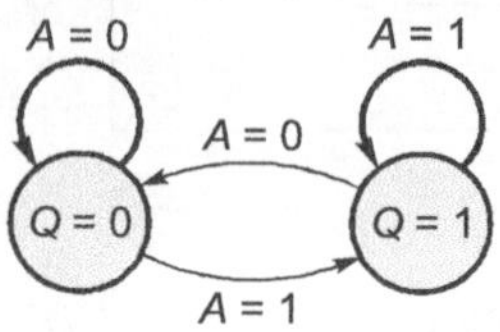

(d) 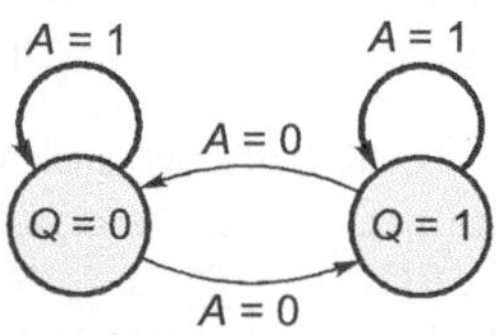

[2012 : 2 M]

4.15 The clock frequency applied to the digital circuit shown in the figure below is 1 kHz. If the initial state of the output Q of the flip-flop is '0', then the frequency of the output waveform Q in kHz is

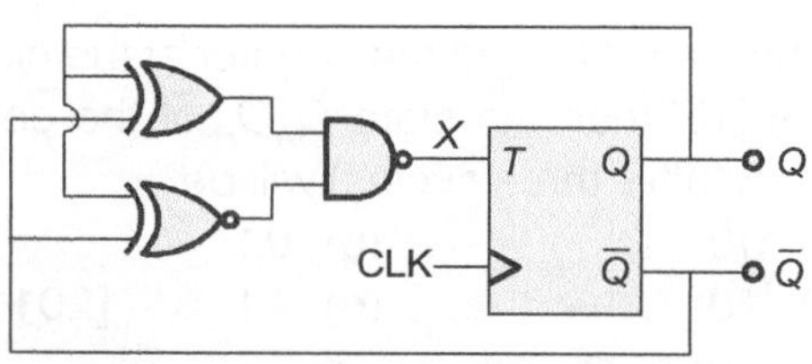

(a) 0.25 (b) 0.5
(c) 1 (d) 2 [2013 : 2 M]

4.16 A cascade of three identical modulo-5 counters has an overall modulus of

(a) 5 (b) 25
(c) 125 (d) 625

[2014 : 1 M, Set-1]

4.17 A JK flip-flop can be implemented by T flip-flops. Identify the correct implementation.

(a) 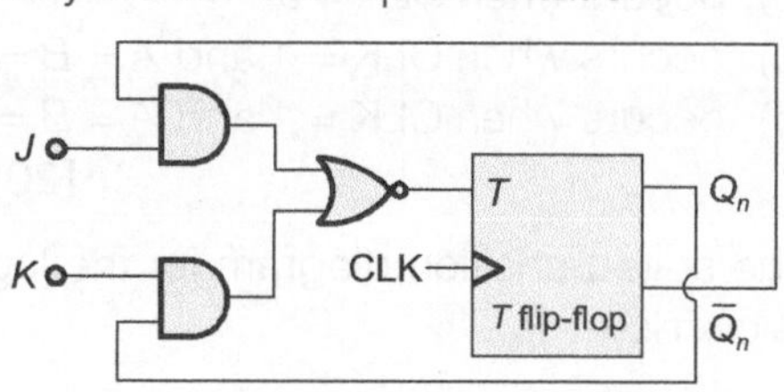

(b) 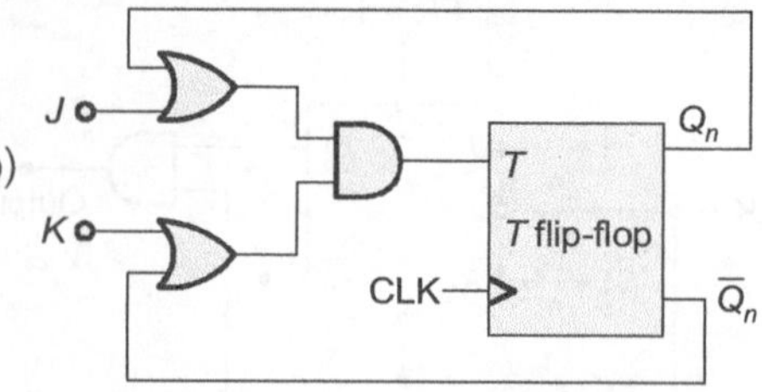

(c) 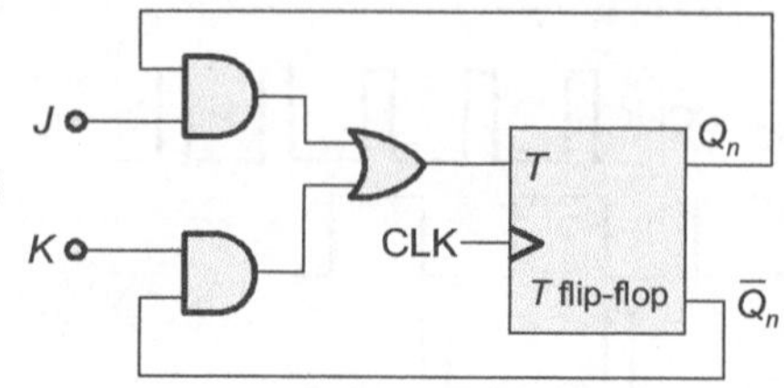

(d) 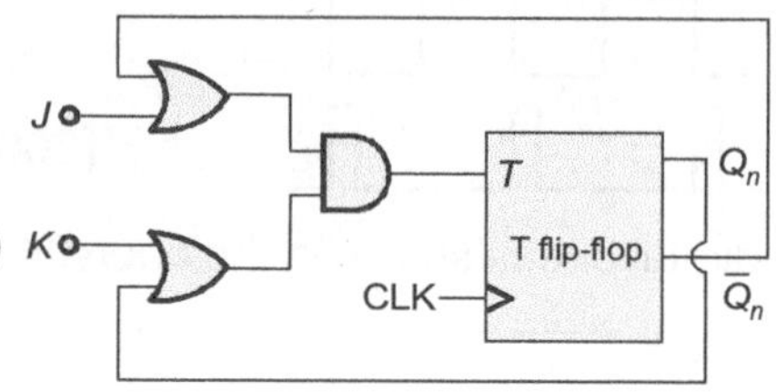

[2014 : 2 M, Set-2]

4.18 A state diagram of a logic which exhibits a delay in the output is shown in the figure, where X is the do not care condition, and Q is the output representing the state.

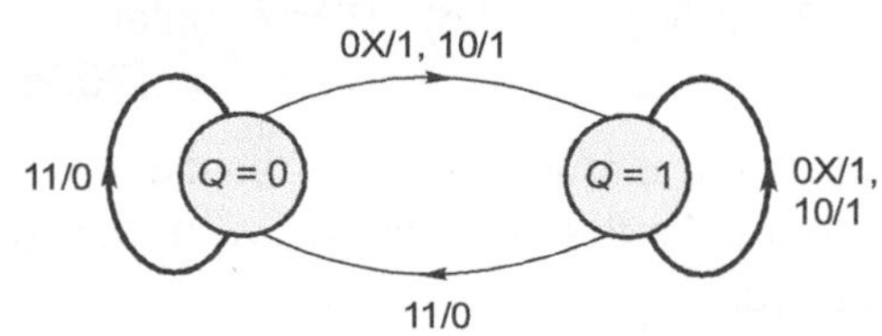

The logic gate represented by the state diagram is

(a) XOR (b) OR
(c) AND (d) NAND

[2014 : 1 M, Set-3]

4.19 The figure shows a digital circuit constructed using negative edge triggered J-K flip flops. Assume a starting state of $Q_2Q_1Q_0$ = 000 will repeat after ______ number of cycles of the clock CLK

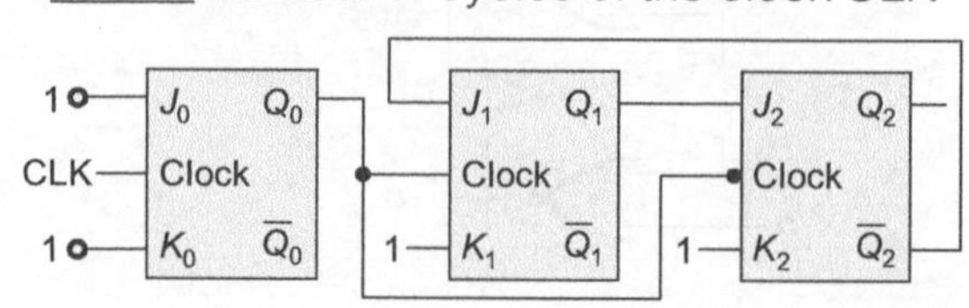

[2015 : 2 M, Set-1]

4.20 In the following sequential circuit, the initial state (before the first clock pulse) of the circuit is Q_1Q_0 = 00. The state (Q_1Q_0), immediately after the 333rd clock pulse is

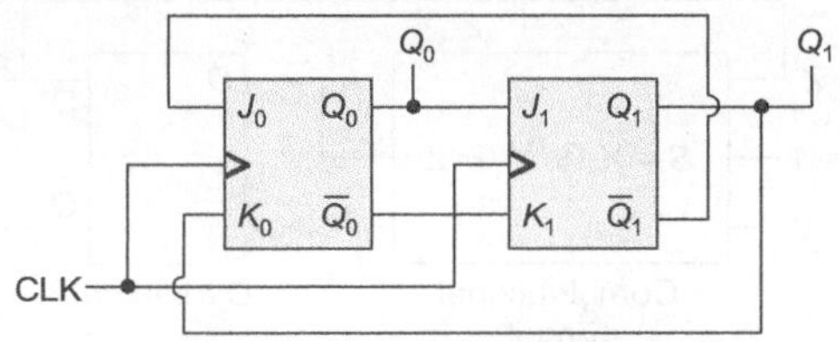

(a) 00 (b) 01
(c) 10 (d) 11 **[2015 : 2 M, Set-2]**

4.21 The current state $Q_A\ Q_B$ of a two JK flip-flop system is 00. Assume that the clock rise-time is much smaller than the delay of the JK flip-flop. The next state of the system is

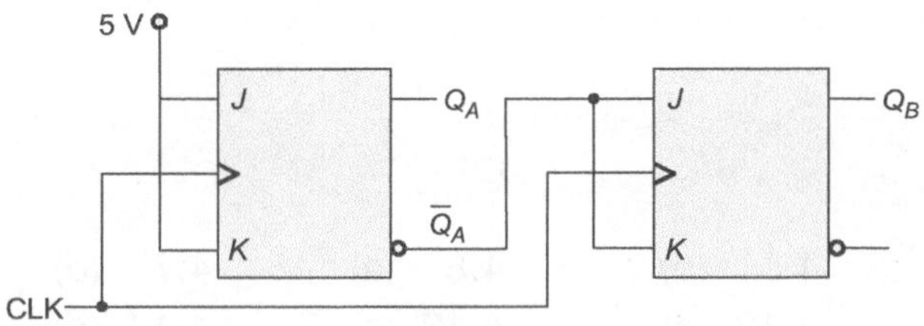

(a) 00 (b) 01
(c) 11 (d) 10 **[2016 : 2 M, Set-1]**

4.22 For the synchronous sequential circuit shown below, the output Z is zero for the initial conditions $Q_A\,Q_B\,Q_C = Q'_A\,Q'_B\,Q'_C = 100$.

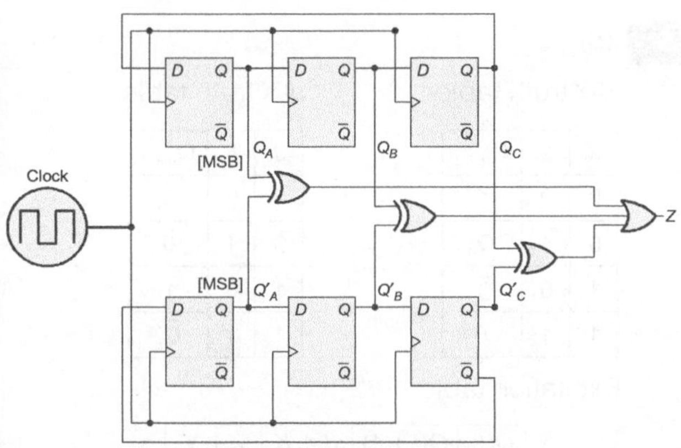

The minimum number of clock cycles after which the output z would again become zero is ______.
[2017 : 2 M, Set-2]

4.23 Which one of the following statements is true about the digital circuit shown in the figure?

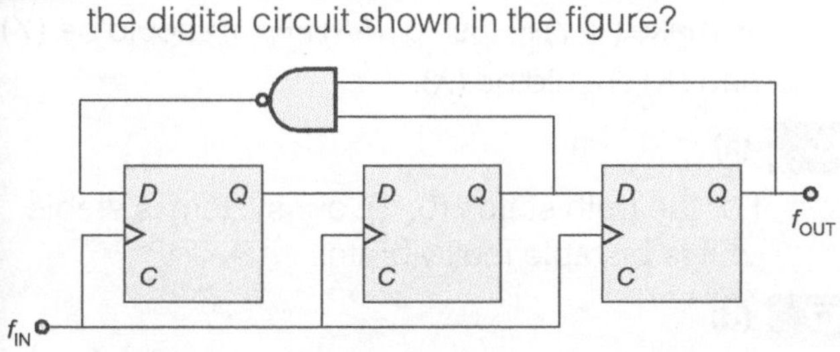

(a) It can be used for dividing the input frequency by 3.
(b) It can be used for dividing input frequency by 5.
(c) It can be used for dividing the input frequency by 7.
(d) It cannot be reliably used as a frequency divider due to disjoint internal cycles.
[2018 : 2 M]

4.24 A sequence detector is designed to detect precisely 3 digital inputs, with overlapping sequences detectable. For the sequence (1, 0, 1) and input data (1,1, 0,1,0,0,1,1,0,1,0,1,1,0): what is the output of this detector?
(a) 0,1,0,0,0,0,0,1,0,1,0,0
(b) 0,1,0,0,0,0,0,0,1,0,0,0
(c) 1,1,0,0,0,0,1,1,0,1,0,0
(d) 0,1,0,0,0,0,0,1,0,1,1,0 **[2020 : 1 M]**

4.25 A counter is constructed with three D flip-flops. The input-output pairs are named (D_0, Q_0), (D_1, Q_1) and (D_2, Q_2), where the subscript 0 denotes the least significant bit. The output sequence is desired to be the Gray-code sequence 000, 001, 011, 010, 110, 111, 101 and 100, repeating periodically. Note that the bits are listed in the $Q_2Q_1Q_0$ format. The combinational logic expression for D_1 is

(a) $Q_2Q_1 + \bar{Q}_2\bar{Q}_1$ (b) $Q_2Q_1Q_0$

(c) $\bar{Q}_2Q_0 + Q_1\bar{Q}_0$ (d) $Q_2Q_0 + Q_1\bar{Q}_0$

[2021 : 2 M]

4.26 A 16-bit synchronous binary up-counter is clocked with a frequency f_{CLK}. The two most significant bits are OR-ed together to form an output Y. Measurements show that Y is periodic, and the duration for which Y remains high in each period is 24 ms. The clock frequency f_{CLK} is ______ MHz. (Round off to 2 decimal places.) **[2021 : 1 M]**

4.27 A MOD-2 and a MOD-5 up-counter when cascaded together results in a MOD ____ counter. (in integer). **[2022 : 1 M]**

4.28 The maximum clock frequency in MHz of a 4-stage ripple counter, utilizing flip-flops, with each flip-flop having a propagation delay of 20 ns, is _____. (round off to one decimal place). **[2022 : 1 M]**

4.29 Neglecting the delays due to the logic gates in the circuit shown in figure, the decimal equivalent of the binary sequence [ABCD] of initial logic states, which will not change with clock, is ______.

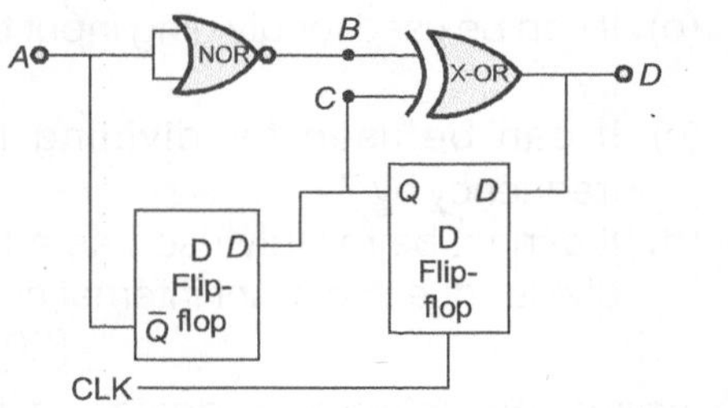

[2023 : 2 M]

4.30 In the circuit, the present value of Z is 1. Neglecting the delay in the combinatorial circuit, the values of S and Z, respectively, after the application of the clock will be

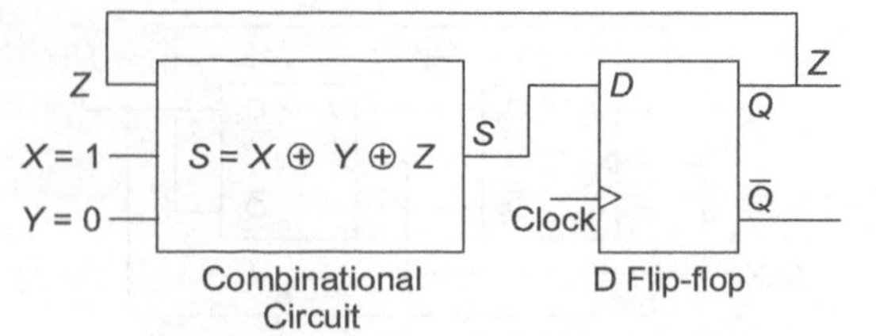

(a) $S = 0, Z = 0$ (b) $S = 1, Z = 1$
(c) $S = 0, Z = 1$ (d) $S = 1, Z = 0$

[2024 : 1 M]

Answers Sequential Logic Circuits

4.1 (0.5)	4.2 (a)	4.3 (d)	4.4 (b)	4.5 (d)	4.6 (a)	4.7 (d)
4.8 (b)	4.9 (c)	4.10 (a)	4.11 (d)	4.12 (c)	4.13 (a)	4.14 (d)
4.15 (b)	4.16 (c)	4.17 (b)	4.18 (d)	4.19 (6)	4.20 (b)	4.21 (c)
4.22 (6)	4.23 (b)	4.24 (a)	4.25 (c)	4.26 (2.05)	4.27 (10)	4.28 (12.5)
4.29 (8)	4.30 (d)					

Explanations Sequential Logic Circuits

4.1 (0.5)

$$f_{out} = \frac{f_{in}}{2} = \frac{1\,\text{MHz}}{2} = 0.5\,\text{MHz}$$

4.3 (d)

In toggle mode,

$$f_{out} = \frac{f_{in}}{2} = \frac{10\,\text{kHz}}{2} = 5\,\text{kHz}$$

4.4 (b)

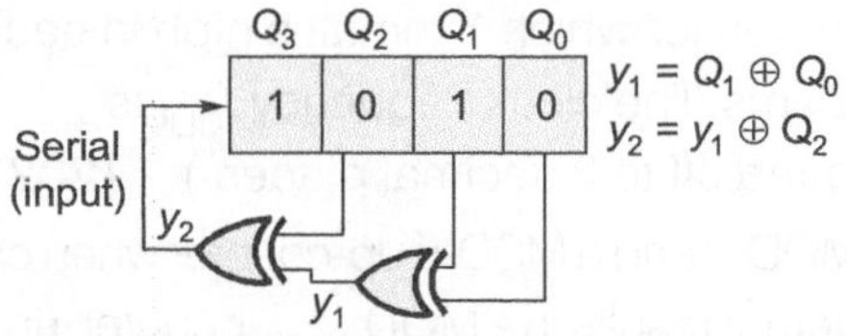

$$Q_3(t+1) = Q_0(t) \oplus Q_1(t) \oplus Q_2(t)$$

CLK pulse		Q_3	Q_2	Q_1	Q_0
	0	1	0	1	0
1	1	1	1	0	1
2	2	0	1	1	0
3	3	0	0	1	1
4	4	0	0	0	1
5	5	1	0	0	0
6	6	0	1	0	0
7	7	1	0	1	0

4.5 (d)

X-Y truth table

X	Y	Q_{n+1}
0	0	1
0	1	Q_n
1	0	$\bar{Q}_n$
1	1	0

J-K truth table

J	K	Q_{n+1}
0	0	Q_n
0	1	0
1	0	1
1	1	$\bar{Q}_n$

Excitation table

$Q(t)$	$Q(t+1)$	J	K	X	Y
0	0	0	x	x	1
0	1	1	x	x	0
1	0	x	1	1	x
1	1	x	0	0	x

To make (X-Y) FF using (J-K) FF, (J) should be ($\bar{Y}$) and (K) should be (X).

4.6 (a)

For the both states (0, 1); our system is stable.
∴ It is bistable multivibrator.

4.7 (d)

For a 4-input X-NOR gate output will be zero if number of 1's will be odd.

We also know that output of XOR gate will be '1' if number of 1's will be odd.

If the inputs will be same then output of XOR gate will be 0 so all inputs to XNOR will be zero so output Y will be '1'. So only in option (d) the inputs are different so Y will be zero.

4.8 (b)

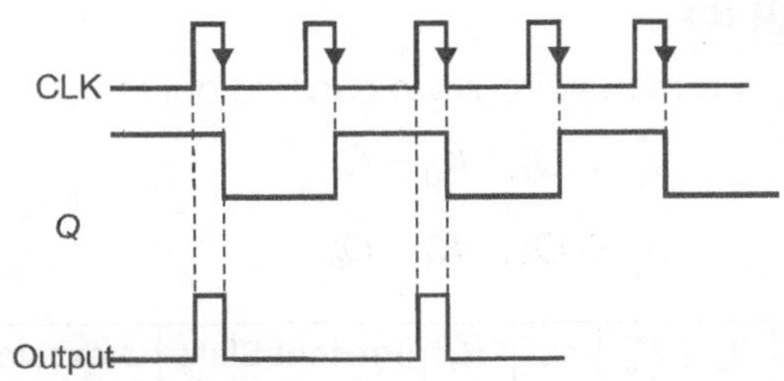

4.9 (c)

Truth table of circuit,

X	Q(t)	Q(t + 1)
0	0	0
0	1	1
1	0	1
1	1	0

(The rows for X = 0 and X = 1 are bracketed together and labelled T-FF.)

4.10 (a)

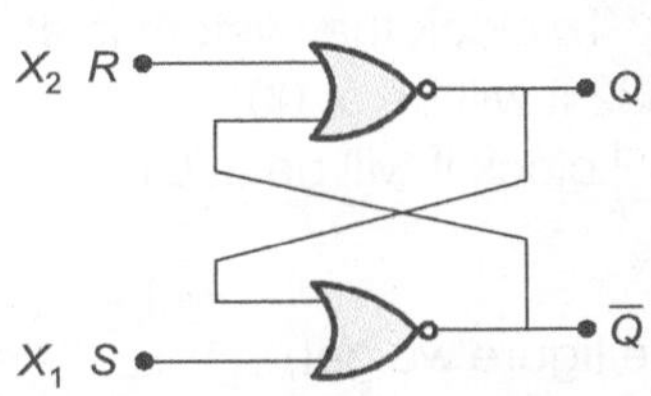

S	R	
X_1	X_2	Q^+
1	0	1
0	0	NC
0	1	0
1	0	1
0	1	0
1	0	1

S	R	Q^+
0	0	NC
0	1	0
1	0	1
1	1	X

4.11 (d)

As no combination of 'Q' with (X_1 and X_2) = 1 output is stable.

It always switches its state from '1' to '0' and from '0' to '1'.

4.12 (c)

	J	K	Q_A	T	Q_B
CLK	1	0	1	1	1
CLK	0	1	0	1	0
CLK	1	0	1	1	0

Q_AQ_B at $t_n + 3$ is '1 0'

4.13 (a)

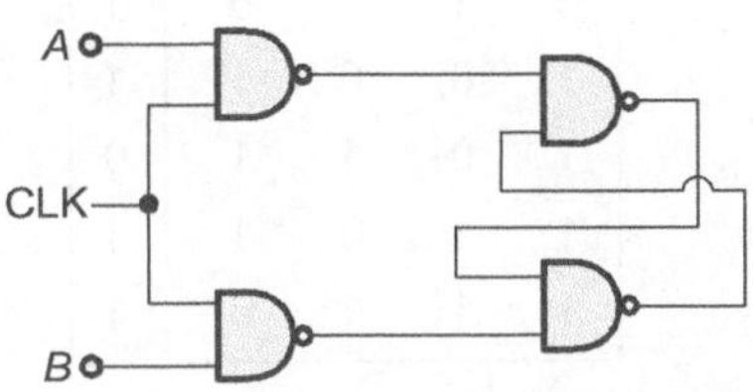

The above circuit is SR flip-flop. The race around condition occurs in J-K flip-flop.

4.14 (d)

- When $A = 1$, then Q will be selected by MUX and feedback to D flipflop which gives output Q again. So, at $A = 1$ it holds its state. When $A = 0$, then $\bar{Q}$ will be selected by MUX and feedback to D flipflop and output will be inverted.
- So, for $A = 1$ it holds the state and for $A = 0$ it interchange the state i.e. if $Q = 0$ then it will go to $Q = 1$ and it $Q = 1$ then it will go to $Q = 0$.

So, option (d) is correct option.

4.15 (b)

$$x = \overline{(Q \oplus \bar{Q}) \cdot (Q \odot \bar{Q})} = \overline{1 \cdot 0} = 1 \text{ (always)}$$

$$\because \quad X = 1 = T$$

$\Rightarrow$ Q always toggles whenever clock triggers.

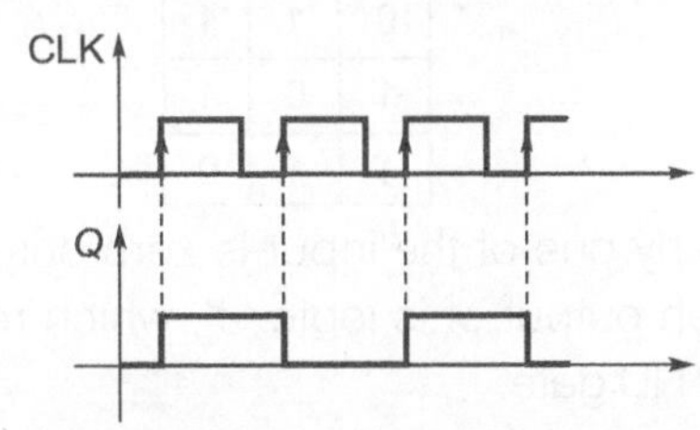

$$\therefore \quad f_Q = \frac{f_{clk}}{2} = \frac{1\,\text{kHz}}{2} = 0.5\,\text{kHz}$$

4.16 (c)

Overall modulus = 5 × 5 × 5 = 125

4.17 (b)

To obtain a JK flip-flop from a T flip-flop, we first construct the characteristic table of JK flip-flop; and then obtain the excitation values for the T flip-flop as shown below:

Characteristic table of JK-flip-flop → ; ← Excitation values for T flip-flop

J	K	Q_n	Q_{n+1}	T
0	0	0	0	0
0	0	1	1	0
0	1	0	0	0
0	1	1	0	1
1	0	0	1	1
1	0	1	1	0
1	1	0	1	1
1	1	1	0	1

Now, assuming T to be an output, we solve it in terms of J, K, Q_n inputs. This gives the definition of the logic to be applied on the T input.
Also, observing the given options, we solve for T using a maxterms map instead of using a minterms map, as shown below:

J \ KQ_n	$K+Q_n$	$K+\bar{Q}_n$	$\bar{K}+\bar{Q}_n$	$\bar{K}+Q_n$
J	0	0		0
$\bar{J}$		0		

$$T = (J+Q_n)\cdot(K+\bar{Q}_n)$$

The circuit corresponding to this expression is given option (b).

4.18 (d)

A	B	Q
0	0	1
0	1	1
1	0	1
1	1	0

If any one of the input is zero output is logic '1' then output of is logic '0', which represents the NAND gate.

4.19 (6)

JK Flip-flop 1 and 2 form a synchronous sequential circuits and they are synchronized with the output of 0th JK Flip-flop.

J_1	K_1	J_2	K_2	Q_2	Q_1	Q_0	
1	1	0	1	0	0	0	T_1
1	1	1	1	0	1	1	T_2
0	1	0	1	1	0	0	T_3
1	1	0	1	0	0	0	

Number of cycles = 3 i.e. equal to 6 clock cycles.

4.20 (b)

From the circuit we can find out that,

$$J_0 = Q_1',\ \ K_0 = Q_1$$

$$J_1 = Q_0,\ \ K_1 = Q_0'$$

J_0	K_0	J_1	K_1	Present State		Next state	
				Q_1	Q_0	Q_1	Q_0
1	0	0	1	0	1	0	1
1	0	1	0	0	1	1	1
0	1	1	0	1	1	1	0
0	1	0	1	1	0	0	0

So, the state diagram is

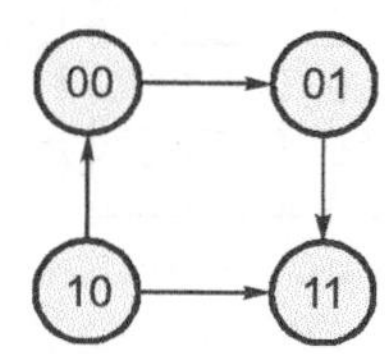

At every $4n$ clock the system is at 00.
So, at 332 it will be at 00.
So, 333rd clock it will be at 01.

4.21 (c)

From the figure we get,

$$J_A = K_A = 1;\ \ J_B = K_B = \bar{Q}_A$$

Clock	J_AK_A	J_BK_B	Q_AQ_B
0	11	11	00
1			11

So, next state will be 11.

4.22 (6)

Clock	Q_A	Q_B	Q_C	Q_A'	Q_B'	Q_C'	$Q_A \oplus Q_A'$	$Q_B \oplus Q_B'$	$Q_C \oplus Q_C'$	Z
0	1	0	0	1	0	0	0	0	0	0
1	0	1	0	1	1	0	1	0	0	1
2	0	0	1	1	1	1	1	1	0	1
3	1	0	0	0	1	1	1	1	1	1
4	0	1	0	0	0	1	0	1	1	1
5	0	0	1	0	0	0	0	0	1	1
6	1	0	0	1	0	0	0	0	0	0

The output Z will again become zero after 6 clock cycles.

4.23 (b)

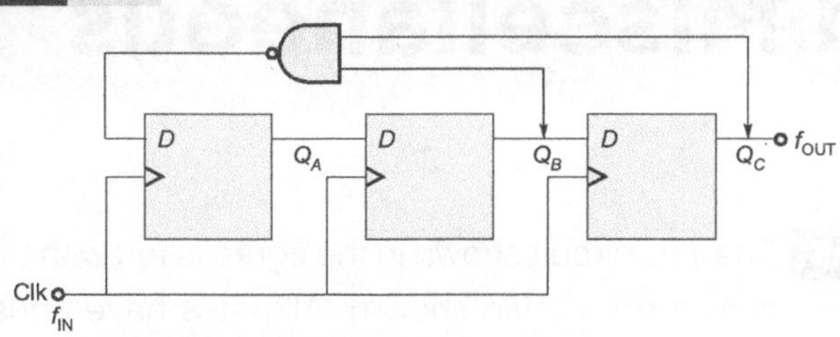

$\overline{Q_B}\cdot Q_C$	Q_A	Q_B				
D_A	D_B	D_C	Q_A	Q_B	Q_C	
			0	0	0	
1	0	0	1	0	0	
1	1	0	1	1	0	Cycle
1	1	1	1	1	1	Cycle
0	1	1	0	1	1	Cycle
0	0	1	0	0	1	Cycle
1	0	0	1	0	0	Cycle
1	1	0	1	1	0	Repeated

mod 5 counter

So, frequency will be divided by 5.

4.24 (a)

Sequence detector problem:

- If consider the case of non-overlapping sequence detector, then the pattern 101 is appearing 2 times in the given bit sequence.
- If we consider the case of overlapping sequence detector, then the pattern 101 is appearing 3 times in the given bit sequence.

The question says that the detector is overlapping, hence answer is (a).

4.25 (c)

Present state Q_2	Q_1	Q_0	Next state Q_2^+	Q_1^+	Q_0^+	D_2	D_1	D_0
0	0	0	0	0	1		0	
0	0	1	0	1	1		1	
0	1	1	0	1	0		1	
0	1	0	1	1	0		1	
1	1	0	1	1	1		1	
1	1	1	1	0	1		0	
1	0	1	1	0	0		0	
1	0	0	0	0	0		0	

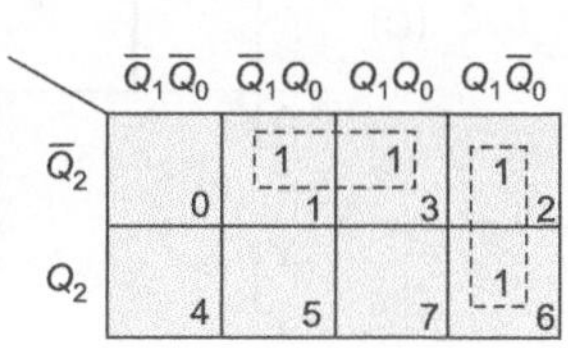

$$D_1 = \bar{Q}_2Q_0 + Q_1\bar{Q}_0$$

4.26 (2.05)

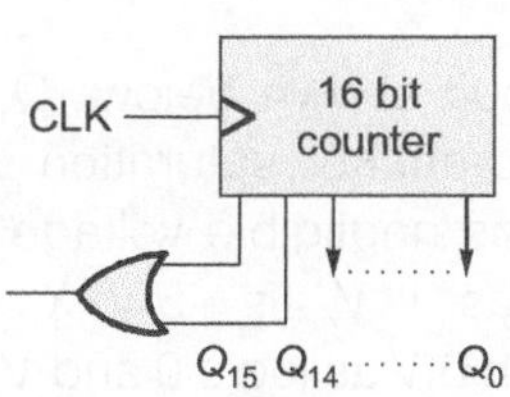

y = 1 for 24 m sec

		Q_{15}	Q_{14}	.....	Q_0
y = 1	Starts at	0100	0000	0000	0000
y = 1	Ends at	1111	1111	1111	1111

Total number of states for which

$$y = 1 \text{ is } 2^{16} - 2^{14}$$

Time is $(2^{16} - 2^{14})T = 24$ msec

$$f = \frac{1}{T} = 2.05 \text{ MHz}$$

4.27 (10)

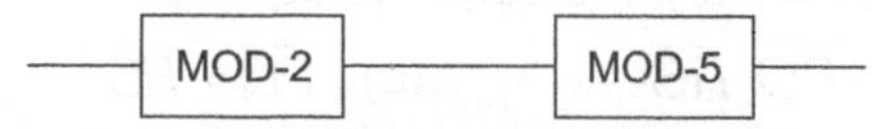

For overall configuration = Mod (2 × 5) = 10

It will be mod 10 counter.

4.28 (12.5)

Given ripple counter 4 flip-flop, t_p = 20 nsec for each flip-flop.

$$T = nt_p = 4 \times 20\text{nsec} = 80 \text{ nsec}$$

$$\text{Clock frequency} = \frac{1}{T}$$

$$= \frac{1}{4\times 20\times 10^{-9}} = 12.5 \text{ MHz}$$

4.29 (8)

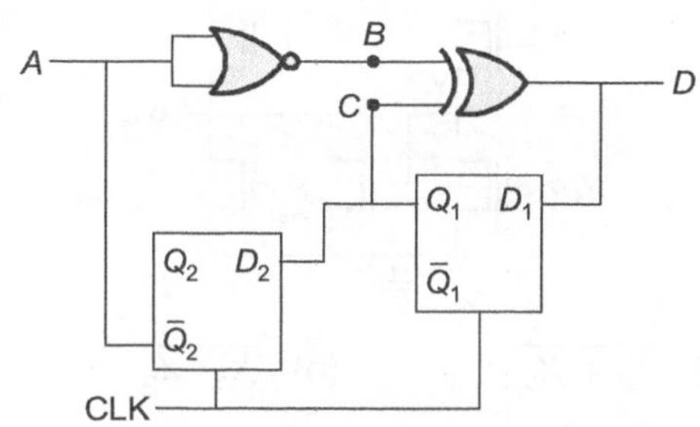

$$B = \bar{A} \quad \text{...(i)}$$

$$A = \bar{Q}_2 = \bar{C} \quad \text{...(ii)}$$

$$C = D \quad \text{...(iii)}$$

$$D = B \oplus C$$

∵ by (i) and (ii), $B = C$

∴ $D = B\oplus B = 0$

∴ $C = 0$; $A = 1$; $B = 0$

$$ABCD = (1000)_2 = (8)_{10}$$

4.30 (d)

Before clock pulse :

$$S = 1\oplus 0\oplus 1 = 0$$

After clock pulse :

$$Z = D = S = 0$$

$$S = 1\oplus 0\oplus 0 = 1$$

5 Logic Families and Miscellaneous

5.1 The open collector outputs of two inputs NAND-gates are connected to a common pull up resistor. If the input to the gates are P, Q and R, S respectively, the output is equal to

(a) $\overline{PQ}\cdot\overline{RS}$ (b) $\overline{PQ}+\overline{RS}$

(c) $PQ + RS$ (d) $PQRS$ [1998 : 1 M]

5.2 In standard *TTL* gates, the totem pole output stage is primarily used to

(a) increase the noise margin of the gate.
(b) decrease the output switching delay.
(c) facilitate a wired OR logic connection.
(d) increase the output impedance of the circuit.

[1998 : 1 M]

5.3 If X_1 and X_2 are the inputs to the circuit shown in the figure, the output Q is

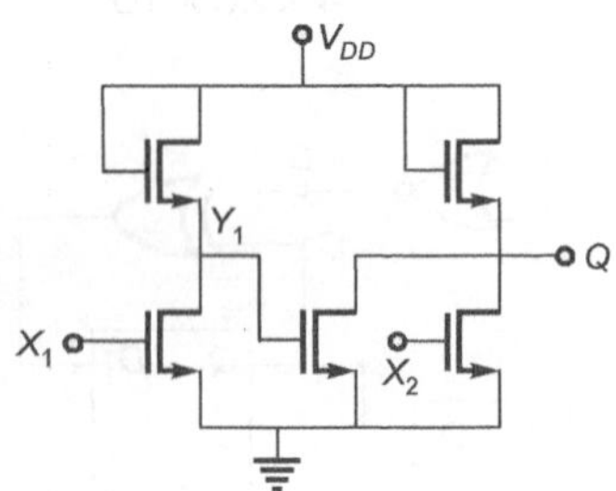

(a) $\overline{X_1 + X_2}$ (b) $\overline{X_1 \cdot X_2}$

(c) $\overline{X}_1 \cdot X_2$ (d) $X_1 \cdot \overline{X}_2$ [2005 : 2 M]

5.4 A TTL NOT gate circuit is shown in the figure. Assuming V_{BE} = 0.7 V of both the transistors, if V_i = 3.0 V, then the states of the two transistors will be

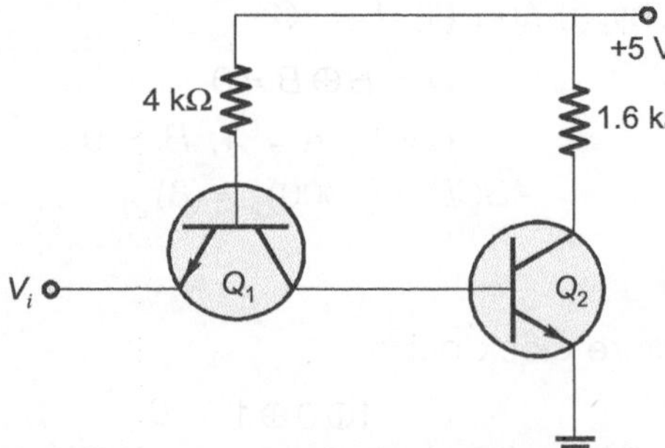

(a) Q_1 ON and Q_2 OFF
(b) Q_1 reverse ON and Q_2 OFF
(c) Q_1 reverse ON and Q_2 ON
(d) Q_1 OFF and Q_2 reverse ON [2006 : 2 M]

5.5 The TTL circuit shown in the figure is fed with the waveform X (also shown). All gates have equal propagation delay of 10 ns. The output Y of the circuit is

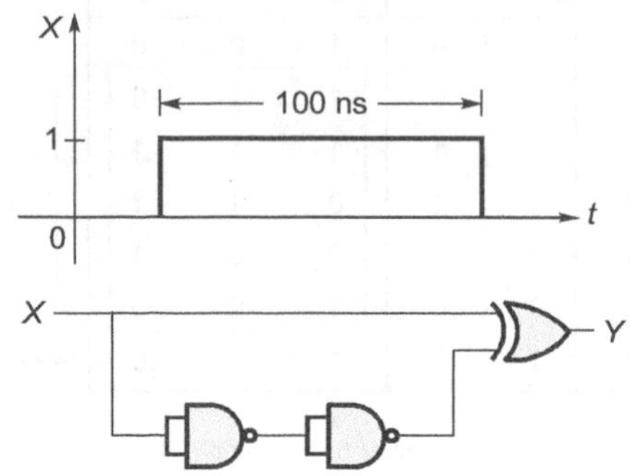

(a)

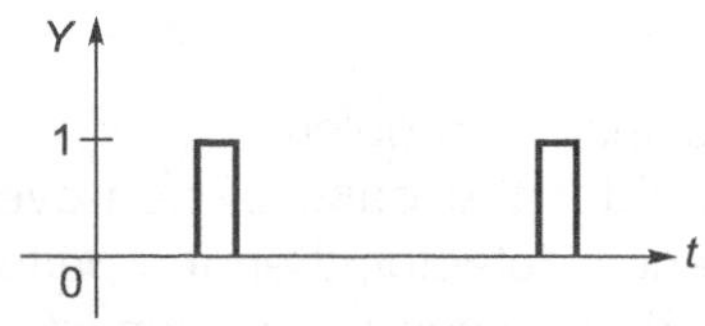

(b)

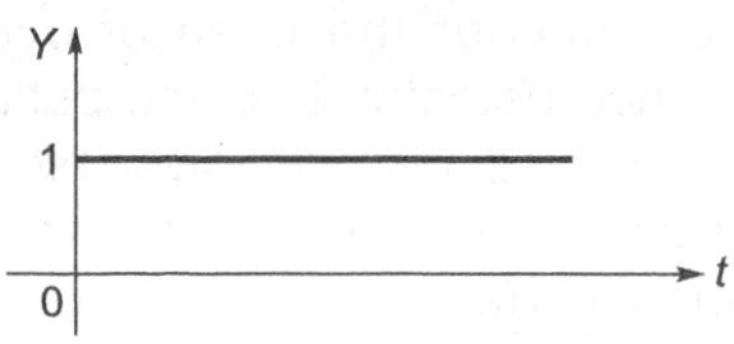

(c)

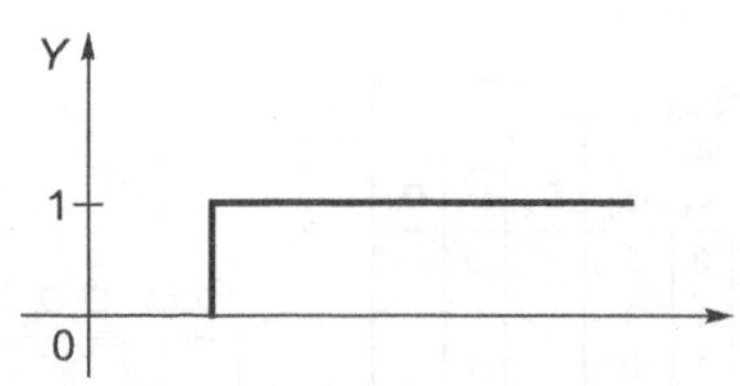

(d)

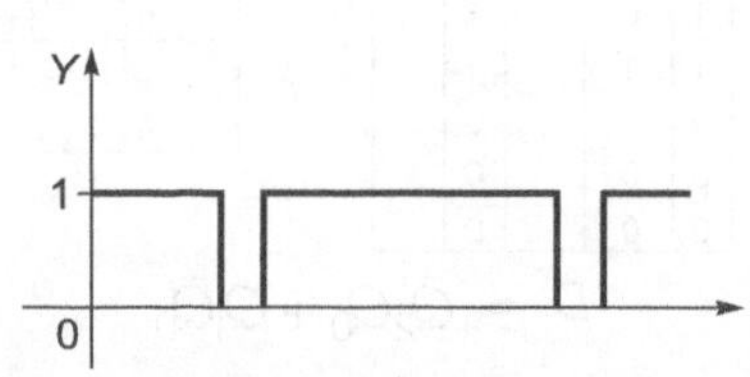

[2010 : 2 M]

5.6 In the circuit shown below, Q_1 has negligible collector-to-emitter saturation voltage and the diode drops negligible voltage across it under forward bias. If V_{cc} is +5 V, X and Y are digital signals with 0 V as logic 0 and V_{cc} as logic 1, the Boolean expression for Z is

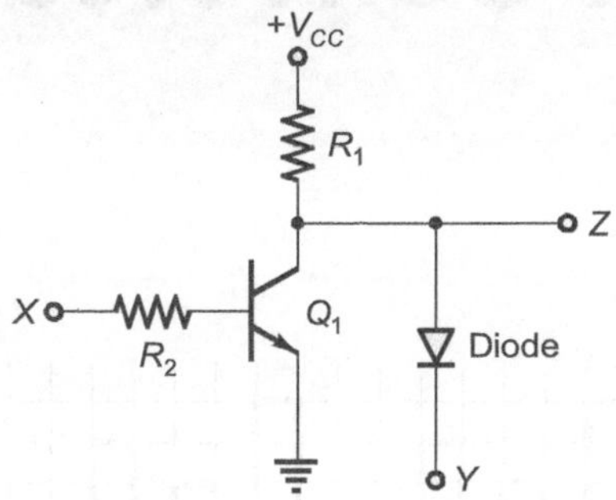

(a) XY (b) $\bar{X}Y$

(c) $X\bar{Y}$ (d) $\overline{XY}$ [2013 : 2 M]

5.7 The logical gate implemented using the circuit shown below where, V_1 and V_2 are inputs (with 0 V as digital 0 and 5 V as digital 1) and V_{OUT} is the output, is

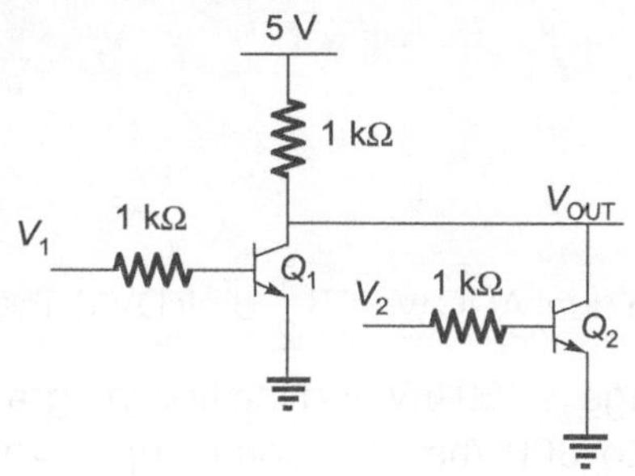

(a) NOT (b) NOR

(c) NAND (d) XOR

[2017 : 2 M, Set-1]

Answers Logic Families and Miscellaneous

5.1 (a) 5.2 (b) 5.3 (d) 5.4 (c) 5.5 (a) 5.6 (b) 5.7 (b)

Explanations Logic Families and Miscellaneous

5.1 (a)

Pull up resistor works as a AND Gate

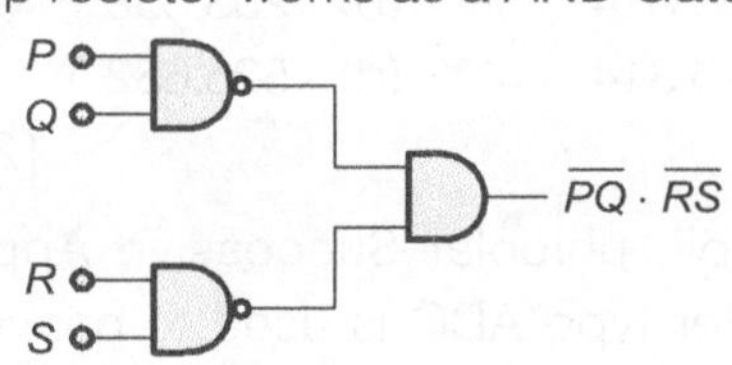

5.3 (d)

$$Y_1 = \bar{X}_1$$

Output, $Q = \overline{Y_1 + X_2} = \overline{(\bar{X}_1 + X_2)} \Rightarrow Q = X_1 \cdot \bar{X}_2$

5.4 (c)

When $V_i = 3$ V then Q_1 will be in reverse active mode i.e. reverse ON and Q_2 will be ON.

5.5 (a)

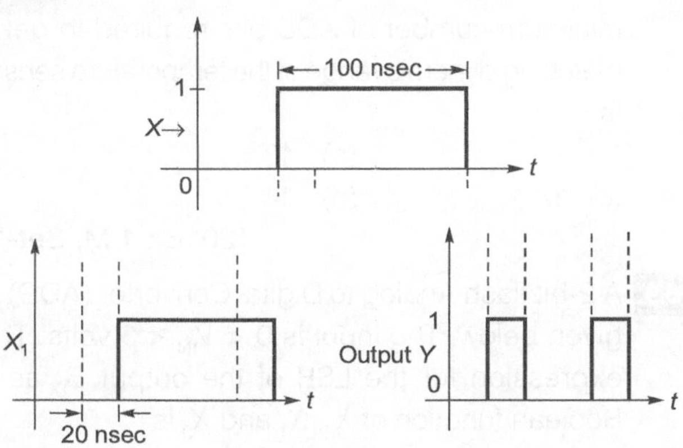

5.6 (b)

X	Y	Q	Z
0	0	OFF	0
0	+5V	OFF	+5V
+5V	0	ON	0
+5V	+5V	ON	0

$\Rightarrow Z = \bar{X}Y$

5.7 (b)

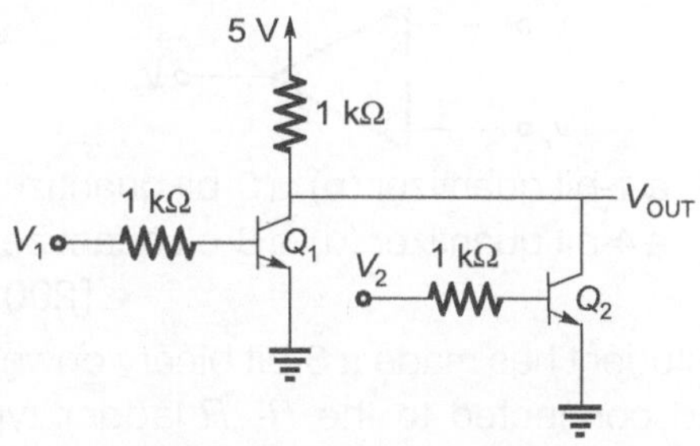

When, V_1 = high 5 V Q_1 on $V_{out} = 0$

V_2 = high 5 V Q_2 on $V_{out} = 0$

Thus when any V_1 or V_2 is high then,

$V_{out} = 0$

$\Rightarrow$

V_1	V_2	V_{out}
0	0	1
0	1	0
1	0	0
1	1	0

Thus, it is a NOR gate.

6 A/D and D/A Converters

6.1 For a dual ADC type $3\frac{1}{2}$ digit DVM, the reference voltage is 100 mV and the first integration time is set to 300 ms. For some input voltage; the 'deintegration' period is 370.2 ms. The DVM will indicate (in mV)

(a) 123.4 (b) 199.9
(c) 100.0 (d) 1.414 **[1999 : 2 M]**

6.2 A dual-slope analog-to-digital converter uses an N-bit counter. When the input signal V_a is being integrated, the counter is allowed to count up to a value

(a) equal to $2^N - 2$ (b) equal to $2^N - 1$
(c) proportional to V_a
(d) inversely proport ional to V_a **[2000 : 2 M]**

6.3 Among the following four, the slowest *ADC* (analog-to-digital converter) is

(a) parallel-comparator (i.e. flash) type
(b) successive approximation type
(c) integrating type
(d) counting type **[2001 : 1 M]**

6.4 The voltage comparator shown in figure can be used in the analog-to-digital conversion as

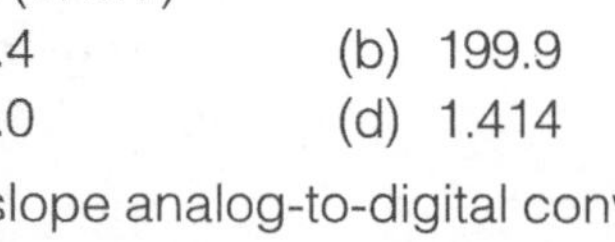

(a) a 1-bit quantizer (b) a 2-bit quantizer
(c) a 4-bit quantizer (d) a 8-bit quantizer
[2004 : 1 M]

6.5 A student has made a 3-bit binary down counter and connected to the R-$2R$ ladder type DAC [Gain = (–1 KΩ/2R) as shown in the figure to generate a staircase waveform. The output achieved is different as shown in the figure. What could be the possible cause of this error?

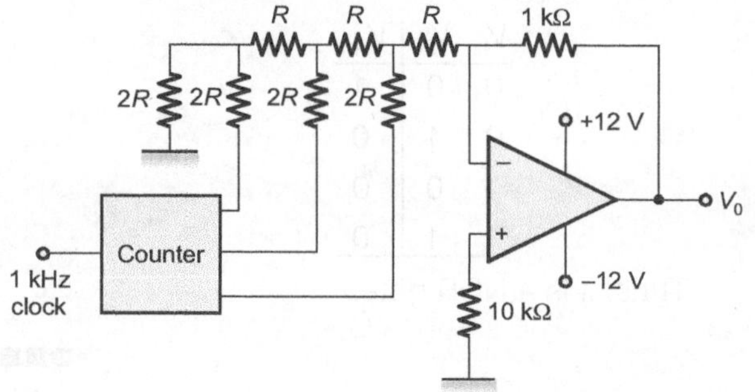

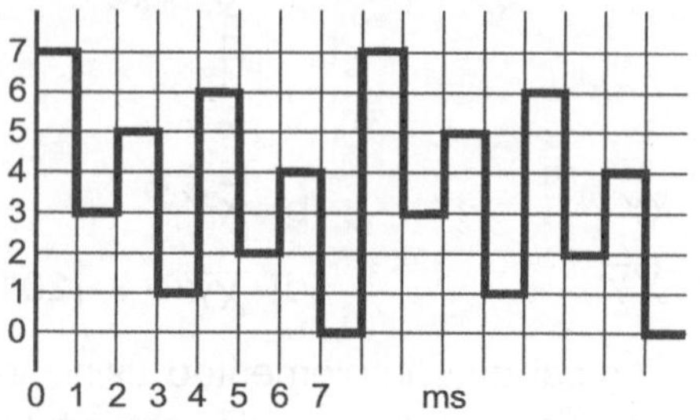

(a) The resistance values are incorrect.
(b) The counter is not working properly.
(c) The connection from the counter to DAC is not proper.
(d) The R and $2R$ resistances are interchanged.
[2006 : 2 M]

6.6 The Octal equivalent of the HEX number *AB.CD* is

(a) 253.314 (b) 253.632
(c) 526.314 (d) 526.632
[2007 : 2 M]

6.7 An 8-bit, unipolar Successive Approximation Register type ADC is used to convert 3.5 V to digital equivalent output.
The reference voltage is +5 V. The output of the ADC at the end of 3rd clock pulse after the start of conversion, is

(a) 1010 0000 (b) 1000 0000
(c) 0000 0001 (d) 0000 0011
[2015 : 2 M, Set-1]

6.8 A temperature in the range of –40°C to 55°C is to be measured with a resolution of 0.1°C. The minimum number of *ADC* bits required to get a matching dynamic range of the temperature sensor is

(a) 8 (b) 10
(c) 12 (d) 14
[2016 : 1 M, Set-1]

6.9 A 2-bit flash Analog to Digital Converter (ADC) is given below. The input is $0 < V_{IN} < 3$ volts. The expression for the LSB of the output B_0 as a Boolean function of X_2, X_1 and X_0 is

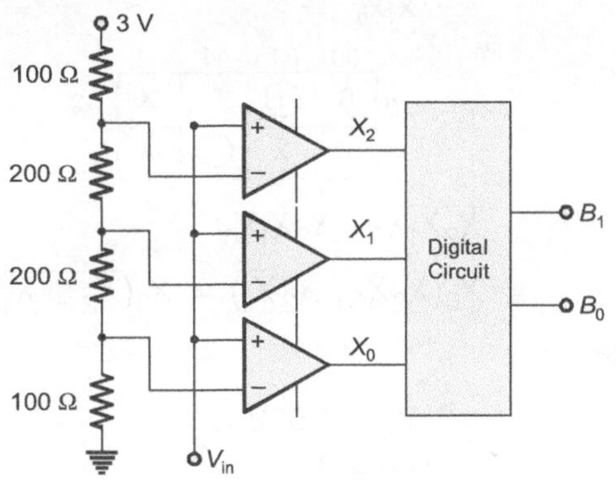

(a) $X_0\left[\overline{X_2 \oplus X_1}\right]$ (b) $\bar{X}_0\left[\overline{X_2 \oplus X_1}\right]$

(c) $X_0[X_2 \oplus X_1]$ (d) $\bar{X}_0[X_2 \oplus X_1]$

[2016 : 2 M, Set-1]

6.10 An 8-bit ADC converts analog voltage in the range of 0 to +5 V to the corresponding digital code as per the conversion characteristics shown in figure. For V_{in} = 1.9922 V, which of the following digital output, given in hex, is true?

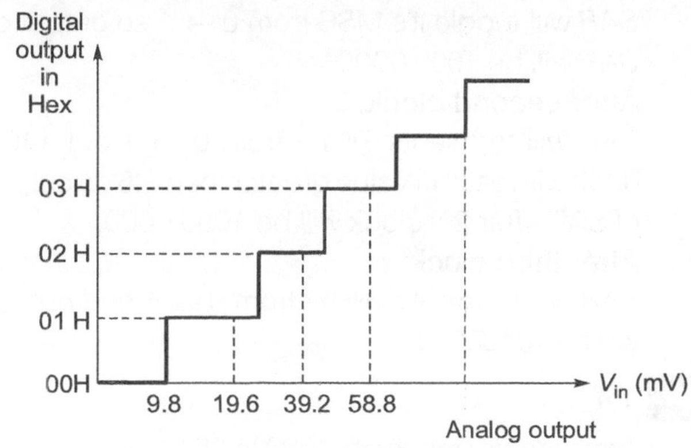

(a) 64H (b) 65H

(c) 66H (d) 67H

[2023 : 2 M]

Answers A/D and D/A Converters

6.1 (a)	6.2 (a)	6.3 (c)	6.4 (a)	6.5 (c)	6.6 (b)	6.7 (a)
6.8 (b)	6.9 (a)	6.10 (c)				

Explanations A/D and D/A Converters

6.1 (a)

$$V_{in} = \frac{100\text{ mV} \times 370.2\text{ ms}}{300\text{ ms}} = 123.33\text{ mV}$$

6.2 (a)

Even when $V_1 > V_2$, the (o/p) 'V_o' is high and for the next case ($V_1 < V_2$) (o/p) is low it is 1 bit quantizer. Since it has two states which can be represented by 1 bit.

6.5 (c)

Initial stage of the counter = $(111)_2 = (7)_{10}$

So output will be equal to 7 V.

Next state of counter = $(110)_2 = (6)_{10}$

So output should be = 6 V

But output is 3 V that means LSB of counter is connected to MSB of DAC and MSB of counter is connected to LSB of DAC.

Similarly next state of counter = $(101)_2 = (5)_{10}$

Input to DAC = $(101)_2 = (5)_{10}$

So output = 5 V

When counter goes to $(100)_2$ then input to DAC = $(001)_2 = (1)_{10}$

So output = 1 V

So connections are not proper.

6.6 (b)

Hex number ($AB.CD$)

$$\underbrace{1010}_{A}\ \underbrace{1011}_{B}\cdot\underbrace{1100}_{C}\ \underbrace{1101}_{D}$$

For finding its octal number, we can add one zero in both extreme and grouping.

$$\underbrace{010}_{2}\ \underbrace{101}_{5}\ \underbrace{011}_{3}.\ \underbrace{110}_{6}\ \underbrace{011}_{3}\ \underbrace{010}_{2}$$

$\therefore$ equivalent octal number : 253.632_8

6.7 (a)

The reference voltage is 5 V.

The number of bits in ADC are 8.

So, the resolution will be = $\dfrac{5}{2^8 - 1} = \dfrac{5}{255}$

The applied input is 3.5 V.

The successive approximation ADC start working from the MSB so.

After one clock:

SAR will toggle it's MSB from 0 → 1 so output of *SAR* will be 1000 0000.

After second clock:

SAR will toggle its 7th bit from 0 → 1 but 1100 0000 will result in value greater than 3.5 so output of SAR after 2nd clock will be 1000 0000.

After third clock:

SAR will toggle it's 6th bit from 0 → 1 and output will be 10100000.

6.8 (b)

Temperature range of –40°C to 55°C.
So total range in 95°C.
Since resolution is 0.1°C.
So number of steps will be 950.
To have 950 steps we need at least 10 bits.

6.9 (a)

The input to digital circuit is X_2, X_1, X_0 and output is B_1B_0

X_2	X_1	X_0	B_1	B_0
0	0	0	0	0
0	0	1	0	1
0	1	1	1	0
1	1	1	1	1

$B_0 = X_2$ \ X_1X_0	00	01	11	11
0	0	(1)	0	X
1	X	X	(1)	X

$$B_0 = \bar{X}_2\bar{X}_1X_0 + X_2X_1X_0$$

$$= X_0(\bar{X}_2\bar{X}_1 + X_2X_1) = X_0(\overline{X_2 \oplus X_1})$$

6.10 (c)

$$V_{in} = 1.992 \text{ V},$$
$$n = 8,$$
$$V_{fs} = 5 \text{ V}$$

We know, from graph

$$\text{Step size} = \frac{V_{fs}}{2^n - 1}$$

$$\text{Step size} = \frac{5}{2^8 - 1} = \frac{5}{255}$$

Analog input = Step size × (decimal equivalent of binary code)

$$1.992 = \frac{5}{255} \times D$$

$$D = \frac{1.992 \times 255}{5} = (101.592)_{10}$$

Rounding off to nearest integer,

$$D \approx 102 \Rightarrow (102)_{10} = (66)_{16} = 66\text{H}$$

7 Microprocessor

7.1 In an 8085 microprocessor, after the execution of XRA *A* instruction
(a) the carry flag is set.
(b) the accumulator contain FFH.
(c) the zero flag is set.
(d) the accumulator contents are shifted left by one bit. **[1995 : 1 M]**

7.2 In a 8085 microprocessor, the following instructions may results in change of accumulator contents and change in status flags. Choose the correct match for each instruction.

List-I
(a) ANA *R*
(b) XRA *R*
(c) CMP *R*

List-II

Contents of *ACC*	*CY* flag	*AC* flag
(P) unchanged	may be SET	may change
(Q) unchanged	SET	SET
(R) unchanged	SET	RESET
(S) may change	RESET	RESET
(T) may change	RESET	SET

[1996 : 3 M]

7.3 In a microprocessor, the address of the next instruction to be executed, is stored in
(a) stack pointer
(b) address latch
(c) program counter
(d) general purpose register **[1997 : 1 M]**

7.4 The range of addresses for which the memory chip shown in figure, will be selected is to

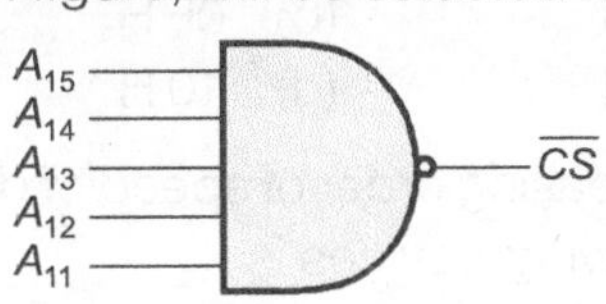

[1997 : 2 M]

7.5 Which of the following is not a vectored interrupt?
(a) TRAP (b) INTR
(c) RST 7.5 (d) RST 3 **[2000 : 1 M]**

7.6 An Intel 8085 processor is executing the program given below.

```
        MVI A, 10 H
        MVI B, 10 H
Back:   NOP
        ADD B
        RLC
        JNCBACK
        HLT
```

The number of times that the operation *NOP* will be executed is equal to
(a) 1 (b) 2
(c) 3 (d) 4 **[2001 : 2 M]**

7.7 The logic circuit used to generate the active low chip select (CS) by an 8085 microprocessor to address a peripheral is shown in figure. The peripheral will respond to addresses in the figure.

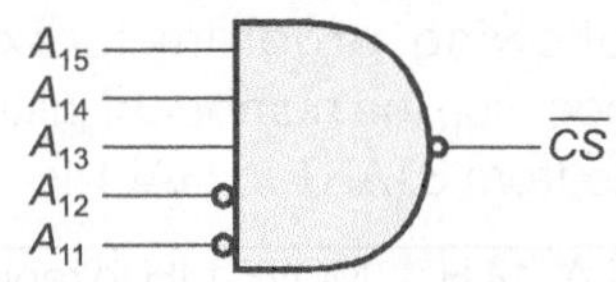

(a) *E*000-*EFFF* (b) 000*E*-*FFFE*
(c) 1000-*FFFF* (d) 0001-*FFF*1
[2002 : 1 M]

7.8 When a program is being executed in an 8085 microprocessor, its program counter contains
(a) the number of instructions in the current program that have already been executed.
(b) the total number of instructions in the program being executed.
(c) the memory address of the instruction that is being currently executed.
(d) the memory address of the instruction that is to be executed next. **[2002 : 1 M]**

7.9 When a program is being executed in an 8085 microprocessor, its Program Counter contains
(a) the number of instructions in the current program that have already been executed.
(b) the total number of instructions in the program being executed.
(c) the memory address of the instruction that is being currently executed.
(d) the memory address of the instruction that is to be executed next. **[2003 : 1 M]**

7.10 The following program is written for an 8085 microprocessor to add two bytes located at memory addresses 1FFE and 1FFF

```
LXI   H, 1FFE
MOV   B, M
INR   L
MOV   A, M
ADD   B
INR   L
MOV   M, A
XRA   A
```

On completion of the execution of the program, the result of addition is found

(a) in the register *A*
(b) at the memory address 1000
(c) at the memory address 1F00
(d) at the memory address 2000 [2003 : 2 M]

7.11 A memory system has a total of 8 memory chips, each with 12 address lines and 4 data lines. The total size of the memory system is

(a) 16 kbytes (b) 32 kbytes
(c) 48 kbytes (d) 64 kbytes [2003 : 2 M]

7.12* If the following program is executed in a microprocessor, the number of instruction cycles it will take from START to HALT is

START MVI A, 14 H ;	MOVE 14H to register A
SHIFT RLC ;	Rotate left without carry
JNZ SHIFT ;	Jump on non-zero to SHIFT
HALT ;	

(a) 4 (b) 8
(c) 13 (d) 16 [2004 : 2 M]

7.13 The 8085 assembly language instruction that stores the content of *H* and *L* registers into the memory locations 2050_H and 2051_H, respectively is

(a) SPHL 2050_H (b) SPHL 2051_H
(c) SHLD 2050_H (d) STAX 2050_H
[2005 : 1 M]

7.14 A software delay subroutine is written as given below:

```
DELAY : MVI   H,   255 D
        MVI   L,   255 D
LOOP :  DCR   L
        JNZ   LOOP
        DCR   H
        JNZ   LOOP
```

How many times DCR *L* instruction will be executed?

(a) 255 (b) 510
(c) 65025 (d) 65279 [2006 : 2 M]

7.15 In an 8085 A microprocessor bases system, it is desired to increment the contents of memory location whose address is available in (*D*, *E*) register pair and store the result in same location. The sequence of instruction is

(a) XCHG
INR M
(b) XCHG
INX H
(c) INX D
XCHG
(d) INR M
XCHG [2006 : 2 M]

7.16 The content of some of the memory location in an 8085A based system are given below.

Address	Content
..	..
26FE	00
26FF	01
2700	02
2701	03
2702	04
..	..

The content of stack (SP), program counter (PC) and (H, L) are 2700 H, 2100 H and 0000 H respectively. When the following sequence of instruction are executed

2100 H : DAD SP
2101 H : PCHL

The content of (SP) and (PC) at the end of execution will be

(a) PC = 2102 H, SP = 2700 H
(b) PC = 2700 H, SP = 2700 H
(c) PC = 2800 H, SP = 26FE H
(d) PC = 2A02 H, SP = 2702 H [2008 : 2 M]

7.17* In an 8085 microprocessor, the contents of the Accumulator, after the following instructions are executed will become

XRA A
MVIB F0H
SUB B

(a) 01 H (b) 0F H
(c) F0 H (d) 10 H [2009 : 2 M]

7.18 The increasing order of speed of data access for the following devices is

i. Cache Memory
ii. CD-ROM
iii. Dynamic RAM
iv. Processor Registers
v. Magnetic Tape

(a) (v), (ii), (iii), (iv), (i)
(b) (v), (ii), (iii), (i), (iv)
(c) (ii), (i), (iii), (iv), (v)
(d) (v), (ii), (i), (iii), (iv) [2009 : 1 M]

7.19 When a "CALL Addr" instruction is executed, the CPU carries out the following sequential operations internally:

Note:
(R) means content of register R
((R)) means content of memory location pointed to by R
PC means Program Counter
SP means Stack Pointer

(a) (SP) incremented
(PC) ← Addr
((SP)) ← (PC)

(b) (PC) ← Addr
((SP)) ← (PC)
(SP) incremented

(c) (PC) ← Addr
(SP) incremented
((SP)) ← (PC)

(d) ((SP)) ← (PC)
(SP) decremented
(PC) ← Addr

[2010 : 2 M]

7.20 A portion of the main program to call a subroutine SUB in an 8085 environment is given below.

```
        :
      LXI    D, DISP
LP:   CALL   SUB
        :
        :
```

It is desired that control be returned to LP + DISP + 3 when the RET instruction is executed in the subroutine. The set of instructions that precede the RET instruction in the subroutine are

(a) POP D
DAD H
PUSH D

(b) POP H
DAD D
INX H
INX H
INX H
PUSH H

(c) POP H
DAD D
PUSH H

(d) XTHL
INX D
INX D
INX D
XTHL

[2011 : 2 M]

7.21 An output device is interfaced with 8-bit microprocessor 8085 A. The interfacing circuit is shown in figure.

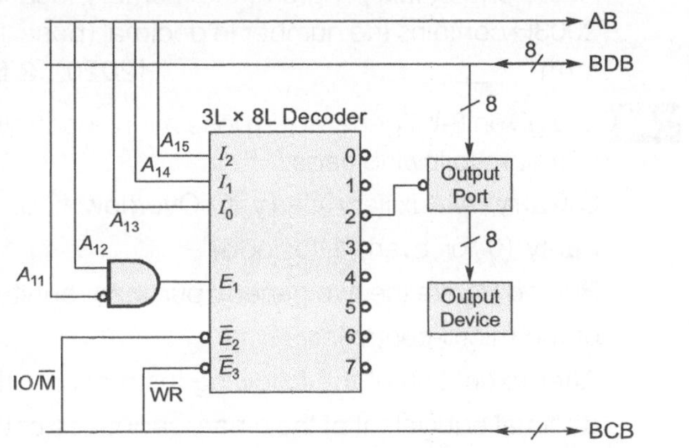

The interfacing circuit makes use of 3 Line to 8 Line decoder having 3 enable lines $E_1, \overline{E}_2, \overline{E}_3$. The address of the device is

(a) 50*H* (b) 5000*H*
(c) A0*H* (d) A000*H*

[2014 : 2 M, Set-1]

7.22 In 8085 microprocessor, the operation performed by the instruction *LHLD* 2100_H is

(a) (H) ← 21_H, (L) ← 00_H
(b) (H) ← M(2100_H), (L) ← M(2101_H)
(c) (H) ← M(2101_H), (L) ← M(2100_H)
(d) (H) ← 00_H, (L) ← 21_H

[2014 : 1 M, Set-3]

7.23 In an 8085 microprocessor, the following program is executed.

Address location-Instruction

2000 H	XRA A
2001 H	MVI B, 04 H
2003 H	MVI A, 03 H
2005 H	RAR
2006 H	DCR B
2007 H	JNZ 2005
200 AH	HLT

At the end of program, register *A* contains

(a) 60 H (b) 30 H
(c) 06 H (d) 03 H

[2014 : 2 M, Set-2]

7.24 An 8085 microprocessor accesses two memory locations (2001 H) and (2002H), that contain 8-bit numbers 98H and B1H respectively. The following program is executed:

```
LXI H,   2001 H
MVIA,    21H
INX H
ADD M
INX H
MOV M,   A
HLT
```

At the end of this program, the memory location 2003H contains the number in decimal (base 10) form ________. [2020 : 2 M]

7.25 In a given 8-bit general purpose micro-controller there are following flags.

C-Carry, A-Auxiliary Carry, O-Overflow flag, P-Parity (0 for even, 1 for odd)

R0 and R1 are the two general purpose registers of the micro-controller.

After execution of the following instructions, the decimal equivalent of the binary sequence of the flag pattern [CAOP] will be ________.

```
MOV R0, +0x60
MOV R1, +0x46
ADD R0, R1
```

[2023 : 2 Marks]

'*' Not in the syllabus GATE-2018.

Answers Microprocessors

7.1 (c)	7.2 (Sol.)	7.3 (c)	7.4 (Sol.)	7.5 (b)	7.6 (c)	7.7 (a)
7.8 (d)	7.9 (d)	7.10 (c)	7.11 (a)	7.12 (d)	7.13 (c)	7.14 (d)
7.15 (a)	7.16 (b)	7.17 (d)	7.18 (b)	7.19 (d)	7.20 (c)	7.21 (b)
7.22 (c)	7.23 (a)	7.24 (210)	7.25 (2)			

Explanations Microprocessors

7.1 (c)

Accumulator gets cleared.

7.2 Sol.

$a \rightarrow T$, $b \rightarrow S$, $c \rightarrow P$.

	Content of A	CY flag	AC flag
ANA R (T)	May change	Reset	Set
XRA R(S)	May change	Reset	Reset
CMP R(P)	Unchanged	May be set	May change

7.3 (c)

The address of next instruction is stored in program counter.

7.4 Sol.

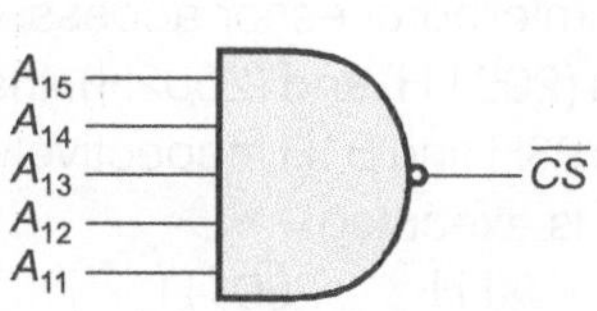

A_{15}	A_{14}	A_{13}	A_{12}	A_{11}	A_{10}	⋯	A_0
1	1	1	1	1	0	⋯	0
1	1	1	1	1	1	⋯	1

F800 to FFFFH.

7.6 (c)

First loop,

ADD B

Accumulator-20H | 0 | 0 | 1 | 0 | 0 | 0 | 0 | 0 |

RLC | 0 | 1 | 0 | 0 | 0 | 0 | 0 | 0 |

Second loop:

```
ACC  0 1 0 0   0 0 0 0
     0 0 0 1   0 0 0 0
     ---------------
     0 1 0 1   0 0 0 0
RLC  1 0 1 0   0 0 0 0
```

Third loop:

```
ACC  1 0 1 0   0 0 0 0   CY [0]
     0 0 0 1   0 0 0 0
     ---------------
     1 0 1 1   0 0 0 0
RLC  1 0 1 1   0 0 0 0   CY [1]
```

Carry is generated. Now program will halt. So NOP instruction is executed 3 times.

7.7 (a)

A_{15}	A_{14}	A_{13}	A_{12}	A_{11}	⋯	A_0
1	1	1	0	0	⋯	0
1	1	1	0	1	⋯	1

$\Rightarrow$ $E000 - EFFF$

7.9 (d)

Program counter contains the memory address of the instruction that is to be executed next.

7.10 (c)

LXI H, IFFE → Load 1FFE in H-L memory
MOV B, M → move content of 1FFE memory location to register B
INR L → Increment the content of 'HL' by 1
1FFE → 1FFF
MOV A, M → move content of 1FFF memory location to accumulator
ADD B → (A → A + B)
INR L → FF + 1 → 00
∴ HL → 1F 00
MOV M, A → move content of accumulator to memory location 1F 00
XRA A → content of accumulator becomes zero

7.11 (a)

Total size of the memory = No. of chip × data lines

$= 4 \times 8 \times 2^{12} = 128$ kbits $= 16$ kbyte

7.13 (c)

Instruction SHLD 2050_H stores the content of L to memory location 2050_H and content of register H to next memory location 2051_H.

7.14 (d)

First of all DCR L will be executed for 255 times till content of L becomes 0.
After that for every decrement of H it will be executed
256 times (first when L is decremented by 1 and become FF_H and then $FF_H = 255_D$ times) and this will be happen till content of H become 1, i.e., 254 times.
So overall it will be executed = 255 + 254 × 256 = 65279.

7.15 (a)

The address of the memory location is stored in *DE* register pair. But INR *M* command will increase the content of memory location *M*. But this command will execute only on *HL* pair. So we have to exchange the address of memory location in *HL* pair from *DE* pair first.

7.16 (b)

Given: (SP) = 2700 H; (PC) = 2100 H
(HL) = 0000 H
2100 H : DAD SP
(SP) + (HL) → (HL)
2700 H + 0000 H = 2700 H stored in HL pair
2101 H : PCHL : the content of HL are transferred to (PC)
So now (PC) = 2700 H and (SP) also unchanged
(PC) = 2700 H; (SP) = 2700 H

7.18 (b)

Access time from register is very less than that from a memory access. So speed of data access is fastest in case of processor registers.
Second highest is cache memory because it's size is small so searching of data takes less time.
So, option (b) is right option satisfying above two.

7.19 (d)

First of all content of PC is loaded into stack i.e. address of next instruction to be executed is loaded onto stack. i.e. SP is decremented then PC is loaded by address given in call instruction.

7.20 (c)

Call takes 3 address locations. RET always returns to **LP + 3** location, this stored in SP.
So to return to LP + DISP + 3.
We have to add DISP to SP.

POP H
DAD D
PUSH H

Normal call operation shown.

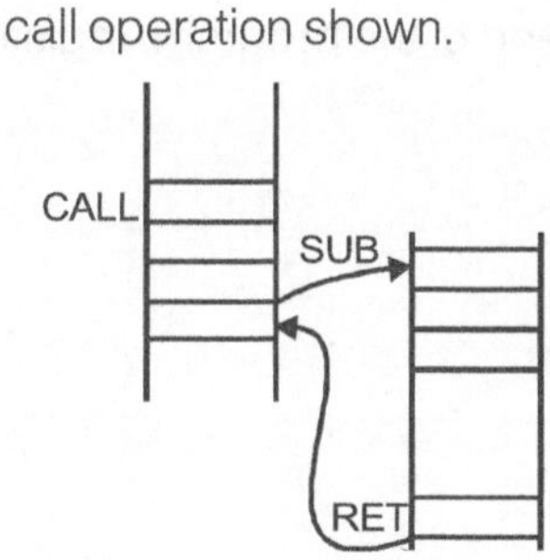

7.21 (b)

For the given decoder,

$$E_1 = A_{12}\bar{A}_{11} = 1\text{(enable)}$$

∴ A_{12} should be HIGH and A_{11} should be LOW.
Also, $E_2 = E_3 = 0$ (disable)
The decoder output depends on $I_2 = A_{15}, I_1 = A_{14}$

and $I_0 = A_{13}$ which should be 010 respectively to enable the output port.

A_{15}	A_{14}	A_{13}	A_{12}	A_{11}	A_{10}	A_9	A_8	A_7	A_6	A_5	A_4	A_3	A_2	A_1	A_0
0	1	0	1	0	0	0	0	0	0	0	0	0	0	0	0
0	1	0	1	0	1	1	1	1	1	1	1	1	1	1	1

The address of the device is 5000H-57FFH.

7.23 (a)

Address location	Instruction	Comment
2000 H	XRA A	$(A) \leftarrow 0000\,H$
2001H	MVI B, 04H	$(B) \leftarrow 04\,H$
2003H	MVI A, 03H	$(A) \leftarrow 03$ H
2005 H	RAR	Rotate the content of accumulator to the right with carry
2006 H	DCR *B*	$(B) \leftarrow (B) - 1$
2007 H	*JNZ* 2005	*Jump* to the address 2005 if the content of B is non zero
200A H	*HLT*	STOP

For the first RAR operation

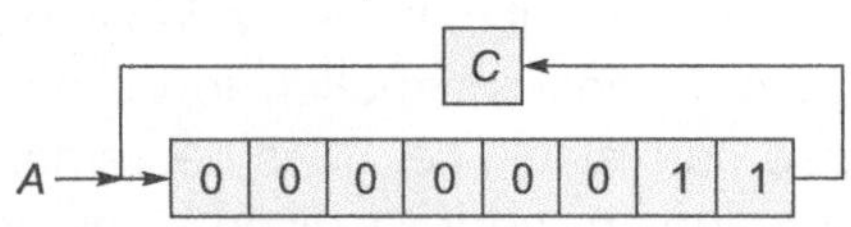

and $B \leftarrow 04$

Similarly for each RAR operation the content of *B* is decremented by '1' till it becomes 00H thus,

$B \leftarrow 04 \quad A \leftarrow 00000011$

$B \leftarrow 03 \quad A \leftarrow 10000001$

$B \leftarrow 02 \quad A \leftarrow 11000000$

$B \leftarrow 01 \quad A \leftarrow 01100000$

$B \leftarrow 00$

$\therefore$ At the end of the program, register '*A*' contains 60 H.

7.24 Sol.

LXI H, 2001 H $\rightarrow$ [H: 20 | L: 01]

MVI A, 21 H $\rightarrow$ [A: 21*H*]

INX H $\rightarrow$ HL + 1 [H: 20 | L: 02]

ADD M $\rightarrow$ [A] + data $\cong$ reference of HL pair

21 H + B1H = D2H $\rightarrow$ [A]

INX H $\rightarrow$ [HL] + 1 $\rightarrow$ 2002H + 1H $\rightarrow$ 2003H

MOV M, A $\rightarrow$ [*A*] to Memory, i.e., $\cong$ reference of HL pair

2003H [*D2*] $\leftarrow$ [A: *D2*]

HLT $\rightarrow$ Stop

$\therefore$ Content in the 2003 H is D2H

Converting in decimal

$D \times 16^1 + 2 \times 16^0 \Rightarrow 13 \times 16 + 2 = (210)_{10}$

7.25 (2)

MOV R_0, 0 × 60 ; $\quad R_0 \leftarrow 60$ H

MOV R_1, +0 × 46 ; $\quad R_1 \leftarrow 46$ H

ADD R_0, R_1; $R_0 \leftarrow [R_0] + [R_1]$

```
 D4   D3
0110 0000
0100 0110
1
---------
1010 0110
```

60H + 46H = A6H, i.e., 10100110

Overflow (O) $\rightarrow$ 1 ;Since if the two 8-bit data were considered as signed data then the result shows negative, i.e., Msb = 1 in A6H but both data bytes are positive.

Parity (P) $\rightarrow$ Even, as there are 'four' binary '1's in result A6H.

$\therefore$ P $\rightarrow$ 0

For Carry Flag (C $\rightarrow$ 0) .... No carry bit out of Mantisa.

For auxiliary carry (AC $\rightarrow$ 0)

No carry from D_3 to D_4 bit.

$\therefore$ [CAOP] $\rightarrow [0010]_2 = (2)_{10}$

Power Electronics

UNIT

IX

CONTENTS

Power Electronics

Syllabus

Static V-I Characteristics and Firing/Gating Circuits for Thyristor, MOSFET, IGBT; DC to DC Conversion: Buck, Boost and Buck-Boost Converters; Single and Three-Phase Configuration of Uncontrolled Rectifiers; Voltage and Current commutated Thyristor based Converters; Bidirectional AC to DC Voltage Source Converters; Magnitude and Phase of Line Current Harmonics for Uncontrolled and Thyristor based Converters; Power Factor and Distortion Factor of AC to DC Converters; Single-Phase and Three-Phase Voltage and Current Source Inverters, Sinusoidal Pulse Width Modulation.

Analysis of Previous GATE Papers

Exam Year	1 Mark Ques.	2 Marks Ques.	3 Marks Ques.	5 Marks Ques.	Total Marks
1995	7	–	–	1	**12**
1996	5	–	–	1	**10**
1997	2	–	–	2	**12**
1998	4	2	–	2	**18**
1999	2	–	–	1	**7**
2000	2	2	–	2	**16**
2001	2	2	–	2	**16**
2002	2	4	–	1	**15**
2003	3	4	–	–	**11**
2004	4	6	–	–	**16**
2005	3	5	–	–	**13**
2006	1	8	–	–	**17**
2007	4	7	–	–	**18**
2008	2	6	–	–	**14**
2009	1	4	–	–	**9**
2010	2	–	–	–	**2**
2011	2	3			**8**
2012	2	3			**8**

Exam Year	1 Mark Ques.	2 Mark Ques.	Total Marks
2013	–	7	**14**
2014 Set-1	2	3	**8**
2014 Set-2	1	3	**7**
2014 Set-3	–	3	**6**
2015 Set-1	2	3	**8**
2015 Set-2	2	3	**8**
2016 Set-1	2	4	**10**
2016 Set-2	3	4	**11**
2017 Set-1	4	2	**8**
2017 Set-2	4	3	**10**
2018	3	3	**9**
2019	2	4	**10**
2020	4	2	**8**
2021	1	3	**7**
2022	3	4	**11**
2023	2	2	**6**
2024	3	4	**11**

1 Power Semiconductor Devices and Commutation Techniques

1.1 The figure show two thyristors, each rated 500 A (continuous) sharing a load current. Current through thyristor y is 120 A. The current through thyristor x will be nearly ______ A

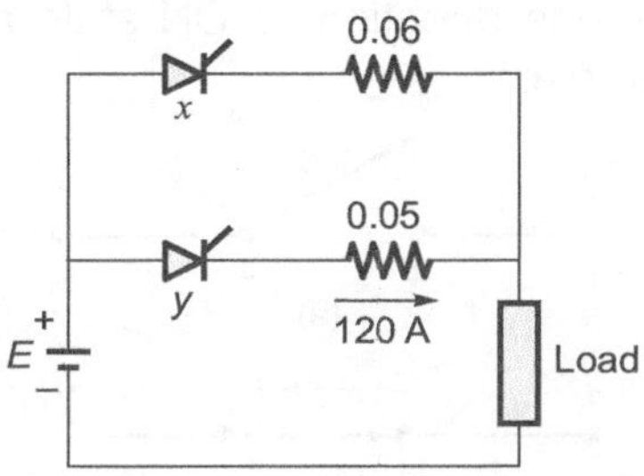

[1995 : 1 M]

1.2 The Triac can be used in
(a) inverter
(b) rectifier
(c) multiquadrant chopper
(d) AC voltage regulator [1996 : 1 M]

1.3 Which semiconductor power device out of the following is not a current triggered device?
(a) Thyristor (b) G.T.O
(c) Triac (d) MOSFET [1996 : 1 M]

1.4 Which of the following does not cause permanent damage of an SCR?
(a) High current
(b) High rate of rise of current
(c) High temperature rise
(d) High rate of rise of voltage [1996 : 1 M]

1.5 The MOSFET switch in its ON state may be considered equivalent to
(a) resistor (b) inductor
(c) capacitor (d) battery [1998 : 1 M]

1.6 The uncontrolled electronic switch employed in power electronic converters is
(a) thyristor
(b) bipolar junction transistor
(c) diode
(d) MOSFET [1998 : 1 M]

1.7 In a commutation circuit employed to turn-off an SCR, satisfactory turn-off is obtained when
(a) Circuit turn-off time < device turn-off time
(b) Circuit turn-off time > device turn-off time
(c) Circuit time constant > device turn-off time
(d) Circuit time constant < device turn-off time
[1998 : 1 M]

1.8 The figure shows a MOSFET with an integral body diode. It is employed as a power switching device in the ON and OFF states through appropriate control. The ON and OFF states of the switch are given on the V_{DS}-I_s plane by

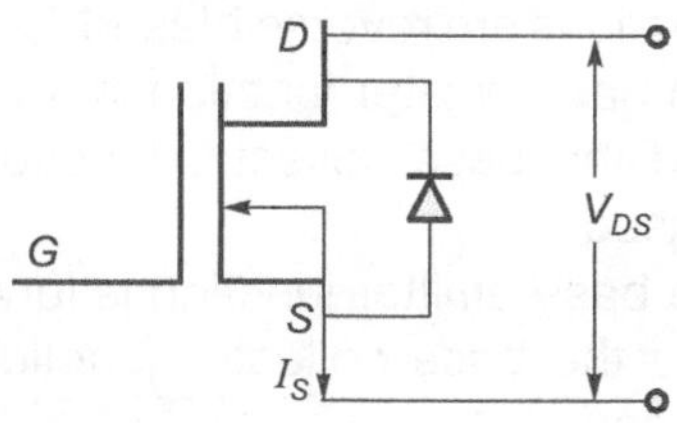

(a)
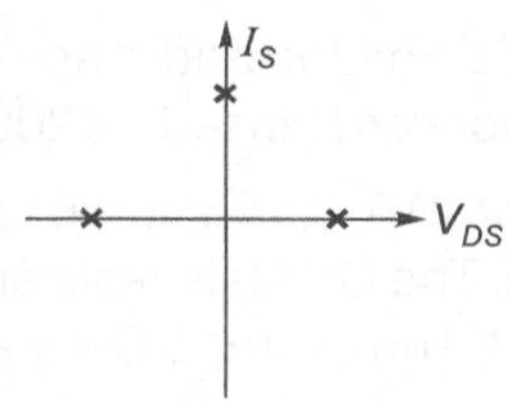

(b)
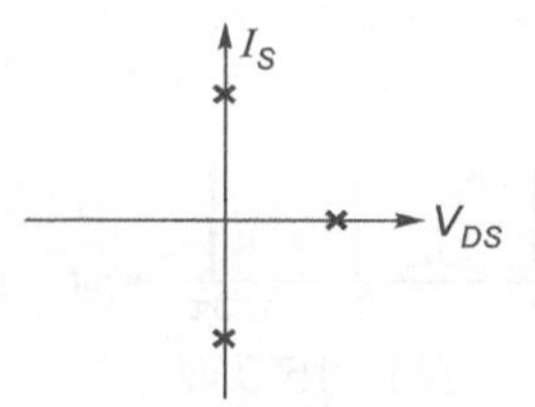

(c)
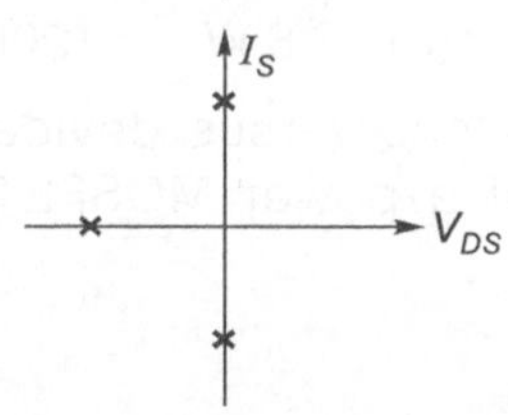

(d)
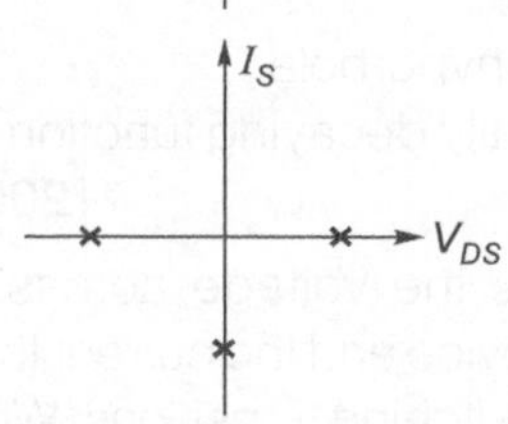

[2003 : 1 M]

1.9 The figure shows a thyristor with the standard terminations of anode (A), cathode (K), gate (G) and the different junctions named J_1, J_2 and J_3. When the thyristor is turned on and conducting.

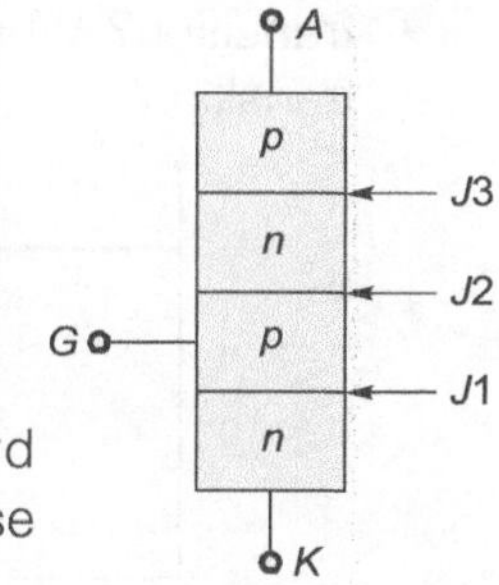

(a) J_1 and J_2 are forward biased and J_3 is reverse biased.

(b) J_1 and J_3 are forward biased and J_2 is reverse biased.
(c) J_1 is forward biased and J_2 and J_3 are reverse biased.
(d) J_1, J_2 and J_3 are all forward biased.

[2003 : 1 M]

1.10 A bipolar junction transistor (BJT) is used as a power control switch by biasing it in the cut-off region (OFF state) or in the saturation region (ON state). In the ON state, for the BJT
(a) both the base-emitter and base-collector junctions are reverse biased.
(b) the base-emitter junction is reverse biased, and the base-collector junction is forward biased.
(c) the base-emitter junction is forward biased, and the base-collector junction is reverse biased.
(d) both the base-emitter and base-collector junctions are forward biased. [2004 : 1 M]

1.11 A MOSFET rated for 10 A, carries a periodic current as shown in figure. The ON state resistance of the MOSFET is 0.15 Ω. The average ON state loss in the MOSFET is

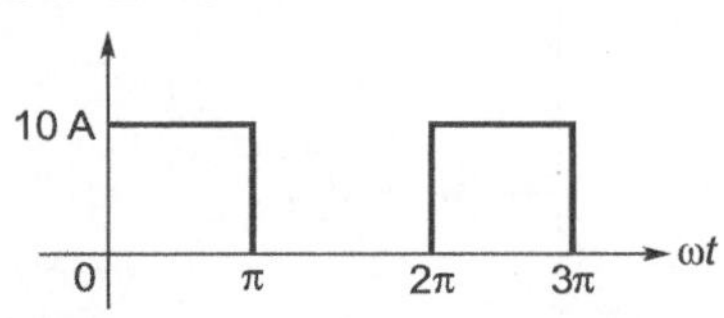

(a) 33.8 W (b) 15.0 W
(c) 7.5 W (d) 3.8 W [2004 : 2 M]

1.12 The conduction loss versus device current characteristic of a power MOSFET is best approximated by
(a) a parabola
(b) a straight line
(c) a rectangular hyperbola
(d) an exponentially decaying function

[2005 : 1 M]

1.13 The figure shows the voltage across a power semiconductor device and the current through the device during a switching transitions. Whether the transition is a turn ON transition or a turn OFF transition? What is the energy lost during the transition?

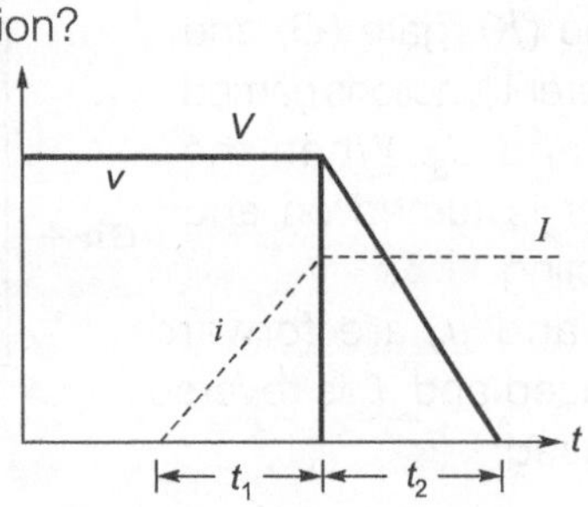

(a) Turn ON, $\frac{VI}{2}(t_1 + t_2)$
(b) Turn OFF, $VI(t_1 + t_2)$
(c) Turn ON, $VI(t_1 + t_2)$
(d) Turn OFF, $\frac{VI}{2}(t_1 + t_2)$ [2005 : 2 M]

1.14 An electronic switch S is required to block voltage of either polarity during its OFF state as shown in the figure (a). This switch is required to conduct in only one direction its ON state as shown in the figure (b).

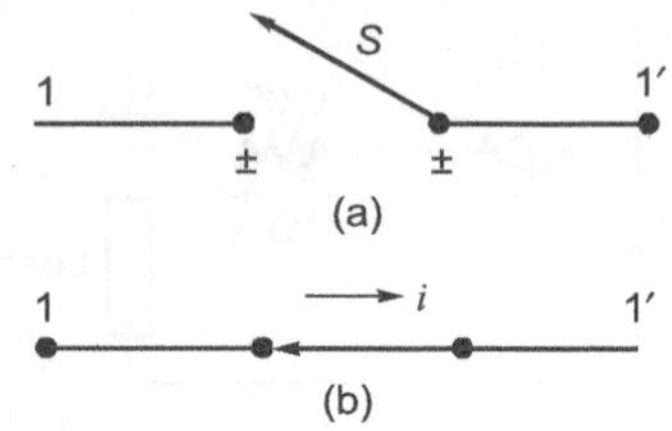

Which of the following are valid realizations of the switch S?

1. 

2. 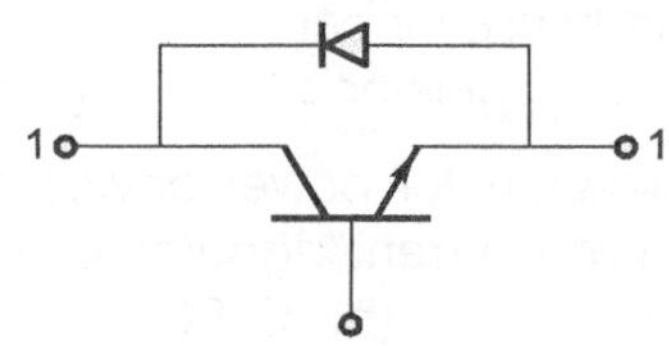

3. 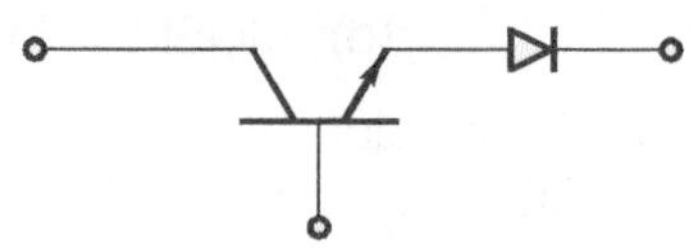

4. 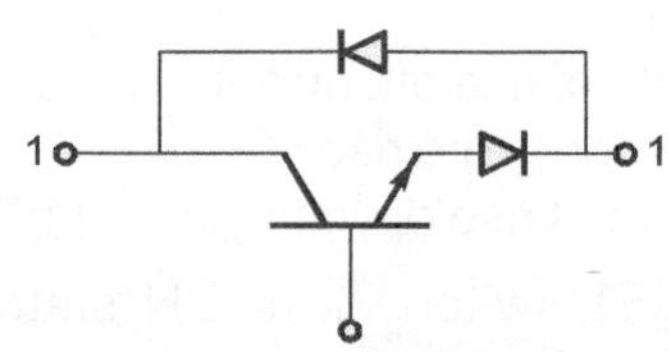

(a) Only 1 (b) 1 and 2
(c) 1 and 3 (d) 3 and 4 [2005 : 2 M]

1.15 A voltage commutation circuit is shown in the figure. If the turn off time of the SCRs is 50 μ sec and a safety margin of 2 is considered, what will be the approximate minimum value of capacitor required for proper commutation?

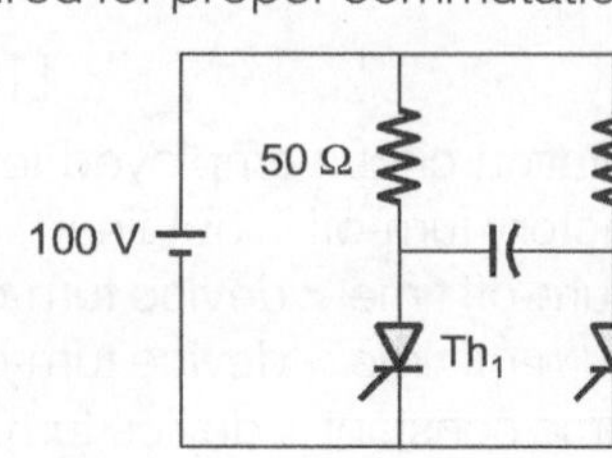

(a) 2.88 μF (b) 1.44 μF
(c) 0.91 μF (d) 0.72 μF [2006 : 2 M]

1.16 An SCR having a turn ON time of 5 μsec, latching current of 50 mA and holding current of 40 mA is triggered by a short duration pulse and is used in the circuit shown in the figure. The minimum pulse width required to turn the SCR ON will be

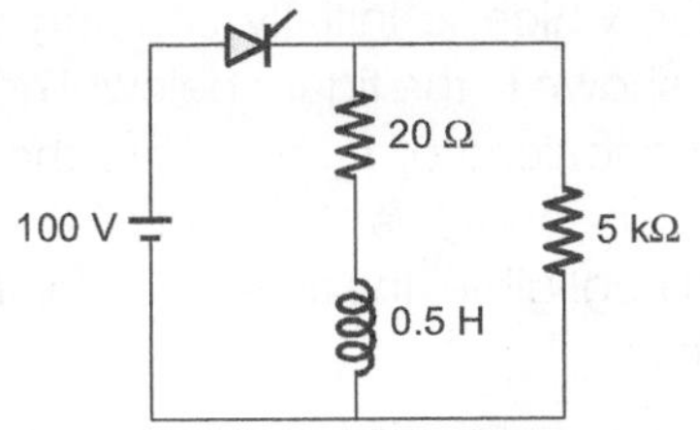

(a) 251 μsec (b) 150 μsec
(c) 100 μsec (d) 5 μsec [2006 : 2 M]

1.17 The circuit in the figure is a current commutated DC-DC chopper where, Th_M is the main SCR and Th_{AUX} is the auxiliary SCR. The load current is constant at 10 A. Th_M is ON. Th_{AUX} is triggered at $t = 0$. T_{hM} is turned OFF between.

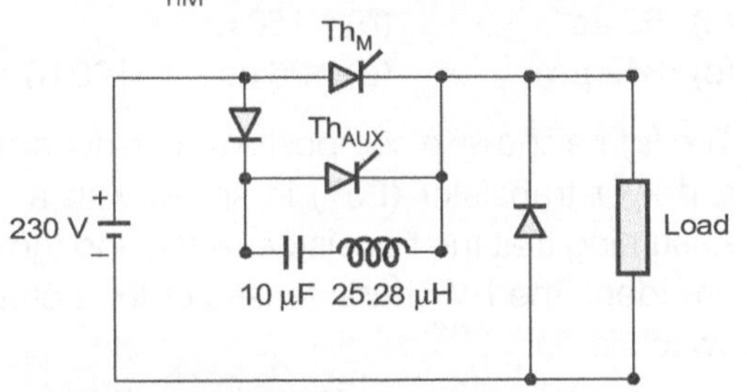

(a) $0\ \mu s < t \leq 25\ \mu s$
(b) $25\ \mu s < t \leq 50\ \mu s$
(c) $50\ \mu s < t \leq 75\ \mu s$
(d) $75\ \mu s < t \leq 100\ \mu s$ [2007 : 2 M]

Common Data for Questions (1.18 and 1.19):

A 1:1 Pulse transformer (PT) is used to trigger the SCR in the figure. The SCR is rated at 1.5 kV, 250 A with I_L = 250 mA, I_H = 150 mA, and I_{Gmax} = 150 mA with I_L =250 mA, I_{Gmin} = 100mA. The SCR is connected to an inductive load, where L =150 mH in series with a small resistance and the supply voltage is 200 V DC. The forward drops of all transistors /diodes and gate-cathode junction during ON state are 1.0 V

1.18 The resistance R should be

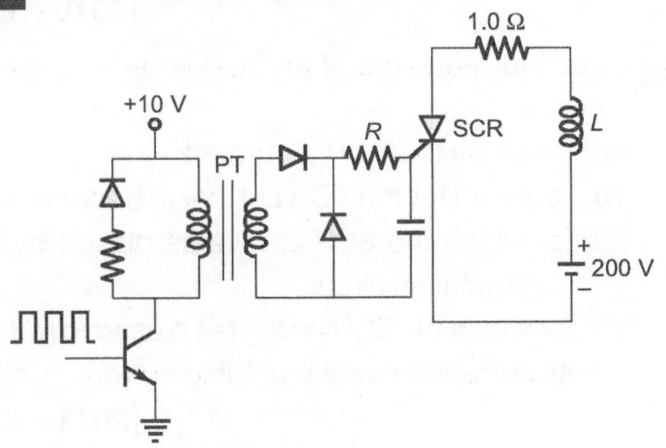

(a) 4.7 kΩ (b) 470 kΩ
(c) 47 Ω (d) 4.7 Ω [2007 : 2 M]

1.19 The minimum approximate volt-second rating of the pulse transformer suitable for triggering the SCR should be: (Volt-second rating is the maximum of product of the voltage and the width of the pulse that may be applied)

(a) 2000 μV-s (b) 200 μV-s
(c) 20 μV-s (d) 2 μV-s [2007 : 2 M]

1.20 In the circuit of adjacent figure the diode connects the AC source to a pure inductance L.

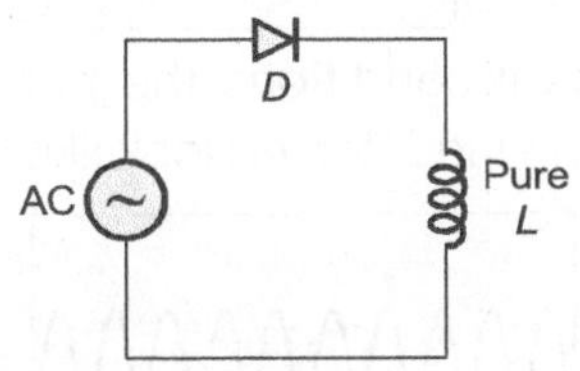

The diode conducts for
(a) 90° (b) 180°
(c) 270° (d) 360° [2007 : 2 M]

1.21 An SCR is considered to be a semi-controlled device because
(a) it can be turned OFF but not ON with a gate pulse.
(b) it conducts only during one half-cycle of an alternating current wave.
(c) it can be turned ON but not OFF with a gate pulse.
(d) it can be turned ON only during one half-cycle of an alternating voltage wave. [2009 : 1 M]

1.22 Match the switch arrangements on the top row to the steady-state V-I characteristics on the lower row. The steady state operating points are shown by large black dots.

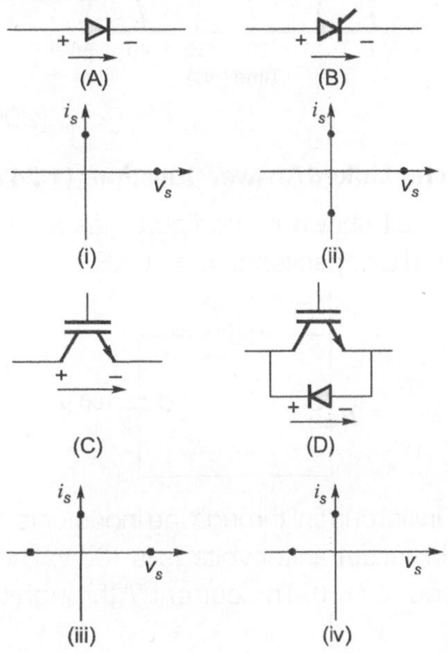

(a) A–i, B-ii, C-iii, D-iv
(b) A–ii, B-iv, C-i, D-iii
(c) A–iv, B-iii, C-i, D-ii
(d) A–iv, B-iii, C-ii, D-i [2009 : 2 M]

1.23 The circuit shows an ideal diode connected to a pure inductor and is connected to a purely sinusoidal 50 Hz voltage source.

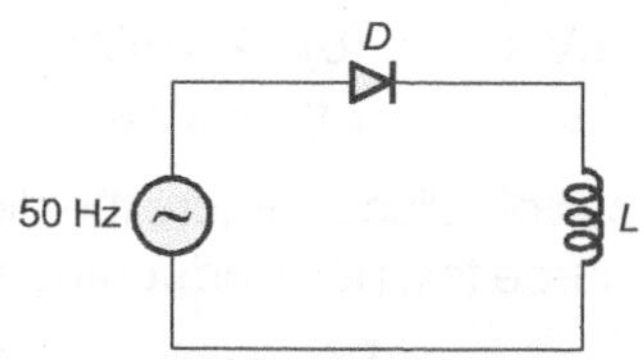

Under ideal conditions the current waveform through the inductor will look like

(a)
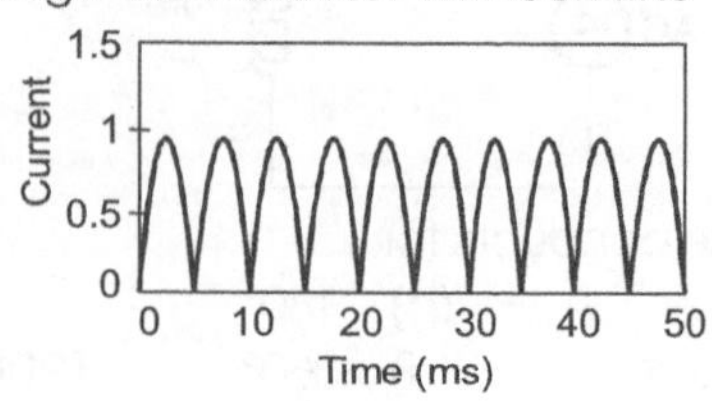

(b)
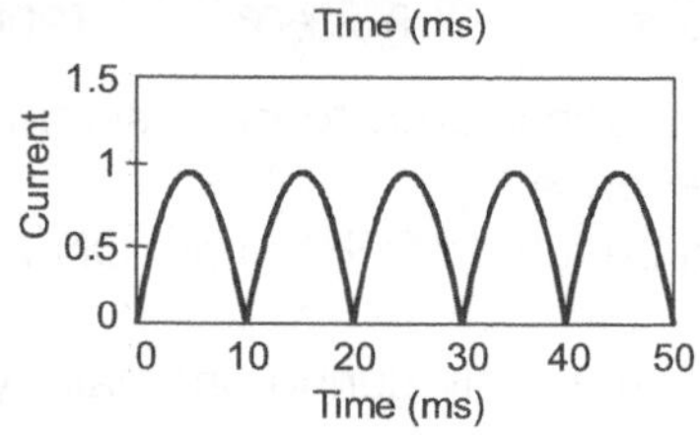

(c)
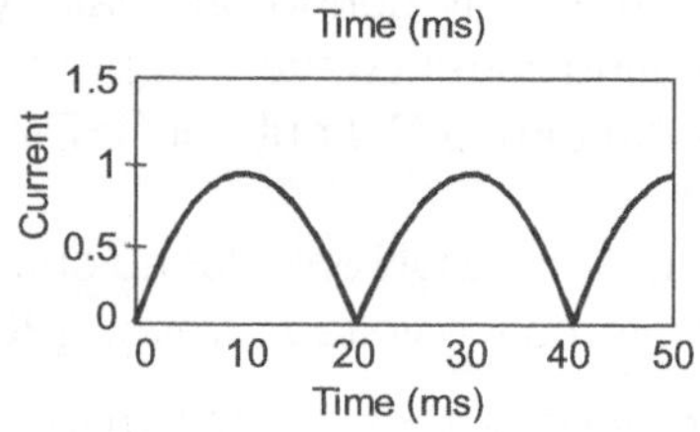

(d)
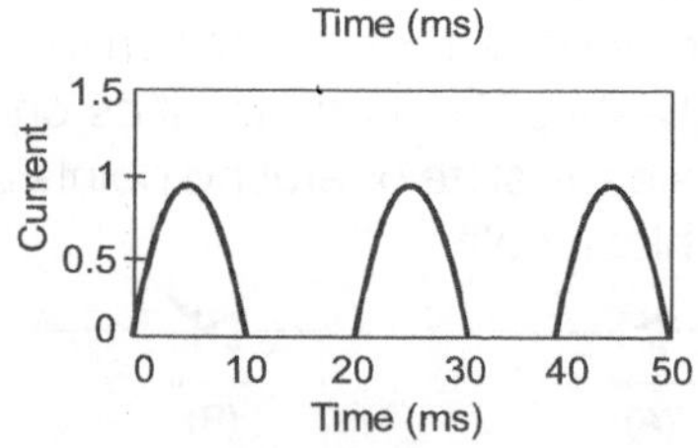

[2009 : 2 M]

Statement for Linked Answer Questions (1.24 and 1.25):

The L-C circuit shown in the figure has an inductance $L = 1$ mH and a capacitance $C = 10$ μF.

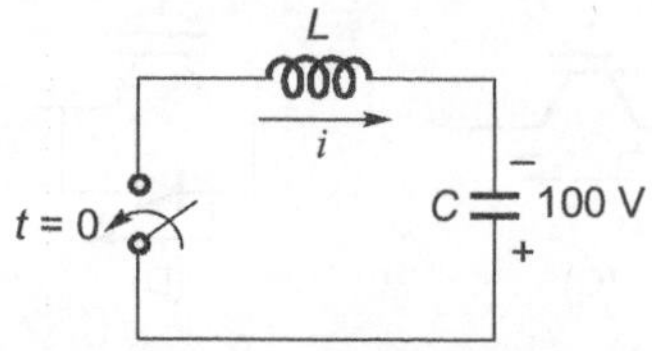

1.24 The initial current through the inductor is zero, while the initial capacitor voltage is 100 V. The switch is closed at $t = 0$. The current 'i' through the circuit is

(a) $5\cos(5 \times 10^3 t)$ A
(b) $5\sin(10^4 t)$ A
(c) $10\cos(5 \times 10^3 t)$ A
(d) $10\sin(10^4 t)$ A [2010 : 2 M]

1.25 The L-C circuit of Q.1.35 is used to commutate a thyristor, which is initially carrying a current of 5 A as shown in the figure below. The values and initial conditions of L and C are the same as in Q.1.35. The switch is closed at $t = 0$. If the forward drop is negligible, the time taken for the device to turn off is

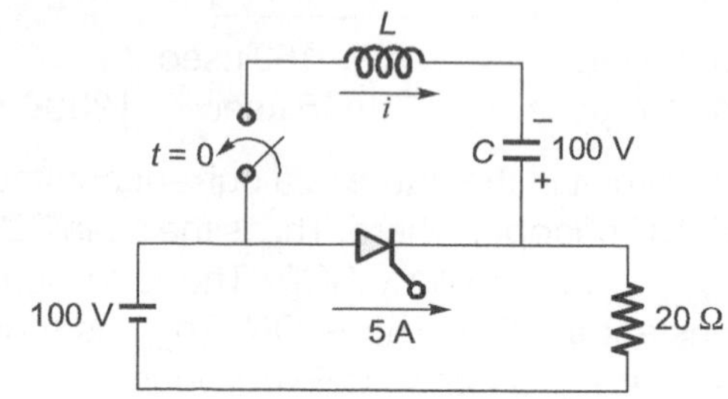

(a) 52 μs (b) 156 μs
(c) 312 μs (d) 26 μs [2010 : 2 M]

1.26 The figure shows a composite switch consisting of a power transistor (BJT) in series with a diode. Assuming that the transistor switch and the diode are ideal, the I-V characteristic of the composite switch is

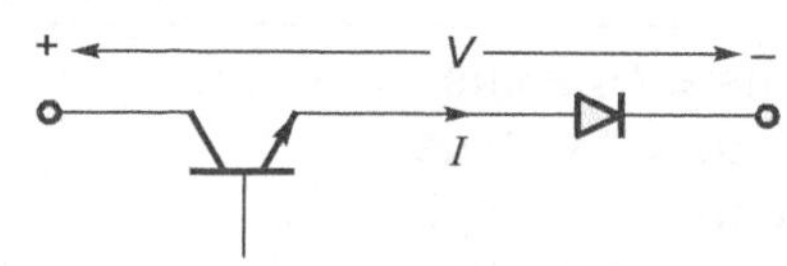

(a)
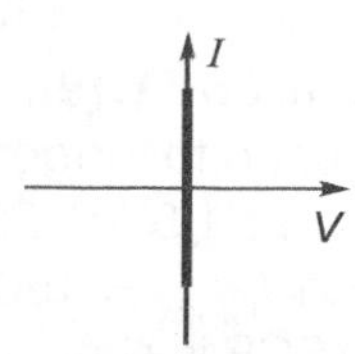

(b)
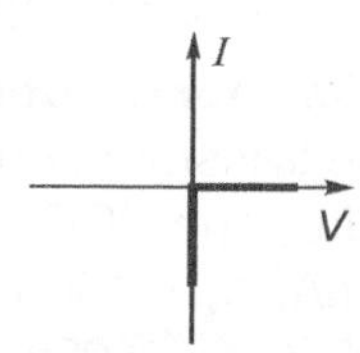

(c)
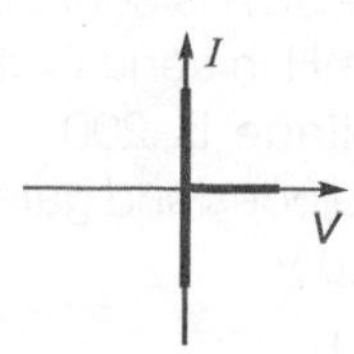

(d)
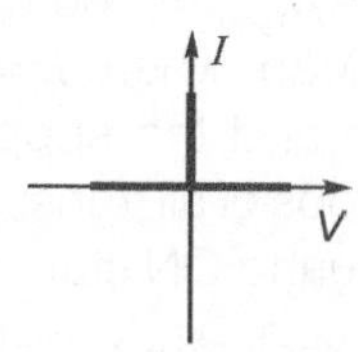

[2010 : 1 M]

1.27 Circuit turn-off time of an SCR is defined as the time

(a) taken by the SCR to turn off
(b) required for the SCR current to become zero
(c) for which the SCR is reverse biased by the commutation circuit
(d) for which the SCR is reverse biased to reduce its current below the holding current

[2011 : 1 M]

1.28 A voltage commuted chopper circuit, operated at 500 Hz, is shown below.

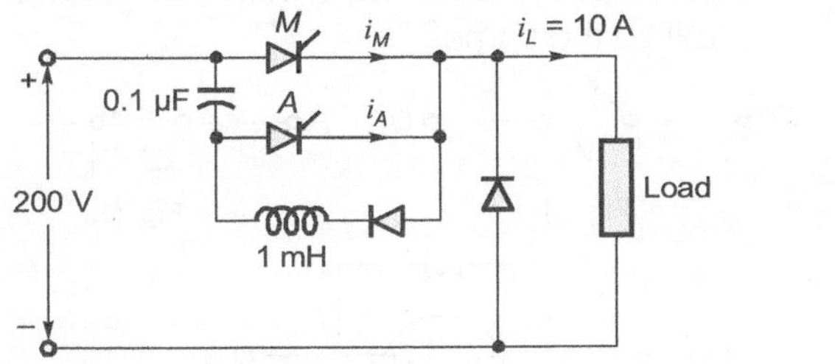

If the maximum value of load current is 10 A, then the maximum current through the main (*M*) and auxiliary (*A*) thyristors will be

(a) $i_{M\,max} = 12$ A and $i_{A\,max} = 10$ A
(b) $i_{M\,max} = 12$ A and $i_{A\,max} = 2$ A
(c) $i_{M\,max} = 10$ A and $i_{A\,max} = 12$ A
(d) $i_{M\,max} = 10$ A and $i_{A\,max} = 8$ A **[2011 : 2 M]**

1.29 The typical ratio of latching current to holding current in a 20 thyristor is

(a) 5.0 (b) 2.0
(c) 1.0 (d) 0.5 **[2012 : 1 M]**

1.30 The figure shows four electronic switches (i), (ii), (iii) and (iv). Which of the switches can block voltages of either polarity (applied between terminals '*a*' and '*b*') when the active device is in the OFF state.

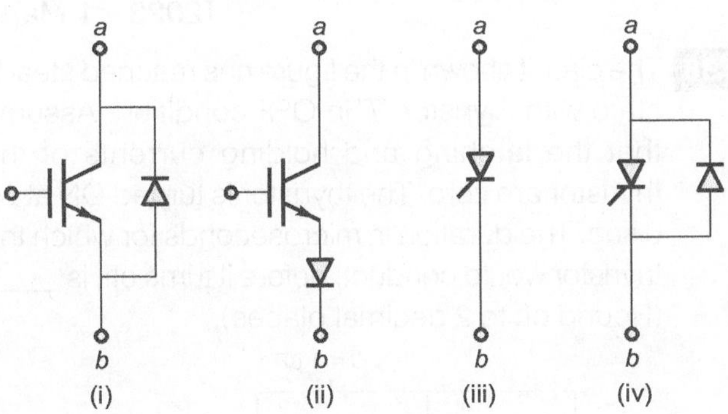

(a) (i), (ii) and (iii) (b) (ii), (iii) and (iv)
(c) (ii) and (iii) (d) (i) and (iv)

[2014 : 1 M, Set-1]

1.31 A steady DC current of 100 A is flowing through a power module (*S*, *D*) as shown in the figure (a). The V-I characteristics of the IGBT (*S*) and the diode (*D*) are shown in the figures (b) and (c), respectively. The conduction power loss in the power module (*S*, *D*), in watts, is ________.

[2016 : 1 M, Set-1]

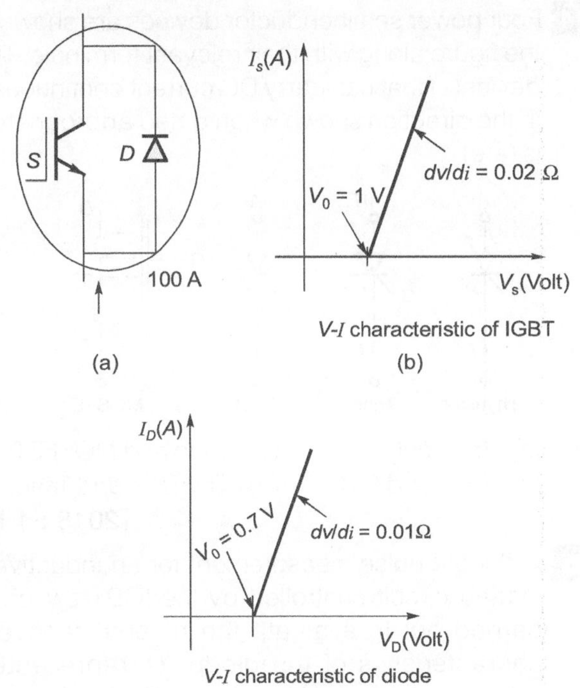

1.32 The voltage (v_s) across and the current (i_s) through a semiconductor switch during a turn-ON transition are shown in the figure. The energy dissipated during the turn-ON transition, in mJ, is

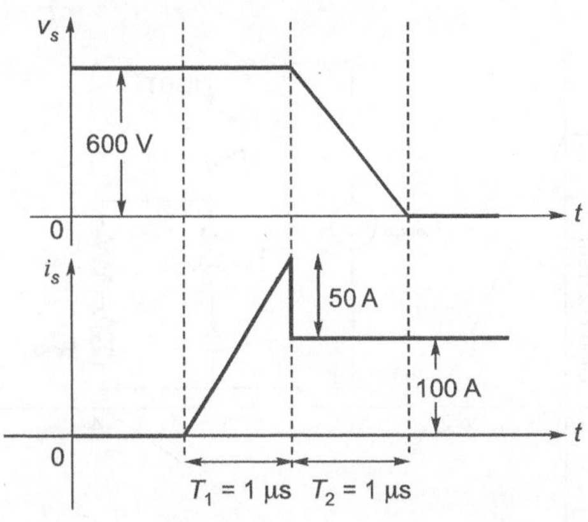

[2016 : 2 M, Set-1]

1.33 For the power semiconductor devices IGBT, MOSFET, diode and Thyristor, which one of the following statements is TRUE?

(a) All the four are majority carrier devices.
(b) All the four are minority carrier devices.
(c) IGBT and MOSFET are majority carrier devices, whereas Diode and Thyristor are minority carrier devices.
(d) MOSFET is majority carrier device, whereas IGBT, Diode, thyristor are minority carrier devices. **[2017 : 1 M, Set-1]**

1.34 Four power semiconductor devices are shown in the figure along with their relevant terminals. The device(s) that can carry DC current continuously in the direction shown when gated appropriately is (are)

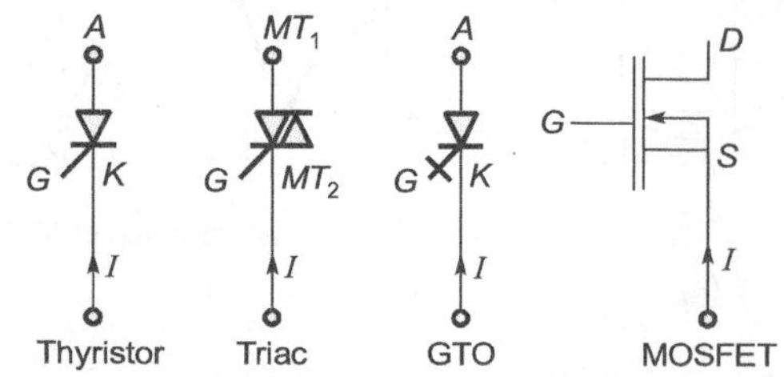

(a) Triac only
(b) Triac and MOSFET
(c) Triac and GTO
(d) Thyristor and Triac

[2018 : 1 M]

1.35 A double pulse measurement for an inductively loaded circuit controlled by the IGBT switch is carried out to evaluate the reverse recovery characteristics of the diode. D, represented approximately as a piecewise linear plot of current vs time at diode turn-off. L_{par} is a parasitic inductance due to the wiring of the circuit, and is in series with the diode. The point on the plot (indicate your choice by entering 1. 2, 3 or 4) at which the IGBT experiences the highest current stress is ______.

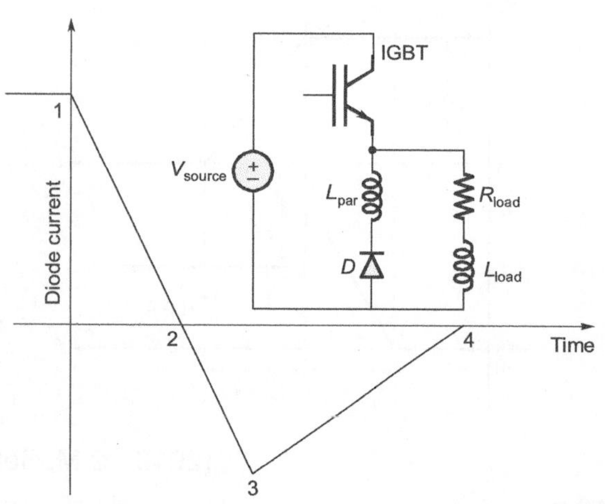

[2020 : 1 M]

1.36 A charger supplies 100 W at 20 V for charging the battery of a laptop. The power devices, used in the converter inside the charger, operate at a switching frequency of 200 kHz. Which power device is best suited for this purpose?

(a) IGBT
(b) Thyristor
(c) MOSFET
(d) BJT

[2022 : 1 M]

1.37 A semiconductor switch needs to block voltage V of only one polarity ($V > 0$) during OFF state as shown in figure (i) and carry current in both directions during ON state as shown in figure (ii). Which of the following switch combination(s) will realize the same?

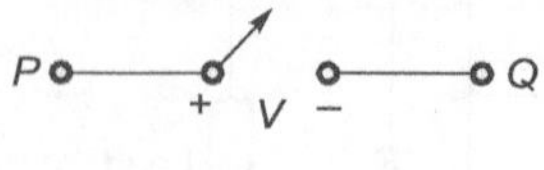

Fig. (i)

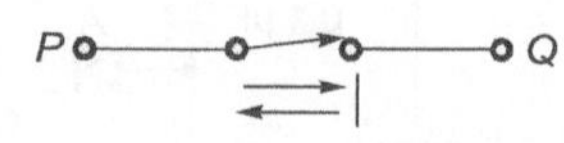

Fig. (ii)

(a) 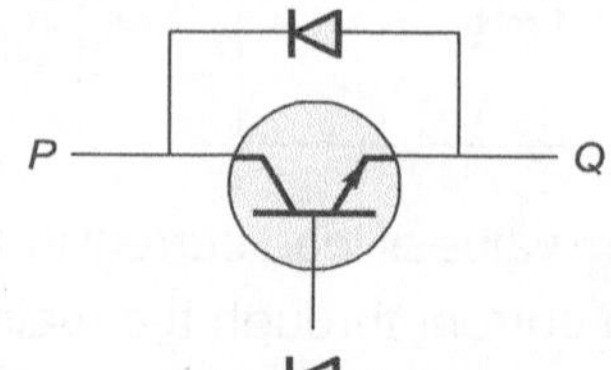

(b) 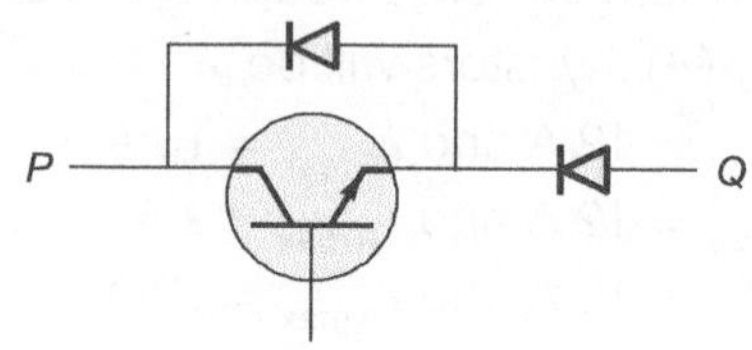

(c) 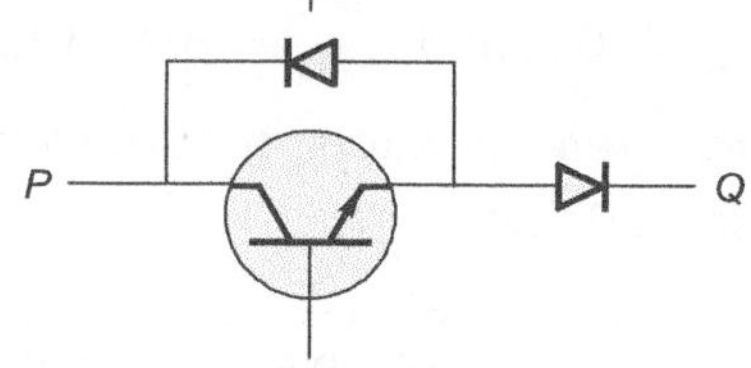

(d) 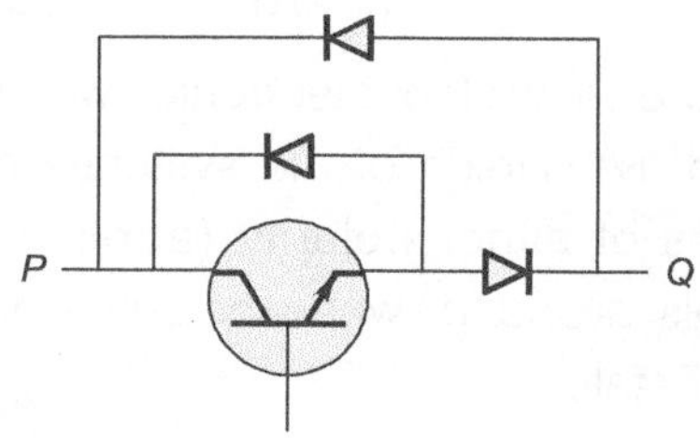

[2023 : 1 Mark]

1.38 The circuit shown in the figure has reached steady state with thyristor 'T' in OFF condition. Assume that the latching and holding currents of the thyristor are zero. The thyristor is turned ON at t = 0 sec. The duration in microseconds for which the thyristor would conduct, before it turns off, is ______ (Round off to 2 decimal places).

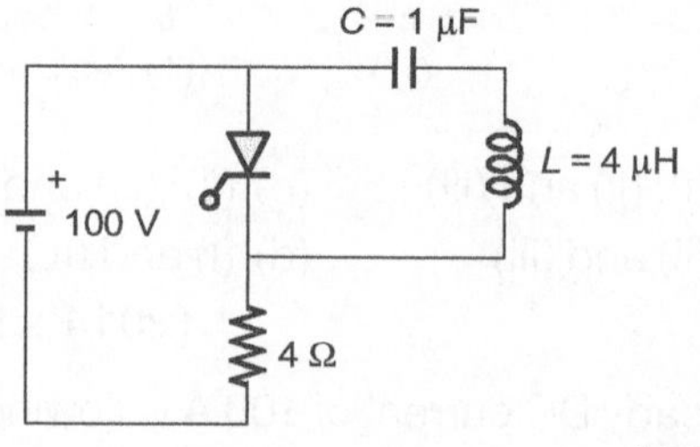

[2023 : 2 M]

1.39 If the following switching devices have similar power ratings, which one of them is the fastest?

(a) SCR
(b) GTO
(c) Power MOSFET
(d) IGBT

[2024 : 1 M]

■■■■

Answers Power Semiconductor Devices & Commutation Techniques

1.1	(100)	1.2	(d)	1.3	(d)	1.4	(d)	1.5	(a)	1.6	(c)	1.7	(b)
1.8	(b)	1.9	(d)	1.10	(d)	1.11	(c)	1.12	(a)	1.13	(a)	1.14	(c)
1.15	(a)	1.16	(b)	1.17	(c)	1.18	(c)	1.19	(a)	1.20	(d)	1.21	(c)
1.22	(c)	1.23	(c)	1.24	(d)	1.25	(a)	1.26	(d)	1.27	(c)	1.28	(a)
1.29	(b)	1.30	(c)	1.31	(170)	1.32	(75)	1.33	(d)	1.34	(b)	1.35	(3)
1.36	(c)	1.37	(a, d)	1.38	(7.33)	1.39	(c)						

Explanations Power Semiconductor Devices & Commutation Techniques

1.1 Sol.

Since both the thyristors are working in parallel, the voltage difference will be same.

$$120 \times 0.05 = I_x \times 0.06 \Rightarrow I_x = \frac{120 \times 5}{6} = 100 \text{ A}$$

1.5 (a)

During on state MOSFET works as resistor.

1.7 (b)

The circuit should not turn-off before making off the thyristor.

1.8 (b)

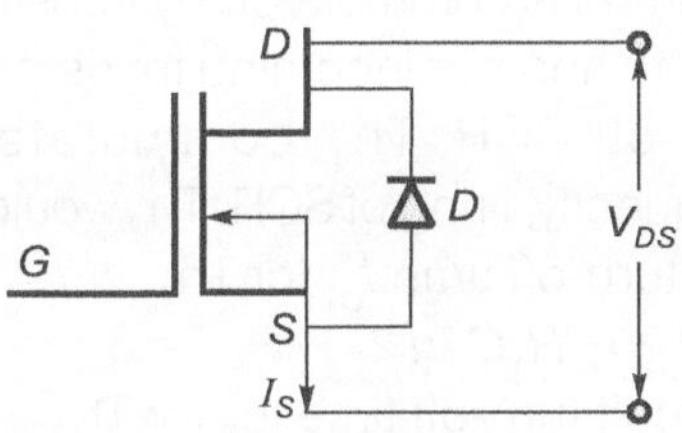

When reverse current flows through diode D.

So, $I_S < 0$ and $V_{DS} = 0$

When MOSFET is in ON state,

$I_S > 0$ and $V_{DS} = 0$

When MOSFET is in OFF state,

$I_s = 0$ and $V_{DS} > 0$

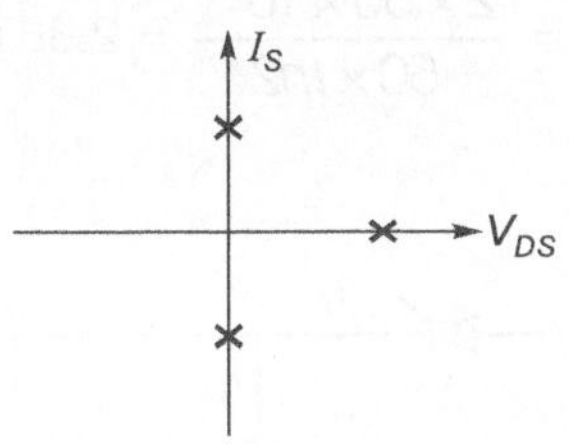

1.10 (d)

$$V_{CB} = V_{CE} - V_{DE} \quad \text{...(i)}$$

Under saturated state, V_{BES} is greater than V_{CES} this means base-emitter junction (BEJ) is forward biased. Further eq. (i) shows that V_{CB} is negative under saturated conditions, therefore base-collector junction (CBJ) is also forward biased.

1.11 (c)

Rated current during on state, $I = 10$

ON state resistance, $R_{ON} = 0.15\ \Omega$

MOSFET is ON

$$0 < \omega t < \pi \Rightarrow 0 < t < \pi/\omega$$

MOSFET is OFF

$$\pi < \omega t < 2\pi \Rightarrow \pi/\omega < t < 2\pi/\omega$$

Average ON state Loss,

$$P_{avg} = \frac{1}{(2\pi/\omega)}\int_0^{\pi/\omega} I^2 R_{ON}\, dt$$

$$= \frac{\omega}{2\pi} \times I^2 R_{ON} \times \frac{\pi}{\omega}$$

$$= \frac{I^2 R_{ON}}{2} = \frac{10^2 \times 0.15}{2} = 7.5 \text{ W}$$

1.12 (a)

Let, I = device current

R_{ON} = ON state resistance of power MOSFET

Conduction loss = $P = I^2 R_{ON}$

Therefore, condition losses versus device current characteristics can be best approximated by a parabola.

1.13 (a)

During interval t_2, voltage starts decreasing and becomes zero and current starts increasing and becomes constant (I), so transition is turn on.

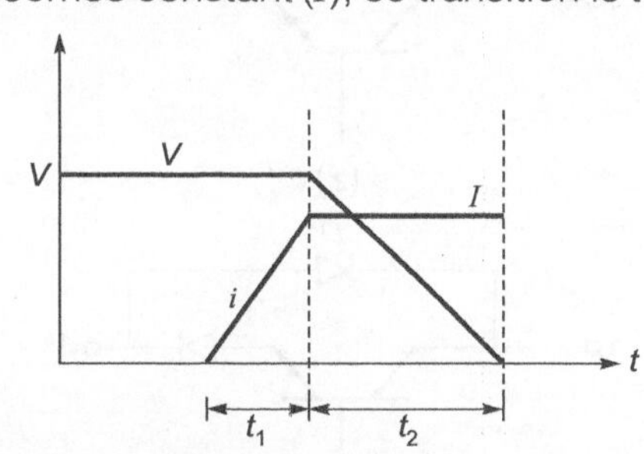

During t_1 interval:

Power loss = vi

$$E_1 = \text{Energy loss} = \int vi\,dt = V\int i\,dt$$

V is constant during this period $v = V$

$\int i\,dt$ represents area under i-t curve

$$\int i dt = \frac{1}{2} \times I \times t_1$$

$$E_1 = V\int i dt = \frac{1}{2} V I t_1 \quad \text{...(i)}$$

During t_2 interval: Power loss = vi

E_2 = Energy loss = $\int v i\,dt = I\int v dt$

i is constant during this period $i = I$

$\int v dt$ represents are under v-t curve

$$\int v dt = \frac{1}{2} V I t_2$$

$$E_2 = I\int v dt = \frac{1}{2} V I t_2 \quad \text{...(ii)}$$

Total energy lost during the transition

$$E = E_1 + E_2 = \frac{1}{2} V I t_1 + \frac{1}{2} V I t_2$$

1.14 (c)

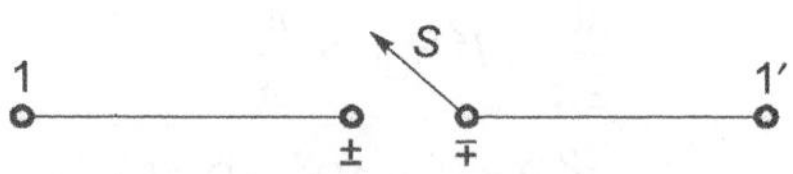

1. Switch s blocks voltage of both polarity, it means s can block forward as well as reverse voltage.

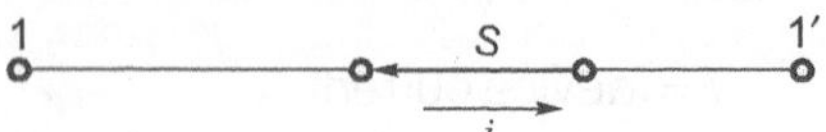

(1)

2. Current through s, flows in forward direction only.

Thyristor blocks voltage in either polarity until gate is triggered. Once the thyristor is triggered current flows in forward direction.

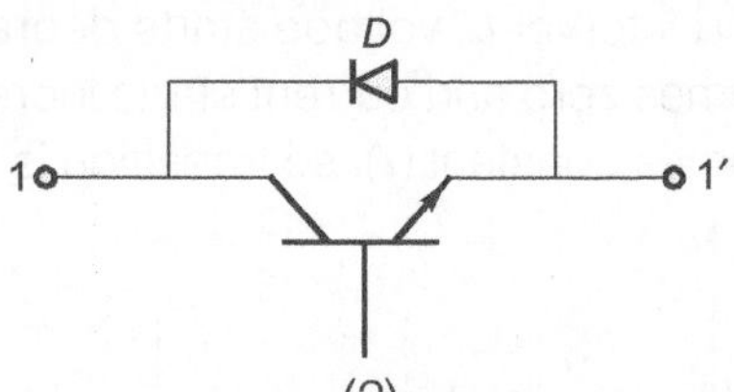

(2)

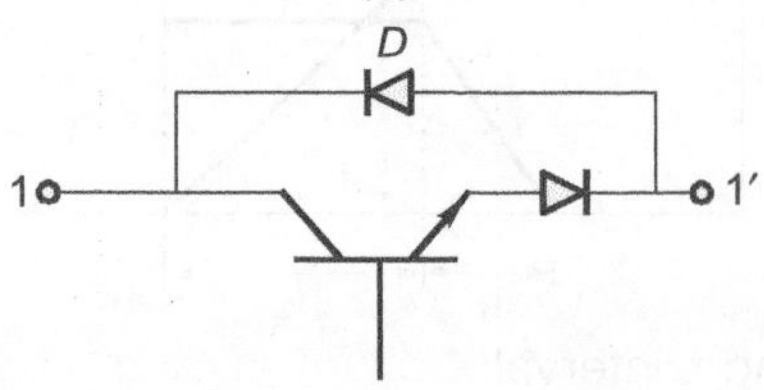

(4)

When the reverse voltage is applied across the devices (2) and (4), current flows through diode indicated with D. So, these device do not satisfy the requirement.

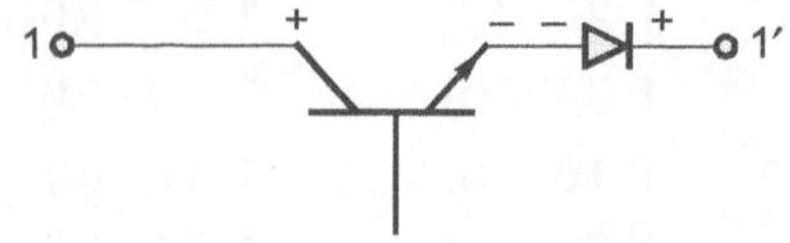

(3)

Transistor blocks positive voltage and diode blocks, negative voltage. Once the base signal is applied, the device conducts in forward direction.

1.15 (a)

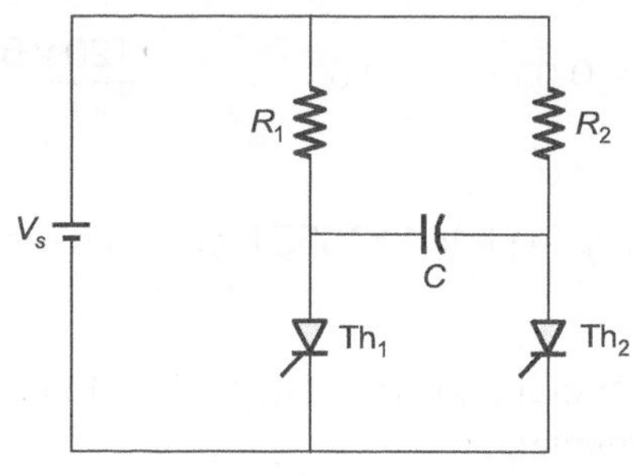

In this type of commutation, a thyristor carrying load current is commutated by transferring its load current to another incoming thyristor.

Firing of SCR Th_1 commutates Th_2 and subsequently, firing of SCR, Th_2 would turn-off Th_1.

Circuit turn-off time t_{c1} for Th_1

$$t_{c1} = R_1 C \ln 2$$

and circuit turn-off time tc_2 for Th_2

$$t_{c2} = R_2 C \ln 2$$

as $R_1 = R_2 = 50\ \Omega$

$$t_{c2} = t_{c1} = R_1 C \ln 2$$

Safety margin = 2

So, $R_1 C \ln 2 = 2\, t_{c1}$

$$C = \frac{2 \times 50 \times 10^{-6}}{50 \times \ln 2} = 2.88\ \mu\text{F}$$

1.16 (b)

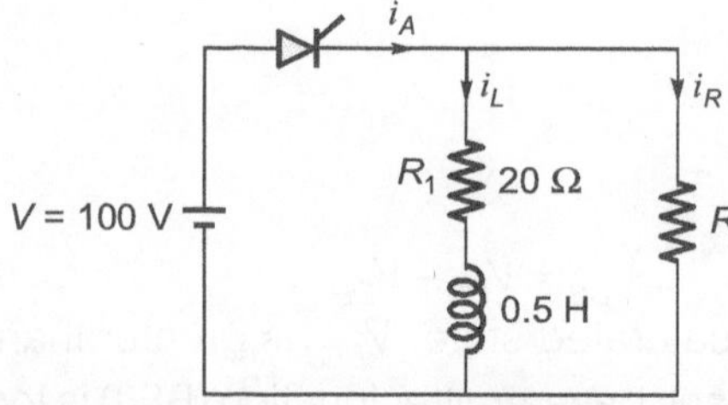

Current through 5 kΩ resistor,

$$i_R = \frac{V}{R_2} = \frac{100}{5 \times 10^3} = 20\text{ mA} = 0.02\text{ A}$$

Current through inductor,

$$i_L = \frac{V}{R_1}(1-e^{-R_1/Lt})$$

$$= \frac{100}{20}(1-e^{20/0.5t}) = 5(1-e^{-40t})$$

Anode current,

$$I_a = i_R + i_L = 0.02 + 5(1-e^{-40t})$$

Let minimum pulse width is T

To turn on $i_a \geq$ latching current

$\Rightarrow$ $0.02 + 5(1-e^{-40t}) = 50 \text{ mA} = 0.5$

$$T = 150\,\mu\text{sec}$$

1.17 (c)

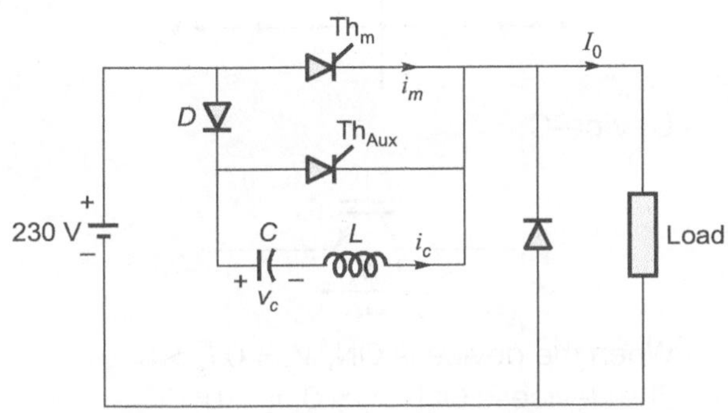

At $t = 0^-$, $v_c = V_s$, $i_c = 0$ and $i_{T1} = I_0$.

At $t = 0$, Th_{aux} is triggered, a resonant current i_c designs to flow from C through Th_{aux}, L and back to C.

This resonant current is given by

$$i_c = -V_s\sqrt{\frac{C}{L}}\sin\omega_0 t$$

$$= -I_P \sin\omega_0 t$$

After half a cycle of $i_c \left\{t_1 = \dfrac{\pi}{\omega_0}\right\}$;

$i_c = 0$, $v_c = -V_s$ and $i_{T1} = I_0$. As i_c tends to reverse, Th_{aux} is turned off.

When $v_c = -V_s$, right hand plate has positive polarity, resonant current i_c now builds up through C, L, D and Th_m. As this current i_c grows opposite to forward thyristor current of Th_m, net forward current $i_m = I_0 - i_c$ begins to decrease. Finally when i_c in the reversed direction attains the value I_0, i_m is reduced to zero and Th_m is turned off.

$$i_m = I_0 - i_c$$

$$= I_0 - I_p \sin\omega_0\Delta t = 0$$

$$\Delta t = \frac{1}{\omega_0}\sin^{-1}\left(\frac{I_0}{I_p}\right)$$

So, Th_m is turned off between

$$t_1 < t < t_1 + \Delta t$$

$$t_1 = \frac{\pi}{\omega_0} = \pi\sqrt{LC}$$

$$= \pi \times \sqrt{10 \times 25.28}$$

$$= 50\,\mu\text{ sec}$$

Option (c) is correct.

Since, commutation of Th_m starts from $t_1 = 50\,\mu$sec.

1.18 (c)

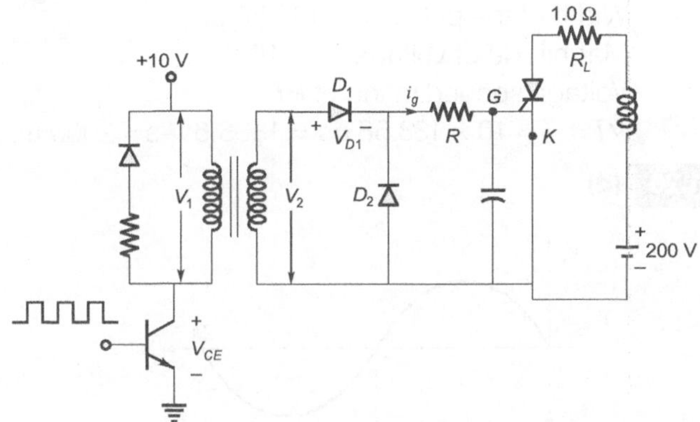

When the pulses are applied to the base of the transistor. Transistor operates in ON state. So, the forward voltage drop in transistor $V_{CE} = 1$ V.

$$V_1 = 10 - V_{CE} = 10 - 1 = 9 \text{ V}$$

$$V_2 = V_1\left(\frac{1}{1}\right) = V_1 = 9 \text{ V} \qquad \text{[turn ratio 1 : 1]}$$

D_1 is forward biased and voltage drop in diode $V_{D1} = 1$ V.

D_2 is reversed biased and acts as open circuit.

Capacitor behaves as open circuit for DC voltage.

Forward voltage drop of gate cathode junction

$$V_{gk} = 1 \text{ V}$$

Voltage drop across resistor R,

$$V_R = V_2 - V_{D1} - V_{gk} = 9 - 1 - 1 = 7 \text{ V}$$

To ensure turn –ON of SCR,

$$R = \frac{V_R}{I_{g(max)}} = \frac{7}{150 \text{ mA}} \approx 47\,\Omega$$

1.19 (a)

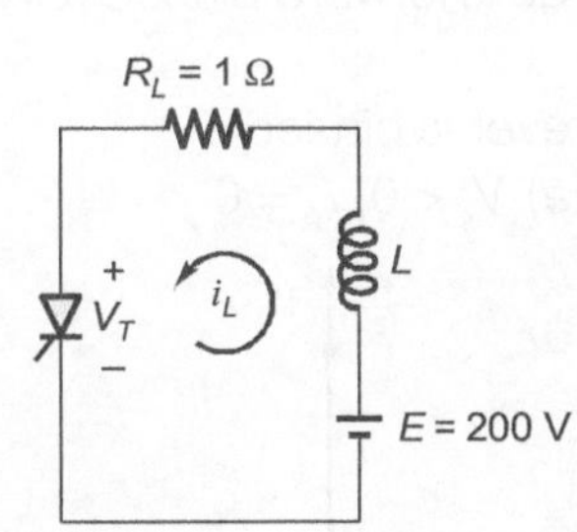

Forward voltage drop of SCR during ON-state

$$V_T = 1 \text{ V}$$

$$E - \frac{Ldi_a}{dt} - Ri_a - V_T = 0$$

$$\Rightarrow \quad 200 - 0.15\frac{di_a}{dt} - i_a - 1 = 0$$

$$\Rightarrow \quad i_a = 199(1 - e^{-t/0.15})$$

Gate pulse width required = time taken by i_a to rise upto $I_L = T$

$$I_L = i_a$$

$$\Rightarrow \quad 250 \times 10^{-3} = 199(1 - e^{-T/0.15})$$

$$T = 188.56\ \mu\text{s}$$

Width of the pulse, $T = 188.56\ \mu$s

Magnitude of voltage, $V = 10$ V

Voltage second rating of PT

$VT = T = 10 \times 188.56\ \mu\text{s} = 1885.6 \text{ V-s} \approx 2000\ \mu\text{s}$

1.20 (d)

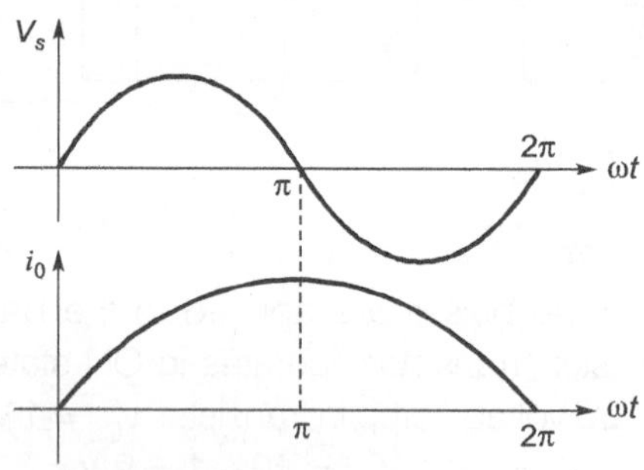

$$i_0 = \frac{V_m}{\omega L}(1 - \cos\omega t)$$

Diode conducts for 360°.

1.21 (c)

During one half cycle, SCR can be in forward blocking mode and by applying gate pulse, the SCR operates in forward conduction mode (ON state). But SCR can not be turned off by applying gate pulse.

1.22 (c)

Device-A:

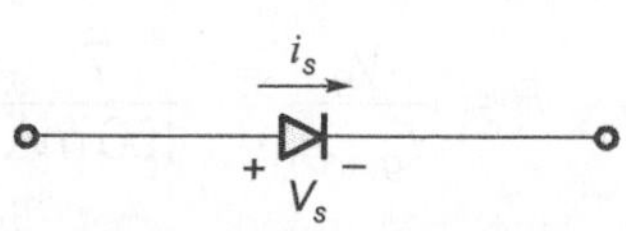

When diode is forward biased (ON state), $V_s = 0$, $i_s > 0$.

Diode is reverse biased

(OFF state) $V_s < 0$, $i_s = 0$

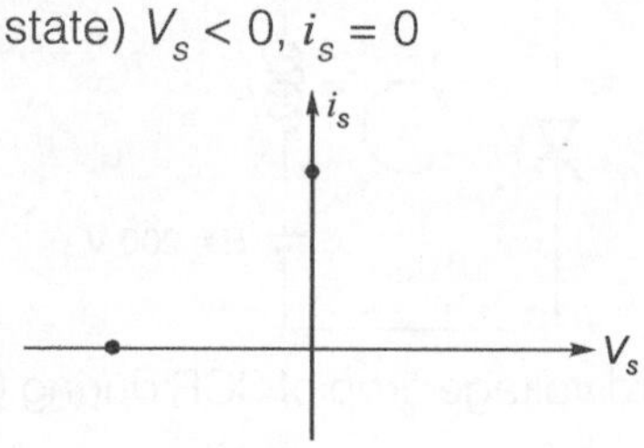

Device-B:

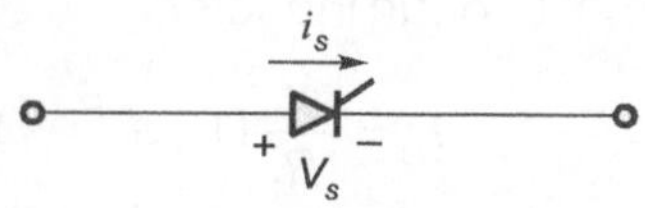

When thyristor is in reverse blocking mode, $V_s < 0$, $i_s = 0$

thyristor is in forward blocking,

$V_s > 0$, $i_s = 0$

Thyristor is in forward conduction mode,

$V_s = 0$, $i_s > 0$

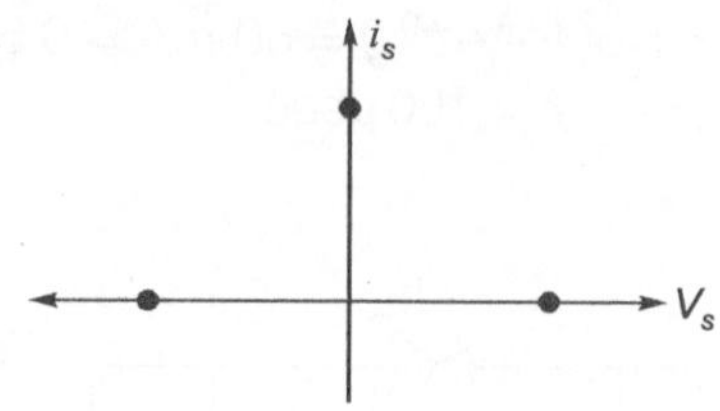

Device-C:

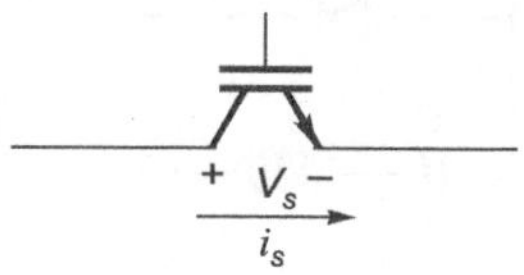

When the device is ON, $V_s = 0$ $i_s > 0$

The device is OFF, $V_s > 0$, $i_s = 0$

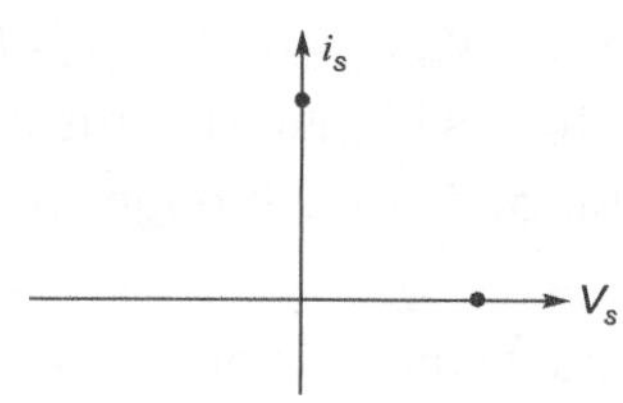

Device-D:

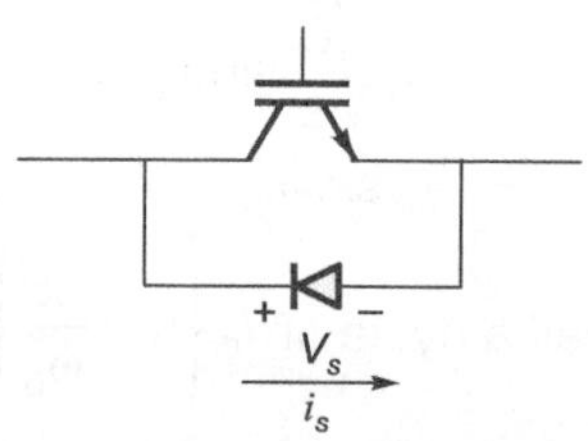

Reverse current can flow through the diode so $i_s < 0$ and $V_s = 0$

During ON state of the device, $i_s > 0$, $V_s = 0$

During OFF state of the device, $i_s = 0$, $V_s = 0$

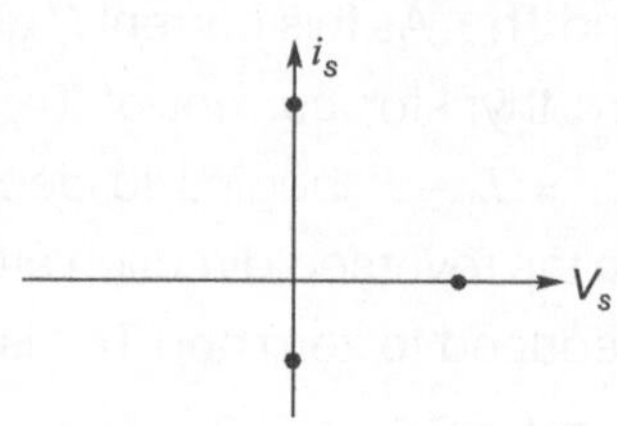

1.23 (c)

Frequency of the voltage source, $f = 50$ Hz

Time period, $T = \frac{1}{f} = \frac{1}{50} = 20$ ms.

During positive half cycle of the source voltage, $0 < t < \frac{T}{2}$, energy is stored in the inductor and current increases.

During negative half cycle of the source voltage, $\frac{T}{2} \le t \le T$, current decreases and energy stored in the inductor is delivered to source.

1.24 (d)

Initial current through the inductor is zero and capacitor voltage is charged upto voltage,

$$V_c(0^-) = 100 \text{ V}$$

As current through inductor and voltage across capacitor can not change abruptly.

So, after closing the switch,

$$i_L(0^+) = i_L(0^-) = 0$$

and $\quad V_c(0^+) = V_c(0^-) = 100 \text{ V}$

The circuit is s-domain,

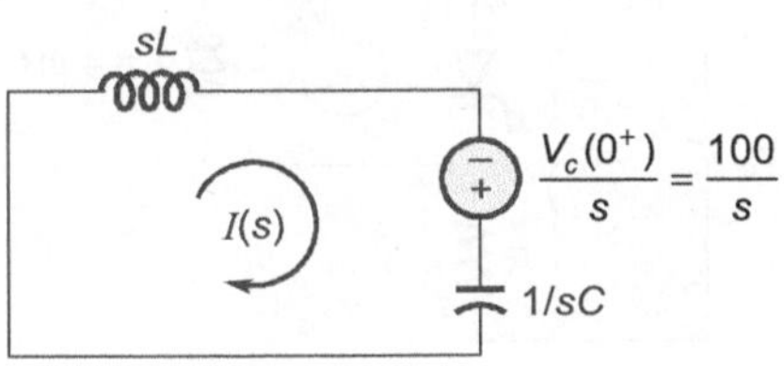

$$I(s) = \frac{100/s}{\left(sL + \frac{1}{sC}\right)} = \frac{100}{L}\left(\frac{1}{s^2 + \frac{1}{LC}}\right)$$

$$= 100\sqrt{\frac{C}{L}}\left(\frac{1/\sqrt{LC}}{s^2 + (1/\sqrt{LC})^2}\right)$$

Taking inverse Laplace transform

$$i(t) = \mathcal{L}^{-1}[I(s)] = 100\sqrt{\frac{C}{L}} \sin\frac{1}{\sqrt{LC}}t$$

$$= 100 \times \sqrt{\frac{10 \times 10^3}{1 \times 10^{-3}}} \times \sin\left(\frac{1}{\sqrt{1 \times 10^{-3} \times 10 \times 10^{-6}}}t\right)$$

$$i(t) = 10 \sin(10^4 t) \text{ A}$$

1.25 (a)

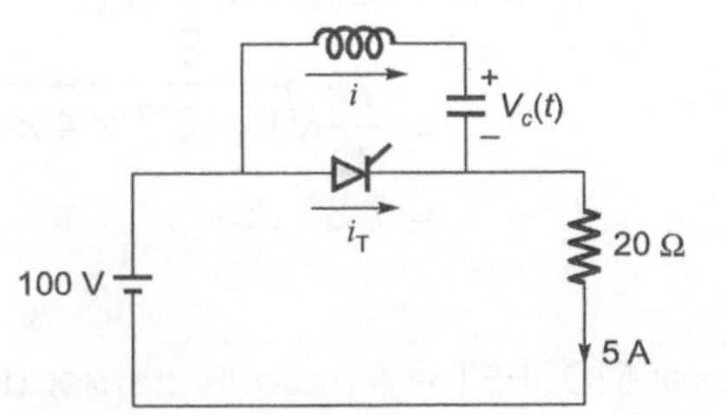

When the switch is closed, i flows through thyristor.

Load current, $\quad I_0 = 5 \text{ A}$

Net current through thyristor,

$$i_T = I_0 - i \Rightarrow i_T = 5 - 10 \sin 10^4 t$$

Let at $t = T$, circuit get turned off and current i_T becomes zero

$$i_T = 5 - 10 \sin 10^4 T = 0$$

$$10 \sin^4 T = 5 \Rightarrow \sin 10^4 T = 0.5$$

$$10^4 T = 30° \text{ or } 0.523 \text{ rad}$$

$$T \approx 52 \text{ μs}$$

1.26 (d)

BJT blocks positive voltage and diode blocks negative voltage and both devices allows positive current.

1.27 (c)

It is defined as the time between the instant anode current becomes zero and the instant reverse voltage is applied across SCR.

1.28 (a)

When main thyristor (M) is turned on, an oscillatory current in the circuit C, M, L and diode is set up and it is given by

$$i_c(t) = V_s\sqrt{\frac{C}{L}} \sin\omega_0 t$$

Peak value of current through capacitor,

$$I_P = V_s\sqrt{\frac{C}{L}} = 200 \times \sqrt{\frac{0.1 \times 10^{-6}}{1 \times 10^{-3}}} = 2 \text{ A}$$

Current through main thyristor,

$$i_m = I_0 + I_c(t) = I_0 + I_P \sin \omega_0 t$$

So, maximum value of

$$i_m = I_0 + I_P = 10 + 2 = 12 \text{ A}$$

When auxiliary thyristor (A) is turned on, capacitor voltage applies a reverse voltage across main thyristor and main thyristor is turned off. The load current is now carried by C and auxiliary thyristor.

Current through auxiliary thyristor,

$$i_A = I_0$$

Maximum value of i_A = maximum value of I_0 = 10 A.

1.29 (b)

For medium power thyristors of rating 6 A to 60 A the ratio of the latching current to holding current is 1.5 to 2.

1.30 (c)

(ii)

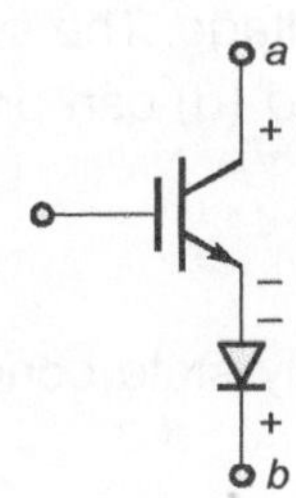

- As given that the active device is in OFF state it means the device can block the voltage when '*a*' is positive with respect to '*b*' and when '*b*' is positive with respect to '*a*' then the diode is reverse biased and voltage is blocked.

(iii) 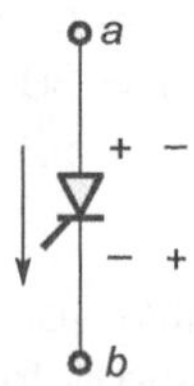

- When '*a*' is positive with respect to '*b*', as given the device is OFF so it will block the voltage.
- When '*b*' is positive with respect to '*a*' it will block the voltage.
- Where as the other two devices will conduct when '*b*' is positive with respect to '*a*'.

1.31 Sol.

No current flows through IGBT. So current flows only in diode. Equivalent circuit of diode is as shown below :

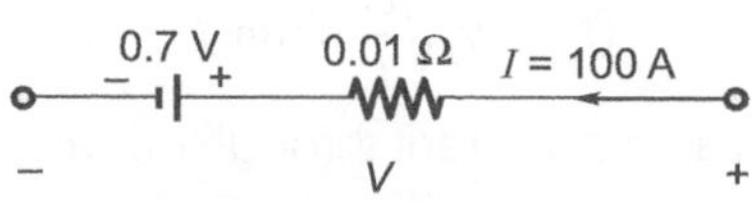

$$V = IR + 0.7 = (100 \times 0.01) + 0.7$$
$$V = 1.7 \text{ V}$$

Power loss during conduction,

$$P = V \times I = 1.7 \times 100 = 170 \text{ W}$$

1.32 Sol.

$$\text{Energy} = \int_0^{T_1} V \cdot i\,dt + \int_0^{T_2} v \cdot i\,dt$$

$$= V\left[\frac{1}{2} I T_1\right] + I\left[\frac{1}{2} V T_2\right]$$

$$= 600\left[\frac{150}{2} \times 1 \times 10^{-6}\right] + 100\left[\frac{1}{2} \times 600 \times 1 \times 10^{-6}\right]$$

Energy = 75 mJ

1.37 (a, d)

From given configuration, the current flows in both direction (Bidirectional). The switch also has on drop voltage. The switch configuration in option (a) and (d) can provide bidirectional current.

1.38 (7.33)

Case-1 : Steady state condition before $t = 0$ sec

The capacitor is charged already with supply voltage $V_S = 100$ V

Case-2 : Now thyristor T is turned on

Mode-1:

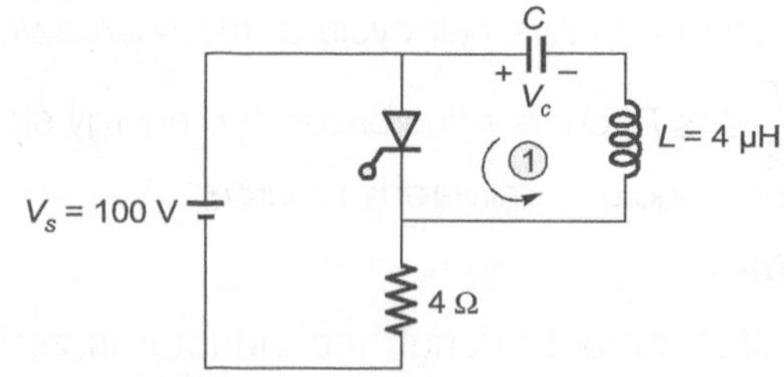

At starting $V_C = 100$ V

The capacitor will discharge through LC circuit at the end, $t_1 = \pi\sqrt{LC}$ sec,

the capacitor voltage will become $V_C = -100$ V (polarity is changed)

Mode-2:

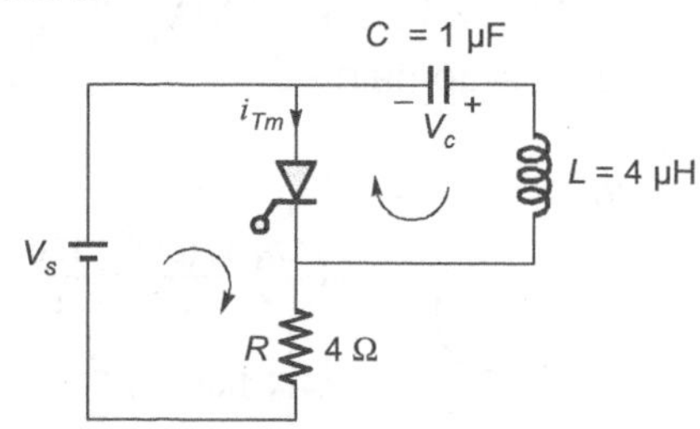

The thyristor current, $i_{Tm} = \dfrac{V_s}{R} - I_p \sin\omega_0 t$

at the end, $I_0 = \dfrac{V_s}{R}$

$i_{Tm} = 0 \qquad (\therefore T_m \to \text{off})$

So, at $t = t_2, \omega_0 t_2 = \sin^{-1}\left(\dfrac{I_0}{I_p}\right)$

$$I_p = V_s\sqrt{\frac{C}{L}} = 100\sqrt{\frac{1}{4}} = 50 \text{ A}$$

and $\quad I_0 = \dfrac{V_s}{R} = \dfrac{100}{4} = 25$ A

So, $\quad t_2 = \sqrt{LC}\sin^{-1}\left(\dfrac{25}{50}\right) = \dfrac{\pi}{6}\sqrt{LC}$

So total time for conduction,

$$t_1 + t_2 = \pi\sqrt{LC} + \frac{\pi}{6}\sqrt{LC}$$
$$= \frac{7\pi}{6}\sqrt{1 \times 10^{-6} \times 4 \times 10^{-6}}$$
$$t_1 + t_2 = 7.33 \text{ μsec}$$

1.39 (c)

Power MOSFET is a majority carrier device and hence it is fastest.

2 Phase Controlled Rectifiers

2.1 A three-phase AC to DC diode bridge rectifier is supplying from a three-phase, 440 V source. The rectifier supplies a purely resistive load. The average DC voltage across the load will be ___ V. **[1995 : 1 M]**

2.2 A single-phase diode bridge rectifier supplies a highly inductive load. The load current can be assumed to be ripple free. The AC supply side current waveforms will be
(a) sinusoidal (b) constant DC
(c) square (d) triangular **[1995 : 1 M]**

2.3 In a 3-phase controlled bridge rectifier, with an increase of overlap angle the output dc voltage
(a) decreases
(b) increases
(c) does not change
(d) depends upon load inductance **[1996 : 1 M]**

2.4 In a dual converter, the circulating current
(a) allows smooth reversal of load current, but increases the response time.
(b) does not allow smooth reversal of load current, but reduces the response time.
(c) allows smooth reversal of load current with improved speed of response.
(d) flows only if there is not interconnecting inductor **[1997 : 1 M]**

2.5 A 3-phase, fully controlled, converter is feeding power into a dc load at a constant current of 150 A. The rms current through each thyristor of the converter is
(a) 50 A (b) 100 A
(c) $\frac{150\sqrt{2}}{\sqrt{3}}$ A (d) $\frac{150}{\sqrt{3}}$ A **[1998 : 2 M]**

2.6 When the firing angle α of a single-phase, fully controlled rectifier feeding a constant DC current into a load is 30°, the displacement power factor of the rectifier is
(a) 1 (b) 0.5
(c) $\frac{1}{\sqrt{3}}$ (d) $\frac{\sqrt{3}}{2}$ **[1998 : 2 M]**

2.7 A thyristor based, three-phase, fully controlled converter feeds a DC load that draws a constant current. Then the input AC line current to the converter has
(a) an rms value equal to the DC load current.
(b) an average value equal to the DC load current.
(c) a peak value equal to the DC load current.
(d) a fundamental frequency component, whose rms value is equal to the DC load current **[2000 : 1 M]**

2.8 AC to DC circulating current dual converters are operated with the following relationship between their triggering angles (α_1 and α_2).
(a) $\alpha_1 + \alpha_2 = 180°$ (b) $\alpha_1 + \alpha_2 = 360°$
(c) $\alpha_1 - \alpha_2 = 180°$ (d) $\alpha_1 + \alpha_2 = 90°$ **[2001 : 1 M]**

2.9 A half wave thyristor converter supplies a purely inductive load as shown in the figure. If triggering angle of the thyristor is 120°, the extinction angle will be

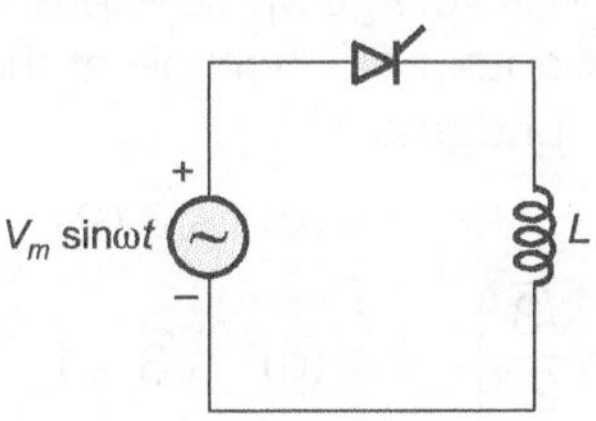

(a) 240° (b) 180°
(c) 200° (d) 120° **[2001 : 2 M]**

2.10 A six pulse thyristor rectifier bridge is connected to a balanced 50 Hz three-phase AC source. Assuming that the DC output current of the rectifier is constant. The lowest frequency harmonics component of the AC source line current is
(a) 100 Hz (b) 150 Hz
(c) 250 Hz (d) 300 Hz **[2002 : 1 M]**

2.11 In the single-phase, diode bridge rectifier shown in figure, the load resistor is $R = 50\ \Omega$. The source voltage is $V = 200 \sin \omega t$, where $\omega = 2\pi \times 50$ radians per second. The power dissipated in the load resistor R is

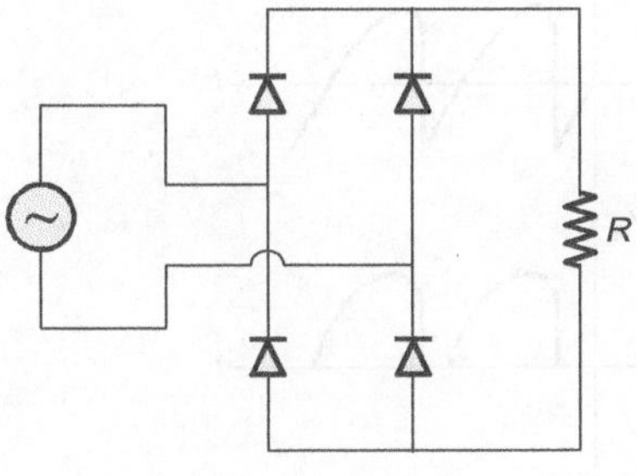

(a) $\frac{3200}{\pi}$ W (b) $\frac{400}{\pi}$ W

(c) 400 W (d) 800 W [2002 : 2 M]

2.12 A three-phase thyristor bridge rectifier is used in a HVDC link. The firing angle α (as measured from the point of natural commutation), is constrained to lie between 5° and 30°. If the DC side current and AC side voltage magnitudes are constant, which of the following statement is true (neglect harmonics in the AC side current and commutation overlap in your analysis).

(a) Reactive power absorbed by the rectifier is maximum when $\alpha = 5°$.

(b) Reactive power absorbed by the rectifier is maximum when $\alpha = 30°$.

(c) Reactive power absorbed by the rectifier is maximum when $\alpha = 15°$.

(d) Reactive power absorbed by the rectifier is maximum when $\alpha = 20°$. [2002 : 2 M]

2.13 A fully controlled natural commutated 3-phase bridge rectifier is operating with a firing angle $\alpha = 30°$. The peak to peak voltage ripple expressed as a ratio of the peak output DC voltage at the output of the converter bridge is

(a) 0.5 (b) $\sqrt{3}/2$

(c) $\left(1-\frac{\sqrt{3}}{2}\right)$ (d) $\sqrt{3}-1$ [2003 : 1 M]

2.14 A phase-controlled half-controlled single-phase converter is shown in the figure. The control angle $\alpha = 30°$.

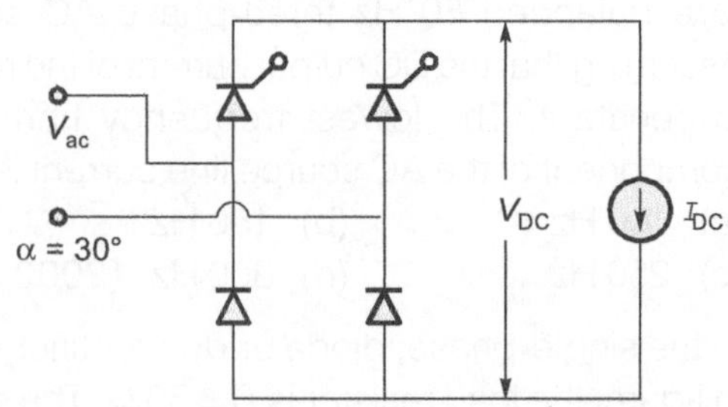

The output DC voltage waveshape will be as shown in

(a)

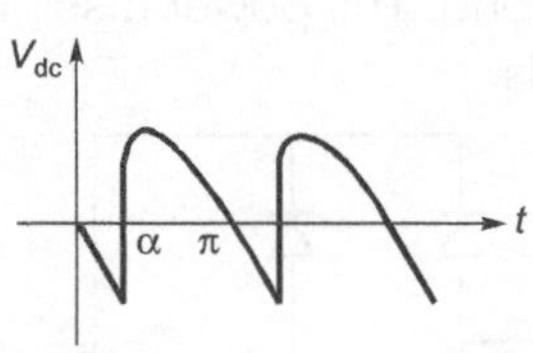

(b)

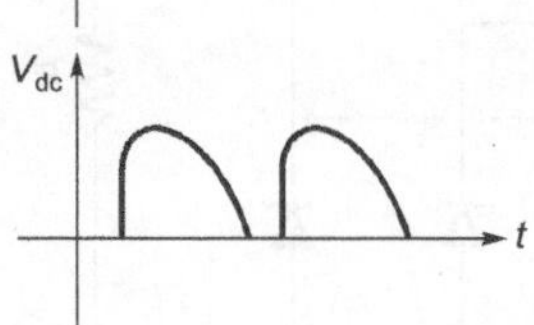

(c)

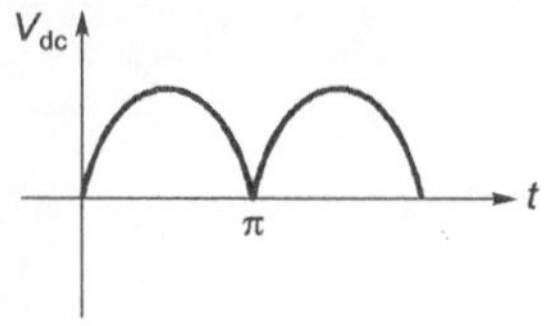

(d)

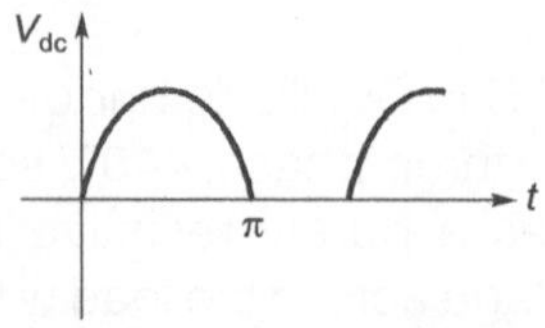

[2003 : 2 M]

2.15 The triggering circuit of a thyristor is shown in the figure. The thyristor requires a gate current of 10 mA, for guaranteed turn-on. The value of R required for the thyristor to turn on reliably under all conditions of V_b variation is

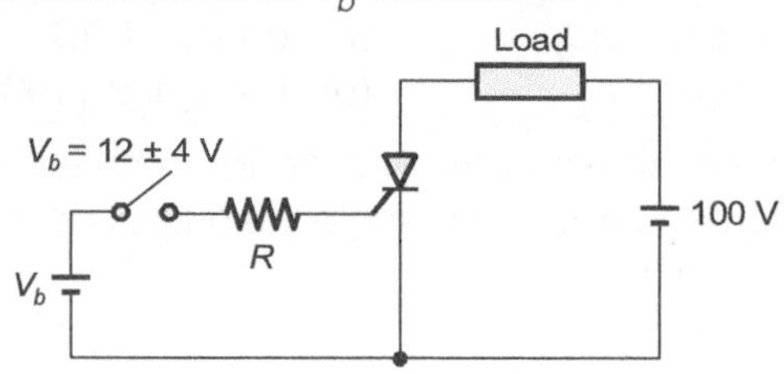

(a) 10000 Ω (b) 1600 Ω

(c) 1200 Ω (d) 800 Ω [2004 : 1 M]

2.16 The circuit in figure shows a full-wave rectifier. The input voltage is 230 V (rms) single-phase AC. The peak reverse voltage across the diodes D_1 and D_2 is

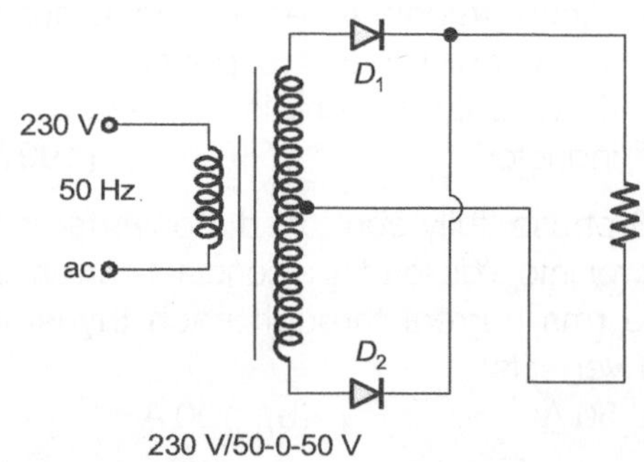

230 V/50-0-50 V

(a) $100\sqrt{2}$ V (b) 100 V

(c) $50\sqrt{2}$ V (d) 50 V [2004 : 1 M]

2.17 The circuit in the figure shows a 3-phase half-wave rectifier. The source is a symmetrical, 3-phase four-wire system. The line-to-line voltage of the source is 100 V. The supply frequency is 400 Hz. The ripple frequency at the output is

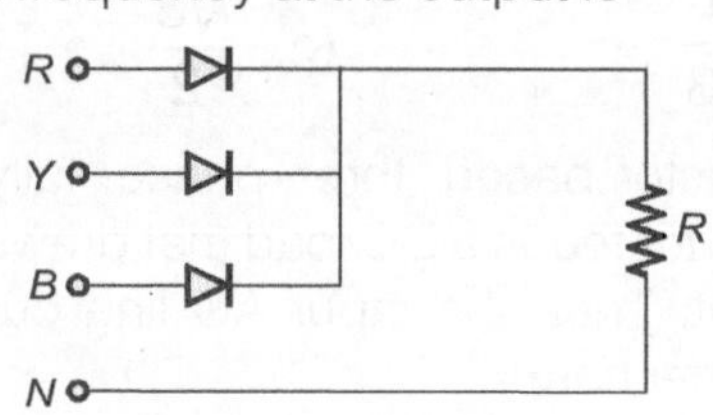

(a) 400 Hz (b) 800 Hz
(c) 1200 Hz (d) 2400 Hz [2004 : 1 M]

2.18 A three-phase diode bridge rectifier is fed from a 400 V RMS, 50 Hz, three-phase AC source. If the load is purely resistive, then peak instantaneous output voltage is equal to

(a) 400 V (b) $400\sqrt{2}$ V
(c) $400\sqrt{\frac{2}{3}}$ V (d) $\frac{400}{\sqrt{3}}$ V [2005 : 1 M]

2.19 Consider a phase-controlled converter shown in the figure. The thyristor is fired at an angle α in every positive half cycle of the input voltage. If the peak value of the instantaneous output voltage equals 230 V, the firing angle α is close to

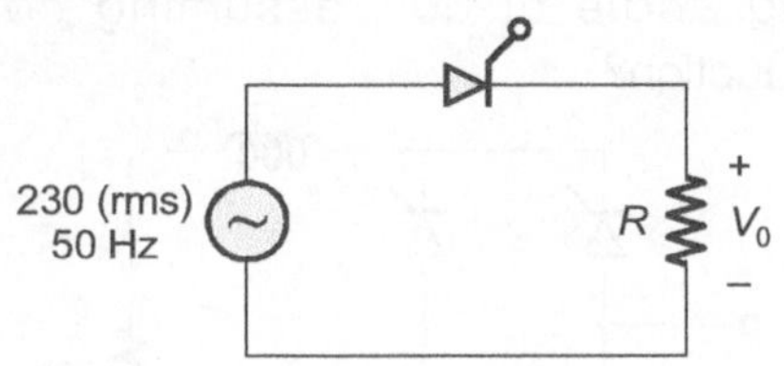

(a) 45° (b) 135°
(c) 90° (d) 83.6° [2005 : 2 M]

2.20 A single-phase half wave uncontrolled converter circuit is shown in the figure. A 2-winding transformer is used at the input for isolation. Assuming the load current to be constant and $v = V_m \sin\omega t$, the current waveform through diode D_2 will be

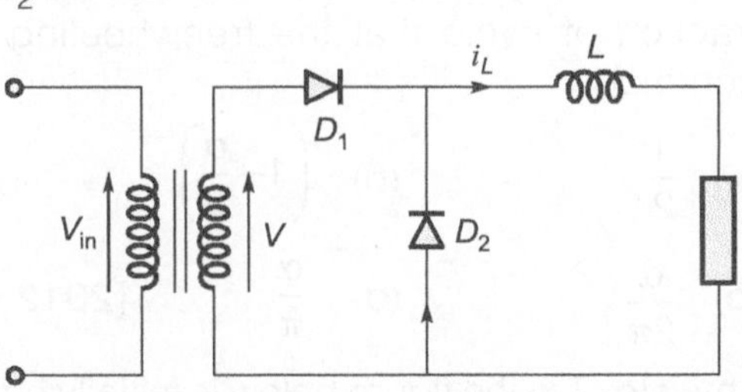

(a)
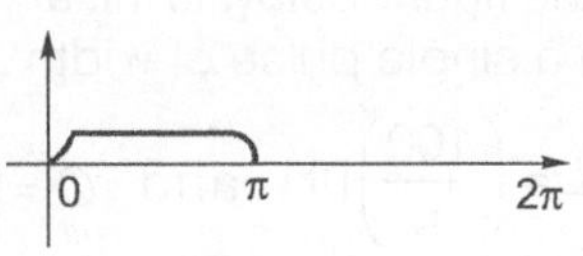

(b)
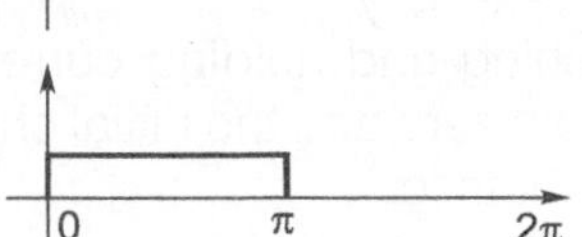

(c)
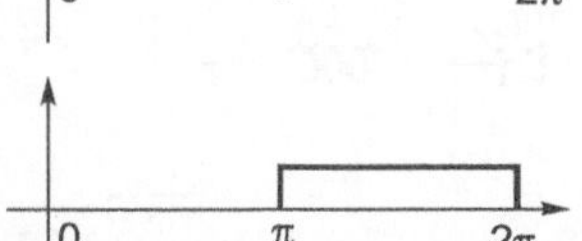

(d)
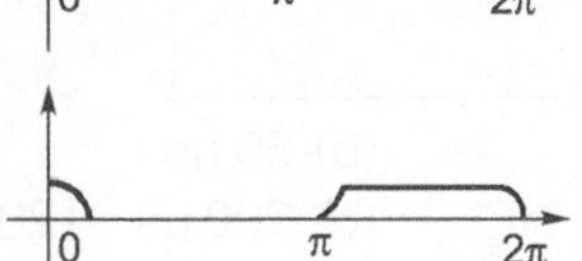

[2006 : 1 M]

2.21 A 3-phase fully controlled bridge converter with free wheeling diode is fed from 400 V, 50 Hz AC source and is operating at a firing angle of 60°. The load current is assumed constant at 10 A due to high load inductance. The input displacement factor (IDF) and the input power factor (IPF) of the converter will be

(a) IDF = 0.867; IPF = 0.828
(b) IDF = 0.867; IPF = 0.552
(c) IDF = 0.5; IPF = 0.478
(d) IDF = 0.5; IPF = 0.318 [2006 : 2 M]

2.22 A single-phase bridge converter is used to charge a battery of 200 V having an internal resistance of 2 Ω as shown in figure. The SCRs are triggered by a constant DC signal. If SCR 2 gets open circuited, what will be the average charging current?

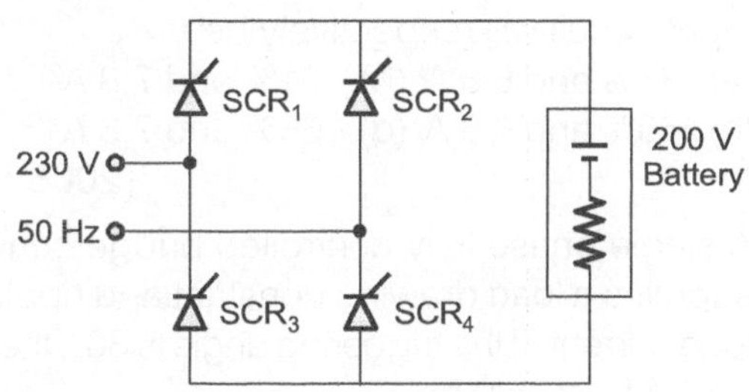

(a) 23.8 A (b) 15 A
(c) 11.9 A (d) 3.54 A [2006 : 2 M]

2.23 A single-phase fully controlled the thyristor bridge AC-DC converter is operating at a firing angle of 25° and on overlap angle of 10° constant DC output current of 20 A. The fundamental power factor (displacement factor) at input AC mains is

(a) 0.78 (b) 0.827
(c) 0.866 (d) 0.9 [2007 : 1 M]

2.24 A three-phase fully-controlled thyristor bridge converter is used as line commutated inverter to feed 50 kW power 420 V DC to a three phase, 415 V (line), 50 Hz AC mains, Consider DC link current to be constant. The rms current of the thyristor is

(a) 119.05 A (b) 79.37 A
(c) 68.73 A (d) 39.68 A [2007 : 1 M]

2.25 A single phase full-wave half-controlled bridge converter feeds an inductive load. The two SCRs in the converter are connected to a common DC bus. The converter has to have a freewheeling diode

(a) because the converter inherently does not provide for free-wheeling.
(b) because the converter does not provide for free-wheeling for high values of triggering angles.

(c) or else the free-wheeling action of the converter will cause shorting of the AC triggering angles.
(d) or else if a gate pulse to one of the SCRs is missed, it will subsequently cause a high load current in the other SCR. [2007 : 1 M]

2.26 A single-phase, 230 V, 50 Hz ac mains fed step down transformer (4:1) is supplying power to a half-wave uncontrolled AC-DC converter used for charging a battery (12 V DC) with the series limiting resistor being 19.04 Ω. The charging current is
(a) 3.43 A (b) 1.65 A
(c) 1.22 A (d) 1.0 A [2007 : 2 M]

2.27 A three phase fully controlled bridge converter is feeding a load drawing a constant and ripple free load current of 10 A at a firing angle of 30°. The approximate Total harmonic Distortion (%THD) and the rms value of fundamental component of the input circuit will respectively be
(a) 31% and 6.8 A (b) 31% and 7.8 A
(c) 66% and 6.8 A (d) 66% and 7.8 A
[2008 : 2 M]

2.28 A single phase fully controlled bridge converter supplies a load drawing constant and ripple free load current. If the triggering angle is 30°, the input power factor will be
(a) 0.65 (b) 0.78
(c) 0.85 (d) 0.866 [2008 : 2 M]

2.29 A single-phase half controlled converter shown in the figure feeding power to highly inductive load. The converter is operating at a firing angle of 60°.

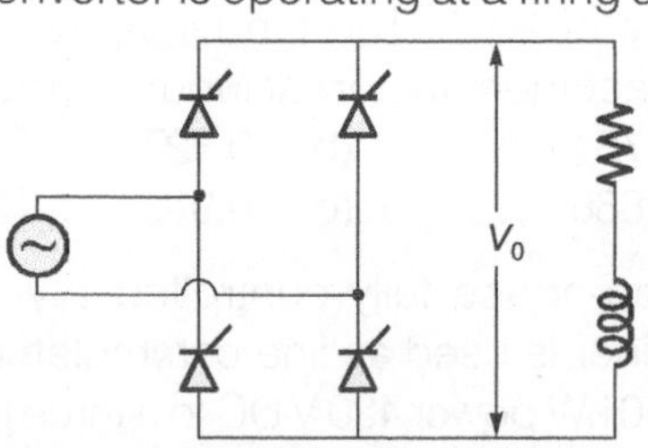

If the firing pulses are suddenly removed, the steady state voltage (V_0) waveform of the converter will become

(a) 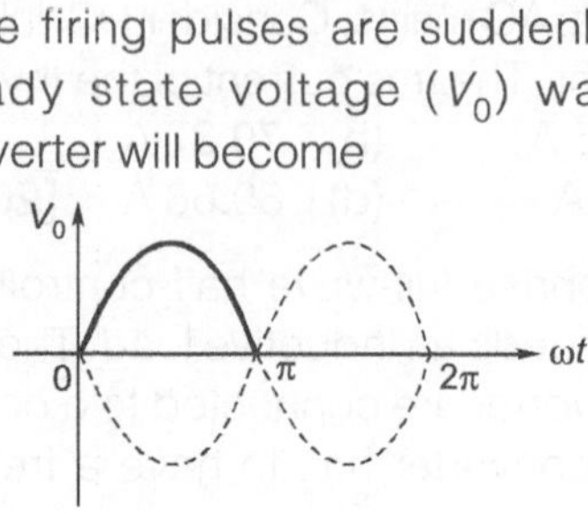

(b) 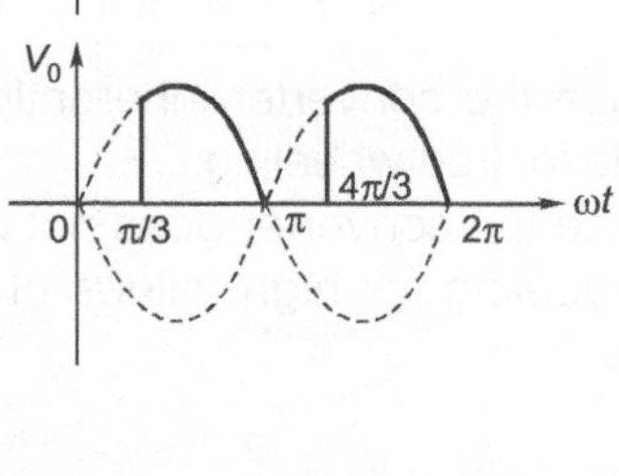

(c) 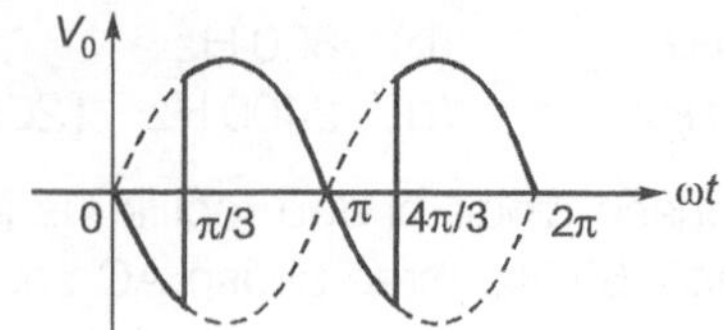

(d) 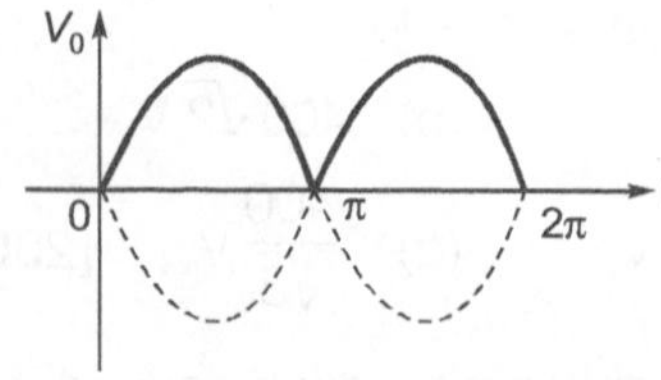

[2008 : 2 M]

2.30 The fully controlled thyristor converter in the figure is fed from a single-phase source. When the firing angle is 0°, the DC output voltage of the converter is 300 V. What will be the output voltage for a firing angle of 60°, assuming continuous conduction?

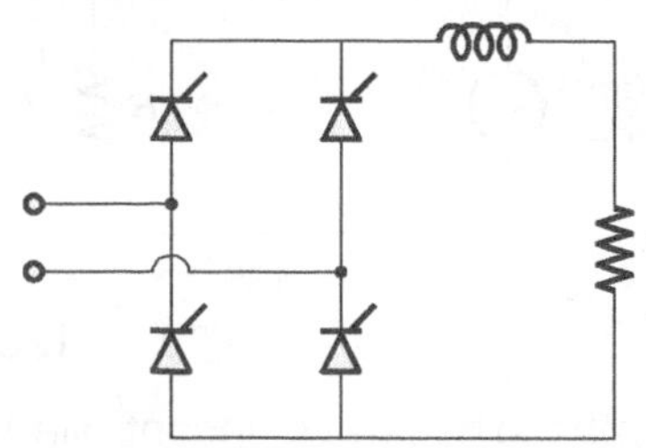

(a) 150 V (b) 210 V
(c) 300 V (d) 100π V [2010 : 1 M]

2.31 A half-controlled single-phase bridge rectifier is supplying an *R-L* load. It is operated at a firing angle α and the load current is continuous. The fraction of cycle that the freewheeling diode conduct is
(a) $\frac{1}{2}$ (b) $\left(1-\frac{\alpha}{\pi}\right)$
(c) $\frac{\alpha}{2\pi}$ (d) $\frac{\alpha}{\pi}$ [2012 : 1 M]

2.32 Thyristor *T* in the figure below is initially off and is triggered with a single pulse of width 10 μs. It is given that $L=\left(\frac{100}{\pi}\right)$μH and $C=\left(\frac{100}{\pi}\right)$μF. Assuming latching and holding currents of the thyristor are both zero and the initial charge on *C* is zero, *T* conducts for

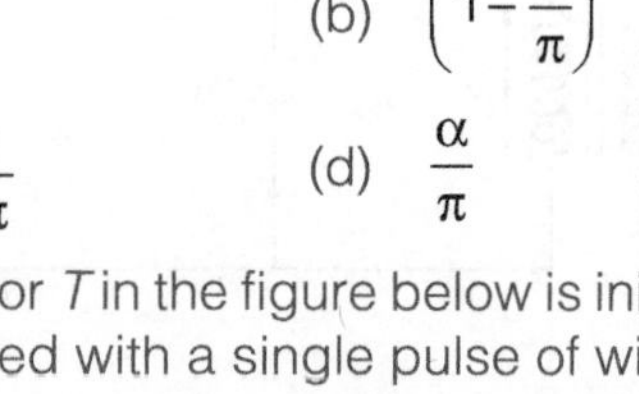

(a) 10 μs (b) 50 μs
(c) 100 μs (d) 200 μs [2013 : 2 M]

2.33 The figure shows the circuit of a rectifier fed from a 230 V (rms), 50 Hz sinusoidal voltage source. If we want to replace the current source with a resistor so that the rms value of the current supplied by the voltage source remains unchanged, the value of the resistance (in ohms) is ______. (Assume diodes to be ideal).

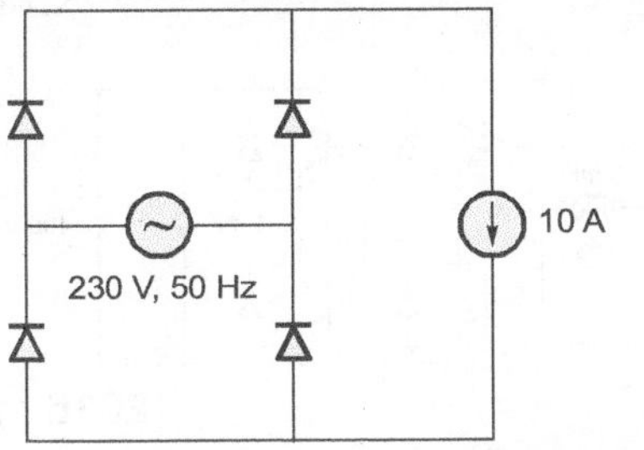

[2014 : 1 M, Set-1]

2.34 The figure shows the circuit diagram of a rectifier. The load consists of a resistance 10 Ω and an inductance 0.05 H connected in series. Assuming ideal thyristor and ideal diode, the thyristor firing angle (in degree) needs to obtain an average load voltage of 70 V is ___.

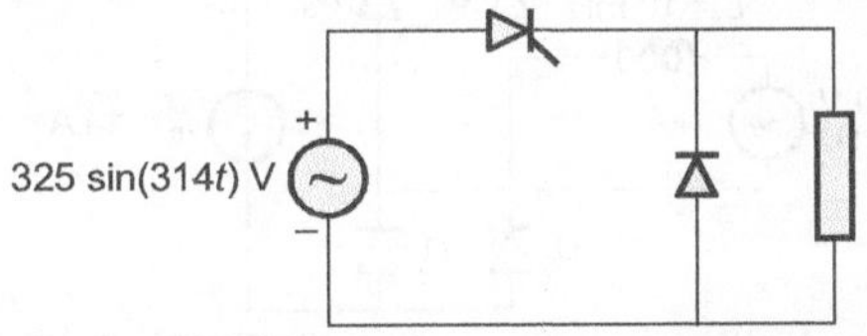

[2014 : 2 M, Set-1]

2.35 A fully controlled converter bridge feeds a highly inductive load with ripple free load current. The input supply (v_s) to the bridge is a sinusoidal source. Triggering angle of the bridge converter is $\alpha = 30°$. The input power factor of the bridge is ______ .

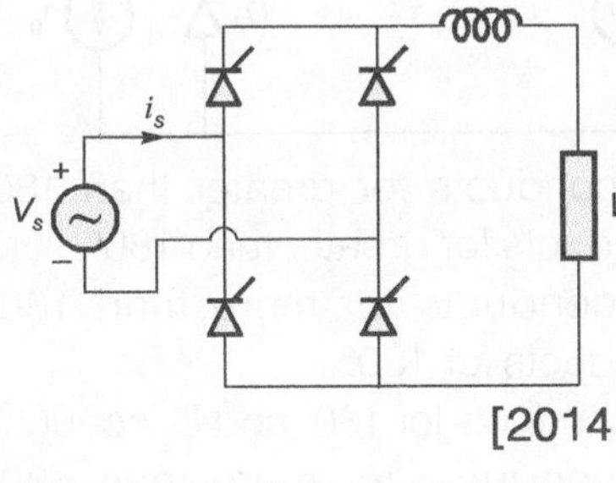

[2014 : 2 M, Set-2]

2.36 A three-phase fully controlled bridge converter is fed through star-delta transformer as shown in the figure.

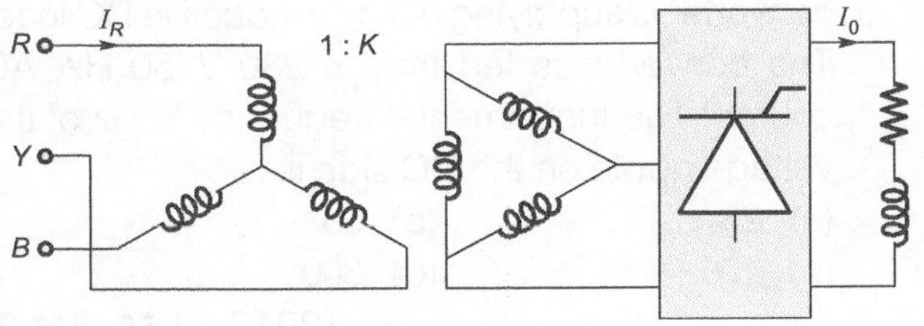

The converter is operated at a firing angle of 30°. Assuming the load current (I_0) to be virtually constant at 1 pu and transformer to be an ideal one, the input phase current waveform is

(a) 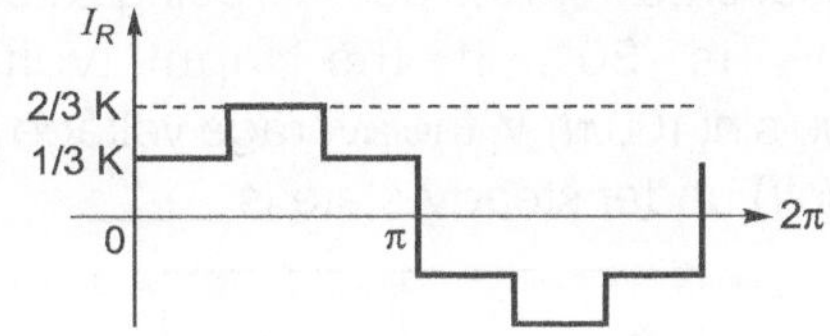

(b) 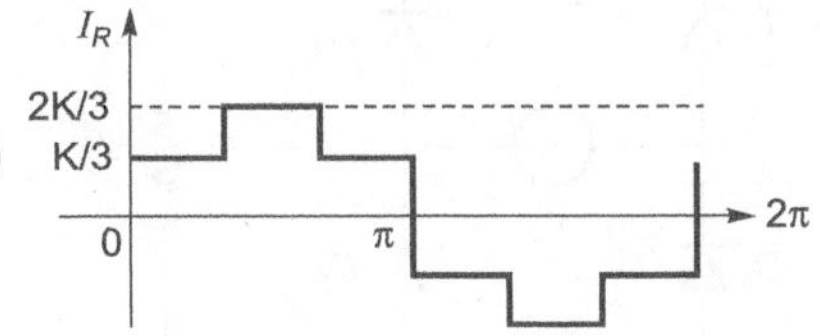

(c) 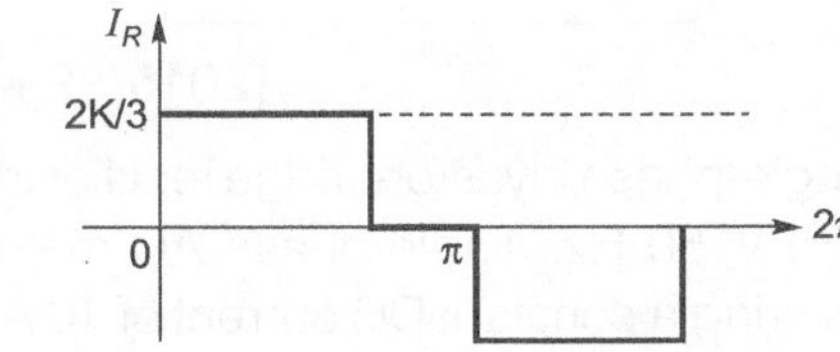

(d) 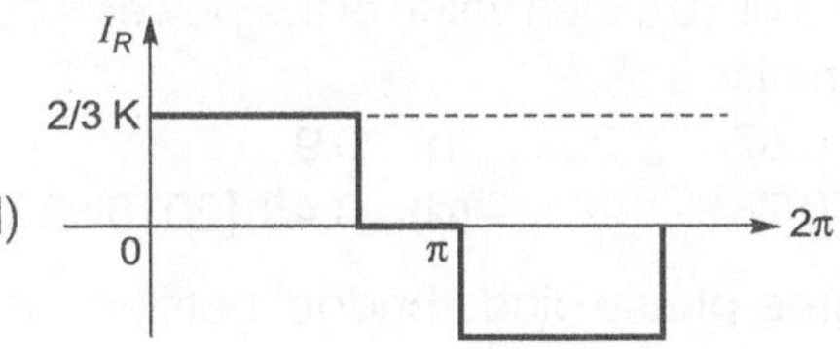

[2014 : 2 M, Set-3]

2.37 A diode circuit feeds an ideal inductor as shown in the figure. Given $v_s = 100 \sin(\omega t)$ V, where $\omega = 100\pi$ rad/s, and $L = 31.83$ mH. The initial value of inductor current is zero. Switch S is closed at $t = 2.5$ ms. The peak value of inductor current i_L (in A) in the first cycle is ______.

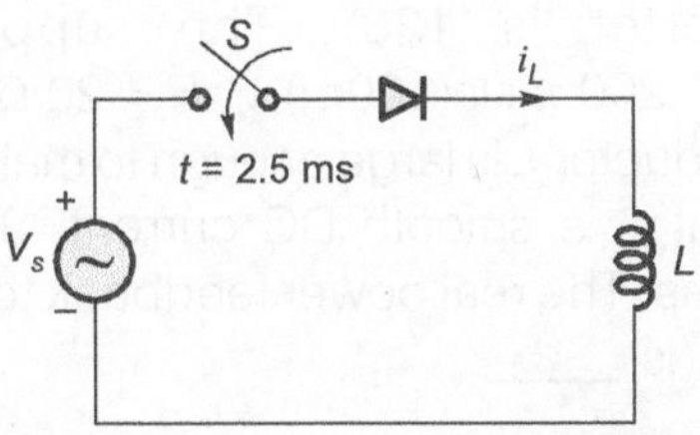

[2014 : 2 M, Set-3]

2.38 In the following circuit, the input voltage V_{in} is $100 \sin(100\pi t)$. For $100\pi RC = 50$, the average voltage across R(in Volts) under steady-state is nearest to

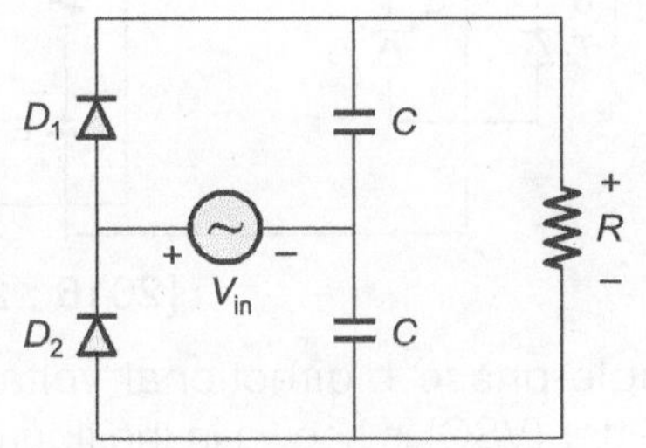

(a) 100 (b) 31.8
(c) 200 (d) 63.6 [2015 : 1 M]

2.39 In the given rectifier, the delay angle of the thyristor T_1 measured from the positive going zero crossing of V_s is 30°. If the input voltage V_s is $100 \sin(100\pi t)$ V, the average voltage across R (in Volt) under steady-state is _____.

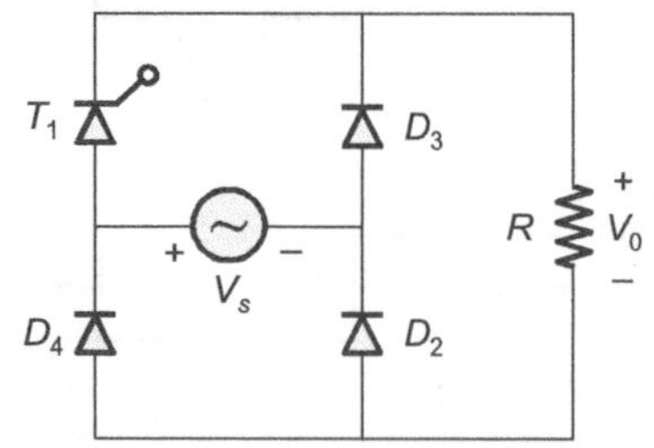

[2015 : 2 M, Set-2]

2.40 A single-phase thyristor-bridge rectifier is fed from a 230 V, 50 Hz, single-phase AC mains. If it is delivering a constant DC current of 10 A, at firing angle of 30°, then value of the power factor at AC mains is
(a) 0.87 (b) 0.9
(c) 0.78 (d) 0.45 [2016 : 2 M, Set-1]

2.41 A three-phase diode bridge rectifier is feeding a constant DC current of 100 A to a highly inductive load. If three-phase, 415 V, 50 Hz AC source is supplying to this bridge rectifier then the rms value of the current in each diode, in Ampere, is _____.
[2016 : 1 M, Set-2]

2.42 A full-bridge converter supplying an RLE load is shown in figure. The firing angle of the bridge converter is 120°. The supply voltage $v_m(t) = 200\pi \sin(100\pi t)$ V, $R = 20\ \Omega$, $E = 800$ V. The inductor L is large enough to make the output current I_L a smooth DC current. Switches are lossless. The real power feedback to the source, in kW, is _____.

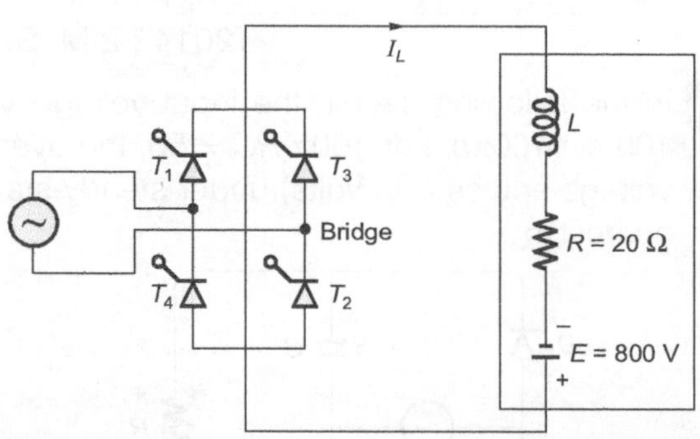

[2016 : 2 M, Set-2]

2.43 A single-phase bidirectional voltage source converter (VSC) is shown in the figure below. All devices are ideal. It is used to charge a battery at 400 V with power of 5 kW from a source $V_s = 220$ V (rms), 50 Hz sinusoidal AC mains at unity p.f. If its AC side interfacing inductor is 5 mH and the switches are operated at 20 kHz, then the phase shift (δ) between AC mains voltage (V_s) and fundamental AC rms VSC voltage (V_{C1}), in degree, is _____.

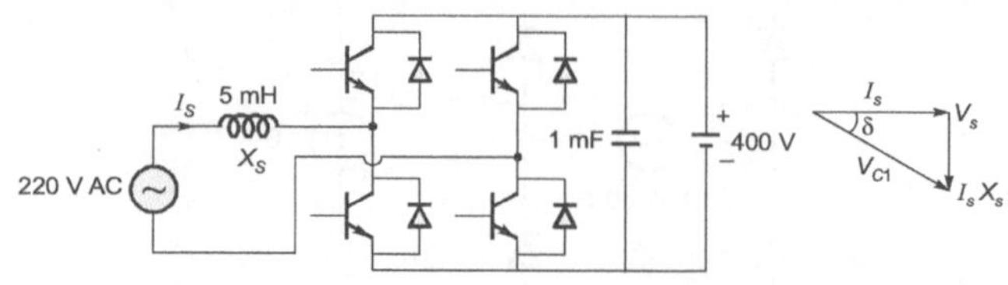

[2016 : 2 M, Set-2]

2.44 The figure below shown an uncontrolled diode bridge rectifier supplied from a 220 V, 50 Hz, 1-phase ac source. The load draws a constant current $I_0 = 14$ A. The conduction angle of the diode, D_1 in degrees (rounded off to two decimal places) is _____.

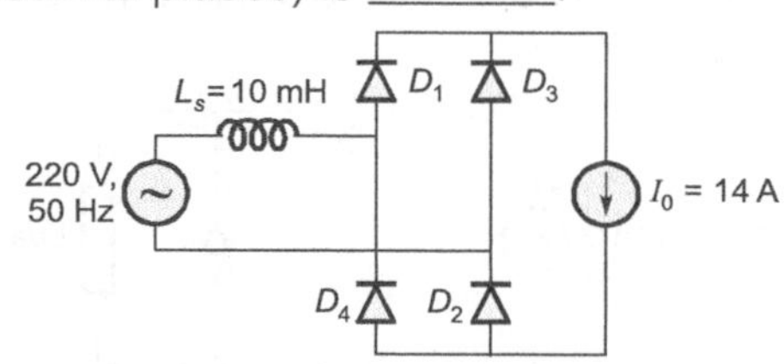

[2017 : 2 M, Set-1]

2.45 In the circuit shown, the diodes are ideal, the inductance is small, and $I_0 \neq 0$. Which one of the following statements is true?

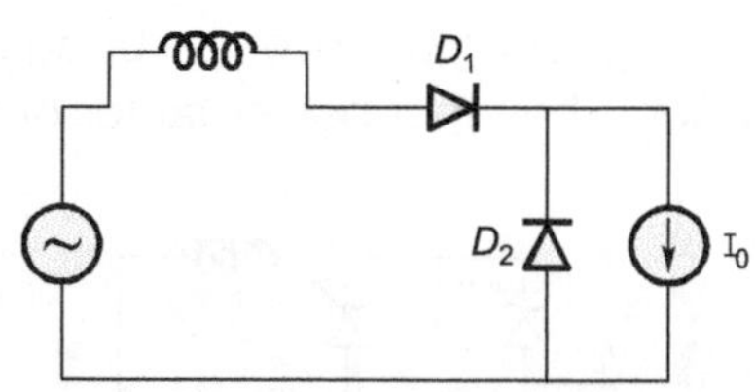

(a) D_1 conducts for greater than 180° and D_2 conducts for greater than 180°.
(b) D_2 conducts for more than 180° and D_1 conducts for 180°.
(c) D_1 conducts for 180° and D_2 conducts for 180°.
(d) D_1 conducts for more than 180° and D_2 conducts for 180°. [2017 : 1 M, Set-2]

2.46 A phase-controlled, single-phase, full-bridge converter is supplying a highly inductive DC load. The converter is fed from a 230 V, 50 Hz, AC source. The fundamental frequency in Hz of the voltage ripple on the DC side is
(a) 25 (b) 50
(c) 100 (d) 300
[2017 : 1 M, Set-2]

2.47 The figure below shows the circuit diagram of a controlled rectifier supplied from a 230 V, 50 Hz, 1-phase voltage source and a 10 : 1 ideal transformer. Assume that all devices are ideal. The firing angles of the thyristors T_1 and T_2 are 90° and 270°, respectively.

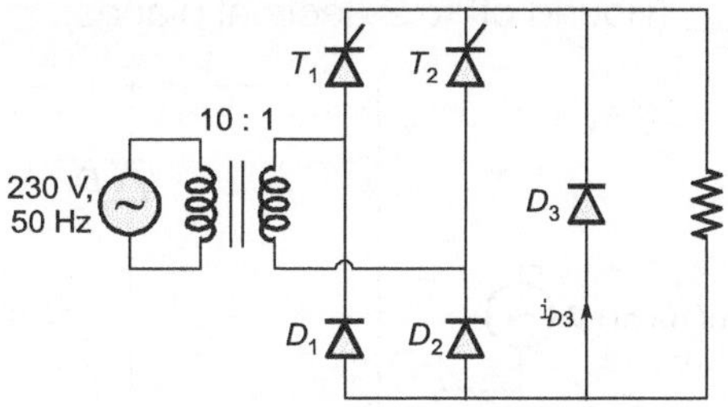

The RMS value of the current through diode D_3 in amperes is _____. [2017 : 1 M, Set-2]

2.48 In the circuit shown in the figure, the diode used is ideal. The input power factor is _____. (Give the answer up to two decimal places.)

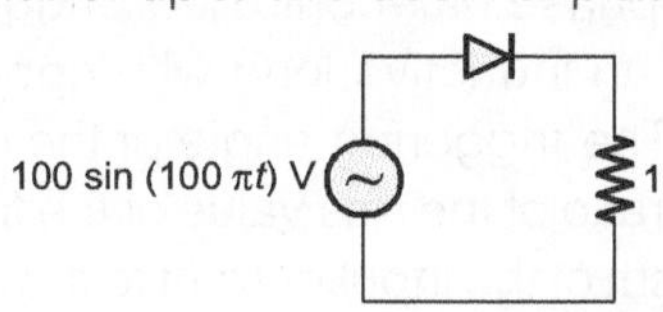

[2017 : 2 M, Set-2]

2.49 The waveform of the current drawn by a semi-converter from a sinusoidal AC voltage source is shown in the figure. If I_0 = 20 A, the rms value of fundamental component of the current is _____ A (up to 2 decimal places).

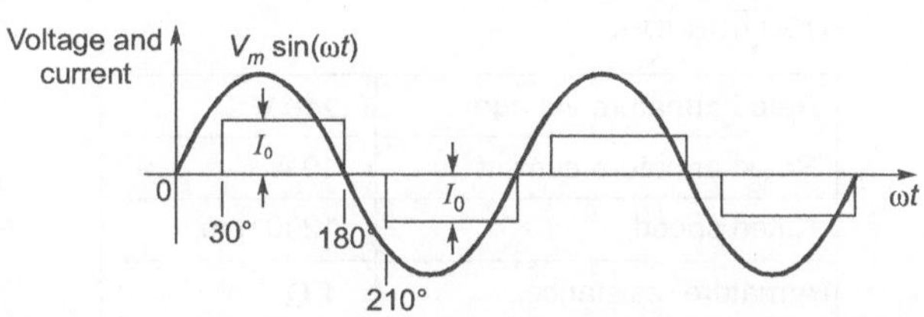

[2018 : 1 M]

2.50 A single-phase fully controlled rectifier is supplying a load with an anti-parallel diode as shown in the figure. All switches and diodes are ideal. Which one of the following is true for instantaneous load voltage and current?

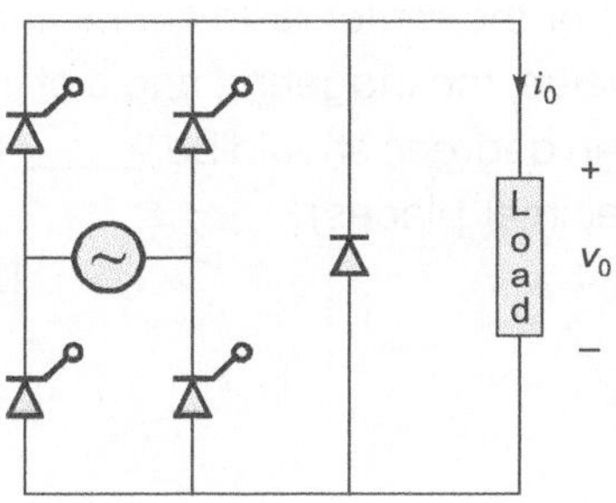

(a) $v_0 \geq 0$ and $i_0 < 0$ (b) $v_0 < 0$ and $i_0 < 0$
(c) $v_0 \geq 0$ and $i_0 \geq 0$ (d) $v_0 < 0$ and $i_0 \geq 0$

[2018 : 1 M]

2.51 A phase controlled single-phase rectifier, supplied by an AC source, feeds power to an R-L-E load as shown in the figure. The rectifier output voltage has an average value given by,

$$V_0 = \frac{V_m}{2\pi}(3 + \cos\alpha)$$

where, $V_m = 80\pi$ volts and α is the firing angle. If the power delivered to the lossless batter is 1600 W. α in degree is ___ (upto 2 decimal places).

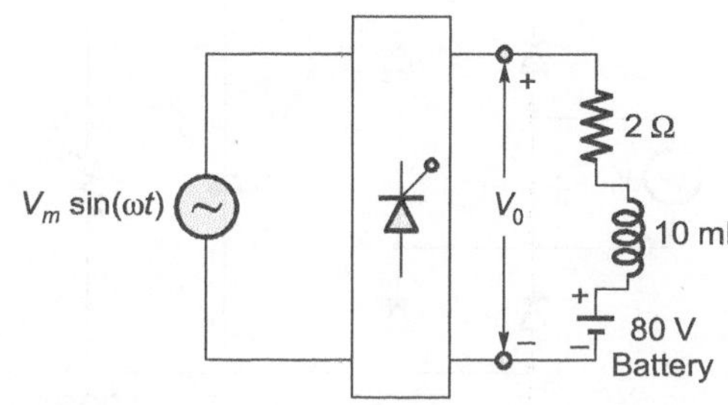

[2018 : 2 M]

2.52 A six-pulse thyristor bridge rectifier is connected to a balanced three-phase 50 Hz AC source. Assuming that the DC output current of the rectifier is constant, the lowest harmonic component in the AC input current is

(a) 250 Hz (b) 100 Hz
(c) 300 Hz (d) 150 Hz [2019 : 1 M]

2.53 A single-phase fully controlled thyristor converter is used to obtain an average voltage of 180 V with 10 A constant current to feed a DC load. It is fed from single-phase AC supply of 230 V, 50 Hz. Neglect the source impedance. The power factor (round off to two decimal places) of AC mains is _____. [2019 : 2 M]

2.54 A fully-controlled three-phase bridge converter is working from a 415 V, 50 Hz AC supply. It is supplying constant current of 100 A at 400 V to a DC load. Assume large inductive smoothing and neglect overlap. The rms value of the AC line current in amperes (round off to two decimal places) is _____. [2019 : 2 M]

2.55 A single-phase, full-bridge diode rectifier fed from a 230 V, 50 Hz sinusoidal source supplies a series combination of finite resistance, *R*, and a very large inductance, *L*, The two most dominant frequency components in the source current are:

(a) 150 Hz, 250 Hz (b) 50 Hz, 150 Hz
(c) 50 Hz, 100 Hz (d) 50 Hz, 0 Hz

[2020 : 1 M]

2.56 A single-phase, full-bridge, fully controlled thyristor rectifier feeds a load comprising a 10 Ω resistance in series with a very large inductance. The rectifier is fed from an ideal 230 V, 50 Hz sinusoidal source through cables which have negligible internal resistance and a total inductance of 2.28 mH. If the thyristors are triggered at an angle α = 45°, the commutation overlap angle in degree (rounded off to 2 decimal places) is_____. [2020 : 2 M]

2.57 In the circuit shown, the input V_i is a sinusoidal AC voltage having an RMS value of 230 V ± 20%. The worst-case peak-inverse voltage seen across any diode is _______V. (Round off to 2 decimal places.)

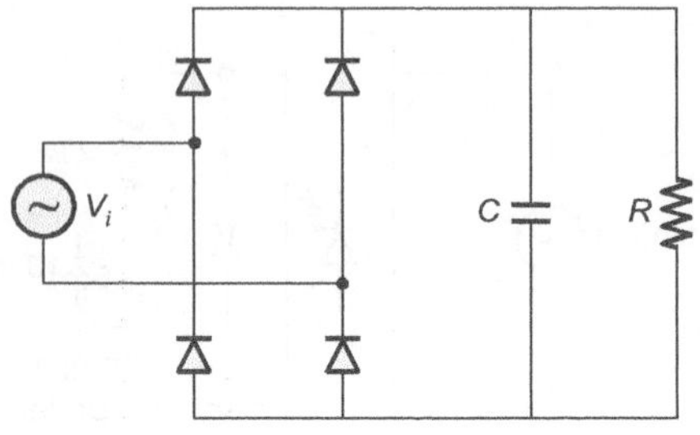

[2021 : 1 M]

2.58 A single-phase full-bridge diode rectifier feeds a resistive load of 50 Ω from a 200 V, 50 Hz single phase AC supply. If the diodes are ideal, then the active power, in watts, drawn by the load is _____. (round off to nearest integer). [2022 : 1 M]

2.59 The voltage at the input of an AC-DC rectifier is given by $v(t) = 230\sqrt{2}\sin\omega t$ where $\omega = 2\pi \times 50$ rad/s. The input current drawn by the rectifier is given by

$$i(t) = 10\sin\left(\omega t - \frac{\pi}{3}\right) + 4\sin\left(3\omega t - \frac{\pi}{6}\right) + 3\sin\left(5\omega t - \frac{\pi}{3}\right)$$

The input power factor, (rounded off to two decimal places), is _____ lag. [2022 : 1 M]

2.60 For the ideal AC-DC rectifier circuit shown in the figure below, the load current magnitude is Idc = 15 A and is ripple free. The thyristors are fired with a delay angle of 45°. The amplitude of the fundamental component of the source current, in amperes, is ___. (round off to 2 decimal places).

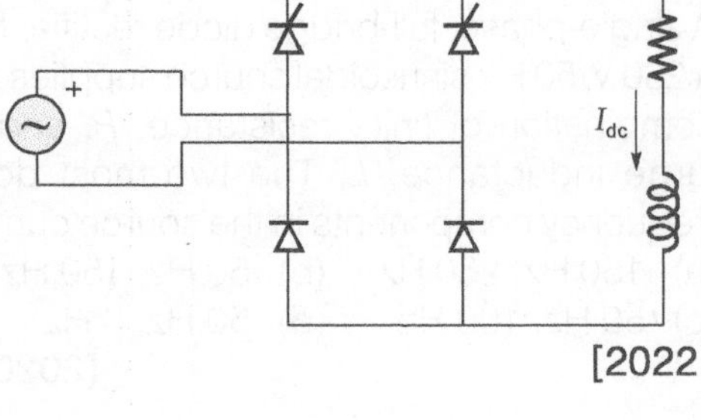

[2022 : 2 M]

2.61 The single phase rectifier consisting of three thyristors T_1, T_2, T_3 and a diode D_1 feed power to a 10 A constant current load. T_1 and T_3 are fired at $\alpha = 60°$ and T_2 is fired at $\alpha = 240°$. The reference for α is the positive zero crossing of V_{in}. The average voltage V_O across the load in volts is _____ (Round off to 2 decimal places).

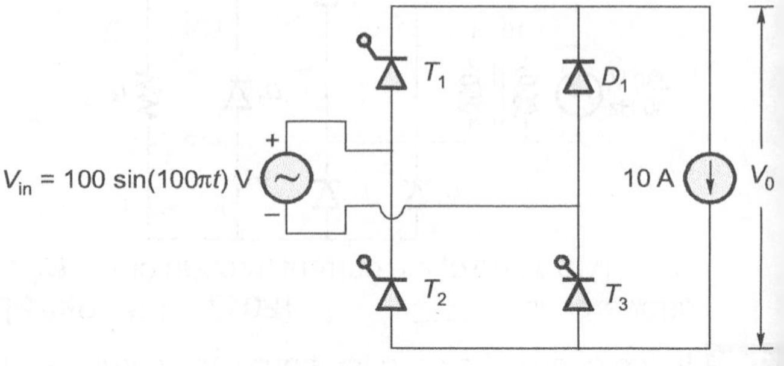

[2023 : 2 M]

2.62 A single-phase half-controlled bridge converter supplies an inductive load with ripple free load current. The triggering angle of the converter is 60°. The ratio of the rms value of the fundamental component of the input current to the rms value of the total input current of the bridge is _________(rounded off to 3 decimal places). [2024 : 2 M]

2.63 A single-phase full-controlled thyristor converter bridge is used for regenerative braking of a separately excited DC motor with the following specifications:

Rated armature voltage	210 V
Rated armature current	10 A
Rated speed	1200 rpm
Armature resistance	1 Ω
Input to the converter bridge	240 V at 50 Hz
The armature of the DC motor is fed from the full-controlled bridge and the field current is kept constant.	

Assume that the motor is running at 600 rpm and the armature terminals of the motor are suitably reversed for regenerative braking. If the armature current of the motor is to be maintained at the rated value, the triggering angle of the converter bridge in degrees should be ______(rounded off to 2 decimal places).

[2024 : 2 M]

■■■■

Answers Phase Controlled Rectifiers

2.1 (594.21)	2.2 (c)	2.3 (a)	2.4 (c)	2.5 (d)	2.6 (d)	2.7 (c)
2.8 (a)	2.9 (a)	2.10 (c)	2.11 (c)	2.12 (b)	2.13 (a)	2.14 (b)
2.15 (d)	2.16 (a)	2.17 (c)	2.18 (b)	2.19 (b)	2.20 (d)	2.21 (c)
2.22 (c)	2.23 (c)	2.24 (c)	2.25 (c)	2.26 (d)	2.27 (b)	2.28 (b)
2.29 (a)	2.30 (a)	2.31 (d)	2.32 (c)	2.33 (23)	2.34 (69°31′)	2.35 (0.78)
2.36 (b)	2.37 (17.07)	2.38 (c)	2.39 (61.53)	2.40 (c)	2.41 (57.7)	2.42 (6)
2.43 (9.21)	2.44 (224.17)	2.45 (a)	2.46 (c)	2.47 (0)	2.48 (0.707)	2.49 (17.39)
2.50 (c)	2.51 (90°)	2.52 (a)	2.53 (0.78)	2.54 (81.65)	2.55 (b)	2.56 (4.80)
2.57 (390.32)	2.58 (800)	2.59 (0.45)	2.60 (17.65)	2.61 (39.78)	2.62 (0.955)	2.63 (114.58)

Explanations Phase Controlled Rectifiers

2.1 Sol.

$$V_a = \frac{3V_{mL}}{\pi} = \frac{3\times 440\sqrt{2}}{\pi} = 594.208 \text{ V}$$

2.2 (c)

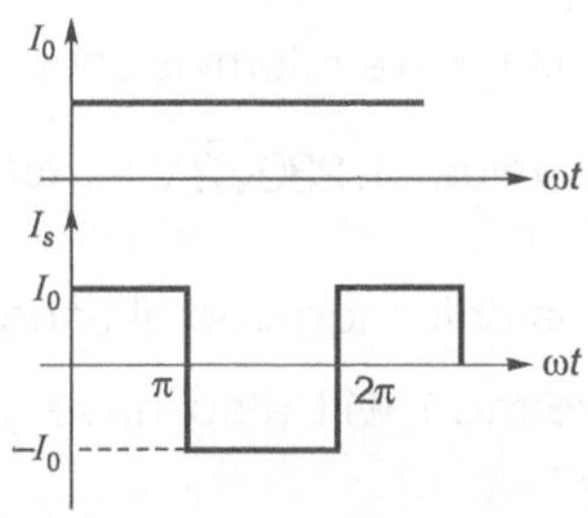

2.5 (d)

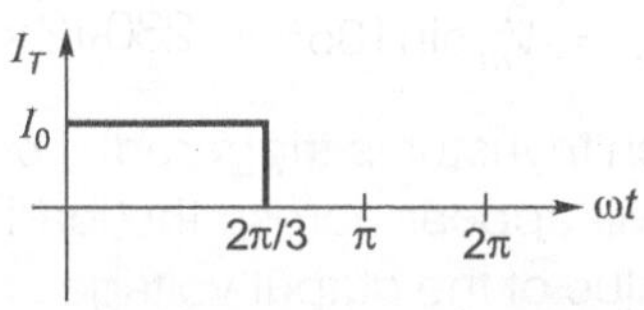

Each thyristor conducts for 120° out of 360°.

$$\therefore\ I_{T_{rms}} = \sqrt{\frac{1}{2\pi}\int_0^{2\pi/3} I_0^2\, d\omega t} = I_0\sqrt{\frac{1}{3}} = \frac{150}{\sqrt{3}} \text{ A}$$

2.6 (d)

$$\text{D.F.} = \cos\alpha = \cos 30° = \frac{\sqrt{3}}{2}$$

2.9 (a)

$$V_m \sin\omega t = L\frac{di}{dt}$$

$$\frac{di}{dt} = \frac{V_m}{L}\sin\omega t$$

$$\int_0^i di = \frac{V_m}{L}\int_\alpha^t \sin\omega t\, dt$$

$$i = \frac{V_m}{\omega L}\cos\omega t\Big|_t^\alpha$$

$$i = \frac{V_m}{\omega L}(\cos\alpha - \cos\omega t) \quad \text{...(i)}$$

For extinction angle, at $\omega t = \beta$, $i = 0$

$$\Rightarrow \quad 0 = \frac{V_m}{\omega L}(\cos\alpha - \cos\beta)$$

which gives, $\cos\alpha = \cos\beta$

$$\beta = 2n\pi \pm \alpha$$

For $n = 1$, $\beta = 240°, 480°$

2.10 (c)

$$i_s \propto \frac{I_a}{n}\cdot\cos\frac{n\pi}{6} \quad \text{where } n \in 1, 3, 5$$

For $n = 3$, $i_s = 0$

For $n = 5$, $i_s \propto -\frac{I_a}{5}$

Lowest harmonic present is fifth harmonic. Its frequency = 50 × 5 = 250 Hz.

2.11 (c)

$$P = \frac{V_{rms}^2}{R} = \frac{(V_m/\sqrt{2})^2}{R} = \frac{(200/\sqrt{2})^2}{50} = 400 \text{ W}$$

2.13 (a)

$$\frac{\text{Peak to peak ripple voltage}}{\text{Peak output DC voltage}} = \frac{V_{ML} - V_{ML}\sin 150°}{V_{ML}} = 0.5$$

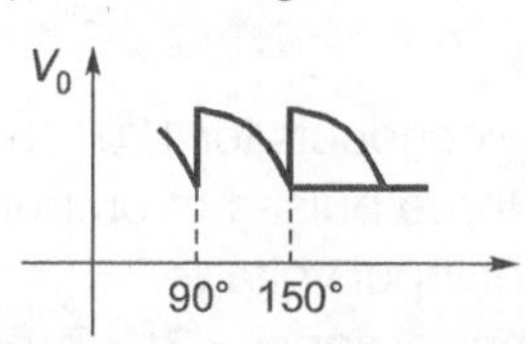

2.14 (b)

Due to free wheeling action, negative voltage is eliminated and power factor is improved in single-phase half controlled rectifier.

2.15 (d)

$$V_b = 12 \pm 4 \text{ V}$$

$$V_{b(\min)} = 8 \text{ V and } V_{b(\max)} = 16 \text{ V}$$

The thyristor must turn on even for minimum value of V_b.

Gate current for guaranteed turn-on, $I_g = 10$ mA, neglecting voltage between gate and cathode.

Therefore, required value of

$$R = \frac{V_{b(\min)}}{I_g} = \frac{8}{10 \text{ mA}} = 800 \ \Omega$$

2.16 (a)

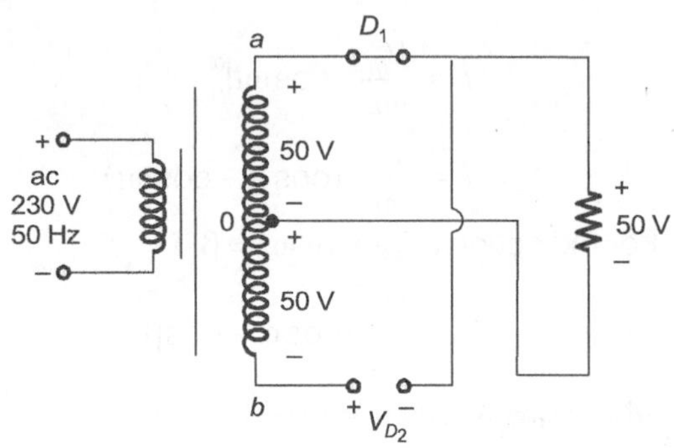

When '*a*' is positive with respect to '*b*' diode D_1 conducts and diode D_2 is reverse biased.

Applying KVL, $V_{D2} = 100$ V (rms)

So, diode D_2 is subjected to reverse voltage of 100 V (rms). Similarly, during negative half cycle, D_1 is subjected to reverse voltage of 100 V (rms). Thus, for diode D_1 and D_2 peak reverse voltage is $100\sqrt{2}$ V (maximum value of the voltage across the diodes).

2.17 (c)

By drawing the (output) voltage.

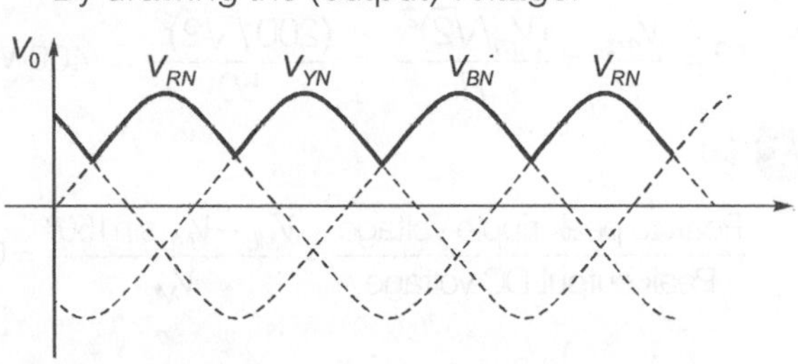

Each diode conducts for 120° only.

There are three pulses of output voltage during one cycle of input voltage.

So, frequency of ripple = $3f = 3 \times 400 = 1200$ Hz

2.18 (b)

Maximum value of input voltage,

$$V_m = 400\sqrt{2} \text{ V}$$

Since load is purely resistive, therefore peak instantaneous output voltage

$$V_m = 400\sqrt{2} \text{ V}$$

2.19 (b)

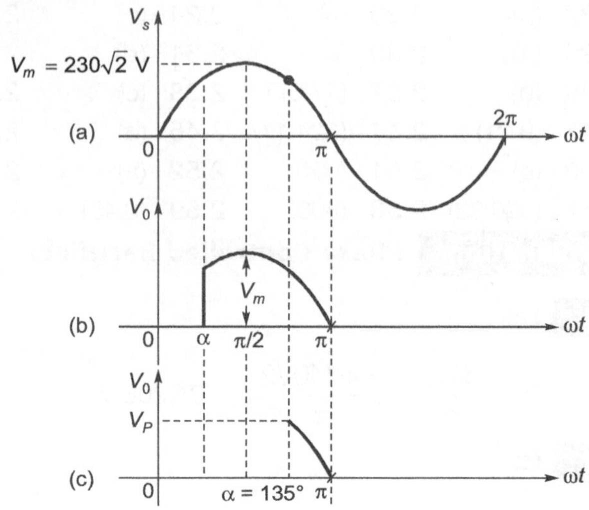

Supply voltage waveform is shown in figure (a) it has peak value of $230\sqrt{2}$ V at $\omega t = \frac{\pi}{2}$.

If the thyristor is triggered at firing angle $\alpha < \frac{\pi}{2}$, output voltage will also have peak value of $230\sqrt{2}$ V.

The supply voltage at $\alpha = 135°$.

$$V_s\big|_{\alpha=135°} = V_m \sin 135° = 230\sqrt{2} \times \frac{1}{\sqrt{2}} = 230 \text{V}$$

So, when thyristor is triggered at $\alpha = 135°$ 230 V will appear across the load, it is also the peak value of the output voltage.

2.20 (d)

Without considering source inductance i.e. (Transformer inductance).

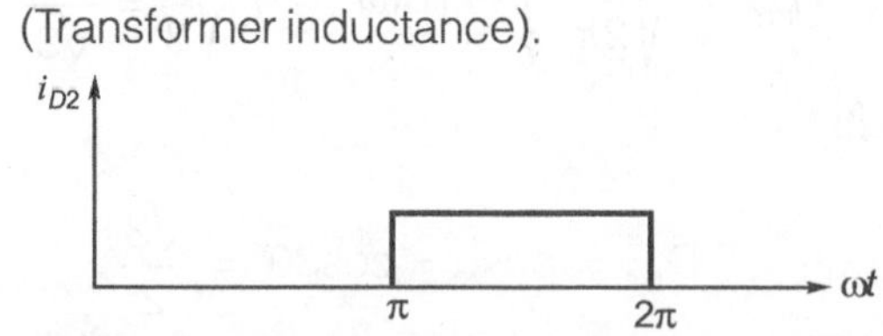

After considering transformer leakage inductance.

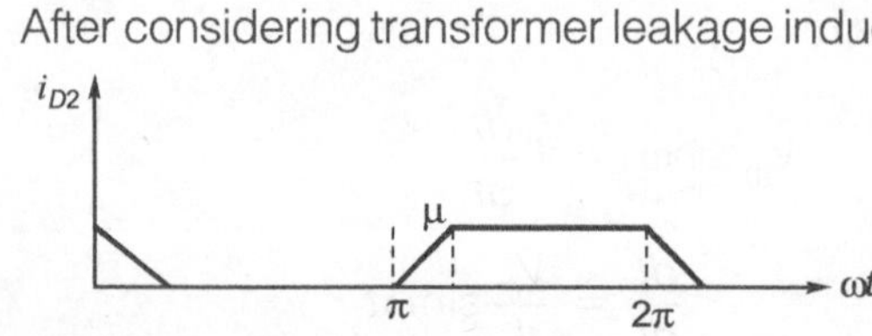

2.21 (c)

Load current is constant,

$$I_0 = 10 \text{ A}$$

In 3-ϕ full converter with free wheeling diode,

Input displacement factor (IDF)

$$= \cos\alpha = \cos 60° = 0.5$$

RMS value of source current,

$$I_s = \sqrt{I_0^2 \times \frac{2\pi}{3} \times \frac{1}{\pi}} = I_0\sqrt{\frac{2}{3}}$$

RMS value of nth harmonic,

$$I_{sh} = \frac{4I_0}{\sqrt{2}n\pi}\sin\frac{n\pi}{3}$$

RMS value of fundamental current,

$$I_{s1} = \frac{2\sqrt{2}}{\pi}I_0\sin 60° = \frac{\sqrt{6}}{\pi}I_0$$

Current distortion factor (CDF)

$$= \frac{I_{s1}}{I_s} = \frac{\sqrt{6}}{\pi}I_0 \times \left(\frac{1}{I_o}\sqrt{\frac{3}{2}}\right) = \frac{3}{\pi} = 0.955$$

Input power factor (IPF)

$$= \text{CDF} \times \text{IDF} = 0.955 \times 0.5 \approx 0.478$$

2.22 (c)

T_1 and T_4 gets forward biased, when

$$V_m \sin\theta_1 \le E$$

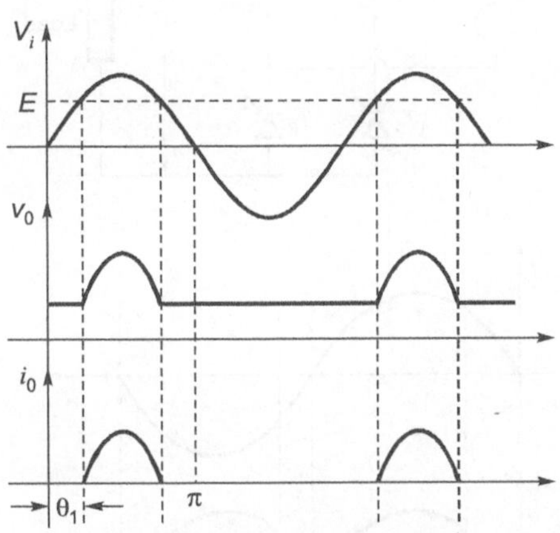

$$I_{avg} = \text{(Average current)}$$

$$= \frac{1}{2\pi R}\int_{\theta_1}^{\pi-\theta_1}(V_m \sin\omega t - E)d\theta$$

$$\therefore I_0\text{(avg)} = \frac{1}{2\pi R}[2V_m\cos\theta - E(\pi - 2\theta_1)]$$

$$= \frac{1}{2\pi \times 2}[2 \times (230 \times \sqrt{2})\cos\theta_1 - 200(\pi - 2\theta_1)]$$

where, $\theta_1 = \sin^{-1}\left(\frac{E}{V_m}\right)$

$$= \sin^{-1}\left(\frac{200}{230 \times \sqrt{2}}\right) = 38° = 0.66 \text{ rad}$$

$$\therefore I_0\text{(avg)} = \frac{1}{2\pi \times 2}[2\sqrt{2} \times 230\cos 38° - 200(\pi - 2 \times 0.66)] = 11.9 \text{ A}$$

2.23 (c)

$$\text{FDF} = \cos\left(\alpha + \frac{\mu}{2}\right) = \cos\left(25° + \frac{10°}{2}\right) = 0.866$$

2.24 (c)

Let DC link current = I_d

DC voltage applied to the inverse,

$$V_d = 420 \text{ V}$$

Power fed to the inverter,

$$P = V_d I_d = 50 \text{ kW}$$

$$\Rightarrow \quad 420 I_d = 50 \times 10^3$$

$$I_d = 119.05 \text{ A}$$

Current through each thyristor flows for period of $2\pi/3$.

So, rms current of thyristor.

$$(I_{th})_{rms} = \sqrt{\frac{1}{2\pi}\int_0^{2\pi/3} I_d^2\, d\omega t}$$

$$= \frac{I_d}{\sqrt{3}} = \frac{119.05}{\sqrt{3}} = 68.73 \text{A}$$

2.25 (c)

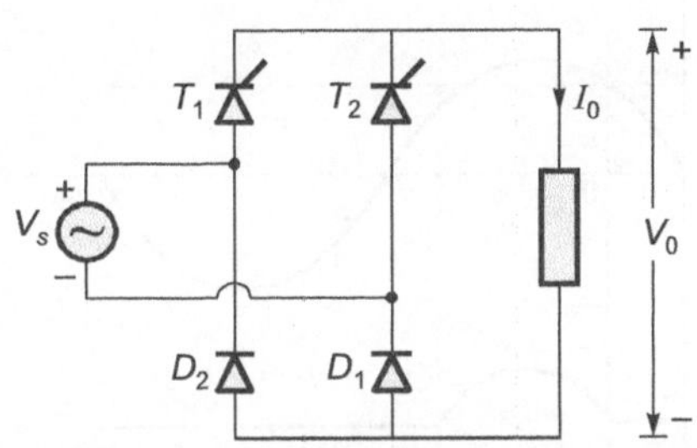

1-ϕ full wave half controlled bridge converter without free wheeling diode is shown in the figure.

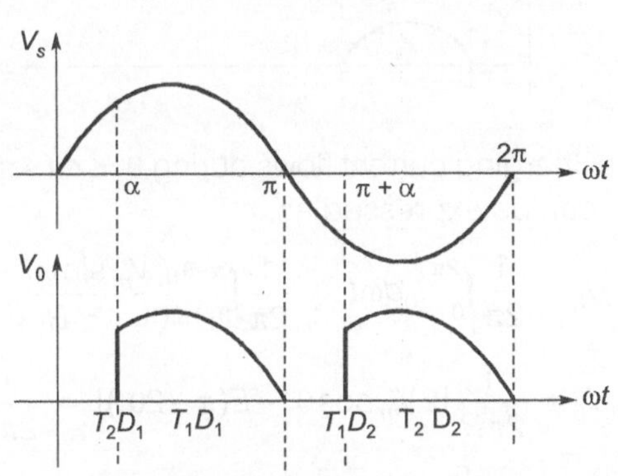

At $\omega t = \alpha$, T_1 is fired and T_1 starts conducting load current flows through T_1D_1 for $\alpha < \omega t < \pi$.

At $\omega t = \pi$, V_s becomes negative and D_1 gets reverse biased and D_2 is forward biased. So during $\pi < \omega t < \pi + \alpha$, free wheeling action takes place through T_1 and D_2 and output voltage becomes zeros.

At $\omega t = \pi + \alpha$, Load T_2 is triggered and load current is transferred from T_1 to T_2. So, during $\pi + \alpha < \omega t < 2\pi$, T_2D_2 conducts.

At $\omega t = \pi + \alpha$, It may be possible that load current is not transferred completely from T_1 to T_2, and T_1 and T_2 may be conducting simultaneously which results in short circuit of the supply for short direction.

2.26 (d)

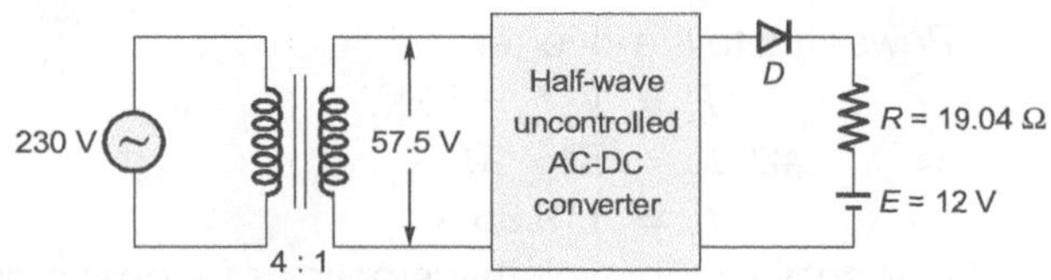

Input to the converter,

$$V_s = \left(\frac{1}{4}\right) 230 = 57.5\text{V}$$

Diode conducts when $V_s \geq E$

$$V_m \sin\theta_1 = E$$

$$57.5\sqrt{2}\sin\theta_1 = 12 \Rightarrow \theta_1 = 8.486° \text{ or } 0.148 \text{ rad}$$

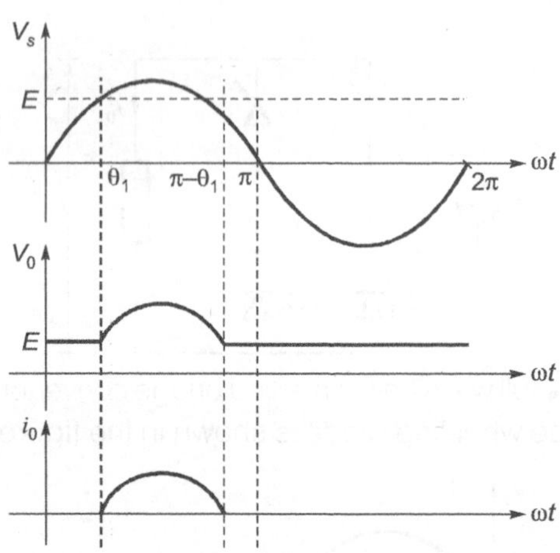

Charging current flows during $\theta_1 \leq \omega t \leq \pi - \theta_1$ and can be expressed as,

$$I_0 = \frac{1}{2\pi}\int_0^{2\pi} i_0 d\omega t = \frac{1}{2\pi}\int_{\theta_1}^{\pi-\theta_1}\left(\frac{V_m \sin\omega t - E}{R}\right) d\omega t$$

$$I_0 = \frac{1}{2\pi R}[2V_m\cos\theta_1 - E(\pi - 2\theta_1)] = \frac{1}{2\pi \times 19.04}$$

$$\left[2 \times 57.5\sqrt{2} \times \cos 8.486° - 12 \times (\pi - 2 \times 0.148)\right)$$

$$= 1.06 \text{ A} \approx 1\text{A}$$

2.27 (b)

Load current, $I_0 = 10$ A

RMS value of total source current

$$I_s = I_0\sqrt{\frac{2}{3}} = 10\sqrt{\frac{2}{3}} = 8.165\text{A}$$

Supply current i_s can be expressed by fourier series,

$$i_s(t) = \sum_{n=1,3,5}^{\infty} \frac{4I_0}{n\pi}\sin\frac{n\pi}{3}\sin(n\omega t - n\alpha)$$

RMS value of fundamental current,

$$I_{s1} = \frac{4I_0}{\sqrt{2}\pi}\sin\frac{\pi}{3} \Rightarrow I_{s1} = \frac{4 \times 10}{\sqrt{2}\pi} \times \frac{\sqrt{3}}{2} = 7.8\text{ A}$$

Total Harmonic distortion

$$= \sqrt{\left(\frac{I_s}{I_{s1}}\right)^2 - 1} \times 100$$

$$= \sqrt{\left(\frac{8.165}{7.8}\right)^2 - 1} \times 100 = 31\%$$

2.28 (b)

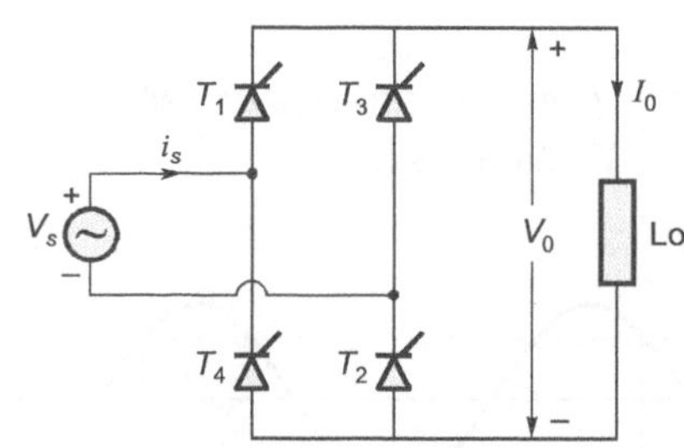

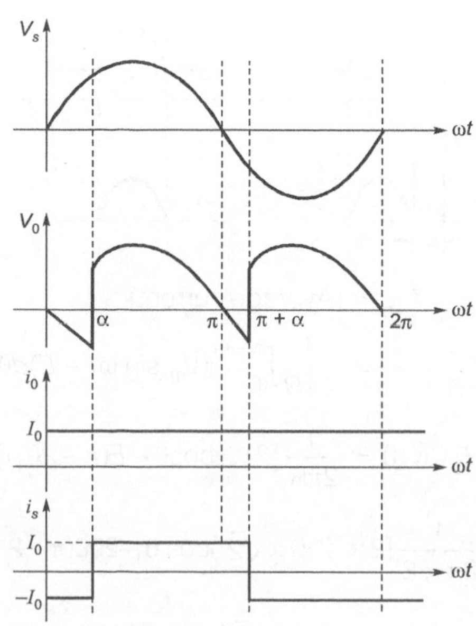

Average output voltage,

$$V_0 = \frac{1}{\pi}\int_{\alpha}^{\pi+\alpha} V_m \sin\omega t \cdot d(\omega t)$$

$$\Rightarrow \quad V_0 = \frac{2V_m}{\pi}\cos\alpha$$

For $\alpha = 30°$, $V_0 = \frac{\sqrt{3}}{\pi}V_m$

Average output current, I_0 (constant)

RMS value of supply current, $I_s = I_0$

RMS value of supply voltage, $V_s = \frac{V_m}{\sqrt{2}}$

$$\text{Input power factor} = \frac{\text{Power delivered to load}}{\text{Input VA}}$$

$$= \frac{V_0 I_0}{V_s I_s} = \frac{(\sqrt{3}/\pi)\,V_m I_0}{V_m/\sqrt{2}\cdot I_0} = \frac{\sqrt{6}}{\pi} = 0.78$$

2.29 (a)

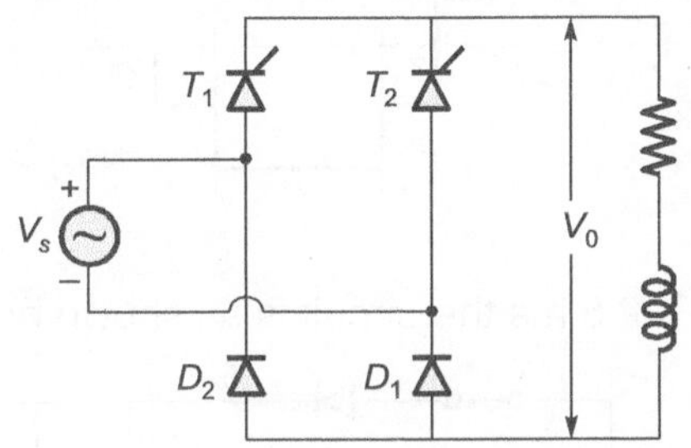

Single phase half controlled converter.
As load is highly inductive, it means load current is continuous and almost constant.

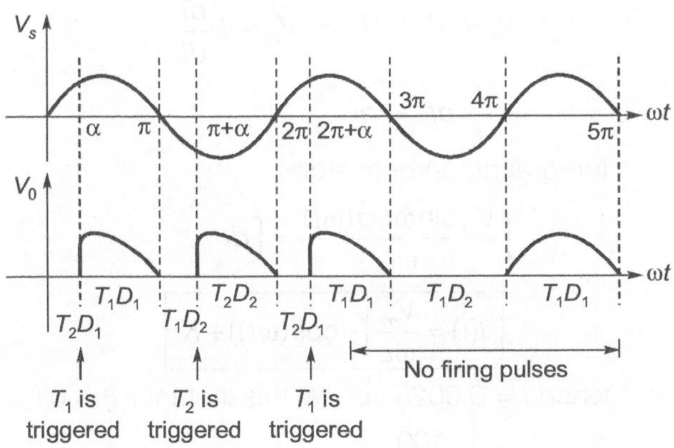

During,

- $0 < \omega t < \alpha$
  freewheeling by T_2D_1, $V_0 = 0$
- $\alpha < \omega t < \pi$
  T_1D_1 conducts $V_0 = V_s$
- $\pi < \omega t < \pi + \alpha$
  Freewheeling by T_1D_2, $V_0 = 0$
- $\pi + \alpha < \omega t < 2\pi$
  T_2D_2 conduct, $V_0 = -V_s$
  T_1 is again triggered at $2\pi + \alpha$

So, during $2\pi + \alpha < \omega t < 3\pi$, T_1D_1 conducts.
Now, if firing pulses are removed after T_1 is triggered at $2\pi + \alpha$.
At $\omega t = 3\pi + \alpha$, no firing pulse will be available to trigger T_2. So, load current will flow through T_1D_2 as load current is continuous. So during $3\pi + \alpha < \omega t < 4\pi$, T_1D_2 will continue to conduct and $V_0 = 0$.
At $\omega t = 4\pi$, D_1 will become forward biased and D_2 will become reverse biased. So, D_1 will conduct and T_1 is already conducting.
So load current will flow through T_1D_1 and $V_0 = V_s$.

2.30 (a)

For firing angle, α, average output voltage of the converter is given by

$$V_0 = \frac{2V_m}{\pi}\cos\alpha$$

when, $\alpha = 0°$

$$V_0 = \frac{2V_m}{\pi}\cos 0° = 300 \Rightarrow \frac{2V_m}{\pi} = 300$$

when, $\alpha = 60°$

$$V_0 = \frac{2V_m}{\pi}\cos 60° = 300\cos 60° = 150 \text{ V}$$

2.31 (d)

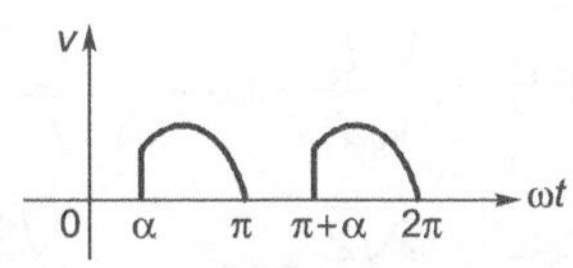

Freewheeling diode conducts for 2α over complete cycle

$$\therefore \quad \text{Fraction} = \frac{2\alpha}{2\pi} = \frac{\alpha}{\pi}$$

2.32 (c)

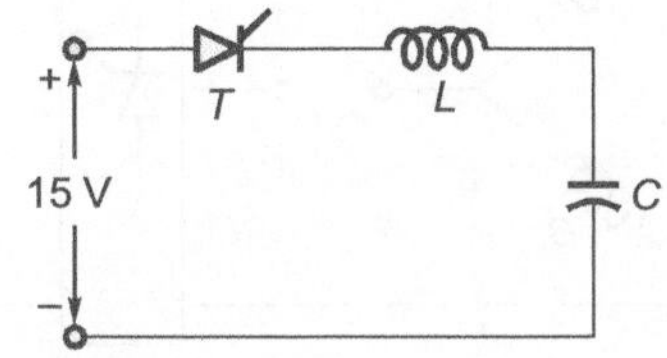

Conduction time = $\pi\sqrt{LC}$

$$= \pi\sqrt{\left(\frac{100}{\pi}\times 10^{-6}\right)\left(\frac{100}{\pi}\times 10^{-6}\right)}$$

$= 100$ μs

2.33 Sol.

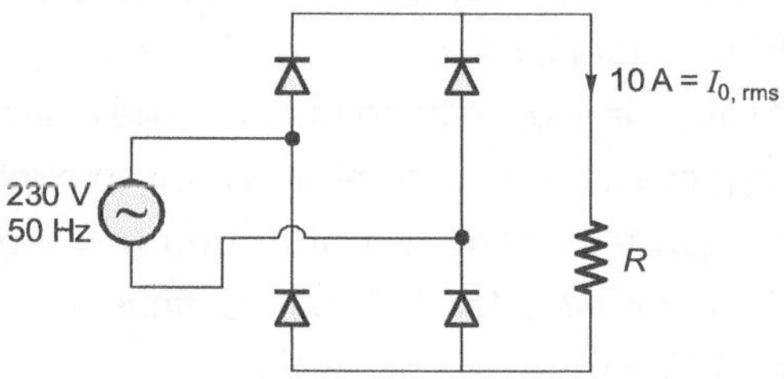

$$I_{0,\text{rms}} = \frac{V_s}{R} \Rightarrow R = \frac{230}{10} = 23\ \Omega$$

2.34 Sol.

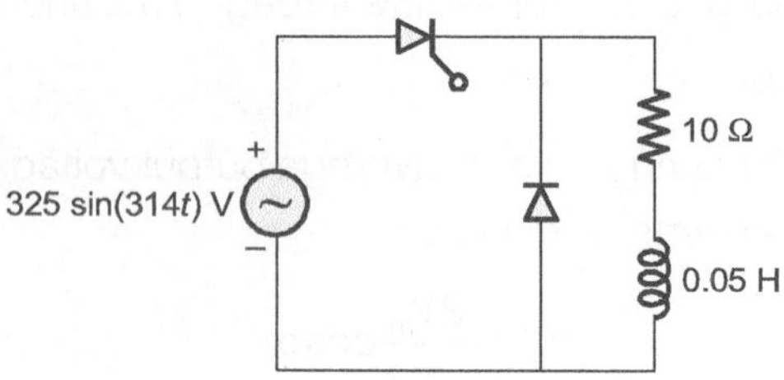

$$\boxed{V_0 = \frac{V_m}{2\pi}(1+\cos\alpha)}$$

$$\frac{70\times 2\pi}{325} - 1 = \cos\alpha \Rightarrow \alpha = \cos^{-1}\left(\frac{70\times 2\pi}{325} - 1\right)$$

$$\boxed{\alpha = 69°31'}$$

2.35 Sol.

The input power factor (PF) is defined as

$$\text{P.F.} = \frac{V_s I_{s1}}{V_s I_s}\cos\phi = \frac{I_{s1}}{I_s}\cos\phi$$

$$\text{P.F.} = \frac{2\sqrt{2}}{\pi}\cos\alpha = \frac{2\sqrt{2}}{\pi}\cos 30°$$

$$= 0.78 \text{ (Lagging)}$$

2.36 (b)

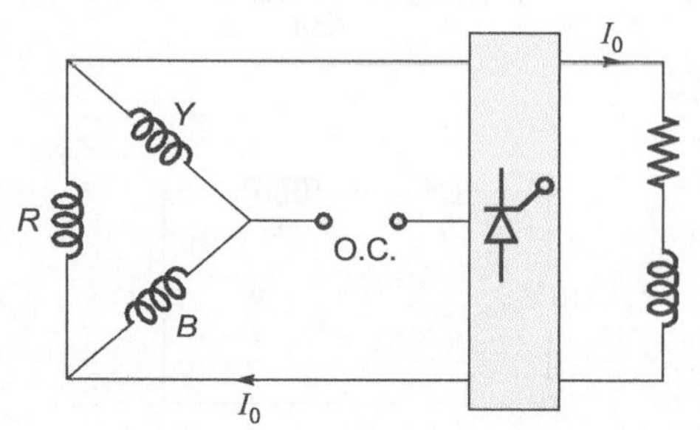

Let us assume all the Impedance are Z and to flow I_0 only closed path is required, so in three phases one phase is open circuited.

According to correct division,

$$I_R = \frac{I_0 2Z}{3Z} = \frac{2I_0}{3} = \frac{2}{3}\text{ p.u.} \quad [\because I_0 = 1 \text{ p.u.}]$$

$$I_{YB} = \frac{I_0 Z}{3Z} = \frac{I_0}{3} = \frac{1}{3}\text{ p.u.}$$

The transformer ratio is given as 1 : K.

$$\therefore \quad I_{R(\text{primary})} = \frac{2K}{3}\text{ p.u. and } I_{YB} = \frac{K}{3}\text{ p.u.}$$

$\therefore$ According 3 phase source waveform,

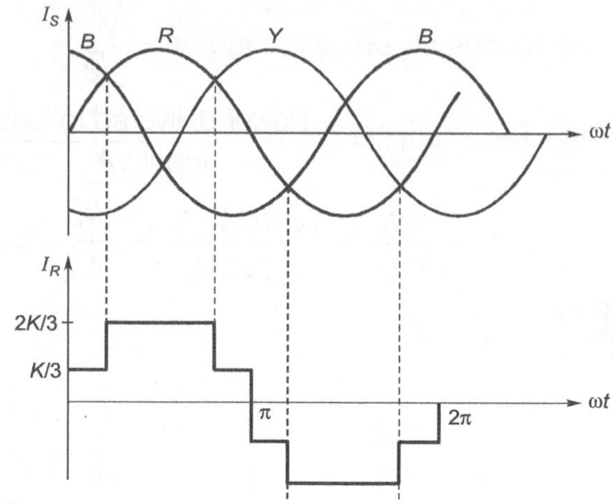

2.37 Sol.

At t = 2.5 ms the circuit is as shown below,

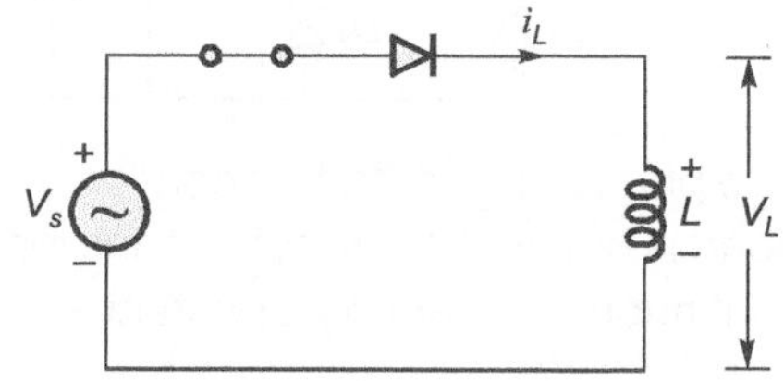

$$-V_S + V_L = 0 \Rightarrow V_S = L\frac{di}{dt}$$

$$\frac{V_S}{L}dt = di$$

Integrating on both sides,

$$\int \frac{V_m \sin\omega t\, d(\omega t)}{\omega L} = \int di$$

$$\boxed{i(t) = \frac{V_m}{\omega L}[-\cos(\omega t)] + K}$$

when t = 0.0025 sec, at this instant $i(t)$ = 0

$$0 = \frac{100}{100\pi\times 31.83\times 10^{-3}}[-\cos(100\pi\times 0.0025] + K$$

$$= -7.07 + K \Rightarrow \boxed{K = 7.07}$$

The peak value of the inductor current will be at $(\omega t = \pi^c)$

$$i(t)_{\max} = \frac{V_m}{\omega L}[-\cos(\pi)] + K$$

$$I_{\max} = \frac{100}{100\pi\times 31.83\times 10^{-3}} + 7.07$$

$$\boxed{I_{\max} = 17.07 \text{ A}}$$

2.38 (c)

Given,

$$100\pi\ RC = 50 \quad \Rightarrow \quad 100\pi \cdot \frac{RC}{2} = 25$$

$$\omega\tau = 25$$

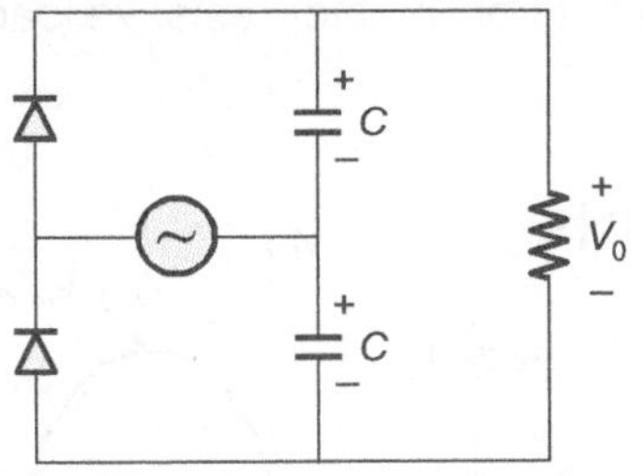

For one cycle $\omega\tau = 2\pi$ rad = 6.28

$[\omega\tau = 25] >> [\omega\tau = 2\pi \text{ rad} = 6.28]$

∴ Each capacitor voltage is approximately equal to

$$V_m = 100 \text{ V}$$

$$\therefore \quad V_0 = 2\ V_m \simeq 200 \text{ V}$$

2.39 Sol.

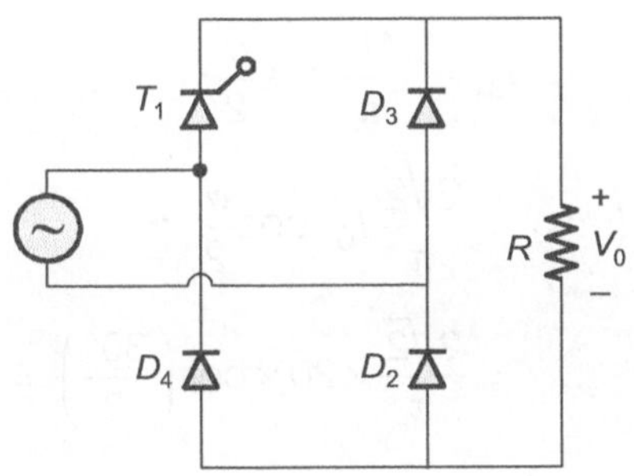

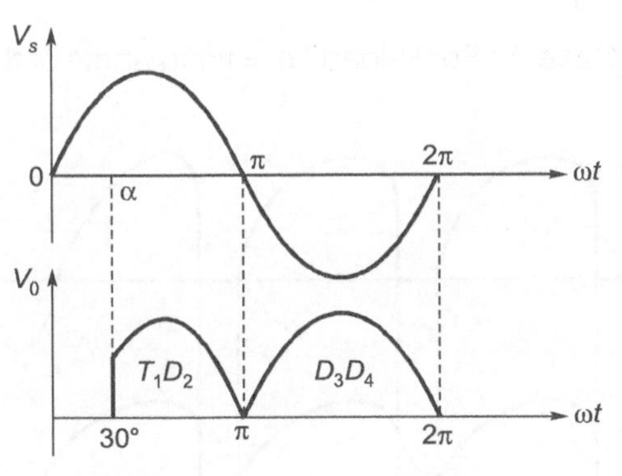

$$V_0 = \frac{1}{2\pi}\left[\int_\alpha^\pi V_m \sin\omega t \cdot d(\omega t) - \int_\pi^{2\pi} V_m \sin\omega t \cdot d(\omega t)\right]$$

$$= \frac{V_m}{2\pi}[3+\cos\alpha] = \frac{100}{2\pi}[3+\cos 30°] = 61.53 \text{ V}$$

2.40 (c)

Input power factor = power factor at ac mains

= C.D.F. × D.F. = $g \cos\alpha$

$$= \frac{2\sqrt{2}}{\pi}\cos\alpha = \frac{2\sqrt{2}}{\pi}\cos 30° = 0.78$$

2.41 Sol.

In the 3-ϕ diode bridge rectifier each diode conducts for 120° for one complete cycle.

$$I_{D\text{rms}} = \sqrt{\frac{1}{2\pi}\int_0^{2\pi/3} I_0^2\, d\omega t} = I_o\sqrt{\frac{2\pi}{2\pi \times 3}}$$

$$= \frac{I_o}{\sqrt{3}} = \frac{100}{\sqrt{3}} = 57.7 \text{ A}$$

2.42 Sol.

$$V_o = 2\frac{V_m}{\pi}\cos\alpha = 2\frac{200\pi}{\pi}\cos 120°$$

$$V_o = -200 \text{ V}$$

$$|V_o| = 200 \text{ V}$$

Power balance equation,

$$EI_o = I_o^2 R + V_o I_o$$

$$800 I_o = I_o^2(20) + 200 I_o \Rightarrow I_o = 30 \text{ A}$$

$$I_o = I_{or}$$

Power fed to source

$$= V_o I_o = 200 \times 30 = 6 \text{ kW}$$

2.43 Sol.

$$P = V_s I_s \text{(I.P.F.)}$$

$$5 \times 10^3 = 220 \times I_s \times 1$$

$$I_s = 22.72 \text{ A}$$

$$\tan\delta = \frac{I_s X_s}{V_s}$$

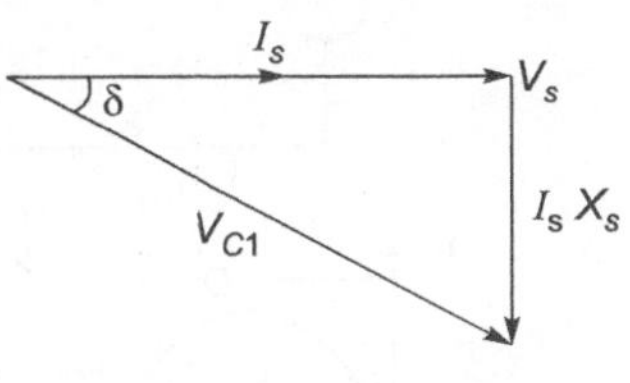

$$\delta = \tan^{-1}\left(\frac{I_s X_s}{V_s}\right)$$

$$\delta = \tan^{-1}\left(\frac{22.72 \times 2\pi \times 50 \times 5 \times 10^{-3}}{220}\right) = 9.21°$$

2.44 Sol.

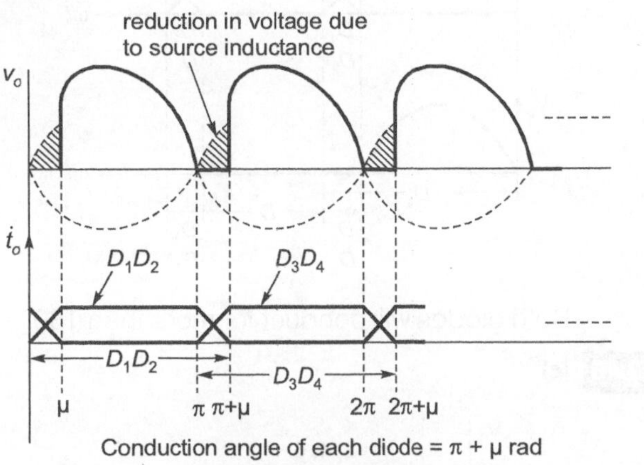

Conduction angle of each diode = π + μ rad

Output Waveform

$$V_{o(\text{avg})} = \frac{2V_m}{\pi} \quad (\because \text{ without } L_s)$$

$$V'_{o(avg)} = \frac{1}{\pi}\int_{\mu}^{\pi} V_m \sin\omega t \cdot d\omega t \quad (\because \text{ with } L_s)$$

$$= \frac{V_m}{\pi}[1+\cos\mu]$$

$$\Delta V_{do} = V_{o(avg)} - V'_{o(avg)} = \frac{V_m}{\pi}[1-\cos\mu]$$

and $\Delta V_{do} = 4fL_sI_o$ [average reduction in voltage due to source inductance]

On equating,

$$\frac{V_m}{\pi}[1-\cos\mu] = 4fL_sI_o$$

$$\frac{220\sqrt{2}}{\pi}(1-\cos\mu) = 4\times 50\times 10\times 10^{-3}\times 14$$

On solving, $\mu = 44.17°$

So, conduction angle of each diode

$\gamma_D = 180° + \mu = 180° + 44.17° = 224.17°$

2.45 (a)

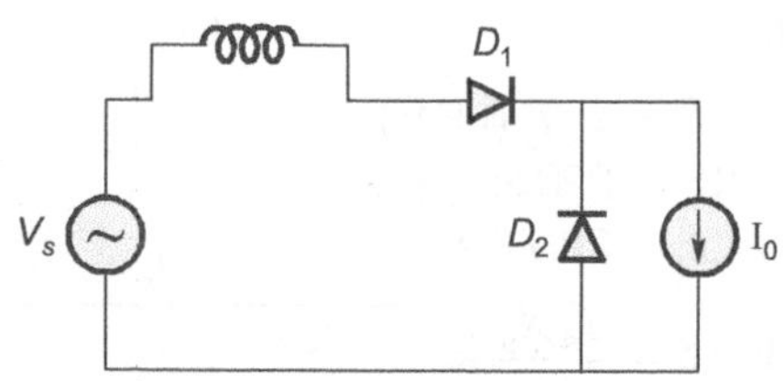

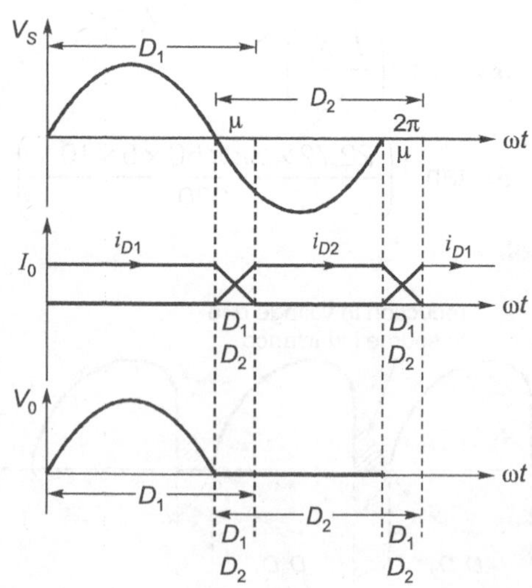

Both diodes will conduct for more than 180°.

2.46 (c)

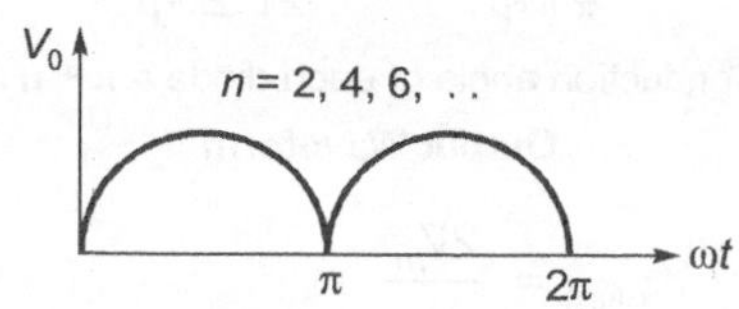

All even harmonics are present.

Fundamental frequency of output voltage

$$= 2f_s = 2\times 50 = 100 \text{ Hz}$$

2.47 Sol.

Diode current, $I_{D3(rms)} = 0$

FD will not conduct for resistive load.

2.48 Sol.

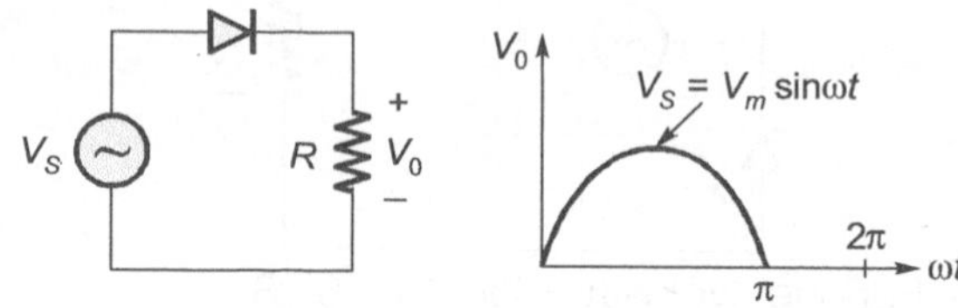

$$V_{or} = \frac{V_m}{2}$$

$$\text{PF} = \frac{V_{or}}{V_{sr}} = \frac{V_m/2}{V_m/\sqrt{2}} = \frac{1}{\sqrt{2}} = 0.707 \quad (\because I_{or} = I_{sr})$$

2.49 Sol.

$$i_{s1} = \frac{4I_0}{\pi}\cos\frac{\alpha}{2}$$

$$I_{s1}(\text{rms}) = \frac{2\sqrt{2}}{\pi}I_0\cdot\cos\frac{\alpha}{2}$$

$$= \frac{2\sqrt{2}}{\pi}\times 20\times\cos\left(\frac{30°}{2}\right) = 17.39 \text{ A}$$

2.50 Sol.

Case 1 : For R-load : α = firing angle of thyristor

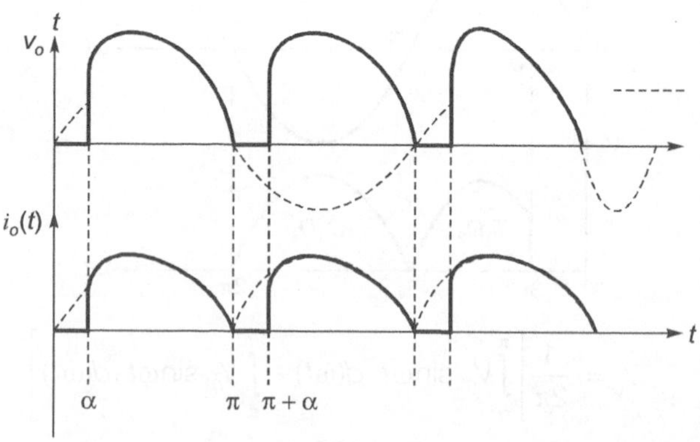

Since load is purely resistive. So, there is no free wheeling action and free wheeling diode will not conduct at all for resistive load.

So, it is clearly seen from the waveforms that $V_o(t) \geq 0$, $i_o(t) \geq 0$.

Case 2 : For R-L load : Assume highly inductive load.

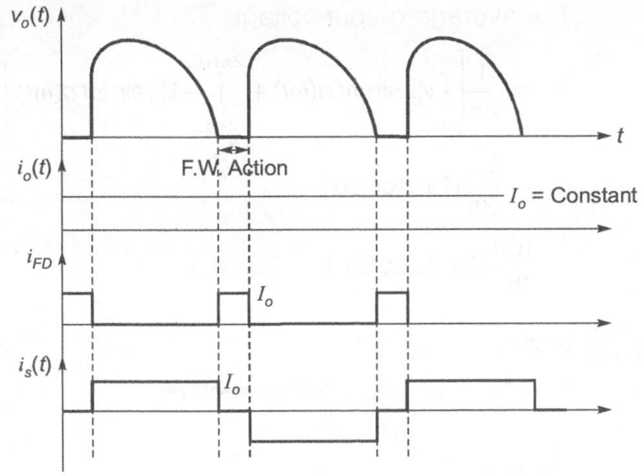

$$V_o(t) \geq 0; \quad i_o(t) \geq 0$$

If F.D. would not present at load side, then $V_o(t)$ can be negative as well, depending on inductive nature of load but $i_o(t)$ will be always positive in this case also.

2.51 Sol.

$$V_0 = \frac{V_m}{2\pi}(3+\cos\alpha)$$

$$E_b I_0 = 1600 \text{ W}$$

$$I_0 = \frac{1600}{80} = 20 \text{ A}$$

$$V_0 = E_b + I_0 R_a$$

$$\frac{V_m}{2\pi}(3+\cos\alpha) = 80 + (20 \times 2)$$

$$\frac{80\pi}{2\pi}(3+\cos\alpha) = 80 + 40$$

$$\alpha = 90°$$

2.52 (a)

For 6 pulse converter, harmonics present in AC current are $6k \pm 1$. ($\because K = 1, 2, 3, 4, ...$)

Lowest order harmonic = 5

Lowest harmonic frequency = 5 × 50 = 250 Hz

2.53 (0.78)

$$V_{sr} I_{sr} \cos\phi = V_0 I_0$$

For single-phase fully controlled converter,

$$I_0 = I_{sr} = 10 \text{ A}$$

$$\cos\phi = \frac{V_0}{V_{sr}} = \frac{180}{230} = 0.78$$

2.54 (81.65)

AC line current rms

$$= (I_s)_{rms} = I_0\sqrt{\frac{2}{3}} = 100 \times \sqrt{\frac{2}{3}} = 81.65 \text{ A}$$

2.55 Sol.

Output waveforms of highly inductive load (without F.W. diode).

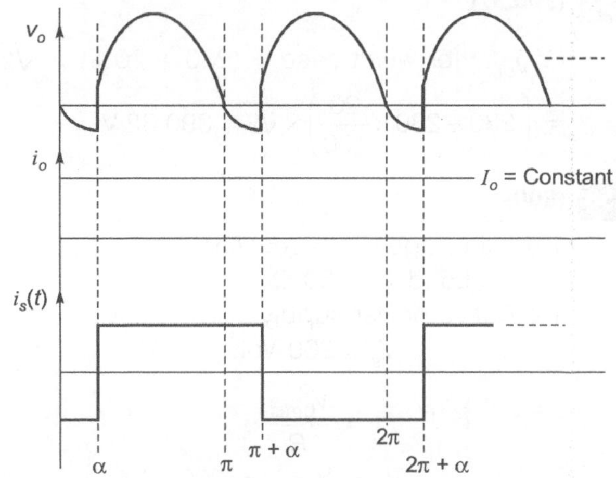

Fourier series of supply current is given as

$$i_s(t) = \sum_{n=1,3,5}^{\infty} \frac{4I_o}{n\pi} \sin n\omega_o t$$

Frequency components present in supply current is

1st, 3rd, 5th, 7th : all odd frequencies.

Two most dominant harmonics are 3rd and 5th, i.e., 150 Hz and 250 Hz.

Two most dominant frequencies are 1st and 3rd, i.e., 50 Hz and 150 Hz.

Except fundamental, all other frequencies are harmonics in supply current.

2.56 Sol.

1-ϕ, SCR bridge rectifier

$$\alpha = 45°, R = 10 \ \Omega$$

supply 230 V, 50 Hz

$$L_s = 2.28 \text{ mH}, \quad \mu = ?$$

$$\Delta V_d = \frac{V_m}{\pi}[\cos\alpha - \cos(\alpha+\mu)] = 4f L_s I_0$$

$$V_0 = \frac{2V_m}{\pi}\cos\alpha - 4f L_s I_0 \text{ (with } L_s)$$

$$I_0 R = \frac{2V_m}{\pi}\cos\alpha - 4f L_s I_0$$

Find I_0

$$I_0 \times 10 = \frac{2\times 230\sqrt{2}}{\pi}\cdot\cos 45 - 4\times 50\times 2.28\times 10^{-3} I_0$$

$$I_0(10 + 0.456) = 146.42$$

$$I_0 = \frac{146.49}{10.456} = 14.0036 \text{ A}$$

$$\Delta V_{d0} = \frac{230\sqrt{2}}{\pi}[\cos 45 - \cos(45+\mu)]$$

$$= 4 \times 50 \times 2.28 \times 10^{-3} \times 14 = 6.384$$

$$\cos 45° - \cos(45° + \mu) = 0.061659$$

$$45 + \mu = 49.80 \Rightarrow \mu = 4.80°$$

2.57 (390.32)

$(V_D)_{peak}$ for worst case $= (230 + 20\%) \times \sqrt{2}$

$$= \left(230 + 230 \times \frac{20}{100}\right) \times \sqrt{2} = 390.32 \text{ V}$$

2.58 (800)

1-φ full bridge diode rectifier,

Load, $R = 50\ \Omega$

1-φ Active power supply

$V_s = 200$ Volts

$$P_{0\,avg} = \frac{V_{0rms}^2}{R}$$

$$V_{0r} = V_{s\,rms} = 200$$

(for 1-φ full bridge rectifier)

$$P_{0\,avg} = \frac{200^2}{50} = 800 \text{ W}$$

2.59 (0.45)

$$g = \frac{I_{s1}}{I_{sr}} = \frac{10/\sqrt{2}}{\sqrt{\left(\frac{10}{\sqrt{2}}\right)^2 + \left(\frac{4}{\sqrt{2}}\right)^2 + \left(\frac{3}{\sqrt{2}}\right)^2}}$$

$$g = 0.89$$

$$pF = g \cdot FDF$$

$$pF = g \cdot \cos\left(\frac{\pi}{3}\right) = 0.447$$

2.60 (17.65)

$$\hat{I}_{sn} = \frac{4 \cdot I_0}{n\pi} \cos\left(\frac{n\alpha}{2}\right)$$

$$\hat{I}_{s1} = \frac{4 \times 15}{\pi} \cos(22.5°) = 17.65 \text{ A}$$

2.61 (39.78)

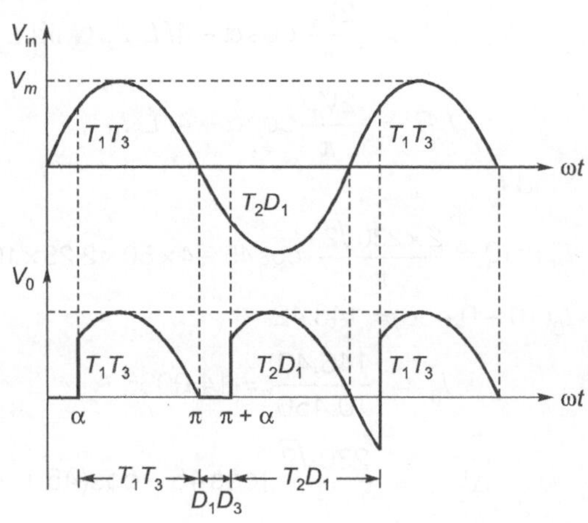

The average output voltage,

$$V_0 = \frac{1}{2\pi}\left[\int_{\alpha}^{\pi} V_m \sin\omega t\, d(\omega t) + \int_{\pi+\alpha}^{2\pi+\alpha} -V_m \sin\omega t\, d(\omega t)\right]$$

$$V_0 = \frac{V_m}{2\pi}[1 + 3\cos\alpha]$$

$$= \frac{100}{2\pi}[1 + 3\cos 60°] = 39.78 \text{ V}$$

2.62 (0.955)

1-φ Half controlled bridge rectifier

$\alpha = 60°$

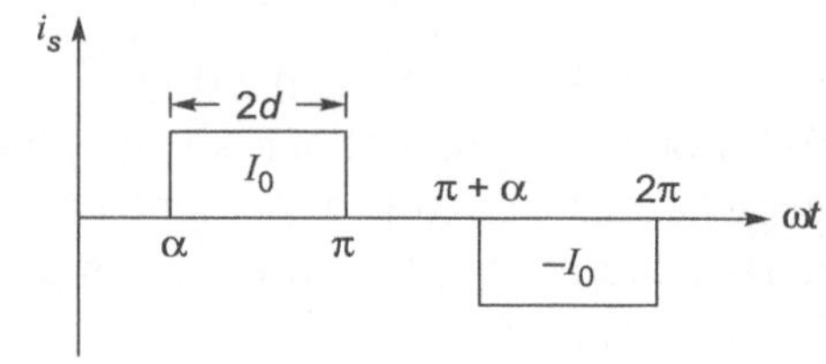

$$g = \frac{I_{s1}}{I_{sr}} = \frac{\frac{2\sqrt{2}}{\pi} I_0 \cdot \cos\frac{\alpha}{2}}{I_0 \left(\frac{\pi - \alpha}{\pi}\right)^{1/2}} = \frac{2\sqrt{2} \cdot \cos\frac{\alpha}{2}}{\sqrt{\pi \cdot (\pi - \alpha)}}$$

$$= \frac{2\sqrt{2} \cdot \cos\left(\frac{60}{2}\right)}{\sqrt{\pi \cdot \left(\pi - \frac{\pi}{3}\right)}} = \frac{3}{\pi} = 0.955$$

Alternate solution:

$$2d = \pi - \alpha = \pi - \frac{3}{\pi}$$

$$2d = \frac{2\pi}{\pi} = 120°$$

$$g = \frac{3}{\pi} = 0.955$$

2.63 (114.58)

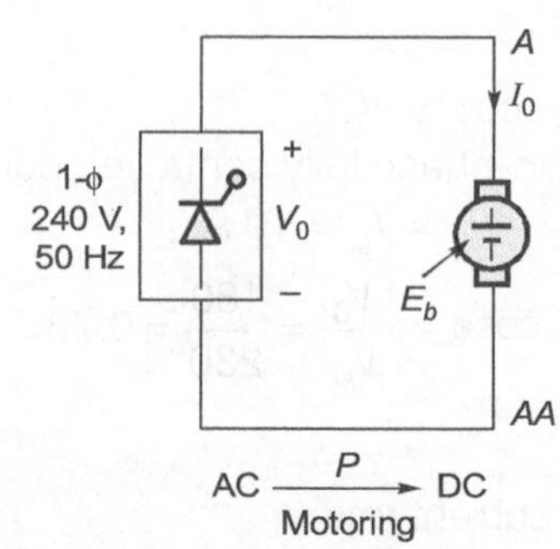

Rating of DC machine

210 V, 10 A, 1200 rpm,

$R_a = 1\ \Omega$

$$I_F \rightarrow \text{constant}$$

$$\therefore \quad \phi \rightarrow \text{constant}$$

$$\therefore \quad E_b = K_m W = K_m \cdot \frac{2\pi \cdot N}{60}$$

$$V_0 = E_b + I_0 R_a$$

$$210 = K_m \cdot \frac{2\pi N}{60} + I_0 R_a$$

$$210 = K_m \cdot \frac{2\pi \cdot 1200}{60} + (10 \times 1)$$

$$K_m = 1.59 \text{ (V-s)/rad}$$

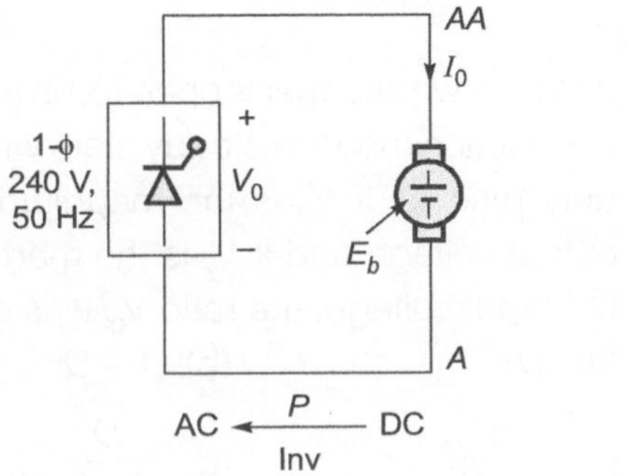

Armature terminals reversed

$$\therefore \quad E_b \rightarrow \text{Negative}$$

Inv for regenerative braking

$$V_0 = -E_b + I_0 R_a$$

$$\frac{2V_m}{\pi} \cos\alpha = -E_b + I_0 R_a$$

$$\frac{2.240\sqrt{2}}{\pi} \cos\alpha = -K_m \cdot \frac{2\pi N}{60} + I_0 R_a$$

$$\frac{2 \cdot 240\sqrt{2}}{\pi} \cdot \cos\alpha = -1.59\left(\frac{2\pi \cdot 600}{60}\right) + 10$$

$$\alpha = 114.58°$$

■■■■

3 Choppers

3.1 Consider the chopper circuit shown in figure. The chopper operates at 400 Hz and 50% duty cycle. The load current remains almost ripple free at 10 A. Assuming the input voltage to be 200 V and the devices to be ideal, the circuit turn-off time available to the thyristor Th M is _____ μs.

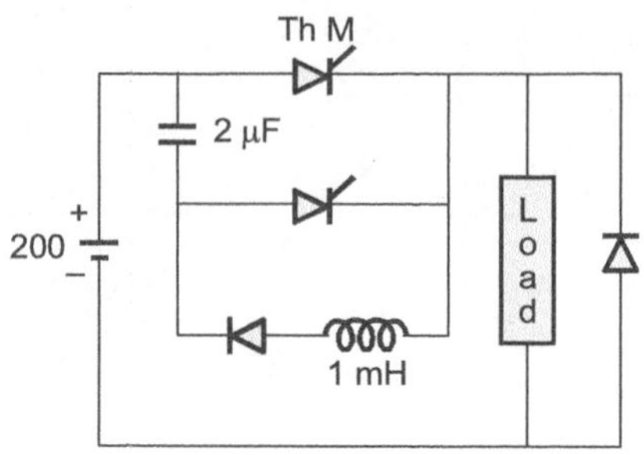

[1995 : 1 M]

3.2 A DC to DC transistor chopper supplied from a fixed voltage DC source feeds a fixed resistive-inductive load and a free-wheeling diode. The chopper operates at 1 kHz and 50% duty cycle. Without changing the value of the average DC current through the load, if it is desired to reduce the ripple constant of load current, the control action needed with be

(a) increase the chopper frequency keeping its duty cycle constant.
(b) increase the chopper frequency and duty cycle in equal ratio.
(c) decrease only the chopper frequency.
(d) decrease only the duty cycle. [1995 : 1 M]

3.3 In a thyristor DC chopper, which type of commutation results in best performance?

(a) Voltage commutation
(b) Current commutation
(c) Load commutation
(d) Supply commutation [1996 : 1 M]

3.4 A step down, chopper operates from a DC voltage source and feeds a DC motor armature with a back e.m.f E_b. From oscilloscope traces, it is found that the current increases for time t_r, falls to zero over time t_f, and remains zero for time t_o in every chopping cycle, then the average DC voltage across the free-wheeling diode is

(a) $\frac{V_s t_r}{(t_r + t_f + t_o)}$
(b) $\frac{(V_s t_r + E_b t_f)}{(t_r + t_f + t_o)}$
(c) $\frac{(V_s t_r + E_b t_o)}{(t_r + t_f + t_o)}$
(d) $\frac{V_s t_r + E_b [t_f + t_o]}{(t_r + t_f + t_o)}$

[2000 : 2 M]

3.5 A step down chopper is operated in the continuous conduction mode in steady state with a constant duty ratio D. If V_0 is the magnitude of the DC output voltage and if V_s is the magnitude of the DC input voltage, the ratio V_0/V_s is given by

(a) D
(b) $1 - D$
(c) $\frac{1}{1-D}$
(d) $\frac{D}{1-D}$

[2002 : 1 M]

3.6 In the chopper circuit shown in figure, the input DC voltage has a constant value V_s. The output voltage V_0 is assumed ripple free. The switch S is operated with a switching time period T and a duty ratio D. What is the value of D at the boundary of continuous and discontinuous conduction of the induction current i_L?

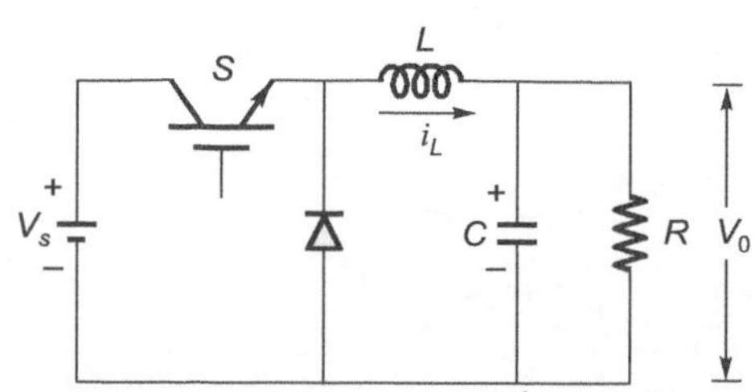

(a) $D = 1 - \frac{V_s}{V_0}$
(b) $D = \frac{2L}{RT}$
(c) $D = 1 - \frac{2L}{RT}$
(d) $D = \frac{RT}{L}$

[2002 : 2 M]

3.7 A chopper is employed to charge a battery as shown in the figure. The charging current is 5 A. The duty ratio is 0.2. The chopper output voltage is also shown in the figure. The peak to peak ripple current in the charging current is

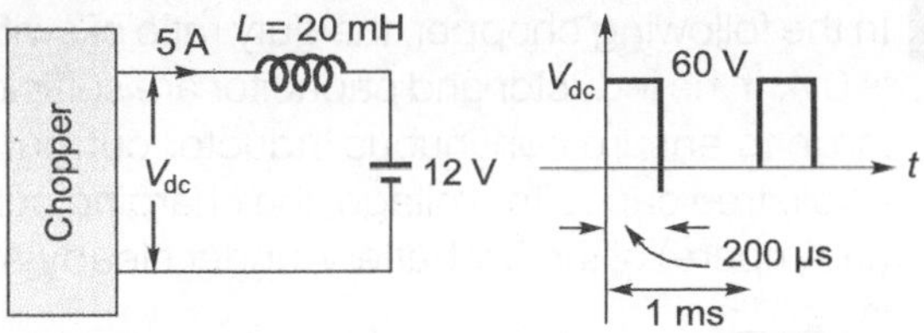

(a) 0.48 A (b) 1.2 A
(c) 2.4 A (d) 1 A [2003 : 2 M]

3.8 The figure shows a chopper operating from a 100 V DC input. The duty ratio of the main switch S is 0.8. The load is sufficiently inductive so that the load current is ripple free. The average current through the diode D under steady state is

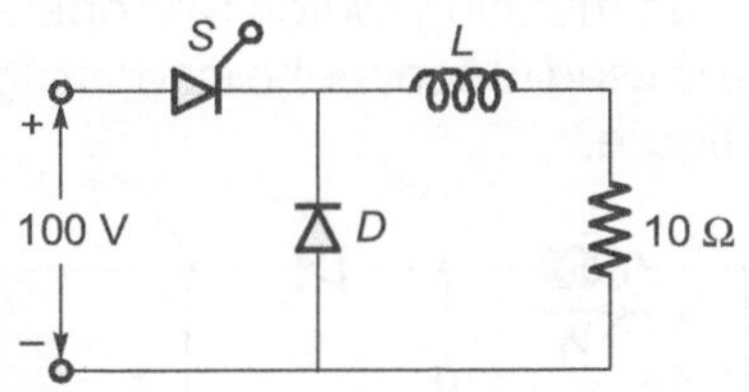

(a) 1.6 A (b) 6.4 A
(c) 8.0 A (d) 10.0 A [2004 : 2 M]

3.9 The figure shows a chopper. The device S_1 is the main switching device. S_2 is the auxiliary commutation device. S_1 is rated for 400 V, 60 A. S_2 is rated for 400 V, 30 A. The load current is 20 A. The main device operates with a duty ratio of 0.5. The peak current through S_1 is

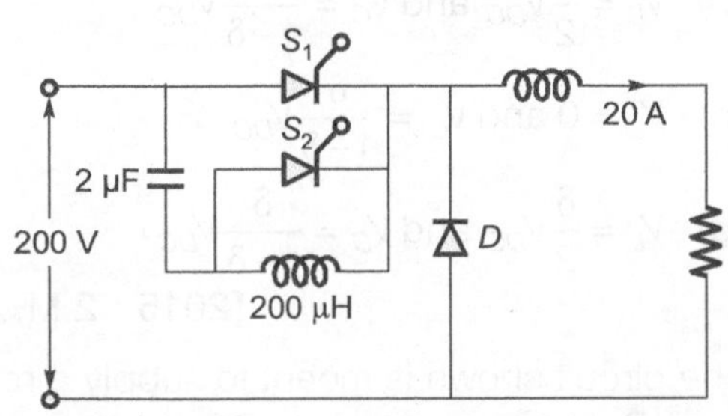

(a) 10 A (b) 20 A
(c) 30 A (d) 40 A [2004 : 2 M]

3.10 The given figure shows a step-down chopper switched at 1 kHz with a duty ratio $D = 0.5$. The peak-peak ripple in the load current is close to

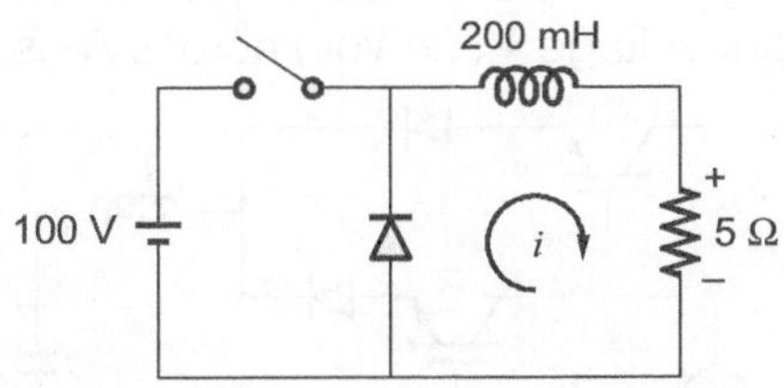

(a) 10 A (b) 0.5 A
(c) 0.125 A (d) 0.25 A [2005 : 2 M]

Statement for Common Data (3.11 and 3.12):

A voltage commutated chopper operating at 1 kHz is used to control the speed of dc motor as shown in figure. The load current is assumed to be constant at 10 A.

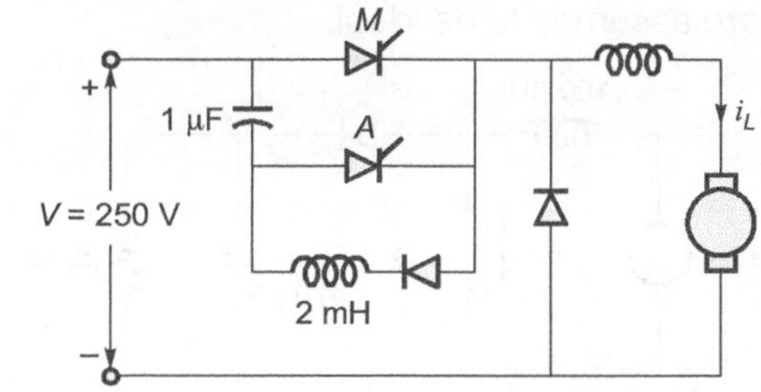

3.11 The minimum time in μsec for which the SCR M should be ON is

(a) 280 μs (b) 140 μs
(c) 70 μs (d) 0 μs [2006 : 2 M]

3.12 The minimum average output voltage of the chopper will be

(a) 70 V (b) 47.5 V
(c) 35 V (d) 0 V [2006 : 2 M]

3.13 In the circuit shown in the figure, the switch is operated at a duty cycle of 0.5. A large capacitor is connected across the load. The inductor current is assumed to be continuous.

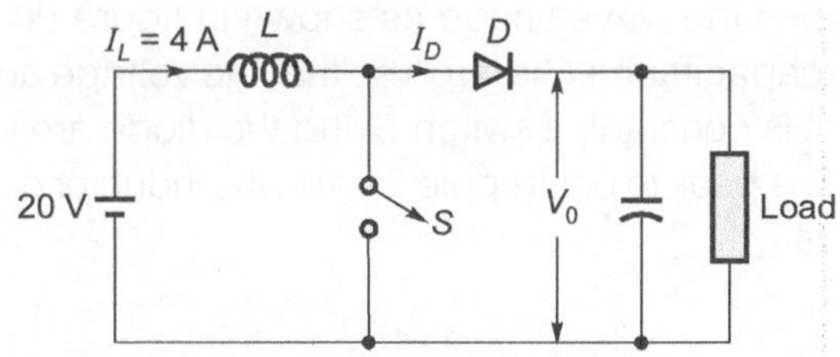

The average voltage across the load and the average current through the diode will respectively be

(a) 10 V, 2 A (b) 10 V, 8 A
(c) 40 V, 2 A (d) 40 V, 8 A [2008 : 2 M]

3.14 In the circuit shown, an ideal switch S is operated at 100 kHz with a duty ratio of 50%. Given that Δi_c is 1.6 A peak-to-peak and I_0 is 5 A DC, the peak current in S is

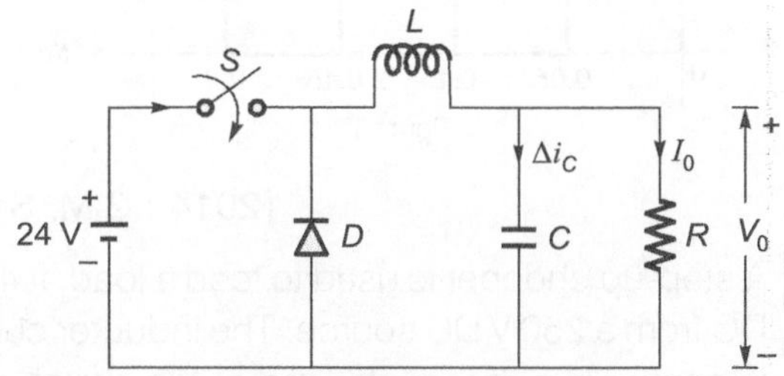

(a) 6.6 A (b) 5.0 A
(c) 5.8 A (d) 4.2 A [2012 : 2 M]

Common Data for Questions (3.15 and 3.16):

In the figure shown below, the chopper feeds a resistive load from a battery source. MOSFET Q is switched at 250 kHz, with a duty ratio of 0.4. All elements of the circuit are assumed to be ideal.

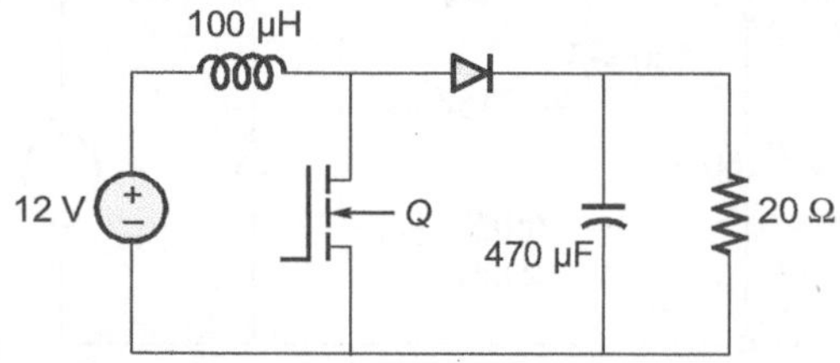

3.15 The average source current (in Amps) in steady-state is

(a) 3/2 (b) 5/3
(c) 5/2 (d) 15/4 [2013 : 2 M]

3.16 The PEAK-TO-PEAK source current ripple in Amps is

(a) 0.96 (b) 0.144
(c) 0.192 (d) 0.288

[2013 : 2 M]

3.17 The figure (i) shows the circuit diagram of a chopper. The switch S in the circuit in figure (i) is switched such that the voltage v_D across the diode has the wave shape as shown in figure (ii). The capacitance C is large so that the voltage across it is constant. If switch S and the diode are ideal, the peak to peak ripple (in A) in the inductor current is ______.

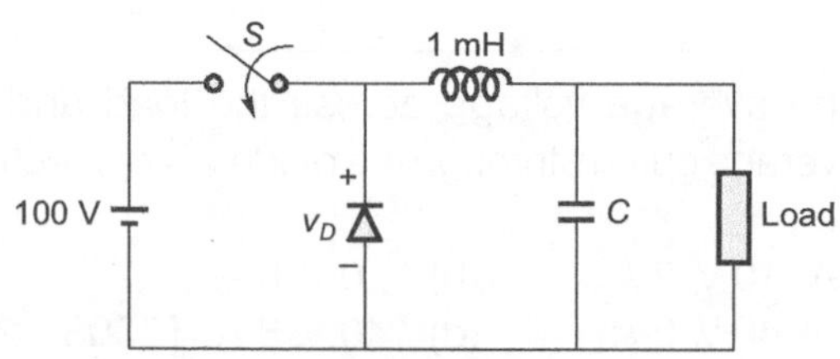

Figure-I

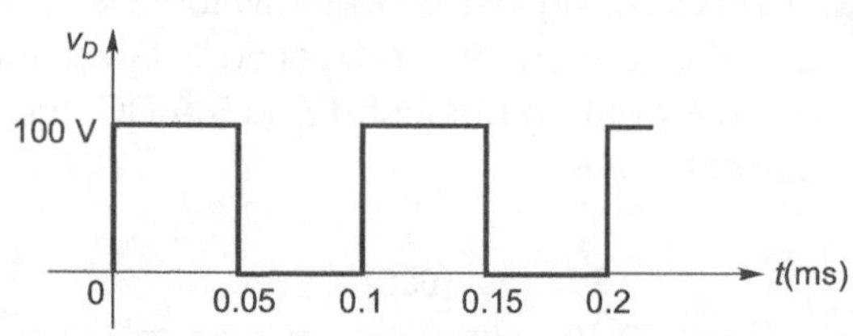

Figure-II

[2014 : 2 M, Set-1]

3.18 A step-up chopper is used to feed a load at 400 V DC from a 250 V DC source. The inductor current is continuous. If the 'off' time of the switch is 20 μs, the switching frequency of the chopper in kHz is _____ . [2014 : 1 M, Set-2]

3.19 In the following chopper, the duty ratio of switch S is 0.4. If the inductor and capacitor are sufficiently large to ensure continuous inductor current and ripple free capacitor voltage, the charging current (in Ampere) of the 5 V battery, under steady-state, is ______.

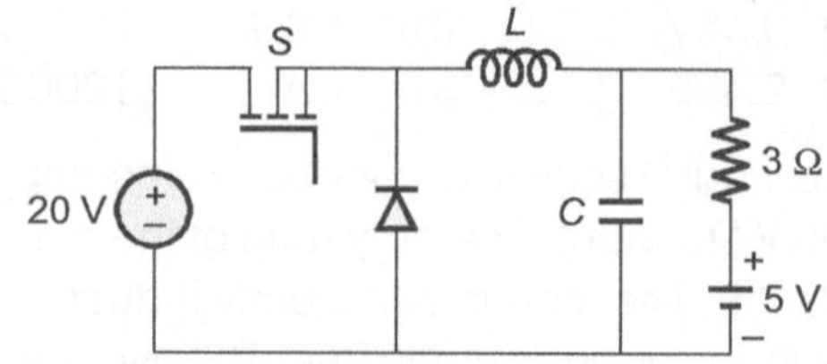

[2015 : 1 M, Set-1]

3.20 A self commutating switch SW, operated at duty cycle δ is used to control the load voltage as shown in the figure.

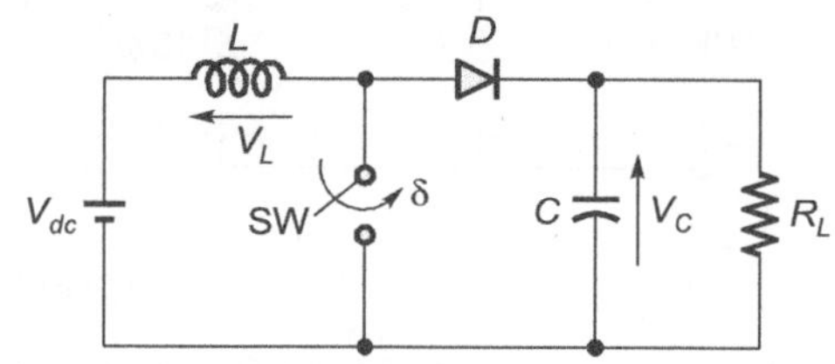

Under steady state operating conditions, the average voltage across the inductor and the capacitor respectively, are

(a) $V_L = 0$ and $V_C = \frac{1}{1-\delta}V_{DC}$

(b) $V_L = \frac{\delta}{2}V_{DC}$ and $V_C = \frac{1}{1-\delta}V_{DC}$

(c) $V_L = 0$ and $V_C = \frac{\delta}{1-\delta}V_{DC}$

(d) $V_L = \frac{\delta}{2}V_{DC}$ and $V_C = \frac{\delta}{1-\delta}V_{DC}$

[2015 : 2 M, Set-1]

3.21 The circuit shown is meant to supply a resistive load R_L from two separate DC voltage sources. The switches $S1$ and $S2$ are controlled so that only one of them is ON at any instant. $S1$ is turned on for 0.2 ms and $S2$ is turned on for 0.3 ms in a 0.5 ms switching cycle time period. Assuming continuous conduction of the inductor current and negligible ripple on the capacitor voltage, the output voltage V_0 (in Volt) across R_L is ______.

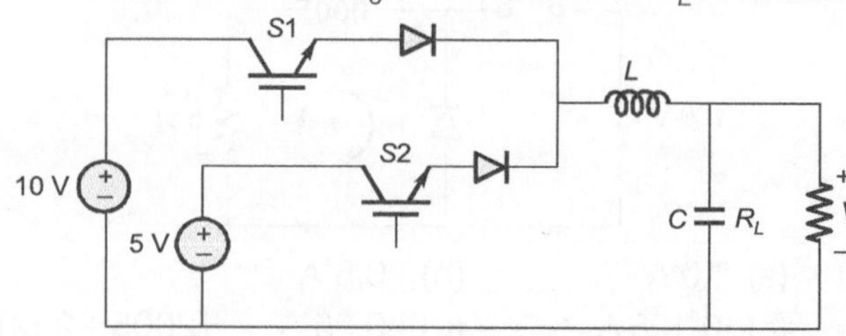

[2015 : 2 M, Set-1]

3.22 For the switching converter shown in the following figure, assume steady-state operation. Also assume that the components are ideal, the inductor current is always positive and continuous and switching period is T_s. If the voltage V_L is as shown, the duty cycle of the switch S is ______.

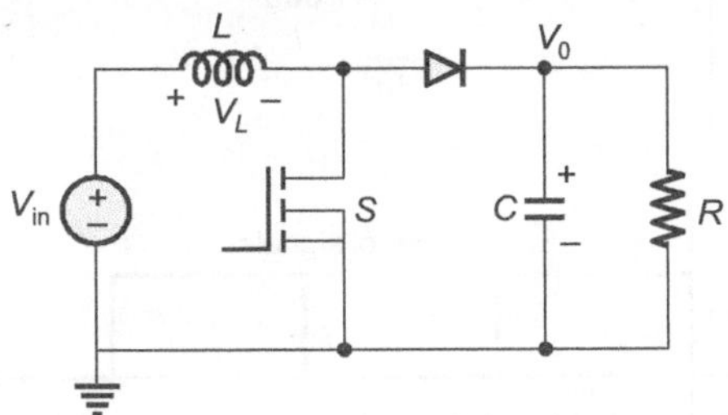

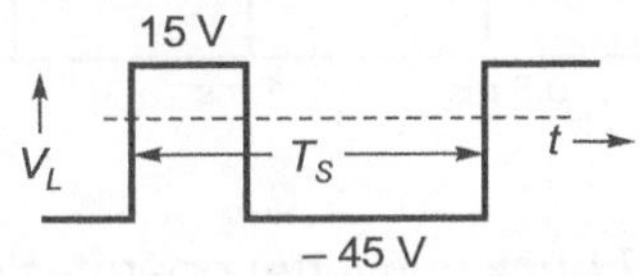

[2015 : 2 M, Set-2]

3.23 A buck converter feeding a variable resistive load is shown in the figure. The switching frequency of the switch S is 100 kHz and the duty ratio is 0.6. The output voltage V_0 is 36 V. Assume that all the components are ideal, and that the output voltage is ripple-free. The value of R (in Ohm) that will make the inductor current (i_L) just continuous is ______.

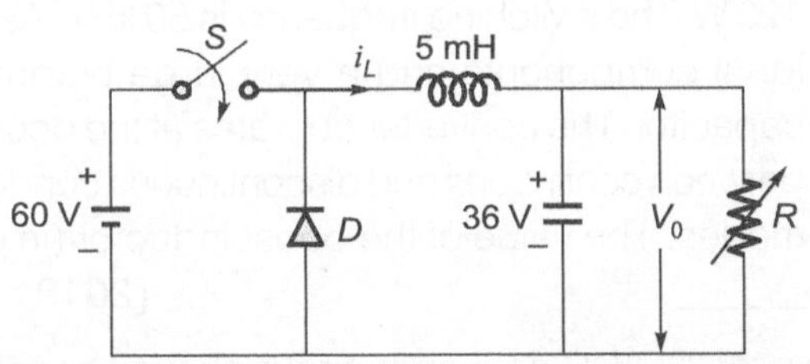

[2015 : 2 M, Set-2]

3.24 A buck converter, as shown in the figure (a) below, is working in steady state. The output voltage and the inductor current can be assumed to be ripple free. Figure (b) shows the inductor voltage V_L during a complete switching interval. Assuming all devices are ideal, the duty cycle of the buck converter is ______.

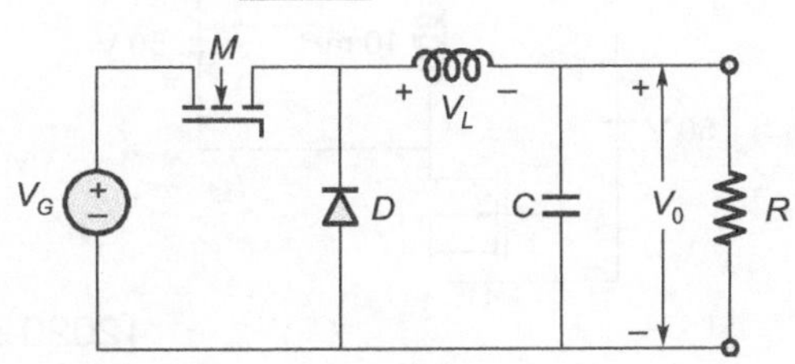

(a)

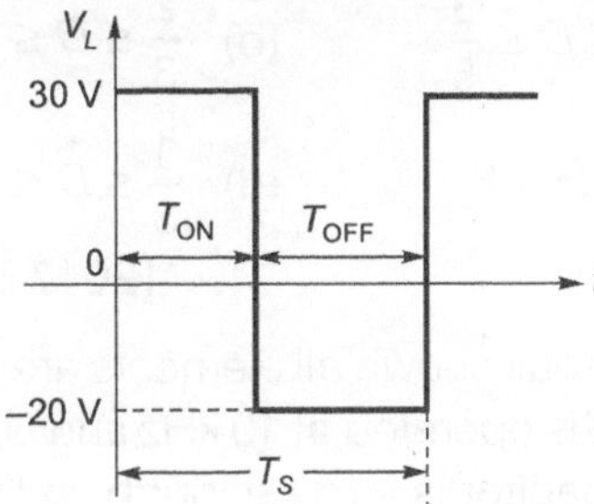

(b)

[2016 : 1 M, Set-1]

3.25 A buck-boost DC-DC converter, shown in the figure below, is used to convert 24 V battery voltage to 36 V DC voltage to feed a load of 72 Ω. It is operated at 20 kHz with an inductor of 2 mH and output capacitor of 1000 μF. All devices are considered to be ideal. The peak voltage across the solid-state switch (S), in volt, is ______.

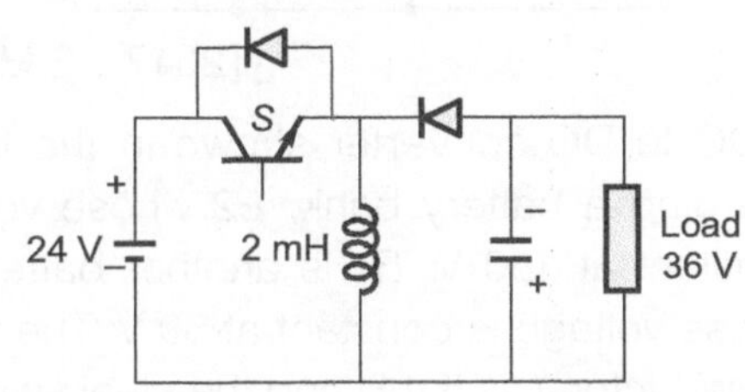

[2016 : 1 M, Set-2]

3.26 A DC-DC boost converter, as shown in the figure below, is used to boost 360V to 400 V, at a power of 4 kW. All devices are ideal. Considering continuous inductor current, the rms current in the solid state switch (S), in ampere, is ______.

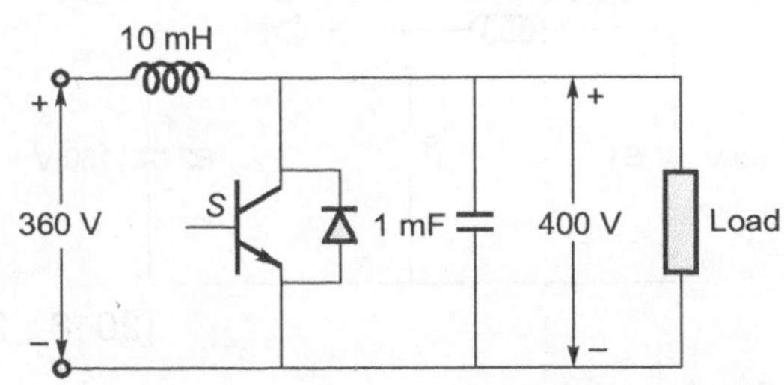

[2016 : 2 M, Set-2]

3.27 The input voltage V_{DC} of the buck-boost converter shown below varies from 32 V to 72 V. Assume that all components are ideal, inductor current is continuous and output voltage is ripple free. The range of duty ratio D of the converter for which the magnitude of the steady-state output voltage remains constant at 48 V is

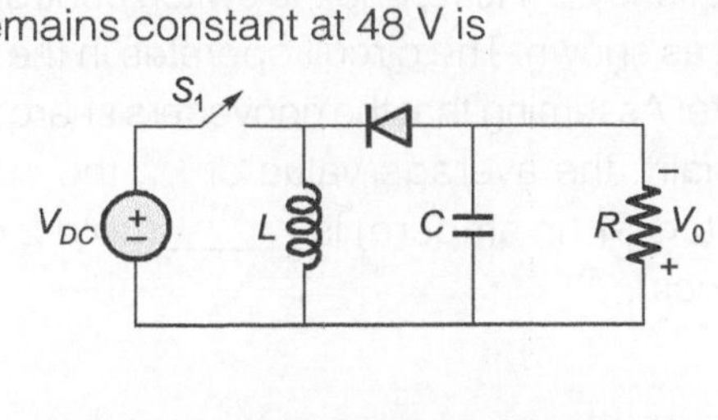

(a) $\frac{2}{5} \le D \le \frac{3}{5}$ (b) $\frac{2}{3} \le D \le \frac{3}{4}$

(c) $0 \le D \le 1$ (d) $\frac{1}{3} \le D \le \frac{2}{3}$

[2017 : 2 M, Set-1]

3.28 In the circuit shown all elements are ideal and the switch S is operated at 10 kHz and 60% duty ratio. The capacitor is large enough so that the ripple across it is negligible and at steady state acquires a voltage as shown. The peak current in amperes drawn from the 50 V DC source is ______. (Give the answer up to one decimal place.)

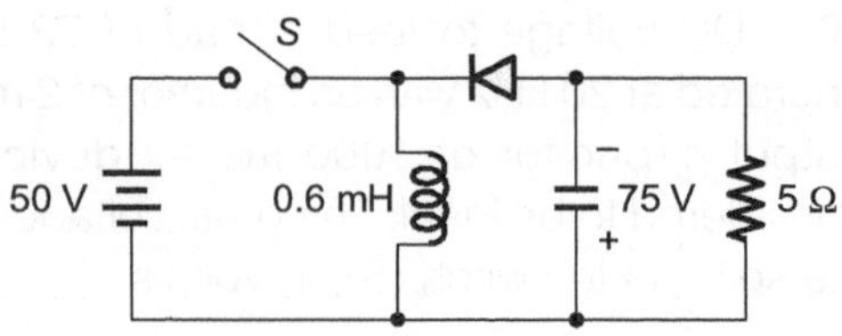

[2017 : 2 M, Set-2]

3.29 A DC to DC converter shown in the figure is charging a battery bank, $B2$ whose voltage is constant at 150 V. $B1$ is another battery bank whose voltage is constant at 50 V. The value of the inductor. L is 5 mH and the ideal switch, S is operated with a switching frequency of 5 kHz with a duty ratio of 0.4. Once the circuit has attained steady state and assuming the diode D to be ideal, the power transferred from $B1$ to $B2$ (in Watt) is ______ (upto 2 decimal places).

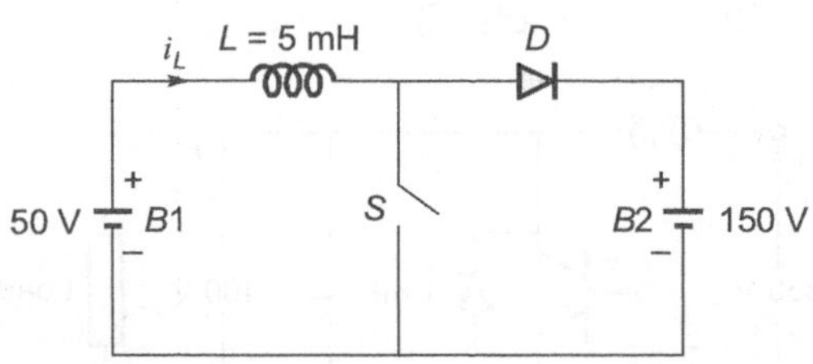

[2018 : 2 M]

3.30 The figure shows two buck converters connected in parallel. The common input dc voltage for the converters has a value of 100 V. The converters have inductors of identical value. The load resistance is 1 Ω. The capacitor voltage has negligible ripple. Both converters operate in the continuous conduction mode. The switching frequency is 1 kHz, and the switch control signals are as shown. The circuit operates in the steady-state. Assuming that the converters share the load equally, the average value of i_{s1}, the current of switch $S1$ (in ampere) is ______ (upto 2 decimal places).

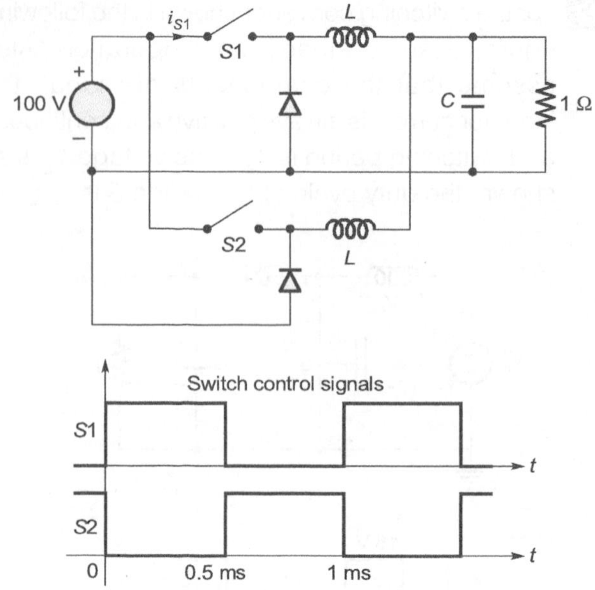

[2018 : 2 M]

3.31 A DC-DC buck converter operates in continuous conduction mode. It has 48 V input voltage, and it feeds a resistive load of 24 Ω. The switching frequency of the converter is 250 Hz. If switch-on duration is 1 ms, the load power is

(a) 48 W (b) 24 W

(c) 6 W (d) 12 W

[2019 : 2 M]

3.32 In a DC-DC boost converter, the duty ratio is controlled to regulate the output voltage at 48 V. The input DC voltage is 24 V. The output power is 120 W. The switching frequency is 50 kHz. Assume ideal components and a very large output filter capacitor. The converter operates at the boundary between continuous and discontinuous conduction modes. The value of the boost inductor (in μH) is ______.

[2019 : 2 M]

3.33 In the DC-DC converter circuit shown, switch Q is switched at a frequency of 10 kHz with a duty ratio of 0.6. All components of the circuit are ideal, and the initial current in the inductor is zero. Energy stored in the inductor in mJ (rounded off to 2 decimal places) at the end of 10 complete switching cycles is ______.

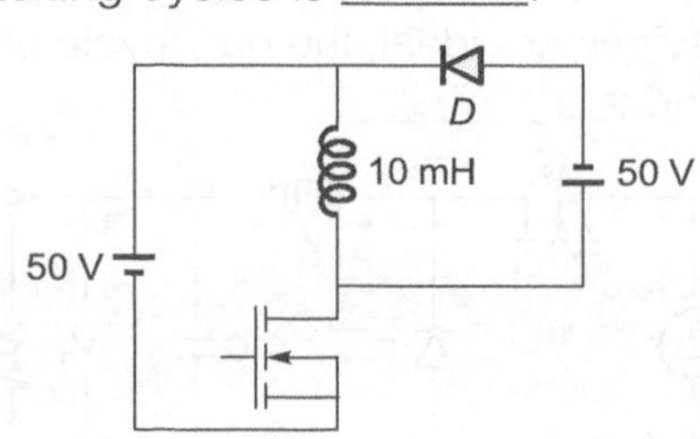

[2020 : 2 M]

3.34 Consider the buck-boost converter shown. Switch Q is operating at 25 kHz and 0.75 duty-cycle. Assume diode and switch to be ideal. Under steady-state condition, the average current flowing through the inductor is ______ A.

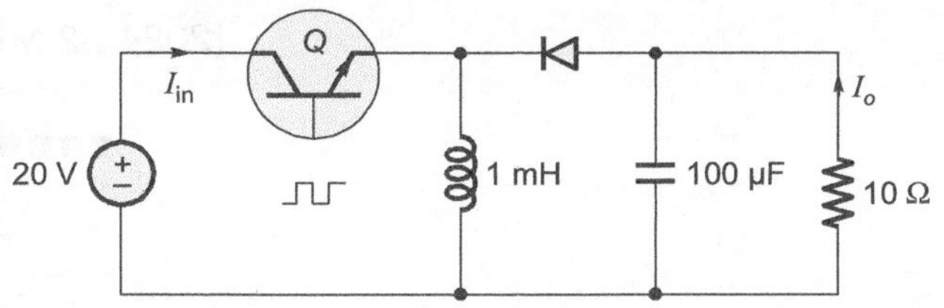

[2021 : 2 M]

3.35 Consider the boost converter shown. Switch Q is operating at 25 kHz with a duty cycle of 0.6. Assume the diode and switch to be ideal. Under steady-state condition, the average resistance R_{in} as seen by the source is ________ Ω. (Round off to 2 decimal places.)

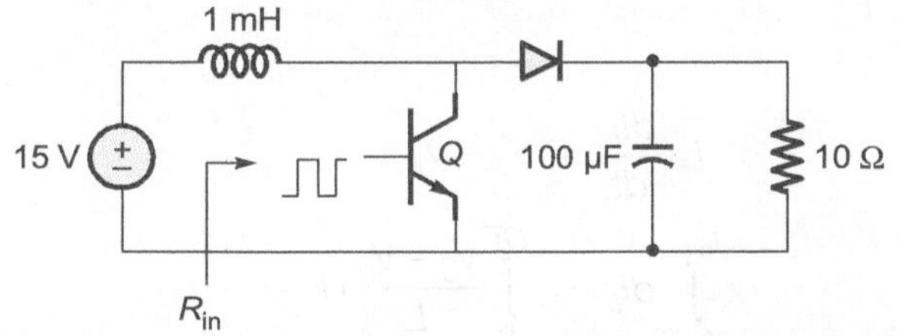

[2021 : 2 M]

3.36 The steady state current flowing through the inductor of a DC-DC buck boost converter is given in the figure below. If the peak-to-peak ripple in the output voltage of the converter is 1 V, then the value of the output capacitor, in μF, is _____. (round off to nearest integer).

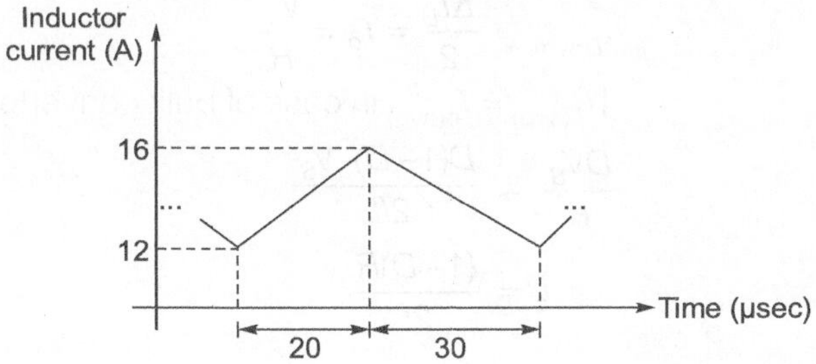

[2022 : 2 M]

3.37 The chopper circuit shown in figure (i) feeds power to a 5 A DC constant current source. The switching frequency of the chopper is 100 kHz. All the components can be assumed to be ideal. The gate signals of switches S_1 and S_2 are shown in figure (ii). Average voltage across the 5 A current source is

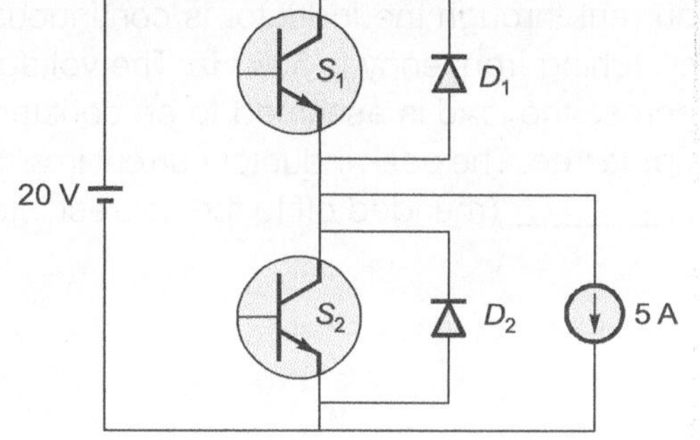

Fig. (i)

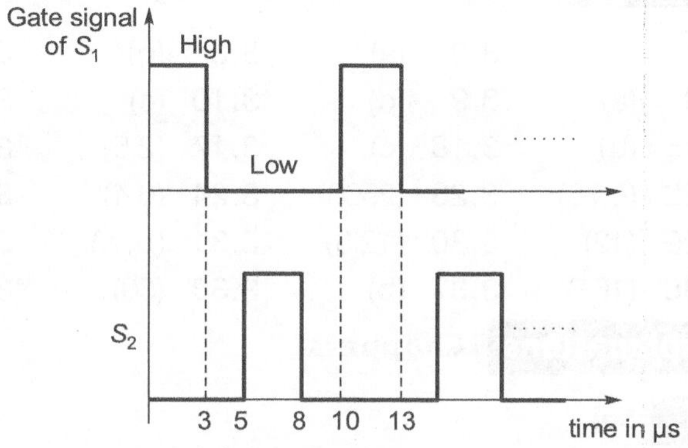

Fig. (ii)

(a) 10 V (b) 6 V
(c) 12 V (d) 20 V

[2023 : 1 M]

3.38 A forced commutated thyristorized step-down chopper is shown in the figure. Neglect the ON-state drop across the power devices. Assume that the capacitor is initially charged to 50 V with the polarity shown in the figure. The load current (I_L) can be assumed to be constant at 10 A. Initially, Th_M is ON and Th_A is OFF. The turn-off time available to Th_M in microseconds, when Th_A is triggered, is ______ (rounded off to the nearest integer).

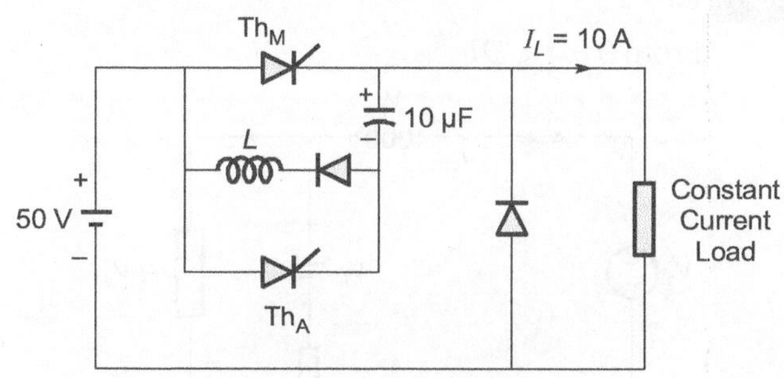

[2024 : 1 M]

3.39 In the DC-DC converter shown in the figure, the current through the inductor is continuous, The switching frequency is 500 Hz. The voltage (V_0) across the load is assumed to be constant and ripple free. The peak inductor current in amperes is ________ (rounded off to the nearest integer).

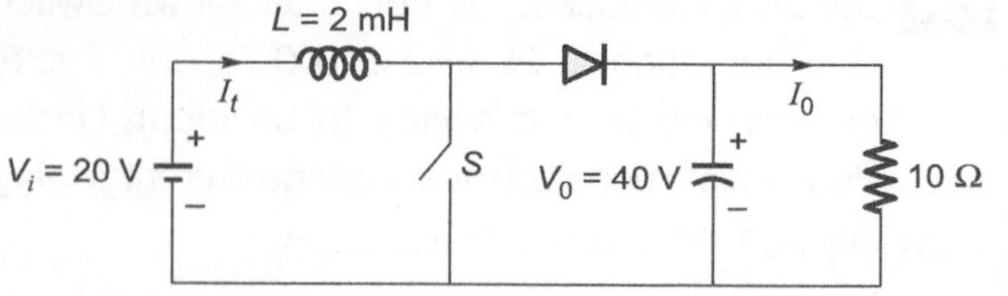

[2024 : 2 M]

Answers Choppers

3.1 (40)	3.2 (a)	3.3 (b)	3.4 (c)	3.5 (a)	3.6 (c)	3.7 (a)
3.8 (a)	3.9 (d)	3.10 (c)	3.11 (b)	3.12 (b)	3.13 (c)	3.14 (c)
3.15 (b)	3.16 (c)	3.17 (2.5)	3.18 (31.2)	3.19 (1)	3.20 (a)	3.21 (7)
3.22 (0.75)	3.23 (2500)	3.24 (0.4)	3.25 (60)	3.26 (3.5)	3.27 (a)	3.28 (40)
3.29 (12)	3.30 (12.5)	3.31 (b, c)	3.32 (24)	3.33 (5)	3.34 (24)	3.35 (1.60)
3.36 (168)	3.37 (b)	3.38 (50)	3.39 (10)			

Explanations Choppers

3.1 Sol.

$$t_c = \frac{CV_s}{I} = \frac{2\times10^{-6}\times200}{10} = 40\,\mu s$$

3.2 (a)

$$\Delta I_c = \frac{V_s}{4fL}$$

3.4 (c)

$$V_{av} = \frac{V_s t_r + E_b t_o}{t_r + t_f + t_o}$$

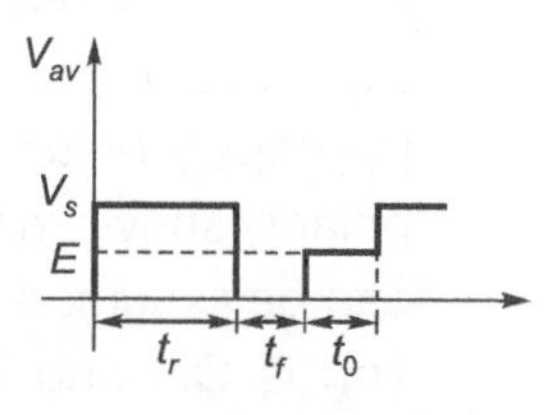

3.5 (a)

$$V_o = DV_s \quad \text{or} \quad \frac{V_o}{V_s} = D$$

3.6 (c)

From $0 < t < DT$

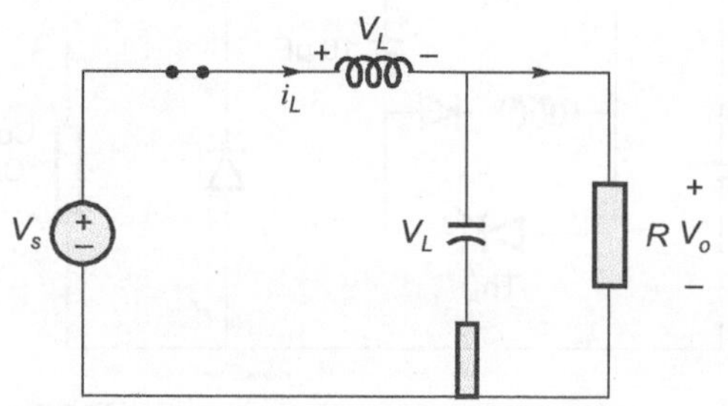

On applying KVL

$-V_S + V_L + V_o = 0 \Rightarrow V_L = V_S - V_o$

$$L\cdot\frac{di_L}{dt} = V_S - V_o$$

$$\int_{I_{min}}^{I_{max}} di_L = \int_0^{DT} \frac{V_S - V_o}{L}dt$$

$$\Delta I_L = \frac{D(1-D)V_S}{fL}$$

At the boundary of continuous and discontinuous conduction,

$$I_{L,min} = 0$$

$$I_{L(avg)} - \frac{\Delta I_L}{2} = 0$$

$$I_{L(avg)} = \frac{\Delta I_L}{2} = I_o = \frac{V_o}{R}$$

[$I_{L(avg)} = I_{o(avg)}$ in case of bulk converter]

$$\frac{DV_S}{R} = \frac{D(1-D)\cdot V_S}{2fL}$$

$$L = \frac{(1-D)R}{2f}$$

$$1 - D = \frac{2L}{RT}$$

$$D = 1 - \frac{2L}{RT}$$

3.7 (a)

$$D = \frac{T_{on}}{T} = \frac{0.2\text{ ms}}{1\text{ ms}} = 0.2$$

From $0 < t < DT$

$$-V_{dc} + V_L + E = 0$$

$$V_L = V_{dc} - E$$

$$L\frac{di}{dt} = V_{dc} - E$$

$$\int_{I_{min}}^{I_{max}} di = \int_0^{DT} \frac{V_{dc} - E}{L} dt$$

$$I_{max} - I_{min} = \frac{(60-12)(DT)}{L}$$

$$\Delta I = \frac{48 \times 0.2 \times 1 \times 10^{-3}}{20 \times 10^{-3}}$$

$$\Delta I = \frac{9.6}{20} = 0.48 \text{ A}$$

So, peak to peak ripples in inductor current is 0.48 A.

3.8 (a)

Since it is buck converter,

$$I_{L(avg)} = I_{o(avg)}$$

$$I_{L(avg)} = \frac{DV_s}{R} = \frac{0.8 \times 100}{10} = 8 \text{ A}$$

and from $DT < t < T$

$$I_D = I_{L(avg)}$$

$$I_{D(avg)} = \frac{I_{L(avg)} \times (T - DT)}{T} = I_{L(avg)}(1 - D)$$

$$I_{D(avg)} = 8 \times (1 - 0.8) = 1.6 \text{ Amp}$$

3.9 (d)

Initially main thyristor (S_1) and auxiliary thyristor (S_2) are off and capacitor is assumed charged to voltage V_s with upper plate positive. When S_1 is turned on, source voltage V_s is applied across load and load current I_0 begins to flow which is assumed to remain constant. With S_1 ON another circuit consisting of C, S_1, and L forms LC oscillation, where the capacitor current is given by

$$i_c = V_s\sqrt{\frac{C}{L}} \sin\omega_0 t$$

So current through S_1,

$$I_{s1} = I_0 + i_c = I_0 + V_s\sqrt{\frac{C}{L}} \sin\omega_0 t$$

Peak value of

$$I_{s1} = I_0 + V_s\sqrt{\frac{C}{L}} = 20 + 200\sqrt{\frac{2 \times 10^{-6}}{200 \times 10^{-6}}} = 40 \text{ A}$$

3.10 (c)

Here, $T = \frac{1}{1 \times 10^3} = 10^{-3}$ sec

$$T_a = \frac{L}{R} = \frac{200 \text{ mH}}{5} = 40 \text{ msec}$$

$$\text{Ripple} = \frac{V_s}{R}\left[\frac{(1 - e^{-\alpha T/T_a})(1 - e^{-(1-\alpha)T/T_a})}{1 - e^{-T/T_a}}\right]$$

$$(\Delta I)_{max} = \frac{V_s}{4fL} = \frac{100}{4 \times 10^3 \times 200 \times 10^{-3}} = 0.125 \text{ A}$$

3.11 (b)

Initially, main thyristor (m) and auxiliary thyristor (A) are off and capacitor is assumed charged to voltage V with upper plate positive. When 'm' is turned on at $t = 0$, source voltage V is applied across load and load current I_0 begins to flow which is assumed to remain constant.

With 'm' ON at $t = 0$, another oscillatory circuit consisting of C, m, L and D is formed where the capacitor current is given by

$$i_c = V\sqrt{\frac{C}{L}} \sin\omega_0 t = I_p \sin\omega_0 t$$

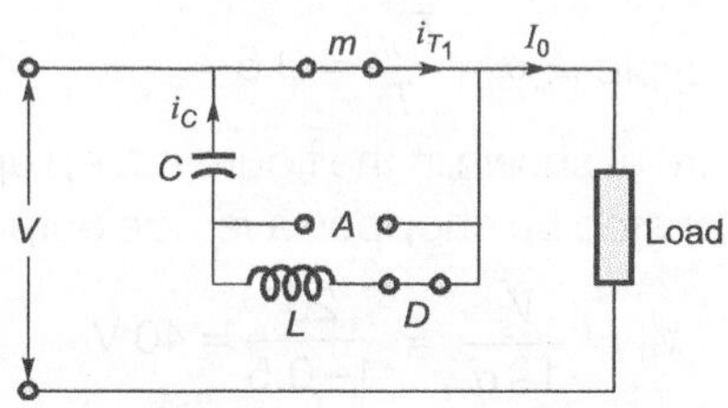

when, $\omega_0 t = \pi$, $i_c = 0$.

Between $0 < t < \frac{\pi}{\omega_0}$, $i_{T1} = I_0 + I_p \sin\omega_0 t$ capacitor voltage charges from $+V_s$ to $-V_s$ sinusoidally and the lower plate becomes positive.

At $\omega_0 t = \pi$, $i_c = 0_i$, $i_{T1} = I_0$ and $V_c = -V_s$.

At t_1, thyristor A is turned on and capacitor voltage V applies a reverse voltage across thyristor in and SCR 'm' is turned off. The load current is now carried by C and SCR A.

So, minimum time for which SCR 'm' should be ON.

$$T_{min} = \frac{\pi}{\omega_0} = \pi\sqrt{LC}$$

$$= \pi\sqrt{2 \times 10^{-3} \times 1 \times 10^{-6}} = 140 \text{ µs}$$

3.12 (b)

$$(V_o) = V_s\left[\frac{T_{ON} + 2t_{cm}}{T}\right]$$

where, $t_{cm} = \frac{CV_s}{I_0} = \frac{1 \times 10^{-6} \times 250}{10}$

$$(V_o)_{min} = V_s[(T_{ON})_{min} + 2t_{cm}]f$$
$$= 250[140 \times 10^{-6} \times (2 \times 25 \times 10^{-6})] \times (1 \times 10^3)$$
$$= 47.5 \text{ V}$$

3.13 (c)

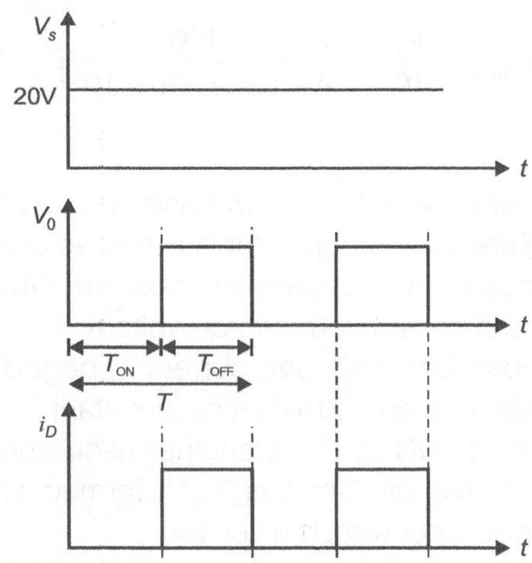

When, switch in ON, $V_0 = 0$ and $i_D = 0$
Switch is OFF, $V_0 = V_s$ and $I_D = I_L$

$$\text{Duty cycle} = \alpha = \frac{T_{ON}}{T} = 0.5$$

The circuit shown in the figure is step up chopper and for step up chopper, average output voltage,

$$V_0 = \frac{V_s}{1-\alpha} = \frac{20}{1-0.5} = 40 \text{ V}$$

Average current through diode

$$I_D = \frac{1}{T}\int_{T_{ON}}^{T} I_L dt = \frac{I_L}{T}(T - T_{ON})$$
$$I_D = I_L(1-\alpha) = (1-0.5) \times 4 = 2 \text{ A}$$

3.14 (c)

$$\text{Peak current} = I_o + \frac{\Delta I_L}{2} = 5 + \frac{1.6}{2} = 5.8 \text{ A}$$

3.15 (b)

In given circuit, it is boot converter. So,
$I_{S(avg)} = I_{L(avg)}$
Because inductor is directly connected to source.

$$V_{o(avg)} = \frac{V_S}{1-D} = \frac{12}{1-0.4} = 20 \text{ V}$$

$$I_{o(avg)} = \frac{V_{o(avg)}}{R} = \frac{20}{20} = 1 \text{ A}$$

From $DT < t < T$,
$$I_o = I_L$$
$$I_{o(avg)} = I_L(1-D)$$
$$I_{L(avg)} = I_{S(avg)} = \frac{I_o}{1-D} = \frac{1}{1-0.4} = \frac{5}{3}\text{A}$$

3.16 (c)

$$I_{S(peak)} = \Delta I_L \Rightarrow \Delta I_L = \frac{V_S D}{f.L}$$
$$= \frac{12 \times 0.4}{250 \times 0.1} = 0.192 \text{ A}$$

The circuit shown is a bulk converter.
From $0 < t < DT$, S/w = 0 N
$V_s = 100$ V, $V_o = DV_s = 0.5 \times 100 = 50$ V
On applying KVL
$$-V_s + V_L + V_o = 0$$
$$L\frac{di}{dt} = 100 - 50 = 50$$
$$\int_{I_{min}}^{I_{max}} di = \int_0^{DT} \frac{50}{L} dt$$
$$\Delta I = \frac{50 \times 0.5 \times 0.1 \times 10^{-3}}{1 \times 10^{-3}}$$
$$= 2.5 \text{ A}$$

3.17 Sol.

The circuit shown is a Buck chopper.

$$\boxed{V_0 = V_s\frac{T_{ON}}{T}}$$

$$V_0 = 100 \times \frac{0.05 \times 10^{-3}}{0.1 \times 10^{-3}}$$
$$= 50 \text{ V}$$

The peak to peak ripple in the inductor current is

$$\boxed{\Delta I = \frac{V_0 T_{OFF}}{L}} = 50 \times \frac{0.05 \times 10^{-3}}{1 \times 10^{-3}}$$

$$\boxed{\Delta I = 2.5 \text{ A}}$$

3.18 Sol.

Given step up chopper,
$$V_0 = 400 \text{ V}; \; V_s = 250 \text{ V}$$
$$T_{OFF} = 2 \times 10^{-6} \text{ sec}$$

$$\boxed{V_0 = V_s \times \frac{T}{T_{OFF}}}$$

$$400 = 250 \times \frac{T}{20 \times 10^{-6}}$$

$$T = \frac{400 \times 20 \times 10^{-6}}{250} = 3.2 \times 10^{-5} \text{ sec}$$

$$f = \frac{1}{T} = \frac{1}{3.2 \times 10^{-5}}$$

$$\therefore \quad \boxed{f = 31.2 \text{ kHz}}$$

3.19 Sol.

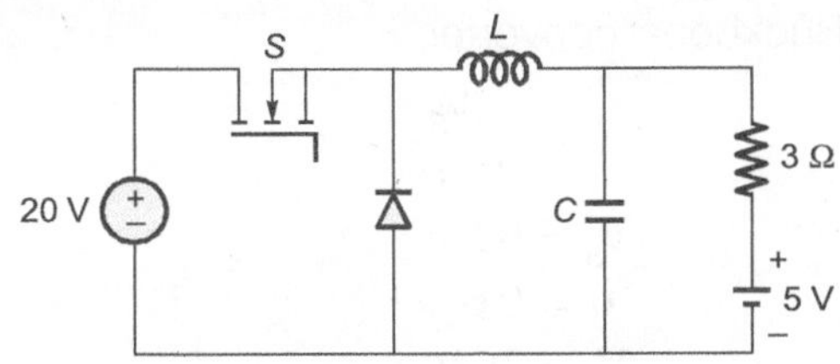

Step down chopper $\Rightarrow V_0 = \alpha V_s$

$$V_0 = I_0 R + E$$

$$I_0 = \frac{V_0 - E}{R} = \frac{\alpha V_s - E}{R} = \frac{(0.4)\,20 - 5}{3} = 1 \text{ A}$$

3.20 (a)

The average voltage across inductor, $V_L = 0$ and hence the given circuit is a boost converter so average voltage across capacitor is equal to average voltage across the load so,

$$V_0 = \left(\frac{1}{1-\delta}\right) V_{dc}$$

3.21 Sol.

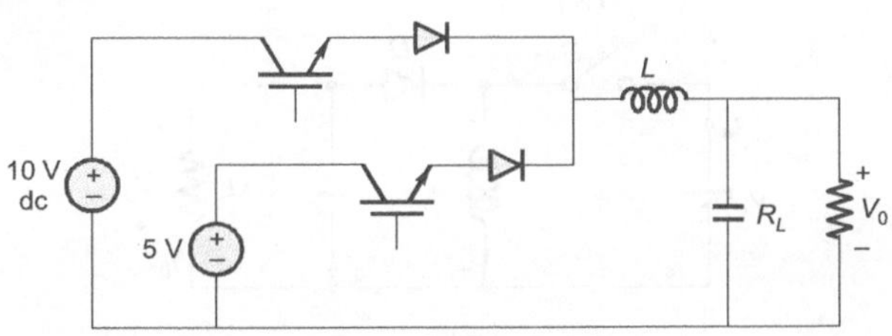

$S1$ is turn ON for 0.2 ms

$S2$ is turn ON for 0.3 ms

Switching cycle time period is 0.5 ms.

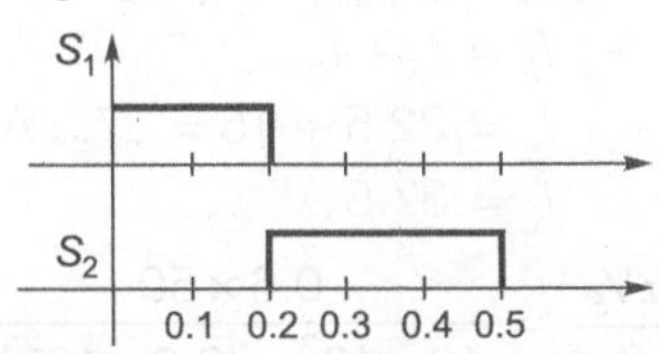

The output voltage V_0 across R_L is

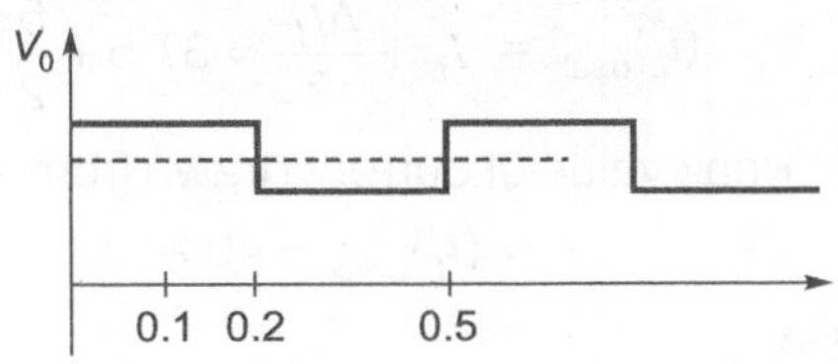

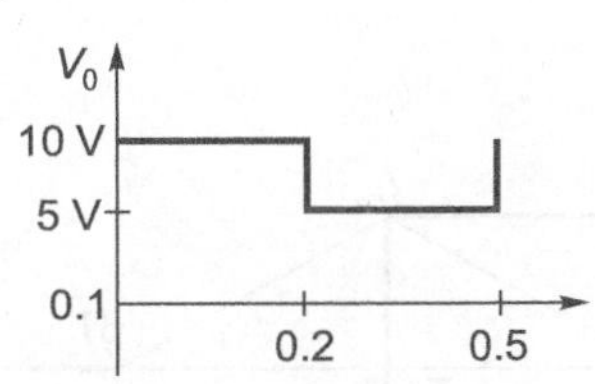

$$V_0 = \frac{1}{0.5\times10^{-3}}\left[\int_0^{0.2} 10\,dt + \int_{0.2}^{0.5} 5\,dt\right]$$

$$= \frac{1}{0.5\times10^{-3}}[10(0.2) + 5(0.5-0.2)]$$

$$= \frac{1}{0.5\times10^{-3}}[2\times10^{-3} + 1.5\times10^{-3}]$$

$$= \frac{3.5\times10^{-3}}{0.5\times10^{-3}} = 7 \text{ V}$$

3.22 Sol.

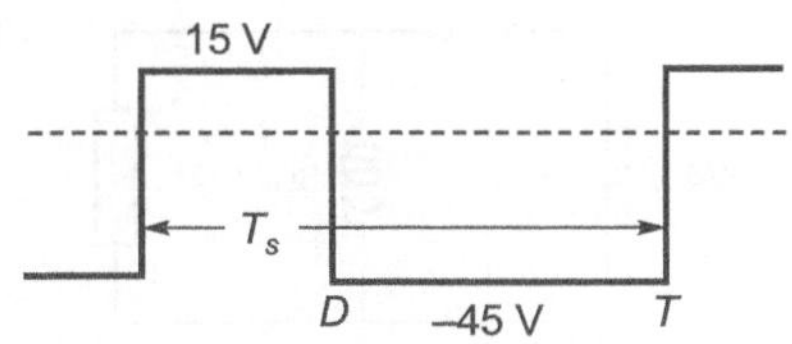

During the interval $0 \le t \le T_{on}$

$$V_S = V_L$$

During the interval $T_{on} \le t \le T$

$$V_L = -(V_0 - V_S)$$

$$V_S = 15 \text{ V}$$

$$-(V_0 - V_S) = -45 \text{ V}$$

i.e., $V_0 = 60$ V

Hence, it is a boost regular,

$$V_0 = \left(\frac{1}{1-\alpha}\right) V_s \Rightarrow 60 = \left(\frac{1}{1-\alpha}\right) 15 \Rightarrow \alpha = 0.75$$

Alternate :

Since inductor current is always continuous, so on applying volt-sec balance equation

$$\int V_L dt = 0$$

$$15DT + (-45)(1 - DT) = 0$$

$$15D = 45(1 - D) \Rightarrow 60D = 45$$

$$D = \frac{45}{60} = \frac{3}{4} = 0.75$$

3.23 (2500)

On the verge of discontinuity

$$L = L_c \text{(critical inductance)}$$

$$I_{L,\min} = 0$$

$$I_{L(avg)} - \frac{\Delta I_L}{2} = 0 \Rightarrow I_{L(avg)} = \frac{\Delta I_L}{2}$$

$$I_{L(avg)} = \frac{D(1-D)\cdot V_s}{2fL} \quad \{\because I_{L(avg)} = I_{o(avg)}\}$$

$$\frac{V_{o(avg)}}{R} = \frac{D(1-D)V_s}{2fL}$$

$$\frac{36}{R} = \frac{60\times0.4\times0.6}{2\times100\times10^3\times5\times10^{-3}}$$

On solving, $R = 2500\ \Omega$

3.24 Sol.

Average voltage across inductor is zero.

$$V_{L(Avg)} = 0$$
$$30(T_{on}) - 20(T_{off}) = 0$$
$$30(DT) = 20(1-D)T$$
$$30a + 2D = 20$$
$$50D = 20 = \frac{2}{5} = 0.4$$

3.25 Sol.

When switch '*S*' is off, diode *D* is on then

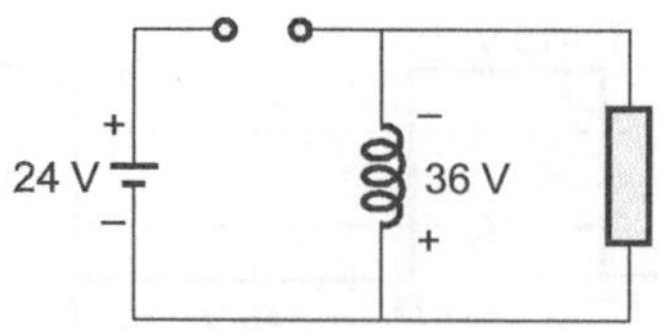

$\therefore$ Peak voltage across switch = 24 + 36 = 60 V

3.26 Sol.

$$\frac{V_o}{V_s} = \frac{1}{1-\alpha}$$

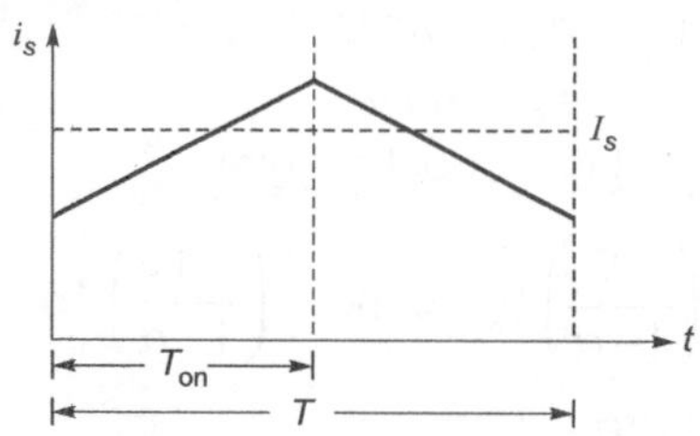

$$\frac{400}{360} = \frac{1}{1-\alpha}$$
$$\alpha = 0.1$$
$$V_s I_s = \text{Power}$$
$$360 I_s = 4000 \Rightarrow I_s = 11.1 \text{ A}$$

Neglecting ripple in i_s,

$$I_{switch\,(rms)} = I_S\left(\frac{T_{on}}{T}\right)^{1/2}$$
$$= I_S\sqrt{\alpha} = 11.1\sqrt{0.1} = 3.5 \text{ A}$$

3.27 (a)

$$V_0 = \frac{\alpha V_s}{1-\alpha}$$

Case 1 :

When $V_s = 32$ V and $V_0 = 48$ V

$$\alpha \le \frac{3}{5}$$

Case 2 :

When, $V_s = 72$ V

and $V_0 = 48$ V

$$\alpha \ge \frac{2}{5}$$

So, $$\frac{2}{5} \le \alpha \le \frac{3}{5}$$

3.28 Sol.

Buckboost converter,

$$V_0 = \frac{\alpha V_S}{1-\alpha}$$
$$V_S = 50 \text{ V}$$
$$\alpha = 0.6$$
$$V_0 = 75 \text{ V}$$
$$\frac{V_0}{V_S} = \frac{I_S}{I_0} = \frac{\alpha}{1-\alpha} = \frac{0.6}{1-0.6} = \frac{0.6}{0.4} = \frac{3}{2}$$
$$I_0 = \frac{V_0}{R} = \frac{75}{5} = 15 \text{ A}$$
$$I_S = \frac{\alpha}{1-\alpha}\cdot I_0 = \frac{3}{2}\times 15 = 22.5 \text{ A}$$

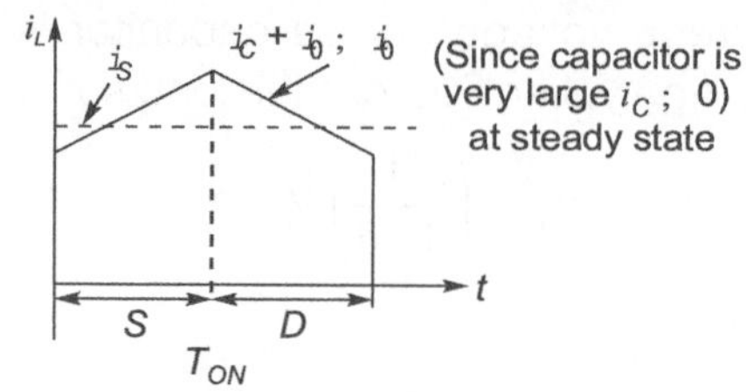

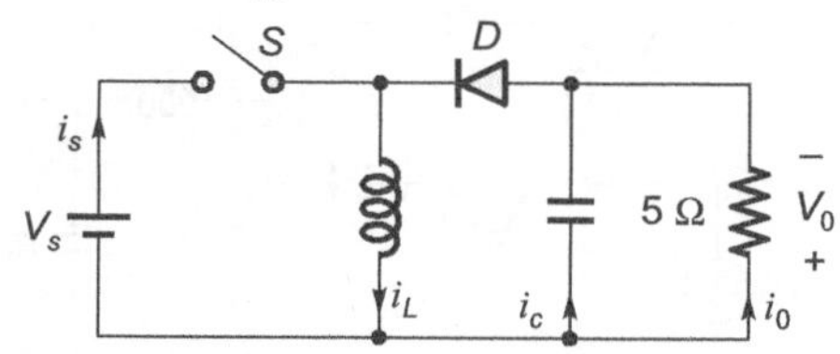

Since capacitor is very large $i_C = 0$ at steady state

$$(i_L)_{avg} = (i_S)_{avg} + (i_0)_{avg}$$
$$I_L = I_S + I_0$$
$$I_L = 22.5 + 15 = 37.5 \text{ A}$$
$$\therefore \quad I_L = 37.5 \text{ A}$$
$$\Delta I_L = \frac{\alpha V_S}{fL} = \frac{0.6\times 50}{10\times 10^3 \times (0.6\times 10^{-3})} = 5 \text{ A}$$
$$(i_L)_{peak} = I_L + \frac{\Delta I_L}{2} = 37.5 + \frac{5}{2} = 40 \text{ A}$$

$\therefore$ Peak value of current drawn from source

$$= (i_L)_{peak} = 40 \text{ A}$$

3.29 Sol.

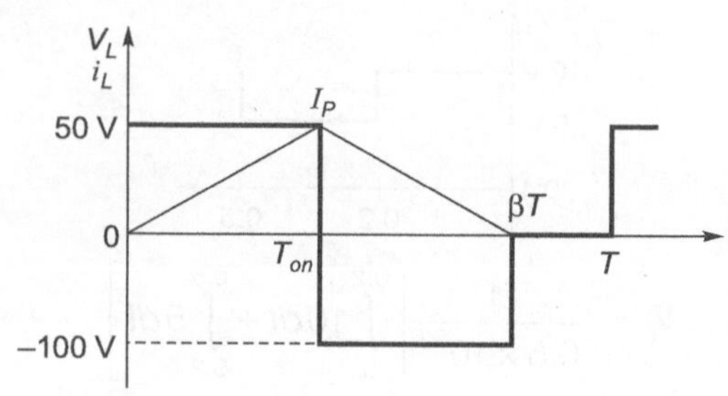

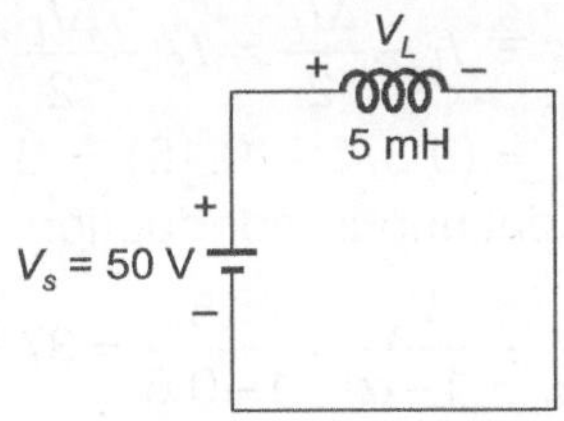

During T_{on} the circuit behaves as,

$$V_s = L\frac{di}{dt} \Rightarrow di = \frac{V_s}{L}dt$$

Integrating on both sides,

$$I_P = \frac{V_s}{L}T_{on}$$

$$T_{on} = \alpha T = \frac{\alpha}{f} = 80 \times 10^{-6} \text{ s}$$

$$I_P = \frac{50}{5 \times 10^{-3}} \times (80 \times 10^{-6}) = 0.8 \text{ A}$$

During T_{off}, it is $T_{on} \le t \le \beta T$

KVL, $V_L = V_s - V_0$

$$(V_L)_{avg} = 0$$

$$V_s T_{on} + (V_s - V_0)(\beta T - T_{on}) = 0$$

$$V_s \beta T = V_0(\beta T - T_{on})$$

$$\frac{V_0}{V_s} = \frac{\beta}{\beta - \alpha}$$

$$\beta = 0.6$$

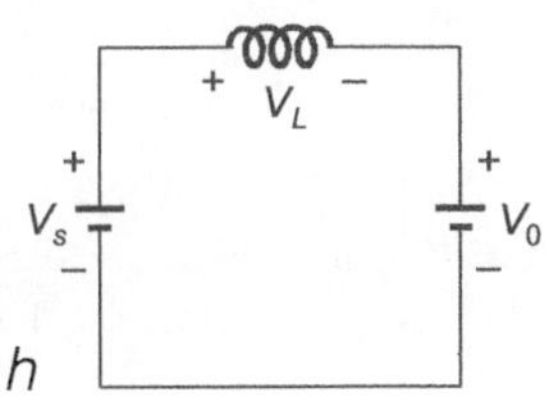

From the graph of I_L,

$$I_{L(avg)} = \frac{\frac{1}{2} \times b \times h}{T}$$

$$= \frac{\frac{1}{2} \times \beta T \times I_P}{T} = \frac{1}{2} \times 0.6 \times 0.8 = 0.24 \text{ A}$$

Power transferred to B_2 is,

$P = V_s \times (I_L)_{avg} = 50 \times 0.24 = 12$ W [where $I_s = I_L$]

3.30 Sol.

Hence it is a buck converter,

$$V_0 = \alpha V_s \Rightarrow V_0 = 0.5 \times 100 = 50 \text{ V}$$

$$I_0 = \frac{V_0}{R} = \frac{50}{1} = 50 \text{ A}$$

$$V_s I_s = V_0 I_0$$

$$I_s = \frac{V_0 I_0}{V_s} = \frac{50 \times 50}{100} = 25 \text{ A}$$

$$I_{s1} = I_{s2} = \frac{I_s}{2} = 12.5 \text{ A}$$

3.31 (b & c)

Given that:

Supply voltage, $V_s = 48$ V

Load resistance, $R_L = 24\ \Omega$

Switch frequency, $f_s = 250$ Hz

On time of switch (T_{ON}) = 1 ms

$$\text{Time period, } T = \frac{1}{f_s} = \frac{1}{250} = 4 \text{ ms}$$

$$\text{Duty cycle, } \alpha = \frac{T_{ON}}{T} = \frac{1 \text{ ms}}{4 \text{ ms}} = 0.25$$

$$\text{Load power} = \frac{(V_0)^2}{R} = \frac{(\alpha V_s)^2}{24}$$

$$= \frac{(0.25 \times 48)^2}{24} = \frac{12^2}{24} = 6 \text{ Watt}$$

Other solution:

$$\text{Load power } (P_{rms}) = \frac{V_{rms}^2}{R} = \frac{(\sqrt{a}V_s)^2}{R}$$

$$= \frac{0.25 \times 48^2}{24} = 24 \text{ W}$$

3.32 Sol.

Given : $P_0 = 120$ W, $V_s = 24$ V, $V_0 = 48$ V

$$V_0 = \frac{V_s}{(1-D)} \Rightarrow 1 - D = \frac{24}{48}$$

$$D = 0.5$$

$$P_0 = V_0 I_0 = 120$$

$$I_0 = \frac{120}{48} = 2.54 \text{ A}$$

$$V_s I_s = V_0 I_0$$

$$I_s = \frac{120}{24} = 5 \text{ A}$$

At boundary of continuous and discontinuous,

$$I_L = I_s = \frac{\Delta I_L}{2}$$

$$\Delta I_L = \frac{DV_s}{fL_c} = 2 \times 5$$

$$L_c = \frac{0.5 \times 24}{50 \times 10^3 \times 10} = 24\ \mu\text{H}$$

3.33 Sol.

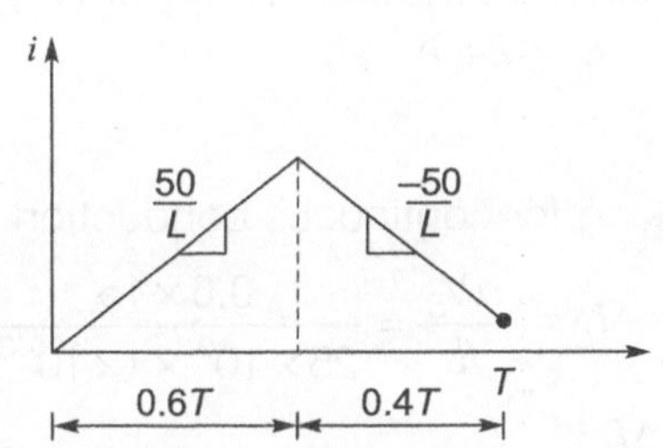

Buck boost converter,

$D = 0.6 \rightarrow$ stores energy

$$D = \frac{T_{ON}}{T} = 0.6$$

$$T_{ON} = 0.6T \rightarrow \text{store energy}$$

$$T_{OFF} = 0.4\,T \rightarrow \text{releasing energy}$$

For one cycle: Rise in current for 0.2T

For 10 cycles: Find rise in current $(0.2T) \times 10 = 2T$

$$i = \frac{50}{L}t = \frac{50}{L}(2T) = \frac{50 \times 2}{LP}$$

$$= \frac{100}{10 \cdot 10^{-3} \times 10 \cdot 10^{3}} = 1\text{A}$$

$\therefore$ Energy stored

$$= \frac{1}{2}Li^2 = \frac{1}{2} \times (10 \cdot 10^{-3}) \cdot (1)^2 = 5 \text{ mJ}$$

3.34 (24)

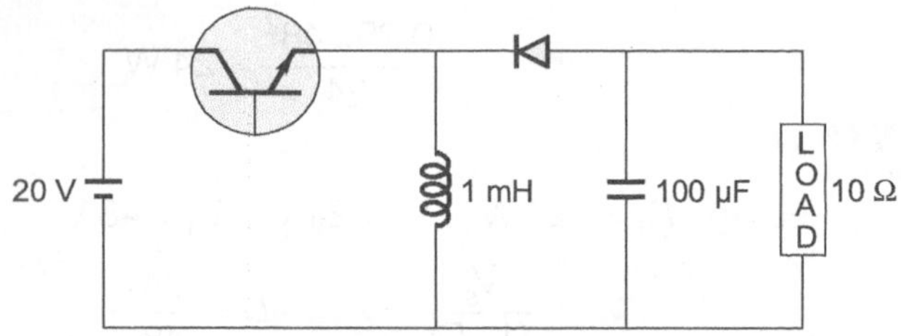

$\alpha = 0.75, \quad f = 25$ kHz

Assume continous conduction:

$$V_0 = \frac{\alpha V_s}{1-\alpha} = \frac{0.75 \times 20}{1-0.75}$$

$$V_0 = 60 \text{ V}$$

$$I_0 = \frac{V_0}{R} = \frac{60}{10} = 6 \text{ A}$$

$$I_L = \frac{I_0}{1-\alpha} = \frac{6}{1-0.75} = 24 \text{ A}$$

$$\Delta I_L = \frac{\alpha V_s}{f_L}$$

$$= \frac{0.75 \times 60}{25 \times 10^3 \times (1 \times 10^{-3})} = 1.8 \text{ A}$$

$$I_{L\,min} = I_L - \frac{\Delta I_L}{2} = 24 - \frac{1.8}{2} = 24 - 0.9$$

$(I_{L\,min} = 23.1 \text{ A}) > 0$

$\therefore$ Continous conduction assumption is correct.

$$I_L = 24 \text{ A}$$

3.35 (1.60)

Checking for continuous conduction mode

$$\Delta I_L = \frac{\alpha V_S}{fL} = \frac{0.6 \times 15}{25 \times 10^3 \times 1 \times 10^{-3}} = 0.36A$$

$$\frac{\Delta I_L}{2} = 0.18\text{A}$$

$$I_{L,min} = I_L - \frac{\Delta I_L}{2} = I_S - \frac{\Delta I_L}{2}$$

$$= (9.375 - 0.18) = 9.195 > 0$$

As it is continuous conduction

$$V_0 = \frac{V_S}{1-\alpha} = \frac{15}{1-0.6} = 37.5V$$

$$I_0 = \frac{V_0}{R} = \frac{37.5}{10} = 3.75V$$

$$\frac{V_0}{V_S} = \frac{I_S}{I_o} = \frac{1}{1-\alpha}$$

$$I_S = \frac{I_o}{1-\alpha} = \frac{3.75}{1-0.6} = 9.375A$$

$$R_{in} = \frac{V_S}{I_S} = \frac{15}{9.375} = 1.6\Omega$$

3.36 (168)

$$I_L = \frac{I_{max} + I_{min}}{2} = \frac{16+12}{2} = 14 \text{ A}$$

$$I_0 = I_L(1-\alpha)$$

$$I_0 = 14 \times \left(\frac{30}{50}\right) = 8.4 \text{ A}$$

$$\Delta V_0 = \Delta V_c = \frac{\alpha \cdot I_0}{f \cdot C}$$

$$1 = \frac{\left(\frac{2}{5}\right) \times 8.4}{\frac{1}{50.10^{-6}} \cdot C} \Rightarrow C = 168 \text{ μF}$$

3.37 (b)

Mode 1 : (t = 0 to 3 μs)

S_1 = ON, $D_1 = D_2$, S_2 = OFF

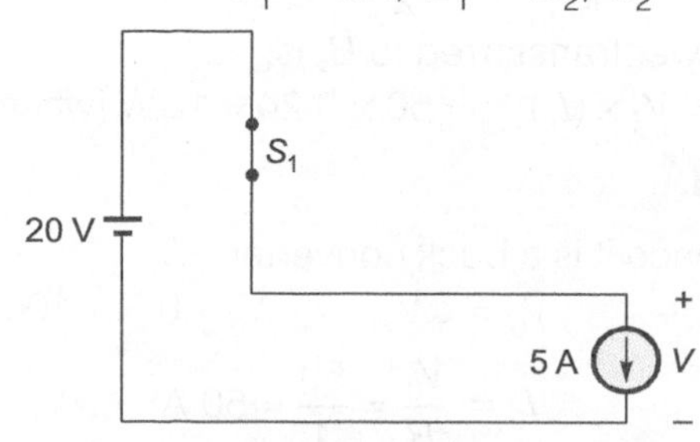

V = 20 Volts

Mode 2 : (t = 3 to 5 μs)

$S_1 = S_2 = D_1$ = OFF; D_2 = ON

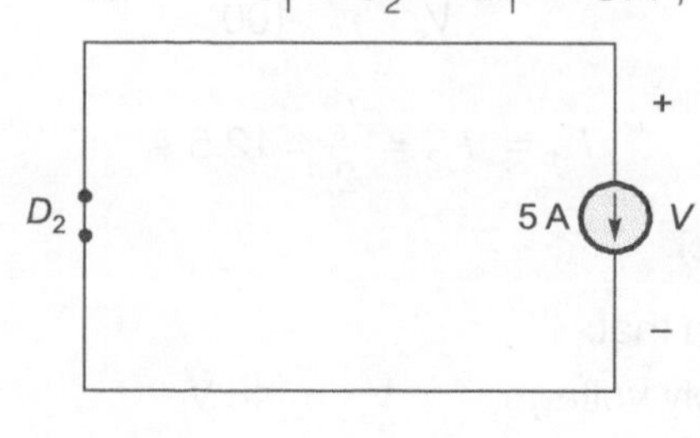

Mode 3 : (t = 5 to 8 μs)

$$S_1 = S_2 = D_1 = \text{OFF}; \quad D_2 = \text{ON}$$

Although gate signal is given to S_2, it still does not turn ON, because two anti-parallel devices cannot conduct simultaneously.

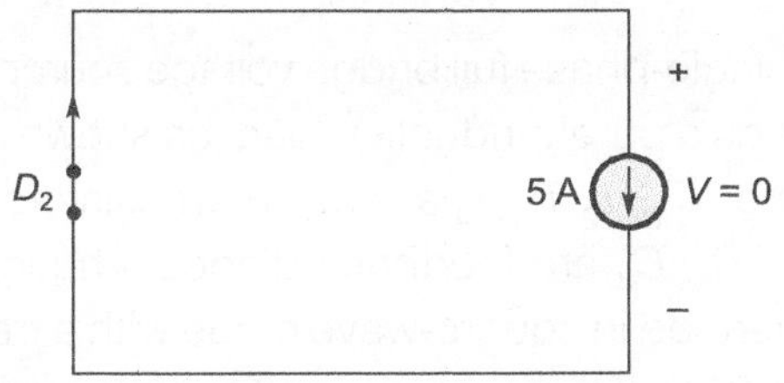

Mode 4 : (t = 8 to 10 μs)

$$S_1 = S_2 = D_1 = \text{OFF}; \quad D_2 = \text{ON}$$

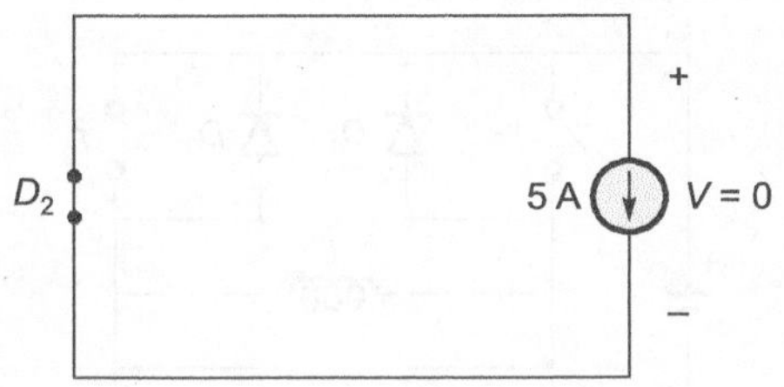

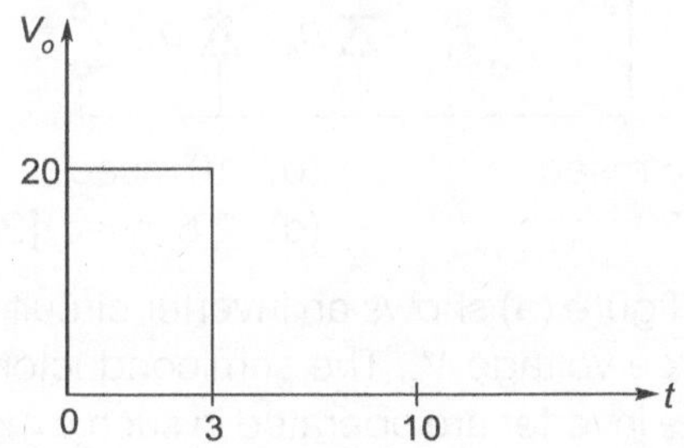

$$V_{(av)} = \frac{20 \times 3}{10} = 6 \text{ Volts}$$

3.38 (50)

$$t_{cm} = \frac{C}{I_o} V_s$$

$$= \frac{10 \cdot 10^{-6}}{10} \cdot 50 = 50 \text{ μs}$$

3.39 (10)

For continuous conduction in BOOST converter,

$$V_0 = \frac{V_S}{1-\alpha}$$

$$40 = \frac{20}{1-\alpha}$$

$$\alpha = 0.5$$

$$I_0 = \frac{V_0}{R} = \frac{40}{10} = 4 \text{ A}$$

Boost converter :

$$I_L = \frac{I_0}{1-\alpha} = \frac{4}{1-0.5} = 8 \text{ A}$$

$$\Delta I_L = \frac{\alpha V_s}{fL} = \frac{0.5 \times 20}{500 \times 2.10^{-3}} = 10$$

$$I_{mx} = (iL)_{peak}$$

$$= I_L + \frac{\Delta I_L}{2} = 8 + \frac{10}{2} = 13 \text{ A}$$

■■■■

4 Inverters

4.1 A single-phase inverter with square wave output voltage will have in its output waveform a fifth harmonic component equal to ______ percentage of the fundamental. **[1995 : 1 M]**

4.2 An inverter capable of supplying a balanced three-phase variable voltage variable frequency output, is feeding a three-phase induction motor rated for 50 Hz and 440 V. The stator winding resistance of the motor are negligibly small during starting. The current inrush can be avoided without sacrificing the starting torque by suitably applying.
(a) Low voltage as rated frequency.
(b) Low voltage keeping the v/f ratio constant.
(c) Rated voltage at low frequency.
(d) Rated voltage at rated frequency.
[1995 : 1 M]

4.3 A PWM switching scheme is used with a three-phase inverter to
(a) reduce the total harmonic distortion with modes filtering.
(b) minimise the load on the DC side.
(c) increase the life of the batteries.
(d) reduce low order harmonics and increase high order harmonics. **[1999 : 1 M]**

4.4 Triangular PWM control, when applied to a three-phase BJT based voltage source inverter, introduces
(a) low order harmonic voltages on the DC side.
(b) very high order harmonic voltage on the DC side.
(c) low order harmonic voltage on the AC side.
(d) very high order harmonic voltage on the AC side. **[2000 : 1 M]**

4.5 A three-phase voltage source inverter supplies a purely inductive three-phase load. Upon magnitude α_h times that of the fundamental frequency component ($\alpha_h < 1$), the load current would then have an n-th order harmonic of magnitude
(a) zero
(b) α_h times the fundamental frequency component.
(c) $n\alpha_h$ times the fundamental frequency component.
(d) α_h/n times the fundamental frequency component. **[2000 : 2 M]**

4.6 A single-phase full bridge voltage source inverter feeds a purely induction load, as shown in figure. When T_1, T_2 T_3, T_4 are power transistors and D_1, D_2, D_3, D_4 are feedback diodes. The inverter is operated in square-wave mode with a frequency of 50 Hz. If the average load current is zero, what is the time duration of conduction of each feedback diode in a cycle?

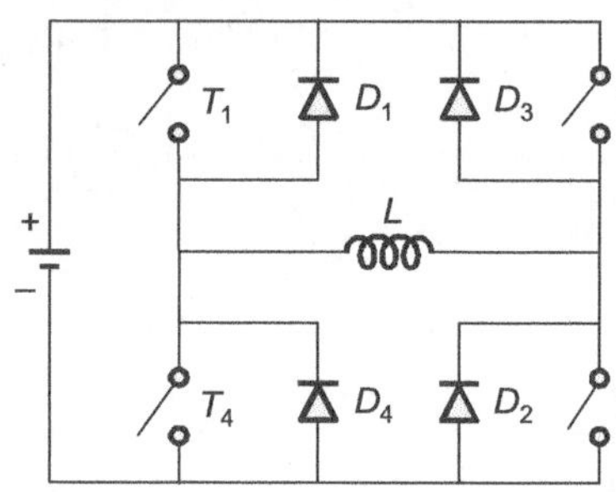

(a) 5 msec (b) 10 msec
(c) 20 msec (d) 2.5 msec **[2001 : 2 M]**

4.7 The figure (a) shows an inverter circuit with a DC source voltage V_s. The semiconductor switches of the inverter are operated in such a manner that the pole voltage V_{10} and V_{20} are as shown in the figure (b) what is the rms value of the pole-to-pole voltage V_{12}?

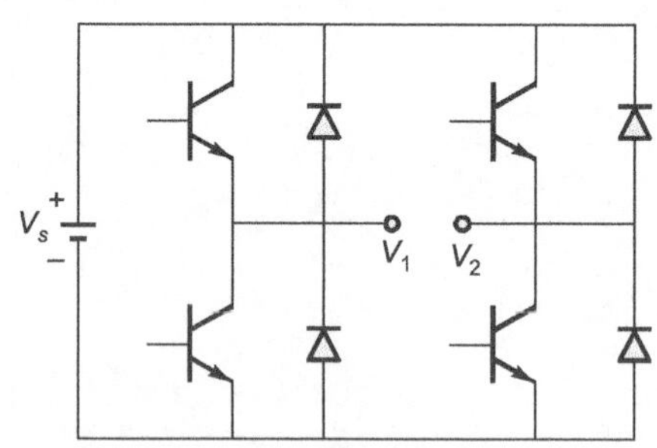

Figure (a)

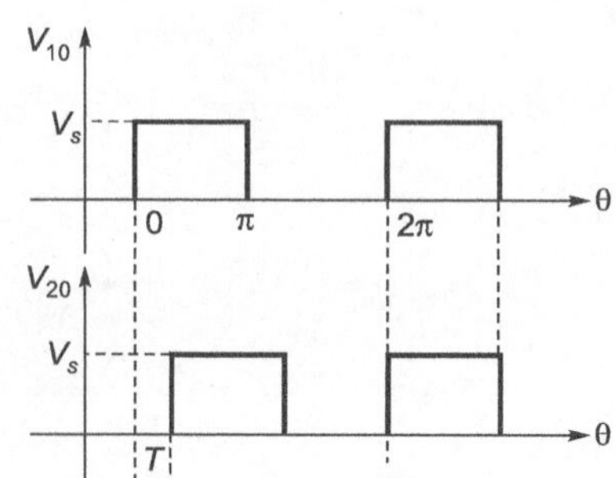

Figure (b)

(a) $\frac{V_s\phi}{\pi/2}$ (b) $V_s\sqrt{\frac{\phi}{\pi}}$

(c) $V_s\sqrt{\frac{\phi}{2\pi}}$ (d) $\frac{V_s}{\pi}$ [2002 : 2 M]

Statement for Common Data (4.8 and 4.9):

An inverter has a periodic output voltage with the output waveform as shown in figure.

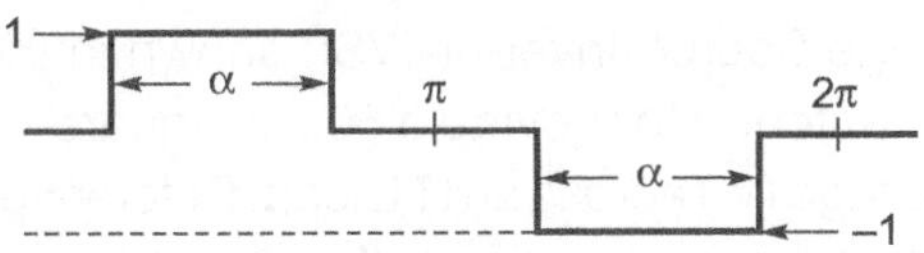

4.8 When the conduction angle α = 120°, the rms fundamental component of the output voltage is

(a) 0.78 V (b) 1.10 V
(c) 0.90 V (d) 1.27 V [2003 : 2 M]

4.9 With reference to the output waveform given in figure, the output of the converter will be free from 5th harmonic when

(a) α = 72° (b) α = 36°
(c) α = 150° (d) α = 120° [2003 : 2 M]

4.10 The output voltage waveform of a three-phase square-wave inverter contains

(a) only even harmonics
(b) both odd and even harmonics
(c) only odd harmonics, excluding triple harmonics
(d) only triple harmonics [2005 : 1 M]

4.11 A single-phase inverter is operated in PWM mode generating a single-pulse of width $2d$ in the centre of each half cycle as shown in figure. It is found that the output voltage is free from 5th harmonic for pulse width 144°. What will be percentage of 3rd harmonic present in the output voltage ($V_{o3}/V_{o1\,max}$)?

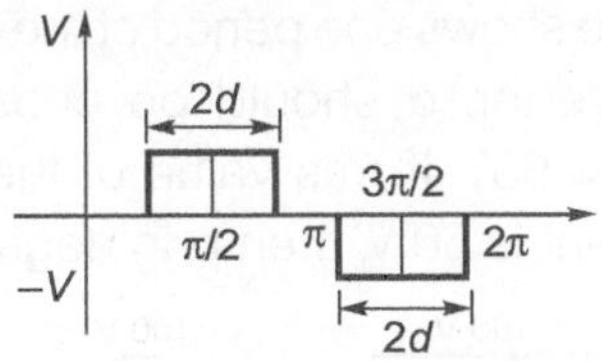

(a) 0.0% (b) 19.6%
(c) 31.7% (d) 53.9% [2006 : 2 M]

4.12 "Six MOSFETs connected in a bridge configuration (having no other power device) MUST be operated as a Voltage Source Inverter (VSI)". This statement is

(a) True, because being majority carrier devices. MOSFETs are voltage driven.
(b) True, because MOSFETs have inherently anti-parallel diodes.
(c) False, because it can be operated both as Current Source Inverter (CSI) or a VSI.
(d) False, because MOSFETs can be operated as excellent constant current sources in the saturation region. [2007 : 1 M]

4.13 A single-phase voltages source inverter is controlled in a single pulse-width modulated mode with a pulse width of 150° in each half cycle. Total harmonic distortion is defined as

$$THD = \frac{\sqrt{V_{rms}^2 - V_1^2}}{V_1} \times 100,$$

where, V_1 is the rms value of the fundamental component of the output voltage. The THD of output AC voltage waveform is

(a) 65.65% (b) 48.42%
(c) 31.83% (d) 30.49% [2007 : 2 M]

4.14 A 3-phase voltage source Inverter is operated in 180° conduction mode. Which one of the following statements is true?

(a) Both pole-voltage and line-voltage will have have 3rd harmonic components.
(b) Pole-voltage will have 3rd harmonic component but line-voltage will be free from 3rd harmonic.
(c) Line-voltage will have 3rd harmonic component but pole-voltage will be free from 3rd harmonic.
(d) Both pole-voltage and line-voltage will be free from 3rd harmonic components. [2008 : 1 M]

4.15 A single phase voltage source inverter is feeding a purely inductive load as shown in the figure

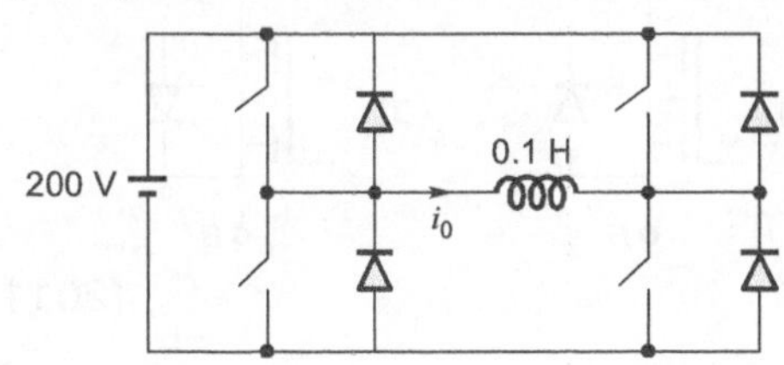

The inverter is operated at 50 Hz in 180° square wave mode. Assume that the load current does not have any DC component. The peak value of the inductor current i_0 will be

(a) 6.37 A (b) 10 A
(c) 20 A (d) 40 A [2008 : 2 M]

4.16 The Current Source Inverter shown in the figure is operated by alternately turning on thyristor pairs (T_1, T_2) and (T_3, T_4). If the load is purely resistive, the theoretical maximum output frequency obtainable will be

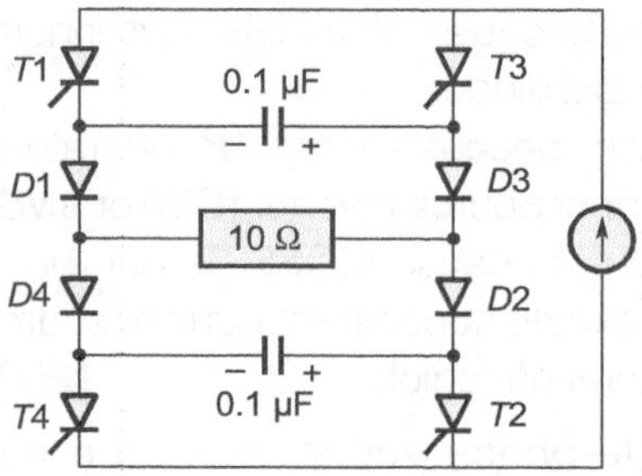

(a) 125 kHz (b) 250 kHz
(c) 500 kHz (d) 50 kHz [2009 : 2 M]

4.17 A three-phase current source inverter used for the speed control of an induction motor is to be realized using MOSFET switches as shown below. Switches S_1 to S_6 are identical switches.

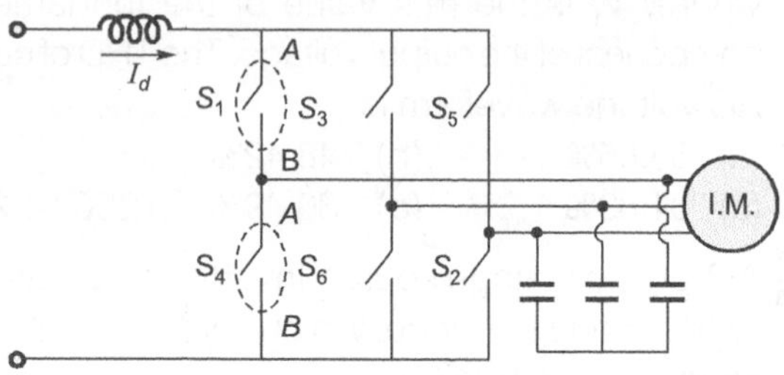

The proper configuration for realizing switches S_1 to S_6 is

(a) 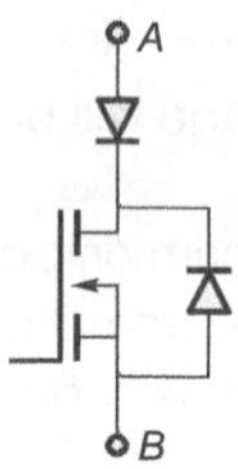

(b) 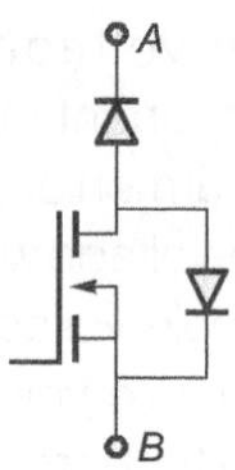

(c) 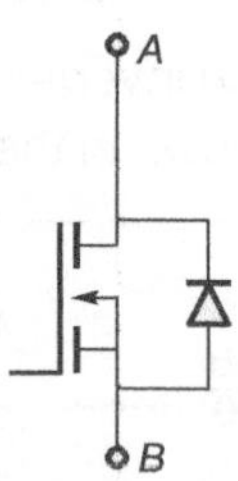

(d) 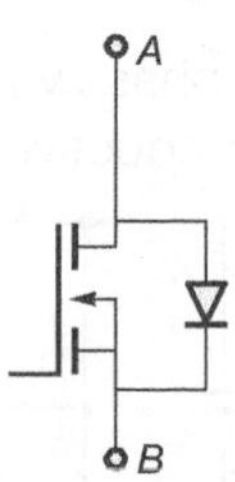

[2011 : 1 M]

Common Data for Questions (4.18 and 4.19):

In the 3-phase inverter circuit shown, the load is balanced and the gating scheme is 180°-conduction mode. All the switching devices are ideal.

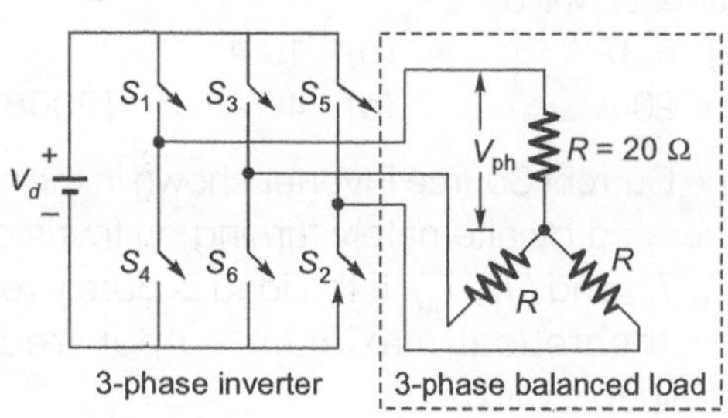

4.18 The rms value of load phase voltage is
(a) 106.1 V (b) 141.4 V
(c) 212.2 V (d) 282.8 V [2012 : 2 M]

4.19 If the DC bus voltage $V_d = 300$ V, the power consumed by 3-phase load is
(a) 1.5 kW (b) 2.0 kW
(c) 2.5 kW (d) 3.0 kW [2012 : 2 M]

Statement for Linked Answer Questions (4.20 and 4.21):

The Voltage Source Inverter (VSI) shown in the figure below is switched to provide a 50 Hz, square-wave AC output voltage (v_o) across an R-L load. Reference polarity of v_o and reference direction of the output current, i_o are indicated in the figure. It is given that $R = 3\ \Omega$, $L = 9.55$ mH.

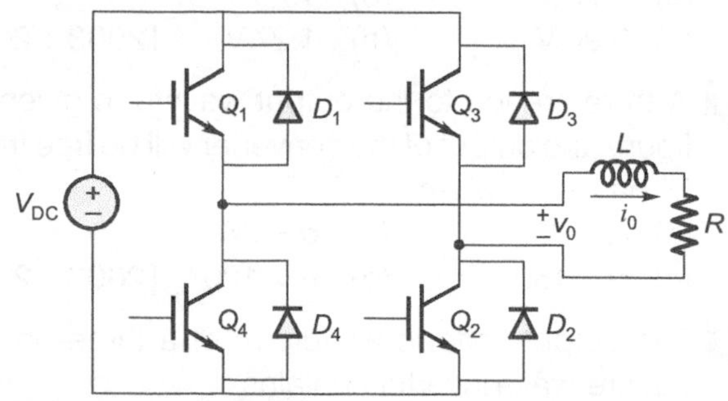

4.20 In the interval when $v_o < 0$ and $i_o > 0$ the pair of devices which conducts the load current is
(a) Q_1, Q_2 (b) Q_3, Q_4
(c) D_1, D_2 (d) D_3, D_4 [2013 : 2 M]

4.21 Appropriate transition, i.e., Zero Voltage Switching (ZVS)/Zero Current Switching (ZCS) of the IGBTs during turn-on/turn-off is
(a) ZVS during turn-off
(b) ZVS during turn-on
(c) ZCS during turn-off
(d) ZCS during turn-on [2013 : 2 M]

4.22 The figure shows one period of the output voltage of an inverter. α should be chosen such that 60° < α < 90°. If rms value of the fundamental component is 50 V, then α in degree is ______.

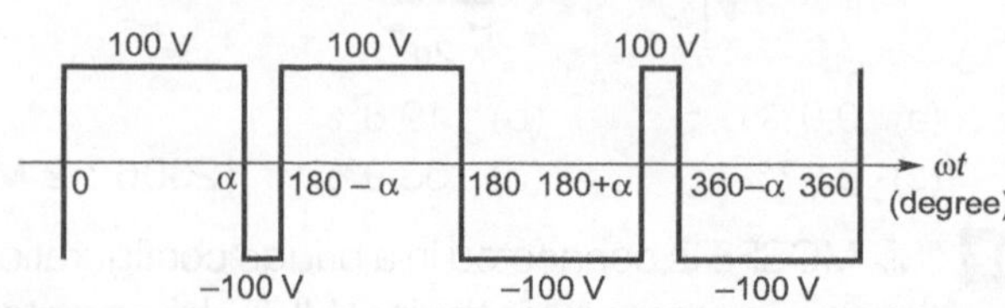

[2014 : 2 M, Set-1]

4.23 A single-phase voltage source inverter shown in the figure is feeding power to a load. The triggering pulses of the devices are also shown in the figure.

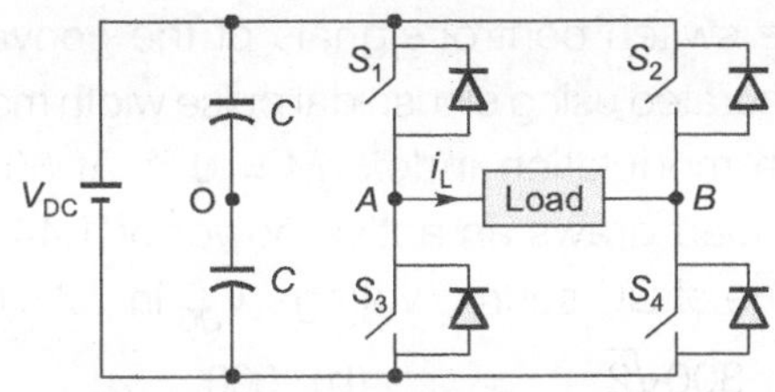

If the load current is sinusoidal and is zero at 0, π, 2π..., the node voltage V_{AO} has the waveform.

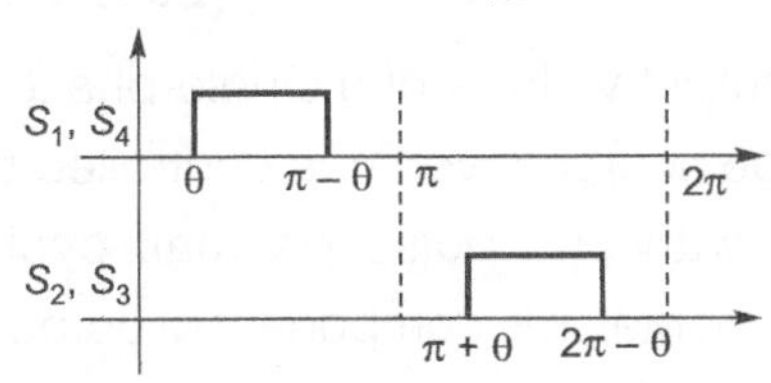

(a) 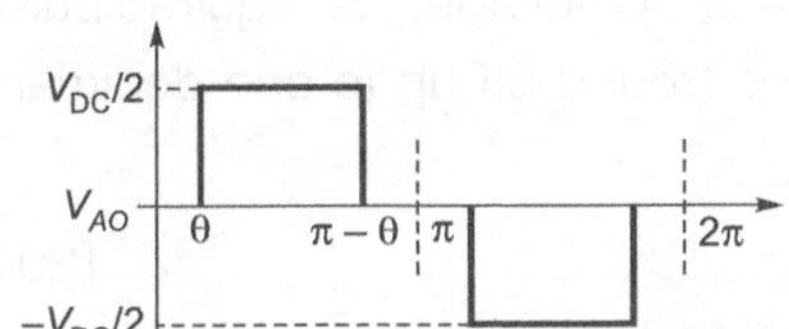

(b) 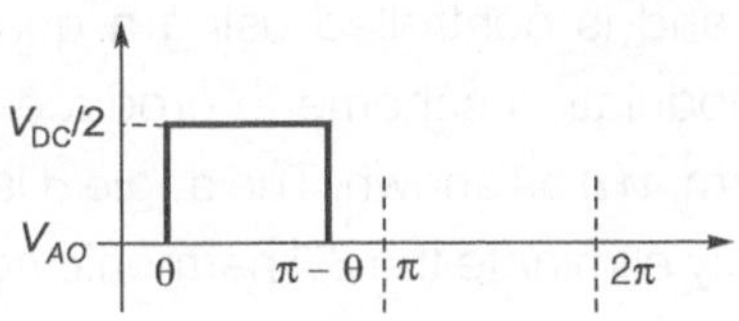

(c) 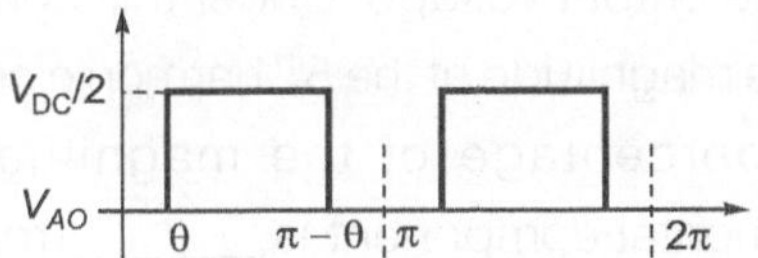

(d) 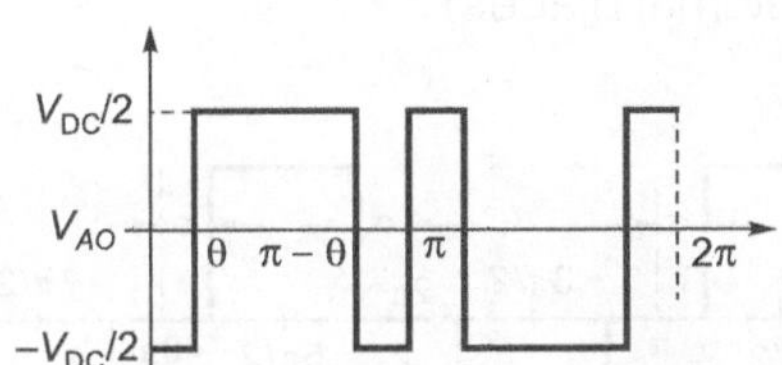

[2014 : 2 M, Set-3]

4.24 The single-phase full-bridge voltage source inverter (VSI), shown in the figure, has an output frequency of 50 Hz. It uses unipolar pulse width modulation with switching frequency of 50 kHz and modulation index of 0.7. For V_{in} = 100 V DC, L = 9.55 mH, C = 63.66 μF and R = 5 Ω, the amplitude of the fundamental component in the output voltage V_0 (in Volt) under steady-state is _____.

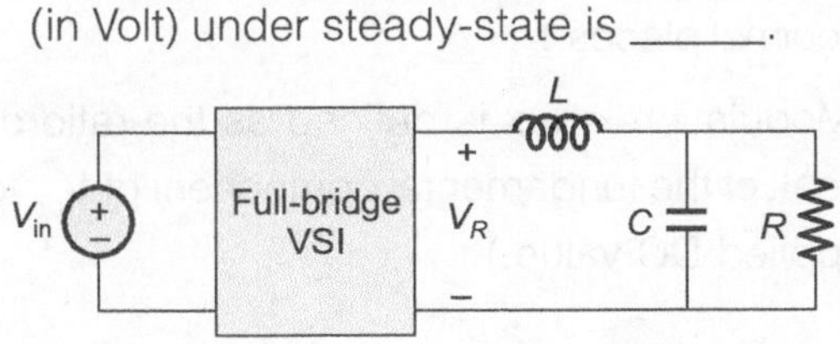

[2015 : 2 M, Set-1]

4.25 The switches T1 and T2 in the figure (a) are switched in a complementary fashion with sinusoidal pulse width modulation technique. The modulating voltage $v_m(t) = 0.8 \sin(200\pi t)$ V and the triangular carrier voltage (v_c) are as shown in the figure (b). The carrier frequency is 5 kHz. The peak value of the 100 Hz component of the load current (i_L), in ampere is _____.

[2016 : 2 M, Set-2]

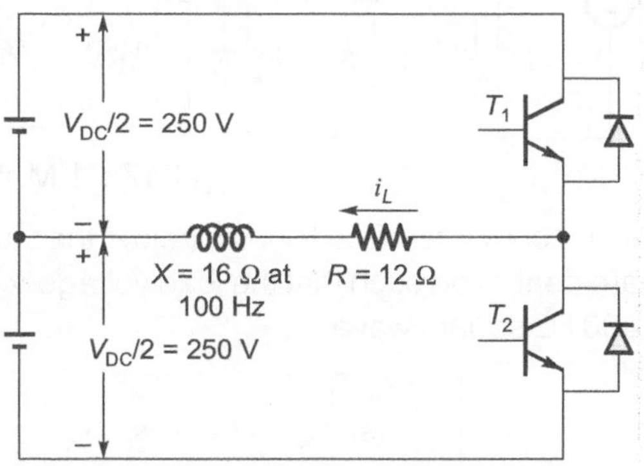

(a)

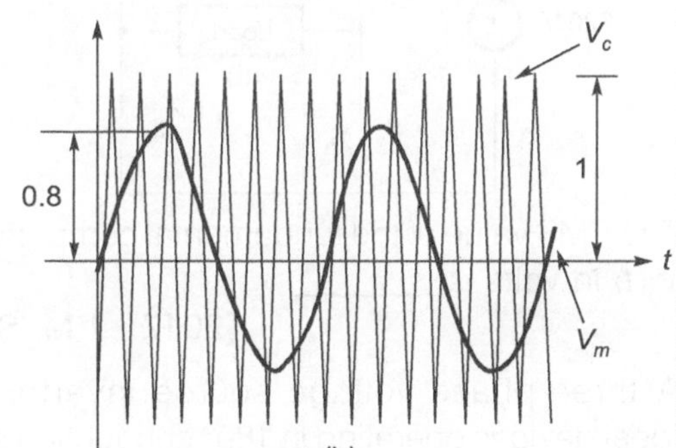

(b)

4.26 A single-phase full-bridge voltage source inverter (VSI) is fed from a 300 V battery. A pulse of 120° duration is used to trigger the appropriate devices in each half-cycle. The rms value of the fundamental component of the output voltage, in volts, is

(a) 234 (b) 245
(c) 300 (d) 331 [2016 : 2 M, Set-1]

4.27 A three-phase Voltage Source Inverter (VSI) as shown in the figure is feeding a delta connected resistive load of 30 Ω/phase. If it is fed from a 600 V battery, with 180° conduction of solid-state devices, the power consumed by the load, in kW, is_____.

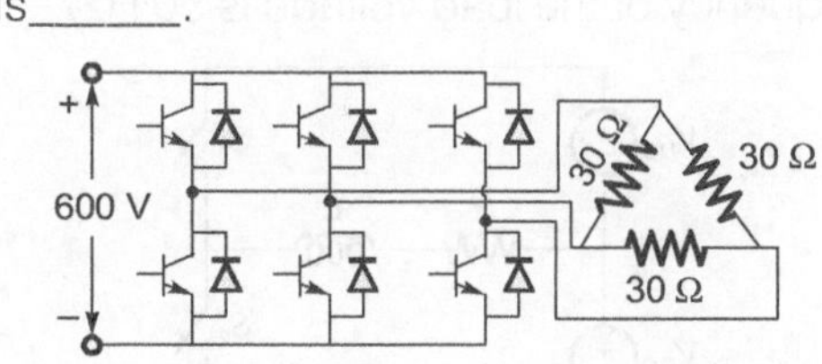

[2016 : 2 M, Set-2]

4.28 A 3-phase voltage source inverter is supplied from a 600 V DC source as shown in the figure below. For a star connected resistive load of 20 Ω per phase, the load power for 120° device conduction. in kW, is ________.

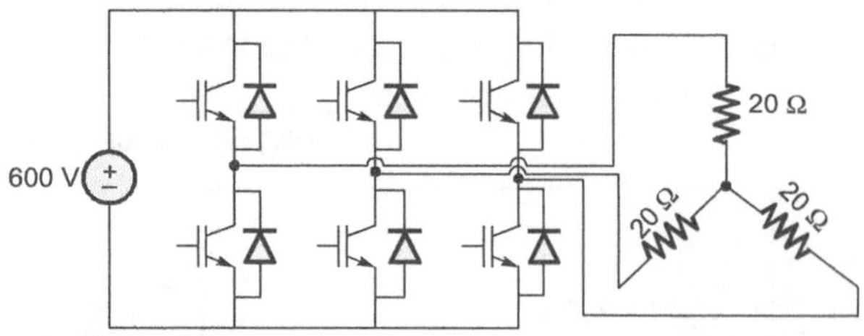

[2017 : 1 M, Set-1]

4.29 In the converter circuit shown below, the switches are controlled such that the load voltage $v_0(t)$ is a 400 Hz square wave.

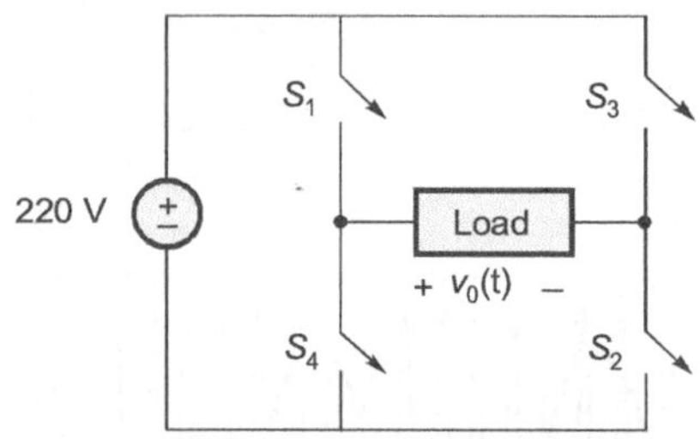

The RMS value of the fundamental component of $v_0(t)$ in volts is ________.

[2017 : 1 M, Set-1]

4.30 A three-phase voltage source inverter with ideal devices operating in 180° conduction mode is feeding a balanced star-connected resistive load. The DC voltage input is V_{dc}. The peak of the fundamental component of the phase voltage is

(a) $\frac{V_{DC}}{\pi}$ (b) $\frac{2V_{DC}}{\pi}$

(c) $\frac{3V_{DC}}{\pi}$ (d) $\frac{4V_{DC}}{\pi}$

[2017 : 1 M, Set-2]

4.31 The figure below shows a half-bridge voltage source inverter supplying an RL-load with $R = 40\ \Omega$ and $L = \left(\frac{0.3}{\pi}\right)H$. The desired fundamental frequency of the load voltage is 50 Hz.

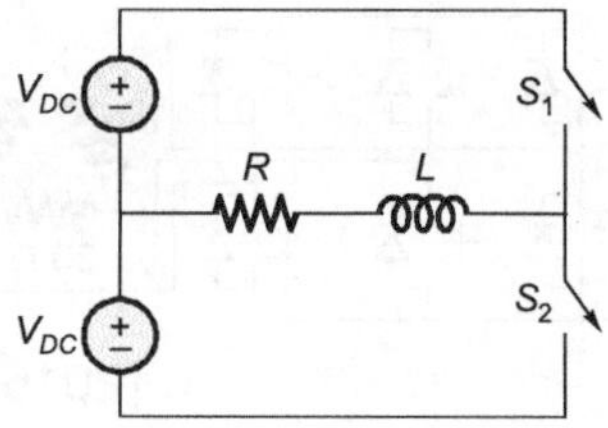

The switch control signals of the converter are generated using sinusoidal pulse width modulation with modulation index, $M = 0.6$. At 50 Hz, the RL-load draws an active power of 1.44 kW. The value of DC source voltage V_{DC} in volts is

(a) $300\sqrt{2}$ (b) 500

(c) $500\sqrt{2}$ (d) $1000\sqrt{2}$

[2017 : 2 M, Set-2]

4.32 The output voltage of a single-phase full bridge voltage source inverter is controlled by unipolar PWM with one pulse per half cycle. For the fundamental rms component of output voltage to be 75% of DC voltage, the required pulse width in degrees (round off up to one decimal place) is ________.

[2019 : 1 M]

4.33 A single-phase inverter is fed from a 100 V dc source and is controlled using a quasi-square wave modulation scheme to produce an output waveform, $v(t)$ as shown. The angle σ is adjusted to entirely eliminate the 3rd harmonic component from the output voltage. Under this condition, for $v(t)$, the magnitude of the 5th harmonic component as a percentage of the magnitude of the fundamental component is ________(rounded off to 2 decimal places).

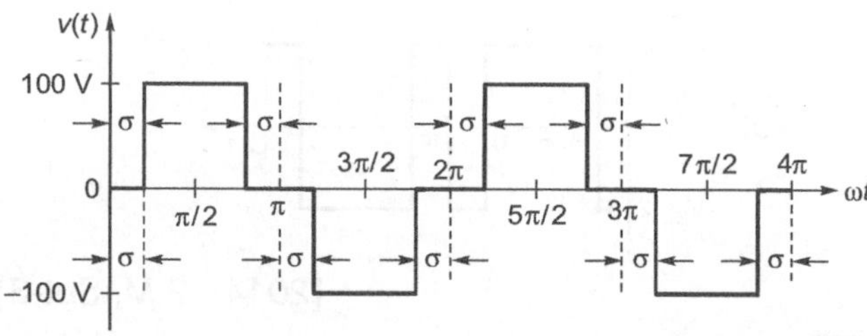

[2020 : 1 M]

4.34 A single-phase full-bridge inverter fed by a 325 V DC produces a symmetric quasi-square waveform across 'ab' as shown. To achieve a modulation index of 0.8. the angle θ expressed in degrees should be ________. (Round off to 2 decimal places,)

(Modulation index is defined as the ratio of the peak of the fundamental component of V_{ab} to the applied DC value.)

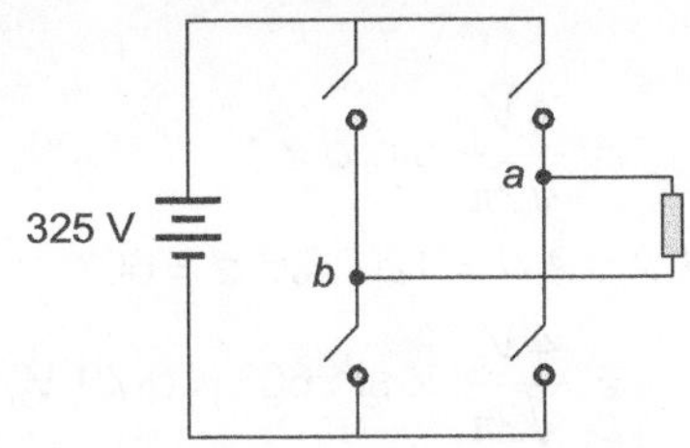

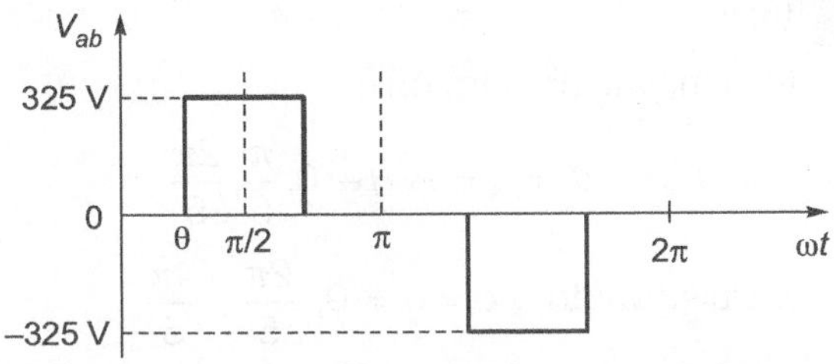

[2021 : 2 M]

4.35 Consider an ideal full-bridge single-phase DC-AC inverter with a DC bus voltage magnitude of 1000 V. The inverter output voltage v(t) shown below, is obtained when diagonal switches of the inverter are switched with 50% duty cycle. The inverter feeds a load with a sinusoidal current given by, $i(t) = 10\sin\left(\omega t - \frac{\pi}{3}\right)$ A, where $\omega = \frac{2\pi}{T}$. The active power, in watts, delivered to the load is _____. (round off to nearest integer).

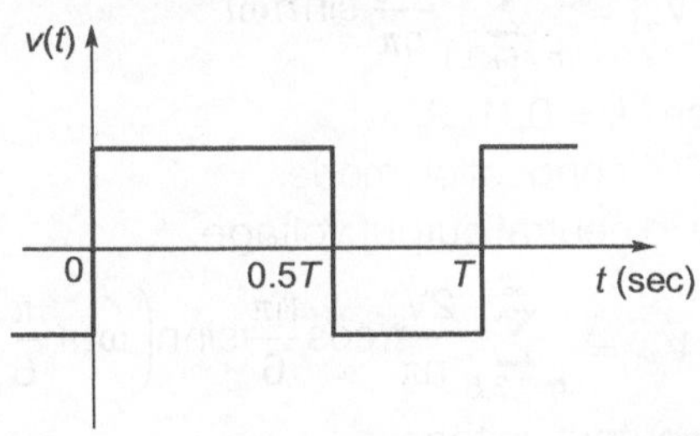

[2022 : 2 M]

4.36 A 3-phase grid-connected voltage source converter with DC link voltage of 1000 V is switched using sinusoidal Pulse Width Modulation (PWM) technique. If the grid phase current is 10 A and the 3-phase complex power supplied by the converter is given by (–4000 – *j*3000) VA, then the modulation index used in sinusoidal PWM is _____. (round off to two decimal places).

[2022 : 2 M]

4.37 A single-phase full bridge voltage source inverter (VSI) feeds a purely inductive load. The inverter output voltage is a square wave in 180° conduction mode. The fundamental frequency of the output voltage is 50 Hz. If the DC input voltage of the inverter is 100 V and the value of the load inductance is 20 mH, the peak-to-peak load current in amperes is _______ (rounded off to the nearest integer).

[2024 : 2 M]

■■■■

Answers Inverters

4.1	(20)	4.2	(b)	4.3	(a)	4.4	(d)	4.5	(d)	4.6	(a)	4.7	(b)
4.8	(a)	4.9	(a)	4.10	(c)	4.11	(b)	4.12	(b)	4.13	(c)	4.14	(b)
4.15	(b)	4.16	(b)	4.17	(a)	4.18	(b)	4.19	(d)	4.20	(d)	4.21	(d)
4.22	(77.15)	4.23	(d)	4.24	(62.75)	4.25	(10)	4.26	(a)	4.27	(24)	4.28	(9)
4.29	(198.06)	4.30	(b)	4.31	(c)	4.32	(112.9)	4.33	(20)	4.34	(51.10)	4.35	(3182)
4.36	(0.47)	4.37	(50)										

Explanations Inverters

4.1 Sol.

For a square wave, fourier series can be given as:

$$V_o = \sum_{n=1,3,5}^{\infty} \frac{4V_s}{n\pi} \sin n\omega t$$

For fundamental component

$$V_1 = \frac{4V_s}{\pi \times \sqrt{2}} \quad \text{...(i)}$$

For fifth harmonics

$$V_5 = \frac{4V_s}{5\pi \times \sqrt{2}} \quad \text{...(ii)}$$

$$\text{Percentage} = \frac{4V_s/5\sqrt{2}\pi}{4V_s/\sqrt{2}\pi} \times 100 = \frac{1}{5} \times 100 = 20\%$$

4.6 (a)

$$\text{Conduction time} = \frac{T}{4} = \frac{20}{4} = 5 \text{ msec}$$

4.7 (b)

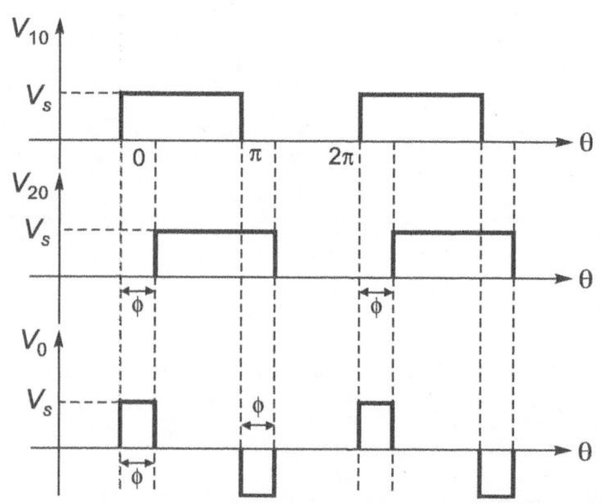

$$V_{12} = V_{10} - V_{20}$$

$$V_{rms} = V_s \sqrt{\frac{\phi}{\pi}}$$

4.8 (a)

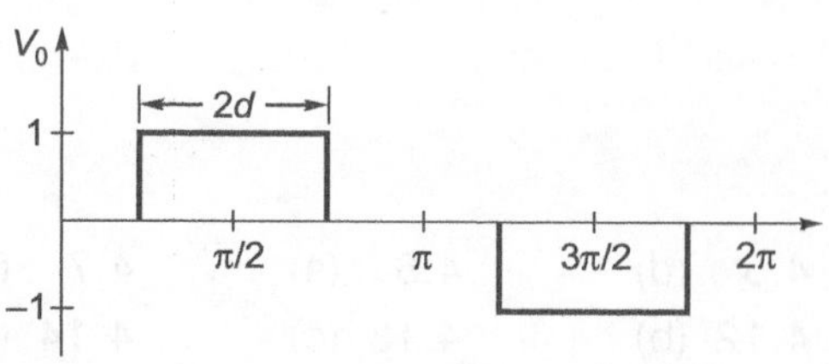

(output) voltage

$$V_0 = \sum_{n=1,3,5\ldots}^{\infty} \frac{4V_s}{n\pi} \sin nd \sin\frac{n\pi}{2} \sin n\omega t$$

∴ R.M.S. value of fundamental component, $n = 1$

$$= \frac{4V_s}{\sqrt{2}\pi} \sin d \times 1$$

$\alpha = 120$, ∴ $2d = 120°$, ∴ $d = 60°$

$$= \frac{4V_s}{\sqrt{2}\pi} \times \sin 60° = 0.78\, V_s = 0.78 \text{ V}$$

4.9 (a)

To eliminate 5th harmonic,

$$5d = 0, \pi, 2\pi \Rightarrow d = 0, \frac{\pi}{5}, \frac{2\pi}{5}$$

∴ Pulse width, $2d = \alpha = 0, \frac{2\pi}{5}, \frac{4\pi}{5}$

$= (0°, 72°, 144°)$

4.10 (c)

The output voltage in both 180° and 120° conduction mode can be described by the fourier series as follows:

In 180° conduction mode,

line output voltage,

$$V_{ab} = \sum_{n=1,3,5}^{\infty} \frac{4V_s}{n\pi} \cos\frac{n\pi}{6} \sin n\left(\omega t + \frac{\pi}{6}\right) \quad \text{...(i)}$$

Line to neutral output voltage,

$$V_{a0} = \sum_{n=6k\pm1}^{\infty} \frac{2V_s}{n\pi} \sin n\omega t \quad \text{...(ii)}$$

Where $k = 0, 1, 2, \ldots$

In 120° conduction mode

Line to neutral output voltage,

$$V_{a0} = \sum_{n=1,3,5}^{\infty} \frac{2V_s}{n\pi} \cos\frac{n\pi}{6} \sin n\left(\omega t + \frac{\pi}{6}\right) \quad \text{...(iii)}$$

Line output voltage,

$$V_{ab} = \sum_{n=6k\pm1}^{\infty} \frac{3V_s}{n\pi} \sin n\left(\omega t + \frac{\pi}{3}\right) \quad \text{...(iv)}$$

Where $k = 0, 1, 2\ldots$

It is clear from equation (i) to (iv), output voltage wave form contains only odd-harmonic.

4.11 (b)

$$\left(\frac{V_{03}}{V_{01}}\right)_{max} = \frac{\frac{4V_s}{3\pi}\sin 3d}{\frac{4V_s}{\pi}} = 19.6\%$$

4.12 (b)

As MOSFET have internally antiparallel diodes, MOSFET can not withstand reverse voltage, due to which MOSFET can not be as current source inverter (CSI).

4.13 (c)

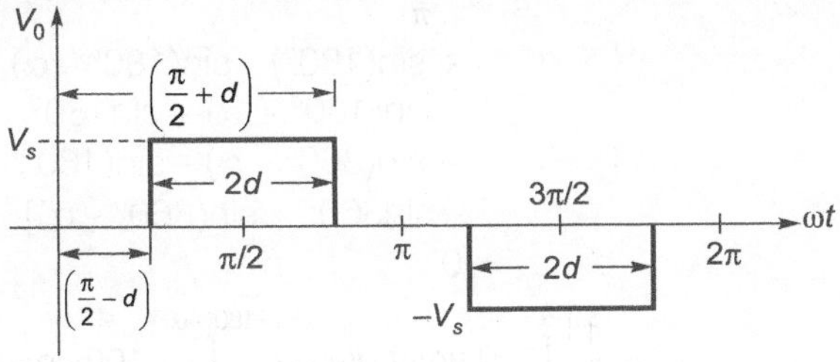

Pulse width, $2d = 150°$

$\Rightarrow \quad d = 75°$

$$V_{rms} = \sqrt{\frac{1}{\pi}\int_{\pi/2-d}^{\pi/2+d} V_s^2 d(\omega t)}$$

$$= V_s\sqrt{\frac{2d}{\pi}} = V_s\sqrt{\frac{5}{6}} = 0.913V_s$$

Output voltage in fourier series.

$$v_0 = \sum_{n=1,3,5}^{\infty} \frac{4V_s}{n\pi}\sin\frac{n\pi}{2}\sin nd \sin n\omega t$$

RMS value of fundamental component of output voltage.

$$V_1 = \frac{1}{\sqrt{2}}\left[\frac{4V_s}{\pi}\sin 75°\right] = 0.87V_s$$

Total Harmonic Distortion,

$$\text{THD} = \sqrt{\frac{V_{rms}^2 - V_1^2}{V_1^2}} \times 100$$

$$= \sqrt{\frac{(0.913V_s)^2 - (0.87V_s)^2}{(0.87V_s)^2}} \times 100$$

$= 31.83\ \%$

4.14 (b)

In 3-phase VSI working in 180° mode:

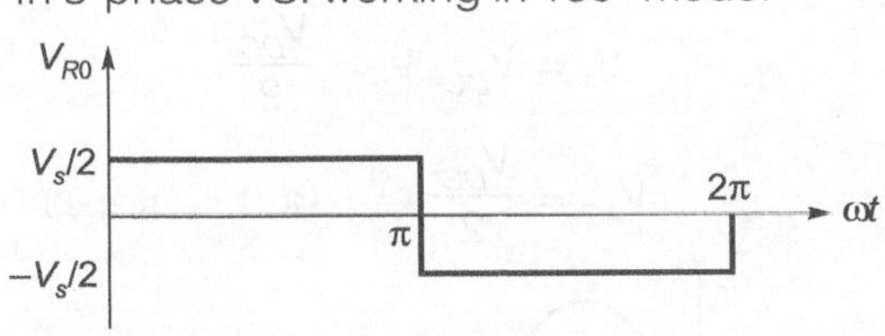

Pole voltage is a square wave.

Therefore 3rd harmonic is present in pole voltage.

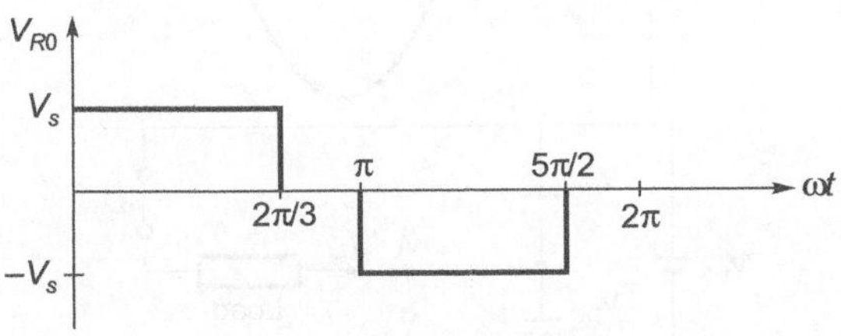

In line voltage waveform even and Tripplen harmonics are not present.

4.15 (b)

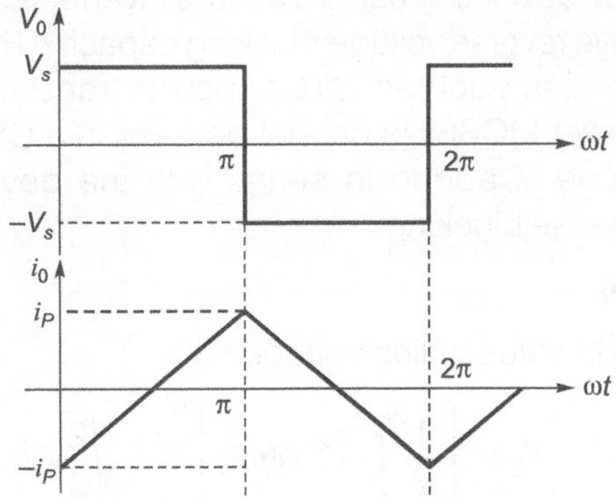

Load current increases linearly from $-I_P$ to I_P during $0 < \omega t \le \pi$. Where I_P = Peak value of i_0

$$\Rightarrow \quad t_1 = \frac{\pi}{\omega} = \frac{\pi}{2\pi 50} = \frac{1}{100}\text{sec}$$

At $\quad t = 0,\ i_0 = -I_P$ and $t = t_1,\ i_0 = I_P$

$$V_0 = L\frac{di}{dt}$$

During $0 < \omega t < \pi$

$$V_0 = V_s = 200\text{ V}$$

$$\Rightarrow \quad 200 = 0.1\left(\frac{I_P - (-I_P)}{t_1 - 0}\right)$$

$$\Rightarrow \quad 2I_P = \frac{200}{0.1}\times t_1$$

$$I_P = \frac{2000}{2}\times\frac{1}{100} = 10\text{ A}$$

4.16 (b)

- The circuit shown in the figure is a single phase bridge auto sequential commutated inverter (1-phase ASCI).
- Thyristor pairs T_1, T_2 and T_3, T_4 are alternatively switches to obtain a nearly square wave load current. Two commutating capacitors, one C_1 in the upper half and the other C_2 in the lower half are connected as shown.
- Diodes D_1 to D_4 are connected in series with each SCR to prevent the commutation capacitors from discharging into the load.
  The inverter output frequency is controlled by adjusting the period T through the triggering circuits of thyristors.
  The theoretical maximum output frequency obtainable

$$f_{max} = \frac{1}{4RC} = \frac{1}{4\times 10\times 0.1\times 10^{-6}} = 250\text{ kHz}$$

4.17 (a)

Device used in current source inverter (CSI) must have reverse voltage blocking capacity. Therefore, devices such as GTOs, power transistors and power MOSFETs cannot be used in a CSI. So, a diode is added in series with the devices for reverse blocking.

4.18 (b)

RMS value of line voltage

$$V_L = \left[\frac{1}{\pi}\int_0^{2\pi/3} V_s^2\, d(\omega t)\right]^{1/2} = \sqrt{\frac{2}{3}}V_s$$

∴ RMS value of load phase voltage,

$$V_p = \frac{V_L}{\sqrt{3}} = \frac{\sqrt{2}}{3}V_s = \frac{300\times\sqrt{2}}{3} = 141.4 \text{ V}$$

4.19 (d)

Power consumed by each resistor,

$$P = \frac{V_p^2}{R} = \frac{(141.4)^2}{20} = 1000 \text{ W}$$

Total power consumed, 3P = 3 kW

4.20 (d)

As $V_o < 0$, (Q_3, D_3) and (Q_4, D_4) should work.

Also $P = v_o i_o$

As $I_o > 0$

$P < 0$

Power is being feedback.

So, D_3 and D_4 are working.

4.22 Sol.

The output voltage V_0 can be represented by Fourier series as under:

$$V_0 = \sum_{n=1,3,5,\ldots}^{\infty} a_n \cos n\omega t + b_n \sin n\omega t$$

where, $$a_n = \frac{2}{T}\int_0^T f(t)\cos n\omega t\, d(\omega t)$$

$$b_n = \frac{2}{T}\int_0^T f(t)\sin n\omega t\, d(\omega t)$$

$$a_n = \frac{1}{\pi}\Bigg[\int_0^{\alpha} 100\cos\omega t\, d(\omega t) + \int_{\alpha}^{180^\circ-\alpha}(-100)\cos\omega t\, d(\omega t) + \int_{180^\circ-\alpha}^{180^\circ} 100\cos\omega t\, d(\omega t) + \int_{180^\circ}^{180^\circ+\alpha} -100\cos\omega t\, d(\omega t) + \int_{180^\circ+\alpha}^{360^\circ-\alpha} 100\cos\omega t\, d(\omega t) + \int_{360^\circ-\alpha}^{360^\circ} -100\cos\omega t\, d(\omega t)\Bigg]$$

$$a_n = \frac{100}{\pi}[\sin\alpha - \sin(180^\circ-\alpha) + \sin\alpha + \sin(180^\circ) - \sin(180^\circ-\alpha) - \sin(180^\circ+\alpha) + \sin 180^\circ + \sin(360^\circ-\alpha) - \sin(180^\circ+\alpha) - \sin 360^\circ + \sin(360^\circ-\alpha)]$$

$$a_n = 0$$

$$b_n = \frac{1}{\pi}\Bigg[\int_0^{\alpha} 100\sin\omega t\, d(\omega t) + \int_{\alpha}^{180^\circ-\alpha} -100\sin\omega t\, d(\omega t) + \int_{180^\circ-\alpha}^{180^\circ} 100\sin\omega t\, d(\omega t) + \int_{180^\circ}^{180^\circ+\alpha} -100\sin\omega t\, d(\omega t) + \int_{180^\circ+\alpha}^{360^\circ-\alpha} 100\sin\omega t\, d(\omega t) + \int_{360^\circ-\alpha}^{360^\circ} -100\sin\omega t\, d(\omega t)\Bigg]$$

$$b_n = \frac{100}{\pi}[-\cos\alpha + 1 + \cos(180^\circ-\alpha) - \cos\alpha - \cos 180^\circ + \cos(180^\circ-\alpha) + \cos(180^\circ+\alpha) - \cos 180^\circ - \cos(360^\circ-\alpha) + \cos(180^\circ+\alpha) + \cos 360^\circ - \cos(360^\circ-\alpha)]$$

$$b_n = \frac{100}{\pi}[4 - 8\cos\alpha] = 50\sqrt{2}$$

$$\cos\alpha = 0.22231 = 77.15^\circ$$

4.23 (d)

$\theta \le \omega t \le \pi\text{-}\theta$ (S_{11} $S_4 \rightarrow$ ON)

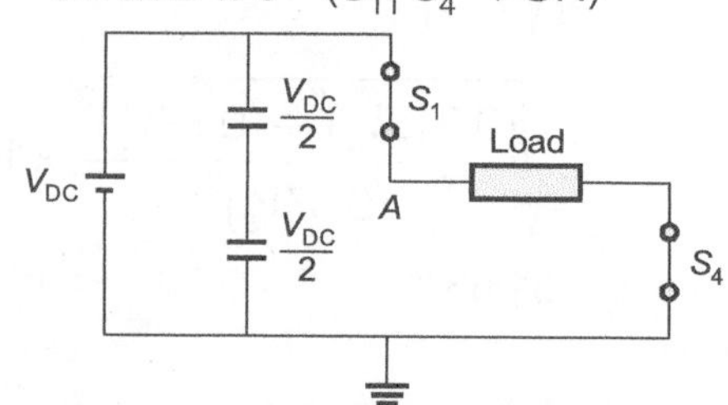

$$V_A = V_{DC},\; V_o = \frac{V_{DC}}{2}$$

$$\Rightarrow \quad V_{AO} = \frac{V_{DC}}{2} \quad (\pi\text{-}\theta < \omega t < z)$$

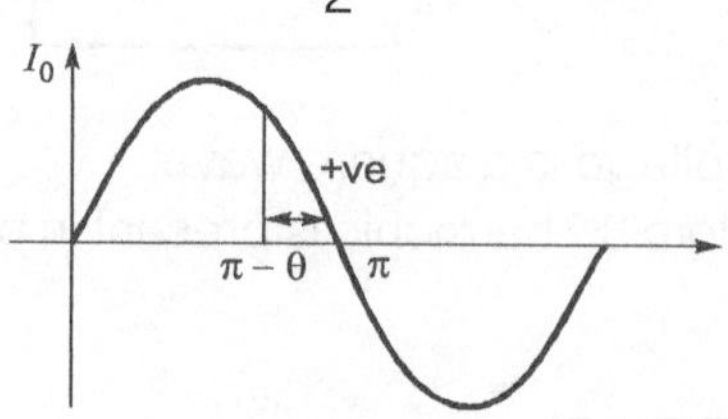

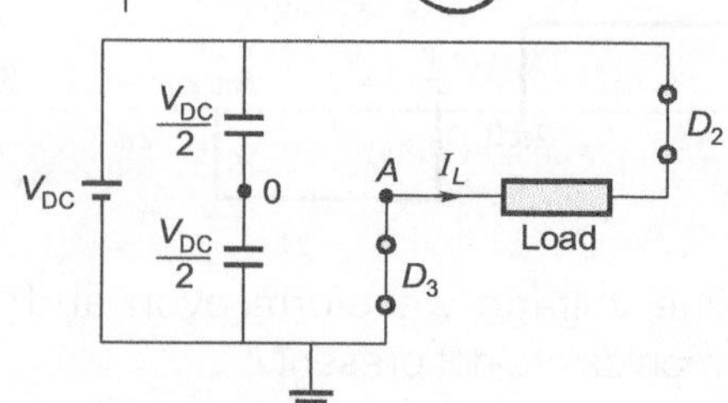

$$V_A = 0$$

$$V_{AO} = V \cdot 0 - \frac{V_{DC}}{2} = -\frac{V_{DC}}{2}$$

So, option (d) is correct.

$$\pi + \theta \leq \omega t \leq 2\pi$$

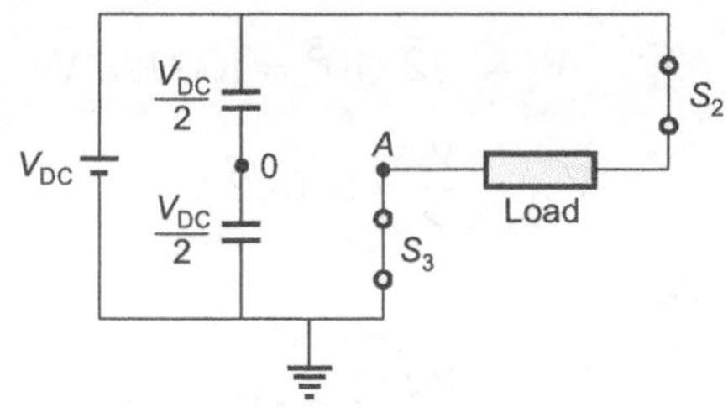

$$V_A = 0, \quad V_0 = -\frac{V_d}{2}$$

4.24 Sol.

The rms value of the fundamental component of output voltage,

$$V_{R1} = \frac{V_s}{\sqrt{2}} \times m_i = \frac{100}{\sqrt{2}} \times 0.7 = 49.49 \text{ V}$$

Amplitude of the value is peak value (i.e.),

$$V_{R_{peak}} = \sqrt{2} \times 49.49 = 70 \text{ V}$$

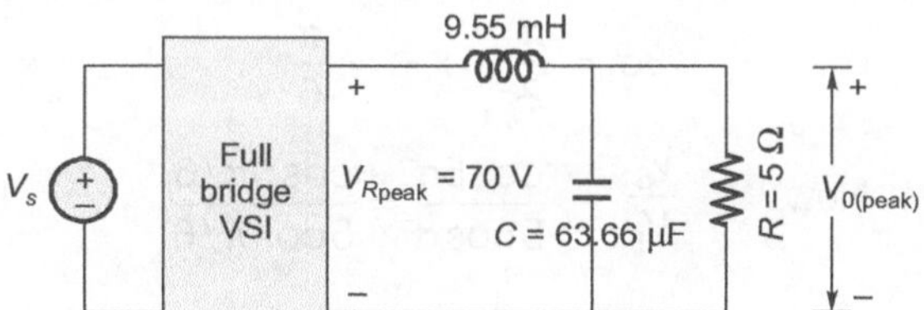

$V_{R_{peak}}$ is the amplitude of fundamental voltage at the VSI. $V_{0(peak)}$ is the amplitude of fundamental voltage at the resistor,

$$Z = \frac{R(-jX_C)}{R - X_C j} = \frac{5(-j50)}{5 - j50} = 4.97\angle -5.71° \ \Omega$$

where, $X_C = \frac{1}{\omega C}$

$$= \frac{1}{2\pi \times 50 \times 63.66 \times 10^{-6}} = 50 \ \Omega$$

$$X_L = \omega L = 2\pi \times 50 \times 9.55 \times 10^{-3} = 3 \ \Omega$$

$$V_{0(peak)} = V_{R_{peak}} \times \frac{4.97\angle -5.71°}{(4.97\angle -5.71°) + j3}$$

$$= 62.75\angle -32.5° \text{ V}$$

∴ The amplitude of the fundamental component in the output voltage, $V_0 = 62.75$ V

4.25 Sol.

$$m_a = 0.8$$

$$(V_{01})_{peak} = m_a \frac{V_d}{2} = 0.8 \times 250 = 200 \text{ V} \quad [m_a \leq 1]$$

$$(I_{01})_{peak} = \frac{(V_{01})_{peak}}{Z_1}$$

$$= \frac{200}{\sqrt{R^2 + (\omega L)^2}} = \frac{200}{\sqrt{12^2 + 16^2}} = 10 \text{ A}$$

4.26 (a)

$$V_{01\,(rms)} = \frac{2\sqrt{2}}{\pi} V_s \cdot \sin d$$

Pulse width where $2d = 120°$

$$d = 60°$$

$$V_{01\,(rms)} = \frac{2\sqrt{2}}{\pi} V_s \cdot \sin 60°$$

$$= \frac{2\sqrt{2}}{\pi} V_s \frac{\sqrt{3}}{2} = 234 \text{ V}$$

4.27 Sol.

$$V_L = V_{ph} = \sqrt{\frac{2}{3}} V_s$$

$$V_{ph} = \sqrt{\frac{2}{3}} \times 600$$

$$P = 3\frac{V_{ph}^2}{R} = \frac{3 \times \frac{2}{3} \times 600^2}{30} = 24 \text{ kW}$$

4.28 Sol.

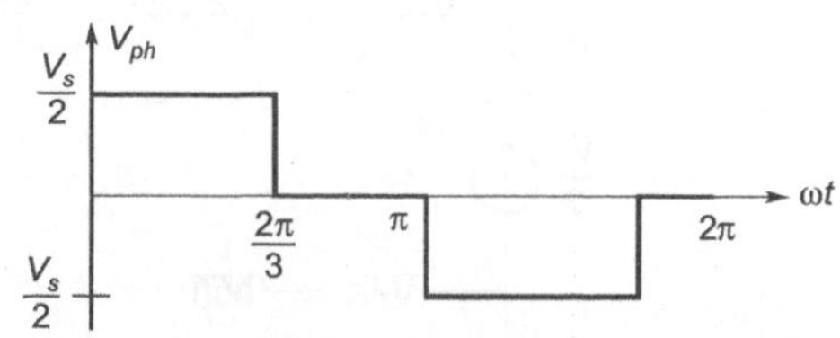

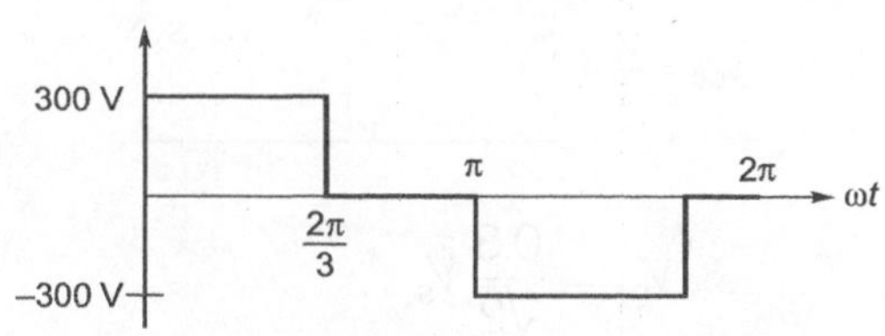

$$V_{rms} = \sqrt{\frac{1}{\pi}(300)^2 \cdot \frac{2\pi}{3}} = 300\sqrt{\frac{2}{3}} = 244.94 \text{ V}$$

$$3\frac{V^2}{R} = \frac{3 \times (244.94)^2}{20} = 9 \text{ kW}$$

4.29 Sol.

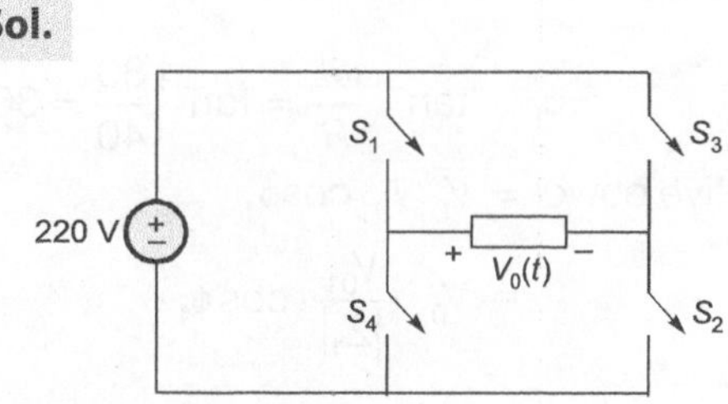

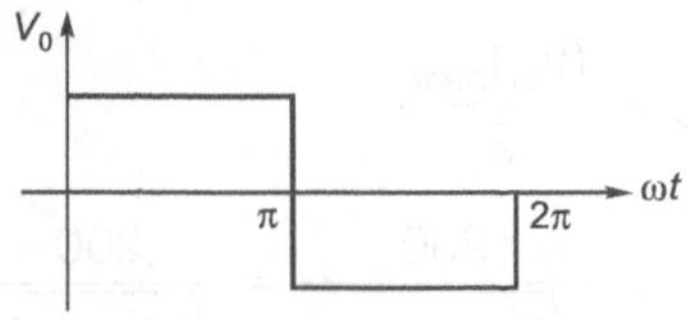

$$V_{01} = \frac{4V_s}{\sqrt{2}\pi} = \frac{4 \times 220}{\sqrt{2}\pi} = 198.06 \text{ V}$$

4.30 (b)

3-ϕ VSI 180° mode:

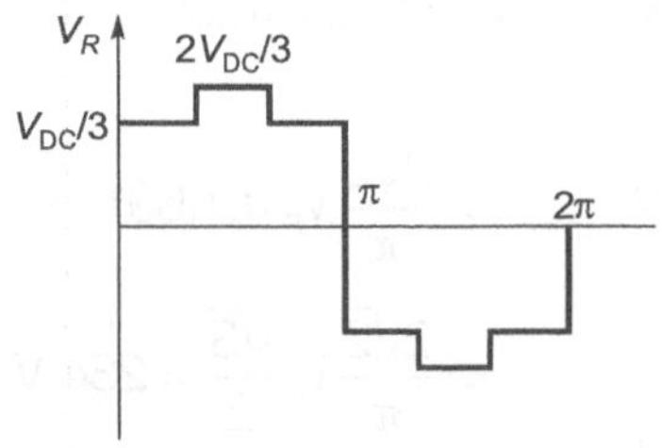

$$V_{Rn} = \frac{6V_{DC}/3}{n\pi}\sin n\omega t = \frac{2 \times V_{DC}}{n\pi}\sin n\omega t$$

$$V_{R1} = \frac{2V_{DC}}{\pi}\sin \omega t$$

4.31 (c)

$$V_{01\text{peak}} = M \cdot \frac{V_s}{2} \quad M \rightarrow \text{Modulation index}$$

$$V_{01\text{rms}} = \frac{M \cdot \frac{V_s}{2}}{\sqrt{2}} = M \cdot \frac{V_s}{2\sqrt{2}} = \frac{0.6\,V_s}{2\sqrt{2}}$$

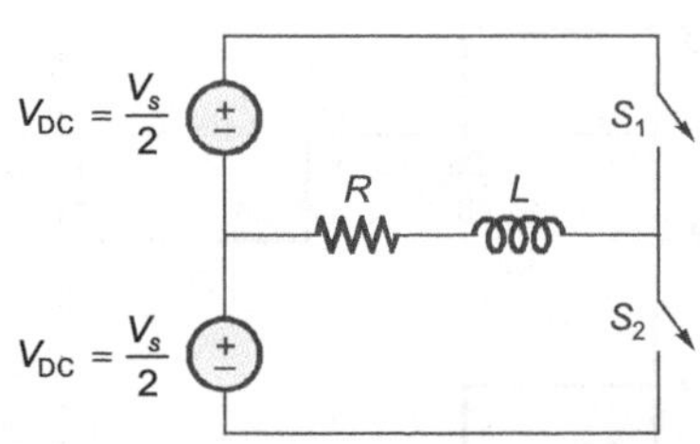

$$V_{01} = \frac{0.3}{\sqrt{2}} V_s$$

$$|Z_1| = \sqrt{R^2 + (\omega L)^2} = \sqrt{40^2 + (2\pi f \cdot L)^2}$$

$$= \sqrt{40^2 + \left(2\pi \times 50 \times \frac{0.3}{\pi}\right)^2} = \sqrt{40^2 + 30^2} = 50$$

$$\phi_1 = \tan^{-1}\frac{\omega L}{R} = \tan^{-1}\frac{30}{40} = 36.869°$$

$$\text{Active power} = V_{01} I_{01} \cos\phi_1$$

$$= V_{01} \cdot \frac{V_{01}}{|Z_1|} \cdot \cos\phi_1$$

$$1.44 \times 10^3 = \left(\frac{0.3}{\sqrt{2}} V_s\right)^2 \cdot \frac{1}{50}(\cos 36.869°)$$

$$\therefore \quad V_s^2 = \frac{1.44 \times 10^3 \times 100}{0.3^2(0.8)} = 20 \times 10^5$$

$$V_s = \sqrt{2} \cdot 10^3 = 1000\sqrt{2} \text{ V}$$

$$V_{DC} = \frac{V_s}{2} = 500\sqrt{2} \text{ V}$$

4.32 (112.9)

$$V_{01,\text{rms}} = \frac{2\sqrt{2}}{\pi}\sin d\, V_s$$

$$0.9 \sin d\, V_s = 0.75\, V_s$$

$$d = 56.44°$$

$$\Rightarrow \quad 2d = 112.88°$$

$$\text{Pulse width} = 112.88°$$

4.33 Sol.

Using result, $V_n = \frac{4V_s}{n\pi}\cos n\sigma$

For $V_3 = 0$, $\cos 3\sigma = 0$

or $\quad 3\sigma = \frac{\pi}{2} \Rightarrow \sigma = \frac{\pi}{6}$

Now, $\quad \frac{V_5}{V_1} = \frac{\cos 5\sigma}{5\cos\sigma} = \frac{\cos 5\pi/6}{5\cos\pi/6}$

$$= -\frac{1}{5}$$

$$\%\left|\frac{V_5}{V_1}\right| = \frac{1}{5} \times 100 = 20\%$$

4.34 (51.10)

Peak value, $\hat{V}_{01} = m_a V_S = 0.8 \times 325 = 260$ V

$$\hat{V}_{01} = \frac{4V_S}{\pi}\sin d = 260$$

$$\frac{4(325)}{\pi}\sin d = 260$$

$$\sin d = \frac{260 \times \pi}{4 \times 325} = 0.628 \Rightarrow d = 38.9$$

$$\therefore \quad \theta = \frac{\pi}{2} - d = 90° - 38.9 = 51.1$$

4.35 (3182)

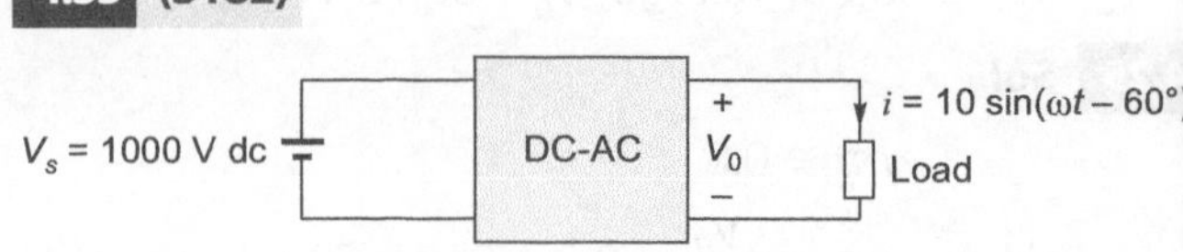

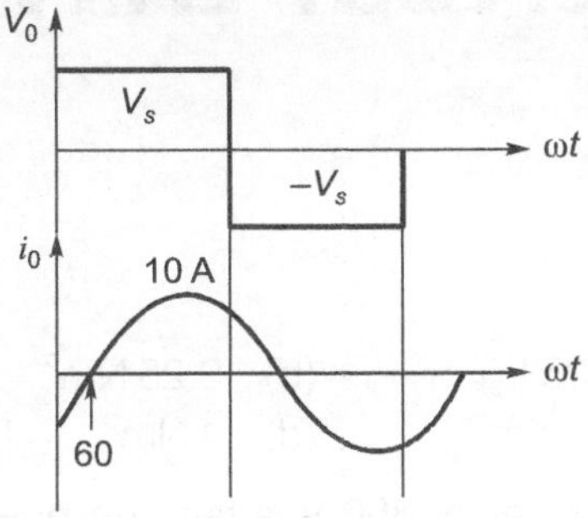

$$P_0 = P_{01} = V_{01} \cdot I_{0r} \cdot \cos(60°)$$

$$= \left(\frac{2\sqrt{2}}{\pi} \cdot (1000)\right) \cdot \left(\frac{10}{\sqrt{2}}\right) \times (0.5) \simeq 3182 \text{ W}$$

4.36 (0.47)

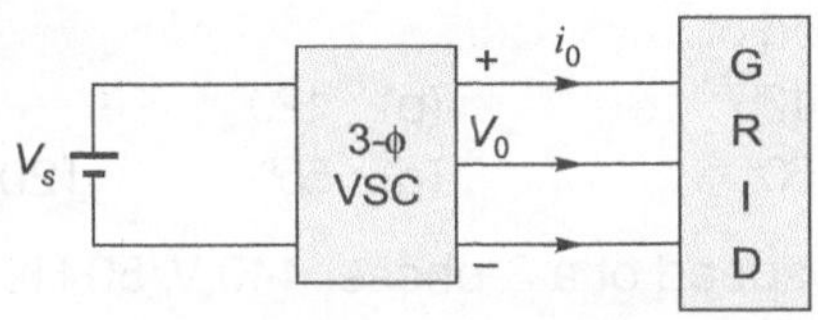

$$S = (-4000 - j3000) \text{ VA}$$

$$P_0 = \sqrt{3} \cdot V_{L1} \cdot I_{0r} \cdot (\text{pF})$$

$$4000 = \sqrt{3} \cdot V_{L1} \cdot 10 \times \left(\frac{4000}{\sqrt{4000^2 + 3000^2}}\right)$$

$$V_{L1\,(rms)} = 288.675 \text{ V}$$

3-φ VSI – SPWM,

$$M_A \le 1$$

$$\hat{V}_{L1} = \sqrt{3} \cdot M_A \cdot \frac{V_s}{2}$$

$$V_{L1\,(rms)} = M_A\left(\frac{\sqrt{3}}{2\sqrt{2}} \times 1000\right)$$

$$288.675 = M_A\left(\frac{\sqrt{3}}{2\sqrt{2}} \times 1000\right)$$

$$M_A = 0.47$$

4.37 (50)

1-φ VSI, $f_1 = f_0 = 50$ Hz

$$V_S = 100 \text{ V}$$

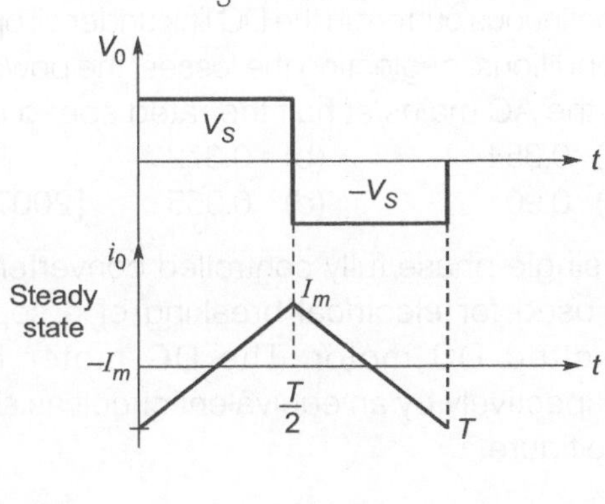

$$V_0 = V_S \quad \left[0 \le t \le \frac{T}{2}\right]$$

$$L \cdot \frac{di}{dt} = V_S$$

$$\frac{di}{dt} = \frac{V_S}{L}$$

$$\int di_0 = \frac{V_S}{L}\int dt$$

$$i_0 = \frac{V_S}{L}t + K$$

$$-I_m = \frac{V_S}{L}(0) + K$$

$$i_0 = \frac{V_S}{L} \cdot t - I_m \qquad \text{At } t = 0,$$

$$i_0 = -I_m$$

$$\therefore \quad K = -I_m$$

$$\text{At } t = \frac{T}{2}, \quad i_0 = I_M$$

$$i_0 = \frac{V_S}{L} \cdot t - I_m$$

$$I_M = \frac{V_S}{L} \cdot \frac{T}{2} - I_m$$

$$2I_m = \frac{V_S}{2fL}$$

$$I_m = \frac{V_S}{4fL}$$

∴ Peak to peak load current,

$$2I_m = \frac{V_S}{2fL} = \frac{100}{2 \times 50(20 \cdot 10^{-3})}$$

$$= 50 \text{ A}$$

■■■■

5 Electrical Drives

5.1 A three-phase semiconverter feeds the armature of separately excited DC motor, supplying a non-zero torque, for steady-state operation, the motor armature current is found to drop to zero at certain instances of time. At such instances, the voltage assumes a value that is
(a) equal is the instantaneous value of the AC phase voltage
(b) equal to the instantaneous value of the motor back emf
(c) arbitrary
(d) zero **[2001 : 1 M]**

5.2 A single-phase half-controlled rectifier is driving a separately excited DC motor. The DC motor has a back emf constant of 0.25 V/rpm. The armature current is 5 A without any ripple. The armature resistance is 2 Ω. The converter is working from a 230 V, single phase AC source with a firing angle of 30°. Under this operating condition, the speed of the motor will be
(a) 346 rpm (b) 359 rpm
(c) 366 rpm (d) 386 rpm **[2004 : 2 M]**

5.3 A variable speed drive rated for 1500 rpm, 40 Nm is reversing under no load. Figure shows the reversing torque and the speed during the transient. The moment of inertia of the drive is

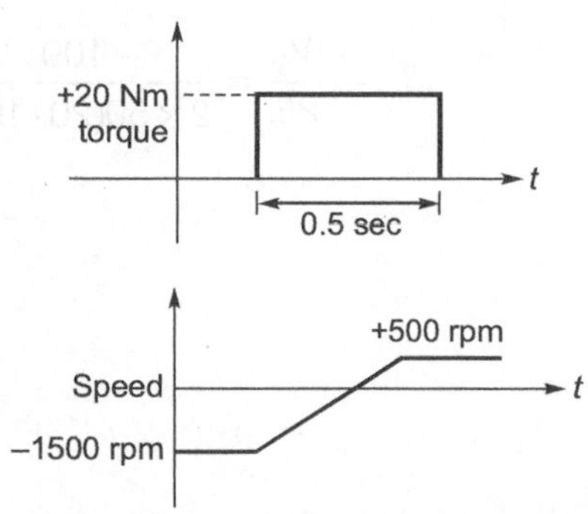

(a) 0.048 kgm^2 (b) 0.064 kgm^2
(c) 0.096 kgm^2 (d) 0.128 kgm^2 **[2004 : 2 M]**

5.4 An electric motor, developing a starting torque of 15 Nm, starts with a load torque of 7 Nm on its shaft. If the acceleration at start is 2 rad/sec^2, the moment of inertia of the systems must be (neglecting viscous and Coulomb friction)
(a) 0.25 kg m^2 (b) 0.25 Nm^2
(c) 4 kg m^2 (d) 4 Nm^2 **[2005 : 2 M]**

5.5 A solar cell of 350 V is feeding power to an AC supply of 440 V, 50 Hz through a 3-phase fully controlled bridge converter. A large inductance is connected in the DC circuit to maintain the DC current at 20 A. If the solar cell resistance is 0.5 Ω, then each thyristor will be reverse biased for a period of
(a) 125° (b) 120°
(c) 60° (d) 55° **[2006 : 2 M]**

5.6 The speed of a 3-phase, 440 V, 50 Hz induction motor is to be controlled over a wide range from zero speed to 1.5 time the rated speed using a 3-phase voltage source inverter. It is desired to keep the flux in the machine constant in the constant torque region by controlling the terminal voltage as the frequency changes. The inverter output voltage vs frequency characteristic should be

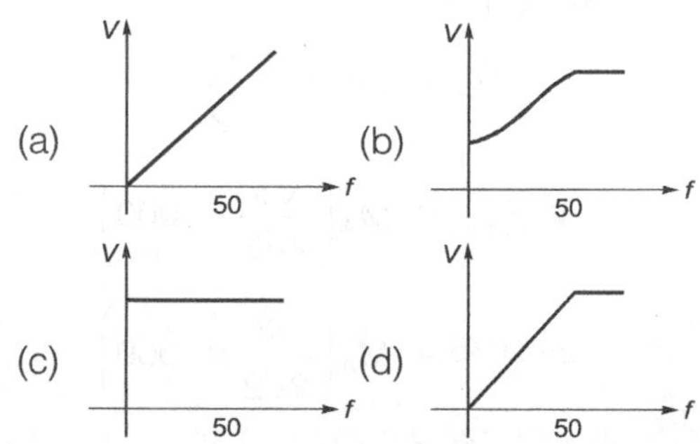

[2006 : 1 M]

5.7 A three-phase, 440 V, 50 Hz ac mains fed thyristor bridge is feeding a 440 V dc, 15 kW, 1500 rpm separately excited DC motor with a ripple free continuous current in the DC link under all operating conditions. neglecting the losses, the power factor of the AC mains at half the rated speed is
(a) 0.354 (b) 0.372
(c) 0.90 (d) 0.955 **[2007 : 2 M]**

5.8 A single phase fully controlled converter bridge is used for electrical breaking of a separately excited DC motor. The DC motor load is respectively by an equivalent circuit as shown in the figure.

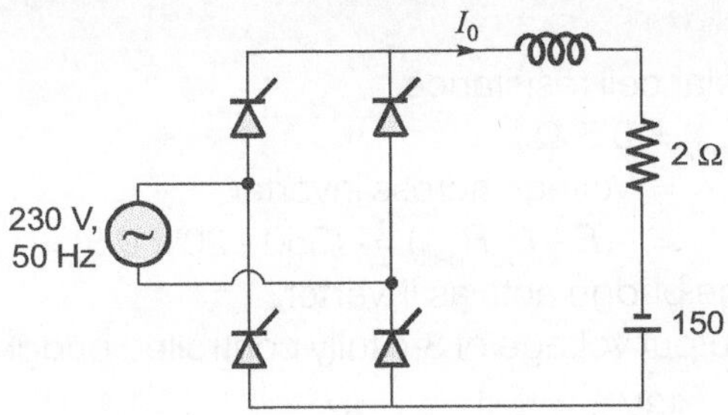

Assume that the load inductance is sufficient to ensure continuous and ripple free load current. The firing angle of the bridge for a load current of I_0 = 10 A will be

(a) 44° (b) 51°
(c) 129° (d) 136° **[2008 : 2 M]**

5.9 A 3-phase squirrel cage induction motor supplied from a balanced 3-phase source drives a mechanical load. The torque-speed characteristics of the motor (solid curve) and of the load (dotted curve) are shown. Of the two equilibrium points *A* and *B*, which of the following options correctly describes the stability of *A* and *B*?

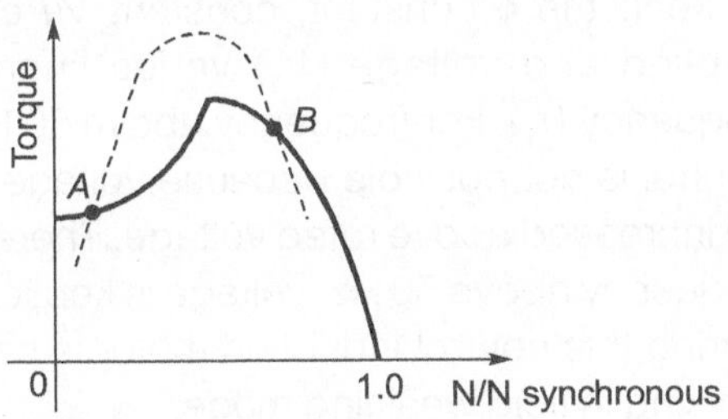

(a) *A* is stable *B* is unstable.
(b) *A* is unstable *B* is stable.
(c) Both are stable.
(d) Both are unstable. **[2009 : 1 M]**

5.10 The separately excited DC motor in the figure below has a rated armature current of 20 A and a rated armature voltage of 150 V. An ideal chopper switching at 5 kHz is used to control the armature voltage. If L_a = 0.1 mH, R_a = 1 Ω, neglecting armature reaction, the duty ratio of the chopper to obtain 50% of the rated torque at the rated speed and the rated field current is

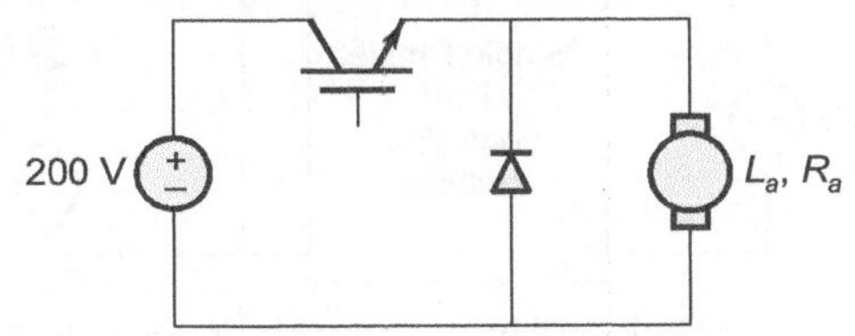

(a) 0.4 (b) 0.5
(c) 0.6 (d) 0.7 **[2013 : 2 M]**

5.11 A shunt-connected DC motor operates at its rated terminal voltage. Its no-load speed is 200 rad/sec. At its rated torque of 500 Nm, its speed is 180 rad/sec. The motor is used to directly drive a load whose load torque T_L depends on its rotational speed ω_r (in radian/second), such that $T_L = 2.78 \times \omega_r$. Neglecting rotational losses, the steady-state speed (in radian/second) of the motor, when it drives this load, is ______.

[2015 : 1 M, Set-2]

■■■■

Answers Electrical Drives

5.1 (b)	5.2 (a)	5.3 (a)	5.4 (c)	5.5 (d)	5.6 (d)	5.7 (a)
5.8 (c)	5.9 (a)	5.10 (d)	5.11 (180)			

Explanations Electrical Drives

5.1 (b)

$$V = E_b + I_a R_a$$
$$I_a = 0; \quad V = E_b$$

5.2 (a)

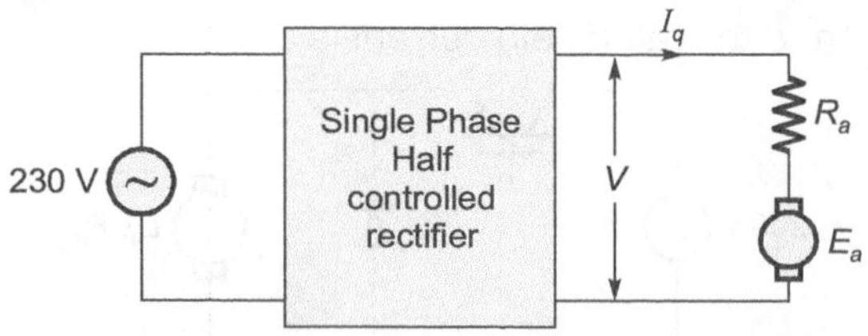

Back emf $= E_a = k\phi N$ or $E_a = k_b N$

where, k_b = Back-emf constant = 0.25 V/rpm

Average output voltage of 1-ϕ half controlled rectifier = V

$$V = \frac{V_m}{\pi}(1+\cos\alpha) = \frac{230\sqrt{2}}{2\pi}(1+\cos 30^\circ)$$

$$\Rightarrow \quad V = 96.6 \text{ V}$$

$$E_a = V - I_a R_a = 96.6 - 5 \times 2 = 86.6 \text{ V}$$

$$\text{Speed} = N = \frac{E_a}{k_b} = \frac{86.6}{0.25} = 346.4 \text{ V}$$

So, option (a) is closer to 346.4 V.

5.3 (a)

Speed changes from –1500 rpm to 500 rpm in 0.5 sec.

So angular acceleration ,

$$\alpha = \frac{500-(-1500)}{0.5} \times \frac{2\pi}{60} \text{rad/sec}^2$$
$$= 418.88 \text{ rad/sec}^2$$

Torque = T = 20 N-m

$$T = I\alpha$$

$$\text{Moment of inertia, } I = \frac{T}{\alpha} = \frac{20}{418.88} = 0.048 \text{ kgm}^2$$

5.4 (c)

T_S = Starting torque developed by the motor = 15 N-m

T_L = Load torque = 7 N-m

T_a = Accelerating torque, $T_S - T_L = 15 - 7 = 8$ N-m

α = Acceleration = 2 rad/sec^2

$T_a = I\alpha$

$$I = \text{Moment of inertia} = \frac{T_a}{\alpha} = \frac{8}{2} = 4 \text{kg m}^2$$

5.5 (d)

Solar cell emf, E = 350 V

DC current, I_{dc} = 20 A

Solar cell resistance,

$R_{cell} = 0.5\ \Omega$

V_0 = Voltage across inverter

$= -(E - I_{dc}R_{cell}) = -(350 - 20 \times 0.5) = -340$ V

The bridge acts as inverter,

Output voltage of 3-ϕ fully controlled bridge

$$V_0 = \frac{3V_{ml}}{\pi}\cos\alpha \Rightarrow \frac{3V_{ml}}{\pi}\cos\alpha = -340$$

$$\Rightarrow \frac{3 \times 440\sqrt{2}\cos\alpha}{\pi} = -340 \Rightarrow \alpha = 125^\circ$$

Therefore, each thyristor will be reverse biased for a period of 55°.

5.6 (d)

For a 3-ϕ induction motor, stator voltage per phase is given by

$$V_1 = \sqrt{2}\pi f_1 N_{pn_1} \phi k_{\omega_1}$$

$$\phi \propto \frac{V_1}{f_1}$$

To keep the ϕ constant, constant V/f control is applied and voltage (V_1) varies linearly with frequency (f_1). For frequency above 50 Hz, such control is not possible because voltage can not be increased above rated voltage. Therefore, for frequency above 50 Hz, voltage is kept constant, During this control induction motor is said to be working in field weakling mode.

5.7 (a)

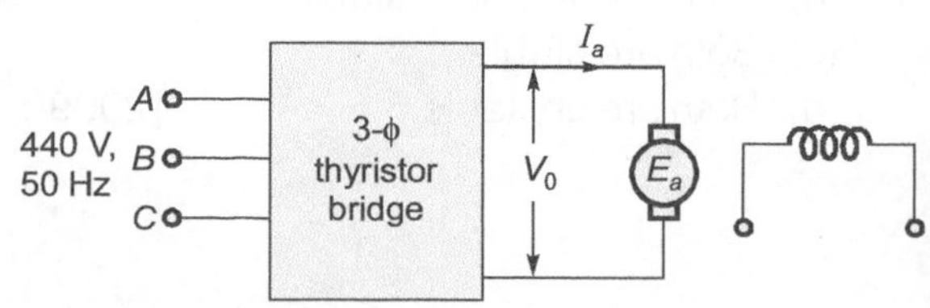

For a separately excited DC motor

Back emf = $E_a = V_0 - I_a R_a$

Since, losses are neglected R_a can be neglected

So, $E_a \approx V_0$

$$V_0 = E_a = k_a \phi N \quad \text{...(i)}$$
$$V_0 \propto N$$

At rated voltage V_0 = 440 V and N = 1500 rpm so,

at half the rated speed. $\left(\frac{N}{2} = 750\text{rpm}\right)$ output voltage of the bridge (V_0) is 220 V.

If I_a is the average value of armature current rms value of supply current will be

$$I_s = I_a\sqrt{\frac{2}{3}}$$

Power delivered to the motor

$$P_0 = V_0 I_a$$

Input VA to the thyristor bridge

$$S_{in} = \sqrt{3} V_s I_s$$

Input power factor

$$= \frac{P_0}{S_{in}} = \frac{V_0 I_a}{\sqrt{3} V_s I_s} = \frac{220 \times I_a}{\sqrt{3} \times 440 \times I_a \sqrt{\frac{2}{3}}} = 0.354$$

5.8 (c)

Average output voltage of the converter,

$$V_0 = \frac{2V_m}{\pi} \cos\alpha$$

Load current, $I_0 = 10$ A

Back emf, $E_b = 150$ V

Armature resistance, $R_a = 2\Omega$

Applying KVL,

$V_0 - 2I_0 + 150 = 0$

$\Rightarrow \quad V_0 = -150 + 2 \times 10 = -130$ V

$$\Rightarrow \quad \frac{2V_m}{\pi} \cos\alpha = -130$$

$$\Rightarrow \quad \frac{2 \times \sqrt{2} \times 230}{\pi} \cos\alpha = -130 \Rightarrow \alpha = 129°$$

5.9 (a)

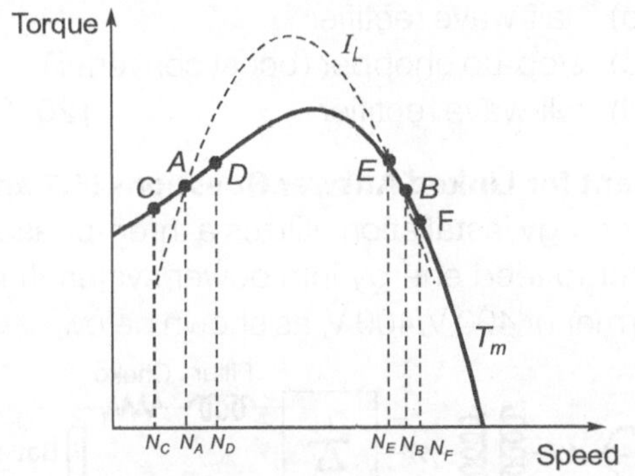

Let torque developed by the motor = T_m

Load torque = T_L

Accelerating torque, $T_a = T_m - T_L$

At point A, $\quad T_m = T_L \Rightarrow T_a = T_m - T_L = 0$

So, no acceleration, hence the motor runs at speed N_A. When a small disturbance decreases speed to N_C.

At point C, $\quad T_m > T_L \Rightarrow T_a > 0$

So, motor will accelerate and comes to point A.

When the disturbance increases speed to N_D

At point D, $\quad T_m < T_L$

So $\quad T_a < 0$

Motor will decelerate and comes to point A

So, motor is stable at point A.

When the motor is operating at point B.

$$T_m = T_L$$

$$\Rightarrow \quad T_a = 0$$

If a small disturbance decreases speed to N_E

$$T_m < T_L$$

$$\Rightarrow \quad T_a < 0$$

Motor will decelerate and motor speed will keep on decreasing.

If the disturbance increases speed to N_F

At point F,

$$T_m > T_L$$

$$\Rightarrow \quad T_a > 0$$

Motor will accelerate and motor speed will keep on increasing. So, motor is unstable at point B.

5.10 (d)

$$V_a = E_b + I_a R_a$$

$$150 = E_b + 20 \times 1 \Rightarrow E_b = 130 \text{ V}$$

For half the rated torque,

$$I_a = 10 \text{ A}$$

$$V_a = 130 + 10 \times 1 = 140$$

$$D = \frac{140}{200} = 0.7$$

5.11 Sol.

$$T_L = 2.78 \times \omega_r$$

After neglecting rotational losses,

$$(T_{sh} = T_a) = T_L$$

At rated load,

$$T_a = 500 \text{ Nm} = 2.78 \times \omega_r$$

$$\omega_r = 179.85$$

$$\omega_r \simeq 180 \text{ rad/sec}$$

6 Miscellaneous

6.1 If a diode is connected in anti-parallel with a thyristor, then
(a) both turn-off power loss and turn-off time decreases.
(b) turn-off power loss decreases but turn-off time increases.
(c) turn-off power loss increases, but turn-off time decreases.
(d) None of the above

[1997 : 1 M]

6.2 Resonant converter's are basically used to
(a) generate large peaky voltage.
(b) reduce the switching losses.
(c) eliminate harmonics.
(d) convert a square wave into a sine wave.

[1999 : 1 M]

6.3 A three phase, wound rotor induction motor is to be operated with slip energy in the constant torque mode, when it delivers an output power P_o at slips. Then theoretically, the maximum power that is available for recovery at the rotor terminals, is equal to
(a) P_o (b) sP_o
(c) $\frac{P_0}{(1-s)}$ (d) $\frac{sP_o}{(1-s)}$

[2000 : 2 M]

6.4 The triac circuit shown in the figure controls the AC output power to the resistive load. The peak power dissipation in the load is

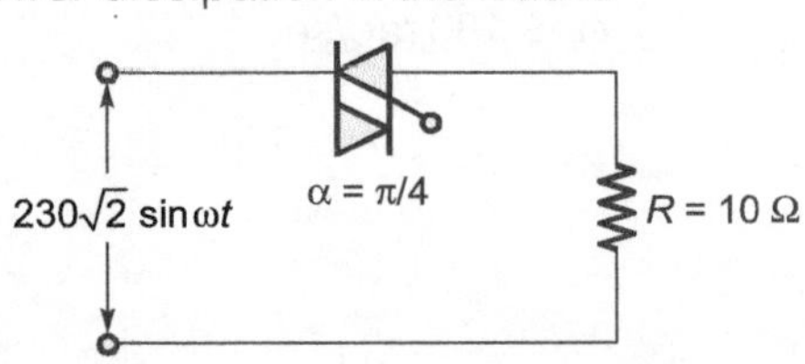

(a) 3968 W (b) 5290 W
(c) 7935 W (d) 10580 W [2004 : 2 M]

6.5 In the single phase voltage controller circuit shown in the figure, for what range of triggering angle (α), the output voltage (V_0) is not controllable?

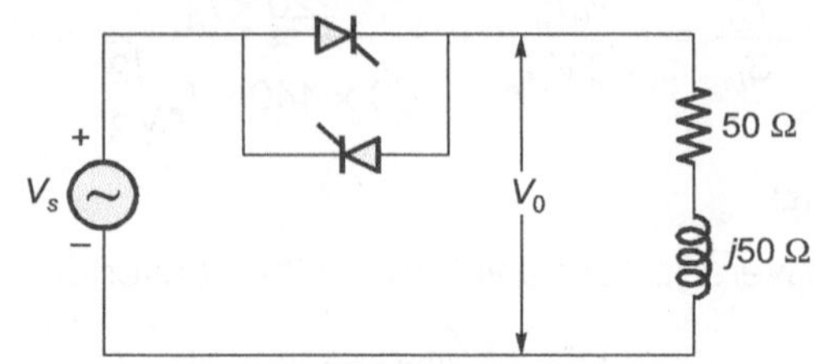

(a) $0° < \alpha < 45°$ (b) $45° < \alpha < 135°$
(c) $90° < \alpha < 180°$ (d) $135° < \alpha < 180°$

[2008 : 1 M]

6.6 The power electronic converter shown in the figure has a single-pole double-throw switch. The pole P of the switch is connected alternately to throws A and B. The converter shown is a

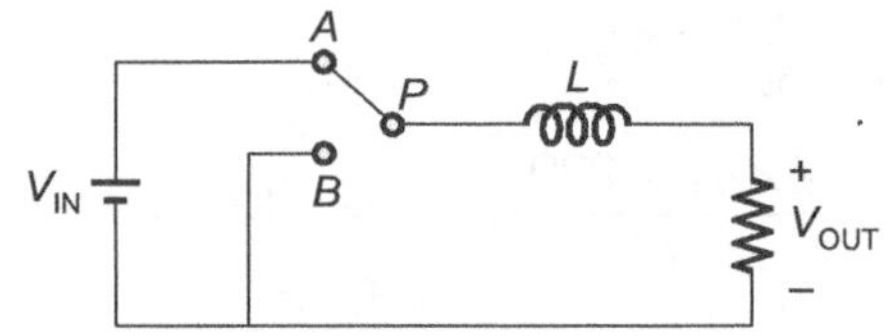

(a) step-down chopper (buck converter)
(b) half-wave rectifier
(c) step-up chopper (boost converter)
(d) full-wave rectifier [2010 : 1 M]

Statement for Linked Answer Questions (6.7 and 6.9):
A solar energy installation utilizes a three-phase bridge converter to feed energy into power system through a transformer of 400 V/400 V, as shown below:

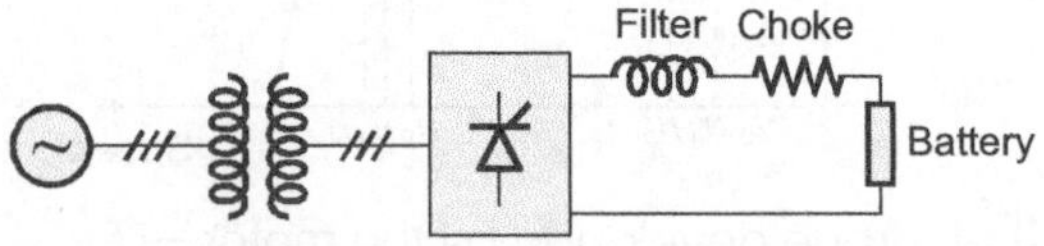

The energy is collected in a bank of 400 V battery and is connected to converter through a large filter choke of resistance 10 Ω.

6.7 The maximum current through the battery will be
(a) 14 A (b) 40 A
(c) 80 A (d) 94 A [2011 : 2 M]

6.8 The kVA rating of the input transformer is
(a) 53.2 kVA (b) 46.0 kVA
(c) 22.6 kVA (d) 19.6 kVA [2011 : 2 M]

6.9 A single-phase SCR based AC regulator is feeding power to a load consisting of 5 Ω resistance and 16 mH inductance. The input supply is 230 V, 50 Hz ac. The maximum firing angle at which the voltage across the device becomes zero all throughout and the rms value of current through SCR, under this operating condition, are

(a) 30° and 46 A (b) 30° and 23 A
(c) 45° and 23 A (d) 45° and 32 A

[2014 : 2 M, Set-2]

6.10 The SCR in the circuit shown has a latching current of 40 mA. A gate pulse of 50 μs is applied to the SCR. The maximum value of R in Ω to ensure successful firing of the SCR is ______.

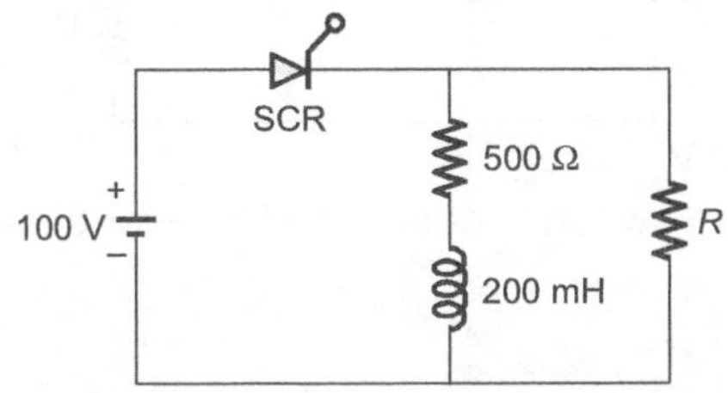

[2014 : 2 M, Set-2]

6.11 Consider a HVDC link which uses thyristor based line-commutated converters as shown in the figure. For a power flow of 750 MW from System 1 to System 2, the voltage at the two ends, and the current are given by: V_1 = 500 kV, V_2 = 485 kV and I = 1.5 kA. If the direction of power flow is to be reversed (that is, from System 2 to System 1) without changing the electrical connections, then which one of the following combinations is feasible?

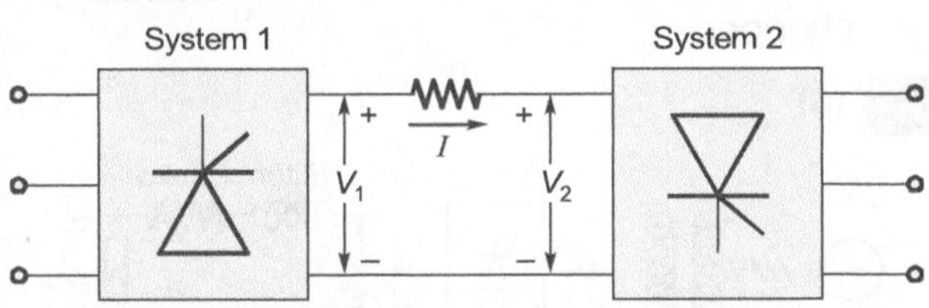

(a) V_1 = –500 kV, V_2 = –485 kV and I = 1.5 kA
(b) V_1 = –485 kV, V_2 = –500 kV and I = 1.5 kA
(c) V_1 = 500 kV, V_2 = 485 kV and I = –1.5 kA
(d) V_1 = –500 kV, V_2 = –485 kV and I = –1.5 kA [2015 : 1 M, Set-1]

Common Data Questions (6.12 and 6.13):

The input voltage given to a converter is

$$v_i = 100\sqrt{2}\sin(100\pi t)\text{ V}$$

The current drawn by the converter is

$$i_1 = (10\sqrt{2}\sin(100\pi t - \pi/3) + 5\sqrt{2}\sin(300\pi t + \pi/4) + 2\sqrt{2}\sin(500\pi t - \pi/6)\ A$$

6.12 The active power drawn by the convert is

(a) 181 W (b) 500 W
(c) 707 W (d) 887 W [2011 : 2 M]

6.13 The input power factor of the converter is

(a) 0.31 (b) 0.44
(c) 0.5 (d) 1 [2011 : 2 M]

6.14 For the circuit shown in the figure below, assume that diodes D_1, D_2 and D_3 are ideal.

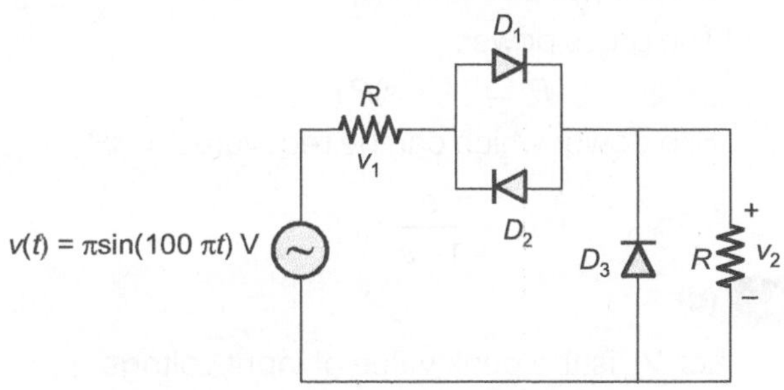

The DC components of voltages v_1 and v_2, respectively are

(a) 0 V and 1 V (b) –0.5 V and 0.5 V
(c) 1 V and 0.5 V (d) 1 V and 1 V

[2017 : 1 M, Set-1]

6.15 Thyristor T_1 is triggered at an angle α (in degree), and T_2 at angle 180° + α, in each cycle of the sinusoidal input voltage. Assume both thyristors to be ideal. To control the load power over the range 0 to 2 kW, the minimum range of variation in α is

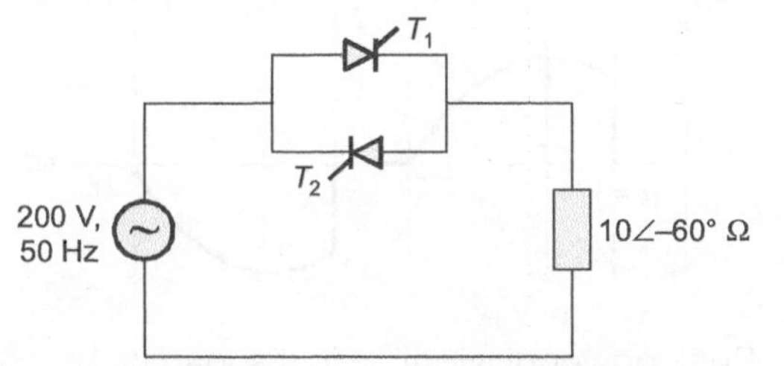

(a) 0° to 60° (b) 60° to 180°
(c) 0° to 120° (d) 60° to 120°

[2020 : 1 M]

6.16 A single-phase triac based AC voltage controller feeds a series RL load. The input AC supply is 230 V, 50 Hz. The values of R and L are 10 Ω and 18.37 mH, respectively. The minimum triggering angle of the triac to obtain controllable output voltage is

(a) 30° (b) 15°
(c) 45° (d) 60°

[2024 : 1 M]

Answers Miscellaneous

6.1	(b)	6.2	(b)	6.3	(d)	6.4	(d)	6.5	(a)	6.6	(a)	6.7	(b)
6.8	(c)	6.9	(c)	6.10	(6060)	6.11	(b)	6.12	(b)	6.13	(b)	6.14	(b)
6.15	(b)	6.16	(a)										

Explanations Miscellaneous

6.3 (d)

If the input power is P_{in},

The gross power,

$$P_0 = (1-s)P_{in}$$

Slip power which can be recovered $= sP_{in}$

$$= \frac{s}{1-s} P_0$$

6.4 (d)

Let V_m is the peak value of input voltage

$$V_m = 230\sqrt{2}\text{ V}$$

As firing angle $= \alpha = \frac{\pi}{4}$, so peak voltage across resistance load is also V_m.

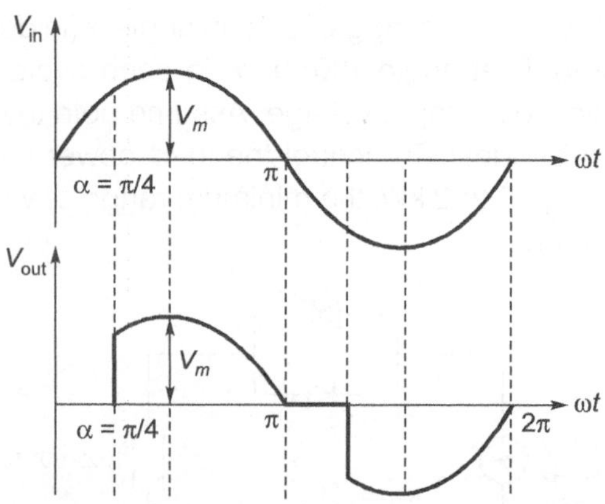

Peak power dissipation in the load

$$P = \frac{V_m^2}{R} = \frac{(230\sqrt{2})^2}{10} = 10580\text{ W}$$

6.5 (a)

Impedance of the load

$$Z_L = R_L + jX_L = 50 + j50\ \Omega = 50\sqrt{2}\angle 45°$$

$$\phi_L = 45°$$

The output voltage (V_0) is controllable when $\alpha \geq \phi_L$, so for $0° < \alpha < 45°$, output voltage is not controllable.

6.6 (a)

When switch is connected to A for time duration T_1

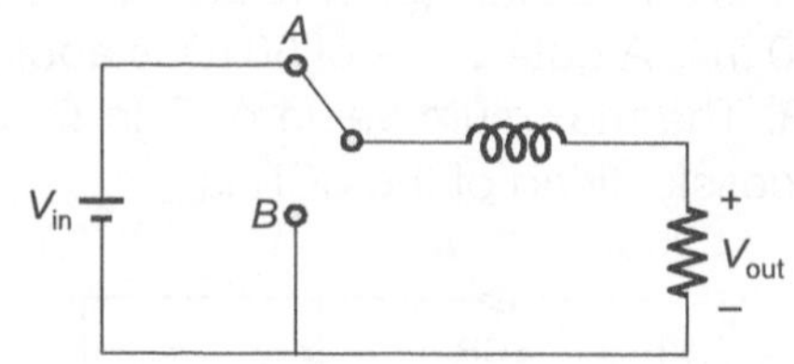

$$V_{out} = V_{in}$$

When switch is connected to B for time duration T_2

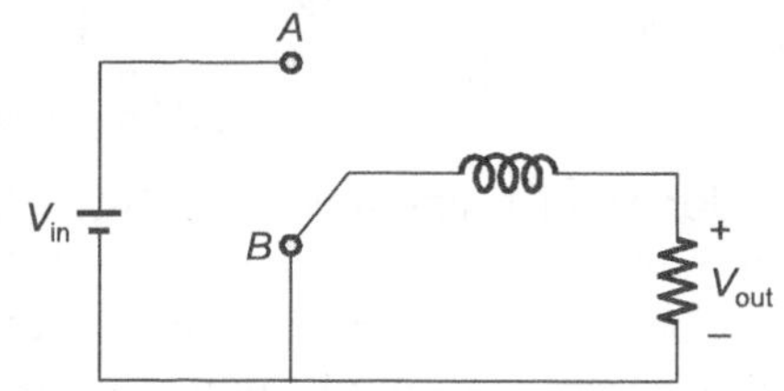

$$\text{Average output voltage} = \frac{V_{in}T_1}{T_1+T_2} = \alpha V_{in}$$

$$\text{Where, } \alpha = \text{duty cycle} = \frac{T_1}{T_1+T_2}$$

Therefore, the converter shown is a step down chopper.

6.7 (b)

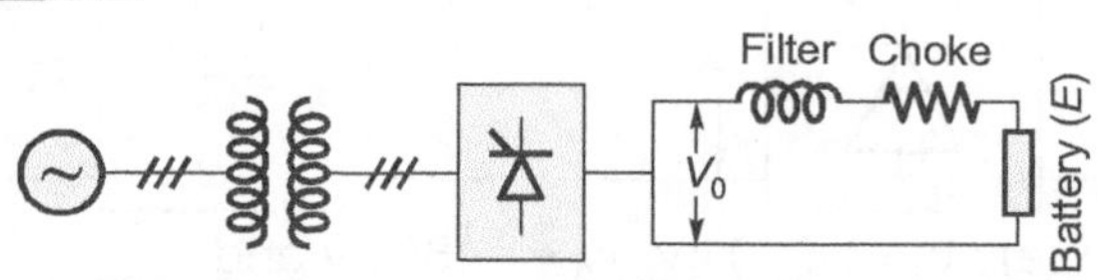

Average output voltage of the converter,

$$V_0 = \frac{3V_{ml}}{\pi}\cos\alpha$$

The converter acts as line commutated inverter and for such mode $\alpha > 90°$ and V_0 is negative. Therefore, battery supplies energy to AC system. So, current through battery

$$I_0 = \frac{400 - V_0}{R}$$

For $V_0 = 0$ or $\alpha = 90°$,

maximum current flow through battery

$$(I_0)_{max} = \frac{400}{10} = 40\text{ A}$$

6.8 (c)

RMS value of supply current in case of 3-ϕ bridge converter

$$I_s = I_0\sqrt{\frac{2}{3}} = 40\sqrt{\frac{2}{3}} = 32.66 \text{ A}$$

kVA rating of the input transformer = $\sqrt{3}V_S I_S$

$= \sqrt{3} \times 400 \times 32.66 \times 10^{-3} \text{kVA} = 22.62 \text{ kVA}$

6.9 (c)

The maximum firing angle at which the voltage across the device becomes 'ϕ' = load angle.

$$\phi = \tan^{-1}\left(\frac{\omega L}{R}\right)$$

$$= \tan^{-1}\left(\frac{2\pi \times 50 \times 16 \times 10^{-3}}{5}\right)$$

$$\boxed{\phi = 45°.15 \simeq 45°}$$

Rms value of current through SCR is

$$I_{T_{rms}} = \sqrt{\left[\frac{1}{2\pi}\int_{\alpha}^{\pi+\alpha}\left\{\frac{V_m}{z}\sin(\omega t - \phi)\right\}^2 d(\omega t)\right]}$$

$$= \sqrt{\frac{V_m^2}{2\pi z^2}\int_{\alpha}^{\pi+\alpha}\left[\frac{1-\cos 2(\omega t - \phi)}{2}\right] d(\omega t)}$$

$$= \sqrt{\frac{V_m^2}{2 \times 2\pi z^2}\left[\pi - \frac{\sin 2(\omega t - \phi)}{2}\Bigg|_{\alpha}^{(\pi+\alpha)}\right]}$$

$$= \frac{V_m}{2z\sqrt{\pi}}\sqrt{\left[\pi + \left(\frac{-\sin 2(\pi + \alpha - \alpha) + \sin 2(\alpha - \alpha)}{2}\right)\right]}$$

$$= \frac{V}{2z\sqrt{\pi}}\sqrt{\pi} = \frac{V_m}{2z}$$

$$I_{T_{rms}} = \frac{230\sqrt{2}}{2 \times \sqrt{5^2(2\pi \times 50 \times 16 \times 10^{-3})^2}}$$

$= 22.93 \simeq 23 \text{ A}$

$$\boxed{I_{T_{rms}} = 23 \text{ A}}$$

6.10 Sol.

Let us assume the SCR is conducting,

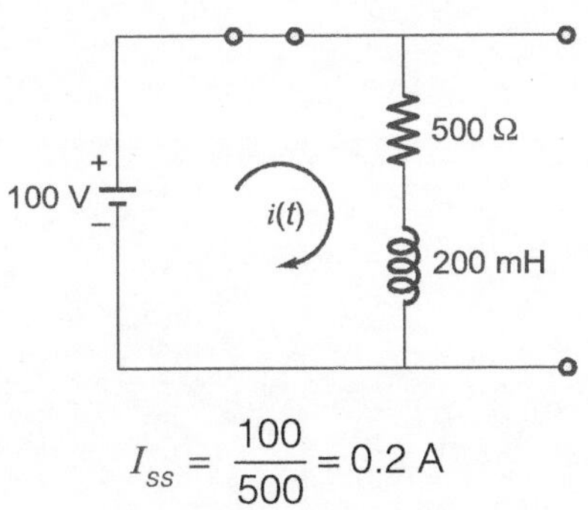

$$I_{ss} = \frac{100}{500} = 0.2 \text{ A}$$

[∵ Inductor will be short circuited in DC]

$$i(t) = I_{ss}(1 - e^{-t/\tau})$$

$$\tau = \frac{L}{R} = \frac{200 \times 10^{-3}}{500} = 4 \times 10^{-4} \text{ sec}$$

Given, $t = 50 \times 10^{-6}$ sec

$$\therefore \quad i(t) = 0.2\left(1 - e^{-\frac{50 \times 10^{-6}}{4 \times 10^{-4}}}\right) = 23.5 \text{ mA}$$

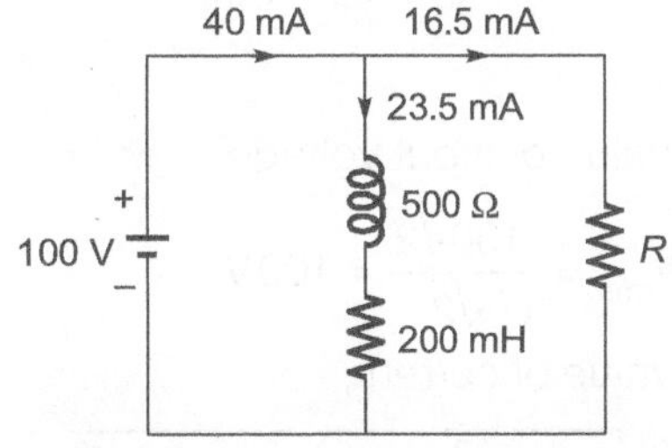

$$V = I \times R$$

$$R = \frac{V}{I} = \frac{100}{16.5 \times 10^{-3}} = 6060\ \Omega$$

6.11 (b)

To maintain the direction of power flow from system 2 to system 1 the voltage V_1 = –485 kV and voltage V_2 = –500 kV and I = 1.5 kA.

Since, current cannot flow in reverse direction. Option (b) is correct answer.

6.12 (b)

$$V_i = 100\sqrt{2}\sin(100\pi t) \text{ V}$$

$$i_i = 10\sqrt{2}\sin\left(100\pi t - \frac{\pi}{3}\right) +$$

$$5\sqrt{2}\sin\left(300\pi t + \frac{\pi}{4}\right) +$$

$$2\sqrt{2}\sin\left(500\pi t - \frac{\pi}{4}\right) \text{ A}$$

Fundamental component of input voltage

$$(V_i)_1 = 100\sqrt{2}\sin(100\pi t)$$

$$(V_i)_{1,\text{rms}} = \frac{100\sqrt{2}}{\sqrt{2}} = 100 \text{ V}$$

Fundamental component of current

$$(i_L)_1 = 10\sqrt{2}\sin\left(100\pi t - \frac{\pi}{3}\right)$$

$$(i_L)_{1,\text{rms}} = \frac{10\sqrt{2}}{\sqrt{2}} = 10$$

Phase difference between these two components

$$\phi_1 = \frac{\pi}{3}, \cos\phi_1 = \cos\frac{\pi}{3} = 0.5$$

Active power due to fundamental components

$$P_1 = (V_i)_{1,\,rms} \times (i_i)_{1,\,rms} \cos\phi_1$$
$$= 100 \times 10 \times 0.5 = 500 \text{ W}$$

Since, 3rd and 5th harmonics are absent in input voltage, there is no active power due to the these components.

Hence, active power drawn by the converter

P_0 = Active power due to fundamental components = 500 W

6.13 (b)

Rms value of input voltage,

$$V_{rms} = \frac{100\sqrt{2}}{\sqrt{2}} = 100\text{V}$$

Rms value of current,

$$I_{rms} = \sqrt{\left(\frac{10\sqrt{2}}{\sqrt{2}}\right)^2 + \left(\frac{5\sqrt{2}}{\sqrt{2}}\right)^2 + \left(\frac{2\sqrt{2}}{\sqrt{2}}\right)^2} = 11.358 \text{ A}$$

Let input power factor $\cos\phi$

$V_{rms} I_{rms} \cos\phi$ = active power drawn by the converter

$$\Rightarrow \quad 100 \times 11.358 \times \cos\phi = 500 \text{ W}$$
$$\Rightarrow \quad \cos\phi = 0.44$$

6.14 (b)

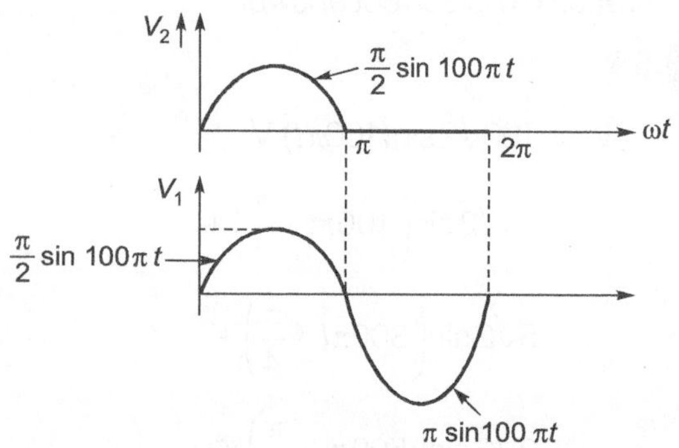

$$V_{2\,avg} = \frac{V_m}{\pi} = \frac{\pi/2}{\pi} = \frac{1}{2} = 0.5 \text{ V}$$

$$V_{1\,avg} = \frac{1}{2\pi}\left(\int_0^{\pi} \frac{\pi}{2} \cdot \sin 100\pi t \cdot d(\omega t) + \int_{\pi}^{2\pi} \pi \sin 100\pi t \cdot d(\omega t)\right)$$
$$= -0.5 \text{ V}$$

6.16 (a)

Output voltage is controlled if

$$\alpha > \theta$$

Given: f = 50 Hz, R = 10Ω, L = 18.37 mH

$$\theta = \tan^{-1}\left(\frac{\omega L}{R}\right)$$

$$\theta = \tan^{-1}\left(\frac{2\pi \times 50 \times 18.37 \times 10^{-3}}{10}\right)$$

$$\theta = 29.989°$$

So, α_{min} can be 30°. Hence answer is (a).

Electromagnetic Theory

UNIT

X

CONTENTS

Electromagnetic Theory

Syllabus

Coulomb's Law, Electric Field Intensity, Electric Flux Density, Gauss's Law, Divergence, Electric Field and Potential due to Point, Line, Plane and Spherical Charge Distributions, Effect of Dielectric Medium, Capacitance of Simple Configurations, Biot Savart's Law, Ampere's Law,Curl, Faraday's Law, Lorentz Force, Inductance, Magnetomotive Force, Reluctance, Magnetic Circuits, Self and Mutual Inductance of Simple Configurations.

Analysis of Previous GATE Papers

Exam Year	1 Mark Ques.	2 Marks Ques.	4 Marks Ques.	5 Marks Ques.	Total Marks
1995	1	–	–	–	**1**
1996	2	–	–	–	**2**
1997	1	3	–	3	**22**
1998	2	–	–	2	**12**
1999	2	–	–	1	**7**
2000	1	–	–	1	**6**
2001	–	3	–	–	**6**
2002	1	–	–	1	**6**
2003	2	3	–	–	**8**
2004	1	–	–	–	**1**
2005	1	–	–	–	**1**
2006	–	1	–	–	**2**
2007	1	4	–	–	**9**
2008	1	3			**7**
2009	–	–			**–**
2010	–	–			**–**
2011	–	1			**2**

Exam Year	1 Mark Ques.	2 Mark Ques.	Total Marks
2012	–	1	**2**
2013	1	1	**3**
2014 Set-1	1	1	**3**
2014 Set-2	1	1	**3**
2014 Set-3	1	1	**3**
2015 Set-1	3	1	**5**
2015 Set-2	3	1	**5**
2016 Set-1	2	1	**4**
2016 Set-2	2	2	**6**
2017 Set-1	2	1	**4**
2017 Set-2	2	1	**4**
2018	1	1	**3**
2019	1	–	**1**
2020	–	3	**6**
2021	2	2	**6**
2022	1	2	**5**
2023	–	2	**4**
2024	–	2	**4**

1 Coordinate Systems and Vector Calculus

1.1 Given a vector field $\vec{F}$, the divergence theorem states that

(a) $\oint_s \vec{F}\cdot d\vec{s} = \int_V \nabla\cdot\vec{F}\,dV$

(b) $\int_s \vec{F}\cdot d\vec{s} = \int_V \nabla\times\vec{F}\,dV$

(c) $\int_s \vec{F}\times d\vec{s} = \int_V \nabla\cdot\vec{F}\,dV$

(d) $\int_s \vec{F}\times d\vec{s} = \int_V \nabla\times\vec{F}\,dV$ **[2002 : 1 M]**

1.2 If $\overline{E}$ is the electric field intensity, $\nabla\cdot(\nabla\times\overline{E})$ is equal to

(a) $\overline{E}$ (b) $|\overline{E}|$
(c) null vector (d) zero **[2005 : 1 M]**

1.3 Divergence of the vector field

$V(x,y,z) = -(x\cos xy + x)\hat{i} + (y\cos xy)\hat{j} + (\sin z^2 + x^2 + y^2)\hat{k}$ is

(a) $2z\cos z^2$ (b) $\sin xy + 2z\cos z^2$
(c) $x\sin xy - \cos z$ (d) None of these
[2007 : 1 M]

1.4 The direction of vector A is radially outward from the origin, with $|A| = kr^n$ where $r^2 = x^2 + y^2 + z^2$ and k is a constant. The value of n for which $\nabla\cdot A = 0$ is

(a) –2 (b) 2
(c) 1 (d) 0 **[2012 : 2 M]**

1.5 Consider a function $\overline{f} = \frac{1}{r^2}\hat{r}$, where r is the distance from the origin and $\hat{r}$ is the unit vector in the radial direction. The divergence of this function over a sphere of radius R, which includes the origin, is

(a) 0 (b) 2π
(c) 4π (d) $R\pi$
[2015 : 1 M, Set-1]

1.6 Match **List-I** with **List-II** the following:

List-I
A. Stoke's Theorem
B. Gauss's Theorem
C. Divergence Theorem
D. Cauchy's Integral Theorem

List-II
1. $\oiint D\cdot ds = Q$
2. $\oint f(z)dz = 0$
3. $\iiint(\nabla\cdot A)dv = \oiint A\cdot ds$
4. $\iint(\nabla\times A)\cdot ds = \oint A\cdot dl$

Codes:

	A	B	C	D
(a)	2	1	4	3
(b)	4	1	3	2
(c)	4	3	1	2
(d)	3	4	2	1

[2015 : 1 M, Set-2]

1.7 In cylindrical coordinate system, the potential produced by a uniform ring charge is given by $f = f(r, z)$, where f is a continuous function of r and z. Let $\vec{E}$ be the resulting electric field. Then the magnitude of $\nabla\times\vec{E}$

(a) increases with r
(b) is 0
(c) is 3
(d) decreases with z **[2016 : 1 M, Set-1]**

1.8 The line integral of the vector field $F = 5xz\hat{i} + (3x^2 + 2y)\hat{j} + x^2z\hat{k}$ along a path from (0, 0, 0) to (1, 1, 1) parameterized by (t, t^2, t) is ________.
[2016 : 2 M, Set-2]

1.9 The figures show diagrammatic representations of vector fields $\vec{X}, \vec{Y}$ and $\vec{Z}$, respectively. Which one of the following choices is true?

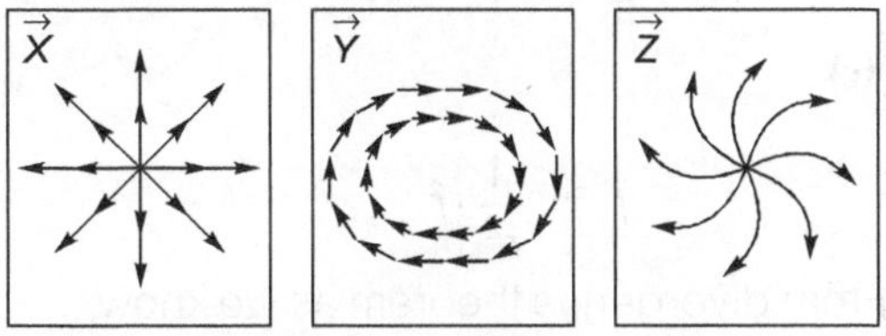

(a) $\nabla\cdot\vec{X} = 0, \nabla\times\vec{Y} \neq 0, \nabla\times\vec{Z} = 0$

(b) $\nabla\cdot\vec{X} \neq 0, \nabla\times\vec{Y} = 0, \nabla\times\vec{Z} \neq 0$

(c) $\nabla\cdot\vec{X} \neq 0, \nabla\times\vec{Y} \neq 0, \nabla\times\vec{Z} \neq 0$

(d) $\nabla\cdot\vec{X} = 0, \nabla\times\vec{Y} = 0, \nabla\times\vec{Z} = 0$

[2017 : 1 M, Set-2]

1.10 Let a_r, a_ϕ and a_z be unit vectors along r, ϕ and z directions, respectively in the cylindrical coordinate system. For the electric flux density given by $D = (a_r 15 + a_\phi 2r - a_z 3rz)$ Coulomb/m^2, the total electric flux, in Coulomb, emanating from the volume enclosed by a solid cylinder of radius 3 m and height 5 m oriented along the z-axis with its base at the origin is:

(a) 108π (b) 54π
(c) 90π (d) 180π [2020 : 2 M]

Answers Coordinate Systems & Vector Calculus

1.1	(a)	1.2	(d)	1.3	(a)	1.4	(a)	1.5	(c)	1.6	(b)	1.7	(b)
1.8	(4.41)	1.9	(c)	1.10	(180)								

Explanations Coordinate Systems & Vector Calculus

1.2 (d)

Divergence of a curl field is always zero,

i.e., $\nabla\cdot(\nabla\times\vec{E}) = 0$

1.3 (a)

$$V(x, y, z) = -(x\cos xy + y)i + (y\cos xy)j \times [\sin(z^2) + (x^2) + (y^2)]\, k$$

$$\text{Divergence} = \nabla\cdot V = \frac{\partial V_x}{\partial x} + \frac{\partial V_y}{\partial y} + \frac{\partial V_z}{\partial z}$$

$$= -\cos xy + x(\sin xy)y + \cos xy - y\sin(xy)x + 2z\cos z^2$$

$$= 2z\cos z^2$$

1.4 (a)

$$\vec{A} = kr^n\,\hat{i}_r$$

$$\nabla\cdot\vec{A} = \frac{1}{r^2}\frac{\partial}{\partial r}(r^2\cdot kr^n)$$

$$= \frac{1}{r^2}\cdot\frac{\partial}{\partial r}(kr^{n+2})$$

$$= (n+2)\,k\cdot\frac{r^{n+1}}{r^2} = (n+2)k\,r^{n-1}$$

For $\nabla\cdot\vec{A} = 0$

$n+2 = 0 \Rightarrow n = -2$

1.5 (c)

$$\bar{f} = \frac{1}{r^2}\cdot\hat{r}$$

From divergence theorem as we know,

$$\int_{\text{vol.}}(\nabla\cdot\bar{f})dV = \oiint_S \bar{f}\cdot d\bar{S}$$

$$\oiint_S \bar{f}\cdot d\bar{S} = \oiint_S\left(\frac{1}{r^2}\cdot\hat{r}\right)\times r^2\sin\theta\cdot d\theta\cdot d\phi\cdot\hat{r}$$

$$= \oiint_S \sin\theta\cdot d\theta\cdot d\phi = 4\pi$$

1.6 (b)

Stokes theorem $\oint\bar{A}\cdot dl = \iint(\nabla\times A)\cdot\hat{n}ds$

Gauss theorem $\iint D\cdot ds = Q$

Divergence theorem

$$\oiint A\cdot\hat{n}ds = \iiint\nabla\cdot\bar{A}\,dV$$

Cauchys integral theorem $\oint_c f(z)dz = 0$

1.7 (b)

V is given as static field in time invariant.

Hence, $\nabla\times E = 0$

1.8 Sol.

$$E = 5xz\bar{i} + (3x^2+2y)\,\bar{j} + x^2z\bar{k}$$

$$= \int_C \bar{F}.\bar{d}r$$

$$= \int_C 5xz\,dx + (3x^2+2y)\,dy + x^2z\,dz$$

$x = t,\quad y = t^2,\quad z = t,\ t = 0$ to 1

$dx = dt$

$dy = 2t\,dt,\quad dz = dt$

$$= \int_0^1 5t^2dt + (3t^2+2t^2)2t\,dt + t^3dt$$

$$= \int_0^1(5t^2+11t^3)dt$$

$$= \left[\frac{5t^3}{3}+\frac{11t^4}{4}\right]_0^1 = \frac{5}{3}+\frac{11}{4} = \frac{53}{12} = 4.41$$

1.9 (c)

$\vec{X}$ is going away so $\vec{\nabla}\cdot\vec{X}\neq 0$.

$\vec{Y}$ is moving circulator direction so $\vec{\nabla}\times\vec{Y}\neq 0$.

$\vec{Z}$ has circular rotation so $\vec{\nabla}\times\vec{Z}\neq 0$.

1.10 Sol.

ψ/crossing closed surface

$$= \oint\!\!\oint \vec{D}\cdot\vec{ds} = \iiint (\vec{\nabla}\cdot\vec{D})\, dv \quad \ldots\text{(i)}$$

$$\vec{\nabla}\cdot\vec{D} = \frac{1}{\rho}\frac{\partial}{\partial\rho}(\rho D_\rho) + \frac{1}{\rho}\frac{\partial D_\phi}{\partial\phi} + \frac{\partial D_z}{\partial z}$$

$$= \frac{1}{\rho}\frac{\partial}{\partial\rho}(\rho 15) + \frac{1}{\rho}\frac{\partial}{\partial\phi}(2\rho) + \frac{\partial}{\partial z}(-3\rho z) = \frac{1}{\rho}15 - 3\rho$$

$$\iiint (\vec{\nabla}\cdot\vec{D})\, dv$$

$$= \iiint \left(\frac{15}{\rho} - 3\rho\right)\rho\, d\rho\, d\phi\, dz$$

$$= \iiint 15\, d\rho\, d\phi\, dz - 3\iiint \rho^2\, d\rho\, d\phi\, dz$$

$$= 15\int_{\rho=0}^{3} d\rho \int_{\phi=0}^{2\pi} d\phi \int_{z=0}^{5} -3\int_{\rho=0}^{3}\rho^2 d\rho \int_{\phi=0}^{2\pi} d\phi \int_{z=0}^{5} dz$$

$$= 15(3-0)(2\pi)(5) - 3\left(\frac{3^3}{3}\right)\times(2\pi)(5)$$

$$= 45(10\pi) - 27(10\pi) = 180\,\pi\text{C}$$

■■■■

2 Electrostatic Fields

2.1 If *V*, *W*, *q* stand for voltage, energy and charge, then *V* can be expressed as

(a) $V = \frac{dq}{dW}$ (b) $V = \frac{dW}{dq}$

(c) $dV = \frac{dW}{dq}$ (d) $dV = \frac{dq}{dW}$

[1996 : 1 M]

2.2 In a uniform electric field, field lines at equipotential surfaces

(a) are parallel to one another
(b) intersect at 45°
(c) intersect at 30°
(d) are orthogonal [1997 : 1 M]

2.3 The capacitance of the arrangement shown in the figure is ____ pF.

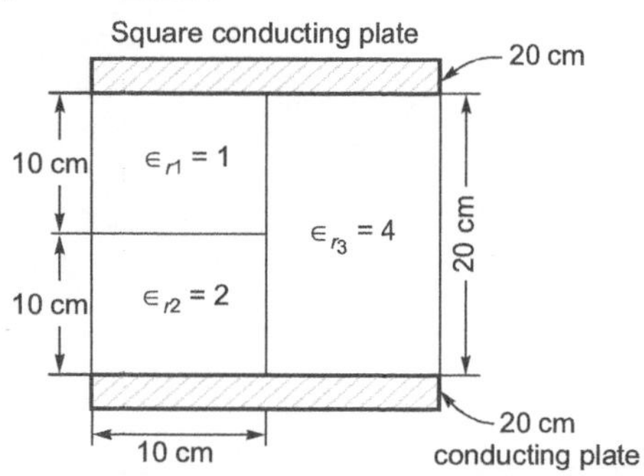

[1997 : 2 M]

2.4 The capacitance of an isolated sphere of radius 10 cm in air is equal to _____ pF. [1997 : 2 M]

2.5 Given the potential function in free space to be $V(x) = (50\,x^2 + 50\,y^2 + 50\,z^2)$ volts, the magnitude (in volts/meter) and the direction of the electric field at a point (1, –1, 1), where the dimensions are in meters, are

(a) $100; (\hat{i} + \hat{j} + \hat{k})$

(b) $\frac{100}{\sqrt{3}}; (\hat{i} - \hat{j} + \hat{k})$

(c) $100\sqrt{3}; [(\hat{i} + \hat{j} - \hat{k})/\sqrt{3}]$

(d) $100\sqrt{3}; [(-\hat{i} + \hat{j} - \hat{k})/\sqrt{3}]$ [2001 : 2 M]

2.6 The electric field $\vec{E}$ (in volts/meter) at the point (1, 1, 0) due to a point charge of +1 μC located at (–1, 1, 1) (co-ordinates in meters) is

(a) $\frac{10^{-6}}{20\sqrt{5}\,\pi\varepsilon_0}(2i - k)$ (b) $\frac{10^{-6}}{20\pi\,\varepsilon_0}(2i - k)$

(c) $\frac{-10^{-6}}{20\sqrt{5}\,\pi\,\varepsilon_0}(2i - k)$ (d) $\frac{-10^{-6}}{20\pi\,\varepsilon_0}(2i - k)$

[2001 : 2 M]

2.7 A point charge of + 1 nC is placed in a space with permittivity of 8.85×10^{-12} F/m as shown in the figure. The potential difference V_{PQ} between two points *P* and *Q* at distances of 40 mm and 20 mm respectively from the point charge is

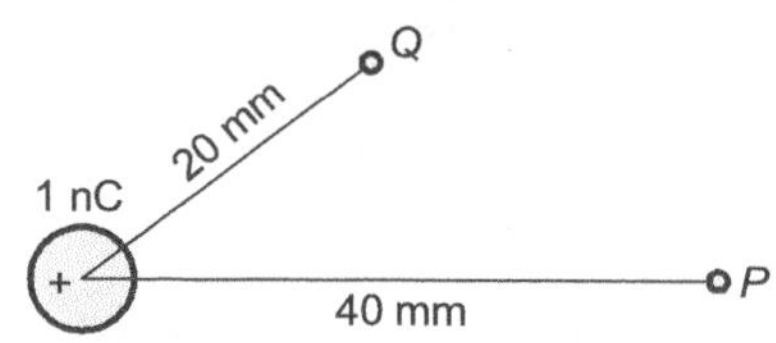

(a) 0.22 kV (b) –225 V
(c) –2.24 kV (d) 15 V [2003 : 2 M]

2.8 A parallel plate capacitor has an electrode area of 100 mm², with spacing of 0.1 mm between the electrodes. The dielectric between the plates is air with a permittivity of 8.85×10^{-12} F/m. The voltage on the capacitor is 100 V. The stored energy in the capacitor is

(a) 8.85 pJ (b) 440 pJ
(c) 22.1 nJ (d) 44.3 nJ [2003 : 2 M]

2.9 A composite parallel plate capacitor is made up of two different dielectric materials with different thickness (t_1 and t_2) as shown in the figure. The two different dielectric materials are separated by a conducting foil *F*. The voltage of the conducting foil is

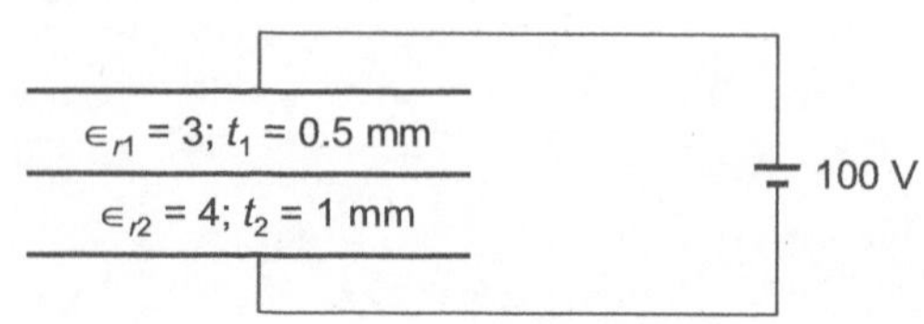

(a) 52 V (b) 60 V
(c) 67 V (d) 33 V [2003 : 2 M]

2.10 A parallel plate capacitor is shown in the figure. It is made of two square metal plates of 400 mm side. The 14 mm space between the plates is filled with two layers of dielectrics of $\varepsilon_r = 4$, 6 mm thick and $\varepsilon_r = 2$, 8 mm thick. Neglecting fringing of fields at the edges the capacitance is

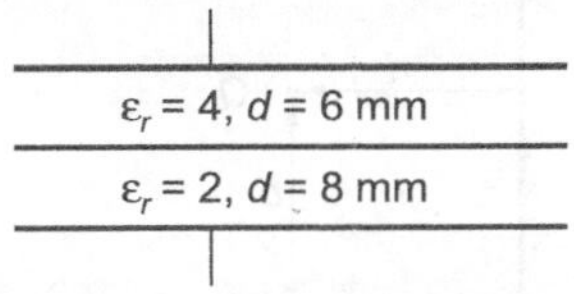

(a) 1298 pF (b) 944 pF
(c) 354 pF (d) 257 pF [2004 : 1 M]

2.11 A solid sphere made of insulating material has a radius R and has a total charge Q distributed uniformly in its volume. What is the magnitude of the electric field intensity, E, at a distance $r(0 < r < R)$ inside the sphere?

(a) $\frac{1}{4\pi\varepsilon_0}\frac{Qr}{R^3}$ (b) $\frac{3}{4\pi\varepsilon_0}\frac{Qr}{R^3}$
(c) $\frac{1}{4\pi\varepsilon_0}\frac{Q}{r^2}$ (d) $\frac{1}{4\pi\varepsilon_0}\frac{QR}{r^3}$ [2007 : 2 M]

2.12 Two point charges $Q_1 = 10\mu C$ and $Q_2 = 20\,\mu C$ are placed at coordinates (1,1,0) and (–1, –1, 0) respectively. The total electric flux passing through a plane $z = 20$ will be
(a) 7.5 μC (b) 13.5 μC
(c) 15.0 μC (d) 22.5 μC [2008 : 2 M]

2.13 A capacitor consists of two metal plates each $500 \times 500\ mm^2$ and spaced 6 mm apart. The space between the metal plates is filled with a glass plate of 4 mm, thickness and a layer of paper of 2 mm thickness. The relative primitivities of the glass and paper are 8 and 2 respectively. Neglecting the fringing effect, the capacitance will be
(Given that $\varepsilon_0 = 8.85 \times 10^{-12}$ F/m)
(a) 983.3 pF (b) 1475 pF
(c) 637.5 pF (d) 9956.25 pF [2008 : 2 M]

2.14 A capacitor is made with a polymeric dielectric having an ε_r of 2.26 and a dielectric breakdown strength of 50 kV/cm. The permittivity of free space is 8.85 pF/m. If the rectangular plates of the capacitor have a width of 20 cm and a length of 40 cm, then the maximum electric charge in the capacitor is
(a) 2 μC (b) 4 μC
(c) 8 μC (d) 10 μC [2011 : 2 M]

2.15 A dielectric slab with 500 mm × 500 mm cross-section is 0.4 m long. The slab is subjected to a uniform electric field of $E = 6\boldsymbol{a}_x + 8\boldsymbol{a}_y$ kV/mm. The relative permittivity of the dielectric material is equal to 2. The value of constant ε_0 is 8.85×10^{-12} F/m. The energy stored in the dielectric in Joules is
(a) 8.85×10^{-11} (b) 8.85×10^{-5}
(c) 88.5 (d) 885 [2013 : 2 M]

2.16 C_0 is capacitance of a parallel plate capacitor with air as dielectric (as in figure (a)). If, half of the entire gap as shown in figure (b) is filled with a dielectric of permittivity ε_r, the expression for the modified capacitance is

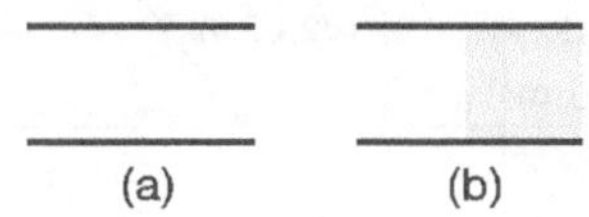

(a) $\frac{C_0}{2}(1+\epsilon_r)$ (b) $(C_0 + \epsilon_r)$
(c) $\frac{C_0}{2}\epsilon_r$ (d) $C_0(1+\epsilon_r)$
[2014 : 1 M, Set-1]

2.17 A parallel plate capacitor consisting two dielectric material is shown in the figure. The middle dielectric slab is placed symmetrically with respect to the plates.

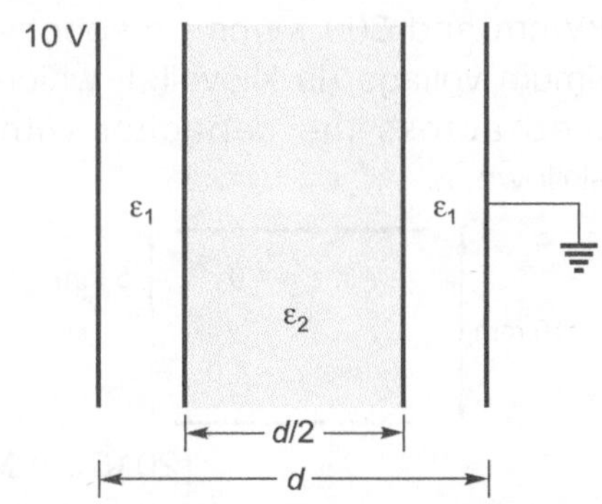

If the potential difference between one of the plates and the nearest surface of dielectric interface is 2 volts, then the ratio $\varepsilon_1 : \varepsilon_2$ is
(a) 1 : 4 (b) 2 : 3
(c) 3 : 2 (d) 4 : 1
[2014 : 1 M, Set-2]

2.18 A hollow metallic sphere of radius r is kept at potential of 1 volt. The total electric flux coming out of the concentric spherical surface of radius $R (> r)$ is
(a) $4\pi\varepsilon_0 r$ (b) $4\pi\varepsilon_0 r^2$
(c) $4\pi\varepsilon_0 R$ (d) $4\pi\varepsilon_0 R^2$
[2014 : 1 M, Set-3]

2.19 A perfectly conducting metal plate is placed in x-y plane in a right handed coordinate system. A charge of $+32\pi\epsilon_0\sqrt{2}$ Coulomb is placed at co-ordinate (0, 0, 2). ϵ_0 is the permittivity of free space. Assume $\hat{i}, \hat{j}, \hat{k}$ to be unit vectors along x,

y and z axes respectively. At the co-ordinate $(\sqrt{2}, \sqrt{2}, 0)$, the electric field vector $\vec{E}$ (Newtons/Coulomb) will be

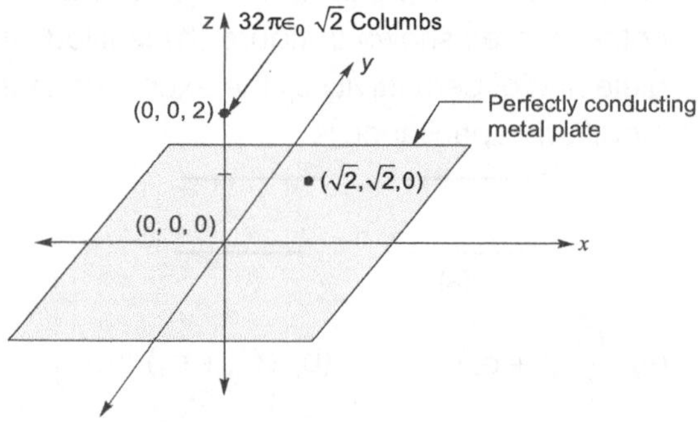

(a) $2\sqrt{2}\,\hat{k}$
(b) $-2\hat{k}$
(c) $2\hat{k}$
(d) $-2\sqrt{2}\,\hat{k}$

[2014 : 2 M, Set-3]

2.20 A parallel plate capacitor is partially filled with glass of dielectric constant 4.0 as shown below. The dielectric strengths of air and glass are 30 kV/cm and 300 kV/cm, respectively. The maximum voltage (in kilovolts), which can be applied across the capacitor without any breakdown, is ____.

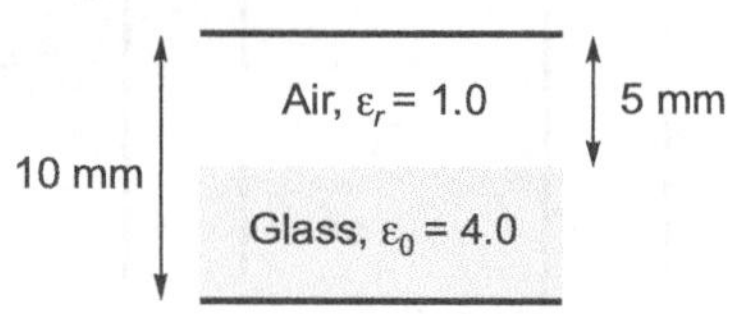

[2015 : 2 M, Set-1]

2.21 Two semi-infinite dielectric regions are separated by a plane boundary at $y = 0$. The dielectric constants of region 1$(y < 0)$ and region 2$(y > 0)$ are 2 and 5, respectively. Region 1 has uniform electric field $\vec{E} = 3\hat{a}_x + 4\hat{a}_y + 2\hat{a}_z$, where $\hat{a}_x, \hat{a}_y$ and $\hat{a}_z$ are unit vectors along the x, y and z axes, respectively. The electric field in region 2 is

(a) $3\hat{a}_x + 1.6\hat{a}_y + 2\hat{a}_z$
(b) $1.2\hat{a}_x + 4\hat{a}_y + 2\hat{a}_z$
(c) $1.2\hat{a}_x + 4\hat{a}_y + 0.8\hat{a}_z$
(d) $3\hat{a}_x + 10\hat{a}_y + 0.8\hat{a}_z$

[2015 : 1 M, Set-2]

2.22 Two semi-infinite conducting sheets are placed at right angles to each other as shown in the figure. A point charge of $+Q$ is placed at a distance of d from both sheets. The net force on the charge is $\frac{Q^2}{4\pi\varepsilon_0}\frac{K}{d^2}$, where K is given by

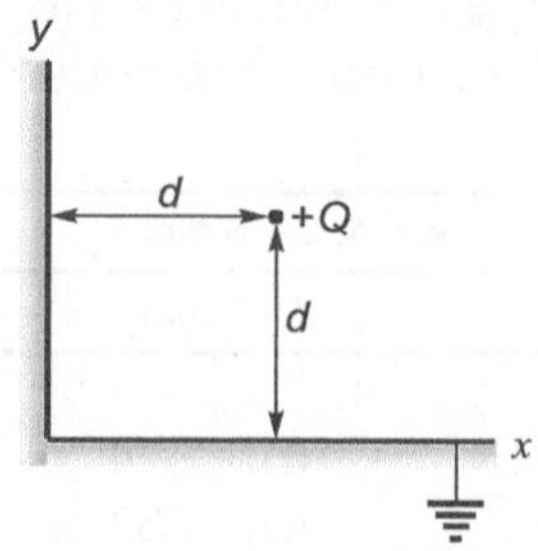

(a) 0
(b) $-\frac{1}{4}\hat{i} - \frac{1}{4}\hat{j}$
(c) $-\frac{1}{8}\hat{i} - \frac{1}{8}\hat{j}$
(d) $\frac{1-2\sqrt{2}}{8\sqrt{2}}\hat{i} + \frac{1-2\sqrt{2}}{8\sqrt{2}}\hat{j}$

[2015 : 2 M, Set-2]

2.23 Two electric charges Q and $-2Q$ are placed at (0,0) and (6,0) on the x-y plane. The equation of the zero equipotential curve in the x-y plane is

(a) $x = -2$
(b) $y = 2$
(c) $x^2 + y^2 = 2$
(d) $(x + 2)^2 + y^2 = 16$

[2016 : 2 M, Set-1]

2.24 A parallel plate capacitor filled with two dielectrics is shown in the figure below. If the electric field in the region A is 4 kV/cm, the electric field in the region B, in kV/cm, is

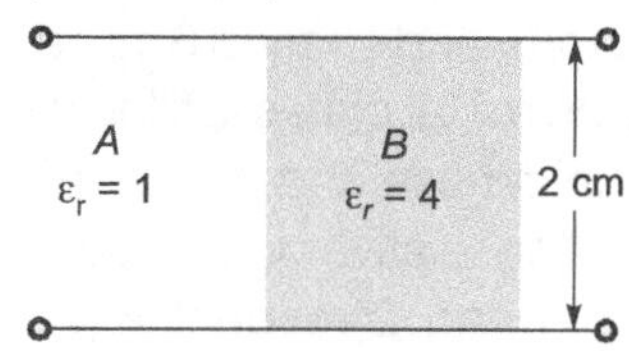

(a) 1
(b) 2
(c) 4
(d) 16

[2016 : 1 M, Set-2]

2.25 Two electrodes, whose cross-sectional view is shown in the figure below, are at the same potential. The maximum electric field will be at the point

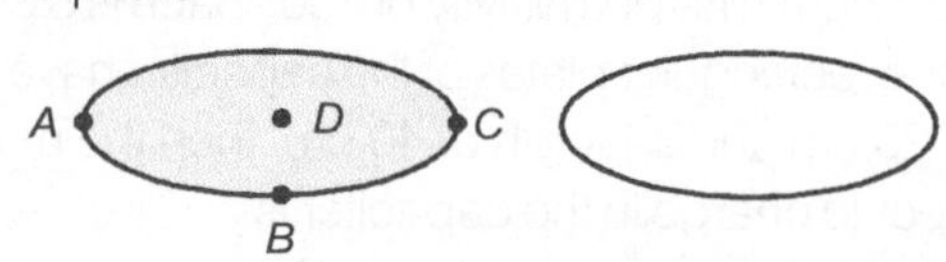

(a) A
(b) B
(c) C
(d) D

[2016 : 1 M, Set-2]

2.26 Consider an electron, a neutron and a proton initially at rest and placed along a straight line such that the neutron is exactly at the center of the line joining the electron and proton. At $t = 0$, the particles are released but are constrained to move along the same straight line. Which of these will collide first?

(a) The particles will never collide
(b) All will collide together
(c) Proton and neutron
(d) Electron and neutron [2017 : 1 M, Set-1]

2.27 Consider a solid sphere of radius 5 cm made of a perfect electric conductor. If one million electrons are added to this sphere, these electrons will be distributed
(a) uniformly over the entire volume of the sphere.
(b) uniformly over the outer surface of the sphere.
(c) concentrated around the centre of the sphere.
(d) along a straight line passing through the centre of the sphere. [2017 : 1 M, Set-2]

2.28 A thin soap bubble of radius R = 1 cm, and thickness a = 3.3 μm ($a << R$), is at a potential of 1 V with respect to a reference point at infinity, The bubble bursts and becomes a single spherical drop of soap (assuming all the soap is contained in the drop) of radius r. The volume of the soap in the thin bubble is $4\pi R^2 a$ and that of the drop is $\frac{4}{3}\pi r^3$. The potential in volts, of the resulting single spherical drop with respect to the same reference point at infinity is ______.
(Give the answer up to two decimal places.)

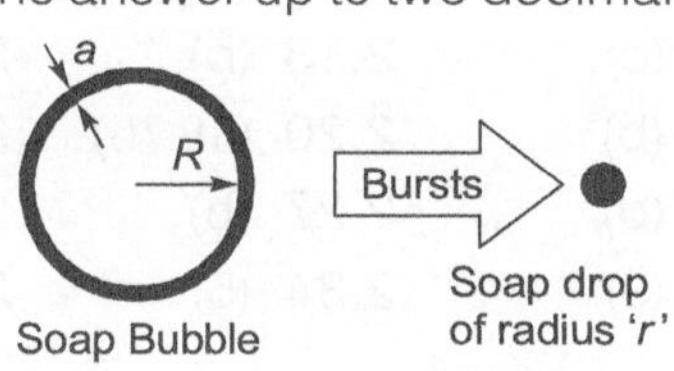

[2017 : 2 M, Set-2]

2.29 A positive charge of 1 nC is placed at (0, 0, 0.2) where all dimensions are in meters. Consider the x-y plane to be a conducting ground plane. Take $\epsilon_0 = 8.85 \times 10^{-12}$ F/m. The Z component of the E field at (0, 0, 0.1) is closed to
(a) 899.18 V/m (b) –899.18 V/m
(c) 999.09 V/m (d) –999.09 V/m
[2018 : 1 M]

2.30 The capacitance of an air-filled parallel-plate capacitor is 60 pF. When a dielectric slab whose thickness is half the distance between the plates, is placed on one of the plates covering it entirely, the capacitance becomes 86 pF. Neglecting the fringing effects, the relative permittivity of the dielectric is ______ (upto 2 decimal places).
[2018 : 2 M]

2.31 A co-axial cylindrical capacitor shown in the figure (i) has dielectric with relative permittivity $\varepsilon_{r1} = 2$. When one-fourth portion of the dielectric is replaced with another dielectric of relative permittivity ε_{r2} as shown in the figure (ii); the capacitance is doubled. The value of ε_{r2} is ______.

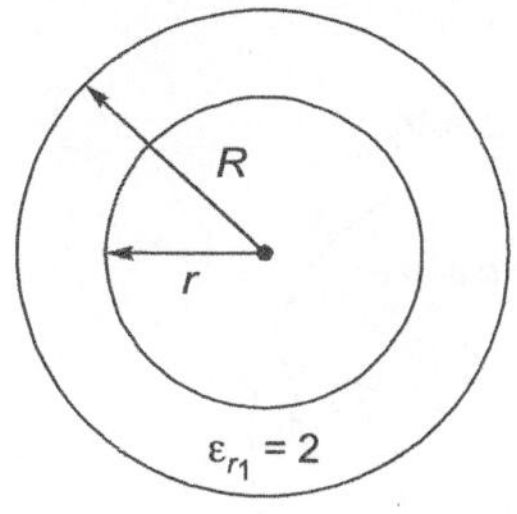

Fig. (i)

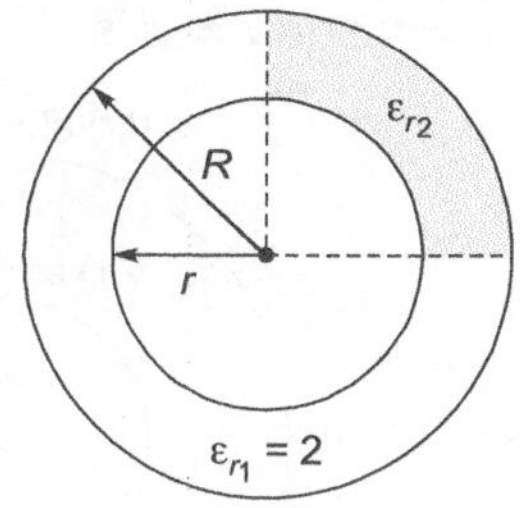

Fig. (ii)

[2019 : 1 M]

2.32 The static electric field inside a dielectric medium with relative permittivity. $\varepsilon_r = 2.25$, expressed in cylindrical coordinate system is given by the following expression

$$E = \hat{a}_r 2r + a_\varphi\left(\frac{3}{r}\right) + \hat{a}_z 6$$

where $\hat{a}_r, a_\varphi, \hat{a}_z$ are unit vectors along r, φ and z directions, respectively. If the above expression represents a valid electrostatic field inside the medium, then the volume charge density associated with this field in terms of free space permittivity, ε_0, in SI units is given by:
(a) $4\,\varepsilon_0$ (b) $5\,\varepsilon_0$
(c) $3\,\varepsilon_0$ (d) $9\,\varepsilon_0$ [2020 : 2 M]

2.33 A 1 μC point charge is held at the origin of a Cartesian coordinate system. If a second point charge of 10 μC is moved from (0, 10, 0) to (5, 5, 5) and subsequently to (5, 0, 0), then the total work done is ______ mJ. (Round off to 2 decimal places.)

Take $\frac{1}{4\pi\varepsilon_0} = 9 \times 10^9$ in SI units. All coordinates are in meters. [2021 : 1 M]

2.34 Consider a large parallel plate capacitor. The gap 'd' between the two plates is filled entirely with a dielectric slab of relative permittivity 5. The plates are initially charged to a potential difference of V volts and then disconnected from the source, If the dielectric slab is pulled out completely, then the ratio of the new electric field E_2 in the gap to the original electric field E_1 is ______.
[2021 : 2 M]

2.35 As shown in the figure below, two concentric conducting spherical shells, centred at $r = 0$ and having radii $r = c$ and $r = d$ are maintained at potentials such that the potential $V(r)$ at $r = c$ is V_1

and $V(r)$ at $r = d$ is V_2. Assume that $V(r)$ depends only on r, where r is the radial distance. The expression for $V(r)$ in the region between $r = c$ and $r = d$ is

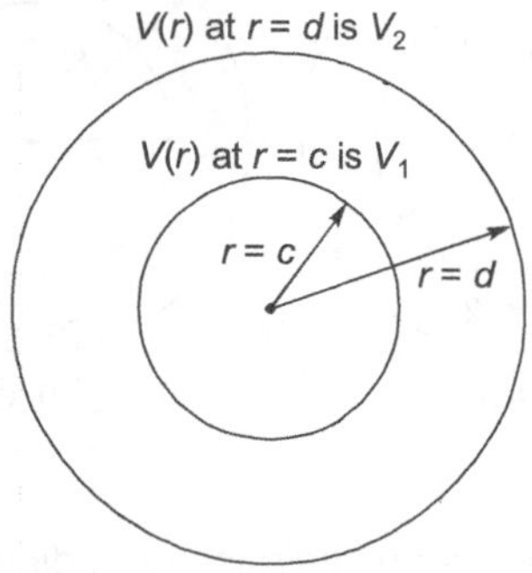

(a) $V(r) = \dfrac{cd(V_2 - V_1)}{(d-c)r} - \dfrac{V_1c + V_2d - 2V_1d}{d-c}$

(b) $V(r) = \dfrac{cd(V_1 - V_2)}{(d-c)r} + \dfrac{V_2d - V_1c}{d-c}$

(c) $V(r) = \dfrac{cd(V_1 - V_2)}{(d-c)r} - \dfrac{V_1c - V_2c}{d-c}$

(d) $V(r) = \dfrac{cd(V_2 - V_1)}{(d-c)r} - \dfrac{V_2c - V_1c}{d-c}$ [2022 : 2 M]

2.36 In the (x, y, z) coordinate system, three point-charges Q, Q. and αQ are located in free space at (–1, 0, 0), (1, 0, 0) and (0, –1, 0) respectively. The value of α for the electric field to be zero at (0, 0.5, 0) is _______ (rounded off to 1 decimal places).

[2024 : 2 M]

Answers Electrostatic Fields

2.1 (b)	2.2 (d)	2.3 (4.72)	2.4 (11.11)	2.5 (d)	2.6 (a)	2.7 (b)
2.8 (d)	2.9 (b)	2.10 (d)	2.11 (a)	2.12 (c)	2.13 (b)	2.14 (c)
2.15 (c)	2.16 (a)	2.17 (c)	2.18 (a)	2.19 (b)	2.20 (18.75)	2.21 (a)
2.22 (d)	2.23 (d)	2.24 (c)	2.25 (c)	2.26 (d)	2.27 (b)	2.28 (10.03)
2.29 (d)	2.30 (2.53)	2.31 (10)	2.32 (d)	2.33 (9)	2.34 (5)	2.35 (b)
2.36 (–1.6)						

Explanations Electrostatic Fields

2.1 (b)

By the fundamental definition of voltage.

$$V = \frac{dW}{dq}$$

2.2 (d)

Field lines are always from higher potential to lower potential point, so at equipotential surfaces field lines are always normal or orthogonal.

2.3 Sol.

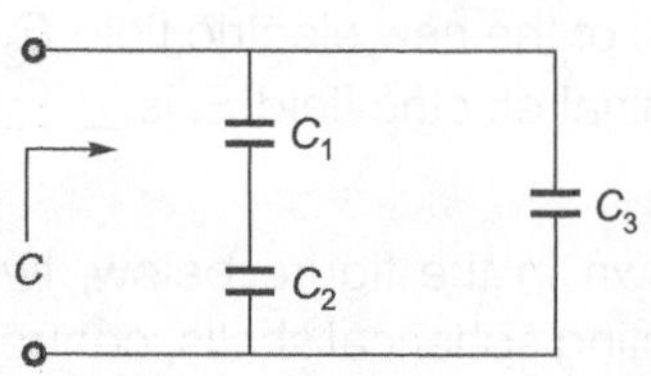

$d = 20$ cm; $A = 20 \times 20 = 400$ cm^2

$$C = \frac{C_1C_2}{C_1 + C_2} + C_3$$

$$= \frac{\left(\dfrac{\epsilon_o \epsilon_{r1}\frac{A}{2}}{\frac{d}{2}}\right)\cdot\left(\dfrac{\epsilon_o \epsilon_{r2}\frac{A}{2}}{\frac{d}{2}}\right)}{\dfrac{\epsilon_o \epsilon_{r1}\frac{A}{2}}{\frac{d}{2}} + \dfrac{\epsilon_o \epsilon_{r2}\frac{A}{2}}{\frac{d}{2}}} + \frac{\epsilon_o \epsilon_{r3}\frac{A}{2}}{d}$$

$$= \epsilon_o \frac{A}{d}\left(\frac{\epsilon_{r1}\epsilon_{r2}}{\epsilon_{r1} + \epsilon_{r2}}\right) + \frac{\epsilon_o \epsilon_{r3} A}{2d}$$

$$= \frac{\epsilon_o A}{d}\left(\frac{\epsilon_{r1}\epsilon_{r2}}{\epsilon_{r1} + \epsilon_{r2}} + \frac{\epsilon_{r3}}{2}\right)$$

On putting the given values,

$$C = \frac{8.85\times10^{-12}\times400\times10^{-4}}{20\times10^{-2}}\left(\frac{1\times2}{1+2}+\frac{4}{2}\right)$$

$C = 4.72 \times 10^{-12}\ F = 4.72$ pF

2.4 Sol.

$$C = 4\pi\epsilon_0 R$$

$$= 4\pi\times8.85\times10^{-12}\times\frac{10}{100}\text{(in air)} = 11.11\text{ pF}$$

2.5 (d)

$$V_{(x,\,y,\,z)} = 50x^2 + 50y^2 + 50z^2$$

$$\vec{E} = -\nabla V$$

$$= -\left[100x\hat{i}+100y\hat{j}+100z\hat{k}\right]\text{V/m}$$

$$\vec{E}(1,-1,1) = -\left[100\hat{i}-100\hat{j}+100\hat{k}\right]\text{V/m}$$

$$|E| = \sqrt{100^2+100^2+100^2} = 100\sqrt{3}$$

$$\hat{a}_E = \frac{\vec{E}}{|E|} = -\frac{1}{\sqrt{3}}(i-j+k)$$

2.6 (a)

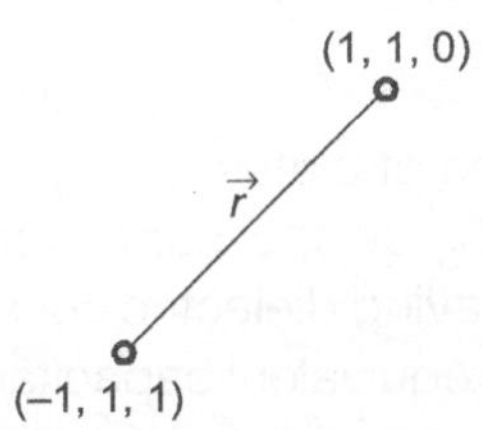

$$\vec{r} = [1-(-1)]\hat{i}+(1-1)\hat{j}+(0-1)\hat{k} = 2\hat{i}-\hat{k}$$

Distance between (1, 1, 0) and (–1, 1, 1) = r

$$= \sqrt{(1-(-1))^2+(1-1)^2+(0-1)^2} = \sqrt{5}$$

$$\vec{E} = \frac{Q}{4\pi\epsilon_o r^3}(\vec{r}) = \frac{10^{-6}}{4\pi\epsilon_o\times(\sqrt{5})^3}(2\hat{i}-\hat{k})$$

$$= \frac{10^{-6}}{20\sqrt{5}\pi\epsilon_o}(2\hat{i}-\hat{k})$$

2.7 (b)

$$V_P - V_Q = \frac{q}{4\pi\varepsilon_0(OP)}-\frac{q}{4\pi\varepsilon_0(OQ)}$$

$$= 9\times10^9\left[\frac{10^{-9}}{40\times10^{-3}}-\frac{10^{-9}}{20\times10^{-3}}\right] = -225\text{ volts}$$

2.8 (d)

$$E = \frac{1}{2}QV = \frac{1}{2}CV^2 = \frac{1}{2}\times\frac{\epsilon_0 A}{d}\times V^2$$

$$= \frac{1}{2}\times\frac{8.85\times10^{-12}\times100\times10^{-6}\times10^4}{10^{-4}} = 44.3\text{ nJ}$$

2.9 (b)

$$C_1 = \frac{\varepsilon_{r1}\,\varepsilon_0\,A}{t_1};\quad C_2 = \frac{\varepsilon_{r2}\,\varepsilon_0\,A}{t_2}$$

Let voltage on the conductor coil is V, then

$$C_1(100-V) = C_2 V$$

[Charges on both capacitors are equal]

$$\frac{\varepsilon_{r1}}{t_1}(100-V) = \frac{\varepsilon_{r2}}{t_2}V$$

$$6(100-V) = 4\text{ V} \Rightarrow V = \frac{600}{10} = 60\text{ Volts}$$

2.10 (d)

When two capacitors formed by two layers of dielectrics are connected in series, then equivalent capacitance,

$$C_{eq} = \frac{C_1\times C_2}{C_1+C_2} = \frac{\frac{\varepsilon_{r1}\,\varepsilon_0\,A}{d_1}\times\frac{\varepsilon_0\,\varepsilon_{r2}\,A}{d_2}}{\frac{\varepsilon_{r1}\,\varepsilon_0\,A}{d_1}+\frac{\varepsilon_0\,\varepsilon_{r2}\,A}{d_2}}$$

$$= \frac{(8.85\times10^{-12})\,(400\times10^{-3})^2\times4\times2}{4\times8\times10^{-3}+6\times2\times10^{-3}} = 257\text{ pF}$$

2.11 (a)

By Gauss's theorem,

$$\int_s \vec{\epsilon}\times\vec{ds} = \frac{Q_{\text{enclosed}}}{\epsilon_0}$$

Where, Q_{enclosed} in radius r

$$\int_s \vec{E}\cdot\vec{dS} = \frac{Q}{\frac{4}{3}\pi R^3}\times\frac{4}{3}\pi r^3\times\frac{1}{\epsilon_0}$$

$$E(4\pi r^2) = \frac{Qr^3}{R^3}\times\frac{1}{\epsilon_0} \Rightarrow E = \frac{Qr}{4\pi\,\epsilon_0\,R^3}$$

2.12 (c)

Net electrical flux passing through plane, $Z = 20$

$$Q = \frac{Q_1+Q_2}{2} = \frac{10+20}{2} = 15\ \mu\text{C}$$

2.13 (b)

$A_1 = A_2 = 500 \times 500\text{ mm}^2 = 25 \times 10^{-2}\text{ m}^2$

$\varepsilon_{r1} = 8$, $\varepsilon_{r2} = 2$, $d_1 = 4$ mm, $d_2 = 2$ mm

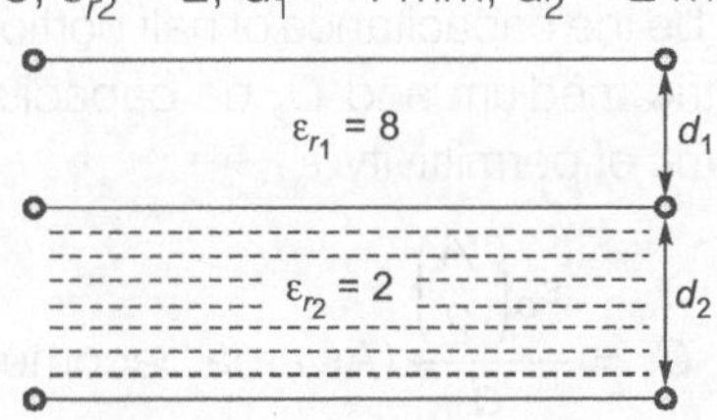

$$C_{eq} = \frac{C_1 \times C_2}{C_1 + C_2} = \frac{\frac{A\,\varepsilon_{r1}\,\varepsilon_0}{d_1} \times \frac{A\,\varepsilon_{r2}\,\varepsilon_0}{d_2}}{\frac{A\,\varepsilon_{r1}\,\varepsilon_0}{d_1} + \frac{A\,\varepsilon_{r2}\,\varepsilon_0}{d_2}}$$

$$C_{eq} = \frac{A\,\varepsilon_0\,\varepsilon_{r1}\,\varepsilon_{r2}}{\varepsilon_{r_1} d_2 + \varepsilon_{r_2} d_1}$$

$$= \frac{500 \times 500 \times 10^{-6} \times 8.85 \times 10^{-12} \times 8 \times 2}{(8 \times 2 + 2 \times 4) \times 10^{-3}}$$

$= 1475$ pF

2.14 (c)

$$Q = CV = \frac{\epsilon_0\,\epsilon_r\,A}{d} \times V$$

$$= \epsilon_0\,\epsilon_r\,A \times \frac{V}{d} = \epsilon_0\,\epsilon_r\,A \times E$$

$$= 8.85 \times 10^{-12} \times 2.26 \times 20 \times 10^{-2} \times 40 \times 10^{-2} \times 5000 \times 10^3 = 8\ \mu\text{C}$$

2.15 (c)

Energy density $= \frac{1}{2} \in |\bar{E}|^2$

Energy = Energy density × Volume

$$\text{Energy} = \frac{1}{2} \times \epsilon_0 \times \epsilon_r (6^2 + 8^2) \times 10^{12} \times 0.5 \times 0.5 \times 0.4$$

$$= \frac{1}{2} \times 8.85 \times 10^{-12} \times 2 \times 100 \times 10^{12} \times 0.1 = 88.5\ \text{J}$$

2.16 (a)

Let A be the area of the parallel plate capacitor and d be the distance between the plates.

⇒ **With air dielectric:**

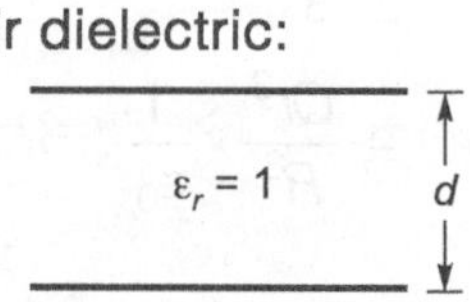

Capacitance, $C_o = \frac{\varepsilon_o A}{d}$ ...(i)

⇒ **With the new arrangement:**

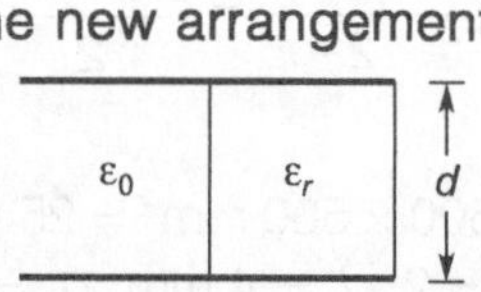

Let C_1 be the capacitance of half portion with air as dielectric medium and C_2 be capacitance with a dielectric of permittivity, ε_r.

Then, $C_1 = \frac{\varepsilon_o \left(\frac{A}{2}\right)}{d}$ (As area becomes half)

and $C_2 = \frac{\varepsilon_o \varepsilon_r \left(\frac{A}{2}\right)}{d}$ (As area becomes half)

Now, these two capacitance will be in parallel if a voltage is applied between the plates as same potential difference will be there between both the capacitances.

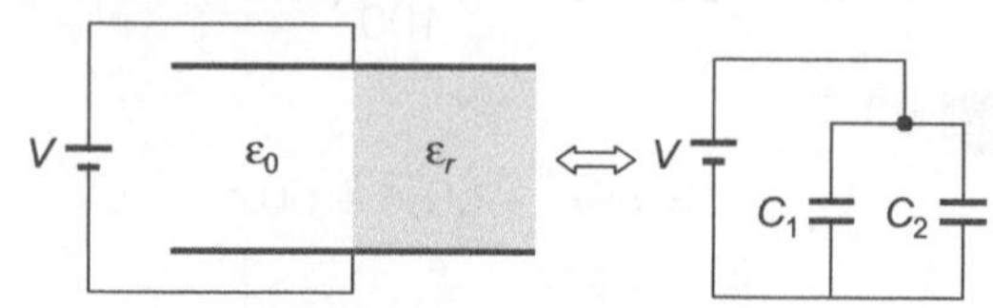

∴ Equivalent capacitance is

$$C_{eqv} = C_1 + C_2 = \frac{\varepsilon_o A}{2d} + \frac{\varepsilon_o \varepsilon_r A}{2d}$$

$$C_{eqv} = \frac{\varepsilon_o A}{2d}(1 + \varepsilon_r) = \frac{C_o}{2}(1 + \varepsilon_r)$$

[Using equation (i)]

Therefore, modified capacitance,

$$C_{eqv} = \frac{C_o}{2}(1 + \varepsilon_r)$$

2.17 (c)

Let A = Area of plates

Let $C_1 = C_3$ be the capacitance formed with dielectric having dielectric constant ε_1.

C_{eqv} be the equivalent capacitance.

C_2 be the capacitance formed with dielectric having dielectric constant ε_{r_2}.

Then, $C_1 = C_3 = \frac{\varepsilon_0\,\varepsilon_1\,A}{\frac{d}{4}} = \frac{4\varepsilon_0\,\varepsilon_1\,A}{d}$

and $C_2 = \frac{\varepsilon_0\,\varepsilon_2\,A}{\frac{d}{2}} = \frac{2\varepsilon_0\,\varepsilon_2\,A}{d}$

Also, equivalent capacitance = C_{eqv}

$$\therefore \quad \frac{1}{C_{eqv}} = \frac{1}{C_1} + \frac{1}{C_2} + \frac{1}{C_3} = \frac{2}{C_1} + \frac{1}{C_2}$$

($\because C_1 = C_3$) (C_1, C_2, C_3 all in series)

$$\text{or,} \quad \frac{1}{C_{eqv}} = \frac{2d}{4\varepsilon_0\,\varepsilon_1 d} + \frac{d}{2\varepsilon_0\,\varepsilon_2\,A}$$

$$\text{or,} \quad C_{eqv} = \frac{2\,\varepsilon_1\,\varepsilon_2\,\varepsilon_0\,A}{d(\varepsilon_1 + \varepsilon_2)}$$

Given, V_{eqv} = Total voltage = 10 Volt,

$V_1 = V_3 = 2$ Volt

We know tht: $C \propto \frac{1}{V}$

$$\therefore \quad \frac{C_{eqv}}{C_1} = \frac{V_1}{V_{eqv}} \text{ or } \frac{C_{eqv}}{C_1} = \frac{1}{5}$$

$$\text{or,} \quad \frac{2\varepsilon_0 \varepsilon_1 \varepsilon_2 A}{d(\varepsilon_1+\varepsilon_2)} \times \frac{d}{4\varepsilon_0 \varepsilon_1 A} = \frac{1}{5}$$

$$\text{or,} \quad \frac{\varepsilon_2}{2(\varepsilon_1+\varepsilon_2)} = \frac{1}{5}$$

$$\text{or,} \quad 5\varepsilon_2 = 2\varepsilon_1 + 2\varepsilon_2 \text{ or, } 2\varepsilon_1 = 3\varepsilon_2$$

$$\text{or,} \quad \boxed{\varepsilon_1 : \varepsilon_2 = 3:2}$$

2.19 (b)

Due to change at (0, 0, 2) and conductor plane there is an image at (0, 0, –2).

Total field is $\vec{E}_1 + \vec{E}_2$

$$\vec{E}_1 = \frac{32\pi\epsilon_0 \sqrt{2}}{4\pi\epsilon_0 (2+2+4)^{3/2}}(\sqrt{2}\hat{a}_x + \sqrt{2}\hat{a}_y - 2\hat{a}_z)$$

$$\vec{E}_2 = \frac{-32\pi\epsilon_0 \sqrt{2}}{4\pi\epsilon_0 (2+2+4)^{3/2}}(\sqrt{2}\hat{a}_x + \sqrt{2}\hat{a}_y + 2\hat{a}_z)$$

$$\vec{E} = -2\hat{a}_z = -2\hat{k}$$

2.20 Sol.

A capacitor is made as shown in figure.

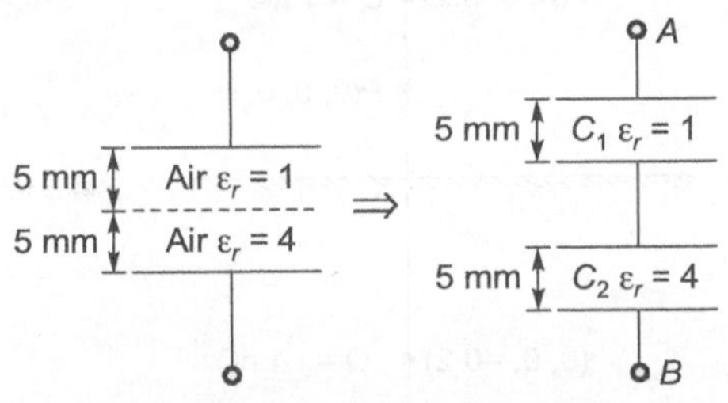

Breakdown of dielectric air is 30 kV/cm and breakdown of dielectric glass is 300 kV/cm.
We need to find maximum value of V_{AB} so that no C_1 or C_2 breaks down.
Since, area and width of C_1 and C_2 is same but $\varepsilon_{r1} = 1$ and $\varepsilon_{r2} = 4$
So, $C_2 = 4C_1$
So, if V_{AB} is applied,

$$V_{C1} = \frac{4}{5}V_{AB} \text{ and } V_{C_2} = \frac{1}{5}V_{AB}$$

For C_1 not breakdown, we need

$$\frac{4}{5}\frac{V_{AB}}{5\text{mm}} < 30 \text{ kV/cm}$$

$$V_{AB} < \frac{30 \times 25 \times 1000}{4} \text{ V}$$

$$V_A < 18.75 \text{ kV}$$

2.21 (a)

Given that,
at the interface ($y = 0$) there is no surface charge.
Region 1 ($y < 0$) has $\varepsilon_r = 2$
Region 2 ($y > 0$) has $\varepsilon_r = 5$
Electric field in region 1 is $3a_x + 4a_y + 2a_z$
Normal component of electric field is $4a_y$ tangential component of electric field is $3a_x + 2a_y$
Now, $E_1 = E_{1t} + E_{1n}$
The tangential component of E_1 = tangential component of E_2

$$E_{1t} = E_{2t} = 3a_x + 2a_z$$

and $\varepsilon_1 E_{1n} = \varepsilon_2 E_{2n}$

$$\text{so,} \quad E_{2n} = \frac{\varepsilon_1 \times 4a_y}{\varepsilon_2} = \frac{2 \times 4a_y}{5} = 1.6\, a_y$$

$$\text{so,} \quad E_2 = E_{2t} + E_{2n} = 3a_x + 1.6a_y + 2a_z$$

2.22 (d)

Due to image of the charge, three new charges will be created as shown in the figure.

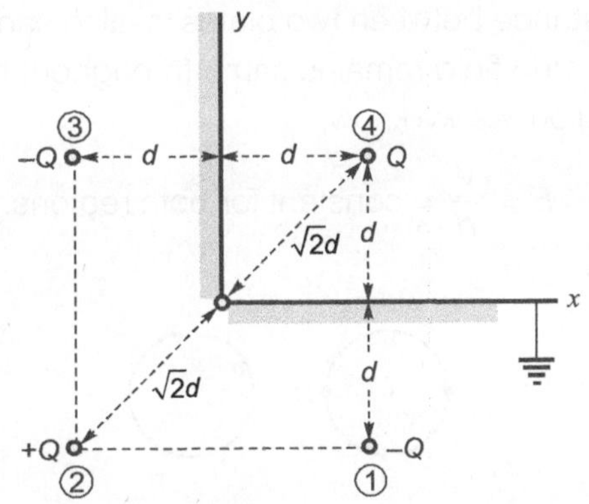

Total force (4) due to (1), (2), (3) will be

$$F_{total} = F_{41} + F_{42} + F_{43}$$

F_{41} = Force as (4) due to (1)

$$= -\frac{Q^2}{4\pi\varepsilon_0}\cdot\frac{1}{(4d^2)}\hat{j}$$

F_{42} = Force as (4) due to (2)

$$= +\frac{Q^2}{4\pi\varepsilon_0}\cdot\frac{1}{8d^2}\left(\frac{\hat{j}}{\sqrt{2}} + \frac{\hat{i}}{\sqrt{2}}\right)$$

F_{43} = Force as (4) due to (1)

$$= -\frac{Q^2}{4\pi\varepsilon_0}\cdot\frac{1}{4d^2}\hat{i}$$

$$F_{total} = \frac{Q^2}{4\pi\varepsilon_0}\left(\frac{1-2\sqrt{2}}{8\sqrt{2}}\hat{i} + \frac{1-2\sqrt{2}}{8\sqrt{2}}\hat{j}\right)$$

2.23 (d)

Charge, Q is located at (0, 0) and -2θ is located at (6, 0). To find V at any point (x, y)

$$V_Q = \frac{Q}{4\pi\in\sqrt{x^2+y^2}}$$

$$V_{-2Q} = \frac{-2Q}{4\pi\in\left(\sqrt{(x-6)^2+y^2}\right)}$$

$$V_{total} = 0$$

$$= \frac{Q}{4\pi\in(\sqrt{x^2+y^2})} + \frac{-2Q}{4\pi\in\left(\sqrt{(x-6)^2+y^2}\right)}$$

$$\sqrt{(x-6)^2+y^2} = 2(\sqrt{x^2+y^2})$$

$$x^2 + 36 - 12x + y^2 = 4x^2 + 4y^2$$

$$3x^2 + 3y^2 + 12x = 36$$

$$x^2 + y^2 + 4x = 12$$

$$(x+2)^2 + y^2 = 16$$

2.24 (c)

As voltage is same across both the regions and distance between two plates is also same, then electric field remains same throughout the both region. As we know,

$$E = \frac{V}{d} = \text{constant for both regions.}$$

2.25 (c)

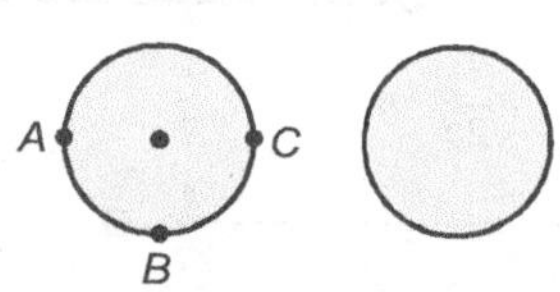

At the point C, $\bar{E}$ (electric field intensity) is maximum being closest to the other plate.

2.26 (d)

Given that electron, neutron and proton are in a straight line.

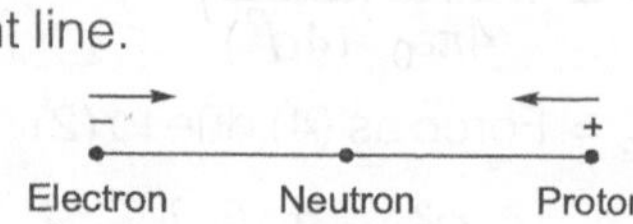

The electron will move towards proton and proton will more towards electron and force will be same

$F = \dfrac{q_1 q_2}{4\pi \in_0 R^2}$. But acceleration of electron will be more than proton as mass of electron < mass of proton. Since neutron are neutral they will not move. Thus electron will hit neutron first.

2.27 (b)

Added charge (one million electrons) to be solid spherical conductor is uniformly distributed over the outer surface of the sphere.

2.28 Sol.

Given: $R = 1$ cm; $a = 3.3$ μm; $V_1 = 1$ V

Now, bubble becomes spherical drop of radius r after burst. So,

$$\frac{4}{3}\pi r^3 = 4\pi R^2 a$$

$$r = \sqrt[3]{3R^2 a} = 9.96 \times 10^{-4} \text{ m}$$

As we know, potential (V) due to point charge is inversely proportional to distance.

$$V \propto \frac{1}{R}$$

$$\frac{V_2}{V_1} = \frac{R}{r}$$

$$V_2 = \frac{1\times1\times10^{-2}}{9.96\times10^{-4}} = 10.04 \text{ V}$$

So, potential of spherical drop becomes 10.04 V.

2.29 (d)

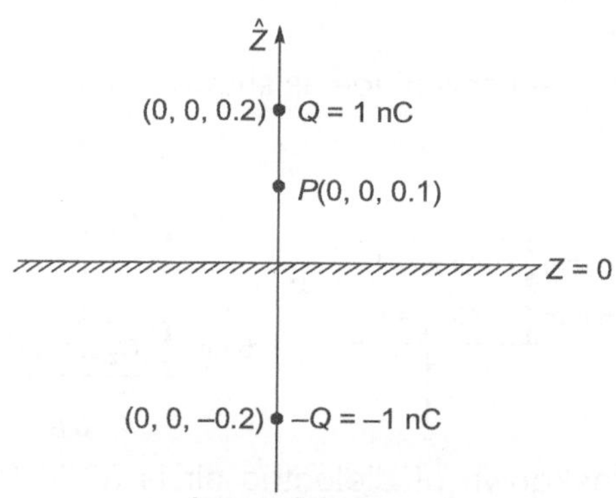

Image change

Net electric field at point P due to charge Q is,

$$\vec{E}_{12} = \frac{Q\vec{R}_{12}}{4\pi\varepsilon_0|\vec{R}_{12}|^3}$$

Total,
$$\vec{E}_{12} = \frac{1\times10^{-9}(-0.1\hat{a}_z)}{4\pi(8.854\times10^{-12})(0.1)^3} + \frac{(-1\times10^{-9})((0.3)\hat{a}_z)}{4\pi(8.854\times10^{-12})(0.3)^3}$$

$$= \left[\frac{-10^5}{4\pi(8.854)} - \frac{10^5}{4\pi(8.854)9}\right]\hat{a}_z$$

$$= (-898.774 - 99.863)\hat{a}_z \cong -999.09\hat{a}_z \text{ V/m}$$

2.30 Sol.

Given, $C = \dfrac{\varepsilon_0 A}{d} = 60 \text{ pF}$

In second case:

Capacitance,

$$C_1 = \frac{\varepsilon_0 A}{d/2} = \frac{2\varepsilon_0 A}{d} = 2 \times (60 \text{ pF}) = 120 \text{ pF}$$

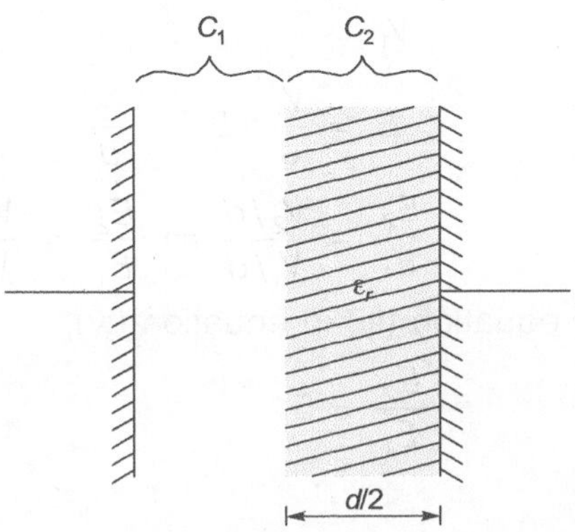

and $C_2 = \dfrac{2\varepsilon_0 \varepsilon_r A}{d} = (2 \times 60)\, \varepsilon_r$ pF

$= 120 \text{ e}_r \text{ pF}$

Now, $C_{eq} = \dfrac{C_1 C_2}{C_1 + C_2} = \dfrac{120 \times 120\varepsilon_r}{(120 + 120\varepsilon_r)}$ pF

$= 86$ pF (given)

or, $86 = \dfrac{120 \times 120 \varepsilon_r}{120(1+\varepsilon_r)}$

$$\frac{86}{120} = \frac{\varepsilon_r}{1+\varepsilon_r} \text{ or } \varepsilon_r = \frac{86}{34} = 2.53$$

2.31 (10)

Co-axial cylindrical capacitor-1

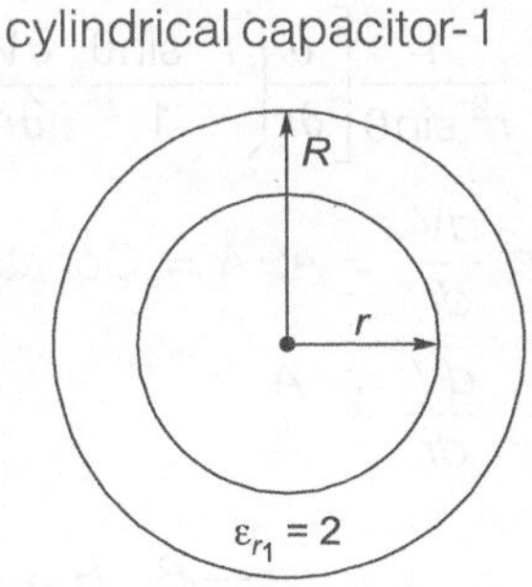

$$C_1 = \frac{2\pi \in_h}{\ln\left(\dfrac{b}{a}\right)} = \frac{2\pi \in_0 (2)h}{\ln\left(\dfrac{R}{r}\right)}$$

$$C_1 = \frac{4\pi \in_0 h}{\ln\left(\dfrac{R}{r}\right)} \quad \text{...(i)}$$

Co-axial cylindrical capacitor-2

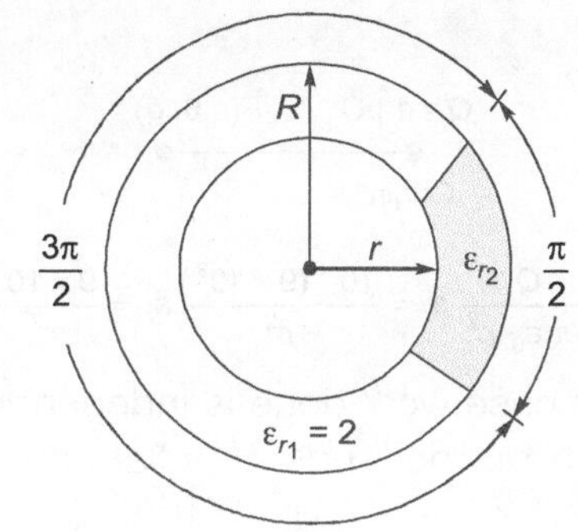

$$C_2 = \frac{\left(\dfrac{3\pi}{2}\right) \in_0 (2)h}{\ln\left(\dfrac{R}{r}\right)} + \frac{\dfrac{\pi}{2} \in_0 \in_{r2} h}{\ln\left(\dfrac{R}{r}\right)}$$

$$C_2 = \frac{\pi \in_0 h}{\ln\left(\dfrac{R}{r}\right)}\left[3 + \frac{\in_{r_2}}{2}\right] \quad \text{...(ii)}$$

Given, $C_2 = 2C_1$ ...(iii)

Put equation (i), (ii) in equation (iii),

$$\frac{\pi \in_0 h}{\ln\left(\dfrac{R}{r}\right)}\left[3 + \frac{\in_{r_2}}{2}\right] = 2\left[\frac{4\pi \in_0 h}{\ln\left(\dfrac{R}{r}\right)}\right]$$

$$3 + \frac{\in_{r_2}}{2} = 8$$

$$\Rightarrow \quad \frac{\in_{r_2}}{2} = 5 \Rightarrow \in_{r_2} = 10$$

2.32 (d)

$$\vec{D} = \in \vec{E} = \in_o \in_r \vec{E}$$

$$\vec{D} = \in_o 2.25\left(2r\hat{a}_r + \frac{3}{r}\hat{a}_\phi + 6\hat{a}_z\right)$$

$$\vec{D} = 4.5 \in_o r\hat{a}_r + \frac{6.75 \in_o}{r}\hat{a}_\phi + 13.5 \in_o \hat{a}_z$$

Volume charge density,

$$\rho_v = \vec{\nabla} \cdot \vec{D}$$

$$\rho_v = \frac{1}{r}\frac{\partial}{\partial r}(rD_r) + \frac{1}{r}\frac{\partial D_\phi}{\partial \phi} + \frac{\partial D_z}{\partial z}$$

$$\rho_v = \frac{1}{r}\frac{\partial}{\partial r}(r 4.5 \in_o r) + \frac{1}{r}\frac{\partial}{\partial \phi}\left(\frac{6.75 \in_o}{r}\right) + \frac{\partial}{\partial z}(13.5 \in_o)$$

$$= \frac{1}{r}\frac{\partial}{\partial r}(4.5 \in_o r^2) + 0 + 0$$

$$= \frac{1}{r}(4.5 \in_o)(2r) = 9 \in_o$$

2.33 (9.00)

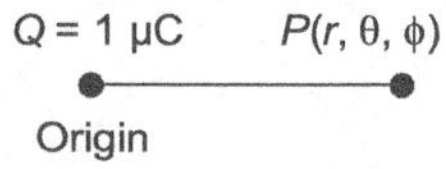

$$\vec{E} = \frac{Q}{4\pi \epsilon_0 r^2}\hat{a}_r = \frac{10^{-6}(9\times10^9)}{r^2}\hat{a}_r = \frac{9\times10^3}{r^2}\hat{a}_r$$

In this case work done is independent of type of path but depends an intial and final point.

$x \quad y \quad z \qquad\qquad r \quad \theta \quad \phi$
$(0, 10, 0) \rightarrow (10, 90°, 90°)$
Initial point

$x \quad y \quad z \qquad\qquad r \quad \theta \quad \phi$
$(5, 5, 5) \rightarrow (5\sqrt{3}, 54.73°, 45°)$
Intermediate point

$x \quad y \quad z \qquad\qquad r \quad \theta \quad \phi$
$(5, 0, 0) \rightarrow (5, 90°, 0°)$ Final point

Work done (by external source) in moving Q-charge (10 μC) in the presence of electric field $\vec{E}$ is

$$W = -Q\int_{\text{initial point}}^{\text{final point}} \vec{E}\cdot d\vec{l}$$

$$= -(10\times10^{-6})\int_{r=10}^{r=5}\frac{9\times10^3}{r^2}\hat{a}_r\cdot dr\,\hat{a}_r$$

$$W = -10\times10^{-6}(9\times10^3)\left(-\frac{1}{r}\right)_{r=10}^{5}$$

$$= 9\times10^{-2}\left[\frac{1}{5}-\frac{1}{10}\right]$$

$$= 9\times10^{-2}\left[\frac{10-5}{50}\right] = 9\times10^{-2}(10^{-1})$$

$$W = 9 \text{ mJ}$$

2.34 (5)

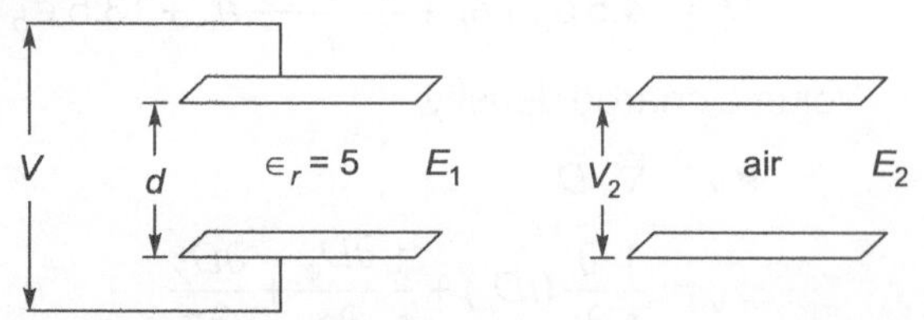

If voltage source is removed then in both cases charge Q is constant.

Case-1: $(Q_1 = Q;\ V_1 = V)$

$$Q_1 = C_1V_1$$

$$Q = \frac{\epsilon_0(5)A}{d}V_1 \Rightarrow Q = 5\left(\frac{\epsilon_0 A}{d}\right)V_1 \qquad ...(i)$$

Case-2: $(Q_2 = Q;\ V_2)$

$$Q_2 = C_2V_2$$

$$Q = \frac{\epsilon_0(1)A}{d}V_2 \Rightarrow Q = \left(\frac{\epsilon_0 A}{d}\right)V_2 \qquad ...(ii)$$

Equation (i) is equal to equation (ii)

$$\Rightarrow 5\left(\frac{\epsilon_0 A}{d}\right)V_1 = \left(\frac{\epsilon_0 A}{d}\right)V_2$$

$$\Rightarrow 5V_1 = V_2$$

$$\frac{V_2}{V_1} = 5 \qquad ...(iii)$$

$$E_1 = \frac{V_1}{d};\ E_2 = \frac{V_2}{d}$$

$$\frac{E_2}{E_1} = \frac{V_2/d}{V_1/d} \Rightarrow \frac{E_2}{E_1} = \frac{V_2}{V_1} \qquad ...(iv)$$

Put equation (iii) in equation (iv),

$$\Rightarrow \frac{E_2}{E_1} = 5$$

2.35 (b)

From Laplace's equation, $\nabla^2 V = 0$

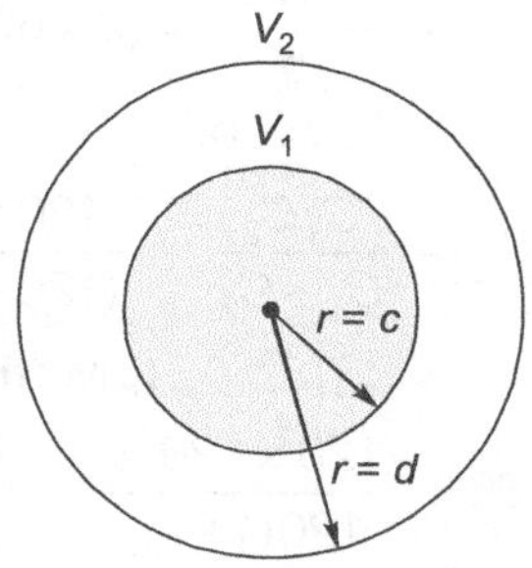

For spherical co-ordinate system,

$h_1 \quad h_2 \quad h_3$
$1 \quad r \quad r\sin\theta$

$$\therefore \nabla^2 V = \frac{1}{r^2\sin\theta}\left[\frac{\partial}{\partial r}\left(\frac{r^2\sin\theta}{1}\cdot\frac{\partial V}{\partial r}\right)\right] = 0$$

$$\Rightarrow r^2\frac{dV}{dr} = A;\ A = \text{Constant}$$

$$\Rightarrow \frac{dV}{dr} = \frac{A}{r^2}$$

$$\Rightarrow V = \frac{-A}{r}+B;\ B = \text{Constant}$$

at $r = c$; $V = V_1$

$$\Rightarrow V_1 = \frac{-A}{c}+B \qquad ...(1)$$

at $r = d$; $V = V_2$

$$\Rightarrow V_2 = \frac{-A}{d}+B \qquad ...(2)$$

(1) – (2), we get

$$V_1 - V_2 = \frac{-A}{c} + \frac{A}{d}$$

$$\Rightarrow \quad V_1 - V_2 = A\left[\frac{c-d}{cd}\right]$$

$$\Rightarrow \quad A = \left(\frac{V_1 - V_2}{c-d}\right)cd$$

From (1),

$$B = V_1 + \frac{A}{c}$$

$$\Rightarrow \quad B = V_1 + \left(\frac{V_1 - V_2}{c-d}\right)d$$

$$\therefore \quad V = \frac{-\left(\frac{V_1 - V_2}{c-d}\right)cd}{r} + V_1 + d\left(\frac{V_1 - V_2}{c-d}\right)$$

$$\Rightarrow \quad V(r) = \frac{(V_1 - V_2)cd}{(d-c)r} + V_1 - d\left(\frac{V_1 - V_2}{d-c}\right)$$

$$\Rightarrow \quad V(r) = \frac{(V_1 - V_2)cd}{(d-c)r} + \frac{V_1 d - V_1 c - V_1 d + V_2 d}{d-c}$$

$$\Rightarrow \quad V(r) = \frac{cd(V_1 - V_2)}{(d-c)r} + \frac{V_2 d - V_1 c}{d-c}$$

2.36 (-1.6)

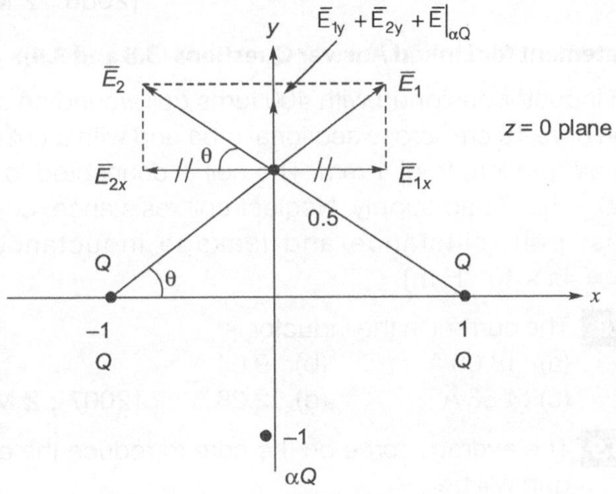

From the figure,

$$\bar{E}_{net} = \bar{E}_{1y} + \bar{E}_{2y} + \bar{E}|_{\alpha Q} = 0$$

[as per question]

Here, $\bar{E}_{1y} = \bar{E}_{2y} = \bar{E}|_Q$ (say)

$$\therefore \quad \bar{E}_{net} = 2\bar{E}|_Q + \bar{E}|_{\alpha Q} = 0$$

Now, $\bar{E}|_Q = \dfrac{kQ}{\left[\sqrt{1+0.5^2}\right]^2}\sin\theta\hat{a}_y; \quad k = \dfrac{1}{4\pi \in_o}$

$$\Rightarrow \bar{E}|_Q = \frac{kQ}{1.25} \times \frac{0.5}{\sqrt{1+0.5^2}}\hat{a}_y = 0.357\,kQ\hat{a}_y$$

Also, $\bar{E}|_{\alpha Q} = \dfrac{k(\alpha Q)}{(1.5)^2}\hat{a}_y = 0.444k(\alpha Q)\hat{a}_y$

$$\therefore \quad \bar{E}_{net} = 2 \times 0.357kQ\hat{a}_y + 0.444k(\alpha Q)\hat{a}_y = 0$$

$$\Rightarrow \quad 0.715kQ = -0.444\,k(\alpha Q)$$

$$\Rightarrow \quad \alpha = \frac{-0.715}{0.444} = -1.61$$

$$\therefore \quad \alpha = -1.61$$

■■■■

3 Magnetostatic Fields

3.1 The energy stored in the magnetic field of solenoid 30 cm long and 3 cm diameter wound with 1000 turns of wire carrying a current of 10 A is

(a) 0.015 J (b) 0.15 J
(c) 0.5 J (d) 1.15 J

[1996 : 1 M]

3.2 Two parallel wires separated by a distance 'd' are carrying a current 'I' in the same direction. The magnetic field along a line running parallel to these wires and midway between them

(a) depends upon 'I'
(b) is zero
(c) depends upon 'd'
(d) depends upon the permeability of medium between the wires

[1999 : 1 M]

3.3 An electron with velocity, $\vec{u}$ is placed in an electric field, $\vec{E}$ and magnetic field, $\vec{B}$. The force experienced by the electron is given by

(a) $-e\vec{E}$ (b) $e\vec{u}\times\vec{B}$
(c) $-e(\vec{u}\times\vec{E}+\vec{B})$ (d) $-e(\vec{E}+\vec{u}\times\vec{B})$

[2000 : 1 M]

3.4 Two conductors are carrying forward and return current of $+I$ and $-I$ as shown in the figure. The magnetic field intensity, $\overline{H}$ at point P is

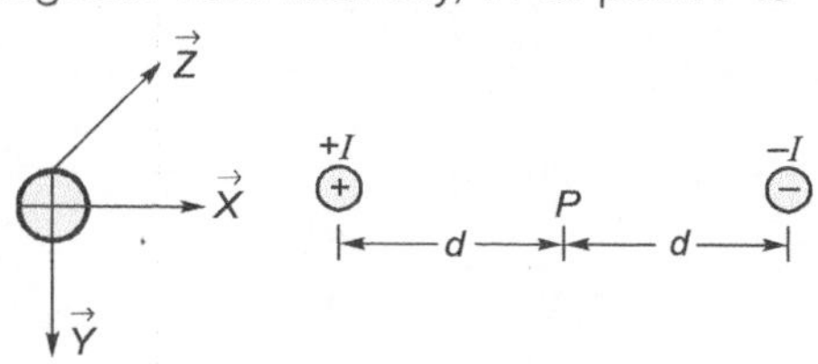

(a) $\frac{I}{\pi d}\overline{y}$ (b) $\frac{I}{\pi d}\overline{x}$
(c) $\frac{I}{2\pi d}\overline{y}$ (d) $\frac{I}{2\pi d}\overline{x}$

[2003 : 1 M]

3.5 Two infinite strips of width w m in X-direction as shown in figure, are carrying forward and return currents of $+I$ and $-I$ in the Z-direction. The strips are separated by a distance of x m. The inductance per unit length of the configuration is measured to be L H/m. If the distance of separation between the strips is now reduced to $x/2$ m, the inductance per unit length of the configuration is

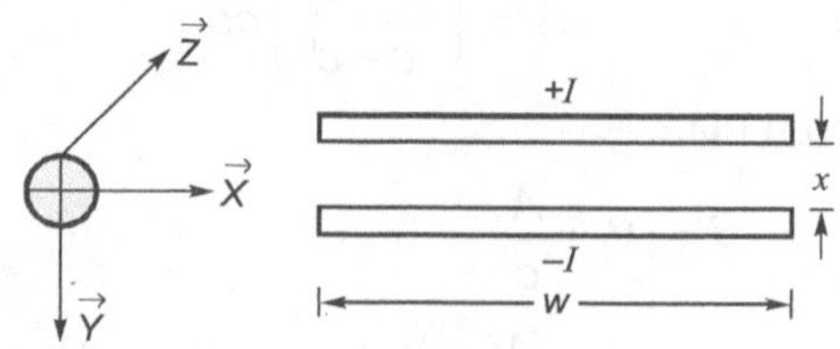

(a) 2L H/m (b) L/4 H/m
(c) L/2 H/m (d) 4L H/m [2003 : 1 M]

3.6 The inductance of a long solenoid of length 1000 mm wound uniformly with 3000 turns on a cylindrical paper tube of 60 mm diameter is

(a) 3.2 μH (b) 3.2 mH
(c) 32.0 mH (d) 3.2 H [2004 : 1 M]

3.7 Which of the following statements holds for the divergence of electric and magnetic flux densities?

(a) Both are zero.
(b) These are zero for static densities but non zero for time varying densities.
(c) It is zero for the electric flux density.
(d) It is zero for the magnetic flux density.

[2006 : 2 M]

Statement for Linked Answer Questions (3.8 and 3.9):

An inductor designed with 400 turns coil wound on an iron core 16 cm^2 cross sectional area and with a cut of an air gap length of 1 mm. The coil is connected to a 230 V, 50 Hz ac supply. Neglect coil resistance, core loss, iron reluctance and leakage inductance, ($\mu_0 = 4\pi \times 10^{-7}$ H/m).

3.8 The current in the inductor is

(a) 18.08 A (b) 9.04 A
(c) 4.56 A (d) 2.28 A [2007 : 2 M]

3.9 The average force on the core to reduce the air gap will be

(a) 832.29 N (b) 1666.22 N
(c) 3332.47 N (d) 6664.84 N [2007 : 2 M]

3.10 A coil of 300 turns is wound on a non-magnetic core having a mean circumference of 300 mm and a cross-sectional area of 300 mm^2. The inductance of the coil corresponding to a magnetizing current of 3 A will be
(Given that $\mu_0 = 4\pi \times 10^{-7}$ H/m)

(a) 37.68 μH (b) 113.04 μH
(c) 3.768 μH (d) 1.1304 μH [2008 : 2 M]

3.11 The flux density at a point in space is given by $B = 4x\mathbf{a}_x + 2ky\mathbf{a}_y + 8\mathbf{a}_z$ Wb/m^2. The value of constant k must be equal to

(a) –2 (b) –0.5
(c) +0.5 (d) +2 [2013 : 1 M]

3.12 The magnitude of magnetic flux density ($\vec{B}$) at a point having normal distance d meters from an infinitely extended wire carrying current of I A is $\frac{\mu_0 I}{2\pi d}$ (in SI units).

An infinitely extended wire is laid along the x-axis and is carrying current of 4 A in the +ve x direction. Another infinitely extended wire is laid along the y-axis and is carrying 2 A current in the +ve y direction. μ_0 is permeability of free space. Assume $\hat{i}, \hat{j}, \hat{k}$ to be unit vectors along x, y and z axes respectively.

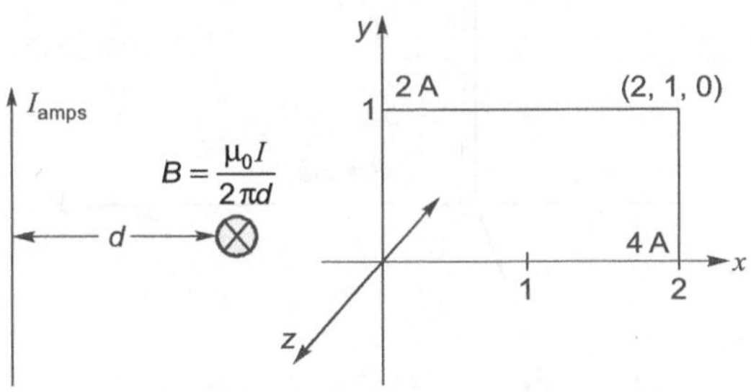

Assuming right handed coordinate system, magnetic field intensity, $\vec{H}$ at coordinate (2, 1, 0) will be

(a) $\frac{3}{2\pi}\hat{k}$ weber/m^2 (b) $\frac{4}{3\pi}\vec{i}$ A/m
(c) $\frac{3}{2\pi}\vec{k}$ A/m (d) 0 A/m

[2014 : 2 M, Set-2]

3.13 The following four vector fields are given in Cartesian co-ordinate system. The vector field which does not satisfy the property of magnetic flux density is

(a) $y^2\hat{a}_x + z^2\hat{a}_y + x^2\hat{a}_z$
(b) $z^2\hat{a}_x + x^2\hat{a}_y + y^2\hat{a}_z$
(c) $x^2\hat{a}_x + y^2\hat{a}_y + z^2\hat{a}_z$
(d) $y^2z^2\hat{a}_x + x^2z^2\hat{a}_y + x^2y^2\hat{a}_z$

[2014 : 2 M, Set-1]

3.14 A steady current I is flowing in the $-x$ direction through each of two infinitely long wires at $y = \pm\frac{L}{2}$ as shown in the figure. The permeability of the medium is μ_0. The $\vec{B}$ field at (0, L, 0) is

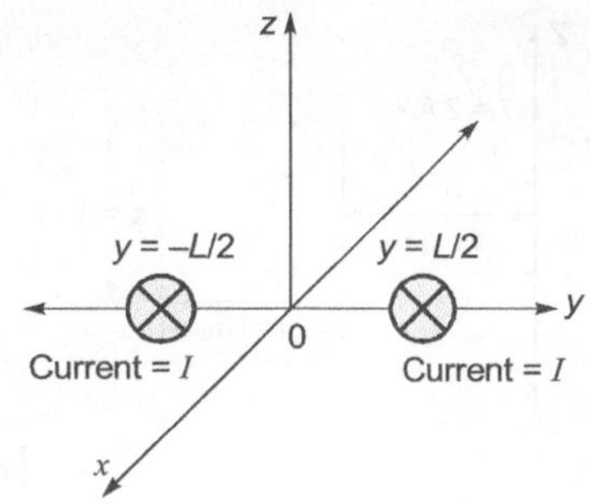

(a) $-\frac{4\mu_0 I}{3\pi L}\hat{z}$ (b) $+\frac{4\mu_0 I}{3\pi L}\hat{z}$
(c) 0 (d) $-\frac{3\mu_0 I}{4\pi L}\hat{z}$

[2015 : 1 M, Set-1]

3.15 A soft-iron toroid is concentric with a long straight conductor carrying a direct current I. If the relative permeability μ_r of soft-iron is 100, the ratio of the magnetic flux densities at two adjacent points located just inside and just outside the toroid, is _____. [2016 : 1 M, Set-1]

3.16 A solid iron cylinder is placed in a region containing a uniform magnetic field such that the cylinder axis is parallel to the magnetic field direction. The magnetic field lines inside the cylinder will

(a) bend closer to the cylinder axis.
(b) bend farther away from the axis.
(c) remain uniform as before.
(d) cease to exist inside the cylinder.

[2017 : 1 M, Set-1]

3.17 The magnitude of magnetic flux density (B) in micro Teslas (μT), at the centre of a loop of wire wound as a regular hexagon of side length 1 m carrying a current (I = 1 A) and placed in vacuum as shown in the figure is _____. (Give the answer up to two decimal places.)

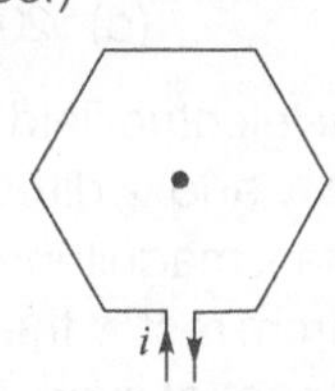

[2017 : 2 M, Set-1]

3.18 A conducting square loop of side length 1 m is placed at a distance of 1 m from a long straight wire carrying a current I = 2 A as shown below. The mutual inductance, in nH (rounded off to 2 decimal places), between the conducting loop and the long wire is _________.

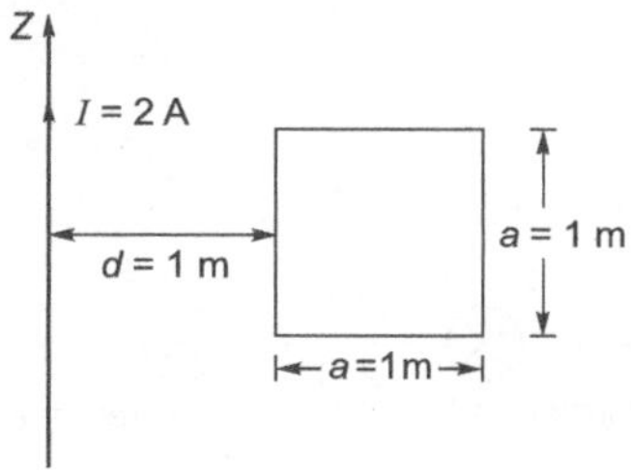

[2020 : 2 M]

3.19 Which one of the following vector functions represents a magnetic field $\vec{B}$?

($\hat{x}$, $\hat{y}$ and $\hat{z}$ are unit vectors along x-axis, y-axis and z-axis, respectively)

(a) $10x\hat{X} + 20y\hat{Y} - 30z\hat{Z}$

(b) $10x\hat{X} - 30z\hat{Y} + 20y\hat{Z}$

(c) $10z\hat{X} + 20y\hat{Y} - 30x\hat{Z}$

(d) $10y\hat{X} + 20x\hat{Y} - 10z\hat{Z}$ [2021 : 1 M]

3.20 A long conducting cylinder having a radius b is placed along the z-axis. The current density is $\vec{J} = J_a r^3 \hat{z}$ for the region $r < b$ where r is the distance in the radial direction. The magnetic field intensity ($\vec{H}$) for the region inside the conductor (i.e., for $r < b$) is

(a) $\frac{J_a}{4} r^4$ (b) $\frac{J_a}{3} r^3$

(c) $\frac{J_a}{5} r^4$ (d) $J_a r^3$ [2022 : 1 M]

3.21 If the magnetic field intensity ($\vec{H}$) in a conducting region is given by the expression, $\vec{H} = x^2\hat{i} + x^2y^2\hat{j} + x^2y^2z^2\hat{k}$ A/m. The magnitude of the current density, in A/m^2, at x = 1 m, y = 2 m and z = 1 m is

(a) 8 (b) 12

(c) 16 (d) 20 [2022 : 2 M]

3.22 In the figure, the electric field E and the magnetic field B point to x and z directions, respectively, and have constant magnitudes. A positive charge 'q' is released from rest at the origin. Which of the following statement(s) is/are true.

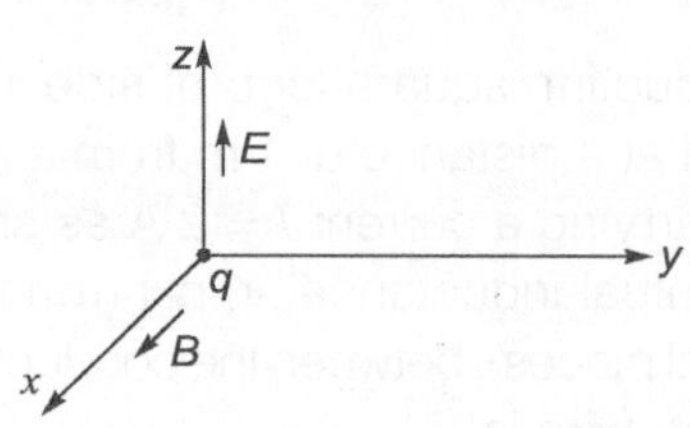

(a) The charge will move in the direction of **z** with constant velocity.

(b) The charge will always move on the **y-z** plane only.

(c) The trajectory of the charge will be a circle.

(d) The charge will progress in the direction of **y**.

[2023 : 2 M]

3.23 An infinite surface of linear current density $\hat{K} = 5\hat{a}_x$ A/m exists on the x-y plane, as shown in the figure. The magnitude of the magnetic field intensity (H) at a point (1,1,1) due to the surface current in Ampere/meter is _______ (Round off to 2 decimal places).

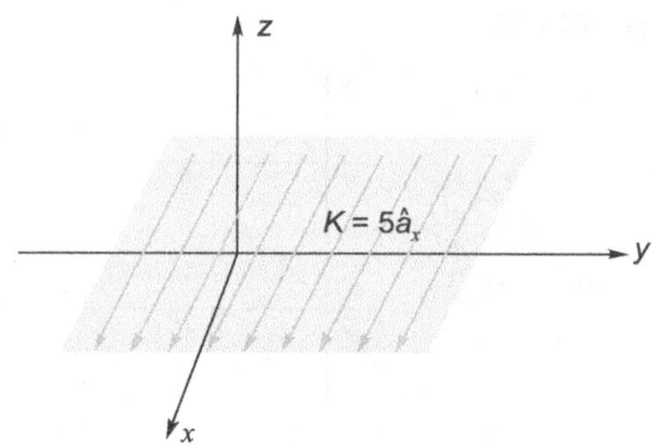

[2023 : 2 M]

3.24 The given equation represents a magnetic field strength $\bar{H}(r,\theta,\phi)$ in the spherical coordinate system, in free space. Here, $\hat{r}$ and $\hat{\theta}$ represent the unit vectors along r and θ, respectively. The value of P in the equation should be _______ (rounded off to the nearest integer).

$$\bar{H}(r,\theta,\phi) = \frac{1}{r^3}(\hat{r}P\cos\theta + \hat{\theta}\sin\theta)$$

[2024 : 2 M]

■■■■

Answers Magnetostatic Fields

3.1 (b)	3.2 (b)	3.3 (d)	3.4 (a)	3.5 (c)	3.6 (c)	3.7 (d)
3.8 (d)	3.9 (a)	3.10 (b)	3.11 (a)	3.12 (c)	3.13 (c)	3.14 (a)
3.15 (100)	3.16 (a)	3.17 (0.69)	3.18 (138.63)	3.19 (a)	3.20 (c)	3.21 (b)
3.22 (*)	3.23 (2.5)	3.24 (2)				

Explanations Magnetostatic Fields

3.1 (b)

$$L = \frac{\mu_0 N^2 A}{l}$$

$$= \frac{4\pi\times10^{-7}\times(1000)^2\times\pi\times\left(\frac{3}{2}\times10^{-2}\right)^2}{30\times10^{-2}}$$

$$= 2.96 \text{ mH}$$

$$E = \frac{1}{2}LI^2 = \frac{1}{2}\times2.96\times10^{-3}\times10^2$$

$$= 0.148 \text{ J} \approx 0.15 \text{ J}$$

3.2 (b)

Use right hand thumb rule.

3.3 (d)

$$F = F_{elect} + F_{mag} = q\vec{E} + q\vec{u}\times\vec{B}$$

$$= q(\vec{E}+\vec{u}\times\vec{B}) \cdots \text{Lorentz force}$$

For an electron $q = -e$

$$\Rightarrow \quad F = -e(\vec{E}+\vec{u}\times\vec{B})$$

3.4 (a)

$$\bar{H} = \bar{H}_1 + \bar{H}_2 = \frac{I}{2\pi d}\hat{y} + \frac{I}{2\pi d}\hat{y} = \frac{I}{\pi d}\hat{y}$$

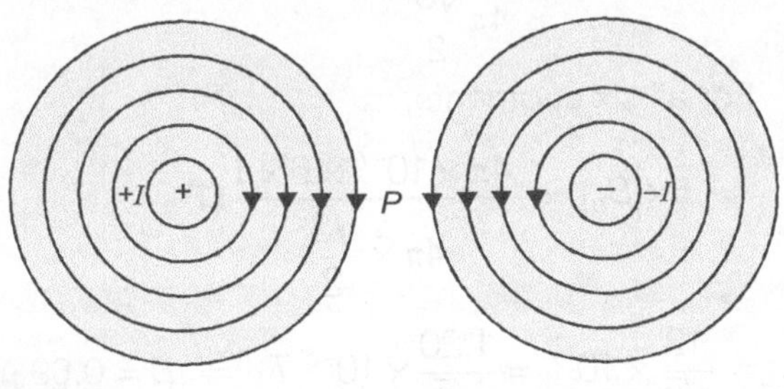

3.5 (c)

Inductance/m is proportional to separation between the strips so when separation is reduced to $x/2$ then inductance will become $L/2$ H/m.

3.6 (c)

Let, $B = \mu_0 nI$

Then, $\phi = B.S = \mu_0 n SI$

Where 'S' is the cross sectional area of solenoid

Flux linkage,

$$a' = n\phi = \mu_0 n^2 SI$$

Hence Inductance/length $= \mu_0 n^2 S$

for l = 1000 mm = 1 m

$\therefore L = 4\pi\times10^{-7}\times3000^2\times\pi\times(30\times10^{-3})^2 = 32$ mH

3.7 (d)

The divergence of magnetic field is always zero because magnetic flux makes always a closed path.

So, $\nabla\cdot\vec{B} = 0$ (Maxwell's equation)

While divergence of electric field, $\nabla\cdot\vec{E} = \frac{\rho_v}{\epsilon}$

3.8 (d)

$$L = \frac{\mu_0 N^2 A}{l}$$

$$= \frac{4\pi\times10^{-7}\times(400)^2\times16\times10^{-4}}{(1\times10^{-3})} = 0.3215$$

$$X_L = 2\pi fL$$

$$\therefore \quad I = \frac{V}{X_L} = \frac{V}{2\pi fL} = \frac{230}{2\pi\times50\times0.3215} = 2.28 \text{ A}$$

3.9 (a)

$$\text{Energy stored} = \frac{1}{2}\times L\times I^2$$

$$= \frac{1}{2}\times0.3215\times2.28^2 = 0.8356 \text{ J}$$

Force to reduce 1 mm air gap $= E/d$

$$= \frac{0.8356}{1\times10^{-3}} = 835.6 \text{ N}$$

3.10 (b)

$n = 300$

(l) circumference $= 2\pi r = 300$ mm

$A = 30 \text{ mm}^2 = 300\times10^{-6} \text{ m}^2$

$\therefore$ Inductance of coil,

$$L = \frac{\mu_0 N^2 A}{I}$$

$$= \frac{4\pi \times 10^{-7} \times 300 \times 300 \times 300 \times 10^{-6}}{300 \times 10^{-3}}$$

$$= 113.04\,\mu\text{H}$$

3.11 (a)

$$\nabla \cdot B = 0$$

$$\left(\frac{\partial}{\partial x}a_x + \frac{\partial}{\partial y}a_y + \frac{\partial}{\partial z}a_z\right)\cdot(4xa_x + 2kya_y + 8a_z) = 0$$

or $4 + 2k = 0 \Rightarrow k = -2$

3.13 (c)

The property of a solid magnetic field is

$\nabla \cdot B = 0$ (i.e. solenoidal property)

When, $B = x^2\hat{a}_x + y^2\hat{a}_y + z^2\hat{a}_z$

$\nabla \cdot B = 2x + 2y + 2z \neq 0$

Hence option (c).

3.14 (a)

$$\vec{B} = \frac{\mu_0 I}{2\pi\left(\frac{L}{2}\right)}(-\hat{z}) + \frac{\mu_0 I}{2\pi\left(\frac{3L}{2}\right)}(-\hat{z})$$

$$= \frac{\mu_0 I(-\hat{z})}{2\pi L}\left(2 + \frac{2}{3}\right) = -\frac{4}{3}\frac{\mu_0 I}{\pi L}\hat{z}$$

3.15 Sol.

Toroid has field, $B \propto \mu$

As $\mu = 100$ (inside field)

Magnetic field density B at any point at a distance at r is

$$B = \frac{\mu I}{2\pi r}$$

Now, $B_{\text{at } r-} = \frac{\mu_0 \mu_r I}{2\pi r^-}$ (just inside toroid)

and $B_{\text{at } r+} = \frac{\mu_0 I}{2\pi r^+}$ (just outside toroid)

$$\frac{B_{\text{at}\,r^-}}{B_{\text{at}\,r^+}} = \mu_r = 100$$

3.16 (a)

Iron being a ferromagnetic material, magnetic lines of force bend closer to cylindrical axis.

3.17 Sol.

$$i = 1\text{ A}$$

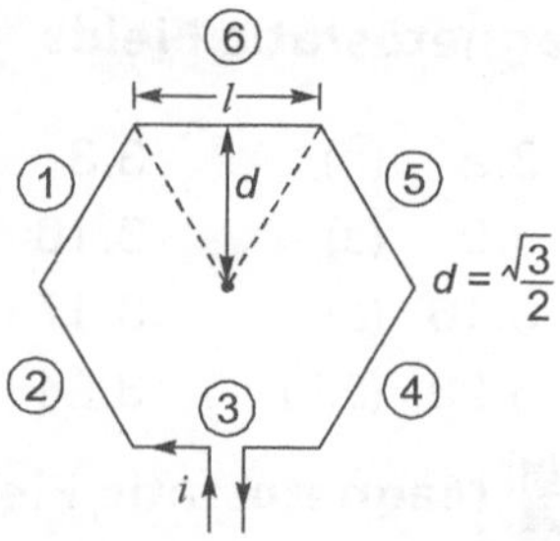

Here, B at point P is

$$B = \frac{\mu_0 I}{4\pi d}(\sin\theta_1 + \sin\theta_2)$$

We know for each segment of hexagon

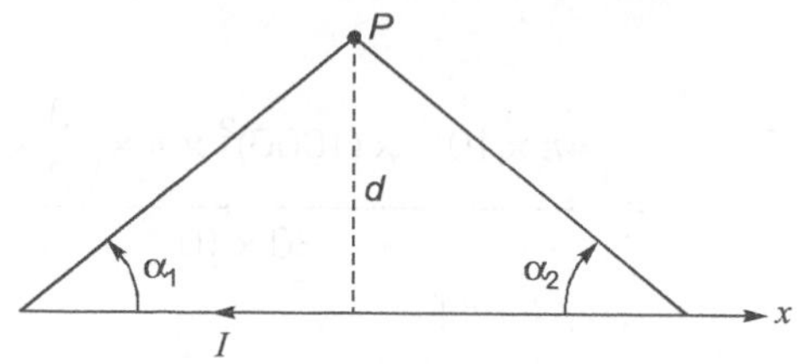

Magnetic field intensity due to elemental length dx,

$$dH = \frac{I\,dx}{4\pi r^2}\,\boldsymbol{i} \times u_r$$

where,

$\boldsymbol{i} \times u_r = (\sin\theta)\hat{k}$ .....Perpendicular to plane of paper

$$H = \int dH = \int \frac{I\sin\theta}{4\pi r^2}(dx)\hat{k} \text{ and } B = \mu_0 H$$

So, $B_p = \frac{\mu_0 I}{4\pi d}(\cos\alpha_1 + \cos\alpha_2)\,\hat{k}$

$d = \frac{\sqrt{3}}{2}l$ ... where $l = 1$ m

and $\alpha_1 = \alpha_2 = 60°$

$$|B_p| = \frac{\mu_0 I}{4\pi\frac{\sqrt{3}}{2}}(\cos 60° + \cos 60°)$$

For all six segments,

$$6 \times |B_p| = \frac{4\pi \times 10^{-7} \times 6 \times 1}{4\pi \times \frac{\sqrt{3}}{2}}(1)$$

$$= \frac{12}{\sqrt{3}} \times 10^{-7} = \frac{1.20}{\sqrt{3}} \times 10^{-6}\,T \Rightarrow B = 0.69\,\mu\text{T}$$

3.18 Sol.

$$\phi \propto I$$

$$\phi = MI$$

$$\vec{B} = \frac{\mu_0 I}{2\pi\rho}\hat{a}_\phi \quad (\vec{B} \text{ due to infinite long line})$$

Magnetic flux crossing square loop is

$$\phi = \iint \vec{B}\cdot\vec{ds}$$

$$= \iint \frac{\mu_o I}{2\pi\rho}\hat{a}_\phi\cdot(d\rho dz)\,\hat{a}_\phi = \frac{\mu_o I}{2\pi}\int_{\rho=1}^{2}\frac{d\rho}{\rho}\int_{z=0}^{1}dz$$

$$\phi = \frac{\mu_o I}{2\pi}(\ln\rho)_{\rho=1}^{2}(z)_{z=0}^{1}$$

$$\phi = \frac{\mu_o I}{2\pi}(\ln 2)$$

$$M = \frac{\phi}{I} = \frac{\mu_o(\ln 2)}{2\pi} = \frac{4\pi\times10^{-7}(\ln_2)}{2\pi}$$

$$M = 1.386\times10^{-7}\text{ Henry} \simeq 138.63\text{ nH}$$

3.19 (a)

If $\vec{B}$ is magnetic flux density then $\vec{\nabla}\cdot\vec{B} = 0$

$$\vec{\nabla}\cdot\vec{B} = \frac{\partial Bx}{\partial x}+\frac{\partial By}{\partial y}+\frac{\partial Bz}{\partial z}$$

$$\frac{\partial}{\partial x}(10x)+\frac{\partial}{\partial y}(20y)+\frac{\partial}{\partial z}(-30z) = \vec{\nabla}\cdot\vec{B}$$

$$\vec{\nabla}\cdot\vec{B} = 10+20-30 = 0$$

3.20 (c)

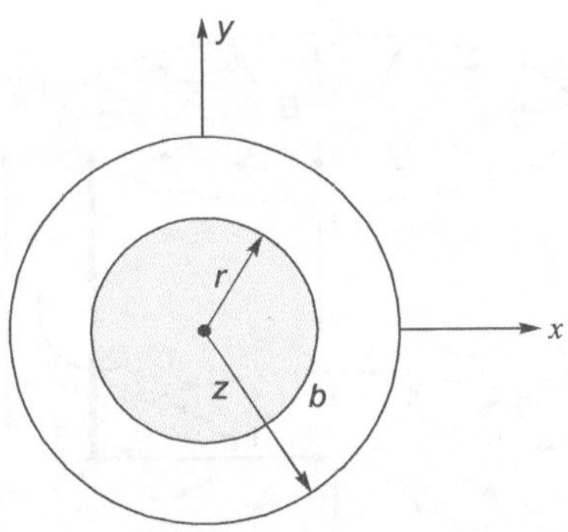

Given : $\vec{J} = J_a r^3\hat{z}$

$$\therefore \quad I = \int_s \vec{J}\cdot\vec{ds};\ \vec{ds} = rdrd\phi\hat{z}$$

$$\Rightarrow I = \int J_a r^3\hat{z}\cdot rdrd\phi\hat{z} = J_a\int_{r=0}^{r}r^4dr\int_{\phi=0}^{2\pi}d\phi$$

$$= J_a\left.\frac{r^5}{5}\right|_0^r\cdot\phi|_0^{2\pi} = \frac{J_a(2\pi)r^5}{5}$$

As $\oint\vec{H}\cdot\vec{dL} = I_{enc} = \int_s\vec{J}\cdot\vec{ds}$

$$\Rightarrow \quad H(2\pi r) = \frac{J_a(2\pi)r^5}{5} \Rightarrow H = \frac{J_a r^4}{5}$$

3.21 (b)

Given: $\vec{H} = x^2\hat{i}+x^2y^2\hat{j}+x^2y^2z^2\hat{k}$ A/m

As $\vec{J} = \nabla\times\vec{H}$

$$\Rightarrow \quad \vec{J} = \begin{vmatrix}\hat{a}_x & \hat{a}_y & \hat{a}_z\\ \frac{\partial}{\partial x} & \frac{\partial}{\partial y} & \frac{\partial}{\partial z}\\ x^2 & x^2y^2 & x^2y^2z^2\end{vmatrix}$$

$$= 2x^2yz^2\hat{a}_x - 2xy^2z^2\hat{a}_y + 2xy^2\hat{a}_z$$

$$\therefore \quad \vec{J}\Big|_{1,2,1} = 2(1)(2)(1)\hat{a}_x - 2(1)(2)^2(1)^2\hat{a}_y + 2(1)(2)^2\hat{a}_z$$

$$\Rightarrow \quad \vec{J} = 4\hat{a}_x - 8\hat{a}_y + 8\hat{a}_z$$

$$\therefore \quad |\vec{J}| = \sqrt{(4)^2+(8)^2+(8)^2} = 12$$

3.23 (2.5)

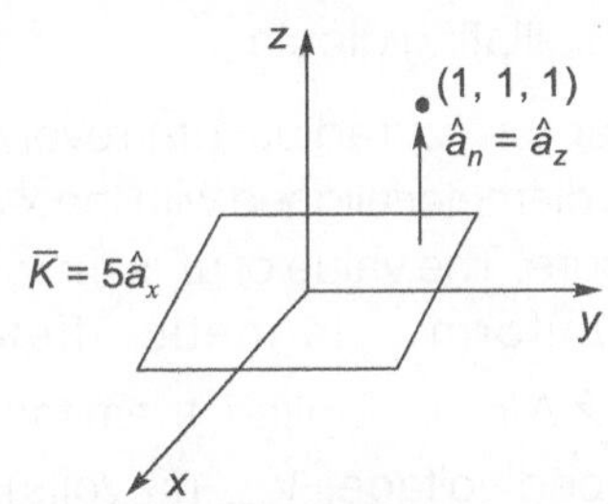

Magnetic field due to sheet current is

$$\bar{H} = \frac{1}{2}(\bar{K}\times\hat{a}_n)\text{ A/m}$$

Here, $\bar{K} = 5\hat{a}_x;\ \hat{a}_n = \hat{a}_z$

$$\therefore \quad \bar{H} = \frac{1}{2}(5\hat{a}_x\times\hat{a}_z) = 2.5(-\hat{a}_y)$$

Here, $\bar{H} = -2.5\hat{a}_y$ A/m

So, $|\bar{H}| = 2.5$ A/m

3.24 (2)

$$\nabla\cdot\bar{B} = 0;\quad \nabla\cdot\bar{H} = 0$$

$$\frac{1}{r^2}\frac{\partial}{\partial r}(r^2H_r)+\frac{1}{r\sin\theta}\frac{\partial}{\partial\theta}(\sin\theta H_\theta)+\frac{1}{r\sin\theta}\frac{\partial H_\phi}{\partial\phi} = 0$$

$$\frac{1}{r^2}\frac{\partial}{\partial r}\left(\frac{r^2}{r^3}P\cos\theta\right)+\frac{1}{r\sin\theta}\frac{\partial}{\partial\theta}\left(\frac{1}{r^3}\sin^2\theta\right)+0 = 0$$

$$\frac{1}{r^2}\frac{\partial}{\partial r}\left(\frac{P\cos\theta}{r}\right)+\frac{1}{r\sin\theta}\frac{\partial}{\partial\theta}\left(\frac{1}{r^3}\sin^2\theta\right)+0 = 0$$

$$\frac{1}{r^2}\cdot\frac{-P\cos\theta}{r^2}+\frac{1}{r\sin\theta}\times\frac{1}{r^3}(2\sin\theta\cos\theta) = 0$$

$$\frac{-P\cos\theta}{r^4}+\frac{2\cos\theta}{r^4} = 0 \Rightarrow P = 2$$

■■■■

4 Time Varying Fields

4.1 The laws of electromagnetic induction (Faraday's and Lenz's law) are summarized in the following equation

(a) $e = iR$ (b) $e = L\frac{di}{dt}$

(c) $e = -\frac{d\Psi}{dt}$ (d) none of these

[1998 : 1 M]

4.2 An electromagnetic field is radiated from
(a) a stationary point charge
(b) a capacitor with a DC voltage
(c) a conductor carrying a DC current
(d) an oscillating dipole [1999 : 1 M]

4.3 A circular turn of radius 1 m revolves at 60 rpm about its diameter aligned with the *x*-axis as shown in the figure. The value of μ_0 is $4\pi \times 10^{-7}$ in SI unit. If a uniform magnetic field intensity $\vec{H} = 10^7 \hat{z}$ A/m is applied, them the peak value if the induced voltage, V_{turn} (in Volts), is ______.

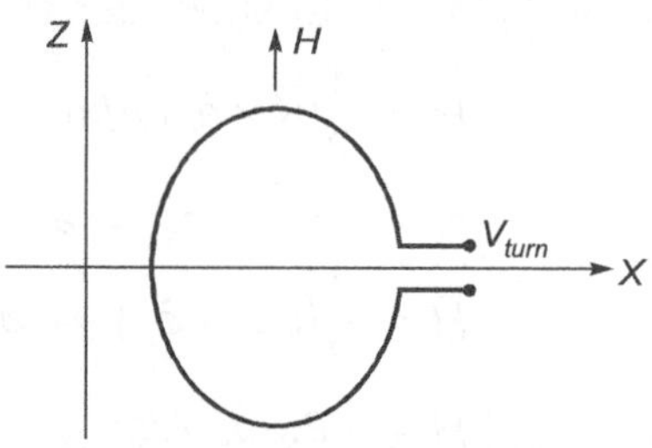

[2015 : 1 M, Set-2]

4.4 Consider a one-turn rectangular loop of wire placed in a uniform magnetic field as shown in the figure. The plane of the loop is perpendicular to the field lines. The resistance of the loop is 0.4 Ω, and its inductance is negligible. The magnetic flux density (in Tesla) is a function of time, and is given by $B(t) = 0.25 \sin\omega t$, where $\omega = 2\pi \times 50$ radian/second. The power absorbed (in Watt) by the loop from the magnetic field is ______.

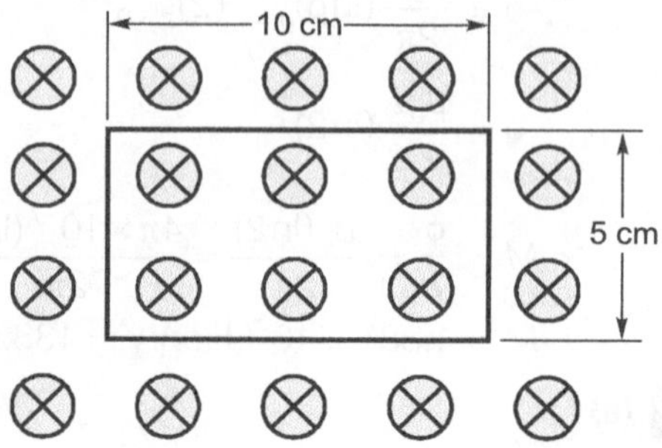

[2015 : 1 M, Set-1]

4.5 A rotating conductor of 1 m length is placed in a radially outward (about the *z*-axis) magnetic flux density (*B*) of 1 Tesla as shown in the figure below. Conductor is parallel to and at 1 m distance from the *z*-axis. The speed of the conductor in r.p.m. required to induce a voltage of 1 V across it, should be ______.

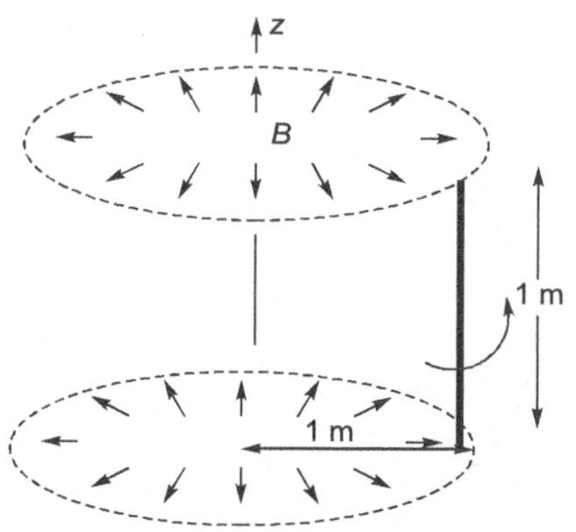

[2016 : 2 M, Set-2]

4.6 One coulomb of point charge moving with a uniform velocity $10\hat{x}$ m/s enters the region $x \geq 0$ having a magnetic flux density.

$\vec{B} = (10y\hat{x} + 10x\hat{y} + 10\hat{z})$ T .

The magnitude of force on the charge at $x = 0^+$ is ________ N.

($\hat{x}$, $\hat{y}$ and $\hat{z}$ are unit vectors along *x*-axis, *y*-axis and *z*-axis, respectively)

[2021 : 2 M]

■■■■

Answers Time Varying Fields

4.1 (c) 4.2 (d) 4.3 (248.05) 4.4 (0.193) 4.5 (9.55) 4.6 (100)

Explanations Time Varying Fields

4.1 (c)

According to Faraday's law $e \propto \frac{d\psi}{dt}$.

Even by considering Lenz's law we have,

$$e = -\frac{d\psi}{dt}$$

4.2 (d)

An oscillating dipole is a dipole (or antenna) through which when ac current is fed, it radiates EM wave.

4.3 Sol.

The circular turn rotate with 60 rpm, let the angle made by ring w.r.t. x-axis θ and $\theta = \omega_0 t$, the turn rotate at 60 rpm,

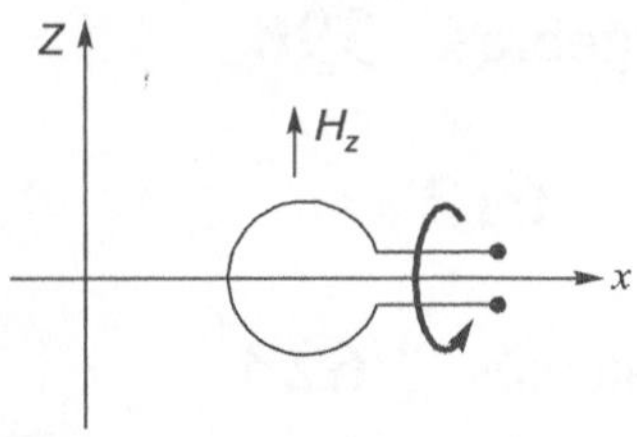

so, $\omega_0 = 2\pi$

So, the flux flowing through the circular turn will be

$$\Psi = (\mu_0 H_z \times \text{Area of turn} \times \cos\omega_0 t)$$
$$\Psi = 4\pi \times 10^{-7} \times 10^7 \text{ A/m} \times \pi \times 1^2 \times \cos\omega_0 t$$

Maximum voltage induced is $\left.\frac{d\Psi}{dt}\right|_{max}$

so, $$\left.\frac{d\Psi}{dt}\right|_{max} = (\omega_0 \times 4\pi \times \pi \sin\omega_0 t)_{max}$$
$$V_{max} = (4\pi^2 \times 2\pi) = 248.05 \text{ volts}$$

4.4 Sol.

$$A = 10 \text{ cm} \times 5 \text{ cm};$$
$$B(t) = 0.25 \sin\omega t,\ R = 0.4\ \Omega$$
$$P = i^2 R$$

$$\because \quad e = -\frac{d\phi}{dt} = -\frac{d}{dt}(B \times A)$$
$$= -\frac{d}{dt}[0.25 \times 50 \times 10^{-4} \sin\omega t]$$
$$e = -0.25 \times 50 \times 10^{-4}\, \omega \cos\omega t$$
$$= -1.25 \times 10^{-3}\, \omega \cos\omega t$$

$$P_{avg.} = \frac{e_{rms}^2}{R}$$

$$P = \int_0^T \frac{(12.5 \times 100\pi \times 10^{-4})^2}{0.4} \times \cos^2 \omega t\, d(\omega t)$$
$$= (12.5 \times \pi \times 10^{-2})^2 \times \frac{1}{2 \times 0.4} = 0.193 \text{ Watt}$$

4.5 Sol.

$$\text{Voltage induced} = \int_0^1 E_m \cdot dl$$

(where E_m is induced electric field)

$$= E_m \times \text{volts}$$

Since,

$$\text{Voltage induced} = 1 \text{ V}$$

So, $$E_m = 1 \text{ V/m}$$

As we know, $$E_m = \bar{V} \times \bar{B}$$

where, V = (radius of path) × (angular velocity)

$$\frac{1 \text{ V}}{m} = (V \times 1 \text{ Tesla})$$
$$v = 1 \text{ m/sec}$$
$$v = r \times \omega = 1 \text{ m/sec}$$

Since, $r = 1$ m, So, $\omega = 1$ rad/sec

Now from this we get,

$$\omega = 2 \times \pi \times \frac{N}{60} = 1 \text{ rad/sec}$$

$$N = \frac{30}{\pi} = 9.55 \text{ revolutions per minute}$$

4.6 (100)

Force on a charge moving with $\vec{v}$ velocity due to magnetic field is

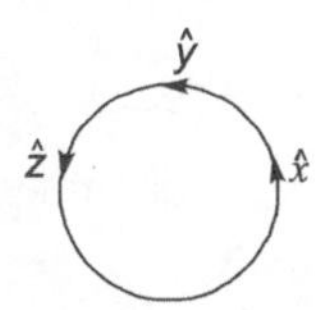

$$\vec{F} = Q(\vec{v} \times \vec{B})$$
$$= 1[10\hat{x} \times (10y\hat{x} + 10x\hat{y} + 10\hat{z}]$$
$$= 10(10x)\hat{z} + 10(10)(-\hat{y})$$
$$= 100x\hat{z} - 100\hat{y}$$
$$\left.\vec{F}\right|_{x=0^+} = -100\hat{y}$$
$$|\vec{F}| = \sqrt{(-100)^2} = 100 \text{ Newton}$$

Engineering Mathematics

UNIT

XI

CONTENTS

Engineering Mathematics

Syllabus

Linear Algebra: Matrix Algebra, Systems of Linear Equations, Eigenvalues, Eigenvectors.

Calculus: Mean Value Theorems, Theorems of Integral Calculus, Evaluation of Definite and Improper Integrals, Partial Derivatives, Maxima and Minima, Multiple Integrals, Fourier Series, Vector Identities, Directional Derivatives, Line Integral, Surface Integral, Volume Integral, Stokes's Theorem, Gauss's Theorem, Divergence Theorem, Green's Theorem.

Differential Equations: First Order Equations (Linear and Non-Linear), Higher Order Linear Differential Equations with Constant Coefficients, Method of Variation of Parameters, Cauchy's Equation, Euler's Equation, Initial and Boundary Value Problems, Partial Differential Equations, Method of Separation of Variables.

Complex Variables: Analytic Functions, Cauchy's Integral Theorem, Cauchy's Integral Formula, Taylor Series, Laurent Series, Residue Theorem, Solution Integrals.

Probability and Statistics: Sampling Theorems, Conditional Probability, Mean, Median, Mode, Standard Deviation, Random Variables, Discrete and Continuous Distributions, Poisson Distribution, Normal Distribution, Binomial Distribution, Correlation Analysis, Regression Analysis.

Analysis of Previous GATE Papers

Exam Year	1 Mark Ques.	2 Marks Ques.	3 Marks Ques.	5 Marks Ques.	Total Marks
1995	1	1	–	–	**3**
1996	1	–	–	–	**1**
1997	1	1	–	–	**3**
1998	3	2	–	–	**7**
1999	–	–	–	–	**–**
2000	–	–	–	–	**–**
2001	–	–	–	–	**–**
2002	2	–	–	–	**2**
2003	–	–	–	–	**–**
2004	–	–	–	–	**–**
2005	4	5	–	–	**14**
2006	–	5	–	–	**10**
2007	1	8	–	–	**17**
2008	3	5	–	–	**13**
2009	1	5	–	–	**11**
2010	2	4	–	–	**10**
2011	3	4	–	–	**11**

Exam Year	1 Mark Ques.	2 Mark Ques.	Total Marks
2012	4	3	**10**
2013	4	4	**12**
2014 Set-1	5	4	**13**
2014 Set-2	5	3	**11**
2014 Set-3	3	3	**9**
2015 Set-1	3	3	**9**
2015 Set-2	2	3	**8**
2016 Set-1	5	3	**11**
2016 Set-2	5	5	**15**
2017 Set-1	2	4	**10**
2017 Set-2	5	3	**11**
2018	5	5	**15**
2019	6	3	**12**
2020	4	4	**12**
2021	4	4	**12**
2022	1	7	**15**
2023	5	4	**13**
2024	4	6	**16**

1 Linear Algebra

1.1 The inverse of the matrix, $S = \begin{bmatrix} 1 & -1 & 0 \\ 1 & 1 & 1 \\ 0 & 0 & 1 \end{bmatrix}$ is

(a) $\begin{bmatrix} 1 & 0 & 1 \\ 0 & 0 & 0 \\ 0 & 1 & 1 \end{bmatrix}$ (b) $\begin{bmatrix} 0 & 1 & 1 \\ -1 & -1 & 1 \\ 1 & 0 & 1 \end{bmatrix}$

(c) $\begin{bmatrix} 2 & 2 & -2 \\ -2 & 2 & -2 \\ 0 & 2 & 2 \end{bmatrix}$ (d) $\begin{bmatrix} \frac{1}{2} & \frac{1}{2} & \frac{-1}{2} \\ \frac{-1}{2} & \frac{1}{2} & \frac{-1}{2} \\ 0 & 0 & 1 \end{bmatrix}$

[1995 : 2 M]

1.2 Given the matrix, $A = \begin{bmatrix} 0 & 1 & 0 \\ 0 & 0 & 1 \\ -6 & -11 & -6 \end{bmatrix}$. Its eigen values are

[1995 : 2 M]

1.3 The rank of the following $(n + 1) \times (n + 1)$ matrix, where 'a' is a real number is

$$\begin{bmatrix} 1 & a & a^2 & \dots & a^n \\ 1 & a & a^2 & \dots & a^n \\ . & & & & \\ . & & & & \\ 1 & a & a^2 & \dots & a^n \end{bmatrix}$$

(a) 1
(b) 2
(c) n
(d) depends on the value of a **[1996 : 1 M]**

1.4 Express the given matrix, $A = \begin{bmatrix} 2 & 1 & 5 \\ 4 & 8 & 13 \\ 6 & 27 & 31 \end{bmatrix}$ as a product of triangular matrices L and U where the diagonal elements of the upper triangular matrices U are unity and L is an lower triangular matrix.

[1997 : 2 M]

1.5 The vector $\begin{bmatrix} 1 \\ 2 \\ -1 \end{bmatrix}$ is an eigen vector of $A = \begin{bmatrix} -2 & 2 & -3 \\ 2 & 1 & -6 \\ -1 & -2 & 0 \end{bmatrix}$ if the eigen value of A is

(a) 1 (b) 2
(c) 5 (d) – 1 **[1998 : 1 M]**

1.6 $A = \begin{bmatrix} 2 & 0 & 0 & -1 \\ 0 & 1 & 0 & 0 \\ 0 & 0 & 3 & 0 \\ -1 & 0 & 0 & 4 \end{bmatrix}$ The sum of the eigen value of the matrix A is

(a) 10 (b) –10
(c) 24 (d) 22 **[1998 : 1 M]**

1.7 If $A = \begin{bmatrix} 5 & 0 & 2 \\ 0 & 3 & 0 \\ 2 & 0 & 1 \end{bmatrix}$ then $A^{-1} =$

(a) $\begin{bmatrix} 1 & 0 & -2 \\ 0 & 1/3 & 0 \\ -2 & 0 & 5 \end{bmatrix}$ (b) $\begin{bmatrix} 5 & 0 & 2 \\ 0 & -1/3 & 0 \\ 2 & 0 & 1 \end{bmatrix}$

(c) $\begin{bmatrix} 1/5 & 0 & 1/2 \\ 0 & 1/3 & 0 \\ 1/2 & 0 & 1 \end{bmatrix}$ (d) $\begin{bmatrix} 1/5 & 0 & -1/2 \\ 0 & 1/3 & 0 \\ -1/2 & 0 & 1 \end{bmatrix}$

[1998 : 2 M]

1.8 A set of linear equation is represented by the matrix equation $AX = B$. The necessary condition for the existence of a solution for this system is

(a) A must be invertible
(b) B must be linearly dependent of columns of A
(c) B must be linearly independent on the columns of A
(d) None **[1998 : 1 M]**

1.9 The determinant of the matrix:

$\begin{bmatrix} 1 & 0 & 0 & 0 \\ 100 & 1 & 0 & 0 \\ 100 & 200 & 1 & 0 \\ 100 & 200 & 300 & 1 \end{bmatrix}$ is

(a) 100 (b) 200
(c) 1 (d) 300

[2002 : 1 M]

1.10 In the matrix equation $Px = q$, which of the following is a necessary condition for the existence of at least one solution for the unknown vector x

(a) Augmented matrix $[Pq]$ must have the same rank as matrix P
(b) Vector q must have only non-zero elements
(c) Matrix P must be singular
(d) Matrix P must be square

[2005 : 1 M]

1.11 For the matrix $p = \begin{bmatrix} 3 & -2 & 2 \\ 0 & -2 & 1 \\ 0 & 0 & 1 \end{bmatrix}$, one of the eigen values is equal to –2. Which of the following is an eigen vector?

(a) $\begin{bmatrix} 3 \\ -2 \\ 1 \end{bmatrix}$ (b) $\begin{bmatrix} -3 \\ 2 \\ -1 \end{bmatrix}$

(c) $\begin{bmatrix} 1 \\ -2 \\ 3 \end{bmatrix}$ (d) $\begin{bmatrix} 2 \\ 5 \\ 0 \end{bmatrix}$

[2005 : 2 M]

1.12 If $R = \begin{bmatrix} 1 & 0 & -1 \\ 2 & 1 & -1 \\ 2 & 3 & 2 \end{bmatrix}$, then top row of R^{-1} is

(a) [5 6 4] (b) [5 –3 1]
(c) [2 0 –1] (d) [2 –1 1/2]

[2005 : 2 M]

Statement for Linked Answer Question (1.13 & 1.14):

$$P = \begin{bmatrix} -10 \\ 1 \\ 3 \end{bmatrix}^T, Q = \begin{bmatrix} -2 \\ -5 \\ 9 \end{bmatrix}^T, R = \begin{bmatrix} 2 \\ -7 \\ 12 \end{bmatrix}^T \text{ are three vectors.}$$

1.13 An orthogonal set of vectors having a span that contains P, Q, R is

(a) $\begin{bmatrix} -6 \\ -3 \\ +6 \end{bmatrix} \begin{bmatrix} 4 \\ -2 \\ 3 \end{bmatrix}$

(b) $\begin{bmatrix} -4 \\ 2 \\ 4 \end{bmatrix} \begin{bmatrix} 5 \\ 7 \\ -11 \end{bmatrix} \begin{bmatrix} 8 \\ 2 \\ -3 \end{bmatrix}$

(c) $\begin{bmatrix} 6 \\ 7 \\ -1 \end{bmatrix} \begin{bmatrix} -3 \\ 2 \\ -2 \end{bmatrix} \begin{bmatrix} 3 \\ 9 \\ -4 \end{bmatrix}$

(d) $\begin{bmatrix} 4 \\ 3 \\ 11 \end{bmatrix} \begin{bmatrix} 1 \\ 31 \\ 3 \end{bmatrix} \begin{bmatrix} 5 \\ 3 \\ 4 \end{bmatrix}$

[2006 : 2 M]

1.14 The following vector is linearly dependent upon the solution to the previous problem

(a) $\begin{bmatrix} 8 \\ 9 \\ 3 \end{bmatrix}$ (b) $\begin{bmatrix} -2 \\ -17 \\ 30 \end{bmatrix}$

(c) $\begin{bmatrix} 4 \\ 4 \\ 5 \end{bmatrix}$ (d) $\begin{bmatrix} 13 \\ 2 \\ -3 \end{bmatrix}$

[2006 : 2 M]

1.15 $X = [x_1 x_2 ... x_n]^T$ is an n-tuple nonzero vector. The $n \times n$ matrix $V = XX^T$

(a) has rank zero (b) has rank 1
(c) is orthogonal (d) has rank n

[2007 : 1 M]

1.16 Let x and y be two vectors in a 3 dimensional space and $<x, y>$ denote their dot product. Then the determinant

$$\det \begin{bmatrix} <x,x> & <x,y> \\ <y,x> & <y,y> \end{bmatrix}$$

(a) is zero when x and y are linearly independent
(b) is positive when x and y are linearly independent
(c) is non-zero for all non-zero x and y
(d) is zero only when either x or y is zero

[2007 : 2 M]

1.17 The linear operation $L(x)$ is defined by the cross product $L(x) = bX_x$, where $b = [0\ 1\ 0]^T$ and $x = [x_1 x_2 x_3]^T$ are three dimensional vectors. The 3×3 matrix M of this operation satisfies

$$L(x) = M \begin{bmatrix} x_1 \\ x_2 \\ x_3 \end{bmatrix}$$

Then the eigenvalues of M are

(a) 0, +1, –1 (b) 1, –1, 1
(c) i, –i, 1 (d) i, –i, 0

[2007 : 2 M]

Statement for Linked Answer Questions (1.18 and 1.19):

Cayley-Hamilton Theorem states that a square matrix satisfies its own characteristic equation. Consider a matrix

$$A = \begin{bmatrix} -3 & 2 \\ -1 & 0 \end{bmatrix}$$

1.18 A satisfies the relation

(a) $A + 3I + 2A^{-1} = 0$
(b) $A^2 + 2A + 2I = 0$
(c) $(A + I)(A + 2I)$
(d) $\exp(A) = 0$

[2007 : 2 M]

1.19 A^9 equals

(a) $511A + 510I$ (b) $309A + 104I$
(c) $154A + 155I$ (d) $\exp(9A)$

[2007 : 2 M]

1.20 The characteristic equation of a (3 × 3) matrix P is defined as

$a(\lambda) = |\lambda I - P| = \lambda^3 + \lambda^2 + 2\lambda + 1 = 0$

If I denotes identity matrix, then the inverse of matrix P will be

(a) $(P^2 + P + 2I)$ (b) $(P^2 + P + I)$
(c) $-(P^2 + P + I)$ (d) $-(P^2 + P + 2I)$

[2008 : 1 M]

1.21 If the rank of a (5 × 6) matrix Q is 4, then which one of the following statements is correct?

(a) Q will have four linearly independent rows and four linearly independent columns.
(b) Q will have four linearly independent rows and five linearly independent columns.
(c) QQ^T will be invertible.
(d) Q^TQ will be invertible.

[2008 : 1 M]

1.22 A is $m \times n$ full rank matrix with $m > n$ and I is an identity matrix. Let matric $A' = (A^TA)^{-1}A^T$, Then, which one of the following statement is FALSE?

(a) $AA'A = A$ (b) $(AA')^2 = A$
(c) $A'A = I$ (d) None of these

[2008 : 2 M]

1.23 Let P be a 2 × 2 real orthogonal matrix and $\vec{x}$ is a real vector $[x_1, x_2]^T$ with length $\|\vec{x}\| = (x_1^2 + x_2^2)^{\frac{1}{2}}$.

Then, which one of the following statements is correct?

(a) $\|P\vec{x}\| \le \|\vec{x}\|$ where at least one vector satisfies $\|P\vec{x}\| < \|\vec{x}\|$
(b) $\|P\vec{x}\| \le \|\vec{x}\|$ for all vectors $\vec{x}$
(c) $\|P\vec{x}\| \ge \|\vec{x}\|$ where at least one vector satisfies $\|P\vec{x}\| > \|\vec{x}\|$
(d) No relationship can be established between $\|\vec{x}\|$ and $\|P\vec{x}\|$

[2008 : 2 M]

1.24 The trace and determinant of a 2 × 2 matrix are known to be –2 and –35 respectively. It eigenvalues are

(a) –30 & –5 (b) –37 & –1
(c) –7 & 5 (d) 17.5 & –2

[2009 : 1 M]

1.25 An eigenvector of $P = \begin{pmatrix} 1 & 1 & 0 \\ 0 & 2 & 2 \\ 0 & 0 & 3 \end{pmatrix}$ is

(a) $[-1 \ \ 1 \ \ 1]^T$ (b) $[1 \ \ 2 \ \ 1]^T$
(c) $[1 \ \ -1 \ \ 2]^T$ (d) $[2 \ \ 1 \ \ -1]^T$

[2010 : 2 M]

1.26 For the set of equations:

$x_1 + 2x_2 + x_3 + 4x_4 = 2$
$3x_1 + 6x_2 + 3x_3 + 12x_4 = 6$

the following statement is true:

(a) Only the trivial solution $x_1 = x_2 = x_3 = x_4 = 0$ exists
(b) There are no solution
(c) A unique non-trivial solution exists
(d) Multiple non-trivial solutions exist

[2010 : 2 M]

1.27 The matrix $[A] = \begin{bmatrix} 2 & 1 \\ 4 & -1 \end{bmatrix}$ is decomposed into a product of a lower triangular matrix $x[L]$ and an upper triangular matrix $[U]$. The properly decomposed $[L]$ and $[U]$ matrices respectively are

(a) $\begin{bmatrix} 1 & 0 \\ 4 & -1 \end{bmatrix}$ and $\begin{bmatrix} 1 & 1 \\ 0 & -2 \end{bmatrix}$
(b) $\begin{bmatrix} 2 & 0 \\ 4 & -1 \end{bmatrix}$ and $\begin{bmatrix} 1 & 1 \\ 0 & 1 \end{bmatrix}$
(c) $\begin{bmatrix} 1 & 0 \\ 4 & 1 \end{bmatrix}$ and $\begin{bmatrix} 2 & 1 \\ 0 & -1 \end{bmatrix}$
(d) $\begin{bmatrix} 2 & 0 \\ 4 & -3 \end{bmatrix}$ and $\begin{bmatrix} 1 & 0.5 \\ 0 & 1 \end{bmatrix}$

[2011 : 2 M]

1.28 Given that, $A = \begin{bmatrix} -5 & -3 \\ 2 & 0 \end{bmatrix}$ and $I = \begin{bmatrix} 1 & 0 \\ 0 & 1 \end{bmatrix}$, the value A^3 is

(a) $15A + 12I$ (b) $19A + 30I$
(c) $17A + 15I$ (d) $17A + 21I$

[2012 : 2 M]

1.29 The equation $\begin{bmatrix} 2 & -2 \\ 1 & -1 \end{bmatrix}\begin{bmatrix} x_1 \\ x_2 \end{bmatrix} = \begin{bmatrix} 0 \\ 0 \end{bmatrix}$ has

(a) no solution

(b) only one solution $\begin{bmatrix} x_1 \\ x_2 \end{bmatrix} = \begin{bmatrix} 0 \\ 0 \end{bmatrix}$

(c) non-zero unique solution

(d) multiple solutions [2013 : 1 M]

1.30 A matrix has eigenvalues –1 and –2. The corresponding eigen vectors are $\begin{bmatrix} 1 \\ -1 \end{bmatrix}$ and $\begin{bmatrix} 1 \\ -2 \end{bmatrix}$ respectively. The matrix is

(a) $\begin{bmatrix} 1 & 1 \\ -1 & -2 \end{bmatrix}$ (b) $\begin{bmatrix} 1 & 2 \\ -2 & -4 \end{bmatrix}$

(c) $\begin{bmatrix} -1 & 0 \\ 0 & -2 \end{bmatrix}$ (d) $\begin{bmatrix} 0 & 1 \\ -2 & -3 \end{bmatrix}$ [2013 : 2 M]

1.31 Given a system of equations:

$$x + 2y + 2z = b_1$$
$$5x + y + 3z = b_2$$

Which of the following is true regarding its solution?

(a) The system has a unique solution for any given b_1 and b_2

(b) The system will have infinitely many solutions for any given b_1 and b_2

(c) Whether or not a solution exists depends on the given b_1 and b_2

(d) The system would have no solution for any values of b_1 and b_2 [2014 : 1 M, Set-1]

1.32 A system matrix is given as follows:

$$A \begin{bmatrix} 0 & 1 & -1 \\ -6 & -11 & 6 \\ -6 & -11 & 5 \end{bmatrix}$$

The absolute value of the ratio of the maximum eigenvalue to the minimum eigenvalue is ______.

[2014 : 2 M, Set-1]

1.33 Which one of the following statements is true for all real symmetric matrices?

(a) All the eigenvalues are real.

(b) All the eigenvalues are positive.

(c) All the eigenvalues are distinct.

(d) Sum of all the eigenvalues is zero.

[2014 : 1 M, Set-2]

1.34 Two matrices A and B are given below:

$$A = \begin{bmatrix} p & q \\ r & s \end{bmatrix}; \; B = \begin{bmatrix} p^2+q^2 & pr+qs \\ pr+qs & r^2+s^2 \end{bmatrix}$$

If the rank of matrix A is N, then the rank of matrix B is

(a) $\frac{N}{2}$ (b) $N-1$

(c) N (d) $2N$ [2014 : 1 M, Set-3]

1.35 If the sum of the diagonal elements of a 2 × 2 symmetric matrix is –6, then the maximum possible value of determinant of the matrix is ____.

[2015 : 1 M, Set-1]

1.36 We have a set of 3 linear equations in 3 unknowns. '$X \equiv Y$' means X and Y are not equivalent statements and '$X \not\equiv Y$' means X and Y are not equivalent statements.

P : There is a unique solution.

Q : The equations are linearly independent.

R : All eigenvalues of the coefficient matrix are nonzero.

S : The determinant of the coefficient matrix is nonzero.

Which one of the following is TRUE?

(a) $P \equiv Q \equiv R \equiv S$ (b) $P \equiv R \not\equiv Q \equiv S$

(c) $P \equiv Q \not\equiv R \equiv S$ (d) $P \not\equiv Q \not\equiv R \not\equiv S$

[2015 : 1 M, Set-2]

1.37 The maximum value of "a" such that the matrix $\begin{pmatrix} -3 & 0 & -2 \\ 1 & -1 & 0 \\ 0 & a & -2 \end{pmatrix}$ has three linearly independent real eigenvectors is

(a) $\frac{2}{3\sqrt{3}}$ (b) $\frac{1}{3\sqrt{3}}$

(c) $\frac{1+2\sqrt{3}}{3\sqrt{3}}$ (d) $\frac{1+\sqrt{3}}{3\sqrt{3}}$ [2015 : 2 M]

1.38 Consider a 3 × 3 matrix with every element being equal to 1. Its only non-zero eigenvalue is ______.

[2016 : 1 M, Set-1]

1.39 Let the eigenvalues of a 2 × 2 matrix A be 1, –2 with eigenvectors x_1 and x_2 respectively. Then the eigenvalues and eigenvectors of the matrix $A^2 - 3A + 4I$ would, respectively, be

(a) 2, 14; x_1, x_2 (b) 2, 14; $x_1 + x_2, x_1 - x_2$

(c) 2, 0; x_1, x_2 (d) 2, 0; $x_1 + x_2, x_1 - x_2$

[2016 : 2 M, Set-1]

1.40 Let A be a 4 × 3 real matrix with rank 2. Which one of the following statement is TRUE?

(a) Rank of A^TA is less than 2.

(b) Rank of A^TA is equal to 2.

(c) Rank of A^TA is greater than 2.

(d) Rank of $A^T A$ can be any number between 1 and 3. [2016 : 2 M, Set-1]

1.41 A 3 × 3 matrix P is such that, $P^3 = P$. Then the eigenvalues of P are
(a) 1, 1, –1
(b) 1, 0.5 + j0.866, 0.5 – j0.866
(c) 1, –0.5 + j0.866, – 0.5 – j0.866
(d) 0, 1, –1
[2016 : 1 M, Set-2]

1.42 Let $P = \begin{bmatrix} 3 & 1 \\ 1 & 3 \end{bmatrix}$. Consider the set S of all vectors $\begin{pmatrix} x \\ y \end{pmatrix}$ such that $a^2 + b^2 = 1$ where $\begin{pmatrix} a \\ b \end{pmatrix} = P\begin{pmatrix} x \\ y \end{pmatrix}$. Then S is
(a) a circle of radius $\sqrt{10}$
(b) a circle of radius $\frac{1}{\sqrt{10}}$
(c) an ellipse with major axis along $\begin{pmatrix} 1 \\ 1 \end{pmatrix}$
(d) an ellipse with minor axis along $\begin{pmatrix} 1 \\ 1 \end{pmatrix}$
[2016 : 2 M, Set-2]

1.43 The matrix, $A = \begin{bmatrix} \frac{3}{2} & 0 & \frac{1}{2} \\ 0 & -1 & 0 \\ \frac{1}{2} & 0 & \frac{3}{2} \end{bmatrix}$ has three distinct eigenvalues and one of its eigenvectors is $\begin{bmatrix} 1 \\ 0 \\ 1 \end{bmatrix}$.

Which one of the following can be another eigenvector of A?
(a) $\begin{bmatrix} 0 \\ 0 \\ -1 \end{bmatrix}$ (b) $\begin{bmatrix} -1 \\ 0 \\ 0 \end{bmatrix}$
(c) $\begin{bmatrix} 1 \\ 0 \\ -1 \end{bmatrix}$ (d) $\begin{bmatrix} 1 \\ -1 \\ 1 \end{bmatrix}$
[2017 : 1 M, Set-1]

1.44 The eigen values of the matrix given below are
$$\begin{bmatrix} 0 & 1 & 0 \\ 0 & 0 & 1 \\ 0 & -3 & -4 \end{bmatrix}$$
(a) (0, –1, –3) (b) (0, –2, –3)
(c) (0, 2, 3) (d) (0, 1, 3)
[2017 : 2 M, Set-2]

1.45 Consider a non-singular 2 × 2 square matrix A. If trace (A) = 4 and trace (A^2) = 5, the determinant of the matrix A is ______ (upto 1 decimal place).
[2018 : 1 M]

1.46 Let $A = \begin{bmatrix} 1 & 0 & -1 \\ -1 & 2 & 0 \\ 0 & 0 & -2 \end{bmatrix}$ and $B = A^3 - A^2 - 4A + 5I$, where I is the 3 × 3 identity matrix. The determinant of B is ______ (upto 1 decimal place).
[2018 : 2 M]

1.47 M is a 2 × 2 matrix with eigenvalues 4 and 9. The eigenvalues of M^2 are
(a) –2 and –3 (b) 2 and 3
(c) 4 and 9 (d) 16 and 81
[2019 : 1 M]

1.48 The rank of the matrix, $M = \begin{bmatrix} 0 & 1 & 1 \\ 1 & 0 & 1 \\ 1 & 1 & 0 \end{bmatrix}$, is ____ .
[2019 : 1 M]

1.49 Consider a 2 × 2 matrix $M = [v_1 \ \ v_2]$, where v_1 and v_2 are the column vectors. Suppose $M^{-1} = \begin{bmatrix} u_1^T \\ u_2^T \end{bmatrix}$, where u_1^T and u_2^T are the row vectors. Consider the following statements:

Statement-1 : $u_1^T v_1 = 1$ and $u_2^T v_2 = 1$

Statement-2 : $u_1^T v_2 = 0$ and $u_2^T v_1 = 0$

Which of the following options is correct?
(a) Statement 2 is true and statement 1 is false.
(b) Both the statements are false.
(c) Both the statements are true.
(d) Statement 1 is true and statement 2 is false.
[2019 : 2 M]

1.50 The number of purely real elements in a lower triangular representation of the given 3 × 3 matrix, obtained through the given decomposition is
$$\begin{bmatrix} 2 & 3 & 3 \\ 3 & 2 & 1 \\ 3 & 1 & 7 \end{bmatrix} = \begin{bmatrix} a_{11} & 0 & 0 \\ a_{12} & a_{22} & 0 \\ a_{13} & a_{23} & a_{33} \end{bmatrix} \begin{bmatrix} a_{11} & 0 & 0 \\ a_{12} & a_{22} & 0 \\ a_{13} & a_{23} & a_{33} \end{bmatrix}^T$$
(a) 6 (b) 5
(c) 8 (d) 9
[2020 : 2 M]

1.51 Let p and q be real numbers such that $p^2 + q^2 = 1$. The eigenvalues of the matrix $\begin{bmatrix} p & q \\ q & -p \end{bmatrix}$ are

(a) pq and $-pq$ (b) 1 and 1
(c) j and $-j$ (d) 1 and −1 [2021 : 1 M]

1.52 Let A be a 10 × 10 matrix such that A^5 is null matrix, and let I be the 10 × 10 identity matrix. The determinant of $A + I$ is ____. [2021 : 1 M]

1.53 Consider a 3 × 3 matrix A whose (i, j)-th element, $a_{i,j} = (i - j)^3$. Then the matrix A will be
(a) symmetric. (b) skew-symmetric.
(c) unitary. (d) null. [2022 : 1 M]

1.54 e^A denotes the exponential of a square matrix A. Suppose λ is an eigen value and v is the corresponding eigen-vector of matrix A.
Consider the following two statements :
Statement 1 : e^{λ} is an eigen value of e^A.
Statement 2 : v is an eigen-vector of e^A.
Which one of the following options is correct?
(a) Statement 1 is true and statement 2 is false.
(b) Statement 1 is false and statement 2 is true.
(c) Both the statements are correct.
(d) Both the statements are false.
[2022 : 2 M]

1.55 Consider a matrix $A = \begin{bmatrix} 1 & 0 & 0 \\ 0 & 4 & -2 \\ 0 & 1 & 1 \end{bmatrix}$. The matrix A satisfies the equation $6A^{-1} = A^2 + cA + dI$, where c and d are scalars and I is the identity matrix. Then $(c + d)$ is equal to
(a) 5 (b) 17
(c) −6 (d) 11 [2022 : 2 M]

1.56 For a given vector $\mathbf{w} = [1\ 2\ 3]^T$, the vector normal to the plane defined by $\mathbf{w}^T\mathbf{x} = 1$ is
(a) $[-2\ \ -2\ \ 2]^T$ (b) $[3\ \ 0\ \ -1]^T$
(c) $[3\ \ 2\ \ 1]^T$ (d) $[1\ \ 2\ \ 3]^T$
[2023 : 1 Mark]

1.57 In the figure, the vectors **u** and **v** are related as: **Au** = **v** by a transformation matrix **A**. The correct choice of **A** is

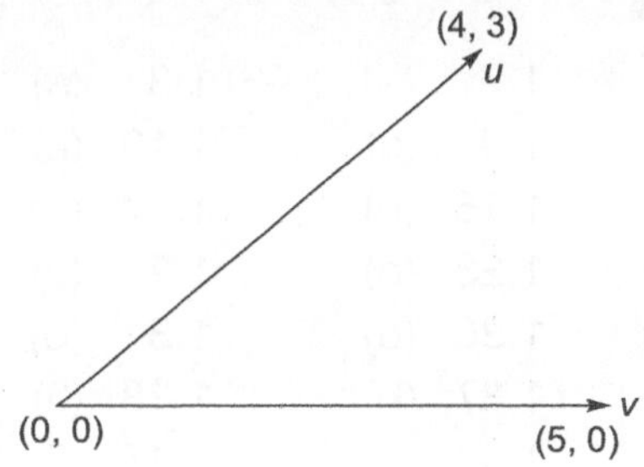

(a) $\begin{bmatrix} \frac{4}{5} & \frac{3}{5} \\ -\frac{3}{5} & \frac{4}{5} \end{bmatrix}$ (b) $\begin{bmatrix} \frac{4}{5} & -\frac{3}{5} \\ \frac{3}{5} & \frac{4}{5} \end{bmatrix}$

(c) $\begin{bmatrix} \frac{4}{5} & \frac{3}{5} \\ \frac{3}{5} & \frac{4}{5} \end{bmatrix}$ (d) $\begin{bmatrix} \frac{4}{5} & -\frac{3}{5} \\ \frac{3}{5} & -\frac{4}{5} \end{bmatrix}$

[2023 : 1 M]

1.58 Which one of the following matrices has an inverse?
(a) $\begin{bmatrix} 1 & 4 & 8 \\ 0 & 4 & 2 \\ 1 & 2 & 4 \end{bmatrix}$ (b) $\begin{bmatrix} 1 & 2 & 3 \\ 2 & 4 & 6 \\ 3 & 2 & 9 \end{bmatrix}$

(c) $\begin{bmatrix} 1 & 4 & 8 \\ 0 & 4 & 2 \\ 0.5 & 2 & 4 \end{bmatrix}$ (d) $\begin{bmatrix} 1 & 4 & 8 \\ 0 & 4 & 2 \\ 3 & 12 & 24 \end{bmatrix}$

[2024 : 1 M]

1.59 The sum of the eigen values of the matrix $A = \begin{bmatrix} 1 & 2 \\ 3 & 4 \end{bmatrix}^2$ is ______ (rounded off to the nearest integer).

[2024 : 1 M]

■■■■

Answers Linear Algebra

1.1 (d)	1.2 (–1,–2,–3)	1.3 (a)	1.4 (Sol.)	1.5 (c)	1.6 (a)	1.7 (a)
1.8 (b)	1.9 (c)	1.10 (a)	1.11 (d)	1.12 (b)	1.13 (a)	1.14 (b)
1.15 (b)	1.16 (b)	1.17 (d)	1.18 (a)	1.19 (a)	1.20 (d)	1.21 (a)
1.22 (b)	1.23 (b)	1.24 (c)	1.25 (b)	1.26 (d)	1.27 (d)	1.28 (b)
1.29 (d)	1.30 (d)	1.31 (b)	1.32 (3)	1.33 (a)	1.34 (c)	1.35 (9)
1.36 (a)	1.37 (b)	1.38 (3)	1.39 (a)	1.40 (b)	1.41 (d)	1.42 (d)
1.43 (c)	1.44 (a)	1.45 (5.5)	1.46 (1)	1.47 (d)	1.48 (3)	1.49 (c)
1.50 (*)	1.51 (d)	1.52 (1)	1.53 (b)	1.54 (c)	1.55 (a)	1.56 (d)
1.57 (a)	1.58 (a)	1.59 (29)				

Explanations Linear Algebra

1.1 (d)

$$|S| = 1(1-0)+1(1-0) = 1+1 = 2$$

$$\text{Adj } S = \begin{bmatrix} 1 & 1 & -1 \\ -1 & 1 & -1 \\ 0 & 0 & 2 \end{bmatrix}$$

$$S^{-1} = \frac{Adj\,S}{|S|} = \begin{bmatrix} 1/2 & 1/2 & -1/2 \\ -1/2 & 1/2 & -1/2 \\ 0 & 0 & 1 \end{bmatrix}$$

1.2 (–1, –2, –3)

The ch. equation is $|A-\lambda I| = 0$

$$\begin{vmatrix} 0-\lambda & 1 & 0 \\ 0 & 0-\lambda & 1 \\ -6 & -11 & -6-\lambda \end{vmatrix} = 0$$

$(-\lambda)\,[6\lambda+\lambda^2+11] - 1\,(0+6) = 0$

$-6\lambda^2 - \lambda^3 - 11\lambda - 6 = 0$

$\lambda^3 + 6\lambda^2 + 11\lambda + 6 = 0$

$\lambda = -1, -2, -3$

1.3 (a)

$R_2 - R_1,\ R_3 - R_1, \ldots\ldots R_{n+1} - R_1$

$$\begin{bmatrix} 1 & a & a^2 & \cdots & a^n \\ 0 & 0 & 0 & \cdots & 0 \\ \vdots & & & & \\ 0 & 0 & 0 & \cdots & 0 \end{bmatrix}$$

Number of non-zero rows = 1

$\therefore$ Rank of $A = 1$

1.4 Sol.

$$A = LU$$

$$\begin{bmatrix} 2 & 1 & 5 \\ 4 & 8 & 13 \\ 6 & 27 & 31 \end{bmatrix} = \begin{bmatrix} l_{11} & 0 & 0 \\ l_{21} & l_{22} & 0 \\ l_{31} & l_{32} & l_{33} \end{bmatrix}\begin{bmatrix} u_{11} & u_{12} & u_{13} \\ 0 & u_{22} & u_{23} \\ 0 & 0 & u_{33} \end{bmatrix}$$

Initially, $u_{11} = u_{22} = u_{33} = 1$

$$= \begin{bmatrix} l_{11} & 0 & 0 \\ l_{21} & l_{22} & 0 \\ l_{31} & l_{32} & l_{33} \end{bmatrix}\begin{bmatrix} 1 & u_{12} & u_{13} \\ 0 & 1 & u_{23} \\ 0 & 0 & 1 \end{bmatrix}$$

$$= \begin{bmatrix} l_{11} & l_{11}u_{12} & l_{11}u_{13} \\ l_{21} & l_{21}u_{12}+l_{22} & l_{21}u_{13}+l_{22}u_{23} \\ l_{31} & l_{31}u_{12}+l_{32} & l_{31}u_{13}+l_{32}u_{23}+l_{33} \end{bmatrix}$$

$$l_{21}u_{23} + l_{22}u_{23} = 13$$

$$(4)\left(\frac{5}{2}\right) + 6u_{23} = 13$$

$$10 + 6u_{23} = 12$$

$$u_{23} = \frac{1}{2}$$

$$l_{31}u_{12} + l_{32} = 27$$

$$6\left(\frac{1}{2}\right) + l_{32} = 27$$

$$l_{32} = 24$$

$$l_{31}u_{13} + l_{32}u_{23} + l_{33} = 31$$

$$6\left(\frac{5}{2}\right) + 24\left(\frac{1}{2}\right) + l_{33} = 31$$

$$15 + 12 + l_{33} = 31$$

$$l_{33} = 4$$

$$L = \begin{bmatrix} 2 & 0 & 0 \\ 4 & 6 & 0 \\ 6 & 24 & 4 \end{bmatrix};\quad U = \begin{bmatrix} 1 & 1/2 & 5/2 \\ 0 & 1 & 1/2 \\ 0 & 0 & 1 \end{bmatrix}$$

1.5 (c)

$AX = \lambda X$

$$AX = \begin{bmatrix} -2 & 2 & -3 \\ 2 & 1 & -6 \\ -1 & -2 & 0 \end{bmatrix}\begin{bmatrix} 1 \\ 2 \\ -1 \end{bmatrix} = \begin{bmatrix} 5 \\ 10 \\ -5 \end{bmatrix}$$

$$= 5\begin{bmatrix}1\\2\\-1\end{bmatrix} = 5X$$

$\therefore$ Eigen value is 5

1.6 (a)

Sum of eigen values = Trace of the matrix
= 2 + 1 + 3 + 4 = 10

1.7 (a)

$$|A| = 15 - 12 = 3$$

$$\text{Adj } A = \begin{bmatrix}3 & 0 & -6\\0 & 1 & 0\\-6 & 0 & 15\end{bmatrix}$$

$$A^{-1} = \frac{Adj\, A}{|A|} = \begin{bmatrix}1 & 0 & -2\\0 & 1/2 & 0\\-2 & 0 & 5\end{bmatrix}$$

1.8 (b)

In $AX = B$, the column matrix B must be linearly dependent of columns of A. Then only $\rho(A)$ will be equal to $\rho(AB)$.

1.9 (c)

The given matrix is a lower triangular matrix
$\therefore |A|$ = Product of its leading diagonal elements
= (1) (1) (1) (1) = 1

1.10 (a)

Rank $[Pq]$ = Rank $[P]$ is necessary for existence of at least one solution to $Px = q$.

1.11 (d)

Since matrix is triangular, the eigen values are the diagonal elements themselves namely $\lambda = 3, -2$ and 1.
Corresponding to eigen value $\lambda = -2$ let us find the eigen vector

$$[A - \lambda I]\hat{x} = 0$$

$$\begin{bmatrix}3-\lambda & -2 & 2\\0 & -2-\lambda & 1\\0 & 0 & 1-\lambda\end{bmatrix}\begin{bmatrix}x_1\\x_2\\x_3\end{bmatrix} = \begin{bmatrix}0\\0\\0\end{bmatrix}$$

Putting, $\lambda = -2$ in above equation we get,

$$\begin{bmatrix}5 & -2 & 2\\0 & 0 & 1\\0 & 0 & 3\end{bmatrix}\begin{bmatrix}x_1\\x_2\\x_3\end{bmatrix} = \begin{bmatrix}0\\0\\0\end{bmatrix}$$

Which gives the equations,

$$5x_1 - 2x_2 + 2x_3 = 0 \quad \text{... (i)}$$
$$x_3 = 0 \quad \text{... (ii)}$$
$$3x_3 = 0 \quad \text{... (iii)}$$

Since eqa (ii) and (iii) are same we have

$$5x_1 - 2x_2 + 2x_3 = 0 \quad \text{... (i)}$$
$$x_3 = 0 \quad \text{... (ii)}$$

Putting, $x_2 = k$, we get
$x_3 = 0$, $x_2 = k$ and $5x_1 - 2k + 2 \times 0 = 0$
$\Rightarrow \quad x_1 = 2/5\, k$
$\therefore$ Eigen vectors are of the form

$$\begin{bmatrix}x_1\\x_2\\x_3\end{bmatrix} = \begin{bmatrix}2/5\,k\\k\\0\end{bmatrix}$$

i.e. $\quad x_1 : x_2 : x_3 = 2/5\,k : k : 0$
$= 2/5 : 1 : 0 = 2 : 5 : 0$

$$\therefore \begin{bmatrix}x_1\\x_2\\x_3\end{bmatrix} = \begin{bmatrix}2\\5\\0\end{bmatrix} \text{ is an eigen vector of matrix } p.$$

1.12 (b)

$$R = \begin{bmatrix}1 & 0 & -1\\2 & 1 & -1\\2 & 3 & 2\end{bmatrix}$$

$$R^{-1} = \frac{adj(R)}{|R|} = \frac{[\text{cofactor }(R)]^T}{|R|}$$

$$|R| = \begin{vmatrix}1 & 0 & -1\\2 & 1 & -1\\2 & 3 & 2\end{vmatrix}$$

$$= 1(2+3) - 0(4+2) - 1(6-2)$$
$$= 5 - 4 = 1$$

Since we need only the top row of R^{-1}, we need to find only first column of (R) which after transpose will become first row of adj A).

$$\text{Co-factor of } (1, 1) = +\begin{vmatrix}1 & -1\\3 & 2\end{vmatrix} = 2 + 3 = 5$$

$$\text{Co-factor of } (2, 1) = +\begin{vmatrix}0 & -1\\1 & -1\end{vmatrix} = -3$$

$$\text{Co-factor of } (3, 1) = +\begin{vmatrix}0 & -1\\1 & -1\end{vmatrix} = +1$$

$$\therefore \text{Co-factor of } (A) = \begin{bmatrix}5 & - & -\\-3 & - & -\\1 & - & -\end{bmatrix}$$

$$\text{Adj }(A) = [\text{cof. }(A)]^T$$

$$= \begin{bmatrix} 5 & -3 & 1 \\ - & - & - \\ - & - & - \end{bmatrix}$$

Dividing by $|R| = 1$ gives

$$R^{-1} = \begin{bmatrix} 5 & -3 & 1 \\ - & - & - \\ - & - & - \end{bmatrix}$$

$\therefore$ Top row of $R^{-1} = [5\ -3\ 1]$

1.15 (b)

The $n \times n$ matrix, $V = XX^T$ has rank 1.

1.17 (d)

The cross product of $b = [0\ \ 1\ \ 0]^T$

and $X = [x_1\ \ x_2\ \ x_3]^T$ can be written as

$$b \times X = \begin{vmatrix} \hat{i} & \hat{j} & \hat{k} \\ 0 & 1 & 0 \\ x_1 & x_2 & x_3 \end{vmatrix}$$

$$= x_3\hat{i} + 0\hat{j} - x_1\hat{k} = [x_3\ \ 0\ \ -x_1]$$

Now, $L(x) = b \times X = M\begin{bmatrix} x_1 \\ x_2 \\ x_3 \end{bmatrix}$

where M is a 3×3 matrix.

Let, $M = \begin{bmatrix} C_1 & C_2 & C_3 \\ C_4 & C_5 & C_6 \\ C_7 & C_8 & C_9 \end{bmatrix}$

Now $M\begin{bmatrix} x_1 \\ x_2 \\ x_3 \end{bmatrix} = b \times X$

$$\Rightarrow \begin{bmatrix} C_1 & C_2 & C_3 \\ C_4 & C_5 & C_6 \\ C_7 & C_8 & C_9 \end{bmatrix}\begin{bmatrix} x_1 \\ x_2 \\ x_3 \end{bmatrix} = \begin{bmatrix} x_3 \\ 0 \\ -x_1 \end{bmatrix}$$

By matching LHS and RHS we get

$$\begin{bmatrix} 0 & 0 & 1 \\ 0 & 0 & 0 \\ -1 & 0 & 0 \end{bmatrix}\begin{bmatrix} x_1 \\ x_2 \\ x_3 \end{bmatrix} = \begin{bmatrix} x_3 \\ 0 \\ -x_1 \end{bmatrix}$$

So, $M = \begin{bmatrix} 0 & 0 & 1 \\ 0 & 0 & 0 \\ -1 & 0 & 0 \end{bmatrix}$

Now we have to find the eigen values of M

$|M - \lambda I| = 0$

$$\Rightarrow \begin{vmatrix} -\lambda & 0 & 1 \\ 0 & -\lambda & 0 \\ -1 & 0 & -\lambda \end{vmatrix} = 0$$

$\Rightarrow -\lambda(\lambda^2 - 0) + 1(0 - \lambda) = 0$

$\Rightarrow \lambda^3 + \lambda = 0$

$\Rightarrow \lambda(\lambda^2 + 1) = 0 \Rightarrow \lambda = 0, \lambda = \pm i$

So, the eigen values of M are i, $-i$ and 0.

1.18 (a)

$$A = \begin{bmatrix} -3 & 2 \\ -1 & 0 \end{bmatrix}$$

$$|A - \lambda I| = 0$$

$$\begin{vmatrix} -3-\lambda & 2 \\ -1 & 0-\lambda \end{vmatrix} = 0$$

$$(-3-\lambda)(-\lambda) + 2 = 0$$

$$\lambda^2 + 3\lambda + 2 = 0$$

A will satisfy this equation according to cayley hamition theorem

i.e., $A^2 + 3A + 2I = 0$

Multiplying by A^{-1} on both sides we get

$A^{-1}A^2 + 3A^{-1}A + 2A^{-1}I = 0$

$A + 3I + 2A^{-1} = 0$

1.19 (a)

To calculate A^9

start from $A^2 + 3A + 2I = 0$ which has been derived above

$\Rightarrow$ $A^2 = -3A - 2I$

$A^4 = A^2 \times A^2 = (-3A - 2I)(-3A - 2I)$

$= 9A^2 + 12A + 4I$

$= 9(-3A - 2I) + 12A + 4I = -15A - 14I$

$A^8 = A^4 \times A^4$

$= (-15A - 14I)(-15A - 14I)$

$= 225A^2 + 420A + 156I$

$= 225(-3A - 2I) + 420A\ 196I$

$= -255A - 254I$

$A^9 = A \times A^8 = A(-255A - 254I)$

$= -255A^2 - 254A$

$= -255(-3A - 2I) - 254A$

$= 511A + 510I$

1.20 (d)

If characteristic equation is

$\lambda^3 + \lambda^2 + 2\lambda + 1 = 0$

Then by cayley - hamilton theorem,

$P^3 + P^2 + 2P + I = 0$

$I = -P^3 - P^2 - 2P$

Multiplying by P^{-1} on both sides,

$P^{-1} = -P^2 - P - 2I = -(P^2 + P + 2I)$

1.21 (a)

If rank of (5 × 6) matrix is 4, then surely it must have exactly 4 linearly independent rows as will as 4 linearly in dependent columns.

1.22 (b)

Choice (a) $AA'A = A$ is correct

Since, $AA'A = A[(A^TA)^{-1}A^T]A = A[(A^TA)^{-1}A^TA]$

Put, $A^TA = P$

Then, $= A[P^{-1}P] = A.I = A$

Choice (c) $A'A = I$ is also correct since

$A'A = (A^TA)^{-1}A^TA = P^{-1}P = I$

1.23 (b)

Consider orthogonal matrix,

$$P = \begin{bmatrix} \cos\theta & -\sin\theta \\ \sin\theta & \cos\theta \end{bmatrix}$$

Properly of orthogonal matrix A,

$AA^T = I$

So, $\vec{P_x} = \begin{bmatrix} x_1\cos\theta - x_2\sin\theta \\ x_1\sin\theta + x_2\cos\theta \end{bmatrix}$

$$||\vec{P_x}|| = \sqrt{(x_1Cos\theta - x_2Sin\theta)^2 + (x_1Sin\theta + x_2Cos\theta)^2}$$

$$||\vec{P_x}|| = \sqrt{x_1^2 + x_2^2}$$

$||\vec{P_x}|| = ||.\bar{x}||$ for any vector $\bar{x}$

1.24 (c)

Trace = Sum of principal diagonal element.

1.25 (b)

Given, $P = \begin{bmatrix} 1 & 1 & 0 \\ 0 & 2 & 2 \\ 0 & 0 & 3 \end{bmatrix}$

P is triangular. So eigen values are the diagonal elements themselves. Eigen values are therefore, $\lambda_1 = 1, \lambda_2 = 2. \lambda_3 = 3.$

Now, the eigen value problem is $[A - \lambda I]\hat{x} = 0$

$$\begin{bmatrix} 1-\lambda & 1 & 0 \\ 0 & 2-\lambda & 2 \\ 0 & 0 & 3-\lambda \end{bmatrix}\begin{bmatrix} x_1 \\ x_2 \\ x_3 \end{bmatrix} = \begin{bmatrix} 0 \\ 0 \\ 0 \end{bmatrix}$$

Putting $\lambda_1 = 1$, we get the eigen vector corresponding to this eigen value,

$$\begin{bmatrix} 0 & 1 & 0 \\ 0 & 1 & 2 \\ 0 & 0 & 2 \end{bmatrix}\begin{bmatrix} x_1 \\ x_2 \\ x_3 \end{bmatrix} = \begin{bmatrix} 0 \\ 0 \\ 0 \end{bmatrix}$$

Which gives the equations,

$$x_2 = 0$$
$$x_2 + 2x_3 = 0$$
$$2x_3 = 0$$

The solution is $x_2 = 0, x_3 = 0, x_1 = k$

So, one eigen vector is $\hat{x}_1 = \begin{bmatrix} k \\ 0 \\ 0 \end{bmatrix}$ i.e., $x_1 : x_2 : x_3 = k : 0 : 0$

Since, none of the eigen vectors given in choices matches with this, ratio we need to proceed further and find the other eigen vectors corresponding to the other Eigen values.

Now, corresponding to $\lambda_2 = 2$, we get by substituting $\lambda = 2$, in the eigen value problem, the following set of equations,

$$\begin{bmatrix} -1 & 1 & 0 \\ 0 & 0 & 2 \\ 0 & 0 & 1 \end{bmatrix}\begin{bmatrix} x_1 \\ x_2 \\ x_3 \end{bmatrix} = \begin{bmatrix} 0 \\ 0 \\ 0 \end{bmatrix}$$

Which gives the equations,

$$-x_1 + x_2 = 0$$
$$2x_3 = 0$$
$$x_3 = 0$$

Solution is $x_3 = 0, x_1 = k, x_2 = k$

$\therefore \quad \hat{X}_2 = \begin{bmatrix} k \\ k \\ 0 \end{bmatrix}$ i.e., $x_1 : x_2 : x_3 = 1 : 1 : 0$

Since none of the eigen vectors given in the choices is of this ratio, we need to proceed further and find 3rd eigen vector also.

By putting $\lambda = 3$ in the eigen valve problem, we get

$$\begin{bmatrix} -2 & 1 & 0 \\ 0 & -1 & 2 \\ 0 & 0 & 0 \end{bmatrix}\begin{bmatrix} x_1 \\ x_2 \\ x_3 \end{bmatrix} = \begin{bmatrix} 0 \\ 0 \\ 0 \end{bmatrix}$$

$$-2x_1 + x_2 = 0$$
$$-x_2 + 2x_3 = 0$$

Putting $x_1 = k$, we get, $x_2 = 2k$ and $x_3 = x_2/2 = k$

$\therefore \quad \hat{X}_3 = \begin{bmatrix} k \\ 2k \\ k \end{bmatrix}$ i.e., $x_1 : x_2 : x_3 = 1 : 2 : 1$

Only the eigen vector given in choice (b) $\begin{bmatrix} 1 \\ 2 \\ 1 \end{bmatrix}$, is in this ratio. So, the correct answer is choice (b).

1.26 (d)

$$x_1 + 2x_2 + x_3 + 4x_4 = 2$$
$$3x_1 + 6x_2 + 3x_3 + 12x_4 = 6$$

The augmental matrix is $\left[\begin{array}{cccc|c} 1 & 2 & 1 & 4 & 2 \\ 3 & 6 & 3 & 12 & 6 \end{array}\right]$.

Performing gauss-elimination on this we get

$$\left[\begin{array}{cccc|c} 1 & 2 & 1 & 4 & 2 \\ 3 & 6 & 3 & 12 & 6 \end{array}\right] \xrightarrow{R_2 - 3R_1} \left[\begin{array}{cccc|c} 1 & 2 & 1 & 4 & 2 \\ 0 & 0 & 0 & 0 & 0 \end{array}\right]$$

rank (A) = rank $(A \mid B) = 1$

So, system is consistent.

Since, system's rank = 1 is less than the number of variables, only infinite (multiple) non-trivial solution exists.

1.27 (d)

Let us try Dolittle's decomposition by putting $l_{11} = 1$ and $l_{22} = 1$

$$\begin{bmatrix} 2 & 1 \\ 4 & -1 \end{bmatrix} = \begin{bmatrix} 1 & 0 \\ l_{21} & 1 \end{bmatrix}\begin{bmatrix} u_{11} & u_{12} \\ 0 & u_{22} \end{bmatrix}$$

$$u_{11} = 2,\ u_{12} = 1$$
$$l_{21}\, u_{11} = 4$$
$$\Rightarrow \quad l_{21} = \frac{4}{2} = 2$$
$$l_{21}\, u_{12} + u_{22} = -1$$
$$\Rightarrow 2 \times 1 + u_{22} = -1$$
$$\Rightarrow \quad u_{22} = -3$$

So, one possible breakdown is

$$\begin{bmatrix} 2 & 1 \\ 4 & -1 \end{bmatrix} = \begin{bmatrix} 1 & 0 \\ 2 & 1 \end{bmatrix}\begin{bmatrix} 2 & 1 \\ 0 & -3 \end{bmatrix}$$

But this is not any of the choices given.

So, let us do Crout's decomposition, by putting $u_{11} = 1$ and $u_{22} = 1$

$$\begin{bmatrix} 2 & 1 \\ 4 & -1 \end{bmatrix} = \begin{bmatrix} l_{11} & 0 \\ l_{21} & l_{22} \end{bmatrix}\begin{bmatrix} 1 & u_{12} \\ 0 & 1 \end{bmatrix}$$

$$l_{11} = 2,\ l_{11}u_{12} = 1$$
$$\Rightarrow \quad u_{12} = \frac{1}{2} = 0.5$$
$$l_{21} = 4,\ l_{21}u_{12} + l_{22} = -1$$
$$\Rightarrow \quad 4 \times \frac{1}{2} + l_{22} = -1$$
$$\Rightarrow \quad l_{22} = -3$$

So $$\begin{bmatrix} 2 & 1 \\ 4 & -1 \end{bmatrix} = \begin{bmatrix} 2 & 0 \\ 4 & -3 \end{bmatrix}\begin{bmatrix} 1 & 0.5 \\ 0 & 1 \end{bmatrix}$$

Which is choice (d).

1.28 (b)

$$A = \begin{bmatrix} -5 & -3 \\ 2 & 0 \end{bmatrix} \text{ and } I = \begin{bmatrix} 1 & 0 \\ 0 & 1 \end{bmatrix}$$

Characteristic equation of A is

$$\begin{vmatrix} -5-\lambda & -3 \\ 2 & 0-\lambda \end{vmatrix} = 0$$

$$(-5-\lambda)(-\lambda) + 6 = 0 \Rightarrow \lambda^2 + 5\lambda + 6 = 0$$

So, $A^2 + 5A + 6I = 0$ (by Cayley Hamilton theorem)

$$\Rightarrow \quad A^2 = -5A - 6I$$

Multiplying by A on both sides, we have,

$$A^3 = -5A^2 - 6A$$
$$\Rightarrow \quad A^3 = -5(-5A - 6I) - 6A = 19A + 30I$$

1.29 (d)

$$\begin{bmatrix} 2 & -2 \\ 1 & -1 \end{bmatrix}\begin{bmatrix} x_1 \\ x_2 \end{bmatrix} = \begin{bmatrix} 0 \\ 0 \end{bmatrix}$$

$$2x_1 - 2x_2 = 0$$
$$x_1 - x_2 = 0 \Rightarrow x_1 = x_2$$

i.e., x_1 and x_2 are having infinite number of solutions. $\Rightarrow$ Multiple solutions are these.

1.30 (d)

$$AX = \lambda X$$

$$\begin{bmatrix} a & b \\ c & d \end{bmatrix}\begin{bmatrix} 1 \\ -1 \end{bmatrix} = (-1)\begin{bmatrix} 1 \\ -1 \end{bmatrix}$$

$$a - b = -1 \quad \text{...(i)}$$
$$c - d = 1 \quad \text{...(ii)}$$

$$\begin{bmatrix} a & b \\ c & d \end{bmatrix}\begin{bmatrix} 1 \\ -2 \end{bmatrix} = (-2)\begin{bmatrix} 1 \\ -2 \end{bmatrix}$$

$$\Rightarrow \quad a - 2b = -2 \quad \text{...(iii)}$$
$$c - 2d = 4 \quad \text{...(iv)}$$

From equation (i) and (iii), $a = 0$ and $b = 1$

From equation (ii) and (iv), $c = -2$ and $d = -3$

$$\therefore \quad A = \begin{bmatrix} a & b \\ c & d \end{bmatrix} = \begin{bmatrix} 0 & 1 \\ -2 & -3 \end{bmatrix}$$

1.32 (3)

Characteristic equation is $|A - \lambda I| = 0$

$$\begin{vmatrix} -\lambda & 1 & -1 \\ -6 & -11-\lambda & 6 \\ -6 & -11 & 5-\lambda \end{vmatrix} = 0$$

$$\Rightarrow -\lambda[-55 + 11\lambda - 5\lambda + \lambda^2 + 66]$$
$$- 1[-30 + 6\lambda + 36] - 1[66 - 66 - \lambda 6] = 0$$
$$\Rightarrow -\lambda(\lambda^2 + 6\lambda + 11) - 1(6\lambda + 6) + 6\lambda = 0$$
$$\Rightarrow -\lambda^3 - 6\lambda^2 - 11\lambda - 6 = 0$$

$\Rightarrow \lambda^3 + 6\lambda^2 + 11\lambda + 6 = 0$

$\Rightarrow (\lambda + 1)(\lambda + 2)(\lambda + 3) = 0$

$\Rightarrow \lambda = -1, -2, -3$

The absolute values of λ are $|\lambda| = 1, 2, 3$.

Ratio of maximum and minimum eigen value is

$$= 3 : 1 = \frac{3}{1} = 3$$

1.34 (c)

$$A = \begin{bmatrix} p & q \\ r & s \end{bmatrix}$$

$$A \times A = A^2 = \begin{bmatrix} p^2+q^2 & pr+qs \\ pr+qs & r^2+s^2 \end{bmatrix} = B$$

$$\boxed{A^2 = B}$$

Rank of matrix does not change when we squaring the matrix, hence rank of B = rank of $A = N$.

1.35 (9)

Consider a symmetric matrix $A = \begin{bmatrix} a & b \\ b & d \end{bmatrix}$.

Given, $a + d = -6$

$|A| = ad - b^2$

Now since b^2 is always non-negative, maximum determinant will come when $b^2 = 0$.

So, we need to maximize

$|A| = ad - 0 = ad = a \times -(6 + a)$

$= -a^2 - 6a$

$$\frac{d|A|}{da} = -2a - 6 = 0$$

$\Rightarrow$ $a = -3$ is the only stationary point

Since, $\left[\frac{d^2|A|}{da}\right]_{a=-3} = -2 < 0$, we have a maximum at $a = -3$.

Since $a + d = -6$, Corresponding value of $d = -3$.

Now, the maximum value of determinant is

$|A| = ad = -3 \times -3 = 9$

1.36 (a)

$a_{11}x_1 + a_{12}x_2 + a_{13}x_3 = b_1$

$a_{21}x_1 + a_{22}x_2 + a_{23}x_3 = b_2$

$a_{31}x_1 + a_{32}x_2 + a_{33}x_3 = b_3$

$$\begin{bmatrix} a_{11} & a_{12} & a_{13} \\ a_{21} & a_{22} & a_{23} \\ a_{31} & a_{32} & a_{33} \end{bmatrix} \begin{bmatrix} x_1 \\ x_2 \\ x_3 \end{bmatrix} = \begin{bmatrix} b_1 \\ b_2 \\ b_3 \end{bmatrix}$$

If $|A| \neq 0$ then $AX = B$ can be written as $X = A^{-1}B$.

It leads unique solutions.

If $|A| \neq 0$ then $\lambda_1 \cdot \lambda_2 \cdot \lambda_3 \neq 0$ each λ_i is non-zero.

If $|A| \neq 0$ then all the row (column) vectors of A are linearly independent.

1.38 (3)

$$A = \begin{bmatrix} 1 & 1 & 1 \\ 1 & 1 & 1 \\ 1 & 1 & 1 \end{bmatrix} \text{ Eigen value are } 0, 0, 3$$

1.39 (a)

Eigen values of $A^2 - 3A + 4I$ are

$= (1)^2 - 3(1) + 4$ and $(-2)^2 - 3(-2) + 4 = 2, 14$

Note: $A^2X = l^2X$

$\Rightarrow X$ is eigen vector for A^2 corresponding to eigen value l^2

X_1 and X_2 are $e\ v$ of A corresponding to 1, –2

Then X_1 and X_2 are $e.v$ of $A^2 - 3 + 4I$ corresponding to 2, 14.

1.40 (b)

Result, Rank $(A^T A)$ = Rank (A)

1.41 (d)

By Cayley Hamilton theorem,

$\lambda^3 = \lambda \Rightarrow \lambda = 0, 1, -1$

1.42 (d)

Method 1: Using quadratic forms

$$P\begin{bmatrix} x \\ y \end{bmatrix} = \begin{bmatrix} a \\ b \end{bmatrix}$$

$$\begin{bmatrix} 3 & 1 \\ 1 & 3 \end{bmatrix}\begin{bmatrix} x \\ y \end{bmatrix} = \begin{bmatrix} a \\ b \end{bmatrix}$$

$3x + y = a$

$x + 3y = b$

$a^2 + b^2 = 1$

$10x^2 + 12xy + 10y^2 = 1$

$$[x \quad y]\begin{bmatrix} 10 & 6 \\ 6 & 10 \end{bmatrix}\begin{bmatrix} x \\ y \end{bmatrix} = 1$$

Quadratic form, $X^T \cdot AX = 1$

If A is real symmetric,

$$A = \begin{bmatrix} 10 & 6 \\ 6 & 10 \end{bmatrix}$$

Find S which orthogonally diagonalize A ($S^TAS = D$)

Eigen values of A are 16, 4.

Unit eigen vector for $\lambda = 16$ is $\begin{bmatrix} 1/\sqrt{2} \\ 1/\sqrt{2} \end{bmatrix}$

Unit eigen vector for $\lambda = 4$ is $\begin{bmatrix} -1/\sqrt{2} \\ 1/\sqrt{2} \end{bmatrix}$

$$\therefore S = \begin{bmatrix} 1/\sqrt{2} & -1/\sqrt{2} \\ 1/\sqrt{2} & 1/\sqrt{2} \end{bmatrix} = \begin{bmatrix} \cos\theta & -\sin\theta \\ \sin\theta & \cos\theta \end{bmatrix}$$

$\Rightarrow \quad \theta = 45°$

$X = SX^{-1}$ transform $10x^2 + 12xy + 10y^2 = 1$ into $16x'^2 + 4y'^2 = 1$

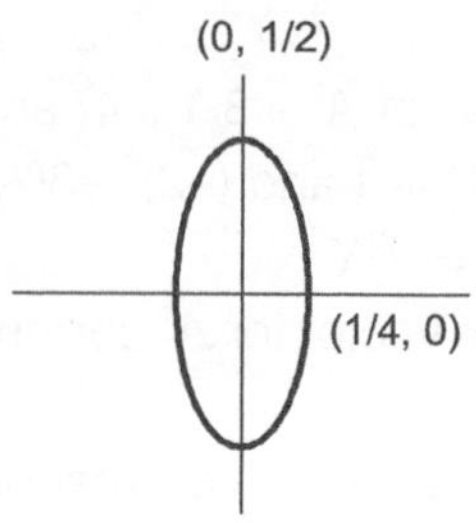

$$\therefore X^T AX = X^T S^T ASX' \qquad (X^T = (SX')^T = X'^T S^T)$$
$$= X'^T \cdot DX'$$
$$= [X' \;\; Y'] \begin{bmatrix} 16 & 0 \\ 0 & 4 \end{bmatrix} \begin{bmatrix} X' \\ Y' \end{bmatrix}$$

$16X'^2 + 4Y'^2 = 1$ in ellipse obtained by rotating original ellipse by 45°.

$\therefore$ Original ellipse is

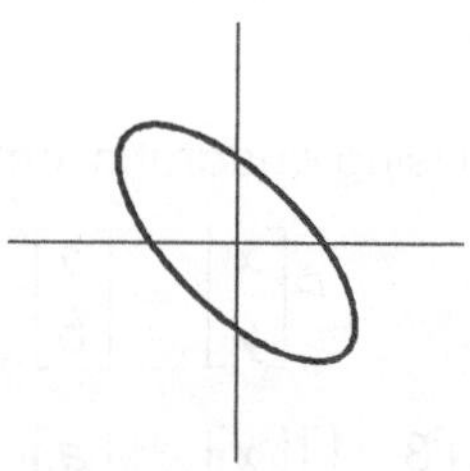

Method 2:

$$P\begin{bmatrix} x \\ y \end{bmatrix} = \begin{bmatrix} a \\ b \end{bmatrix}$$

$$\Rightarrow \begin{bmatrix} 3 & 1 \\ 1 & 3 \end{bmatrix}\begin{bmatrix} x \\ y \end{bmatrix} = \begin{bmatrix} a \\ b \end{bmatrix}$$

$\Rightarrow \quad 3x + y = a$

$\quad x + 3y = y$

$\because \quad a^2 + b^2 = 1$

$\Rightarrow 10x^2 + 10y^2 + 12xy = 1 \qquad$ ...(i)

To transform the equation

$Ax^2 + Bxy + Cy^2 + Dx + Ey + F = 0$

into an equation in x' and y' with no $x'y'$ term, find θ such that,

$$\cot 2\theta = \frac{A-C}{B} \qquad 0 < \theta < 90°$$

$\therefore \quad A = 10 \;;\; B = 12 \;;\; C = 10$

$$\cot 2\theta = \frac{A-C}{B} = 0 \Rightarrow 2\theta = 90 \Rightarrow \theta = 45°$$

Rotation formulas:

$$x = x'\cos\theta - y'\sin\theta$$
$$x = \frac{x'}{\sqrt{2}} - \frac{y'}{\sqrt{2}}$$
$$y = x'\sin\theta + y'\cos\theta$$
$$y = \frac{x'}{\sqrt{2}} + \frac{y'}{\sqrt{2}}$$

Substitute in (1)

$$10\left(\frac{x'-y'}{\sqrt{2}}\right)^2 + 12\left(\frac{x'-y'}{\sqrt{2}}\right)\left(\frac{x'+y'}{\sqrt{2}}\right) + 10\left(\frac{x'+y'}{\sqrt{2}}\right)^2 = 1$$

$\Rightarrow \quad 16x'^2 + 4y'^2 = 1$

$$\Rightarrow \quad \left(\frac{x'}{1/4}\right)^2 + \left(\frac{y'}{1/2}\right)^2 = 1$$

Which is ellipse

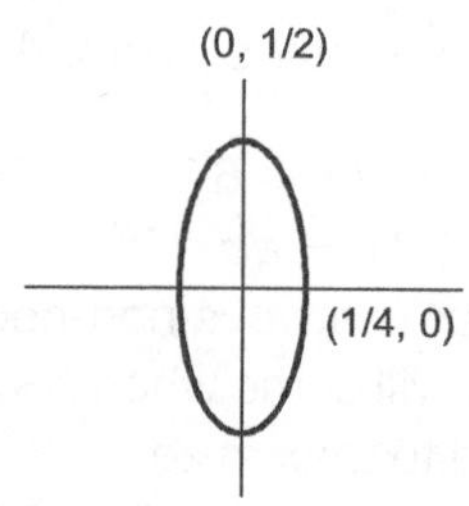

$\therefore$ Original ellipse is

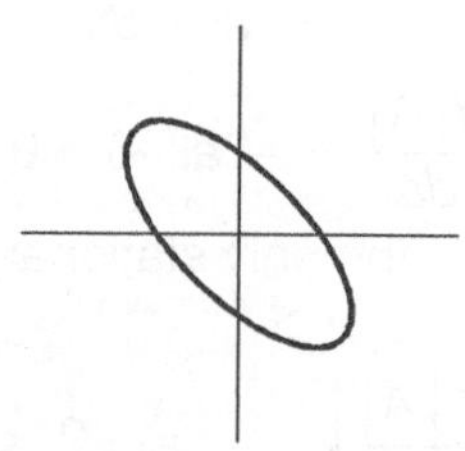

and minor axis is along $\begin{bmatrix} 1 \\ 1 \end{bmatrix}$.

1.43 (c)

The given matrix is symmetric and all its eigen values are distinct. Hence all its eigen vectors are orthogonal one of the eigen vector is

$$x_1 = \begin{bmatrix} 1 \\ 0 \\ 1 \end{bmatrix}$$

The corresponding orthogonal vector in the given option is C. i.e., $x_2 = \begin{bmatrix} 1 \\ 0 \\ -1 \end{bmatrix}$.

$$x_1^T \cdot x_2 = [1 \ 0 \ 1]\begin{bmatrix}1\\0\\-1\end{bmatrix} = 1 + 0 - 1 = 0$$

1.44 (a)

The characteristics equation is $|A - \lambda| i = 0$.

$$\begin{vmatrix}0-\lambda & 1 & 0\\0 & 0-\lambda & 1\\0 & -3 & -4-\lambda\end{vmatrix} = 0$$

$$-\lambda(4\lambda + \lambda^2 + 3) - 1(0 - 0) = 0$$
$$-\lambda(\lambda^2 + 4\lambda + 3) = 0$$
$$\lambda = 0, (\lambda + 1)(\lambda + 3) = 0$$
$$\lambda = -1, -3$$
$$\lambda = (0, -1, -3)$$

1.45 (5.5)

A is 2×2 matrix,

$$Tr\,A = 4$$
$$\lambda_1 + \lambda_2 = 4$$
$$tr(A^2) = 5$$
$$\lambda_1^2 + \lambda_2^2 = 5$$
$$(\lambda_1 + \lambda^2)^2 = \lambda_1^2 + \lambda_2^2 + 2\lambda_1\lambda_2$$
$$16 = 5 + 2\lambda_1\lambda_2$$
$$2\lambda_1\lambda_2 = 11$$
$$\lambda_1\lambda_2 = \frac{11}{2}$$
$$|A| = \frac{11}{2} = 5.5$$

1.46 (1)

$$A = \begin{bmatrix}1 & 0 & -1\\-1 & 2 & 0\\0 & 0 & -2\end{bmatrix}$$

$$|A - \lambda I| = 0$$

$$\begin{vmatrix}1-\lambda & 0 & -1\\-1 & 2-\lambda & 0\\0 & 0 & -2-\lambda\end{vmatrix} = 0$$

$(1 - \lambda)((2 - \lambda)(-2 - \lambda)) - 1(0 - 0) = 0$

$\lambda = 1, 2, -2$

Eigen values of A are	1, 2, – 2
Eigen values of A^3 are	1, 8, –8
A^2 are	1, 4, 4
$4A$ are	4, 8, –8
$5I$ are	5, 5, 5
$A^3 - A^2 - 4A + 5I$ are	1, 1, 1

$\therefore\ |B| = (1)(1)(1) = 1$

1.47 (d)

M is a 2×2 matrix with eigen values 4 and 9. The eigen values of M^2 are 16 and 81.

1.48 (3)

$$M = \begin{bmatrix}0 & 1 & 1\\1 & 0 & 1\\1 & 1 & 0\end{bmatrix}$$

$R_1 \leftrightarrow R_2$

$$\begin{bmatrix}1 & 0 & 1\\0 & 1 & 1\\1 & 1 & 0\end{bmatrix}$$

$R_3 \to R_3 - R_1$

$$= \begin{bmatrix}1 & 0 & 1\\0 & 1 & 1\\0 & 1 & -1\end{bmatrix}$$

$R_3 \to R_3 - R_2$

$$\begin{bmatrix}1 & 0 & 1\\0 & 1 & 1\\0 & 0 & -2\end{bmatrix}$$

Which is in echelon form

$\therefore\quad \rho(A) = 3$

1.49 (c)

$$M^{-1}M = I$$

$$\Rightarrow \begin{bmatrix}u_1^T\\u_2^T\end{bmatrix}[v_1 \ v_2] = \begin{bmatrix}1 & 0\\0 & 1\end{bmatrix}$$

$$\begin{bmatrix}u_1^T v_1 & u_1^T v_2\\u_2^T v_1 & u_2^T v_2\end{bmatrix} = \begin{bmatrix}1 & 0\\0 & 1\end{bmatrix}$$

$$u_1^T v_1 = 1; \quad u_1^T v_2 = 0$$
$$u_2^T v_1 = 0; \quad u_2^T v_2 = 1$$

Statement (1) and (2) are both correct. Option (c) is correct.

1.50 (*)

$$\begin{bmatrix}2 & 3 & 3\\3 & 2 & 1\\3 & 1 & 7\end{bmatrix} = \begin{bmatrix}l_{11} & 0 & 0\\l_{21} & l_{22} & 0\\l_{31} & l_{32} & l_{33}\end{bmatrix}\begin{bmatrix}u_{11} & u_{12} & u_{13}\\0 & u_{22} & u_{23}\\0 & 0 & u_{33}\end{bmatrix}$$

Consider, $u_{11} = u_{22} = u_{33} = 1$

$$\begin{bmatrix}2 & 3 & 3\\3 & 2 & 1\\3 & 1 & 7\end{bmatrix} = \begin{bmatrix}l_{11} & 0 & 0\\l_{21} & l_{22} & 0\\l_{31} & l_{32} & l_{33}\end{bmatrix}\begin{bmatrix}1 & u_{12} & u_{13}\\0 & 1 & u_{23}\\0 & 0 & 1\end{bmatrix}$$

$$\begin{bmatrix}2&3&3\\3&2&1\\3&1&7\end{bmatrix} = \begin{bmatrix}l_{11} & l_{11}u_{12} & l_{11}u_{13}\\ l_{21} & l_{21}u_{12}+l_{22} & l_{21}u_{13}+l_{22}u_{23}\\ l_{31} & l_{31}u_{12}+l_{32} & l_{31}u_{13}+l_{32}u_{23}+l_{33}\end{bmatrix}$$

$l_{11} = 2 \quad l_{11}u_{12} = 3 \quad l_{11}u_{13} = 3$

$l_{21} = 3 \quad 2u_{12} = 3 \quad 2u_{13} = 3$

$l_{31} = 3 \quad u_{12} = \frac{3}{2} \quad u_{13} = \frac{3}{2}$

$l_{21}u_{12} + l_{22} = 2 \quad l_{21}u_{13} + l_{22}u_{23} = 1$

$(3)\left(\frac{3}{2}\right) + l_{22} = 2 \quad (3)\left(\frac{3}{2}\right) + \left(-\frac{5}{2}\right)u_{23} = 1$

$l_{22} = -\frac{5}{2} \quad u_{23} = \frac{7}{5}$

$l_{31}u_{12} + l_{32} = 1 \quad l_{31}u_{13} + l_{32}u_{23} + l_{33} = 7$

$(3)\left(\frac{3}{2}\right) + l_{32} = 1 \quad (3)\left(\frac{3}{2}\right) + \left(-\frac{7}{2}\right)\left(\frac{7}{5}\right) + l_{33} = 7$

$l_{32} = -\frac{7}{2} \quad l_{33} = \frac{74}{10}$

$$L = \begin{bmatrix}2 & 0 & 0\\ 3 & -5/2 & 0\\ 3 & -7/2 & 74/10\end{bmatrix}$$

The number of purely real elements of lower triangular matrix are 9.

[Marks to all was awarded for this question].

1.51 (d)

Characteristic equation of A

$|A-\lambda I| = (-1)^2\lambda^2 + (-1)^1\text{Tr}(A)\lambda + |A| = 0$

$\lambda^2 - (p-p)\lambda + (-p^2 - q^2) = 0$

$\Rightarrow \quad \lambda^2 - 1 = 0 \quad \Rightarrow \lambda = \pm 1$

1.52 (1)

Given: $A^5 = 0$

$Ax = \lambda x$

$\Rightarrow \quad A^5x = \lambda^5 x \quad (\because x \neq 0)$

$\Rightarrow \quad \lambda^5 = 0 \quad \Rightarrow \lambda = 0$

Eigen values of $A + I$ is $\lambda + 1$

$\because$ Eigen values of $A + I = 0 + 1 = 1$

Hence, $|A+I|$ = Product of eigen values of $A + I = 1 \times 1 \times 1 \times$ ... 10 times = 1.

1.53 (b)

$A = [a_{ij}]_{3\times3},\ a_{ij} = (i-j)^3$

for $i = j \Rightarrow a_{ii} = (i-i)^3 = 0 \ \forall i$

for $i \neq j \Rightarrow a_{ij} = (i-j)^3 = (-(j-i))^3$

$= -(j-i)^3 = -a_{ji}$

$\therefore A_{3\times3}$ is a skew-symmetric matrix.

1.54 (c)

Given λ is an eigen value of A.

Then eigen value of $\left(e^A = 1 + A + \frac{A^2}{2!} + \frac{A^3}{3!} + ...\right)$

is $\left(1 + \lambda + \frac{\lambda^2}{2!} + ... = e^\lambda\right)$.

$\therefore$ Statement (1) is true.

We know that eigen vector of A and polynomial matrix in A is same.

$\Rightarrow$ Eigen vector of A and e^A is same.

$\therefore$ Statement (2) is true.

1.55 (a)

$$A = \begin{bmatrix}1&0&0\\0&4&-2\\0&1&1\end{bmatrix} \Rightarrow \lambda_A = 1$$

Given : $6A^{-1} = A^2 + cA + dI$

$\Rightarrow \quad 6\lambda^{-1} = \lambda^2 + c\lambda + d$

Sub. $\lambda = 1 \Rightarrow \frac{6}{1} = 1 + c + d \Rightarrow c + d = 5$

1.56 (d)

$$w = \begin{bmatrix}1\\2\\3\end{bmatrix}$$

$\phi = W^T x = 1$

Let $X = \begin{bmatrix}x\\y\\z\end{bmatrix}$

$$\phi = \begin{bmatrix}1 & 2 & 3\end{bmatrix}\begin{bmatrix}x\\y\\z\end{bmatrix} = 1$$

$\phi = x + 2y + 3z = 1$

Normal vector to ϕ is given by $\vec{\nabla}\phi$

$N = \vec{\nabla}\phi = i\frac{\partial\phi}{\partial x} + j\frac{\partial\phi}{\partial y} + k\frac{\partial\phi}{\partial z}$

$N = \vec{\nabla}\phi = i(1) + j(2) + k(3)$

$$N = \begin{bmatrix}1\\2\\3\end{bmatrix}$$

1.57 (a)

By verifying $Au = V$ with given options, we get

$$A = \begin{bmatrix} \frac{4}{5} & \frac{3}{5} \\ -\frac{3}{5} & \frac{4}{5} \end{bmatrix}$$

1.58 (a)

Option (b) : det = 0

Cannot have inverse.

Option (c) : det = 0

Cannot have inverse.

Option (d) : det = 0

It cannot have inverse.

Option (a) : $|A| \neq 0$

$\therefore$ It has inverse.

1.59 (29)

$$A = \begin{bmatrix} 1 & 2 \\ 3 & 4 \end{bmatrix}\begin{bmatrix} 1 & 2 \\ 3 & 4 \end{bmatrix} = \begin{bmatrix} 7 & 10 \\ 15 & 22 \end{bmatrix}$$

$$\lambda_1 + \lambda_2 = \text{Trace (A)} = 22 + 7 = 29$$

■■■■

2 Calculus

2.1 If $S = \int_1^\infty x^{-3}\, dx$, then S has the value

(a) $\frac{-1}{3}$ (b) $\frac{1}{4}$

(c) $\frac{1}{2}$ (d) 1 [2005 : 1 M]

2.2 For the function, $f(x) = x^2 e^{-x}$, the maximum occurs when x is equal to

(a) 2 (b) 1

(c) 0 (d) –1 [2005 : 2 M]

2.3 For the scalar field, $u = \frac{x^2}{2} + \frac{y^2}{3}$, magnitude of the gradient at the point (1, 3) is

(a) $\sqrt{\frac{13}{9}}$ (b) $\sqrt{\frac{9}{2}}$

(c) $\sqrt{5}$ (d) $\frac{9}{2}$ [2005 : 2 M]

2.4 The expression $V = \int_0^H \pi R^2 (1 - h/H)^2\, dh$ for the volume of a cone is equal to

(a) $\int_0^R \pi R^2 (1 - h/H)^2\, dr$

(b) $\int_0^R \pi R^2 (1 - h/H)^2\, dh$

(c) $\int_0^H 2\pi r H(1 - r/R)\, dh$

(d) $4\int_0^R 2\pi r H\left(1 - \frac{r}{R}\right)^2 dr$ [2006 : 2 M]

2.5 A surface $S(x, y) = 2x + 5y - 3$ is integrated once over a path consisting of the points that satisfy $(x + 1)^2 + (y - 1)^2 = 2$. The integral evaluates to

(a) $17\sqrt{2}$ (b) $17/\sqrt{2}$

(c) $\sqrt{2}/17$ (d) 0 [2006 : 2 M]

2.6 The integral $\frac{1}{2\pi}\int_0^{2\pi} \sin(t - \tau)\cos\tau\, d\tau$ equals

(a) $\sin t \cos t$ (b) 0

(c) $(1/2)\cos t$ (d) $(1/2)\sin t$ [2007 : 2 M]

2.7 Consider function $f(x) = (x^2 - 4)^2$ where x is a real number. Then the function has

(a) only one minimum

(b) only two minima

(c) three minima

(d) three maxima [2008 : 2 M]

2.8 $f(x, y)$ is a continuous function defined over $(x, y) \in [0, 1] \times [0, 1]$. Given the two constraints, $x > y^2$ and $y > x^2$, the volume under $f(x, y)$ is

(a) $\int_{y=0}^{y=1}\int_{x=y^2}^{x=\sqrt{y}} f(x,y)\,dx\,dy$

(b) $\int_{y=x^2}^{y=1}\int_{x=y^2}^{x=1} f(x,y)\,dx\,dy$

(c) $\int_{y=0}^{y=1}\int_{x=0}^{x=1} f(x,y)\,dx\,dy$

(d) $\int_{y=0}^{y=\sqrt{x}}\int_{x=0}^{x=\sqrt{y}} f(x,y)\,dx\,dy$ [2009 : 2 M]

2.9 A cubic polynomial with real coefficients

(a) can possibly have no extrema and no zero crossings

(b) may have up to three extrema and upto 2 zero crossings

(c) cannot have more than two extrema and more than three zero crossings

(d) will always have an equal number of extrema and zero crossings [2009 : 2 M]

2.10 The value of the quantity P, where $P = \int_0^1 xe^x\,dx$, is equal to

(a) 0 (b) 1

(c) e (d) $1/e$ [2010 : 1 M]

2.11 At $t = 0$, the function $f(t) = \frac{\sin t}{t}$ has

(a) a minimum

(b) a discontinuity

(c) a point of inflection

(d) a maximum [2010 : 2 M]

2.12 Roots of the algebraic equation $x^3 + x^2 + x + 1 = 0$ are

(a) $(+1, +j, -j)$ (b) $(+1, -1, +1)$

(c) $(0, 0, 0)$ (d) $(-1, +j, -j)$ [2011 : 1 M]

2.13 The function $f(x) = 2x - x^2 + 3$ has
(a) a maxima at $x = 1$ and a minima at $x = 5$
(b) a maxima at $x = 1$ and a minima at $x = -5$
(c) only a maxima at $x = 1$
(d) only a minima at $x = 1$ [2011 : 2 M]

2.14 The two vectors [1, 1, 1] and [1, a, a^2], where $a = \left(-\frac{1}{2} + j\frac{\sqrt{3}}{2}\right)$, are
(a) orthonormal (b) orthogonal
(c) parallel (d) collinear [2011 : 2 M]

2.15 The maximum value of $f(x) = x^3 - 9x^2 + 24x + 5$ in the interval [1, 6] is
(a) 21 (b) 25
(c) 41 (d) 46 [2012 : 2 M]

2.16 A function $y = 5x^2 + 10x$ is defined over an open interval $x = (1, 2)$. At least at one point in this interval, $\frac{dy}{dx}$ is exactly
(a) 20 (b) 25
(c) 30 (d) 35 [2013 : 2 M]

2.17 Let $f(x) = x\,e^{-x}$. The maximum value of the function in the interval $(0, \infty)$ is
(a) e^{-1} (b) e
(c) $1 - e^{-1}$ (d) $1 + e^{-1}$
[2014 : 1 M, Set-1]

2.18 The line integral of function $F = yzi$, in the counter clockwise direction, along the circle $x^2 + y^2 = 1$ at $z = 1$ is
(a) -2π (b) $-\pi$
(c) π (d) 2π [2014 : 2 M, Set-1]

2.19 Minimum of the real valued function, $f(x) = (x-1)^{2/3}$ occurs at x equal to
(a) $-\infty$ (b) 0
(c) 1 (d) ∞ [2014 : 1 M, Set-2]

2.20 To evaluate the double integral $\int_0^8\left(\int_{y/2}^{(y/2)+1}\left(\frac{2x-y}{2}\right)dx\right)dy$, we make the substitution $u = \left(\frac{2x-y}{2}\right)$ and $v = \frac{y}{2}$. The integral will reduce to
(a) $\int_0^4\left(\int_0^2 2u\,du\right)dv$ (b) $\int_0^4\left(\int_0^1 2u\,du\right)dv$
(c) $\int_0^4\left(\int_0^1 u\,du\right)dv$ (d) $\int_0^4\left(\int_0^2 u\,du\right)dv$
[2014 : 2 M, Set-2]

2.21 The minimum value of the function,
$$f(x) = x^3 - 3x^2 - 24x + 100$$
in the interval [–3, 3] is
(a) 20 (b) 28
(c) 16 (d) 32 [2014 : 2 M, Set-2]

2.22 A particle, starting from origin at $t = 0$ s, is traveling along x-axis with velocity
$$v = \frac{\pi}{2}\cos\left(\frac{\pi}{2}t\right) \text{ m/s}$$
At $t = 3$ s, the difference between the distance covered by the particle and the magnitude of displacement from the origin is ______.
[2014 : 1 M, Set-3]

2.23 The mean thickness and variance of silicon steel laminations are 0.2 mm and 0.02 respectively. The varnish insulation is applied on both the sides of the laminations. The mean thickness of one side insulation and its variance are 0.1 mm and 0.01 respectively. If the transformer core is made using 100 such varnish coated laminations, the mean thickness and variance of the core respectively are
(a) 30 mm and 0.22 (b) 30 mm and 2.44
(c) 40 mm and 2.44 (d) 40 mm and 0.24
[2014 : 2 M, Set-3]

2.24 If a continuous function $f(x)$ does not have a root in the interval $[a, b]$, then which one of the following statements is TRUE?
(a) $f(a) \cdot f(b) = 0$ (b) $f(a) \cdot f(b) < 0$
(c) $f(a) \cdot f(b) > 0$ (d) $f(a)/f(b) \le 0$
[2015 : 1 M, Set-1]

2.25 The volume enclosed by the surface, $f(x, y) = e^x$ over the triangle bounded by the lines $x = y$; $x = 0$; $y = 1$ in the xy plane is ____. [2015 : 2 M, Set-2]

2.26 The maximum value attained by the function, $f(x) = x(x-1)(x-2)$ in the interval [1, 2] is ______.
[2016 : 1 M, Set-1]

2.27 Let, $I = c\iint_R xy^2\,dxdy$, where R is the region shown in the figure and $c = 6 \times 10^{-4}$. The value of I equals ______.
(Give the answer up to two decimal places.)

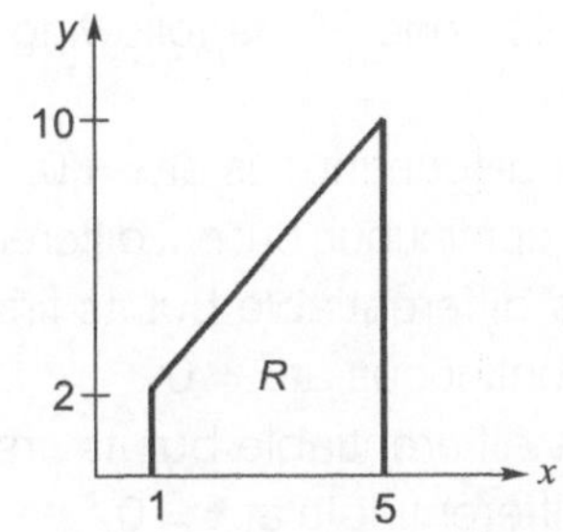

[2017 : 1 M, Set-1]

2.28 Consider a function $f(x, y, z)$ given by

$$f(x, y, z) = (x^2 + y^2 - 2z^2)(y^2 + z^2)$$

The partial derivative of this function with respect to x at the point, $x = 2$, $y = 1$ and $z = 3$ is ____

[2017 : 1 M, Set-2]

2.29 Let $g(x) = \begin{cases} -x, & x \le 1 \\ x+1, & x \ge 1 \end{cases}$ and $f(x) = \begin{cases} 1-x, & x \le 0 \\ x^2, & x > 0 \end{cases}$.

Consider the composition of f and g i.e. (fog)$(x) = f(g(x))$. The number of discontinuities in (fog)(x) present in the interval $(-\infty, 0)$ is:

(a) 0 (b) 1
(c) 2 (d) 4 [2017 : 2 M, Set-2]

2.30 A function $f(x)$ is defined as,

$$f(x) = \begin{cases} e^x & x < 1 \\ \ln x + ax^2 + bx, & x \ge 1 \end{cases}$$

where $x \in R$ which one of the following statements is TRUE?

(a) $f(x)$ is **NOT** differentiable at $x = 1$ for any values of a and b.
(b) $f(x)$ is differentiable at $x = 1$ for the unique values of a and b.
(c) $f(x)$ is differentiable at $x = 1$ for all the values of a and b such that $a + b = e$.
(d) $f(x)$ is differentiable at $x = 1$ for all values of a and b. [2017 : 2 M, Set-1]

2.31 The value of the directional derivative of the function $\phi(x, y, z) = xy^2 + yz^2 + zx^2$ at the point $(2, -1, 1)$ in the direction of the vector $p = i + 2j + 2k$ is

(a) 1 (b) 0.95
(c) 0.93 (d) 0.9 [2018 : 1 M]

2.32 Let f be a real valued function of a real variable defined as $f(x) = x - [x]$, where $[x]$ denotes the largest integer less than or equal to x. The value of $\int_{0.25}^{1.25} f(x)\, dx$ is ____(up to 2 decimal places).

[2018 : 1 M]

2.33 Let f be a real-valued function of a real variable defined as $f(x) = x^2$ for $x \ge 0$, and $f(x) = -x^2$ for $x < 0$. Which one of the following statements is true?

(a) $f(x)$ is discontinuous at $x = 0$.
(b) $f(x)$ is continuous but not differentiable at $x = 0$.
(c) $f(x)$ is differentiable but its first derivative is not continuous at $x = 0$.
(d) $f(x)$ is differentiable but its first derivative is not differentiable at $x = 0$. [2018 : 1 M]

2.34 Let $f(x) = 3x^3 - 7x^2 + 5x + 6$. The maximum value of $f(x)$ over the interval $[0, 2]$ is ____ (upto 1 decimal place).

[2018 : 2 M]

2.35 Which of the following is true for all possible non-zero choices of integers m, n; $m \ne n$. or all possible non-zero choices of real numbers p, q; $p \ne q$, as applicable?

(a) $\frac{1}{\pi}\int_0^{\pi} \sin m\theta \sin n\theta\, d\theta = 0$

(b) $\frac{1}{2\pi}\int_{-\pi}^{\pi} \sin p\theta \cos q\theta\, d\theta = 0$

(c) $\frac{1}{2\pi}\int_{-\pi/2}^{\pi/2} \sin p\theta \sin q\theta\, d\theta = 0$

(d) $\lim_{\alpha \to \infty} \frac{1}{2a}\int_{-\alpha}^{\alpha} \sin p\theta \sin q\theta\, d\theta = 0$ [2020 : 1 M]

2.36 $ax^3 + bx^2 + cx + d$ is a polynomial on real x over real coefficients a, b, c, d wherein $a \ne 0$. Which of the following statements is true?

(a) No choice of coefficients can make all roots identical.
(b) a, b, c, d can be chosen to ensure that all roots are complex.
(c) d can be chosen to ensure that $x = 0$ is a root for any given set a, b, c.
(d) c alone cannot ensure that all roots are real.

[2020 : 1 M]

2.37 Let a_x and a_y be unit vectors along x and y directions, respectively. A vector function is given by $F = \mathbf{a}_x y - \mathbf{a}_y x$.

The line integral of the above function, $\int_C F \cdot dl$ along the curve C, which follows the parabola $y = x^2$ as shown below is ______ (rounded off to 2 decimal places).

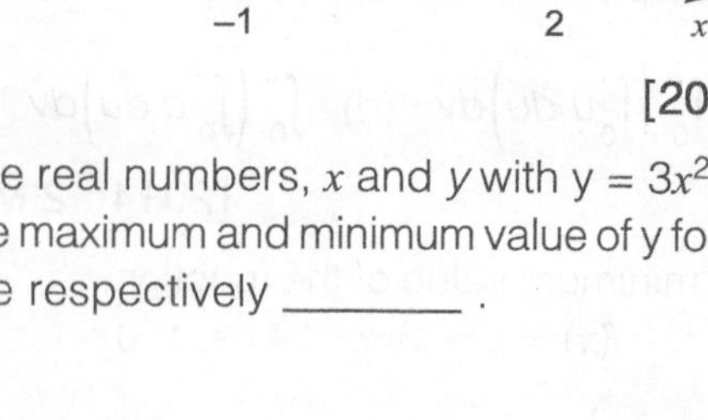

[2020 : 2 M]

2.38 The real numbers, x and y with $y = 3x^2 + 3x + 1$, the maximum and minimum value of y for $x \in [-2, 0]$ are respectively ______.

(a) 7 and $\frac{1}{4}$ (b) 7 and 1

(c) −2 and $\frac{-1}{2}$ (d) 1 and $\frac{1}{4}$ [2020 : 2 M]

2.39 Suppose the circle $x^2 + y^2 = 1$ and $(x-1)^2 + (y-1)^2 = r^2$ intersect each other orthogonally at the point (u, v). Then $u + u$ = _____. [2021 : 1 M]

2.40 In the open interval (0, 1), the polynomial $p(x) = x^4 - 4x^3 + 2$ has

(a) three real roots (b) two real roots

(c) one real root (d) no real roots

[2021 : 2 M]

2.41 Let $f(x)$ be a real-valued function such that $f'(x_0) = 0$ for some $x_0 \in (0, 1)$ and $f''(x) > 0$ for all $x \in (0, 1)$. Then $f(x)$ has

(a) two distinct local minima in (0, 1)

(b) exactly one local minimum in (0, 1)

(c) one local maximum in (0, 1)

(d) no local minimum in (0, 1) [2021 : 1 M]

2.42 Let $f(x) = \int_0^x e^t(t-1)(t-2)dt$. Then $f(x)$ decreases in the interval.

(a) $x \in (1, 2)$ (b) $x \in (2, 3)$

(c) $x \in (0, 1)$ (d) $x \in (0.5, 1)$

[2022 : 2 M]

2.43 Let R be a region in the first quadrant of the xy plane enclosed by a closed curve C considered in counter-clockwise direction. Which of the following expressions does not represent the area of the region R?

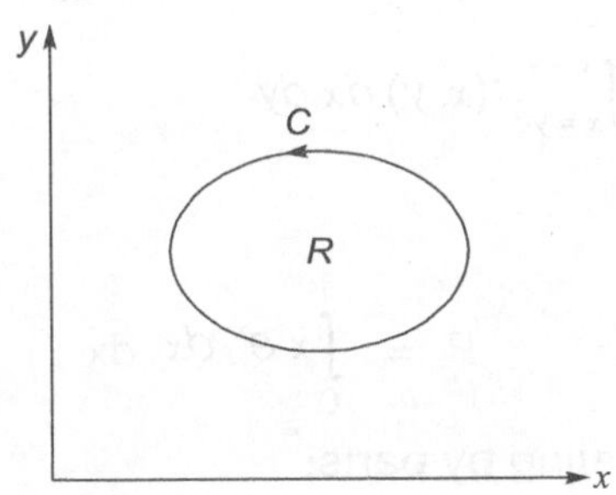

(a) $\iint_R dxdy$ (b) $\oint_C xdy$

(c) $\oint_C ydx$ (d) $\frac{1}{2}\oint_C (xdy - ydx)$

[2022 : 2 M]

2.44 Consider the following equation in a 2-D real-space.

$$|x_1|^p + |x_2|^p = 1 \text{ for } p > 0$$

Which of the following statement(s) is/are true.

(a) When $p = 2$, the area enclosed by the curve is π.

(b) When p tends to ∞, the area enclosed by the curve tends to 4.

(c) When p tends to 0, the area enclosed by the curve is 1.

(d) When $p = 1$, the area enclosed by the curve is 2. [2023 : 2 M]

2.45 The closed curve shown in the figure is described by $r = 1 + \cos\theta$, where $r = \sqrt{x^2 + y^2}$; $x = r\cos\theta$, $y = r\sin\theta$. The magnitude of the line integral of the vector field $F = -y\hat{i} + x\hat{j}$ around the closed curve is __ (Round off to 2 decimal places).

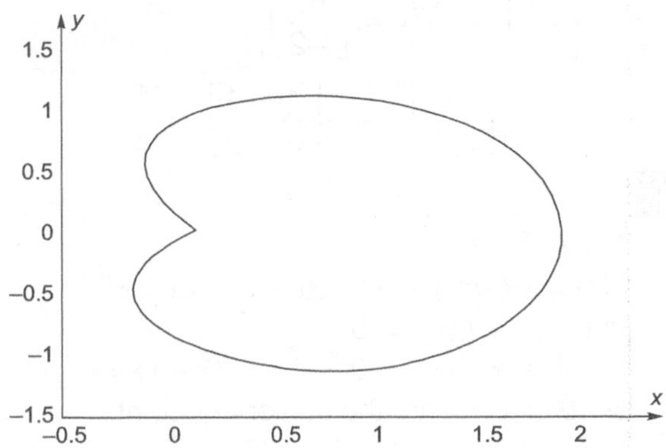

[2023 : 2 M]

2.46 A quadratic function of two variables is given as $f(x_1, x_2) = x_1^2 + 2x_2^2 + 3x_1 + 3x_2 + x_1x_2 + 1$. The magnitude of the maximum rate of change of the function at the point (1,1) is ________ (Round off to the nearest integer). [2023 : 2 M]

2.47 Let $f(t)$ be a real-valued function whose second derivative is positive for $-\infty < t < \infty$. Which of the following statements is/are always true?

(a) $f(t)$ cannot have two distinct local minima.

(b) The minimum value of $f(t)$ cannot be negative.

(c) $f(t)$ has at least one local maximum.

(d) $f(t)$ has at least one local minimum.

[2024 : 2 M]

2.48 Consider the function $f(t) = (\max(0, t))^2$ for $-\infty < t < \infty$, where max(a, b) denotes the maximum of a and b. Which of the following statements is/are true?

(a) $f(t)$ and its derivative are differentiable.

(b) $f(t)$ is differentiable but its derivative is not continuous.

(c) $f(t)$ is not differentiable.

(d) $f(t)$ is differentiable and its derivative is continuous. [2024 : 2 M]

■■■■

Answers Calculus

2.1 (c)	2.2 (a)	2.3 (c)	2.4 (d)	2.5 (d)	2.6 (b)	2.7 (b)
2.8 (a)	2.9 (c)	2.10 (b)	2.11 (d)	2.12 (d)	2.13 (c)	2.14 (b)
2.15 (c)	2.16 (b)	2.17 (a)	2.18 (b)	2.19 (c)	2.20 (b)	2.21 (b)
2.22 (2)	2.23 (d)	2.24 (c)	2.25 (0.718)	2.26 (0)	2.27 (0.99)	2.28 (40)
2.29 (a)	2.30 (b)	2.31 (a)	2.32 (0.5)	2.33 (d)	2.34 (12)	2.35 (b)
2.36 (c, d)	2.37 (–3)	2.38 (a)	2.39 (1)	2.40 (c)	2.41 (b)	2.42 (a)
2.43 (c)	2.44 (a, b, d)	2.45 (9.42)	2.46 (10)	2.47 (a)	2.48 (d)	

Explanations Calculus

2.1 (c)

$$\delta = \int_1^{\infty} x^{-3} dx = \left[\frac{x^{-2}}{-2}\right]_1^{\infty}$$

$$= -\left[\frac{1}{2x^2}\right]_1^{\infty} = -\left[\frac{1}{\infty} - \frac{1}{2}\right] = \frac{1}{2}$$

2.2 (a)

$f(x) = x^2 e^{-x}$

$f'(x) = x^2(-e^{-x}) + e^{-x} \times 2x = e^{-x}(2x - x^2)$

Putting, $f'(x) = 0$

$e^{-x}(2x - x^2) = 0 \Rightarrow e^{-x} x(2 - x) = 0$

$x = 0$ or $x = 2$ are the stationary points.

Now, $f''(x) = e^{-x}(2 - 2x) + (2x - x^2)(-e^{-x})$

$= e^{-x}(2 - 2x - (2x - x^2)) = e^{-x}(x^2 - 4x + 2)$

At $x = 0$, $f''(0) = e^{-0}(0 - 0 + 2) = 2$

Since $f''(x) = 2$ is > 0 at $x = 0$ we have a minima.

Now at, $x = 2$

$f''(2) = e^{-2}(2^2 - 4 \times 2 + 2)$

$= e^{-2}(4 - 8 + 2) = -2\,e^{-2} < 0$

$\therefore$ At $x = 2$ we have a maxima.

2.3 (c)

$$\text{Grad } u = i\frac{\partial u}{\partial x} + \frac{\partial u}{\partial y} = x\hat{i} + \tfrac{2}{3}y\hat{j}$$

$$\text{At}(1,3) \text{ Grad } u = \sqrt{1^2 + \left[\left(\tfrac{2}{3}\right) \times 3\right]^2} = \sqrt{5}$$

2.4 (d)

We consider options (a) and (d) only because which contains variable r. By integrating (d), we get $1/3\,\pi r^2 H$, which is volume of cone.

2.6 (b)

By property of definite integral,

$$\int_0^{2\pi} f(x) = \int_0^{2\pi} t(2\pi - x) = -f(x) = 0$$

2.7 (b)

$f(x) = (x^2 - 4)^2$

$f'(x) = 2(x^2 - 4) \times 2x = 4x(x^2 - 4) = 0$

$x = 0$, $x = 2$ and $x = -2$ are the stationary points

$f''(x) = 4[x(2x) + (x^2 - 4) \times 1]$

$= 4[2x^2 + x^2 - 4] = 4[3x^2 - 4] = 12x^2 - 16$

$f''(0) = -16 <$ – maxima at $x = 0$

$f''(2) = (12)2^2 - 16 = 32 > 0$ minima at $x = 2$

$f''(-2) = 12(-2)^2 - 16 = 32 > 0$ minima at $x = -2$

$\therefore$ There is only one maxima and only two minima for this function.

2.8 (a)

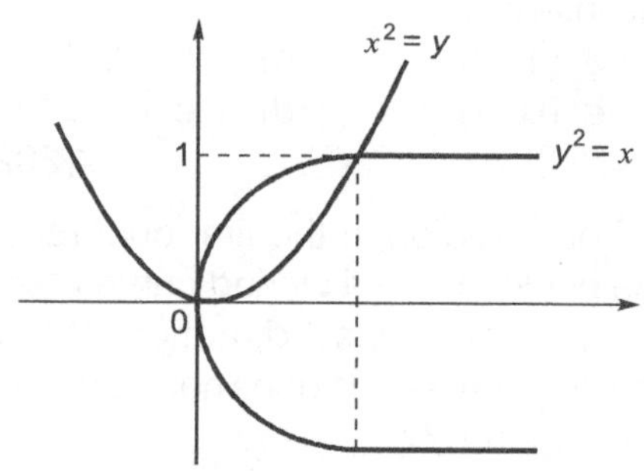

$$\int_{y=0}^{y=1} \int_{x=y^2}^{x=y} f(x, y)\, dx\, dy$$

2.10 (b)

$$P = \int_0^1 x\, e^x\, dx \; dx$$

Integrating by parts:

Let $u = x$, $dv = e^x\, dx$, $du = dx$,

$$v = \int e^x\, dx = e^x$$

Now, $\int u\,dv = uv - \int v\, du$

$$\therefore \int x\, e^x\, dx = x\, e^x - \int e^x\, dx$$

$$= x\, e^x - e^x + c$$

$$\int_0^1 x\, e^x\, dx = [x\, e^x - e^x]_0^1$$

$= (1 \times e^1 - e^1) - (0 \times e^0 - e^0) = 0 - (-1) = 1$

2.11 (d)

$$f(t) = \frac{\sin t}{t}$$

$$f(t) = \frac{t - \frac{t^3}{3!} + \frac{t^5}{5!} - \dots}{t}$$

$$f(t) = 1 - \frac{t^2}{3!} + \frac{t^4}{5!} - \dots$$

$$f'(t) = -\frac{2t}{3!} + \frac{4t^3}{5!} - \dots$$

$$f''(t) = -\frac{2t}{3!} + \frac{4t^3}{5!} - \dots$$

At $t = 0$, $f'(t) = 0$, $f''(t) < 0$. $\therefore$ $f(t)$ attains maxima.

2.12 (d)

-1 is one of the roots since

$(-1)^3 + (-1)^2 + (-1) + 1 = 0$

By polynomial division

$$\frac{x^3 + x^2 + x + 1}{\{x - (-1)\}} = x^2 + 1$$

$\Rightarrow$ $x^3 + x^2 + x + 1 = (x^2 + 1)(x + 1)$

So, roots are $(-1, +j, -j)$

2.13 (c)

$$f(x) = 2x - x^2 + 3$$
$$f'(x) = 2 - 2x = 0$$

$\Rightarrow$ $x = 1$ is the stationary point

$$f''(x) = -2 \Rightarrow f''(1) = -2 < 0$$

So at $x = 1$ we have a relative maxima.

2.14 (b)

Given $[1, 1, 1]$ and $[1, a, a^2]$

hence $a = \omega = -\frac{1}{2} + j\frac{\sqrt{3}}{2}$

$$a^2 = \omega^2$$

So the vectors we

$u = [1, 1, 1]$ and $v = [1, \omega, \omega^2]$

Now, $u.v = 1.1 + 1.\omega + 1.\omega^2$

$= 1 + \omega + \omega^2 = 0$

So u and v are orthogonal.

2.15 (c)

We need absolute maximum of

$f(x) = x^3 - 9x^2 + 24x + 5$ in the interval $[1, 6]$

First find local maximum if any by putting $f'(x) = 0$.

i.e. $f'(x) = 3x^2 - 18x + 24 = 0$

i.e. $x^2 - 6x + 8 = 0$

$x = 2, 4$

Now, $f''(x) = 6x - 18$

$f''(2) = 12 - 18 = -6 < 0$

(So $x = 2$ is a point of local maximum)

and $f''(4) = 24 - 18 = +6 > 0$

(So $x = 4$ is a point of local minimum)

Now tabulate the values of f at end point of interval and at local maximum point, to find absolute maximum in given range, as shown below:

x	$f(x)$
1	21
2	25
6	41

Clearly the absolute maxima is at $x = 6$ and absolute maximum value is 41.

2.16 (b)

$$\frac{dy}{dx} = 10x + 10$$

$$\left.\frac{dy}{dx}\right|_{x=1} = 20$$

$$\left.\frac{dy}{dx}\right|_{x=2} = 30$$

$\because$ x is defined open interval $x = (1, 2)$

$\therefore$ $1 < x < 2$

$\therefore$ $20 < \frac{dy}{dx} < 30$

2.17 (a)

$$f(x) = xe^{-x}$$
$$f'(x) = e^{-x} - x\,e^{-x} = 0$$
$$e^{-x}(1 - x) = 0$$
$$x = 1$$
$$f''(x) = -e^{-x} - e^{-x} + xe^{-x}$$
$$= e^{-x}[x - 2]$$
$$f''(1) = e^{-1}(-1) = -e^{-1} < 0$$

Hence, $f(x)$ has maximum value at $x = 1$

$$f(1) = 1 \cdot e^{-1} = e^{-1}$$

2.18 (b)

$$\bar{F} = yz\,\hat{i}$$

$$\nabla \times \bar{F} = \begin{vmatrix} \hat{i} & \hat{j} & \hat{k} \\ \frac{\partial}{\partial x} & \frac{\partial}{\partial y} & \frac{\partial}{\partial z} \\ yz & 0 & 0 \end{vmatrix}$$

$$= \hat{i}(0) - \hat{j}(0 - y) + \hat{k}(0 - z) = 0\hat{i} + y\hat{j} - z\hat{k}$$

By Stokes theorem,

$$\int_C \bar{F} \cdot d\bar{r} = \int_S (\nabla \times \bar{F}) \cdot \hat{n}ds = \int_S (y\hat{j} - z\hat{k}) \cdot \hat{k}\,ds$$

$$= \int_S -z\, ds \qquad \text{Since } z = 1$$

$$= \int_S -1 ds = (-1)S = (-1)\pi = -\pi$$

where $\underline{S}$ is surface area of $x^2 + y^2 = 1$

$\therefore \quad S = \pi(1)^2 = \pi$

2.19 (c)

$$f(x) = (x-1)^{2/3} = (\sqrt[3]{x-1})^2$$

As $f(x)$ is square of $\sqrt[3]{x-1}$ hence, its minimum value be 0 where at $x = 1$.

2.20 (b)

$$\int_0^8 \left(\int_{y/2}^{(y/2)+1} \left(\frac{2x-y}{2} \right) dx \right) dy$$

$$\frac{2x-y}{2} = u$$

$$x - \frac{y}{2} = u$$

$$dx = du$$

At $\quad x = \dfrac{y}{2}$

$$u = \frac{\frac{2y}{2} - y}{2} = 0$$

At $\quad x = \dfrac{y}{2} + 1$

$$u = \frac{2\left(\frac{y}{2}+1\right) - y}{2} = \frac{y+2-y}{2} = 1$$

Thus, integral becomes $\displaystyle\int_0^8 \left[\int_0^1 u du \right] dy$

$$v = \frac{y}{2}$$

$$dv = \frac{dy}{2} \quad \Rightarrow \quad dy = 2\, dv$$

$$y = 0$$

$\Rightarrow \quad v = 0, \quad y = 8$

$$\Rightarrow \quad v = 4 \quad = \int_0^4 \left[\int_0^1 u du \right] \times 2 dv = \int_0^4 \left[\int_0^1 2u\, du \right] dv$$

2.21 (b)

$$f(x) = x^3 - 3x^2 - 24x + 100 \quad x \in [-3, 3]$$

$$f'(x) = 3x^2 - 6x - 24$$

$$f'(x) = 0 \quad \text{at } x = 4, -2$$

Critical points are $\{-3, -2, 3\}$

$$f(-3) = -27 - 27 + 72 + 100 = 118$$

$$f(-2) = -8 - 12 + 48 + 100 = 128$$

$$f(3) = 27 - 27 - 72 + 100 = 28$$

Hence, $f(x)$ has minimum value at $x = 3$ which is 28.

2.24 (c)

Intermediate value theorem states that if a function is continuous and $f(a) \cdot f(b) < 0$, then surely there is a root in (a, b). The contrapositive of this theorem is that if a function is continuous and has no root in (a, b) then surely $f(a) \cdot f(b) \geq 0$. But since it is given that there is no root in the closed interval $[a, b]$ it means $f(a) \cdot f(b) \neq 0$.

So, surely $f(a) \cdot f(b) > 0$ which is choice (c).

2.25 (0.718)

$$\text{Volume} = \iint f(x,y) dx\, dy = \int_0^1 \int_0^y e^x\, dx\, dy$$

$$= \int_0^1 [e^x]_0^y dy = \int_0^1 (e^y - 1)\, dy = (e^y - y)\Big|_0^1$$

$$= (e-1) - (1-0) = e - 1 - 1 = e - 2 = 0.71828$$

2.26 (0)

$$f(x) = x^3 - 3x^2 + 2x \quad [1, 2]$$

$$f'(x) = 3x^2 - 6x + 2$$

$$f'(x) = 0 \text{ for stationary point}$$

stationary points are $1 \pm \dfrac{1}{\sqrt{3}}$

only $1 + \dfrac{1}{\sqrt{3}}$ lies in [1, 2]

$$f(1) = 0; \quad f(2) = 0$$

$$f\left(1 + \frac{1}{\sqrt{3}}\right) = -\frac{2}{3\sqrt{3}}$$

Maximum value is 0.

2.27 (0.99)

$$I = C \iint xy^2 dx\, dy = C \int_{x=1}^{5} \int_{y=0}^{2x} xy^2 dy\, dx$$

$$= C \int_1^5 x \cdot \left(\frac{y^3}{3} \right) \Bigg|_0^{2x} dx$$

$$= C \int_1^5 x \cdot \frac{8x^3}{3} dx = C \int_1^5 8 \frac{x^4}{3} dx$$

$$= C \cdot \frac{8}{3} \left(\frac{x^5}{5} \right) \Bigg|_1^5$$

$$= \frac{8C}{3}\left(5^4 - \frac{1}{5}\right) = \frac{8C}{3}(625 - 0.2)$$

$$= \frac{8}{3}(6 \times 10^{-4})(625 - 0.2) = 0.99968$$

2.28 (40)

$$f(x, y, z) = (x^2 + y^2 - 2z^2)(y^2 + z^2)$$

$$\frac{\partial f}{\partial x} = (x^2 + y^2 - 2z^2)(0) + (y^2 + z^2)(2x)$$

$$= 0 + (y^2 + z^2)(2x)$$

$$\left.\frac{\partial f}{\partial x}\right|_{\substack{x=2\\y=1\\z=3}} = (1 + 9)2(2) = 40$$

2.29 (a)

$f(x) = 1 - x;\quad x < 0$

$g(x) = -x;\quad x < 0$

(Both are continuous for $x < 0$)

$\therefore$ log(x) is continuous for $x < 0$

The composite function of two continuous function is always continuous. Therefore the number of discontinuities are zero.

2.30 (b)

$$f(x) = \begin{vmatrix} e^x & x < 1 \\ \ln x + ax^2 + bx & x \geq 1 \end{vmatrix}$$

At $x = 1$,

L.H.L = R.H.L

$e = a + b$ ...(i)

At $x = 1$, $f'(x)$ exists

$$f'(x) = \begin{vmatrix} e^x & x < 1 \\ \frac{1}{x} + 20x + b & x \geq 1 \end{vmatrix}$$

At $x = 1$,

L.H.D = R.H.D

$e = 1 + 2a + b$

$2a + b = e - 1$ ...(ii)

From equation (i) and (ii),

$a = -1,\ b = e + 1$

$f(x)$ is differentiable at $x = 1$ for unique values of a and b.

2.31 (a)

$$\phi = xy^2 + yz^2 + zx^2$$

$$\nabla\phi = \bar{i}\frac{\partial\phi}{\partial x} + \bar{j}\frac{\partial\phi}{\partial y} + \bar{k}\frac{\partial\phi}{\partial z}$$

$$= \bar{i}(y^2 + 2xz) + \bar{j}(2xy + z^2) + \bar{k}(2yz + x^2)$$

$$\nabla\phi_{(2,-1,1)} = \bar{i}(1+4) + \bar{j}(-4+1) + \bar{k}(-2+4)$$

$$= 5\bar{i} - 3\bar{j} + 2\bar{k}$$

$$\bar{P} = \bar{i} + 2\bar{j} + 2\bar{k}$$

$$|\bar{P}| = \sqrt{1+4+4} = 3$$

The directional derivative of f(x, y, z) at (2, –1, 1) in the direction of $\bar{P}$ is $\nabla\phi_{\text{at }P} \cdot \frac{\bar{P}}{|\bar{P}|}$

$$= (5\bar{i} - 3\bar{j} + 2\bar{k}) \cdot \left(\frac{\bar{i} + 2\bar{j} + 2\bar{k}}{3}\right) = \frac{5 - 6 + 4}{3} = 1$$

2.32 (0.5)

$$\int_{0.25}^{1.25} f(x)\,dx = \int_{0.25}^{1} (x - [x])dx + \int_{1}^{1.25} (x - [x])\,dx$$

$$= \int_{0.25}^{1.25} (x\,dx) - \left[\int_{0.25}^{1} [x]dx + \int_{1}^{1.25} [x]dx\right]$$

$$= \left(\frac{x^2}{2}\right)\Bigg|_{0.25}^{1.25} - \left(\int_{0.25}^{1} 0\,dx + \int_{1}^{1.25} 1\,dx\right)$$

$$= \frac{(1.25)^2}{2} - \left(\frac{(0.25)^2}{2}\right) - (0 + 0.25)$$

$$= \frac{1}{2}[(1.5625 - 0.0625)] - 0.25 = 0.5$$

2.33 (d)

$$f(x) = \begin{cases} x^2 & x \geq 0 \\ -x^2 & x < 0 \end{cases}$$

$$f'(x) = \begin{cases} 2x & x \geq 0 \\ -2x & x < 0 \end{cases}$$

$$f''(x) = \begin{cases} 2 & x \geq 0 \\ -2 & x < 0 \end{cases}$$

The first derivation of f (i.e) $f'(x)$ is not derivable at $x = 0$.

2.34 (12)

$f(x) = 3x^3 - 7x^2 + 5x + 6 \quad$ in[0, 2]

$f'(x) = 9x^2 - 14x + 5$

$f''(x) = 18x - 14$

$f'(x) = 0$

$9x^2 - 14x + 5 = 0$

$x = 1, 0.55$

$x = 1$

$f''(1) = 18 - 14 = 4 > 0$ minima

$x = 0.55$

$f''(0.55) = -4.1 < 0$ maxima

Maximum {f(0), f(0.55), f(2)}

Maximum {6, 7.13, 12} = 12

2.35 (b)

1. $$\int_{-l}^{l} \cos\left(\frac{n\pi x}{l}\right)\cos\left(\frac{m\pi x}{l}\right)dx = \begin{cases} 2l & \text{if } n=m=0 \\ l & \text{if } n=m\neq 0 \\ 0 & \text{if } m\neq n \end{cases}$$

2. $$\int_{0}^{l} \cos\left(\frac{n\pi x}{l}\right)\cos\left(\frac{m\pi x}{l}\right)dx = \begin{cases} l & \text{if } n=m=0 \\ l/2 & \text{if } n=m\neq 0 \\ 0 & \text{if } m\neq n \end{cases}$$

3. $$\int_{-l}^{l} \sin\left(\frac{n\pi x}{l}\right)\sin\left(\frac{m\pi x}{l}\right)dx = \begin{cases} l & \text{if } m=n \\ 0 & \text{if } m\neq n \end{cases}$$

4. $$\int_{0}^{l} \sin\left(\frac{n\pi x}{l}\right)\sin\left(\frac{m\pi x}{l}\right)dx = \begin{cases} l/2 & \text{if } n=m \\ 0 & \text{if } n\neq m \end{cases}$$

5. $$\int_{-l}^{l} \sin\left(\frac{n\pi x}{l}\right)\cos\left(\frac{m\pi x}{l}\right)dx = 0$$

Given option (b),

$$\frac{1}{2\pi}\int_{-\pi}^{\pi} \sin p\theta \cos q\theta \, d\theta = 0$$

Put $l = \pi$ in rule 5

$$\int_{-\pi}^{\pi} \sin\left(\frac{n\pi x}{\pi}\right)\cos\left(\frac{m\pi x}{\pi}\right)dx$$

Given that, $m \neq n$

$$\int_{-\pi}^{\pi} \sin nx \cos mx \, dx = 0$$

2.36 (c, d)

$ax^3 + bx^2 + cx + d = 0$; $a \neq 0$

$x = 0$ is the root for any values of a, b, c only when $d = 0$.

c alone cannot ensure that all roots are real.

2.37 Sol.

$$\vec{F} = y\hat{a}_x - x\hat{a}_y$$

$$\vec{r} = x\hat{i} + y\hat{j}$$

$$\vec{F} = y\hat{i} - x\hat{j}$$

$$d\vec{r} = dx\hat{i} + dy\hat{j}$$

$$= \int_C F\cdot d\vec{r} = \int_C F_1\,dx + F_2\,dy$$

$$= \int_C y\,dx - x\,dy$$

where C is, $y = x^2$

$dy = 2x\,dx$

x varies from -1 to 2,

$$\int_C \vec{F}\,dl = \int_{-1}^{2} x^2dx - x\cdot 2x\,dx = \int_{-1}^{2}(x^2 - 2x)\,dx$$

$$= \int_{-1}^{2} -x^2dx = -\left.\frac{x^3}{3}\right|_{-1}^{2} = -\frac{8}{3} - \frac{1}{3} = -\frac{9}{3} = -3$$

2.38 (a)

$y = 3x^2 + 3x + 1$ in $[-2, 0]$

$$\frac{\partial y}{\partial x} = 6x + 3, \quad \frac{\partial^2 y}{\partial x^2} = 6$$

$$\frac{dy}{dx} = 0$$

$$6x + 3 = 0$$

$$x = \frac{-1}{2}$$

$$\frac{d^2y}{dx^2} = 6 > 0 \text{ minimum}$$

Maximum value of y in $[-2, 0]$ is maximum $\{f(-2), f(0)\}$

max{7, 1} = 7

Minimum value of y in $[-2, 0]$

$$\min\left\{\begin{matrix} f(-2) & f(0) & f\left(-\frac{1}{2}\right) \\ \downarrow & \downarrow & \downarrow \\ 7 & 1 & \frac{1}{4} \end{matrix}\right\} + = \frac{1}{4}$$

Maximum value 7, minimum value $\frac{1}{4}$.

2.39 (1)

If two curves cut orthogonally then product of slopes = −1

$$x^2 + y^2 = 1$$

$$2x + 2y\frac{dy}{dx} = 0$$

$$(M_1)_{u,v} = \frac{dy}{dx} = \frac{-x}{y} = \frac{-u}{v}$$

$$(x-1)^2 + (y-1)^2 = r^2$$

$$(M_2)_{u,v} = \left(\frac{dy}{dx}\right) = \left(\frac{-2(x-1)}{2(y-1)}\right)_{(u,v)} = \frac{1-u}{v-1}$$

$\because$ $M_1M_2 = -1$

$$\frac{-u}{v}\times\frac{1-u}{v-1} = -1$$

$$-u + u^2 = -v^2 + v$$
$$u^2 + v^2 = v + u$$
$\because$ $x^2 + y^2 = 1$

$\therefore$ $u + v = 1$

2.40 (c)

Given : $P(x) = x^4 - 4x^3 + 2$

For root, $x^4 - 4x^3 + 2 = 0$

$$x^4 + 2 = 4x^3$$

We can write it into two functions,

$$f_1(x) = x^4 + 2 \text{ and } f_2(x) = 4x^3$$

It is clear that point of intersection of these functions is solution (or) root of $p(x) = 0$

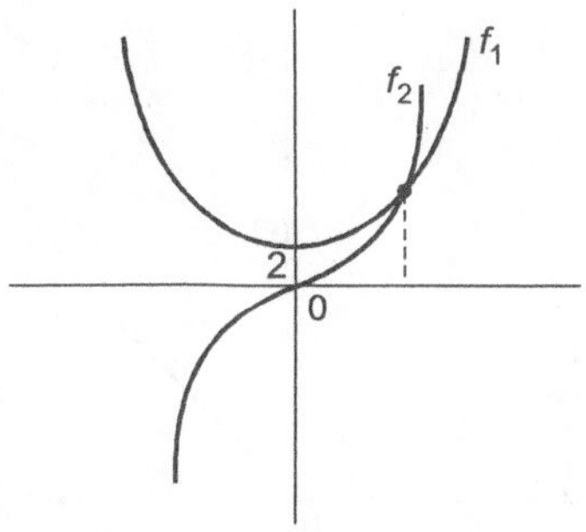

According to intermediate value theorem $P(0)$ and $P(1)$ are having opposite signs

$\therefore$ Root of $p(x) = 0$ lies in (0, 1)

and also from graph, there is only one point of intersection. Hence exactly one real root exists in (0, 1).

2.41 (b)

$x_0 \in (0, 1)$, where $f'(x) = 0$ is stationary point and $f''(x) > 0 \quad \forall x \in (0,1)$

So, for $f'(x) = 0$ and $f''(0) > 0$, where $x \in (0, 1)$

Hence, $f(x)$ has exactly one local minima in (0, 1).

2.42 (a)

$f(x) = \int_0^x e^t(t-1)(t-2)dt$ is decreasing if $f'(x) = 0$.

By Leibnitz rule :

$$f'(x) = \frac{d}{dx}f(x) < 0$$

$$= e^x(x-1)(x-2)\frac{d}{dx}(x) - [e^0(0-1)(0-2)] \times \left[\frac{d}{dx}(0)\right]$$

$$f'(x) = [e^x(x-1)(x-2)] - 0 < 0$$
$$= (x-1)(x-2) < 0$$
$$= x > 1 \text{ and } x < 2$$
$$= x \in (1, 2)$$

2.43 (c)

(a) $\iint_R dxdy$ = Area of region R bounded by C

By Green's

(b) $\oint_C xdy = \iint_R (1-0)dxdy$ = Area of R bounded by C

By Green's

(c) $\oint_C ydx = \iint_R (0-1)dxdy$ = –[Area of R]

By Green's

(d) $\frac{1}{2}\oint_C xdy - ydx = \frac{1}{2}\iint_R [1-(-1)]dxdy$ = Area of R

2.44 (a, b, d)

Given space $|x_1|^P + |x_2|^P = 1$, for $P > 0$

(a) For $P = 2$,

Eqn. : $|x_1|^2 + |x_2|^2 = 1$ is a unit circle.

$\therefore$ Area $= \pi(1)^2 = \pi$

$\therefore$ It is true.

As P tends to infinity, the curve $x_1^P + x_2^P = 1$ approaches a square with sides of length 2 centred at the origin.

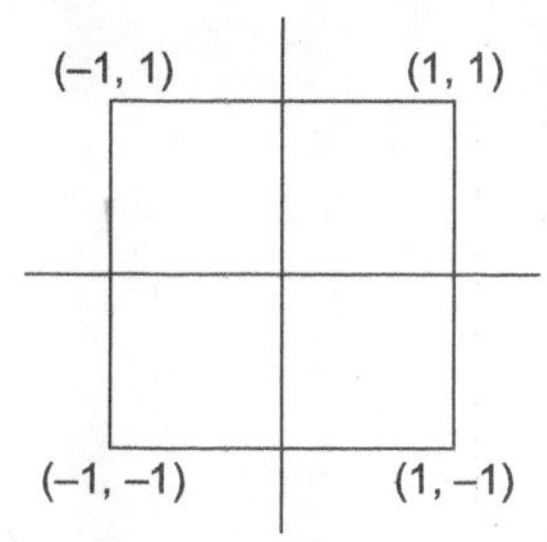

(d) For $P = 1$,

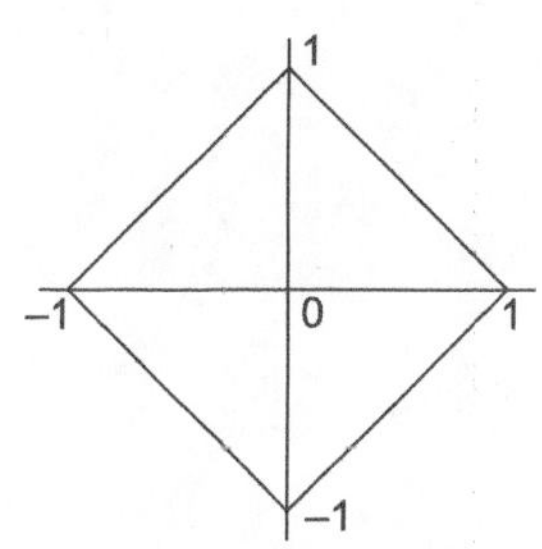

Eqn. : $|x_1| + |x_2| = 1$ is a square.

$$\text{Area of square} = \frac{d^2}{2} = \frac{2^2}{2} = 2$$

2.45 (9.42)

$$C : r = (1 + \cos\theta)$$

$$\oint_C \bar{F}\cdot\overline{dr} = \oint_C -ydx + xdy$$

By Green's Theorem

$= 2\iint_R dydx$ = 2[Area of region bounded by curve 'C']

$$= 2\times\frac{1}{2}\int_{\theta=0}^{2\pi} r^2 d\theta = \int_0^{2\pi}(1+\cos\theta)^2 d\theta$$

$$= \int_0^{2\pi}(1+\cos^2\theta + 2\cos\theta)d\theta$$

$$= (\theta)_0^{2\pi} + 4\int_0^{\pi/2}\cos^2\theta d\theta + 4\int_0^{\pi}\cos\theta d\theta$$

$$= 2\pi + 4\times\frac{1}{2}\times\frac{\pi}{2} + 0 = 2\pi + \pi$$

$$= 3\pi = 3 \times 3.14 = 9.42$$

2.46 (10)

Let, $\phi = x_1^2 + 2x_2^2 + 3x_1 + 3x_2 + x_1x_2 + 1$

$\vec{\nabla}\phi = i\,\phi_{x1} + j\phi_{x2}$

$= i(2x_1 + 3 + x_2) + j(4x_2 + 3 + x_1)$

$$\vec{\nabla}\phi\Big|_{(1,1)} = 6\hat{i} + 8\hat{j}$$

$$\left|\vec{\nabla}\phi\right| = \sqrt{6^2+8^2} = 10$$

2.47 (a)

Since $f''(t) > 0$

If $t \in (-\infty, \infty)$

$\Rightarrow f(t)$ is a parabola open upwards. So, it has only one minima at its stationary point.

$\therefore f(t)$ cannot have two distinct local minima.

2.48 (d)

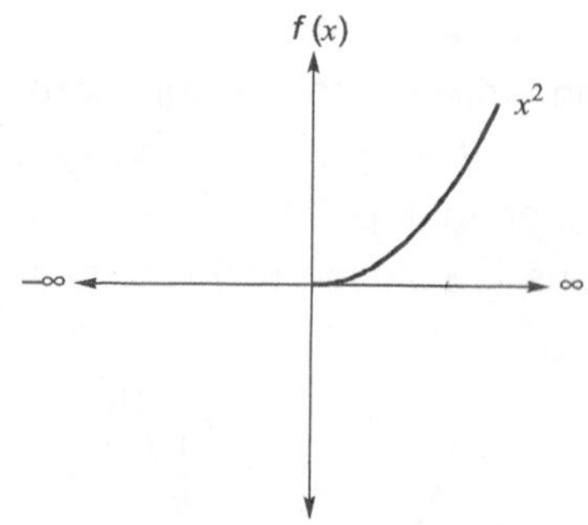

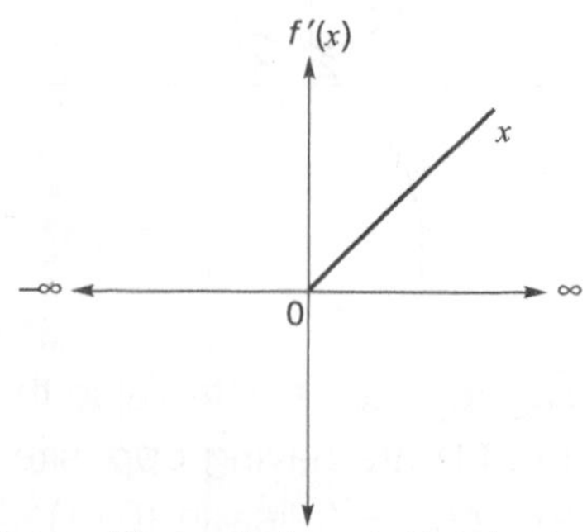

$\therefore f(x)$ is differential and $f'(x)$ is continuous but $f''(x)$ is not continuous.

■■■■

3 Vector Calculus

3.1 Given a vector field $\bar{F}$, the divergence theorem states that

(a) $\int_S \bar{F}\cdot d\bar{s} = \int_V \bar{\nabla}\cdot\bar{F}\,dv$

(b) $\int_S \bar{F}\cdot d\bar{s} = \int_V \bar{\nabla}\times\bar{F}\,dv$

(c) $\int_S \bar{F}\times d\bar{s} = \int_V \bar{\nabla}\cdot\bar{F}\,dv$

(d) $\int_S \bar{F}\times d\bar{s} = \int_V \bar{\nabla}\times\bar{F}\,dv$ **[2002 : 1 M]**

3.2 $F(x, y) = (x^2 + xy)\hat{a}_x + (y^2 + xy)\hat{a}_y$. It's line integral over the straight line from $(x, y) = (0, 2)$ to $(x, y) = (2, 0)$ evaluates to

(a) –8 (b) 4
(c) 8 (d) 0 **[2009 : 2 M]**

3.3 Divergence of the three-dimensional radial vector field $\vec{r}$ is

(a) 3 (b) $1/r$
(c) $\hat{i}+\hat{j}+\hat{k}$ (d) $3(\hat{i}+\hat{j}+\hat{k})$ **[2010 : 1 M]**

3.4 The curl of the gradient of the scalar field defined by $V = 2x^2y + 3y^2z + 4z^2x$ is

(a) $4xya_x + 6yza_y + 8zxa_z$
(b) $4a_x + 6a_y + 8a_z$
(c) $(4xy + 4z^2)a_x + (2x^2 + 6yz)a_y + (3y^2 + 8zx)a_z$
(d) 0 **[2013 : 1 M]**

3.5 Given a vector field $\vec{F} = y^2x\hat{a}_x - yz\hat{a}_y - x^2\hat{a}_z$, the line integral $\int \vec{F}\cdot\vec{dl}$ evaluated along a segment on the x-axis from $x = 1$ to $x = 2$ is

(a) –2.33 (b) 0
(c) 2.33 (d) 7 **[2013 : 1 M]**

3.6 Let $\nabla\cdot(f\vec{v}) = x^2y + y^2z + z^2x$, where f and v are scalar and vector fields respectively. If $\vec{v} = y\vec{i} + z\vec{j} + x\vec{k}$, then $\vec{v}\cdot\nabla f$ is

(a) $x^2y + y^2z + z^2x$ (b) $2xy + 2yz + 2zx$
(c) $x + y + z$ (d) 0 **[2014 : 1 M, Set-3]**

3.7 The value of the line integral

$$\int_c (2xy^2dx + 2x^2ydy + dz)$$

along a path joining the origin (0, 0, 0) and the point (1, 1, 1) is

(a) 0 (b) 2
(c) 4 (d) 6 **[2016 : 1 M, Set-2]**

3.8 The line integral of the vector field $F = 5xz\hat{i} + (3x^2 + 2y)\hat{j} + x^2z\hat{k}$ along a path from (0, 0, 0) to (1, 1, 1) parameterized by (t, t^2, t) is ______. **[2016 : 2 M, Set-2]**

3.9 As shown in the figure, C is the arc from the point (3, 0) to the point (0, 3) on the circle $x^2 + y^2 = 9$. The value of the integral

$$\int_C (y^2 + 2yx)\,dx + (2xy + x^2)\,dy$$

is _____ (upto 2 decimal places).

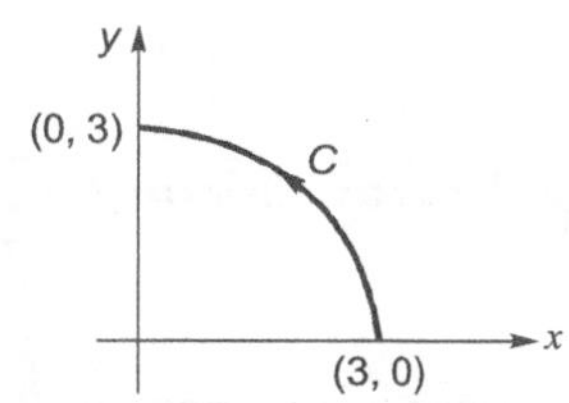

[2018 : 2 M]

3.10 If $f = 2x^3 + 3y^2 + 4z$, the value of line integral $\int_c$ grad $f.dr$ evaluated over contour C formed by the segments $(-3, -3, 2) \to (2, -3, 2) \to (2, 6, 2) \to (2, 6, -1)$ is ______. **[2019 : 1 M]**

3.11 If $A = 2xi + 3yj + 4zk$ and $u = x^2 + y^2 + z^2$, then div(uA) at (1, 1, 1) is _____. **[2019 : 2 M]**

3.12 The vector function expressed by

$F = a_x(5y - k_1z) + a_y(3z + k_2x) + a_z(k_3y - 4x)$

represents a conservative field, where a_x, a_y, a_z are unit vectors along x, y and z directions, respectively. The values of constants k_1, k_2, k_3 are given by:

(a) $k_1 = 3, k_2 = 3, k_3 = 7$
(b) $k_1 = 4, k_2 = 5, k_3 = 3$
(c) $k_1 = 3, k_2 = 8, k_3 = 5$
(d) $k_1 = 0, k_2 = 0, k_3 = 0$ **[2020 : 2 M]**

3.13 Let $f(x, y, z) = 4x^2 + 7xy + 3xz^2$. The direction in which the function $f(x, y, z)$ increases most rapidly at point $P = (1, 0, 2)$ is

(a) $20\hat{i} + 7\hat{j}$ (b) $20\hat{i} + 7\hat{j} + 12\hat{k}$
(c) $20\hat{i} + 12\hat{k}$ (d) $20\hat{i}$ **[2022 : 2 M]**

3.14 Let $\vec{E}(x,y,z) = 2x^2\hat{i} + 5y\hat{j} + 3z\hat{k}$. The value of $\iiint_V (\vec{\nabla}\cdot\vec{E})dV$, where V is the volume enclosed by the unit cube defined by $0 \le x \le 1$, $0 \le y \le 1$, and $0 \le z \le 1$, is

(a) 3 (b) 8
(c) 10 (d) 5 [2022 : 2 M]

3.15 Consider a vector $\bar{u} = 2\hat{x} + \hat{y} + 2\hat{z}$, where $\hat{x}, \hat{y}, \hat{z}$ represent unit vector along the coordinate axes x, y, z respectively. The directional derivative of the function $f(x, y, z) = 2\ln(xy) + \ln(yz) + 3\ln(xz)$ at the point $(x, y, z) = (1, 1, 1)$ in the direction of $\bar{u}$ is

(a) 21 (b) 7
(c) $\dfrac{7}{5\sqrt{2}}$ (d) 0

[2024 : 2 M]

■■■■

Answers Vector Calculus

3.1	(a)	3.2	(d)	3.3	(a)	3.4	(d)	3.5	(b)	3.6	(a)	3.7	(b)
3.8	(4.41)	3.9	(0)	3.10	(139)	3.11	(45)	3.12	(b)	3.13	(b)	3.14	(c)
3.15	(b)												

Explanations Vector Calculus

3.1 (a)

By Gauss divergence theorem,

$$\oint_S \bar{F}\cdot\hat{n}\bar{d}s = \int_V div\,\bar{F}dv \;\Rightarrow\; \oint_S \bar{F}\cdot\hat{n}\bar{d}s = \int_V \nabla\cdot\bar{F}dv$$

3.2 (d)

$$\bar{F} = (x^2 + xy)\,\hat{a}_x + (y^2 + xy)\,\hat{a}_y$$

$$(x_1, y_1) = (0, 2);\quad (x_2, y_2) = (2, 0)$$

Equation of starting line $\dfrac{y - y_1}{x - x_1}$

$$= \frac{y_2 - y_1}{x_2 - x_1} \Rightarrow \boxed{y = 2 - x}\quad dy = -dx$$

$$\bar{F}\cdot\overline{dl} = (x^2 + xy)\,dx + (y^2 + xy)\,dy$$

Putting $y = 2 - x$ and $dy = -dx$

$$\int \bar{F}\cdot\overline{dl} = \int_0^2 2x dx - \int_0^2 (4 - 2x)dx = 0$$

3.3 (a)

$$\vec{r} = x\hat{i} + y\hat{j} + z\hat{k}$$

$$\text{div}\,\vec{r} = \nabla\cdot\vec{r}$$

$$= \left(\hat{i}\frac{\partial}{\partial x} + \hat{j}\frac{\partial}{\partial y} + \hat{k}\frac{\partial}{\partial z}\right)\cdot(x\hat{i} + y\hat{j} + z\hat{k})$$

$$= \frac{\partial x}{\partial x} + \frac{\partial y}{\partial y} + \frac{\partial z}{\partial z} = 1 + 1 + 1 = 3$$

3.4 (d)

Curl of gradient of a scalar field is always zero.

$$\nabla \times \nabla V = 0$$

3.5 (b)

To find: $\int \vec{F}\cdot\overrightarrow{dl}$ along a segment on the x-axis from $x = 1$ to $x = 2$.

i.e. $y = 0$, $z = 0$, $dy = 0$ and $dz = 0$

$$\int \vec{F}\cdot\overrightarrow{dl} = \int (y^2 \times \hat{a}_x - yz\hat{a}_y - x^2\hat{a}_z)\cdot(\hat{a}_x dx + \hat{a}_y dy + \hat{a}_z dz)$$

$$= \int y^2 x dx - yz dy - x^2 dz$$

Putting, $y = 0$, $z = 0$, $dy = 0$ and $dz = 0$

We get, $\int \vec{F}\cdot\overrightarrow{dl} = 0$

3.6 (a)

$$\vec{v} = y\hat{i} + z\hat{j} + x\hat{k}$$

$$\hat{i}\frac{\partial(fV)}{\partial x} + \hat{j}\frac{\partial(fV)}{\partial y} + \hat{k}\frac{\partial(fV)}{\partial z} = x^2y + y^2z + z^2x$$

$$y\frac{\partial f}{\partial x} + z\frac{\partial f}{\partial y} + x\frac{\partial f}{\partial z} = x^2y + y^2z + z^2x \quad \text{...(i)}$$

$$\vec{v}\,\Delta f = (y\hat{i} + z\hat{j} + x\hat{k})\left(\frac{\hat{i}\partial}{\partial x} + \frac{\hat{j}\partial}{\partial y} + \frac{\hat{k}\partial}{\partial z}\right)f$$

$$\vec{v}\,\Delta f = \frac{y\partial f}{\partial x} + \frac{z\partial f}{\partial y} + \frac{x\partial f}{\partial z} \quad \text{...(ii)}$$

From equations (i) and (ii)

$$\vec{v}\cdot\nabla f = x^2y + y^2z + z^2x$$

3.7 (b)

$$\int_C \bar{F}\cdot\bar{d}r$$

where, $\bar{F} = xy^2\bar{i} + 2x^2y\bar{j} + \bar{k}$

$$\nabla\times\vec{F} = \vec{O}$$

($\vec{F}$ is irrotational $\Rightarrow$ $\vec{F}$ is conservative)

$\vec{F} = \nabla\phi$ (ϕ is scalar rotational function)

$\phi_x = 2xy^2;\ \phi_y = 2x^2y;\ \phi_z = 1$

$\Rightarrow\quad \phi = x^2y^2 + z + C$

where, $\vec{F}$ is conservative

$$\int_C \bar{F}\cdot\bar{d}r = \int_{(0,0,0)}^{(1,1,1)} d\phi = \left[x^2y^2 + z\right]_{(0,0,0)}^{(1,1,1)} = 2$$

3.8 (4.41)

$$E = 5xz\bar{i} + (3x^2+2y)\bar{j} + x^2z\bar{k} = \int_C \bar{F}.\bar{d}r$$

$$= \int_C 5xz\,dx + (3x^2+2y)dy + x^2z dz$$

$x = t,\quad y = t^2,\quad z = t,\ t = 0$ to 1

$dx = dt$

$dy = 2t\,dy,\quad dz = dt$

$$= \int_0^1 5t^2dt + (3t^2+2t^2)2t\,dt + t^3dt$$

$$= \int_0^1 (5t^2 + 11t^3)\,dt$$

$$= \left[\frac{5t^3}{3} + \frac{11t^4}{4}\right]_0^1 = \frac{5}{3} + \frac{11}{4} = \frac{53}{12} = 4.41$$

3.9 (0)

$x^2 + y^2 = 9$

$x = 3\cos\theta$

$y = 3\sin\theta$

$dx = -3\sin\theta\,d\theta$

$dy = 3\cos\theta\,d\theta$

θ varies from 0 to $\frac{\pi}{2}$

$$\int (y^2+2xy)\,dx + (2xy+x^2)\,dy$$

$$= \int_0^{\pi/2} (9\sin^2\theta + 18\sin\theta\cos\theta)\,(-3\sin\theta d\theta)$$

$$+(18\sin\theta\cos\theta + 9\cos^2\theta)\,(3\cos\theta)\,d\theta$$

$$= \int_0^{\pi/2} (-27\sin^3\theta - 54\sin^2\theta\cos\theta + 54\sin\theta\cos^2\theta + 27\cos^3\theta)\,d\theta$$

$= 0$

3.10 (139)

$$f = 2x^3 + 3y^2 + 4z$$

$$\Delta f = 6x^2\hat{i} + 6y\hat{j} + 4\hat{k}$$

$$\text{curl}\,(\Delta f) = \begin{vmatrix} \hat{i} & \hat{j} & \hat{k} \\ \partial/\partial x & \partial/\partial y & \partial/\partial z \\ 6x^2 & 6y & 4 \end{vmatrix}$$

$$\therefore\quad \int_c \nabla f\cdot dr = \int_c d(2x^3+3y^2+4z) = 0$$

$$= \int_{(-3,-3,2)}^{(2,-3,2)} d(2x^3+3y^2+4z)$$

$$+ \int_{(2,-3,2)}^{(2,6,2)} d(2x^3+3y^2+4z)$$

$$+ \int_{(2,6,2)}^{(2,6,-1)} d(2x^3+2x^2+4z)$$

$$= 70 + 81 + (-12) = 139$$

3.11 (45)

$$\nabla\cdot(uA) = u(\nabla\cdot A) + (\nabla A)\,F$$

$$= (x^2+y^2+z^2)\,[2+3+4]$$

$$+ (2x\hat{i} + 2y\hat{j} + 2z\hat{k})\cdot(2x\hat{i} + 3y\hat{j} + 4z\hat{k})$$

$$= 9(x^2+y^2+z^2) + (4x^2+6y^2+8z^2)$$

$$\text{At}(1,1,1) = 9(3) + [4+6+8]$$

$$= 27 + 18 = 45$$

3.12 (b)

$$\bar{F} = (5y - k_1z)\hat{i} + (3z + k_2x)\hat{j} + (k_3y - 4x)\hat{k}$$

is conservative field.

$\bar{F}$ is irrotational,

$$\nabla\times\bar{F} = 0$$

$$\begin{vmatrix} \hat{i} & \hat{j} & \hat{k} \\ \dfrac{\partial}{\partial x} & \dfrac{\partial}{\partial y} & \dfrac{\partial}{\partial z} \\ 5y-k_1z & 3z+k_2x & k_3y-4x \end{vmatrix} = 0$$

$$\hat{i}(k_3-3) - \hat{j}(-4+k_1) + \hat{k}(k_2-5) = 0$$

$k_3 - 3 = 0$	$4 - k_1 = 0$	$k_2 - 5 = 0$
$k_3 = 3$	$k_1 = 4$	$k_2 = 5$
$k_1 = 4$	$k_2 = 5$	$k_3 = 3$

3.13 (b)

Scalar function $f(x, y, z)$ rapidly increases in the direction of its normal vector at P.

$$\therefore\quad \bar{a} = \nabla f$$

$$= \left(i\frac{\partial}{\partial x} + j\frac{\partial}{\partial y} + k\frac{\partial}{\partial z}\right)f$$

$$= i(8x + 7y + 3z^2) + j(7x) + k(6xz)$$

$$(\nabla f)_{(1,0,2)} = 20i + 7j + 12k$$

3.14 (c)

$$\bar{\nabla}\cdot\bar{E} = \left(i\frac{\partial}{\partial x} + j\frac{\partial}{\partial y} + k\frac{\partial}{\partial z}\right)\cdot\left(2x^2 i + 5yj + 3zk\right)$$

$$= 4x + 5 + 3$$

$$\therefore \quad \iiint_V \bar{\nabla}\cdot\bar{E}\,dV = \int_{x=0}^{1}\int_{y=0}^{1}\int_{z=0}^{1}(4x+5+3)dzdydx$$

$$= \int_0^1\int_0^1 (4x+5+3)dzdydx = \int_0^1\int_0^1 (4xz+8z)_0^1 dydx$$

$$= \int_0^1 (4xy+8y)_0^1 dx = (2x^2 + 8x)_0^1 = 10$$

3.15 (b)

We know that: directional derivative of '*f*' at *P*(1, 1, 1) in the direction of $\vec{a}$ is given by

$$\text{Directional derivative} = \nabla f\cdot\frac{\hat{a}}{|\bar{a}|}$$

$$\nabla f = \hat{i}\frac{\partial f}{\partial x} + \hat{j}\frac{\partial f}{\partial y} + \hat{k}\frac{\partial f}{\partial z}$$

$$\nabla f = i\frac{5}{x} + j\frac{3}{y} + k\frac{4}{z}$$

$$(\nabla f)_{(1,1,1)} = 5i + 3j + 4k$$

$$\therefore \text{Directional derivative} = \overline{\nabla f}\cdot\frac{\bar{a}}{|\bar{a}|}$$

$$= (5i+3j+4k)\cdot\frac{(2i+j+2k)}{\sqrt{2^2+1^2+2^2}} = \frac{21}{3} = 7$$

■■■■

4 Differential Equations

4.1 The solution of the first order differential equation $x'(t) = -3x(t)$, $x(0) = x_0$ is
(a) $x(t) = x_0 e^{-3t}$ (b) $x(t) = x_0 e^{-3}$
(c) $x(t) = x_0 e^{-1/3}$ (d) $x(t) = x_0 e^{-1}$
[2005 : 1 M]

4.2 For the equation $x''(t) + 3x'(t) + 2x(t) = 5$, the solution $x(t)$ approaches which of the following values as $t \rightarrow \infty$?
(a) 0 (b) $\frac{5}{2}$
(c) 5 (d) 10 **[2005 : 2 M]**

4.3 A differential equation $dx/dt = e^{-2t} u(t)$ has to be solved using trapezoidal rule of integration with a step size, $h = 0.01$ s. Function $u(t)$ indicates a unit step function. If $x(0^-) = 0$, then value of x at $t = 0.01$s will be given by
(a) 0.00099 (b) 0.00495
(c) 0.0099 (d) 0.0198 **[2008 : 2 M]**

4.4 For the differential equation $\frac{d^2x}{dt^2} + 6\frac{dx}{dt} + 8x = 0$ with initial conditions $x(0) = 1$ and $\left.\frac{dx}{dt}\right|_{t=0} = 0$, the solution is
(a) $x(t) = 2e^{-6t} - e^{-2t}$
(b) $x(t) = 2e^{-2t} - e^{-4t}$
(c) $x(t) = -e^{-6t} + 2e^{-4t}$
(d) $x(t) = e^{-2t} + 2e^{-4t}$ **[2010 : 2 M]**

4.5 With K as a constant the possible for the first order differential equation $\frac{dy}{dx} = e^{-3x}$ is
(a) $-\frac{1}{3}e^{-3x} + K$ (b) $-\frac{1}{3}e^{3x} + K$
(c) $-3e^{-3x} + K$ (d) $-3e^{-x} + K$ **[2011 : 1 M]**

4.6 With initial condition $x(1) = 0.5$, the solution of the differential equation $t\frac{dx}{dt} + x = t$ is
(a) $x = t - \frac{1}{2}$ (b) $x = t^2 - \frac{1}{2}$
(c) $x = \frac{t^2}{2}$ (d) $x = \frac{t}{2}$ **[2012 : 1 M]**

4.7 The solution for the differential equation:

$$\frac{d^2x}{dt^2} = -9x$$

with initial conditions $x(0) = 1$ and $\left.\frac{dx}{dt}\right|_{t=0} = 1$, is
(a) $t^2 + t + 1$
(b) $\sin 3t + \frac{1}{3}\cos 3t + \frac{2}{3}$
(c) $\frac{1}{3}\sin 3t + \cos 3t$
(d) $\cos 3t + t$ **[2014 : 1 M, Set-1]**

4.8 Consider the differential equation $x^2\frac{d^2y}{dx^2} + x\frac{dy}{dx} - y = 0$. Which of the following is a solution to this differential equation for $x > 0$?
(a) e^x (b) x^2
(c) $1/x$ (d) $\ln x$ **[2014 : 1 M, Set-2]**

4.9 A solution of the ordinary differential equation $\frac{d^2y}{dt^2} + 5\frac{dy}{dt} + 6y = 0$ is such that $y(0) = 2$ and $y(1) = -\frac{1-3e}{e^3}$. The value of $\frac{dy}{dt}(0)$ is ______.
[2015 : 2 M, Set-1]

4.10 Let $y(x)$ be the solution of the differential equation $\frac{d^2y}{dx^2} - 4\frac{dy}{dx} + 4y = 0$ with initial conditions $y(0) = 0$ and $\left.\frac{dy}{dx}\right|_{x=0} = 1$. Then the value of $y(1)$ is ____.
[2016 : 2 M, Set-2]

4.11 A function $y(t)$, such that $y(0) = 1$ and $y(1) = 3e^{-1}$, is a solution of the differential equation $\frac{d^2y}{dt^2} + 2\frac{dy}{dt} + y = 0$. Then $y(2)$ is
(a) $5e^{-1}$ (b) $5e^{-2}$
(c) $7e^{-1}$ (d) $7e^{-2}$
[2016 : 1 M, Set-1]

4.12 Consider the differential equation:

$$(t^2 - 81)\frac{dy}{dt} + 5t\,y = \sin(t) \text{ with } y(1) = 2\pi$$

There exists a unique solution for this differential equation when t belongs to the interval

(a) (−2, 2) (b) (−10, 10)
(c) (−10, 2) (d) (0, 10)

[2017 : 2 M, Set-1]

4.13 Consider a system governed by the following equations:

$$\frac{dx_1(t)}{dt} = x_2(t) - x_1(t)$$

$$\frac{dx_2(t)}{dt} = x_1(t) - x_2(t)$$

The initial conditions are such that $x_1(0) < x_2(0) < \infty$. Let $x_{1f} = \lim_{t\to\infty} x_1(t)$ and $x_{2f} = \lim_{t\to\infty} x_2(t)$. Which one of the following is true?

(a) $x_{1f} < x_{2f} < \infty$ (b) $x_{2f} < x_{1f} < \infty$
(c) $x_{1f} = x_{2f} < \infty$ (d) $x_{1f} = x_{2f} = \infty$

[2018 : 2 M]

Partial Differential Equations

4.14 The partial differential equation:

$$\frac{\partial^2 u}{\partial t^2} - C^2\left(\frac{\partial^2 u}{\partial x^2} + \frac{\partial^2 u}{\partial y^2}\right) = 0; \quad \text{where } C \neq 0$$

is known as,

(a) Poisson's equation
(b) Wave equation
(c) Heat equation
(d) Laplace equation

[2019 : 1 M]

4.15 Consider the initial value problem below. The value of y at $x = \ln 2$. (rounded off to 3 decimal places) is ________ .

$$\frac{dy}{dx} = 2x - y, \; y(0) = 1$$

[2020 : 1 M]

4.16 In the following differential equation, the numerically obtained value of $y(t)$, at $t = 1$, is ________ (Round off to 2 decimal places).

$$\frac{dy}{dt} = \frac{e^{-\alpha t}}{2 + \alpha t}, \; \alpha = 0.01 \text{ and } y(0) = 0$$

[2023 : 1 M]

4.17 Which of the following differential equations is/are nonlinear?

(a) $x(t) + e^{\left(\frac{dx(t)}{dt}\right)} = 1, \; x(0) = 0$

(b) $t\,x(t) + \left(\frac{dx(t)}{dt}\right) = t^2 e^t, \; x(0) = 0$

(c) $\frac{1}{2}e^t + x(t)\left(\frac{dx(t)}{dt}\right) = 0, \; x(0) = 0$

(d) $x(t)\cos t - \frac{dx(t)}{dt}\sin t = 1, \; x(0) = 0$

[2024 : 2 M]

■■■■

Answers Differential Equations

4.1 (a)	4.2 (b)	4.3 (c)	4.4 (b)	4.5 (a)	4.6 (d)	4.7 (c)
4.8 (c)	4.9 (–3)	4.10 (7.38)	4.11 (b)	4.12 (a)	4.13 (c)	4.14 (b)
4.15 (0.886)	4.16 (0.50)	4.17 (a, c)				

Explanations Differential Equations

4.1 (a)

Given: $\dot{x}(t) = -3\,x(t)$, i.e., $\frac{dx}{dt} = -3\,x$

$$\frac{dx}{x} = -3\,dt$$

$$\int \frac{dx}{x} = \int -3\,dt$$

$\Rightarrow$ $\ln x = -3\,t + C$

$\Rightarrow$ $x = e^{-3t + C} = e^C \times e^{-3t}$

Putting, $e^C = C_1$

$$x = C_1 \times e^{-3t}$$

Now putting initial condition $x(0) = x_0$

$$x_0 = C_1\, e^0 = C_1$$

$\therefore$ $C_1 = x_0$

$\therefore$ Solution is $x = x_0\, e^{-3t}$, i.e., $x(t) = x_0\, e^{-3t}$

4.2 (b)

$x''(t) + 3x'(t) + 2x(t) = 5$

By taking Laplace transform,

$$(s^2 + 3s + 2)X(s) = \frac{5}{s}$$

$$\therefore \quad X(s) = \frac{5/s}{s^2+3s+2}$$

$$\lim_{t\to\infty} = \lim_{s\to 0} s\,X(s) = \frac{5}{2}$$

4.3 (c)

$$\frac{dx}{dt} = e^{-2t}u(t)$$

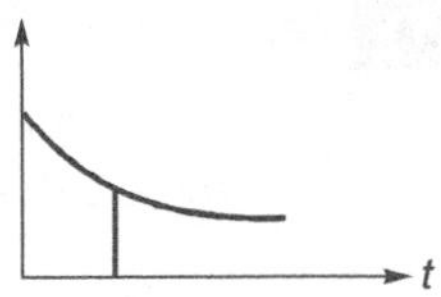

$$x = \int e^{-2t}u(t)dt$$

$$= \int f(t)\,dt$$

At $t = 0.015$, x = Area of trapezoidal

$$= \frac{h}{2}[f(0) + f(0.01)]$$

$$= \frac{0.01}{2}[1 + 3^{-0.02}] = 0.0099$$

4.4 (b)

Given: $\dfrac{d^2x}{dt^2} + 6\dfrac{dx}{dt} + 8x = 0$

$$x(0) = 1 \text{ and } \left.\frac{dx}{dt}\right|_{t=0} = 0$$

$$D^2 + 6D + 8 = 0$$

$$(D+4)(D+2) = 0$$

$$D = -2 \;\&\; D = -4$$

$\therefore$ Solution is

$$x = C_1 e^{-2t} + C_2 e^{-4t}$$

Since, $x(0) = 1$

We have $C_1 + C_2 = 1$ ... (i)

$$\frac{dx}{dt} = -2C_1 e^{-2t} - 4C_2 e^{-4t}$$

Since, $\left[\dfrac{dx}{dt}\right]_{t=0} = 0$, we have

$-2C_1 - 4C_2 = 0$ ... (ii)

Solving (i) and (ii) we have, $C_1 = 2$ and $C_2 = -1$

So, the solution is

$$x(t) = 2e^{-2t} - e^{-4t}$$

4.5 (a)

$$\frac{dy}{dx} = e^{-3x}$$

$$\int dy = \int e^{-3x}dx$$

$$y = \frac{e^{-3x}}{-3} + K = -\frac{1}{3}e^{-3x} + K$$

4.6 (d)

The given differential equation is

$\dfrac{t\,dx}{dt} + x = t$ with initial condition $x(1) = \dfrac{1}{2}$ which is same as

$$\frac{dx}{dt} + \frac{x}{t} = 1$$

Which is a linear differential equation,

$$\frac{dx}{dt} + Px = Q$$

Where, $P = \dfrac{1}{t}$ and $Q = 1$

Integrating factor $= e^{\int P dt} = e^{\int \frac{1}{t}dt} = e^{\log_e t} = t$

Solution is,

$$x\cdot(IF) = \int Q\cdot(IF)dt + C$$

$$x\cdot t = \int 1\cdot t\cdot dt + C$$

$$xt = \frac{t^2}{2} + C$$

$$x = \frac{t}{2} + \frac{C}{t}$$

Put, $x(1) = \dfrac{1}{2}$

$$\Rightarrow \quad \frac{1}{2} + \frac{C}{1} = \frac{1}{2}$$

$$\Rightarrow \quad C = 0$$

So, $\boxed{x = \dfrac{t}{2}}$ is the solution.

4.7 (c)

$$\frac{d^2x}{dt^2} = -9x \qquad \frac{d}{dt} = D$$

$$\frac{d^2x}{dt^2} + 9x = 0 \qquad (D^2 + 9)x = 0$$

Auxiliary equation is $m^2 + 9 = 0$

$$m = \pm 3i$$

$x = C_1\cos 3t + C_2 \sin 3t$ ...(i)

$x(0) = 1$, i.e., $x \to 1$ when $t \to 0$

$$\boxed{1 = C_1}$$

$\dfrac{dx}{dt} = -3C_1 \sin 3t + 3C_2\cos 3t$ ...(ii)

$x'(0) = 1$ i.e. $x' \to 1$ when $t \to 0$

$$1 = 3C_2 \quad \boxed{C_2 = \frac{1}{3}}$$

$$\therefore \quad x = \cos 3t + \frac{1}{3}\sin 3t$$

4.8 (c)

$$x^2\frac{d^2y}{dx^2}+x\frac{dy}{dx}-y=0$$

Let, $x = e^z \longleftrightarrow z = \log x$

$$x\frac{d}{dx} = xD = \theta = \frac{d}{dz}$$

$$x^2 D^2 = \theta(\theta-1)$$

$$(x^2 D^2 + xD - 1)y = 0$$

$$[\theta(\theta-1)+\theta-1]y = 0$$

$$(\theta^2-\theta+\theta-1) = 0$$

$$(\theta^2-1)y = 0$$

Auxilliary equation is $m^2 - 1 = 0$

$$m = \pm 1$$

CF is $C_1 e^{-z} + C_2 e^z$

Solution is $y = C_1 e^{-z} + C_2 e^z$

$$y = C_1 x^{-1} + C_2 x$$

$$y = C_1\frac{1}{x}+C_2 x$$

One independent solution is $\frac{1}{x}$

Another independent solution is x.

4.9 (–3)

$$D^2 + 5D + 6 = 0$$

$$D = -2, -3$$

$$y(t) = c_1 e^{-2t} + c_2 e^{-3t}$$

Given, $y(0) = 2$

$$\Rightarrow \quad c_1 + c_2 = 2 \quad ...(i)$$

$$y(1) = -\left(\frac{1-3e}{e^3}\right)$$

$$\Rightarrow \quad \frac{c_1}{e^2}+\frac{c_2}{e^3} = -\left(\frac{1-3e}{e^3}\right)$$

$$\Rightarrow \quad ec_1 + c_2 = 3e-1 \quad ...(ii)$$

Now solving equation (i) and (ii), we get

$$c_1 = 3; \quad c_2 = -1$$

Substituting in $y(t)$, we get

$$y(t) = 3e^{-2t} - e^{-3t}$$

Now, $$\frac{dy}{dt} = -6e^{-2t} + 3e^{-3t}$$

$$\left(\frac{dy}{dt}\right)_{t=0} = -6+3 = -3$$

4.10 (7.38)

A.E. $m^2 - 4m + 4 = 0$

$$m = 2, 2$$

$$y = (C_1 + C_2 x)e^{2x}$$

$$y(0) = 0 \Rightarrow C_1 = 0$$

$$y = C_2 x e^{2x}$$

$$y' = C_2 e^{2x} + 2C_2 x e^{2x}$$

$$y'(0) = 1$$

$$\Rightarrow \quad C_2 = 1$$

$$y = x e^{2x}$$

$$y(1) = e^2 = 7.38$$

4.11 (b)

Auxiliary equation,

$$m^2 + 2m + 1 = 0$$

$$m = -1, -1$$

$$y = (c_1 + c_2 t)e^{-t}$$

$$y(0) = 1$$

$$\Rightarrow \quad c_1 = 1$$

$$y = (1 + c_2 t)e^{-t}$$

$$y(1) = 3e^{-1}$$

$$\Rightarrow \quad (1+(c_2)e^{-3} = 3e^{-3}$$

$$c_2 = 2$$

$$y = (1+2t)e^{-t}$$

$$y(2) = 5e^{-2}$$

4.12 (a)

The differential equation,

$$(t^2-81)\frac{dy}{dt}+5+y = \sin t$$

$$\frac{dy}{dt}+\frac{5t}{t^2-81}y = \frac{\sin t}{t^2-81}$$

$$P = \frac{5t}{t^2-81}$$

$$Q = \frac{\sin t}{t^2-81}$$

$$I.F = e^{\int p dt} = e^{\int \frac{5t}{t^2-81}dt}$$

$$= e^{\frac{5}{2}\int\frac{2t}{t^2-81}dt} = e^{\frac{5}{2}ln(t^2-81)}$$

$$= e^{ln(t^2-81)^{5/2}} = (t^2-81)^{5/2}$$

Solution is,

$$y(t^2-81)^{5/2} = \int\frac{\sin t}{t^2-81}\cdot(t^2-81)^{5/2}dt$$

$$y = \frac{\int(\sin t)(t^2-81)^{3/2}dt}{(t^2-81)^{5/2}}+\frac{C}{(t^2-81)^{5/2}}$$

The solution exists for $t \neq -9$.

$$t \neq +9$$

Hence option (a) is correct.

Because the remaining option are involving either 9 or (–9).

4.13 (c)

$$(D+1)\,x_1 = x_2$$
$$\Rightarrow \quad (D+1)\,x_1 - x_2 = 0$$
$$(D+1)\,x_2 = x_1$$
$$\Rightarrow \quad -x_1 + (D+1)\,x_2 = 0$$
$$x_{1f} = \lim_{t\to\infty} x_1(t)$$
$$x_{2f} = \lim_{t\to\infty} x_2(t)$$
$$(D+1)\,x_1 - x_2 = 0$$
$$-(D+1)\,x_1 + (D+1)^2 x_2 = 0$$
$$((D+1)^2 - 1)\,x_2 = 0$$
$$(D^2 + 2D)\,x_2 = 0$$
$$D^2 + 2D = 0$$
$$D(D+2) = 0$$
$$\therefore \quad x^2 = C_1 + C_2 e^{-2t} \quad ...(1)$$
$$D = 0, -2$$
$$x_{2f} = \lim_{t\to\infty} C_1 + C_2 e^{-2t}$$
$$= C_1$$
$$(D+1)x_1 - (D+1)\,x_2 = 0$$
$$-x_1 + (D+1)\,x_2 = 0$$
$$((D+1)^2 - 1)\,x_1 = 0$$
$$(D_2 + 2D)\,x_1 = 0$$
$$\therefore \quad x_{1f} = \lim_{t\to\infty} C_1 + C_2 e^{-2t}$$
$$= C_1$$
$$\therefore \quad x_{1f} = C_1$$

4.15 Sol.

$$\frac{dy}{dx} = 2x - y;$$
$y(0) = 1$, y at $x = \ln 2$
$$\frac{dy}{dx} + y = 2x$$
$$P = 1, \; Q = 2x$$
$$I.F. = e^{\int P dx} = e^{\int 1 dx} = e^x$$
Solution, $y(I.F) = \int Q(I.F.)dx$
$$ye^x = \int 2x \cdot e^x dx = 2(xe^x - e^x) + C$$
$$y = 2x - 2 + ce^{-x}$$
$$y(0) = 1$$
$$1 = 0 - 2 + C$$
$$C = 3$$
$$\therefore \quad y = 2x - 2 + 3e^{-x}$$
At $x = \ln 2$
$$y = 2(\ln 2) - 2 + 3e^{-\ln 2}$$
$$= 1.386 - 2 + \frac{3}{2} = 0.886$$

4.16 (0.50)

Given : $\dfrac{dy}{dt} = \dfrac{e^{-\alpha t}}{2+\alpha t}$, $\alpha = 0.01$, $y(0) = 0$

$$y'' = \frac{d^2y}{dt^2} = \frac{(2+\alpha t)e^{-\alpha t} \times (-\alpha) - (e^{-\alpha t} \times \alpha)}{(2+\alpha t)^2}$$
$$= \frac{-\alpha e^{-\alpha t}(3+\alpha t)}{(2+\alpha t)^2} \Rightarrow y''(0) = \frac{-\alpha \times 3}{4}$$

By Taylor's series :
$$y(x) = y(0) + xy'(0) + \frac{x^2}{2!}y''(0) + ...$$
$$= 0 + x\left(\frac{e^{-\alpha(0)}}{2+\alpha(0)}\right) + \frac{x^2}{2}\left(\frac{-3\alpha}{4}\right)$$
$$y(x) = \frac{x}{2} - \frac{3}{8}ax^2$$
$$y(1) = \frac{1}{2} - \frac{3}{8}(0.01)^2$$
$$y(1) = 0.5 - 0.00375 = 0.496$$
$$y(1) = 0.49 \text{ or } 0.50$$

4.17 (a, c)

Option (a) and (c) do not satisfy the criteria of linear differential equation.

■■■■

5 Complex Variables

5.1 e^z is a periodic with a period of

(a) 2π (b) $2\pi i$

(c) π (d) $i\pi$ [1997 : 1 M]

5.2 The value of $\oint_C \frac{dz}{(1+z^2)}$ where C is the contour $|z-i/2| = 1$ is

(a) $2\pi i$ (b) π

(c) $\tan^{-1} z$ (d) $\pi i \tan^{-1} z$ [2007 : 2 M]

5.3 A point z has been plotted in the complex plane, as shown in figure below.

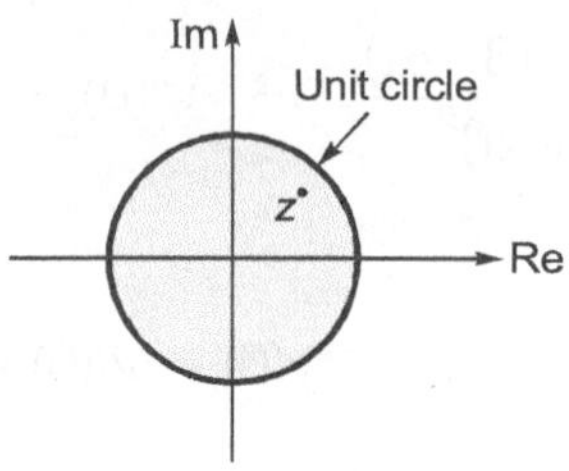

Image of $1/z$ is

(a)

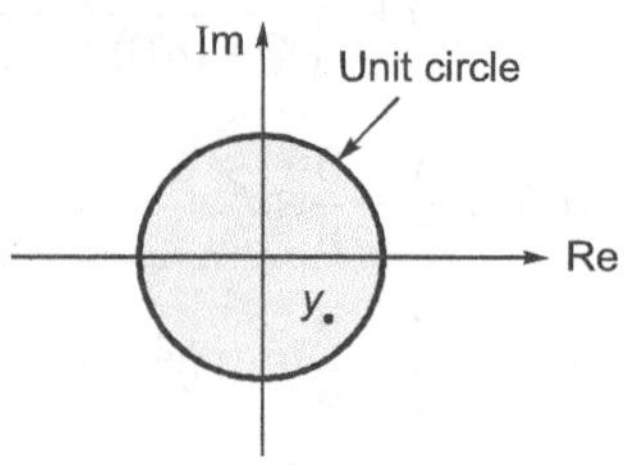

(b)

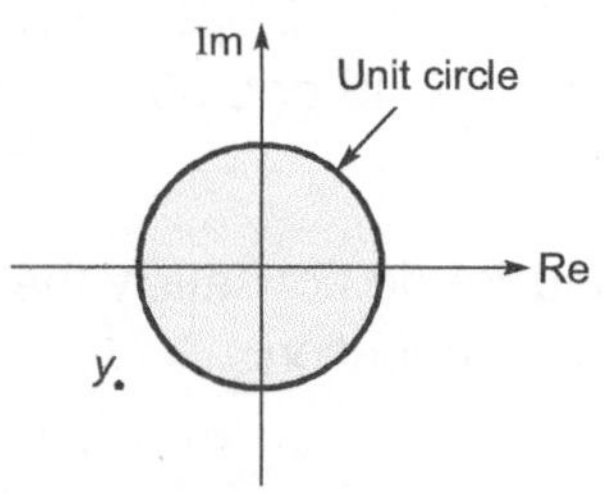

(c)

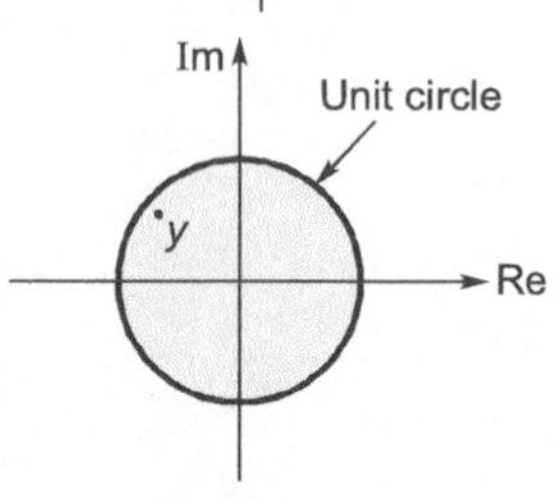

(d)

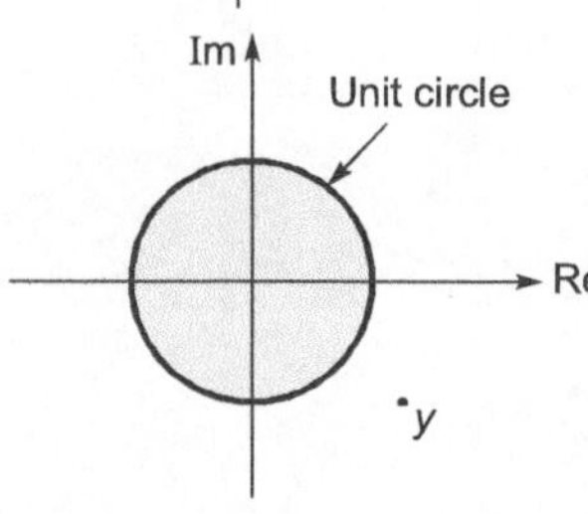

[2011 : 1 M]

5.4 If $x = \sqrt{-1}$, then the value of x^x is

(a) $e^{-\pi/2}$ (b) $e^{\pi/2}$

(c) x (d) 1 [2012 : 1 M]

5.5 Given: $f(z) = \frac{1}{z+1} - \frac{2}{z+3}$. If C is a counterclockwise path in the z-plane such that $|z+1| = 1$, the value of $\frac{1}{2\pi j}\oint_C f(z)\,dz$ is

(a) -2 (b) -1

(c) 1 (d) 2 [2012 : 1 M]

5.6 Square roots of $-i$, where $i = \sqrt{-1}$, are

(a) $i, -i$

(b) $\cos\left(-\frac{\pi}{4}\right) + i\sin\left(-\frac{\pi}{4}\right), \cos\left(\frac{3\pi}{4}\right) + i\sin\left(\frac{3\pi}{4}\right)$

(c) $\cos\left(\frac{\pi}{4}\right) + i\sin\left(\frac{3\pi}{4}\right), \cos\left(\frac{3\pi}{4}\right) + i\sin\left(\frac{\pi}{4}\right)$

(d) $\cos\left(\frac{3\pi}{4}\right) + i\sin\left(-\frac{3\pi}{4}\right), \cos\left(-\frac{3\pi}{4}\right) + i\sin\left(\frac{3\pi}{4}\right)$

[2013 : 1 M]

5.7 $\oint \frac{z^2-4}{z^2+4} dz$ evaluated anticlockwise around the circle $|z-i| = 2$, where $i = \sqrt{-1}$, is

(a) -4π (b) 0

(c) $2+\pi$ (d) $2+2i$ [2013 : 2 M]

5.8 Let S be the set of points in the complex plane corresponding to the unit circle.

(That is, $S = \{z : |z| = 1\}$. Consider the function $f(z) = zz^*$ where $z*$ denotes the complex conjugate of z. The $f(z)$ maps S to which one of the following in the complex plane:

(a) unit circle

(b) horizontal axis line segment from origin to (1, 0)

(c) the point (1, 0)

(d) none of these [2014 : 1 M, Set-1]

5.9 All the values of the multi-valued complex function 1^i, where $i = \sqrt{-1}$, are

(a) purely imaginary

(b) real and non-negative

(c) on the unit circle

(d) equal in real and imaginary parts

[2014 : 1 M, Set-2]

5.10 Integration of the complex function $f(z) = \frac{z^2}{z^2 - 1}$, in the counterclockwise direction, around $|z - 1| = 1$, is

(a) $-\pi i$ (b) 0
(c) πi (d) $2\pi i$ [2014 : 2 M, Set-3]

5.11 Given $f(z) = g(z) + h(z)$, where f, g, h are complex valued functions of a complex variable z. Which one of the following statements is TRUE?

(a) If $f(z)$ is differentiable at z_0, then $g(z)$ and $h(z)$ are also differentiable at z_0.
(b) If $g(z)$ and $h(z)$ are differentiable at z_0, then $f(z)$ is also differentiable at z_0.
(c) If $f(z)$ is continuous at z_0, then it is differentiable at z_0.
(d) If $f(z)$ is differentiable at z_0, then so are its real and imaginary parts.

[2015 : 1 M, Set-2]

5.12 Consider the function $f(z) = z + z^*$ where z is a complex variable and z^* denotes its complex conjugate. Which one of the following is TRUE?

(a) $f(z)$ is both continuous and analytic.
(b) $f(z)$ is continuous but not analytic.
(c) $f(z)$ is not continuous but is analytic.
(d) $f(z)$ is neither continuous nor analytic.

[2016 : 1 M, Set-2]

5.13 The value of the integral

$$\oint \frac{2z+5}{\left(z-\frac{1}{2}\right)(z^2-4z+5)} dz$$

over the contour $|z| = 1$, taken in the anti-clockwise direction, would be

(a) $\frac{24\pi i}{13}$ (b) $\frac{48\pi i}{13}$
(c) $\frac{24}{13}$ (d) $\frac{12}{13}$

[2016 : 1 M, Set-1]

5.14 For a complex number z, $\lim_{z \to i} \frac{z^2+1}{z^3+2z-i(z^2+2)}$ is

(a) $-2i$ (b) $-i$
(c) i (d) $2i$

[2017 : 1 M, Set-1]

5.15 The value of the contour integral in the complex-plane

$$\oint \frac{z^3-2z+3}{z-2} dz$$

along the contour $|z| = 3$, taken counter-clockwise is

(a) $-18\pi i$ (b) 0
(c) $14\pi i$ (d) $48\pi i$

[2017 : 2 M, Set-2]

5.16 Consider the line integral $I = \int_c (x^2 + iy^2) dz$, where $z = x + iy$. The line c is shown in the figure below.

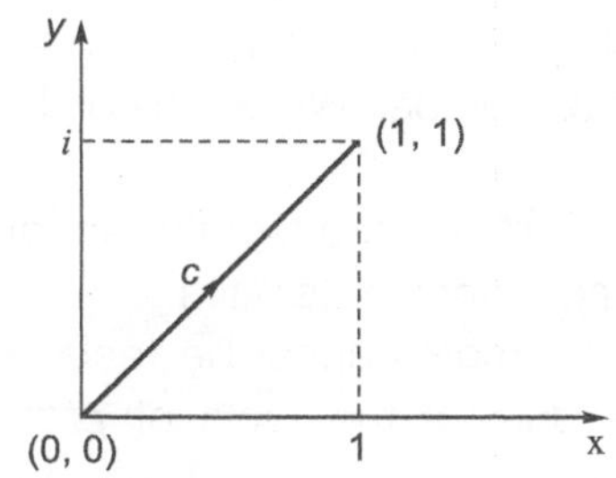

The value of I is

(a) $\frac{1}{2}i$ (b) $\frac{2}{3}i$
(c) $\frac{3}{4}i$ (d) $\frac{4}{5}i$

[2017 : 2 M, Set-1]

5.17 The value of the integral $\oint \frac{z+1}{z^2-4} dz$ in counter clockwise direction around a circle C of radius 1 with center at the point $z = -2$ is

(a) $\frac{\pi i}{2}$ (b) $2\pi i$
(c) $-\frac{\pi i}{2}$ (d) $-2\pi i$ [2018 : 1 M]

5.18 If C is a circle $|z| = 4$ and $f(z) = \frac{z^2}{(z^2-3z+2)^2}$, then $\oint_C f(z)\,dz$ is

(a) 1 (b) 0
(c) -1 (d) -2 [2018 : 2 M]

5.19 Which one of the following functions is analytic in the region $|z| \le 1$?

(a) $\frac{z^2-1}{z+2}$ (b) $\frac{z^2-1}{z+j0.5}$
(c) $\frac{z^2-1}{z}$ (d) $\frac{z^2-1}{z-0.5}$ [2019 : 1 M]

5.20 The closed-loop line integral $\oint_{|z|=5} \frac{z^3+z^2+8}{z+2} dz$ evaluated counter-clockwise, is

(a) $+8j\pi$ (b) $+4j\pi$
(c) $-8j\pi$ (d) $-4j\pi$ [2019 : 2 M]

5.21 The value of the following complex integral, with C representing the unit circle centered at origin in the counterclockwise sense, is

$$\int_C \frac{z^2+1}{z^2-2z}dz$$

(a) $-\pi i$ (b) πi
(c) $8\pi i$ (d) $-8\pi i$ [2020 : 1 M]

5.22 Let $p(z) = z^3 + (1+j)z^2 + (2+j)z + 3$, where z is a complex number. Which one of the following is true?

(a) The complex roots of the equation $p(z) = 0$ come in conjugate pairs.
(b) All the roots cannot be real.
(c) The sum of the roots of $p(z) = 0$ is real number
(d) Conjugate $\{p(z)\} = p\{\text{conjugate}(z)\}$ for all z [2021 : 1 M]

5.23 Let $(-1, -j)$, $(3, -j)$, $(3, +j)$ and $(-1, +j)$ be the vertices of rectangle C in the complex plane. Assuming that C is traversed in counter-clockwise direction, the value of the contour integral

$$\oint_C \frac{dz}{z^2(z-4)}$$ is

(a) 0 (b) $\frac{-j\pi}{8}$
(c) $\frac{j\pi}{2}$ (d) $\frac{j\pi}{16}$ [2021 : 2 M]

5.24 Which of the following complex functions is/are analytic on the complex plane?

(a) $f(z) = \text{Im}(z)$ (b) $f(z) = z^2 - z$
(c) $f(z) = e^{|z|}$ (d) $f(z) = j\,\text{Re}(z)$ [2024 : 1 M]

5.25 Consider the complex function $f(z) = \cos z + e^{z^2}$. The coefficient of z^5 in the Taylor series expansion of $f(z)$ about the origin is ______ (rounded off to 1 decimal place).

[2024 : 1 M]

Answers Complex Variables

5.1 (b)	5.2 (b)	5.3 (d)	5.4 (a)	5.5 (c)	5.6 (b)	5.7 (a)
5.8 (c)	5.9 (b)	5.10 (c)	5.11 (b)	5.12 (b)	5.13 (b)	5.14 (d)
5.15 (c)	5.16 (b)	5.17 (a)	5.18 (b)	5.19 (a)	5.20 (a)	5.21 (a)
5.22 (b)	5.23 (b)	5.24 (b)	5.25 (0)			

Explanations Complex Variables

5.1 (b)

A function $f(t)$ is said to be periodic function with period T if $f(t+T) = f(t)$ for all t

Now, $e^{z+2\pi i} = e^{z\cdot 2\pi i} = e^z\cdot(\cos 2\pi + i\sin 2\pi)$
$= e^z(1 + i(0)) = e^z$
$\cos 2\pi = 1$
$\sin 2\pi = 0$
$\therefore$ e^z is a periodic function with period $2\pi i$.

5.3 (d)

Let $Z = a + bi$

Since Z is shown inside the unit circle in I quadrant, a and b are both +ve and $0 < \sqrt{a^2+b^2} < 1$

Now, $\frac{1}{Z} = \frac{1}{a+bi}$

$$\frac{a-bi}{a^2+b^2} = \frac{a}{a^2+b^2} - \frac{b}{a^2+b^2}i$$

Since $a, b > 0$,

$$\frac{a}{\sqrt{a^2+b^2}} > 0$$

$\frac{-b}{a^2+b^2} < 0$. So, $\frac{1}{Z}$ is in IV quadrant.

$$\left|\frac{1}{Z}\right| = \sqrt{\left(\frac{a}{a^2+b^2}\right)^2 + \left(\frac{-b}{a^2+b^2}\right)^2}$$

$$= \sqrt{\frac{1}{a^2+b^2}} = \frac{1}{\sqrt{a^2+b^2}}$$

Since, $0 < \sqrt{a^2+b^2} < 1$

$$\frac{1}{\sqrt{a^2+b^2}} > 1$$

So, $\frac{1}{Z}$ is outside the unit circle is IV quadrant.

5.4 (a)

$x = i$, then in polar coordinates,

$$x = \cos\frac{\pi}{2} + i\sin\frac{\pi}{2} = e^{\frac{\pi}{2}i}$$

Now, $x^x = i^i = (e^{\pi/2 i})^i = e^{i^2\pi/2} = e^{-\pi/2}$

5.5 (c)

Given:

$$f(z) = \frac{1}{z+1} - \frac{2}{z+3}$$

$$= \frac{(z+3)-2(z+1)}{(z+1)(z+3)} = \frac{-z+1}{(z+1)(z+3)}$$

Poles are at –1 and –3, i.e., (–1, 0) and (–3, 0).
From figure below of $|z+1| = 1$,
We see that (–1, 0) is inside the circle and (–3, 0) is outside the circle.

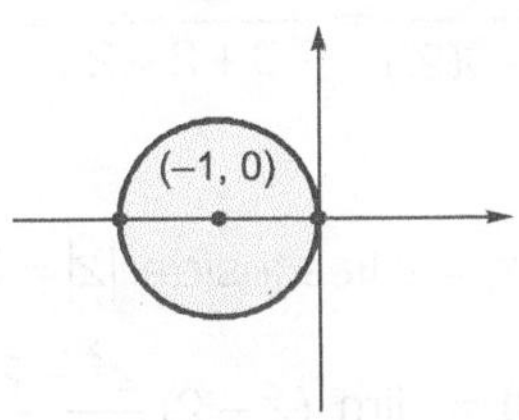

Residue theorem says
$$\frac{1}{2\pi j}\oint_C f(z)\,dz = \text{Residue of those poles which are inside } C.$$

So, the required integral $\frac{1}{2\pi j}\oint_C f(z)\,dz$ is given by the residue of function at pole (–1, 0) (which is inside the circle).

This residue is $= \frac{-(-1)+1}{(-1+3)} = \frac{2}{2} = 1$

5.6 (b)

$$-i = \cos\left(-\frac{\pi}{2}\right) + i\sin\left(-\frac{\pi}{2}\right)$$

$$= \cos\left(\frac{3\pi}{2}\right) + i\sin\left(\frac{3\pi}{2}\right)$$

$$(-i)^{1/2} = \left[\cos\left(-\frac{\pi}{2}\right) + i\sin\left(-\frac{\pi}{2}\right)\right]^{1/2}$$

$$= \cos\left(-\frac{\pi}{4}\right) + \sin\left(-\frac{\pi}{4}\right)$$

$$= \cos\left(\frac{3\pi}{4}\right) + i\sin\left(\frac{3\pi}{4}\right)$$

5.7 (a)

$$\frac{Z^2-4}{Z^2+4} = \frac{Z^2-4}{(Z+2i)(Z-2i)}$$

Poles at $2i$ and $-2i$, i.e., $(0, 2i)$ and $(0, -2i)$

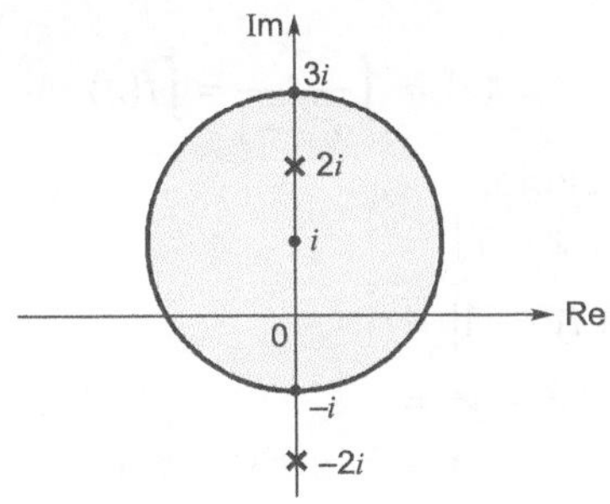

From figure of $|Z-i| = 2$, we see that pole, is inside C,
While pole, $-2i$ is outside C.

$$\therefore \oint \frac{Z^2-4}{Z^2+4}dz = 2\pi i \times \text{Res. F(z)}$$

$$= 2\pi i \cdot \left.\frac{(Z-2i)(Z^2-4)}{(Z+2i)(Z-2i)}\right|_{Z=2i}$$

$$= 2\pi i\,\frac{\left[(2i)^2-4\right]}{(2i+2i)} = -4\pi$$

5.8 (c)

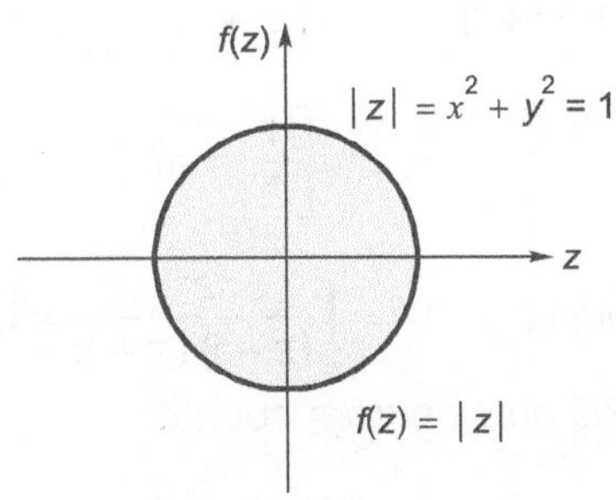

zz^*

$\Rightarrow$

$$z = x + iy$$
$$z^* = x - iy$$
$$zz^* = (x+iy)(x-iy) = x^2 + y^2$$

Which is equal to (1) always as given

$$|z| = 1$$
$$zz^* = x^2 + y^2$$

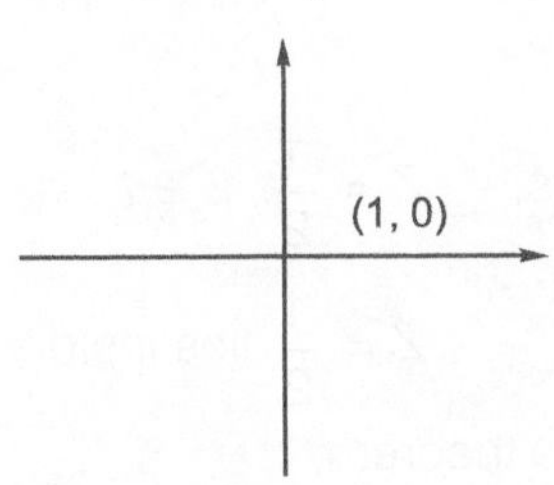

5.9 (b)

Let, $z = 1^i = 1^{e^{i(4n+1)\pi/2}} \quad n \in I$

$z = 1$ which is purely real and non negative.

5.10 (c)

$$f(z) = \int \frac{z^2}{z^2 - 1} = \int f(z)$$

Given circle,

$$|z - 1| = 1$$

$$\Rightarrow |(x + iy) - 1| = 1$$

$$(x - 1)^2 + y^2 = 1$$

$$x = 1, y = 0, r = 1$$

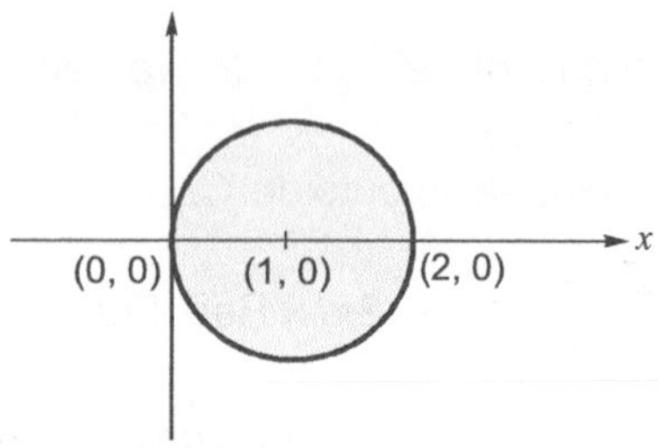

Poles of $f(z)$

$$z^2 - 1 = 0$$

$$[z = +1, 1]$$

So, $-1 \rightarrow$ Outside circle

$+1 \rightarrow$ Inside circle

$$\int \frac{z^2}{(z-1)(z+1)} = 2\pi i \left[\frac{z^2}{z+1}\right]_{z=+1}$$

$$= 2\pi i \left[\frac{1}{2}\right] = \pi i$$

For pole $(z = -1) = \int \frac{z^2}{(z-1)(z+1)} = 0$

As it lies outside from counter.

5.12 (b)

$$f(z) = z + z^*$$

$f(z) = 2x$ is continuous (polynomial)

$u = 2x \qquad v = 0$

$u_x = 2 \qquad u_y = 0$

$v_x = 0 \qquad v_y = 0$

C.R. equation not satisfied. ∴ No where analytic.

5.13 (b)

Singlarilies, $Z = \frac{1}{2}, 2 \pm i$

only, $Z = \frac{1}{2}$ lies inside C

By residue theorem,

$$\oint_C = 2\pi i(R) = \frac{48\pi i}{13}$$

Residue at $\frac{1}{2}$

$$= R_{1/2} = \lim_{Z \to 1/2}\left[\left(Z - \frac{1}{2}\right) \cdot \frac{2Z + 5}{\left(Z - \frac{1}{2}\right)(Z^2 + 4Z + 5)}\right]$$

$$= \frac{24}{13}$$

5.14 (d)

$$\lim_{z \to i} = \frac{z^2 + 1}{z^3 + 2z - i(z^2 + 2)} \quad \left(\frac{0}{0} \text{ from}\right)$$

$$\lim_{z \to i} = \frac{2z}{3z^2 + 2 - i(2z)}$$

$$= \frac{2i}{3i^2 + 2 - i(2i)} = \frac{2i}{-3 + 2 + 2} = \frac{2i}{-3 + 4} = 2i$$

5.15 (c)

Pole, $z = 2$ lies inside $|z| = 3$

$$\text{Res } f(z) = \lim_{z \to 2} (z - 2)\frac{z^2 - 2z + 3}{z - 2}$$

$z = 2, \quad = 8 - 4 + 3 = 7$

By Cauche residue theorem, $I = 2\pi i(7) = 14\pi i$

5.16 (b)

From the diagram C is $y = x$

$$I = \int_C (x^2 + iy^2)\, dz = \int_C (x^2 + iy^2)(dx + idy)$$

$$= \int_C (x^2 + ix^2)(dx + idx)$$

$$= \int x^2 dx + ix^2 dx + ix^2 dx - x^2 dx$$

$$= 2i\int_0^1 x^2 dx = 2i\left(\frac{x^3}{3}\right)\Big|_0^1 = \frac{2i}{3}$$

5.17 (a)

$$\int \frac{z+1}{z^2 - 4} dz$$

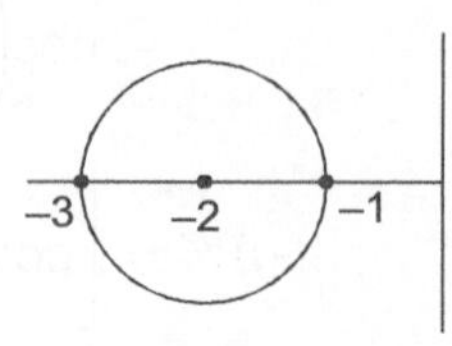

$$\int \frac{z+1}{(z-2)(z+2)} dz$$

$$\int \frac{\left(\frac{z+1}{z-2}\right)}{(z+2)} dz \qquad \text{where, } f(z) = \frac{z+1}{z-2}$$

$$= 2\pi i\, f(-2) = 2\pi i\left(\frac{-2+1}{-2-2}\right) = 2\pi i\left(\frac{-1}{-4}\right) = \frac{\pi i}{2}$$

5.18 (b)

$$\int \frac{z^2}{(z^2-3z+2)^2}\,dz$$

$$\int \frac{z^2}{(z-1)^2(z-2)^2}\,dz$$

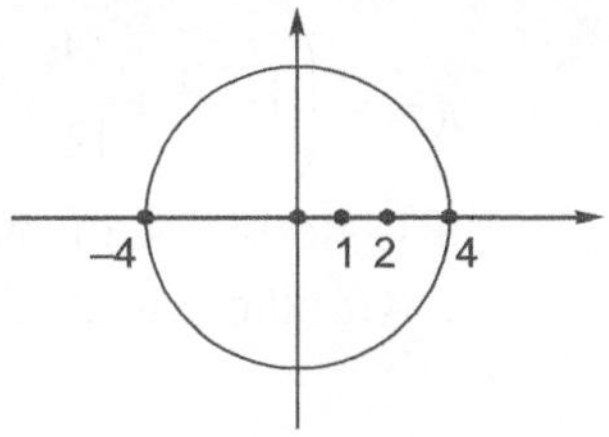

$$\underset{z=1}{\text{Res.}} f(z) = \lim_{z\to 1}\frac{1}{1!}\frac{d}{dz}\left((z-1)^2\cdot\frac{z^2}{(z-1)^2(z-2)^2}\right)$$

$$= \lim_{z\to 1}\left(\frac{2z(z-2)^2-2z^2(z-2)}{(z-2)^4}\right)$$

$$= \lim_{z\to 1}\left(\frac{2z(z-2)-2z^2}{(z-2)^3}\right) = \frac{-4}{-1} = 4$$

$$\underset{z=2}{\text{Res.}} f(z) = \lim_{z\to 2}\frac{1}{1!}\frac{d}{dz}\left((z-2)^2\cdot\frac{z^2}{(z-1)^2(z-2)^2}\right)$$

$$= \lim_{z\to 2}\left(\frac{(z-1)^2\cdot 2z - z^2 2(z-1)}{(z-1)^4}\right)$$

$$= \lim_{z\to 2}\left(\frac{2z(z-1)-2z^2}{(z-1)^3}\right) = \frac{4-8}{1} = -4$$

By residue theorem, $I = 2\pi i\,(4-4) = 0$

5.19 (a)

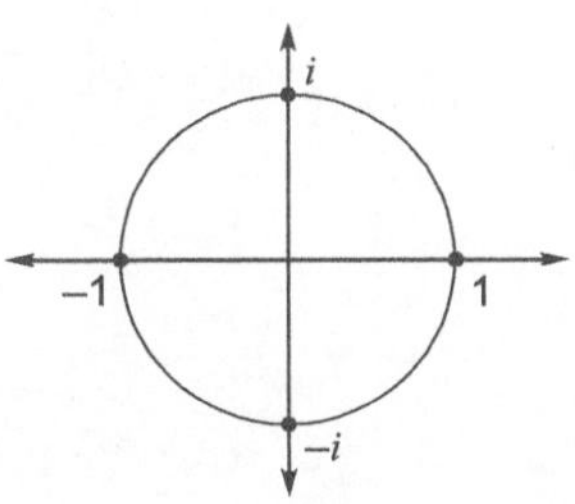

By Cauchy integral theorem, $\int \frac{z^2-1}{z+2}\,dz = 0$

Therefore, $\frac{z^2-1}{z+2}$ is analytic in the region $|z| \le 1$.

5.20 (a)

$$\oint_{|z|=5} \frac{z^3+z^2+8}{2+2}\,dz = 2\pi j\,(\text{sum of residues})$$

$$= 2\pi j \times \left[\lim_{z\to -2}\frac{(z+2)(z^3+z^2+8)}{(z+2)}\right]$$

$$= 2\pi j \times \left[\frac{-8+4+8}{1}\right] = 8\pi j$$

5.21 (a)

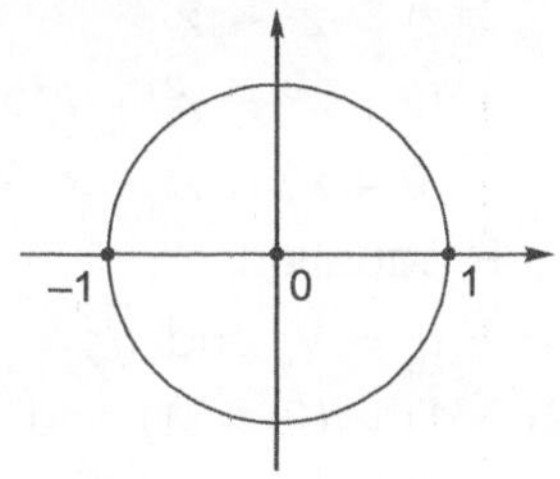

$\int \frac{z^2+1}{z^2-2z}\,dz$ where C is $|z| = 1$

$$= \int \frac{z^2+1}{z(z-2)}\,dz = \int \frac{\frac{z^2+1}{z-2}}{z}\,dz$$

By integral formula, $2\pi i\, f(0)$ where,

$$f(z) = \frac{z^2+1}{z-2} = 2\pi i\left(\frac{1}{-2}\right) = -\pi i$$

5.22 (b)

Since sum of the roots is a complex number.

$\Rightarrow$ Atleast one root is complex.

So all the roots cannot be real.

5.23 (b)

Given : $\oint_C \frac{dz}{z^2(z-4)}$

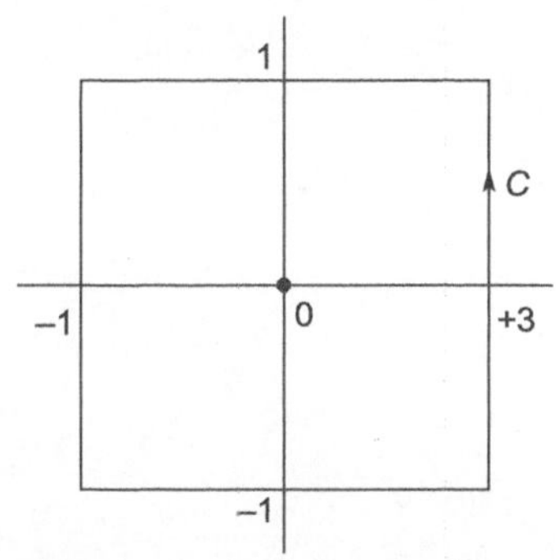

Singularities are given by $z^2(z-4) = 0$

$\Rightarrow \quad z = 0, 4$

$z = 0$ is pole of order m = 2 lies inside contour 'c'

$z = 4$ is pole of order m = 1 lies outside 'c'

Residue at $x = 0$

$$= \frac{1}{(2-1)!}\lim_{z\to 0}\frac{d^{2-1}}{dz^{2-1}}\left[(z-0)^2\frac{1}{z^2(z-4)}\right]$$

$$= \frac{-1}{(0-4)^2} = \frac{-1}{16}$$

By Cauchy residue theorem,

$$\oint_C f(z)\,dz = 2\pi i\,\text{Res} = 2\pi i\left[\frac{-1}{16}\right] = \frac{-\pi i}{8}$$

5.24 (b)

$$f(z) = z^2 - z$$

$$u + iv = x^2 - y^2 + i(2xy) - x - iy$$

$$\Rightarrow \quad u = x^2 - y^2 - x,\ V = 2xy - y$$

Now by C–R equation,

$$u_x = V_y \text{ and } u_y = -V_x$$

$$\Rightarrow \quad (2x - 1) = (2x - 1) \text{ and } -2y = -(2y)$$

i.e., C-R equation are satisfied.

5.25 (0)

$$f(z) = \left(1 + z^2 + \frac{z^4}{\lfloor 2} + \frac{z^6}{\lfloor 3} + ...\right) + \left(1 - \frac{z^2}{\lfloor 2} + \frac{z^4}{\lfloor 4} + ...\right)$$

It is series is of even powers.

$\therefore$ Coefficient of $z^5 = 0$.

■■■■

6 Probability & Statistics

6.1 If P and Q are two random events, then the following is TRUE

(a) Independence of P and Q implies that probability $(P \cap Q) = 0$

(b) Probability $(P \cup Q) \geq$ Probability (P) + Probability (Q)

(c) If P and Q are mutually exclusive, then they must be independent

(d) Probability $(P \cap Q) \leq$ Probability (P)

[2005 : 1 Mark]

6.2 A fair coin is tossed three times in succession. If the first toss produces a head, then the probability of getting exactly two heads in three tosses is

(a) $\frac{1}{8}$ (b) $\frac{1}{2}$

(c) $\frac{3}{8}$ (d) $\frac{3}{4}$ **[2005 : 2 M]**

6.3 Two fair dice are rolled and the sum r of the numbers turned up is considered

(a) $Pr(r > 6) = \frac{1}{6}$

(b) $Pr(r/3 \text{ is an integer}) = \frac{5}{6}$

(c) $Pr(r = 8 \mid r/4 \text{ is an integer}) = \frac{5}{9}$

(d) $Pr(r = 6 \mid r/5 \text{ is an integer}) = \frac{1}{18}$

[2006 : 2 M]

6.4 A loaded dice has following probability distribution of occurrences

Dice Value	1	2	3	4	5	6
Probability	1/4	1/8	1/8	1/8	1/8	1/4

If three identical dice as the above are thrown, the probability of occurrence of values 1, 5 and 6 on the three dice is

(a) same as that of occurrence of 3, 4, 5

(b) same as that of occurrence of 1, 2, 5

(c) 1/128

(d) 5/8 **[2007 : 2 M]**

6.5 X is a uniformly distributed random variable that takes values between 0 and 1. The value of $E\{X^3\}$ will be

(a) 0 (b) 1/8

(c) 1/4 (d) 1/2 **[2008 : 1 M]**

6.6 Assume for simplicity that N people, all born in April (a month of 30 days), are collected in a room. Consider the event of at least two people in the room being born on the same date of the month, even if in different years, e.g. 1980 and 1985. What is the smallest N so that the probability of this event exceeds 0.5?

(a) 20 (b) 7

(c) 15 (d) 16 **[2009 : 2 M]**

6.7 A box contains 4 white balls and 3 red balls. In succession, two balls are randomly selected and removed from the box. Give that the first removed ball is white, the probability that the second removed ball is red is

(a) 1/3 (b) 3/7

(c) 1/2 (d) 4/7 **[2010 : 2 M]**

6.8 Two independent random variables X and Y are uniformly distributed in the interval $[-1, 1]$. The probability that max $[X, Y]$ is less than 1/2 is

(a) 3/4 (b) 9/16

(c) 1/4 (d) 2/3 **[2012 : 1 M]**

6.9 A fair coin is tossed till a head appears for the first time. The probability that the number of required tosses is odd, is

(a) 1/3 (b) 1/2

(c) 2/3 (d) 3/4 **[2012 : 2 M]**

6.10 A continuous random variable X has a probability density function, $f(x) = e^{-x}$, $0 < x < \infty$. Then $P\{X > 1\}$ is

(a) 0.368 (b) 0.5

(c) 0.632 (d) 1.0 **[2013 : 1 M]**

6.11 A fair coin is tossed n times. The probability that the difference between the number of heads and tails is $(n - 3)$ is

(a) 2^{-n} (b) 0

(c) ${}^{n}C_{n-3}\, 2^{-n}$ (d) 2^{-n+3}

[2014 : 2 M, Set-1]

6.12 Consider a dice with the property that the probability of a face with n dots showing up is proportional to n. The probability of the face with three dots showing up is ____. **[2014 : 1 M, Set-2]**

6.13 Let X be a random variable with probability density function

$$f(x) = \begin{cases} 0.2, & \text{for} |x| \le 1 \\ 0.1, & \text{for } 1 < |x| \le 4 \\ 0, & \text{otherwise} \end{cases}$$

The probability $P(0.5 < X < 5)$ is ______.

[2014 : 2 M, Set-2]

6.14 Lifetime of an electric bulb is a random variable with density $f(x) = kx^2$, where x is measured in years. If the minimum and maximum lifetimes of bulb are 1 and 2 years respectively, then the value of k is ______.

[2014 : 1 M, Set-3]

6.15 A random variable X has probability density function $f(x)$ as given below:

$$f(x) = \begin{cases} a + bx & \text{for } 0 < x < 1 \\ 0 & \text{otherwise} \end{cases}$$

If the expected value $E[X] = 2/3$, then $Pr[X < 0.5]$ is ______.

[2015 : 1 M, Set-1]

6.16 Two players, A and B, alternately keep rolling a fair dice. The person to get a six first wins the game. Given that player A starts the game, the probability that A wins the game is

(a) $\frac{5}{11}$ (b) $\frac{1}{2}$

(c) $\frac{7}{13}$ (d) $\frac{6}{11}$ [2015 : 2 M, Set-1]

6.17 Two coins R and S are tossed. The 4 joint events H_RH_S, T_RT_S, H_RT_S, T_RH_S have probabilities 0.28, 0.18, 0.30, 0.24, respectively, where H represents head and T represents tail. Which one of the following is TRUE?

(a) The coin tosses are independent.
(b) R is fair, S is not
(c) S is fair, R is not
(d) The coin tosses are dependent

[2015 : 2 M, Set-2]

6.18 Candidates were asked to come to an interview with 3 pens each. Black, blue, green and red were the permitted pen colours that the candidate could bring. The probability that a candidate comes with all 3 pens having the same colour is ______.

[2016 : 1 M, Set-2]

6.19 Let the probability density function of a random variable , X, be given as:

$$f_x(x) = \frac{3}{2}e^{-3x}u(x) + ae^{4x}u(-x)$$

where $u(x)$ is the unit step function.

Then the value of 'a' and Prob $\{X \le 0\}$, respectively, are

(a) $2, \frac{1}{2}$ (b) $4, \frac{1}{2}$

(c) $2, \frac{1}{4}$ (d) $4, \frac{1}{4}$

[2016 : 2 M, Set-2]

6.20 An urn contains 5 red balls and 5 black balls. In the first draw, one ball is picked at random and discarded without noticing its colour. The probability to get a red ball in the second draw is

(a) $\frac{1}{2}$ (b) $\frac{4}{9}$

(c) $\frac{5}{9}$ (d) $\frac{6}{9}$ [2017 : 1 M, Set-2]

6.21 A person decides to toss a fair coin repeatedly until he gets a head. He will make at most 3 tosses. Let the random variable Y denote the number of heads. The value of var$\{Y\}$, where var$\{\cdot\}$ denotes the variance, equals

(a) $\frac{7}{8}$ (b) $\frac{49}{64}$

(c) $\frac{7}{64}$ (d) $\frac{105}{64}$

[2017 : 2 M, Set-2]

6.22 Assume that in a traffic junction, the cycle of the traffic signal lights is 2 minutes of green (vehicle does not stop) and 3 minutes of red (vehicle stops). Consider that the arrival time of vehicles at the junction is uniformly distributed over 5 minute cycle. The expected waiting time (in minutes) for the vehicle at the junction is ______.

[2017 : 1 M, Set-2]

6.23 The mean-square of a zero-mean random process is $\frac{kT}{C}$, where k is Boltzmann's constant, T is the absolute temperature, and C is a capacitance. The standard deviation of the random process is,

(a) $\sqrt{\frac{kT}{C}}$ (b) $\frac{C}{kT}$

(c) $\frac{\sqrt{kT}}{C}$ (d) $\frac{kT}{C}$ [2019 : 1 M]

6.24 The probability of a resistor being defective is 0.02. There are 50 such resistors in a circuit. The probability of two or more defective resistors in the circuit (round off to two decimal places) is ______.

[2019 : 2 M]

6.25 Suppose the probability that a coin toss shows "head" is p, where $0 < p < 1$. The coin is tossed repeatedly until the first "head" appears. The expected number of tosses required is

(a) $\frac{(1-p)}{p}$ (b) $\frac{1}{p}$

(c) $\frac{1}{p^2}$ (d) $\frac{p}{(1-p)}$ [2021 : 2 M]

6.26 Let the probability density function of a random variable x be given as

$$f(x) = ae^{-2|x|}$$

The value of a is _____. [2022 : 2 M]

6.27 One million random numbers are generated from a statistically stationary process with a Gaussian distribution with mean zero and standard deviation σ_o.

The σ_o is estimated by randomly drawing out 10,000 numbers of samples (x_n). The estimates $\hat{\sigma}_1, \hat{\sigma}_2$ are computed in the following two ways.

$$\hat{\sigma}_1^2 = \frac{1}{10000}\sum_{n=1}^{10000} x_n^2$$

$$\hat{\sigma}_2^2 = \frac{1}{9999}\sum_{n=1}^{10000} x_n^2$$

Which of the following statements is true?

(a) $E(\hat{\sigma}_2^2) = \sigma_o^2$ (b) $E(\hat{\sigma}_2) = \sigma_o$

(c) $E(\hat{\sigma}_1^2) = \sigma_o^2$ (d) $E(\hat{\sigma}_1) = E(\hat{\sigma}_2)$

[2023 : 1 M]

6.28 Three points in the x-y plane are (–1, 0.8), (0, 2.2) and (1, 2.8). The value of the slope of the best fit straight line in the least square sense is _________ (Round off to 2 decimal places).

[2023 : 1 M]

6.29 The expected number of trials for first occurrence of a "head" in a biased coin is known to be 4. The probability of first occurrence of a "head" in the second trial is ____________ (Round off to 3 decimal places).

[2023 : 2 M]

6.30 Let X be a discrete random variable that is uniformly distributed over the set {–10, –9, ..., 0, ..., 9, 10}. Which of the following random variables is/are uniformly distributed?

(a) X^2 (b) $(X + 10)^2$

(c) X^3 (d) $(X - 5)^2$

[2024 : 2 M]

■■■■

Answers Probability & Statistics

6.1 (d)	6.2 (b)	6.3 (c)	6.4 (c)	6.5 (c)	6.6 (b)	6.7 (c)
6.8 (b)	6.9 (c)	6.10 (a)	6.11 (b)	6.12 (0.142)	6.13 (0.4)	6.14 (0.42)
6.15 (0.25)	6.16 (d)	6.17 (d)	6.18 (0.0625)	6.19 (a)	6.20 (a)	6.21 (c)
6.22 (0.9)	6.23 (a)	6.24 (0.26)	6.25 (b)	6.26 (1)	6.27 (c)	6.28 (1)
6.29 (0.188)	6.30 (b, c)					

Explanations Probability & Statistics

6.1 (d)

(a) is false since of P & Q are independent

$$Pr(P \cap Q) = Pr(P) \times Pr(Q)$$

which need not be zero.

(b) is false since

$Pr(P \cup Q) = Pr(P) + Pr(Q) - Pr(P \cap Q)$

$\therefore Pr(P \cup Q) \le Pr(P) + Pr(Q)$

(c) is false since independence and mutually exclusion are unrelated properties.

(d) is true

Since $P \cap Q \subseteq P$

$\Rightarrow n(P \cap Q) \le n(P)$

$\Rightarrow Pr(P \cap Q) \le Pr(P)$

6.2 (b)

$$P(A \mid B) = \frac{P(A \cap B)}{P(B)}$$

$\therefore$ P(2 heads in 3 tosses | first toss is head)

$$= \frac{P(\text{2 heads in 3 tosses and first toss is head})}{P(\text{first toss is head})}$$

$= P(\text{first toss is head}) = \frac{1}{2}$

P(2 head in 3 tosses and first toss is heard)

$$= P(HHT) + P(HTH)$$

$$= \frac{1}{2}\times\frac{1}{2}\times\frac{1}{2}+\frac{1}{2}\times\frac{1}{2}\times\frac{1}{2}=\frac{1}{4}$$

$\therefore$ Required probability $= \frac{1/4}{1/2} = \frac{1}{2}$

6.3 (c)

If two fair dices are rolles the probability distribution of r where r is the sum of the numbers on each die is given by

r	2	3	4	5	6	7	8	9	10	11	12
$P(r)$	$\frac{1}{36}$	$\frac{2}{36}$	$\frac{3}{36}$	$\frac{4}{36}$	$\frac{5}{36}$	$\frac{6}{36}$	$\frac{5}{36}$	$\frac{4}{36}$	$\frac{3}{36}$	$\frac{2}{36}$	$\frac{1}{36}$

The above table has been obtained by taking all different ways of obtaining a particular sum. For example, a sum of 5 can be obtained by (1, 4), (2, 3), (3, 2) and (4, 1).

$\therefore \quad P(x = 5) = 4/36$

Now let us consider choice (a)

$$Pr(r > 6) = Pr(r \geq 7)$$

$$= \frac{6}{36}+\frac{5}{36}+\frac{4}{36}+\frac{3}{36}+\frac{2}{36}+\frac{1}{36}$$

$$= \frac{21}{36} = \frac{7}{12}$$

$\therefore$ Choice (a) $Pr(r > 6) = 1/6$ is wrong.

Consider choice (b):

$Pr(r/3$ is an integer)

$= Pr(r = 3) + Pr(r = 6) + Pr(r = 9) + Pr(r = 12)$

$$= \frac{2}{36}+\frac{5}{36}+\frac{4}{36}+\frac{1}{36}=\frac{12}{36}=\frac{1}{3}$$

$\therefore$ choice (b) $Pr\ r/3$ is an integer = 5/6 is wrong

Consider choice (c):

$Pr(r = 8 \mid r/4 \text{ is an integer}) = \frac{1}{36}$

Now, $Pr(r/4$ is an integer)

$= Pr(r = 4) + Pr(r = 8) + Pr(r = 12)$

$$= \frac{3}{36}+\frac{5}{36}+\frac{1}{36}=\frac{9}{36}=\frac{1}{4}$$

$Pr(r = 8$ and $r/4$ is an integer)

$$= Pr(r = 8) = \frac{5}{36}$$

$\therefore Pr(r = 8 \mid r/4 \text{ is an integer})$

$$= \frac{5/36}{1/4} = \frac{20}{36} = \frac{5}{9}$$

$\therefore$ Choice (c) is correct.

6.4 (c)

Dice value	1	2	3	4	5	6
Probability	$\frac{1}{4}$	$\frac{1}{8}$	$\frac{1}{8}$	$\frac{1}{8}$	$\frac{1}{8}$	$\frac{1}{4}$

$$P(1, 5, 6) = \frac{1}{4}\times\frac{1}{8}\times\frac{1}{4} = \frac{1}{128}$$

6.5 (c)

x is uniformly distributes in [0, 1]

$\therefore$ Probability density function

$$f(x) = \frac{1}{b-a} = \frac{1}{1-0} = 1$$

$$\therefore \quad f(x) = 1 \quad 0 < x < 1$$
$$= 0 \quad \text{elsewhere}$$

Now, $E(x^3) = \int_0^1 x^3 f(x)\,dx$

$$= \int_0^1 x^3 \times 1 \times dx = \left[\frac{x^4}{4}\right]_0^1 = \frac{1}{4}$$

6.7 (c)

$P(II$ is red $\mid I$ is white)

$$= \frac{P(II \text{ is red and } I \text{ is white})}{P(I \text{ is white})}$$

$$= \frac{P(I \text{ is white and } II \text{ is red})}{P(I \text{ is white})} = \frac{\frac{4}{7}\times\frac{3}{6}}{\frac{4}{7}} = \frac{3}{6} = \frac{1}{2}$$

6.8 (b)

$-1 \leq x \leq 1$ and $-1 \leq y \leq 1$ is the entire rectangle. The region in which maximum of $\{x, y\}$ is less than $\frac{1}{2}$ is shown below as shaded region inside this rectangle.

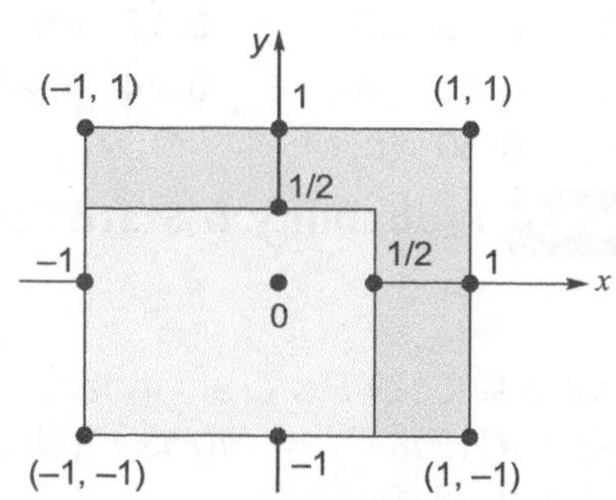

$$P\left(\max\{x, y\} < \frac{1}{2}\right) = \frac{\text{Area of shaded region}}{\text{Area of entire rectangle}}$$

$$= \frac{\frac{3}{2}\times\frac{3}{2}}{2\times 2} = \frac{9}{16}$$

6.9 (c)

P(number of tosses is odd)
= P(number of tosses is 1, 3, 5, 7...)

P(no. of toss is 1) = P(Head in first toss) = $\frac{1}{2}$

P(no. of toss is 3) = P(tail in first toss, tail in second toss and head in third toss)

$$= \frac{1}{2}\times\frac{1}{2}\times\frac{1}{2} = \frac{1}{8}$$

P(no. of toss is 5) = P(T, T, T, T, H)

$$= \left(\frac{1}{2}\right)^5 = \frac{1}{32} \text{ ... etc.}$$

So P(no. of tosses is odd) = $\frac{1}{2}+\frac{1}{8}+\frac{1}{32}+\cdots$

Sum of infinite geometric series with

$$a = \frac{1}{2} \text{ and } r = \frac{1}{4} \Rightarrow \frac{\frac{1}{2}}{1-\frac{1}{4}} = \frac{\frac{1}{2}}{\frac{3}{4}} = \frac{2}{3}$$

6.10 (a)

$$P = \int_1^{\infty} f(x)\,dx = \int_1^{\infty} e^{-x}dx = e^{-x}\Big|_{\infty}^{1} = e^{-1} = 0.368$$

6.11 (b)

$$(x + y)^n = \sum_{r=0}^{n} {}^nC_r\, x^{n-r} y^n$$

when, n = Probability of occurrence of head
y = Probability of occurrence of tail

Let number of head is P.
Number of tail is q

$$P + q = n \quad \text{...(i)}$$

Total number of tails
Given: $|P - q| = n - 3$

$$|P - (n - P)| = n - 3$$

$n = n - 3$ which is not possible.

Here required probability is zero.

6.12 (0.142)

Let probability of occurrence of one dot is P.
So, writing total probability
$P + 2P + 3P + 4P + 5P + 6P = 1$

$$P = \frac{1}{21}$$

hence problem of occurrence of 3 dot is

$$= 3P = \frac{3}{21} = \frac{1}{7} = 0.142$$

6.13 (0.4)

$$\text{Probability } (0.5 < n < 5) = \int_{0.5}^{5} f(x)\,dx$$

$$= \int_{0.5}^{1} 0.2\,dx + \int_{1}^{4} 0.1\,dx + \int_{4}^{5} 0\,dx$$

$= 0.2[1 - 0.5] + 0.1[4 - 1] + 0[5 - 4]$
$= 0.2 \times 0.5 + 0.1 \times 3 = 0.1 + 0.3 = 0.4$

6.14 (0.42)

Life time of an electric bulb with density

$$f(x) = Kx^2$$

If minimum and maximum lifetimes of bulb are 1 and 2 years respectively then

$$\int_1^2 Kx^2\,dx = 1$$

$$K\frac{x^3}{3}\Big|_1^2 = 1$$

$$K\left(\frac{8}{3}-\frac{1}{3}\right) = 1$$

$$\frac{7K}{3} = 1 \Rightarrow K = \frac{3}{7} = 0.42$$

6.15 (0.25)

$$f(x) = \begin{cases} a + bx & \text{for } 0 < x < 1 \\ 0 & \text{otherwise} \end{cases}$$

Now given, $E(X) = 2/3$

$$\Rightarrow \int_0^1 xf(x)\,dx = \frac{2}{3}$$

$$\Rightarrow \int_0^1 x(a + bx)\,dx = \frac{2}{3}$$

$$\Rightarrow a\left(\frac{x^2}{2}\right)_0^1 + b\left(\frac{x^3}{3}\right)_0^1 = \frac{2}{3}$$

$$a\left(\frac{1}{2}\right) + b\left(\frac{1}{3}\right) = \frac{2}{3}$$

$$\Rightarrow 3a + 2b = 4 \quad \text{...(i)}$$

$$\text{Now, } \int_0^1 f(x)\,dx = 1$$

(Total probability is always equal to 1)

$$\Rightarrow \int_0^1 (a + bx)\,dx = \left(ax + \frac{bx^2}{2}\right)_0^1 = 1$$

$$\Rightarrow a + \frac{b}{2} = 1 \Rightarrow 2a + b = 2 \quad \text{...(ii)}$$

Now solving (i) and (ii), we get

$$a = 0,\ b = 2$$

$$\text{So } f(x) = \begin{cases} 2x & \text{for } 0 < x < 1 \\ 0 & \text{otherwise} \end{cases}$$

Now we need,

$$p(x < 0.5) = \int_0^{0.5} 2x\,dx = 2\left(\frac{x^2}{2}\right)_0^{0.5} = 0.5^2 - 0^2 = 0.25$$

6.16 (d)

$$P(\text{A wins}) = p(6 \text{ in first throw by } A) + p(A \text{ not } 6, B \text{ not } 6, A\ 6) + \ldots$$

$$= \frac{1}{6} + \frac{5}{6}\cdot\frac{5}{6}\cdot\frac{1}{6} + \cdots$$

$$= \frac{1}{6}\left(1 + \left(\frac{5}{6}\right)^2 + \left(\frac{5}{6}\right)^4 + \cdots\right) = \frac{1}{6}\cdot\frac{1}{1-\left(\frac{5}{6}\right)^2} = \frac{6}{11}$$

6.17 (d)

From the given information, we can create a joint probability table as follows:

S \ R	H_R	T_R	
H_S	0.28	0.24	0.52
T_S	0.30	0.18	0.48
	0.58	0.42	1

From the table, we can get

$P(H_R) = 0.58$, $P(T_R) = 0.42$, $P(H_S) = 0.52$

$P(T_S) = 0.48$

So, Coins *R* and *S* are biased (not fair). So choices (*b*) and (*c*) are both false.

The coin tosses are not independent since their probability of heads and tails is not 0.5.

R and *S* are dependent.

If *R* and *S* were independent then all the joint probabilities will be equal to the product of the marginal probabilities.

For example

$$P(H_R \cap H_S) = 0.28$$

$$P(H_R)\cdot P(H_S) = 0.58 \times 0.52 = 0.3016$$

Clearly $P(H_R \cap H_S) \neq P(H_R)\cdot P(H_S)$

So *R* and *S* are not independent.

i.e. *R* and *S* are dependent. So, choice (a) is false and choice (d) is true.

6.19 (a)

$$f_x(x) = \begin{cases} ae^{4x} & x < 0 \\ \frac{3}{2}e^{-3x} & x \geq 0 \end{cases}$$

$$\int_{-\infty}^{\infty} f_x(x) = 1$$

$$\int_{-\infty}^{0} ae^{4x}dx + \int_0^{\infty} \frac{3}{2}e^{-3x}dx = 1$$

$$\left[\frac{ae^{4x}}{4}\right]_{-\infty}^{0} + \left[\frac{\frac{3}{2}e^{-3x}}{-3}\right]_0^{\infty} = 1$$

$$\frac{a}{4} + \frac{3}{6} = 1$$

$$a = 2$$

$$P(x \leq 0) = \int_{-\infty}^{0} 2e^{4x}dx = \left[\frac{e^{4x}}{2}\right]_{-\infty}^{0} = \frac{1}{2}$$

6.20 (a)

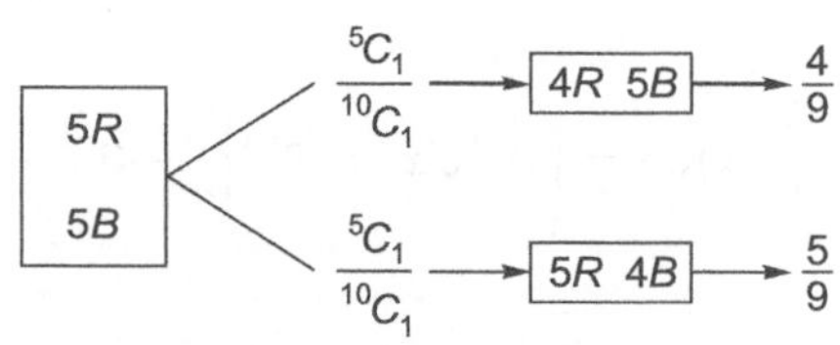

$$P(\text{red}) = \frac{5}{10}\cdot\frac{4}{9} + \frac{5}{10}\cdot\frac{5}{9} = \frac{45}{90} = 0.5$$

6.21 (c)

y: No. of Heads	0	1
$P(y)$	1/8	7/8

$$E(y) = 0 \times \frac{1}{8} + 1 \times \frac{7}{8} = \frac{7}{8}$$

$$E(y^2) = 0^2\left(\frac{1}{8}\right) + 1^2 \times \frac{7}{8} = \frac{7}{8}$$

$$\text{Variance}(y) = E(y^2) - (E(y))^2$$

$$= \frac{7}{8} - \frac{49}{64} = \frac{56-49}{64} = \frac{7}{64}$$

6.22 (0.9)

t be arrival time of vehicles of the junction is uniformally distributed in [0, 5].

Let *y* be the waiting time of the junction.

$$\text{Then,} \quad y = \begin{cases} 0 & t < 2 \\ 5 - t & 2 \leq t < 5 \end{cases}$$

$$y \rightarrow [0, 5]$$

$$f(y) = \frac{1}{5-0} = \frac{1}{5}$$

$$E(y) = \int_{-\infty}^{0} y(y)\,dy = \int_{0}^{5} y f(y) dy$$

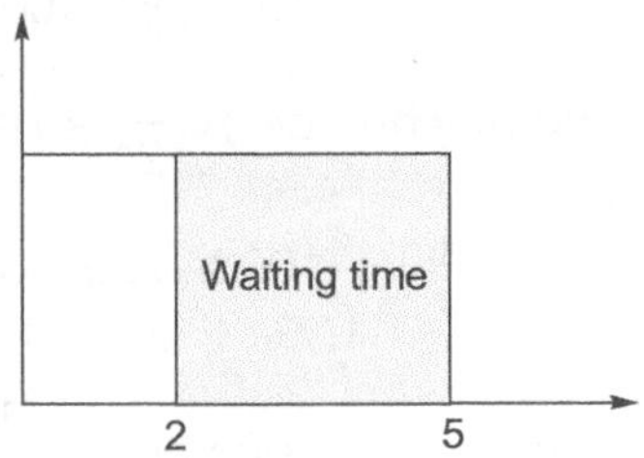

$$= \int_{2}^{5} y\left(\frac{1}{5}\right) dy = \frac{1}{5}\int_{2}^{5}(5-t)\,dt$$

$$= \frac{1}{5}\left(5t - \frac{t^2}{2}\right)\Bigg|_{2}^{5}$$

$$= \frac{1}{5}\left\{\left(25 - \frac{25}{2}\right) - \left(10 - \frac{4}{2}\right)\right\}$$

$$= \frac{1}{5}\left(\frac{25}{2} - 8\right) = \frac{1}{5}\left(\frac{9}{2}\right) = 0.9$$

6.23 (a)

Given that, $E(x^2) = \frac{kT}{C}$

$$E(x) = 0 = E(x^2) - (E(x))^2$$

$$\text{Var}(x) = \frac{kT}{C} - 0 = \frac{kT}{C}$$

$$\text{Standard deviation} = \sqrt{\frac{kT}{C}}$$

6.24 (0.26)

$$p = 0.02; \quad n = 50$$

$$\lambda = np = 50(0.02) = 1$$

$$p(x \geq 2) = 1 - p(x < 2) = 1 - (p(x=0) + p(x=1))$$

$$= 1 - \left(\frac{e^{-\lambda}\lambda^0}{0!} + \frac{e^{-\lambda}\lambda^1}{1!}\right)$$

$$= 1 - e^{-\lambda}(1 + \lambda) = 1 - e^{-1}(1+1) = 0.26$$

6.25 (b)

$P(H) = p$, let x = Number of tosses

x	1	2	3	4	5...
$P(x)$	p	$(1-p)p$	$(1-p)^2 p$		

$$E(x) = \Sigma x_i P(x)$$

$$= p + 2(1-p)p + 3(p)(1-p)^2 + \ldots$$

$$= p\left[1 + 2(1-p) + 3(1-p)^2 + \ldots\right]$$

$$= p[1-(1-p)]^{-2} = \frac{1}{p}$$

6.26 (1)

$f(x)$ is a p.d.f. if $\int_{-\infty}^{\infty} f(x)dx = 1$.

$$\Rightarrow \int_{-\infty}^{\infty} ae^{-2|x|}dx = 1 \Rightarrow \int_{-\infty}^{0} ae^{2x}dx + \int_{0}^{\infty} ae^{-2(x)}dx = 1$$

$$= a\frac{e^{2x}}{2}\Bigg]_{-\infty}^{0} + a\left(\frac{e^{-2x}}{-2}\right)_{0}^{\infty} = 1$$

$$\Rightarrow \frac{a}{2} + \frac{a}{2} = 1 \Rightarrow a = 1$$

6.27 (c)

σ_o^2 is the variance of population of 1000 samples given by

$$\sigma_o^2 = \frac{1}{10000}\sum_{n=1}^{10000}\left(X - \bar{X}\right)^2$$

Given $\bar{X} = 0$

$$\sigma_o^2 = \frac{1}{10000}\sum X_n^2$$

We know that $E\left[\sigma_1^2\right] = \sigma_1^2 = \sigma_0^2$

Hence, option (c) is the correct answer.

6.28 (1)

Given that :

X	–1	0	1
Y	0.8	2.2	2.8

Best fit of straight line ($y = ax + b$) by least square.

Approximation given by normal equations

$$\Sigma y_i = a\,\Sigma x_i + b\,n \qquad \ldots\text{(i)}$$

$$\Sigma x_i y_i = a\,\Sigma x_i^2 + b\,\Sigma x_i \qquad \ldots\text{(ii)}$$

x	y	x_i^2	xy
–1	0.8	1	–0.8
0	2.2	0	0
1	2.8	1	2.8
$\Sigma x_i = 0$	$\Sigma y_i = 5.8$	$\Sigma x^2 = 2$	$\Sigma xy = 2$

Substituting values in eqn. (1) and (2)

$$5.8 = a(0) + 3b \qquad \ldots\text{(iii)}$$

$$2 = a(2) + 0(b) \qquad \ldots\text{(iv)}$$

Solving we get, $a = 1$, $b = \frac{5.8}{3} = 1.93$

$\therefore$ Slope of best fit $y = ax + b$ is 'a' = 1.

6.29 (0.188)

By Geometrical distribution,

$$E(x) = \frac{1}{P} \Rightarrow E(x) = 4 = \frac{1}{P}$$

$$q = 1 - \frac{1}{4} = \frac{3}{4}$$

$$P(H) = \frac{1}{4},\quad P(T) = \frac{3}{4}$$

Required probability, $P(E) = P(T\,H)$

$$= \frac{3}{4} \times \frac{1}{4} = \frac{3}{16} = 0.1875$$

6.30 (b, c)

$X^2 = \{100, 81, 64, ..., 4, 1, 0, 1, 4, 9, ..., 81, 100\}$

$$P(\text{Choosing } 0) = \frac{1}{21},\ P\ (\text{choosing } 1) = \frac{2}{21}$$

So, X^2 is not uniformly distributed.

$(X - 5)^2 = \{225, 199, ..., 4, 1, 0, 1, 4, 9, 16, 25\}$

Again probability of choosing any number is not equal so

$(X - 5)^2$ is also not uniformly - distributed.

$(X + 10)^2 = \{0, 1, 4, 9, 16, ... 81, 100, 121, .., 400\}$

$P(\text{Choosing any number}) = \frac{1}{21}$ so is it uniformly-distributed

Now, $X^3 = \{-1000, -729, -512, ..., -8, -1, 0, 1, 8, ... 729, 1000\}$

Again $P(\text{Choosing any number}) = \frac{1}{21} = \text{Constant}$

So, it is also uniformly-distributed.

■■■■

7 Numerical Methods

7.1 The differential equation $\frac{dx}{dt} = \frac{1-x}{\tau}$ is discretised using Euler's numerical integration method with a time step $\Delta T > 0$. What is the maximum permissible value of ΔT to ensure stability of the solution of the corresponding discrete time equation?
(a) 1 (b) $\tau/2$
(c) τ (d) 2τ **[2007 : 2 M]**

7.2 Equation $e^x - 1 = 0$ is required to be solved using Newton's method with an initial guess $x_0 = -1$. Then, after one step of Newton's method, estimate x_1 of the solution will be given by
(a) 0.71828 (b) 0.36784
(c) 0.20587 (d) 0.00000 **[2008 : 2 M]**

7.3 Let $x^2 - 117 = 0$. The iterative steps for the solution using Newton-Raphson's method is given by

(a) $x_{k+1} = \frac{1}{2}\left(x_k + \frac{117}{x_k}\right)$

(b) $x_{k+1} = x_k - \frac{117}{x_k}$

(c) $x_{k+1} = x_k - \frac{x_k}{117}$

(d) $x_{k+1} = x_k - \frac{1}{2}\left(x_k + \frac{117}{x_k}\right)$ **[2009 : 2 M]**

7.4 Solution of the variables x_1 and x_2 for the following equations is to be obtained by employing the Newton-Raphson iterative method
equation (i): $10x_2 \sin x_1 - 0.8 = 0$
equation (ii): $10x_2^2 - 10x_2 \cos x_1 - 0.6 = 0$
Assuming the initial values, $x_1 = 0.0$ and $x_2 = 1.0$, the Jacobian matrix is

(a) $\begin{bmatrix} 10 & -0.8 \\ 0 & -0.6 \end{bmatrix}$ (b) $\begin{bmatrix} 10 & 0 \\ 0 & 10 \end{bmatrix}$

(c) $\begin{bmatrix} 0 & -0.8 \\ 10 & -0.6 \end{bmatrix}$ (d) $\begin{bmatrix} 10 & 0 \\ 10 & -10 \end{bmatrix}$ **[2011 : 2 M]**

7.5 When the Newton-Raphson method is applied to solve the equation, $f(x) = x^3 + 2x - 1 = 0$, the solution at the end of the first iteration with the initial guess value as $x_0 = 1.2$ is
(a) –0.82 (b) 0.49
(c) 0.705 (d) 1.69 **[2013 : 2 M]**

7.6 The function $f(x) = e^x - 1$ is to be solved using Newton-Raphson method. If the initial value of x_0 is taken as 1.0, then the absolute error observed at 2nd iteration is ______. **[2014 : 2 M, Set-3]**

7.7 A differential equation $\frac{di}{dt} - 0.2i = 0$ is applicable over $-10 < t < 10$. If $i(4) = 10$, then $i(-5)$ is ______. **[2015 : 2 M, Set-2]**

7.8 Only one of the real roots of $f(x) = x^6 - x - 1$ lies in the interval $1 \le x \le 2$ and bisection method is used to find its value. For achieving an accuracy of 0.001, the required minimum number of iterations is ____. (Give the answer up to two decimal places.) **[2017 : 2 M, Set-1]**

7.9 Let x and y be integers satisfying the following equations:
$$2x^2 + y^2 = 34$$
$$x + 2y = 11$$
The value of $(x + y)$ is ____. **[2017 : 1 M, Set-2]**

7.10 Let, $y^2 - 2y + 1 = x$ and $\sqrt{x} + y = 5$. The value of $x + \sqrt{y}$ equals______.
(Give the answer up to three decimal places). **[2017 : 1 M, Set-2]**

Answers Numerical Methods

7.1	(d)	7.2	(a)	7.3	(a)	7.4	(b)	7.5	(c)	7.6	(0.06)	7.7	(1.652)
7.8	(10)	7.9	(7)	7.10	(5.732)								

Explanations Numerical Methods

7.2 (a)

Here, $f(x) = e^x - 1;\ f'(x) = e^x$

The newton raphson iterative equation is

$$x_{i+1} = x_i - \frac{f(x_i)}{f'(x_i)}$$

$$f(x_i) = e^{xi} - 1;\ f'(x_i) = e^{xi}$$

$$\therefore \quad x_{i+1} = x_i - \frac{e^{x_i} - 1}{e^{x_i}}$$

i.e. $$x_{i+1} = \frac{x_i e^{x_i} - (e^{x_i} - 1)}{e^{x_i}} = \frac{e^{x_i}(x_i - 1) + 1}{e^{x_i}}$$

Now put $i = 0$

$$x_1 = \frac{e^{x_0}(x_0 - 1) + 1}{e^{x_0}}$$

Put, $x_0 = -1$ as given,

$$x_1 = [e^{-1}(-2) + 1]/e^{-1} = 0.71828$$

7.3 (a)

$$x_{K+1} = x_K - \frac{f(x_K)}{f'(x_K)} = x_K - \frac{x_K^2 - 117}{2x_K} = \frac{1}{2}\left[x_K + \frac{117}{x_K}\right]$$

7.4 (b)

$$u(x_1, x_2) = 10x_2 \sin x_1 - 0.8 = 0$$

$$v(x_1, x_2) = 10x_2^2 - 10x_2 \cos x_1 - 0.6 = 0$$

The Jacobian matrix is

$$\begin{bmatrix} \frac{\partial u}{\partial x_1} & \frac{\partial u}{\partial x_2} \\ \frac{\partial v}{\partial x_1} & \frac{\partial v}{\partial x_2} \end{bmatrix} = \begin{bmatrix} 10x_2 \cos x_1 & 10\sin x_1 \\ 10x_2 \sin x_1 & 20x_2 - 10\cos x_1 \end{bmatrix} = \begin{bmatrix} 10 & 0 \\ 0 & 10 \end{bmatrix}$$

7.5 (c)

$$f'(x) = 3x^2 + 2$$

$$f'(x_o) = 3(1.2)^2 + 2 = 6.32$$

$$f(x_o) = (1.2)^3 + 2 \times 1.2 - 1 = 3.128$$

$$f'(x_1) = x_o - \frac{f(x_o)}{f'(x_1)} = 1.2 - \frac{3.728}{6.32} = 0.705$$

7.6 (0.06)

Given, $f(x) = e^x - 1$

or, $f(x_K) = (e^{x_K} - 1)$ and $x_o = 1$

In Newton-Raphson method, we have:

$$x_{K+1} = \left[x_K - \frac{f(x_K)}{f'(x_K)}\right]$$

$$\therefore \quad x_1 = \left[x_o - \frac{f(x_o)}{f'(x_o)}\right] \quad \text{...(i)}$$

Now, $f(x_0) = e^{x_o} - 1 = e^1 - 1 = (e - 1)$

and $f'(x) = e^x$

$\therefore$ $f'(x_0) = e^1 = e$

Putting the values, we get:

$$x_1 = \left[1 - \frac{(e-1)}{e}\right] = \left[\frac{e - e + 1}{e}\right] = e^{-1}$$

Also, $$x_2 = \left[x_1 - \frac{f(x_1)}{f'(x_1)}\right] \quad \text{...(ii)}$$

Now, $x_1 = \frac{1}{e} = e^{-1}$ and $f(x_1) = (e^{e^{-1}} - 1)$,

$$f'(x_1) = e^{e^{-1}}$$

Putting the values, we get:

$$x_2 = \left[e^{-1} - \frac{(e^{e^{-1}} - 1)}{e^{e^{-1}}}\right] = \left[e^{-1} - \frac{(e^{0.37} - 1)}{e^{0.37}}\right] = 0.06$$

Therefore, the absolute error observed at second iteration = 0.06.

7.7 (1.652)

$$\frac{di}{dt} = 0.2i$$

$$\frac{di}{i} = 0.2\, dt$$

$$\int \frac{di}{i} = \int 0.2\, dt$$

$$\log i = 0.2t + \log C$$

$$\log i - \log C = 0.2t$$

$$\log\left(\frac{i}{C}\right) = 0.2t$$

$$\frac{i}{C} = e^{0.2t}$$

$$i = Ce^{0.2t} \quad \text{...(i)}$$

$i(4) = 10$ i.e. $i = 10$ when $t = 4$

$$10 = Ce^{(0.2)4} \Rightarrow 10 = C(2.225)$$

$$\therefore \quad C = 4.493$$

$$i = (4.493)\, e^{0.2t} \quad \text{...(ii)}$$

when, $i = -5$

$$i = (4.493)\, e^{(0.2)(-5)} = 1.652$$

7.8 (10)

$$f(x) = x^6 - x - 1$$

$$a = 1, \quad b = 2, \quad \in = 0.01 = 10^{-3}$$

The minimum number of iterations by Bisection method is given by

$$\frac{|b-a|}{2^n} < \in$$

$$\frac{2-1}{2^n} < 10^{-3}$$

$$\frac{1}{2^n} < \frac{1}{10^3}$$

$$2^n > 10^3 \Rightarrow 2^n > 1000$$

$$\frac{n}{n_2} > In1000 \Rightarrow n > \frac{In1000}{In2}$$

$$n > 9.96$$

$$\therefore \quad n = 10$$

7.9 (7)

$$x + 2y = 1,$$

$$x = 11 - 2y$$

$$2x^2 + y^2 = 34$$

$$2(11 - 2y)^2 + y^2 = 34$$

$$24^2 + 8y^2 - 88y + y^2 - 34 = 0$$

$$9y^2 - 88y + 208 = 0$$

$$y = 5.77, 4$$

$$x = -0.54,$$

$$x = 11 - 2(4) = 3$$

$$x = 3, y = 4: \quad x + y = 3 + 4 = 7$$

7.10 (5.732)

$$y^2 - 2y + 1 = x$$

$$(y-1)^2 = x \qquad \sqrt{x} + y = 5$$

$$y - 1 = \sqrt{x} \qquad \sqrt{x} + 1 + 5x = 5$$

$$y = 1 + \sqrt{x} \qquad 2\sqrt{x} = 4$$

$$y = 1 + \sqrt{4} = 3 = 3 \qquad \sqrt{x} = 2$$

$$\therefore \quad = 3 \qquad x = 4$$

$$x + \sqrt{y} = 4 + \sqrt{3} = 5.732$$

8 Laplace Transform

8.1 The Laplace transform of $f(t)$ is $F(s)$.

Given $F(s) = \dfrac{\omega}{s^2+\omega^2}$, the final value of $f(t)$ is __

(a) initially (b) zero
(c) one (d) none [1995 : 1 M]

8.2 The Laplace transform of $(t^2 - 2t)\,u(t-1)$ is

(a) $\frac{2}{s^3}e^{-s} - \frac{2}{s^2}e^{-s}$ (b) $\frac{2}{s^3}e^{-2s} - \frac{2}{s^2}e^{-s}$

(c) $\frac{2}{s^3}e^{-s} - \frac{2}{s}e^{-s}$ (d) None [1998: 2 M]

8.3 Let $Y(s)$ be the Laplace transform of function $y(t)$, then the final value of the function is _____.

(a) $\lim\limits_{s\to 0} Y(s)$ (b) $\lim\limits_{s\to\infty} Y(s)$

(c) $\lim\limits_{s\to 0} sY(s)$ (d) $\lim\limits_{s\to\infty} sY(s)$

[2002 : 1 M]

8.4 Let $X(s) = \dfrac{3s+5}{s^2+10s+21}$ be the Laplace Transform of a signal $x(t)$. Then, $x(0^+)$ is

(a) 0 (b) 3
(c) 5 (d) 21 [2014 : 1 M, Set-1]

8.5 The Laplace Transform of $f(t) = e^{2t}\sin(5t)\,u(t)$ is

(a) $\dfrac{5}{s^2-4s+29}$ (b) $\dfrac{5}{s^2+5}$

(c) $\dfrac{s-2}{s^2-4s+29}$ (d) $\dfrac{5}{s+5}$

[2016 : 1 M, Set-1]

8.6 The solution of the differential equation, for $t > 0$, $y''(t) + 2y'(t) + y(t) = 0$ with initial conditions $y(0) = 0$ and $y'(0) = 1$, is ($u(t)$ denotes the unit step function),

(a) $te^{-t}u(t)$ (b) $(e^{-t} - te^{-t})u(t)$
(c) $(-e^{-t} + te^{-t})u(t)$ (d) $e^{-t}u(t)$

[2016 : 1 M, Set-2]

Fourier Series

8.7 The period of the signal

$$x(t) = 8\sin\left(0.8\pi t + \frac{\pi}{4}\right) \text{ is}$$

(a) 0.4π s (b) 0.8π s
(c) 1.25 s (d) 2.5 s [2010 : 1 M]

■■■■

Answers Laplace Transform

8.1 (d) 8.2 (d) 8.3 (c) 8.4 (b) 8.5 (a) 8.6 (a) 8.7 (d)

Explanations Laplace Transform

8.1 (d)

$$F(s) = \frac{\omega}{s^2+\omega^2}$$

$$\therefore \quad f(t) = \sin\omega t$$

$$\underset{t\to\infty}{Lt} f(t) = \underset{t\to\infty}{Lt} \sin\omega t = \text{Lies between } -1 \text{ and } 1$$

8.2 (d)

$L[(t^2 - 2t)\, u(t-1)]$

$L(t^2 - 2t + 1 - 1)\, u(t-1)]$

$$L((t-1)^2 u(t-1) - L(u(t-1)) = \frac{2}{s^3} e^{-s} - \frac{1}{s} e^{-s}$$

8.3 (c)

As per final value theorem,

$$\underset{t\to\infty}{Lt}\, y(t) = \underset{s\to 0}{Lt}\, s\cdot Y(s)$$

8.4 (b)

Given, $X(s) = \left[\frac{3s+5}{s^2+10s+21}\right]$

Using initial value theorem,

$$x(0^+) = \lim_{s\to\infty} [sX(s)]$$

$$\therefore \quad x(0^+) = \lim_{s\to\infty}\left[\frac{s(3s+5)}{s^2+10s+21}\right]$$

$$= \lim_{s\to\infty}\left[\frac{3+\frac{5}{s}}{1+\frac{10}{s}+\frac{21}{s^2}}\right] = \frac{3}{1} = 3$$

8.5 (a)

Laplace transform of $\sin 5t\, u(t) \to \frac{5}{s^2+25}$

$$e^{2t}\sin 5t\, u(t) \to \frac{5}{(s-2)^2+25} = \frac{5}{s^2-4s+29}$$

8.6 (a)

The differential equation is

$y''(t) + 2y'(t) + y(t) = 0$

So, $(s^2Y(s) - sy(0) - y'(0)) + 2[sY(s) - y(0)] + Y(s)$

So, $$Y(s) = \frac{sy(0)+y'(0)+2y(0)}{(s^2+2s+1)}$$

Given that, $y'(0) = 1,\ y(0) = 0$

So, $Y(s) = \frac{1}{(s+1)^2}$. So, $y(t) = te^{-t}\, u(t)$

8.7 (d)

If period = T, then $x(t+T) = x(t)$

Here, $x(t) = 8\sin(0.8\pi t + \pi/4)$

Putting $x(t+T) = x(t)$ we get,

$8\sin(0.8\pi(t+T) + \pi/4) = 8\sin(0.8\pi t + \pi/4)$

$\sin(0.8\pi(t+T) + \pi/4) = \sin(0.8\pi t + \pi/4)$

$\sin(\theta_1) = \sin(\theta_2) \Rightarrow \theta_1 = \theta_2 + 2\pi$

$\therefore 0.8\pi(t+T) + \pi/4 = 0.8\pi t + \pi/4 + 2\pi$

$$0.8\pi T = 2\pi \Rightarrow T = \frac{2}{0.8} = 2.5\ s$$

■■■■

General Aptitude

UNIT

XII

SYLLABUS

Numerical Ability : Numerical Computation, Numerical Estimation, Numerical Reasoning and Data Interpretation.

Verbal Ability : English Grammar, Sentence Completion, Verbal Analogies, Word Groups, Instructions, Critical Reasoning and Verbal Deduction.

CONTENTS

Analysis of Previous GATE Papers

Exam Year	1 Mark Ques.	2 Mark Ques.	Total Marks
2010	5	5	**15**
2011	5	5	**15**
2012	5	5	**15**
2013	5	5	**15**
2014 Set-1	5	5	**15**
2014 Set-2	5	5	**15**
2014 Set-3	5	5	**15**
2014 Set-1	5	5	**15**
2015 Set-2	5	5	**15**
2016 Set-1	5	5	**15**
2016 Set-2	5	5	**15**

Exam Year	1 Mark Ques.	2 Mark Ques.	Total Marks
2017 Set-1	10	–	**10**
2017 Set-2	10	–	**10**
2018	5	5	**15**
2019	6	4	**14**
2020	5	5	**15**
2021	5	5	**15**
2022	5	5	**15**
2023	5	5	**15**
2024	5	5	**15**

1 Numerical Ability

1.1 25 persons are in a room. 15 of them play hockey, 17 of them play football and 10 of them play both hockey and football. Then the number of persons playing neither hockey nor football is

(a) 2 (b) 17
(c) 13 (d) 3 **[2010 : 1 M]**

1.2 If 137 + 276 = 435 how much is 731 + 672?

(a) 534 (b) 1403
(c) 1623 (d) 1531 **[2010 : 2 M]**

1.3 5 skilled workers can build a wall in 20 days; 8 semiskilled workers can build a wall in 25 days; 10 unskilled workers can build a wall in 30 days. If a team has 2 skilled, 6 semiskilled and 5 unskilled workers, how long will it take to build the wall?

(a) 20 days (b) 18 days
(c) 16 days (d) 15 days **[2010 : 2 M]**

1.4 Given digits 2, 2, 3, 3, 3, 4, 4, 4, 4 how many distinct 4 digit numbers greater than 3000 can be formed?

(a) 50 (b) 51
(c) 52 (d) 54 **[2010 : 2 M]**

1.5 Hari (H), Gita (G), Irfan (I) and Saira (S) are siblings (i.e. brothers and sisters). All were born on I^st January. The age difference between any two successive siblings (that is born one after another) is less than 3 years. Given the following facts:

1. Hari's age + Gita's age > Irfan's age + Saira's age.
2. The age difference between Gita and Saira is 1 year. However, Gita is not the oldest and Saira is not the youngest.
3. There are no twins.

In what order were they born (oldest first)?

(a) HSIG (b) SGHI
(c) IGSH (d) IHSG **[2010 : 2 M]**

1.6 There are two candidates *P* and *Q* in an election. During the campaign 40% of the voters promised to vote for *P*, and rest for *Q*. However, on the day of election 15% of the voters went back on their promise to vote for *P* and instead voted for *Q*. 25% of the voters went back on their promise to vote for *Q* and instead voted for P. Suppose, *P* lost by 2 votes, then what was the total number of voters?

(a) 100 (b) 110
(c) 90 (d) 95 **[2011 : 1 M]**

1.7 The fuel consumed by a motorcycle during a journey while travelling at various speeds is indicated in the graph below

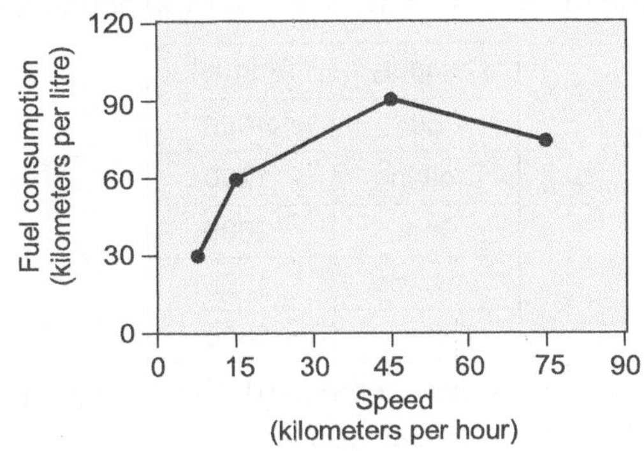

The distance covered during four laps of the journey are listed in the table below.

Lap	Distance (kilometers)	Average speed (kilometers per hour)
P	15	15
Q	75	45
R	40	75
S	10	10

From the given data, we can conclude that the fuel consumed per kilometre was least during the lap

(a) *P* (b) *Q*
(c) *R* (d) *S* **[2011 : 2 M]**

1.8 Three friends, *R*, *S* and *T* shared toffee from a bowl. *R* took 1/3rd of the toffees, but returned to the bowl. S took 1/4th of what was left but returned three toffees to the bowl. *T* took half of the remainder but returned two back into the bowl. If the bowl had 17 toffees left, how many toffees were originally there in the bowl?

(a) 38 (b) 31
(c) 48 (d) 41 **[2011 : 2 M]**

1.9 Given that: $f(y) = |y|\ y$, and q is any non-zero real number, the value of $|f(q) - f(-q)|$ is

(a) 0 (b) –1
(c) 1 (d) 2 **[2011 : 2 M]**

1.10 The sum of n terms of the series 4 + 44 + 444 + ... is

(a) $(4/81)[10^{n+1} - 9n - 1]$
(b) $(4/81)[10^{n-1} - 9n - 1]$
(c) $(4/81)[10^{n+1} - 9n - 10]$
(d) $(4/81)[10^{n} - 9n - 10]$ [2011 : 2 M]

1.11 If $(1.001)^{1259} = 3.52$ $(1.001)^{2062} = 7.85$, then $(1.001)^{3321}$

(a) 2.23 (b) 4.33
(c) 11.37 (d) 27.64 [2012 : 1 M]

1.12 The data given in the following table summarizes the monthly budget of an average household.

Category	Amount
Food	4000
Clothing	1200
Rent	2000
Savings	1500
Others	1800

The approximate percentage of the monthly budget **NOT** spent on savings is

(a) 10% (b) 14%
(c) 81% (d) 86% [2012 : 2 M]

1.13 A and B friends. They decide to meet between 1 PM and 2 PM on a given day. There is a condition that whoever arrives first will not wait for the other for more than 15 minutes. The probability that they will meet on that day is

(a) 1/4 (b) 1/16
(c) 7/16 (d) 9/16 [2012 : 2 M]

1.14 Raju has 14 currency notes in his pocket consisting of only Rs. 20 notes and Rs. 10 notes. The total money value of the notes is Rs. 230. The number of Rs. 10 notes in Raju's pocket is

(a) 5 (b) 6
(c) 9 (d) 10 [2012 : 2 M]

1.15 There are eight bags of rice looking alike, seven of which have equal and one is slightly heavier. The weighing balance is of unlimited capacity. Using this balance the minimum number of weighting required to identify the heavier bag is

(a) 2 (b) 3
(c) 4 (d) 8 [2012 : 2 M]

1.16 In the summer of 2012, in New Delhi, the mean temperature of Monday to Wednesday was 41°C and of Tuesday to Thursday was 43°C. If the temperature on Thursday was 15% higher than that of Monday, then the temperature in °C on Thursday was

(a) 40 (b) 43
(c) 46 (d) 49 [2013 : 1 M]

1.17 A car travels 8 km in the first quarter of an hour, 6 km in the second quarter and 16 km in the third quarter. The average speed of the car in km per hour over the entire journey is

(a) 30 (b) 36
(c) 40 (d) 24 [2013 : 2 M]

1.18 Find the sum to n terms of the series 10 + 84 + 734 +....

(a) $\frac{9(9^n+1)}{10}+1$ (b) $\frac{9(9^n-1)}{8}+1$
(c) $\frac{9(9^n-1)}{8}+n$ (d) $\frac{9(9^n-1)}{8}+n^2$

[2013 : 2 M]

1.19 The set of values of p for which the roots of the equation $3x^2 + 2x + p(p-1) = 0$ are of opposite sign is

(a) $(-\infty, 0)$ (b) $(0, 1)$
(c) $(1, \infty)$ (d) $(0, \infty)$ [2013 : 2 M]

1.20 What is the change that a leap year, selected at random, will contain 53 Saturdays?

(a) $\frac{2}{7}$ (b) $\frac{3}{7}$
(c) $\frac{1}{7}$ (d) $\frac{5}{7}$ [2013 : 2 M]

1.21 If $\left(z+\frac{1}{z}\right)^2 = 98$, compute $\left(z^2+\frac{1}{z^2}\right)$.

[2014 : 1 M, Set-1]

1.22 The roots of $ax^2 + bx + c = 0$ are real and positive a, b and c are real.
Then $ax^2 + b|x| + c = 0$ has

(a) no roots (b) 2 real roots
(c) 3 real roots (d) 4 real roots

[2014 : 1 M, Set-1]

1.23 Round-trip tickets to a tourist destination are eligible for a discount of 10% on the total fare. In addition, groups of 4 or more get a discount of 5% on the total fare. If the one way single person fare is Rs. 100, a group of 5 tourists purchasing round trip tickets will be charged Rs. ______.

[2014 : 2 M, Set-1]

1.24 In a survey, 300 respondents were asked whether they own a vehicle or not. If yes, they were further asked to mention whether they own a car or scooter or both. Their responses are tabulated below. What percent of respondents do not own a scooter.

		Men	Women
Own vehicle	Car	40	34
	Scooter	30	20
	Both	60	46
Do not own vehicle		20	50

[2014 : 2 M, Set-1]

1.25 When a point inside of a tetrahedron (a solid with four triangular surfaces) is connected by straight lines to corners, how many (new) internal planes are created with these lines? ________.

[2014 : 2 M, Set-1]

1.26 What is the average of all multiples of 10 from 2 to 198?

(a) 90 (b) 100
(c) 110 (d) 120 [2014 : 1 M, Set-2]

1.27 The value of $\sqrt{12+\sqrt{12+\sqrt{12+...}}}$ is

(a) 3.464 (b) 3.932
(c) 4.000 (d) 4.444

[2014 : 1 M, Set-2]

1.28 If x is real and $|x^2 - 2x + 3| = 11$, then possible values of $|-x^3 + x^2 - x|$ include

(a) 2, 4 (b) 2, 14
(c) 4, 52 (d) 14, 52

[2014 : 2 M, Set-2]

1.29 The ratio of male to female students in a college for five years is plotted in the following line graph. If the number of female students doubled in 2009, by what percent did the number of male students increase in 2009?

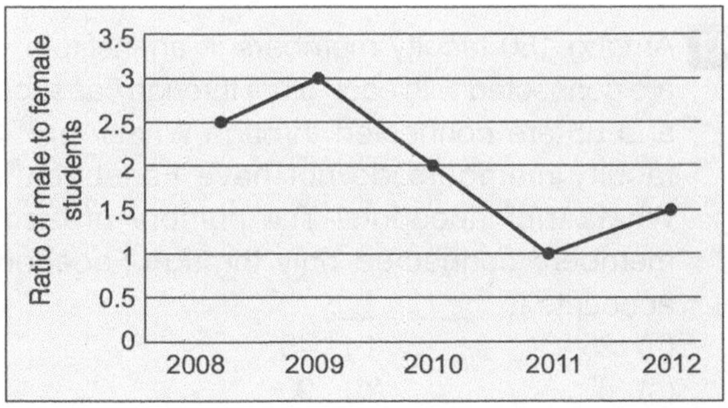

[2014 : 2 M, Set-2]

1.30 At what time between 6 a.m. and 7 a.m. will the minute hand and hour hand of a clock make an angle closest to 60°?

(a) 6:22 a.m. (b) 6:27 a.m.
(c) 6:38 a.m. (d) 6:45 a.m.

[2014 : 2 M, Set-2]

1.31 Which number does not belong in the series below?
2, 5, 10, 17, 26, 37, 50, 64

(a) 17 (b) 37
(c) 64 (d) 26 [2014 : 1 M, Set-3]

1.32 The table below has question wise data on the performance of students in an examination. The marks for each question are also listed. There is no negative or partial marking in the examination.

Q.No.	Marks	Answered correctly	Answered wrongly	Not attemped
1.	2	21	17	6
2.	3	15	27	2
3.	2	23	18	3

What is the average of the marks obtained by the class in the examination?

(a) 1.34 (b) 1.74
(c) 3.02 (d) 3.91 [2014 : 1 M, Set-3]

1.33 The Gross Domestic Product (GDP) in Rupees grew at 7% during 2012-2013. For international comparison, the GDP is compared in US Dollars (USD) after conversion based on the market exchange rate. During the period 2012-2013 the exchange rate for the USD increased from Rs. 50/USD to Rs.60/USD. India's GDP in USD during the period 2012-2013.

(a) Increased by 5%
(b) Decreased by 13%
(c) Decreased by 20%
(d) Decreased by 11% [2014 : 2 M, Set-3]

1.34 The ratio of male to female students in a college for five years is plotted in the following line graph. If the number of female students in 2011 and 2012 is equal, what is the ratio of male students in 2012 to male students in 2011?

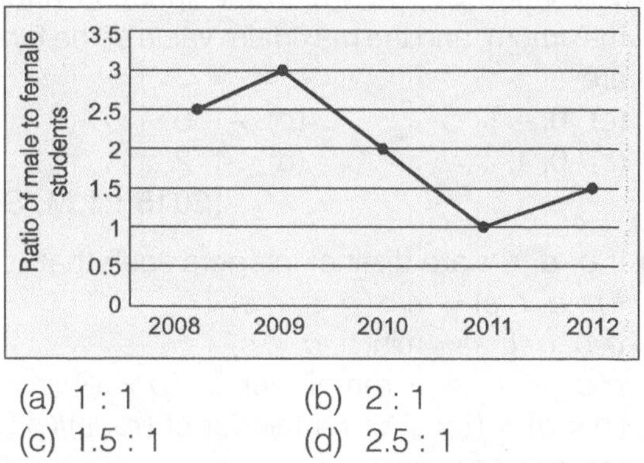

(a) 1 : 1 (b) 2 : 1
(c) 1.5 : 1 (d) 2.5 : 1

[2014 : 2 M, Set-3]

1.35 Consider the equation: $(7526)_8 - (Y)_8 = (4364)_8$, where $(X)_N$ stands for X to the base N. Find Y.

(a) 1634 (b) 1737
(c) 3142 (d) 3162 [2014 : 2 M, Set-3]

1.36 The probabilities that a student passes in Mathematics, Physics and Chemistry are *m*, *p*, and *c* respectively. Of these subjects, the student has 75% chance of passing in at least one, a 50% chance of passing in at least two and a 40% chance of passing in exactly two. Following relations are drawn in *m*, *p*, *c*:

(I) $p + m + c = 27/20$

(II) $p + m + c = 13/20$

(III) $(p) \times (m) \times (c) = 1/10$

(a) Only relation I is true

(b) Only relation II is true

(c) Relations II & III are true

(d) Relations I & III are true **[2015 : 2 M, Set-1]**

1.37 The number of students in a class who have answered correctly, wrongly, or not attempted each question in an exam, are listed in the table below. The marks for each question are also listed. There is no negative or partial marking.

Q No.	Marks	Answered correctly	Answered wrongly	Not attempted
1	2	21	17	6
2	3	15	27	2
3	1	11	29	4
4	2	23	18	3
5	5	31	12	1

What is the average of the marks obtained by the class in the examination?

(a) 2.290 (b) 2.970

(c) 6.795 (d) 8.795

[2015 : 2 M, Set-1]

1.38 Consider a function $f(x) = 1 - |x|$ on $-1 \le x \le 1$. The value of x at which the function attains a maximum, and the maximum value of the function are:

(a) 0, – 1 (b) –1, 0

(c) 0, 1 (d) –1, 2

[2015 : 1 M, Set-2]

1.39 If *p*, *q*, *r*, *s* are distinct integers such that

$f(p, q, r, s) = \max(p, q, r, s)$

$g(p, q, r, s) = \min(p, q, r, s)$

$h(p, q, r, s)$ = remainder of $(p \times q)/(r \times s)$ if $(p \times q) > (r \times s)$ or remainder of $(r \times s)/(p \times q)$ if $(r \times s) > (p \times q)$

Also a function $fgh(p, q, r, s) = f(p, q, r, s) \times g(p, q, r, s) \times h(p, q, r, s)$

Also the same operations are valid with two variable function of the form $f(p, q)$

What is the value of $fg(h(2, 5, 7, 3), 4, 6, 8)$?

[2015 : 2 M, Set-2]

1.40 Four branches of a company are located at *M*, *N*, *O*, and *P*. *M* is north of *N* at a distance of 4 km; *P* is south of *O* at a distance of 2 km; *N* is southeast of *O* by 1 km. What is the distance between *M* and *P* in km?

(a) 5.34 (b) 6.74

(c) 28.5 (d) 45.49

[2015 : 2 M, Set-2]

1.41 In a triangle PQR, PS is the angle bisector of ∠QPR and ∠QPS = 60°. What is the length of PS?

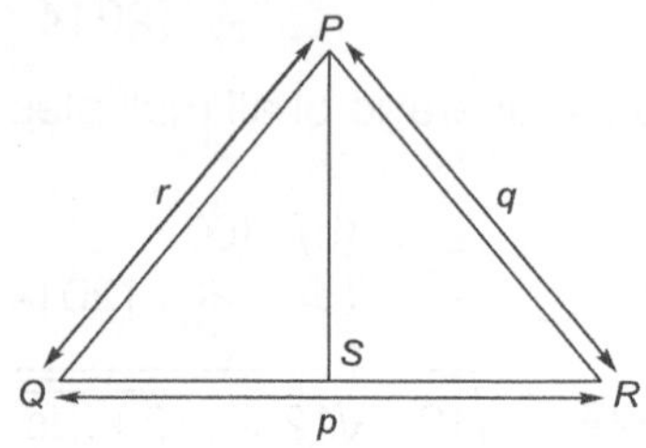

(a) $\dfrac{(q+r)}{qr}$ (b) $\dfrac{qr}{(q+r)}$

(c) $\sqrt{(q^2+r^2)}$ (d) $\dfrac{(q+r)^2}{qr}$

[2015 : 2 M, Set-2]

1.42 In a quadratic function, the value of the product of the roots (α, β) is 4. Find the value of

$$\frac{\alpha^n + \beta^n}{\alpha^{-n} + \beta^{-n}}$$

(a) n^4 (b) 4^n

(c) 2^{2n-1} (d) 4^{n-1}

[2016 : 1 M, Set-1]

1.43 Among 150 faculty members in an institute, 55 are connected with each other through Facebook® and 85 are connected through WhatsApp®. 30 faculty members do not have Facebook® or WhatsApp® accounts. The number of faculty members connected only through Facebook® accounts is ________.

(a) 35 (b) 45

(c) 65 (d) 90

[2016 : 2 M, Set-1]

1.44 In a 2 × 4 rectangle grid shown below, each cell is a rectangle. How many rectangles can be observed in the grid?

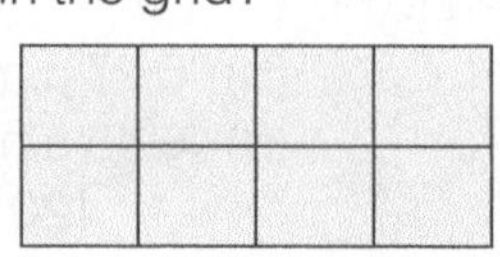

(a) 21 (b) 27

(c) 30 (d) 36 **[2016 : 2 M, Set-1]**

1.45 Choose the correct expression for $f(x)$ given in the graph.

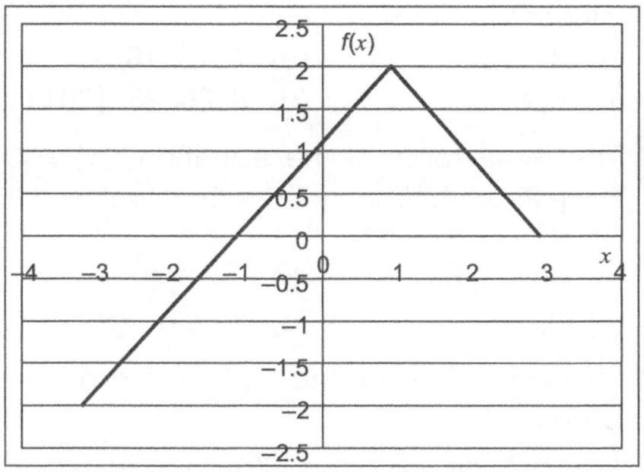

(a) $f(x) = 1 - |x - 1|$
(b) $f(x) = 1 + |x - 1|$
(c) $f(x) = 2 - |x - 1|$
(d) $f(x) = 2 + |x - 1|$ **[2016 : 2 M, Set-1]**

1.46 Pick the odd one out in the following:
13, 23, 33, 43, 53
(a) 23 (b) 33
(c) 43 (d) 53 **[2016 : 1 M, Set-2]**

1.47 If $|9y - 6| = 3$, then $y^2 - \frac{4y}{3}$ is________.
(a) 0 (b) $+\frac{1}{3}$
(c) $-\frac{1}{3}$ (d) undefined
[2016 : 1 M, Set-2]

1.48 Shaquille O' Neal is a 60% career free throw shooter, meaning that he successfully makes 60 free throws out of 100 attempts on average. What is the probability that he will successfully make exactly 6 free throws in 10 attempts?
(a) 0.2508 (b) 0.2816
(c) 0.2934 (d) 0.6000
[2016 : 2 M, Set-2]

1.49 The numeral in the units position of $211^{870} + 146^{127} \times 3^{424}$ is ___. **[2016 : 2 M, Set-2]**

1.50 The probability that a k-digit number does NOT contain the digits 0, 5 or 9 is
(a) 0.3^k (b) 0.6^k
(c) 0.7^k (d) 0.9^k
[2017 : 1 M, Set-1]

1.51 Find the smallest number y such that $y \times 162$ is a perfect cube.
(a) 24 (b) 27
(c) 32 (d) 36
[2017 : 1 M, Set-1]

1.52 After Rajendra Chola returned from his voyage to Indonesia, he ________ to visit the temple in Thanjavur.
(a) was wishing (b) is wishing
(c) wished (d) had wished
[2017 : 1 M, Set-1]

1.53 A contour line joins locations having the same height above the mean sea level. The following is a contour plot of a geographical region. Contour lines are shown at 25 m intervals in this plot. If in a flood, the water level rises to 525 m, which of the villages P, Q, R, S, T get submerged?

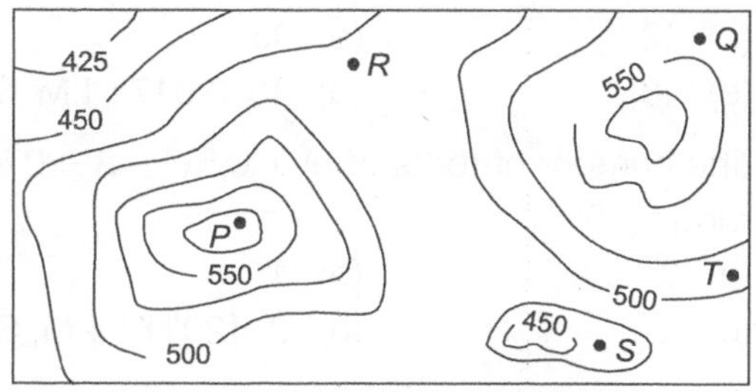

(a) P, Q (b) P, Q, T
(c) R, S, T (d) Q, R, S
[2017 : 1 M, Set-1]

1.54 The expression $\frac{(x+y) - |x-y|}{2}$ is equal to
(a) the maximum of x and y
(b) the minimum of x and y
(c) 1
(d) none of the above
[2017 : 1 M, Set-1]

1.55 Arun, Gulab, Neel and Shweta must choose one shirt each from a pile of four shirts coloured red, pink, blue and white respectively. Arun dislikes the colour red and Shweta dislikes the colour white. Gulab and Neel like all the colours. In how many different ways can they choose the shirts so that no one has a shirt with a colour he or she dislikes?
(a) 21 (b) 18
(c) 16 (d) 14
[2017 : 1 M, Set-1]

1.56 Six people are seated around a circular table. There are at least two men and two women. There are at least three right-handed persons. Every woman has a left-handed person to her immediate right. None of the women are right-handed. The number of women at the table is
(a) 2 (b) 3
(c) 4 (d) Can't be determined
[2017 : 1 M, Set-1]

1.57 There are 3 red socks, 4 green socks and 3 blue socks. You choose 2 socks. The probability that they are of the same colour is

(a) 1/5 (b) 7/30
(c) 1/4 (d) 4/15

[2017 : 1 M, Set-2]

1.58 A test has twenty questions worth 100 marks in total. There are two types of questions. Multiple choice questions are worth 3 marks each and essay questions are worth 11 marks each. How many multiple choice questions does the exam have?

(a) 12 (b) 15
(c) 18 (d) 19 [2017 : 1 M, Set-2]

1.59 The number of roots of $e^x + 0.5x^2 - 2 = 0$ in the range [–5, 5] is

(a) 0 (b) 1
(c) 2 (d) 3 [2017 : 1 M, Set-2]

1.60 An air pressure contour line joins locations in a region having the same atmospheric pressure. The following is an air pressure contour plot of a geographical region. Contour lines are shown at 0.05 bar intervals in this plot.

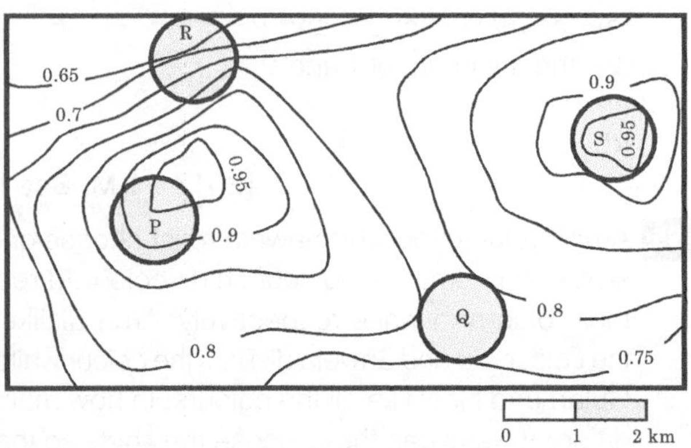

If the possibility of a thunderstorm is given by how fast air pressure rises or drops over a region, which of the following regions is most likely to have a thunderstorm?

(a) *P* (b) Q
(c) *R* (d) *S* [2017 : 1 M, Set-2]

1.61 *X* is a 30 digit number starting with the digit 4 followed by the digit 7. Then the number X^3 will have

(a) 90 digits (b) 91 digits
(c) 92 digits (d) 93 digits

[2017 : 1 M, Set-2]

1.62 For what values of k given below is $\frac{(k+2)^2}{k-3}$ an integer?

(a) 4, 8, 18 (b) 4, 10, 16
(c) 4, 8, 28 (d) 8, 26, 28 [2018 : 1 M]

1.63 The three roots of the equation $f(x) = 0$ are $x = \{-2, 0, 3\}$. What are the three values of x for which $f(x - 3) = 0$?

(a) –5, –3, 0 (b) –2, 0, 3
(c) 0, 6, 8 (d) 1, 3, 6 [2018 : 1 M]

1.64 Functions, *F*(*a*, *b*) and *G*(*a*, *b*) are defined as follows:

$F(a, b) = (a - b)^2$ and $G(a, b) = |a - b|$, where $|x|$ represents the absolute value of x. What would be the value of $G(F(1, 3), G(1, 3))$?

(a) 2 (b) 4
(c) 6 (d) 36 [2018 : 1 M]

1.65 The missing number in the given sequence 343, 1331, _______, 4913 is

(a) 4096 (b) 2744
(c) 3375 (d) 2197 [2019 : 1 M]

1.66 It takes two hours for a person *X* to mow the lawn. *Y* can mow the same lawn in four hours. How long (in minutes) will it take *X* and *Y*, if they work together to mow the lawn?

(a) 90 (b) 60
(c) 80 (d) 120 [2019 : 1 M]

1.67 Given two sets *X* = {1,2,3} and *Y* = {2, 3, 4}, we construct a set *Z* of all possible fractions where the numerators belong to set *X* and the denominators belong to set *Y*. The product of elements having minimum and maximum values in the set *Z* is ______.

(a) $\frac{1}{6}$ (b) $\frac{1}{12}$
(c) $\frac{3}{8}$ (d) $\frac{1}{8}$ [2019 : 2 M]

1.68 How many integers are there between 100 and 1000 all of whose digits are even?

(a) 90 (b) 60
(c) 100 (d) 80 [2019 : 2 M]

1.69 The ratio of the number of boys and girls who participated in an examination is 4 : 3. The total percentage of candidates who passed the examination is 80 and the percentage of girls who passed is 90. The percentage of boys who passed is _______.

(a) 72.50 (b) 90.00
(c) 80.50 (d) 55.50 [2019 : 2 M]

1.70 Select the next element of the series: Z, WV, RQP, ______ .

(a) JIHG (b) KJIH
(c) NMLK (d) LKJI [2020 : 1 M]

1.71 If P, Q, R, S are four individuals. How many teams of size exceeding one can be formed, with Q as a member?

(a) 8 (b) 6
(c) 5 (d) 7 [2020 : 1 M]

1.72 In four-digit integer numbers from 1001 to 9999, the digit group "37" (in the same sequence) appears ______ times.

(a) 299 (b) 270
(c) 280 (d) 279 [2020 : 1 M]

1.73 Non-performing Assets (NPAs) of a bank in India is defined as an asset, which remains unpaid by a borrower for a certain period of time in terms of interest, principal, or both. Reserve Bank of India (RBI) has changed the definition of NPA thrice during 1993-2004. in terms of the holding period of loans. The holding period was reduced by one quarter each time. In 1993, the holding period was four quarters (360 days). Based on the above paragraph, the holding period of loans in 2004 after the third revision was ______ days.

(a) 45 (b) 135
(c) 90 (d) 180 [2020 : 1 M]

1.74 Given a semicircle with O as the centre; as shown in the figure, the ratio $\frac{\overline{AC}+\overline{CB}}{\overline{AB}}$ is ________ .

Where $\overline{AC}$, $\overline{CB}$ and $\overline{AB}$ are chords.

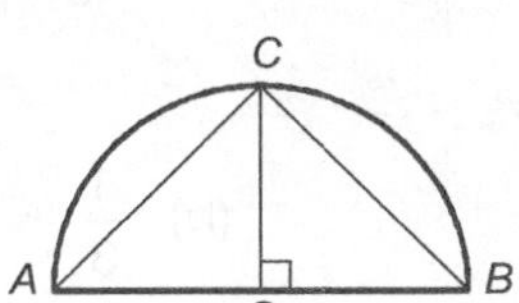

(a) $\sqrt{2}$ (b) $\sqrt{3}$
(c) 2 (d) 3 [2020 : 1 M]

1.75 The revenue and expenditure of four different companies P, Q, R and S in 2015 are shown in the figure. If the revenue of company Q in 2015 was 20% more than that in 2014, and company Q had earned a profit of 10% on expenditure in 2014, then its expenditure (in million rupees) in 2014 was ______.

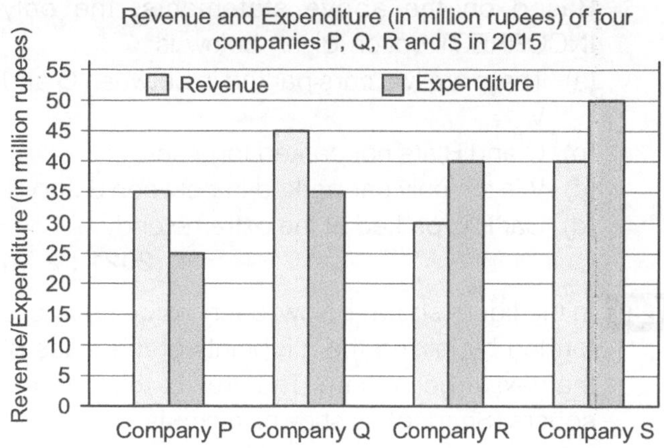

(a) 34.1 (b) 32.7
(c) 33.7 (d) 35.1 [2020 : 1 M]

1.76 Which one of the following numbers is exactly divisible by $(11^{13} + 1)$?

(a) $11^{33} + 1$ (b) $11^{52} - 1$
(c) $11^{26} + 1$ (d) $11^{39} + 1$ [2021 : 1 M]

1.77 For a regular polygon having 10 sides, the interior angle between the sides of the polygon, in degrees, is :

(a) 396° (b) 144°
(c) 324° (d) 216° [2021 : 1 M]

1.78 A transparent square sheet above is folded along the dotted line. The folded sheet will look like ______.

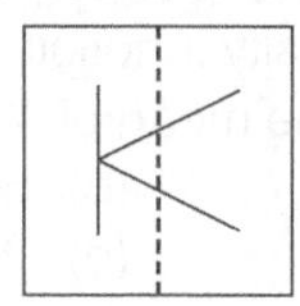

(a) 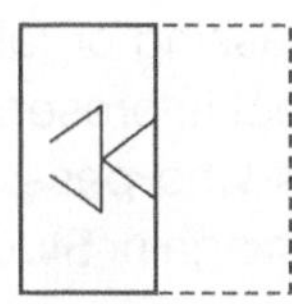 (b) 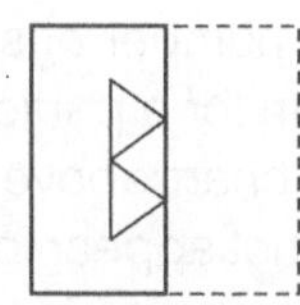

(c) 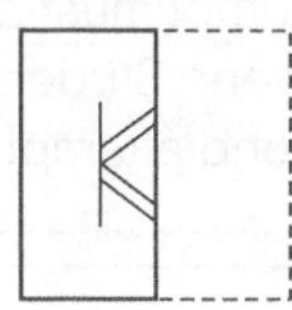 (d) 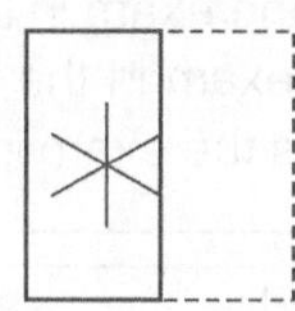

[2021 : 1 M]

1.79 Seven cars P, Q, R, S, T, U and V are parked in a row not necessarily in that order. The cars T and U should be parked next to each other. The cars S and V also should be parked next to each other, whereas P and Q cannot be parked next to each other. Q and S must be parked next to each other. R is parked to the immediate right of V. T is parked to the left of U.

Based on the above statements, the only INCORRECT option given below is:

(a) There are two cars parked in between Q and V.
(b) Q and R are not parked together.
(c) V is the only car parked in between S and R
(d) Car P is parked at the extreme end.

[2021 : 2 M]

1.80 In the figure shown below, each inside square is formed by joining the midpoints of the sides of the next larger square. The area of the smallest square (Shaded) as shown, in cm^2 is

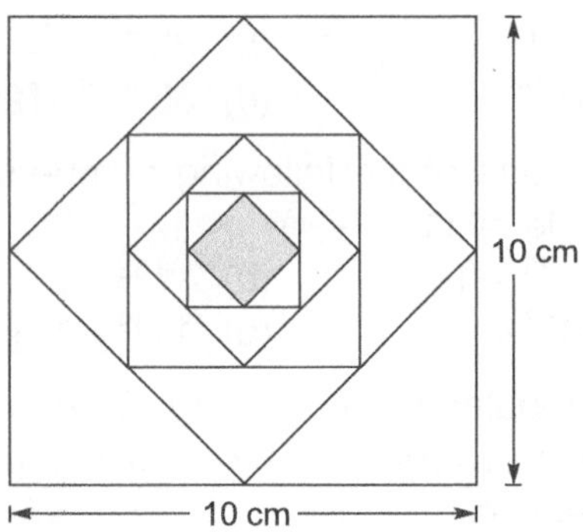

(a) 3.125 (b) 12.50
(c) 1.5625 (d) 6.25 [2021 : 2 M]

1.81 Let X be a continuous random variable denoting the temperature measured. The range of temperature is (0, 100) degree Celsius and let the probability density function of X be $f(x) = 0.01$ for $0 \le X \le 100$. The mean of X is _______ .

(a) 5.0 (b) 50.0
(c) 25.0 (d) 2.5 [2021 : 2 M]

1.82 The number of students passing or failing in an exam for a particular subject is presented in the bar chart above. Students who pass the exam cannot appear for the exam again. Students who fail the exam in the first attempt must appear for the exam in the following year. Students always pass the exam in their second attempt.

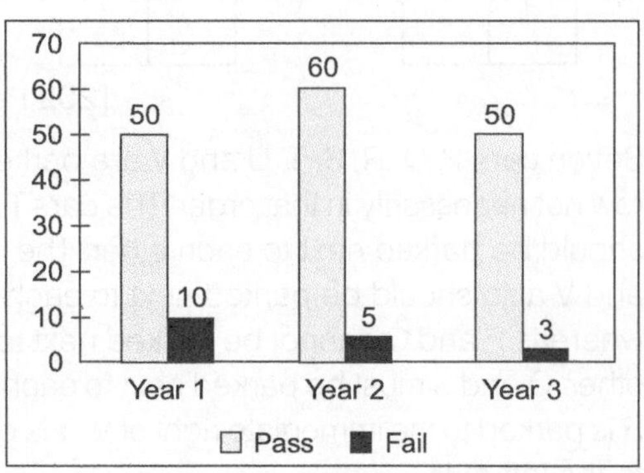

The number of students who took the exam for the first time in the year 2 and the year 3 respectively, are _______ .

(a) 65 & 53 (b) 55 & 53
(c) 60 & 50 (d) 55 & 48 [2021 : 2 M]

1.83 In a 500 m race, P and Q have speeds in the ratio of 3 : 4. Q starts the race when P has already covered 140 m.

What is the distance between P and Q (in m) when P wins the race?

(a) 20 (b) 40
(c) 60 (d) 140 [2022 : 1 M]

1.84 Three bells P, Q and R are rung periodically in a school. P is rung every 20 minutes; Q is rung every 30 minutes and R is rung every 50 minutes.

If all the three bells are rung at 12:00 PM, when will the three bells ring together again the next time?

(a) 5:00 PM (b) 5:30 PM
(c) 6:00 PM (d) 6:30 PM [2022 : 1 M]

1.85 The figure below shows the front and rear view of a disc, which is shaded with identical patterns. The disc is flipped once with respect to any one of the fixed axes 1–1, 2–2 or 3–3 chosen uniformly at random.

What is the probability that the disc DOES NOT retain the same front and rear views after the flipping operation?

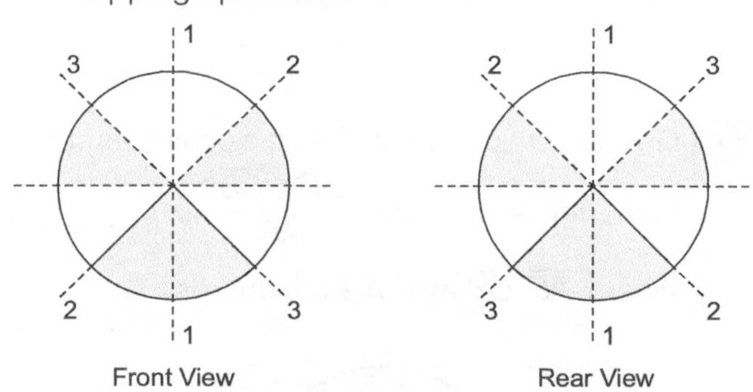

(a) 0 (b) $\frac{1}{3}$

(c) $\frac{2}{3}$ (d) 1 [2022 : 1 M]

1.86 There are two identical dice with a single letter on each of the faces. The following six letters : Q, R, S, T, U and V, one on each of the faces. Any of the six outcomes are equally likely.

The two dice are thrown once independently at random.

What is the probability that the outcomes on the dice were composed only of any combination of the following possible outcomes : Q, U and V?

(a) $\frac{1}{4}$ (b) $\frac{3}{4}$

(c) $\frac{1}{6}$ (d) $\frac{5}{36}$ [2022 : 2 M]

1.87 The price of an item is 10% cheaper in an online store S compared to the price at another online store M. Store S charges Rs. 150 for delivery. There are no delivery charges for orders from the store M. A person bought the item from the store S and saved Rs. 100.

What is the price of the item at the online store S (in Rs.) if there are no other charges than what is described above?

(a) 2500 (b) 2250

(c) 1750 (d) 1500

[2022 : 2 M]

1.88 The letters P, Q, R, S, T and U are to be placed one per vertex on a regular convex hexagon, but not necessarily in the same order.

Consider the following statements :

- The line segment joint R and S is longer than the line segment joining P and Q.
- The line segment joining R and S is perpendicular to the line segment joining P and Q.
- The line segment joining R and U is parallel to the line segment joining T and Q.

Based on the above statements, which one of the following options is CORRECT?

(a) The line segment joining R and T is parallel to the line segment joining Q and S.

(b) The line segment joining T and Q is parallel to the line joining P and U.

(c) The line segment joining R and P is perpendicular to the line segment joining U and Q.

(d) The line segment joining Q and S is perpendicular to the line segment joining R and P. [2022 : 2 M]

1.89 An ant is at the bottom-left corner of a grid (point P) as shown below in the figure. It aims to move to the top-right corner of the grid. The ant moves only along the lines marked in the grid such that the current distance to the top-right corner strictly decreases.

Which one of the following is a part of a possible trajectory of the ant during the movement?

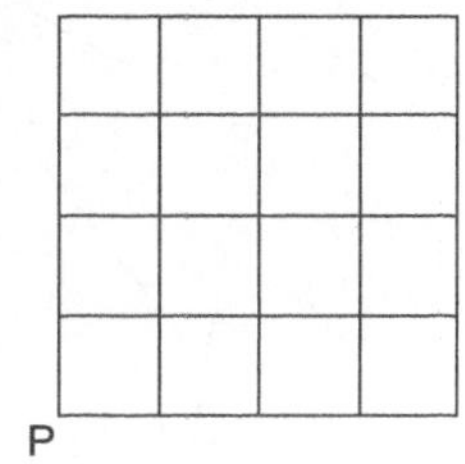

(a) 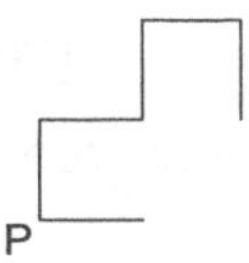 (b) 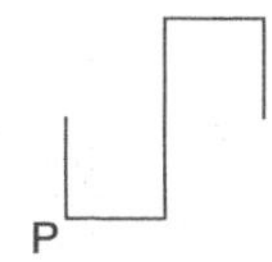

(c) 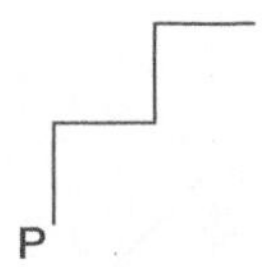 (d) 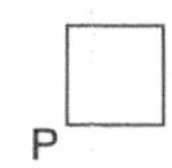

[2022 : 2 M]

1.90 Given a fair six-faced dice where the faces are labelled '1', '2', '3', '4', '5' and '6', what is the probability of getting a '1' on the first roll of the dice and a '4' on the second roll?

(a) $\frac{1}{36}$ (b) $\frac{1}{6}$

(c) $\frac{5}{6}$ (d) $\frac{1}{3}$

[2023 : 1 M]

1.91 A recent survey shows that 65% of tobacco users were advised to stop consuming tobacco. The survey also shows that 3 out of 10 tobacco users attempted to stop using tobacco.

Based only on the information in the above passage, which one of the following options can be logically inferred with *certainty*?

(a) A majority of tobacco users who were advised to stop consuming tobacco made an attempt to do so.

(b) A majority of tobacco users who were advised to stop consuming tobacco did not attempt to do so.

(c) Approximately 30% of tobacco users successfully stopped consuming tobacco.

(d) Approximately 65% of tobacco users successfully stopped consuming tobacco.

[2023 : 1 M]

1.92 How many triangles are present in the given figure?

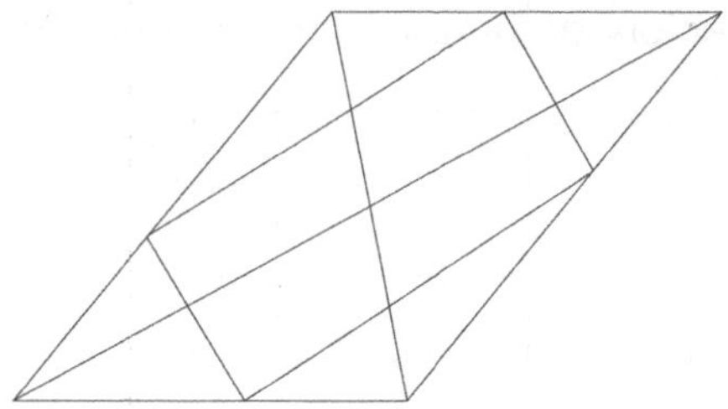

(a) 12 (b) 16
(c) 20 (d) 24

[2023 : 1 M]

1.93 Which one of the following options represents the given graph?

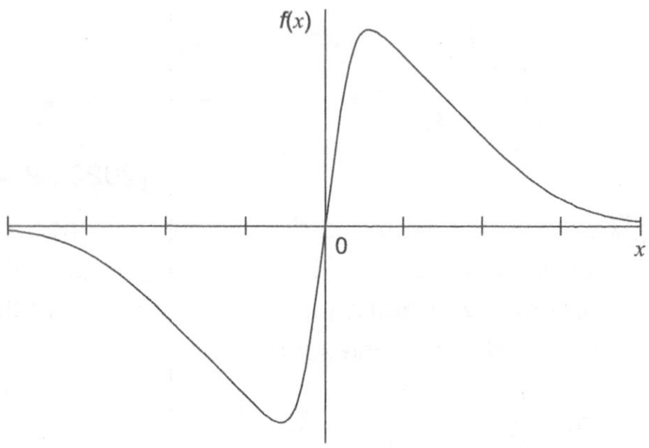

(a) $f(x) = x^2 2^{-|x|}$ (b) $f(x) = x2^{-|x|}$
(c) $f(x) = |x|2^{-x}$ (d) $f(x) = x2^{-x}$

[2023 : 2 M]

1.94 The digit in the unit's place of the product $3^{999} \times 7^{1000}$ is _____.

(a) 7 (b) 1
(c) 3 (d) 9

[2023 : 2 M]

1.95 A square with sides of length 6 cm is given. The boundary of the shaded region is defined by two semi-circles whose diameters are the sides of the square, as shown.

The area of the shaded region is _____ cm^2.

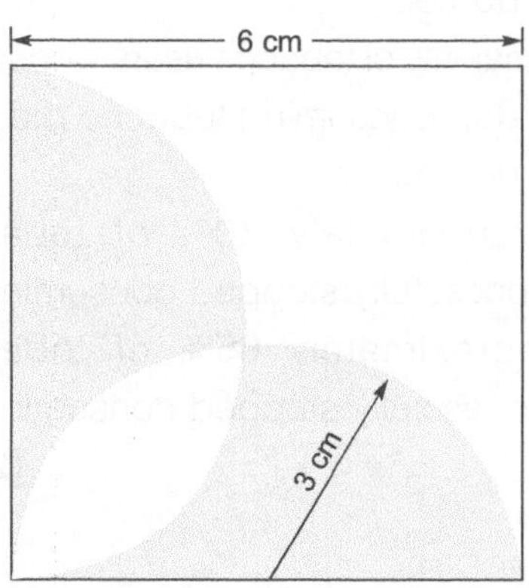

(a) 6π (b) 18
(c) 20 (d) 9π

[2023 : 2 M]

1.96 The asembly shown below has three teethed circular objects (Pinions) and two teethed flat objects (Racks), which are perfectly mating with each other. Pinions can only rotate clockwise or anti-clockwise staying at its own center. Racks can translate towards the left (←) and the right (→) direction.

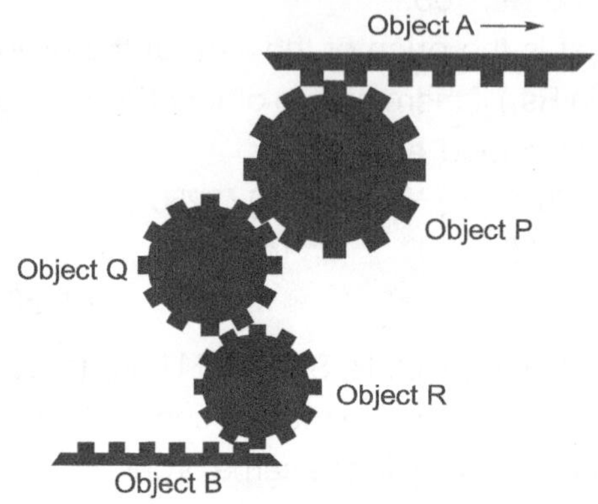

If the object A (Rack) is translating towards the right (→) direction, the correct statement among the following is

(a) Object B translates towards the right direction.
(b) Object Q rotates in the clockwise direction.
(c) Object B translates towards the left direction.
(d) Object R rotates in the anti-clockwise direction.

[2024 : 2 M]

1.97 If, for non-zero real variables x, y and real parameter $a > 1$,

$$x : y = (a + 1) : (a - 1),$$

then, the ratio $(x^2 - y^2) : (x^2 + y^2)$ is

(a) $a : (a^2 - 1)$ (b) $a : (a^2 + 1)$
(c) $2a : (a^2 + 1)$ (d) $2a : (a^2 - 1)$

[2024 : 1 M]

1.98 The decimal number system uses the characters 0, 1, 2, ..., 8, 9, and the octal number system uses the characters 0, 1, 2, ..., 6, 7.

For example, the decimal number 12 ($= 1 \times 10^1 + 2 \times 10^0$) is expressed as 14 ($= 1 \times 8^1 + 4 \times 8^0$) in the octal number system.

The decimal number 108 in the octal number system is

(a) 150 (b) 168
(c) 108 (d) 154

[2024 : 1 M]

1.99 In the following figure,

$$CD = 5 \text{ cm}, BE = 10 \text{ cm}, AE = 12 \text{ cm}$$
$$\angle DAB = \angle DCB \text{ and } \angle DAE = \angle DBC = 90°$$

Points AFCD create a rhombus.

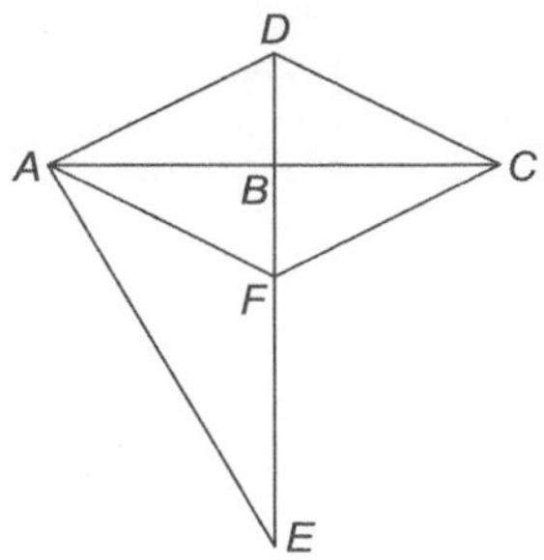

The length of BF (in cm) is

(a) 2 (b) 6
(c) 4 (d) 3 [2024 : 2 M]

1.100 The chart below shows the data of the number of cars bought by Millennials and Gen X people in a country from the year 2010 to 2020 as well as the yearly fuel consumption of the country (in Million liters).

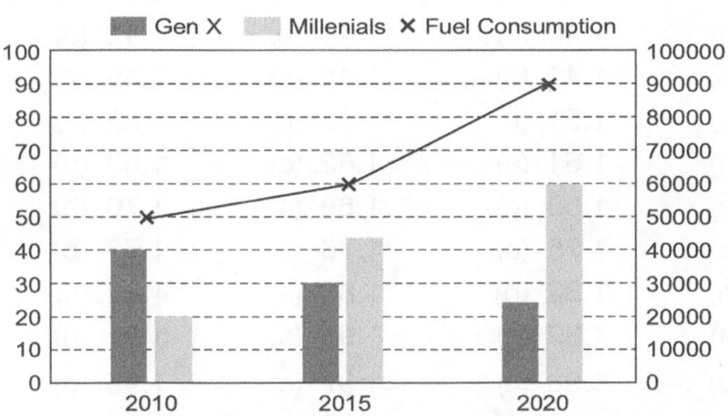

Considering the data presented in the chart, which one of the following options is true?

(a) The increase in the number of Millennial car buyers from 2015 to 2020 is less than the decrease in the number of Gen X car buyers from 2010 to 2015.
(b) The decrease in the number of Gen X car buyers from 2015 to 2020 is more than the increase in the number of Millennial car buyers from 2010 to 2015.
(c) The increase in the number of Millennial car buyers from 2010 to 2015 is more than the decrease in the number of Gen X car buyers from 2010 to 2015.
(d) The percentage increase in fuel consumption from 2010 to 2015 is more than the percentage increase in fuel consumption from 2015 to 2020. [2024 : 2 M]

1.101 A surveyor has to measure the horizontal distance from her position to a distant reference point C. Using her position as the center, a 200 m horizontal line segment is drawn with the two endpoints A and B. Points A, B and C are not collinear. Each of the angles ∠CAB and ∠CBA are measured as 87.8°. The distance (in m) of the reference point C from her position is nearest to

(a) 2606 (b) 2306
(c) 2603 (d) 2063 [2024 : 2 M]

1.102 P and Q have been allotted a hostel room with two beds, a study table, and an almirah. P is an avid bird-watcher and wants to sit at the table and watch birds outside the window. Q does not mind that as long as his bed is close to the ceiling fan. Which one of the following arrangements suits them the most?

(a)
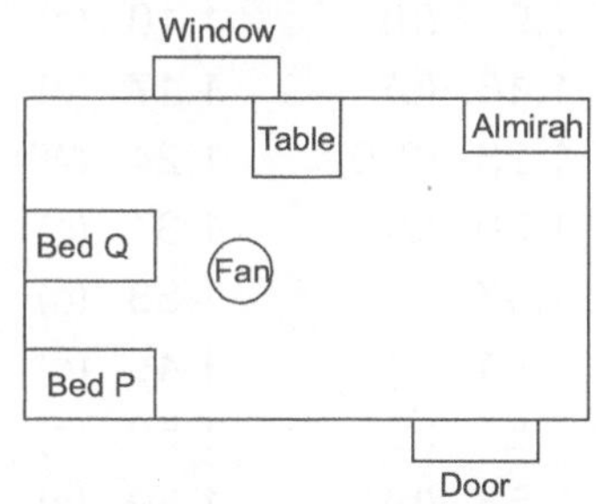

(b)
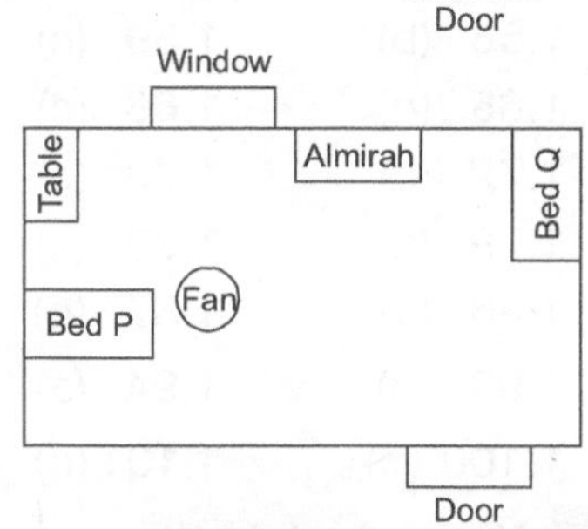

(c)
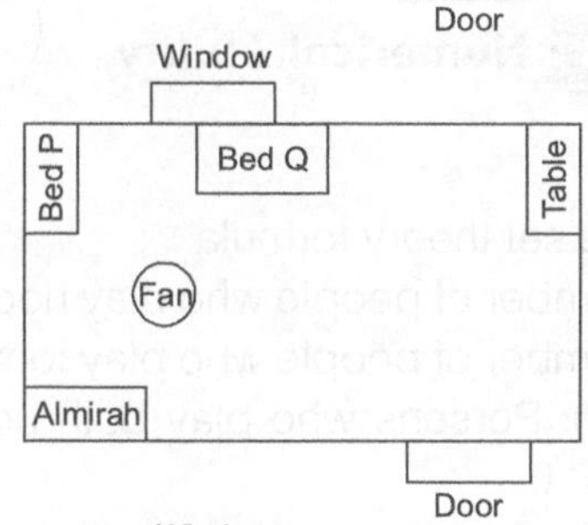

(d)
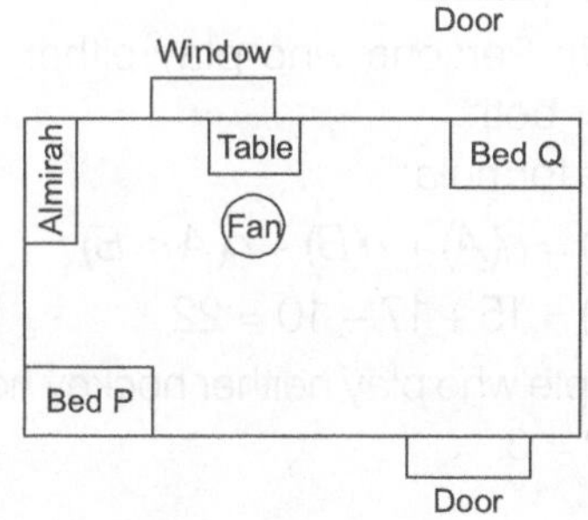

[2024 : 1 M]

1.103 A shopkeeper buys shirts from a producer and sells them at 20% profit. A customer has to pay ₹3,186.00 including 18% taxes, per shirt. At what price did the shopkeeper buy each shirt from the producer?

(a) ₹ 2,250.00 (b) ₹ 1,975.40
(c) ₹ 2,500.00 (d) ₹ 2,548.00

[2024 : 1 M]

Answers Numerical Ability

1.1 (d)	1.2 (c)	1.3 (d)	1.4 (b)	1.5 (b)	1.6 (a)	1.7 (b)
1.8 (c)	1.9 (d)	1.10 (c)	1.11 (d)	1.12 (d)	1.13 (c)	1.14 (a)
1.15 (a)	1.16 (c)	1.17 (c)	1.18 (d)	1.19 (b)	1.20 (a)	1.21 (96)
1.22 (d)	1.23 (850)	1.24 (48)	1.25 (6)	1.26 (b)	1.27 (c)	1.28 (d)
1.29 (140)	1.30 (a)	1.31 (c)	1.32 (c)	1.33 (d)	1.34 (c)	1.35 (c)
1.36 (c)	1.37 (c)	1.38 (c)	1.39 (8)	1.40 (a)	1.41 (b)	1.42 (a)
1.43 (a)	1.44 (c)	1.45 (c)	1.46 (b)	1.47 (c)	1.48 (a)	1.49 (7)
1.50 (c)	1.51 (d)	1.52 (c)	1.53 (c)	1.54 (b)	1.55 (d)	1.56 (a)
1.57 (d)	1.58 (b)	1.59 (c)	1.60 (c)	1.61 (a)	1.62 (c)	1.63 (d)
1.64 (a)	1.65 (d)	1.66 (c)	1.67 (c)	1.68 (c)	1.69 (a)	1.70 (b)
1.71 (d)	1.72 (c)	1.73 (c)	1.74 (a)	1.75 (a)	1.76 (b)	1.77 (b)
1.78 (b)	1.79 (a)	1.80 (a)	1.81 (b)	1.82 (d)	1.83 (a)	1.84 (a)
1.85 (c)	1.86 (a)	1.87 (b)	1.88 (a)	1.89 (c)	1.90 (a)	1.91 (b)
1.92 (c)	1.93 (b)	1.94 (a)	1.95 (b)	1.96 (c)	1.97 (c)	1.98 (d)
1.99 (*)	1.100 (c)	1.101 (c)	1.102 (a)	1.103 (a)		

Explanations Numerical Ability

1.1 (d)

Using the set theory formula

$n(A)$: Number of people who play hockey = 15

$n(B)$: Number of people who play football = 17

$n(A \cap B)$: Persons who play both hockey and football = 10

$n(A \cup B)$: Persons who play either hockey or football or both.

Using the formula,

$$n(A \cup B) = n(A) + n(B) - n(A \cap B)$$

$$n(A \cup B) = 15 + 17 - 10 = 22$$

Thus people who play neither hockey nor football = 25 − 22 = 3

Alternative Method:

Refer to Venn diagram given below:

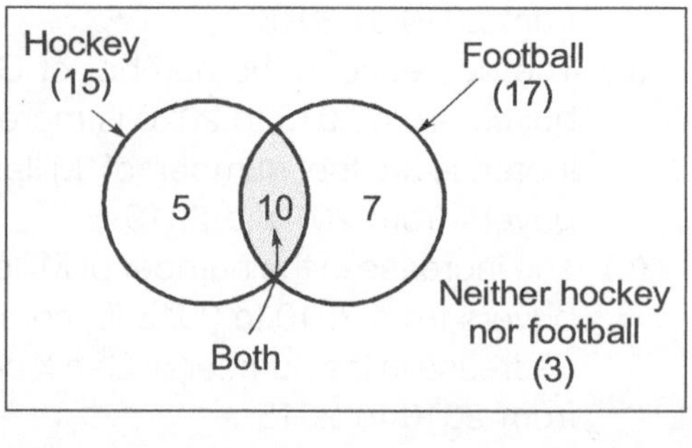

1.2 (c)

7 and 6 added is becoming five means the given two numbers are added on base 8.

$$\begin{array}{r}(137)_8\\+(276)_8\\\hline(435)_8\end{array}$$

Hence we have to add the another two given set of numbers also on base 8.

$$\begin{array}{r}(731)_8\\+(672)_8\\\hline(1623)_8\end{array}$$

Hence the overall problem was based on identifying base which was 8 and adding number on base 8.

1.3 (d)

$\Rightarrow$ Per day work or rate of 5 skilled workers $= \frac{1}{20}$

$\Rightarrow$ Per day work or rate of one skill worker

$= \frac{1}{5\times 20} = \frac{1}{100}$.

Similarly Per day work or rate of 8 semiskilled workers $= \frac{1}{25}$

$\Rightarrow$ Per day work or rate of one semi-skill worker

$= \frac{1}{8\times 25} = \frac{1}{200}$

And per day work or rate of 10 unskilled workers $= \frac{1}{30}$

$\Rightarrow$ Per day work or rate of one semi-skill worker

$= \frac{1}{10\times 30} = \frac{1}{300}$

Thus total per day work of 2 skilled, 6 semiskilled and 5 unskilled workers

$= \frac{2}{100} + \frac{6}{200} + \frac{5}{300} = \frac{12+18+10}{600}$

$= \frac{40}{600} = \frac{1}{15}$

Thus, time to complete the work is 15 days.

Alternative Method:

Let one day work of skilled semi-skilled and unskilled worker be a, b, c units respectively.

$5a\times 20 = 8b + 25 = 10c\times 30$ = Total unit of work

$100a = 200b = 300c \Rightarrow a = 2b = 3c$

$\Rightarrow \quad b = \frac{a}{2}$ and $c = \frac{a}{3}$

Given that: 2 skilled, 6 semi-skilled and 5 unskilled workers are working. Let they finish the work in 'x' days.

$(2a + 6b + 5c)\,x = 5a + 20$ = Total units of work

x = 15 days

1.4 (b)

We have to make 4 digit numbers, so the number should be start with 3 or 4, two cases possible;

Case (1) Thousands digit is 3.

Now other three digits may be any of 2, 2, 3, 3, 4, 4, 4, 4.

(a) Using 2, 2, 3

$\Rightarrow$ 223, 232, 322 -------

$\left(\frac{3!}{2!} = 3 \text{ numbers are possible}\right)$

(b) Using 2, 2, 4 $\Rightarrow$ 224, 242, 422

$\left(\frac{3!}{2!} = 3 \text{ numbers are possible}\right)$

(c) Using 2, 3, 3 $\Rightarrow$ 233, 323, 332

$\left(\frac{3!}{2!} = 3 \text{ numbers are possible}\right)$

(d) Using 2, 3, 4 $\Rightarrow$ 234, 243, 324, 342, 423, 432

(3! = 6 numbers are possible)

(e) Using 2, 4, 4 $\Rightarrow$ 244, 424, 442

$\left(\frac{3!}{2!} = 3 \text{ numbers are possible}\right)$

(f) Using 3, 3, 4 $\Rightarrow$ 334, 343, 433

$\left(\frac{3!}{2!} = 3 \text{ numbers are possible}\right)$

(g) Using 3, 4, 4 $\Rightarrow$ 344, 434, 443

$\left(\frac{3!}{2!} = 3 \text{ numbers are possible}\right)$

(h) Using 4, 4, 4 $\Rightarrow$ 444

$\left(\frac{3!}{3!} = 1 \text{ numbers are possible}\right)$

Total 4 digit numbers in case 1 = 3 + 3 + 3 + 6 + 3 + 3 + 3 + 1 = 25

Case (2) : Thousands digit is 4 ; Now other three digits may be any of 2, 2, 3, 3, 3, 4, 4, 4.

(a) Using 2, 2, 3 $\Rightarrow$ 223, 232, 322

$\left(\frac{3!}{2!} = 3 \text{ numbers are possible}\right)$

(b) Using 2, 2, 4 $\Rightarrow$ 224, 242, 422

$\left(\frac{3!}{2!} = 3 \text{ numbers are possible}\right)$

(c) Using 2, 3, 3 $\Rightarrow$ 233, 323, 332

$\left(\frac{3!}{2!} = 3 \text{ numbers are possible}\right)$

(d) Using 2, 3, 4 $\Rightarrow$ 234, 243, 324, 342, 423, 432

$\left(\frac{3!}{2!} = 3 \text{ numbers are possible}\right)$

(e) Using 2, 4, 4 $\Rightarrow$ 244, 424, 442

$\left(\frac{3!}{2!} = 3 \text{ numbers are possible}\right)$

(f) Using 3, 3, 3 $\Rightarrow$ 333

$\left(\frac{3!}{3!} = 1 \text{ numbers are possible}\right)$

(g) Using 3, 3, 4 $\Rightarrow$ 334, 343, 433

$\left(\frac{3!}{2!} = 3 \text{ numbers are possible}\right)$

(h) Using 3, 4, 4 $\Rightarrow$ 344, 434, 443

$\left(\frac{3!}{2!} = 3 \text{ numbers are possible}\right)$

(i) Using 4, 4, 4 $\Rightarrow$ 444

$\left(\frac{3!}{3!} = 1 \text{ numbers are possible}\right)$

Total 4 digit numbers in case 2 = 3 + 3 + 3 + 6 + 3 + 3 = 1 + 3 + 1 = 26.

Thus total 4 digits numbers using case (1) and case (2) = 25 + 26 = 51.

Alternative Method / Shortcut method:

As the number is greater than 3000. So thousand's place can be either 3 or 4.

Let's consider the following two cases

Case (I) When thousand's place is 3.

3 a b c

If there is no restriction on number of two's, three's and four's. Then each of a, b, c can be filled with 2 or 3 or 4 each in 3 ways.

So 3 × 3 × 3 = 27 numbers are there. Out of which 3222, 3333 are invalid as 2 can be used twice and three thrice only so number of such valid numbers beginning with 3 are 27 – 2 = 25.

Case (II) When thousand's place is 4

4 a b c

Without restriction on number of 2's, 3's and 4's *a*, *b*, *c* (as explained in case I) can be filled in 27 ways.

Out of these 27 numbers, 4 2 2 2 is only invalid as two have to be used twice only.

So valid numbers are 27 – 1 = 26.

Total numbers from Case (I) & Case (II)

25 + 26 = 51.

1.5 (b)

Suppose: Hari's age : *H*, Gita's age : *G*, Saira's age : *S*, Irfan's age : I

- $H + G > I + S$
- Using Statement (2) both $G - S = 1$ or $S - G = 1$; *G* can't be oldest and *S* can't be youngest.
- There are no twins thus using statement (2) either *GS* or *SG* possible.

(A) **HSIG:** not possible as there is I between *S* and *G* which is not possible using statement (3)

(B) **SGHI:** *SG* order is possible,
$S > G > H > I$ and $G + H > S + I$ (possible)
Because if $\{S = G + 1$; and $G = H + 1$ and $H = I + 2$ then $G + (I + 2) > (G + 1) + I\}$

(C) **IGSH:** according to this $I > G$ and $S > H$ thus adding these both inequalities we get $I + S > G + H$ which is opposite of statement (2) thus not possible.

(D) **IHSG:** according to this $I > H$ and $S > G$ thus adding both inequalities $I + S > H + G$ which is opposite of statement (2) thus not possible.

1.6 (a)

Let there be overall 'x' candidates. Initially those who decided to vote for *P* and *Q* are. $0.4x$ and $0.6x$ respectively. On the day of election 15% of $0.4x = 0.06x$ went from P's side to Q's side and 25% of $0.6x = 0.15x$ went from Q's side to P's side.

Now, after transfer '*p*' has $0.4x - 0.06x + 0.15x = 0.49x$

and after transfer '*q*' has $0.6x - 0.15x + 0.06x = 0.51x$

Given, in the question that *P* lost by 2 votes,

$$Q - P = 2 \text{ votes}$$

$$0.51x - 0.49 = 0.02x = 2 \text{ votes}$$

Hence, $(x = 100)$

Total number of votes are 100.

Alternate Solution:

Using options:

Lets take option (a) 100 as total number of voters. Initially *P* and *Q* has 40 and 60 voters on their sides respectively.

15% went from *P*'s side to *Q* and 25% went from *Q*'s side to *P*.

P	Q
(40)	(60)
(–) 15% of 40	– 25% of (60)
+ 25% of 60	+ 15% of (40)
40 – 6 + 15	60 – 15 + 6
= 49	= 51

Hence, Q and P has 51 and 49 respectively difference in number of votes = 2.
Hence answer is (a).

1.7 (b)

Fuel consumed per km will we least will be when mileage (kilometres per litres) mentioned on 'y' axis of graph will be maximum irrespective of number of kilometres travelled.
From the graph ('y' axis) we can observe that mileage (kilometres per litres) is maximum when vehicle is driven at 45 kilometres per hour. Hence the stretch which 'Q' covered at 45 kmph, mileage was highest and fuel consumption per kilometre was lowest.

1.8 (c)

Shortcut method:
The question can be done orally as R took 1/3rd of toffees initially. Hence total number of toffees have to be multiple of 3. There is only one option (c) as 48 which is multiple of 3.
Alternate Solution:
Using options:
If we use options for option (c), if total number of toffees are 48. R took 1/3rd if 48 and returned 4. So R has 16 – 4 = 12 toffees. Out of left over 36 toffees S took 1/4th of 36 and returned 3. So, S has 9 – 3 = 6 and remaining 30 will be left in bowl. Now T took 1/2 of 30 and returned two back. So, the bowl has 17 left.
*Always solve such questions by options.

1.9 (d)

$$f(y) = \frac{|y|}{y}$$

As we know, $$|y| = \begin{cases} y & y \geq 0 \\ -y & y < 0 \end{cases} \quad ...(i)$$

So from (i), $$f(y) = \frac{y}{y} = 1 \quad \text{If } y \geq 0$$

$$f(y) = \frac{-y}{y} = -1 \quad \text{If } y < 0$$

Hence, $|f(q) - f(-q)| = |1 - (-1)| = 2$

Alternate Solution:

Otherwise if $f(y) = \dfrac{|y|}{y}$

$$f(1) = \frac{|1|}{1} = 1$$

$$f(-1) = \frac{|-1|}{-1} = -1$$

For calculating $|f(q) - f(-q)|$

Let, $q = 1$, $|f(q) - f(-q)|$

$$|f(1) - f(-1)| = 1 - (-1) = 2$$

1.10 (c)

4 + 44 + 444............
If we see first term of series = 4
Hence, sum upto 1st term is also = 4
Put, $n = 1$ (first term) only option (c) satisfies.

$$\frac{4}{81}[10^{n+1} - 9n - 10] = \frac{4}{81}[10^{1+1} - 9 \times 1 - 10]$$

$$= \frac{4}{81} \times 81 = 4$$

Hence, option (c) is correct answer. Always solve these questions by putting values.

Alternative Solution:

4 + 44 + 444......................

$$\frac{4}{9}[9 + 99 + 999.............]$$

$$= \frac{4}{9}[(10-1) + (10^2 - 1) + (10^3 - 1).......]$$

$$= \frac{4}{9}[(10 + 10^2 + 10^3....10^n) - (1 + 1 + 1.....)]$$

$$= \frac{4}{81}[10^{n+1} - 9n - 10]$$

1.11 (d)

$(1.001)^{1259} = 3.52$ and $(1.001)^{2062} = 7.85$

$$(1.001)^{1259} \times (1.001)$$
$$\downarrow \qquad \downarrow$$
$$3.52 \times 7.85$$

$= (1.001)^{3321} = (1.001)^{3321}$ (as $a^m \times a^n = a^m + a^n$)
Hence, $(1.001)^{3321} = 27.632 \approx 27.64$

1.12 (d)

Category	Amount
Food	4000
Clothing	1200
Rent	2000
Savings	1500
Others	1800
Total	10500

So, total expenses are 10,500.
Expenses excluding savings are 9,000.
Hence, required percentage

$$= \frac{9{,}000}{10{,}500} \times 100 = 85.714\% \approx 86\%$$

1.13 (c)

Probability that A and B will meet will be given by the graphical representation where shaded region represents favorable area

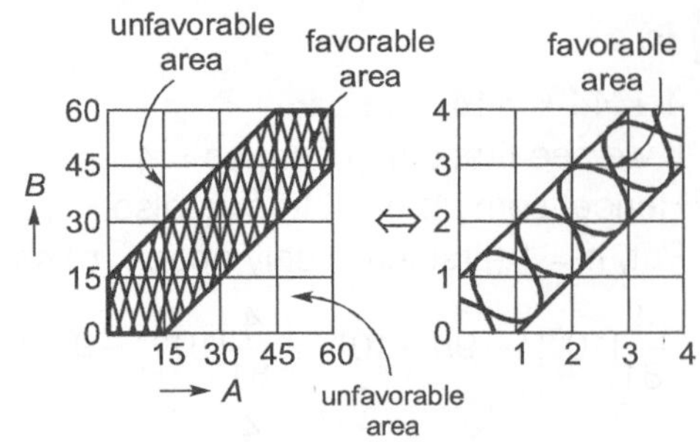

$$\text{Required probability} = \frac{\text{Fabourable area}}{\text{Total area}} = \frac{7}{16}$$

1.14 (a)

Let the number of Rs. 20 notes be 'a' and Rs. 10 notes be 'b' respectively.

So, $a + b = 14$

$20a + 10b = 230$

on solving, $a = 9, b = 5$

Hence, number of Rs.10 notes are 5.

Alternative Solution:

If we use option (a),

Number of ten Rs. notes are 5.

$\text{Rs } 10 \times 5 = 50 \text{ Rs.}$

The remaining nine notes have to be 20 Rs. notes

$\text{Rs. } 20 \times 9 = 180$

and hence $50 + 180 = 230$ (total money value)

Hence, option (a) number of ten Rs. notes are 5.

1.15 (a)

Divide 8 bags into three parts.

3, 3 and 2 respectively. Now two cases arises.

Case (i):

Firstly, if we compare the 3, 3 bags on pans of a balance. We can identify on which side is the lighter bag placed. And then we will need only one more weighing for identifying faulty bag. So only two weighings are required.

Case (ii):

Secondly, if the 3, 3 compared bags are equal, then to be sure faulty bag is one among the remaining two bags. So one more weighing is required to find out faulty bag from these two bags. Hence overall two weighings are required. Thus in either cases only two weighings are required.

Alternative Solution:

In case of a weighing balance (i.e. beam balance) the following model can be observed

$1 - 3 \rightarrow$ 1 weighing required

$4 - 9 \rightarrow$ 2 weighing required

$10 - 27 \rightarrow$ 3 weighing required

$28 - 81 \rightarrow$ 4 weighing required

and the process will go so on.

As there are eight objects. Minimum number of weighings required will be only 2.

1.16 (c)

$$\frac{\text{M} + \text{Tue} + \text{W}}{3} = 41$$

$$\Rightarrow \quad \text{M} + \text{Tue} + \text{W} = 123 \quad \text{...(i)}$$

$$\frac{\text{Tue} + \text{W} + \text{Th}}{3} = 43$$

$$\Rightarrow \quad \text{Tue} + \text{W} + \text{Th} = 129 \quad \text{...(ii)}$$

From equation (ii) and (i),

$$\text{Th} - \text{M} = 6$$

Also $\text{Th} = 1.15\,\text{M}$

$$1.15\,\text{M} - \text{M} = 6 \Rightarrow 15\,\text{M} = 6$$

$$\Rightarrow \quad \text{M} = 40$$

$$\text{Th} = 1.15\,\text{m} \Rightarrow \text{Th} = 1.15 \times 40$$

$$\Rightarrow \quad \text{Th} = 46$$

1.17 (c)

$$\text{Average speed} = \frac{\text{Total distance}}{\text{Total time}}$$

$$\frac{\text{Total distance}}{\text{Total time}} = \frac{(8+6+16)\text{km}}{\left(\frac{1}{4}+\frac{1}{4}+\frac{1}{4}\right)\text{hr}}$$

$$\frac{30\,\text{km}}{\frac{3}{4}\text{hr}} = 40 \text{ km/hr}$$

1.18 (d)

$S = 10 + 84 + 73 + \ldots$

$S = (9 + 1) + (9^2 + 3) + (9^3 + 5)\ldots$

$S = \{9 + 9^2 + 9^3 + \ldots 9^n\} + \{1 + 3 + 5 \ldots (2n - 1)\}$

$$S = \frac{9(9^n - 1)}{(9 - 1)} + n^2$$

$$\left(\begin{array}{l}\text{As } a + ar + ar^2 \ldots\ldots\ldots ar^{n-1} \\ = \dfrac{a(r^n - 1)}{(r - 1)} \quad \text{for} \quad r > 1\end{array}\right)$$

$$\left(\begin{array}{l}\text{and } 1 + 3 + 5 \ldots\ldots\ldots (2n - 1) = n^2 \\ \text{Sum of odd natural numbers} = n^2\end{array}\right)$$

$$\text{So, } S = \frac{9(9^n - 1)}{8} + n^2$$

Alternate Method:

Put $n = 2$ in all options and check whether sum $= 10 + 84 = 94$ is in which option.

1.19 (b)

$3x^2 + 2x + p(p - 1) = 0$

If roots are of opposite signs product of the roots will be negative.

$$\frac{p(p-1)}{3} < 0$$
$$p(p-1) < 0$$
$$0 < p < 1$$
$$P \in (0, 1)$$

1.20 (a)

Every leap year will have 366 days.

1 leap year = 366 days = 52 × 7 + 2 days

Every leap year will have 52 complete weeks. For 53rd Saturday there are two favourable cases from 7 possible cases.

Fri	Sat
Sat	Sun
Sun	Mon
Mon	Tue
Tue	Wed
Wed	Thu
Thu	Fri

(Fri Sat, Sat Sun) → 2 favourable cases

Required probability = $\frac{2}{7}$

1.21 Sol.

$$\left(z + \frac{1}{z}\right)^2 = 98$$

So, $$z^2 + \frac{1}{z^2} = 96$$

1.22 (d)

Given: $ax^2 + bx + c = 0$

Roots are real and positive

For e.g.

$x^2 - 5x + 6 = 0$ roots are 2, 3

Then roots of $ax^2 + b|x| + c = 0$

Then roots of $x^2 - 5|x| + 6 = 0$

$\therefore |x|^2 - 5|x| + 6 = 0$

$\left.\begin{matrix} \text{if } x > 0 \\ |x| = x \end{matrix}\right\}$ $x^2 - 5x + 6 = 0$

$(x - 2)(x - 3) = 0$ $x = 2, 3$ are roots

$\left.\begin{matrix} \text{If } x < 0 \\ |x| = -x \end{matrix}\right\}$ $x^2 + 5x + 6 = 0$

$(x + 2)(x + 3) = 0$ $x = -2, -3$ are roots

So, if $x^2 - 5x + 6 = 0$ has two real roots then, $|x|^2 - 5|x| + 6 = 0$ has four real roots

$ax^2 + bx + c = 0$ has real roots then, $ax^2 + b|x| + c = 0$ has 4 real roots.

1.23 Sol.

One way fare of single person = 100

Undiscounted round trip fare of singe person = 200

Undiscounted round trip fare of five persons = 1000

Round trip will result in a discount of 10%. Also additional discount of 5% on total fare for 4 or more person.

So, a term of 5 person will be eligible for overall discount of (10% + 5%) = 15% discount

Discounted round trip for 5 persons

1000 × 0.85 = 850 Rs.

1.24 Sol.

Following table can be redrawn:

		Men	Women	Total
Own vehicle	Car	40	34	74
	Scooter	30	20	50
	Both	60	46	106
Do not own vehicle		20	50	70

Now following venn-diagram can be drawn

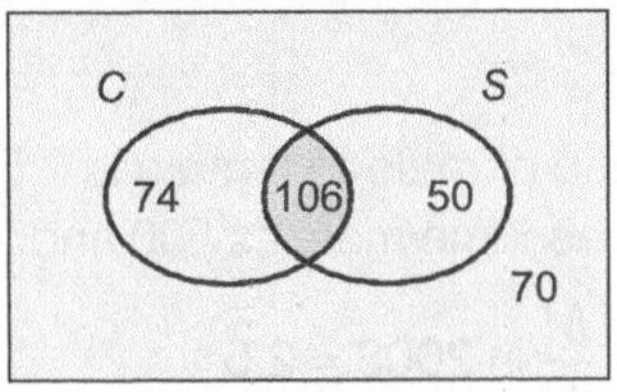

So, % of people not owing a scooter

$$\left[\left(\frac{74+70}{300}\right) \times 100\right] = \frac{144}{300} \times 100 = 48\%$$

1.25 Sol.

Answer is 6.

1.26 (b)

There are 19 multiples of 10 from 2 to 198 typically 10, 20, 30 ··· 190.

Average of these number will be

$$= \frac{10 + 20 + 30 + \cdots + 190}{19}$$

$$= 10\frac{[1 + 2 + 3 + \cdots + 19]}{19}$$

$$= 10\frac{[19 \times 20]}{19 \times 2} = 100$$

1.27 (c)

Let, $y = \sqrt{12+\sqrt{12+\sqrt{12+\cdots}}}$

So, $y = \sqrt{12+y}$

Squaring, $y^2 = 12 + y$

$y^2 - y - 12 = 0 \Rightarrow y-4)(y+3) = 0$

$y = 4$

Option (c) → 4.000

1.28 (d)

Given x is real.

$$|x^2 - 2x + 3| = 11$$
$$x^2 - 2x + 3 = 11$$
$$x^2 - 2x - 8 = 0$$
$$(x-4)(x+2) = 0$$
$$x = 4, -2$$

or
$$x^2 - 2x + 3 = -11$$
$$x^2 - 2x + 14 = 0$$

Will not give real roots as $D = b^2 - 4ac < 0$

Now, $|-x^3 + x^2 - x|$ can attach

If $x = 4$,

$$|-x^3 + x^2 - x| = |-64+16-4| = 52$$

If $x = -2$,

$$|-x^3 + x^2 - x| = |8+4+2| = 14$$

1.29 Sol.

Given ratio of male to female is 2.5 in year 2008.
So let us assume male as 500 and female as 200.

So, ratio $\frac{M}{F}$ in 2008 = 2.5

Now female doubles up in 2009

Number of female in 2009 = 2 × 200 = 400

$\frac{M}{F}$ ratio in 2009 is 3

$$\frac{M}{400} = 3 \Rightarrow M = 1200$$

So number of males in 2009 is 1200.

% increase in number of males from 2008 to 2009

$$\% \text{ increase} = \frac{1200-500}{500} \times 100 = 140\%$$

1.30 (a)

Any angle and corresponding time can be driven by

$$\left[5x - \left(\frac{D^0}{6}\right)\right] \times \frac{12}{11}$$

Between 6 and 7 pm angle have to be 60°

Put $x = 6$ and $D = 60°$

$$\left[5\times 6 - \frac{60°}{6}\right] \times \frac{12}{11} = 20 \times \frac{12}{11} = 21\frac{9}{11}$$

So, at $4:21\frac{9}{11}$ angle between minutes and hour hand will be 60°.

In given question angle have to close to 60° nearest option is 6 : 22.

1.31 (c)

2, 5, 10, 17, 26, 37, 50, 64.

$2 = 1^2 + 1,\quad 5 = 2^2 + 1,\quad 10 = 3^2 + 1,$

$17 = 4^2 + 1,\quad 26 = 5^2 + 1,\quad 37 = 6^2 + 1,$

$50 = 7^2 + 1$, ***64 = 8²***

So 64 is the incorrect term.

1.32 (c)

Q.No.	Marks	Answered correctly	Answered wrongly	Not attemped	Total students
1.	2	21	17	6	44
2.	3	15	27	2	44
3.	2	23	18	3	44

$$\text{Average} = \frac{\text{Total marks scored}}{\text{Total number of students}}$$

$$= \frac{21\times 2 + 15\times 3 + 23\times 2}{44} = 3.022727 \simeq 3.02$$

1.33 (d)

Let GDP before year 2012-2013 be 100 Rs.

During 2012-2013 GDP increased by 7%.

So, GDP in year 2013-2013 = 107 Rs.

Now exchange rate for USD change from Rs. 50/USD to Rs. 60/USD.

i.e., value of Rupee reduced to $\frac{5}{6}$ of original value.

So, new GDP (in USD) = $107 \times \frac{5}{6} = 89.16$

Effective change is from 100 to 89.16

= 10.83% decrease ≃ 11% decrease

1.34 (c)

Ratio of male to female 2001 is given.

Let there be 100 males and 100 female in 2011.

Now number of female in 2012 is same as number of females in 2012 as given in question.

So female in 2012 is also 100.

Ratio of male of female in 2012 is 1.5

$$\frac{M}{F} = 1.5 \Rightarrow M = 150$$

So, there are 150 males n 2012.

Ratio of male in 2012 to male in 2011

$$= \frac{150}{100} = \frac{1.5}{1}$$

1.35 (c)

$$(7526) - (Y)_8 = (4364)_8$$

So, $Y = (7526)_8 - (4364)_8$

$$= (7526 - 4364)_8 = (31420)_8$$

1.36 (c)

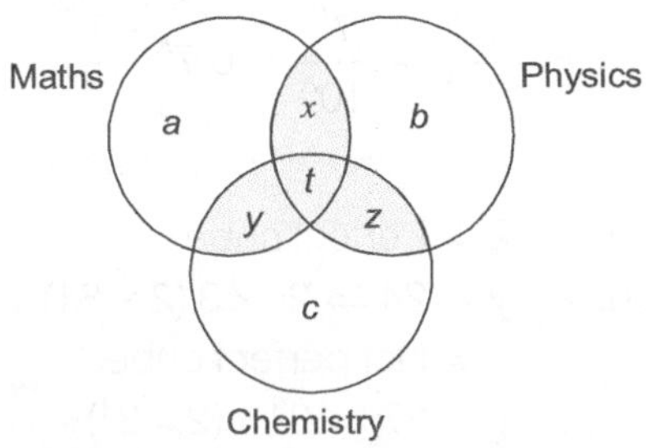

According to question,

$$75\% = a + b + c + (x + y + z) + t$$

Also, $x + y + z + t = 50\%$

$x + y + z = 40\%$

so, $t = 10\%$

so, (*iii*) is correct.

and $a + b + c = 25\%$

Also, $m + c + ch = (a + x + y + t) + (b + x + z + t) + (c + y + z + t)$

$$m + c + ch = (a + b + c) + 2(x + y + z) + 3t\, 25 + 2(40) + 3 \times 10$$

$$= 135\% = \frac{27}{30}$$

1.37 (c)

Marks are allotted only for correct answer

$$\text{Average marks} = \frac{\text{Total marks}}{\text{No. of students}}$$

$$= \frac{21\times 2 + 15\times 3 + 11\times 1 + 23\times 2 + 31\times 5}{21 + 17 + 6}$$

$$= \frac{299}{44} = 6.795$$

1.38 (c)

$f(x) = 1 - |x|$, use options

Put $x = 0$, $f(x) = 1$ is maximum value

1.39 Sol.

$fg(h\,(2, 5, 7, 3), 4, 6, 8)$

$$fg\left[\left(\text{Re}\,\frac{7\times 3}{2\times 5}\right), 4, 6, 8\right] = fg[1, 4, 6, 8]$$

$$= f[1, 4, 6, 8] \times g[1, 4, 6, 8] = 8 \times 1 = 8$$

1.40 (a)

Refer to the figure below.

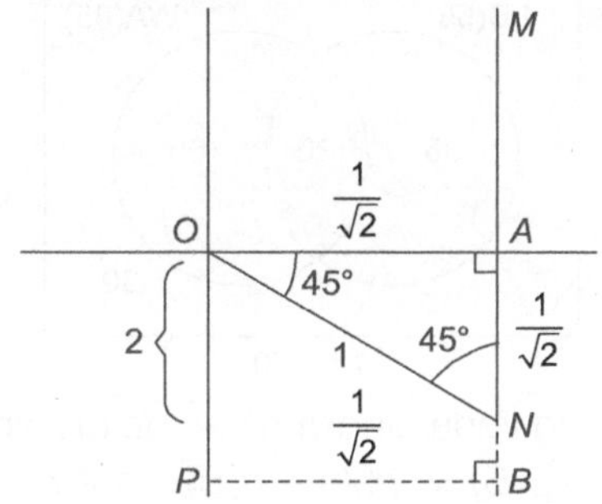

As $ON = 1$, $OA = AN = \frac{1}{\sqrt{2}} = PB$

So, $MB = \left(4 + 2 - \frac{1}{\sqrt{2}}\right)$ and $PB = \left(\frac{1}{\sqrt{2}}\right)$

$$PM = \sqrt{PB^2 + BM^2} = 5.34$$

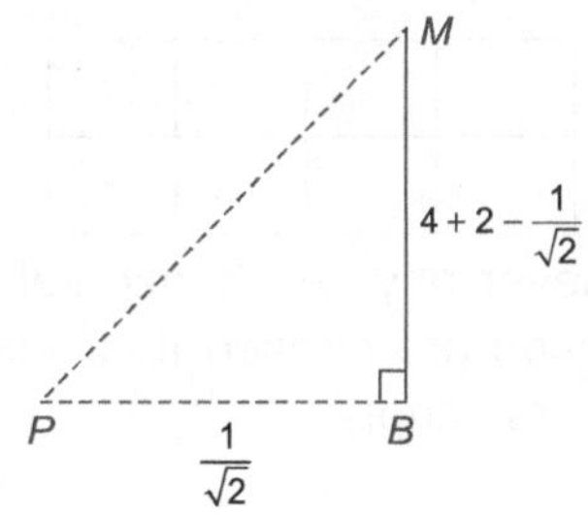

1.41 (b)

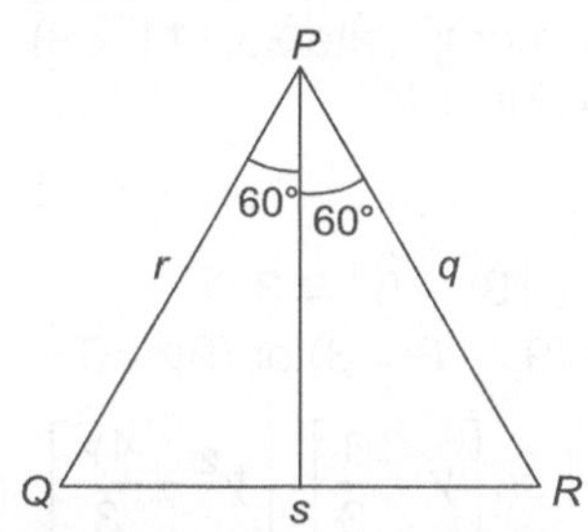

In $\Delta PQR \Rightarrow Ar\Delta PQR = Ar(\Delta QPS) + Ar(\Delta PSR)$

$$\frac{1}{2} r \times q \times \sin 120° = \frac{1}{2} r \times PS \times \sin 60° + \frac{1}{2} q \times PS \sin 60°$$

$$\frac{1}{2} r \times q \times \sin(180 - 60°) = \frac{1}{2} PS \sin 60° (r + q)$$

$$\frac{1}{2} r \times q \times \sin 60° = \frac{1}{2} PS \sin 60° (r + q)$$

$$PS = \left[\frac{r \times q}{r + q}\right]$$

1.42 (a)

$$\frac{\alpha^n + \beta^n}{\alpha^{-n} + \beta^{-n}} = \frac{\alpha^n + \beta^n}{\frac{1}{\alpha^n} + \frac{1}{\beta^n}} = \frac{\alpha^n + \beta^n}{\left(\frac{\beta^n + \alpha^n}{\alpha^n \beta^n}\right)} = (\alpha\beta)^n$$

Given : $\alpha \times \beta = 4$. Hence, $(\alpha\beta)^n = 4^n$

1.43 (a)

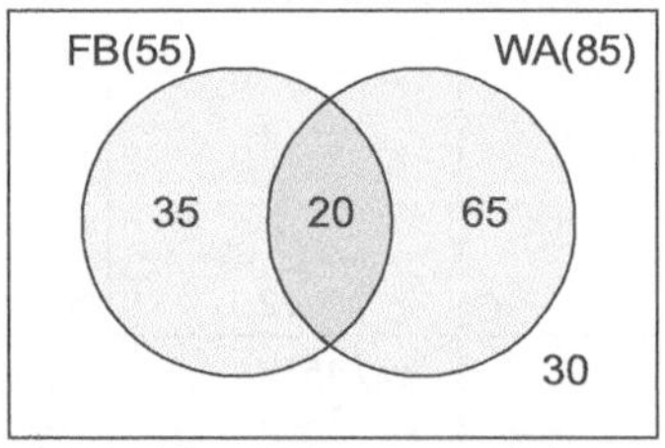

$T = 150$

Following venn diagram can be drawn as

$$n(A \cup B) = n(A) + n(B) - n(A \cap B)$$
$$120 = 55 + 85 - n(A \cap B)$$
$$n(A \cap B) = 20. \text{ So, only facebook 35(A).}$$

1.44 (c)

Number of rectangles will be

$${}^5C_2 \times {}^3C_2$$
$$10 \times 3 = 30$$

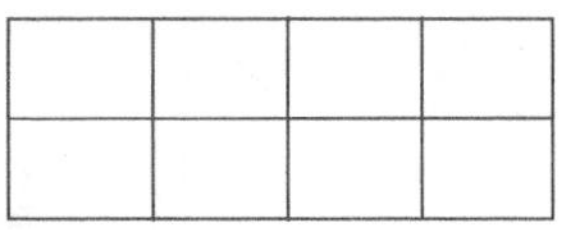

As whenever any two horizontal and any two vertical lines are chosen their intersection will produce a rectangle.

1.46 (b)

13, 23, 43, 53 are all prime numbers.
Only 33 is composite 33, (11 × 3).
Odd one out is 33.

1.47 (c)

$$|9y - 6| = 3$$
$$\Rightarrow \text{Either } (9y - 6 = 3) \text{ or } (9y - 6 = -3)$$
$$\Rightarrow [y = 1] \text{ or } \left[y = \frac{1}{3}\right] \left[y^2 - \frac{4y}{3}\right]$$

Put $y = \frac{1}{3}$, $\left(\frac{1}{3}\right)^2 - \frac{4}{3} \times \frac{1}{3} = -\frac{1}{3}$ or $y^2 - \frac{4y}{3}$

Put $y = 1$, $1^2 - \frac{4}{3} = -\frac{1}{3}$

1.48 (a)

Probability of free throw $= \frac{60}{100} = 0.6$

Probability of NOT free throw = 1 – 0.6 = 0.4

So, required probability of exactly 6 throws in 10 attempts will be given by

$${}^{10}C_6 (0.6)^6 \times (0.4)^4 = 0.2508$$

1.49 Sol.

Unit digit of $211^{870} + 146^{127} \times 3^{424}$ is $1 + 6 \times 1 = 7$

1.50 (c)

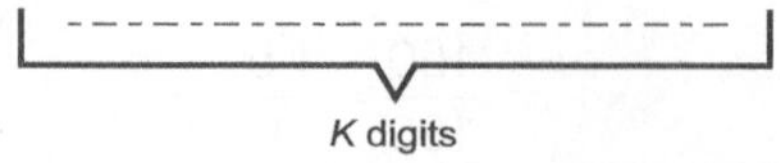

Each digit can be filled in 7 ways as 0, 5 and 9 is not allowed, so each of these places can be filled by 1, 2, 3, 4, 6, 7, 8 so required probability is $\left(\frac{7}{10}\right)^K$ or 0.7^K

$$\frac{fC}{TC} = \frac{7^K}{10^K} = 0.7^K$$

1.51 (d)

$$y \times 162 = \text{Perfect cube}$$

Option, $y = 24 \Rightarrow 2^3 \times 3\,(2 \times 81)$
$\neq$ Not perfect cube
$y = 27 \Rightarrow 3^3 \times (2 \times 3^4)$
$\neq$ Not perfect cube
$y = 32 \Rightarrow 2^5 \times 2 \times 3^4$
$\neq$ Not perfect cube
$y = 36 \Rightarrow 2^2 \times 3^2 \times 2 \times 3^4 = 2^3 \times 3^6$
$= (2 \times 3^2)^3$ is a perfect cube

Hence, the answer is $y = 36$

1.52 (c)

Correct option is; Wished
After Rajender Chola **returned** from his voyage to Indonesia, he **wished** to visit the temple in Thanjavur.
Both are events of past. Use of past perfect form is unwarranted as it reflects part of past.

1.53 (c)

P will not get submerged as it is @ 550. So *P* is not present in correct options. Hence (i) and (ii) options are incorrect. Now compare *Q* and *T*. As *T* is between 500 and 525. T will get submerged. Hence, ans (c) as among option *C* and *D*. *T* is present in only option (c).

1.54 (b)

$$\frac{(x + y) - |x - y|}{2} \quad \ldots(i)$$

If $x > y\,|x - y| = x - y$;
If $x < y\,|x - y| = y - x$
Now, if $x > y$, above expression (i) becomes

$$\frac{(x + y) - (x - y)}{2} = \frac{2y}{2} = y = \text{minimum of } (x, y)$$

as $x > y$

Now if $x < y$;

$$\frac{x+y-(y-x)}{2} = \frac{2x}{2} = x = \text{minimum of } (x, y)$$
as $x < y$

Therefore correct answer is option (b).

Alternate solution:

Use easy values, $x = 1$ and $y = -2$

Now, $\frac{(1-2)-|1-(-2)|}{2} = -2$ or $x = 2$ and $y = -1$

$$\frac{(2-1)-|2-(-1)|}{2} = \frac{1-3}{2} = -1$$

Which is minimum of (x, y).

Therefore, correct answer is option (b).

1.55 (d)

As there are 4 people A, G, N, S and four colours so without any restriction total ways have to be $4 \times 4 = 16$.

Now, Arun $\rightarrow$ dislikes Red and Shweta dislikes white so $16 - 2 = 14$ ways.

Therefore correct answer should be option (d).

Alternate solution:

Only one option is less than 16.

Therefore correct answer should be option (d).

1.56 (a)

Out of six people. 3 place definitely occupied by right handed people as at least 2 women are there so these two will sit adjacently. Now as only on seat is left it will be occupied by a left handed man because on right side of this seat is sitting an right handed man.

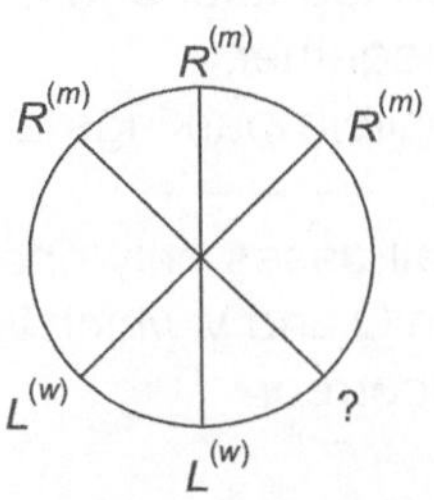

? have to be a left handed man.

Here, $R^{(m)}$ indicates right handed man and $L^{(w)}$ indicates left handed women.

Therefore, answer should be only 2 women.

1.62 (c)

$$K = 4 \Rightarrow \frac{(4+2)^2}{4-3} = 36$$
$$K = 8 \Rightarrow \frac{(8+2)^2}{8-3} = \frac{100}{5} = 20$$
$$K = 28 \Rightarrow \frac{(28+2)^2}{28-3} = \frac{900}{25} = 36$$
} Integers

So, option (c) $\Rightarrow$ 4, 8, 28

1.63 (d)

$$f(x) = 0$$
$$x = \{-2, 0, 3\}$$
$$f(-2) = 0,\ f(0) = 0,\ f(3) = 0$$
$$f(x-3) = 0$$
$$1, 3, 6 : f(1-3) = f(-2) = 0$$
$$f(3-3) = f(0) = 0$$
$$f(6-3) = f(3) = 0$$

1.64 (a)

$$F(a, b) = (a-b)^2$$
$$G(a, b) = |a-b|$$
$$G(F(1,3)\ G(1,3)) = G((1-3)^2, |1-3|)$$
$$= G(4, 2) = |4-2| = 2$$

1.65 (d)

$7^3, 11^3,$ _____, 17^3

$\because$ 7, 11, 13 and 17 are all prime numbers.

The series is cube of these numbers.

1.66 (c)

Work done by X in 1 hour $= \frac{1}{2}$

Work done by Y in 1 hour $= \frac{1}{4}$

Work done by X and Y together in 1 hour

$$= \frac{1}{2} + \frac{1}{4} = \frac{3}{4}$$

$\therefore$ Total time to complete mowing of lawn

$$= \frac{4}{3}\text{hour} = \frac{4}{3} \times 60 = 80 \text{ min}$$

1.67 (c)

$$X = \{1, 2, 3\}; \quad Y = \{2, 3, 4\}$$
$$Z = \left\{\frac{1}{2}, \frac{1}{3}, \frac{1}{4}, \frac{2}{3}, \frac{2}{4}, \frac{3}{2}, \frac{3}{4}\right\}$$

Minimum value in $\{Z\} = \frac{1}{4}$

Maximum value in $\{Z\} = \frac{3}{2}$

$$\text{Product} = \frac{1}{4} \times \frac{3}{2} = \frac{3}{8}$$

1.68 (c)

All numbers between 100 and 1000 can be 3 digit numbers only

For units and tens digits = 5 integers can be filled (0, 2, 4, 6, 8)

For hundreds digit = 0 can't be filled only 4 digits can be filled (2, 4, 6, 8)

$\therefore$ Total choices $= 4 \times 5 \times 5 = 100$ numbers

1.69 (a)

Let, Number of boys = $4x$
and Number of girls = $3x$

Total passed candidates = $\frac{80}{100} \times 7x = \frac{28}{5}x$

Number of girls candidates who passed

$$= \frac{90}{100} \times 3x = \frac{27}{10}x$$

Now total number of candidates passed = Number of girls who passed + Number of boys who passed
⇒ Number of boys who passed

$$= \left(\frac{28}{5} - \frac{27}{10}\right)x = \frac{56-27}{10}x = \frac{29}{10}x$$

% of boys = $\frac{29}{10 \times 4x} x \times 100 = 72.5\%$

Option (a) is correct.

1.70 (b)

Z,	M V,	R Q P	K J I H
26,	23,22	18,17,16	11,10,9,8

1.71 (d)

$^3C_1 + {}^3C_2 + {}^3C_3$
$3 + 3 + 1 = 7$
Possible combinations,
PQ, RQ, SQ, PRQ, PSQ, RSQ, PQRS

1.72 (c)

$10 \times 10 + 9 \times 10 + 9 \times 10 = 280$
Question is asking 37 appears how many times and not how many numbers.

1.73 (c)

As given in question holding period was reduced by one quarter each time. Therefore, after third division holding period remains 90 days.

1.74 (a)

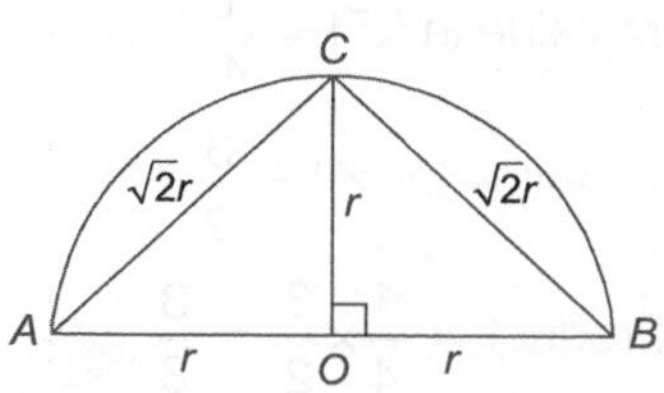

The ratio,

$$\frac{\overline{AC} + \overline{CB}}{\overline{AB}} = \frac{\sqrt{2}r + \sqrt{2}r}{2r} = \frac{2\sqrt{2}}{2} = \sqrt{2}$$

1.75 (a)

Let the revenue of company Q in 2010 = x in million rupees. Then the revenue of company Q in 2015 = 1.2 x in million rupees.

Given in graph,
$1.2x = 45 \Rightarrow x = 37.5$ in million rupees
As given in question Q had earned a profit of 10%. ∵ Expenditure of company Q in 2014

$$= \frac{37.5}{1.1} = 34.09 \approx 34.1$$

1.76 (b)

Consider expression $\frac{x^n - a^n}{x+a}$.

Going by options and trying to convert it in the above format seeing option (b)
$11^{52} - 1 = (11^{13})^4 - (1)^4$

$$\frac{(11^{13})^4 - (1)^4}{11^{13}+1} = \frac{((11)^{13})^2 - (1)^2)((11^{13})^2 + (1)^2)}{11^{13}+1}$$
$$= [(11^{13} - 1)][(11^{13})^2 + (1)^2]$$

Hence, $11^{52} - 1$ is divisible by $11^{13} + 1$.

1.77 (b)

Sum of the interior angle = $(n-2) \times 180° = 1440°$.
The interior angle between two sides of polygon is $\frac{1440°}{10} = 144°$.

1.78 (b)

Upon rotation, we get the below figure.

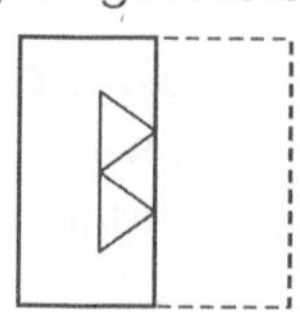

1.79 (a)

∵ S and V must be parked together and Q and S also must be parked together.
These two condition possiible parkings are:
Q S V – – – –; V S Q – – – –
We can observe that in all cases only one car can be parked in between Q and V which is car S. Hence option (a) is incorrect.

1.80 (a)

Side of outer first square = 10 cm

Side of outer second square = $\frac{10}{\sqrt{2}}$ cm

Likewise,

Side of outer third square = $\left(\frac{10}{(\sqrt{2})^2}\right)$

Hence, side of smallest square = $\frac{10}{(\sqrt{2})^5}$

Area of smallest square

$$= \left(\frac{10}{(\sqrt{2})^5}\right)^2 = \frac{100}{2^5} = 3.125$$

1.81 (b)

Given: pdf $f(x) = 10^{-2}$, $0 < x < 100$

$$E(x) = \int_0^{100} xf(x)dx = \int_0^{100} x10^{-2}dx$$

$$= \frac{1}{100}\left.\frac{x^2}{2}\right]_0^{100} = \frac{100}{2} = 50$$

Alternate Solution:

n is a continuous random variable for continuous random variable

$$\therefore \quad f(x) = \begin{cases} \frac{1}{\beta - \alpha} & \alpha \le x \le \beta \\ 0 & \text{otherwise} \end{cases}$$

From here, $\alpha = 0$, $\beta = 100$

For continuous probability function mean

$$= \frac{\alpha+\beta}{2} = \frac{0+100}{2} = 50$$

1.82 (d)

Total number of students in the year-2
= 60 + 5 = 65

Students who failed in the year-1 and appeared in year-2 = 10

So the students who appeared first time in year-2 = 65 − 10 = 55

Similarly, total no. of students in year 3 = 53

Students who failed in the year-2 and appeared in year 3 = 5

So, students who appeared first time in year 3
= 53 − 5 = 48

1.83 (a)

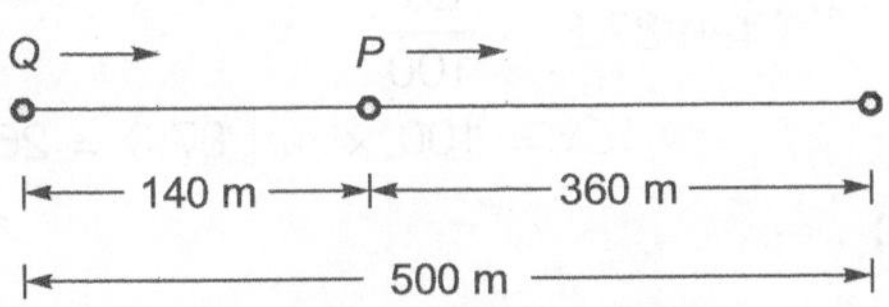

$$P : Q = 3 : 4$$

$$P = 3x \; ; \; Q = 4x$$

So, $3x = 360 \Rightarrow x = 120$ m

$$Q = 4x = 4 \times 120 = 480 \text{ m}$$

When P reaches 500 metres and Q reaches only 480 metres. $\therefore$ P wins by 20 m.

1.84 (a)

Three bells P, Q, R.

P = 20 minutes; Q = 30 minutes; R = 50 minutes

L.C.M. of 20, 30, 50 = 300 minutes = 5 hours

12:00 PM + 5 hours = 5:00 P.M.

So, three bells ring together again at 5:00 P.M.

1.85 (c)

n = Total = 3

m = Favourable = disc does not retain the same front and rear views after the flipping = 2

$$\text{Probability} = \frac{m}{n} = \frac{2}{3}$$

1.86 (a)

Given two dice.

$$P(E) = \frac{\text{Favourable}}{\text{Total}} \Rightarrow P(E) = \frac{3}{6} \times \frac{3}{6} = \frac{1}{4}$$

1.87 (b)

Given online stores S and M.

Cap price on M= Rs. x

Price on S = $(x - 0.1x) = 0.9x$

Delivery charges on S = Rs. 100

$\therefore x - (0.9x + 150) = 100$

$0.1x - 150 = 100 \Rightarrow x = 2500$

Price on store S= 0.9 × 2500 = 2250

1.90 (a)

$$P(1) = \frac{1}{6}; \quad P(4) = \frac{1}{6}$$

$$\text{Required probability} = P(1) \times P(4) = \frac{1}{6} \times \frac{1}{6} = \frac{1}{36}$$

1.91 (b)

Tobacco users advised to stop consumption = 65%

Tobacco users attempted to stop consumption

$$= \frac{3}{10} = 30\%$$

$\because$ Majority of users who were advised to stop consumption did not attempt to do so.

1.92 (c)

Total number of triangles is 20.

1.93 (b)

The function shown in figure follows ODD symmetry.

$$f(-x) = -f(x)$$

Only function in option (b) follows ODD symmetry.

$f(x) = x2^{-|x|}$

$f(-x) = -x2^{-|-x|} = -x2^{-|x|} \Rightarrow f(-x) = -f(x)$

1.94 (a)

$\Rightarrow 3^{999} \times 7^{1000} = 3^{4(249)} \times 3^3 \times 7^{4(250)}$

$= 3^0 \times 3^3 \times 7^0 = 1 \times 7 \times 1 = 7$

Unit place = 7

1.95 (b)

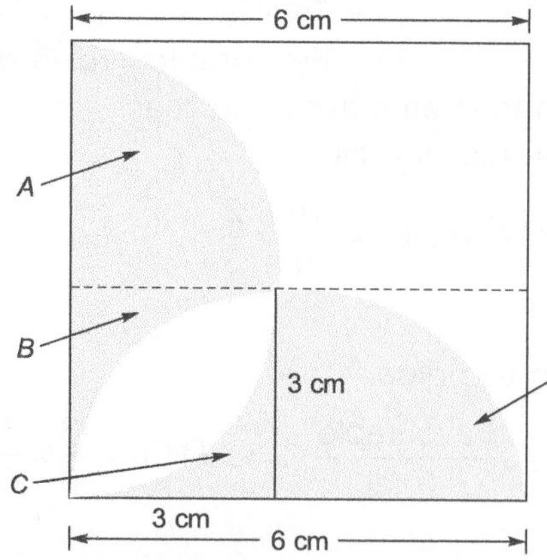

$$A = D = \frac{\pi r^2}{4} = \frac{\pi \times 3^2}{4} \Rightarrow A = D = \frac{9\pi}{4} \quad ...(i)$$

$$B = C = r^2 - \frac{\pi r^2}{4} = 9 - \frac{9\pi}{4} \quad ...(ii)$$

Shaded area = $A + B + C + D$

$$= 2 \times \frac{9\pi}{4} + 2 \times \left(9 - \frac{9\pi}{4}\right) = 2 \times 9 = 18$$

1.96 (c)

Object A : Right (→)
Object P : Clockwise (↻)
Object Q : Anti-clockwise (↺)
Object R : Clockwise (↻)
Object B : Left (←)
Object B translates towards the left direction.

1.97 (c)

$$x : y = (a+1) : (a-1)$$

$$\frac{x}{y} = \frac{(a+1)}{(a-1)}$$

$$\frac{x^2 - y^2}{x^2 + y^2} = \frac{\left(\frac{x}{y}\right)^2 - 1}{\left(\frac{x}{y}\right)^2 + 1} = \frac{\left(\frac{(a+1)}{(a-1)}\right)^2 - 1}{\left(\frac{(a+1)}{(a-1)}\right)^2 + 1}$$

$$= \frac{(a+1)^2 - (a-1)^2}{(a+1)^2 + (a-1)^2} = \frac{4a}{2a^2 + 2} = \frac{2a}{a^2 + 1}$$

1.98 (d)

		Remainder
8	108	
8	13	4 (LSB)
8	1	5
8	0	1 (MSB)

$$(108)_{10} = (154)_8$$

1.99 (*)

$\because$ AFCD is a rhombus

$\therefore \quad AD = CD = 5$ cm

$\therefore \quad \angle DAE = 90°$

$\therefore \quad AD^2 + AE^2 = DE^2$

$25 + 144 = DE^2 \Rightarrow DE = 13$ cm

$DB + BE = DE = 13$ cm

$DB = BF = 13 - 10 = 3$ cm

There is data inconsistency because the values given in question are not practical and such figures can not be made.

1.101 (c)

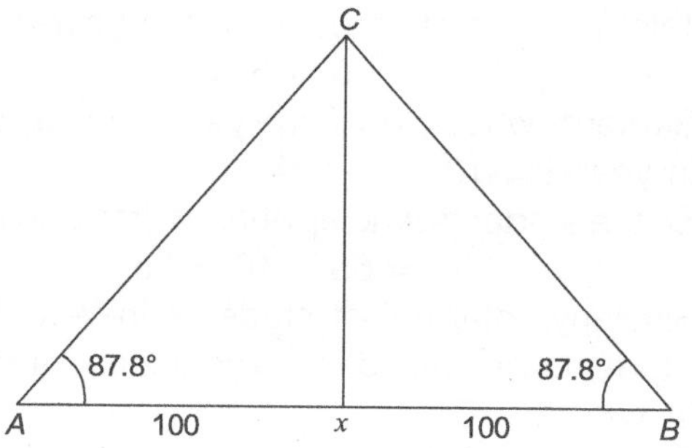

$$\tan\theta = \frac{Cx}{Ax}$$

$$\tan 87.8 = \frac{Cx}{100}$$

$$Cx = 100 \times \tan 87.8 = 2603$$

1.103 (a)

Let Cost price = x

$$x \times \frac{120}{100} \times \frac{118}{100} = 3186$$

$$x = \frac{100 \times 100 \times 3186}{120 \times 118} = 2250$$

2 Verbal Ability

2.1 Which of the following options is the closest in meaning to the word below?

Circuitous

(a) Cyclic (b) Indirect
(c) Confusing (d) Crooked [2010 : 1 M]

2.2 The question below consists of a pair of related words followed by four pairs of words. Select the pair that best expresses the relation in the original pair.

Unemployed: Worker

(a) Fallow : Land (b) Unaware : Sleeper
(c) Wit : Jester (d) Renovated : House

[2010 : 1 M]

2.3 Choose the most appropriate word from the options given below to complete the following sentence:

If we manage to .... our natural resources, we would leave a better planet for our children.

(a) uphold (b) restrain
(c) cherish (d) conserve [2010 : 1 M]

2.4 Choose the most appropriate word from the options given below to complete the following sentence:

His rather casual remarks on politics .... his lack of seriousness about the subject.

(a) masked (b) belied
(c) betrayed (d) suppressed

[2010 : 1 M]

2.5 Modern warfare has changed from large scale clashes of armies to suppression of civilian populations. Chemical agents that do their work silently appear to be suited to such warfare; and regretfully, there exist people in military establishments who think that chemical agents are useful tools for their cause.

Which of the following statements best sums up the meaning of the above passage?

(a) Modern warfare has resulted in civil strife.
(b) Chemical agents are useful in modern warfare.
(c) Use of chemical agents in warfare would be undesirable.
(d) People in military establishments like to use chemical agents in war. [2010 : 2 M]

2.6 The question below consists of a pair of related words followed by four pairs of words. Select the pair that best expresses the relation in the original pair:

Gladiator : Arena

(a) dancer : stage
(b) commuter : train
(c) teacher : classroom
(d) lawyer : courtroom [2011 : 1 M]

2.7 Choose the most appropriate word from the options given below to complete the following sentence:

It was her view that the country's problems had been ____ by foreign technocrats, so that to invite them to come back would be counter-productive.

(a) identified (b) ascertained
(c) exacerbated (d) analysed [2011 : 1 M]

2.8 Choose the word from the options given below that is most nearly opposite in meaning to the given word:

Frequency

(a) periodicity (b) rarity
(c) gradualness (d) persistency

[2011 : 1 M]

2.9 Choose the most appropriate word from the options given below to complete the following sentence:

Under ethical guidelines recently adopted by the Indian Medical Association, human genes are to be manipulated only to correct diseases for which ______ treatments are unsatisfactory.

(a) similar (b) most
(c) uncommon (d) available [2011 : 1 M]

2.10 The horse has played a little known but very important role in the field of medicine. Horses were injected with toxins of diseases until their blood built up immunities. Then a serum was made from their blood. Serums to fight with diphtheria and tetanus were developed this way.

It can be inferred from the passage, that horses were

(a) given immunity to diseases
(b) generally quite immune to diseases
(c) given medicines to fight toxins
(d) given diphtheria and tetanus serums

[2011 : 2 M]

2.11 Which one of the following options is the closest in meaning to the word given below?

Latitude

(a) Eligibility (b) Freedom
(c) Coercion (d) Meticulousness

[2012 : 1 M]

2.12 One of the parts (A, B, C, D) in the sentence given below contains an ERROR. Which one of the following is INCORRECT?

I requested that he should be given the driving test today instead of tomorrow.

(a) requested that
(b) should be given
(c) the driving test
(d) instead of tomorrow [2012 : 1 M]

2.13 Choose the most appropriate alternative from the options given below to complete the following sentence:

If the tired soldier wanted to lie down, he ____ the mattress out on the balcony.

(a) should take
(b) shall take
(c) should have taken
(d) will have taken [2012 : 1 M]

2.14 Choose the most appropriate word from the options given below to complete the following sentence:

Given the seriousness of the situation that he had to face, his ___ was impressive.

(a) beggary (b) nomenclature
(c) jealousy (d) nonchalance

[2012 : 1 M]

2.15 One of the legacies of the Roman legions was discipline. In the legions, military law prevailed and discipline was brutal. Discipline on the battlefield kept units obedient, intact and fighting, even when the odds and conditions were against them.

Which one of the following statements best sums up the meaning of the above passage?

(a) Thorough regimentation was the main reason for the efficiency of the Roman legions even in adverse circumstances.
(b) The legions were treated inhumanly as if the men were animals.
(c) Discipline was the armies inheritance from their seniors.
(d) The harsh discipline to which the legions were subjected to led to the odds and conditions being against them. [2012 : 2 M]

2.16 Choose the grammatically **CORRECT** sentence:

(a) Two and two add four.
(b) Two and two become four.
(c) Two and two are four.
(d) Two and two make four. [2013 : 1 M]

2.17 **Statement:** You can always give me a ring whenever you need.

Which one of the following is the best inference from the above statement?

(a) Because I have a nice caller tune.
(b) Because I have a better telephone facility.
(c) Because a friend in need is a friend indeed.
(d) Because you need not pay towards the telephone bills when you give me a ring.

[2013 : 1 M]

2.18 Complete the sentence: Dare ______ mistakes.

(a) commit (b) to commit
(c) committed (d) committing

[2013 : 1 M]

2.19 They were requested not to **quarrel** with others. Which one of the following options is the closest in meaning to the word **quarrel**?

(a) make out (b) call out
(c) dig out (d) fall out [2013 : 1 M]

2.20 **Statement:** There were different streams of freedom movements in colonial India carried out by the moderates, liberals, radicals, socialists, and so on.

Which one of the following is the best inference from the above statement?

(a) The emergence of nationalism in colonial India led to our Independence.
(b) Nationalism in India emerged in the context of colonialism.
(c) Nationalism in India is homogeneous.
(d) Nationalism in India is heterogeneous.

[2013 : 2 M]

2.21 Which of the following options is the closest in meaning to the phrase underlined in the sentence below?

It is fascinating to see life forms cope with varied environment conditions.

(a) adopt to (b) adapt to
(c) adept in (d) accept with

[2014 : 1 M, Set-1]

2.22 Choose the most appropriate word from the options given below to complete the following sentence.

He could not understand the judges awarding her

the first prize, because he thought that her performance was quite ______.

(a) superb (b) medium
(c) mediocre (d) exhilarating

[2014 : 1 M, Set-1]

2.23 In a press meet on the recent scam, the minister said, "The buck stops here". What did the minister convey by the statement?

(a) He wants all the money
(b) He will return the money
(c) He will assume final responsibility
(d) He will resist all enquiries [2014 : 1 M, Set-1]

2.24 The Palghat Gap (or Palakkad Gap), a region about 30 km wide in the southern part of the Western Ghats in India, is lower than the hilly terrain to its north and south. The exact reasons for the formation of this gap are not clear. It result in the neighbouring regions of Tamil Nadu getting more rainfall from the South West monsoon and the neighbouring regions of Kerala having higher summer temperatures. What can be inferred from this passage?

(a) The Palghat gap is caused by high temperatures in southern Tamil Nadu and Kerala.
(b) The regions in Tamil Nadu and Kerala that are near the Palghat Gap are low lying.
(c) The low terrain of the Palghat Gap has a significant impact on weather patterns in neighbouring parts of Tamil Nadu and Kerala.
(d) Higher summer temperatures result in higher rainfall near the Palghat Gap area.

[2014 : 2 M, Set-1]

2.25 Geneticists say that are very close to confirming the genetic roots of psychiatric illnesses such as depression and schizophrenia, and consequently, that doctors will be able to eradicate these diseases through early identification and gene therapy.
On which of the following assumptions dose the statement above rely?

(a) Strategies are now available for eliminating psychiatric illness.
(b) Certain psychiatric illness have a genetic basis.
(c) All human disease can be traced back to genes and how they are expressed.
(d) In the future, genetics will become the only relevant field for identifying psychiatric illness.

[2014 : 2 M, Set-1]

2.26 Choose the most appropriate phrase from the options given below to complete the following sentence.
India is a post colonial country because

(a) It was a former British colony
(b) Indian Information Technology processional have colonized the world
(c) India does not follow any colonial practices
(d) India has helped other countries gain freedom

[2014 : 1 M, Set-2]

2.27 Who ________ was coming to see us this evening?

(a) you said (b) did you say
(c) did you say that (d) had you said

[2014 : 1 M, Set-2]

2.28 Match the columns:

	Column-1		Column-2
1.	eradicate	P.	misrepresent
2.	distort	Q.	soak completely
3.	saturate	R.	use
4.	utilize	S.	destroy utterly

(a) 1-S, 2-P, 3-Q, 4-R
(b) 1-P, 2-Q, 3-R, 4-S
(c) 1-Q, 2-R, 3-S, 4-P
(d) 1-S, 2-P, 3-R, 4-Q [2014 : 1 M, Set-2]

2.29 The old city of Koenigsberg, which had a German majority population before World War 2, is now called Kaliningrad. After the events of the war, Kaliningrad is now a Russian territory and has a predominantly Russian population. It is bordered by the Baltic Sea on the north and the countries of Poland to the South and West and Lithuania to the East respectively. Which of the statements below can be inferred from this passage.

(a) Kaliningrad was historically Russian in its ethnic make up
(b) Kaliningrad is a part of Russia despite it not being contiguous with the rest of Russia
(c) Koenigsberg was renamed Kaliningrad, as that was its original Russian name
(d) Poland and Lithuania are on the route from Kaliningrad to the rest of Russia

[2014 : 2 M, Set-2]

2.30 The number of people diagnosed with dengue fever (contracted from the bite of a mosquito) in North India is twice the number diagnosed last year. Municipal authorities have concluded that measures to control the mosquito population have failed in this region.
Which one of the following statements, if true, does not contradict this conclusion?

(a) A high proportion of the affected population has returned from neighbouring countries where dengue is prevalent.
(b) More cases of dengue are now reported because of an increase in the Municipal Office's administrative efficiency.
(c) Many more cases of dengue are being diagnosed this year since the introduction of a new effective diagnostic test.
(d) The number of people with malarial fever (also contracted from mosquito bites) has increased this year. [2014 : 2 M, Set-2]

2.31 While trying to collect (I) an envelope from under the table (II), Mr. X fell down and (III) and was losing consciousness (IV).
Which one of the above underlined parts of the sentence is NOT appropriate?
(a) I (b) II
(c) III (d) IV [2014 : 1 M, Set-3]

2.32 If she ______ how to calibrate the instrument, she ______ done the experiment.
(a) knows, will have
(b) knew, had
(c) had known, could have
(d) should have known, would have
[2014 : 1 M, Set-3]

2.33 Choose the word that is opposite in meaning to the word "coherent"
(a) sticky (b) well-connected
(c) rambling (d) friendly
[2014 : 1 M, Set-3]

2.34 A dance programme is scheduled for 10.00 a.m. Some students are participating in the programme and they need to come an hour earlier than the start of the event. These students should be accompanied by a parent. Other students and parents should come in time for the programme. The instruction you think that is appropriate for this is
(a) students should come at 9.00 am and parents should come at 10.00 am.
(b) participating students should come at 9.00 am accompanied by a parent, and other parents and students should come by 10.00 am.
(c) students who are not participating should come by 10.00 am and they should not bring their parent. Participating students should come at 9.00 am.
(d) participating students should come before 9.00 am. Parents who accompany them should come at 9.00 am. All other should come at 10.00 am. [2014 : 2 M, Set-3]

2.35 By the beginning of the 20th century, several hypotheses were being proposed, suggesting a paradigm shift in our understanding of the universe. However, the clinching evidence was provided by experimental measurements of the position of a star which was directly behind our sun.
Which of the following inference(s) may be drawn from the above passage?
(*i*) Out understanding of the universe changes based on the positions of stars.
(*ii*) Paradigm shifts usually occur at the beginning of centuries.
(*iii*) Stars are important objects in the universe.
(*iv*) Experimental evidence was important in confirming this paradigm shift.
(a) (*i*), (*ii*) and (*iv*) (b) (*iii*) only
(c) (*i*) and (*iv*) (d) (*iv*) only
[2014 : 2 M, Set-3]

2.36 Which one of the following combinations is incorrect?
(a) Acquiescence - Submission
(b) Wheedle - Roundabout
(c) Flippancy - Lightness
(d) Profligate - Extravagant [2015 : 1 M, Set-1]

2.37 Didn't you buy ______ when you went shopping?
(a) any paper (b) much paper
(c) no paper (d) a few paper
[2015 : 1 M, Set-1]

2.38 Base on the given statements, select the most appropriate option to solve the given question. If two floors in a certain building are 9 feet apart, how many steps are there in a set of stairs that extends from the first floor to the second floor of the building?
Statements:
(I) Each step is 3/4 foot high.
(II) Each step is 1 foot wide.
(a) Statement-I alone is sufficient, but statement-II alone is not sufficient.
(b) Statement-II alone is sufficient, but statement I alone is not sufficient.
(c) Both statements together are sufficient, but neither statement alone is sufficient.

(d) Statement I and II together are not sufficient. [2015 : 1 M, Set-1]

2.39 Given Set A = {2, 3, 4, 5} Set B = {11, 12, 13, 14, 15}, two numbers are randomly selected, one from each set. What is the probability that the sum of the two numbers are equals 16?

(a) 0.20 (b) 0.25
(c) 0.30 (d) 0.33

[2015 : 1 M, Set-1]

2.40 Which of the following options is the closest in meaning to the sentence below?

She enjoyed herself immensely at the party.

(a) She had a terrible time at the party.
(b) She had a horrible time at the party.
(c) She had a terrific time at the party.
(d) She had a terrifying time at the party.

[2015 : 1 M, Set-1]

2.41 The pie chart below has the breakup of the number of students from different departments in an engineering college for the year 2012. The proportion of male to female students in each department is 5 : 4. There are 40 males in Electrical Engineering. What is the difference between the numbers of female students in the Civil department and the female students in the Mechanical department?

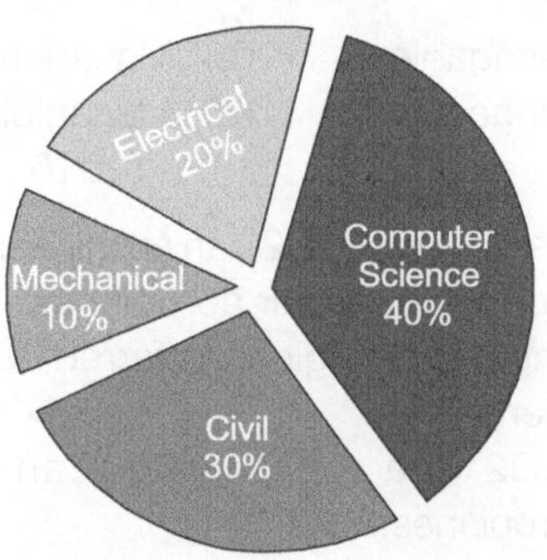

[2015 : 2 M, Set-1]

2.42 Select the alternative meaning of the underlined part of the sentence.

The chain snatchers <u>took to their heels</u> when the police party arrived.

(a) took shelter in a thick jungle
(b) open indiscriminate fire
(c) took to flight
(d) unconditionally surrendered

[2015 : 2 M, Set-1]

2.43 The given statement is followed by some courses of action. Assuming the statement to be true, decide the correct option.

Statement: There has been a significant drop in the water level in the lakes supplying water to the city.

Course of action:

(I) The water supply authority should impose a partial cut in supply to tackle the situation.
(II) The government should appeal to the all the residents through mass media for minimal use of water.
(III) The government should ban the water supply in lower areas.

(a) Statements I and II follow
(b) Statements I and III follow
(c) Statements II and III follow
(d) All statements follow [2015 : 2 M, Set-1]

2.44 Base on the given statements select the most appropriate option to solve the given question. What will be the total weight of 10 poles each of same weight?

Statements:

(I) One fourth of the weight of a pole is 5 kg.
(II) The total weight of these poles is 160 kg more than the total weight of two poles.

(a) Statement I alone is not sufficient
(b) Statement II alone is not sufficient
(c) Either I or II alone is sufficient
(d) Both statement I and II together are not sufficient [2015 : 1 M, Set-2]

2.45 We ______ our friend's birthday and we _____ how to make it up to him

(a) completely forgot - - - don't just know
(b) forgot completely - - - don't just know
(c) completely forgot - - - just don't know
(d) forgot completely - - - just don't know

[2015 : 1 M, Set-2]

2.46 A generic term that includes various items of clothing such as a skirt, a pair of trousers and a shirt is

(a) fabric (b) textile
(c) fibre (d) apparel

[2015 : 1 M, Set-2]

2.47 Choose the statement where underlined word is used correctly.

(a) The industrialist had a <u>personnel</u> jet.
(b) I write my experience in my <u>personnel</u> diary.
(c) All <u>personnel</u> are being given the day off.
(d) Being religious is a <u>personnel</u> aspect.

2.48 If the list of letters, *P, R, S, T, U* is an arithmetic sequence, which of the following are also in arithmetic sequence?

(I) $2P, 2R, 2S, 2T, 2U$
(II) P–3, R–3, S–3, T–3, U–3
(II) P^2, R^2, S^2, T^2, U^2
(a) I only (b) I an II
(c) II and III (d) I and III
[2015 : 1 M, Set-2]

2.49 Out of the following four sentences, select the most suitable sentence with respect to grammar and usage.
(a) Since the report lacked needed information, it was of no use to them.
(b) The report was useless to them because there were no needed information in it.
(c) Since the report did not contain the needed information, it was not real useful to them.
(d) Since the report lacked needed information, it would not had been useful to them.
[2015 : 2 M, Set-2]

2.50 The man who is now Municipal Commissioner worked as _______.
(a) the security guard at a university
(b) a security guard at the university
(c) a security guard at university
(d) the security guard at the university
[2016 : 1 M, Set-1]

2.51 Nobody knows how the Indian cricket team is going to cope with the difficult and seamer-friendly wickets in Australia. Choose the option which is closest in meaning to the underlined phrase in the above sentence.
(a) put up with (b) put in with
(c) put down to (d) put up against
[2016 : 1 M, Set-2]

2.52 Find the odd one in the following group of words. mock, deride, praise, jeer
(a) mock (b) deride
(c) praise (d) jeer [2016 : 1 M, Set-2]

2.53 Pick the odd one from the following options.
(a) CADBE (b) JHKIL
(c) XVYWZ (d) ONPMQ
[2016 : 1 M, Set-2]

2.54 Computers were invented for performing only high-end useful computations. However, it is no understatement that they have taken over our world today. The internet, for example, is ubiquitous. Many believe that the internet itself is an unintended consequence of the original invention. With the advent of mobile computing on our phones, a whole new dimension is now enabled. One is left wondering if all these developments are good or, more importantly, required.
Which of the statement(s) below is/are logically valid and can be inferred from the above paragraph?
(i) The author believes that computers are not good for us.
(ii) Mobile computers and the internet are both intended inventions
(a) (i) only (b) (ii) only
(c) both (i) and (ii) (d) neither (i) nor (ii)
[2016 : 2 M, Set-2]

2.55 All hill-stations have a lake. Ooty has two lakes. Which of the statement(s) below is/are logically valid and can be inferred from the above sentences?
(i) Ooty is not a hill-station.
(ii) No hill-station can have more than one lake.
(a) (i) only (b) (ii) only
(c) Both (i) and (ii) (d) Neither (i) nor (ii)
[2016 : 2 M, Set-2]

2.56 The chairman requested the aggrieved shareholders to _________ him.
(a) bare with (b) bore with
(c) bear with (d) bare
[2016 : 1 M, Set-2]

2.57 Identify the correct spelling out of the given options:
(a) Managable (b) Manageable
(c) Mangaeble (d) Managible
[2016 : 1 M, Set-2]

2.58 R2D2 is a robot. R2D2 can repair aeroplanes. No other robot can repair aeroplanes. Which of the following can be logically inferred from the above statements?
(a) R2D2 is a robot which can only repair aeroplanes.
(b) R2D2 is the only robot which can repair aeroplanes.
(c) R2D2 is a robot which can repair only aeroplanes.
(d) Only R2D2 is a robot. [2016 : 1 M, Set-2]

2.59 The following graph represents the installed capacity for cement production (in tonnes) and the actual production (in tonnes) of nine cement plants of a cement company. Capacity utilization of a plant is defined as ratio of actual production of cement to installed capacity. A plant with installed capacity of at least 200 tonnes is called a large plant and a plant with lesser capacity is called a small plant. The difference between total production of large plants and small plants, in tonnes is ____.

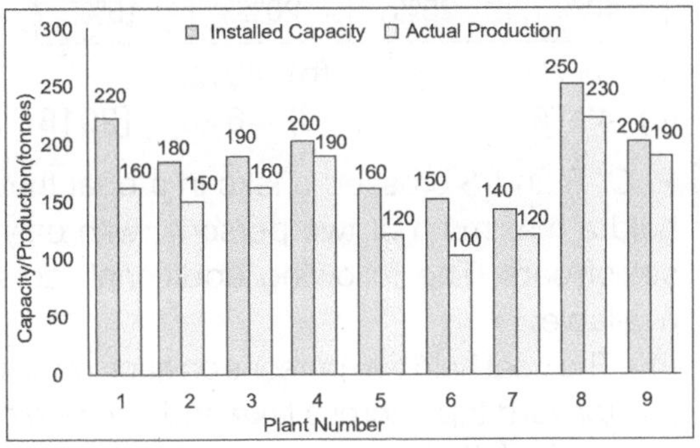

[2016 : 2 M, Set-2]

2.60 A poll of students appearing for masters in engineering indicated that 60 % of the students believed that mechanical engineering is a profession unsuitable for women. A research study on women with masters or higher degrees in mechanical engineering found that 99 % of such women were successful in their professions.
Which of the following can be logically inferred from the above paragraph?
(a) Many students have misconceptions regarding various engineering disciplines.
(b) Men with advanced degrees in mechanical engineering believe women are well suited to be mechanical engineers.
(c) Mechanical engineering is a profession well suited for women with masters or higher degrees in mechanical engineering.
(d) The number of women pursuing higher degrees in mechanical engineering is small.
[2016 : 2 M, Set-2]

2.61 Sourya committee had proposed the establishment of Sourya Institutes of Technology (SITs) in line with Indian Institutes of Technology (IITs) to cater to the technological and industrial needs of a developing country.
Which of the following can be logically inferred from the above sentence?
Based on the proposal,
(i) In the initial years, SIT students will get degrees from IIT.
(ii) SITs will have a distinct national objective.
(iii) SIT like institutions can only be established in consultation with IIT.
(iv) SITs will serve technological needs of a developing country.
(a) (iii) and (iv) only. (b) (i) and (iv) only.
(c) (ii) and (iv) only (d) (ii) and (iii) only
[2016 : 2 M, Set-2]

2.62 Research in the workplace reveal that people work for many reasons ______ .
(a) money beside (b) beside money
(c) money besides (d) besides money
[2017 : 1 M, Set-1]

2.63 Rahul, Murali, Srinivas and Arul are seated around a square table. Rahul is sitting to the left of Murali. Srinivas is sitting to the right of Arul. Which of the following pairs are seated opposite each other?
(a) Rahul and Murali
(b) Srinivas and Arul
(c) Srinivas and Murali
(d) Srinivas and Rahul [2017 : 1 M, Set-1]

2.64 "The hold of the nationalist imagination on our colonial past is such that anything inadequately or improperly nationalist is just not history."
Which of the following statements best reflects the author's opinion?
(a) Nationalists are highly imaginative.
(b) History is viewed through the filter of nationalism.
(c) Our colonial past never happened.
(d) Nationalism has to be both adequately and properly imagined. [2017 : 1 M, Set-1]

2.65 Choose the option with words that are not synonyms.
(a) aversion, dislike (b) luminous, radiant
(c) plunder, loot (d) yielding, resistant
[2017 : 1 M, Set-2]

2.66 There are five buildings called *V*, *W*, *X*, *Y* and *Z* in a row (not necessarily in that order). *V* is to the West of *W*. *Z* is to the East of *X* and the West of *V*. *W* is to the West of *Y*. Which is the building in the middle?
(a) *V* (b) *W*
(c) *X* (d) *Y* [2017 : 1 M, Set-2]

2.67 Saturn is ______ to be seen on a clear night with the naked eye.
(a) enough bright (b) bright enough
(c) as enough bright (d) bright as enough
[2017 : 1 M, Set-2]

2.68 "We lived in a culture that denied any merit to literary works, considering them important only when they were handmaidens to something seemingly more urgent-namely ideology. This was a country where all gestures, even the most private, were interpreted in political terms,". The author's belief that ideology is not as important as literature is revealed by the word:

(a) 'culture' (b) 'seemingly'
(c) 'urgent' (d) 'political'
[2017 : 1 M, Set-2]

2.69 There are three boxes. One contains apples, another contains oranges and the last one contains both apples and oranges. All three are known to be incorrectly labelled. If you are permitted to open just one box and then pull out and inspect only one fruit, which box would you open to determine the contents of all three boxes?
(a) The box labelled 'Apples'
(b) The box labelled 'Apples and Oranges'
(c) The box labelled 'Oranges'
(d) Cannot be determined [2017 : 1 M, Set-2]

2.70 "Since you have gone off the _____, the ______ sand is likely to damage the car". The words that best fill the blanks in the above sentence are
(a) course, coarse (b) course, course
(c) coarse, course (d) coarse, coarse
[2018 : 1 M]

2.71 "A common misconception among writers is that sentence structure mirrors thought; the more ____ the structure, the more complicated the ideas". The word that best fills the blank in the above sentence is
(a) detailed (b) simple
(c) clear (d) convoluted
[2018 : 1 M]

2.72 In a certain code AMCF is written as EQGJ and NKUF is written as ROYJ. How will DHLP be written in the code?
(a) RSTN (b) TLPH
(c) HLPT (d) XSVR [2018 : 2 M]

2.73 An e-mail password must contain three characters. The password has to contain one numeral from 0 to 9, one upper case and one lower case character from the English alphabet. How many distinct passwords are possible?
(a) 6,760 (b) 13,520
(c) 40,560 (d) 1,05,456 [2018 : 2 M]

2.74 A designer uses marbles of four different colours for his designs. The cost of each marble is the same, irrespective of the colour. The table below shows the percentage of marbles of each colour used in the current design. The cost of each marble increased by 25%. Therefore, the designer decided to reduce equal numbers of marbles of each colour to keep the total cost unchanged. What is the percentage of blue marbles in the new design?

Blue	Black	Red	Yellow
40%	25%	20%	15%

(a) 35.75 (b) 40.25
(c) 43.75 (d) 46.25 [2018 : 2 M]

2.75 *P*, *Q*, *R* and *S* crossed a lake in a boat that can hold a maximum of two persons, with only one set of oars. The following additional facts are available:
(i) The boat held two persons on each of the three forward trips across lake and one person on each of the two return trips.
(ii) *P* is unable to row when someone else is in the boat.
(iii) *Q* is unable to row with anyone else except *R*.
(iv) Each person rowed for at least one trip.
(v) Only one person can row during a trip.
Who rowed twice?
(a) *P* (b) *Q*
(c) *R* (d) *S* [2018 : 2 M]

2.76 A class of twelve children has two more boys than girls. A group of three children are randomly picked from this class to accompany the teacher on a field trip. What is the probability that the group accompanying the teacher contains more girls than boys?
(a) 0 (b) $\frac{325}{864}$
(c) $\frac{525}{864}$ (d) $\frac{5}{12}$ [2018 : 2 M]

2.77 Newspapers are a constant source of delight and recreation for me. The _______ trouble is that I read _______ many of them.
(a) even, quite (b) only, too
(c) even, too (d) only, quite
[2019 : 1 M]

2.78 The passengers were angry _______ the airline staff about the delay.
(a) towards (b) on
(c) about (d) with [2019 : 1 M]

2.79 I am not sure if the bus that has been booked will be able to _______ all the students.
(a) fill (b) sit
(c) accommodate (d) deteriorate
[2019 : 1 M]

2.80 An award-winning study by a group of researchers suggests that men are as prone to buying on impulse as women but women feel more guilty about shopping.

Which of the following statements can be inferred from the given text?
(a) Many men and women indulge in buying on impulse.
(b) Few men and women indulge in buying on impulse.
(c) All men and women indulge in buying on impulse.
(d) Some men and women indulge in buying on impulse. [2019 : 2 M]

2.81 Consider five people - Mita, Ganga, Rekha, Lakshmi and Sana. Ganga is taller than both Rekha and Lakshmi. Lakshmi is taller than Sana. Mita is taller than Ganga. Which of the following conclusions are true?
1. Lakshmi is taller than Rekha
2. Rekha is shorter than Mita
3. Rekha is taller than Sana
4. Sana is shorter than Ganga

(a) 1 and 3 (b) 3 only
(c) 1 only (d) 2 and 4 [2019 : 2 M]

2.82 Stock markets ________ at the news of the coup.
(a) plunged (b) plugged
(c) poised (d) probed [2020 : 1 M]

2.83 People were prohibited _______ their vehicles near the entrance of the main administrative building.
(a) to park (b) to have parked
(c) from parking (d) parking [2020 : 1 M]

2.84 Select the word that fits the analogy:
Do : Undo : Trust : ______
(a) Untrust (b) Intrust
(c) Distrust (d) Entrust [2020 : 1 M]

2.85 This book, including all its chapters ______ interesting. The students as well as the instructor ______ in agreement about it.
(a) are, are (b) is, was
(c) is, are (d) were, was
[2020 : 1 M]

2.86 The people _______ were at the demonstration were from all section of society.
(a) whose (b) who
(c) whom (d) which [2021 : 1 M]

2.87 Oasis is to sand as island is to ______ .
Which one of the following options maintains a similar logical relation in the above sentence?
(a) Mountain (b) Water
(c) Land (d) Stone [2021 : 1 M]

2.88 The importance of sleep is often overlooked by students when they are preparing for exams. Research has consistently shown that sleep deprivation greatly reduces the ability to recall the material learn. Hence, cutting down on sleep to study longer hours can be counterproductive.
Which one of the following statements is the CORRECT inference from the above passage?
(a) Sleeping well alone is enough to prepare for an exam. Studying has lesser benefit.
(b) To do well in an exam, adequate sleep must be part of the preparation.
(c) Students are efficient and are not wrong in thinking that sleep is a waste of time.
(d) If a student is extremely well prepared for an exam, he needs little or no sleep.
[2021 : 2 M]

2.89 As you grow older, an injury to your _____ may take longer to _____.
(a) heel / heel (b) heal / heel
(c) heal / heal (d) heel / heal
[2022 : 1 M]

2.90 Given below are two statements and four conclusions drawn based on the statements.
Statement I : Some bottles are cups.
Statement II : All cups are knives.
Conclusion I : Some bottles are knives.
Conclusion II : Some knives are cups.
Conclusion III : All cups are bottles.
Conclusion IV : All knives are cups.
Which one of the following options can be logically inferred?
(a) Only conclusion I and conclusion II are correct.
(b) Only conclusion II and conclusion III are correct.
(c) Only conclusion II and conclusion IV are correct.
(d) Only conclusion III and conclusion IV are correct. [2022 : 1 M]

2.91 Altruism is the human concern for the well being of others. Altruism has been shown to be motivated more by social bonding, familiarity and identification of belongingness to a group. The notion that altruism may be attributed to empathy or guilt has now been rejected.
Which one of the following is the CORRECT logical inference based on the information in the above passage?

(a) Humans engage in altruism due to guilt but not empathy.
(b) Humans engage in altruism due to empathy but not guilt.
(c) Humans engage in altruism due to group identification but not empathy.
(d) Humans engage in altruism due to empathy but not familiarity. [2022 : 2 M]

2.92 Rafi told Mary, "I am thinking of watching a film this weekend."
The following reports the above statement in indirect speech :
Rafi told Mary that he _____ of watching a film that weekend.

(a) thought (b) is thinking
(c) am thinking (d) was thinking

[2023 : 1 M]

2.93 Permit : ____ :: Enforce : Relax (by word meaning)

(a) Allow (b) Forbid
(c) License (d) Reinforce

[2023 : 1 M]

2.94 Students of all the departments of a college who have successfully completed the registration process are eligible to vote in the upcoming college elections. However, by the time the due date for registration was over, it was found that surprisingly none of the students from the Department of Human Sciences had completed the registration process.
Based only on the information provided above, which one of the following sets of statement(s) can be logically inferred with *certainty*?

(i) All those students who would not be eligible to vote in the college elections would certainly belong to the Department of Human Sciences.
(ii) None of the students from departments other than Human Sciences failed to complete the registration process within the due time.
(iii) All the eligible voters would certainly be students who are not from the Department of Human Sciences.

(a) (i) and (ii) (b) (i) and (iii)
(c) Only (i) (d) Only (iii)

[2023 : 2 M]

2.95 Which one of the options does NOT describe the passage below or follow from it?
We tend to think of cancer as a 'modern' illness because its metaphors are so modern. It is a disease of overproduction, of sudden growth, a growth that is unstoppable, tipped into the abyss of no control. Modern cell biology encourages us to imagine the cell as a molecular machine. Cancer is that machine unable to quench its initial command (to grow) and thus transform into an indestructible, self-propelled automation.
(Adapted from *The Emperor of All Maladies* by Siddhartha Mukherjee)

(a) It is a reflection of why cancer seems so modern to most of us.
(b) It tells us that modern cell biology uses and promotes metaphors of machinery.
(c) Modern cell biology encourages metaphors of machinery, and cancer is often imagined as a machine.
(d) Modern cell biology never uses figurative language, such as metaphors, to describe or explain anything.

[2023 : 2 M]

2.96 If '→' denotes increasing order of intensity, then the meaning of the words [talk → shout → scream] is analogous to [please → __________ → pander]. Which one of the given options is appropriate to fill the blank?

(a) flutter (b) fritter
(c) frizzle (d) flatter

[2024 : 1 M]

2.97 In the given text, the blanks are numbered (i)-(iv). Select the best match for all the blanks.
Following a row __(i)__ the shopkeeper __(ii)__ the price of a frying pan, the cook stood __(iii)__ a row to withdraw cash __(iv)__ the ATM booth.

(a) (i) with; (ii) over; (iii) in; (iv) at
(b) (i) with; (ii) over; (iii) at; (iv) with
(c) (i) over; (ii) with; (iii) over; (iv) at
(d) (i) at; (ii) over; (iii) over; (iv) in

[2024 : 2 M]

Answers Verbal Ability

2.1 (b)	2.2 (a)	2.3 (d)	2.4 (c)	2.5 (d)	2.6 (a)	2.7 (c)
2.8 (c)	2.9 (c)	2.10 (c)	2.11 (b)	2.12 (a)	2.13 (c)	2.14 (d)
2.15 (a)	2.16 (d)	2.17 (c)	2.18 (a)	2.19 (d)	2.20 (d)	2.21 (b)
2.22 (c)	2.23 (c)	2.24 (c)	2.25 (b)	2.26 (a)	2.27 (b)	2.28 (a)
2.29 (b)	2.30 (d)	2.31 (d)	2.32 (c)	2.33 (c)	2.34 (b)	2.35 (d)
2.36 (b)	2.37 (a)	2.38 (a)	2.39 (a)	2.40 (c)	2.41 (32)	2.42 (c)
2.43 (a)	2.44 (c)	2.45 (c)	2.46 (d)	2.47 (c)	2.48 (b)	2.49 (a)
2.50 (b)	2.51 (a)	2.52 (c)	2.53 (d)	2.54 (d)	2.55 (d)	2.56 (c)
2.57 (b)	2.58 (b)	2.59 (120)	2.60 (a)	2.61 (c)	2.62 (d)	2.63 (c)
2.64 (b)	2.65 (d)	2.66 (a)	2.67 (b)	2.68 (b)	2.69 (b)	2.70 (a)
2.71 (d)	2.72 (c)	2.73 (c)	2.74 (c)	2.75 (c)	2.76 (b)	2.77 (b)
2.78 (d)	2.79 (c)	2.80 (d)	2.81 (d)	2.82 (a)	2.83 (c)	2.84 (c)
2.85 (c)	2.86 (b)	2.87 (b)	2.88 (b)	2.89 (d)	2.90 (a)	2.91 (c)
2.92 (d)	2.93 (b)	2.94 (d)	2.95 (d)	2.96 (d)	2.97 (a)	

Explanations Verbal Ability

2.1 (b)

Circuitous: Deviating from a straight course ⇒ Indirect

(A) Cyclic: Recurring in cycle
(B) Indirect: Not leading by straight line
(C) Confusing: Lacking clarity
(D) Crooked: For shapes (irregular in shape)

2.2 (a)

Unemployed: Worker ⇒ Here one is opposite to other.

(A) Fallow: Land ⇒ Fallow means undeveloped land.
(B) Unaware: sleeper ⇒ Both are same unaware or asleep.
(C) Wit: Jester ⇒ Wit means ability to make jokes and jester is a joker.
(D) Renovated : House ⇒ Renovate means to make better and house can be renovated.

2.3 (d)

(A) Uphold: cause to remain ⇒ not appropriate
(B) Restrain: keep under control ⇒ not appropriate
(C) Cherish: be fond of ⇒ not related
(D) Conserve: Keep in safety and protect from harm, decay, loss, or destruction ⇒ most appropriate.

2.4 (c)

(A) Masked: Hide under a false appearance ⇒ opposite
(B) Belied: Be in contradiction with ⇒ not appropriate
(C) Betrayed: Reveal unintentionally⇒ most appropriate
(D) Suppressed: To put down by force or authority ⇒ irrelevant

2.5 (d)

(A) Modern warfare has resulted in civil strife: There is no direct consequence of warfare given, so it is not appropriate.
(B) Chemical agents are useful in modern warfare: Passage does not say whether chemical agents are useful or not, so not appropriate.
(C) Use of chemical agents in warfare would be undesirable: Given that people in military think these are useful, undesirable is wrong
(D) People in military establishments like to use chemical agents in war; Correct choice as last statement tells that military people think that chemical agents are useful tools for their cause (work silently in warfare).

2.11 (b)

Latitude refers to freedom of action, freedom of expression from restrictions etc. For example, he allowed his children a fair amount of latitude. Coercion refers to force which is an opposite of the word Latitude. Meticulousness refers to being extremely careful and conscientious.

2.12 (a)

'Requested that' part should be modified. The sentence is in Reported speech so 'should be given' part is correct. Instead of Request, words like 'suggested', 'recommended' can be introduced.

2.13 (c)

If the tired soldier wanted to lie down, he should have taken the mattress out on the balcony.

2.14 (d)

Given the seriousness of the situation that he had to face, his nonchalance was impressive.
Nonchalance refers to casual / non-serious/ indifferent attitude. The author wants to convey that in spite of the seriousness of the situation, he was very casual.
No other word fits in the blank. Beggary is word that is related to extreme poverty (Beggar). Nomenclature refers to the art of naming.

2.15 (a)

Thorough regimentation was the main reason for the efficiency of the Roman legions even in adverse circumstances. As the author says, 'Discipline on the battlefield kept units obedient, intact and fighting, even when the odds and conditions were against them.' So, choice (A) emerges as the best answer option.
We have been asked to summarize the paragraph in the best possible manner. Choice (C) presents one-sided picture of the paragraph. Summary is the essence or the gist of the paragraph.
Choice (B) and Choice (D) are wrong as per the information mentioned in the paragraph.

2.16 (d)

This is a famous idiom in English.

2.17 (c)

'You can always give me a ring whenever you need' implies that I am available even at a single call (whenever you need me). 'Ring' has been used as a metaphor here and not literally.

2.20 (d)

As the given argument states that there were different streams of freedom movement during colonialism, it clearly implies that nationalism in India is heterogeneous. Option (c) is ruled out as it states the contrary. Options (a) and (b) fail to establish a direct link with the argument.

2.38 (a)

From statements A, if height of each step is known, we can get the number of steps as height of building is given in the question.

2.39 (a)

$A = \{2, 3, 4, 5\}$, $B = [11, 12, 13, 14, 15]$
Sample space will contain $4 \times 5 = 20$ events favourable events [2, 14] [3, 13] [4, 12] [5, 11] for sum 16.

$$\text{Required probability} = \frac{4}{20} = \frac{1}{5} = 0.2$$

2.41 Sol.

Let the total number of students in electrical be T_e. Male students in electrical is 40.

$$\text{So, } \frac{5}{9}T_e = 40 \Rightarrow T_e = 72$$

Also, let total number of students (overall) = T.
Electrical is 20% of T
$\Rightarrow$ 20% of $T = 72 \Rightarrow T = 360$
Now, number of female students in Civil

$$\Rightarrow \quad 30\% \text{ of } 360 \times \frac{4}{9} = 48$$

and number of female students in Mechanical

$$\Rightarrow \quad 10\% \text{ of } 360 \times \frac{4}{9} = 16$$

Difference = 48 – 16 = 32

2.48 (b)

If subtract or multiply every term of an A.P. by a constant resulting sequence will be also A.P. as common difference will still be constant.

2.53 (d)

$$\underbrace{C \quad A \quad D \quad B \quad E}_{-2,\ \ +3,\ \ -2,\ \ 3}$$

Same pattern follows in all options A, B, C only (d) doesn't follow this pattern , Hence odd number out (D).

2.59 Sol.

According to information in Question
Large plants are 1, 4, 8, 9 which are having installed capacity of at least 200 tonnes.
Total production of large plant

[160 + 190 + 230 + 190] = 770

Remaining plant number 2, 3, 5, 6, 7 all are small

plants with capacity less than 200 tonnes.
Total production of small plants

$= 150 + 160 + 120 + 100 + 120 = 650$

Difference $= 770 - 650 = 120$

2.62 (d)

Besides money
Research in the work place reveals the people warks for many reasons **besides money**.
Besides conveys the meaning of 'in addition'
Beside means 'next to'

2.63 (c)

Following seating arrangement can be drawn

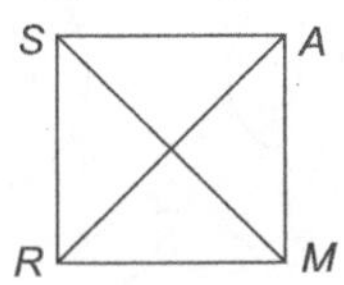

Therefore correct option is (c).

2.64 (b)

History is viewed through the filter of rationalism.

2.67 (b)

Enough is an adverb. It is usually positioned after the adjective/adverb it modifies.

2.70 (a)

- Going off the course - not following the intended route.
- Coarse sand - harsh in texture

2.72 (c)

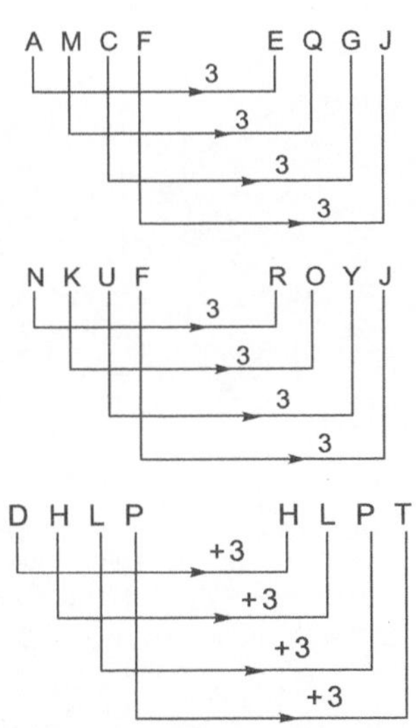

2.73 (c)

Numeral can be selected in 10 ways (0 - 9).
Each of upper case and lower case alphabets can be done in 26 ways each.

All three chosen (1 numeral and 2 alphabet) can be arrange in 3! ways.
So, total number of ways will be $10 \times 26 \times 26 \times 3!$
$= 40560$ ways

2.74 (c)

Assume number of marbles = 100
Cost of marbles increased = 25

New cost = Rs. 125

Number of marbles in Rs.100

$$= \frac{100}{125} \times 100 = 80 \text{ marbles}$$

Now, $(40 - x) + (25 - x) + (20 - x) + (15 - x) = 80$

$$100 - 4x = 80 \Rightarrow x = 5$$

% blue marbles in new design

$$= \frac{(40-5)}{80} \times 100 = 35 \times \frac{5}{4} = 43.75\%$$

2.75 (c)

(i) Q and R moves first.
In forward trip Q rowed.
In return trip R rowed.

(ii) P and R moves in second trip
R rowed in forward trip
P rowed in return trip

(iii) P and S moves in last trip
S rowed in forward trip.
R rowed twice.

2.76 (b)

$B + G = 12$

$$B = G + 2 \Rightarrow B = 7;\ G = 5$$

7 Boys and 5 Girls are there is 12 students. Among 3 students selected boys have to be more than girls. So only two cases arise.

G	B
2	1
3	0

As case (i),

$$GGG \rightarrow \frac{5}{12} \times \frac{5}{12} \times \frac{5}{12}$$

Case (ii),

$$\left.\begin{matrix} G & G & B \\ G & B & G \\ B & G & G \end{matrix}\right\} \rightarrow 3 \times \frac{5}{12} \times \frac{5}{12} \times \frac{7}{12}$$

$$\text{Required prob.} = \left(\frac{5}{12}\right)^3 + 3\left(\frac{5}{12}\right)^2 \times \left(\frac{7}{12}\right)$$

$$= \frac{650}{1728} = \frac{325}{864} = 0.3761574 \text{ » } 0.376$$

Alternative Method:

There are 7*B* and 5*G*. Through as question is stating 3 students are taken at random.

This can be a possible way to approach it.

G	B
2	1
3	0

$$= \frac{{}^5C_2 + {}^7C_1 + {}^5C_3 + {}^7C_0}{{}^{12}C_3}$$

$$= \frac{10 \times 7 + 10 \times 1}{\frac{12 \times 10 \times 11}{6}} = \frac{80}{220} = \frac{4}{11} \approx 0.3636$$

2.78 (d)

Angry on ... an issue

angry with ... someone.

2.81 (d)

If '>' Implies taller than

1. Ganga > Rekha, Ganga > Lakshmi
2. Lakshmi > Sana
3. Mita > Ganga

$\Rightarrow$ Mita > Ganga > Lakshmi > Sana

$\Rightarrow$ Mita > Ganga > Rekha

Statement 2 is correct. Statement 4 is correct.

2.84 (c)

Do : Undo : Trust : Distrust

2.86 (b)

We cannot use 'which' for people, 'who' is the only word which acts as a subject for the verb.

2.87 (b)

Oasis is surrounded by sand as Island is surrounded by water.

2.88 (b)

First statement is ridiculous, third and fourth are contradictory to what is given in the passage.

2.90 (a)

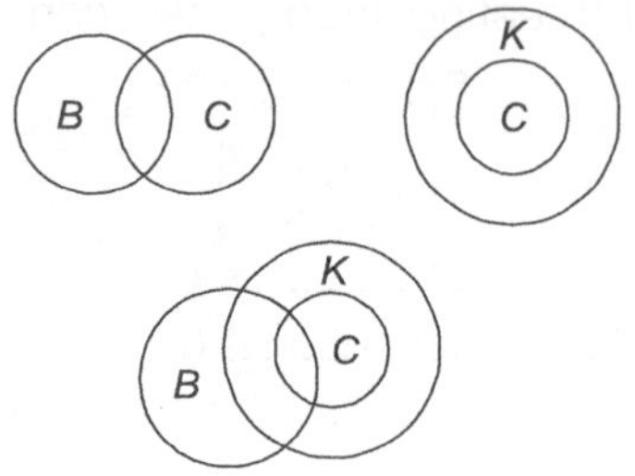

Conclusions I and II are correct.

2.93 (b)

FORBID is antonym of PERMIT.

■■■■